A TEXTBOOK OF

Machine Design

(S.I. UNITS)

A TEXTBOOK OF

Machine Design

(S.I. UNITS)

[A Textbook for the Students of B.E. / B.Tech., U.P.S.C. (Engg. Services); Section 'B' of A.M.I.E. (I)]

R.S. KHURMI

J.K. GUPTA

EURASIA PUBLISHING HOUSE (PVT.) LTD.

S Chand And Company Limited

(ISO 9001 Certified Company)

Head Office: D-92, Sector–2, Noida – 201301, U.P. (India), Ph. 91-120-4682700

Registered Office: A-27, 2nd Floor, Mohan Co-operative Industrial Estate, New Delhi – 110 044, Phone: 011-49731800

www.**schandpublishing.com**; e-mail: **info@schandpublishing.com**

Marketing Offices:

Chennai : Ph: 23632120; chennai@schandpublishing.com
Guwahati : Ph: 2738811, 2735640; guwahati@schandpublishing.com
Hyderabad : Ph: 40186018; hyderabad@schandpublishing.com
Jalandhar : Ph: 4645630; jalandhar@schandpublishing.com
Kolkata : Ph: 23357458, 23353914; kolkata@schandpublishing.com
Lucknow : Ph: 4003633; lucknow@schandpublishing.com
Mumbai : Ph: 25000297; mumbai@schandpublishing.com
Patna : Ph: 2260011; patna@schandpublishing.com

First Edition 1980
Subsequent Editions and Reprints 1980, 81, 83, 84, 85, 86, 87, 88, (Twice), 89, 90, (Twice), 91, 92, 93, 94, 95, 96, (Twice), 97, 98 (Twice), 2000, 2001, 2002, 2003 (Thrice), 2004 (Twice), 2005, First Multicolour Revised & Updated Edition 2005, Reprints 2006, 2007, 2008, 2010 (Twice), 2011 (Twice), 2012, 2013, 2014 (Twice), 2015 (Twice), 2016 (Twice), 2018, 2019 (Twice)

Low Priced Students' Paperback Edition (LPSPE), 2022

Reprint 2023 (Twice)

ISBN: 978-93-5501-083-4 **Product Code:** H3MDG62MACD10ENAA21L

PRINTED IN INDIA

By Vikas Publishing House Private Limited, Plot 20/4, Site-IV, Industrial Area Sahibabad, Ghaziabad – 201 010 and Published by S Chand And Company Limited, A-27, 2nd Floor, Mohan Co-operative Industrial Estate, New Delhi – 110 044.

Preface to the Fourteenth Edition

We feel satisfied in presenting the new edition of this popular treatise. The favourable and warm reception which the previous editions and reprints of this book have enjoyed all over India and abroad, is a matter of great satisfaction for us.

The present edition has been thoroughly revised and brought up-to-date. Pictures have been added to enhance the content value and to give the students an idea of what he will be dealing in reality, and to bridge the gap between theory and practice. This book has already been included in the 'Suggested Reading' for the A.M.I.E. (India) examinations. The mistakes which had crept in, have been eliminated. We wish to express our sincere thanks to numerous professors and students, both at home and abroad, for sending their valuable suggestions and recommending the book to their students and friends. We hope, that they will continue to patronise this book in the future also.

Any errors, omissions and suggestions, for the improvement of this volume brought to our notice, will be thankfully acknowledged and incorporated in the next edition.

R.S. KHURMI
J.K. GUPTA

Preface

We take an opportunity to present this standard treatise entitled as 'A TEXTBOOK OF MACHINE DESIGN' to the students of Degree, Diploma and A.M.I.E. (India) classes in M.K.S. and S.I. units. The objective of this book is to present the subject matter in a most concise, compact, to the point and lucid manner.

While writing the book, we have continuously kept in mind the examination requirement of the students preparing for U.P.S.C. (Engg. Services) and A.M.I.E. (India) examinations. In order to make this volume more useful for them, complete solutions of their examination papers upto 1977 have also been included. Every care has been taken to make this treatise as self-explanatory as possible. The subject matter has been amply illustrated by incorporating a good number of solved, unsolved and well graded examples of almost every variety. Most of these examples are taken from the recent examination papers of Indian and foreign universities as well as professional examining bodies, to make the students familiar with the type of questions, usually, set in their examinations. At the end of each chapter, a few exercises have been added for the students to solve them independently. Answers to these problems have been provided, but it is too much to hope that these are entirely free from errors. In short, it is earnestly hoped that the book will earn appreciation of all the teachers and students alike.

Although every care has been taken to check mistakes and misprints, yet it is difficult to claim perfection. Any errors, omissions and suggestions for the improvement of this treatise, brought to our notice, will be thankfully acknowledged and incorporated in the next edition.

R.S. KHURMI
J.K. GUPTA

CONTENTS

CHP
N827BG

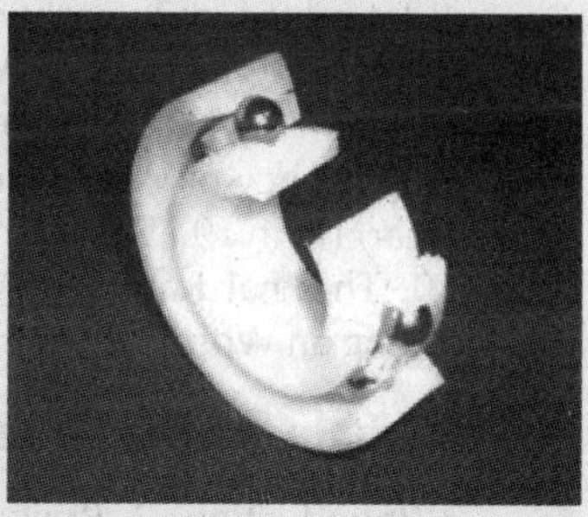

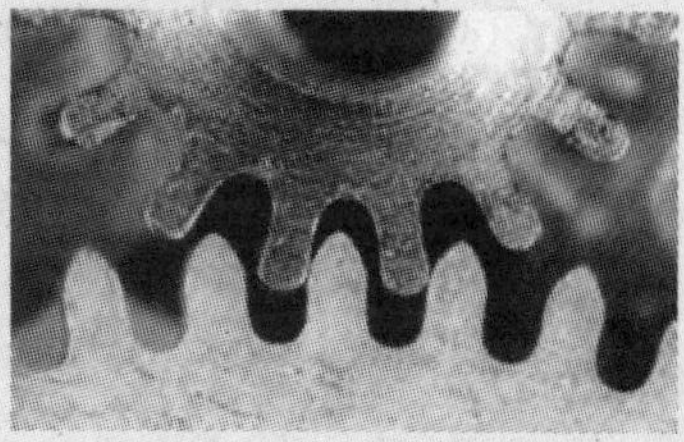

CHAPTER 1

Introduction

1.1 Definition

The subject Machine Design is the creation of new and better machines and improving the existing ones. A new or better machine is one which is more economical in the overall cost of production and operation. The process of design is a long and time consuming one. From the study of existing ideas, a new idea has to be conceived. The idea is then studied keeping in mind its commercial success and given shape and form in the form of drawings. In the preparation of these drawings, care must be taken of the availability of resources in money, in men and in materials required for the successful completion of the new idea into an actual reality. In designing a machine component, it is necessary to have a good knowledge of many subjects such as Mathematics, Engineering Mechanics, Strength of Materials, Theory of Machines, Workshop Processes and Engineering Drawing.

1.2 Classifications of Machine Design

The machine design may be classified as follows :

1. *Adaptive design.* In most cases, the designer's work is concerned with adaptation of existing designs. This type of design needs no special knowledge or skill and can be attempted by designers of ordinary technical training. The designer only makes minor alternation or modification in the existing designs of the product.

2. *Development design.* This type of design needs considerable scientific training and design ability in order to modify the existing designs into a new idea by adopting a new material or different method of manufacture. In this case, though the designer starts from the existing design, but the final product may differ quite markedly from the original product.

3. *New design.* This type of design needs lot of research, technical ability and creative thinking. Only those designers who have personal qualities of a sufficiently high order can take up the work of a new design.

The designs, depending upon the methods used, may be classified as follows :

(*a*) ***Rational design.*** This type of design depends upon mathematical formulae of principle of mechanics.

(*b*) ***Empirical design.*** This type of design depends upon empirical formulae based on the practice and past experience.

(*c*) ***Industrial design.*** This type of design depends upon the production aspects to manufacture any machine component in the industry.

(*d*) ***Optimum design.*** It is the best design for the given objective function under the specified constraints. It may be achieved by minimising the undesirable effects.

(*e*) ***System design.*** It is the design of any complex mechanical system like a motor car.

(*f*) ***Element design.*** It is the design of any element of the mechanical system like piston, crankshaft, connecting rod, etc.

(*g*) ***Computer aided design.*** This type of design depends upon the use of computer systems to assist in the creation, modification, analysis and optimisation of a design.

1.3 General Considerations in Machine Design

Following are the general considerations in designing a machine component :

1. *Type of load and stresses caused by the load.* The load, on a machine component, may act in several ways due to which the internal stresses are set up. The various types of load and stresses are discussed in chapters 4 and 5.

2. *Motion of the parts or kinematics of the machine.* The successful operation of any machine depends largely upon the simplest arrangement of the parts which will give the motion required. The motion of the parts may be :

(*a*) Rectilinear motion which includes unidirectional and reciprocating motions.

(*b*) Curvilinear motion which includes rotary, oscillatory and simple harmonic.

(*c*) Constant velocity.

(*d*) Constant or variable acceleration.

3. *Selection of materials.* It is essential that a designer should have a thorough knowledge of the properties of the materials and their behaviour under working conditions. Some of the important characteristics of materials are : strength, durability, flexibility, weight, resistance to heat and corrosion, ability to cast, welded or hardened, machinability, electrical conductivity, etc. The various types of engineering materials and their properties are discussed in chapter 2.

4. ***Form and size of the parts.*** The form and size are based on judgement. The smallest practicable cross-section may be used, but it may be checked that the stresses induced in the designed cross-section are reasonably safe. In order to design any machine part for form and size, it is necessary to know the forces which the part must sustain. It is also important to anticipate any suddenly applied or impact load which may cause failure.

5. ***Frictional resistance and lubrication.*** There is always a loss of power due to frictional resistance and it should be noted that the friction of starting is higher than that of running friction. It is, therefore, essential that a careful attention must be given to the matter of lubrication of all surfaces which move in contact with others, whether in rotating, sliding, or rolling bearings.

6. ***Convenient and economical features.*** In designing, the operating features of the machine should be carefully studied. The starting, controlling and stopping levers should be located on the basis of convenient handling. The adjustment for wear must be provided employing the various take-up devices and arranging them so that the alignment of parts is preserved. If parts are to be changed for different products or replaced on account of wear or breakage, easy access should be provided and the necessity of removing other parts to accomplish this should be avoided if possible.

The economical operation of a machine which is to be used for production, or for the processing of material should be studied, in order to learn whether it has the maximum capacity consistent with the production of good work.

Design considerations play important role in the successful production of machines.

7. ***Use of standard parts.*** The use of standard parts is closely related to cost, because the cost of standard or stock parts is only a fraction of the cost of similar parts made to order.

The standard or stock parts should be used whenever possible ; parts for which patterns are already in existence such as gears, pulleys and bearings and parts which may be selected from regular shop stock such as screws, nuts and pins. Bolts and studs should be as few as possible to avoid the delay caused by changing drills, reamers and taps and also to decrease the number of wrenches required.

8. ***Safety of operation.*** Some machines are dangerous to operate, especially those which are speeded up to insure production at a maximum rate. Therefore, any moving part of a machine which is within the zone of a worker is considered an accident hazard and may be the cause of an injury. It is, therefore, necessary that a designer should always provide safety devices for the safety of the operator. The safety appliances should in no way interfere with operation of the machine.

9. ***Workshop facilities.*** A design engineer should be familiar with the limitations of his employer's workshop, in order to avoid the necessity of having work done in some other workshop. It is sometimes necessary to plan and supervise the workshop operations and to draft methods for casting, handling and machining special parts.

10. ***Number of machines to be manufactured.*** The number of articles or machines to be manufactured affects the design in a number of ways. The engineering and shop costs which are called fixed charges or overhead expenses are distributed over the number of articles to be manufactured. If only a few articles are to be made, extra expenses are not justified unless the machine is large or of some special design. An order calling for small number of the product will not permit any undue expense in the workshop processes, so that the designer should restrict his specification to standard parts as much as possible.

11. *Cost of construction.* The cost of construction of an article is the most important consideration involved in design. In some cases, it is quite possible that the high cost of an article may immediately bar it from further considerations. If an article has been invented and tests of hand made samples have shown that it has commercial value, it is then possible to justify the expenditure of a considerable sum of money in the design and development of automatic machines to produce the article, especially if it can be sold in large numbers. The aim of design engineer under all conditions, should be to reduce the manufacturing cost to the minimum.

Car assembly line.

12. *Assembling.* Every machine or structure must be assembled as a unit before it can function. Large units must often be assembled in the shop, tested and then taken to be transported to their place of service. The final location of any machine is important and the design engineer must anticipate the exact location and the local facilities for erection.

1.4 General Procedure in Machine Design

In designing a machine component, there is no rigid rule. The problem may be attempted in several ways. However, the general procedure to solve a design problem is as follows :

1. *Recognition of need.* First of all, make a complete statement of the problem, indicating the need, aim or purpose for which the machine is to be designed.

2. *Synthesis (Mechanisms).* Select the possible mechanism or group of mechanisms which will give the desired motion.

3. *Analysis of forces.* Find the forces acting on each member of the machine and the energy transmitted by each member.

4. *Material selection.* Select the material best suited for each member of the machine.

5. *Design of elements (Size and Stresses).* Find the size of each member of the machine by considering the force acting on the member and the permissible stresses for the material used. It should be kept in mind that each member should not deflect or deform than the permissible limit.

6. *Modification.* Modify the size of the member to agree with the past experience and judgment to facilitate manufacture. The modification may also be necessary by consideration of manufacturing to reduce overall cost.

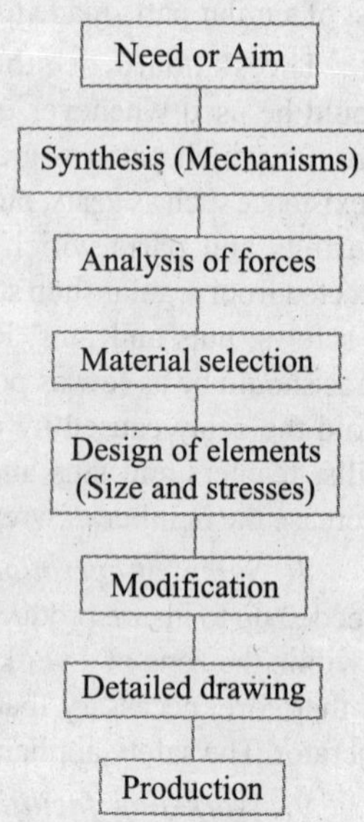

Fig. 1.1. General procedure in Machine Design.

7. *Detailed drawing.* Draw the detailed drawing of each component and the assembly of the machine with complete specification for the manufacturing processes suggested.

8. *Production.* The component, as per the drawing, is manufactured in the workshop.

The flow chart for the general procedure in machine design is shown in Fig. 1.1.

Note : When there are number of components in the market having the same qualities of efficiency, durability and cost, then the customer will naturally attract towards the most appealing product. The aesthetic and ergonomics are very important features which gives grace and lustre to product and dominates the market.

1.5 Fundamental Units

The measurement of physical quantities is one of the most important operations in engineering. Every quantity is measured in terms of some arbitrary, but internationally accepted units, called ***fundamental units.***

1.6 Derived Units

Some units are expressed in terms of other units, which are derived from fundamental units, are known as ***derived units*** *e.g.* the unit of area, velocity, acceleration, pressure, etc.

1.7 System of Units

There are only four systems of units, which are commonly used and universally recognised. These are known as :

1. C.G.S. units, 2. F.P.S. units, 3. M.K.S. units, and 4. S.I. units.

Since the present course of studies are conducted in S.I. system of units, therefore, we shall discuss this system of unit only.

1.8 S.I. Units (International System of Units)

The 11th General Conference* of Weights and Measures have recommended a unified and systematically constituted system of fundamental and derived units for international use. This system is now being used in many countries. In India, the standards of Weights and Measures Act 1956 (vide which we switched over to M.K.S. units) has been revised to recognise all the S.I. units in industry and commerce.

In this system of units, there are seven fundamental units and two supplementary units, which cover the entire field of science and engineering. These units are shown in Table 1.1

Table 1.1. Fundamental and supplementary units.

S.No.	*Physical quantity*	*Unit*
	Fundamental units	
1.	Length (l)	Metre (m)
2.	Mass (m)	Kilogram (kg)
3.	Time (t)	Second (s)
4.	Temperature (T)	Kelvin (K)
5.	Electric current (I)	Ampere (A)
6.	Luminous intensity(Iv)	Candela (cd)
7.	Amount of substance (n)	Mole (mol)
	Supplementary units	
1.	Plane angle (α, β, θ, ϕ)	Radian (rad)
2.	Solid angle (Ω)	Steradian (sr)

* It is known as General Conference of Weights and Measures (G.C.W.M). It is an international organisation of which most of the advanced and developing countries (including India) are members. The conference has been entrusted with the task of prescribing definitions for various units of weights and measures, which are the very basics of science and technology today.

The derived units, which will be commonly used in this book, are given in Table 1.2.

Table 1.2. Derived units.

S.No.	Quantity	Symbol	Units
1.	Linear velocity	V	m/s
2.	Linear acceleration	a	m/s^2
3.	Angular velocity	ω	rad/s
4.	Angular acceleration	α	rad/s^2
5.	Mass density	ρ	kg/m^3
6.	Force, Weight	F, W	N ; 1N = 1kg-m/s^2
7.	Pressure	P	N/m^2
8.	Work, Energy, Enthalpy	W, E, H	J ; 1J = 1N-m
9.	Power	P	W ; 1W = 1J/s
10.	Absolute or dynamic viscosity	μ	N-s/m^2
11.	Kinematic viscosity	ν	m^2/s
12.	Frequency	f	Hz ; 1Hz = 1cycle/s
13.	Gas constant	R	J/kg K
14.	Thermal conductance	h	W/m^2 K
15.	Thermal conductivity	k	W/m K
16.	Specific heat	c	J/kg K
17.	Molar mass or Molecular mass	M	kg/mol

1.9 Metre

The metre is defined as the length equal to 1 650 763.73 wavelengths in vacuum of the radiation corresponding to the transition between the levels $2\,p_{10}$ and $5\,d_5$ of the Krypton– 86 atom.

1.10 Kilogram

The kilogram is defined as the mass of international prototype (standard block of platinum-iridium alloy) of the kilogram, kept at the International Bureau of Weights and Measures at Sevres near Paris.

1.11 Second

The second is defined as the duration of 9 192 631 770 periods of the radiation corresponding to the transition between the two hyperfine levels of the ground state of the caesium – 133 atom.

1.12 Presentation of Units and their Values

The frequent changes in the present day life are facilitated by an international body known as International Standard Organisation (ISO) which makes recommendations regarding international standard procedures. The implementation of ISO recommendations, in a country, is assisted by its organisation appointed for the purpose. In India, Bureau of Indian Standards (BIS), has been created for this purpose. We have already discussed that the fundamental units in S.I. units for length, mass and time is metre, kilogram and second respectively. But in actual practice, it is not necessary to express all lengths in metres, all masses in kilograms and all times in seconds. We shall, sometimes, use the convenient units, which are multiples or divisions of our basic units in tens. As a typical example, although the metre is the unit of length, yet a smaller length of one-thousandth of a metre proves to be more convenient unit, especially in the dimensioning of drawings. Such convenient units are formed by using a prefix in the basic units to indicate the multiplier. The full list of these prefixes is given in the following table :

Table 1.3. Prefixes used in basic units.

Factor by which the unit is multiplied	*Standard form*	*Prefix*	*Abbreviation*
1 000 000 000 000	10^{12}	tera	T
1 000 000 000	10^{9}	giga	G
1 000 000	10^{6}	mega	M
1000	10^{3}	kilo	k
100	10^{2}	hecto*	h
10	10^{1}	deca*	da
0.1	10^{-1}	deci*	d
0.01	10^{-2}	centi*	c
0.001	10^{-3}	milli	m
0.000 001	10^{-6}	micro	μ
0.000 000 001	10^{-9}	nano	n
0.000 000 000 001	10^{-12}	pico	p

1.13 Rules for S.I. Units

The eleventh General Conference of Weights and Measures recommended only the fundamental and derived units of S.I. units. But it did not elaborate the rules for the usage of the units. Later on many scientists and engineers held a number of meetings for the style and usage of S.I. units. Some of the decisions of the meeting are :

1. For numbers having five or more digits, the digits should be placed in groups of three separated by spaces (instead of commas)** counting both to the left and right of the decimal point.
2. In a four*** digit number, the space is not required unless the four digit number is used in a column of numbers with five or more digits.
3. A dash is to be used to separate units that are multiplied together. For example, newton × metre is written as N-m. It should not be confused with mN, which stands for milli newton.
4. Plurals are never used with symbols. For example, metre or metres are written as m.
5. All symbols are written in small letters except the symbol derived from the proper names. For example, N for newton and W for watt.
6. The units with names of the scientists should not start with capital letter when written in full. For example, 90 newton and not 90 Newton.

At the time of writing this book, the authors sought the advice of various international authorities, regarding the use of units and their values. Keeping in view the international reputation of the authors, as well as international popularity of their books, it was decided to present **** units and

---

* These prefixes are generally becoming obsolete, probably due to possible confusion. Moreover it is becoming a conventional practice to use only those power of ten which conform to 10^{3x}, where x is a positive or negative whole number.

** In certain countries, comma is still used as the decimal mark

*** In certain countries, a space is used even in a four digit number.

**** In some of the question papers of the universities and other examining bodies standard values are not used. The authors have tried to avoid such questions in the text of the book. However, at certain places the questions with sub-standard values have to be included, keeping in view the merits of the question from the reader's angle.

their values as per recommendations of ISO and BIS. It was decided to use :

4500	not	4 500	or	4,500
75 890 000	not	75890000	or	7,58,90,000
0.012 55	not	0.01255	or	.01255
30×10^6	not	3,00,00,000	or	3×10^7

The above mentioned figures are meant for numerical values only. Now let us discuss about the units. We know that the fundamental units in S.I. system of units for length, mass and time are metre, kilogram and second respectively. While expressing these quantities, we find it time consuming to write the units such as metres, kilograms and seconds, in full, every time we use them. As a result of this, we find it quite convenient to use some standard abbreviations :

We shall use :

m	for metre or metres
km	for kilometre or kilometres
kg	for kilogram or kilograms
t	for tonne or tonnes
s	for second or seconds
min	for minute or minutes
N-m	for netwon × metres (*e.g.* work done)
kN-m	for kilonewton × metres
rev	for revolution or revolutions
rad	for radian or radians

1.14 Mass and Weight

Sometimes much confusion and misunderstanding is created, while using the various systems of units in the measurements of force and mass. This happens because of the lack of clear understanding of the difference between the mass and weight. The following definitions of mass and weight should be clearly understood :

Mass. It is the amount of matter contained in a given body and does not vary with the change in its position on the earth's surface. The mass of a body is measured by direct comparison with a standard mass by using a lever balance.

Weight. It is the amount of pull, which the earth exerts upon a given body. Since the pull varies with the distance of the body from the centre of the earth, therefore, the weight of the body will vary with its position on the earth's surface (say latitude and elevation). It is thus obvious, that the weight is a force.

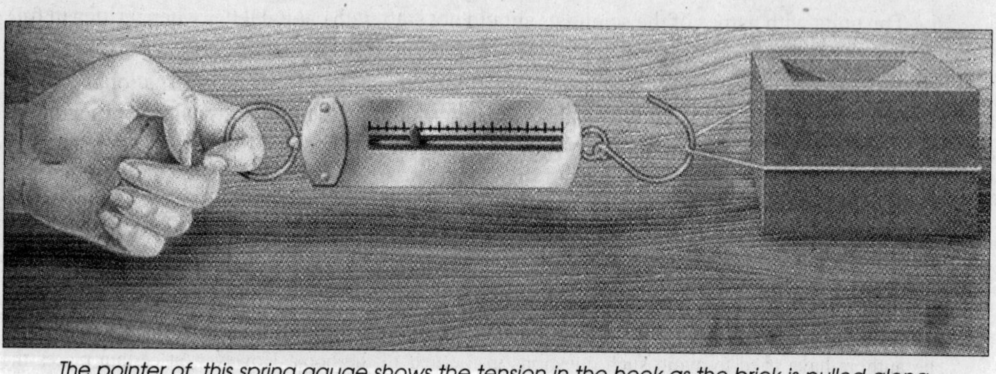

The pointer of this spring gauge shows the tension in the hook as the brick is pulled along.

The earth's pull in metric units at sea level and 45° latitude has been adopted as one force unit and named as one kilogram of force. Thus, it is a definite amount of force. But, unfortunately, has the same name as the unit of mass.

The weight of a body is measured by the use of a spring balance, which indicates the varying tension in the spring as the body is moved from place to place.

Note : The confusion in the units of mass and weight is eliminated to a great extent, in S.I units . In this system, the mass is taken in kg and the weight in newtons. The relation between mass (m) and weight (W) of a body is

$$W = m.g \quad \text{or} \quad m = W / g$$

where W is in newtons, m in kg and g is the acceleration due to gravity in m/s^2.

1.15 Inertia

It is that property of a matter, by virtue of which a body cannot move of itself nor change the motion imparted to it.

1.16 Laws of Motion

Newton has formulated three laws of motion, which are the basic postulates or assumptions on which the whole system of dynamics is based. Like other scientific laws, these are also justified as the results, so obtained, agree with the actual observations. Following are the three laws of motion :

1. ***Newton's First Law of Motion.*** It states, "*Every body continues in its state of rest or of uniform motion in a straight line, unless acted upon by some external force*". This is also known as *Law of Inertia.*

2. ***Newton's Second Law of Motion.*** It states, "*The rate of change of momentum is directly proportional to the impressed force and takes place in the same direction in which the force acts*".

3. ***Newton's Third Law of Motion.*** It states, "*To every action, there is always an equal and opposite reaction*".

1.17 Force

It is an important factor in the field of Engineering science, which may be defined as **an agent, which produces or tends to produce, destroy or tends to destroy motion.**

According to Newton's Second Law of Motion, the applied force or impressed force is directly proportional to the rate of change of momentum. We know that

Momentum = Mass × Velocity

Let m = Mass of the body,

u = Initial velocity of the body,

v = Final velocity of the body,

a = Constant acceleration, and

t = Time required to change velocity from u to v.

$\therefore$ Change of momentum $= mv - mu$

and rate of change of momentum

$$= \frac{mv - mu}{t} = \frac{m(v-u)}{t} = m.a \quad \ldots \left(\because \frac{v-u}{t} = a\right)$$

or Force, $F \propto ma$ or $F = k\,m\,a$

where k is a constant of proportionality.

For the sake of convenience, the unit of force adopted is such that it produces a unit acceleration to a body of unit mass.

$$\therefore \quad F = m.a = \text{Mass} \times \text{Acceleration}$$

In S.I. system of units, the unit of force is called **newton** (briefly written as N). A ***newton may be defined as the force, while acting upon a mass of one kg, produces an acceleration of 1 m/s² in the direction in which it acts.*** Thus

$$1\text{N} = 1\text{kg} \times 1 \text{ m/s}^2 = 1\text{kg-m/s}^2$$

Far away from Earth's gravity and its frictional forces, a spacecraft shows Newton's three laws of motion at work.

1.18 Absolute and Gravitational Units of Force

We have already discussed, that when a body of mass 1 kg is moving with an acceleration of 1 m/s^2, the force acting on the body is one newton (briefly written as 1 N). Therefore, when the same body is moving with an acceleration of 9.81 m/s^2, the force acting on the body is 9.81N. But we denote 1 kg mass, attracted towards the earth with an acceleration of 9.81 m/s^2 as 1 kilogram force (briefly written as kgf) or 1 kilogram weight (briefly written as kg-wt). It is thus obvious that

$$1\text{kgf} = 1\text{kg} \times 9.81 \text{ m/s}^2 = 9.81 \text{ kg-m/s}^2 = 9.81 \text{ N} \quad \ldots (\because 1\text{N} = 1\text{kg-m/s}^2)$$

The above unit of force *i.e.* kilogram force (kgf) is called ***gravitational*** or ***engineer's unit of force,*** whereas netwon is the ***absolute*** or ***scientific*** or ***S.I. unit of force.*** It is thus obvious, that the gravitational units are 'g' times the unit of force in the absolute or S. I. units.

It will be interesting to know that ***the mass of a body in absolute units is numerically equal to the weight of the same body in gravitational units.***

For example, consider a body whose mass, $m = 100$ kg.

∴ The force, with which it will be attracted towards the centre of the earth,

$$F = m.a = m.g = 100 \times 9.81 = 981 \text{ N}$$

Now, as per definition, we know that the weight of a body is the force, by which it is attracted towards the centre of the earth.

∴ Weight of the body,

$$W = 981 \text{ N} = \frac{981}{9.81} = 100 \text{ kgf} \quad \ldots (\because 1 \text{ kgf} = 9.81 \text{ N})$$

In brief, the weight of a body of mass m kg at a place where gravitational acceleration is 'g' m/s^2 is $m.g$ newtons.

1.19 Moment of Force

It is the turning effect produced by a force, on the body, on which it acts. The moment of a force is equal to the product of the force and the perpendicular distance of the point, about which the moment is required, and the line of action of the force. Mathematically,

$$\text{Moment of a force} = F \times l$$

where F = Force acting on the body, and

l = Perpendicular distance of the point and the line of action of the force (F) as shown in Fig. 1.2.

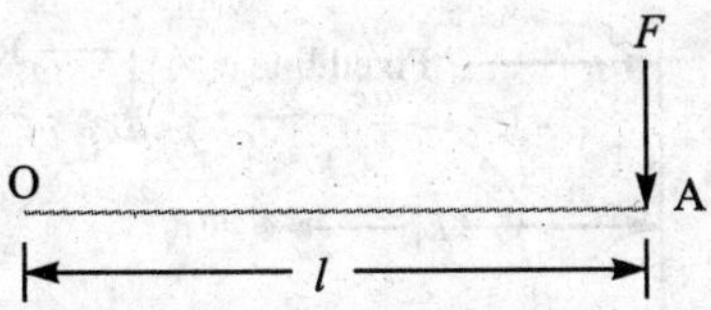

Fig. 1.2. Moment of a force.

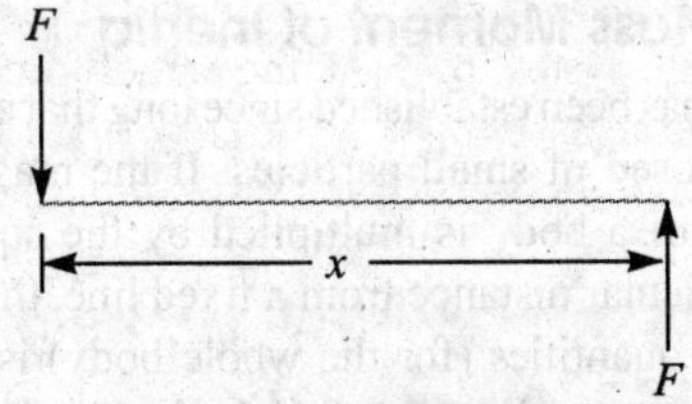

Fig. 1.3. Couple.

1.20 Couple

The two equal and opposite parallel forces, whose lines of action are different form a couple, as shown in Fig. 1.3.

The perpendicular distance (x) between the lines of action of two equal and opposite parallel forces is known as ***arm of the couple.*** The magnitude of the couple (*i.e.* moment of a couple) is the product of one of the forces and the arm of the couple. Mathematically,

$$\text{Moment of a couple} = F \times x$$

A little consideration will show, that a couple does not produce any translatory motion (*i.e.* motion in a straight line). But, a couple produces a motion of rotation of the body on which it acts.

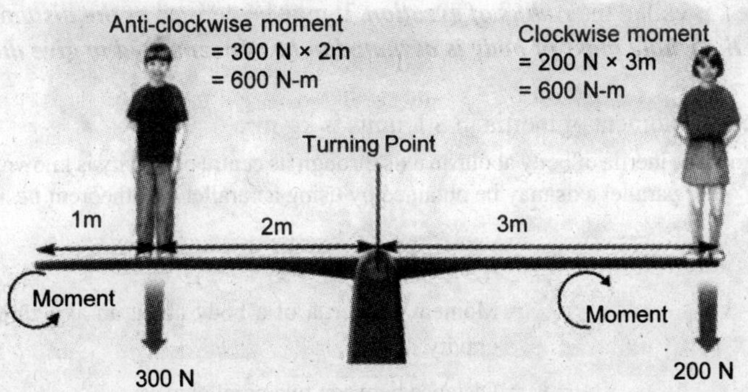

A see saw is balanced when the clockwise moment equals the anti-clockwise moment. The boy's weight is 300 newtons (300 N) and he stands 2 metres (2 m) from the pivot. He causes the anti-clockwise moment of 600 newton-metres (N-m). The girl is lighter (200 N) but she stands further from the pivot (3m). She causes a clockwise moment of 600 N-m, so the seesaw is balanced.

1.21 Mass Density

The mass density of the material is the mass per unit volume. The following table shows the mass densities of some common materials used in practice.

Table 1.4. Mass density of commonly used materials.

Material	*Mass density (kg/m³)*	*Material*	*Mass density (kg/m³)*
Cast iron	7250	Zinc	7200
Wrought iron	7780	Lead	11 400
Steel	7850	Tin	7400
Brass	8450	Aluminium	2700
Copper	8900	Nickel	8900
Cobalt	8850	Monel metal	8600
Bronze	8730	Molybdenum	10 200
Tungsten	19 300	Vanadium	6000

1.22 Mass Moment of Inertia

It has been established since long that a rigid body is composed of small particles. If the mass of every particle of a body is multiplied by the square of its perpendicular distance from a fixed line, then the sum of these quantities (for the whole body) is known as ***mass moment of inertia*** of the body. It is denoted by *I*.

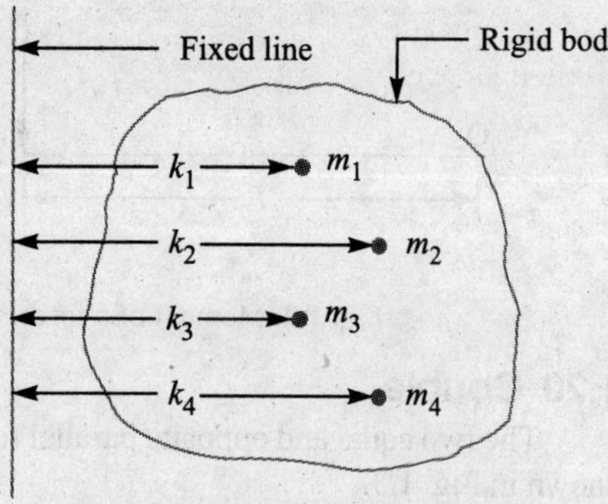

Fig. 1.4. Mass moment of inertia.

Consider a body of total mass *m*. Let it be composed of small particles of masses m_1, m_2, m_3, m_4, etc. If k_1, k_2, k_3, k_4, etc., are the distances from a fixed line, as shown in Fig. 1.4, then the mass moment of inertia of the whole body is given by

$$I = m_1 (k_1)^2 + m_2 (k_2)^2 + m_3 (k_3)^2 + m_4 (k_4)^2 + ...$$

If the total mass of a body may be assumed to concentrate at one point (known as centre of mass or centre of gravity), at a distance *k* from the given axis, such that

$$mk^2 = m_1 (k_1)^2 + m_2 (k_2)^2 + m_3 (k_3)^2 + m_4 (k_4)^2 + ...$$

then

$$I = m\, k^2$$

The distance *k* is called the ***radius of gyration.*** It may be defined ***as the distance, from a given reference, where the whole mass of body is assumed to be concentrated to give the same value of I.***

The unit of mass moment of inertia in S.I. units is kg-m².

Notes : 1. If the moment of inertia of body about an axis through its centre of gravity is known, then the moment of inertia about any other parallel axis may be obtained by using a parallel axis theorem *i.e.* moment of inertia about a parallel axis,

$$I_p = I_G + mh^2$$

where I_G = Moment of inertia of a body about an axis through its centre of gravity, and

h = Distance between two parallel axes.

2. The following are the values of *I* for simple cases :

(*a*) The moment of inertia of a thin disc of radius *r*, about an axis through its centre of gravity and perpendicular to the plane of the disc is,

$$I = mr^2/2 = 0.5\, mr^2$$

and moment of inertia about a diameter,

$$I = mr^2/4 = 0.25\, mr^2$$

(*b*) The moment of inertia of a thin rod of length *l,* about an axis through its centre of gravity and perpendicular to its length,

$$I_G = ml^2/12$$

and moment of inertia about a parallel axis through one end of a rod,

$$I_P = ml^2/3$$

3. The moment of inertia of a solid cylinder of radius *r* and length *l,* about the longitudinal axis or polar axis

$$= mr^2/2 = 0.5\, mr^2$$

and moment of inertia through its centre perpendicular to the longitudinal axis

$$= m\left(\frac{r^2}{4} + \frac{l^2}{12}\right)$$

1.23 Angular Momentum

It is the product of the mass moment of inertia and the angular velocity of the body. Mathematically,

$$\text{Angular momentum} = I.\omega$$

where I = Mass moment of inertia, and

ω = Angular velocity of the body.

1.24 Torque

It may be defined as the product of force and the perpendicular distance of its line of action from the given point or axis. A little consideration will show that the torque is equivalent to a couple acting upon a body.

The Newton's second law of motion when applied to rotating bodies states, the ***torque is directly proportional to the rate of change of angular momentum.*** Mathematically,

$$\text{Torque, } T \propto \frac{d(I\omega)}{dt}$$

Since I is constant, therefore,

$$T = I \times \frac{d\omega}{dt} = I.\alpha$$

$$\dots \left[\because \frac{d\omega}{dt} = \text{Angular acceleration } (\alpha) \right]$$

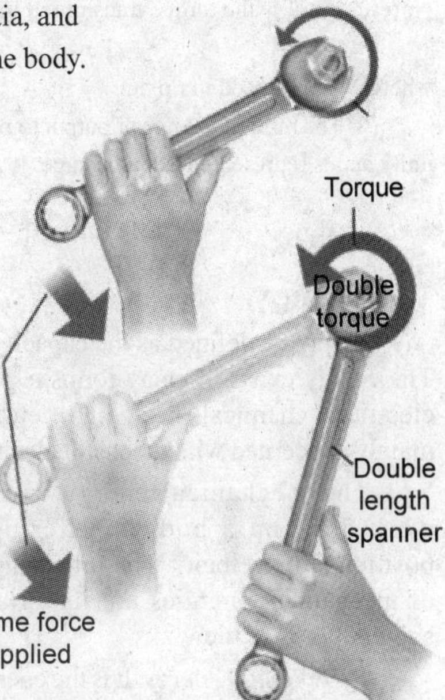

Same force applied at double the length, doubles the torque.

1.25 Work

Whenever a force acts on a body and the body undergoes a displacement in the direction of the force, then work is said to be done. For example, if a force F acting on a body causes a displacement x of the body in the direction of the force, then

$$\text{Work done} = \text{Force} \times \text{Displacement} = F \times x$$

If the force varies linearly from zero to a maximum value of F, then

$$\text{Work done} = \frac{0 + F}{2} \times x = \frac{F}{2} \times x$$

When a couple or torque (T) acting on a body causes the angular displacement (θ) about an axis perpendicular to the plane of the couple, then

$$\text{Work done} = \text{Torque} \times \text{Angular displacement} = T.\theta$$

The unit of work depends upon the units of force and displacement. In S. I. system of units, the practical unit of work is N-m. It is the work done by a force of 1 newton, when it displaces a body through 1 metre. The work of 1 N-m is known as joule (briefly written as J), such that 1 N-m = 1 J.

Note : While writing the unit of work, it is a general practice to put the units of force first followed by the units of displacement (*e.g.* N-m).

1.26 Power

It may be defined as the rate of doing work or work done per unit time. Mathematically,

$$\text{Power, } P = \frac{\text{Work done}}{\text{Time taken}}$$

In S.I system of units, the unit of power is watt (briefly written as W) which is equal to 1 J/s or 1N-m/s. Thus, the power developed by a force of F (in newtons) moving with a velocity v m/s is $F.v$ watt. Generally, a bigger unit of power called kilowatt (briefly written as kW) is used which is equal to 1000 W

Notes : 1. If T is the torque transmitted in N-m or J and ω is angular speed in rad/s, then

$$\text{Power, } P = T.\omega = T \times 2\,\pi\,N/60 \text{ watts} \qquad \ldots (\because \omega = 2\,\pi\,N/60)$$

where N is the speed in r.p.m.

2. The ratio of the power output to power input is known as ***efficiency*** of a machine. It is always less than unity and is represented as percentage. It is denoted by a Greek letter eta (η). Mathematically,

$$\text{Efficiency, } \eta = \frac{\text{Power output}}{\text{Power input}}$$

1.27 Energy

It may be defined as the capacity to do work. The energy exists in many forms *e.g.* mechanical, electrical, chemical, heat, light, etc. But we are mainly concerned with mechanical energy.

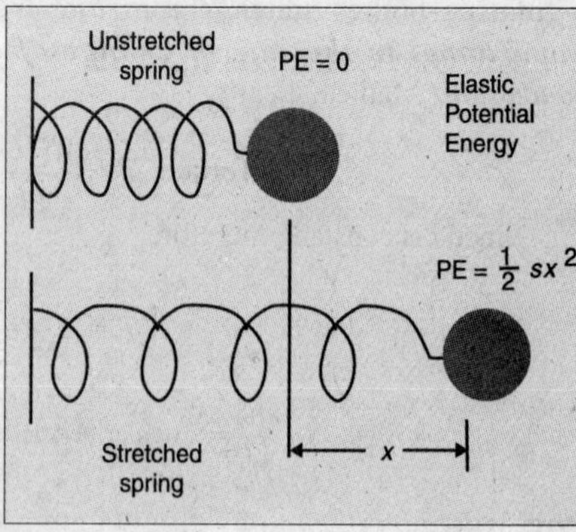

The mechanical energy is equal to the work done on a body in altering either its position or its velocity. The following three types of mechanical energies are important from the subject point of view :

1. ***Potential energy.*** It is the energy possessed by a body, for doing work, by virtue of its position. For example, a body raised to some height above the ground level possesses potential energy, because it can do some work by falling on earth's surface.

Let W = Weight of the body,

m = Mass of the body, and

h = Distance through which the body falls.

∴ Potential energy,

$$\text{P.E.} = W.\,h = m.g.h$$

It may be noted that

(*a*) When W is in newtons and h in metres, then potential energy will be in N-m.

(*b*) When m is in kg and h in metres, then the potential energy will also be in N-m as discussed below :

We know that potential energy

$$= m.g.h = \text{kg} \times \frac{\text{m}}{\text{s}^2} \times \text{m} = \text{N-m} \qquad \ldots \left(\because 1\text{N} = \frac{1 \text{ kg-m}}{\text{s}^2}\right)$$

2. ***Strain energy.*** It is the potential energy stored by an elastic body when deformed. A compressed spring possesses this type of energy, because it can do some work in recovering its original shape. Thus, if a compressed spring of stiffness (s) N per unit deformation (*i.e.* extension or compression) is deformed through a distance x by a weight W, then

$$\text{Strain energy} = \text{Work done} = \frac{1}{2}\,W.x = \frac{1}{2}\,s.x^2 \qquad \ldots (\because W = s.x)$$

In case of a torsional spring of stiffness (q) N-m per unit angular deformation when twisted through an angle θ radians, then

$$\text{Strain energy} = \text{Work done} = \frac{1}{2}\, q.\theta^2$$

3. ***Kinetic energy.*** It is the energy possessed by a body, for doing work, by virtue of its mass and velocity of motion. If a body of mass m attains a velocity v from rest in time t, under the influence of a force F and moves a distance s, then

$$\text{Work done} = F.s = m.a.s \qquad ...(\because F = m.a)$$

$\therefore$ Kinetic energy of the body or the kinetic energy of translation,

$$\text{K.E.} = m.a.s = m \times a \times \frac{{}^{*}v^2}{2a} = \frac{1}{2}\, mv^2$$

It may be noted that when m is in kg and v in m/s, then kinetic energy will be in N-m as discussed below :

We know that kinetic energy,

$$\text{K.E.} = \frac{1}{2}\, mv^2 = \text{kg} \times \frac{\text{m}^2}{\text{s}^2} = \frac{\text{kg - m}}{\text{s}^2} \times \text{m} = \text{N-m} \; ... \left(\because 1\text{N} = \frac{1\text{ kg-m}}{\text{s}^2}\right)$$

Notes : 1. When a body of mass moment of inertia I (about a given axis) is rotated about that axis, with an angular velocity ω, then it possesses some kinetic energy. In this case,

$$\text{Kinetic energy of rotation} = \frac{1}{2}\, I.\omega^2$$

2. When a body has both linear and angular motions, *e.g.* wheels of a moving car, then the total kinetic energy of the body is equal to the sum of linear and angular kinetic energies.

$$\therefore \qquad \text{Total kinetic energy} = \frac{1}{2}\, m.v^2 + \frac{1}{2}\, I.\omega^2$$

3. The energy can neither be created nor destroyed, though it can be transformed from one form into any of the forms, in which energy can exist. This statement is known as **'*Law of Conservation of Energy*'**.

4. The loss of energy in any one form is always accompanied by an equivalent increase in another form. When work is done on a rigid body, the work is converted into kinetic or potential energy or is used in overcoming friction. If the body is elastic, some of the work will also be stored as strain energy.

---

* We know that $v^2 - u^2 = 2\, a.s$

Since the body starts from rest (*i.e.* $u = 0$), therefore,

$$v^2 = 2\, a.s \quad \text{or} \quad s = v^2 / 2a$$

CHAPTER 2

Engineering Materials and their Properties

2.1 Introduction

The knowledge of materials and their properties is of great significance for a design engineer. The machine elements should be made of such a material which has properties suitable for the conditions of operation. In addition to this, a design engineer must be familiar with the effects which the manufacturing processes and heat treatment have on the properties of the materials. In this chapter, we shall discuss the commonly used engineering materials and their properties in Machine Design.

2.2 Classification of Engineering Materials

The engineering materials are mainly classified as :

1. Metals and their alloys, such as iron, steel, copper, aluminium, etc.
2. Non-metals, such as glass, rubber, plastic, etc.

The metals may be further classified as :

(*a*) Ferrous metals, and (*b*) Non-ferrous metals.

The **ferrous metals* are those which have the iron as their main constituent, such as cast iron, wrought iron and steel.

The ***non-ferrous*** metals are those which have a metal other than iron as their main constituent, such as copper, aluminium, brass, tin, zinc, etc.

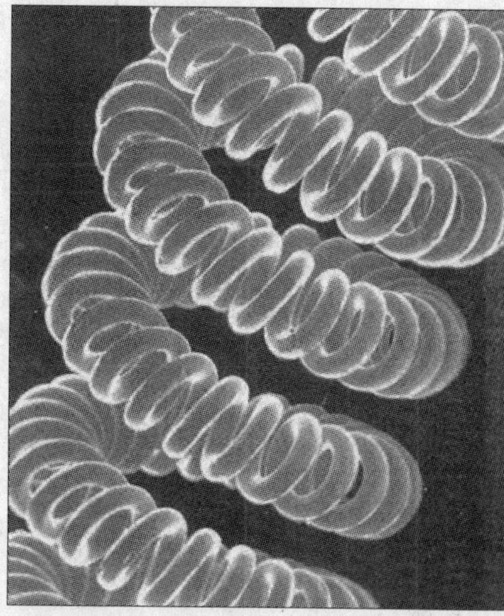

A filament of bulb needs a material like tungsten which can withstand high temperatures without undergoing deformation.

2.3 Selection of Materials for Engineering Purposes

The selection of a proper material, for engineering purposes, is one of the most difficult problem for the designer. The best material is one which serve the desired objective at the minimum cost. The following factors should be considered while selecting the material :

1. Availability of the materials,
2. Suitability of the materials for the working conditions in service, and
3. The cost of the materials.

The important properties, which determine the utility of the material are physical, chemical and mechanical properties. We shall now discuss the physical and mechanical properties of the material in the following articles.

Valuable Metals

2.4 Physical Properties of Metals

The physical properties of the metals include luster, colour, size and shape, density, electric and thermal conductivity, and melting point. The following table shows the important physical properties of some pure metals.

---

* The word 'ferrous' is derived from a latin word 'ferrum' which means iron.

Table 2.1. Physical properties of metals.

Metal	*Density* (*kg/m*^{3})	*Melting point* (°C)	*Thermal conductivity* (*W/m°C*)	*Coefficient of linear expansion at 20°C (μm/m/°C)*
Aluminium	2700	660	220	23.0
Brass	8450	950	130	16.7
Bronze	8730	1040	67	17.3
Cast iron	7250	1300	54.5	9.0
Copper	8900	1083	393.5	16.7
Lead	11 400	327	33.5	29.1
Monel metal	8600	1350	25.2	14.0
Nickel	8900	1453	63.2	12.8
Silver	10 500	960	420	18.9
Steel	7850	1510	50.2	11.1
Tin	7400	232	67	21.4
Tungsten	19 300	3410	201	4.5
Zinc	7200	419	113	33.0
Cobalt	8850	1490	69.2	12.4
Molybdenum	10 200	2650	13	4.8
Vanadium	6000	1750	—	7.75

2.5 Mechanical Properties of Metals

The mechanical properties of the metals are those which are associated with the ability of the material to resist mechanical forces and load. These mechanical properties of the metal include strength, stiffness, elasticity, plasticity, ductility, brittleness, malleability, toughness, resilience, creep and hardness. We shall now discuss these properties as follows:

1. ***Strength.*** It is the ability of a material to resist the externally applied forces without breaking or yielding. The internal resistance offered by a part to an externally applied force is called *stress.

2. ***Stiffness.*** It is the ability of a material to resist deformation under stress. The modulus of elasticity is the measure of stiffness.

3. ***Elasticity.*** It is the property of a material to regain its original shape after deformation when the external forces are removed. This property is desirable for materials used in tools and machines. It may be noted that steel is more elastic than rubber.

4. ***Plasticity.*** It is property of a material which retains the deformation produced under load permanently. This property of the material is necessary for forgings, in stamping images on coins and in ornamental work.

5. ***Ductility.*** It is the property of a material enabling it to be drawn into wire with the application of a tensile force. A ductile material must be both strong and plastic. The ductility is usually measured by the terms, percentage elongation and percentage reduction in area. The ductile material commonly used in engineering practice (in order of diminishing ductility) are mild steel, copper, aluminium, nickel, zinc, tin and lead.

Note : The ductility of a material is commonly measured by means of percentage elongation and percentage reduction in area in a tensile test. (Refer Chapter 4, Art. 4.11).

* For further details, refer Chapter 4 on Simple Stresses in Machine Parts.

6. ***Brittleness.*** It is the property of a material opposite to ductility. It is the property of breaking of a material with little permanent distortion. Brittle materials when subjected to tensile loads, snap off without giving any sensible elongation. Cast iron is a brittle material.

7. ***Malleability.*** It is a special case of ductility which permits materials to be rolled or hammered into thin sheets. A malleable material should be plastic but it is not essential to be so strong. The malleable materials commonly used in engineering practice (in order of diminishing malleability) are lead, soft steel, wrought iron, copper and aluminium.

8. ***Toughness.*** It is the property of a material to resist fracture due to high impact loads like hammer blows. The toughness of the material decreases when it is heated. It is measured by the amount of energy that a unit volume of the material has absorbed after being stressed upto the point of fracture. This property is desirable in parts subjected to shock and impact loads.

9. ***Machinability.*** It is the property of a material which refers to a relative case with which a material can be cut. The machinability of a material can be measured in a number of ways such as comparing the tool life for cutting different materials or thrust required to remove the material at some given rate or the energy required to remove a unit volume of the material. It may be noted that brass can be easily machined than steel.

10. ***Resilience.*** It is the property of a material to absorb energy and to resist shock and impact loads. It is measured by the amount of energy absorbed per unit volume within elastic limit. This property is essential for spring materials.

11. ***Creep.*** When a part is subjected to a constant stress at high temperature for a long period of time, it will undergo a slow and permanent deformation called ***creep.*** This property is considered in designing internal combustion engines, boilers and turbines.

12. ***Fatigue.*** When a material is subjected to repeated stresses, it fails at stresses below the yield point stresses. Such type of failure of a material is known as ****fatigue***. The failure is caused by means of a progressive crack formation which are usually fine and of microscopic size. This property is considered in designing shafts, connecting rods, springs, gears, etc.

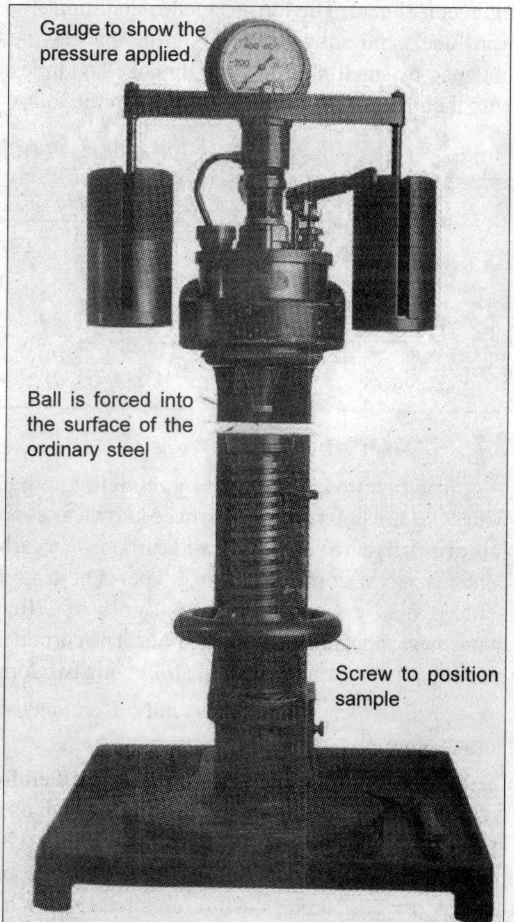

Brinell Tester : Hardness can be defined as the resistance of a metal to attempts to deform it. This machine invented by the Swedish metallurgist Johann August Brinell (1849-1925), measure hardness precisely.

13. ***Hardness.*** It is a very important property of the metals and has a wide variety of meanings. It embraces many different properties such as resistance to wear, scratching, deformation and machinability etc. It also means the ability of a metal to cut another metal. The hardness is usually

* For further details, refer Chapter 6 (Art. 6.3) on Variable Stresses in Machine Parts.

expressed in numbers which are dependent on the method of making the test. The hardness of a metal may be determined by the following tests :

(*a*) Brinell hardness test,

(*b*) Rockwell hardness test,

(*c*) Vickers hardness (also called Diamond Pyramid) test, and

(*d*) Shore scleroscope.

2.6 Ferrous Metals

We have already discussed in Art. 2.2 that the ferrous metals are those which have iron as their main constituent. The ferrous metals commonly used in engineering practice are cast iron, wrought iron, steels and alloy steels. The principal raw material for all ferrous metals is pig iron which is obtained by smelting iron ore with coke and limestone, in the blast furnace. The principal iron ores with their metallic contents are shown in the following table :

Table 2.2. Principal iron ores.

Iron ore	*Chemical formula*	*Colour*	*Iron content (%)*
Magnetite	Fe_2O_3	Black	72
Haematite	Fe_3O_4	Red	70
Limonite	$FeCO_3$	Brown	60–65
Siderite	Fe_2O_3 (H_2O)	Brown	48

2.7 Cast Iron

The cast iron is obtained by re-melting pig iron with coke and limestone in a furnace known as cupola. It is primarily an alloy of iron and carbon. The carbon contents in cast iron varies from 1.7 per cent to 4.5 per cent. It also contains small amounts of silicon, manganese, phosphorous and sulphur. The carbon in a cast iron is present in either of the following two forms:

1. Free carbon or graphite, and 2. Combined carbon or cementite.

Since the cast iron is a brittle material, therefore, it cannot be used in those parts of machines which are subjected to shocks. The properties of cast iron which make it a valuable material for engineering purposes are its low cost, good casting characteristics, high compressive strength, wear resistance and excellent machinability. The compressive strength of cast iron is much greater than the tensile strength. Following are the values of ultimate strength of cast iron :

Tensile strength = 100 to 200 MPa*

Compressive strength = 400 to 1000 MPa

Shear strength = 120 MPa

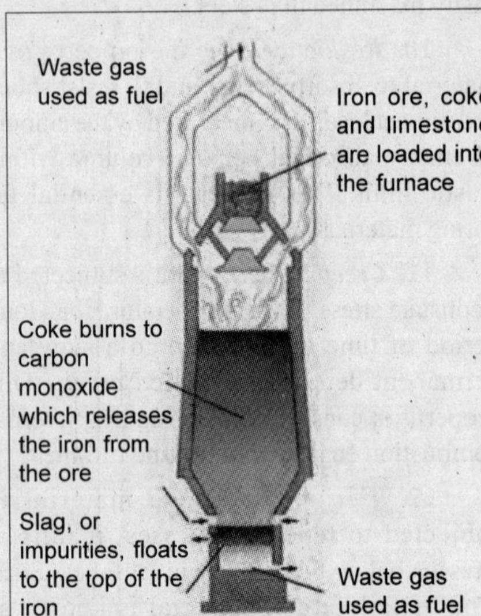

Smelting : Ores consist of non-metallic elements like oxygen or sulphur combined with the wanted metal. Iron is separated from the oxygen in its ore heating it with carbon monoxide derived from coke (a form of carbon made from coal). Limestone is added to keep impurities liquid so that the iron can separate from them.

* $1 MPa = 1 MN/m^2 = 1 \times 10^6 N/m^2 = 1 N/mm^2$

2.8 Types of Cast Iron

The various types of cast iron in use are discussed as follows :

Haematite is an ore of iron. It often forms kidney-shaped lumps, These give the ore its nickname of kidney ore.

1. ***Grey cast iron.*** It is an ordinary commercial iron having the following compositions :

Carbon = 3 to 3.5%; Silicon = 1 to 2.75%; Manganese = 0.40 to 1.0%; Phosphorous = 0.15 to 1% ; Sulphur = 0.02 to 0.15% ; and the remaining is iron.

The grey colour is due to the fact that the carbon is present in the form of *free graphite. It has a low tensile strength, high compressive strength and no ductility. It can be easily machined. A very good property of grey cast iron is that the free graphite in its structure acts as a lubricant. Due to this reason, it is very suitable for those parts where sliding action is desired. The grey iron castings are widely used for machine tool bodies, automotive cylinder blocks, heads, housings, fly-wheels, pipes and pipe fittings and agricultural implements.

Table 2.3. Grey iron castings, as per IS : 210 – 1993.

IS Designation	*Tensile strength (MPa or N/mm²)*	*Brinell hardness number (B.H.N.)*
FG 150	150	130 to 180
FG 200	200	160 to 220
FG 220	220	180 to 220
FG 260	260	180 to 230
FG 300	300	180 to 230
FG 350	350	207 to 241
FG 400	400	207 to 270

According to Indian standard specifications (IS: 210 – 1993), the grey cast iron is designated by the alphabets 'FG' followed by a figure indicating the minimum tensile strength in MPa or N/mm^2. For example, 'FG 150' means grey cast iron with 150 MPa or N/mm^2 as minimum tensile strength. The seven recommended grades of grey cast iron with their tensile strength and Brinell hardness number (B.H.N) are given in Table 2.3.

2. ***White cast iron.*** The white cast iron shows a white fracture and has the following approximate compositions :

Carbon = 1.75 to 2.3% ; Silicon = 0.85 to 1.2% ; Manganese = less than 0.4% ; Phosphorus = less than 0.2% ; Sulphur = less than 0.12%, and the remaining is iron.

The white colour is due to fact that it has no graphite and whole of the carbon is in the form of carbide (known as cementite) which is the hardest constituent of iron. The white cast iron has a high tensile strength and a low compressive strength. Since it is hard, therefore, it cannot be machined with ordinary cutting tools but requires grinding as shaping process. The white cast iron may be produced by casting against metal chills or by regulating analysis. The chills are used when a hard, wear resisting surface is desired for such products as for car wheels, rolls for crushing grains and jaw crusher plates.

3. ***Chilled cast iron.*** It is a white cast iron produced by quick cooling of molten iron. The quick cooling is generally called chilling and the cast iron so produced is called chilled cast iron. All castings

* When filing or machining cast iron makes our hands black, then it shows that free graphite is present in cast iron.

are chilled at their outer skin by contact of the molten iron with the cool sand in the mould. But on most castings, this hardness penetrates to a very small depth (less than 1 mm). Sometimes, a casting is chilled intentionally and sometimes chilled becomes accidently to a considerable depth. The intentional chilling is carried out by putting inserts of iron or steel (chills) into the mould. When the molten metal comes into contact with the chill, its heat is readily conducted away and the hard surface is formed. Chills are used on any faces of a casting which are required to be hard to withstand wear and friction.

4. *Mottled cast iron.* It is a product in between grey and white cast iron in composition, colour and general properties. It is obtained in castings where certain wearing surfaces have been chilled.

5. *Malleable cast iron.* The malleable iron is a cast iron-carbon alloy which solidifies in the as-cast condition in a graphite free structure, *i.e.* total carbon content is present in its combined form as cementite (Fe_3C).

It is ductile and may be bent without breaking or fracturing the section. The tensile strength of the malleable cast iron is usually higher than that of grey cast iron and has excellent machining qualities. It is used for machine parts for which the steel forgings would be too expensive and in which the metal should have a fair degree of accuracy, *e.g.* hubs of wagon wheels, small fittings for railway rolling stock, brake supports, parts of agricultural machinery, pipe fittings, door hinges, locks etc.

In order to obtain a malleable iron castings, it is first cast into moulds of white cast iron. Then by a suitable heat treatment (*i.e.* annealing), the combined carbon of the white cast iron is separated into nodules of graphite. The following two methods are used for this purpose :

1. Whiteheart process, and **2.** Blackheart process.

In a ***whiteheart process,*** the white iron castings are packed in iron or steel boxes surrounded by a mixture of new and used haematite ore. The boxes are slowly heated to a temperature of 900 to 950°C and maintained at this temperature for several days. During this period, some of the carbon is oxidised out of the castings and the remaining carbon is dispersed in small specks throughout the structure. The heating process is followed by the cooling process which takes several more days. The result of this heat treatment is a casting which is tough and will stand heat treatment without fracture.

In a ***blackheart process,*** the castings used contain less carbon and sulphur. They are packed in a neutral substance like sand and the reduction of sulphur helps to accelerate the process. The castings are heated to a temperature of 850 to 900°C and maintained at that temperature for 3 to 4 days. The carbon in this process transforms into globules, unlike whiteheart process. The castings produced by this process are more malleable.

Notes : (*a*) According to Indian standard specifications (*IS : 14329 – 1995), the malleable cast iron may be either whiteheart, blackheart or pearlitic, according to the chemical composition, temperature and time cycle of annealing process.

(*b*) The ***whiteheart malleable cast iron*** obtained after annealing in a decarburizing atmosphere have a silvery-grey fracture with a heart dark grey to black. The microstructure developed in a section depends upon the size of the section. In castings of small sections, it is mainly ferritic with certain amount of pearlite. In large sections, microstructure varies from the surface to the core as follows :

Core and intermediate zone : Pearlite + ferrite + temper carbon

Surface zone : Ferrite.

The microstructure shall not contain flake graphite.

---

* This standard (IS : 14329-1995) supersedes the previous three standards, *i.e.*

(*a*) IS : 2107–1977 for white heart malleable iron casting,

(*b*) IS : 2108–1977 for black heart malleable iron casting, and

(*c*) IS : 2640–1977 for pearlitic malleable iron casting.

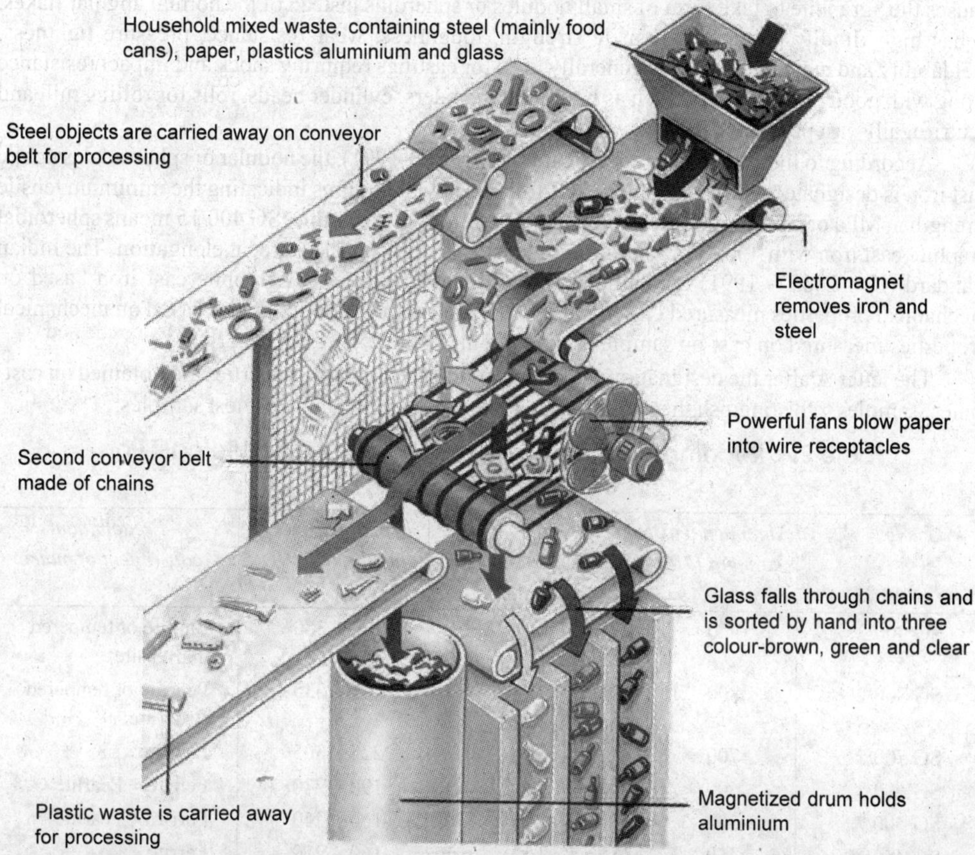

In a modern materials recovery plant, mixed waste (but no organic matter) is passed along a conveyor belt and sorted into reusable materials-steel, aluminium, paper, glass. Such recycling plants are expensive, but will become essential as vital resources become scarce.

Note : This picture is given as additional information and is not a direct example of the current chapter.

(*c*) The ***blackheart malleable cast iron*** obtained after annealing in an inert atmosphere have a black fracture. The microstructure developed in the castings has a matrix essentially of ferrite with temper carbon and shall not contain flake graphite.

(*d*) The ***pearlitic malleable cast iron*** obtained after heat-treatment have a homogeneous matrix essentially of pearlite or other transformation products of austenite. The graphite is present in the form of temper carbon nodules. The microstructure shall not contain flake graphite.

(*e*) According to IS: 14329 – 1995, the whiteheart, blackheart and pearlitic malleable cast irons are designated by the alphabets WM, BM and PM respectively. These designations are followed by a figure indicating the minimum tensile strength in MPa or N/mm^2. For example 'WM 350' denotes whiteheart malleable cast iron with 350 MPa as minimum tensile strength. The following are the different grades of malleable cast iron :

Whiteheart malleable cast iron — WM 350 and WM 400

Blackheart malleable cast iron — BM 300 ; BM 320 and BM 350

Pearlitic malleable cast iron — PM 450 ; PM 500 ; PM 550 ; PM 600 and PM 700

6. ***Nodular or spheroidal graphite cast iron.*** The nodular or spheroidal graphite cast iron is also called ***ductile cast iron*** or ***high strength cast iron***. This type of cast iron is obtained by adding small amounts of magnesium (0.1 to 0.8%) to the molten grey iron. The addition of magnesium causes the *graphite to take form of small nodules or spheroids instead of the normal angular flakes. It has high fluidity, castability, tensile strength, toughness, wear resistance, pressure tightness, weldability and machinability. It is generally used for castings requiring shock and impact resistance along with good machinability, such as hydraulic cylinders, cylinder heads, rolls for rolling mill and centrifugally cast products.

According to Indian standard specification (IS : 1865-1991), the nodular or spheroidal graphite cast iron is designated by the alphabets 'SG' followed by the figures indicating the minimum tensile strength in MPa or N/mm^2 and the percentage elongation. For example, SG 400/15 means spheroidal graphite cast iron with 400 MPa as minimum tensile strength and 15 percent elongation. The Indian standard (IS : 1865 – 1991) recommends nine grades of spheroidal graphite cast iron based on mechanical properties measured on separately-cast test samples and six grades based on mechanical properties measured on cast-on sample as given in the Table 2.4.

The letter A after the designation of the grade indicates that the properties are obtained on cast-on test samples to distinguish them from those obtained on separately-cast test samples.

Table 2.4. Recommended grades of spheroidal graphite cast iron as per IS : 1865–1991.

Grade	*Minimum tensile strength (MPa)*	*Minimum percentage elongation*	*Brinell hardness number (BHN)*	*Predominant constituent of matrix*
SG 900/2	900	2	280 – 360	Bainite or tempered martensite
SG 800/2	800	2	245 – 335	Pearlite or tempered structure
SG 700/2	700	2	225 – 305	Pearlite
SG 600/3	600	3	190 – 270	Ferrite + Pearlite
SG 500/7	500	7	160 – 240	Ferrite + Pearlite
SG 450/10	450	10	160 – 210	Ferrite
SG 400/15	400	15	130 – 180	Ferrite
SG 400/18	400	18	130 – 180	Ferrite
SG 350/22	350	22	≤ 150	Ferrite
SG 700/2A	700	2	220 – 320	Pearlite
SG 600/3A	600	2	180 – 270	Pearlite + Ferrite
SG 500/7A	450	7	170 – 240	Pearlite + Ferrite
SG 400/15A	390	15	130 – 180	Ferrite
SG 400/18A	390	15	130 – 180	Ferrite
SG 350/22A	330	18	≤ 150	Ferrite

2.9 Alloy Cast Iron

The cast irons as discussed in Art. 2.8 contain small percentages of other constituents like silicon, manganese, sulphur and phosphorus. These cast irons may be called as **plain cast irons.** The alloy cast iron is produced by adding alloying elements like nickel, chromium, molybdenum, copper and manganese in sufficient quantities. These alloying elements give more strength and result in improvement of properties. The alloy cast iron has special properties like increased strength, high wear resistance, corrosion resistance or heat resistance. The alloy cast irons are extensively used for

* The graphite flakes in cast iron act as discontinuities in the matrix and thus lower its mechanical properties. The sharp corners of the flakes also act as stress raisers. The weakening effect of the graphite can be reduced by changing its form from a flake to a spheroidal form.

gears, automobile parts like cylinders, pistons, piston rings, crank cases, crankshafts, camshafts, sprockets, wheels, pulleys, brake drums and shoes, parts of crushing and grinding machinery etc.

2.10 Effect of Impurities on Cast Iron

We have discussed in the previous articles that the cast iron contains small percentages of silicon, sulphur, manganese and phosphorous. The effect of these impurities on the cast iron are as follows:

1. ***Silicon.*** It may be present in cast iron upto 4%. It provides the formation of free graphite which makes the iron soft and easily machinable. It also produces sound castings free from blow-holes, because of its high affinity for oxygen.

2. ***Sulphur.*** It makes the cast iron hard and brittle. Since too much sulphur gives unsound casting, therefore, it should be kept well below 0.1% for most foundry purposes.

3. ***Manganese.*** It makes the cast iron white and hard. It is often kept below 0.75%. It helps to exert a controlling influence over the harmful effect of sulphur.

4. ***Phosphorus.*** It aids fusibility and fluidity in cast iron, but induces brittleness. It is rarely allowed to exceed 1%. Phosphoric irons are useful for casting of intricate design and for many light engineering castings when cheapness is essential.

Phosphorus is a non-metallic element. It must be stored underwater (above), since it catches fire when exposed to air, forming a compound.

2.11 Wrought Iron

It is the purest iron which contains at least 99.5% iron but may contain upto 99.9% iron. The typical composition of a wrought iron is

Carbon = 0.020%, Silicon = 0.120%, Sulphur = 0.018%, Phosphorus = 0.020%, Slag = 0.070%, and the remaining is iron.

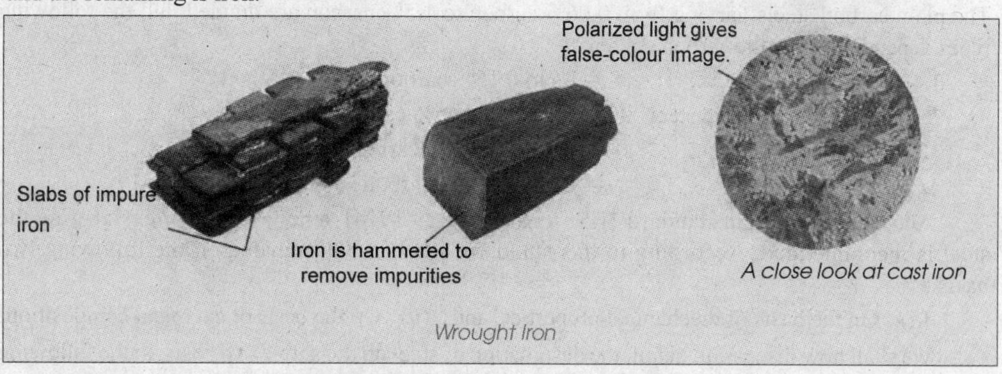

Wrought Iron

The wrought iron is produced from pig iron by remelting it in the puddling furnace of reverberatory type. The molten metal free from impurities is removed from the furnace as a pasty mass of iron and slag. The balls of this pasty mass, each about 45 to 65 kg are formed. These balls are then mechanically worked both to squeeze out the slag and to form it into some commercial shape.

The wrought iron is a tough, malleable and ductile material. It cannot stand sudden and excessive shocks. Its ultimate tensile strength is 250 MPa to 500 MPa and the ultimate compressive strength is 300 MPa.

It can be easily forged or welded. It is used for chains, crane hooks, railway couplings, water and steam pipes.

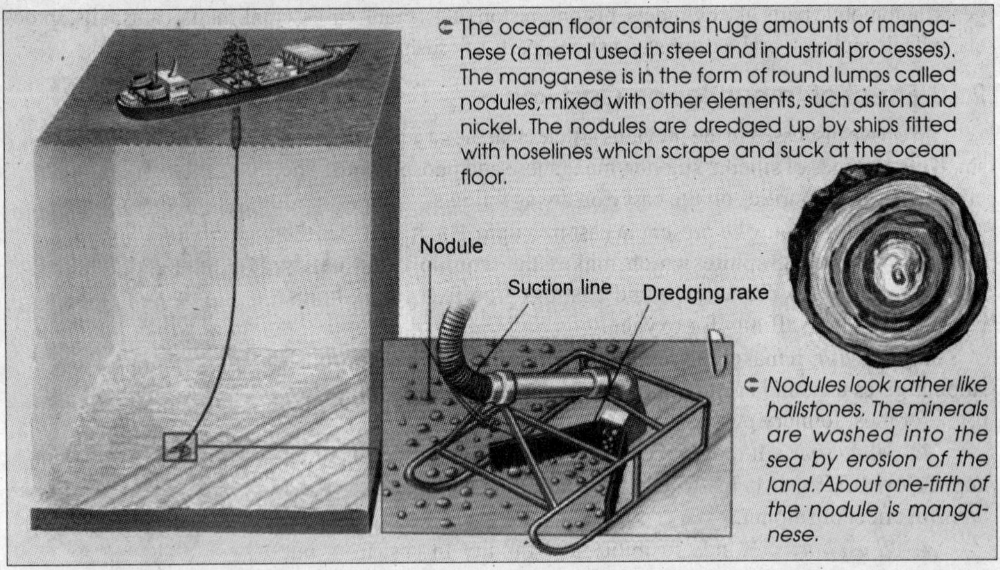

Note : This picture is given as additional information and is not a direct example of the current chapter.

2.12 Steel

It is an alloy of iron and carbon, with carbon content up to a maximum of 1.5%. The carbon occurs in the form of iron carbide, because of its ability to increase the hardness and strength of the steel. Other elements *e.g.* silicon, sulphur, phosphorus and manganese are also present to greater or lesser amount to impart certain desired properties to it. Most of the steel produced now-a-days is ***plain carbon steel*** or simply ***carbon steel.*** A carbon steel is defined as a steel which has its properties mainly due to its carbon content and does not contain more than 0.5% of silicon and 1.5% of manganese. The plain carbon steels varying from 0.06% carbon to 1.5% carbon are divided into the following types depending upon the carbon content.

1. Dead mild steel — up to 0.15% carbon
2. Low carbon or mild steel — 0.15% to 0.45% carbon
3. Medium carbon steel — 0.45% to 0.8% carbon
4. High carbon steel — 0.8% to 1.5% carbon

According to Indian standard *[IS : 1762 (Part-I)–1974], a new system of designating the steel is recommended. According to this standard, steels are designated on the following two basis :

(*a*) On the basis of mechanical properties, and (*b*) On the basis of chemical composition.

We shall now discuss, in detail, the designation of steel on the above two basis, in the following pages.

2.13 Steels Designated on the Basis of Mechanical Properties

These steels are carbon and low alloy steels where the main criterion in the selection and inspection of steel is the tensile strength or yield stress. According to Indian standard **IS: 1570 (Part–I)-1978 (Reaffirmed 1993), these steels are designated by a symbol 'Fe' or 'Fe E' depending on whether

* This standard was reaffirmed in 1993 and covers the code designation of wrought steel based on letter symbols.

** The Indian standard IS : 1570-1978 (Reaffirmed 1993) on wrought steels for general engineering purposes has been revised on the basis of experience gained in the production and use of steels. This standard is now available in seven parts.

the steel has been specified on the basis of minimum tensile strength or yield strength, followed by the figure indicating the minimum tensile strength or yield stress in N/mm^2. For example 'Fe 290' means a steel having minimum tensile strength of 290 N/mm^2 and 'Fe E 220' means a steel having yield strength of 220 N/mm^2.

Table 2.5 shows the tensile and yield properties of standard steels with their uses according to IS : 1570 (Part I)-1978 (Reaffirmed 1993).

Table 2.5. Indian standard designation of steel according to IS : 1570 (Part I)-1978 (Reaffirmed 1993).

Indian standard designation	*Tensile strength (Minimum) N/mm^2*	*Yield stress (Minimum) N/mm^2*	*Minimum percentage elongation*	*Uses as per IS : 1871 (Part I)–1987 (Reaffirmed 1993)*
Fe 290 Fe E 220	290 290	170 220	27 27	It is used for plain drawn or enamelled parts, tubes for oil well casing, steam, water and air passage, cycle, motor cycle and automobile tubes, rivet bars and wire.
Fe 310 Fe E 230 Fe 330 Fe E 250	310 310 330 330	180 230 200 250	26 26 26 26	These steels are used for locomotive carriages and car structures, screw stock and other general engineering purposes.
Fe 360 Fe E 270	360 360	220 270	25 25	It is used for chemical pressure vessels and other general engineering purposes.
Fe 410 Fe E 310	410 410	250 310	23 23	It is used for bridges and building construction, railway rolling stock, screw spikes, oil well casing, tube piles, and other general engineering purposes.
Fe 490 Fe E 370	490 490	290 370	21 21	It is used for mines, forgings for marine engines, sheet piling and machine parts.
Fe 540 Fe E 400	540 540	320 400	20 20	It is used for locomotive, carriage, wagon and tramway axles, arches for mines, bolts, seamless and welded tubes.
Fe 620 Fe E 460	620 620	380 460	15 15	It is used for tramway axles and seamless tubes.
Fe 690 Fe E 520	690 690	410 520	12 12	It is used for locomotive, carriage and wagon wheels and tyres, arches for mines, seamless oil well casing and drill tubes, and machine parts for heavy loading.
Fe 770 Fe E 580	770 770	460 580	10 10	It is used for locomotive, carriage and wagon wheels and tyres, and machine parts for heavy loading.
Fe 870 Fe E 650	870 870	520 650	8 8	It is used for locomotive, carriage and wagon wheels and tyres.

Notes : 1. The steels from grades Fe 290 to Fe 490 are general structural steels and are available in the form of bars, sections, tubes, plates, sheets and strips.

2. The steels of grades Fe 540 and Fe 620 are medium tensile structural steels.

3. The steels of grades Fe 690, Fe 770 and Fe 870 are high tensile steels.

2.14 Steels Designated on the Basis of Chemical Composition

According to Indian standard, IS : 1570 (Part II/Sec I)-1979 (Reaffirmed 1991), the carbon steels are designated in the following order :

(*a*) Figure indicating 100 times the average percentage of carbon content,

(*b*) Letter 'C', and

(*c*) Figure indicating 10 times the average percentage of manganese content. The figure after multiplying shall be rounded off to the nearest integer.

For example 20C8 means a carbon steel containing 0.15 to 0.25 per cent (0.2 per cent on an average) carbon and 0.60 to 0.90 per cent (0.75 per cent rounded off to 0.8 per cent on an average) manganese.

Table 2.6 shows the Indian standard designation of carbon steel with composition and their uses.

Table 2.6. Indian standard designation of carbon steel according to IS : 1570 (Part II/Sec 1) – 1979 (Reaffirmed 1991).

Indian standard designation	*Composition in percentages*		*Uses as per IS : 1871 (Part II)–1987 (Reaffirmed 1993)*
	Carbon (C)	*Manganese (Mn)*	
4C2	0.08 Max.	0.40 Max.	It is a dead soft steel generally used in electrical industry.
5C4 7C4 10C4	0.10 Max. 0.12 Max. 0.15 Max.	0.50 Max. 0.50 Max. 0.30 – 0.60	These steels are used where cold formability is the primary requirement. In the rimming quality, they are used as sheet, strip, rod and wire especially where excellent surface finish or good drawing qualities are required, such as automobile body, and fender stock, hoods, lamps, oil pans and a multiple of deep drawn and formed products. They are also used for cold heading wire and rivets and low carbon wire products. The killed steel is used for forging and heat treating applications.
10C4 14C6	0.15 Max. 0.10 – 0.18	0.30 – 0.60 0.40 – 0.70	The case hardening steels are used for making camshafts, cams, light duty gears, worms, gudgeon pins, spindles, pawls, ratchets, chain wheels, tappets, etc.
15C4	0.20 Max.	0.30 – 0.60	It is used for lightly stressed parts. The material, although easily machinable, is not designed specifically for rapid cutting, but is suitable where cold web, such as bending and riveting may be necessary.

Indian standard designation	*Composition in percentages*		*Uses as per IS : 1871 (Part II)–1987 (Reaffirmed 1993)*
	Carbon (C)	*Manganese (Mn)*	
15C8	0.10 – 0.20	0.60 – 0.90	These steels are general purposes steels used for low stressed components.
20C8	0.15 – 0.25	0.60 – 0.90	
25C4	0.20 – 0.30	0.30 – 0.60	
25C8	0.20 – 0.30	0.60 – 0.90	
30C8	0.25 – 0.35	0.60 – 0.90	It is used for making cold formed parts such as shift and brake levers. After suitable case hardening or hardening and tempering, this steel is used for making sprockets, tie rods, shaft fork and rear hub, 2 and 3 wheeler scooter parts such as sprocket, lever, hubs for forks, cams, rocket arms and bushes. Tubes for aircraft, automobile, bicycle and furniture are also made of this steel.
35C4	0.30 – 0.40	0.30 – 0.60	It is used for low stressed parts, automobile tubes and fasteners.
35C8	0.30 – 0.40	0.60 – 0.90	It is used for low stressed parts in machine structures, cycle and motor cycle tubes, fish plates for rails and fasteners.
40C8	0.35 – 0.45	0.60 – 0.90	It is used for crankshafts, shafts, spindles, push rods, automobile axle beams, connecting rods, studs, bolts, lightly stressed gears, chain parts, umbrella ribs, washers, etc.
45C8	0.40 – 0.50	0.60 – 0.90	It is used for spindles of machine tools, bigger gears, bolts, lead screws, feed rods, shafts and rocks.
50C4	0.45 – 0.55	0.30 – 0.60	It is used for keys, crankshafts, cylinders and machine parts requiring moderate wear resistance. In surface hardened condition, it is also suitable for large pitch worms and gears.
50C12	0.45 – 0.55	1.1 – 1.50	It is a rail steel. It is also used for making spike bolts, gear shafts, rocking levers and cylinder liners.
55C4	0.50 – 0.60	0.30 – 0.60	These steels are used for making gears, coil springs, cylinders, cams, keys, crankshafts, sprockets and machine parts requiring moderate wear resistance for which toughness is not of primary importance. It is also used for cycle and industrial chains, spring, can opener, umbrella ribs, parts of camera and typewriter.
55C8	0.50 – 0.60	0.60 – 0.90	
60C4	0.55 – 0.65	0.30 – 0.60	It is used for making clutch springs, hardened screws and nuts, machine tool spindles, couplings, crankshafts, axles and pinions.
65C9	0.60 – 0.70	0.50 – 0.80	It is a high tensile structural steel used for making locomotive carriage and wagon tyres. It is also used for engine valve springs, small washers and thin stamped parts.

Indian standard designation	*Composition in percentages*		*Uses as per IS : 1871 (Part II)–1987 (Reaffirmed 1993)*
	Carbon (C)	*Manganese (Mn)*	
70C6	0.65 – 0.75	0.50 – 0.80	It is used for making baffle springs, shock absorbers, springs for seat cushions for road vehicles. It is also used for making rail tyres, unhardened gears and worms, washers, wood working saw, textile and jute machinery parts and clutch plates, etc.
75C6	0.70 – 0.80	0.50 – 0.80	It is used for making light flat springs formed from annealed stock. Because of good wear properties when properly heat treated, it is used for making shear blades, rack teeth, scrappers and cutlivators' shovels.
80C6 85C6	0.75 – 0.85 0.80 – 0.90	0.50 – 0.80 0.50 – 0.80	These steels are used for making flat and coil springs for automobile and railway vehicles. It is also used for girder rails. The valve spring wire and music wire are special applications of steel 85 C6. After suitable heat treatment, these steels are also used for making scraper blades, discs and spring tooth harrows. It is also used for clutch parts, wood working saw, band saw and textile and jute machinery parts.
98C6 113C6	0.90 – 1.05 1.05 – 1.20	0.50 – 0.80 0.50 – 0.80	These steels in the oil hardened and tempered condition are used for coil or spiral springs. It is also used for pen nibs, volute spring, spring cutlery, knitting needle and hacksaw blades.

2.15 Effect of Impurities on Steel

The following are the effects of impurities like silicon, sulphur, manganese and phosphorus on steel.

1. ***Silicon.*** The amount of silicon in the finished steel usually ranges from 0.05 to 0.30%. Silicon is added in low carbon steels to prevent them from becoming porous. It removes the gases and oxides, prevent blow holes and thereby makes the steel tougher and harder.

2. ***Sulphur.*** It occurs in steel either as iron sulphide or manganese sulphide. Iron sulphide because of its low melting point produces red shortness, whereas manganese sulphide does not effect so much. Therefore, manganese sulphide is less objectionable in steel than iron sulphide.

3. ***Manganese.*** It serves as a valuable deoxidising and purifying agent in steel. Manganese also combines with sulphur and thereby decreases the harmful effect of this element remaining in the steel. When used in ordinary low carbon steels, manganese makes the metal ductile and of good bending qualities. In high speed steels, it is used to toughen the metal and to increase its critical temperature.

4. ***Phosphorus.*** It makes the steel brittle. It also produces cold shortness in steel. In low carbon steels, it raises the yield point and improves the resistance to atmospheric corrosion. The sum of carbon and phosphorus usually does not exceed 0.25%.

2.16 Free Cutting Steels

The free cutting steels contain sulphur and phosphorus. These steels have higher sulphur content than other carbon steels. In general, the carbon content of such steels vary from 0.1 to 0.45 per cent and sulphur from 0.08 to 0.3 per cent. These steels are used where rapid machining is the prime requirement. It may be noted that the presence of sulphur and phosphorus causes long chips in machining to be easily broken and thus prevent clogging of machines. Now a days, lead is used from 0.05 to 0.2 per cent instead of sulphur, because lead also greatly improves the machinability of steel without the loss of toughness.

According to Indian standard, IS : 1570 (Part III)-1979 (Reaffirmed 1993), carbon and carbon manganese free cutting steels are designated in the following order :

1. Figure indicating 100 times the average percentage of carbon,
2. Letter 'C',
3. Figure indicating 10 times the average percentage of manganese, and
4. Symbol 'S' followed by the figure indicating the 100 times the average content of sulphur. If instead of sulphur, lead (Pb) is added to make the steel free cutting, then symbol 'Pb' may be used.

Table 2.7 shows the composition and uses of carbon and carbon-manganese free cutting steels, as per IS : 1570 (Part III)-1979 (Reaffirmed 1993).

2.17 Alloy Steel

An alloy steel may be defined as a steel to which elements other than carbon are added in sufficient amount to produce an improvement in properties. The alloying is done for specific purposes to increase wearing resistance, corrosion resistance and to improve electrical and magnetic properties, which cannot be obtained in plain carbon steels. The chief alloying elements used in steel are nickel, chromium, molybdenum, cobalt, vanadium, manganese, silicon and tungsten. Each of these elements confer certain qualities upon the steel to which it is added. These elements may be used separately or in combination to produce the desired characteristic in steel. Following are the effects of alloying elements on steel:

1. *Nickel.* It increases the strength and toughness of the steel. These steels contain 2 to 5% nickel and from 0.1 to 0.5% carbon. In this range, nickel contributes great strength and hardness with high elastic limit, good ductility and good resistance to corrosion. An alloy containing 25% nickel possesses maximum toughness and offers the greatest resistance to rusting, corrosion and burning at high temperature. It has proved to be of advantage in the manufacture of boiler tubes, valves for use with superheated steam, valves for I.C. engines and spark plugs for petrol engines. A nickel steel alloy containing 36% of nickel is known as ***invar.*** It has nearly zero coefficient of expansion. So it is in great demand for measuring instruments and standards of lengths for everyday use.

2. *Chromium.* It is used in steels as an alloying element to combine hardness with high strength and high elastic limit. It also imparts corrosion-resisting properties to steel. The most common chrome steels contains from 0.5 to 2% chromium and 0.1 to 1.5% carbon. The chrome steel is used for balls, rollers and races for bearings. A ***nickel chrome steel*** containing 3.25% nickel, 1.5% chromium and 0.25% carbon is much used for armour plates. Chrome nickel steel is extensively used for motor car crankshafts, axles and gears requiring great strength and hardness.

3. *Tungsten.* It prohibits grain growth, increases the depth of hardening of quenched steel and confers the property of remaining hard even when heated to red colour. It is usually used in conjuction with other elements. Steel containing 3 to 18% tungsten and 0.2 to 1.5% carbon is used for cutting tools. The principal uses of tungsten steels are for cutting tools, dies, valves, taps and permanent magnets.

Table 2.7. Indian standard designation of carbon and carbon-manganese free cutting steels according to IS:1570 (Part III) – 1979 (Reaffirmed 1993).

Indian standard designation	*Composition in percentages*					*Uses as per IS : 1871 (Part III)–1987 (Reaffirmed 1993)*
	Carbon (C)	*Silicon (Si)*	*Manganese (Mn)*	*Sulphur (S)*	*Phosphorus (P) Max*	
10C8S10	0.15 Max.	0.05 – 0.30	0.60 – 0.90	0.08 – 0.13	0.06	It is used for small parts to be cyanided or carbonitrided.
14C14S14	0.10 – 0.18	0.05 – 0.30	1.20 – 1.50	0.1 – 0.18	0.06	It is used for parts where good machinability and finish are important.
25C12S14	0.20 – 0.30	0.25 Max.	1.00 – 1.50	0.10 – 0.18	0.06	It is used for bolts, studs and other heat treated parts of small section. It is suitable in either cold drawn, normalised or heat treated condition for moderately stressed parts requiring more strength than mild steel.
40C10S18	0.35 – 0.45	0.25 Max.	0.80 – 1.20	0.14 – 0.22	0.06	It is used for heat treated bolts, engine shafts, connecting rods, miscellaneous gun carriage, and small arms parts not subjected to high stresses and severe wear.
11C10S25	0.08 – 0.15	0.10 Max.	0.80 – 1.20	0.20 – 0.30	0.06	It is used for lightly stressed components not subjected to shock (nuts, studs, etc.) and suitable for production on automatic lathes. It is not recommended for general case hardening work but should be used when ease of machining is the deciding factor.
40C15S12	0.35 – 0.45	0.25 Max.	1.30 – 1.70	0.08 – 0.15	0.06	It is used for heat treated axles, shafts, small crankshafts and other vehicle parts. It is not recommended for forgings in which transverse properties are important.

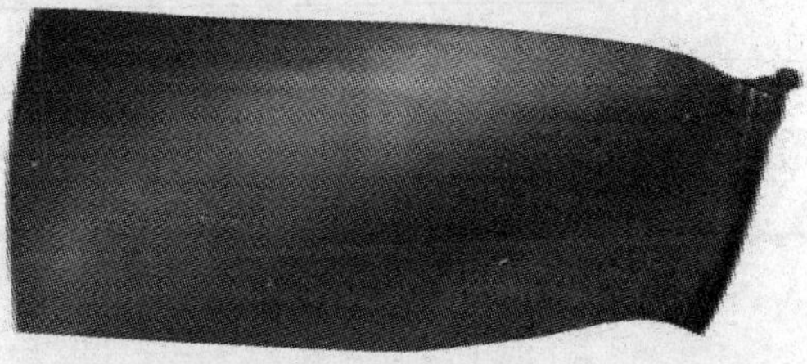

This is a fan blade from a jumbo jet engine. On take-off, the stress on the metal is immense, so to prevent the fan from flying apart, the blades must be both light and very strong. Titanium, though expensive, is the only suitable metal.

4. ***Vanadium.*** It aids in obtaining a fine grain structure in tool steel. The addition of a very small amount of vanadium (less than 0.2%) produces a marked increase in tensile strength and elastic limit in low and medium carbon steels without a loss of ductility. The ***chrome-vanadium steel*** containing about 0.5 to 1.5% chromium, 0.15 to 0.3% vanadium and 0.13 to 1.1% carbon have extremely good tensile strength, elastic limit, endurance limit and ductility. These steels are frequently used for parts such as springs, shafts, gears, pins and many drop forged parts.

5. ***Manganese.*** It improves the strength of the steel in both the hot rolled and heat treated condition. The manganese alloy steels containing over 1.5% manganese with a carbon range of 0.40 to 0.55% are used extensively in gears, axles, shafts and other parts where high strength combined with fair ductility is required. The principal uses of manganese steel is in machinery parts subjected to severe wear. These steels are all cast and ground to finish.

6. ***Silicon.*** The silicon steels behave like nickel steels. These steels have a high elastic limit as compared to ordinary carbon steel. Silicon steels containing from 1 to 2% silicon and 0.1 to 0.4% carbon and other alloying elements are used for electrical machinery, valves in I.C. engines, springs and corrosion resisting materials.

7. ***Cobalt.*** It gives red hardness by retention of hard carbides at high temperatures. It tends to decarburise steel during heat-treatment. It increases hardness and strength and also residual magnetism and coercive magnetic force in steel for magnets.

8. ***Molybdenum.*** A very small quantity (0.15 to 0.30%) of molybdenum is generally used with chromium and manganese (0.5 to 0.8%) to make molybdenum steel. These steels possess extra tensile strength and are used for air-plane fuselage and automobile parts. It can replace tungsten in high speed steels.

2.18 Indian Standard Designation of Low and Medium Alloy Steels

According to Indian standard, IS : 1762 (Part I)-1974 (Reaffirmed 1993), low and medium alloy steels shall be designated in the following order :

1. Figure indicating 100 times the average percentage carbon.
2. Chemical symbol for alloying elements each followed by the figure for its average percentage content multiplied by a factor as given below :

Element	*Multiplying factor*
Cr, Co, Ni, Mn, Si and W	4
Al, Be, V, Pb, Cu, Nb, Ti, Ta, Zr and Mo	10
P, S and N	100

For example 40 Cr 4 Mo 2 means alloy steel having average 0.4% carbon, 1% chromium and 0.25% molybdenum.

Notes : 1. The figure after multiplying shall be rounded off to the nearest integer.

2. Symbol 'Mn' for manganese shall be included in case manganese content is equal to or greater than 1 per cent.

3. The chemical symbols and their figures shall be listed in the designation in the order of decreasing content.

Table 2.8 shows the composition and uses of some low and medium alloy steels according to Indian standard IS : 1570-1961 (Reaffirmed 1993).

Table 2.8. Composition and uses of alloy steels according to IS : 1570-1961 (Reaffirmed 1993).

Indian standard designation	*Composition in percentages*						*Uses as per IS : 1871–1965*
	Carbon (C)	*Silicon (Si)*	*Manganese (Mn)*	*Nickel (Ni)*	*Chromium (Cr)*	*Molybdenum (Mo)*	
11Mn2	0.16 Max.	0.10 – 0.35	1.30 – 1.70	–	–	–	It is a notch ductile steel for general purposes. It is also used in making filler rods, colliery cage suspension gear tub, mine car draw gear, couplings and rope sockets.
20Mn2	0.16 – 0.24	0.10 – 0.35	1.30 – 1.70	–	–	–	These are used for welded structures, crankshafts, steering levers, shafting spindles, etc.
27Mn2	0.22 – 0.32	0.10 – 0.35	1.30 – 1.70	–	–	–	
37Mn2	0.32 – 0.42	0.10 – 0.35	1.30 – 1.70	–	–	–	It is used for making axles, shafts, crankshafts, connecting rods, etc.
47Mn2	0.42 – 0.52	0.10 – 0.35	1.30 – 1.70	–	–	–	It is used for tram rails and similar other structural purposes.
40Cr1	0.35 – 0.45	0.10 – 0.35	0.60 – 0.09	–	0.90 – 1.20	–	It is used for making gears, connecting rods, stub axles, steering arms, wear resistant plates for earth moving and concrete handling equipment, etc.
50Cr1	0.45 – 0.55	0.10 – 0.35	0.60 – .90	–	0.90 – 1.20	–	It is spring steel. It is used in a helical automobile front suspension springs.
35Mn2Mo28	0.30 – 0.40	0.10 – 0.35	1.30 – 1.80	–	–	0.20 – 0.35	These are used for making general engineering components such as crankshafts, bolts, wheel studs, axle shafts, levers and connecting rods.
35Mn2Mo45	0.30 – 0.40	0.10 – 0.35	1.30 – 1.80	–	–	0.35 – 0.55	
40Cr1Mo28	0.35 – 0.45	0.10 – 0.35	0.50 – 0.80	–	0.90 – 1.20	0.20 – 0.35	It is used for making axle shafts, crankshafts, connecting rods, gears, high tensile bolts and studs, propeller shaft joints, etc.

Contd..

Indian standard designation	*Composition in percentages*						*Uses as per IS : 1871–1965*
	Carbon (C)	*Silicon (Si)*	*Manganese (Mn)*	*Nickel (Ni)*	*Chromium (Cr)*	*Molybdenum (Mo)*	
15Cr3Mo55 25Cr3Mo55	0.10 – 0.20 0.20 – 0.30	0.10 – 0.35 0.10 – 0.35	0.40 – 0.70 0.40 – 0.70	0.30 Max. 0.30 Max.	2.90 – 3.40 2.90 – 3.40	0.45 – 0.65 0.45 – 0.65	These are used for components requiring medium to high tensile properties. In the nitrided condition, it is used for crank-shafts, cylinder liners for aero and automobile engines, gears, and machine parts requiring high surface hardness and wear resistance.
40Ni3	0.35 – 0.45	0.10 – 0.35	0.50 – 0.80	3.20 – 3.60	0.30 Max.	–	It is used for parts requiring excessively high toughness. In particular, it is used for components working at low temperatures (in refrigerators, compressors, locomotives and aircraft) and for heavy forgings, turbine blades, severely stressed screws, bolts and nuts.
30Ni4Crl	0.26 – 0.34	0.10 – 0.35	0.40 – 0.70	3.90 – 4.30	1.10 – 1.40	–	It is used for highly stressed gears and other components requiring high tensile strength of the order of 16 N/mm^2 and where minimum distortion in heat treatment is essential.
35NilCr60	0.30 – 0.40	0.10 – 0.35	0.60 – 0.90	1.00 – 1.50	0.45 – 0.75	–	It is used in the construction of aircraft and heavy vehicles for crankshafts, connecting rods, gear shafts, chain parts, clutches, flexible shafts for plenary gears, camshafts, etc.
40Ni2CrlMo28	0.35 – 0.45	0.10 – 0.35	0.40 – 70	1.25 – 1.75	0.90 – 1.30	0.20 – 0.35	It is used for high strength machine parts collets, spindles, screws, high tensile bolts and studs, gears, pinions, axle shafts, tappets, crankshafts, connecting rods, boring bars, arbours, etc.

2.19 Stainless Steel

It is defined as that steel which when correctly heat treated and finished, resists oxidation and corrosive attack from most corrosive media. The different types of stainless steels are discussed below :

1. ***Martensitic stainless steel.*** The chromium steels containing 12 to 14 per cent chromium and 0.12 to 0.35 per cent carbon are the first stainless steels developed. Since these steels possess martensitic structure, therefore, they are called ***martensitic stainless steels.*** These steels are magnetic and may be hardened by suitable heat treatment and the hardness obtainable depends upon the carbon content. These steels can be easily welded and machined. When formability, softness, etc. are required in fabrication, steel having 0.12 per cent maximum carbon is often used in soft condition. With increasing carbon, it is possible by hardening and tempering to obtain tensile strength in the range of 600 to 900 N/mm^2, combined with reasonable toughness and ductility. In this condition, these steels find many useful general applications where mild corrosion resistance is required. Also, with the higher carbon range in the hardened and lightly tempered condition, tensile strength of about 1600 N/mm^2 may be developed with lowered ductility.

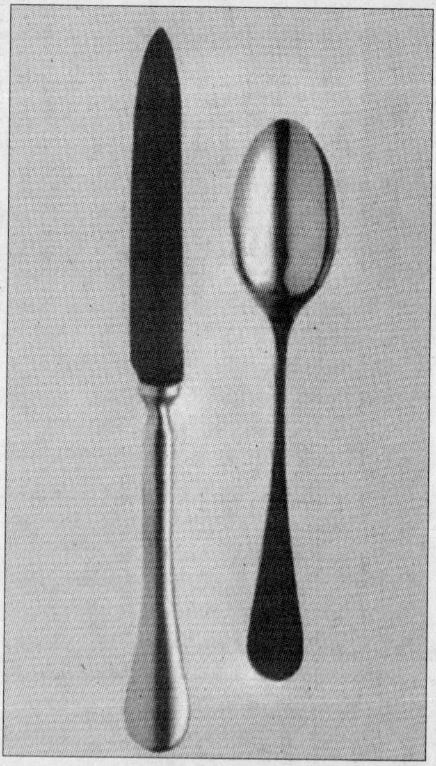

Stainless steel was invented in 1913 by British metallurgist Harry Brearley (1871-1948). He made a steel containing 13 per cent chromium. The new alloy proved to be highly resistant to corrosion: chromium reacts with oxygen in the air to form a tough, protective film which renews itself if the metal is scratched.

These steels may be used where the corrosion conditions are not too severe, such as for hydraulic, steam and oil pumps, valves and other engineering components. However, these steels are not suitable for shafts and parts working in contact with non-ferrous metals (*i.e.* brass, bronze or gun metal bearings) and with graphite packings, because electrolytic corrosion is likely to occur. After hardening and light tempering, these steels develop good cutting properties. Therefore, they are used for cutlery, springs, surgical and dental instruments.

Note: The presence of chromium provides good resistance to scaling upto a temperature of about 750°C, but it is not suitable where mechanical strength in the temperature range of 600 to 750°C is required. In fact, creep resistance of these steels at this temperature is not superior to that of mild steel. But at temperature below 600°C, the strength of these steels is better than that of carbon steels and upto 480°C is even better than that of austenitic steels.

2. ***Ferritic stainless steel.*** The steels containing greater amount of chromium (from 16 to 18 per cent) and about 0.12 per cent carbon are called ***ferritic stainless steels.*** These steels have better corrosion resistant property than martensitic stainless steels. But, such steels have little capacity for hardening by heat treatment. However, in the softened condition, they possess good ductility and are mainly used as sheet or strip for cold forming and pressing operations for purposes where moderate corrosion resistance is required. They may be cold worked or hot worked. They are ferro-magnetic, usually undergo excessive grain growth during prolonged exposure to elevated temperatures, and may develop brittleness after electric arc resistance or gas welding. These steels have lower strength

at elevated temperatures than martensitic steels. However, resistance to scaling and corrosion at elevated temperatures are usually better. The machinability is good and they show no tendency to intercrystalline corrosion.

Note: When nickel from 1.5 to 2.5 per cent is added to 16 to 18 per cent chromium steel, it not only makes more resistant to corrosion than martensitic steel but also makes it hardenable by heat treatment. Such a steel has good resistance to electrolytic corrosion when in contact with non-ferrous metals and graphite packings. Thus it is widely used for pump shafts, spindles and valves as well as for many other fittings where a good combination of mechanical and corrosion properties are required.

3. *Austenitic stainless steel.* The steel containing high content of both chromium and nickel are called ***austenitic stainless steels.*** There are many variations in chemical composition of these steels, but the most widely used steel contain 18 per cent chromium and 8 per cent nickel with carbon content as low as possible. Such a steel is commonly known as ***18/8 steel.*** These steels cannot be hardened by quenching, in fact they are softened by rapid cooling from about 1000°C. They are non-magnetic and possess greatest resistance to corrosion and good mechanical properties at elevated temperature.

These steels are very tough and can be forged and rolled but offer great difficulty in machining. They can be easily welded, but after welding, it is susceptible to corrosive attack in an area adjacent to the weld. This susceptibility to corrosion (called intercrystalline corrosion or weld decay) may be removed by softening after welding by heating to about 1100°C and cooling rapidly. These steels are used in the manufacture of pump shafts, rail road car frames and sheathing, screws, nuts and bolts and small springs. Since 18/8 steel provide excellent resistance to attack by many chemicals, therefore, it is extensively used in chemical, food, paper making and dyeing industries.

Note : When increased corrosion resistance properties are required, for some purposes, then molybdenum from 2 to 3 per cent may be added.

2.20 Heat Resisting Steels

The steels which can resist creep and oxidation at high temperatures and retain sufficient strength are called ***heat resisting steels.*** A number of heat resisting steels have been developed as discussed below :

1. *Low alloy steels.* These steels contain 0.5 per cent molybdenum. The main application of these steels are for superheater tubes and pipes in steam plants, where service temperatures are in the range of 400°C to 500°C.

2. *Valve steels.* The chromium-silicon steels such as ***silchrome*** (0.4% C, 8% Cr, 3.5% Si) and ***Volmax*** (0.5% C, 8% Cr, 3.5% Si, 0.5% Mo) are used for automobile valves. They possess good resistance to scaling at dull red heat, although their strength at elevated temperatures is relatively low. For aeroplane engines and marine diesel engine valves, 13/13/3 nickel-chromium-tungsten valve steel is usually used.

3. *Plain chromium steel.* The plain chromium steel consists of

(*a*) Martensitic chromium steel with 12–13% Cr, and

(*b*) Ferritic chromium steels with 18–30% Cr.

These steels are very good for oxidation resistance at high temperatures as compared to their strength which is not high at such conditions. The maximum operating temperature for martensitic steels is about 750°C, whereas for ferritic steels it is about 1000 – 1150°C.

4. *Austenitic chromium-nickel steels.* These steels have good mechanical properties at high temperatures with good scaling resistance. These alloys contain a minimum of 18 per cent chromium and 8 per cent nickel stabilised with titanium or niobium. Other carbide forming elements such as molybdenum or tungsten may also be added in order to improve creep strength. Such alloys are suitable for use upto 1100°C and are used for gas turbine discs and blades.

Table 2.9. Indian standard designation of high alloy steels (stainless steel and heat resisting steels) according to IS : 1570 (Part V)-1985 (Reaffirmed 1991).

Indian standard designation	Composition in percentages						Uses as per IS : 1871–1965
	Carbon (C)	Silicon (Si)	Manganese (Mn)	Nickel (Ni)	Chromium (Cr)	Molybdenum (Mo)	
30Cr13	0.26 – 0.35	1.0 Max.	1.0 Max.	1.0 Max.	12.0 – 14.0	–	It is used for structural parts with high strength and kitchen utensils.
15Cr16Ni2	0.10 – 0.20	1.0 Max.	1.0 Max.	1.5 – 3.0	15.0 – 18.0	–	It is used for aircraft fittings, wind shield wiper arms, bolting materials, paper machinery etc.
07Cr18Ni9	0.12 Max.	1.0 Max.	2.0 Max.	8.0 – 10.0	17.0 – 19.0	–	It is used for aircraft fire walls and cawlings, radar and microwaves antennae, jewellery, household novelties, automotive wheel covers, refrigerator trays, kitchen utensils, railway passenger car bodies, ice making equipment, tubular furniture, screen door and storm window frames, electric switch parts, flexible couplings etc.
04Cr17Ni12 Mo2	0.08 Max.	1.0 Max.	2.0 Max.	10.5 – 14.0	16.0 – 18.5	2.0 – 3.0	It is used for high temperature chemical handling equipment for rayon, rubber and marine industries, photographic developing equipment, pulp handling equipment, steam jacketed kettles, coke plant equipment, food processing equipment, edible oil storage tanks.
45Cr9Si4	0.40 – 0.50	3.25 – 3.75	0.30 – 0.60	0.05 Max.	7.50 – 9.50	–	It is used for heat resisting outlet valves in oil engines, lorries and cars.
80Cr20Si2 Nil	0.75 – 0.85	1.75 – 2.25	0.20 – 0.60	1.20 – 1.70	19.0 – 21.0	–	It is used for highly stressed outlet valves in high speed carburetors and heavy oil engines.

2.21 Indian Standard Designation of High Alloy Steels (Stainless Steel and Heat Resisting Steel)

According to Indian standard, IS : 1762 (Part I)-1974 (Reaffirmed 1993), the high alloy steels (*i.e.* stainless steel and heat resisting steel) are designated in the following order:

1. Letter 'X'.
2. Figure indicating 100 times the percentage of carbon content.
3. Chemical symbol for alloying elements each followed by a figure for its average percentage content rounded off to the nearest integer.
4. Chemical symbol to indicate specially added element to allow the desired properties.

For example, X 10 Cr 18 Ni 9 means alloy steel with average carbon 0.10 per cent, chromium 18 per cent and nickel 9 per cent.

Table 2.9 shows the composition and uses of some types of the stainless steels and heat resisting steels according to Indian standard IS : 1570 (Part V)-1985 (Reaffirmed 1991).

2.22 High Speed Tool Steels

These steels are used for cutting metals at a much higher cutting speed than ordinary carbon tool steels. The carbon steel cutting tools do not retain their sharp cutting edges under heavier loads and higher speeds. This is due to the fact that at high speeds, sufficient heat may be developed during the cutting operation and causes the temperature of the cutting edge of the tool to reach a red heat. This temperature would soften the carbon tool steel and thus the tool will not work efficiently for a longer period. The high speed steels have the valuable property of retaining their hardness even when heated to red heat. Most of the high speed steels contain tungsten as the chief alloying element, but other elements like cobalt, chromium, vanadium, etc. may be present in some proportion. Following are the different types of high speed steels:

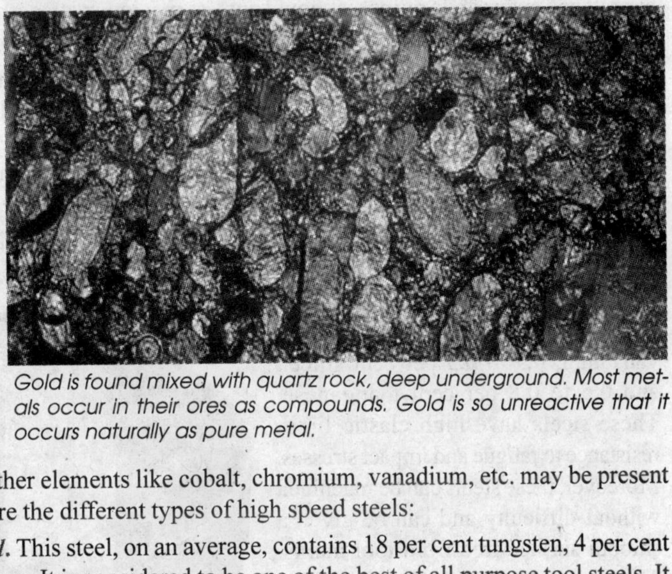

Gold is found mixed with quartz rock, deep underground. Most metals occur in their ores as compounds. Gold is so unreactive that it occurs naturally as pure metal.

1. ***18-4-1 High speed steel.*** This steel, on an average, contains 18 per cent tungsten, 4 per cent chromium and 1 per cent vanadium. It is considered to be one of the best of all purpose tool steels. It is widely used for drills, lathe, planer and shaper tools, milling cutters, reamers, broaches, threading dies, punches, etc.

2. ***Molybdenum high speed steel.*** This steel, on an average, contains 6 per cent tungsten, 6 per cent molybdenum, 4 per cent chromium and 2 per cent vanadium. It has excellent toughness and cutting ability. The molybdenum high speed steels are better and cheaper than other types of steels. It is particularly used for drilling and tapping operations.

3. ***Super high speed steel.*** This steel is also called ***cobalt high speed steel*** because cobalt is added from 2 to 15 per cent, in order to increase the cutting efficiency especially at high temperatures. This steel, on an average, contains 20 per cent tungsten, 4 per cent chromium, 2 per cent vanadium and 12 per cent cobalt. Since the cost of this steel is more, therefore, it is principally used for heavy cutting operations which impose high pressure and temperatures on the tool.

2.23 Indian Standard Designation of High Speed Tool Steel

According to Indian standard, IS : 1762 (Part I)-1974 (Reaffirmed 1993), the high speed tool steels are designated in the following order :

1. Letter 'XT'.
2. Figure indicating 100 times the percentage of carbon content.
3. Chemical symbol for alloying elements each followed by the figure for its average percentage content rounded off to the nearest integer, and
4. Chemical symbol to indicate specially added element to attain the desired properties.

For example, XT 75 W 18 Cr 4 V 1 means a tool steel with average carbon content 0.75 per cent, tungsten 18 per cent, chromium 4 per cent and vanadium 1 per cent.

Table 2.10 shows the composition of high speed tool steels as per Indian standard, IS : 7291-1981 (Reaffirmed 1993).

2.24 Spring Steels

The most suitable material for springs are those which can store up the maximum amount of work or energy in a given weight or volume of spring material, without permanent deformation. These steels should have a high elastic limit as well as high deflection value. The spring steel, for aircraft and automobile purposes should possess maximum strength against fatigue effects and shocks. The steels most commonly used for making springs are as follows:

1. ***High carbon steels.*** These steels contain 0.6 to 1.1 per cent carbon, 0.2 to 0.5 per cent silicon and 0.6 to 1 per cent manganese. These steels are heated to 780 – 850°C according to the composition and quenched in oil or water. It is then tempered at 200 – 500°C to suit the particular application. These steels are used for laminated springs for locomotives, carriages, wagons, and for heavy road vehicles. The higher carbon content oil hardening steels are used for volute, spiral and conical springs and for certain types of petrol engine inlet valve springs.

2. ***Chrome-vanadium steels.*** These are high quality spring steels and contain 0.45 to 0.55 per cent carbon, 0.9 to 1.2 per cent chromium, 0.15 to 0.20 per cent vanadium, 0.3 to 0.5 per cent silicon and 0.5 to 0.8 per cent manganese. These steels have high elastic limit, resistance to fatigue and impact stresses. Moreover, these steels can be machined without difficulty and can be given a smooth surface free from tool marks. These are hardened by oil quenching at 850 – 870°C and tempered at 470 – 510°C for vehicle and other spring purposes. These steels are used for motor car laminated and coil springs for suspension purposes, automobile and aircraft engine valve springs.

Sodium is in Group I. Although it is a metal, it is so soft that a knife can cut easily through a piece. Sodium is stored in oil to stop air or moisture reacting with it.

3. ***Silicon-manganese steels.*** These steels contain 1.8 to 2.0 per cent silicon, 0.5 to 0.6 per cent carbon and 0.8 to 1 per cent manganese. These steels have high fatigue strength, resistance and toughness. These are hardened by quenching in oil at 850 – 900°C and tempered at 475 – 525°C. These are the usual standard quality modern spring materials and are much used for many engineering purposes.

Table 2.10. Indian standard designation of high speed tool steel according to IS : 7291-1981 (Reaffirmed 1993).

Indian standard designation	Chemical composition in percentages								Brinell hardness in annealed condition (HB) Max.
	Carbon (C)	Silicon (Si)	Manganese (Mn)	Chromium (Cr)	Molybdenum (Mo)	Vanadium (V)	Tungsten (W)	Cobalt (Co)	
XT 72 W 18 Cr 4 V 1	0.65 – 0.80	0.15 – 0.40	0.20 – 0.40	3.75 – 4.50	–	1.00 – 1.25	17.50 – 19.0	–	255
XT 75 W 18 Co 5 Cr 4 Mo V 1	0.70 – 0.80	0.15 – 0.40	0.20 – 0.40	3.75 – 4.50	0.40 – 1.00	1.00 – 1.25	17.50 – 19.0	4.50 – 5.50	269
XT 80 W 20 Co 12 Cr 4 V 2 Mo 1	0.75 – 0.85	0.15 – 0.40	0.20 – 0.40	4.00 – 4.75	0.40 – 1.00	1.25 – 1.75	19.50 – 21.0	11.00 – 12.50	302
XT 125 W Co 10 Cr Mo 4 V 3	1.20 – 1.30	0.15 – 0.40	0.20 – 0.40	3.75 – 4.75	3.00 – 4.00	2.80 – 3.50	8.80 – 10.70	8.80 – 10.70	269
XT 87 W 6 Mo 5 Cr 4 V 2	0.82 – 0.92	0.15 – 0.40	0.15 – 0.40	3.75 – 4.75	4.75 – 5.50	1.75 – 2.05	5.75 – 6.75	–	248
XT 90 W 6 Co Mo 5 Cr 4 V 2	0.85 – 0.95	0.15 – 0.40	0.20 – 0.40	3.75 – 4.75	4.75 – 5.50	1.70 – 2.20	5.75 – 6.75	4.75 – 5.25	269
XT 110 Mo 10 Co 8 Cr 4 W 2	1.05 – 1.15	0.15 – 0.40	0.15 – 0.40	3.50 – 4.50	9.0 – 10.0	0.95 – 1.35	1.15 – 1.85	7.75 – 8.75	269

Notes:
1. For all steels, sulphur (*S*) and phosphorus (*P*) is 0.030 per cent Max.
2. If sulphur is added to give free machining properties, then it shall be between 0.09 and 0.15 per cent.

2.25 Heat Treatment of Steels

The term heat treatment may be defined as an operation or a combination of operations, involving the heating and cooling of a metal or an alloy in the solid state for the purpose of obtaining certain desirable conditions or properties without change in chemical composition. The aim of heat treatment is to achieve one or more of the following objects :

1. To increase the hardness of metals.
2. To relieve the stresses set up in the material after hot or cold working.
3. To improve machinability.
4. To soften the metal.
5. To modify the structure of the material to improve its electrical and magnetic properties.
6. To change the grain size.
7. To increase the qualities of a metal to provide better resistance to heat, corrosion and wear.

Following are the various heat treatment processes commonly employed in engineering practice:

1. *Normalising.* The main objects of normalising are :
   1. To refine the grain structure of the steel to improve machinability, tensile strength and structure of weld.
   2. To remove strains caused by cold working processes like hammering, rolling, bending, etc., which makes the metal brittle and unreliable.
   3. To remove dislocations caused in the internal structure of the steel due to hot working.
   4. To improve certain mechanical and electrical properties.

The process of normalising consists of heating the steel from 30 to 50°C above its upper critical temperature (for hypoeutectoid steels) or Acm line (for hypereutectoid steels). It is held at this temperature for about fifteen minutes and then allowed to cool down in still air.

This process provides a homogeneous structure consisting of ferrite and pearlite for hypoeutectoid steels, and pearlite and cementite for hypereutectoid steels. The homogeneous structure provides a higher yield point, ultimate tensile strength and impact strength with lower ductility to steels. The process of normalising is frequently applied to castings and forgings, etc. The alloy steels may also be normalised but they should be held for two hours at a specified temperature and then cooling in the furnace.

Notes : (*a*) The upper critical temperature for a steel depends upon its carbon content. It is 900°C for pure iron, 860°C for steels with 2.2% carbon, 723°C for steel with 0.8% carbon and 1130°C for steel with 1.8% carbon.

(*b*) Steel containing 0.8% carbon is known as ***eutectoid steel***, steel containing less than 0.8% carbon is called ***hypoeutectoid steel*** and steel containing above 0.8% carbon is called ***hypereutectoid steel.***

2. *Annealing.* The main objects of annealing are :
   1. To soften the steel so that it may be easily machined or cold worked.
   2. To refine the grain size and structure to improve mechanical properties like strength and ductility.
   3. To relieve internal stresses which may have been caused by hot or cold working or by unequal contraction in casting.
   4. To alter electrical, magnetic or other physical properties.
   5. To remove gases trapped in the metal during initial casting.

The annealing process is of the following two types :

(*a*) ***Full annealing.*** The purpose of full annealing is to soften the metal to refine the grain structure, to relieve the stresses and to remove trapped gases in the metal. The process consists of (*i*) heating the steel from 30 to 50°C above the upper critical temperature for hypoeutectoid steel and by the same temperature above the lower critical temperature *i.e.* 723°C for hypereutectoid steels.

(*ii*) holding it at this temperature for sometime to enable the internal changes to take place. The time allowed is approximately 3 to 4 minutes for each millimetre of thickness of the largest section, and

(*iii*) cooling slowly in the furnace. The rate of cooling varies from 30 to 200°C per hour depending upon the composition of steel.

In order to avoid decarburisation of the steel during annealing, the steel is packed in a cast iron box containing a mixture of cast iron borings, charcoal, lime, sand or ground mica. The box along with its contents is allowed to cool slowly in the furnace after proper heating has been completed.

The following table shows the approximate temperatures for annealing depending upon the carbon contents in steel.

Table 2.11. Annealing temperatures.

S.No.	*Carbon content, per cent*	*Annealing temperature, °C*
1.	Less than 0.12 (Dead mild steel)	875 – 925
2.	0.12 to 0.45 (Mild steel)	840 – 970
3.	0.45 to 0.50 (Medium carbon steel)	815 – 840
4.	0.50 to 0.80 (Medium carbon steel)	780 – 810
5.	0.80 to 1.50 (High carbon or tool steel)	760 – 780

(b) Process annealing. The process annealing is used for relieving the internal stresses previously set up in the metal and for increasing the machinability of the steel. In this process, steel is heated to a temperature below or close to the lower critical temperature, held at this temperature for sometime and then cooled slowly. This causes complete recrystallisation in steels which have been severely cold worked and a new grain structure is formed. The process annealing is commonly used in the sheet and wire industries.

3. ***Spheroidising.*** It is another form of annealing in which cementite in the granular form is produced in the structure of steel. This is usually applied to high carbon tool steels which are difficult to machine. The operation consists of heating the steel to a temperature slightly above the lower critical temperature (730 to 770°C). It is held at this temperature for some time and then cooled slowly to a temperature of 600°C. The rate of cooling is from 25 to 30°C per hour.

The spheroidising improves the machinability of steels, but lowers the hardness and tensile strength. These steels have better elongation properties than the normally annealed steel.

4. ***Hardening.*** The main objects of hardening are :

1. To increase the hardness of the metal so that it can resist wear.
2. To enable it to cut other metals *i.e.* to make it suitable for cutting tools.

Clay can be hardened by heat. Bricks and ceramic items are made by firing soft clay objects in a kiln.

The process of hardening consists of

(*a*) heating the metal to a temperature from 30 to 50°C above the upper critical point for hypoeutectoid steels and by the same temperature above the lower critical point for hypereutectoid steels.

(*b*) keeping the metal at this temperature for a considerable time, depending upon its thickness.

(*c*) quenching (cooling suddenly) in a suitable cooling medium like water, oil or brine.

It may be noted that the low carbon steels cannot be hardened appreciably, because of the presence of ferrite which is soft and is not changed by the treatment. As the carbon content goes on increasing, the possible obtainable hardness also increases.

Notes : 1. The greater the rate of quenching, the harder is the resulting structure of steel.

2. For hardening alloy steels and high speed steels, they are heated from 1100°C to 1300°C followed by cooling in a current of air.

5. ***Tempering.*** The steel hardened by rapid quenching is very hard and brittle. It also contains internal stresses which are severe and unequally distributed to cause cracks or even rupture of hardened steel. The tempering (also known as ***drawing***) is, therefore, done for the following reasons :

1. To reduce brittleness of the hardened steel and thus to increase ductility.
2. To remove the internal stresses caused by rapid cooling of steel.
3. To make steel tough to resist shock and fatigue.

The tempering process consists of reheating the hardened steel to some temperature below the lower critical temperature, followed by any desired rate of cooling. The exact tempering temperature depends upon the purpose for which the article or tool is to be used.

6. ***Surface hardening or case hardening.*** In many engineering applications, it is desirable that a steel being used should have a hardened surface to resist wear and tear. At the same time, it should have soft and tough interior or core so that it is able to absorb any shocks, etc. This is achieved by hardening the surface layers of the article while the rest of it is left as such. This type of treatment is applied to gears, ball bearings, railway wheels, etc.

Following are the various *surface or case hardening processes by means of which the surface layer is hardened:

1. Carburising, 2. Cyaniding, 3. Nitriding, 4. Induction hardening, and 5. Flame hardening.

2.26 Non-ferrous Metals

We have already discussed that the non-ferrous metals are those which contain a metal other than iron as their chief constituent. The non-ferrous metals are usually employed in industry due to the following characteristics :

1. Ease of fabrication (casting, rolling, forging, welding and machining),
2. Resistance to corrosion,
3. Electrical and thermal conductivity, and
4. Weight.

The various non-ferrous metals used in engineering practice are aluminium, copper, lead, tin, zinc, nickel, etc. and their alloys. We shall now discuss these non-ferrous metals and their alloys in detail, in the following pages.

2.27 Aluminium

It is white metal produced by electrical processes from its oxide (alumina), which is prepared from a clayey mineral called ***bauxite.*** It is a light metal having specific gravity 2.7 and melting point 658°C. The tensile strength of the metal varies from 90 MPa to 150 MPa.

* For complete details, please refer authors' popular book **'A Text Book of Workshop Technology'.**

In its pure state, the metal would be weak and soft for most purposes, but when mixed with small amounts of other alloys, it becomes hard and rigid. So, it may be blanked, formed, drawn, turned, cast, forged and die cast. Its good electrical conductivity is an important property and is widely used for overhead cables. The high resistance to corrosion and its non-toxicity makes it a useful metal for cooking utensils under ordinary condition and thin foils are used for wrapping food items. It is extensively used in aircraft and automobile components where saving of weight is an advantage.

2.28 Aluminium Alloys

The aluminium may be alloyed with one or more other elements like copper, magnesium, manganese, silicon and nickel. The addition of small quantities of alloying elements converts the soft and weak metal into hard and strong metal, while still retaining its light weight. The main aluminium alloys are discussed below:

1. ***Duralumin.*** It is an important and interesting wrought alloy. Its composition is as follows:

Copper = 3.5 – 4.5%; Manganese = 0.4 – 0.7%; Magnesium = 0.4 – 0.7%, and the remainder is aluminium.

This alloy possesses maximum tensile strength (upto 400 MPa) after heat treatment and age hardening. After working, if the metal is allowed to age for 3 or 4 days, it will be hardened. This phenomenon is known as ***age hardening.***

It is widely used in wrought conditions for forging, stamping, bars, sheets, tubes and rivets. It can be worked in hot condition at a temperature of 500°C. However, after forging and annealing, it can also be cold worked. Due to its high strength and light weight, this alloy may be used in automobile and aircraft components. It is also used in manufacturing connecting rods, bars, rivets, pulleys, etc.

2. ***Y-alloy.*** It is also called copper-aluminium alloy. The addition of copper to pure aluminium increases its strength and machinability. The composition of this alloy is as follows :

Copper = 3.5 – 4.5%; Manganese = 1.2 – 1.7%; Nickel = 1.8 – 2.3%; Silicon, Magnesium, Iron = 0.6% each; and the remainder is aluminium.

This alloy is heat treated and age hardened like duralumin. The ageing process is carried out at room temperature for about five days.

It is mainly used for cast purposes, but it can also be used for forged components like duralumin. Since *Y*-alloy has better strength (than duralumin) at high temperature, therefore, it is much used in aircraft engines for cylinder heads and pistons.

3. ***Magnalium.*** It is made by melting the aluminium with 2 to 10% magnesium in a vacuum and then cooling it in a vacuum or under a pressure of 100 to 200 atmospheres. It also contains about 1.75% copper. Due to its light weight and good mechanical properties, it is mainly used for aircraft and automobile components.

4. ***Hindalium.*** It is an alloy of aluminium and magnesium with a small quantity of chromium. It is the trade name of aluminium alloy produced by Hindustan Aluminium Corporation Ltd, Renukoot (U.P.). It is produced as a rolled product in 16 gauge, mainly for anodized utensil manufacture.

2.29 Copper

It is one of the most widely used non-ferrous metals in industry. It is a soft, malleable and ductile material with a reddish-brown appearance. Its specific gravity is 8.9 and melting point is 1083°C. The tensile strength varies from 150 MPa to 400 MPa under different conditions. It is a good conductor of electricity. It is largely used in making electric cables and wires for electric machinery and appliances, in electrotyping and electroplating, in making coins and household utensils.

It may be cast, forged, rolled and drawn into wires. It is non-corrosive under ordinary conditions and resists weather very effectively. Copper in the form of tubes is used widely in mechanical engineering. It is also used for making ammunitions. It is used for making useful alloys with tin, zinc, nickel and aluminium.

Malachite is an ore of copper. Its dramatic bands of dark green make it popular in jewellery.

2.30 Copper Alloys

The copper alloys are broadly classified into the following two groups :

1. ***Copper-zinc alloys (Brass).*** The most widely used copper-zinc alloy is brass. There are various types of brasses, depending upon the proportions of copper and zinc. This is fundamentally a binary alloy of copper with zinc each 50%. By adding small quantities of other elements, the properties of brass may be greatly changed. For example, the addition of lead (1 to 2%) improves the machining quality of brass. It has a greater strength than that of copper, but have a lower thermal and electrical conductivity. Brasses are very resistant to atmospheric corrosion and can be easily soldered. They can be easily fabricated by processes like spinning and can also be electroplated with metals like nickel and chromium. The following table shows the composition of various types of brasses according to Indian standards.

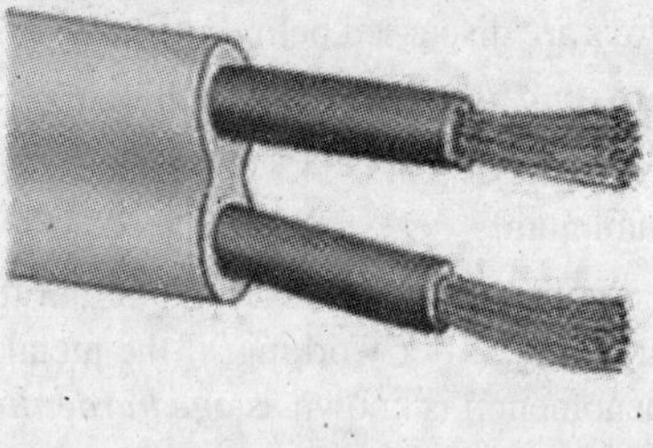
Electrical cables often consist of fine strands of copper wire woven together and encased in a plastic sleeve.

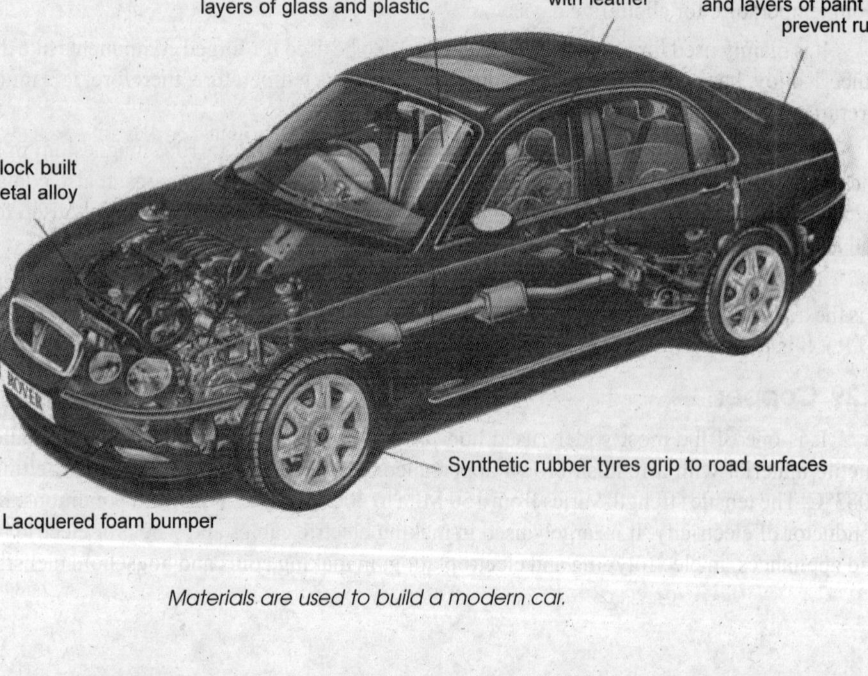

Materials are used to build a modern car.

Table 2.12. Composition and uses of brasses.

Indian standard designation	*Composition in percentages*	*Uses*
Cartridge brass	Copper = 70 Zinc = 30	It is a cold working brass used for cold rolled sheets, wire drawing, deep drawing, pressing and tube manufacture.
Yellow brass (Muntz metal)	Copper = 60 Zinc = 40	It is suitable for hot working by rolling, extrusion and stamping.
Leaded brass	Copper = 62.5 Zinc = 36 Lead = 1.5	
Admiralty brass	Copper = 70 Zinc = 29 Tin = 1	These are used for plates, tubes, etc.
Naval brass	Copper = 59 Zinc = 40 Tin = 1	It is used for marine castings.
Nickel brass (German silver or Nickel silver)	Copper = 60 – 45 Zinc = 35 – 20 Nickel = 5 – 35	It is used for valves, plumbing fittings, automobile fitting, type writer parts and musical instruments.

2. ***Copper-tin alloys (Bronze).*** The alloys of copper and tin are usually termed as bronzes. The useful range of composition is 75 to 95% copper and 5 to 25% tin. The metal is comparatively hard, resists surface wear and can be shaped or rolled into wires, rods and sheets very easily. In corrosion resistant properties, bronzes are superior to brasses. Some of the common types of bronzes are as follows:

(*a*) ***Phosphor bronze.*** When bronze contains phosphorus, it is called phosphor bronze. Phosphorus increases the strength, ductility and soundness of castings. The tensile strength of this alloy when cast varies from 215 MPa to 280 MPa but increases upto 2300 MPa when rolled or drawn. This alloy possesses good wearing qualities and high elasticity. The metal is resistant to salt water corrosion. The composition of the metal varies according to whether it is to be forged, wrought or made into castings. A common type of phosphor bronze has the following composition according to Indian standards :

Copper = 87–90%, Tin = 9–10%, and Phosphorus = 0.1–3%.

It is used for bearings, worm wheels, gears, nuts for machine lead screws, pump parts, linings and for many other purposes. It is also suitable for making springs.

(*b*) ***Silicon bronze.*** It contains 96% copper, 3% silicon and 1% manganese or zinc. It has good general corrosion resistance of copper combined with higher strength. It can be cast, rolled, stamped, forged and pressed either hot or cold and it can be welded by all the usual methods.

It is widely used for boilers, tanks, stoves or where high strength and good corrosion resistance is required.

(*c*) ***Beryllium bronze.*** It is a copper base alloy containing about 97.75% copper and 2.25% beryllium. It has high yield point, high fatigue limit and excellent cold and hot corrosion resistance. It is particularly suitable material for springs, heavy duty electrical switches, cams and bushings. Since the wear resistance of beryllium copper is five times that of phosphor bronze, therefore, it may be used as a bearing metal in place of phosphor bronze.

It has a film forming and a soft lubricating property, which makes it more suitable as a bearing metal.

(*d*) ***Manganese bronze.*** It is an alloy of copper, zinc and little percentage of manganese. The usual composition of this bronze is as follows:

Copper = 60%, Zinc = 35%, and Manganese = 5%

This metal is highly resistant to corrosion. It is harder and stronger than phosphor bronze. It is generally used for bushes, plungers, feed pumps, rods etc. Worm gears are frequently made from this bronze.

(*e*) ***Aluminium bronze.*** It is an alloy of copper and aluminium. The aluminium bronze with 6–8% aluminium has valuable cold working properties. The maximum tensile strength of this alloy is 450 MPa with 11% of aluminium. They are most suitable for making components exposed to severe corrosion conditions. When iron is added to these bronzes, the mechanical properties are improved by refining the grain size and improving the ductility.

Aluminium bronzes are widely used for making gears, propellers, condenser bolts, pump components, tubes, air pumps, slide valves and bushings, etc. Cams and rollers are also made from this alloy. The 6% aluminium alloy has a fine gold colour which is used for imitation jewellery and decorative purposes.

2.31 Gun Metal

It is an alloy of copper, tin and zinc. It usually contains 88% copper, 10% tin and 2% zinc. This metal is also known as ***Admiralty gun metal.*** The zinc is added to clean the metal and to increase its fluidity.

It is not suitable for being worked in the cold state but may be forged when at about 600°C. The metal is very strong and resistant to corrosion by water and atmosphere. Originally, it was made for casting guns. It is extensively used for casting boiler fittings, bushes, bearings, glands, etc.

2.32 Lead

It is a bluish grey metal having specific gravity 11.36 and melting point 326°C. It is so soft that it can be cut with a knife. It has no tenacity. It is extensively used for making solders, as a lining for acid tanks, cisterns, water pipes, and as coating for electrical cables.

The lead base alloys are employed where a cheap and corrosion resistant material is required. An alloy containing 83% lead, 15% antimony, 1.5% tin and 0.5% copper is used for large bearings subjected to light service.

2.33 Tin

It is brightly shining white metal. It is soft, malleable and ductile. It can be rolled into very thin sheets. It is used for making important alloys, fine solder, as a protective coating for iron and steel sheets and for making tin foil used as moisture proof packing.

A tin base alloy containing 88% tin, 8% antimony and 4% copper is called ***babbit metal.*** It is a soft material with a low coefficient of friction and has little strength. It is the most common bearing metal used with cast iron boxes where the bearings are subjected to high pressure and load.

Note : Those alloys in which lead and tin are predominating are designated as ***white metal bearing alloys.*** This alloy is used for lining bearings subjected to high speeds like the bearings of aero-engines.

2.34 Bearing Metals

The following are the widely used bearing metals :

1. Copper-base alloys, **2.** Lead-base alloys, **3.** Tin-base alloys, and **4.** Cadmium-base alloys

The **copper base alloys** are the most important bearing alloys. These alloys are harder and stronger than the white metals (lead base and tin base alloys) and are used for bearings subjected to heavy pressures. These include brasses and bronzes which are discussed in Art 2.30. The **lead base** and **tin base alloys** are discussed in Art. 2.32 and 2.33. The **cadmium base alloys** contain 95% cadmium and 5% silver. It is used for medium loaded bearings subjected to high temperature.

The selection of a particular type of bearing metal depends upon the conditions under which it is to be used. It involves factors relating to bearing pressures, rubbing speeds, temperatures, lubrication, etc. A bearing material should have the following properties:

1. It should have low coefficient of friction.
2. It should have good wearing qualities.
3. It should have ability to withstand bearing pressures.
4. It should have ability to operate satisfactorily with suitable lubrication means at the maximum rubbing speeds.
5. It should have a sufficient melting point.
6. It should have high thermal conductivity.
7. It should have good casting qualities.
8. It should have minimum shrinkage after casting.
9. It should have non-corrosive properties.
10. It should be economical in cost.

2.35 Zinc Base Alloys

The most of the die castings are produced from zinc base alloys. These alloys can be casted easily with a good finish at fairly low temperatures. They have also considerable strength and are low in cost. The usual alloying elements for zinc are aluminium, copper and magnesium and they are all held in close limits.

The composition of two standard die casting zinc alloys are as follows :

1. Aluminium 4.1%, copper 0.1%, magnesium 0.04% and the remainder is zinc.
2. Aluminium 4.1%, copper 1%, magnesium 0.04% and the remainder is zinc.

Aluminium improves the mechanical properties and also reduces the tendency of zinc to dissolve iron. Copper increases the tensile strength, hardness and ductility. Magnesium has the beneficial effect of making the castings permanently stable. These alloys are widely used in the automotive industry and for other high production markets such as washing machines, oil burners, refrigerators, radios, photographs, television, business machines, etc.

This copper statue, believed to be the world's oldest metal sculpture, is an image of Egyptian pharaoh Pepi I. This old kingdom pharaoh reigned from 2289 to 2244 BC.

2.36 Nickel Base Alloys

The nickel base alloys are widely used in engineering industry on account of their high mechanical strength properties, corrosion resistance, etc. The most important nickel base alloys are discussed below:

1. *Monel metal.* It is an important alloy of nickel and copper. It contains 68% nickel, 29% copper and 3% other constituents like iron, manganese, silicon and carbon. Its specific gravity is 8.87 and melting point 1360°C. It has a tensile strength from 390 MPa to 460 MPa. It resembles nickel in appearance and is strong, ductile and tough. It is superior to brass and bronze in corrosion resisting properties. It is used for making propellers, pump fittings, condenser tubes, steam turbine blades, sea water exposed parts, tanks and chemical and food handling plants.

2. *Inconel.* It consists of 80% nickel, 14% chromium, and 6% iron. Its specific gravity is 8.55 and melting point 1395°C. This alloy has excellent mechanical properties at ordinary and elevated temperatures. It can be cast, rolled and cold drawn. It is used for making springs which have to withstand high temperatures and are exposed to corrosive action. It is also used for exhaust manifolds of aircraft engines.

3. *Nichrome.* It consists of 65% nickel, 15% chromium and 20% iron. It has high heat and oxidation resistance. It is used in making electrical resistance wire for electric furnaces and heating elements.

4. *Nimonic.* It consists of 80% nickel and 20% chromium. It has high strength and ability to operate under intermittent heating and cooling conditions. It is widely used in gas turbine engines.

2.37 Non-metallic Materials

The non-metallic materials are used in engineering practice due to their low density, low cost, flexibility, resistant to heat and electricity. Though there are many non-metallic materials, yet the following are important from the subject point of view.

1. *Plastics.* The plastics are synthetic materials which are moulded into shape under pressure with or without the application of heat. These can also be cast, rolled, extruded, laminated and machined. Following are the two types of plastics :

(*a*) Thermosetting plastics, and

(*b*) Thermoplastic.

The ***thermosetting plastics*** are those which are formed into shape under heat and pressure and results in a permanently hard product. The heat first softens the material, but as additional heat and pressure is applied, it becomes hard by a chemical change known as phenol-formaldehyde (Bakelite), phenol-furfural (Durite), urea-formaldehyde (Plaskon), etc.

Reinforced plastic with fibreglass makes the material to withstand high compressive as well as tensile stresses.

The ***thermoplastic*** materials do not become hard with the application of heat and pressure and no chemical change occurs. They remain soft at elevated temperatures until they are hardened by cooling. These can be remelted repeatedly by successive application of heat. Some of the common thermoplastics are cellulose nitrate (Celluloid), polythene, polyvinyl acetate, polyvinyl chloride (P.V.C.), etc.

The plastics are extremely resistant to corrosion and have a high dimensional stability. They are mostly used in the manufacture of aeroplane and automobile parts. They are also used for making safety glasses, laminated gears, pulleys, self-lubricating bearing, etc. due to their resilience and strength.

2. *Rubber.* It is one of the most important natural plastics. It resists abrasion, heat, strong alkalis and fairly strong acids. Soft rubber is used for electrical insulations. It is also used for power transmission belting, being applied to woven cotton or cotton cords as a base. The hard rubber is used for piping and as lining for pickling tanks.

3. *Leather.* It is very flexible and can withstand considerable wear under suitable conditions. It is extensively used for power transmission belting and as a packing or as washers.

4. *Ferrodo.* It is a trade name given to asbestos lined with lead oxide. It is generally used as a friction lining for clutches and brakes.

QUESTIONS

1. How do you classify materials for engineering use?
2. What are the factors to be considered for the selection of materials for the design of machine elements? Discuss.
3. Enumerate the most commonly used engineering materials and state at least one important property and one application of each.
4. Why are metals in their pure form unsuitable for industrial use?
5. Define 'mechanical property' of an engineering material. State any six mechanical properties, give their definitions and one example of the material possessing the properties.
6. Define the following properties of a material :
   (*i*) Ductility, (*ii*) Toughness, (*iii*) Hardness, and (*iv*) Creep.
7. Distinguish clearly amongst cast iron, wrought iron and steel regarding their constituents and properties.
8. How cast iron is obtained? Classify and explain different types of cast irons.
9. How is grey cast iron designated in Indian standards?
10. Discuss the effect of silicon, manganese, sulphur and phosphorus on cast iron.
11. Define plain carbon steel. How it is designated according to Indian standards?
12. Define alloy steel. Discuss the effects of nickel, chromium and manganese on steel.
13. What are the common materials used in Mechanical Engineering Design? How can the properties of steel be improved?
14. State the alloying elements added to steel to get alloy steels and the effect they produce. Give at least one example of each.
15. Give the composition of 35 Mn 2 Mo 45 steel. List its main uses.
16. Write short notes on free cutting steel, and stainless steel.
17. Select suitable material for the following cases, indicating the reason;
   1. A shaft subjected to variable torsional and bending load ; 2. Spring used in a spring loaded safety valve; 3. Nut of a heavy duty screw jack; and 4. Low speed line-shaft coupling.
18. Select suitable materials for the following parts stating the special property which makes it most suitable for use in manufacturing:
   1. Turbine blade, 2. Bush bearing, 3. Dies, 4. Carburetor body, 5. Keys (used for fastening), 6. Cams, 7. Heavy duty machine tool beds, 8. Ball bearing, 9. Automobile cylinder block, 10. Helical springs.
19. Suggest suitable materials for the following parts stating the special property which makes it more suitable for use in manufacturing:
   1. Diesel engine crankshaft ; 2. Automobile tyres ; 3. Roller bearings ; 4. High pressure steam pipes ; 5. Stay bar of boilers ; 6. Worm and worm gear ; 7. Dies; 8. Tramway axle ; 9. Cam follower ; 10. Hydraulic brake piston.
20. Write short notes on high speed tool steel and spring steel.
21. Explain the following heat treatment processes:
   1. Normalising; 2. Hardening; and 3. Tempering.
22. Write short note on the type of bearing metals.
23. Discuss the important non-metallic materials of construction used in engineering practice.

OBJECTIVE TYPE QUESTIONS

1. Which of the following material has the maximum ductility?
   (*a*) Mild steel (*b*) Copper
   (*c*) Zinc (*d*) Aluminium
2. According to Indian standard specifications, a grey cast iron designated by 'FG 200' means that the
   (*a*) carbon content is 2%
   (*b*) maximum compressive strength is 200 N/mm^2
   (*c*) minimum tensile strength is 200 N/mm^2
   (*d*) maximum shear strength is 200 N/mm^2

3. Steel containing upto 0.15% carbon is known as
   (*a*) mild steel (*b*) dead mild steel
   (*c*) medium carbon steel (*d*) high carbon steel
4. According to Indian standard specifications, a plain carbon steel designated by 40C8 means that
   (*a*) carbon content is 0.04 per cent and manganese is 0.08 per cent
   (*b*) carbon content is 0.4 per cent and manganese is 0.8 per cent
   (*c*) carbon content is 0.35 to 0.45 per cent and manganese is 0.60 to 0.90 per cent
   (*d*) carbon content is 0.60 to 0.80 per cent and manganese is 0.8 to 1.2 per cent
5. The material commonly used for machine tool bodies is
   (*a*) mild steel (*b*) aluminium
   (*c*) brass (*d*) cast iron
6. The material commonly used for crane hooks is
   (*a*) cast iron (*b*) wrought iron
   (*c*) mild steel (*d*) aluminium
7. Shock resistance of steel is increased by adding
   (*a*) nickel (*b*) chromium
   (*c*) nickel and chromium (*d*) sulphur, lead and phosphorus
8. The steel widely used for motor car crankshafts is
   (*a*) nickel steel (*b*) chrome steel
   (*c*) nickel-chrome steel (*d*) silicon steel
9. A steel with 0.8 per cent carbon is known as
   (*a*) eutectoid steel (*b*) hypereutectoid steel
   (*c*) hypoeutectoid steel (*d*) none of these
10. 18/8 steel contains
    (*a*) 18 per cent nickel and 8 per cent chromium
    (*b*) 18 per cent chromium and 8 per cent nickel
    (*c*) 18 per cent nickel and 8 per cent vanadium
    (*d*) 18 per cent vanadium and 8 per cent nickel
11. Ball bearing are usually made from
    (*a*) low carbon steel (*b*) high carbon steel
    (*c*) medium carbon steel (*d*) high speed steel
12. The process which improves the machinability of steels, but lower the hardness and tensile strength is
    (*a*) normalising (*b*) full annealing
    (*c*) process annealing (*d*) spheroidising
13. The metal suitable for bearings subjected to heavy loads is
    (*a*) silicon bronze (*b*) white metal
    (*c*) monel metal (*d*) phosphor bronze
14. The metal suitable for bearings subjected to light loads is
    (*a*) silicon bronze (*b*) white metal
    (*c*) monel metal (*d*) phosphor bronze
15. Thermoplastic materials are those materials which
    (*a*) are formed into shape under heat and pressure and results in a permanently hard product
    (*b*) do not become hard with the application of heat and pressure and no chemical change occurs
    (*c*) are flexible and can withstand considerable wear under suitable conditions
    (*d*) are used as a friction lining for clutches and brakes

ANSWERS

1. (*a*)	2. (*c*)	3. (*b*)	4. (*c*)	5. (*d*)
6. (*b*)	7. (*c*)	8. (*b*)	9. (*a*)	10. (*b*)
11. (*c*)	12. (*d*)	13. (*b*)	14. (*d*)	15. (*b*)

CHAPTER

3

Manufacturing Considerations in Machine Design

3.1 Introduction

In the previous chapter, we have only discussed about the composition, properties and uses of various materials used in Mechanical Engineering. We shall now discuss in this chapter a few of the manufacturing processes, limits and fits, etc.

3.2 Manufacturing Processes

The knowledge of manufacturing processes is of great importance for a design engineer. The following are the various manufacturing processes used in Mechanical Engineering.

1. *Primary shaping processes.* The processes used for the preliminary shaping of the machine component are known as primary shaping processes. The common operations used for this process are casting, forging, extruding, rolling, drawing, bending, shearing, spinning, powder metal forming, squeezing, etc.

2. ***Machining processes.*** The processes used for giving final shape to the machine component, according to planned dimensions are known as machining processes. The common operations used for this process are turning, planning, shaping, drilling, boring, reaming, sawing, broaching, milling, grinding, hobbing, etc.

3. ***Surface finishing processes.*** The processes used to provide a good surface finish for the machine component are known as surface finishing processes. The common operations used for this process are polishing, buffing, honing, lapping, abrasive belt grinding, barrel tumbling, electroplating, superfinishing, sheradizing, etc.

4. ***Joining processes.*** The processes used for joining machine components are known as joining processes. The common operations used for this process are welding, riveting, soldering, brazing, screw fastening, pressing, sintering, etc.

5. ***Processes effecting change in properties.*** These processes are used to impart certain specific properties to the machine components so as to make them suitable for particular operations or uses. Such processes are heat treatment, hot-working, cold-working and shot peening.

To discuss in detail all these processes is beyond the scope of this book, but a few of them which are important from the subject point of view will be discussed in the following pages.

3.3 Casting

It is one of the most important manufacturing process used in Mechanical Engineering. The castings are obtained by remelting of ingots* in a cupola or some other foundry furnace and then pouring this molten metal into metal or sand moulds. The various important casting processes are as follows:

1. ***Sand mould casting.*** The casting produced by pouring molten metal in sand mould is called sand mould casting. It is particularly used for parts of larger sizes.

2. ***Permanent mould casting.*** The casting produced by pouring molten metal in a metallic mould is called permanent mould casting. It is used for casting aluminium pistons, electric iron parts, cooking utensils, gears, etc. The permanent mould castings have the following advantages:

1. Shaping the Sand : A wooden pattern cut to the shape of one half of the casting is positioned in an iron box and surrounded by tightly packed moist sand.

2. Ready for the Metal : After the wooden patterns have been removed, the two halves of the mould are clamped together. Molten iron is poured into opening called the runner.

* Most of the metals used in industry are obtained from ores. These ores are subjected to suitable reducing or refining process which gives the metal in a molten form. This molten metal is poured into moulds to give commercial castings, called **ingots.**

(*a*) It has more favourable fine grained structure.

(*b*) The dimensions may be obtained with close tolerances.

(*c*) The holes up to 6.35 mm diameter may be easily cast with metal cores.

3. ***Slush casting.*** It is a special application of permanent metal mould casting. This method is used for production of hollow castings without the use of cores.

4. ***Die casting.*** The casting produced by forcing molten metal under pressure into a permanent metal mould (known as die) is called die casting. A die is usually made in two halves and when closed it forms a cavity similar to the casting desired. One half of the die that remains stationary is known as ***cover die*** and the other movable half is called ***ejector die.*** The die casting method is mostly used for castings of non-ferrous metals of comparatively low fusion temperature. This process is cheaper and quicker than permanent or sand mould casting. Most of the automobile parts like fuel pump, carburettor bodies, horn, heaters, wipers, brackets, steering wheels, hubs and crank cases are made with this process. Following are the advantages and disadvantages of die casting :

Aluminium die casting component

Advantages

(*a*) The production rate is high, ranging up to 700 castings per hour.

(*b*) It gives better surface smoothness.

(*c*) The dimensions may be obtained within tolerances.

(*d*) The die retains its trueness and life for longer periods. For example, the life of a die for zinc base castings is upto one million castings, for copper base alloys upto 75 000 castings and for aluminium base alloys upto 500 000 castings.

Sand Casting

Investment Casting

(*e*) It requires less floor area for equivalent production by other casting methods.

(*f*) By die casting, thin and complex shapes can be easily produced.

(*g*) The holes up to 0.8 mm can be cast.

Disadvantages

(*a*) The die casting units are costly.

(*b*) Only non-ferrous alloys are casted more economically.

(*c*) It requires special skill for maintenance and operation of a die casting machine.

5. ***Centrifugal casting.*** The casting produced by a process in which molten metal is poured and allowed to solidify while the mould is kept revolving, is known as centrifugal casting. The metal thus poured is subjected to centrifugal force due to which it flows in the mould cavities. This results in the production of high density castings with promoted directional solidification. The examples of centrifugal castings are pipes, cylinder liners and sleeves, rolls, bushes, bearings, gears, flywheels, gun barrels, piston rings, brake drums, etc.

3.4 Casting Design

An engineer must know how to design the castings so that they can effectively and efficiently render the desired service and can be produced easily and economically. In order to design a casting, the following factors must be taken into consideration :

1. The function to be performed by the casting,
2. Soundness of the casting,
3. Strength of the casting,
4. Ease in its production,
5. Consideration for safety, and
6. Economy in production.

In order to meet these requirements, a design engineer should have a thorough knowledge of production methods including pattern making, moulding, core making, melting and pouring, etc. The best designs will be achieved only when one is able to make a proper selection out of the various available methods. However, a few rules for designing castings are given below to serve as a guide:

1. The sharp corners and frequent use of fillets should be avoided in order to avoid concentration of stresses.
2. All sections in a casting should be designed of uniform thickness, as far as possible. If, however, variation is unavoidable, it should be done gradually.
3. An abrupt change of an extremely thick section into a very thin section should always be avoided.
4. The casting should be designed as simple as possible, but with a good appearance.
5. Large flat surfaces on the casting should be avoided because it is difficult to obtain true surfaces on large castings.
6. In designing a casting, the various allowances must be provided in making a pattern.
7. The ability to withstand contraction stresses of some members of the casting may be improved by providing the curved shapes *e.g.*, the arms of pulleys and wheels.
8. The stiffening members such as webs and ribs used on a casting should be minimum possible in number, as they may give rise to various defects like hot tears and shrinkage, etc.
9. The casting should be designed in such a way that it will require a simpler pattern and its moulding is easier.
10. In order to design cores for casting, due consideration should be given to provide them adequate support in the mould.

11. The deep and narrow pockets in the casting should invariably be avoided to reduce cleaning costs.
12. The use of metal inserts in the casting should be kept minimum.
13. The markings such as names or numbers, etc., should never be provided on vertical surfaces because they provide a hindrance in the withdrawl of pattern.
14. A tolerance of ± 1.6 mm on small castings (below 300 mm) should be provided. In case more dimensional accuracy is desired, a tolerance of ± 0.8 mm may be provided.

3.5 Forging

It is the process of heating a metal to a desired temperature in order to acquire sufficient plasticity, followed by operations like hammering, bending and pressing, etc. to give it a desired shape. The various forging processes are :

1. Smith forging or hand forging
2. Power forging,
3. Machine forging or upset forging, and
4. Drop forging or stamping

The **smith** or **hand forging** is done by means of hand tools and it is usually employed for small jobs. When the forging is done by means of power hammers, it is then known as ***power forging.*** It is used for medium size and large articles requiring very heavy blows. The **machine forging** is done by means of forging machines. The **drop forging** is carried out with the help of drop hammers and is particularly suitable for mass production of identical parts. The forging process has the following advantages :

1. It refines the structure of the metal.
2. It renders the metal stronger by setting the direction of grains.
3. It effects considerable saving in time, labour and material as compared to the production of a similar item by cutting from a solid stock and then shaping it.
4. The reasonable degree of accuracy may be obtained by forging.
5. The forgings may be welded.

It may be noted that wrought iron and various types of steels and steel alloys are the common raw material for forging work. Low carbon steels respond better to forging work than the high carbon steels. The common non-ferrous metals and alloys used in forging work are brass, bronze, copper, aluminium and magnesium alloys. The following table shows the temperature ranges for forging some common metals.

Table 3.1. Temperature ranges for forging.

Material	*Forging temperature (°C)*	*Material*	*Forging temperature (°C)*
Wrought iron	900 – 1300	Stainless steel	940 – 1180
Mild steel	750 – 1300	Aluminium and magnesium alloys	350 – 500
Medium carbon steel	750 – 1250		
High carbon and alloy steel	800 – 1150	Copper, brass and bronze	600 – 950

3.6 Forging Design

In designing a forging, the following points should always be considered.

1. The forged components should ultimately be able to achieve a radial flow of grains or fibres.
2. The forgings which are likely to carry flash, such as drop and press forgings, should preferably have the parting line in such a way that the same will divide them in two equal halves.
3. The parting line of a forging should lie, as far as possible, in one plane.
4. Sufficient draft on surfaces should be provided to facilitate easy removal of forgings from dies.
5. The sharp corners should always be avoided in order to prevent concentration of stress and to facilitate ease in forging.
6. The pockets and recesses in forgings should be minimum in order to avoid increased die wear.
7. The ribs should not be high and thin.
8. Too thin sections should be avoided to facilitate easy flow of metal.

3.7 Mechanical Working of Metals

The mechanical working of metals is defined as an intentional deformation of metals plastically under the action of externally applied forces.

The mechanical working of metal is described as hot working and cold working depending upon whether the metal is worked above or below the recrystallisation temperature. The metal is subjected to mechanical working for the following purposes :

1. To reduce the original block or ingot into desired shapes,
2. To refine grain size, and
3. To control the direction of flow lines.

3.8 Hot Working

The working of metals above the *recrystallisation temperature is called ***hot working.*** This temperature should not be too high to reach the solidus temperature, otherwise the metal will burn and become unsuitable for use. The hot working of metals has the following advantages and disadvantages :

Advantages

1. The porosity of the metal is largely eliminated.
2. The grain structure of the metal is refined.
3. The impurities like slag are squeezed into fibres and distributed throughout the metal.
4. The mechanical properties such as toughness, ductility, percentage elongation, percentage reduction in area, and resistance to shock and vibration are improved due to the refinement of grains.

Disadvantages

1. It requires expensive tools.
2. It produces poor surface finish, due to the rapid oxidation and scale formation on the metal surface.
3. Due to the poor surface finish, close tolerance cannot be maintained.

---

* The temperature at which the new grains are formed in the metal is known as **recrystallisation temperature.**

3.9 Hot Working Processes

The various *hot working processes are described as below :

1. *Hot rolling.* The hot rolling process is the most rapid method of converting large sections into desired shapes. It consists of passing the hot ingot through two rolls rotating in opposite directions at the same speed. The space between the rolls is adjusted to conform to the desired thickness of the rolled section. The rolls, thus, squeeze the passing ingot to reduce its cross-section and increase its length. The forming of bars, plates, sheets, rails, angles, I-beam and other structural sections are made by hot rolling.

Hot Rolling : When steel is heated until it glows bright red, it becomes soft enough to form into elabrate shapes.

2. *Hot forging.* It consists of heating the metal to plastic state and then the pressure is applied to form it into desired shapes and sizes. The pressure applied in this is not continuous as for hot rolling, but intermittent. The pressure may be applied by hand hammers, power hammers or by forging machines.

3. *Hot spinning.* It consists of heating the metal to forging temperature and then forming it into the desired shape on a spinning lathe. The parts of circular cross-section which are symmetrical about the axis of rotation, are made by this process.

4. *Hot extrusion.* It consists of pressing a metal inside a chamber to force it out by high pressure through an orifice which is shaped to provide the desired form of the finished part. Most commercial metals and their alloys such as steel, copper, aluminium and nickel are directly extruded at elevated temperatures. The rods, tubes, structural shapes, flooring strips and lead covered cables, etc., are the typical products of extrusion.

5. *Hot drawing or cupping.* It is mostly used for the production of thick walled seamless tubes and cylinders. It is usually performed in two stages. The first stage consists of drawing a cup out of a hot circular plate with the help of a die and punch. The second stage consists of reheating the drawn cup and drawing it further to the desired length having the required wall thickness. The second drawing operation is performed through a number of dies, which are arranged in a descending order of their diameters, so that the reduction of wall thickness is gradual in various stages.

Cold Rolled Steel : Many modern products are made from easily shaped sheet metal.

6. *Hot piercing.* This process is used for the manufacture of seamless tubes. In its operation, the heated cylindrical billets of steel are passed between two conical shaped rolls operating in the same direction. A mandrel is provided between these rolls which assist in piercing and controls the size of the hole, as the billet is forced over it.

* For complete details, please refer to Authors' popular book **'A Text Book of Workshop Technology'.**

3.10 Cold Working

The working of metals below their recrystallisation temperature is known as ***cold working.*** Most of the cold working processes are performed at room temperature. The cold working distorts the grain structure and does not provide an appreciable reduction in size. It requires much higher pressures than hot working. The extent to which a metal can be cold worked depends upon its ductility. The higher the ductility of the metal, the more it can be cold worked. During cold working, severe stresses known as residual stresses are set up. Since the presence of these stresses is undesirable, therefore, a suitable heat treatment may be employed to neutralise the effect of these stresses. The cold working is usually used as finishing operation, following the shaping of the metal by hot working. It also increases tensile strength, yield strength and hardness of steel but lowers its ductility. The increase in hardness due to cold working is called ***work-hardening.***

In general, cold working produces the following effects :

1. The stresses are set up in the metal which remain in the metal, unless they are removed by subsequent heat treatment.
2. A distortion of the grain structure is created.
3. The strength and hardness of the metal are increased with a corresponding loss in ductility.
4. The recrystalline temperature for steel is increased.
5. The surface finish is improved.
6. The close dimensional tolerance can be maintained.

3.11 Cold Working Processes

The various cold working processes are discussed below:

1. *Cold rolling*. It is generally employed for bars of all shapes, rods, sheets and strips, in order to provide a smooth and bright surface finish. It is also used to finish the hot rolled components to close tolerances and improve their toughness and hardness. The hot rolled articles are first immersed in an acid to remove the scale and washed in water, and then dried. This process of cleaning the articles is known as ***pickling.*** These cleaned articles are then passed through rolling mills. The rolling mills are similar to that used in hot rolling.

Gallium arsenide (GaAs) is now being manufactured as an alternative to silicon for microchips. This combination of elements is a semiconductor like silicon, but is electronically faster and therefore better for microprocessors.

Note : This picture is given as additional information and is not a direct example of the current chapter.

2. ***Cold forging.*** The cold forging is also called ***swaging.*** During this method of cold working, the metal is allowed to flow in some pre-determined shape according to the design of dies, by a compressive force or impact. It is widely used in forming ductile metals. Following are the three, commonly used cold forging processes :

(*a*) ***Sizing.*** It is the simplest form of cold forging. It is the operation of slightly compressing a forging, casting or steel assembly to obtain close tolerance and a flat surface. The metal is confined only in a vertical direction.

(*b*) ***Cold heading.*** This process is extensively used for making bolts, rivets and other similar headed parts. This is usually done on a cold header machine. Since the cold header is made from unheated material, therefore, the equipment must be able to withstand the high pressures that develop. The rod is fed to the machine where it is cut off and moved into the header die. The operation may be either single or double and upon completion, the part is ejected from the dies.

After making the bolt head, the threads are produced on a thread rolling machine. This is also a cold working process. The process consists of pressing the blank between two rotating rolls which have the thread form cut in their surface.

(*c*) ***Rotary swaging.*** This method is used for reducing the diameter of round bars and tubes by rotating dies which open and close rapidly on the work. The end of rod is tapered or reduced in size by a combination of pressure and impact.

3. ***Cold spinning.*** The process of cold spinning is similar to hot spinning except that the metal is worked at room temperature. The process of cold spinning is best suited for aluminium and other soft metals. The commonly used spun articles out of aluminum and its alloys are processing kettles, cooking utensils, liquid containers, and light reflectors, etc.

4. ***Cold extrusion.*** The principle of cold extrusion is exactly similar to hot extrusion. The most common cold extrusion process is ***impact extrusion.*** The operation of cold extrusion is performed with the help of a punch and die. The work material is placed in position into a die and struck from top

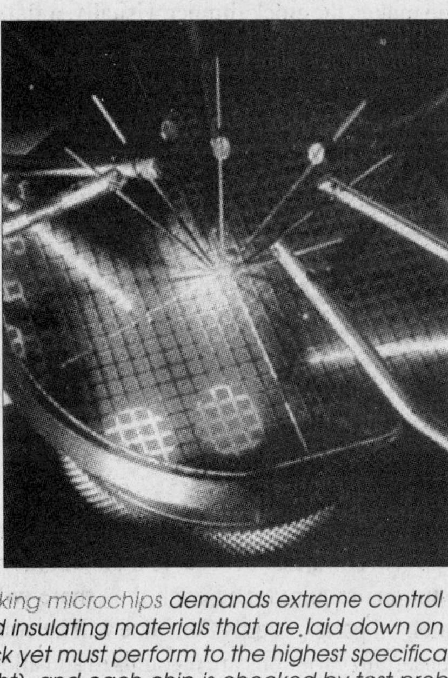

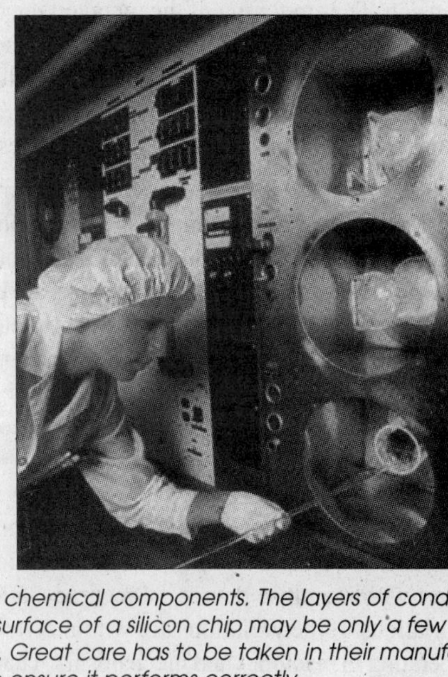

Making microchips demands extreme control over chemical components. The layers of conducting and insulating materials that are laid down on the surface of a silicon chip may be only a few atoms thick yet must perform to the highest specifications. Great care has to be taken in their manufacture (right), and each chip is checked by test probes to ensure it performs correctly.

Note : This picture is given as additional information and is not a direct example of the current chapter.

by a punch operating at high pressure and speed. The metal flows up along the surface of the punch forming a cup-shaped component. When the punch moves up, compressed air is used to separate the component from the punch. The thickness of the side wall is determined by the amount of clearance between the punch and die. The process of impact extrusion is limited to soft and ductile materials such as lead, tin, aluminium, zinc and some of their alloys. The various items of daily use such as tubes for shaving creams and tooth pastes and such other thin walled products are made by impact extrusion.

5. ***Cold drawing.*** It is generally employed for bars, rods, wires, etc. The important cold drawing processes are as follows:

(*a*) ***Bar or rod drawing.*** In bar drawing, the hot drawn bars or rods from the mills are first pickled, washed and coated to prevent oxidation. A draw bench, is employed for cold drawing. One end of the bar is reduced in diameter by the swaging operation to permit it to enter a drawing die. This end of bar is inserted through the die and gripped by the jaws of the carriage fastened to the chain of the draw bench. The length of bars which can be drawn is limited by the maximum travel of the carriage, which may be from 15 metres to 30 metres. A high surface finish and dimensional accuracy is obtained by cold drawing. The products may be used directly without requiring any machining.

(*b*) ***Wire drawing.*** In wire drawing, the rolled bars from the mills are first pickled, washed and coated to prevent oxidation. They are then passed through several dies of decreasing diameter to provide the desired reduction in size. The dies are usually made of carbide materials.

(*c*) ***Tube drawing.*** The tube drawing is similar to bar drawing and in most cases it is accomplished with the use of a draw bench.

6. ***Cold bending.*** The bars, wires, tubes, structural shapes and sheet metal may be bent to many shapes in cold condition through dies. A little consideration will show that when the metal is bend beyond the elastic limit, the inside of the bend will be under compression while the outside will be under tension. The stretching of the metal on the outside makes the stock thinner. Usually, a flat strip of metal is bend by ***roll forming.*** The materials commonly used for roll forming are carbon steel, stainless steel, bronze, copper, brass, zinc and aluminium. Some of its products are metal windows, screen frame parts, bicycle wheel rims, trolley rails, etc. Most of the tubing is now-a-days are roll formed in cold conditions and then welded by resistance welding.

7. ***Cold peening.*** This process is used to improve the fatigue resistance of the metal by setting up compressive stresses in its surface. This is done by blasting or hurling a rain of small shot at high velocity against the surface to be peened. The shot peening is done by air blast or by some mechanical means. As the shot strikes, small indentations are produced, causing a slight plastic flow of the surface metal to a depth of a few hundreds of a centimetre. This stretching of the outer fibres is resisted by those underneath, which tend to return them to their original length, thus producing an outer layer having a compressive stress while those below are in tension. In addition, the surface is slightly hardened and strengthened by the cold working operation.

3.12 Interchangeability

The term interchangeability is normally employed for the mass production of indentical items within the prescribed limits of sizes. A little consideration will show that in order to maintain the sizes of the part within a close degree of accuracy, a lot of time is required. But even then there will be small variations. If the variations are within certain limits, all parts of equivalent size will be equally fit for operating in machines and mechanisms. Therefore, certain variations are recognised and allowed in the sizes of the mating parts to give the required fitting. This facilitates to select at random from a large number of parts for an assembly and results in a considerable saving in the cost of production. In order to control the size of finished part, with due allowance for error, for interchangeable parts is called ***limit system.***

It may be noted that when an assembly is made of two parts, the part which enters into the other, is known as ***enveloped surface*** (or **shaft** for cylindrical part) and the other in which one enters is called ***enveloping surface*** (or **hole** for cylindrical part).

Notes: 1. The term ***shaft*** refers not only to the diameter of a circular shaft, but it is also used to designate any external dimension of a part.

2. The term ***hole*** refers not only to the diameter of a circular hole, but it is also used to designate any internal dimension of a part.

3.13 Important Terms used in Limit System

The following terms used in limit system (or interchangeable system) are important from the subject point of view:

1. ***Nominal size.*** It is the size of a part specified in the drawing as a matter of convenience.

2. ***Basic size.*** It is the size of a part to which all limits of variation (*i.e.* tolerances) are applied to arrive at final dimensioning of the mating parts. The nominal or basic size of a part is often the same.

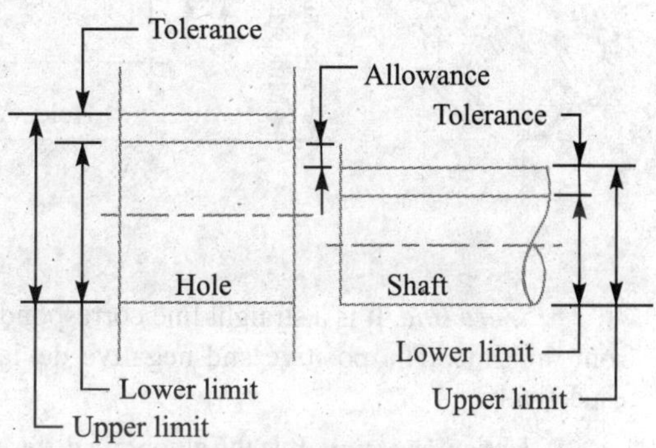

Fig. 3.1. Limits of sizes.

3. ***Actual size.*** It is the actual measured dimension of the part. The difference between the basic size and the actual size should not exceed a certain limit, otherwise it will interfere with the interchangeability of the mating parts.

4. ***Limits of sizes.*** There are two extreme permissible sizes for a dimension of the part as shown in Fig. 3.1. The largest permissible size for a dimension of the part is called ***upper*** or ***high*** or ***maximum limit,*** whereas the smallest size of the part is known as ***lower*** or ***minimum limit.***

5. ***Allowance.*** It is the difference between the basic dimensions of the mating parts. The allowance may be ***positive*** or ***negative***. When the shaft size is less than the hole size, then the allowance is ***positive*** and when the shaft size is greater than the hole size, then the allowance is ***negative.***

6. ***Tolerance.*** It is the difference between the upper limit and lower limit of a dimension. In other words, it is the maximum permissible variation in a dimension. The tolerance may be ***unilateral*** or ***bilateral.*** When all the tolerance is allowed on one side of the nominal size, *e.g.* $20^{+0.000}_{-0.004}$, then it is said to be ***unilateral system of tolerance.*** The unilateral system is mostly used in industries as it permits changing the tolerance value while still retaining the same allowance or type of fit.

(*a*) Unilateral tolerance. (*b*) Bilateral tolerance.

Fig. 3.2. Method of assigning tolerances.

When the tolerance is allowed on both sides of the nominal size, *e.g.* $20^{+0.002}_{-0.002}$, then it is said to be ***bilateral system of tolerance.*** In this case + 0.002 is the upper limit and – 0.002 is the lower limit.

The method of assigning unilateral and bilateral tolerance is shown in Fig. 3.2 (*a*) and (*b*) respectively.

7. ***Tolerance zone.*** It is the zone between the maximum and minimum limit size, as shown in Fig. 3.3.

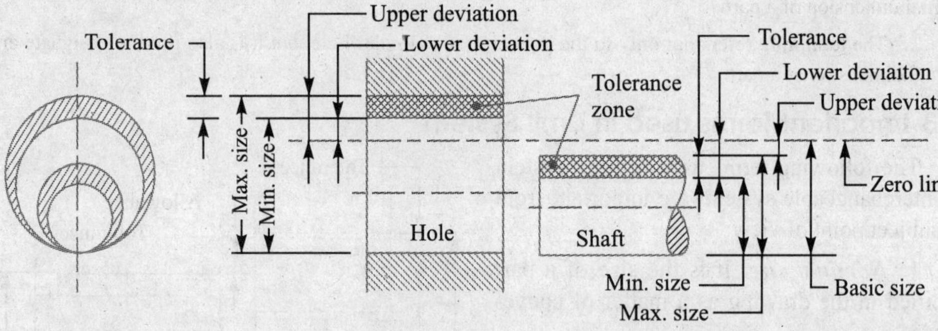

Fig. 3.3. Tolerance zone.

8. ***Zero line.*** It is a straight line corresponding to the basic size. The deviations are measured from this line. The positive and negative deviations are shown above and below the zero line respectively.

9. ***Upper deviation.*** It is the algebraic difference between the maximum size and the basic size. The upper deviation of a hole is represented by a symbol *ES* (Ecart Superior) and of a shaft, it is represented by *es*.

10. ***Lower deviation.*** It is the algebraic difference between the minimum size and the basic size. The lower deviation of a hole is represented by a symbol *EI* (Ecart Inferior) and of a shaft, it is represented by *ei*.

11. ***Actual deviation.*** It is the algebraic difference between an actual size and the corresponding basic size.

12. ***Mean deviation.*** It is the arithmetical mean between the upper and lower deviations.

13. ***Fundamental deviation.*** It is one of the two deviations which is conventionally chosen to define the position of the tolerance zone in relation to zero line, as shown in Fig. 3.4.

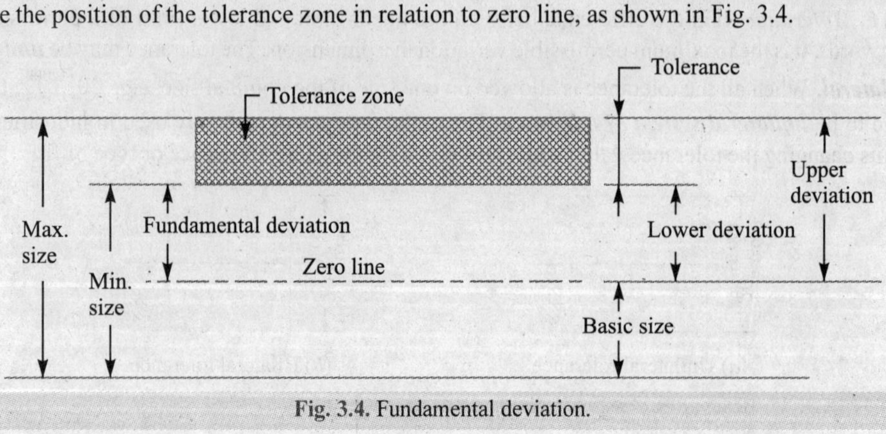

Fig. 3.4. Fundamental deviation.

3.14 Fits

The degree of tightness or looseness between the two mating parts is known as a *fit* of the parts. The nature of fit is characterised by the presence and size of clearance and interference.

The ***clearance*** is the amount by which the actual size of the shaft is less than the actual size of the mating hole in an assembly as shown in Fig. 3.5 (*a*). In other words, the clearance is the difference between the sizes of the hole and the shaft before assembly. The difference must be ***positive.***

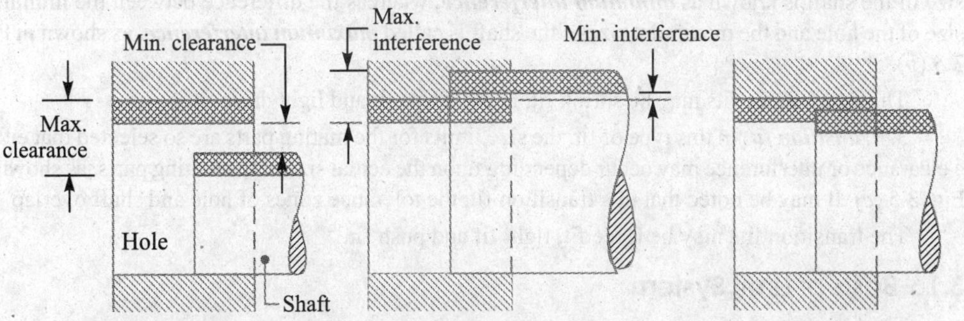

(*a*) Clearance fit. (*b*) Interference fit. (*c*) Transition fit.

Fig. 3.5. Types of fits.

The ***interference*** is the amount by which the actual size of a shaft is larger than the actual finished size of the mating hole in an assembly as shown in Fig. 3.5 (*b*). In other words, the interference is the arithmetical difference between the sizes of the hole and the shaft, before assembly. The difference must be ***negative.***

3.15 Types of Fits

According to Indian standards, the fits are classified into the following three groups :

1. ***Clearance fit.*** In this type of fit, the size limits for mating parts are so selected that clearance between them always occur, as shown in Fig. 3.5 (*a*). It may be noted that in a clearance fit, the tolerance zone of the hole is entirely above the tolerance zone of the shaft.

In a clearance fit, the difference between the minimum size of the hole and the maximum size of the shaft is known as ***minimum clearance*** whereas the difference between the maximum size of the hole and minimum size of the shaft is called ***maximum clearance*** as shown in Fig. 3.5 (*a*).

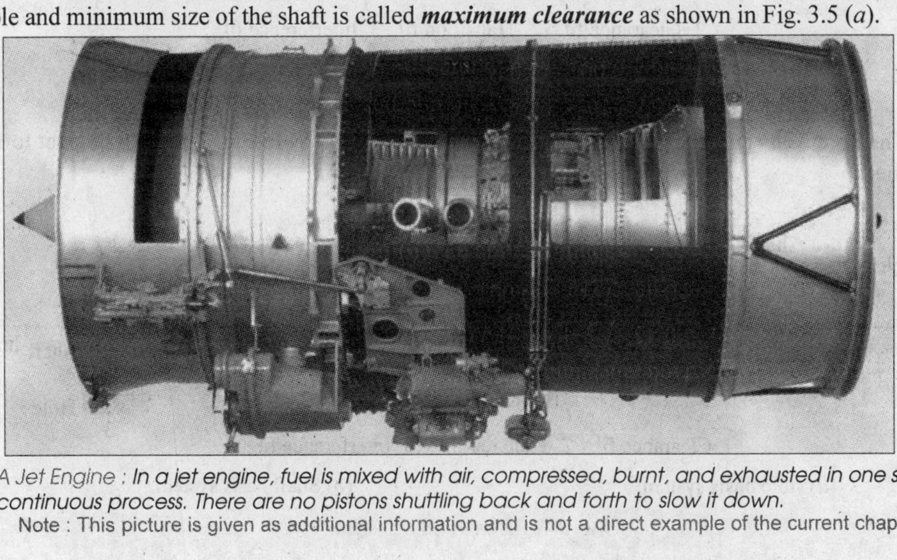

A Jet Engine : In a jet engine, fuel is mixed with air, compressed, burnt, and exhausted in one smooth, continuous process. There are no pistons shuttling back and forth to slow it down.

Note : This picture is given as additional information and is not a direct example of the current chapter.

The clearance fits may be slide fit, easy sliding fit, running fit, slack running fit and loose running fit.

2. ***Interference fit.*** In this type of fit, the size limits for the mating parts are so selected that interference between them always occur, as shown in Fig. 3.5 (*b*). It may be noted that in an interference fit, the tolerance zone of the hole is entirely below the tolerance zone of the shaft.

In an interference fit, the difference between the maximum size of the hole and the minimum size of the shaft is known as ***minimum interference,*** whereas the difference between the minimum size of the hole and the maximum size of the shaft is called ***maximum interference,*** as shown in Fig. 3.5 (*b*).

The interference fits may be shrink fit, heavy drive fit and light drive fit.

3. ***Transition fit.*** In this type of fit, the size limits for the mating parts are so selected that either a clearance or interference may occur depending upon the actual size of the mating parts, as shown in Fig. 3.5 (*c*). It may be noted that in a transition fit, the tolerance zones of hole and shaft overlap.

The transition fits may be force fit, tight fit and push fit.

3.16 Basis of Limit System

The following are two bases of limit system:

1. ***Hole basis system.*** When the hole is kept as a constant member (*i.e.* when the lower deviation of the hole is zero) and different fits are obtained by varying the shaft size, as shown in Fig. 3.6 (*a*), then the limit system is said to be on a hole basis.

2. ***Shaft basis system.*** When the shaft is kept as a constant member (*i.e.* when the upper deviation of the shaft is zero) and different fits are obtained by varying the hole size, as shown in Fig. 3.6 (*b*), then the limit system is said to be on a shaft basis.

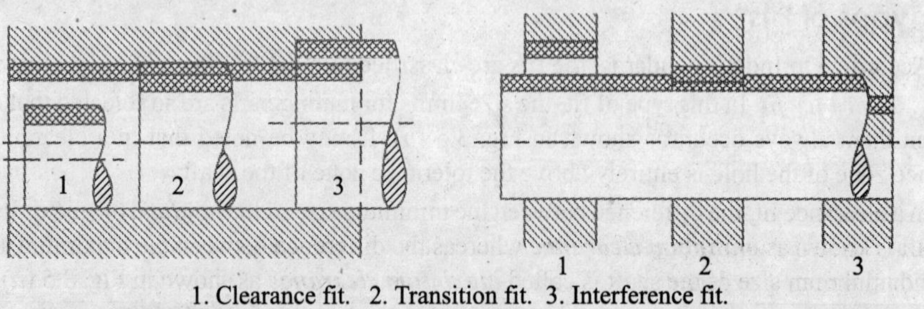

1. Clearance fit. 2. Transition fit. 3. Interference fit.

(*a*) Hole basis system. (*b*) Shaft basis system.

Fig. 3.6. Bases of limit system.

The hole basis and shaft basis system may also be shown as in Fig. 3.7, with respect to the zero line.

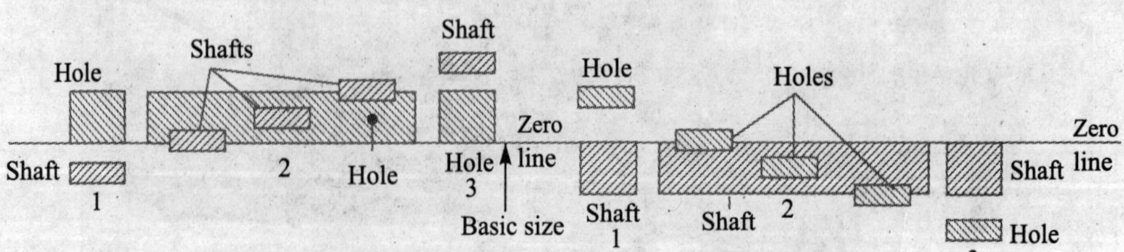

1. Clearance fit. 2. Transition fit. 3. Interference fit.

(*a*) Hole basis system. (*b*) Shaft basis system.

Fig. 3.7. Bases of limit system.

It may be noted that from the manufacturing point of view, a hole basis system is always preferred. This is because the holes are usually produced and finished by standard tooling like drill, reamers, etc., whose size is not adjustable easily. On the other hand, the size of the shaft (which is to go into the hole) can be easily adjusted and is obtained by turning or grinding operations.

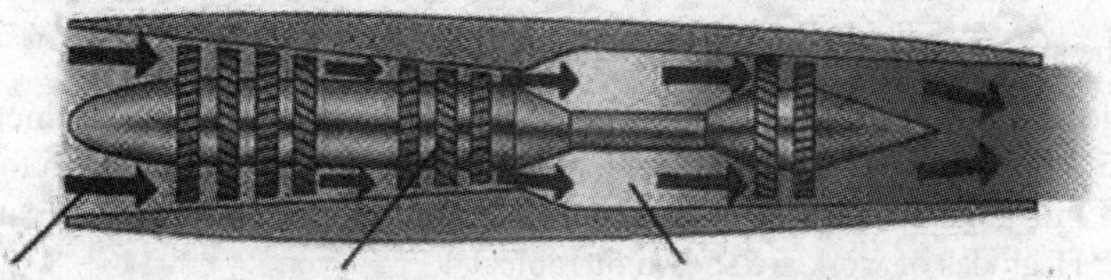

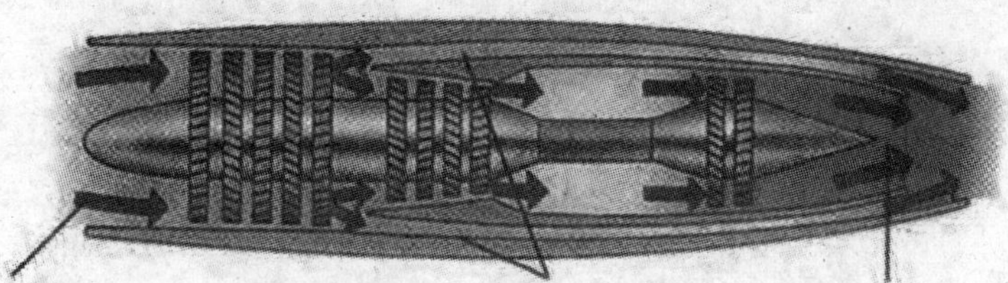

Turbofan engines are quieter and more efficient than simple turbojet engines. Turbofans drive air around the combustion engine as well as through it.

Note : This picture is given as additional information and is not a direct example of the current chapter.

3.17 Indian Standard System of Limits and Fits

According to Indian standard [IS : 919 (Part I)-1993], the system of limits and fits comprises 18 grades of fundamental tolerances *i.e.* grades of accuracy of manufacture and 25 types of fundamental deviations indicated by letter symbols for both holes and shafts (capital letter *A* to *ZC* for holes and small letters *a* to *zc* for shafts) in diameter steps ranging from 1 to 500 mm. A unilateral hole basis system is recommended but if necessary a unilateral or bilateral shaft basis system may also be used. The 18 tolerance grades are designated as IT 01, IT 0 and IT 1 to IT 16. These are called ***standard tolerances.*** The standard tolerances for grades IT 5 to IT 7 are determined in terms of standard tolerance unit (i) in microns, where

i (microns) = $0.45 \sqrt[3]{D} + 0.001 D$, where D is the size or geometric mean diameter in mm.

The following table shows the relative magnitude for grades between IT 5 and IT 16.

Table 3.2. Relative magnitude of tolerance grades.

Tolerance grade	IT 5	IT 6	IT 7	IT 8	IT 9	IT 10	IT 11	IT 12	IT 13	IT 14	IT 15	IT 16
Magnitude	7 i	10 i	16 i	25 i	40 i	64 i	100 i	160 i	250 i	400 i	640 i	1000 i

The values of standard tolerances corresponding to grades IT 01, IT 0 and IT 1 are as given below:

For IT 01, i (microns) $= 0.3 + 0.008\,D$,

For IT 0, i (microns) $= 0.5 + 0.012\,D$, and

For IT 1, i (microns) $= 0.8 + 0.020\,D$,

where D is the size or geometric mean diameter in mm.

The tolerance values of grades IT 2 to IT 4 are scaled approximately geometrically between IT 1 and IT 5. The fundamental tolerances of grades IT 01, IT 0 and IT 1 to IT 16 for diameter steps ranging from 1 to 500 mm are given in Table 3.3. The manufacturing processes capable of producing the particular IT grades of work are shown in Table 3.4.

The alphabetical representation of fundamental deviations for basic shaft and basic hole system is shown in Fig. 3.8.

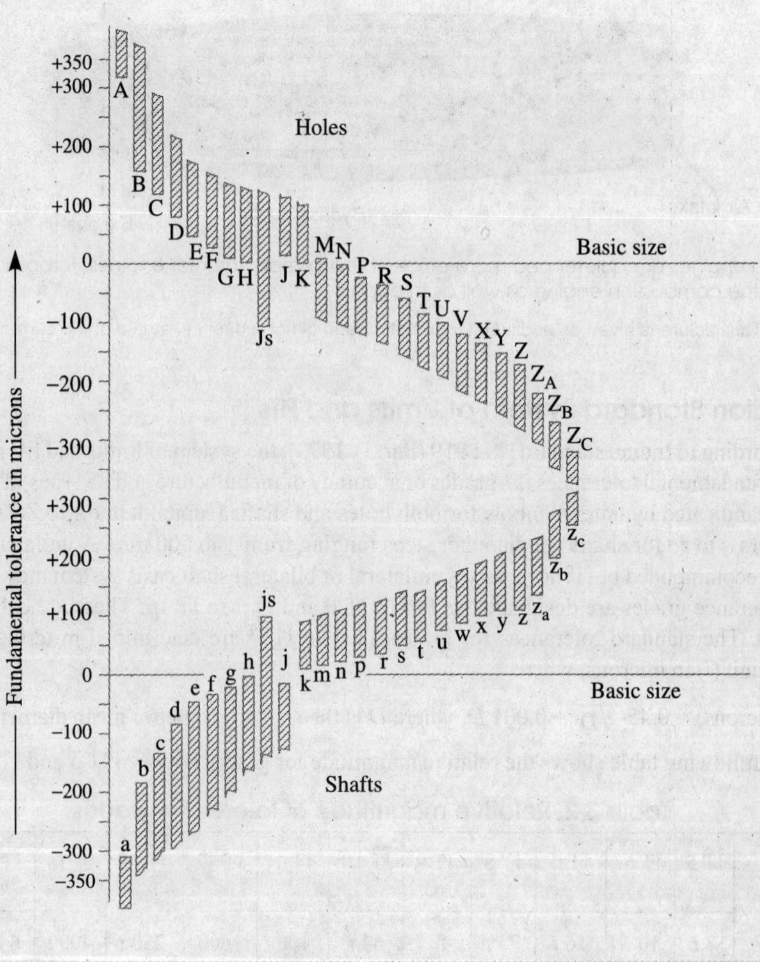

Fig. 3.8. Fundamental deviations for shafts and holes.

Table 3.3. Fundamental tolerances of grades IT01, IT0 and IT1 to IT16, according to IS : 919 (Part I) – 1993.

Basic size (Diameter steps) in mm		*Standard tolerance grades, in micron (1 micron = 0.001 mm)*																	
		IT01	IT0	IT1	IT2	IT3	IT4	IT5	IT6	IT7	IT8	IT9	IT10	IT11	IT12	IT13	IT14	IT15	IT16
Over To and inc.	1 3	0.3	0.5	0.8	1.2	2	3	4	6	10	14	25	40	60	100	140	250	400	600
Over To and inc.	3 6	0.4	0.6	1	1.5	2.5	4	5	8	12	18	30	48	75	120	180	300	480	750
Over To and inc.	6 10	0.4	0.6	1	1.5	2.5	4	6	9	15	22	36	58	90	150	220	360	580	900
Over To and inc.	10 18	0.5	0.8	1.2	2	3	5	8	11	18	27	43	70	110	180	270	430	700	1100
Over To and inc.	18 30	0.6	1	1.5	2.5	4	6	9	13	21	33	52	84	130	210	330	520	840	1300
Over To and inc.	30 50	0.6	1	1.5	2.5	4	7	11	16	25	39	62	100	160	250	390	620	1000	1600
Over To and inc.	50 80	0.8	1.2	2	3	5	8	13	19	30	46	74	120	190	300	460	740	1200	1900
Over To and inc.	80 120	1	1.5	2.5	4	6	10	15	22	35	54	87	140	220	350	540	870	1400	2200
Over To and inc.	120 180	1.2	2	3.5	5	8	12	18	25	40	63	100	160	250	400	630	1000	1600	2500
Over To and inc.	180 250	2	3	4.5	7	10	14	20	29	46	72	115	185	290	460	720	1150	1850	2900
Over To and inc.	250 315	2.5	4	6	8	12	16	23	32	52	81	130	210	320	520	810	1300	2100	3200
Over To and inc.	315 400	3	5	7	9	13	18	25	36	57	89	140	230	360	570	890	1400	2300	3800
Over To and inc.	400 500	4	6	8	10	15	20	27	40	63	97	155	250	400	630	970	1550	2500	4000

Table 3.4. Manufacturing processes and IT grades produced.

S.No.	*Manufacturing process*	*IT grade produced*	*S.No.*	*Manufacturing process*	*IT grade produced*
1.	Lapping	4 and 5	9.	Extrusion	8 to 10
2.	Honing	4 and 5	10	Boring	8 to 13
3.	Cylindrical grinding	5 to 7	11.	Milling	10 to 13
4.	Surface grinding	5 to 8	12.	Planing and shaping	10 to 13
5.	Broaching	5 to 8	13.	Drilling	10 to 13
6.	Reaming	6 to 10	14.	Die casting	12 to 14
7.	Turning	7 to 13	15.	Sand casting	14 to 16
8.	Hot rolling	8 to 10	16.	Forging	14 to 16

For hole, *H* stands for a dimension whose lower deviation refers to the basic size. The hole *H* for which the lower deviation is zero is called a ***basic hole.*** Similarly, for shafts, *h* stands for a dimension whose upper deviation refers to the basic size. The shaft *h* for which the upper deviation is zero is called a ***basic shaft.***

This view along the deck of a liquefied natural gas (LNG) carrier shows the tops of its large, insulated steel tanks. The tanks contain liquefied gas at-162°C.

A fit is designated by its basic size followed by symbols representing the limits of each of its two components, the hole being quoted first. For example, 100 *H*6/*g*5 means basic size is 100 mm and the tolerance grade for the hole is 6 and for the shaft is 5. Some of the fits commonly used in engineering practice, for holes and shafts are shown in Tables 3.5 and 3.6 respectively according to IS : 2709 – 1982 (Reaffirmed 1993).

Table 3.5. Commonly used fits for holes according to IS : 2709 – 1982 (Reaffirmed 1993).

Type of fit	*Class of shaft*	*With holes*				*Remarks and uses*
		H6	*H7*	*H8*	*H11*	
Clearance fit	*a*	—	—	—	*a*11	Large clearance fit and widely used.
	b	—	—	—	*b*11	
	c	—	*c*8	**c* 9	*c* 11	Slack running fit.
	d	—	*d*8	**d* 8 *d* 9, *d*10	*d* 11	Loose running fit—used for plummer block bearings and loose pulleys.
	e	*e*7	*e*8	**e* 8-*e* 9	—	Easy running fit—used for properly lubricated bearings requiring appreciable clearance. In the finer grades, it may be used on large electric motor and turbogenerator bearings according to the working condition.
	f	**f*6	*f* 7	**f*8	—	Normal running fit—widely used for grease lubricated or oil lubricated bearings where no substantial temperature differences are encountered—Typical applications are gear box shaft bearings and the bearings of small electric motors, pumps, etc.
	g	**g* 5	**g* 6	*g* 7	—	Close running fit or sliding fit—Also fine spigot and location fit—used for bearings for accurate link work and for piston and slide valves.
	h	**h* 5	**h* 6	**h* 7–*h* 8	**h*11	Precision sliding fit. Also fine spigot and location fit—widely used for non-running parts.
Transition fit	*j*	**j*5	**j*6	**j*7	—	Push fit for very accurate location with easy assembly and dismantling—Typical applications are coupling, spigots and recesses, gear rings clamped to steel hubs, etc.
	k	**k* 5	**k* 6	*k* 7	—	True transition fit (light keying fit)—used for keyed shaft, non-running locked pins, etc.
	m	**m* 5	**m* 6	*m* 7	—	Medium keying fit.
	n	*n* 5	**n* 6	*n*7	—	Heavy keying fit—used for tight assembly of mating parts.

* Second preference fits.

Type of fit	*Class of shaft*	*With holes*				*Remarks and uses*
		H6	*H7*	*H8*	*H11*	
Interference fit	*p*	*p5*	**p6*	—	—	Light press fit with easy dismantling for non-ferrous parts. Standard press fit with easy dismantling for ferrous and non-ferrous parts assembly.
	r	*r5*	**r6*	—	—	Medium drive fit with easy dismantling for ferrous parts assembly. Light drive fit with easy dismantling for non-ferrous parts assembly.
	s	*s5*	**s6*	*s7*	—	Heavy drive fit on ferrous parts for permanent or semi-permanent assembly. Standard press fit for non-ferrous parts.
	t	*t5*	*t6*	**t7*	—	Force fit on ferrous parts for permanent assembly.
	u	*u5*	*u6*	**u7*	—	Heavy force fit or shrink fit.
	v, x	—	—	—	—	Very large interference fits — not recommended for use
	y, z	—	—	—	—	

Table 3.6. Commonly used fits for shafts according to IS : 2709 – 1982 (Reaffirmed 1993).

Type of fit	*Class of hole*	*With shafts*						*Remarks and uses*
		**h5*	*h6*	*h7*	**h8*	*h9*	*h11*	
Clearance fit	*A*	—	—	—	—	—	*A11*	Large clearance fit and widely used.
	B	—	—	—	—	—	*B11*	
	C	—	—	—	—	—	*C11*	Slack running fit.
	D	—	**D9*	—	*D10*	*D10*	**D11*	Loose running fit.
	E	—	**E8*	—	*E8**	*E9*	—	Easy running fit.
	F	—	**F7*	—	*F8*	**F8*	—	Normal running fit.
	G	**G6*	*G7*	—	—	—	—	Close running fit or sliding fit, also spigot and location fit.
	H	**H6*	*H7*	*H8*	*H8*	*H8, H9*	*H11*	Precision sliding fit. Also fine spigot and location fit.
	Js	**Js6*	*Js7*	**Js8*	—	—	—	Push fit for very accurate location with easy assembly and disassembly.

* Second preference fits.

Type of fit	*Class of hole*	*With shafts*						*Remarks and uses*
		**h5*	*h6*	*h7*	**h8*	*h9*	*h11*	
Transition fit	*K*	**K6*	*K7*	**K8*	—	—	—	Light keying fit (true transition) for keyed shafts, non-running locked pins, etc.
	M	**M6*	**M7*	**M8*	—	—	—	Medium keying fit.
	N	**N6*	*N7*	**N8*	—	—	—	Heavy keying fit (for tight assembly of mating surfaces).
Interference fit	*P*	**P6*	*P7*	—	—	—	—	Light press fit with easy dismantling for non-ferrous parts. Standard press fit with easy dismantling for ferrous and non-ferrous parts assembly.
	R	**R6*	*R7*	—	—	—	—	Medium drive fit with easy dismantling for ferrous parts assembly. Light drive fit with easy dismantling for non-ferrous parts assembly.
	S	**S6*	*S7*	—	—	—	—	Heavy drive fit for ferrous parts permanent or semi- permanent assembly, standard press fit for non-ferrous parts.
	T	**T6*	*T7*	—	—	—	—	Force fit on ferrous parts for permanent assembly.

3.18 Calculation of Fundamental Deviation for Shafts

Computer simulation of stresses on a jet engine blades.

We have already discussed that for holes, the upper deviation is denoted by *ES* and the lower deviation by *EI*. Similarly for shafts, the upper deviation is represented by *es* and the lower deviation by *ei*. According to Indian standards, for each letter symbol, the magnitude and sign for one of the two deviations (*i.e.* either upper or lower deviation), which is known as fundamental deviation, have been determined by means of formulae given in Table 3.7. The other deviation may be calculated by using the absolute value of the standard tolerance (*IT*) from the following relation:

$$ei = es - IT \qquad \text{or} \qquad es = ei + IT$$

It may be noted for shafts *a* to *h*, the upper deviations (*es*) are considered whereas for shafts *j* to *Zc*, the lower deviation (*ei*) is to be considered.

* Second preference fits.

The fundamental deviation for Indian standard shafts for diameter steps from 1 to 200 mm may be taken directly from Table 3.10 (page 76).

Table 3.7. Formulae for fundamental shaft deviations.

Upper deviation (es)		*Lower deviation (ei)*	
Shaft designation	*In microns (for D in mm)*	*Shaft designation*	*In microns (for D in mm)*
a	$= -(265 + 1.3\ D)$	*J* 5 to *j* 8	No formula
	for $D \leq 120$	*k* 4 to *k* 7	$= +0.6\ \sqrt[3]{D}$
	$= -3.5\ D$		
	for $D > 120$	*k* for grades	$= 0$
		≤ 3 and ≤ 8	
b	$= -(140 + 0.85\ D)$	*m*	$= +(IT\ 7 - IT\ 6)$
	for $D \leq 160$		
	$= -1.8\ D$	*n*	$= +5\ (D)^{0.34}$
	for $D > 160$	*p*	$= +IT\ 7 + 0$ to 5
c	$= -52\ (D)^{0.2}$	*r*	= Geometric mean of values of *ei*
	for $D \leq 40$		for shaft *p* and *s*
	$= -(95 + 0.8\ D)$	*s*	$= +(IT\ 8 + 1$ to $4)$ for $D \leq 50$
	for $D > 40$		$= +(IT\ 7 + 0.4\ D)$ for $D > 50$
d	$= -16\ (D)^{0.44}$	*t*	$= +(IT\ 7 + 0.63\ D)$
e	$= -11\ (D)^{0.41}$	*u*	$= +(IT\ 7 + D)$
f	$= -5.5\ (D)^{0.41}$	*v*	$= +(IT\ 7 + 1.25\ D)$
		x	$= +(IT\ 7 + 1.6\ D)$
g	$= -2.5\ (D)^{0.34}$	*y*	$= +(IT\ 7 + 2\ D)$
		z	$= +(IT\ 7 + 2.5\ D)$
h	$= 0$	*za*	$= +(IT\ 8 + 3.15\ D)$
		zb	$= +(IT\ 9 + 4\ D)$
		zc	$= +(IT\ 10 + 5\ D)$

For *js*, the two deviations are equal to $\pm IT/2$.

3.19 Calculation of Fundamental Deviation for Holes

The fundamental deviation for holes for those of the corresponding shafts, are derived by using the rule as given in Table 3.8.

Table 3.8. Rules for fundamental deviation for holes.

<table>
<tr><td colspan="3">All deviation except those below</td><td>General rule
Hole limits are identical with the shaft limits of the same symbol (letter and grade) but disposed on the other side of the zero line.
EI = Upper deviation es of the shaft of the same letter symbol but of opposite sign.</td></tr>
<tr><td rowspan="3">For sizes above 3 mm</td><td>N</td><td>9 and coarser grades</td><td>ES = 0</td></tr>
<tr><td>J, K, M and N</td><td>Upto grade 8 inclusive</td><td rowspan="2">Special rule
ES = Lower deviation ei of the shaft of the same letter symbol but one grade finer and of opposite sign increased by the difference between the tolerances of the two grades in question.</td></tr>
<tr><td>P to ZC</td><td>upto grade 7 inclusive</td></tr>
</table>

The fundamental deviation for Indian standard holes for diameter steps from 1 to 200 mm may be taken directly from the following table.

Table 3.9. Indian standard 'H' Hole
Limits for H5 to H13 over the range 1 to 200 mm as per IS : 919 (Part II) -1993.

<table>
<tr><td colspan="2" rowspan="2">Diameter steps in mm</td><td colspan="10">Deviations in micron (1 micron = 0.001 mm)</td></tr>
<tr><td>H5</td><td>H6</td><td>H7</td><td>H8</td><td>H9</td><td>H10</td><td>H11</td><td>H12</td><td>H13</td><td>H5 – H13</td></tr>
<tr><td>Over</td><td>To</td><td>High +</td><td>High +</td><td>High +</td><td>High +</td><td>High +</td><td>High +</td><td>High +</td><td>High +</td><td>High +</td><td>Low</td></tr>
<tr><td>1</td><td>3</td><td>5</td><td>7</td><td>9</td><td>14</td><td>25</td><td>40</td><td>60</td><td>90</td><td>140</td><td>0</td></tr>
<tr><td>3</td><td>6</td><td>5</td><td>8</td><td>12</td><td>18</td><td>30</td><td>48</td><td>75</td><td>120</td><td>180</td><td>0</td></tr>
<tr><td>6</td><td>10</td><td>6</td><td>9</td><td>15</td><td>22</td><td>36</td><td>58</td><td>90</td><td>150</td><td>220</td><td>0</td></tr>
<tr><td>10</td><td>14</td><td rowspan="2">8</td><td rowspan="2">11</td><td rowspan="2">18</td><td rowspan="2">27</td><td rowspan="2">43</td><td rowspan="2">70</td><td rowspan="2">110</td><td rowspan="2">180</td><td rowspan="2">270</td><td rowspan="2">0</td></tr>
<tr><td>14</td><td>18</td></tr>
<tr><td>18</td><td>24</td><td rowspan="2">9</td><td rowspan="2">13</td><td rowspan="2">21</td><td rowspan="2">33</td><td rowspan="2">52</td><td rowspan="2">84</td><td rowspan="2">130</td><td rowspan="2">210</td><td rowspan="2">330</td><td rowspan="2">0</td></tr>
<tr><td>24</td><td>30</td></tr>
<tr><td>30</td><td>40</td><td rowspan="2">11</td><td rowspan="2">16</td><td rowspan="2">25</td><td rowspan="2">39</td><td rowspan="2">62</td><td rowspan="2">100</td><td rowspan="2">160</td><td rowspan="2">250</td><td rowspan="2">460</td><td rowspan="2">0</td></tr>
<tr><td>40</td><td>50</td></tr>
<tr><td>50</td><td>65</td><td rowspan="2">13</td><td rowspan="2">19</td><td rowspan="2">30</td><td rowspan="2">46</td><td rowspan="2">74</td><td rowspan="2">120</td><td rowspan="2">190</td><td rowspan="2">300</td><td rowspan="2">390</td><td rowspan="2">0</td></tr>
<tr><td>65</td><td>80</td></tr>
<tr><td>80</td><td>100</td><td rowspan="2">15</td><td rowspan="2">22</td><td rowspan="2">35</td><td rowspan="2">54</td><td rowspan="2">87</td><td rowspan="2">140</td><td rowspan="2">220</td><td rowspan="2">350</td><td rowspan="2">540</td><td rowspan="2">0</td></tr>
<tr><td>100</td><td>120</td></tr>
<tr><td>120</td><td>140</td><td rowspan="2">18</td><td rowspan="2">25</td><td rowspan="2">40</td><td rowspan="2">63</td><td rowspan="2">100</td><td rowspan="2">160</td><td rowspan="2">250</td><td rowspan="2">400</td><td rowspan="2">630</td><td rowspan="2">0</td></tr>
<tr><td>140</td><td>160</td></tr>
<tr><td>160</td><td>180</td><td rowspan="2">20</td><td rowspan="2">29</td><td rowspan="2">46</td><td rowspan="2">72</td><td rowspan="2">115</td><td rowspan="2">185</td><td rowspan="2">290</td><td rowspan="2">460</td><td rowspan="2">720</td><td rowspan="2">0</td></tr>
<tr><td>180</td><td>200</td></tr>
</table>

Table 3.10. Indian standard shafts for common use as per IS : 919 (Part II)–1993.

Shaft	*Limit*	*Values of deviations in microns for diameter steps 1 to 200 mm (1 micron = 0.001 mm)*																
		1 to 3	3 to 6	6 to 10	10 to 14	14 to 18	18 to 24	24 to 30	30 to 40	40 to 50	50 to 65	65 to 80	80 to 100	100 to 120	120 to 140	140 to 160	160 to 180	180 to 200
s6	High+	22	37	32	39	39	48	48	59	59	72	78	93	101	117	125	133	151
s7	High+	24	31	38	46	46	56	66	68	68	83	89	106	114	132	140	148	168
s6 & s7	Low+	15	19	23	28	28	35	35	43	43	53	59	71	79	92	100	108	122
p9	High+	16	20	24	29	29	35	35	42	42	51	51	59	59	68	68	68	79
	Low+	9	12	15	18	18	22	22	26	26	32	32	37	37	43	43	43	50
k6	High+	—	—	10	12	12	15	15	18	18	21	21	25	25	28	28	28	33
k7	High+	—	—	16	19	19	23	23	27	27	32	32	38	38	43	43	43	50
k6 & k7	Low+	—	—	1	1	1	2	2	2	2	2	2	3	3	3	3	3	4
	High+	7	9	10	12	12	13	13	15	15	18	18	20	20	22	22	22	25
	Low–	2	3	5	6	6	8	8	10	10	12	12	15	15	18	18	18	21
h6	High–	0	0	0	0	0	0	0	0	0	0	0	0	0	0	0	0	0

Shaft	Limit	Values of deviations in microns for diameter steps 1 to 200 mm (1 micron = 0.001 mm)																
		1 to 3	3 to 6	6 to 10	10 to 14	14 to 18	18 to 24	24 to 30	30 to 40	40 to 50	50 to 65	65 to 80	80 to 100	100 to 120	120 to 140	140 to 160	160 to 180	180 to 200
*h*7	High–	0	0	0	0	0	0	0	0	0	0	0	0	0	0	0	0	0
*h*6	Low–	7	8	9	11	11	13	13	16	16	19	19	22	22	25	25	26	29
&		&	&	&	&	&	&	&	&	&	&	&	&	&	&	&	&	&
*h*7		9	12	15	18	18	21	21	25	25	30	30	35	35	40	40	40	46
g 6	High–	3	4	5	6	6	7	7	9	9	10	10	12	12	14	14	14	15
	Low–	10	12	14	17	17	20	20	25	25	29	29	34	34	39	39	39	44
f 7	High–	7	10	13	16	16	20	20	25	25	30	30	36	36	43	43	43	50
f 8	High–	7	10	13	16	16	20	20	25	25	30	30	36	36	43	43	43	50
f 7	Low–	16	22	28	34	34	41	41	50	50	60	60	71	71	83	83	83	96
&		&	&	&	&	&	&	&	&	&	&	&	&	&	&	&	&	&
f 8		21	28	35	43	43	53	53	64	64	76	76	90	90	106	106	106	122
e 8	High–	14	20	25	32	32	40	40	50	50	60	60	72	72	85	85	85	100
e 9	High–	14	20	25	32	32	40	40	50	50	60	60	72	72	85	85	85	100
e 8	Low–	28	38	47	59	58	73	73	89	89	106	106	126	126	148	148	148	172
&		&	&	&	&	&	&	&	&	&	&	&	&	&	&	&	&	&
e 9		39	50	61	75	75	92	92	112	112	134	134	158	158	185	185	185	215
d 9	High–	20	30	40	50	50	65	65	80	80	100	100	120	120	145	145	145	170
	Low–	45	60	76	93	93	117	117	142	142	174	174	207	207	245	245	245	285
c 9	High–	60	70	80	95	95	110	110	120	130	140	150	170	180	200	210	230	240
	Low–	85	100	116	138	138	162	162	182	192	214	224	257	267	300	310	330	355
b 9	High–	140	140	150	150	150	160	160	170	180	190	200	220	240	260	280	310	340
	Low–	165	170	186	193	193	212	212	232	242	264	274	307	327	360	380	410	545

Example 3.1. *The dimensions of the mating parts, according to basic hole system, are given as follows :*

Hole : 25.00 mm *Shaft : 24.97 mm*

25.02 mm *24.95 mm*

Find the hole tolerance, shaft tolerance and allowance.

Solution. Given : Lower limit of hole = 25 mm ; Upper limit of hole = 25.02 mm ; Upper limit of shaft = 24.97 mm ; Lower limit of shaft = 24.95 mm

Hole tolerance

We know that hole tolerance

= Upper limit of hole – Lower limit of hole

= 25.02 – 25 = 0.02 mm **Ans.**

Shaft tolerance

We know that shaft tolerance

= Upper limit of shaft – Lower limit of shaft

= 24.97 – 24.95 = 0.02 mm **Ans.**

Allowance

We know that allowance

= Lower limit of hole – Upper limit of shaft

= 25.00 – 24.97 = 0.03 mm **Ans.**

Example 3.2. *Calculate the tolerances, fundamental deviations and limits of sizes for the shaft designated as 40 H8 / f7.*

Solution. Given: Shaft designation = 40 *H*8 / *f* 7

The shaft designation 40 *H*8 / *f* 7 means that the basic size is 40 mm and the tolerance grade for the hole is 8 (*i.e. I T* 8) and for the shaft is 7 (*i.e. I T* 7).

Tolerances

Since 40 mm lies in the diameter steps of 30 to 50 mm, therefore the geometric mean diameter,

$$D = \sqrt{30 \times 50} = 38.73 \text{ mm}$$

We know that standard tolerance unit,

$$i = 0.45 \sqrt[3]{D} + 0.001 D$$

$$= 0.45 \sqrt[3]{38.73} + 0.001 \times 38.73$$

$$= 0.45 \times 3.38 + 0.03873 = 1.559\ 73 \quad \text{or} \quad 1.56 \text{ microns}$$

$$= 1.56 \times 0.001 = 0.001\ 56 \text{ mm} \qquad ...(\because 1 \text{ micron} = 0.001 \text{ mm})$$

From Table 3.2, we find that standard tolerance for the hole of grade 8 (*I T* 8)

= 25 *i* = 25 × 0.001 56 = 0.039 mm **Ans.**

and standard tolerance for the shaft of grade 7 (*I T* 7)

= 16 *i* = 16 × 0.001 56 = 0.025 mm **Ans.**

Note : The value of *I T* 8 and *I T* 7 may be directly seen from Table 3.3.

Fundamental deviation

We know that fundamental deviation (lower deviation) for hole H,

$$EI = 0$$

From Table 3.7, we find that fundamental deviation (upper deviation) for shaft f,

$$es = -5.5\,(D)^{0.41}$$
$$= -5.5\,(38.73)^{0.41} = -24.63 \text{ or } -25 \text{ microns}$$
$$= -25 \times 0.001 = -0.025 \text{ mm } \textbf{Ans.}$$

∴ Fundamental deviation (lower deviation) for shaft f,

$$ei = es - IT = -0.025 - 0.025 = -0.050 \text{ mm } \textbf{Ans.}$$

The –ve sign indicates that fundamental deviation lies below the zero line.

Limits of sizes

We know that lower limit for hole

= Basic size = 40 mm **Ans.**

Upper limit for hole = Lower limit for hole + Tolerance for hole

= 40 + 0.039 = 40.039 mm **Ans.**

Upper limit for shaft = Lower limit for hole or Basic size – Fundamental deviation (upper deviation) ...(∵ Shaft f lies below the zero line)

= 40 – 0.025 = 39.975 mm **Ans.**

and lower limit for shaft = Upper limit for shaft – Tolerance for shaft

= 39.975 – 0.025 = 39.95 mm **Ans.**

Example 3.3. *Give the dimensions for the hole and shaft for the following:*

(*a*) *A 12 mm electric motor sleeve bearing;*

(*b*) *A medium force fit on a 200 mm shaft; and*

(*c*) *A 50 mm sleeve bearing on the elevating mechanism of a road grader.*

Solution.

(*a*) ***Dimensions for the hole and shaft for a 12 mm electric motor sleeve bearing***

From Table 3.5, we find that for an electric motor sleeve bearing, a shaft e 8 should be used with H 8 hole.

Since 12 mm size lies in the diameter steps of 10 to 18 mm, therefore the geometric mean diameter,

$$D = \sqrt{10 \times 18} = 13.4 \text{ mm}$$

We know that standard tolerance unit,

$$i = 0.45\sqrt[3]{D} + 0.001\,D$$
$$= 0.45\sqrt[3]{13.4} + 0.001 \times 13.4 = 1.07 + 0.0134 = 1.0834 \text{ microns}$$

∴ *Standard tolerance for shaft and hole of grade 8 (IT 8)

$= 25\,i$ ...(From Table 3.2)

$= 25 \times 1.0834 = 27$ microns

$= 27 \times 0.001 = 0.027$ mm ...(∵ 1 micron = 0.001 mm)

From Table 3.7, we find that upper deviation for shaft 'e',

$$es = -11(D)^{0.41} = -11\,(13.4)^{0.41} = -32 \text{ microns}$$
$$= -32 \times 0.001 = -0.032 \text{ mm}$$

* The tolerance values may be taken directly from Table 3.3.

We know that lower deviation for shaft 'e',

$$ei = es - IT = -0.032 - 0.027 = -0.059 \text{ mm}$$

∴ Dimensions for the hole (H 8)

$$= 12^{+0.027}_{+0.000} \textbf{ Ans.}$$

and dimension for the shaft (e 8)

$$= 12^{-0.032}_{-0.059} \textbf{ Ans.}$$

(*b*) ***Dimensions for the hole and shaft for a medium force fit on a 200 mm shaft***

From Table 3.5, we find that shaft r 6 with hole H 7 gives the desired fit.

Since 200 mm lies in the diameter steps of 180 mm to 250 mm, therefore the geometric mean diameter,

$$D = \sqrt{180 \times 250} = 212 \text{ mm}$$

We know that standard tolerance unit,

$$i = 0.45 \sqrt[3]{D} + 0.001\,D$$

$$= 0.45 \sqrt[3]{212} + 0.001 \times 212 = 2.68 + 0.212 = 2.892 \text{ microns}$$

∴ Standard tolerance for the shaft of grade 6 ($IT6$) from Table 3.2

$$= 10\,i = 10 \times 2.892 = 28.92 \text{ microns}$$

$$= 28.92 \times 0.001 = 0.02892 \text{ or } 0.029 \text{ mm}$$

and standard tolerance for the hole of grade 7 (IT 7)

$$= 16\,i = 16 \times 2.892 = 46 \text{ microns}$$

$$= 46 \times 0.001 = 0.046 \text{ mm}$$

We know that lower deviation for shaft 'r' from Table 3.7

$$ei = \frac{1}{2}[(IT\,7 + 0.4\,D) + (IT\,7 + 0 \text{ to } 5)]$$

$$= \frac{1}{2}[(46 + 0.4 \times 212) + (46 + 3)] = 90 \text{ microns}$$

$$= 90 \times 0.001 = 0.09 \text{ mm}$$

and upper deviation for the shaft r,

$$es = ei + IT = 0.09 + 0.029 = 0.119 \text{ mm}$$

∴ Dimension for the hole H 7

$$= 200^{+0.046}_{+0.00} \textbf{ Ans.}$$

and dimension for the shaft r 6

$$= 200^{+0.119}_{+0.09} \textbf{ Ans.}$$

(*c*) ***Dimensions for the hole and shaft for a 50 mm sleeve bearing on the elevating mechanism of a road grader***

From Table 3.5, we find that for a sleeve bearing, a loose running fit will be suitable and a shaft d 9 should be used with hole H 8.

Since 50 mm size lies in the diameter steps of 30 to 50 mm or 50 to 80 mm, therefore the geometric mean diameter,

$$D = \sqrt{30 \times 50} = 38.73 \text{ mm}$$

We know that standard tolerance unit,

$$i = 0.45\ \sqrt[3]{D} + 0.001\ D$$

$$= 0.45\ \sqrt[3]{38.73} + 0.001 \times 38.73$$

$$= 1.522 + 0.03873 = 1.56073 \quad \text{or} \quad 1.56 \text{ microns}$$

∴ Standard tolerance for the shaft of grade 9 (*IT* 9) from Table 3.2

$$= 40\ i = 40 \times 1.56 = 62.4 \text{ microns}$$

$$= 62.4 \times 0.001 = 0.0624 \quad \text{or} \quad 0.062 \text{ mm}$$

and standard tolerance for the hole of grade 8 (*IT* 8)

$$= 25\ i = 25 \times 1.56 = 39 \text{ microns}$$

$$= 39 \times 0.001 = 0.039 \text{ mm}$$

We know that upper deviation for the shaft *d*, from Table 3.7

$$es = -16\ (D)^{0.44} = -16\ (38.73)^{0.44} = -80 \text{ microns}$$

$$= -80 \times 0.001 = -0.08 \text{ mm}$$

and lower deviation for the shaft *d*,

$$ei = es - IT = -0.08 - 0.062 = -0.142 \text{ mm}$$

∴ Dimension for the hole *H* 8

$$= 50^{+0.039}_{+0.000} \quad \textbf{Ans.}$$

and dimension for the shaft *d* 9

$$= 50^{-0.08}_{-0.142} \quad \textbf{Ans.}$$

Example 3.4. *A journal of nominal or basic size of 75 mm runs in a bearing with close running fit. Find the limits of shaft and bearing. What is the maximum and minimum clearance?*

Solution. Given: Nominal or basic size = 75 mm

From Table 3.5, we find that the close running fit is represented by *H* 8/*g* 7, *i.e.* a shaft *g* 7 should be used with *H* 8 hole.

Since 75 mm lies in the diameter steps of 50 to 80 mm, therefore the geometric mean diameter,

$$D = \sqrt{50 \times 80} = 63 \text{ mm}$$

We know that standard tolerance unit,

$$i = 0.45\ \sqrt[3]{D} + 0.001\ D = 0.45\ \sqrt[3]{63} + 0.001 \times 63$$

$$= 1.79 + 0.063 = 1.853 \text{ micron}$$

$$= 1.853 \times 0.001 = 0.001\ 853 \text{ mm}$$

∴ Standard tolerance for hole '*H*' of grade 8 (*IT* 8)

$$= 25\ i = 25 \times 0.001\ 853 = 0.046 \text{ mm}$$

and standard tolerance for shaft '*g*' of grade 7 (*IT* 7)

$$= 16\ i = 16 \times 0.001\ 853 = 0.03 \text{ mm}$$

From Table 3.7, we find that upper deviation for shaft *g*,

$$es = -2.5\ (D)^{0.34} = -2.5\ (63)^{0.34} = -10 \text{ micron}$$

$$= -10 \times 0.001 = -0.01 \text{ mm}$$

∴ Lower deviation for shaft *g*,

$$ei = es - IT = -0.01 - 0.03 = -0.04 \text{ mm}$$

We know that lower limit for hole

= Basic size = 75 mm

Upper limit for hole = Lower limit for hole + Tolerance for hole

= 75 + 0.046 = 75.046 mm

Upper limit for shaft = Lower limit for hole – Upper deviation for shaft

...(∵ Shaft *g* lies below zero line)

= 75 – 0.01 = 74.99 mm

and lower limit for shaft = Upper limit for shaft – Tolerance for shaft

= 74.99 – 0.03 = 74.96 mm

We know that maximum clearance

= Upper limit for hole – Lower limit for shaft

= 75.046 – 74.96 = 0.086 mm **Ans.**

and minimum clearance = Lower limit for hole – Upper limit for shaft

= 75 – 74.99 = 0.01 mm **Ans.**

3.20 Surface Roughness and its Measurement

A little consideration will show that surfaces produced by different machining operations (*e.g.* turning, milling, shaping, planing, grinding and superfinishing) are of different characteristics. They show marked variations when compared with each other. The variation is judged by the degree of smoothness. A surface produced by superfinishing is the smoothest, while that by planing is the roughest. In the assembly of two mating parts, it becomes absolutely necessary to describe the surface finish in quantitative terms which is measure of micro-irregularities of the surface and expressed in microns. In order to prevent stress concentrations and proper functioning, it may be necessary to avoid or to have certain surface roughness.

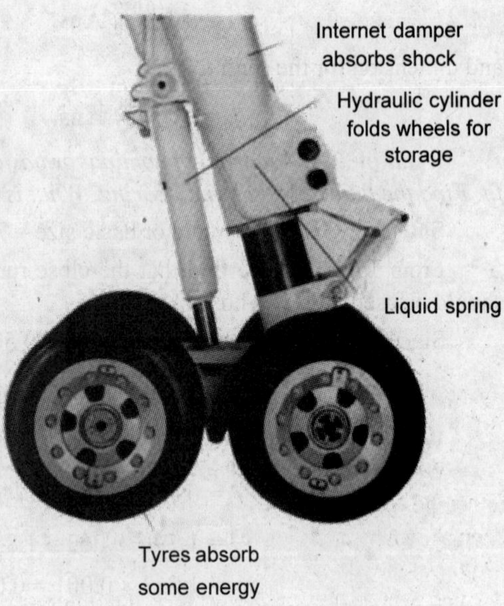

Landing Gear : When an aircraft comes in to land, it has to lose a lot of energy in a very short time. the landing gear deals with this and prevents disaster. First, mechanical or liquid springs absorb energy rapidly by being compressed. As the springs relax, this energy will be released again, but in a slow controlled manner in a damper-the second energy absorber. Finally, the tyres absorb energy, getting hot in the process.

There are many ways of expressing the surface roughness numerically, but the following two methods are commonly used :

1. Centre line average method (briefly known as CLA method), and
2. Root mean square method (briefly known as RMS method).

The ***centre line average method*** is defined as the average value of the ordinates between the surface and the mean line, measured on both sides of it. According to Indian standards, the surface finish is measured in terms of 'CLA' value and it is denoted by *Ra*.

$$\text{CLA value or } Ra \text{ (in microns)} = \frac{y_1 + y_2 + y_3 + ... y_n}{n}$$

where, $y_1, y_2, ...y_n$ are the ordinates measured on both sides of the mean line and n are the number of ordinates.

The ***root mean square method*** is defined as the square root of the arithmetic mean of the squares of the ordinates. Mathematically,

$$\text{R.M.S. value (in microns)} = \sqrt{\frac{y_1^2 + y_2^2 + y_3^2 + ... y_n^2}{n}}$$

According to Indian standards, following symbols are used to denote the various degrees of surface roughness :

Symbol	Surface roughness (Ra) in microns
∇	8 to 25
∇ ∇	1.6 to 8
∇ ∇ ∇	0.025 to 1.6
∇ ∇ ∇ ∇	Less than 0.025

The following table shows the range of surface roughness that can be produced by various manufacturing processes.

Table 3.11. Range of surface roughness.

S.No.	*Manufacturing process*	*Surface roughness in microns*	*S.No.*	*Manufacturing process*	*Surface roughness in microns*
1.	Lapping	0.012 to 0.016	9	Extrusion	0.16 to 5
2.	Honing	0.025 to 0.40	10.	Boring	0.40 to 6.3
3.	Cylindrical grinding	0.063 to 5	11.	Milling	0.32 to 25
4.	Surface grinding	0.063 to 5	12.	Planing and shaping	1.6 to 25
5.	Broaching	0.40 to 3.2	13.	Drilling	1.6 to 20
6.	Reaming	0.40 to 3.2	14.	Sand casting	5 to 50
7.	Turning	0.32 to 25	15.	Die casting	0.80 to 3.20
8.	Hot rolling	2.5 to 50	16.	Forging	1.60 to 2.5

3.21 Preferred Numbers

When a machine is to be made in several sizes with different powers or capacities, it is necessary to decide what capacities will cover a certain range efficiently with minimum number of sizes. It has been shown by experience that a certain range can be covered efficiently when it follows a geometrical progression with a constant ratio. The preferred numbers are the conventionally rounded off values derived from geometric series including the integral powers of 10 and having as common ratio of the following factors:

$$\sqrt[5]{10},\ \sqrt[10]{10},\ \sqrt[20]{10} \text{ and } \sqrt[40]{10}$$

These ratios are approximately equal to 1.58, 1.26, 1.12 and 1.06. The series of preferred numbers are designated as *R5, R10, R20 and R40 respectively. These four series are called ***basic series.*** The other series called ***derived series*** may be obtained by simply multiplying or dividing the basic sizes by 10, 100, etc. The preferred numbers in the series R5 are 1, 1.6, 2.5, 4.0 and 6.3. Table 3.12 shows basic series of preferred numbers according to IS : 1076 (Part I) – 1985 (Reaffirmed 1990).

* The symbol R is used as a tribute to Captain Charles Renard, the first man to use preferred numbers.

Notes : **1.** The standard sizes (in mm) for wrought metal products are shown in Table 3.13 according to IS : 1136 – 1990. The standard G.P. series used correspond to R10, R20 and R40.

2. The hoisting capacities (in tonnes) of cranes are in R10 series, while the hydraulic cylinder diameters are in R40 series and hydraulic cylinder capacities are in R5 series.

3. The basic thickness of sheet metals and diameter of wires are based on R10, R20 and R40 series. Wire diameter of helical springs are in R20 series.

Table 3.12. Preferred numbers of the basic series, according to IS : 1076 (Part I)–1985 (Reaffirmed 1990).

Basic series	*Preferred numbers*
R5	1.00, 1.60, 2.50, 4.00, 6.30, 10.00
R10	1.00, 1.25, 1.60, 2.00, 2.50, 3.15, 4.00, 5.00, 6.30, 8.00, 10.00
R20	1.00, 1.12, 1.25, 1.40, 1.60, 1.80, 2.00, 2.24, 2.50, 2.80, 3.15, 3.55, 4.00, 4.50, 5.00, 5.60, 6.30, 7.10, 8.00, 9.00, 10.00
R40	1.00, 1.06, 1.12, 1.18, 1.25, 1.32, 1.40, 1.50, 1.60, 1.70, 1.80, 1.90, 2.00, 2.12, 2.24, 2.36, 2.50, 2.65, 2.80, 3.00, 3.15, 3.35, 3.55, 3.75, 4.00, 4.25, 4.50, 4.75, 5.00, 5.30, 5.60, 6.00, 6.30, 6.70, 7.10, 7.50, 8.00, 8.50, 9.00, 9.50, 10.00

Table 3.13. Preferred sizes for wrought metal products according to IS : 1136 – 1990.

Size range	*Preferred sizes (mm)*
0.01 – 0.10 mm	0.02, 0.025, 0.030, 0.04, 0.05, 0.06, 0.08 and 0.10
0.10 – 1 mm	0.10, 0.11, 0.12, 0.14, 0.16, 0.18, 0.20, 0.22, 0.25, 0.28, 0.30, 0.32, 0.35, 0.36, 0.40, 0.45, 0.50, 0.55, 0.60, 0.63, 0.70, 0.80, 0.90 and 1
1 – 10 mm	1, 1.1, 1.2, 1.4, 1.5, 1.6, 1.8, 2.22, 2.5, 2.8, 3, 3.2, 3.5, 3.6, 4, 4.5, 5, 5.5, 5.6, 6, 6.3, 7, 8, 9 and 10
10 – 100 mm	10 to 25 (in steps of 1 mm), 28, 30, 32, 34, 35, 36, 38, 40, 42, 44, 45, 46, 48,50, 52, 53, 55, 56, 58, 60, 62, 63, 65, 67, 68, 70, 71, 72, 75, 78, 80, 82, 85, 88, 90, 92, 95, 98 and 100
100 – 1000 mm	100 to 200 (in steps of 5 mm), 200 to 310 (in steps of 10 mm), 315, 320, 330, 340, 350, 355, 360, 370, 375, 380 to 500 (in steps of 10 mm), 520, 530,550, 560, 580, 600, 630, 650, 670, 700, 710 and 750 – 1000 (in steps of 50 mm)
1000 – 10 000 mm	1000, 1100, 1200, 1250, 1400, 1500, 1600, 1800, 2000, 2200, 2500, 2800, 3000, 3200, 3500, 3600, 4000, 4500, 5000, 5500, 5600, 6000, 6300, 7000, 7100, 8000, 9000 and 10 000

EXERCISES

1. A journal of basic size of 75 mm rotates in a bearing. The tolerance for both the shaft and bearing is 0.075 mm and the required allowance is 0.10 mm. Find the dimensions of the shaft and the bearing bore. **[Ans. For shaft : 74.90 mm, 74.825 mm ; For hole : 75.075 mm, 75 mm]**

2. A medium force fit on a 75 mm shaft requires a hole tolerance and shaft tolerance each equal to 0.225 mm and average interference of 0.0375 mm. Find the hole and shaft dimensions. **[Ans. 75 mm, 75.225 mm ; 75.2625 mm, 75.4875 mm]**

3. Calculate the tolerances, fundamental deviations and limits of size for hole and shaft in the following cases of fits :

   (*a*) 25 $H\,8\,/\,d\,9$; and (*b*) 60 $H\,7\,/\,m\,6$

   [Ans. (*a*) 0.033 mm, 0.052 mm; 0, – 0.064 mm, – 0.116 mm; 25 mm, 25.033 mm, 24.936 mm, 24.884 mm (*b*) 0.03 mm, 0.019 mm; 0.011 mm, – 0.008 mm; 60 mm, 60.03 mm, 59.989 mm, 59.97 mm]

4. Find the extreme diameters of shaft and hole for a transition fit *H*7/*n*6, if the nominal or basic diameter is 12 mm. What is the value of clearance and interference?

   [Ans. 12.023 mm, 12.018 mm; 0.006 mm, – 0.023 mm]

5. A gear has to be shrunk on a shaft of basic size 120 mm. An interference fit *H*7/*u*6 is being selected. Determine the minimum and maximum diameter of the shaft and interference.

   [Ans. 120.144 mm, 120.166 mm; 0.109 mm, 0.166 mm]

QUESTIONS

1. Enumerate the various manufacturing methods of machine parts which a designer should know.
2. Explain briefly the different casting processes.
3. Write a brief note on the design of castings?
4. State and illustrate two principal design rules for casting design.
5. List the main advantages of forged components.
6. What are the salient features used in the design of forgings? Explain.
7. What do you understand by 'hot working' and 'cold working' processes? Explain with examples.
8. State the advantages and disadvantages of hot working of metals. Discuss any two hot working processes.
9. What do you understand by cold working of metals? Describe briefly the various cold working processes.
10. What are fits and tolerances? How are they designated?
11. What do you understand by the nominal size and basic size?
12. Write short notes on the following :

    (*a*) Interchangeability; (*b*) Tolerance; (*c*) Allowance; and (*d*) Fits.
13. What is the difference in the type of assembly generally used in running fits and interference fits?
14. State briefly unilateral system of tolerances covering the points of definition, application and advantages over the bilateral system.
15. What is meant by 'hole basis system' and 'shaft basis system'? Which one is preferred and why?
16. Discuss the Indian standard system of limits and fits.
17. What are the commonly used fits according to Indian standards?
18. What do you understand by preferred numbers? Explain fully.

OBJECTIVE TYPE QUESTIONS

1. The castings produced by forcing molten metal under pressure into a permanent metal mould is known as

   (*a*) permanent mould casting (*b*) slush casting

   (*c*) die casting (*d*) centrifugal casting
2. The metal is subjected to mechanical working for

   (*a*) refining grain size (*b*) reducing original block into desired shape

   (*c*) controlling the direction of flow lines (*d*) all of these

3. The temperature at which the new grains are formed in the metal is called
   (*a*) lower critical temperature (*b*) upper critical temperature
   (*c*) eutectic temperature (*d*) recrystallisation temperature
4. The hot working of metals is carried out
   (*a*) at the recrystallisation temperature (*b*) below the recrystallisation temperature
   (*c*) above the recrystallisation temperature (*d*) at any temperature
5. During hot working of metals
   (*a*) porosity of the metal is largely eliminated
   (*b*) grain structure of the metal is refined
   (*c*) mechanical properties are improved due to refinement of grains
   (*d*) all of the above
6. The parts of circular cross-section which are symmetrical about the axis of rotation are made by
   (*a*) hot forging (*b*) hot spinning
   (*c*) hot extrusion (*d*) hot drawing
7. The cold working of metals is carried out .... the recrystallisation temperature.
   (*a*) above (*b*) below
8. The process extensively used for making bolts and nuts is
   (*a*) hot piercing (*b*) extrusion
   (*c*) cold peening (*d*) cold heading
9. In a unilateral system of tolerance, the tolerance is allowed on
   (*a*) one side of the actual size (*b*) one side of the nominal size
   (*c*) both sides of the actual size (*d*) both sides of the nominal size
10. The algebraic difference between the maximum limit and the basic size is called
   (*a*) actual deviation (*b*) upper deviation
   (*c*) lower deviation (*d*) fundamental deviation
11. A basic shaft is one whose
   (*a*) lower deviation is zero (*b*) upper deviation is zero
   (*c*) lower and upper deviations are zero (*d*) none of these
12. A basic hole is one whose
   (*a*) lower deviation is zero (*b*) upper deviation is zero
   (*c*) lower and upper deviations are zero (*d*) none of these
13. According to Indian standard specifications, 100 $H\,6\,/\,g\,5$ means that the
   (*a*) actual size is 100 mm
   (*b*) basic size is 100 mm
   (*c*) difference between the actual size and basic size is 100 mm
   (*d*) none of the above
14. According to Indian standards, total number of tolerance grades are
   (*a*) 8 (*b*) 12
   (*c*) 18 (*d*) 20
15. According to Indian standard specification, 100 $H6/g5$ means that
   (*a*) tolerance grade for the hole is 6 and for the shaft is 5
   (*b*) tolerance grade for the shaft is 6 and for the hole is 5
   (*c*) tolerance grade for the shaft is 4 to 8 and for the hole is 3 to 7
   (*d*) tolerance grade for the hole is 4 to 8 and for the shaft is 3 to 7

ANSWERS

1. (*c*)	2. (*d*)	3. (*d*)	4. (*c*)	5. (*d*)
6. (*b*)	7. (*b*)	8. (*d*)	9. (*b*)	10. (*b*)
11. (*b*)	12. (*a*)	13. (*b*)	14. (*c*)	15. (*a*)

CHAPTER

4

Simple Stresses in Machine Parts

4.1 Introduction

In engineering practice, the machine parts are subjected to various forces which may be due to either one or more of the following:

1. Energy transmitted,
2. Weight of machine,
3. Frictional resistances,
4. Inertia of reciprocating parts,
5. Change of temperature, and
6. Lack of balance of moving parts.

The different forces acting on a machine part produces various types of stresses, which will be discussed in this chapter.

4.2 Load

It is defined as **any external force acting upon a machine part.** The following four types of the load are important from the subject point of view:

1. ***Dead or steady load.*** A load is said to be a dead or steady load, when it does not change in magnitude or direction.

2. ***Live or variable load.*** A load is said to be a live or variable load, when it changes continually.

3. ***Suddenly applied or shock loads.*** A load is said to be a suddenly applied or shock load, when it is suddenly applied or removed.

4. ***Impact load.*** A load is said to be an impact load, when it is applied with some initial velocity.

Note: A machine part resists a dead load more easily than a live load and a live load more easily than a shock load.

4.3 Stress

When some external system of forces or loads act on a body, the internal forces (equal and opposite) are set up at various sections of the body, which resist the external forces. This internal force per unit area at any section of the body is known as ***unit stress*** or simply a ***stress.*** It is denoted by a Greek letter sigma (σ). Mathematically,

$$\text{Stress, } \sigma = P/A$$

where P = Force or load acting on a body, and

A = Cross-sectional area of the body.

In S.I. units, the stress is usually expressed in Pascal (Pa) such that 1 Pa = 1 N/m^2. In actual practice, we use bigger units of stress *i.e.* megapascal (MPa) and gigapascal (GPa), such that

$$1 \text{ MPa} = 1 \times 10^6 \text{ N/m}^2 = 1 \text{ N/mm}^2$$

and

$$1 \text{ GPa} = 1 \times 10^9 \text{ N/m}^2 = 1 \text{ kN/mm}^2$$

4.4 Strain

When a system of forces or loads act on a body, it undergoes some deformation. This deformation per unit length is known as ***unit strain*** or simply a ***strain.*** It is denoted by a Greek letter epsilon (ε). Mathematically,

$$\text{Strain, } \varepsilon = \delta l / l \quad \text{or} \quad \delta l = \varepsilon . l$$

where δl = Change in length of the body, and

l = Original length of the body.

4.5 Tensile Stress and Strain

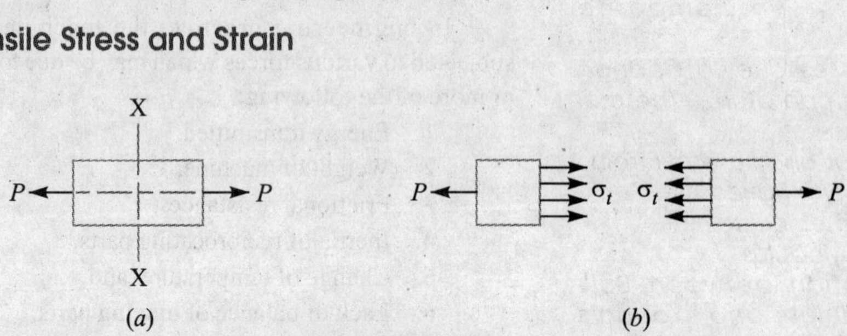

Fig. 4.1. Tensile stress and strain.

When a body is subjected to two equal and opposite axial pulls P (also called tensile load) as shown in Fig. 4.1 (*a*), then the stress induced at any section of the body is known as ***tensile stress*** as shown in Fig. 4.1 (*b*). A little consideration will show that due to the tensile load, there will be a decrease in cross-sectional area and an increase in length of the body. The ratio of the increase in length to the original length is known as ***tensile strain.***

Let P = Axial tensile force acting on the body,

A = Cross-sectional area of the body,

l = Original length, and

δl = Increase in length.

∴ Tensile stress, $\sigma_t = P/A$

and tensile strain, $\varepsilon_t = \delta l / l$

4.6 Compressive Stress and Strain

Shock absorber of a motorcycle absorbs stresses.

Note : This picture is given as additional information and is not a direct example of the current chapter.

When a body is subjected to two equal and opposite axial pushes P (also called compressive load) as shown in Fig. 4.2 (*a*), then the stress induced at any section of the body is known as ***compressive stress*** as shown in Fig. 4.2 (*b*). A little consideration will show that due to the compressive load, there will be an increase in cross-sectional area and a decrease in length of the body. The ratio of the decrease in length to the original length is known as ***compressive strain.***

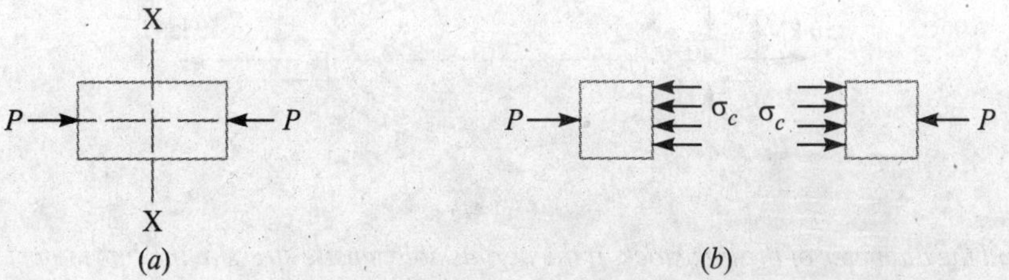

Fig. 4.2. Compressive stress and strain.

Let P = Axial compressive force acting on the body,

A = Cross-sectional area of the body,

l = Original length, and

δl = Decrease in length.

∴ Compressive stress, $\sigma_c = P/A$

and compressive strain, $\varepsilon_c = \delta l / l$

Note : In case of tension or compression, the area involved is at right angles to the external force applied.

4.7 Young's Modulus or Modulus of Elasticity

Hooke's law* states that when a material is loaded within elastic limit, the stress is directly proportional to strain, *i.e.*

$$\sigma \propto \varepsilon \quad \text{or} \quad \sigma = E.\varepsilon$$

$$\therefore \quad E = \frac{\sigma}{\varepsilon} = \frac{P \times l}{A \times \delta l}$$

* It is named after Robert Hooke, who first established it by experiments in 1678.

where E is a constant of proportionality known as ***Young's modulus*** or ***modulus of elasticity.*** In S.I. units, it is usually expressed in GPa *i.e.* GN/m^2 or kN/mm^2. It may be noted that Hooke's law holds good for tension as well as compression.

The following table shows the values of modulus of elasticity or Young's modulus (E) for the materials commonly used in engineering practice.

Table 4.1. Values of E for the commonly used engineering materials.

Material	*Modulus of elasticity (E) in GPa i.e. GN/m² or kN/mm²*
Steel and Nickel	200 to 220
Wrought iron	190 to 200
Cast iron	100 to 160
Copper	90 to 110
Brass	80 to 90
Aluminium	60 to 80
Timber	10

Example 4.1. *A coil chain of a crane required to carry a maximum load of 50 kN, is shown in Fig. 4.3.*

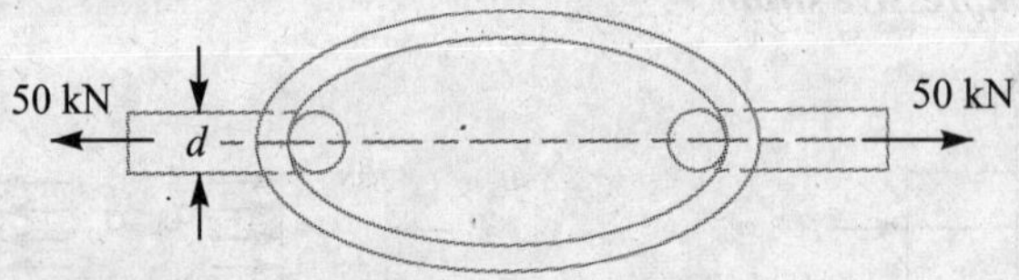

Fig. 4.3

Find the diameter of the link stock, if the permissible tensile stress in the link material is not to exceed 75 MPa.

Solution. Given : $P = 50 \text{ kN} = 50 \times 10^3 \text{ N}$; $\sigma_t = 75 \text{ MPa} = 75 \text{ N/mm}^2$

Let d = Diameter of the link stock in mm.

$\therefore$ Area, $A = \dfrac{\pi}{4} \times d^2 = 0.7854\, d^2$

We know that the maximum load (P),

$$50 \times 10^3 = \sigma_t \cdot A = 75 \times 0.7854\, d^2 = 58.9\, d^2$$

$\therefore$ $d^2 = 50 \times 10^3 / 58.9 = 850$ or $d = 29.13$ say 30 mm **Ans.**

Example 4.2. *A cast iron link, as shown in Fig. 4.4, is required to transmit a steady tensile load of 45 kN. Find the tensile stress induced in the link material at sections A-A and B-B.*

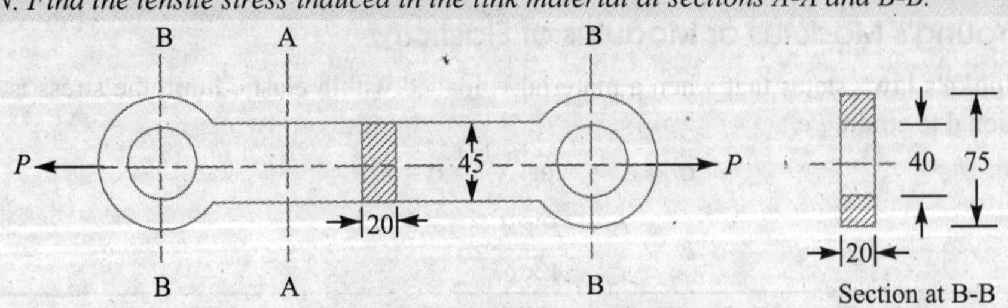

Fig. 4.4. All dimensions in mm.

Solution. Given : $P = 45 \text{ kN} = 45 \times 10^3 \text{ N}$

Tensile stress induced at section A-A

We know that the cross-sectional area of link at section *A-A*,

$$A_1 = 45 \times 20 = 900 \text{ mm}^2$$

∴ Tensile stress induced at section *A-A*,

$$\sigma_{t1} = \frac{P}{A_1} = \frac{45 \times 10^3}{900} = 50 \text{ N/mm}^2 = 50 \text{ MPa} \textbf{ Ans.}$$

Tensile stress induced at section B-B

We know that the cross-sectional area of link at section B-B,

$$A_2 = 20\,(75 - 40) = 700 \text{ mm}^2$$

∴ Tensile stress induced at section *B-B*,

$$\sigma_{t2} = \frac{P}{A_2} = \frac{45 \times 10^3}{700} = 64.3 \text{ N/mm}^2 = 64.3 \text{ MPa} \textbf{ Ans.}$$

Example 4.3. *A hydraulic press exerts a total load of 3.5 MN. This load is carried by two steel rods, supporting the upper head of the press. If the safe stress is 85 MPa and E = 210 kN/mm², find : 1. diameter of the rods, and 2. extension in each rod in a length of 2.5 m.*

Solution. Given : $P = 3.5 \text{ MN} = 3.5 \times 10^6 \text{ N}$; $\sigma_t = 85 \text{ MPa} = 85 \text{ N/mm}^2$; $E = 210 \text{ kN/mm}^2 = 210 \times 10^3 \text{ N/mm}^2$; $l = 2.5 \text{ m} = 2.5 \times 10^3 \text{ mm}$

1. ***Diameter of the rods***

Let d = Diameter of the rods in mm.

∴ Area, $A = \frac{\pi}{4} \times d^2 = 0.7854\, d^2$

Since the load *P* is carried by two rods, therefore load carried by each rod,

$$P_1 = \frac{P}{2} = \frac{3.5 \times 10^6}{2} = 1.75 \times 10^6 \text{ N}$$

We know that load carried by each rod (P_1),

$$1.75 \times 10^6 = \sigma_t \,.\, A = 85 \times 0.7854\, d^2 = 66.76\, d^2$$

∴ $d^2 = 1.75 \times 10^6 / 66.76 = 26\,213$ or $d = 162 \text{ mm}$ **Ans.**

2. ***Extension in each rod***

Let δl = Extension in each rod.

We know that Young's modulus (*E*),

$$210 \times 10^3 = \frac{P_1 \times l}{A \times \delta l} = \frac{\sigma_t \times l}{\delta l} = \frac{85 \times 2.5 \times 10^3}{\delta l} = \frac{212.5 \times 10^3}{\delta l} \quad \ldots \left(\because \frac{P_1}{A} = \sigma_t\right)$$

∴ $\delta l = 212.5 \times 10^3 / (210 \times 10^3) = 1.012 \text{ mm}$ **Ans.**

Example 4.4. *A rectangular base plate is fixed at each of its four corners by a 20 mm diameter bolt and nut as shown in Fig. 4.5. The plate rests on washers of 22 mm internal diameter and 50 mm external diameter. Copper washers which are placed between the nut and the plate are of 22 mm internal diameter and 44 mm external diameter.*

If the base plate carries a load of 120 kN (including self-weight, which is equally distributed on the four corners), calculate the stress on the lower washers before the nuts are tightened.

What could be the stress in the upper and lower washers, when the nuts are tightened so as to produce a tension of 5 kN on each bolt?

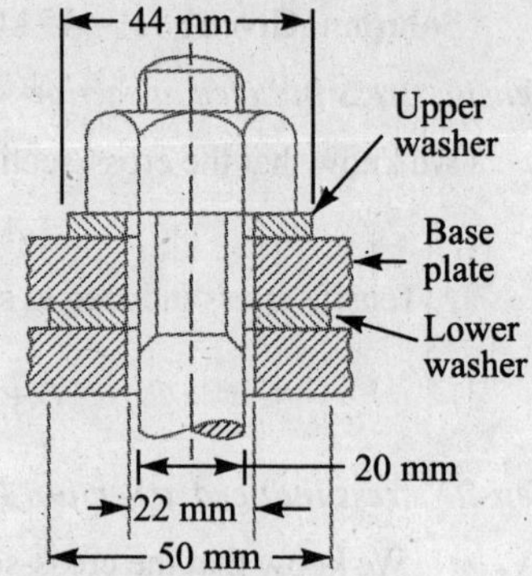

Fig. 4.5

Solution. Given : $d = 20$ mm ; $d_1 = 22$ mm ; $d_2 = 50$ mm ; $d_3 = 22$ mm ; $d_4 = 44$ mm ; $P_1 = 120$ kN ; $P_2 = 5$ kN

Stress on the lower washers before the nuts are tightened

We know that area of lower washers,

$$A_1 = \frac{\pi}{4}\left[(d_2)^2 - (d_1)^2\right] = \frac{\pi}{4}\left[(50)^2 - (22)^2\right] = 1583 \text{ mm}^2$$

and area of upper washers,

$$A_2 = \frac{\pi}{4}\left[(d_4)^2 - (d_3)^2\right] = \frac{\pi}{4}\left[(44)^2 - (22)^2\right] = 1140 \text{ mm}^2$$

Since the load of 120 kN on the four washers is equally distributed, therefore load on each lower washer before the nuts are tightened,

$$P_1 = \frac{120}{4} = 30 \text{ kN} = 30\,000 \text{ N}$$

We know that stress on the lower washers before the nuts are tightened,

$$\sigma_{c1} = \frac{P_1}{A_1} = \frac{30\,000}{1583} = 18.95 \text{ N/mm}^2 = 18.95 \text{ MPa} \quad \textbf{Ans.}$$

Stress on the upper washers when the nuts are tightened

Tension on each bolt when the nut is tightened,

$$P_2 = 5 \text{ kN} = 5000 \text{ N}$$

∴ Stress on the upper washers when the nut is tightened,

$$\sigma_{c2} = \frac{P_2}{A_2} = \frac{5000}{1140} = 4.38 \text{ N/mm}^2 = 4.38 \text{ MPa} \quad \textbf{Ans.}$$

Stress on the lower washers when the nuts are tightened

We know that the stress on the lower washers when the nuts are tightened,

$$\sigma_{c3} = \frac{P_1 + P_2}{A_1} = \frac{30\,000 + 5000}{1583} = 22.11 \text{ N/mm}^2 = 22.11 \text{ MPa} \quad \textbf{Ans.}$$

Example 4.5. *The piston rod of a steam engine is 50 mm in diameter and 600 mm long. The diameter of the piston is 400 mm and the maximum steam pressure is 0.9 N/mm². Find the compression of the piston rod if the Young's modulus for the material of the piston rod is 210 kN/mm².*

Solution. Given : $d = 50$ mm ; $l = 600$ mm ; $D = 400$ mm ; $p = 0.9$ N/mm² ; $E = 210$ kN/mm² $= 210 \times 10^3$ N/mm²

Let δl = Compression of the piston rod.

We know that cross-sectional area of piston,

$$= \frac{\pi}{4} \times D^2 = \frac{\pi}{4}(400)^2 = 125\,680 \text{ mm}^2$$

∴ Maximum load acting on the piston due to steam,

$$P = \text{Cross-sectional area of piston} \times \text{Steam pressure}$$

$$= 125\,680 \times 0.9 = 113\,110 \text{ N}$$

We also know that cross-sectional area of piston rod,

$$A = \frac{\pi}{4} \times d^2 = \frac{\pi}{4} (50)^2$$

$$= 1964 \text{ mm}^2$$

and Young's modulus (E),

$$210 \times 10^3 = \frac{P \times l}{A \times \delta l}$$

$$= \frac{113\,110 \times 600}{1964 \times \delta l} = \frac{34\,555}{\delta l}$$

$$\therefore \quad \delta l = 34\,555 / (210 \times 10^3)$$

$$= 0.165 \text{ mm} \quad \textbf{Ans.}$$

This picture shows a jet engine being tested for bearing high stresses.

4.8 Shear Stress and Strain

When a body is subjected to two equal and opposite forces acting tangentially across the resisting section, as a result of which the body tends to shear off the section, then the stress induced is called ***shear stress.***

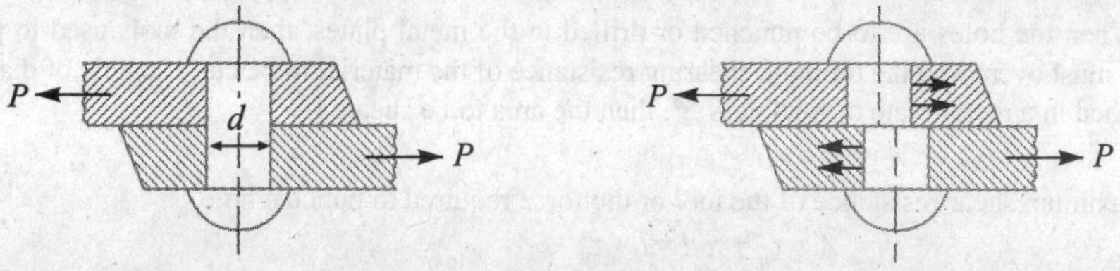

Fig. 4.6. Single shearing of a riveted joint.

The corresponding strain is known as ***shear strain*** and it is measured by the angular deformation accompanying the shear stress. The shear stress and shear strain are denoted by the Greek letters tau (τ) and phi (ϕ) respectively. Mathematically,

$$\text{Shear stress, } \tau = \frac{\text{Tangential force}}{\text{Resisting area}}$$

Consider a body consisting of two plates connected by a rivet as shown in Fig. 4.6 (*a*). In this case, the tangential force P tends to shear off the rivet at one cross-section as shown in Fig. 4.6 (*b*). It may be noted that when the tangential force is resisted by one cross-section of the rivet (or when shearing takes place at one cross-section of the rivet), then the rivets are said to be in ***single shear.*** In such a case, the area resisting the shear off the rivet,

$$A = \frac{\pi}{4} \times d^2$$

and shear stress on the rivet cross-section,

$$\tau = \frac{P}{A} = \frac{P}{\frac{\pi}{4} \times d^2} = \frac{4P}{\pi\, d^2}$$

Now let us consider two plates connected by the two cover plates as shown in Fig. 4.7 (*a*). In this case, the tangential force P tends to shear off the rivet at two cross-sections as shown in Fig. 4.7 (*b*). It may be noted that when the tangential force is resisted by two cross-sections of the rivet (or when the shearing takes place at two cross-sections of the rivet), then the rivets are said to be in ***double shear.*** In such a case, the area resisting the shear off the rivet,

$$A = 2 \times \frac{\pi}{4} \times d^2 \qquad \text{... (For double shear)}$$

and shear stress on the rivet cross-section,

$$\tau = \frac{P}{A} = \frac{P}{2 \times \frac{\pi}{4} \times d^2} = \frac{2P}{\pi d^2}$$

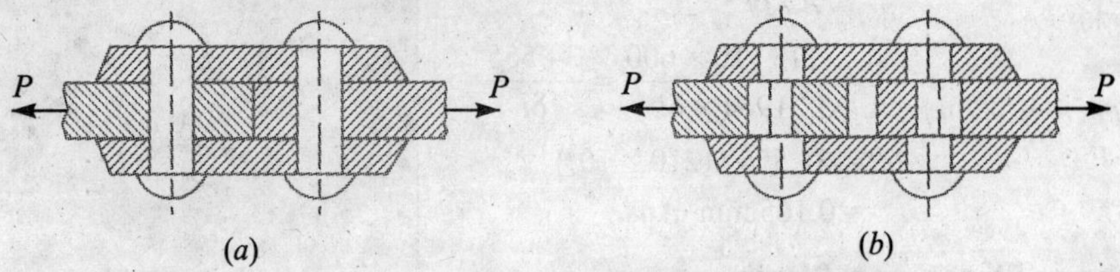

Fig. 4.7. Double shearing of a riveted joint.

Notes : 1. All lap joints and single cover butt joints are in single shear, while the butt joints with double cover plates are in double shear.

2. In case of shear, the area involved is parallel to the external force applied.

3. When the holes are to be punched or drilled in the metal plates, then the tools used to perform the operations must overcome the ultimate shearing resistance of the material to be cut. If a hole of diameter '*d*' is to be punched in a metal plate of thickness '*t*', then the area to be sheared,

$$A = \pi\, d \times t$$

and the maximum shear resistance of the tool or the force required to punch a hole,

$$P = A \times \tau_u = \pi\, d \times t \times \tau_u$$

where τ_u = Ultimate shear strength of the material of the plate.

4.9 Shear Modulus or Modulus of Rigidity

It has been found experimentally that within the elastic limit, the shear stress is directly proportional to shear strain. Mathematically

$$\tau \propto \phi \quad \text{or} \quad \tau = C \,.\, \phi \quad \text{or} \quad \tau / \phi = C$$

where τ = Shear stress,

ϕ = Shear strain, and

C = Constant of proportionality, known as shear modulus or modulus of rigidity. It is also denoted by *N* or *G*.

The following table shows the values of modulus of rigidity (*C*) for the materials in every day use:

Table 4.2. Values of *C* for the commonly used materials.

Material	*Modulus of rigidity (C) in GPa i.e. GN/m² or kN/mm²*
Steel	80 to 100
Wrought iron	80 to 90
Cast iron	40 to 50
Copper	30 to 50
Brass	30 to 50
Timber	10

Example 4.6. *Calculate the force required to punch a circular blank of 60 mm diameter in a plate of 5 mm thick. The ultimate shear stress of the plate is 350 N/mm².*

Solution. Given: $d = 60$ mm ; $t = 5$ mm ; $\tau_u = 350$ N/mm²

We know that area under shear,

$$A = \pi\, d \times t = \pi \times 60 \times 5 = 942.6 \text{ mm}^2$$

and force required to punch a hole,

$$P = A \times \tau_u = 942.6 \times 350 = 329\,910 \text{ N} = 329.91 \text{ kN} \quad \textbf{Ans.}$$

Example 4.7. *A pull of 80 kN is transmitted from a bar X to the bar Y through a pin as shown in Fig. 4.8.*

If the maximum permissible tensile stress in the bars is 100 N/mm² and the permissible shear stress in the pin is 80 N/mm², find the diameter of bars and of the pin.

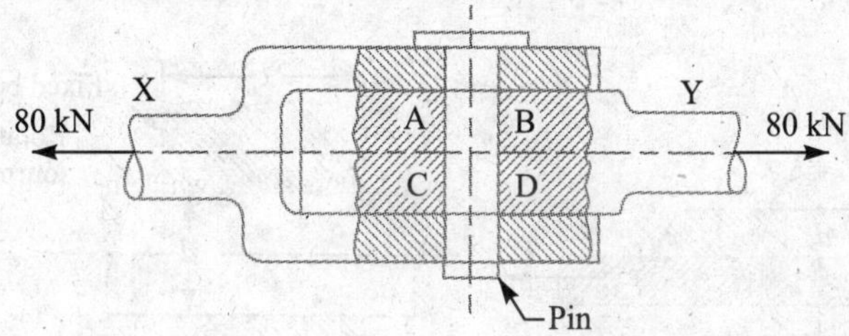

Fig. 4.8

Solution. Given : $P = 80$ kN $= 80 \times 10^3$ N; $\sigma_t = 100$ N/mm² ; $\tau = 80$ N/mm²

Diameter of the bars

Let D_b = Diameter of the bars in mm.

$$\therefore \text{ Area, } A_b = \frac{\pi}{4}(D_b)^2 = 0.7854\,(D_b)^2$$

We know that permissible tensile stress in the bar (σ_t),

$$100 = \frac{P}{A_b} = \frac{80 \times 10^3}{0.7854\,(D_b)^2} = \frac{101\,846}{(D_b)^2}$$

$$\therefore \quad (D_b)^2 = 101\,846 / 100 = 1018.46$$

or $D_b = 32$ mm **Ans.**

High force injection moulding machine.

Note : This picture is given as additional information and is not a direct example of the current chapter.

Diameter of the pin

Let D_p = Diameter of the pin in mm.

Since the tensile load P tends to shear off the pin at two sections *i.e.* at AB and CD, therefore the pin is in double shear.

$\therefore$ Resisting area,

$$A_p = 2 \times \frac{\pi}{4}(D_p)^2 = 1.571\,(D_p)^2$$

We know that permissible shear stress in the pin (τ),

$$80 = \frac{P}{A_p} = \frac{80 \times 10^3}{1.571\,(D_p)^2} = \frac{50.9 \times 10^3}{(D_p)^2}$$

$$\therefore \quad (D_p)^2 = 50.9 \times 10^3/80 = 636.5 \text{ or } D_p = 25.2 \text{ mm} \quad \textbf{Ans.}$$

4.10 Bearing Stress

A localised compressive stress at the surface of contact between two members of a machine part, that are relatively at rest is known as ***bearing stress*** or ***crushing stress.*** The bearing stress is taken into account in the design of riveted joints, cotter joints, knuckle joints, etc. Let us consider a riveted joint subjected to a load P as shown in Fig. 4.9. In such a case, the bearing stress or crushing stress (stress at the surface of contact between the rivet and a plate),

$$\sigma_b \text{ (or } \sigma_c) = \frac{P}{d.t.n}$$

where d = Diameter of the rivet,

t = Thickness of the plate,

$d.t$ = Projected area of the rivet, and

n = Number of rivets per pitch length in bearing or crushing.

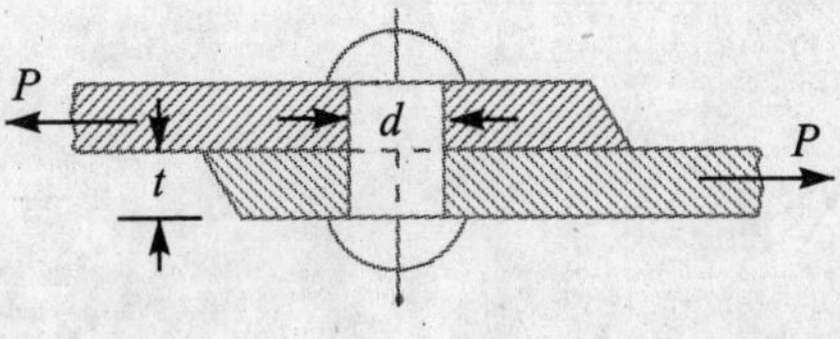

Fig. 4.9. Bearing stress in a riveted joint.

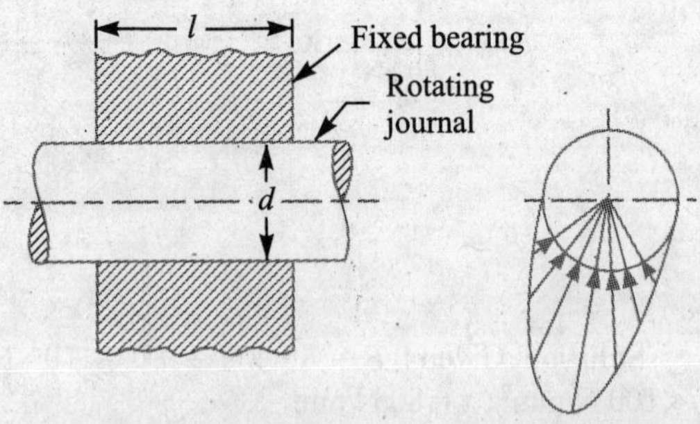

(*a*) Journal supported in a bearing. (*b*) Distribution of bearing pressure.

Fig. 4.10. Bearing pressure in a journal supported in a bearing.

It may be noted that the local compression which exists at the surface of contact between two members of a machine part that are in relative motion, is called ***bearing pressure*** (not the bearing stress). This term is commonly used in the design of a journal supported in a bearing, pins for levers, crank pins, clutch lining, etc. Let us consider a journal rotating in a fixed bearing as shown in Fig. 4.10 (*a*). The journal exerts a bearing pressure on the curved surfaces of the brasses immediately below it. The distribution of this bearing pressure will not be uniform, but it will be in accordance with the shape of the surfaces in contact and deformation characteristics of the two materials. The distribution of bearing pressure will be similar to that as shown in Fig. 4.10 (*b*). Since the actual bearing pressure is difficult to determine, therefore the average bearing pressure is usually calculated by dividing the load to the projected area of the curved surfaces in contact. Thus, the average bearing pressure for a journal supported in a bearing is given by

$$p_b = \frac{P}{l.d}$$

where p_b = Average bearing pressure,

P = Radial load on the journal,

l = Length of the journal in contact, and

d = Diameter of the journal.

Example 4.8. *Two plates 16 mm thick are joined by a double riveted lap joint as shown in Fig. 4.11. The rivets are 25 mm in diameter.*

Find the crushing stress induced between the plates and the rivet, if the maximum tensile load on the joint is 48 kN.

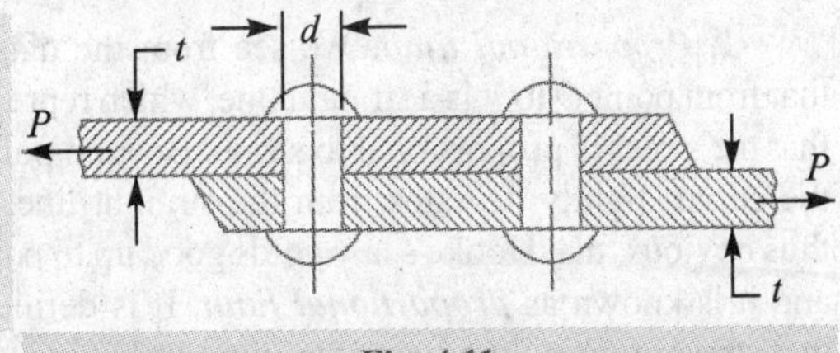

Fig. 4.11

Solution. Given : $t = 16$ mm ; $d = 25$ mm ; $P = 48$ kN $= 48 \times 10^3$ N

Since the joint is double riveted, therefore, strength of two rivets in bearing (or crushing) is taken. We know that crushing stress induced between the plates and the rivets,

$$\sigma_c = \frac{P}{d.t.n} = \frac{48 \times 10^3}{25 \times 16 \times 2} = 60 \text{ N/mm}^2 \textbf{ Ans.}$$

Example 4.9. *A journal 25 mm in diameter supported in sliding bearings has a maximum end reaction of 2500 N. Assuming an allowable bearing pressure of 5 N/mm², find the length of the sliding bearing.*

Solution. Given : $d = 25$ mm ; $P = 2500$ N ; $p_b = 5$ N/mm²

Let l = Length of the sliding bearing in mm.

We know that the projected area of the bearing,

$$A = l \times d = l \times 25 = 25\, l \text{ mm}^2$$

∴ Bearing pressure (p_b),

$$5 = \frac{P}{A} = \frac{2500}{25\, l} = \frac{100}{l} \quad \text{or} \quad l = \frac{100}{5} = 20 \text{ mm } \textbf{Ans.}$$

4.11 Stress-strain Diagram

In designing various parts of a machine, it is necessary to know how the material will function in service. For this, certain characteristics or properties of the material should be known. The mechanical properties mostly used in mechanical engineering practice are commonly determined from a standard tensile test. This test consists of gradually loading a standard specimen of a material and noting the corresponding values of load and elongation until the specimen fractures. The load is applied and measured by a testing machine. The stress is determined by dividing the load values by the original cross-sectional area of the specimen. The elongation is measured by determining the amounts that two reference points on the specimen are moved apart by the action of the machine. The original distance between the two reference points is known as ***gauge length.*** The strain is determined by dividing the elongation values by the gauge length.

In addition to bearing the stresses, some machine parts are made of stainless steel to make them corrosion resistant.

Note : This picture is given as additional information and is not a direct example of the current chapter.

The values of the stress and corresponding strain are used to draw the stress-strain diagram of the material tested. A stress-strain diagram for a mild steel under tensile test is shown in Fig. 4.12 (*a*). The various properties of the material are discussed below :

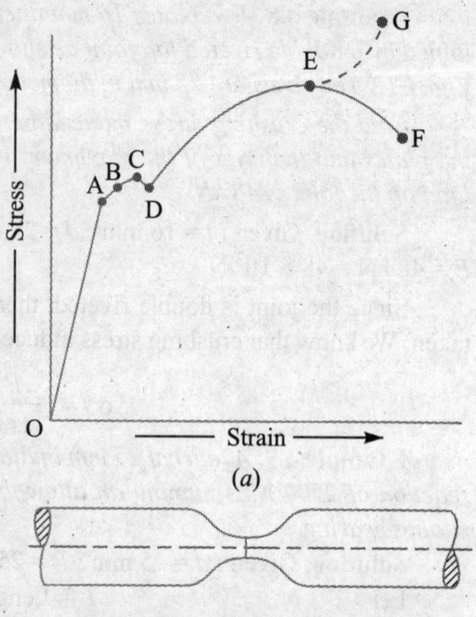

(b) Shape of specimen after elongation.

Fig. 4.12. Stress-strain diagram for a mild steel.

1. ***Proportional limit.*** We see from the diagram that from point *O* to *A* is a straight line, which represents that the stress is proportional to strain. Beyond point *A*, the curve slightly deviates from the straight line. It is thus obvious, that Hooke's law holds good up to point *A* and it is known as ***proportional limit***. It is defined as that stress at which the stress-strain curve begins to deviate from the straight line.

2. ***Elastic limit.*** It may be noted that even if the load is increased beyond point *A* upto the point *B*, the material will regain its shape and size when the load is removed. This means that the material has elastic properties up to the point *B*. This point is known as ***elastic limit.*** It is defined as the stress developed in the material without any permanent set.

Note: Since the above two limits are very close to each other, therefore, for all practical purposes these are taken to be equal.

3. ***Yield point.*** If the material is stressed beyond point *B*, the plastic stage will reach *i.e.* on the removal of the load, the material will not be able to recover its original size and shape. A little consideration will show that beyond point *B*, the strain increases at a faster rate with any increase in the stress until the point *C* is reached. At this point, the material yields before the load and there is an appreciable strain without any increase in stress. In case of mild steel, it will be seen that a small load drops to *D*, immediately after yielding commences. Hence there are two yield points *C* and *D*. The points *C* and *D* are called the ***upper*** and ***lower yield points*** respectively. The stress corresponding to yield point is known as ***yield point stress.***

4. ***Ultimate stress.*** At *D*, the specimen regains some strength and higher values of stresses are required for higher strains, than those between *A* and *D*. The stress (or load) goes on increasing till the

A crane used on a ship.

Note : This picture is given as additional information and is not a direct example of the current chapter.

point E is reached. The gradual increase in the strain (or length) of the specimen is followed with the uniform reduction of its cross-sectional area. The work done, during stretching the specimen, is transformed largely into heat and the specimen becomes hot. At E, the stress, which attains its maximum value is known as ***ultimate stress.*** It is defined as the largest stress obtained by dividing the largest value of the load reached in a test to the original cross-sectional area of the test piece.

A recovery truck with crane.

Note : This picture is given as additional information and is not a direct example of the current chapter.

5. ***Breaking stress.*** After the specimen has reached the ultimate stress, a neck is formed, which decreases the cross-sectional area of the specimen, as shown in Fig. 4.12 (*b*). A little consideration will show that the stress (or load) necessary to break away the specimen, is less than the maximum stress. The stress is, therefore, reduced until the specimen breaks away at point F. The stress corresponding to point F is known as ***breaking stress.***

Note : The breaking stress (*i.e.* stress at F which is less than at E) appears to be somewhat misleading. As the formation of a neck takes place at E which reduces the cross-sectional area, it causes the specimen suddenly to fail at F. If for each value of the strain between E and F, the tensile load is divided by the reduced cross-sectional area at the narrowest part of the neck, then the true stress-strain curve will follow the dotted line EG. However, it is an established practice, to calculate strains on the basis of original cross-sectional area of the specimen.

6. ***Percentage reduction in area.*** It is the difference between the original cross-sectional area and cross-sectional area at the neck (*i.e.* where the fracture takes place). This difference is expressed as percentage of the original cross-sectional area.

Let A = Original cross-sectional area, and

a = Cross-sectional area at the neck.

Then reduction in area $= A - a$

and percentage reduction in area $= \dfrac{A - a}{A} \times 100$

7. ***Percentage elongation.*** It is the percentage increase in the standard gauge length (*i.e.* original length) obtained by measuring the fractured specimen after bringing the broken parts together.

Let l = Gauge length or original length, and

L = Length of specimen after fracture or final length.

$\therefore$ Elongation $= L - l$

and percentage elongation $= \dfrac{L - l}{l} \times 100$

Note : The percentage elongation gives a measure of ductility of the metal under test. The amount of local extensions depends upon the material and also on the transverse dimensions of the test piece. Since the specimens are to be made from bars, strips, sheets, wires, forgings, castings, etc., therefore it is not possible to make all specimens of one standard size. Since the dimensions of the specimen influence the result, therefore some standard means of comparison of results are necessary.

As a result of series of experiments, Barba estabilished a law that in tension, similar test pieces deform similarly and two test pieces are said to be similar if they have the same value of $\frac{l}{\sqrt{A}}$, where l is the gauge length and A is the cross-sectional area. A little consideration will show that the same material will give the same percentage elongation and percentage reduction in area.

It has been found experimentally by Unwin that the general extension (up to the maximum load) is proportional to the gauge length of the test piece and that the local extension (from maximum load to the breaking load) is proportional to the square root of the cross-sectional area. According to Unwin's formula, the increase in length,

$$\delta l = b.l + C\sqrt{A}$$

and percentage elongation $= \frac{\delta l}{l} \times 100$

where l = Gauge length,

A = Cross-sectional area, and

b and C = Constants depending upon the quality of the material.

The values of b and C are determined by finding the values of δl for two test pieces of known length (l) and area (A).

Example 4.10. *A mild steel rod of 12 mm diameter was tested for tensile strength with the gauge length of 60 mm. Following observations were recorded :*

Final length = 80 mm; Final diameter = 7 mm; Yield load = 3.4 kN and Ultimate load = 6.1 kN.

Calculate : 1. yield stress, 2. ultimate tensile stress, 3. percentage reduction in area, and 4. percentage elongation.

Solution. Given : D = 12 mm ; l = 60 mm ; L = 80 mm ; d = 7 mm ; W_y = 3.4 kN = 3400 N; W_u = 6.1 kN = 6100 N

We know that original area of the rod,

$$A = \frac{\pi}{4} \times D^2 = \frac{\pi}{4}(12)^2 = 113 \text{ mm}^2$$

and final area of the rod,

$$a = \frac{\pi}{4} \times d^2 = \frac{\pi}{4}(7)^2 = 38.5 \text{ mm}^2$$

1. *Yield stress*

We know that yield stress

$$= \frac{W_y}{A} = \frac{3400}{113} = 30.1 \text{ N/mm}^2 = 30.1 \text{ MPa} \quad \textbf{Ans.}$$

2. *Ultimate tensile stress*

We know the ultimate tensile stress

$$= \frac{W_u}{A} = \frac{6100}{113} = 54 \text{ N/mm}^2 = 54 \text{ MPa} \quad \textbf{Ans.}$$

3. *Percentage reduction in area*

We know that percentage reduction in area

$$= \frac{A - a}{A} = \frac{113 - 38.5}{113} = 0.66 \text{ or } 66\% \quad \textbf{Ans.}$$

4. *Percentage elongation*

We know that percentage elongation

$$= \frac{L - l}{L} = \frac{80 - 60}{80} = 0.25 \text{ or } 25\% \textbf{ Ans.}$$

4.12 Working Stress

When designing machine parts, it is desirable to keep the stress lower than the maximum or ultimate stress at which failure of the material takes place. This stress is known as the ***working stress*** or ***design stress.*** It is also known as ***safe*** or ***allowable stress.***

Note : By failure it is not meant actual breaking of the material. Some machine parts are said to fail when they have plastic deformation set in them, and they no more perform their function satisfactory.

4.13 Factor of Safety

It is defined, in general, as the **ratio of the maximum stress to the working stress.** Mathematically,

$$\text{Factor of safety} = \frac{\text{Maximum stress}}{\text{Working or design stress}}$$

In case of ductile materials *e.g.* mild steel, where the yield point is clearly defined, the factor of safety is based upon the yield point stress. In such cases,

$$\text{Factor of safety} = \frac{\text{Yield point stress}}{\text{Working or design stress}}$$

In case of brittle materials *e.g.* cast iron, the yield point is not well defined as for ductile materials. Therefore, the factor of safety for brittle materials is based on ultimate stress.

$$\therefore \quad \text{Factor of safety} = \frac{\text{Ultimate stress}}{\text{Working or design stress}}$$

This relation may also be used for ductile materials.

Note: The above relations for factor of safety are for static loading.

4.14 Selection of Factor of Safety

The selection of a proper factor of safety to be used in designing any machine component depends upon a number of considerations, such as the material, mode of manufacture, type of stress, general service conditions and shape of the parts. Before selecting a proper factor of safety, a design engineer should consider the following points :

1. The reliability of the properties of the material and change of these properties during service ;
2. The reliability of test results and accuracy of application of these results to actual machine parts ;
3. The reliability of applied load ;
4. The certainty as to exact mode of failure ;
5. The extent of simplifying assumptions ;
6. The extent of localised stresses ;
7. The extent of initial stresses set up during manufacture ;
8. The extent of loss of life if failure occurs ; and
9. The extent of loss of property if failure occurs.

Each of the above factors must be carefully considered and evaluated. The high factor of safety results in unnecessary risk of failure. The values of factor of safety based on ultimate strength for different materials and type of load are given in the following table:

Table 4.3. Values of factor of safety.

Material	*Steady load*	*Live load*	*Shock load*
Cast iron	5 to 6	8 to 12	16 to 20
Wrought iron	4	7	10 to 15
Steel	4	8	12 to 16
Soft materials and alloys	6	9	15
Leather	9	12	15
Timber	7	10 to 15	20

4.15 Stresses in Composite Bars

A composite bar may be defined as a bar made up of two or more different materials, joined together, in such a manner that the system extends or contracts as one unit, equally, when subjected to tension or compression. In case of composite bars, the following points should be kept in view:

1. The extension or contraction of the bar being equal, the strain *i.e.* deformation per unit length is also equal.
2. The total external load on the bar is equal to the sum of the loads carried by different materials.

Consider a composite bar made up of two different materials as shown in Fig. 4.13.

Let P_1 = Load carried by bar 1,

A_1 = Cross-sectional area of bar 1,

σ_1 = Stress produced in bar 1,

E_1 = Young's modulus of bar 1,

P_2, A_2, σ_2, E_2 = Corresponding values of bar 2,

P = Total load on the composite bar,

l = Length of the composite bar, and

δl = Elongation of the composite bar.

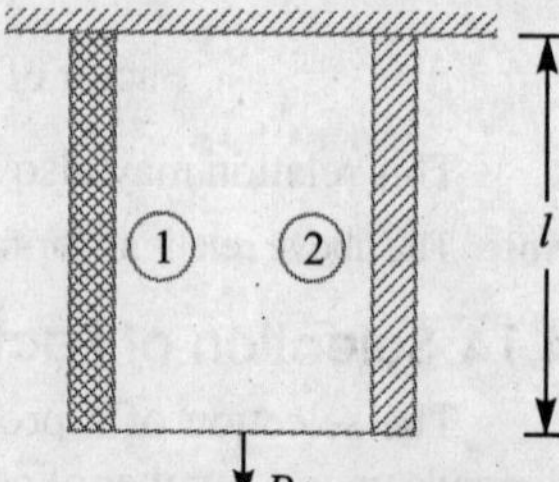

Fig. 4.13. Stresses in composite bars.

We know that $P = P_1 + P_2$ ...(*i*)

Stress in bar 1, $\sigma_1 = \dfrac{P_1}{A_1}$

and strain in bar 1, $\varepsilon = \dfrac{\sigma_1}{E_1} = \dfrac{P_1}{A_1 . E_1}$

∴ Elongation of bar 1,

$$\delta l_1 = \frac{P_1 . l}{A_1 . E_1}$$

Similarly, elongation of bar 2,

$$\delta l_2 = \frac{P_2 . l}{A_2 . E_2}$$

Since $\delta l_1 = \delta l_2$

A Material handling system

Note : This picture is given as additional information and is not a direct example of the current chapter.

Therefore, $\dfrac{P_1.l}{A_1.E_1} = \dfrac{P_2.l}{A_2.E_2}$ or $P_1 = P_2 \times \dfrac{A_1.E_1}{A_2.E_2}$ ...(*ii*)

But $$P = P_1 + P_2 = P_2 \times \frac{A_1.E_1}{A_2.E_2} + P_2 = P_2\left(\frac{A_1.E_1}{A_2.E_2} + 1\right)$$

$$= P_2\left(\frac{A_1.E_1 + A_2.E_2}{A_2.E_2}\right)$$

or $$P_2 = P \times \frac{A_2.E_2}{A_1.E_1 + A_2.E_2} \quad ...(iii)$$

Similarly $$P_1 = P \times \frac{A_1.E_1}{A_1.E_1 + A_2.E_2} \quad \text{...[From equation (ii)]} \quad ...(iv)$$

We know that

$$\frac{P_1.l}{A_1.E_1} = \frac{P_2.l}{A_2.E_2}$$

$\therefore$ $$\frac{\sigma_1}{E_1} = \frac{\sigma_2}{E_2}$$

or $$\sigma_1 = \frac{E_1}{E_2} \times \sigma_2 \quad ...(v)$$

Similarly, $$\sigma_2 = \frac{E_2}{E_1} \times \sigma_1 \quad ...(vi)$$

From the above equations, we can find out the stresses produced in the different bars. We also know that

$$P = P_1 + P_2 = \sigma_1.A_1 + \sigma_2.A_2$$

From this equation, we can also find out the stresses produced in different bars.

Note : The ratio E_1 / E_2 is known as ***modular ratio*** of the two materials.

Example 4.11. *A bar 3 m long is made of two bars, one of copper having E = 105 GN/m² and the other of steel having E = 210 GN/m². Each bar is 25 mm broad and 12.5 mm thick. This compound bar is stretched by a load of 50 kN. Find the increase in length of the compound bar and the stress produced in the steel and copper. The length of copper as well as of steel bar is 3 m each.*

Solution. Given : $l_c = l_s = 3$ m $= 3 \times 10^3$ mm ; $E_c = 105$ GN/m² $= 105$ kN/mm² ; $E_s = 210$ GN/m² $= 210$ kN/mm² ; $b = 25$ mm ; $t = 12.5$ mm ; $P = 50$ kN

Increase in length of the compound bar

Let δl = Increase in length of the compound bar.

The compound bar is shown in Fig. 4.14. We know that cross-sectional area of each bar,

$$A_c = A_s = b \times t = 25 \times 12.5 = 312.5 \text{ mm}^2$$

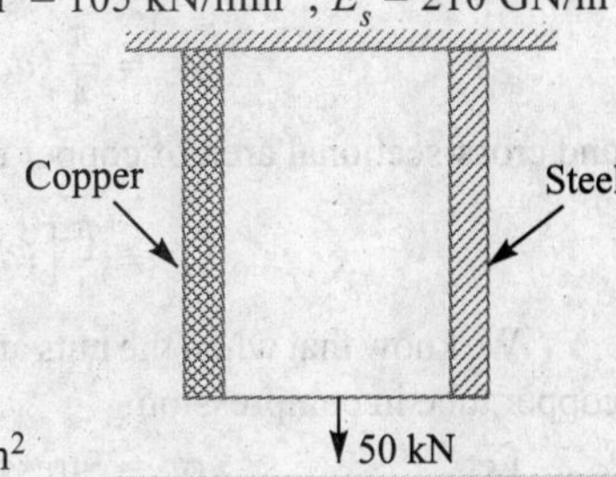

Fig. 4.14

∴ Load shared by the copper bar,

$$P_c = P \times \frac{A_c . E_c}{A_c . E_c + A_s . E_s} = P \times \frac{E_c}{E_c + E_s} \qquad ...(\because A_c = A_s)$$

$$= 50 \times \frac{105}{105 + 210} = 16.67 \text{ kN}$$

and load shared by the steel bar,

$$P_s = P - P_c = 50 - 16.67 = 33.33 \text{ kN}$$

Since the elongation of both the bars is equal, therefore

$$\delta l = \frac{P_c . l_c}{A_c . E_c} = \frac{P_s . l_s}{A_s . E_s} = \frac{16.67 \times 3 \times 10^3}{312.5 \times 105} = 1.52 \text{ mm } \textbf{Ans.}$$

Stress produced in the steel and copper bar

We know that stress produced in the steel bar,

$$\sigma_s = \frac{E_s}{E_c} \times \sigma_c = \frac{210}{105} \times \sigma_c = 2\, \sigma_c$$

and total load, $P = P_s + P_c = \sigma_s . A_s + \sigma_c . A_c$

∴ $50 = 2\, \sigma_c \times 312.5 + \sigma_c \times 312.5 = 937.5\, \sigma_c$

or $\sigma_c = 50 / 937.5 = 0.053 \text{ kN/mm}^2 = 53 \text{ N/mm}^2 = 53 \text{ MPa}$ **Ans.**

and $\sigma_s = 2\, \sigma_c = 2 \times 53 = 106 \text{ N/mm}^2 = 106 \text{ MPa}$ **Ans.**

Example 4.12. *A central steel rod 18 mm diameter passes through a copper tube 24 mm inside and 40 mm outside diameter, as shown in Fig. 4.15. It is provided with nuts and washers at each end. The nuts are tightened until a stress of 10 MPa is set up in the steel.*

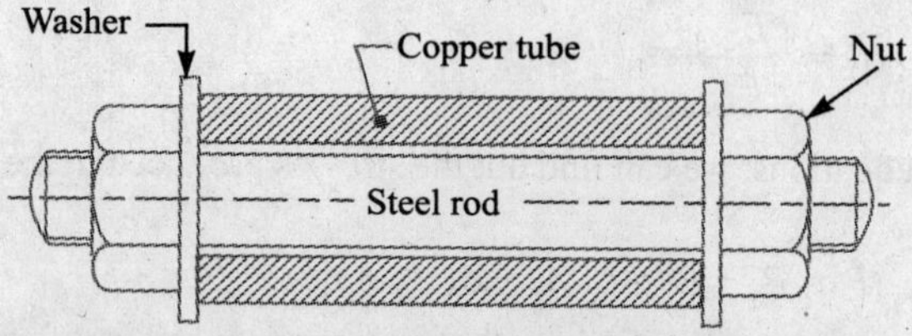

Fig. 4.15

The whole assembly is then placed in a lathe and turned along half the length of the tube removing the copper to a depth of 1.5 mm. Calculate the stress now existing in the steel. Take $E_s = 2E_c$.

Solution. Given : $d_s = 18$ mm ; $d_{c1} = 24$ mm ; $d_{c2} = 40$ mm ; $\sigma_s = 10$ MPa $= 10$ N/mm²

We know that cross-sectional area of steel rod,

$$A_s = \frac{\pi}{4}(d_s)^2 = \frac{\pi}{4}(18)^2 = 254.5 \text{ mm}^2$$

and cross-sectional area of copper tube,

$$A_c = \frac{\pi}{4}\left[(d_{c2})^2 - (d_{c1})^2\right] = \frac{\pi}{4}\left[(40)^2 - (24)^2\right] = 804.4 \text{ mm}^2$$

We know that when the nuts are tightened on the tube, the steel rod will be under tension and the copper tube in compression.

Let σ_c = Stress in the copper tube.

Since the tensile load on the steel rod is equal to the compressive load on the copper tube, therefore

$$\sigma_s \times A_s = \sigma_c \times A_c$$

$$10 \times 254.5 = \sigma_c \times 804.4$$

$$\therefore \quad \sigma_c = \frac{10 \times 254.5}{804.4} = 3.16 \text{ N/mm}^2$$

When the copper tube is reduced in the area for half of its length, then outside diameter of copper tube,

$$= 40 - 2 \times 1.5 = 37 \text{ mm}$$

∴ Cross-sectional area of the half length of copper tube,

$$A_{c1} = \frac{\pi}{4}(37^2 - 24^2) = 623 \text{ mm}^2$$

The cross-sectional area of the other half remains same. If A_{c2} be the area of the remainder, then

$$A_{c2} = A_c = 804.4 \text{ mm}^2$$

Let σ_{c1} = Compressive stress in the reduced section,

σ_{c2} = Compressive stress in the remainder, and

σ_{s1} = Stress in the rod after turning.

Since the load on the copper tube is equal to the load on the steel rod, therefore

$$A_{c1} \times \sigma_{c1} = A_{c2} \times \sigma_{c2} = A_s \times \sigma_{s1}$$

$$\therefore \quad \sigma_{c1} = \frac{A_s}{A_{c1}} \times \sigma_{s1} = \frac{254.5}{623} \times \sigma_{s1} = 0.41\, \sigma_{s1} \quad \text{...(i)}$$

and

$$\sigma_{c2} = \frac{A_s}{A_{c2}} \times \sigma_{s1} = \frac{254.5}{804.4} \times \sigma_{s1} = 0.32\, \sigma_{s1} \quad \text{...(ii)}$$

Let δl = Change in length of the steel rod before and after turning,

l = Length of the steel rod and copper tube between nuts,

δl_1 = Change in length of the reduced section (*i.e.* $l/2$) before and after turning, and

δl_2 = Change in length of the remainder section (*i.e.* $l/2$) before and after turning.

Since $\delta l = \delta l_1 + \delta l_2$

$$\therefore \quad \frac{\sigma_s - \sigma_{s1}}{E_s} \times l = \frac{\sigma_{c1} - \sigma_c}{E_c} \times \frac{l}{2} + \frac{\sigma_{c2} - \sigma_c}{E_c} \times \frac{l}{2}$$

or

$$\frac{10 - \sigma_{s1}}{2E_c} = \frac{0.41\,\sigma_{s1} - 3.16}{2E_c} + \frac{0.32\,\sigma_{s1} - 3.16}{2E_c} \quad \text{...(Cancelling } l \text{ throughout)}$$

$$\therefore \quad \sigma_{s1} = 9.43 \text{ N/mm}^2 = 9.43 \text{ MPa} \textbf{ Ans.}$$

4.16 Stresses due to Change in Temperature—Thermal Stresses

Whenever there is some increase or decrease in the temperature of a body, it causes the body to expand or contract. A little consideration will show that if the body is allowed to expand or contract freely, with the rise or fall of the temperature, no stresses are induced in the body. But, if the deformation of the body is prevented, some stresses are induced in the body. Such stresses are known as ***thermal stresses.***

Let l = Original length of the body,

t = Rise or fall of temperature, and

α = Coefficient of thermal expansion,

∴ Increase or decrease in length,

$$\delta l = l.\,\alpha.t$$

If the ends of the body are fixed to rigid supports, so that its expansion is prevented, then compressive strain induced in the body,

$$\varepsilon_c = \frac{\delta l}{l} = \frac{l.\alpha.t}{l} = \alpha.t$$

∴ Thermal stress, $\sigma_{th} = \varepsilon_c.E = \alpha.t.E$

Notes : 1. When a body is composed of two or different materials having different coefficient of thermal expansions, then due to the rise in temperature, the material with higher coefficient of thermal expansion will be subjected to compressive stress whereas the material with low coefficient of expansion will be subjected to tensile stress.

2. When a thin tyre is shrunk on to a wheel of diameter D, its internal diameter d is a little less than the wheel diameter. When the tyre is heated, its circumferance $\pi\, d$ will increase to $\pi\, D$. In this condition, it is slipped on to the wheel. When it cools, it wants to return to its original circumference $\pi\, d$, but the wheel if it is assumed to be rigid, prevents it from doing so.

$$\therefore \qquad \text{Strain, } \varepsilon = \frac{\pi D - \pi d}{\pi d} = \frac{D - d}{d}$$

This strain is known as ***circumferential*** or ***hoop strain.***

∴ Circumferential or hoop stress,

$$\sigma = E.\varepsilon = \frac{E\,(D - d)}{d}$$

Steel tyres of a locomotive.

Example 4.13. *A thin steel tyre is shrunk on to a locomotive wheel of 1.2 m diameter. Find the internal diameter of the tyre if after shrinking on, the hoop stress in the tyre is 100 MPa. Assume $E = 200$ kN/mm^2. Find also the least temperature to which the tyre must be heated above that of the wheel before it could be slipped on. The coefficient of linear expansion for the tyre is 6.5×10^{-6} per °C.*

Solution. Given : $D = 1.2$ m = 1200 mm ; $\sigma = 100$ MPa = 100 N/mm^2 ; $E = 200$ kN/mm^2 = 200×10^3 N/mm^2 ; $\alpha = 6.5 \times 10^{-6}$ per °C

Internal diameter of the tyre

Let d = Internal diameter of the tyre.

We know that hoop stress (σ),

$$100 = \frac{E\,(D-d)}{d} = \frac{200 \times 10^3\,(D-d)}{d}$$

$$\therefore \quad \frac{D-d}{d} = \frac{100}{200 \times 10^3} = \frac{1}{2 \times 10^3} \qquad ...(i)$$

$$\frac{D}{d} = 1 + \frac{1}{2 \times 10^3} = 1.0005$$

$$\therefore \quad d = \frac{D}{1.0005} = \frac{1200}{1.0005} = 1199.4 \text{ mm} = 1.1994 \text{ m} \quad \textbf{Ans.}$$

Least temperature to which the tyre must be heated

Let t = Least temperature to which the tyre must be heated.

We know that

$$\pi D = \pi d + \pi d\,.\,\alpha.t = \pi d\,(1 + \alpha.t)$$

$$\alpha.t = \frac{\pi D}{\pi d} - 1 = \frac{D-d}{d} = \frac{1}{2 \times 10^3} \qquad ...[\text{From equation } (i)]$$

$$\therefore \quad t = \frac{1}{\alpha \times 2 \times 10^3} = \frac{1}{6.5 \times 10^{-6} \times 2 \times 10^3} = 77°\text{C} \quad \textbf{Ans.}$$

Example 4.14. *A composite bar made of aluminium and steel is held between the supports as shown in Fig. 4.16. The bars are stress free at a temperature of 37°C. What will be the stress in the two bars when the temperature is 20°C, if (a) the supports are unyielding; and (b) the supports yield and come nearer to each other by 0.10 mm?*

It can be assumed that the change of temperature is uniform all along the length of the bar. Take E_s = 210 GPa ; E_a = 74 GPa ; $\alpha_s = 11.7 \times 10^{-6}$ / °C ; and $\alpha_a = 23.4 \times 10^{-6}$ / °C.

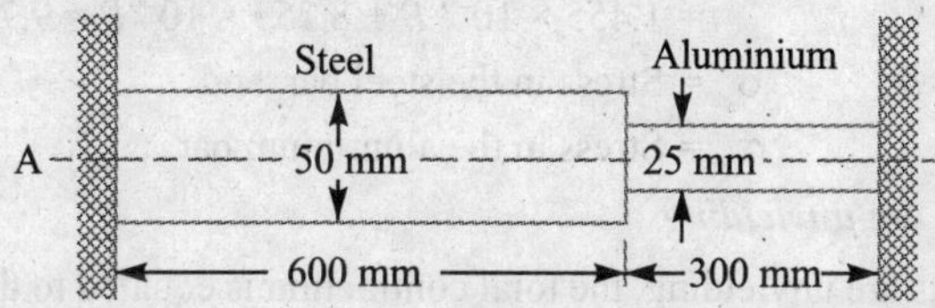

Fig. 4.16

Solution. Given : t_1 = 37°C ; t_2 = 20°C ; E_s = 210 GPa = 210 × 10^9 N/m^2 ; E_a = 74 GPa = 74 × 10^9 N/m^2 ; $\alpha_s = 11.7 \times 10^{-6}$ / °C ; $\alpha_a = 23.4 \times 10^{-6}$ / °C , d_s = 50 mm = 0.05 m ; d_a = 25 mm = 0.025 m ; l_s = 600 mm = 0.6 m ; l_a = 300 mm = 0.3 m

Let us assume that the right support at *B* is removed and the bar is allowed to contract freely due to the fall in temperature. We know that the fall in temperature,

$$t = t_1 - t_2 = 37 - 20 = 17°\text{C}$$

∴ Contraction in steel bar

$$= \alpha_s\,.\,l_s\,.\,t = 11.7 \times 10^{-6} \times 600 \times 17 = 0.12 \text{ mm}$$

and contraction in aluminium bar

$$= \alpha_a\,.\,l_a\,.\,t = 23.4 \times 10^{-6} \times 300 \times 17 = 0.12 \text{ mm}$$

$$\text{Total contraction} = 0.12 + 0.12 = 0.24 \text{ mm} = 0.24 \times 10^{-3} \text{ m}$$

It may be noted that even after this contraction (*i.e.* 0.24 mm) in length, the bar is still stress free as the right hand end was assumed free.

Let an axial force P is applied to the right end till this end is brought in contact with the right hand support at B, as shown in Fig. 4.17.

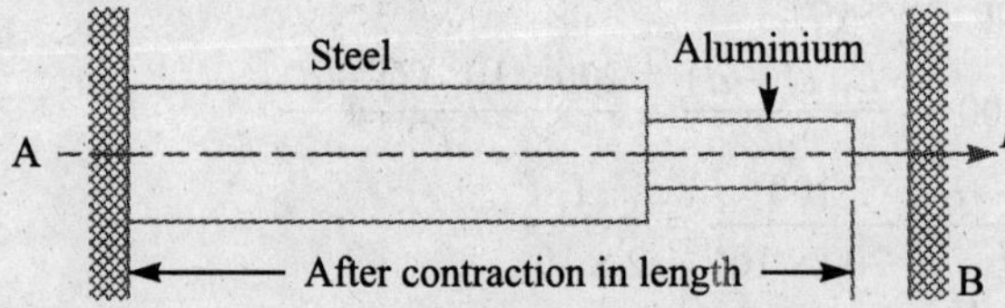

Fig. 4.17

We know that cross-sectional area of the steel bar,

$$A_s = \frac{\pi}{4}(d_s)^2 = \frac{\pi}{4}(0.05)^2 = 1.964 \times 10^{-3} \text{ m}^2$$

and cross-sectional area of the aluminium bar,

$$A_a = \frac{\pi}{4}(d_a)^2 = \frac{\pi}{4}(0.025)^2 = 0.491 \times 10^{-3} \text{ m}^2$$

We know that elongation of the steel bar,

$$\delta l_s = \frac{P \times l_s}{A_s \times E_s} = \frac{P \times 0.6}{1.964 \times 10^{-3} \times 210 \times 10^9} = \frac{0.6P}{412.44 \times 10^6} \text{ m}$$

$$= 1.455 \times 10^{-9}\, P \text{ m}$$

and elongation of the aluminium bar,

$$\delta l_a = \frac{P \times l_a}{A_a \times E_a} = \frac{P \times 0.3}{0.491 \times 10^{-3} \times 74 \times 10^9} = \frac{0.3\,P}{36.334 \times 10^6} \text{ m}$$

$$= 8.257 \times 10^{-9}\, P \text{ m}$$

$\therefore$ Total elongation, $\delta l = \delta l_s + \delta l_a$

$$= 1.455 \times 10^{-9}\, P + 8.257 \times 10^{-9} P = 9.712 \times 10^{-9}\, P \text{ m}$$

Let σ_s = Stress in the steel bar, and

σ_a = Stress in the aluminium bar.

(a) When the supports are unyielding

When the supports are unyielding, the total contraction is equated to the total elongation, *i.e.*

$$0.24 \times 10^{-3} = 9.712 \times 10^{-9} P \quad \text{or} \quad P = 24\,712 \text{ N}$$

$\therefore$ Stress in the steel bar,

$$\sigma_s = P/A_s = 24\,712 / (1.964 \times 10^{-3}) = 12\,582 \times 10^3 \text{ N/m}^2$$

$$= 12.582 \text{ MPa} \quad \textbf{Ans.}$$

and stress in the aluminium bar,

$$\sigma_a = P/A_a = 24\,712 / (0.491 \times 10^{-3}) = 50\,328 \times 10^3 \text{ N/m}^2$$

$$= 50.328 \text{ MPa} \quad \textbf{Ans.}$$

(b) When the supports yield by 0.1 mm

When the supports yield and come nearer to each other by 0.10 mm, the net contraction in length

$$= 0.24 - 0.1 = 0.14 \text{ mm} = 0.14 \times 10^{-3} \text{ m}$$

Equating this net contraction to the total elongation, we have

$$0.14 \times 10^{-3} = 9.712 \times 10^{-9} P \quad \text{or} \quad P = 14\,415 \text{ N}$$

∴ Stress in the steel bar,

$$\sigma_s = P/A_s = 14\,415 / (1.964 \times 10^{-3}) = 7340 \times 10^3 \text{ N/m}^2$$
$$= 7.34 \text{ MPa} \textbf{ Ans.}$$

and stress in the aluminium bar,

$$\sigma_a = P/A_a = 14\,415 / (0.491 \times 10^{-3}) = 29\,360 \times 10^3 \text{ N/m}^2$$
$$= 29.36 \text{ MPa} \textbf{ Ans.}$$

Example 4.15. *A copper bar 50 mm in diameter is placed within a steel tube 75 mm external diameter and 50 mm internal diameter of exactly the same length. The two pieces are rigidly fixed together by two pins 18 mm in diameter, one at each end passing through the bar and tube. Calculate the stress induced in the copper bar, steel tube and pins if the temperature of the combination is raised by 50°C. Take E_s = 210 GN/m²; E_c = 105 GN/m²; $\alpha_s = 11.5 \times 10^{-6}$/°C and $\alpha_c = 17 \times 10^{-6}$/°C.*

Solution. Given: d_c = 50 mm ; d_{se} = 75 mm ; d_{si} = 50 mm ; d_p = 18 mm = 0.018 m ; t = 50°C; E_s = 210 GN/m² = 210 × 10^9 N/m² ; E_c = 105 GN/m² = 105 × 10^9 N/m² ; $\alpha_s = 11.5 \times 10^{-6}$/°C ; $\alpha_c = 17 \times 10^{-6}$/°C

The copper bar in a steel tube is shown in Fig. 4.18.

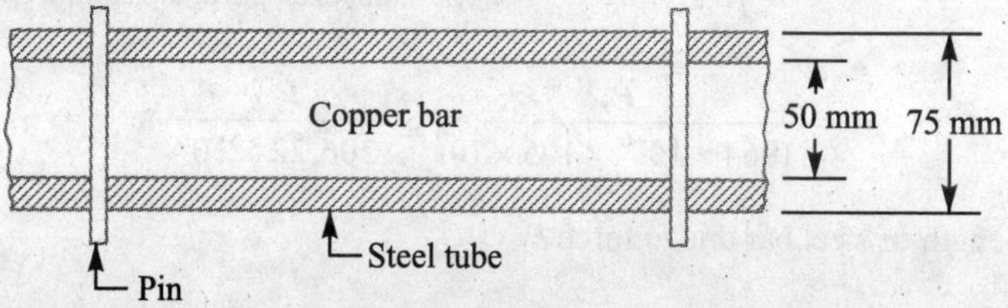

Fig. 4.18

We know that cross-sectional area of the copper bar,

$$A_c = \frac{\pi}{4}(d_c)^2 = \frac{\pi}{4}(50)^2 = 1964 \text{ mm}^2 = 1964 \times 10^{-6} \text{ m}^2$$

and cross-sectional area of the steel tube,

$$A_s = \frac{\pi}{4}\left[(d_{se})^2 - (d_{si})^2\right] = \frac{\pi}{4}\left[(75)^2 - (50)^2\right] = 2455 \text{ mm}^2$$
$$= 2455 \times 10^{-6} \text{ m}^2$$

Let l = Length of the copper bar and steel tube.

We know that free expansion of copper bar

$$= \alpha_c \,.\, l \,.\, t = 17 \times 10^{-6} \times l \times 50 = 850 \times 10^{-6}\, l$$

and free expansion of steel tube

$$= \alpha_s \,.\, l \,.\, t = 11.5 \times 10^{-6} \times l \times 50 = 575 \times 10^{-6}\, l$$

∴ Difference in free expansion

$$= 850 \times 10^{-6}\, l - 575 \times 10^{-6}\, l = 275 \times 10^{-6}\, l \qquad ...(i)$$

Since the free expansion of the copper bar is more than the free expansion of the steel tube, therefore the copper bar is subjected to a *compressive stress, while the steel tube is subjected to a tensile stress.

Let a compressive force P newton on the copper bar opposes the extra expansion of the copper bar and an equal tensile force P on the steel tube pulls the steel tube so that the net effect of reduction in length of copper bar and the increase in length of steel tube equalises the difference in free expansion of the two.

Main wheels on the undercarraige of an airliner. Air plane landing gears and wheels need to bear high stresses and shocks.

Note : This picture is given as additional information and is not a direct example of the current chapter.

$\therefore$ Reduction in length of copper bar due to force P

$$= \frac{P.l}{A_c . E_c}$$

$$= \frac{P.l}{1964 \times 10^{-6} \times 105 \times 10^9} = \frac{P.l}{206.22 \times 10^6} \text{ m}$$

and increase in length of steel bar due to force P

$$= \frac{P.l}{A_s . E_s} = \frac{P.l}{2455 \times 10^{-6} \times 210 \times 10^9} = \frac{P.l}{515.55 \times 10^6} \text{ m}$$

$$\therefore \text{Net effect in length} = \frac{P.l}{206.22 \times 10^6} + \frac{P.l}{515.55 \times 10^6}$$

$$= 4.85 \times 10^{-9}\, P.l + 1.94 \times 10^{-9}\, P.l = 6.79 \times 10^{-9}\, P.l$$

Equating this net effect in length to the difference in free expansion, we have

$$6.79 \times 10^{-9}\, P.l = 275 \times 10^{-6}\, l \quad \text{or} \quad P = 40\,500 \text{ N}$$

Stress induced in the copper bar, steel tube and pins

We know that stress induced in the copper bar,

$$\sigma_c = P / A_c = 40\,500 / (1964 \times 10^{-6}) = 20.62 \times 10^6 \text{ N/m}^2 = 20.62 \text{ MPa } \textbf{Ans.}$$

Stress induced in the steel tube,

$$\sigma_s = P / A_s = 40\,500 / (2455 \times 10^{-6}) = 16.5 \times 10^6 \text{ N / m}^2 = 16.5 \text{ MPa } \textbf{Ans.}$$

* In other words, we can also say that since the coefficient of thermal expansion for copper (α_c) is more than the coefficient of thermal expansion for steel (α_s), therefore the copper bar will be subjected to compressive stress and the steel tube will be subjected to tensile stress.

and shear stress induced in the pins,

$$\tau_p = \frac{P}{2\,A_p} = \frac{40\,500}{2 \times \frac{\pi}{4}(0.018)^2} = 79.57 \times 10^6 \text{ N/m}^2 = 79.57 \text{ MPa} \quad \textbf{Ans.}$$

...($\because$ The pin is in double shear)

4.17 Linear and Lateral Strain

Consider a circular bar of diameter *d* and length *l*, subjected to a tensile force *P* as shown in Fig. 4.19 (*a*).

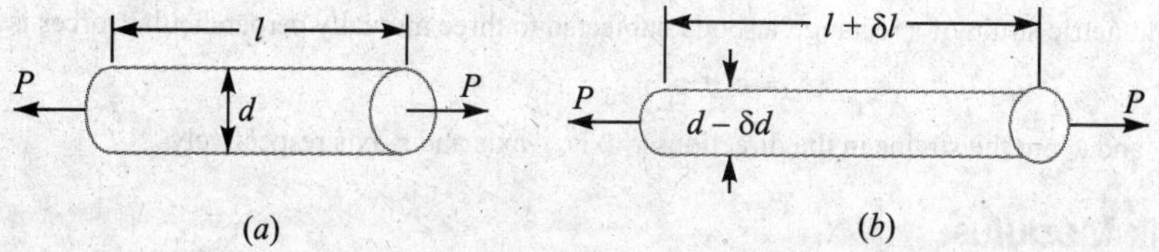

Fig. 4.19. Linear and lateral strain.

A little consideration will show that due to tensile force, the length of the bar increases by an amount δl and the diameter decreases by an amount δd, as shown in Fig. 4.19 (*b*). Similarly, if the bar is subjected to a compressive force, the length of bar will decrease which will be followed by increase in diameter.

It is thus obvious, that every direct stress is accompanied by a strain in its own direction which is known as ***linear strain*** and an opposite kind of strain in every direction, at right angles to it, is known as ***lateral strain.***

4.18 Poisson's Ratio

It has been found experimentally that when a body is stressed within elastic limit, the lateral strain bears a constant ratio to the linear strain, Mathematically,

$$\frac{\text{Lateral strain}}{\text{Linear strain}} = \text{Constant}$$

This constant is known as **Poisson's ratio** and is denoted by $1/m$ or μ.

Following are the values of Poisson's ratio for some of the materials commonly used in engineering practice.

Table 4.4. Values of Poisson's ratio for commonly used materials.

S.No.	*Material*	*Poisson's ratio* ($1/m$ *or* μ)
1	Steel	0.25 to 0.33
2	Cast iron	0.23 to 0.27
3	Copper	0.31 to 0.34
4	Brass	0.32 to 0.42
5	Aluminium	0.32 to 0.36
6	Concrete	0.08 to 0.18
7	Rubber	0.45 to 0.50

4.19 Volumetric Strain

When a body is subjected to a system of forces, it undergoes some changes in its dimensions. In other words, the volume of the body is changed. The ratio of the change in volume to the original volume is known as ***volumetric strain.*** Mathematically, volumetric strain,

$$\varepsilon_v = \delta V / V$$

where δV = Change in volume, and V = Original volume.

Notes : **1.** Volumetric strain of a rectangular body subjected to an axial force is given as

$$\varepsilon_v = \frac{\delta V}{V} = \varepsilon\left(1 - \frac{2}{m}\right); \text{ where } \varepsilon = \text{Linear strain.}$$

2. Volumetric strain of a rectangular body subjected to three mutually perpendicular forces is given by

$$\varepsilon_v = \varepsilon_x + \varepsilon_y + \varepsilon_z$$

where ε_x, ε_y and ε_z are the strains in the directions *x*-axis, *y*-axis and *z*-axis respectively.

4.20 Bulk Modulus

When a body is subjected to three mutually perpendicular stresses, of equal intensity, then the ratio of the direct stress to the corresponding volumetric strain is known as ***bulk modulus.*** It is usually denoted by *K*. Mathematically, bulk modulus,

$$K = \frac{\text{Direct stress}}{\text{Volumetric strain}} = \frac{\sigma}{\delta V / V}$$

4.21 Relation Between Bulk Modulus and Young's Modulus

The bulk modulus (*K*) and Young's modulus (*E*) are related by the following relation,

$$K = \frac{m.E}{3\,(m - 2)} = \frac{E}{3\,(1 - 2\mu)}$$

4.22 Relation Between Young's Modulus and Modulus of Rigidity

The Young's modulus (*E*) and modulus of rigidity (*G*) are related by the following relation,

$$G = \frac{m.E}{2\,(m + 1)} = \frac{E}{2\,(1 + \mu)}$$

Example 4.16. *A mild steel rod supports a tensile load of 50 kN. If the stress in the rod is limited to 100 MPa, find the size of the rod when the cross-section is 1. circular, 2. square, and 3. rectangular with width = 3 × thickness.*

Solution. Given : $P = 50$ kN $= 50 \times 10^3$ N ; $\sigma_t = 100$ MPa $= 100$ N/mm²

1. *Size of the rod when it is circular*

Let d = Diameter of the rod in mm.

$\therefore$ Area, $A = \frac{\pi}{4} \times d^2 = 0.7854\, d^2$

We know that tensile load (*P*),

$$50 \times 10^3 = \sigma_t \times A = 100 \times 0.7854\, d^2 = 78.54\, d^2$$

$\therefore$ $d^2 = 50 \times 10^3 / 78.54 = 636.6$ or $d = 25.23$ mm **Ans.**

2. ***Size of the rod when it is square***

Let x = Each side of the square rod in mm.

∴ Area, $A = x \times x = x^2$

We know that tensile load (P),

$$50 \times 10^3 = \sigma_t \times A = 100 \times x^2$$

∴ $x^2 = 50 \times 10^3/100 = 500$ or $x = 22.4$ mm **Ans.**

3. ***Size of the rod when it is rectangular***

Let t = Thickness of the rod in mm, and

b = Width of the rod in mm = $3\,t$ ...(Given)

∴ Area, $A = b \times t = 3t \times t = 3\,t^2$

We know that tensile load (P),

$$50 \times 10^3 = \sigma_t \times A = 100 \times 3\,t^2 = 300\,t^2$$

∴ $t^2 = 50 \times 10^3 / 300 = 166.7$ or $t = 12.9$ mm **Ans.**

and $b = 3t = 3 \times 12.9 = 38.7$ mm **Ans.**

Example 4.17. *A steel bar 2.4 m long and 30 mm square is elongated by a load of 500 kN. If poisson's ratio is 0.25, find the increase in volume. Take E = 0.2 × 10⁶ N/mm².*

Solution. Given : $l = 2.4$ m = 2400 mm ; $A = 30 \times 30 = 900$ mm^2 ; $P = 500$ kN = 500×10^3 N ; $1/m = 0.25$; $E = 0.2 \times 10^6$ N/mm^2

Let δV = Increase in volume.

We know that volume of the rod,

$$V = \text{Area} \times \text{length} = 900 \times 2400 = 2160 \times 10^3 \text{ mm}^3$$

and Young's modulus,

$$E = \frac{\text{Stress}}{\text{Strain}} = \frac{P/A}{\varepsilon}$$

$$\therefore \quad \varepsilon = \frac{P}{A.E} = \frac{500 \times 10^3}{900 \times 0.2 \times 10^6} = 2.8 \times 10^{-3}$$

We know that volumetric strain,

$$\frac{\delta V}{V} = \varepsilon\left(1 - \frac{2}{m}\right) = 2.8 \times 10^{-3}\ (1 - 2 \times 0.25) = 1.4 \times 10^3$$

$$\therefore \quad \delta V = V \times 1.4 \times 10^{-3} = 2160 \times 10^3 \times 1.4 \times 10^{-3} = 3024 \text{ mm}^3 \textbf{ Ans.}$$

4.23 Impact Stress

Sometimes, machine members are subjected to the load with impact. The stress produced in the member due to the falling load is known as ***impact stress.***

Consider a bar carrying a load W at a height h and falling on the collar provided at the lower end, as shown in Fig. 4.20.

Let A = Cross-sectional area of the bar,

E = Young's modulus of the material of the bar,

l = Length of the bar,

δl = Deformation of the bar,

P = Force at which the deflection δl is produced,

σ_i = Stress induced in the bar due to the application of impact load, and

h = Height through which the load falls.

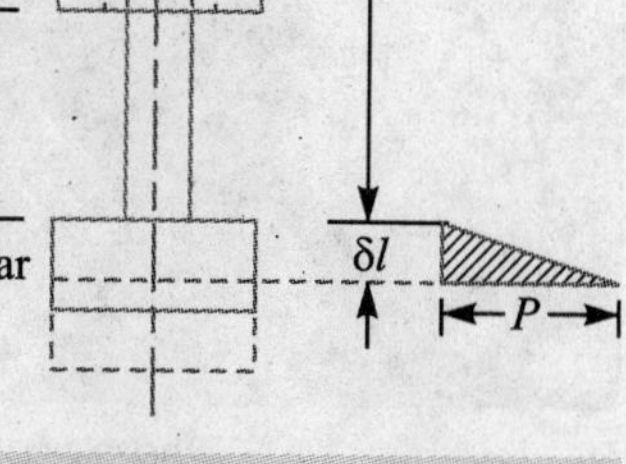

Fig. 4.20. Impact stress.

We know that energy gained by the system in the form of strain energy

$$= \frac{1}{2} \times P \times \delta l$$

and potential energy lost by the weight

$$= W(h + \delta l)$$

Since the energy gained by the system is equal to the potential energy lost by the weight, therefore

$$\frac{1}{2} \times P \times \delta l = W(h + \delta l)$$

$$\frac{1}{2} \sigma_i \times A \times \frac{\sigma_i \times l}{E} = W\left(h + \frac{\sigma_i \times l}{E}\right) \quad \ldots \left[\because P = \sigma_i \times A, \text{ and } \delta l = \frac{\sigma_i \times l}{E}\right]$$

$$\therefore \quad \frac{Al}{2E}(\sigma_i)^2 - \frac{Wl}{E}(\sigma_i) - Wh = 0$$

From this quadratic equation, we find that

$$\sigma_i = \frac{W}{A}\left(1 + \sqrt{1 + \frac{2hAE}{Wl}}\right) \quad \ldots \text{[Taking +ve sign for maximum value]}$$

Note : When $h = 0$, then $\sigma_i = 2W/A$. This means that the stress in the bar when the load in applied suddenly is double of the stress induced due to gradually applied load.

Example 4.18. *An unknown weight falls through 10 mm on a collar rigidly attached to the lower end of a vertical bar 3 m long and 600 mm² in section. If the maximum instantaneous extension is known to be 2 mm, what is the corresponding stress and the value of unknown weight? Take E = 200 kN/mm².*

Solution. Given : $h = 10$ mm ; $l = 3$ m $= 3000$ mm ; $A = 600$ mm^2 ; $\delta l = 2$ mm ; $E = 200$ kN/mm$^2 = 200 \times 10^3$ N/mm^2

Stress in the bar

Let σ = Stress in the bar.

We know that Young's modulus,

$$E = \frac{\text{Stress}}{\text{Strain}} = \frac{\sigma}{\varepsilon} = \frac{\sigma . l}{\delta l}$$

$$\therefore \quad \sigma = \frac{E.\delta l}{l} = \frac{200 \times 10^3 \times 2}{3000} = \frac{400}{3} = 133.3 \text{ N/mm}^2 \quad \textbf{Ans.}$$

These bridge shoes are made to bear high compressive stresses.

Note : This picture is given as additional information and is not a direct example of the current chapter.

Value of the unknown weight

Let W = Value of the unknown weight.

We know that $\sigma = \frac{W}{A}\left[1+\sqrt{1+\frac{2hAE}{Wl}}\right]$

$$\frac{400}{3} = \frac{W}{600}\left[1+\sqrt{1+\frac{2\times 10\times 600\times 200\times 10^3}{W\times 3000}}\right]$$

$$\frac{400\times 600}{3\,W} = 1+\sqrt{1+\frac{800\ 000}{W}}$$

$$\frac{80\ 000}{W} - 1 = \sqrt{1+\frac{800\ 000}{W}}$$

Squaring both sides,

$$\frac{6400\times 10^6}{W^2} + 1 - \frac{160000}{W} = 1 + \frac{800000}{W}$$

$$\frac{6400\times 10^2}{W} - 16 = 80 \quad \text{or} \quad \frac{6400\times 10^2}{W} = 96$$

$\therefore$ $W = 6400 \times 10^2 / 96 = 6666.7$ N **Ans.**

4.24 Resilience

When a body is loaded within elastic limit, it changes its dimensions and on the removal of the load, it regains its original dimensions. So long as it remains loaded, it has stored energy in itself. On removing the load, the energy stored is given off as in the case of a spring. This energy, which is absorbed in a body when strained within elastic limit, is known as ***strain energy.*** The strain energy is always capable of doing some work.

The strain energy stored in a body due to external loading, within elastic limit, is known as ***resilience*** and the maximum energy which can be stored in a body up to the elastic limit is called ***proof resilience***. The proof resilience per unit volume of a material is known as ***modulus of resilience.*** It is an important property of a material and gives capacity of the material to bear impact or shocks. Mathematically, strain energy stored in a body due to tensile or compressive load or resilience,

$$U = \frac{\sigma^2 \times V}{2E}$$

and Modulus of resilience $= \frac{\sigma^2}{2E}$

where σ = Tensile or compressive stress,

V = Volume of the body, and

E = Young's modulus of the material of the body.

Notes : 1. When a body is subjected to a shear load, then modulus of resilience (shear)

$$= \frac{\tau^2}{2C}$$

where τ = Shear stress, and

C = Modulus of rigidity.

2. When the body is subjected to torsion, then modulus of resilience

$$= \frac{\tau^2}{4C}$$

Example 4.19. *A wrought iron bar 50 mm in diameter and 2.5 m long transmits a shock energy of 100 N-m. Find the maximum instantaneous stress and the elongation. Take E = 200 GN/m².*

Solution. Given : $d = 50$ mm ; $l = 2.5$ m = 2500 mm ; $U = 100$ N-m = 100×10^3 N-mm ; $E = 200$ GN/m² = 200×10^3 N/mm²

Maximum instantaneous stress

Let σ = Maximum instantaneous stress.

We know that volume of the bar,

$$V = \frac{\pi}{4} \times d^2 \times l = \frac{\pi}{4}(50)^2 \times 2500 = 4.9 \times 10^6 \text{ mm}^3$$

We also know that shock or strain energy stored in the body (U),

$$100 \times 10^3 = \frac{\sigma^2 \times V}{2E} = \frac{\sigma^2 \times 4.9 \times 10^6}{2 \times 200 \times 10^3} = 12.25\,\sigma^2$$

$$\therefore \quad \sigma^2 = 100 \times 10^3 / 12.25 = 8163 \quad \text{or} \quad \sigma = 90.3 \text{ N/mm}^2 \text{ Ans.}$$

Elongation produced

Let δl = Elongation produced.

We know that Young's modulus,

$$E = \frac{\text{Stress}}{\text{Strain}} = \frac{\sigma}{\varepsilon} = \frac{\sigma}{\delta l / l}$$

$$\therefore \quad \delta l = \frac{\sigma \times l}{E} = \frac{90.3 \times 2500}{200 \times 10^3} = 1.13 \text{ mm} \quad \textbf{Ans.}$$

A double-decker train.

EXERCISES

1. A reciprocating steam engine connecting rod is subjected to a maximum load of 65 kN. Find the diameter of the connecting rod at its thinnest part, if the permissible tensile stress is 35 N/mm². **[Ans. 50 mm]**

2. The maximum tension in the lower link of a Porter governor is 580 N and the maximum stress in the link is 30 N/mm². If the link is of circular cross-section, determine its diameter. **[Ans. 5 mm]**

3. A wrought iron rod is under a compressive load of 350 kN. If the permissible stress for the material is 52.5 N/mm^2, calculate the diameter of the rod. **[Ans. 95 mm]**

4. A load of 5 kN is to be raised by means of a steel wire. Find the minimum diameter required, if the stress in the wire is not to exceed 100 N/mm^2. **[Ans. 8 mm]**

5. A square tie bar 20 mm × 20 mm in section carries a load. It is attached to a bracket by means of 6 bolts. Calculate the diameter of the bolt if the maximum stress in the tie bar is 150 N/mm^2 and in the bolts is 75 N/mm^2. **[Ans. 13 mm]**

6. The diameter of a piston of the steam engine is 300 mm and the maximum steam pressure is 0.7 N/mm^2. If the maximum permissible compressive stress for the piston rod material is 40 N/mm^2, find the size of the piston rod. **[Ans. 40 mm]**

7. Two circular rods of 50 mm diameter are connected by a knuckle joint, as shown in Fig. 4.21, by a pin of 40 mm in diameter. If a pull of 120 kN acts at each end, find the tensile stress in the rod and shear stress in the pin. **[Ans. 61 N/mm^2; 48 N/mm^2]**

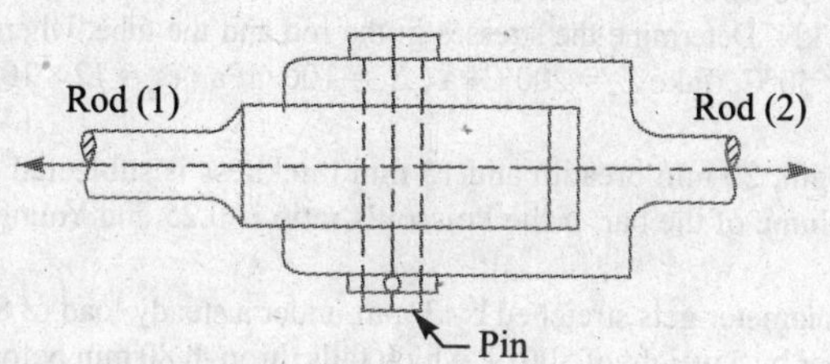

Fig. 4.21

8. Find the minimum size of a hole that can be punched in a 20 mm thick mild steel plate having an ultimate shear strength of 300 N/mm^2. The maximum permissible compressive stress in the punch material is 1200 N/mm^2. **[Ans. 20 mm]**

9. The crankpin of an engine sustains a maximum load of 35 kN due to steam pressure. If the allowable bearing pressure is 7 N/mm^2, find the dimensions of the pin. Assume the length of the pin equal to 1.2 times the diameter of the pin. **[Ans. 64.5 mm; 80 mm]**

10. The following results were obtained in a tensile test on a mild steel specimen of original diameter 20 mm and gauge length 40 mm.

Load at limit of proportionality	=	80 kN
Extension at 80 kN load	=	0.048 mm
Load at yield point	=	85 kN
Maximum load	=	150 kN

When the two parts were fitted together after being broken, the length between gauge length was found to be 55.6 mm and the diameter at the neck was 15.8 mm.

Calculate Young's modulus, yield stress, ultimate tensile stress, percentage elongation and percentage reduction in area. **[Ans. 213 kN/mm^2; 270 N/mm^2; 478 N/mm^2; 39%; 38%]**

11. A steel rod of 25 mm diameter is fitted inside a brass tube of 25 mm internal diameter and 375 mm external diameter. The projecting ends of the steel rod are provided with nuts and washers. The nuts are tightened up so as to produce a pull of 5 kN in the rod. The compound is then placed in a lathe and the brass is turned down to 4 mm thickness. Calculate the stresses in the two materials.

[Ans. 7 N/mm^2, 7.8 N/mm^2]

12. A composite bar made up of aluminium bar and steel bar, is firmly held between two unyielding supports as shown in Fig. 4.22.

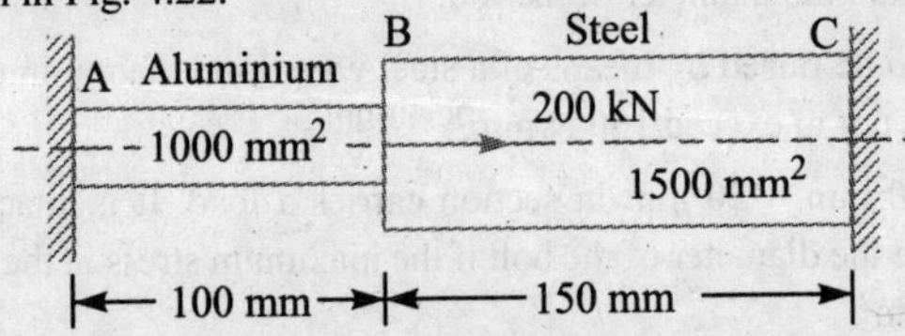

Fig. 4.22

An axial load of 200 kN is applied at *B* at 47°C. Find the stresses in each material, when the temperature is 97°C. Take E_a = 70 GPa ; E_s = 210 GPa ; $\alpha_a = 24 \times 10^{-6}$/°C and $\alpha_s = 12 \times 10^{6}$/°C. **[Ans. 60.3 MPa; 173.5 MPa]**

13. A steel rod of 20 mm diameter passes centrally through a copper tube of external diameter 40 mm and internal diameter 20 mm. The tube is closed at each end with the help of rigid washers (of negligible thickness) which are screwed by the nuts. The nuts are tightened until the compressive load on the copper tube is 50 kN. Determine the stresses in the rod and the tube, when the temperature of whole assembly falls by 50°C. Take E_s = 200 GPa ; E_c = 100 GPa ; $\alpha_s = 12 \times 10^{-6}$/°C and $\alpha_c = 18 \times 10^{6}$/°C. **[Ans. 99.6 MPa; 19.8 MPa]**

14. A bar of 2 m length, 20 mm breadth and 15 mm thickness is subjected to a tensile load of 30 kN. Find the final volume of the bar, if the Poisson's ratio is 0.25 and Young's modulus is 200 GN/m^2. **[Ans. 600 150 mm^3]**

15. A bar of 12 mm diameter gets stretched by 3 mm under a steady load of 8 kN. What stress would be produced in the bar by a weight of 800 N, which falls through 80 mm before commencing the stretching of the rod, which is initially unstressed. Take *E* = 200 kN/mm^2. **[Ans. 170.6 N/mm^2]**

QUESTIONS

1. Define the terms load , stress and strain. Discuss the various types of stresses and strain.
2. What is the difference between modulus of elasticity and modulus of rigidity?
3. Explain clearly the bearing stress developed at the area of contact between two members.
4. What useful informations are obtained from the tensile test of a ductile material?
5. What do you mean by factor of safety?
6. List the important factors that influence the magnitude of factor of safety.
7. What is meant by working stress and how it is calculated from the ultimate stress or yield stress of a material? What will be the factor of safety in each case for different types of loading?
8. Describe the procedure for finding out the stresses in a composite bar.
9. Explain the difference between linear and lateral strain.
10. Define the following :
    (*a*) Poisson's ratio, (*b*) Volumetric strain, and (*c*) Bulk modulus
11. Derive an expression for the impact stress induced due to a falling load.
12. Write short notes on :
    (*a*) Resilience (*b*) Proof resilience, and (*c*) Modulus of resilience

OBJECTIVE TYPE QUESTIONS

1. Hooke's law holds good upto
   (*a*) yield point (*b*) elastic limit
   (*c*) plastic limit (*d*) breaking point
2. The ratio of linear stress to linear strain is called
   (*a*) Modulus of elasticity (*b*) Modulus of rigidity
   (*c*) Bulk modulus (*d*) Poisson's ratio

3. The modulus of elasticity for mild steel is approximately equal to

(a) 80 kN/mm² (b) 100 kN/mm²
(c) 110 kN/mm² (d) 210 kN/mm²

4. When the material is loaded within elastic limit, then the stress is .... to strain.

(a) equal (b) directly proportional (c) inversely proportional

5. When a hole of diameter '*d*' is punched in a metal of thickness '*t*', then the force required to punch a hole is equal to

(a) $d.t.\tau_u$ (b) $\pi\, d.t.\tau_u$

(c) $\frac{\pi}{4} \times d^2\ \tau_u$ (d) $\frac{\pi}{4} \times d^2.t.\tau_u$

where τ_u = Ultimate shear strength of the material of the plate.

6. The ratio of the ultimate stress to the design stress is known as

(a) elastic limit (b) strain
(c) factor of safety (d) bulk modulus

7. The factor of safety for steel and for steady load is

(a) 2 (b) 4
(c) 6 (d) 8

8. An aluminium member is designed based on

(a) yield stress (b) elastic limit stress
(c) proof stress (d) ultimate stress

9. In a body, a thermal stress is one which arises because of the existence of

(a) latent heat (b) temperature gradient
(c) total heat (d) specific heat

10. A localised compressive stress at the area of contact between two members is known as

(a) tensile stress (b) bending stress
(c) bearing stress (d) shear stress

11. The Poisson's ratio for steel varies from

(a) 0.21 to 0.25 (b) 0.25 to 0.33
(c) 0.33 to 0.38 (d) 0.38 to 0.45

12. The stress in the bar when load is applied suddenly is .... as compared to the stress induced due to gradually applied load.

(a) same (b) double
(c) three times (d) four times

13. The energy stored in a body when strained within elastic limit is known as

(a) resilience (b) proof resilience
(c) strain energy (d) impact energy

14. The maximum energy that can be stored in a body due to external loading upto the elastic limit is called

(a) resilience (b) proof resilience
(c) strain energy (d) modulus of resilience

15. The strain energy stored in a body, when suddenly loaded, is .... the strain energy stored when same load is applied gradually.

(a) equal to (b) one-half
(c) twice (d) four times

ANSWERS

1. (b)	2. (a)	3. (d)	4. (b)	5. (b)
6. (c)	7. (b)	8. (a)	9. (b)	10. (c)
11. (b)	12. (b)	13. (c)	14. (b)	15. (d)

CHAPTER

5

Torsional and Bending Stresses in Machine Parts

5.1 Introduction

Sometimes machine parts are subjected to pure torsion or bending or combination of both torsion and bending stresses. We shall now discuss these stresses in detail in the following pages.

5.2 Torsional Shear Stress

When a machine member is subjected to the action of two equal and opposite couples acting in parallel planes (or torque or twisting moment), then the machine member is said to be subjected to ***torsion***. The stress set up by torsion is known as ***torsional shear stress***. It is zero at the centroidal axis and maximum at the outer surface.

Consider a shaft fixed at one end and subjected to a torque (*T*) at the other end as shown in Fig. 5.1. As a result of this torque, every cross-section of the shaft is subjected to torsional shear stress. We have discussed above that the torsional shear stress is zero at the centroidal axis and maximum at the outer surface. The maximum torsional shear stress at the outer surface of the shaft may be obtained from the following equation:

$$\frac{\tau}{r} = \frac{T}{J} = \frac{C \cdot \theta}{l} \quad ...(i)$$

where τ = Torsional shear stress induced at the outer surface of the shaft or maximum shear stress,

r = Radius of the shaft,

T = Torque or twisting moment,

J = Second moment of area of the section about its polar axis or polar moment of inertia,

C = Modulus of rigidity for the shaft material,

l = Length of the shaft, and

θ = Angle of twist in radians on a length l.

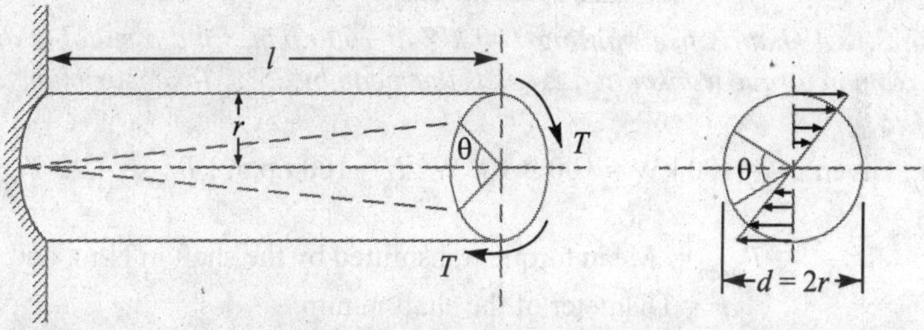

Fig. 5.1. Torsional shear stress.

The equation (*i*) is known as ***torsion equation.*** It is based on the following assumptions:

1. The material of the shaft is uniform throughout.
2. The twist along the length of the shaft is uniform.
3. The normal cross-sections of the shaft, which were plane and circular before twist, remain plane and circular after twist.
4. All diameters of the normal cross-section which were straight before twist, remain straight with their magnitude unchanged, after twist.
5. The maximum shear stress induced in the shaft due to the twisting moment does not exceed its elastic limit value.

Notes : 1. Since the torsional shear stress on any cross-section normal to the axis is directly proportional to the distance from the centre of the axis, therefore the torsional shear stress at a distance x from the centre of the shaft is given by

$$\frac{\tau_x}{x} = \frac{\tau}{r}$$

2. From equation (*i*), we know that

$$\frac{T}{J} = \frac{\tau}{r} \quad \text{or} \quad T = \tau \times \frac{J}{r}$$

For a solid shaft of diameter (d), the polar moment of inertia,

$$J = I_{XX} + I_{YY} = \frac{\pi}{64} \times d^4 + \frac{\pi}{64} \times d^4 = \frac{\pi}{32} \times d^4$$

$$\therefore \quad T = \tau \times \frac{\pi}{32} \times d^4 \times \frac{2}{d} = \frac{\pi}{16} \times \tau \times d^3$$

In case of a hollow shaft with external diameter (d_o) and internal diameter (d_i), the polar moment of inertia,

$$J = \frac{\pi}{32}[(d_o)^4 - (d_i)^4] \text{ and } r = \frac{d_o}{2}$$

$$\therefore \quad T = \tau \times \frac{\pi}{32}[(d_o)^4 - (d_4)^4] \times \frac{2}{d_o} = \frac{\pi}{16} \times \tau \left[\frac{(d_o)^4 - (d_i)^4}{d_o}\right]$$

$$= \frac{\pi}{16} \times \tau\, (d_o)^3\, (1 - k^4) \qquad \ldots \left(\text{Substituting, } k = \frac{d_i}{d_o}\right)$$

3. The expression ($C \times J$) is called ***torsional rigidity*** of the shaft.

4. The strength of the shaft means the maximum torque transmitted by it. Therefore, in order to design a shaft for strength, the above equations are used. The power transmitted by the shaft (in watts) is given by

$$P = \frac{2\,\pi\, N \,.\, T}{60} = T.\omega \qquad \ldots \left(\because \omega = \frac{2\,\pi\, N}{60}\right)$$

where T = Torque transmitted in N-m, and

ω = Angular speed in rad/s.

Example 5.1. *A shaft is transmitting 100 kW at 160 r.p.m. Find a suitable diameter for the shaft, if the maximum torque transmitted exceeds the mean by 25%. Take maximum allowable shear stress as 70 MPa.*

Solution. Given : P = 100 kW = 100 × 10³ W ; N = 160 r.p.m ; $T_{max} = 1.25\, T_{mean}$; τ = 70 MPa = 70 N/mm²

Let T_{mean} = Mean torque transmitted by the shaft in N-m, and

d = Diameter of the shaft in mm.

We know that the power transmitted (P),

$$100 \times 10^3 = \frac{2\,\pi\, N \,.\, T_{mean}}{60} = \frac{2\pi \times 160 \times T_{mean}}{60} = 16.76\, T_{mean}$$

$$\therefore \quad T_{mean} = 100 \times 10^3/16.76 = 5966.6 \text{ N-m}$$

A Helicopter propeller shaft has to bear torsional, tensile, as well as bending stresses.

Note : This picture is given as additional information and is not a direct example of the current chapter.

and maximum torque transmitted,

$$T_{max} = 1.25 \times 5966.6 = 7458 \text{ N-m} = 7458 \times 10^3 \text{ N-mm}$$

We know that maximum torque (T_{max}),

$$7458 \times 10^3 = \frac{\pi}{16} \times \tau \times d^3 = \frac{\pi}{16} \times 70 \times d^3 = 13.75\, d^3$$

$$\therefore \quad d^3 = 7458 \times 10^3/13.75 = 542.4 \times 10^3 \text{ or } d = 81.5 \text{ mm } \textbf{Ans.}$$

Example 5.2. *A steel shaft 35 mm in diameter and 1.2 m long held rigidly at one end has a hand wheel 500 mm in diameter keyed to the other end. The modulus of rigidity of steel is 80 GPa.*

1. What load applied to tangent to the rim of the wheel produce a torsional shear of 60 MPa?

2. How many degrees will the wheel turn when this load is applied?

Solution. Given : d = 35 mm or r = 17.5 mm ; l = 1.2 m = 1200 mm ; D = 500 mm or R = 250 mm ; C = 80 GPa = 80 kN/mm^2 = 80 × 10^3 N/mm^2 ; τ = 60 MPa = 60 N/mm^2

1. *Load applied to the tangent to the rim of the wheel*

Let W = Load applied (in newton) to tangent to the rim of the wheel.

We know that torque applied to the hand wheel,

$$T = W.R = W \times 250 = 250\, W \text{ N-mm}$$

and polar moment of inertia of the shaft,

$$J = \frac{\pi}{32} \times d^4 = \frac{\pi}{32} (35)^4 = 147.34 \times 10^3 \text{ mm}^4$$

We know that $\frac{T}{J} = \frac{\tau}{r}$

$$\therefore \quad \frac{250\, W}{147.34 \times 10^3} = \frac{60}{17.5} \quad \text{or} \quad W = \frac{60 \times 147.34 \times 10^3}{17.5 \times 250} = 2020 \text{ N } \textbf{Ans.}$$

2. *Number of degrees which the wheel will turn when load W = 2020 N is applied*

Let θ = Required number of degrees.

We know that $\frac{T}{J} = \frac{C.\theta}{l}$

$$\therefore \quad \theta = \frac{T.l}{C.J} = \frac{250 \times 2020 \times 1200}{80 \times 10^3 \times 147.34 \times 10^3} = 0.05° \textbf{ Ans.}$$

Example 5.3. *A shaft is transmitting 97.5 kW at 180 r.p.m. If the allowable shear stress in the material is 60 MPa, find the suitable diameter for the shaft. The shaft is not to twist more that 1° in a length of 3 metres. Take C = 80 GPa.*

Solution. Given : P = 97.5 kW = 97.5 × 10^3 W ; N = 180 r.p.m. ; τ = 60 MPa = 60 N/mm^2 ; θ = 1° = π / 180 = 0.0174 rad ; l = 3 m = 3000 mm ; C = 80 GPa = 80 × 10^9 N/m^2 = 80 × 10^3 N/mm^2

Let T = Torque transmitted by the shaft in N-m, and

d = Diameter of the shaft in mm.

We know that the power transmitted by the shaft (P),

$$97.5 \times 10^3 = \frac{2\,\pi\, N.T}{60} = \frac{2\pi \times 180 \times T}{60} = 18.852\, T$$

$$\therefore \quad T = 97.5 \times 10^3/18.852 = 5172 \text{ N-m} = 5172 \times 10^3 \text{ N-mm}$$

Now let us find the diameter of the shaft based on the strength and stiffness.

1. *Considering strength of the shaft*

We know that the torque transmitted (T),

$$5172 \times 10^3 = \frac{\pi}{16} \times \tau \times d^3 = \frac{\pi}{16} \times 60 \times d^3 = 11.78\ d^3$$

$$\therefore \quad d^3 = 5172 \times 10^3/11.78 = 439 \times 10^3 \text{ or } d = 76 \text{ mm} \quad ...(i)$$

2. *Considering stiffness of the shaft*

Polar moment of inertia of the shaft,

$$J = \frac{\pi}{32} \times d^4 = 0.0982\ d^4$$

We know that $\dfrac{T}{J} = \dfrac{C.\theta}{l}$

$$\frac{5172 \times 10^3}{0.0982\ d^4} = \frac{80 \times 10^3 \times 0.0174}{3000} \text{ or } \frac{52.7 \times 10^6}{d^4} = 0.464$$

$$\therefore \quad d^4 = 52.7 \times 10^6/0.464 = 113.6 \times 10^6 \text{ or } d = 103 \text{ mm} \quad ...(ii)$$

Taking larger of the two values, we shall provide $d = 103$ say 105 mm **Ans.**

Example 5.4. *A hollow shaft is required to transmit 600 kW at 110 r.p.m., the maximum torque being 20% greater than the mean. The shear stress is not to exceed 63 MPa and twist in a length of 3 metres not to exceed 1.4 degrees. Find the external diameter of the shaft, if the internal diameter to the external diameter is 3/8. Take modulus of rigidity as 84 GPa.*

Solution. Given : $P = 600$ kW $= 600 \times 10^3$ W ; $N = 110$ r.p.m. ; $T_{max} = 1.2\ T_{mean}$; $\tau = 63$ MPa $= 63$ N/mm^2 ; $l = 3$ m $= 3000$ mm ; $\theta = 1.4 \times \pi / 180 = 0.024$ rad ; $k = d_i / d_o = 3/8$; $C = 84$ GPa $= 84 \times 10^9$ N/m^2 $= 84 \times 10^3$ N/mm^2

Let T_{mean} = Mean torque transmitted by the shaft,

d_o = External diameter of the shaft, and

d_i = Internal diameter of the shaft.

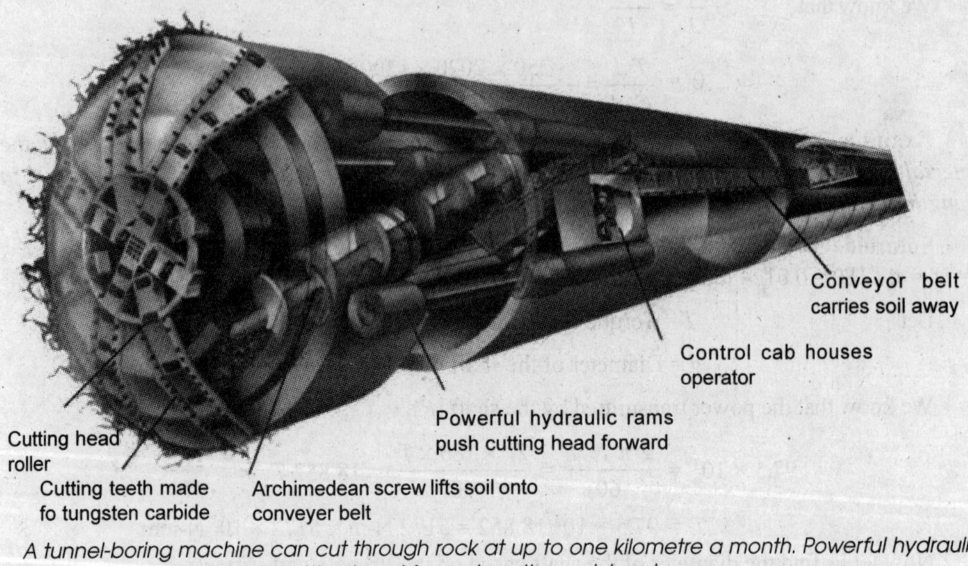

A tunnel-boring machine can cut through rock at up to one kilometre a month. Powerful hydraulic rams force the machine's cutting head fowards as the rock is cut away.

Note : This picture is given as additional information and is not a direct example of the current chapter.

We know that power transmitted by the shaft (P),

$$600 \times 10^3 = \frac{2\,\pi\,N.T_{mean}}{60} = \frac{2\pi \times 110 \times T_{mean}}{60} = 11.52\ T_{mean}$$

$$\therefore \quad T_{mean} = 600 \times 10^3/11.52 = 52 \times 10^3 \text{ N-m} = 52 \times 10^6 \text{ N-mm}$$

and maximum torque transmitted by the shaft,

$$T_{max} = 1.2\ T_{mean} = 1.2 \times 52 \times 10^6 = 62.4 \times 10^6 \text{ N-mm}$$

Now let us find the diameter of the shaft considering strength and stiffness.

1. ***Considering strength of the shaft***

We know that maximum torque transmitted by the shaft,

$$T_{max} = \frac{\pi}{16} \times \tau\,(d_o)^3\,(1 - k^4)$$

$$62.4 \times 10^6 = \frac{\pi}{16} \times 63 \times (d_o)^3 \left[1 - \left(\frac{3}{8}\right)^4\right] = 12.12\,(d_o)^3$$

$$\therefore \quad (d_o)^3 = 62.4 \times 10^6/12.12 = 5.15 \times 10^6 \text{ or } d_o = 172.7 \text{ mm} \qquad ...(i)$$

2. ***Considering stiffness of the shaft***

We know that polar moment of inertia of a hollow circular section,

$$J = \frac{\pi}{32}\left[(d_o)^4 - (d_i)^4\right] = \frac{\pi}{32}(d_o)^4\left[1 - \left(\frac{d_i}{d_o}\right)^4\right]$$

$$= \frac{\pi}{32}(d_o)^4\,(1 - k^4) = \frac{\pi}{32}(d_o)^4\left[1 - \left(\frac{3}{8}\right)^4\right] = 0.0962\,(d_o)^4$$

We also know that

$$\frac{T}{J} = \frac{C.\theta}{l}$$

$$\frac{62.4 \times 10^6}{0.0962\,(d_o)^4} = \frac{84 \times 10^3 \times 0.024}{3000} \quad \text{or} \quad \frac{648.6 \times 10^6}{(d_o)^4} = 0.672$$

$$\therefore \quad (d_o)^4 = 648.6 \times 10^6/0.672 = 964 \times 10^6 \text{ or } d_o = 176.2 \text{ mm} \qquad ...(ii)$$

Taking larger of the two values, we shall provide

$$d_o = 176.2 \text{ say } 180 \text{ mm } \textbf{Ans.}$$

5.3 Shafts in Series and Parallel

When two shafts of different diameters are connected together to form one shaft, it is then known as ***composite shaft.*** If the driving torque is applied at one end and the resisting torque at the other end, then the shafts are said to be connected in series as shown in Fig. 5.2 (*a*). In such cases, each shaft transmits the same torque and the total angle of twist is equal to the sum of the angle of twists of the two shafts.

Mathematically, total angle of twist,

$$\theta = \theta_1 + \theta_2 = \frac{T.l_1}{C_1\,J_1} + \frac{T.l_2}{C_2\,J_2}$$

If the shafts are made of the same material, then $C_1 = C_2 = C$.

$$\therefore \quad \theta = \frac{T.l_1}{CJ_1} + \frac{T.l_2}{CJ_2} = \frac{T}{C}\left[\frac{l_1}{J_1} + \frac{l_2}{J_2}\right]$$

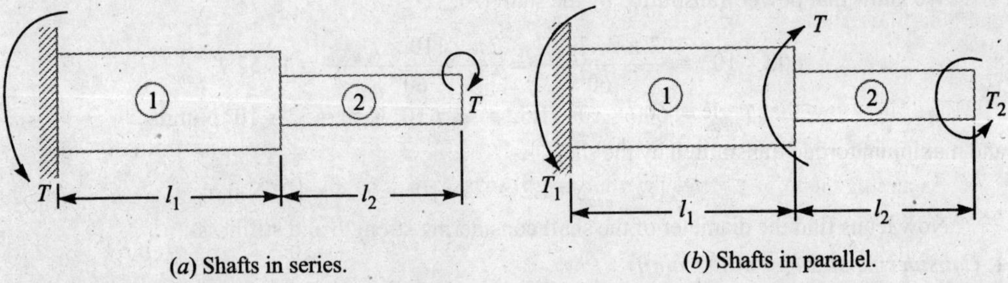

Fig. 5.2. Shafts in series and parallel.

When the driving torque (T) is applied at the junction of the two shafts, and the resisting torques T_1 and T_2 at the other ends of the shafts, then the shafts are said to be connected in parallel, as shown in Fig. 5.2 (*b*). In such cases, the angle of twist is same for both the shafts, *i.e.*

$$\theta_1 = \theta_2$$

or
$$\frac{T_1 l_1}{C_1 J_1} = \frac{T_2 l_2}{C_2 J_2} \quad \text{or} \quad \frac{T_1}{T_2} = \frac{l_2}{l_1} \times \frac{C_1}{C_2} \times \frac{J_1}{J_2}$$

and
$$T = T_1 + T_2$$

If the shafts are made of the same material, then $C_1 = C_2$.

$$\therefore \quad \frac{T_1}{T_2} = \frac{l_2}{l_1} \times \frac{J_1}{J_2}$$

Example 5.5. *A steel shaft ABCD having a total length of 3.5 m consists of three lengths having different sections as follows:*

AB is hollow having outside and inside diameters of 100 mm and 62.5 mm respectively, and BC and CD are solid. BC has a diameter of 100 mm and CD has a diameter of 87.5 mm. If the angle of twist is the same for each section, determine the length of each section. Find the value of the applied torque and the total angle of twist, if the maximum shear stress in the hollow portion is 47.5 MPa and shear modulus, C = 82.5 GPa.

Solution. Given: $L = 3.5$ m ; $d_o = 100$ mm ; $d_i = 62.5$ mm ; $d_2 = 100$ mm ; $d_3 = 87.5$ mm ; $\tau = 47.5$ MPa $= 47.5$ N/mm^2 ; $C = 82.5$ GPa $= 82.5 \times 10^3$ N/mm^2

The shaft *ABCD* is shown in Fig. 5.3.

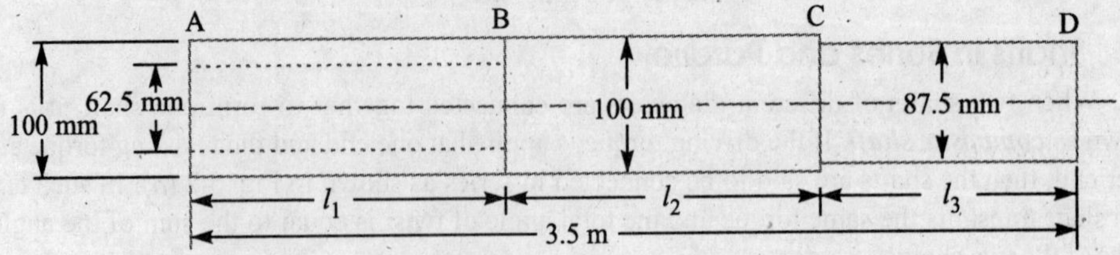

Fig. 5.3

Length of each section

Let l_1, l_2 and l_3 = Length of sections *AB*, *BC* and *CD* respectively.

We know that polar moment of inertia of the hollow shaft *AB*,

$$J_1 = \frac{\pi}{32}\left[(d_o)^4 - (d_i)^4\right] = \frac{\pi}{32}\left[(100)^4 - (62.5)^4\right] = 8.32 \times 10^6 \text{ mm}^4$$

Polar moment of inertia of the solid shaft *BC*,

$$J_2 = \frac{\pi}{32}(d_2)^4 = \frac{\pi}{32}(100)^4 = 9.82 \times 10^6 \text{ mm}^4$$

and polar moment of inertia of the solid shaft *CD*,

$$J_3 = \frac{\pi}{32}(d_3)^4 = \frac{\pi}{32}(87.5)^4 = 5.75 \times 10^6 \text{ mm}^4$$

We also know that angle of twist,

$$\theta = T \cdot l / C \cdot J$$

Assuming the torque *T* and shear modulus *C* to be same for all the sections, we have

Angle of twist for hollow shaft *AB*,

$$\theta_1 = T \cdot l_1 / C \cdot J_1$$

Similarly, angle of twist for solid shaft *BC*,

$$\theta_2 = T \cdot l_2 / C \cdot J_2$$

and angle of twist for solid shaft *CD*,

$$\theta_3 = T \cdot l_3 / C \cdot J_3$$

Machine part of a jet engine.

Note : This picture is given as additional information and is not a direct example of the current chapter.

Since the angle of twist is same for each section, therefore

$$\theta_1 = \theta_2$$

$$\frac{T \cdot l_1}{C \cdot J_1} = \frac{T \cdot l_2}{C \cdot J_2} \text{ or } \frac{l_1}{l_2} = \frac{J_1}{J_2} = \frac{8.32 \times 10^6}{9.82 \times 10^6} = 0.847 \qquad ...(i)$$

Also $\qquad \theta_1 = \theta_3$

$$\frac{T \cdot l_1}{C \cdot J_1} = \frac{T \cdot l_3}{C \cdot J_3} \text{ or } \frac{l_1}{l_3} = \frac{J_1}{J_3} = \frac{8.32 \times 10^6}{5.75 \times 10^6} = 1.447 \qquad ...(ii)$$

We know that $l_1 + l_2 + l_3 = L = 3.5 \text{ m} = 3500 \text{ mm}$

$$l_1\left(1 + \frac{l_2}{l_1} + \frac{l_3}{l_1}\right) = 3500$$

$$l_1\left(1 + \frac{1}{0.847} + \frac{1}{1.447}\right) = 3500$$

$$l_1 \times 2.8717 = 3500 \text{ or } l_1 = 3500 / 2.8717 = 1218.8 \text{ mm } \textbf{Ans.}$$

From equation (*i*),

$$l_2 = l_1 / 0.847 = 1218.8 / 0.847 = 1439 \text{ mm } \textbf{Ans.}$$

and from equation (*ii*), $\qquad l_3 = l_1 / 1.447 = 1218.8 / 1.447 = 842.2 \text{ mm } \textbf{Ans.}$

Value of the applied torque

We know that the maximum shear stress in the hollow portion,

$$\tau = 47.5 \text{ MPa} = 47.5 \text{ N/mm}^2$$

For a hollow shaft, the applied torque,

$$T = \frac{\pi}{16} \times \tau \left[\frac{(d_o)^4 - (d_i)^4}{d_o}\right] = \frac{\pi}{16} \times 47.5 \left[\frac{(100)^4 - (62.5)^4}{100}\right]$$

$$= 7.9 \times 10^6 \text{ N-mm} = 7900 \text{ N-m } \textbf{Ans.}$$

Total angle of twist

When the shafts are connected in series, the total angle of twist is equal to the sum of angle of twists of the individual shafts. Mathematically, the total angle of twist,

$$\theta = \theta_1 + \theta_2 + \theta_3$$

$$= \frac{T \cdot l_1}{C \cdot J_1} + \frac{T \cdot l_2}{C \cdot J_2} + \frac{T \cdot l_3}{C \cdot J_3} = \frac{T}{C}\left[\frac{l_1}{J_1} + \frac{l_2}{J_2} + \frac{l_3}{J_3}\right]$$

$$= \frac{7.9 \times 10^6}{82.5 \times 10^3}\left[\frac{1218.8}{8.32 \times 10^6} + \frac{1439}{9.82 \times 10^6} + \frac{842.2}{5.75 \times 10^6}\right]$$

$$= \frac{7.9 \times 10^6}{82.5 \times 10^3 \times 10^6}[146.5 + 146.5 + 146.5] = 0.042 \text{ rad}$$

$= 0.042 \times 180 / \pi = 2.406°$ **Ans.**

5.4 Bending Stress in Straight Beams

In engineering practice, the machine parts of structural members may be subjected to static or dynamic loads which cause bending stress in the sections besides other types of stresses such as tensile, compressive and shearing stresses.

Consider a straight beam subjected to a bending moment M as shown in Fig. 5.4. The following assumptions are usually made while deriving the bending formula.

1. The material of the beam is perfectly homogeneous (*i.e.* of the same material throughout) and isotropic (*i.e.* of equal elastic properties in all directions).
2. The material of the beam obeys Hooke's law.
3. The transverse sections (*i.e.* BC or GH) which were plane before bending, remain plane after bending also.
4. Each layer of the beam is free to expand or contract, independently, of the layer, above or below it.
5. The Young's modulus (E) is the same in tension and compression.
6. The loads are applied in the plane of bending.

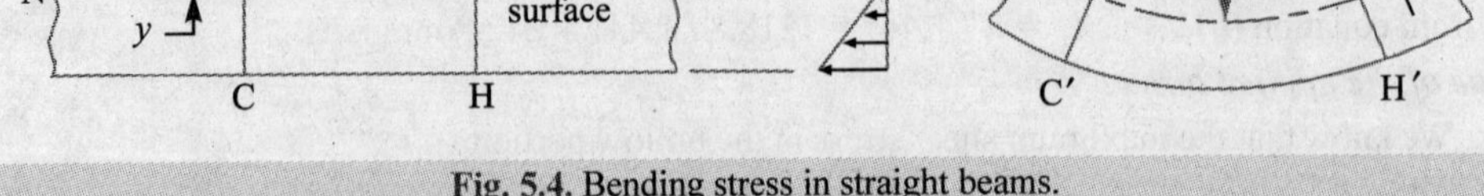

Fig. 5.4. Bending stress in straight beams.

A little consideration will show that when a beam is subjected to the bending moment, the fibres on the upper side of the beam will be shortened due to compression and those on the lower side will be elongated due to tension. It may be seen that somewhere between the top and bottom fibres there is a surface at which the fibres are neither shortened nor lengthened. Such a surface is called ***neutral surface.*** The intersection of the neutral surface with any normal cross-section of the beam is known *as* ***neutral axis.*** The stress distribution of a beam is shown in Fig. 5.4. The bending equation is given by

$$\frac{M}{I} = \frac{\sigma}{y} = \frac{E}{R}$$

where M = Bending moment acting at the given section,

σ = Bending stress,

I = Moment of inertia of the cross-section about the neutral axis,
y = Distance from the neutral axis to the extreme fibre,
E = Young's modulus of the material of the beam, and
R = Radius of curvature of the beam.

From the above equation, the bending stress is given by

$$\sigma = y \times \frac{E}{R}$$

Since E and R are constant, therefore within elastic limit, the stress at any point is directly proportional to y, *i.e.* the distance of the point from the neutral axis.

Also from the above equation, the bending stress,

$$\sigma = \frac{M}{I} \times y = \frac{M}{I/y} = \frac{M}{Z}$$

The ratio I/y is known as ***section modulus*** and is denoted by Z.

Notes : 1. The neutral axis of a section always passes through its centroid.

2. In case of symmetrical sections such as circular, square or rectangular, the neutral axis passes through its geometrical centre and the distance of extreme fibre from the neutral axis is $y = d / 2$, where d is the diameter in case of circular section or depth in case of square or rectangular section.

3. In case of unsymmetrical sections such as L-section or T-section, the neutral axis does not pass through its geometrical centre. In such cases, first of all the centroid of the section is calculated and then the distance of the extreme fibres for both lower and upper side of the section is obtained. Out of these two values, the bigger value is used in bending equation.

Parts in a machine.

Table 5.1 (from pages 130 to 134) shows the properties of some common cross-sections.

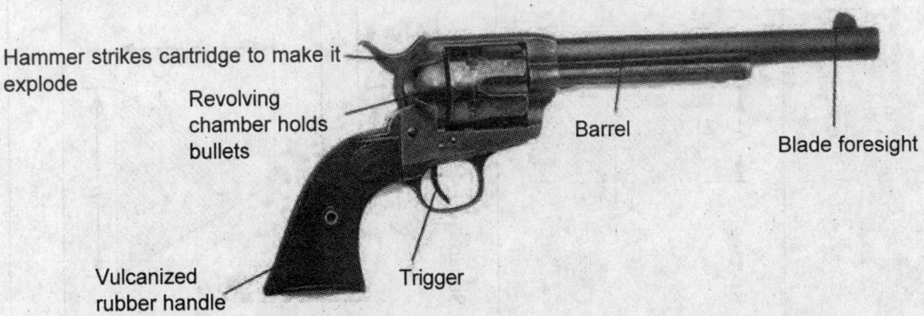

This is the first revolver produced in a production line using interchangeable parts.

Note : This picture is given as additional information and is not a direct example of the current chapter.

Table 5.1. Properties of commonly used cross-sections.

Section	Area (A)	Moment of inertia (I)	*Distance from the neutral axis to the extreme fibre (y)	Section modulus $\left[Z = \frac{I}{y}\right]$	Radius of gyration $\left[k = \sqrt{\frac{I}{A}}\right]$
1. Rectangle 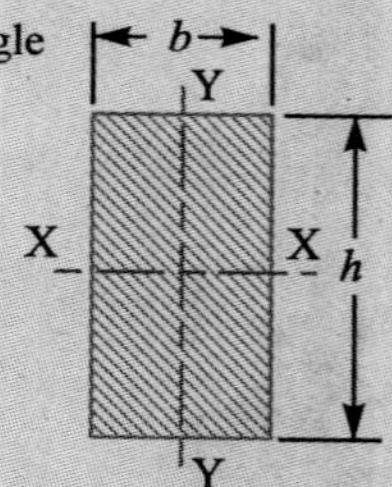	bh	$I_{xx} = \frac{b.h^3}{12}$ $I_{yy} = \frac{h.b^3}{12}$	$\frac{h}{2}$ $\frac{b}{2}$	$Z_{xx} = \frac{b.h^2}{6}$ $Z_{yy} = \frac{h.b^2}{6}$	$k_{xx} = 0.289\ h$ $k_{yy} = 0.289\ b$
2. Square 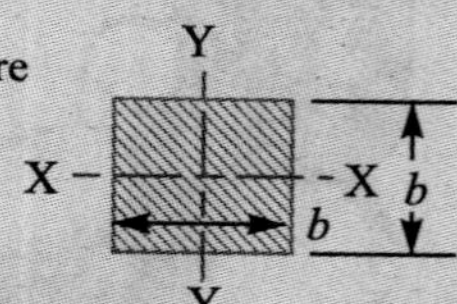	b^2	$I_{xx} = I_{yy} = \frac{b^4}{12}$	$\frac{b}{2}$	$Z_{xx} = Z_{yy} = \frac{b^3}{6}$	$k_{xx} = k_{yy} = 0.289\ b$
3. Triangle 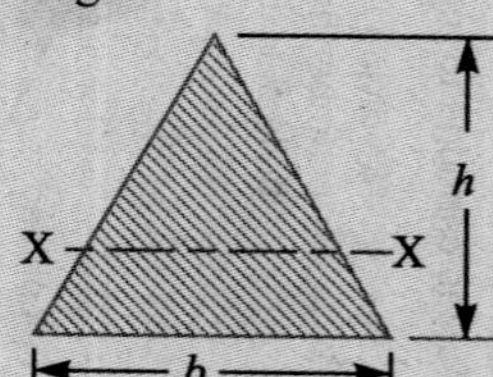	$\frac{bh}{2}$	$I_{xx} = \frac{b.h^3}{36}$	$\frac{h}{3}$	$Z_{xx} = \frac{bh^2}{12}$	$k_{xx} = 0.2358\ h$

* The distances from the neutral axis to the bottom extreme fibre is taken into consideration.

Section	(A)	(I)	(y)	$Z = \frac{I}{y}$	$k = \sqrt{\frac{I}{A}}$
4. Hollow rectangle 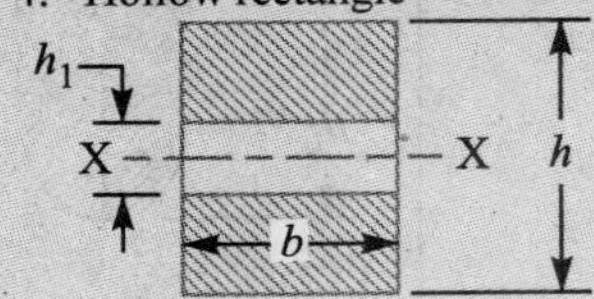	$b(h - h_1)$	$I_{xx} = \frac{b}{12}(h^3 - h_1^3)$	$\frac{h}{2}$	$Z_{xx} = \frac{b}{6}\left(\frac{h^3 - h_1^3}{h}\right)$	$k_{xx} = 0.289\sqrt{\frac{h^3 - h_1^3}{h - h_1}}$
5. Hollow square 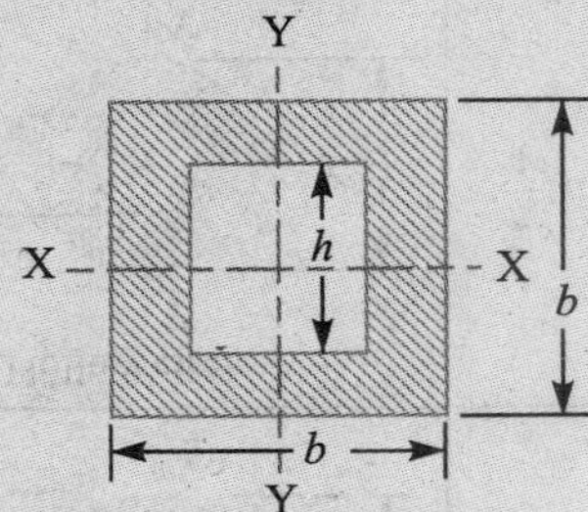	$b^2 - h^2$	$I_{xx} = I_{yy} = \frac{b^4 - h^4}{12}$	$\frac{b}{2}$	$Z_{xx} = Z_{yy} = \frac{b^4 - h^4}{6b}$	$0.289\sqrt{b^2 + h^2}$
6. Trapezoidal 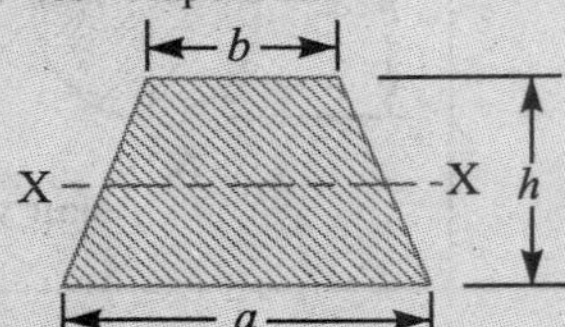	$\frac{a + b}{2} \times h$	$I_{xx} = \frac{h^2(a^2 + 4ab + b^2)}{36(a + b)}$	$\frac{a + 2b}{3(a + b)} \times h$	$Z_{xx} = \frac{a^2 + 4ab + b^2}{12(a + 2b)}$	$\frac{0.236}{a+b}\sqrt{h(a^2 + 4ab + b^2)}$

Section	(A)	(I)	(y)	$Z = \frac{I}{y}$	$k = \sqrt{\frac{I}{A}}$
7. Circle 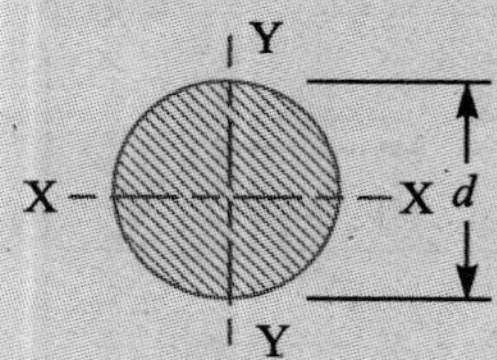	$\frac{\pi}{4} \times d^2$	$I_{xx} = I_{yy} = \frac{\pi d^4}{64}$	$\frac{d}{2}$	$Z_{xx} = Z_{yy} = \frac{\pi d^3}{32}$	$k_{xx} = k_{yy} = \frac{d}{2}$
8. Hollow circle 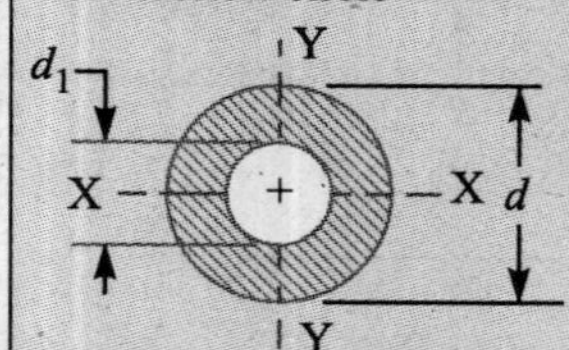	$\frac{\pi}{4}(d^2 - d_1^2)$	$I_{xx} = I_{yy} = \frac{\pi}{64}(d^4 - d_1^4)$	$\frac{d}{2}$	$Z_{xx} = Z_{yy} = \frac{\pi}{32}\left(\frac{d^4 - d_1^4}{d}\right)$	$k_{xx} = k_{yy} = \frac{\sqrt{d^2 + d_1^2}}{4}$
9. Elliptical 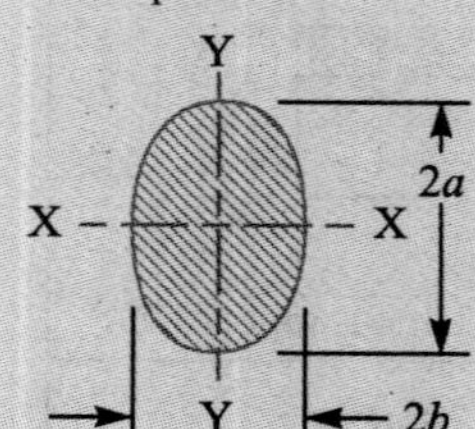	πab	$I_{xx} = \frac{\pi}{4} \times a^3 b$ $I_{yy} = \frac{\pi}{4} \times ab^3$	a b	$Z_{xx} = \frac{\pi}{4} \times a^2 b$ $Z_{yy} = \frac{\pi}{4} \times ab^2$	$k_{xx} = 0.5a$ $k_{yy} = 0.5b$

Section	(A)	(I)	(y)	$Z = \frac{I}{y}$	$k = \sqrt{\frac{I}{A}}$
10. Hollow elliptical 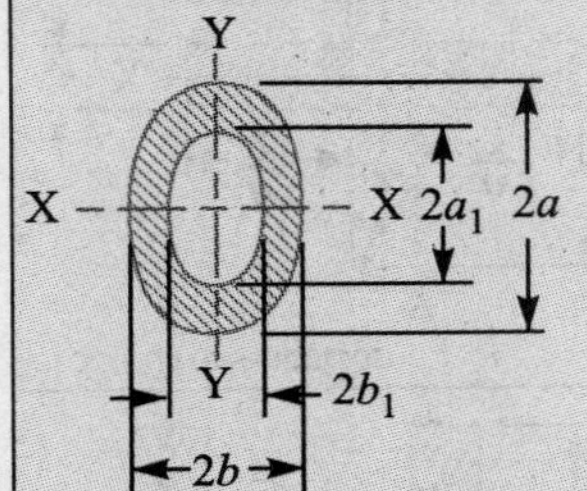	$\pi(ab - a_1 b_1)$	$I_{xx} = \frac{\pi}{4}(ba^3 - b_1 a_1^3)$ $I_{yy} = \frac{\pi}{4}(ab^3 - a_1 b_1^3)$	a b	$Z_{xx} = \frac{\pi}{4a}(ba^3 - b_1 a_1^3)$ $Z_{yy} = \frac{\pi}{4b}(ab^3 - a_1 b_1^3)$	$k_{xx} = \frac{1}{2}\sqrt{\frac{ba^3 - b_1 a_1^3}{ab - a_1 b_1}}$ $k_{yy} = \frac{1}{2}\sqrt{\frac{ab^3 - a_1 b_1^3}{ab - a_1 b_1}}$
11. I-section 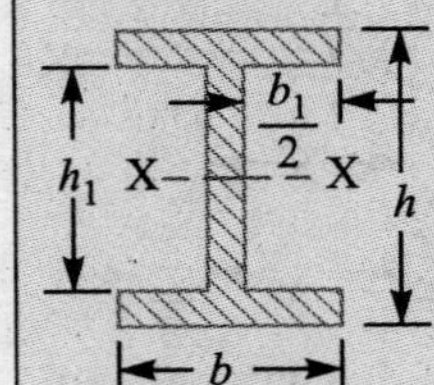	$bh - b_1 h_1$	$I_{xx} = \frac{bh^3 - b_1 h_1^3}{12}$	$\frac{h}{2}$	$Z_{xx} = \frac{bh^3 - b_1 h_1^3}{6h}$	$k_{xx} = 0.289\sqrt{\frac{bh^3 - b_1 h_1^3}{bh - b_1 h_1}}$
12. T-section 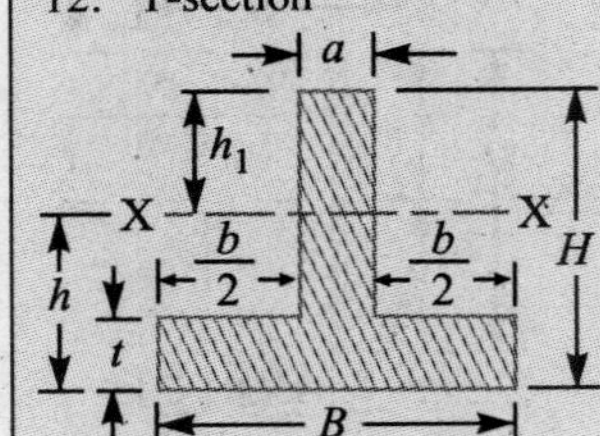	$Bt + (H - t)\,a$	$I_{xx} = \frac{Bh^3 - b(h-t)^3 + ah_1^3}{3}$	$h = H - h_1$ $= \frac{aH^2 + bt^2}{2(aH + bt)}$	$Z_{xx} = \frac{2I_{xx}(aH + bt)}{aH^2 + bt^2}$	$k_{xx} = \sqrt{\frac{I_{xx}}{Bt + (H - t)a}}$

Section	(A)	(I)	(y)	$Z = \frac{I}{y}$	$k = \sqrt{\frac{I}{A}}$
13. Channel Section 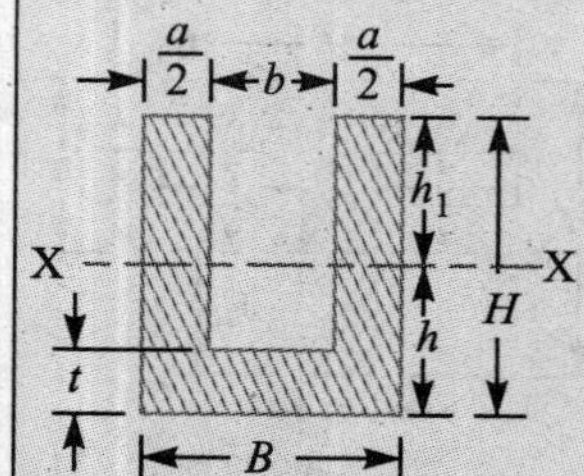	$Bt + (H - t)\,a$	$I_{xx} = \frac{Bh^3 - b(h-t)^3 + ah_1^3}{3}$	$h = H - h_1 = \frac{aH^2 + bt^2}{2(aH + bt)}$	$Z_{xx} = \frac{2I_{xx}(aH + bt)}{aH^2 + bt^2}$	$k_{xx} = \sqrt{\frac{I_{xx}}{Bt + (H-t)a}}$
14. H-Section 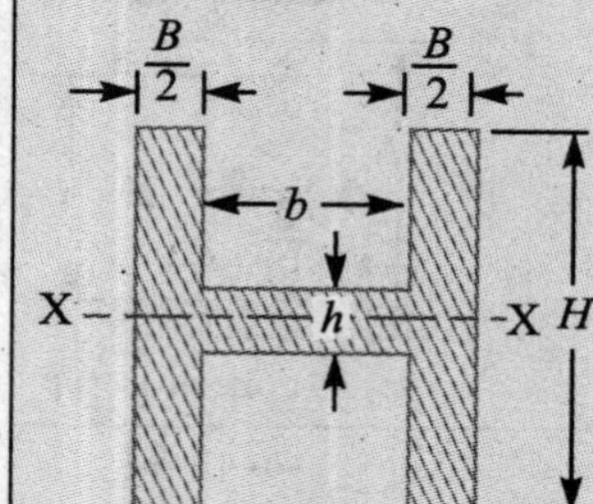	$BH + bh$	$I_{xx} = \frac{BH^3 + bh^3}{12}$	$\frac{H}{2}$	$Z_{xx} = \frac{BH^3 + bh^3}{6H}$	$k_{xx} = 0.289\sqrt{\frac{BH^3 + bh^3}{BH + bh}}$
15. Cross-section 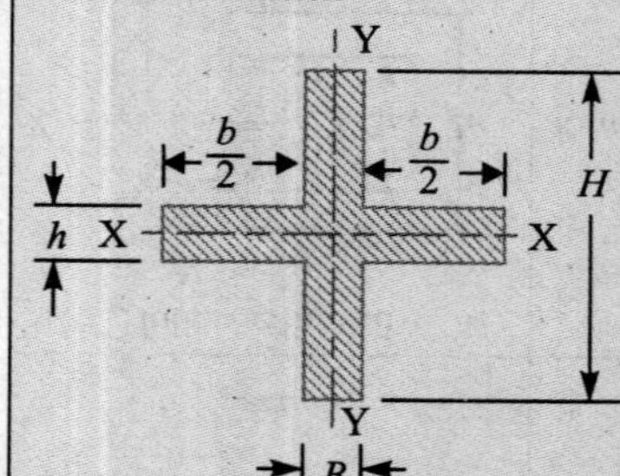	$BH + bh$	$I_{xx} = \frac{Bh^3 + bh^3}{12}$	$\frac{H}{2}$	$Z_{xx} = \frac{BH^3 + bh^3}{6H}$	$k_{xx} = 0.289\sqrt{\frac{BH^3 + bh^3}{BH + bh}}$

Example 5.6. *A pump lever rocking shaft is shown in Fig. 5.5. The pump lever exerts forces of 25 kN and 35 kN concentrated at 150 mm and 200 mm from the left and right hand bearing respectively. Find the diameter of the central portion of the shaft, if the stress is not to exceed 100 MPa.*

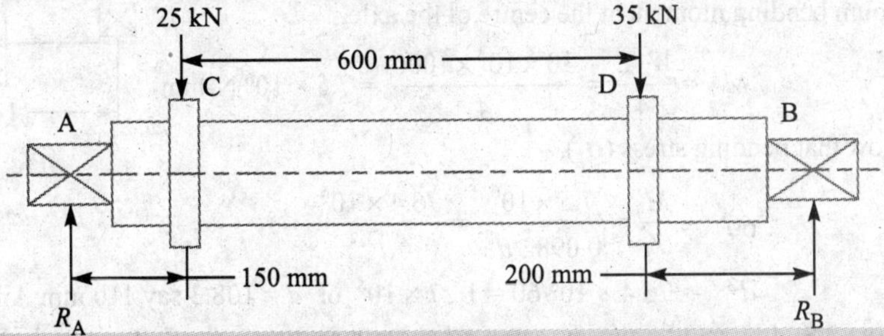

Fig. 5.5

Solution. Given : $\sigma_b = 100$ MPa = 100 N/mm^2

Let R_A and R_B = Reactions at *A* and *B* respectively.

Taking moments about *A*, we have

$$R_B \times 950 = 35 \times 750 + 25 \times 150 = 30\ 000$$

$$\therefore \quad R_B = 30\ 000 / 950 = 31.58 \text{ kN} = 31.58 \times 10^3 \text{ N}$$

and $$R_A = (25 + 35) - 31.58 = 28.42 \text{ kN} = 28.42 \times 10^3 \text{ N}$$

$\therefore$ Bending moment at *C*

$$= R_A \times 150 = 28.42 \times 10^3 \times 150 = 4.263 \times 10^6 \text{ N-mm}$$

and bending moment at $D = R_B \times 200 = 31.58 \times 10^3 \times 200 = 6.316 \times 10^6$ N-mm

We see that the maximum bending moment is at *D*, therefore maximum bending moment, $M = 6.316 \times 10^6$ N-mm.

Let d = Diameter of the shaft.

$\therefore$ Section modulus,

$$Z = \frac{\pi}{32} \times d^3 = 0.0982\, d^3$$

We know that bending stress (σ_b),

$$100 = \frac{M}{Z} = \frac{6.316 \times 10^6}{0.0982\, d^3} = \frac{64.32 \times 10^6}{d^3}$$

$\therefore \quad d^3 = 64.32 \times 10^6/100 = 643.2 \times 10^3$ or $d = 86.3$ say 90 mm **Ans.**

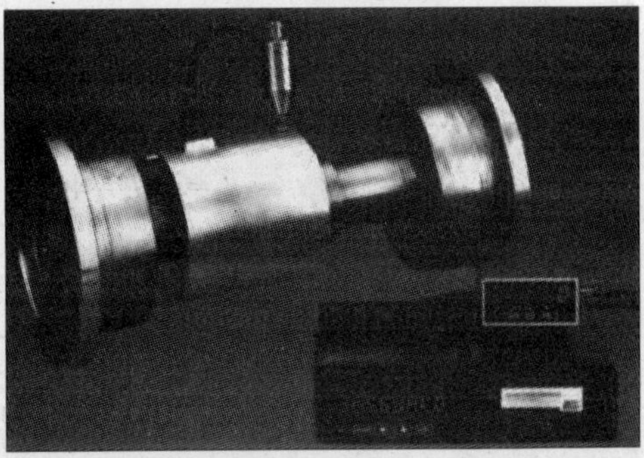

The picture shows a method where sensors are used to measure torsion

Note : This picture is given as additional information and is not a direct example of the current chapter.

Example 5.7. *An axle 1 metre long supported in bearings at its ends carries a fly wheel weighing 30 kN at the centre. If the stress (bending) is not to exceed 60 MPa, find the diameter of the axle.*

Solution. Given : $L = 1$ m = 1000 mm ; $W = 30$ kN = 30×10^3 N ; $\sigma_b = 60$ MPa = 60 N/mm^2

The axle with a flywheel is shown in Fig. 5.6.

Let d = Diameter of the axle in mm.

∴ Section modulus,

$$Z = \frac{\pi}{32} \times d^3 = 0.0982\, d^3$$

Maximum bending moment at the centre of the axle,

$$M = \frac{W.L}{4} = \frac{30 \times 10^3 \times 1000}{4} = 7.5 \times 10^6 \text{ N-mm}$$

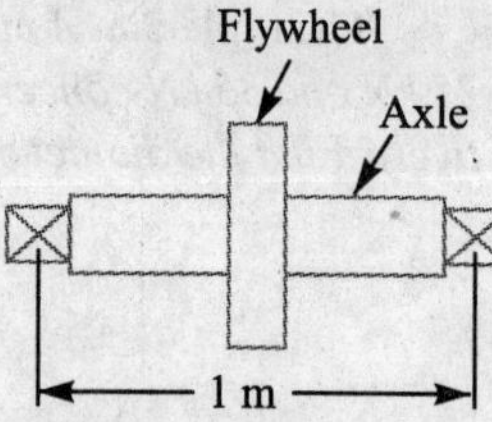

Fig. 5.6

We know that bending stress (σ_b),

$$60 = \frac{M}{Z} = \frac{7.5 \times 10^6}{0.0982\, d^3} = \frac{76.4 \times 10^6}{d^3}$$

∴ $d^3 = 76.4 \times 10^6/60 = 1.27 \times 10^6$ or $d = 108.3$ say 110 mm **Ans.**

Example 5.8. *A beam of uniform rectangular cross-section is fixed at one end and carries an electric motor weighing 400 N at a distance of 300 mm from the fixed end. The maximum bending stress in the beam is 40 MPa. Find the width and depth of the beam, if depth is twice that of width.*

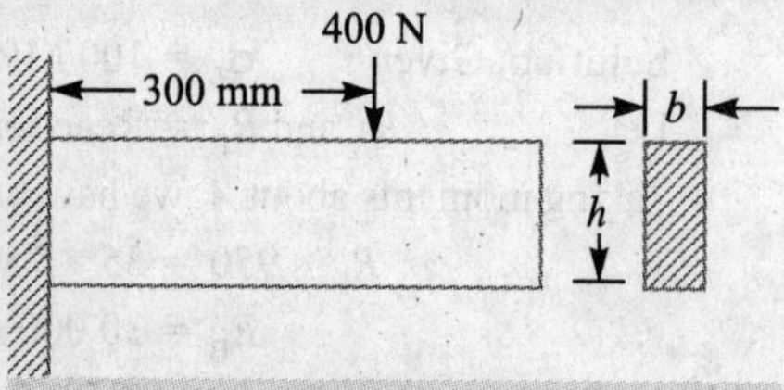

Fig. 5.7

Solution. Given: $W = 400$ N ; $L = 300$ mm ; $\sigma_b = 40$ MPa = 40 N/mm^2 ; $h = 2b$

The beam is shown in Fig. 5.7.

Let b = Width of the beam in mm, and

h = Depth of the beam in mm.

∴ Section modulus,

$$Z = \frac{b \cdot h^2}{6} = \frac{b\,(2b)^2}{6} = \frac{2\,b^3}{3} \text{ mm}^3$$

Maximum bending moment (at the fixed end),

$$M = W.L = 400 \times 300 = 120 \times 10^3 \text{ N-mm}$$

We know that bending stress (σ_b),

$$40 = \frac{M}{Z} = \frac{120 \times 10^3 \times 3}{2\,b^3} = \frac{180 \times 10^3}{b^3}$$

∴ $b^3 = 180 \times 10^3/40 = 4.5 \times 10^3$ or $b = 16.5$ mm **Ans.**

and $h = 2b = 2 \times 16.5 = 33$ mm **Ans.**

Example 5.9. *A cast iron pulley transmits 10 kW at 400 r.p.m. The diameter of the pulley is 1.2 metre and it has four straight arms of elliptical cross-section, in which the major axis is twice the minor axis. Determine the dimensions of the arm if the allowable bending stress is 15 MPa.*

Solution. Given : $P = 10$ kW $= 10 \times 10^3$ W ; $N = 400$ r.p.m ; $D = 1.2$ m = 1200 mm or $R = 600$ mm ; $\sigma_b = 15$ MPa = 15 N/mm^2

Let T = Torque transmitted by the pulley.

We know that the power transmitted by the pulley (P),

$$10 \times 10^3 = \frac{2\,\pi\,N.T}{60} = \frac{2\pi \times 400 \times T}{60} = 42\,T$$

∴ $T = 10 \times 10^3/42 = 238$ N-m $= 238 \times 10^3$ N-mm

Since the torque transmitted is the product of the tangential load and the radius of the pulley, therefore tangential load acting on the pulley

$$= \frac{T}{R} = \frac{238 \times 10^3}{600} = 396.7 \text{ N}$$

Since the pulley has four arms, therefore tangential load on each arm,

$$W = 396.7/4 = 99.2 \text{ N}$$

and maximum bending moment on the arm,

$$M = W \times R = 99.2 \times 600 = 59\ 520 \text{ N-mm}$$

Let $2b$ = Minor axis in mm, and

$2a$ = Major axis in mm = $2 \times 2b = 4b$ ...(Given)

∴ Section modulus for an elliptical cross-section,

$$Z = \frac{\pi}{4} \times a^2 b = \frac{\pi}{4} (2b)^2 \times b = \pi\, b^3 \text{ mm}^3$$

We know that bending stress (σ_b),

$$15 = \frac{M}{Z} = \frac{59\ 520}{\pi\, b^3} = \frac{18\ 943}{b^3}$$

or $b^3 = 18\ 943/15 = 1263$ or $b = 10.8$ mm

∴ Minor axis, $2b = 2 \times 10.8 = 21.6$ mm **Ans.**

and major axis, $2a = 2 \times 2b = 4 \times 10.8 = 43.2$ mm **Ans.**

5.5 Bending Stress in Curved Beams

We have seen in the previous article that for the straight beams, the neutral axis of the section coincides with its centroidal axis and the stress distribution in the beam is linear. But in case of curved beams, the neutral axis of the cross-section is shifted towards the centre of curvature of the beam causing a non-linear (hyperbolic) distribution of stress, as shown in Fig. 5.8. It may be noted that the neutral axis lies between the centroidal axis and the centre of curvature and always occurs within the curved beams. The application of curved beam principle is used in crane hooks, chain links and frames of punches, presses, planers etc.

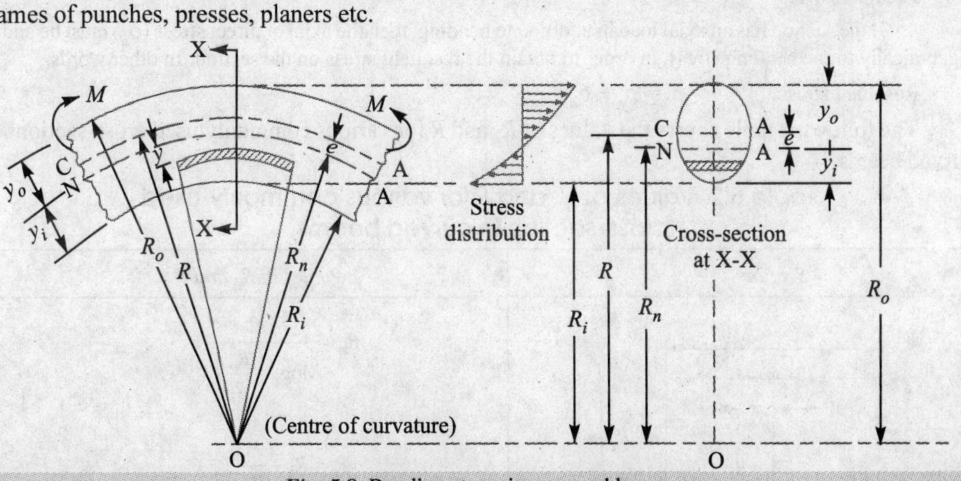

Fig. 5.8. Bending stress in a curved beam.

Consider a curved beam subjected to a bending moment M, as shown in Fig. 5.8. In finding the bending stress in curved beams, the same assumptions are used as for straight beams. The general expression for the bending stress (σ_b) in a curved beam at any fibre at a distance y from the neutral axis, is given by

$$\sigma_b = \frac{M}{A \cdot e}\left(\frac{y}{R_n - y}\right)$$

where M = Bending moment acting at the given section about the centroidal axis,

A = Area of cross-section,

e = Distance from the centroidal axis to the neutral axis = $R - R_n$,

R = Radius of curvature of the centroidal axis,

R_n = Radius of curvature of the neutral axis, and

y = Distance from the neutral axis to the fibre under consideration. It is positive for the distances towards the centre of curvature and negative for the distances away from the centre of curvature.

Notes : 1. The bending stress in the curved beam is zero at a point other than at the centroidal axis.

2. If the section is symmetrical such as a circle, rectangle, I-beam with equal flanges, then the maximum bending stress will always occur at the inside fibre.

3. If the section is unsymmetrical, then the maximum bending stress may occur at either the inside fibre or the outside fibre. The maximum bending stress at the inside fibre is given by

$$\sigma_{bi} = \frac{M \cdot y_i}{A \cdot e \cdot R_i}$$

where y_i = Distance from the neutral axis to the inside fibre = $R_n - R_i$, and

R_i = Radius of curvature of the inside fibre.

The maximum bending stress at the outside fibre is given by

$$\sigma_{bo} = \frac{M \cdot y_o}{A \cdot e \cdot R_o}$$

where y_o = Distance from the neutral axis to the outside fibre = $R_o - R_n$, and

R_o = Radius of curvature of the outside fibre.

It may be noted that the bending stress at the inside fibre is ***tensile*** while the bending stress at the outside fibre is ***compressive.***

4. If the section has an axial load in addition to bending, then the axial or direct stress (σ_d) must be added algebraically to the bending stress, in order to obtain the resultant stress on the section. In other words,

Resultant stress, $\sigma = \sigma_d \pm \sigma_b$

The following table shows the values of R_n and R for various commonly used cross-sections in curved beams.

Table 5.2. Values of R_n and R for various commonly used cross-section in curved beams.

Section	*Values of R_n and R*
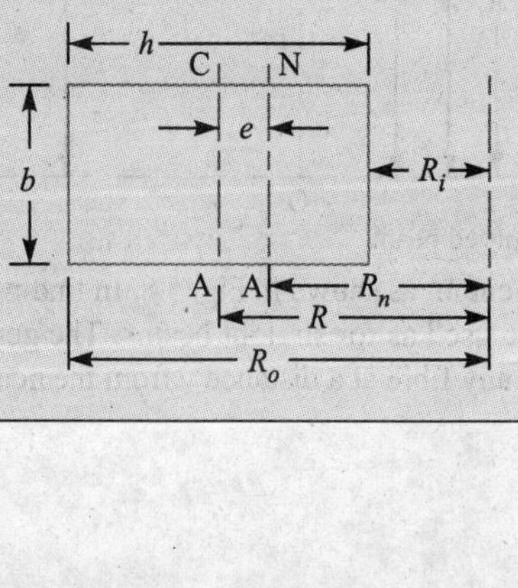	$R_n = \dfrac{h}{\log_e\left(\dfrac{R_o}{R_i}\right)}$ $R = R_i + \dfrac{h}{2}$

Section	*Values of R_n and R*
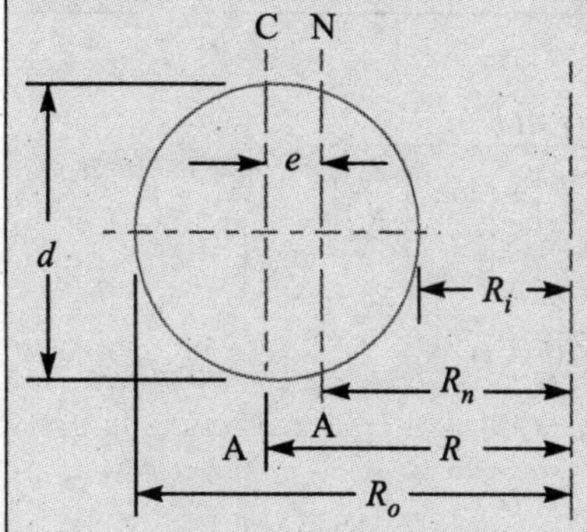	$R_n = \frac{\left[\sqrt{R_o} + \sqrt{R_i}\right]^2}{4}$ $R = R_i + \frac{d}{2}$
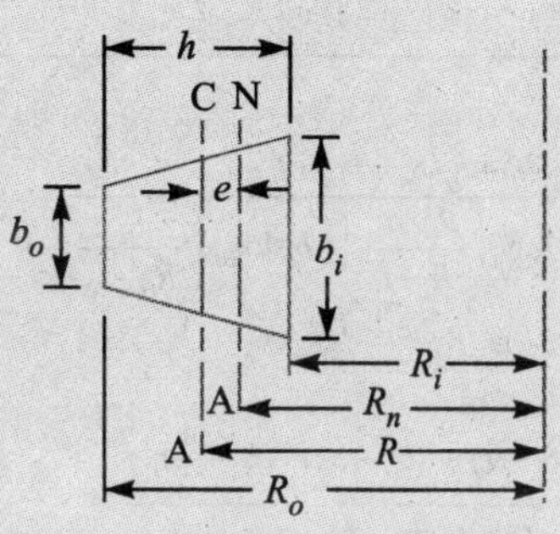	$R_n = \frac{\left(\frac{b_i + b_o}{2}\right) h}{\left(\frac{b_i R_o - b_o R_i}{h}\right) \log_e \left(\frac{R_o}{R_i}\right) - (b_i - b_o)}$ $R = R_i + \frac{h\,(b_i + 2b_o)}{3\,(b_i + b_o)}$
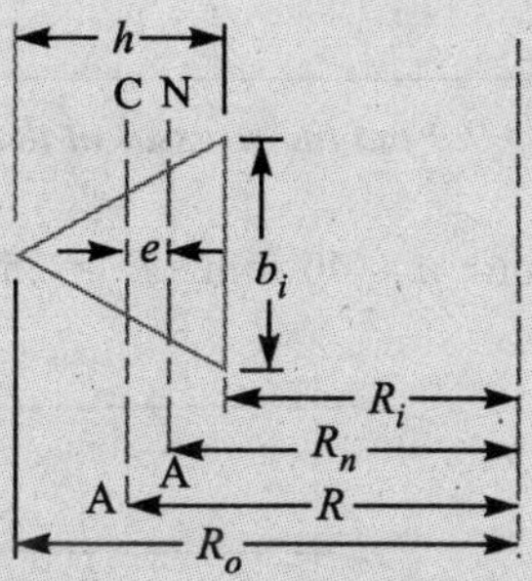	$R_n = \frac{\frac{1}{2} b_i \times h}{\frac{b_i R_o}{h} \log_e \left(\frac{R_o}{R_i}\right) - b_i}$ $R = R_i + \frac{h}{3}$
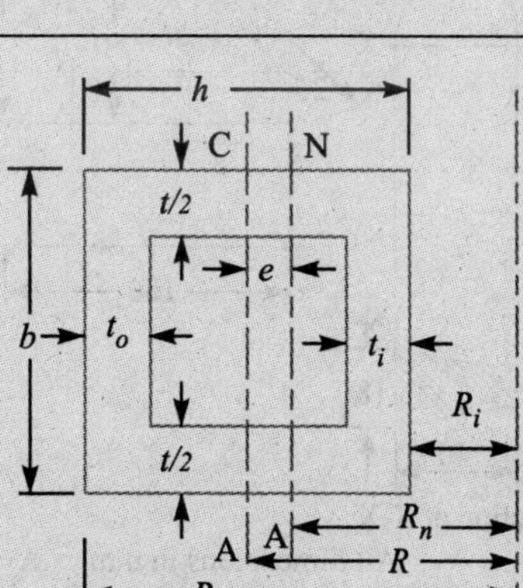	$R_n = \frac{(b - t)(t_i + t_o) + t.h}{b\left[\log_e\left(\frac{R_i + t_i}{R_i}\right) + \log_e\left(\frac{R_o}{R_o - t_o}\right)\right] + t.\log_e\left(\frac{R_o - t_o}{R_i + t_i}\right)}$ $R = R_i + \frac{\frac{1}{2}h^2.t + \frac{1}{2}t_i^2(b - t) + (b - t)\,t_o\,(h - \frac{1}{2}t_o)}{h.t + (b - t)(t_i + t_o)}$

Section	*Values of R_n and R*
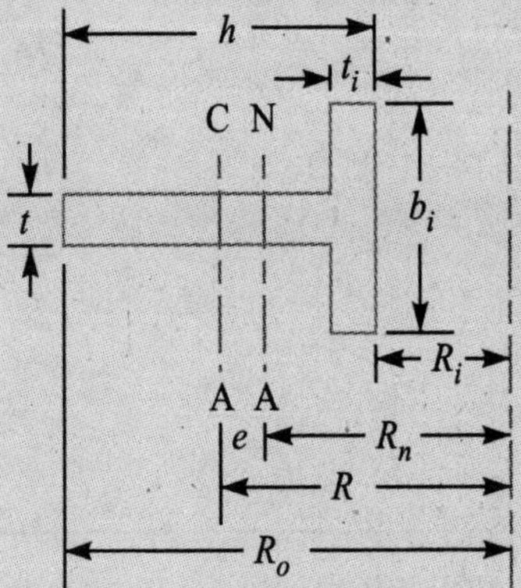	$R_n = \dfrac{t_i(b_i - t) + t.h}{(b_i - t)\log_e\left(\dfrac{R_i + t_i}{R_i}\right) + t.\log_e\left(\dfrac{R_o}{R_i}\right)}$ $R = R_i + \dfrac{\frac{1}{2}h^2 t + \frac{1}{2}t_i^2 (b_i - t)}{h.t + t_i(b_i - t)}$
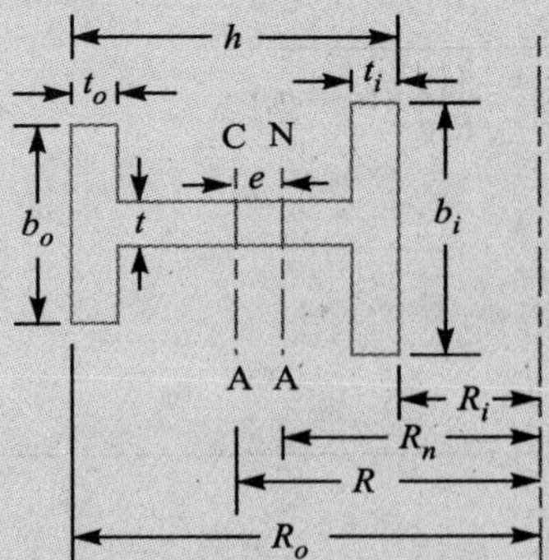	$R_n = \dfrac{t_i(b_i - t) + t_o(b_o - t) + t.h}{b_i\log_e\left(\dfrac{R_i + t_i}{R_i}\right) + t\log_e\left(\dfrac{R_o - t_o}{R_i + t_i}\right) + b_o\log_e\left(\dfrac{R_o}{R_o - t_o}\right)}$ $R = R_i + \dfrac{\frac{1}{2}h^2 t + \frac{1}{2}t_i^2(b_i - t) + (b_o - t)\,t_o\left(h - \frac{1}{2}t_o\right)}{t_i(b_i - t) + t_o(b_o - t) + t.h}$

Example 5.10. *The frame of a punch press is shown in Fig. 5.9. Find the stresses at the inner and outer surface at section X-X of the frame, if W = 5000 N.*

Solution. Given : $W = 5000$ N ; $b_i = 18$ mm ; $b_o = 6$ mm ; $h = 40$ mm ; $R_i = 25$ mm ; $R_o = 25 + 40 = 65$ mm

We know that area of section at X-X,

$$A = \frac{1}{2}(18 + 6)\,40 = 480 \text{ mm}^2$$

The various distances are shown in Fig. 5.10.

We know that radius of curvature of the neutral axis,

$$R_n = \frac{\left(\dfrac{b_i + b_o}{2}\right)h}{\left(\dfrac{b_i R_o - b_o R_i}{h}\right)\log_e\left(\dfrac{R_o}{R_i}\right) - (b_i - b_o)}$$

$$= \frac{\left(\dfrac{18 + 6}{2}\right) \times 40}{\left(\dfrac{18 \times 65 - 6 \times 25}{40}\right)\log_e\left(\dfrac{65}{25}\right) - (18 - 6)}$$

$$= \frac{480}{(25.5 \times 0.9555) - 12} = 38.83 \text{ mm}$$

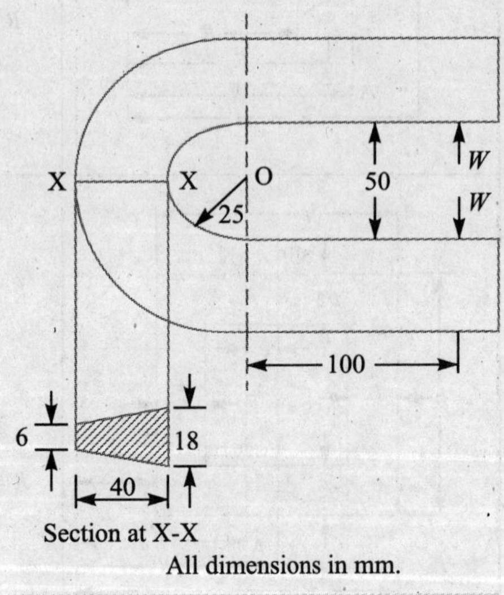

Fig. 5.9

and radius of curvature of the centroidal axis,

$$R = R_i + \frac{h\,(b_i + 2b_o)}{3\,(b_i + b_o)} = 25 + \frac{40\,(18 + 2 \times 6)}{3\,(18 + 6)} \text{ mm}$$

$$= 25 + 16.67 = 41.67 \text{ mm}$$

Distance between the centroidal axis and neutral axis,

$$e = R - R_n = 41.67 - 38.83 = 2.84 \text{ mm}$$

and the distance between the load and centroidal axis,

$$x = 100 + R = 100 + 41.67 = 141.67 \text{ mm}$$

∴ Bending moment about the centroidal axis,

$$M = W.x = 5000 \times 141.67 = 708\,350 \text{ N-mm}$$

The section at X-X is subjected to a direct tensile load of $W = 5000$ N and a bending moment of $M = 708\,350$ N-mm. We know that direct tensile stress at section X-X,

$$\sigma_t = \frac{W}{A} = \frac{5000}{480} = 10.42 \text{ N/mm}^2 = 10.42 \text{ MPa}$$

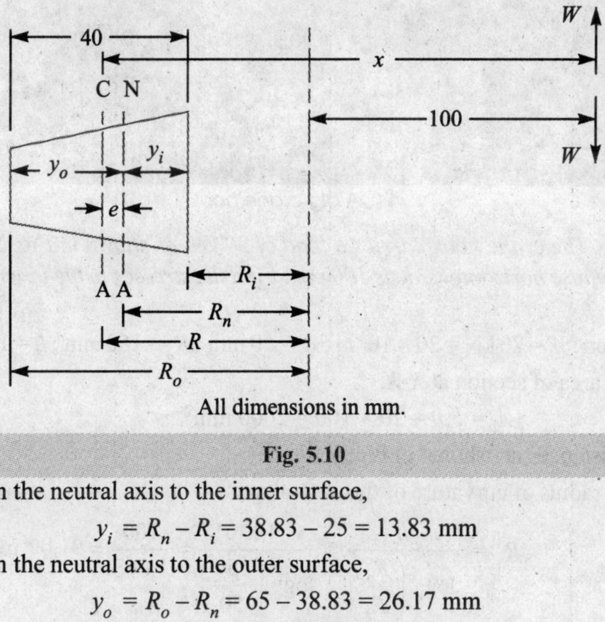

All dimensions in mm.

Fig. 5.10

Distance from the neutral axis to the inner surface,

$$y_i = R_n - R_i = 38.83 - 25 = 13.83 \text{ mm}$$

Distance from the neutral axis to the outer surface,

$$y_o = R_o - R_n = 65 - 38.83 = 26.17 \text{ mm}$$

We know that maximum bending stress at the inner surface,

$$\sigma_{bi} = \frac{M\,.\,y_i}{A\,.\,e\,.\,R_i} = \frac{708\,350 \times 13.83}{480 \times 2.84 \times 25} = 287.4 \text{ N/mm}^2$$

$$= 287.4 \text{ MPa (tensile)}$$

and maximum bending stress at the outer surface,

$$\sigma_{b0} = \frac{M\,.\,y_o}{A\,.\,e\,.\,R_o} = \frac{708\,350 \times 26.17}{480 \times 2.84 \times 65} = 209.2 \text{ N/mm}^2$$

$$= 209.2 \text{ MPa (compressive)}$$

∴ Resultant stress on the inner surface

$$= \sigma_t + \sigma_{bi} = 10.42 + 287.4 = 297.82 \text{ MPa (tensile) } \textbf{Ans.}$$

and resultant stress on the outer surface,

$$= \sigma_t - \sigma_{bo} = 10.42 - 209.2 = -198.78 \text{ MPa}$$

$$= 198.78 \text{ MPa (compressive) } \textbf{Ans.}$$

A big crane hook

Example 5.11. *The crane hook carries a load of 20 kN as shown in Fig. 5.11. The section at X-X is rectangular whose horizontal side is 100 mm. Find the stresses in the inner and outer fibres at the given section.*

Solution. Given : $W = 20 \text{ kN} = 20 \times 10^3 \text{ N}$; $R_i = 50$ mm ; $R_o = 150$ mm ; $h = 100$ mm ; $b = 20$ mm

We know that area of section at *X-X*,

$$A = b.h = 20 \times 100 = 2000 \text{ mm}^2$$

The various distances are shown in Fig. 5.12.

We know that radius of curvature of the neutral axis,

$$R_n = \frac{h}{\log_e\left(\frac{R_o}{R_i}\right)} = \frac{100}{\log_e\left(\frac{150}{50}\right)} = \frac{100}{1.098} = 91.07 \text{ mm}$$

and radius of curvature of the centroidal axis,

$$R = R_i + \frac{h}{2} = 50 + \frac{100}{2} = 100 \text{ mm}$$

∴ Distance between the centroidal axis and neutral axis,

$$e = R - R_n = 100 - 91.07 = 8.93 \text{ mm}$$

and distance between the load and the centroidal axis,

$$x = R = 100 \text{ mm}$$

∴ Bending moment about the centroidal axis,

$$M = W \times x = 20 \times 10^3 \times 100 = 2 \times 10^6 \text{ N-mm}$$

The section at *X-X* is subjected to a direct tensile load of $W = 20 \times 10^3$ N and a bending moment of $M = 2 \times 10^6$ N-mm. We know that direct tensile stress at section *X-X*,

$$\sigma_t = \frac{W}{A} = \frac{20 \times 10^3}{2000} = 10 \text{ N/mm}^2 = 10 \text{ MPa}$$

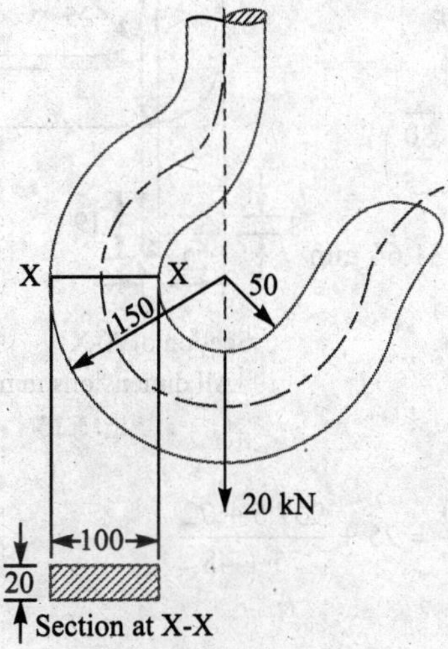

Fig. 5.11

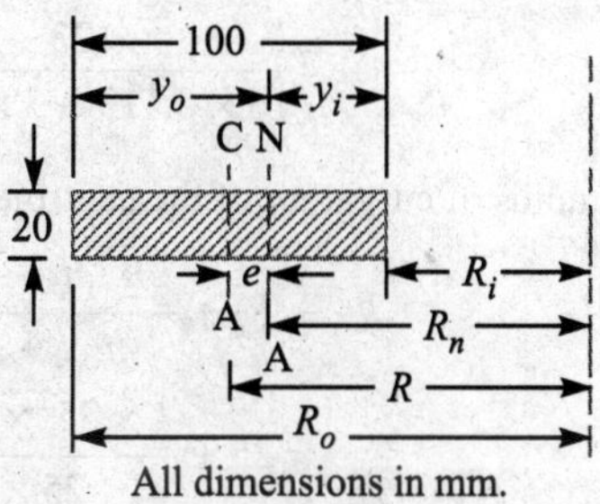

Fig. 5.12

We know that the distance from the neutral axis to the inside fibre,

$$y_i = R_n - R_i = 91.07 - 50 = 41.07 \text{ mm}$$

and distance from the neutral axis to outside fibre,

$$y_o = R_o - R_n = 150 - 91.07 = 58.93 \text{ mm}$$

∴ Maximum bending stress at the inside fibre,

$$\sigma_{bi} = \frac{M \,.\, y_i}{A \,.\, e \,.\, R_i} = \frac{2 \times 10^6 \times 41.07}{2000 \times 8.93 \times 50} = 92 \text{ N/mm}^2 = 92 \text{ MPa (tensile)}$$

and maximum bending stress at the outside fibre,

$$\sigma_{bo} = \frac{M \,.\, y_o}{A \,.\, e \,.\, R_o} = \frac{2 \times 10^6 \times 58.93}{2000 \times 8.93 \times 150} = 44 \text{ N/mm}^2$$

$$= 44 \text{ MPa (compressive)}$$

∴ Resultant stress at the inside fibre

$$= \sigma_t + \sigma_{bi} = 10 + 92 = 102 \text{ MPa (tensile)} \textbf{ Ans.}$$

and resultant stress at the outside fibre

$$= \sigma_t - \sigma_{bo} = 10 - 44 = -34 \text{ MPa} = 34 \text{ MPa (compressive)} \textbf{ Ans.}$$

Example 5.12. *A C-clamp is subjected to a maximum load of W, as shown in Fig. 5.13. If the maximum tensile stress in the clamp is limited to 140 MPa, find the value of load W.*

Solution. Given : $\sigma_{t(max)} = 140$ MPa $= 140$ N/mm² ; $R_i = 25$ mm ; $R_o = 25 + 25 = 50$ mm ; $b_i = 19$ mm ; $t_i = 3$ mm ; $t = 3$ mm ; $h = 25$ mm

We know that area of section at *X-X*,

$$A = 3 \times 22 + 3 \times 19 = 123 \text{ mm}^2$$

The various distances are shown in Fig. 5.14. We know that radius of curvature of the neutral axis,

$$R_n = \frac{t_i\,(b_i - t) + t\,.\,h}{(b_i - t)\log_e\left(\frac{R_i + t_i}{R_i}\right) + t\log_e\left(\frac{R_o}{R_i}\right)}$$

$$= \frac{3\,(19 - 3) + 3 \times 25}{(19 - 3)\log_e\left(\frac{25 + 3}{25}\right) + 3\log_e\left(\frac{50}{25}\right)}$$

$$= \frac{123}{16 \times 0.113 + 3 \times 0.693} = \frac{123}{3.887} = 31.64 \text{ mm}$$

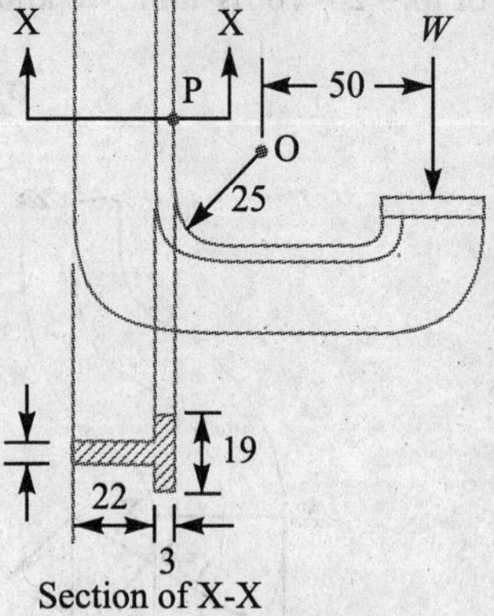

All dimensions in mm.

Fig. 5.13

and radius of curvature of the centroidal axis,

$$R = R_i + \frac{\frac{1}{2}h^2\,.\,t + \frac{1}{2}t_i^2\,(b_i - t)}{h\,.\,t + t_i\,(b_i - t)}$$

$$= 25 + \frac{\frac{1}{2} \times 25^2 \times 3 + \frac{1}{2} \times 3^2\,(19 - 3)}{25 \times 3 + 3\,(19 - 3)} = 25 + \frac{937.5 + 72}{75 + 48}$$

$$= 25 + 8.2 = 33.2 \text{ mm}$$

Distance between the centroidal axis and neutral axis,

$$e = R - R_n = 33.2 - 31.64 = 1.56 \text{ mm}$$

and distance between the load W and the centroidal axis,

$$x = 50 + R = 50 + 33.2 = 83.2 \text{ mm}$$

∴ Bending moment about the centroidal axis,

$$M = W.x = W \times 83.2 = 83.2\,W \text{ N-mm}$$

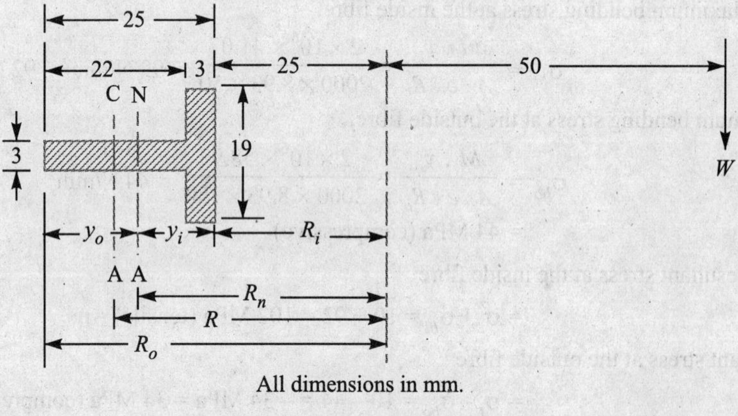

All dimensions in mm.

Fig. 5.14

The section at X-X is subjected to a direct tensile load of W and a bending moment of 83.2 W. The maximum tensile stress will occur at point P (*i.e.* at the inner fibre of the section).

Distance from the neutral axis to the point P,

$$y_i = R_n - R_i = 31.64 - 25 = 6.64 \text{ mm}$$

Direct tensile stress at section X-X,

$$\sigma_t = \frac{W}{A} = \frac{W}{123} = 0.008\ W\ \text{N/mm}^2$$

and maximum bending stress at point P,

$$\sigma_{bi} = \frac{M\,.\,y_i}{A\,.\,e\,.\,R_i} = \frac{83.2\ W \times 6.64}{123 \times 1.56 \times 25} = 0.115\ W\ \text{N/mm}^2$$

We know that the maximum tensile stress $\sigma_{t(max)}$,

$$140 = \sigma_t + \sigma_{bi} = 0.008\ W + 0.115\ W = 0.123\ W$$

$$\therefore \quad W = 140/0.123 = 1138\ \text{N}\ \textbf{Ans.}$$

Note : We know that distance from the neutral axis to the outer fibre,

$$y_o = R_o - R_n = 50 - 31.64 = 18.36\ \text{mm}$$

$\therefore$ Maximum bending stress at the outer fibre,

$$\sigma_{bo} = \frac{M\,.\,y_o}{A\,.\,e\,.\,R_o} = \frac{83.2\ W \times 18.36}{123 \times 1.56 \times 50} = 0.16\ W$$

and maximum stress at the outer fibre,

$$= \sigma_t - \sigma_{bo} = 0.008\ W - 0.16\ W = -0.152\ W\ \text{N/mm}^2$$

$$= 0.152\ W\ \text{N/mm}^2\ \text{(compressive)}$$

From above we see that stress at the outer fibre is larger in this case than at the inner fibre, but this stress at outer fibre is compressive.

5.6 Principal Stresses and Principal Planes

In the previous chapter, we have discussed about the direct tensile and compressive stress as well as simple shear. Also we have always referred the stress in a plane which is at right angles to the line of action of the force. But it has been observed that at any point in a strained material, there are three planes, mutually perpendicular to each other which carry direct stresses only and no shear stress. It may be noted that out of these three direct stresses, one will be maximum and the other will be minimum. These perpendicular planes which have no shear stress are known as ***principal planes*** and the direct stresses along these planes are known as ***principal stresses.*** The planes on which the maximum shear stress act are known as planes of maximum shear.

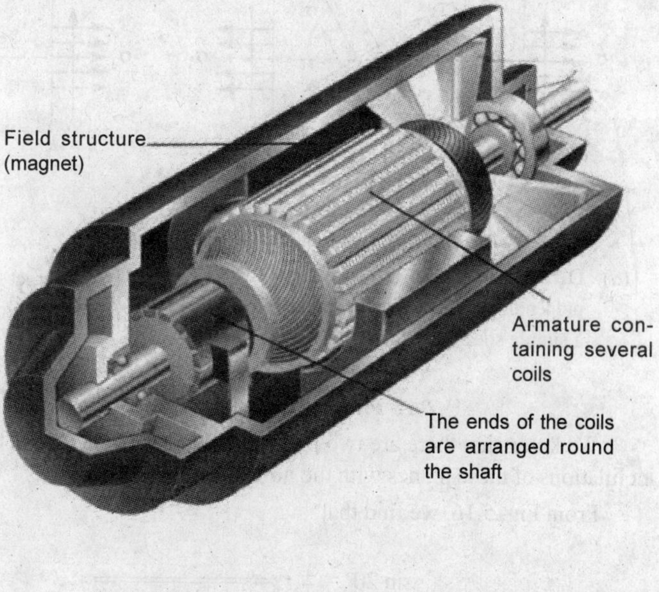

Big electric generators undergo high torsional stresses.

5.7 Determination of Principal Stresses for a Member Subjected to Bi-axial Stress

When a member is subjected to bi-axial stress (*i.e.* direct stress in two mutually perpendicular planes accompanied by a simple shear stress), then the normal and shear stresses are obtained as discussed below:

Consider a rectangular body *ABCD* of uniform cross-sectional area and unit thickness subjected to normal stresses σ_1 and σ_2 as shown in Fig. 5.15 (*a*). In addition to these normal stresses, a shear stress τ also acts.

It has been shown in books on ***'Strength of Materials'*** that the normal stress across any oblique section such as *EF* inclined at an angle θ with the direction of σ_2, as shown in Fig. 5.15 (*a*), is given by

$$\sigma_t = \frac{\sigma_1 + \sigma_2}{2} + \frac{\sigma_1 + \sigma_2}{2} \cos 2\theta + \tau \sin 2\theta \qquad \text{...(i)}$$

and tangential stress (*i.e.* shear stress) across the section *EF*,

$$\tau_1 = \frac{1}{2}(\sigma_1 - \sigma_2) \sin 2\theta - \tau \cos 2\theta \qquad \text{...(ii)}$$

Since the planes of maximum and minimum normal stress (*i.e.* principal planes) have no shear stress, therefore the inclination of principal planes is obtained by equating $\tau_1 = 0$ in the above equation (*ii*), *i.e.*

$$\frac{1}{2}(\sigma_1 - \sigma_2) \sin 2\theta - \tau \cos 2\theta = 0$$

$$\therefore \qquad \tan 2\theta = \frac{2\,\tau}{\sigma_1 - \sigma_2} \qquad \text{...(iii)}$$

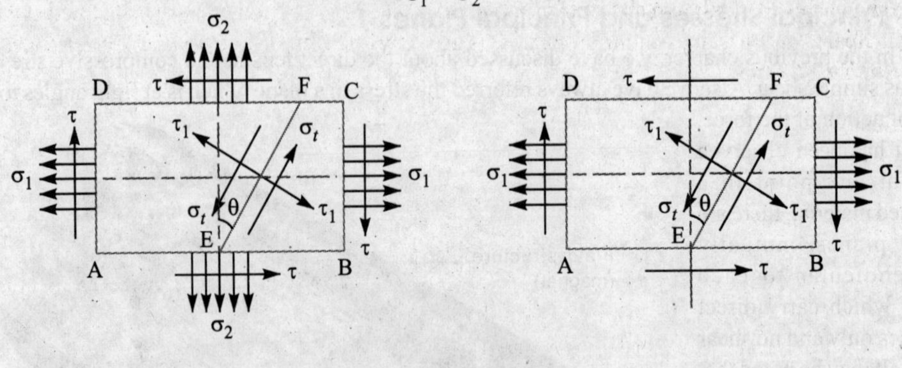

(*a*) Direct stress in two mutually prependicular planes accompanied by a simple shear stress.

(*b*) Direct stress in one plane accompanied by a simple shear stress.

Fig. 5.15. Principal stresses for a member subjected to bi-axial stress.

We know that there are two principal planes at right angles to each other. Let θ_1 and θ_2 be the inclinations of these planes with the normal cross-section.

From Fig. 5.16, we find that

$$\sin 2\theta = \pm \frac{2\tau}{\sqrt{(\sigma_1 - \sigma_2)^2 + 4\,\tau^2}}$$

$$\therefore \quad \sin 2\theta_1 = + \frac{2\tau}{\sqrt{(\sigma_1 - \sigma_2)^2 + 4\tau^2}}$$

and
$$\sin 2\theta_2 = - \frac{2\tau}{\sqrt{(\sigma_1 - \sigma_2)^2 + 4\tau^2}}$$

Also
$$\cos 2\theta = \pm \frac{\sigma_1 - \sigma_2}{\sqrt{(\sigma_1 - \sigma_2)^2 + 4\tau^2}}$$

$$\therefore \quad \cos 2\theta_1 = + \frac{\sigma_1 - \sigma_2}{\sqrt{(\sigma_1 - \sigma_2)^2 + 4\tau^2}}$$

and
$$\cos 2\theta_2 = - \frac{\sigma_1 - \sigma_2}{\sqrt{(\sigma_1 - \sigma_2)^2 + 4\tau^2}}$$

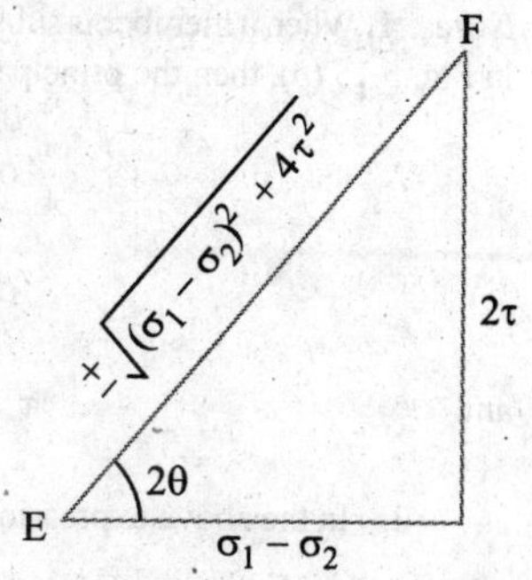

Fig. 5.16

The maximum and minimum principal stresses may now be obtained by substituting the values of sin 2θ and cos 2θ in equation (*i*).

∴ Maximum principal (or normal) stress,

$$\sigma_{t1} = \frac{\sigma_1 + \sigma_2}{2} + \frac{1}{2}\sqrt{(\sigma_1 - \sigma_2)^2 + 4\tau^2} \quad \text{...(iv)}$$

and minimum principal (or normal) stress,

$$\sigma_{t2} = \frac{\sigma_1 + \sigma_2}{2} - \frac{1}{2}\sqrt{(\sigma_1 - \sigma_2)^2 + 4\tau^2} \quad \text{...(v)}$$

The planes of maximum shear stress are at right angles to each other and are inclined at 45° to the principal planes. The maximum shear stress is given by ***one-half the algebraic difference between the principal stresses,*** *i.e.*

$$\tau_{max} = \frac{\sigma_{t1} - \sigma_{t2}}{2} = \frac{1}{2}\sqrt{(\sigma_1 - \sigma_2)^2 + 4\tau^2} \quad \text{...(vi)}$$

A Boring mill.

Note : This picture is given as additional information and is not a direct example of the current chapter.

Notes: 1. When a member is subjected to direct stress in one plane accompanied by a simple shear stress as shown in Fig. 5.15 (*b*), then the principal stresses are obtained by substituting $\sigma_2 = 0$ in equation (*iv*), (*v*) and (*vi*).

$$\therefore \quad \sigma_{t1} = \frac{\sigma_1}{2} + \frac{1}{2}\left[\sqrt{(\sigma_1)^2 + 4\,\tau^2}\right]$$

$$\sigma_{t2} = \frac{\sigma_1}{2} - \frac{1}{2}\left[\sqrt{(\sigma_1)^2 + 4\,\tau^2}\right]$$

and

$$\tau_{max} = \frac{1}{2}\left[\sqrt{(\sigma_1)^2 + 4\,\tau^2}\right]$$

2. In the above expression of σ_{t2}, the value of $\frac{1}{2}\left[\sqrt{(\sigma_1)^2 + 4\,\tau^2}\right]$ is more than $\frac{\sigma_1}{2}$. Therefore the nature of σ_{t2} will be opposite to that of σ_{t1}, *i.e.* if σ_{t1} is tensile then σ_{t2} will be compressive and *vice-versa*.

5.8 Application of Principal Stresses in Designing Machine Members

There are many cases in practice, in which machine members are subjected to combined stresses due to simultaneous action of either tensile or compressive stresses combined with shear stresses. In many shafts such as propeller shafts, C-frames etc., there are direct tensile or compressive stresses due to the external force and shear stress due to torsion, which acts normal to direct tensile or compressive stresses. The shafts like crank shafts, are subjected simultaneously to torsion and bending. In such cases, the maximum principal stresses, due to the combination of tensile or compressive stresses with shear stresses may be obtained.

The results obtained in the previous article may be written as follows:

1. Maximum tensile stress,

$$\sigma_{t(max)} = \frac{\sigma_t}{2} + \frac{1}{2}\left[\sqrt{(\sigma_t)^2 + 4\,\tau^2}\right]$$

2. Maximum compressive stress,

$$\sigma_{c(max)} = \frac{\sigma_c}{2} + \frac{1}{2}\left[\sqrt{(\sigma_c)^2 + 4\,\tau^2}\right]$$

3. Maximum shear stress,

$$\tau_{max} = \frac{1}{2}\left[\sqrt{(\sigma_t)^2 + 4\,\tau^2}\right]$$

where σ_t = Tensile stress due to direct load and bending,

σ_c = Compressive stress, and

τ = Shear stress due to torsion.

Notes : 1. When $\tau = 0$ as in the case of thin cylindrical shell subjected in internal fluid pressure, then

$$\sigma_{t\,(max)} = \sigma_t$$

2. When the shaft is subjected to an axial load (*P*) in addition to bending and twisting moments as in the propeller shafts of ship and shafts for driving worm gears, then the stress due to axial load must be added to the bending stress (σ_b). This will give the resultant tensile stress or compressive stress (σ_t or σ_c) depending upon the type of axial load (*i.e.* pull or push).

Example 5.13. *A hollow shaft of 40 mm outer diameter and 25 mm inner diameter is subjected to a twisting moment of 120 N-m, simultaneously, it is subjected to an axial thrust of 10 kN and a bending moment of 80 N-m. Calculate the maximum compressive and shear stresses.*

Solution. Given: d_o = 40 mm ; d_i = 25 mm ; T = 120 N-m = 120 × 10³ N-mm ; P = 10 kN = 10 × 10³ N ; M = 80 N-m = 80 × 10³ N-mm

We know that cross-sectional area of the shaft,

$$A = \frac{\pi}{4}\left[(d_o)^2 - (d_i)^2\right] = \frac{\pi}{4}\left[(40)^2 - (25)^2\right] = 766 \text{ mm}^2$$

∴ Direct compressive stress due to axial thrust,

$$\sigma_o = \frac{P}{A} = \frac{10 \times 10^3}{766} = 13.05 \text{ N/mm}^2 = 13.05 \text{ MPa}$$

Section modulus of the shaft,

$$Z = \frac{\pi}{32}\left[\frac{(d_o)^4 - (d_i)^4}{d_o}\right] = \frac{\pi}{32}\left[\frac{(40)^4 - (25)^4}{40}\right] = 5325 \text{ mm}^3$$

∴ Bending stress due to bending moment,

$$\sigma_b = \frac{M}{Z} = \frac{80 \times 10^3}{5325} = 15.02 \text{ N/mm}^2 = 15.02 \text{ MPa (compressive)}$$

and resultant compressive stress,

$$\sigma_c = \sigma_b + \sigma_o = 15.02 + 13.05 = 28.07 \text{ N/mm}^2 = 28.07 \text{ MPa}$$

We know that twisting moment (T),

$$120 \times 10^3 = \frac{\pi}{16} \times \tau \left[\frac{(d_o)^4 - (d_i)^4}{d_o}\right] = \frac{\pi}{16} \times \tau \left[\frac{(40)^4 - (25)^4}{40}\right] = 10\,650\,\tau$$

$$\therefore \quad \tau = 120 \times 10^3/10\,650 = 11.27 \text{ N/mm}^2 = 11.27 \text{ MPa}$$

Maximum compressive stress

We know that maximum compressive stress,

$$\sigma_{c(max)} = \frac{\sigma_c}{2} + \frac{1}{2}\left[\sqrt{(\sigma_c)^2 + 4\,\tau^2}\right]$$

$$= \frac{28.07}{2} + \frac{1}{2}\left[\sqrt{(28.07)^2 + 4\,(11.27)^2}\right]$$

$$= 14.035 + 18 = 32.035 \text{ MPa } \textbf{Ans.}$$

Maximum shear stress

We know that maximum shear stress,

$$\tau_{max} = \tfrac{1}{2}\left[\sqrt{(\sigma_c)^2 + 4\tau^2}\right] = \tfrac{1}{2}\left[\sqrt{(28.07)^2 + 4(11.27)^2}\right] = 18 \text{ MPa } \textbf{Ans.}$$

Example 5.14. *A shaft, as shown in Fig. 5.17, is subjected to a bending load of 3 kN, pure torque of 1000 N-m and an axial pulling force of 15 kN.*

Calculate the stresses at A and B.

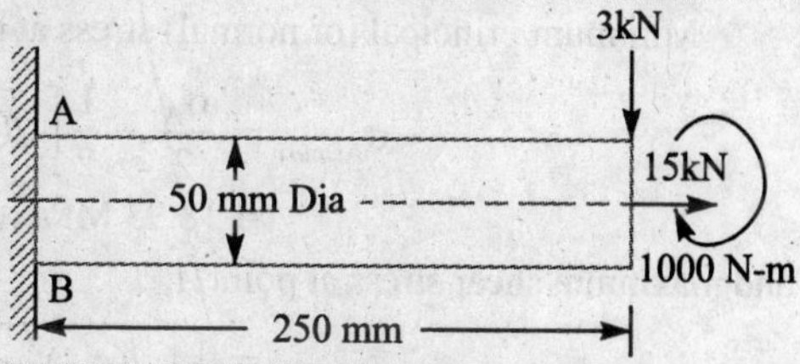

Fig. 5.17

Solution. Given : W = 3 kN = 3000 N ; T = 1000 N-m = 1 × 10^6 N-mm ; P = 15 kN = 15 × 10^3 N ; d = 50 mm; x = 250 mm

We know that cross-sectional area of the shaft,

$$A = \frac{\pi}{4} \times d^2$$

$$= \frac{\pi}{4}(50)^2 = 1964 \text{ mm}^2$$

∴ Tensile stress due to axial pulling at points A and B,

$$\sigma_o = \frac{P}{A} = \frac{15 \times 10^3}{1964} = 7.64 \text{ N/mm}^2 = 7.64 \text{ MPa}$$

Bending moment at points A and B,

$$M = W.x = 3000 \times 250 = 750 \times 10^3 \text{ N-mm}$$

Section modulus for the shaft,

$$Z = \frac{\pi}{32} \times d^3 = \frac{\pi}{32} (50)^3$$

$$= 12.27 \times 10^3 \text{ mm}^3$$

∴ Bending stress at points *A* and *B*,

$$\sigma_b = \frac{M}{Z} = \frac{750 \times 10^3}{12.27 \times 10^3}$$

$$= 61.1 \text{ N/mm}^2 = 61.1 \text{ MPa}$$

This picture shows a machine component inside a crane

Note : This picture is given as additional information and is not a direct example of the current chapter.

This bending stress is tensile at point *A* and compressive at point *B*.

∴ Resultant tensile stress at point *A*,

$$\sigma_A = \sigma_b + \sigma_o = 61.1 + 7.64$$
$$= 68.74 \text{ MPa}$$

and resultant compressive stress at point *B*,

$$\sigma_B = \sigma_b - \sigma_o = 61.1 - 7.64 = 53.46 \text{ MPa}$$

We know that the shear stress at points *A* and *B* due to the torque transmitted,

$$\tau = \frac{16\,T}{\pi\, d^3} = \frac{16 \times 1 \times 10^6}{\pi\,(50)^3} = 40.74 \text{ N/mm}^2 = 40.74 \text{ MPa} \qquad \ldots\left(\because\ T = \frac{\pi}{16} \times \tau \times d^3\right)$$

Stresses at point A

We know that maximum principal (or normal) stress at point *A*,

$$\sigma_{A(max)} = \frac{\sigma_A}{2} + \frac{1}{2}\left[\sqrt{(\sigma_A)^2 + 4\,\tau^2}\right]$$

$$= \frac{68.74}{2} + \frac{1}{2}\left[\sqrt{(68.74)^2 + 4\,(40.74)^2}\right]$$

$$= 34.37 + 53.3 = 87.67 \text{ MPa (tensile)}$$ **Ans.**

Minimum principal (or normal) stress at point *A*,

$$\sigma_{A(min)} = \frac{\sigma_A}{2} - \frac{1}{2}\left[\sqrt{(\sigma_A)^2 + 4\,\tau^2}\right] = 34.37 - 53.3 = -18.93 \text{ MPa}$$

$$= 18.93 \text{ MPa (compressive)}$$ **Ans.**

and maximum shear stress at point *A*,

$$\tau_{A(max)} = \frac{1}{2}\left[\sqrt{(\sigma_A)^2 + 4\,\tau^2}\right] = \frac{1}{2}\left[\sqrt{(68.74)^2 + 4\,(40.74)^2}\right]$$

$$= 53.3 \text{ MPa}$$ **Ans.**

Stresses at point B

We know that maximum principal (or normal) stress at point *B*,

$$\sigma_{B(max)} = \frac{\sigma_B}{2} + \frac{1}{2}\left[\sqrt{(\sigma_B)^2 + 4\,\tau^2}\right]$$

$$= \frac{53.46}{2} + \frac{1}{2}\left[\sqrt{(53.46)^2 + 4\,(40.74)^2}\right]$$

$$= 26.73 + 48.73 = 75.46 \text{ MPa (compressive)}$$ **Ans.**

Minimum principal (or normal) stress at point *B*,

$$\sigma_{B(min)} = \frac{\sigma_B}{2} - \frac{1}{2}\left[\sqrt{(\sigma_B)^2 + 4\tau^2}\right]$$

$$= 26.73 - 48.73 = -22 \text{ MPa}$$

$$= 22 \text{ MPa (tensile) } \textbf{Ans.}$$

and maximum shear stress at point *B*,

$$\tau_{B(max)} = \frac{1}{2}\left[\sqrt{(\sigma_B)^2 + 4\,\tau^2}\right] = \frac{1}{2}\left[\sqrt{(53.46)^2 + 4\,(40.74)^2}\right]$$

$$= 48.73 \text{ MPa } \textbf{Ans.}$$

Example 5.15. *An overhang crank with pin and shaft is shown in Fig. 5.18. A tangential load of 15 kN acts on the crank pin. Determine the maximum principal stress and the maximum shear stress at the centre of the crankshaft bearing.*

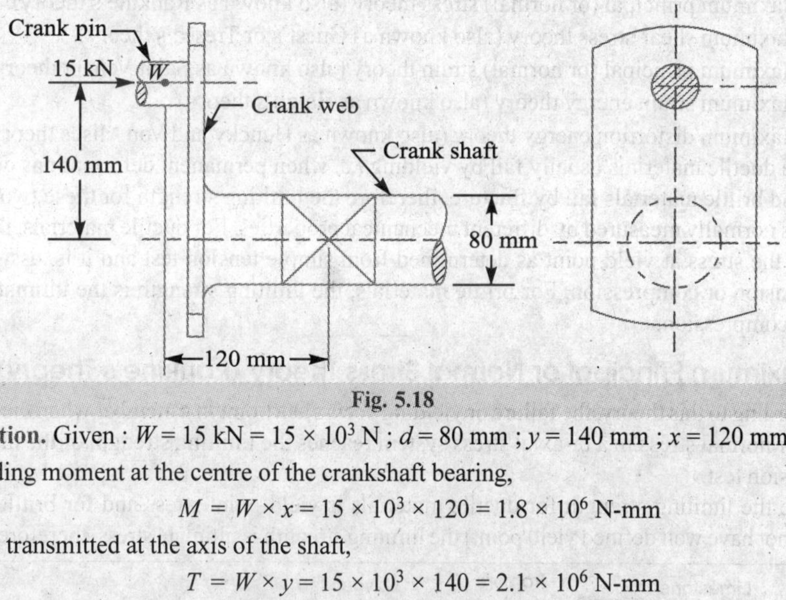

Fig. 5.18

Solution. Given : $W = 15$ kN $= 15 \times 10^3$ N ; $d = 80$ mm ; $y = 140$ mm ; $x = 120$ mm

Bending moment at the centre of the crankshaft bearing,

$$M = W \times x = 15 \times 10^3 \times 120 = 1.8 \times 10^6 \text{ N-mm}$$

and torque transmitted at the axis of the shaft,

$$T = W \times y = 15 \times 10^3 \times 140 = 2.1 \times 10^6 \text{ N-mm}$$

We know that bending stress due to the bending moment,

$$\sigma_b = \frac{M}{Z} = \frac{32\,M}{\pi\,d^3} \qquad \ldots\left(\because Z = \frac{\pi}{32} \times d^3\right)$$

$$= \frac{32 \times 1.8 \times 10^6}{\pi\,(80)^3} = 35.8 \text{ N/mm}^2 = 35.8 \text{ MPa}$$

and shear stress due to the torque transmitted,

$$\tau = \frac{16\,T}{\pi\,d^3} = \frac{16 \times 2.1 \times 10^6}{\pi\,(80)^3} = 20.9 \text{ N/mm}^2 = 20.9 \text{ MPa}$$

Maximum principal stress

We know that maximum principal stress,

$$\sigma_{t(max)} = \frac{\sigma_t}{2} + \frac{1}{2}\left[\sqrt{(\sigma_t)^2 + 4\,\tau^2}\right]$$

$$= \frac{35.8}{2} + \frac{1}{2}\left[\sqrt{(35.8)^2 + 4\,(20.9)^2}\right] \qquad \ldots (\text{Substituting } \sigma_t = \sigma_b)$$

$$= 17.9 + 27.5 = 45.4 \text{ MPa } \textbf{Ans.}$$

Maximum shear stress

We know that maximum shear stress,

$$\tau_{max} = \frac{1}{2}\left[\sqrt{(\sigma_t)^2 + 4\,\tau^2}\right] = \frac{1}{2}\left[\sqrt{(35.8)^2 + 4\,(20.9)^2}\right]$$

$$= 27.5 \text{ MPa } \textbf{Ans.}$$

5.9 Theories of Failure Under Static Load

It has already been discussed in the previous chapter that strength of machine members is based upon the mechanical properties of the materials used. Since these properties are usually determined from simple tension or compression tests, therefore, predicting failure in members subjected to uni-axial stress is both simple and straight-forward. But the problem of predicting the failure stresses for members subjected to bi-axial or tri-axial stresses is much more complicated. In fact, the problem is so complicated that a large number of different theories have been formulated. The principal theories of failure for a member subjected to bi-axial stress are as follows:

1. Maximum principal (or normal) stress theory (also known as Rankine's theory).
2. Maximum shear stress theory (also known as Guest's or Tresca's theory).
3. Maximum principal (or normal) strain theory (also known as Saint Venant theory).
4. Maximum strain energy theory (also known as Haigh's theory).
5. Maximum distortion energy theory (also known as Hencky and Von Mises theory).

Since ductile materials usually fail by yielding *i.e.* when permanent deformations occur in the material and brittle materials fail by fracture, therefore the limiting strength for these two classes of materials is normally measured by different mechanical properties. For ductile materials, the limiting strength is the stress at yield point as determined from simple tension test and it is, assumed to be equal in tension or compression. For brittle materials, the limiting strength is the ultimate stress in tension or compression.

5.10 Maximum Principal or Normal Stress Theory (Rankine's Theory)

According to this theory, the failure or yielding occurs at a point in a member when the maximum principal or normal stress in a bi-axial stress system reaches the limiting strength of the material in a simple tension test.

Since the limiting strength for ductile materials is yield point stress and for brittle materials (which do not have well defined yield point) the limiting strength is ultimate stress, therefore according

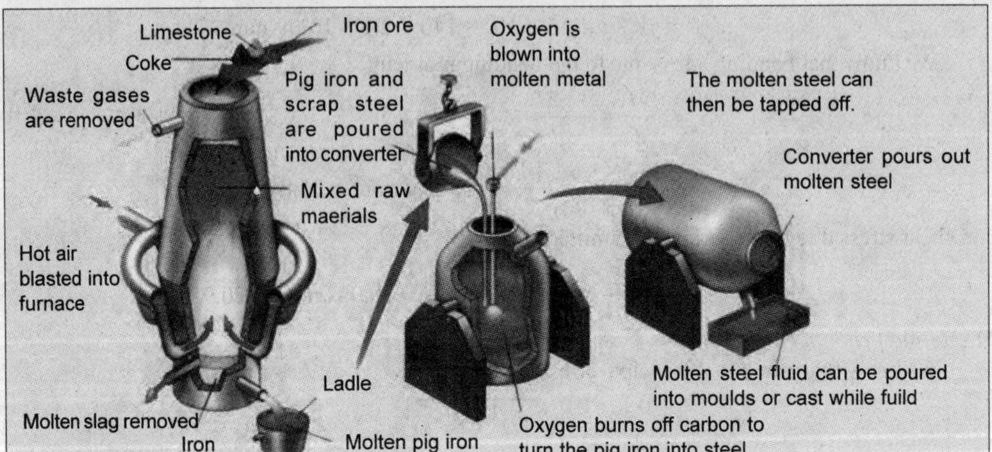

Pig iron is made from iron ore in a blast furnace. It is a brittle form of iron that contains 4-5 per cent carbon.

Note : This picture is given as additional information and is not a direct example of the current chapter.

to the above theory, taking factor of safety (*F.S.*) into consideration, the maximum principal or normal stress (σ_{t1}) in a bi-axial stress system is given by

$$\sigma_{t1} = \frac{\sigma_{yt}}{F.S.}, \text{ for ductile materials}$$

$$= \frac{\sigma_u}{F.S.}, \text{ for brittle materials}$$

where σ_{yt} = Yield point stress in tension as determined from simple tension test, and

σ_u = Ultimate stress.

Since the maximum principal or normal stress theory is based on failure in tension or compression and ignores the possibility of failure due to shearing stress, therefore it is not used for ductile materials. However, for brittle materials which are relatively strong in shear but weak in tension or compression, this theory is generally used.

Note : The value of maximum principal stress (σ_{t1}) for a member subjected to bi-axial stress system may be determined as discussed in Art. 5.7.

5.11 Maximum Shear Stress Theory (Guest's or Tresca's Theory)

According to this theory, the failure or yielding occurs at a point in a member when the maximum shear stress in a bi-axial stress system reaches a value equal to the shear stress at yield point in a simple tension test. Mathematically,

$$\tau_{max} = \tau_{yt}/F.S. \quad \text{...(i)}$$

where τ_{max} = Maximum shear stress in a bi-axial stress system,

τ_{yt} = Shear stress at yield point as determined from simple tension test, and

$F.S.$ = Factor of safety.

Since the shear stress at yield point in a simple tension test is equal to one-half the yield stress in tension, therefore the equation (*i*) may be written as

$$\tau_{max} = \frac{\sigma_{yt}}{2 \times F.S.}$$

This theory is mostly used for designing members of ductile materials.

Note: The value of maximum shear stress in a bi-axial stress system (τ_{max}) may be determined as discussed in Art. 5.7.

5.12 Maximum Principal Strain Theory (Saint Venant's Theory)

According to this theory, the failure or yielding occurs at a point in a member when the maximum principal (or normal) strain in a bi-axial stress system reaches the limiting value of strain (*i.e.* strain at yield point) as determined from a simple tensile test. The maximum principal (or normal) strain in a bi-axial stress system is given by

$$\varepsilon_{max} = \frac{\sigma_{t1}}{E} - \frac{\sigma_{t2}}{m \cdot E}$$

∴ According to the above theory,

$$\varepsilon_{max} = \frac{\sigma_{t1}}{E} - \frac{\sigma_{t2}}{m \cdot E} = \varepsilon = \frac{\sigma_{yt}}{E \times F.S.} \quad \text{...(i)}$$

where σ_{t1} and σ_{t2} = Maximum and minimum principal stresses in a bi-axial stress system,

ε = Strain at yield point as determined from simple tension test,

$1/m$ = Poisson's ratio,

E = Young's modulus, and

$F.S.$ = Factor of safety.

From equation (*i*), we may write that

$$\sigma_{t1} - \frac{\sigma_{t2}}{m} = \frac{\sigma_{yt}}{F.S.}$$

This theory is not used, in general, because it only gives reliable results in particular cases.

5.13 Maximum Strain Energy Theory (Haigh's Theory)

According to this theory, the failure or yielding occurs at a point in a member when the strain energy per unit volume in a bi-axial stress system reaches the limiting strain energy (*i.e.* strain energy at the yield point) per unit volume as determined from simple tension test.

This double-decker A 380 has a passenger capacity of 555. Its engines and parts should be robust which can bear high torsional and variable stresses.

We know that strain energy per unit volume in a bi-axial stress system,

$$U_1 = \frac{1}{2E}\left[(\sigma_{t1})^2 + (\sigma_{t2})^2 - \frac{2\,\sigma_{t1} \times \sigma_{t2}}{m}\right]$$

and limiting strain energy per unit volume for yielding as determined from simple tension test,

$$U_2 = \frac{1}{2E}\left(\frac{\sigma_{yt}}{F.S.}\right)^2$$

According to the above theory, $U_1 = U_2$.

$$\therefore \quad \frac{1}{2E}\left[(\sigma_{t1})^2 + (\sigma_{t2})^2 - \frac{2\,\sigma_{t1} \times \sigma_{t2}}{m}\right] = \frac{1}{2E}\left(\frac{\sigma_{yt}}{F.S.}\right)^2$$

or

$$(\sigma_{t1})^2 + (\sigma_{t2})^2 - \frac{2\,\sigma_{t1} \times \sigma_{t2}}{m} = \left(\frac{\sigma_{yt}}{F.S.}\right)^2$$

This theory may be used for ductile materials.

5.14 Maximum Distortion Energy Theory (Hencky and Von Mises Theory)

According to this theory, the failure or yielding occurs at a point in a member when the distortion strain energy (also called shear strain energy) per unit volume in a bi-axial stress system reaches the limiting distortion energy (*i.e.* distortion energy at yield point) per unit volume as determined from a simple tension test. Mathematically, the maximum distortion energy theory for yielding is expressed as

$$(\sigma_{t1})^2 + (\sigma_{t2})^2 - 2\sigma_{t1} \times \sigma_{t2} = \left(\frac{\sigma_{yt}}{F.S.}\right)^2$$

This theory is mostly used for ductile materials in place of maximum strain energy theory.

Note: The maximum distortion energy is the difference between the total strain energy and the strain energy due to uniform stress.

Example 5.16. *The load on a bolt consists of an axial pull of 10 kN together with a transverse shear force of 5 kN. Find the diameter of bolt required according to*

1. Maximum principal stress theory; 2. Maximum shear stress theory; 3. Maximum principal strain theory; 4. Maximum strain energy theory; and 5. Maximum distortion energy theory.

Take permissible tensile stress at elastic limit = 100 MPa and poisson's ratio = 0.3.

Solution. Given : P_{t1} = 10 kN ; P_s = 5 kN ; $\sigma_{t(el)}$ = 100 MPa = 100 N/mm² ; $1/m$ = 0.3

Let d = Diameter of the bolt in mm.

∴ Cross-sectional area of the bolt,

$$A = \frac{\pi}{4} \times d^2 = 0.7854\, d^2 \text{ mm}^2$$

We know that axial tensile stress,

$$\sigma_1 = \frac{P_{t1}}{A} = \frac{10}{0.7854\, d^2} = \frac{12.73}{d^2} \text{ kN/mm}^2$$

and transverse shear stress,

$$\tau = \frac{P_s}{A} = \frac{5}{0.7854\, d^2} = \frac{6.365}{d^2} \text{ kN/mm}^2$$

1. ***According to maximum principal stress theory***

We know that maximum principal stress,

$$\sigma_{t1} = \frac{\sigma_1 + \sigma_2}{2} + \frac{1}{2}\left[\sqrt{(\sigma_1 - \sigma_2)^2 + 4\,\tau^2}\right]$$

$$= \frac{\sigma_1}{2} + \frac{1}{2}\left[\sqrt{(\sigma_1)^2 + 4\,\tau^2}\right] \qquad \text{...}(\because \sigma_2 = 0)$$

$$= \frac{12.73}{2\,d^2} + \frac{1}{2}\left[\sqrt{\left(\frac{12.73}{d^2}\right)^2 + 4\left(\frac{6.365}{d^2}\right)^2}\right]$$

$$= \frac{6.365}{d^2} + \frac{1}{2} \times \frac{6.365}{d^2}\left[\sqrt{4+4}\right]$$

$$= \frac{6.365}{d^2}\left[1 + \frac{1}{2}\sqrt{4+4}\right] = \frac{15.365}{d^2} \text{ kN/mm}^2 = \frac{15\,365}{d^2} \text{ N/mm}^2$$

According to maximum principal stress theory,

$$\sigma_{t1} = \sigma_{t(el)} \quad \text{or} \quad \frac{15\,365}{d^2} = 100$$

$$\therefore \quad d^2 = 15\,365/100 = 153.65 \quad \text{or} \quad d = 12.4 \text{ mm } \textbf{Ans.}$$

2. ***According to maximum shear stress theory***

We know that maximum shear stress,

$$\tau_{max} = \tfrac{1}{2}\left[\sqrt{(\sigma_1 - \sigma_2)^2 + 4\,\tau^2}\right] = \tfrac{1}{2}\left[\sqrt{(\sigma_1)^2 + 4\,\tau^2}\right] \qquad \text{...}(\because \sigma_2 = 0)$$

$$= \frac{1}{2}\left[\sqrt{\left(\frac{12.73}{d^2}\right)^2 + 4\left(\frac{6.365}{d^2}\right)^2}\right] = \frac{1}{2} \times \frac{6.365}{d^2}\left[\sqrt{4+4}\right]$$

$$= \frac{9}{d^2} \text{ kN/mm}^2 = \frac{9000}{d^2} \text{ N/mm}^2$$

According to maximum shear stress theory,

$$\tau_{max} = \frac{\sigma_{t(el)}}{2} \quad \text{or} \quad \frac{9000}{d^2} = \frac{100}{2} = 50$$

$$\therefore \quad d^2 = 9000 / 50 = 180 \quad \text{or} \quad d = 13.42 \text{ mm } \textbf{Ans.}$$

3. ***According to maximum principal strain theory***

We know that maximum principal stress,

$$\sigma_{t1} = \frac{\sigma_1}{2} + \frac{1}{2}\left[\sqrt{(\sigma_1)^2 + 4\,\tau^2}\right] = \frac{15\,365}{d^2} \quad \text{...(As calculated before)}$$

and minimum principal stress,

$$\sigma_{t2} = \frac{\sigma_1}{2} - \frac{1}{2}\left[\sqrt{(\sigma_1)^2 + 4\,\tau^2}\right]$$

$$= \frac{12.73}{2\,d^2} - \frac{1}{2}\left[\sqrt{\left(\frac{12.73}{d^2}\right)^2 + 4\left(\frac{6.365}{d^2}\right)^2}\right]$$

$$= \frac{6.365}{d^2} - \frac{1}{2} \times \frac{6.365}{d^2}\left[\sqrt{4+4}\right]$$

$$= \frac{6.365}{d^2}\left[1 - \sqrt{2}\right] = \frac{-\,2.635}{d^2}\ \text{kN/mm}^2$$

$$= \frac{-\,2635}{d^2}\ \text{N/mm}^2$$

Front view of a jet engine. The rotors undergo high torsional and bending stresses.

We know that according to maximum principal strain theory,

$$\frac{\sigma_{t1}}{E} - \frac{\sigma_{t2}}{mE} = \frac{\sigma_{t(el)}}{E} \text{ or } \sigma_{t1} - \frac{\sigma_{t2}}{m} = \sigma_{t(el)}$$

$$\therefore \quad \frac{15\,365}{d^2} + \frac{2635 \times 0.3}{d^2} = 100 \quad \text{or} \quad \frac{16\,156}{d^2} = 100$$

$d^2 = 16\,156 / 100 = 161.56$ or $d = 12.7$ mm **Ans.**

4. ***According to maximum strain energy theory***

We know that according to maximum strain energy theory,

$$(\sigma_{t1})^2 + (\sigma_{t2})^2 - \frac{2\,\sigma_{t1} \times \sigma_{t2}}{m} = [\sigma_{t(el)}]^2$$

$$\left[\frac{15\,365}{d^2}\right]^2 + \left[\frac{-\,2635}{d^2}\right]^2 - 2 \times \frac{15\,365}{d^2} \times \frac{-\,2635}{d^2} \times 0.3 = (100)^2$$

$$\frac{236 \times 10^6}{d^4} + \frac{6.94 \times 10^6}{d^4} + \frac{24.3 \times 10^6}{d^4} = 10 \times 10^3$$

$$\frac{23\,600}{d^4} + \frac{694}{d^4} + \frac{2430}{d^4} = 1 \quad \text{or} \quad \frac{26\,724}{d^4} = 1$$

$\therefore \quad d^4 = 26\,724$ or $d = 12.78$ mm **Ans.**

5. ***According to maximum distortion energy theory***

According to maximum distortion energy theory,

$$(\sigma_{t1})^2 + (\sigma_{t2})^2 - 2\sigma_{t1} \times \sigma_{t2} = [\sigma_{t(el)}]^2$$

$$\left[\frac{15\,365}{d^2}\right]^2 + \left[\frac{-\,2635}{d^2}\right]^2 - 2 \times \frac{15\,365}{d^2} \times \frac{-\,2635}{d^2} = (100)^2$$

$$\frac{236 \times 10^6}{d^4} + \frac{6.94 \times 10^6}{d^4} + \frac{80.97 \times 10^6}{d^4} = 10 \times 10^3$$

$$\frac{23\,600}{d^4} + \frac{694}{d^4} + \frac{8097}{d^4} = 1 \quad \text{or} \quad \frac{32\,391}{d^4} = 1$$

$\therefore \quad d^4 = 32\,391$ or $d = 13.4$ mm **Ans.**

Example 5.17. *A cylindrical shaft made of steel of yield strength 700 MPa is subjected to static loads consisting of bending moment 10 kN-m and a torsional moment 30 kN-m. Determine the diameter of the shaft using two different theories of failure, and assuming a factor of safety of 2. Take E = 210 GPa and poisson's ratio = 0.25.*

Solution. Given : $\sigma_{yt} = 700$ MPa $= 700$ N/mm^2 ; $M = 10$ kN-m $= 10 \times 10^6$ N-mm ; $T = 30$ kN-m $= 30 \times 10^6$ N-mm ; $F.S. = 2$; $E = 210$ GPa $= 210 \times 10^3$ N/mm^2; $1/m = 0.25$

Let d = Diameter of the shaft in mm.

First of all, let us find the maximum and minimum principal stresses.

We know that section modulus of the shaft

$$Z = \frac{\pi}{32} \times d^3 = 0.0982\, d^3 \text{ mm}^3$$

∴ Bending (tensile) stress due to the bending moment,

$$\sigma_1 = \frac{M}{Z} = \frac{10 \times 10^6}{0.0982\, d^3} = \frac{101.8 \times 10^6}{d^3} \text{ N/mm}^2$$

and shear stress due to torsional moment,

$$\tau = \frac{16\, T}{\pi\, d^3} = \frac{16 \times 30 \times 10^6}{\pi\, d^3} = \frac{152.8 \times 10^6}{d^3} \text{ N/mm}^2$$

We know that maximum principal stress,

$$\sigma_{t1} = \frac{\sigma_1 + \sigma_2}{2} + \frac{1}{2}\left[\sqrt{(\sigma_1 - \sigma_2)^2 + 4\,\tau^2}\right]$$

$$= \frac{\sigma_1}{2} + \frac{1}{2}\left[\sqrt{(\sigma_1)^2 + 4\,\tau^2}\right] \qquad \text{...}(\because \sigma_2 = 0)$$

$$= \frac{101.8 \times 10^6}{2d^3} + \frac{1}{2}\left[\sqrt{\left(\frac{101.8 \times 10^6}{d^3}\right)^2 + 4\left(\frac{152.8 \times 10^6}{d^3}\right)^2}\right]$$

$$= \frac{50.9 \times 10^6}{d^3} + \frac{1}{2} \times \frac{10^6}{d^3}\left[\sqrt{(101.8)^2 + 4\,(152.8)^2}\right]$$

$$= \frac{50.9 \times 10^6}{d^3} + \frac{161 \times 10^6}{d^3} = \frac{211.9 \times 10^6}{d^3} \text{ N/mm}^2$$

and minimum principal stress,

$$\sigma_{t2} = \frac{\sigma_1 + \sigma_2}{2} - \frac{1}{2}\left[\sqrt{(\sigma_1 - \sigma_2)^2 + 4\,\tau^2}\right]$$

$$= \frac{\sigma_1}{2} - \frac{1}{2}\left[\sqrt{(\sigma_1)^2 + 4\,\tau^2}\right] \qquad \text{...}(\because \sigma_2 = 0)$$

$$= \frac{50.9 \times 10^6}{d^3} - \frac{161 \times 10^6}{d^3} = \frac{-110.1 \times 10^6}{d^3} \text{ N/mm}^2$$

Let us now find out the diameter of shaft (d) by considering the maximum shear stress theory and maximum strain energy theory.

1. ***According to maximum shear stress theory***

We know that maximum shear stress,

$$\tau_{max} = \frac{\sigma_{t1} - \sigma_{t2}}{2} = \frac{1}{2}\left[\frac{211.9 \times 10^6}{d^3} + \frac{110.1 \times 10^6}{d^3}\right] = \frac{161 \times 10^6}{d^3}$$

We also know that according to maximum shear stress theory,

$$\tau_{max} = \frac{\sigma_{yt}}{2\,F.S.} \quad \text{or} \quad \frac{161 \times 10^6}{d^3} = \frac{700}{2 \times 2} = 175$$

$$\therefore \quad d^3 = 161 \times 10^6 / 175 = 920 \times 10^3 \text{ or } d = 97.2 \text{ mm } \textbf{Ans.}$$

Note: The value of maximum shear stress (τ_{max}) may also be obtained by using the relation,

$$\tau_{max} = \frac{1}{2}\left[\sqrt{(\sigma_1)^2 + 4\,\tau^2}\right]$$

$$= \frac{1}{2}\left[\sqrt{\left(\frac{101.8\times10^6}{d^3}\right)^2 + 4\left(\frac{152.8\times10^6}{d^3}\right)^2}\right]$$

$$= \frac{1}{2}\times\frac{10^6}{d^3}\left[\sqrt{(101.8)^2 + 4\,(152.8)^2}\right]$$

$$= \frac{1}{2}\times\frac{10^6}{d^3}\times 322 = \frac{161\times10^6}{d^3}\text{ N/mm}^2 \qquad \text{...(Same as before)}$$

2. *According to maximum strain energy theory*

We know that according to maximum strain energy theory,

$$\frac{1}{2E}\left[(\sigma_{t1})^2 + (\sigma_{t2})^2 - \frac{2\,\sigma_{t1}\times\sigma_{t2}}{m}\right] = \frac{1}{2E}\left(\frac{\sigma_{yt}}{F.S.}\right)^2$$

or
$$(\sigma_{t1})^2 + (\sigma_{t2})^2 - \frac{2\,\sigma_{t1}\times\sigma_{t2}}{m} = \left(\frac{\sigma_{yt}}{F.S.}\right)^2$$

$$\left[\frac{211.9\times10^6}{d^3}\right]^2 + \left[\frac{-110.1\times10^6}{d^3}\right]^2 - 2\times\frac{211.9\times10^6}{d^3}\times\frac{-110.1\times10^6}{d^3}\times 0.25 = \left(\frac{700}{2}\right)^2$$

or
$$\frac{44\,902\times10^{12}}{d^6} + \frac{12\,122\times10^{12}}{d^6} + \frac{11\,665\times10^{12}}{d^6} = 122\,500$$

$$\frac{68\,689\times10^{12}}{d^6} = 122\,500$$

$\therefore$ $d^6 = 68\,689 \times 10^{12}/122\,500 = 0.5607 \times 10^{12}$ or $d = 90.8$ mm **Ans.**

Example 5.18. *A mild steel shaft of 50 mm diameter is subjected to a bending moment of 2000 N-m and a torque T. If the yield point of the steel in tension is 200 MPa, find the maximum value of this torque without causing yielding of the shaft according to 1. the maximum principal stress; 2. the maximum shear stress; and 3. the maximum distortion strain energy theory of yielding.*

Solution. Given: $d = 50$ mm ; $M = 2000$ N-m $= 2 \times 10^6$ N-mm ; $\sigma_{yt} = 200$ MPa $= 200$ N/mm^2

Let T = Maximum torque without causing yielding of the shaft, in N-mm.

1. *According to maximum principal stress theory*

We know that section modulus of the shaft,

$$Z = \frac{\pi}{32}\times d^3 = \frac{\pi}{32}(50)^3 = 12\,273\text{ mm}^3$$

$\therefore$ Bending stress due to the bending moment,

$$\sigma_1 = \frac{M}{Z} = \frac{2\times10^6}{12\,273} = 163\text{ N/mm}^2$$

and shear stress due to the torque,

$$\tau = \frac{16\,T}{\pi\,d^3} = \frac{16\,T}{\pi\,(50)^3} = 0.0407\times10^{-3}\,T\text{ N/mm}^2 \qquad \text{...}\left[\because T = \frac{\pi}{16}\times\tau\times d^3\right]$$

We know that maximum principal stress,

$$\sigma_{t1} = \frac{\sigma_1}{2} + \frac{1}{2}\left[\sqrt{(\sigma_1)^2 + 4\,\tau^2}\right]$$

$$= \frac{163}{2} + \frac{1}{2}\left[\sqrt{(163)^2 + 4\,(0.0407\times10^{-3}\,T)^2}\right]$$

$$= 81.5 + \sqrt{6642.5 + 1.65 \times 10^{-9}\, T^2} \text{ N/mm}^2$$

Minimum principal stress,

$$\sigma_{t2} = \frac{\sigma_1}{2} - \frac{1}{2}\left[\sqrt{(\sigma_1)^2 + 4\,\tau^2}\right]$$

$$= \frac{163}{2} - \frac{1}{2}\left[\sqrt{(163)^2 + 4\,(0.0407 \times 10^{-3}\, T)^2}\right]$$

$$= 81.5 - \sqrt{6642.5 + 1.65 \times 10^{-9}\, T^2} \text{ N/mm}^2$$

and maximum shear stress,

$$\tau_{max} = \tfrac{1}{2}\left[\sqrt{(\sigma_1)^2 + 4\,\tau^2}\right] = \tfrac{1}{2}\left[\sqrt{(163)^2 + 4\,(0.0407 \times 10^{-3}\, T)^2}\right]$$

$$= \sqrt{6642.5 + 1.65 \times 10^{-9}\, T^2} \text{ N/mm}^2$$

We know that according to maximum principal stress theory,

$$\sigma_{t1} = \sigma_{yt} \qquad \text{...(Taking } F.S. = 1)$$

$$\therefore \quad 81.5 + \sqrt{6642.5 + 1.65 \times 10^{-9}\, T^2} = 200$$

$$6642.5 + 1.65 + 10^{-9}\, T^2 = (200 - 81.5)^2 = 14\,042$$

$$T^2 = \frac{14\,042 - 6642.5}{1.65 \times 10^{-9}} = 4485 \times 10^9$$

or $T = 2118 \times 10^3$ N-mm = 2118 N-m **Ans.**

2. ***According to maximum shear stress theory***

We know that according to maximum shear stress theory,

$$\tau_{max} = \tau_{yt} = \frac{\sigma_{yt}}{2}$$

$$\therefore \quad \sqrt{6642.5 + 1.65 \times 10^{-9}\, T^2} = \frac{200}{2} = 100$$

$$6642.5 + 1.65 \times 10^{-9}\, T^2 = (100)^2 = 10\,000$$

$$T^2 = \frac{10\,000 - 6642.5}{1.65 \times 10^{-9}} = 2035 \times 10^9$$

$\therefore$ $T = 1426 \times 10^3$ N-mm = 1426 N-m **Ans.**

3. ***According to maximum distortion strain energy theory***

We know that according to maximum distortion strain energy theory

$$(\sigma_{t1})^2 + (\sigma_{t2})^2 - \sigma_{t1} \times \sigma_{t2} = (\sigma_{yt})^2$$

$$\left[81.5 + \sqrt{6642.5 + 1.65 \times 10^{-9}\, T^2}\right]^2 + \left[81.5 - \sqrt{6642.5 + 1.65 \times 10^{-9}\, T^2}\right]^2$$

$$- \left[81.5 + \sqrt{6642.5 + 1.65 \times 10^{-9}\, T^2}\right]\left[81.5 - \sqrt{6642.5 + 1.65 \times 10^{-9}\, T^2}\right] = (200)^2$$

$$2\left[(81.5)^2 + 6642.5 + 1.65 \times 10^{-9}\, T^2\right] - \left[(81.5)^2 - 6642.5 + 1.65 \times 10^{-9}\, T^2\right] = (200)^2$$

$$(81.5)^2 + 3 \times 6642.5 + 3 \times 1.65 \times 10^{-9}\, T^2 = (200)^2$$

$$26\,570 + 4.95 \times 10^{-9}\, T^2 = 40\,000$$

$$T^2 = \frac{40\,000 - 26\,570}{4.95 \times 10^{-9}} = 2713 \times 10^9$$

$\therefore$ $T = 1647 \times 10^3$ N-mm = 1647 N-m **Ans.**

5.15 Eccentric Loading - Direct and Bending Stresses Combined

An external load, whose line of action is parallel but does not coincide with the centroidal axis of the machine component, is known as an ***eccentric load.*** The distance between the centroidal axis of the machine component and the eccentric load is called ***eccentricity*** and is generally denoted by *e*. The examples of eccentric loading, from the subject point of view, are *C*-clamps, punching machines, brackets, offset connecting links etc.

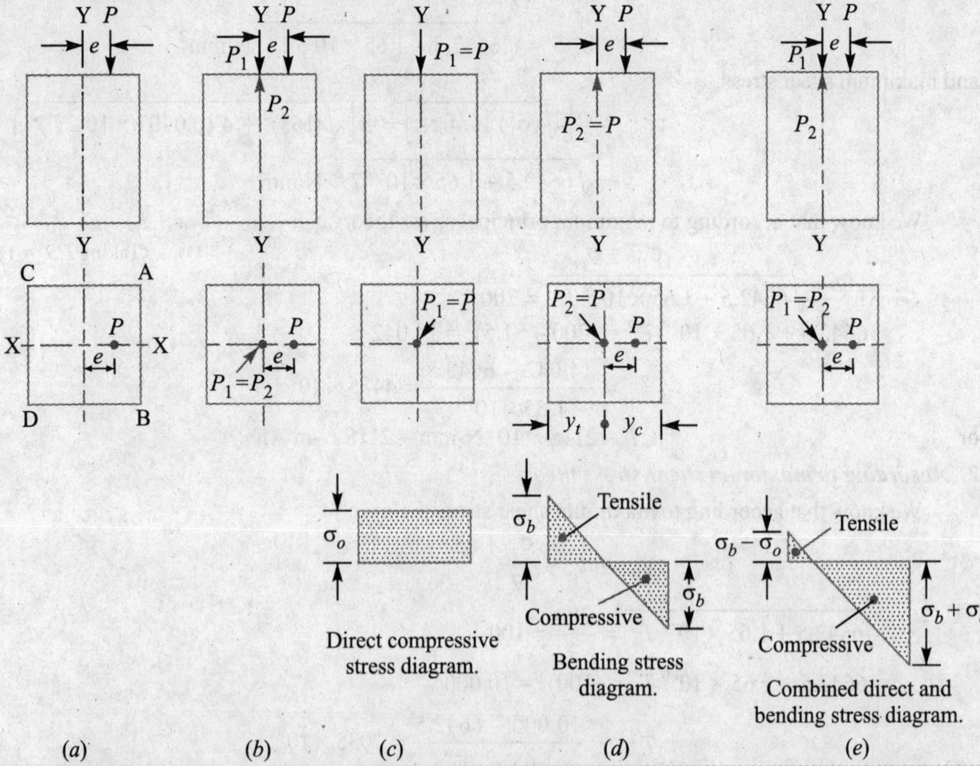

Fig. 5.19. Eccentric loading.

Consider a short prismatic bar subjected to a compressive load *P* acting at an eccentricity of *e* as shown in Fig. 5.19 (*a*).

Let us introduce two forces P_1 and P_2 along the centre line or neutral axis equal in magnitude to *P*, without altering the equilibrium of the bar as shown in Fig. 5.19 (*b*). A little consideration will show that the force P_1 will induce a direct compressive stress over the entire cross-section of the bar, as shown in Fig. 5.19 (*c*).

The magnitude of this direct compressive stress is given by

$$\sigma_o = \frac{P_1}{A} \text{ or } \frac{P}{A}, \text{ where } A \text{ is the cross-sectional area of the bar.}$$

The forces P_1 and P_2 will form a couple equal to $P \times e$ which will cause bending stress. This bending stress is compressive at the edge *AB* and tensile at the edge *CD*, as shown in Fig. 5.19 (*d*). The magnitude of bending stress at the edge *AB* is given by

$$\sigma_b = \frac{P \,.\, e \,.\, y_c}{I} \text{ (compressive)}$$

and bending stress at the edge *CD*,

$$\sigma_b = \frac{P \,.\, e \,.\, y_t}{I} \text{ (tensile)}$$

where y_c and y_t = Distances of the extreme fibres on the compressive and tensile sides, from the neutral axis respectively, and

I = Second moment of area of the section about the neutral axis *i.e.* Y-axis.

According to the principle of superposition, the maximum or the resultant compressive stress at the edge AB,

$$\sigma_c = \frac{P \cdot e \cdot y_c}{I} + \frac{P}{A} = {}^*\frac{M}{Z} + \frac{P}{A} = \sigma_b + \sigma_o$$

and the maximum or resultant tensile stress at the edge CD,

$$\sigma_t = \frac{P \cdot e \cdot y_t}{I} - \frac{P}{A} = \frac{M}{Z} - \frac{P}{A} = \sigma_b - \sigma_o$$

The resultant compressive and tensile stress diagram is shown in Fig. 5.19 (*e*).

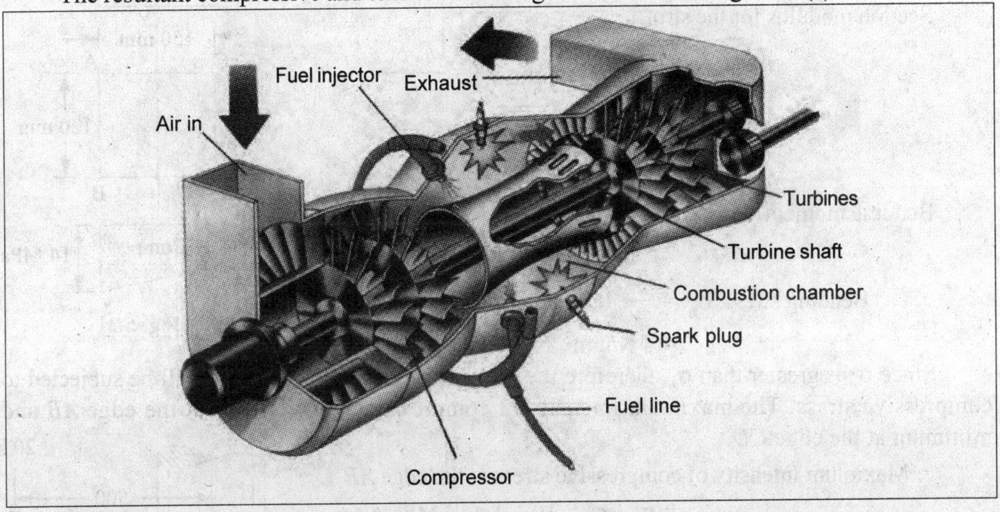

In a gas-turbine system, a compressor forces air into a combustion chamber. There, it mixes with fuel. The mixture is ignited by a spark. Hot gases are produced when the fuel burns. They expand and drive a series of fan blades called a turbine.

Note : This picture is given as additional information and is not a direct example of the current chapter.

Notes: 1. When the member is subjected to a tensile load, then the above equations may be used by interchanging the subscripts *c* and *t*.

2. When the direct stress σ_o is greater than or equal to bending stress σ_b, then the compressive stress shall be present all over the cross-section.

3. When the direct stress σ_o is less than the bending stress σ_b, then the tensile stress will occur in the left hand portion of the cross-section and compressive stress on the right hand portion of the cross-section. In Fig. 5.19, the stress diagrams are drawn by taking σ_o less than σ_b.

In case the eccentric load acts with eccentricity about two axes, as shown in Fig. 5.20, then the total stress at the extreme fibre

$$= \frac{P}{A} \pm \frac{P \cdot e_x \cdot x}{I_{XX}} \pm \frac{P \cdot e_y \cdot y}{I_{YY}}$$

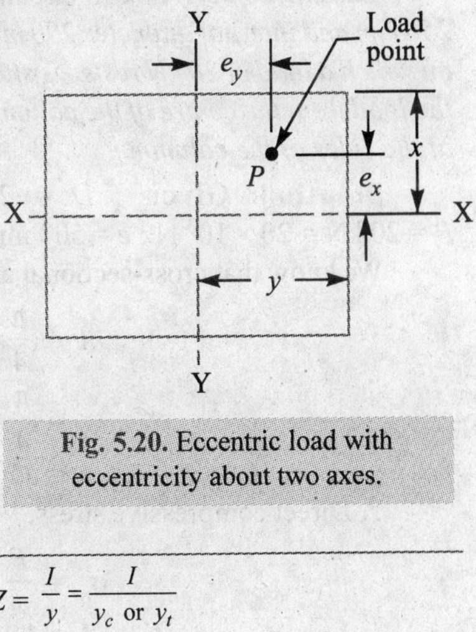

Fig. 5.20. Eccentric load with eccentricity about two axes.

* We know that bending moment, $M = P.e$ and section modulus, $Z = \frac{I}{y} = \frac{I}{y_c \text{ or } y_t}$

∴ Bending stress, $\sigma_b = M / Z$

Example 5.19. *A rectangular strut is 150 mm wide and 120 mm thick. It carries a load of 180 kN at an eccentricity of 10 mm in a plane bisecting the thickness as shown in Fig. 5.21. Find the maximum and minimum intensities of stress in the section.*

Solution. Given : b = 150 mm ; d = 120 mm ; P = 180 kN = 180×10^3 N ; e = 10 mm

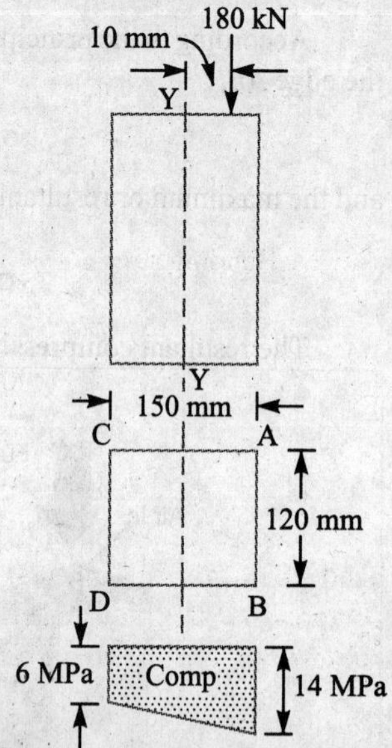

Fig. 5.21

We know that cross-sectional area of the strut,

$$A = b.d = 150 \times 120 = 18 \times 10^3 \text{ mm}^2$$

∴ Direct compressive stress,

$$\sigma_o = \frac{P}{A} = \frac{180 \times 10^3}{18 \times 10^3} = 10 \text{ N/mm}^2 = 10 \text{ MPa}$$

Section modulus for the strut,

$$Z = \frac{I_{YY}}{y} = \frac{d \cdot b^3/12}{b/2} = \frac{d \cdot b^2}{6} = \frac{120\,(150)^2}{6} = 450 \times 10^3 \text{ mm}^3$$

Bending moment, $M = P.e = 180 \times 10^3 \times 10 = 1.8 \times 10^6$ N-mm

∴ Bending stress, $\sigma_b = \frac{M}{Z} = \frac{1.8 \times 10^6}{450 \times 10^3} = 4 \text{ N/mm}^2 = 4 \text{ MPa}$

Since σ_o is greater than σ_b, therefore the entire cross-section of the strut will be subjected to compressive stress. The maximum intensity of compressive stress will be at the edge *AB* and minimum at the edge *CD*.

∴ Maximum intensity of compressive stress at the edge *AB*

$$= \sigma_o + \sigma_b = 10 + 4 = 14 \text{ MPa } \textbf{Ans.}$$

and minimum intensity of compressive stress at the edge *CD*

$$= \sigma_o - \sigma_b = 10 - 4 = 6 \text{ MPa } \textbf{Ans.}$$

Example 5.20. *A hollow circular column of external diameter 250 mm and internal diameter 200 mm, carries a projecting bracket on which a load of 20 kN rests, as shown in Fig. 5.22. The centre of the load from the centre of the column is 500 mm. Find the stresses at the sides of the column.*

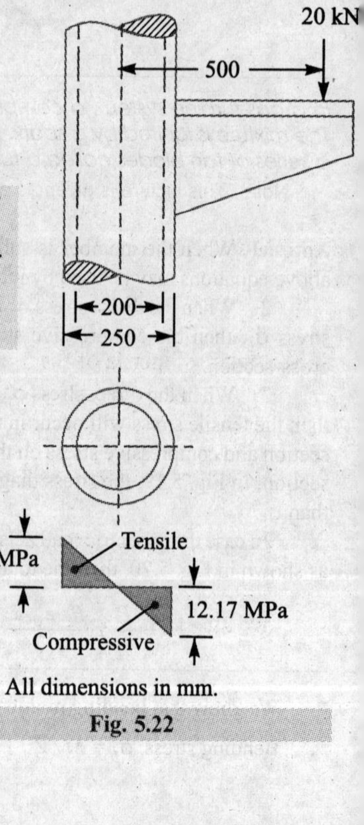

All dimensions in mm.

Fig. 5.22

Solution. Given : D = 250 mm ; d = 200 mm ; P = 20 kN = 20×10^3 N ; e = 500 mm

We know that cross-sectional area of column,

$$A = \frac{\pi}{4}(D^2 - d^2) = \frac{\pi}{4}[(250)^2 - (200)^2] = 17\,674 \text{ mm}^2$$

∴ Direct compressive stress,

$$\sigma_o = \frac{P}{A} = \frac{20 \times 10^3}{17\,674} = 1.13 \text{ N/mm}^2 = 1.13 \text{ MPa}$$

Section modulus for the column,

$$Z = \frac{I}{y} = \frac{\frac{\pi}{64}\left[D^4 - d^4\right]}{D/2} = \frac{\frac{\pi}{64}\left[(250)^4 - (200)^4\right]}{250/2}$$

$$= 905.8 \times 10^3 \text{ mm}^3$$

Bending moment,

$$M = P.e$$
$$= 20 \times 10^3 \times 500$$
$$= 10 \times 10^6 \text{ N-mm}$$

∴ Bending stress,

$$\sigma_b = \frac{M}{Z} = \frac{10 \times 10^6}{905.8 \times 10^3}$$
$$= 11.04 \text{ N/mm}^2$$
$$= 11.04 \text{ MPa}$$

Since σ_o is less than σ_b, therefore right hand side of the column will be subjected to compressive stress and the left hand side of the column will be subjected to tensile stress.

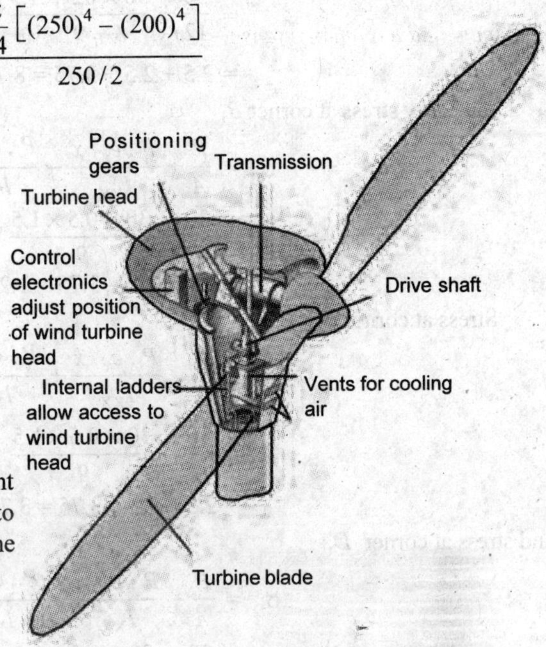

Wind turbine.

Note : This picture is given as additional information and is not a direct example of the current chapter.

∴ Maximum compressive stress,

$$\sigma_c = \sigma_b + \sigma_o = 11.04 + 1.13$$
$$= 12.17 \text{ MPa } \textbf{Ans.}$$

and maximum tensile stress,

$$\sigma_t = \sigma_b - \sigma_o = 11.04 - 1.13 = 9.91 \text{ MPa } \textbf{Ans.}$$

Example 5.21. *A masonry pier of width 4 m and thickness 3 m, supports a load of 30 kN as shown in Fig. 5.23. Find the stresses developed at each corner of the pier.*

Solution. Given: $b = 4$ m ; $d = 3$ m ; $P = 30$ kN ; $e_x = 0.5$ m ; $e_y = 1$ m

We know that cross-sectional area of the pier,

$$A = b \times d = 4 \times 3 = 12 \text{ m}^2$$

Moment of inertia of the pier about X-axis,

$$I_{XX} = \frac{b \cdot d^3}{12} = \frac{4 \times 3^3}{12} = 9 \text{ m}^4$$

and moment of inertia of the pier about Y-axis,

$$I_{YY} = \frac{d \cdot b^3}{12} = \frac{3 \times 4^3}{12} = 16 \text{ m}^4$$

Distance between X-axis and the corners A and B,

$$x = 3/2 = 1.5 \text{ m}$$

Distance between Y-axis and the corners A and C,

$$y = 4/2 = 2 \text{ m}$$

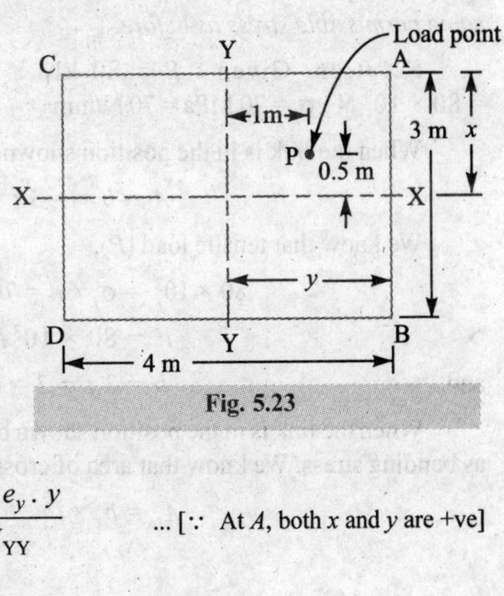

Fig. 5.23

We know that stress at corner A,

$$\sigma_A = \frac{P}{A} + \frac{P \cdot e_x \cdot x}{I_{XX}} + \frac{P \cdot e_y \cdot y}{I_{YY}} \quad \text{... [∵ At } A\text{, both } x \text{ and } y \text{ are +ve]}$$

$$= \frac{30}{12} + \frac{30 \times 0.5 \times 1.5}{9} + \frac{30 \times 1 \times 2}{16}$$

$$= 2.5 + 2.5 + 3.75 = 8.75 \text{ kN/m}^2 \textbf{ Ans.}$$

Similarly stress at corner *B*,

$$\sigma_B = \frac{P}{A} + \frac{P \cdot e_x \cdot x}{I_{XX}} - \frac{P \cdot e_y \cdot y}{I_{YY}} \quad \text{... [} \because \text{ At } B, x \text{ is +ve and } y \text{ is –ve]}$$

$$= \frac{30}{12} + \frac{30 \times 0.5 \times 1.5}{9} - \frac{30 \times 1 \times 2}{16}$$

$$= 2.5 + 2.5 - 3.75 = 1.25 \text{ kN/m}^2 \textbf{ Ans.}$$

Stress at corner *C*,

$$\sigma_C = \frac{P}{A} - \frac{P \cdot e_x \cdot x}{I_{XX}} + \frac{P \cdot e_y \cdot y}{I_{YY}} \quad \text{... [At } C, x \text{ is –ve and } y \text{ is +ve]}$$

$$= \frac{30}{12} - \frac{30 \times 0.5 \times 1.5}{9} + \frac{30 \times 1 \times 2}{16}$$

$$= 2.5 - 2.5 + 3.75 = 3.75 \text{ kN/m}^2 \textbf{ Ans.}$$

and stress at corner *D*,

$$\sigma_D = \frac{P}{A} - \frac{P \cdot e_x \cdot x}{I_{XX}} - \frac{P \cdot e_y \cdot y}{I_{YY}} \quad \text{... [At } D, \text{ both } x \text{ and } y \text{ are – ve]}$$

$$= \frac{30}{12} - \frac{30 \times 0.5 \times 1.5}{9} - \frac{30 \times 1 \times 2}{16}$$

$$= 2.5 - 2.5 - 3.75 = -3.75 \text{ kN/m}^2 = 3.75 \text{ kN/m}^2 \text{ (tensile) } \textbf{Ans.}$$

Example 5.22. *A mild steel link, as shown in Fig. 5.24 by full lines, transmits a pull of 80 kN. Find the dimensions b and t if b = 3t. Assume the permissible tensile stress as 70 MPa. If the original link is replaced by an unsymmetrical one, as shown by dotted lines in Fig. 5.24, having the same thickness t, find the depth b_1, using the same permissible stress as before.*

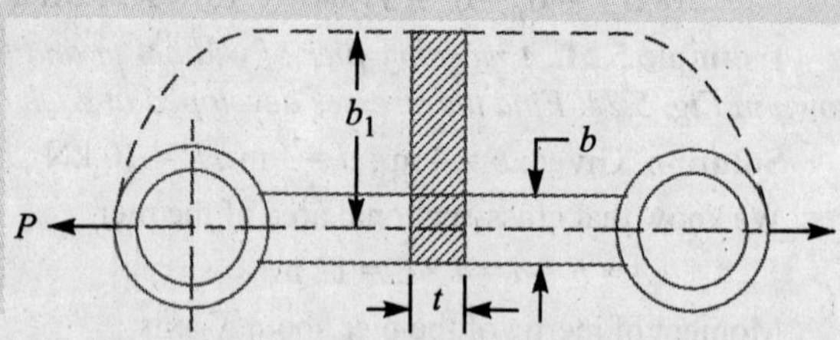

Fig. 5.24

Solution. Given : P = 80 kN = 80 × 10^3 N ; σ_t = 70 MPa = 70 N/mm²

When the link is in the position shown by full lines in Fig. 5.24, the area of cross-section,

$$A = b \times t = 3\,t \times t = 3\,t^2 \quad \text{...(} \because b = 3\,t\text{)}$$

We know that tensile load (P),

$$80 \times 10^3 = \sigma_t \times A = 70 \times 3\,t^2 = 210\,t^2$$

$$\therefore \quad t^2 = 80 \times 10^3 / 210 = 381 \text{ or } t = 19.5 \text{ say } 20 \text{ mm } \textbf{Ans.}$$

and $\quad b = 3\,t = 3 \times 20 = 60$ mm **Ans.**

When the link is in the position shown by dotted lines, it will be subjected to direct stress as well as bending stress. We know that area of cross-section,

$$A_1 = b_1 \times t$$

∴ Direct tensile stress,

$$\sigma_o = \frac{P}{A} = \frac{P}{b_1 \times t}$$

and bending stress, $$\sigma_b = \frac{M}{Z} = \frac{P \cdot e}{Z} = \frac{6\,P \cdot e}{t\,(b_1)^2} \qquad \ldots \left(\because Z = \frac{t\,(b_1)^2}{6}\right)$$

∴ Total stress due to eccentric loading

$$= \sigma_b + \sigma_o = \frac{6\,P \cdot e}{t\,(b_1)^2} + \frac{P}{b_1 \times t} = \frac{P}{t \cdot b_1}\left(\frac{6\,e}{b_1} + 1\right)$$

Since the permissible tensile stress is the same as 70 N/mm², therefore

$$70 = \frac{80 \times 10^3}{20\,b_1}\left(\frac{6 \times b_1}{b_1 \times 2} + 1\right) = \frac{16 \times 10^3}{b_1} \qquad \ldots \left(\because \text{Eccentricity, } e = \frac{b_1}{2}\right)$$

∴ $b_1 = 16 \times 10^3 / 70 = 228.6$ say 230 mm **Ans.**

Example 5.23. *A cast-iron link, as shown in Fig. 5.25, is to carry a load of 20 kN. If the tensile and compressive stresses in the link are not to exceed 25 MPa and 80 MPa respectively, obtain the dimensions of the cross-section of the link at the middle of its length.*

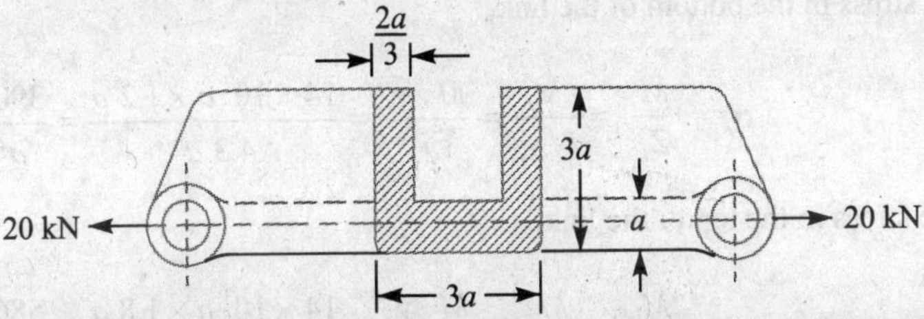

Fig. 5.25

Solution. Given : $P = 20$ kN $= 20 \times 10^3$ N ; $\sigma_{t(max)} = 25$ MPa $= 25$ N/mm² ; $\sigma_{c(max)} = 80$ MPa $= 80$ N/mm²

Since the link is subjected to eccentric loading, therefore there will be direct tensile stress as well as bending stress. The bending stress at the bottom of the link is tensile and in the upper portion is compressive.

We know that cross-sectional area of the link,

$$A = 3a \times a + 2 \times \frac{2a}{3} \times 2a$$

$$= 5.67\,a^2 \text{ mm}^2$$

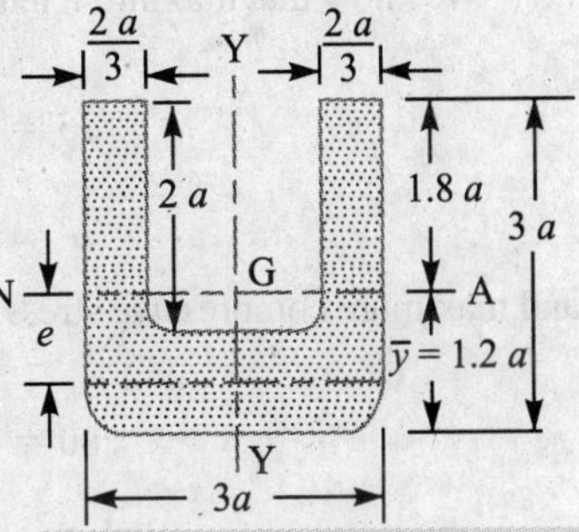

Fig. 5.26

∴ Direct tensile stress,

$$\sigma_o = \frac{P}{A} = \frac{20 \times 10^3}{5.67\,a^2} = \frac{3530}{a^2} \text{ N/mm}^2$$

Now let us find the position of centre of gravity (or neutral axis) in order to find the bending stresses.

Let $\bar{y}$ = Distance of neutral axis (N.A.) from the bottom of the link as shown in Fig. 5.26.

∴ $$\bar{y} = \frac{3a^2 \times \frac{a}{2} + 2 \times \frac{4a^2}{3} \times 2a}{5.67\,a^2} = 1.2\,a \text{ mm}$$

Moment of inertia about N.A.,

$$I = \left[\frac{3a \times a^3}{12} + 3a^2 (1.2a - 0.5a)^2\right] + 2\left[\frac{\frac{2}{3} a \times (2a)^3}{12} + \frac{4a^2}{3} (2a - 1.2a)^2\right]$$

$$= (0.25\, a^4 + 1.47\, a^4) + 2\,(0.44a^4 + 0.85\, a^4) = 4.3\, a^4 \text{ mm}^4$$

Distance of N.A. from the bottom of the link,

$$y_t = \bar{y} = 1.2\, a \text{ mm}$$

Distance of N.A. from the top of the link,

$$y_c = 3\,a - 1.2\,a = 1.8\,a \text{ mm}$$

Eccentricity of the load (*i.e.* distance of N.A. from the point of application of the load),

$$e = 1.2\,a - 0.5\,a = 0.7\,a \text{ mm}$$

We know that bending moment exerted on the section,

$$M = P.e = 20 \times 10^3 \times 0.7\,a = 14 \times 10^3\,a \text{ N-mm}$$

∴ Tensile stress in the bottom of the link,

$$\sigma_t = \frac{M}{Z_t} = \frac{M}{I / y_t} = \frac{M\,.\,y_t}{I} = \frac{14 \times 10^3\,a \times 1.2\,a}{4.3\,a^4} = \frac{3907}{a^2}$$

and compressive stress in the top of the link,

$$\sigma_c = \frac{M}{Z_c} = \frac{M}{I / y_c} = \frac{M\,.\,y_c}{I} = \frac{14 \times 10^3\,a \times 1.8\,a}{4.3\,a^4} = \frac{5860}{a^2}$$

We know that maximum tensile stress $[\sigma_{t\,(max)}]$,

$$25 = \sigma_t + \sigma_c = \frac{3907}{a^2} + \frac{5860}{a^2} = \frac{9767}{a^2}$$

∴ $a^2 = 9767 / 25 = 390.7$ or $a = 19.76$ mm ...(*i*)

and maximum compressive stress $[\sigma_{c(max)}]$,

$$80 = \sigma_c - \sigma_0 = \frac{5860}{a^2} - \frac{3530}{a^2} = \frac{2330}{a^2}$$

∴ $a^2 = 2330 / 80 = 29.12$ or $a = 5.4$ mm ...(*ii*)

We shall take the larger of the two values, *i.e.*

$$a = 19.76 \text{ mm } \textbf{Ans.}$$

Example 5.24. *A horizontal pull P = 5 kN is exerted by the belting on one of the cast iron wall brackets which carry a factory shafting. At a point 75 mm from the wall, the bracket has a T-section as shown in Fig. 5.27. Calculate the maximum stresses in the flange and web of the bracket due to the pull.*

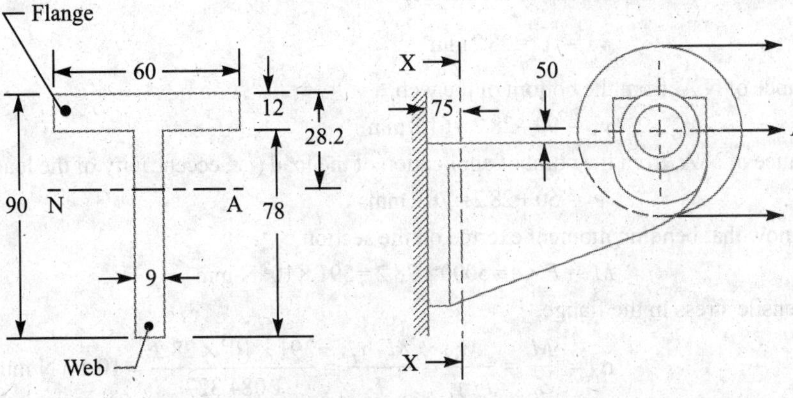

All dimensions in mm.

Fig. 5.27

Solution. Given : Horizontal pull, $P = 5$ kN = 5000 N

Since the section is subjected to eccentric loading, therefore there will be direct tensile stress as well as bending stress. The bending stress at the flange is tensile and in the web is compressive.

We know that cross-sectional area of the section,

$$A = 60 \times 12 + (90 - 12)9 = 720 + 702 = 1422 \text{ mm}^2$$

$$\therefore \text{ Direct tensile stress, } \sigma_0 = \frac{P}{A} = \frac{5000}{1422} = 3.51 \text{ N/mm}^2 = 3.51 \text{ MPa}$$

Now let us find the position of neutral axis in order to determine the bending stresses. The neutral axis passes through the centre of gravity of the section.

Let $\bar{y}$ = Distance of centre of gravity (*i.e.* neutral axis) from top of the flange.

$$\therefore \quad \bar{y} = \frac{60 \times 12 \times \frac{12}{2} + 78 \times 9 \left(12 + \frac{78}{2}\right)}{720 + 702} = 28.2 \text{ mm}$$

Moment of inertia of the section about N.A.,

$$I = \left[\frac{60\,(12)^3}{12} + 720\,(28.2 - 6)^2\right] + \left[\frac{9\,(78)^3}{12} + 702\,(51 - 28.2)^2\right]$$

$$= (8640 + 354\,845) + (355\,914 + 364\,928) = 1\,084\,327 \text{ mm}^4$$

This picture shows a reconnoissance helicopter of air force. Its dark complexion absorbs light that falls on its surface. The flat and sharp edges deflect radar waves and they do not return back to the radar. These factors make it difficult to detect the helicopter.

Note : This picture is given as additional information and is not a direct example of the current chapter.

Distance of N.A. from the top of the flange,

$$y_t = \bar{y} = 28.2 \text{ mm}$$

Distance of N.A. from the bottom of the web,

$$y_c = 90 - 28.2 = 61.8 \text{ mm}$$

Distance of N.A. from the point of application of the load (*i.e.* eccentricity of the load),

$$e = 50 + 28.2 = 78.2 \text{ mm}$$

We know that bending moment exerted on the section,

$$M = P \times e = 5000 \times 78.2 = 391 \times 10^3 \text{ N-mm}$$

∴ Tensile stress in the flange,

$$\sigma_t = \frac{M}{Z_t} = \frac{M}{I/y_t} = \frac{M \cdot y_t}{I} = \frac{391 \times 10^3 \times 28.2}{1\,084\,327} = 10.17 \text{ N/mm}^2$$
$$= 10.17 \text{ MPa}$$

and compressive stress in the web,

$$\sigma_c = \frac{M}{Z_c} = \frac{M}{I/y_c} = \frac{M \cdot y_c}{I} = \frac{391 \times 10^3 \times 61.8}{1\,084\,327} = 22.28 \text{ N/mm}^2$$
$$= 22.28 \text{ MPa}$$

We know that maximum tensile stress in the flange,

$$\sigma_{t(max)} = \sigma_b + \sigma_o = \sigma_t + \sigma_o = 10.17 + 3.51 = 13.68 \text{ MPa} \quad \textbf{Ans.}$$

and maximum compressive stress in the flange,

$$\sigma_{c(max)} = \sigma_b - \sigma_o = \sigma_c - \sigma_o = 22.28 - 3.51 = 18.77 \text{ MPa} \quad \textbf{Ans.}$$

Example 5.25. *A mild steel bracket as shown in Fig. 5.28, is subjected to a pull of 6000 N acting at 45° to its horizontal axis. The bracket has a rectangular section whose depth is twice the thickness. Find the cross-sectional dimensions of the bracket, if the permissible stress in the material of the bracket is limited to 60 MPa.*

Solution. Given : $P = 6000$ N ; $\theta = 45°$; $\sigma = 60$ MPa $= 60$ N/mm²

Let t = Thickness of the section in mm, and

b = Depth or width of the section = $2\,t$ ...(Given)

We know that area of cross-section,

$$A = b \times t = 2\,t \times t = 2\,t^2 \text{ mm}^2$$

and section modulus,

$$Z = \frac{t \times b^2}{6} = \frac{t\,(2t)^2}{6} = \frac{4\,t^3}{6} \text{ mm}^3$$

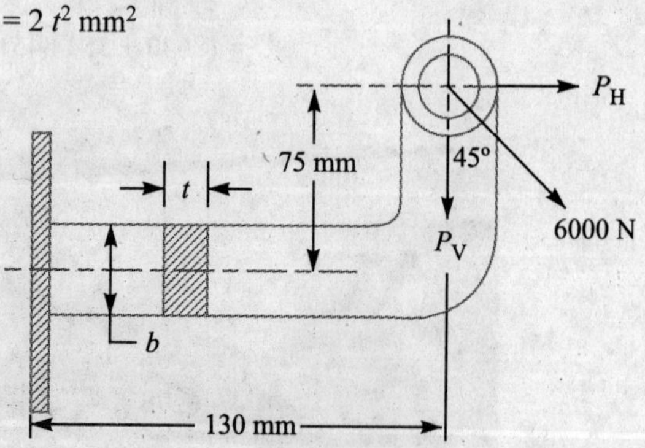

Fig. 5.28

Horizontal component of the load,

$$P_H = 6000 \cos 45° = 6000 \times 0.707 = 4242 \text{ N}$$

∴ Bending moment due to horizontal component of the load,

$$M_H = P_H \times 75 = 4242 \times 75 = 318\,150 \text{ N-mm}$$

A little consideration will show that the bending moment due to the horizontal component of the load induces tensile stress on the upper surface of the bracket and compressive stress on the lower surface of the bracket.

∴ Maximum bending stress on the upper surface due to horizontal component,

$$\sigma_{bH} = \frac{M_H}{Z} = \frac{318\ 150 \times 6}{4\ t^3}$$

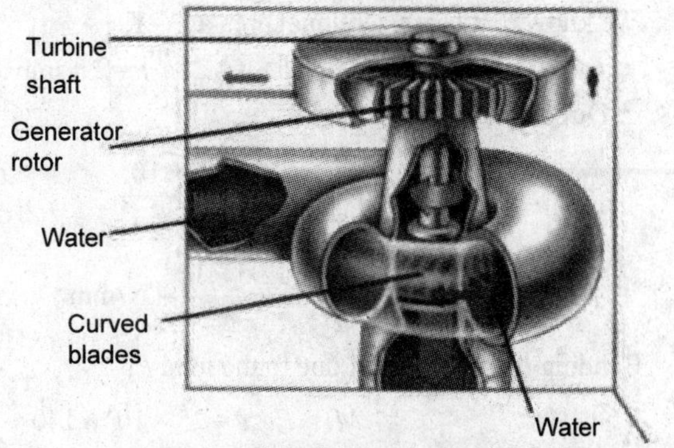

Schematic of a hydel turbine.

Note : This picture is given as additional information and is not a direct example of the current chapter.

$$= \frac{477\ 225}{t^3} \text{ N/mm}^2 \text{ (tensile)}$$

Vertical component of the load,

$$P_V = 6000 \sin 45° = 6000 \times 0.707 = 4242 \text{ N}$$

∴ Direct stress due to vertical component,

$$\sigma_{oV} = \frac{P_V}{A} = \frac{4242}{2t^2} = \frac{2121}{t^2} \text{ N/mm}^2 \text{ (tensile)}$$

Bending moment due to vertical component of the load,

$$M_V = P_V \times 130 = 4242 \times 130 = 551\ 460 \text{ N-mm}$$

This bending moment induces tensile stress on the upper surface and compressive stress on the lower surface of the bracket.

∴ Maximum bending stress on the upper surface due to vertical component,

$$\sigma_{bV} = \frac{M_V}{Z} = \frac{551\ 460 \times 6}{4\ t^3} = \frac{827\ 190}{t^3} \text{ N/mm}^2 \text{ (tensile)}$$

and total tensile stress on the upper surface of the bracket,

$$\sigma = \frac{477\ 225}{t^3} + \frac{2121}{t^2} + \frac{827\ 190}{t^3} = \frac{1\ 304\ 415}{t^3} + \frac{2121}{t^2}$$

Since the permissible stress (σ) is 60 N/mm², therefore

$$\frac{1\ 304\ 415}{t^3} + \frac{2121}{t^2} = 60 \text{ or } \frac{21\ 740}{t^3} + \frac{35.4}{t^2} = 1$$

∴ $t = 28.4$ mm **Ans.** ... (By hit and trial)

and $b = 2t = 2 \times 28.4 = 56.8$ mm **Ans.**

Example 5.26. *A C-clamp as shown in Fig. 5.29, carries a load P = 25 kN. The cross-section of the clamp at X-X is rectangular having width equal to twice thickness. Assuming that the clamp is made of steel casting with an allowable stress of 100 MPa, find its dimensions. Also determine the stresses at sections Y-Y and Z-Z.*

Solution. Given : $P = 25$ kN $= 25 \times 10^3$ N ; $\sigma_{t(max)} = 100$ MPa $= 100$ N/mm²

Dimensions at X-X

Let t = Thickness of the section at X-X in mm, and

b = Width of the section at X-X in mm = $2t$ ...(Given)

We know that cross-sectional area at *X-X*,

$$A = b \times t = 2\,t \times t = 2\,t^2 \text{ mm}^2$$

∴ Direct tensile stress at *X-X*,

$$\sigma_o = \frac{P}{A} = \frac{25 \times 10^3}{2\,t^2}$$

$$= \frac{12.5 \times 10^3}{t^3} \text{ N/mm}^2$$

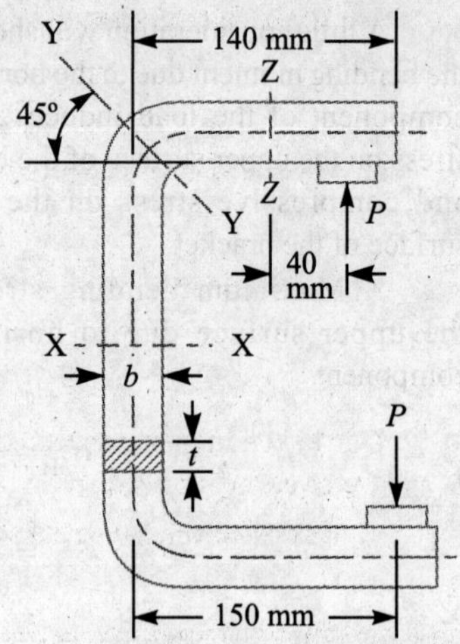

Fig. 5.29

Bending moment at *X-X* due to the load *P*,

$$M = P \times e = 25 \times 10^3 \times 140$$

$$= 3.5 \times 10^6 \text{ N-mm}$$

Section modulus,
$$Z = \frac{t\,.\,b^2}{6} = \frac{t\,(2t)^2}{6} = \frac{4\,t^3}{6} \text{ mm}^3 \quad ...(\because b = 2t)$$

∴ Bending stress at *X-X*,

$$\sigma_b = \frac{M}{Z} = \frac{3.5 \times 10^6 \times 6}{4\,t^3} = \frac{5.25 \times 10^6}{t^3} \text{ N/mm}^2 \text{ (tensile)}$$

We know that the maximum tensile stress $[\sigma_{t(max)}]$,

$$100 = \sigma_o + \sigma_b = \frac{12.5 \times 10^3}{t^2} + \frac{5.25 \times 10^6}{t^3}$$

or
$$\frac{125}{t^2} + \frac{52.5 \times 10^3}{t^3} - 1 = 0$$

∴ $t = 38.5$ mm **Ans.** ...(By hit and trial)

and $b = 2\,t = 2 \times 38.5 = 77$ mm **Ans.**

Stresses at section Y-Y

Since the cross-section of frame is uniform throughout, therefore cross-sectional area of the frame at section *Y-Y*,

$$A = b \sec 45° \times t = 77 \times 1.414 \times 38.5 = 4192 \text{ mm}^2$$

Component of the load perpendicular to the section

$$= P \cos 45° = 25 \times 10^3 \times 0.707 = 17\,675 \text{ N}$$

This component of the load produces uniform tensile stress over the section.

∴ Uniform tensile stress over the section,

$$\sigma = 17\,675 / 4192 = 4.2 \text{ N/mm}^2 = 4.2 \text{ MPa}$$

Component of the load parallel to the section

$$= P \sin 45° = 25 \times 10^3 \times 0.707 = 17\,675 \text{ N}$$

This component of the load produces uniform shear stress over the section.

∴ Uniform shear stress over the section,

$$\tau = 17\,675 / 4192 = 4.2 \text{ N/mm}^2 = 4.2 \text{ MPa}$$

We know that section modulus,

$$Z = \frac{t\,(b \sec 45°)^2}{6} = \frac{38.5\,(77 \times 1.414)^2}{6} = 76 \times 10^3 \text{ mm}^3$$

Bending moment due to load (P) over the section Y-Y,

$$M = 25 \times 10^3 \times 140 = 3.5 \times 10^6 \text{ N-mm}$$

∴ Bending stress over the section,

$$\sigma_b = \frac{M}{Z} = \frac{3.5 \times 10^6}{76 \times 10^3} = 46 \text{ N/mm}^2 = 46 \text{ MPa}$$

Due to bending, maximum tensile stress at the inner corner and the maximum compressive stress at the outer corner is produced.

∴ Maximum tensile stress at the inner corner,

$$\sigma_t = \sigma_b + \sigma_o = 46 + 4.2 = 50.2 \text{ MPa}$$

and maximum compressive stress at the outer corner,

$$\sigma_c = \sigma_b - \sigma_o = 46 - 4.2 = 41.8 \text{ MPa}$$

Since the shear stress acts at right angles to the tensile and compressive stresses, therefore maximum principal stress (tensile) on the section Y-Y at the inner corner

$$= \frac{\sigma_t}{2} + \frac{1}{2}\left[\sqrt{(\sigma_t)^2 + 4\,\tau^2}\right] = \frac{50.2}{2} + \frac{1}{2}\left[\sqrt{(50.2)^2 + 4 \times (4.2)^2}\right] \text{ MPa}$$

$$= 25.1 + 25.4 = 50.5 \text{ MPa } \textbf{Ans.}$$

and maximum principal stress (compressive) on section Y-Y at outer corner

$$= \frac{\sigma_c}{2} + \frac{1}{2}\left[\sqrt{(\sigma_c)^2 + 4\,\tau^2}\right] = \frac{41.8}{2} + \frac{1}{2}\left[\sqrt{(41.8)^2 + 4 \times (4.2)^2}\right] \text{ MPa}$$

$$= 20.9 + 21.3 = 42.2 \text{ MPa } \textbf{Ans.}$$

Maximum shear stress $= \frac{1}{2}\left[\sqrt{(\sigma_t)^2 + 4\,\tau^2}\right] = \frac{1}{2}\left[\sqrt{(50.2)^2 + 4 \times (4.2)^2}\right] = 25.4 \text{ MPa}$ **Ans.**

Stresses at section Z-Z

We know that bending moment at section Z-Z,

$$= 25 \times 10^3 \times 40 = 1 \times 10^6 \text{ N-mm}$$

and section modulus,

$$Z = \frac{t\,.\,b^2}{6} = \frac{38.5\,(77)^2}{6} = 38 \times 10^3 \text{ mm}^3$$

∴ Bending stress at section Z-Z,

$$\sigma_b = \frac{M}{Z} = \frac{1 \times 10^6}{38 \times 10^3} = 26.3 \text{ N/mm}^2 = 26.3 \text{ MPa } \textbf{Ans.}$$

The bending stress is tensile at the inner edge and compressive at the outer edge. The magnitude of both these stresses is 26.3 MPa. At the neutral axis, there is only transverse shear stress. The shear stress at the inner and outer edges will be zero.

We know that *maximum transverse shear stress,

$$\tau_{max} = 1.5 \times \text{Average shear stress} = 1.5 \times \frac{P}{b.t} = 1.5 \times \frac{25 \times 10^3}{77 \times 38.5}$$

$$= 12.65 \text{ N/mm}^2 = 12.65 \text{ MPa } \textbf{Ans.}$$

* Refer Art. 5.16

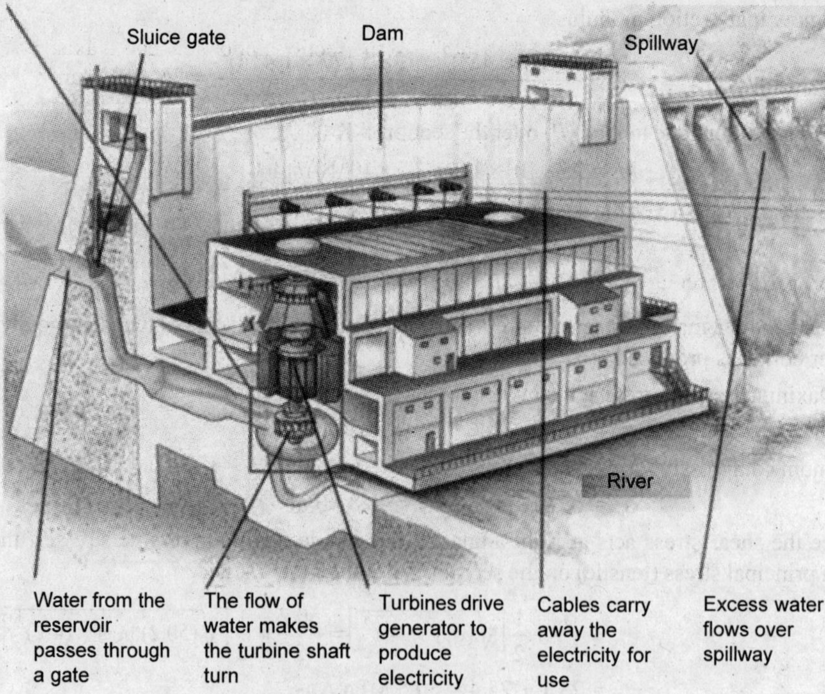

General layout of a hydroelectric plant.

Note : This picture is given as additional information and is not a direct example of the current chapter.

5.16 Shear Stresses in Beams

In the previous article, we have assumed that no shear force is acting on the section. But, in actual practice, when a beam is loaded, the shear force at a section always comes into play along with the bending moment. It has been observed that the effect of the shear stress, as compared to the bending stress, is quite negligible and is of not much importance. But, sometimes, the shear stress at a section is of much importance in the design. It may be noted that the shear stress in a beam is not uniformly distributed over the cross-section but varies from zero at the outer fibres to a maximum at the neutral surface as shown in Fig. 5.30 and Fig. 5.31.

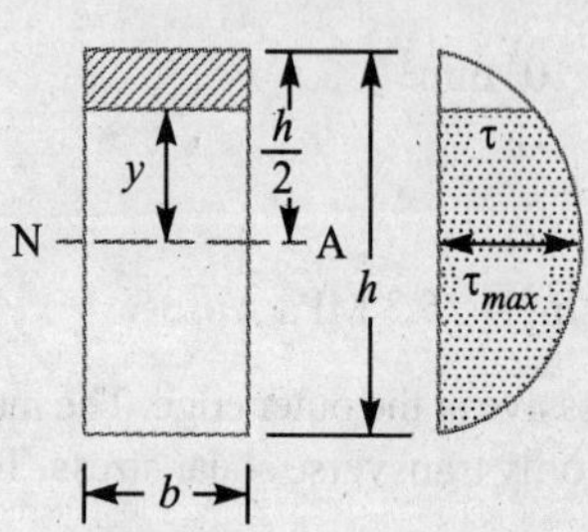

Fig. 5.30. Shear stress in a rectangular beam.

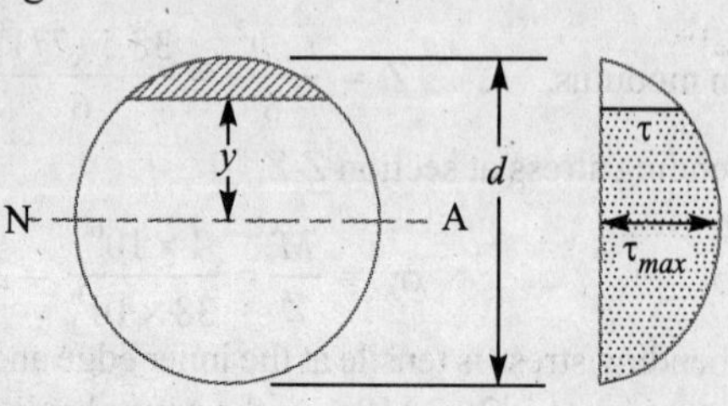

Fig. 5.31. Shear stress in a circular beam.

The shear stress at any section acts in a plane at right angle to the plane of the bending stress and its value is given by

$$\tau = \frac{F}{I\,.\,b} \times A.\bar{y}$$

where F = Vertical shear force acting on the section,

I = Moment of inertia of the section about the neutral axis,

b = Width of the section under consideration,

A = Area of the beam above neutral axis, and

$\bar{y}$ = Distance between the C.G. of the area and the neutral axis.

The following values of maximum shear stress for different cross-section of beams may be noted :

1. For a beam of rectangular section, as shown in Fig. 5.30, the shear stress at a distance y from neutral axis is given by

$$\tau = \frac{F}{2I}\left(\frac{h^2}{4} - y^2\right) = \frac{3F}{2b \cdot h^3}(h^2 - 4y^2) \qquad \ldots\left[\because I = \frac{b \cdot h^3}{12}\right]$$

and maximum shear stress,

$$\tau_{max} = \frac{3F}{2b \cdot h} \qquad \ldots\left(\text{Substituting } y = \frac{h}{2}\right)$$

$$= 1.5\,\tau_{(average)} \qquad \ldots\left[\because \tau_{(average)} = \frac{F}{\text{Area}} = \frac{F}{b.h}\right]$$

The distribution of stress is shown in Fig. 5.30.

2. For a beam of circular section as shown in Fig. 5.31, the shear stress at a distance y from neutral axis is given by

$$\tau = \frac{F}{3I}\left(\frac{d^2}{4} - y^2\right) = \frac{16\,F}{3\,\pi\,d^4}(d^2 - 4y^2)$$

and the maximum shear stress,

$$\tau_{max} = \frac{4F}{3 \times \frac{\pi}{4} d^2} \qquad \ldots\left(\text{Substituting } y = \frac{d}{2}\right)$$

$$= \frac{4}{3}\,\tau_{(average)} \qquad \ldots\left[\because \tau_{(average)} = \frac{F}{\text{Area}} = \frac{F}{\frac{\pi}{4} d^2}\right]$$

The distribution of stress is shown in Fig. 5.31.

3. For a beam of *I*-section as shown in Fig. 5.32, the maximum shear stress occurs at the neutral axis and is given by

$$\tau_{max} = \frac{F}{I \cdot b}\left[\frac{B}{8}(H^2 - h^2) + \frac{b.h^2}{8}\right]$$

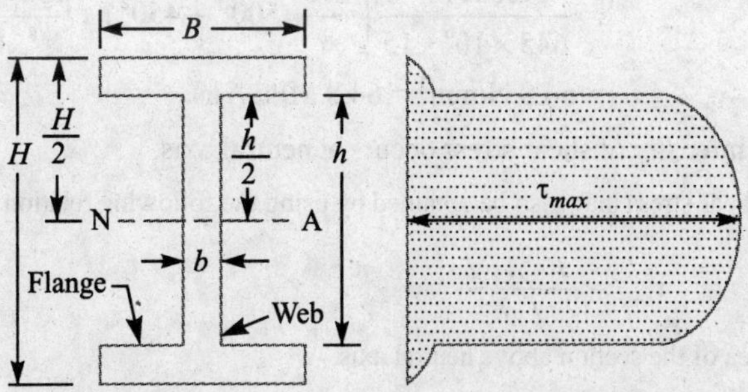

Fig. 5.32

Shear stress at the joint of the web and the flange

$$= \frac{F}{8I}(H^2 - h^2)$$

and shear stress at the junction of the top of the web and bottom of the flange

$$= \frac{F}{8I} \times \frac{B}{b}(H^2 - h^2)$$

The distribution of stress is shown in Fig. 5.32.

Example 5.27. *A beam of I-section 500 mm deep and 200 mm wide has flanges 25 mm thick and web 15 mm thick, as shown in Fig. 5.33 (a). It carries a shearing force of 400 kN. Find the maximum intensity of shear stress in the section, assuming the moment of inertia to be 645×10^6 mm^4. Also find the shear stress at the joint and at the junction of the top of the web and bottom of the flange.*

Solution. Given : $H = 500$ mm ; $B = 200$ mm ; $h = 500 - 2 \times 25 = 450$ mm ; $b = 15$ mm ; $F = 400$ kN $= 400 \times 10^3$ N ; $I = 645 \times 10^6$ mm^4

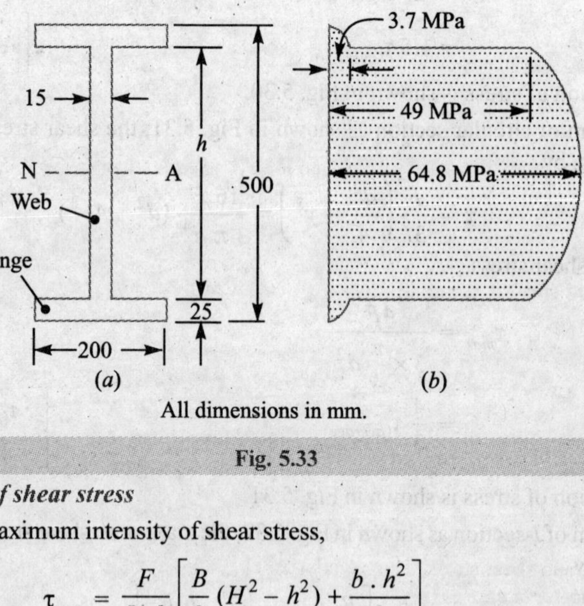

All dimensions in mm.

Fig. 5.33

Maximum intensity of shear stress

We know that maximum intensity of shear stress,

$$\tau_{max} = \frac{F}{I \,.\, b}\left[\frac{B}{8}(H^2 - h^2) + \frac{b \,.\, h^2}{8}\right]$$

$$= \frac{400 \times 10^3}{645 \times 10^6 \times 15}\left[\frac{200}{8}(500^2 - 450^2) + \frac{15 \times 450^2}{8}\right] \text{N/mm}^2$$

$$= 64.8 \text{ N/mm}^2 = 64.8 \text{ MPa } \textbf{Ans.}$$

The maximum intensity of shear stress occurs at neutral axis.

Note : The maximum shear stress may also be obtained by using the following relation :

$$\tau_{max} = \frac{F \,.\, A \,.\, \bar{y}}{I \,.\, b}$$

We know that area of the section above neutral axis,

$$A = 200 \times 25 + \frac{450}{2} \times 15 = 8375 \text{ mm}^2$$

Distance between the centre of gravity of the area and neutral axis,

$$\bar{y} = \frac{200 \times 25\,(225 + 12.5) + 225 \times 15 \times 112.5}{8375} = 187 \text{ mm}$$

$$\therefore \quad \tau_{max} = \frac{400 \times 10^3 \times 8375 \times 187}{645 \times 10^6 \times 15} = 64.8 \text{ N/mm}^2 = 64.8 \text{ MPa } \textbf{Ans.}$$

Shear stress at the joint of the web and the flange

We know that shear stress at the joint of the web and the flange

$$= \frac{F}{8I}(H^2 - h^2) = \frac{400 \times 10^3}{8 \times 645 \times 10^6}\left[(500)^2 - (450)^2\right] \text{N/mm}^2$$

$$= 3.7 \text{ N/mm}^2 = 3.7 \text{ MPa } \textbf{Ans.}$$

Shear stress at the junction of the top of the web and bottom of the flange

We know that shear stress at junction of the top of the web and bottom of the flange

$$= \frac{F}{8I} \times \frac{B}{b}(H^2 - h^2) = \frac{400 \times 10^3}{8 \times 645 \times 10^6} \times \frac{200}{15}\left[(500)^2 - (450)^2\right] \text{N/mm}^2$$

$$= 49 \text{ N/mm}^2 = 49 \text{ MPa } \textbf{Ans.}$$

The stress distribution is shown in Fig. 5.33 (*b*)

EXERCISES

1. A steel shaft 50 mm diameter and 500 mm long is subjected to a twisting moment of 1100 N-m, the total angle of twist being 0.6°. Find the maximum shearing stress developed in the shzaft and modulus of rigidity. **[Ans. 44.8 MPa; 85.6 kN/m^2]**
2. A shaft is transmitting 100 kW at 180 r.p.m. If the allowable stress in the material is 60 MPa, find the suitable diameter for the shaft. The shaft is not to twist more than 1° in a length of 3 metres. Take C = 80 GPa. **[Ans. 105 mm]**
3. Design a suitable diameter for a circular shaft required to transmit 90 kW at 180 r.p.m. The shear stress in the shaft is not to exceed 70 MPa and the maximum torque exceeds the mean by 40%. Also find the angle of twist in a length of 2 metres. Take C = 90 GPa. **[Ans. 80 mm; 2.116°]**
4. Design a hollow shaft required to transmit 11.2 MW at a speed of 300 r.p.m. The maximum shear stress allowed in the shaft is 80 MPa and the ratio of the inner diameter to outer diameter is 3/4. **[Ans. 240 mm; 320 mm]**
5. Compare the weights of equal lengths of hollow shaft and solid shaft to transmit a given torque for the same maximum shear stress. The material for both the shafts is same and inside diameter is 2/3 of outside diameter in case of hollow shaft. **[Ans. 0.56]**
6. A spindle as shown in Fig. 5.34, is a part of an industrial brake and is loaded as shown. Each load P is equal to 4 kN and is applied at the mid point of its bearing. Find the diameter of the spindle, if the maximum bending stress is 120 MPa. **[Ans. 22 mm]**

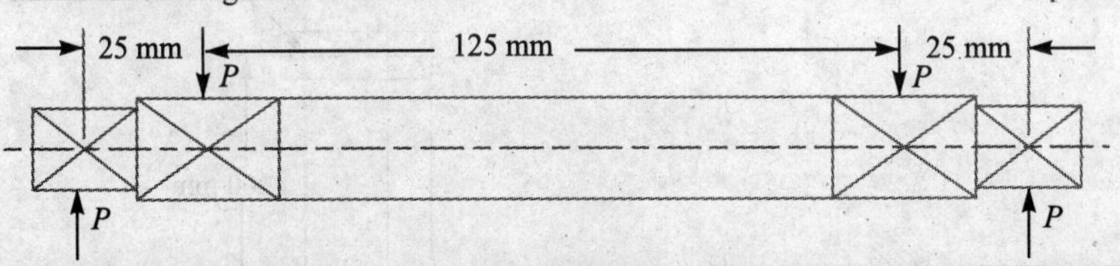

Fig. 5.34

7. A cast iron pulley transmits 20 kW at 300 r.p.m. The diameter of the pulley is 550 mm and has four straight arms of elliptical cross-section in which the major axis is twice the minor axis. Find the dimensions of the arm, if the allowable bending stress is 15 MPa. **[Ans. 60 mm; 30 mm]**

8. A shaft is supported in bearings, the distance between their centres being 1 metre. It carries a pulley in the centre and it weighs 1 kN. Find the diameter of the shaft, if the permissible bending stress for the shaft material is 40 MPa. **[Ans. 40 mm]**

9. A punch press, used for stamping sheet metal, has a punching capacity of 50 kN. The section of the frame is as shown in Fig. 5.35. Find the resultant stress at the inner and outer fibre of the section. **[Ans. 28.3 MPa (tensile); 17.7 MPa (compressive)]**

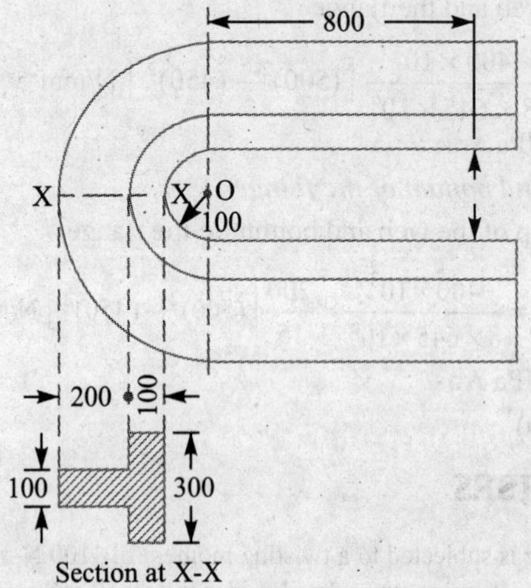

All dimensions in mm.

Fig. 5.35

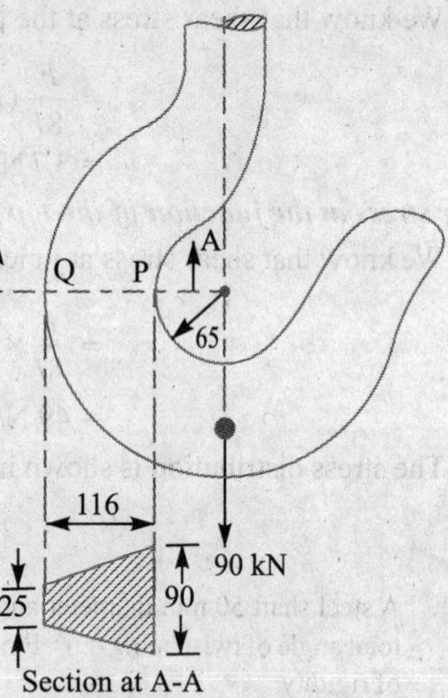

All dimensions in mm.

Fig. 5.36

10. A crane hook has a trapezoidal section at *A-A* as shown in Fig. 5.36. Find the maximum stress at points *P* and *Q*. **[Ans. 118 MPa (tensile); 62 MPa (compressive)]**

11. A rotating shaft of 16 mm diameter is made of plain carbon steel. It is subjected to axial load of 5000 N, a steady torque of 50 N-m and maximum bending moment of 75 N-m. Calculate the factor of safety available based on 1. Maximum normal stress theory; and 2. Maximum shear stress theory.

Assume yield strength as 400 MPa for plain carbon steel. If all other data remaining same, what maximum yield strength of shaft material would be necessary using factor of safety of 1.686 and maximum distortion energy theory of failure. Comment on the result you get. **[Ans. 1.752; 400 MPa]**

12. A hand cranking lever, as shown in Fig. 5.37, is used to start a truck engine by applying a force $F = 400$ N. The material of the cranking lever is 30C8 for which yield strength = 320 MPa; Ultimate tensile strength = 500 MPa ; Young's modulus = 205 GPa ; Modulus of rigidity = 84 GPa and poisson's ratio = 0.3.

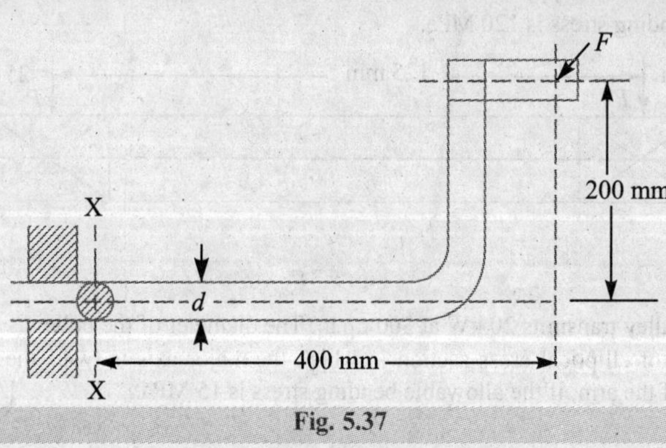

Fig. 5.37

Assuming factor of safety to be 4 based on yield strength, design the diameter 'd' of the lever at section X-X near the guide bush using : 1. Maximum distortion energy theory; and 2. Maximum shear stress theory. **[Ans. 28.2 mm; 28.34 mm]**

13. An offset bar is loaded as shown in Fig. 5.38. The weight of the bar may be neglected. Find the maximum offset (*i.e.*, the dimension x) if allowable stress in tension is limited to 70 MPa. **[Ans. 418 mm]**

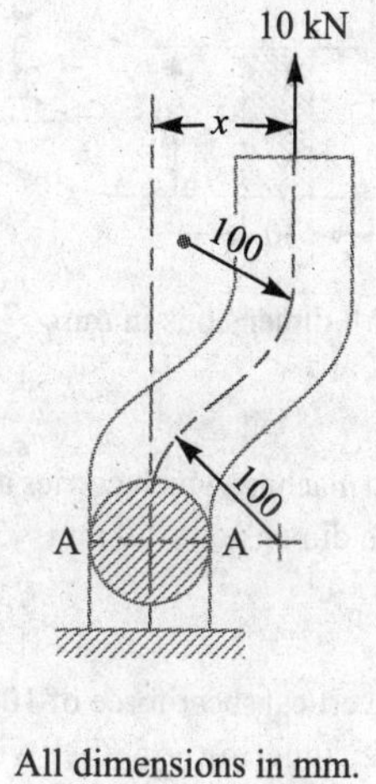

Fig. 5.38

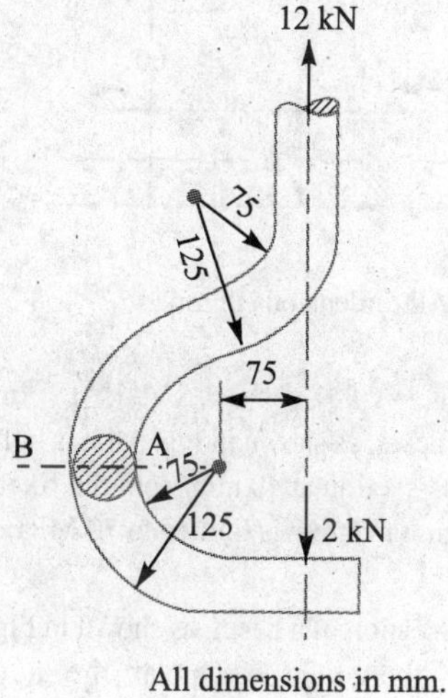

Fig. 5.39

14. A crane hook made from a 50 mm diameter bar is shown in Fig. 5.39. Find the maximum tensile stress and specify its location. **[Ans. 35.72 MPa at A]**

15. An overhang crank, as shown in Fig. 5.40 carries a tangential load of 10 kN at the centre of the crankpin. Find the maximum principal stress and the maximum shear stress at the centre of the crankshaft bearing. **[Ans. 29.45 MPa; 18.6 MPa]**

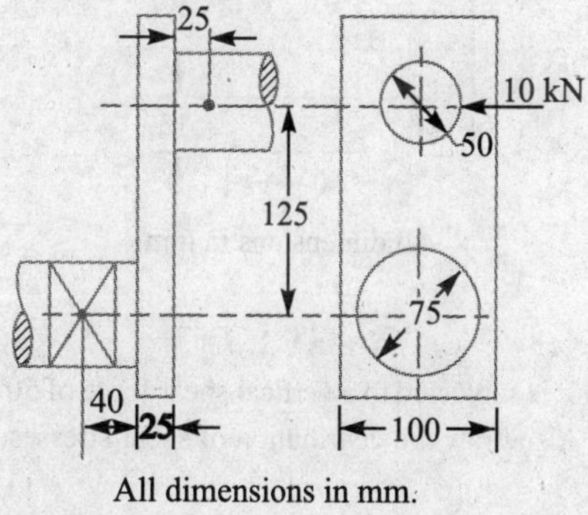

Fig. 5.40

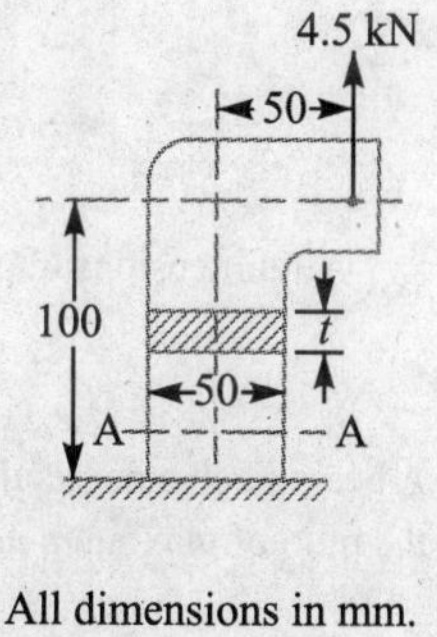

Fig. 5.41

16. A steel bracket is subjected to a load of 4.5 kN, as shown in Fig. 5.41. Determine the required thickness of the section at A-A in order to limit the tensile stress to 70 MPa. **[Ans. 9 mm]**

17. A wall bracket, as shown in Fig. 5.42, is subjected to a pull of $P = 5$ kN, at 60° to the vertical. The cross-section of bracket is rectangular having $b = 3t$. Determine the dimensions b and t if the stress in the material of the bracket is limited to 28 MPa. **[Ans. 75 mm; 25 mm]**

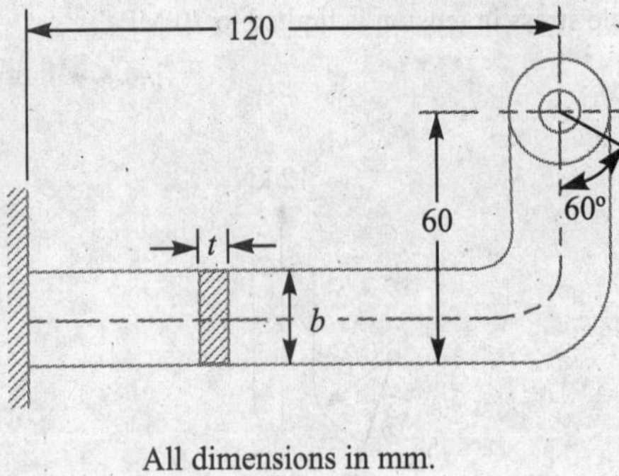

Fig. 5.42

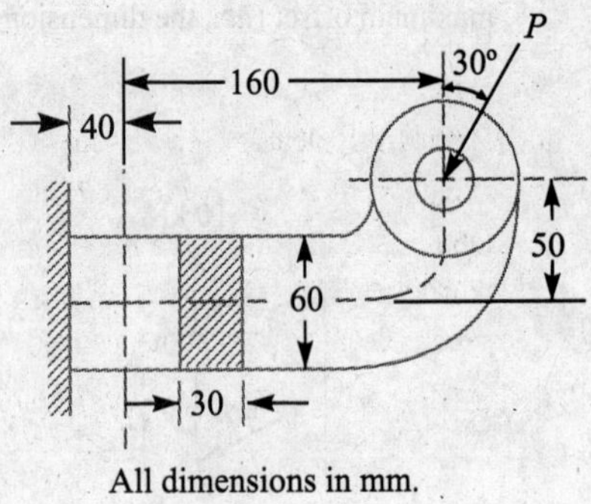

Fig. 5.43

18. A bracket, as shown in Fig. 5.43, is bolted to the framework of a machine which carries a load P. The cross-section at 40 mm from the fixed end is rectangular with dimensions, 60 mm × 30 mm. If the maximum stress is limited to 70 MPa, find the value of P.

[Ans. 3000 N]

19. A T-section of a beam, as shown in Fig. 5.44, is subjected to a vertical shear force of 100 kN. Calculate the shear stress at the neutral axis and at the junction of the web and the flange. The moment of inertia at the neutral axis is 113.4×10^6 mm^4.

[Ans. 11.64 MPa; 11 MPa; 2.76 MPa]

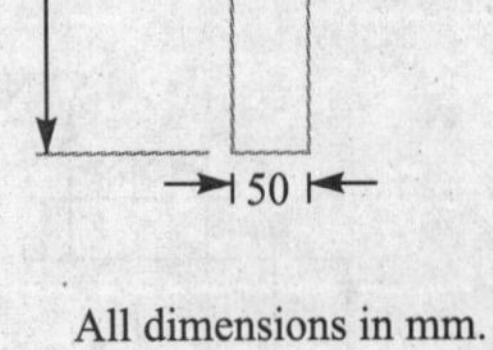

Fig. 5.44

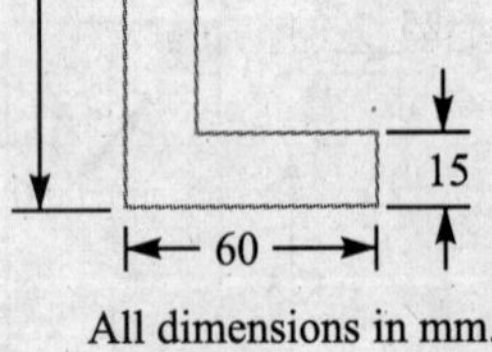

Fig. 5.45

20. A beam of channel section, as shown in Fig. 5.45, is subjected to a vertical shear force of 50 kN. Find the ratio of maximum and mean shear stresses. Also draw the distribution of shear stresses.

[Ans. 2.22]

QUESTIONS

1. Derive a relation for the shear stress developed in a shaft, when it is subjected to torsion.
2. State the assumptions made in deriving a bending formula.

3. Prove the relation: $M/I = \sigma/y = E/R$

   where M = Bending moment; I = Moment of inertia; σ = Bending stress in a fibre at a distance y from the neutral axis; E = Young's modulus; and R = Radius of curvature.
4. Write the relations used for maximum stress when a machine member is subjected to tensile or compressive stresses along with shearing stresses.
5. Write short note on maximum shear stress theory *verses* maximum strain energy theory.
6. Distinguish clearly between direct stress and bending stress.
7. What is meant by eccentric loading and eccentricity?
8. Obtain a relation for the maximum and minimum stresses at the base of a symmetrical column, when it is subjected to

   (*a*) an eccentric load about one axis, and (*b*) an eccentric load about two axes.

OBJECTIVE TYPE QUESTIONS

1. When a machine member is subjected to torsion, the torsional shear stress set up in the member is
   (*a*) zero at both the centroidal axis and outer surface of the member
   (*b*) Maximum at both the centroidal axis and outer surface of the member
   (*c*) zero at the centroidal axis and maximum at the outer surface of the member
   (*d*) none of the above
2. The torsional shear stress on any cross-section normal to the axis is .... the distance from the centre of the axis.
   (*a*) directly proportional to   (*b*) inversely proportional to
3. The neutral axis of a beam is subjected to
   (*a*) zero stress   (*b*) maximum tensile stress
   (*c*) maximum compressive stress   (*d*) maximum shear stress
4. At the neutral axis of a beam,
   (*a*) the layers are subjected to maximum bending stress
   (*b*) the layers are subjected to tension   (*c*) the layers are subjected to compression
   (*d*) the layers do not undergo any strain
5. The bending stress in a curved beam is
   (*a*) zero at the centroidal axis   (*b*) zero at the point other than centroidal axis
   (*c*) maximum at the neutral axis   (*d*) none of the above
6. The maximum bending stress, in a curved beam having symmetrical section, always occur, at the
   (*a*) centroidal axis   (*b*) neutral axis
   (*c*) inside fibre   (*d*) outside fibre
7. If d = diameter of solid shaft and τ = permissible stress in shear for the shaft material, then torsional strength of shaft is written as
   (*a*) $\frac{\pi}{32} d^4 \tau$ (*b*) $d \log_e \tau$
   (*c*) $\frac{\pi}{16} d^3 \tau$ (*d*) $\frac{\pi}{32} d^3 \tau$
8. If d_i and d_o are the inner and outer diameters of a hollow shaft, then its polar moment of inertia is
   (*a*) $\frac{\pi}{32}\left[(d_o)^4 - (d_i)^4\right]$ (*b*) $\frac{\pi}{32}\left[(d_o)^3 - (d_i)^3\right]$
   (*c*) $\frac{\pi}{32}\left[(d_o)^2 - (d_i)^2\right]$ (*d*) $\frac{\pi}{32}(d_o - d_i)$

9. Two shafts under pure torsion are of identical length and identical weight and are made of same material. The shaft A is solid and the shaft B is hollow. We can say that
   (*a*) shaft B is better than shaft A
   (*b*) shaft A is better than shaft B
   (*c*) both the shafts are equally good
10. A solid shaft transmits a torque T. The allowable shear stress is τ. The diameter of the shaft is
   (*a*) $\sqrt[3]{\dfrac{16\,T}{\pi\,\tau}}$ (*b*) $\sqrt[3]{\dfrac{32\,T}{\pi\,\tau}}$
   (*c*) $\sqrt[3]{\dfrac{64\,T}{\pi\,\tau}}$ (*d*) $\sqrt[3]{\dfrac{16\,T}{\tau}}$
11. When a machine member is subjected to a tensile stress (σ_t) due to direct load or bending and a shear stress (τ) due to torsion, then the maximum shear stress induced in the member will be
   (*a*) $\frac{1}{2}\left[\sqrt{(\sigma_t)^2 + 4\,\tau^2}\right]$ (*b*) $\frac{1}{2}\left[\sqrt{(\sigma_t)^2 - 4\,\tau^2}\right]$
   (*c*) $\left[\sqrt{(\sigma_t)^2 + 4\,\tau^2}\right]$ (*d*) $(\sigma_t)^2 + 4\,\tau^2$
12. Rankine's theory is used for
   (*a*) brittle materials (*b*) ductile materials
   (*c*) elastic materials (*d*) plastic materials
13. Guest's theory is used for
   (*a*) brittle materials (*b*) ductile materials
   (*c*) elastic materials (*d*) plastic materials
14. At the neutral axis of a beam, the shear stress is
   (*a*) zero (*b*) maximum
   (*c*) minimum
15. The maximum shear stress developed in a beam of rectangular section is .... the average shear stress.
   (*a*) equal to (*b*) $\frac{4}{3}$ times
   (*c*) 1.5 times

ANSWERS

1. (*b*)	2. (*a*)	3. (*a*)	4. (*d*)	5. (*b*)
6. (*c*)	7. (*c*)	8. (*a*)	9. (*a*)	10. (*a*)
11. (*a*)	12. (*a*)	13. (*b*)	14. (*b*)	15. (*c*)

C H A P T E R

6

Variable Stresses in Machine Parts

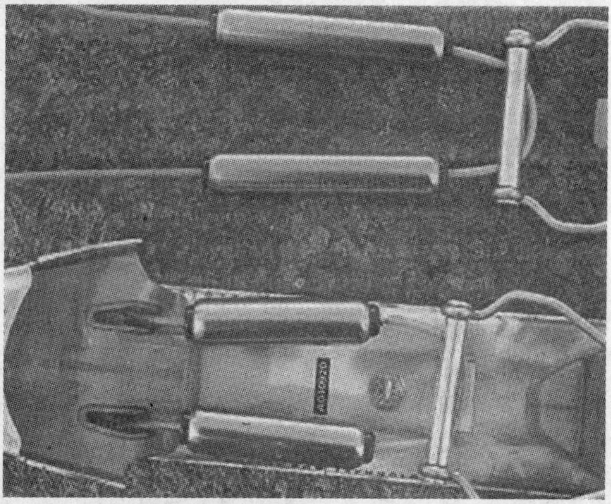

6.1 Introduction

We have discussed, in the previous chapter, the stresses due to static loading only. But only a few machine parts are subjected to static loading. Since many of the machine parts (such as axles, shafts, crankshafts, connecting rods, springs, pinion teeth etc.) are subjected to variable or alternating loads (also known as fluctuating or fatigue loads), therefore we shall discuss, in this chapter, the variable or alternating stresses.

6.2 Completely Reversed or Cyclic Stresses

Consider a rotating beam of circular cross-section and carrying a load W, as shown in Fig. 6.1. This load induces stresses in the beam which are cyclic in nature. A little consideration will show that the upper fibres of the beam (*i.e.* at point A) are under compressive stress and the lower fibres (*i.e.* at point B) are under tensile stress. After half a revolution, the point *B* occupies the position of point *A* and the point *A* occupies the position of point *B*. Thus the point *B* is now under compressive stress and the point *A* under tensile stress. The speed of variation of these stresses depends upon the speed of the beam.

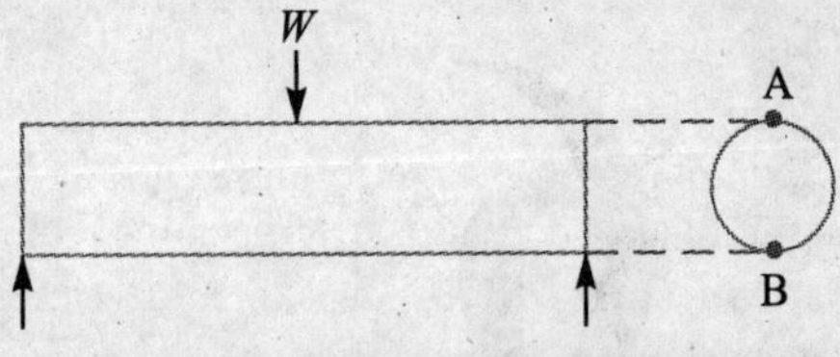

Fig. 6.1. Reversed or cyclic stresses.

From above we see that for each revolution of the beam, the stresses are reversed from compressive to tensile. The stresses which vary from one value of compressive to the same value of tensile or ***vice versa***, are known as ***completely reversed*** or ***cyclic stresses.***

Notes: 1. The stresses which vary from a minimum value to a maximum value of the same nature, (*i.e.* tensile or compressive) are called ***fluctuating stresses.***

2. The stresses which vary from zero to a certain maximum value are called ***repeated stresses.***

3. The stresses which vary from a minimum value to a maximum value of the opposite nature (*i.e.* from a certain minimum compressive to a certain maximum tensile or from a minimum tensile to a maximum compressive) are called ***alternating stresses.***

6.3 Fatigue and Endurance Limit

It has been found experimentally that when a material is subjected to repeated stresses, it fails at stresses below the yield point stresses. Such type of failure of a material is known as **fatigue.** The failure is caused by means of a progressive crack formation which are usually fine and of microscopic size. The failure may occur even without any prior indication. The fatigue of material is effected by the size of the component, relative magnitude of static and fluctuating loads and the number of load reversals.

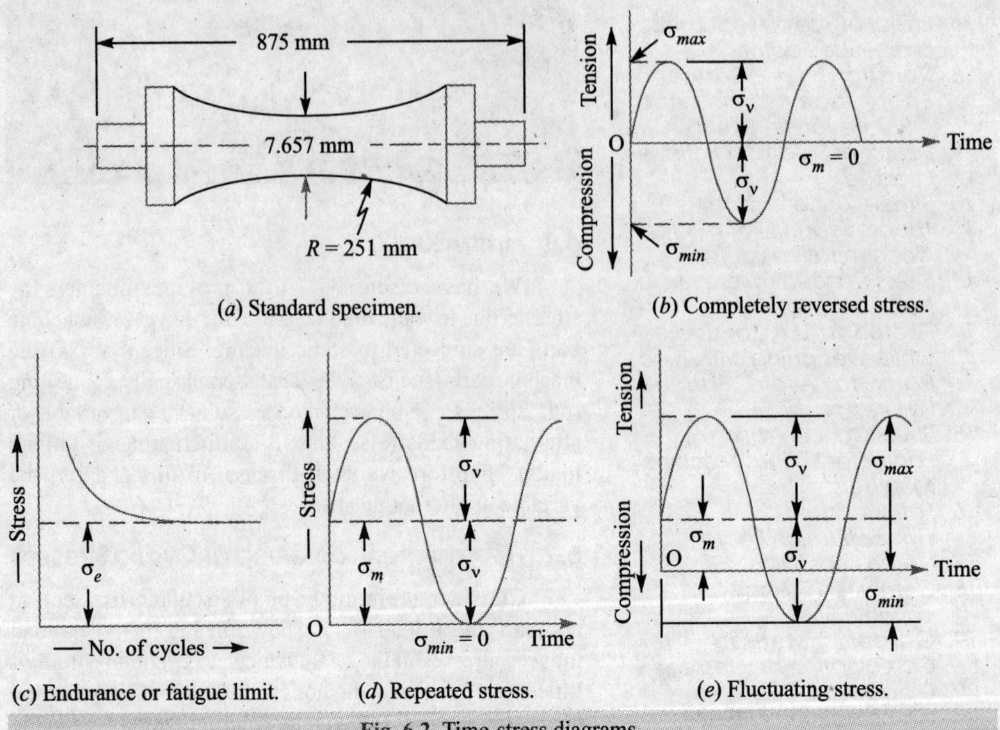

(*a*) Standard specimen. (*b*) Completely reversed stress.

(*c*) Endurance or fatigue limit. (*d*) Repeated stress. (*e*) Fluctuating stress.

Fig. 6.2. Time-stress diagrams.

In order to study the effect of fatigue of a material, a rotating mirror beam method is used. In this method, a standard mirror polished specimen, as shown in Fig. 6.2 (*a*), is rotated in a fatigue

A machine part is being turned on a Lathe.

testing machine while the specimen is loaded in bending. As the specimen rotates, the bending stress at the upper fibres varies from maximum compressive to maximum tensile while the bending stress at the lower fibres varies from maximum tensile to maximum compressive. In other words, the specimen is subjected to a completely reversed stress cycle. This is represented by a time-stress diagram as shown in Fig. 6.2 (*b*). A record is kept of the number of cycles required to produce failure at a given stress, and the results are plotted in stress-cycle curve as shown in Fig. 6.2 (*c*). A little consideration will show that if the stress is kept below a certain value as shown by dotted line in Fig. 6.2 (*c*), the material will not fail whatever may be the number of cycles. This stress, as represented by dotted line, is known as ***endurance*** or ***fatigue limit*** (σ_e). It is defined as maximum value of the completely reversed bending stress which a polished standard specimen can withstand without failure, for infinite number of cycles (usually 10^7 cycles).

It may be noted that the term endurance limit is used for reversed bending only while for other types of loading, the term ***endurance strength*** may be used when referring the fatigue strength of the material. It may be defined as the safe maximum stress which can be applied to the machine part working under actual conditions.

We have seen that when a machine member is subjected to a completely reversed stress, the maximum stress in tension is equal to the maximum stress in compression as shown in Fig. 6.2 (*b*). In actual practice, many machine members undergo different range of stress than the completely reversed stress.

The stress ***verses*** time diagram for fluctuating stress having values σ_{min} and σ_{max} is shown in Fig. 6.2 (*e*). The variable stress, in general, may be considered as a combination of steady (or mean or average) stress and a completely reversed stress component σ_v. The following relations are derived from Fig. 6.2 (*e*):

1. Mean or average stress,

$$\sigma_m = \frac{\sigma_{max} + \sigma_{min}}{2}$$

2. Reversed stress component or alternating or variable stress,

$$\sigma_v = \frac{\sigma_{max} - \sigma_{min}}{2}$$

Note: For repeated loading, the stress varies from maximum to zero (*i.e.* $\sigma_{min} = 0$) in each cycle as shown in Fig. 6.2 (*d*).

$$\therefore \qquad \sigma_m = \sigma_v = \frac{\sigma_{max}}{2}$$

3. Stress ratio, $R = \dfrac{\sigma_{max}}{\sigma_{min}}$. For completely reversed stresses, $R = -1$ and for repeated stresses, $R = 0$. It may be noted that R cannot be greater than unity.

4. The following relation between endurance limit and stress ratio may be used

$$\sigma'_e = \frac{3\sigma_e}{2 - R}$$

where σ'_e = Endurance limit for any stress range represented by R.

σ_e = Endurance limit for completely reversed stresses, and

R = Stress ratio.

6.4 Effect of Loading on Endurance Limit—Load Factor

The endurance limit (σ_e) of a material as determined by the rotating beam method is for reversed bending load. There are many machine members which are subjected to loads other than reversed bending loads. Thus the endurance limit will also be different for different types of loading. The endurance limit depending upon the type of loading may be modified as discussed below:

Shaft drive.

Let K_b = Load correction factor for the reversed or rotating bending load. Its value is usually taken as unity.

K_a = Load correction factor for the reversed axial load. Its value may be taken as 0.8.

K_s = Load correction factor for the reversed torsional or shear load. Its value may be taken as 0.55 for ductile materials and 0.8 for brittle materials.

∴ Endurance limit for reversed bending load, $\sigma_{eb} = \sigma_e.K_b = \sigma_e$ ...(∵ $K_b = 1$)

Endurance limit for reversed axial load, $\sigma_{ea} = \sigma_e.K_a$

and endurance limit for reversed torsional or shear load, $\tau_e = \sigma_e.K_s$

6.5 Effect of Surface Finish on Endurance Limit—Surface Finish Factor

When a machine member is subjected to variable loads, the endurance limit of the material for that member depends upon the surface conditions. Fig. 6.3 shows the values of surface finish factor for the various surface conditions and ultimate tensile strength.

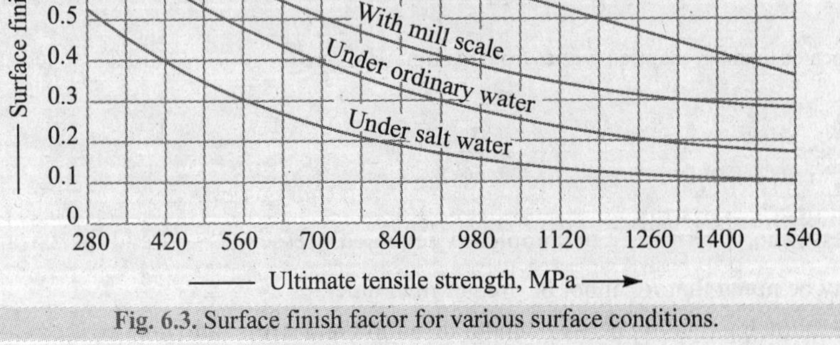

Fig. 6.3. Surface finish factor for various surface conditions.

When the surface finish factor is known, then the endurance limit for the material of the machine member may be obtained by multiplying the endurance limit and the surface finish factor. We see that for a mirror polished material, the surface finish factor is unity. In other words, the endurance limit for mirror polished material is maximum and it goes on reducing due to surface condition.

Let K_{sur} = Surface finish factor.

∴ Endurance limit,

$$\sigma_{e1} = \sigma_{eb}.K_{sur} = \sigma_e.K_b.K_{sur} = \sigma_e.K_{sur} \quad ...(\because K_b = 1)$$

...(For reversed bending load)

$$= \sigma_{ea}.K_{sur} = \sigma_e.K_a.K_{sur} \quad \text{...(For reversed axial load)}$$

$$= \tau_e.K_{sur} = \sigma_e.K_s.K_{sur} \quad \text{...(For reversed torsional or shear load)}$$

Note : The surface finish factor for non-ferrous metals may be taken as unity.

6.6 Effect of Size on Endurance Limit—Size Factor

A little consideration will show that if the size of the standard specimen as shown in Fig. 6.2 (*a*) is increased, then the endurance limit of the material will decrease. This is due to the fact that a longer specimen will have more defects than a smaller one.

Let K_{sz} = Size factor.

∴ Endurance limit,

$$\sigma_{e2} = \sigma_{e1} \times K_{sz} \quad \text{...(Considering surface finish factor also)}$$

$$= \sigma_{eb}.K_{sur}.K_{sz} = \sigma_e.K_b.K_{sur}.K_{sz} = \sigma_e.K_{sur}.K_{sz} \quad (\because K_b = 1)$$

$$= \sigma_{ea}.K_{sur}.K_{sz} = \sigma_e.K_a.K_{sur}.K_{sz} \quad \text{...(For reversed axial load)}$$

$$= \tau_e.K_{sur}.K_{sz} = \sigma_e.K_s.K_{sur}.K_{sz} \quad \text{... (For reversed torsional or shear load)}$$

Notes: 1. The value of size factor is taken as unity for the standard specimen having nominal diameter of 7.657 mm.

2. When the nominal diameter of the specimen is more than 7.657 mm but less than 50 mm, the value of size factor may be taken as 0.85.

3. When the nominal diameter of the specimen is more than 50 mm, then the value of size factor may be taken as 0.75.

6.7 Effect of Miscellaneous Factors on Endurance Limit

In addition to the surface finish factor (K_{sur}), size factor (K_{sz}) and load factors K_b, K_a and K_s, there are many other factors such as reliability factor (K_r), temperature factor (K_t), impact factor (K_i) etc. which has effect on the endurance limit of a material. Considering all these factors, the endurance limit may be determined by using the following expressions :

1. For the reversed bending load, endurance limit,

$$\sigma'_e = \sigma_{eb}.K_{sur}.K_{sz}.K_r.K_t.K_i$$

2. For the reversed axial load, endurance limit,

$$\sigma'_e = \sigma_{ea}.K_{sur}.K_{sz}.K_r.K_t.K_i$$

3. For the reversed torsional or shear load, endurance limit,

$$\sigma'_e = \tau_e.K_{sur}.K_{sz}.K_r.K_t.K_i$$

In solving problems, if the value of any of the above factors is not known, it may be taken as unity.

In addition to shear, tensile, compressive and torsional stresses, temperature can add its own stress (Ref. Chapter 4)

Note : This picture is given as additional information and is not a direct example of the current chapter.

6.8 Relation Between Endurance Limit and Ultimate Tensile Strength

It has been found experimentally that endurance limit (σ_e) of a material subjected to fatigue loading is a function of ultimate tensile strength (σ_u). Fig. 6.4 shows the endurance limit of steel corresponding to ultimate tensile strength for different surface conditions. Following are some empirical relations commonly used in practice :

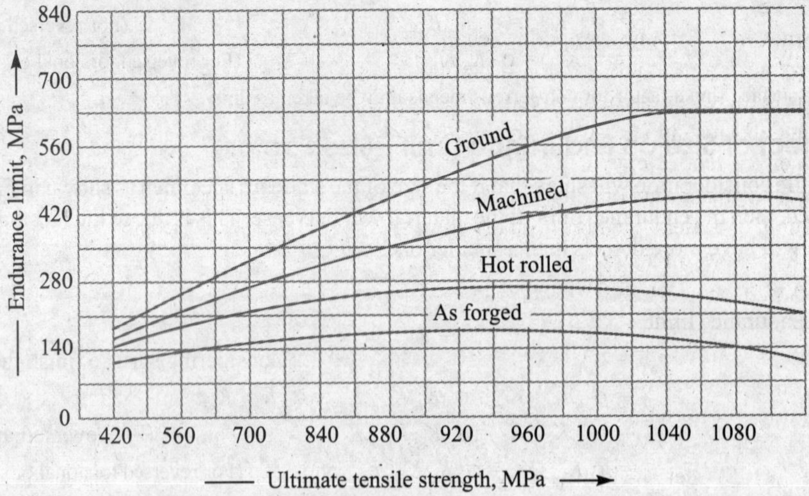

Fig. 6.4. Endurance limit of steel corresponding to ultimate tensile strength.

For steel, $\sigma_e = 0.5\ \sigma_u$;

For cast steel, $\sigma_e = 0.4\ \sigma_u$;

For cast iron, $\sigma_e = 0.35\ \sigma_u$;

For non-ferrous metals and alloys, $\sigma_e = 0.3\ \sigma_u$

6.9 Factor of Safety for Fatigue Loading

When a component is subjected to fatigue loading, the endurance limit is the criterion for faliure. Therefore, the factor of safety should be based on endurance limit. Mathematically,

$$\text{Factor of safety } (F.S.) = \frac{\text{Endurance limit stress}}{\text{Design or working stress}} = \frac{\sigma_e}{\sigma_d}$$

Note: For steel, $\sigma_e = 0.8 \text{ to } 0.9\ \sigma_y$

where σ_e = Endurance limit stress for completely reversed stress cycle, and

σ_y = Yield point stress.

Example 6.1. *Determine the design stress for a piston rod where the load is completely reversed. The surface of the rod is ground and the surface finish factor is 0.9. There is no stress concentration. The load is predictable and the factor of safety is 2.*

Solution. Given : $K_{sur} = 0.9$; $F.S. = 2$

The piston rod is subjected to reversed axial loading. We know that for reversed axial loading, the load correction factor (K_a) is 0.8.

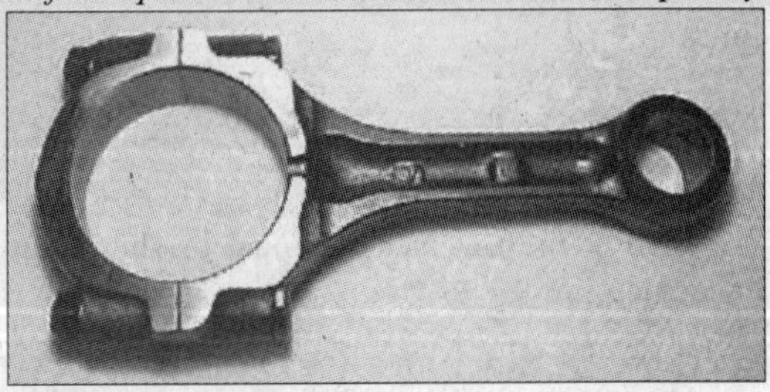

Piston rod

If σ_e is the endurance limit for reversed bending load, then endurance limit for reversed axial load,

$$\sigma_{ea} = \sigma_e \times K_a \times K_{sur} = \sigma_e \times 0.8 \times 0.9 = 0.72\,\sigma_e$$

We know that design stress,

$$\sigma_d = \frac{\sigma_{ea}}{F.S.} = \frac{0.72\,\sigma_e}{2} = 0.36\,\sigma_e \textbf{ Ans.}$$

6.10 Stress Concentration

Whenever a machine component changes the shape of its cross-section, the simple stress distribution no longer holds good and the neighbourhood of the discontinuity is different. This irregularity in the stress distribution caused by abrupt changes of form is called ***stress concentration.*** It occurs for all kinds of stresses in the presence of fillets, notches, holes, keyways, splines, surface roughness or scratches etc.

In order to understand fully the idea of stress concentration, consider a member with different cross-section under a tensile load as shown in Fig. 6.5. A little consideration will show that the nominal stress in the right and left hand sides will be uniform but in the region where the cross-section is changing, a re-distribution of the force within the member must take place. The material near the edges is stressed considerably higher than the average value. The maximum stress occurs at some point on the fillet and is directed parallel to the boundary at that point.

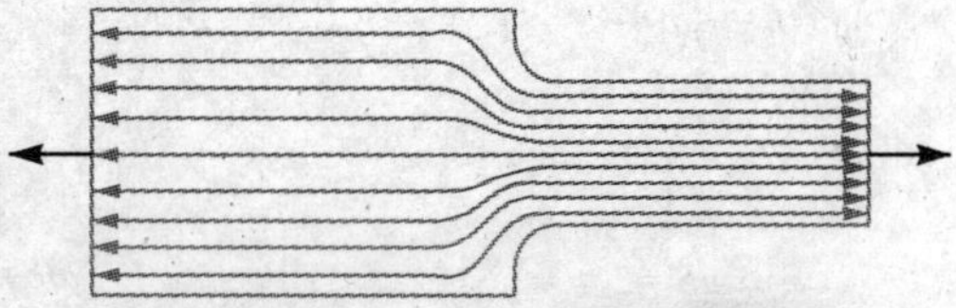

Fig. 6.5. Stress concentration.

6.11 Theoretical or Form Stress Concentration Factor

The theoretical or form stress concentration factor is defined as the ratio of the maximum stress in a member (at a notch or a fillet) to the nominal stress at the same section based upon net area. Mathematically, theoretical or form stress concentration factor,

$$K_t = \frac{\text{Maximum stress}}{\text{Nominal stress}}$$

The value of K_t depends upon the material and geometry of the part.

Notes: 1. In static loading, stress concentration in ductile materials is not so serious as in brittle materials, because in ductile materials local deformation or yielding takes place which reduces the concentration. In brittle materials, cracks may appear at these local concentrations of stress which will increase the stress over the rest of the section. It is, therefore, necessary that in designing parts of brittle materials such as castings, care should be taken. In order to avoid failure due to stress concentration, fillets at the changes of section must be provided.

2. In cyclic loading, stress concentration in ductile materials is always serious because the ductility of the material is not effective in relieving the concentration of stress caused by cracks, flaws, surface roughness, or any sharp discontinuity in the geometrical form of the member. If the stress at any point in a member is above the endurance limit of the material, a crack may develop under the action of repeated load and the crack will lead to failure of the member.

6.12 Stress Concentration due to Holes and Notches

Consider a plate with transverse elliptical hole and subjected to a tensile load as shown in Fig. 6.6 (*a*). We see from the stress-distribution that the stress at the point away from the hole is practically uniform and the maximum stress will be induced at the edge of the hole. The maximum stress is given by

$$\sigma_{max} = \sigma\left(1 + \frac{2a}{b}\right)$$

and the theoretical stress concentration factor,

$$K_t = \frac{\sigma_{max}}{\sigma} = \left(1 + \frac{2a}{b}\right)$$

When a/b is large, the ellipse approaches a crack transverse to the load and the value of K_t becomes very large. When a/b is small, the ellipse approaches a longitudinal slit [as shown in Fig. 6.6 (*b*)] and the increase in stress is small. When the hole is circular as shown in Fig. 6.6 (*c*), then $a/b = 1$ and the maximum stress is three times the nominal value.

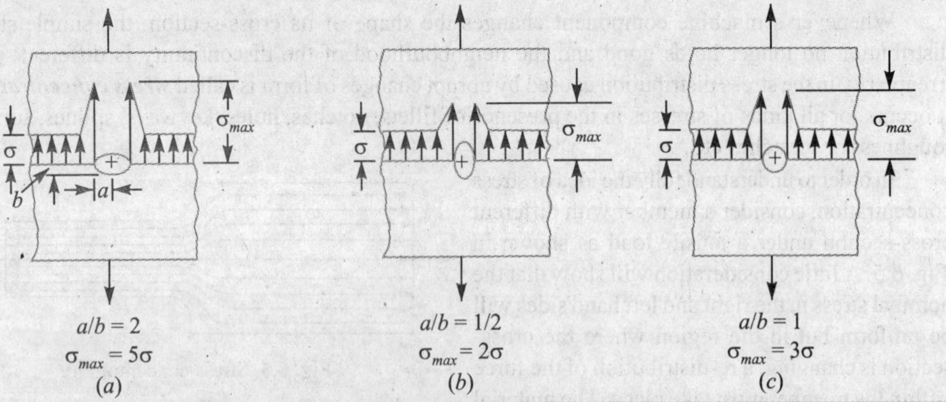

Fig. 6.6. Stress concentration due to holes.

The stress concentration in the notched tension member, as shown in Fig. 6.7, is influenced by the depth a of the notch and radius r at the bottom of the notch. The maximum stress, which applies to members having notches that are small in comparison with the width of the plate, may be obtained by the following equation,

$$\sigma_{max} = \sigma\left(1 + \frac{2a}{r}\right)$$

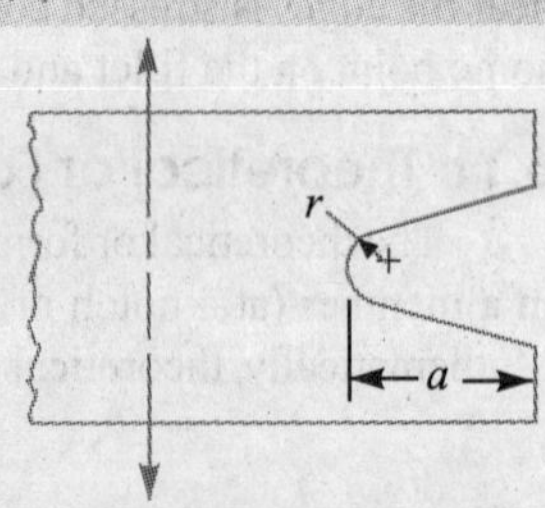

Fig. 6.7. Stress concentration due to notches.

6.13 Methods of Reducing Stress Concentration

We have already discussed in Art 6.10 that whenever there is a change in cross-section, such as shoulders, holes, notches or keyways and where there is an interference fit between a hub or bearing race and a shaft, then stress concentration results. The presence of

Crankshaft

stress concentration can not be totally eliminated but it may be reduced to some extent. A device or concept that is useful in assisting a design engineer to visualize the presence of stress concentration and how it may be mitigated is that of stress flow lines, as shown in Fig. 6.8. The mitigation of stress concentration means that the stress flow lines shall maintain their spacing as far as possible.

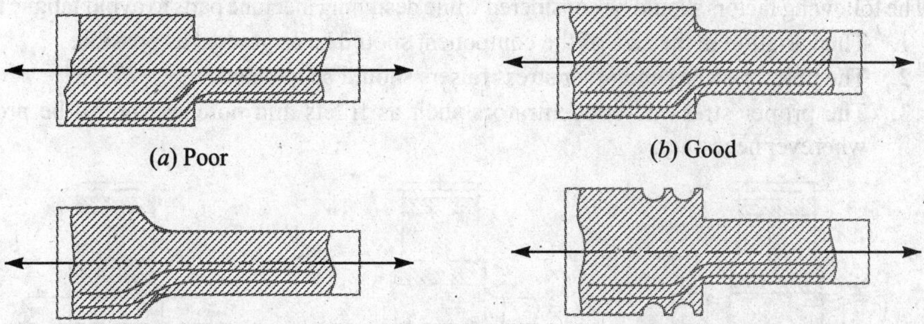

(*a*) Poor (*b*) Good (*c*) Preferred (*d*) Preferred

Fig. 6.8

In Fig. 6.8 (*a*) we see that stress lines tend to bunch up and cut very close to the sharp re-entrant corner. In order to improve the situation, fillets may be provided, as shown in Fig. 6.8 (*b*) and (*c*) to give more equally spaced flow lines.

Figs. 6.9 to 6.11 show the several ways of reducing the stress concentration in shafts and other cylindrical members with shoulders, holes and threads respectively. It may be noted that it is not practicable to use large radius fillets as in case of ball and roller bearing mountings. In such cases, notches may be cut as shown in Fig. 6.8 (*d*) and Fig. 6.9 (*b*) and (*c*).

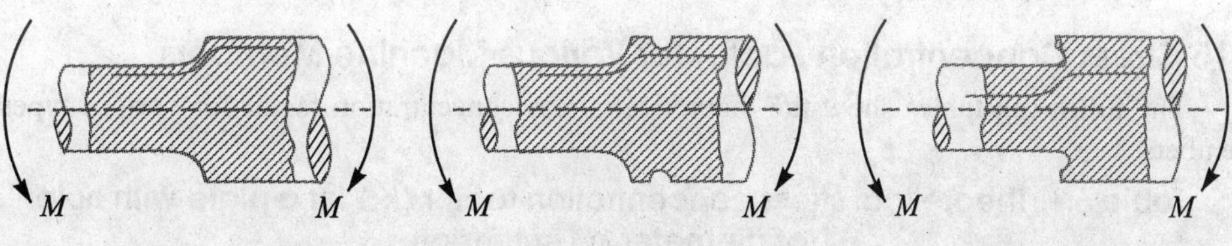

(*a*) Poor (*b*) Good (*c*) Preferred

Fig. 6.9. Methods of reducing stress concentration in cylindrical members with shoulders.

(*a*) Poor (*b*) Preferred

Fig. 6.10. Methods of reducing stress concentration in cylindrical members with holes.

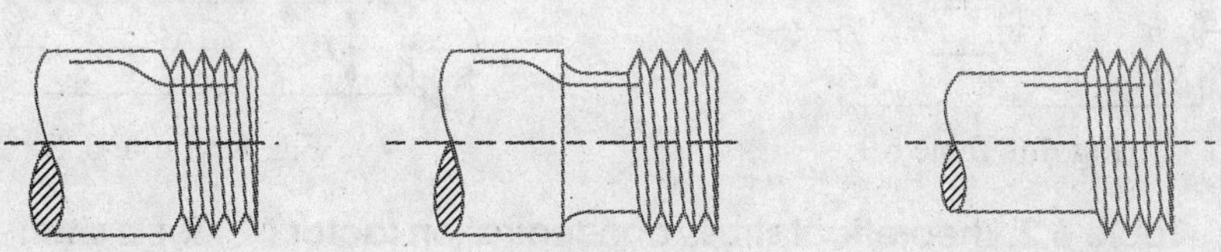

(*a*) Poor (*b*) Good (*c*) Preferred

Fig. 6.11. Methods of reducing stress concentration in cylindrical members with holes.

The stress concentration effects of a press fit may be reduced by making more gradual transition from the rigid to the more flexible shaft. The various ways of reducing stress concentration for such cases are shown in Fig. 6.12 (*a*), (*b*) and (*c*).

6.14 Factors to be Considered while Designing Machine Parts to Avoid Fatigue Failure

The following factors should be considered while designing machine parts to avoid fatigue failure:

1. The variation in the size of the component should be as gradual as possible.
2. The holes, notches and other stress raisers should be avoided.
3. The proper stress de-concentrators such as fillets and notches should be provided wherever necessary.

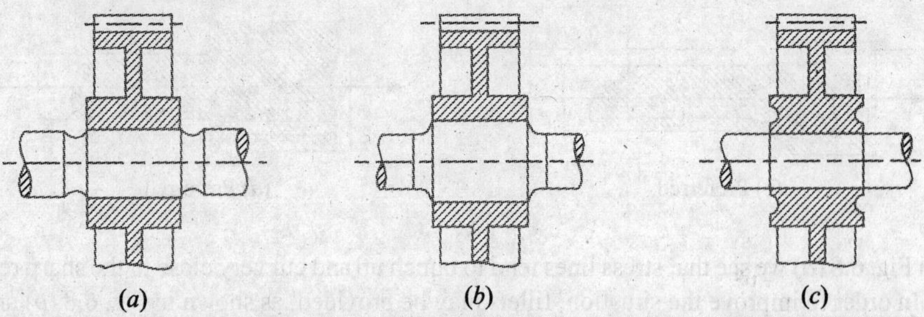

Fig. 6.12. Methods of reducing stress concentration of a press fit.

4. The parts should be protected from corrosive atmosphere.
5. A smooth finish of outer surface of the component increases the fatigue life.
6. The material with high fatigue strength should be selected.
7. The residual compressive stresses over the parts surface increases its fatigue strength.

6.15 Stress Concentration Factor for Various Machine Members

The following tables show the theoretical stress concentration factor for various types of members.

Table 6.1. Theoretical stress concentration factor (K_t) for a plate with hole (of diameter d) in tension.

$\frac{d}{b}$	0.05	0.1	0.15	0.20	0.25	0.30	0.35	0.40	0.45	0.50	0.55
K_t	2.83	2.69	2.59	2.50	2.43	2.37	2.32	2.26	2.22	2.17	2.13

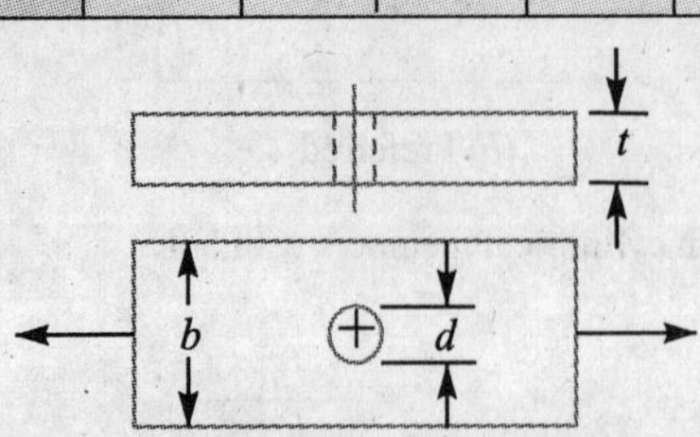

Fig. for Table 6.1

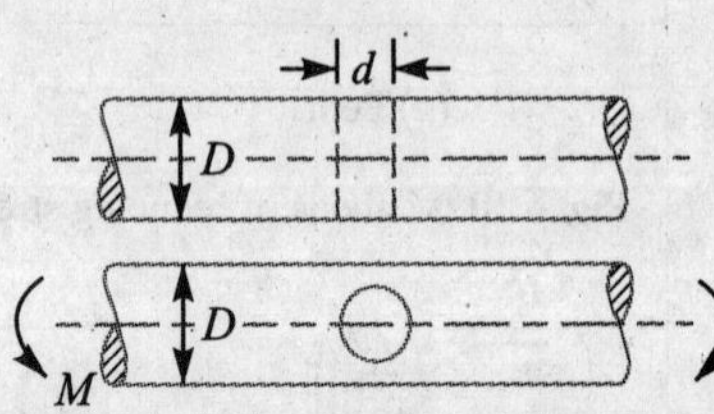

Fig. for Table 6.2

Table 6.2. Theoretical stress concentration factor (K_t) for a shaft with transverse hole (of diameter d) in bending.

$\frac{d}{D}$	0.02	0.04	0.08	0.10	0.12	0.16	0.20	0.24	0.28	0.30
K_t	2.70	2.52	2.33	2.26	2.20	2.11	2.03	1.96	1.92	1.90

Table 6.3. Theoretical stress concentration factor (K_t) for stepped shaft with a shoulder fillet (of radius r) in tension.

$\frac{D}{d}$	*Theoretical stress concentration factor (K_t)*									
	r/d									
	0.08	*0.10*	*0.12*	*0.16*	*0.18*	*0.20*	*0.22*	*0.24*	*0.28*	*0.30*
1.01	1.27	1.24	1.21	1.17	1.16	1.15	1.15	1.14	1.13	1.13
1.02	1.38	1.34	1.30	1.26	1.24	1.23	1.22	1.21	1.19	1.19
1.05	1.53	1.46	1.42	1.36	1.34	1.32	1.30	1.28	1.26	1.25
1.10	1.65	1.56	1.50	1.43	1.39	1.37	1.34	1.33	1.30	1.28
1.15	1.73	1.63	1.56	1.46	1.43	1.40	1.37	1.35	1.32	1.31
1.20	1.82	1.68	1.62	1.51	1.47	1.44	1.41	1.38	1.35	1.34
1.50	2.03	1.84	1.80	1.66	1.60	1.56	1.53	1.50	1.46	1.44
2.00	2.14	1.94	1.89	1.74	1.68	1.64	1.59	1.56	1.50	1.47

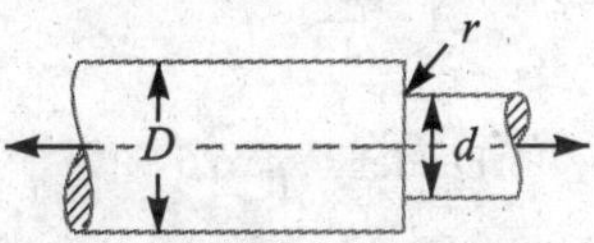

$$A = \frac{\pi}{4} \times d^2$$

Fig. for Table 6.3

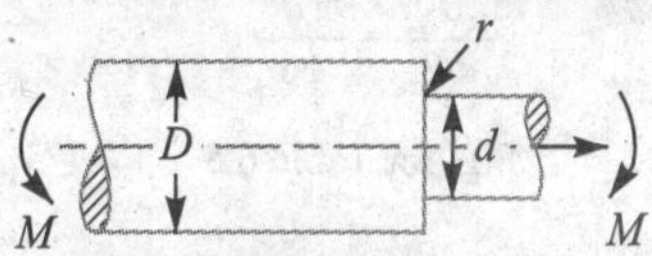

$$Z = \frac{\pi}{32} \times d^3$$

Fig. for Table 6.4

Table 6.4. Theoretical stress concentration factor (K_t) for a stepped shaft with a shoulder fillet (of radius r) in bending.

$\frac{D}{d}$	*Theoretical stress concentration factor (K_t)*									
	r/d									
	0.02	*0.04*	*0.08*	*0.10*	*0.12*	*0.16*	*0.20*	*0.24*	*0.28*	*0.30*
1.01	1.85	1.61	1.42	1.36	1.32	1.24	1.20	1.17	1.15	1.14
1.02	1.97	1.72	1.50	1.44	1.40	1.32	1.27	1.23	1.21	1.20
1.05	2.20	1.88	1.60	1.53	1.48	1.40	1.34	1.30	1.27	1.25
1.10	2.36	1.99	1.66	1.58	1.53	1.44	1.38	1.33	1.28	1.27
1.20	2.52	2.10	1.72	1.62	1.56	1.46	1.39	1.34	1.29	1.28
1.50	2.75	2.20	1.78	1.68	1.60	1.50	1.42	1.36	1.31	1.29
2.00	2.86	2.32	1.87	1.74	1.64	1.53	1.43	1.37	1.32	1.30
3.00	3.00	2.45	1.95	1.80	1.69	1.56	1.46	1.38	1.34	1.32
6.00	3.04	2.58	2.04	1.87	1.76	1.60	1.49	1.41	1.35	1.33

Table 6.5. Theoretical stress concentration factor (K_t) for a stepped shaft with a shoulder fillet (of radius r) in torsion.

$\frac{D}{d}$	Theoretical stress concentration factor (K_t)									
	r/d									
	0.02	0.04	0.08	0.10	0.12	0.16	0.20	0.24	0.28	0.30
1.09	1.54	1.32	1.19	1.16	1.15	1.12	1.11	1.10	1.09	1.09
1.20	1.98	1.67	1.40	1.33	1.28	1.22	1.18	1.15	1.13	1.13
1.33	2.14	1.79	1.48	1.41	1.35	1.28	1.22	1.19	1.17	1.16
2.00	2.27	1.84	1.53	1.46	1.40	1.32	1.26	1.22	1.19	1.18

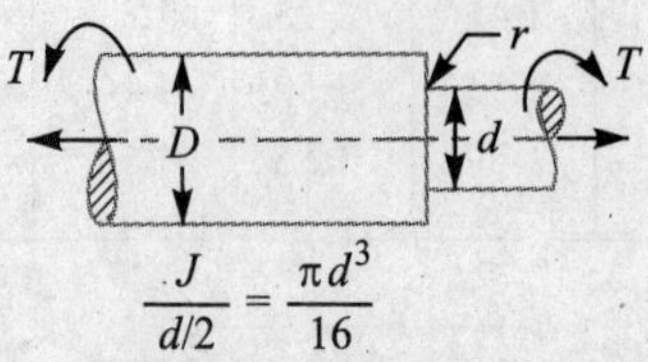

Fig. for Table 6.5

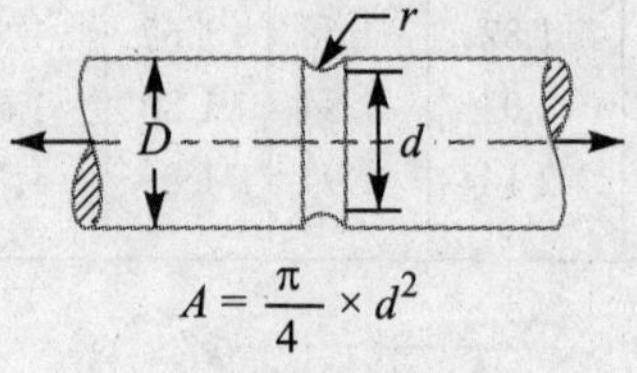

Fig. for Table 6.6

Table 6.6. Theoretical stress concentration factor (K_t) for a grooved shaft in tension.

$\frac{D}{d}$	Theoretical stress concentration (K_t)									
	r/d									
	0.02	0.04	0.08	0.10	0.12	0.16	0.20	0.24	0.28	0.30
1.01	1.98	1.71	1.47	1.42	1.38	1.33	1.28	1.25	1.23	1.22
1.02	2.30	1.94	1.66	1.59	1.54	1.45	1.40	1.36	1.33	1.31
1.03	2.60	2.14	1.77	1.69	1.63	1.53	1.46	1.41	1.37	1.36
1.05	2.85	2.36	1.94	1.81	1.73	1.61	1.54	1.47	1.43	1.41
1.10	..	2.70	2.16	2.01	1.90	1.75	1.70	1.57	1.50	1.47
1.20	..	2.90	2.36	2.17	2.04	1.86	1.74	1.64	1.56	1.54
1.30	..	..	2.46	2.26	2.11	1.91	1.77	1.67	1.59	1.56
1.50	..	..	2.54	2.33	2.16	1.94	1.79	1.69	1.61	1.57
2.00	..	..	2.61	2.38	2.22	1.98	1.83	1.72	1.63	1.59
∞	..	..	2.69	2.44	2.26	2.03	1.86	1.74	1.65	1.61

Table 6.7. Theoretical stress concentration factor (K_t) of a grooved shaft in bending.

$\frac{D}{d}$	Theoretical stress concentration factor (K_t)									
	r/d									
	0.02	0.04	0.08	0.10	0.12	0.16	0.20	0.24	0.28	0.30
1.01	1.74	1.68	1.47	1.41	1.38	1.32	1.27	1.23	1.22	1.20
1.02	2.28	1.89	1.64	1.53	1.48	1.40	1.34	1.30	1.26	1.25
1.03	2.46	2.04	1.68	1.61	1.55	1.47	1.40	1.35	1.31	1.28
1.05	2.75	2.22	1.80	1.70	1.63	1.53	1.46	1.40	1.35	1.33
1.12	3.20	2.50	1.97	1.83	1.75	1.62	1.52	1.45	1.38	1.34
1.30	3.40	2.70	2.04	1.91	1.82	1.67	1.57	1.48	1.42	1.38
1.50	3.48	2.74	2.11	1.95	1.84	1.69	1.58	1.49	1.43	1.40
2.00	3.55	2.78	2.14	1.97	1.86	1.71	1.59	1.55	1.44	1.41
∞	3.60	2.85	2.17	1.98	1.88	1.71	1.60	1.51	1.45	1.42

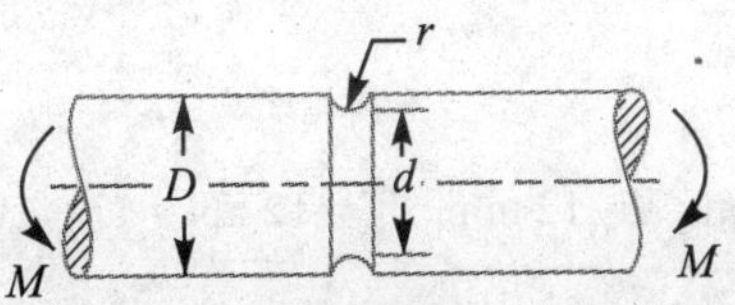

$$Z = \frac{\pi}{32} \times d^3$$

Fig. for Table 6.7

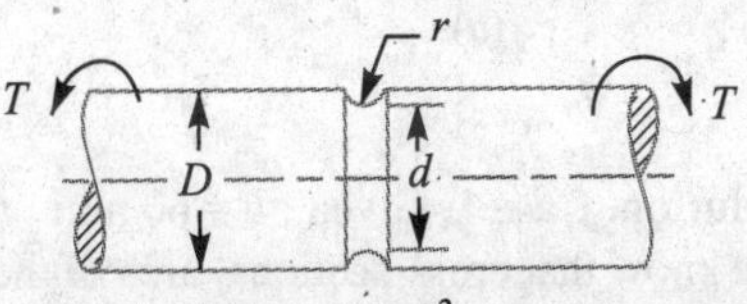

$$\frac{J}{d/2} = \frac{\pi d^3}{16}$$

Fig. for Table 6.8

Table 6.8. Theoretical stress concentration factor (K_t) for a grooved shaft in torsion.

$\frac{D}{d}$	Theoretical stress concentration factor (K_{ts})									
	r/d									
	0.02	0.04	0.08	0.10	0.12	0.16	0.20	0.24	0.28	0.30
1.01	1.50	1.03	1.22	1.20	1.18	1.16	1.13	1.12	1.12	1.12
1.02	1.62	1.45	1.31	1.27	1.23	1.20	1.18	1.16	1.15	1.16
1.05	1.88	1.61	1.40	1.35	1.32	1.26	1.22	1.20	1.18	1.17
1.10	2.05	1.73	1.47	1.41	1.37	1.31	1.26	1.24	1.21	1.20
1.20	2.26	1.83	1.53	1.46	1.41	1.34	1.27	1.25	1.22	1.21
1.30	2.32	1.89	1.55	1.48	1.43	1.35	1.30	1.26	—	—
2.00	2.40	1.93	1.58	1.50	1.45	1.36	1.31	1.26	—	—
∞	2.50	1.96	1.60	1.51	1.46	1.38	1.32	1.27	1.24	1.23

Example 6.2. *Find the maximum stress induced in the following cases taking stress concentration into account:*

1. A rectangular plate 60 mm × 10 mm with a hole 12 diameter as shown in Fig. 6.13 (a) and subjected to a tensile load of 12 kN.

2. A stepped shaft as shown in Fig. 6.13 (b) and carrying a tensile load of 12 kN.

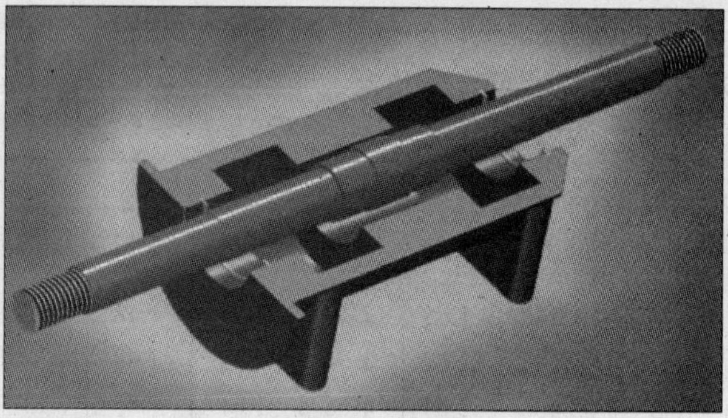

Stepped shaft

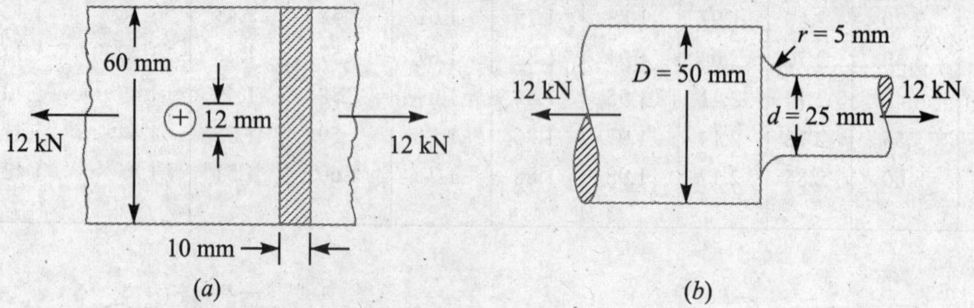

Fig. 6.13

Solution. Case 1. Given : $b = 60$ mm ; $t = 10$ mm ; $d = 12$ mm ; $W = 12$ kN $= 12 \times 10^3$ N

We know that cross-sectional area of the plate,

$$A = (b - d)\,t = (60 - 12)\,10 = 480 \text{ mm}^2$$

$$\therefore \quad \text{Nominal stress} = \frac{W}{A} = \frac{12 \times 10^3}{480} = 25 \text{ N/mm}^2 = 25 \text{ MPa}$$

Ratio of diameter of hole to width of plate,

$$\frac{d}{b} = \frac{12}{60} = 0.2$$

From Table 6.1, we find that for $d / b = 0.2$, theoretical stress concentration factor,

$$K_t = 2.5$$

$$\therefore \quad \text{Maximum stress} = K_t \times \text{Nominal stress} = 2.5 \times 25 = 62.5 \text{ MPa } \textbf{Ans.}$$

Case 2. Given : $D = 50$ mm ; $d = 25$ mm ; $r = 5$ mm ; $W = 12$ kN $= 12 \times 10^3$ N

We know that cross-sectional area for the stepped shaft,

$$A = \frac{\pi}{4} \times d^2 = \frac{\pi}{4} (25)^2 = 491 \text{ mm}^2$$

$$\therefore \quad \text{Nominal stress} = \frac{W}{A} = \frac{12 \times 10^3}{491} = 24.4 \text{ N/mm}^2 = 24.4 \text{ MPa}$$

Ratio of maximum diameter to minimum diameter,

$$D/d = 50/25 = 2$$

Ratio of radius of fillet to minimum diameter,

$$r/d = 5/25 = 0.2$$

From Table 6.3, we find that for $D/d = 2$ and $r/d = 0.2$, theoretical stress concentration factor, $K_t = 1.64$.

$$\therefore \quad \text{Maximum stress} = K_t \times \text{Nominal stress} = 1.64 \times 24.4 = 40 \text{ MPa } \textbf{Ans.}$$

6.16 Fatigue Stress Concentration Factor

When a machine member is subjected to cyclic or fatigue loading, the value of fatigue stress concentration factor shall be applied instead of theoretical stress concentration factor. Since the determination of fatigue stress concentration factor is not an easy task, therefore from experimental tests it is defined as

Fatigue stress concentration factor,

$$K_f = \frac{\text{Endurance limit without stress concentration}}{\text{Endurance limit with stress concentration}}$$

6.17 Notch Sensitivity

In cyclic loading, the effect of the notch or the fillet is usually less than predicted by the use of the theoretical factors as discussed before. The difference depends upon the stress gradient in the region of the stress concentration and on the hardness of the material. The term ***notch sensitivity*** is applied to this behaviour. It may be defined as the degree to which the theoretical effect of stress concentration is actually reached. The stress gradient depends mainly on the radius of the notch, hole or fillet and on the grain size of the material. Since the extensive data for estimating the notch sensitivity factor (q) is not available, therefore the curves, as shown in Fig. 6.14, may be used for determining the values of q for two steels.

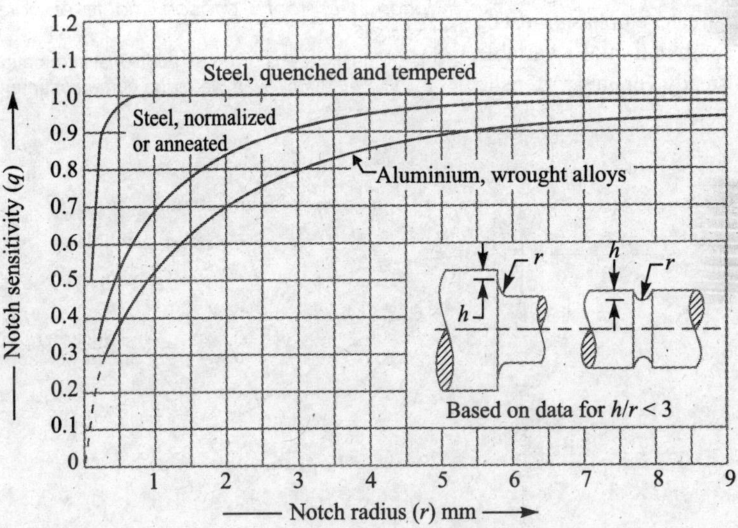

Fig. 6.14. Notch sensitivity.

When the notch sensitivity factor q is used in cyclic loading, then fatigue stress concentration factor may be obtained from the following relations:

$$q = \frac{K_f - 1}{K_t - 1}$$

or $$K_f = 1 + q\,(K_t - 1) \quad \text{...[For tensile or bending stress]}$$

and $$K_{fs} = 1 + q\,(K_{ts} - 1) \quad \text{...[For shear stress]}$$

where K_t = Theoretical stress concentration factor for axial or bending loading, and

K_{ts} = Theoretical stress concentration factor for torsional or shear loading.

6.18 Combined Steady and Variable Stress

The failure points from fatigue tests made with different steels and combinations of mean and variable stresses are plotted in Fig. 6.15 as functions of variable stress (σ_v) and mean stress (σ_m). The most significant observation is that, in general, the failure point is little related to the mean stress when it is compressive but is very much a function of the mean stress when it is tensile. In practice, this means that fatigue failures are rare when the mean stress is compressive (or negative). Therefore, the greater emphasis must be given to the combination of a variable stress and a steady (or mean) tensile stress.

Protective colour coatings are added to make components it corrosion resistant. Corrosion if not taken care can magnify other stresses.

Note : This picture is given as additional information and is not a direct example of the current chapter.

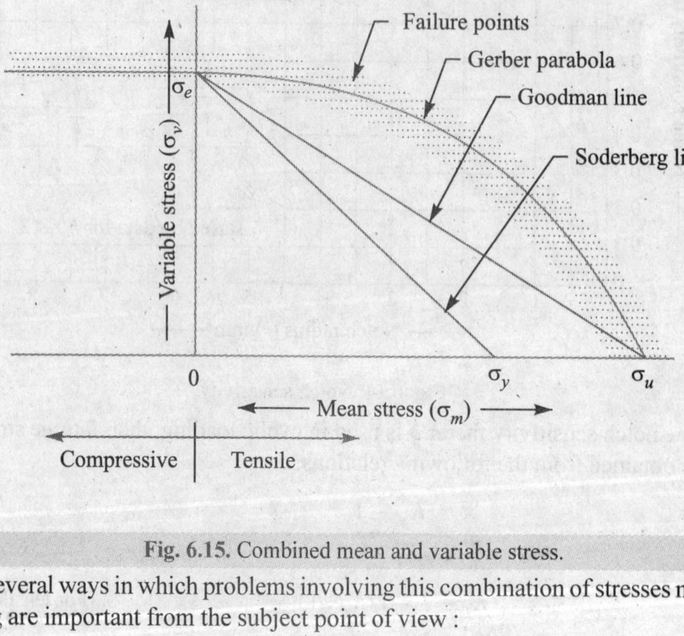

Fig. 6.15. Combined mean and variable stress.

There are several ways in which problems involving this combination of stresses may be solved, but the following are important from the subject point of view :

1. Gerber method, **2.** Goodman method, and **3.** Soderberg method.

We shall now discuss these methods, in detail, in the following pages.

6.19 Gerber Method for Combination of Stresses

The relationship between variable stress (σ_v) and mean stress (σ_m) for axial and bending loading for ductile materials are shown in Fig. 6.15. The point σ_e represents the fatigue strength corresponding to the case of complete reversal ($\sigma_m = 0$) and the point σ_u represents the static ultimate strength corresponding to $\sigma_v = 0$.

A parabolic curve drawn between the endurance limit (σ_e) and ultimate tensile strength (σ_u) was proposed by Gerber in 1874. Generally, the test data for ductile material fall closer to Gerber parabola as shown in Fig. 6.15, but because of scatter in the test points, a straight line relationship (*i.e.* Goodman line and Soderberg line) is usually preferred in designing machine parts.

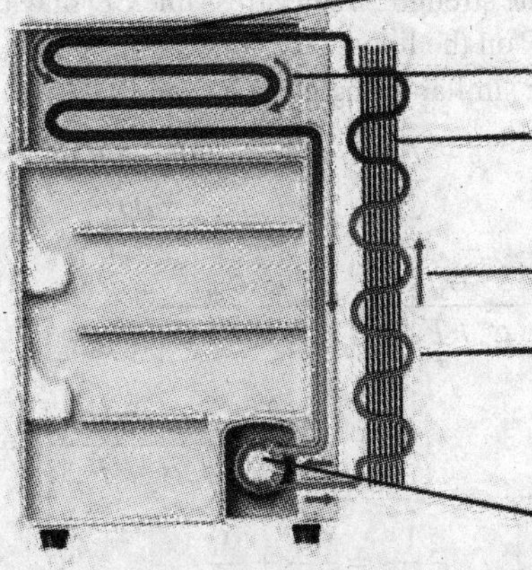

Liquid refrigerant absorbs heat as it vaporizes inside the evaporator coil of a refrigerator. The heat is released when a compressor turns the refrigerant back to liquid.

Note : This picture is given as additional information and is not a direct example of the current chapter.

According to Gerber, variable stress,

$$\sigma_v = \sigma_e \left[\frac{1}{F.S.} - \left(\frac{\sigma_m}{\sigma_u} \right)^2 F.S. \right]$$

or

$$\frac{1}{F.S.} = \left(\frac{\sigma_m}{\sigma_u} \right)^2 F.S. + \frac{\sigma_v}{\sigma_e} \quad \text{...}(i)$$

where $F.S.$ = Factor of safety,

σ_m = Mean stress (tensile or compressive),

σ_u = Ultimate stress (tensile or compressive), and

σ_e = Endurance limit for reversal loading.

Considering the fatigue stress concentration factor (K_f), the equation (*i*) may be written as

$$\frac{1}{F.S.} = \left(\frac{\sigma_m}{\sigma_u} \right)^2 F.S. + \frac{\sigma_v \times K_f}{\sigma_e}$$

6.20 Goodman Method for Combination of Stresses

A straight line connecting the endurance limit (σ_e) and the ultimate strength (σ_u), as shown by line AB in Fig. 6.16, follows the suggestion of Goodman. A Goodman line is used when the design is based on ultimate strength and may be used for ductile or brittle materials.

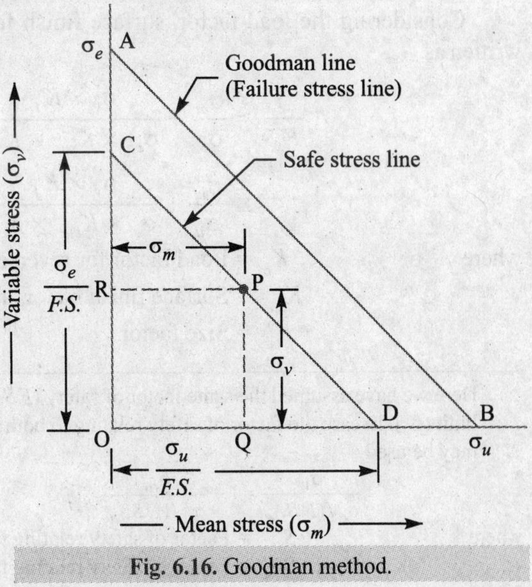

Fig. 6.16. Goodman method.

In Fig. 6.16, line AB connecting σ_e and

σ_u is called ***Goodman's failure stress line.*** If a suitable factor of safety (*F.S.*) is applied to endurance limit and ultimate strength, a safe stress line *CD* may be drawn parallel to the line *AB*. Let us consider a design point *P* on the line *CD*.

Now from similar triangles *COD* and *PQD*,

$$\frac{PQ}{CO} = \frac{QD}{OD} = \frac{OD - OQ}{OD} = 1 - \frac{OQ}{OD} \qquad ...(\because QD = OD - OQ)$$

$$\therefore \quad \frac{^*\sigma_v}{\sigma_e / F.S.} = 1 - \frac{\sigma_m}{\sigma_u / F.S.}$$

$$\sigma_v = \frac{\sigma_e}{F.S.}\left[1 - \frac{\sigma_m}{\sigma_u / F.S.}\right] = \sigma_e\left[\frac{1}{F.S.} - \frac{\sigma_m}{\sigma_u}\right]$$

$$\text{or} \quad \frac{1}{F.S.} = \frac{\sigma_m}{\sigma_u} + \frac{\sigma_v}{\sigma_e} \qquad ...(i)$$

This expression does not include the effect of stress concentration. It may be noted that for ductile materials, the stress concentration may be ignored under steady loads.

Since many machine and structural parts that are subjected to fatigue loads contain regions of high stress concentration, therefore equation (*i*) must be altered to include this effect. In such cases, the fatigue stress concentration factor (K_f) is used to multiply the variable stress (σ_v). The equation (*i*) may now be written as

$$\frac{1}{F.S.} = \frac{\sigma_m}{\sigma_u} + \frac{\sigma_v \times K_f}{\sigma_e} \qquad ...(ii)$$

where $F.S.$ = Factor of safety,

σ_m = Mean stress,

σ_u = Ultimate stress,

σ_v = Variable stress,

σ_e = Endurance limit for reversed loading, and

K_f = Fatigue stress concentration factor.

Considering the load factor, surface finish factor and size factor, the equation (*ii*) may be written as

$$\frac{1}{F.S.} = \frac{\sigma_m}{\sigma_u} + \frac{\sigma_v \times K_f}{\sigma_{eb} \times K_{sur} \times K_{sz}} = \frac{\sigma_m}{\sigma_u} + \frac{\sigma_v \times K_f}{\sigma_e \times K_b \times K_{sur} \times K_{sz}} \qquad ...(iii)$$

$$= \frac{\sigma_m}{\sigma_u} + \frac{\sigma_v \times K_f}{\sigma_e \times K_{sur} \times K_{sz}} \qquad ...(\because \sigma_{eb} = \sigma_e \times K_b \text{ and } K_b = 1)$$

where K_b = Load factor for reversed bending load,

K_{sur} = Surface finish factor, and

K_{sz} = Size factor.

---

* Here we have assumed the same factor of safety (*F.S.*) for the ultimate tensile strength (σ_u) and endurance limit (σ_e). In case the factor of safety relating to both these stresses is different, then the following relation may be used :

$$\frac{\sigma_v}{\sigma_e / (F.S.)_e} = 1 - \frac{\sigma_m}{\sigma_u / (F.S.)_u}$$

where $(F.S.)_e$ = Factor of safety relating to endurance limit, and

$(F.S.)_u$ = Factor of safety relating to ultimate tensile strength.

Notes : 1. The equation (*iii*) is applicable to ductile materials subjected to reversed bending loads (tensile or compressive). For brittle materials, the theoretical stress concentration factor (K_t) should be applied to the mean stress and fatigue stress concentration factor (K_f) to the variable stress. Thus for brittle materials, the equation (*iii*) may be written as

$$\frac{1}{F.S.} = \frac{\sigma_m \times K_t}{\sigma_u} + \frac{\sigma_v \times K_f}{\sigma_{eb} \times K_{sur} \times K_{sz}} \quad ...(iv)$$

2. When a machine component is subjected to a load other than reversed bending, then the endurance limit for that type of loading should be taken into consideration. Thus for reversed axial loading (tensile or compressive), the equations (*iii*) and (*iv*) may be written as

$$\frac{1}{F.S.} = \frac{\sigma_m}{\sigma_u} + \frac{\sigma_v \times K_f}{\sigma_{ea} \times K_{sur} \times K_{sz}} \quad ...\text{(For ductile materials)}$$

and

$$\frac{1}{F.S.} = \frac{\sigma_m \times K_t}{\sigma_u} + \frac{\sigma_v \times K_f}{\sigma_{ea} \times K_{sur} \times K_{sz}} \quad ...\text{(For brittle materials)}$$

Similarly, for reversed torsional or shear loading,

$$\frac{1}{F.S.} = \frac{\tau_m}{\tau_u} + \frac{\tau_v \times K_{fs}}{\tau_e \times K_{sur} \times K_{sz}} \quad ...\text{(For ductile materials)}$$

and

$$\frac{1}{F.S.} = \frac{\tau_m \times K_{ts}}{\tau_u} + \frac{\tau_v \times K_{fs}}{\tau_e \times K_{sur} \times K_{sz}} \quad ...\text{(For brittle materials)}$$

where suffix '*s*' denotes for shear.

For reversed torsional or shear loading, the values of ultimate shear strength (τ_u) and endurance shear strength (τ_e) may be taken as follows:

$$\tau_u = 0.8\,\sigma_u; \text{ and } \tau_e = 0.8\,\sigma_e$$

6.21 Soderberg Method for Combination of Stresses

A straight line connecting the endurance limit (σ_e) and the yield strength (σ_y), as shown by the line *AB* in Fig. 6.17, follows the suggestion of Soderberg line. This line is used when the design is based on yield strength.

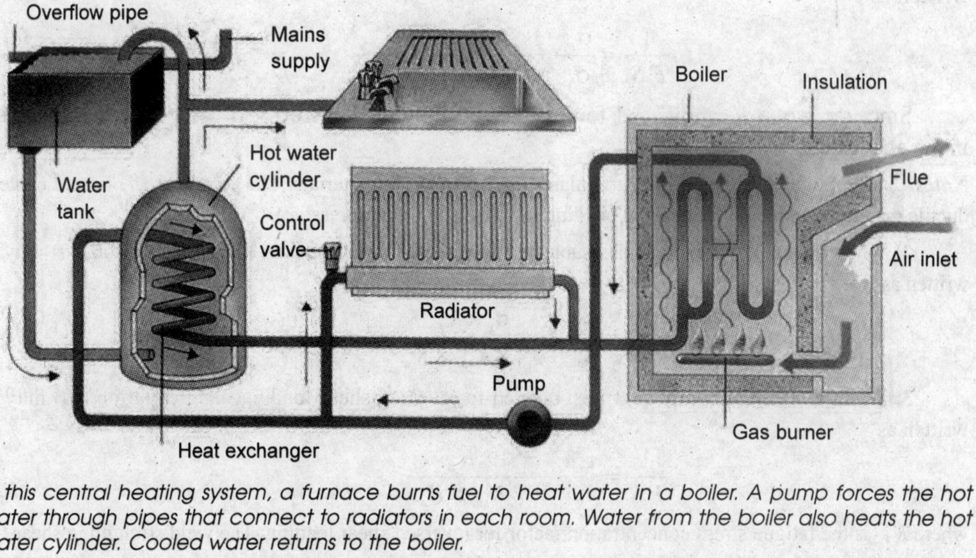

In this central heating system, a furnace burns fuel to heat water in a boiler. A pump forces the hot water through pipes that connect to radiators in each room. Water from the boiler also heats the hot water cylinder. Cooled water returns to the boiler.

Note : This picture is given as additional information and is not a direct example of the current chapter.

Proceeding in the same way as discussed in Art 6.20, the line AB connecting σ_e and σ_y, as shown in Fig. 6.17, is called ***Soderberg's failure stress line***. If a suitable factor of safety (*F.S.*) is applied to the endurance limit and yield strength, a safe stress line CD may be drawn parallel to the line AB. Let us consider a design point P on the line CD. Now from similar triangles COD and PQD,

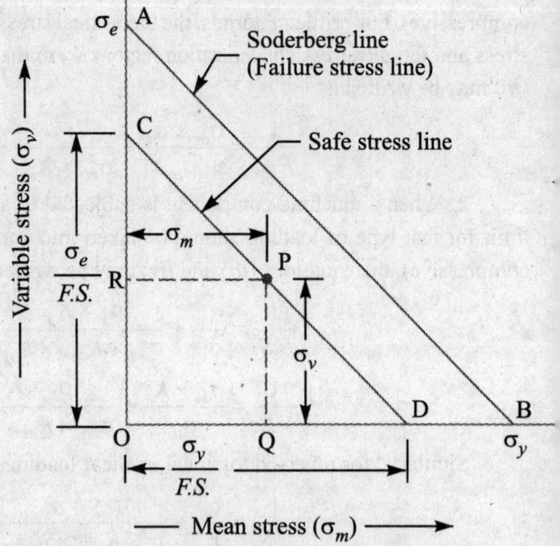

Fig. 6.17. Soderberg method.

$$\frac{PQ}{CO} = \frac{QD}{OD} = \frac{OD - OQ}{OD}$$

$$= 1 - \frac{OQ}{OD}$$

$$\ldots (\because QD = OD - OQ)$$

$$\therefore \quad \frac{\sigma_v}{\sigma_e / F.S.} = 1 - \frac{\sigma_m}{\sigma_y / F.S.}$$

or

$$\sigma_v = \frac{\sigma_e}{F.S.}\left[1 - \frac{\sigma_m}{\sigma_y / F.S.}\right] = \sigma_e\left[\frac{1}{F.S.} - \frac{\sigma_m}{\sigma_y}\right]$$

$$\therefore \quad \frac{1}{F.S.} = \frac{\sigma_m}{\sigma_y} + \frac{\sigma_v}{\sigma_e} \qquad \ldots (i)$$

For machine parts subjected to fatigue loading, the fatigue stress concentration factor (K_f) should be applied to only variable stress (σ_v). Thus the equations (*i*) may be written as

$$\frac{1}{F.S.} = \frac{\sigma_m}{\sigma_y} + \frac{\sigma_v \times K_f}{\sigma_e} \qquad \ldots (ii)$$

Considering the load factor, surface finish factor and size factor, the equation (*ii*) may be written as

$$\frac{1}{F.S.} = \frac{\sigma_m}{\sigma_y} + \frac{\sigma_v \times K_f}{\sigma_{eb} \times K_{sur} \times K_{sz}} \qquad \ldots (iii)$$

Since $\sigma_{eb} = \sigma_e \times K_b$ and $K_b = 1$ for reversed bending load, therefore $\sigma_{eb} = \sigma_e$ may be substituted in the above equation.

Notes: 1. The Soderberg method is particularly used for ductile materials. The equation (*iii*) is applicable to ductile materials subjected to reversed bending load (tensile or compressive).

2. When a machine component is subjected to reversed axial loading, then the equation (*iii*) may be written as

$$\frac{1}{F.S.} = \frac{\sigma_m}{\sigma_y} + \frac{\sigma_v \times K_f}{\sigma_{ea} \times K_{sur} \times K_{sz}}$$

3. When a machine component is subjected to reversed shear loading, then equation (*iii*) may be written as

$$\frac{1}{F.S.} = \frac{\tau_m}{\tau_y} + \frac{\tau_v \times K_{fs}}{\tau_e \times K_{sur} \times K_{sz}}$$

where K_{fs} is the fatigue stress concentration factor for reversed shear loading. The yield strength in shear (τ_y) may be taken as one-half the yield strength in reversed bending (σ_y).

Example 6.3. *A machine component is subjected to a flexural stress which fluctuates between + 300 MN/m² and – 150 MN/m². Determine the value of minimum ultimate strength according to 1. Gerber relation; 2. Modified Goodman relation; and 3. Soderberg relation.*

Take yield strength = 0.55 Ultimate strength; Endurance strength = 0.5 Ultimate strength; and factor of safety = 2.

Springs often undergo variable stresses.

Solution. Given : $\sigma_1 = 300$ MN/m² ; $\sigma_2 = -150$ MN/m² ; $\sigma_y = 0.55\ \sigma_u$; $\sigma_e = 0.5\ \sigma_u$; $F.S. = 2$

Let σ_u = Minimum ultimate strength in MN/m².

We know that the mean or average stress,

$$\sigma_m = \frac{\sigma_1 + \sigma_2}{2} = \frac{300 + (-150)}{2} = 75 \text{ MN/m}^2$$

and variable stress,

$$\sigma_v = \frac{\sigma_1 - \sigma_2}{2} = \frac{300 - (-150)}{2} = 225 \text{ MN/m}^2$$

1. ***According to Gerber relation***

We know that according to Gerber relation,

$$\frac{1}{F.S.} = \left(\frac{\sigma_m}{\sigma_u}\right)^2 F.S. + \frac{\sigma_v}{\sigma_e}$$

$$\frac{1}{2} = \left(\frac{75}{\sigma_u}\right)^2 2 + \frac{225}{0.5\sigma_u} = \frac{11\ 250}{(\sigma_u)^2} + \frac{450}{\sigma_u} = \frac{11\ 250 + 450\ \sigma_u}{(\sigma_u)^2}$$

$$(\sigma_u)^2 = 22\ 500 + 900\ \sigma_u$$

or

$$(\sigma_u)^2 - 900\ \sigma_u - 22\ 500 = 0$$

$$\therefore \quad \sigma_u = \frac{900 \pm \sqrt{(900)^2 + 4 \times 1 \times 22\ 500}}{2 \times 1} = \frac{900 \pm 948.7}{2}$$

$= 924.35$ MN/m² **Ans.** ...(Taking +ve sign)

2. ***According to modified Goodman relation***

We know that according to modified Goodman relation,

$$\frac{1}{F.S.} = \frac{\sigma_m}{\sigma_u} + \frac{\sigma_v}{\sigma_e}$$

or

$$\frac{1}{2} = \frac{75}{\sigma_u} + \frac{225}{0.5\ \sigma_u} = \frac{525}{\sigma_u}$$

$\therefore \quad \sigma_u = 2 \times 525 = 1050$ MN/m² **Ans.**

3. ***According to Soderberg relation***

We know that according to Soderberg relation,

$$\frac{1}{F.S.} = \frac{\sigma_m}{\sigma_y} + \frac{\sigma_v}{\sigma_e}$$

or

$$\frac{1}{2} = \frac{75}{0.55\ \sigma_u} + \frac{255}{0.5\ \sigma_u} = \frac{586.36}{\sigma_u}$$

$\therefore \quad \sigma_u = 2 \times 586.36 = 1172.72$ MN/m² **Ans.**

Example 6.4. *A bar of circular cross-section is subjected to alternating tensile forces varying from a minimum of 200 kN to a maximum of 500 kN. It is to be manufactured of a material with an ultimate tensile strength of 900 MPa and an endurance limit of 700 MPa. Determine the diameter of bar using safety factors of 3.5 related to ultimate tensile strength and 4 related to endurance limit and a stress concentration factor of 1.65 for fatigue load. Use Goodman straight line as basis for design.*

Solution. Given : W_{min} = 200 kN ; W_{max} = 500 kN ; σ_u = 900 MPa = 900 N/mm^2 ; σ_e = 700 MPa = 700 N/mm^2 ; $(F.S.)_u$ = 3.5 ; $(F.S.)_e$ = 4 ; K_f = 1.65

Let d = Diameter of bar in mm.

$$\therefore \quad \text{Area, } A = \frac{\pi}{4} \times d^2 = 0.7854\, d^2 \text{ mm}^2$$

We know that mean or average force,

$$W_m = \frac{W_{max} + W_{min}}{2} = \frac{500 + 200}{2} = 350 \text{ kN} = 350 \times 10^3 \text{ N}$$

$$\therefore \quad \text{Mean stress, } \sigma_m = \frac{W_m}{A} = \frac{350 \times 10^3}{0.7854\, d^2} = \frac{446 \times 10^3}{d^2} \text{ N/mm}^2$$

$$\text{Variable force, } W_v = \frac{W_{max} - W_{min}}{2} = \frac{500 - 200}{2} = 150 \text{ kN} = 150 \times 10^3 \text{ N}$$

$$\therefore \quad \text{Variable stress, } \sigma_v = \frac{W_v}{A} = \frac{150 \times 10^3}{0.7854\, d^2} = \frac{191 \times 10^3}{d^2} \text{ N/mm}^2$$

We know that according to Goodman's formula,

$$\frac{\sigma_v}{\sigma_e / (F.S.)_e} = 1 - \frac{\sigma_m . K_f}{\sigma_u / (F.S.)_u}$$

$$\frac{\dfrac{191 \times 10^3}{d^2}}{700/4} = 1 - \frac{\dfrac{446 \times 10^3}{d^2} \times 1.65}{900/3.5}$$

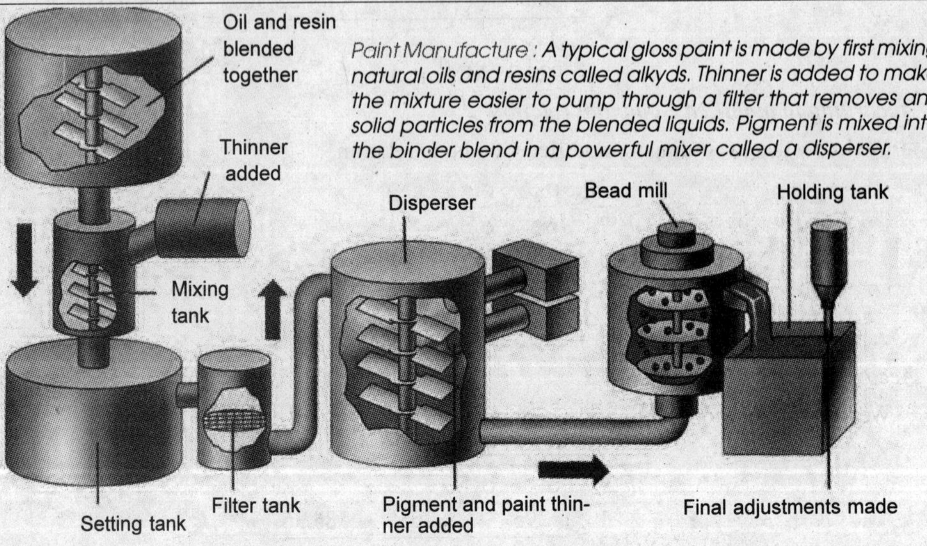

Paint Manufacture : A typical gloss paint is made by first mixing natural oils and resins called alkyds. Thinner is added to make the mixture easier to pump through a filter that removes any solid particles from the blended liquids. Pigment is mixed into the binder blend in a powerful mixer called a disperser.

Note : This picture is given as additional information and is not a direct example of the current chapter.

$$\frac{1100}{d^2} = 1 - \frac{2860}{d^2} \quad \text{or} \quad \frac{1100 + 2860}{d^2} = 1$$

$\therefore \quad d^2 = 3960$ or $d = 62.9$ say 63 mm **Ans.**

Example 6.5. *Determine the thickness of a 120 mm wide uniform plate for safe continuous operation if the plate is to be subjected to a tensile load that has a maximum value of 250 kN and a minimum value of 100 kN. The properties of the plate material are as follows:*

Endurance limit stress = 225 MPa, and Yield point stress = 300 MPa.

The factor of safety based on yield point may be taken as 1.5.

Solution. Given : $b = 120$ mm ; $W_{max} = 250$ kN; $W_{min} = 100$ kN ; $\sigma_e = 225$ MPa = 225 N/mm^2 ; $\sigma_y = 300$ MPa = 300 N/mm^2; $F.S. = 1.5$

Let $\quad t$ = Thickness of the plate in mm.

$\therefore \quad$ Area, $A = b \times t = 120\, t$ mm^2

We know that mean or average load,

$$W_m = \frac{W_{max} + W_{min}}{2} = \frac{250 + 100}{2} = 175 \text{ kN} = 175 \times 10^3 \text{ N}$$

$$\therefore \quad \text{Mean stress, } \sigma_m = \frac{W_m}{A} = \frac{175 \times 10^3}{120t} \text{ N/mm}^2$$

$$\text{Variable load, } W_v = \frac{W_{max} - W_{min}}{2} = \frac{250 - 100}{2} = 75 \text{ kN} = 75 \times 10^3 \text{ N}$$

$$\therefore \quad \text{Variable stress, } \sigma_v = \frac{W_v}{A} = \frac{75 \times 10^3}{120t} \text{ N/mm}^2$$

According to Soderberg's formula,

$$\frac{1}{F.S.} = \frac{\sigma_m}{\sigma_y} + \frac{\sigma_v}{\sigma_e}$$

$$\frac{1}{1.5} = \frac{175 \times 10^3}{120t \times 300} + \frac{75 \times 10^3}{120t \times 225} = \frac{4.86}{t} + \frac{2.78}{t} = \frac{7.64}{t}$$

$\therefore \quad t = 7.64 \times 1.5 = 11.46$ say 11.5 mm **Ans.**

Example 6.6. *Determine the diameter of a circular rod made of ductile material with a fatigue strength (complete stress reversal), σ_e = 265 MPa and a tensile yield strength of 350 MPa. The member is subjected to a varying axial load from $W_{min} = -300 \times 10^3$ N to $W_{max} = 700 \times 10^3$ N and has a stress concentration factor = 1.8. Use factor of safety as 2.0.*

Solution. Given : $\sigma_e = 265$ MPa = 265 N/mm^2 ; $\sigma_y = 350$ MPa = 350 N/mm^2 ; $W_{min} = -300 \times 10^3$ N ; $W_{max} = 700 \times 10^3$ N ; $K_f = 1.8$; $F.S. = 2$

Let $\quad d$ = Diameter of the circular rod in mm.

$$\therefore \quad \text{Area, } A = \frac{\pi}{4} \times d^2 = 0.7854\, d^2 \text{ mm}^2$$

We know that the mean or average load,

$$W_m = \frac{W_{max} + W_{min}}{2} = \frac{700 \times 10^3 + (-300 \times 10^3)}{2} = 200 \times 10^3 \text{ N}$$

$$\therefore \quad \text{Mean stress, } \sigma_m = \frac{W_m}{A} = \frac{200 \times 10^3}{0.7854\, d^2} = \frac{254.6 \times 10^3}{d^2} \text{ N/mm}^2$$

Variable load, $W_v = \frac{W_{max} - W_{min}}{2} = \frac{700 \times 10^3 - (-300 \times 10^3)}{2} = 500 \times 10^3$ N

$\therefore$ Variable stress, $\sigma_v = \frac{W_v}{A} = \frac{500 \times 10^3}{0.7854\ d^2} = \frac{636.5 \times 10^3}{d^2}$ N/mm^2

We know that according to Soderberg's formula,

$$\frac{1}{F.S.} = \frac{\sigma_m}{\sigma_y} + \frac{\sigma_v \times K_f}{\sigma_e}$$

$$\frac{1}{2} = \frac{254.6 \times 10^3}{d^2 \times 350} + \frac{636.5 \times 10^3 \times 1.8}{d^2 \times 265} = \frac{727}{d^2} + \frac{4323}{d^2} = \frac{5050}{d^2}$$

$\therefore \quad d^2 = 5050 \times 2 = 10\,100$ or $d = 100.5$ mm **Ans.**

Example 6.7. *A steel rod is subjected to a reversed axial load of 180 kN. Find the diameter of the rod for a factor of safety of 2. Neglect column action. The material has an ultimate tensile strength of 1070 MPa and yield strength of 910 MPa. The endurance limit in reversed bending may be assumed to be one-half of the ultimate tensile strength. Other correction factors may be taken as follows:*

For axial loading = 0.7; For machined surface = 0.8 ; For size = 0.85 ; For stress concentration = 1.0.

Solution. Given : W_{max} = 180 kN ; W_{min} = – 180 kN ; $F.S.$ = 2 ; σ_u = 1070 MPa = 1070 N/mm^2; σ_y = 910 MPa = 910 N/mm^2 ; $\sigma_e = 0.5\ \sigma_u$; K_a = 0.7 ; K_{sur} = 0.8 ; K_{sz} = 0.85 ; K_f = 1

Let $\quad d$ = Diameter of the rod in mm.

$\therefore \quad$ Area, $A = \frac{\pi}{4} \times d^2 = 0.7854\ d^2$ mm^2

We know that the mean or average load,

$$W_m = \frac{W_{max} + W_{min}}{2} = \frac{180 + (-180)}{2} = 0$$

$\therefore$ Mean stress, $\sigma_m = \frac{W_m}{A} = 0$

Variable load, $W_v = \frac{W_{max} - W_{min}}{2} = \frac{180 - (-180)}{2} = 180$ kN $= 180 \times 10^3$ N

$\therefore$ Variable stress, $\sigma_v = \frac{W_v}{A} = \frac{180 \times 10^3}{0.7854\ d^2} = \frac{229 \times 10^3}{d^2}$ N/mm^2

Endurance limit in reversed axial loading,

$$\sigma_{ea} = \sigma_e \times K_a = 0.5\ \sigma_u \times 0.7 = 0.35\ \sigma_u \quad ...(\because \sigma_e = 0.5\ \sigma_u)$$

$$= 0.35 \times 1070 = 374.5 \text{ N/mm}^2$$

We know that according to Soderberg's formula for reversed axial loading,

$$\frac{1}{F.S.} = \frac{\sigma_m}{\sigma_y} + \frac{\sigma_v \times K_f}{\sigma_{ea} \times K_{sur} \times K_{sz}}$$

$$\frac{1}{2} = 0 + \frac{229 \times 10^3 \times 1}{d^2 \times 374.5 \times 0.8 \times 0.85} = \frac{900}{d^2}$$

$\therefore \quad d^2 = 900 \times 2 = 1800$ or $d = 42.4$ mm **Ans.**

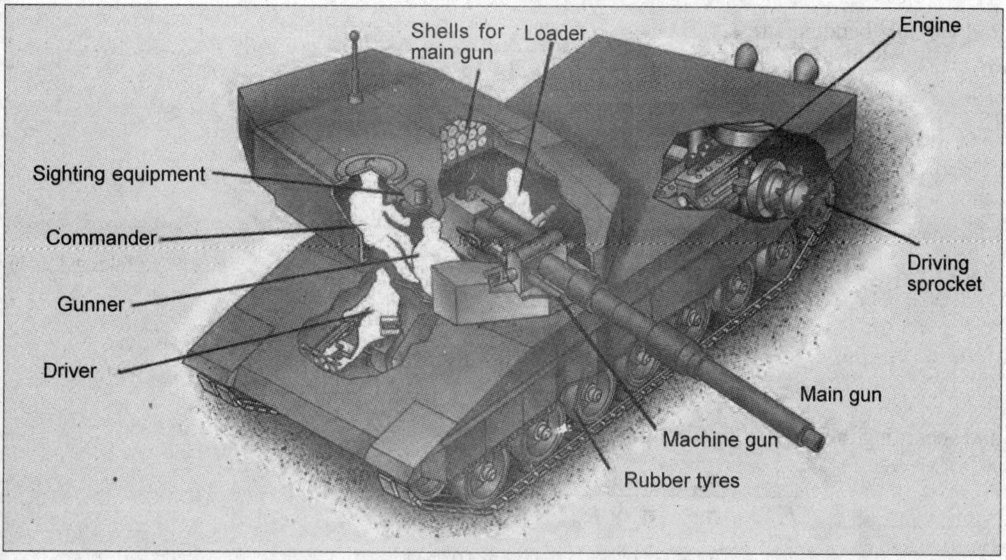

Layout of a military tank.

Note : This picture is given as additional information and is not a direct example of the current chapter.

Example 6.8. *A circular bar of 500 mm length is supported freely at its two ends. It is acted upon by a central concentrated cyclic load having a minimum value of 20 kN and a maximum value of 50 kN. Determine the diameter of bar by taking a factor of safety of 1.5, size effect of 0.85, surface finish factor of 0.9. The material properties of bar are given by : ultimate strength of 650 MPa, yield strength of 500 MPa and endurance strength of 350 MPa.*

Solution. Given : $l = 500$ mm ; $W_{min} = 20$ kN $= 20 \times 10^3$ N ; $W_{max} = 50$ kN $= 50 \times 10^3$ N ; F.S. = 1.5 ; $K_{sz} = 0.85$; $K_{sur} = 0.9$; $\sigma_u = 650$ MPa $= 650$ N/mm^2 ; $\sigma_y = 500$ MPa $= 500$ N/mm^2 ; $\sigma_e = 350$ MPa $= 350$ N/mm^2

Let d = Diameter of the bar in mm.

We know that the maximum bending moment,

$$M_{max} = \frac{W_{max} \times l}{4} = \frac{50 \times 10^3 \times 500}{4} = 6250 \times 10^3 \text{ N-mm}$$

and minimum bending moment,

$$M_{min} = \frac{W_{min} \times l}{4} = \frac{20 \times 10^3 \times 500}{4} = 2550 \times 10^3 \text{ N-mm}$$

∴ Mean or average bending moment,

$$M_m = \frac{M_{max} + M_{min}}{2} = \frac{6250 \times 10^3 + 2500 \times 10^3}{2} = 4375 \times 10^3 \text{ N-mm}$$

and variable bending moment,

$$M_v = \frac{M_{max} - M_{min}}{2} = \frac{6250 \times 10^3 - 2500 \times 10^3}{2} = 1875 \times 10^3 \text{ N-mm}$$

Section modulus of the bar,

$$Z = \frac{\pi}{32} \times d^3 = 0.0982\, d^3 \text{ mm}^3$$

∴ Mean or average bending stress,

$$\sigma_m = \frac{M_m}{Z} = \frac{4375 \times 10^3}{0.0982\, d^3} = \frac{44.5 \times 10^6}{d^3} \text{ N/mm}^2$$

and variable bending stress,

$$\sigma_v = \frac{M_v}{Z} = \frac{1875 \times 10^3}{0.0982\,d^3} = \frac{19.1 \times 10^6}{d^3} \text{ N/mm}^2$$

We know that according to Goodman's formula,

$$\frac{1}{F.S.} = \frac{\sigma_m}{\sigma_u} + \frac{\sigma_v \times K_f}{\sigma_e \times K_{sur} \times K_{sz}}$$

$$\frac{1}{1.5} = \frac{44.5 \times 10^6}{d^3 \times 650} + \frac{19.1 \times 10^6 \times 1}{d^3 \times 350 \times 0.9 \times 0.85} \qquad \text{...(Taking } K_f = 1)$$

$$= \frac{68 \times 10^3}{d^3} + \frac{71 \times 10^3}{d^3} = \frac{139 \times 10^3}{d^3}$$

$\therefore \quad d^3 = 139 \times 10^3 \times 1.5 = 209 \times 10^3$ or $d = 59.3$ mm

and according to Soderberg's formula,

$$\frac{1}{F.S.} = \frac{\sigma_m}{\sigma_y} + \frac{\sigma_v \times K_f}{\sigma_e \times K_{sur} \times K_{sz}}$$

$$\frac{1}{1.5} = \frac{44.5 \times 10^6}{d^3 \times 500} + \frac{19.1 \times 10^6 \times 1}{d^3 \times 350 \times 0.9 \times 0.85} \qquad \text{...(Taking } K_f = 1)$$

$$= \frac{89 \times 10^3}{d^3} + \frac{71 \times 10^3}{d^3} = \frac{160 \times 10^3}{d^3}$$

$\therefore \quad d^3 = 160 \times 10^3 \times 1.5 = 240 \times 10^3$ or $d = 62.1$ mm

Taking larger of the two values, we have $d = 62.1$ mm **Ans.**

Example 6.9. *A 50 mm diameter shaft is made from carbon steel having ultimate tensile strength of 630 MPa. It is subjected to a torque which fluctuates between 2000 N-m to – 800 N-m. Using Soderberg method, calculate the factor of safety. Assume suitable values for any other data needed.*

Solution. Given : $d = 50$ mm ; $\sigma_u = 630$ MPa $= 630$ N/mm² ; $T_{max} = 2000$ N-m ; $T_{min} = -800$ N-m

We know that the mean or average torque,

$$T_m = \frac{T_{max} + T_{min}}{2} = \frac{2000 + (-800)}{2} = 600 \text{ N-m} = 600 \times 10^3 \text{ N-mm}$$

$\therefore$ Mean or average shear stress,

$$\tau_m = \frac{16\,T_m}{\pi d^3} = \frac{16 \times 600 \times 10^3}{\pi (50)^3} = 24.4 \text{ N/mm}^2 \qquad \text{...}\left(\because T = \frac{\pi}{16} \times \tau \times d^3\right)$$

Variable torque,

$$T_v = \frac{T_{max} - T_{min}}{2} = \frac{2000 - (-800)}{2} = 1400 \text{ N-m} = 1400 \times 10^3 \text{ N-mm}$$

$$\therefore \text{ Variable shear stress, } \tau_v = \frac{16\,T_v}{\pi d^3} = \frac{16 \times 1400 \times 10^3}{\pi (50)^3} = 57 \text{ N/mm}^2$$

Since the endurance limit in reversed bending (σ_e) is taken as one-half the ultimate tensile strength (*i.e.* $\sigma_e = 0.5\,\sigma_u$) and the endurance limit in shear (τ_e) is taken as 0.55 σ_e, therefore

$$\tau_e = 0.55\,\sigma_e = 0.55 \times 0.5\,\sigma_u = 0.275\,\sigma_u$$

$$= 0.275 \times 630 = 173.25 \text{ N/mm}^2$$

Assume the yield stress (σ_y) for carbon steel in reversed bending as 510 N/mm², surface finish factor (K_{sur}) as 0.87, size factor (K_{sz}) as 0.85 and fatigue stress concentration factor (K_{fs}) as 1.

Army Tank

Note : This picture is given as additional information and is not a direct example of the current chapter.

Since the yield stress in shear (τ_y) for shear loading is taken as one-half the yield stress in reversed bending (σ_y), therefore

$$\tau_y = 0.5\ \sigma_y = 0.5 \times 510 = 255 \text{ N/mm}^2$$

Let $\quad F.S.$ = Factor of safety.

We know that according to Soderberg's formula,

$$\frac{1}{F.S.} = \frac{\tau_m}{\tau_y} + \frac{\tau_v \times K_{fs}}{\tau_e \times K_{sur} \times K_{sz}} = \frac{24.4}{255} + \frac{57 \times 1}{173.25 \times 0.87 \times 0.85}$$

$$= 0.096 + 0.445 = 0.541$$

$\therefore \quad F.S. = 1 / 0.541 = 1.85$ **Ans.**

Example 6.10. *A cantilever beam made of cold drawn carbon steel of circular cross-section as shown in Fig. 6.18, is subjected to a load which varies from – F to 3 F. Determine the maximum load that this member can withstand for an indefinite life using a factor of safety as 2. The theoretical stress concentration factor is 1.42 and the notch sensitivity is 0.9. Assume the following values :*

Ultimate stress = 550 MPa
Yield stress = 470 MPa
Endurance limit = 275 MPa
Size factor = 0.85
Surface finish factor = 0.89

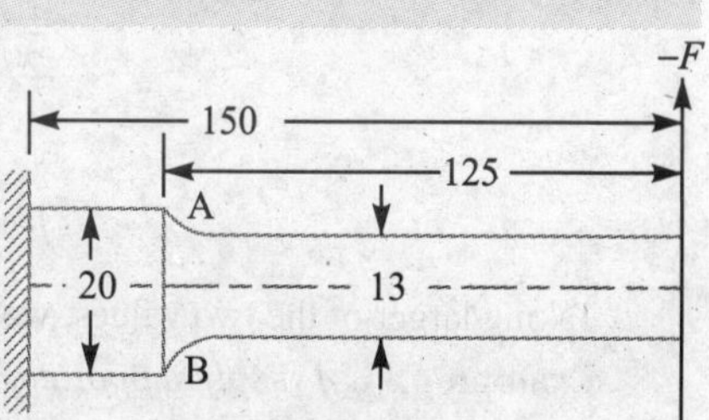

All dimensions in mm.

Fig. 6.18

Solution. Given : $W_{min} = -F$; $W_{max} = 3F$; $F.S. = 2$; $K_t = 1.42$; $q = 0.9$; $\sigma_u = 550$ MPa $= 550$ N/mm^2 ; $\sigma_y = 470$ MPa $= 470$ N/mm^2 ; $\sigma_e = 275$ MPa $= 275$ N/mm^2 ; $K_{sz} = 0.85$; $K_{sur} = 0.89$

The beam as shown in Fig. 6.18 is subjected to a reversed bending load only. Since the point *A* at the change of cross section is critical, therefore we shall find the bending moment at point *A*.

We know that maximum bending moment at point *A*,

$$M_{max} = W_{max} \times 125 = 3F \times 125 = 375\, F \text{ N-mm}$$

and minimum bending moment at point *A*,

$$M_{min} = W_{min} \times 125 = -F \times 125 = -125\, F \text{ N-mm}$$

∴ Mean or average bending moment,

$$M_m = \frac{M_{max} + M_{min}}{2} = \frac{375\,F + (-125\,F)}{2} = 125\,F \text{ N-mm}$$

and variable bending moment,

$$M_v = \frac{M_{max} - M_{min}}{2} = \frac{375\,F - (-125\,F)}{2} = 250\,F \text{ N-mm}$$

Section modulus, $Z = \frac{\pi}{32} \times d^3 = \frac{\pi}{32}(13)^3 = 215.7 \text{ mm}^3$ ...(∵ d = 13 mm)

∴ Mean bending stress, $\sigma_m = \frac{M_m}{Z} = \frac{125\,F}{215.7} = 0.58\,F \text{ N/mm}^2$

and variable bending stress, $\sigma_v = \frac{M_v}{Z} = \frac{250\,F}{215.7} = 1.16\,F \text{ N/mm}^2$

Fatigue stress concentration factor, $K_f = 1 + q\,(K_t - 1) = 1 + 0.9\,(1.42 - 1) = 1.378$

We know that according to Goodman's formula

$$\frac{1}{F.S.} = \frac{\sigma_m}{\sigma_u} + \frac{\sigma_v \times K_f}{\sigma_e \times K_{sur} \times K_{sz}}$$

$$\frac{1}{2} = \frac{0.58\,F}{550} + \frac{1.16\,F \times 1.378}{275 \times 0.89 \times 0.85}$$

$$= 0.001\,05\,F + 0.007\,68\,F = 0.008\,73\,F$$

$$\therefore \quad F = \frac{1}{2 \times 0.008\,73} = 57.3 \text{ N}$$

and according to Soderberg's formula,

$$\frac{1}{F.S.} = \frac{\sigma_m}{\sigma_y} + \frac{\sigma_v \times K_f}{\sigma_e \times K_{sur} \times K_{sz}}$$

$$\frac{1}{2} = \frac{0.58\,F}{470} + \frac{1.16\,F \times 1.378}{275 \times 0.89 \times 0.85}$$

$$= 0.001\,23\,F + 0.007\,68\,F = 0.008\,91\,F$$

$$\therefore \quad F = \frac{1}{2 \times 0.008\,91} = 56 \text{ N}$$

Taking larger of the two values, we have F = 57.3 N **Ans.**

Example 6.11. *A simply supported beam has a concentrated load at the centre which fluctuates from a value of P to 4 P. The span of the beam is 500 mm and its cross-section is circular with a diameter of 60 mm. Taking for the beam material an ultimate stress of 700 MPa, a yield stress of 500 MPa, endurance limit of 330 MPa for reversed bending, and a factor of safety of 1.3, calculate the maximum value of P. Take a size factor of 0.85 and a surface finish factor of 0.9.*

Solution. Given : $W_{min} = P$; $W_{max} = 4P$; L = 500 mm; d = 60 mm ; σ_u = 700 MPa = 700 N/mm^2 ; σ_y = 500 MPa = 500 N/mm^2 ; σ_e = 330 MPa = 330 N/mm^2 ; $F.S.$ = 1.3 ; K_{sz} = 0.85 ; K_{sur} = 0.9

We know that maximum bending moment,

$$M_{max} = \frac{W_{max} \times L}{4} = \frac{4P \times 500}{4} = 500 P \text{ N-mm}$$

and minimum bending moment,

$$M_{min} = \frac{W_{min} \times L}{4} = \frac{P \times 500}{4} = 125 P \text{ N-mm}$$

∴ Mean or average bending moment,

$$M_m = \frac{M_{max} + M_{min}}{2} = \frac{500 P + 125 P}{2} = 312.5 P \text{ N-mm}$$

and variable bending moment,

$$M_v = \frac{M_{max} - M_{min}}{2} = \frac{500 P - 125 P}{2} = 187.5 P \text{ N-mm}$$

Section modulus, $Z = \frac{\pi}{32} \times d^3 = \frac{\pi}{32} (60)^3 = 21.21 \times 10^3 \text{ mm}^3$

∴ Mean bending stress,

$$\sigma_m = \frac{M_m}{Z} = \frac{312.5 P}{21.21 \times 10^3} = 0.0147 P \text{ N/mm}^2$$

and variable bending stress,

$$\sigma_v = \frac{M_v}{Z} = \frac{187.5 P}{21.21 \times 10^3} = 0.0088 P \text{ N/mm}^2$$

We know that according to Goodman's formula,

$$\frac{1}{F.S.} = \frac{\sigma_m}{\sigma_u} + \frac{\sigma_v \times K_f}{\sigma_e \times K_{sur} \times K_{sz}}$$

$$\frac{1}{1.3} = \frac{0.0147 P}{700} + \frac{0.0088 P \times 1}{330 \times 0.9 \times 0.85} \quad \text{...(Taking } K_f = 1)$$

$$= \frac{21 P}{10^6} + \frac{34.8 P}{10^6} = \frac{55.8 P}{10^6}$$

$$\therefore \quad P = \frac{1}{1.3} \times \frac{10^6}{55.8} = 13\ 785 \text{ N} = 13.785 \text{ kN}$$

and according to Soderberg's formula,

$$\frac{1}{F.S.} = \frac{\sigma_m}{\sigma_y} + \frac{\sigma_v \times K_f}{\sigma_e \times K_{sur} \times K_{sz}}$$

$$\frac{1}{1.3} = \frac{0.0147 P}{500} + \frac{0.0088 P \times 1}{330 \times 0.9 \times 0.85} = \frac{29.4 P}{10^6} + \frac{34.8 P}{10^6} = \frac{64.2 P}{10^6}$$

$$\therefore \quad P = \frac{1}{1.3} \times \frac{10^6}{64.2} = 11\ 982 \text{ N} = 11.982 \text{ kN}$$

From the above, we find that maximum value of $P = 13.785$ kN **Ans.**

6.22 Combined Variable Normal Stress and Variable Shear Stress

When a machine part is subjected to both variable normal stress and a variable shear stress; then it is designed by using the following two theories of combined stresses :

1. Maximum shear stress theory, and **2.** Maximum normal stress theory.

We have discussed in Art. 6.21, that according to Soderberg's formula,

$$\frac{1}{F.S.} = \frac{\sigma_m}{\sigma_y} + \frac{\sigma_v \times K_{fb}}{\sigma_{eb} \times K_{sur} \times K_{sz}} \qquad \text{...(For reversed bending load)}$$

Multiplying throughout by σ_y, we get

$$\frac{\sigma_y}{F.S.} = \sigma_m + \frac{\sigma_v \times \sigma_y \times K_{fb}}{\sigma_{eb} \times K_{sur} \times K_{sz}}$$

The term on the right hand side of the above expression is known as ***equivalent normal stress*** due to reversed bending.

∴ Equivalent normal stress due to reversed bending,

$$\sigma_{neb} = \sigma_m + \frac{\sigma_v \times \sigma_y \times K_{fb}}{\sigma_{eb} \times K_{sur} \times K_{sz}} \qquad \text{...}(i)$$

Similarly, equivalent normal stress due to reversed axial loading,

$$\sigma_{nea} = \sigma_m + \frac{\sigma_v \times \sigma_y \times K_{fa}}{\sigma_{ea} \times K_{sur} \times K_{sz}} \qquad \text{...}(ii)$$

and total equivalent normal stress,

$$\sigma_{ne} = \sigma_{neb} + \sigma_{nea} = \frac{\sigma_y}{F.S.} \qquad \text{...}(iii)$$

We have also discussed in Art. 6.21, that for reversed torsional or shear loading,

$$\frac{1}{F.S.} = \frac{\tau_m}{\tau_y} + \frac{\tau_v \times K_{fs}}{\tau_e \times K_{sur} \times K_{sz}}$$

Multiplying throughout by τ_y, we get

$$\frac{\tau_y}{F.S.} = \tau_m + \frac{\tau_v \times \tau_y \times K_{fs}}{\tau_e \times K_{sur} \times K_{sz}}$$

The term on the right hand side of the above expression is known as ***equivalent shear stress.***

∴ Equivalent shear stress due to reversed torsional or shear loading,

$$\tau_{es} = \tau_m + \frac{\tau_v \times \tau_y \times K_{fs}}{\tau_e \times K_{sur} \times K_{sz}} \qquad \text{...}(iv)$$

The maximum shear stress theory is used in designing machine parts of ductile materials. According to this theory, maximum equivalent shear stress,

$$\tau_{es(max)} = \frac{1}{2}\sqrt{(\sigma_{ne})^2 + 4\,(\tau_{es})^2} = \frac{\tau_y}{F.S.}$$

The maximum normal stress theory is used in designing machine parts of brittle materials. According to this theory, maximum equivalent normal stress,

$$\sigma_{ne(max)} = \frac{1}{2}(\sigma_{ne}) + \frac{1}{2}\sqrt{(\sigma_{ne})^2 + 4\,(\tau_{es})^2} = \frac{\sigma_y}{F.S.}$$

Example 6.12. *A steel cantilever is 200 mm long. It is subjected to an axial load which varies from 150 N (compression) to 450 N (tension) and also a transverse load at its free end which varies from 80 N up to 120 N down. The cantilever is of circular cross-section. It is of diameter 2d for the first 50 mm and of diameter d for the remaining length. Determine its diameter taking a factor of safety of 2. Assume the following values :*

Yield stress	= *330 MPa*
Endurance limit in reversed loading	= *300 MPa*
Correction factors	= *0.7 in reversed axial loading*
	= *1.0 in reversed bending*

Stress concentration factor	$= 1.44$ *for bending*
	$= 1.64$ *for axial loading*
Size effect factor	$= 0.85$
Surface effect factor	$= 0.90$
Notch sensitivity index	$= 0.90$

Solution. Given : $l = 200$ mm; $W_{a(max)} = 450$ N; $W_{a(min)} = -150$ N ; $W_{t(max)} = 120$ N ; $W_{t(min)} = -80$ N; *F.S.* $=2$; $\sigma_y = 330$ MPa $= 330$ N/mm^2 ; $\sigma_e = 300$ MPa $= 300$ N/mm^2 ; $K_a = 0.7$; $K_b = 1$; $K_{tb} = 1.44$; $K_{ta} = 1.64$; $K_{sz} = 0.85$; $K_{sur} = 0.90$; $q = 0.90$

First of all, let us find the equivalent normal stress for point *A* which is critical as shown in Fig. 6.19. It is assumed that the equivalent normal stress at this point will be the algebraic sum of the equivalent normal stress due to axial loading and equivalent normal stress due to bending (*i.e.* due to transverse load acting at the free end).

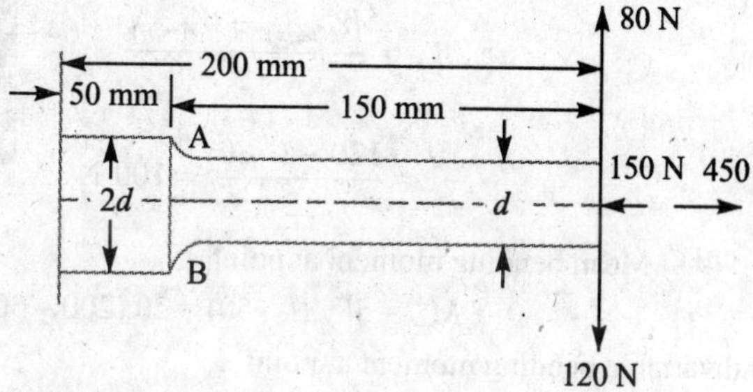

Fig. 6.19

Let us first consider the reversed axial loading. We know that mean or average axial load,

$$W_m = \frac{W_{a(max)} + W_{a(min)}}{2} = \frac{450 + (-150)}{2} = 150 \text{ N}$$

and variable axial load,

$$W_v = \frac{W_{a(max)} - W_{a(min)}}{2} = \frac{450 - (-150)}{2} = 300 \text{ N}$$

∴ Mean or average axial stress,

$$\sigma_m = \frac{W_m}{A} = \frac{150 \times 4}{\pi d^2} = \frac{191}{d^2} \text{ N/mm}^2 \quad \ldots\left(\because A = \frac{\pi}{4} \times d^2\right)$$

and variable axial stress,

$$\sigma_v = \frac{W_v}{A} = \frac{300 \times 4}{\pi d^2} = \frac{382}{d^2} \text{ N/mm}^2$$

We know that fatigue stress concentration factor for reversed axial loading,

$$K_{fa} = 1 + q\,(K_{ta} - 1) = 1 + 0.9\,(1.64 - 1) = 1.576$$

and endurance limit stress for reversed axial loading,

$$\sigma_{ea} = \sigma_e \times K_a = 300 \times 0.7 = 210 \text{ N/mm}^2$$

We know that equivalent normal stress at point *A* due to axial loading,

$$\sigma_{nea} = \sigma_m + \frac{\sigma_v \times \sigma_y \times K_{fa}}{\sigma_{ea} \times K_{sur} \times K_{sz}} = \frac{191}{d^2} + \frac{382 \times 330 \times 1.576}{d^2 \times 210 \times 0.9 \times 0.85}$$

$$= \frac{191}{d^2} + \frac{1237}{d^2} = \frac{1428}{d^2} \text{ N/mm}^2$$

Machine transporter

Now let us consider the reversed bending due to transverse load. We know that mean or average bending load,

$$W_m = \frac{W_{t(max)} + W_{t(min)}}{2}$$

$$= \frac{120 + (-80)}{2} = 20 \text{ N}$$

and variable bending load,

$$W_v = \frac{W_{t(max)} - W_{t(min)}}{2}$$

$$= \frac{120 - (-80)}{2} = 100 \text{ N}$$

∴ Mean bending moment at point A,

$$M_m = W_m (l - 50) = 20\,(200 - 50) = 3000 \text{ N-mm}$$

and variable bending moment at point A,

$$M_v = W_v (l - 50) = 100\,(200 - 50) = 15\,000 \text{ N-mm}$$

We know that section modulus,

$$Z = \frac{\pi}{32} \times d^3 = 0.0982\, d^3 \text{ mm}^3$$

∴ Mean or average bending stress,

$$\sigma_m = \frac{M_m}{Z} = \frac{3000}{0.0982\, d^3} = \frac{30\,550}{d^3} \text{ N/mm}^2$$

and variable bending stress,

$$\sigma_v = \frac{M_v}{Z} = \frac{15\,000}{0.0982\, d^3} = \frac{152\,750}{d^3} \text{ N/mm}^2$$

We know that fatigue stress concentration factor for reversed bending,

$$K_{fb} = 1 + q\,(K_{tb} - 1) = 1 + 0.9\,(1.44 - 1) = 1.396$$

Since the correction factor for reversed bending load is 1 (*i.e.* $K_b = 1$), therefore the endurance limit for reversed bending load,

$$\sigma_{eb} = \sigma_e \,.\, K_b = \sigma_e = 300 \text{ N/mm}^2$$

We know that the equivalent normal stress at point A due to bending,

$$\sigma_{neb} = \sigma_m + \frac{\sigma_v \times \sigma_y \times K_{fb}}{\sigma_{eb} \times K_{sur} \times K_{sz}} = \frac{30\,550}{d^3} + \frac{152\,750 \times 330 \times 1.396}{d^3 \times 300 \times 0.9 \times 0.85}$$

$$= \frac{30\,550}{d^3} + \frac{306\,618}{d^3} = \frac{337\,168}{d^3} \text{ N/mm}^2$$

∴ Total equivalent normal stress at point A,

$$\sigma_{ne} = \sigma_{neb} + \sigma_{nea} = \frac{337\,168}{d^3} + \frac{1428}{d^2} \text{ N/mm}^2 \qquad ...(i)$$

We know that equivalent normal stress at point A,

$$\sigma_{ne} = \frac{\sigma_y}{F.S.} = \frac{330}{2} = 165 \text{ N/mm}^2 \qquad ...(ii)$$

Equating equations (*i*) and (*ii*), we have

$$\frac{337\,168}{d^3} + \frac{1428}{d^2} = 165 \quad \text{or} \quad 337\,168 + 1428\,d = 165\,d^3$$

$$\therefore \quad 236.1 + d = 0.116\,d^3 \text{ or } d = 12.9 \text{ mm} \quad \textbf{Ans.} \qquad ...\text{(By hit and trial)}$$

Example 6.13. *A hot rolled steel shaft is subjected to a torsional moment that varies from 330 N-m clockwise to 110 N-m counterclockwise and an applied bending moment at a critical section varies from 440 N-m to – 220 N-m. The shaft is of uniform cross-section and no keyway is present at the critical section. Determine the required shaft diameter. The material has an ultimate strength of 550 MN/m² and a yield strength of 410 MN/m². Take the endurance limit as half the ultimate strength, factor of safety of 2, size factor of 0.85 and a surface finish factor of 0.62.*

Solution. Given : T_{max} = 330 N-m (clockwise) ; T_{min} = 110 N-m (counterclockwise) = – 110 N-m (clockwise) ; M_{max} = 440 N-m ; M_{min} = – 220 N-m ; $\sigma_u = 550 \text{ MN/m}^2 = 550 \times 10^6 \text{ N/m}^2$; $\sigma_y = 410 \text{ MN/m}^2 = 410 \times 10^6 \text{ N/m}^2$; $\sigma_e = \frac{1}{2}\sigma_u = 275 \times 10^6 \text{ N/m}^2$; $F.S. = 2$; $K_{sz} = 0.85$; $K_{sur} = 0.62$

Let d = Required shaft diameter in metres.

We know that mean torque,

$$T_m = \frac{T_{max} + T_{min}}{2} = \frac{330 + (-110)}{2} = 110 \text{ N-m}$$

and variable torque,

$$T_v = \frac{T_{max} - T_{min}}{2} = \frac{330 - (-110)}{2} = 220 \text{ N-m}$$

$\therefore$ Mean shear stress,

$$\tau_m = \frac{16\,T_m}{\pi\,d^3} = \frac{16 \times 110}{\pi d^3} = \frac{560}{d^3} \text{ N/m}^2$$

and variable shear stress,

$$\tau_v = \frac{16\,T_v}{\pi\,d^3} = \frac{16 \times 220}{\pi d^3} = \frac{1120}{d^3} \text{ N/m}^2$$

Since the endurance limit in shear (τ_e) is 0.55 σ_e, and yield strength in shear (τ_y) is 0.5 σ_y, therefore

$$\tau_e = 0.55 \times 275 \times 10^6 = 151.25 \times 10^6 \text{ N/m}^2$$

and

$$\tau_y = 0.5 \times 410 \times 10^6 = 205 \times 10^6 \text{ N/m}^2$$

We know that equivalent shear stress,

$$\tau_{es} = \tau_m + \frac{\tau_v \times \tau_y\,K_{fs}}{\tau_e \times K_{sur} \times K_{sz}}$$

$$= \frac{560}{d^3} + \frac{1120 \times 205 \times 10^6 \times 1}{d^3 \times 151.25 \times 10^6 \times 0.62 \times 0.85} \qquad ...(\text{Taking } K_{fs} = 1)$$

$$= \frac{560}{d^3} + \frac{2880}{d^3} = \frac{3440}{d^3} \text{ N/m}^2$$

Mean or average bending moment,

$$M_m = \frac{M_{max} + M_{min}}{2} = \frac{440 + (-220)}{2} = 110 \text{ N-m}$$

and variable bending moment,

$$M_v = \frac{M_{max} - M_{min}}{2} = \frac{440 - (-220)}{2} = 330 \text{ N-m}$$

Section modulus, $Z = \frac{\pi}{32} \times d^3 = 0.0982\ d^3 \text{ m}^3$

$\therefore$ Mean bending stress,

$$\sigma_m = \frac{M_m}{Z} = \frac{110}{0.0982\ d^3} = \frac{1120}{d^3} \text{ N/m}^2$$

and variable bending stress,

$$\sigma_v = \frac{M_v}{Z} = \frac{330}{0.0982\ d^3} = \frac{3360}{d^3} \text{ N/m}^2$$

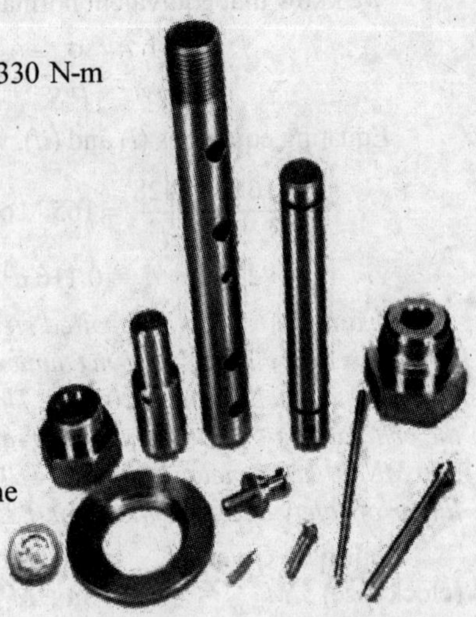

Machine parts are often made of alloys to improve their mechanical properties.

Since there is no reversed axial loading, therefore the equivalent normal stress due to reversed bending load,

$$\sigma_{neb} = \sigma_{ne} = \sigma_m + \frac{\sigma_v \times \sigma_y \times K_{fb}}{\sigma_{eb} \times K_{sur} \times K_{sz}}$$

$$= \frac{1120}{d^3} + \frac{3360 \times 410 \times 10^6 \times 1}{d^3 \times 275 \times 10^6 \times 0.62 \times 0.85}$$

...(Taking $K_{fb} = 1$ and $\sigma_{eb} = \sigma_e$)

$$= \frac{1120}{d^3} + \frac{9506}{d^3} = \frac{10\,626}{d^3} \text{ N/m}^2$$

We know that the maximum equivalent shear stress,

$$\tau_{es(max)} = \frac{\tau_y}{F.S.} = \frac{1}{2}\sqrt{(\sigma_{ne})^2 + 4\,(\tau_{es})^2}$$

$$\frac{205 \times 10^6}{2} = \frac{1}{2}\sqrt{\left(\frac{10\,625}{d^3}\right)^2 + 4\left(\frac{3440}{d^3}\right)^2}$$

$$205 \times 10^6 \times d^3 = \sqrt{113 \times 10^6 + 4 \times 11.84 \times 10^6} = 12.66 \times 10^3$$

$$\therefore \quad d^3 = \frac{12.66 \times 10^3}{205 \times 10^6} = \frac{0.0617}{10^3}$$

or $\quad d = \frac{0.395}{10} = 0.0395 \text{ m} = 39.5$ say 40 mm **Ans.**

Example 6.14. *A pulley is keyed to a shaft midway between two bearings. The shaft is made of cold drawn steel for which the ultimate strength is 550 MPa and the yield strength is 400 MPa. The bending moment at the pulley varies from – 150 N-m to + 400 N-m as the torque on the shaft varies from – 50 N-m to + 150 N-m. Obtain the diameter of the shaft for an indefinite life. The stress concentration factors for the keyway at the pulley in bending and in torsion are 1.6 and 1.3 respectively. Take the following values:*

Factor of safety	= *1.5*
Load correction factors	= *1.0 in bending, and 0.6 in torsion*
Size effect factor	= *0.85*
Surface effect factor	= *0.88*

Solution. Given : σ_u = 550 MPa = 550 N/mm^2 ; σ_y = 400 MPa = 400 N/mm^2 ; M_{min} = – 150 N-m; M_{max} = 400 N-m ; T_{min} = – 50 N-m ; T_{max} = 150 N-m ; K_{fb} = 1.6 ; K_{fs} = 1.3 ; *F.S.* = 1.5 ; K_b = 1 ; K_s = 0.6 ; K_{sz} = 0.85 ; K_{sur} = 0.88

Let d = Diameter of the shaft in mm.

First of all, let us find the equivalent normal stress due to bending.

We know that the mean or average bending moment,

$$M_m = \frac{M_{max} + M_{min}}{2} = \frac{400 + (-150)}{2} = 125 \text{ N-m} = 125 \times 10^3 \text{ N-mm}$$

and variable bending moment,

$$M_v = \frac{M_{max} - M_{min}}{2} = \frac{400 - (-150)}{2} = 275 \text{ N-m} = 275 \times 10^3 \text{ N-mm}$$

Section modulus, $Z = \frac{\pi}{32} \times d^3 = 0.0982\, d^3 \text{ mm}^3$

∴ Mean bending stress,

$$\sigma_m = \frac{M_m}{Z} = \frac{125 \times 10^3}{0.0982\, d^3} = \frac{1273 \times 10^3}{d^3} \text{ N/mm}^2$$

and variable bending stress,

$$\sigma_v = \frac{M_v}{Z} = \frac{275 \times 10^3}{0.0982\, d^3} = \frac{2800 \times 10^3}{d^3} \text{ N/mm}^2$$

Assuming the endurance limit in reversed bending as one-half the ultimate strength and since the load correction factor for reversed bending is 1 (*i.e.* $K_b = 1$), therefore endurance limit in reversed bending,

$$\sigma_{eb} = \sigma_e = \frac{\sigma_u}{2} = \frac{550}{2} = 275 \text{ N/mm}^2$$

Since there is no reversed axial loading, therefore equivalent normal stress due to bending,

$$\sigma_{neb} = \sigma_{ne} = \sigma_m + \frac{\sigma_v \times \sigma_y \times K_{fb}}{\sigma_{eb} \times K_{sur} \times K_{sz}}$$

$$= \frac{1273 \times 10^3}{d^3} + \frac{2800 \times 10^3 \times 400 \times 1.6}{d^3 \times 275 \times 0.88 \times 0.85}$$

$$= \frac{1273 \times 10^3}{d^3} + \frac{8712 \times 10^3}{d^3} = \frac{9985 \times 10^3}{d^3} \text{ N/mm}^2$$

Now let us find the equivalent shear stress due to torsional moment. We know that the mean torque,

$$T_m = \frac{T_{max} + T_{min}}{2} = \frac{150 + (-50)}{2} = 50 \text{ N-m} = 50 \times 10^3 \text{ N-mm}$$

and variable torque, $$T_v = \frac{T_{max} - T_{min}}{2} = \frac{150 - (-50)}{2} = 100 \text{ N-m} = 100 \times 10^3 \text{ N-mm}$$

∴ Mean shear stress,

$$\tau_m = \frac{16\, T_m}{\pi d^3} = \frac{16 \times 50 \times 10^3}{\pi d^3} = \frac{255 \times 10^3}{d^3} \text{ N/mm}^2$$

and variable shear stress,

$$\tau_v = \frac{16\,T_v}{\pi d^3} = \frac{16 \times 100 \times 10^3}{\pi d^3} = \frac{510 \times 10^3}{d^3} \text{ N/mm}^2$$

Endurance limit stress for reversed torsional or shear loading,

$$\tau_e = \sigma_e \times K_s = 275 \times 0.6 = 165 \text{ N/mm}^2$$

Assuming yield strength in shear,

$$\tau_y = 0.5\,\sigma_y = 0.5 \times 400 = 200 \text{ N/mm}^2$$

We know that equivalent shear stress,

$$\tau_{es} = \tau_m + \frac{\tau_v \times \tau_y \times K_{fs}}{\tau_e \times K_{sur} \times K_{sz}}$$

$$= \frac{255 \times 10^3}{d^3} + \frac{510 \times 10^3 \times 200 \times 1.3}{d^3 \times 165 \times 0.88 \times 0.85}$$

$$= \frac{255 \times 10^3}{d^3} + \frac{1074 \times 10^3}{d^3} = \frac{1329 \times 10^3}{d^3} \text{ N/mm}^2$$

and maximum equivalent shear stress,

$$\tau_{es(max)} = \frac{\tau_y}{F.S.} = \frac{1}{2}\sqrt{(\sigma_{ne})^2 + 4\,(\tau_{es})^2}$$

$$\frac{200}{1.5} = \frac{1}{2}\sqrt{\left(\frac{9985 \times 10^3}{d^3}\right)^2 + 4\left(\frac{1329 \times 10^3}{d^3}\right)^2} = \frac{5165 \times 10^3}{d^3}$$

$$\therefore \quad d^3 = \frac{5165 \times 10^3 \times 1.5}{200} = 38\,740 \text{ or } d = 33.84 \text{ say } 35 \text{ mm} \quad \textbf{Ans.}$$

6.23 Application of Soderberg's Equation

We have seen in Art. 6.21 that according to Soderberg's equation,

$$\frac{1}{F.S.} = \frac{\sigma_m}{\sigma_y} + \frac{\sigma_v \times K_f}{\sigma_e} \qquad ...(i)$$

This equation may also be written as

$$\frac{1}{F.S.} = \frac{\sigma_m \times \sigma_e + \sigma_v \times \sigma_y \times K_f}{\sigma_y \times \sigma_e}$$

or

$$F.S. = \frac{\sigma_y \times \sigma_e}{\sigma_m \times \sigma_e + \sigma_v \times \sigma_y \times K_f} = \frac{\sigma_y}{\sigma_m + \left(\dfrac{\sigma_y}{\sigma_e}\right) K_f \times \sigma_v} \qquad ...(ii)$$

Since the factor of safety based on yield strength is the ratio of the yield point stress to the working or design stress, therefore from equation (*ii*), we may write

Working or design stress

$$= \sigma_m + \left(\frac{\sigma_y}{\sigma_e}\right) K_f \times \sigma_v \qquad ...(iii)$$

Let us now consider the use of Soderberg's equation to a ductile material under the following loading conditions.

1. ***Axial loading***

In case of axial loading, we know that the mean or average stress,

$$\sigma_m = W_m / A$$

and variable stress, $\sigma_v = W_v / A$

where W_m = Mean or average load,

W_v = Variable load, and

A = Cross-sectional area.

The equation (*iii*) may now be written as follows :

Working or design stress,

$$= \frac{W_m}{A} + \left(\frac{\sigma_y}{\sigma_e}\right) K_f \times \frac{W_v}{A} = \frac{W_m + \left(\frac{\sigma_y}{\sigma_e}\right) K_f \times W_v}{A}$$

$$\therefore \quad F.S. = \frac{\sigma_y \times A}{W_m + \left(\frac{\sigma_y}{\sigma_e}\right) K_f \times W_v}$$

2. *Simple bending*

In case of simple bending, we know that the bending stress,

$$\sigma_b = \frac{M.y}{I} = \frac{M}{Z} \qquad \ldots \left(\because Z = \frac{I}{y}\right)$$

∴ Mean or average bending stress,

$$\sigma_m = M_m / Z$$

and variable bending stress,

$$\sigma_v = M_v / Z$$

where M_m = Mean bending moment,

M_v = Variable bending moment, and

Z = Section modulus.

The equation (*iii*) may now be written as follows :

Working or design bending stress,

$$\sigma_b = \frac{M_m}{Z} + \left(\frac{\sigma_y}{\sigma_e}\right) K_f \times \frac{M_v}{Z}$$

$$= \frac{M_m + \left(\frac{\sigma_y}{\sigma_e}\right) K_f \times M_v}{Z}$$

$$= \frac{32}{\pi d^3}\left[M_m + \left(\frac{\sigma_y}{\sigma_e}\right) K_f \times M_v\right] \qquad \ldots \left(\because \text{For circular shafts, } Z = \frac{\pi}{32} \times d^3\right)$$

$$\therefore \quad F.S. = \frac{\sigma_y}{\frac{32}{\pi d^3}\left[M_m + \left(\frac{\sigma_y}{\sigma_e}\right) K_f \times M_v\right]}$$

A large disc- shaped electromagnet hangs from jib of this scrapyard crane. Steel and iron objects fly towards the magnet when the current is switched on. In this way, iron and steel can be separated for recycling.

Note : This picture is given as additional information and is not a direct example of the current chapter.

3. *Simple torsion of circular shafts*

In case of simple torsion, we know that the torque,

$$T = \frac{\pi}{16} \times \tau \times d^3 \text{ or } \tau = \frac{16\,T}{\pi\,d^3}$$

∴ Mean or average shear stress,

$$\tau_m = \frac{16\,T_m}{\pi\,d^3}$$

and variable shear stress, $\tau_v = \dfrac{16\,T_v}{\pi\,d^3}$

where T_m = Mean or average torque,

T_v = Variable torque, and

d = Diameter of the shaft.

The equation *(iii)* may now be written as follows :

Working or design shear stress,

$$\tau = \frac{16\,T_m}{\pi d^3} + \left(\frac{\tau_y}{\tau_e}\right) K_{fs} \times \frac{16\,T_v}{\pi d^3} = \frac{16}{\pi d^3}\left[T_m + \left(\frac{\tau_y}{\tau_e}\right) K_{fs} \times T_v\right]$$

$$\therefore \quad F.S. = \frac{\tau_y}{\dfrac{16}{\pi d^3}\left[T_m + \left(\dfrac{\tau_y}{\tau_e}\right) K_{fs} \times T_v\right]}$$

where K_{fs} = Fatigue stress concentration factor for torsional or shear loading.

Note : For shafts made of ductile material, $\tau_y = 0.5\ \sigma_y$, and $\tau_e = 0.5\ \sigma_e$ may be taken.

4. *Combined bending and torsion of circular shafts*

In case of combined bending and torsion of circular shafts, the maximum shear stress theory may be used. According to this theory, maximum shear stress,

$$\tau_{max} = \frac{\tau_y}{F.S.} = \frac{1}{2}\sqrt{(\sigma_b)^2 + 4\tau^2}$$

$$= \frac{1}{2}\sqrt{\left[\frac{32}{\pi d^3}\left\{M_m + \left(\frac{\sigma_y}{\sigma_e}\right) K_f \times M_v\right\}\right]^2 + 4\left[\frac{16}{\pi d^3}\left\{T_m + \left(\frac{\tau_y}{\tau_e}\right) K_{fs} \times T_v\right\}\right]^2}$$

$$= \frac{16}{\pi d^3}\sqrt{\left[M_m + \left(\frac{\sigma_y}{\sigma_e}\right) K_f \times M_v\right]^2 + \left[T_m + \left(\frac{\tau_y}{\tau_e}\right) K_{fs} \times T_v\right]^2}$$

The majority of rotating shafts carry a steady torque and the loads remain fixed in space in both direction and magnitude. Thus during each revolution every fibre on the surface of the shaft undergoes a complete reversal of stress due to bending moment. Therefore for the usual case when $M_m = 0$, $M_v = M$, $T_m = T$ and $T_v = 0$, the above equation may be written as

$$\frac{\tau_y}{F.S.} = \frac{16}{\pi d^3}\sqrt{\left[\left(\frac{\sigma_y}{\sigma_e}\right) K_f \times M\right]^2 + T^2}$$

Note: The above relations apply to a solid shaft. For hollow shaft, the left hand side of the above equations must be multiplied by $(1 - k^4)$, where k is the ratio of inner diameter to outer diameter.

Example 6.15. *A centrifugal blower rotates at 600 r.p.m. A belt drive is used to connect the blower to a 15 kW and 1750 r.p.m. electric motor. The belt forces a torque of 250 N-m and a force of 2500 N on the shaft. Fig. 6.20 shows the location of bearings, the steps in the shaft and the plane in which the resultant belt force and torque act. The ratio of the journal diameter to the overhung shaft diameter is 1.2 and the radius of the fillet is 1/10th of overhung shaft diameter. Find the shaft diameter, journal diameter and radius of fillet to have a factor of safety 3. The blower shaft is to be machined from hot rolled steel having the following values of stresses:*

Endurance limit = 180 MPa; Yield point stress = 300 MPa; Ultimate tensile stress = 450 MPa.

Solution. Given: $^*N_B = 600$ r.p.m. ; $^*P = 15$ kW; $^*N_M = 1750$ r.p.m. ; $T = 250$ N-m $= 250 \times 10^3$ N-mm; $F = 2500$ N ; $F.S. = 3$; $\sigma_e = 180$ MPa $= 180$ N/mm^2 ; $\sigma_y = 300$ MPa $= 300$ N/mm^2 ; $\sigma_u = 450$ MPa $= 450$ N/mm^2

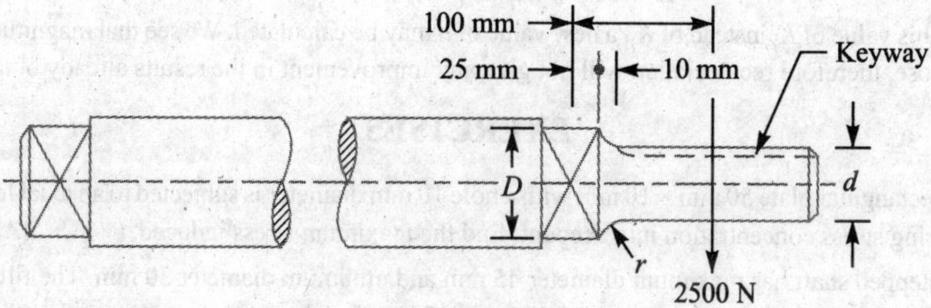

Fig. 6.20

Let D = Journal diameter,

d = Shaft diameter, and r = Fillet radius.

∴ Ratio of journal diameter to shaft diameter,

$D/d = 1.2$ ...(Given)

and radius of the fillet, $r = 1/10 \times$ Shaft diameter $(d) = 0.1\,d$

∴ $r/d = 0.1$ ...(Given)

From Table 6.3, for $D/d = 1.2$ and $r/d = 0.1$, the theoretical stress concentration factor,

$K_t = 1.62$

The two points at which failure may occur are at the end of the keyway and at the shoulder fillet. The critical section will be the one with larger product of $K_f \times M$. Since the notch sensitivity factor q is dependent upon the unknown dimensions of the notch and since the curves for notch sensitivity factor (Fig. 6.14) are not applicable to keyways, therefore the product $K_t \times M$ shall be the basis of comparison for the two sections.

∴ Bending moment at the end of the keyway,

$K_t \times M = 1.6 \times 2500\,[100 - (25 + 10)] = 260 \times 10^3$ N-mm

...(∵ K_t for key ways = 1.6)

and bending moment at the shoulder fillet,

$K_t \times M = 1.62 \times 2500\,(100 - 25) = 303\,750$ N-mm

Since $K_t \times M$ at the shoulder fillet is large, therefore considering the shoulder fillet as the critical section. We know that

$$\frac{\tau_y}{F.S.} = \frac{16}{\pi d^3}\sqrt{\left[\left(\frac{\sigma_y}{\sigma_e}\right) K_f \times M\right]^2 + T^2}$$

* Superfluous data

$$\frac{0.5 \times 300}{3} = \frac{16}{\pi d^3} \sqrt{\left(\frac{300}{180} \times 303750\right)^2 + (250 \times 10^3)^2}$$

... (Substituting, $\tau_y = 0.5\ \sigma_y$)

$$50 = \frac{16}{\pi d^3} \times 565 \times 10^3 = \frac{2877 \times 10^3}{d^3}$$

$\therefore$ $d^3 = 2877 \times 10^3/50 = 57\ 540$ or $d = 38.6$ say 40 mm **Ans.**

Note: Since r is known (because $r/d = 0.1$ or $r = 0.1d = 4$ mm), therefore from Fig. 6.14, the notch sensitivity factor (q) may be obtained. For $r = 4$ mm, we have $q = 0.93$.

$\therefore$ Fatigue stress concentration factor,

$$K_f = 1 + q\,(K_t - 1) = 1 + 0.93\,(1.62 - 1) = 1.58$$

Using this value of K_f instead of K_t, a new value of d may be calculated. We see that magnitudes of K_f and K_t are very close, therefore recalculation will not give any improvement in the results already obtained.

EXERCISES

1. A rectangular plate 50 mm × 10 mm with a hole 10 mm diameter is subjected to an axial load of 10 kN. Taking stress concentration into account, find the maximum stress induced. **[Ans. 50 MPa]**
2. A stepped shaft has maximum diameter 45 mm and minimum diameter 30 mm. The fillet radius is 6 mm. If the shaft is subjected to an axial load of 10 kN, find the maximum stress induced, taking stress concentration into account. **[Ans. 22 MPa]**
3. A leaf spring in an automobile is subjected to cyclic stresses. The average stress = 150 MPa; variable stress = 500 MPa; ultimate stress = 630 MPa; yield point stress = 350 MPa and endurance limit = 150 MPa. Estimate, under what factor of safety the spring is working, by Goodman and Soderberg formulae. **[Ans. 1.75, 1.3]**
4. Determine the design stress for bolts in a cylinder cover where the load is fluctuating due to gas pressure. The maximum load on the bolt is 50 kN and the minimum is 30 kN. The load is unpredictable and factor of safety is 3. The surface of the bolt is hot rolled and the surface finish factor is 0.9. During a simple tension test and rotating beam test on ductile materials (40 C 8 steel annealed), the following results were obtained :

   Diameter of specimen = 12.5 mm; Yield strength = 240 MPa; Ultimate strength = 450 MPa; Endurance limit = 180 MPa. **[Ans. 65.4 MPa]**
5. Determine the diameter of a tensile member of a circular cross-section. The following data is given : Maximum tensile load = 10 kN; Maximum compressive load = 5 kN; Ultimate tensile strength = 600 MPa; Yield point = 380 MPa; Endurance limit = 290 MPa; Factor of safety = 4; Stress concentration factor = 2.2 **[Ans. 24 mm]**
6. Determine the size of a piston rod subjected to a total load of having cyclic fluctuations from 15 kN in compression to 25 kN in tension. The endurance limit is 360 MPa and yield strength is 400 MPa. Take impact factor = 1.25, factor of safety = 1.5, surface finish factor = 0.88 and stress concentration factor = 2.25. **[Ans. 35.3 mm]**
7. A steel connecting rod is subjected to a completely reversed axial load of 160 kN. Suggest the suitable diameter of the rod using a factor of safety 2. The ultimate tensile strength of the material is 1100 MPa, and yield strength 930 MPa. Neglect column action and the effect of stress concentration. **[Ans. 30.4 mm]**
8. Find the diameter of a shaft made of 37 Mn 2 steel having the ultimate tensile strength as 600 MPa and yield stress as 440 MPa. The shaft is subjected to completely reversed axial load of 200 kN. Neglect stress concentration factor and assume surface finish factor as 0.8. The factor of safety may be taken as 1.5. **[Ans. 51.7 mm]**

9. Find the diameter of a shaft to transmit twisting moments varying from 800 N-m to 1600 N-m. The ultimate tensile strength for the material is 600 MPa and yield stress is 450 MPa. Assume the stress concentration factor = 1.2, surface finish factor = 0.8 and size factor = 0.85. **[Ans. 27.7 mm]**

10. A simply supported shaft between bearings carries a steady load of 10 kN at the centre. The length of shaft between bearings is 450 mm. Neglecting the effect of stress concentration, find the minimum diameter of shaft. Given that

    Endurance limit = 600 MPa; surface finish factor = 0.87; size factor = 0.85; and factor of safety = 1.6. **[Ans. 35 mm]**

11. Determine the diameter of a circular rod made of ductile material with a fatigue strength (complete stress reversal) σ_e = 280 MPa and a tensile yield strength of 350 MPa. The member is subjected to a varying axial load from 700 kN to – 300 kN. Assume K_t = 1.8 and *F.S.* = 2. **[Ans. 80 mm]**

12. A cold drawn steel rod of circular cross-section is subjected to a variable bending moment of 565 N-m to 1130 N-m as the axial load varies from 4500 N to 13 500 N. The maximum bending moment occurs at the same instant that the axial load is maximum. Determine the required diameter of the rod for a factor of safety 2. Neglect any stress concentration and column effect. Assume the following values:

Ultimate strength	=	550 MPa
Yield strength	=	470 MPa
Size factor	=	0.85
Surface finish factor	=	0.89
Correction factors	=	1.0 for bending
	=	0.7 for axial load

The endurance limit in reversed bending may be taken as one-half the ultimate strength. **[Ans. 41 mm]**

13. A steel cantilever beam, as shown in Fig. 6.21, is subjected to a transverse load at its end that varies from 45 N up to 135 N down as the axial load varies from 110 N (compression) to 450 N (tension). Determine the required diameter at the change of section for infinite life using a factor of safety of 2. The strength properties are as follows:

Ultimate strength	=	550 MPa
Yield strength	=	470 MPa
Endurance limit	=	275 MPa

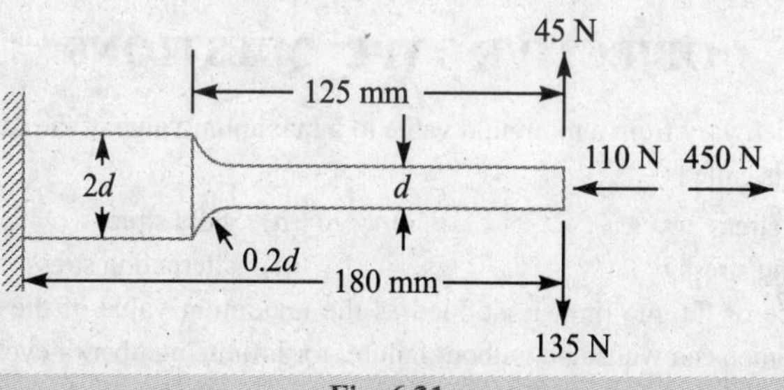

Fig. 6.21

The stress concentration factors for bending and axial loads are 1.44 and 1.63 respectively, at the change of cross-section. Take size factor = 0.85 and surface finish factor = 0.9. **[Ans. 12.5 mm]**

14. A steel shaft is subjected to completely reversed bending moment of 800 N-m and a cyclic twisting moment of 500 N-m which varies over a range of ± 40%. Determine the diameter of shaft if a reduction factor of 1.2 is applied to the variable component of bending stress and shearing stress. Assume

(a) that the maximum bending and shearing stresses are in phase;

(b) that the tensile yield point is the limiting stress for steady state component;

(c) that the maximum shear strength theory can be applied; and

(d) that the Goodman relation is valid.

Take the following material properties:

Yield strength = 500 MPa ; Ultimate strength = 800 MPa ; Endurance limit = ± 400 MPa.

[Ans. 40 mm]

15. A pulley is keyed to a shaft midway between two anti-friction bearings. The bending moment at the pulley varies from – 170 N-m to 510 N-m and the torsional moment in the shaft varies from 55 N-m to 165 N-m. The frequency of the variation of the loads is the same as the shaft speed. The shaft is made of cold drawn steel having an ultimate strength of 540 MPa and a yield strength of 400 MPa. Determine the required diameter for an indefinite life. The stress concentration factor for the keyway in bending and torsion may be taken as 1.6 and 1.3 respectively. The factor of safety is 1.5. Take size factor = 0.85 and surface finish factor = 0.88. **[Ans. 36.5 mm]**

[**Hint.** Assume $\sigma_e = 0.5\ \sigma_u$; $\tau_y = 0.5\ \sigma_y$; $\tau_e = 0.55\ \sigma_e$]

QUESTIONS

1. Explain the following terms in connection with design of machine members subjected to variable loads:

   (a) Endurance limit, (b) Size factor,

   (c) Surface finish factor, and (d) Notch sensitivity.

2. What is meant by endurance strength of a material? How do the size and surface condition of a component and type of load affect such strength?
3. Write a note on the influence of various factors of the endurance limit of a ductile material.
4. What is meant by 'stress concentration'? How do you take it into consideration in case of a component subjected to dynamic loading?
5. Illustrate how the stress concentration in a component can be reduced.
6. Explain how the factor of safety is determined under steady and varying loading by different methods.
7. Write Soderberg's equation and state its application to different type of loadings.
8. What information do you obtain from Soderberg diagram?

OBJECTIVE TYPE QUESTIONS

1. The stress which vary from a minimum value to a maximum value of the same nature (*i.e.* tensile or compressive) is called

   (a) repeated stress (b) yield stress

   (c) fluctuating stress (d) alternating stress

2. The endurance or fatigue limit is defined as the maximum value of the stress which a polished standard specimen can withstand without failure, for infinite number of cycles, when subjected to

   (a) static load (b) dynamic load

   (c) static as well as dynamic load (d) completely reversed load

3. Failure of a material is called fatigue when it fails

   (a) at the elastic limit (b) below the elastic limit

   (c) at the yield point (d) below the yield point

4. The resistance to fatigue of a material is measured by
   (*a*) elastic limit (*b*) Young's modulus
   (*c*) ultimate tensile strength (*d*) endurance limit
5. The yield point in static loading is .... as compared to fatigue loading.
   (*a*) higher (*b*) lower (*c*) same
6. Factor of safety for fatigue loading is the ratio of
   (*a*) elastic limit to the working stress
   (*b*) Young's modulus to the ultimate tensile strength
   (*c*) endurance limit to the working stress
   (*d*) elastic limit to the yield point
7. When a material is subjected to fatigue loading, the ratio of the endurance limit to the ultimate tensile strength is
   (*a*) 0.20 (*b*) 0.35
   (*c*) 0.50 (*d*) 0.65
8. The ratio of endurance limit in shear to the endurance limit in flexure is
   (*a*) 0.25 (*b*) 0.40
   (*c*) 0.55 (*d*) 0.70
9. If the size of a standard specimen for a fatigue testing machine is increased, the endurance limit for the material will
   (*a*) have same value as that of standard specimen (*b*) increase (*c*) decrease
10. The residential compressive stress by way of surface treatment of a machine member subjected to fatigue loading
    (*a*) improves the fatigue life (*b*) deteriorates the fatigue life
    (*c*) does not affect the fatigue life (*d*) immediately fractures the specimen
11. The surface finish factor for a mirror polished material is
    (*a*) 0.45 (*b*) 0.65
    (*c*) 0.85 (*d*) 1
12. Stress concentration factor is defined as the ratio of
    (*a*) maximum stress to the endurance limit (*b*) nominal stress to the endurance limit
    (*c*) maximum stress to the nominal stress (*d*) nominal stress to the maximum stress
13. In static loading, stress concentration is more serious in
    (*a*) brittle materials (*b*) ductile materials
    (*c*) brittle as well as ductile materials (*d*) elastic materials
14. In cyclic loading, stress concentration is more serious in
    (*a*) brittle materials (*b*) ductile materials
    (*c*) brittle as well as ductile materials (*d*) elastic materials
15. The notch sensitivity q is expressed in terms of fatigue stress concentration factor K_f and theoretical stress concentration factor K_t, as

    (*a*) $\frac{K_f + 1}{K_t + 1}$ (*b*) $\frac{K_f - 1}{K_t - 1}$

    (*c*) $\frac{K_t + 1}{K_f + 1}$ (*d*) $\frac{K_t - 1}{K_f - 1}$

ANSWERS

1. (*c*)	2. (*d*)	3. (*d*)	4. (*d*)	5. (*a*)
6. (*c*)	7. (*c*)	8. (*c*)	9. (*c*)	10. (*a*)
11. (*d*)	12. (*c*)	13. (*a*)	14. (*b*)	15. (*b*)

CHAPTER

7

Pressure Vessels

1. Introduction.
2. Classification of Pressure Vessels.
3. Stresses in a Thin Cylindrical Shell due to an Internal Pressure.
4. Circumferential or Hoop Stress.
5. Longitudinal Stress.
6. Change in Dimensions of a Thin Cylindrical Shell due to an Internal Pressure.
7. Thin Spherical Shells Subjected to an Internal Pressure.
8. Change in Dimensions of a Thin Spherical Shell due to an Internal Pressure.
9. Thick Cylindrical Shell Subjected to an Internal Pressure.
10. Compound Cylindrical Shells.
11. Stresses in Compound Cylindrical Shells.
12. Cylinder Heads and Cover Plates.

7.1 Introduction

The pressure vessels (*i.e.* cylinders or tanks) are used to store fluids under pressure. The fluid being stored may undergo a change of state inside the pressure vessel as in case of steam boilers or it may combine with other reagents as in a chemical plant. The pressure vessels are designed with great care because rupture of a pressure vessel means an explosion which may cause loss of life and property. The material of pressure vessels may be brittle such as cast iron, or ductile such as mild steel.

7.2 Classification of Pressure Vessels

The pressure vessels may be classified as follows:

1. ***According to the dimensions.*** The pressure vessels, according to their dimensions, may be classified as ***thin shell*** or ***thick shell.*** If the wall thickness of the shell (t) is less than 1/10 of the diameter of the shell (d), then it is called a ***thin shell.*** On the other hand, if the wall thickness of the shell is greater than 1/10 of the diameter of the shell, then it is said to be a ***thick shell.*** Thin shells are used in boilers, tanks and pipes, whereas thick shells are used in high pressure cylinders, tanks, gun barrels etc.

Note: Another criterion to classify the pressure vessels as thin shell or thick shell is the internal fluid pressure (p) and the allowable stress (σ_t). If the internal fluid pressure (p) is less than 1/6 of the allowable stress, then it is called a ***thin shell.*** On the other hand, if the internal fluid pressure is greater than 1/6 of the allowable stress, then it is said to be a ***thick shell.***

Pressure vessels.

2. *According to the end construction.* The pressure vessels, according to the end construction, may be classified as ***open end*** or ***closed end.*** A simple cylinder with a piston, such as cylinder of a press is an example of an open end vessel, whereas a tank is an example of a closed end vessel. In case of vessels having open ends, the circumferential or hoop stresses are induced by the fluid pressure, whereas in case of closed ends, longitudinal stresses in addition to circumferential stresses are induced.

7.3 Stresses in a Thin Cylindrical Shell due to an Internal Pressure

The analysis of stresses induced in a thin cylindrical shell are made on the following assumptions:

1. The effect of curvature of the cylinder wall is neglected.
2. The tensile stresses are uniformly distributed over the section of the walls.
3. The effect of the restraining action of the heads at the end of the pressure vessel is neglected.

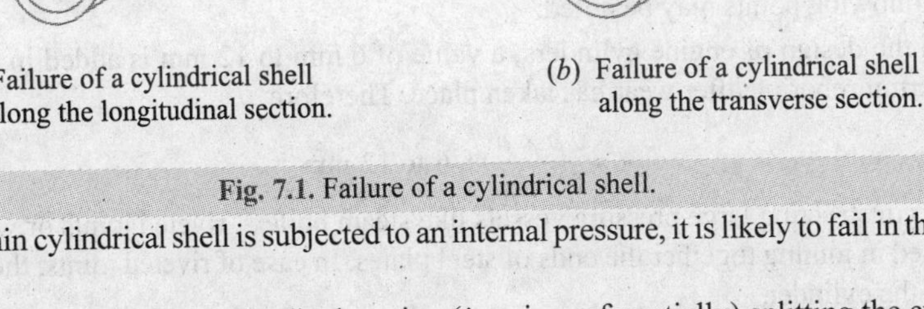

(*a*) Failure of a cylindrical shell along the longitudinal section.

(*b*) Failure of a cylindrical shell along the transverse section.

Fig. 7.1. Failure of a cylindrical shell.

When a thin cylindrical shell is subjected to an internal pressure, it is likely to fail in the following two ways:

1. It may fail along the longitudinal section (*i.e.* circumferentially) splitting the cylinder into two troughs, as shown in Fig. 7.1 (*a*).
2. It may fail across the transverse section (*i.e.* longitudinally) splitting the cylinder into two cylindrical shells, as shown in Fig. 7.1 (*b*).

Thus the wall of a cylindrical shell subjected to an internal pressure has to withstand tensile stresses of the following two types:

(*a*) Circumferential or hoop stress, and (*b*) Longitudinal stress.

These stresses are discussed, in detail, in the following articles.

7.4 Circumferential or Hoop Stress

Consider a thin cylindrical shell subjected to an internal pressure as shown in Fig. 7.2 (*a*) and (*b*). A tensile stress acting in a direction tangential to the circumference is called ***circumferential*** or ***hoop stress.*** In other words, it is a tensile stress on *longitudinal section (or on the cylindrical walls).

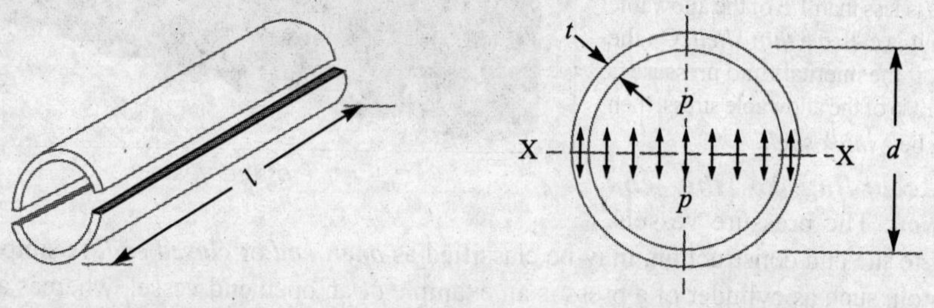

(*a*) View of shell. (*b*) Cross-section of shell.

Fig. 7.2. Circumferential or hoop stress.

Let p = Intensity of internal pressure,

d = Internal diameter of the cylindrical shell,

l = Length of the cylindrical shell,

t = Thickness of the cylindrical shell, and

σ_{t1} = Circumferential or hoop stress for the material of the cylindrical shell.

We know that the total force acting on a longitudinal section (*i.e.* along the diameter *X-X*) of the shell

$$= \text{Intensity of pressure} \times \text{Projected area} = p \times d \times l \qquad ...(i)$$

and the total resisting force acting on the cylinder walls

$$= \sigma_{t1} \times 2t \times l \qquad ...(\because \text{ of two sections}) \qquad ...(ii)$$

From equations (*i*) and (*ii*), we have

$$\sigma_{t1} \times 2t \times l = p \times d \times l \quad \text{or} \quad \sigma_{t1} = \frac{p \times d}{2t} \quad \text{or} \quad t = \frac{p \times d}{2\,\sigma_{t1}} \qquad ...(iii)$$

The following points may be noted:

1. In the design of engine cylinders, a value of 6 mm to 12 mm is added in equation (*iii*) to permit reboring after wear has taken place. Therefore

$$t = \frac{p \times d}{2\,\sigma_{t1}} + 6 \text{ to } 12 \text{ mm}$$

2. In constructing large pressure vessels like steam boilers, riveted joints or welded joints are used in joining together the ends of steel plates. In case of riveted joints, the wall thickness of the cylinder,

$$t = \frac{p \times d}{2\sigma_{t1} \times \eta_l}$$

where η_l = Efficiency of the longitudinal riveted joint.

* A section cut from a cylinder by a plane that contains the axis is called longitudinal section.

3. In case of cylinders of ductile material, the value of circumferential stress (σ_{t1}) may be taken 0.8 times the yield point stress (σ_y) and for brittle materials, σ_{t1} may be taken as 0.125 times the ultimate tensile stress (σ_u).
4. In designing steam boilers, the wall thickness calculated by the above equation may be compared with the minimum plate thickness as provided in boiler code as given in the following table.

Table 7.1. Minimum plate thickness for steam boilers.

Boiler diameter	*Minimum plate thickness (t)*
0.9 m or less	6 mm
Above 0.9 m and upto 1.35 m	7.5 mm
Above 1.35 m and upto 1.8 m	9 mm
Over 1.8 m	12 mm

Note: If the calculated value of *t* is less than the code requirement, then the latter should be taken, otherwise the calculated value may be used.

The boiler code also provides that the factor of safety shall be at least 5 and the steel of the plates and rivets shall have as a minimum the following ultimate stresses.

Tensile stress, $\sigma_t = 385$ MPa

Compressive stress, $\sigma_c = 665$ MPa

Shear stress, $\tau = 308$ MPa

7.5 Longitudinal Stress

Consider a closed thin cylindrical shell subjected to an internal pressure as shown in Fig. 7.3 (*a*) and (*b*). A tensile stress acting in the direction of the axis is called ***longitudinal stress.*** In other words, it is a tensile stress acting on the *transverse or circumferential section *Y-Y* (or on the ends of the vessel).

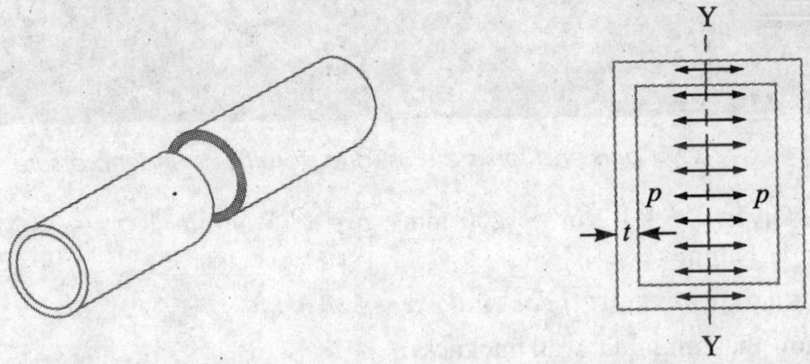

(*a*) View of shell. (*b*) Cross-section of shell.

Fig. 7.3. Longitudinal stress.

Let σ_{t2} = Longitudinal stress.

In this case, the total force acting on the transverse section (*i.e.* along *Y-Y*)

= Intensity of pressure × Cross-sectional area

$$= p \times \frac{\pi}{4}(d)^2 \qquad \text{...(i)}$$

and total resisting force $= \sigma_{t2} \times \pi\, d.t$ ...(*ii*)

* A section cut from a cylinder by a plane at right angles to the axis of the cylinder is called transverse section.

From equations (*i*) and (*ii*), we have

$$\sigma_{t2} \times \pi\, d.t = p \times \frac{\pi}{4} (d)^2$$

$$\therefore \qquad \sigma_{t2} = \frac{p \times d}{4\, t} \qquad \text{or} \qquad t = \frac{p \times d}{4\, \sigma_{t2}}$$

If η_c is the efficiency of the circumferential joint, then

$$t = \frac{p \times d}{4 \sigma_{t2} \times \eta_c}$$

From above we see that the longitudinal stress is half of the circumferential or hoop stress. Therefore, the design of a pressure vessel must be based on the maximum stress *i.e.* hoop stress.

Example 7.1. *A thin cylindrical pressure vessel of 1.2 m diameter generates steam at a pressure of 1.75 N/mm². Find the minimum wall thickness, if (a) the longitudinal stress does not exceed 28 MPa; and (b) the circumferential stress does not exceed 42 MPa.*

Cylinders and tanks are used to store fluids under pressure.

Solution. Given : $d = 1.2$ m $= 1200$ mm ; $p = 1.75$ N/mm² ; $\sigma_{t2} = 28$ MPa $= 28$ N/mm² ; $\sigma_{t1} = 42$ MPa $= 42$ N/mm²

(a) When longitudinal stress (σ_{t2}) does not exceed 28 MPa

We know that minimum wall thickness,

$$t = \frac{p\,.\,d}{4\,\sigma_{t2}} = \frac{1.75 \times 1200}{4 \times 28} = 18.75 \text{ say } 20 \text{ mm} \quad \textbf{Ans.}$$

(b) When circumferential stress (σ_{t1}) does not exceed 42 MPa

We know that minimum wall thickness,

$$t = \frac{p\,.\,d}{2\,\sigma_{t1}} = \frac{1.75 \times 1200}{2 \times 42} = 25 \text{ mm} \quad \textbf{Ans.}$$

Example 7.2. *A thin cylindrical pressure vessel of 500 mm diameter is subjected to an internal pressure of 2 N/mm². If the thickness of the vessel is 20 mm, find the hoop stress, longitudinal stress and the maximum shear stress.*

Solution. Given : $d = 500$ mm ; $p = 2$ N/mm² ; $t = 20$ mm

Hoop stress

We know that hoop stress,

$$\sigma_{t1} = \frac{p \cdot d}{2\,t} = \frac{2 \times 500}{2 \times 20} = 25 \text{ N/mm}^2 = 25 \text{ MPa } \textbf{Ans.}$$

Longitudinal stress

We know that longitudinal stress,

$$\sigma_{t2} = \frac{p \cdot d}{4\,t} = \frac{2 \times 500}{4 \times 20} = 12.5 \text{ N/mm}^2 = 12.5 \text{ MPa } \textbf{Ans.}$$

Maximum shear stress

We know that according to maximum shear stress theory, the maximum shear stress is one-half the algebraic difference of the maximum and minimum principal stress. Since the maximum principal stress is the hoop stress (σ_{t1}) and minimum principal stress is the longitudinal stress (σ_{t2}), therefore maximum shear stress,

$$\tau_{max} = \frac{\sigma_{t1} - \sigma_{t2}}{2} = \frac{25 - 12.5}{2} = 6.25 \text{ N/mm}^2 = 6.25 \text{ MPa } \textbf{Ans.}$$

Example 7.3. *An hydraulic control for a straight line motion, as shown in Fig. 7.4, utilises a spherical pressure tank 'A' connected to a working cylinder B. The pump maintains a pressure of 3 N/mm² in the tank.*

1. If the diameter of pressure tank is 800 mm, determine its thickness for 100% efficiency of the joint. Assume the allowable tensile stress as 50 MPa.

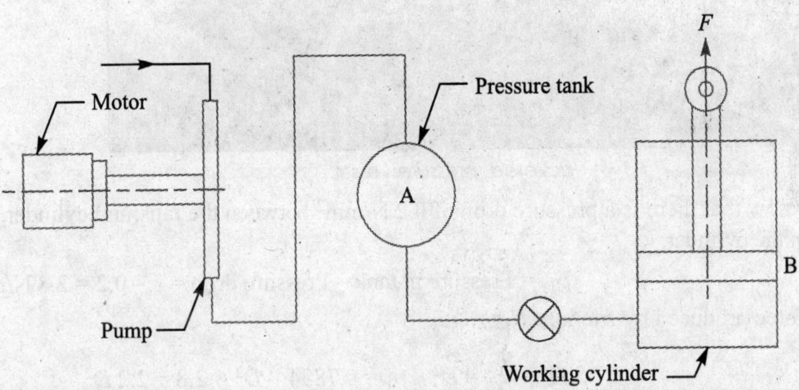

Fig. 7.4

2. Determine the diameter of a cast iron cylinder and its thickness to produce an operating force F = 25 kN. Assume (i) an allowance of 10 per cent of operating force F for friction in the cylinder and packing, and (ii) a pressure drop of 0.2 N/mm² between the tank and cylinder. Take safe stress for cast iron as 30 MPa.

3. Determine the power output of the cylinder, if the stroke of the piston is 450 mm and the time required for the working stroke is 5 seconds.

4. Find the power of the motor, if the working cycle repeats after every 30 seconds and the efficiency of the hydraulic control is 80 percent and that of pump 60 percent.

Solution. Given : $p = 3$ N/mm² ; $d = 800$ mm ; $\eta = 100\% = 1$; $\sigma_{t1} = 50$ MPa $= 50$ N/mm² ; $F = 25$ kN $= 25 \times 10^3$ N ; $\sigma_{tc} = 30$ MPa $= 30$ N/mm² : $\eta_H = 80\% = 0.8$; $\eta_P = 60\% = 0.6$

1. *Thickness of pressure tank*

We know that thickness of pressure tank,

$$t = \frac{p \cdot d}{2\sigma_{t1} \cdot \eta} = \frac{3 \times 800}{2 \times 50 \times 1} = 24 \text{ mm} \quad \textbf{Ans.}$$

2. *Diameter and thickness of cylinder*

Let D = Diameter of cylinder, and

t_1 = Thickness of cylinder.

Since an allowance of 10 per cent of operating force F is provided for friction in the cylinder and packing, therefore total force to be produced by friction,

$$F_1 = F + \frac{10}{100} F = 1.1\,F = 1.1 \times 25 \times 10^3 = 27\,500 \text{ N}$$

Jacketed pressure vessel.

We know that there is a pressure drop of 0.2 N/mm² between the tank and cylinder, therefore pressure in the cylinder,

$$p_1 = \text{Pressure in tank} - \text{Pressure drop} = 3 - 0.2 = 2.8 \text{ N/mm}^2$$

and total force produced by friction (F_1),

$$27\,500 = \frac{\pi}{4} \times D^2 \times p_1 = 0.7854 \times D^2 \times 2.8 = 2.2\,D^2$$

$$\therefore \quad D^2 = 27\,500 / 2.2 = 12\,500 \quad \text{or} \quad D = 112 \text{ mm} \quad \textbf{Ans.}$$

We know that thickness of cylinder,

$$t_1 = \frac{p_1 \cdot D}{2\,\sigma_{tc}} = \frac{2.8 \times 112}{2 \times 30} = 5.2 \text{ mm} \quad \textbf{Ans.}$$

3. *Power output of the cylinder*

We know that stroke of the piston

= 450 mm = 0.45 m ...(Given)

and time required for working stroke

= 5 s ...(Given)

∴ Distance moved by the piston per second

$$= \frac{0.45}{5} = 0.09 \text{ m}$$

We know that work done per second

$= \text{Force} \times \text{Distance moved per second}$

$= 27\,500 \times 0.09 = 2475$ N-m

∴ Power output of the cylinder

$= 2475$ W $= 2.475$ kW **Ans.** ...($\because$ 1 N-m/s = 1 W)

4. ***Power of the motor***

Since the working cycle repeats after every 30 seconds, therefore the power which is to be produced by the cylinder in 5 seconds is to be provided by the motor in 30 seconds.

∴ Power of the motor

$$= \frac{\text{Power of the cylinder}}{\eta_H \times \eta_P} \times \frac{5}{30} = \frac{2.475}{0.8 \times 0.6} \times \frac{5}{30} = 0.86 \text{ kW } \textbf{Ans.}$$

7.6 Change in Dimensions of a Thin Cylindrical Shell due to an Internal Pressure

When a thin cylindrical shell is subjected to an internal pressure, there will be an increase in the diameter as well as the length of the shell.

Let l = Length of the cylindrical shell,

d = Diameter of the cylindrical shell,

t = Thickness of the cylindrical shell,

p = Intensity of internal pressure,

E = Young's modulus for the material of the cylindrical shell, and

μ = Poisson's ratio.

The increase in diameter of the shell due to an internal pressure is given by,

$$\delta d = \frac{p.d^2}{2\,t.E}\left(1 - \frac{\mu}{2}\right)$$

The increase in length of the shell due to an internal pressure is given by,

$$\delta l = \frac{p.d.l}{2\,t.E}\left(\frac{1}{2} - \mu\right)$$

It may be noted that the increase in diameter and length of the shell will also increase its volume. The increase in volume of the shell due to an internal pressure is given by

$$\delta V = \text{Final volume} - \text{Original volume} = \frac{\pi}{4}(d + \delta d)^2 (l + \delta l) - \frac{\pi}{4} \times d^2.l$$

$$= \frac{\pi}{4}(d^2.\delta l + 2\,d.l.\delta d) \qquad \text{...(Neglecting small quantities)}$$

Example 7.4. *Find the thickness for a tube of internal diameter 100 mm subjected to an internal pressure which is 5/8 of the value of the maximum permissible circumferential stress. Also find the increase in internal diameter of such a tube when the internal pressure is 90 N/mm². Take E = 205 kN/mm² and μ = 0.29. Neglect longitudinal strain.*

Solution. Given : $p = 5/8 \times \sigma_{t1} = 0.625\,\sigma_{t1}$; $d = 100$ mm ; $p_1 = 90$ N/mm² ; $E = 205$ kN/mm² $= 205 \times 10^3$ N/mm² ; $\mu = 0.29$

Thickness of a tube

We know that thickness of a tube,

$$t = \frac{p\,.\,d}{2\,\sigma_{t1}} = \frac{0.625\,\sigma_{t1} \times 100}{2\,\sigma_{t1}} = 31.25 \text{ mm } \textbf{Ans.}$$

Increase in diameter of a tube

We know that increase in diameter of a tube,

$$\delta d = \frac{p_1 d^2}{2\,t.E}\left(1 - \frac{\mu}{2}\right) = \frac{90\,(100)^2}{2 \times 31.25 \times 205 \times 10^3}\left[1 - \frac{0.29}{2}\right] \text{ mm}$$

$$= 0.07\,(1 - 0.145) = 0.06 \text{ mm } \textbf{Ans.}$$

7.7 Thin Spherical Shells Subjected to an Internal Pressure

Consider a thin spherical shell subjected to an internal pressure as shown in Fig. 7.5.

Let V = Storage capacity of the shell,

p = Intensity of internal pressure,

d = Diameter of the shell,

t = Thickness of the shell,

σ_t = Permissible tensile stress for the shell material.

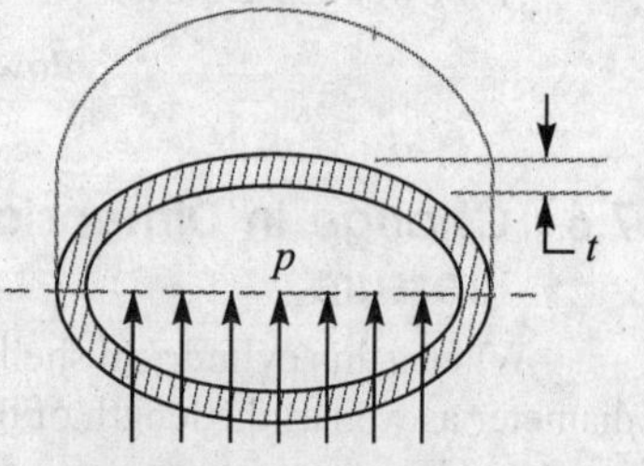

Fig. 7.5. Thin spherical shell.

In designing thin spherical shells, we have to determine 1. Diameter of the shell, and 2. Thickness of the shell.

1. ***Diameter of the shell***

We know that the storage capacity of the shell,

$$V = \frac{4}{3} \times \pi\, r^3 = \frac{\pi}{6} \times d^3 \quad \text{or} \quad d = \left(\frac{6\,V}{\pi}\right)^{1/3}$$

2. ***Thickness of the shell***

As a result of the internal pressure, the shell is likely to rupture along the centre of the sphere. Therefore force tending to rupture the shell along the centre of the sphere or bursting force,

$$= \text{Pressure} \times \text{Area} = p \times \frac{\pi}{4} \times d^2 \qquad \ldots(i)$$

and resisting force of the shell

$$= \text{Stress} \times \text{Resisting area} = \sigma_t \times \pi\, d.t \qquad \ldots(ii)$$

Equating equations (*i*) and (*ii*), we have

$$p \times \frac{\pi}{4} \times d^2 = \sigma_t \times \pi\, d.t$$

or

$$t = \frac{p.d}{4\,\sigma_t}$$

If η is the efficiency of the circumferential joints of the spherical shell, then

$$t = \frac{p.d}{4\,\sigma_t.\eta}$$

The Trans-Alaska Pipeline carries crude oil 1, 284 kilometres through Alaska. The pipeline is 1.2 metres in diameter and can transport 318 million litres of crude oil a day.

Example 7.5. *A spherical vessel 3 metre diameter is subjected to an internal pressure of 1.5 N/mm². Find the thickness of the vessel required if the maximum stress is not to exceed 90 MPa. Take efficiency of the joint as 75%.*

Solution. Given: d = 3 m = 3000 mm ; p = 1.5 N/mm^2 ; σ_t = 90 MPa = 90 N/mm^2 ; η = 75% = 0.75

We know that thickness of the vessel,

$$t = \frac{p.d}{4\,\sigma_t\,.\,\eta} = \frac{1.5 \times 3000}{4 \times 90 \times 0.75} = 16.7 \text{ say } 18 \text{ mm } \textbf{Ans.}$$

7.8 Change in Dimensions of a Thin Spherical Shell due to an Internal Pressure

Consider a thin spherical shell subjected to an internal pressure as shown in Fig. 7.5.

Let d = Diameter of the spherical shell,

t = Thickness of the spherical shell,

p = Intensity of internal pressure,

E = Young's modulus for the material of the spherical shell, and

μ = Poisson's ratio.

Increase in diameter of the spherical shell due to an internal pressure is given by,

$$\delta d = \frac{p.d^2}{4\,t.E}(1 - \mu) \quad ...(i)$$

and increase in volume of the spherical shell due to an internal pressure is given by,

$$\delta V = \text{Final volume} - \text{Original volume} = \frac{\pi}{6}(d + \delta d)^3 - \frac{\pi}{6} \times d^3$$

$$= \frac{\pi}{6}(3\,d^2 \times \delta d) \quad \text{...(Neglecting higher terms)}$$

Substituting the value of δd from equation (*i*), we have

$$\delta V = \frac{3\,\pi\,d^2}{6}\left[\frac{p.d^2}{4\,t.E}(1 - \mu)\right] = \frac{\pi\,p\,d^4}{8\,t.E}(1 - \mu)$$

Example 7.6. *A seamless spherical shell, 900 mm in diameter and 10 mm thick is being filled with a fluid under pressure until its volume increases by 150 × 10³ mm³. Calculate the pressure exerted by the fluid on the shell, taking modulus of elasticity for the material of the shell as 200 kN/mm² and Poisson's ratio as 0.3.*

Solution. Given : d = 900 mm ; t = 10 mm ; $\delta V = 150 \times 10^3$ mm³ ; E = 200 kN/mm² $= 200 \times 10^3$ N/mm² ; $\mu = 0.3$

Let p = Pressure exerted by the fluid on the shell.

We know that the increase in volume of the spherical shell (δV),

$$150 \times 10^3 = \frac{\pi\,p\,d^4}{8\,t\,E}(1 - \mu) = \frac{\pi\,p\,(900)^4}{8 \times 10 \times 200 \times 10^3}(1 - 0.3) = 90\,190\,p$$

$\therefore$ $p = 150 \times 10^3 / 90\,190 = 1.66$ N/mm² **Ans.**

7.9 Thick Cylindrical Shells Subjected to an Internal Pressure

When a cylindrical shell of a pressure vessel, hydraulic cylinder, gunbarrel and a pipe is subjected to a very high internal fluid pressure, then the walls of the cylinder must be made extremely heavy or thick.

In thin cylindrical shells, we have assumed that the tensile stresses are uniformly distributed over the section of the walls. But in the case of thick wall cylinders as shown in Fig. 7.6 (*a*), the stress over the section of the walls cannot be assumed to be uniformly distributed. They develop both tangential and radial stresses with values which are dependent upon the radius of the element under consideration. The distribution of stress in a thick cylindrical shell is shown in Fig. 7.6 (*b*) and (*c*). We see that the tangential stress is maximum at the inner surface and minimum at the outer surface of the shell. The radial stress is maximum at the inner surface and zero at the outer surface of the shell.

In the design of thick cylindrical shells, the following equations are mostly used:

1. Lame's equation; **2.** Birnie's equation; **3.** Clavarino's equation; and **4.** Barlow's equation.

The use of these equations depends upon the type of material used and the end construction.

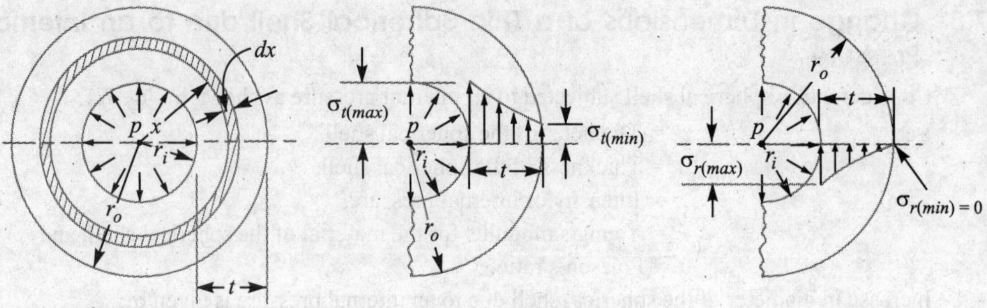

(*a*) Thick cylindrical shell. (*b*) Tangential stress distribution. (*c*) Radial stress distribution.

Fig. 7.6. Stress distribution in thick cylindrical shells subjected to internal pressure.

Let r_o = Outer radius of cylindrical shell,

r_i = Inner radius of cylindrical shell,

t = Thickness of cylindrical shell = $r_o - r_i$,

p = Intensity of internal pressure,

μ = Poisson's ratio,

σ_t = Tangential stress, and

σ_r = Radial stress.

All the above mentioned equations are now discussed, in detail, as below:

1. ***Lame's equation.*** Assuming that the longitudinal fibres of the cylindrical shell are equally strained, Lame has shown that the tangential stress at any radius x is,

$$\sigma_t = \frac{p_i (r_i)^2 - p_o (r_o)^2}{(r_o)^2 - (r_i)^2} + \frac{(r_i)^2 (r_o)^2}{x^2} \left[\frac{p_i - p_o}{(r_o)^2 - (r_i)^2} \right]$$

While designing a tanker, the pressure added by movement of the vehicle also should be considered.

and radial stress at any radius x,

$$\sigma_r = \frac{p_i (r_i)^2 - p_o (r_o)^2}{(r_o)^2 - (r_i)^2} - \frac{(r_i)^2 (r_o)^2}{x^2} \left[\frac{p_i - p_o}{(r_o)^2 - (r_i)^2} \right]$$

Since we are concerned with the internal pressure ($p_i = p$) only, therefore substituting the value of external pressure, $p_o = 0$.

∴ Tangential stress at any radius x,

$$\sigma_t = \frac{p (r_i)^2}{(r_o)^2 - (r_i)^2} \left[1 + \frac{(r_o)^2}{x^2} \right] \quad ...(i)$$

and radial stress at any radius x,

$$\sigma_r = \frac{p (r_i)^2}{(r_o)^2 - (r_i)^2} \left[1 - \frac{(r_o)^2}{x^2} \right] \quad ...(ii)$$

We see that the tangential stress is always a tensile stress whereas the radial stress is a compressive stress. We know that the tangential stress is maximum at the inner surface of the shell (*i.e.* when $x = r_i$) and it is minimum at the outer surface of the shell (*i.e.* when $x = r_o$). Substituting the value of $x = r_i$ and $x = r_o$ in equation (*i*), we find that the *maximum tangential stress at the inner surface of the shell,

$$\sigma_{t(max)} = \frac{p [(r_o)^2 + (r_i)^2]}{(r_o)^2 - (r_i)^2}$$

and minimum tangential stress at the outer surface of the shell,

$$\sigma_{t(min)} = \frac{2 p (r_i)^2}{(r_o)^2 - (r_i)^2}$$

We also know that the radial stress is maximum at the inner surface of the shell and zero at the outer surface of the shell. Substituting the value of $x = r_i$ and $x = r_o$ in equation (*ii*), we find that maximum radial stress at the inner surface of the shell,

$$\sigma_{r(max)} = -p \text{ (compressive)}$$

and minimum radial stress at the outer surface of the shell,

$$\sigma_{r(min)} = 0$$

In designing a thick cylindrical shell of brittle material (*e.g.* cast iron, hard steel and cast aluminium) with closed or open ends and in accordance with the maximum normal stress theory failure, the tangential stress induced in the cylinder wall,

$$\sigma_t = \sigma_{t(max)} = \frac{p [(r_o)^2 + (r_i)^2]}{(r_o)^2 - (r_i)^2}$$

Since $r_o = r_i + t$, therefore substituting this value of r_o in the above expression, we get

$$\sigma_t = \frac{p [(r_i + t)^2 + (r_i)^2]}{(r_i + t)^2 - (r_i)^2}$$

$$\sigma_t (r_i + t)^2 - \sigma_t (r_i)^2 = p (r_i + t)^2 + p (r_i)^2$$

$$(r_i + t)^2 (\sigma_t - p) = (r_i)^2 (\sigma_t + p)$$

$$\frac{(r_i + t)^2}{(r_i)^2} = \frac{\sigma_t + p}{\sigma_t - p}$$

---

* The maximum tangential stress is always greater than the internal pressure acting on the shell.

$$\frac{r_i + t}{r_i} = \sqrt{\frac{\sigma_t + p}{\sigma_t - p}} \quad \text{or} \quad 1 + \frac{t}{r_i} = \sqrt{\frac{\sigma_t + p}{\sigma_t - p}}$$

$$\therefore \quad \frac{t}{r_i} = \sqrt{\frac{\sigma_t + p}{\sigma_t - p}} - 1 \quad \text{or} \quad t = r_i \left[\sqrt{\frac{\sigma_t + p}{\sigma_t - p}} - 1 \right] \qquad \text{...(iii)}$$

The value of σ_t for brittle materials may be taken as 0.125 times the ultimate tensile strength (σ_u).

We have discussed above the design of a thick cylindrical shell of brittle materials. In case of cylinders made of ductile material, Lame's equation is modified according to maximum shear stress theory.

According to this theory, the maximum shear stress at any point in a strained body is equal to one-half the algebraic difference of the maximum and minimum principal stresses at that point. We know that for a thick cylindrical shell,

Maximum principal stress at the inner surface,

$$\sigma_{t\,(max)} = \frac{p\,[(r_o)^2 + (r_i)^2]}{(r_o)^2 - (r_i)^2}$$

and minimum principal stress at the outer surface,

$$\sigma_{t(min)} = -p$$

$\therefore$ Maximum shear stress,

$$\tau = \tau_{max} = \frac{\sigma_{t(max)} - \sigma_{t(min)}}{2} = \frac{\dfrac{p\,[(r_o)^2 + (r_i)^2]}{(r_o)^2 - (r_i)^2} - (-p)}{2}$$

$$= \frac{p\,[(r_o)^2 + (r_i)^2] + p\,[(r_o)^2 - (r_i)^2]}{2[(r_o)^2 - (r_i)^2]} = \frac{2\,p\,(r_o)^2}{2[(r_o)^2 - (r_i)^2]}$$

$$= \frac{p\,(r_i + t)^2}{(r_i + t)^2 - (r_i)^2} \qquad \text{... } (\because r_o = r_i + t)$$

or
$$\tau(r_i + t)^2 - \tau(r_i)^2 = p(r_i + t)^2$$

$$(r_i + t)^2\,(\tau - p) = \tau(r_i)^2$$

$$\frac{(r_i + t)^2}{(r_i)^2} = \frac{\tau}{\tau - p}$$

$$\frac{r_i + t}{r_i} = \sqrt{\frac{\tau}{\tau - p}} \quad \text{or} \quad 1 + \frac{t}{r_i} = \sqrt{\frac{\tau}{\tau - p}}$$

$$\therefore \quad \frac{t}{r_i} = \sqrt{\frac{\tau}{\tau - p}} - 1 \quad \text{or} \quad t = r_i \left[\sqrt{\frac{\tau}{\tau - p}} - 1 \right] \qquad \text{...(iv)}$$

The value of shear stress (τ) is usually taken as one-half the tensile stress (σ_t). Therefore the above expression may be written as

$$t = r_i \left[\sqrt{\frac{\sigma_t}{\sigma_t - 2p}} - 1 \right] \qquad \text{...(v)}$$

From the above expression, we see that if the internal pressure (p) is equal to or greater than the allowable working stress (σ_t or τ), then no thickness of the cylinder wall will prevent failure. Thus, it is impossible to design a cylinder to withstand fluid pressure greater than the allowable working stress for a given material. This difficulty is overcome by using compound cylinders (See Art. 7.10).

Oil is frequently transported by ships called tankers. The larger tankers, such as this Acrco Alaska oil transporter, are known as supertankers. They can be hundreds of metres long.

2. *Birnie's equation.* In case of open-end cylinders (such as pump cylinders, rams, gun barrels etc.) made of ductile material (*i.e.* low carbon steel, brass, bronze, and aluminium alloys), the allowable stresses cannot be determined by means of maximum-stress theory of failure. In such cases, the maximum-strain theory is used. According to this theory, the failure occurs when the strain reaches a limiting value and Birnie's equation for the wall thickness of a cylinder is

$$t = r_i \left[\sqrt{\frac{\sigma_t + (1 - \mu)\, p}{\sigma_t - (1 + \mu)\, p}} - 1 \right]$$

The value of σ_t may be taken as 0.8 times the yield point stress (σ_y).

3. *Clavarino's equation.* This equation is also based on the maximum-strain theory of failure, but it is applied to closed-end cylinders (or cylinders fitted with heads) made of ductile material. According to this equation, the thickness of a cylinder,

$$t = r_i \left[\sqrt{\frac{\sigma_t + (1 - 2\mu)\, p}{\sigma_t - (1 + \mu)\, p}} - 1 \right]$$

In this case also, the value of σ_t may be taken as 0.8 σ_y.

4. *Barlow's equation.* This equation is generally used for high pressure oil and gas pipes. According to this equation, the thickness of a cylinder,

$$t = p.r_o / \sigma_t$$

For ductile materials, $\sigma_t = 0.8\, \sigma_y$ and for brittle materials $\sigma_t = 0.125\, \sigma_u$, where σ_u is the ultimate stress.

Example 7.7. *A cast iron cylinder of internal diameter 200 mm and thickness 50 mm is subjected to a pressure of 5 N/mm². Calculate the tangential and radial stresses at the inner, middle (radius = 125 mm) and outer surfaces.*

Solution. Given : d_i = 200 mm or r_i = 100 mm ; t = 50 mm ; p = 5 N/mm²

We know that outer radius of the cylinder,

$$r_o = r_i + t = 100 + 50 = 150 \text{ mm}$$

Tangential stresses at the inner, middle and outer surfaces

We know that the tangential stress at any radius x,

$$\sigma_t = \frac{p\,(r_i)^2}{(r_o)^2 - (r_i)^2}\left[1 + \frac{(r_o)^2}{x^2}\right]$$

∴ Tangential stress at the inner surface (*i.e.* when $x = r_i = 100$ mm),

$$\sigma_{t(inner)} = \frac{p\,[(r_o)^2 + (r_i)^2]}{(r_o)^2 - (r_i)^2} = \frac{5\,[(150)^2 + (100)^2]}{(150)^2 - (100)^2} = 13 \text{ N/mm}^2 = 13 \text{ MPa } \textbf{Ans.}$$

Tangential stress at the middle surface (*i.e.* when $x = 125$ mm),

$$\sigma_{t(middle)} = \frac{5\,(100)^2}{(150)^2 - (100)^2}\left[1 + \frac{(150)^2}{(125)^2}\right] = 9.76 \text{ N/mm}^2 = 9.76 \text{ MPa } \textbf{Ans.}$$

and tangential stress at the outer surface (*i.e.* when $x = r_o = 150$ mm),

$$\sigma_{t(outer)} = \frac{2\,p\,(r_i)^2}{(r_o)^2 - (r_i)^2} = \frac{2 \times 5\,(100)^2}{(150)^2 - (100)^2} = 8 \text{ N/mm}^2 = 8 \text{ MPa } \textbf{Ans.}$$

Radial stresses at the inner, middle and outer surfaces

We know that the radial stress at any radius x,

$$\sigma_r = \frac{p\,(r_i)^2}{(r_o)^2 - (r_i)^2}\left[1 - \frac{(r_o)^2}{x^2}\right]$$

∴ Radial stress at the inner surface (*i.e.* when $x = r_i = 100$ mm),

$$\sigma_{r(inner)} = -p = -5 \text{ N/mm}^2 = 5 \text{ MPa (compressive) } \textbf{Ans.}$$

Radial stress at the middle surface (*i.e.* when $x = 125$ mm)

$$\sigma_{r(middle)} = \frac{5\,(100)^2}{(150)^2 - (100)^2}\left[1 - \frac{(150)^2}{(125)^2}\right] = -1.76 \text{ N/mm}^2 = -1.76 \text{ MPa}$$

$$= 1.76 \text{ MPa (compressive) } \textbf{Ans.}$$

and radial stress at the outer surface (*i.e.* when $x = r_o = 150$ mm),

$$\sigma_{r(outer)} = 0 \textbf{ Ans.}$$

Example 7.8. *A hydraulic press has a maximum capacity of 1000 kN. The piston diameter is 250 mm. Calculate the wall thickness if the cylinder is made of material for which the permissible strength may be taken as 80 MPa. This material may be assumed as a brittle material.*

Hydraulic Press

Solution. Given : $W = 1000$ kN $= 1000 \times 10^3$ N ; $d = 250$ mm ; $\sigma_t = 80$ MPa $= 80$ N/mm²

First of all, let us find the pressure inside the cylinder (p). We know that load on the hydraulic press (W),

$$1000 \times 10^3 = \frac{\pi}{4} \times d^2 \times p = \frac{\pi}{4}(250)^2\,p = 49.1 \times 10^3 p$$

$$\therefore \quad p = 1000 \times 10^3 / 49.1 \times 10^3 = 20.37 \text{ N/mm}^2$$

Let r_i = Inside radius of the cylinder $= d/2 = 125$ mm

We know that wall thickness of the cylinder,

$$t = r_i\left[\sqrt{\frac{\sigma_t + p}{\sigma_t - p}} - 1\right] = 125\left[\sqrt{\frac{80 + 20.37}{80 - 20.37}} - 1\right] \text{ mm}$$

$$= 125\ (1.297 - 1) = 37 \text{ mm } \textbf{Ans.}$$

Example 7.9. *A closed-ended cast iron cylinder of 200 mm inside diameter is to carry an internal pressure of 10 N/mm² with a permissible stress of 18 MPa. Determine the wall thickness by means of Lame's and the maximum shear stress equations. What result would you use? Give reason for your conclusion.*

Solution. Given : $d_i = 200$ mm or $r_i = 100$ mm ; $p = 10$ N/mm² ; $\sigma_t = 18$ MPa = 18 N/mm²

According to Lame's equation, wall thickness of a cylinder,

$$t = r_i\left[\sqrt{\frac{\sigma_t + p}{\sigma_t - p}} - 1\right] = 100\left[\sqrt{\frac{80 + 10}{80 - 10}} - 1\right] = 87 \text{ mm}$$

According to maximum shear stress equation, wall thickness of a cylinder,

$$t = r_i\left[\sqrt{\frac{\tau}{\tau - p}} - 1\right]$$

We have discussed in Art. 7.9 [equation (*iv*)], that the shear stress (τ) is usually taken one-half the tensile stress (σ_t). In the present case, $\tau = \sigma_t / 2 = 18/2 = 9$ N/mm². Since τ is less than the internal pressure ($p = 10$ N/mm²), therefore the expression under the square root will be negative. Thus no thickness can prevent failure of the cylinder. Hence it is impossible to design a cylinder to withstand fluid pressure greater than the allowable working stress for the given material. This difficulty is overcome by using compound cylinders as discussed in Art. 7.10.

Thus, we shall use a cylinder of wall thickness, $t = 87$ mm **Ans.**

Example 7.10. *The cylinder of a portable hydraulic riveter is 220 mm in diameter. The pressure of the fluid is 14 N/mm² by gauge. Determine suitable thickness of the cylinder wall assuming that the maximum permissible tensile stress is not to exceed 105 MPa.*

Solution. Given : $d_i = 220$ mm or $r_i = 110$ mm ; $p = 14$ N/mm² ; $\sigma_t = 105$ MPa = 105 N/mm²

Since the pressure of the fluid is high, therefore thick cylinder equation is used.

Assuming the material of the cylinder as steel, the thickness of the cylinder wall (t) may be obtained by using Birnie's equation. We know that

$$t = r_i\left[\sqrt{\frac{\sigma_t + (1 - \mu)\,p}{\sigma_t - (1 + \mu)\,p}} - 1\right]$$

$$= 110\left[\sqrt{\frac{105 + (1 - 0.3)\,14}{105 - (1 + 0.3)\,14}} - 1\right] = 16.5 \text{ mm } \textbf{Ans.}$$

...(Taking Poisson's ratio for steel, $\mu = 0.3$)

Example 7.11. *The hydraulic cylinder 400 mm bore operates at a maximum pressure of 5 N/mm². The piston rod is connected to the load and the cylinder to the frame through hinged joints. Design: 1. cylinder, 2. piston rod, 3. hinge pin, and 4. flat end cover.*

The allowable tensile stress for cast steel cylinder and end cover is 80 MPa and for piston rod is 60 MPa.

Draw the hydraulic cylinder with piston, piston rod, end cover and O-ring.

Solution. Given : $d_i = 400$ mm or $r_i = 200$ mm ; $p = 5$ N/mm² ; $\sigma_t = 80$ MPa = 80 N/mm² ; $\sigma_{tp} = 60$ MPa = 60 N/mm²

1. ***Design of cylinder***

Let d_o = Outer diameter of the cylinder.

We know that thickness of cylinder,

$$t = r_i \left[\sqrt{\frac{\sigma_t + p}{\sigma_t - p}} - 1 \right] = 200 \left[\sqrt{\frac{80 + 5}{80 - 5}} - 1 \right] \text{ mm}$$

$$= 200\,(1.06 - 1) = 12 \text{ mm } \textbf{Ans.}$$

∴ Outer diameter of the cylinder,

$$d_o = d_i + 2t = 400 + 2 \times 12 = 424 \text{ mm } \textbf{Ans.}$$

2. ***Design of piston rod***

Let d_p = Diameter of the piston rod.

We know that the force acting on the piston rod,

$$F = \frac{\pi}{4}\,(d_i)^2\,p = \frac{\pi}{4}\,(400)^2\,5 = 628\ 400 \text{ N} \qquad ...(i)$$

We also know that the force acting on the piston rod,

$$F = \frac{\pi}{4}\,(d_i)^2\,\sigma_{tp} = \frac{\pi}{4}\,(d_p)^2\,60 = 47.13\,(d_p)^2 \text{ N} \qquad ...(ii)$$

From equations (*i*) and (*ii*), we have

$$(d_p)^2 = 628\ 400/47.13 = 13\ 333.33 \quad \text{or} \quad d_p = 115.5 \text{ say } 116 \text{ mm } \textbf{Ans.}$$

3. ***Design of the hinge pin***

Let d_h = Diameter of the hinge pin of the piston rod.

Since the load on the pin is equal to the force acting on the piston rod, and the hinge pin is in double shear, therefore

$$F = 2 \times \frac{\pi}{4}\,(d_h)^2\,\tau$$

$$628\ 400 = 2 \times \frac{\pi}{4}\,(d_h)^2\,45 = 70.7\,(d_h)^2 \qquad \text{...(Taking } \tau = 45 \text{ N/mm}^2\text{)}$$

$$\therefore \quad (d_h)^2 = 628\ 400 / 70.7 = 8888.3 \quad \text{or} \quad d_h = 94.3 \text{ say } 95 \text{ mm } \textbf{Ans.}$$

When the cover is hinged to the cylinder, we can use two hinge pins only diametrically opposite to each other. Thus the diameter of the hinge pins for cover,

$$d_{hc} = \frac{d_h}{2} = \frac{95}{2} = 47.5 \text{ mm } \textbf{Ans.}$$

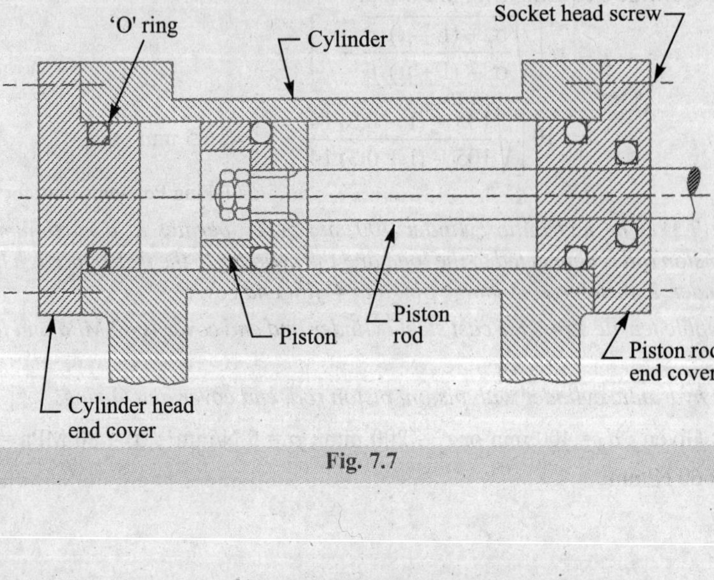

Fig. 7.7

4. *Design of the flat end cover*

Let t_c = Thickness of the end cover.

We know that force on the end cover,

$$F = d_i \times t_c \times \sigma_t$$

$$628\ 400 = 400 \times t_c \times 80 = 32 \times 10^3\ t_c$$

$$\therefore \quad t_c = 628\ 400 / 32 \times 10^3 = 19.64 \text{ say } 20 \text{ mm } \textbf{Ans.}$$

The hydraulic cylinder with piston, piston rod, end cover and *O*-ring is shown in Fig. 7.7.

7.10 Compound Cylindrical Shells

According to Lame's equation, the thickness of a cylindrical shell is given by

$$t = r_i\left(\sqrt{\frac{\sigma_t + p}{\sigma_t - p}} - 1\right)$$

From this equation, we see that if the internal pressure (p) acting on the shell is equal to or greater than the allowable working stress (σ_t) for the material of the shell, then no thickness of the shell will prevent failure. Thus it is impossible to design a cylinder to withstand internal pressure equal to or greater than the allowable working stress.

This difficulty is overcome by inducing an initial compressive stress on the wall of the cylindrical shell. This may be done by the following two methods:

1. By using compound cylindrical shells, and
2. By using the theory of plasticity.

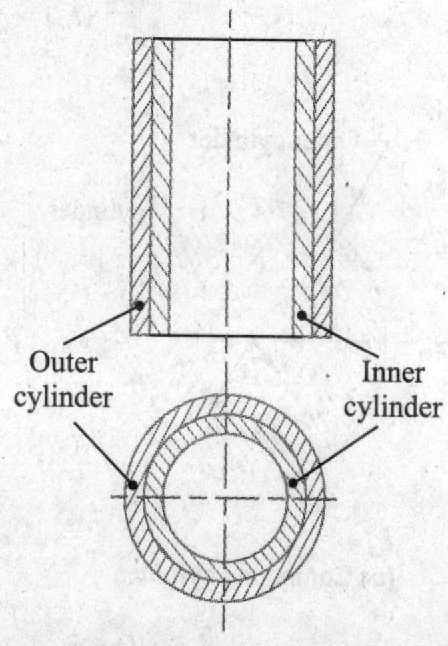

Fig. 7.8. Compound cylindrical shell.

In a compound cylindrical shell, as shown in Fig. 7.8, the outer cylinder (having inside diameter smaller than the outside diameter of the inner cylinder) is shrunk fit over the inner cylinder by heating and cooling. On cooling, the contact pressure is developed at the junction of the two cylinders, which induces compressive tangential stress in the material of the inner cylinder and tensile tangential stress in the material of the outer cylinder. When the cylinder is loaded, the compressive stresses are first relieved and then tensile stresses are induced. Thus, a compound cylinder is effective in resisting higher internal pressure than a single cylinder with the same overall dimensions. The principle of compound cylinder is used in the design of gun tubes.

In the theory of plasticity, a temporary high internal pressure is applied till the plastic stage is reached near the inside of the cylinder wall. This results in a residual compressive stress upon the removal of the internal pressure, thereby making the cylinder more effective to withstand a higher internal pressure.

7.11 Stresses in Compound Cylindrical Shells

Fig. 7.9 (*a*) shows a compound cylindrical shell assembled with a shrink fit. We have discussed in the previous article that when the outer cylinder is shrunk fit over the inner cylinder, a contact pressure (p) is developed at junction of the two cylinders (*i.e.* at radius r_2) as shown in Fig. 7.9 (*b*) and (*c*). The stresses resulting from this pressure may be easily determined by using Lame's equation.

According to this equation (See Art. 7.9), the tangential stress at any radius x is

$$\sigma_t = \frac{p_i (r_i)^2 - p_o (r_o)^2}{(r_o)^2 - (r_i)^2} + \frac{(r_i)^2 (r_o)^2}{x^2}\left[\frac{p_i - p_o}{(r_o)^2 - (r_i)^2}\right] \quad \text{...}(i)$$

and radial stress at any radius x,

$$\sigma_r = \frac{p_i (r_i)^2 - p_o (r_o)^2}{(r_o)^2 - (r_i)^2} - \frac{(r_i)^2 (r_o)^2}{x^2} \left[\frac{p_i - p_o}{(r_o)^2 - (r_i)^2} \right] \quad \text{...(ii)}$$

Considering the external pressure only,

$$\sigma_t = \frac{- p_o (r_o)^2}{(r_o)^2 - (r_i)^2} \left[1 + \frac{(r_i)^2}{x^2} \right] \quad \text{...(iii)}$$

...[Substituting $p_i = 0$ in equation (*i*)]

and $$\sigma_r = \frac{- p_o (r_o)^2}{(r_o)^2 - (r_i)^2} \left[1 - \frac{(r_i)^2}{x^2} \right] \quad \text{...(iv)}$$

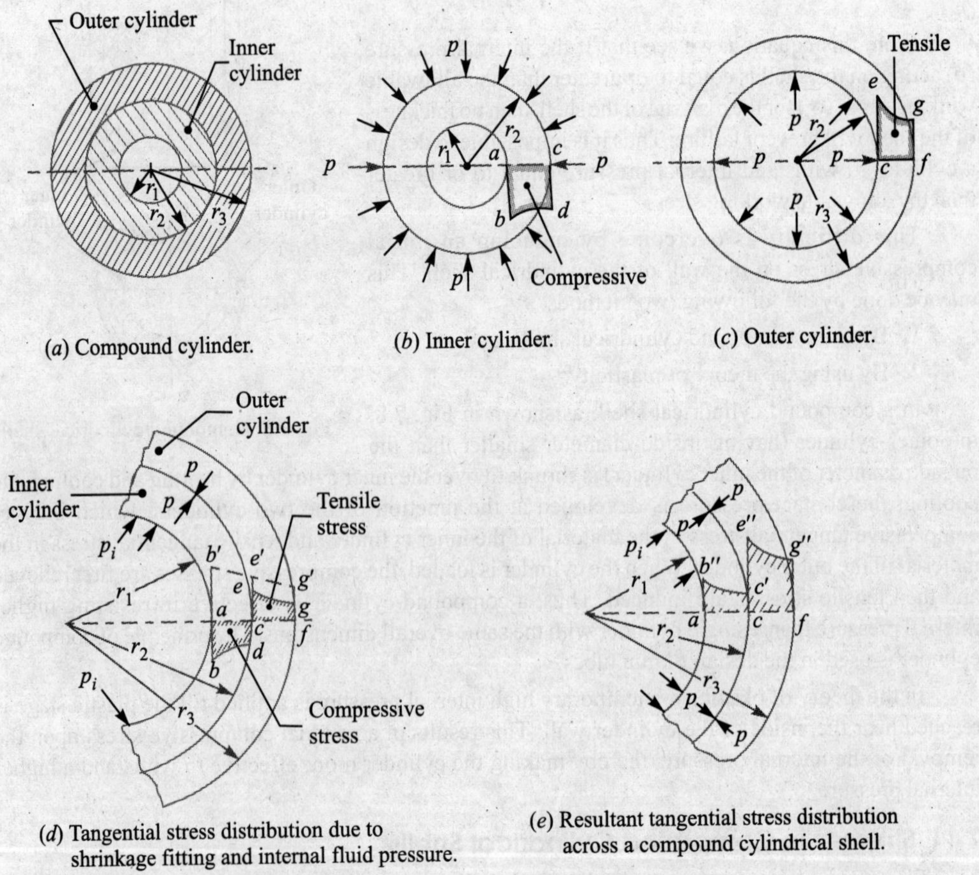

Fig. 7.9. Stresses in compound cylindrical shells.

Considering the internal pressure only,

$$\sigma_t = \frac{p_i (r_i)^2}{(r_o)^2 - (r_i)^2} \left[1 + \frac{(r_o)^2}{x^2} \right] \quad \text{...(v)}$$

...[Substituting $p_o = 0$ in equation (*i*)]

and $$\sigma_r = \frac{p_i (r_i)^2}{(r_o)^2 - (r_i)^2}\left[1 - \frac{(r_o)^2}{x^2}\right] \qquad ...(vi)$$

Since the inner cylinder is subjected to an external pressure (p) caused by the shrink fit and the outer cylinder is subjected to internal pressure (p), therefore from equation (*iii*), we find that the tangential stress at the inner surface of the inner cylinder,

$$\sigma_{t1} = \frac{-p (r_2)^2}{(r_2)^2 - (r_1)^2}\left[1 + \frac{(r_1)^2}{(r_1)^2}\right] = \frac{-2p (r_2)^2}{(r_2)^2 - (r_1)^2} \text{ (compressive)} \qquad ...(vii)$$

... [Substituting $p_o = p, x = r_1, r_o = r_2$ and $r_i = r_1$]

This stress is compressive and is shown by *ab* in Fig. 7.9 (*b*).

Radial stress at the inner surface of the inner cylinder,

$$\sigma_{r1} = \frac{-p (r_2)^2}{(r_2)^2 - (r_1)^2}\left[1 - \frac{(r_1)^2}{(r_1)^2}\right] = 0$$

...[From equation (*iv*)]

Similarly from equation (*iii*), we find that tangential stress at the outer surface of the inner cylinder,

$$\sigma_{t2} = \frac{-p (r_2)^2}{(r_2)^2 - (r_1)^2}\left[1 + \frac{(r_1)^2}{(r_2)^2}\right] = \frac{-p\,[(r_2)^2 + (r_1)^2]}{(r_2)^2 - (r_1)^2} \text{ (compressive)} \qquad ...(viii)$$

...[Substituting $p_o = p, x = r_2, r_o = r_2$ and $r_i = r_1$]

This stress is compressive and is shown by *cd* in Fig. 7.9 (*b*).

Radial stress at the outer surface of the inner cylinder,

$$\sigma_{r2} = \frac{-p (r_2)^2}{(r_2)^2 - (r_1)^2}\left[1 - \frac{(r_1)^2}{(r_2)^2}\right] = -p$$

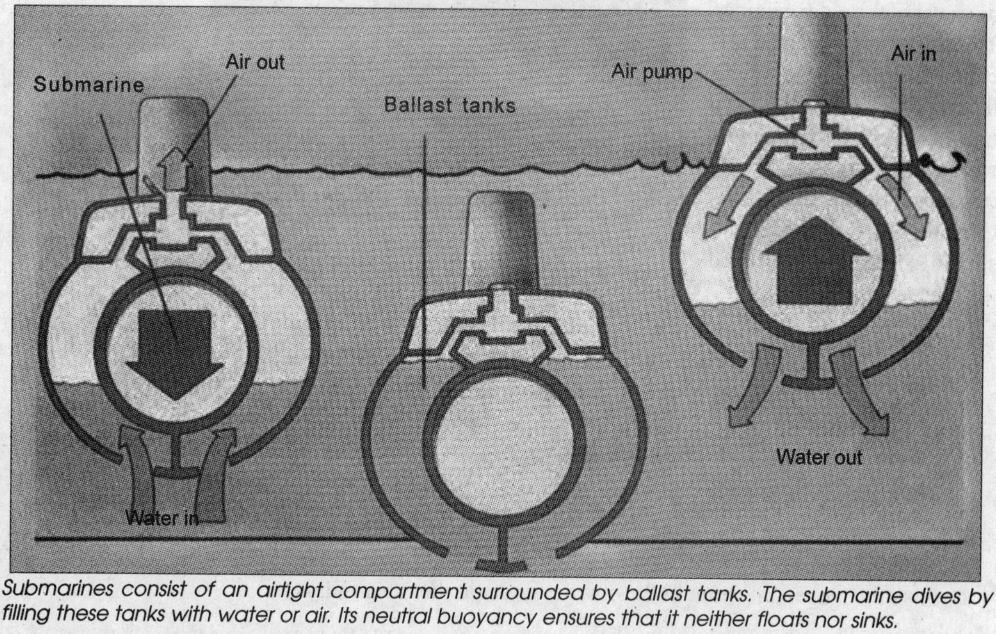

Submarines consist of an airtight compartment surrounded by ballast tanks. The submarine dives by filling these tanks with water or air. Its neutral buoyancy ensures that it neither floats nor sinks.

Note : This picture is given as additional information and is not a direct example of the current chapter.

Now let us consider the outer cylinder subjected to internal pressure (p). From equation (v), we find that the tangential stress at the inner surface of the outer cylinder,

$$\sigma_{t3} = \frac{p\,(r_2)^2}{(r_3)^2 - (r_2)^2}\left[1 + \frac{(r_3)^2}{(r_2)^2}\right] = \frac{p\,[(r_3)^2 + (r_2)^2]}{(r_3)^2 - (r_2)^2} \text{ (tensile)} \qquad ...(ix)$$

...[Substituting $p_i = p$, $x = r_2$, $r_o = r_3$ and $r_i = r_2$]

This stress is tensile and is shown by ce in Fig. 7.9 (c).

Radial stress at the inner surface of the outer cylinder,

$$\sigma_{r3} = \frac{p\,(r_2)^2}{(r_3)^2 - (r_2)^2}\left[1 - \frac{(r_3)^2}{(r_2)^2}\right] = -\,p \qquad \text{...[From equation } (vi)]$$

Similarly from equation (v), we find that the tangential stress at the outer surface of the outer cylinder,

$$\sigma_{t4} = \frac{p\,(r_2)^2}{(r_3)^2 - (r_2)^2}\left[1 + \frac{(r_3)^2}{(r_3)^2}\right] = \frac{2\,p\,(r_2)^2}{(r_3)^2 - (r_2)^2} \text{ (tensile)} \qquad ...(x)$$

...[Substituting $p_i = p$, $x = r_3$, $r_o = r_3$ and $r_i = r_2$]

This stress is tensile and is shown by fg in Fig. 7.9 (c).

Radial stress at the outer surface of the outer cylinder,

$$\sigma_{r4} = \frac{p\,(r_2)^2}{(r_3)^2 - (r_2)^2}\left[1 - \frac{(r_3)^2}{(r_3)^2}\right] = 0$$

The equations (vii) to (x) cannot be solved until the contact pressure (p) is known. In obtaining a shrink fit, the outside diameter of the inner cylinder is made larger than the inside diameter of the outer cylinder. This difference in diameters is called the **interference** and is the deformation which the two cylinders must experience. Since the diameters of the cylinders are usually known, therefore the deformation should be calculated to find the contact pressure.

Submarine is akin a to pressure vessel. CAD and CAM were used to design and manufacture this French submarine.

Let δ_o = Increase in inner radius of the outer cylinder,

δ_i = Decrease in outer radius of the inner cylinder,

E_o = Young's modulus for the material of the outer cylinder,

E_i = Young's modulus for the material of the inner cylinder, and

μ = Poisson's ratio.

We know that the tangential strain in the outer cylinder at the inner radius (r_2),

$$\varepsilon_{to} = \frac{\text{Change in circumference}}{\text{Original circumference}} = \frac{2\pi\,(r_2 + \delta_o) - 2\pi\,r_2}{2\pi\,r_2} = \frac{\delta_o}{r_2} \qquad ...(xi)$$

Also the tangential strain in the outer cylinder at the inner radius (r_2),

$$\varepsilon_{to} = \frac{\sigma_{to}}{E_o} - \frac{\mu.\sigma_{ro}}{E_o} \qquad ...(xii)$$

We have discussed above that the tangential stress at the inner surface of the outer cylinder (or at the contact surfaces),

$$\sigma_{to} = \sigma_{t3} = \frac{p\,[(r_3)^2 + (r_2)^2]}{(r_3)^2 - (r_2)^2} \qquad \text{...[From equation } (ix)]$$

and radial stress at the inner surface of the outer cylinder (or at the contact surfaces),

$$\sigma_{ro} = \sigma_{r3} = -p$$

Substituting the value of σ_{to} and σ_{ro} in equation (*xii*), we get

$$\varepsilon_{to} = \frac{p\,[(r_3)^2 + (r_2)^2]}{E_o\,[(r_3)^2 - (r_2)^2]} + \frac{\mu.p}{E_o} = \frac{p}{E_o}\left[\frac{(r_3)^2 + (r_2)^2}{(r_3)^2 - (r_2)^2} + \mu\right] \qquad \text{...}(xiii)$$

From equations (*xi*) and (*xiii*),

$$\delta_o = \frac{p.r_2}{E_o}\left[\frac{(r_3)^2 + (r_2)^2}{(r_3)^2 - (r_2)^2} + \mu\right] \qquad \text{...}(xiv)$$

Similarly, we may find that the decrease in the outer radius of the inner cylinder,

$$\delta_i = \frac{-\,p.r_2}{E_i}\left[\frac{(r_2)^2 + (r_1)^2}{(r_2)^2 - (r_1)^2} - \mu\right] \qquad \text{...}(xv)$$

∴ Difference in radius,

$$\delta_r = \delta_o - \delta_i = \frac{p.r_2}{E_o}\left[\frac{(r_3)^2 + (r_2)^2}{(r_3)^2 - (r_2)^2} + \mu\right] + \frac{p.r_2}{E_i}\left[\frac{(r_2)^2 + (r_1)^2}{(r_2)^2 - (r_1)^2} - \mu\right]$$

If both the cylinders are of the same material, then $E_o = E_i = E$. Thus the above expression may be written as

$$\delta_r = \frac{p.r_2}{E}\left[\frac{(r_3)^2 + (r_2)^2}{(r_3)^2 - (r_2)^2} + \frac{(r_2)^2 + (r_1)^2}{(r_2)^2 - (r_1)^2}\right]$$

$$= \frac{p.r_2}{E}\left[\frac{[(r_3)^2 + (r_2)^2]\,[(r_2)^2 - (r_1)^2] + [(r_2)^2 + (r_1)^2]\,[(r_3)^2 - (r_2)^2]}{[(r_3)^2 - (r_2)^2]\,[(r_2)^2 - (r_1)^2]}\right]$$

$$= \frac{p.r_2}{E}\left[\frac{2(r_2)^2\,[(r_3)^2 - (r_1)^2]}{[(r_3)^2 - (r_2)^2]\,[(r_2)^2 - (r_1)^2]}\right]$$

or

$$p = \frac{E.\delta_r}{r_2}\left[\frac{[(r_3)^2 - (r_2)^2]\,[(r_2)^2 - (r_1)^2]}{2(r_2)^2\,[(r_3)^2 - (r_1)^2]}\right]$$

Substituting this value of p in equations (*vii*) to (*x*), we may obtain the tangential stresses at the various surfaces of the compound cylinder.

Now let us consider the compound cylinder subjected to an internal fluid pressure (p_i). We have discussed above that when the compound cylinder is subjected to internal pressure (p_i), then the tangential stress at any radius (x) is given by

$$\sigma_t = \frac{p_i\,(r_i)^2}{(r_o)^2 - (r_i)^2}\left[1 + \frac{(r_o)^2}{x^2}\right]$$

∴ Tangential stress at the inner surface of the inner cylinder,

$$\sigma_{t5} = \frac{p_i\,(r_1)^2}{(r_3)^2 - (r_1)^2}\left[1 + \frac{(r_3)^2}{(r_1)^2}\right] = \frac{p_i\,[(r_3)^2 + (r_1)^2]}{(r_3)^2 - (r_1)^2} \text{ (tensile)}$$

... [Substituting $x = r_1$, $r_o = r_3$ and $r_i = r_1$]

This stress is tensile and is shown by *ab'* in Fig. 7.9 (*d*).

Tangential stress at the outer surface of the inner cylinder or inner surface of the outer cylinder,

$$\sigma_{t6} = \frac{p_i (r_1)^2}{(r_3)^2 - (r_1)^2}\left[1 + \frac{(r_3)^2}{(r_2)^2}\right] = \frac{p_i (r_1)^2}{(r_2)^2}\left[\frac{(r_3)^2 + (r_2)^2}{(r_3)^2 - (r_1)^2}\right] \text{(tensile)}$$

... [Substituting $x = r_2$, $r_o = r_3$ and $r_i = r_1$]

This stress is tensile and is shown by *ce'* in Fig. 7.9 (*d*),

and tangential stress at the outer surface of the outer cylinder,

$$\sigma_{t7} = \frac{p_i (r_1)^2}{(r_3)^2 - (r_1)^2}\left[1 + \frac{(r_3)^2}{(r_3)^2}\right] = \frac{2\,p_i (r_1)^2}{(r_3)^2 - (r_1)^2} \text{(tensile)}$$

...[Substituting $x = r_3$, $r_o = r_3$ and $r_i = r_1$]

This stress is tensile and is shown by *fg'* in Fig. 7.9 (*d*).

Now the resultant stress at the inner surface of the compound cylinder,

$$\sigma_{ti} = \sigma_{t1} + \sigma_{t5} \quad \text{or} \quad ab' - ab$$

This stress is tensile and is shown by *ab"* in Fig. 7.9 (*e*).

Resultant stress at the outer surface of the inner cylinder

$$= \sigma_{t2} + \sigma_{t6} \quad \text{or} \quad ce' - cd \text{ or } cc'$$

Resultant stress at the inner surface of the outer cylinder

$$= \sigma_{t3} + \sigma_{t6} \quad \text{or} \quad ce + ce' \text{ or } c'e''$$

∴ Total resultant stress at the mating or contact surface,

$$\sigma_{tm} = \sigma_{t2} + \sigma_{t6} + \sigma_{t3} + \sigma_{t6}$$

This stress is tensile and is shown by *ce"* in Fig. 7.9 (*e*),

and resultant stress at the outer surface of the outer cylinder,

$$\sigma_{to} = \sigma_{t4} + \sigma_{t7} \quad \text{or} \quad fg + fg'$$

This stress is tensile and is shown by *fg"* in Fig. 7.9 (*e*).

Example 7.12. *The hydraulic press, having a working pressure of water as 16 N/mm² and exerting a force of 80 kN is required to press materials upto a maximum size of 800 mm × 800 mm and 800 mm high, the stroke length is 80 mm. Design and draw the following parts of the press : 1. Design of ram; 2. Cylinder; 3. Pillars; and 4. Gland.*

Solution. Given: $p = 16$ N/mm² ; $F = 80$ kN $= 80 \times 10^3$ N

The hydraulic press is shown in Fig. 7.10.

1. ***Design of ram***

Let d_r = Diameter of ram.

We know that the maximum force to be exerted by the ram (F),

$$80 \times 10^3 = \frac{\pi}{4}(d_r)^2 p = \frac{\pi}{4}(d_r)^2\, 16 = 12.57\,(d_r)^2$$

$$\therefore \quad (d_r)^2 = 80 \times 10^3 / 12.57 = 6364 \quad \text{or} \quad d_r = 79.8 \text{ say } 80 \text{ mm}$$ **Ans.**

In case the ram is made hollow in order to reduce its weight, then it can be designed as a thick cylinder subjected to external pressure. We have already discussed in Art. 7.11 that according to Lame's equation, maximum tangential stress (considering external pressure only) is

$$\sigma_{t(max)} = \frac{-p_o (d_{ro})^2}{(d_{ro})^2 - (d_{ri})^2}\left[1 + \frac{(d_{ri})^2}{(d_{ro})^2}\right] = -p_o\left[\frac{(d_{ro})^2 + (d_{ri})^2}{(d_{ro})^2 - (d_{ri})^2}\right] \text{ (compressive)}$$

and maximum radial stress,

$$\sigma_{r(max)} = -p_o \text{ (compressive)}$$

where d_{ro} = Outer diameter of ram = d_r = 80 mm

d_{ri} = Inner diameter of ram, and

p_o = External pressure = p = 16 N/mm² ...(Given)

Now according to maximum shear stress theory for ductile materials, maximum shear stress is

$$\tau_{max} = \frac{\sigma_{t(max)} - \sigma_{r(max)}}{2} = \frac{-p_o\left[\frac{(d_{ro})^2 + (d_{ri})^2}{(d_{ro})^2 - (d_{ri})^2}\right] - (-p_o)}{2}$$

$$= -p_o\left[\frac{(d_{ri})^2}{(d_{ro})^2 - (d_{ri})^2}\right]$$

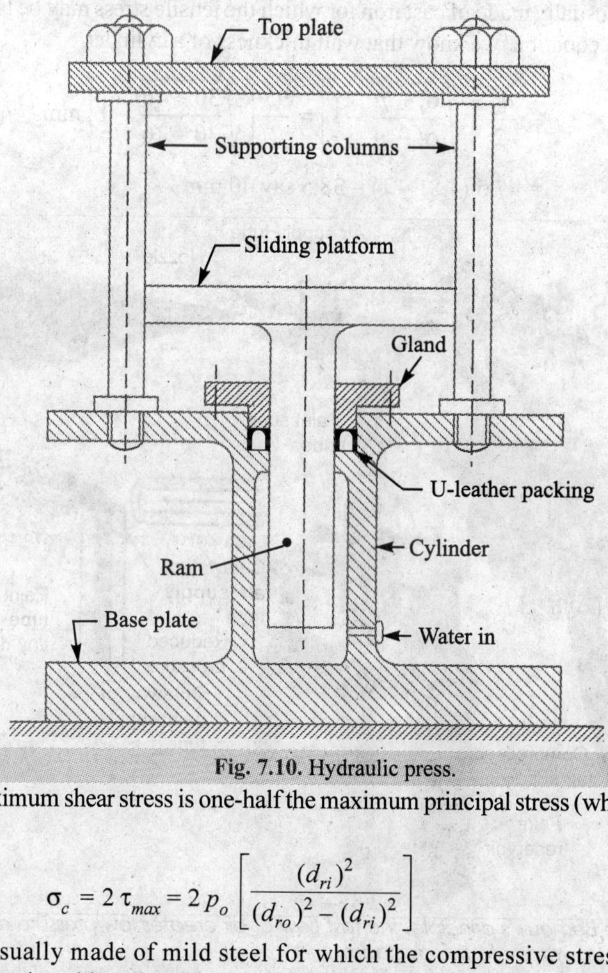

Fig. 7.10. Hydraulic press.

Since the maximum shear stress is one-half the maximum principal stress (which is compressive), therefore

$$\sigma_c = 2\,\tau_{max} = 2\,p_o\left[\frac{(d_{ri})^2}{(d_{ro})^2 - (d_{ri})^2}\right]$$

The ram is usually made of mild steel for which the compressive stress may be taken as 75 N/mm². Substituting this value of stress in the above expression, we get

$$75 = 2 \times 16 \left[\frac{(d_{ri})^2}{(80)^2 - (d_{ri})^2} \right] = \frac{32\,(d_{ri})^2}{6400 - (d_{ri})^2}$$

or
$$\frac{(d_{ri})^2}{6400 - (d_{ri})^2} = \frac{75}{32} = 2.34$$

$$(d_{ri})^2 = 2.34\,[6400 - (d_{ri})^2] = 14\,976 - 2.34\,(d_{ri})^2$$

$$3.34\,(d_{ri})^2 = 14\,976 \quad \text{or} \quad (d_{ri})^2 = 14\,976/3.34 = 4484$$

$\therefore$ $d_{ri} = 67$ mm **Ans.**

and $d_{ro} = d_r = 80$ mm **Ans.**

2. ***Design of cylinder***

Let d_{ci} = Inner diameter of cylinder, and

d_{co} = Outer diameter of cylinder.

Assuming a clearance of 15 mm between the ram and the cylinder bore, therefore inner diameter of the cylinder,

$$d_{ci} = d_{ro} + \text{Clearance} = 80 + 15 = 95 \text{ mm } \textbf{Ans.}$$

The cylinder is usually made of cast iron for which the tensile stress may be taken as 30 N/mm^2. According to Lame's equation, we know that wall thickness of a cylinder,

$$t = \frac{d_{ci}}{2} \left[\sqrt{\frac{\sigma_t + p}{\sigma_t - p}} - 1 \right] = \frac{95}{2} \left[\sqrt{\frac{30 + 16}{30 - 16}} - 1 \right] \text{ mm}$$

$$= 47.5\,(1.81 - 1) = 38.5 \text{ say } 40 \text{ mm}$$

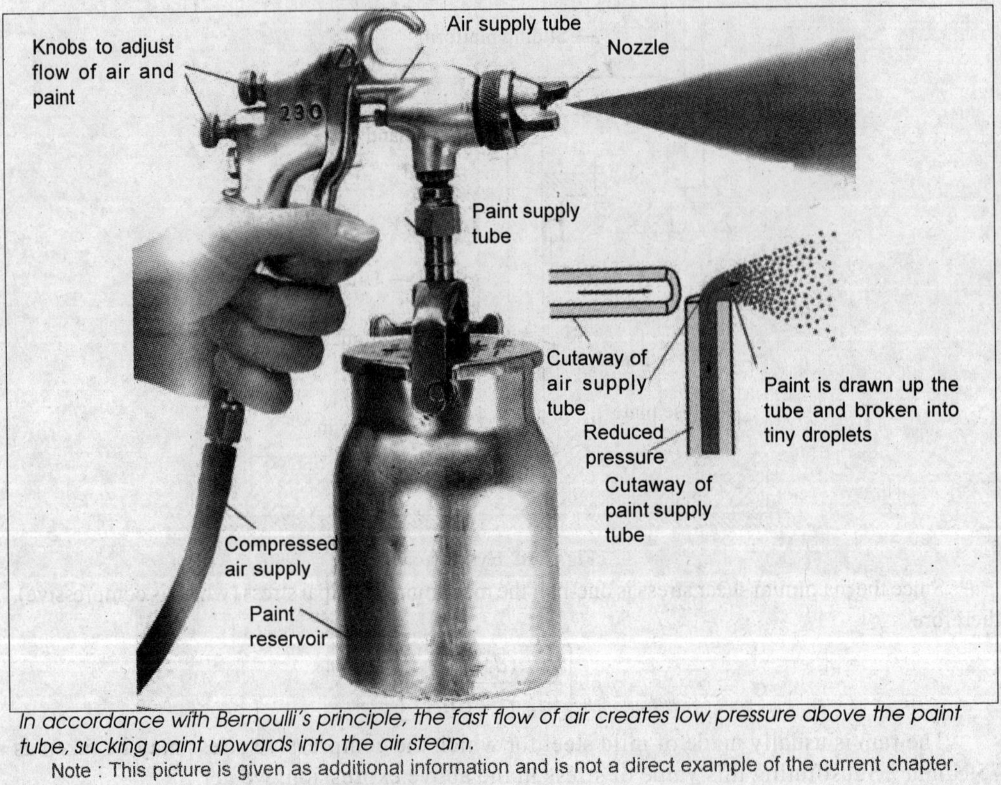

In accordance with Bernoulli's principle, the fast flow of air creates low pressure above the paint tube, sucking paint upwards into the air steam.

Note : This picture is given as additional information and is not a direct example of the current chapter.

and outside diameter of the cylinder,

$$d_{co} = d_{ci} + 2\,t = 95 + 2 \times 40 = 175 \text{ mm } \textbf{Ans.}$$

3. ***Design of pillars***

Let d_p = Diameter of the pillar.

The function of the pillars is to support the top plate and to guide the sliding plate. When the material is being pressed, the pillars will be under direct tension. Let there are four pillars and the load is equally shared by these pillars.

∴ Load on each pillar

$$= 80 \times 10^3/4 = 20 \times 10^3 \text{ N} \qquad ...(i)$$

We know that load on each pillar

$$= \frac{\pi}{4}\,(d_p)^2\,\sigma_t = \frac{\pi}{4}\,(d_p)^2\,75 = 58.9\,(d_p)^2 \qquad ...(ii)$$

From equations (*i*) and (*ii*),

$$(d_p)^2 = 20 \times 10^3/58.9 = 340 \quad \text{or} \quad d_p = 18.4 \text{ mm}$$

From fine series of metric threads, let us adopt the threads on pillars as M 20 × 1.5 having major diameter as 20 mm and core diameter as 18.16 mm. **Ans.**

4. ***Design of gland***

The gland is shown in Fig 7.11. The width (w) of the U-leather packing for a ram is given empirically as $2\sqrt{d_r}$ to $2.5\sqrt{d_r}$, where d_r is the diameter (outer) of the ram in mm. Let us take width of the packing as $2.2\sqrt{d_r}$.

∴ Width of packing,

$$w = 2.2\sqrt{80} = 19.7 \text{ say } 20 \text{ mm } \textbf{Ans.}$$

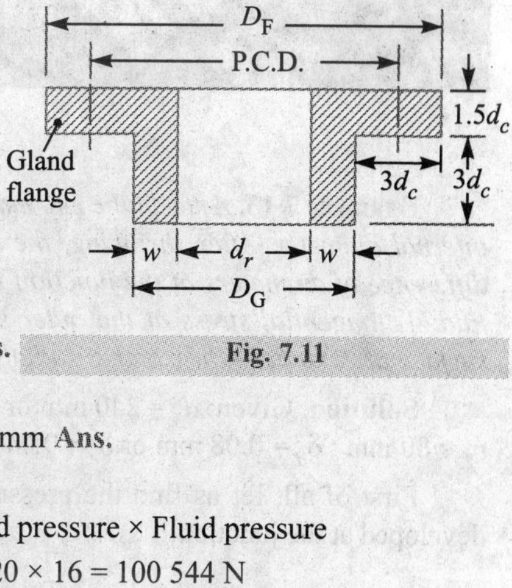

Fig. 7.11

and outer diameter of gland,

$$D_G = d_r + 2\,w = 80 + 2 \times 20 = 120 \text{ mm } \textbf{Ans.}$$

We know that total upward load on the gland

= Area of gland exposed to fluid pressure × Fluid pressure

$$= \pi\,(d_r + w)\,w.p = \pi\,(80 + 20)\,20 \times 16 = 100\,544 \text{ N}$$

Let us assume that 8 studs equally spaced on the pitch circle of the gland flange are used for holding down the gland.

∴ Load on each stud = 100 544 / 8 = 12 568 N

If d_c is the core diameter of the stud and σ_t is the permissible tensile stress for the stud material, then

Load on each stud,

$$12\,568 = \frac{\pi}{4}\,(d_c)^2\,\sigma_t = \frac{\pi}{4}\,(d_c)^2\,75 = 58.9\,(d_c)^2 \qquad ... (\text{Taking } \sigma_t = 75 \text{ N/mm}^2)$$

$$\therefore \quad (d_c)^2 = 12\,568 / 58.9 = 213.4 \quad \text{or} \quad d_c = 14.6 \text{ mm}$$

From fine series of metric threads, let us adopt the studs of size M 18 × 1.5 having major diameter as 18 mm and core diameter (d_c) as 16.16 mm. **Ans.**

The other dimensions for the gland are taken as follows:

Pitch circle diameter of the gland flange,

$$\text{P.C.D.} = D_G + 3\,d_c = 120 + 3 \times 16.16 = 168.48 \text{ or } 168.5 \text{ mm } \textbf{Ans.}$$

Outer diameter of the gland flange,

$$D_F = D_G + 6\,d_c = 120 + 6 \times 16.16 = 216.96 \text{ or } 217 \text{ mm } \textbf{Ans.}$$

and thickness of the gland flange $= 1.5\,d_c = 1.5 \times 16.16 = 24.24$ or 24.5 mm **Ans.**

A long oil tank.

Example 7.13. *A steel tube 240 mm external diameter is shrunk on another steel tube of 80 mm internal diameter. After shrinking, the diameter at the junction is 160 mm. Before shrinking, the difference of diameters at the junction was 0.08 mm. If the Young's modulus for steel is 200 GPa, find: 1. tangential stress at the outer surface of the inner tube; 2. tangential stress at the inner surface of the outer tube ; and 3. radial stress at the junction.*

Solution. Given: $d_3 = 240$ mm or $r_3 = 120$ mm ; $d_1 = 80$ mm or $r_1 = 40$ mm ; $d_2 = 160$ mm or $r_2 = 80$ mm ; $\delta_d = 0.08$ mm or $\delta_r = 0.04$ mm ; $E = 200$ GPa $= 200$ kN/mm² $= 200 \times 10^3$ N/mm²

First of all, let us find the pressure developed at the junction. We know that the pressure developed at the junction,

$$p = \frac{E.\delta_r}{r_2}\left[\frac{[(r_3)^2 - (r_2)^2]\,[(r_2)^2 - (r_1)^2]}{2(r_2)^2\,[(r_3)^2 - (r_1)^2]}\right]$$

$$= \frac{200 \times 10^3 \times 0.04}{80}\left[\frac{[(120)^2 - (80)^2]\,[(80)^2 - (40)^2]}{2 \times (80)^2\,[(120)^2 - (40)^2]}\right]$$

$$= 100 \times 0.234 = 23.4 \text{ N/mm}^2 \textbf{ Ans.}$$

1. ***Tangential stress at the outer surface of the inner tube***

We know that the tangential stress at the outer surface of the inner tube,

$$\sigma_{ti} = \frac{-\,p\,[(r_2)^2 + (r_1)^2]}{(r_2)^2 - (r_1)^2} = \frac{-\,23.4\,[(80)^2 + (40)^2]}{(80)^2 - (40)^2} = -\,39 \text{ N/mm}^2$$

$$= 39 \text{ MPa (compressive) } \textbf{Ans.}$$

2. ***Tangential stress at the inner surface of the outer tube***

We know that the tangential stress at the inner surface of the outer tube,

$$\sigma_{to} = \frac{-p\,[(r_3)^2 + (r_2)^2]}{(r_3)^2 - (r_2)^2} = \frac{23.4\,[(120)^2 + (80)^2]}{(120)^2 - (80)^2} = 60.84 \text{ N/mm}^2$$

$$= 60.84 \text{ MPa } \textbf{Ans.}$$

3. ***Radial stress at the junction***

We know that the radial stress at the junction, (*i.e.* at the inner radius of the outer tube),

$$\sigma_{ro} = -p = -23.4 \text{ N/mm}^2 = 23.4 \text{ MPa (compressive) } \textbf{Ans.}$$

Example 7.14. *A shrink fit assembly, formed by shrinking one tube over another, is subjected to an internal pressure of 60 N/mm². Before the fluid is admitted, the internal and the external diameters of the assembly are 120 mm and 200 mm and the diameter at the junction is 160 mm. If after shrinking on, the contact pressure at the junction is 8 N/mm², determine using Lame's equations, the stresses at the inner, mating and outer surfaces of the assembly after the fluid has been admitted.*

Solution. Given : $p_i = 60$ N/mm² ; $d_1 = 120$ mm or $r_1 = 60$ mm ; $d_3 = 200$ mm or $r_3 = 100$ mm ; $d_2 = 160$ mm or $r_2 = 80$ mm ; $p = 8$ N/mm²

First of all, let us find out the stresses induced in the assembly due to contact pressure at the junction (p).

We know that the tangential stress at the inner surface of the inner tube,

$$\sigma_{t1} = \frac{-2p\,(r_2)^2}{(r_2)^2 - (r_1)^2} = \frac{-2 \times 8\,(80)^2}{(80)^2 - (60)^2} = -36.6 \text{ N/mm}^2$$

$$= 36.6 \text{ MPa (compressive)}$$

Tangential stress at the outer surface of the inner tube,

$$\sigma_{t2} = \frac{-p\,[(r_2)^2 + (r_1)^2]}{(r_2)^2 - (r_1)^2} = \frac{-8\,[(80)^2 + (60)^2]}{(80)^2 - (60)^2} = -28.6 \text{ N/mm}^2$$

$$= 28.6 \text{ MPa (compressive)}$$

Tangential stress at the inner surface of the outer tube,

$$\sigma_{t3} = \frac{p\,[(r_3)^2 + (r_2)^2]}{(r_3)^2 - (r_2)^2} = \frac{8\,[(100)^2 + (80)^2]}{(100)^2 - (80)^2} = 36.4 \text{ N/mm}^2$$

$$= 36.4 \text{ MPa (tensile)}$$

and tangential stress at the outer surface of the outer tube,

$$\sigma_{t4} = \frac{2p\,(r_2)^2}{(r_3)^2 - (r_2)^2} = \frac{2 \times 8\,(80)^2}{(100)^2 - (80)^2} = 28.4 \text{ N/mm}^2$$

$$= 28.4 \text{ MPa (tensile)}$$

Now let us find out the stresses induced in the assembly due to internal fluid pressure (p_i).

We know that the tangential stress at the inner surface of the inner tube,

$$\sigma_{t5} = \frac{p_i\,[(r_3)^2 + (r_1)^2]}{(r_3)^2 - (r_1)^2} = \frac{60\,[(100)^2 + (60)^2]}{(100)^2 - (60)^2} = 127.5 \text{ N/mm}^2$$

$$= 127.5 \text{ MPa (tensile)}$$

Tangential stress at the outer surface of the inner tube or inner surface of the outer tube (*i.e.*, mating surface),

$$\sigma_{t6} = \frac{p_i (r_1)^2}{(r_2)^2}\left[\frac{(r_3)^2 + (r_2)^2}{(r_3)^2 - (r_1)^2}\right] = \frac{60\,(60)^2}{(80)^2}\left[\frac{(100)^2 + (80)^2}{(100)^2 - (60)^2}\right] = 86.5 \text{ N/mm}^2$$

$$= 86.5 \text{ MPa (tensile)}$$

and tangential stress at the outer surface of the outer tube,

$$\sigma_{t7} = \frac{2\,p_i (r_1)^2}{(r_3)^2 - (r_1)^2} = \frac{2 \times 60\,(60)^2}{(100)^2 - (60)^2} = 67.5 \text{ N/mm}^2 = 67.5 \text{ MPa (tensile)}$$

We know that resultant stress at the inner surface of the assembly

$$\sigma_{ti} = \sigma_{t1} + \sigma_{t5} = -36.6 + 127.5 = 90.9 \text{ N/mm}^2 = 90.9 \text{ MPa (tensile)}$$ **Ans.**

Resultant stress at the outer surface of the inner tube

$$= \sigma_{t2} + \sigma_{t6} = -28.6 + 86.5 = 57.9 \text{ N/mm}^2 = 57.9 \text{ MPa (tensile)}$$

Resultant stress at the inner surface of the outer tube

$$= \sigma_{t3} + \sigma_{t6} = 36.4 + 86.5 = 122.9 \text{ N/mm}^2 = 122.9 \text{ MPa (tensile)}$$

∴ Total resultant stress at the mating surface of the assembly,

$$\sigma_{tm} = 57.9 + 122.9 = 180.8 \text{ N/mm}^2 = 180.8 \text{ MPa (tensile)}$$ **Ans.**

and resultant stress at the outer surface of the assembly,

$$\sigma_{to} = \sigma_{t4} + \sigma_{t7} = 28.4 + 67.5 = 95.9 \text{ N/mm}^2 = 95.9 \text{ MPa (tensile)}$$ **Ans.**

7.12 Cylinder Heads and Cover Plates

The heads of cylindrical pressure vessels and the sides of rectangular or square tanks may have flat plates or slightly dished plates. The plates may either be cast integrally with the cylinder walls or fixed by means of bolts, rivets or welds. The design of flat plates forming the heads depend upon the following two factors:

(*a*) Type of connection between the head and the cylindrical wall, (*i.e.* freely supported or rigidly fixed); and

(*b*) Nature of loading (*i.e.* uniformly distributed or concentrated).

This 2500-ton hydraulic press is used to forge machine parts a high temperature.

Note : This picture is given as additional information and is not a direct example of the current chapter.

Since the stress distribution in the cylinder heads and cover plates are of complex nature, therefore empirical relations based on the work of Grashof and Bach are used in the design of flat plates. Let us consider the following cases:

1. ***Circular flat plate with uniformly distributed load.*** The thickness (t_1) of a plate with a diameter (d) supported at the circumference and subjected to a pressure (p) uniformly distributed over the area is given by

$$t_1 = k_1.d\sqrt{\frac{p}{\sigma_t}}$$

where σ_t = Allowable design stress.

The coefficient k_1 depends upon the material of the plate and the method of holding the edges. The values of k_1 for the cast iron and mild steel are given in Table 7.2.

2. ***Circular flat plate loaded centrally.*** The thickness (t_1) of a flat cast iron plate supported freely at the circumference with a diameter (d) and subjected to a load (F) distributed uniformly over an area $\frac{\pi}{4}(d_0)^2$, is given by

$$t_1 = 3\sqrt{\left(1 - \frac{0.67\, d_0}{d}\right)\frac{F}{\sigma_t}}$$

If the plate with the above given type of loading is fixed rigidly around the circumference, then

$$t_1 = 1.65\sqrt{\frac{F}{\sigma_t}\log_e\left(\frac{d}{d_0}\right)}$$

3. ***Rectangular flat plate with uniformly distributed load.*** The thickness (t_1) of a rectangular plate subjected to a pressure (p) uniformly distributed over the total area is given by

$$t_1 = a.b.k_2\sqrt{\frac{p}{\sigma_t(a^2+b^2)}}$$

where a = Length of the plate; and

b = Width of the plate.

The values of the coefficient k_2 are given in Table 7.2.

Table 7.2. Values of coefficients k_1, k_2, k_3 and k_4.

Material of the cover plate	*Type of connection*	*Circular plate*	*Rectangular plate*		*Elliptical plate*
		k_1	k_2	k_3	k_4
Cast iron	Freely supported	0.54	0.75	4.3	1.5
	Fixed	0.44	0.62	4.0	1.2
Mild Steel	Freely supported	0.42	0.60	3.45	1.2
	Fixed	0.35	0.49	3.0	0.9

4. ***Rectangular flat plate with concentrated load.*** The thickness (t_1) of a rectangular plate subjected to a load (F) at the intersection of the diagonals is given by

$$t_1 = k_3\sqrt{\frac{a.b.F}{\sigma_t(a^2+b^2)}}$$

The values of coefficient k_3 are given in Table 7.2.

5. ***Elliptical plate with uniformly distributed load.*** The thickness (t_1) of an elliptical plate subjected to a pressure (p) uniformly distributed over the total area, is given by

$$t_1 = a.b.k_4\sqrt{\frac{p}{\sigma_t(a^2+b^2)}}$$

where a and b = Major and minor axes respectively.

The values of coefficient k_4 are given in Table 7.2.

6. ***Dished head with uniformly distributed load.*** Let us consider the following cases of dished head:

(*a*) *Riveted or welded dished head.* When the cylinder head has a dished plate, then the thickness of such a plate that is riveted or welded as shown in Fig. 7.12 (*a*), is given by

$$t_1 = \frac{4.16\ p.R}{\sigma_u}$$

where p = Pressure inside the cylinder,

R = Inside radius of curvature of the plate, and

σ_u = Ultimate strength for the material of the plate.

When there is an opening or manhole in the head, then the thickness of the dished plate is given by

$$t_1 = \frac{4.8\ p.R}{\sigma_u}$$

It may be noted that the inside radius of curvature of the dished plate (*R*) should not be greater than the inside diameter of the cylinder (*d*).

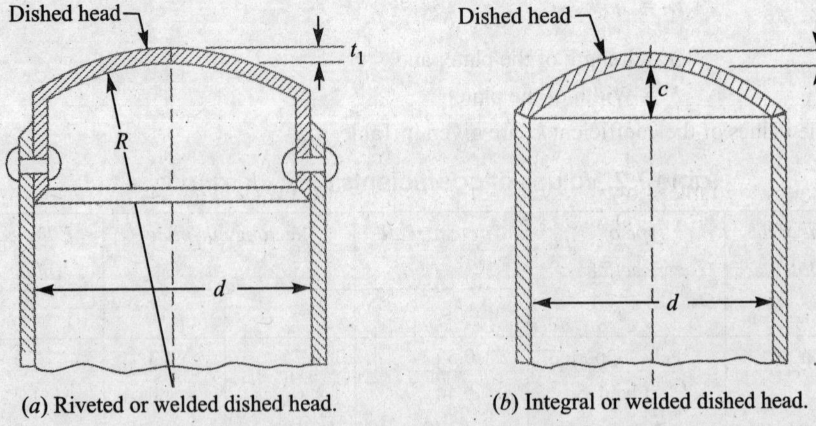

Fig. 7.12. Dished plate with uniformly distributed load.

(*b*) *Integral or welded dished head.* When the dished plate is fixed integrally or welded to the cylinder as shown in Fig. 7.12 (*b*), then the thickness of the dished plate is given by

$$t_1 = \frac{p\,(d^2 + 4c^2)}{16\ \sigma_t \times c}$$

where c = Camber or radius of the dished plate.

Mostly the cylindrical shells are provided with hemispherical heads. Thus for hemispherical heads, $c = \frac{d}{2}$. Substituting the value of *c* in the above expression, we find that the thickness of the hemispherical head (fixed integrally or welded),

$$t_1 = \frac{p\left(d^2 + 4 \times \frac{d^2}{4}\right)}{16\ \sigma_t \times \frac{d}{2}} = \frac{p.d}{4\ \sigma_t}$$ ...(Same as for thin spherical shells)

7. ***Unstayed flat plate with uniformly distributed load.*** The minimum thickness (t_1) of an unstayed steel flat head or cover plate is given by

$$t_1 = d\sqrt{\frac{k.p}{\sigma_t}}$$

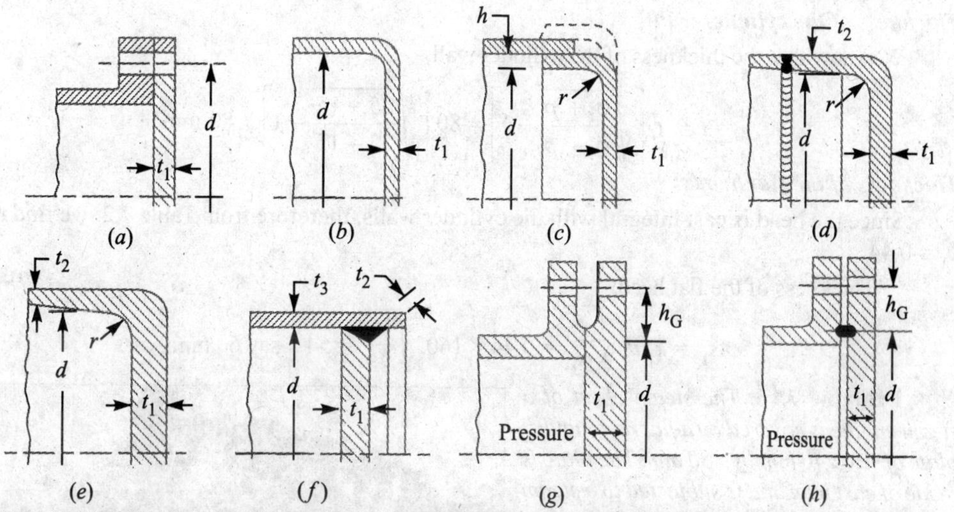

Fig. 7.13. Types of unstayed flat head and covers.

The following table shows the value of the empirical coefficient (k) for the various types of plate (or head) connection as shown in Fig. 7.13.

Table 7.3. Values of an empirical coefficient (k).

S.No.	*Particulars of plate connection*	*Value of 'k'*
1.	Plate riveted or bolted rigidly to the shell flange, as shown in Fig. 7.13 (*a*).	0.162
2.	Integral flat head as shown in Fig. 7.13 (*b*), $d \le 600$ mm, $t_1 \ge 0.05\ d$.	0.162
3.	Flanged plate attached to the shell by a lap joint as shown in Fig. 7.13 (*c*), $r \ge 3t_1$.	0.30
4.	Plate butt welded as shown in Fig. 7.13 (*d*), $r \ge 3\ t_2$	0.25
5.	Integral forged plate as shown in Fig. 7.13 (*e*), $r \ge 3\ t_2$	0.25
6.	Plate fusion welded with fillet weld as shown in Fig. 7.13 (*f*), $t_2 \ge 1.25\ t_3$.	0.50
7.	Bolts tend to dish the plate as shown in Fig. 7.13 (*g*) and (*h*).	$0.3 + \frac{1.04\ W.h_G}{H.d}$, W = Total bolt load, and H = Total load on area bounded by the outside diameter of the gasket.

Example 7.15. *A cast iron cylinder of inside diameter 160 mm is subjected to a pressure of 15 N/mm². The permissible working stress for the cast iron may be taken as 25 MPa. If the cylinder is closed by a flat head cast integral with the cylinder walls, find the thickness of the cylinder wall and the flat head.*

Solution. Given : d_i = 160 mm or r_i = 80 mm ; p = 15 N/mm² ; σ_t = 25 MPa = 25 N/mm²

Thickness of the cylinder wall

We know that the thickness of the cylinder wall,

$$t = r_i \left[\sqrt{\frac{\sigma_t + p}{\sigma_t - p}} - 1\right] = 80\left[\sqrt{\frac{25+15}{25-15}} - 1\right] = 80 \text{ mm } \textbf{Ans.}$$

Thickness of the flat head

Since the head is cast integral with the cylinder walls, therefore from Table 7.2, we find that k_1 = 0.44.

∴ Thickness of the flat head,

$$t_1 = k_1.d\sqrt{\frac{p}{\sigma_t}} = 0.44 \times 160\sqrt{\frac{15}{25}} = 54.5 \text{ say } 60 \text{ mm } \textbf{Ans.}$$

Example 7.16. *The steam chest of a steam engine is covered by a flat rectangular plate of size 240 mm by 380 mm. The plate is made of cast iron and is subjected to a steam pressure of 1.2 N/mm². If the plate is assumed to be uniformly loaded and freely supported at the edges, find the thickness of plate for an allowable stress of 35 N/mm².*

Steam engine.

Solution. Given: b = 240 m ; a = 380 mm ; p = 1.2 N/mm² ; σ_t = 35 N/mm²

From Table 7.2, we find that for a rectangular plate freely supported, the coefficient k_2 = 0.75.

We know that the thickness of a rectangular plate,

$$t_1 = a.b.k_2\sqrt{\frac{p}{\sigma_t(a^2+b^2)}} = 380 \times 240 \times 0.75\sqrt{\frac{1.2}{35[(380)^2+(240)^2]}}$$

$$= 68\,400 \times 0.412 \times 10^{-3} = 28.2 \text{ say } 30 \text{ mm } \textbf{Ans.}$$

Example 7.17. *Determine the wall thickness and the head thickness required for a 500 mm fusion-welded steel drum that is to contain ammonia at 6 N/mm² pressure. The radius of curvature of the head is to be 450 mm.*

Solution. Given: d = 500 mm ; p = 6 N/mm² ; R = 450 mm

Wall thickness for a steel drum

For the chemical pressure vessels, steel Fe 360 is used. The ultimate tensile strength (σ_u) of the steel is 360 N/mm². Assuming a factor of safety (*F.S.*) as 6, the allowable tensile strength,

$$\sigma_t = \frac{\sigma_u}{F.S.} = \frac{360}{6} = 60 \text{ N/mm}^2$$

We know that the wall thickness,

$$t = \frac{p.d}{2\sigma_t} = \frac{6 \times 500}{2 \times 60} = 25 \text{ mm } \textbf{Ans.}$$

Head thickness for a steel drum

We know that the head thickness,

$$t_1 = \frac{4.16\ p.R}{\sigma_u} = \frac{4.16 \times 6 \times 450}{360} = 31.2 \text{ say } 32 \text{ mm } \textbf{Ans.}$$

Example 7.18. *A pressure vessel consists of a cylinder of 1 metre inside diameter and is closed by hemispherical ends. The pressure intensity of the fluid inside the vessel is not to exceed 2 N/mm². The material of the vessel is steel, whose ultimate strength in tension is 420 MPa. Calculate the required wall thickness of the cylinder and the thickness of the hemispherical ends, considering a factor of safety of 6. Neglect localised effects at the junction of the cylinder and the hemisphere.*

Solution. Given: $d = 1$ m $= 1000$ mm ; $p = 2$ N/mm^2 ; $\sigma_u = 420$ MPa $= 420$ N/mm^2 ; $F.S. = 6$

We know that allowable tensile stress,

$$\sigma_t = \frac{\sigma_u}{F.S.} = \frac{420}{6} = 70 \text{ N/mm}^2$$

Wall thickness of the cylinder

We know that wall thickness of the cylinder,

$$t = \frac{p.d}{2\ \sigma_t} = \frac{2 \times 1000}{2 \times 70} = 14.3 \text{ say } 15 \text{ mm } \textbf{Ans.}$$

Thickness of hemispherical ends

We know that the thickness of hemispherical ends,

$$t_1 = \frac{p.d}{4\ \sigma_t} = \frac{2 \times 1000}{4 \times 70} = 7.15 \text{ say } 8 \text{ mm } \textbf{Ans.}$$

Example 7.19. *A cast steel cylinder of 350 mm inside diameter is to contain liquid at a pressure of 13.5 N/mm². It is closed at both ends by flat cover plates which are made of alloy steel and are attached by bolts.*

1. *Determine the wall thickness of the cylinder if the maximum hoop stress in the material is limited to 55 MPa.*
2. *Calculate the minimum thickness necessary of the cover plates if the working stress is not to exceed 65 MPa.*

Steel drums.

Solution. Given : $d_i = 350$ mm or $r_i = 175$ mm ; $p = 13.5$ N/mm^2 ; $\sigma_t = 55$ MPa $= 55$ N/mm^2 ; $\sigma_{t1} = 65$ MPa $= 65$ N/mm^2

1. ***Wall thickness of the cylinder***

We know that the wall thickness of the cylinder,

$$t = r_i \left[\sqrt{\frac{\sigma_t + p}{\sigma_t - p}} - 1\right] = 175 \left[\sqrt{\frac{55 + 13.5}{55 - 13.5}} - 1\right] = 49.8 \text{ say } 50 \text{ mm } \textbf{Ans.}$$

2. ***Minimum thickness of the cover plates***

From Table 7.3, we find that for a flat cover plate bolted to the shell flange, the value of coefficient $k = 0.162$. Therefore, minimum thickness of the cover plates

$$t_1 = d_i \sqrt{\frac{k.p}{\sigma_{t1}}} = 350 \sqrt{\frac{0.162 \times 13.5}{64}} = 64.2 \text{ say } 65 \text{ mm } \textbf{Ans.}$$

EXERCISES

1. A steel cylinder of 1 metre diameter is carrying a fluid under a pressure of 10 N/mm^2. Calculate the necessary wall thickness, if the tensile stress is not to exceed 100 MPa. **[Ans. 50 mm]**
2. A steam boiler, 1.2 metre in diameter, generates steam at a gauge pressure of 0.7 N/mm^2. Assuming the efficiency of the riveted joints as 75%, find the thickness of the shell. Given that ultimate tensile stress = 385 MPa and factor of safety = 5. **[Ans. 7.3 mm]**
3. Find the thickness of a cast iron cylinder 250 mm in diameter to carry a pressure of 0.7 N/mm^2. Take maximum tensile stress for cast iron as 14 MPa. **[Ans. 6.25 mm]**
4. A pressure vessel has an internal diameter of 1 m and is to be subjected to an internal pressure of 2.75 N/mm^2 above the atmospheric pressure. Considering it as a thin cylinder and assuming the efficiency of its riveted joint to be 79%, calculate the plate thickness if the tensile stress in the material is not to exceed 88 MPa. **[Ans. 20 mm]**
5. A spherical shell of 800 mm diameter is subjected to an internal pressure of 2 N/mm^2. Find the thickness required for the shell if the safe stress is not to exceed 100 MPa. **[Ans. 4 mm]**
6. A bronze spherical shell of thickness 15 mm is installed in a chemical plant. The shell is subjected to an internal pressure of 1 N/mm^2. Find the diameter of the shell, if the permissible stress for the bronze is 55 MPa. The efficiency may be taken as 80%. **[Ans. 2.64 m]**
7. The pressure within the cylinder of a hydraulic press is 8.4 N/mm^2. The inside diameter of the cylinder is 25.4 mm. Determine the thickness of the cylinder wall, if the allowable tensile stress is 17.5 MPa. **[Ans. 8.7 mm]**
8. A thick cylindrical shell of internal diameter 150 mm has to withstand an internal fluid pressure of 50 N/mm^2. Determine its thickness so that the maximum stress in the section does not exceed 150 MPa. **[Ans. 31 mm]**
9. A steel tank for shipping gas is to have an inside diameter of 30 mm and a length of 1.2 metres. The gas pressure is 15 N/mm^2. The permissible stress is to be 57.5 MPa. **[Ans. 4.5 mm]**
10. The ram of a hydraulic press 200 mm internal diameter is subjected to an internal pressure of 10 N/mm^2. If the maximum stress in the material of the wall is not to exceed 28 MPa, find the external diameter. **[Ans. 265 mm]**
11. The maximum force exerted by a small hydraulic press is 500 kN. The working pressure of the fluid is 20 N/mm^2. Determine the diameter of the plunger, operating the table. Also suggest the suitable thickness for the cast steel cylinder in which the plunger operates, if the permissible stress for cast steel is 100 MPa. **[Ans. 180 mm ; 20 mm]**
12. Find the thickness of the flat end cover plates for a 1 N/mm^2 boiler that has a diameter of 600 mm. The limiting tensile stress in the boiler shell is 40 MPa. **[Ans. 38 mm]**

This vessel holds oil at high pressure.

QUESTIONS

1. What is the pressure vessel ?
2. Make out a systematic classification of pressure vessels and discuss the role of statutory regulations.
3. How do you distinguish between a thick and thin cylinder?
4. What are the important points to be considered while designing a pressure vessel ?
5. Distinguish between circumferential stress and longitudinal stress in a cylindrical shell, when subjected to an internal pressure.
6. Show that in case of a thin cylindrical shell subjected to an internal fluid pressure, the tendency to burst lengthwise is twice as great as at a transverse section.
7. When a thin cylinder is subjected to an internal pressure *p*, the tangential stress should be the criterion for determining the cylinder wall thickness. Explain.
8. Derive a formula for the thickness of a thin spherical tank subjected to an internal fluid pressure.
9. Compare the stress distribution in a thin and thick walled pressure vessels.
10. When the wall thickness of a pressure vessel is relatively large, the usual assumptions valid in thin cylinders do not hold good for its analysis. Enumerate the important violations. List any two theories suggested for the analysis of thick cylinders.
11. Discuss the design procedure for pressure vessels subjected to higher external pressure.
12. Explain the various types of ends used for pressure vessel giving practical applications of each.

OBJECTIVE TYPE QUESTIONS

1. A pressure vessel is said to be a thin cylindrical shell, if the ratio of the wall thickness of the shell to its diameter is

   (*a*) equal to 1/10 (*b*) less than 1/10

   (*c*) more than 1/10 (*d*) none of these

2. In case of pressure vessels having open ends, the fluid pressure induces

   (*a*) longitudinal stress (*b*) circumferential stress

   (*c*) shear stress (*d*) none of these

3. The longitudinal stress is ...... of the circumferential stress.

   (*a*) one-half (*b*) two-third

   (*c*) three-fourth

4. The design of the pressure vessel is based on

   (*a*) longitudinal stress (*b*) hoop stress

   (*c*) longitudinal and hoop stress (*d*) none of these

5. A thin spherical shell of internal diameter *d* is subjected to an internal pressure *p*. If σ_t is the tensile stress for the shell material, then thickness of the shell (t) is equal to

   (*a*) $\frac{p.d}{\sigma_t}$ (*b*) $\frac{p.d}{2\,\sigma_t}$

   (*c*) $\frac{p.d}{3\,\sigma_t}$ (*d*) $\frac{p.d}{4\,\sigma_t}$

6. In case of thick cylinders, the tangential stress across the thickness of cylinder is
   (*a*) maximum at the outer surface and minimum at the inner surface
   (*b*) maximum at the inner surface and minimum at the outer surface
   (*c*) maximum at the inner surface and zero at the outer surface
   (*d*) maximum at the outer surface and zero at the inner surface
7. According to Lame's equation, the thickness of a cylinder is equal to
   (*a*) $r_i\left[\sqrt{\dfrac{\sigma_t + (1-2\mu)\,p}{\sigma_t - (1-2\mu)\,p}} - 1\right]$ (*b*) $r_i\left[\sqrt{\dfrac{\sigma_t + (1-\mu)\,p}{\sigma_t - (1-\mu)\,p}} - 1\right]$
   (*c*) $r_i\left[\sqrt{\dfrac{\sigma_t + p}{\sigma_t - p}} - 1\right]$ (*d*) $r_i\left[\sqrt{\dfrac{\sigma_t}{\sigma_t - 2p}} - 1\right]$

   where
   r_i = Internal radius of the cylinder,
   σ_t = Allowable tensile stress,
   p = Internal fluid pressure, and
   μ = Poisson's ratio.
8. In a thick cylindrical shell, the maximum radial stress at the outer surfaces of the shell is
   (*a*) zero (*b*) p
   (*c*) $-p$ (*d*) $2p$
9. For high pressure oil and gas cylinders, the thickness of the cylinder is determined by
   (*a*) Lame's equation (*b*) Clavarino's equation
   (*c*) Barlow's equation (*d*) Birnie's equation
10. The thickness of a dished head that is riveted or welded to the cylindrical wall is
   (*a*) $\dfrac{4.16\ p.R}{\sigma_u}$ (*b*) $\dfrac{5.36\ p.R}{\sigma_u}$
   (*c*) $\dfrac{6.72\ p.R}{\sigma_u}$ (*d*) $\dfrac{8.33\ p.R}{\sigma_u}$

   where
   p = Internal pressure,
   R = Inside radius of curvature of the dished plate, and
   σ_u = Ultimate strength for the material of the plate.

ANSWERS

1. (*b*)	**2.** (*b*)	**3.** (*a*)	**4.** (*b*)	**5.** (*b*)
6. (*b*)	**7.** (*c*)	**8.** (*a*)	**9.** (*c*)	**10.** (*a*)

CHAPTER

8

Pipes and Pipe Joints

8.1 Introduction

The pipes are used for transporting various fluids like water, steam, different types of gases, oil and other chemicals with or without pressure from one place to another. Cast iron, wrought iron, steel and brass are the materials generally used for pipes in engineering practice. The use of cast iron pipes is limited to pressures of about 0.7 N/mm^2 because of its low resistance to shocks which may be created due to the action of water hammer. These pipes are best suited for water and sewage systems. The wrought iron and steel pipes are used chiefly for conveying steam, air and oil. Brass pipes, in small sizes, finds use in pressure lubrication systems on prime movers. These are made up and threaded to the same standards as wrought iron and steel pipes. Brass pipe is not liable to corrosion. The pipes used in petroleum industry are generally seamless pipes made of heat-resistant chrome-molybdenum alloy steel. Such type of pipes can resist pressures more than 4 N/mm^2 and temperatures greater than 440°C.

8.2 Stresses in Pipes

The stresses in pipes due to the internal fluid pressure are determined by Lame's equation as discussed in the previous chapter (Art. 7.9). According to Lame's equation, tangential stress at any radius *x*,

$$\sigma_t = \frac{p\,(r_i)^2}{(r_o)^2 - (r_i)^2}\left[1 + \frac{(r_o)^2}{x^2}\right] \qquad ...(i)$$

and radial stress at any radius *x*,

$$\sigma_r = \frac{p\,(r_i)^2}{(r_o)^2 - (r_i)^2}\left[1 - \frac{(r_o)^2}{x^2}\right] \qquad ...(ii)$$

where p = Internal fluid pressure in the pipe,

r_i = Inner radius of the pipe, and

r_o = Outer radius of the pipe.

The tangential stress is maximum at the inner surface (when $x = r_i$) of the pipe and minimum at the outer surface (when $x = r_o$) of the pipe.

Substituting the values of $x = r_i$ and $x = r_o$ in equation (*i*), we find that the maximum tangential stress at the inner surface of the pipe,

$$\sigma_{t(max)} = \frac{p\,[(r_o)^2 + (r_i)^2]}{(r_o)^2 - (r_i)^2}$$

and minimum tangential stress at the outer surface of the pipe,

$$\sigma_{t(min)} = \frac{2\,p\,(r_i)^2}{(r_o)^2 - (r_i)^2}$$

The radial stress is maximum at the inner surface of the pipe and zero at the outer surface of the pipe. Substituting the values of $x = r_i$ and $x = r_o$ in equation (*ii*), we find that maximum radial stress at the inner surface,

$$\sigma_{r(max)} = -p \text{ (compressive)}$$

Cast iron pipes.

and minimum radial stress at the outer surface of the pipe,

$$\sigma_{r(min)} = 0$$

The thick cylindrical formula may be applied when

(*a*) the variation of stress across the thickness of the pipe is taken into account,

(*b*) the internal diameter of the pipe (*D*) is less than twenty times its wall thickness (t), *i.e.* $D/t < 20$, and

(*c*) the allowable stress (σ_t) is less than six times the pressure inside the pipe (p) *i.e.* $\sigma_t / p < 6$.

According to thick cylindrical formula (Lame's equation), wall thickness of pipe,

$$t = R\left[\sqrt{\frac{\sigma_t + p}{\sigma_t - p}} - 1\right]$$

where R = Internal radius of the pipe.

The following table shows the values of allowable tensile stress (σ_t) to be used in the above relations:

Table 8.1. Values of allowable tensile stress for pipes of different materials.

S.No.	*Pipes*	*Allowable tensile stress (σ_t) in MPa or N/mm²*
1.	Cast iron steam or water pipes	14
2.	Cast iron steam engine cylinders	12.5
3.	Lap welded wrought iron tubes	60
4.	Solid drawn steel tubes	140
5.	Copper steam pipes	25
6.	Lead pipes	1.6

Example 8.1. *A cast iron pipe of internal diameter 200 mm and thickness 50 mm carries water under a pressure of 5 N/mm². Calculate the tangential and radial stresses at radius (r) = 100 mm ; 110 mm ; 120 mm ; 130 mm ; 140 mm and 150 mm. Sketch the stress distribution curves.*

Solution. Given : $d_i = 200$ mm or $r_i = 100$ mm ; $t = 50$ mm ; $p = 5$ N/mm²

We know that outer radius of the pipe,

$$r_o = r_i + t = 100 + 50 = 150 \text{ mm}$$

Tangential stresses at radius 100 mm, 110 mm, 120 mm, 130 mm, 140 mm and 150 mm

We know that tangential stress at any radius x,

$$\sigma_t = \frac{p\,(r_i)^2}{(r_o)^2 - (r_i)^2}\left[1 + \frac{(r_o)^2}{x^2}\right] = \frac{5\,(100)^2}{(150)^2 - (100)^2}\left[1 + \frac{(r_o)^2}{x^2}\right]$$

$$= 4\left[1 + \frac{(r_o)^2}{x^2}\right] \text{ N/mm}^2 \text{ or MPa}$$

∴ Tangential stress at radius 100 mm (*i.e.* when $x = 100$ mm),

$$\sigma_{t1} = 4\left[1 + \frac{(150)^2}{(100)^2}\right] = 4 \times 3.25 = 13 \text{ MPa} \quad \textbf{Ans.}$$

Tangential stress at radius 110 mm (*i.e.* when $x = 110$ mm),

$$\sigma_{t2} = 4\left[1 + \frac{(150)^2}{(110)^2}\right] = 4 \times 2.86 = 11.44 \text{ MPa} \quad \textbf{Ans.}$$

Tangential stress at radius 120 mm (*i.e.* when $x = 120$ mm),

$$\sigma_{t3} = 4\left[1 + \frac{(150)^2}{(120)^2}\right] = 4 \times 2.56 = 10.24 \text{ MPa} \quad \textbf{Ans.}$$

Tangential stress at radius 130 mm (*i.e.* when $x = 130$ mm),

$$\sigma_{t4} = 4\left[1 + \frac{(150)^2}{(130)^2}\right] = 4 \times 2.33 = 9.32 \text{ MPa} \quad \textbf{Ans.}$$

Tangential stress at radius 140 mm (*i.e.* when $x = 140$ mm),

$$\sigma_{t5} = 4\left[1 + \frac{(150)^2}{(140)^2}\right] = 4 \times 2.15 = 8.6 \text{ MPa} \quad \textbf{Ans.}$$

and tangential stress at radius 150 mm (*i.e.* when $x = 150$ mm),

$$\sigma_{t6} = 4\left[1 + \frac{(150)^2}{(150)^2}\right] = 4 \times 2 = 8 \text{ MPa} \quad \textbf{Ans.}$$

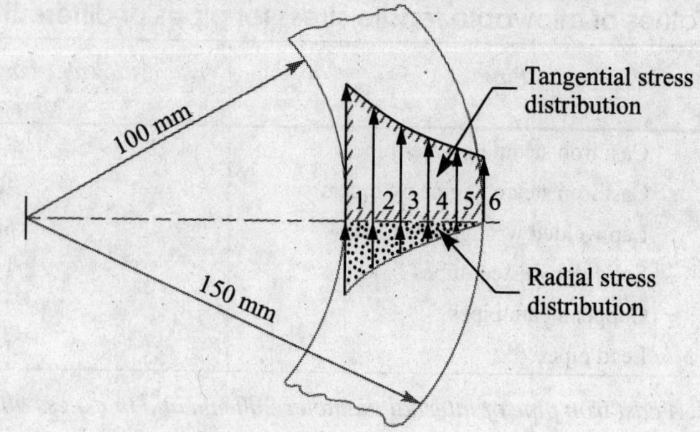

Fig. 8.1

Radial stresses at radius 100 mm, 110 mm, 120 mm, 130 mm, 140 mm and 150 mm

We know that radial stress at any radius *x*,

$$\sigma_r = \frac{p\,(r_i)^2}{(r_o)^2 - (r_i)^2}\left[1 - \frac{(r_o)^2}{x^2}\right] = \frac{5\,(100)^2}{(150)^2 - (100)^2}\left[1 - \frac{(r_o)^2}{x^2}\right]$$

$$= 4\left[1 - \frac{(r_o)^2}{x^2}\right] \text{ N/mm}^2 \text{ or MPa}$$

∴ Radial stress at radius 100 mm (*i.e.* when x = 100 mm),

$$\sigma_{r1} = 4\left[1 - \frac{(150)^2}{(100)^2}\right] = 4 \times -1.25 = -5 \text{ MPa} \quad \textbf{Ans.}$$

Radial stress at radius 110 mm (*i.e.*, when x = 110 mm),

$$\sigma_{r2} = 4\left[1 - \frac{(150)^2}{(110)^2}\right] = 4 \times -0.86 = -3.44 \text{ MPa} \quad \textbf{Ans.}$$

Radial stress at radius 120 mm (*i.e.* when x = 120 mm),

$$\sigma_{r3} = 4\left[1 - \frac{(150)^2}{(120)^2}\right] = 4 \times -0.56 = -2.24 \text{ MPa} \quad \textbf{Ans.}$$

Radial stress at radius 130 mm (*i.e.* when x = 130 mm),

$$\sigma_{r4} = 4\left[1 - \frac{(150)^2}{(130)^2}\right] = 4 \times -0.33 = -1.32 \text{ MPa} \quad \textbf{Ans.}$$

Radial stress at radius 140 mm (*i.e.* when x = 140 mm),

$$\sigma_{r5} = 4\left[1 - \frac{(150)^2}{(140)^2}\right] = 4 \times -0.15 = -0.6 \text{ MPa} \quad \textbf{Ans.}$$

Radial stress at radius 150 mm (*i.e.* when x = 150 mm),

$$\sigma_{r6} = 4\left[1 - \frac{(150)^2}{(150)^2}\right] = 0 \quad \textbf{Ans.}$$

The stress distribution curves for tangential and radial stresses are shown in Fig. 8.1.

8.3 Design of Pipes

The design of a pipe involves the determination of inside diameter of the pipe and its wall thickness as discussed below:

1. ***Inside diameter of the pipe.*** The inside diameter of the pipe depends upon the quantity of fluid to be delivered.

Let D = Inside diameter of the pipe,

v = Velocity of fluid flowing per minute, and

Q = Quantity of fluid carried per minute.

We know that the quantity of fluid flowing per minute,

$$Q = \text{Area} \times \text{Velocity} = \frac{\pi}{4} \times D^2 \times v$$

$$\therefore \quad D = \sqrt{\frac{4}{\pi} \times \frac{Q}{v}} = 1.13\sqrt{\frac{Q}{v}}$$

2. ***Wall thickness of the pipe.*** After deciding upon the inside diameter of the pipe, the thickness of the wall (t) in order to withstand the internal fluid pressure (p) may be obtained by using thin cylindrical or thick cylindrical formula.

The thin cylindrical formula may be applied when

(*a*) the stress across the section of the pipe is uniform,

(*b*) the internal diameter of the pipe (D) is more than twenty times its wall thickness (t), *i.e.* $D/t > 20$, and

(*c*) the allowable stress (σ_t) is more than six times the pressure inside the pipe (p), *i.e.* $\sigma_t / p > 6$.

Pipe Joint

According to thin cylindrical formula, wall thickness of pipe,

$$t = \frac{p.D}{2\sigma_t} \quad \text{or} \quad \frac{p.D}{2\sigma_t \, \eta_l}$$

where η_l = Efficiency of longitudinal joint.

A little consideration will show that the thickness of wall as obtained by the above relation is too small. Therefore for the design of pipes, a certain constant is added to the above relation. Now the relation may be written as

$$t = \frac{p.D}{2\sigma_t} + C$$

The value of constant 'C', according to Weisback, are given in the following table.

Table 8.2. Values of constant 'C'.

Material	*Cast iron*	*Mild steel*	*Zinc and Copper*	*Lead*
Constant (C) in mm	9	3	4	5

Example 8.2. *A seamless pipe carries 2400 m³ of steam per hour at a pressure of 1.4 N/mm². The velocity of flow is 30 m/s. Assuming the tensile stress as 40 MPa, find the inside diameter of the pipe and its wall thickness.*

Solution. Given : $Q = 2400$ m³/h = 40 m³/min ; $p = 1.4$ N/mm²; $v = 30$ m/s = 1800 m/min ; $\sigma_t = 40$ MPa = 40 N/mm²

Inside diameter of the pipe

We know that inside diameter of the pipe,

$$D = 1.13\sqrt{\frac{Q}{v}} = 1.13\sqrt{\frac{40}{1800}} = 0.17 \text{ m} = 170 \text{ mm} \quad \textbf{Ans.}$$

Wall thickness of the pipe

From Table 8.2, we find that for a steel pipe, $C = 3$ mm. Therefore wall thickness of the pipe,

$$t = \frac{p.D}{2\sigma_t} + C = \frac{1.4 \times 170}{2 \times 40} + 3 = 6 \text{ mm} \quad \textbf{Ans.}$$

8.4 Pipe Joints

The pipes are usually connected to vessels from which they transport the fluid. Since the length of pipes available are limited, therefore various lengths of pipes have to be joined to suit any particular installation. There are various forms of pipe joints used in practice, but most common of them are discussed below.

1. ***Socket or a coupler joint.*** The most common method of joining pipes is by means of a socket or a coupler as shown in Fig. 8.2. A socket is a small piece of pipe threaded inside. It is screwed on half way on the threaded end of one pipe and the other pipe is then screwed in the remaining half of socket. In order to prevent leakage, jute or hemp is wound around the threads at the end of each pipe. This type of joint is mostly used for pipes carrying water at low pressure and where the overall smallness of size is most essential.

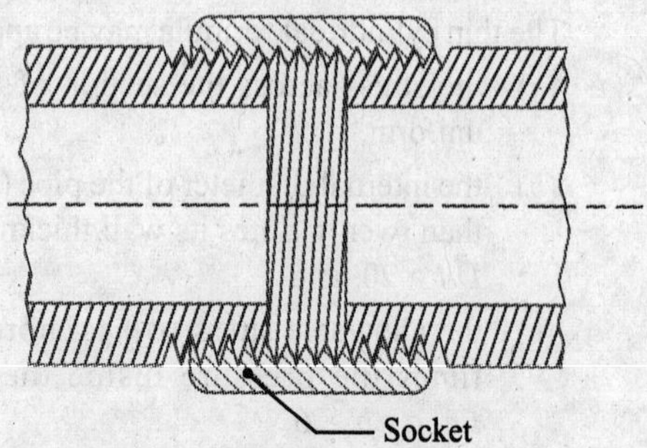

Fig. 8.2. Socket or coupler joint.

2. ***Nipple joint.*** In this type of joint, a nipple which is a small piece of pipe threaded outside is screwed in the internally threaded end of each pipe, as shown in Fig. 8.3. The disadvantage of this joint is that it reduces the area of flow.

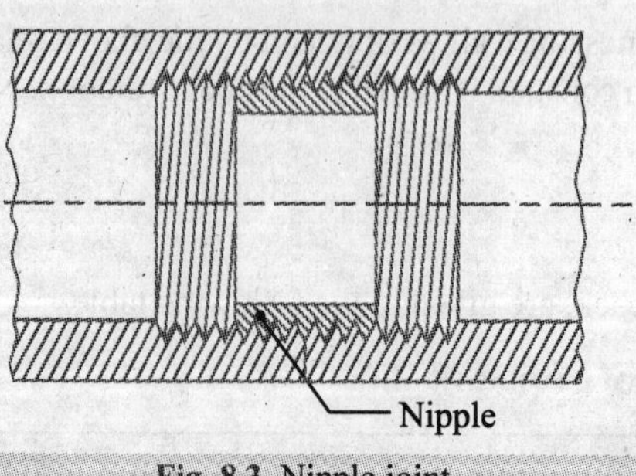

Fig. 8.3. Nipple joint.

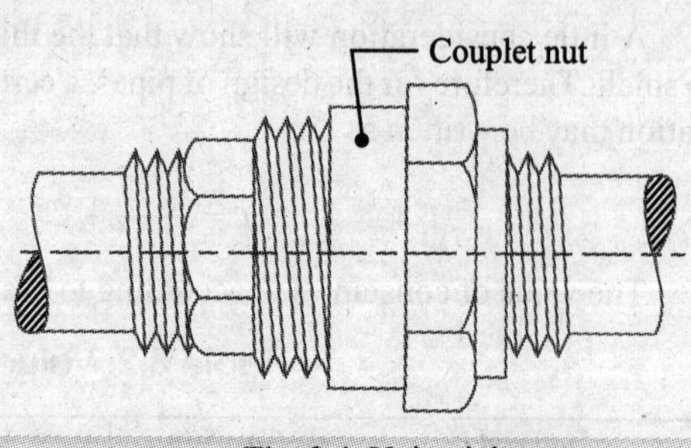

Fig. 8.4. Union joint.

3. ***Union joint.*** In order to disengage pipes joined by a socket, it is necessary to unscrew pipe from one end. This is sometimes inconvenient when pipes are long.

The union joint, as shown in Fig. 8.4, provide the facility of disengaging the pipes by simply unscrewing a coupler nut.

4. ***Spigot and socket joint.*** A spigot and socket joint as shown in Fig. 8.5, is chiefly used for pipes which are buried in the earth. Some pipe lines are laid straight as far as possible. One of the important features of this joint is its flexibility as it adopts itself to small changes in level due to settlement of earth which takes place due to climate and other conditions.

In this type of joint, the spigot end of one pipe fits into the socket end of the other pipe. The remaining space between the two is filled with a jute rope and a ring of lead. When the lead solidifies, it is caulked-in tightly.

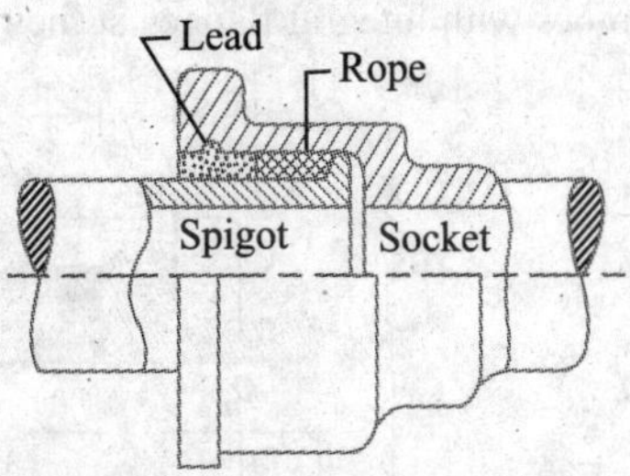

Fig. 8.5. Spigot and socket joint.

5. ***Expansion joint.*** The pipes carrying steam at high pressures are usually joined by means of expansion joint. This joint is used in steam pipes to take up expansion and contraction of pipe line due to change of temperature.

In order to allow for change in length, steam pipes are not rigidly clamped but supported on rollers. The rollers may be arranged on wall bracket, hangers or floor stands. The expansion bends, as shown in Fig. 8.6 (*a*) and (*b*), are useful in a long pipe line. These pipe bends will spring in either direction and readily accommodate themselves to small movements of the actual pipe ends to which they are attached.

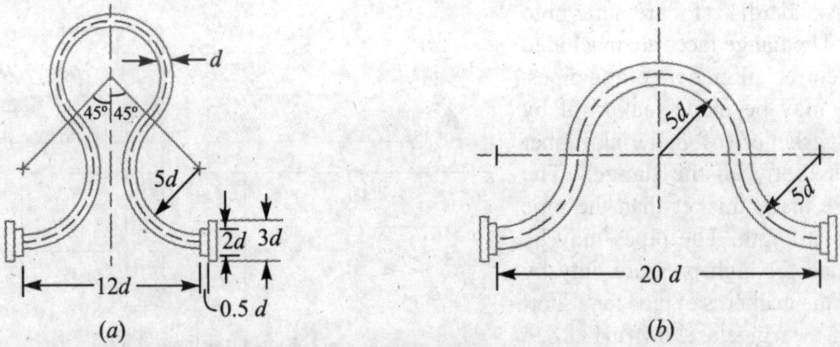

Fig. 8.6. Expansion bends.

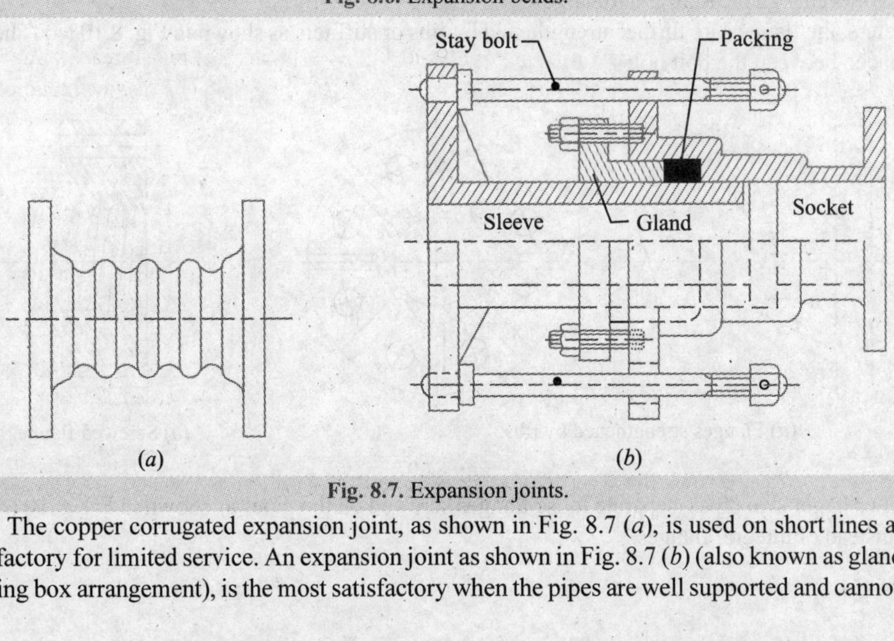

Fig. 8.7. Expansion joints.

The copper corrugated expansion joint, as shown in Fig. 8.7 (*a*), is used on short lines and is satisfactory for limited service. An expansion joint as shown in Fig. 8.7 (*b*) (also known as gland and stuffing box arrangement), is the most satisfactory when the pipes are well supported and cannot sag.

6. ***Flanged joint.*** It is one of the most widely used pipe joint. A flanged joint may be made with flanges cast integral with the pipes or loose flanges welded or screwed. Fig. 8.8 shows two cast iron pipes with integral flanges at their ends. The flanges are connected by means of bolts. The flanges have seen standardised for pressures upto 2 N/mm^2. The flange faces are machined to ensure correct alignment of the pipes. The joint may be made leakproof by placing a gasket of soft material, rubber or convass between the flanges. The flanges are made thicker than the pipe walls, for strength. The pipes may be strengthened for high pressure duty by increasing the thickness of pipe for a short length from the flange, as shown in Fig. 8.9.

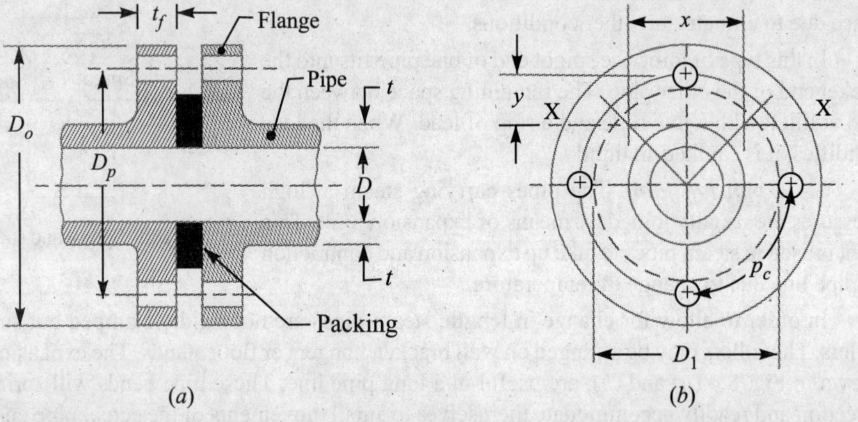

Fig. 8.8. Flanged joint.

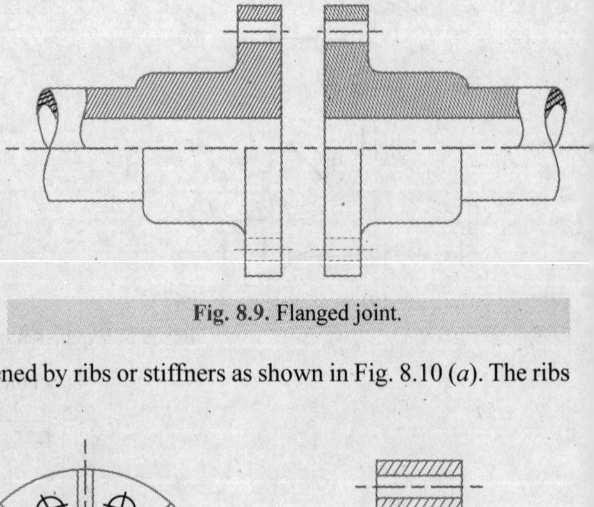
Fig. 8.9. Flanged joint.

For even high pressure and for large diameters, the flanges are further strengthened by ribs or stiffners as shown in Fig. 8.10 (*a*). The ribs are placed between the bolt holes.

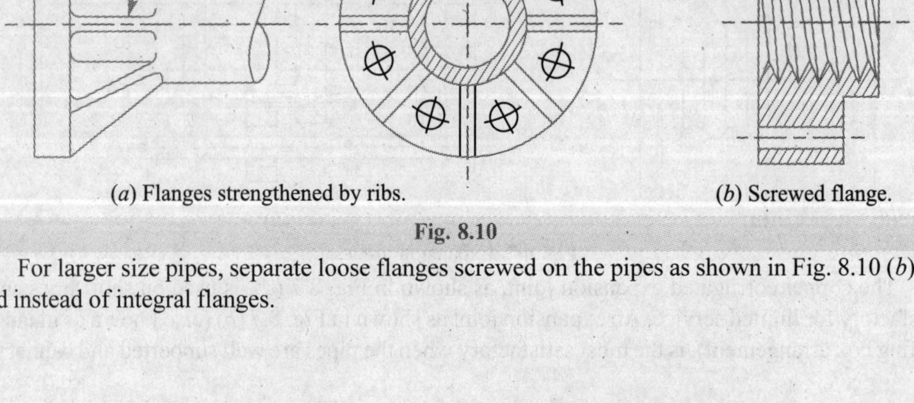

(*a*) Flanges strengthened by ribs. (*b*) Screwed flange.

Fig. 8.10

For larger size pipes, separate loose flanges screwed on the pipes as shown in Fig. 8.10 (*b*) are used instead of integral flanges.

7. ***Hydraulic pipe joint.*** This type of joint has oval flanges and are fastened by means of two bolts, as shown in Fig. 8.11. The oval flanges are usually used for small pipes, upto 175 mm diameter. The flanges are generally cast integral with the pipe ends. Such joints are used to carry fluid pressure varying from 5 to 14 N/mm². Such a high pressure is found in hydraulic applications like riveting, pressing, lifts etc. The hydraulic machines used in these installations are pumps, accumulators, intensifiers etc.

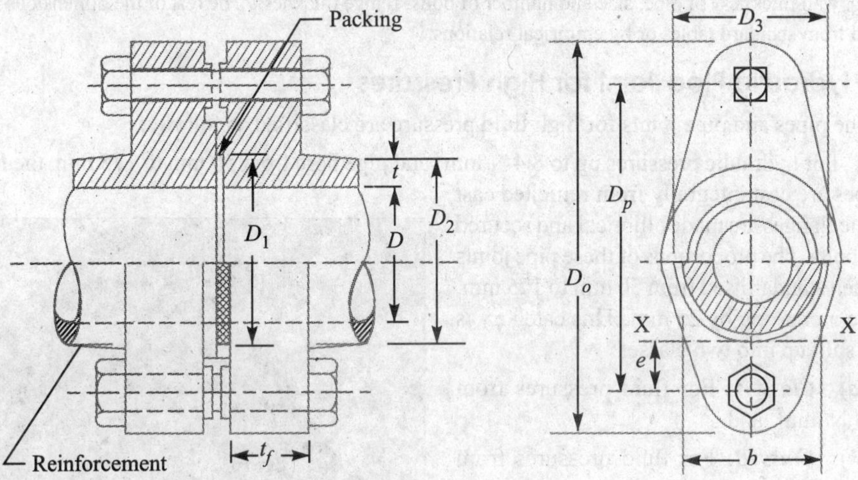

Fig. 8.11. Hydraulic pipe joint.

8.5 Standard Pipe Flanges for Steam

The Indian boiler regulations (I.B.R.) 1950 (revised 1961) have standardised all dimensions of pipe and flanges based upon steam pressure. They have been divided into five classes as follows:

Class I : For steam pressures up to 0.35 N/mm² and water pressures up to 1.4 N/mm². This is not suitable for feed pipes and shocks.

Class II : For steam pressures over 0.35 N/mm² but not exceeding 0.7 N/mm².

Class III : For steam pressures over 0.7 N/mm² but not exceeding 1.05 N/mm².

Class IV : For steam pressures over 1.05 N/mm² but not exceeding 1.75 N/mm².

Class V : For steam pressures from 1.75 N/mm² to 2.45 N/mm².

The Trans-Alaska Pipeline was built to carry oil across the frozen sub-Arctic landscape of North America.

According to I.B.R., it is desirable that for classes II, III, IV and V, the diameter of flanges, diameter of bolt circles and number of bolts should be identical and that difference should consist in variations of the thickness of flanges and diameter of bolts only. The I.B.R. also recommends that all nuts should be chamfered on the side bearing on the flange and that the bearing surfaces of the flanges, heads and nuts should be true. The number of bolts in all cases should be a multiple of four. The I.B.R. recommends that for 12.5 mm and 15 mm bolts, the bolt holes should be 1.5 mm larger and for higher sizes of bolts, the bolt holes should be 3 mm larger. All dimensions for pipe flanges having internal diameters 1.25 mm to 600 mm are standardised for the above mentioned classes (I to V). The flanged tees, bends are also standardised.

Note: As soon as the size of pipe is determined, the rest of the dimensions for the flanges, bolts, bolt holes, thickness of pipe may be fixed from standard tables. In practice, dimensions are not calculated on a rational basis. The standards are evolved on the basis of long practical experience, suitability and interchangeability. The calculated dimensions as discussed in the previous articles do not agree with the standards. It is of academic interest only that the students should know how to use fundamental principles in determining various dimensions *e.g.* wall thickness of pipe, size and number of bolts, flange thickness. The rest of the dimensions may be obtained from standard tables or by empirical relations.

8.6 Hydraulic Pipe Joint for High Pressures

The pipes and pipe joints for high fluid pressure are classified as follows:

1. For hydraulic pressures up to 8.4 N/mm^2 and pipe bore from 50 mm to 175 mm, the flanges and pipes are cast integrally from remelted cast iron. The flanges are made elliptical and secured by two bolts. The proportions of these pipe joints have been standardised from 50 mm to 175 mm, the bore increasing by 25 mm. This category is further split up into two classes:

(*a*) ***Class A***: For fluid pressures from 5 to 6.3 N/mm^2, and

(*b*) ***Class B***: For fluid pressures from 6.3 to 8.4 N/mm^2.

The flanges in each of the above classes may be of two types. Type I is suitable for pipes of 50 to 100 mm bore in class *A*, and for 50 to 175 mm bore in class *B*. The flanges of type II are stronger than those of Type I and are usually set well back on the pipe.

Hydraulic pipe joints use two or four bolts which is a great advantage while assembling the joint especially in narrow space.

2. For pressures above 8.4 N/mm^2 with bores of 50 mm or below, the piping is of wrought steel, solid drawn, seamless or rolled. The flanges may be of cast iron, steel mixture or forged steel. These are screwed or welded on to the pipe and are square in elevation secured by four bolts. These joints are made for pipe bores 12.5 mm to 50 mm rising in increment of 3 mm from 12.5 to 17.5 mm and by 6 mm from 17.5 to 50 mm. The flanges and pipes in this category are strong enough for service under pressures ranging up to 47.5 N/mm^2.

In all the above classes, the joint is of the spigot and socket type made with a jointing ring of gutta-percha.

Notes: The hydraulic pipe joints for high pressures differ from those used for low or medium pressure in the following ways:

1. The flanges used for high pressure hydraulic pipe joints are heavy oval or square in form, They use two or four bolts which is a great advantage while assembling and disassembling the joint especially in narrow space.

2. The bolt holes are made square with sufficient clearance to accomodate square bolt heads and to allow for small movements due to setting of the joint.

3. The surfaces forming the joint make contact only through a gutta-percha ring on the small area provided by the spigot and recess. The tightening up of the bolts squeezes the ring into a triangular shape and makes a perfectly tight joint capable of withstanding pressure up to 47.5 N/mm^2.

4. In case of oval and square flanged pipe joints, the condition of bending is very clearly defined due to the flanges being set back on the pipe and thickness of the flange may be accurately determined to withstand the bending action due to tightening of bolts.

8.7 Design of Circular Flanged Pipe Joint

Consider a circular flanged pipe joint as shown in Fig. 8.8. In designing such joints, it is assumed that the fluid pressure acts in between the flanges and tends to separate them with a pressure existing at the point of leaking. The bolts are required to take up tensile stress in order to keep the flanges together.

The effective diameter on which the fluid pressure acts, just at the point of leaking, is the diameter of a circle touching the bolt holes. Let this diameter be D_1. If d_1 is the diameter of bolt hole and D_p is the pitch circle diameter, then

$$D_1 = D_p - d_1$$

∴ Force trying to separate the two flanges,

$$F = \frac{\pi}{4}(D_1)^2\, p \qquad \text{...(i)}$$

Let n = Number of bolts,

d_c = Core diameter of the bolts, and

σ_t = Permissible stress for the material of the bolts.

∴ Resistance to tearing of bolts

$$= \frac{\pi}{4}(d_c)^2\, \sigma_t \times n \qquad \text{...(ii)}$$

Assuming the value of d_c, the value of n may be obtained from equations (*i*) and (*ii*). The number of bolts should be even because of the symmetry of the section.

The circumferential pitch of the bolts is given by

$$p_c = \frac{\pi D_p}{n}$$

In order to make the joint leakproof, the value of p_c should be between $20\sqrt{d_1}$ to $30\sqrt{d_1}$, where d_1 is the diameter of the bolt hole. Also a bolt of less than 16 mm diameter should never be used to make the joint leakproof.

The thickness of the flange is obtained by considering a segment of the flange as shown in Fig. 8.8 (*b*).

In this it is assumed that each of the bolt supports one segment. The effect of joining of these segments on the stresses induced is neglected. The bending moment is taken about the section *X-X*, which is tangential to the outside of the pipe. Let the width of this segment is x and the distance of this section from the centre of the bolt is y.

∴ Bending moment on each bolt due to the force F

$$= \frac{F}{n} \times y \qquad \text{...(iii)}$$

and resisting moment on the flange

$$= \sigma_b \times Z \qquad \text{...(iv)}$$

where σ_b = Bending or tensile stress for the flange material, and

Z = Section modulus of the cross-section of the flange $= \frac{1}{6} \times x\,(t_f)^2$

Equating equations (*iii*) and (*iv*), the value of t_f may be obtained.

The dimensions of the flange may be fixed as follows:

Nominal diameter of bolts, $d = 0.75\, t + 10$ mm

Number of bolts, $n = 0.0275\, D + 1.6$ ...(D is in mm)

Thickness of flange, $t_f = 1.5\,t + 3$ mm

Width of flange, $B = 2.3\,d$

Outside diameter of flange,

$$D_o = D + 2t + 2B$$

Pitch circle diameter of bolts,

$$D_p = D + 2t + 2d + 12 \text{ mm}$$

The pipes may be strengthened by providing greater thickness near the flanges $\left(\text{equal to } \dfrac{t + t_f}{2}\right)$ as shown in Fig. 8.9. The flanges may be strengthened by providing ribs equal to thickness of $\dfrac{t + t_f}{2}$, as shown in Fig. 8.10 (*a*).

Example 8.3. *Find out the dimensions of a flanged joint for a cast iron pipe 250 mm diameter to carry a pressure of 0.7 N/mm².*

Solution. Given: $D = 250$ mm ; $p = 0.7$ N/mm²

From Table 8.1, we find that for cast iron, allowable tensile stress, $\sigma_t = 14$ N/mm² and from Table 8.2, $C = 9$ mm. Therefore thickness of the pipe,

$$t = \frac{p.D}{2\sigma_t} + C = \frac{0.7 \times 250}{2 \times 14} + 9 = 15.3 \text{ say } 16 \text{ mm} \quad \textbf{Ans.}$$

Other dimensions of a flanged joint for a cast iron pipe may be fixed as follows:

Nominal diameter of the bolts,

$$d = 0.75\,t + 10 \text{ mm} = 0.75 \times 16 + 10 = 22 \text{ mm} \quad \textbf{Ans.}$$

Number of bolts, $n = 0.0275\,D + 1.6 = 0.0275 \times 250 + 1.6 = 8.475$ say 10 **Ans.**

Thickness of the flanges, $t_f = 1.5\,t + 3 \text{ mm} = 1.5 \times 16 + 3 = 27$ mm **Ans.**

Width of the flange, $B = 2.3\,d = 2.3 \times 22 = 50.6$ say 52 mm **Ans.**

Outside diameter of the flange,

$$D_o = D + 2t + 2B = 250 + 2 \times 16 + 2 \times 52 = 386 \text{ mm} \quad \textbf{Ans.}$$

Pitch circle diameter of the bolts,

$$D_p = D + 2t + 2d + 12 \text{ mm} = 250 + 2 \times 16 + 2 \times 22 + 12 \text{ mm}$$
$$= 338 \text{ mm} \quad \textbf{Ans.}$$

Circumferential pitch of the bolts,

$$p_c = \frac{\pi \times D_p}{n} = \frac{\pi \times 338}{10} = 106.2 \text{ mm} \quad \textbf{Ans.}$$

In order to make the joint leak proof, the value of p_c should be between $20\sqrt{d_1}$ to $30\sqrt{d_1}$ where d_1 is the diameter of bolt hole.

Let us take $d_1 = d + 3 \text{ mm} = 22 + 3 = 25$ mm

$$\therefore \quad 20\sqrt{d_1} = 20\sqrt{25} = 100 \text{ mm}$$

and

$$30\sqrt{d_1} = 30\sqrt{25} = 150 \text{ mm}$$

Since the circumferential pitch as obtained above (*i.e.* 106.2 mm) is within $20\sqrt{d_1}$ to $30\sqrt{d_1}$, therefore the design is satisfactory.

Example 8.4. *A flanged pipe with internal diameter as 200 mm is subjected to a fluid pressure of 0.35 N/mm². The elevation of the flange is shown in Fig. 8.12. The flange is connected by means of eight M 16 bolts. The pitch circle diameter of the bolts is 290 mm. If the thickness of the flange is 20 mm, find the working stress in the flange.*

Solution. Given : $D = 200$ mm ; $p = 0.35$ N/mm² ; $n = 8$;* $d = 16$ mm ; $D_p = 290$ mm ; $t_f = 20$ mm

First of all, let us find the thickness of the pipe. Assuming the pipe to be of cast iron, we find from Table 8.1 that the allowable tensile stress for cast iron, $\sigma_t = 14$ N/mm² and from Table 8.2, $C = 9$ mm.

∴ Thickness of the pipe,

$$t = \frac{p.D}{2\,\sigma_t} + C = \frac{0.35 \times 200}{2 \times 14} + 9 = 11.5 \text{ say } 12 \text{ mm}$$

Since the diameter of the bolt holes (d_1) is taken larger than the nominal diameter of the bolts (d), therefore let us take diameter of the bolt holes,

$$d_1 = d + 2 \text{ mm} = 16 + 2 = 18 \text{ mm}$$

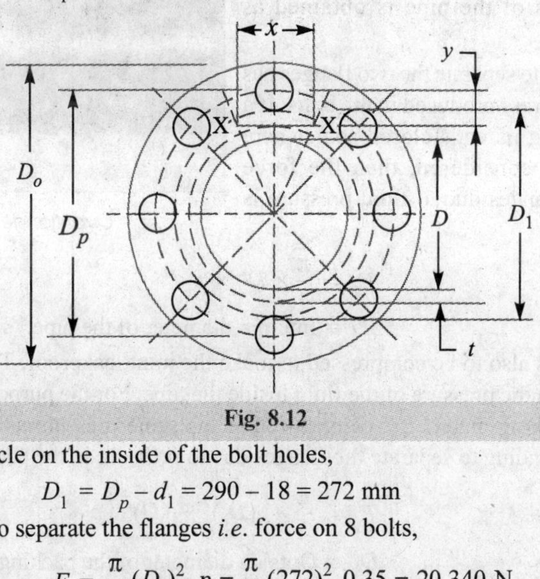

Fig. 8.12

and diameter of the circle on the inside of the bolt holes,

$$D_1 = D_p - d_1 = 290 - 18 = 272 \text{ mm}$$

∴ Force trying to separate the flanges *i.e.* force on 8 bolts,

$$F = \frac{\pi}{4}(D_1)^2\; p = \frac{\pi}{4}(272)^2\; 0.35 = 20\,340 \text{ N}$$

Now let us find the bending moment about the section *X-X* which is tangential to the outside of the pipe. The width of the segment is obtained by measuring the distance from the drawing. On measuring, we get

$$x = 90 \text{ mm}$$

and distance of the section *X-X* from the centre of the bolt,

$$y = \frac{D_p}{2} - \left(\frac{D}{2} + t\right) = \frac{290}{2} - \left(\frac{200}{2} + 12\right) = 33 \text{ mm}$$

Let σ_b = Working stress in the flange.

We know that bending moment on each bolt due to force F

$$= \frac{F}{n} \times y = \frac{20340}{8} \times 33 = 83\,900 \text{ N-mm} \qquad \text{...}(i)$$

* M16 bolt means that the nominal diameter of the bolt (d) is 16 mm.

and resisting moment on the flange

$$= \sigma_b \times Z = \sigma_b \times \frac{1}{6} \times x\,(t_f)^2$$

$$= \sigma_b \times \frac{1}{6} \times 90\,(20)^2 = 6000\,\sigma_b \text{ N-mm} \quad ...(ii)$$

From equations (*i*) and (*ii*), we have

$$\sigma_b = 83\,900 / 6000$$

$$= 13.98 \text{ N/mm}^2 = 13.98 \text{ MPa} \quad \textbf{Ans.}$$

8.8 Design of Oval Flanged Pipe Joint

Consider an oval flanged pipe joint as shown in Fig. 8.11. A spigot and socket is provided for locating the pipe bore in a straight line. A packing of trapezoidal section is used to make the joint leak proof. The thickness of the pipe is obtained as discussed previously.

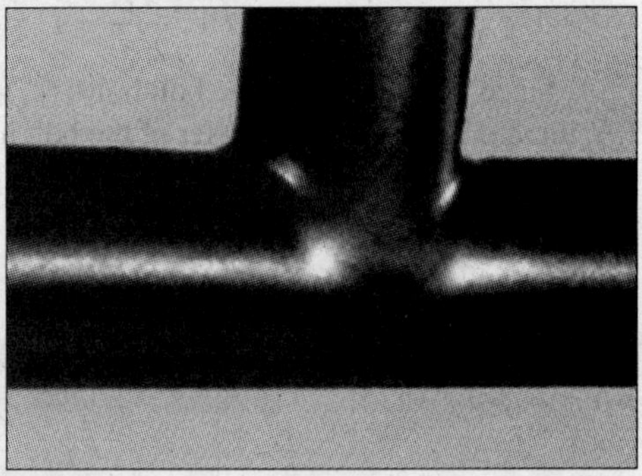

Oval flanged pipe joint.

The force trying to separate the two flanges has to be resisted by the stress produced in the bolts. If a length of pipe, having its ends closed somewhere along its length, be considered, then the force separating the two flanges due to fluid pressure is given by

$$F_1 = \frac{\pi}{4} \times D^2 \times p$$

where D = Internal diameter of the pipe.

The packing has also to be compressed to make the joint leakproof. The intensity of pressure should be greater than the pressure of the fluid inside the pipe. For the purposes of calculations, it is assumed that the packing material is compressed to the same pressure as that of inside the pipe. Therefore the force tending to separate the flanges due to pressure in the packing is given by

$$F_2 = \frac{\pi}{4} \times \left[(D_1)^2 - (D)^2\right] p$$

where D_1 = Outside diameter of the packing.

∴ Total force trying to separate the two flanges,

$$F = F_1 + F_2$$

$$= \frac{\pi}{4} \times D^2 \times p + \frac{\pi}{4}\left[(D_1)^2 - (D)^2\right] p = \frac{\pi}{4}(D_1)^2\, p$$

Since an oval flange is fastened by means of two bolts, therefore load taken up by each bolt is $F_b = F/2$. If d_c is the core diameter of the bolts, then

$$F_b = \frac{\pi}{4}(d_c)^2\, \sigma_{tb}$$

where σ_{tb} is the allowable tensile stress for the bolt material. The value of σ_{tb} is usually kept low to allow for initial tightening stress in the bolts. After the core diameter is obtained, then the nominal diameter of the bolts is chosen from *tables. It may be noted that bolts of less than 12 mm diameter

* In the absence of tables, nominal diameter = $\dfrac{\text{Core diameter}}{0.84}$

should never be used for hydraulic pipes, because very heavy initial tightening stresses may be induced in smaller bolts. The bolt centres should be as near the centre of the pipe as possible to avoid bending of the flange. But sufficient clearance between the bolt head and pipe surface must be provided for the tightening of the bolts without damaging the pipe material.

The thickness of the flange is obtained by considering the flange to be under bending stresses due to the forces acting in one bolt. The maximum bending stress will be induced at the section *X-X*. The bending moment at this section is given by

$$M_{xx} = F_b \times e = \frac{F}{2} \times e$$

and section modulus,
$$Z = \frac{1}{6} \times b\,(t_f)^2$$

where b = Width of the flange at the section *X-X*, and

t_f = Thickness of the flange.

Using the bending equation, we have

$$M_{xx} = \sigma_b . Z$$

or
$$F_b \times e = \sigma_b \times \frac{1}{6} \times b\,(t_f)^2$$

where σ_b = Permissible bending stress for the flange material.

From the above expression, the value of t_f may be obtained, if b is known. The width of the flange is estimated from the lay out of the flange. The hydraulic joints with oval flanges are known as ***Armstrong's pipe joints.*** The various dimensions for a hydraulic joint may be obtained by using the following empirical relations:

Nominal diameter of bolts, $d = 0.75\,t + 10$ mm

Thickness of the flange, $t_f = 1.5\,t + 3$ mm

Outer diameter of the flange,

$$D_o = D + 2t + 4.6\,d$$

Pitch circle diameter, $D_p = D_o - (3\,t + 20 \text{ mm})$

Example 8.5. *Design and draw an oval flanged pipe joint for a pipe having 50 mm bore. It is subjected to an internal fluid pressure of 7 N/mm². The maximum tensile stress in the pipe material is not to exceed 20 MPa and in the bolts 60 MPa.*

Solution. Given: D = 50 mm or R = 25 mm ; p = 7 N/mm² ; σ_t = 20 MPa = 20 N/mm² ; σ_{tb} = 60 MPa = 60 N/mm²

First of all let us find the thickness of the pipe (t). According to Lame's equation, we know that thickness of the pipe,

$$t = R\left[\sqrt{\frac{\sigma_t + p}{\sigma_t - p}} - 1\right] = 25\left[\sqrt{\frac{20+7}{20-7}} - 1\right] = 11.03 \text{ say } 12 \text{ mm } \textbf{Ans.}$$

Assuming the width of packing as 10 mm, therefore outside diameter of the packing,

$$D_1 = D + 2 \times \text{Width of packing} = 50 + 2 \times 10 = 70 \text{ mm}$$

∴ Force trying to separate the flanges,

$$F = \frac{\pi}{4}(D_1)^2\, p = \frac{\pi}{4}(70)^2\, 7 = 26\,943 \text{ N}$$

Since the flange is secured by means of two bolts, therefore load on each bolt,

$$F_b = F/2 = 26\,943/2 = 13\,471.5 \text{ N}$$

Let d_c = Core diameter of bolts.

We know that load on each bolt (F_b),

$$13\,471.5 = \frac{\pi}{4}(d_c)^2\,\sigma_{tb} = \frac{\pi}{4}(d_c)^2\,60 = 47.2\,(d_c)^2$$

$\therefore$ $(d_c)^2 = 13\,471.5 / 47.2 = 285.4$ or $d_c = 16.9$ say 17 mm

and nominal diameter of bolts,

$$d = \frac{d_c}{0.84} = \frac{17}{0.84} = 20.2 \text{ say } 22 \text{ mm} \quad \textbf{Ans.}$$

Outer diameter of the flange,

$$D_o = D + 2t + 4.6\,d = 50 + 2 \times 12 + 4.6 \times 22$$
$$= 175.2 \text{ say } 180 \text{ mm} \quad \textbf{Ans.}$$

and pitch circle diameter of the bolts,

$$D_p = D_o - (3t + 20 \text{ mm}) = 180 - (3 \times 12 + 20) = 124 \text{ mm}$$

The elevation of the flange as shown in Fig. 8.13 (which is an ellipse) may now be drawn by taking major axis as D_o (*i.e.* 180 mm) and minor axis as $(D_p - d)$ *i.e.* 124 – 22 = 102 mm. In order to find thickness of the flange (t_f), consider the section *X-X*. By measurement, we find that the width of the flange at the section *X-X*,

$$b = 89 \text{ mm}$$

and the distance of the section *X-X* from the centre line of the bolt,

$$e = 33 \text{ mm}$$

$\therefore$ Bending moment at the section *X-X*,

$$M_{xx} = F_b \times e = 13\,471.5 \times 33 \text{ N-mm}$$
$$= 444\,560 \text{ N-mm}$$

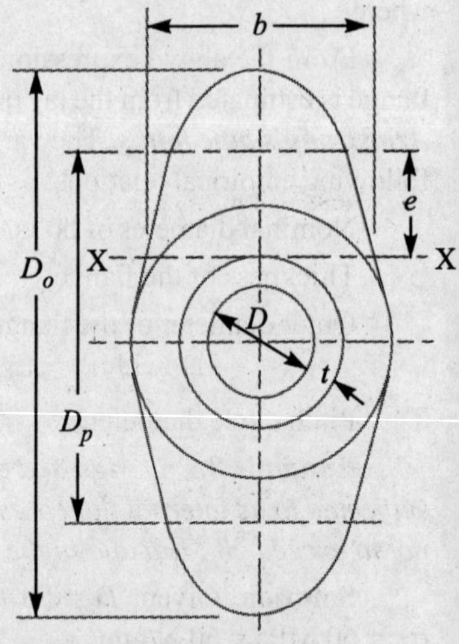

Fig. 8.13

and section modulus, $Z = \frac{1}{6}\,b\,(t_f)^2 = \frac{1}{6} \times 89\,(t_f)^2$

$$= 14.83\,(t_f)^2$$

We know that $M_{xx} = \sigma_b \times Z$

$$444\,560 = 20 \times 14.83\,(t_f)^2 = 296.6\,(t_f)^2$$

$\therefore$ $(t_f)^2 = 444\,560 / 296.6 = 1500$

or $t_f = 38.7$ say 40 mm **Ans.**

8.9 Design of Square Flanged Pipe Joint

The design of a square flanged pipe joint, as shown in Fig. 8.14, is similar to that of an oval flanged pipe joint except that the load has to be divided into four bolts. The thickness of the flange may be obtained by considering the bending of the flange about one of the sections *A-A*, *B-B*, or *C-C*.

A little consideration will show that the flange is weakest in bending about section *A-A*. Therefore the thickness of the flange is calculated by considering the bending of the flange, about section *A-A*.

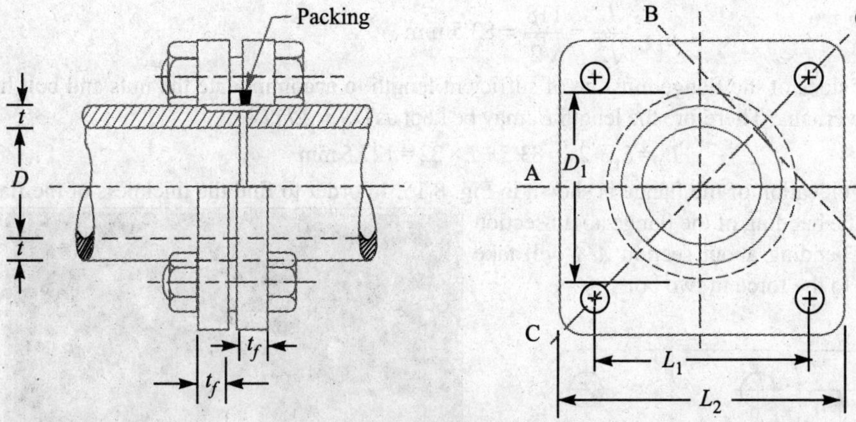

Fig. 8.14. Square flanged pipe joint.

Example 8.6. *Design a square flanged pipe joint for pipes of internal diameter 50 mm subjected to an internal fluid pressure of 7 N/mm². The maximum tensile stress in the pipe material is not to exceed 21 MPa and in the bolts 28 MPa.*

Solution. Given: D = 50 mm or R = 25 mm ; p = 7 N/mm² ; σ_t = 21 MPa = 21 N/mm² ; σ_{tb} = 28 MPa = 28 N/mm²

First of all, let us find the thickness of the pipe. According to Lame's equation, we know that thickness of the pipe,

$$t = R\left[\sqrt{\frac{\sigma_t + p}{\sigma_t - p}} - 1\right] = 25\left[\sqrt{\frac{21+7}{21-7}} - 1\right] = 10.35 \text{ say } 12 \text{ mm}$$

Assuming the width of packing as 10 mm, therefore outside diameter of the packing,

$$D_1 = 50 + 2 \times \text{Width of packing} = 50 + 2 \times 10 = 70 \text{ mm}$$

∴ Force trying to separate the flanges,

$$F = \frac{\pi}{4}(D_1)^2\ p = \frac{\pi}{4}(70)^2\ 7 = 26\ 943 \text{ N}$$

Since this force is to be resisted by four bolts, therefore force on each bolt,

$$F_b = F/4 = 26\ 943/4 = 6735.8 \text{ N}$$

Let d_c = Core diameter of the bolts.

We know that force on each bolt (F_b),

$$6735.8 = \frac{\pi}{4}(d_c)^2\ \sigma_{tb} = \frac{\pi}{4}(d_c)^2\ 28 = 22\ (d_c)^2$$

$$\therefore \quad (d_c)^2 = 6735.8/22 = 306 \quad \text{or} \quad d_c = 17.5 \text{ mm}$$

and nominal diameter of the bolts,

$$d = \frac{d_c}{0.84} = \frac{17.5}{0.84} = 20.9 \text{ say } 22 \text{ mm} \quad \textbf{Ans.}$$

The axes of the bolts are arranged at the corners of a square of such size that the corners of the nut clear the outside of the pipe.

∴ Minimum length of a diagonal for this square,

$$L = \text{Outside diameter of pipe} + 2 \times \text{Dia. of bolt} = D + 2t + 2d$$

$$= 50 + (2 \times 12) + (2 \times 22) = 118 \text{ mm}$$

and side of this square,

$$L_1 = \frac{L}{\sqrt{2}} = \frac{118}{\sqrt{2}} = 83.5 \text{ mm}$$

The sides of the flange must be of sufficient length to accommodate the nuts and bolt heads without overhang. Therefore the length L_2 may be kept as $(L_1 + 2d)$ *i.e.*

$$L_2 = L_1 + 2d = 83.5 + 2 \times 22 = 127.5 \text{ mm}$$

The elevation of the flange is shown in Fig. 8.15. In order to find the thickness of the flange, consider the bending of the flange about section *A-A*. The bending about section *A-A* will take place due to the force in two bolts.

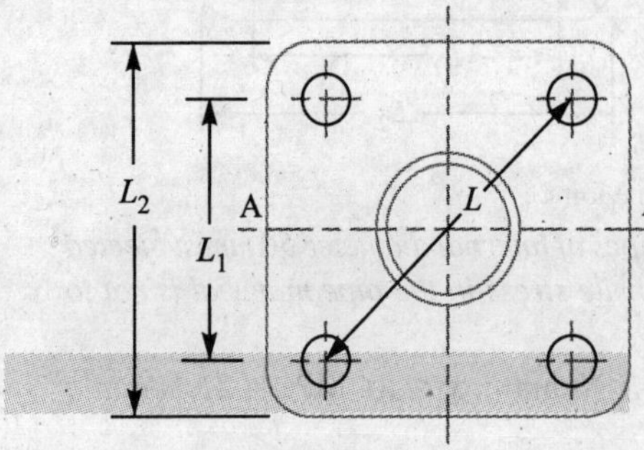

Fig. 8.15

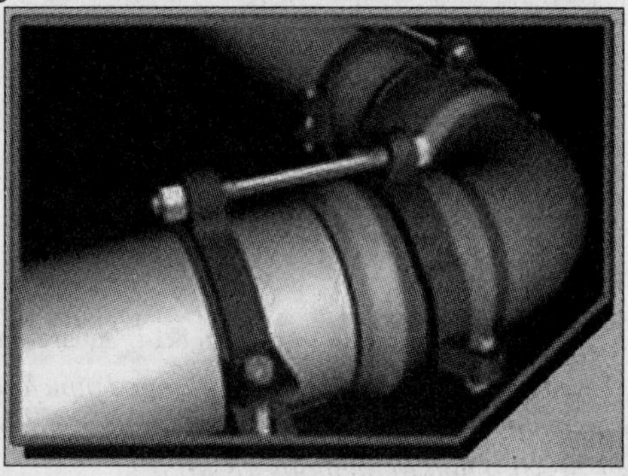

Square flanged pipe joint.

∴ Bending moment due to the force in two bolts (*i.e.* due to $2F_b$),

$$M_1 = 2F_b \times \frac{L_1}{2} = 2 \times 6735.8 \times \frac{83.5}{2} = 562\ 440 \text{ N-mm}$$

Water pressure acting on half the flange

$$= 2F_b = 2 \times 6735.8 = 13\ 472 \text{ N}$$

The flanges are screwed with pipe having metric threads of 4.4 threads in 10 mm (*i.e.* pitch of the threads is 10/4.4 = 2.28 mm).

Nominal or major diameter of the threads

$$= \text{Outside diameter of the pipe} = D + 2t = 50 + 2 \times 12 = 74 \text{ mm}$$

∴ Nominal radius of the threads

$$= 74/2 = 37 \text{ mm}$$

Depth of the threads $= 0.64 \times \text{Pitch of threads} = 0.64 \times 2.28 = 1.46 \text{ mm}$

∴ Core or minor radius of the threads

$$= 37 - 1.46 = 35.54 \text{ mm}$$

∴ Mean radius of the arc from *A-A* over which the load due to fluid pressure may be taken to be concentrated

$$= \frac{1}{2}(37 + 35.54) = 36.27 \text{ mm}$$

The centroid of this arc from *A-A*

$$= 0.6366 \times \text{Mean radius} = 0.6366 \times 36.27 = 23.1 \text{ mm}$$

∴ Bending moment due to the water pressure,

$$M_2 = 2F_b \times 23.1 = 2 \times 6735.8 \times 23.1 = 311\ 194 \text{ N-mm}$$

Since the bending moments M_1 and M_2 are in opposite directions, therefore

Net resultant bending moment on the flange about section *A-A*,

$$M = M_1 - M_2 = 562\,440 - 311\,194 = 251\,246 \text{ N-mm}$$

Width of the flange at the section *A-A*,

$$b = L_2 - \text{Outside diameter of pipe} = 127.5 - 74 = 53.5 \text{ mm}$$

Let t_f = Thickness of the flange in mm.

∴ Section modulus,

$$Z = \frac{1}{6} \times b\,(t_f)^2 = \frac{1}{6} \times 53.5\,(t_f)^2 = 8.9\,(t_f)^2 \text{ mm}^3$$

We know that net resultant bending moment (*M*),

$$251\,246 = \sigma_b . Z = 21 \times 8.9\,(t_f)^2 = 187\,(t_f)^2$$

$$\therefore \quad (t_f)^2 = 251\,246 / 187 = 1344 \text{ or } t_f = 36.6 \text{ say } 38 \text{ mm} \quad \textbf{Ans.}$$

EXERCISES

1. A cast iron pipe of internal diameter 200 mm and thickness 50 mm carries water under a pressure of 5 N/mm^2. Calculate the tangential and radial stresses at the inner, middle (radius = 125 mm) and outer surfaces. **[Ans. 13 MPa, 9.76 MPa, 8 MPa ; – 5 MPa, – 1.76 MPa, 0]**
2. A cast iron pipe is to carry 60 m^3 of compressed air per minute at a pressure of 1 N/mm^2. The velocity of air in the pipe is limited to 10 m/s and the permissible tensile stress for the material of the pipe is 14 MPa. Find the diameter of the pipe and its wall thickness. **[Ans. 360 mm ; 22 mm]**
3. A seamless steel pipe carries 2000 m^3 of steam per hour at a pressure of 1.2 N/mm^2. The velocity of flow is 28 m/s. Assuming the tensile stress as 40 MPa, find the inside diameter of the pipe and its wall thickness. **[Ans. 160 mm ; 5.4 mm]**
4. Compute the dimensions of a flanged cast iron pipe 200 mm in diameter to carry a pressure of 0.7 N/mm^2.

   [Ans. t = 20 mm ; d = 16 mm ; n = 8 ; t_f = 33 mm ; B = 37 mm ; D_o = 314 mm ; D_p = 284 mm]
5. Design an oval flanged pipe joint for pipes of internal diameter 50 mm subjected to a fluid pressure of 7 N/mm^2. The maximum tensile stress in the pipe material is not to exceed 21 MPa and in the bolts 28 MPa. **[Ans. t = 12 mm ; d = 30 mm ; t_f = 38 mm]**

QUESTIONS

1. Discuss how the pipes are designed.
2. Describe with sketches, the various types of pipe joints commonly used in engineering practice.
3. Explain the procedure for design of a circular flanged pipe point.
4. Describe the procedure for designing an oval flanged pipe joint.

OBJECTIVE TYPE QUESTIONS

1. Cast iron pipes are mainly used
   (*a*) for conveying steam
   (*b*) in water and sewage systems
   (*c*) in pressure lubrication systems on prime movers
   (*d*) all of the above

2. The diameter of a pipe carrying steam Q m³/min at a velocity v m/min is

(*a*) $\frac{Q}{v}$ (*b*) $\sqrt{\frac{Q}{v}}$

(*c*) $\frac{\pi}{4}\sqrt{\frac{Q}{v}}$ (*d*) $1.13\sqrt{\frac{Q}{v}}$

3. When the internal diameter of the pipe exceeds twenty times its wall thickness, then .... cylindrical shell formula may be applied.

(*a*) thin (*b*) thick

4. Which of the following joint is commonly used for joining pipes carrying water at low pressure?

(*a*) union joint (*b*) spigot and socket joint

(*c*) socket or a coupler joint (*d*) nipple joint

5. The pipes which are burried in the earth should be joined with

(*a*) union joint (*b*) spigot and socket joint

(*c*) coupler joint (*d*) nipple joint

6. The expansion joint is mostly used for pipes which carry steam at .... pressures.

(*a*) low (*b*) high

7. The pipes carrying fluid pressure varying from 5 to 14 N/mm² should have

(*a*) square flanged joint (*b*) circular flanged joint

(*c*) oval flanged joint (*d*) spigot and socket joint

8. An oval type flange is fastened by means of

(*a*) two bolts (*b*) four bolts

(*c*) six bolts (*d*) eight bolts

9. A flanged pipe joint will be a leakproof, if the circumferential pitch of the bolts is

(*a*) less then $20\sqrt{d}$ (*b*) greater than $30\sqrt{d}$

(*c*) between $20\sqrt{d}$ and $30\sqrt{d}$ (*d*) equal to one-third of inside diameter of pipe

where d = Diameter of the bolt hole.

10. The flanges in a circular flanged pipe joint are strengthened by providing ribs between the bolt holes. The thickness of such ribs is taken as

(*a*) t (*b*) t_f

(*c*) $\frac{t - t_f}{2}$ (*d*) $\frac{t + t_f}{2}$

where t = Thickness of pipe, and

t_f = Thickness of flange.

ANSWERS

1. (*b*)	**2.** (*d*)	**3.** (*a*)	**4.** (*c*)	**5.** (*b*)
6. (*a*)	**7.** (*c*)	**8.** (*a*)	**9.** (*c*)	**10.** (*d*)

CHAPTER

9

Riveted Joints

9.1 Introduction

A rivet is a short cylindrical bar with a head integral to it. The cylindrical portion of the rivet is called ***shank*** or ***body*** and lower portion of shank is known as ***tail***, as shown in Fig. 9.1. The rivets are used to make permanent fastening between the plates such as in structural work, ship building, bridges, tanks and boiler shells. The riveted joints are widely used for joining light metals.

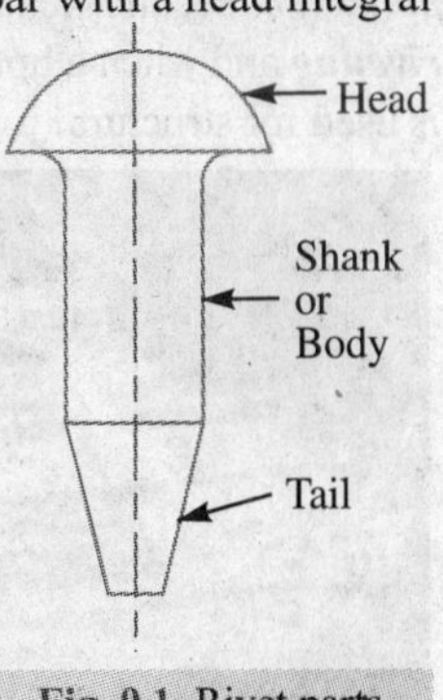

Fig. 9.1. Rivet parts.

The fastenings (*i.e.* joints) may be classified into the following two groups :

1. Permanent fastenings, and
2. Temporary or detachable fastenings.

The ***permanent fastenings*** are those fastenings which can not be disassembled without destroying the connecting components. The examples of permanent fastenings in order of strength are soldered, brazed, welded and riveted joints.

The ***temporary*** or ***detachable fastenings*** are those fastenings which can be disassembled without destroying the connecting components. The examples of temporary fastenings are screwed, keys, cotters, pins and splined joints.

9.2 Methods of Riveting

The function of rivets in a joint is to make a connection that has strength and tightness. The strength is necessary to prevent failure of the joint. The tightness is necessary in order to contribute to strength and to prevent leakage as in a boiler or in a ship hull.

When two plates are to be fastened together by a rivet as shown in Fig. 9.2 (*a*), the holes in the plates are punched and reamed or drilled. Punching is the cheapest method and is used for relatively thin plates and in structural work. Since punching injures the material around the hole, therefore drilling is used in most pressure-vessel work. In structural and pressure vessel riveting, the diameter of the rivet hole is usually 1.5 mm larger than the nominal diameter of the rivet.

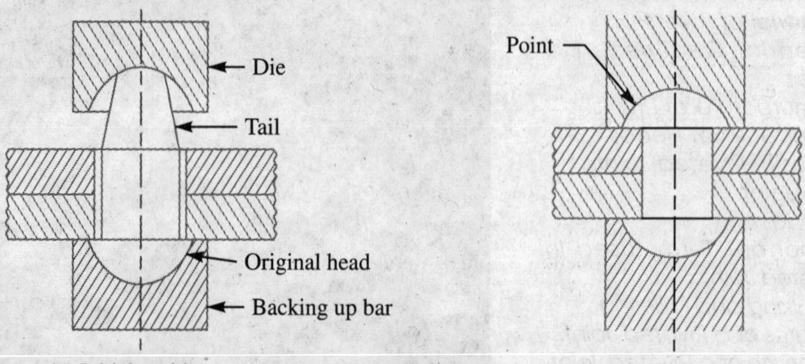

(*a*) Initial position.

(*b*) Final position.

Fig. 9.2. Methods of riveting.

The plates are drilled together and then separated to remove any burrs or chips so as to have a tight flush joint between the plates. A cold rivet or a red hot rivet is introduced into the plates and the ***point*** (*i.e.* second head) is then formed. When a cold rivet is used, the process is known as ***cold riveting*** and when a hot rivet is used, the process is known as ***hot riveting.*** The cold riveting process is used for structural joints while hot riveting is used to make leak proof joints.

A ship's body is a combination of riveted, screwed and welded joints.

The riveting may be done by hand or by a riveting machine. In hand riveting, the original rivet head is backed up by a hammer or heavy bar and then the die or set, as shown in Fig. 9.2 (*a*), is placed against the end to be headed and the blows are applied by a hammer. This causes the shank to expand thus filling the hole and the tail is converted into a ***point*** as shown in Fig. 9.2 (*b*). As the rivet cools, it tends to contract. The lateral contraction will be slight, but there will be a longitudinal tension introduced in the rivet which holds the plates firmly together.

In machine riveting, the die is a part of the hammer which is operated by air, hydraulic or steam pressure.

Notes : **1.** For steel rivets upto 12 mm diameter, the cold riveting process may be used while for larger diameter rivets, hot riveting process is used.

2. In case of long rivets, only the tail is heated and not the whole shank.

9.3 Material of Rivets

The material of the rivets must be tough and ductile. They are usually made of steel (low carbon steel or nickel steel), brass, aluminium or copper, but when strength and a fluid tight joint is the main consideration, then the steel rivets are used.

The rivets for general purposes shall be manufactured from steel conforming to the following Indian Standards :

(*a*) IS : 1148–1982 (Reaffirmed 1992) – Specification for hot rolled rivet bars (up to 40 mm diameter) for structural purposes; or

(*b*) IS : 1149–1982 (Reaffirmed 1992) – Specification for high tensile steel rivet bars for structural purposes.

The rivets for boiler work shall be manufactured from material conforming to IS : 1990 – 1973 (Reaffirmed 1992) – Specification for steel rivets and stay bars for boilers.

Note : The steel for boiler construction should conform to IS : 2100 – 1970 (Reaffirmed 1992) – Specification for steel billets, bars and sections for boilers.

9.4 Essential Qualities of a Rivet

According to Indian standard, IS : 2998 – 1982 (Reaffirmed 1992), the material of a rivet must have a tensile strength not less than 40 N/mm^2 and elongation not less than 26 percent. The material must be of such quality that when in cold condition, the shank shall be bent on itself through 180° without cracking and after being heated to 650°C and quenched, it must pass the same test. The rivet when hot must flatten without cracking to a diameter 2.5 times the diameter of shank.

9.5 Manufacture of Rivets

According to Indian standard specifications, the rivets may be made either by cold heading or by hot forging. If rivets are made by the cold heading process, they shall subsequently be adequately heat treated so that the stresses set up in the cold heading process are eliminated. If they are made by hot forging process, care shall be taken to see that the finished rivets cool gradually.

9.6 Types of Rivet Heads

According to Indian standard specifications, the rivet heads are classified into the following three types :

1. Rivet heads for general purposes (below 12 mm diameter) as shown in Fig. 9.3, according to IS : 2155 – 1982 (Reaffirmed 1996).

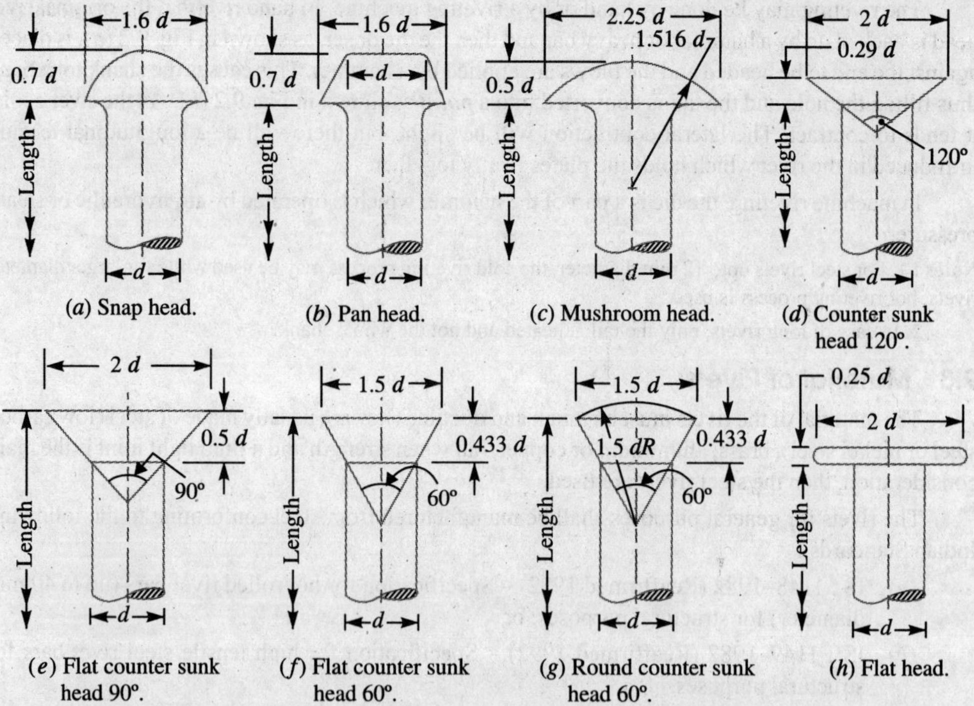

Fig. 9.3. Rivet heads for general purposes (below 12 mm diameter).

2. Rivet heads for general purposes (From 12 mm to 48 mm diameter) as shown in Fig. 9.4, according to IS : 1929 – 1982 (Reaffirmed 1996).

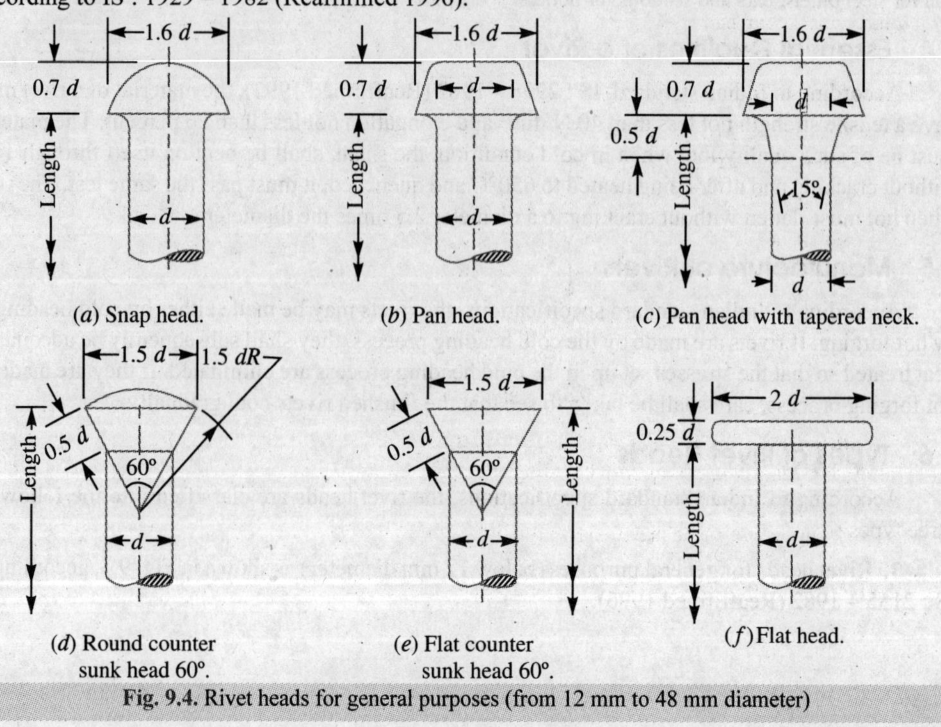

Fig. 9.4. Rivet heads for general purposes (from 12 mm to 48 mm diameter)

3. Rivet heads for boiler work (from 12 mm to 48 mm diameter, as shown in Fig. 9.5, according to IS : 1928 – 1961 (Reaffirmed 1996).

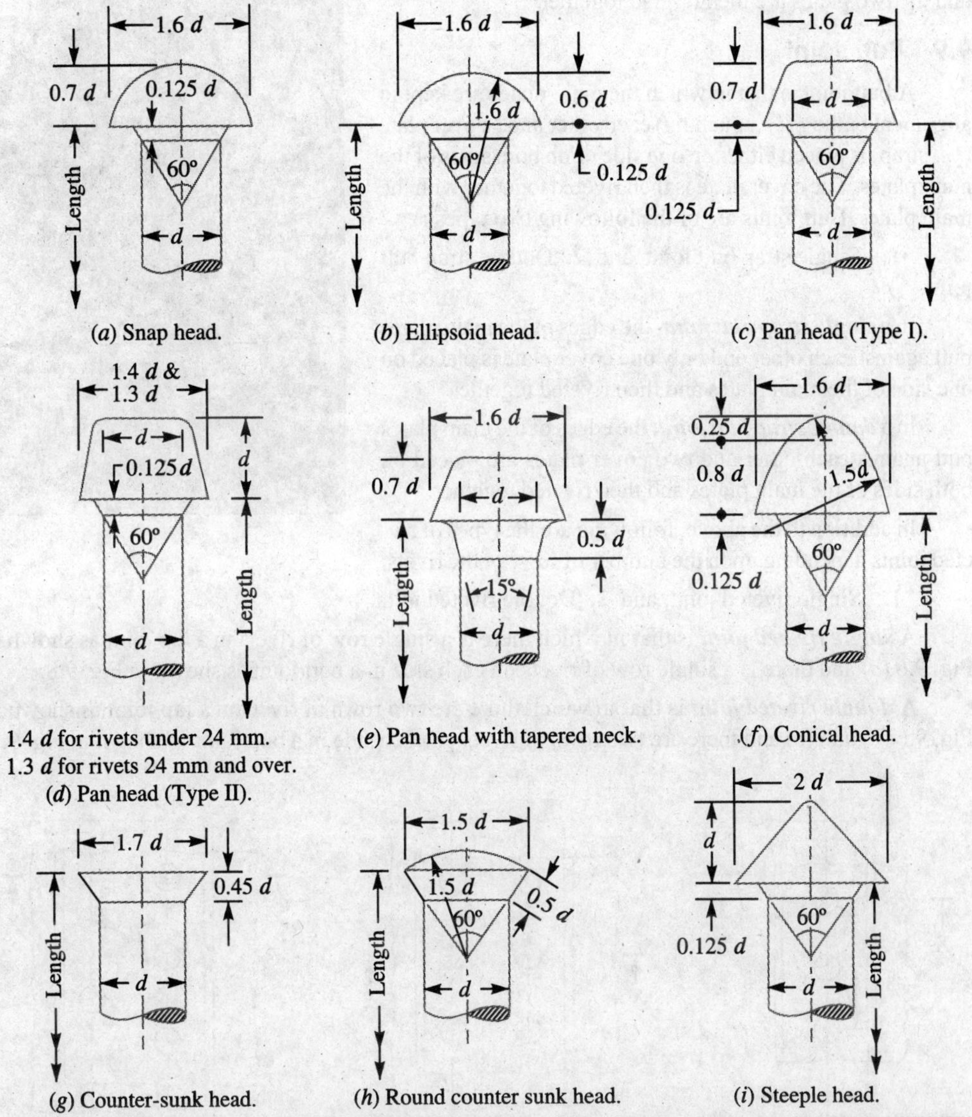

Fig. 9.5. Rivet heads for boiler work.

The **snap heads** are usually employed for structural work and machine riveting. The **counter sunk heads** are mainly used for ship building where flush surfaces are necessary. The **conical heads** (also known as **conoidal heads)** are mainly used in case of hand hammering. The **pan heads** have maximum strength, but these are difficult to shape.

9.7 Types of Riveted Joints

Following are the two types of riveted joints, depending upon the way in which the plates are connected.

1. Lap joint, and 2. Butt joint.

9.8 Lap Joint

A lap joint is that in which one plate overlaps the other and the two plates are then riveted together.

9.9 Butt Joint

A butt joint is that in which the main plates are kept in alignment butting (*i.e.* touching) each other and a cover plate (*i.e.* strap) is placed either on one side or on both sides of the main plates. The cover plate is then riveted together with the main plates. Butt joints are of the following two types :

1. Single strap butt joint, and 2. Double strap butt joint.

In a ***single strap butt joint***, the edges of the main plates butt against each other and only one cover plate is placed on one side of the main plates and then riveted together.

In a ***double strap butt joint***, the edges of the main plates butt against each other and two cover plates are placed on both sides of the main plates and then riveted together.

In addition to the above, following are the types of riveted joints depending upon the number of rows of the rivets.

1. Single riveted joint, and 2. Double riveted joint.

A ***single riveted joint*** is that in which there is a single row of rivets in a lap joint as shown in Fig. 9.6 (*a*) and there is a single row of rivets on each side in a butt joint as shown in Fig. 9.8.

A ***double riveted joint*** is that in which there are two rows of rivets in a lap joint as shown in Fig. 9.6 (*b*) and (*c*) and there are two rows of rivets on each side in a butt joint as shown in Fig. 9.9.

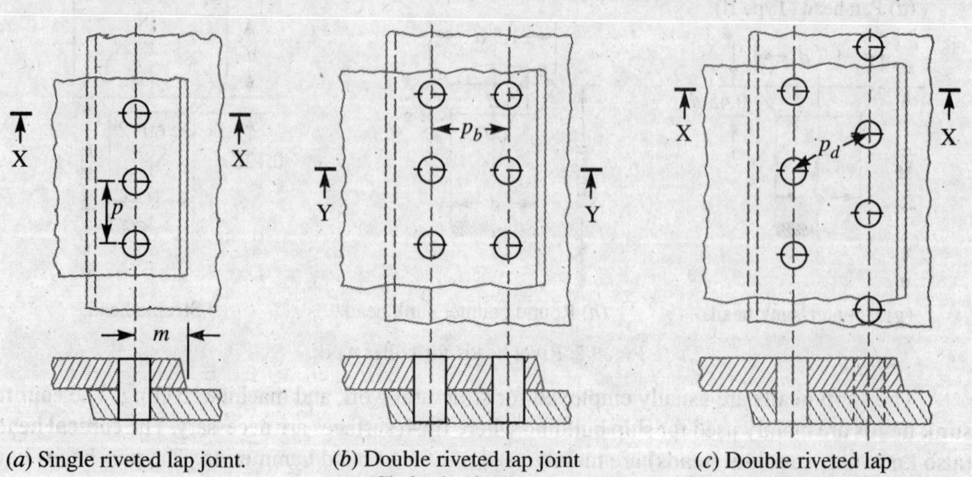

(*a*) Single riveted lap joint. (*b*) Double riveted lap joint (Chain riveting). (*c*) Double riveted lap joint (Zig-zag riveting).

Fig. 9.6. Single and double riveted lap joints.

Similarly the joints may be ***triple riveted*** or ***quadruple riveted.***

Notes : 1. When the rivets in the various rows are opposite to each other, as shown in Fig. 9.6 (*b*), then the joint is said to be ***chain riveted.*** On the other hand, if the rivets in the adjacent rows are staggered in such a way that every rivet is in the middle of the two rivets of the opposite row as shown in Fig. 9.6 (*c*), then the joint is said to be ***zig-zag riveted.***

2. Since the plates overlap in lap joints, therefore the force *P*, *P* acting on the plates [See Fig. 9.15 (*a*)] are not in the same straight line but they are at a distance equal to the thickness of the plate. These forces will form a couple which may bend the joint. Hence the lap joints may be used only where small loads are to be transmitted. On the other hand, the forces *P*, *P* in a butt joint [See Fig. 9.15 (*b*)] act in the same straight line, therefore there will be no couple. Hence the butt joints are used where heavy loads are to be transmitted.

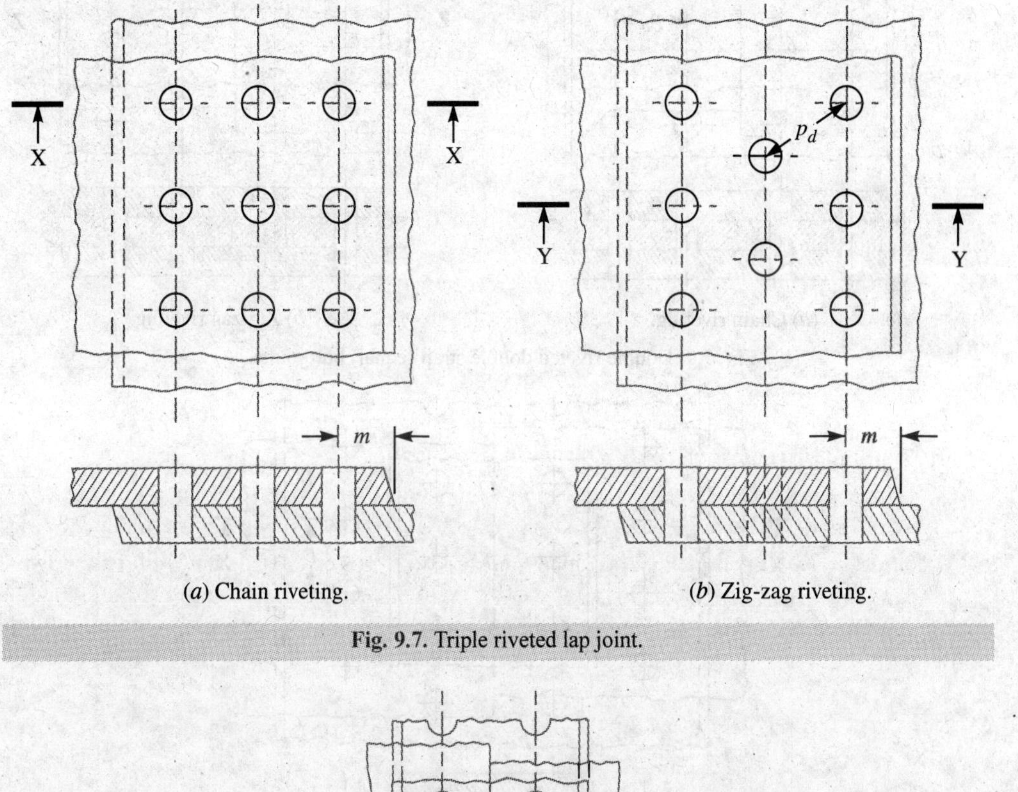

(*a*) Chain riveting. (*b*) Zig-zag riveting.

Fig. 9.7. Triple riveted lap joint.

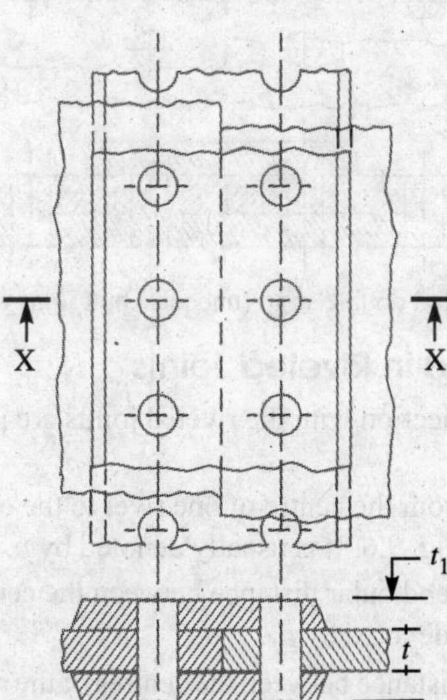

Fig. 9.8. Single riveted double strap butt joint.

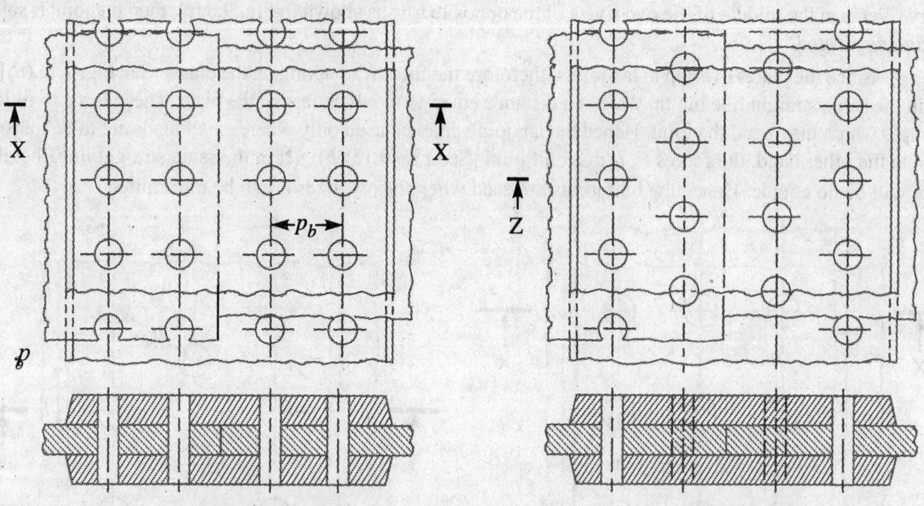

(*a*) Chain riveting. (*b*) Zig-zag riveting.

Fig. 9.9. Double riveted double strap (equal) butt joints.

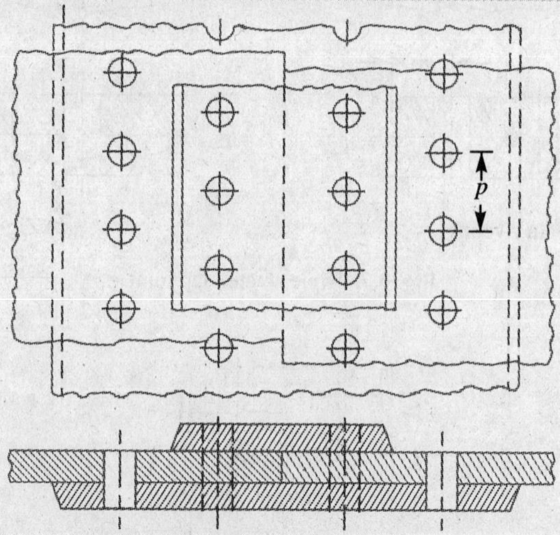

Fig. 9.10. Double riveted double strap (unequal) butt joint with zig-zag riveting.

9.10 Important Terms Used in Riveted Joints

The following terms in connection with the riveted joints are important from the subject point of view :

1. ***Pitch.*** It is the distance from the centre of one rivet to the centre of the next rivet measured parallel to the seam as shown in Fig. 9.6. It is usually denoted by p.

2. ***Back pitch.*** It is the perpendicular distance between the centre lines of the successive rows as shown in Fig. 9.6. It is usually denoted by p_b.

3. ***Diagonal pitch.*** It is the distance between the centres of the rivets in adjacent rows of zig-zag riveted joint as shown in Fig. 9.6. It is usually denoted by p_d.

4. ***Margin or marginal pitch.*** It is the distance between the centre of rivet hole to the nearest edge of the plate as shown in Fig. 9.6. It is usually denoted by m.

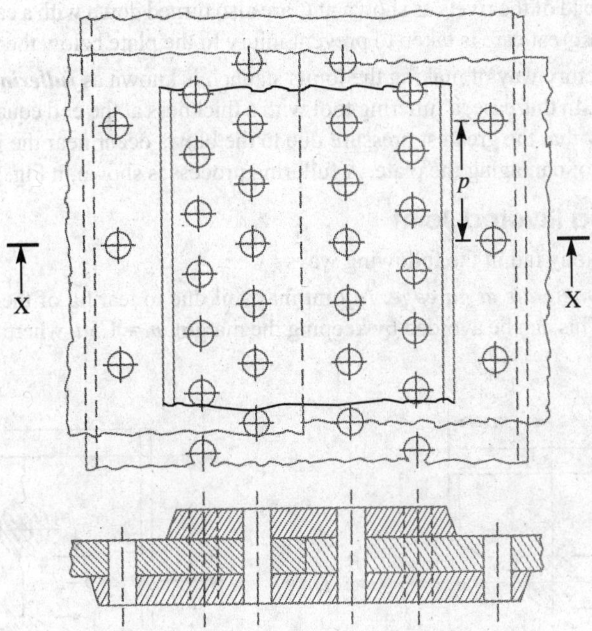

Fig. 9.11. Triple riveted double strap (unequal) butt joint.

9.11 Caulking and Fullering

Caulking process is employed to make the joints leak proofs or fluid tight jn steam boiler.

In order to make the joints leak proof or fluid tight in pressure vessels like steam boilers, air receivers and tanks etc. a process known as ***caulking*** is employed. In this process, a narrow blunt tool called caulking tool, about 5 mm thick and 38 mm in breadth, is used. The edge of the tool is ground to an angle of 80°. The tool is moved after each blow along the edge of the plate, which is planed to a bevel of 75° to 80° to facilitate the forcing down of edge. It is seen that the tool burrs down the plate at *A* in Fig. 9.12 (*a*) forming a metal to metal joint. In actual practice, both the edges at *A* and

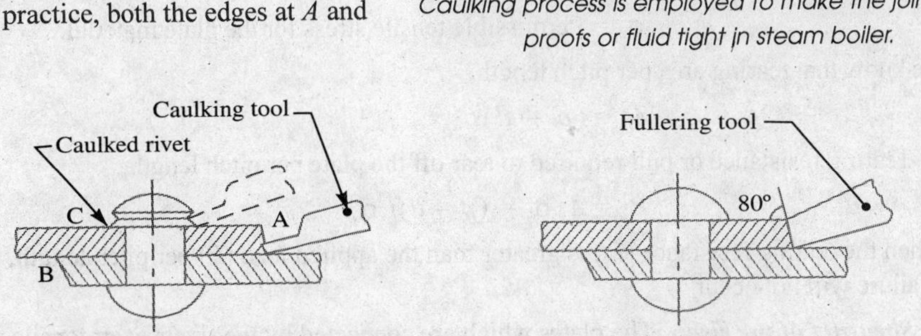

(*a*) Caulking. (*b*) Fullering.

Fig. 9.12. Caulking and fullering.

B are caulked. The head of the rivets as shown at *C* are also turned down with a caulking tool to make a joint steam tight. A great care is taken to prevent injury to the plate below the tool.

A more satisfactory way of making the joints staunch is known as ***fullering*** which has largely superseded caulking. In this case, a fullering tool with a thickness at the end equal to that of the plate is used in such a way that the greatest pressure due to the blows occur near the joint, giving a clean finish, with less risk of damaging the plate. A fullering process is shown in Fig. 9.12 (*b*).

9.12 Failures of a Riveted Joint

A riveted joint may fail in the following ways :

1. ***Tearing of the plate at an edge.*** A joint may fail due to tearing of the plate at an edge as shown in Fig. 9.13. This can be avoided by keeping the margin, $m = 1.5d$, where d is the diameter of the rivet hole.

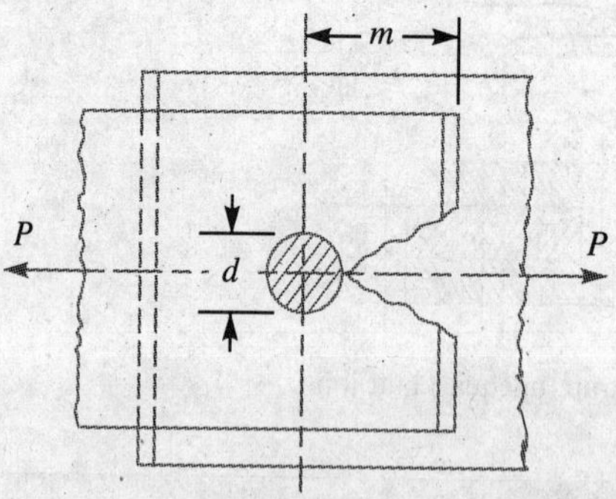

Fig. 9.13. Tearing of the plate at an edge.

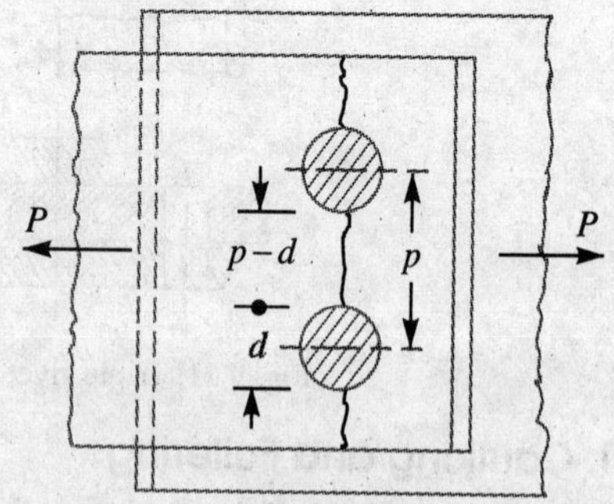

Fig. 9.14. Tearing of the plate across the rows of rivets.

2. ***Tearing of the plate across a row of rivets.*** Due to the tensile stresses in the main plates, the main plate or cover plates may tear off across a row of rivets as shown in Fig. 9.14. In such cases, we consider only one pitch length of the plate, since every rivet is responsible for that much length of the plate only.

The resistance offered by the plate against tearing is known as ***tearing resistance*** or ***tearing strength*** or ***tearing value*** of the plate.

Let p = Pitch of the rivets,

d = Diameter of the rivet hole,

t = Thickness of the plate, and

σ_t = Permissible tensile stress for the plate material.

We know that tearing area per pitch length,

$$A_t = (p - d)t$$

∴ Tearing resistance or pull required to tear off the plate per pitch length,

$$P_t = A_t.\sigma_t = (p - d)t.\sigma_t$$

When the tearing resistance (P_t) is greater than the applied load (P) per pitch length, then this type of failure will not occur.

3. ***Shearing of the rivets.*** The plates which are connected by the rivets exert tensile stress on the rivets, and if the rivets are unable to resist the stress, they are sheared off as shown in Fig. 9.15.

It may be noted that the rivets are in *single shear in a lap joint and in a single cover butt joint, as shown in Fig. 9.15. But the rivets are in double shear in a double cover butt joint as shown in Fig. 9.16. The resistance offered by a rivet to be sheared off is known as ***shearing resistance*** or ***shearing strength*** or ***shearing value*** of the rivet.

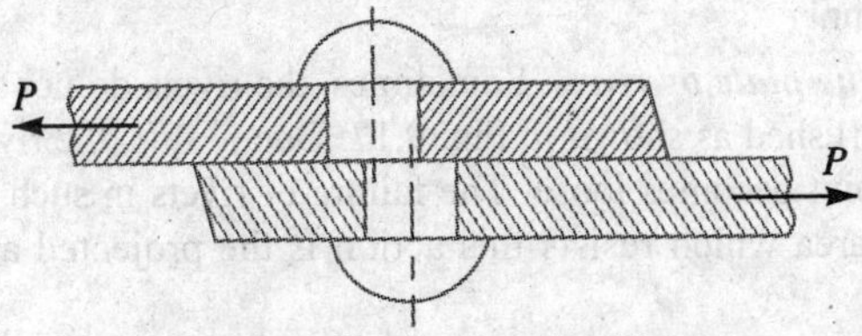

(*a*) Shearing off a rivet in a lap joint.

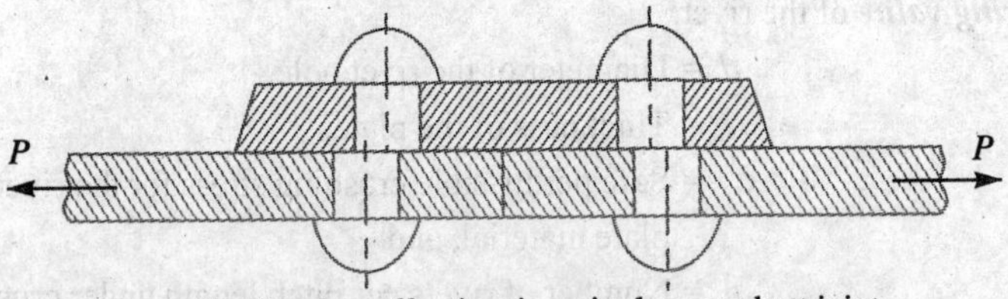

(*b*) Shearing off a rivet in a single cover butt joint.

Fig. 9.15. Shearing of rivets.

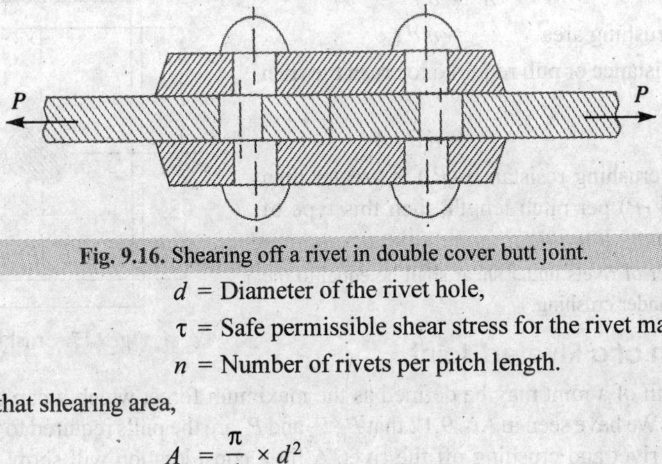

Fig. 9.16. Shearing off a rivet in double cover butt joint.

Let d = Diameter of the rivet hole,

τ = Safe permissible shear stress for the rivet material, and

n = Number of rivets per pitch length.

We know that shearing area,

$$A_s = \frac{\pi}{4} \times d^2 \quad \text{...(In single shear)}$$

$$= 2 \times \frac{\pi}{4} \times d^2 \quad \text{...(Theoretically, in double shear)}$$

$$= 1.875 \times \frac{\pi}{4} \times d^2 \quad \text{...(In double shear, according to Indian Boiler Regulations)}$$

∴ Shearing resistance or pull required to shear off the rivet per pitch length,

$$P_s = n \times \frac{\pi}{4} \times d^2 \times \tau \quad \text{...(In single shear)}$$

$$= n \times 2 \times \frac{\pi}{4} \times d^2 \times \tau \quad \text{...(Theoretically, in double shear)}$$

* We have already discussed in Chapter 4 (Art. 4.8) that when the shearing takes place at one cross-section of the rivet, then the rivets are said to be in ***single shear.*** Similarly, when the shearing takes place at two cross-sections of the rivet, then the rivets are said to be in ***double shear.***

$$= n \times 1.875 \times \frac{\pi}{4} \times d^2 \times \tau$$ ...(In double shear, according to Indian Boiler Regulations)

When the shearing resistance (P_s) is greater than the applied load (P) per pitch length, then this type of failure will occur.

4. ***Crushing of the plate or rivets.*** Sometimes, the rivets do not actually shear off under the tensile stress, but are crushed as shown in Fig. 9.17. Due to this, the rivet hole becomes of an oval shape and hence the joint becomes loose. The failure of rivets in such a manner is also known as ***bearing failure.*** The area which resists this action is the projected area of the hole or rivet on diametral plane.

The resistance offered by a rivet to be crushed is known as ***crushing resistance*** or ***crushing strength*** or ***bearing value*** of the rivet.

Let d = Diameter of the rivet hole,

t = Thickness of the plate,

σ_c = Safe permissible crushing stress for the rivet or plate material, and

n = Number of rivets per pitch length under crushing.

We know that crushing area per rivet (*i.e.* projected area per rivet),

$$A_c = d.t$$

$\therefore$ Total crushing area $= n.d.t$

and crushing resistance or pull required to crush the rivet per pitch length,

$$P_c = n.d.t.\sigma_c$$

When the crushing resistance (P_c) is greater than the applied load (P) per pitch length, then this type of failure will occur.

Note : The number of rivets under shear shall be equal to the number of rivets under crushing.

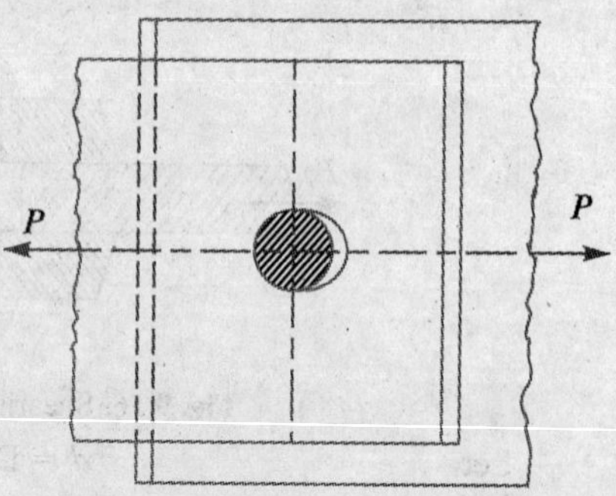

Fig. 9.17. Crushing of a rivet.

9.13 Strength of a Riveted Joint

The strength of a joint may be defined as the maximum force, which it can transmit, without causing it to fail. We have seen in Art. 9.12 that P_t, P_s and P_c are the pulls required to tear off the plate, shearing off the rivet and crushing off the rivet. A little consideration will show that if we go on increasing the pull on a riveted joint, it will fail when the least of these three pulls is reached, because a higher value of the other pulls will never reach since the joint has failed, either by tearing off the plate, shearing off the rivet or crushing off the rivet.

If the joint is ***continuous*** as in case of boilers, the strength is calculated ***per pitch length.*** But if the joint is ***small,*** the strength is calculated for the ***whole length*** of the plate.

9.14 Efficiency of a Riveted Joint

The efficiency of a riveted joint is defined as the ratio of the strength of riveted joint to the strength of the un-riveted or solid plate.

We have already discussed that strength of the riveted joint

= Least of P_t, P_s and P_c

Strength of the un-riveted or solid plate per pitch length,

$$P = p \times t \times \sigma_t$$

∴ Efficiency of the riveted joint,

$$\eta = \frac{\text{Least of } P_t, P_s \text{ and } P_c}{p \times t \times \sigma_t}$$

where p = Pitch of the rivets,

t = Thickness of the plate, and

σ_t = Permissible tensile stress of the plate material.

Example 9.1. *A double riveted lap joint is made between 15 mm thick plates. The rivet diameter and pitch are 25 mm and 75 mm respectively. If the ultimate stresses are 400 MPa in tension, 320 MPa in shear and 640 MPa in crushing, find the minimum force per pitch which will rupture the joint.*

If the above joint is subjected to a load such that the factor of safety is 4, find out the actual stresses developed in the plates and the rivets.

Solution. Given : t = 15 mm ; d = 25 mm ; p = 75 mm ; σ_{tu} = 400 MPa = 400 N/mm^2 ; τ_u = 320 MPa = 320 N/mm^2 ; σ_{cu} = 640 MPa = 640 N/mm^2

Minimum force per pitch which will rupture the joint

Since the ultimate stresses are given, therefore we shall find the ultimate values of the resistances of the joint. We know that ultimate tearing resistance of the plate per pitch,

$$P_{tu} = (p - d)t \times \sigma_{tu} = (75 - 25)15 \times 400 = 300\ 000 \text{ N}$$

Ultimate shearing resistance of the rivets per pitch,

$$P_{su} = n \times \frac{\pi}{4} \times d^2 \times \tau_u = 2 \times \frac{\pi}{4}(25)^2\, 320 = 314\ 200 \text{ N} \quad ...(\because n = 2)$$

and ultimate crushing resistance of the rivets per pitch,

$$P_{cu} = n \times d \times t \times \sigma_{cu} = 2 \times 25 \times 15 \times 640 = 480\ 000 \text{ N}$$

From above we see that the minimum force per pitch which will rupture the joint is 300 000 N or 300 kN. **Ans.**

Actual stresses produced in the plates and rivets

Since the factor of safety is 4, therefore safe load per pitch length of the joint

$$= 300\ 000/4 = 75\ 000 \text{ N}$$

Let σ_{ta}, τ_a and σ_{ca} be the actual tearing, shearing and crushing stresses produced with a safe load of 75 000 N in tearing, shearing and crushing.

We know that actual tearing resistance of the plates (P_{ta}),

$$75\ 000 = (p - d)\,t \times \sigma_{ta} = (75 - 25)15 \times \sigma_{ta} = 750\ \sigma_{ta}$$

$$\therefore \quad \sigma_{ta} = 75\ 000 / 750 = 100 \text{ N/mm}^2 = 100 \text{ MPa} \quad \textbf{Ans.}$$

Actual shearing resistance of the rivets (P_{sa}),

$$75\ 000 = n \times \frac{\pi}{4} \times d^2 \times \tau_a = 2 \times \frac{\pi}{4}(25)^2\,\tau_a = 982\ \tau_a$$

$$\therefore \quad \tau_a = 75000 / 982 = 76.4 \text{ N/mm}^2 = 76.4 \text{ MPa} \quad \textbf{Ans.}$$

and actual crushing resistance of the rivets (P_{ca}),

$$75\ 000 = n \times d \times t \times \sigma_{ca} = 2 \times 25 \times 15 \times \sigma_{ca} = 750\ \sigma_{ca}$$

$$\therefore \quad \sigma_{ca} = 75000 / 750 = 100 \text{ N/mm}^2 = 100 \text{ MPa} \quad \textbf{Ans.}$$

Example 9.2. *Find the efficiency of the following riveted joints :*

1. *Single riveted lap joint of 6 mm plates with 20 mm diameter rivets having a pitch of 50 mm.*
2. *Double riveted lap joint of 6 mm plates with 20 mm diameter rivets having a pitch of 65 mm.*

Assume

Permissible tensile stress in plate = 120 MPa
Permissible shearing stress in rivets = 90 MPa
Permissible crushing stress in rivets = 180 MPa

Solution. Given : $t = 6$ mm ; $d = 20$ mm ; $\sigma_t = 120$ MPa $= 120$ N/mm^2 ; $\tau = 90$ MPa $= 90$ N/mm^2 ; $\sigma_c = 180$ MPa $= 180$ N/mm^2

1. ***Efficiency of the first joint***

Pitch, $p = 50$ mm ...(Given)

First of all, let us find the tearing resistance of the plate, shearing and crushing resistances of the rivets.

(i) Tearing resistance of the plate

We know that the tearing resistance of the plate per pitch length,

$$P_t = (p - d)\, t \times \sigma_t = (50 - 20)\, 6 \times 120 = 21\,600 \text{ N}$$

(ii) Shearing resistance of the rivet

Since the joint is a single riveted lap joint, therefore the strength of one rivet in single shear is taken. We know that shearing resistance of one rivet,

$$P_s = \frac{\pi}{4} \times d^2 \times \tau = \frac{\pi}{4}\,(20)^2\, 90 = 28\,278 \text{ N}$$

(iii) Crushing resistance of the rivet

Since the joint is a single riveted, therefore strength of one rivet is taken. We know that crushing resistance of one rivet,

$$P_c = d \times t \times \sigma_c = 20 \times 6 \times 180 = 21\,600 \text{ N}$$

∴ Strength of the joint

$$= \text{Least of } P_t, P_s \text{ and } P_c = 21\,600 \text{ N}$$

We know that strength of the unriveted or solid plate,

$$P = p \times t \times \sigma_t = 50 \times 6 \times 120 = 36\,000 \text{ N}$$

∴ Efficiency of the joint,

$$\eta = \frac{\text{Least of } P_t, P_s \text{ and } P_c}{P} = \frac{21\,600}{36\,000} = 0.60 \text{ or } 60\% \quad \textbf{Ans.}$$

2. ***Efficiency of the second joint***

Pitch, $p = 65$ mm ...(Given)

(i) Tearing resistance of the plate,

We know that the tearing resistance of the plate per pitch length,

$$P_t = (p - d)\, t \times \sigma_t = (65 - 20)\, 6 \times 120 = 32\,400 \text{ N}$$

(ii) Shearing resistance of the rivets

Since the joint is double riveted lap joint, therefore strength of two rivets in single shear is taken. We know that shearing resistance of the rivets,

$$P_s = n \times \frac{\pi}{4} \times d^2 \times \tau = 2 \times \frac{\pi}{4}\,(20)^2\, 90 = 56\,556 \text{ N}$$

(iii) Crushing resistance of the rivet

Since the joint is double riveted, therefore strength of two rivets is taken. We know that crushing resistance of rivets,

$$P_c = n \times d \times t \times \sigma_c = 2 \times 20 \times 6 \times 180 = 43\,200 \text{ N}$$

∴ Strength of the joint

$$= \text{Least of } P_t, P_s \text{ and } P_c = 32\,400 \text{ N}$$

We know that the strength of the unriveted or solid plate,

$$P = p \times t \times \sigma_t = 65 \times 6 \times 120 = 46\,800 \text{ N}$$

∴ Efficiency of the joint,

$$\eta = \frac{\text{Least of } P_t,\, P_s \text{ and } P_c}{P} = \frac{32\,400}{46\,800} = 0.692 \text{ or } 69.2\% \quad \textbf{Ans.}$$

Example 9.3. *A double riveted double cover butt joint in plates 20 mm thick is made with 25 mm diameter rivets at 100 mm pitch. The permissible stresses are :*

$\sigma_t = 120$ *MPa;* $\tau = 100$ *MPa;* $\sigma_c = 150$ *MPa*

Find the efficiency of joint, taking the strength of the rivet in double shear as twice than that of single shear.

Solution. Given : $t = 20$ mm ; $d = 25$ mm ; $p = 100$ mm ; $\sigma_t = 120$ MPa = 120 N/mm^2 ; $\tau = 100$ MPa = 100 N/mm^2 ; $\sigma_c = 150$ MPa = 150 N/mm^2

First of all, let us find the tearing resistance of the plate, shearing resistance and crushing resistance of the rivet.

(i) ***Tearing resistance of the plate***

We know that tearing resistance of the plate per pitch length,

$$P_t = (p - d)\, t \times \sigma_t = (100 - 25)\, 20 \times 120 = 180\,000 \text{ N}$$

(ii) ***Shearing resistance of the rivets***

Since the joint is double riveted butt joint, therefore the strength of two rivets in double shear is taken. We know that shearing resistance of the rivets,

$$P_s = n \times 2 \times \frac{\pi}{4} \times d^2 \times \tau = 2 \times 2 \times \frac{\pi}{4} (25)^2\, 100 = 196\,375 \text{ N}$$

(iii) ***Crushing resistance of the rivets***

Since the joint is double riveted, therefore the strength of two rivets is taken. We know that crushing resistance of the rivets,

$$P_c = n \times d \times t \times \sigma_c = 2 \times 25 \times 20 \times 150 = 150\,000 \text{ N}$$

∴ Strength of the joint

= Least of P_t, P_s and P_c

= 150 000 N

Efficiency of the joint

We know that the strength of the unriveted or solid plate,

$$P = p \times t \times \sigma_t = 100 \times 20 \times 120$$

= 240 000 N

∴ Efficiency of the joint

$$= \frac{\text{Least of } P_t,\, P_s \text{ and } P_c}{P} = \frac{150\,000}{240\,000}$$

= 0.625 or 62.5% **Ans.**

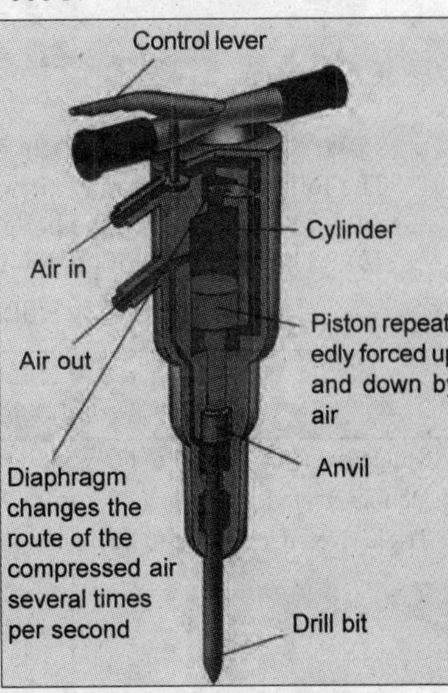

Preumatic drill uses compressed air.

9.15 Design of Boiler Joints

The boiler has a longitudinal joint as well as circumferential joint. The ***longitudinal joint*** is used to join the ends of the plate to get the required diameter of a boiler. For this purpose, a butt joint with two cover plates is used. The

circumferential joint is used to get the required length of the boiler. For this purpose, a lap joint with one ring overlapping the other alternately is used.

Since a boiler is made up of number of rings, therefore the longitudinal joints are staggered for convenience of connecting rings at places where both longitudinal and circumferential joints occur.

9.16 Assumptions in Designing Boiler Joints

The following assumptions are made while designing a joint for boilers :

1. The load on the joint is equally shared by all the rivets. The assumption implies that the shell and plate are rigid and that all the deformation of the joint takes place in the rivets themselves.
2. The tensile stress is equally distributed over the section of metal between the rivets.
3. The shearing stress in all the rivets is uniform.
4. The crushing stress is uniform.
5. There is no bending stress in the rivets.
6. The holes into which the rivets are driven do not weaken the member.
7. The rivet fills the hole after it is driven.
8. The friction between the surfaces of the plate is neglected.

9.17 Design of Longitudinal Butt Joint for a Boiler

According to Indian Boiler Regulations (I.B.R), the following procedure should be adopted for the design of longitudinal butt joint for a boiler.

1. ***Thickness of boiler shell.*** First of all, the thickness of the boiler shell is determined by using the thin cylindrical formula, *i.e.*

$$t = \frac{P.D}{2\,\sigma_t \times \eta_l} + 1 \text{ mm as corrosion allowance}$$

where
t = Thickness of the boiler shell,
P = Steam pressure in boiler,
D = Internal diameter of boiler shell,
σ_t = Permissible tensile stress, and
η_l = Efficiency of the longitudinal joint.

The following points may be noted :

(*a*) The thickness of the boiler shell should not be less than 7 mm.

(*b*) The efficiency of the joint may be taken from the following table.

Table 9.1. Efficiencies of commercial boiler joints.

Lap joints	*Efficiency (%)*	**Maximum efficiency*	*Butt joints (Double strap)*	*Efficiency (%)*	**Maximum efficiency*
Single riveted	45 to 60	63.3	Single riveted	55 to 60	63.3
Double riveted	63 to 70	77.5	Double riveted	70 to 83	86.6
Triple riveted	72 to 80	86.6	Triple riveted (5 rivets per pitch with unequal width of straps)	80 to 90	95.0
			Quadruple riveted	85 to 94	98.1

* The maximum efficiencies are valid for ideal equistrength joints with tensile stress = 77 MPa, shear stress = 62 MPa and crushing stress = 133 MPa.

Indian Boiler Regulations (I.B.R.) allow a maximum efficiency of 85% for the best joint.

(*c*) According to I.B.R., the factor of safety should not be less than 4. The following table shows the values of factor of safety for various kind of joints in boilers.

Table 9.2. Factor of safety for boiler joints.

Type of joint	*Factor of safety*	
	Hand riveting	*Machine riveting*
Lap joint	4.75	4.5
Single strap butt joint	4.75	4.5
Single riveted butt joint with two equal cover straps	4.75	4.5
Double riveted butt joint with two equal cover straps	4.25	4.0

2. ***Diameter of rivets.*** After finding out the thickness of the boiler shell (t), the diameter of the rivet hole (d) may be determined by using Unwin's empirical formula, *i.e.*

$$d = 6\sqrt{t} \qquad \text{(when } t \text{ is greater than 8 mm)}$$

But if the thickness of plate is less than 8 mm, then the diameter of the rivet hole may be calculated by equating the shearing resistance of the rivets to crushing resistance. In no case, the diameter of rivet hole should not be less than the thickness of the plate, because there will be danger of punch crushing. The following table gives the rivet diameter corresponding to the diameter of rivet hole as per IS : 1928 – 1961 (Reaffirmed 1996).

Table 9.3. Size of rivet diameters for rivet hole diameter as per IS : 1928 - 1961 (Reaffirmed 1996).

Basic size of rivet mm	12	14	16	18	20	22	24	27	30	33	36	39	42	48
Rivet hole diameter (min) mm	13	15	17	19	21	23	25	28.5	31.5	34.5	37.5	41	44	50

According to IS : 1928 – 1961 (Reaffirmed 1996), the table on the next page (Table 9.4) gives the preferred length and diameter combination for rivets.

3. ***Pitch of rivets.*** The pitch of the rivets is obtained by equating the tearing resistance of the plate to the shearing resistance of the rivets. It may noted that

(*a*) The pitch of the rivets should not be less than $2d$, which is necessary for the formation of head.

(*b*) The maximum value of the pitch of rivets for a longitudinal joint of a boiler as per I.B.R. is

$$p_{max} = C \times t + 41.28 \text{ mm}$$

where t = Thickness of the shell plate in mm, and

C = Constant.

The value of the constant C is given in Table 9.5.

Table 9.4. Preferred length and diameter combinations for rivets used in boilers as per IS : 1928-1961 (Reaffirmed 1996).

(All dimensions in mm)

	Diameter													
Length	12	14	16	18	20	22	24	27	30	33	36	39	42	48
28	×	–	–	–	–	–	–	–	–	–	–	–	–	–
31.5	×	×	–	–	–	–	–	–	–	–	–	–	–	–
35.5	×	×	×	–	–	–	–	–	–	–	–	–	–	–
40	×	×	×	×	–	–	–	–	–	–	–	–	–	–
45	×	×	×	×	×	–	–	–	–	–	–	–	–	–
50	×	×	×	×	×	×	–	–	–	–	–	–	–	–
56	×	×	×	×	×	×	×	–	–	–	–	–	–	–
63	×	×	×	×	×	×	×	×	–	–	–	–	–	–
71	×	×	×	×	×	×	×	×	×	–	–	–	–	–
80	×	×	×	×	×	×	×	×	×	–	–	–	–	–
85	–	×	×	×	×	×	×	×	×	×	–	–	–	–
90	–	×	×	×	×	×	×	×	×	×	–	–	–	–
95	–	×	×	×	×	×	×	×	×	×	×	–	–	–
100	–	–	×	×	×	×	×	×	×	×	×	–	–	–
106	–	–	×	×	×	×	×	×	×	×	×	×	–	–
112	–	–	×	×	×	×	×	×	×	×	×	×	–	–
118	–	–	–	×	×	×	×	×	×	×	×	×	×	–
125	–	–	–	–	×	×	×	×	×	×	×	×	×	×
132	–	–	–	–	–	×	×	×	×	×	×	×	×	×
140	–	–	–	–	–	×	×	×	×	×	×	×	×	×
150	–	–	–	–	–	–	×	×	×	×	×	×	×	×
160	–	–	–	–	–	–	×	×	×	×	×	×	×	×
180	–	–	–	–	–	–	–	×	×	×	×	×	×	×
200	–	–	–	–	–	–	–	–	×	×	×	×	×	×
224	–	–	–	–	–	–	–	–	–	×	×	×	×	×
250	–	–	–	–	–	–	–	–	–	–	–	–	×	×

Preferred numbers are indicated by ×.

Table 9.5. Values of constant *C*.

Number of rivets per pitch length	*Lap joint*	*Butt joint (single strap)*	*Butt joint (double strap)*
1	1.31	1.53	1.75
2	2.62	3.06	3.50
3	3.47	4.05	4.63
4	4.17	–	5.52
5	–	–	6.00

Note : If the pitch of rivets as obtained by equating the tearing resistance to the shearing resistance is more than p_{max}, then the value of p_{max} is taken.

4. ***Distance between the rows of rivets.*** The distance between the rows of rivets as specified by Indian Boiler Regulations is as follows :

(a) For equal number of rivets in more than one row for lap joint or butt joint, the distance between the rows of rivets (p_b) should not be less than

$0.33\,p + 0.67\,d$, for zig-zig riveting, and

$2\,d$, for chain riveting.

(b) For joints in which the number of rivets in outer rows is ***half*** the number of rivets in inner rows and if the inner rows are chain riveted, the distance between the outer rows and the next rows should not be less than

$0.33\,p + 0.67$ or $2\,d$, whichever is greater.

The distance between the rows in which there are full number of rivets shall not be less than $2d$.

(c) For joints in which the number of rivets in outer rows is ***half*** the number of rivets in inner rows and if the inner rows are zig-zig riveted, the distance between the outer rows and the next rows shall not be less than $0.2\,p + 1.15\,d$. The distance between the rows in which there are full number of rivets (zig-zag) shall not be less than $0.165\,p + 0.67\,d$.

Note : In the above discussion, p is the pitch of the rivets in the outer rows.

5. ***Thickness of butt strap.*** According to I.B.R., the thicknesses for butt strap (t_1) are as given below :

(a) The thickness of butt strap, in no case, shall be less than 10 mm.

(b) $t_1 = 1.125\,t$, for ordinary (chain riveting) single butt strap.

$t_1 = 1.125\,t\left(\dfrac{p-d}{p-2d}\right)$, for single butt straps, every alternate rivet in outer rows being omitted.

$t_1 = 0.625\,t$, for double butt-straps of equal width having ordinary riveting (chain riveting).

$t_1 = 0.625\,t\left(\dfrac{p-d}{p-2d}\right)$, for double butt straps of equal width having every alternate rivet in the outer rows being omitted.

(c) For unequal width of butt straps, the thicknesses of butt strap are

$t_1 = 0.75\,t$, for wide strap on the inside, and

$t_2 = 0.625\,t$, for narrow strap on the outside.

6. ***Margin.*** The margin (m) is taken as $1.5\,d$.

Note : The above procedure may also be applied to ordinary riveted joints.

9.18 Design of Circumferential Lap Joint for a Boiler

The following procedure is adopted for the design of circumferential lap joint for a boiler.

1. ***Thickness of the shell and diameter of rivets.*** The thickness of the boiler shell and the diameter of the rivet will be same as for longitudinal joint.

2. ***Number of rivets.*** Since it is a lap joint, therefore the rivets will be in single shear.

∴ Shearing resistance of the rivets,

$$P_s = n \times \frac{\pi}{4} \times d^2 \times \tau \qquad ...(i)$$

where n = Total number of rivets.

Knowing the inner diameter of the boiler shell (D), and the pressure of steam (P), the total shearing load acting on the circumferential joint,

$$W_s = \frac{\pi}{4} \times D^2 \times P \qquad ...(ii)$$

From equations (i) and (ii), we get

$$n \times \frac{\pi}{4} \times d^2 \times \tau = \frac{\pi}{4} \times D^2 \times P$$

$$\therefore \quad n = \left(\frac{D}{d}\right)^2 \frac{P}{\tau}$$

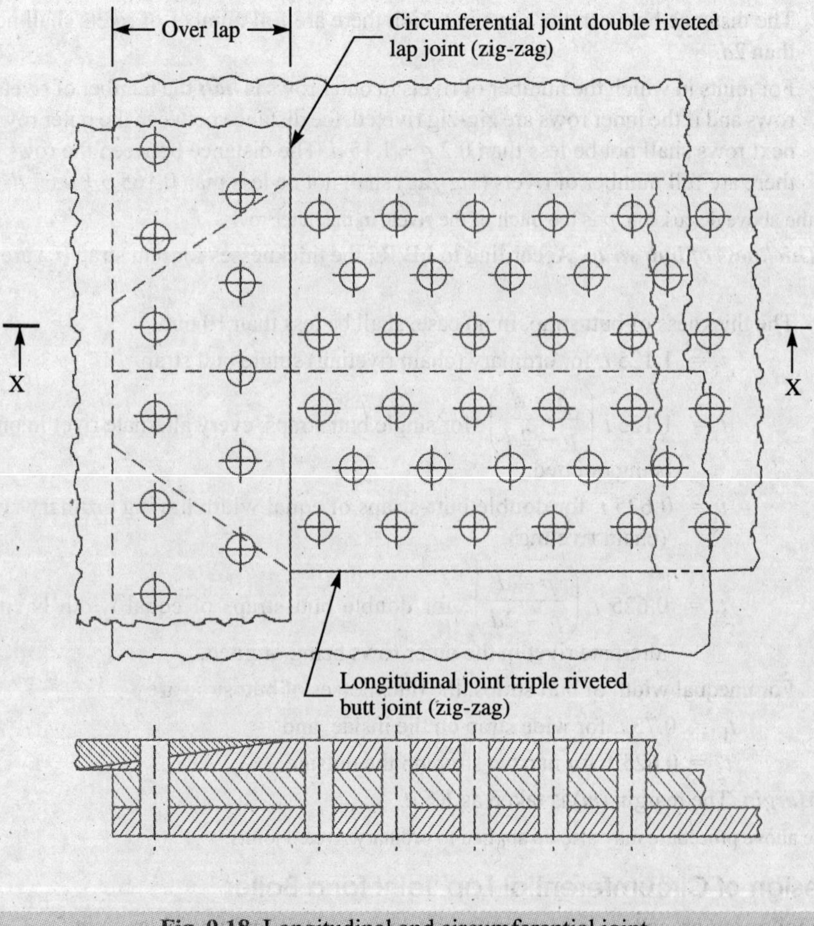

Fig. 9.18. Longitudinal and circumferential joint.

3. *Pitch of rivets.* If the efficiency of the longitudinal joint is known, then the efficiency of the circumferential joint may be obtained. It is generally taken as 50% of tearing efficiency in longitudinal joint, but if more than one circumferential joints is used, then it is 62% for the intermediate joints. Knowing the efficiency of the circumferential lap joint (η_c), the pitch of the rivets for the lap joint

(p_1) may be obtained by using the relation :

$$\eta_c = \frac{p_1 - d}{p_1}$$

4. *Number of rows.* The number of rows of rivets for the circumferential joint may be obtained from the following relation :

$$\text{Number of rows} = \frac{\text{Total number of rivets}}{\text{Number of rivets in one row}}$$

and the number of rivets in one row

$$= \frac{\pi (D + t)}{p_1}$$

where D = Inner diameter of shell.

5. After finding out the number of rows, the type of the joint (*i.e.* single riveted or double riveted etc.) may be decided. Then the number of rivets in a row and pitch may be re-adjusted. In order to have a leak-proof joint, the pitch for the joint should be checked from Indian Boiler Regulations.

6. The distance between the rows of rivets (*i.e.* back pitch) is calculated by using the relations as discussed in the previous article.

7. After knowing the distance between the rows of rivets (p_b), the overlap of the plate may be fixed by using the relation,

$$\text{Overlap} = (\text{No. of rows of rivets} - 1)\, p_b + m$$

where m = Margin.

There are several ways of joining the longitudinal joint and the circumferential joint. One of the methods of joining the longitudinal and circumferential joint is shown in Fig. 9.18.

9.19 Recommended Joints for Pressure Vessels

The following table shows the recommended joints for pressure vessels.

Table 9.6. Recommended joints for pressure vessels.

Diameter of shell (metres)	*Thickness of shell (mm)*	*Type of joint*
0.6 to 1.8	6 to 13	Double riveted
0.9 to 2.1	13 to 25	Triple riveted
1.5 to 2.7	19 to 40	Quadruple riveted

Example 9.4. *A double riveted lap joint with zig-zag riveting is to be designed for 13 mm thick plates. Assume*

σ_t = 80 MPa ; τ = 60 MPa ; and σ_c = 120 MPa

State how the joint will fail and find the efficiency of the joint.

Solution. Given : t = 13 mm ; σ_t = 80 MPa = 80 N/mm^2 ; τ = 60 MPa = 60 N/mm^2 ; σ_c = 120 MPa = 120 N/mm^2

1. *Diameter of rivet*

Since the thickness of plate is greater than 8 mm, therefore diameter of rivet hole,

$$d = 6\sqrt{t} = 6\sqrt{13} = 21.6 \text{ mm}$$

From Table 9.3, we find that according to IS : 1928 – 1961 (Reaffirmed 1996), the standard size of the rivet hole (d) is 23 mm and the corresponding diameter of the rivet is 22 mm. **Ans.**

2. *Pitch of rivets*

Let p = Pitch of the rivets.

Since the joint is a double riveted lap joint with zig-zag riveting [See Fig. 9.6 (*c*)], therefore there are two rivets per pitch length, *i.e.* $n = 2$. Also, in a lap joint, the rivets are in single shear.

We know that tearing resistance of the plate,

$$P_t = (p - d)t \times \sigma_t = (p - 23)\ 13 \times 80 = (p - 23)\ 1040 \text{ N} \quad ...(i)$$

and shearing resistance of the rivets,

$$P_s = n \times \frac{\pi}{4} \times d^2 \times \tau = 2 \times \frac{\pi}{4}\ (23)^2\ 60 = 49\ 864 \text{ N} \quad ...(ii)$$

...($\because$ There are two rivets in single shear)

From equations (*i*) and (*ii*), we get

$$p - 23 = 49864 / 1040 = 48 \quad \text{or} \quad p = 48 + 23 = 71 \text{ mm}$$

The maximum pitch is given by,

$$p_{max} = C \times t + 41.28 \text{ mm}$$

From Table 9.5, we find that for 2 rivets per pitch length, the value of C is 2.62.

$$\therefore \quad p_{max} = 2.62 \times 13 + 41.28 = 75.28 \text{ mm}$$

Since p_{max} is more than p, therefore we shall adopt

$$p = 71 \text{ mm} \quad \textbf{Ans.}$$

3. *Distance between the rows of rivets*

We know that the distance between the rows of rivets (for zig-zag riveting),

$$p_b = 0.33\ p + 0.67\ d = 0.33 \times 71 + 0.67 \times 23 \text{ mm}$$
$$= 38.8 \text{ say } 40 \text{ mm} \quad \textbf{Ans.}$$

4. *Margin*

We know that the margin,

$$m = 1.5\ d = 1.5 \times 23 = 34.5 \text{ say } 35 \text{ mm} \quad \textbf{Ans.}$$

Failure of the joint

Now let us find the tearing resistance of the plate, shearing resistance and crushing resistance of the rivets.

We know that tearing resistance of the plate,

$$P_t = (p - d)\ t \times \sigma_t = (71 - 23) 13 \times 80 = 49\ 920 \text{ N}$$

Shearing resistance of the rivets,

$$P_s = n \times \frac{\pi}{4} \times d^2 \times \tau = 2 \times \frac{\pi}{4}\ (23)^2\ 60 = 49\ 864 \text{ N}$$

and crushing resistance of the rivets,

$$P_c = n \times d \times t \times \sigma_c = 2 \times 23 \times 13 \times 120 = 71\ 760 \text{ N}$$

The least of P_t, P_s and P_c is P_s = 49 864 N. Hence the joint will fail due to shearing of the rivets. **Ans.**

Efficiency of the joint

We know that strength of the unriveted or solid plate,

$$P = p \times t \times \sigma_t = 71 \times 13 \times 80 = 73\ 840 \text{ N}$$

$\therefore$ Efficiency of the joint,

$$\eta = \frac{P_s}{P} = \frac{49\ 864}{73\ 840} = 0.675 \text{ or } 67.5\% \quad \textbf{Ans.}$$

Example 9.5. *Two plates of 7 mm thick are connected by a triple riveted lap joint of zig-zag pattern. Calculate the rivet diameter, rivet pitch and distance between rows of rivets for the joint. Also state the mode of failure of the joint. The safe working stresses are as follows :*

σ_t *= 90 MPa ;* τ *= 60 MPa ; and* σ_c *= 120 MPa.*

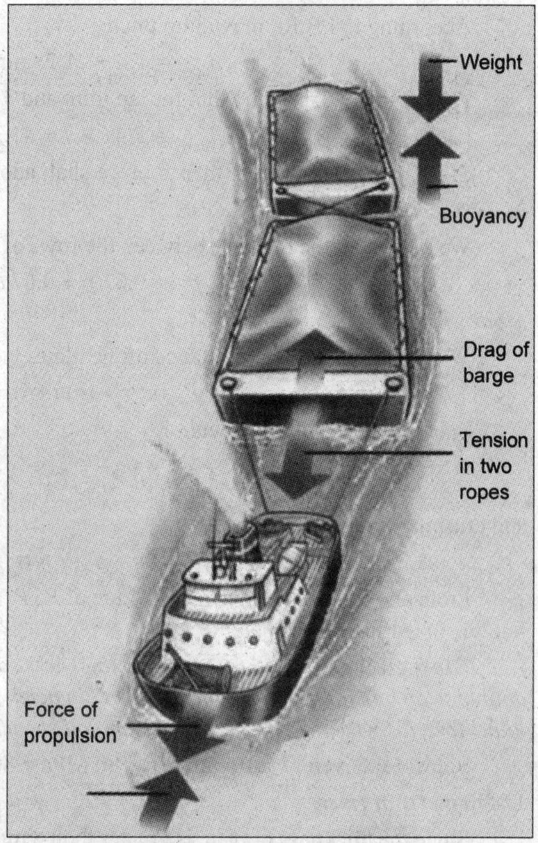

Forces on a ship as shown above need to be consider while designing various joints

Note : This picture is given as additional information and is not a direct example of the current chapter.

Solution. Given : $t = 7$ mm ; $\sigma_t = 90$ MPa = 90 N/mm^2 ; $\tau = 60$ MPa = 60 N/mm^2 ; $\sigma_c = 120$ MPa = 120 N/mm^2

1. ***Diameter of rivet***

Since the thickness of plate is less than 8 mm, therefore diameter of the rivet hole (d) is obtained by equating the shearing resistance (P_s) to the crushing resistance (P_c) of the rivets. The triple riveted lap joint of zig-zag pattern is shown in Fig. 9.7 (*b*). We see that there are three rivets per pitch length (*i.e.* $n = 3$). Also, the rivets in lap joint are in single shear.

We know that shearing resistance of the rivets,

$$P_s = n \times \frac{\pi}{4} \times d^2 \times \tau$$

$$= 3 \times \frac{\pi}{4} \times d^2 \times 60 = 141.4\, d^2 \text{ N} \quad ...(i)$$

$$...(\because n = 3)$$

and crushing resistance of the rivets,

$$P_c = n \times d \times t \times \sigma_c = 3 \times d \times 7 \times 120 = 2520\, d \text{ N} \quad ...(ii)$$

From equations (*i*) and (*ii*), we get

$$141.4\, d^2 = 2520\, d \quad \text{or} \quad d = 2520 / 141.4 = 17.8 \text{ mm}$$

From Table 9.3, we see that according to IS : 1928 – 1961 (Reaffirmed 1996), the standard diameter of rivet hole (d) is 19 mm and the corresponding diameter of rivet is 18 mm. **Ans.**

2. ***Pitch of rivets***

Let p = Pitch of rivets.

We know that tearing resistance of the plate,

$$P_t = (p - d)\, t \times \sigma_t = (p - 19)\, 7 \times 90 = 630\, (p - 19) \text{ N} \quad ...(iii)$$

and shearing resistance of the rivets,

$$P_s = 141.4\, d^2 = 141.4\, (19)^2 = 51\,045 \text{ N} \quad ...[\text{From equation } (i)] \quad ...(iv)$$

Equating equations (*iii*) and (*iv*), we get

$$630\, (p - 19) = 51\,045$$

$$p - 19 = 51\,045 / 630 = 81 \text{ or } p = 81 + 19 = 100 \text{ mm}$$

According to I.B.R., maximum pitch,

$$p_{max} = C.t + 41.28 \text{ mm}$$

From Table 9.5, we find that for lap joint and 3 rivets per pitch length, the value of C is 3.47.

$$\therefore \quad p_{max} = 3.47 \times 7 + 41.28 = 65.57 \text{ say } 66 \text{ mm}$$

Since p_{max} is less than p, therefore we shall adopt $p = p_{max} = 66$ mm **Ans.**

3. *Distance between rows of rivets*

We know that the distance between the rows of rivets for zig-zag riveting,

$$p_b = 0.33\,p + 0.67\,d = 0.33 \times 66 + 0.67 \times 19 = 34.5 \text{ mm} \quad \textbf{Ans.}$$

Mode of failure of the joint

We know that tearing resistance of the plate,

$$P_t = (p - d)\,t \times \sigma_t = (66 - 19)\,7 \times 90 = 29\,610 \text{ N}$$

Shearing resistance of rivets,

$$P_s = n \times \frac{\pi}{4} \times d^2 \times \tau = 3 \times \frac{\pi}{4}\,(19)^2\,60 = 51\,045 \text{ N}$$

and crushing resistance of rivets.

$$P_c = n \times d \times t \times \sigma_c = 3 \times 19 \times 7 \times 120 = 47\,880 \text{ N}$$

From above we see that the least value of P_t, P_s and P_c is $P_t = 29\,610$ N. Therefore the joint will fail due to tearing off the plate.

Example 9.6. *Two plates of 10 mm thickness each are to be joined by means of a single riveted double strap butt joint. Determine the rivet diameter, rivet pitch, strap thickness and efficiency of the joint. Take the working stresses in tension and shearing as 80 MPa and 60 MPa respectively.*

Solution. Given : $t = 10$ mm ; $\sigma_t = 80$ MPa $= 80$ N/mm^2 ; $\tau = 60$ MPa $= 60$ N/mm^2

1. *Diameter of rivet*

Since the thickness of plate is greater than 8 mm, therefore diameter of rivet hole,

$$d = 6\sqrt{t} = 6\sqrt{10} = 18.97 \text{ mm}$$

From Table 9.3, we see that according to IS : 1928 – 1961 (Reaffirmed 1996), the standard diameter of rivet hole (d) is 19 mm and the corresponding diameter of the rivet is 18 mm. **Ans.**

2. *Pitch of rivets*

Let p = Pitch of rivets.

Since the joint is a single riveted double strap butt joint as shown in Fig. 9.8, therefore there is one rivet per pitch length (*i.e.* $n = 1$) and the rivets are in double shear.

We know that tearing resistance of the plate,

$$P_t = (p - d)\,t \times \sigma_t = (p - 19)\,10 \times 80 = 800\,(p - 19) \text{ N} \quad ...(i)$$

and shearing resistance of the rivets,

$$P_s = n \times 1.875 \times \frac{\pi}{4} \times d^2 \times \tau \quad ...(\because \text{Rivets are in double shear})$$

$$= 1 \times 1.875 \times \frac{\pi}{4}\,(19)^2\,60 = 31\,900 \text{ N} \quad ...(\because n = 1) \quad ...(ii)$$

From equations (*i*) and (*ii*), we get

$$800\,(p - 19) = 31\,900$$

$$\therefore \quad p - 19 = 31\,900 / 800 = 39.87 \text{ or } p = 39.87 + 19 = 58.87 \text{ say } 60 \text{ mm}$$

According to I.B.R., the maximum pitch of rivets,

$$p_{max} = C.t + 41.28 \text{ mm}$$

From Table 9.5, we find that for double strap butt joint and 1 rivet per pitch length, the value of C is 1.75.

$\therefore \quad p_{max} = 1.75 \times 10 + 41.28 = 58.78$ say 60 mm

From above we see that $p = p_{max} = 60$ mm **Ans.**

3. ***Thickness of cover plates***

We know that thickness of cover plates,

$$t_1 = 0.625\, t = 0.625 \times 10 = 6.25 \text{ mm} \quad \textbf{Ans.}$$

Efficiency of the joint

We know that tearing resistance of the plate,

$$P_t = (p - d)\, t \times \sigma_t = (60 - 19)\, 10 \times 80 = 32\,800 \text{ N}$$

and shearing resistance of the rivets,

$$P_s = n \times 1.875 \times \frac{\pi}{4} \times d^2 \times \tau = 1 \times 1.875 \times \frac{\pi}{4}\, (19)^2\, 60 = 31\,900 \text{ N}$$

$\therefore$ Strength of the joint

$$= \text{Least of } P_t \text{ and } P_s = 31\,900 \text{ N}$$

Strength of the unriveted plate per pitch length

$$P = p \times t \times \sigma_t = 60 \times 10 \times 80 = 48\,000 \text{ N}$$

$\therefore$ Efficiency of the joint,

$$\eta = \frac{\text{Least of } P_t \text{ and } P_s}{P} = \frac{31\,900}{48\,000} = 0.665 \text{ or } 66.5\% \quad \textbf{Ans.}$$

Example 9.7. *Design a double riveted butt joint with two cover plates for the longitudinal seam of a boiler shell 1.5 m in diameter subjected to a steam pressure of 0.95 N/mm². Assume joint efficiency as 75%, allowable tensile stress in the plate 90 MPa ; compressive stress 140 MPa ; and shear stress in the rivet 56 MPa.*

Solution. Given : $D = 1.5$ m $= 1500$ mm ; $P = 0.95$ N/mm² ; $\eta_l = 75\% = 0.75$; $\sigma_t = 90$ MPa $= 90$ N/mm² ; $\sigma_c = 140$ MPa $= 140$ N/mm² ; $\tau = 56$ MPa $= 56$ N/mm²

1. ***Thickness of boiler shell plate***

We know that thickness of boiler shell plate,

$$t = \frac{P.D}{2\sigma_t \times \eta_l} + 1 \text{ mm} = \frac{0.95 \times 1500}{2 \times 90 \times 0.75} + 1 = 11.6 \text{ say } 12 \text{ mm} \quad \textbf{Ans.}$$

2. ***Diameter of rivet***

Since the thickness of the plate is greater than 8 mm, therefore the diameter of the rivet hole,

$$d = 6\sqrt{t} = 6\sqrt{12} = 20.8 \text{ mm}$$

From Table 9.3, we see that according to IS : 1928 – 1961 (Reaffirmed 1996), the standard diameter of the rivet hole (d) is 21 mm and the corresponding diameter of the rivet is 20 mm. **Ans.**

3. ***Pitch of rivets***

Let $\quad p$ = Pitch of rivets.

The pitch of the rivets is obtained by equating the tearing resistance of the plate to the shearing resistance of the rivets.

We know that tearing resistance of the plate,

$$P_t = (p - d)\, t \times \sigma_t = (p - 21)12 \times 90 = 1080\, (p - 21)\text{N} \quad ...(i)$$

Since the joint is double riveted double strap butt joint, as shown in Fig. 9.9, therefore there are two rivets per pitch length (*i.e.* $n = 2$) and the rivets are in double shear. Assuming that the rivets in double shear are 1.875 times stronger than in single shear, we have

Shearing strength of the rivets,

$$P_s = n \times 1.875 \times \frac{\pi}{4} \times d^2 \times \tau = 2 \times 1.875 \times \frac{\pi}{4} (21)^2 \times 56 \text{ N}$$

$$= 72\ 745 \text{ N} \quad \ldots(ii)$$

From equations (*i*) and (*ii*), we get

$$1080\,(p - 21) = 72\ 745$$

$$\therefore \quad p - 21 = 72\ 745 / 1080 = 67.35 \text{ or } p = 67.35 + 21 = 88.35 \text{ say } 90 \text{ mm}$$

According to I.B.R., the maximum pitch of rivets for longitudinal joint of a boiler is given by

$$p_{max} = C \times t + 41.28 \text{ mm}$$

From Table 9.5, we find that for a double riveted double strap butt joint and two rivets per pitch length, the value of C is 3.50.

$$\therefore \quad p_{max} = 3.5 \times 12 + 41.28 = 83.28 \text{ say } 84 \text{ mm}$$

Since the value of p is more than p_{max}, therefore we shall adopt pitch of the rivets,

$$p = p_{max} = 84 \text{ mm} \quad \textbf{Ans.}$$

4. ***Distance between rows of rivets***

Assuming zig-zag riveting, the distance between the rows of the rivets (according to I.B.R.),

$$p_b = 0.33\,p + 0.67\,d = 0.33 \times 84 + 0.67 \times 21 = 41.8 \text{ say } 42 \text{ mm} \quad \textbf{Ans.}$$

5. ***Thickness of cover plates***

According to I.B.R., the thickness of each cover plate of equal width is

$$t_1 = 0.625\,t = 0.625 \times 12 = 7.5 \text{ mm} \quad \textbf{Ans.}$$

6. ***Margin***

We know that the margin,

$$m = 1.5\,d = 1.5 \times 21 = 31.5 \text{ say } 32 \text{ mm} \quad \textbf{Ans.}$$

Let us now find the efficiency for the designed joint.

Tearing resistance of the plate,

$$P_t = (p - d)\,t \times \sigma_t = (84 - 21)12 \times 90 = 68\ 040 \text{ N}$$

Shearing resistance of the rivets,

$$P_s = n \times 1.875 \times \frac{\pi}{4} \times d^2 \times \tau = 2 \times 1.875 \times \frac{\pi}{4} (21)^2 \times 56 = 72\ 745 \text{ N}$$

and crushing resistance of the rivets,

$$P_c = n \times d \times t \times \sigma_c = 2 \times 21 \times 12 \times 140 = 70\ 560 \text{ N}$$

Since the strength of riveted joint is the least value of P_t, P_s or P_c, therefore strength of the riveted joint,

$$P_t = 68\ 040 \text{ N}$$

We know that strength of the un-riveted plate,

$$P = p \times t \times \sigma_t = 84 \times 12 \times 90 = 90\ 720 \text{ N}$$

∴ Efficiency of the designed joint,

$$\eta = \frac{P_t}{P} = \frac{68\ 040}{90\ 720} = 0.75 \text{ or } 75\% \quad \textbf{Ans.}$$

Since the efficiency of the designed joint is equal to the given efficiency of 75%, therefore the design is satisfactory.

Example 9.8. *A pressure vessel has an internal diameter of 1 m and is to be subjected to an internal pressure of 2.75 N/mm² above the atmospheric pressure. Considering it as a thin cylinder and assuming efficiency of its riveted joint to be 79%, calculate the plate thickness if the tensile stress in the material is not to exceed 88 MPa.*

Design a longitudinal double riveted double strap butt joint with equal straps for this vessel. The pitch of the rivets in the outer row is to be double the pitch in the inner row and zig-zag riveting is proposed. The maximum allowable shear stress in the rivets is 64 MPa. You may assume that the rivets in double shear are 1.8 times stronger than in single shear and the joint does not fail by crushing.

Make a sketch of the joint showing all calculated values. Calculate the efficiency of the joint.

Pressure vessel.

Solution. Given : D = 1 m = 1000 mm ; P = 2.75 N/mm² ; η_l = 79% = 0.79 ; σ_t = 88 MPa = 88 N/mm²; τ = 64 MPa = 64 N/mm²

1. ***Thickness of plate***

We know that the thickness of plate,

$$t = \frac{P.D}{2\,\sigma_t \times \eta_l} + 1\text{ mm} = \frac{2.75 \times 1000}{2 \times 88 \times 0.79} + 1\text{ mm}$$

$$= 20.8 \text{ say } 21 \text{ mm} \quad \textbf{Ans.}$$

2. ***Diameter of rivet***

Since the thickness of plate is more than 8 mm, therefore diameter of rivet hole,

$$d = 6\sqrt{t} = 6\sqrt{21} = 27.5 \text{ mm}$$

From Table 9.3, we see that according to IS : 1928 – 1961 (Reaffirmed 1996), the standard diameter of the rivet hole (d) is 28.5 mm and the corresponding diameter of the rivet is 27 mm. **Ans.**

3. ***Pitch of rivets***

Let p = Pitch in the outer row.

The pitch of the rivets is obtained by equating the tearing resistance of the plate to the shearing resistance of the rivets.

We know that the tearing resistance of the plate per pitch length,

$$P_t = (p - d)\,t \times \sigma_t = (p - 28.5)\,21 \times 88 = 1848\,(p - 28.5)\text{ N} \quad ...(i)$$

Since the pitch in the outer row is twice the pitch of the inner row and the joint is double riveted, therefore for one pitch length there will be three rivets in double shear (*i.e.* n = 3). It is given that the strength of rivets in double shear is 1.8 times that of single shear, therefore

Shearing strength of the rivets per pitch length,

$$P_s = n \times 1.8 \times \frac{\pi}{4} \times d^2 \times \tau = 3 \times 1.8 \times \frac{\pi}{4}\,(28.5)^2\,64\text{ N}$$

$$= 220\,500\text{ N} \quad ...(ii)$$

From equations (*i*) and (*ii*), we get

$$1848\,(p - 28.5) = 220\,500$$

$$\therefore \quad p - 28.5 = 220\,500 / 1848 = 119.3$$

or $$p = 119.3 + 28.5 = 147.8 \text{ mm}$$

According to I.B.R., the maximum pitch,

$$p_{max} = C \times t + 41.28 \text{ mm}$$

From Table 9.5, we find that for 3 rivets per pitch length and for double strap butt joint, the value of C is 4.63.

$$\therefore \quad p_{max} = 4.63 \times 21 + 41.28 = 138.5 \text{ say } 140 \text{ mm}$$

Since the value of p_{max} is less than p, therefore we shall adopt the value of

$$p = p_{max} = 140 \text{ mm} \quad \textbf{Ans.}$$

∴ Pitch in the inner row

$$= 140 / 2 = 70 \text{ mm} \quad \textbf{Ans.}$$

4. ***Distance between the rows of rivets***

According to I.B.R., the distance between the rows of rivets,

$$p_b = 0.2\,p + 1.15\,d = 0.2 \times 140 + 1.15 \times 28.5 = 61 \text{ mm} \quad \textbf{Ans.}$$

5. ***Thickness of butt strap***

According to I.B.R., the thickness of double butt straps of equal width,

$$t_1 = 0.625\,t\left(\frac{p-d}{p-2d}\right) = 0.625 \times 21\left(\frac{140-28.5}{140-2\times 28.5}\right) \text{ mm}$$

$$= 17.6 \text{ say } 18 \text{ mm} \quad \textbf{Ans.}$$

6. ***Margin***

We know that the margin,

$$m = 1.5\,d = 1.5 \times 28.5 = 43 \text{ mm} \quad \textbf{Ans.}$$

Efficiency of the joint

We know that tearing resistance of the plate,

$$P_t = (p-d)\,t \times \sigma_t = (140 - 28.5)\,21 \times 88 = 206\,050 \text{ N}$$

Shearing resistance of the rivets,

$$P_s = n \times 1.8 \times \frac{\pi}{4} \times d^2 \times \tau = 3 \times 1.8 \times \frac{\pi}{4}\,(28.5)^2\,64 = 220\,500 \text{ N}$$

Strength of the solid plate,

$$= p \times t \times \sigma_t = 140 \times 21 \times 88 = 258\,720 \text{ N}$$

∴ Efficiency of the joint

$$= \frac{\text{Least of } P_t \text{ and } P_s}{\text{Strength of solid plate}} = \frac{206\,050}{258\,720} = 0.796 \text{ or } 79.6\% \quad \textbf{Ans.}$$

Since the efficiency of the designed joint is more than the given efficiency, therefore the design is satisfactory.

Example 9.9. *Design the longitudinal joint for a 1.25 m diameter steam boiler to carry a steam pressure of 2.5 N/mm². The ultimate strength of the boiler plate may be assumed as 420 MPa, crushing strength as 650 MPa and shear strength as 300 MPa. Take the joint efficiency as 80%. Sketch the joint with all the dimensions. Adopt the suitable factor of safety.*

Solution. Given : $D = 1.25$ m = 1250 mm; $P = 2.5$ N/mm² ; $\sigma_{tu} = 420$ MPa = 420 N/mm²; $\sigma_{cu} = 650$ MPa = 650 N/mm² ; $\tau_u = 300$ MPa = 300 N/mm² ; $\eta_l = 80\% = 0.8$

Assuming a factor of safety (*F.S.*) as 5, the allowable stresses are as follows :

$$\sigma_t = \frac{\sigma_{tu}}{F.S.} = \frac{420}{5} = 84 \text{ N/mm}^2$$

$$\sigma_c = \frac{\sigma_{cu}}{F.S.} = \frac{650}{5} = 130 \text{ N/mm}^2$$

and $$\tau = \frac{\tau_u}{F.S.} = \frac{300}{5} = 60 \text{ N/mm}^2$$

1. ***Thickness of plate***

We know that thickness of plate,

$$t = \frac{P.D}{2\,\sigma_t \times \eta_l} + 1 \text{ mm} = \frac{2.5 \times 1250}{2 \times 84 \times 0.8} + 1 \text{ mm}$$

$$= 24.3 \text{ say } 25 \text{ mm} \quad \textbf{Ans.}$$

2. ***Diameter of rivet***

Since the thickness of the plate is more than 8 mm, therefore diameter of the rivet hole,

$$d = 6\sqrt{t} = 6\sqrt{25} = 30 \text{ mm}$$

From Table 9.3, we see that according to IS : 1928 – 1961 (Reaffirmed 1996), the standard diameter of the rivet hole is 31.5 mm and the corresponding diameter of the rivet is 30 mm. **Ans.**

3. ***Pitch of rivets***

Assume a triple riveted double strap butt joint with unequal straps, as shown in Fig. 9.11.

Let p = Pitch of the rivets in the outer most row.

∴ Tearing strength of the plate per pitch length,

$$P_t = (p - d)\,t \times \sigma_t = (p - 31.5)\,25 \times 84 = 2100\,(p - 31.5) \text{ N} \quad ...(i)$$

Since the joint is triple riveted with two unequal cover straps, therefore there are 5 rivets per pitch length. Out of these five rivets, four rivets are in double shear and one is in single shear. Assuming the strength of the rivets in double shear as 1.875 times that of single shear, therefore

Shearing resistance of the rivets per pitch length,

$$P_s = 4 \times 1.875 \times \frac{\pi}{4} \times d^2 \times \tau + \frac{\pi}{4} \times d^2 \times \tau = 8.5 \times \frac{\pi}{4} \times d^2 \times \tau$$

$$= 8.5 \times \frac{\pi}{4}\,(31.5)^2\,60 = 397\,500 \text{ N} \quad ...(ii)$$

From equations (*i*) and (*ii*), we get

$$2100\,(p - 31.5) = 397\,500$$

$$\therefore \quad p - 31.5 = 397\,500 / 2100 = 189.3 \text{ or } p = 31.5 + 189.3 = 220.8 \text{ mm}$$

According to I.B.R., maximum pitch,

$$p_{max} = C \times t + 41.28 \text{ mm}$$

From Table 9.5, we find that for double strap butt joint with 5 rivets per pitch length, the value of C is 6.

$$\therefore \quad p_{max} = 6 \times 25 + 41.28 = 191.28 \text{ say } 196 \text{ mm} \quad \textbf{Ans.}$$

Since p_{max} is less than p, therefore we shall adopt $p = p_{max}$ = 196 mm **Ans.**

∴ Pitch of rivets in the inner row,

$$p' = 196 / 2 = 98 \text{ mm} \quad \textbf{Ans.}$$

4. ***Distance between the rows of rivets***

According to I.B.R., the distance between the outer row and the next row,

$$= 0.2\,p + 1.15\,d = 0.2 \times 196 + 1.15 \times 31.5 \text{ mm}$$

$$= 75.4 \text{ say } 76 \text{ mm} \quad \textbf{Ans.}$$

and the distance between the inner rows for zig-zag riveting

$$= 0.165\,p + 0.67\,d = 0.165 \times 196 + 0.67 \times 31.5 \text{ mm}$$

$$= 53.4 \text{ say } 54 \text{ mm} \quad \textbf{Ans.}$$

5. *Thickness of butt straps*

We know that for unequal width of butt straps, the thicknesses are as follows :

For wide butt strap, $t_1 = 0.75\ t = 0.75 \times 25 = 18.75$ say 20 mm **Ans.**

and for narrow butt strap, $t_2 = 0.625\ t = 0.625 \times 25 = 15.6$ say 16 mm **Ans.**

It may be noted that wide and narrow butt straps are placed on the inside and outside of the shell respectively.

6. *Margin*

We know that the margin,

$$m = 1.5\ d = 1.5 \times 31.5 = 47.25 \text{ say } 47.5 \text{ mm} \quad \textbf{Ans.}$$

Let us now check the efficiency of the designed joint.

Tearing resistance of the plate in the outer row,

$$P_t = (p - d)\ t \times \sigma_t = (196 - 31.5)\ 25 \times 84 = 345\ 450 \text{ N}$$

Shearing resistance of the rivets,

$$P_s = 4 \times 1.875 \times \frac{\pi}{4} \times d^2 \times \tau + \frac{\pi}{4} \times d^2 \times \tau = 8.5 \times \frac{\pi}{4} \times d^2 \times \tau$$

$$= 8.5 \times \frac{\pi}{4}\ (31.5)^2 \times 60 = 397\ 500 \text{ N}$$

and crushing resistance of the rivets,

$$P_c = n \times d \times t \times \sigma_c = 5 \times 31.5 \times 25 \times 130 = 511\ 875 \text{ N} \quad ...(\because n = 5)$$

The joint may also fail by tearing off the plate between the rivets in the second row. This is only possible if the rivets in the outermost row gives way (*i.e.* shears). Since there are two rivet holes per pitch length in the second row and one rivet is in the outer most row, therefore combined tearing and shearing resistance

$$= (p - 2d)\ t \times \sigma_t + \frac{\pi}{4} \times d^2 \times \tau$$

$$= (196 - 2 \times 31.5)\ 25 \times 84 + \frac{\pi}{4}\ (31.5)^2\ 60 = 326\ 065 \text{ N}$$

From above, we see that strength of the joint

$$= 326\ 065 \text{ N}$$

Strength of the unriveted or solid plate,

$$P = p \times t \times \sigma_t = 196 \times 25 \times 84 = 411\ 600 \text{ N}$$

∴ Efficiency of the joint,

$$\eta = 326\ 065 / 411\ 600 = 0.792 \text{ or } 79.2\%$$

Since the efficiency of the designed joint is nearly equal to the given efficiency, therefore the design is satisfactory.

Example 9.10. *A steam boiler is to be designed for a working pressure of 2.5 N/mm² with its inside diameter 1.6 m. Give the design calculations for the longitudinal and circumferential joints for the following working stresses for steel plates and rivets :*

In tension = 75 MPa ; In shear = 60 MPa; In crushing = 125 MPa.

Draw the joints to a suitable scale.

Solution. Given : $P = 2.5$ N/mm² ; $D = 1.6$ m $= 1600$ mm ; $\sigma_t = 75$ MPa $= 75$ N/mm² ; $\tau = 60$ MPa $= 60$ N/mm² ; $\sigma_c = 125$ MPa $= 125$ N/mm²

Design of longitudinal joint

The longitudinal joint for a steam boiler may be designed as follows :

1. ***Thickness of boiler shell***

We know that the thickness of boiler shell,

$$t = \frac{P.D}{2\,\sigma_t} + 1 \text{ mm} = \frac{2.5 \times 1600}{2 \times 75} + 1 \text{ mm}$$

$$= 27.6 \text{ say } 28 \text{ mm} \quad \textbf{Ans.}$$

2. ***Diameter of rivet***

Since the thickness of the plate is more than 8 mm, therefore diameter of rivet hole,

$$d = 6\sqrt{t} = 6\sqrt{28} = 31.75 \text{ mm}$$

From Table 9.3, we see that according to IS : 1928 – 1961 (Reaffirmed 1996), the standard diameter of rivet hole (d) is 34.5 mm and the corresponding diameter of the rivet is 33 mm. **Ans.**

3. ***Pitch of rivets***

Assume the joint to be triple riveted double strap butt joint with unequal cover straps, as shown in Fig. 9.11.

Let p = Pitch of the rivet in the outer most row.

∴ Tearing resistance of the plate per pitch length,

$$P_t = (p - d)\,t \times \sigma_t = (p - 34.5)\, 28 \times 75 \text{ N}$$

$$= 2100\,(p - 34.5) \text{ N} \qquad ...(i)$$

Since the joint is triple riveted with two unequal cover straps, therefore there are 5 rivets per pitch length. Out of these five rivets, four are in double shear and one is in single shear. Assuming the strength of rivets in double shear as 1.875 times that of single shear, therefore

Shearing resistance of the rivets per pitch length,

$$P_s = 4 \times 1.875 \times \frac{\pi}{4} \times d^2 \times \tau + \frac{\pi}{4} \times d^2 \times \tau$$

$$= 8.5 \times \frac{\pi}{4} \times d^2 \times \tau$$

$$= 8.5 \times \frac{\pi}{4}\,(34.5)^2\, 60 = 476\,820 \text{ N} \qquad ...(ii)$$

Equating equations (*i*) and (*ii*), we get

$$2100\,(p - 34.5) = 476\,820$$

$$\therefore \quad p - 34.5 = 476\,820 / 2100 = 227 \text{ or } p = 227 + 34.5 = 261.5 \text{ mm}$$

According to I.B.R., the maximum pitch,

$$p_{max} = C.t + 41.28 \text{ mm}$$

From Table 9.5, we find that for double strap butt joint with 5 rivets per pitch length, the value of C is 6.

$$\therefore \quad p_{max} = 6 \times 28 + 41.28 = 209.28 \text{ say } 220 \text{ mm}$$

Since p_{max} is less than p, therefore we shall adopt

$$p = p_{max} = 220 \text{ mm} \quad \textbf{Ans.}$$

∴ Pitch of rivets in the inner row,

$$p' = 220 / 2 = 110 \text{ mm } \textbf{Ans.}$$

4. ***Distance between the rows of rivets***

According to I.B.R., the distance between the outer row and the next row

$$= 0.2\,p + 1.15\,d = 0.2 \times 220 + 1.15 \times 34.5 \text{ mm}$$

$$= 83.7 \text{ say } 85 \text{ mm} \quad \textbf{Ans.}$$

and the distance between the inner rows for zig-zig riveting

$$= 0.165\,p + 0.67\,d = 0.165 \times 220 + 0.67 \times 34.5 \text{ mm}$$
$$= 59.4 \text{ say } 60 \text{ mm} \quad \textbf{Ans.}$$

5. ***Thickness of butt straps***

We know that for unequal width of butt straps, the thicknesses are :

For wide butt strap, $t_1 = 0.75\,t = 0.75 \times 28 = 21$ mm **Ans.**

and for narrow butt strap, $t_2 = 0.625\,t = 0.625 \times 28 = 17.5$ say 18 mm **Ans.**

It may be noted that the wide and narrow butt straps are placed on the inside and outside of the shell respectively.

6. ***Margin***

We know that the margin,

$$m = 1.5\,d = 1.5 \times 34.5 = 51.75 \text{ say } 52 \text{ mm} \quad \textbf{Ans.}$$

Let us now check the efficiency of the designed joint.

Tearing resistance of the plate in the outer row,

$$P_t = (p - d)\,t \times \sigma_t = (220 - 34.5)\,28 \times 75 = 389\,550 \text{ N}$$

Shearing resistance of the rivets,

$$P_s = 4 \times 1.875 \times \frac{\pi}{4} \times d^2 \times \tau + \frac{\pi}{4} \times d^2 \times \tau = 8.5 \times \frac{\pi}{4} \times d^2 \times \tau$$
$$= 8.5 \times \frac{\pi}{4}\,(34.5)^2\,60 = 476\,820 \text{ N}$$

and crushing resistance of the rivets,

$$P_c = n \times d \times t \times \sigma_c = 5 \times 34.5 \times 28 \times 125 = 603\,750 \text{ N}$$

The joint may also fail by tearing off the plate between the rivets in the second row. This is only possible if the rivets in the outermost row gives way (*i.e.* shears). Since there are two rivet holes per pitch length in the second row and one rivet in the outermost row, therefore

Combined tearing and shearing resistance

$$= (p - 2d)\,t \times \sigma_t + \frac{\pi}{4} \times d^2 \times \tau$$
$$= (220 - 2 \times 34.5)\,28 \times 75 + \frac{\pi}{4}\,(34.5)^2\,60$$
$$= 317\,100 + 56\,096 = 373\,196 \text{ N}$$

From above, we see that the strength of the joint

$$= 373\,196 \text{ N}$$

Strength of the unriveted or solid plate,

$$P = p \times t \times \sigma_t = 220 \times 28 \times 75 = 462\,000 \text{ N}$$

∴ Efficiency of the designed joint,

$$\eta = \frac{373\,196}{462\,000} = 0.808 \text{ or } 80.8\% \quad \textbf{Ans.}$$

Design of circumferential joint

The circumferential joint for a steam boiler may be designed as follows :

1. The thickness of the boiler shell (t) and diameter of rivet hole (d) will be same as for longitudinal joint, *i.e.*

$$t = 28 \text{ mm} \; ; \text{ and } d = 34.5 \text{ mm}$$

2. ***Number of rivets***

Let n = Number of rivets.

We know that shearing resistance of the rivets

$$= n \times \frac{\pi}{4} \times d^2 \times \tau \quad \text{...}(i)$$

and total shearing load acting on the circumferential joint

$$= \frac{\pi}{4} \times D^2 \times P \quad \text{...}(ii)$$

From equations (*i*) and (*ii*), we get

$$n \times \frac{\pi}{4} \times d^2 \times \tau = \frac{\pi}{4} \times D^2 \times P$$

$$\therefore \quad n = \frac{D^2 \times P}{d^2 \times \tau} = \frac{(1600)^2 \; 2.5}{(34.5)^2 \; 60} = 89.6 \text{ say } 90 \quad \textbf{Ans.}$$

3. ***Pitch of rivets***

Assuming the joint to be double riveted lap joint with zig-zag riveting, therefore number of rivets per row

$$= 90 / 2 = 45$$

We know that the pitch of the rivets,

$$p_1 = \frac{\pi (D + t)}{\text{Number of rivets per row}} = \frac{\pi (1600 + 28)}{45} = 113.7 \text{ mm}$$

Let us take pitch of the rivets, p_1 = 140 mm **Ans.**

4. ***Efficiency of the joint***

We know that the efficiency of the circumferential joint,

$$\eta_c = \frac{p_1 - d}{p_1} = \frac{140 - 34.5}{140} = 0.753 \text{ or } 75.3\%$$

5. ***Distance between the rows of rivets***

We know that the distance between the rows of rivets for zig-zag riveting,

$$= 0.33\, p_1 + 0.67\, d = 0.33 \times 140 + 0.67 \times 34.5 \text{ mm}$$

$$= 69.3 \text{ say } 70 \text{ mm} \quad \textbf{Ans.}$$

6. ***Margin***

We know that the margin,

$$m = 1.5\, d = 1.5 \times 34.5$$

$$= 51.75 \text{ say } 52 \text{ mm} \quad \textbf{Ans.}$$

9.20 Riveted Joint for Structural Use–Joints of Uniform Strength (Lozenge Joint)

A riveted joint known as ***Lozenge joint*** used for roof, bridge work or girders etc. is shown in Fig. 9.19. In such a joint, *diamond riveting is employed so that the joint is made of uniform strength.

Fig. 9.19 shows a triple riveted double strap butt joint.

Riveted joints are used for roofs, bridge work and girders.

* In diamond riveting, the number of rivets increases as we proceed from the outermost row to the innermost row.

Let b = Width of the plate,

t = Thickness of the plate, and

d = Diameter of the rivet hole.

In designing a Lozenge joint, the following procedure is adopted.

1. *Diameter of rivet*

The diameter of the rivet hole is obtained by using Unwin's formula, *i.e.*

$$d = 6\sqrt{t}$$

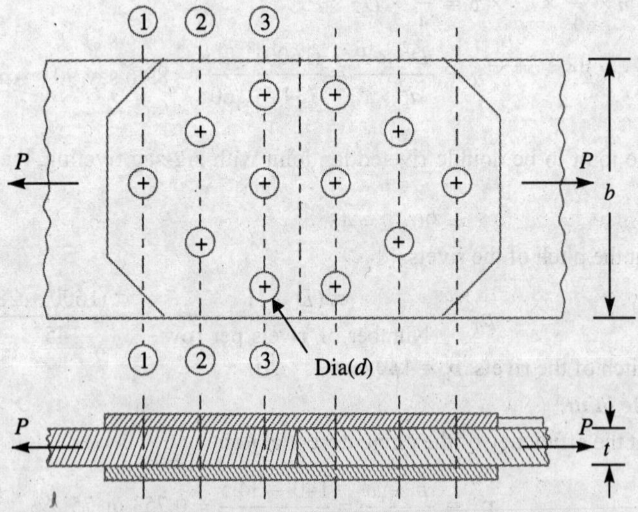

Fig. 9.19. Riveted joint for structural use.

According to IS : 1929–1982 (Reaffirmed 1996), the sizes of rivets for general purposes are given in the following table.

Table 9.7. Sizes of rivets for general purposes, according to IS : 1929 - 1982 (Reaffirmed 1996).

Diameter of rivet hole (mm)	13.5	15.5	17.5	19.5	21.5	23.5	25.5	29	32	35	38	41	44	50
Diameter of rivet (mm)	12	14	16	18	20	22	24	27	30	33	36	39	42	48

2. *Number of rivets*

The number of rivets required for the joint may be obtained by the shearing or crushing resistance of the rivets.

Let P_t = Maximum pull acting on the joint. This is the tearing resistance of the plate at the outer row which has only one rivet.

$= (b - d)\, t \times \sigma_t$

and n = Number of rivets.

Since the joint is double strap butt joint, therefore the rivets are in double shear. It is assumed that resistance of a rivet in double shear is 1.75 times than in single shear in order to allow for possible eccentricity of load and defective workmanship.

∴ Shearing resistance of one rivet,

$$P_s = 1.75 \times \frac{\pi}{4} \times d^2 \times \tau$$

and crushing resistance of one rivet,

$$P_c = d \times t \times \sigma_c$$

∴ Number of rivets required for the joint,

$$n = \frac{P_t}{\text{Least of } P_s \text{ or } P_c}$$

3. From the number of rivets, the number of rows and the number of rivets in each row is decided.

4. *Thickness of the butt straps*

The thickness of the butt strap,

$$t_1 = 1.25\, t, \text{ for single cover strap}$$
$$= 0.75\, t, \text{ for double cover strap}$$

5. *Efficiency of the joint*

First of all, calculate the resistances along the sections 1-1, 2-2 and 3-3.

At section 1-1, there is only one rivet hole.

∴ Resistance of the joint in tearing along 1-1,

$$P_{t1} = (b - d)\, t \times \sigma_t$$

At section 2-2, there are two rivet holes.

∴ Resistance of the joint in tearing along 2-2,

$$P_{t2} = (b - 2d)\, t \times \sigma_t + \text{Strength of one rivet in front of section 2-2}$$

(This is due to the fact that for tearing off the plate at section 2-2, the rivet in front of section 2-2 *i.e.* at section 1-1 must first fracture).

Similarly at section 3-3 there are three rivet holes.

∴ Resistance of the joint in tearing along 3-3,

$$P_{t3} = (b - 3d)\, t \times \sigma_t + \text{Strength of 3 rivets in front of section 3-3}$$

The least value of P_{t1}, P_{t2}, P_{t3}, P_s or P_c is the strength of the joint.

We know that the strength of unriveted plate,

$$P = b \times t \times \sigma_t$$

∴ Efficiency of the joint,

$$\eta = \frac{\text{Least of } P_{t1}, P_{t2}, P_{t3}, P_s \text{ or } P_c}{P}$$

Note : The permissible stresses employed in structural joints are higher than those used in design of pressure vessels. The following values are usually adopted.

For plates in tension	... 140 MPa
For rivets in shear	... 105 MPa
For crushing of rivets and plates	
Single shear	... 224 MPa
Double shear	... 280 MPa

6. The pitch of the rivets is obtained by equating the strength of the joint in tension to the strength of the rivets in shear. The pitches allowed in structural joints are larger than those of pressure vessels. The following table shows the values of pitch due to Rotscher.

Table 9.8. Pitch of rivets for structural joints.

Thickness of plate (mm)	*Diameter of rivet hole (mm)*	*Diameter of rivet (mm)*	*Pitch of rivet p = 3d + 5mm*	*Marginal pitch (mm)*
2	8.4	8	29	16
3	9.5	9	32	17
4	11	10	35	17
5–6	13	12	38	18
6–8	15	14	47	21
8–12	17	16	56	25
11–15	21	20	65	30

7. The marginal pitch (m) should not be less than 1.5 d.
8. The distance between the rows of rivets is 2.5 d to 3 d.

Example 9.11. *Two lengths of mild steel tie rod having width 200 mm and thickness 12.5 mm are to be connected by means of a butt joint with double cover plates. Design the joint if the permissible stresses are 80 MPa in tension, 65 MPa in shear and 160 MPa in crushing. Make a sketch of the joint.*

Solution. Given : b = 200 mm ; t = 12.5 mm ; σ_t = 80 MPa = 80 N/mm^2 ; τ = 65 MPa = 65 N/mm^2 ; σ_c = 160 MPa = 160 N/mm^2

1. ***Diameter of rivet***

We know that the diameter of rivet hole,

$$d = 6\sqrt{t} = 6\sqrt{12.5} = 21.2 \text{ mm}$$

From Table 9.7, we see that according to IS : 1929 – 1982 (Reaffirmed 1996), the standard diameter of the rivet hole (d) is 21.5 mm and the corresponding diameter of rivet is 20 mm. **Ans.**

2. ***Number of rivets***

Let n = Number of rivets.

We know that maximum pull acting on the joint,

$$P_t = (b - d)\, t \times \sigma_t = (200 - 21.5)\, 12.5 \times 80 = 178\ 500 \text{ N}$$

Since the joint is a butt joint with double cover plates as shown in Fig. 9.20, therefore the rivets are in double shear. Assume that the resistance of the rivet in double shear is 1.75 times than in single shear.

∴ Shearing resistance of one rivet,

$$P_s = 1.75 \times \frac{\pi}{4} \times d^2 \times \tau = 1.75 \times \frac{\pi}{4} (21.5)^2\, 65 = 41\ 300 \text{ N}$$

and crushing resistance of one rivet,

$$P_c = d \times t \times \sigma_c = 21.5 \times 12.5 \times 160 = 43\ 000 \text{ N}$$

Since the shearing resistance is less than the crushing resistance, therefore number of rivets required for the joint,

$$n = \frac{P_t}{P_s} = \frac{178\ 500}{41\ 300} = 4.32 \text{ say } 5 \quad \textbf{Ans.}$$

3. The arrangement of the rivets is shown in Fig. 9.20.

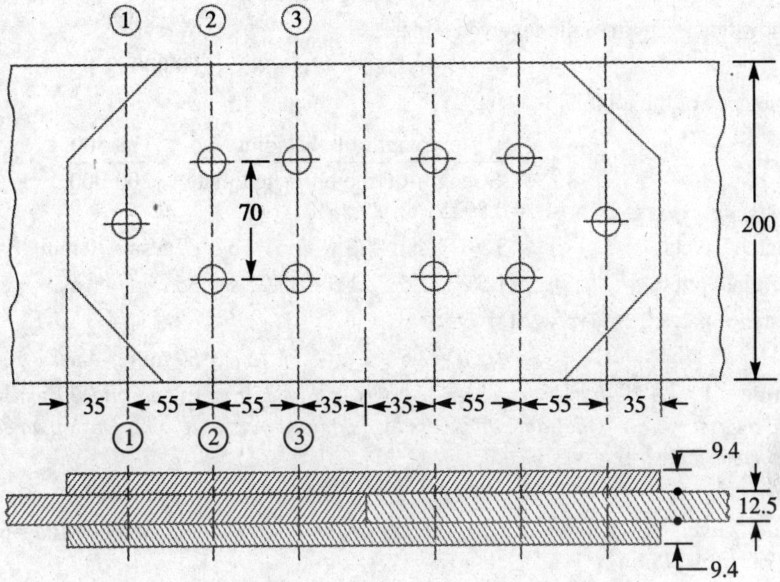

Fig. 9.20. All dimensions in mm.

4. ***Thickness of butt straps***

We know that thickness of butt straps,

$$t_1 = 0.75\, t = 0.75 \times 12.5 = 9.375 \text{ say } 9.4 \text{ mm} \quad \textbf{Ans.}$$

5. ***Efficiency of the joint***

First of all, let us find the resistances along the sections 1-1, 2-2 and 3-3.

At section 1-1, there is only one rivet hole.

∴ Resistance of the joint in tearing along section 1-1,

$$P_{t1} = (b - d)\, t \times \sigma_t = (200 - 21.5)\, 12.5 \times 80 = 178\ 500 \text{ N}$$

At section 2-2, there are two rivet holes. In this case, the tearing of the plate will only take place if the rivet at section 1-1 (in front of section 2-2) gives way (*i.e.* shears).

∴ Resistance of the joint in tearing along section 2-2,

$$P_{t2} = (b - 2d)\, t \times \sigma_t + \text{Shearing resistance of one rivet}$$
$$= (200 - 2 \times 21.5)\, 12.5 \times 80 + 41\ 300 = 198\ 300 \text{ N}$$

At section 3-3, there are two rivet holes. The tearing of the plate will only take place if one rivet at section 1-1 and two rivets at section 2-2 gives way (*i.e.* shears).

∴ Resistance of the joint in tearing along section 3-3,

$$P_{t3} = (b - 2d)\, t \times \sigma_t + \text{Shearing resistance of 3 rivets}$$
$$= (200 - 2 \times 21.5)\, 12.5 \times 80 + 3 \times 41\ 300 = 280\ 900 \text{ N}$$

Shearing resistance of all the 5 rivets

$$P_s = 5 \times 41\ 300 = 206\ 500 \text{ N}$$

and crushing resistance of all the 5 rivets,

$$P_c = 5 \times 43\ 000 = 215\ 000 \text{ N}$$

Since the strength of the joint is the least value of P_{t1}, P_{t2}, P_{t3}, P_s and P_c, therefore strength of the joint

$$= 178\ 500 \text{ N along section 1-1}$$

We know that strength of the un-riveted plate,

$$= b \times t \times \sigma_t = 20 \times 12.5 \times 80 = 200\ 000 \text{ N}$$

∴ Efficiency of the joint,

$$\eta = \frac{\text{Strength of the joint}}{\text{Strength of the unriveted plate}} = \frac{178\ 500}{200\ 000}$$

$$= 0.8925 \text{ or } 89.25\% \textbf{ Ans.}$$

6. Pitch of rivets, $p = 3\,d + 5 \text{ mm} = (3 \times 21.5) + 5 = 69.5$ say 70 mm **Ans.**

7. Marginal pitch, $m = 1.5\,d = 1.5 \times 21.5 = 33.25$ say 35 mm **Ans.**

8. Distance between the rows of rivets

$$= 2.5\,d = 2.5 \times 21.5 = 53.75 \text{ say } 55 \text{ mm} \quad \textbf{Ans.}$$

Example 9.12. *A tie-bar in a bridge consists of flat 350 mm wide and 20 mm thick. It is connected to a gusset plate of the same thickness by a double cover butt joint. Design an economical joint if the permissible stresses are :*

$$\sigma_t = 90 \text{ MPa}, \tau = 60 \text{ MPa and } \sigma_c = 150 \text{ MPa}$$

Solution. Given : $b = 350$ mm ; $t = 20$ mm ; $\sigma_t = 90$ MPa $= 90$ N/mm^2 ; $\tau = 60$ MPa $= 60$ N/mm^2 ; $\sigma_c = 150$ MPa $= 150$ N/mm^2

Riveted, screwed and welded joints are employed in bridges.

1. ***Diameter of rivet***

We know that the diameter of rivet hole,

$$d = 6\sqrt{t} = 6\sqrt{20} = 26.8 \text{ mm}$$

From Table 9.7, we see that according to IS : 1929–1982 (Reaffirmed 1996), the standard diameter of rivet hole (d) is 29 mm and the corresponding diameter of rivet is 27 mm. **Ans.**

2. ***Number of rivets***

Let n = Number of rivets.

We know that the maximum pull acting on the joint,

$$P_t = (b - d)\, t \times \sigma_t = (350 - 29)\, 20 \times 90 = 577\,800 \text{ N}$$

Since the joint is double strap butt joint, therefore the rivets are in double shear. Assume that the resistance of the rivet in double shear is 1.75 times than in single shear.

∴ Shearing resistance of one rivet,

$$P_s = 1.75 \times \frac{\pi}{4} \times d^2 \times \tau = 1.75 \times \frac{\pi}{4}\,(29)^2\, 60 = 69\,360 \text{ N}$$

and crushing resistance of one rivet,

$$P_c = d \times t \times \sigma_c = 29 \times 20 \times 150 = 87\,000 \text{ N}$$

Since the shearing resistance is less than crushing resistance, therefore number of rivets required for the joint,

$$n = \frac{P_t}{P_s} = \frac{577\,800}{69\,360} = 8.33 \text{ say } 9 \quad \textbf{Ans.}$$

3. The arrangement of rivets is shown in Fig. 9.21.

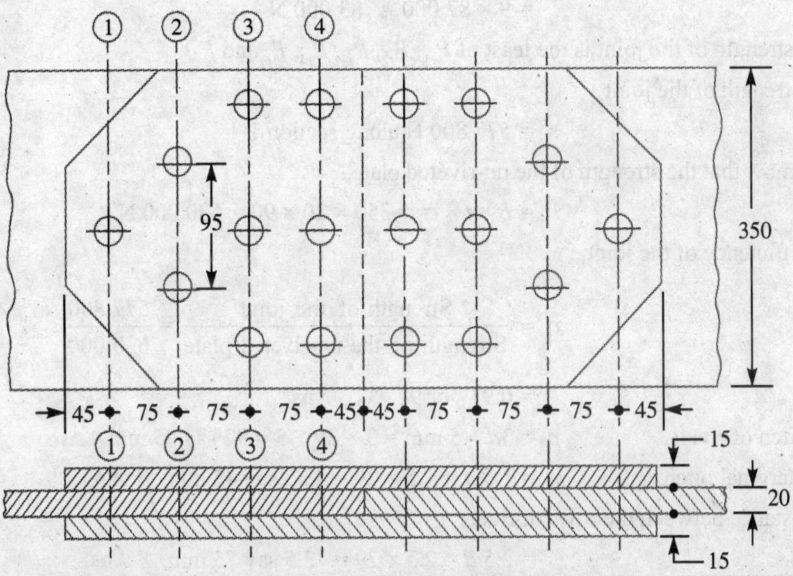

Fig. 9.21. All dimensions in mm.

4. ***Thickness of butt straps***

We know that the thickness of butt straps,

$$t_1 = 0.75\, t = 0.75 \times 20 = 15 \text{ mm} \quad \textbf{Ans.}$$

5. *Efficiency of the johint*

First of all, let us find the resistances along the sections 1-1, 2-2, 3-3 and 4-4.

At section 1-1, there is only one rivet hole.

∴ Resistance of the joint in tearing along 1-1,

$$P_{t1} = (b - d)\,t \times \sigma_t = (350 - 29)\,20 \times 90 = 577\,800 \text{ N}$$

At section 2-2, there are two rivet holes. In this case the tearing of the plate will only take place if the rivet at section 1-1 (in front of section 2-2) gives way.

∴ Resistance of the joint in tearing along 2-2,

$$P_{t2} = (b - 2d)\,t \times \sigma_t + \text{Shearing strength of one rivet in front}$$
$$= (350 - 2 \times 29)\,20 \times 90 + 69\,360 = 594\,960 \text{ N}$$

At section 3-3, there are three rivet holes. The tearing of the plate will only take place if one rivet at section 1-1 and two rivets at section 2-2 gives way.

∴ Resistance of the joint in tearing along 3-3,

$$P_{t3} = (b - 3d)\,t \times \sigma_t + \text{Shearing strength of 3 rivets in front.}$$
$$= (350 - 3 \times 29)\,20 \times 90 + 3 \times 69\,360 = 681\,480 \text{ N}$$

Similarly, resistance of the joint in tearing along 4-4,

$$P_{t4} = (b - 3d)\,t \times \sigma_t + \text{Shearing strength of 6 rivets in front}$$
$$= (350 - 3 \times 29)\,20 \times 90 + 6 \times 69\,360 = 889\,560 \text{ N}$$

Shearing resistance of all the 9 rivets,

$$P_s = 9 \times 69\,360 = 624\,240 \text{ N}$$

and crushing resistance of all the 9 rivets,

$$P_c = 9 \times 87\,000 = 783\,000 \text{ N}$$

The strength of the joint is the least of P_{t1}, P_{t2}, P_{t3}, P_{t4}, P_s and P_c.

∴ Strength of the joint

$$= 577\,800 \text{ N along section 1-1}$$

We know that the strength of the un-riveted plate,

$$P = b \times t \times \sigma_t = 350 \times 20 \times 90 = 630\,000 \text{ N}$$

∴ Efficiency of the joint,

$$\eta = \frac{\text{Strength of the joint}}{\text{Strength of the un-riveted plate}} = \frac{577\,800}{630\,000}$$
$$= 0.917 \text{ or } 91.7\% \quad \textbf{Ans.}$$

6. Pitch of rivets, $p = 3d + 5 \text{ mm} = 3 \times 29 + 5 = 92 \text{ say } 95 \text{ mm}$ **Ans.**

7. Marginal pitch, $m = 1.5\,d = 1.5 \times 29 = 43.5 \text{ say } 45 \text{ mm}$ **Ans.**

8. Distance between the rows of rivets

$$= 2.5\,d = 2.5 \times 29 = 72.5 \text{ say } 75 \text{ mm} \quad \textbf{Ans.}$$

Note : If chain riveting with three rows of three rivets in each is used instead of diamond riveting, then

Least strength of the joint

$$= (b - 3\,d)\,t \times \sigma_t = (350 - 3 \times 29)\,20 \times 90 = 473\,400 \text{ N}$$

∴ Efficiency of the joint $= \dfrac{473\,400}{630\,000} = 0.752 \text{ or } 75.2\%$

Thus we see that with the use of diamond riveting, efficiency of the joint is increased.

Example 9.13. *Design a lap joint for a mild steel flat tie-bar 200 mm × 10 mm thick, using 24 mm diameter rivets. Assume allowable stresses in tension and compression of the plate material as 112 MPa and 200 MPa respectively and shear stress of the rivets as 84 MPa. Show the disposition of the rivets for maximum joint efficiency and determine the joint efficiency. Take diameter of rivet hole as 25.5 mm for a 24 mm diameter rivet.*

Solution. Given : $b = 200$ mm; $t = 10$ mm ; $\sigma_t = 112$ MPa = 112 N/mm^2 ; $\sigma_c = 200$ MPa = 200 N/mm^2 ; $\tau = 84$ MPa = 84 N/mm^2 ; $d = 25.5$ mm ; $d_1 = 24$ mm

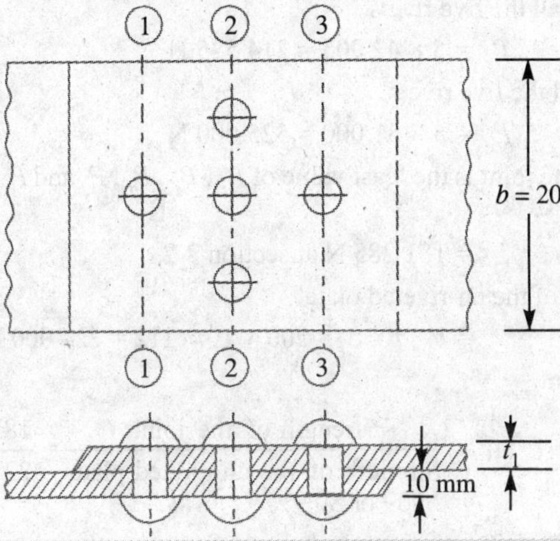

Fig. 9.22

1. *Number of rivets*

Let n = Number of rivets.

We know that the maximum pull acting on the joint,

$$P_t = (b - d)\, t \times \sigma_t = (200 - 25.5)\, 10 \times 112 = 195\,440 \text{ N}$$

Since the joint is a lap joint, therefore shearing resistance of one rivet,

$$P_s = \frac{\pi}{4} \times d^2 \times \tau = \frac{\pi}{4}\,(25.5)^2\, 84 = 42\,905 \text{ N}$$

and crushing resistance of one rivet,

$$P_c = d \times t \times \sigma_c = 25.5 \times 10 \times 200 = 51\,000 \text{ N}$$

Since the shearing resistance is less than the crushing resistance, therefore number of rivets required for the joint,

$$n = \frac{P_t}{P_s} = \frac{195\,440}{42\,905} = 4.56 \text{ say } 5 \quad \textbf{Ans.}$$

2. The arrangement of the rivets is shown in Fig. 9.22.

3. *Thickness of the cover plate*

We know that the thickness of a cover plate for lap joint,

$$t_1 = 1.25\, t = 1.25 \times 10 = 12.5 \text{ mm} \quad \textbf{Ans.}$$

4. *Efficiency of the joint*

First of all, let us find the resistances along the sections 1-1, 2-2 and 3-3. At section 1-1, there is only one rivet hole.

∴ Resistance of the joint in tearing along section 1-1,

$$P_{t1} = (b - d)\, t \times \sigma_t = (200 - 25.5)\, 10 \times 112 = 195\,440 \text{ N}$$

At section 2-2, there are three rivet holes. In this case, the tearing of the plate will only take place if the rivet at section 1-1 (in front of section 2-2) gives way (*i.e.* shears).

∴ Resistance of the joint in tearing along section 2-2,

$$P_{t2} = (b - 3d)\, t \times \sigma_t + \text{Shearing resistance of one rivet}$$
$$= (200 - 3 \times 25.5)\, 10 \times 112 + 42\,905 = 181\,285 \text{ N}$$

At section 3-3, there is only one rivet hole. The resistance of the joint in tearing along section 3-3 will be same as at section 1-1.

$$P_{t3} = P_{t1} = 195\,440 \text{ N}$$

Shearing resistance of all the five rivets,

$$P_s = 5 \times 42\,905 = 214\,525 \text{ N}$$

and crushing resistance of all the five rivets,

$$P_c = 5 \times 51\,000 = 525\,000 \text{ N}$$

Since the strength of the joint is the least value of P_{t1}, P_{t2}, P_{t3}, P_s and P_c, therefore strength of the joint

$$= 181\,285 \text{ N at section 2-2}$$

We know that strength of the un-riveted plate

$$= b \times t \times \sigma_t = 200 \times 10 \times 112 = 224\,000 \text{ N}$$

∴ Efficiency of the joint,

$$\eta = \frac{\text{Strength of the joint}}{\text{Strength of the un-riveted plate}} = \frac{181\,225}{224\,000}$$
$$= 0.809 \text{ or } 80.9\% \quad \textbf{Ans.}$$

9.21 Eccentric Loaded Riveted Joint

When the line of action of the load does not pass through the centroid of the rivet system and thus all rivets are not equally loaded, then the joint is said to be an ***eccentric loaded riveted joint,*** as shown in Fig. 9.23 (*a*). The eccentric loading results in secondary shear caused by the tendency of force to twist the joint about the centre of gravity in addition to direct shear or primary shear.

Let P = Eccentric load on the joint, and

e = Eccentricity of the load *i.e.* the distance between the line of action of the load and the centroid of the rivet system *i.e.* G.

The following procedure is adopted for the design of an eccentrically loaded riveted joint.

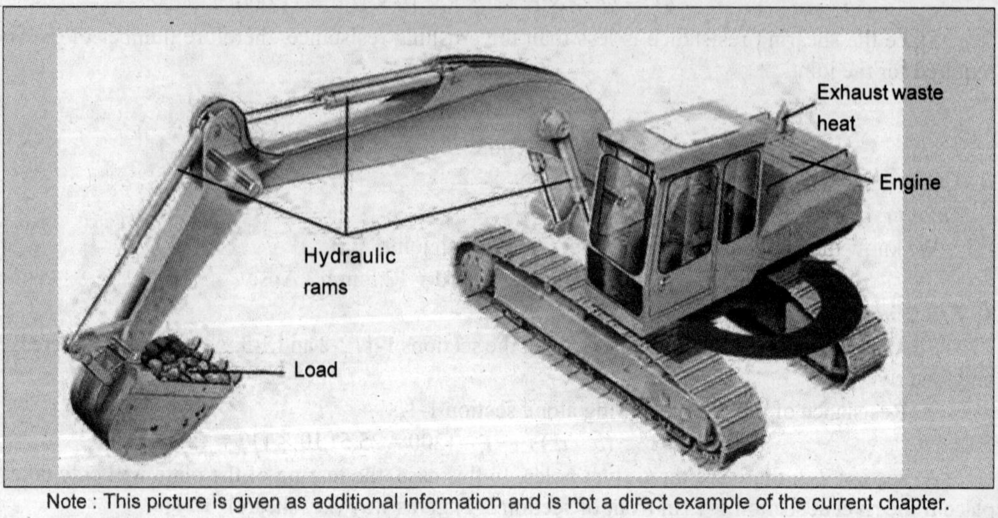

Note : This picture is given as additional information and is not a direct example of the current chapter.

1. First of all, find the centre of gravity G of the rivet system.

Let A = Cross-sectional area of each rivet,

x_1, x_2, x_3 etc. = Distances of rivets from OY, and

y_1, y_2, y_3 etc. = Distances of rivets from OX.

We know that
$$\bar{x} = \frac{A_1 x_1 + A_2 x_2 + A_3 x_3 + \ldots}{A_1 + A_2 + A_3 + \ldots} = \frac{A x_1 + A x_2 + A x_3 + \ldots}{n.A}$$
$$= \frac{x_1 + x_2 + x_3 + \ldots}{n} \quad \text{...(where } n = \text{Number of rivets)}$$

Similarly,
$$\bar{y} = \frac{y_1 + y_2 + y_3 + \ldots}{n}$$

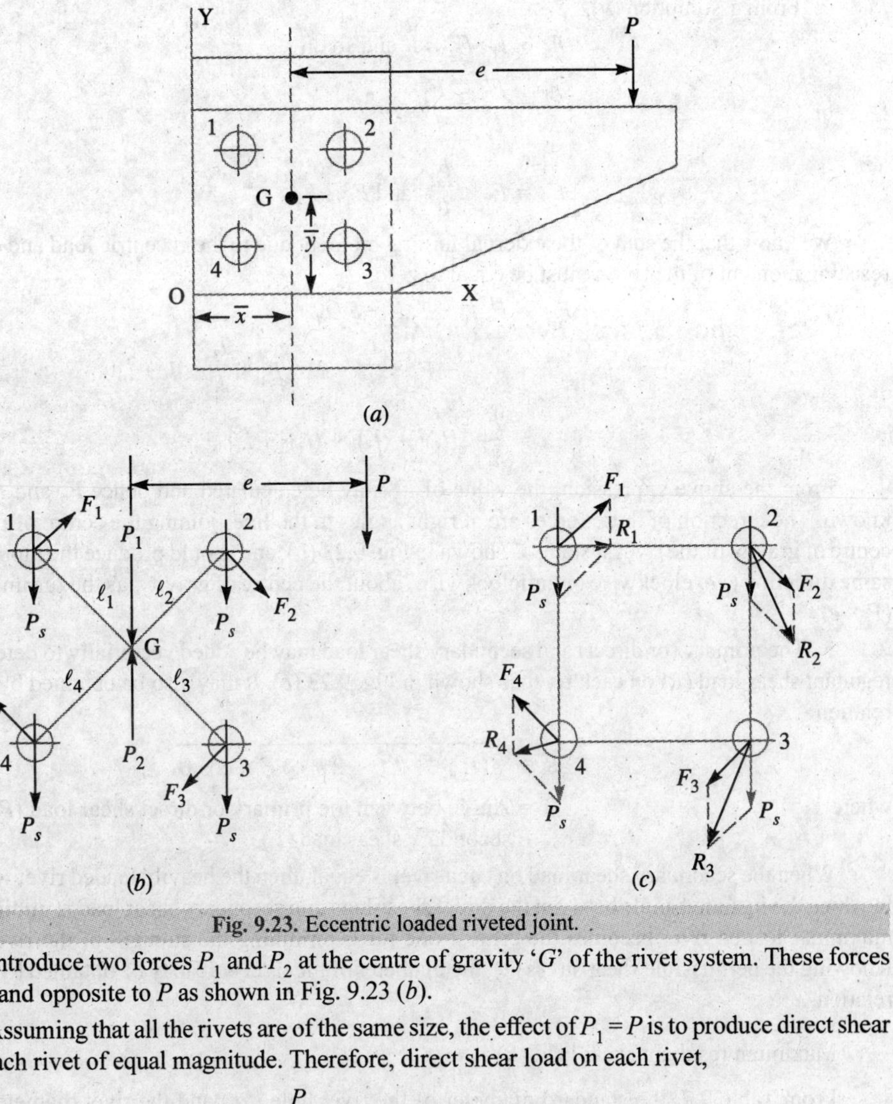

Fig. 9.23. Eccentric loaded riveted joint.

2. Introduce two forces P_1 and P_2 at the centre of gravity 'G' of the rivet system. These forces are equal and opposite to P as shown in Fig. 9.23 (*b*).

3. Assuming that all the rivets are of the same size, the effect of $P_1 = P$ is to produce direct shear load on each rivet of equal magnitude. Therefore, direct shear load on each rivet,

$$P_s = \frac{P}{n}, \text{ acting parallel to the load } P.$$

4. The effect of $P_2 = P$ is to produce a turning moment of magnitude $P \times e$ which tends to rotate the joint about the centre of gravity 'G' of the rivet system in a clockwise direction. Due to the turning moment, secondary shear load on each rivet is produced. In order to find the secondary shear load, the following two assumptions are made :

(*a*) The secondary shear load is proportional to the radial distance of the rivet under consideration from the centre of gravity of the rivet system.

(*b*) The direction of secondary shear load is perpendicular to the line joining the centre of the rivet to the centre of gravity of the rivet system..

Let $F_1, F_2, F_3 ...$ = Secondary shear loads on the rivets 1, 2, 3...etc.

$l_1, l_2, l_3 ...$ = Radial distance of the rivets 1, 2, 3 ...etc. from the centre of gravity 'G' of the rivet system.

∴ From assumption (*a*),

$$F_1 \propto l_1 ; F_2 \propto l_2 \text{ and so on}$$

or

$$\frac{F_1}{l_1} = \frac{F_2}{l_2} = \frac{F_3}{l_3} = ...$$

∴

$$F_2 = F_1 \times \frac{l_2}{l_1}, \text{ and } F_3 = F_1 \times \frac{l_3}{l_1}$$

We know that the sum of the external turning moment due to the eccentric load and of internal resisting moment of the rivets must be equal to zero.

∴

$$P.e = F_1.l_1 + F_2.l_2 + F_3.l_3 + ...$$

$$= F_1.l_1 + F_1 \times \frac{l_2}{l_1} \times l_2 + F_1 \times \frac{l_3}{l_1} \times l_3 + ...$$

$$= \frac{F_1}{l_1} [(l_1)^2 + (l_2)^2 + (l_3)^2 + ...]$$

From the above expression, the value of F_1 may be calculated and hence F_2 and F_3 etc. are known. The direction of these forces are at right angles to the lines joining the centre of rivet to the centre of gravity of the rivet system, as shown in Fig. 9.23 (*b*), and should produce the moment in the same direction (*i.e.* clockwise or anticlockwise) about the centre of gravity, as the turning moment ($P \times e$).

5. The primary (or direct) and secondary shear load may be added vectorially to determine the resultant shear load (R) on each rivet as shown in Fig. 9.23 (*c*). It may also be obtained by using the relation

$$R = \sqrt{(P_s)^2 + F^2 + 2 P_s \times F \times \cos \theta}$$

where θ = Angle between the primary or direct shear load (P_s) and secondary shear load (F).

When the secondary shear load on each rivet is equal, then the heavily loaded rivet will be one in which the included angle between the direct shear load and secondary shear load is minimum. The maximum loaded rivet becomes the critical one for determining the strength of the riveted joint. Knowing the permissible shear stress (τ), the diameter of the rivet hole may be obtained by using the relation,

$$\text{Maximum resultant shear load } (R) = \frac{\pi}{4} \times d^2 \times \tau$$

From Table 9.7, the standard diameter of the rivet hole (d) and the rivet diameter may be specified, according to IS : 1929 – 1982 (Reaffirmed 1996).

Notes : 1. In the solution of a problem, the primary and shear loads may be laid off approximately to scale and generally the rivet having the maximum resultant shear load will be apparent by inspection. The values of the load for that rivet may then be calculated.

2. When the thickness of the plate is given, then the diameter of the rivet hole may be checked against crushing.

3. When the eccentric load P is inclined at some angle, then the same procedure as discussed above may be followed to find the size of rivet (See Example 9.18).

Example 9.14. *An eccentrically loaded lap riveted joint is to be designed for a steel bracket as shown in Fig. 9.24.*

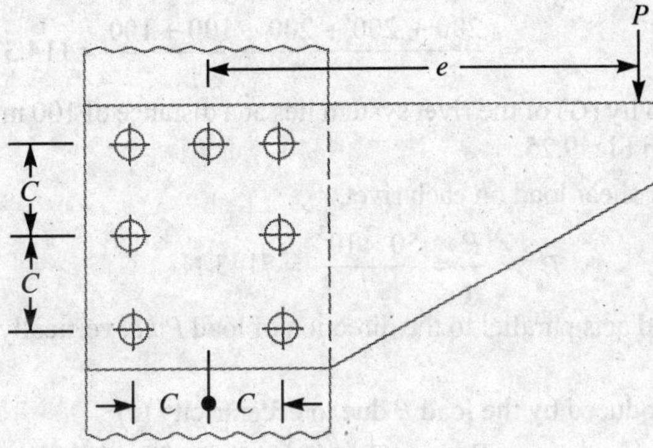

Fig. 9.24

The bracket plate is 25 mm thick. All rivets are to be of the same size. Load on the bracket, P = 50 kN ; rivet spacing, C = 100 mm; load arm, e = 400 mm.

Permissible shear stress is 65 MPa and crushing stress is 120 MPa. Determine the size of the rivets to be used for the joint.

Solution. Given : t = 25 mm ; P = 50 kN = 50 × 10^3 N ; e = 400 mm ; n = 7 ; τ = 65 MPa = 65 N/mm^2 ; σ_c = 120 MPa = 120 N/mm^2

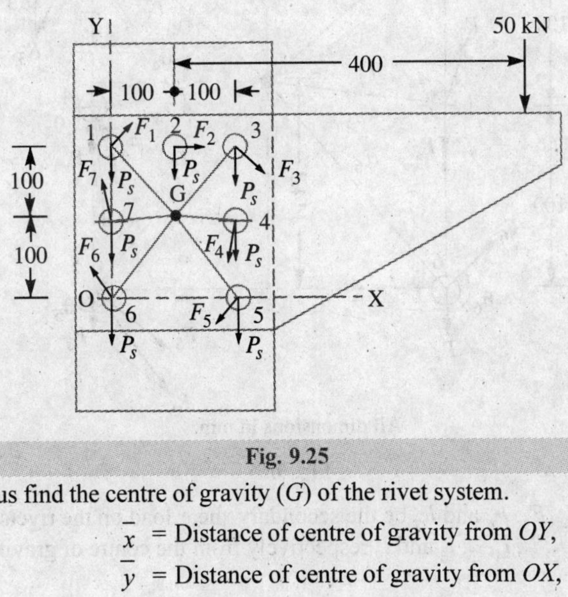

Fig. 9.25

First of all, let us find the centre of gravity (G) of the rivet system.

Let $\bar{x}$ = Distance of centre of gravity from OY,

$\bar{y}$ = Distance of centre of gravity from OX,

$x_1, x_2, x_3...$ = Distances of centre of gravity of each rivet from OY, and

$y_1, y_2, y_3...$ = Distances of centre of gravity of each rivet from OX.

We know that
$$\bar{x} = \frac{x_1 + x_2 + x_3 + x_4 + x_5 + x_6 + x_7}{n}$$
$$= \frac{100 + 200 + 200 + 200}{7} = 100 \text{ mm} \quad ...(\because x_1 = x_6 = x_7 = 0)$$

and
$$\bar{y} = \frac{y_1 + y_2 + y_3 + y_4 + y_5 + y_6 + y_7}{n}$$
$$= \frac{200 + 200 + 200 + 100 + 100}{7} = 114.3 \text{ mm} \quad ...(\because y_5 = y_6 = 0)$$

∴ The centre of gravity (G) of the rivet system lies at a distance of 100 mm from OY and 114.3 mm from OX, as shown in Fig. 9.25.

We know that direct shear load on each rivet,
$$P_s = \frac{P}{n} = \frac{50 \times 10^3}{7} = 7143 \text{ N}$$

The direct shear load acts parallel to the direction of load P *i.e.* vertically downward as shown in Fig. 9.25.

Turning moment produced by the load P due to eccentricity (e)
$$= P \times e = 50 \times 10^3 \times 400 = 20 \times 10^6 \text{ N-mm}$$

This turning moment is resisted by seven rivets as shown in Fig. 9.25.

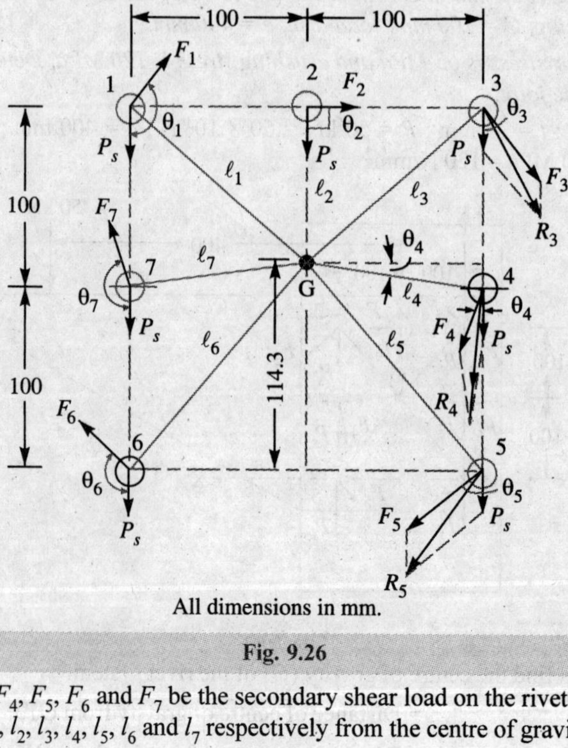

All dimensions in mm.

Fig. 9.26

Let $F_1, F_2, F_3, F_4, F_5, F_6$ and F_7 be the secondary shear load on the rivets 1, 2, 3, 4, 5, 6 and 7 placed at distances $l_1, l_2, l_3, l_4, l_5, l_6$ and l_7 respectively from the centre of gravity of the rivet system as shown in Fig. 9.26.

From the geometry of the figure, we find that

$$l_1 = l_3 = \sqrt{(100)^2 + (200 - 114.3)^2} = 131.7 \text{ mm}$$

$$l_2 = 200 - 114.3 = 85.7 \text{ mm}$$

$$l_4 = l_7 = \sqrt{(100)^2 + (114.3 - 100)^2} = 101 \text{ mm}$$

and
$$l_5 = l_6 = \sqrt{(100)^2 + (114.3)^2} = 152 \text{ mm}$$

Now equating the turning moment due to eccentricity of the load to the resisting moment of the rivets, we have

$$P \times e = \frac{F_1}{l_1}\left[(l_1)^2 + (l_2)^2 + (l_3)^2 + (l_4)^2 + (l_5)^2 + (l_6)^2 + (l_7)^2\right]$$

$$= \frac{F_1}{l_1}\left[2(l_1)^2 + (l_2)^2 + 2(l_4)^2 + 2(l_5)^2\right]$$

....($\because$ $l_1 = l_3$; $l_4 = l_7$ and $l_5 = l_6$)

$$50 \times 10^3 \times 400 = \frac{F_1}{131.7}\left[2(131.7)^2 + (85.7)^2 + 2(101)^2 + 2(152)^2\right]$$

$$20 \times 10^6 \times 131.7 = F_1(34\,690 + 7345 + 20\,402 + 46\,208) = 108\,645\, F_1$$

$$\therefore \quad F_1 = 20 \times 10^6 \times 131.7 / 108\,645 = 24\,244 \text{ N}$$

Since the secondary shear loads are proportional to their radial distances from the centre of gravity, therefore

$$F_2 = F_1 \times \frac{l_2}{l_1} = 24\,244 \times \frac{85.7}{131.7} = 15\,776 \text{ N}$$

$$F_3 = F_1 \times \frac{l_3}{l_1} = F_1 = 24\,244 \text{ N} \qquad ...(\because l_1 = l_3)$$

$$F_4 = F_1 \times \frac{l_4}{l_1} = 24\,244 \times \frac{101}{131.7} = 18\,593 \text{ N}$$

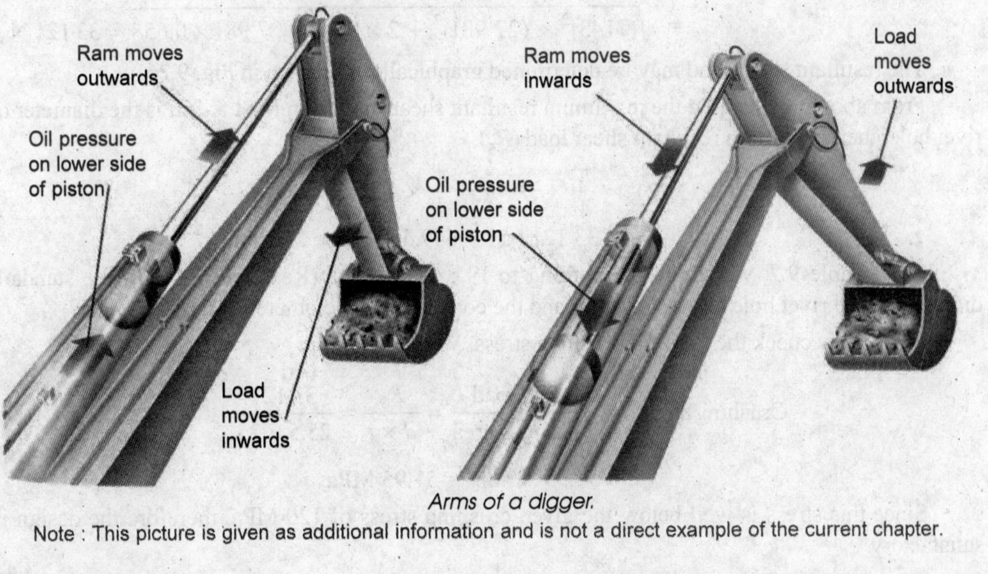

Arms of a digger.

Note : This picture is given as additional information and is not a direct example of the current chapter.

$$F_5 = F_1 \times \frac{l_5}{l_1} = 24\ 244 \times \frac{152}{131.7} = 27\ 981 \text{ N}$$

$$F_6 = F_1 \times \frac{l_6}{l_1} = F_5 = 27\ 981 \text{ N} \qquad ...(\because l_6 = l_5)$$

$$F_7 = F_1 \times \frac{l_7}{l_1} = F_4 = 18\ 593 \text{ N} \qquad ...(\because l_7 = l_4)$$

By drawing the direct and secondary shear loads on each rivet, we see that the rivets 3, 4 and 5 are heavily loaded. Let us now find the angles between the direct and secondary shear load for these three rivets. From the geometry of Fig. 9.26, we find that

$$\cos \theta_3 = \frac{100}{l_3} = \frac{100}{131.7} = 0.76$$

$$\cos \theta_4 = \frac{100}{l_4} = \frac{100}{101} = 0.99$$

and $$\cos \theta_5 = \frac{100}{l_5} = \frac{100}{152} = 0.658$$

Now resultant shear load on rivet 3,

$$R_3 = \sqrt{(P_s)^2 + (F_3)^2 + 2 P_s \times F_3 \times \cos \theta_3}$$

$$= \sqrt{(7143)^2 + (24\ 244)^2 + 2 \times 7143 \times 24\ 244 \times 0.76} = 30\ 033 \text{ N}$$

Resultant shear load on rivet 4,

$$R_4 = \sqrt{(P_s)^2 + (F_4)^2 + 2\ P_s \times F_4 \times \cos \theta_4}$$

$$= \sqrt{(7143)^2 + (18\ 593)^2 + 2 \times 7143 \times 18\ 593 \times 0.99} = 25\ 684 \text{ N}$$

and resultant shear load on rivet 5,

$$R_5 = \sqrt{(P_s)^2 + (F_5)^2 + 2\ P_s \times F_5 \times \cos \theta_5}$$

$$= \sqrt{(7143)^2 + (27\ 981)^2 + 2 \times 7143 \times 27\ 981 \times 0.658} = 33\ 121 \text{ N}$$

The resultant shear load may be determined graphically, as shown in Fig. 9.26.

From above we see that the maximum resultant shear load is on rivet 5. If d is the diameter of rivet hole, then maximum resultant shear load (R_5),

$$33\ 121 = \frac{\pi}{4} \times d^2 \times \tau = \frac{\pi}{4} \times d^2 \times 65 = 51\ d^2$$

$$\therefore \qquad d^2 = 33\ 121 / 51 = 649.4 \quad \text{or} \quad d = 25.5 \text{ mm}$$

From Table 9.7, we see that according to IS : 1929–1982 (Reaffirmed 1996), the standard diameter of the rivet hole (d) is 25.5 mm and the corresponding diameter of rivet is 24 mm.

Let us now check the joint for crushing stress. We know that

$$\text{Crushing stress} = \frac{\text{Max. load}}{\text{Crushing area}} = \frac{R_5}{d \times t} = \frac{33\ 121}{25.5 \times 25}$$

$$= 51.95 \text{ N/mm}^2 = 51.95 \text{ MPa}$$

Since this stress is well below the given crushing stress of 120 MPa, therefore the design is satisfactory.

Example 9.15. *The bracket as shown in Fig. 9.27, is to carry a load of 45 kN. Determine the size of the rivet if the shear stress is not to exceed 40 MPa. Assume all rivets of the same size.*

Solution. Given : $P = 45$ kN $= 45 \times 10^3$ N ; $\tau = 40$ MPa $= 40$ N/mm^2 ; $e = 500$ mm; $n = 9$

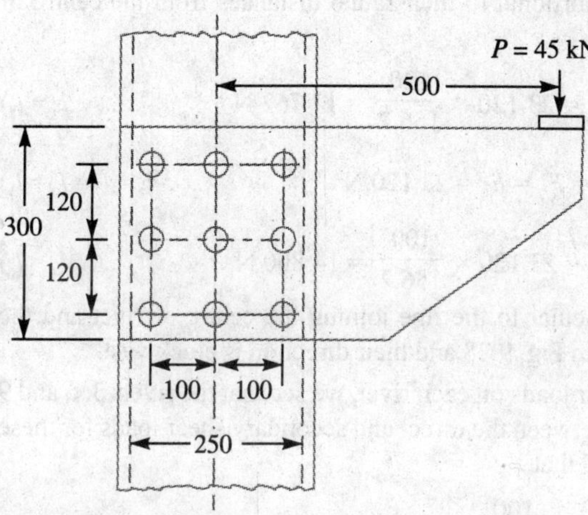

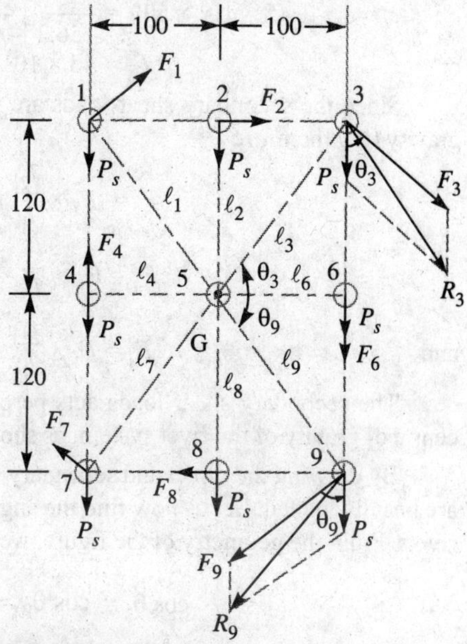

All dimensions in mm.

Fig. 9.27 **Fig. 9.28**

First of all, let us find the centre of gravity of the rivet system.

Since all the rivets are of same size and placed symmetrically, therefore the centre of gravity of the rivet system lies at G (rivet 5) as shown in Fig. 9.28.

We know that direct shear load on each rivet,

$$P_s = P / n = 45 \times 10^3 / 9 = 5000 \text{ N}$$

The direct shear load acts parallel to the direction of load P, *i.e.* vertically downward as shown in the figure.

Turning moment produced by the load P due to eccentricity e

$$= P.e = 45 \times 10^3 \times 500 = 22.5 \times 10^6 \text{ N-mm}$$

This turning moment tends to rotate the joint about the centre of gravity (G) of the rivet system in a clockwise direction. Due to this turning moment, secondary shear load on each rivet is produced. It may be noted that rivet 5 does not resist any moment.

Let F_1, F_2, F_3, F_4, F_6, F_7, F_8 and F_9 be the secondary shear load on rivets 1, 2, 3, 4, 6, 7, 8 and 9 at distances l_1, l_2, l_3, l_4, l_6, l_7, l_8 and l_9 from the centre of gravity (G) of the rivet system as shown in Fig. 9.28. From the symmetry of the figure, we find that

$$l_1 = l_3 = l_7 = l_9 = \sqrt{(100)^2 + (120)^2} = 156.2 \text{ mm}$$

Now equating the turning moment due to eccentricity of the load to the resisting moments of the rivets, we have

$$P \times e = \frac{F_1}{l_1}\left[(l_1)^2 + (l_2)^2 + (l_3)^2 + (l_4)^2 + (l_6)^2 + (l_7)^2 + (l_8)^2 + (l_9)^2\right]$$

$$= \frac{F_1}{l_1}\left[4(l_1)^2 + 2(l_2)^2 + 2(l_4)^2\right] \quad ....(\because l_1 = l_3 = l_7 = l_9;\ l_2 = l_8 \text{ and } l_4 = l_6)$$

$$\therefore \quad 45 \times 10^3 \times 500 = \frac{F_1}{156.2}\left[4(156.2)^2 + 2(120)^2 + 2(100)^2\right] = 973.2\ F_1$$

or $$F_1 = 45 \times 10^3 \times 500 / 973.2 = 23\ 120 \text{ N}$$

Since the secondary shear loads are proportional to their radial distances from the centre of gravity (G), therefore

$$F_2 = F_1 \times \frac{l_2}{l_1} = F_8 = 23\ 120 \times \frac{120}{156.2} = 17\ 762 \text{ N} \quad ...(\because l_2 = l_8)$$

$$F_3 = F_1 \times \frac{l_3}{l_1} = F_1 = F_7 = F_9 = 23\ 120 \text{ N} \quad ...(\because l_3 = l_7 = l_9 = l_1)$$

and $$F_4 = F_1 \times \frac{l_4}{l_1} = F_6 = 23\ 120 \times \frac{100}{156.2} = 14\ 800 \text{ N} \quad ...(\because l_4 = l_6)$$

The secondary shear loads acts perpendicular to the line joining the centre of rivet and the centre of gravity of the rivet system, as shown in Fig. 9.28 and their direction is clockwise.

By drawing the direct and secondary shear loads on each rivet, we see that the rivets 3, 6 and 9 are heavily loaded. Let us now find the angle between the direct and secondary shear loads for these rivets. From the geometry of the figure, we find that

$$\cos\theta_3 = \cos\theta_9 = \frac{100}{l_3} = \frac{100}{156.2} = 0.64$$

$\therefore$ Resultant shear load on rivets 3 and 9,

$$R_3 = R_9 = \sqrt{(P_s)^2 + (F_3)^3 + 2\,P_s \times F_3 \times \cos\theta_3}$$

$$= \sqrt{(5000)^2 + (23\ 120)^2 + 2 \times 5000 \times 23\ 120 \times 0.64} = 26\ 600 \text{ N}$$

$$...(\because F_3 = F_9 \text{ and } \cos\theta_3 = \cos\theta_9)$$

and resultant shear load on rivet 6,

$$R_6 = P_s + F_6 = 5000 + 14\ 800 = 19\ 800 \text{ N}$$

The resultant shear load (R_3 or R_9) may be determined graphically as shown in Fig. 9.28.

From above we see that the maximum resultant shear load is on rivets 3 and 9.

If d is the diameter of the rivet hole, then maximum resultant shear load (R_3),

$$26\ 600 = \frac{\pi}{4} \times d^2 \times \tau = \frac{\pi}{4} \times d^2 \times 40 = 31.42\ d^2$$

$$\therefore \quad d^2 = 26\ 600 / 31.42 = 846 \quad \text{or} \quad d = 29 \text{ mm}$$

From Table 9.7, we see that according to IS : 1929 – 1982 (Reaffirmed 1996), the standard diameter of the rivet hole (d) is 29 mm and the corresponding diameter of the rivet is 27 mm. **Ans.**

Example 9.16. *Find the value of P for the joint shown in Fig. 9.29 based on a working shear stress of 100 MPa for the rivets. The four rivets are equal, each of 20 mm diameter.*

Solution. Given : $\tau = 100$ MPa $= 100$ N/mm^2 ; $n = 4$; $d = 20$ mm

We know that the direct shear load on each rivet,

$$P_s = \frac{P}{n} = \frac{P}{4} = 0.25\ P$$

The direct shear load on each rivet acts in the direction of the load P, as shown in Fig. 9.30. The centre of gravity of the rivet group will lie at E (because of symmetry). From Fig. 9.30, we find that the perpendicular distance from the centre of gravity E to the line of action of the load (or eccentricity),

$$EC = e = 100 \text{ mm}$$

∴ Turning moment produced by the load at the centre of gravity (E) of the rivet system due to eccentricity

$$= P.e = P \times 100 \text{ N-mm (anticlockwise)}$$

This turning moment is resisted by four rivets as shown in Fig. 9.30. Let F_A, F_B, F_C and F_D be the secondary shear load on the rivets, A, B, C, and D placed at distances l_A, l_B, l_C and l_D respectively from the centre of gravity of the rivet system.

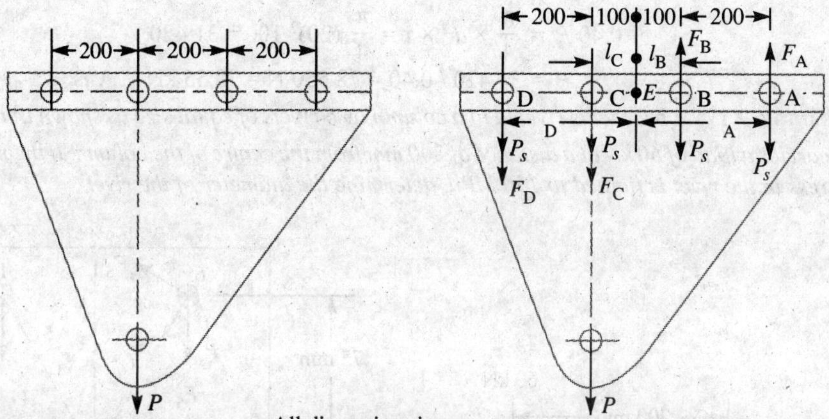

All dimensions in mm.

Fig. 9.29 Fig. 9.30

From Fig. 9.30, we find that

$$l_A = l_D = 200 + 100 = 300 \text{ mm} \text{ ; and } l_B = l_C = 100 \text{ mm}$$

We know that

$$P \times e = \frac{F_A}{l_A}\left[(l_A)^2 + (l_B)^2 + (l_C)^2 + (l_D)^2\right] = \frac{F_A}{l_A}\left[2(l_A)^2 + 2(l_B)^2\right]$$

$$\ldots(\because l_A = l_D \text{ and } l_B = l_C)$$

$$P \times 100 = \frac{F_A}{300}\left[2\,(300)^2 + 2\,(100)^2\right] = \frac{2000}{3} \times F_A$$

$$\therefore \quad F_A = P \times 100 \times 3 / 2000 = 3\,P/20 = 0.15\,P \text{ N}$$

Since the secondary shear loads are proportional to their radial distances from the centre of gravity, therefore

$$F_B = F_A \times \frac{l_B}{l_A} = \frac{3\,P}{20} \times \frac{100}{300} = 0.05\,P\text{N}$$

$$F_C = F_A \times \frac{l_C}{l_A} = \frac{3\,P}{20} \times \frac{100}{300} = 0.05\,P\text{N}$$

and

$$F_D = F_A \times \frac{l_D}{l_A} = \frac{3\,P}{20} \times \frac{300}{300} = 0.15\,P\text{N}$$

The secondary shear loads on each rivet act at right angles to the lines joining the centre of the rivet to the centre of gravity of the rivet system as shown in Fig. 9.30.

Now let us find out the resultant shear load on each rivet. From Fig. 9.30, we find that

Resultant load on rivet A,

$$R_A = P_s - F_A = 0.25\,P - 0.15\,P = 0.10\,P$$

Resultant load on rivet *B*,

$$R_B = P_s - F_B = 0.25\,P - 0.05\,P = 0.20\,P$$

Resultant load on rivet *C*,

$$R_C = P_s + F_C = 0.25\,P + 0.05\,P = 0.30\,P$$

and resultant load on rivet *D*,

$$R_D = P_s + F_D = 0.25\,P + 0.15\,P = 0.40\,P$$

From above we see that the maximum shear load is on rivet *D*. We know that the maximum shear load (R_D),

$$0.40\,P = \frac{\pi}{4} \times d^2 \times \tau = \frac{\pi}{4}\,(20)^2\,100 = 31\,420$$

$$\therefore \quad P = 31\,420 / 0.40 = 78\,550 \text{ N} = 78.55 \text{ kN} \quad \textbf{Ans.}$$

Example 9.17. *A bracket is riveted to a column by 6 rivets of equal size as shown in Fig. 9.31. It carries a load of 60 kN at a distance of 200 mm from the centre of the column. If the maximum shear stress in the rivet is limited to 150 MPa, determine the diameter of the rivet.*

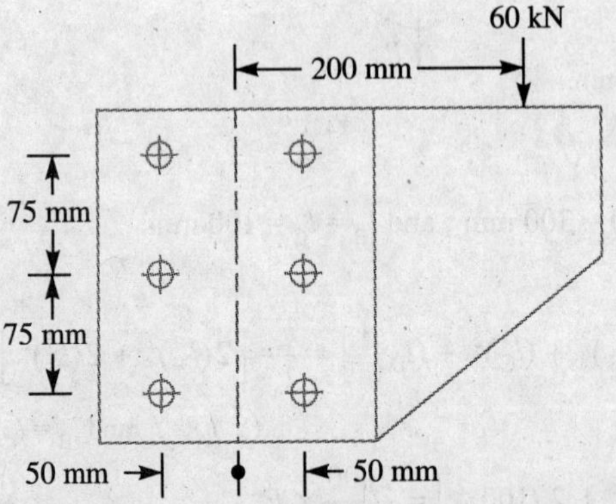

Fig. 9.31

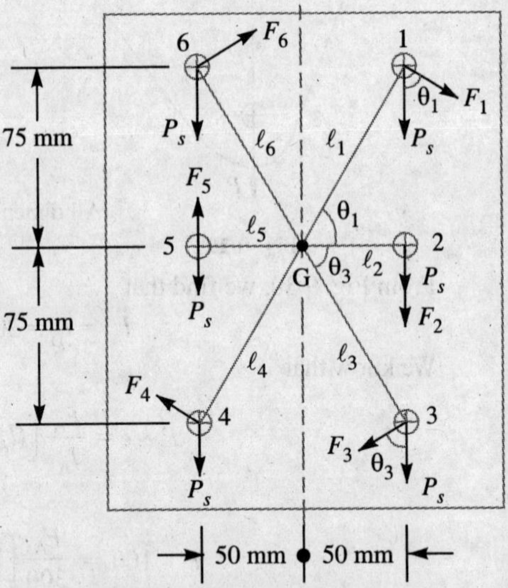

Fig. 9.32

Solution. Given : $n = 6$; $P = 60 \text{ kN} = 60 \times 10^3$ N ; $e = 200$ mm ; $\tau = 150 \text{ MPa} = 150 \text{ N/mm}^2$

Since the rivets are of equal size and placed symmetrically, therefore the centre of gravity of the rivet system lies at *G* as shown in Fig. 9.32. We know that ditect shear load on each rivet,

$$P_s = \frac{P}{n} = \frac{60 \times 10^3}{6} = 10\,000 \text{ N}$$

Let F_1, F_2, F_3, F_4, F_5 and F_6 be the secondary shear load on the rivets 1, 2, 3, 4, 5 and 6 at distances l_1, l_2, l_3, l_4, l_5 and l_6 from the centre of gravity (*G*) of the rivet system. From the symmetry of the figure, we find that

$$l_1 = l_3 = l_4 = l_6 = \sqrt{(75)^2 + (50)^2} = 90.1 \text{ mm}$$

and $$l_2 = l_5 = 50 \text{ mm}$$

Now equating the turning moment due to eccentricity of the load to the resisting moments of the rivets, we have

$$P \times e = \frac{F_1}{l_1}\left[(l_1)^2 + (l_2)^2 + (l_3)^2 + (l_4)^2 + (l_5)^2 + (l_6)^2\right]$$

$$= \frac{F_1}{l_1}\left[4(l_1)^2 + 2(l_2)^2\right]$$

$$\therefore \quad 60 \times 10^3 \times 200 = \frac{F_1}{90.1}\,[4(90.1)^2 + 2(50)^2] = 416\,F_1$$

or
$$F_1 = 60 \times 10^3 \times 200 / 416 = 28\,846 \text{ N}$$

Since the secondary shear loads are proportional to the radial distances from the centre of gravity, therefore

$$F_2 = F_1 \times \frac{l_2}{l_1} = 28\,846 \times \frac{50}{90.1} = 16\,008 \text{ N}$$

$$F_3 = F_1 \times \frac{l_3}{l_1} = F_1 = 28\,846 \text{ N} \qquad \ldots(\because l_3 = l_1)$$

$$F_4 = F_1 \times \frac{l_4}{l_1} = F_1 = 28\,846 \text{ N} \qquad \ldots(\because l_4 = l_1)$$

$$F_5 = F_1 \times \frac{l_5}{l_1} = F_2 = 16\,008 \text{ N} \qquad \ldots(\because l_5 = l_2)$$

and
$$F_6 = F_1 \times \frac{l_6}{l_1} = F_1 = 28\,846 \text{ N} \qquad \ldots(\because l_6 = l_1)$$

By drawing the direct and secondary shear loads on each rivet, we see that the rivets 1, 2 and 3 are heavily loaded. Let us now find the angles between the direct and secondary shear loads for these three rivets. From the geometry of the figure, we find that

$$\cos\theta_1 = \cos\theta_3 = \frac{50}{l_1} = \frac{50}{90.1} = 0.555$$

Excavator in action

∴ Resultant shear load on rivets 1 and 3,

$$R_1 = R_3 = \sqrt{(P_s)^2 + (F_1)^2 + 2\, P_s \times F_1 \times \cos\theta_1}$$

...($\because F_1 = F_3$ and $\cos\theta_1 = \cos\theta_3$)

$$= \sqrt{(10\ 000)^2 + (28\ 846)^2 + 2 \times 10\ 000 \times 28\ 846 \times 0.555}$$

$$= \sqrt{100 \times 10^6 + 832 \times 10^6 + 320 \times 10^6} = 35\ 348 \text{ N}$$

and resultant shear load on rivet 2,

$$R_2 = P_s + F_2 = 10\ 000 + 16\ 008 = 26\ 008 \text{ N}$$

From above we see that the maximum resultant shear load is on rivets 1 and 3. If d is the diameter of rivet hole, then maximum resultant shear load (R_1 or R_3),

$$35\ 384 = \frac{\pi}{4} \times d^2 \times \tau = \frac{\pi}{4} \times d^2 \times 150 = 117.8\, d^2$$

$$\therefore \quad d^2 = 35\ 384 / 117.8 = 300.4 \quad \text{or} \quad d = 17.33 \text{ mm}$$

From Table 9.7, we see that according to IS : 1929 – 1982 (Reaffirmed 1996), the standard diameter of the rivet hole (d) is 19.5 mm and the corresponding diameter of the rivet is 18 mm. **Ans.**

Example 9.18. *A bracket in the form of a plate is fitted to a column by means of four rivets A, B, C and D in the same vertical line, as shown in Fig. 9.33. AB = BC = CD = 60 mm. E is the mid-point of BC. A load of 100 kN is applied to the bracket at a point F which is at a horizontal distance of 150 m from E. The load acts at an angle of 30° to the horizontal. Determine the diameter of the rivets which are made of steel having a yield stress in shear of 240 MPa. Take a factor of safety of 1.5.*

What would be the thickness of the plate taking an allowable bending stress of 125 MPa for the plate, assuming its total width at section ABCD as 240 mm?

Solution. Given : $n = 4$; $AB = BC = CD = 60$ mm ; $P = 100$ kN $= 100 \times 10^3$ N; $EF = 150$ mm; $\theta = 30°$; $\tau_y = 240$ MPa $= 240$ N/mm² ; $F.S. = 1.5$; $\sigma_b = 125$ MPa $= 125$ N/mm² ; $b = 240$ mm

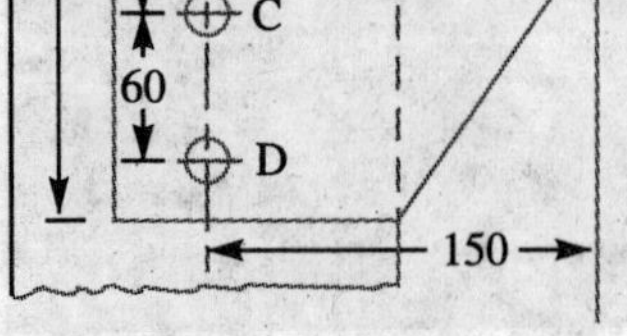

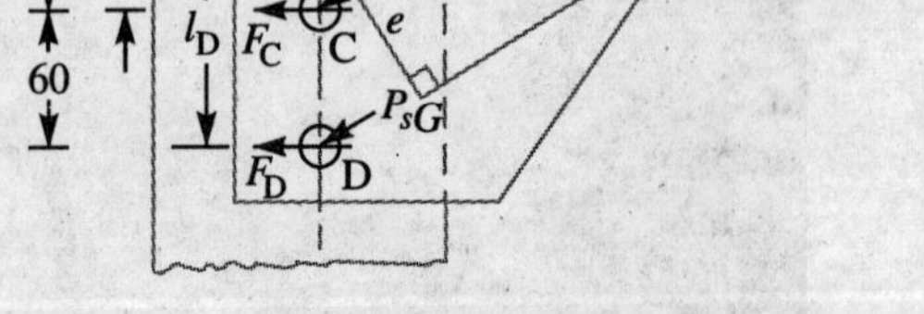

All dimensions in mm.

Fig. 9.33

Fig. 9.34

Diameter of rivets

Let d = Diameter of rivets.

We know that direct shear load on each rivet,

$$P_s = \frac{P}{n} = \frac{100 \times 10^3}{4} = 25\ 000 \text{ N}$$

The direct shear load on each rivet acts in the direction of 100 kN load (*i.e.* at 30° to the horizontal) as shown in Fig. 9.34. The centre of gravity of the rivet group lies at *E*. From Fig. 9.34, we find that the perpendicular distance from the centre of gravity *E* to the line of action of the load (or eccentricity of the load) is

$$EG = e = EF \sin 30° = 150 \times \frac{1}{2} = 75 \text{ mm}$$

∴ Turning moment produced by the load *P* due to eccentricity

$$= P.e = 100 \times 10^3 \times 75 = 7500 \times 10^3 \text{ N-mm}$$

This turning moment is resisted by four bolts, as shown in Fig. 9.34. Let F_A, F_B, F_C and F_D be the secondary shear load on the rivets, *A*, *B*, *C*, and *D* placed at distances l_A, l_B, l_C and l_D respectively from the centre of gravity of the rivet system.

From Fig. 9.34, we find that

$$l_A = l_D = 60 + 30 = 90 \text{ mm and } l_B = l_C = 30 \text{ mm}$$

We know that

$$P \times e = \frac{F_A}{l_A}[(l_A)^2 + (l_B)^2 + (l_C)^2 + (l_D)^2] = \frac{F_A}{l_A}\left[2(l_A)^2 + 2(l_B)^2\right]$$

...($\because l_A = l_D$ and $l_B = l_C$)

$$7500 \times 10^3 = \frac{F_A}{90}\left[2(90)^2 + 2(30)^2\right] = 200\ F_A$$

$$\therefore \quad F_A = 7500 \times 10^3 / 200 = 37\ 500 \text{ N}$$

Since the secondary shear loads are proportional to their radial distances from the centre of gravity, therefore,

$$F_B = F_A \times \frac{l_B}{l_A} = 37\ 500 \times \frac{30}{90} = 12\ 500 \text{ N}$$

$$F_C = F_A \times \frac{l_C}{l_A} = 37\ 500 \times \frac{30}{90} = 12\ 500 \text{ N}$$

and

$$F_D = F_A \times \frac{l_D}{l_A} = 37\ 500 \times \frac{90}{90} = 37\ 500 \text{ N}$$

Now let us find the resultant shear load on each rivet.

From Fig. 9.34, we find that angle between F_A and $P_s = \theta_A = 150°$

Angle between F_B and $P_s = \theta_B = 150°$

Angle between F_C and $P_s = \theta_C = 30°$

Angle between F_D and $P_s = \theta_D = 30°$

∴ Resultant load on rivet *A*,

$$R_A = \sqrt{(P_s)^2 + (F_A)^2 + 2P_s \times F_A \times \cos\theta_A}$$

$$= \sqrt{(25\ 000)^2 + (37\ 500)^2 + 2 \times 25\ 000 \times 37\ 500 \times \cos 150°}$$

$$= \sqrt{625 \times 10^6 + 1406 \times 10^6 - 1623.8 \times 10^6} = 15\ 492 \text{ N}$$

Resultant shear load on rivet *B*,

$$R_B = \sqrt{(P_s)^2 + (F_B)^2 + 2P_s \times F_B \times \cos\theta_B}$$

$$= \sqrt{(25\ 000)^2 + (12\ 500)^2 + 2 \times 25\ 000 \times 12\ 500 \times \cos 150°}$$

$$= \sqrt{625 \times 10^6 + 156.25 \times 10^6 - 541.25 \times 10^6} = 15\ 492 \text{ N}$$

Resultant shear load on rivet *C*,

$$R_C = \sqrt{(P_s)^2 + (F_C)^2 + 2P_s \times F_C \times \cos\theta_C}$$

$$= \sqrt{(25\ 000)^2 + (12\ 500)^2 + 2 \times 25\ 000 \times 12\ 500 \times \cos 30°}$$

$$= \sqrt{625 \times 10^6 + 156.25 \times 10^6 + 541.25 \times 10^6} = 36\ 366 \text{ N}$$

and resultant shear load on rivet *D*,

$$R_D = \sqrt{(P_s)^2 + (F_D)^2 + 2P_s \times F_D \times \cos\theta_D}$$

$$= \sqrt{(25\ 000)^2 + (37\ 500)^2 + 2 \times 25\ 000 \times 37\ 500 \times \cos 30°}$$

$$= \sqrt{625 \times 10^6 + 1406 \times 10^6 + 1623.8 \times 10^6} = 60\ 455 \text{ N}$$

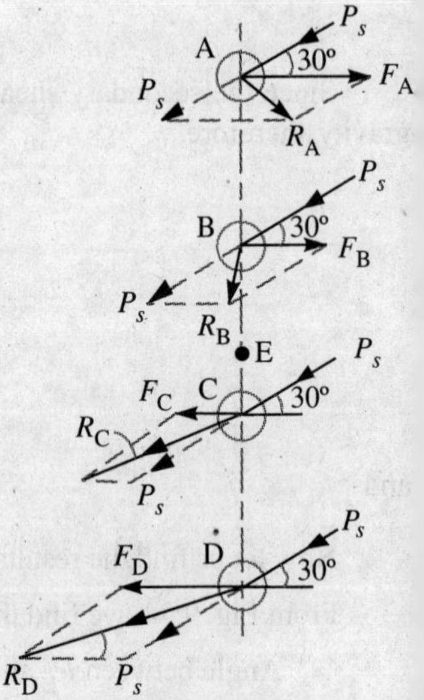

Fig. 9.35

The resultant shear load on each rivet may be determined graphically as shown in Fig. 9.35.

From above we see that the maximum resultant shear load is on rivet *D*. We know that maximum resultant shear load (R_D),

$$60\ 455 = \frac{\pi}{4} \times d^2 \times \tau = \frac{\pi}{4} \times d^2 \times \frac{\tau_y}{F.S.}$$

$$= \frac{\pi}{4} \times d^2 \times \frac{240}{1.5} = 125.7\ d^2$$

$\therefore$ $$d^2 = 60\ 455 / 125.7 = 481$$

or $$d = 21.9 \text{ mm}$$

From Table 9.7, we see that the standard diameter of the rivet hole (d) is 23.5 mm and the corresponding diameter of rivet is 22 mm. **Ans.**

Thickness of the plate

Let t = Thickness of the plate in mm,

σ_b = Allowable bending stress for the plate

= 125 MPa = 125 N/mm² ...(Given)

b = Width of the plate = 240 mm ...(Given)

Consider the weakest section of the plate (*i.e.* the section where it receives four rivet holes of diameter 23.5 mm and thickness *t* mm) as shown in Fig. 9.36. We know that moment of inertia of the plate about *X-X*,

$$I_{XX} = \text{M.I. of solid plate about } X\text{-}X - {}^*\text{M.I. of 4 rivet holes about } X\text{-}X$$

* M.I. of four rivet holes about *X-X*

= M.I. of four rivet holes about their centroidal axis + $2A(h_1)^2 + 2A(h_2)^2$

where *A* = Area of rivet hole.

$$= \frac{1}{12} \times t\,(240)^3 - \left[4 \times \frac{1}{12} \times t\,(23.5)^3 + 2 \times t \times 23.5\,(30^2 + 90^2)\right]$$

$$= 1152 \times 10^3\, t - [4326\, t + 423 \times 10^3\, t] = 724\,674\, t \text{ mm}^4$$

Bending moment,

$$M = P \times e = 100 \times 10^3 \times 75$$

$$= 7500 \times 10^3 \text{ N-mm}$$

Distance of neutral axis (*X–X*) from the top most fibre of the plate,

$$y = \frac{b}{2} = \frac{240}{2} = 120 \text{ mm}$$

We know that $\dfrac{M}{I} = \dfrac{\sigma_b}{y}$

or $\dfrac{7500 \times 10^3}{724\,674\, t} = \dfrac{125}{120}$

$\therefore \dfrac{10.35}{t} = 1.04$ or $t = \dfrac{10.35}{1.04} = 9.95$ say 10 mm **Ans.**

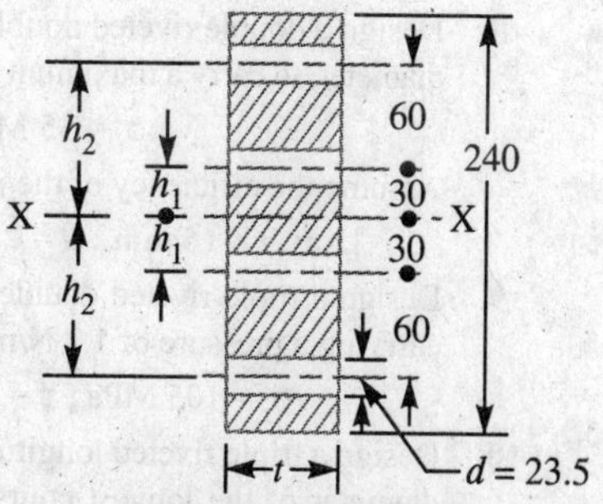

All dimensions in mm.

Fig. 9.36

EXERCISES

1. A single riveted lap joint is made in 15 mm thick plates with 20 mm diameter rivets. Determine the strength of the joint, if the pitch of rivets is 60 mm. Take $\sigma_t = 120$ MPa; $\tau = 90$ MPa and $\sigma_c = 160$ MPa. **[Ans. 28 280 N]**
2. Two plates 16 mm thick are joined by a double riveted lap joint. The pitch of each row of rivets is 90 mm. The rivets are 25 mm in diameter. The permissible stresses are as follows :

   $\sigma_t = 140$ MPa ; $\tau = 110$ MPa and $\sigma_c = 240$ MPa

   Find the efficiency of the joint. **[Ans. 53.5%]**
3. A single riveted double cover butt joint is made in 10 mm thick plates with 20 mm diameter rivets with a pitch of 60 mm. Calculate the efficiency of the joint, if

   $\sigma_t = 100$ MPa ; $\tau = 80$ MPa and $\sigma_c = 160$ MPa. **[Ans. 53.8%]**
4. A double riveted double cover butt joint is made in 12 mm thick plates with 18 mm diameter rivets. Find the efficiency of the joint for a pitch of 80 mm, if

   $\sigma_t = 115$ MPa ; $\tau = 80$ MPa and $\sigma_c = 160$ MPa. **[Ans. 62.6%]**
5. A double riveted lap joint with chain riveting is to be made for joining two plates 10 mm thick. The allowable stresses are : $\sigma_t = 60$ MPa ; $\tau = 50$ MPa and $\sigma_c = 80$ MPa. Find the rivet diameter, pitch of rivets and distance between rows of rivets. Also find the efficiency of the joint.

   [Ans. $d = 20$ mm ; $p = 73$ mm; $p_b = 38$ mm; $\eta = 71.7\%$]
6. A triple riveted lap joint with zig-zag riveting is to be designed to connect two plates of 6 mm thickness. Determine the dia. of rivet, pitch of rivets and distance between the rows of rivet. Indicate how the joint will fail. Assume : $\sigma_t = 120$ MPa ; $\tau = 100$ MPa and $\sigma_c = 150$ MPa.

   [Ans. $d = 14$ mm ; $p = 78$ mm; $p_b = 35.2$ mm]
7. A double riveted butt joint, in which the pitch of the rivets in the outer rows is twice that in the inner rows, connects two 16 mm thick plates with two cover plates each 12 mm thick. The diameter of rivets is 22 mm. Determine the pitches of the rivets in the two rows if the working stresses are not to exceed the following limits:

Tensile stress in plates = 100 MPa ; Shear stress in rivets = 75 MPa; and bearing stress in rivets and plates = 150 MPa.

Make a fully dimensioned sketch of the joint by showing at least two views.

[Ans. 107 mm, 53.5 mm]

8. Design a double riveted double strap butt joint for the longitudinal seam of a boiler shell, 750 mm in diameter, to carry a maximum steam pressure of 1.05 N/mm^2 gauge. The allowable stresses are :s

$$\sigma_t = 35 \text{ MPa; } \tau = 28 \text{ MPa and } \sigma_c = 52.5 \text{ MPa}$$

Assume the efficiency of the joint as 75%.

[Ans. t = 16 mm ; d = 25 mm ; p = 63 mm ; p_b = 37.5 mm ; $t_1 = t_2$ = 10 mm ; m = 37.5 mm]

9. Design a triple riveted double strap butt joint with chain riveting for a boiler of 1.5 m diameter and carrying a pressure of 1.2 N/mm^2. The allowable stresses are :

$\sigma_t = 105$ MPa ; $\tau = 77$ MPa and $\sigma_c = 162.5$ MPa **[Ans. d = 20 mm; p = 50 mm]**

10. Design a triple riveted longitudinal double strap butt joint with unequal straps for a boiler. The inside diameter of the longest course of the drum is 1.3 metres. The joint is to be designed for a steam pressure of 2.4 N/mm^2. The working stresses to be used are :

$$\sigma_t = 77 \text{ MPa; } \tau = 62 \text{ MPa and } \sigma_c = 120 \text{ MPa}$$

Assume the efficiency of the joint as 81%.

[Ans. t = 26 mm; d = 31.5 mm ; p = 200 mm ; t_1 = 19.5 mm ; t_2 = 16.5 mm ; m = 47.5 mm]

11. Design the longitudinal and circumferential joint for a boiler whose diameter is 2.4 metres and is subjected to a pressure of 1 N/mm^2. The longitudinal joint is a triple riveted butt joint with an efficiency of about 85% and the circumferential joint is a double riveted lap joint with an efficiency of about 70%. The pitch in the outer rows of the rivets is to be double than in the inner rows and the width of the cover plates is unequal. The allowable stresses are :

$$\sigma_t = 77 \text{ MPa ; } \tau = 56 \text{ MPa and } \sigma_c = 120 \text{ MPa}$$

Assume that the resistance of rivets in double shear is 1.875 times that of single shear. Draw the complete joint.

12. A triple riveted butt joint with equal double cover plates (zig-zag riveting) is used for the longitudinal joint of a Lancashire boiler of 2.5 m internal diameter. The working steam pressure is 1.12 N/mm^2 and the efficiency of the joint is 85 per cent. Calculate the plate thickness for mild steel of 460 MPa ultimate tensile strength. Assume ratio of tensile to shear stresses as 7/6 and factor of safety 4. The resistance of the rivets in double shear is to be taken as 1.875 times that of single shear. Design a suitable circumferential joint also.

13. Two lengths of mild steel flat tie bars 200 mm × 10 mm are to be connected by a double riveted double cover butt joint, using 24 mm diameter rivets. Design the joint, if the allowable working stresses are 112 MPa in tension, 84 MPa in shear and 200 MPa in crushing.

[Ans. n = 5; η = 88%]

14. Two mild steel tie bars for a bridge structure are to be joined by a double cover butt joint. The thickness of the tie bar is 20 mm and carries a tensile load of 400 kN. Design the joint if the allowable stresses are : $\sigma_t = 90$ MPa ; $\tau = 75$ MPa and $\sigma_c = 150$ MPa.

Assume the strength of rivet in double shear to be 1.75 times that of in single shear.

[Ans. b = 150 mm ; d = 27 mm ; n = 6 ; η = 90%]

15. Two lengths of mild steel tie rod having width 200 mm are to be connected by means of Lozenge joint with two cover plates to withstand a tensile load of 180 kN. Completely design the joint, if the permissible stresses are 80 MPa in tension; 65 MPa in shear and 160 MPa in crushing. Draw a neat sketch of the joint.

[Ans. t = 13 mm ; d = 22 mm ; n = 5 ; η = 86.5%]

16. A bracket is supported by means of 4 rivets of same size, as shown in Fig. 9.37. Determine the diameter of the rivet if the maximum shear stress is 140 MPa. **[Ans. 16 mm]**

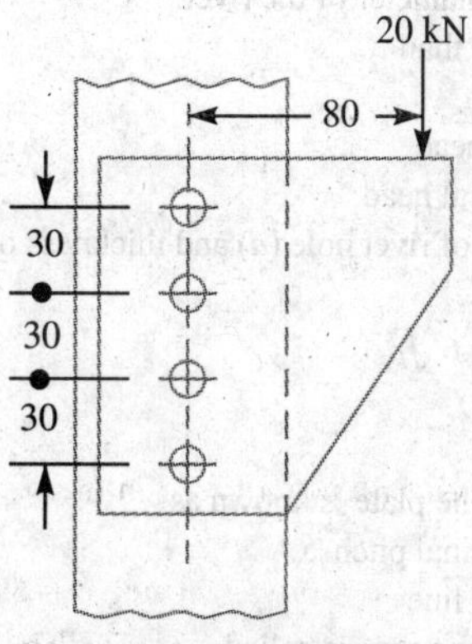

Fig. 9.37

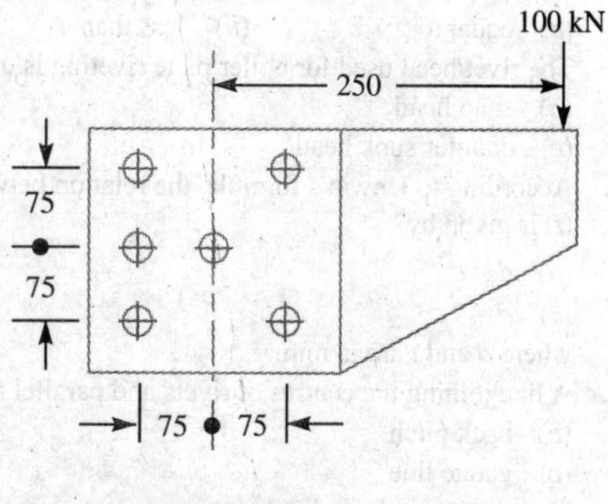

Fig. 9.38

17. A bracket is riveted to a columm by 6 rivets of equal size as shown in Fig. 9.38.
It carries a load of 100 kN at a distance of 250 mm from the column. If the maximum shear stress in the rivet is limited to 63 MPa, find the diameter of the rivet. **[Ans. 41 mm]**

18. A bracket in the form of a plate is fitted to a column by means of four rivets of the same size, as shown in Fig. 9.39. A load of 100 kN is applied to the bracket at an angle of 60° to the horizontal and the line of action of the load passes through the centre of the bottom rivet. If the maximum shear stress for the material of the rivet is 70 MPa, find the diameter of rivets. What will be the thickness of the plate if the crushing stress is 100 MPa? **[Ans. 29 mm; 1.5 mm]**

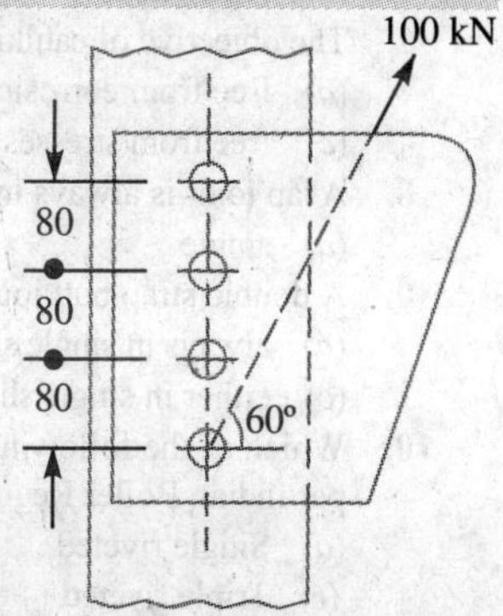

Fig. 9.39

QUESTIONS

1. What do you understand by the term riveted joint? Explain the necessity of such a joint.
2. What are the various permanent and detachable fastenings? Give a complete list with the different types of each category.
3. Classify the rivet heads according to Indian standard specifications.
4. What is the material used for rivets?
5. Enumerate the different types of riveted joints and rivets.
6. What is an economical joint and where does it find applications?
7. What is the difference between caulking and fullering? Explain with the help of neat sketches.
8. Show by neat sketches the various ways in which a riveted joint may fail.
9. What do you understand by the term 'efficiency of a riveted joint'? According to I.B.R., what is the highest efficiency required of a riveted joint?
10. Explain the procedure for designing a longitudinal and circumferential joint for a boiler.
11. Describe the procedure for designing a lozenge joint.
12. What is an eccentric riveted joint? Explain the method adopted for designing such a joint?

OBJECTIVE TYPE QUESTIONS

1. A rivet is specified by

(*a*) shank diameter	(*b*) length of rivet
(*c*) type of head	(*d*) length of tail

2. The diameter of the rivet hole is usually .... the nominal diameter of the rivet.
(a) equal to (b) less than (c) more than
3. The rivet head used for boiler plate riveting is usually
(a) snap head (b) pan head
(c) counter sunk head (d) conical head
4. According to Unwin's formula, the relation between diameter of rivet hole (d) and thickness of plate (t) is given by
(a) $d = t$ (b) $d = 1.6\sqrt{t}$
(c) $d = 2\,t$ (d) $d = 6\,t$
where d and t are in mm.
5. A line joining the centres of rivets and parallel to the edge of the plate is known as
(a) back pitch (b) marginal pitch
(c) gauge line (d) pitch line
6. The centre to centre distance between two consecutive rivets in a row, is called
(a) margin (b) pitch
(c) back pitch (d) diagonal pitch
7. The objective of caulking in a riveted joint is to make the joint
(a) free from corrosion (b) stronger in tension
(c) free from stresses (d) leak-proof
8. A lap joint is always in .....shear.
(a) single (b) double
9. A double strap butt joint (with equal straps) is
(a) always in single shear (b) always in double shear
(c) either in single shear or double shear (d) any one of these
10. Which of the following riveted butt joints with double straps should have the highest efficiency as per Indian Boiler Regulations?
(a) Single riveted (b) Double riveted
(c) Triple riveted (d) Quadruple riveted
11. If the tearing efficiency of a riveted joint is 50%, then ratio of diameter of rivet hole to the pitch of rivets is
(a) 0.20 (b) 0.30
(c) 0.50 (d) 0.60
12. The strength of the unriveted or solid plate per pitch length is equal to
(a) $p \times d \times \sigma_t$ (b) $p \times t \times \sigma_t$
(c) $(p - t)\,d \times \sigma_t$ (d) $(p - d)\,t \times \sigma_t$
13. The longitudinal joint in boilers is used to get the required
(a) length of boiler (b) diameter of boiler
(c) length and diameter of boiler (d) efficiency of boiler
14. For longitudinal joint in boilers, the type of joint used is
(a) lap joint with one ring overlapping the other (b) butt joint with single cover plate
(c) butt joint with double cover plates (d) any one of these
15. According to Indian standards, the diameter of rivet hole for a 24 mm diameter of rivet, should be
(a) 23 mm (b) 24 mm
(c) 25 mm (d) 26 mm

ANSWERS

1. (a)	2. (c)	3. (a)	4. (d)	5. (b)
6. (b)	7. (d)	8. (a)	9. (b)	10. (d)
11. (c)	12. (b)	13. (b)	14. (c)	15. (c)

C H A P T E R

10

Welded Joints

1. *Introduction.*
2. *Advantages and Disadvantages of Welded Joints over Riveted Joints.*
3. *Welding Processes.*
4. *Fusion Welding.*
5. *Thermit Welding.*
6. *Gas Welding.*
7. *Electric Arc Welding.*
8. *Forge Welding.*
9. *Types of Welded Joints.*
10. *Lap Joint.*
11. *Butt Joint.*
12. *Basic Weld Symbols.*
13. *Supplementary Weld Symbols.*
14. *Elements of a Weld Symbol.*
15. *Standard Location of Elements of a Welding Symbol.*
16. *Strength of Transverse Fillet Welded Joints.*
17. *Strength of Parallel Fillet Welded Joints.*
18. *Special Cases of Fillet Welded Joints.*
19. *Strength of Butt Joints.*
20. *Stresses for Welded Joints.*
21. *Stress Concentration Factor for Welded Joints.*
22. *Axially Loaded Unsymmetrical Welded Sections.*
23. *Eccentrically Loaded Welded Joints.*
24. *Polar Moment of Inertia and Section Modulus of Welds.*

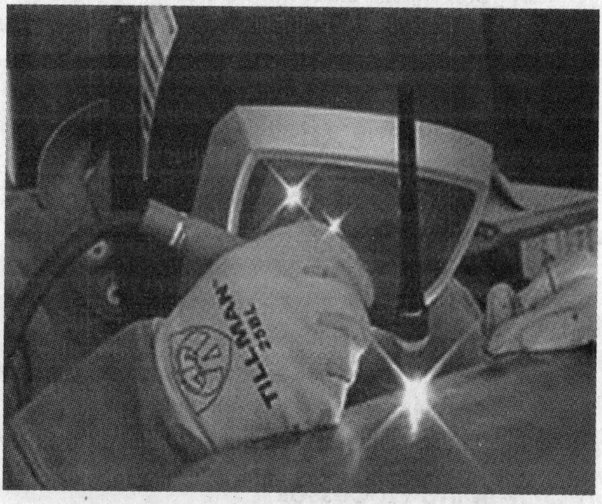

10.1 Introduction

A welded joint is a permanent joint which is obtained by the fusion of the edges of the two parts to be joined together, with or without the application of pressure and a filler material. The heat required for the fusion of the material may be obtained by burning of gas (in case of gas welding) or by an electric arc (in case of electric arc welding). The latter method is extensively used because of greater speed of welding.

Welding is extensively used in fabrication as an alternative method for casting or forging and as a replacement for bolted and riveted joints. It is also used as a repair medium *e.g.* to reunite metal at a crack, to build up a small part that has broken off such as gear tooth or to repair a worn surface such as a bearing surface.

10.2 Advantages and Disadvantages of Welded Joints over Riveted Joints

Following are the advantages and disadvantages of welded joints over riveted joints.

Advantages

1. The welded structures are usually lighter than riveted structures. This is due to the reason, that in welding, gussets or other connecting components are not used.
2. The welded joints provide maximum efficiency (may be 100%) which is not possible in case of riveted joints.
3. Alterations and additions can be easily made in the existing structures.
4. As the welded structure is smooth in appearance, therefore it looks pleasing.
5. In welded connections, the tension members are not weakened as in the case of riveted joints.
6. A welded joint has a great strength. Often a welded joint has the strength of the parent metal itself.
7. Sometimes, the members are of such a shape (*i.e.* circular steel pipes) that they afford difficulty for riveting. But they can be easily welded.
8. The welding provides very rigid joints. This is in line with the modern trend of providing rigid frames.
9. It is possible to weld any part of a structure at any point. But riveting requires enough clearance.
10. The process of welding takes less time than the riveting.

Disadvantages

1. Since there is an uneven heating and cooling during fabrication, therefore the members may get distorted or additional stresses may develop.
2. It requires a highly skilled labour and supervision.
3. Since no provision is kept for expansion and contraction in the frame, therefore there is a possibility of cracks developing in it.
4. The inspection of welding work is more difficult than riveting work.

10.3 Welding Processes

The welding processes may be broadly classified into the following two groups:

1. Welding processes that use heat alone *e.g.* fusion welding.
2. Welding processes that use a combination of heat and pressure *e.g.* forge welding.

These processes are discussed in detail, in the following pages.

10.4 Fusion Welding

In case of fusion welding, the parts to be jointed are held in position while the molten metal is supplied to the joint. The molten metal may come from the parts themselves (*i.e.* parent metal) or filler metal which normally have the composition of the parent metal. The joint surface become plastic or even molten because of the heat

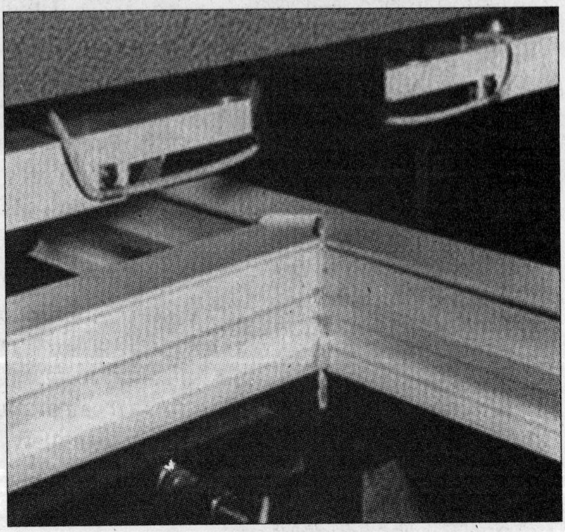

Fusion welding at 245°C produces permanent molecular bonds between sections.

from the molten filler metal or other source. Thus, when the molten metal solidifies or fuses, the joint is formed.

The fusion welding, according to the method of heat generated, may be classified as:

1. Thermit welding, 2. Gas welding, and 3. Electric arc welding.

10.5 Thermit Welding

In thermit welding, a mixture of iron oxide and aluminium called ***thermit*** is ignited and the iron oxide is reduced to molten iron. The molten iron is poured into a mould made around the joint and fuses with the parts to be welded. A major advantage of the thermit welding is that all parts of weld section are molten at the same time and the weld cools almost uniformly. This results in a minimum problem with residual stresses. It is fundamentally a melting and casting process.

The thermit welding is often used in joining iron and steel parts that are too large to be manufactured in one piece, such as rails, truck frames, locomotive frames, other large sections used on steam and rail roads, for stern frames, rudder frames etc. In steel mills, thermit electric welding is employed to replace broken gear teeth, to weld new necks on rolls and pinions, and to repair broken shears.

10.6 Gas Welding

A gas welding is made by applying the flame of an oxy-acetylene or hydrogen gas from a welding torch upon the surfaces of the prepared joint. The intense heat at the white cone of the flame heats up the local surfaces to fusion point while the operator manipulates a welding rod to supply the metal for the weld. A flux is being used to remove the slag. Since the heating rate in gas welding is slow, therefore it can be used on thinner materials.

10.7 Electric Arc Welding

In electric arc welding, the work is prepared in the same manner as for gas welding. In this case the filler metal is supplied by metal welding electrode. The operator, with his eyes and face protected, strikes an arc by touching the work of base metal with the electrode. The base metal in the path of the arc stream is melted, forming a pool of molten metal, which seems to be forced out of the pool by the blast from the arc, as shown in Fig. 10.1. A small depression is formed in the base metal and the molten metal is deposited around the edge of this depression, which is called the ***arc crater.*** The slag is brushed off after the joint has cooled.

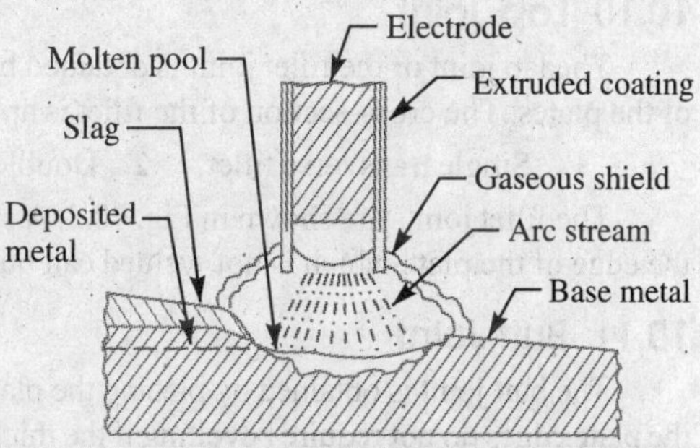

Fig. 10.1. Shielded electric arc welding.

The arc welding does not require the metal to be preheated and since the temperature of the arc is quite high, therefore the fusion of the metal is almost instantaneous. There are two kinds of arc weldings depending upon the type of electrode.

1. Un-shielded arc welding, and
2. Shielded arc welding.

When a large electrode or filler rod is used for welding, it is then said to be ***un-shielded arc welding.*** In this case, the deposited weld metal while it is hot will absorb oxygen and nitrogen from the atmosphere. This decreases the strength of weld metal and lower its ductility and resistance to corrosion.

In ***shielded arc welding,*** the welding rods coated with solid material are used, as shown in Fig. 10.1. The resulting projection of coating focuses a concentrated arc stream, which protects the globules of metal from the air and prevents the absorption of large amounts of harmful oxygen and nitrogen.

10.8 Forge Welding

In forge welding, the parts to be jointed are first heated to a proper temperature in a furnace or forge and then hammered. This method of welding is rarely used now-a-days. An ***electric-resistance welding*** is an example of forge welding.

In this case, the parts to be joined are pressed together and an electric current is passed from one part to the other until the metal is heated to the fusion temperature of the joint. The principle of applying heat and pressure, either sequentially or simultaneously, is widely used in the processes known as ****spot, seam, projection, upset and flash welding.***

Forge welding.

10.9 Types of Welded Joints

Following two types of welded joints are important from the subject point of view:

1. Lap joint or fillet joint, and 2. Butt joint.

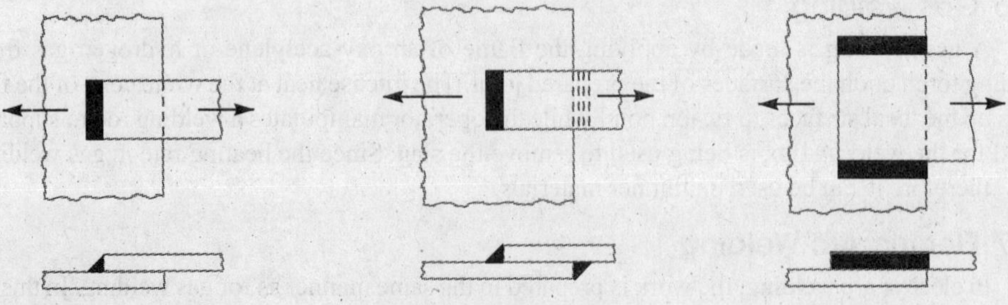

(*a*) Single transverse. (*b*) Double transverse. (*c*) Parallel fillet.

Fig. 10.2. Types of lap or fillet joints.

10.10 Lap Joint

The lap joint or the fillet joint is obtained by overlapping the plates and then welding the edges of the plates. The cross-section of the fillet is approximately triangular. The fillet joints may be

1. Single transverse fillet, 2. Double transverse fillet, and 3. Parallel fillet joints.

The fillet joints are shown in Fig. 10.2. A single transverse fillet joint has the disadvantage that the edge of the plate which is not welded can buckle or warp out of shape.

10.11 Butt Joint

The butt joint is obtained by placing the plates edge to edge as shown in Fig. 10.3. In butt welds, the plate edges do not require bevelling if the thickness of plate is less than 5 mm. On the other hand, if the plate thickness is 5 mm to 12.5 mm, the edges should be bevelled to V or U-groove on both sides.

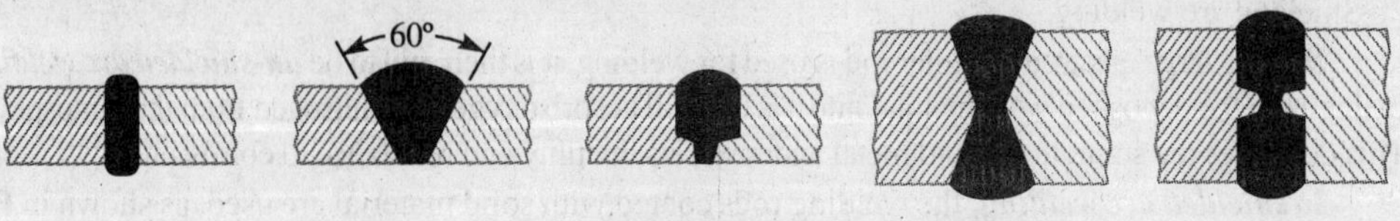

(*a*) Square butt joint. (*b*) Single *V*-butt joint. (*c*) Single *U*-butt joint. (*d*) Double *V*-butt joint. (*e*) Double *U*-butt joint.

Fig. 10.3. Types of butt joints.

* For further details, refer author's popular book **'A Textbook of Workshop Technology'**.

The butt joints may be

1. Square butt joint,
2. Single V-butt joint
3. Single U-butt joint,
4. Double V-butt joint, and
5. Double U-butt joint.

These joints are shown in Fig. 10.3.

The other type of welded joints are corner joint, edge joint and T-joint as shown in Fig. 10.4.

(*a*) Corner joint. (*b*) Edge joint. (*c*) *T*-joint.

Fig. 10.4. Other types of welded joints.

The main considerations involved in the selection of weld type are:

1. The shape of the welded component required,
2. The thickness of the plates to be welded, and
3. The direction of the forces applied.

10.12 Basic Weld Symbols

The basic weld symbols according to IS : 813 – 1961 (Reaffirmed 1991) are shown in the following table.

Table 10.1. Basic weld symbols.

S. No.	*Form of weld*	*Sectional representation*	*Symbol*
1.	Fillet		
2.	Square butt		
3.	Single-*V* butt		
4.	Double-*V* butt		
5.	Single-*U* butt		
6.	Double-*U* butt	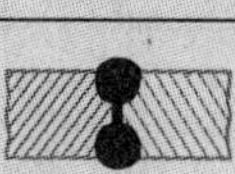	
7.	Single bevel butt	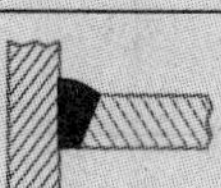	
8.	Double bevel butt	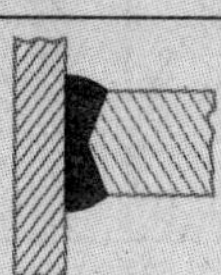	

S. No.	Form of weld	Sectional representation	Symbol
9.	Single-*J* butt	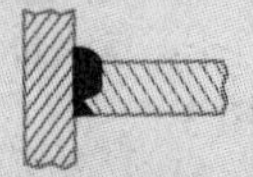	
10.	Double-*J* butt	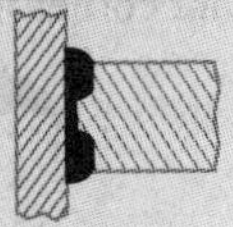	
11.	Bead (edge or seal)		
12.	Stud	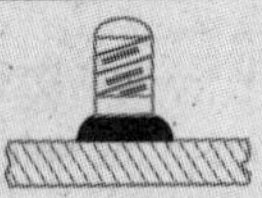	
13.	Sealing run		
14.	Spot	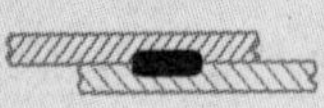	
15.	Seam	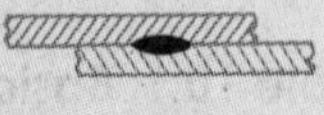	
16.	Mashed seam	 Before  After	
17.	Plug	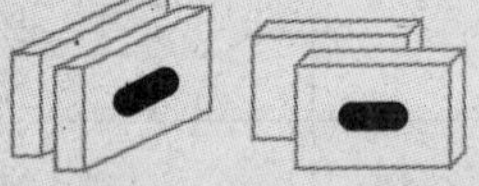	
18.	Backing strip		
19.	Stitch	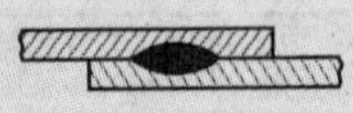	
20.	Projection	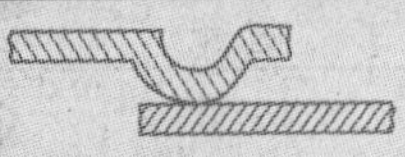 Before 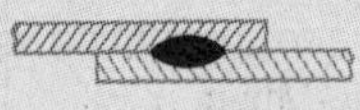 After	
21.	Flash	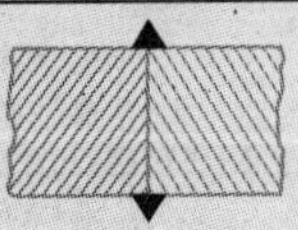 Rod or bar 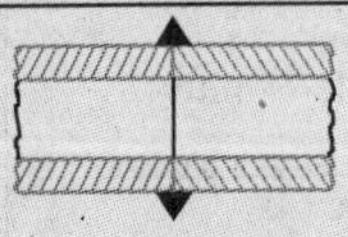 Tube	
22.	Butt resistance or pressure (upset)	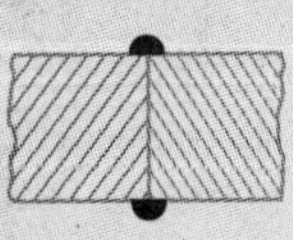 Rod or bar 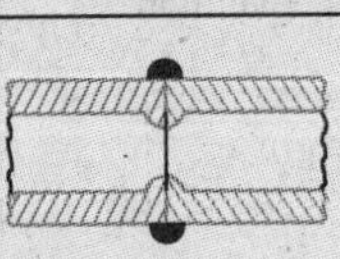 Tube	

10.13 Supplementary Weld Symbols

In addition to the above symbols, some supplementary symbols, according to IS:813 – 1961 (Reaffirmed 1991), are also used as shown in the following table.

Table 10.2. Supplementary weld symbols.

S. No.	*Particulars*	*Drawing representation*	*Symbol*
1.	Weld all round		○
2.	Field weld		●
3.	Flush contour	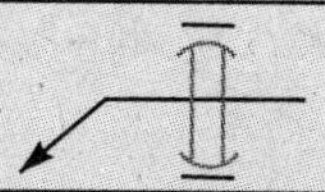	—
4.	Convex contour		⌒
5.	Concave contour	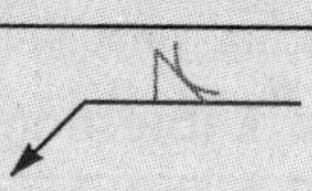	◡
6.	Grinding finish	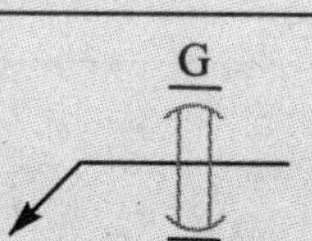	G
7.	Machining finish	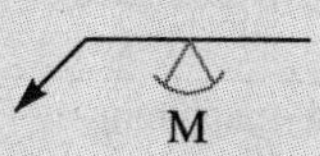	M
8.	Chipping finish		C

10.14 Elements of a Welding Symbol

A welding symbol consists of the following eight elements:

1. Reference line,
2. Arrow,
3. Basic weld symbols,
4. Dimensions and other data,
5. Supplementary symbols,
6. Finish symbols,
7. Tail, and
8. Specification, process or other references.

10.15 Standard Location of Elements of a Welding Symbol

According to Indian Standards, IS: 813 – 1961 (Reaffirmed 1991), the elements of a welding symbol shall have standard locations with respect to each other.

The arrow points to the location of weld, the basic symbols with dimensions are located on one or both sides of reference line. The specification if any is placed in the tail of arrow. Fig. 10.5 shows the standard locations of welding symbols represented on drawing.

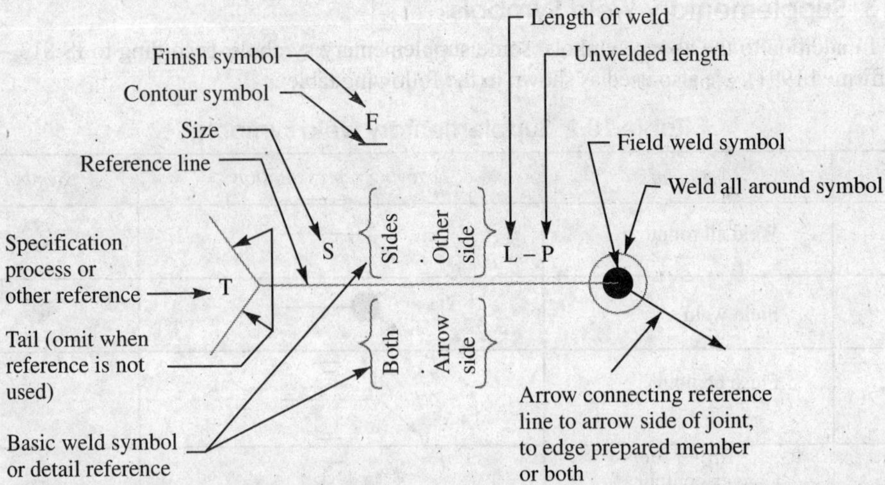

Fig. 10.5. Standard location of welding symbols.

Some of the examples of welding symbols represented on drawing are shown in the following table.

Table 10.3. Representation of welding symbols.

S. No.	*Desired weld*	*Representation on drawing*
1.	Fillet-weld each side of Tee- convex contour	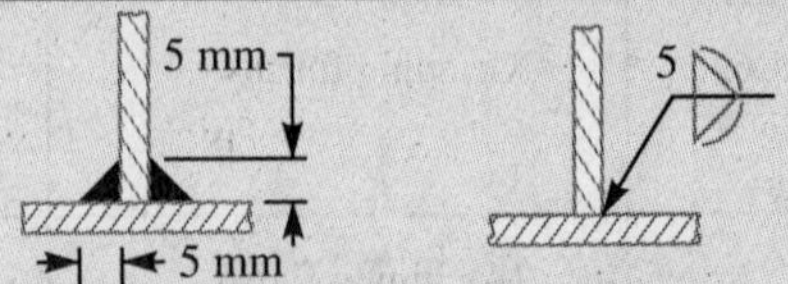
2.	Single V-butt weld -machining finish	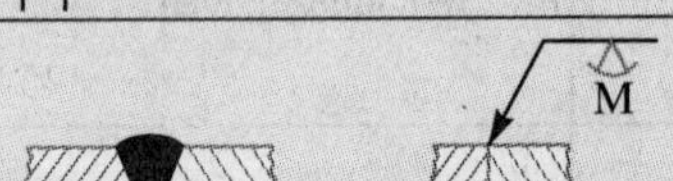
3.	Double V- butt weld	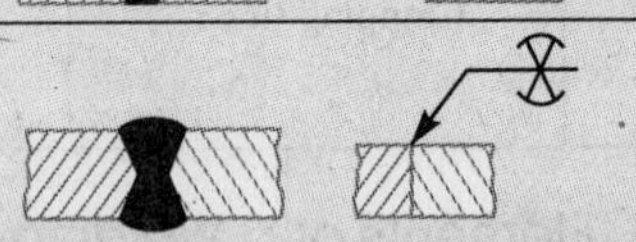
4.	Plug weld - 30° Groove-angle-flush contour	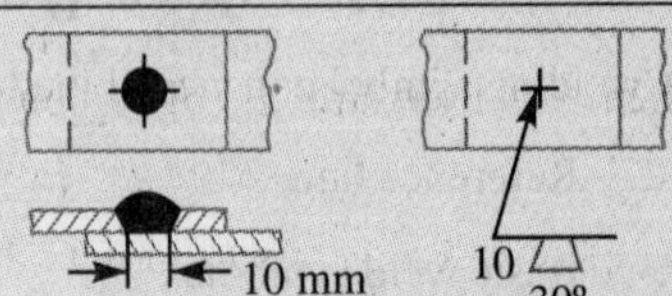
5.	Staggered intermittent fillet welds	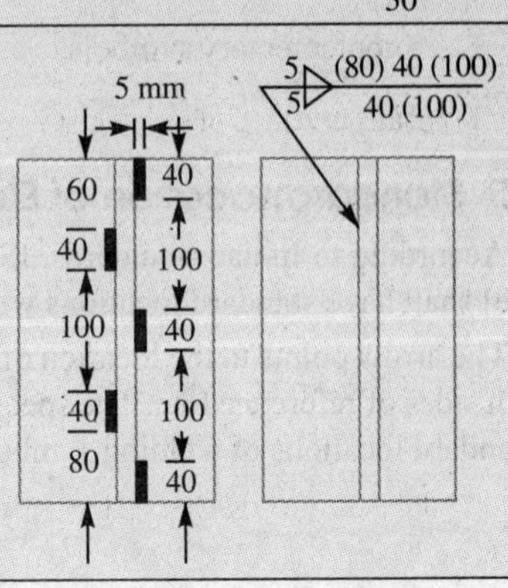

10.16 Strength of Transverse Fillet Welded Joints

We have already discussed that the fillet or lap joint is obtained by overlapping the plates and then welding the edges of the plates. The transverse fillet welds are designed for tensile strength. Let us consider a single and double transverse fillet welds as shown in Fig. 10.6 (*a*) and (*b*) respectively.

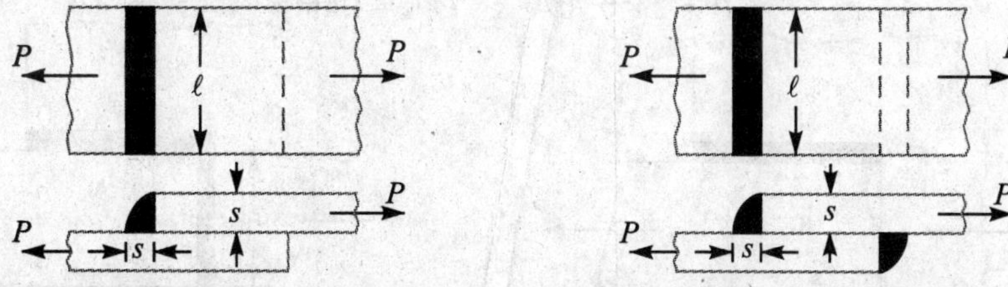

(*a*) Single transverse fillet weld. (*b*) Double transverse fillet weld.

Fig. 10.6. Transverse fillet welds.

In order to determine the strength of the fillet joint, it is assumed that the section of fillet is a right angled triangle *ABC* with hypotenuse *AC* making equal angles with other two sides *AB* and *BC*. The enlarged view of the fillet is shown in Fig. 10.7. The length of each side is known as ***leg*** or ***size of the weld*** and the perpendicular distance of the hypotenuse from the intersection of legs (*i.e. BD*) is known as ***throat thickness.*** The minimum area of the weld is obtained at the throat *BD*, which is given by the product of the throat thickness and length of weld.

Let t = Throat thickness (*BD*),

s = Leg or size of weld,

= Thickness of plate, and

l = Length of weld,

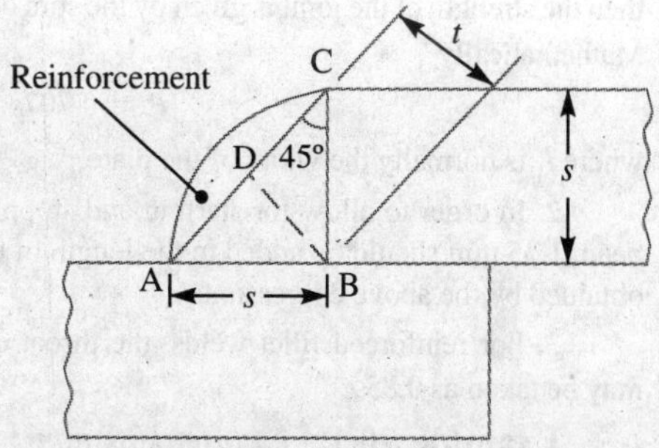

Fig. 10.7. Enlarged view of a fillet weld.

From Fig. 10.7, we find that the throat thickness,

$$t = s \times \sin 45° = 0.707\, s$$

∴ *Minimum area of the weld or throat area,

$$A = \text{Throat thickness} \times \text{Length of weld} = t \times l = 0.707\, s \times l$$

If σ_t is the allowable tensile stress for the weld metal, then the tensile strength of the joint for single fillet weld,

$$P = \text{Throat area} \times \text{Allowable tensile stress} = 0.707\, s \times l \times \sigma_t$$

and tensile strength of the joint for double fillet weld,

$$P = 2 \times 0.707\, s \times l \times \sigma_t = 1.414\, s \times l \times \sigma_t$$

Note: Since the weld is weaker than the plate due to slag and blow holes, therefore the weld is given a reinforcement which may be taken as 10% of the plate thickness.

10.17 Strength of Parallel Fillet Welded Joints

The parallel fillet welded joints are designed for shear strength. Consider a double parallel fillet welded joint as shown in Fig. 10.8 (*a*). We have already discussed in the previous article, that the minimum area of weld or the throat area,

$$A = 0.707\, s \times l$$

* The minimum area of the weld is taken because the stress is maximum at the minimum area.

If τ is the allowable shear stress for the weld metal, then the shear strength of the joint for single parallel fillet weld,

$$P = \text{Throat area} \times \text{Allowable shear stress} = 0.707\ s \times l \times \tau$$

and shear strength of the joint for double parallel fillet weld,

$$P = 2 \times 0.707 \times s \times l \times \tau = 1.414\ s \times l \times \tau$$

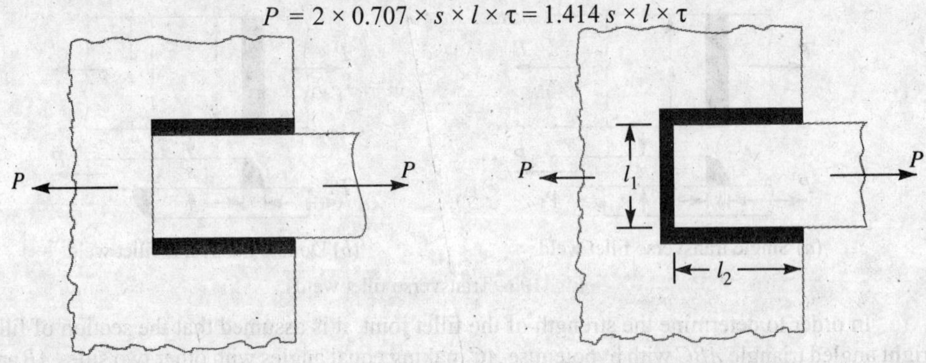

(a) Double parallel fillet weld.

(b) Combination of transverse and parallel fillet weld.

Fig. 10.8

Notes: 1. If there is a combination of single transverse and double parallel fillet welds as shown in Fig. 10.8 (*b*), then the strength of the joint is given by the sum of strengths of single transverse and double parallel fillet welds. Mathematically,

$$P = 0.707s \times l_1 \times \sigma_t + 1.414\ s \times l_2 \times \tau$$

where l_1 is normally the width of the plate.

2. In order to allow for starting and stopping of the bead, 12.5 mm should be added to the length of each weld obtained by the above expression.

3. For reinforced fillet welds, the throat dimension may be taken as 0.85 *t*.

Electric arc welding

Example 10.1. *A plate 100 mm wide and 10 mm thick is to be welded to another plate by means of double parallel fillets. The plates are subjected to a static load of 80 kN. Find the length of weld if the permissible shear stress in the weld does not exceed 55 MPa.*

Solution. Given: *Width = 100 mm ; Thickness = 10 mm ; P = 80 kN = 80×10^3 N ; τ = 55 MPa = 55 N/mm²

Let l = Length of weld, and

s = Size of weld = Plate thickness = 10 mm ... (Given)

We know that maximum load which the plates can carry for double parallel fillet weld (P),

$$80 \times 10^3 = 1.414 \times s \times l \times \tau = 1.414 \times 10 \times l \times 55 = 778\ l$$

$$\therefore \quad l = 80 \times 10^3 / 778 = 103 \text{ mm}$$

Adding 12.5 mm for starting and stopping of weld run, we have

$$l = 103 + 12.5 = 115.5 \text{ mm} \quad \textbf{Ans.}$$

* Superfluous data.

10.18 Special Cases of Fillet Welded Joints

The following cases of fillet welded joints are important from the subject point of view.

1. *Circular fillet weld subjected to torsion.* Consider a circular rod connected to a rigid plate by a fillet weld as shown in Fig. 10.9.

Let d = Diameter of rod,

r = Radius of rod,

T = Torque acting on the rod,

s = Size (or leg) of weld,

t = Throat thickness,

$^*J$ = Polar moment of inertia of the weld section $= \dfrac{\pi t d^3}{4}$

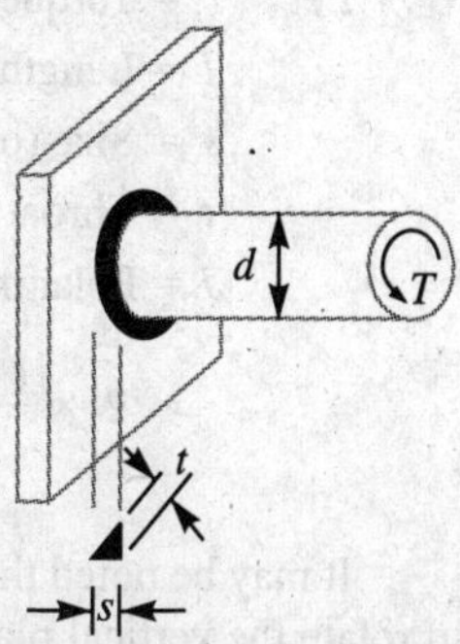

Fig. 10.9. Circular fillet weld subjected to torsion.

We know that shear stress for the material,

$$\tau = \frac{T.r}{J} = \frac{T \times d/2}{J}$$

$$= \frac{T \times d/2}{\pi t d^3/4} = \frac{2T}{\pi t d^2} \quad \ldots \left(\because \frac{T}{J} = \frac{\tau}{r}\right)$$

This shear stress occurs in a horizontal plane along a leg of the fillet weld. The maximum shear occurs on the throat of weld which is inclined at 45° to the horizontal plane.

$\therefore$ Length of throat, $t = s \sin 45° = 0.707\, s$

and maximum shear stress,

$$\tau_{max} = \frac{2T}{\pi \times 0.707\, s \times d^2} = \frac{2.83\, T}{\pi\, s\, d^2}$$

2. *Circular fillet weld subjected to bending moment.* Consider a circular rod connected to a rigid plate by a fillet weld as shown in Fig. 10.10.

Let d = Diameter of rod,

M = Bending moment acting on the rod,

s = Size (or leg) of weld,

t = Throat thickness,

$^{**}Z$ = Section modulus of the weld section $= \dfrac{\pi t d^2}{4}$

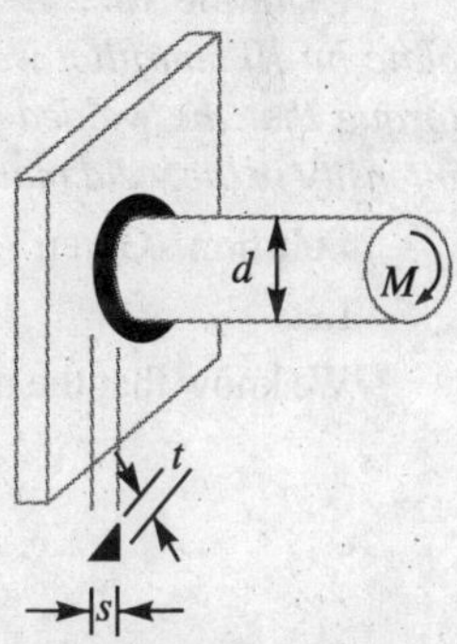

Fig. 10.10. Circular fillet weld subjected to bending moment.

We know that the bending stress,

$$\sigma_b = \frac{M}{Z} = \frac{M}{\pi t d^2/4} = \frac{4M}{\pi t d^2}$$

This bending stress occurs in a horizontal plane along a leg of the fillet weld. The maximum bending stress occurs on the throat of the weld which is inclined at 45° to the horizontal plane.

$\therefore$ Length of throat, $t = s \sin 45° = 0.707\, s$

and maximum bending stress,

$$\sigma_{b(max)} = \frac{4\, M}{\pi \times 0.707\, s \times d^2} = \frac{5.66\, M}{\pi\, s\, d^2}$$

---

* See Art, 10.24.

** See Art, 10.24.

3. ***Long fillet weld subjected to torsion.*** Consider a vertical plate attached to a horizontal plate by two identical fillet welds as shown in Fig. 10.11.

Let T = Torque acting on the vertical plate,

l = Length of weld,

s = Size (or leg) of weld,

t = Throat thickness, and

J = Polar moment of inertia of the weld section

$$= 2 \times \frac{t \times l^3}{12} = \frac{t \times l^3}{6} \quad ...$$

($\because$ of both sides weld)

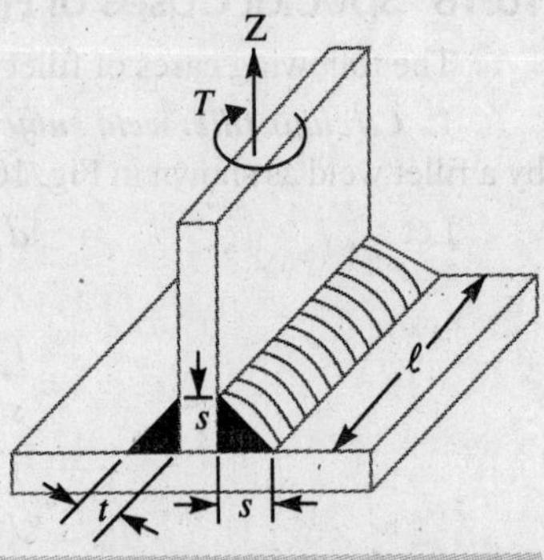

Fig. 10.11. Long fillet weld subjected to torsion.

It may be noted that the effect of the applied torque is to rotate the vertical plate about the Z-axis through its mid point. This rotation is resisted by shearing stresses developed between two fillet welds and the horizontal plate. It is assumed that these horizontal shearing stresses vary from zero at the Z-axis and maximum at the ends of the plate. This variation of shearing stress is analogous to the variation of normal stress over the depth (l) of a beam subjected to pure bending.

$$\therefore \text{ Shear stress,} \quad \tau = \frac{T \times l/2}{t \times l^3/6} = \frac{3\,T}{t \times l^2}$$

The maximum shear stress occurs at the throat and is given by

$$\tau_{max} = \frac{3\,T}{0.707\,s \times l^2} = \frac{4.242\,T}{s \times l^2}$$

Example 10.2. *A 50 mm diameter solid shaft is welded to a flat plate by 10 mm fillet weld as shown in Fig. 10.12. Find the maximum torque that the welded joint can sustain if the maximum shear stress intensity in the weld material is not to exceed 80 MPa.*

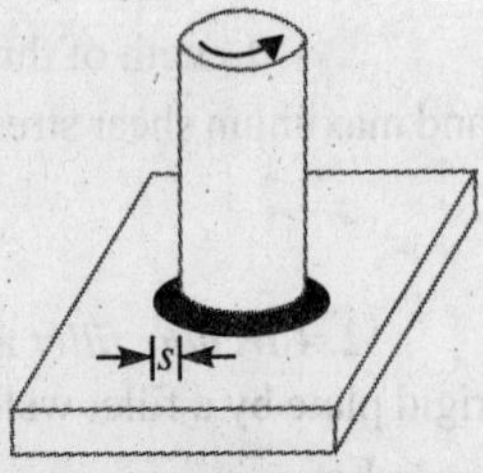

Fig. 10.12

Solution. Given : d = 50 mm ; s = 10 mm ; τ_{max} = 80 MPa = 80 N/mm^2

Let T = Maximum torque that the welded joint can sustain.

We know that the maximum shear stress (τ_{max}),

$$80 = \frac{2.83\,T}{\pi s \times d^2} = \frac{2.83\,T}{\pi \times 10\,(50)^2} = \frac{2.83\,T}{78\,550}$$

$$\therefore \quad T = 80 \times 78\,550/2.83$$

$$= 2.22 \times 10^6 \text{ N-mm} = 2.22 \text{ kN-m} \quad \textbf{Ans.}$$

Example 10.3. *A plate 1 m long, 60 mm thick is welded to another plate at right angles to each other by 15 mm fillet weld, as shown in Fig. 10.13. Find the maximum torque that the welded joint can sustain if the permissible shear stress intensity in the weld material is not to exceed 80 MPa.*

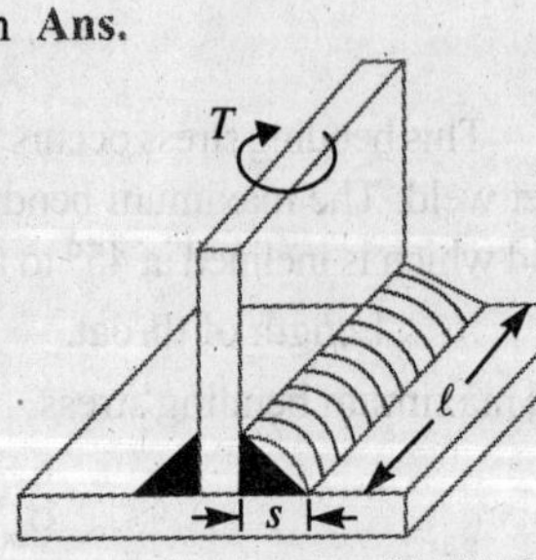

Fig. 10.13

Solution. Given: l = 1m = 1000 mm ; Thickness = 60 mm ; s = 15 mm ; τ_{max} = 80 MPa = 80 N/mm^2

Let T = Maximum torque that the welded joint can sustain.

We know that the maximum shear stress (τ_{max}),

$$80 = \frac{4.242\ T}{s \times l^2} = \frac{4.242\ T}{15\ (1000)^2} = \frac{0.283\ T}{10^6}$$

$$\therefore \quad T = 80 \times 10^6 / 0.283 = 283 \times 10^6 \text{ N-mm} = 283 \text{ kN-m} \quad \textbf{Ans.}$$

10.19 Strength of Butt Joints

The butt joints are designed for tension or compression. Consider a single V-butt joint as shown in Fig. 10.14 (*a*).

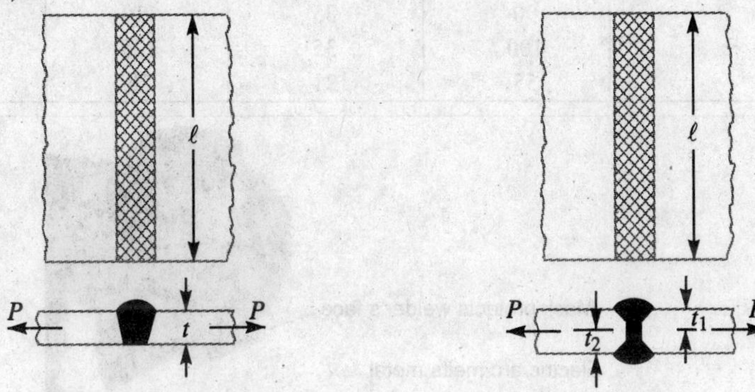

(*a*) Single *V*-butt joint. (*b*) Double *V*-butt joint.

Fig. 10.14. Butt joints.

In case of butt joint, the length of leg or size of weld is equal to the throat thickness which is equal to thickness of plates.

∴ Tensile strength of the butt joint (single-*V* or square butt joint),

$$P = t \times l \times \sigma_t$$

where l = Length of weld. It is generally equal to the width of plate.

and tensile strength for double-*V* butt joint as shown in Fig. 10.14 (*b*) is given by

$$P = (t_1 + t_2)\ l \times \sigma_t$$

where t_1 = Throat thickness at the top, and

t_2 = Throat thickness at the bottom.

It may be noted that size of the weld should be greater than the thickness of the plate, but it may be less. The following table shows recommended minimum size of the welds.

Table 10.4. Recommended minimum size of welds.

Thickness of plate (mm)	3 – 5	6 – 8	10 – 16	18 – 24	26 – 55	Over 58
Minimum size of weld (mm)	3	5	6	10	14	20

10.20 Stresses for Welded Joints

The stresses in welded joints are difficult to determine because of the variable and unpredictable parameters like homogenuity of the weld metal, thermal stresses in the welds, changes of physical properties due to high rate of cooling etc. The stresses are obtained, on the following assumptions:

1. The load is distributed uniformly along the entire length of the weld, and
2. The stress is spread uniformly over its effective section.

The following table shows the stresses for welded joints for joining ferrous metals with mild steel electrode under steady and fatigue or reversed load.

Table 10.5. Stresses for welded joints.

Type of weld	*Bare electrode*		*Coated electrode*	
	Steady load (MPa)	*Fatigue load (MPa)*	*Steady load (MPa)*	*Fatigue load (MPa)*
1. Fillet welds (All types)	80	21	98	35
2. Butt welds				
Tension	90	35	110	55
Compression	100	35	125	55
Shear	55	21	70	35

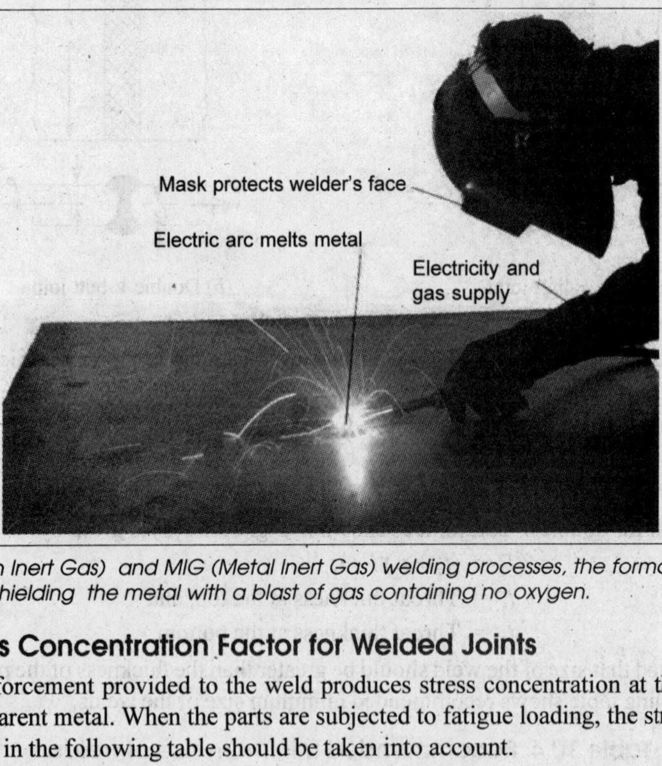

In TIG (Tungsten Inert Gas) and MIG (Metal Inert Gas) welding processes, the formation of oxide is prevented by shielding the metal with a blast of gas containing no oxygen.

10.21 Stress Concentration Factor for Welded Joints

The reinforcement provided to the weld produces stress concentration at the junction of the weld and the parent metal. When the parts are subjected to fatigue loading, the stress concentration factor as given in the following table should be taken into account.

Table 10.6. Stress concentration factor for welded joints.

Type of joint	*Stress concentration factor*
1. Reinforced butt welds	1.2
2. Toe of transverse fillet welds	1.5
3. End of parallel fillet weld	2.7
4. T-butt joint with sharp corner	2.0

Note : For static loading and any type of joint, stress concentration factor is 1.0.

Example 10.4. *A plate 100 mm wide and 12.5 mm thick is to be welded to another plate by means of parallel fillet welds. The plates are subjected to a load of 50 kN. Find the length of the weld so that the maximum stress does not exceed 56 MPa. Consider the joint first under static loading and then under fatigue loading.*

Solution. Given: *Width = 100 mm ; Thickness = 12.5 mm ; P = 50 kN = 50×10^3N ; τ = 56 MPa = 56 N/mm^2

Length of weld for static loading

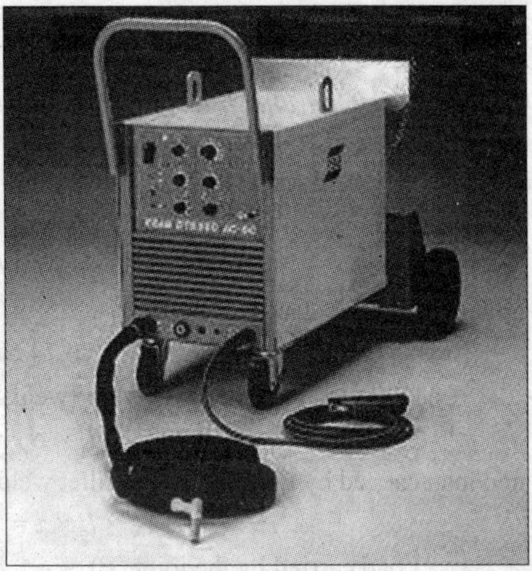

TIG (Tungsten Inert Gas) welding Machine

Let l = Length of weld, and

s = Size of weld = Plate thickness

= 12.5 mm ... (Given)

We know that the maximum load which the plates can carry for double parallel fillet welds (P),

$$50 \times 10^3 = 1.414\, s \times l \times \tau$$

$$= 1.414 \times 12.5 \times l \times 56 = 990\, l$$

$$\therefore \quad l = 50 \times 10^3 / 990 = 50.5 \text{ mm}$$

Adding 12.5 mm for starting and stopping of weld run, we have

l = 50.5 + 12.5 = 63 mm **Ans.**

Length of weld for fatigue loading

From Table 10.6, we find that the stress concentration factor for parallel fillet welding is 2.7.

$\therefore$ Permissible shear stress,

$$\tau = 56 / 2.7 = 20.74 \text{ N/mm}^2$$

We know that the maximum load which the plates can carry for double parallel fillet welds (P),

$$50 \times 10^3 = 1.414\, s \times l \times \tau = 1.414 \times 12.5 \times l \times 20.74 = 367\, l$$

$$\therefore \quad l = 50 \times 10^3 / 367 = 136.2 \text{ mm}$$

Adding 12.5 for starting and stopping of weld run, we have

l = 136.2 + 12.5 = 148.7 mm **Ans.**

Example 10.5. *A plate 75 mm wide and 12.5 mm thick is joined with another plate by a single transverse weld and a double parallel fillet weld as shown in Fig. 10.15. The maximum tensile and shear stresses are 70 MPa and 56 MPa respectively.*

Find the length of each parallel fillet weld, if the joint is subjected to both static and fatigue loading.

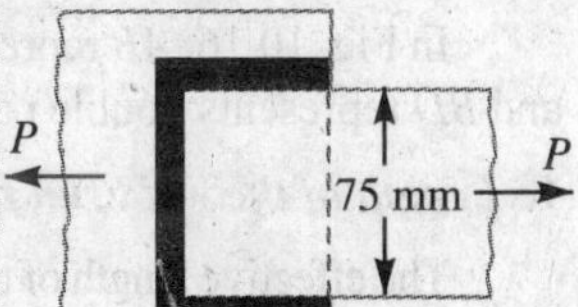

Fig. 10.15

Solution. Given : Width = 75 mm ; Thickness = 12.5 mm ; σ_τ = 70 MPa = 70 N/mm^2 ; τ = 56 MPa = 56 N/mm^2.

The effective length of weld (l_1) for the transverse weld may be obtained by subtracting 12.5 mm from the width of the plate.

$$\therefore \quad l_1 = 75 - 12.5 = 62.5 \text{ mm}$$

Length of each parallel fillet for static loading

Let l_2 = Length of each parallel fillet.

We know that the maximum load which the plate can carry is

$$P = \text{Area} \times \text{Stress} = 75 \times 12.5 \times 70 = 65\,625 \text{ N}$$

Load carried by single transverse weld,

$$P_1 = 0.707\, s \times l_1 \times \sigma_t = 0.707 \times 12.5 \times 62.5 \times 70 = 38\,664 \text{ N}$$

and the load carried by double parallel fillet weld,

$$P_2 = 1.414\, s \times l_2 \times \tau = 1.414 \times 12.5 \times l_2 \times 56 = 990\, l_2 \text{ N}$$

* Superfluous data.

∴ Load carried by the joint (P),

$$65\ 625 = P_1 + P_2 = 38\ 664 + 990\ l_2 \quad \text{or} \quad l_2 = 27.2 \text{ mm}$$

Adding 12.5 mm for starting and stopping of weld run, we have

$$l_2 = 27.2 + 12.5 = 39.7 \text{ say } 40 \text{ mm } \textbf{Ans.}$$

Length of each parallel fillet for fatigue loading

From Table 10.6, we find that the stress concentration factor for transverse welds is 1.5 and for parallel fillet welds is 2.7.

∴ Permissible tensile stress,

$$\sigma_t = 70 / 1.5 = 46.7 \text{ N/mm}^2$$

and permissible shear stress,

$$\tau = 56 / 2.7 = 20.74 \text{ N/mm}^2$$

Load carried by single transverse weld,

$$P_1 = 0.707\ s \times l_1 \times \sigma_t = 0.707 \times 12.5 \times 62.5 \times 46.7 = 25\ 795 \text{ N}$$

and load carried by double parallel fillet weld,

$$P_2 = 1.414\ s \times l_2 \times \tau = 1.414 \times 12.5\ l_2 \times 20.74 = 366\ l_2 \text{ N}$$

∴ Load carried by the joint (P),

$$65\ 625 = P_1 + P_2 = 25\ 795 + 366\ l_2 \quad \text{or} \quad l_2 = 108.8 \text{ mm}$$

Adding 12.5 mm for starting and stopping of weld run, we have

$$l_2 = 108.8 + 12.5 = 121.3 \text{ mm } \textbf{Ans.}$$

Example 10.6. *Determine the length of the weld run for a plate of size 120 mm wide and 15 mm thick to be welded to another plate by means of*

1. A single transverse weld; and

2. Double parallel fillet welds when the joint is subjected to variable loads.

Solution. Given : Width = 120 mm ; Thickness = 15 mm

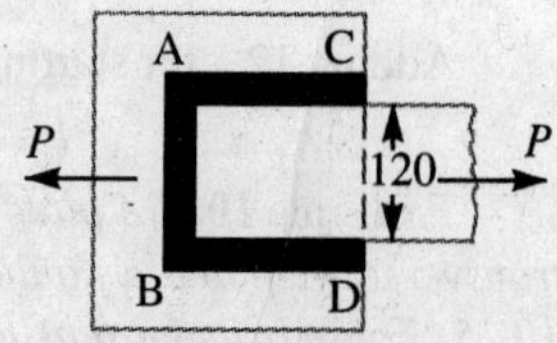

Fig. 10.16

In Fig. 10.16, AB represents the single transverse weld and AC and BD represents double parallel fillet welds.

1. ***Length of the weld run for a single transverse weld***

The effective length of the weld run (l_1) for a single transverse weld may be obtained by subtracting 12.5 mm from the width of the plate.

$$\therefore \quad l_1 = 120 - 12.5 = 107.5 \text{ mm } \textbf{Ans.}$$

2. ***Length of the weld run for a double parallel fillet weld subjected to variable loads***

Let l_2 = Length of weld run for each parallel fillet, and

s = Size of weld = Thickness of plate = 15 mm

Assuming the tensile stress as 70 MPa or N/mm² and shear stress as 56 MPa or N/mm² for static loading. We know that the maximum load which the plate can carry is

$$P = \text{Area} \times \text{Stress} = 120 \times 15 \times 70 = 126 \times 10^3 \text{ N}$$

From Table 10.6, we find that the stress concentration factor for transverse weld is 1.5 and for parallel fillet welds is 2.7.

∴ Permissible tensile stress,

$$\sigma_t = 70 / 1.5 = 46.7 \text{ N/mm}^2$$

and permissible shear stress,

$$\tau = 56 / 2.7 = 20.74 \text{ N/mm}^2$$

∴ Load carried by single transverse weld,

$$P_1 = 0.707\, s \times l_1 \times \sigma_t = 0.707 \times 15 \times 107.5 \times 46.7 = 53\,240 \text{ N}$$

and load carried by double parallel fillet weld,

$$P_2 = 1.414\, s \times l_2 \times \tau = 1.414 \times 15 \times l_2 \times 20.74 = 440\, l_2 \text{ N}$$

∴ Load carried by the joint (P),

$$126 \times 10^3 = P_1 + P_2 = 53\,240 + 440\, l_2 \quad \text{or} \quad l_2 = 165.4 \text{ mm}$$

Adding 12.5 mm for starting and stopping of weld run, we have

$$l_2 = 165.4 + 12.5 = 177.9 \text{ say } 178 \text{ mm} \qquad \textbf{Ans.}$$

Example 10.7. *The fillet welds of equal legs are used to fabricate a 'T' as shown in Fig. 10.17 (a) and (b), where s is the leg size and l is the length of weld.*

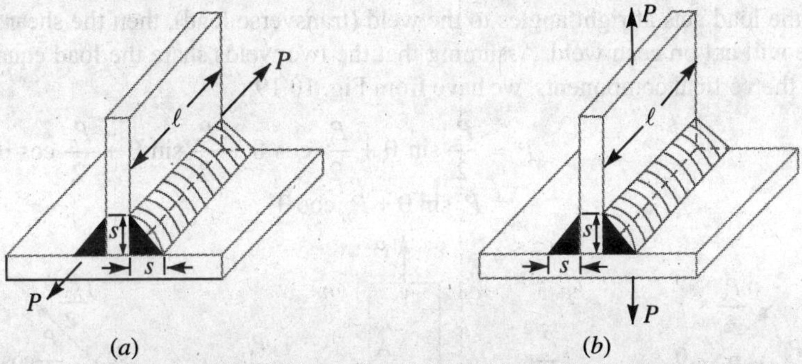

Fig. 10.17

Locate the plane of maximum shear stress in each of the following loading patterns:

1. Load parallel to the weld (neglect eccentricity), and

2. Load at right angles to the weld (transverse load).

Find the ratio of these limiting loads.

Solution. Given : Leg size = s ; Length of weld = l

1. ***Plane of maximum shear stress when load acts parallel to the weld (neglecting eccentricity)***

Let θ = Angle of plane of maximum shear stress, and

t = Throat thickness BD.

From the geometry of Fig. 10.18, we find that

$$BC = BE + EC$$

$$= BE + DE \qquad ...(\because EC = DE)$$

or

$$s = BD\cos\theta + BD\sin\theta$$

$$= t\cos\theta + t\sin\theta$$

$$= t(\cos\theta + \sin\theta)$$

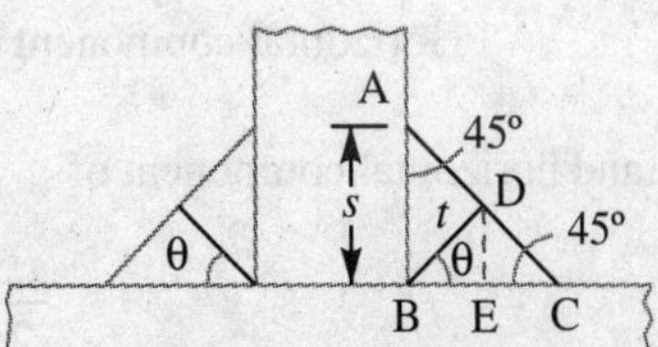

Fig. 10.18

$$\therefore \quad t = \frac{s}{\cos\theta + \sin\theta}$$

We know that the minimum area of the weld or throat area,

$$A = 2\,t \times l = \frac{2\,s \times l}{(\cos\theta + \sin\theta)} \qquad ...(\because \text{ of double fillet weld})$$

and shear stress,
$$\tau = \frac{P}{A} = \frac{P(\cos\theta + \sin\theta)}{2s \times l} \quad ...(i)$$

For maximum shear stress, differentiate the above expression with respect to θ and equate to zero.

$$\therefore \quad \frac{d\tau}{d\theta} = \frac{P}{2s \times l}(-\sin\theta + \cos\theta) = 0$$

or
$$\sin\theta = \cos\theta \text{ or } \theta = 45°$$

Substituting the value of θ = 45° in equation (*i*), we have maximum shear stress,

$$\tau_{max} = \frac{P(\cos 45° + \sin 45°)}{2s \times l} = \frac{1.414\,P}{2s \times l}$$

or
$$P = \frac{2s \times l \times \tau_{max}}{1.414} = 1.414\,s \times l \times \tau_{max} \textbf{ Ans.}$$

2. *Plane of maximum shear stress when load acts at right angles to the weld*

When the load acts at right angles to the weld (transverse load), then the shear force and the normal force will act on each weld. Assuming that the two welds share the load equally, therefore summing up the vertical components, we have from Fig. 10.19,

$$P = \frac{P_s}{2}\sin\theta + \frac{P_n}{2}\cos\theta + \frac{P_s}{2}\sin\theta + \frac{P_n}{2}\cos\theta$$
$$= P_s \sin\theta + P_n \cos\theta \quad ...(i)$$

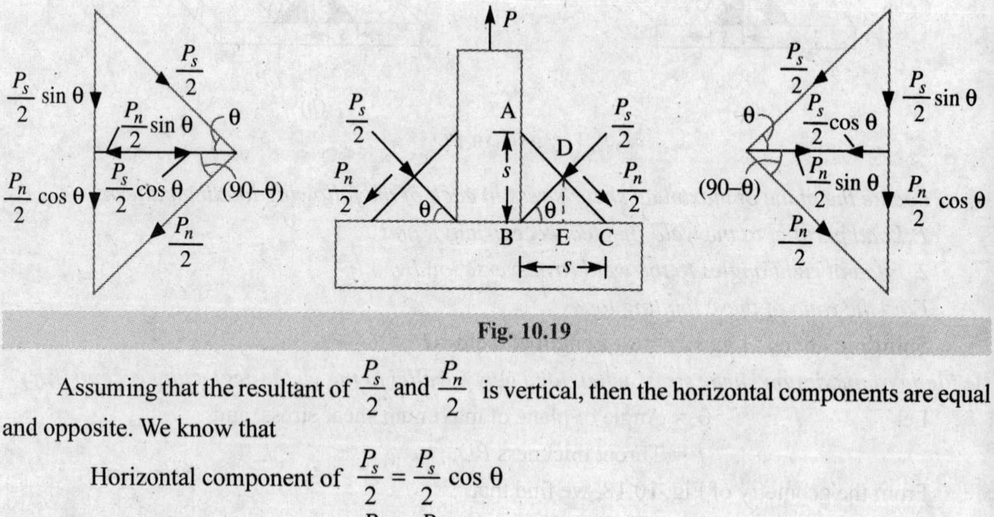

Fig. 10.19

Assuming that the resultant of $\frac{P_s}{2}$ and $\frac{P_n}{2}$ is vertical, then the horizontal components are equal and opposite. We know that

$$\text{Horizontal component of } \frac{P_s}{2} = \frac{P_s}{2}\cos\theta$$

$$\text{and horizontal component of } \frac{P_n}{2} = \frac{P_n}{2}\sin\theta$$

$$\therefore \quad \frac{P_s}{2}\cos\theta = \frac{P_n}{2}\sin\theta \text{ or } P_n = \frac{P_s\cos\theta}{\sin\theta}$$

Substituting the value of P_n in equation (*i*), we have

$$P = P_s\sin\theta + \frac{P_s\cos\theta \times \cos\theta}{\sin\theta}$$

Multiplying throughout by sin θ, we have

$$P\sin\theta = P_s\sin^2\theta + P_s\cos^2\theta$$
$$= P_s(\sin^2\theta + \cos^2\theta) = P_s \quad ...(ii)$$

From the geometry of Fig. 10.19, we have

$$BC = BE + EC = BE + DE \quad ...(\because EC = DE)$$

or $$s = t\cos\theta + t\sin\theta = t(\cos\theta + \sin\theta)$$

∴ Throat thickness, $$t = \frac{s}{\cos\theta + \sin\theta}$$

and minimum area of the weld or throat area,

$$A = 2t \times l \quad ...(\because \text{ of double fillet weld})$$

$$= 2 \times \frac{s}{\cos\theta + \sin\theta} \times l = \frac{2s \times l}{\cos\theta + \sin\theta}$$

∴ Shear stress, $$\tau = \frac{P_s}{A} = \frac{P\sin\theta(\cos\theta + \sin\theta)}{2s \times l} \quad ...[\text{From equation } (ii)] \quad ...(iii)$$

For maximum shear stress, differentiate the above expression with respect to θ and equate to zero.

∴ $$\frac{d\tau}{d\theta} = \frac{P}{2\,s\,l}[\sin\theta(-\sin\theta + \cos\theta) + (\cos\theta + \sin\theta)\cos\theta] = 0$$

$$...\left(\because \frac{d(u.v)}{d\theta} = u\frac{dv}{d\theta} + v\frac{du}{d\theta}\right)$$

or $$-\sin^2\theta + \sin\theta\cos\theta + \cos^2\theta + \sin\theta\cos\theta = 0$$

$$\cos^2\theta - \sin^2\theta + 2\sin\theta\cos\theta = 0$$

Since $\cos^2\theta - \sin^2\theta = \cos 2\theta$ and $2\sin\theta\cos\theta = \sin 2\theta$, therefore,

$$\cos 2\theta + \sin 2\theta = 0$$

or $$\sin 2\theta = -\cos 2\theta$$

$$\frac{\sin 2\theta}{\cos 2\theta} = -1 \quad \text{or} \quad \tan 2\theta = -1$$

∴ $$2\theta = 135° \quad \text{or} \quad \theta = 67.5° \quad \textbf{Ans.}$$

Substituting the value of θ = 67.5° in equation (*iii*), we have maximum shear stress,

$$\tau_{max} = \frac{P\sin 67.5°(\cos 67.5° + \sin 67.5°)}{2\,s \times l}$$

$$= \frac{P \times 0.9239\,(0.3827 + 0.9229)}{2s \times l} = \frac{1.21\,P}{2\,s \times l}$$

and $$P = \frac{2s \times l \times \tau_{max}}{1.21} = 1.65\,s \times l \times \tau_{max} \quad \textbf{Ans.}$$

Ratio of the limiting loads

We know that the ratio of the limiting (or maximum) loads

$$= \frac{1.414\,s \times l \times \tau_{max}}{1.65\,s \times l \times \tau_{max}} = 0.857 \quad \textbf{Ans.}$$

10.22 Axially Loaded Unsymmetrical Welded Sections

Sometimes unsymmetrical sections such as angles, channels, *T*-sections etc., welded on the flange edges are loaded axially as shown in Fig. 10.20. In such cases, the lengths of weld should be proportioned in such a way that the sum of resisting moments of the welds about the gravity axis is zero. Consider an angle section as shown in Fig. 10.20.

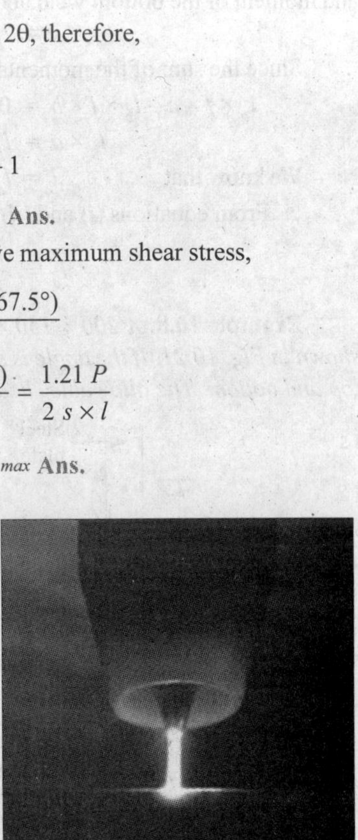

Plasma arc welding

Let l_a = Length of weld at the top,

l_b = Length of weld at the bottom,

l = Total length of weld = $l_a + l_b$

P = Axial load,

a = Distance of top weld from gravity axis,

b = Distance of bottom weld from gravity axis, and

f = Resistance offered by the weld per unit length.

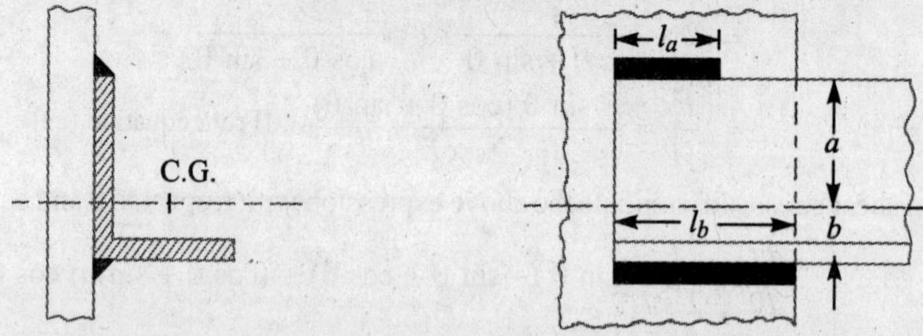

Fig. 10.20. Axially loaded unsymmetrical welded section.

∴ Moment of the top weld about gravity axis

$$= l_a \times f \times a$$

and moment of the bottom weld about gravity axis

$$= l_b \times f \times b$$

Since the sum of the moments of the weld about the gravity axis must be zero, therefore,

$$l_a \times f \times a - l_b \times f \times b = 0$$

or $$l_a \times a = l_b \times b \quad \text{...(i)}$$

We know that $$l = l_a + l_b \quad \text{...(ii)}$$

∴ From equations (*i*) and (*ii*), we have

$$l_a = \frac{l \times b}{a + b}, \quad \text{and} \quad l_b = \frac{l \times a}{a + b}$$

Example 10.8. *A 200 × 150 × 10 mm angle is to be welded to a steel plate by fillet welds as shown in Fig. 10.21. If the angle is subjected to a static load of 200 kN, find the length of weld at the top and bottom. The allowable shear stress for static loading may be taken as 75 MPa.*

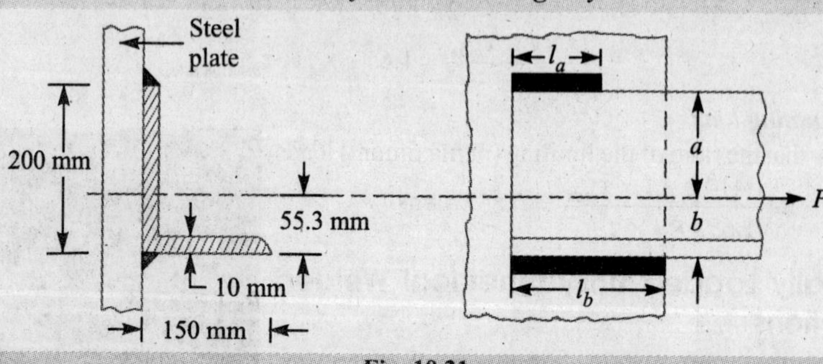

Fig. 10.21

Solution. Given : $a + b$ = 200 mm ; P = 200 kN = 200 × 10^3 N ; τ = 75 MPa = 75 N/mm^2

Let l_a = Length of weld at the top,

l_b = Length of weld at the bottom, and

l = Total length of the weld = $l_a + l_b$

Since the thickness of the angle is 10 mm, therefore size of weld,

$$s = 10 \text{ mm}$$

We know that for a single parallel fillet weld, the maximum load (P),

$$200 \times 10^3 = 0.707\, s \times l \times \tau = 0.707 \times 10 \times l \times 75 = 530.25\, l$$

$$\therefore \quad l = 200 \times 10^3 / 530.25 = 377 \text{ mm}$$

or $$l_a + l_b = 377 \text{ mm}$$

Now let us find out the position of the centroidal axis.

Let b = Distance of centroidal axis from the bottom of the angle.

$$\therefore \quad b = \frac{(200 - 10)\, 10 \times 95 + 150 \times 10 \times 5}{190 \times 10 + 150 \times 10} = 55.3 \text{ mm}$$

and $$a = 200 - 55.3 = 144.7 \text{ mm}$$

We know that $$l_a = \frac{l \times b}{a + b} = \frac{377 \times 55.3}{200} = 104.2 \text{ mm} \quad \textbf{Ans.}$$

and $$l_b = l - l_a = 377 - 104.2 = 272.8 \text{ mm} \quad \textbf{Ans.}$$

10.23 Eccentrically Loaded Welded Joints

An eccentric load may be imposed on welded joints in many ways. The stresses induced on the joint may be of different nature or of the same nature. The induced stresses are combined depending upon the nature of stresses. When the shear and bending stresses are simultaneously present in a joint (see case 1), then maximum stresses are as follows:

Maximum normal stress,

$$\sigma_{t(max)} = \frac{\sigma_b}{2} + \frac{1}{2}\sqrt{(\sigma_b)^2 + 4\,\tau^2}$$

and maximum shear stress,

$$\tau_{max} = \frac{1}{2}\sqrt{(\sigma_b)^2 + 4\,\tau^2}$$

where σ_b = Bending stress, and

τ = Shear stress.

When the stresses are of the same nature, these may be combined vectorially (see case 2).

We shall now discuss the two cases of eccentric loading as follows:

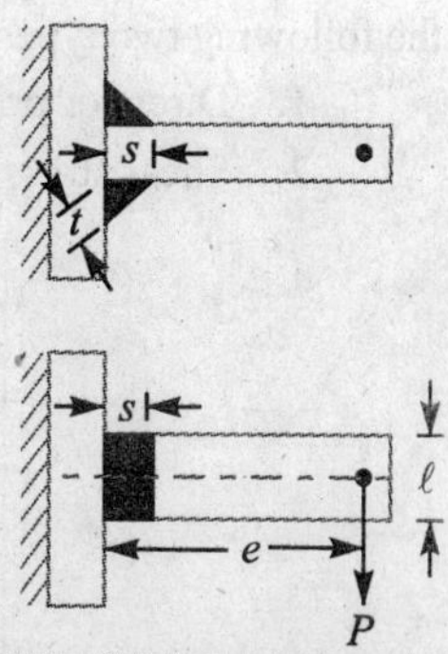

Fig. 10.22. Eccentrically loaded welded joint.

Case 1

Consider a *T*-joint fixed at one end and subjected to an eccentric load P at a distance e as shown in Fig. 10.22.

Let s = Size of weld,

l = Length of weld, and

t = Throat thickness.

The joint will be subjected to the following two types of stresses:

1. Direct shear stress due to the shear force P acting at the welds, and
2. Bending stress due to the bending moment $P \times e$.

We know that area at the throat,

$$A = \text{Throat thickness} \times \text{Length of weld}$$

$$= t \times l \times 2 = 2\,t \times l \quad \text{... (For double fillet weld)}$$

$$= 2 \times 0.707\, s \times l = 1.414\, s \times l \quad \text{... } (\because t = s \cos 45^\circ = 0.707\, s)$$

∴ Shear stress in the weld (assuming uniformly distributed),

$$\tau = \frac{P}{A} = \frac{P}{1.414\, s \times l}$$

Section modulus of the weld metal through the throat,

$$Z = \frac{t \times l^2}{6} \times 2 \quad \text{...(For both sides weld)}$$

$$= \frac{0.707\, s \times l^2}{6} \times 2 = \frac{s \times l^2}{4.242}$$

Bending moment, $M = P \times e$

$$\therefore \text{ Bending stress, } \sigma_b = \frac{M}{Z} = \frac{P \times e \times 4.242}{s \times l^2} = \frac{4.242\, P \times e}{s \times l^2}$$

We know that the maximum normal stress,

$$\sigma_{t(max)} = \frac{1}{2}\sigma_b + \frac{1}{2}\sqrt{(\sigma_b)^2 + 4\,\tau^2}$$

and maximum shear stress,

$$\tau_{max} = \frac{1}{2}\sqrt{(\sigma_b)^2 + 4\,\tau^2}$$

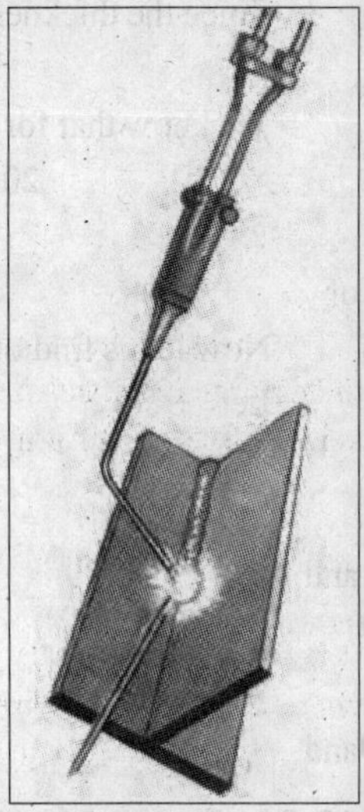

Soldering is done by melting a metal which melts at a lower temperature than the metal that is soldered.

Case 2

When a welded joint is loaded eccentrically as shown in Fig. 10.23, the following two types of the stresses are induced:

1. Direct or primary shear stress, and
2. Shear stress due to turning moment.

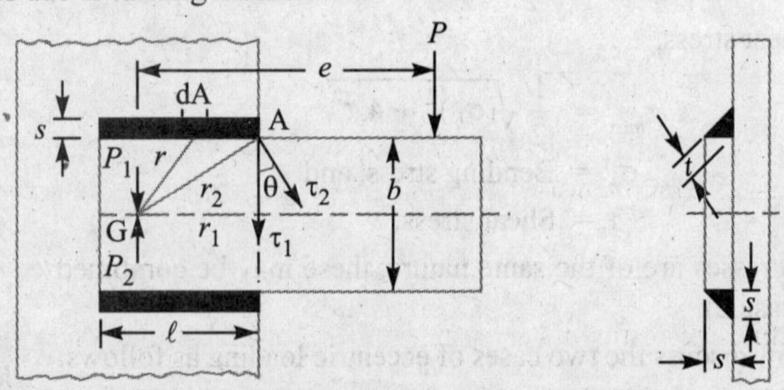

Fig. 10.23. Eccentrically loaded welded joint.

Let P = Eccentric load,

e = Eccentricity *i.e.* perpendicular distance between the line of action of load and centre of gravity (G) of the throat section or fillets,

l = Length of single weld,

s = Size or leg of weld, and

t = Throat thickness.

Let two loads P_1 and P_2 (each equal to P) are introduced at the centre of gravity 'G' of the weld system. The effect of load $P_1 = P$ is to produce direct shear stress which is assumed to be uniform over the entire weld length. The effect of load $P_2 = P$ is to produce a turning moment of magnitude $P \times e$ which tends of rotate the joint about the centre of gravity 'G' of the weld system. Due to the turning moment, secondary shear stress is induced.

We know that the direct or primary shear stress,

$$\tau_1 = \frac{\text{Load}}{\text{Throat area}} = \frac{P}{A} = \frac{P}{2\,t \times l}$$

$$= \frac{P}{2 \times 0.707\ s \times l} = \frac{P}{1.414\ s \times l}$$

... ($\because$ Throat area for single fillet weld = $t \times l = 0.707\ s \times l$)

Since the shear stress produced due to the turning moment ($T = P \times e$) at any section is proportional to its radial distance from G, therefore stress due to $P \times e$ at the point A is proportional to AG (r_2) and is in a direction at right angles to AG. In other words,

$$\frac{\tau_2}{r_2} = \frac{\tau}{r} = \text{Constant}$$

or
$$\tau = \frac{\tau_2}{r_2} \times r \qquad ...(i)$$

where τ_2 is the shear stress at the maximum distance (r_2) and τ is the shear stress at any distance r.

Consider a small section of the weld having area dA at a distance r from G.

$\therefore$ Shear force on this small section

$$= \tau \times dA$$

and turning moment of this shear force about G,

$$dT = \tau \times dA \times r = \frac{\tau_2}{r_2} \times dA \times r^2 \qquad \text{... [From equation (i)]}$$

$\therefore$ Total turning moment over the whole weld area,

$$T = P \times e = \int \frac{\tau_2}{r_2} \times dA \times r^2 = \frac{\tau_2}{r_2} \int dA \times r^2$$

$$= \frac{\tau_2}{r_2} \times J \qquad \left(\because J = \int dA \times r^2\right)$$

where J = Polar moment of inertia of the throat area about G.

$\therefore$ Shear stress due to the turning moment *i.e.* secondary shear stress,

$$\tau_2 = \frac{T \times r_2}{J} = \frac{P \times e \times r_2}{J}$$

In order to find the resultant stress, the primary and secondary shear stresses are combined vectorially.

$\therefore$ Resultant shear stress at A,

$$\tau_A = \sqrt{(\tau_1)^2 + (\tau_2)^2 + 2\tau_1 \times \tau_2 \times \cos\theta}$$

where θ = Angle between τ_1 and τ_2, and

$$\cos\theta = r_1 / r_2$$

Note: The polar moment of inertia of the throat area (A) about the centre of gravity (G) is obtained by the parallel axis theorem, *i.e.*

$$J = 2\,[I_{xx} + A \times x^2] \qquad \text{... ($\because$ of double fillet weld)}$$

$$= 2\left[\frac{A \times l^2}{12} + A \times x^2\right] = 2A\left(\frac{l_2}{12} + x^2\right)$$

where A = Throat area = $t \times l = 0.707\ s \times l$,

l = Length of weld, and

x = Perpendicular distance between the two parallel axes.

10.24 Polar Moment of Inertia and Section Modulus of Welds

The following table shows the values of polar moment of inertia of the throat area about the centre of gravity *'G'* and section modulus for some important types of welds which may be used for eccentric loading.

Table 10.7. Polar moment of inertia and section modulus of welds.

S.No	*Type of weld*	*Polar moment of inertia (J)*	*Section modulus (Z)*
1.	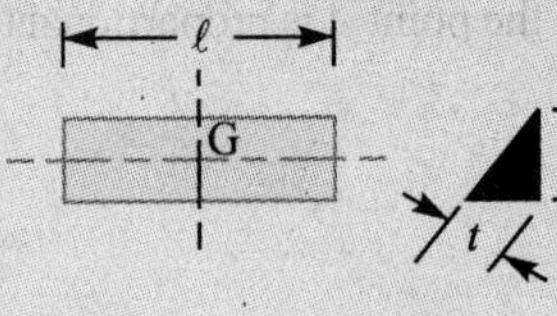	$\frac{t.l^3}{12}$	—
2.	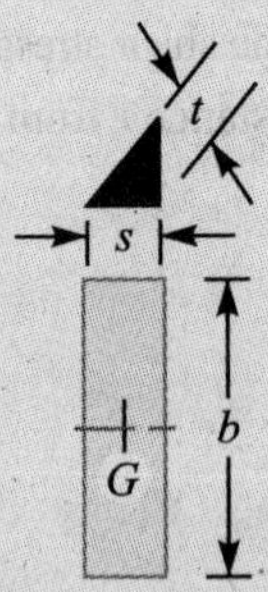	$\frac{t.b^3}{12}$	$\frac{t.b^2}{6}$
3.	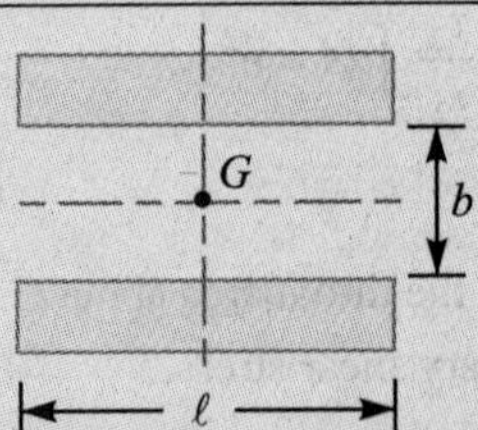	$\frac{t.l(3b^2 + l^2)}{6}$	$t.b.l$
4.	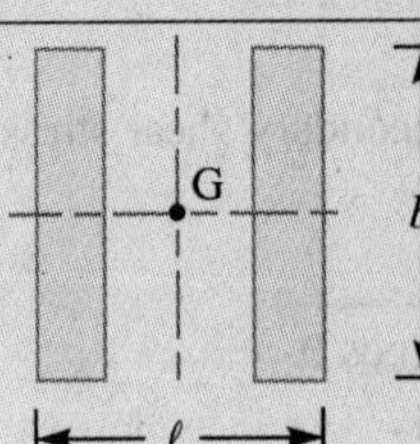	$\frac{t.b(b^2 + 3l^2)}{6}$	$\frac{t.b^2}{3}$
5.	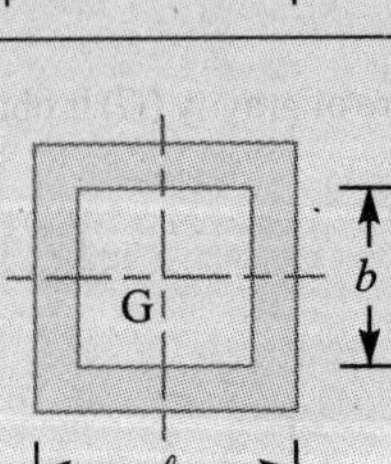	$\frac{t(b + l)^3}{6}$	$t\left(b.l + \frac{b^2}{3}\right)$

S.No	Type of weld	Polar moment of inertia (J)	Section modulus (Z)
6.	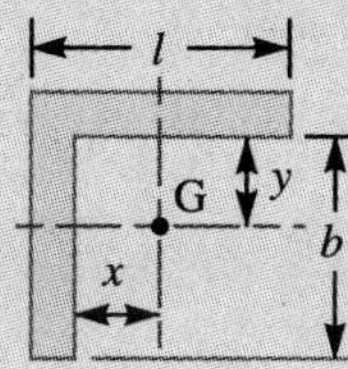 $x = \frac{l^2}{2(l+b)}, y = \frac{b^2}{2(l+b)}$	$t\left[\frac{(b+l)^4 - 6b^2l^2}{12(l+b)}\right]$	$t\left(\frac{4l.b + b^2}{6}\right)$ (Top) $t\left[\frac{b^2(4lb+b)}{6(2l+b)}\right]$ (Bottom)
7.	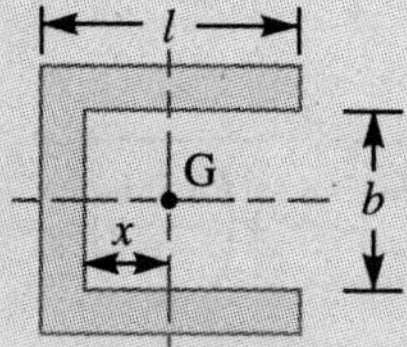 $x = \frac{l^2}{2l+b}$	$t\left[\frac{(b+2l)^3}{12} - \frac{l^2(b+l)^2}{b+2l}\right]$	$t\left(l.b + \frac{b^2}{6}\right)$
8.	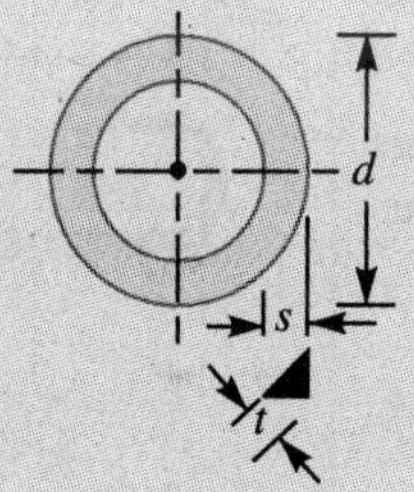	$\frac{\pi t d^3}{4}$	$\frac{\pi t d^2}{4}$

Note: In the above expressions, t is the throat thickness and s is the size of weld. It has already been discussed that $t = 0.707\ s$.

Example 10.9. *A welded joint as shown in Fig. 10.24, is subjected to an eccentric load of 2 kN. Find the size of weld, if the maximum shear stress in the weld is 25 MPa.*

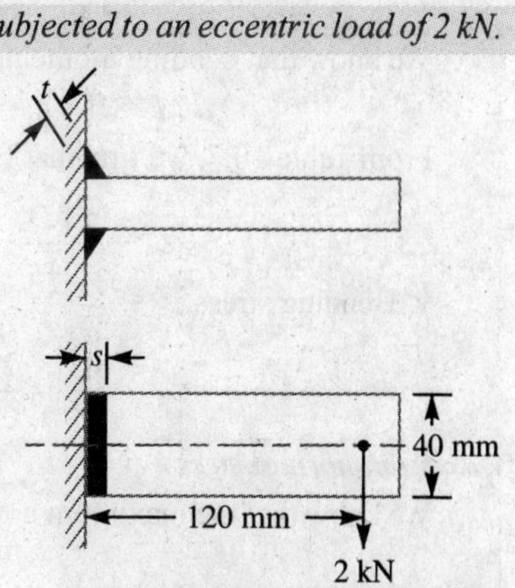

Fig. 10.24

Solution. Given: P = 2kN = 2000 N ; e = 120 mm ; l = 40 mm ; τ_{max} = 25 MPa = 25 N/mm²

Let s = Size of weld in mm, and

t = Throat thickness.

The joint, as shown in Fig. 10.24, will be subjected to direct shear stress due to the shear force, P = 2000 N and bending stress due to the bending moment of $P \times e$.

We know that area at the throat,

$$A = 2t \times l = 2 \times 0.707\ s \times l$$

$$= 1.414\ s \times l$$

$$= 1.414\ s \times 40 = 56.56 \times s \text{ mm}^2$$

$\therefore$ Shear stress, $\tau = \frac{P}{A} = \frac{2000}{56.56 \times s} = \frac{35.4}{s}$ N/mm^2

Bending moment, $M = P \times e = 2000 \times 120 = 240 \times 10^3$ N-mm

Section modulus of the weld through the throat,

$$Z = \frac{s \times l^2}{4.242} = \frac{s\,(40)^2}{4.242} = 377 \times s \text{ mm}^3$$

$$\therefore \text{ Bending stress, } \sigma_b = \frac{M}{Z} = \frac{240 \times 10^3}{377 \times s} = \frac{636.6}{s} \text{ N/mm}^2$$

We know that maximum shear stress (τ_{max}),

$$25 = \frac{1}{2}\sqrt{(\sigma_b)^2 + 4\,\tau^2} = \frac{1}{2}\sqrt{\left(\frac{636.6}{s}\right)^2 + 4\left(\frac{35.4}{s}\right)^2} = \frac{320.3}{s}$$

$\therefore$ $s = 320.3 / 25 = 12.8$ mm **Ans.**

Example 10.10. *A 50 mm diameter solid shaft is welded to a flat plate as shown in Fig. 10.25. If the size of the weld is 15 mm, find the maximum normal and shear stress in the weld.*

Solution. Given : $D = 50$ mm ; $s = 15$ mm ; $P = 10$ kN $= 10\,000$ N ; $e = 200$ mm

Let t = Throat thickness.

The joint, as shown in Fig. 10.25, is subjected to direct shear stress and the bending stress. We know that the throat area for a circular fillet weld,

$$A = t \times \pi D = 0.707\, s \times \pi D$$
$$= 0.707 \times 15 \times \pi \times 50$$
$$= 1666 \text{ mm}^2$$

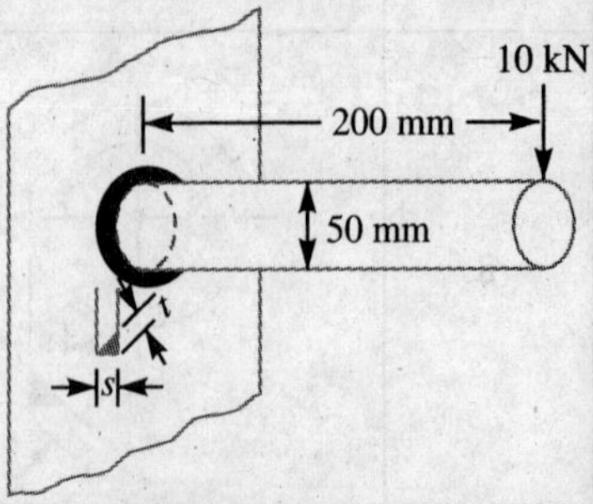

Fig. 10.25

$\therefore$ Direct shear stress,

$$\tau = \frac{P}{A} = \frac{10\,000}{1666} = 6 \text{ N/mm}^2 = 6 \text{ MPa}$$

We know that bending moment,

$$M = P \times e = 10\,000 \times 200 = 2 \times 10^6 \text{ N-mm}$$

From Table 10.7, we find that for a circular section, section modulus,

$$Z = \frac{\pi\, t\, D^2}{4} = \frac{\pi \times 0.707\, s \times D^2}{4} = \frac{\pi \times 0.707 \times 15\,(50)^2}{4} = 20\,825 \text{ mm}^3$$

$\therefore$ Bending stress,

$$\sigma_b = \frac{M}{Z} = \frac{2 \times 10^6}{20\,825} = 96 \text{ N/mm}^2 = 96 \text{ MPa}$$

Maximum normal stress

We know that the maximum normal stress,

$$\sigma_{t(max)} = \frac{1}{2}\,\sigma_b + \frac{1}{2}\sqrt{(\sigma_b)^2 + 4\,\tau^2} = \frac{1}{2} \times 96 + \frac{1}{2}\sqrt{(96)^2 + 4 \times 6^2}$$
$$= 48 + 48.4 = 96.4 \text{ MPa } \textbf{Ans.}$$

Maximum shear stress

We know that the maximum shear stress,

$$\tau_{max} = \frac{1}{2}\sqrt{(\sigma_b)^2 + 4\,\tau^2} = \frac{1}{2}\sqrt{(96)^2 + 4 \times 6^2} = 48.4 \text{ MPa} \quad \textbf{Ans.}$$

Example 10.11. *A rectangular cross-section bar is welded to a support by means of fillet welds as shown in Fig. 10.26.*

Determine the size of the welds, if the permissible shear stress in the weld is limited to 75 MPa.

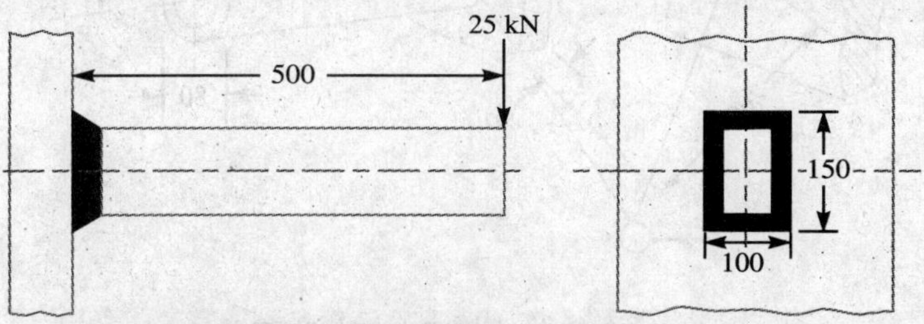

All dimensions in mm

Fig. 10.26

Solution. Given : P = 25 kN = 25 × 10³N ; τ_{max} = 75 MPa = 75N/mm² ; l = 100 mm ; b = 150 mm ; e = 500 mm

Let s = Size of the weld, and

t = Throat thickness.

The joint, as shown in Fig. 10.26, is subjected to direct shear stress and the bending stress. We know that the throat area for a rectangular fillet weld,

$$A = t\,(2b + 2l) = 0.707\,s\,(2b + 2l)$$

$$= 0.707s\,(2 \times 150 + 2 \times 100) = 353.5\,s \text{ mm}^2 \qquad \text{... } (\because t = 0.707s)$$

$$\therefore \text{ Direct shear stress, } \tau = \frac{P}{A} = \frac{25 \times 10^3}{353.5\,s} = \frac{70.72}{s} \text{ N/mm}^2$$

We know that bending moment,

$$M = P \times e = 25 \times 10^3 \times 500 = 12.5 \times 10^6 \text{ N-mm}$$

From Table 10.7, we find that for a rectangular section, section modulus,

$$Z = t\left(b.l + \frac{b^2}{3}\right) = 0.707\,s\left[150 \times 100 + \frac{(150)^2}{3}\right] = 15\,907.5\,s \text{ mm}^3$$

$$\therefore \text{ Bending stress, } \sigma_b = \frac{M}{Z} = \frac{12.5 \times 10^6}{15\,907.5\,s} = \frac{785.8}{s} \text{ N/mm}^2$$

We know that maximum shear stress (τ_{max}),

$$75 = \frac{1}{2}\sqrt{(\sigma_b)^2 + 4\,\tau^2} = \frac{1}{2}\sqrt{\left(\frac{785.8}{s}\right)^2 + 4\left(\frac{70.72}{s}\right)^2} = \frac{399.2}{s}$$

$$\therefore \quad s = 399.2 / 75 = 5.32 \text{ mm} \quad \textbf{Ans.}$$

Example 10.12. *An arm A is welded to a hollow shaft at section '1'. The hollow shaft is welded to a plate C at section '2'. The arrangement is shown in Fig. 10.27, along with dimensions. A force P = 15 kN acts at arm A perpendicular to the axis of the arm.*

Calculate the size of weld at section '1' and '2'. The permissible shear stress in the weld is 120 MPa.

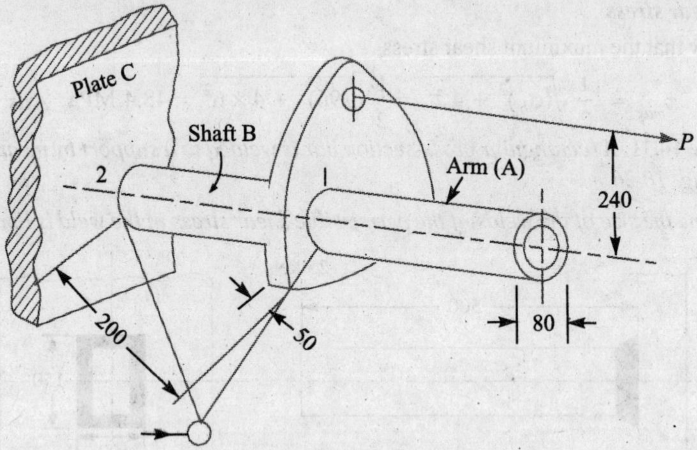

Fig. 10.27. All dimensions in mm.

Solution. Given : $P = 15$ kN $= 15 \times 10^3$ N ; $\tau_{max} = 120$ MPa $= 120$ N/mm^2 ; $d = 80$ mm

Let s = Size of the weld.

The welded joint, as shown in Fig. 10.27, is subjected to twisting moment or torque (T) as well as bending moment (M).

We know that the torque acting on the shaft,

$$T = 15 \times 10^3 \times 240 = 3600 \times 10^3 \text{ N-mm}$$

$\therefore$ Shear stress,

$$\tau = \frac{2.83\, T}{\pi\, s\, d^2} = \frac{2.83 \times 3600 \times 10^3}{\pi \times s\,(80)^2} = \frac{506.6}{s} \text{ N/mm}^2$$

Bending moment,

$$M = 15 \times 10^3 \left(200 - \frac{50}{2}\right) = 2625 \times 10^3 \text{ N-mm}$$

$\therefore$ Bending stress,

$$\sigma_b = \frac{5.66\, M}{\pi\, s\, d^2} = \frac{5.66 \times 26.25 \times 10^3}{\pi\, s\,(80)^2} = \frac{738.8}{s} \text{ N/mm}^2$$

We know that maximum shear stress (τ_{max}),

$$120 = \frac{1}{2}\sqrt{(\sigma_b)^2 + 4\,\tau^2} = \frac{1}{2}\sqrt{\left(\frac{738.8}{s}\right)^2 + 4\left(\frac{506.6}{s}\right)^2} = \frac{627}{s}$$

$\therefore$ $s = 627/120 = 5.2$ mm **Ans.**

Example 10.13. *A bracket carrying a load of 15 kN is to be welded as shown in Fig. 10.28. Find the size of weld required if the allowable shear stress is not to exceed 80 MPa.*

Solution. Given : $P = 15$ kN $= 15 \times 10^3$ N ; $\tau = 80$ MPa $= 80$ N/mm^2 ; $b = 80$ mm ; $l = 50$ mm; $e = 125$ mm

Let s = Size of weld in mm, and

t = Throat thickness.

We know that the throat area,

$$A = 2 \times t \times l = 2 \times 0.707\, s \times l$$
$$= 1.414\, s \times l = 1.414 \times s \times 50 = 70.7\, s \text{ mm}^2$$

$\therefore$ Direct or primary shear stress,

$$\tau_1 = \frac{P}{A} = \frac{15 \times 10^3}{70.7\, s} = \frac{212}{s} \text{ N/mm}^2$$

From Table 10.7, we find that for such a section, the polar moment of inertia of the throat area of the weld about G is

$$J = \frac{t.l\,(3b^2 + l^2)}{6} = \frac{0.707\ s \times 50\,[3\,(80)^2 + (50)^2]}{6}\ \text{mm}^4$$

$$= 127\ 850\ s\ \text{mm}^4 \qquad \ldots (\because\ t = 0.707\ s)$$

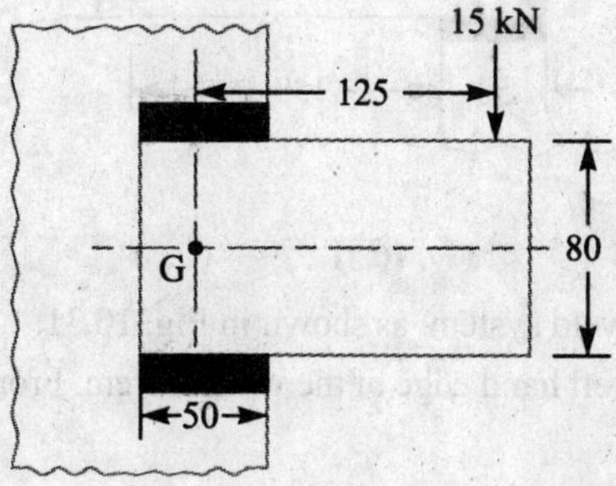

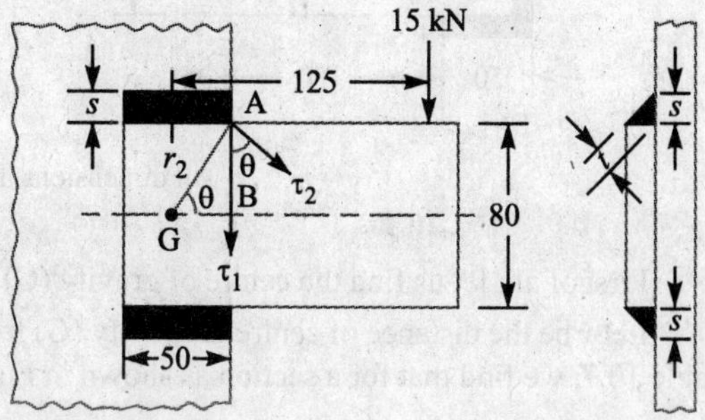

All dimensions in mm.

Fig. 10.28 **Fig. 10.29**

From Fig. 10.29, we find that AB = 40 mm and $BG = r_1$ = 25 mm.

∴ Maximum radius of the weld,

$$r_2 = \sqrt{(AB)^2 + (BG)^2} = \sqrt{(40)^2 + (25)^2} = 47\ \text{mm}$$

Shear stress due to the turning moment *i.e.* secondary shear stress,

$$\tau_2 = \frac{P \times e \times r_2}{J} = \frac{15 \times 10^3 \times 125 \times 47}{127\ 850\ s} = \frac{689.3}{s}\ \text{N/mm}^2$$

and $$\cos\theta = \frac{r_1}{r_2} = \frac{25}{47} = 0.532$$

We know that resultant shear stress,

$$\tau = \sqrt{(\tau_1)^2 + (\tau_2)^2 + 2\,\tau_1 \times \tau_2\,\cos\theta}$$

$$80 = \sqrt{\left(\frac{212}{s}\right)^2 + \left(\frac{689.3}{s}\right)^2 + 2 \times \frac{212}{s} \times \frac{689.3}{s} \times 0.532} = \frac{822}{s}$$

∴ $s = 822 / 80 = 10.3$ mm **Ans.**

Example 10.14. *A rectangular steel plate is welded as a cantilever to a vertical column and supports a single concentrated load P, as shown in Fig. 10.30.*

Determine the weld size if shear stress in the same is not to exceed 140 MPa.

Solution. Given : $P = 60$ kN $= 60 \times 10^3$ N ; $b = 100$ mm ; $l = 50$ mm ; $\tau = 140$ MPa $= 140$ N/mm²

Let s = Weld size, and

t = Throat thickness.

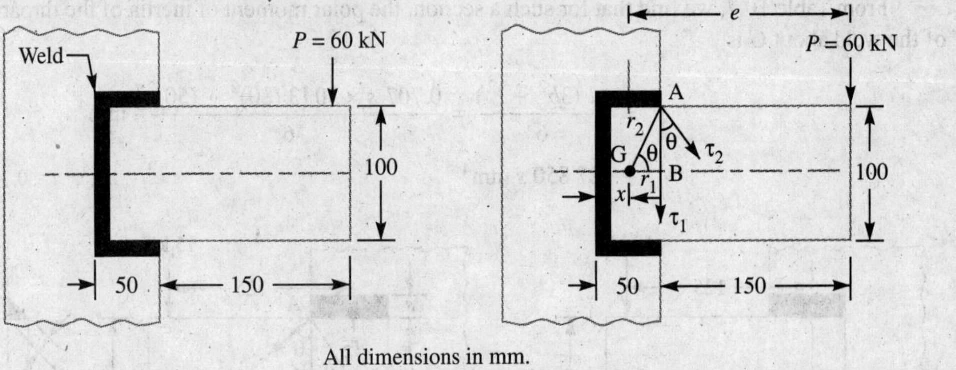

All dimensions in mm.

Fig. 10.30 **Fig. 10.31**

First of all, let us find the centre of gravity (*G*) of the weld system, as shown in Fig. 10.31.

Let *x* be the distance of centre of gravity (*G*) from the left hand edge of the weld system. From Table 10.7, we find that for a section as shown in Fig. 10.31,

$$x = \frac{l^2}{2l + b} = \frac{(50)^2}{2 \times 50 + 100} = 12.5 \text{ mm}$$

and polar moment of inertia of the throat area of the weld system about *G*,

$$J = t\left[\frac{(b + 2l)^3}{12} - \frac{l^2\,(b + l)^2}{b + 2l}\right]$$

$$= 0.707\,s\left[\frac{(100 + 2 \times 50)^3}{12} - \frac{(50)^2\,(100 + 50)^2}{100 + 2 \times 50}\right] \quad \ldots (\because t = 0.707\,s)$$

$$= 0.707s\,[670 \times 10^3 - 281 \times 10^3] = 275 \times 10^3\,s \text{ mm}^4$$

Distance of load from the centre of gravity (*G*) *i.e.* eccentricity,

$$e = 150 + 50 - 12.5 = 187.5 \text{ mm}$$

$$r_1 = BG = 50 - x = 50 - 12.5 = 37.5 \text{ mm}$$

$$AB = 100 / 2 = 50 \text{ mm}$$

We know that maximum radius of the weld,

$$r_2 = \sqrt{(AB)^2 + (BG)^2} = \sqrt{(50)^2 + (37.5)^2} = 62.5 \text{ mm}$$

$$\therefore \quad \cos\theta = \frac{r_1}{r_2} = \frac{37.5}{62.5} = 0.6$$

We know that throat area of the weld system,

$$A = 2 \times 0.707s \times l + 0.707s \times b = 0.707\,s\,(2l + b)$$

$$= 0.707s\,(2 \times 50 + 100) = 141.4\,s \text{ mm}^2$$

∴ Direct or primary shear stress,

$$\tau_1 = \frac{P}{A} = \frac{60 \times 10^3}{141.4\,s} = \frac{424}{s} \text{ N/mm}^2$$

and shear stress due to the turning moment or secondary shear stress,

$$\tau_2 = \frac{P \times e \times r_2}{J} = \frac{60 \times 10^3 \times 187.5 \times 62.5}{275 \times 10^3\,s} = \frac{2557}{s} \text{ N/mm}^2$$

We know that the resultant shear stress,

$$\tau = \sqrt{(\tau_1)^2 + (\tau_2)^2 + 2\,\tau_1 \times \tau_2 \times \cos\theta}$$

$$140 = \sqrt{\left(\frac{424}{s}\right)^2 + \left(\frac{2557}{s}\right)^2 + 2 \times \frac{424}{s} \times \frac{2557}{s} \times 0.6} = \frac{2832}{s}$$

$\therefore \quad s = 2832 / 140 = 20.23$ mm **Ans.**

Example 10.15. *Find the maximum shear stress induced in the weld of 6 mm size when a channel, as shown in Fig. 10.32, is welded to a plate and loaded with 20 kN force at a distance of 200 mm.*

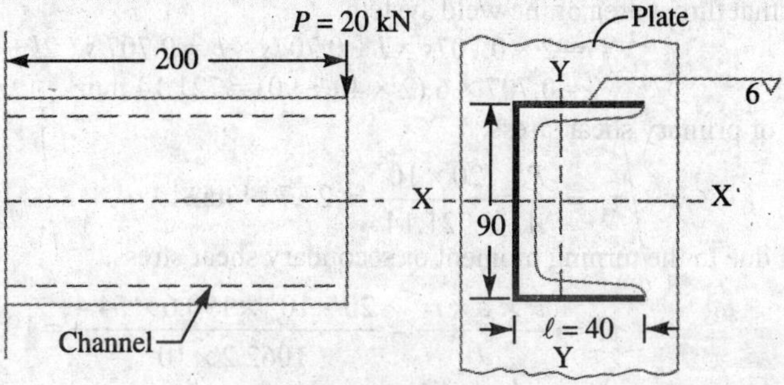

All dimensions in mm.

Fig. 10.32

Solution. Given : $s = 6$ mm ; $P = 20$ kN $= 20 \times 10^3$ N ; $l = 40$ mm ; $b = 90$ mm

Let $\quad t$ = Throat thickness.

First of all, let us find the centre of gravity (*G*) of the weld system as shown in Fig. 10.33. Let *x* be the distance of centre of gravity from the left hand edge of the weld system. From Table 10.7, we find that for a section as shown in Fig. 10.33,

$$x = \frac{l^2}{2l + b} = \frac{(40)^2}{2 \times 40 + 90} = 9.4 \text{ mm}$$

and polar moment of inertia of the throat area of the weld system about *G*,

$$J = t\left[\frac{(b + 2l)^3}{12} - \frac{l^2\,(b + l)^2}{b + 2l}\right]$$

$$= 0.707\,s\left[\frac{(90 + 2 \times 40)^3}{12} - \frac{(40)^2\,(90 + 40)^2}{90 + 2 \times 40}\right] \quad \text{... } (\because t = 0.707\,s)$$

$$= 0.707 \times 6\,[409.4 \times 10^3 - 159 \times 10^3] = 1062.2 \times 10^3 \text{ mm}^4$$

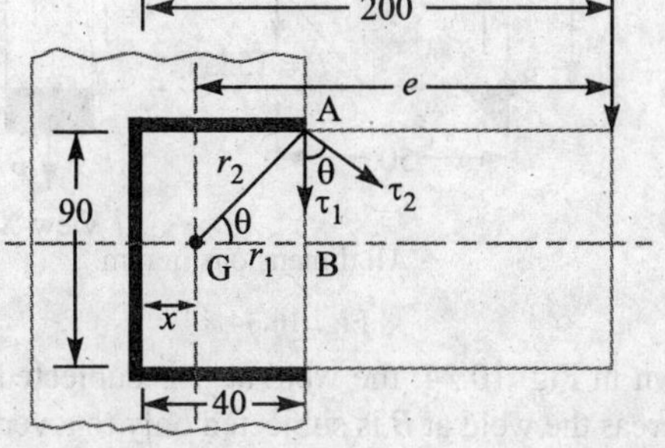

Fig. 10.33

Distance of load from the centre of gravity (G), *i.e.* eccentricity,

$$e = 200 - x = 200 - 9.4 = 190.6 \text{ mm}$$

$$r_1 = BG = 40 - x = 40 - 9.4 = 30.6 \text{ mm}$$

$$AB = 90/2 = 45 \text{ mm}$$

We know that maximum radius of the weld,

$$r_2 = \sqrt{(AB)^2 + (BG)^2} = \sqrt{(45)^2 + (30.6)^2} = 54.4 \text{ mm}$$

$$\therefore \quad \cos\theta = \frac{r_1}{r_2} = \frac{30.6}{54.4} = 0.5625$$

We know that throat area of the weld system,

$$A = 2 \times 0.707s \times l + 0.707s \times b = 0.707\, s\,(2l + b)$$

$$= 0.707 \times 6\,(2 \times 40 + 90) = 721.14 \text{ mm}^2$$

$\therefore$ Direct or primary shear stress,

$$\tau_1 = \frac{P}{A} = \frac{20 \times 10^3}{721.14} = 27.7 \text{ N/mm}^2$$

and shear stress due to the turning moment or secondary shear stress,

$$\tau_2 = \frac{P \times e \times r_2}{J} = \frac{20 \times 10^3 \times 190.6 \times 54.4}{1062.2 \times 10^3} = 195.2 \text{ N/mm}^2$$

We know that resultant or maximum shear stress,

$$\tau = \sqrt{(\tau_1)^2 + (\tau_2)^2 + 2\tau_1 \times \tau_2 \times \cos\theta}$$

$$= \sqrt{(27.7)^2 + (195.2)^2 + 2 \times 27.7 \times 195.2 \times 0.5625}$$

$$= 212 \text{ N/mm}^2 = 212 \text{ MPa} \quad \textbf{Ans.}$$

Example 10.16. *The bracket, as shown in Fig. 10.34, is designed to carry a dead weight of $P = 15$ kN.*

What sizes of the fillet welds are required at the top and bottom of the bracket? Assume the forces act through the points A and B. The welds are produced by shielded arc welding process with a permissible strength of 150 MPa.

Solution. Given : $P = 15$ kN ; $\tau = 150$ MPa $= 150$ N/mm^2 ; $l = 25$ mm

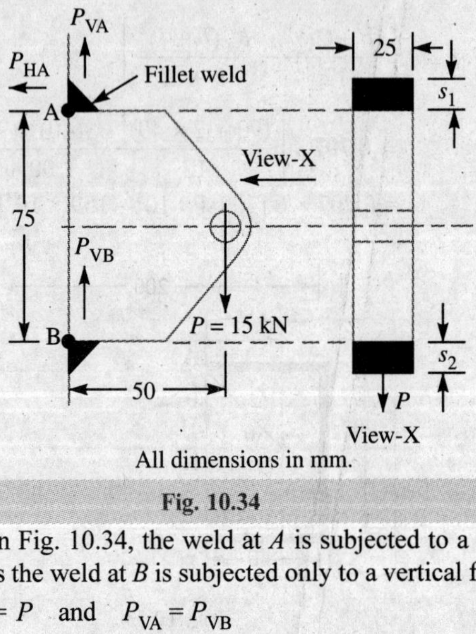

All dimensions in mm.

Fig. 10.34

In the joint, as shown in Fig. 10.34, the weld at A is subjected to a vertical force P_{VA} and a horizontal force P_{HA}, whereas the weld at B is subjected only to a vertical force P_{VB}. We know that

$$P_{VA} + P_{VB} = P \quad \text{and} \quad P_{VA} = P_{VB}$$

∴ Vertical force at A and B,

$$P_{VA} = P_{VB} = P/2 = 15/2 = 7.5 \text{ kN} = 7500 \text{ N}$$

The horizontal force at A may be obtained by taking moments about point B.

$$\therefore \quad P_{HA} \times 75 = 15 \times 50 = 750$$

or $$P_{HA} = 750/75 = 10 \text{ kN}$$

Size of the fillet weld at the top of the bracket

Let s_1 = Size of the fillet weld at the top of the bracket in mm.

We know that the resultant force at A,

$$P_A = \sqrt{(P_{VA})^2 + (P_{HA})^2} = \sqrt{(7.5)^2 + (10)^2} = 12.5 \text{ kN} = 12\,500 \text{ N} \quad ...(i)$$

We also know that the resultant force at A,

$$P_A = \text{Throat area} \times \text{Permissible stress}$$
$$= 0.707\, s_1 \times l \times \tau = 0.707\, s_1 \times 25 \times 150 = 2650\, s_1 \quad ...(ii)$$

From equations (*i*) and (*ii*), we get

$$s_1 = 12\,500/2650 = 4.7 \text{ mm} \quad \textbf{Ans.}$$

Size of fillet weld at the bottom of the bracket

Let s_2 = Size of the fillet weld at the bottom of the bracket.

The fillet weld at the bottom of the bracket is designed for the vertical force (P_{VB}) only. We know that

$$P_{VB} = 0.707\, s_2 \times l \times \tau$$
$$7500 = 0.707\, s_2 \times 25 \times 150 = 2650\, s_2$$
$$\therefore \quad s_2 = 7500/2650 = 2.83 \text{ mm} \quad \textbf{Ans.}$$

EXERCISES

1. A plate 100 mm wide and 10 mm thick is to be welded with another plate by means of transverse welds at the ends. If the plates are subjected to a load of 70 kN, find the size of weld for static as well as fatigue load. The permissible tensile stress should not exceed 70 MPa. **[Ans. 83.2 mm; 118.5 mm]**
2. If the plates in Ex. 1, are joined by double parallel fillets and the shear stress is not to exceed 56 MPa, find the length of weld for (*a*) Static loading, and (*b*) Dynamic loading. **[Ans. 91 mm; 259 mm]**
3. A 125 × 95 × 10 mm angle is joined to a frame by two parallel fillet welds along the edges of 150 mm leg. The angle is subjected to a tensile load of 180 kN. Find the lengths of weld if the permissible static load per mm weld length is 430 N. **[Ans. 137 mm and 307 mm]**
4. A circular steel bar 50 mm diameter and 200 mm long is welded perpendicularly to a steel plate to form a cantilever to be loaded with 5 kN at the free end. Determine the size of the weld, assuming the allowable stress in the weld as 100 MPa. **[Ans. 7.2 mm]**

   $$\left[\textbf{Hint:}\ \sigma_{b(max)} = \frac{5.66\, M}{\pi s d^2}\right]$$

5. A 65 mm diameter solid shaft is to be welded to a flat plate by a fillet weld around the circumference of the shaft. Determine the size of the weld if the torque on the shaft is 3 kN-m. The allowable shear stress in the weld is 70 MPa. **[Ans. 10 mm]**

   $$\left[\textbf{Hint:}\ \tau_{(max)} = \frac{2.83\, T}{\pi s d^2}\right]$$

6. A solid rectangular shaft of cross-section 80 mm × 50 mm is welded by a 5 mm fillet weld on all sides to a flat plate with axis perpendicular to the plate surface. Find the maximum torque that can be applied to the shaft, if the shear stress in the weld is not to exceed 85 MPa.

$$\left[\textbf{Hint}: \tau_{(max)} = \frac{4.242\,T}{s \times l^2}\right]$$ **[Ans. 32.07 kN-m]**

7. A low carbon steel plate of 0.7 m width welded to a structure of similar material by means of two parallel fillet welds of 0.112 m length (each) is subjected to an eccentric load of 4000 N, the line of action of which has a distance of 1.5 m from the centre of gravity of the weld group. Design the required thickness of the plate when the allowable stress of the weld metal is 60 MPa and that of the plate is 40 MPa. **[Ans. 2 mm]**

8. A 125 × 95 × 10 mm angle is welded to a frame by two 10 mm fillet welds, as shown in Fig. 10.35. A load of 16 kN is apsplied normal to the gràvity axis at a distance of 300 mm from the centre of gravity of welds. Find maximum shear stress in the welds, assuming each weld to be 100 mm long and parallel to the axis of the angle. **[Ans. 45.5 MPa]**

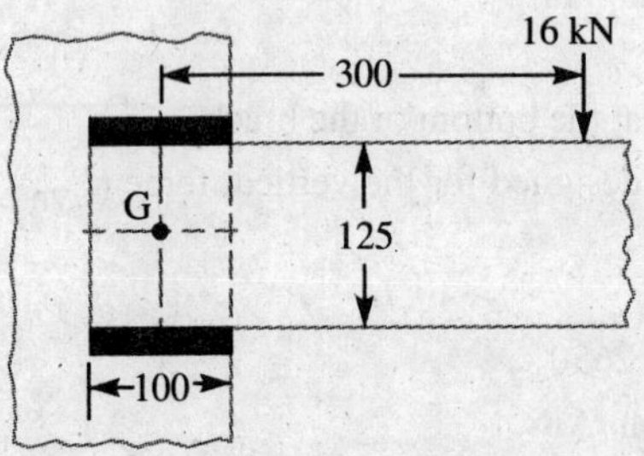

All dimensions in mm.

Fig. 10.35

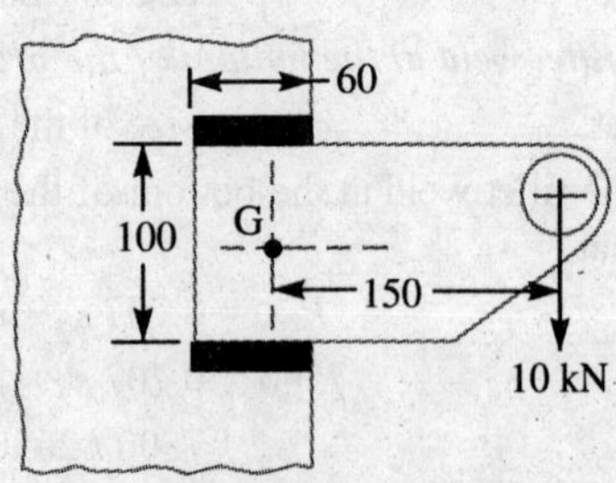

All dimensions in mm.

Fig. 10.36

9. A bracket, as shown in Fig. 10.36, carries a load of 10 kN. Find the size of the weld if the allowable shear stress is not to exceed 80 MPa. **[Ans. 10.83 mm]**

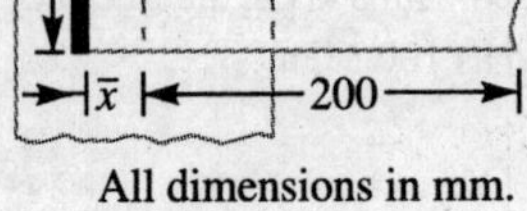

All dimensions in mm.

Fig. 10.37

10. Fig. 10.37 shows a welded joint subjected to an eccentric load of 20 kN. The welding is only on one side. Determine the uniform size of the weld on the entire length of two legs. Take permissible shear stress for the weld material as 80 MPa. **[Ans. 8.9 mm]**

11. A bracket is welded to the side of a column and carries a vertical load P, as shown in Fig. 10.38. Evaluate P so that the maximum shear stress in the 10 mm fillet welds is 80 MPa.

[Ans. 50.7 kN]

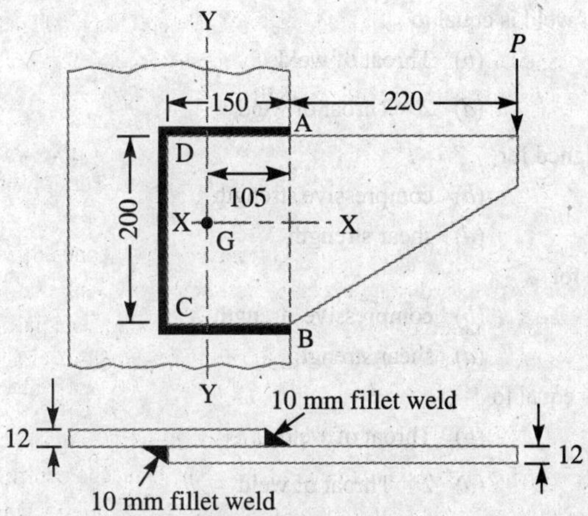

All dimensions in mm.

Fig. 10.38

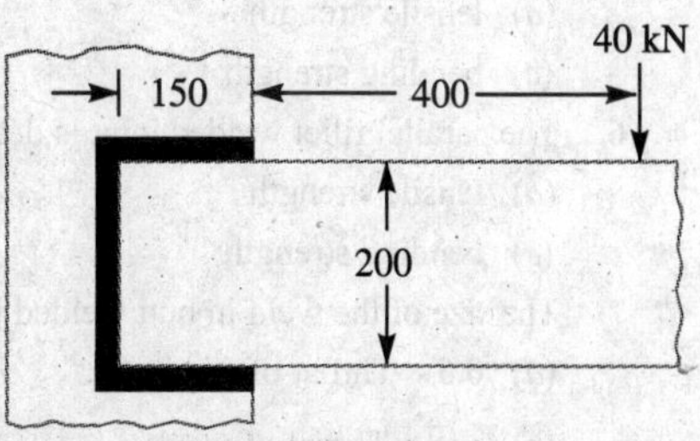

All dimensions in mm.

Fig. 10.39

12. A bracket, as shown in Fig. 10.39, carries a load of 40 kN. Calculate the size of weld, if the allowable shear stress is not to exceed 80 MPa. **[Ans. 7 mm]**

QUESTIONS

1. What do you understand by the term welded joint? How it differs from riveted joint?
2. Sketch and discuss the various types of welded joints used in pressure vessels. What are the considerations involved?
3. State the basic difference between manual welding, semi-automatic welding and automatic welding.
4. What are the assumptions made in the design of welded joint?
5. Explain joint preparation with particular reference to butt welding of plates by arc welding.
6. Discuss the standard location of elements of a welding symbol.
7. Explain the procedure for designing an axially loaded unsymmetrical welded section.
8. What is an eccentric loaded welded joint ? Discuss the procedure for designing such a joint.
9. Show that the normal stress in case of an annular fillet weld subjected to bending is given by

$$\sigma = \frac{5.66\,M}{\pi s d^2}$$

where M = Bending moment; s = Weld size and d = Diameter of cylindrical element welded to flat surface.

OBJECTIVE TYPE QUESTIONS

1. In a fusion welding process,

   (*a*) only heat is used (*b*) only pressure is used

   (*c*) combination of heat and pressure is used (*d*) none of these

2. The electric arc welding is a type of .... welding.

   (*a*) forge (*b*) fusion

3. The principle of applying heat and pressure is widely used in

   (*a*) spot welding (*b*) seam welding

   (*c*) projection welding (*d*) all of these

4. In transverse fillet welded joint, the size of weld is equal to
   (*a*) 0.5 × Throat of weld (*b*) Throat of weld
   (*c*) $\sqrt{2}$ × Throat of weld (*d*) 2 × Throat of weld
5. The transverse fillet welded joints are designed for
   (*a*) tensile strength (*b*) compressive strength
   (*c*) bending strength (*d*) shear strength
6. The parallel fillet welded joint is designed for
   (*a*) tensile strength (*b*) compressive strength
   (*c*) bending strength (*d*) shear strength
7. The size of the weld in butt welded joint is equal to
   (*a*) 0.5 × Throat of weld (*b*) Throat of weld
   (*c*) $\sqrt{2}$ × Throat of weld (*d*) 2 × Throat of weld
8. A double fillet welded joint with parallel fillet weld of length *l* and leg *s* is subjected to a tensile force *P*. Assuming uniform stress distribution, the shear stress in the weld is given by
   (*a*) $\frac{\sqrt{2}\,P}{s.l}$ (*b*) $\frac{P}{2s.l}$
   (*c*) $\frac{P}{\sqrt{2}\,s.l}$ (*d*) $\frac{2P}{s.l}$
9. When a circular rod welded to a rigid plate by a circular fillet weld is subjected to a twisting moment *T*, then the maximum shear stress is given by
   (*a*) $\frac{2.83\,T}{\pi s d^2}$ (*b*) $\frac{4.242\,T}{\pi s d^2}$
   (*c*) $\frac{5.66\,T}{\pi s d^2}$ (*d*) none of these
10. For a parallel load on a fillet weld of equal legs, the plane of maximum shear occurs at
   (*a*) 22.5° (*b*) 30°
   (*c*) 45° (*d*) 60°

ANSWERS

1. (*a*)	2. (*b*)	3. (*d*)	4. (*c*)	5. (*a*)
6. (*d*)	7. (*b*)	8. (*c*)	9. (*a*)	10. (*c*)

CHAPTER

11

Screwed Joints

11.1 Introduction

A screw thread is formed by cutting a continuous helical groove on a cylindrical surface. A screw made by cutting a single helical groove on the cylinder is known as ***single threaded*** (or single-start) screw and if a second thread is cut in the space between the grooves of the first, a ***double threaded*** (or double-start) screw is formed. Similarly, triple and quadruple (*i.e.* multiple-start) threads may be formed. The helical grooves may be cut either ***right hand*** or ***left hand.***

A screwed joint is mainly composed of two elements *i.e.* a bolt and nut. The screwed joints are widely used where the machine parts are required to be readily connected or disconnected without damage to the machine or the fastening. This may be for the purpose of holding or adjustment in assembly or service inspection, repair, or replacement or it may be for the manufacturing or assembly reasons.

The parts may be rigidly connected or provisions may be made for predetermined relative motion.

11.2 Advantages and Disadvantages of Screwed Joints

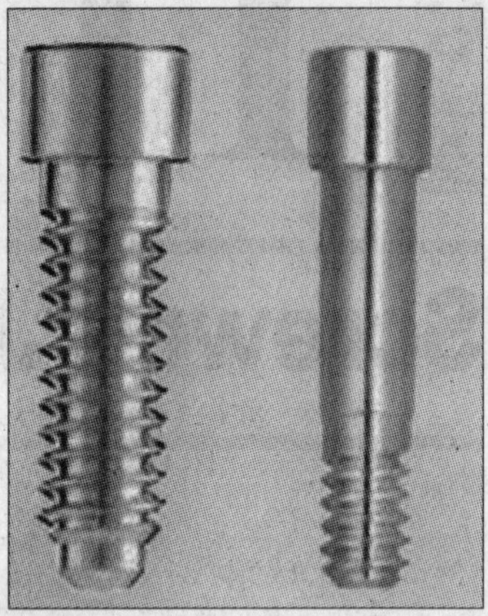

Following are the advantages and disadvantages of the screwed joints.

Advantages

1. Screwed joints are highly reliable in operation.
2. Screwed joints are convenient to assemble and disassemble.
3. A wide range of screwed joints may be adopted to various operating conditions.
4. Screws are relatively cheap to produce due to standardisation and highly efficient manufacturing processes.

Disadvantages

The main disadvantage of the screwed joints is the stress concentration in the threaded portions which are vulnerable points under variable load conditions.

Note : The strength of the screwed joints is not comparable with that of riveted or welded joints.

11.3 Important Terms Used in Screw Threads

The following terms used in screw threads, as shown in Fig. 11.1, are important from the subject point of view :

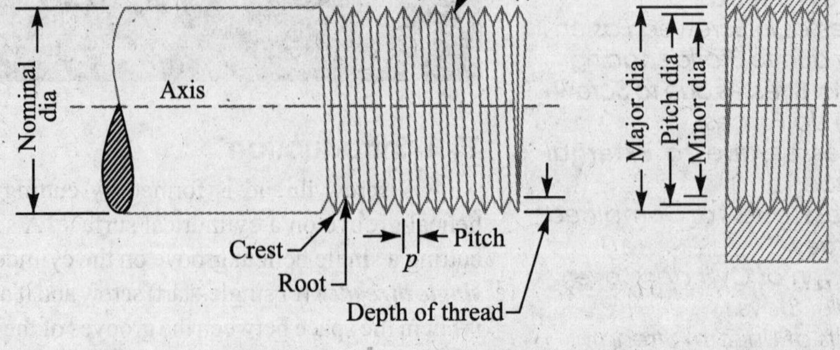

Fig. 11.1. Terms used in screw threads.

1. *Major diameter.* It is the largest diameter of an external or internal screw thread. The screw is specified by this diameter. It is also known as ***outside*** or ***nominal diameter.***

2. *Minor diameter.* It is the smallest diameter of an external or internal screw thread. It is also known as ***core*** or ***root diameter.***

3. *Pitch diameter.* It is the diameter of an imaginary cylinder, on a cylindrical screw thread, the surface of which would pass through the thread at such points as to make equal the width of the thread and the width of the spaces between the threads. It is also called an ***effective diameter.*** In a nut and bolt assembly, it is the diameter at which the ridges on the bolt are in complete touch with the ridges of the corresponding nut.

4. ***Pitch.*** It is the distance from a point on one thread to the corresponding point on the next. This is measured in an axial direction between corresponding points in the same axial plane. Mathematically,

$$\text{Pitch} = \frac{1}{\text{No. of threads per unit length of screw}}$$

5. ***Lead.*** It is the distance between two corresponding points on the same helix. It may also be defined as the distance which a screw thread advances axially in one rotation of the nut. Lead is equal to the pitch in case of single start threads, it is twice the pitch in double start, thrice the pitch in triple start and so on.

6. ***Crest.*** It is the top surface of the thread.

7. ***Root.*** It is the bottom surface created by the two adjacent flanks of the thread.

8. ***Depth of thread.*** It is the perpendicular distance between the crest and root.

9. ***Flank.*** It is the surface joining the crest and root.

10. ***Angle of thread.*** It is the angle included by the flanks of the thread.

11. ***Slope.*** It is half the pitch of the thread.

11.4 Forms of Screw Threads

The following are the various forms of screw threads.

1. ***British standard whitworth (B.S.W.) thread.*** This is a British standard thread profile and has coarse pitches. It is a symmetrical *V*-thread in which the angle between the flankes, measured in an axial plane, is 55°. These threads are found on bolts and screwed fastenings for special purposes. The various proportions of B.S.W. threads are shown in Fig. 11.2.

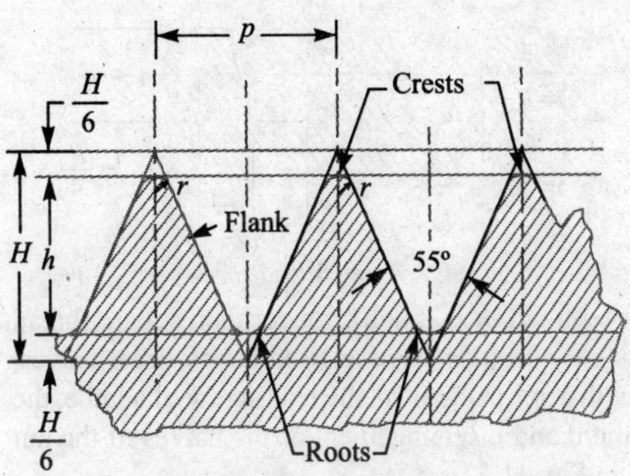

$H = 0.96\,p$; $h = 0.64\,p$; $r = 0.1373\,p$

Fig. 11.2. British standard whitworth (B.S.W) thread.

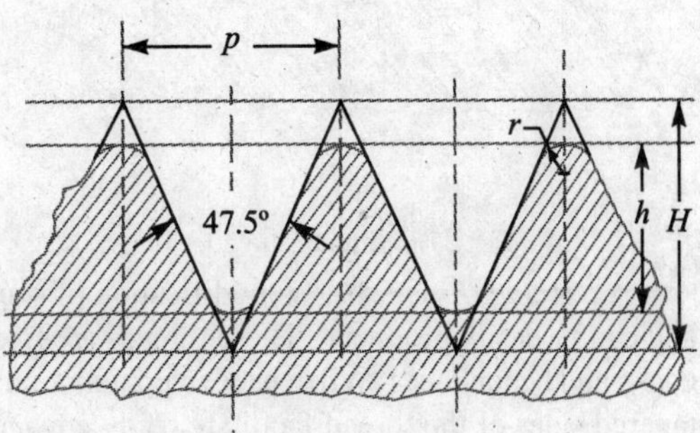

$H = 1.13634\,p$; $h = 0.6\,p$; $r = 0.18083\,p$

Fig. 11.3. British association (B.A.) thread.

The British standard threads with fine pitches (B.S.F.) are used where great strength at the root is required. These threads are also used for line adjustments and where the connected parts are subjected to increased vibrations as in aero and automobile work.

The British standard pipe (B.S.P.) threads with fine pitches are used for steel and iron pipes and tubes carrying fluids. In external pipe threading, the threads are specified by the bore of the pipe.

2. ***British association (B.A.) thread.*** This is a B.S.W. thread with fine pitches. The proportions of the B.A. thread are shown in Fig. 11.3. These threads are used for instruments and other precision works.

3. ***American national standard thread.*** The American national standard or U.S. or Seller's thread has flat crests and roots. The flat crest can withstand more rough usage than sharp *V*-threads. These threads are used for general purposes *e.g.* on bolts, nuts, screws and tapped holes. The various

proportions are shown in Fig. 11.4.

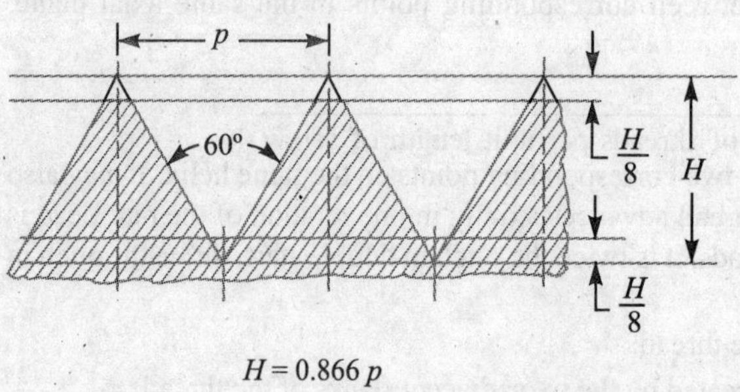

Fig. 11.4. American national standard thread.

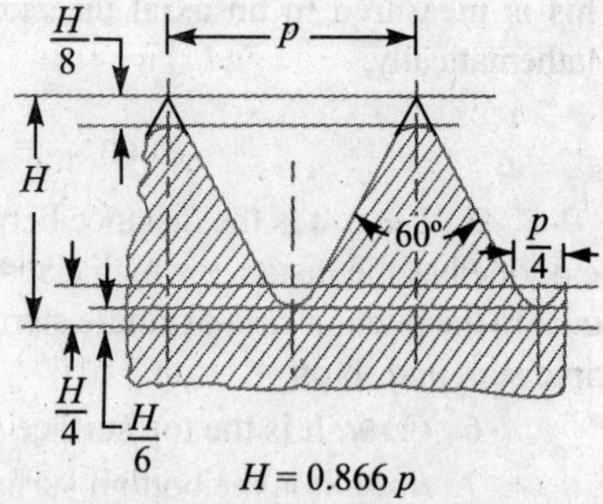

Fig. 11.5. Unified standard thread.

4. ***Unified standard thread.*** The three countries *i.e.*, Great Britain, Canada and United States came to an agreement for a common screw thread system with the included angle of 60°, in order to facilitate the exchange of machinery. The thread has rounded crests and roots, as shown in Fig. 11.5.

5. ***Square thread.*** The square threads, because of their high efficiency, are widely used for transmission of power in either direction. Such type of threads are usually found on the feed mechanisms of machine tools, valves, spindles, screw jacks etc. The square threads are not so strong as V-threads but they offer less frictional resistance to motion than Whitworth threads. The pitch of the square thread is often taken twice that of a B.S.W. thread of the same diameter. The proportions of the thread are shown in Fig. 11.6.

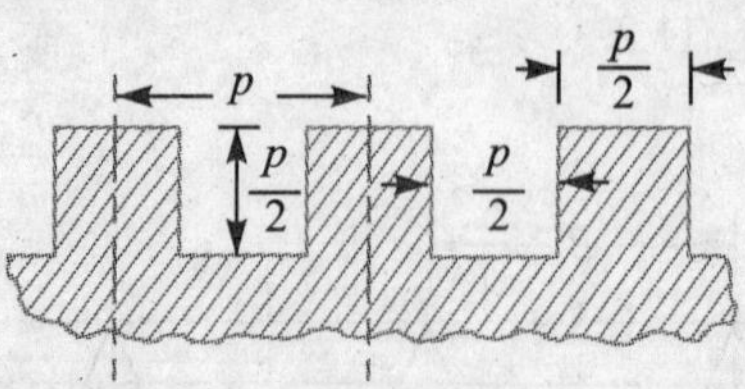

Fig. 11.6. Square thread.

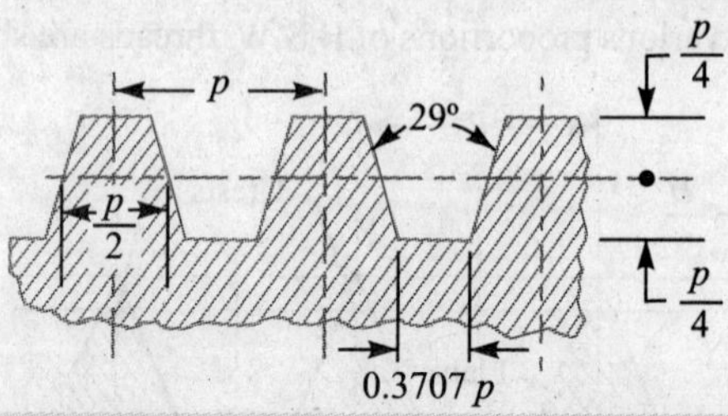

Fig. 11.7. Acme thread.

6. ***Acme thread.*** It is a modification of square thread. It is much stronger than square thread and can be easily produced. These threads are frequently used on screw cutting lathes, brass valves, cocks and bench vices. When used in conjunction with a split nut, as on the lead screw of a lathe, the tapered sides of the thread facilitate ready engagement and disengagement of the halves of the nut when required. The various proportions are shown in Fig. 11.7.

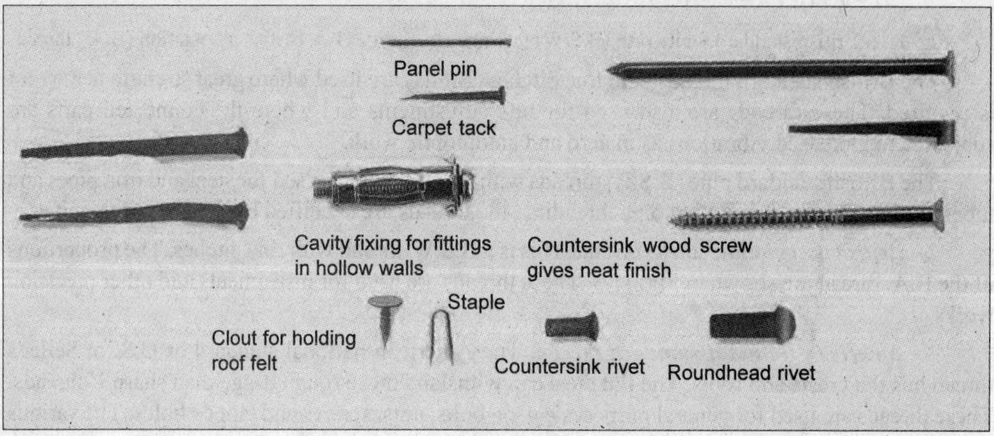

7. ***Knuckle thread.*** It is also a modification of square thread. It has rounded top and bottom. It can be cast or rolled easily and can not economically be made on a machine. These threads are used for rough and ready work. They are usually found on railway carriage couplings, hydrants, necks of glass bottles and large moulded insulators used in electrical trade.

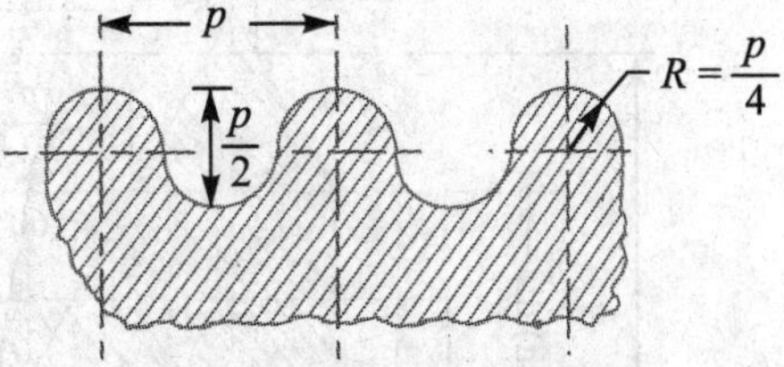

Fig. 11.8. Knuckle thread.

8. ***Buttress thread.*** It is used for transmission of power in one direction only. The force is transmitted almost parallel to the axis. This thread units the advantage of both square and V-threads. It has a low frictional resistance characteristics of the square thread and have the same strength as that of V-thread. The spindles of bench vices are usually provided with buttress thread. The various proportions of buttress thread are shown in Fig. 11.9.

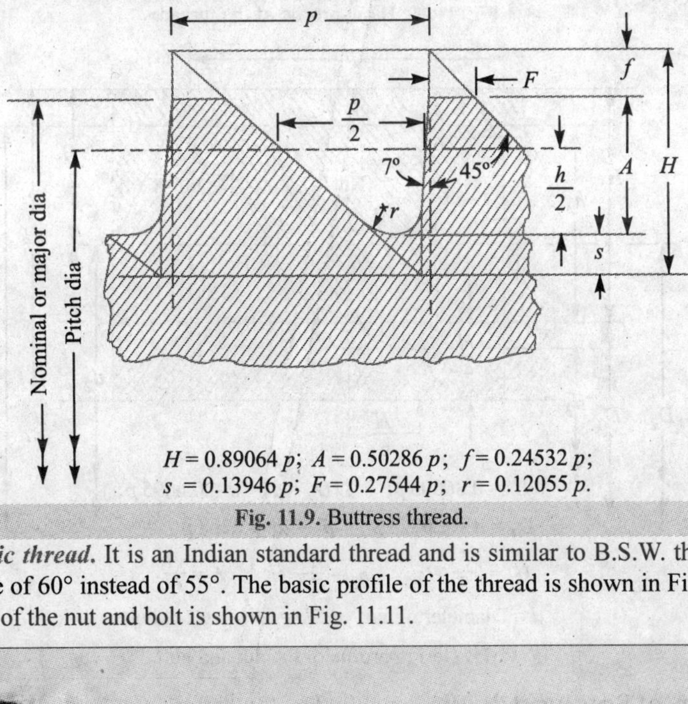

$H = 0.89064\,p;\ A = 0.50286\,p;\ f = 0.24532\,p;$
$s = 0.13946\,p;\ F = 0.27544\,p;\ r = 0.12055\,p.$

Fig. 11.9. Buttress thread.

9. ***Metric thread.*** It is an Indian standard thread and is similar to B.S.W. threads. It has an included angle of 60° instead of 55°. The basic profile of the thread is shown in Fig. 11.10 and the design profile of the nut and bolt is shown in Fig. 11.11.

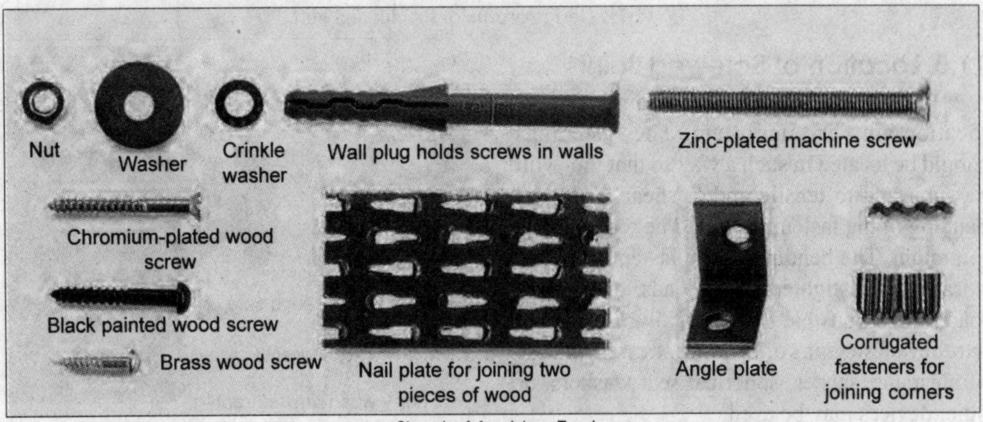

Simple Machine Tools.

Note : This picture is given as additional information and is not a direct example of the current chapter.

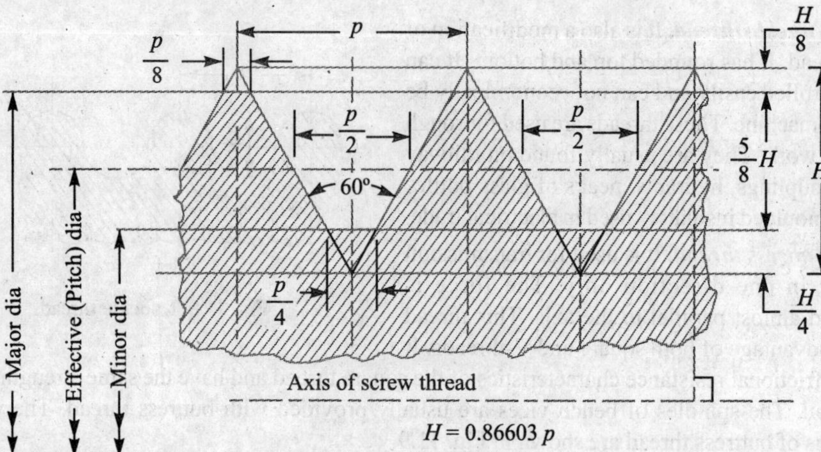

Fig. 11.10. Basic profile of the thread.

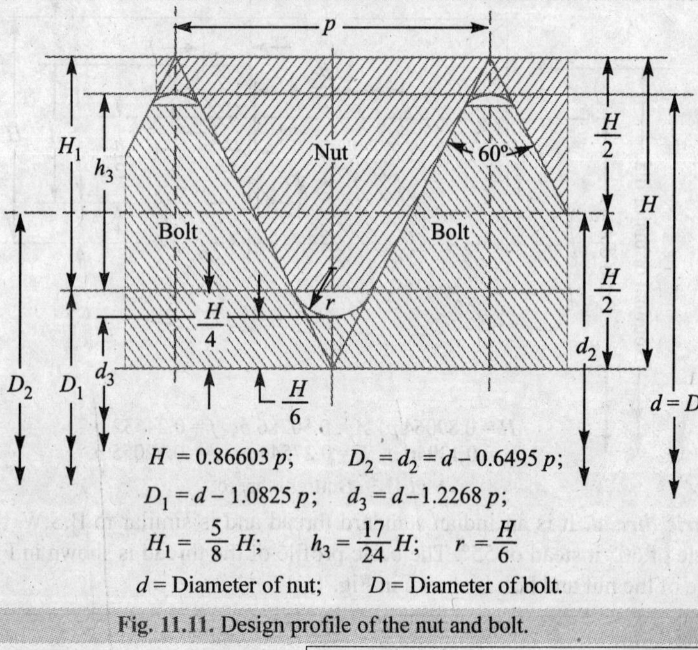

$H = 0.86603\,p;\quad D_2 = d_2 = d - 0.6495\,p;$

$D_1 = d - 1.0825\,p;\quad d_3 = d - 1.2268\,p;$

$H_1 = \frac{5}{8}H;\quad h_3 = \frac{17}{24}H;\quad r = \frac{H}{6}$

d = Diameter of nut; D = Diameter of bolt.

Fig. 11.11. Design profile of the nut and bolt.

11.5 Location of Screwed Joints

The choice of type of fastenings and its location are very important. The fastenings should be located in such a way so that they will be subjected to tensile and/or shear loads and bending of the fastening should be reduced to a minimum. The bending of the fastening due to misalignment, tightening up loads, or external loads are responsible for many failures. In order to relieve fastenings of bending stresses, the use of clearance spaces, spherical seat washers, or other devices may be used.

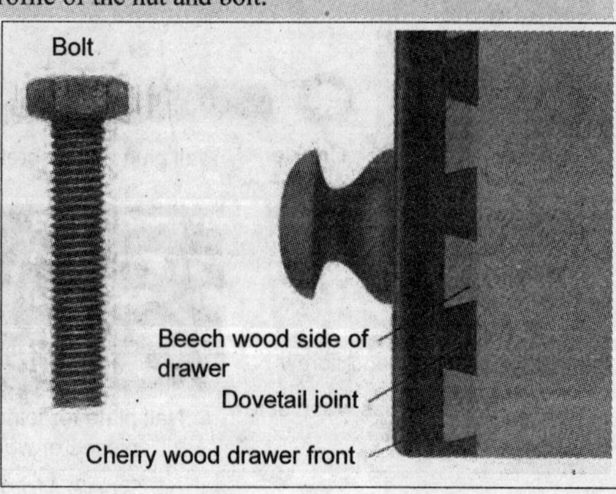

11.6 Common Types of Screw Fastenings

Following are the common types of screw fastenings :

1. ***Through bolts.*** A through bolt (or simply a bolt) is shown in Fig. 11.12 (*a*). It is a cylindrical bar with threads for the nut at one end and head at the other end. The cylindrical part of the bolt is known as **shank.** It is passed through drilled holes in the two parts to be fastened together and clamped them securely to each other as the nut is screwed on to the threaded end. The through bolts may or may not have a machined finish and are made with either hexagonal or square heads. A through bolt should pass easily in the holes, when put under tension by a load along its axis. If the load acts perpendicular to the axis, tending to slide one of the connected parts along the other end thus subjecting it to shear, the holes should be reamed so that the bolt shank fits snugly there in. The through bolts according to their usage may be known as **machine bolts, carriage bolts, automobile bolts, eye bolts etc.**

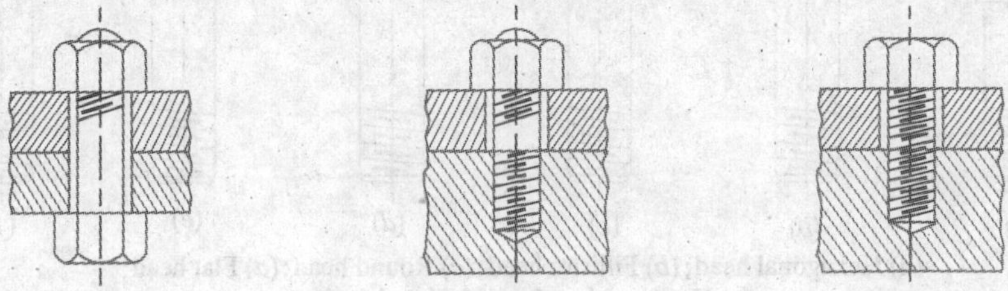

(*a*) Through bolt. (*b*) Tap bolt. (*c*) Stud.

Fig. 11.12

2. ***Tap bolts.*** A tap bolt or screw differs from a bolt. It is screwed into a tapped hole of one of the parts to be fastened without the nut, as shown in Fig. 11.12 (*b*).

3. ***Studs.*** A stud is a round bar threaded at both ends. One end of the stud is screwed into a tapped hole of the parts to be fastened, while the other end receives a nut on it, as shown in Fig. 11.12 (*c*). Studs are chiefly used instead of tap bolts for securing various kinds of covers *e.g.* covers of engine and pump cylinders, valves, chests etc.

Deck-handler crane is used on ships to move loads

Note : This picture is given as additional information and is not a direct example of the current chapter.

This is due to the fact that when tap bolts are unscrewed or replaced, they have a tendency to break the threads in the hole. This disadvantage is overcome by the use of studs.

4. *Cap screws.* The cap screws are similar to tap bolts except that they are of small size and a variety of shapes of heads are available as shown in Fig. 11.13.

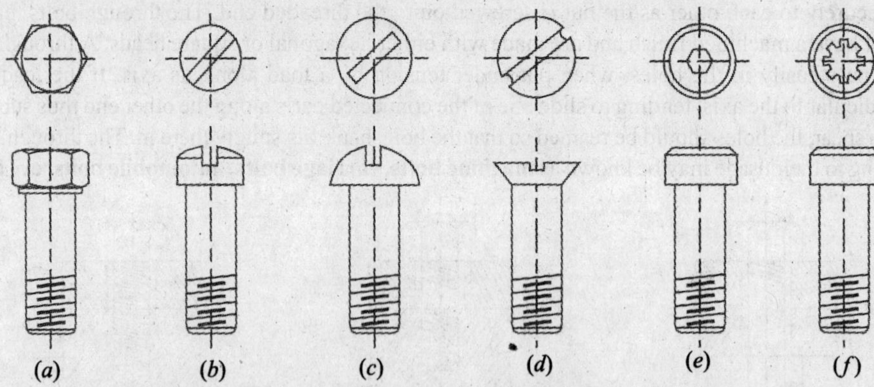

(*a*) Hexagonal head; (*b*) Fillister head; (*c*) Round head; (*d*) Flat head; (*e*) Hexagonal socket; (*f*) Fluted socket.

Fig. 11.13. Types of cap screws.

5. *Machine screws.* These are similar to cap screws with the head slotted for a screw driver. These are generally used with a nut.

6. *Set screws.* The set screws are shown in Fig. 11.14. These are used to prevent relative motion between the two parts. A set screw is screwed through a threaded hole in one part so that its point (*i.e.* end of the screw) presses against the other part. This resists the relative motion between the two parts by means of friction between the point of the screw and one of the parts. They may be used instead of key to prevent relative motion between a hub and a shaft in light power transmission members. They may also be used in connection with a key, where they prevent relative axial motion of the shaft, key and hub assembly.

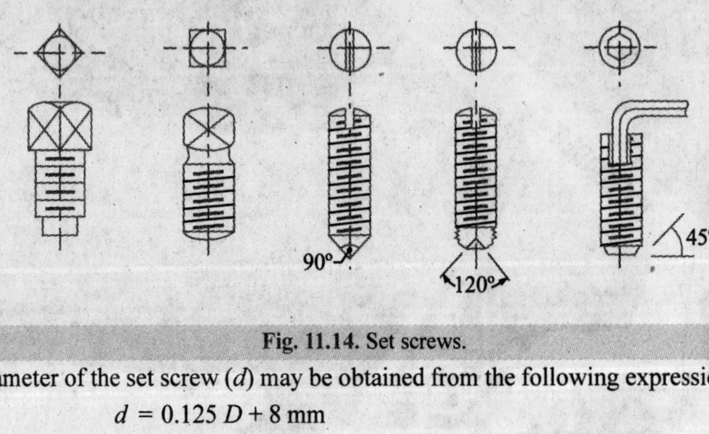

Fig. 11.14. Set screws.

The diameter of the set screw (d) may be obtained from the following expression:

$$d = 0.125\, D + 8 \text{ mm}$$

where D is the diameter of the shaft (in mm) on which the set screw is pressed.

The tangential force (in newtons) at the surface of the shaft is given by

$$F = 6.6\, (d)^{2.3}$$

∴ Torque transmitted by a set screw,

$$T = F \times \frac{D}{2} \text{ N-m} \qquad \text{... (D is in metres)}$$

and power transmitted (in watts), $P = \frac{2\pi N.T}{60}$, where N is the speed in r.p.m.

11.7 Locking Devices

Ordinary thread fastenings, generally, remain tight under static loads, but many of these fastenings become loose under the action of variable loads or when machine is subjected to vibrations. The loosening of fastening is very dangerous and must be prevented. In order to prevent this, a large number of locking devices are available, some of which are discussed below :

1. ***Jam nut or lock nut.*** A most common locking device is a jam, lock or check nut. It has about one-half to two-third thickness of the standard nut. The thin lock nut is first tightened down with ordinary force, and then the upper nut (*i.e.* thicker nut) is tightened down upon it, as shown in Fig. 11.15 (*a*). The upper nut is then held tightly while the lower one is slackened back against it.

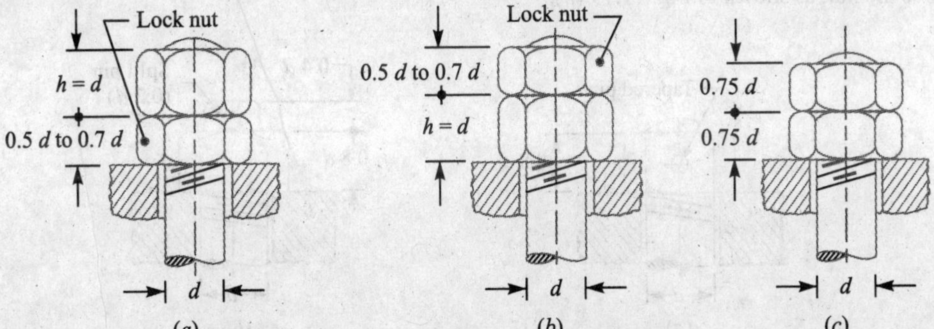

Fig. 11.15. Jam nut or lock nut.

In slackening back the lock nut, a thin spanner is required which is difficult to find in many shops. Therefore to overcome this difficulty, a thin nut is placed on the top as shown in Fig. 11.15 (*b*).

If the nuts are really tightened down as they should be, the upper nut carries a greater tensile load than the bottom one. Therefore, the top nut should be thicker one with a thin nut below it because it is desirable to put whole of the load on the thin nut. In order to overcome both the difficulties, both the nuts are made of the same thickness as shown in Fig. 11.15 (*c*).

2. ***Castle nut.*** It consists of a hexagonal portion with a cylindrical upper part which is slotted in line with the centre of each face, as shown in Fig. 11.16. The split pin passes through two slots in the nut and a hole in the bolt, so that a positive lock is obtained unless the pin shears. It is extensively used on jobs subjected to sudden shocks and considerable vibration such as in automobile industry.

3. ***Sawn nut.*** It has a slot sawed about half way through, as shown in Fig. 11.17. After the nut is screwed down, the small screw is tightened which produces more friction between the nut and the bolt. This prevents the loosening of nut.

4. ***Penn, ring or grooved nut.*** It has a upper portion hexagonal and a lower part cylindrical as shown in Fig. 11.18. It is largely used where bolts pass through connected pieces reasonably near their edges such as in marine type connecting rod ends. The bottom portion is cylindrical and is recessed to receive the tip of the locking set screw. The bolt hole requires counter-boring to receive the cylindrical portion of the nut. In order to prevent bruising of the latter by the case hardened tip of the set screw, it is recessed.

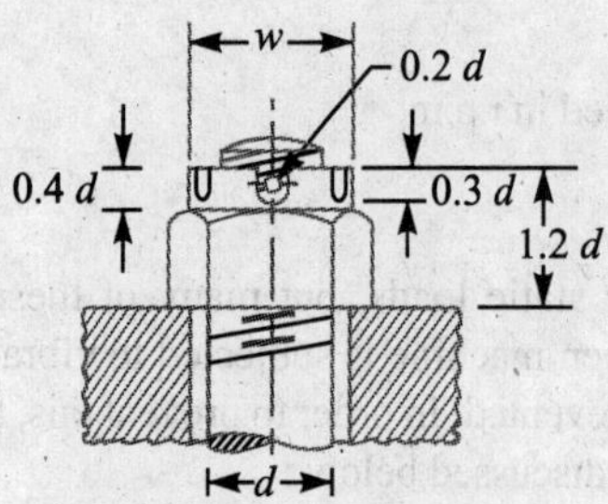

Fig. 11.16. Castle nut.

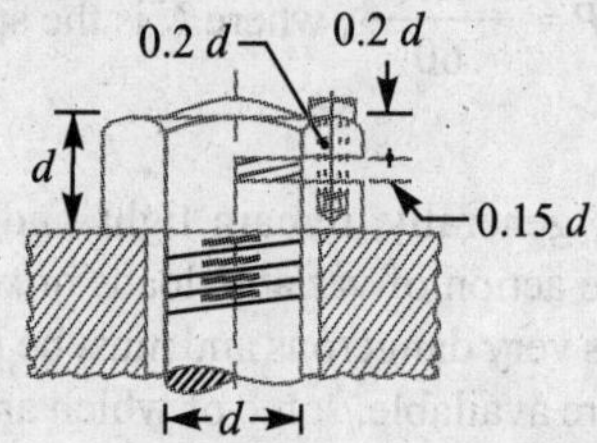

Fig. 11.17. Sawn nut.

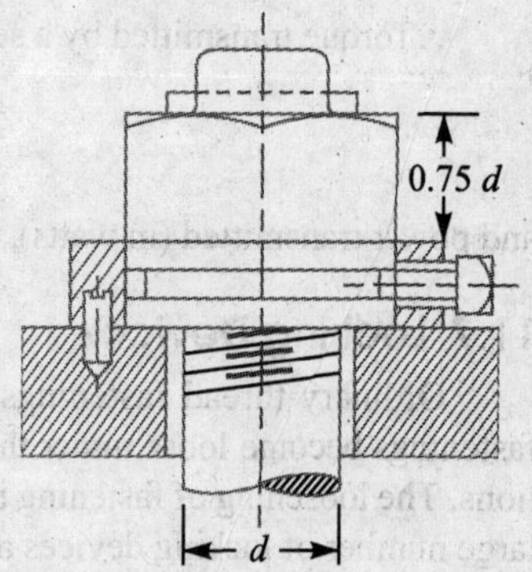

Fig. 11.18. Penn, ring or grooved nut.

5. ***Locking with pin.*** The nuts may be locked by means of a taper pin or cotter pin passing through the middle of the nut as shown in Fig. 11.19 (*a*). But a split pin is often driven through the bolt above the nut, as shown in Fig. 11.19 (*b*).

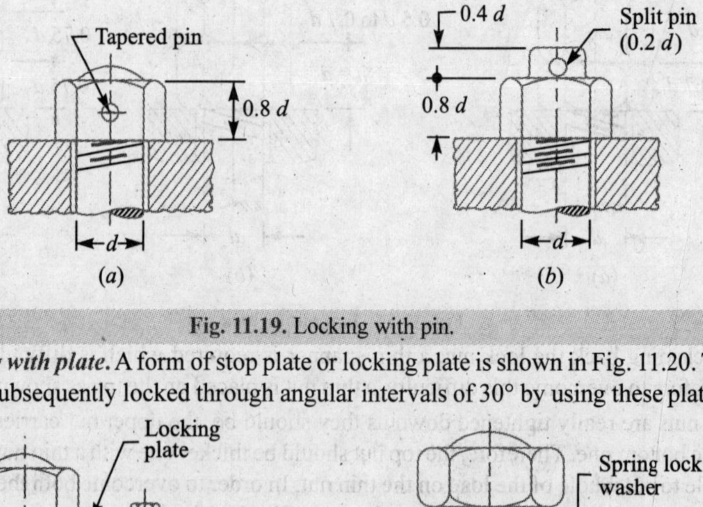

Fig. 11.19. Locking with pin.

6. ***Locking with plate.*** A form of stop plate or locking plate is shown in Fig. 11.20. The nut can be adjusted and subsequently locked through angular intervals of 30° by using these plates.

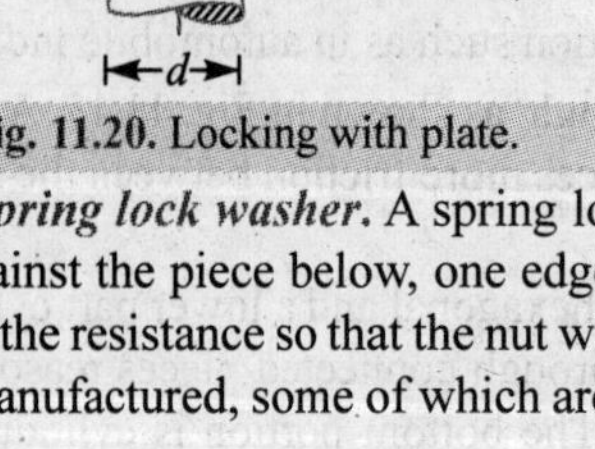

Fig. 11.20. Locking with plate.

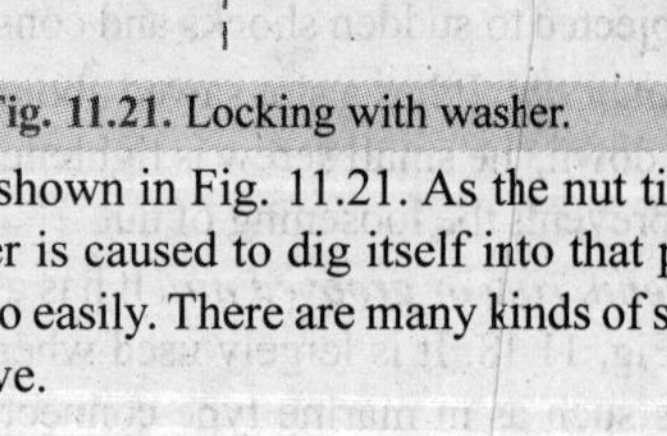

Fig. 11.21. Locking with washer.

7. ***Spring lock washer.*** A spring lock washer is shown in Fig. 11.21. As the nut tightens the washer against the piece below, one edge of the washer is caused to dig itself into that piece, thus increasing the resistance so that the nut will not loosen so easily. There are many kinds of spring lock washers manufactured, some of which are fairly effective.

11.8 Designation of Screw Threads

According to Indian standards, IS : 4218 (Part IV) 1976 (Reaffirmed 1996), the complete designation of the screw thread shall include

1. ***Size designation.*** The size of the screw thread is designated by the letter '*M*' followed by the diameter and pitch, the two being separated by the sign ×. When there is no indication of the pitch, it shall mean that a coarse pitch is implied.

2. ***Tolerance designation.*** This shall include

(*a*) A figure designating tolerance grade as indicated below:
'7' for fine grade, '8' for normal (medium) grade, and '9' for coarse grade.

(*b*) A letter designating the tolerance position as indicated below :
'*H*' for unit thread, '*d*' for bolt thread with allowance, and '*h*' for bolt thread without allowance.

For example, A bolt thread of 6 mm size of coarse pitch and with allowance on the threads and normal (medium) tolerance grade is designated as *M*6-8*d*.

11.9 Standard Dimensions of Screw Threads

The design dimensions of I.S.O. screw threads for screws, bolts and nuts of coarse and fine series are shown in Table 11.1.

Table 11.1. Design dimensions of screw threads, bolts and nuts according to IS : 4218 (Part III) 1976 (Reaffirmed 1996) (Refer Fig. 11.1)

Designation	*Pitch mm*	*Major or nominal diameter Nut and Bolt (d = D) mm*	*Effective or pitch diameter Nut and Bolt (d_p) mm*	*Minor or core diameter (d_c) mm*		*Depth of thread (bolt) mm*	*Stress area mm²*
				Bolt	*Nut*		
(1)	(2)	(3)	(4)	(5)	(6)	(7)	(8)
Coarse series							
M 0.4	0.1	0.400	0.335	0.277	0.292	0.061	0.074
M 0.6	0.15	0.600	0.503	0.416	0.438	0.092	0.166
M 0.8	0.2	0.800	0.670	0.555	0.584	0.123	0.295
M 1	0.25	1.000	0.838	0.693	0.729	0.153	0.460
M 1.2	0.25	1.200	1.038	0.893	0.929	0.158	0.732
M 1.4	0.3	1.400	1.205	1.032	1.075	0.184	0.983
M 1.6	0.35	1.600	1.373	1.171	1.221	0.215	1.27
M 1.8	0.35	1.800	1.573	1.371	1.421	0.215	1.70
M 2	0.4	2.000	1.740	1.509	1.567	0.245	2.07
M 2.2	0.45	2.200	1.908	1.648	1.713	0.276	2.48
M 2.5	0.45	2.500	2.208	1.948	2.013	0.276	3.39
M 3	0.5	3.000	2.675	2.387	2.459	0.307	5.03
M 3.5	0.6	3.500	3.110	2.764	2.850	0.368	6.78
M 4	0.7	4.000	3.545	3.141	3.242	0.429	8.78
M 4.5	0.75	4.500	4.013	3.580	3.688	0.460	11.3
M 5	0.8	5.000	4.480	4.019	4.134	0.491	14.2
M 6	1	6.000	5.350	4.773	4.918	0.613	20.1

(1)	(2)	(3)	(4)	(5)	(6)	(7)	(8)
M 7	1	7.000	6.350	5.773	5.918	0.613	28.9
M 8	1.25	8.000	7.188	6.466	6.647	0.767	36.6
M 10	1.5	10.000	9.026	8.160	8.876	0.920	58.3
M 12	1.75	12.000	10.863	9.858	10.106	1.074	84.0
M 14	2	14.000	12.701	11.546	11.835	1.227	115
M 16	2	16.000	14.701	13.546	13.835	1.227	157
M 18	2.5	18.000	16.376	14.933	15.294	1.534	192
M 20	2.5	20.000	18.376	16.933	17.294	1.534	245
M 22	2.5	22.000	20.376	18.933	19.294	1.534	303
M 24	3	24.000	22.051	20.320	20.752	1.840	353
M 27	3	27.000	25.051	23.320	23.752	1.840	459
M 30	3.5	30.000	27.727	25.706	26.211	2.147	561
M 33	3.5	33.000	30.727	28.706	29.211	2.147	694
M 36	4	36.000	33.402	31.093	31.670	2.454	817
M 39	4	39.000	36.402	34.093	34.670	2.454	976
M 42	4.5	42.000	39.077	36.416	37.129	2.760	1104
M 45	4.5	45.000	42.077	39.416	40.129	2.760	1300
M 48	5	48.000	44.752	41.795	42.587	3.067	1465
M 52	5	52.000	48.752	45.795	46.587	3.067	1755
M 56	5.5	56.000	52.428	49.177	50.046	3.067	2022
M 60	5.5	60.000	56.428	53.177	54.046	3.374	2360
Fine series							
M 8 × 1	1	8.000	7.350	6.773	6.918	0.613	39.2
M 10 × 1.25	1.25	10.000	9.188	8.466	8.647	0.767	61.6
M 12 × 1.25	1.25	12.000	11.184	10.466	10.647	0.767	92.1
M 14 × 1.5	1.5	14.000	13.026	12.160	12.376	0.920	125
M 16 × 1.5	1.5	16.000	15.026	14.160	14.376	0.920	167
M 18 × 1.5	1.5	18.000	17.026	16.160	16.376	0.920	216
M 20 × 1.5	1.5	20.000	19.026	18.160	18.376	0.920	272
M 22 × 1.5	1.5	22.000	21.026	20.160	20.376	0.920	333
M 24 × 2	2	24.000	22.701	21.546	21.835	1.227	384
M 27 × 2	2	27.000	25.701	24.546	24.835	1.227	496
M 30 × 2	2	30.000	28.701	27.546	27.835	1.227	621
M 33 × 2	2	33.000	31.701	30.546	30.835	1.227	761
M 36 × 3	3	36.000	34.051	32.319	32.752	1.840	865
M 39 × 3	3	39.000	37.051	35.319	35.752	1.840	1028

Note : In case the table is not available, then the core diameter (d_c) may be taken as 0.84 d, where d is the major diameter.

11.10 Stresses in Screwed Fastening due to Static Loading

The following stresses in screwed fastening due to static loading are important from the subject point of view :

1. Internal stresses due to screwing up forces,
2. Stresses due to external forces, and
3. Stress due to combination of stresses at (1) and (2).

We shall now discuss these stresses, in detail, in the following articles.

11.11 Initial Stresses due to Screwing up Forces

The following stresses are induced in a bolt, screw or stud when it is screwed up tightly.

1. ***Tensile stress due to stretching of bolt.*** Since none of the above mentioned stresses are accurately determined, therefore bolts are designed on the basis of direct tensile stress with a large factor of safety in order to account for the indeterminate stresses. The initial tension in a bolt, based on experiments, may be found by the relation

$$P_i = 2840\, d \text{ N}$$

where P_i = Initial tension in a bolt, and

d = Nominal diameter of bolt, in mm.

The above relation is used for making a joint fluid tight like steam engine cylinder cover joints etc. When the joint is not required as tight as fluid-tight joint, then the initial tension in a bolt may be reduced to half of the above value. In such cases

$$P_i = 1420\, d \text{ N}$$

The small diameter bolts may fail during tightening, therefore bolts of smaller diameter (less than M 16 or M 18) are not permitted in making fluid tight joints.

If the bolt is not initially stressed, then the maximum safe axial load which may be applied to it, is given by

P = Permissible stress × Cross-sectional area at bottom of the thread (*i.e.* stress area)

The stress area may be obtained from Table 11.1 or it may be found by using the relation

$$\text{Stress area} = \frac{\pi}{4}\left(\frac{d_p + d_c}{2}\right)^2$$

where d_p = Pitch diameter, and

d_c = Core or minor diameter.

Simple machine tools.

Note : This picture is given as additional information and is not a direct example of the current chapter.

2. ***Torsional shear stress caused by the frictional resistance of the threads during its tightening.*** The torsional shear stress caused by the frictional resistance of the threads during its tightening may be obtained by using the torsion equation. We know that

$$\frac{T}{J} = \frac{\tau}{r}$$

$$\therefore \qquad \tau = \frac{T}{J} \times r = \frac{T}{\frac{\pi}{32}(d_c)^4} \times \frac{d_c}{2} = \frac{16\,T}{\pi\,(d_c)^3}$$

where
τ = Torsional shear stress,
T = Torque applied, and
d_c = Minor or core diameter of the thread.

It has been shown during experiments that due to repeated unscrewing and tightening of the nut, there is a gradual scoring of the threads, which increases the torsional twisting moment (T).

3. ***Shear stress across the threads.*** The average thread shearing stress for the screw (τ_s) is obtained by using the relation :

$$\tau_s = \frac{P}{\pi\, d_c \times b \times n}$$

where b = Width of the thread section at the root.

The average thread shearing stress for the nut is

$$\tau_n = \frac{P}{\pi\, d \times b \times n}$$

where d = Major diameter.

4. ***Compression or crushing stress on threads.*** The compression or crushing stress between the threads (σ_c) may be obtained by using the relation :

$$\sigma_c = \frac{P}{\pi\,[d^2 - (d_c)^2]\, n}$$

where
d = Major diameter,
d_c = Minor diameter, and
n = Number of threads in engagement.

5. ***Bending stress if the surfaces under the head or nut are not perfectly parallel to the bolt axis.*** When the outside surfaces of the parts to be connected are not parallel to each other, then the bolt will be subjected to bending action. The bending stress (σ_b) induced in the shank of the bolt is given by

$$\sigma_b = \frac{x \,.\, E}{2l}$$

where
x = Difference in height between the extreme corners of the nut or head,
l = Length of the shank of the bolt, and
E = Young's modulus for the material of the bolt.

Example 11.1. *Determine the safe tensile load for a bolt of M 30, assuming a safe tensile stress of 42 MPa.*

Solution. Given : d = 30 mm ; σ_t = 42 MPa = 42 N/mm^2

From Table 11.1 (coarse series), we find that the stress area *i.e.* cross-sectional area at the bottom of the thread corresponding to M 30 is 561 mm².

$\therefore$ Safe tensile load = Stress area × σ_t = 561 × 42 = 23 562 N = 23.562 kN **Ans.**

Note: In the above example, we have assumed that the bolt is not initially stressed.

Example 11.2. *Two machine parts are fastened together tightly by means of a 24 mm tap bolt. If the load tending to separate these parts is neglected, find the stress that is set up in the bolt by the initial tightening.*

Solution. Given : d = 24 mm

From Table 11.1 (coarse series), we find that the core diameter of the thread corresponding to M 24 is d_c = 20.32 mm.

Let σ_t = Stress set up in the bolt.

We know that initial tension in the bolt,

$$P = 2840\, d = 2840 \times 24 = 68\,160 \text{ N}$$

We also know that initial tension in the bolt (P),

$$68\,160 = \frac{\pi}{4}(d_c)^2\,\sigma_t = \frac{\pi}{4}(20.30)^2\,\sigma_t = 324\,\sigma_t$$

$$\therefore \quad \sigma_t = 68\,160 / 324 = 210 \text{ N/mm}^2 = 210 \text{ MPa } \textbf{Ans.}$$

11.12 Stresses due to External Forces

The following stresses are induced in a bolt when it is subjected to an external load.

1. ***Tensile stress.*** The bolts, studs and screws usually carry a load in the direction of the bolt axis which induces a tensile stress in the bolt.

Let d_c = Root or core diameter of the thread, and

σ_t = Permissible tensile stress for the bolt material.

We know that external load applied,

$$P = \frac{\pi}{4}(d_c)^2\,\sigma_t \quad \text{or} \quad d_c = \sqrt{\frac{4\,P}{\pi\,\sigma_t}}$$

Now from Table 11.1, the value of the nominal diameter of bolt corresponding to the value of d_c may be obtained or stress area $\left[\frac{\pi}{4}(d_c)^2\right]$ may be fixed.

Notes: (*a*) If the external load is taken up by a number of bolts, then

$$P = \frac{\pi}{4}(d_c)^2\,\sigma_t \times n$$

(*b*) In case the standard table is not available, then for coarse threads, d_c = 0.84 d, where d is the nominal diameter of bolt.

Simple machine tools.

Note : This picture is given as additional information and is not a direct example of the current chapter.

2. ***Shear stress.*** Sometimes, the bolts are used to prevent the relative movement of two or more parts, as in case of flange coupling, then the shear stress is induced in the bolts. The shear stresses should be avoided as far as possible. It should be noted that when the bolts are subjected to direct shearing loads, they should be located in such a way that the shearing load comes upon the body (*i.e.* shank) of the bolt and not upon the threaded portion. In some cases, the bolts may be relieved of shear load by using shear pins. When a number of bolts are used to share the shearing load, the finished bolts should be fitted to the reamed holes.

Let d = Major diameter of the bolt, and

n = Number of bolts.

∴ Shearing load carried by the bolts,

$$P_s = \frac{\pi}{4} \times d^2 \times \tau \times n \quad \text{or} \quad d = \sqrt{\frac{4\,P_s}{\pi \tau n}}$$

3. ***Combined tension and shear stress.*** When the bolt is subjected to both tension and shear loads, as in case of coupling bolts or bearing, then the diameter of the shank of the bolt is obtained from the shear load and that of threaded part from the tensile load. A diameter slightly larger than that required for either shear or tension may be assumed and stresses due to combined load should be checked for the following principal stresses.

Maximum principal shear stress,

$$\tau_{max} = \frac{1}{2}\sqrt{(\sigma_t)^2 + 4\tau^2}$$

and maximum principal tensile stress,

$$\sigma_{t(max)} = \frac{\sigma_t}{2} + \frac{1}{2}\sqrt{(\sigma_t)^2 + 4\tau^2}$$

These stresses should not exceed the safe permissible values of stresses.

Example 11.3. *An eye bolt is to be used for lifting a load of 60 kN. Find the nominal diameter of the bolt, if the tensile stress is not to exceed 100 MPa. Assume coarse threads.*

Solution. Given : P = 60 kN = 60 × 10³ N ; σ_t = 100 MPa = 100 N/mm²

An eye bolt for lifting a load is shown in Fig. 11.22.

Let d = Nominal diameter of the bolt, and

d_c = Core diameter of the bolt.

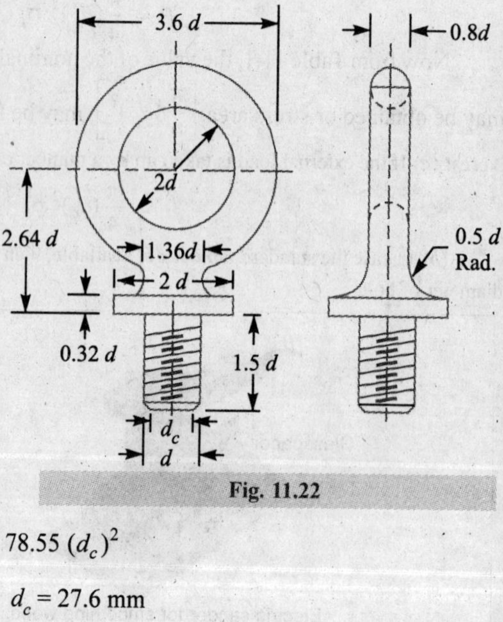

Fig. 11.22

We know that load on the bolt (P),

$$60 \times 10^3 = \frac{\pi}{4}(d_c)^2 \sigma_t = \frac{\pi}{4}(d_c)^2\, 100 = 78.55\,(d_c)^2$$

∴ $(d_c)^2 = 600 \times 10^3 / 78.55 = 764$ or $d_c = 27.6$ mm

From Table 11.1 (coarse series), we find that the standard core diameter (d_c) is 28.706 mm and the corresponding nominal diameter (d) is 33 mm. **Ans.**

Note : A lifting eye bolt, as shown in Fig. 11.22, is used for lifting and transporting heavy machines. It consists of a ring of circular cross-section at the head and provided with threads at the lower portion for screwing inside a threaded hole on the top of the machine.

Example 11.4. *Two shafts are connected by means of a flange coupling to transmit torque of 25 N-m. The flanges of the coupling are fastened by four bolts of the same material at a radius of 30 mm. Find the size of the bolts if the allowable shear stress for the bolt material is 30 MPa.*

Solution. Given : $T = 25$ N-m $= 25 \times 10^3$ N-mm ; $n = 4$; $R_p = 30$ mm ; $\tau = 30$ MPa $= 30$ N/mm²

We know that the shearing load carried by flange coupling,

$$P_s = \frac{T}{R_p} = \frac{25 \times 10^3}{30} = 833.3 \text{ N} \quad ...(i)$$

Let d_c = Core diameter of the bolt.

∴ Resisting load on the bolts

$$= \frac{\pi}{4}(d_c)^2 \tau \times n = \frac{\pi}{4}(d_c)^2\ 30 \times 4 = 94{\cdot}26\ (d_c)^2 \quad ...(ii)$$

From equations (*i*) and (*ii*), we get

$$(d_c)^2 = 833.3 / 94.26 = 8.84 \quad \text{or} \quad d_c = 2.97 \text{ mm}$$

From Table 11.1 (coarse series), we find that the standard core diameter of the bolt is 3.141 mm and the corresponding size of the bolt is M 4. **Ans.**

Example 11.5. *A lever loaded safety valve has a diameter of 100 mm and the blow off pressure is 1.6 N/mm². The fulcrum of the lever is screwed into the cast iron body of the cover. Find the diameter of the threaded part of the fulcrum if the permissible tensile stress is limited to 50 MPa and the leverage ratio is 8.*

Solution. Given : $D = 100$ mm ; $p = 1.6$ N/mm²; $\sigma_t = 50$ MPa $= 50$ N/mm²

We know that the load acting on the valve,

$$F = \text{Area} \times \text{pressure} = \frac{\pi}{4} \times D^2 \times p = \frac{\pi}{4}\ (100)^2\ 1.6 = 12\ 568 \text{ N}$$

Since the leverage is 8, therefore load at the end of the lever,

$$W = \frac{12\ 568}{8} = 1571 \text{ N}$$

∴ Load on the fulcrum,

$$P = F - W = 12\ 568 - 1571 = 10\ 997 \text{ N} \quad ...(i)$$

Let d_c = Core diameter of the threaded part.

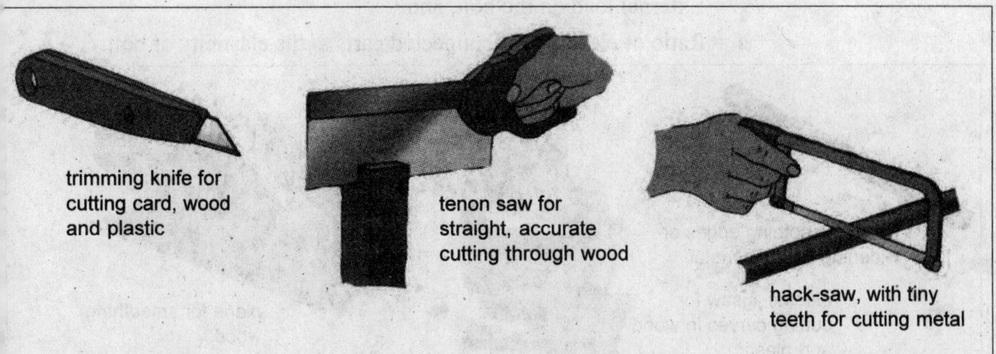

Simple machine tools.

Note : This picture is given as additional information and is not a direct example of the current chapter.

∴ Resisting load on the threaded part of the fulcrum,

$$P = \frac{\pi}{4}(d_c)^2 \sigma_t = \frac{\pi}{4}(d_c)^2\, 50 = 39.3\,(d_c)^2 \qquad ...(ii)$$

From equations (*i*) and (*ii*), we get

$$(d_c)^2 = 10\,997 / 39.3 = 280 \qquad \text{or} \qquad d_c = 16.7 \text{ mm}$$

From Table 11.1 (fine series), we find that the standard core diameter is 18.376 mm and the corresponding size of the bolt is M 20 × 1.5. **Ans.**

11.13 Stress due to Combined Forces

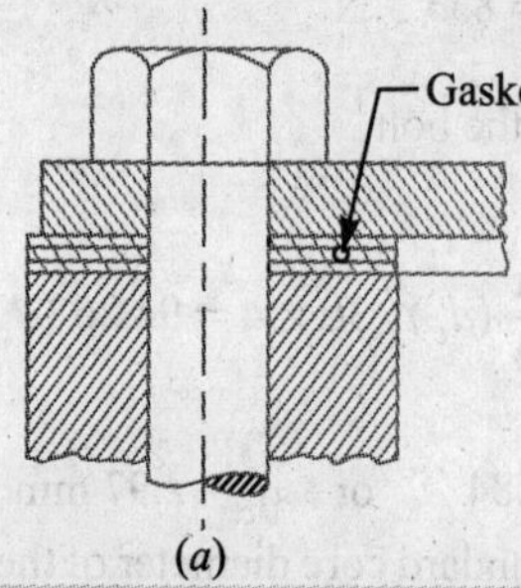

Fig. 11.23

The resultant axial load on a bolt depends upon the following factors :

1. The initial tension due to tightening of the bolt,
2. The extenal load, and
3. The relative elastic yielding (springiness) of the bolt and the connected members.

When the connected members are very yielding as compared with the bolt, which is a soft gasket, as shown in Fig. 11.23 (*a*), then the resultant load on the bolt is approximately equal to the sum of the initial tension and the external load. On the other hand, if the bolt is very yielding as compared with the connected members, as shown in Fig. 11.23 (*b*), then the resultant load will be either the initial tension or the external load, whichever is greater. The actual conditions usually lie between the two extremes. In order to determine the resultant axial load (*P*) on the bolt, the following equation may be used :

$$P = P_1 + \frac{a}{1+a} \times P_2 = P_1 + K.P_2 \qquad ...\left(\text{Substituting } \frac{a}{1+a} = K\right)$$

where P_1 = Initial tension due to tightening of the bolt,

P_2 = External load on the bolt, and

a = Ratio of elasticity of connected parts to the elasticity of bolt.

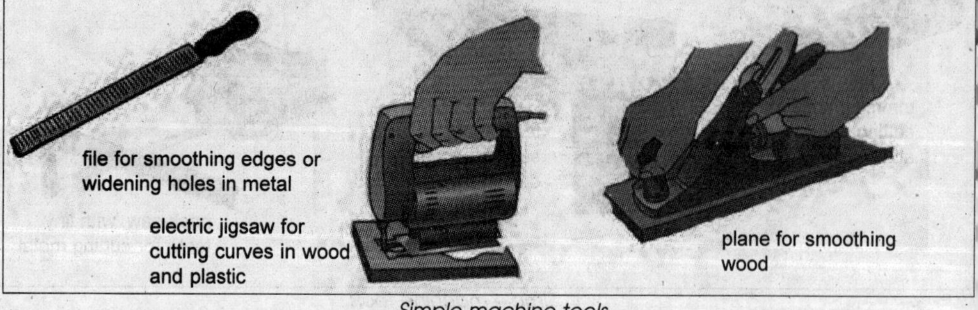

Simple machine tools.

Note : This picture is given as additional information and is not a direct example of the current chapter.

For soft gaskets and large bolts, the value of a is high and the value of $\frac{a}{1+a}$ is approximately equal to unity, so that the resultant load is equal to the sum of the initial tension and the external load.

For hard gaskets or metal to metal contact surfaces and with small bolts, the value of a is small and the resultant load is mainly due to the initial tension (or external load, in rare case it is greater than initial tension).

The value of 'a' may be estimated by the designer to obtain an approximate value for the resultant load. The values of $\frac{a}{1+a}$ (*i.e.* K) for various type of joints are shown in Table 11.2. The designer thus has control over the influence on the resultant load on a bolt by proportioning the sizes of the connected parts and bolts and by specifying initial tension in the bolt.

Table 11.2. Values of *K* for various types of joints.

Type of joint	$K = \frac{a}{1+a}$
Metal to metal joint with through bolts	0.00 to 0.10
Hard copper gasket with long through bolts	0.25 to 0.50
Soft copper gasket with long through bolts	0.50 to 0.75
Soft packing with through bolts	0.75 to 1.00
Soft packing with studs	1.00

11.14 Design of Cylinder Covers

The cylinder covers may be secured by means of bolts or studs, but studs are preferred. The possible arrangement of securing the cover with bolts and studs is shown in Fig. 11.24 (*a*) and (*b*) respectively. The bolts or studs, cylinder cover plate and cylinder flange may be designed as discussed below:

1. *Design of bolts or studs*

In order to find the size and number of bolts or studs, the following procedure may be adopted.

Let D = Diameter of the cylinder,

p = Pressure in the cylinder,

d_c = Core diameter of the bolts or studs,

n = Number of bolts or studs, and

σ_{tb} = Permissible tensile stress for the bolt or stud material.

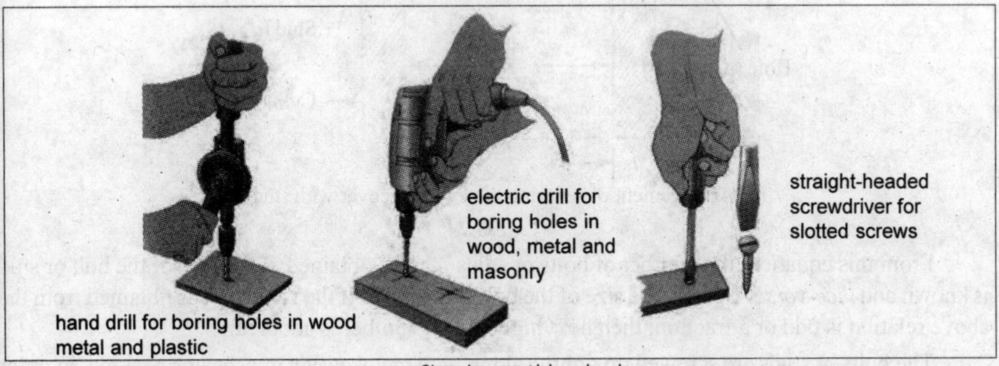

Simple machine tools.

Note : This picture is given as additional information and is not a direct example of the current chapter.

We know that upward force acting on the cylinder cover,

$$P = \frac{\pi}{4}(D^2)\,p \qquad \text{...}(i)$$

This force is resisted by *n* number of bolts or studs provided on the cover.

∴ Resisting force offered by *n* number of bolts or studs,

$$P = \frac{\pi}{4}(d_c)^2\,\sigma_{tb} \times n \qquad \text{...}(ii)$$

From equations (*i*) and (*ii*), we have

$$\frac{\pi}{4}(D^2)\,p = \frac{\pi}{4}(d_c)^2\,\sigma_{tb} \times n \qquad \text{...}(ii)$$

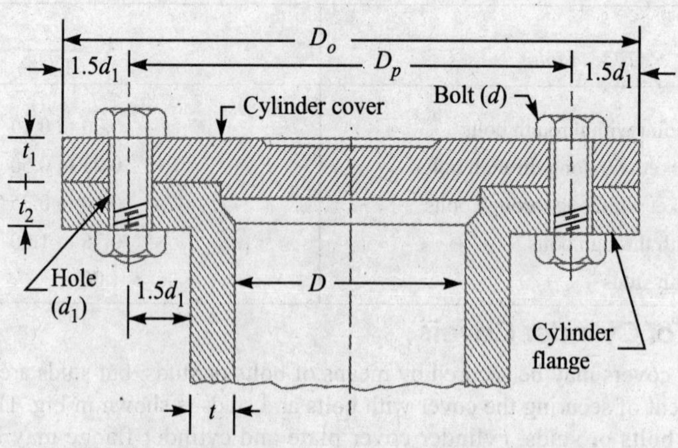

(*a*) Arrangement of securing the cylinder cover with bolts.

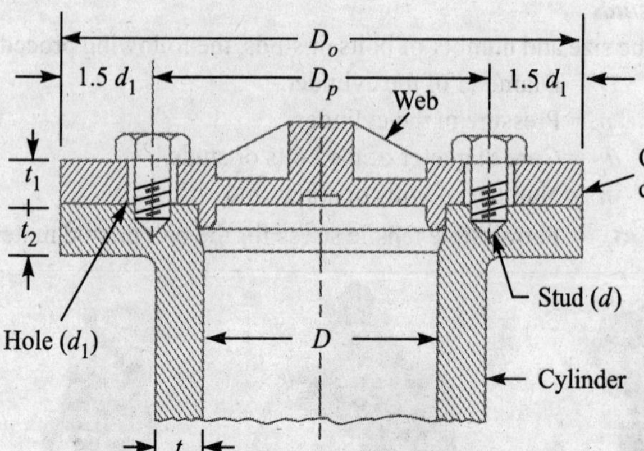

(*b*) Arrangement of securing the cylinder cover with studs.

Fig. 11.24

From this equation, the number of bolts or studs may be obtained, if the size of the bolt or stud is known and *vice-versa*. Usually the size of the bolt is assumed. If the value of *n* as obtained from the above relation is odd or a fraction, then next higher even number is adopted.

The bolts or studs are screwed up tightly, along with metal gasket or asbestos packing, in order to provide a leak proof joint. We have already discussed that due to the tightening of bolts, sufficient tensile stress is produced in the bolts or studs. This may break the bolts or studs, even before any load due to internal pressure acts upon them. Therefore a bolt or a stud less than 16 mm diameter should never be used.

The tightness of the joint also depends upon the circumferential pitch of the bolts or studs. The circumferential pitch should be between $20\sqrt{d_1}$ and $30\sqrt{d_1}$, where d_1 is the diameter of the hole in mm for bolt or stud. The pitch circle diameter (D_p) is usually taken as $D + 2t + 3d_1$ and outside diameter of the cover is kept as

$$D_o = D_p + 3d_1 = D + 2t + 6d_1$$

where t = Thickness of the cylinder wall.

2. *Design of cylinder cover plate*

The thickness of the cylinder cover plate (t_1) and the thickness of the cylinder flange (t_2) may be determined as discussed below:

Let us consider the semi-cover plate as shown in Fig. 11.25. The internal pressure in the cylinder tries to lift the cylinder cover while the bolts or studs try to retain it in its position. But the centres of pressure of these two loads do not coincide. Hence, the cover plate is subjected to bending stress. The point X is the centre of pressure for bolt load and the point Y is the centre of internal pressure.

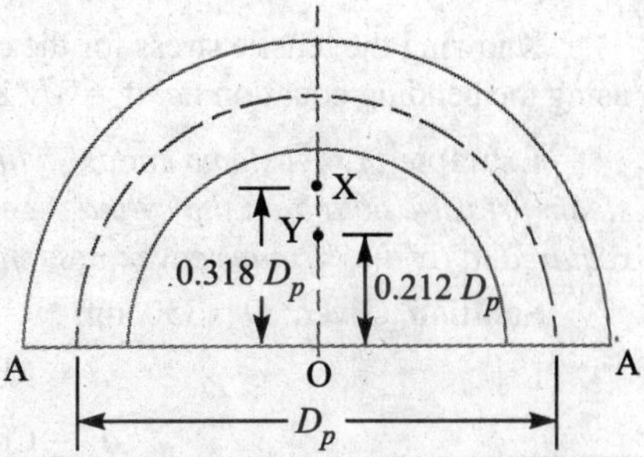

Fig. 11.25. Semi-cover plate of a cylinder.

We know that the bending moment at A-A,

$$M = \frac{\text{Total bolt load}}{2}(OX - OY) = \frac{P}{2}(0.318\ D_p - 0.212\ D_p)$$

$$= \frac{P}{2} \times 0.106\ D_p = 0.053\ P \times D_p$$

Section modulus, $$Z = \frac{1}{6} w\ (t_1)^2$$

where w = Width of plate

= Outside dia. of cover plate – 2 × dia. of bolt hole

$= D_o - 2d_1$

Knowing the tensile stress for the cover plate material, the value of t_1 may be determined by using the bending equation, *i.e.*, $\sigma_t = M / Z$.

3. *Design of cylinder flange*

The thickness of the cylinder flange (t_2) may be determined from bending consideration. A portion of the cylinder flange under the influence of one bolt is shown in Fig. 11.26.

The load in the bolt produces bending stress in the section X-X. From the geometry of the figure, we find that eccentricity of the load from section X-X is

e = Pitch circle radius – (Radius of bolt hole + Thickness of cylinder wall)

$$= \frac{D_p}{2} - \left(\frac{d_1}{2} + t\right)$$

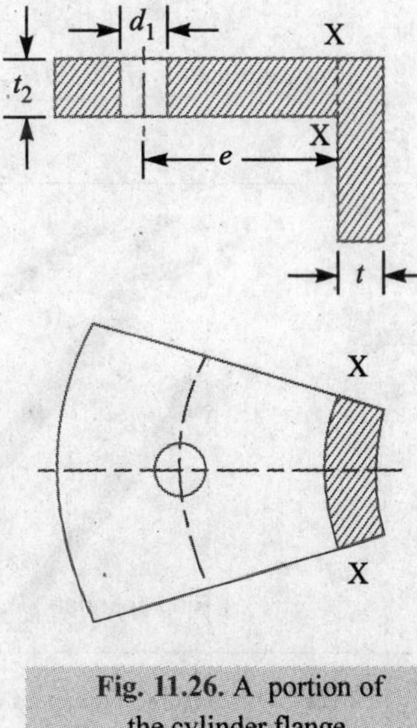

Fig. 11.26. A portion of the cylinder flange.

∴ Bending moment, M = Load on each bolt × $e = \frac{P}{n} \times e$

Radius of the section X-X,

$$R = \text{Cylinder radius} + \text{Thickness of cylinder wall} = \frac{D}{2} + t$$

Width of the section X-X,

$$w = \frac{2\pi R}{n}, \text{ where } n \text{ is the number of bolts.}$$

Section modulus,

$$Z = \frac{1}{6} w (t_2)^2$$

Knowing the tensile stress for the cylinder flange material, the value of t_2 may be obtained by using the bending equation *i.e.* $\sigma_t = M / Z$.

Example 11.6. *A steam engine cylinder has an effective diameter of 350 mm and the maximum steam pressure acting on the cylinder cover is 1.25 N/mm². Calculate the number and size of studs required to fix the cylinder cover, assuming the permissible stress in the studs as 33 MPa.*

Solution. Given: D = 350 mm ; p = 1.25 N/mm² ; σ_t = 33 MPa = 33 N/mm²

Let d = Nominal diameter of studs,

d_c = Core diameter of studs, and

n = Number of studs.

We know that the upward force acting on the cylinder cover,

$$P = \frac{\pi}{4} \times D^2 \times p = \frac{\pi}{4} (350)^2 \, 1.25 = 120\ 265 \text{ N} \qquad ...(i)$$

Assume that the studs of nominal diameter 24 mm are used. From Table 11.1 (coarse series), we find that the corresponding core diameter (d_c) of the stud is 20.32 mm.

∴ Resisting force offered by n number of studs,

$$P = \frac{\pi}{4} \times (d_c)^2 \, \sigma_t \times n = \frac{\pi}{4} (20.32)^2 \; 33 \times n = 10\ 700 \, n \text{N} \qquad ...(ii)$$

From equations (*i*) and (*ii*), we get

$$n = 120\ 265 / 10\ 700 = 11.24 \text{ say } 12 \textbf{ Ans.}$$

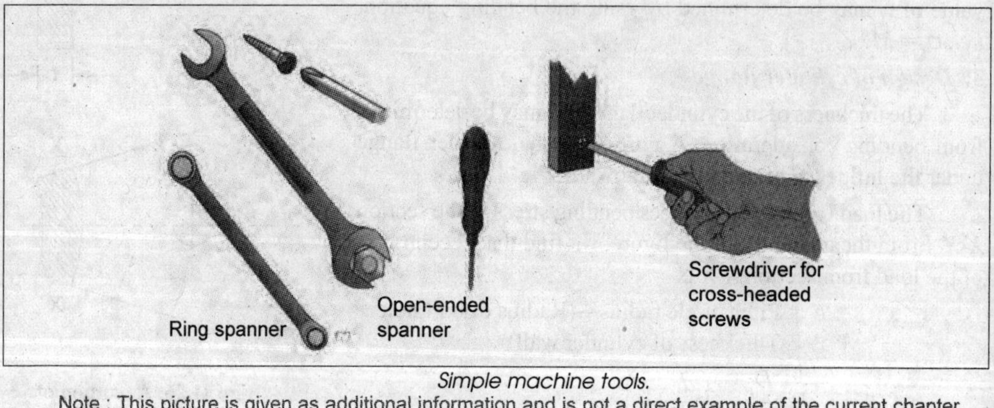

Simple machine tools.

Note : This picture is given as additional information and is not a direct example of the current chapter.

Taking the diameter of the stud hole (d_1) as 25 mm, we have pitch circle diameter of the studs,

$$D_p = D + 2t + 3d_1 = 350 + 2 \times 10 + 3 \times 25 = 445 \text{ mm}$$

...(Assuming t = 10 mm)

∴ *Circumferential pitch of the studs

$$= \frac{\pi \times D_p}{n} = \frac{\pi \times 445}{12} = 116.5 \text{ mm}$$

We know that for a leak-proof joint, the circumferential pitch of the studs should be between $20\sqrt{d_1}$ to $30\sqrt{d_1}$, where d_1 is the diameter of stud hole in mm.

∴ Minimum circumferential pitch of the studs

$$= 20\sqrt{d_1} = 20\sqrt{25} = 100 \text{ mm}$$

and maximum circumferential pitch of the studs

$$= 30\sqrt{d_1} = 30\sqrt{25} = 150 \text{ mm}$$

Since the circumferential pitch of the studs obtained above lies within 100 mm to 150 mm, therefore the size of the stud chosen is satisfactory.

∴ Size of the stud = M 24 **Ans.**

Example 11.7. *A mild steel cover plate is to be designed for an inspection hole in the shell of a pressure vessel. The hole is 120 mm in diameter and the pressure inside the vessel is 6 N/mm². Design the cover plate along with the bolts. Assume allowable tensile stress for mild steel as 60 MPa and for bolt material as 40 MPa.*

Solution. Given : D = 120 mm or r = 60 mm ; p = 6 N/mm² ; σ_t = 60 MPa = 60 N/mm² ; σ_{tb} = 40 MPa = 40 N/mm²

First for all, let us find the thickness of the pressure vessel. According to Lame's equation, thickness of the pressure vessel,

$$t = r\left[\sqrt{\frac{\sigma_t + p}{\sigma_t - p}} - 1\right] = 60\left[\sqrt{\frac{60 + 6}{60 - 6}} - 1\right] = 6 \text{ mm}$$

Let us adopt t = 10 mm

Design of bolts

Let d = Nominal diameter of the bolts,

d_c = Core diameter of the bolts, and

n = Number of bolts.

We know that the total upward force acting on the cover plate (or on the bolts),

$$P = \frac{\pi}{4}(D)^2\, p = \frac{\pi}{4}(120)^2 6 = 67\,860 \text{ N} \quad \text{...(i)}$$

Let the nominal diameter of the bolt is 24 mm. From Table 11.1 (coarse series), we find that the corresponding core diameter (d_c) of the bolt is 20.32 mm.

∴ Resisting force offered by n number of bolts,

$$P = \frac{\pi}{4}(d_c)^2\, \sigma_{tb} \times n = \frac{\pi}{4}(20.32)^2\, 40 \times n = 12\,973\, n \text{ N} \quad \text{...(ii)}$$

* The circumferential pitch of the studs can not be measured and marked on the cylinder cover. The centres of the holes are usually marked by angular distribution of the pitch circle into n number of equal parts. In the present case, the angular displacement of the stud hole centre will be 360°/12 = 30°.

From equations (*i*) and (*ii*), we get

$$n = 67\,860 / 12\,973 = 5.23 \text{ say } 6$$

Taking the diameter of the bolt hole (d_1) as 25 mm, we have pitch circle diameter of bolts,

$$D_p = D + 2t + 3d_1 = 120 + 2 \times 10 + 3 \times 25 = 215 \text{ mm}$$

∴ Circumferential pitch of the bolts

$$= \frac{\pi \times D_p}{n} = \frac{\pi \times 215}{6} = 112.6 \text{ mm}$$

We know that for a leak proof joint, the circumferential pitch of the bolts should lie between $20\sqrt{d_1}$ to $30\sqrt{d_1}$, where d_1 is the diameter of the bolt hole in mm.

∴ Minimum circumferential pitch of the bolts

$$= 20\sqrt{d_1} = 20\sqrt{25} = 100 \text{ mm}$$

and maximum circumferential pitch of the bolts

$$= 30\sqrt{d_1} = 30\sqrt{25} = 150 \text{ mm}$$

Since the circumferential pitch of the bolts obtained above is within 100 mm and 150 mm, therefore size of the bolt chosen is satisfactory.

∴ Size of the bolt = M 24 **Ans.**

Design of cover plate

Let t_1 = Thickness of the cover plate.

The semi-cover plate is shown in Fig. 11.27.

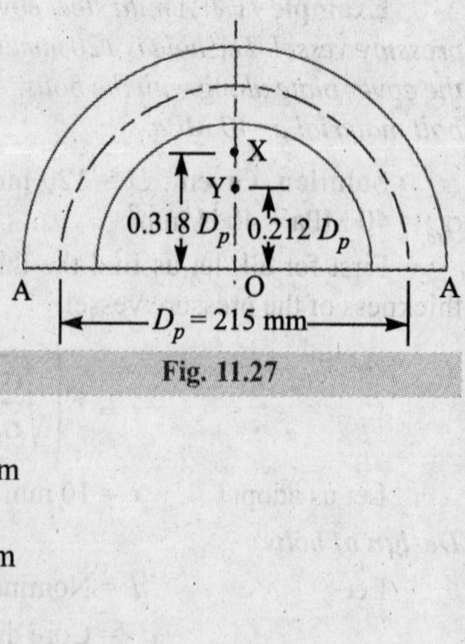

Fig. 11.27

We know that the bending moment at *A-A*,

$$M = 0.053\, P \times D_p$$
$$= 0.053 \times 67\,860 \times 215$$
$$= 773\,265 \text{ N-mm}$$

Outside diameter of the cover plate,

$$D_o = D_p + 3d_1 = 215 + 3 \times 25 = 290 \text{ mm}$$

Width of the plate,

$$w = D_o - 2d_1 = 290 - 2 \times 25 = 240 \text{ mm}$$

∴ Section modulus,

$$Z = \frac{1}{6} w (t_1)^2 = \frac{1}{6} \times 240\,(t_1)^2 = 40\,(t_1)^2 \text{ mm}^3$$

We know that bending (tensile) stress,

$$\sigma_t = M/Z \quad \text{or} \quad 60 = 773\,265 / 40\,(t_1)^2$$

∴ $(t_1)^2 = 773\,265 / 40 \times 60 = 322$ or $t_1 = 18$ mm **Ans.**

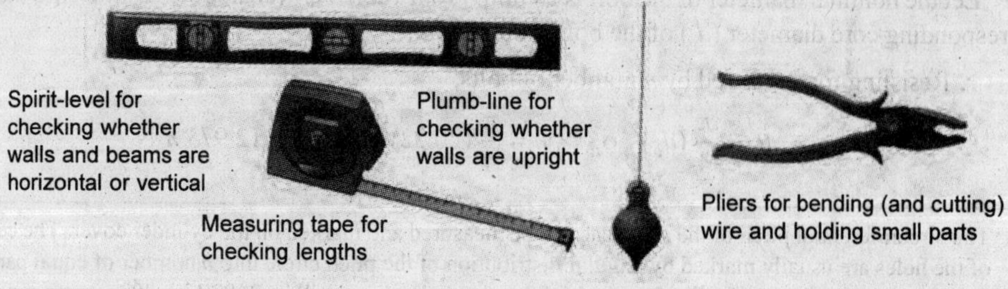

Simple machine tools.

Note : This picture is given as additional information and is not a direct example of the current chapter.

Example 11.8. *The cylinder head of a steam engine is subjected to a steam pressure of 0.7 N/mm². It is held in position by means of 12 bolts. A soft copper gasket is used to make the joint leak-proof. The effective diameter of cylinder is 300 mm. Find the size of the bolts so that the stress in the bolts is not to exceed 100 MPa.*

Solution. Given: $p = 0.7$ N/mm² ; $n = 12$; $D = 300$ mm ; $\sigma_t = 100$ MPa = 100 N/mm²

We know that the total force (or the external load) acting on the cylinder head *i.e.* on 12 bolts,

$$= \frac{\pi}{4}(D)^2 \; p = \frac{\pi}{4}(300)^2 \; 0.7 = 49\,490 \text{ N}$$

∴ External load on the cylinder head per bolt,

$$P_2 = 49\,490 / 12 = 4124 \text{ N}$$

Let d = Nominal diameter of the bolt, and

d_c = Core diameter of the bolt.

We know that initial tension due to tightening of bolt,

$$P_1 = 2840 \, d \text{ N} \qquad \text{... (where } d \text{ is in mm)}$$

From Table 11.2, we find that for soft copper gasket with long through bolts, the minimum value of $K = 0.5$.

∴ Resultant axial load on the bolt,

$$P = P_1 + K \, . \, P_2 = 2840 \, d + 0.5 \times 4124 = (2840 \, d + 2062) \text{ N}$$

We know that load on the bolt (P),

$$2840 \, d + 2062 = \frac{\pi}{4}(d_c)^2 \; \sigma_t = \frac{\pi}{4}(0.84d)^2 \; 100 = 55.4 \, d^2 \qquad \text{...(Taking } d_c = 0.84 \, d)$$

$$\therefore \quad 55.4 \, d^2 - 2840d - 2062 = 0$$

or

$$d^2 - 51.3d - 37.2 = 0$$

$$\therefore \quad d = \frac{51.3 \pm \sqrt{(51.3)^2 + 4 \times 37.2}}{2} = \frac{51.3 \pm 52.7}{2} = 52 \text{ mm}$$

..(Taking + ve sign)

Thus, we shall use a bolt of size M 52. **Ans.**

Example 11.9. *A steam engine of effective diameter 300 mm is subjected to a steam pressure of 1.5 N/mm². The cylinder head is connected by 8 bolts having yield point 330 MPa and endurance limit at 240 MPa. The bolts are tightened with an initial preload of 1.5 times the steam load. A soft copper gasket is used to make the joint leak-proof. Assuming a factor of safety 2, find the size of bolt required. The stiffness factor for copper gasket may be taken as 0.5.*

Solution. Given : $D = 300$ mm ; $p = 1.5$ N/mm² ; $n = 8$; $\sigma_y = 330$ MPa = 330 N/mm²; $\sigma_e = 240$ MPa = 240 N/mm² ; $P_1 = 1.5 \, P_2$; $F.S. = 2$; $K = 0.5$

We know that steam load acting on the cylinder head,

$$P_2 = \frac{\pi}{4}(D)^2 \; p = \frac{\pi}{4}(300)^2 \; 1.5 = 106\,040 \text{ N}$$

∴ Initial pre-load,

$$P_1 = 1.5 \, P_2 = 1.5 \times 106\,040 = 159\,060 \text{ N}$$

We know that the resultant load (or the maximum load) on the cylinder head,

$$P_{max} = P_1 + K.P_2 = 159\,060 + 0.5 \times 106\,040 = 212\,080 \text{ N}$$

This load is shared by 8 bolts, therefore maximum load on each bolt,

$$P_{max} = 212\,080 / 8 = 26\,510 \text{ N}$$

and minimum load on each bolt,

$$P_{min} = P_1 / n = 159\,060/8 = 19\,882 \text{ N}$$

We know that mean or average load on the bolt,

$$P_m = \frac{P_{max} + P_{min}}{2} = \frac{26\,510 + 19\,882}{2} = 23\,196 \text{ N}$$

and the variable load on the bolt,

$$P_v = \frac{P_{max} - P_{min}}{2} = \frac{26\,510 - 19\,882}{2} = 3314 \text{ N}$$

Let d_c = Core diameter of the bolt in mm.

∴ Stress area of the bolt,

$$A_s = \frac{\pi}{4}(d_c)^2 = 0.7854\,(d_c)^2 \text{ mm}^2$$

We know that mean or average stress on the bolt,

$$\sigma_m = \frac{P_m}{A_s} = \frac{23\,196}{0.7854\,(d_c)^2} = \frac{29\,534}{(d_c)^2} \text{ N/mm}^2$$

and variable stress on the bolt,

$$\sigma_v = \frac{P_v}{A_s} = \frac{3314}{0.7854\,(d_c)^2} = \frac{4220}{(d_c)^2} \text{ N/mm}^2$$

According to *Soderberg's formula, the variable stress,

$$\sigma_v = \sigma_e\left(\frac{1}{F.S} - \frac{\sigma_m}{\sigma_y}\right)$$

$$\frac{4220}{(d_c)^2} = 240\left(\frac{1}{2} - \frac{29\,534}{(d_c)^2\,330}\right) = 120 - \frac{21\,480}{(d_c)^2}$$

or $$\frac{4220}{(d_c)^2} + \frac{21\,480}{(d_c)^2} = 120 \quad \text{or} \quad \frac{25\,700}{(d_c)^2} = 120$$

∴ $(d_c)^2 = 25\,700 / 120 = 214$ or $d_c = 14.6$ mm

From Table 11.1 (coarse series), the standard core diameter is d_c = 14.933 mm and the corresponding size of the bolt is M18. **Ans.**

11.15 Boiler Stays

In steam boilers, flat or slightly curved plates are supported by stays. The stays are used in order to increase strength and stiffness of the plate and to reduce distortion. The principal types of stays are:

Vice for holding wood or metal being worked on

G-clamp to hold parts together for glueing

Simple machine tools.

Note : This picture is given as additional information and is not a direct example of the current chapter.

* See Chapter 6, Art. 6.20.

1. ***Direct stays.*** These stays are usually screwed round bars placed at right angles to the plates supported by them.

2. ***Diagonal and gusset stays.*** These stays are used for supporting one plate by trying it to another at right angles to it.

3. ***Girder stays.*** These stays are placed edgewise on the plate to be supported and bolted to it at intervals.

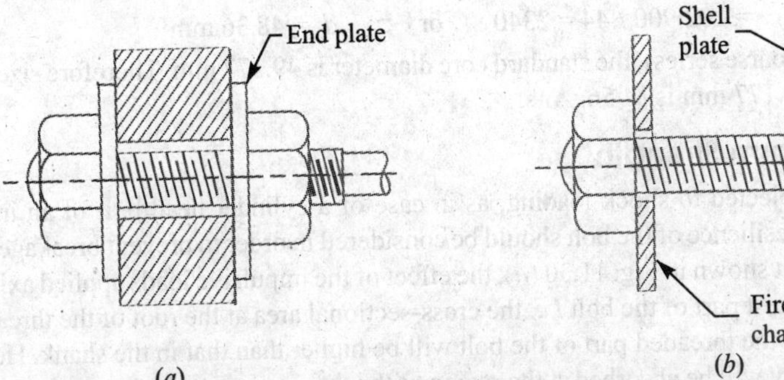

Fig. 11.28. Boiler stays.

Here we are mainly concerned with the direct stays. The direct stays may be ***bar stays*** or ***screwed stays.*** A bar stay for supporting one end plate of a boiler shell from the other end plate is shown in Fig. 11.28 (*a*). The ends of the bar are screwed to receive two nuts between which the end plate is locked. The bar stays are not screwed into the plates.

The fire boxes or combustion chambers of locomotive and marine boilers are supported by screwed stays as shown in Fig. 11.28 (*b*). These stays are called screwed stays, because they are screwed into the plates which they support. The size of the bar or screwed stays may be obtained as discussed below :

Consider a short boiler having longitudinal bar stays as shown in Fig. 11.29.

Let p = Pressure of steam in a boiler,

x = Pitch of the stays,

A = Area of the plate supported by each stay $= x \times x = x^2$

d_c = Core diameter of the stays, and

σ_t = Permissible tensile stress for the material of the stays.

We know that force acting on the stay,

$$P = \text{Pressure} \times \text{Area} = p.A = p.x^2$$

Knowing the force P, we may determine the core diameter of the stays by using the following relation,

$$P = \frac{\pi}{4}(d_c)^2\,\sigma_t$$

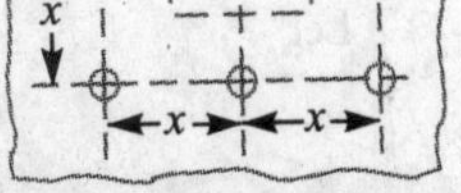

Fig. 11.29. Longitudinal bar stay.

From the core diameter, the standard size of the stay may be fixed from Table 11.1.

Example 11.10. *The longitudinal bar stays of a short boiler are pitched at 350 mm horizontally and vertically as shown in Fig. 11.29. The steam pressure is 0.84 N/mm². Find the size of mild steel bolts having tensile stress as 56 MPa.*

Solution. Given : $p = 0.84$ N/mm^2 ; $\sigma_t = 56$ MPa $= 56$ N/mm^2

Since the pitch of the stays is 350 mm, therefore area of the plate supported by each stay,

$$A = 350 \times 350 = 122\,500 \text{ mm}^2$$

We know that force acting on each stay,

$$P = A \times p = 122\,500 \times 0.84 = 102\,900 \text{ N}$$

Let d_c = Core diameter of the bolts.

We know that the resisting force on the bolts (P),

$$102\,900 = \frac{\pi}{4}(d_c)^2 \sigma_t = \frac{\pi}{4}(d_c)^2 56 = 44\,(d_c)^2$$

$$\therefore \quad (d_c)^2 = 102\,900 / 44 = 2340 \quad \text{or} \quad d_c = 48.36 \text{ mm}$$

From Table 11.1 (coarse series), the standard core diameter is 49.177 mm. Therefore size of the bolt corresponding to 49.177 mm is M 56. **Ans.**

11.16 Bolts of Uniform Strength

When a bolt is subjected to shock loading, as in case of a cylinder head bolt of an internal combustion engine, the resilience of the bolt should be considered in order to prevent breakage at the thread. In an ordinary bolt shown in Fig. 11.30 (*a*), the effect of the impulsive loads applied axially is concentrated on the weakest part of the bolt *i.e.* the cross-sectional area at the root of the threads. In other words, the stress in the threaded part of the bolt will be higher than that in the shank. Hence a great portion of the energy will be absorbed at the region of the threaded part which may fracture the threaded portion because of its small length.

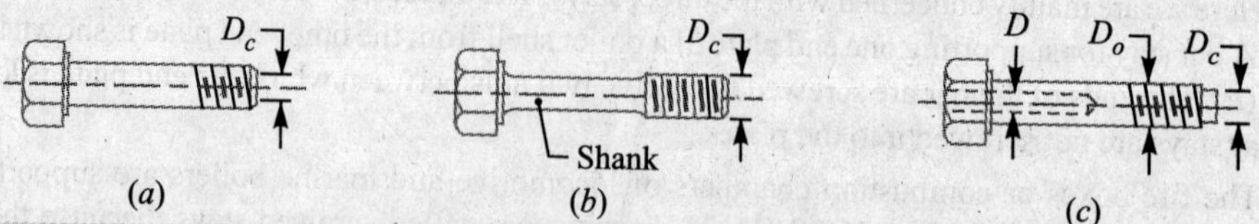

Fig. 11.30. Bolts of uniform strength.

If the shank of the bolt is turned down to a diameter equal or even slightly less than the core diameter of the thread (D_c) as shown in Fig. 11.30 (*b*), then shank of the bolt will undergo a higher stress. This means that a shank will absorb a large portion of the energy, thus relieving the material at the sections near the thread. The bolt, in this way, becomes stronger and lighter and it increases the shock absorbing capacity of the bolt because of an increased modulus of resilience. This gives us ***bolts of uniform strength***. The resilience of a bolt may also be increased by increasing its length.

A second alternative method of obtaining the bolts of uniform strength is shown in Fig. 11.30 (*c*). In this method, an axial hole is drilled through the head as far as the thread portion such that the area of the shank becomes equal to the root area of the thread.

Let D = Diameter of the hole.

D_o = Outer diameter of the thread, and

D_c = Root or core diameter of the thread.

$$\therefore \quad \frac{\pi}{4}D^2 = \frac{\pi}{4}\left[(D_o)^2 - (D_c)^2\right]$$

$$\text{or} \quad D^2 = (D_o)^2 - (D_c)^2$$

$$\therefore \quad D = \sqrt{(D_o)^2 - (D_c)^2}$$

Example 11.11. *Determine the diameter of the hole that must be drilled in a M 48 bolt such that the bolt becomes of uniform strength.*

Solution. Given : D_o = 48 mm

From Table 11.1 (coarse series), we find that the core diameter of the thread (corresponding to D_o = 48 mm) is D_c = 41.795 mm.

We know that for bolts of uniform strength, the diameter of the hole,

$$D = \sqrt{(D_o)^2 - (D_c)^2} = \sqrt{(48)^2 - (41.795)^2} = 23.64 \text{ mm } \textbf{Ans.}$$

11.17 Design of a Nut

When a bolt and nut is made of mild steel, then the effective height of nut is made equal to the nominal diameter of the bolt. If the nut is made of weaker material than the bolt, then the height of nut should be larger, such as 1.5 *d* for gun metal, 2 *d* for cast iron and 2.5 *d* for aluminium alloys (where *d* is the nominal diameter of the bolt). In case cast iron or aluminium nut is used, then *V*-threads are permissible only for permanent fastenings, because threads in these materials are damaged due to repeated screwing and unscrewing. When these materials are to be used for parts frequently removed and fastened, a screw in steel bushing for cast iron and cast-in-bronze or monel metal insert should be used for aluminium and should be drilled and tapped in place.

11.18 Bolted Joints under Eccentric Loading

There are many applications of the bolted joints which are subjected to eccentric loading such as a wall bracket, pillar crane, etc. The eccentric load may be

1. Parallel to the axis of the bolts,
2. Perpendicular to the axis of the bolts, and
3. In the plane containing the bolts.

We shall now discuss the above cases, in detail, in the following articles.

11.19 Eccentric Load Acting Parallel to the Axis of Bolts

Consider a bracket having a rectangular base bolted to a wall by means of four bolts as shown in Fig. 11.31. A little consideration will show that each bolt is subjected to a direct tensile load of

$W_{t1} = \dfrac{W}{n}$, where n is the number of bolts.

Fig. 11.31. Eccentric load acting parallel to the axis of bolts.

Further the load *W* tends to rotate the bracket about the edge *A-A*. Due to this, each bolt is stretched by an amount that depends upon its distance from the tilting edge. Since the stress is a function of *elongation, therefore each bolt will experience a different load which also depends upon the distance from the tilting edge. For convenience, all the bolts are made of same size. In case the flange is heavy, it may be considered as a rigid body.

Let *w* be the load in a bolt per unit distance due to the turning effect of the bracket and let W_1 and W_2 be the loads on each of the bolts at distances L_1 and L_2 from the tilting edge.

* We know that elongation is proportional to strain which in turn is proportional to stress within elastic limits.

∴ Load on each bolt at distance L_1,

$$W_1 = w.L_1$$

and moment of this load about the tilting edge

$$= w_1.L_1 \times L_1 = w\,(L_1)^2$$

Similarly, load on each bolt at distance L_2,

$$W_2 = w.L_2$$

and moment of this load about the tilting edge

$$= w.L_2 \times L_2 = w\,(L_2)^2$$

∴ Total moment of the load on the bolts about the tilting edge

$$= 2w\,(L_1)^2 + 2w\,(L_2)^2 \quad ...(i)$$

... ($\because$ There are two bolts each at distance of L_1 and L_2)

Also the moment due to load W about the tilting edge

$$= W.L \quad ...(ii)$$

From equations (*i*) and (*ii*), we have

$$W.L = 2w\,(L_1)^2 + 2w(L_2)^2 \quad \text{or} \quad w = \frac{W.L}{2\,[(L_1)^2 + (L_2)^2]} \quad ...(iii)$$

It may be noted that the most heavily loaded bolts are those which are situated at the greatest distance from the tilting edge. In the case discussed above, the bolts at distance L_2 are heavily loaded.

∴ Tensile load on each bolt at distance L_2,

$$W_{t2} = W_2 = w.L_2 = \frac{W.L.L_2}{2[(L_1)^2 + (L_2)^2]} \quad \text{... [From equation (iii)]}$$

and the total tensile load on the most heavily loaded bolt,

$$W_t = W_{t1} + W_{t2} \quad ...(iv)$$

If d_c is the core diameter of the bolt and σ_t is the tensile stress for the bolt material, then total tensile load,

$$W_t = \frac{\pi}{4}\,(d_c)^2\,\sigma_t \quad ...(v)$$

From equations (*iv*) and (*v*), the value of d_c may be obtained.

Example 11.12. *A bracket, as shown in Fig. 11.31, supports a load of 30 kN. Determine the size of bolts, if the maximum allowable tensile stress in the bolt material is 60 MPa. The distances are :*

$L_1 = 80$ mm, $L_2 = 250$ mm, and $L = 500$ mm.

Solution. Given : $W = 30$ kN ; $\sigma_t = 60$ MPa $= 60$ N/mm^2 ; $L_1 = 80$ mm ; $L_2 = 250$ mm ; $L = 500$ mm

We know that the direct tensile load carried by each bolt,

$$W_{t1} = \frac{W}{n} = \frac{30}{4} = 7.5 \text{ kN}$$

and load in a bolt per unit distance,

$$w = \frac{W.L}{2\,[(L_1)^2 + (L_2)^2]} = \frac{30 \times 500}{2\,[(80)^2 + (250)^2]} = 0.109 \text{ kN/mm}$$

Since the heavily loaded bolt is at a distance of L_2 mm from the tilting edge, therefore load on the heavily loaded bolt,

$$W_{t2} = w.L_2 = 0.109 \times 250 = 27.25 \text{ kN}$$

∴ Maximum tensile load on the heavily loaded bolt,

$$W_t = W_{t1} + W_{t2} = 7.5 + 27.25 = 34.75 \text{ kN} = 34\,750 \text{ N}$$

Let $\quad d_c$ = Core diameter of the bolts.

We know that the maximum tensile load on the bolt (W_t),

$$34\,750 = \frac{\pi}{4}(d_c)^2\,\sigma_t = \frac{\pi}{4}(d_c)^2\,60 = 47\,(d_c)^2$$

$$\therefore \quad (d_c)^2 = 34\,750 / 47 = 740$$

or $\quad d_c = 27.2$ mm

From Table 11.1 (coarse series), we find that the standard core diameter of the bolt is 28.706 mm and the corresponding size of the bolt is M 33. **Ans.**

Example 11.13. *A crane runway bracket is shown in Fig. 11.32. Determine the tensile and compressive stresses produced in the section X-X when the magnitude of the wheel load is 15 kN.*

Also find the maximum stress produced in the bolts used for fastening the bracket to the roof truss.

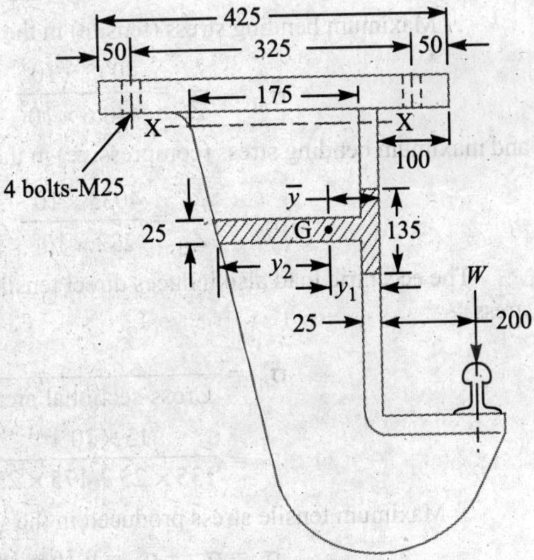

Fig. 11.32

Solution. Given : $W = 15$ kN $= 15 \times 10^3$ N

First of all, let us find the distance of centre of gravity of the section at *X–X*.

Let $\quad \bar{y}$ = Distance of centre of gravity (*G*) from the top of the flange.

$$\therefore \quad \bar{y} = \frac{135 \times 25 \times \frac{25}{2} + 175 \times 25\left(25 + \frac{175}{2}\right)}{135 \times 25 + 175 \times 25} = 69 \text{ mm}$$

Moment of inertia about an axis passing through the centre of gravity of the section,

$$I_{GG} = \left[\frac{135(25)^3}{12} + 135 \times 25\left(69 - \frac{25}{2}\right)^2\right] + \left[\frac{25(175)^3}{12} + 175 \times 25\left(200 - 69 - \frac{175}{2}\right)^2\right]$$

$$= 30.4 \times 10^6 \text{ mm}^4$$

Distance of C.G. from the top of the flange,

$$y_1 = \bar{y} = 69 \text{ mm}$$

and distance of C.G. from the bottom of the web,

$$y_2 = 175 + 25 - 69 = 131 \text{ mm}$$

Due to the tilting action of the load *W*, the cross-section of the bracket *X-X* will be under bending stress. The upper fibres of the top flange will be under maximum tension and the lower fibres of the web will be under maximum compression.

∴ Section modulus for the maximum tensile stress,

$$Z_1 = \frac{I_{GG}}{y_1} = \frac{30.4 \times 10^6}{69} = 440.6 \times 10^3 \text{ mm}^3$$

and section modulus for the maximum compressive stress,

$$Z_2 = \frac{I_{GG}}{y_2} = \frac{30.4 \times 10^6}{131} = 232 \times 10^3 \text{ mm}^3$$

We know that bending moment exerted on the section,

$$M = 15 \times 10^3 (200 + 69) = 4035 \times 10^3 \text{ N-mm}$$

∴ Maximum bending stress (tensile) in the flange,

$$\sigma_{b1} = \frac{M}{Z_1} = \frac{4035 \times 10^3}{440.6 \times 10^3} = 9.16 \text{ N/mm}^2$$

and maximum bending stress (compressive) in the web,

$$\sigma_{b2} = \frac{M}{Z_2} = \frac{4035 \times 10^3}{232 \times 10^3} = 17.4 \text{ N/mm}^2$$

The eccentric load also induces direct tensile stress in the bracket. We know that direct tensile stress,

$$\sigma_{t1} = \frac{\text{Load}}{\text{Cross-sectional area of the bracket at } X - X}$$

$$= \frac{15 \times 10^3}{135 \times 25 + 175 \times 25} = 1.94 \text{ N/mm}^2$$

∴ Maximum tensile stress produced in the section at X–X (*i.e.* in the flange),

$$\sigma_t = \sigma_{b1} + \sigma_{t1} = 9.16 + 1.94 = 11.1 \text{ N/mm}^2 = 11.1 \text{ MPa } \textbf{Ans.}$$

and maximum compressive stress produced in the section at X–X (*i.e.* in the web),

$$\sigma_c = \sigma_{b2} - \sigma_{t1} = 17.4 - 1.94 = 15.46 \text{ N/mm}^2 = 15.46 \text{ MPa } \textbf{Ans.}$$

Let σ_{tb} = Maximum stress produced in bolts,

n = Number of bolts = 4, and ...(Given)

d = Major diameter of the bolts = 25 mm ...(Given)

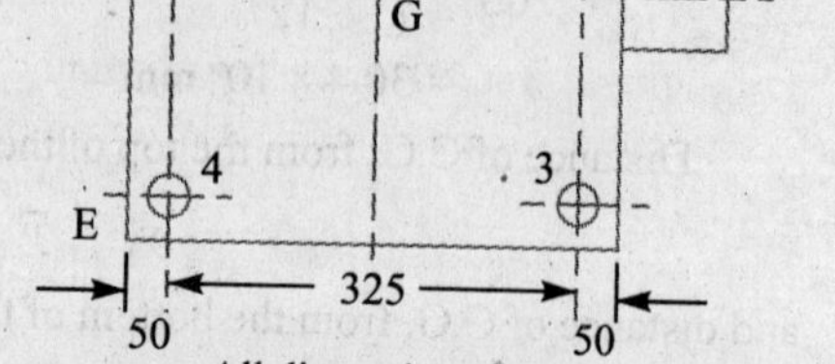

Fig. 11.33

The plan of the bracket is shown in Fig. 11.33. Due to the eccentric load W, the bracket has a tendency to tilt about the edge EE. Since the load is acting parallel to the axis of bolts, therefore direct tensile load on each bolt,

$$W_{t1} = \frac{W}{n} = \frac{15 \times 10^3}{4} = 3750 \text{ N}$$

Let w = Load in each bolt per mm distance from the edge EE due to the turning effect of the bracket,

L_1 = Distance of bolts 1 and 4 from the tilting edge EE = 50 mm, and

L_2 = Distance of bolts 2 and 3 from the tilting edge EE = 50 + 325 = 375 mm

We know that

$$w = \frac{W.L}{2\,[(L_1)^2 + (L_2)^2]} = \frac{15 \times 10^3\ (100 + 50 + 325 + 50)}{2\,[(50)^2 + (375)^2]} = 27.5 \text{ N/mm}$$

Since the heavily loaded bolts are those which lie at greater distance from the tilting edge, therefore the bolts 2 and 3 will be heavily loaded.

∴ Maximum tensile load on each of bolts 2 and 3,

$$W_{t2} = w \times L_2 = 27.5 \times 375 = 10\,312 \text{ N}$$

and the total tensile load on each of the bolts 2 and 3,

$$W_t = W_{t1} + W_{t2} = 3750 + 10\,312 = 14\,062 \text{ N}$$

We know that tensile load on the bolt (W_t),

$$14\,062 = \frac{\pi}{4}(d_c)^2 \sigma_{tb} = \frac{\pi}{4}(0.84 \times 25)^2 \sigma_{tb} = 346.4\,\sigma_{tb}$$

... (Taking, $d_c = 0.84\,d$)

$$\therefore \quad \sigma_{tb} = 14\,062/346.4 = 40.6 \text{ N/mm}^2 = 40.6 \text{ MPa } \textbf{Ans.}$$

11.20 Eccentric Load Acting Perpendicular to the Axis of Bolts

A wall bracket carrying an eccentric load perpendicular to the axis of the bolts is shown in Fig. 11.34.

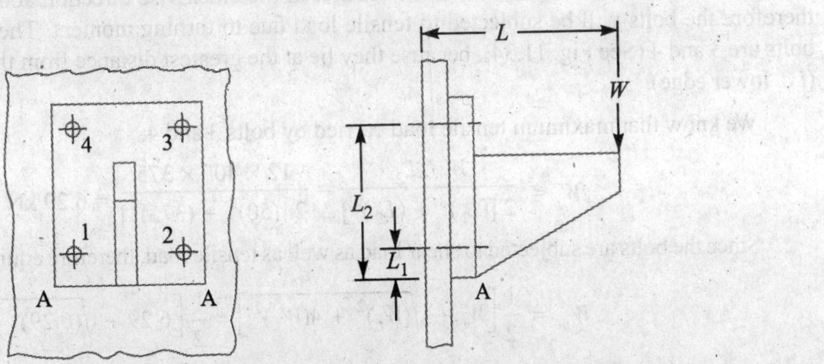

Fig. 11.34. Eccentric load perpendicular to the axis of bolts.

In this case, the bolts are subjected to direct shearing load which is equally shared by all the bolts. Therefore direct shear load on each bolts,

$$W_s = W/n, \text{ where } n \text{ is number of bolts.}$$

A little consideration will show that the eccentric load W will try to tilt the bracket in the clockwise direction about the edge A-A. As discussed earlier, the bolts will be subjected to tensile stress due to the turning moment. The maximum tensile load on a heavily loaded bolt (W_t) may be obtained in the similar manner as discussed in the previous article. In this case, bolts 3 and 4 are heavily loaded.

∴ Maximum tensile load on bolt 3 or 4,

$$W_{t2} = W_t = \frac{W.L.L_2}{2\,[(L_1)^2 + (L_2)^2]}$$

When the bolts are subjected to shear as well as tensile loads, then the equivalent loads may be determined by the following relations :

Equivalent tensile load,

$$W_{te} = \frac{1}{2}\left[W_t + \sqrt{(W_t)^2 + 4(W_s)^2}\right]$$

and equivalent shear load,

$$W_{se} = \frac{1}{2}\left[\sqrt{(W_t)^2 + 4(W_s)^2}\right]$$

Knowing the value of equivalent loads, the size of the bolt may be determined for the given allowable stresses.

Example 11.14. *For supporting the travelling crane in a workshop, the brackets are fixed on steel columns as shown in Fig. 11.35. The maximum load that comes on the bracket is 12 kN acting vertically at a distance of 400 mm from the face of the column. The vertical face of the bracket is secured to a column by four bolts, in two rows (two in each row) at a distance of 50 mm from the lower edge of the bracket. Determine the size of the bolts if the permissible value of the tensile stress for the bolt material is 84 MPa. Also find the cross-section of the arm of the bracket which is rectangular.*

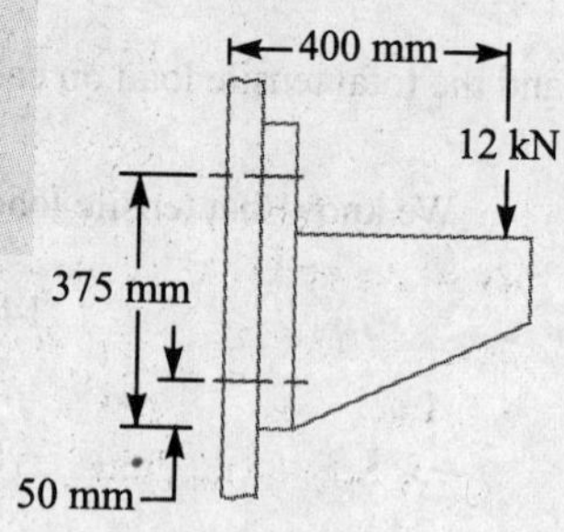

Fig. 11.35

Solution. Given : $W = 12$ kN $= 12 \times 10^3$ N ; $L = 400$ mm ; $L_1 = 50$ mm ; $L_2 = 375$ mm ; $\sigma_t = 84$ MPa $= 84$ N/mm² ; $n = 4$

We know that direct shear load on each bolt,

$$W_s = \frac{W}{n} = \frac{12}{4} = 3 \text{ kN}$$

Since the load W will try to tilt the bracket in the clockwise direction about the lower edge, therefore the bolts will be subjected to tensile load due to turning moment. The maximum loaded bolts are 3 and 4 (See Fig. 11.34), because they lie at the greatest distance from the tilting edge A–A (*i.e.* lower edge).

We know that maximum tensile load carried by bolts 3 and 4,

$$W_t = \frac{W.L.L_2}{2\,[(L_1)^2 + (L_2)^2]} = \frac{12 \times 400 \times 375}{2\,[(50)^2 + (375)^2]} = 6.29 \text{ kN}$$

Since the bolts are subjected to shear load as well as tensile load, therefore equivalent tensile load,

$$W_{te} = \frac{1}{2}\left[W_t + \sqrt{(W_t)^2 + 4(W_s)^2}\right] = \frac{1}{2}\left[6.29 + \sqrt{(6.29)^2 + 4 \times 3^2}\right] \text{ kN}$$

$$= \frac{1}{2}(6.29 + 8.69) = 7.49 \text{ kN} = 7490 \text{ N}$$

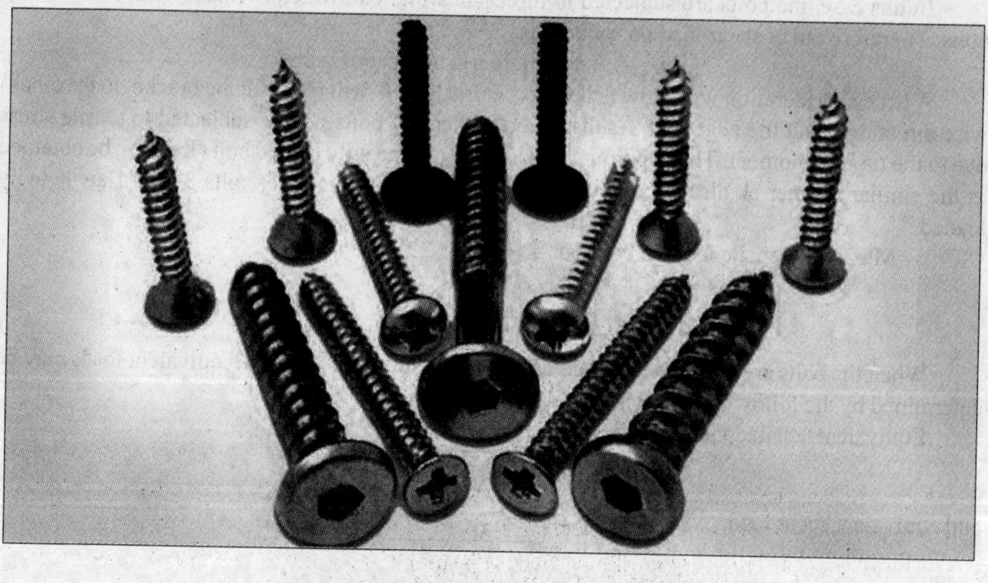

Size of the bolt

Let d_c = Core diameter of the bolt.

We know that the equivalent tensile load (W_{te}),

$$7490 = \frac{\pi}{4}(d_c)^2\,\sigma_t = \frac{\pi}{4}(d_c)^2\,84 = 66\,(d_c)^2$$

$\therefore$ $(d_c)^2 = 7490 / 66 = 113.5$ or $d_c = 10.65$ mm

From Table 11.1 (coarse series), the standard core diameter is 11.546 mm and the corresponding size of the bolt is M 14. **Ans.**

Cross-section of the arm of the bracket

Let t and b = Thickness and depth of arm of the bracket respectively.

$\therefore$ Section modulus,

$$Z = \frac{1}{6}\,t.b^2$$

Assume that the arm of the bracket extends upto the face of the steel column. This assumption gives stronger section for the arm of the bracket.

$\therefore$ Maximum bending moment on the bracket,

$$M = 12 \times 10^3 \times 400 = 4.8 \times 10^6 \text{ N-mm}$$

We know that the bending (tensile) stress (σ_t),

$$84 = \frac{M}{Z} = \frac{4.8 \times 10^6 \times 6}{t.b^2} = \frac{28.8 \times 10^6}{t.b^2}$$

$\therefore$ $t.b^2 = 28.8 \times 10^6 / 84 = 343 \times 10^3$ or $t = 343 \times 10^3 / b^2$

Assuming depth of arm of the bracket, b = 250 mm, we have

$$t = 343 \times 10^3 / (250)^2 = 5.5 \text{ mm } \textbf{Ans.}$$

Example 11.15. *Determine the size of the bolts and the thickness of the arm for the bracket as shown in Fig. 11.36, if it carries a load of 40 kN at an angle of 60° to the vertical.*

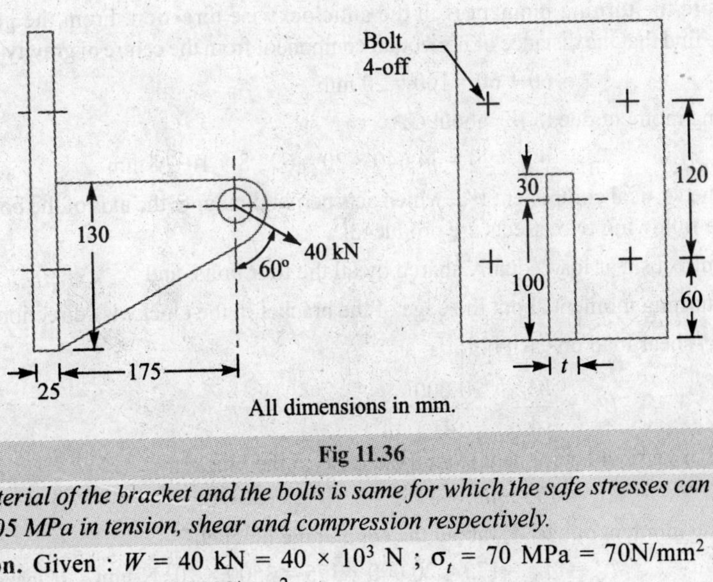

All dimensions in mm.

Fig 11.36

The material of the bracket and the bolts is same for which the safe stresses can be assumed as 70, 50 and 105 MPa in tension, shear and compression respectively.

Solution. Given : $W = 40$ kN $= 40 \times 10^3$ N ; $\sigma_t = 70$ MPa $= 70$N/mm^2 ; $\tau = 50$ MPa $= 50$ N/mm^2; $\sigma_c = 105$ MPa $= 105$ N/mm^2

Since the load $W = 40$ kN is inclined at an angle of 60° to the vertical, therefore resolving it into horizontal and vertical components. We know that horizontal component of 40 kN,

$$W_H = 40 \times \sin 60° = 40 \times 0.866 = 34.64 \text{ kN} = 34\ 640 \text{ N}$$

and vertical component of 40 kN,

$$W_V = 40 \times \cos 60° = 40 \times 0.5 = 20 \text{ kN} = 20\ 000 \text{ N}$$

Due to the horizontal component (W_H), which acts parallel to the axis of the bolts as shown in Fig. 11.37, the following two effects are produced :

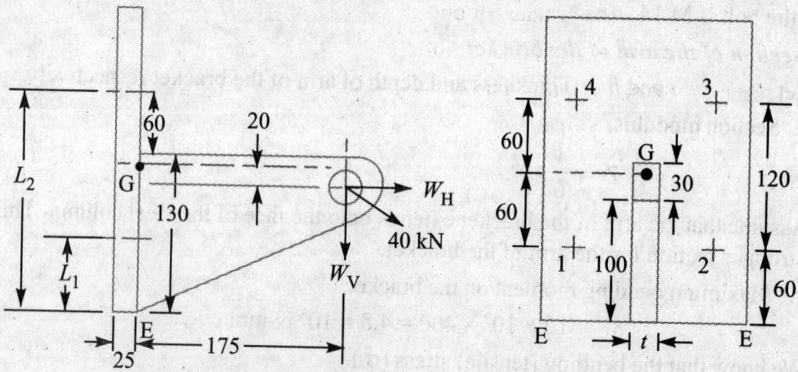

All dimensions in mm.

Fig. 11.37

1. A direct tensile load equally shared by all the four bolts, and
2. A turning moment about the centre of gravity of the bolts, in the anticlockwise direction.

∴ Direct tensile load on each bolt,

$$W_{t1} = \frac{W_H}{4} = \frac{34\ 640}{4} = 8660 \text{ N}$$

Since the centre of gravity of all the four bolts lies in the centre at G (because of symmetrical bolts), therefore the turning moment is in the anticlockwise direction. From the geometry of the Fig. 11.37, we find that the distance of horizontal component from the centre of gravity (G) of the bolts

$$= 60 + 60 - 100 = 20 \text{ mm}$$

∴ Turning moment due to W_H about G,

$$T_H = W_H \times 20 = 34\ 640 \times 20 = 692.8 \times 10^3 \text{ N-mm} \quad \text{...(Anticlockwise)}$$

Due to the vertical component W_V, which acts perpendicular to the axis of the bolts as shown in Fig. 11.37, the following two effects are produced:

1. A direct shear load equally shared by all the four bolts, and
2. A turning moment about the edge of the bracket in the clockwise direction.

∴ Direct shear load on each bolt,

$$W_s = \frac{W_V}{4} = \frac{20\ 000}{4} = 5000 \text{ N}$$

Distance of vertical component from the edge E of the bracket,

$$= 175 \text{ mm}$$

∴ Turning moment due to W_V about the edge of the bracket,

$$T_V = W_V \times 175 = 20\ 000 \times 175 = 3500 \times 10^3 \text{ N-mm} \quad \text{(Clockwise)}$$

From above, we see that the clockwise moment is greater than the anticlockwise moment, therefore,

Net turning moment $= 3500 \times 10^3 - 692.8 \times 10^3 = 2807.2 \times 10^3$ N-mm (Clockwise) ...(*i*)

Due to this clockwise moment, the bracket tends to tilt about the lower edge E.

Let w = Load on each bolt per mm distance from the edge E due to the turning effect of the bracket,

L_1 = Distance of bolts 1 and 2 from the tilting edge E = 60 mm, and

L_2 = Distance of bolts 3 and 4 from the tilting edge E

$= 60 + 120 = 180$ mm

$\therefore$ Total moment of the load on the bolts about the tilting edge E

$$= 2\,(w.L_1)\,L_1 + 2\,(w.L_2)\,L_2$$

...($\because$ There are two bolts each at distance L_1 and L_2.)

$$= 2w\,(L_1)^2 + 2w(L_2)^2 = 2w\,(60)^2 + 2w(180)^2$$

$$= 72\,000\;w \text{ N-mm} \qquad ...(ii)$$

From equations (*i*) and (*ii*),

$$w = 2807.2 \times 10^3 / 72\,000 = 39 \text{ N/mm}$$

Since the heavily loaded bolts are those which lie at a greater distance from the tilting edge, therefore the upper bolts 3 and 4 will be heavily loaded. Thus the diameter of the bolt should be based on the load on the upper bolts. We know that the maximum tensile load on each upper bolt,

$$W_{t2} = w.L_2 = 39 \times 180 = 7020 \text{ N}$$

$\therefore$ Total tensile load on each of the upper bolt,

$$W_t = W_{t1} + W_{t2} = 8660 + 7020 = 15\,680 \text{ N}$$

Since each upper bolt is subjected to a tensile load (W_t = 15 680 N) and a shear load (W_s = 5000 N), therefore equivalent tensile load,

$$W_{te} = \frac{1}{2}\left[W_t + \sqrt{(W_t)^2 + 4(W_s)^2}\right]$$

$$= \frac{1}{2}\left[15\,680 + \sqrt{(15\,680)^2 + 4(5000)^2}\right] \text{ N}$$

$$= \frac{1}{2}[15\,680 + 18\,600] = 17\,140 \text{ N} \qquad ...(iii)$$

Size of the bolts

Let d_c = Core diameter of the bolts.

We know that tensile load on each bolt

$$= \frac{\pi}{2}(d_c)^2\,\sigma_t = \frac{\pi}{4}(d_c)^2\,70 = 55\,(d_c)^2 \text{ N} \qquad ...(iv)$$

From equations (*iii*) and (*iv*), we get

$$(d_c)^2 = 17\,140 / 55 = 311.64 \quad \text{or} \quad d_c = 17.65 \text{ mm}$$

From Table 11.1 (coarse series), we find that the standard core diameter is 18.933 mm and corresponding size of the bolt is M 22. **Ans.**

Thickness of the arm of the bracket

Let t = Thickness of the arm of the bracket in mm, and

b = Depth of the arm of the bracket = 130 mm ...(Given)

We know that cross-sectional area of the arm,

$$A = b \times t = 130\,t \text{ mm}^2$$

and section modulus of the arm,

$$Z = \frac{1}{6}\,t\,(b)^2 = \frac{1}{6} \times t\,(130)^2 = 2817\,t \text{ mm}^3$$

Due to the horizontal component W_H, the following two stresses are induced in the arm :

1. Direct tensile stress,

$$\sigma_{t1} = \frac{W_H}{A} = \frac{34\,640}{130\,t} = \frac{266.5}{t} \text{ N/mm}^2$$

2. Bending stress causing tensile in the upper most fibres of the arm and compressive in the lower most fibres of the arm. We know that the bending moment of W_H about the centre of gravity of the arm,

$$M_H = W_H\left(100 - \frac{130}{2}\right) = 34\,640 \times 35 = 1212.4 \times 10^3 \text{ N-mm}$$

∴ Bending stress, $\sigma_{t2} = \frac{M_H}{Z} = \frac{1212.4 \times 10^3}{2817\,t} = \frac{430{\cdot}4}{t} \text{ N/mm}^2$

Due to the vertical component W_V, the following two stresses are induced in the arm :

1. Direct shear stress,

$$\tau = \frac{W_V}{A} = \frac{20\,000}{130\,t} = \frac{154}{t} \text{ N/mm}^2$$

2. Bending stress causing tensile stress in the upper most fibres of the arm and compressive in the lower most fibres of the arm.

Assuming that the arm extends upto the plate used for fixing the bracket to the structure. This assumption gives stronger section for the arm of the bracket.

∴ Bending moment due to W_V,

$$M_V = W_V\,(175 + 25) = 20\,000 \times 200 = 4 \times 10^6 \text{ N-mm}$$

and bending stress, $\sigma_{t3} = \frac{M_V}{Z} = \frac{4 \times 10^6}{2817\,t} = \frac{1420}{t} \text{ N/mm}^2$

Net tensile stress induced in the upper most fibres of the arm of the bracket,

$$\sigma_t = \sigma_{t1} + \sigma_{t2} + \sigma_{t3} = \frac{266.5}{t} + \frac{430.4}{t} + \frac{1420}{t} = \frac{2116.9}{t} \text{ N/mm}^2 \quad \text{...}(v)$$

We know that maximum tensile stress $[\sigma_{t(max)}]$,

$$70 = \frac{1}{2}\,\sigma_t + \frac{1}{2}\sqrt{(\sigma_t)^2 + 4\tau^2}$$

$$= \frac{1}{2} \times \frac{2116.9}{t} + \frac{1}{2}\sqrt{\left(\frac{2116.9}{t}\right)^2 + 4\left(\frac{154}{t}\right)^2}$$

$$= \frac{1058.45}{t} + \frac{1069.6}{t} = \frac{2128.05}{t}$$

∴ $t = 2128.05 / 70 = 30.4$ say 31 mm **Ans.**

Let us now check the shear stress induced in the arm. We know that maximum shear stress,

$$\tau_{max} = \frac{1}{2}\sqrt{(\sigma_t)^2 + 4\tau^2} = \frac{1}{2}\sqrt{\left(\frac{2116.9}{t}\right)^2 + 4\left(\frac{154}{t}\right)^2}$$

$$= \frac{1069.6}{t} = \frac{1069.6}{31} = 34.5 \text{ N/mm}^2 = 34.5 \text{ MPa}$$

Since the induced shear stress is less than the permissible stress (50 MPa), therefore the design is safe.

Notes : **1.** The value of '*t*' may be obtained as discussed below :

Since the shear stress at the upper most fibres of the arm of the bracket is zero, therefore equating equation (*v*) to the given safe tensile stress (*i.e.* 70 MPa), we have

$$\frac{2116.9}{t} = 70 \quad \text{or} \quad t = 2116.9 / 70 = 30.2 \text{ say } 31 \text{ mm} \quad \textbf{Ans.}$$

2. If the compressive stress in the lower most fibres of the arm is taken into consideration, then the net compressive stress induced in the lower most fibres of the arm,

$$\sigma_c = \sigma_{c1} + \sigma_{c2} + \sigma_{c3}$$

$$= -\sigma_{t1} + \sigma_{t2} + \sigma_{t3}$$

... (∵ The magnitude of tensile and compressive stresses is same.)

$$= -\frac{266.5}{t} + \frac{430.4}{t} + \frac{1420}{t} = \frac{1583.9}{t} \text{ N/mm}^2$$

Since the safe compressive stress is 105 N/mm^2, therefore

$$105 = \frac{1583.9}{t} \quad \text{or} \quad t = 1583.9 / 105 = 15.1 \text{ mm}$$

This value of thickness is low as compared to 31 mm as calculated above. Since the higher value is taken, therefore

$$t = 31 \text{ mm} \quad \textbf{Ans.}$$

Example 11.16. *An offset bracket, having arm of I-cross-section is fixed to a vertical steel column by means of four standard bolts as shown in Fig. 11.38. An inclined pull of 10 kN is acting on the bracket at an angle of 60° to the vertical.*

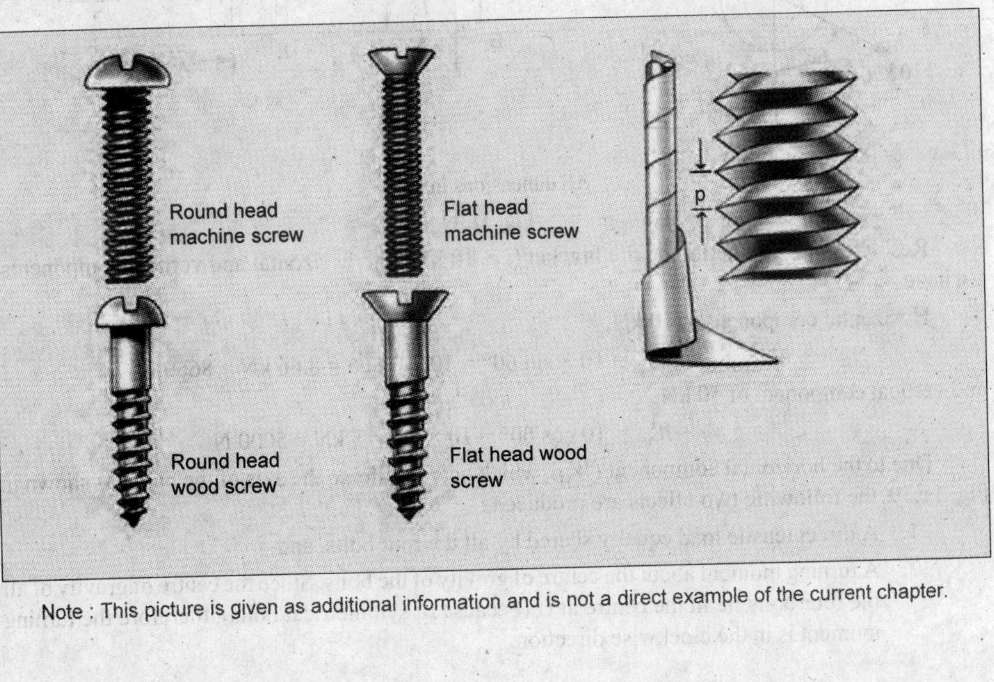

Note : This picture is given as additional information and is not a direct example of the current chapter.

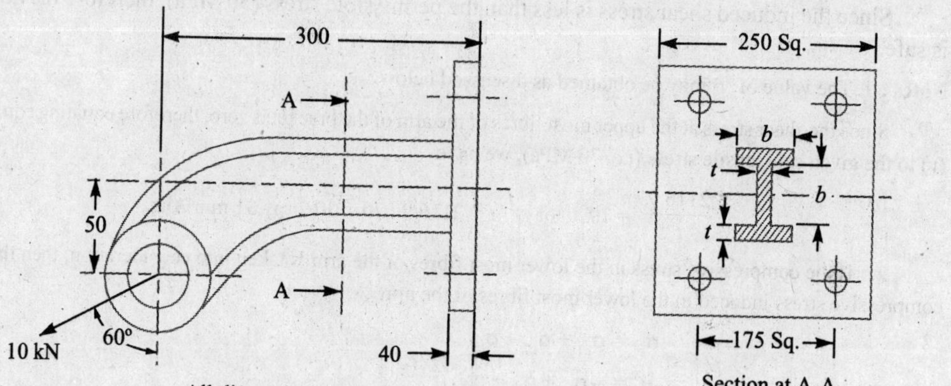

Fig. 11.38

Determine : (a) the diameter of the fixing bolts, and (b) the dimensions of the arm of the bracket if the ratio between b and t is 3 : 1.

For all parts, assume safe working stresses of 100 MPa in tension and 60 MPa in shear.

Solution. Given : $W = 10$ kN ; $\theta = 60°$; $\sigma_1 = 100$ MPa $= 100$ N/mm^2 ; $\tau = 60$ MPa $= 60$ N/mm^2

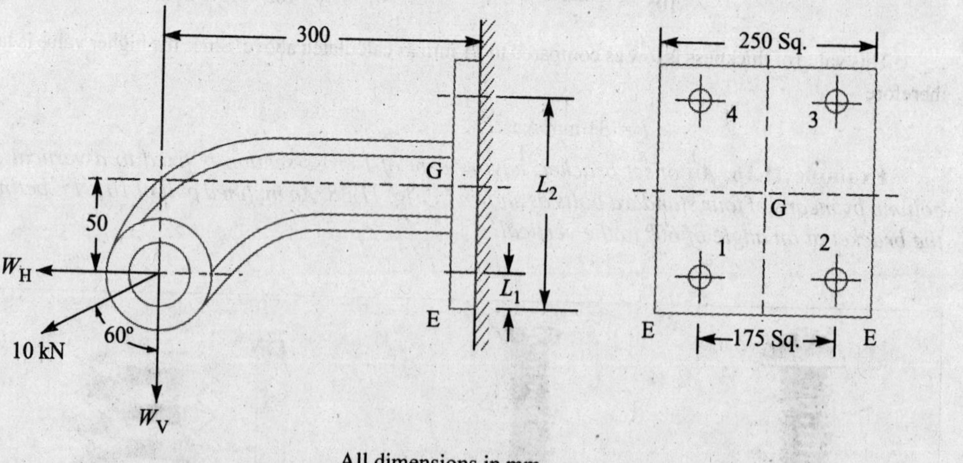

Fig. 11.39

Resolving the pull acting on the bracket (*i.e.* 10 kN) into horizontal and vertical components, we have

Horizontal component of 10 kN,

$$W_H = 10 \times \sin 60° = 10 \times 0.866 = 8.66 \text{ kN} = 8660 \text{ N}$$

and vertical component of 10 kN,

$$W_V = 10 \cos 60° = 10 \times 0.5 = 5 \text{ kN} = 5000 \text{ N}$$

Due to the horizontal component (W_H), which acts parallel to the axis of the bolts, as shown in Fig. 11.39, the following two effects are produced :

1. A direct tensile load equally shared by all the four bolts, and
2. A turning moment about the centre of gravity of the bolts. Since the centre of gravity of all the four bolts lie in the centre at *G* (because of symmetrical bolts), therefore the turning moment is in the clockwise direction.

∴ Direct tensile load on each bolt,

$$W_{t1} = \frac{W_H}{4} = \frac{8660}{4} = 2165 \text{ N}$$

Distance of horizontal component from the centre of gravity (*G*) of the bolts

$$= 50 \text{ mm} = 0.05 \text{ m}$$

∴ Turning moment due to W_H about *G*,

$$T_H = W_H \times 0.05 = 8660 \times 0.05 = 433 \text{ N-m (Clockwise)}$$

Due to the vertical component (W_V), which acts perpendicular to the axis of the bolts, as shown in Fig. 11.39, the following two effects are produced :

1. A direct shear load equally shared by all the four bolts, and
2. A turning moment about the edge of the bracket, in the anticlockwise direction.

∴ Direct shear load on each bolt,

$$W_s = \frac{W_V}{4} = \frac{5000}{4} = 1250 \text{ N}$$

Distance of vertical component from the edge of the bracket

$$= 300 \text{ mm} = 0.3 \text{ m}$$

∴ Turning moment about the edge of the bracket,

$$T_V = W_V \times 0.3 = 5000 \times 0.3 = 1500 \text{ N-m (Anticlockwise)}$$

From above, we see that the anticlockwise moment is greater than the clockwise moment, therefore

Net turning moment

$$= 1500 - 433 = 1067 \text{ N-m (Anticlockwise)} \quad \text{...}(i)$$

Due to this anticlockwise moment, the bracket tends to tilt about the edge *E*.

Let w = Load in each bolt per metre distance from the edge *E*, due to the turning effect of the bracket,

L_1 = Distance of bolts 1 and 2 from the tilting edge *E*

$$= \frac{250 - 175}{2} = 37.5 \text{ mm} = 0.0375 \text{ m}$$

L_3 = Distance of bolts 3 and 4 from the tilting edge

$$= L_1 + 175 \text{ mm} = 37.5 + 175 = 212.5 \text{ mm} = 0.2125 \text{ m}$$

∴ Total moment of the load on the bolts about the tilting edge *E*

$$= 2\,(w.L_1)\,L_1 + 2\,(w.L_2)\,L_2 = 2w\,(L_1)^2 + 2w\,(L_2)^2$$

...(∵ There are two bolts each at distance L_1 and L_2.)

$$= 2w\,(0.0375)^2 + 2w\,(0.2125)^2 = 0.093\,w \text{ N-m} \quad \text{...}(ii)$$

From equations (*i*) and (*ii*), we have

$$w = 1067 / 0.093 = 11\,470 \text{ N/m}$$

Since the heavily loaded bolts are those which lie at a greater distance from the tilting edge, therefore the upper bolts 3 and 4 will be heavily loaded.

∴ Maximum tensile load on each upper bolt,

$$W_{t2} = w.L_2 = 11\,470 \times 0.2125 = 2435 \text{ N}$$

and total tensile load on each of the upper bolt,

$$W_t = W_{t1} + W_{t2} = 2165 + 2435 = 4600 \text{ N}$$

Since each upper bolt is subjected to a total tensile load (W_t = 4600 N) and a shear load (W_s = 1250 N), therefore equivalent tensile load,

$$W_{te} = \frac{1}{2}\left[W_t + \sqrt{(W_t)^2 + 4(W_s)^2}\right] = \frac{1}{2}\left[4600 + \sqrt{(4600)^2 + 4(1250)^2}\right]$$

$$= \frac{1}{2}(4600 + 5240) = 4920 \text{ N}$$

(a) Diameter of the fixing bolts

Let d_c = Core diameter of the fixing bolts.

We know that the equivalent tensile load (W_{te}),

$$4920 = \frac{\pi}{4}(d_c)^2 \sigma_t = \frac{\pi}{4}(d_c)^2 100 = 78.55\,(d_c)^2$$

$$\therefore \quad (d_c)^2 = 4920 / 78.55 = 62.6 \quad \text{or} \quad d_c = 7.9 \text{ mm}$$

From Table 11.1 (coarse series), we find that standard core diameter is 8.18 mm and the corresponding size of the bolt is M 10. **Ans.**

Dimensions of the arm of the bracket

Let t = Thickness of the flanges and web in mm, and

b = Width of the flanges in mm = $3t$ ... (Given)

∴ Cross-sectional area of the *I*-section of the arms,

$$A = 3\,b.t = 3 \times 3\,t \times t = 9\,t^2 \text{ mm}^2$$

and moment of inertia of the *I*-section of the arm about an axis passing through the centre of gravity of the arm,

$$I = \frac{b\,(2t + b)^3}{12} - \frac{(b - t)\,b^3}{12}$$

$$= \frac{3t\,(2t + 3t)^3}{12} - \frac{(3t - t)\,(3t)^3}{12} = \frac{375\,t^4}{12} - \frac{54\,t^4}{12} = \frac{321\,t^4}{12}$$

∴ Section modulus of *I*-section of the arm,

$$Z = \frac{I}{t + b/2} = \frac{321\,t^4}{12\,(t + 3t/2)} = 10.7\,t^3 \text{ mm}^3$$

Due to the horizontal component W_H, the following two stresses are induced in the arm:

1. Direct tensile stress,

$$\sigma_{t1} = \frac{W_H}{A} = \frac{8660}{9t^2} = \frac{962}{t^2} \text{ N/mm}^2$$

2. Bending stress causing tensile in the lower most fibres of the bottom flange and compressive in the upper most fibres of the top flange.

We know that bending moment of W_H about the centre of gravity of the arm,

$$M_H = W_H \times 0.05 = 8660 \times 0.05 = 433 \text{ N-m} = 433 \times 10^3 \text{ N-mm}$$

∴ Bending stress,

$$\sigma_{t2} = \frac{M_H}{Z} = \frac{433 \times 10^3}{10.7\,t^3} = \frac{40.5 \times 10^3}{t^3} \text{ N/mm}^2$$

Due to the vertical component W_V, the following two stresses are induced the arm:

1. Direct shear stress,

$$\tau = \frac{W_V}{A} = \frac{5000}{9t^2} = 556 \text{ N/mm}^2$$

2. Bending stress causing tensile in the upper most fibres of the top flange and compressive in lower most fibres of the bottom flange.

Assuming that the arm extends upto the plate used for fixing the bracket to the structure.

We know that bending moment due to W_V,

$$M_V = W_V \times 0.3 = 5000 \times 0.3 = 1500 \text{ N-m} = 1500 \times 10^3 \text{ N-mm}$$

∴ Bending stress,

$$\sigma_{t3} = \frac{M_V}{Z} = \frac{1500 \times 10^3}{10.7\, t^3} = \frac{140.2 \times 10^3}{t^3} \text{ N/mm}^2$$

Retaining screws on a lamp.

Considering the upper most fibres of the top flange.

Net tensile stress induced in the arm of the bracket

$$= \sigma_{t1} - \sigma_{t2} + \sigma_{t3}$$

$$= \frac{962}{t^2} - \frac{40.5 \times 10^3}{t^3} + \frac{140.2 \times 10^3}{t^3}$$

$$= \frac{962}{t^2} + \frac{99.7 \times 10^3}{t^3}$$

Since the shear stress at the top most fibres is zero, therefore equating the above expression, equal to the given safe tensile stress of 100 N/mm^2, we have

$$\frac{962}{t^2} + \frac{99.7 \times 10^3}{t^3} = 100$$

By hit and trial method, we find that

$$t = 10.4 \text{ mm } \textbf{Ans.}$$

and $$b = 3\,t = 3 \times 10.4 = 31.2 \text{ mm } \textbf{Ans.}$$

11.21 Eccentric Load on a Bracket with Circular Base

Sometimes the base of a bracket is made circular as in case of a flanged bearing of a heavy machine tool and pillar crane etc. Consider a round flange bearing of a machine tool having four bolts as shown in Fig. 11.40.

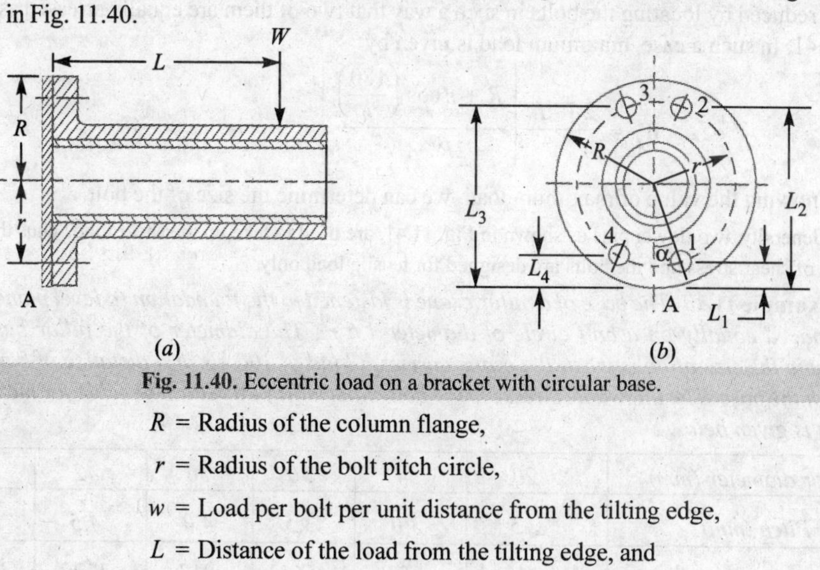

Fig. 11.40. Eccentric load on a bracket with circular base.

Let R = Radius of the column flange,

r = Radius of the bolt pitch circle,

w = Load per bolt per unit distance from the tilting edge,

L = Distance of the load from the tilting edge, and

$L_1, L_2, L_3,$ and L_4 = Distance of bolt centres from the tilting edge A.

As discussed in the previous article, equating the external moment $W \times L$ to the sum of the resisting moments of all the bolts, we have,

$$W.L = w\,[(L_1)^2 + (L_2)^2 + (L_3)^2 + (L_4)^2]$$

$$\therefore \quad w = \frac{W.L}{(L_1)^2 + (L_2)^2 + (L_3)^2 + (L_4)^2} \quad ...(i)$$

Now from the geometry of the Fig. 11.40 (*b*), we find that

$$L_1 = R - r\cos\alpha \qquad L_2 = R + r\sin\alpha$$

$$L_3 = R + r\cos\alpha \quad \text{and} \quad L_4 = R - r\sin\alpha$$

Substituting these values in equation (*i*), we get

$$w = \frac{W.L}{4\,R^2 + 2\,r^2}$$

$$\therefore \text{ Load in the bolt situated at } 1 = w.L_1 = \frac{W.L.L_1}{4\,R^2 + 2\,r^2} = \frac{W.L\,(R - r\cos\alpha)}{4\,R^2 + 2\,r^2}$$

This load will be maximum when $\cos\alpha$ is minimum *i.e.* when $\cos\alpha = -1$ or $\alpha = 180°$.

$\therefore$ Maximum load in a bolt

$$= \frac{W.L\,(R + r)}{4\,R^2 + 2\,r^2}$$

In general, if there are *n* number of bolts, then load in a bolt

$$= \frac{2W.L\,(R - r\cos\alpha)}{n\,(2R^2 + r^2)}$$

and maximum load in a bolt,

$$W_t = \frac{2\,W.L\,(R + r)}{n\,(2R^2 + r^2)}$$

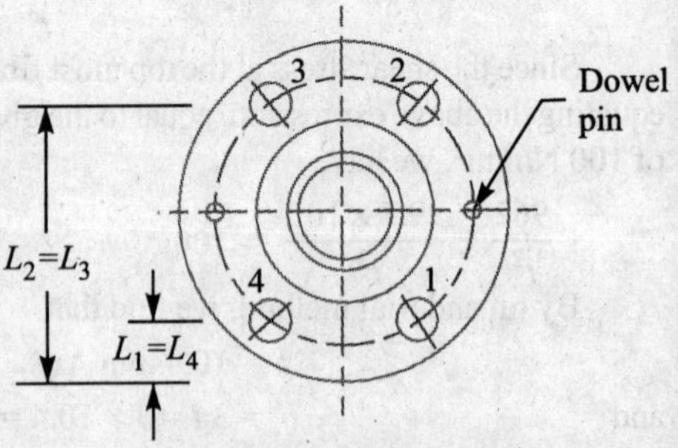

Fig. 11.41

The above relation is used when the direction of the load *W* changes with relation to the bolts as in the case of pillar crane. But if the direction of load is fixed, then the maximum load on the bolts may be reduced by locating the bolts in such a way that two of them are equally stressed as shown in Fig. 11.41. In such a case, maximum load is given by

$$W_t = \frac{2\,W.L}{n}\left[\frac{R + r\cos\left(\dfrac{180}{n}\right)}{2R^2 + r^2}\right]$$

Knowing the value of maximum load, we can determine the size of the bolt.

Note : Generally, two dowel pins as shown in Fig. 11.41, are used to take up the shear load. Thus the bolts are relieved of shear stress and the bolts are designed for tensile load only.

Example 11.17. *The base of a pillar crane is fastened to the foundation (a level plane) by eight bolts spaced equally on a bolt circle of diameter 1.6 m. The diameter of the pillar base is 2 m. Determine the size of bolts when the crane carries a load of 100 kN at a distance of 5 m from the centre of the base. The allowable stress for the bolt material is 100 MPa. The table for metric coarse threads is given below :*

Major diameter (mm)	*20*	*24*	*30*	*36*	*42*	*48*
Pitch (mm)	*2.5*	*3.0*	*3.5*	*4.0*	*4.5*	*5.0*
Stress area (mm^2)	*245*	*353*	*561*	*817*	*1120*	*1472*

Solution. Given : $n = 8$; $d = 1.6$ m or $r = 0.8$ m ; $D = 2$m or $R = 1$m ; $W = 100$ kN $= 100 \times 10^3$ N ; $e = 5$ m; $\sigma_t = 100$ MPa $= 100$ N/mm²

The pillar crane is shown in Fig. 11.42.

We know that the distance of the load from the tilting edge *A-A*,

$$L = e - R = 5 - 1 = 4 \text{ m}$$

Let d_c = Core diameter of the bolts.

We know that maximum load on a bolt,

$$W_t = \frac{2\,W.L\,(R + r)}{n\,(2R^2 + r^2)} = \frac{2 \times 100 \times 10^3 \times 4\,(1 + 0.8)}{8\,[2 \times 1^2 + (0.8)^2]}$$

$$= \frac{1440 \times 10^3}{21.12} = 68.18 \times 10^3 \text{ N}$$

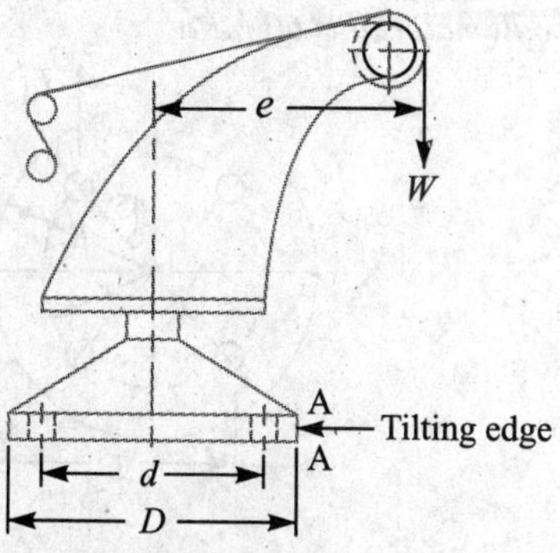

Fig. 11.42

We also know that maximum load on a bolt (W_t),

$$68.18 \times 10^3 = \frac{\pi}{4}\,(d_c)^2\,\sigma_t = \frac{\pi}{4}\,(d_c)^2\,100 = 78.54\,(d_c)^2$$

$$\therefore \quad (d_c)^2 = 68.18 \times 10^3 / 78.54 = 868 \quad \text{or} \quad d_c = 29.5 \text{ mm}$$

From Table 11.1 (coarse series), we find that the standard core diameter of the bolt is 31.093 mm and the corresponding size of the bolt is M 36. **Ans.**

Example 11.18. *A flanged bearing, as shown in Fig. 11.40, is fastened to a frame by means of four bolts spaced equally on 500 mm bolt circle. The diameter of bearing flange is 650 mm and a load of 400 kN acts at a distance of 250 mm from the frame. Determine the size of the bolts, taking safe tensile stress as 60 MPa for the material of the bolts.*

Solution. Given : $n = 4$; $d = 500$ mm or $r = 250$ mm ; $D = 650$ mm or $R = 325$ mm ; $W = 400$ kN $= 400 \times 10^3$ N ; $L = 250$ mm ; $\sigma_t = 60$ MPa $= 60$ N/mm²

Let d_c = Core diameter of the bolts.

We know that when the bolts are equally spaced, the maximum load on the bolt,

$$W_t = \frac{2W.L}{n}\left[\frac{R + r\cos\left(\frac{180}{n}\right)}{2R^2 + r^2}\right]$$

$$= \frac{2 \times 400 \times 10^3 \times 250}{4}\left[\frac{325 + 250\cos\left(\frac{180}{4}\right)}{2\,(325)^2 + (250)^2}\right] = 91\,643 \text{ N}$$

We also know that maximum load on the bolt (W_t),

$$91\,643 = \frac{\pi}{4}\,(d_c)^2\,\sigma_t = \frac{\pi}{4}\,(d_c)^2\,60 = 47.13\,(d_c)^2$$

$$\therefore \quad (d_c)^2 = 91\,643 / 47.13 = 1945 \quad \text{or} \quad d_c = 44 \text{ mm}$$

From Table 11.1, we find that the standard core diameter of the bolt is 45.795 mm and corresponding size of the bolt is M 52. **Ans.**

Example 11.19. *A pillar crane having a circular base of 600 mm diameter is fixed to the foundation of concrete base by means of four bolts. The bolts are of size 30 mm and are equally spaced on a bolt circle diameter of 500 mm.*

Determine : 1. The distance of the load from the centre of the pillar along a line X-X as shown in Fig. 11.43 (a). The load lifted by the pillar crane is 60 kN and the allowable tensile stress for the bolt material is 60 MPa.

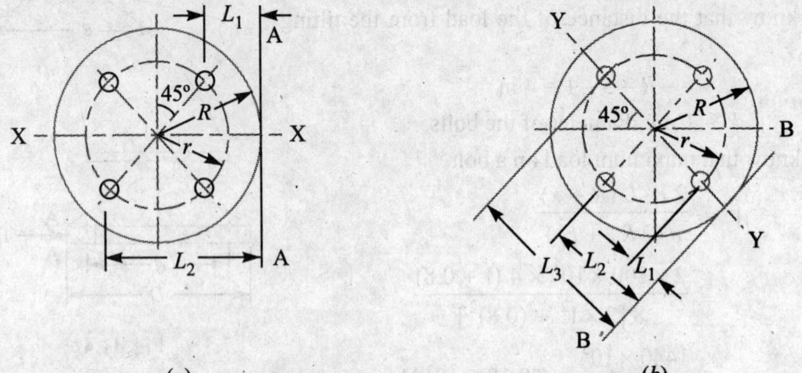

Fig. 11.43

2. The maximum stress induced in the bolts if the load is applied along a line Y-Y of the foundation as shown in Fig. 11.43 (b) at the same distance as in part (1).

Solution. Given : D = 600 mm or R = 300 mm ; n = 4; d_b = 30 mm ; d = 500 mm or r = 250 mm ; W = 60 kN ; σ_t = 60 MPa = 60 N/mm²

Since the size of bolt (*i.e.* d_b = 30 mm), is given therefore from Table 11.1, we find that the stress area corresponding to M 30 is 561 mm².

We know that the maximum load carried by each bolt

$$= \text{Stress area} \times \sigma_t = 561 \times 60 = 33\ 660 \text{ N} = 33.66 \text{ kN}$$

and direct tensile load carried by each bolt

$$= \frac{W}{n} = \frac{60}{4} = 15 \text{ kN}$$

∴ Total load carried by each bolt at distance L_2 from the tilting edge *A-A*

$$= 33.66 + 15 = 48.66 \text{ kN} \quad \text{...}(i)$$

From Fig. 11.43 (*a*), we find that

$$L_1 = R - r\cos 45° = 300 - 250 \times 0.707 = 123 \text{ mm} = 0.123 \text{ m}$$

and $$L_2 = R + r\cos 45° = 300 + 250 \times 0.707 = 477 \text{ mm} = 0.477 \text{ m}$$

Let w = Load (in kN) per bolt per unit distance.

∴ Total load carried by each bolt at distance L_2 from the tilting edge *A-A*

$$= w.L_2 = w \times 0.477 \text{ kN} \quad \text{...}(ii)$$

From equations (*i*) and (*ii*), we have

$$w = 48.66 / 0.477 = 102 \text{ kN/m}$$

∴ Resisting moment of all the bolts about the outer (*i.e.* tilting) edge of the flange along the tangent *A-A*

$$= 2w\,[(L_1)^2 + (L_2)^2] = 2 \times 102\,[(0.123)^2 + (0.477)^2] = 49.4 \text{ kN-m}$$

1. *Distance of the load from the centre of the pillar*

Let e = Distance of the load from the centre of the pillar or eccentricity of the load, and

L = Distance of the load from the tilting edge *A-A* = $e - R = e - 0.3$

We know that turning moment due to load W, about the tilting edge A-A of the flange

$$= W.L = 60\,(e - 0.3) \text{ kN-m}$$

Now equating the turning moment to the resisting moment of all the bolts, we have

$$60\,(e - 0.3) = 49.4$$

$$\therefore \quad e - 0.3 = 49.4 / 60 = 0.823 \quad \text{or} \quad e = 0.823 + 0.3 = 1.123 \text{ m } \textbf{Ans.}$$

2. *Maximum stress induced in the bolt*

Since the load is applied along a line Y-Y as shown in Fig. 11.43 (*b*), and at the same distance as in part (*1*) *i.e.* at $L = e - 0.3 = 1.123 - 0.3 = 0.823$ m from the tilting edge B-B, therefore

Turning moment due to load W about the tilting edge B–B

$$= W.L = 60 \times 0.823 = 49.4 \text{ kN-m}$$

From Fig. 11.43 (*b*), we find that

$$L_1 = R - r = 300 - 250 = 50 \text{ mm} = 0.05 \text{ m}$$

$$L_2 = R = 300 \text{ mm} = 0.3 \text{ m}$$

and

$$L_3 = R + r = 300 + 250 = 550 \text{ mm} = 0.55 \text{ m}$$

$\therefore$ Resisting moment of all the bolts about B–B

$$= w\,[(L_1)^2 + 2(L_2)^2 + (L_3)^2] = w[(0.05)^2 + 2(0.3)^2 + (0.55)^2] \text{ kN-m}$$

$$= 0.485\, w \text{ kN-m}$$

Equating resisting moment of all the bolts to the turning moment, we have

$$0.485\, w = 49.4$$

or

$$w = 49.4 / 0.485 = 102 \text{ kN/m}$$

Since the bolt at a distance of L_3 is heavily loaded, therefore load carried by this bolt

$$= w.L_3 = 102 \times 0.55 = 56.1 \text{ kN}$$

Harvesting machine

Note : This picture is given as additional information and is not a direct example of the current chapter.

and net force taken by the bolt

$$= w.L_3 - \frac{W}{n} = 56.1 - \frac{60}{4} = 41.1 \text{ kN} = 41\ 100 \text{ N}$$

∴ Maximum stress induced in the bolt

$$= \frac{\text{Force}}{\text{Stress area}} = \frac{41\ 000}{516}$$

$$= 79.65 \text{ N/mm}^2 = 79.65 \text{ MPa } \textbf{Ans.}$$

11.22 Eccentric Load Acting in the Plane Containing the Bolts

When the eccentric load acts in the plane containing the bolts, as shown in Fig. 11.44, then the same procedure may be followed as discussed for eccentric loaded riveted joints.

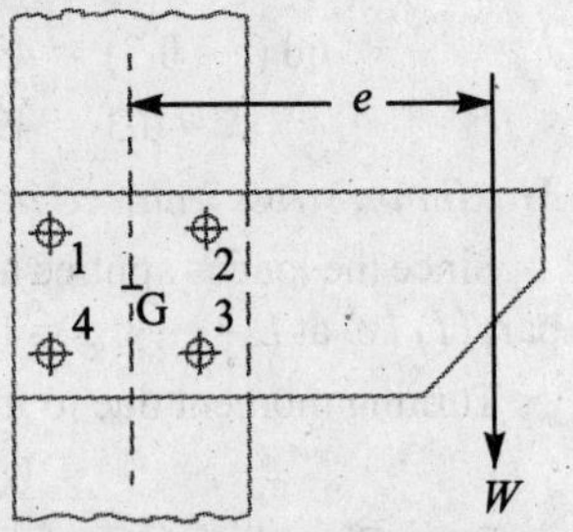

Fig. 11.44. Eccentric load in the plane containing the bolts.

Example 11.20. *Fig. 11.45 shows a solid forged bracket to carry a vertical load of 13.5 kN applied through the centre of hole. The square flange is secured to the flat side of a vertical stanchion through four bolts. Calculate suitable diameter D and d for the arms of the bracket, if the permissible stresses are 110 MPa in tension and 65 MPa in shear.*

Estimate also the tensile load on each top bolt and the maximum shearing force on each bolt.

Solution. Given : $W = 13.5$ kN = 13 500 N ; $\sigma_t = 110$ MPa = 110 N/mm^2 ; $\tau = 65$ MPa = 65 N/mm^2

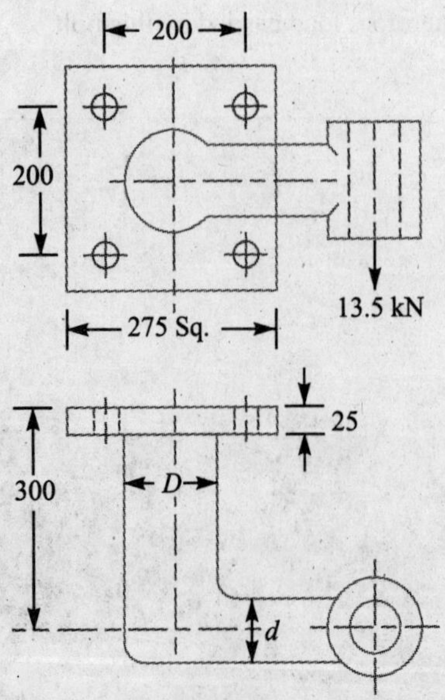

Fig. 11.45

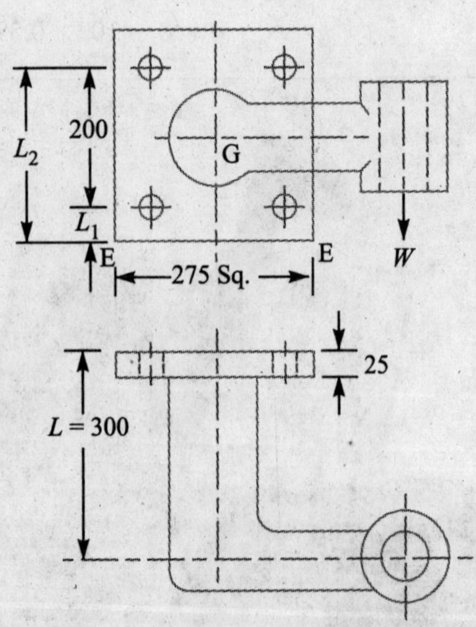

Fig. 11.46

Diameter D for the arm of the bracket

The section of the arm having *D* as the diameter is subjected to bending moment as well as twisting moment. We know that bending moment,

$$M = 13\ 500 \times (300 - 25) = 3712.5 \times 10^3 \text{ N-mm}$$

and twisting moment, $T = 13\,500 \times 250 = 3375 \times 10^3$ N-mm

∴ Equivalent twisting moment,

$$T_e = \sqrt{M^2 + T^2} = \sqrt{(3712.5 \times 10^3)^2 + (3375 \times 10^3)^2} \text{ N-mm}$$
$$= 5017 \times 10^3 \text{ N-mm}$$

We know that equivalent twisting moment (T_e),

$$5017 \times 10^3 = \frac{\pi}{16} \times \tau \times D^3 = \frac{\pi}{16} \times 65 \times D^3 = 12.76\, D^3$$

∴ $D^3 = 5017 \times 10^3 / 12.76 = 393 \times 10^3$

or $D = 73.24$ say 75 mm **Ans.**

Diameter (d) for the arm of the bracket

The section of the arm having d as the diameter is subjected to bending moment only. We know that bending moment,

$$M = 13\,500 \left(250 - \frac{75}{2}\right) = 2868.8 \times 10^3 \text{ N-mm}$$

and section modulus, $Z = \dfrac{\pi}{32} \times d^3 = 0.0982\, d^3$

We know that bending (tensile) stress (σ_t),

$$110 = \frac{M}{Z} = \frac{2868.8 \times 10^3}{0.0982\, d^3} = \frac{29.2 \times 10^6}{d^3}$$

∴ $d^3 = 29.2 \times 10^6 / 110 = 265.5 \times 10^3$ or $d = 64.3$ say 65 mm **Ans.**

Tensile load on each top bolt

Due to the eccentric load W, the bracket has a tendency to tilt about the edge $E–E$, as shown in Fig. 11.46.

Let w = Load on each bolt per mm distance from the tilting edge due to the tilting effect of the bracket.

Since there are two bolts each at distance L_1 and L_2 as shown in Fig. 11.46, therefore total moment of the load on the bolts about the tilting edge $E–E$

$$= 2\,(w.L_1)\,L_1 + 2(w.L_2)\,L_2 = 2w\,[(L_1)^2 + (L_2)^2]$$
$$= 2w\,[(37.5)^2 + (237.5)^2] = 115\,625\,w \text{ N-mm} \quad ...(i)$$

...(∵ $L_1 = 37.5$ mm and $L_2 = 237.5$ mm)

and turning moment of the load about the tilting edge

$$= W.L = 13\,500 \times 300 = 4050 \times 10^3 \text{ N-mm} \quad ...(ii)$$

From equations (*i*) and (*ii*), we have

$$w = 4050 \times 10^3 / 115\,625 = 35.03 \text{ N/mm}$$

∴ Tensile load on each top bolt

$$= w.L_2 = 35.03 \times 237.5 = 8320 \text{ N}$$ **Ans.**

Maximum shearing force on each bolt

We know that primary shear load on each bolt acting vertically downwards,

$$W_{s1} = \frac{W}{n} = \frac{13\,500}{4} = 3375 \text{ N}$$ ...(∵ No. of bolts, $n = 4$)

Since all the bolts are at equal distances from the centre of gravity of the four bolts (G), therefore the secondary shear load on each bolt is same.

Distance of each bolt from the centre of gravity (*G*) of the bolts,

$$l_1 = l_2 = l_3 = l_4 = \sqrt{(100)^2 + (100)^2} = 141.4 \text{ mm}$$

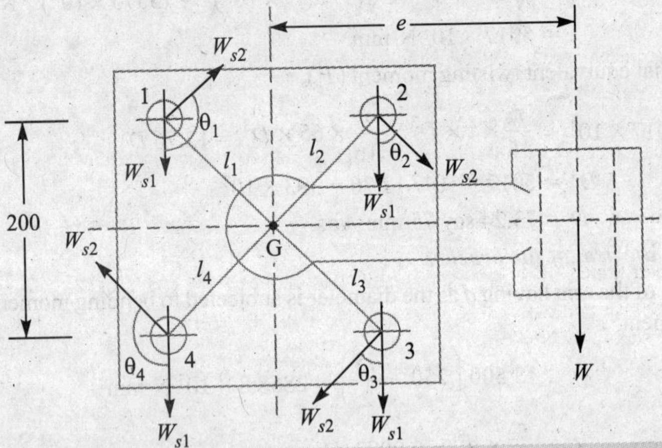

Fig. 11.47

∴ Secondary shear load on each bolt,

$$W_{s2} = \frac{W.e.l_1}{(l_1)^2 + (l_2)^2 + (l_3)^3 + (l_4)^2} = \frac{13\ 500 \times 250 \times 141.4}{4\ (141.4)^2} = 5967 \text{ N}$$

Since the secondary shear load acts at right angles to the line joining the centre of gravity of the bolt group to the centre of the bolt as shown in Fig. 11.47, therefore the resultant of the primary and secondary shear load on each bolt gives the maximum shearing force on each bolt.

From the geometry of the Fig. 11.47, we find that

$$\theta_1 = \theta_4 = 135°, \text{ and } \theta_2 = \theta_3 = 45°$$

∴ Maximum shearing force on the bolts 1 and 4

$$= \sqrt{(W_{s1})^2 + (W_{s2})^2 + 2\, W_{s1} \times W_{s2} \times \cos 135°}$$

$$= \sqrt{(3375)^2 + (5967)^2 - 2 \times 3375 \times 5967 \times 0.7071} = 4303 \text{ N \textbf{Ans.}}$$

and maximum shearing force on the bolts 2 and 3

$$= \sqrt{(W_{s1})^2 + (W_{s2})^2 + 2\, W_{s1} \times W_{s2} \times \cos 45°}$$

$$= \sqrt{(3375)^2 + (5967)^2 + 2 \times 3375 \times 5967 \times 0.7071} = 8687 \text{ N \textbf{Ans.}}$$

EXERCISES

1. Determine the safe tensile load for bolts of M 20 and M 36. Assume that the bolts are not initiall stressed and take the safe tensile stress as 200 MPa. **[Ans. 49 kN; 16.43 kN**

2. An eye bolt carries a tensile load of 20 kN. Find the size of the bolt, if the tensile stress is not to excee 100 MPa. Draw a neat proportioned figure for the bolt. **[Ans. M 20**

3. An engine cylinder is 300 mm in diameter and the steam pressure is 0.7 N/mm². If the cylinder hea is held by 12 studs, find the size. Assume safe tensile stress as 28 MPa. **[Ans. M 24**

4. Find the size of 14 bolts required for a C.I. steam engine cylinder head. The diameter of the cylinder is 400 mm and the steam pressure is 0.12 N/mm^2. Take the permissible tensile stress as 35 MPa. **[Ans. M 24]**

5. The cylinder head of a steam engine is subjected to a pressure of 1 N/mm^2. It is held in position by means of 12 bolts. The effective diameter of the cylinder is 300 mm. A soft copper gasket is used to make the joint leak proof. Determine the size of the bolts so that the stress in the bolts does not exceed 100 MPa. **[Ans. M 36]**

6. A steam engine cylinder of 300 mm diameter is supplied with steam at 1.5 N/mm^2. The cylinder cover is fastened by means of 8 bolts of size M 20. The joint is made leak proof by means of suitable gaskets. Find the stress produced in the bolts. **[Ans. 249 MPa]**

7. The effective diameter of the cylinder is 400 mm. The maximum pressure of steam acting on the cylinder cover is 1.12 N/mm^2. Find the number and size of studs required to fix the cover. Draw a neat proportioned sketch for the elevation of the cylinder cover. **[Ans. 14; M 24]**

8. Specify the size and number of studs required to fasten the head of a 400 mm diameter cylinder containing steam at 2 N/mm^2. A hard gasket (gasket constant = 0.3) is used in making the joint. Draw a neat sketch of the joint also. Other data may be assumed. **[Ans. M 30; 12]**

9. A steam engine cylinder has an effective diameter of 200 mm. It is subjected to a maximum steam pressure of 1.75 N/mm^2. Calculate the number and size of studs required to fix the cylinder cover onto the cylinder flange assuming the permissible stress in the studs as 30 MPa. Take the pitch circle diameter of the studs as 320 mm and the total load on the studs as 20% higher than the external load on the joint. Also check the circumferential pitch of the studs so as to give a leak proof joint. **[Ans. 16; M 16]**

10. A steam engine cylinder of size 300 mm × 400 mm operates at 1.5 N/mm^2 pressure. The cylinder head is connected by means of 8 bolts having yield point stress of 350 MPa and endurance limit of 240 MPa. The bolts are tightened with an initial preload of 1.8 times the steam lead. The joint is made leak-proof by using soft copper gasket which renderes the effect of external load to be half. Determine the size of bolts, if factor of safety is 2 and stress concentration factor is 3. **[Ans. M 20]**

11. The cylinder head of a 200 mm × 350 mm compressor is secured by means of 12 studs of rolled mild steel. The gas pressure is 1.5 N/mm^2 gauge. The initial tension in the bolts, assumed to be equally loaded such that a cylinder pressure of 3 N/mm^2 gauge is required for the joint to be on the point of opening. Suggest the suitable size of the studs in accordance with Soderberg's equation assuming the equivalent diameter of the compressed parts to be twice the bolt size and factor of safety 2. The stress concentration factor may be taken as 2.8 and the value of endurance strength for reversed axial loading is half the value of ultimate strength. **[Ans. M 12]**

12. Find the diameter of screwed boiler stays, each stay supports an area equal to 200 mm × 150 mm. The steam pressure is 1 N/mm^2. The permissible tensile stress for the stay material is 34 MPa. **[Ans. M 36]**

13. What size of hole must be drilled in a M 42 bolt so as to make the bolt of uniform strength? **[Ans. 18.4 mm]**

14. A mounting plate for a drive unit is fixed to the support by means of four M 12 bolts as shown in Fig. 11.48. The core diameter of the bolts can be considered as 9.858 mm. Determine the maximum value of 'W' if the allowable tensile stress in bolt material is 60 MPa. **[Ans. 12.212 kN]**

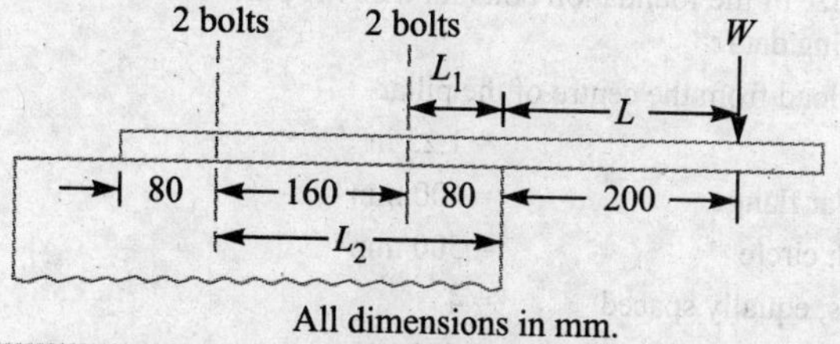

Fig. 11.48

15. A pulley bracket, as shown in Fig. 11.49, is supported by 4 bolts, two at *A-A* and two at *B-B*. Determine the size of bolts using an allowable shear stress of 25 MPa for the material of the bolts.
[Ans. M 27]

16. A wall bracket, as shown in Fig. 11.50, is fixed to a wall by means of four bolts. Find the size of the bolts and the width of bracket. The safe stress in tension for the bolt and bracket may be assumed as 70 MPa. **[Ans. M 30 ; 320 mm]**

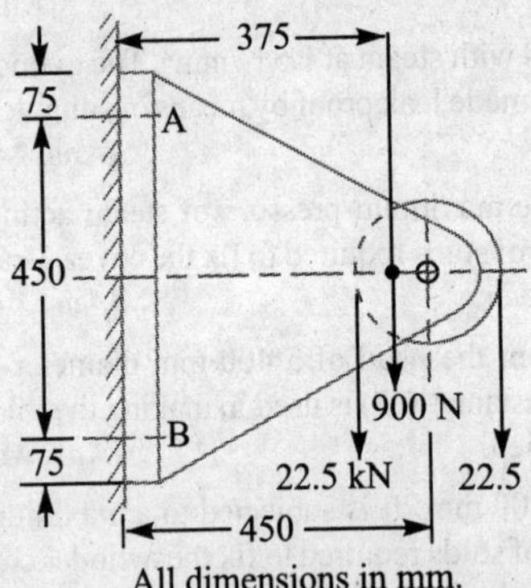

Fig. 11.49

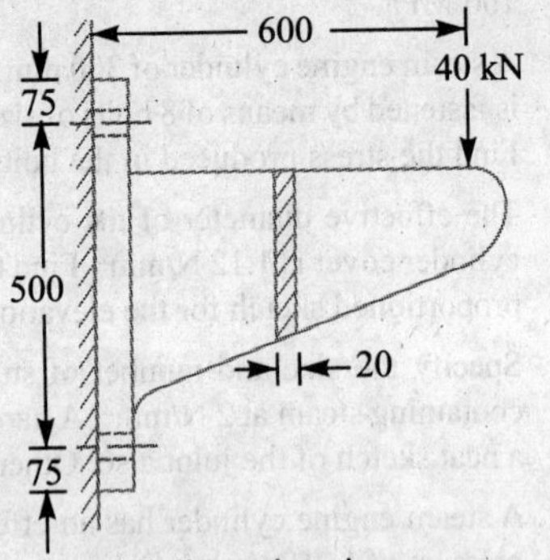

Fig. 11.50

17. A bracket is bolted to a column by 6 bolts of equal size as shown in Fig. 11.51. It carries a load of 5C kN at a distance of 150 mm from the centre of column. If the maximum stress in the bolts is to be limited to 150 MPa, determine the diameter of bolt. **[Ans. 14 mm]**

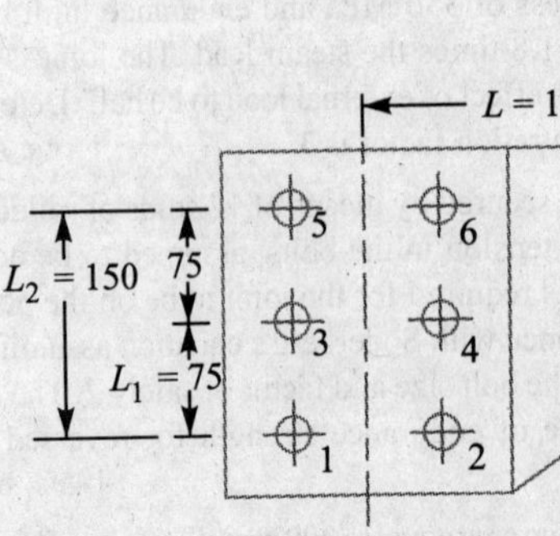

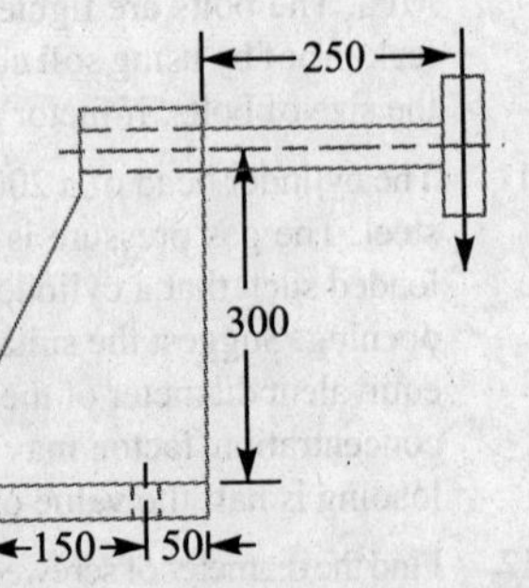

All dimensions in mm.

Fig. 11.51

Fig. 11.52

18. A cast iron bracket to carry a shaft and a belt pulley is shown in Fig. 11.52. The bracket is fixed to th main body by means of four standard bolts. The tensions in the slack and tight sides of the belt are 2. kN and 4.25 kN respectively. Find the size of the bolts, if the safe tensile stress for bolts is 50 MPa
[Ans. M 1

19. Determine the size of the foundation bolts for a 60 kN pillar crane as shown in Fig. 11.42 (page 42 from the following data :

Distance of the load from the centre of the pillar = 1.25 m

Diameter of pillar flange = 600 mm

Diameter of bolt circle = 500 mm

Number of bolts, equally spaced = 4

Allowable tensile stress for bolts = 600 MPa **[Ans. M 3**

20. A bracket, as shown in Fig. 11.53, is fixed to a vertical steel column by means of five standard bolts. Determine : (*a*) The diameter of the fixing bolts, and (*b*) The thickness of the arm of the bracket. Assume safe working stresses of 70 MPa in tension and 50 MPa in shear.

[Ans. M 18; 50 mm]

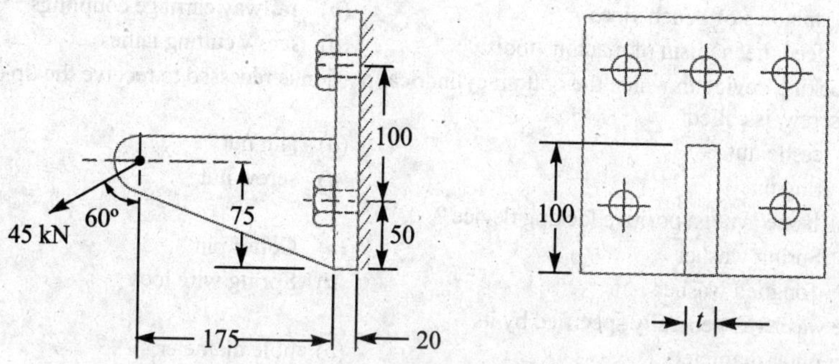

All dimensions in mm.

Fig. 11.53

QUESTIONS

1. What do you understand by the single start and double start threads ?
2. Define the following terms :

   (*a*) Major diameter, (*b*) Minor diameter, (*c*) Pitch, and (*d*) Lead.
3. Write short note on nut locking devices covering the necessity and various types. Your answer should be illustrated with neat sketches.
4. Discuss the significance of the initial tightening load and the applied load so far as bolts are concerned. Explain which of the above loads must be greater for a properly designed bolted joint and show how each affects the total load on the bolt.
5. Discuss on bolts of uniform strength giving examples of practical applications of such bolts.
6. Bolts less than M 16 should normally be used in pre loaded joints. Comment.
7. How the core diameter of the bolt is determined when a bracket having a rectangular base is bolted to a wall by four bolts and carries an eccentric load parallel to the axis of the bolt?
8. Derive an expression for the maximum load in a bolt when a bracket with circular base is bolted to a wall by means of four bolts.
9. Explain the method of determining the size of the bolt when the bracket carries an eccentric load perpendicular to the axis of the bolt.

OBJECTIVE TYPE QUESTIONS

1. The largest diameter of an external or internal screw thread is known as

   (*a*) minor diameter (*b*) major diameter

   (*c*) pitch diameter (*d*) none of these
2. The pitch diameter is the .... diameter of an external or internal screw thread.

   (*a*) effective (*b*) smallest (*c*) largest
3. A screw is specified by its

   (*a*) major diameter (*b*) minor diameter

(c) pitch diameter (d) pitch

4. The railway carriage coupling have
   (a) square threads (b) acme threads
   (c) knuckle threads (d) buttress threads
5. The square threads are usually found on
   (a) spindles of bench vices (b) railway carriage couplings
   (c) feed mechanism of machine tools (d) screw cutting lathes
6. A locking device in which the bottom cylindrical portion is recessed to receive the tip of the locking set screw, is called
   (a) castle nut (b) jam nut
   (c) ring nut (d) screw nut
7. Which one is not a positive locking device ?
   (a) Spring washer (b) Cotter pin
   (c) Tongued washer (d) Spring wire lock
8. The washer is generally specified by its
   (a) outer diameter (b) hole diameter
   (c) thickness (d) mean diameter
9. A locking device extensively used in automobile industry is a
   (a) jam nut (b) castle nut
   (c) screw nut (d) ring nut
10. A bolt of M 24 × 2 means that
    (a) the pitch of the thread is 24 mm and depth is 2 mm
    (b) the cross-sectional area of the threads is 24 mm^2
    (c) the nominal diameter of bolt is 24 mm and the pitch is 2 mm
    (d) the effective diameter of the bolt is 24 mm and there are two threads per cm
11. When a nut is tightened by placing a washer below it, the bolt will be subjected to
    (a) tensile stress (b) compressive stress
    (c) shear stress (d) none of these
12. The eye bolts are used for
    (a) transmission of power (b) locking devices
    (c) lifting and transporting heavy machines (d) absorbing shocks and vibrations
13. The shock absorbing capacity of a bolt may be increased by
    (a) increasing its shank diameter
    (b) decreasing its shank diameter
    (c) tightening the bolt properly
    (d) making the shank diameter equal to the core diameter of the thread.
14. The resilience of a bolt may be increased by
    (a) increasing its shank diameter (b) increasing its length
    (c) decreasing its shank diameter (d) decreasing its length
15. A bolt of uniform strength can be developed by
    (a) keeping the core diameter of threads equal to the diameter of unthreaded portion of the bolt
    (b) keeping the core diameter of threads smaller than the diameter of unthreaded portion of the bolt
    (c) keeping the nominal diameter of threads equal to the diameter of unthreaded portion of bolt
    (d) none of the above

ANSWERS

1. (b)	2. (a)	3. (a)	4. (d)	5. (c)
6. (c)	7. (a)	8. (b)	9. (b)	10. (c)
11. (a)	12. (c)	13. (b)	14. (b)	15. (a)

C H A P T E R

12

Cotter and Knuckle Joints

1. Introduction.
2. Types of Cotter Joints.
3. Socket and Spigot Cotter Joint.
4. Design of Socket and Spigot Cotter Joint.
5. Sleeve and Cotter Joint.
6. Design of Sleeve and Cotter Joint.
7. Gib and Cotter Joint.
8. Design of Gib and Cotter Joint for Strap End of a Connecting Rod.
9. Design of Gib and Cotter Joint for Square Rods.
10. Design of Cotter Joint to Connect Piston Rod and Crosshead.
11. Design of Cotter Foundation Bolt.
12. Knuckle Joint.
13. Dimensions of Various Parts of the Knuckle Joint.
14. Methods of Failure of Knuckle Joint.
15. Design Procedure of Knuckle Joint.
16. Adjustable Screwed Joint for Round Rods (Turn Buckle).
17. Design of Turn Buckle.

12.1 Introduction

A cotter is a flat wedge shaped piece of rectangular cross-section and its width is tapered (either on one side or both sides) from one end to another for an easy adjustment. The taper varies from 1 in 48 to 1 in 24 and it may be increased up to 1 in 8, if a locking device is provided. The locking device may be a taper pin or a set screw used on the lower end of the cotter. The cotter is usually made of mild steel or wrought iron. A cotter joint is a temporary fastening and is used to connect rigidly two co-axial rods or bars which are subjected to axial tensile or compressive forces. It is usually used in connecting a piston rod to the cross-head of a reciprocating steam engine, a piston rod and its extension as a tail or pump rod, strap end of connecting rod etc.

12.2 Types of Cotter Joints

Following are the three commonly used cotter joints to connect two rods by a cotter :

1. Socket and spigot cotter joint, 2. Sleeve and cotter joint, and 3. Gib and cotter joint.

The design of these types of joints are discussed, in detail, in the following pages.

12.3 Socket and Spigot Cotter Joint

In a socket and spigot cotter joint, one end of the rods (say *A*) is provided with a socket type of end as shown in Fig. 12.1 and the other end of the other rod (say *B*) is inserted into a socket. The end of the rod which goes into a socket is also called ***spigot.*** *A* rectangular hole is made in the socket and spigot. *A* cotter is then driven tightly through a hole in order to make the temporary connection between the two rods. The load is usually acting axially, but it changes its direction and hence the cotter joint must be designed to carry both the tensile and compressive loads. The compressive load is taken up by the collar on the spigot.

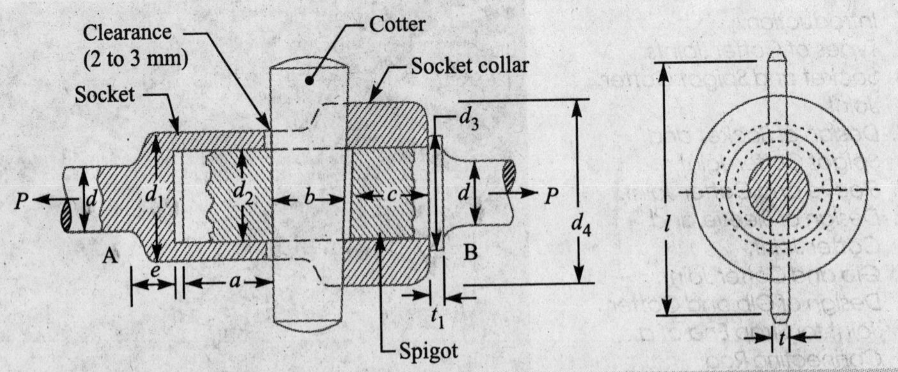

Fig. 12.1. Socket and spigot cotter joint.

12.4 Design of Socket and Spigot Cotter Joint

The socket and spigot cotter joint is shown in Fig. 12.1.

Let P = Load carried by the rods,

d = Diameter of the rods,

d_1 = Outside diameter of socket,

d_2 = Diameter of spigot or inside diameter of socket,

d_3 = Outside diameter of spigot collar,

t_1 = Thickness of spigot collar,

d_4 = Diameter of socket collar,

c = Thickness of socket collar,

b = Mean width of cotter,

t = Thickness of cotter,

l = Length of cotter,

a = Distance from the end of the slot to the end of rod,

σ_t = Permissible tensile stress for the rods material,

τ = Permissible shear stress for the cotter material, and

σ_c = Permissible crushing stress for the cotter material.

The dimensions for a socket and spigot cotter joint may be obtained by considering the various modes of failure as discussed below :

1. ***Failure of the rods in tension***

The rods may fail in tension due to the tensile load P. We know that

Area resisting tearing

$$= \frac{\pi}{4} \times d^2$$

∴ Tearing strength of the rods,

$$= \frac{\pi}{4} \times d^2 \times \sigma_t$$

Equating this to load (P), we have

$$P = \frac{\pi}{4} \times d^2 \times \sigma_t$$

From this equation, diameter of the rods (d) may be determined.

Fork lift is used to move goods from one place to the other within the factory.

2. ***Failure of spigot in tension across the weakest section (or slot)***

Since the weakest section of the spigot is that section which has a slot in it for the cotter, as shown in Fig. 12.2, therefore

Area resisting tearing of the spigot across the slot

$$= \frac{\pi}{4} (d_2)^2 - d_2 \times t$$

and tearing strength of the spigot across the slot

$$= \left[\frac{\pi}{4} (d_2)^2 - d_2 \times t\right] \sigma_t$$

Equating this to load (P), we have

$$P = \left[\frac{\pi}{4} (d_2)^2 - d_2 \times t\right] \sigma_t$$

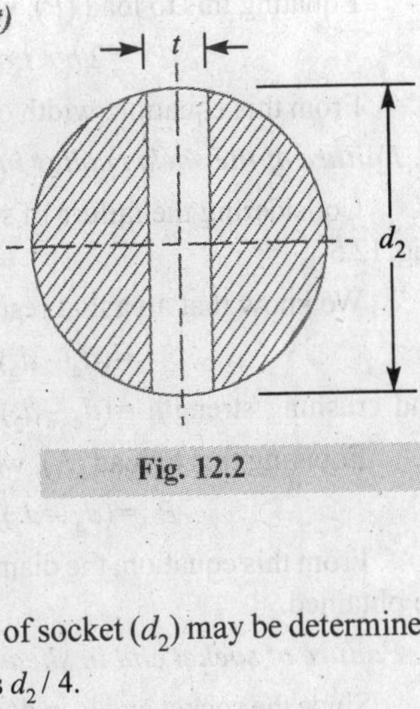

Fig. 12.2

From this equation, the diameter of spigot or inside diameter of socket (d_2) may be determined.

Note : In actual practice, the thickness of cotter is usually taken as $d_2 / 4$.

3. ***Failure of the rod or cotter in crushing***

We know that the area that resists crushing of a rod or cotter

$$= d_2 \times t$$

∴ Crushing strength $= d_2 \times t \times \sigma_c$

Equating this to load (P), we have

$$P = d_2 \times t \times \sigma_c$$

From this equation, the induced crushing stress may be checked.

4. *Failure of the socket in tension across the slot*

We know that the resisting area of the socket across the slot, as shown in Fig. 12.3

$$= \frac{\pi}{4}\left[(d_1)^2 - (d_2)^2\right] - (d_1 - d_2)\, t$$

∴ Tearing strength of the socket across the slot

$$= \left\{\frac{\pi}{4}[(d_1)^2 - (d_2)^2] - (d_1 - d_2)\, t\right\} \sigma_t$$

Equating this to load (P), we have

$$P = \left\{\frac{\pi}{4}[(d_1)^2 - (d_2)^2] - (d_1 - d_2)\, t\right\} \sigma_t$$

From this equation, outside diameter of socket (d_1) may be determined.

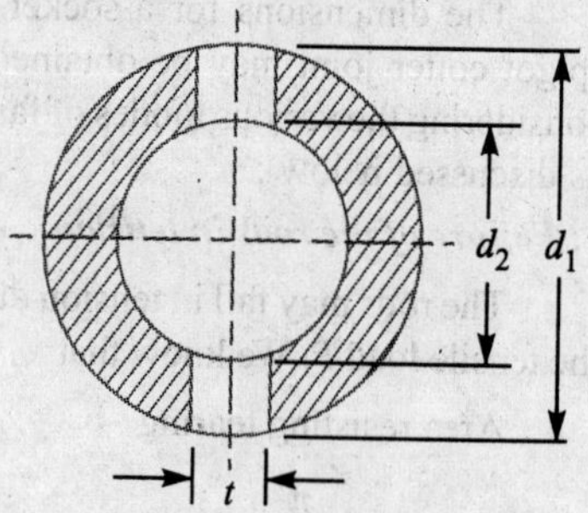

Fig. 12.3

5. *Failure of cotter in shear*

Considering the failure of cotter in shear as shown in Fig. 12.4. Since the cotter is in double shear, therefore shearing area of the cotter

$$= 2\, b \times t$$

and shearing strength of the cotter

$$= 2\, b \times t \times \tau$$

Equating this to load (P), we have

$$P = 2\, b \times t \times \tau$$

From this equation, width of cotter (b) is determined.

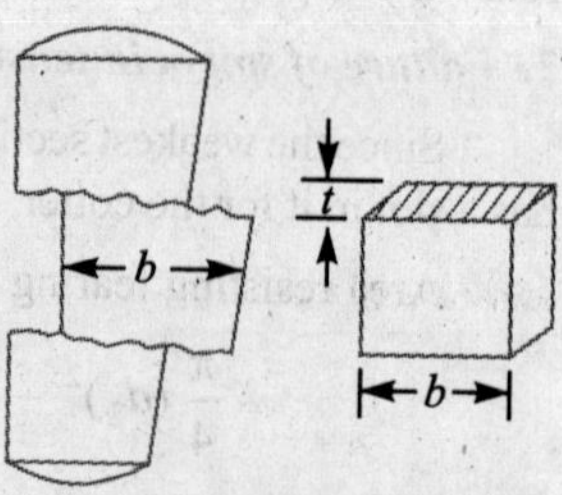

Fig. 12.4

6. *Failure of the socket collar in crushing*

Considering the failure of socket collar in crushing as shown in Fig. 12.5.

We know that area that resists crushing of socket collar

$$= (d_4 - d_2)\, t$$

and crushing strength $= (d_4 - d_2)\, t \times \sigma_c$

Equating this to load (P), we have

$$P = (d_4 - d_2)\, t \times \sigma_c$$

From this equation, the diameter of socket collar (d_4) may be obtained.

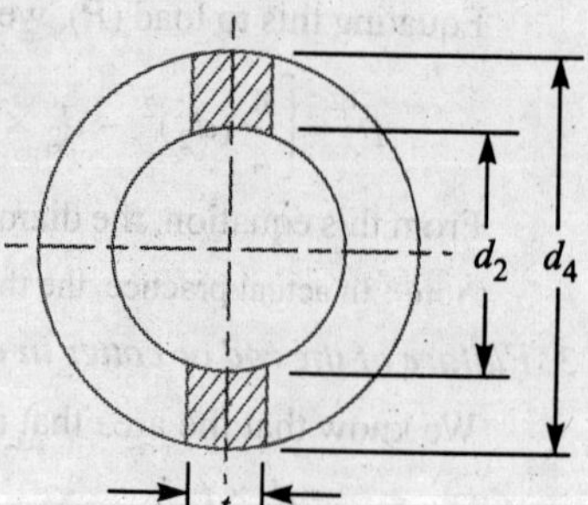

Fig. 12.5

7. *Failure of socket end in shearing*

Since the socket end is in double shear, therefore area that resists shearing of socket collar

$$= 2\,(d_4 - d_2)\, c$$

and shearing strength of socket collar

$$= 2\,(d_4 - d_2)\, c \times \tau$$

Equating this to load (P), we have

$$P = 2\,(d_4 - d_2)\, c \times \tau$$

From this equation, the thickness of socket collar (c) may be obtained.

8. *Failure of rod end in shear*

Since the rod end is in double shear, therefore the area resisting shear of the rod end

$$= 2\,a \times d_2$$

and shear strength of the rod end

$$= 2\,a \times d_2 \times \tau$$

Equating this to load (P), we have

$$P = 2\,a \times d_2 \times \tau$$

From this equation, the distance from the end of the slot to the end of the rod (a) may be obtained.

9. *Failure of spigot collar in crushing*

Considering the failure of the spigot collar in crushing as shown in Fig. 12.6. We know that area that resists crushing of the collar

$$= \frac{\pi}{4}\left[(d_3)^2 - (d_2)^2\right]$$

and crushing strength of the collar

$$= \frac{\pi}{4}\left[(d_3)^2 - (d_2)^2\right]\sigma_c$$

Equating this to load (P), we have

$$P = \frac{\pi}{4}\left[(d_3)^2 - (d_2)^2\right]\sigma_c$$

From this equation, the diameter of the spigot collar (d_3) may be obtained.

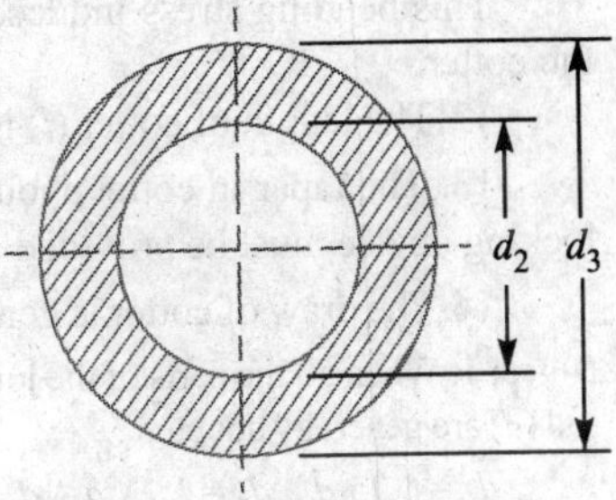

Fig. 12.6

10. *Failure of the spigot collar in shearing*

Considering the failure of the spigot collar in shearing as shown in Fig. 12.7. We know that area that resists shearing of the collar

$$= \pi\,d_2 \times t_1$$

and shearing strength of the collar,

$$= \pi\,d_2 \times t_1 \times \tau$$

Equating this to load (P) we have

$$P = \pi\,d_2 \times t_1 \times \tau$$

From this equation, the thickness of spigot collar (t_1) may be obtained.

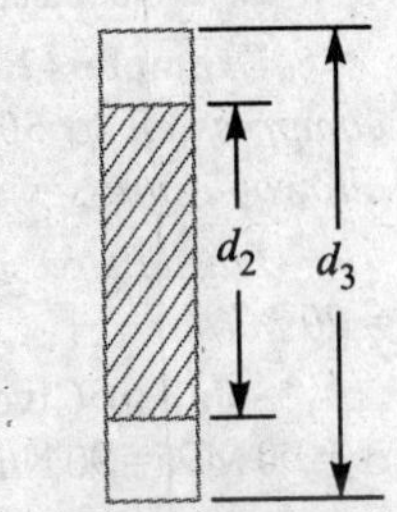

Fig. 12.7

11. *Failure of cotter in bending*

In all the above relations, it is assumed that the load is uniformly distributed over the various cross-sections of the joint. But in actual practice, this does not happen and the cotter is subjected to bending. In order to find out the bending stress induced, it is assumed that the load on the cotter in the rod end is uniformly distributed while in the socket end it varies from zero at the outer diameter (d_4) and maximum at the inner diameter (d_2), as shown in Fig. 12.8.

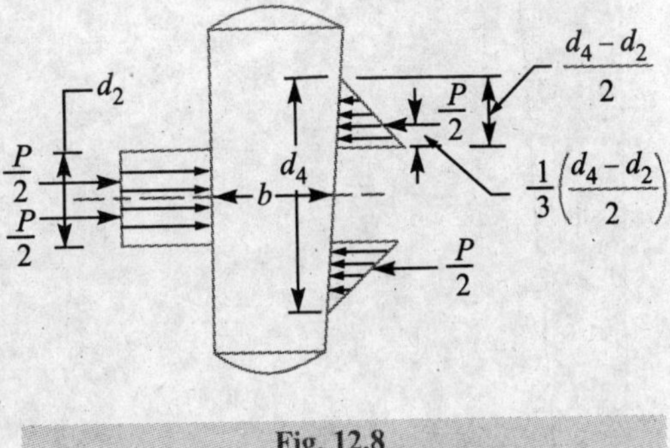

Fig. 12.8

The maximum bending moment occurs at the centre of the cotter and is given by

$$M_{max} = \frac{P}{2}\left(\frac{1}{3} \times \frac{d_4 - d_2}{2} + \frac{d_2}{2}\right) - \frac{P}{2} \times \frac{d_2}{4}$$

$$= \frac{P}{2}\left(\frac{d_4 - d_2}{6} + \frac{d_2}{2} - \frac{d_2}{4}\right) = \frac{P}{2}\left(\frac{d_4 - d_2}{6} + \frac{d_2}{4}\right)$$

We know that section modulus of the cotter,

$$Z = t \times b^2 / 6$$

∴ Bending stress induced in the cotter,

$$\sigma_b = \frac{M_{max}}{Z} = \frac{\frac{P}{2}\left(\frac{d_4 - d_2}{6} + \frac{d_2}{4}\right)}{t \times b^2 / 6} = \frac{P\,(d_4 + 0.5\,d_2)}{2\,t \times b^2}$$

This bending stress induced in the cotter should be less than the allowable bending stress of the cotter.

12. The length of cotter (l) is taken as 4 d.

13. The taper in cotter should not exceed 1 in 24. In case the greater taper is required, then a locking device must be provided.

14. The draw of cotter is generally taken as 2 to 3 mm.

Notes: 1. When all the parts of the joint are made of steel, the following proportions in terms of diameter of the rod (d) are generally adopted :

$d_1 = 1.75\,d$, $d_2 = 1.21\,d$, $d_3 = 1.5\,d$, $d_4 = 2.4\,d$, $a = c = 0.75\,d$, $b = 1.3\,d$, $l = 4\,d$, $t = 0.31\,d$, $t_1 = 0.45\,d$, $e = 1.2\,d$.

Taper of cotter = 1 in 25, and draw of cotter = 2 to 3 mm.

2. If the rod and cotter are made of steel or wrought iron, then $\tau = 0.8\,\sigma_t$ and $\sigma_c = 2\,\sigma_t$ may be taken.

Example 12.1. *Design and draw a cotter joint to support a load varying from 30 kN in compression to 30 kN in tension. The material used is carbon steel for which the following allowable stresses may be used. The load is applied statically.*

Tensile stress = compressive stress = 50 MPa ; shear stress = 35 MPa and crushing stress = 90 MPa.

Solution. Given : $P = 30\text{ kN} = 30 \times 10^3\text{ N}$; $\sigma_t = 50\text{ MPa} = 50\text{ N/mm}^2$; $\tau = 35\text{ MPa} = 35\text{ N/mm}^2$; $\sigma_c = 90\text{ MPa} = 90\text{ N/mm}^2$

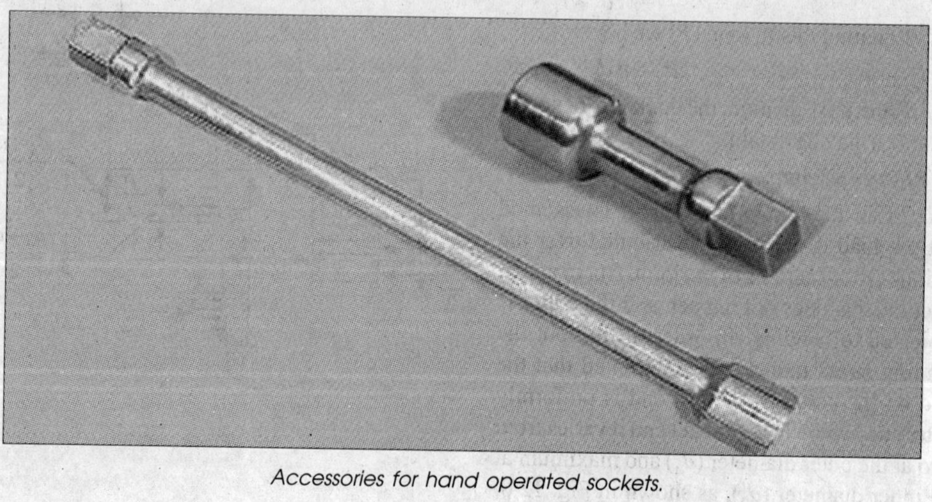

Accessories for hand operated sockets.

The cotter joint is shown in Fig. 12.1. The joint is designed as discussed below :

1. ***Diameter of the rods***

Let d = Diameter of the rods.

Considering the failure of the rod in tension. We know that load (P),

$$30 \times 10^3 = \frac{\pi}{4} \times d^2 \times \sigma_t = \frac{\pi}{4} \times d^2 \times 50 = 39.3\, d^2$$

$$\therefore \quad d^2 = 30 \times 10^3 / 39.3 = 763 \quad \text{or} \quad d = 27.6 \text{ say } 28 \text{ mm} \quad \textbf{Ans.}$$

2. ***Diameter of spigot and thickness of cotter***

Let d_2 = Diameter of spigot or inside diameter of socket, and

t = Thickness of cotter. It may be taken as $d_2 / 4$.

Considering the failure of spigot in tension across the weakest section. We know that load (P),

$$30 \times 10^3 = \left[\frac{\pi}{4}(d_2)^2 - d_2 \times t\right]\sigma_t = \left[\frac{\pi}{4}(d_2)^2 - d_2 \times \frac{d_2}{4}\right] 50 = 26.8\,(d_2)^2$$

$$\therefore \quad (d_2)^2 = 30 \times 10^3 / 26.8 = 1119.4 \quad \text{or} \quad d_2 = 33.4 \text{ say } 34 \text{ mm}$$

and thickness of cotter, $t = \dfrac{d_2}{4} = \dfrac{34}{4} = 8.5$ mm

Let us now check the induced crushing stress. We know that load (P),

$$30 \times 10^3 = d_2 \times t \times \sigma_c = 34 \times 8.5 \times \sigma_c = 289\, \sigma_c$$

$$\therefore \quad \sigma_c = 30 \times 10^3 / 289 = 103.8 \text{ N/mm}^2$$

Since this value of σ_c is more than the given value of $\sigma_c = 90$ N/mm², therefore the dimensions $d_2 = 34$ mm and $t = 8.5$ mm are not safe. Now let us find the values of d_2 and t by substituting the value of $\sigma_c = 90$ N/mm² in the above expression, *i.e.*

$$30 \times 10^3 = d_2 \times \frac{d_2}{4} \times 90 = 22.5\,(d_2)^2$$

$$\therefore \quad (d_2)^2 = 30 \times 10^3 / 22.5 = 1333 \quad \text{or} \quad d_2 = 36.5 \text{ say } 40 \text{ mm} \quad \textbf{Ans.}$$

and $\quad t = d_2 / 4 = 40 / 4 = 10$ mm **Ans.**

3. ***Outside diameter of socket***

Let d_1 = Outside diameter of socket.

Considering the failure of the socket in tension across the slot. We know that load (P),

$$30 \times 10^3 = \left[\frac{\pi}{4}\{(d_1)^2 - (d_2)^2\} - (d_1 - d_2)\, t\right]\sigma_t$$

$$= \left[\frac{\pi}{4}\{(d_1)^2 - (40)^2\} - (d_1 - 40)\, 10\right] 50$$

$$30 \times 10^3 / 50 = 0.7854\,(d_1)^2 - 1256.6 - 10\, d_1 + 400$$

or $(d_1)^2 - 12.7\, d_1 - 1854.6 = 0$

$$\therefore \quad d_1 = \frac{12.7 \pm \sqrt{(12.7)^2 + 4 \times 1854.6}}{2} = \frac{12.7 \pm 87.1}{2}$$

$= 49.9$ say 50 mm **Ans.** ...(Taking +ve sign)

4. ***Width of cotter***

Let b = Width of cotter.

Considering the failure of the cotter in shear. Since the cotter is in double shear, therefore load (P),

$$30 \times 10^3 = 2\,b \times t \times \tau = 2\,b \times 10 \times 35 = 700\,b$$

$$\therefore \quad b = 30 \times 10^3 / 700 = 43 \text{ mm } \textbf{Ans.}$$

5. ***Diameter of socket collar***

Let d_4 = Diameter of socket collar.

Considering the failure of the socket collar and cotter in crushing. We know that load (P),

$$30 \times 10^3 = (d_4 - d_2)\,t \times \sigma_c = (d_4 - 40)10 \times 90 = (d_4 - 40)\,900$$

$$\therefore \quad d_4 - 40 = 30 \times 10^3 / 900 = 33.3 \text{ or } d_4 = 33.3 + 40 = 73.3 \text{ say } 75 \text{ mm } \textbf{Ans.}$$

6. ***Thickness of socket collar***

Let c = Thickness of socket collar.

Considering the failure of the socket end in shearing. Since the socket end is in double shear, therefore load (P),

$$30 \times 10^3 = 2(d_4 - d_2)\,c \times \tau = 2\,(75 - 40\,)\,c \times 35 = 2450\,c$$

$$\therefore \quad c = 30 \times 10^3 / 2450 = 12 \text{ mm } \textbf{Ans.}$$

7. ***Distance from the end of the slot to the end of the rod***

Let a = Distance from the end of slot to the end of the rod.

Considering the failure of the rod end in shear. Since the rod end is in double shear, therefore load (P),

$$30 \times 10^3 = 2\,a \times d_2 \times \tau = 2a \times 40 \times 35 = 2800\,a$$

$$\therefore \quad a = 30 \times 10^3 / 2800 = 10.7 \text{ say } 11 \text{ mm } \textbf{Ans.}$$

8. ***Diameter of spigot collar***

Let d_3 = Diameter of spigot collar.

Considering the failure of spigot collar in crushing. We know that load (P),

$$30 \times 10^3 = \frac{\pi}{4}\left[(d_3)^2 - (d_2)^2\right]\sigma_c = \frac{\pi}{4}\left[(d_3)^2 - (40)^2\right]90$$

or

$$(d_3)^2 - (40)^2 = \frac{30 \times 10^3 \times 4}{90 \times \pi} = 424$$

$$\therefore \quad (d_3)^2 = 424 + (40)^2 = 2024 \text{ or } d_3 = 45 \text{ mm } \textbf{Ans.}$$

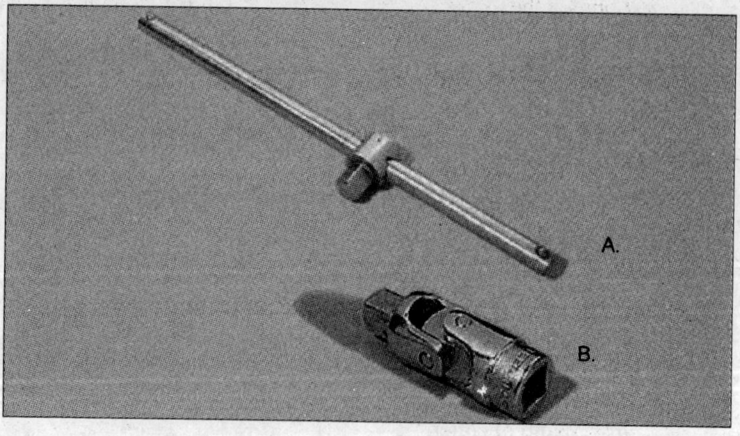

A. T. Handle, B. Universal Joint

9. ***Thickness of spigot collar***

Let t_1 = Thickness of spigot collar.

Considering the failure of spigot collar in shearing. We know that load (P),

$$30 \times 10^3 = \pi\, d_2 \times t_1 \times \tau = \pi \times 40 \times t_1 \times 35 = 4400\, t_1$$

$$\therefore \quad t_1 = 30 \times 10^3 / 4400 = 6.8 \text{ say } 8 \text{ mm } \textbf{Ans.}$$

10. The length of cotter (l) is taken as $4\,d$.

$$\therefore \quad l = 4\,d = 4 \times 28 = 112 \text{ mm } \textbf{Ans.}$$

11. The dimension e is taken as $1.2\,d$.

$$\therefore \quad e = 1.2 \times 28 = 33.6 \text{ say } 34 \text{ mm } \textbf{Ans.}$$

12.5 Sleeve and Cotter Joint

Sometimes, a sleeve and cotter joint as shown in Fig. 12.9, is used to connect two round rods or bars. In this type of joint, a sleeve or muff is used over the two rods and then two cotters (one on each rod end) are inserted in the holes provided for them in the sleeve and rods. The taper of cotter is usually 1 in 24. It may be noted that the taper sides of the two cotters should face each other as shown in Fig. 12.9. The clearance is so adjusted that when the cotters are driven in, the two rods come closer to each other thus making the joint tight.

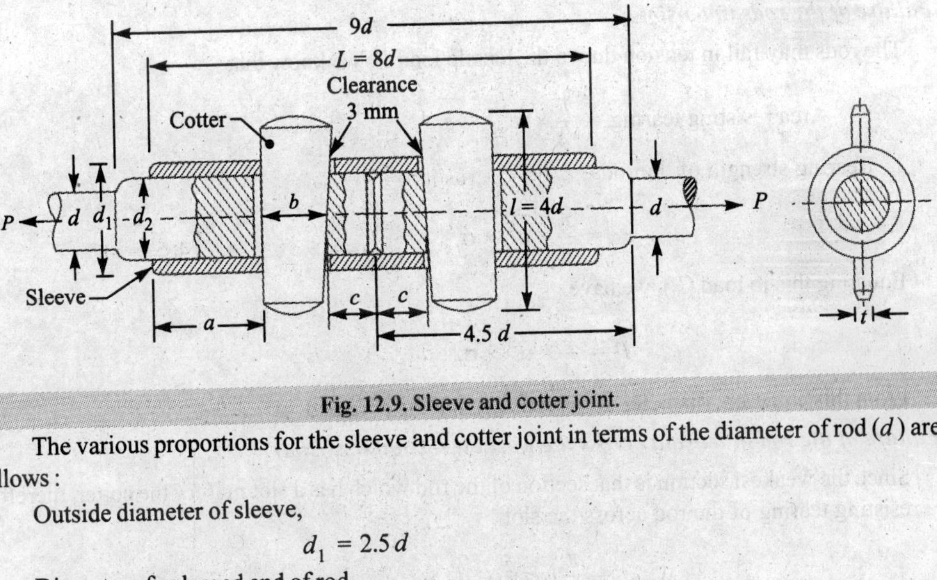

Fig. 12.9. Sleeve and cotter joint.

The various proportions for the sleeve and cotter joint in terms of the diameter of rod (d) are as follows :

Outside diameter of sleeve,

$$d_1 = 2.5\,d$$

Diameter of enlarged end of rod,

$$d_2 = \text{Inside diameter of sleeve} = 1.25\,d$$

Length of sleeve, $L = 8\,d$

Thickness of cotter, $t = d_2/4$ or $0.31\,d$

Width of cotter, $b = 1.25\,d$

Length of cotter, $l = 4\,d$

Distance of the rod end (a) from the beginning to the cotter hole (inside the sleeve end)

= Distance of the rod end (c) from its end to the cotter hole

$= 1.25\,d$

12.6 Design of Sleeve and Cotter Joint

The sleeve and cotter joint is shown in Fig. 12.9.

Let

P = Load carried by the rods,

d = Diameter of the rods,

d_1 = Outside diameter of sleeve,

d_2 = Diameter of the enlarged end of rod,

t = Thickness of cotter,

l = Length of cotter,

b = Width of cotter,

a = Distance of the rod end from the beginning to the cotter hole (inside the sleeve end),

c = Distance of the rod end from its end to the cotter hole,

σ_t, τ and σ_c = Permissible tensile, shear and crushing stresses respectively for the material of the rods and cotter.

The dimensions for a sleeve and cotter joint may be obtained by considering the various modes of failure as discussed below :

1. *Failure of the rods in tension*

The rods may fail in tension due to the tensile load P. We know that

$$\text{Area resisting tearing} = \frac{\pi}{4} \times d^2$$

∴ Tearing strength of the rods

$$= \frac{\pi}{4} \times d^2 \times \sigma_t$$

Equating this to load (P), we have

$$P = \frac{\pi}{4} \times d^2 \times \sigma_t$$

From this equation, diameter of the rods (d) may be obtained.

2. *Failure of the rod in tension across the weakest section (i.e. slot)*

Since the weakest section is that section of the rod which has a slot in it for the cotter, therefore area resisting tearing of the rod across the slot

$$= \frac{\pi}{4}(d_2)^2 - d_2 \times t$$

and tearing strength of the rod across the slot

$$= \left[\frac{\pi}{4}(d_2)^2 - d_2 \times t\right]\sigma_t$$

Equating this to load (P), we have

$$P = \left[\frac{\pi}{4}(d_2)^2 - d_2 \times t\right]\sigma_t$$

From this equation, the diameter of enlarged end of the rod (d_2) may be obtained.

Note: The thickness of cotter is usually taken as $d_2 / 4$.

3. *Failure of the rod or cotter in crushing*

We know that the area that resists crushing of a rod or cotter

$$= d_2 \times t$$

$$\therefore \quad \text{Crushing strength} = d_2 \times t \times \sigma_c$$

Equating this to load (P), we have

$$P = d_2 \times t \times \sigma_c$$

From this equation, the induced crushing stress may be checked.

4. *Failure of sleeve in tension across the slot*

We know that the resisting area of sleeve across the slot

$$= \frac{\pi}{4}\left[(d_1)^2 - (d_2)^2 \right] - (d_1 - d_2)\, t$$

$\therefore$ Tearing strength of the sleeve across the slot

$$= \left[\frac{\pi}{4} [(d_1)^2 - (d_2)^2] - (d_1 - d_2)\, t \right] \sigma_t$$

Equating this to load (P), we have

$$P = \left[\frac{\pi}{4} [(d_1)^2 - (d_2)^2] - (d_1 - d_2)\, t \right] \sigma_t$$

From this equation, the outside diameter of sleeve (d_1) may be obtained.

5. *Failure of cotter in shear*

Since the cotter is in double shear, therefore shearing area of the cotter

$$= 2b \times t$$

and shear strength of the cotter

$$= 2b \times t \times \tau$$

Equating this to load (P), we have

$$P = 2b \times t \times \tau$$

From this equation, width of cotter (b) may be determined.

6. *Failure of rod end in shear*

Since the rod end is in double shear, therefore area resisting shear of the rod end

$$= 2\, a \times d_2$$

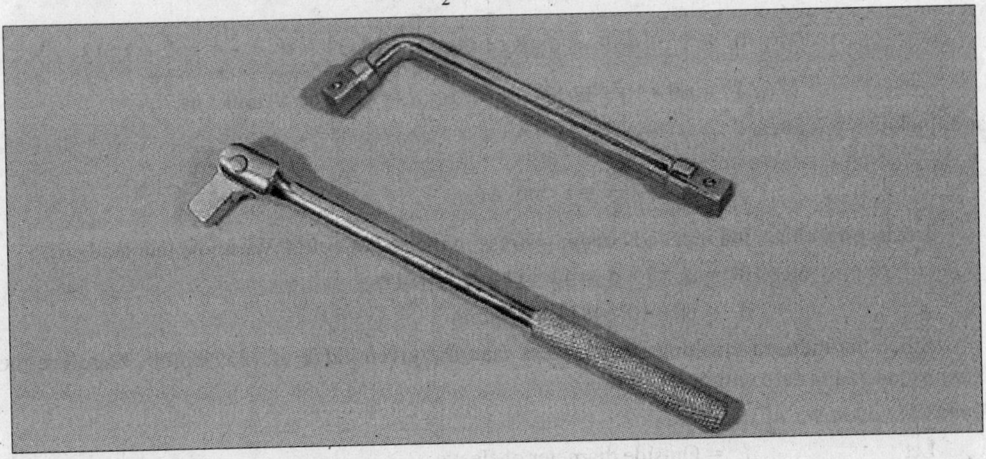

Offset handles.

and shear strength of the rod end

$$= 2\,a \times d_2 \times \tau$$

Equating this to load (P), we have

$$P = 2\,a \times d_2 \times \tau$$

From this equation, distance (a) may be determined.

7. ***Failure of sleeve end in shear***

Since the sleeve end is in double shear, therefore the area resisting shear of the sleeve end

$$= 2\,(d_1 - d_2)\,c$$

and shear strength of the sleeve end

$$= 2\,(d_1 - d_2)\,c \times \tau$$

Equating this to load (P), we have

$$P = 2\,(d_1 - d_2)\,c \times \tau$$

From this equation, distance (c) may be determined.

Example 12.2. *Design a sleeve and cotter joint to resist a tensile load of 60 kN. All parts of the joint are made of the same material with the following allowable stresses :*

σ_t = 60 MPa ; τ = 70 MPa ; and σ_c = 125 MPa.

Solution. Given : $P = 60\text{ kN} = 60 \times 10^3\text{ N}$; $\sigma_t = 60\text{ MPa} = 60\text{ N/mm}^2$; $\tau = 70\text{ MPa} = 70\text{ N/mm}^2$; $\sigma_c = 125\text{ MPa} = 125\text{ N/mm}^2$

1. ***Diameter of the rods***

Let d = Diameter of the rods.

Considering the failure of the rods in tension. We know that load (P),

$$60 \times 10^3 = \frac{\pi}{4} \times d^2 \times \sigma_t = \frac{\pi}{4} \times d^2 \times 60 = 47.13\,d^2$$

$$\therefore \quad d^2 = 60 \times 10^3 / 47.13 = 1273 \quad \text{or} \quad d = 35.7 \text{ say } 36 \text{ mm } \textbf{Ans.}$$

2. ***Diameter of enlarged end of rod and thickness of cotter***

Let d_2 = Diameter of enlarged end of rod, and

t = Thickness of cotter. It may be taken as $d_2 / 4$.

Considering the failure of the rod in tension across the weakest section (*i.e.* slot). We know that load (P),

$$60 \times 10^3 = \left[\frac{\pi}{4}(d_2)^2 - d_2 \times t\right]\sigma_t = \left[\frac{\pi}{4}(d_2)^2 - d_2 \times \frac{d_2}{4}\right]60 = 32.13\,(d_2)^2$$

$$\therefore \quad (d_2)^2 = 60 \times 10^3 / 32.13 = 1867 \quad \text{or} \quad d_2 = 43.2 \text{ say } 44 \text{ mm } \textbf{Ans.}$$

and thickness of cotter,

$$t = \frac{d_2}{4} = \frac{44}{4} = 11 \text{ mm } \textbf{Ans.}$$

Let us now check the induced crushing stress in the rod or cotter. We know that load (P),

$$60 \times 10^3 = d_2 \times t \times \sigma_c = 44 \times 11 \times \sigma_c = 484\,\sigma_c$$

$$\therefore \quad \sigma_c = 60 \times 10^3 / 484 = 124 \text{ N/mm}^2$$

Since the induced crushing stress is less than the given value of 125 N/mm^2, therefore the dimensions d_2 and t are within safe limits.

3. ***Outside diameter of sleeve***

Let d_1 = Outside diameter of sleeve.

Considering the failure of sleeve in tension across the slot. We know that load (P)

$$60 \times 10^3 = \left[\frac{\pi}{4}[(d_1)^2 - (d_2)^2] - (d_1 - d_2)\,t\right]\sigma_t$$

$$= \left[\frac{\pi}{4}[(d_1)^2 - (44)^2] - (d_1 - 44)\,11\right]60$$

$\therefore \quad 60 \times 10^3/60 = 0.7854\,(d_1)^2 - 1520.7 - 11\,d_1 + 484$

or $\quad (d_1)^2 - 14\,d_1 - 2593 = 0$

$$\therefore \quad d_1 = \frac{14 \pm \sqrt{(14)^2 + 4 \times 2593}}{2} = \frac{14 \pm 102.8}{2}$$

$= 58.4$ say 60 mm **Ans.** ...(Taking +ve sign)

4. ***Width of cotter***

Let b = Width of cotter.

Considering the failure of cotter in shear. Since the cotter is in double shear, therefore load (P),

$$60 \times 10^3 = 2\,b \times t \times \tau = 2 \times b \times 11 \times 70 = 1540\,b$$

$\therefore \quad b = 60 \times 10^3/1540 = 38.96$ say 40 mm **Ans.**

5. ***Distance of the rod from the beginning to the cotter hole (inside the sleeve end)***

Let a = Required distance.

Considering the failure of the rod end in shear. Since the rod end is in double shear, therefore load (P),

$$60 \times 10^3 = 2\,a \times d_2 \times \tau = 2\,a \times 44 \times 70 = 6160\,a$$

$\therefore \quad a = 60 \times 10^3/6160 = 9.74$ say 10 mm **Ans.**

6. ***Distance of the rod end from its end to the cotter hole***

Let c = Required distance.

Considering the failure of the sleeve end in shear. Since the sleeve end is in double shear, therefore load (P),

$$60 \times 10^3 = 2\,(d_1 - d_2)\,c \times \tau = 2\,(60 - 44)\,c \times 70 = 2240\,c$$

$\therefore \quad c = 60 \times 10^3/2240 = 26.78$ say 28 mm **Ans.**

12.7 Gib and Cotter Joint

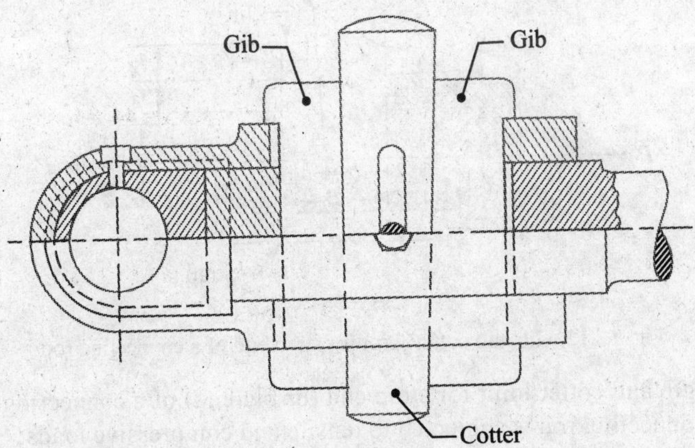

Fig. 12.10. Gib and cotter joint for strap end of a connecting rod.

A *gib and cotter joint is usually used in strap end (or big end) of a connecting rod as shown in Fig. 12.10. In such cases, when the cotter alone (*i.e.* without gib) is driven, the friction between its ends and the inside of the slots in the strap tends to cause the sides of the strap to spring open (or spread) outwards as shown dotted in Fig. 12.11 (*a*). In order to prevent this, gibs as shown in Fig. 12.11 (*b*) and (*c*), are used which hold together the ends of the strap. Moreover, gibs provide a larger bearing surface for the cotter to slide on, due to the increased holding power. Thus, the tendency of cotter to slacken back owing to friction is considerably decreased. The jib, also, enables parallel holes to be used.

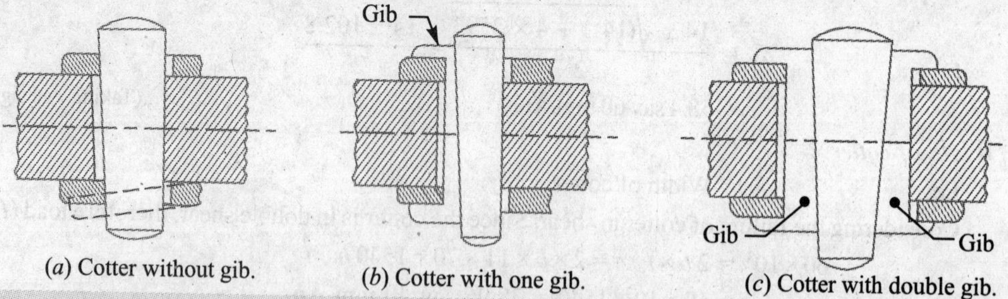

Fig. 12.11. Gib and cotter Joints.

Notes : 1. When one gib is used, the cotter with one side tapered is provided and the gib is always on the outside as shown in Fig. 12.11 (*b*).

2. When two jibs are used, the cotter with both sides tapered is provided.

3. Sometimes to prevent loosening of cotter, a small set screw is used through the rod jamming against the cotter.

12.8 Design of a Gib and Cotter Joint for Strap End of a Connecting Rod

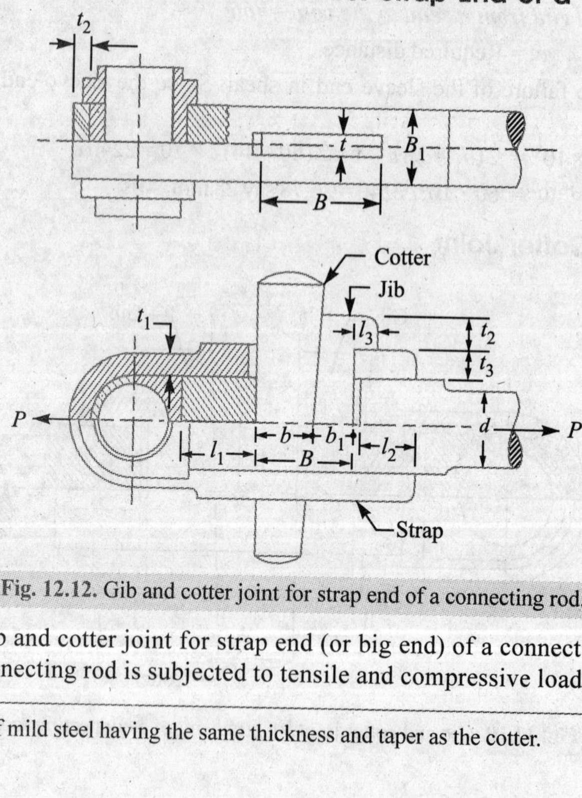

Fig. 12.12. Gib and cotter joint for strap end of a connecting rod.

Consider a gib and cotter joint for strap end (or big end) of a connecting rod as shown in Fig. 12.12. The connecting rod is subjected to tensile and compressive loads.

---

* A gib is a piece of mild steel having the same thickness and taper as the cotter.

Let P = Maximum thrust or pull in the connecting rod,

d = Diameter of the adjacent end of the round part of the rod,

B_1 = Width of the strap,

B = Total width of gib and cotter,

t = Thickness of cotter,

t_1 = Thickness of the strap at the thinnest part,

σ_t = Permissible tensile stress for the material of the strap, and

τ = Permissible shear stress for the material of the cotter and gib.

The width of strap (B_1) is generally taken equal to the diameter of the adjacent end of the round part of the rod (d). The other dimensions may be fixed as follows :

Thickness of cotter,

$$t = \frac{\text{Width of strap}}{4} = \frac{B_1}{4}$$

Thickness of gib = Thickness of cotter (t)

Height (t_2) and length of gib head (l_3)

= Thickness of cotter (t)

In designing the gib and cotter joint for strap end of a connecting rod, the following modes of failure are considered.

1. ***Failure of the strap in tension***

Assuming that no hole is provided for lubrication, the area that resists the failure of the strap due to tearing $= 2\, B_1 \times t_1$

∴ Tearing strength of the strap

$$= 2\, B_1 \times t_1 \times \sigma_t$$

Equating this to the load (P), we get

$$P = 2\, B_1 \times t_1 \times \sigma_t$$

From this equation, the thickness of the strap at the thinnest part (t_1) may be obtained. When an oil hole is provided in the strap, then its weakening effect should be considered.

The thickness of the strap at the cotter (t_3) is increased such that the area of cross-section of the strap at the cotter hole is not less than the area of the strap at the thinnest part. In other words

$$2\, t_3\, (B_1 - t) = 2\, t_1 \times B_1$$

From this expression, the value of t_3 may be obtained.

(a) Hand operated sqaure drive sockets (b) Machine operated sockets.

Note : This picture is given as additional information and is not a direct example of the current chapter.

2. *Failure of the gib and cotter in shearing*

Since the gib and cotter are in double shear, therefore area resisting failure

$$= 2\,B \times t$$

and resisting strength $= 2\,B \times t \times \tau$

Equating this to the load (P), we get

$$P = 2\,B \times t \times \tau$$

From this equation, the total width of gib and cotter (B) may be obtained. In the joint, as shown in Fig. 12.12, one gib is used, the proportions of which are

Width of gib, $b_1 = 0.55\,B$; and width of cotter, $b = 0.45\,B$

The other dimensions may be fixed as follows :

Thickness of the strap at the crown,

$$t_4 = 1.15\,t_1 \text{ to } 1.5\,t_1$$

$$l_1 = 2\,t_1;\ \text{and } l_2 = 2.5\,t_1$$

Example 12.3. *The big end of a connecting rod, as shown in Fig. 12.12, is subjected to a maximum load of 50 kN. The diameter of the circular part of the rod adjacent to the strap end is 75 mm. Design the joint, assuming permissible tensile stress for the material of the strap as 25 MPa and permissible shear stress for the material of cotter and gib as 20 MPa.*

Solution. Given : $P = 50$ kN $= 50 \times 10^3$ N ; $d = 75$ mm ; $\sigma_t = 25$ MPa $= 25$ N/mm^2 ; $\tau = 20$ MPa $= 20$ N/mm^2

1. *Width of the strap*

Let B_1 = Width of the strap.

The width of the strap is generally made equal to the diameter of the adjacent end of the round part of the rod (d).

$\therefore$ $B_1 = d = 75$ mm **Ans.**

Other dimensions are fixed as follows :

Thickness of the cotter

$$t = \frac{B_1}{4} = \frac{75}{4} = 18.75 \text{ say } 20 \text{ mm} \textbf{ Ans.}$$

Thickness of gib = Thickness of cotter = 20 mm **Ans.**

Height (t_2) and length of gib head (l_3)

= Thickness of cotter = 20 mm **Ans.**

2. *Thickness of the strap at the thinnest part*

Let t_1 = Thickness of the strap at the thinnest part.

Considering the failure of the strap in tension. We know that load (P),

$$50 \times 10^3 = 2\,B_1 \times t_1 \times \sigma_t = 2 \times 75 \times t_1 \times 25 = 3750\,t_1$$

$\therefore$ $t_1 = 50 \times 10^3 / 3750 = 13.3$ say 15 mm **Ans.**

3. *Thickness of the strap at the cotter*

Let t_3 = Thickness of the strap at the cotter.

The thickness of the strap at the cotter is increased such that the area of the cross-section of the strap at the cotter hole is not less than the area of the strap at the thinnest part. In other words,

$$2\,t_3\,(B_1 - t) = 2\,t_1 \times B_1$$

$$2\,t_3\,(75 - 20) = 2 \times 15 \times 75 \quad \text{or} \quad 110\,t_3 = 2250$$

$\therefore$ $t_3 = 2250 / 110 = 20.45$ say 21 mm **Ans.**

4. ***Total width of gib and cotter***

Let B = Total width of gib and cotter.

Considering the failure of gib and cotter in double shear. We know that load (P),

$$50 \times 10^3 = 2\,B \times t \times \tau = 2\,B \times 20 \times 20 = 800\,B$$

$$\therefore \quad B = 50 \times 10^3 / 800 = 62.5 \text{ say } 65 \text{ mm } \textbf{Ans.}$$

Since one gib is used, therefore width of gib,

$$b_1 = 0.55\,B = 0.55 \times 65 = 35.75 \text{ say } 36 \text{ mm } \textbf{Ans.}$$

and width of cotter, $b = 0.45\,B = 0.45 \times 65 = 29.25$ say 30 mm **Ans.**

The other dimensions are fixed as follows :

$$t_4 = 1.25\,t_1 = 1.25 \times 15 = 18.75 \text{ say } 20 \text{ mm } \textbf{Ans.}$$

$$l_1 = 2\,t_1 = 2 \times 15 = 30 \text{ mm } \textbf{Ans.}$$

and $l_2 = 2.5\,t_1 = 2.5 \times 15 = 37.5$ say 40 mm **Ans.**

12.9 Design of Gib and Cotter Joint for Square Rods

Consider a gib and cotter joint for square rods as shown in Fig. 12.13. The rods may be subjected to a tensile or compressive load. All components of the joint are assumed to be of the same material.

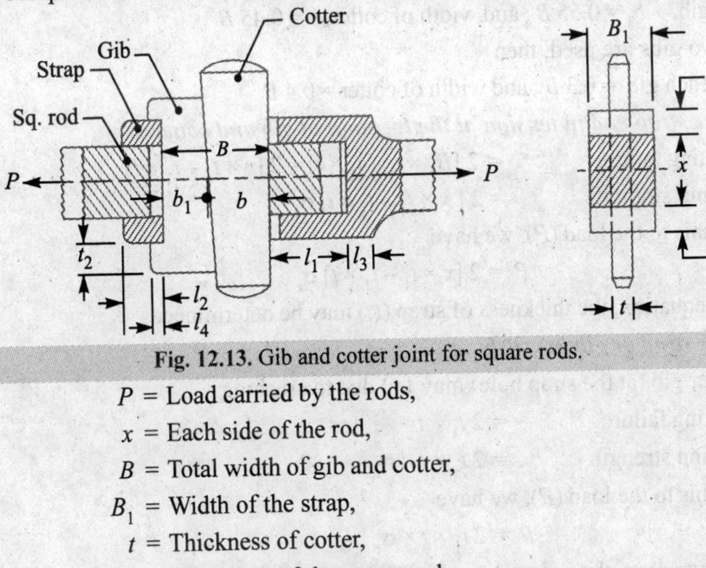

Fig. 12.13. Gib and cotter joint for square rods.

Let P = Load carried by the rods,

x = Each side of the rod,

B = Total width of gib and cotter,

B_1 = Width of the strap,

t = Thickness of cotter,

t_1 = Thickness of the strap, and

σ_t, τ and σ_c = Permissible tensile, shear and crushing stresses.

In designing a gib and cotter joint, the following modes of failure are considered.

1. ***Failure of the rod in tension***

The rod may fail in tension due to the tensile load P. We know that

Area resisting tearing $= x \times x = x^2$

$\therefore$ Tearing strength of the rod

$$= x^2 \times \sigma_t$$

Equating this to the load (P), we have

$$P = x^2 \times \sigma_t$$

From this equation, the side of the square rod (x) may be determined. The other dimensions are fixed as under :

Width of strap, B_1 = Side of the square rod = x

Thickness of cotter, $t = \frac{1}{4}$ width of strap $= \frac{B_1}{4}$

Thickness of gib = Thickness of cotter (t)

Height (t_2) and length of gib head (l_4)

= Thickness of cotter (t)

2. ***Failure of the gib and cotter in shearing***

Since the gib and cotter are in double shear, therefore,

Area resisting failure $= 2\,B \times t$

and resisting strength $= 2\,B \times t \times \tau$

Equating this to the load (P), we have

$$P = 2B \times t \times \tau$$

From this equation, the width of gib and cotter (B) may be obtained. In the joint, as shown in Fig. 12.13, one gib is used, the proportions of which are

Width of gib, $b_1 = 0.55\,B$; and width of cotter, $b = 0.45\,B$

In case two gibs are used, then

Width of each gib $= 0.3\,B$; and width of cotter $= 0.4\,B$

3. ***Failure of the strap end in tension at the location of gib and cotter***

Area resisting failure $= 2\,[B_1 \times t_1 - t_1 \times t] = 2\,[x \times t_1 - t_1 \times t]$ ... ($\because B_1 = x$)

$\therefore$ Resisting strength $= 2\,[x \times t_1 - t_1 \times t]\,\sigma_t$

Equating this to the load (P), we have

$$P = 2\,[x \times t_1 - t_1 \times t]\,\sigma_t$$

From this equation, the thickness of strap (t_1) may be determined.

4. ***Failure of the strap or gib in crushing***

The strap or gib (at the strap hole) may fail due to crushing.

Area resisting failure $= 2\,t_1 \times t$

$\therefore$ Resisting strength $= 2\,t_1 \times t \times \sigma_c$

Equating this to the load (P), we have

$$P = 2\,t_1 \times t \times \sigma_c$$

From this equation, the induced crushing stress may be checked.

5. ***Failure of the rod end in shearing***

Since the rod is in double shear, therefore

Area resisting failure $= 2\,l_1 \times x$

$\therefore$ Resisting strength $= 2\,l_1 \times x \times \tau$

Equating this to the load (P), we have

$$P = 2\,l_1 \times x \times \tau$$

From this equation, the dimension l_1 may be determined.

6. ***Failure of the strap end in shearing***

Since the length of rod (l_2) is in double shearing, therefore

Area resisting failure $= 2 \times 2\,l_2 \times t_1$

∴ Resisting strength $= 2 \times 2\, l_2 \times t_1 \times \tau$

Equating this to the load (P), we have

$$P = 2 \times 2\, l_2 \times t_1 \times \tau$$

From this equation, the length of rod (l_2) may be determined. The length l_3 of the strap end is proportioned as $\frac{2}{3}$ rd of side of the rod. The clearance is usually kept 3 mm. The length of cotter is generally taken as 4 times the side of the rod.

Example 12.4. *Design a gib and cottor joint as shown in Fig. 12.13, to carry a maximum load of 35 kN. Assuming that the gib, cotter and rod are of same material and have the following allowable stresses :*

$$\sigma_t = 20 \text{ MPa} \; ; \; \tau = 15 \text{ MPa} \; ; \text{ and } \sigma_c = 50 \text{ MPa}$$

Solution. Given : $P = 35$ kN $= 35\,000$ N ; $\sigma_t = 20$ MPa $= 20$ N/mm^2 ; $\tau = 15$ MPa $= 15$ N/mm^2 ; $\sigma_c = 50$ MPa $= 50$ N/mm^2

1. ***Side of the square rod***

Let x = Each side of the square rod.

Considering the failure of the rod in tension. We know that load (P),

$$35\,000 = x^2 \times \sigma_t = x^2 \times 20 = 20\, x^2$$

$$\therefore \quad x^2 = 35\,000 / 20 = 1750 \text{ or } x = 41.8 \text{ say } 42 \text{ mm } \textbf{Ans.}$$

Other dimensions are fixed as follows :

Width of strap, $B_1 = x = 42$ mm **Ans.**

Thickness of cotter, $t = \dfrac{B_1}{4} = \dfrac{42}{4} = 10.5$ say 12 mm **Ans.**

Thickness of gib = Thickness of cotter = 12 mm **Ans.**

Height (t_2) and length of gib head (l_4)

= Thickness of cotter = 12 mm **Ans.**

2. ***Width of gib and cotter***

Let B = Width of gib and cotter.

Considering the failure of the gib and cotter in double shear. We know that load (P),

$$35\,000 = 2\, B \times t \times \tau = 2\, B \times 12 \times 15 = 360\, B$$

$$\therefore \quad B = 35\,000 / 360 = 97.2 \text{ say } 100 \text{ mm } \textbf{Ans.}$$

Since one gib is used, therefore

Width of gib, $b_1 = 0.55\, B = 0.55 \times 100 = 55$ mm **Ans.**

and width of cotter, $b = 0.45\, B = 0.45 \times 100 = 45$ mm **Ans.**

3. ***Thickness of strap***

Let t_1 = Thickness of strap.

Considering the failure of the strap end in tension at the location of the gib and cotter. We know that load (P),

$$35\,000 = 2\,(x \times t_1 - t_1 \times t)\, \sigma_t = 2\,(42 \times t_1 - t_1 \times 12)\, 20 = 1200\, t_1$$

$$\therefore \quad t_1 = 35\,000 / 1200 = 29.1 \text{ say } 30 \text{ mm } \textbf{Ans.}$$

Now the induced crushing stress may be checked by considering the failure of the strap or gib in crushing. We know that load (P),

$$35\,000 = 2\, t_1 \times t \times \sigma_c = 2 \times 30 \times 12 \times \sigma_c = 720\, \sigma_c$$

$$\therefore \quad \sigma_c = 35\,000 / 720 = 48.6 \text{ N/mm}^2$$

Since the induced crushing stress is less than the given crushing stress, therefore the joint is safe.

4. ***Length (l_1) of the rod***

Considering the failure of the rod end in shearing. Since the rod is in double shear, therefore load (*P*),

$$35\ 000 = 2\,l_1 \times x \times \tau = 2\,l_1 \times 42 \times 15 = 1260\,l_1$$

$$\therefore \quad l_1 = 35\ 000 / 1260 = 27.7 \text{ say } 28 \text{ mm } \textbf{Ans.}$$

5. ***Length (l_2) of the rod***

Considering the failure of the strap end in shearing. Since the length of the rod (l_2) is in double shear, therefore load (*P*),

$$35\ 000 = 2 \times 2\,l_2 \times t_1 \times \tau = 2 \times 2\,l_2 \times 30 \times 15 = 1800\,l_2$$

$$\therefore \quad l_2 = 35\ 000 / 1800 = 19.4 \text{ say } 20 \text{ mm } \textbf{Ans.}$$

Length (l_3) of the strap end

$$= \frac{2}{3} \times x = \frac{2}{3} \times 42 = 28 \text{ mm } \textbf{Ans.}$$

and length of cotter $= 4x = 4 \times 42 = 168$ mm **Ans.**

12.10 Design of Cotter Joint to Connect Piston Rod and Crosshead

The cotter joint to connect piston rod and crosshead is shown in Fig. 12.14. In such a type of joint, the piston rod is tapered in order to resist the thrust instead of being provided with a collar for the purpose. The taper may be from 1 in 24 to 1 in 12.

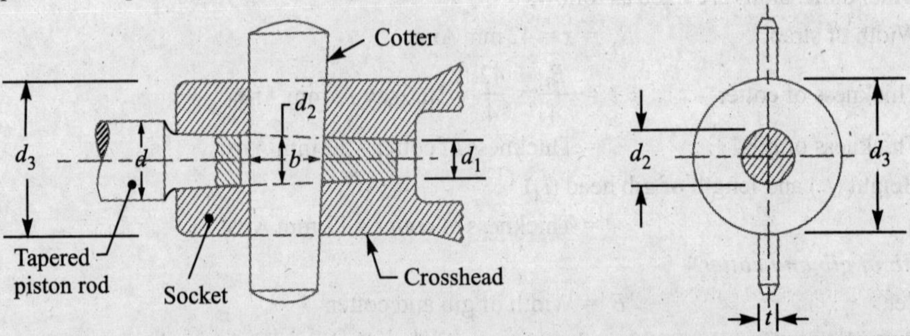

Fig. 12.14. Cotter joint to connect piston rod and crosshead.

Let d = Diameter of parallel part of the piston rod,

d_1 = Diameter at tapered end of the piston,

d_2 = Diameter of piston rod at the cotter,

d_3 = Diameter of socket through the cotter hole,

b = Width of cotter at the centre,

t = Thickness of cotter,

σ_t, τ and σ_c = Permissible stresses in tension, shear and crushing respectively.

We know that maximum load on the piston,

$$P = \frac{\pi}{4} \times D^2 \times p$$

where D = Diameter of the piston, and

p = Effective steam pressure on the piston.

Let us now consider the various failures of the joint as discussed below :

1. ***Failure of piston rod in tension at cotter***

The piston rod may fail in tension at cotter due to the maximum load on the piston. We know that area resisting tearing at the cotter

$$= \frac{\pi}{4}(d_2)^2 - d_2 \times t$$

∴ Tearing strength of the piston rod at the cotter

$$= \left[\frac{\pi}{4}(d_2)^2 - d_2 \times t\right]\sigma_t$$

Equating this to maximum load (P), we have

$$P = \left[\frac{\pi}{4}(d_2)^2 - d_2 \times t\right]\sigma_t$$

From this equation, the diameter of piston rod at the cotter (d_2) may be determined.

Note: The thickness of cotter (t) is taken as 0.3 d_2.

2. ***Failure of cotter in shear***

Since the cotter is in double shear, therefore shearing area of the cotter

$$= 2\,b \times t$$

and shearing strength of the cotter

$$= 2\,b \times t \times \tau$$

Equating this to maximum load (P), we have

$$P = 2\,b \times t \times \tau$$

From this equation, width of cotter (b) is obtained.

3. ***Failure of the socket in tension at cotter***

We know that area that resists tearing of socket at cotter

$$= \frac{\pi}{4}\left[(d_3)^2 - (d_2)^2\right] - (d_3 - d_2)\,t$$

and tearing strength of socket at cotter

$$= \left[\frac{\pi}{4}\left\{(d_3)^2 - (d_2)^2\right\} - (d_3 - d_2)\,t\right]\sigma_t$$

Equating this to maximum load (P), we have

$$P = \left[\frac{\pi}{4}\left\{(d_3)^2 - (d_2)^2\right\} - (d_3 - d_2)\,t\right]\sigma_t$$

From this equation, diameter of socket (d_3) is obtained.

4. ***Failure of socket in crushing***

We know that area that resists crushing of socket

$$= (d_3 - d_2)\,t$$

and crushing strength of socket

$$= (d_3 - d_2)\,t \times \sigma_c$$

Equating this to maximum load (P), we have

$$P = (d_3 - d_2)\,t \times \sigma_c$$

From this equation, the induced crushing stress in the socket may be checked.

The length of the tapered portion of the piston rod (L) is taken as 2.2 d_2. The diameter of the parallel part of the piston rod (d) and diameter of the piston rod at the tapered end (d_1) may be obtained as follows :

$$d = d_2 + \frac{L}{2} \times \text{taper} \text{ ; and } d_1 = d_2 - \frac{L}{2} \times \text{taper}$$

Note: The taper on the piston rod is usually taken as 1 in 20.

Example 12.5. *Design a cotter joint to connect piston rod to the crosshead of a double acting steam engine. The diameter of the cylinder is 300 mm and the steam pressure is 1 N/mm². The allowable stresses for the material of cotter and piston rod are as follows :*

$$\sigma_t = 50 \text{ MPa} \text{ ; } \tau = 40 \text{ MPa} \text{ ; and } \sigma_c = 84 \text{ MPa}$$

Solution. Given : $D = 300$ mm ; $p = 1$ N/mm² ; $\sigma_t = 50$ MPa $= 50$ N/mm² ; $\tau = 40$ MPa $= 40$ N/mm² ; $\sigma_c = 84$ MPa $= 84$ N/mm²

We know that maximum load on the piston rod,

$$P = \frac{\pi}{4} \times D^2 \times p = \frac{\pi}{4} (300)^2\ 1 = 70\ 695 \text{ N}$$

The various dimensions for the cotter joint are obtained by considering the different modes of failure as discussed below :

1. *Diameter of piston rod at cotter*

Let d_2 = Diameter of piston rod at cotter, and

t = Thickness of cotter. It may be taken as $0.3\ d_2$.

Considering the failure of piston rod in tension at cotter. We know that load (P),

$$70\ 695 = \left[\frac{\pi}{4} (d_2)^2 - d_2 \times t\right] \sigma_t = \left[\frac{\pi}{4} (d_2)^2 - 0.3\ (d_2)^2\right] 50 = 24.27\ (d_2)^2$$

$\therefore$ $(d_2)^2 = 70\ 695 / 24.27 = 2913$ or $d_2 = 53.97$ say 55 mm **Ans.**

and $t = 0.3\ d_2 = 0.3 \times 55 = 16.5$ mm **Ans.**

2. *Width of cotter*

Let b = Width of cotter.

Considering the failure of cotter in shear. Since the cotter is in double shear, therefore load (P),

$$70\ 695 = 2\ b \times t \times \tau = 2\ b \times 16.5 \times 40 = 1320\ b$$

$\therefore$ $b = 70\ 695 / 1320 = 53.5$ say 54 mm **Ans.**

3. *Diameter of socket*

Let d_3 = Diameter of socket.

Considering the failure of socket in tension at cotter. We know that load (P),

$$70\ 695 = \left\{\frac{\pi}{4}\left[(d_3)^2 - (d_2)^2\right] - (d_3 - d_2)\ t\right\} \sigma_t$$

$$= \left\{\frac{\pi}{4}\left[(d_3)^2 - (55)^2\right] - (d_3 - 55)\ 16.5\right\} 50$$

$$= 39.27\ (d_3)^2 - 118\ 792 - 825\ d_3 + 45\ 375$$

or $(d_3)^2 - 21\ d_3 - 3670 = 0$

$$\therefore \quad d_3 = \frac{21 \pm \sqrt{(21)^2 + 4 \times 3670}}{2} = \frac{21 \pm 123}{2} = 72 \text{ mm} \quad \text{...(Taking + ve sign)}$$

Let us now check the induced crushing stress in the socket. We know that load (P),

$$70\ 695 = (d_3 - d_2)\ t \times \sigma_c = (72 - 55)\ 16.5 \times \sigma_c = 280.5\ \sigma_c$$

$\therefore$ $\sigma_c = 70\ 695 / 280.5 = 252$ N/mm²

Since the induced crushing is greater than the permissible value of 84 N/mm², therefore let us find the value of d_3 by substituting σ_c = 84 N/mm^2 in the above expression, *i.e.*

$$70\ 695 = (d_3 - 55)\ 16.5 \times 84 = (d_3 - 55)\ 1386$$

$$\therefore \quad d_3 - 55 = 70\ 695 / 1386 = 51$$

or $$d_3 = 55 + 51 = 106 \text{ mm } \textbf{Ans.}$$

We know the tapered length of the piston rod,

$$L = 2.2\, d_2 = 2.2 \times 55 = 121 \text{ mm } \textbf{Ans.}$$

Assuming the taper of the piston rod as 1 in 20, therefore the diameter of the parallel part of the piston rod,

$$d = d_2 + \frac{L}{2} \times \frac{1}{20} = 55 + \frac{121}{2} \times \frac{1}{20} = 58 \text{ mm } \textbf{Ans.}$$

and diameter of the piston rod at the tapered end,

$$d_1 = d_2 - \frac{L}{2} \times \frac{1}{20} = 55 - \frac{121}{2} \times \frac{1}{20} = 52 \text{ mm } \textbf{Ans.}$$

12.11 Design of Cotter Foundation Bolt

The cotter foundation bolt is mostly used in conjunction with foundation and holding down bolts to fasten heavy machinery to foundations. It is generally used where an ordinary bolt or stud cannot be conveniently used. Fig. 12.15 shows the two views of the application of such a cotter foundation bolt. In this case, the bolt is dropped down from above and the cotter is driven in from the side. Now this assembly is tightened by screwing down the nut. It may be noted that two base plates (one under the nut and the other under the cotter) are used to provide more bearing area in order to take up the tightening load on the bolt as well as to distribute the same uniformly over the large surface.

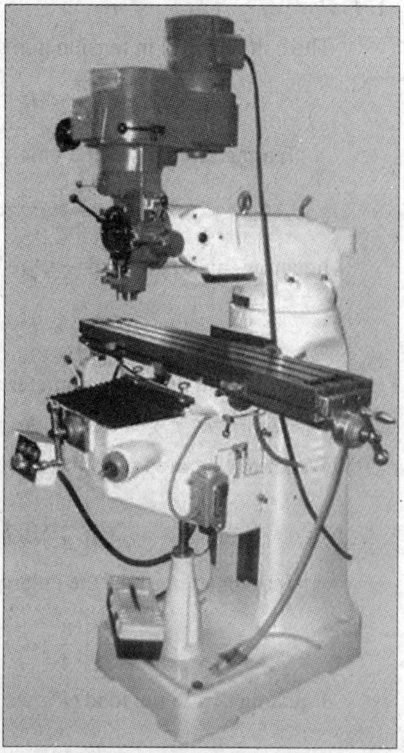

Variable speed Knee-type milling machine.

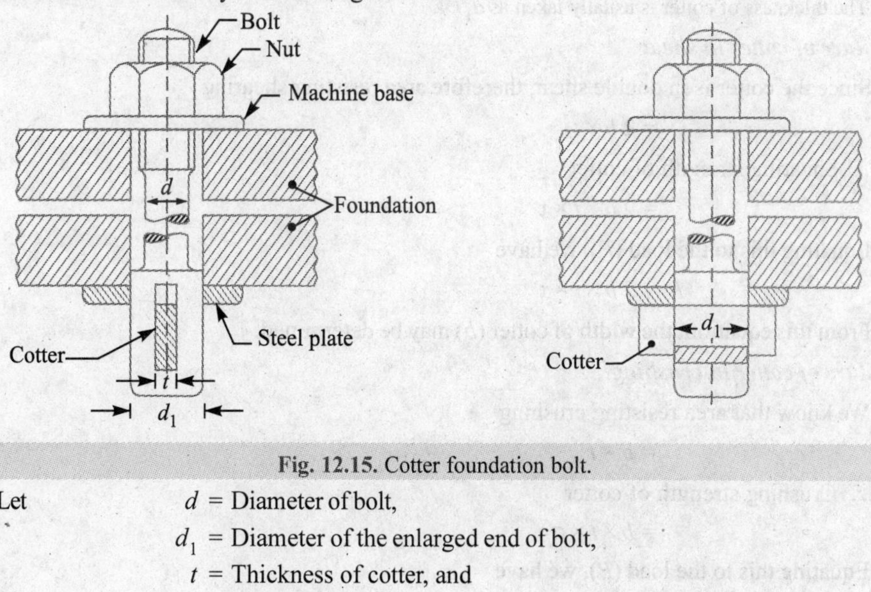

Fig. 12.15. Cotter foundation bolt.

Let d = Diameter of bolt,

d_1 = Diameter of the enlarged end of bolt,

t = Thickness of cotter, and

b = Width of cotter.

The various modes of failure of the cotter foundation bolt are discussed as below:

1. *Failure of bolt in tension*

The bolt may fail in tension due to the load (*P*). We know that area resisting tearing

$$= \frac{\pi}{4} \times d^2$$

∴ Tearing strength of the bolt

$$= \frac{\pi}{4} \times d^2 \times \sigma_t$$

Equating this to the load (*P*), we have

$$P = \frac{\pi}{4} \times d^2 \times \sigma_t$$

From this equation, the diameter of bolt (*d*) may be determined.

2. *Failure of the enlarged end of the bolt in tension at the cotter*

We know that area resisting tearing

$$= \left[\frac{\pi}{4}(d_1)^2 - d_1 \times t\right]$$

∴ Tearing strength of the enlarged end of the bolt

$$= \left[\frac{\pi}{4}(d_1)^2 - d_1 \times t\right]\sigma_t$$

Equating this to the load (*P*), we have

$$P = \left[\frac{\pi}{4}(d_1)^2 - d_1 \times t\right]\sigma_t$$

From this equation, the diameter of the enlarged end of the bolt (d_1) may be determined.

Note: The thickness of cotter is usually taken as $d_1 / 4$.

3. *Failure of cotter in shear*

Since the cotter is in double shear, therefore area resisting shearing

$$= 2\,b \times t$$

∴ Shearing strength of cotter

$$= 2\,b \times t \times \tau$$

Equating this to the load (*P*), we have

$$P = 2\,b \times t \times \tau$$

From this equation, the width of cotter (*b*) may be determined.

4. *Failure of cotter in crushing*

We know that area resisting crushing

$$= b \times t$$

∴ Crushing strength of cotter

$$= b \times t \times \sigma_c$$

Equating this to the load (*P*), we have

$$P = b \times t \times \sigma_c$$

From this equation, the induced crushing stress in the cotter may be checked.

Example 12.6. *Design and draw a cottered foundation bolt which is subjected to a maximum pull of 50 kN. The allowable stresses are :*

$\sigma_t = 80$ *MPa* ; $\tau = 50$ *MPa* ; *and* $\sigma_c = 100$ *MPa*

Solution: Given: $P = 50$ kN $= 50 \times 10^3$ N ; $\sigma_t = 80$ MPa $= 80$ N/mm^2 ; $\tau = 50$ MPa $= 50$ N/mm^2 ; $\sigma_c = 100$ MPa $= 100$ N/mm^2

1. ***Diameter of bolt***

Let d = Diameter of bolt.

Considering the failure of the bolt in tension. We know that load (P),

$$50 \times 10^3 = \frac{\pi}{4} \times d^2 \times \sigma_t = \frac{\pi}{4} \times d^2 \times 80 = 62.84\, d^2$$

$$\therefore \quad d^2 = 50 \times 10^3 / 62.84 = 795.7 \text{ or } d = 28.2 \text{ say } 30 \text{ mm } \textbf{Ans.}$$

2. ***Diameter of enlarged end of the bolt and thickness of cotter***

Let d_1 = Diameter of enlarged end of the bolt, and

t = Thickness of cotter. It may be taken as $d_1 / 4$.

Considering the failure of the enlarged end of the bolt in tension at the cotter. We know that load (P),

$$50 \times 10^3 = \left[\frac{\pi}{4}(d_1)^2 - d_1 \times t\right]\sigma_t = \left[\frac{\pi}{4}(d_1)^2 - d_1 \times \frac{d_1}{4}\right] 80 = 42.84\,(d_1)^2$$

$$\therefore \quad (d_1)^2 = 50 \times 10^3 / 42.84 = 1167 \text{ or } d_1 = 34 \text{ say } 36 \text{ mm } \textbf{Ans.}$$

and

$$t = \frac{d_1}{4} = \frac{36}{4} = 9 \text{ mm } \textbf{Ans.}$$

3. ***Width of cotter***

Let b = Width of cotter.

Considering the failure of cotter in shear. Since the cotter is in double shear, therefore load (P),

$$50 \times 10^3 = 2\, b \times t \times \tau = 2\, b \times 9 \times 50 = 900\, b$$

$$\therefore \quad b = 50 \times 10^3 / 900 = 55.5 \text{ mm say } 60 \text{ mm } \textbf{Ans.}$$

Let us now check the crushing stress induced in the cotter. Considering the failure of cotter in crushing. We know that load (P),

$$50 \times 10^3 = b \times t \times \sigma_c = 60 \times 9 \times \sigma_c = 540\, \sigma_c$$

$$\therefore \quad \sigma_c = 50 \times 10^3 / 540 = 92.5 \text{ N/mm}^2$$

Since the induced crushing stress is less than the permissible value of 100 N/mm^2, therefore the design is safe.

12.12 Knuckle Joint

A knuckle joint is used to connect two rods which are under the action of tensile loads. However, if the joint is guided, the rods may support a compressive load. A knuckle joint may be readily disconnected for adjustments or repairs. Its use may be found in the link of a cycle chain, tie rod joint for roof truss, valve rod joint with eccentric rod, pump rod joint, tension link in bridge structure and lever and rod connections of various types.

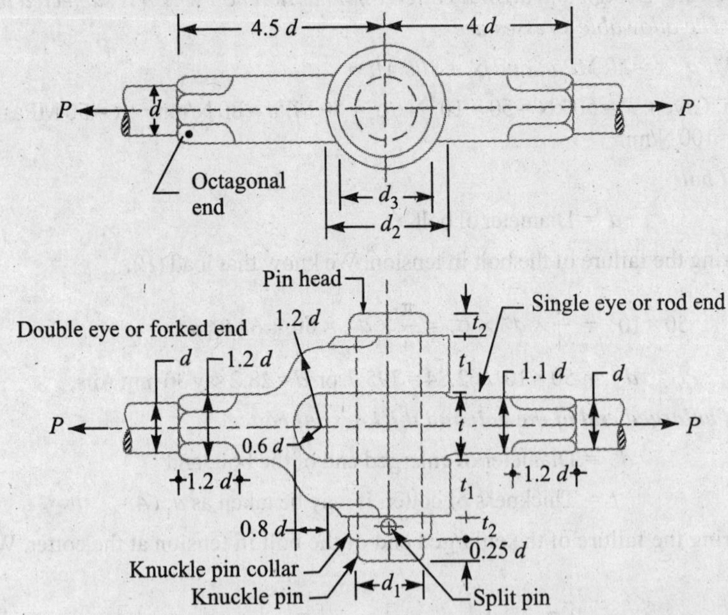

Fig. 12.16. Kunckle joint.

In knuckle joint (the two views of which are shown in Fig. 12.16), one end of one of the rods is made into an eye and the end of the other rod is formed into a fork with an eye in each of the fork leg. The knuckle pin passes through both the eye hole and the fork holes and may be secured by means of a collar and taper pin or spilt pin. The knuckle pin may be prevented from rotating in the fork by means of a small stop, pin, peg or snug. In order to get a better quality of joint, the sides of the fork and eye are machined, the hole is accurately drilled and pin turned. The material used for the joint may be steel or wrought iron.

12.13 Dimensions of Various Parts of the Knuckle Joint

The dimensions of various parts of the knuckle joint are fixed by empirical relations as given below. It may be noted that all the parts should be made of the same material *i.e.* mild steel or wrought iron.

If d is the diameter of rod, then diameter of pin,

$$d_1 = d$$

Outer diameter of eye,

$$d_2 = 2\,d$$

Submersibles like this can work at much greater ocean depths and high pressures where divers cannot reach.

Note : This picture is given as additional information and is not a direct example of the current chapter.

Diameter of knuckle pin head and collar,

$$d_3 = 1.5\,d$$

Thickness of single eye or rod end,

$$t = 1.25\,d$$

Thickness of fork, $t_1 = 0.75\,d$

Thickness of pin head, $t_2 = 0.5\,d$

Other dimensions of the joint are shown in Fig. 12.16.

12.14 Methods of Failure of Knuckle Joint

Consider a knuckle joint as shown in Fig. 12.16.

Let P = Tensile load acting on the rod,

d = Diameter of the rod,

d_1 = Diameter of the pin,

d_2 = Outer diameter of eye,

t = Thickness of single eye,

t_1 = Thickness of fork.

σ_t, τ and σ_c = Permissible stresses for the joint material in tension, shear and crushing respectively.

In determining the strength of the joint for the various methods of failure, it is assumed that

1. There is no stress concentration, and
2. The load is uniformly distributed over each part of the joint.

Due to these assumptions, the strengths are approximate, however they serve to indicate a well proportioned joint. Following are the various methods of failure of the joint :

1. ***Failure of the solid rod in tension***

Since the rods are subjected to direct tensile load, therefore tensile strength of the rod,

$$= \frac{\pi}{4} \times d^2 \times \sigma_t$$

Equating this to the load (P) acting on the rod, we have

$$P = \frac{\pi}{4} \times d^2 \times \sigma_t$$

From this equation, diameter of the rod (d) is obtained.

2. ***Failure of the knuckle pin in shear***

Since the pin is in double shear, therefore cross-sectional area of the pin under shearing

$$= 2 \times \frac{\pi}{4}\,(d_1)^2$$

and the shear strength of the pin

$$= 2 \times \frac{\pi}{4}\,(d_1)^2\,\tau$$

Equating this to the load (P) acting on the rod, we have

$$P = 2 \times \frac{\pi}{4}\,(d_1)^2\,\tau$$

From this equation, diameter of the knuckle pin (d_1) is obtained. This assumes that there is no slack and clearance between the pin and the fork and hence there is no bending of the pin. But, in actual practice, the knuckle pin is loose in forks in order to permit angular movement of one with respect to the other, therefore the pin is subjected to bending in addition to shearing. By making the diameter of knuckle pin equal to the diameter of the rod (*i.e.*, $d_1 = d$), a margin of strength is provided to allow for the bending of the pin.

In case, the stress due to bending is taken into account, it is assumed that the load on the pin is uniformly distributed along the middle portion (*i.e.* the eye end) and varies uniformly over the forks as shown in Fig. 12.17. Thus in the forks, a load $P/2$ acts through a distance of $t_1/3$ from the inner edge and the bending moment will be maximum at the centre of the pin. The value of maximum bending moment is given by

$$M = \frac{P}{2}\left(\frac{t_1}{3} + \frac{t}{2}\right) - \frac{P}{2} \times \frac{t}{4}$$

$$= \frac{P}{2}\left(\frac{t_1}{3} + \frac{t}{2} - \frac{t}{4}\right)$$

$$= \frac{P}{2}\left(\frac{t_1}{3} + \frac{t}{4}\right)$$

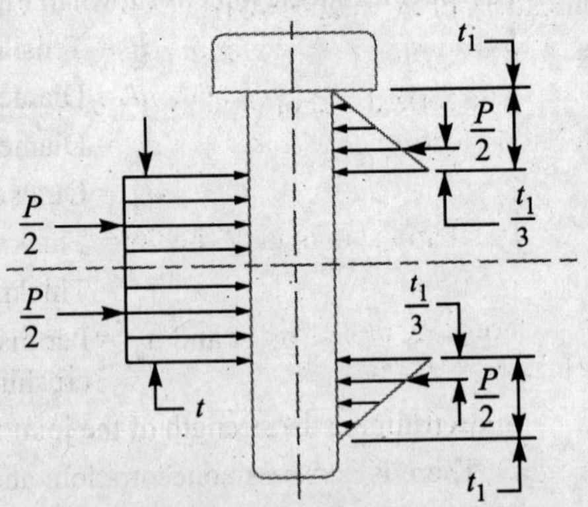

Fig. 12.17. Distribution of load on the pin.

and section modulus, $Z = \frac{\pi}{32}(d_1)^3$

∴ Maximum bending (tensile) stress,

$$\sigma_t = \frac{M}{Z} = \frac{\frac{P}{2}\left(\frac{t_1}{3} + \frac{t}{4}\right)}{\frac{\pi}{32}(d_1)^3}$$

From this expression, the value of d_1 may be obtained.

3. *Failure of the single eye or rod end in tension*

The single eye or rod end may tear off due to the tensile load. We know that area resisting tearing $= (d_2 - d_1)\, t$

∴ Tearing strength of single eye or rod end

$$= (d_2 - d_1)\, t \times \sigma_t$$

Equating this to the load (P) we have

$$P = (d_2 - d_1)\, t \times \sigma_t$$

From this equation, the induced tensile stress (σ_t) for the single eye or rod end may be checked. In case the induced tensile stress is more than the allowable working stress, then increase the outer diameter of the eye (d_2).

4. *Failure of the single eye or rod end in shearing*

The single eye or rod end may fail in shearing due to tensile load. We know that area resisting shearing $= (d_2 - d_1)\, t$

∴ Shearing strength of single eye or rod end

$$= (d_2 - d_1)\, t \times \tau$$

Equating this to the load (P), we have

$$P = (d_2 - d_1)\, t \times \tau$$

From this equation, the induced shear stress (τ) for the single eye or rod end may be checked.

5. *Failure of the single eye or rod end in crushing*

The single eye or pin may fail in crushing due to the tensile load. We know that area resisting crushing

$$= d_1 \times t$$

∴ Crushing strength of single eye or rod end

$$= d_1 \times t \times \sigma_c$$

Equating this to the load (P), we have

$$\therefore \quad P = d_1 \times t \times \sigma_c$$

From this equation, the induced crushing stress (σ_c) for the single eye or pin may be checked. In case the induced crushing stress in more than the allowable working stress, then increase the thickness of the single eye (t).

6. *Failure of the forked end in tension*

The forked end or double eye may fail in tension due to the tensile load. We know that area resisting tearing

$$= (d_2 - d_1) \times 2\,t_1$$

∴ Tearing strength of the forked end

$$= (d_2 - d_1) \times 2\,t_1 \times \sigma_t$$

Equating this to the load (P), we have

$$P = (d_2 - d_1) \times 2t_1 \times \sigma_t$$

From this equation, the induced tensile stress for the forked end may be checked.

7. *Failure of the forked end in shear*

The forked end may fail in shearing due to the tensile load. We know that area resisting shearing

$$= (d_2 - d_1) \times 2t_1$$

∴ Shearing strength of the forked end

$$= (d_2 - d_1) \times 2t_1 \times \tau$$

Equating this to the load (P), we have

$$P = (d_2 - d_1) \times 2t_1 \times \tau$$

From this equation, the induced shear stress for the forked end may be checked. In case, the induced shear stress is more than the allowable working stress, then thickness of the fork (t_1) is increased.

8. *Failure of the forked end in crushing*

The forked end or pin may fail in crushing due to the tensile load. We know that area resisting crushing

$$= d_1 \times 2\,t_1$$

∴ Crushing strength of the forked end

$$= d_1 \times 2\,t_1 \times \sigma_c$$

Equating this to the load (P), we have

$$P = d_1 \times 2\,t_1 \times \sigma_c$$

From this equation, the induced crushing stress for the forked end may be checked.

Note: From the above failures of the joint, we see that the thickness of fork (t_1) should be equal to half the thickness of single eye ($t/2$). But, in actual practice $t_1 > t/2$ in order to prevent deflection or spreading of the forks which would introduce excessive bending of pin.

12.15 Design Procedure of Knuckle Joint

The empirical dimensions as discussed in Art. 12.13 have been formulated after wide experience on a particular service. These dimensions are of more practical value than the theoretical analysis. Thus, a designer should consider the empirical relations in designing a knuckle joint. The following procedure may be adopted :

1. First of all, find the diameter of the rod by considering the failure of the rod in tension. We know that tensile load acting on the rod,

$$P = \frac{\pi}{4} \times d^2 \times \sigma_t$$

where d = Diameter of the rod, and

σ_t = Permissible tensile stress for the material of the rod.

2. After determining the diameter of the rod, the diameter of pin (d_1) may be determined by considering the failure of the pin in shear. We know that load,

$$P = 2 \times \frac{\pi}{4} (d_1)^2 \tau$$

A little consideration will show that the value of d_1 as obtained by the above relation is less than the specified value (*i.e.* the diameter of rod). So fix the diameter of the pin equal to the diameter of the rod.

3. Other dimensions of the joint are fixed by empirical relations as discussed in Art. 12.13.

4. The induced stresses are obtained by substituting the empirical dimensions in the relations as discussed in Art. 12.14.

In case the induced stress is more than the allowable stress, then the corresponding dimension may be increased.

Example 12.7. *Design a knuckle joint to transmit 150 kN. The design stresses may be taken as 75 MPa in tension, 60 MPa in shear and 150 MPa in compression.*

Solution. Given : $P = 150$ kN $= 150 \times 10^3$ N ; $\sigma_t = 75$ MPa $= 75$ N/mm^2 ; $\tau = 60$ MPa $= 60$ N/mm^2 ; $\sigma_c = 150$ MPa $= 150$ N/mm^2

The knuckle joint is shown in Fig. 12.16. The joint is designed by considering the various methods of failure as discussed below :

1. Failure of the solid rod in tension

Let d = Diameter of the rod.

We know that the load transmitted (P),

$$150 \times 10^3 = \frac{\pi}{4} \times d^2 \times \sigma_t = \frac{\pi}{4} \times d^2 \times 75 = 59\, d^2$$

$$\therefore \quad d^2 = 150 \times 10^3 / 59 = 2540 \quad \text{or} \quad d = 50.4 \text{ say } 52 \text{ mm } \textbf{Ans.}$$

Now the various dimensions are fixed as follows :

Diameter of knuckle pin,

$$d_1 = d = 52 \text{ mm}$$

Outer diameter of eye, $d_2 = 2\,d = 2 \times 52 = 104$ mm

Diameter of knuckle pin head and collar,

$$d_3 = 1.5\,d = 1.5 \times 52 = 78 \text{ mm}$$

Thickness of single eye or rod end,

$$t = 1.25\,d = 1.25 \times 52 = 65 \text{ mm}$$

Thickness of fork, $t_1 = 0.75\,d = 0.75 \times 52 = 39$ say 40 mm

Thickness of pin head, $t_2 = 0.5\,d = 0.5 \times 52 = 26$ mm

2. Failure of the knuckle pin in shear

Since the knuckle pin is in double shear, therefore load (P),

$$150 \times 10^3 = 2 \times \frac{\pi}{4} \times (d_1)^2\, \tau = 2 \times \frac{\pi}{4} \times (52)^2\, \tau = 4248\, \tau$$

$\therefore \quad \tau = 150 \times 10^3 / 4248 = 35.3 \text{ N/mm}^2 = 35.3 \text{ MPa}$

3. *Failure of the single eye or rod end in tension*

The single eye or rod end may fail in tension due to the load. We know that load (*P*),

$$150 \times 10^3 = (d_2 - d_1)\, t \times \sigma_t = (104 - 52)\, 65 \times \sigma_t = 3380\, \sigma_t$$

$$\therefore \quad \sigma_t = 150 \times 10^3 / 3380 = 44.4 \text{ N/mm}^2 = 44.4 \text{ MPa}$$

4. *Failure of the single eye or rod end in shearing*

The single eye or rod end may fail in shearing due to the load. We know that load (*P*),

$$150 \times 10^3 = (d_2 - d_1)\, t \times \tau = (104 - 52)\, 65 \times \tau = 3380\, \tau$$

$$\therefore \quad \tau = 150 \times 10^3 / 3380 = 44.4 \text{ N/mm}^2 = 44.4 \text{ MPa}$$

5. *Failure of the single eye or rod end in crushing*

The single eye or rod end may fail in crushing due to the load. We know that load (*P*),

$$150 \times 10^3 = d_1 \times t \times \sigma_c = 52 \times 65 \times \sigma_c = 3380\, \sigma_c$$

$$\therefore \quad \sigma_c = 150 \times 10^3 / 3380 = 44.4 \text{ N/mm}^2 = 44.4 \text{ MPa}$$

6. *Failure of the forked end in tension*

The forked end may fail in tension due to the load. We know that load (*P*),

$$150 \times 10^3 = (d_2 - d_1)\, 2\, t_1 \times \sigma_t = (104 - 52)\, 2 \times 40 \times \sigma_t = 4160\, \sigma_t$$

$$\therefore \quad \sigma_t = 150 \times 10^3 / 4160 = 36 \text{ N/mm}^2 = 36 \text{ MPa}$$

7. *Failure of the forked end in shear*

The forked end may fail in shearing due to the load. We know that load (*P*),

$$150 \times 10^3 = (d_2 - d_1)\, 2\, t_1 \times \tau = (104 - 52)\, 2 \times 40 \times \tau = 4160\, \tau$$

$$\therefore \quad \tau = 150 \times 10^3 / 4160 = 36 \text{ N/mm}^2 = 36 \text{ MPa}$$

8. *Failure of the forked end in crushing*

The forked end may fail in crushing due to the load. We know that load (*P*),

$$150 \times 10^3 = d_1 \times 2\, t_1 \times \sigma_c = 52 \times 2 \times 40 \times \sigma_c = 4160\, \sigma_c$$

$$\therefore \quad \sigma_c = 150 \times 10^3 / 4180 = 36 \text{ N/mm}^2 = 36 \text{ MPa}$$

From above, we see that the induced stresses are less than the given design stresses, therefore the joint is safe.

Example 12.8. *Design a knuckle joint for a tie rod of a circular section to sustain a maximum pull of 70 kN. The ultimate strength of the material of the rod against tearing is 420 MPa. The ultimate tensile and shearing strength of the pin material are 510 MPa and 396 MPa respectively. Determine the tie rod section and pin section. Take factor of safety = 6.*

Solution. Given : $P = 70 \text{ kN} = 70\,000 \text{ N}$; σ_{tu} for rod = 420 MPa ; *σ_{tu} for pin = 510 MPa ; $\tau_u = 396$ MPa ; *F.S.* = 6

We know that the permissible tensile stress for the rod material,

$$\sigma_t = \frac{\sigma_{tu} \text{ for rod}}{F.S.} = \frac{420}{6} = 70 \text{ MPa} = 70 \text{ N/mm}^2$$

and permissible shear stress for the pin material,

$$\tau = \frac{\tau_u}{F.S.} = \frac{396}{6} = 66 \text{ MPa} = 66 \text{ N/mm}^2$$

* Superfluous data.

We shall now consider the various methods of failure of the joint as discussed below:

1. ***Failure of the rod in tension***

Let d = Diameter of the rod.

We know that the load (P),

$$70\,000 = \frac{\pi}{4} \times d^2 \times \sigma_t = \frac{\pi}{4} \times d^2 \times 70 = 55\, d^2$$

$$\therefore \quad d^2 = 70\,000 / 55 = 1273 \text{ or } d = 35.7 \text{ say } 36 \text{ mm } \textbf{Ans.}$$

The other dimensions of the joint are fixed as given below :

Diameter of the knuckle pin,

$$d_1 = d = 36 \text{ mm}$$

Outer diameter of the eye,

$$d_2 = 2\,d = 2 \times 36 = 72 \text{ mm}$$

Diameter of knuckle pin head and collar,

$$d_3 = 1.5\,d = 1.5 \times 36 = 54 \text{ mm}$$

Thickness of single eye or rod end,

$$t = 1.25\,d = 1.25 \times 36 = 45 \text{ mm}$$

Thickness of fork, $t_1 = 0.75\,d = 0.75 \times 36 = 27$ mm

Now we shall check for the induced streses as discussed below :

2. ***Failure of the knuckle pin in shear***

Since the knuckle pin is in double shear, therefore load (P),

$$70\,000 = 2 \times \frac{\pi}{4}\,(d_1)^2\,\tau = 2 \times \frac{\pi}{4}\,(36)^2\,\tau = 2036\,\tau$$

$$\therefore \quad \tau = 70\,000 / 2036 = 34.4 \text{ N/mm}^2$$

3. ***Failure of the single eye or rod end in tension***

The single eye or rod end may fail in tension due to the load. We know that load (P),

$$70\,000 = (d_2 - d_1)\,t \times \sigma_t = (72 - 36)\,45\,\sigma_t = 1620\,\sigma_t$$

$$\therefore \quad \sigma_t = 70\,000 / 1620 = 43.2 \text{ N/mm}^2$$

4. ***Failure of the forked end in tension***

The forked end may fail in tension due to the load. We know that load (P),

$$70\,000 = (d_2 - d_1)\,2\,t_1 \times \sigma_t = (72 - 36) \times 2 \times 27 \times \sigma_t = 1944\,\sigma_t$$

$$\therefore \quad \sigma_t = 70\,000 / 1944 = 36 \text{ N/mm}^2$$

From above we see that the induced stresses are less than given permissible stresses, therefore the joint is safe.

12.16 Adjustable Screwed Joint for Round Rods (Turnbuckle)

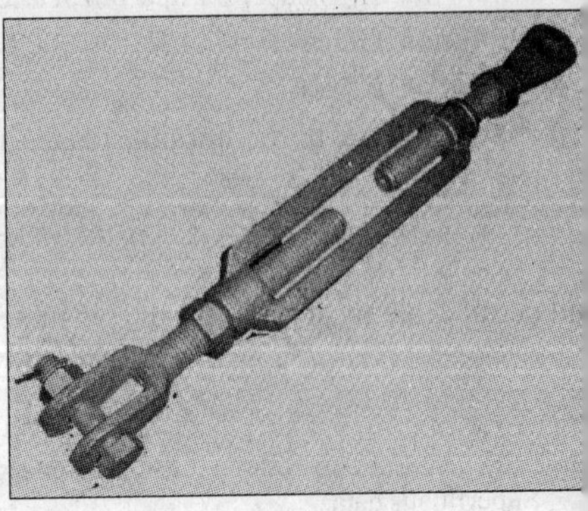

Turnbuckle.

Sometimes, two round tie rods, as shown in Fig. 12.18, are connected by means of a coupling known as a ***turnbuckle.*** In this type of joint, one of the rods has right hand threads and the other rod has left hand threads. The rods are screwed to a coupler which has a threaded hole. The coupler is of hexagonal or rectangular shape in the centre and round at both the ends in order to facilitate the rods to tighten or loosen with the help of a spanner when required. Sometimes instead of a spanner, a round iron rod may be used. The iron rod is inserted in a hole in the coupler as shown dotted in Fig. 12.18.

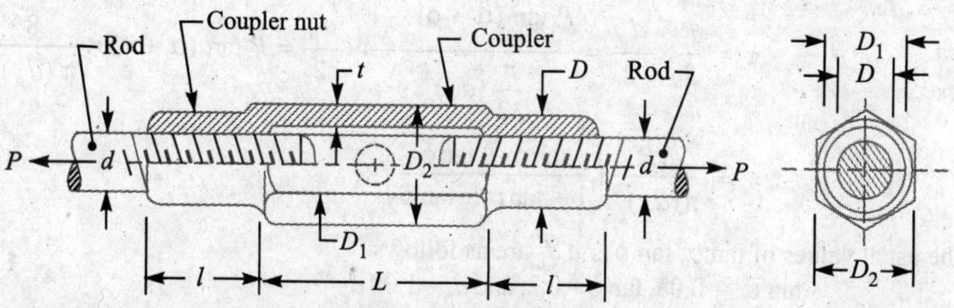

Fig. 12.18. Turnbuckle.

A turnbuckle commonly used in engineering practice (mostly in aeroplanes) is shown in Fig. 12.19. This type of turnbuckle is made hollow in the middle to reduce its weight. In this case, the two ends of the rods may also be seen. It is not necessary that the material of the rods and the turnbuckle may be same or different. It depends upon the pull acting on the joint.

12.17 Design of Turnbuckle

Consider a turnbuckle, subjected to an axial load P, as shown in Fig. 12.19. Due to this load, the threaded rod will be subjected to tensile stress whose magnitude is given by

$$\sigma_t = \frac{P}{A} = \frac{P}{\frac{\pi}{4}(d_c)^2}$$

where $\quad d_c$ = Core diameter of the threaded rod.

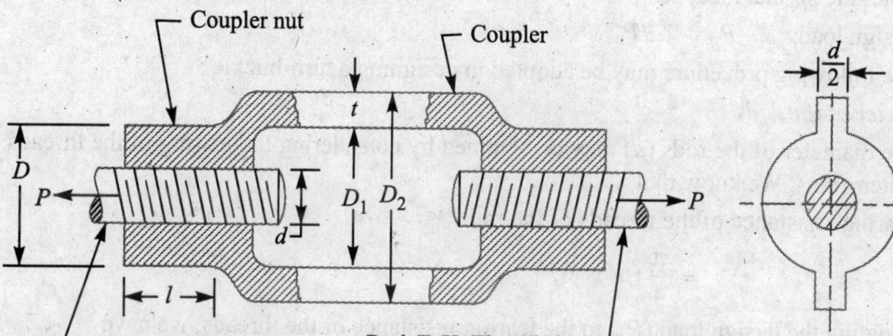

Fig. 12.19. Turnbuckle.

In order to drive the rods, the torque required is given by

$$T = P \tan(\alpha + \phi)\frac{d_p}{2}$$

where $\quad \alpha$ = Helix angle,

$\tan\phi$ = Coefficient of friction between the threaded rod and the coupler nut, and

d_p = Pitch diameter or mean diameter of the threaded rod.

∴ Shear stress produced by the torque,

$$\tau = \frac{T}{J} \times \frac{d_p}{2} = \frac{P \tan(\alpha + \phi) \frac{d_p}{2}}{\frac{\pi}{32}(d_p)^4} \times \frac{d_p}{2} = P \tan(\alpha + \phi) \times \frac{8}{\pi (d_p)^2}$$

$$= \frac{8P}{\pi(d_p)^2}\left(\frac{\tan\alpha + \tan\phi}{1 - \tan\alpha \times \tan\phi}\right)$$

The usual values of tan α, tan ϕ and d_p are as follows :

$$\tan\alpha = 0.03, \tan\phi = 0.2, \text{ and } d_p = 1.08\, d_c$$

Substituting these values in the above expression, we get

$$\tau = \frac{8P}{\pi(1.08\, d_c)^2}\left[\frac{0.03 + 0.2}{1 - 0.03 \times 0.2}\right] = \frac{8P}{4\pi(d_c)^2} = \frac{P}{2A} = \frac{\sigma_t}{2}$$

$$\ldots \left[\because A = \frac{\pi}{4}(d_c)^2\right]$$

Since the threaded rod is subjected to tensile stress as well as shear stress, therefore maximum principal stress,

$$\sigma_{t(max)} = \frac{\sigma_t}{2} + \frac{1}{2}\sqrt{(\sigma_t)^2 + 4\tau^2} = \frac{\sigma_t}{2} + \frac{1}{2}\sqrt{(\sigma_t)^2 + (\sigma_t)^2} \qquad \ldots\left(\because \tau = \frac{\sigma_t}{2}\right)$$

$$= 0.5\,\sigma_t + 0.707\,\sigma_t = 1.207\,\sigma_t = 1.207\, P/A$$

Giving a margin for higher coefficient of friction, the maximum principal stress may be taken as 1.3 times the normal stress. Therefore for designing a threaded section, we shall take the design load as 1.3 times the normal load, *i.e.*

Design load, $P_d = 1.3\,P$

The following procedure may be adopted in designing a turn-buckle :

1. ***Diameter of the rods***

The diameter of the rods (d) may be obtained by considering the tearing of the threads of the rods at their roots. We know that

Tearing resistance of the threads of the rod

$$= \frac{\pi}{4}(d_c)^2\, \sigma_t$$

Equating the design load (P_d) to the tearing resistance of the threads, we have

$$P_d = \frac{\pi}{4}(d_c)^2\, \sigma_t$$

where d_c = Core diameter of the threads of the rod, and

σ_t = Permissible tensile stress for the material of the rod.

From the above expression, the core diameter of the threads may be obtained. The nominal diameter of the threads (or diameter of the rod) may be found from Table 11.1, corresponding to the core diameter, assuming coarse threads.

2. ***Length of the coupler nut***

The length of the coupler nut (l) is obtained by considering the shearing of the threads at their roots in the coupler nut. We know that

Shearing resistance of the threads of the coupler nut

$$= (\pi\, d_c \times l)\,\tau$$

where τ = Shear stress for the material of the coupler nut.

Equating the design load to the shearing resistance of the threads in the coupler nut, we have

$$P_d = (\pi \, d_c \times l) \, \tau$$

From this expression, the value of l may be calculated. In actual practice, the length of coupler nut (l) is taken d to 1.25 d for steel nuts and 1.5 d to 2 d for cast iron and softer material nut. The length of the coupler nut may also be checked for crushing of threads. We know that

Crushing resistance of the threads in the coupler nut

$$= \frac{\pi}{4}\left[(d)^2 - (d_c)^2\right] n \times l \times \sigma_c$$

where σ_c = Crushing stress induced in the coupler nut, and

n = Number of threads per mm length.

Equating the design load to the crushing resistance of the threads, we have

$$P_d = \frac{\pi}{4}\left[(d)^2 - (d_c)^2\right] n \times l \times \sigma_c$$

From this expression, the induced σ_c may be checked.

3. *Outside diameter of the coupler nut*

The outside diameter of the coupler nut (D) may be obtained by considering the tearing at the coupler nut. We know that

Tearing resistance at the coupler nut

$$= \frac{\pi}{4}(D^2 - d^2)\,\sigma_t$$

where σ_t = Permissible tensile stress for the material of the coupler nut.

Equating the axial load to the tearing resistance at the coupler nut, we have

$$P = \frac{\pi}{4}(D^2 - d^2)\,\sigma_t$$

From this expression, the value of D may be calculated. In actual practice, the diameter of the coupler nut (D) is taken from 1.25 d to 1.5 d.

4. *Outside diameter of the coupler*

The outside diameter of the coupler (D_2) may be obtained by considering the tearing of the coupler. We know that

Tearing resistance of the coupler

$$= \frac{\pi}{4}\left[(D_2)^2 - (D_1)^2\right]\sigma_t$$

where D_1 = Inside diameter of the coupler. It is generally taken as (d + 6 mm), and

σ_t = Permissible tensile stress for the material of the coupler.

Equating the axial load to the tearing resistance of the coupler, we have

$$P = \frac{\pi}{4}\left[(D_2)^2 - (D_1)^2\right]\sigma_t$$

From this expression, the value of D_2 may be calculated. In actual practice, the outside diameter of the coupler (D_2) is taken as 1.5 d to 1.7 d. If the section of the coupler is to be made hexagonal or rectangular to fit the spanner, it may be circumscribed over the circle of outside diameter D_2.

5. The length of the coupler between the nuts (L) depends upon the amount of adjustment required. It is usually taken as 6 d.

6. The thickness of the coupler is usually taken as t = 0.75 d, and thickness of the coupler nut, t_1 = 0.5 d.

Example 12.9. *The pull in the tie rod of an iron roof truss is 50 kN. Design a suitable adjustable screwed joint. The permissible stresses are 75 MPa in tension, 37.5 MPa in shear and 90 MPa in crushing.*

Solution. Given : $P = 50 \text{ kN} = 50 \times 10^3 \text{ N}$; $\sigma_t = 75 \text{ MPa} = 75 \text{ N/mm}^2$; $\tau = 37.5 \text{ MPa} = 37.5 \text{ N/mm}^2$

We know that the design load for the threaded section,

$$P_d = 1.3\,P = 1.3 \times 50 \times 10^3 = 65 \times 10^3 \text{ N}$$

An adjustable screwed joint, as shown in Fig. 12.19, is suitable for the given purpose. The various dimensions for the joint are determined as discussed below :

1. ***Diameter of the tie rod***

Let d = Diameter of the tie rod, and

d_c = Core diameter of threads on the tie rod.

Considering tearing of the threads on the tie rod at their roots.

We know that design load (P_d),

$$65 \times 10^3 = \frac{\pi}{4}(d_c)^2\,\sigma_t = \frac{\pi}{4}(d_c)^2\,75 = 59\,(d_c)^2$$

$$\therefore \quad (d_c)^2 = 65 \times 10^3 / 59 = 1100 \quad \text{or} \quad d_c = 33.2 \text{ mm}$$

From Table 11.1 for coarse series, we find that the standard core diameter is 34.093 mm and the corresponding nominal diameter of the threads or diameter of tie rod,

$$d = 39 \text{ mm} \quad \textbf{Ans.}$$

2. ***Length of the coupler nut***

Let l = Length of the coupler nut.

Considering the shearing of threads at their roots in the coupler nut. We know that design load (P_d),

$$65 \times 10^3 = (\pi\, d_c . l)\,\tau = \pi \times 34.093 \times l \times 37.5 = 4107\, l$$

$$\therefore \quad l = 65 \times 10^3 / 4017 = 16.2 \text{ mm}$$

Since the length of the coupler nut is taken from d to 1.25 d, therefore we shall take

$$l = d = 39 \text{ mm} \quad \textbf{Ans.}$$

We shall now check the length of the coupler nut for crushing of threads.

From Table 11.1 for coarse series, we find that the pitch of the threads is 4 mm. Therefore the number of threads per mm length,

$$n = 1/4 = 0.25$$

We know that design load (P_d),

$$65 \times 10^3 = \frac{\pi}{4}\left[(d)^2 - (d_c)^2\right] n \times l \times \sigma_c$$

$$= \frac{\pi}{4}\left[(39)^2 - (34.093)^2\right] 0.25 \times 39 \times \sigma_c = 2750\,\sigma_c$$

$$\therefore \quad \sigma_c = 65 \times 10^3 / 2750 = 23.6 \text{ N/mm}^2 = 23.6 \text{ MPa}$$

Since the induced crushing stress in the threads of the coupler nut is less than the permissible stress, therefore the design is satisfactory.

3. ***Outside diameter of the coupler nut***

Let D = Outside diameter of the coupler nut

Considering tearing of the coupler nut. We know that axial load (P),

$$50 \times 10^3 = \frac{\pi}{4}(D^2 - d^2)\sigma_t$$

$$= \frac{\pi}{4}\left[D^2 - (39)^2\right] 75 = 59\left[D^2 - (39)^2\right]$$

or $$D^2 - (39)^2 = 50 \times 10^3 / 59 = 848$$

$\therefore$ $$D^2 = 848 + (39)^2 = 2369 \quad \text{or} \quad D = 48.7 \text{ say } 50 \text{ mm} \quad \textbf{Ans.}$$

Since the minimum outside diameter of coupler nut is taken as 1.25 *d* (*i.e.* 1.25 × 39 = 48.75 mm), therefore the above value of *D* is satisfactory.

4. *Outside diameter of the coupler*

Let D_2 = Outside diameter of the coupler, and

D_1 = Inside diameter of the coupler = d + 6 mm = 39 + 6 = 45 mm

Considering tearing of the coupler. We know that axial load (*P*),

$$50 \times 10^3 = \frac{\pi}{4}\left[(D_2)^2 - (D_1)^2\right]\sigma_t = \frac{\pi}{4}\left[(D_2)^2 - (45)^2\right] 75 = 59\left[(D_2)^2 - (45)^2\right]$$

$\therefore$ $$(D_2)^2 = 50 \times 10^3 / 59 + (45)^2 = 2873 \text{ or } D_2 = 53.6 \text{ mm}$$

Since the minimum outside diameter of the coupler is taken as 1.5 *d* (*i.e.* 1.5 × 39 = 58.5 say 60 mm), therefore we shall take

$$D_2 = 60 \text{ mm} \quad \textbf{Ans.}$$

5. Length of the coupler between nuts,

$$L = 6\,d = 6 \times 39 = 234 \text{ mm} \quad \textbf{Ans.}$$

6. Thickness of the coupler,

$$t_1 = 0.75\,d = 0.75 \times 39 = 29.25 \text{ say } 30 \text{ mm} \quad \textbf{Ans.}$$

and thickness of the coupler nut,

$$t = 0.5\,d = 0.5 \times 39 = 19.5 \text{ say } 20 \text{ mm} \quad \textbf{Ans.}$$

EXERCISES

1. Design a cotter joint to connect two mild steel rods for a pull of 30 kN. The maximum permissible stresses are 55 MPa in tension ; 40 MPa in shear and 70 MPa in crushing. Draw a neat sketch of the joint designed.

   [Ans. d = 22 mm; d_2 = 32 mm ; t = 14 mm ; d_1 = 44 mm ; b = 30 mm ; a = 12 mm ; d_4 = 65 mm ; c = 12 mm ; d_3 = 40 mm ; t_1 = 8 mm]

2. Two rod ends of a pump are joined by means of a cotter and spigot and socket at the ends. Design the joint for an axial load of 100 kN which alternately changes from tensile to compressive. The allowable stresses for the material used are 50 MPa in tension, 40 MPa in shear and 100 MPa in crushing.

   [Ans. d = 51 mm ; d_2 = 62 mm ; t = 16 mm ; d_1 = 72 mm ; b = 78 mm ; a = 20 mm ; d_3 = 83 mm ; d_4 = 125 mm ; c = 16 mm ; t_1 = 13 mm]

3. Two mild steel rods 40 mm diameter are to be connected by a cotter joint. The thickness of the cotter is 12 mm. Calculate the dimensions of the joint, if the maximum permissible stresses are: 46 MPa in tension ; 35 MPa in shear and 70 MPa in crushing.

   [Ans. d_2 = 30 mm ; d_1 = 48 mm ; b = 70 mm ; a = 27.5 mm ; d_4 = 100 mm ; c = 12 mm ; d_3 = 44.2 mm ; t = 35 mm ; t_1 = 13.5 mm]

4. The big end of a connecting rod is subjected to a load of 40 kN. The diameter of the circular part adjacent to the strap is 50 mm.

Design the joint assuming the permissible tensile stress in the strap as 30 MPa and permissible shear stress in the cotter and gib as 20 MPa.

[Ans. B_1 = 50 mm ; t = 15 mm ; t_1 = 15 mm ; t_3 = 22 mm ; B = 70 mm]

5. Design a cotter joint to connect a piston rod to the crosshead. The maximum steam pressure on the piston rod is 35 kN. Assuming that all the parts are made of the same material having the following permissible stresses :

$$\sigma_1 = 50 \text{ MPa} ; \tau = 60 \text{ MPa and } \sigma_c = 90 \text{ MPa.}$$

[Ans. d_2 = 40 mm ; t = 12 mm ; d_3 = 75 mm ; L = 88 mm ; d = 44 mm ; d_1 = 38 mm]

6. Design and draw a cotter foundation bolt to take a load of 90 kN. Assume the permissible stresses as follows :

$$\sigma_t = 50 \text{ MPa}, \tau = 60 \text{ MPa and } \sigma_c = 100 \text{ MPa.}$$

[Ans. d = 50 mm ; d_1 = 60 mm ; t = 15 mm ; b = 60 mm]

7. Design a knuckle joint to connect two mild steel bars under a tensile load of 25 kN. The allowable stresses are 65 MPa in tension, 50 MPa in shear and 83 MPa in crushing.

[Ans. $d = d_1$ = 23 mm ; d_2 = 46 mm ; d_3 = 35 mm ; t = 29 mm ; t_1 = 18 mm]

8. A knuckle joint is required to withstand a tensile load of 25 kN. Design the joint if the permissible stresses are :

$$\sigma_t = 56 \text{ MPa} ; \tau = 40 \text{ MPa and } \sigma_c = 70 \text{ MPa.}$$

[Ans. $d = d_1$ = 28 mm ; d_2 = 56 mm ; d_3 = 42 mm ; t_1 = 21 mm]

9. The pull in the tie rod of a roof truss is 44 kN. Design a suitable adjustable screw joint. The permissible tensile and shear stresses are 75 MPa and 37.5 MPa respectively. Draw full size two suitable views of the joint. **[Ans. d = 36 mm ; l = 11 mm ; D = 45 mm ; D_2 = 58 mm]**

QUESTIONS

1. What is a cotter joint? Explain with the help of a neat sketch, how a cotter joint is made ?
2. What are the applications of a cottered joint ?
3. Discuss the design procedure of spigot and socket cotter joint.
4. Why gibs are used in a cotter joint? Explain with the help of a neat sketch the use of single and double gib.
5. Describe the design procedure of a gib and cotter joint.
6. Distinguish between cotter joint and knuckle joint.
7. Sketch two views of a knuckle joint and write the equations showing the strength of joint for the most probable modes of failure.
8. Explain the purpose of a turn buckle. Describe its design procedure.

OBJECTIVE TYPE QUESTIONS

1. A cotter joint is used to transmit

   (*a*) axial tensile load only (*b*) axial compressive load only

   (*c*) combined axial and twisting loads (*d*) axial tensile or compressive loads

2. The taper on cotter varies from

   (*a*) 1 in 15 to 1 in 10 (*b*) 1 in 24 to 1 in 20

   (*c*) 1 in 32 to 1 in 24 (*d*) 1 in 48 to 1 in 24

3. Which of the following cotter joint is used to connect strap end of a connecting rod ?
   (*a*) Socket and spigot cotter joint
   (*b*) Sleeve and cotter joint
   (*c*) Gib and cotter joint
   (*d*) none of these
4. In designing a sleeve and cotter joint, the outside diameter of the sleeve is taken as
   (*a*) 1.5 *d*
   (*b*) 2.5 *d*
   (*c*) 3 *d*
   (*d*) 4 *d*

   where *d* = Diameter of the rod.
5. The length of cotter, in a sleeve and cotter joint, is taken as
   (*a*) 1.5 *d*
   (*b*) 2.5 *d*
   (*c*) 3 *d*
   (*d*) 4 *d*
6. In a gib and cotter joint, the thickness of gib is ....thickness of cotter.
   (*a*) more than
   (*b*) less than
   (*c*) equal to
7. When one gib is used in a gib and cotter joint, then the width of gib should be taken as
   (*a*) 0.45 *B*
   (*b*) 0.55 *B*
   (*c*) 0.65 *B*
   (*d*) 0.75 *B*

   where *B* = Total width of gib and cotter.
8. In a steam engine, the piston rod is usually connected to the crosshead by means of a
   (*a*) knuckle joint
   (*b*) universal joint
   (*c*) flange coupling
   (*d*) cotter joint
9. In a steam engine, the valve rod is connected to an eccentric by means of a
   (*a*) knuckle joint
   (*b*) universal joint
   (*c*) flange coupling
   (*d*) cotter joint
10. In a turn buckle, if one of the rods has left hand threads, then the other rod will have
    (*a*) right hand threads
    (*b*) left hand threads
    (*c*) pointed threads
    (*d*) multiple threads

ANSWERS

1. (*d*)	2. (*d*)	3. (*c*)	4. (*b*)	5. (*d*)
6. (*c*)	7. (*b*)	8. (*d*)	9. (*a*)	10. (*a*)

CHAPTER

13

Keys and Coupling

13.1 Introduction

A key is a piece of mild steel inserted between the shaft and hub or boss of the pulley to connect these together in order to prevent relative motion between them. It is always inserted parallel to the axis of the shaft. Keys are used as temporary fastenings and are subjected to considerable crushing and shearing stresses. A keyway is a slot or recess in a shaft and hub of the pulley to accommodate a key.

13.2 Types of Keys

The following types of keys are important from the subject point of view :

1. Sunk keys, **2.** Saddle keys, **3.** Tangent keys, **4.** Round keys, and **5.** Splines.

We shall now discuss the above types of keys, in detail, in the following pages.

13.3 Sunk Keys

The sunk keys are provided half in the keyway of the shaft and half in the keyway of the hub or boss of the pulley. The sunk keys are of the following types :

1. ***Rectangular sunk key.*** A rectangular sunk key is shown in Fig. 13.1. The usual proportions of this key are :

Width of key, $w = d / 4$; and thickness of key, $t = 2w / 3 = d / 6$

where d = Diameter of the shaft or diameter of the hole in the hub.

The key has taper 1 in 100 on the top side only.

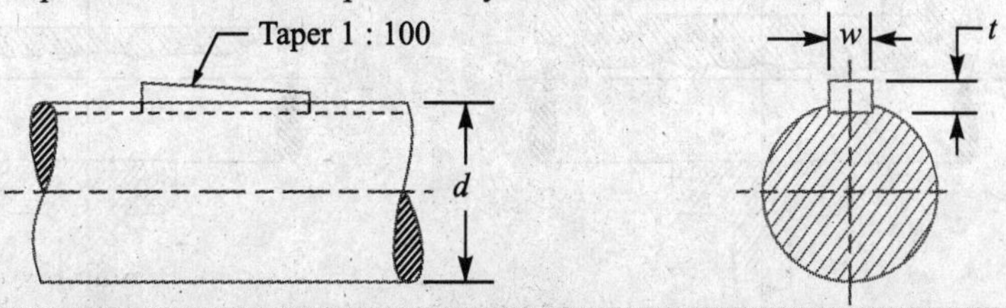

Fig. 13.1. Rectangular sunk key.

2. ***Square sunk key.*** The only difference between a rectangular sunk key and a square sunk key is that its width and thickness are equal, *i.e.*

$w = t = d / 4$

3. ***Parallel sunk key.*** The parallel sunk keys may be of rectangular or square section uniform in width and thickness throughout. It may be noted that a parallel key is a taperless and is used where the pulley, gear or other mating piece is required to slide along the shaft.

4. ***Gib-head key.*** It is a rectangular sunk key with a head at one end known as ***gib head.*** It is usually provided to facilitate the removal of key. A gib head key is shown in Fig. 13.2 (*a*) and its use in shown in Fig. 13.2 (*b*).

Helicopter driveline couplings.

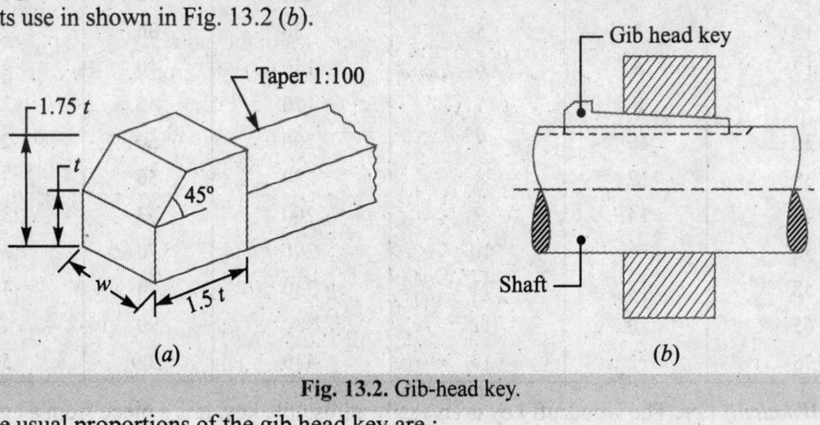

Fig. 13.2. Gib-head key.

The usual proportions of the gib head key are :

Width, $w = d / 4$;

and thickness at large end, $t = 2w / 3 = d / 6$

5. ***Feather key.*** A key attached to one member of a pair and which permits relative axial movement is known as ***feather key.*** It is a special type of parallel key which transmits a turning moment and also permits axial movement. It is fastened either to the shaft or hub, the key being a sliding fit in the key way of the moving piece.

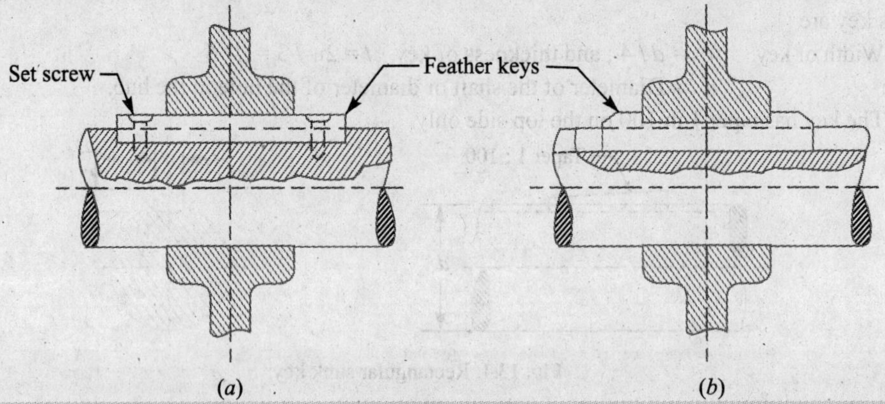

Fig. 13.3. Feather key.

The feather key may be screwed to the shaft as shown in Fig. 13.3 (*a*) or it may have double gib heads as shown in Fig. 13.3 (*b*). The various proportions of a feather key are same as that of rectangular sunk key and gib head key.

The following table shows the proportions of standard parallel, tapered and gib head keys, according to IS : 2292 and 2293-1974 (Reaffirmed 1992).

Table 13.1. Proportions of standard parallel, tapered and gib head keys.

Shaft diameter (mm) upto and including	*Key cross-section*		*Shaft diameter (mm) upto and including*	*Key cross-section*	
	Width (mm)	*Thickness (mm)*		*Width (mm)*	*Thickness (mm)*
6	2	2	85	25	14
8	3	3	95	28	16
10	4	4	110	32	18
12	5	5	130	36	20
17	6	6	150	40	22
22	8	7	170	45	25
30	10	8	200	50	28
38	12	8	230	56	32
44	14	9	260	63	32
50	16	10	290	70	36
58	18	11	330	80	40
65	20	12	380	90	45
75	22	14	440	100	50

6. ***Woodruff key.*** The woodruff key is an easily adjustable key. It is a piece from a cylindrical disc having segmental cross-section in front view as shown in Fig. 13.4. A woodruff key is capable of tilting in a recess milled out in the shaft by a cutter having the same curvature as the disc from which the key is made. This key is largely used in machine tool and automobile construction.

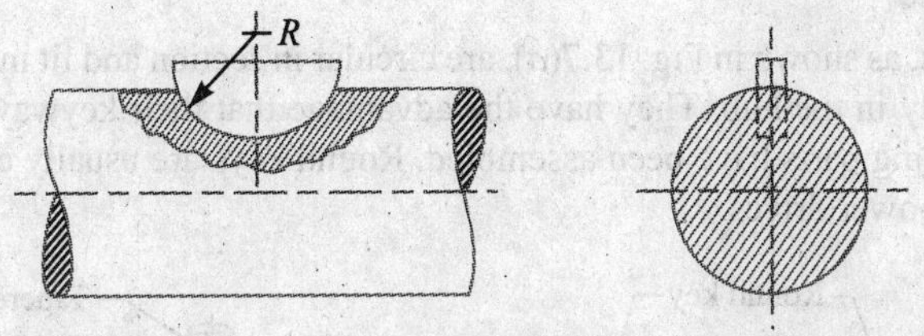

Fig. 13.4. Woodruff key.

The main advantages of a woodruff key are as follows :

1. It accommodates itself to any taper in the hub or boss of the mating piece.
2. It is useful on tapering shaft ends. Its extra depth in the shaft *prevents any tendency to turn over in its keyway.

The disadvantages are :

1. The depth of the keyway weakens the shaft.
2. It can not be used as a feather.

13.4 Saddle keys

The saddle keys are of the following two types :

1. Flat saddle key, and 2. Hollow saddle key.

A ***flat saddle key*** is a taper key which fits in a keyway in the hub and is flat on the shaft as shown in Fig. 13.5. It is likely to slip round the shaft under load. Therefore it is used for comparatively light loads.

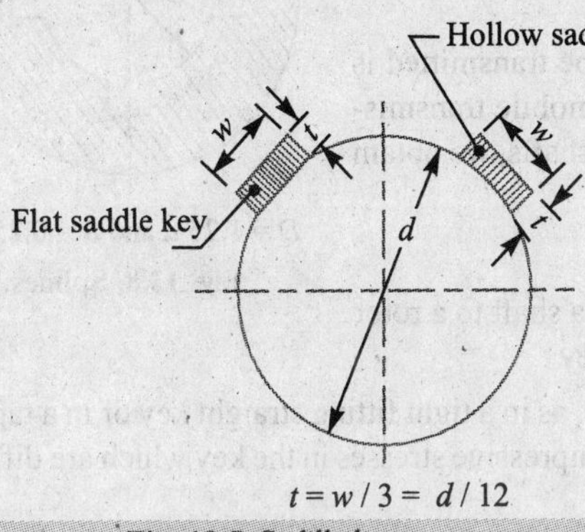

Fig. 13.5. Saddle key.

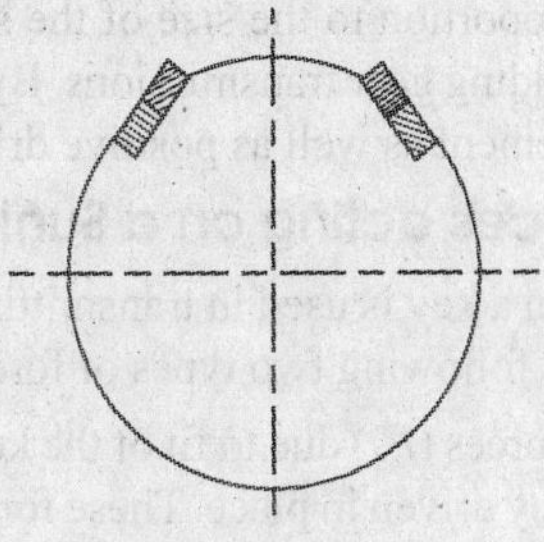

Fig. 13.6. Tangent key.

A ***hollow saddle key*** is a taper key which fits in a keyway in the hub and the bottom of the key is shaped to fit the curved surface of the shaft. Since hollow saddle keys hold on by friction, therefore these are suitable for light loads. It is usually used as a temporary fastening in fixing and setting eccentrics, cams etc.

13.5 Tangent Keys

The tangent keys are fitted in pair at right angles as shown in Fig. 13.6. Each key is to withstand torsion in one direction only. These are used in large heavy duty shafts.

---

* The usual form of rectangular sunk key is very likely to turn over in its keyway unless well fitted as its sides.

13.6 Round Keys

The round keys, as shown in Fig. 13.7(a), are circular in section and fit into holes drilled partly in the shaft and partly in the hub. They have the advantage that their keyways may be drilled and reamed after the mating parts have been assembled. Round keys are usually considered to be most appropriate for low power drives.

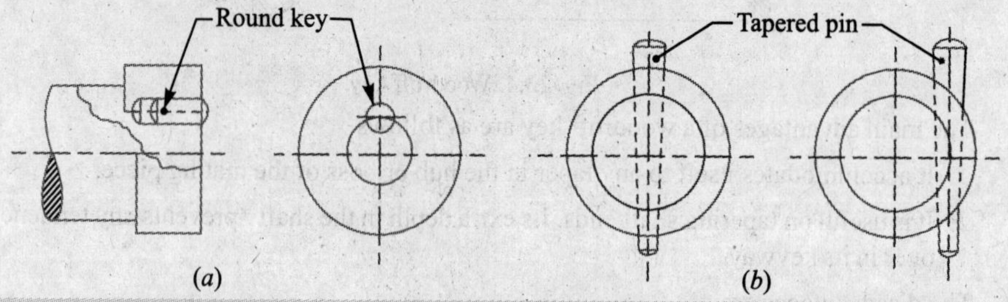

Fig. 13.7. Round keys.

Sometimes the tapered pin, as shown in Fig. 13.7 (b), is held in place by the friction between the pin and the reamed tapered holes.

13.7 Splines

Sometimes, keys are made integral with the shaft which fits in the keyways broached in the hub. Such shafts are known as ***splined shafts*** as shown in Fig. 13.8. These shafts usually have four, six, ten or sixteen splines. The splined shafts are relatively stronger than shafts having a single keyway.

The splined shafts are used when the force to be transmitted is large in proportion to the size of the shaft as in automobile transmission and sliding gear transmissions. By using splined shafts, we obtain axial movement as well as positive drive is obtained.

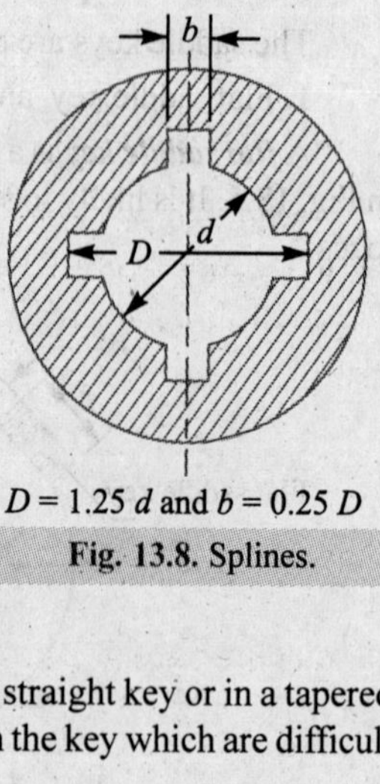

$D = 1.25\ d$ and $b = 0.25\ D$

Fig. 13.8. Splines.

13.8 Forces acting on a Sunk Key

When a key is used in transmitting torque from a shaft to a rotor or hub, the following two types of forces act on the key :

1. Forces (F_1) due to fit of the key in its keyway, as in a tight fitting straight key or in a tapered key driven in place. These forces produce compressive stresses in the key which are difficult to determine in magnitude.

2. Forces (F) due to the torque transmitted by the shaft. These forces produce shearing and compressive (or crushing) stresses in the key.

The distribution of the forces along the length of the key is not uniform because the forces are concentrated near the torque-input end. The non-uniformity of distribution is caused by the twisting of the shaft within the hub.

The forces acting on a key for a clockwise torque being transmitted from a shaft to a hub are shown in Fig. 13.9.

In designing a key, forces due to fit of the key are neglected and it is assumed that the distribution of forces along the length of key is uniform.

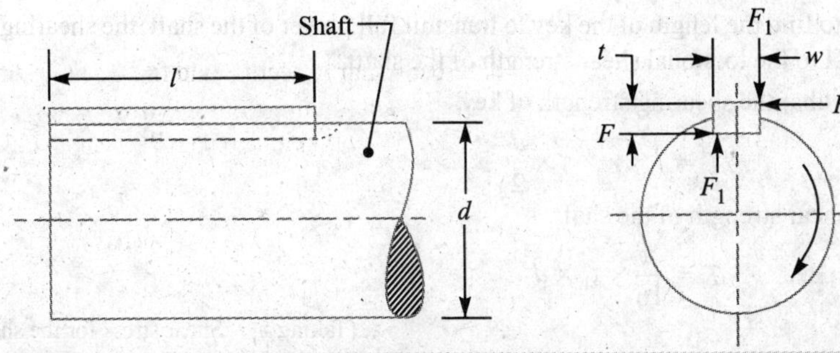

Fig. 13.9. Forces acting on a sunk key.

13.9 Strength of a Sunk Key

A key connecting the shaft and hub is shown in Fig. 13.9.

Let T = Torque transmitted by the shaft,

F = Tangential force acting at the circumference of the shaft,

d = Diameter of shaft,

l = Length of key,

w = Width of key.

t = Thickness of key, and

τ and σ_c = Shear and crushing stresses for the material of key.

A little consideration will show that due to the power transmitted by the shaft, the key may fail due to shearing or crushing.

Considering shearing of the key, the tangential shearing force acting at the circumference of the shaft,

$$F = \text{Area resisting shearing} \times \text{Shear stress} = l \times w \times \tau$$

∴ Torque transmitted by the shaft,

$$T = F \times \frac{d}{2} = l \times w \times \tau \times \frac{d}{2} \qquad ...(i)$$

Considering crushing of the key, the tangential crushing force acting at the circumference of the shaft,

$$F = \text{Area resisting crushing} \times \text{Crushing stress} = l \times \frac{t}{2} \times \sigma_c$$

∴ Torque transmitted by the shaft,

$$T = F \times \frac{d}{2} = l \times \frac{t}{2} \times \sigma_c \times \frac{d}{2} \qquad ...(ii)$$

The key is equally strong in shearing and crushing, if

$$l \times w \times \tau \times \frac{d}{2} = l \times \frac{t}{2} \times \sigma_c \times \frac{d}{2} \qquad \text{...[Equating equations } (i) \text{ and } (ii)]$$

or

$$\frac{w}{t} = \frac{\sigma_c}{2\tau} \qquad ...(iii)$$

The permissible crushing stress for the usual key material is atleast twice the permissible shearing stress. Therefore from equation *(iii)*, we have $w = t$. In other words, a square key is equally strong in shearing and crushing.

In order to find the length of the key to transmit full power of the shaft, the shearing strength of the key is equal to the torsional shear strength of the shaft.

We know that the shearing strength of key,

$$T = l \times w \times \tau \times \frac{d}{2} \quad \text{...}(iv)$$

and torsional shear strength of the shaft,

$$T = \frac{\pi}{16} \times \tau_1 \times d^3 \quad \text{...}(v)$$

...(Taking τ_1 = Shear stress for the shaft material)

From equations (*iv*) and (*v*), we have

$$l \times w \times \tau \times \frac{d}{2} = \frac{\pi}{16} \times \tau_1 \times d^3$$

$$\therefore \quad l = \frac{\pi}{8} \times \frac{\tau_1 d^2}{w \times \tau} = \frac{\pi d}{2} \times \frac{\tau_1}{\tau} = 1.571\, d \times \frac{\tau_1}{\tau} \quad \text{... (Taking } w = d/4) \quad \text{...}(vi)$$

When the key material is same as that of the shaft, then $\tau = \tau_1$.

$$\therefore \quad l = 1.571\, d \quad \text{... [From equation } (vi)]$$

Example 13.1. *Design the rectangular key for a shaft of 50 mm diameter. The shearing and crushing stresses for the key material are 42 MPa and 70 MPa.*

Solution. Given : $d = 50$ mm ; $\tau = 42$ MPa $= 42$ N/mm^2 ; $\sigma_c = 70$ MPa $= 70$ N/mm^2

The rectangular key is designed as discussed below:

From Table 13.1, we find that for a shaft of 50 mm diameter,

Width of key, $w = 16$ mm **Ans.**

and thickness of key, $t = 10$ mm **Ans.**

The length of key is obtained by considering the key in shearing and crushing.

Let l = Length of key.

Considering shearing of the key. We know that shearing strength (or torque transmitted) of the key,

$$T = l \times w \times \tau \times \frac{d}{2} = l \times 16 \times 42 \times \frac{50}{2} = 16\,800\, l \text{ N-mm} \quad \text{...}(i)$$

and torsional shearing strength (or torque transmitted) of the shaft,

$$T = \frac{\pi}{16} \times \tau \times d^3 = \frac{\pi}{16} \times 42\,(50)^3 = 1.03 \times 10^6 \text{ N-mm} \quad \text{...}(ii)$$

From equations (*i*) and (*ii*), we have

$$l = 1.03 \times 10^6 / 16\,800 = 61.31 \text{ mm}$$

Now considering crushing of the key. We know that shearing strength (or torque transmitted) of the key,

$$T = l \times \frac{t}{2} \times \sigma_c \times \frac{d}{2} = l \times \frac{10}{2} \times 70 \times \frac{50}{2} = 8750\, l \text{ N-mm} \quad \text{...}(iii)$$

From equations (*ii*) and (*iii*), we have

$$l = 1.03 \times 10^6 / 8750 = 117.7 \text{ mm}$$

Taking larger of the two values, we have length of key,

$l = 117.7$ say 120 mm **Ans.**

Example 13.2. *A 45 mm diameter shaft is made of steel with a yield strength of 400 MPa. A parallel key of size 14 mm wide and 9 mm thick made of steel with a yield strength of 340 MPa is to be used. Find the required length of key, if the shaft is loaded to transmit the maximum permissible torque. Use maximum shear stress theory and assume a factor of safety of 2.*

Solution. Given : $d = 45$ mm ; σ_{yt} for shaft = 400 MPa = 400 N/mm² ; $w = 14$ mm ; $t = 9$ mm ; σ_{yt} for key = 340 MPa = 340 N/mm²; $F.S. = 2$

Let l = Length of key.

According to maximum shear stress theory (See Art. 5.10), the maximum shear stress for the shaft,

$$\tau_{max} = \frac{\sigma_{yt}}{2 \times F.S.} = \frac{400}{2 \times 2} = 100 \text{ N/mm}^2$$

and maximum shear stress for the key,

$$\tau_k = \frac{\sigma_{yt}}{2 \times F.S.} = \frac{340}{2 \times 2} = 85 \text{ N/mm}^2$$

We know that the maximum torque transmitted by the shaft and key,

$$T = \frac{\pi}{16} \times \tau_{max} \times d^3 = \frac{\pi}{16} \times 100\ (45)^3 = 1.8 \times 10^6 \text{ N-mm}$$

First of all, let us consider the failure of key due to shearing. We know that the maximum torque transmitted (T),

$$1.8 \times 10^6 = l \times w \times \tau_k \times \frac{d}{2} = l \times 14 \times 85 \times \frac{45}{2} = 26\ 775\ l$$

$$\therefore \quad l = 1.8 \times 10^6 / 26\ 775 = 67.2 \text{ mm}$$

Now considering the failure of key due to crushing. We know that the maximum torque transmitted by the shaft and key (T),

$$1.8 \times 10^6 = l \times \frac{t}{2} \times \sigma_{ck} \times \frac{d}{2} = l \times \frac{9}{2} \times \frac{340}{2} \times \frac{45}{2} = 17\ 213\ l \quad \ldots\left(\text{Taking } \sigma_{ck} = \frac{\sigma_{yt}}{F.S.}\right)$$

$$\therefore \quad l = 1.8 \times 10^6 / 17\ 213 = 104.6 \text{ mm}$$

Taking the larger of the two values, we have

l = 104.6 say 105 mm **Ans.**

13.10 Effect of Keyways

A little consideration will show that the keyway cut into the shaft reduces the load carrying capacity of the shaft. This is due to the stress concentration near the corners of the keyway and reduction in the cross-sectional area of the shaft. It other words, the torsional strength of the shaft is reduced. The following relation for the weakening effect of the keyway is based on the experimental results by H.F. Moore.

$$e = 1 - 0.2\left(\frac{w}{d}\right) - 1.1\left(\frac{h}{d}\right)$$

where e = Shaft strength factor. It is the ratio of the strength of the shaft with keyway to the strength of the same shaft without keyway,

w = Width of keyway,

d = Diameter of shaft, and

$$h = \text{Depth of keyway} = \frac{\text{Thickness of key } (t)}{2}$$

It is usually assumed that the strength of the keyed shaft is 75% of the solid shaft, which is somewhat higher than the value obtained by the above relation.

In case the keyway is too long and the key is of sliding type, then the angle of twist is increased in the ratio k_θ as given by the following relation :

$$k_\theta = 1 + 0.4\left(\frac{w}{d}\right) + 0.7\left(\frac{h}{d}\right)$$

where k_θ = Reduction factor for angular twist.

Example 13.3. *A 15 kW, 960 r.p.m. motor has a mild steel shaft of 40 mm diameter and the extension being 75 mm. The permissible shear and crushing stresses for the mild steel key are 56 MPa and 112 MPa. Design the keyway in the motor shaft extension. Check the shear strength of the key against the normal strength of the shaft.*

Solution. Given : P = 15 kW = 15 × 10³ W ; N = 960 r.p.m. ; d = 40 mm ; l = 75 mm ; τ = 56 MPa = 56 N/mm² ; σ_c = 112 MPa = 112 N/mm²

We know that the torque transmitted by the motor,

$$T = \frac{P \times 60}{2\,\pi N} = \frac{15 \times 10^3 \times 60}{2\,\pi \times 960} = 149 \text{ N-m} = 149 \times 10^3 \text{ N-mm}$$

Let w = Width of keyway or key.

Considering the key in shearing. We know that the torque transmitted (T),

$$149 \times 10^3 = l \times w \times \tau \times \frac{d}{2} = 75 \times w \times 56 \times \frac{40}{2} = 84 \times 10^3\, w$$

$$\therefore \quad w = 149 \times 10^3 / 84 \times 10^3 = 1.8 \text{ mm}$$

This width of keyway is too small. The width of keyway should be at least d / 4.

$$\therefore \quad w = \frac{d}{4} = \frac{40}{4} = 10 \text{ mm } \textbf{Ans.}$$

Since $\sigma_c = 2\tau$, therefore a square key of w = 10 mm and t = 10 mm is adopted.

According to H.F. Moore, the shaft strength factor,

$$e = 1 - 0.2\left(\frac{w}{d}\right) - 1.1\left(\frac{h}{d}\right) = 1 - 0.2\left(\frac{w}{d}\right) - 1.1\left(\frac{t}{2d}\right) \qquad \text{... } (\because h = t/2)$$

$$= 1 - 0.2\left(\frac{10}{20}\right) - \left(\frac{10}{2 \times 40}\right) = 0.8125$$

∴ Strength of the shaft with keyway,

$$= \frac{\pi}{16} \times \tau \times d^3 \times e = \frac{\pi}{16} \times 56\,(40)^3\ 0.8125 = 571\ 844 \text{ N}$$

and shear strength of the key

$$= l \times w \times \tau \times \frac{d}{2} = 75 \times 10 \times 56 \times \frac{40}{2} = 840\ 000 \text{ N}$$

$$\therefore \quad \frac{\text{Shear strength of the key}}{\text{Normal strength of the shaft}} = \frac{840\ 000}{571\ 844} = 1.47 \textbf{ Ans.}$$

13.11 Shaft Coupling

Shafts are usually available up to 7 metres length due to inconvenience in transport. In order to have a greater length, it becomes necessary to join two or more pieces of the shaft by means of a coupling.

Shaft couplings are used in machinery for several purposes, the most common of which are the following :

1. To provide for the connection of shafts of units that are manufactured separately such as a motor and generator and to provide for disconnection for repairs or alternations.
2. To provide for misalignment of the shafts or to introduce mechanical flexibility.
3. To reduce the transmission of shock loads from one shaft to another.
4. To introduce protection against overloads.
5. It should have no projecting parts.

Couplings

Note : A coupling is termed as a device used to make permanent or semi-permanent connection where as a clutch permits rapid connection or disconnection at the will of the operator.

13.12 Requirements of a Good Shaft Coupling

A good shaft coupling should have the following requirements :

1. It should be easy to connect or disconnect.
2. It should transmit the full power from one shaft to the other shaft without losses.
3. It should hold the shafts in perfect alignment.
4. It should reduce the transmission of shock loads from one shaft to another shaft.
5. It should have no projecting parts.

13.13 Types of Shafts Couplings

Shaft couplings are divided into two main groups as follows :

1. ***Rigid coupling.*** It is used to connect two shafts which are perfectly aligned. Following types of rigid coupling are important from the subject point of view :

   (*a*) Sleeve or muff coupling.

   (*b*) Clamp or split-muff or compression coupling, and

   (*c*) Flange coupling.

2. ***Flexible coupling.*** It is used to connect two shafts having both lateral and angular misalignment. Following types of flexible coupling are important from the subject point of view :

   (*a*) Bushed pin type coupling,

   (*b*) Universal coupling, and

   (*c*) Oldham coupling.

We shall now discuss the above types of couplings, in detail, in the following pages.

Flexible PVC (non-metallic) coupling

Note : This picture is given as additional information and is not a direct example of the current chapter.

13.14 Sleeve or Muff-coupling

It is the simplest type of rigid coupling, made of cast iron. It consists of a hollow cylinder whose inner diameter is the same as that of the shaft. It is fitted over the ends of the two shafts by means of a gib head key, as shown in Fig. 13.10. The power is transmitted from one shaft to the other shaft by means of a key and a sleeve. It is, therefore, necessary that all the elements must be strong enough to transmit the torque. The usual proportions of a cast iron sleeve coupling are as follows :

Outer diameter of the sleeve, $D = 2d + 13$ mm

and length of the sleeve, $L = 3.5\ d$

where d is the diameter of the shaft.

In designing a sleeve or muff-coupling, the following procedure may be adopted.

1. ***Design for sleeve***

The sleeve is designed by considering it as a hollow shaft.

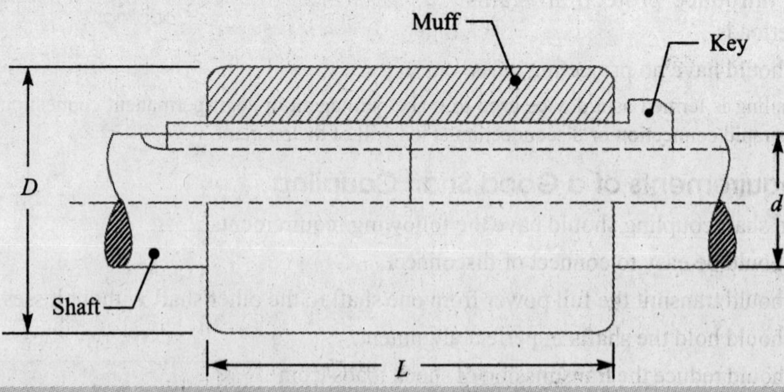

Fig. 13.10. Sleeve or muff coupling.

Let T = Torque to be transmitted by the coupling, and

τ_c = Permissible shear stress for the material of the sleeve which is cast rion. The safe value of shear stress for cast iron may be taken as 14 MPa.

We know that torque transmitted by a hollow section,

$$T = \frac{\pi}{16} \times \tau_c \left(\frac{D^4 - d^4}{D}\right) = \frac{\pi}{16} \times \tau_c \times D^3\ (1 - k^4) \qquad \ldots (\because\ k = d/D)$$

From this expression, the induced shear stress in the sleeve may be checked.

2. ***Design for key***

The key for the coupling may be designed in the similar way as discussed in Art. 13.9. The width and thickness of the coupling key is obtained from the proportions.

The length of the coupling key is atleast equal to the length of the sleeve (*i.e.* 3.5 d). The coupling key is usually made into two parts so that the length of the key in each shaft,

$$l = \frac{L}{2} = \frac{3.5\ d}{2}$$

After fixing the length of key in each shaft, the induced shearing and crushing stresses may be checked. We know that torque transmitted,

$$T = l \times w \times \tau \times \frac{d}{2} \qquad \ldots \text{(Considering shearing of the key)}$$

$$= l \times \frac{t}{2} \times \sigma_c \times \frac{d}{2} \qquad \ldots \text{(Considering crushing of the key)}$$

Note: The depth of the keyway in each of the shafts to be connected should be exactly the same and the diameters should also be same. If these conditions are not satisfied, then the key will be bedded on one shaft while in the other it will be loose. In order to prevent this, the key is made in two parts which may be driven from the same end for each shaft or they may be driven from opposite ends.

Example 13.4. *Design and make a neat dimensioned sketch of a muff coupling which is used to connect two steel shafts transmitting 40 kW at 350 r.p.m. The material for the shafts and key is plain carbon steel for which allowable shear and crushing stresses may be taken as 40 MPa and 80 MPa respectively. The material for the muff is cast iron for which the allowable shear stress may be assumed as 15 MPa.*

Solution. Given : $P = 40$ kW $= 40 \times 10^3$ W; $N = 350$ r.p.m.; $\tau_s = 40$ MPa $= 40$ N/mm^2; $\sigma_{cs} = 80$ MPa $= 80$ N/mm^2; $\tau_c = 15$ MPa $= 15$ N/mm^2

A type of muff couplings.

Note : This picture is given as additional information and is not a direct example of the current chapter.

The muff coupling is shown in Fig. 13.10. It is designed as discussed below :

1. *Design for shaft*

Let d = Diameter of the shaft.

We know that the torque transmitted by the shaft, key and muff,

$$T = \frac{P \times 60}{2\pi N} = \frac{40 \times 10^3 \times 60}{2\pi \times 350} = 1100 \text{ N-m}$$

$$= 1100 \times 10^3 \text{ N-mm}$$

We also know that the torque transmitted (T),

$$1100 \times 10^3 = \frac{\pi}{16} \times \tau_s \times d^3 = \frac{\pi}{16} \times 40 \times d^3 = 7.86\, d^3$$

$\therefore \quad d^3 = 1100 \times 10^3/7.86 = 140 \times 10^3$ or $d = 52$ say 55 mm **Ans.**

2. *Design for sleeve*

We know that outer diameter of the muff,

$$D = 2d + 13 \text{ mm} = 2 \times 55 + 13 = 123 \text{ say } 125 \text{ mm } \textbf{Ans.}$$

and length of the muff,

$$L = 3.5\, d = 3.5 \times 55 = 192.5 \text{ say } 195 \text{ mm } \textbf{Ans.}$$

Let us now check the induced shear stress in the muff. Let τ_c be the induced shear stress in the muff which is made of cast iron. Since the muff is considered to be a hollow shaft, therefore the torque transmitted (T),

$$1100 \times 10^3 = \frac{\pi}{16} \times \tau_c \left(\frac{D^4 - d^4}{D}\right) = \frac{\pi}{16} \times \tau_c \left[\frac{(125)^4 - (55)^4}{125}\right]$$

$$= 370 \times 10^3\, \tau_c$$

$\therefore \quad \tau_c = 1100 \times 10^3/370 \times 10^3 = 2.97$ N/mm^2

Since the induced shear stress in the muff (cast iron) is less than the permissible shear stress of 15 N/mm^2, therefore the design of muff is safe.

3. *Design for key*

From Table 13.1, we find that for a shaft of 55 mm diameter,

Width of key, $w = 18$ mm **Ans.**

Since the crushing stress for the key material is twice the shearing stress, therefore a square key may be used.

∴ Thickness of key, $t = w = 18$ mm **Ans.**

We know that length of key in each shaft,

$$l = L / 2 = 195 / 2 = 97.5 \text{ mm } \textbf{Ans.}$$

Let us now check the induced shear and crushing stresses in the key. First of all, let us consider shearing of the key. We know that torque transmitted (*T*),

$$1100 \times 10^3 = l \times w \times \tau_s \times \frac{d}{2} = 97.5 \times 18 \times \tau_s \times \frac{55}{2} = 48.2 \times 10^3 \, \tau_s$$

$$\therefore \quad \tau_s = 1100 \times 10^3 / 48.2 \times 10^3 = 22.8 \text{ N/mm}^2$$

Now considering crushing of the key. We know that torque transmitted (*T*),

$$1100 \times 10^3 = l \times \frac{t}{2} \times \sigma_{cs} \times \frac{d}{2} = 97.5 \times \frac{18}{2} \times \sigma_{cs} \times \frac{55}{2} = 24.1 \times 10^3 \, \sigma_{cs}$$

$$\therefore \quad \sigma_{cs} = 1100 \times 10^3 / 24.1 \times 10^3 = 45.6 \text{ N/mm}^2$$

Since the induced shear and crushing stresses are less than the permissible stresses, therefore the design of key is safe.

13.15 Clamp or Compression Coupling

It is also known as ***split muff coupling.*** In this case, the muff or sleeve is made into two halves and are bolted together as shown in Fig. 13.11. The halves of the muff are made of cast iron. The shaft ends are made to abutt each other and a single key is fitted directly in the keyways of both the shafts. One-half of the muff is fixed from below and the other half is placed from above. Both the halves are held together by means of mild steel studs or bolts and nuts. The number of bolts may be two, four or six. The nuts are recessed into the bodies of the muff castings. This coupling may be used for heavy duty and moderate speeds. The advantage of this coupling is that the position of the shafts need not be changed for assembling or disassembling of the

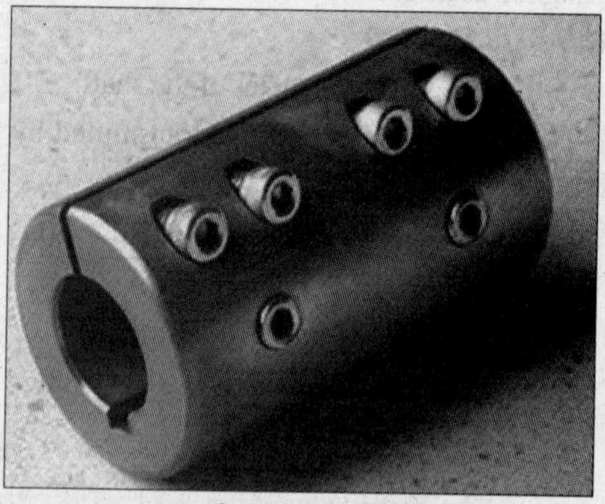

Spilt-sleeve coupling.

Note : This picture is given as additional information and is not a direct example of the current chapter.

(a) Heavy duty flex-flex coupling.

(b) Heavy duty flex-rigid coupling.

coupling. The usual proportions of the muff for the clamp or compression coupling are :

Diameter of the muff or sleeve, $D = 2d + 13$ mm

Length of the muff or sleeve, $L = 3.5\ d$

where d = Diameter of the shaft.

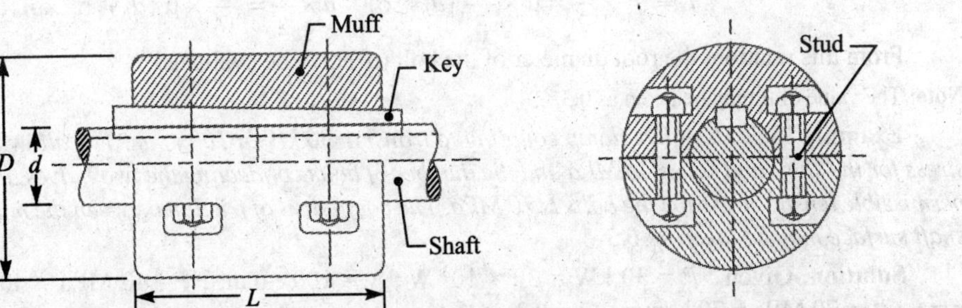

Fig. 13.11. Clamp or compression coupling.

In the clamp or compression coupling, the power is transmitted from one shaft to the other by means of key and the friction between the muff and shaft. In designing this type of coupling, the following procedure may be adopted.

1. *Design of muff and key*

The muff and key are designed in the similar way as discussed in muff coupling (Art. 13.14).

2. *Design of clamping bolts*

Let T = Torque transmited by the shaft,

d = Diameter of shaft,

d_b = Root or effective diameter of bolt,

n = Number of bolts,

σ_t = Permissible tensile stress for bolt material,

μ = Coefficient of friction between the muff and shaft, and

L = Length of muff.

We know that the force exerted by each bolt

$$= \frac{\pi}{4}(d_b)^2\sigma_t$$

∴ Force exerted by the bolts on each side of the shaft

$$= \frac{\pi}{4}(d_b)^2\sigma_t \times \frac{n}{2}$$

Let p be the pressure on the shaft and the muff surface due to the force, then for uniform pressure distribution over the surface,

$$p = \frac{\text{Force}}{\text{Projected area}} = \frac{\frac{\pi}{4}(d_b)^2\sigma_t \times \frac{n}{2}}{\frac{1}{2}L \times d}$$

∴ Frictional force between each shaft and muff,

$$F = \mu \times \text{pressure} \times \text{area} = \mu \times p \times \frac{1}{2} \times \pi d \times L$$

$$= \mu \times \frac{\frac{\pi}{4}(d_b)^2\sigma_t \times \frac{n}{2}}{\frac{1}{2}L \times d} \times \frac{1}{2}\pi d \times L$$

$$= \mu \times \frac{\pi}{4}(d_b)^2 \sigma_t \times \frac{n}{2} \times \pi = \mu \times \frac{\pi^2}{8}(d_b)^2 \sigma_t \times n$$

and the torque that can be transmitted by the coupling,

$$T = F \times \frac{d}{2} = \mu \times \frac{\pi^2}{8}(d_b)^2 \sigma_t \times n \times \frac{d}{2} = \frac{\pi^2}{16} \times \mu\,(d_b)^2 \sigma_t \times n \times d$$

From this relation, the root diameter of the bolt (d_b) may be evaluated.

Note: The value of μ may be taken as 0.3.

Example 13.5. *Design a clamp coupling to transmit 30 kW at 100 r.p.m. The allowable shear stress for the shaft and key is 40 MPa and the number of bolts connecting the two halves are six. The permissible tensile stress for the bolts is 70 MPa. The coefficient of friction between the muff and the shaft surface may be taken as 0.3.*

Solution. Given : $P = 30$ kW $= 30 \times 10^3$ W ; $N = 100$ r.p.m. ; $\tau = 40$ MPa $= 40$ N/mm^2 ; $n = 6$; $\sigma_t = 70$ MPa $= 70$ N/mm^2 ; $\mu = 0.3$

1. *Design for shaft*

Let d = Diameter of shaft.

We know that the torque transmitted by the shaft,

$$T = \frac{P \times 60}{2\,\pi N} = \frac{30 \times 10^3 \times 60}{2\,\pi \times 100} = 2865 \text{ N-m} = 2865 \times 10^3 \text{ N-mm}$$

We also know that the torque transmitted by the shaft (T),

$$2865 \times 10^3 = \frac{\pi}{16} \times \tau \times d^3 = \frac{\pi}{16} \times 40 \times d^3 = 7.86\, d^3$$

$$\therefore \quad d^3 = 2865 \times 10^3 / 7.86 = 365 \times 10^3 \text{ or } d = 71.4 \text{ say } 75 \text{ mm } \textbf{Ans.}$$

2. *Design for muff*

We know that diameter of muff,

$$D = 2d + 13 \text{ mm} = 2 \times 75 + 13 = 163 \text{ say } 165 \text{ mm } \textbf{Ans.}$$

and total length of the muff,

$$L = 3.5\, d = 3.5 \times 75 = 262.5 \text{ mm } \textbf{Ans.}$$

3. *Design for key*

The width and thickness of the key for a shaft diameter of 75 mm (from Table 13.1) are as follows :

Width of key, $w = 22$ mm **Ans.**

Thickness of key, $t = 14$ mm **Ans.**

and length of key = Total length of muff = 262.5 mm **Ans.**

4. *Design for bolts*

Let d_b = Root or core diameter of bolt.

We know that the torque transmitted (T),

$$2865 \times 10^3 = \frac{\pi^2}{16} \times \mu (d_b)^2 \sigma_t \times n \times d = \frac{\pi^2}{16} \times 0.3\,(d_b)^2\; 70 \times 6 \times 75 = 5830 (d_b)^2$$

$$\therefore \quad (d_b)^2 = 2865 \times 10^3 / 5830 = 492 \quad \text{or} \quad d_b = 22.2 \text{ mm}$$

From Table 11.1, we find that the standard core diameter of the bolt for coarse series is 23.32 mm and the nominal diameter of the bolt is 27 mm (M 27). **Ans.**

13.16 Flange Coupling

A flange coupling usually applies to a coupling having two separate cast iron flanges. Each flange is mounted on the shaft end and keyed to it. The faces are turned up at right angle to the axis of the shaft. One of the flange has a projected portion and the other flange has a corresponding recess.

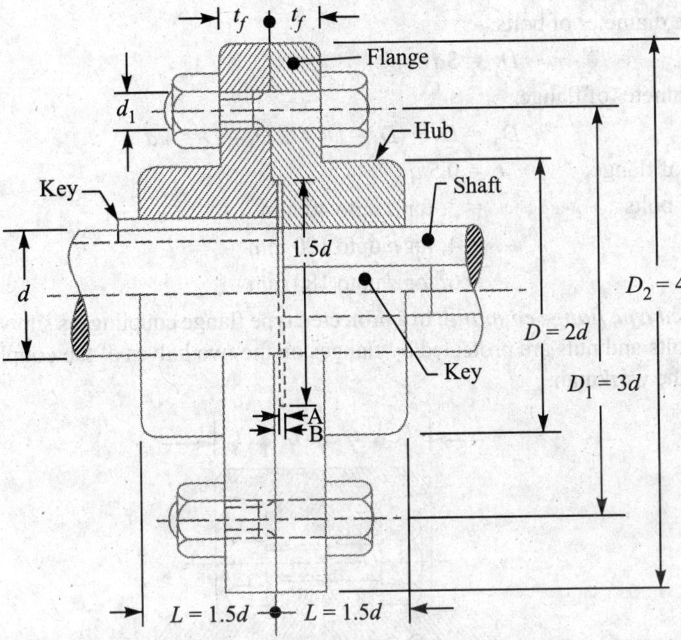

Fig. 13.12. Unprotected type flange coupling.

This helps to bring the shafts into line and to maintain alignment. The two flanges are coupled together by means of bolts and nuts. The flange coupling is adopted to heavy loads and hence it is used on large shafting. The flange couplings are of the following three types :

1. ***Unprotected type flange coupling.*** In an unprotected type flange coupling, as shown in Fig. 13.12, each shaft is keyed to the boss of a flange with a counter sunk key and the flanges are coupled together by means of bolts. Generally, three, four or six bolts are used. The keys are staggered at right angle along the circumference of the shafts in order to divide the weakening effect caused by keyways.

Flange Couplings.

The usual proportions for an unprotected type cast iron flange couplings, as shown in Fig. 13.12, are as follows :

If d is the diameter of the shaft or inner diameter of the hub, then

Outside diameter of hub,

$$D = 2\,d$$

Length of hub, $L = 1.5\ d$

Pitch circle diameter of bolts,

$$D_1 = 3d$$

Outside diameter of flange,

$$D_2 = D_1 + (D_1 - D) = 2\ D_1 - D = 4\ d$$

Thickness of flange, $t_f = 0.5\ d$

Number of bolts = 3, for d upto 40 mm

= 4, for d upto 100 mm

= 6, for d upto 180 mm

2. ***Protected type flange coupling.*** In a protected type flange coupling, as shown in Fig. 13.13, the protruding bolts and nuts are protected by flanges on the two halves of the coupling, in order to avoid danger to the workman.

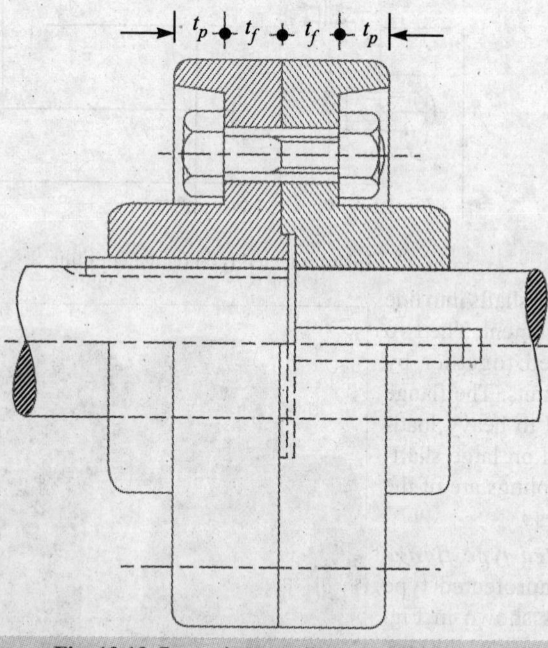

Fig. 13.13. Protective type flange coupling.

The thickness of the protective circumferential flange (t_p) is taken as 0.25 d. The other proportions of the coupling are same as for unprotected type flange coupling.

3. ***Marine type flange coupling.*** In a marine type flange coupling, the flanges are forged integral with the shafts as shown in Fig. 13.14. The flanges are held together by means of tapered headless bolts, numbering from four to twelve depending upon the diameter of shaft.

The number of bolts may be choosen from the following table.

Table 13.2. Number of bolts for marine type flange coupling.
(According to IS : 3653 – 1966 (Reaffirmed 1990))

Shaft diameter (mm)	35 to 55	56 to 150	151 to 230	231 to 390	Above 390
No. of bolts	4	6	8	10	12

The other proportions for the marine type flange coupling are taken as follows :

Thickness of flange $= d/3$

Taper of bolt = 1 in 20 to 1 in 40

Pitch circle diameter of bolts, $D_1 = 1.6\, d$

Outside diameter of flange, $D_2 = 2.2\, d$

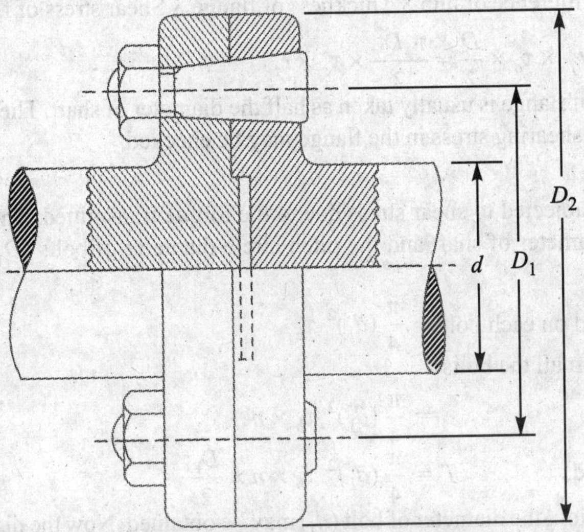

Fig. 13.14. Marine type flange coupling.

13.17 Design of Flange Coupling

Consider a flange coupling as shown in Fig. 13.12 and Fig. 13.13.

Let d = Diameter of shaft or inner diameter of hub,

D = Outer diameter of hub,

d_1 = Nominal or outside diameter of bolt,

D_1 = Diameter of bolt circle,

n = Number of bolts,

t_f = Thickness of flange,

τ_s, τ_b and τ_k = Allowable shear stress for shaft, bolt and key material respectively

τ_c = Allowable shear stress for the flange material *i.e.* cast iron,

σ_{cb}, and σ_{ck} = Allowable crushing stress for bolt and key material respectively.

The flange coupling is designed as discussed below :

1. *Design for hub*

The hub is designed by considering it as a hollow shaft, transmitting the same torque (T) as that of a solid shaft.

$$\therefore \quad T = \frac{\pi}{16} \times \tau_c \left(\frac{D^4 - d^4}{D} \right)$$

The outer diameter of hub is usually taken as twice the diameter of shaft. Therefore from the above relation, the induced shearing stress in the hub may be checked.

The length of hub (L) is taken as 1.5 d.

2. *Design for key*

The key is designed with usual proportions and then checked for shearing and crushing stresses.

The material of key is usually the same as that of shaft. The length of key is taken equal to the length of hub.

3. *Design for flange*

The flange at the junction of the hub is under shear while transmitting the torque. Therefore, the troque transmitted,

$$T = \text{Circumference of hub} \times \text{Thickness of flange} \times \text{Shear stress of flange} \times \text{Radius of hub}$$

$$= \pi\, D \times t_f \times \tau_c \times \frac{D}{2} = \frac{\pi\, D^2}{2} \times \tau_c \times t_f$$

The thickness of flange is usually taken as half the diameter of shaft. Therefore from the above relation, the induced shearing stress in the flange may be checked.

4. *Design for bolts*

The bolts are subjected to shear stress due to the torque transmitted. The number of bolts (n) depends upon the diameter of shaft and the pitch circle diameter of bolts (D_1) is taken as 3 d. We know that

$$\text{Load on each bolt} = \frac{\pi}{4}(d_1)^2\, \tau_b$$

∴ Total load on all the bolts

$$= \frac{\pi}{4}(d_1)^2\, \tau_b \times n$$

and torque transmitted, $T = \frac{\pi}{4}(d_1)^2\, \tau_b \times n \times \frac{D_1}{2}$

From this equation, the diameter of bolt (d_1) may be obtained. Now the diameter of bolt may be checked in crushing.

We know that area resisting crushing of all the bolts

$$= n \times d_1 \times t_f$$

and crushing strength of all the bolts

$$= (n \times d_1 \times t_f)\, \sigma_{cb}$$

∴ Torque, $T = (n \times d_1 \times t_f \times \sigma_{cb}) \frac{D_1}{2}$

From this equation, the induced crushing stress in the bolts may be checked.

Example 13.6. *Design a cast iron protective type flange coupling to transmit 15 kW at 900 r.p.m. from an electric motor to a compressor. The service factor may be assumed as 1.35. The following permissible stresses may be used :*

Shear stress for shaft, bolt and key material	=	*40 MPa*
Crushing stress for bolt and key	=	*80 MPa*
Shear stress for cast iron	=	*8 MPa*

Draw a neat sketch of the coupling.

Solution. Given : $P = 15$ kW $= 15 \times 10^3$ W ; $N = 900$ r.p.m. ; Service factor = 1.35 ; $\tau_s = \tau_b = \tau_k = 40$ MPa $= 40$ N/mm^2 ; $\sigma_{cb} = \sigma_{ck} = 80$ MPa $= 80$ N/mm^2 ; $\tau_c = 8$ MPa $= 8$ N/mm^2

The protective type flange coupling is designed as discussed below :

1. *Design for hub*

First of all, let us find the diameter of the shaft (d). We know that the torque transmitted by the shaft,

$$T = \frac{P \times 60}{2\,\pi N} = \frac{15 \times 10^3 \times 60}{2\,\pi \times 900} = 159.13 \text{ N-m}$$

Since the service factor is 1.35, therefore the maximum torque transmitted by the shaft,

$$T_{max} = 1.35 \times 159.13 = 215 \text{ N-m} = 215 \times 10^3 \text{ N-mm}$$

We know that the torque transmitted by the shaft (T),

$$215 \times 10^3 = \frac{\pi}{16} \times \tau_s \times d^3 = \frac{\pi}{16} \times 40 \times d^3 = 7.86\, d^3$$

$$\therefore \quad d^3 = 215 \times 10^3 / 7.86 = 27.4 \times 10^3 \quad \text{or} \quad d = 30.1 \text{ say } 35 \text{ mm } \textbf{Ans.}$$

We know that outer diameter of the hub,

$$D = 2d = 2 \times 35 = 70 \text{ mm } \textbf{Ans.}$$

and length of hub, $L = 1.5\, d = 1.5 \times 35 = 52.5$ mm **Ans.**

Let us now check the induced shear stress for the hub material which is cast iron. Considering the hub as a hollow shaft. We know that the maximum torque transmitted (T_{max}).

$$215 \times 10^3 = \frac{\pi}{16} \times \tau_c \left[\frac{D^4 - d^4}{D}\right] = \frac{\pi}{16} \times \tau_c \left[\frac{(70)^4 - (35)^4}{70}\right] = 63\,147\, \tau_c$$

$$\therefore \quad \tau_c = 215 \times 10^3 / 63\,147 = 3.4 \text{ N/mm}^2 = 3.4 \text{ MPa}$$

Since the induced shear stress for the hub material (*i.e.* cast iron) is less than the permissible value of 8 MPa, therefore the design of hub is safe.

2. ***Design for key***

Since the crushing stress for the key material is twice its shear stress (*i.e.* $\sigma_{ck} = 2\tau_k$), therefore a square key may be used. From Table 13.1, we find that for a shaft of 35 mm diameter,

Width of key, $w = 12$ mm **Ans.**

and thickness of key, $t = w = 12$ mm **Ans.**

The length of key (l) is taken equal to the length of hub.

$$\therefore \quad l = L = 52.5 \text{ mm } \textbf{Ans.}$$

Let us now check the induced stresses in the key by considering it in shearing and crushing.

Considering the key in shearing. We know that the maximum torque transmitted (T_{max}),

$$215 \times 10^3 = l \times w \times \tau_k \times \frac{d}{2} = 52.5 \times 12 \times \tau_k \times \frac{35}{2} = 11\,025\, \tau_k$$

$$\therefore \quad \tau_k = 215 \times 10^3 / 11\,025 = 19.5 \text{ N/mm}^2 = 19.5 \text{ MPa}$$

Considering the key in crushing. We know that the maximum torque transmitted (T_{max}),

$$215 \times 10^3 = l \times \frac{t}{2} \times \sigma_{ck} \times \frac{d}{2} = 52.5 \times \frac{12}{2} \times \sigma_{ck} \times \frac{35}{2} = 5512.5\, \sigma_{ck}$$

$$\therefore \quad \sigma_{ck} = 215 \times 10^3 / 5512.5 = 39 \text{ N/mm}^2 = 39 \text{ MPa}$$

Since the induced shear and crushing stresses in the key are less than the permissible stresses, therefore the design for key is safe.

3. ***Design for flange***

The thickness of flange (t_f) is taken as 0.5 d.

$$\therefore \quad t_f = 0.5\, d = 0.5 \times 35 = 17.5 \text{ mm } \textbf{Ans.}$$

Let us now check the induced shearing stress in the flange by considering the flange at the junction of the hub in shear.

We know that the maximum torque transmitted (T_{max}),

$$215 \times 10^3 = \frac{\pi D^2}{2} \times \tau_c \times t_f = \frac{\pi (70)^2}{2} \times \tau_c \times 17.5 = 134\ 713\ \tau_c$$

$$\therefore \quad \tau_c = 215 \times 10^3 / 134\ 713 = 1.6 \text{ N/mm}^2 = 1.6 \text{ MPa}$$

Since the induced shear stress in the flange is less than 8 MPa, therefore the design of flange is safe.

4. *Design for bolts*

Let d_1 = Nominal diameter of bolts.

Since the diameter of the shaft is 35 mm, therefore let us take the number of bolts,

$$n = 3$$

and pitch circle diameter of bolts,

$$D_1 = 3d = 3 \times 35 = 105 \text{ mm}$$

The bolts are subjected to shear stress due to the torque transmitted. We know that the maximum torque transmitted (T_{max}),

$$215 \times 10^3 = \frac{\pi}{4} (d_1)^2\ \tau_b \times n \times \frac{D_1}{2} = \frac{\pi}{4} (d_1)^2\ 40 \times 3 \times \frac{105}{2} = 4950\ (d_1)^2$$

$$\therefore \quad (d_1)^2 = 215 \times 10^3 / 4950 = 43.43 \text{ or } d_1 = 6.6 \text{ mm}$$

Assuming coarse threads, the nearest standard size of bolt is M 8. **Ans.**

Other proportions of the flange are taken as follows :

Outer diameter of the flange,

$$D_2 = 4\ d = 4 \times 35 = 140 \text{ mm } \textbf{Ans.}$$

Thickness of the protective circumferential flange,

$$t_p = 0.25\ d = 0.25 \times 35 = 8.75 \text{ say } 10 \text{ mm } \textbf{Ans.}$$

Example 13.7. *Design and draw a protective type of cast iron flange coupling for a steel shaft transmitting 15 kW at 200 r.p.m. and having an allowable shear stress of 40 MPa. The working stress in the bolts should not exceed 30 MPa. Assume that the same material is used for shaft and key and that the crushing stress is twice the value of its shear stress. The maximum torque is 25% greater than the full load torque. The shear stress for cast iron is 14 MPa.*

Solution. Given : P = 15 kW = 15 × 10³ W ; N = 200 r.p.m. ; τ_s = 40 MPa = 40 N/mm² ; τ_b = 30 MPa = 30 N/mm² ; $\sigma_{ck} = 2\tau_k$; $T_{max} = 1.25\ T_{mean}$; τ_c = 14 MPa = 14 N/mm²

The protective type of cast iron flange coupling is designed as discussed below :

1. *Design for hub*

First of all, let us find the diameter of shaft (d). We know that the full load or mean torque transmitted by the shaft,

$$T_{mean} = \frac{P \times 60}{2\ \pi N} = \frac{15 \times 10^3 \times 60}{2\ \pi \times 200} = 716 \text{ N-m} = 716 \times 10^3 \text{ N-mm}$$

and maximum torque transmitted,

$$T_{max} = 1.25\ T_{mean} = 1.25 \times 716 \times 10^3 = 895 \times 10^3 \text{ N-mm}$$

We also know that maximum torque transmitted (T_{max}),

$$895 \times 10^3 = \frac{\pi}{16} \times \tau_s \times d^3 = \frac{\pi}{16} \times 40 \times d^3 = 7.86\ d^3$$

$\therefore \quad d^3 = 895 \times 10^3 / 7.86 = 113\ 868$ or $d = 48.4$ say 50 mm **Ans.**

We know that the outer diameter of the hub,

$$D = 2\,d = 2 \times 50 = 100 \text{ mm } \textbf{Ans.}$$

and length of the hub, $L = 1.5\,d = 1.5 \times 50 = 75$ mm **Ans.**

Let us now check the induced shear stress for the hub material which is cast iron, by considering it as a hollow shaft. We know that the maximum torque transmitted (T_{max}),

$$895 \times 10^3 = \frac{\pi}{16} \times \tau_c \left(\frac{D^4 - d^4}{D}\right) = \frac{\pi}{16} \times \tau_c \left(\frac{(100)^4 - (50)^2}{100}\right) = 184\ 100\ \tau_c$$

$$\therefore \quad \tau_c = 895 \times 10^3 / 184\ 100 = 4.86 \text{ N/mm}^2 = 4.86 \text{ MPa}$$

Since the induced shear stress in the hub is less than the permissible value of 14 MPa, therefore the design for hub is safe.

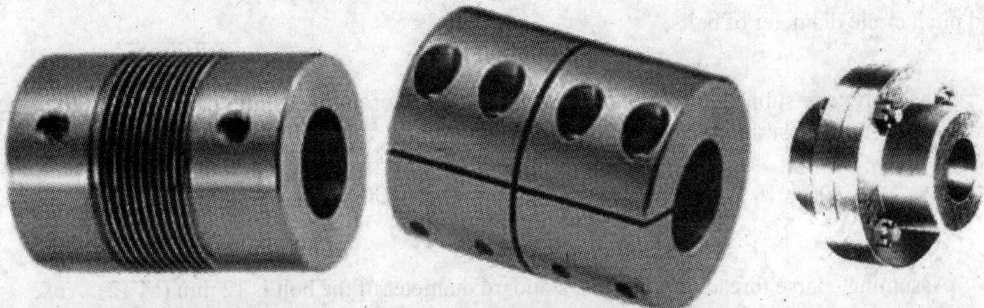

(a) Miniature flexible coupling (b) Miniature rigid coupling (c) Rigid coupling.

2. *Design for key*

Since the crushing stress for the key material is twice its shear stress, therefore a square key may be used.

From Table 13.1, we find that for a 50 mm diameter shaft,

Width of key, $w = 16$ mm **Ans.**

and thickness of key, $t = w = 16$ mm **Ans.**

The length of key (l) is taken equal to the length of hub.

$\therefore \quad l = L = 75$ mm **Ans.**

Let us now check the induced stresses in the key by considering it in shearing and crushing. Considering the key in shearing. We know that the maximum torque transmitted (T_{max}),

$$895 \times 10^3 = l \times w \times \tau_k \times \frac{d}{2} = 75 \times 16 \times \tau_k \times \frac{50}{2} = 30 \times 10^3\ \tau_k$$

$$\therefore \quad \tau_k = 895 \times 10^3 / 30 \times 10^3 = 29.8 \text{ N/mm}^2 = 29.8 \text{ MPa}$$

Considering the key in crushing. We know that the maximum torque transmitted (T_{max}),

$$895 \times 10^3 = l \times \frac{t}{2} \times \sigma_{ck} \times \frac{d}{2} = 75 \times \frac{16}{2} \times \sigma_{ck} \times \frac{50}{2} = 15 \times 10^3\ \sigma_{ck}$$

$$\therefore \quad \sigma_{ck} = 895 \times 10^3 / 15 \times 10^3 = 59.6 \text{ N/mm}^2 = 59.6 \text{ MPa}$$

Since the induced shear and crushing stresses in key are less than the permissible stresses, therefore the design for key is safe.

3. *Design for flange*

The thickness of the flange (t_f) is taken as 0.5 d.

$\therefore \quad t_f = 0.5 \times 50 = 25$ mm **Ans.**

Let us now check the induced shear stress in the flange, by considering the flange at the junction of the hub in shear. We know that the maximum torque transmitted (T_{max}),

$$895 \times 10^3 = \frac{\pi D^2}{2} \times \tau_c \times t_f = \frac{\pi (100)^2}{2} \times \tau_c \times 25 = 392\ 750\ \tau_c$$

$$\therefore \quad \tau_c = 895 \times 10^3 / 392\ 750 = 2.5 \text{ N/mm}^2 = 2.5 \text{ MPa}$$

Since the induced shear stress in the flange is less than the permissible value of 14 MPa, therefore the design for flange is safe.

4. Design for bolts

Let $\quad d_1$ = Nominal diameter of bolts.

Since the diameter of shaft is 50 mm, therefore let us take the number of bolts,

$$n = 4$$

and pitch circle diameter of bolts,

$$D_1 = 3\, d = 3 \times 50 = 150 \text{ mm}$$

The bolts are subjected to shear stress due to the torque transmitted. We know that the maximum torque transmitted (T_{max}),

$$895 \times 10^3 = \frac{\pi}{4} (d_1)^2 \tau_b \times n \times \frac{D_1}{2} = \frac{\pi}{4} (d_1)^2\, 30 \times 4 \times \frac{150}{4} = 7070\, (d_1)^2$$

$$\therefore \quad (d_1)^2 = 895 \times 10^3 / 7070 = 126.6 \quad \text{or} \quad d_1 = 11.25 \text{ mm}$$

Assuming coarse threads, the nearest standard diameter of the bolt is 12 mm (M 12). **Ans.**

Other proportions of the flange are taken as follows :

Outer diameter of the flange,

$$D_2 = 4\, d = 4 \times 50 = 200 \text{ mm } \textbf{Ans.}$$

Thickness of the protective circumferential flange,

$$t_p = 0.25\, d = 0.25 \times 50 = 12.5 \text{ mm } \textbf{Ans.}$$

Example 13.8. *Design and draw a cast iron flange coupling for a mild steel shaft transmitting 90 kW at 250 r.p.m. The allowable shear stress in the shaft is 40 MPa and the angle of twist is not to exceed 1° in a length of 20 diameters. The allowable shear stress in the coupling bolts is 30 MPa.*

Solution. Given : $P = 90$ kW $= 90 \times 10^3$ W ; $N = 250$ r.p.m. ; $\tau_s = 40$ MPa $= 40$ N/mm² ; $\theta = 1° = \pi / 180 = 0.0175$ rad ; $\tau_b = 30$ MPa $= 30$ N/mm²

First of all, let us find the diameter of the shaft (d). We know that the torque transmitted by the shaft,

$$T = \frac{P \times 60}{2\pi N} = \frac{90 \times 10^3 \times 60}{2\pi \times 250} = 3440 \text{ N-m} = 3440 \times 10^3 \text{ N-mm}$$

Considering strength of the shaft, we know that

$$\frac{T}{J} = \frac{\tau_s}{d/2}$$

$$\frac{3440 \times 10^3}{\frac{\pi}{32} \times d^4} = \frac{40}{d/2} \quad \text{or} \quad \frac{35 \times 10^6}{d^4} = \frac{80}{d} \quad \ldots \left(\because J = \frac{\pi}{32} \times d^4\right)$$

$$\therefore \quad d^3 = 35 \times 10^6 / 80 = 0.438 \times 10^6 \quad \text{or} \quad d = 76 \text{ mm}$$

Considering rigidity of the shaft, we know that

$$\frac{T}{J} = \frac{C \times \theta}{l}$$

$$\frac{3440 \times 10^3}{\frac{\pi}{32} \times d^4} = \frac{84 \times 10^3 \times 0.0175}{20\,d} \quad \text{or} \quad \frac{35 \times 10^6}{d^4} = \frac{73.5}{d} \quad \text{... (Taking } C = 84 \text{ kN/mm}^2\text{)}$$

$$\therefore \quad d^3 = 35 \times 10^6 / 73.5 = 0.476 \times 10^6 \text{ or } d = 78 \text{ mm}$$

Taking the larger of the two values, we have

$$d = 78 \text{ say } 80 \text{ mm } \textbf{Ans.}$$

Let us now design the cast iron flange coupling of the protective type as discussed below :

1. *Design for hub*

We know that the outer diameter of hub,

$$D = 2d = 2 \times 80 = 160 \text{ mm } \textbf{Ans.}$$

and length of hub, $L = 1.5\,d = 1.5 \times 80 = 120$ mm **Ans.**

Let us now check the induced shear stress in the hub by considering it as a hollow shaft. The shear stress for the hub material (which is cast iron) is usually 14 MPa. We know that the torque transmitted (T),

$$3440 \times 10^3 = \frac{\pi}{16} \times \tau_c \left[\frac{D^4 - d^4}{D}\right] = \frac{\pi}{16} \times \tau_c \left[\frac{(160)^4 - (80)^4}{160}\right] = 754 \times 10^3\, \tau_c$$

$$\therefore \quad \tau_c = 3440 \times 10^3 / 754 \times 10^3 = 4.56 \text{ N/mm}^2 = 4.56 \text{ MPa}$$

Since the induced shear stress for the hub material is less than 14 MPa, therefore the design for hub is safe.

2. *Design for key*

From Table 13.1, we find that the proportions of key for a 80 mm diameter shaft are :

Width of key, $w = 25$ mm **Ans.**

and thickness of key, $t = 14$ mm **Ans.**

The length of key (l) is taken equal to the length of hub (L).

$$\therefore \quad l = L = 120 \text{ mm } \textbf{Ans.}$$

Assuming that the shaft and key are of the same material. Let us now check the induced shear stress in key. We know that the torque transmitted (T),

$$3440 \times 10^3 = l \times w \times \tau_k \times \frac{d}{2} = 120 \times 25 \times \tau_k \times \frac{80}{2} = 120 \times 10^3\, \tau_k$$

$$\tau_k = 3440 \times 10^3 / 120 \times 10^3 = 28.7 \text{ N/mm}^2 = 28.7 \text{ MPa}$$

Since the induced shear stress in the key is less than 40 MPa, therefore the design for key is safe.

3. *Design for flange*

The thickness of the flange (t_f) is taken as 0.5 d.

$$\therefore \quad t_f = 0.5\,d = 0.5 \times 80 = 40 \text{ mm } \textbf{Ans.}$$

Let us now check the induced shear stress in the cast iron flange by considering the flange at the junction of the hub under shear. We know that the torque transmitted (T),

$$3440 \times 10^3 = \frac{\pi\, D^2}{2} \times t_f \times \tau_c = \frac{\pi (160)^2}{2} \times 40 \times \tau_c = 1608 \times 10^3\, \tau_c$$

$$\therefore \quad \tau_c = 3440 \times 10^3 / 1608 \times 10^3 = 2.14 \text{ N/mm}^2 = 2.14 \text{ MPa}$$

Since the induced shear stress in the flange is less than 14 MPa, therefore the design for flange is safe.

4. ***Design for bolts***

Let d_1 = Nominal diameter of bolts.

Since the diameter of the shaft is 80 mm, therefore let us take number of bolts,

$$n = 4$$

and pitch circle diameter of bolts,

$$D_1 = 3\,d = 3 \times 80 = 240 \text{ mm}$$

The bolts are subjected to shear stress due to the torque transmitted. We know that torque transmitted (T),

$$3440 \times 10^3 = \frac{\pi}{4}(d_1)^2\, n \times \tau_b \times \frac{D_1}{2} = \frac{\pi}{4}(d_1)^2 \times 4 \times 30 \times \frac{240}{2} = 11\,311\,(d_1)^2$$

$$\therefore \quad (d_1)^2 = 3440 \times 10^3/11\,311 = 304 \text{ or } d_1 = 17.4 \text{ mm}$$

Assuming coarse threads, the standard nominal diameter of bolt is 18 mm. **Ans.**

The other proportions are taken as follows :

Outer diameter of the flange,

$$D_2 = 4\,d = 4 \times 80 = 320 \text{ mm } \textbf{Ans.}$$

Thickness of protective circumferential flange,

$$t_p = 0.25\,d = 0.25 \times 80 = 20 \text{ mm } \textbf{Ans.}$$

Example 13.9. *Design a rigid flange coupling to transmit a torque of 250 N-m between two co-axial shafts. The shaft is made of alloy steel, flanges out of cast iron and bolts out of steel. Four bolts are used to couple the flanges. The shafts are keyed to the flange hub. The permissible stresses are given below:*

Shear stress on shaft	*= 100 MPa*
Bearing or crushing stress on shaft	*= 250 MPa*
Shear stress on keys	*= 100 MPa*
Bearing stress on keys	*= 250 MPa*
Shearing stress on cast iron	*= 200 MPa*
Shear stress on bolts	*= 100 MPa*

After designing the various elements, make a neat sketch of the assembly indicating the important dimensions. The stresses developed in the various members may be checked if thumb rules are used for fixing the dimensions.

Soution. Given : T = 250 N-m = 250 × 10^3 N-mm ; n = 4; τ_s = 100 MPa = 100 N/mm^2 ; σ_{cs} = 250 MPa = 250 N/mm^2 ; τ_k = 100 MPa = 100 N/mm^2 ; σ_{ck} = 250 MPa = 250 N/mm^2 ; τ_c = 200 MPa = 200 N/mm^2 ; τ_b = 100 MPa = 100 N/mm^2

The cast iron flange coupling of the protective type is designed as discussed below :

1. ***Design for hub***

First of all, let us find the diameter of the shaft (d). We know that the torque transmitted by the shaft (T),

$$250 \times 10^3 = \frac{\pi}{16} \times \tau_s \times d^3 = \frac{\pi}{16} \times 100 \times d^3 = 19.64\,d^3$$

$$\therefore \quad d^3 = 250 \times 10^3 / 19.64 = 12\,729 \text{ or } d = 23.35 \text{ say } 25 \text{ mm } \textbf{Ans.}$$

We know that the outer diameter of the hub,

$$D = 2\,d = 2 \times 25 = 50 \text{ mm}$$

and length of hub, $L = 1.5\,d = 1.5 \times 25 = 37.5$ mm

Let us now check the induced shear stress in the hub by considering it as a hollow shaft. We know that the torque transmitted (T),

$$250 \times 10^3 = \frac{\pi}{16} \times \tau_c \left(\frac{D^4 - d^4}{D}\right) = \frac{\pi}{16} \times \tau_c \left[\frac{(50)^4 - (25)^4}{50}\right] = 23\,013\ \tau_c$$

$$\therefore \quad \tau_c = 250 \times 10^3/23\,013 = 10.86\ \text{N/mm}^2 = 10.86\ \text{MPa}$$

Since the induced shear stress for the hub material (*i.e.* cast iron) is less than 200 MPa, therefore the design for hub is safe.

2. ***Design for key***

From Table 13.1, we find that the proportions of key for a 25 mm diameter shaft are :

Width of key, $w = 10$ mm **Ans.**

and thickness of key, $t = 8$ mm **Ans.**

The length of key (l) is taken equal to the length of hub,

$$\therefore \quad l = L = 37.5\ \text{mm} \quad \textbf{Ans.}$$

Let us now check the induced shear and crushing stresses in the key. Considering the key in shearing. We know that the torque transmitted (T),

$$250 \times 10^3 = l \times w \times \tau_k \times \frac{d}{2} = 37.5 \times 10 \times \tau_k \times \frac{25}{2} = 4688\ \tau_k$$

$$\therefore \quad \tau_k = 250 \times 10^3 / 4688 = 53.3\ \text{N/mm}^2 = 53.3\ \text{MPa}$$

Considering the key in crushing. We know that the torque transmitted (T),

$$250 \times 10^3 = l \times \frac{t}{2} \times \sigma_{ck} \times \frac{d}{2} = 37.5 \times \frac{8}{2} \times \sigma_{ck} \times \frac{25}{2} = 1875\ \sigma_{ck}$$

$$\therefore \quad \sigma_{ck} = 250 \times 10^3 / 1875 = 133.3\ \text{N/mm}^2 = 133.3\ \text{MPa}$$

Since the induced shear and crushing stresses in the key are less than the given stresses, therefore the design of key is safe.

3. ***Design for flange***

The thickness of the flange (t_f) is taken as 0.5 d.

$$\therefore \quad t_f = 0.5\,d = 0.5 \times 25 = 12.5\ \text{mm} \quad \textbf{Ans.}$$

Let us now check the induced shear stress in the flange by considering the flange at the junction of the hub in shear. We know that the torque transmitted (T),

$$250 \times 10^3 = \frac{\pi D^2}{2} \times \tau_c \times t_f = \frac{\pi (50)^2}{2} \times \tau_c \times 12.5 = 49\,094\ \tau_c$$

$$\therefore \quad \tau_c = 250 \times 10^3 / 49\,094 = 5.1\ \text{N/mm}^2 = 5.1\ \text{MPa}$$

Since the induced shear stress in the flange of cast iron is less than 200 MPa, therefore design of flange is safe.

4. ***Design for bolts***

Let d_1 = Nominal diameter of bolts.

We know that the pitch circle diameter of bolts,

$$\therefore \quad D_1 = 3\,d = 3 \times 25 = 75\ \text{mm} \quad \textbf{Ans.}$$

The bolts are subjected to shear stress due to the torque transmitted. We know that torque transmitted (T),

$$250 \times 10^3 = \frac{\pi}{4} (d_1)^2\, \tau_b \times n \times \frac{D_1}{2} = \frac{\pi}{4} (d_1)^2\, 100 \times 4 \times \frac{75}{2} = 11\,780\ (d_1)^2$$

$\therefore \quad (d_1)^2 = 250 \times 10^3 / 11\ 780 = 21.22 \quad$ or $\quad d_1 = 4.6$ mm

Assuming coarse threads, the nearest standard size of the bolt is M 6. **Ans.**

Other proportions of the flange are taken as follows :

Outer diameter of the flange,

$$D_2 = 4\,d = 4 \times 25 = 100 \text{ mm } \textbf{Ans.}$$

Thickness of the protective circumferential flange,

$$t_p = 0.25\,d = 0.25 \times 25 = 6.25 \text{ mm } \textbf{Ans.}$$

Example 13.10. *Two 35 mm shafts are connected by a flanged coupling. The flanges are fitted with 6 bolts on 125 mm bolt circle. The shafts transmit a torque of 800 N-m at 350 r.p.m. For the safe stresses mentioned below, calculate 1. diameter of bolts ; 2. thickness of flanges ; 3. key dimensions ; 4. hub length; and 5. power transmitted.*

Safe shear stress for shaft material = 63 MPa
Safe stress for bolt material = 56 MPa
Safe stress for cast iron coupling = 10 MPa
Safe stress for key material = 46 MPa

Solution. Given : $d = 35$ mm ; $n = 6$; $D_1 = 125$ mm ; $T = 800$ N-m $= 800 \times 10^3$ N-mm ; $N = 350$ r.p.m.; $\tau_s = 63$ MPa $= 63$ N/mm² ; $\tau_b = 56$ MPa $= 56$ N/mm² ; $\tau_c = 10$ MPa $= 10$ N/mm² ; $\tau_k = 46$ MPa $= 46$ N/mm²

1. ***Diameter of bolts***

Let $\quad d_1$ = Nominal or outside diameter of bolt.

We know that the torque transmitted (T),

$$800 \times 10^3 = \frac{\pi}{4}(d_1)^2\,\tau_b \times n \times \frac{D_1}{2} = \frac{\pi}{4}(d_1)^2\,56 \times 6 \times \frac{125}{2} = 16\ 495\,(d_1)^2$$

$$\therefore \quad (d_1)^2 = 800 \times 10^3 / 16\ 495 = 48.5 \text{ or } d_1 = 6.96 \text{ say } 8 \text{ mm } \textbf{Ans.}$$

2. ***Thickness of flanges***

Let $\quad t_f$ = Thickness of flanges.

We know that the torque transmitted (T),

$$800 \times 10^3 = \frac{\pi D^2}{2} \times \tau_c \times t_f = \frac{\pi\,(2 \times 35)^2}{2} \times 10 \times t_f = 76\ 980\ t_f \quad \text{... } (\because D = 2d)$$

$$\therefore \quad t_f = 800 \times 10^3 / 76\ 980 = 10.4 \text{ say } 12 \text{ mm } \textbf{Ans.}$$

3. ***Key dimensions***

From Table 13.1, we find that the proportions of key for a 35 mm diameter shaft are :

Width of key, $\quad w = 12$ mm **Ans.**

and thickness of key, $\quad t = 8$ mm **Ans.**

The length of key (l) is taken equal to the length of hub (L).

$$\therefore \quad l = L = 1.5\,d = 1.5 \times 35 = 52.5 \text{ mm}$$

Let us now check the induced shear stress in the key. We know that the torque transmitted (T),

$$800 \times 10^3 = l \times w \times \tau_k \times \frac{d}{2} = 52.5 \times 12 \times \tau_k \times \frac{35}{2} = 11\ 025\ \tau_k$$

$$\therefore \quad \tau_k = 800 \times 10^3 / 11\ 025 = 72.5 \text{ N/mm}^2$$

Since the induced shear stress in the key is more than the given safe stress (46 MPa), therefore let us find the length of key by substituting the value of $\tau_k = 46$ MPa in the above equation, *i.e.*

$$800 \times 10^3 = l \times 12 \times 46 \times \frac{35}{2} = 9660\, l$$

$$\therefore \quad l = 800 \times 10^3 / 9660 = 82.8 \text{ say } 85 \text{ mm } \textbf{Ans.}$$

4. *Hub length*

Since the length of key is taken equal to the length of hub, therefore we shall take hub length,

$$L = l = 85 \text{ mm } \textbf{Ans.}$$

5. *Power transmitted*

We know that the power transmitted,

$$P = \frac{T \times 2\pi N}{60} = \frac{800 \times 2\pi \times 350}{60} = 29\,325 \text{ W} = 29.325 \text{ kW } \textbf{Ans.}$$

Example 13.11. *The shaft and the flange of a marine engine are to be designed for flange coupling, in which the flange is forged on the end of the shaft. The following particulars are to be considered in the design :*

Power of the engine	*= 3 MW*
Speed of the engine	*= 100 r.p.m.*
Permissible shear stress in bolts and shaft	*= 60 MPa*
Number of bolts used	*= 8*
Pitch circle diameter of bolts	*= 1.6 × Diameter of shaft*

Find : 1. diameter of shaft ; 2. diameter of bolts ; 3. thickness of flange ; and 4. diameter of flange.

Solution. Given : $P = 3$ MW $= 3 \times 10^6$ W ; $N = 100$ r.p.m. ; $\tau_b = \tau_s = 60$ MPa $= 60$ N/mm^2 ; $n = 8$; $D_1 = 1.6\, d$

1. *Diameter of shaft*

Let $\quad d$ = Diameter of shaft.

We know that the torque transmitted by the shaft,

$$T = \frac{P \times 60}{2\pi N} = \frac{3 \times 10^6 \times 60}{2\pi \times 100} = 286 \times 10^3 \text{ N-m} = 286 \times 10^6 \text{ N-mm}$$

We also know that torque transmitted by the shaft (T),

$$286 \times 10^6 = \frac{\pi}{16} \times \tau_s \times d^3 = \frac{\pi}{16} \times 60 \times d^3 = 11.78\, d^3$$

$$\therefore \quad d^3 = 286 \times 10^6 / 11.78 = 24.3 \times 10^6$$

or $\quad d = 2.89 \times 10^2 = 289$ say 300 mm **Ans.**

2. *Diameter of bolts*

Let $\quad d_1$ = Nominal diameter of bolts.

The bolts are subjected to shear stress due to the torque transmitted. We know that torque transmitted (T),

$$286 \times 10^6 = \frac{\pi}{4} (d_1)^2 \tau_b \times n \times \frac{D_1}{2} = \frac{\pi}{4} \times (d_1)^2\, 60 \times 8 \times \frac{1.6 \times 300}{2}$$

$$= 90\,490\, (d_1)^2 \quad \ldots (\because D_1 = 1.6\, d)$$

$$\therefore \quad (d_1)^2 = 286 \times 10^6 / 90\,490 = 3160 \quad \text{or} \quad d_1 = 56.2 \text{ mm}$$

Assuming coarse threads, the standard diameter of the bolt is 60 mm (M 60). The taper on the bolt may be taken from 1 in 20 to 1 in 40. **Ans.**

3. ***Thickness of flange***

The thickness of flange (t_f) is taken as d / 3.

$\therefore \qquad t_f = d / 3 = 300/3 = 100$ mm **Ans.**

Let us now check the induced shear stress in the flange by considering the flange at the junction of the shaft in shear. We know that the torque transmitted (T),

$$286 \times 10^6 = \frac{\pi d^2}{2} \times \tau_s \times t_f = \frac{\pi (300)^2}{2} \times \tau_s \times 100 = 14.14 \times 10^6 \, \tau_s$$

$\therefore \qquad \tau_s = 286 \times 10^6 / 14.14 \times 10^6 = 20.2 \text{ N/mm}^2 = 20.2 \text{ MPa}$

Since the induced shear stress in the *flange is less than the permissible shear stress of 60 MPa, therefore the thickness of flange (t_f = 100 mm) is safe.

4. ***Diameter of flange***

The diameter of flange (D_2) is taken as 2.2 d.

$\therefore \qquad D_2 = 2.2\, d = 2.2 \times 300 = 660$ mm **Ans.**

13.18 Flexible Coupling

We have already discussed that a flexible coupling is used to join the abutting ends of shafts

(a) Bellows coupling , (b) Elastomeric coupling, (c) Flanged coupling , (d) Flexible coupling

Note : This picture is given as additional information and is not a direct example of the current chapter.

when they are not in exact alignment. In the case of a direct coupled drive from a prime mover to an electric generator, we should have four bearings at a comparatively close distance. In such a case and in many others, as in a direct electric drive from an electric motor to a machine tool, a flexible coupling is used so as to permit an axial misalignemnt of the shaft without undue absorption of the power which the shaft are transmitting. Following are the different types of flexible couplings :

1. Bushed pin flexible coupling, **2.** Oldham's coupling, and **3.** Universal coupling.

We shall now discuss these types of couplings, in detail, in the following articles.

---

* The flange material in case of marine flange coupling is same as that of shaft.

13.19 Bushed-pin Flexible Coupling

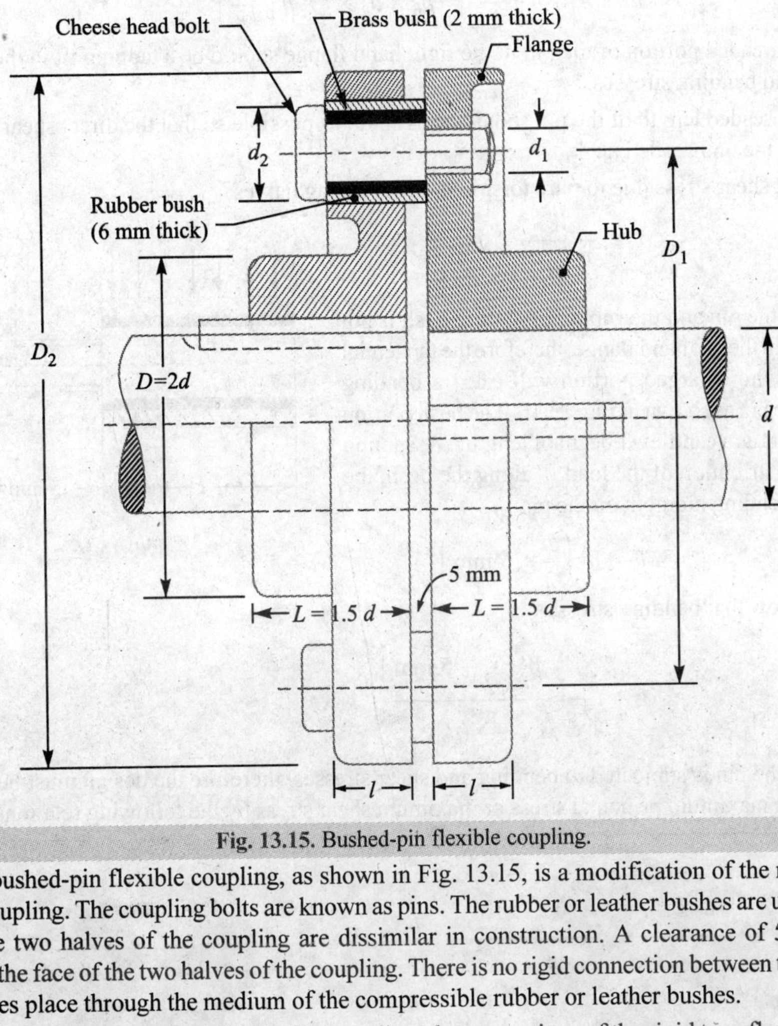

Fig. 13.15. Bushed-pin flexible coupling.

A bushed-pin flexible coupling, as shown in Fig. 13.15, is a modification of the rigid type of flange coupling. The coupling bolts are known as pins. The rubber or leather bushes are used over the pins. The two halves of the coupling are dissimilar in construction. A clearance of 5 mm is left between the face of the two halves of the coupling. There is no rigid connection between them and the drive takes place through the medium of the compressible rubber or leather bushes.

In designing the bushed-pin flexible coupling, the proportions of the rigid type flange coupling are modified. The main modification is to reduce the bearing pressure on the rubber or leather bushes and it should not exceed 0.5 N/mm². In order to keep the low bearing pressure, the pitch circle diameter and the pin size is increased.

Let l = Length of bush in the flange,

d_2 = Diameter of bush,

p_b = Bearing pressure on the bush or pin,

n = Number of pins, and

D_1 = Diameter of pitch circle of the pins.

We know that bearing load acting on each pin,

$$W = p_b \times d_2 \times l$$

∴ Total bearing load on the bush or pins

$$= W \times n = p_b \times d_2 \times l \times n$$

and the torque transmitted by the coupling,

$$T = W \times n \left(\frac{D_1}{2}\right) = p_b \times d_2 \times l \times n \left(\frac{D_1}{2}\right)$$

The threaded portion of the pin in the right hand flange should be a tapping fit in the coupling hole to avoid bending stresses.

The threaded length of the pin should be as small as possible so that the direct shear stress can be taken by the unthreaded neck.

Direct shear stress due to pure torsion in the coupling halves,

$$\tau = \frac{W}{\frac{\pi}{4}(d_1)^2}$$

Since the pin and the rubber or leather bush is not rigidly held in the left hand flange, therefore the tangential load (W) at the enlarged portion will exert a bending action on the pin as shown in Fig. 13.16. The bush portion of the pin acts as a cantilever beam of length l. Assuming a uniform distribution of the load W along the bush, the maximum bending moment on the pin,

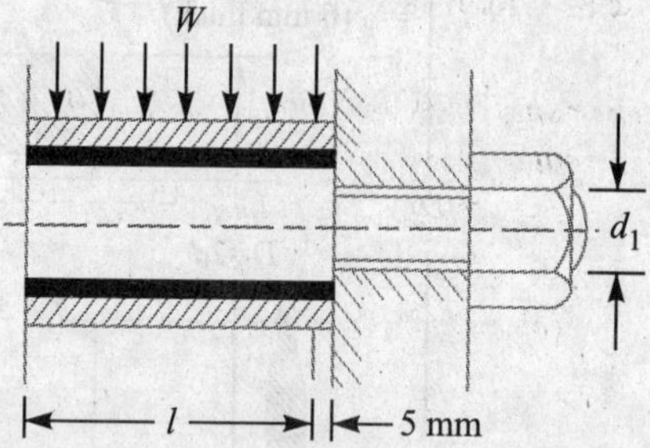

Fig. 13.16

$$M = W\left(\frac{l}{2} + 5 \text{ mm}\right)$$

We know that bending stress,

$$\sigma = \frac{M}{Z} = \frac{W\left(\frac{l}{2} + 5 \text{ mm}\right)}{\frac{\pi}{32}(d_1)^3}$$

Since the pin is subjected to bending and shear stresses, therefore the design must be checked either for the maximum principal stress or maximum shear stress by the following relations :

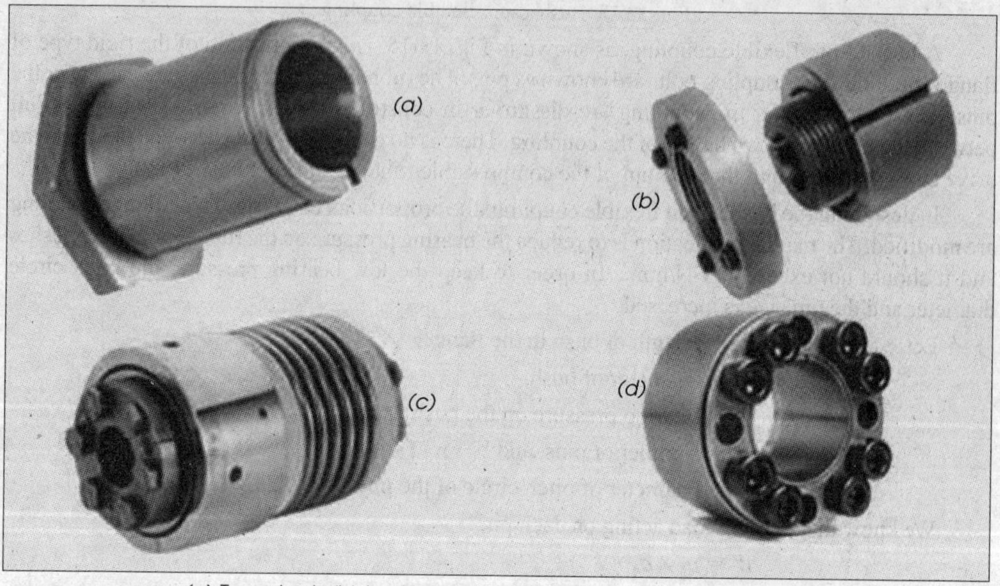

(a) Taper bush (b) Locking-assembly (shaft or bush connectors) (c) Friction joint bushing (d) Safety overload coupling.

Maximum principal stress

$$= \frac{1}{2}\left[\sigma + \sqrt{\sigma^2 + 4\tau^2}\right]$$

and the maximum shear stress on the pin

$$= \frac{1}{2}\sqrt{\sigma^2 + 4\tau^2}$$

The value of maximum principal stress varies from 28 to 42 MPa.

Note: After designing the pins and rubber bush, the hub, key and flange may be designed in the similar way as discussed for flange coupling.

Example 13.12. *Design a bushed-pin type of flexible coupling to connect a pump shaft to a motor shaft transmitting 32 kW at 960 r.p.m. The overall torque is 20 percent more than mean torque. The material properties are as follows :*

(*a*) *The allowable shear and crushing stress for shaft and key material is 40 MPa and 80 MPa respectively.*

(*b*) *The allowable shear stress for cast iron is 15 MPa.*

(*c*) *The allowable bearing pressure for rubber bush is 0.8 N/mm².*

(*d*) *The material of the pin is same as that of shaft and key.*

Draw neat sketch of the coupling.

Solution. Given : $P = 32$ kW $= 32 \times 10^3$ W ; $N = 960$ r.p.m. ; $T_{max} = 1.2\ T_{mean}$; $\tau_s = \tau_k = 40$ MPa $= 40$ N/mm² ; $\sigma_{cs} = \sigma_{ck} = 80$ MPa $= 80$ N/mm² ; $\tau_c = 15$ MPa $= 15$ N/mm² ; $p_b = 0.8$ N/mm²

The bushed-pin flexible coupling is designed as discussed below :

1. ***Design for pins and rubber bush***

First of all, let us find the diameter of the shaft (d). We know that the mean torque transmitted by the shaft,

$$T_{mean} = \frac{P \times 60}{2\pi N} = \frac{32 \times 10^3 \times 60}{2\pi \times 960} = 318.3 \text{ N-m}$$

and the maximum or overall torque transmitted,

$$T_{max} = 1.2\ T_{mean} = 1.2 \times 318.3 = 382 \text{ N-m} = 382 \times 10^3 \text{ N-mm}$$

We also know that the maximum torque transmitted by the shaft (T_{max}),

$$382 \times 10^3 = \frac{\pi}{16} \times \tau_s \times d^3 = \frac{\pi}{16} \times 40 \times d^3 = 7.86\ d^3$$

$$\therefore \quad d^3 = 382 \times 10^3 / 7.86 = 48.6 \times 10^3 \text{ or } d = 36.5 \text{ say } 40 \text{ mm}$$

We have discussed in rigid type of flange coupling that the number of bolts for 40 mm diameter shaft are 3. In the flexible coupling, we shall use the number of pins (n) as 6.

$$\therefore \quad \text{Diameter of pins, } d_1 = \frac{0.5\,d}{\sqrt{n}} = \frac{0.5 \times 40}{\sqrt{6}} = 8.2 \text{ mm}$$

In order to allow for the bending stress induced due to the compressibility of the rubber bush, the diameter of the pin (d_1) may be taken as 20 mm. **Ans.**

The length of the pin of least diameter *i.e.* $d_1 = 20$ mm is threaded and secured in the right hand coupling half by a standard nut and washer. The enlarged portion of the pin which is in the left hand coupling half is made of 24 mm diameter. On the enlarged portion, a brass bush of thickness 2 mm is pressed. A brass bush carries a rubber bush. Assume the thickness of rubber bush as 6 mm.

∴ Overall diameter of rubber bush,

$$d_2 = 24 + 2 \times 2 + 2 \times 6 = 40 \text{ mm } \textbf{Ans.}$$

and diameter of the pitch circle of the pins,

$$D_1 = 2d + d_2 + 2 \times 6 = 2 \times 40 + 40 + 12 = 132 \text{ mm } \textbf{Ans.}$$

Let l = Length of the bush in the flange.

We know that the bearing load acting on each pin,

$$W = p_b \times d_2 \times l = 0.8 \times 40 \times l = 32\, l \text{ N}$$

and the maximum torque transmitted by the coupling (T_{max}),

$$382 \times 10^3 = W \times n \times \frac{D_1}{2} = 32\, l \times 6 \times \frac{132}{2} = 12\,672\, l$$

$$\therefore \quad l = 382 \times 10^3 / 12\,672 = 30.1 \text{ say } 32 \text{ mm}$$

and $W = 32\, l = 32 \times 32 = 1024$ N

∴ Direct stress due to pure torsion in the coupling halves,

$$\tau = \frac{W}{\frac{\pi}{4}(d_1)^2} = \frac{1024}{\frac{\pi}{4}(20)^2} = 3.26 \text{ N/mm}^2$$

Since the pin and the rubber bush are not rigidly held in the left hand flange, therefore the tangential load (W) at the enlarged portion will exert a bending action on the pin. Assuming a uniform distribution of load (W) along the bush, the maximum bending moment on the pin,

$$M = W\left(\frac{l}{2} + 5\right) = 1024\left(\frac{32}{2} + 5\right) = 21\,504 \text{ N-mm}$$

and section modulus, $Z = \frac{\pi}{32}(d_1)^3 = \frac{\pi}{32}(20)^3 = 785.5 \text{ mm}^3$

We know that bending stress,

$$\sigma = \frac{M}{Z} = \frac{21\,504}{785.5} = 27.4 \text{ N/mm}^2$$

∴ Maximum principal stress

$$= \frac{1}{2}\left[\sigma + \sqrt{\sigma^2 + 4\tau^2}\right] = \frac{1}{2}\left[27.4 + \sqrt{(27.4)^2 + 4\,(3.26)^2}\right]$$

$$= 13.7 + 14.1 = 27.8 \text{ N/mm}^2$$

and maximum shear stress

$$= \frac{1}{2}\left[\sqrt{\sigma^2 + 4\tau^2}\right] = \frac{1}{2}\left[\sqrt{(27.4)^2 + 4\,(3.26)^2}\right] = 14.1 \text{ N/mm}^2$$

Since the maximum principal stress and maximum shear stress are within limits, therefore the design is safe.

2. *Design for hub*

We know that the outer diameter of the hub,

$$D = 2\,d = 2 \times 40 = 80 \text{ mm}$$

and length of hub, $L = 1.5\,d = 1.5 \times 40 = 60$ mm

Let us now check the induced shear stress for the hub material which is cast iron. Considering the hub as a hollow shaft. We know that the maximum torque transmitted (T_{max}),

$$382 \times 10^3 = \frac{\pi}{16} \times \tau_c \left[\frac{D^4 - d^4}{D}\right] = \frac{\pi}{16} \times \tau_c \left[\frac{(80)^4 - (40)^4}{80}\right] = 94.26 \times 10^3\, \tau_c$$

$\therefore \quad \tau_c = 382 \times 10^3 / 94.26 \times 10^3 = 4.05 \text{ N/mm}^2 = 4.05 \text{ MPa}$

Since the induced shear stress for the hub material (*i.e.* cast iron) is less than the permissible value of 15 MPa, therefore the design of hub is safe.

3. *Design for key*

Since the crushing stress for the key material is twice its shear stress (*i.e.* $\sigma_{ck} = 2\,\tau_k$), therefore a square key may be used. From Table 13.1, we find that for a shaft of 40 mm diameter,

Width of key, $w = 14$ mm **Ans.**

and thickness of key, $t = w = 14$ mm **Ans.**

The length of key (L) is taken equal to the length of hub, *i.e.*

$$L = 1.5\,d = 1.5 \times 40 = 60 \text{ mm}$$

Let us now check the induced stresses in the key by considering it in shearing and crushing.

Considering the key in shearing. We know that the maximum torque transmitted (T_{max}),

$$382 \times 10^3 = L \times w \times \tau_k \times \frac{d}{2} = 60 \times 14 \times \tau_k \times \frac{40}{2} = 16\,800\,\tau_k$$

$$\therefore \quad \tau_k = 382 \times 10^3 / 16\,800 = 22.74 \text{ N/mm}^2 = 22.74 \text{ MPa}$$

Considering the key in crushing. We know that the maximum torque transmitted (T_{max}),

$$382 \times 10^3 = L \times \frac{t}{2} \times \sigma_{ck} \times \frac{d}{2} = 60 \times \frac{14}{2} \times \sigma_{ck} \times \frac{40}{2} = 8400\,\sigma_{ck}$$

$$\therefore \quad \sigma_{ck} = 382 \times 10^3 / 8400 = 45.48 \text{ N/mm}^2 = 45.48 \text{ MPa}$$

Since the induced shear and crushing stress in the key are less than the permissible stresses of 40 MPa and 80 MPa respectively, therefore the design for key is safe.

4. *Design for flange*

The thickness of flange (t_f) is taken as 0.5 d.

$$\therefore \quad t_f = 0.5\,d = 0.5 \times 40 = 20 \text{ mm}$$

Let us now check the induced shear stress in the flange by considering the flange at the junction of the hub in shear.

We know that the maximum torque transmitted (T_{max}),

$$382 \times 10^3 = \frac{\pi\,D^2}{2} \times \tau_c \times t_f = \frac{\pi (80)^2}{2} \times \tau_c \times 20 = 201 \times 10^3\,\tau_c$$

$$\therefore \quad \tau_c = 382 \times 10^3 / 201 \times 10^3 = 1.9 \text{ N/mm}^2 = 1.9 \text{ MPa}$$

Since the induced shear stress in the flange of cast iron is less than 15 MPa, therefore the design of flange is safe.

13.20 Oldham Coupling

It is used to join two shafts which have lateral mis-alignment. It consists of two flanges A and B with slots and a central floating part E with two tongues T_1 and T_2 at right angles as shown in Fig. 13.17. The central floating part is held by means of a pin passing through the flanges and the floating part. The tongue T_1 fits into the slot of flange A and allows for 'to and fro' relative motion of the shafts, while the tongue T_2 fits into the slot of the flange B and allows for vertical relative motion of the parts. The resultant of these two components of motion will accommodate lateral misalignment of the shaft as they rotate.

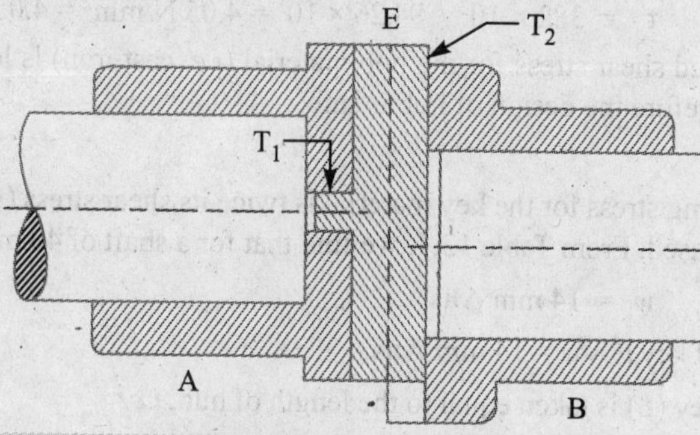

Fig. 13.17. Oldham coupling.

13.21 Universal (or Hooke's) Coupling

A universal or Hooke's coupling is used to connect two shafts whose axes intersect at a small angle. The inclination of the two shafts may be constant, but in actual practice, it varies when the motion is transmitted from one shaft to another. The main application of the universal or Hooke's coupling is found in the transmission from the gear box to the differential or back axle of the automobiles. In such a case, we use two Hooke's coupling, one at each end of the propeller shaft, connecting the gear box at one end and the differential on the other end. A Hooke's coupling is also used for transmission of power to different spindles of multiple drilling machine. It is used as a knee joint in milling machines.

In designing a universal coupling, the shaft diameter and the pin diameter is obtained as discussed below. The other dimensions of the coupling are fixed by proportions as shown in Fig. 13.18.

Let d = Diameter of shaft,

d_p = Diameter of pin, and

τ and τ_1 = Allowable shear stress for the material of the shaft and pin respectively.

We know that torque transmitted by the shafts,

$$T = \frac{\pi}{16} \times \tau \times d^3$$

A type of universal joint.

From this relation, the diameter of shafts may be determined.

Since the pin is in double shear, therefore the torque transmitted,

$$T = 2 \times \frac{\pi}{4} (d_p)^2 \tau_1 \times d$$

From this relation, the diameter of pin may be determined.

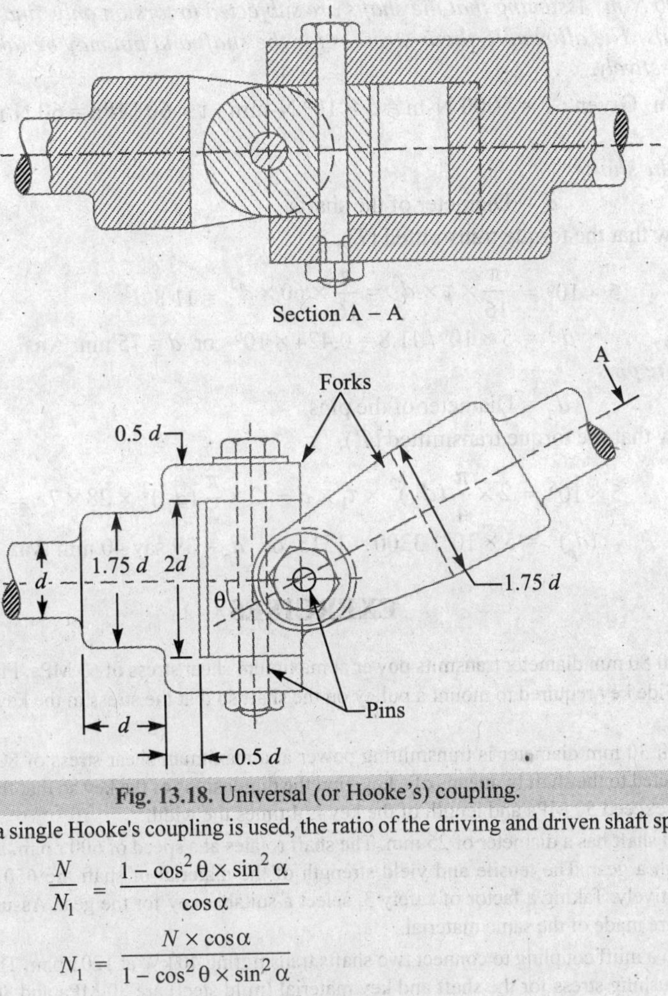

Fig. 13.18. Universal (or Hooke's) coupling.

Note: When a single Hooke's coupling is used, the ratio of the driving and driven shaft speeds is given by

$$\frac{N}{N_1} = \frac{1 - \cos^2 \theta \times \sin^2 \alpha}{\cos \alpha}$$

$$\therefore \quad N_1 = \frac{N \times \cos \alpha}{1 - \cos^2 \theta \times \sin^2 \alpha}$$

where N = Speed of the driving shaft in r.p.m.,

N_1 = Speed of the driven shaft in r.p.m.,

α = Angle of inclination of the shafts, and

θ = Angle of the driving shaft from the position where the pins of the driving shaft fork are in the plane of the two shafts.

We know that maximum speed of the driven shaft,

$$^*N_{1\,(max)} = \frac{N}{\cos \alpha}$$

and minimum speed of the driven shaft,

$$^*N_{1\,(min)} = N \cos \alpha$$

* For further details, please refer to authors' popular book on **'Theory of Machines'**.

From above we see that for a single Hooke's coupling, the speed of the driven shaft is not constant but varies from maximum to minimum. In order to have constant velocity ratio of the driving and driven shafts, an intermediate shaft with a Hooke's coupling at each end (known as double Hooke's coupling) is used.

Example 13.13. *An universal coupling is used to connect two mild steel shafts transmitting a torque of 5000 N-m. Assuming that the shafts are subjected to torsion only, find the diameter of the shafts and pins. The allowable shear stresses for the shaft and pin may be taken as 60 MPa and 28 MPa respectively.*

Solution. Given : $T = 5000$ N-m $= 5 \times 10^6$ N-mm ; $\tau = 60$ MPa $= 60$ N/mm^2 ; $\tau_1 = 28$ MPa $= 28$ N/mm^2

Diameter of the shafts

Let d = Diameter of the shafts.

We know that the torque transmitted (T),

$$5 \times 10^6 = \frac{\pi}{16} \times \tau \times d^3 = \frac{\pi}{16} \times 60 \times d^3 = 11.8\, d^3$$

$$\therefore \quad d^3 = 5 \times 10^6 / 11.8 = 0.424 \times 10^6 \quad \text{or} \quad d = 75 \text{ mm } \textbf{Ans.}$$

Diameter of the pins

Let d_p = Diameter of the pins.

We know that the torque transmitted (T),

$$5 \times 10^6 = 2 \times \frac{\pi}{4} (d_p)^2 \times \tau_1 \times d = 2 \times \frac{\pi}{4} (d_p)^2 \times 28 \times 75 = 3300\, (d_p)^2$$

$$\therefore \quad (d_p)^2 = 5 \times 10^6 / 3300 = 1515 \quad \text{or} \quad d_p = 39 \text{ say } 40 \text{ mm } \textbf{Ans.}$$

EXERCISES

1. A shaft 80 mm diameter transmits power at maximum shear stress of 63 MPa. Find the length of a 20 mm wide key required to mount a pulley on the shaft so that the stress in the key does not exceed 42 MPa. **[Ans. 152 mm]**
2. A shaft 30 mm diameter is transmitting power at a maximum shear stress of 80 MPa. If a pulley is connected to the shaft by means of a key, find the dimensions of the key so that the stress in the key is not to exceed 50 MPa and length of the key is 4 times the width. **[Ans. l = 126 mm]**
3. A steel shaft has a diameter of 25 mm. The shaft rotates at a speed of 600 r.p.m. and transmits 30 kW through a gear. The tensile and yield strength of the material of shaft are 650 MPa and 353 MPa respectively. Taking a factor of safety 3, select a suitable key for the gear. Assume that the key and shaft are made of the same material. **[Ans. l = 102 mm]**
4. Design a muff coupling to connect two shafts transmitting 40 kW at 120 r.p.m. The permissible shear and crushing stress for the shaft and key material (mild steel) are 30 MPa and 80 MPa respectively. The material of muff is cast iron with permissible shear stress of 15 MPa. Assume that the maximum torque transmitted is 25 per cent greater than the mean torque.
   [Ans. d = 90 mm ; w = 28 mm, t = 16 mm, l = 157.5 mm ; D = 195 mm, L = 315 mm]
5. Design a compression coupling for a shaft to transmit 1300 N-m. The allowable shear stress for the shaft and key is 40 MPa and the number of bolts connecting the two halves are 4. The permissible tensile stress for the bolts material is 70 MPa. The coefficient of friction between the muff and the shaft surface may be taken as 0.3. **[Ans. d = 55 mm ; D = 125 mm ; L = 192.5 mm ; d_b = 24 mm]**
6. Design a cast iron protective flange coupling to connect two shafts in order to transmit 7.5 kW at 720 r.p.m. The following permissible stresses may be used :
   Permissible shear stress for shaft, bolt and key material = 33 MPa
   Permissible crushing stress for bolt and key material = 60 MPa
   Permissible shear stress for the cast iron = 15 MPa **[Ans. d = 25 mm; D = 50 mm]**

7. Two shafts made of plain carbon steel are connected by a rigid protective type flange coupling. The shafts are running at 500 r.p.m. and transmit 25 kW power. Design the coupling completely for overload capacity 25 per cent in excess of mean transmitted torque capacity.
   Assume the following permissible stresses for the coupling components :
   Shaft — Permissible tensile stress = 60 MPa; Permissible shear stress = 35 MPa
   Keys — Rectangular formed end sunk key having permissible compressive strength = 60 MPa
   Bolts — Six numbers made of steel having permissible shear stress = 28 MPa
   Flanges — Cast iron having permissible shear stress = 12 MPa
   Draw two views of the coupling you have designed. **[Ans. d = 45 mm ; D = 90 mm]**
8. Design a shaft and flange for a Diesel engine in which protected type of flange coupling is to be adopted for power transmission. The following data is available for design :
   Power of engine = 75 kW; speed of engine = 200 r.p.m.; maximum permissible stress in shaft = 40 MPa; maximum permissible twist in shaft = 1° in length of shaft equal to 30 times the diameter of shaft; maximum torque = 1.25 × mean torque; pitch circle diameter of bolts = 3 × diameter of shaft; maximum permissible stress in bolts = 20 MPa.
   Find out : 1. Diameter of shaft, 2. number of bolts, and 3. diameter of bolts.
   [Ans. 100 mm ; 4 ; 22 mm]
9. A flanged protective type coupling is required to transmit 50 kW at 2000 r.p.m.. Find :
   (*a*) Shaft diameters if the driving shaft is hollow with d_i / d_0 = 0.6 and driven shaft is a solid shaft. Take τ = 100 MPa.
   (*b*) Diameter of bolts, if the coupling uses four bolts. Take $\sigma_c = \sigma_t$ = 70 MPa and τ = 25 MPa. Assume pitch circle diameter as about 3 times the outside diameter of the hollow shaft.
   (*c*) Thickness of the flange and diameter of the hub. Assume σ_c = 100 MPa and τ = 125 MPa.
   (*d*) Make a neat free hand sketch of the assembled coupling showing a longitudinal sectional elevation with the main dimensions. The other dimensions may be assumed suitably.
10. A marine type flange coupling is used to transmit 3.75 MW at 150 r.p.m. The allowable shear stress in the shaft and bolts may be taken as 50 MPa. Determine the shaft diameter and the diameter of the bolts. **[Ans. 300 mm ; 56 mm]**
11. Design a bushed-pin type flexible coupling for connecting a motor shaft to a pump shaft for the following service conditions :
    Power to be transmitted = 40 kW ; speed of the motor shaft = 1000 r.p.m. ; diameter of the motor shaft = 50 mm ; diameter of the pump shaft = 45 mm.
    The bearing pressure in the rubber bush and allowable stress in the pins are to be limited to 0.45 N/mm^2 and 25 MPa respectively. **[Ans. d_1 = 20 mm; n = 6; d_2 = 40 mm ; l = 152 mm]**
12. An universal coupling is used to connect two mild steel shafts transmitting a torque of 6000 N-m. Assuming that the shafts are subjected to torsion only, find the diameter of the shaft and the pin. The allowable shear stresses for the shaft and pin may be taken as 55 MPa and 30 MPa respectively.
    [Ans. d = 85 mm ; d_p = 40 mm]

QUESTIONS

1. What is a key ? State its function.
2. How are the keys classified? Draw neat sketches of different types of keys and state their applications.
3. What are the considerations in the design of dimensions of formed and parallel key having rectangular cross-section ?
4. Write short note on the splined shaft covering the points of application, different types and method of manufacture.
5. What is the effect of keyway cut into the shaft ?
6. Discuss the function of a coupling. Give at least three practical applications.
7. Describe, with the help of neat sketches, the types of various shaft couplings mentioning the uses of each type.
8. How does the working of a clamp coupling differ from that of a muff coupling ? Explain.

9. Sketch a protective type flange coupling and indicate there on its leading dimensions for shaft size of '*d*'.
10. What are flexible couplings and what are their applications ? Illustrate your answer with suitable examples and sketches.
11. Write short note on universal coupling.
12. Why are two universal joints often used when there is angular misalignment between two shafts ?

OBJECTIVE TYPE QUESTIONS

1. The taper on a rectangular sunk key is
   (*a*) 1 in 16 (*b*) 1 in 32
   (*c*) 1 in 48 (*d*) 1 in 100
2. The usual proportion for the width of key is
   (*a*) *d*/8 (*b*) *d*/6
   (*c*) *d*/4 (*d*) *d*/2
   where *d* = Diameter of shaft.
3. When a pulley or other mating piece is required to slide along the shaft, a .... sunk key is used.
   (*a*) rectangular (*b*) square (*c*) parallel
4. A key made from a cylindrical disc having segmental cross-section, is known as
   (*a*) feather key (*b*) gib head key
   (*c*) woodruff key (*d*) flat saddle key
5. A feather key is generally
   (*a*) loose in shaft and tight in hub (*b*) tight in shaft and loose in hub
   (*c*) tight in both shaft and hub (*d*) loose in both shaft and hub.
6. The type of stresses developed in the key is/are
   (*a*) shear stress alone (*b*) bearing stress alone
   (*c*) both shear and bearing stresses (*d*) shearing, bearing and bending stresses
7. For a square key made of mild steel, the shear and crushing strengths are related as
   (*a*) shear strength = crushing strength (*b*) shear strength > crushing strength
   (*c*) shear strength < crushing strength (*d*) none of the above
8. A keyway lowers
   (*a*) the strength of the shaft (*b*) the rigidity of the shaft
   (*c*) both the strength and rigidity of the shaft (*d*) the ductility of the material of the shaft
9. The sleeve or muff coupling is designed as a
   (*a*) thin cylinder (*b*) thick cylinder
   (*c*) solid shaft (*d*) hollow shaft
10. Oldham coupling is used to connect two shafts
    (*a*) which are perfectly aligned (*b*) which are not in exact alignment
    (*c*) which have lateral misalignment (*d*) whose axes intersect at a small angle

ANSWERS

1. (*d*)	2. (*c*)	3. (*c*)	4. (*d*)	5. (*b*)
6. (*c*)	7. (*a*)	8. (*c*)	9. (*d*)	10. (*c*)

CHAPTER

14

Shafts

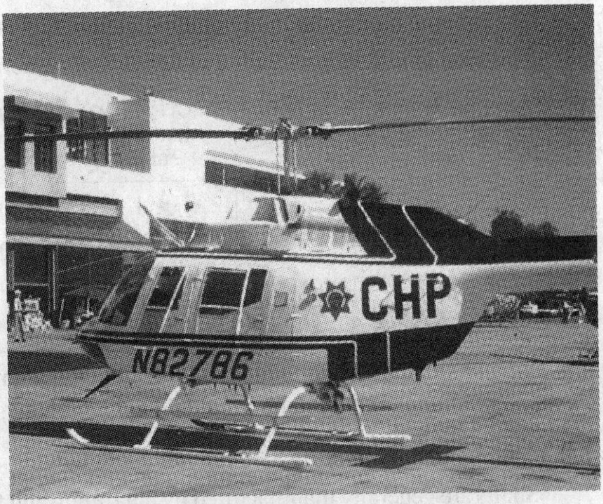

14.1 Introduction

A shaft is a rotating machine element which is used to transmit power from one place to another. The power is delivered to the shaft by some tangential force and the resultant torque (or twisting moment) set up within the shaft permits the power to be transferred to various machines linked up to the shaft. In order to transfer the power from one shaft to another, the various members such as pulleys, gears etc., are mounted on it. These members along with the forces exerted upon them causes the shaft to bending. In other words, we may say that a shaft is used for the transmission of torque and bending moment. The various members are mounted on the shaft by means of keys or splines.

Notes: 1. The shafts are usually cylindrical, but may be square or cross-shaped in section. They are solid in cross-section but sometimes hollow shafts are also used.

2. An ***axle***, though similar in shape to the shaft, is a stationary machine element and is used for the transmission of bending moment only. It simply acts as a support for some rotating body such as hoisting drum, a car wheel or a rope sheave.

3. A ***spindle*** is a short shaft that imparts motion either to a cutting tool (*e.g.* drill press spindles) or to a work piece (*e.g.* lathe spindles).

14.2 Material Used for Shafts

The material used for shafts should have the following properties :

1. It should have high strength.
2. It should have good machinability.
3. It should have low notch sensitivity factor.
4. It should have good heat treatment properties.
5. It should have high wear resistant properties.

The material used for ordinary shafts is carbon steel of grades 40 C 8, 45 C 8, 50 C 4 and 50 C 12. The mechanical properties of these grades of carbon steel are given in the following table.

Table 14.1. Mechanical properties of steels used for shafts.

Indian standard designation	*Ultimate tensile strength, MPa*	*Yield strength, MPa*
40 C 8	560 - 670	320
45 C 8	610 - 700	350
50 C 4	640 - 760	370
50 C 12	700 Min.	390

When a shaft of high strength is required, then an alloy steel such as nickel, nickel-chromium or chrome-vanadium steel is used.

14.3 Manufacturing of Shafts

Shafts are generally manufactured by hot rolling and finished to size by cold drawing or turning and grinding. The cold rolled shafts are stronger than hot rolled shafts but with higher residual stresses. The residual stresses may cause distortion of the shaft when it is machined, especially when slots or keyways are cut. Shafts of larger diameter are usually forged and turned to size in a lathe.

14.4 Types of Shafts

The following two types of shafts are important from the subject point of view :

1. ***Transmission shafts.*** These shafts transmit power between the source and the machines absorbing power. The counter shafts, line shafts, over head shafts and all factory shafts are transmission shafts. Since these shafts carry machine parts such as pulleys, gears etc., therefore they are subjected to bending in addition to twisting.

2. ***Machine shafts.*** These shafts form an integral part of the machine itself. The crank shaft is an example of machine shaft.

14.5 Standard Sizes of Transmission Shafts

The standard sizes of transmission shafts are :

25 mm to 60 mm with 5 mm steps; 60 mm to 110 mm with 10 mm steps ; 110 mm to 140 mm with 15 mm steps ; and 140 mm to 500 mm with 20 mm steps.

The standard length of the shafts are 5 m, 6 m and 7 m.

14.6 Stresses in Shafts

The following stresses are induced in the shafts :

1. Shear stresses due to the transmission of torque (*i.e.* due to torsional load).

2. Bending stresses (tensile or compressive) due to the forces acting upon machine elements like gears, pulleys etc. as well as due to the weight of the shaft itself.

3. Stresses due to combined torsional and bending loads.

14.7 Maximum Permissible Working Stresses for Transmission Shafts

According to American Society of Mechanical Engineers (ASME) code for the design of transmission shafts, the maximum permissible working stresses in tension or compression may be taken as

(*a*) 112 MPa for shafts without allowance for keyways.

(*b*) 84 MPa for shafts with allowance for keyways.

For shafts purchased under definite physical specifications, the permissible tensile stress (σ_t) may be taken as 60 per cent of the elastic limit in tension (σ_{el}), but not more than 36 per cent of the ultimate tensile strength (σ_u). In other words, the permissible tensile stress,

$$\sigma_t = 0.6\ \sigma_{el} \text{ or } 0.36\ \sigma_u, \text{ whichever is less.}$$

The maximum permissible shear stress may be taken as

(*a*) 56 MPa for shafts without allowance for key ways.

(*b*) 42 MPa for shafts with allowance for keyways.

For shafts purchased under definite physical specifications, the permissible shear stress (τ) may be taken as 30 per cent of the elastic limit in tension (σ_{el}) but not more than 18 per cent of the ultimate tensile strength (σ_u). In other words, the permissible shear stress,

$$\tau = 0.3\ \sigma_{el} \text{ or } 0.18\ \sigma_u, \text{ whichever is less.}$$

14.8 Design of Shafts

The shafts may be designed on the basis of

1. Strength, and 2. Rigidity and stiffness.

In designing shafts on the basis of strength, the following cases may be considered :

(*a*) Shafts subjected to twisting moment or torque only,

(*b*) Shafts subjected to bending moment only,

(*c*) Shafts subjected to combined twisting and bending moments, and

(*d*) Shafts subjected to axial loads in addition to combined torsional and bending loads.

We shall now discuss the above cases, in detail, in the following pages.

14.9 Shafts Subjected to Twisting Moment Only

When the shaft is subjected to a twisting moment (or torque) only, then the diameter of the shaft may be obtained by using the torsion equation. We know that

$$\frac{T}{J} = \frac{\tau}{r} \quad ...(i)$$

where T = Twisting moment (or torque) acting upon the shaft,

J = Polar moment of inertia of the shaft about the axis of rotation,

τ = Torsional shear stress, and

r = Distance from neutral axis to the outer most fibre
$= d / 2$; where d is the diameter of the shaft.

We know that for round solid shaft, polar moment of inertia,

$$J = \frac{\pi}{32} \times d^4$$

The equation (*i*) may now be written as

$$\frac{T}{\frac{\pi}{32} \times d^4} = \frac{\tau}{\frac{d}{2}} \quad \text{or} \quad T = \frac{\pi}{16} \times \tau \times d^3 \qquad \text{...(ii)}$$

From this equation, we may determine the diameter of round solid shaft (d).

We also know that for hollow shaft, polar moment of inertia,

$$J = \frac{\pi}{32}\left[(d_o)^4 - (d_i)^4\right]$$

where d_o and d_i = Outside and inside diameter of the shaft, and $r = d_o / 2$.

Substituting these values in equation (*i*), we have

$$\frac{T}{\frac{\pi}{32}\left[(d_o)^4 - (d_i)^4\right]} = \frac{\tau}{\frac{d_o}{2}} \quad \text{or} \quad T = \frac{\pi}{16} \times \tau \left[\frac{(d_o)^4 - (d_i)^4}{d_o}\right] \qquad \text{...(iii)}$$

Let k = Ratio of inside diameter and outside diameter of the shaft
$= d_i / d_o$

Now the equation (*iii*) may be written as

$$T = \frac{\pi}{16} \times \tau \times \frac{(d_o)^4}{d_o}\left[1 - \left(\frac{d_i}{d_o}\right)^4\right] = \frac{\pi}{16} \times \tau\,(d_o)^3\,(1 - k^4) \qquad \text{...(iv)}$$

Shafts inside generators and motors are made to bear high torsional stresses.

From the equations *(iii)* or *(iv)*, the outside and inside diameter of a hollow shaft may be determined.

It may be noted that

1. The hollow shafts are usually used in marine work. These shafts are stronger per kg of material and they may be forged on a mandrel, thus making the material more homogeneous than would be possible for a solid shaft.

When a hollow shaft is to be made equal in strength to a solid shaft, the twisting moment of both the shafts must be same. In other words, for the same material of both the shafts,

$$T = \frac{\pi}{16} \times \tau \left[\frac{(d_o)^4 - (d_i)^4}{d_o}\right] = \frac{\pi}{16} \times \tau \times d^3$$

$$\therefore \quad \frac{(d_o)^4 - (d_i)^4}{d_o} = d^3 \quad \text{or} \quad (d_o)^3 (1 - k^4) = d^3$$

2. The twisting moment (T) may be obtained by using the following relation :

We know that the power transmitted (in watts) by the shaft,

$$P = \frac{2\pi N \times T}{60} \quad \text{or} \quad T = \frac{P \times 60}{2\pi N}$$

where T = Twisting moment in N-m, and

N = Speed of the shaft in r.p.m.

3. In case of belt drives, the twisting moment (T) is given by

$$T = (T_1 - T_2)\, R$$

where T_1 and T_2 = Tensions in the tight side and slack side of the belt respectively, and

R = Radius of the pulley.

Example 14.1. *A line shaft rotating at 200 r.p.m. is to transmit 20 kW. The shaft may be assumed to be made of mild steel with an allowable shear stress of 42 MPa. Determine the diameter of the shaft, neglecting the bending moment on the shaft.*

Solution. Given : N = 200 r.p.m. ; P = 20 kW = 20 × 10³ W; τ = 42 MPa = 42 N/mm²

Let d = Diameter of the shaft.

We know that torque transmitted by the shaft,

$$T = \frac{P \times 60}{2\pi N} = \frac{20 \times 10^3 \times 60}{2\pi \times 200} = 955 \text{ N-m} = 955 \times 10^3 \text{ N-mm}$$

We also know that torque transmitted by the shaft (T),

$$955 \times 10^3 = \frac{\pi}{16} \times \tau \times d^3 = \frac{\pi}{16} \times 42 \times d^3 = 8.25\, d^3$$

$$\therefore \quad d^3 = 955 \times 10^3 / 8.25 = 115\,733 \quad \text{or} \quad d = 48.7 \text{ say } 50 \text{ mm}$$ **Ans.**

Example 14.2. *A solid shaft is transmitting 1 MW at 240 r.p.m. Determine the diameter of the shaft if the maximum torque transmitted exceeds the mean torque by 20%. Take the maximum allowable shear stress as 60 MPa.*

Solution. Given : P = 1 MW = 1 × 10⁶ W ; N = 240 r.p.m. ; $T_{max} = 1.2\, T_{mean}$; τ = 60 MPa = 60 N/mm²

Let d = Diameter of the shaft.

We know that mean torque transmitted by the shaft,

$$T_{mean} = \frac{P \times 60}{2\pi N} = \frac{1 \times 10^6 \times 60}{2\pi \times 240} = 39\,784 \text{ N-m} = 39\,784 \times 10^3 \text{ N-mm}$$

∴ Maximum torque transmitted,

$$T_{max} = 1.2\, T_{mean} = 1.2 \times 39\,784 \times 10^3 = 47\,741 \times 10^3 \text{ N-mm}$$

We know that maximum torque transmitted (T_{max}),

$$47\,741 \times 10^3 = \frac{\pi}{16} \times \tau \times d^3 = \frac{\pi}{16} \times 60 \times d^3 = 11.78\, d^3$$

∴ $d^3 = 47\,741 \times 10^3 / 11.78 = 4053 \times 10^3$

or $d = 159.4$ say 160 mm **Ans.**

Example 14.3. *Find the diameter of a solid steel shaft to transmit 20 kW at 200 r.p.m. The ultimate shear stress for the steel may be taken as 360 MPa and a factor of safety as 8.*

If a hollow shaft is to be used in place of the solid shaft, find the inside and outside diameter when the ratio of inside to outside diameters is 0.5.

Solution. Given : $P = 20$ kW $= 20 \times 10^3$ W ; $N = 200$ r.p.m. ; $\tau_u = 360$ MPa $= 360$ N/mm^2 ; $F.S. = 8$; $k = d_i / d_o = 0.5$

We know that the allowable shear stress,

$$\tau = \frac{\tau_u}{F.S.} = \frac{360}{8} = 45 \text{ N/mm}^2$$

Diameter of the solid shaft

Let d = Diameter of the solid shaft.

We know that torque transmitted by the shaft,

$$T = \frac{P \times 60}{2\pi N} = \frac{20 \times 10^3 \times 60}{2\pi \times 200} = 955 \text{ N-m} = 955 \times 10^3 \text{ N-mm}$$

We also know that torque transmitted by the solid shaft (T),

$$955 \times 10^3 = \frac{\pi}{16} \times \tau \times d^3 = \frac{\pi}{16} \times 45 \times d^3 = 8.84\, d^3$$

∴ $d^3 = 955 \times 10^3 / 8.84 = 108\,032$ or $d = 47.6$ say 50 mm **Ans.**

Diameter of hollow shaft

Let d_i = Inside diameter, and

d_o = Outside diameter.

We know that the torque transmitted by the hollow shaft (T),

$$955 \times 10^3 = \frac{\pi}{16} \times \tau\, (d_o)^3\, (1 - k^4)$$

$$= \frac{\pi}{16} \times 45\, (d_o)^3\, [1 - (0.5)^4] = 8.3\, (d_o)^3$$

∴ $(d_o)^3 = 955 \times 10^3 / 8.3 = 115\,060$ or $d_o = 48.6$ say 50 mm **Ans.**

and $d_i = 0.5\, d_o = 0.5 \times 50 = 25$ mm **Ans.**

14.10 Shafts Subjected to Bending Moment Only

When the shaft is subjected to a bending moment only, then the maximum stress (tensile or compressive) is given by the bending equation. We know that

$$\frac{M}{I} = \frac{\sigma_b}{y} \quad \text{...(}$$

where M = Bending moment,

I = Moment of inertia of cross-sectional area of the shaft about the axis of rotation,

σ_b = Bending stress, and

y = Distance from neutral axis to the outer-most fibre.

We know that for a round solid shaft, moment of inertia,

$$I = \frac{\pi}{64} \times d^4 \quad \text{and} \quad y = \frac{d}{2}$$

Substituting these values in equation (*i*), we have

$$\frac{M}{\frac{\pi}{64} \times d^4} = \frac{\sigma_b}{\frac{d}{2}} \quad \text{or} \quad M = \frac{\pi}{32} \times \sigma_b \times d^3$$

From this equation, diameter of the solid shaft (*d*) may be obtained.

We also know that for a hollow shaft, moment of inertia,

$$I = \frac{\pi}{64}\left[(d_o)^4 - (d_i)^4\right] = \frac{\pi}{64}(d_o)^4 (1 - k^4) \quad \text{...(where } k = d_i / d_o)$$

and $y = d_o / 2$

Again substituting these values in equation (*i*), we have

$$\frac{M}{\frac{\pi}{64}(d_o)^4 (1 - k^4)} = \frac{\sigma_b}{\frac{d_o}{2}} \quad \text{or} \quad M = \frac{\pi}{32} \times \sigma_b (d_o)^3 (1 - k^4)$$

From this equation, the outside diameter of the shaft (d_o) may be obtained.

Note: We have already discussed in Art. 14.1 that the axles are used to transmit bending moment only. Thus, axles are designed on the basis of bending moment only, in the similar way as discussed above.

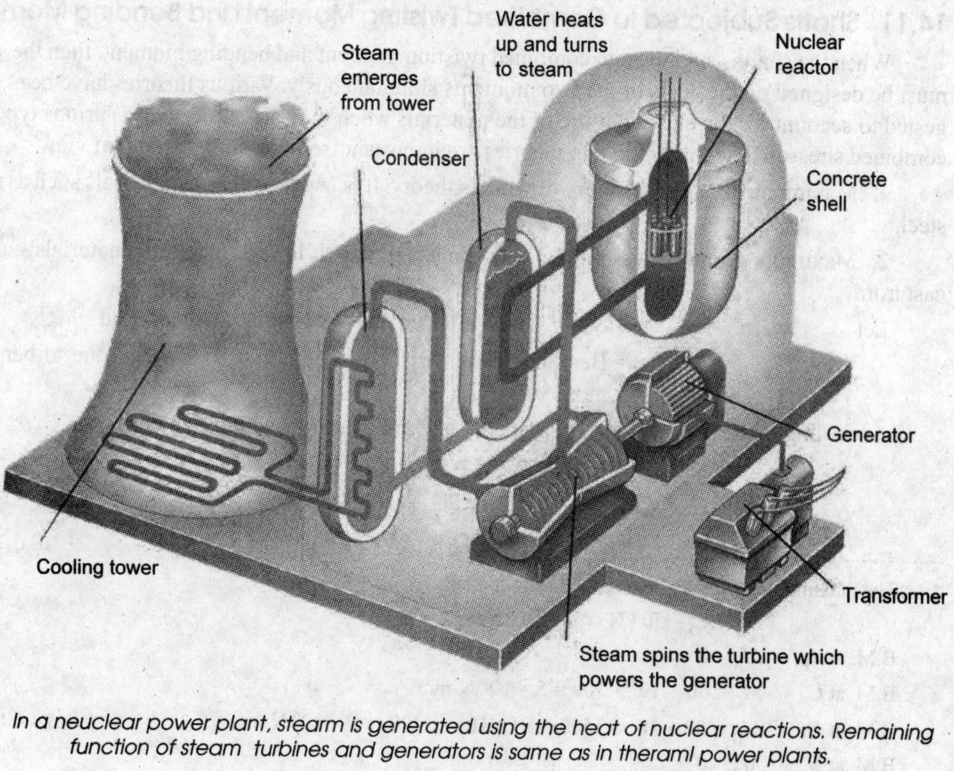

In a neuclear power plant, stearm is generated using the heat of nuclear reactions. Remaining function of steam turbines and generators is same as in theraml power plants.

Example 14.4. *A pair of wheels of a railway wagon carries a load of 50 kN on each axle box, acting at a distance of 100 mm outside the wheel base. The gauge of the rails is 1.4 m. Find the diameter of the axle between the wheels, if the stress is not to exceed 100 MPa.*

Solution. Given : $W = 50$ kN $= 50 \times 10^3$ N ; $L = 100$ mm ; $x = 1.4$ m ; $\sigma_b = 100$ MPa $= 100$ N/mm^2

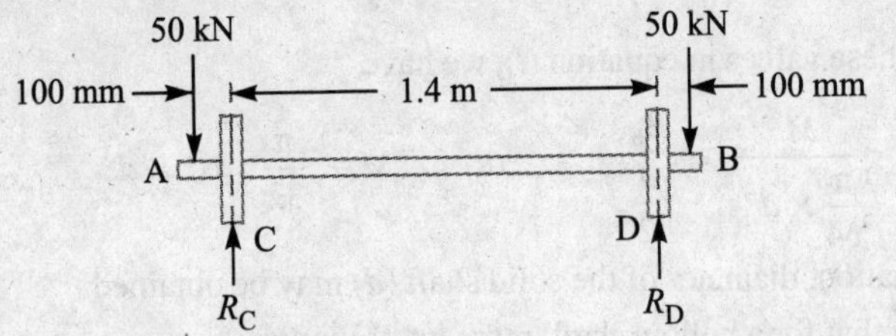

Fig. 14.1

The axle with wheels is shown in Fig. 14.1.

A little consideration will show that the maximum bending moment acts on the wheels at *C* and *D*. Therefore maximum bending moment,

$$*M = W.L = 50 \times 10^3 \times 100 = 5 \times 10^6 \text{ N-mm}$$

Let $\quad d$ = Diameter of the axle.

We know that the maximum bending moment (*M*),

$$5 \times 10^6 = \frac{\pi}{32} \times \sigma_b \times d^3 = \frac{\pi}{32} \times 100 \times d^3 = 9.82\, d^3$$

$$\therefore \quad d^3 = 5 \times 10^6 / 9.82 = 0.51 \times 10^6 \text{ or } d = 79.8 \text{ say } 80 \text{ mm } \textbf{Ans.}$$

14.11 Shafts Subjected to Combined Twisting Moment and Bending Moment

When the shaft is subjected to combined twisting moment and bending moment, then the shaft must be designed on the basis of the two moments simultaneously. Various theories have been suggested to account for the elastic failure of the materials when they are subjected to various types of combined stresses. The following two theories are important from the subject point of view :

1. Maximum shear stress theory or Guest's theory. It is used for ductile materials such as mild steel.

2. Maximum normal stress theory or Rankine's theory. It is used for brittle materials such as cast iron.

Let $\quad \tau$ = Shear stress induced due to twisting moment, and

σ_b = Bending stress (tensile or compressive) induced due to bending moment.

According to maximum shear stress theory, the maximum shear stress in the shaft,

$$\tau_{max} = \frac{1}{2} \sqrt{(\sigma_b)^2 + 4\tau^2}$$

---

* The maximum B.M. may be obtained as follows :

$R_C = R_D = 50$ kN $= 50 \times 10^3$ N

B.M. at *A*, $\quad M_A = 0$

B.M. at *C*, $\quad M_C = 50 \times 10^3 \times 100 = 5 \times 10^6$ N-mm

B.M. at *D*, $\quad M_D = 50 \times 10^3 \times 1500 - 50 \times 10^3 \times 1400 = 5 \times 10^6$ N-mm

B.M. at *B*, $\quad M_B = 0$

Substituting the values of τ and σ_b from Art. 14.9 and Art. 14.10, we have

$$\tau_{max} = \frac{1}{2}\sqrt{\left(\frac{32\,M}{\pi d^3}\right)^2 + 4\left(\frac{16\,T}{\pi d^3}\right)^2} = \frac{16}{\pi d^3}\left[\sqrt{M^2 + T^2}\right]$$

or $$\frac{\pi}{16} \times \tau_{max} \times d^3 = \sqrt{M^2 + T^2} \qquad \text{...}(i)$$

The expression $\sqrt{M^2 + T^2}$ is known as ***equivalent twisting moment*** and is denoted by T_e. The equivalent twisting moment may be defined as that twisting moment, which when acting alone, produces the same shear stress (τ) as the actual twisting moment. By limiting the maximum shear stress (τ_{max}) equal to the allowable shear stress (τ) for the material, the equation (*i*) may be written as

$$T_e = \sqrt{M^2 + T^2} = \frac{\pi}{16} \times \tau \times d^3 \qquad \text{...}(ii)$$

From this expression, diameter of the shaft (d) may be evaluated.

Now according to maximum normal stress theory, the maximum normal stress in the shaft,

$$\sigma_{b(max)} = \frac{1}{2}\sigma_b + \frac{1}{2}\sqrt{(\sigma_b)^2 + 4\tau^2} \qquad \text{...}(iii)$$

$$= \frac{1}{2} \times \frac{32\,M}{\pi d^3} + \frac{1}{2}\sqrt{\left(\frac{32\,M}{\pi d^3}\right)^2 + 4\left(\frac{16\,T}{\pi d^3}\right)^2}$$

$$= \frac{32}{\pi d^3}\left[\frac{1}{2}\left(M + \sqrt{M^2 + T^2}\right)\right]$$

or $$\frac{\pi}{32} \times \sigma_{b\,(max)} \times d^3 = \frac{1}{2}\left[M + \sqrt{M^2 + T^2}\right] \qquad \text{...}(iv)$$

The expression $\frac{1}{2}\left[\left(M + \sqrt{M^2 + T^2}\right)\right]$ is known as ***equivalent bending moment*** and is denoted by M_e. The equivalent bending moment may be defined as **that moment which when acting alone produces the same tensile or compressive stress (σ_b) as the actual bending moment.** By limiting the maximum normal stress [$\sigma_{b(max)}$] equal to the allowable bending stress (σ_b), then the equation (*iv*) may be written as

$$M_e = \frac{1}{2}\left[M + \sqrt{M^2 + T^2}\right] = \frac{\pi}{32} \times \sigma_b \times d^3 \qquad \text{...}(v)$$

From this expression, diameter of the shaft (d) may be evaluated.

Notes: 1. In case of a hollow shaft, the equations (*ii*) and (*v*) may be written as

$$T_e = \sqrt{M^2 + T^2} = \frac{\pi}{16} \times \tau\,(d_o)^3\,(1 - k^4)$$

and $$M_e = \tfrac{1}{2}\left(M + \sqrt{M^2 + T^2}\right) = \frac{\pi}{32} \times \sigma_b\,(d_o)^3\,(1 - k^4)$$

2. It is suggested that diameter of the shaft may be obtained by using both the theories and the larger of the two values is adopted.

Example 14.5. *A solid circular shaft is subjected to a bending moment of 3000 N-m and a torque of 10 000 N-m. The shaft is made of 45 C 8 steel having ultimate tensile stress of 700 MPa and a ultimate shear stress of 500 MPa. Assuming a factor of safety as 6, determine the diameter of the shaft.*

Solution. Given : M = 3000 N-m = 3 × 10^6 N-mm ; T = 10 000 N-m = 10 × 10^6 N-mm ; σ_{tu} = 700 MPa = 700 N/mm^2 ; τ_u = 500 MPa = 500 N/mm^2

We know that the allowable tensile stress,

$$\sigma_t \text{ or } \sigma_b = \frac{\sigma_{tu}}{F.S.} = \frac{700}{6} = 116.7 \text{ N/mm}^2$$

and allowable shear stress,

$$\tau = \frac{\tau_u}{F.S.} = \frac{500}{6} = 83.3 \text{ N/mm}^2$$

Let d = Diameter of the shaft in mm.

According to maximum shear stress theory, equivalent twisting moment,

$$T_e = \sqrt{M^2 + T^2} = \sqrt{(3 \times 10^6)^2 + (10 \times 10^6)^2} = 10.44 \times 10^6 \text{ N-mm}$$

We also know that equivalent twisting moment (T_e),

$$10.44 \times 10^6 = \frac{\pi}{16} \times \tau \times d^3 = \frac{\pi}{16} \times 83.3 \times d^3 = 16.36\ d^3$$

$$\therefore \quad d^3 = 10.44 \times 10^6 / 16.36 = 0.636 \times 10^6 \text{ or } d = 86 \text{ mm}$$

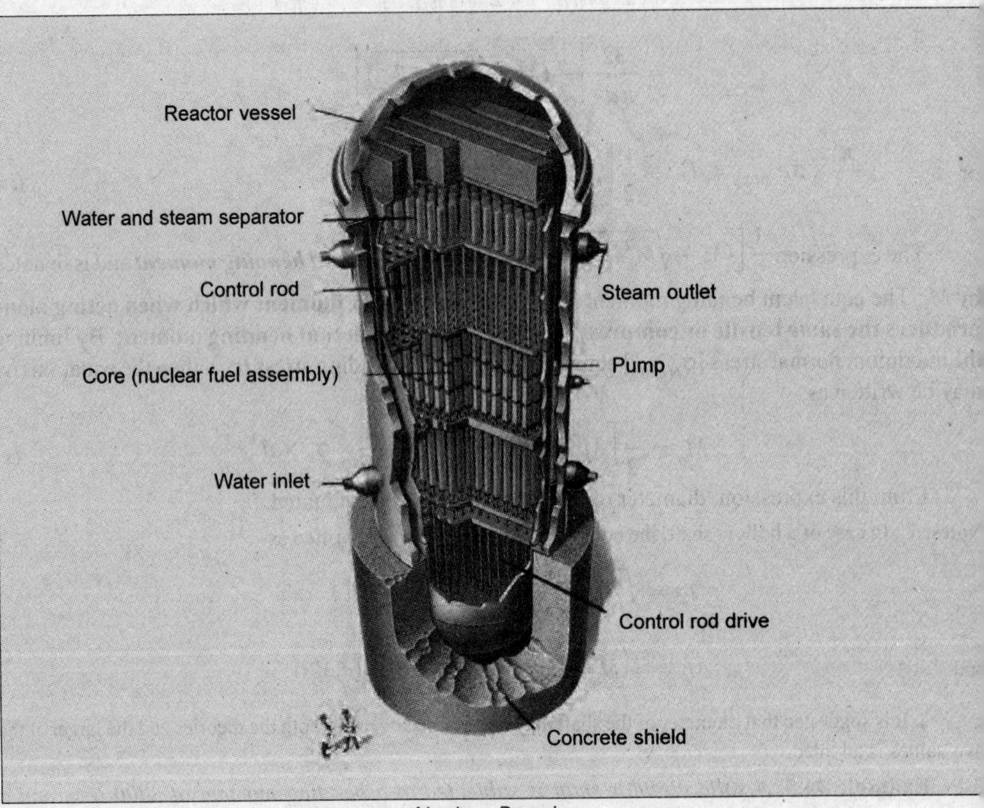

Nuclear Reactor

Note : This picture is given as additional information and is not a direct example of the current chapter.

According to maximum normal stress theory, equivalent bending moment,

$$M_e = \frac{1}{2}\left(M + \sqrt{M^2 + T^2}\right) = \frac{1}{2}(M + T_e)$$

$$= \frac{1}{2}(3 \times 10^6 + 10.44 \times 10^6) = 6.72 \times 10^6 \text{ N-mm}$$

We also know that the equivalent bending moment (M_e),

$$6.72 \times 10^6 = \frac{\pi}{32} \times \sigma_b \times d^3 = \frac{\pi}{32} \times 116.7 \times d^3 = 11.46\, d^3$$

$$\therefore \quad d^3 = 6.72 \times 10^6 / 11.46 = 0.586 \times 10^6 \text{ or } d = 83.7 \text{ mm}$$

Taking the larger of the two values, we have

d = 86 say 90 mm **Ans.**

Example 14.6. *A shaft supported at the ends in ball bearings carries a straight tooth spur gear at its mid span and is to transmit 7.5 kW at 300 r.p.m. The pitch circle diameter of the gear is 150 mm. The distances between the centre line of bearings and gear are 100 mm each. If the shaft is made of steel and the allowable shear stress is 45 MPa, determine the diameter of the shaft. Show in a sketch how the gear will be mounted on the shaft; also indicate the ends where the bearings will be mounted?. The pressure angle of the gear may be taken as 20°.*

Solution. Given : P = 7.5 kW = 7500 W ; N = 300 r.p.m. ; D = 150 mm = 0.15 m ; L = 200 mm = 0.2 m ; τ = 45 MPa = 45 N/mm² ; α = 20°

Fig. 14.2 shows a shaft with a gear mounted on the bearings.

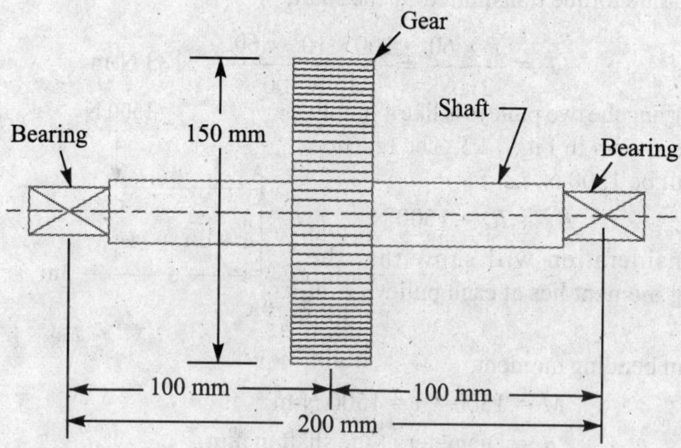

Fig. 14.2

We know that torque transmitted by the shaft,

$$T = \frac{P \times 60}{2\pi N} = \frac{7500 \times 60}{2\pi \times 300} = 238.7 \text{ N-m}$$

$\therefore$ Tangential force on the gear,

$$F_t = \frac{2T}{D} = \frac{2 \times 238.7}{0.15} = 3182.7 \text{ N}$$

and the normal load acting on the tooth of the gear,

$$W = \frac{F_t}{\cos \alpha} = \frac{3182.7}{\cos 20°} = \frac{3182.7}{0.9397} = 3387 \text{ N}$$

Since the gear is mounted at the middle of the shaft, therefore maximum bending moment at the centre of the gear,

$$M = \frac{W.L}{4} = \frac{3387 \times 0.2}{4} = 169.4 \text{ N-m}$$

Let d = Diameter of the shaft.

We know that equivalent twisting moment,

$$T_e = \sqrt{M^2 + T^2} = \sqrt{(169.4)^2 + (238.7)^2} = 292.7 \text{ N-m}$$
$$= 292.7 \times 10^3 \text{ N-mm}$$

We also know that equivalent twisting moment (T_e),

$$292.7 \times 10^3 = \frac{\pi}{16} \times \tau \times d^3 = \frac{\pi}{16} \times 45 \times d^3 = 8.84\ d^3$$

$\therefore$ $d^3 = 292.7 \times 10^3 / 8.84 = 33 \times 10^3$ or $d = 32$ say 35 mm **Ans.**

Example 14.7. *A shaft made of mild steel is required to transmit 100 kW at 300 r.p.m. The supported length of the shaft is 3 metres. It carries two pulleys each weighing 1500 N supported at a distance of 1 metre from the ends respectively. Assuming the safe value of stress, determine the diameter of the shaft.*

Solution. Given : $P = 100$ kW $= 100 \times 10^3$ W ; $N = 300$ r.p.m. ; $L = 3$ m ; $W = 1500$ N

We know that the torque transmitted by the shaft,

$$T = \frac{P \times 60}{2\pi N} = \frac{100 \times 10^3 \times 60}{2\pi \times 300} = 3183 \text{ N-m}$$

The shaft carrying the two pulleys is like a simply supported beam as shown in Fig. 14.3. The reaction at each support will be 1500 N, *i.e.*

$$R_A = R_B = 1500 \text{ N}$$

A little consideration will show that the maximum bending moment lies at each pulley *i.e.* at *C* and *D*.

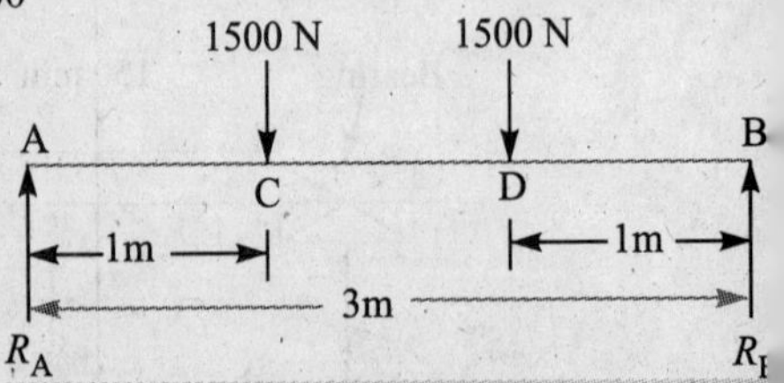

Fig. 14.3

$\therefore$ Maximum bending moment,

$$M = 1500 \times 1 = 1500 \text{ N-m}$$

Let d = Diameter of the shaft in mm.

We know that equivalent twisting moment,

$$T_e = \sqrt{M^2 + T^2} = \sqrt{(1500)^2 + (3183)^2} = 3519 \text{ N-m}$$
$$= 3519 \times 10^3 \text{ N-mm}$$

We also know that equivalent twisting moment (T_e),

$$3519 \times 10^3 = \frac{\pi}{16} \times \tau \times d^3 = \frac{\pi}{16} \times 60 \times d^3 = 11.8\ d^3 \quad \text{...(Assuming } \tau = 60 \text{ N/mm}^2\text{)}$$

$\therefore$ $d^3 = 3519 \times 10^3 / 11.8 = 298 \times 10^3$ or $d = 66.8$ say 70 mm **Ans.**

Example 14.8. *A line shaft is driven by means of a motor placed vertically below it. The pulley on the line shaft is 1.5 metre in diameter and has belt tensions 5.4 kN and 1.8 kN on the tight side and slack side of the belt respectively. Both these tensions may be assumed to be vertical. If the pulley b* *overhang from the shaft, the distance of the centre line of the pulley from the centre line of the bearing being 400 mm, find the diameter of the shaft. Assuming maximum allowable shear stress of 42 MPa.*

Solution . Given : $D = 1.5$ m or $R = 0.75$ m; $T_1 = 5.4$ kN = 5400 N ; $T_2 = 1.8$ kN = 1800 N ; $L = 400$ mm ; $\tau = 42$ MPa = 42 N/mm²

A line shaft with a pulley is shown in Fig 14.4.

We know that torque transmitted by the shaft,

$$T = (T_1 - T_2)\,R = (5400 - 1800)\,0.75 = 2700 \text{ N-m}$$
$$= 2700 \times 10^3 \text{ N-mm}$$

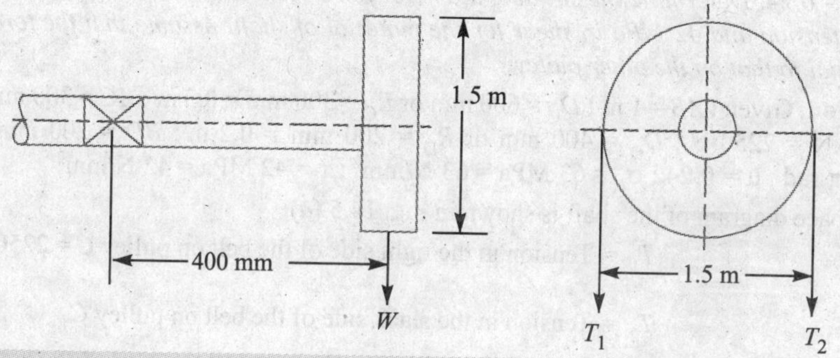

Fig. 14.4

Neglecting the weight of shaft, total vertical load acting on the pulley,

$$W = T_1 + T_2 = 5400 + 1800 = 7200 \text{ N}$$

∴ Bending moment, $M = W \times L = 7200 \times 400 = 2880 \times 10^3$ N-mm

Let d = Diameter of the shaft in mm.

We know that the equivalent twisting moment,

$$T_e = \sqrt{M^2 + T^2} = \sqrt{(2880 \times 10^3)^2 + (2700 \times 10^3)^2}$$
$$= 3950 \times 10^3 \text{ N-mm}$$

Steel shaft

We also know that equivalent twisting moment (T_e),

$$3950 \times 10^3 = \frac{\pi}{16} \times \tau \times d^3 = \frac{\pi}{16} \times 42 \times d^3 = 8.25\ d^3$$

$$\therefore \quad d^3 = 3950 \times 10^3/8.25 = 479 \times 10^3 \text{ or } d = 78 \text{ say } 80 \text{ mm } \textbf{Ans.}$$

Example 14.9. *A shaft is supported by two bearings placed 1 m apart. A 600 mm diameter pulley is mounted at a distance of 300 mm to the right of left hand bearing and this drives a pulley directly below it with the help of belt having maximum tension of 2.25 kN. Another pulley 400 mm diameter is placed 200 mm to the left of right hand bearing and is driven with the help of electric motor and belt, which is placed horizontally to the right. The angle of contact for both the pulleys is 180° and μ = 0.24. Determine the suitable diameter for a solid shaft, allowing working stress of 63 MPa in tension and 42 MPa in shear for the material of shaft. Assume that the torque on one pulley is equal to that on the other pulley.*

Solution. Given : $AB = 1$ m ; $D_C = 600$ mm or $R_C = 300$ mm $= 0.3$ m ; $AC = 300$ mm $= 0.3$ m ; $T_1 = 2.25$ kN $= 2250$ N ; $D_D = 400$ mm or $R_D = 200$ mm $= 0.2$ m ; $BD = 200$ mm $= 0.2$ m ; $\theta = 180° = \pi$ rad ; $\mu = 0.24$; $\sigma_b = 63$ MPa $= 63$ N/mm^2 ; $\tau = 42$ MPa $= 42$ N/mm^2

The space diagram of the shaft is shown in Fig. 14.5 (*a*).

Let T_1 = Tension in the tight side of the belt on pulley $C = 2250$ N ...(Given)

T_2 = Tension in the slack side of the belt on pulley C.

We know that

$$2.3 \log \left(\frac{T_1}{T_2}\right) = \mu.\theta = 0.24 \times \pi = 0.754$$

$$\therefore \quad \log \left(\frac{T_1}{T_2}\right) = \frac{0.754}{2.3} = 0.3278 \text{ or } \frac{T_1}{T_2} = 2.127 \quad \text{...(Taking antilog of 0.3278)}$$

and

$$T_2 = \frac{T_1}{2.127} = \frac{2250}{2.127} = 1058 \text{ N}$$

∴ Vertical load acting on the shaft at C,

$$W_C = T_1 + T_2 = 2250 + 1058 = 3308 \text{ N}$$

and vertical load on the shaft at D

$$= 0$$

The vertical load diagram is shown in Fig. 14.5 (*c*).

We know that torque acting on the pulley C,

$$T = (T_1 - T_2)\ R_C = (2250 - 1058)\ 0.3 = 357.6 \text{ N-m}$$

The torque diagram is shown in Fig. 14.5 (*b*).

Let T_3 = Tension in the tight side of the belt on pulley D, and

T_4 = Tension in the slack side of the belt on pulley D.

Since the torque on both the pulleys (*i.e.* C and D) is same, therefore

$$(T_3 - T_4)\ R_D = T = 357.6 \text{ N-m or } T_3 - T_4 = \frac{357.6}{R_D} = \frac{357.6}{0.2} = 1788 \text{ N} \quad \text{...(}$$

We know that

$$\frac{T_3}{T_4} = \frac{T_1}{T_2} = 2.127 \text{ or } T_3 = 2.127\ T_4 \quad \text{...(}$$

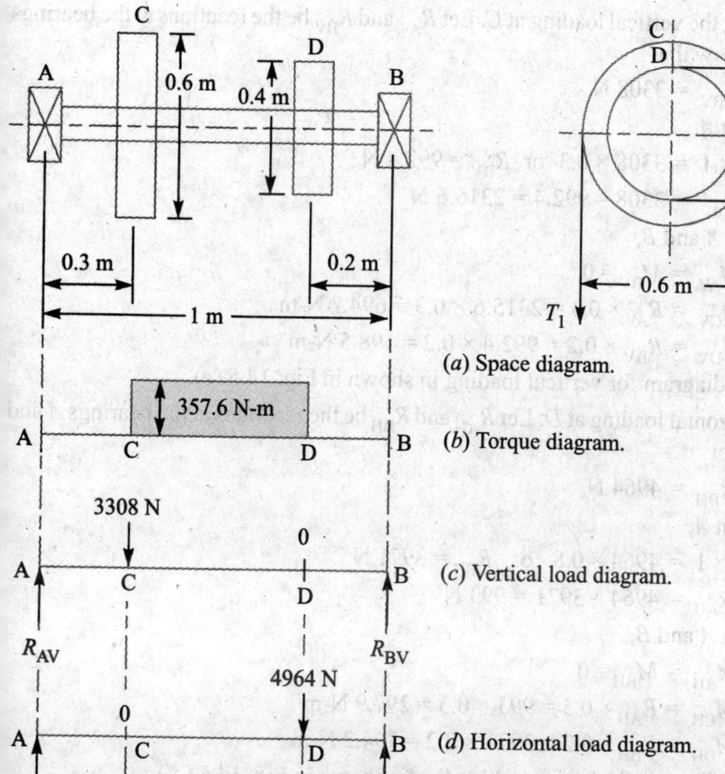

Fig. 14.5

From equations (*i*) and (*ii*), we find that

$$T_3 = 3376 \text{ N, and } T_4 = 1588 \text{ N}$$

∴ Horizontal load acting on the shaft at *D*,

$$W_D = T_3 + T_4 = 3376 + 1588 = 4964 \text{ N}$$

and horizontal load on the shaft at $C = 0$

The horizontal load diagram is shown in Fig. 14.5 (*d*).

Now let us find the maximum bending moment for vertical and horizontal loading.

First of all, considering the vertical loading at C. Let R_{AV} and R_{BV} be the reactions at the bearings A and B respectively. We know that

$$R_{AV} + R_{BV} = 3308 \text{ N}$$

Taking moments about A,

$$R_{BV} \times 1 = 3308 \times 0.3 \text{ or } R_{BV} = 992.4 \text{ N}$$

and $$R_{AV} = 3308 - 992.4 = 2315.6 \text{ N}$$

We know that B.M. at A and B,

$$M_{AV} = M_{BV} = 0$$

B.M. at C, $$M_{CV} = R_{AV} \times 0.3 = 2315.6 \times 0.3 = 694.7 \text{ N-m}$$

B.M. at D, $$M_{DV} = R_{BV} \times 0.2 = 992.4 \times 0.2 = 198.5 \text{ N-m}$$

The bending moment diagram for vertical loading in shown in Fig. 14.5 (*e*).

Now considering horizontal loading at D. Let R_{AH} and R_{BH} be the reactions at the bearings A and B respectively. We know that

$$R_{AH} + R_{BH} = 4964 \text{ N}$$

Taking moments about A,

$$R_{BH} \times 1 = 4964 \times 0.8 \text{ or } R_{BH} = 3971 \text{ N}$$

and $$R_{AH} = 4964 - 3971 = 993 \text{ N}$$

We know that B.M. at A and B,

$$M_{AH} = M_{BH} = 0$$

B.M. at C, $$M_{CH} = R_{AH} \times 0.3 = 993 \times 0.3 = 297.9 \text{ N-m}$$

B.M. at D, $$M_{DH} = R_{BH} \times 0.2 = 3971 \times 0.2 = 794.2 \text{ N-m}$$

The bending moment diagram for horizontal loading is shown in Fig. 14.5 (*f*).

Resultant B.M. at C,

$$M_C = \sqrt{(M_{CV})^2 + (M_{CH})^2} = \sqrt{(694.7)^2 + (297.9)^2} = 756 \text{ N-m}$$

and resultant B.M. at D,

$$M_D = \sqrt{(M_{DV})^2 + (M_{DH})^2} = \sqrt{(198.5)^2 + (794.2)^2} = 819.2 \text{ N-m}$$

The resultant bending moment diagram is shown in Fig. 14.5 (*g*).

We see that bending moment is maximum at D.

∴ Maximum bending moment,

$$M = M_D = 819.2 \text{ N-m}$$

Let d = Diameter of the shaft.

We know that equivalent twisting moment,

$$T_e = \sqrt{M^2 + T^2} = \sqrt{(819.2)^2 + (357.6)^2} = 894 \text{ N-m}$$
$$= 894 \times 10^3 \text{ N-mm}$$

We also know that equivalent twisting moment (T_e),

$$894 \times 10^3 = \frac{\pi}{16} \times \tau \times d^3 = \frac{\pi}{16} \times 42 \times d^3 = 8.25\, d^3$$

$$\therefore \quad d^3 = 894 \times 10^3 / 8.25 = 108 \times 10^3 \text{ or } d = 47.6 \text{ mm}$$

Again we know that equivalent bending moment,

$$M_e = \frac{1}{2}\left(M + \sqrt{M^2 + T^2}\right) = \frac{1}{2}(M + T_e)$$
$$= \frac{1}{2}(819.2 + 894) = 856.6 \text{ N-m} = 856.6 \times 10^3 \text{ N-mm}$$

We also know that equivalent bending moment (M_e),

$$856.6 \times 10^3 = \frac{\pi}{32} \times \sigma_b \times d^3 = \frac{\pi}{32} \times 63 \times d^3 = 6.2\, d^3$$

$$\therefore \quad d^3 = 856.6 \times 10^3 / 6.2 = 138.2 \times 10^3 \text{ or } d = 51.7 \text{ mm}$$

Taking larger of the two values, we have

$$d = 51.7 \text{ say } 55 \text{ mm } \textbf{Ans.}$$

Example 14.10. *A shaft is supported on bearings A and B, 800 mm between centres. A 20° straight tooth spur gear having 600 mm pitch diameter, is located 200 mm to the right of the left hand bearing A, and a 700 mm diameter pulley is mounted 250 mm towards the left of bearing B. The gear is driven by a pinion with a downward tangential force while the pulley drives a horizontal belt having 180° angle of wrap. The pulley also serves as a flywheel and weighs 2000 N. The maximum belt tension is 3000 N and the tension ratio is 3 : 1. Determine the maximum bending moment and the necessary shaft diameter if the allowable shear stress of the material is 40 MPa.*

Solution. Given : $AB = 800$ mm ; $\alpha_C = 20°$; $D_C = 600$ mm or $R_C = 300$ mm ; $AC = 200$ mm ; $D_D = 700$ mm or $R_D = 350$ mm ; $DB = 250$ mm ; $\theta = 180° = \pi$ rad ; $W = 2000$ N ; $T_1 = 3000$ N ; $T_1/T_2 = 3$; $\tau = 40$ MPa $= 40$ N/mm²

The space diagram of the shaft is shown in Fig. 14.6 (*a*).

We know that the torque acting on the shaft at *D*,

$$T = (T_1 - T_2)\, R_D = T_1 \left(1 - \frac{T_2}{T_1}\right) R_D$$

$$= 3000 \left(1 - \frac{1}{3}\right) 350 = 700 \times 10^3 \text{ N-mm} \qquad ...(\because T_1/T_2 = 3)$$

The torque diagram is shown in Fig. 14.6 (*b*).

Assuming that the torque at *D* is equal to the torque at *C*, therefore the tangential force acting on the gear *C*,

$$F_{tc} = \frac{T}{R_C} = \frac{700 \times 10^3}{300} = 2333 \text{ N}$$

and the normal load acting on the tooth of gear *C*,

$$W_C = \frac{F_{tc}}{\cos \alpha_C} = \frac{2333}{\cos 20°} = \frac{2333}{0.9397} = 2483 \text{ N}$$

The normal load acts at 20° to the vertical as shown in Fig. 14.7. Resolving the normal load vertically and horizontally, we get

Vertical component of W_C *i.e.* the vertical load acting on the shaft at *C*,

$$W_{CV} = W_C \cos 20°$$
$$= 2483 \times 0.9397 = 2333 \text{ N}$$

and horizontal component of W_C *i.e.* the horizontal load acting on the shaft at *C*,

$$W_{CH} = W_C \sin 20°$$
$$= 2483 \times 0.342 = 849 \text{ N}$$

Since $T_1 / T_2 = 3$ and $T_1 = 3000$ N, therefore

$$T_2 = T_1 / 3 = 3000 / 3 = 1000 \text{ N}$$

Camshaft

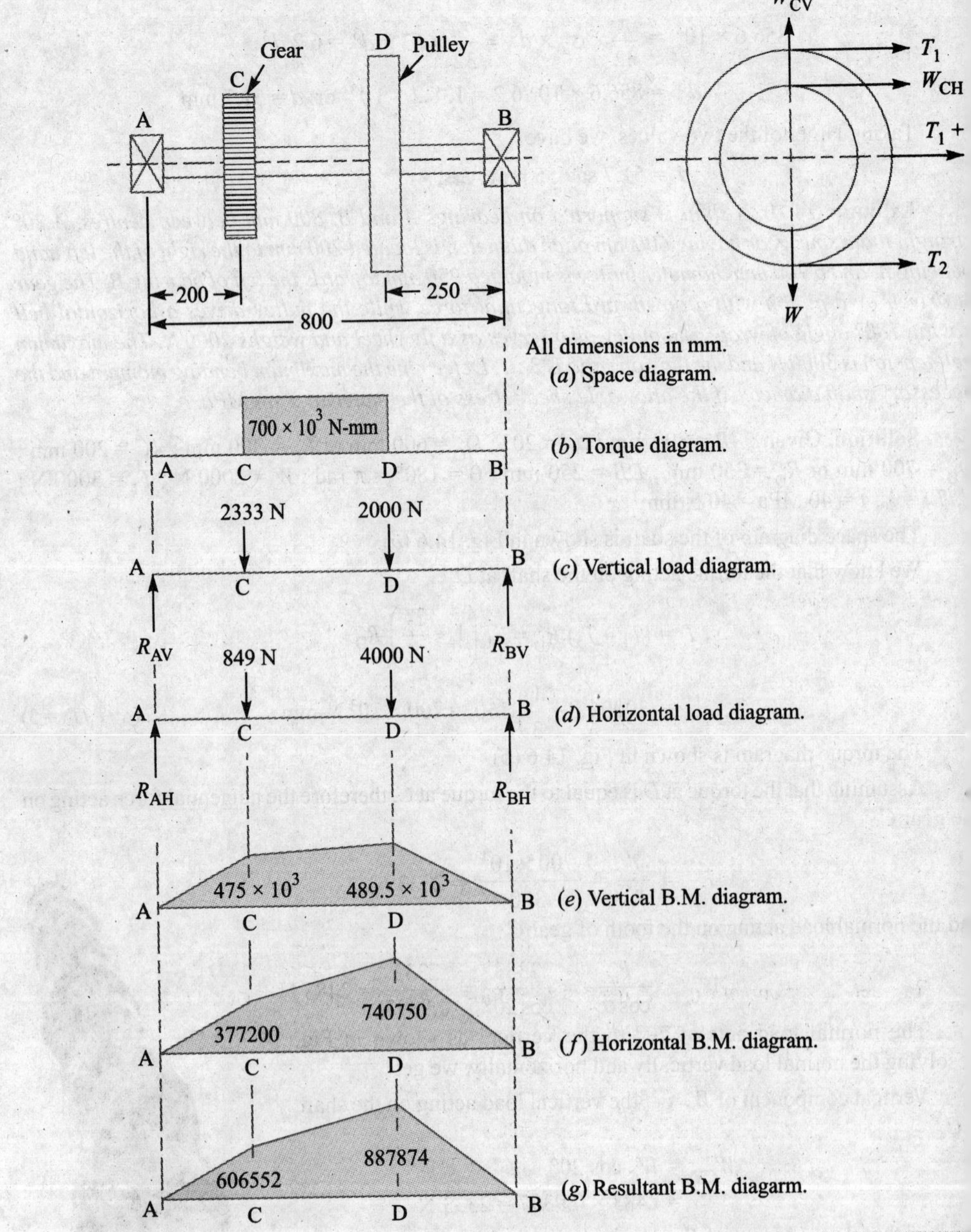

Fig. 14.6

∴ Horizontal load acting on the shaft at *D*,

$$W_{DH} = T_1 + T_2 = 3000 + 1000 = 4000 \text{ N}$$

and vertical load acting on the shaft at *D*,

$$W_{DV} = W = 2000 \text{ N}$$

The vertical and horizontal load diagram at *C* and *D* is shown in Fig. 14.6 (*c*) and (*d*) respectively.

Now let us find the maximum bending moment for vertical and horizontal loading.

First of all considering the vertical loading at *C* and *D*. Let R_{AV} and R_{BV} be the reactions at the bearings *A* and *B* respectively. We know that

$$R_{AV} + R_{BV} = 2333 + 2000 = 4333 \text{ N}$$

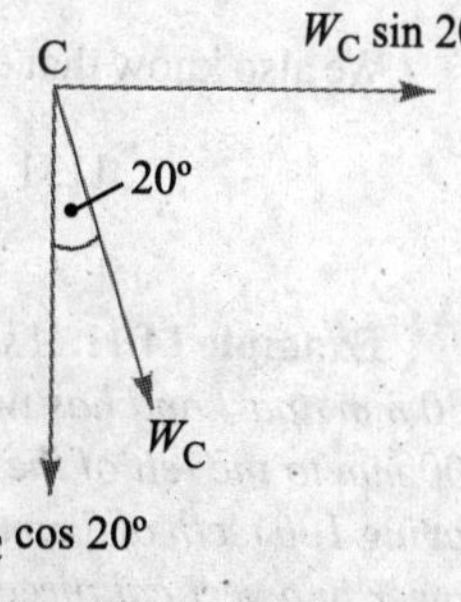

Fig. 14.7

Taking moments about *A*, we get

$$R_{BV} \times 800 = 2000\ (800 - 250) + 2333 \times 200 = 1\ 566\ 600$$

$$\therefore \quad R_{BV} = 1\ 566\ 600 / 800 = 1958 \text{ N}$$

and $$R_{AV} = 4333 - 1958 = 2375 \text{ N}$$

We know that B.M. at *A* and *B*,

$$M_{AV} = M_{BV} = 0$$

B.M. at *C*, $$M_{CV} = R_{AV} \times 200 = 2375 \times 200 = 475 \times 10^3 \text{ N-mm}$$

B.M. at D, $$M_{DV} = R_{BV} \times 250 = 1958 \times 250 = 489.5 \times 10^3 \text{ N-mm}$$

The bending moment diagram for vertical loading is shown in Fig. 14.6 (*e*).

Now consider the horizontal loading at *C* and *D*. Let R_{AH} and R_{BH} be the reactions at the bearings and *B* respectively. We know that

$$R_{AH} + R_{BH} = 849 + 4000 = 4849 \text{ N}$$

Taking moments about *A*, we get

$$R_{BH} \times 800 = 4000\ (800 - 250) + 849 \times 200 = 2\ 369\ 800$$

$$\therefore \quad R_{BH} = 2\ 369\ 800 / 800 = 2963 \text{ N}$$

and $$R_{AH} = 4849 - 2963 = 1886 \text{ N}$$

We know that B.M. at *A* and *B*,

$$M_{AH} = M_{BH} = 0$$

B.M. at *C*, $$M_{CH} = R_{AH} \times 200 = 1886 \times 200 = 377\ 200 \text{ N-mm}$$

B.M. at *D*, $$M_{DH} = R_{BH} \times 250 = 2963 \times 250 = 740\ 750 \text{ N-mm}$$

The bending moment diagram for horizontal loading is shown in Fig. 14.6 (*f*).

We know that resultant B.M. at *C*,

$$M_C = \sqrt{(M_{CV})^2 + (M_{CH})^2} = \sqrt{(475 \times 10^3)^2 + (377\ 200)^2} = 606\ 552 \text{ N-mm}$$

d resultant B.M. at *D*,

$$M_D = \sqrt{(M_{DV})^2 + (M_{DH})^2} = \sqrt{(489.5 \times 10^3)^2 + (740\ 750)^2} = 887\ 874 \text{ N-mm}$$

aximum bending moment

The resultant B.M. diagram is shown in Fig. 14.6 (*g*). We see that the bending moment is aximum at *D*, therefore

Maximum B.M., $M = M_D = 887\ 874$ N-mm **Ans.**

Diameter of the shaft

Let d = Diameter of the shaft.

We know that the equivalent twisting moment,

$$T_e = \sqrt{M^2 + T^2} = \sqrt{(887\ 874)^2 + (700 \times 10^3)^2} = 1131 \times 10^3 \text{ N-mm}$$

We also know that equivalent twisting moment (T_e),

$$1131 \times 10^3 = \frac{\pi}{16} \times \tau \times d^3 = \frac{\pi}{16} \times 40 \times d^3 = 7.86\ d^3$$

$$\therefore \quad d^3 = 1131 \times 10^3 / 7.86 = 144 \times 10^3 \quad \text{or} \quad d = 52.4 \text{ say } 55 \text{ mm } \textbf{Ans.}$$

Example 14.11. *A steel solid shaft transmitting 15 kW at 200 r.p.m. is supported on two bearing 750 mm apart and has two gears keyed to it. The pinion having 30 teeth of 5 mm module is locate 100 mm to the left of the right hand bearing and delivers power horizontally to the right. The gea having 100 teeth of 5 mm module is located 150 mm to the right of the left hand bearing and receive power in a vertical direction from below. Using an allowable stress of 54 MPa in shear, determin the diameter of the shaft.*

Solution. Given : P = 15 kW = 15 × 10³ W ; N = 200 r.p.m. ; AB = 750 mm ; T_D = 30 m_D = 5 mm ; BD = 100 mm ; T_C = 100 ; m_C = 5 mm ; AC = 150 mm ; τ = 54 MPa = 54 N/mm²

The space diagram of the shaft is shown in Fig. 14.8 (*a*).

We know that the torque transmitted by the shaft,

$$T = \frac{P \times 60}{2\pi N} = \frac{15 \times 10^3 \times 60}{2\pi \times 200} = 716 \text{ N-m} = 716 \times 10^3 \text{ N-mm}$$

The torque diagram is shown in Fig. 14.8 (*b*).

We know that diameter of gear

= No. of teeth on the gear × module

∴ Radius of gear C,

$$R_C = \frac{T_C \times m_C}{2} = \frac{100 \times 5}{2} = 250 \text{ mm}$$

and radius of pinion D,

$$R_D = \frac{T_D \times m_D}{2} = \frac{30 \times 5}{2} = 75 \text{ mm}$$

Assuming that the torque at C and D is same (*i.e.* 716 × 10³ N-mm), therefore tangential for on the gear C, acting downward,

$$F_{tC} = \frac{T}{R_C} = \frac{716 \times 10^3}{250} = 2870 \text{ N}$$

and tangential force on the pinion D, acting horizontally,

$$F_{tD} = \frac{T}{R_D} = \frac{716 \times 10^3}{75} = 9550 \text{ N}$$

The vertical and horizontal load diagram is shown in Fig. 14.8 (*c*) and (*d*) respectively.

Now let us find the maximum bending moment for vertical and horizontal loading.

First of all, considering the vertical loading at C. Let R_{AV} and R_{BV} be the reactions at the bearin A and B respectively. We know that

$$R_{AV} + R_{BV} = 2870 \text{ N}$$

Taking moments about A, we get

$$R_{BV} \times 750 = 2870 \times 150$$

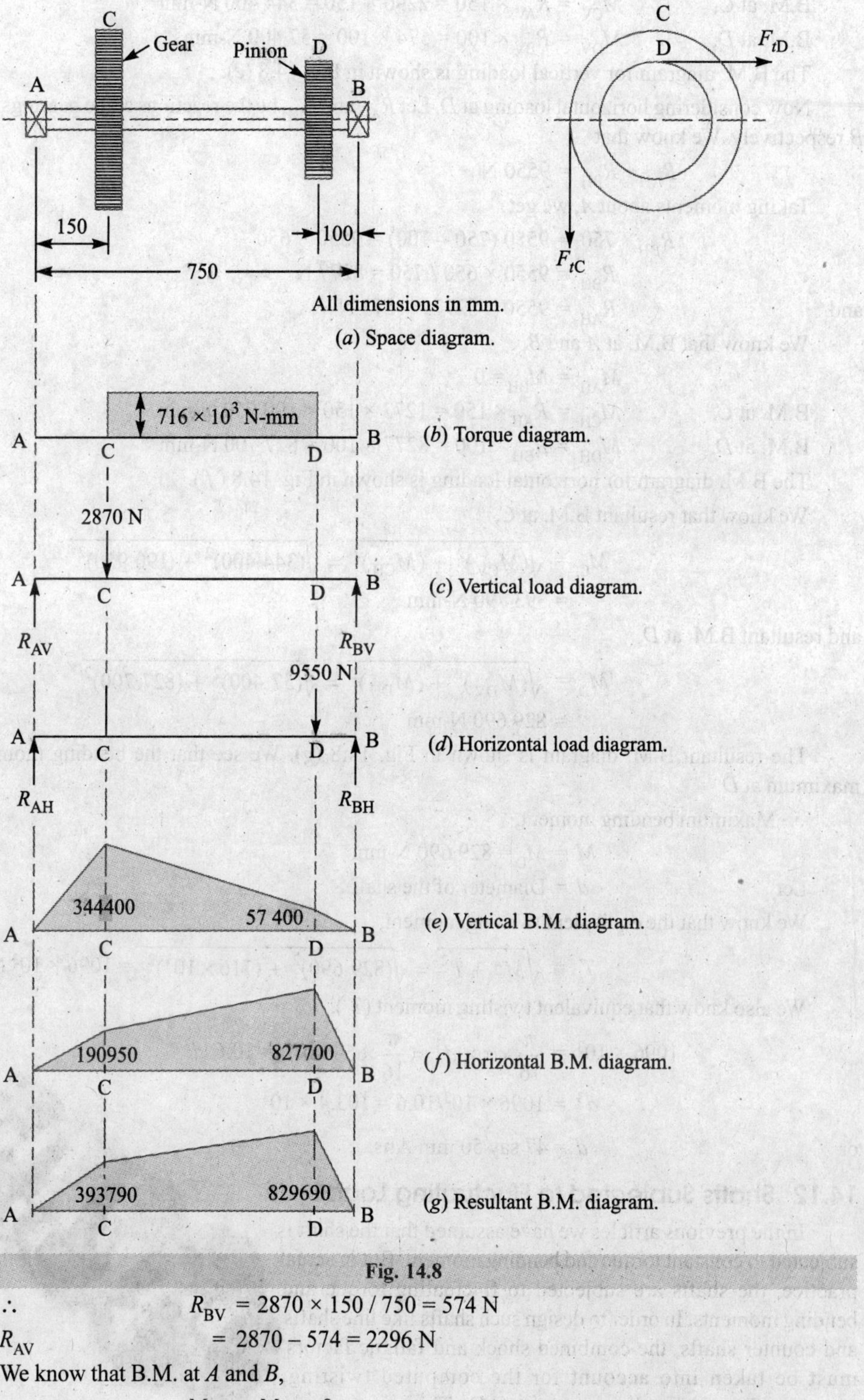

Fig. 14.8

$\therefore$ $R_{BV} = 2870 \times 150 / 750 = 574$ N

and $R_{AV} = 2870 - 574 = 2296$ N

We know that B.M. at A and B,

$$M_{AV} = M_{BV} = 0$$

B.M. at *C*, $M_{CV} = R_{AV} \times 150 = 2296 \times 150 = 344\ 400$ N-mm

B.M. at *D*, $M_{DV} = R_{BV} \times 100 = 574 \times 100 = 57\ 400$ N-mm

The B.M. diagram for vertical loading is shown in Fig. 14.8 (*e*).

Now considering horizontal loading at *D*. Let R_{AH} and R_{BH} be the reactions at the bearings *A* and *B* respectively. We know that

$$R_{AH} + R_{BH} = 9550 \text{ N}$$

Taking moments about *A*, we get

$$R_{BH} \times 750 = 9550\ (750 - 100) = 9550 \times 650$$

$\therefore$ $R_{BH} = 9550 \times 650 / 750 = 8277$ N

and $R_{AH} = 9550 - 8277 = 1273$ N

We know that B.M. at *A* and *B*,

$$M_{AH} = M_{BH} = 0$$

B.M. at *C*, $M_{CH} = R_{AH} \times 150 = 1273 \times 150 = 190\ 950$ N-mm

B.M. at *D*, $M_{DH} = R_{BH} \times 100 = 8277 \times 100 = 827\ 700$ N-mm

The B.M. diagram for horizontal loading is shown in Fig. 14.8 (*f*).

We know that resultant B.M. at *C*,

$$M_C = \sqrt{(M_{CV})^2 + (M_{CH})^2} = \sqrt{(344\ 400)^2 + (190\ 950)^2}$$
$$= 393\ 790 \text{ N-mm}$$

and resultant B.M. at *D*,

$$M_D = \sqrt{(M_{DV})^2 + (M_{DH})^2} = \sqrt{(57\ 400)^2 + (827\ 700)^2}$$
$$= 829\ 690 \text{ N-mm}$$

The resultant B.M. diagram is shown in Fig. 14.8 (g). We see that the bending moment is maximum at *D*.

$\therefore$ Maximum bending moment,

$$M = M_D = 829\ 690 \text{ N-mm}$$

Let d = Diameter of the shaft.

We know that the equivalent twisting moment,

$$T_e = \sqrt{M^2 + T^2} = \sqrt{(829\ 690)^2 + (716 \times 10^3)^2} = 1096 \times 10^3 \text{ N-mm}$$

We also know that equivalent twisting moment (T_e),

$$1096 \times 10^3 = \frac{\pi}{16} \times \tau \times d^3 = \frac{\pi}{16} \times 54 \times d^3 = 10.6\ d^3$$

$\therefore$ $d^3 = 1096 \times 10^3 / 10.6 = 103.4 \times 10^3$

or $d = 47$ say 50 mm **Ans.**

14.12 Shafts Subjected to Fluctuating Loads

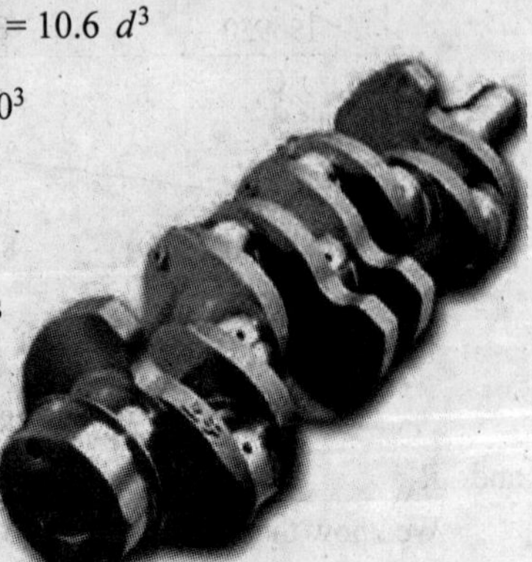

Crankshaft

In the previous articles we have assumed that the shaft is subjected to constant torque and bending moment. But in actual practice, the shafts are subjected to fluctuating torque and bending moments. In order to design such shafts like line shafts and counter shafts, the combined shock and fatigue factors must be taken into account for the computed twisting moment (T) and bending moment (M). Thus for a shaft subjected to combined bending and torsion, the equivalent twisting moment,

$$T_e = \sqrt{(K_m \times M)^2 + (K_t + T)^2}$$

and equivalent bending moment,

$$M_e = \frac{1}{2}\left[K_m \times M + \sqrt{(K_m \times M)^2 + (K_t \times T)^2}\right]$$

where K_m = Combined shock and fatigue factor for bending, and

K_t = Combined shock and fatigue factor for torsion.

The following table shows the recommended values for K_m and K_t.

Table 14.2. Recommended values for K_m and K_t.

Nature of load	K_m	K_t
1. Stationary shafts		
(*a*) Gradually applied load	1.0	1.0
(*b*) Suddenly applied load	1.5 to 2.0	1.5 to 2.0
2. Rotating shafts		
(*a*) Gradually applied or steady load	1.5	1.0
(*b*) Suddenly applied load with minor shocks only	1.5 to 2.0	1.5 to 2.0
(*c*) Suddenly applied load with heavy shocks	2.0 to 3.0	1.5 to 3.0

Example 14.12. *A mild steel shaft transmits 20 kW at 200 r.p.m. It carries a central load of 900 N and is simply supported between the bearings 2.5 metres apart. Determine the size of the shaft, if the allowable shear stress is 42 MPa and the maximum tensile or compressive stress is not to exceed 56 MPa. What size of the shaft will be required, if it is subjected to gradually applied loads?*

Solution. Given : P = 20 kW = 20 × 10^3 W ; N = 200 r.p.m. ; W = 900 N ; L = 2.5 m ; τ = 42 MPa = 42 N/mm^2 ; σ_b = 56 MPa = 56 N/mm^2

Size of the shaft

Let d = Diameter of the shaft, in mm.

We know that torque transmitted by the shaft,

$$T = \frac{P \times 60}{2\pi N} = \frac{20 \times 10^3 \times 60}{2\pi \times 200} = 955 \text{ N-m} = 955 \times 10^3 \text{ N-mm}$$

and maximum bending moment of a simply supported shaft carrying a central load,

$$M = \frac{W \times L}{4} = \frac{900 \times 2.5}{4} = 562.5 \text{ N-m} = 562.5 \times 10^3 \text{ N-mm}$$

We know that the equivalent twisting moment,

$$T_e = \sqrt{M^2 + T^2} = \sqrt{(562.5 \times 10^3)^2 + (955 \times 10^3)^2}$$
$$= 1108 \times 10^3 \text{ N-mm}$$

We also know that equivalent twisting moment (T_e),

$$1108 \times 10^3 = \frac{\pi}{16} \times \tau \times d^3 = \frac{\pi}{16} \times 42 \times d^3 = 8.25\ d^3$$

$$\therefore \quad d^3 = 1108 \times 10^3 / 8.25 = 134.3 \times 10^3 \text{ or } d = 51.2 \text{ mm}$$

We know that the equivalent bending moment,

$$M_e = \frac{1}{2}\left[M + \sqrt{M^2 + T^2}\right] = \frac{1}{2}(M + T_e)$$

$$= \frac{1}{2}(562.5 \times 10^3 + 1108 \times 10^3) = 835.25 \times 10^3 \text{ N-mm}$$

We also know that equivalent bending moment (M_e),

$$835.25 \times 10^3 = \frac{\pi}{32} \times \sigma_b \times d^3 = \frac{\pi}{32} \times 56 \times d^3 = 5.5\, d^3$$

$$\therefore \quad d^3 = 835.25 \times 10^3 / 5.5 = 152 \times 10^3 \text{ or } d = 53.4 \text{ mm}$$

Taking the larger of the two values, we have

$$d = 53.4 \text{ say } 55 \text{ mm } \textbf{Ans.}$$

Size of the shaft when subjected to gradually applied load

Let d = Diameter of the shaft.

From Table 14.2, for rotating shafts with gradually applied loads,

$$K_m = 1.5 \text{ and } K_t = 1$$

We know that equivalent twisting moment,

$$T_e = \sqrt{(K_m \times M)^2 + (K_t \times T)^2}$$

$$= \sqrt{(1.5 \times 562.5 \times 10^3)^2 + (1 \times 955 \times 10^3)^2} = 1274 \times 10^3 \text{ N-mm}$$

We also know that equivalent twisting moment (T_e),

$$1274 \times 10^3 = \frac{\pi}{16} \times \tau \times d^3 = \frac{\pi}{16} \times 42 \times d^3 = 8.25\, d^3$$

$$\therefore \quad d^3 = 1274 \times 10^3 / 8.25 = 154.6 \times 10^3 \text{ or } d = 53.6 \text{ mm}$$

We know that the equivalent bending moment,

$$M_e = \frac{1}{2}\left[K_m \times M + \sqrt{(K_m \times M)^2 + (K_t \times T)^2}\right] = \frac{1}{2}\left[K_m \times M + T_e\right]$$

$$= \frac{1}{2}\left[1.5 \times 562.5 \times 10^3 + 1274 \times 10^3\right] = 1059 \times 10^3 \text{ N-mm}$$

We also know that equivalent bending moment (M_e),

$$1059 \times 10^3 = \frac{\pi}{32} \times \sigma_b \times d^3 = \frac{\pi}{32} \times 56 \times d^3 = 5.5\, d^3$$

$$\therefore \quad d^3 = 1059 \times 10^3 / 5.5 = 192.5 \times 10^3 = 57.7 \text{ mm}$$

Taking the larger of the two values, we have

$$d = 57.7 \text{ say } 60 \text{ mm } \textbf{Ans.}$$

Example 14.13. *Design a shaft to transmit power from an electric motor to a lathe head stock through a pulley by means of a belt drive. The pulley weighs 200 N and is located at 300 mm from the centre of the bearing. The diameter of the pulley is 200 mm and the maximum power transmitted is 1 kW at 120 r.p.m. The angle of lap of the belt is 180° and coefficient of friction between the belt and the pulley is 0.3. The shock and fatigue factors for bending and twisting are 1.5 and 2.0 respectively. The allowable shear stress in the shaft may be taken as 35 MPa.*

Solution. Given : W = 200 N ; L = 300 mm ; D = 200 mm or R = 100 mm ; P = 1 kW = 1000 W ; N = 120 r.p.m. ; θ = 180° = π rad ; μ = 0.3 ; K_m = 1.5 ; K_t = 2 ; τ = 35 MPa = 35 N/mm^2

The shaft with pulley is shown in Fig. 14.9.

We know that torque transmitted by the shaft,

$$T = \frac{P \times 60}{2\pi N} = \frac{1000 \times 60}{2\pi \times 120} = 79.6 \text{ N-m} = 79.6 \times 10^3 \text{ N-mm}$$

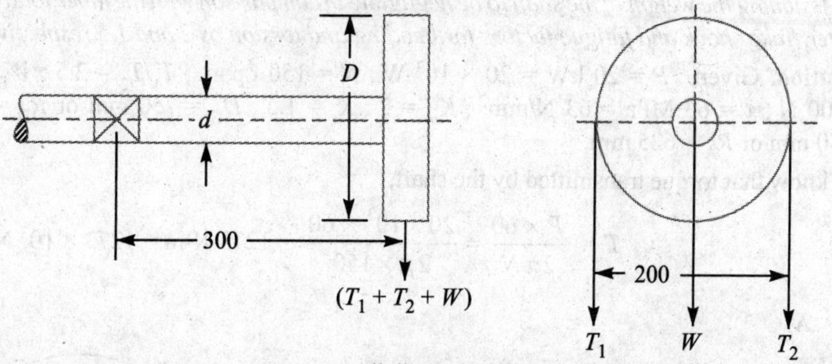

All dimensions in mm.

Fig. 14.9

Let T_1 and T_2 = Tensions in the tight side and slack side of the belt respectively in newtons.

∴ Torque transmitted (*T*),

$$79.6 \times 10^3 = (T_1 - T_2)\, R = (T_1 - T_2)\, 100$$

∴ $$T_1 - T_2 = 79.6 \times 10^3 / 100 = 796 \text{ N} \qquad ...(i)$$

We know that

$$2.3 \log\left(\frac{T_1}{T_2}\right) = \mu.\theta = 0.3\,\pi = 0.9426$$

∴ $$\log\left(\frac{T_1}{T_2}\right) = \frac{0.9426}{2.3} = 0.4098 \text{ or } \frac{T_1}{T_2} = 2.57 \qquad ...(ii)$$

...(Taking antilog of 0.4098)

From equations (*i*) and (*ii*), we get,

$$T_1 = 1303 \text{ N, and } T_2 = 507 \text{ N}$$

We know that the total vertical load acting on the pulley,

$$W_T = T_1 + T_2 + W = 1303 + 507 + 200 = 2010 \text{ N}$$

∴ Bending moment acting on the shaft,

$$M = W_T \times L = 2010 \times 300 = 603 \times 10^3 \text{ N-mm}$$

Let d = Diameter of the shaft.

We know that equivalent twisting moment,

$$T_e = \sqrt{(K_m \times M)^2 + (K_t + T)^2}$$

$$= \sqrt{(1.5 \times 603 \times 10^3)^2 + (2 \times 79.6 \times 10^3)^2} = 918 \times 10^3 \text{ N-mm}$$

We also know that equivalent twisting moment (T_e),

$$918 \times 10^3 = \frac{\pi}{16} \times \tau \times d^3 = \frac{\pi}{16} \times 35 \times d^3 = 6.87\, d^3$$

∴ $d^3 = 918 \times 10^3 / 6.87 = 133.6 \times 10^3$ or $d = 51.1$ say 55 mm **Ans.**

Example 14.14. *Fig. 14.10 shows a shaft carrying a pulley A and a gear B and supported in two bearings C and D. The shaft transmits 20 kW at 150 r.p.m. The tangential force F_t on the gear B acts vertically upwards as shown.*

The pulley delivers the power through a belt to another pulley of equal diameter vertically below the pulley A. The ratio of tensions T_1 / T_2 is equal to 2.5. The gear and the pulley weigh 900 N and 2700 N respectively. The permissible shear stress for the material of the shaft may be taken as 63 MPa. Assuming the weight of the shaft to be negligible in comparison with the other loads, determine its diameter. Take shock and fatigue factors for bending and torsion as 2 and 1.5 respectively.

Solution. Given : $P = 20$ kW $= 20 \times 10^3$ W ; $N = 150$ r.p.m. ; $T_1/T_2 = 2.5$; $W_B = 900$ N ; $W_A = 2700$ N ; $\tau = 63$ MPa $= 63$ N/mm^2 ; $K_m = 2$; $K_t = 1.5$; $D_B = 750$ mm or $R_B = 375$ mm ; $D_A = 1250$ mm or $R_A = 625$ mm.

We know that torque transmitted by the shaft,

$$T = \frac{P \times 60}{2\pi N} = \frac{20 \times 10^3 \times 60}{2\pi \times 150} = 1273 \text{ N-m} = 1273 \times 10^3 \text{ N-mm}$$

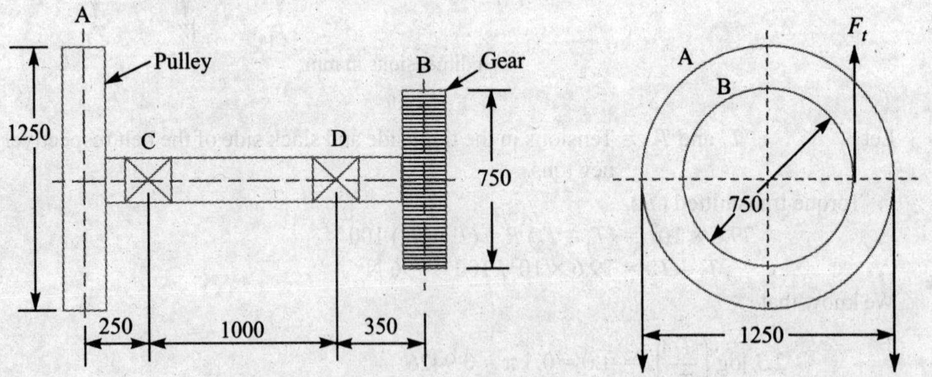

All dimensions in mm.

Fig. 14.10

Let T_1 and T_2 = Tensions in the tight side and slack side of the belt on pulley A.

Since the torque on the pulley is same as that of shaft (*i.e.* 1273×10^3 N-mm), therefore

$$(T_1 - T_2) R_A = 1273 \times 10^3 \quad \text{or} \quad T_1 - T_2 = 1273 \times 10^3 / 625 = 2037 \text{ N} \qquad ...(i)$$

Since $\quad T_1 / T_2 = 2.5 \quad$ or $\quad T_1 = 2.5\, T_2$, therefore

$$2.5\, T_2 - T_2 = 2037 \quad \text{or} \quad T_2 = 2037/1.5 = 1358 \text{ N} \qquad ...\text{[From equation } (i)]$$

and $\quad T_1 = 2.5 \times 1358 = 3395$ N

∴ Total vertical load acting downward on the shaft at A

$$= T_1 + T_2 + W_A = 3395 + 1358 + 2700 = 7453 \text{ N}$$

Assuming that the torque on the gear B is same as that of the shaft, therefore the tangential force acting vertically upward on the gear B,

$$F_t = \frac{T}{R_B} = \frac{1273 \times 10^3}{375} = 3395 \text{ N}$$

Since the weight of gear B ($W_B = 900$ N) acts vertically downward, therefore the total vertical load acting upward on the shaft at B

$$= F_t - W_B = 3395 - 900 = 2495 \text{ N}$$

Now let us find the reactions at the bearings C and D. Let R_C and R_D be the reactions at C and D respectively. A little consideration will show that the reaction R_C will act upward while the reaction R_D act downward as shown in Fig. 14.11.

Taking moments about D, we get

$$R_C \times 1000 = 7453 \times 1250 + 2495 \times 350 = 10.2 \times 10^6$$

$$\therefore \quad R_C = 10.2 \times 10^6 / 1000 = 10\,200 \text{ N}$$

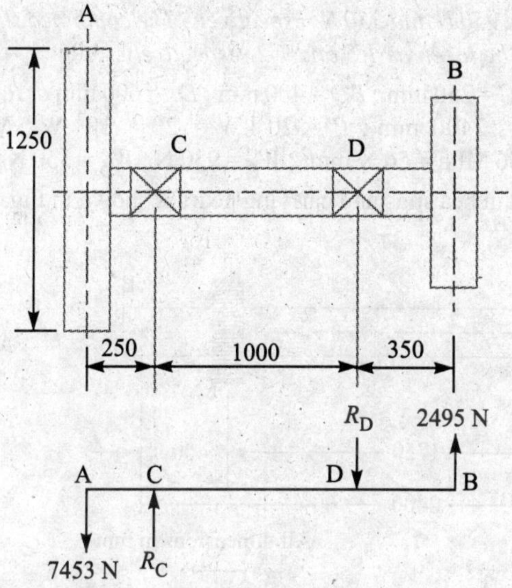

Fig. 14.11

For the equilibrium of the shaft,

$$R_D + 7453 = R_C + 2495 = 10\,200 + 2495 = 12\,695$$

$$\therefore \quad R_D = 12\,695 - 7453 = 5242 \text{ N}$$

We Know that B.M. at A and B

$$= 0$$

B.M. at C $= 7453 \times 250 = 1863 \times 10^3$ N-mm

B.M. at D $= 2495 \times 350 = 873 \times 10^3$ N-mm

We see that the bending moment is maximum at C.

$\therefore$ Maximum B.M. $= M = M_C = 1863 \times 10^3$ N-mm

We know that the equivalent twisting moment,

$$T_e = \sqrt{(K_m \times M)^2 + (K_t \times T)^2}$$

$$= \sqrt{(2 \times 1863 \times 10^3)^2 + (1.5 \times 1273 \times 10^3)^2}$$

$$= 4187 \times 10^3 \text{ N-mm}$$

We also know that equivalent twisting moment (T_e),

$$4187 \times 10^3 = \frac{\pi}{16} \times \tau \times d^3 = \frac{\pi}{16} \times 63 \times d^3 = 12.37\, d^3.$$

$$\therefore \quad d^3 = 4187 \times 10^3 / 12.37 = 338 \times 10^3$$

or $\quad d = 69.6$ say 70 mm **Ans.**

Example 14.15. *A horizontal nickel steel shaft rests on two bearings, A at the left and B at the right end and carries two gears C and D located at distances of 250 mm and 400 mm respectively from the centre line of the left and right bearings. The pitch diameter of the gear C is 600 mm and that of gear D is 200 mm. The distance between the centre line of the bearings is 2400 mm. The shaft* *transmits 20 kW at 120 r.p.m. The power is delivered to the shaft at gear C and is taken out at gear D in such a manner that the tooth pressure F_{tC} of the gear C and F_{tD} of the gear D act vertically downwards.*

Find the diameter of the shaft, if the working stress is 100 MPa in tension and 56 MPa in shear. The gears C and D weighs 950 N and 350 N respectively. The combined shock and fatigue factors for bending and torsion may be taken as 1.5 and 1.2 respectively.

Solution. Given : $AC = 250$ mm ; $BD = 400$ mm ; $D_C = 600$ mm or $R_C = 300$ mm ; $D_D = 200$ mm or $R_D = 100$ mm ; $AB = 2400$ mm ; $P = 20$ kW $= 20 \times 10^3$ W ; $N = 120$ r.p.m ; $\sigma_t = 100$ MPa $= 100$ N/mm^2 ; $\tau = 56$ MPa $= 56$ N/mm^2 ; $W_C = 950$ N ; $W_D = 350$ N ; $K_m = 1.5$; $K_t = 1.2$

The shaft supported in bearings and carrying gears is shown in Fig. 14.12.

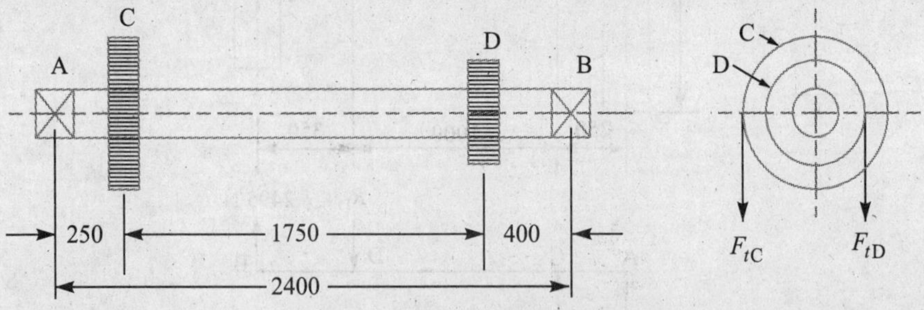

All dimensions in mm.

Fig. 14.12

We know that the torque transmitted by the shaft,

$$T = \frac{P \times 60}{2\pi N} = \frac{20 \times 10^3 \times 60}{2\pi \times 120} = 1590 \text{ N-m} = 1590 \times 10^3 \text{ N-mm}$$

Since the torque acting at gears C and D is same as that of the shaft, therefore the tangential force acting at gear C,

$$F_{tC} = \frac{T}{R_C} = \frac{1590 \times 10^3}{300} = 5300 \text{ N}$$

Car rear axle.

and total load acting downwards on the shaft at *C*

$$= F_{tC} + W_C = 5300 + 950 = 6250 \text{ N}$$

Similarly tangential force acting at gear *D*,

$$F_{tD} = \frac{T}{R_D} = \frac{1590 \times 10^3}{100} = 15\,900 \text{ N}$$

and total load acting downwards on the shaft at *D*

$$= F_{tD} + W_D = 15\,900 + 350 = 16\,250 \text{ N}$$

Now assuming the shaft as a simply supported beam as shown in Fig. 14.13, the maximum bending moment may be obtained as discussed below :

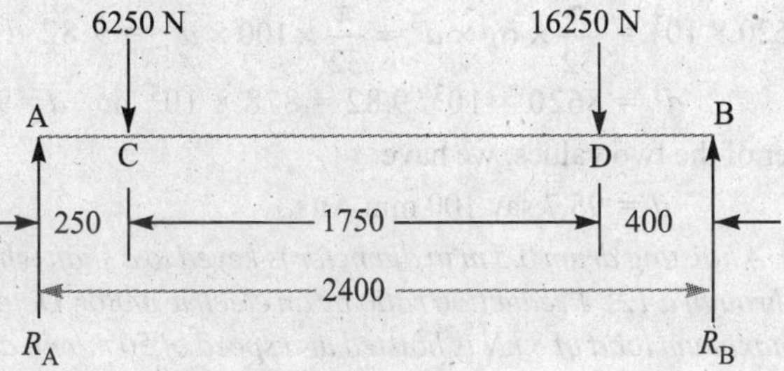

Fig. 14.13

Let R_A and R_B = Reactions at *A* and *B* respectively.

$\therefore$ $R_A + R_B$ = Total load acting downwards at *C* and *D*

$= 6250 + 16\,250 = 22\,500$ N

Now taking moments about *A*,

$$R_B \times 2400 = 16\,250 \times 2000 + 6250 \times 250 = 34\,062.5 \times 10^3$$

$$\therefore \quad R_B = 34\,062.5 \times 10^3 / 2400 = 14\,190 \text{ N}$$

and $R_A = 22\,500 - 14\,190 = 8310$ N

A little consideration will show that the maximum bending moment will be either at *C* or *D*.

We know that bending moment at *C*,

$$M_C = R_A \times 250 = 8310 \times 250 = 2077.5 \times 10^3 \text{ N-mm}$$

Bending moment at *D*,

$${}^*M_D = R_B \times 400 = 14\,190 \times 400 = 5676 \times 10^3 \text{ N-mm}$$

$\therefore$ Maximum bending moment transmitted by the shaft,

$$M = M_D = 5676 \times 10^3 \text{ N-mm}$$

Let d = Diameter of the shaft.

We know that the equivalent twisting moment,

$$T_e = \sqrt{(K_m \times M)^2 + (K_t \times T)^2}$$

$$= \sqrt{(1.5 \times 5676 \times 10^3)^2 + (1.2 \times 1590 \times 10^3)^2}$$

$$= 8725 \times 10^3 \text{ N-mm}$$

---

* The bending moment at *D* may also be calculated as follows :

$M_D = R_A \times 2000 - (\text{Total load at } C)\ 1750$

We also know that the equivalent twisting moment (T_e),

$$8725 \times 10^3 = \frac{\pi}{16} \times \tau \times d^3 = \frac{\pi}{16} \times 56 \times d^3 = 11\ d^3$$

$$\therefore \quad d^3 = 8725 \times 10^3 / 11 = 793 \times 10^3 \text{ or } d = 92.5 \text{ mm}$$

Again we know that the equivalent bending moment,

$$M_e = \frac{1}{2}\left[K_m \times M + \sqrt{(K_m \times M)^2 + (K_t \times T)^2}\right] = \frac{1}{2}(K_m \times M + T_e)$$

$$= \frac{1}{2}\left[1.5 \times 5676 \times 10^3 + 8725 \times 10^3\right] = 8620 \times 10^3 \text{ N-mm}$$

We also know that the equivalent bending moment (M_e),

$$8620 \times 10^3 = \frac{\pi}{32} \times \sigma_b \times d^3 = \frac{\pi}{32} \times 100 \times d^3 = 9.82\ d^3 \quad \text{...(Taking } \sigma_b = \sigma_t\text{)}$$

$$\therefore \quad d^3 = 8620 \times 10^3 / 9.82 = 878 \times 10^3 \text{ or } d = 95.7 \text{ mm}$$

Taking the larger of the two values, we have

d = 95.7 say 100 mm **Ans.**

Example 14.16. *A hoisting drum 0.5 m in diameter is keyed to a shaft which is supported in two bearings and driven through a 12 : 1 reduction ratio by an electric motor. Determine the power of the driving motor, if the maximum load of 8 kN is hoisted at a speed of 50 m/min and the efficiency of the drive is 80%. Also determine the torque on the drum shaft and the speed of the motor in r.p.m. Determine also the diameter of the shaft made of machinery steel, the working stresses of which are 115 MPa in tension and 50 MPa in shear. The drive gear whose diameter is 450 mm is mounted at the end of the shaft such that it overhangs the nearest bearing by 150 mm. The combined shock and fatigue factors for bending and torsion may be taken as 2 and 1.5 respectively.*

Solution. Given : D = 0.5 m or R = 0.25 m ; Reduction ratio = 12 : 1 ; W = 8 kN = 8000 N ; v = 50 m/min ; η = 80% = 0.8 ; σ_t = 115 MPa = 115 N/mm^2 ; τ = 50 MPa = 50 N/mm^2 ; D_1 = 450 mm or R_1 = 225 mm = 0.225 m ; Overhang = 150 mm = 0.15 m ; K_m = 2 ; K_t = 1.5

Power of the driving motor

We know that the energy supplied to the hoisting drum per minute

$$= W \times v = 8000 \times 50 = 400 \times 10^3 \text{ N-m/min}$$

$\therefore$ Power supplied to the hoisting drum

$$= \frac{400 \times 10^3}{60} = 6670 \text{ W} = 6.67 \text{ kW} \quad \text{...}(\because 1 \text{ N-m/s} = 1 \text{ W})$$

Since the efficiency of the drive is 0.8, therefore power of the driving motor

$$= \frac{6.67}{0.8} = 8.33 \text{ kW } \textbf{Ans.}$$

Torque on the drum shaft

We know that the torque on the drum shaft,

$$T = W.R = 8000 \times 0.25 = 2000 \text{ N-m } \textbf{Ans.}$$

Speed of the motor

Let N = Speed of the motor in r.p.m.

We know that angular speed of the hoisting drum

$$= \frac{\text{Linear speed}}{\text{Radius of the drum}} = \frac{v}{R} = \frac{50}{0.25} = 200 \text{ rad / min}$$

Since the reduction ratio is 12 : 1, therefore the angular speed of the electric motor,

$$\omega = 200 \times 12 = 2400 \text{ rad/min}$$

and speed of the motor in r.p.m.,

$$N = \frac{\omega}{2\pi} = \frac{2400}{2\pi} = 382 \text{ r.p.m. } \textbf{Ans.}$$

Diameter of the shaft

Let d = Diameter of the shaft.

Since the torque on the drum shaft is 2000 N-m, therefore the tangential tooth load on the drive gear,

$$F_t = \frac{T}{R_1} = \frac{2000}{0.225} = 8900 \text{ N}$$

Assuming that the pressure angle of the drive gear in 20°, therefore the maximum bending load on the shaft due to tooth load

$$= \frac{F_t}{\cos 20°} = \frac{8900}{0.9397} = 9470 \text{ N}$$

Since the overhang of the shaft is 150 mm = 0.15 m, therefore bending moment at the bearing,

$$M = 9470 \times 0.15 = 1420 \text{ N-m}$$

We know that the equivalent twisting moment,

$$T_e = \sqrt{(K_m \times M)^2 + (K_t \times T)^2}$$

$$= \sqrt{(2 \times 1420)^2 + (1.5 \times 2000)^2} = 4130 \text{ N-m} = 4130 \times 10^3 \text{ N-mm}$$

We also know that equivalent twisting moment (T_e),

$$4130 \times 10^3 = \frac{\pi}{16} \times \tau \times d^3 = \frac{\pi}{16} \times 50 \times d^3 = 9.82\, d^3$$

$$\therefore \quad d^3 = 4130 \times 10^3 / 9.82 = 420.6 \times 10^3 \quad \text{or} \quad d = 75 \text{ mm}$$

Again we know that the equivalent bending moment,

$$M_e = \tfrac{1}{2}\left[K_m \times M + \sqrt{(K_m \times M)^2 + (K_t \times T)^2}\right] = \tfrac{1}{2}(K_m \times M + T_e)$$

$$= \tfrac{1}{2}(2 \times 1420 + 4130) = 3485 \text{ N-m} = 3485 \times 10^3 \text{ N-mm}$$

We also know that equivalent bending moment (M_e),

$$3485 \times 10^3 = \frac{\pi}{32} \times \sigma_b \times d^3 = \frac{\pi}{32} \times 115 \times d^3 = 11.3\, d^3$$

$$\therefore \quad d^3 = 3485 \times 10^3 / 11.3 = 308.4 \times 10^3 \quad \text{or} \quad d = 67.5 \text{ mm}$$

Taking the larger of the two values, we have

$$d = 75 \text{ mm } \textbf{Ans.}$$

Example 14.17. *A solid steel shaft is supported on two bearings 1.8 m apart and rotates at 250 r.p.m. A 20° involute gear D, 300 mm diameter is keyed to the shaft at a distance of 150 mm to the left on the right hand bearing. Two pulleys B and C are located on the shaft at distances of 600 mm and 1350 mm respectively to the right of the left hand bearing. The diameters of the pulleys B and C are 750 mm and 600 mm respectively. 30 kW is supplied to the gear, out of which 18.75 kW is taken off at the pulley C and 11.25 kW from pulley B. The drive from B is vertically downward while from C the drive is downward at an angle of 60° to the horizontal. In both cases the belt tension ratio is 2 and the angle of lap is 180°. The combined fatigue and shock factors for torsion and bending may be taken as 1.5 and 2 respectively.*

Design a suitable shaft taking working stress to be 42 MPa in shear and 84 MPa in tension.

Solution. Given : $PQ = 1.8$ m ; $N = 250$ r.p.m ; $\alpha_D = 20°$; $D_D = 300$ mm or $R_D = 150$ mm $= 0.15$ m ; $QD = 150$ mm $= 0.15$ m ; $PB = 600$ mm $= 0.6$ m ; $PC = 1350$ mm $= 1.35$ m ; $D_B = 750$ mm or $R_B = 375$ mm $= 0.375$ m ; $D_C = 600$ mm or $R_C = 300$ mm $= 0.3$ m ; $P_D = 30$ kW $= 30 \times 10^3$ W ; $P_C = 18.75$ kW $= 18.75 \times 10^3$ W ; $P_B = 11.25$ kW $= 11.25 \times 10^3$ W ; $T_{B1}/T_{B2} = T_{C1}/T_{C2} = 2$; $\theta = 180° = \pi$ rad ; $K_t = 1.5$; $K_m = 2$; $\tau = 42$ MPa $= 42$ N/mm² ; $\sigma_t = 84$ MPa $= 84$ N/mm²

First of all, let us find the total loads acting on the gear D and pulleys C and B respectively.

For gear D

We know that torque transmitted by the gear D,

$$T_D = \frac{P_D \times 60}{2\pi N} = \frac{30 \times 10^3 \times 60}{2\pi \times 250} = 1146 \text{ N-m}$$

∴ Tangential force acting on the gear D,

$$F_{tD} = \frac{T_D}{R_D} = \frac{1146}{0.15} = 7640 \text{ N}$$

and the normal load acting on the gear tooth,

$$W_D = \frac{F_{tD}}{\cos 20°} = \frac{7640}{0.9397} = 8130 \text{ N}$$

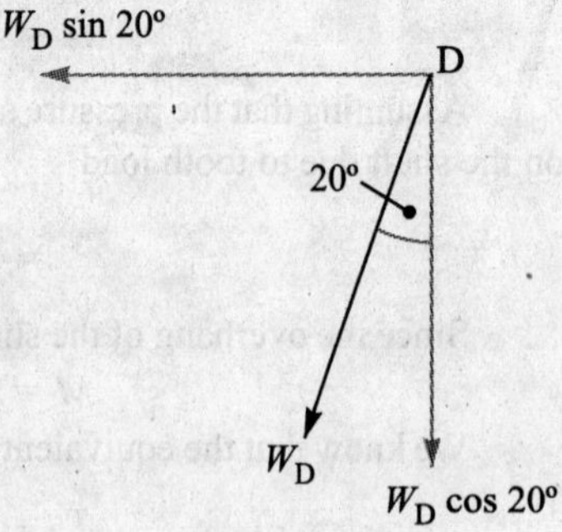

Fig. 14.14

The normal load acts at 20° to the vertical as shown in Fig. 14.14. Resolving the normal load vertically and horizontally, we have

Vertical component of W_D

$$= W_D \cos 20° = 8130 \times 0.9397 = 7640 \text{ N}$$

Horizontal component of W_D

$$= W_D \sin 20° = 8130 \times 0.342 = 2780 \text{ N}$$

For pulley C

We know that torque transmitted by pulley C,

$$T_C = \frac{P_C \times 60}{2\pi N} = \frac{18.75 \times 10^3 \times 60}{2\pi \times 250} = 716 \text{ N-m}$$

Let T_{C1} and T_{C2} = Tensions in the tight side and slack side of the belt for pulley C.

We know that torque transmitted by pulley C (T_C),

$$716 = (T_{C1} - T_{C2}) R_C = (T_{C1} - T_{C2})\, 0.3$$

$$\therefore \quad T_{C1} - T_{C2} = 716 / 0.3 = 2387 \text{ N} \quad \text{...(i)}$$

Since $T_{C1} / T_{C2} = 2$ or $T_{C1} = 2\, T_{C2}$, therefore from equation (*i*), we have

$$T_{C2} = 2387 \text{ N ; and } T_{C1} = 4774 \text{ N}$$

∴ Total load acting on pulley C,

$$W_C = T_{C1} + T_{C2} = 4774 + 2387 = 7161 \text{ N}$$

...(Neglecting weight of pulley C)

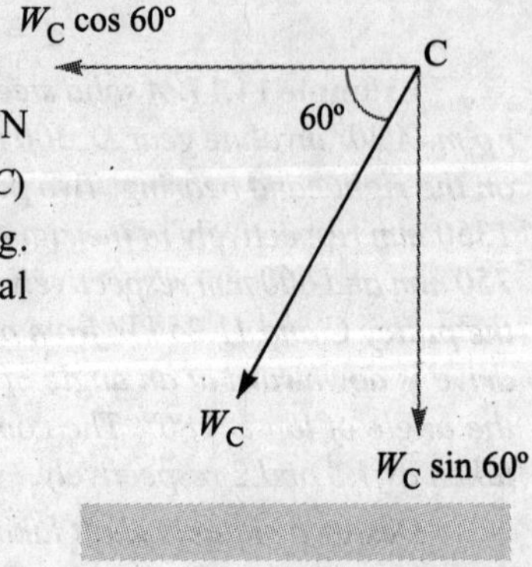

This load acts at 60° to the horizontal as shown in Fig. 14.15. Resolving the load W_C into vertical and horizontal components, we have

Vertical component of W_C

$$= W_C \sin 60° = 7161 \times 0.866$$

$$= 6200 \text{ N}$$

Trainwheels and Axles

and horizontal component of W_C

$$= W_C \cos 60° = 7161 \times 0.5$$

$$= 3580 \text{ N}$$

For pulley B

We know that torque transmitted by pulley *B*,

$$T_B = \frac{P_B \times 60}{2\pi N} = \frac{11.25 \times 10^3 \times 60}{2\pi \times 250} = 430 \text{ N-m}$$

Let T_{B1} and T_{B2} = Tensions in the tight side and slack side of the belt for pulley *B*.

We know that torque transmitted by pulley *B* (T_B),

$$430 = (T_{B1} - T_{B2})\, R_B = (T_{B1} - T_{B2})\, 0.375$$

$\therefore$ $T_{B1} - T_{B2} = 430 / 0.375 = 1147$ N ...(*ii*)

Since $T_{B1} / T_{B2} = 2$ or $T_{B1} = 2T_{B2}$, therefore from equation (*ii*), we have

$$T_{B2} = 1147 \text{ N, and } T_{B1} = 2294 \text{ N}$$

$\therefore$ Total load acting on pulley *B*,

$$W_B = T_{B1} + T_{B2} = 2294 + 1147 = 3441 \text{ N}$$

This load acts vertically downwards.

From above, we may say that the shaft is subjected to the vertical and horizontal loads as follows :

Type of loading	*Load in N*		
	At D	*At C*	*At B*
Vertical	7640	6200	3441
Horizontal	2780	3580	0

The vertical and horizontal load diagrams are shown in Fig. 14.16 (*c*) and (*d*).

First of all considering vertical loading on the shaft. Let R_{PV} and R_{QV} be the reactions at bearings *P* and *Q* respectively for vertical loading. We know that

$$R_{PV} + R_{QV} = 7640 + 6200 + 3441 = 17\,281 \text{ N}$$

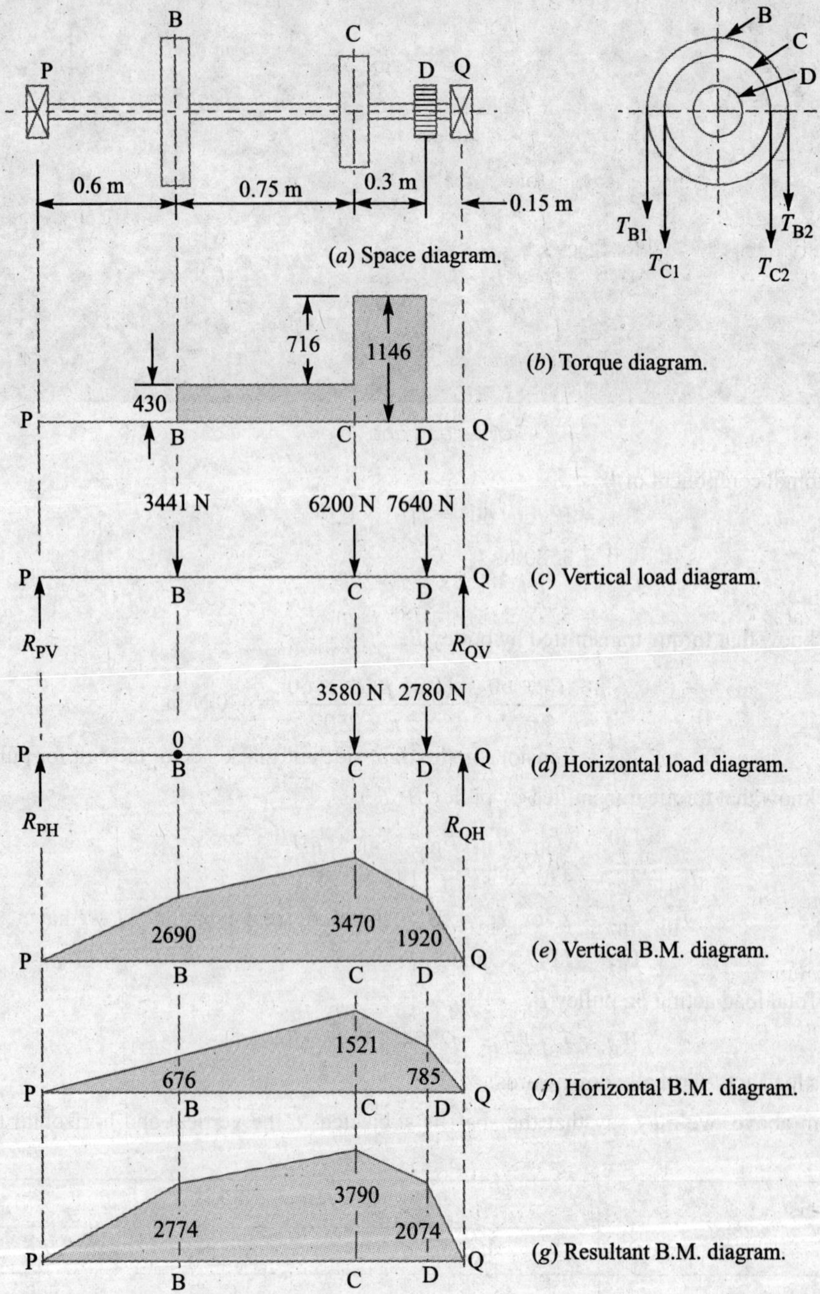

Fig. 14.16

Taking moments about P, we get

$$R_{QV} \times 1.8 = 7640 \times 1.65 + 6200 \times 1.35 + 3441 \times 0.6 = 23\ 041$$

$$\therefore \quad R_{QV} = 23\ 041 / 1.8 = 12\ 800 \text{ N}$$

and $$R_{PV} = 17\ 281 - 12\ 800 = 4481 \text{ N}$$

We know that B.M. at P and Q,

$$M_{PV} = M_{QV} = 0$$

B.M. at B, $$M_{BV} = 4481 \times 0.6 = 2690 \text{ N-m}$$

B.M. at C, $$M_{CV} = 4481 \times 1.35 - 3441 \times 0.75 = 3470 \text{ N-m}$$

and B.M. at D, $$M_{DV} = 12\ 800 \times 0.15 = 1920 \text{ N-m}$$

The bending moment diagram for vertical loading is shown in Fig. 14.16 (*e*).

Now considering horizontal loading. Let R_{PH} and R_{QH} be the reactions at the bearings P and Q respectively for horizontal loading. We know that

$$R_{PH} + R_{QH} = 2780 + 3580 = 6360 \text{ N}$$

Taking moments about P, we get

$$R_{QH} \times 1.8 = 2780 \times 1.65 + 3580 \times 1.35 = 9420 \text{ N}$$

$$\therefore \quad R_{QH} = 9420 / 1.8 = 5233 \text{ N}$$

and $$R_{PH} = 6360 - 5233 = 1127 \text{ N}$$

We know that B.M. at P and Q,

$$M_{PH} = M_{QH} = 0$$

B.M. at B, $$M_{BH} = 1127 \times 0.6 = 676 \text{ N-m}$$

B.M. at C, $$M_{CH} = 1127 \times 1.35 = 1521 \text{ N-m}$$

and B.M. at D, $$M_{DH} = 5233 \times 0.15 = 785 \text{ N-m}$$

The bending moment diagram for horizontal loading is shown in Fig. 14.16 (*f*).

The resultant bending moments for the points B, C and D are as follows :

$$\text{Resultant B.M. at } B = \sqrt{(M_{BV})^2 + (M_{BH})^2} = \sqrt{(2690)^2 + (676)^2} = 2774 \text{ N-m}$$

$$\text{Resultant B.M. at } C = \sqrt{(M_{CV})^2 + (M_{CH})^2} = \sqrt{(3470)^2 + (1521)^2} = 3790 \text{ N-m}$$

$$\text{Resultant B.M. at } D = \sqrt{(M_{DV})^2 + (M_{DH})^2} = \sqrt{(1920)^2 + (785)^2} = 2074 \text{ N-m}$$

From above we see that the resultant bending moment is maximum at C.

$$\therefore \quad M = M_C = 3790 \text{ N-m}$$

and maximum torque at C,

$$T = \text{Torque corresponding to 30 kW} = T_D = 1146 \text{ N-m}$$

Let d = Diameter of the shaft in mm.

We know that equivalent twisting moment,

$$T_e = \sqrt{(K_m \times M)^2 + (K_t \times T)^2} = \sqrt{(2 \times 3790)^2 + (1.5 \times 1146)^2}$$

$$= 7772 \text{ N-m} = 7772 \times 10^3 \text{ N-mm}$$

We also know that the equivalent twisting moment (T_e),

$$7772 \times 10^3 = \frac{\pi}{16} \times \tau \times d^3 = \frac{\pi}{16} \times 42 \times d^3 = 8.25\ d^3$$

$$\therefore \quad d^3 = 7772 \times 10^3 / 8.25 = 942 \times 10^3 \text{ or } d = 98 \text{ mm}$$

Again, we know that equivalent bending moment,

$$M_e = \frac{1}{2}\left[K_m \times M + \sqrt{(K_m \times M)^2 + (K_t \times T)^2}\right] = \frac{1}{2}(K_m \times M + T_e)$$

$$= \frac{1}{2}(2 \times 3790 + 7772) = 7676 \text{ N-m} = 7676 \times 10^3 \text{ N-mm}$$

We also know that the equivalent bending moment (M_e),

$$7676 \times 10^3 = \frac{\pi}{32} \times \sigma_b \times d^3 = \frac{\pi}{32} \times 84 \times d^3 = 8.25\, d^3$$

$$\therefore \quad d^3 = 7676 \times 10^3 / 8.25 = 930 \times 10^3 \text{ or } d = 97.6 \text{ mm}$$

Taking the larger of the two values, we have

$$d = 98 \text{ say } 100 \text{ mm } \textbf{Ans.}$$

14.13 Shafts Subjected to Axial Load in addition to Combined Torsion and Bending Loads

When the shaft is subjected to an axial load (F) in addition to torsion and bending loads as in propeller shafts of ships and shafts for driving worm gears, then the stress due to axial load must be added to the bending stress (σ_b). We know that bending equation is

$$\frac{M}{I} = \frac{\sigma_b}{y} \text{ or } \sigma_b = \frac{M.y}{I} = \frac{M \times d/2}{\frac{\pi}{64} \times d^4} = \frac{32M}{\pi d^3}$$

and stress due to axial load

$$= \frac{F}{\frac{\pi}{4} \times d^2} = \frac{4F}{\pi d^2} \quad \text{...(For round solid shaft)}$$

$$= \frac{F}{\frac{\pi}{4}\left[(d_o)^2 - (d_i)^2\right]} = \frac{4F}{\pi\left[(d_o)^2 - (d_i)^2\right]} \quad \text{...(For hollow shaft)}$$

$$= \frac{F}{\pi (d_o)^2 (1 - k^2)} \quad \text{... } (\because k = d_i/d_o)$$

$\therefore$ Resultant stress (tensile or compressive) for solid shaft,

$$\sigma_1 = \frac{32M}{\pi d^3} + \frac{4F}{\pi d^2} = \frac{32}{\pi d^3}\left(M + \frac{F \times d}{8}\right) \quad \text{...}(i)$$

$$= \frac{32 M_1}{\pi d^3} \quad \text{...}\left(\text{Substituting } M_1 = M + \frac{F \times d}{8}\right)$$

In case of a hollow shaft, the resultant stress,

$$\sigma_1 = \frac{32M}{\pi (d_o)^3 (1 - k^4)} + \frac{4F}{\pi (d_o)^2 (1 - k^2)}$$

$$= \frac{32}{\pi (d_o)^3 (1 - k^4)}\left[M + \frac{F d_o (1 + k^2)}{8}\right] = \frac{32 M_1}{\pi (d_o)^3 (1 - k^4)}$$

$$\text{...}\left[\text{Substituting for hollow shaft, } M_1 = M + \frac{F d_o (1 + k^2)}{8}\right]$$

In case of long shafts (slender shafts) subjected to compressive loads, a factor known as ***column factor*** (α) must be introduced to take the column effect into account.

$\therefore$ Stress due to the compressive load,

$$\sigma_c = \frac{\alpha \times 4F}{\pi d^2} \quad \text{...(For round solid shaft)}$$

$$= \frac{\alpha \times 4F}{\pi (d_o)^2 \, (1 - k^2)}$$ ...(For hollow shaft)

The value of column factor (α) for compressive loads* may be obtained from the following relation :

$$\text{Column factor,} \quad \alpha = \frac{1}{1 - 0.0044 \, (L / K)}$$

This expression is used when the slenderness ratio (L / K) is less than 115. When the slenderness ratio (L / K) is more than 115, then the value of column factor may be obtained from the following relation :

$$\text{**Column factor,} \quad \alpha = \frac{\sigma_y \, (L / K)^2}{C \pi^2 E}$$

where L = Length of shaft between the bearings,

K = Least radius of gyration,

σ_y = Compressive yield point stress of shaft material, and

C = Coefficient in Euler's formula depending upon the end conditions.

The following are the different values of C depending upon the end conditions.

C = 1, for hinged ends,

= 2.25, for fixed ends,

= 1.6, for ends that are partly restrained as in bearings.

Note: In general, for a hollow shaft subjected to fluctuating torsional and bending load, along with an axial load, the equations for equivalent twisting moment (T_e) and equivalent bending moment (M_e) may be written as

$$T_e = \sqrt{\left[K_m \times M + \frac{\alpha F \, d_o \, (1 + k^2)}{8}\right]^2 + (K_t \times T)^2}$$

$$= \frac{\pi}{16} \times \tau \, (d_o)^3 \, (1 - k^4)$$

and

$$M_e = \frac{1}{2}\left[K_m \times M + \frac{\alpha F d_o \, (1 + k^2)}{8} + \sqrt{\left\{K_m \times M + \frac{\alpha F d_o \, (1 + k^2)}{8}\right\}^2 + (K_t \times T)^2}\right]$$

$$= \frac{\pi}{32} \times \sigma_b \, (d_o)^3 \, (1 - k^4)$$

It may be noted that for a solid shaft, $k = 0$ and $d_0 = d$. When the shaft carries no axial load, then $F = 0$ and when the shaft carries axial tensile load, then $\alpha = 1$.

Example 14.18. *A hollow shaft is subjected to a maximum torque of 1.5 kN-m and a maximum bending moment of 3 kN-m. It is subjected, at the same time, to an axial load of 10 kN. Assume that the load is applied gradually and the ratio of the inner diameter to the outer diameter is 0.5. If the outer diameter of the shaft is 80 mm, find the shear stress induced in the shaft.*

Solution. Given : T = 1.5 kN-m = 1.5 × 10³ N-m ; M = 3 kN-m = 3 × 10³ N-m ; F = 10 kN = 10 × 10³ N ; $k = d_i / d_o = 0.5$; d_o = 80 mm = 0.08 m

Let τ = Shear stress induced in the shaft.

Since the load is applied gradually, therefore from Table 14.2, we find that

K_m = 1.5 ; and K_t = 1.0

---

* The value of column factor (α) for tensile load is unity.

** It is an Euler's formula for long columns.

We know that the equivalent twisting moment for a hollow shaft,

$$T_e = \sqrt{\left[K_m \times M + \frac{\alpha F d_o (1+k^2)}{8}\right]^2 + (K_t \times T)^2}$$

$$= \sqrt{\left[1.5 \times 3 \times 10^3 + \frac{1 \times 10 \times 10^3 \times 0.08 (1+0.5^2)}{8}\right]^2 + (1 \times 1.5 \times 10^3)^2}$$

... ($\because \alpha = 1$, for axial tensile loading)

$$= \sqrt{(4500 + 125)^2 + (1500)^2} = 4862 \text{ N-m} = 4862 \times 10^3 \text{ N-mm}$$

We also know that the equivalent twisting moment for a hollow shaft (T_e),

$$4862 \times 10^3 = \frac{\pi}{16} \times \tau (d_o)^3 (1 - k^4) = \frac{\pi}{16} \times \tau (80)^3 (1 - 0.5^4) = 94\,260\,\tau$$

$\therefore \quad \tau = 4862 \times 10^3 / 94\,260 = 51.6 \text{ N/mm}^2 = 51.6$ MPa **Ans.**

Crankshaft inside the crank-case

Example 14.19. *A hollow shaft of 0.5 m outside diameter and 0.3 m inside diameter is used to drive a propeller of a marine vessel. The shaft is mounted on bearings 6 metre apart and it transmits 5600 kW at 150 r.p.m. The maximum axial propeller thrust is 500 kN and the shaft weighs 70 kN. Determine :*

1. The maximum shear stress developed in the shaft, and

2. The angular twist between the bearings.

Solution. Given : $d_o = 0.5$ m ; $d_i = 0.3$ m ; $P = 5600$ kW $= 5600 \times 10^3$ W ; $L = 6$ m ; $N = 150$ r.p.m. ; $F = 500$ kN $= 500 \times 10^3$ N ; $W = 70$ kN $= 70 \times 10^3$ N

1. *Maximum shear stress developed in the shaft*

Let τ = Maximum shear stress developed in the shaft.

We know that the torque transmitted by the shaft,

$$T = \frac{P \times 60}{2 \pi N} = \frac{5600 \times 10^3 \times 60}{2 \pi \times 150} = 356\,460 \text{ N-m}$$

and the maximum bending moment,

$$M = \frac{W \times L}{8} = \frac{70 \times 10^3 \times 6}{8} = 52\,500 \text{ N-m}$$

Now let us find out the column factor α. We know that least radius of gyration,

$$K = \sqrt{\frac{I}{A}} = \sqrt{\frac{\frac{\pi}{64}\left[(d_o)^4 - (d_i)^4\right]}{\frac{\pi}{4}\left[(d_o)^2 - (d_i)^2\right]}}$$

$$= \sqrt{\frac{[(d_o)^2 + (d_i)^2]\,[(d_o)^2 - (d_i)^2]}{16\,[(d_o)^2 - (d_i)^2]}}$$

$$= \frac{1}{4}\sqrt{(d_o)^2 + (d_i)^2} = \frac{1}{4}\sqrt{(0.5)^2 + (0.3)^2} = 0.1458 \text{ m}$$

∴ Slenderness ratio,

$$L / K = 6 / 0.1458 = 41.15$$

and column factor,

$$\alpha = \frac{1}{1 - 0.0044\left(\frac{L}{K}\right)} \qquad \ldots\left(\because \frac{L}{K} < 115\right)$$

$$= \frac{1}{1 - 0.0044 \times 41.15} = \frac{1}{1 - 0.18} = 1.22$$

Assuming that the load is applied gradually, therefore from Table 14.2, we find that

$$K_m = 1.5 \text{ and } K_t = 1.0$$

Also $\quad k = d_i / d_o = 0.3 / 0.5 = 0.6$

We know that the equivalent twisting moment for a hollow shaft,

$$T_e = \sqrt{\left[K_m \times M + \frac{\alpha F d_o (1 + k^2)}{8}\right]^2 + (K_t \times T)^2}$$

$$= \sqrt{\left[1.5 \times 52\,500 + \frac{1.22 \times 500 \times 10^3 \times 0.5\,(1 + 0.6^2)}{8}\right]^2 + (1 \times 356\,460)^2}$$

$$= \sqrt{(78\,750 + 51\,850)^2 + (356\,460)^2} = 380 \times 10^3 \text{ N-m}$$

We also know that the equivalent twisting moment for a hollow shaft (T_e),

$$380 \times 10^3 = \frac{\pi}{16} \times \tau\,(d_o)^3\,(1 - k^4) = \frac{\pi}{16} \times \tau\,(0.5)^3\,[1 - (0.6)^4] = 0.02\,\tau$$

$$\therefore \quad \tau = 380 \times 10^3 / 0.02 = 19 \times 10^6 \text{ N/m}^2 = 19 \text{ MPa } \textbf{Ans.}$$

2. ***Angular twist between the bearings***

Let $\quad \theta$ = Angular twist between the bearings in radians.

We know that the polar moment of inertia for a hollow shaft,

$$J = \frac{\pi}{32}\,[(d_o)^4 - (d_i)^4] = \frac{\pi}{32}\,[(0.5)^4 - (0.3)^4] = 0.005\,34 \text{ m}^4$$

From the torsion equation,

$$\frac{T}{J} = \frac{G \times \theta}{L}, \text{ we have}$$

$$\theta = \frac{T \times L}{G \times J} = \frac{356\,460 \times 6}{84 \times 10^9 \times 0.00\,534} = 0.0048 \text{ rad}$$

... (Taking $G = 84$ GPa $= 84 \times 10^9$ N/m²)

$$= 0.0048 \times \frac{180}{\pi} = 0.275° \textbf{ Ans.}$$

Example 14.20. *A hollow steel shaft is to transmit 20 kW at 300 r.p.m. The loading is such that the maximum bending moment is 1000 N-m, the maximum torsional moment is 500 N-m and axial compressive load is 15 kN. The shaft is supported on rigid bearings 1.5 m apart. The maximum permissible shear stress on the shaft is 40 MPa. The inside diameter is 0.8 times the outside diameter. The load is cyclic in nature and applied with shocks. The values for the shock factors are K_t = 1.5 and K_m = 1.6.*

Solution. Given : *P = 20 kW ; *N = 300 r.p.m. ; M = 1000 N-m = 1000 × 10^3 N-mm ; T = 500 N-m = 500 × 10^3 N-mm ; F = 15 kN = 15 000 N ; L = 1.5 m = 1500 mm ; τ = 40 MPa = 40 N/mm^2 ; $d_i = 0.8\, d_o$ or $k = d_i/d_o = 0.8$; $K_t = 1.5$; $K_m = 1.6$

Let d_o = Outside diameter of the shaft, and

d_i = Inside diameter of the shaft = 0.8 d_o ...(Given)

We know that moment of inertia of a hollow shaft,

$$I = \frac{\pi}{64}\left[(d_o)^4 - (d_i)^4\right]$$

and cross-sectional area of the hollow shaft,

$$A = \frac{\pi}{4}\left[(d_o)^2 - (d_i)^2\right]$$

∴ Radius of gyration of the hollow shaft,

$$K = \sqrt{\frac{I}{A}} = \sqrt{\frac{\frac{\pi}{64}[(d_o)^4 - (d_i)^4]}{\frac{\pi}{4}[(d_o)^2 - (d_i)^2]}}$$

$$= \sqrt{\frac{[(d_o)^2 + (d_i)^2]\,[(d_o)^2 - (d_i)^2]}{16\,[(d_o)^2 - (d_i)^2]}} = \sqrt{\frac{(d_o)^2 + (d_i)^2}{16}}$$

$$= \frac{d_o}{4}\sqrt{1 + \left(\frac{d_i}{d_o}\right)^2} = \frac{d_o}{4}\sqrt{1 + (0.8)^2} = 0.32\, d_o$$

and column factor for compressive loads,

$$\alpha = \frac{1}{1 - 0.0044\,(L/K)} = \frac{1}{1 - 0.0044\,(1500/0.32\, d_o)}$$

$$= \frac{1}{1 - 20.6/d_o} = \frac{d_o}{d_o - 20.6}$$

We know that equivalent twisting moment for a hollow shaft,

$$T_e = \sqrt{\left[K_m \times M + \frac{\alpha F\, d_o\,(1 + k^2)}{8}\right]^2 + (K_t \times T)^2}$$

$$= \sqrt{\left[1.6 \times 1000 \times 10^3 + \frac{\left(\frac{d_o}{d_o - 20.6}\right) 15000 \times d_o\,(1 + 0.8^2)}{8}\right]^2 + (1.5 \times 500 \times 10^3)^2}$$

$$= \sqrt{\left[1600 \times 10^3 + \frac{3075\,(d_o)^2}{d_o - 20.6}\right]^2 + (750 \times 10^3)^2} \quad ...(i)$$

* Superfluous data.

We also know that equivalent twisting moment for a hollow shaft,

$$T_e = \frac{\pi}{16} \times \tau (d_o)^3 (1 - k^4)$$

$$= \frac{\pi}{16} \times 40 (d_o)^3 (1 - 0.8^4) = 4.65 (d_o)^3 \quad ...(ii)$$

Equating equations (*i*) and (*ii*), we have

$$4.65 (d_o)^3 = \sqrt{\left[1600 \times 10^3 + \frac{3075 (d_o)^2}{d_o - 20.6}\right]^2 + (750 \times 10^3)^2} \quad ...(iii)$$

Solving this expression by hit and trial method, we find that

$$d_o = 76.32 \text{ say } 80 \text{ mm } \textbf{Ans.}$$

and $$d_i = 0.8\, d_o = 0.8 \times 80 = 64 \text{ mm } \textbf{Ans.}$$

Note : In order to find the minimum value of d_o to be used for the hit and trial method, determine the equivalent twisting moment without considering the axial compressive load. We know that equivalent twisting moment,

$$T_e = \sqrt{(K_m \times M)^2 + (K_t \times T)^2} = \sqrt{(1.6 \times 1000 \times 10^3)^2 + (1.5 \times 500 \times 10^3)^2} \quad ... (iv)$$

$$= 1767 \times 10^3 \text{ N-mm}$$

Equating equations (*ii*) and (*iv*),

$$4.65(d_o)^3 = 1767 \times 10^3 \text{ or } (d_o)^3 = 1767 \times 10^3/4.65 = 380 \times 10^3$$

$$\therefore \quad d_o = 72.4 \text{ mm}$$

Thus the value of d_o to be substituted in equation (*iii*) must be greater than 72.4 mm.

14.14 Design of Shafts on the basis of Rigidity

Sometimes the shafts are to be designed on the basis of rigidity. We shall consider the following two types of rigidity.

1. ***Torsional rigidity.*** The torsional rigidity is important in the case of camshaft of an I.C. engine where the timing of the valves would be effected. The permissible amount of twist should not exceed 0.25° per metre length of such shafts. For line shafts or transmission shafts, deflections 2.5 to 3 degree per metre length may be used as limiting value. The widely used deflection for the shafts is limited to 1 degree in a length equal to twenty times the diameter of the shaft.

The torsional deflection may be obtained by using the torsion equation,

$$\frac{T}{J} = \frac{G.\theta}{L} \text{ or } \theta = \frac{T.L}{J.G}$$

where θ = Torsional deflection or angle of twist in radians,

T = Twisting moment or torque on the shaft,

J = Polar moment of inertia of the cross-sectional area about the axis of rotation,

$$= \frac{\pi}{32} \times d^4 \quad ...(\text{For solid shaft})$$

$$= \frac{\pi}{32}\left[(d_o)^4 - (d_i)^4\right] \quad ...(\text{For hollow shaft})$$

G = Modulus of rigidity for the shaft material, and

L = Length of the shaft.

2. ***Lateral rigidity.*** It is important in case of transmission shafting and shafts running at high speed, where small lateral deflection would cause huge out-of-balance forces. The lateral rigidity is also important for maintaining proper bearing clearances and for correct gear teeth alignment. If the shaft is of uniform cross-section, then the lateral deflection of a shaft may be obtained by using the deflection formulae as in Strength of Materials. But when the shaft is of variable cross-section, then

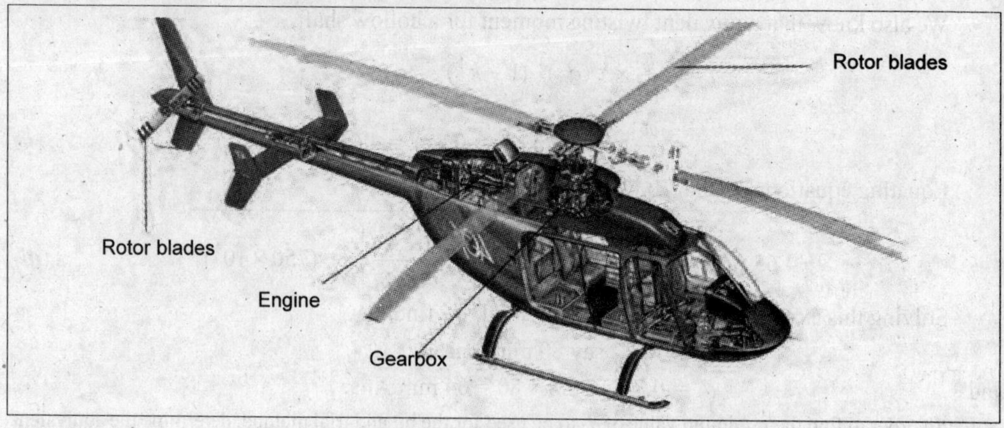

Air acclerating downwards, pushed by the rotating blades, produced an upwards reaction that lifts the helicopter.

Note : This picture is given as additional information and is not a direct example of the current chapter.

the lateral deflection may be determined from the fundamental equation for the elastic curve of a beam, *i.e.*

$$\frac{d^2 y}{dx^2} = \frac{M}{EI}$$

Example 14.21. *A steel spindle transmits 4 kW at 800 r.p.m. The angular deflection should not exceed 0.25° per metre of the spindle. If the modulus of rigidity for the material of the spindle is 84 GPa, find the diameter of the spindle and the shear stress induced in the spindle.*

Solution. Given : $P = 4$ kW = 4000 W ; $N = 800$ r.p.m. ; $\theta = 0.25° = 0.25 \times \frac{\pi}{180} = 0.0044$ rad ; $L = 1$ m = 1000 mm ; $G = 84$ GPa = 84×10^9 N/m^2 = 84×10^3 N/mm^2

Diameter of the spindle

Let d = Diameter of the spindle in mm.

We know that the torque transmitted by the spindle,

$$T = \frac{P \times 60}{2\pi N} = \frac{4000 \times 60}{2\pi \times 800} = 47.74 \text{ N-m} = 47\ 740 \text{ N-mm}$$

We also know that $\frac{T}{J} = \frac{G \times \theta}{L}$ or $J = \frac{T \times l}{G \times \theta}$

or $$\frac{\pi}{32} \times d^4 = \frac{47\ 740 \times 1000}{84 \times 10^3 \times 0.0044} = 129\ 167$$

$\therefore$ $d^4 = 129\ 167 \times 32 / \pi = 1.3 \times 10^6$ or $d = 33.87$ say 35 mm **Ans.**

Shear stress induced in the spindle

Let τ = Shear stress induced in the spindle.

We know that the torque transmitted by the spindle (T),

$$47\ 740 = \frac{\pi}{16} \times \tau \times d^3 = \frac{\pi}{16} \times \tau\ (35)^3 = 8420\ \tau$$

$\therefore$ $\tau = 47\ 740 / 8420 = 5.67$ N/mm^2 = 5.67 MPa **Ans.**

Example 14.22. *Compare the weight, strength and stiffness of a hollow shaft of the same external diameter as that of solid shaft. The inside diameter of the hollow shaft being half the external diameter. Both the shafts have the same material and length.*

Solution. Given : $d_o = d$; $d_i = d_o / 2$ or $k = d_i / d_o = 1/2 = 0.5$

Comparison of weight

We know that weight of a hollow shaft,

$$W_H = \text{Cross-sectional area} \times \text{Length} \times \text{Density}$$

$$= \frac{\pi}{4}\left[(d_o)^2 - (d_i)^2\right] \times \text{Length} \times \text{Density} \quad ...(i)$$

and weight of the solid shaft,

$$W_S = \frac{\pi}{4} \times d^2 \times \text{Length} \times \text{Density} \quad ...(ii)$$

Since both the shafts have the same material and length, therefore by dividing equation (*i*) by equation (*ii*), we get

$$\frac{W_H}{W_S} = \frac{(d_o)^2 - (d_i)^2}{d^2} = \frac{(d_o)^2 - (d_i)^2}{(d_o)^2} \quad ...(\because d = d_o)$$

$$= 1 - \frac{(d_i)^2}{(d_o)_2} = 1 - k^2 = 1 - (0.5)^2 = 0.75 \textbf{ Ans.}$$

Comparison of strength

We know that strength of the hollow shaft,

$$T_H = \frac{\pi}{16} \times \tau\,(d_o)^3\,(1 - k^4) \quad ...(iii)$$

and strength of the solid shaft,

$$T_S = \frac{\pi}{16} \times \tau \times d^3 \quad ...(iv)$$

Dividing equation (*iii*) by equation (*iv*), we get

$$\frac{T_H}{T_S} = \frac{(d_o)^3\,(1 - k^4)}{d^3} = \frac{(d_o)^3\,(1 - k^4)}{(d_o)^3} = 1 - k^4 \quad ...(\because d = d_o)$$

$$= 1 - (0.5)^4 = 0.9375 \textbf{ Ans.}$$

Comparison of stiffness

We know that stiffness

$$= \frac{T}{\theta} = \frac{G \times J}{L}$$

The propeller shaft of this heavy duty helicopter is subjected to very high torsion.

∴ Stiffness of a hollow shaft,

$$S_H = \frac{G}{L} \times \frac{\pi}{32}\left[(d_o)^4 - (d_i)^4\right] \quad ...(v)$$

and stiffness of a solid shaft,

$$S_S = \frac{G}{L} \times \frac{\pi}{32} \times d^4 \quad ...(vi)$$

Dividing equation (*v*) by equation (*vi*), we get

$$\frac{S_H}{S_S} = \frac{(d_o)^4 - (d_i)^4}{d^4} = \frac{(d_o)^4 - (d_i)^4}{(d_o)^4} = 1 - \frac{(d_i)^4}{(d_o)^4} \quad ...(\because d = d_o)$$

$$= 1 - k^4 = 1 - (0.5)^4 = 0.9375 \textbf{ Ans.}$$

EXERCISES

1. A shaft running at 400 r.p.m. transmits 10 kW. Assuming allowable shear stress in shaft as 40 MPa, find the diameter of the shaft. **[Ans. 35 mm]**
2. A hollow steel shaft transmits 600 kW at 500 r.p.m. The maximum shear stress is 62.4 MPa. Find the outside and inside diameter of the shaft, if the outer diameter is twice of inside diameter, assuming that the maximum torque is 20% greater than the mean torque. **[Ans. 100 mm ; 50 mm]**
3. A hollow shaft for a rotary compressor is to be designed to transmit a maximum torque of 4750 N-m. The shear stress in the shaft is limited to 50 MPa. Determine the inside and outside diameters of the shaft, if the ratio of the inside to the outside diameter is 0.4. **[Ans. 35 mm ; 90 mm]**
4. A motor car shaft consists of a steel tube 30 mm internal diameter and 4 mm thick. The engine develops 10 kW at 2000 r.p.m. Find the maximum shear stress in the tube when the power is transmitted through a 4 : 1 gearing. **[Ans. 30 MPa]**
5. A cylindrical shaft made of steel of yield strength 700 MPa is subjected to static loads consisting of a bending moment of 10 kN-m and a torsional moment of 30 kN-m. Determine the diameter of the shaft using two different theories of failure and assuming a factor of safety of 2. **[Ans. 100 mm]**
6. A line shaft rotating at 200 r.p.m. is to transmit 20 kW. The allowable shear stress for the material of the shaft is 42 MPa. If the shaft carries a central load of 900 N and is simply supported between bearing 3 metre apart, determine the diameter of the shaft. The maximum tensile or compressive stress is not to exceed 56 MPa. **[Ans. 50 mm]**
7. Two 400 mm diameter pulleys are keyed to a simply supported shaft 500 mm apart. Each pulley is 100 mm from its support and has horizontal belts, tension ratio being 2.5. If the shear stress is to be limited to 80 MPa while transmitting 45 kW at 900 r.p.m., find the shaft diameter if it is to be used for the input-output belts being on the same or opposite sides. **[Ans. 40 mm]**
8. A cast gear wheel is driven by a pinion and transmits 100 kW at 375 r.p.m. The gear has 200 machine cut teeth having 20° pressure angle and is mounted at the centre of a 0.4 m long shaft. The gear weighs 2000 N and its pitch circle diameter is 1.2 m. Design the gear shaft. Assume that the axes of the gear and pinion lie in the same horizontal plane. **[Ans. 80 mm]**
9. Fig. 14.17 shows a shaft from a hand-operated machine. The frictional torque in the journal bearings at *A* and *B* is 15 N-m each. Find the diameter (*d*) of the shaft (on which the pulley is mounted) using maximum distortion energy criterion. The shaft material is 40 C 8 steel for which the yield stress in tension is 380 MPa and the factor of safety is 1.5. **[Ans. 20 mm]**

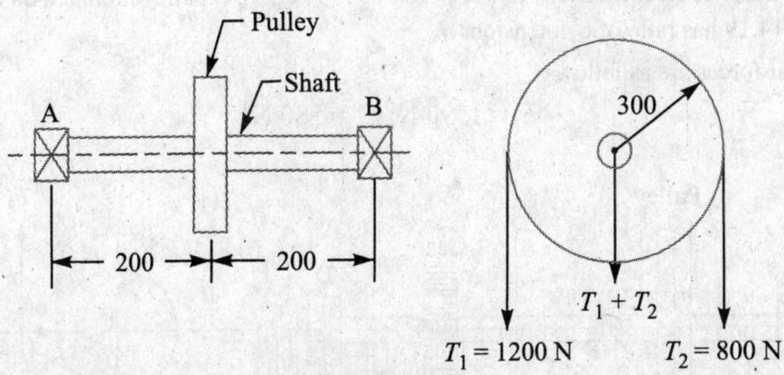

All dimensions in mm.

Fig. 14.17

10. A line shaft is to transmit 30 kW at 160 r.p.m. It is driven by a motor placed directly under it by means of a belt running on a 1 m diameter pulley keyed to the end of the shaft. The tension in the tight side of the belt is 2.5 times that in the slack side and the centre of the pulley over-hangs 150 mm beyond the centre line of the end bearing. Determine the diameter of the shaft, if the allowable shear stress is 56 MPa and the pulley weighs 1600 N. **[Ans. 60 mm]**

11. Determine the diameter of hollow shaft having inside diameter 0.5 times the outside diameter. The permissible shear stress is limited to 200 MPa. The shaft carries a 900 mm diameter cast iron pulley. This pulley is driven by another pulley mounted on the shaft placed below it. The belt ends are parallel and vertical. The ratio of tensions in the belt is 3. The pulley on the hollow shaft weighs 800 N and overhangs the nearest bearing by 250 mm. The pulley is to transmit 35 kW at 400 r.p.m.

[Ans. $d_o = 40$ mm, $d_i = 20$ mm]

12. A horizontal shaft *AD* supported in bearings at *A* and *B* and carrying pulleys at *C* and *D* is to transmit 75 kW at 500 r.p.m. from drive pulley *D* to off-take pulley *C*, as shown in Fig. 14.18.

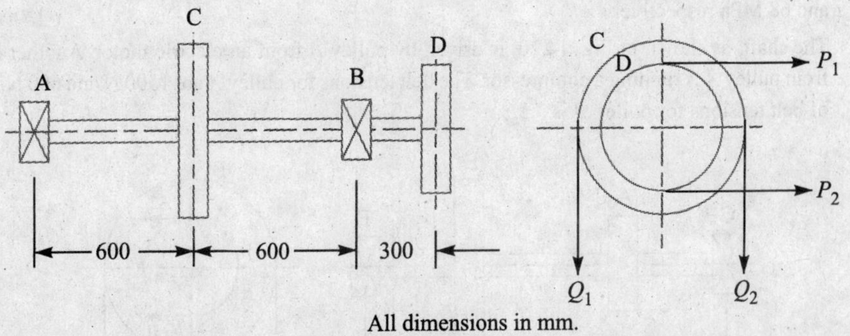

All dimensions in mm.

Fig. 14.18

Calculate the diameter of shaft. The data given is : $P_1 = 2\ P_2$ (both horizontal), $Q_1 = 2\ Q_2$ (both vertical), radius of pulley $C = 220$ mm, radius of pulley $D = 160$ mm, allowable shear stress = 45 MPa.

[Ans. 100 mm]

13. A line shaft *ABCD*, 9 metres long, has four pulleys *A*, *B*, *C* and *D* at equal distance apart. Power of 45 kW is being supplied to the shaft through the pulley *C* while the power is being taken off equally from the pulleys *A*, *B* and *D*. The shaft runs at 630 r.p.m.

Calculate the most economical diameters for the various portions of the shaft so that the shear stress does not exceed 55 MPa. If the shear modulus is 85 GPa, determine the twist of the pulley *D* with respect to the pulley *A*. **[Ans. 28 mm, 36 mm, 28 mm ; 0.0985°]**

14. A shaft made of steel receives 7.5 kW power at 1500 r.p.m. A pulley mounted on the shaft as shown in Fig. 14.19 has ratio of belt tensions 4.

The gear forces are as follows :

$$F_t = 1590 \text{ N}; F_r = 580 \text{ N}$$

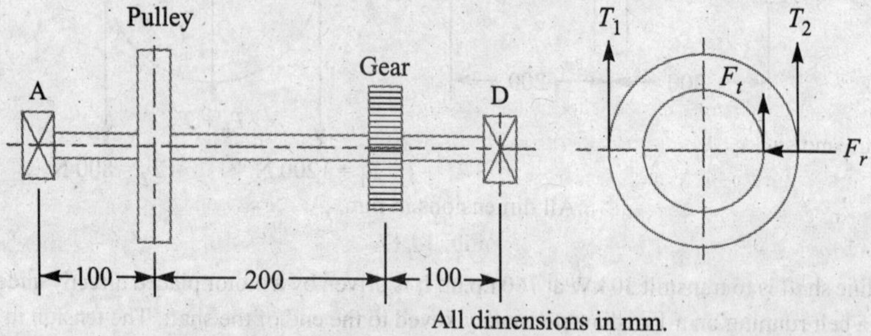

Fig. 14.19

Design the shaft diameter by maximum shear stress theory. The shaft material has the following properties :

Ultimate tensile strength = 720 MPa; Yield strength = 380 MPa; Factor of safety = 1.5.

[Ans. 20 mm]

15. An overhang hollow shaft carries a 900 mm diameter pulley, whose centre is 250 mm from the centre of the nearest bearing. The weight of the pulley is 600 N and the angle of lap is 180°. The pulley is driven by a motor vertically below it. If permissible tension in the belt is 2650 N and if coefficient of friction between the belt and pulley surface is 0.3, estimate, diameters of shaft, when the internal diameter is 0.6 of the external.

Neglect centrifugal tension and assume permissible tensile and shear stresses in the shaft as 84 MPa and 68 MPa respectively.

[Ans. 65 mm]

16. The shaft, as shown in Fig. 14.20, is driven by pulley *B* from an electric motor. Another belt drive from pulley *A* is running a compressor. The belt tensions for pulley *A* are 1500 N and 600 N. The ratio of belt tensions for pulley *B* is 3.5.

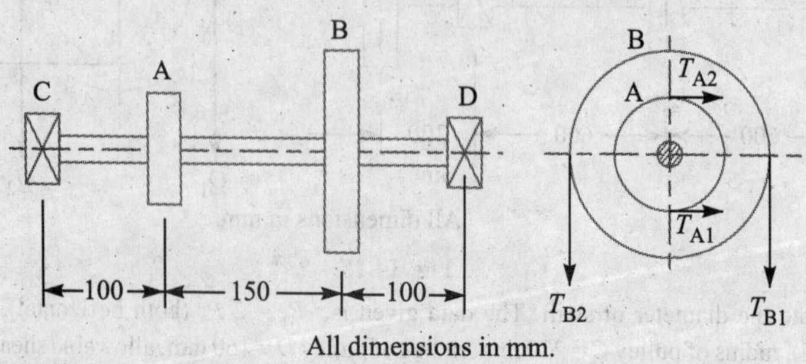

Fig. 14.20

The diameter of pulley *A* is 150 mm and the diameter of pulley *B* is 480 mm. The allowable tensile stress for the shaft material is 170 MPa and the allowable shear stress is 85 MPa. Taking torsion and bending factors as 1.25 and 1.75 respectively, find the shaft diameter.

Also find out the dimensions for a hollow shaft with outside diameter limited to 30 mm. Compare the weights of the two shafts. **[Ans. 30 mm ; 24 mm ; 1.82]**

17. A mild steel shaft transmits 15 kW at 210 r.p.m. It is supported on two bearings 750 mm apart and has two gears keyed to it. The pinion having 24 teeth of 6 mm module is located 100 mm to the left of the right hand bearing and delivers the power horizontally to the right. The gear having 80 teeth of 6 mm module is located 15 mm to the right of the left hand bearing and receives power in a vertical direction from below. Assuming an allowable working shear stress as 53 MPa, and a combined shock and fatigue factor of 1.5 in bending as well as in torsion, determine the diameter of the shaft.

[Ans. 60 mm]

18. A steel shaft 800 mm long transmitting 15 kW at 400 r.p.m. is supported at two bearings at the two ends. A gear wheel having 80 teeth and 500 mm pitch circle diameter is mounted at 200 mm from the left hand side bearing and receives power from a pinion meshing with it. The axis of pinion and gear lie in the horizontal plane. A pulley of 300 mm diameter is mounted at 200 mm from right hand side bearing and is used for transmitting power by a belt. The belt drive is inclined at 30° to the vertical in the forward direction. The belt lap angle is 180 degrees. The coefficient of friction between belt and pulley is 0.3. Design and sketch the arrangement of the shaft assuming the values of safe stresses as : $\tau = 55$ MPa; $\sigma_t = 80$ MPa. Take torsion and bending factor 1.5 and 2 respectively. **[Ans. 120 mm]**

19. A machine shaft, supported on bearings having their centres 750 mm apart, transmitted 185 kW at 600 r.p.m. A gear of 200 mm and 20° tooth profile is located 250 mm to the right of left hand bearing and a 450 mm diameter pulley is mounted at 200 mm to right of right hand bearing. The gear is driven by a pinion with a downward tangential force while the pulley drives a horizontal belt having 180° angle of contact. The pulley weighs 1000 N and tension ratio is 3. Find the diameter of the shaft, if the allowable shear stress of the material is 63 MPa. **[Ans. 80 mm]**

20. If in the above Exercise 19, the belt drive is at an angle of 60° to the horizontal and a combined shock and fatigue factor is 1.5 for bending and 1.0 for torque, find the diameter of the shaft. **[Ans. 90 mm]**

21. A shaft made of 40 C 8 steel is used to drive a machine. It rotates at 1500 r.p.m. The pulleys *A*, *B* and the bearings *C*, *D* are located as shown in Fig. 14.21. The belt tensions are also shown in the figure.

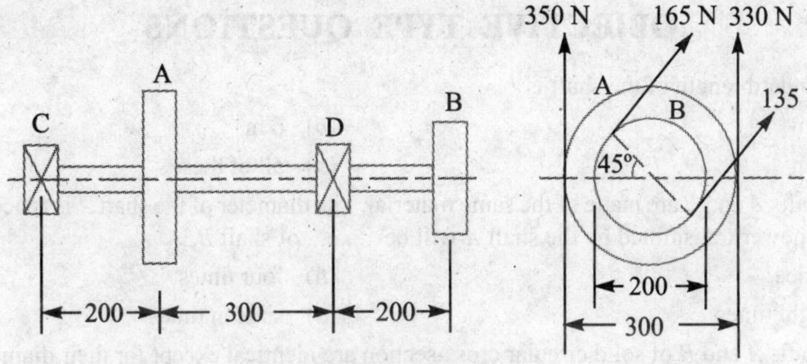

All dimensions in mm.

Fig. 14.21

Determine the diameter of the shaft. The permissible shear stress for the shaft material is 100 MPa. The combined shock and fatigue factor applied to bending and torsion are 1.5 and 1.2 respectively.

[Ans. 25 mm]

22. The engine of a ship develops 440 kW and transmits the power by a horizontal propeller shaft which runs at 120 r.p.m. It is proposed to design a hollow propeller shaft with inner diameter as 0.6 of the outer diameter. Considering torsion alone, calculate the diameter of the propeller shaft if stress in the material is not to exceed 63 MPa and also the angular twist over a length of 2.5 m is not to be more than 1°. The modulus of rigidity of the shaft material is 80 GPa. **[Ans. 30 mm ; 18 mm]**

23. A shaft is required to transmit 1 MW power at 240 r.p.m. The shaft must not twist more than 1 degree on a length of 15 diameters. If the modulus of rigidity for material of the shaft is 80 GPa, find the diameter of the shaft and shear stress induced. **[Ans. 165 mm ; 46.5 MPa]**

24. The internal diameter of a hollow shaft is 2/3 rd of its external diameter. Compare the strength and stiffness of the shaft with that of a solid shaft of the same material. **[Ans. 1.93 ; 2.6]**

25. The shaft of an axial flow rotary compressor is subjected to a maximum torque of 2000 N-m and a maximum bending moment of 4000 N-m. The combined shock and fatigue factor in torsion is 1.5 and that in bending is 2. Design the diameter of the shaft, if the shear stress in the shaft is 50 MPa. Design a hollow shaft for the above compressor taking the ratio of outer diameter to the inner diameter as 2. What is the percentage saving in material ? Also compare the stiffness.

[Ans. 96 mm ; 98 mm, 49 mm ; 21.84%; 1.018]

QUESTIONS

1. Distinguish clearly, giving examples between pin, axle and shaft.
2. How the shafts are formed ?
3. Discuss the various types of shafts and the standard sizes of transmissions shafts.
4. What type of stresses are induced in shafts ?
5. How the shaft is designed when it is subjected to twisting moment only ?
6. Define equivalent twisting moment and equivalent bending moment. State when these two terms are used in design of shafts.
7. When the shaft is subjected to fluctuating loads, what will be the equivalent twisting moment and equivalent bending moment ?
8. What do you understand by torsional rigidity and lateral rigidity.
9. A hollow shaft has greater strength and stiffness than solid shaft of equal weight. Explain.
10. Under what circumstances are hollow shafts preferred over solid shafts ? Give any two examples where hollow shafts are used. How are they generally manufactured ?

OBJECTIVE TYPE QUESTIONS

1. The standard length of the shaft is

   (*a*) 5 m (*b*) 6 m

   (*c*) 7 m (*d*) all of these

2. Two shafts *A* and *B* are made of the same material. The diameter of the shaft *A* is twice as that of shaft *B*. The power transmitted by the shaft *A* will be .... of shaft *B*.

   (*a*) twice (*b*) four times

   (*c*) eight times (*d*) sixteen times

3. Two shafts *A* and *B* of solid circular cross-section are identical except for their diameters d_A and d_B. The ratio of power transmitted by the shaft *A* to that of shaft *B* is

   (*a*) $\frac{d_A}{d_B}$ (*b*) $\frac{(d_A)^2}{(d_B)^2}$

   (*c*) $\frac{(d_A)^3}{(d_B)^3}$ (*d*) $\frac{(d_A)^4}{(d_B)^4}$

4. Two shafts will have equal strength, if

   (*a*) diameter of both the shafts is same

   (*b*) angle of twist of both the shafts is same

   (*c*) material of both the shafts is same

   (*d*) twisting moment of both the shafts is same

5. A transmission shaft subjected to bending loads must be designed on the basis of
   (*a*) maximum normal stress theory
   (*b*) maximum shear stress theory
   (*c*) maximum normal stress and maximum shear stress theories
   (*d*) fatigue strength
6. Which of the following loading is considered for the design of axles ?
   (*a*) Bending moment only
   (*b*) Twisting moment only
   (*c*) Combined bending moment and torsion
   (*d*) Combined action of bending moment, twisting moment and axial thrust
7. When a shaft is subjected to a bending moment *M* and a twisting moment *T*, then the equivalent twisting moment is equal to
   (*a*) $M + T$ (*b*) $M^2 + T^2$
   (*c*) $\sqrt{M^2 + T^2}$ (*d*) $\sqrt{M^2 - T^2}$
8. The maximum shear stress theory is used for
   (*a*) brittle materials (*b*) ductile materials
   (*c*) plastic materials (*d*) non-ferrous materials
9. The maximum normal stress theory is used for
   (*a*) brittle materials (*b*) ductile materials
   (*c*) plastic materials (*d*) non-ferrous materials
10. The design of shafts made of brittle materials is based on
   (*a*) Guest's theory (*b*) Rankine's theory
   (*c*) St. Venant's theory (*d*) Von Mises Theory

ANSWERS

1. (*d*)	2. (*c*)	3. (*c*)	4. (*d*)	5. (*a*)
6. (*a*)	7. (*c*)	8. (*b*)	9. (*a*)	10. (*b*)

CHAPTER

15

Levers

15.1 Introduction

A lever is a rigid rod or bar capable of turning about a fixed point called ***fulcrum***. It is used as a machine to lift a load by the application of a small effort. The ratio of load lifted to the effort applied is called ***mechanical advantage***. Sometimes, a lever is merely used to facilitate the application of force in a desired direction. A lever may be ***straight*** or ***curved*** and the forces applied on the lever (or by the lever) may be parallel or inclined to one another. The principle on which the lever works is same as that of moments.

Consider a straight lever with parallel forces acting in the same plane as shown in Fig 15.1. The points *A* and *B* through which the load and effort is applied are known as load and effort points respectively. *F* is the fulcrum about which the lever is capable of turning. The perpendicular distance between the load point and fulcrum (l_1) is known as ***load arm*** and the perpendicular distance between the effort point and fulcrum (l_2) is called ***effort arm***. According to the principle of moments,

$$W \times l_1 = P \times l_2 \quad \text{or} \quad \frac{W}{P} = \frac{l_2}{l_1}$$

i.e. Mechanical advantage,

$$M.A. = \frac{W}{P} = \frac{l_2}{l_1}$$

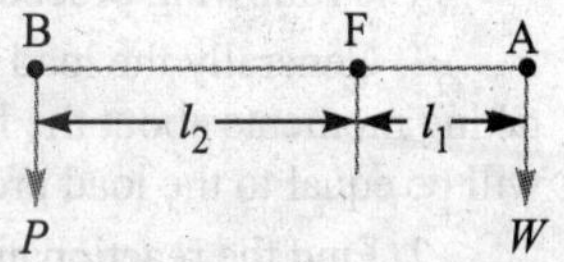

Fig. 15.1. Straight lever.

The ratio of the effort arm to the load arm *i.e.* l_2 / l_1 is called ***leverage***.

A little consideration will show that if a large load is to be lifted by a small effort, then the effort arm should be much greater than the load arm. In some cases, it may not be possible to provide a lever with large effort arm due to space limitations. Therefore in order to obtain a great leverage, ***compound levers*** may be used. The compound levers may be made of straight pieces, which may be attached to one another with pin joints. The bell cranked levers may be used instead of a number of jointed levers. In a compound lever, the leverage is the product of leverages of various levers.

15.2 Application of Levers in Engineering Practice

The load W and the effort P may be applied to the lever in three different ways as shown in Fig. 15.2. The levers shown at (*a*), (*b*) and (*c*) in Fig. 15.2 are called ***first type***, ***second type*** and ***third type*** of levers respectively.

In the ***first type*** of levers, the fulcrum is in between the load and effort. In this case, the effort arm is greater than load arm, therefore mechanical advantage obtained is more than one. Such type of levers are commonly found in bell cranked levers used in railway signalling arrangement, rocker arm in internal combustion engines, handle of a hand pump, hand wheel of a punching press, beam of a balance, foot lever etc.

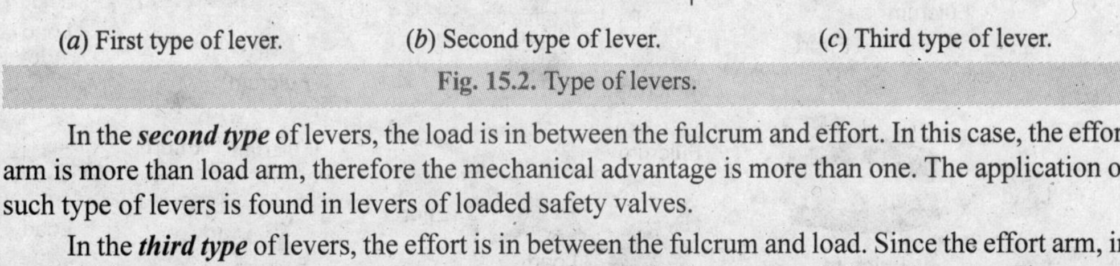

(*a*) First type of lever. (*b*) Second type of lever. (*c*) Third type of lever.

Fig. 15.2. Type of levers.

In the ***second type*** of levers, the load is in between the fulcrum and effort. In this case, the effort arm is more than load arm, therefore the mechanical advantage is more than one. The application of such type of levers is found in levers of loaded safety valves.

In the ***third type*** of levers, the effort is in between the fulcrum and load. Since the effort arm, in this case, is less than the load arm, therefore the mechanical advantage is less that one. The use of such type of levers is not recommended in engineering practice. However a pair of tongs, the treadle of a sewing machine etc. are examples of this type of lever.

15.3 Design of a Lever

The design of a lever consists in determining the physical dimensions of a lever when forces acting on the lever are given. The forces acting on the lever are

1. Load (W), **2.** Effort (P), and **3.** Reaction at the fulcrum F (R_F).

The load and effort cause moments in opposite directions about the fulcrum.

The following procedure is usually adopted in the design of a lever :

1. Generally the load W is given. Find the value of the effort (P) required to resist this load by taking moments about the fulcrum. When the load arm is equal to the effort arm, the effort required will be equal to the load provided the friction at bearings is neglected.

2. Find the reaction at the fulcrum (R_F), as discussed below :

(*i*) When W and P are parallel and their direction is same as shown in Fig. 15.2 (*a*), then

$$R_F = W + P$$

The direction of R_F will be opposite to that of W and P.

(*ii*) When W and P are parallel and acts in opposite directions as shown in Fig. 15.2 (*b*) and (*c*), then R_F will be the difference of W and P. For load positions as shown in Fig. 15.2 (*b*),

$$R_F = W - P$$

and for load positions as shown in Fig. 15.2 (*c*),

$$R_F = P - W$$

The direction of R_F will be opposite to that of W or P whichever is greater.

(*iii*) When W and P are inclined to each other as shown in Fig. 15.3 (*a*), then R_F, which is equal to the resultant of W and P, is determined by parallelogram law of forces. The line of action of R_F passes through the intersection of W and P and also through F. The direction of R_F depends upon the direction of W and P.

(*iv*) When W and P acts at right angles and the arms are inclined at an angle θ as shown in Fig. 15.3 (*b*), then R_F is determined by using the following relation :

$$R_F = \sqrt{W^2 + P^2 - 2W \times P \cos \theta}$$

In case the arms are at right angles as shown in Fig. 15.3 (*c*), then

$$R_F = \sqrt{W^2 + P^2}$$

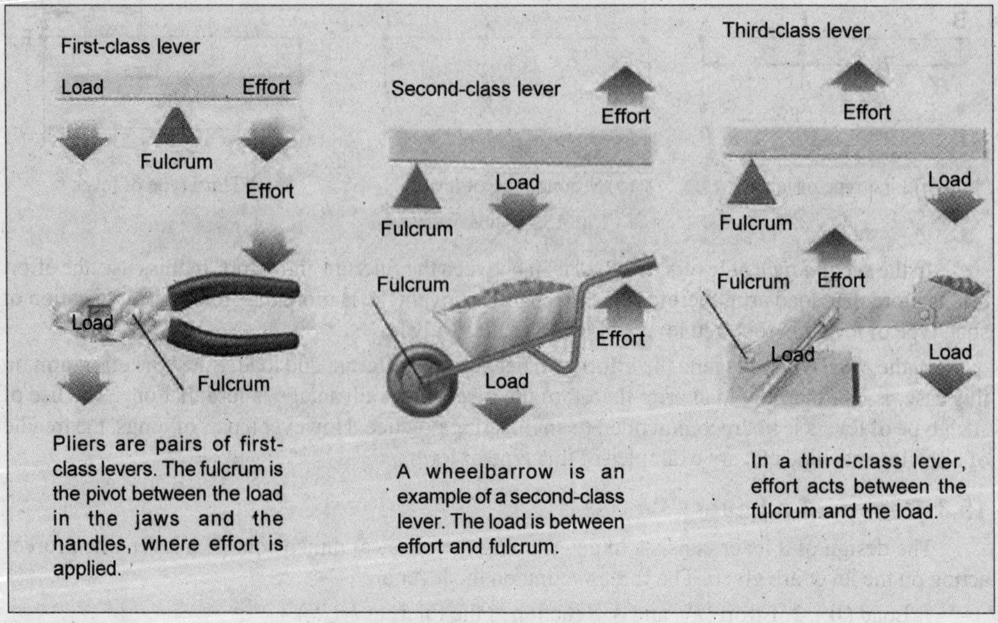

There are three classes of levers.

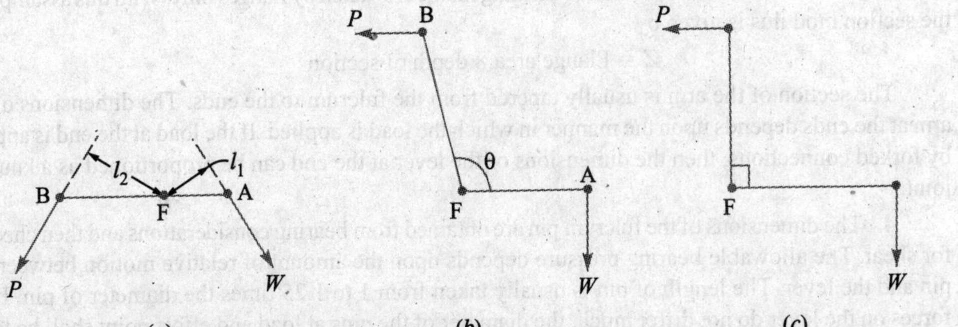

(a) (b) (c)

Fig. 15.3

3. Knowing the forces acting on the lever, the cross-section of the arm may be determined by considering the section of the lever at which the maximum bending moment occurs. In case of levers having two arms as shown in Fig. 15.4 (*a*) and cranked levers, the maximum bending moment occurs at the boss. The cross-section of the arm may be rectangular, elliptical or *I*-section as shown in Fig. 15.4 (*b*). We know that section modulus for rectangular section,

$$Z = \frac{1}{6} \times t \times h^2$$

where t = Breadth or thickness of the lever, and

h = Depth or height of the lever.

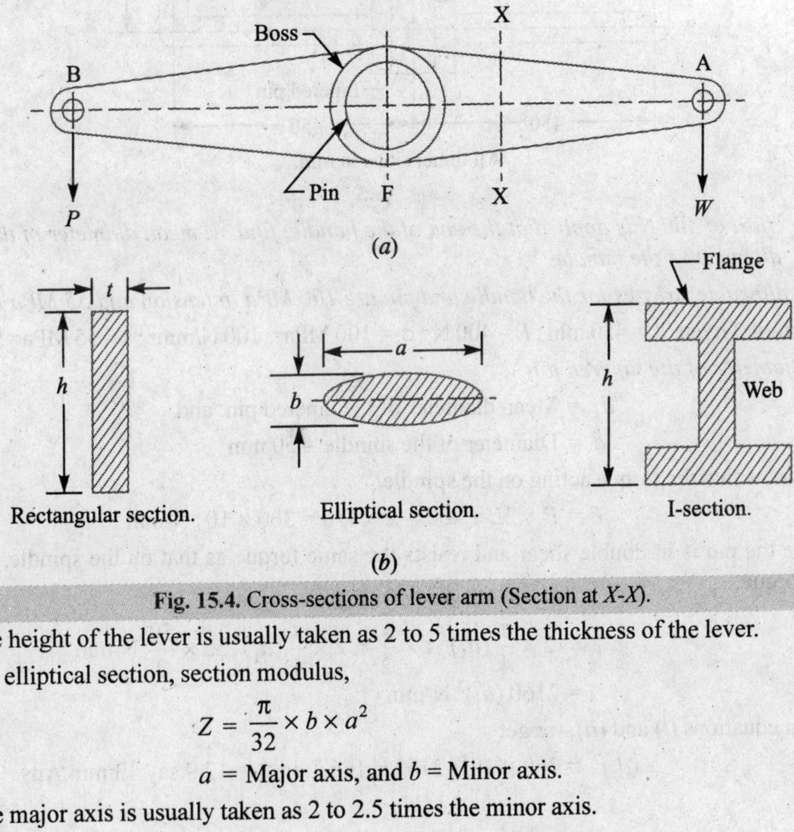

(b)

Fig. 15.4. Cross-sections of lever arm (Section at *X-X*).

The height of the lever is usually taken as 2 to 5 times the thickness of the lever.

For elliptical section, section modulus,

$$Z = \frac{\pi}{32} \times b \times a^2$$

where a = Major axis, and b = Minor axis.

The major axis is usually taken as 2 to 2.5 times the minor axis.

For *I*-section, it is assumed that the bending moment is taken by flanges only. With this assumption, the section modulus is given by

$$Z = \text{Flange area} \times \text{depth of section}$$

The section of the arm is usually tapered from the fulcrum to the ends. The dimensions of the arm at the ends depends upon the manner in which the load is applied. If the load at the end is applied by forked connections, then the dimensions of the lever at the end can be proportioned as a knuckle joint.

4. The dimensions of the fulcrum pin are obtained from bearing considerations and then checked for shear. The allowable bearing pressure depends upon the amount of relative motion between the pin and the lever. The length of pin is usually taken from 1 to 1.25 times the diameter of pin. If the forces on the lever do not differ much, the diameter of the pins at load and effort point shall be taken equal to the diameter of the fulcrum pin so that the spares are reduced. Instead of choosing a thick lever, the pins are provided with a boss in order to provide sufficient bearing length.

5. The diameter of the boss is taken twice the diameter of pin and length of the boss equal to the length of pin. The boss is usually provided with a 3 mm thick phosphor bronze bush with a dust proof lubricating arrangement in order to reduce wear and to increase the life of lever.

Example 15.1. *A handle for turning the spindle of a large valve is shown in Fig. 15.5. The length of the handle from the centre of the spindle is 450 mm. The handle is attached to the spindle by means of a round tapered pin.*

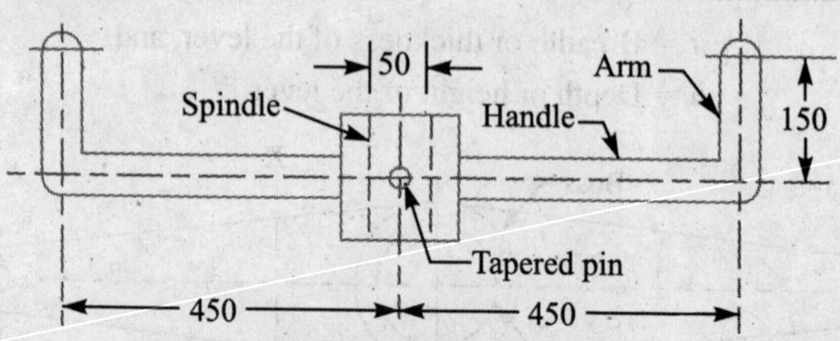

All dimensions in mm.

Fig. 15.5

If an effort of 400 N is applied at the end of the handle, find: 1. mean diameter of the tapered pin, and 2. diameter of the handle.

The allowable stresses for the handle and pin are 100 MPa in tension and 55 MPa in shear.

Solution. Given : $L = 450$ mm ; $P = 400$ N ; $\sigma_t = 100$ MPa $= 100$ N/mm^2 ; $\tau = 55$ MPa $= 55$ N/mm^2

1. *Mean diameter of the tapered pin*

Let d_1 = Mean diameter of the tapered pin, and

d = Diameter of the spindle = 50 mm ...(Given)

We know that the torque acting on the spindle,

$$T = P \times 2L = 400 \times 2 \times 450 = 360 \times 10^3 \text{ N-mm} \quad ...(i)$$

Since the pin is in double shear and resists the same torque as that on the spindle, therefore resisting torque,

$$T = 2 \times \frac{\pi}{4} (d_1)^2 \tau \times \frac{d}{2} = 2 \times \frac{\pi}{4} (d_1)^2 \, 55 \times \frac{50}{2} \text{ N-mm}$$

$$= 2160 \, (d_1)^2 \text{ N-mm} \quad ...(ii)$$

From equations (*i*) and (*ii*), we get

$$(d_1)^2 = 360 \times 10^3 / 2160 = 166.7 \text{ or } d_1 = 12.9 \text{ say } 13 \text{ mm} \textbf{ Ans.}$$

2. *Diameter of the handle*

Let D = Diameter of the handle.

Since the handle is subjected to both bending moment and twisting moment, therefore the design will be based on either equivalent twisting moment or equivalent bending moment. We know that bending moment,

$$M = P \times L = 400 \times 450 = 180 \times 10^3 \text{ N-mm}$$

The twisting moment depends upon the point of application of the effort. Assuming that the effort acts at a distance 100 mm from the end of the handle, we have twisting moment,

$$T = 400 \times 100 = 40 \times 10^3 \text{ N-mm}$$

We know that equivalent twisting moment,

$$T_e = \sqrt{M^2 + T^2} = \sqrt{(180 \times 10^3)^2 + (40 \times 10^3)^2} = 184.4 \times 10^3 \text{ N-mm}$$

We also know that equivalent twisting moment (T_e),

$$184.4 \times 10^3 = \frac{\pi}{16} \times \tau \times D^3 = \frac{\pi}{16} \times 55 \times D^3 = 10.8\, D^3$$

$$\therefore \quad D^3 = 184.4 \times 10^3 / 10.8 = 17.1 \times 10^3 \quad \text{or} \quad D = 25.7 \text{ mm}$$

Again we know that equivalent bending moment,

$$M_e = \frac{1}{2}\left[M + \sqrt{M^2 + T^2}\right] = \frac{1}{2}(M + T_e)$$

$$= \frac{1}{2}(180 \times 10^3 + 184.4 \times 10^3) = 182.2 \times 10^3 \text{ N-mm}$$

We also know that equivalent bending moment (M_e),

$$182.2 \times 10^3 = \frac{\pi}{32} \times \sigma_b \times D^3 = \frac{\pi}{32} \times 100 \times D^3 = 9.82\, D^3 \qquad ...(\because \sigma_b = \sigma_t)$$

$$\therefore \quad D^3 = 182.2 \times 10^3 / 9.82 = 18.6 \times 10^3 \quad \text{or} \quad D = 26.5 \text{ mm}$$

Taking larger of the two values, we have

D = 26.5 mm **Ans.**

Example 15.2. *A vertical lever PQR, 15 mm thick is attached by a fulcrum pin at R and to a horizontal rod at Q, as shown in Fig. 15.6.*

An operating force of 900 N is applied horizontally at P. Find :

1. *Reactions at Q and R,*
2. *Tensile stress in 12 mm diameter tie rod at Q*
3. *Shear stress in 12 mm diameter pins at P, Q and R, and*
4. *Bearing stress on the lever at Q.*

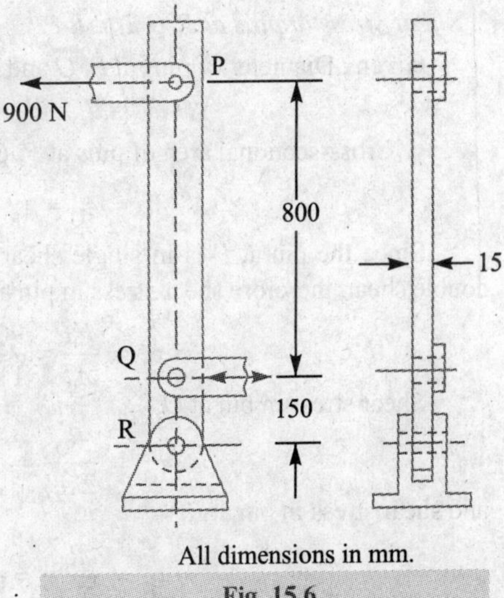

All dimensions in mm.

Fig. 15.6

Solution. Given : t = 15 mm ; F_P = 900 N

1. *Reactions at Q and R*

Let R_Q = Reaction at Q, and

R_R = Reaction at R,

Taking moments about R, we have

$$R_Q \times 150 = 900 \times 950 = 855\,000$$

$$\therefore \quad R_Q = 855\,000 / 150 = 5700 \text{ N} \quad \textbf{Ans.}$$

These levers are used to change railway tracks.

Since the forces at P and Q are parallel and opposite as shown in Fig. 15.7, therefore reaction at R,

$$R_R = R_Q - 900 = 5700 - 900 = 4800 \text{ N} \quad \textbf{Ans.}$$

2. ***Tensile stress in the tie rod at Q***

Let d_t = Diameter of tie rod = 12 mm ...(Given)

$$\therefore \quad \text{Area, } A_t = \frac{\pi}{4}(12)^2 = 113 \text{ mm}^2$$

We know that tensile stress in the tie rod,

$$\sigma_t = \frac{\text{Force at } Q\ (R_Q)}{\text{Cross - sectional area } (A_t)} = \frac{5700}{113}$$

$$= 50.4 \text{ N/mm}^2 = 50.4 \text{ MPa} \quad \textbf{Ans.}$$

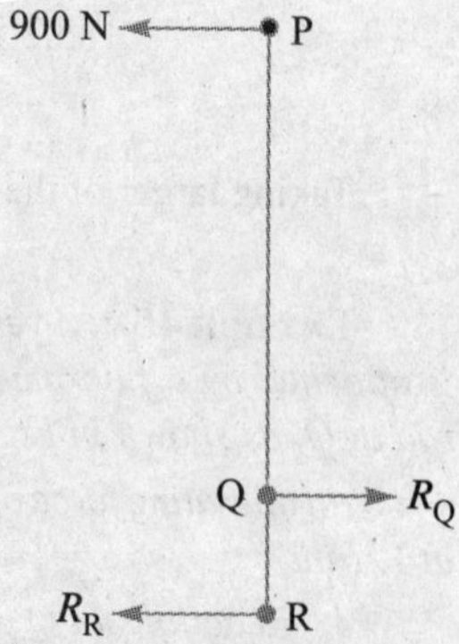

Fig. 15.7

3. ***Shear stress in pins at P, Q and R***

Given : Diameter of pins at P, Q and R,

$$d_P = d_Q = d_R = 12 \text{ mm}$$

$\therefore$ Cross-sectional area of pins at P, Q and R,

$$A_P = A_Q = A_R = \frac{\pi}{4}(12)^2 = 113 \text{ mm}^2$$

Since the pin at P is in single shear and pins at Q and R are in double shear, therefore shear stress in pin at P,

$$\tau_P = \frac{F_P}{A_P} = \frac{900}{113} = 7.96 \text{ N/mm}^2 = 7.96 \text{ MPa} \quad \textbf{Ans.}$$

Shear stress in pin at Q,

$$\tau_Q = \frac{R_Q}{2\,A_Q} = \frac{5700}{2 \times 113} = 25.2 \text{ N/mm}^2 = 25.2 \text{ MPa} \quad \textbf{Ans.}$$

and shear stress in pin at R,

$$\tau_R = \frac{R_R}{2\,A_R} = \frac{4800}{2 \times 113} = 21.2 \text{ N/mm}^2 = 21.2 \text{ MPa} \quad \textbf{Ans.}$$

4. ***Bearing stress on the lever at Q***

Bearing area of the lever at the pin Q,

$$A_b = \text{Thickness of lever} \times \text{Diameter of pin} = 15 \times 12 = 180 \text{ mm}^2$$

∴ Bearing stress on the lever at Q,

$$\sigma_b = \frac{R_Q}{A_b} = \frac{5700}{180} = 31.7 \text{ N/mm}^2 = 31.7 \text{ MPa} \quad \textbf{Ans.}$$

15.4 Hand Levers

A hand lever with suitable dimensions and proportions is shown in Fig. 15.8.

Let P = Force applied at the handle,

L = Effective length of the lever,

σ_t = Permissible tensile stress, and

τ = Permissible shear stress.

For wrought iron, σ_t may be taken as 70 MPa and τ as 60 MPa.

In designing hand levers, the following procedure may be followed :

1. The diameter of the shaft (d) is obtained by considering the shaft under pure torsion. We know that twisting moment on the shaft,

$$T = P \times L$$

and resisting torque,

$$T = \frac{\pi}{16} \times \tau \times d^3$$

From this relation, the diameter of the shaft (d) may be obtained.

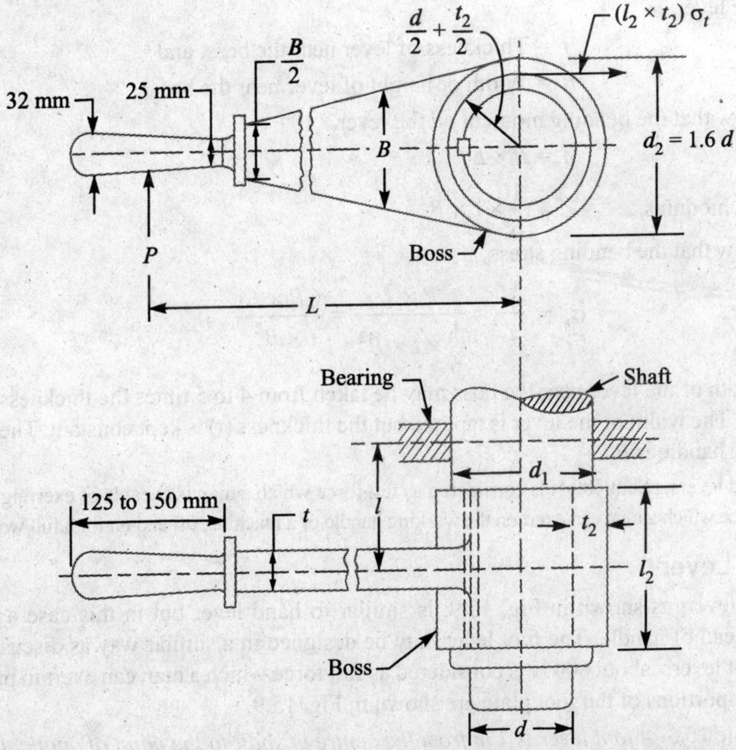

Fig. 15.8. Hand lever.

2. The diameter of the boss (d_2) is taken as 1.6 d and thickness of the boss (t_2) as 0.3 d.

3. The length of the boss (l_2) may be taken from d to 1.25 d. It may be checked for a trial thickness t_2 by taking moments about the axis. Equating the twisting moment ($P \times L$) to the moment

of resistance to tearing parallel to the axis, we get

$$P \times L = l_2 t_2 \sigma_t \left(\frac{d + t_2}{2}\right) \quad \text{or} \quad l_2 = \frac{2P \times L}{t_2 \sigma_t (d + t_2)}$$

4. The diameter of the shaft at the centre of the bearing (d_1) is obtained by considering the shaft in combined bending and twisting.

We know that bending moment on the shaft,

$$M = P \times l$$

and twisting moment, $T = P \times L$

∴ Equivalent twisting moment,

$$T_e = \sqrt{M^2 + T^2} = \sqrt{(P \times l)^2 + (P \times L)^2} = P\sqrt{l^2 + L^2}$$

We also know that equivalent twisting moment,

$$T_e = \frac{\pi}{16} \times \tau (d_1)^3 \quad \text{or} \quad P\sqrt{l^2 + L^2} = \frac{\pi}{16} \times \tau (d_1)^3$$

The length l may be taken as $2\ l_2$.

From the above expression, the value of d_1 may be determined.

5. The key for the shaft is designed as usual for transmitting a torque of $P \times L$.

6. The cross-section of the lever near the boss may be determined by considering the lever in bending. It is assumed that the lever extends to the centre of the shaft which results in a stronger section of the lever.

Let t = Thickness of lever near the boss, and

B = Width or height of lever near the boss.

We know that the bending moment on the lever,

$$M = P \times L$$

Section modulus, $Z = \frac{1}{6} \times t \times B^2$

We know that the bending stress,

$$\sigma_b = \frac{M}{Z} = \frac{P \times L}{\frac{1}{6} \times t \times B^2} = \frac{6P \times L}{t \times B^2}$$

The width of the lever near the boss may be taken from 4 to 5 times the thickness of lever, *i.e.* $B = 4\,t$ to $5\,t$. The width of the lever is tapered but the thickness (t) is kept constant. The width of the lever near the handle is $B/2$.

Note: For hand levers, about 400 N is considered as full force which a man is capable of exerting. About 100 N is the mean force which a man can exert on the working handle of a machine, off and on for a full working day.

15.5 Foot Lever

A foot lever, as shown in Fig. 15.9, is similar to hand lever but in this case a foot plate is provided instead of handle. The foot lever may be designed in a similar way as discussed for hand lever. For foot levers, about 800 N is considered as full force which a man can exert in pushing a foot lever. The proportions of the foot plate are shown in Fig. 15.9.

Example 15.3. *A foot lever is 1 m from the centre of shaft to the point of application of 800 N load. Find :*

1. Diameter of the shaft, 2. Dimensions of the key, and 3. Dimensions of rectangular arm of the foot lever at 60 mm from the centre of shaft assuming width of the arm as 3 times thickness.

The allowable tensile stress may be taken as 73 MPa and allowable shear stress as 70 MPa.

Solution. Given : $L = 1$ m $= 1000$ mm ; $P = 800$ N ; $\sigma_t = 73$ MPa $= 73$ N/mm^2 ; $\tau = 70$ MPa $= 70$ N/mm^2

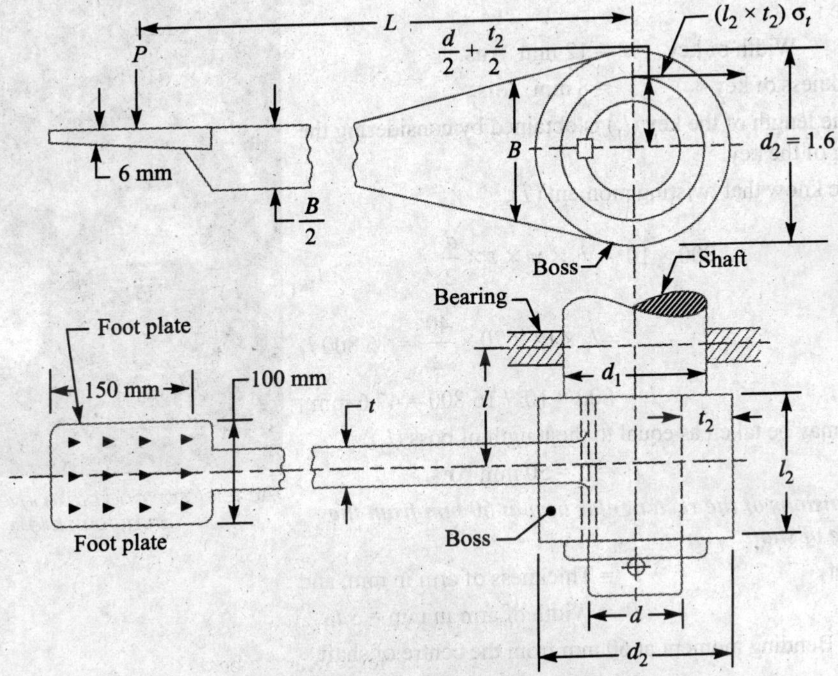

Fig. 15.9. Foot lever.

1. ***Diameter of the shaft***

Let d = Diameter of the shaft.

We know that the twisting moment on the shaft,

$$T = P \times L = 800 \times 1000 = 800 \times 10^3 \text{ N-mm}$$

We also know that the twisting moment on the shaft (T),

$$800 \times 10^3 = \frac{\pi}{16} \times \tau \times d^3 = \frac{\pi}{16} \times 70 \times d^3 = 13.75\, d^3$$

$\therefore$ $d^3 = 800 \times 10^3 / 13.75 = 58.2 \times 10^3$

or $d = 38.8$ say 40 mm **Ans.**

We know that diameter of the boss,

$$d_2 = 1.6\, d = 1.6 \times 40 = 64 \text{ mm}$$

Thickness of the boss,

$$t_2 = 0.3\, d = 0.3 \times 40 = 12 \text{ mm}$$

and length of the boss, $l_2 = 1.25\, d = 1.25 \times 40 = 50$ mm

Now considering the shaft under combined bending and twisting, the diameter of the shaft at the centre of the bearing (d_1) is given by the relation

$$\frac{\pi}{16} \times \tau\, (d_1)^3 = P\sqrt{l^2 + L^2}$$

$$\frac{\pi}{16} \times 70 \times (d_1)^3 = 800\sqrt{(100)^2 + (1000)^2} \quad \text{...(Taking } l = 2\, l_2\text{)}$$

or $13.75\,(d_1)^3 = 804 \times 10^3$

$\therefore$ $(d_1)^3 = 804 \times 10^3 / 13.75 = 58.5 \times 10^3$ or $d_1 = 38.8$ say 40 mm **Ans.**

2. ***Dimensions of the key***

The standard dimensions of the key for a 40 mm diameter shaft are :

Width of key, $w = 12$ mm **Ans.**

and thickness of key $= 8$ mm **Ans.**

The length of the key (l_1) is obtained by considering the shearing of the key.

We know that twisting moment (T),

$$800 \times 10^3 = l_1 \times w \times \tau \times \frac{d}{2}$$

$$= l_1 \times 12 \times 70 \times \frac{40}{2} = 16\ 800\ l_1$$

$\therefore$ $l_1 = 800 \times 10^3 / 16\ 800 = 47.6$ mm

It may be taken as equal to the length of boss (l_2).

$\therefore$ $l_1 = l_2 = 50$ mm **Ans.**

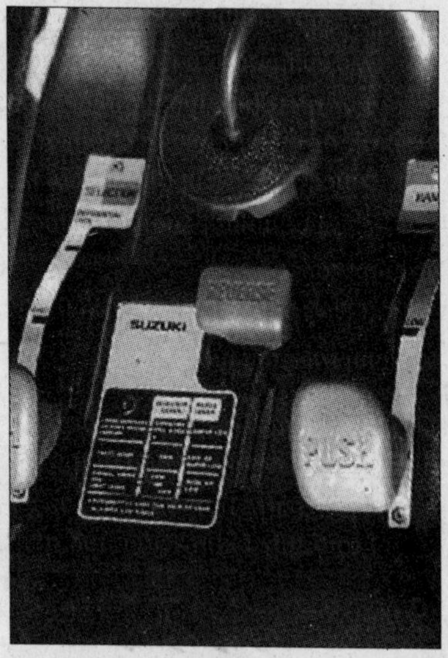

Accelerator and brake levers inside an automobile.

3. ***Dimensions of the rectangular arm at 60 mm from the centre of shaft***

Let t = Thickness of arm in mm, and

B = Width of arm in mm = $3\ t$ ...(Given)

$\therefore$ Bending moment at 60 mm from the centre of shaft,

$$M = 800\ (1000 - 60) = 752 \times 10^3 \text{ N-mm}$$

and section modulus,

$$Z = \frac{1}{6} \times t \times B^2 = \frac{1}{6} \times t\ (3t)^2 = 1.5\ t^3 \text{ mm}^3$$

We know that the tensile bending stress (σ_t),

$$73 = \frac{M}{Z} = \frac{752 \times 10^3}{1.5\ t^3} = \frac{501.3 \times 10^3}{t^3}$$

$\therefore$ $t^3 = 501.3 \times 10^3 / 73 = 6.87 \times 10^3$

or $t = 19$ say 20 mm **Ans.**

and $B = 3\ t = 3 \times 20 = 60$ mm **Ans.**

The width of the arm is tapered while the thickness is kept constant throughout. The width of the arm on the foot plate side,

$$B_1 = B / 2 = 30 \text{ mm } \textbf{Ans.}$$

15.6 Cranked Lever

A cranked lever, as shown in Fig. 15.10, is a hand lever commonly used for operating hoisting winches.

The lever can be operated either by a single person or by two persons. The maximum force in order to operate the lever may be taken as 400 N and the length of handle as 300 mm. In case the lever is operated by two persons, the maximum force of operation will be doubled and length of handle may be taken as 500 mm. The handle is covered in a pipe to prevent hand scoring. The end of the shaft is usually squared so that the lever may be easily fixed and removed. The length (L) is usually from 400 to 450 mm and the height of the shaft centre line from the ground is usually one metre. In order to design such levers, the following procedure may be adopted :

1. The diameter of the handle (d) is obtained from bending considerations. It is assumed that the effort (P) applied on the handle acts at $\frac{2}{3}$rd of its length (l).

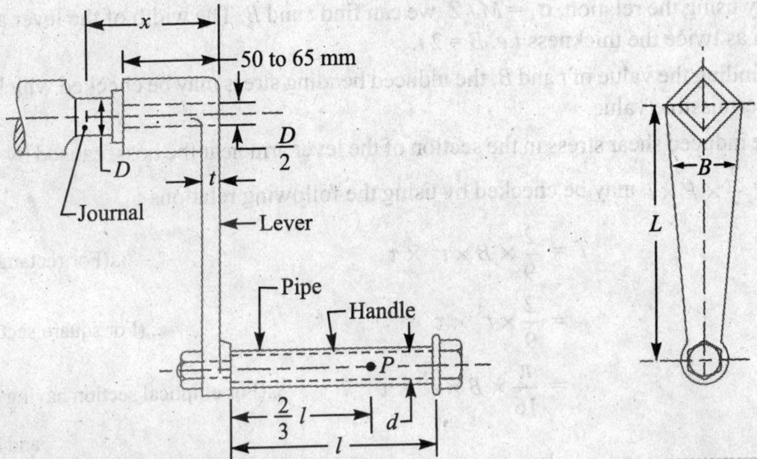

Fig. 15.10. Cranked lever.

∴ Maximum bending moment,

$$M = P \times \frac{2l}{3} = \frac{2}{3} \times P \times l$$

and section modulus,

$$Z = \frac{\pi}{32} \times d^3$$

∴ Resisting moment $= \sigma_b \times Z = \sigma_b \times \frac{\pi}{32} \times d^3$

where σ_b = Permissible bending stress for the material of the handle.

Equating resisting moment to the maximum bending moment, we have

$$\sigma_b \times \frac{\pi}{32} \times d^3 = \frac{2}{3} \times P \times l$$

From this expression, the diameter of the handle (d) may be evaluated. The diameter of the handle is usually proportioned as 25 mm for single person and 40 mm for two persons.

2. The cross-section of the lever arm is usually rectangular having uniform thickness throughout. The width of the lever arm is tapered from the boss to the handle. The arm is subjected to constant twisting moment, $T = \frac{2}{3} \times P \times l$ and a varying bending moment which is maximum near the boss. It is assumed that the arm of the lever extends upto the centre of shaft, which results in a slightly stronger lever.

∴ Maximum bending moment $= P \times L$

Since, at present time, there is insufficient information on the subject of combined bending and twisting of rectangular sections to enable us to find equivalent bending or twisting, with sufficient accuracy, therefore the indirect procedure is adopted.

We shall design the lever arm for 25% more bending moment.

∴ Maximum bending moment

$$M = 1.25\, P \times L$$

Let t = Thickness of the lever arm, and

B = Width of the lever arm near the boss.

∴ Section modulus for the lever arm,

$$Z = \frac{1}{6} \times t \times B^2$$

Now by using the relation, $\sigma_b = M / Z$, we can find t and B. The width of the lever arm near the boss is taken as twice the thickness *i.e.* $B = 2\,t$.

After finding the value of t and B, the induced bending stress may be checked which should not exceed the permissible value.

3. The induced shear stress in the section of the lever arm near the boss, caused by the twisting moment, $T = \frac{2}{3} \times P \times l$ may be checked by using the following relations :

$$T = \frac{2}{9} \times B \times t^2 \times \tau \qquad \text{...(For rectangular section)}$$

$$= \frac{2}{9} \times t^3 \times \tau \qquad \text{...(For square section of side } t\text{)}$$

$$= \frac{\pi}{16} \times B \times t^2 \times \tau \qquad \text{...(For elliptical section having major axis } B \text{ and minor axis } t\text{)}$$

4. Knowing the values of σ_b and τ, the maximum principal or shear stress induced may be checked by using the following relations :

Maximum principal stress,

$$\sigma_{b(max)} = \frac{1}{2}\left[\sigma_b + \sqrt{(\sigma_b)^2 + 4\tau^2}\right]$$

Maximum shear stress,

$$\tau_{max} = \frac{1}{2}\sqrt{(\sigma_b)^2 + 4\tau^2}$$

5. Since the journal of the shaft is subjected to twisting moment and bending moment, therefore its diameter is obtained from equivalent twisting moment.

We know that twisting moment on the journal of the shaft,

$$T = P \times L$$

and bending moment on the journal of the shaft,

$$M = P\left(\frac{2l}{3} + x\right)$$

where x = Distance from the end of boss to the centre of journal.

∴ Equivalent twisting moment,

$$T_e = \sqrt{M^2 + T^2} = P\sqrt{\left(\frac{2l}{3} + x\right)^2 + L^2}$$

We know that equivalent twisting moment,

$$T_e = \frac{\pi}{16} \times \tau \times D^3$$

From this expression, we can find the diameter (D) of the journal.

The diameter of the journal is usually taken as

$$D = 30 \text{ to } 40 \text{ mm, for single person}$$
$$= 40 \text{ to } 45 \text{ mm, for two persons.}$$

Note: The above procedure may be used in the design of overhung cranks of engines.

Example 15.4. *A cranked lever, as shown in 15.10, has the following dimensions :*

Length of the handle = 300 *mm*

Length of the lever arm = 400 *mm*

Overhang of the journal = 100 *mm*

If the lever is operated by a single person exerting a maximum force of 400 N at a distance of $\frac{1}{3}$ *rd length of the handle from its free end, find : 1. Diameter of the handle, 2. Cross-section of the lever arm, and 3. Diameter of the journal.*

The permissible bending stress for the lever material may be taken as 50 MPa and shear stress for shaft material as 40 MPa.

Solution. Given : $l = 300$ mm ; $L = 400$ mm ; $x = 100$ mm ; $P = 400$ N ; $\sigma_b = 50$ MPa $= 50$ N/mm^2 ; $\tau = 40$ MPa $= 40$ N/mm^2

1. ***Diameter of the handle***

Let d = Diameter of the handle in mm.

Since the force applied acts at a distance of 1/3 rd length of the handle from its free end,therefore maximum bending moment,

$$M = \left(1 - \frac{1}{3}\right) P \times l = \frac{2}{3} \times P \times l = \frac{2}{3} \times 400 \times 300 \text{ N-mm}$$

$$= 80 \times 10^3 \text{ N-mm} \quad ...(i)$$

Section modulus, $Z = \frac{\pi}{32} \times d^3 = 0.0982\ d^3$

$\therefore$ Resisting bending moment,

$$M = \sigma_b \times Z = 50 \times 0.0982\ d^3 = 4.91\ d^3 \text{ N-mm} \quad ..(ii)$$

From equations (*i*) and (*ii*), we get

$$d^3 = 80 \times 10^3 / 4.91 = 16.3 \times 10^3 \quad \text{or} \quad d = 25.4 \text{ mm } \textbf{Ans.}$$

2. ***Cross-section of the lever arm***

Let t = Thickness of the lever arm in mm, and

B = Width of the lever arm near the boss, in mm.

Since the lever arm is designed for 25% more bending moment, therefore maximum bending moment,

$$M = 1.25\ P \times L = 1.25 \times 400 \times 400 = 200 \times 10^3 \text{ N-mm}$$

Section modulus, $Z = \frac{1}{6} \times t \times B^2 = \frac{1}{6} \times t\ (2t)^2 = 0.667\ t^3$ ...(Assuming $B = 2t$)

We know that bending stress (σ_b),

$$50 = \frac{M}{Z} = \frac{200 \times 10^3}{0.667\ t^3} = \frac{300 \times 10^3}{t^3}$$

$\therefore$ $t^3 = 300 \times 10^3/50 = 6 \times 10^3$ or $t = 18.2$ say 20 mm **Ans.**

and $B = 2\,t = 2 \times 20 = 40$ mm **Ans.**

Let us now check the lever arm for induced bending and shear stresses.

Bending moment on the lever arm near the boss (assuming that the length of the arm extends upto the centre of shaft) is given by

$$M = P \times L = 400 \times 400 = 160 \times 10^3 \text{ N-mm}$$

and section modulus, $Z = \frac{1}{6} \times t \times B^2 = \frac{1}{6} \times 20\ (40)^2 = 5333 \text{ mm}^3$

∴ Induced bending stress,

$$\sigma_b = \frac{M}{Z} = \frac{160 \times 10^3}{5333} = 30 \text{ N/mm}^2 = 30 \text{ MPa}$$

The induced bending stress is within safe limits.

We know that the twisting moment,

$$T = \frac{2}{3} \times P \times l = \frac{2}{3} \times 400 \times 300 = 80 \times 10^3 \text{ N-mm}$$

We also know that the twisting moment (T),

$$80 \times 10^3 = \frac{2}{9} \times B \times t^2 \times \tau = \frac{2}{9} \times 40\ (20)^2\ \tau = 3556\ \tau$$

$$\therefore \quad \tau = 80 \times 10^3 / 3556 = 22.5 \text{ N/mm}^2 = 22.5 \text{ MPa}$$

The induced shear stress is also within safe limits.

Let us now check the cross-section of lever arm for maximum principal or shear stress.

We know that maximum principal stress,

$$\sigma_{b\,(max)} = \frac{1}{2}\left[\sigma_b + \sqrt{(\sigma_b)^2 + 4\tau^2}\right] = \frac{1}{2}\left[30 + \sqrt{(30)^2 + 4\ (22.5)^2}\right]$$

$$= \frac{1}{2}\ (30 + 54) = 42 \text{ N/mm}^2 = 42 \text{ MPa}$$

and maximum shear stress,

$$\tau_{max} = \frac{1}{2}\sqrt{(\sigma_b)^2 + 4\tau^2} = \frac{1}{2}\sqrt{(30)^2 + 4\ (22.5)^2} = 27 \text{ N/mm}^2 = 27 \text{ MPa}$$

The maximum principal and shear stresses are also within safe limits.

3. *Diameter of the journal*

Let D = Diameter of the journal.

Since the journal of the shaft is subjected to twisting moment and bending moment, therefore its diameter is obtained from equivalent twisting moment.

We know that equivalent twisting moment,

$$T_e = P\sqrt{\left(\frac{2l}{3} + x\right)^2 + L^2} = 400\sqrt{\left(\frac{2 \times 300}{3} + 100\right)^2 + (400)^2}$$

$$= 200 \times 10^3 \text{ N-mm}$$

We know that equivalent twisting moment (T_e),

$$200 \times 10^3 = \frac{\pi}{16} \times \tau \times D^3 = \frac{\pi}{16} \times 40 \times D^3 = 7.86\ D^3$$

$$\therefore \quad D^3 = 200 \times 10^3 / 7.86 = 25.4 \times 10^3 \text{ or } D = 29.4 \text{ say } 30 \text{ mm} \textbf{ Ans.}$$

15.7 Lever for a Lever Safety Valve

A lever safety valve is shown in Fig. 15.11. It is used to maintain a constant safe pressure inside the boiler. When the pressure inside the boiler increases the safe value, the excess steam blows off through the valve automatically. The valve rests over the gunmetal seat which is secured to a casing fixed upon the boiler. One end of the lever is pivoted at the fulcrum *F* by a pin to the toggle, while the other end carries the weights. The valve is held on its seat against the upward steam pressure by the force *P* provided by the weights at *B*. The weights and its distance from the fulcrum are so adjusted that when the steam pressure acting upward on the valve exceeds the normal limit, it lifts the valve and the lever with its weights. The excess steam thus escapes until the pressure falls to the required limit.

The lever may be designed in the similar way as discussed earlier. The maximum steam load (W), at which the valve blows off, is given by

$$W = \frac{\pi}{4} \times D^2 \times p$$

where D = Diameter of the valve, and

p = Steam pressure.

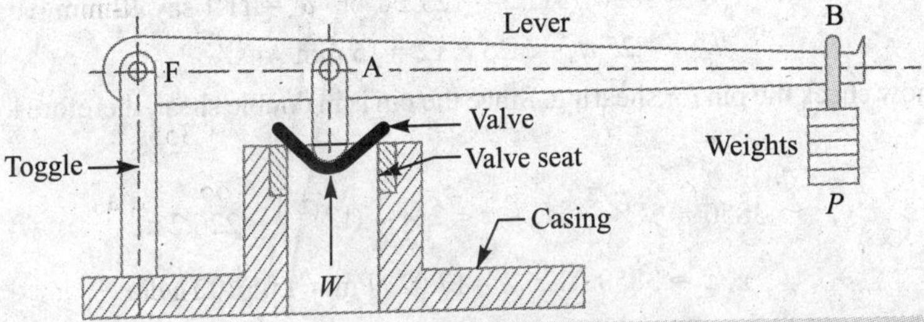

Fig. 15.11. Lever safety valve.

Example 15.5. *A lever loaded safety valve is 70 mm in diameter and is to be designed for a boiler to blow-off at pressure of 1 N/mm² gauge. Design a suitable mild steel lever of rectangular cross-section using the following permissible stresses :*

Tensile stress = 70 MPa; Shear stress = 50 MPa; Bearing pressure intensity = 25 N/mm².

The pin is also made of mild steel. The distance from the fulcrum to the weight of the lever is 880 mm and the distance between the fulcrum and pin connecting the valve spindle links to the lever is 80 mm.

Solution. Given : D = 70 mm ; p = 1 N/mm² ; σ_t = 70 MPa = 70 N/mm² ; τ = 50 MPa = 50 N/mm² ; p_b = 25 N/mm² ; FB = 880 mm ; FA = 80 mm

We know that the maximum steam load at which the valve blows off,

$$W = \frac{\pi}{4} \times D^2 \times p = \frac{\pi}{4}(70)^2 \times 1 = 3850 \text{ N} \qquad ...(i)$$

Taking moments about the fulcrum F, we have

$$P \times 880 = 3850 \times 80 = 308 \times 10^3 \quad \text{or} \quad P = 308 \times 10^3 / 880 = 350 \text{ N}$$

Since the load (W) and the effort (P) in the form of dead weight are parallel and opposite, therefore reaction at F,

$$R_F = W - P = 3850 - 350 = 3500 \text{ N}$$

This rection will act vertically downward as shown in Fig. 15.12.

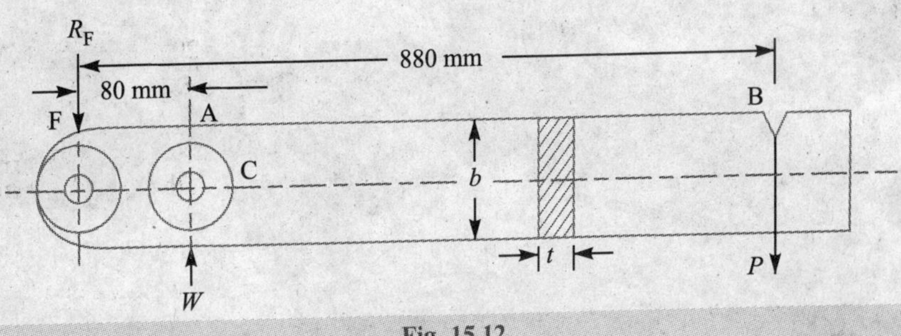

Fig. 15.12

First of all, let us find the diameter of the pin at A from bearing considerations.

Let d_p = Diameter of the pin at A, and

l_p = Length of the pin at A.

∴ Bearing area of the pin at A

$$= d_p \times l_p = 1.25\,(d_p)^2 \qquad \text{...(Assuming } l_p = 1.25\, d_p\text{)}$$

and load on the pin at A

$$= \text{Bearing area} \times \text{Bearing pressure}$$

$$= 1.25\,(d_p)^2\, p_b = 1.25\,(d_p)^2\, 25 = 31.25\,(d_p)^2 \qquad \text{...(ii)}$$

Since the load acting on the pin at A is $W = 3850$ N, therefore from equations (*i*) and (*ii*), we get

$$(d_p)^2 = 3850 / 31.25 = 123.2 \quad \text{or} \quad d_p = 11.1 \text{ say } 12 \text{ mm } \textbf{Ans.}$$

and

$$l_p = 1.25\, d_p = 1.25 \times 12 = 15 \text{ mm } \textbf{Ans.}$$

Let us now check the pin for shearing. Since the pin is in double shear, therefore load on the pin at A (W),

$$3850 = 2 \times \frac{\pi}{4}(d_p)^2\,\tau = 2 \times \frac{\pi}{4}(12)^2\,\tau = 226.2\,\tau$$

$$\therefore \quad \tau = 3850 / 226.2 = 17.02 \text{ N/mm}^2 = 17.02 \text{ MPa}$$

This value of shear stress is less than the permissible value of 50 MPa, therefore the design for pin at A is safe. Since the load at F does not very much differ with the load at A, therefore the same diameter of pin may be used at F, in order to facilitate the interchangeability of parts.

∴ Diameter of the fulcrum pin at F

$$= 12 \text{ mm}$$

A gun metal bush of 2 mm thickness is provided in the pin holes at A and F in order to reduce wear and to increase the life of lever.

∴ Diameter of hole at A and F

$$= 12 + 2 \times 2 = 16 \text{ mm}$$

and outside diameter of the boss

$$= 2 \times \text{Dia. of hole} = 2 \times 16 = 32 \text{ mm}$$

Power clamp of an excavator.

Note : This picture is given as additional information and is not a direct example of the current chapter.

Now let us find out the cross-section of the lever considering the bending moment near the boss at *A*.

Let t = Thickness of the lever, and

b = Width of the lever.

Bending moment near the boss at *A i.e.* at point *C*,

$$M = P \times BC = P(BF - AF - AC) = 350\left(880 - 80 - \frac{16}{2}\right) \text{ N-mm}$$

$$= 277\ 200 \text{ N-mm}$$

and section modulus, $Z = \frac{1}{6} \times t.b^2 = \frac{1}{6} \times t\,(4t)^2 = 2.67\,t^3$ ...(Assuming $b = 4\,t$)

We know that the bending stress (σ_b)

$$70 = \frac{M}{Z} = \frac{277\ 200}{2.67\,t^3} = \frac{104 \times 10^3}{t^3} \quad ...(\because \sigma_b = \sigma_t)$$

$\therefore$ $t^3 = 104 \times 10^3 / 70 = 1.5 \times 10^3$ or $t = 11.4$ say 12 mm **Ans.**

and $b = 4\,t = 4 \times 12 = 48$ mm **Ans.**

Now let us check for the maximum shear stress induced in the lever. From the shear force diagram as shown in Fig. 15.13 (*a*), we see that the maximum shear force on the lever is $(W - P)$ *i.e.* 3500 N.

$\therefore$ Maximum shear stress induced,

$$\tau_{max} = \frac{\text{Maximum shear force}}{\text{Cross-sectional area of the lever}} = \frac{3500}{12 \times 48}$$

$$= 6.07 \text{ N/mm}^2 = 6.07 \text{ MPa}$$

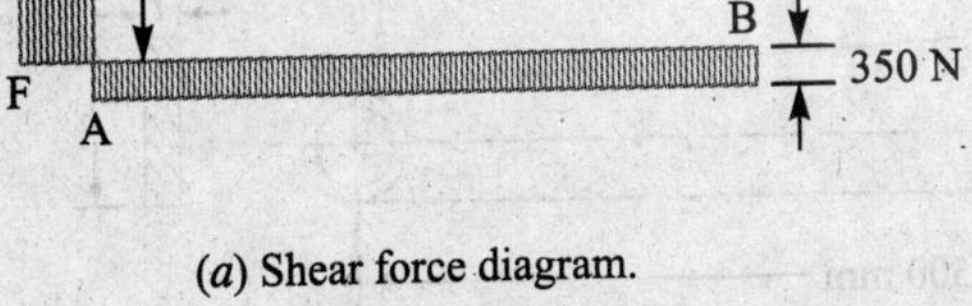

(*a*) Shear force diagram.

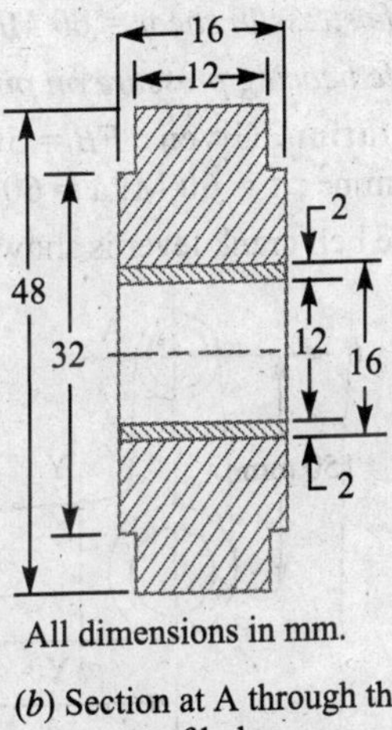

All dimensions in mm.

(*b*) Section at A through the centre of hole.

Fig. 15.13

Since this value of maximum shear stress is much below the permissible shear stress of 50 MPa therefore the design for lever is safe.

Again checking for the bending stress induced at the section passing through the centre of hole at *A*. The section at *A* through the centre of the hole is shown in Fig. 15.13 (*b*).

$\therefore$ Maximum bending moment at the centre of hole at *A*,

$$M = 350\,(880 - 80) = 280 \times 10^3 \text{ N-mm}$$

Section modulus,

$$Z = \frac{\frac{1}{12} \times 12 \left[(48)^3 - (16)^3\right] + 2 \times \frac{1}{12} \times 2 \left[(32)^3 - (16)^3\right]}{48/2}$$

$$= \frac{106\ 496 + 9557}{24} = 4836 \text{ mm}^3$$

∴ Maximum bending stress induced,

$$\sigma_t = \frac{M}{Z} = \frac{280 \times 10^3}{4836} = 58 \text{ N/mm}^2 = 58 \text{ MPa}$$

Since this maximum stress is below the permissible value of 70 MPa, therefore the design in safe.

15.8 Bell Crank Lever

In a bell crank lever, the two arms of the lever are at right angles. Such type of levers are used in railway signalling, governors of Hartnell type, the drive for the air pump of condensors etc. The bell crank lever is designed in a similar way as discussed earlier. The arms of the bell crank lever may be assumed of rectangular, elliptical or I-section. The complete design procedure for the bell crank lever is given in the following example.

Example 15.6. *Design a right angled bell crank lever. The horizontal arm is 500 mm long and a load of 4.5 kN acts vertically downward through a pin in the forked end of this arm. At the end of the 150 mm long arm which is perpendicular to the 500 mm long arm, a force P act at right angles to the axis of 150 mm arm through a pin into a forked end. The lever consists of forged steel material and a pin at the fulcrum. Take the following data for both the pins and lever material:*

Safe stress in tension = 75 MPa

Safe stress in shear = 60 MPa

Safe bearing pressure on pins = 10 N/mm²

Solution. Given : FB = 500 mm ; W = 4.5 kN = 4500 N ; FA = 150 mm ; σ_t = 75 MPa = 75 N/mm² ; τ = 60 MPa = 60 N/mm² ; p_b = 10 N/mm²

The bell crank lever is shown in Fig. 15.14.

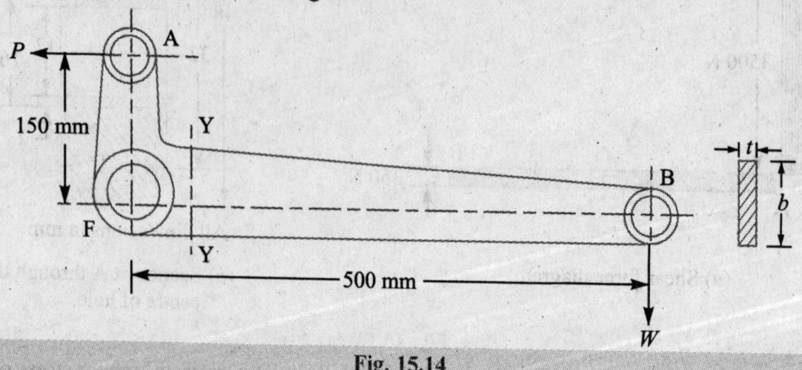

Fig. 15.14

First of all, let us find the effort (P) required to raise the load (W). Taking moments about the fulcrum F, we have

$$W \times 500 = P \times 150$$

$$\therefore \quad P = \frac{W \times 500}{150} = \frac{4500 \times 500}{150} = 15\ 000 \text{ N}$$

and reaction at the fulcrum pin at F,

$$R_F = \sqrt{W^2 + P^2} = \sqrt{(4500)^2 + (15\,000)^2} = 15\,660 \text{ N}$$

1. *Design for fulcrum pin*

Let d = Diameter of the fulcrum pin, and

l = Length of the fulcrum pin.

Considering the fulcrum pin in bearing. We know that load on the fulcrum pin (R_F),

$$15\,660 = d \times l \times p_b = d \times 1.25\,d \times 10 = 12.5\,d^2 \quad \text{...(Assuming } l = 1.25\,d)$$

$\therefore$ $d^2 = 15\,660 / 12.5 = 1253$ or $d = 35.4$ say 36 mm **Ans.**

and $l = 1.25\,d = 1.25 \times 36 = 45$ mm **Ans.**

Let us now check for the shear stress induced in the fulcrum pin. Since the pin is in double shear, therefore load on the fulcrum pin (R_F),

$$15\,660 = 2 \times \frac{\pi}{4} \times d^2 \times \tau = 2 \times \frac{\pi}{4}(36)^2\,\tau = 2036\,\tau$$

$\therefore$ $\tau = 15\,660/2036 = 7.7 \text{ N/mm}^2 = 7.7$ MPa

Since the shear stress induced in the fulcrum pin is less than the given value of 60 MPa, therefore design for the fulcrum pin is safe.

A brass bush of 3 mm thickness is pressed into the boss of fulcrum as a bearing so that the renewal become simple when wear occurs.

$\therefore$ Diameter of hole in the lever

$= d + 2 \times 3$

$= 36 + 6 = 42$ mm

and diameter of boss at fulcrum

$= 2\,d = 2 \times 36 = 72$ mm

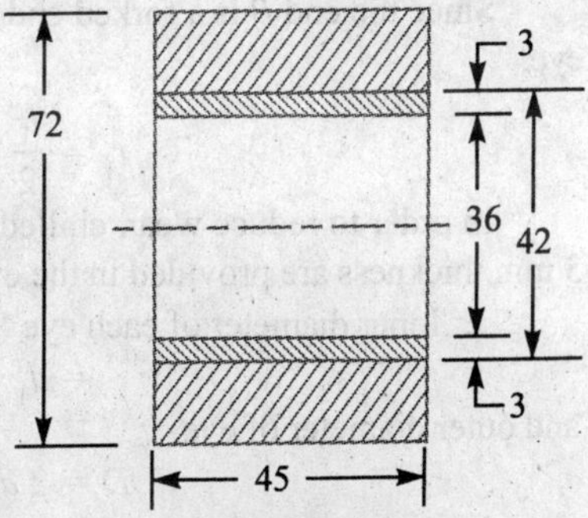

All dimensions in mm.

Fig. 15.15

Now let us check the bending stress induced in the lever arm at the fulcrum. The section of the fulcrum is shown in Fig. 15.15.

Bending moment at the fulcrum

$$M = W \times FB = 4500 \times 500 = 2250 \times 10^3 \text{ N-mm}$$

Section modulus,

$$Z = \frac{\frac{1}{12} \times 45\left[(72)^3 - (42)^3\right]}{72/2} = 311\,625 \text{ mm}^3$$

$\therefore$ Bending stress,

$$\sigma_b = \frac{M}{Z} = \frac{2250 \times 10^3}{311\,625} = 7.22 \text{ N/mm}^2 = 7.22 \text{ MPa}$$

Since the bending stress induced in the lever arm at the fulcrum is less than the given value of 85 MPa, therefore it is safe.

2. Design *for pin at A*

Since the effort at A (which is 15 000 N), is not very much different from the reaction at fulcrum (which is 15 660 N), therefore the same dimensions for the pin and boss may be used as for fulcrum pin to reduce spares.

$\therefore$ Diameter of pin at A = 36 mm **Ans.**

Length of pin at A = 45 mm **Ans.**

and diameter of boss at A = 72 mm **Ans.**

3. *Design for pin at B*

Let d_1 = Diameter of the pin at *B*, and

l_1 = Length of the pin at *B*.

Considering the bearing of the pin at *B*. We know that load on the pin at *B* (*W*),

$$4500 = d_1 \times l_1 \times p_b = d_1 \times 1.25\, d_1 \times 10 = 12.5\,(d_1)^2$$

... (Assuming $l_1 = 1.25\, d_1$)

$\therefore$ $(d_1)^2 = 4500 / 12.5 = 360$ or $d_1 = 18.97$ say 20 mm **Ans.**

and $l_1 = 1.25\, d_1 = 1.25 \times 20 = 25$ mm **Ans.**

Let us now check for the shear stress induced in the pin at *B*. Since the pin is in double shear, therefore load on the pin at *B* (*W*),

$$4500 = 2 \times \frac{\pi}{4}(d_1)^2\,\tau = 2 \times \frac{\pi}{4}(20)^2\,\tau = 628.4\,\tau$$

$$\therefore \quad \tau = 4500 / 628.4 = 7.16 \text{ N/mm}^2 = 7.16 \text{ MPa}$$

Since the shear stress induced in the pin at *B* is within permissible limits, therefore the design is safe.

Since the end *B* is a forked end, therefore thickness of each eye,

$$t_1 = \frac{l_1}{2} = \frac{25}{2} = 12.5 \text{ mm}$$

In order to reduce wear, chilled phosphor bronze bushes of 3 mm thickness are provided in the eyes.

$\therefore$ Inner diameter of each eye

$$= d_1 + 2 \times 3 = 20 + 6 = 26 \text{ mm}$$

and outer diameter of eye,

$$D = 2\, d_1 = 2 \times 20 = 40 \text{ mm}$$

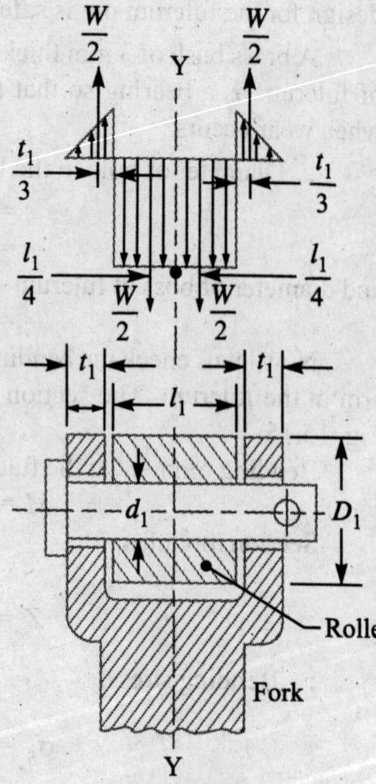

Fig. 15.16

Let us now check the induced bending stress in the pin. The pin is neither simply supported nor rigidly fixed at its ends. Therefore the common practice is to assume the load distribution as shown in Fig. 15.16. The maximum bending moment will occur at *Y-Y.*

$\therefore$ Maximum bending moment at *Y-Y*,

$$M = \frac{W}{2}\left(\frac{l_1}{2} + \frac{t_1}{3}\right) - \frac{W}{2} \times \frac{l_1}{4}$$

$$= \frac{5}{24} W \times l_1 \qquad \ldots(\because t_1 = l_1/2)$$

$$= \frac{5}{24} \times 4500 \times 25 = 23\,438 \text{ N-mm}$$

and section modulus,

$$Z = \frac{\pi}{32}(d_1)^3 = \frac{\pi}{32}(20)^3 = 786 \text{ mm}^3$$

$\therefore$ Bending stress induced,

$$\sigma_b = \frac{M}{Z} = \frac{23\,438}{786} = 29.8 \text{ N/mm}^2 = 29.8 \text{ MPa}$$

This induced bending stress is within safe limits.

4. *Design of lever*

It is assumed that the lever extends upto the centre of the fulcrum from the point of application of the load. This assumption is commonly made and results in a slightly stronger section. Considering the weakest section of failure at *Y-Y*.

Let t = Thickness of the lever at *Y-Y*, and

b = Width or depth of the lever at *Y-Y*.

Taking distance from the centre of the fulcrum to *Y-Y* as 50 mm, therefore maximum bending moment at *Y-Y*,

$$= 4500\,(500 - 50) = 2025 \times 10^3 \text{ N-mm}$$

and section modulus, $Z = \frac{1}{6} \times t \times b^2 = \frac{1}{6} \times t\,(3t)^2 = 1.5\,t^3$ ...(Assuming $b = 3\,t$)

We know that the bending stress (σ_b),

$$75 = \frac{M}{Z} = \frac{2025 \times 10^3}{1.5 t^3} = \frac{1350 \times 10^3}{t^3}$$

∴ $t^3 = 1350 \times 10^3 / 75 = 18 \times 10^3$ or $t = 26$ mm **Ans.**

and $b = 3\,t = 3 \times 26 = 78$ mm **Ans.**

Bucket of a bulldozer.

Note : This picture is given as additional information and is not a direct example of the current chapter.

Example 15.7. *In a Hartnell governor, the length of the ball arm is 190 mm, that of the sleeve arm is 140 mm, and the mass of each ball is 2.7 kg. The distance of the pivot of each bell crank lever from the axis of rotation is 170 mm and the speed when the ball arm is vertical, is 300 r.p.m. The speed is to increase 0.6 per cent for a lift of 12 mm of the sleeve.*

(*a*) *Find the necessary stiffness of the spring.*

(*b*) *Design the bell crank lever. The permissible tensile stress for the material of the lever may be taken as 80 MPa and the allowable bearing pressure at the pins is 8 N/mm².*

Solution. Given : $x = 190$ mm ; $y = 140$ mm ; $m = 2.7$ kg ; $r_2 = 170$ mm $= 0.17$ m ; $N_2 = 300$ r.p.m. ; $h = 12$ mm ; $\sigma_t = 80$ MPa $= 80$ N/mm² ; $p_b = 8$ N/mm²

A Hartnell governor is shown in Fig. 15.17.

(a) Stiffness of the spring

Let s_1 = Stiffness of the spring.

We know that minimum angular speed of the ball arm (*i.e.* when the ball arm is vertical),

$$\omega_2 = \frac{2\pi N_2}{60} = \frac{2\pi \times 300}{60} = 31.42 \text{ rad/s}$$

Since the increase in speed is 0.6 per cent, therefore maximum angular speed of the ball arm,

$$\omega_1 = \omega_2 + \frac{0.6}{100} \times \omega_2 = 1.006\ \omega_2 = 1.006 \times 31.42 = 31.6 \text{ rad/s}$$

We know that radius of rotation at the maximum speed,

$$r_1 = r_2 + h \times \frac{x}{y} = 170 + 12 \times \frac{190}{140} = 186.3 \text{ mm} = 0.1863 \text{ m}$$

$$\quad ...\left[\because h = (r_1 - r_2)\frac{y}{x}\right]$$

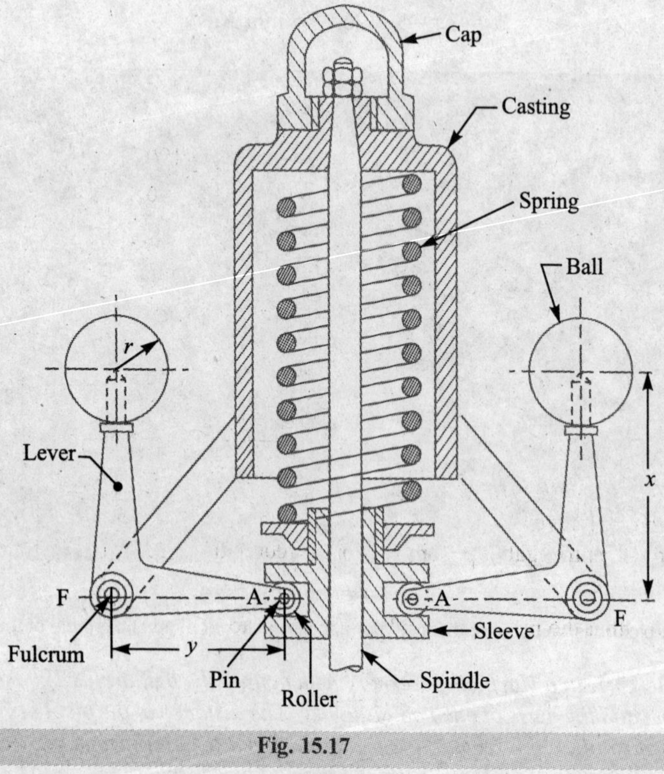

Fig. 15.17

The minimum and maximum position of the ball arm and sleeve arm is shown in Fig. 15.18 (*a*) and (*b*) respectively.

Let F_{C1} = Centrifugal force at the maximum speed = $m\,(\omega_1)^2\, r_1$,

F_{C2} = Centrifugal force at the minimum speed = $m\,(\omega_2)^2\, r_2$,

S_1 = Spring force at the maximum speed (ω_1), and

S_2 = Spring force at the minimum speed (ω_2).

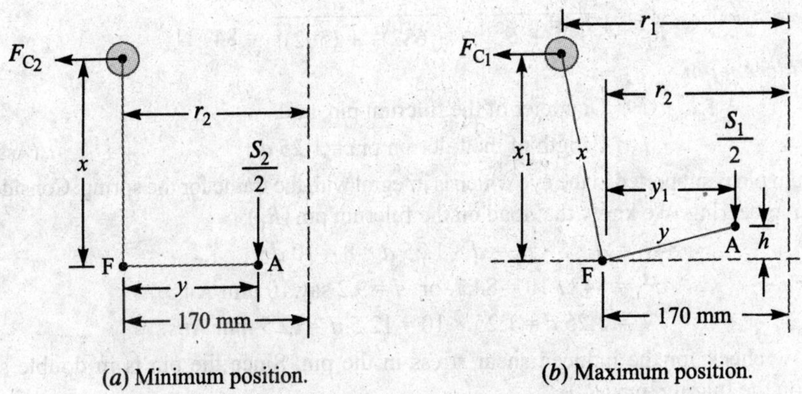

Fig. 15.18

Taking moments about the fulcrum F of the bell crank lever, neglecting the obliquity effect of the arms (*i.e.* taking $x_1 = x$ and $y_1 = y$) and the moment due to mass of the balls, we have for *maximum position,

$$S_1 = 2F_{C1} \times \frac{x}{y} = 2m\,(\omega_1)^2\, r_1 \times \frac{x}{y} \qquad \ldots\left(\because \frac{S_1}{2} \times y = F_{C1} \times x\right)$$

$$= 2 \times 2.7\,(31.6)^2\; 0.1863 \times \frac{190}{140} = 1364 \text{ N}$$

Similarly

$$S_2 = 2F_{C2} \times \frac{x}{y} = 2m\,(\omega_2)^2\, r_2 \times \frac{x}{y}$$

$$= 2 \times 2.7\,(31.42)^2\; 0.17 \times \frac{190}{140} = 1230 \text{ N}$$

We know that

$$S_1 - S_2 = h \times s_1$$

$$\therefore \quad s_1 = \frac{S_1 - S_2}{h} = \frac{1364 - 1230}{12} = 11.16 \text{ N/mm} \quad \textbf{Ans.}$$

(b) Design of bell crank lever

The bell crank lever is shown in Fig. 15.19. First of all, let us find the centrifugal force (or the effort P) required at the ball end to resist the load at A.

We know that the maximum load on the roller arm at A,

$$W = \frac{S_1}{2} = \frac{1364}{2} = 682 \text{ N}$$

Taking moments about F, we have

$$P \times x = W \times y$$

$$\therefore \quad P = \frac{W \times y}{x} = \frac{682 \times 140}{190}$$

$$= 502 \text{ N}$$

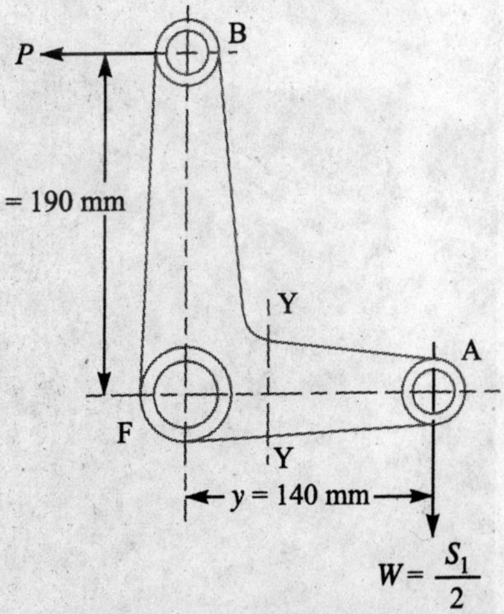

Fig. 15.19

* For further details, please refer chapter on 'Governors' of authors' popular book on **'Theory of Machines'**.

We know that reaction at the fulcrum F,

$$R_F = \sqrt{W^2 + P^2} = \sqrt{(682)^2 + (502)^2} = 847 \text{ N}$$

1. ***Design for fulcrum pin***

Let d = Diameter of the fulcrum pin, and

l = Length of the fulcrum pin = 1.25 d ...(Assume)

The fulcrum pin is supported in the eye which is integral with the frame for the spring. Considering the fulcrum pin in bearing. We know that load on the fulcrum pin (R_F),

$$847 = d \times l \times p_b = d \times 1.25\, d \times 8 = 10\, d^2$$

$$\therefore \quad d^2 = 847 / 10 = 84.7 \text{ or } d = 9.2 \text{ say } 10 \text{ mm } \textbf{Ans.}$$

and $$l = 1.25\, d = 1.25 \times 10 = 12.5\, d = 12.5 \text{ mm } \textbf{Ans.}$$

Let us now check for the induced shear stress in the pin. Since the pin is in double shear, therefore load on the fulcrum pin (R_F),

$$847 = 2 \times \frac{\pi}{4} \times d^2 \times \tau = 2 \times \frac{\pi}{4} (10)^2\, \tau = 157.1\, \tau$$

$$\therefore \quad \tau = 847 / 157.1 = 5.4 \text{ N/mm}^2 = 5.4 \text{ MPa}$$

This induced shear stress is very much within safe limits.

A brass bush of 3 mm thick may be pressed into the boss. Therefore diameter of hole in the lever or inner diameter of boss

$$= 10 + 2 \times 3 = 16 \text{ mm}$$

and outer diameter of boss

$$= 2\, d = 2 \times 10 = 20 \text{ mm}$$

2. ***Design for lever***

The cross-section of the lever is obtained by considering the lever in bending. It is assumed that the lever arm extends upto the centre of the fulcrum from the point of application of load. This assumption results in a slightly stronger lever. Considering the weakest section of failure at Y-Y (40 mm from the centre of the fulcrum).

Lapping is a surface finishing process for finishing gears, etc.

Note : This picture is given as additional information and is not a direct example of the current chapter.

∴ Maximum bending moment at *Y-Y*,

$$= 682\,(140 - 40) = 68\,200 \text{ N-mm}$$

Let t = Thickness of the lever, and

B = Depth or width of the lever.

∴ Section modulus,

$$Z = \frac{1}{6} \times t \times B^2 = \frac{1}{6} \times t\,(3t)^2 = 1.5\,t^3 \quad \text{...(Assuming } B = 3\,t\text{)}$$

We know that bending stress (σ_b),

$$80 = \frac{M}{Z} = \frac{68\,200}{1.5t^3} = \frac{45\,467}{t^3}$$

∴ $t^3 = 45\,467 / 80 = 568$ or $t = 8.28$ say 10 mm **Ans.**

and $B = 3\,t = 3 \times 10 = 30$ mm **Ans.**

3. ***Design for ball***

Let r = Radius of the ball.

The balls are made of cast iron, whose density is 7200 kg/m^3. We know that mass of the ball (m),

$$2.7 = \text{Volume} \times \text{density} = \frac{4}{3}\pi r^3 \times 7200 = 30\,163\,r^3$$

∴ $r^3 = 2.7 / 30\,163 = 0.089/10^3$

or $r = 0.0447$ m $= 44.7$ say 45 mm **Ans.**

The ball is screwed to the end of the lever. The screwed length of lever will be equal to the radius of ball.

∴ Maximum bending moment on the screwed end of the lever,

$$M = P \times r = 502 \times 45 = 22\,590 \text{ N-mm}$$

Let d_c = Core diameter of the screwed length of the lever.

∴ Section modulus,

$$Z = \frac{\pi}{32}\,(d_c)^3 = 0.0982\,(d_c)^3$$

We know that bending stress (σ_b),

$$80 = \frac{M}{Z} = \frac{22\,590}{0.0982\,(d_c)^3} = \frac{230 \times 10^3}{(d_c)^3}$$

∴ $(d_c)^3 = 230 \times 10^3 / 80 = 2876$ or $d_c = 14.2$ mm

We shall take nominal diameter of the screwed length of lever as 16 mm. **Ans.**

4. ***Design for roller end A***

Let d_1 = Diameter of the pin at *A*, and

l_1 = Length of the pin at $A = 1.25\,d_1$ ...(Assume)

We know that the maximum load on the roller at *A*,

$$W = S_1 / 2 = 1364 / 2 = 682 \text{ N}$$

Considering the pin in bearing. We know that load on the pin at *A* (W),

$$682 = d_1.l_1.p_b = d_1 \times 1.25\,d_1 \times 8 = 10\,(d_1)^2$$

∴ $(d_1)^2 = 682 / 10 = 68.2$ or $d_1 = 8.26$ say 10 mm **Ans.**

and $l_1 = 1.25\,d_1 = 1.25 \times 10 = 12.5$ mm **Ans.**

Let us now check the pin for induced shear stress. Since the pin is in double shear, therefore load on the pin at A (W),

$$682 = 2 \times \frac{\pi}{4} (d_1)^2 \tau = 2 \times \frac{\pi}{4} (10)^2 \tau = 157.1 \tau$$

$$\therefore \quad \tau = 682 / 157.1 = 4.35 \text{ N/mm}^2 = 4.35 \text{ MPa}$$

This induced stress is very much within safe limits.

The roller pin is fixed in the forked end of the bell crank lever and the roller moves freely on the pin. Let us now check the pin for induced bending stress. We know that maximum bending moment,

$$M = \frac{5}{24} \times W \times l_1 = \frac{5}{24} \times 682 \times 12.5 = 1776 \text{ N-mm}$$

and section modulus of the pin,

$$Z = \frac{\pi}{32} (d_1)^3 = \frac{\pi}{32} (10)^3 = 98.2 \text{ mm}^3$$

$\therefore$ Bending stress induced

$$= \frac{M}{Z} = \frac{1776}{98.2} = 18.1 \text{ N/mm}^2 = 18.1 \text{ MPa}$$

This induced bending stress is within safe limits.

We know that the thickness of each eye of the fork,

$$t_1 = \frac{l_1}{2} = \frac{12.5}{2} = 6.25 \text{ mm}$$

and outer diameter of the eye,

$$D = 2 \, d_1 = 2 \times 10 = 20 \text{ mm}$$

The outer diameter of the roller is taken slightly larger (at least 3 mm more) than the outer diameter of the eye. In the present case, 23 mm outer diameter of the roller will be sufficient. The roller is not provided with bush because after sufficient service, the roller has to be replaced due to wear on the profile. A clearance of 1.5 mm is provided between the roller and fork on either side of roller.

$\therefore$ Total length of the pin,

$$l_2 = l_1 + 2 \, t_1 + 2 \times 1.5 = 12.5 + 2 \times 6.25 + 3 = 28 \text{ mm} \textbf{ Ans.}$$

15.9 Rocker Arm for Exhaust Valve

A rocker arm for operating the exhaust valve is shown in Fig. 15.20. In designing a rocker arm, the following procedure may be followed :

1. The rocker arm is usually of I-section. Due to the load on the valve, it is subjected to bending moment. In order to find the bending moment, it is assumed that the arm of the lever extends from the point of application of the load to the centre of the pivot which acts as a fulcrum of the rocker arm. This assumption results in a slightly stronger lever near the boss.
2. The ratio of the length to the diameter of the fulcrum and roller pin is taken as 1.25. The permissible bearing pressure on this pin is taken from 3.5 to 6 N/mm^2.
3. The outside diameter of the boss at fulcrum is usually taken as twice the diameter of the pin at fulcrum. The boss is provided with a 3 mm thick phosphor bronze bush to take up wear.
4. One end of the rocker arm has a forked end to receive the roller. The roller is carried on a pin and is free to revolve in an eye to reduce wear. The pin or roller is not provided with a bush because after sufficient service the roller has to be discarded due to wear at the profile.
5. The outside diameter of the eye at the forked end is also taken as twice the diameter of pin. The diameter of the roller is taken slightly larger (at least 3 mm more) than the diameter of

eye at the forked end. The radial thickness of each eye of the forked end is taken as half the diameter of pin. Some clearance, about 1.5 mm, must be provided between the roller and eye at the forked end so that the roller can move freely. The pin should, therefore, be checked for bending.

6. The other end of the rocker arm (*i.e.* tappet end) is made circular to receive the tappet which is a stud with a lock nut. The outside diameter of the circular arm is taken as twice the diameter of the stud. The depth of the section is also taken equal to twice the diameter of the stud.

Example 15.8. *For operating the exhaust valve of a petrol engine, the maximum load required on the valve is 5000 N. The rocker arm oscillates around a pin whose centre line is 250 mm away from the valve axis. The two arms of the rocker are equal and make an included angle of 160°. Design the rocker arm with the fulcrum if the tensile stress is 70 MPa and the bearing pressure is 7 N/mm². Assume the cross-section of the rocker arm as rectangular.*

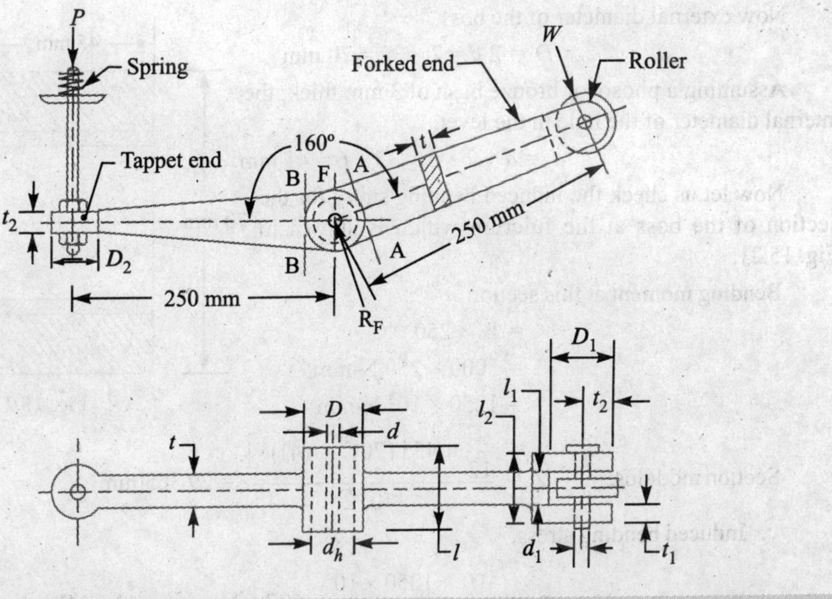

Fig. 15.20

Solution. Given : $W = 5000$ N ; $\theta = 160°$; $\sigma_t = 70$ MPa $= 70$ N/mm²; $p_b = 7$N/mm²

A rocker arm for operating the exhaust valve is shown in Fig. 15.20.

First of all, let us find out the reaction at the fulcrum pin.

Let R_F = Reaction at the fulcrum pin.

Since the two arms of the rocker are equal, therefore the load at the two ends of the arm are equal *i.e.* $W = P = 5000$ N.

We know that

$$R_F = \sqrt{W^2 + P^2 - 2W \times P \times \cos\theta}$$

$$= \sqrt{(5000)^2 + (5000)^2 - 2 \times 5000 \times 5000 \times \cos 160°}$$

$$= \sqrt{25 \times 10^6 + 25 \times 10^6 + 47 \times 10^6} = 9850 \text{ N}$$

Design of fulcrum

Let d = Diameter of the fulcrum pin, and

l = Length of the fulcrum pin = 1.25 d ...(Assume)

Considering the bearing of the fulcrum pin. We know that load on the fulcrum pin (R_F),

$$9850 = d \times l \times p_b = d \times 1.25\, d \times 7 = 8.75\, d^2$$

$\therefore$ $d^2 = 9850 / 8.75 = 1126$ or $d = 33.6$ say 35 mm **Ans.**

and $l = 1.25\, d = 1.25 \times 35 = 43.75$ say 45 mm **Ans.**

Now let us check the average shear stress induced in the pin. Since the pin is in double shear, therefore load on the fulcrum pin (R_F),

$$9850 = 2 \times \frac{\pi}{4} \times d^2 \times \tau = 2 \times \frac{\pi}{4} (35)^2\, \tau = 1924.5\, \tau$$

$\therefore$ $\tau = 9850 / 1924.5 = 5.12$ N/mm^2 = 5.12 MPa

The induced shear stess is quite safe.

Now external diameter of the boss,

$$D = 2\,d = 2 \times 35 = 70 \text{ mm}$$

Assuming a phosphor bronze bush of 3 mm thick, the internal diameter of the hole in the lever,

$$d_h = d + 2 \times 3 = 35 + 6 = 41 \text{ mm}$$

Now let us check the induced bending stress for the section of the boss at the fulcrum which is shown in Fig. 15.21.

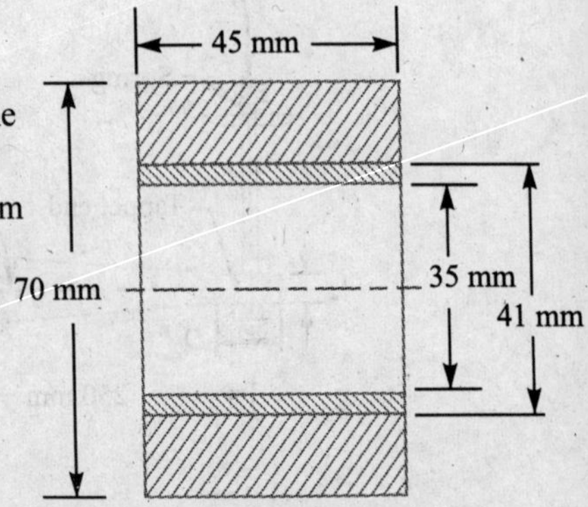

Fig. 15.21

Bending moment at this section

$= W \times 250$

$= 5000 \times 250$ N-mm

$= 1250 \times 10^3$ N-mm

Section modulus, $$Z = \frac{\frac{1}{12} \times 45 \left[(70)^3 - (41)^3\right]}{70/2} = 29\,365 \text{ mm}^3$$

$\therefore$ Induced bending stress,

$$\sigma_b = \frac{M}{Z} = \frac{1250 \times 10^3}{29\,365} = 42.6 \text{ N/mm}^2 = 42.6 \text{ MPa}$$

Since the induced bending stress is less than the permissible value of 70 MPa, therefore it is safe.

Design for forked end

Let d_1 = Diameter of the roller pin, and

l_1 = Length of the roller pin = 1.25 d_1 ...(Assume)

Considering bearing of the roller pin. We know that load on the roller pin (W),

$$5000 = d_1 \times l_1 \times p_b = d_1 \times 1.25\, d_1 \times 7 = 8.75\, (d_1)^2$$

$\therefore$ $(d_1)^2 = 5000 / 8.75 = 571.4$ or $d_1 = 24$ mm **Ans.**

and $l_1 = 1.25\, d_1 = 1.25 \times 24 = 30$ mm **Ans.**

Let us now check the roller pin for induced shearing stress. Since the pin is in double shear, therefore load on the roller pin (W),

$$5000 = 2 \times \frac{\pi}{4} (d_1)^2\, \tau = 2 \times \frac{\pi}{4} (24)^2\, \tau = 905\, \tau$$

$\therefore \quad \tau = 5000/905 = 5.5 \text{ N/mm}^2 = 5.5 \text{ MPa}$

This induced shear stress is quite safe.

The roller pin is fixed in eye and the thickness of each eye is taken as half the length of the roller pin.

∴ Thickness of each eye,

$$t_1 = \frac{l_1}{2} = \frac{30}{2} = 15 \text{ mm}$$

Let us now check the induced bending stress in the roller pin. The pin is neither simply supported in fork nor rigidly fixed at the end. Therefore the common practice is to assume the load distribution as shown in Fig. 15.22.

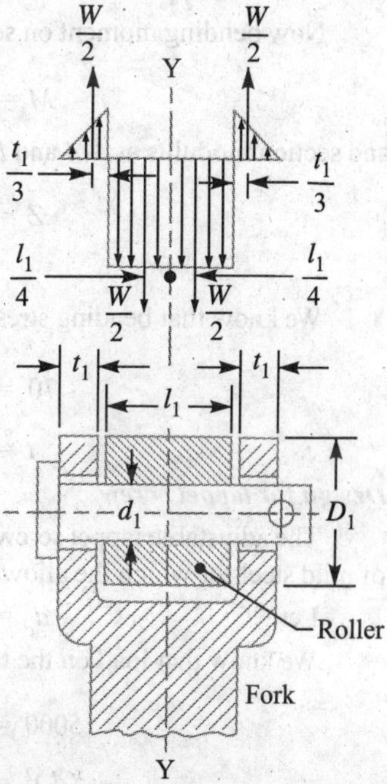

Fig. 15.22

The maximum bending moment will occur at Y-Y.

Neglecting the effect of clearance, we have

Maximum bending moment at Y–Y,

$$M = \frac{W}{2}\left(\frac{l_1}{2} + \frac{t_1}{3}\right) - \frac{W}{2} \times \frac{l_1}{4}$$

$$= \frac{W}{2}\left(\frac{l_1}{2} + \frac{l_1}{6}\right) - \frac{W}{2} \times \frac{l_1}{4}$$

$$\ldots(\because t_1 = l_1/2)$$

$$= \frac{5}{24} W \times l_1 = \frac{5}{24} \times 5000 \times 30 \text{ N-mm}$$

$$= 31\,250 \text{ N-mm}$$

and section modulus of the pin,

$$Z = \frac{\pi}{32}(d_1)^3 = \frac{\pi}{32}(24)^3 = 1357 \text{ mm}^3$$

∴ Bending stress induced in the pin

$$= \frac{M}{Z} = \frac{31\,250}{1357} = 23 \text{ N/mm}^2 = 23 \text{ MPa}$$

The bending stress induced in the pin is within permissible limit of 70 MPa.

Since the radial thickness of eye (t_2) is taken as $d_1/2$, therefore overall diameter of the eye,

$$D_1 = 2d_1 = 2 \times 24 = 48 \text{ mm}$$

The outer diameter of the roller is taken slightly larger (at least 3 mm more) than the outer diameter of the eye.

In the present case, 54 mm outer diameter of the roller will be sufficient.

Providing a clearance of 1.5 mm between the roller and the fork on either side of the roller, we have

$$l_2 = l_1 + 2 \times \frac{t_1}{2} + 2 \times 1.5 = 30 + 2 \times \frac{15}{2} + 3 = 48 \text{ mm}$$

Design of lever arm

The cross-section of the lever arm is obtained by considering the bending of the sections just near the boss of fulcrum on both sides, such as section A-A and B-B.

Let t = Thickness of the lever arm which is uniform throughout.

B = Width or depth of the lever arm which varies from boss diameter of fulcrum to outside diameter of the eye (for the forked end side) and from boss diameter of fulcrum to thickness t_2 (for the tappet or stud end side).

Now bending moment on sections *A-A* and *B-B*,

$$M = 5000\left(250 - \frac{D}{2}\right) = 5000\left(250 - \frac{70}{2}\right) = 1075 \times 10^3 \text{ N-mm}$$

and section modulus at *A-A* and *B-B*,

$$Z = \frac{1}{6} \times t \times B^2 = \frac{1}{6} \times t \times D^2 = \frac{1}{6} \times t\,(70)^2 = 817\,t \text{ mm}^3$$

...(At sections *A-A* and *B-B*, $B = D$)

We know that bending stress (σ_b),

$$70 = \frac{M}{Z} = \frac{1075 \times 10^3}{817\,t} = \frac{1316}{t}$$

$\therefore$ $t = 1316 / 70 = 18.8$ say 20 mm **Ans.**

Design for tappet screw

The adjustable tappet screw carries a compressive load of 5000 N. Assuming the screw is made of mild steel for which the allowable compressive stress (σ_c) may be taken as 50 N/mm².

Let d_c = Core diameter of the screw.

We know that load on the tappet screw (W),

$$5000 = \frac{\pi}{4}(d_c)^2 \sigma_c = \frac{\pi}{4}(d_c)^2\, 50 = 39.3\,(d_c)^2$$

$\therefore$ $(d_c)^2 = 5000 / 39.3 = 127$ or $d_c = 11.3$ mm

and outer or nominal diameter of the screw,

$$d = d_c / 0.84 = 11.3 / 0.84$$
$$= 13.5 \text{ say } 14 \text{ mm } \textbf{Ans.}$$

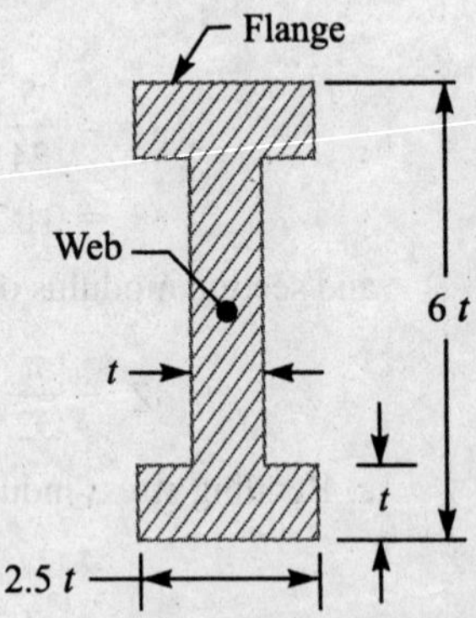

Fig. 15.23

We shall use 14 mm stud and it is provided with a lock nut. The diameter of the circular end of the lever arm (D_2) and its depth (t_2) is taken twice the diameter of stud.

$\therefore$ $D_2 = 2 \times 14 = 28$ mm

and $t_2 = 2 \times 14 = 28$ mm

If the lever arm is assumed to be of *I*-section with proportions as shown in Fig. 15.23 at *A-A* and *B-B*, then section modulus,

$$Z = \frac{\frac{1}{12}\left[2.5\,t\,(6t)^3 - 1.5\,t\,(4t)^3\right]}{6t/2} = \frac{37\,t^4}{3t} = 12.3\,t^3$$

We know that the maximum bending moment at *A-A* and *B-B*,

$$M = 5000\left(250 - \frac{70}{2}\right) = 1075 \times 10^3 \text{ N-mm}$$

$\therefore$ Bending stress (σ_b),

$$70 = \frac{M}{Z} = \frac{1075 \times 10^3}{12.3\,t^3} = \frac{87.4 \times 10^3}{t^3}$$

$\therefore$ $t^3 = 87.4 \times 10^3 / 70 = 1248$ or $t = 10.77$ say 12 mm

We have assumed that width of the flange

$= 2.5\,t = 2.5 \times 12 = 30$ mm **Ans.**

Depth of the web $= 4\,t = 4 \times 12 = 48$ mm **Ans.**

and depth of the section $= 6\,t = 6 \times 12 = 72$ mm **Ans.**

Normally thickness of the flange and web is constant throughout whereas the width and the depth is tapered.

15.10 Miscellaneous Levers

In the previous articles, we have discussed the design of various types of levers used in engineering practice. Some more types of levers designed on the same principle are discussed in the following examples.

Example 15.9. *A pressure vessel as shown in Fig. 15.24, is used as a digester in a chemical process. It is designed to withstand a pressure of 0.2 N/mm² by gauge. The diameter of the pressure vessel is 600 mm. The vessel and its cover are made of cast iron. All other parts are made of steel. The cover is held tightly against the vessel by a screw B which is turned down through the tapped hole in the beam A, so that the end of the screw presses against the cover. The beam A is of rectangular section in which* $b_1 = 2\,t_1$.

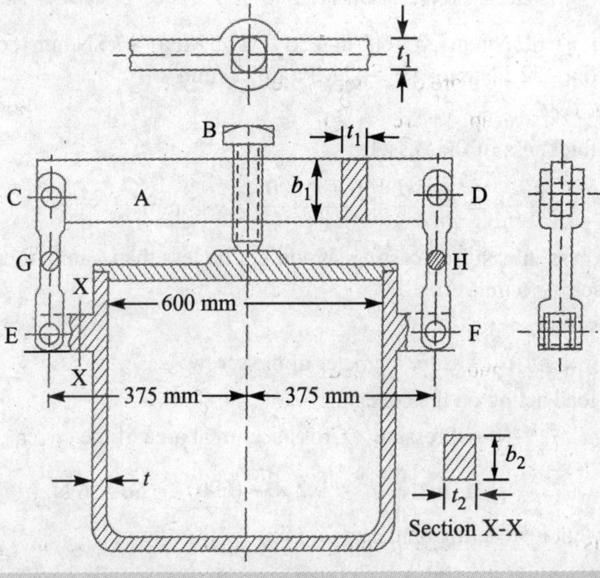

Fig. 15.24

The rectangular section is opened up at the centre to take the tapped hole as shown in the figure. The beam is attached by pins C and D to the links G and H which are secured by pins E and F to the extensions cast on the vessel.

Assume allowable stresses as under :

Material	*Tension*	*Compression*	*Shear*
Cast iron	*17.5 MPa*	—	—
Steel	*52.5 MPa*	*52.5 MPa*	*42 MPa*

Find: 1. Thickness of the vessel; 2. Diameter of the screw; 3. Cross-section of beam A; 3. Diameter of pins C and D; 5. Diameter of pins E and F; 6. Diameter of pins G and H; and 7. Cross-section of the supports of pins E and F.

This massive crane is used for construction work.

Note : This picture is given as additional information and is not a direct example of the current chapter.

Solution. Given : $p = 0.2$ N/mm^2 ; $d = 600$ mm ; $\sigma_{tc} = 17.5$ MPa $= 17.5$ N/mm^2 ; $\sigma_{ts} = 52.5$ MPa $= 52.5$ N/mm^2 ; $\sigma_{cs} = 52.5$ MPa $= 52.5$ N/mm^2 ; $\tau_s = 42$ MPa $= 42$ N/mm^2

1. *Thickness of the vessel*

We know that thickness of the vessel,

$$t = \frac{p \times d}{2\,\sigma_{tc}} = \frac{0.2 \times 600}{2 \times 17.5} = 3.43 \text{ mm}$$

Since the thickness of cast iron casting should not be less than 6 mm, therefore we shall take thickness of the vessel, $t = 6$ mm. **Ans.**

2. *Diameter of the screw*

Let d_c = Core diameter of the screw.

We know that load acting on the cover,

$$W = \text{Pressure} \times \text{Cross-sectional area of the cover}$$

$$= p \times \frac{\pi}{4} d^2 = 0.2 \times \frac{\pi}{4} (600)^2 = 56\,556 \text{ N} \quad ...(i)$$

We also know that load acting on the cover (W),

$$56\,556 = \frac{\pi}{4} (d_c)^2 \sigma_{ts} = \frac{\pi}{4} (d_c)^2\, 52.5 = 41.24\,(d_c)^2 \quad ...(ii)$$

From equations (*i*) and (*ii*), we have

$$\therefore \quad (d_c)^2 = 56\,556 / 41.24 = 1372 \text{ or } d_c = 37 \text{ mm}$$

We shall use a standard screw of size M 48 with core diameter 41.5 mm and outer diameter 48 mm. **Ans.**

3. *Cross-section of the beam A*

Let t_1 = Thickness of the beam, and

b_1 = Width of the beam = $2\,t_1$ ...(Given)

Since it is a simply supported beam supported at C and D and the load W acts in the centre, therefore the reactions at C and D (R_C and R_D) will be $W/2$.

$$\therefore \quad R_C = R_D = \frac{W}{2} = \frac{56\,556}{2} = 28\,278 \text{ N}$$

Maximum bending moment at the centre of beam,

$$M = \frac{W}{2} \times \frac{l}{2} = \frac{56\ 556}{2} \times \frac{750}{2} = 10.6 \times 10^6 \text{ N-mm}$$

and section modulus of the beam,

$$Z = \frac{1}{6} \times t_1\ (b_1)^2 = \frac{1}{6} \times t_1\ (2t_1)^2 = \frac{2}{3}\ (t_1)^3 \qquad ...(\because b_1 = 2\ t_1)$$

We know that bending stress (σ_b),

$$52.5 = \frac{M}{Z} = \frac{10.6 \times 10^6 \times 3}{2\ (t_1)^3} = \frac{15.9 \times 10^6}{(t_1)^3} \qquad ...(\text{Substituting } \sigma_b = \sigma_{ts})$$

$\therefore$ $(t_1)^3 = 15.9 \times 10^6 / 52.5 = 303 \times 10^3$ or $t_1 = 67.5$ mm **Ans.**

and $b = 2\ t = 2 \times 67.5 = 135$ mm **Ans.**

4. Diameter of pins C and D

Let d_1 = Diameter of pins *C* and *D*.

The load acting on the pins *C* and *D* are reactions at *C* and *D* due to the load acting on the beam. Since the pins at *C* and *D* are in double shear, therefore load acting on the pins (R_C or R_D).

$$28\ 278 = 2 \times \frac{\pi}{4}\ (d_1)^2\ \tau_s = 2 \times \frac{\pi}{4}\ (d_1)^2\ 42 = 66\ (d_1)^2$$

$\therefore$ $(d_1)^2 = 28\ 278 / 66 = 428.5$ or $d_1 = 20.7$ say 21 mm **Ans.**

5. Diameter of pins E and F

Since the load on pins *E* and *F* is same as that of *C* and *D*, therefore diameter of pins *E* and *F* will be of same diameter *i.e.* 21 mm. **Ans.**

6. Diameter of links G and H

Let d_2 = Diameter of links *G* and *H*.

A little consideration will show that the links are in tension and the load acting on each link

$$= \frac{W}{2} = \frac{56\ 556}{2} = 28\ 278 \text{ N}$$

We also know that load acting on each link,

$$28\ 278 = \frac{\pi}{4}\ (d_2)^2\ \sigma_{ts} = \frac{\pi}{4}\ (d_2)^2\ 52.5 = 41(d_2)^2$$

$\therefore$ $(d_2)^2 = 28\ 278 / 41 = 689.7$ or $d_2 = 26.3$ mm **Ans.**

7. Cross-section of the supports of pins E and F

Let t_2 = Thickness of the support, and

b_2 = Width of the support.

The supports are a part of casting with a vessel and acts as a cantilever, therefore maximum bending moment at the support,

$$M = R_C \times x = R_C\ [375 - (300 + t)]$$
$$= 28\ 278\ [375 - (300 + 6)] = 1.95 \times 10^6 \text{ N-mm}$$

and section modulus,

$$Z = \frac{1}{6} \times t_2\ (b_2)^2 = \frac{1}{6} \times t_2\ (2t_2)^2 = \frac{2}{3}\ (t_2)^3 \qquad ...(\text{Assuming } b_2 = 2t_2)$$

We know that bending stress (σ_b),

$$17.5 = \frac{M}{Z} = \frac{1.95 \times 10^6 \times 3}{2\ (t_2)^3} = \frac{2.9 \times 10^6}{(t_2)^3} \qquad ...(\text{Substituting } \sigma_b = \sigma_{tc})$$

$\therefore$ $(t_2)^3 = 2.9 \times 10^6 / 17.5 = 165.7 \times 10^3$ or $t_2 = 55$ mm **Ans.**

and $b_2 = 2\ t_2 = 2 \times 55 = 110$ mm **Ans.**

Example 15.10. *A cross-lever to operate a double cylinder double acting pump is shown in Fig. 15.25. Find*

1. *Dimension of pins at L, M, N and Q,*
2. *Cross-section for the vertical arm of the lever, and*
3. *Cross-section for the horizontal arm of the lever.*

The permissible shear stress for the material of the pin is 40 MPa. The bearing pressure on the pins should not exeed 17.5 N/mm².

The permissible bending stress for the material of the lever should not exceed 70 MPa.

Solution. Given : $W_L = 3$ kN ; $W_N = 5$ kN ; $\tau = 40$ MPa $= 40$ N/mm² ; $p_b = 17.5$ N/mm² ; $\sigma_b = 70$ MPa $= 70$ N/mm²

First of all, let us find the effort P applied at Q. Taking moments about the fulcrum M, we have

$$P \times 800 = 5 \times 300 + 3 \times 300 = 2400 \quad \text{or} \quad P = 2400 / 800 = 3 \text{ kN}$$

When both sides of pump operate, then load on the fulcrum pin M,

$$W_M = 5 - 3 = 2 \text{ kN}$$

∴ Resultant force on the fulcrum pin M,

$$R_M = \sqrt{(W_M)^2 + P^2} = \sqrt{2^2 + 3^2} = 3.6 \text{ kN}$$

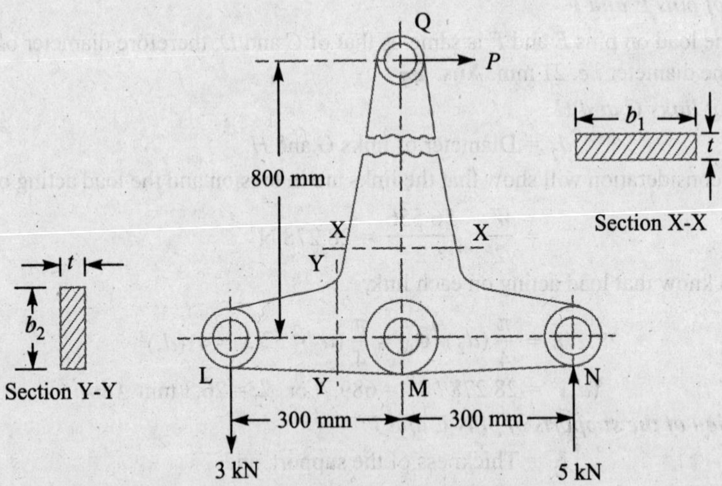

Fig. 15.25

The worst condition arises when one side of the pump does not work. At that time, the effort required increases. Taking moments about M, we get

$$P \times 800 = 5 \times 300 = 1500 \quad \text{or} \quad P = 1500 / 800 = 1.875 \text{ kN}$$

∴ In worst condition, the resultant force on the fulcrum pin M,

$$R_{M1} = \sqrt{(1.875)^2 + 5^2} = 5.34 \text{ kN} = 5340 \text{ N}$$

Therefore the fulcrum pin M will be designed for a maximum load of 5.34 kN.

A little consideration will show that the load on the pins L and Q is 3 kN each; therefore the pins L and Q will be of the same size. Since the load on pin N (5 kN) do not differ much with the maximum load on pin M *i.e.* 5.34 kN, therefore the pins at N and M may be taken of the same size.

1. ***Dimension of pins at L, M, N and Q***

First of all, let us find the diameter of pins at *M* and *N*. These pins will be designed for a maximum load of 5.34 kN or 5340 N.

Let d = Diameter of pins at *M* and *N*, and

l = Length of pins at *M* and *N* = 1.25 d ...(Assume)

Considering the bearing of the pins. We know that load on the pins,

$$5340 = d \times l \times p_b = d \times 1.25\, d \times 17.5 = 21.87\, d^2$$

$\therefore$ $d^2 = 5340 / 21.87 = 244$ or $d = 15.6$ say 16 mm **Ans.**

and $l = 1.25\, d = 1.25 \times 16 = 20$ mm **Ans.**

Let us check the pin for induced shear stress. Since the pin is in double shear, therefore load on the pin,

$$5340 = 2 \times \frac{\pi}{4}\, d^2 \times \tau = 2 \times \frac{\pi}{4}\,(16)^2\, \tau = 402\, \tau$$

$\therefore$ $\tau = 5340 / 402 = 13.3 \text{ N/mm}^2 = 13.3$ MPa

The induced shear stress is within safe limits.

A 3 mm thick bush may be inserted so that the diameter of hole in the lever is 22 mm. The outside diameter of the boss may be taken as twice the diameter of hole.

$\therefore$ Outside diameter of boss = 2 × 22 = 44 mm

Let us now check the section of the lever for induced bending stress. The section at the fulcrum is shown in Fig. 15.26.

We know the maximum bending moment,

$$M = \frac{5}{24} \times W \times l = \frac{5}{24} \times 5340 \times 20 = 22\,250 \text{ N}$$

and section modulus,

$$Z = \frac{\frac{1}{12} \times 20\left[(44)^3 - (22)^3\right]}{44/2} = 5647 \text{ mm}^3$$

$\therefore$ Bending stress induced,

$$\sigma_b = \frac{M}{Z} = \frac{22\,250}{5647} = 3.94 \text{ N/mm}^2 = 3.94 \text{ MPa}$$

The bending stress induced is very much within safe limits.

It should be remembered that the direction of the load will be reversed, consequently the loads will be changed. Hence the pins at *L* and *N* must be identical. We shall provide the pin at *Q* of the same size as for *L*, *M* and *N* in order to avoid extra storage. Thus the diameter of pins at *L*, *M*, *N* and *Q* is 16 mm and length 20 mm. **Ans.**

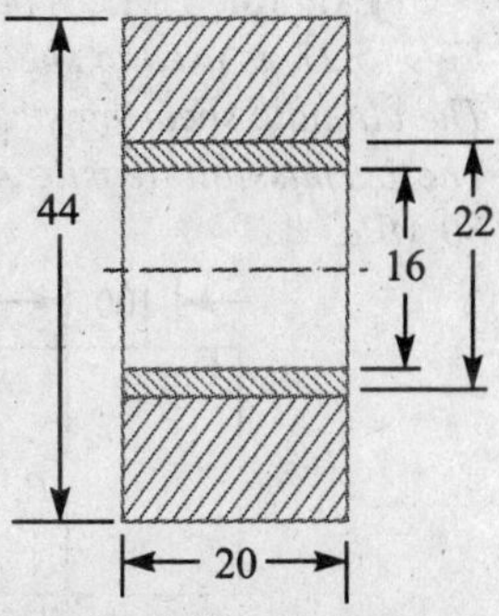

All dimensions in mm.

Fig. 15.26

2. ***Cross-section for the vertical arm of the lever***

Considering the cross-section of the vertical arm at *X-X*,

Let t = Thickness of the arm, and

b_1 = Width of the arm = 3 t ... (Assume)

It is assumed that the length of the arm extends upto the centre of the fulcrum. This assumption results in a slightly stronger arm.

$\therefore$ Maximum bending moment,

$$M = P \times 800 = 3 \times 800 = 2400 \text{ kN-mm} = 2.4 \times 10^6 \text{ N-mm}$$

and section modulus,

$$Z = \frac{1}{6} \times t\,(b_1)^2 = \frac{1}{6} \times t\,(3t)^2 = 1.5\,t^3 \text{ mm}^3$$

We know that bending stress (σ_b),

$$70 = \frac{M}{Z} = \frac{2.4 \times 10^6}{1.5\,t^3} = \frac{1.6 \times 10^6}{t^3}$$

$$\therefore \quad t^3 = 1.6 \times 10^6 / 70 = 23 \times 10^3 \quad \text{or} \quad t = 28.4 \text{ say } 30 \text{ mm } \textbf{Ans.}$$

and $\quad b_1 = 3\,t = 3 \times 30 = 90$ mm **Ans.**

3. ***Cross-section of horizontal arm of the lever***

Considering the cross-section of the arm at *Y-Y*.

Let t = Thickness of the arm. The thickness of the horizontal arm will be same as that of vertical arm.

b_2 = Width of the arm.

Again, assuming that the length of arm extends upto the centre of the fulcrum, therefore maximum bending moment,

$$M = W_N \times 300 = 5 \times 300$$
$$= 1500 \text{ kN-mm} = 1.5 \times 10^6 \text{ N-mm}$$

and section modulus,

$$Z = \frac{1}{6} \times t\,(b_2)^2 = \frac{1}{6} \times 30\,(b_2)^2 = 5\,(b_2)^2$$

We know that bending stress (σ_b),

$$70 = \frac{M}{Z} = \frac{1.5 \times 10^6}{5\,(b_2)^2} = \frac{0.3 \times 10^6}{(b_2)^2}$$

$$\therefore \quad (b_2)^2 = 0.3 \times 10^6 / 70 = 43 \times 10^2$$

or $\quad b_2 = 65.5$ say 66 mm **Ans.**

Lasers supported by computer controls can cut the metal very accurately

Note : This picture is given as additional information and is not a direct example of the current chapter.

Example 15.11. *A bench shearing machine as shown in Fig. 15.27, is used to shear mild steel bars of 5 mm × 3 mm. The ultimate shearing strength of the mild steel is 400 MPa. The permissible tensile stress for pins, links and lever is 80 MPa.*

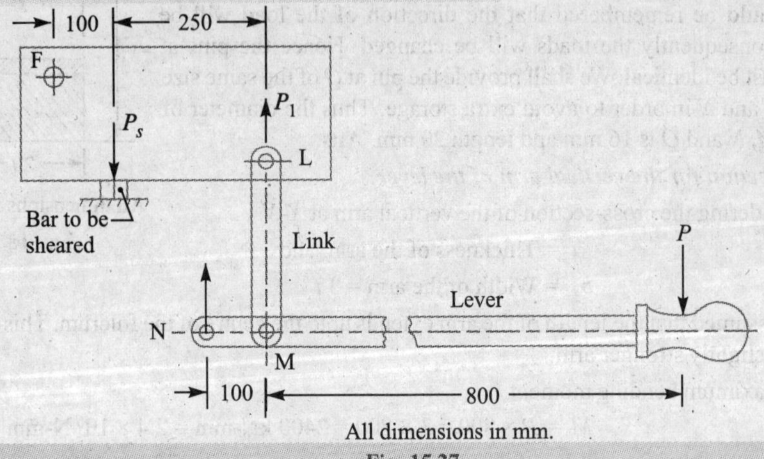

All dimensions in mm.

Fig. 15.27

The allowable bearing pressure on pins may be taken as 20 N/mm². Design the pins at L, M and N; the link and the lever.

Solution. Given : $A_s = 5 \times 3 = 15$ mm² ; $\tau_u = 400$ MPa $= 400$ N/mm² ; $\sigma_t = 80$ MPa $= 80$ N/mm² ; $p_b = 20$ N/mm²

We know that maximum shearing force required,

$$P_s = \text{Area sheared} \times \text{Ultimate shearing strength}$$

$$= A_s \times \tau_u = 15 \times 400 = 6000 \text{ N}$$

Let P_1 = Force in the link *LM*.

Taking moments about *F*, we have

$$P_1 \times 350 = P_s \times 100 = 6000 \times 100 = 600 \times 10^3$$

$$\therefore \quad P_1 = 600 \times 10^3/350 = 1715 \text{ N}$$

Again taking moments about *N* to find the force *P* required to operate the handle, we have

$$P \times 900 = P_1 \times 100 = 1715 \times 100 = 171\,500$$

$$\therefore \quad P = 171\,500/900 = 191 \text{ N}$$

and force in the pin at $N = P_1 + P = 1715 + 191 = 1906$ N

Design of pins at L, M and N

We see that the force in the pins at *L* and *M* is equal to 1715 N and force in the pin at *N* is 1906 N. Since the forces in pins at *L*, *M* and *N* do not differ very much, therefore the same size of pins may be used. These pins will be designed for a maximum load of 1906 N.

Let d = Diameter of the pins at *L*, *M* and *N*, and

l = Length of the pins = 1.25 d ...(Assume)

Considering the pins in bearing. We know that load on the pins,

$$1906 = d \times l \times p_b = d \times 1.25\, d \times 20 = 25\, d^2$$

$$\therefore \quad d^2 = 1906 / 25 = 76.2 \text{ or } d = 8.73 \text{ say } 10 \text{ mm } \textbf{Ans.}$$

and $l = 1.25\, d = 1.25 \times 10 = 12.5$ mm **Ans.**

Let us check the pins for induced shear stress. Since the pins are in double shear, therefore load on the pins,

$$1906 = 2 \times \frac{\pi}{4}\, d^2 \times \tau = 2 \times \frac{\pi}{4}\, (10)^2\, \tau = 157.1\, \tau$$

$$\therefore \quad \tau = 1906 / 157.1 = 12.1 \text{ N/mm}^2 = 12.1 \text{ MPa}$$

The induced shear stress is within safe limits.

A 3 mm thick bush in inserted in the hole. Therefore, diameter of the hole in the link and lever

$$= d + 2 \times 3 = 10 + 6 = 16 \text{ mm}$$

The diameter of the boss may be taken as twice the diameter of hole.

$\therefore$ Diameter of the boss

$$= 2 \times 16 = 32 \text{ mm}$$

Let us now check the induced bending stress for the cross-section of the lever at *N*. The cross-section at *N* is shown in Fig. 15.28.

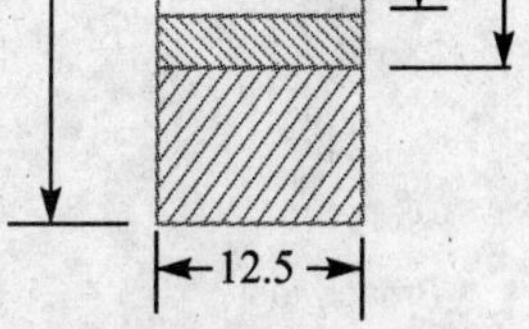

All dimensions in mm.

Fig. 15.28

We know that maximum bending moment,

$$M = \frac{5}{24} \times W \times l$$

$$= \frac{5}{24} \times 1906 \times 12.5 = 4964 \text{ N-mm}$$

and section modulus,

$$Z = \frac{\frac{1}{12} \times 12.5 \left[(32)^3 - (16)^3 \right]}{32/2} = 1867 \text{ mm}^3$$

∴ Induced bending stress

$$= \frac{M}{Z} = \frac{4964}{1867} = 2.66 \text{ N/mm}^2 = 2.66 \text{ MPa}$$

The induced bending stress is very much within safe limits.

Design for link

The link is of circular cross-section with ends forked.

Let d_1 = Diameter of the link.

The link is designed for a maximum load of 1906 N. Since the link is under tension, therefore load on the link,

$$1906 = \frac{\pi}{4} (d_1)^2 \sigma_t = \frac{\pi}{4} (d_1)^2 \, 80 = 62.84 \, (d_1)^2$$

$$\therefore \quad (d_1)^2 = 1906 / 62.84 = 30.3 \quad \text{or} \quad d_1 = 5.5 \text{ mm}$$

We shall provide the diameter of links (d_1) as 10 mm because forks are to be made at each end. **Ans.**

Design for lever

Assuming the lever to be rectangular.

Let t = Thickness of the lever. The thickness of the lever will be same as that of length of pin *i.e.* 12.5 mm.

B = Width of the lever.

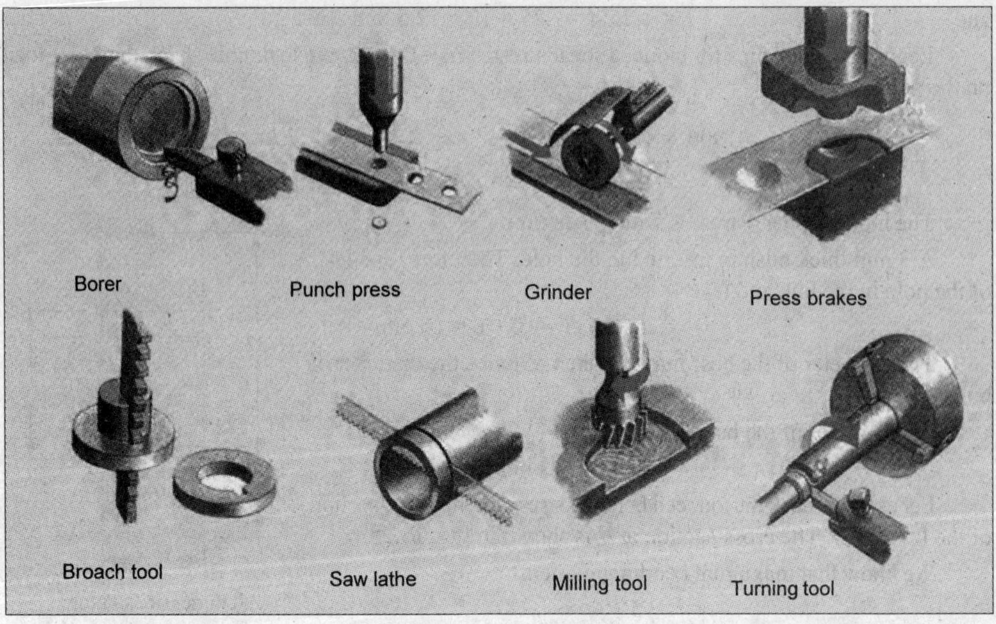

Common machine tools.

Note : This picture is given as additional information and is not a direct example of the current chapter.

We know that maximum bending moment on the lever,

$$M = 1906 \times 100 = 190\,600 \text{ N-mm}$$

Section modulus, $\quad Z = \frac{1}{6} \times t\, B^2 = \frac{1}{6} \times 12.5\, B^2 = 2.1\, B^2$

We know that bending stress (σ_b),

$$80 = \frac{M}{Z} = \frac{190\,600}{2.1\, B^2} = \frac{90\,762}{B^2}$$

$\therefore \quad B^2 = 90\,762 / 80 = 1134$ or $B = 33.7$ say 34 mm **Ans.**

The handle at the end of the lever is made 125 mm long with maximum diameter 32 mm and minimum diameter 25 mm. **Ans.**

EXERCISES

1. The spindle of a large valve is turned by a handle as shown in Fig. 15.5. The length of the handle from the centre of the spindle is 500 mm. The handle is attached to the spindle by means of a round tapered pin. If the spindle diameter is 60 mm and an effort of 300 N is applied at the end of the handle, find the dimensions for the tapered pin and the handle. The grip length of the handle may be taken as 200 mm. The allowable stresses for the handle and key are 100 MPa and 55 MPa in shear.

   [Ans. d_1 = 12 mm ; D = 25 mm]

2. A vertical lever *PQR* of length 1 m is attached by a fulcrum pin at *R* and to a horizontal rod at *Q*. An operating force of 700 N is applied horizontally at *P*. The distance of the horizontal rod *Q* from the fulcrum pin *R* is 140 mm. If the permissible stresses are 52.5 MPa in tension and compression and 32 MPa in shear; find the diameter of the pins, tie rod at *Q* and thickness of the lever. The bearing pressure on the pins may be taken as 22 N/mm^2.

   [Ans. 5.4 mm ; 14 mm ; 13 mm ; 11 mm ; 16.5 mm]

3. A hand lever for a brake is 0.8 m long from the centre of gravity of the spindle to the point of application of the pull of 300 N. The effective overhang from the nearest bearing is 100 mm. If the permissible stress in tension, shear and crushing is not to exceed 66 MPa, design the spindle, key and lever. Assume the arm of the lever to be rectangular having width twice of its thickness.

   [Ans. d = 45 mm ; l_1 = 45 mm ; t = 13 mm]

4. Design a foot brake lever from the following data :

   Length of lever from the centre of gravity of the spindle to the point of application of load = 1 metre
   Maximum load on the foot plate = 800 N
   Overhang from the nearest bearing = 100 mm
   Permissible tensile and shear stress = 70 MPa **[Ans. d = 40 mm ; l_1 = 40 mm ; t = 20 mm]**

5. Design a cranked lever for the following dimensions :

   Length of the handle = 320 mm
   Length of the lever arm = 450 mm
   Overhang of the journal = 120 mm

   The lever is operated by a single person exerting a maximum force of 400 N at a distance of 1/3rd length of the handle from its free end. The permissible stresses may be taken as 50 MPa for lever material and 40 MPa for shaft material.

   [Ans. d = 42 mm ; t = 20 mm ; B = 40 mm ; D = 32 mm]

6. A lever safety valve is 75 mm in diameter. It is required to blow off at 1.3 N/mm^2. Design the mild steel lever of rectangular cross-section if the permissible stresses are 70 MPa in tension, 52.5 MPa in shear and 24.5 MPa in bearing. The pin is made of the same material as that of the lever. The distance from the fulcrum to the dead weight of the lever is 800 mm and the distance between the fulcrum pin and the valve spindle link pin is 80 mm. **[Ans. t = 7.25 mm ; b = 21.75 mm]**

7. The line sketch of a lever of loaded safety valve is shown in Fig. 15.29. The maximum force at which the valve blows is 4000 N. The weight at the end of the lever is 300 N and the distance between the fulcrum point and the line of action of valve force is 'a'. The following permissible values may be used :

   For lever : $\sigma_t = \sigma_c = 40$ MPa and $\tau = 25$ MPa

   For pins : $\sigma_t = \sigma_c = 60$ MPa and $\tau = 35$ MPa

   Find the distance 'a'. Design the lever and make a neat sketch of the lever. Take lever cross-section as ($t \times 3\,t$). The permissible bearing stress is 20 MPa. **[Ans. 73.5 mm]**

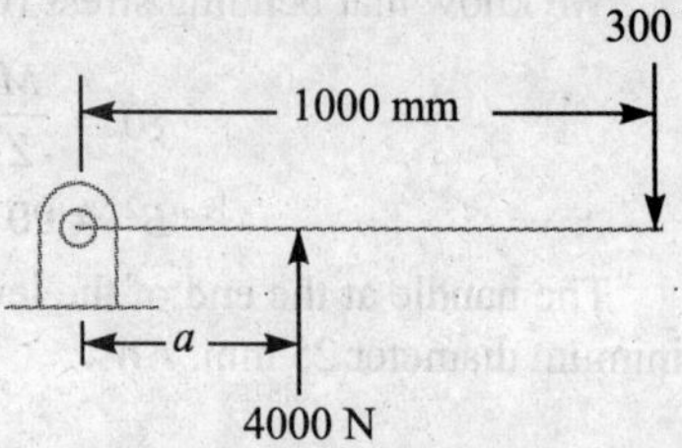

Fig. 15.29

8. Design a right angled bell crank lever having one arm 500 mm and the other 150 mm long. The load of 5 kN is to be raised acting on a pin at the end of 500 mm arm and the effort is applied at the end of 150 mm arm. The lever consists of a steel forgings, turning on a point at the fulcrum. The permissible stresses for the pin and lever are 84 MPa in tension and compression and 70 MPa in shear. The bearing pressure on the pin is not to exceed 10 N/mm^2. **[Ans. t = 27 mm ; b = 81 mm]**

9. Design a bell crank lever to apply a load of 5 kN (vertical) at the end A of an horizontal arm of length 400 mm. The end of the vertical arm C and the fulcrum B are to be fixed with the help of pins inside forked shaped supports. The end A is itself forked. Determine the cross-section of the arms and the dimensions of the pins. The lever is to have mechanical advantage of 4 with a shorter vertical arm BC. The ultimate stresses in shear and tension for the lever and pins are 400 MPa and 500 MPa respectively. The allowable bearing pressure for the pins is 12 N/mm^2. Make a sketch of the lever to scale and give all the dimensions. **[Ans. t = 22 mm ; b = 66 mm]**

   [**Hint.** Assume a factor of safety as 4 and the cross-section of the lever as rectangular with depth (b) as three times the thickness (t).]

10. A Hartnell type governor as shown in Fig. 15.17 has the ball arm length 120 mm and sleeve arm length 90 mm. The maximum and minimum distances of the balls from the axis of governor are 150 mm and 75 mm. The mass of each ball is 2.2 kg. The speed of the governor fluctuates between 310 r.p.m. and 290 r.p.m. Design the cast iron ball and mild steel lever. The permissible tensile stress for the lever material may be taken as 100 MPa. The bearing pressure for the roller and pin should not exceed 7 N/mm^2. **[Ans. r = 42 mm ; t = 12 mm ; b = 36 mm]**

11. The pivots of the bell crank levers of a spring loaded governor of Hartnell type are fixed at 100 mm radius from the spindle axis. The length of the ball arm of each lever is 150 mm, the length of the sleeve arm is 75 mm and the two arms are at right angles. The mass of each ball is 2 kg. The equilibrium speed in the lowest position of the governor is 300 r.p.m. when the radius of rotation of the ball path is 82 mm. The speed is to be limited to 6% more than the lowest equilibrium speed. The lift of the sleeve, for the operating speed range, is 15 mm. Design and draw a bell crank lever for the governor. **[Ans. t = 7 mm ; b = 21 mm]**

12. The maximum load at the roller end of a rocker arm is 2000 N. The distance between the centre of boss and the load line is 200 mm. Suggest suitable I-section of the rocker arm, if the permissible normal stress is limited to 70 MPa. **[Ans. t = 10 mm]**

   [**Hint.** The dimensions of I-section may be taken as follows :

   Top and bottom flanges = $2.5\,t \times t$ and Web = $4\,t \times t$]

QUESTIONS

1. What is a lever ? Explain the principle on which it works.
2. What do you understand by leverage ?
3. Why are levers usually tapered ?
4. (*a*) Why are bushes of softer material inserted in the eyes of levers ?
   (*b*) Why is a boss generally needed at the fulcrum of the levers.
5. State the application of hand and foot levers. Discuss the procedure for designing a hand or foot lever.
6. A lever is to be designed for a hoisting winch. Write the procedure for designing a lever for such operation.
7. Explain the design procedure of a lever for a lever safety valve.
8. Discuss the design procedure of a rocker arm for operating the exhaust valve.

OBJECTIVE TYPE QUESTIONS

1. In levers, the leverage is the ratio of
   (*a*) load lifted to the effort applied
   (*b*) mechanical advantage to the velocity ratio
   (*c*) load arm to the effort arm
   (*d*) effort arm to the load arm
2. In the levers of first type, the mechanical advantage is................ one.
   (*a*) less than
   (*b*) equal to
   (*c*) more than
3. The bell crank levers used in railway signalling arrangement are of
   (*a*) first type of levers
   (*b*) second type of levers
   (*c*) third type of levers
4. The rocker arm in internal combustion engines are of .... type of levers.
   (*a*) first
   (*b*) second
   (*c*) third
5. The cross-section of the arm of a bell crank lever is
   (*a*) rectangular
   (*b*) elliptical
   (*c*) I-section
   (*d*) any one of these
6. All the types of levers are subjected to
   (*a*) twisting moment
   (*b*) bending moment
   (*c*) direct axial load
   (*d*) combined twisting and bending moment
7. The method of manufacturing usually adopted for levers is
   (*a*) casting
   (*b*) fabrication
   (*c*) forging
   (*d*) machining
8. An *I*-section is more suitable for a
   (*a*) rocker arm
   (*b*) cranked lever
   (*c*) foot lever
   (*d*) lever of lever safety valve
9. The design of the pin of a rocker arm of an I.C. Engine is based on
   (*a*) tensile, creep and bearing failure
   (*b*) creep, bearing and shearing failure
   (*c*) bearing, shearing and bending failure
   (*d*) none of these
10. In designing a rocker arm for operating the exhaust valve, the ratio of the length to the diameter of the fulcrum and roller pin is taken as
    (*a*) 1.25 (*b*) 1.5 (*c*) 1.75 (*d*) 2

ANSWERS

1. (*d*)	2. (*c*)	3. (*c*)	4. (*a*)	5. (*d*)
6. (*b*)	7. (*c*)	8. (*a*)	9. (*c*)	10. (*a*)

CHAPTER

16

Columns and Struts

16.1 Introduction

A machine part subjected to an axial compressive force is called a ***strut.*** A strut may be horizontal, inclined or even vertical. But a vertical strut is known as a ***column, pillar*** or ***stanchion.*** The machine members that must be investigated for column action are piston rods, valve push rods, connecting rods, screw jack, side links of toggle jack etc. In this chapter, we shall discuss the design of piston rods, valve push rods and connecting rods.

Note: The design of screw jack and toggle jack is discussed in the next chapter on 'Power screws'.

16.2 Failure of a Column or Strut

It has been observed that when a column or a strut is subjected to a compressive load and the load is gradually increased, a stage will reach when the column will be subjected to ultimate load. Beyond this, the column will fail by crushing and the load will be known as ***crushing load.***

It has also been experienced, that sometimes, a compression member does not fail entirely by crushing, but also by bending *i.e.* buckling. This happens in the case of long columns. It has also been observed, that all the *short columns fail due to their crushing. But, if a **long column is subjected to a compressive load, it is subjected to a compressive stress. If the load is gradually increased, the column will reach a stage, when it will start buckling. The load, at which the column tends to have lateral displacement or tends to buckle is called ***buckling load, critical load,*** or ***crippling load*** and the column is said to have developed an elastic instability. The buckling takes place about the axis having minimum radius of gyration or least moment of inertia. It may be noted that for a long column, the value of buckling load will be less than the crushing load. Moreover, the value of buckling load is low for long columns, and relatively high for short columns.

Depending on the end conditions, different columns have different crippling loads

16.3 Types of End Conditions of Columns

In actual practice, there are a number of end conditions for columns. But we shall study the Euler's column theory on the following four types of end conditions which are important from the subject point of view:

1. Both the ends hinged or pin jointed as shown in Fig. 16.1 (*a*),
2. Both the ends fixed as shown in Fig. 16.1 (*b*),
3. One end is fixed and the other hinged as shown in Fig. 16.1 (*c*), and
4. One end is fixed and the other free as shown in Fig. 16.1 (*d*).

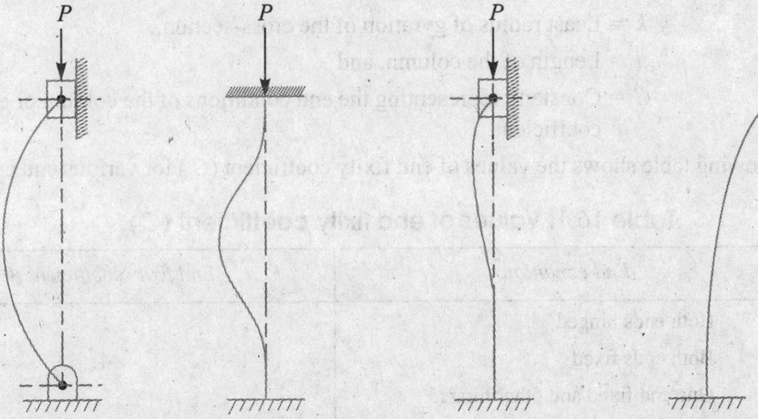

Fig. 16.1. Types of end conditions of columns.

16.4 Euler's Column Theory

The first rational attempt, to study the stability of long columns, was made by Mr. Euler. He

* The columns which have lengths less than 8 times their diameter, are called ***short columns*** (see also Art 16.8).

** The columns which have lengths more than 30 times their diameter are called ***long columns.***

derived an equation, for the buckling load of long columns based on the bending stress. While deriving this equation, the effect of direct stress is neglected. This may be justified with the statement, that the direct stress induced in a long column is negligible as compared to the bending stress. It may be noted that Euler's formula cannot be used in the case of short columns, because the direct stress is considerable, and hence cannot be neglected.

16.5 Assumptions in Euler's Column Theory

The following simplifying assumptions are made in Euler's column theory :

1. Initially the column is perfectly straight, and the load applied is truly axial.
2. The cross-section of the column is uniform throughout its length.
3. The column material is perfectly elastic, homogeneous and isotropic, and thus obeys Hooke's law.
4. The length of column is very large as compared to its cross-sectional dimensions.
5. The shortening of column, due to direct compression (being very small) is neglected.
6. The failure of column occurs due to buckling alone.
7. The weight of the column itself is neglected.

16.6 Euler's Formula

According to Euler's theory, the crippling or buckling load (W_{cr}) under various end conditions is represented by a general equation,

$$W_{cr} = \frac{C \pi^2 E I}{l^2} = \frac{C \pi^2 E A k^2}{l^2} \qquad \ldots (\because I = A.k^2)$$

$$= \frac{C \pi^2 E A}{(l/k)^2}$$

where
E = Modulus of elasticity or Young's modulus for the material of the column,
A = Area of cross-section,
k = Least radius of gyration of the cross-section,
l = Length of the column, and
C = Constant, representing the end conditions of the column or end fixity coefficient.

The following table shows the values of end fixity coefficient (C) for various end conditions.

Table 16.1. Values of end fixity coefficient (*C*).

S. No.	*End conditions*	*End fixity coefficient (C)*
1.	Both ends hinged	1
2.	Both ends fixed	4
3.	One end fixed and other hinged	2
4.	One end fixed and other end free	0.25

Notes : **1.** The vertical column will have two moment of inertias (*viz.* I_{xx} and I_{yy}). Since the column will tend to buckle in the direction of least moment of inertia, therefore the least value of the two moment of inertias is to be used in the relation.

2. In the above formula for crippling load, we have not taken into account the direct stresses induced in the material due to the load which increases gradually from zero to the crippling value. As a matter of fact, the combined stresses (due to the direct load and slight bending), reaches its allowable value at a load lower than that required for buckling and therefore this will be the limiting value of the safe load.

16.7 Slenderness Ratio

In Euler's formula, the ratio l / k is known as ***slenderness ratio.*** It may be defined as the ratio of the effective length of the column to the least radius of gyration of the section.

It may be noted that the formula for crippling load, in the previous article is based on the assumption that the slenderness ratio l / k is so large, that the failure of the column occurs only due to bending, the effect of direct stress (*i.e.* W / A) being negligible.

This equipment is used to determine the crippling load for axially loaded long struts.

16.8 Limitations of Euler's Formula

We have discussed in Art. 16.6 that the general equation for the crippling load is

$$W_{cr} = \frac{C \pi^2 E A}{(l / k)^2}$$

∴ Crippling stress,

$$\sigma_{cr} = \frac{W_{cr}}{A} = \frac{C \pi^2 E}{(l / k)^2}$$

A little consideration will show that the crippling stress will be high, when the slenderness ratio is small. We know that the crippling stress for a column cannot be more than the crushing stress of the column material. It is thus obvious that the Euler's fromula will give the value of crippling stress of the column (equal to the crushing stress of the column material) corresponding to the slenderness ratio. Now consider a mild steel column. We know that the crushing stress for mild steel is 330 N/mm^2 and Young's modulus for mild steel is 0.21×10^6 N/mm^2.

Now equating the crippling stress to the crushing stress, we have

$$\frac{C \pi^2 E}{(l / k)^2} = 330$$

$$\frac{1 \times 9.87 \times 0.21 \times 10^6}{(l / k)^2} = 330 \qquad \text{... (Taking } C = 1\text{)}$$

or $$(l / k)^2 = 6281$$

$\therefore$ $$l / k = 79.25 \text{ say } 80$$

Hence if the slenderness ratio is less than 80, Euler's formula for a mild steel column is not valid.

Sometimes, the columns whose slenderness ratio is more than 80, are known as ***long columns,*** and those whose slenderness ratio is less than 80 are known as ***short columns.*** It is thus obvious that the Euler's formula holds good only for long columns.

16.9 Equivalent Length of a Column

Sometimes, the crippling load according to Euler's formula may be written as

$$W_{cr} = \frac{\pi^2 E I}{L^2}$$

where L is the equivalent length or effective length of the column. The equivalent length of a given column with given end conditions is the length of an equivalent column of the same material and cross-section with hinged ends to that of the given column. The relation between the equivalent length and actual length for the given end conditions is shown in the following table.

Table 16.2. Relation between equivalent length (L) and actual length (l).

S.No.	*End Conditions*	*Relation between equivalent length (L) and actual length (l)*
1.	Both ends hinged	$L = l$
2.	Both ends fixed	$L = \frac{l}{2}$
3.	One end fixed and other end hinged	$L = \frac{l}{\sqrt{2}}$
4.	One end fixed and other end free	$L = 2l$

Example 16.1. *A T-section 150 mm × 120 mm × 20 mm is used as a strut of 4 m long hinged at both ends. Calculate the crippling load, if Young's modulus for the material of the section is 200 kN/mm².*

Solution. Given : $l = 4$ m = 4000 mm ; $E = 200$ kN/mm² = 200×10^3 N/mm²

First of all, let us find the centre of gravity (G) of the T-section as shown in Fig. 16.2.

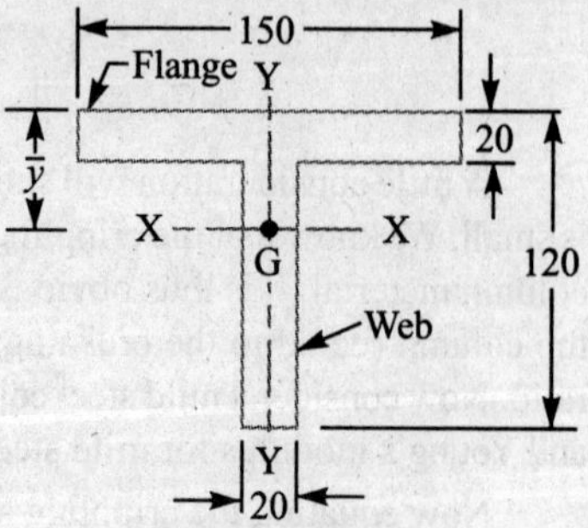

All dimensions in mm.

Fig. 16.2

Let $\bar{y}$ be the distance between the centre of gravity (G) and top of the flange,

We know that the area of flange,

$$a_1 = 150 \times 20 = 3000 \text{ mm}^2$$

Its distance of centre of gravity from top of the flange,

$$y_1 = 20 / 2 = 10 \text{ mm}$$

Area of web, $$a_2 = (120 - 20)\, 20 = 2000 \text{ mm}^2$$

Its distance of centre of gravity from top of the flange,

$$y_2 = 20 + 100 / 2 = 70 \text{ mm}$$

$$\therefore \quad \bar{y} = \frac{a_1 y_1 + a_2 y_2}{a_1 + a_2} = \frac{3000 \times 10 + 2000 \times 70}{3000 + 2000} = 34 \text{ mm}$$

We know that the moment of inertia of the section about *X-X*,

$$I_{XX} = \left[\frac{150\,(20)^3}{12} + 3000\,(34 - 10)^2 + \frac{20\,(100)^3}{12} + 2000\,(70 - 34)^2 \right]$$
$$= 6.1 \times 10^6 \text{ mm}^4$$

and
$$I_{YY} = \frac{20\,(150)^3}{12} + \frac{100\,(20)^3}{12} = 5.7 \times 10^6 \text{ mm}^4$$

Since I_{YY} is less than I_{XX}, therefore the column will tend to buckle in *Y-Y* direction. Thus we shall take the value of *I* as $I_{YY} = 5.7 \times 10^6$ mm^4.

Moreover as the column is hinged at its both ends, therefore equivalent length,

$$L = l = 4000 \text{ mm}$$

We know that the crippling load,

$$W_{cr} = \frac{\pi^2 E\, I}{L^2} = \frac{9.87 \times 200 \times 10^3 \times 5.7 \times 10^6}{(4000)^2} = 703 \times 10^3 \text{ N} = 703 \text{ kN } \textbf{Ans.}$$

Example 16.2. *An I-section 400 mm × 200 mm × 10 mm and 6 m long is used as a strut with both ends fixed. Find Euler's crippling load. Take Young's modulus for the material of the section as 200 kN/mm².*

Solution. Given : $D = 400$ mm ; $B = 200$ mm ; $t = 10$ mm ; $l = 6$ m $= 6000$ mm ; $E = 200$ kN/mm^2 $= 200 \times 10^3$ N/mm^2

The *I*-section is shown in Fig. 16.3.

Crippling load.

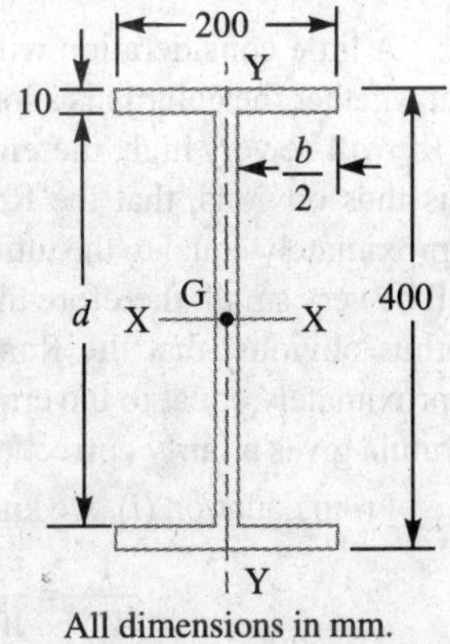

All dimensions in mm.

Fig. 16.3

We know that the moment of inertia of the *I*-section about *X-X*,

$$I_{XX} = \frac{B.D^3}{12} - \frac{b.d^3}{12}$$
$$= \frac{200\,(400)^3}{12} - \frac{(200 - 10)\,(400 - 20)^3}{12}$$
$$= 200 \times 10^6 \text{ mm}^4$$

and moment of inertia of the *I*-section about *Y-Y*,

$$I_{YY} = 2\left(\frac{t.B^3}{12}\right) + \frac{d.t^3}{12}$$
$$= 2\left[\frac{10\,(200)^3}{12}\right] + \frac{(400 - 20)\,10^3}{12}$$
$$= 13.36 \times 10^6 \text{ mm}^4$$

Since I_{YY} is less than I_{XX}, therefore the section will tend to buckle about *Y-Y* axis. Thus we shall take *I* as $I_{YY} = 13.36 \times 10^4$ mm^4.

Since the column is fixed at its both ends, therefore equivalent length,

$$L = l/2 = 6000/2 = 3000 \text{ mm}$$

We know that the crippling load,

$$W_{cr} = \frac{\pi^2 E I}{L^2} = \frac{9.87 \times 200 \times 10^3 \times 13.36 \times 10^6}{(3000)^2} = 2.93 \times 10^6 \text{ N}$$

$$= 2930 \text{ kN} \quad \textbf{Ans.}$$

16.10 Rankine's Formula for Columns

We have already discussed that Euler's formula gives correct results only for very long columns. Though this formula is applicable for columns, ranging from very long to short ones, yet it does not give reliable results. Prof. Rankine, after a number of experiments, gave the following empirical formula for columns.

$$\frac{1}{W_{cr}} = \frac{1}{W_C} + \frac{1}{W_E} \qquad \text{...}(i)$$

where
W_{cr} = Crippling load by Rankine's formula,
W_C = Ultimate crushing load for the column = $\sigma_c \times A$,
W_E = Crippling load, obtained by Euler's formula = $\dfrac{\pi^2 E I}{L^2}$

A little consideration will show, that the value of W_C will remain constant irrespective of the fact whether the column is a long one or short one. Moreover, in the case of short columns, the value of W_E will be very high, therefore the value of $1/W_E$ will be quite negligible as compared to $1/W_C$. It is thus obvious, that the Rankine's formula will give the value of its crippling load (*i.e.* W_{cr}) approximately equal to the ultimate crushing load (*i.e.* W_C). In case of long columns, the value of W_E will be very small, therefore the value of $1/W_E$ will be quite considerable as compared to $1/W_C$. It is thus obvious, that the Rankine's formula will give the value of its crippling load (*i.e.* W_{cr}) approximately equal to the crippling load by Euler's formula (*i.e.* W_E). Thus, we see that Rankine's formula gives a fairly correct result for all cases of columns, ranging from short to long columns.

From equation (*i*), we know that

$$\frac{1}{W_{cr}} = \frac{1}{W_C} + \frac{1}{W_E} = \frac{W_E + W_C}{W_C \times W_E}$$

$$\therefore \quad W_{cr} = \frac{W_C \times W_E}{W_C + W_E} = \frac{W_C}{1 + \dfrac{W_C}{W_E}}$$

Now substituting the value of W_C and W_E in the above equation, we have

$$W_{cr} = \frac{\sigma_c \times A}{1 + \dfrac{\sigma_c \times A \times L^2}{\pi^2 E I}} = \frac{\sigma_c \times A}{1 + \dfrac{\sigma_c}{\pi^2 E} \times \dfrac{A.L^2}{A.k^2}} \qquad \text{... } (\because I = A.k^2)$$

$$= \frac{\sigma_c \times A}{1 + a\left(\dfrac{L}{k}\right)^2} = \frac{\text{Crushing load}}{1 + a\left(\dfrac{L}{k}\right)^2}$$

where
σ_c = Crushing stress or yield stress in compression,
A = Cross-sectional area of the column,
a = Rankine's constant = $\dfrac{\sigma_c}{\pi^2 E}$,

L = Equivalent length of the column, and

k = Least radius of gyration.

The following table gives the values of crushing stress and Rankine's constant for various materials.

Table 16.3. Values of crushing stress (σ_c) and Rankine's constant (a) for various materials.

S.No.	Material	σ_c in MPa	$a = \frac{\sigma_c}{\pi^2 E}$
1.	Wrought iron	250	$\frac{1}{9000}$
2.	Cast iron	550	$\frac{1}{1600}$
3.	Mild steel	320	$\frac{1}{7500}$
4.	Timber	50	$\frac{1}{750}$

16.11 Johnson's Formulae for Columns

Prof. J.B. Johnson proposed the following two formula for short columns.

1. ***Straight line formula.*** According to straight line formula proposed by Johnson, the critical or crippling load is

$$W_{cr} = A\left[\sigma_y - \frac{2\,\sigma_y}{3\pi}\left(\frac{L}{k}\right)\sqrt{\frac{\sigma_y}{3C \times E}}\right] = A\left[\sigma_y - C_1\left(\frac{L}{k}\right)\right]$$

where A = Cross-sectional area of column,

σ_y = Yield point stress,

$C_1 = \frac{2\,\sigma_y}{3\pi}\sqrt{\frac{\sigma_y}{3C.E}}$

= A constant, whose value depends upon the type of material as well as the type of ends, and

$\frac{L}{k}$ = Slenderness ratio.

If the safe stress (W_{cr} / A) is plotted against slenderness ratio (L / k), it works out to be a straight line, so it is known as straight line formula.

2. ***Parabolic formula.*** Prof. Johnson after proposing the straight line formula found that the results obtained by this formula are very approximate. He then proposed another formula, according to which the critical or crippling load,

$$W_{cr} = A \times \sigma_y\left[1 - \frac{\sigma_y}{4\,C\,\pi^2\,E}\left(\frac{L}{k}\right)^2\right] \text{ with usual notations.}$$

If a curve of safe stress (W_{cr} / A) is plotted against (L / k), it works out to be a parabolic, so it is known as parabolic formula.

Fig. 16.4 shows the relationship of safe stress (W_{cr} / A) and the slenderness ratio (L / k) as given by Johnson's formula and Euler's formula for a column made of mild steel with both ends hinged (*i.e.* $C = 1$), having a yield strength, σ_y = 210 MPa. We see from the figure that point A (the point of tangency between the Johnson's straight line formula and Euler's formula) describes the use of two formulae. In other words, Johnson's straight line formula may be used when $L / k < 180$ and the Euler's formula is used when $L / k > 180$.

Similarly, the point B (the point of tangency between the Johnson's parabolic formula and Euler's formula) describes the use of two formulae. In other words, Johnson's parabolic formula is used when $L / k < 140$ and the Euler's formula is used when $L / k > 140$.

Note : For short columns made of ductile materials, the Johnson's parabolic formula is used.

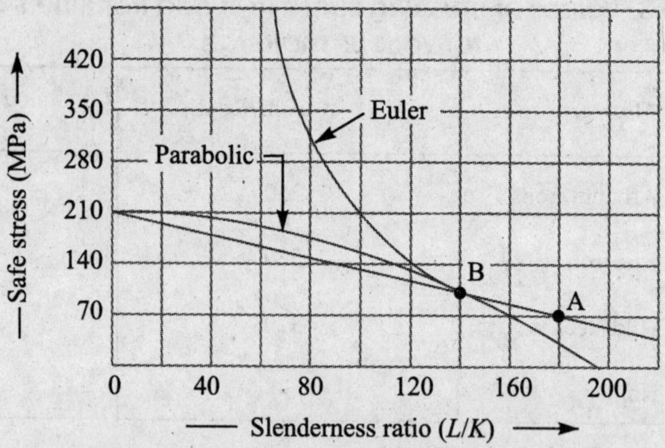

Fig. 16.4. Relation between slendeness ratio and safe stress.

16.12 Long Columns Subjected to Eccentric Loading

In the previous articles, we have discussed the effect of loading on long columns. We have always referred the cases when the load acts axially on the column (*i.e.* the line of action of the load coincides with the axis of the column). But in actual practice it is not always possible to have an axial load on the column, and eccentric loading takes place. Here we shall discuss the effect of eccentric loading on the Rankine's and Euler's formula for long columns.

Consider a long column hinged at both ends and subjected to an eccentric load as shown in Fig. 16.5.

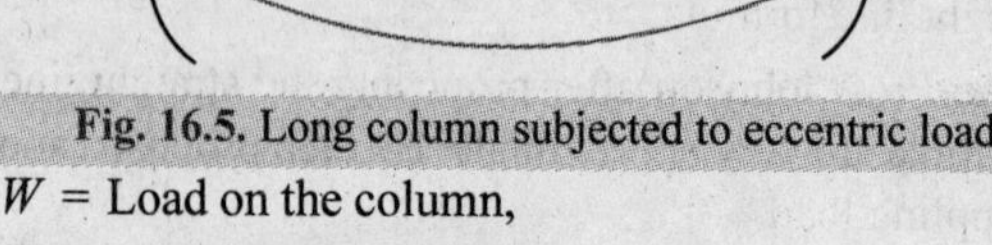

Fig. 16.5. Long column subjected to eccentric loading.

Let W = Load on the column,

A = Area of cross-section,

e = Eccentricity of the load,

Z = Section modulus,

y_c = Distance of the extreme fibre (on compression side) from the axis of the column,

k = Least radius of gyration,

I = Moment of inertia = $A.k^2$,

E = Young's modulus, and

l = Length of the column.

We have already discussed that when a column is subjected to an eccentric load, the maximum intensity of compressive stress is given by the relation

$$\sigma_{max} = \frac{W}{A} + \frac{M}{Z}$$

The maximum bending moment for a column hinged at both ends and with eccentric loading is given by

$$M = W.e.\sec\frac{l}{2}\sqrt{\frac{W}{E.I}} = W.e.\sec\frac{l}{2\,k}\sqrt{\frac{W}{E.A}} \qquad \ldots (\because I = A.k^2)$$

$$\therefore \qquad \sigma_{max} = \frac{W}{A} + \frac{W.e.\sec\dfrac{l}{2\,k}\sqrt{\dfrac{W}{E.A}}}{Z}$$

$$= \frac{W}{A} + \frac{W.e.y_c.\sec\dfrac{l}{2\,k}\sqrt{\dfrac{W}{E.A}}}{A.k^2} \qquad \ldots (\because Z = I/y_c = A.k^2/y_c)$$

$$= \frac{W}{A}\left[1 + \frac{e.y_c}{k^2}\sec\frac{l}{2\,k}\sqrt{\frac{W}{E.A}}\right]$$

$$= {}^{*}\frac{W}{A}\left[1 + \frac{e.y_c}{k^2}\sec\frac{L}{2\,k}\sqrt{\frac{W}{E.A}}\right]$$

... (Substituting $l = L$, equivalent length for both ends hinged).

16.13 Design of Piston Rod

Since a piston rod moves forward and backward in the engine cylinder, therefore it is subjected to alternate tensile and compressive forces. It is usually made of mild steel. One end of the piston rod is secured to the piston by means of tapered rod provided with nut. The other end of the piston rod is joined to crosshead by means of a cotter.

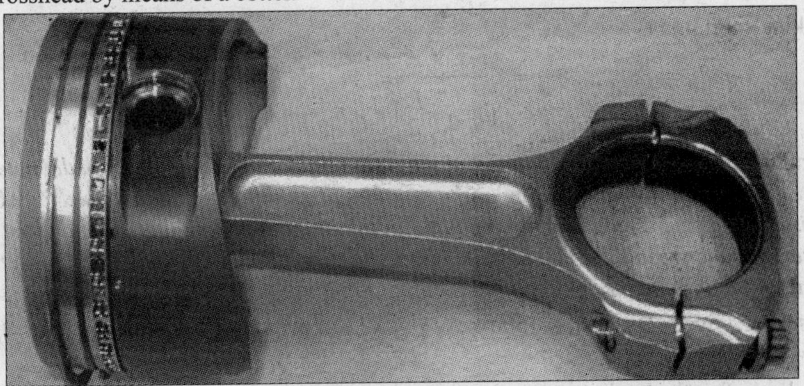

Piston rod is made of mild steel.

---

* The expression $\sigma_{max} = \frac{W}{A}\left[1 + \frac{e.y_c}{k^2}\sec\frac{L}{2k}\sqrt{\frac{W}{E.A}}\right]$ may also be written as follows:

$$\sigma_{max} = \frac{W}{A} + \frac{W}{A} \times \frac{e.y_c}{\frac{I}{A}}\sec\frac{L}{2k}\sqrt{\frac{W}{E \times \frac{I}{k^2}}} \qquad \ldots \left(\text{Substituting } k^2 = \frac{I}{A} \text{ and } A = \frac{I}{k^2}\right)$$

$$= \frac{W}{A} + \frac{W.e}{Z}\sec\frac{L}{2}\sqrt{\frac{W}{E.I}}$$

Let
p = Pressure acting on the piston,
D = Diameter of the piston,
d = Diameter of the piston rod,
W = Load acting on the piston rod,
W_{cr} = Buckling or crippling load = $W \times$ Factor of safety,
σ_t = Allowable tensile stress for the material of rod,
σ_c = Compressive yield stress,
A = Cross-sectional area of the rod,
l = Length of the rod, and
k = Least radius of gyration of the rod section.

The diameter of the piston rod is obtained as discussed below:

1. When the length of the piston rod is small *i.e.* when slenderness ratio (l / k) is less than 40, then the diameter of piston rod may be obtained by equating the load acting on the piston rod to its tensile strength, *i.e.*

$$W = \frac{\pi}{4} \times d^2 \times \sigma_t$$

or

$$\frac{\pi}{4} \times D^2 \times p = \frac{\pi}{4} \times d^2 \times \sigma_t$$

$$\therefore \quad d = D\sqrt{\frac{p}{\sigma_t}}$$

2. When the length of the piston rod is large, then the diameter of the piston rod is obtained by using Euler's formula or Rankine's formula. Since the piston rod is securely fastened to the piston and cross head, therefore it may be considered as fixed ends. The Euler's formula is

$$W_{cr} = \frac{\pi^2 E I}{L^2}$$

and Rankine's formula is,

$$W_{cr} = \frac{\sigma_c \times A}{1 + a\left(\frac{L}{k}\right)^2}$$

Example 16.3. *Calculate the diameter of a piston rod for a cylinder of 1.5 m diameter in which the greatest difference of steam pressure on the two sides of the piston may be assumed to be 0.2 N/mm². The rod is made of mild steel and is secured to the piston by a tapered rod and nut and to the crosshead by a cotter. Assume modulus of elasticity as 200 kN/mm² and factor of safety as 8. The length of rod may be assumed as 3 metres.*

Solution. Given : D = 1.5 m = 1500 mm ; p = 0.2 N/mm² ; E = 200 kN/mm² = 200 × 10³ N/mm² ; l = 3 m = 3000 mm

We know that the load acting on the piston,

$$W = \frac{\pi}{4} \times D^2 \times p = \frac{\pi}{4}(1500)^2 \times 0.2 = 353\,475 \text{ N}$$

∴ Buckling load on the piston rod,

$$W_{cr} = W \times \text{Factor of safety} = 353\,475 \times 8 = 2.83 \times 10^6 \text{ N}$$

Since the piston rod is considered to have both ends fixed, therefore from Table 16.2, the equivalent length of the piston rod,

$$L = \frac{l}{2} = \frac{3000}{2} = 1500 \text{ mm}$$

Let d = Diameter of piston rod in mm, and

I = Moment of inertia of the cross-section of the rod = $\frac{\pi}{64} \times d^4$

According to Euler's formula, buckling load (W_{cr}),

$$2.83 \times 10^6 = \frac{\pi^2 E I}{L^2} = \frac{9.87 \times 200 \times 10^3 \times \pi d^4}{(1500)^2 \times 64} = 0.043\, d^4$$

$$\therefore \quad d^4 = 2.83 \times 10^6 / 0.043 = 65.8 \times 10^6 \quad \text{or} \quad d = 90 \text{ mm}$$

According to Rankine's formula, buckling load,

$$W_{cr} = \frac{\sigma_c \times A}{1 + a\left(\frac{L}{k}\right)^2} \qquad \text{...(i)}$$

We know that for mild steel, the crushing stress,

$$\sigma_c = 320 \text{ MPa} = 320 \text{ N/mm}^2, \text{ and } a = \frac{1}{7500}$$

and least radius of gyration for the piston rod section,

$$k = \sqrt{\frac{I}{A}} = \sqrt{\frac{\pi d^4}{64} \times \frac{4}{\pi d^2}} = \frac{d}{4}$$

Substituting these values in the above equation (*i*), we have

$$2.83 \times 10^6 = \frac{320 \times \frac{\pi d^2}{4}}{1 + \frac{1}{7500}\left(\frac{1500 \times 4}{d}\right)^2} = \frac{251.4\, d^2}{1 + \frac{4800}{d^2}} = \frac{251.4\, d^4}{d^2 + 4800}$$

$$251.4\, d^4 - 2.83 \times 10^6\, d^2 - 2.83 \times 10^6 \times 4800 = 0$$

or

$$d^4 - 11\,257\, d^2 - 54 \times 10^6 = 0$$

$$\therefore \quad d^2 = \frac{11\,250 \pm \sqrt{(11\,257)^2 + 4 \times 1 \times 54 \times 10^6}}{2} = \frac{11\,257 \pm 18\,512}{2}$$

$$= 14\,885 \qquad \text{... (Taking +ve sign)}$$

or

$$d = 122 \text{ mm}$$

Taking larger of the two values, we have

d = 122 mm **Ans.**

16.14 Design of Push Rods

The push rods are used in overhead valve and side valve engines. Since these are designed as long columns, therefore Euler's formula should be used. The push rods may be treated as pin end columns because they use spherical seated bearings.

These rods are used in overhead valve and side valve engines.

Let W = Load acting on the push rod,

D = Diameter of the push rod,

d = Diameter of the hole through the push rod,

I = Moment of inertia of the push rod,

$= \frac{\pi}{64} \times D^4$, for solid rod

$= \frac{\pi}{64}(D^4 - d^4)$, for tubular section

l = Length of the push rod, and

E = Young's modulus for the material of push rod.

If m is the factor of safety for the long columns, then the critical or crippling load on the rod is given by

$$W_{cr} = m \times W$$

Now using Euler's formula, $W_{cr} = \frac{\pi^2 EI}{L^2}$, the diameter of the push rod (D) can be obtained.

Notes: 1. Generally the diameter of the hole through the push rod is 0.8 times the diameter of push rod, *i.e.*

$$d = 0.8\,D$$

2. Since the push rods are treated as pin end columns, therefore the equivalent length of the rod (L) is equal to the actual length of the rod (l).

Example 16.4. *The maximum load on a petrol engine push rod 300 mm long is 1400 N. It is hollow having the outer diameter 1.25 times the inner diameter. Spherical seated bearings are used for the push rod. The modulus of elasticity for the material of the push rod is 210 kN/mm². Find a suitable size for the push rod, taking a factor of safety of 2.5.*

Solution. Given : l = 300 mm ; W = 1400 N ; D = 1.25 d ; E = 210 kN/mm² = 210 × 10³ N/mm² ; m = 2.5

Let d = Inner diameter of push rod in mm, and

D = Outer diameter of the push rod in mm = 1.25 d ...(Given)

∴ Moment of inertia of the push rod section,

$$I = \frac{\pi}{64}(D^4 - d^4) = \frac{\pi}{64}[(1.25\,d)^4 - d^4] = 0.07\,d^4\ \text{mm}^4$$

We know that the crippling load on the push rod,

$$W_{cr} = m \times W = 2.5 \times 1400 = 3500\ \text{N}$$

Now according to Euler's formula, crippling load (W_{cr}),

$$3500 = \frac{\pi^2 E\,I}{L^2} = \frac{9.87 \times 210 \times 10^3 \times 0.07\,d^4}{(300)^2} = 1.6\,d^4 \quad ...(\because L = l)$$

∴ $d^4 = 3500 / 1.6 = 2188$ or $d = 6.84$ mm **Ans.**

and $D = 1.25\,d = 1.25 \times 6.84 = 8.55$ mm **Ans.**

16.15 Design of Connecting Rod

A connecting rod is a machine member which is subjected to alternating direct compressive and tensile forces. Since the compressive forces are much higher than the tensile forces, therefore the cross-section of the connecting rod is designed as a strut and the Rankine's formula is used.

A connecting rod subjected to an axial load W may buckle with X-axis as neutral axis (*i.e.* in the plane of motion of the connecting rod) or Y-axis as neutral axis (*i.e.* in the plane perpendicular to the plane of motion). The connecting rod is considered like both ends hinged for buckling about X-axis and both ends fixed for buckling about Y-axis. A connecting rod should be equally strong in buckling about either axes.

Let A = Cross-sectional area of the connecting rod,

l = Length of the connecting rod,

σ_c = Compressive yield stress,

W_{cr} = Crippling or buckling load,

I_{xx} and I_{yy} = Moment of inertia of the section about X-axis and Y-axis respectively, and

k_{xx} and k_{yy} = Radius of gyration of the section about X-axis and Y-axis respectively.

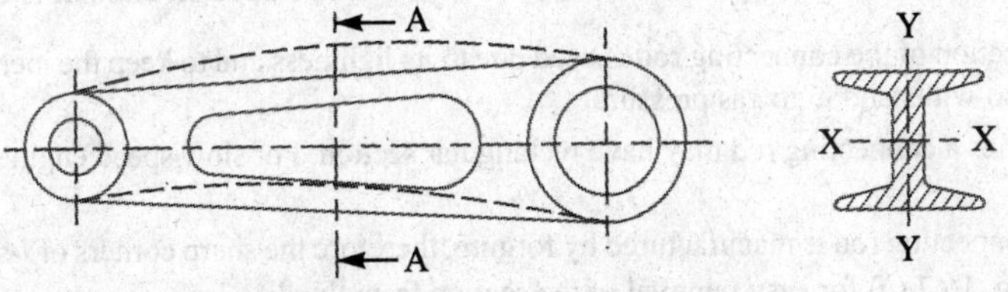

Fig. 16.6. Buckling of connecting rod.

According to Rankine's formula,

$$W_{cr} \text{ about } X\text{-axis} = \frac{\sigma_c \times A}{1 + a\left(\frac{L}{k_{xx}}\right)^2} = \frac{\sigma_c \times A}{1 + a\left(\frac{l}{k_{xx}}\right)^2} \quad \text{... } (\because \text{ For both ends hinged, } L = l)$$

and

$$W_{cr} \text{ about } Y\text{-axis} = \frac{\sigma_c \times A}{1 + a\left(\frac{L}{k_{yy}}\right)^2} = \frac{\sigma_c \times A}{1 + a\left(\frac{l}{2k_{yy}}\right)^2} \quad \text{... } \left(\because \text{ For both ends fixed, } L = \frac{l}{2}\right)$$

In order to have a connecting rod equally strong in buckling about both the axes, the buckling loads must be equal, *i.e.*

$$\frac{\sigma_c \times A}{1 + a\left(\frac{l}{k_{xx}}\right)^2} = \frac{\sigma_c \times A}{1 + a\left(\frac{l}{2\,k_{yy}}\right)^2} \quad \text{or} \quad \left(\frac{l}{k_{xx}}\right)^2 = \left(\frac{l}{2\,k_{yy}}\right)^2$$

$$\therefore \quad k^2_{xx} = 4\,k^2_{yy} \quad \text{or} \quad I_{xx} = 4\,I_{yy} \quad \text{... } (\because I = A \times k^2)$$

This shows that the connecting rod is four times strong in buckling about Y-axis than about X-axis. If $I_{xx} > 4\,I_{yy}$, then buckling will occur about Y-axis and if $I_{xx} < 4\,I_{yy}$, buckling will occur about X-axis. In actual practice, I_{xx} is kept slightly less than $4\,I_{yy}$. It is usually taken between 3 and 3.5 and the connecting rod is designed for buckling about X-axis. The design will alwyas be satisfactory for buckling about Y-axis.

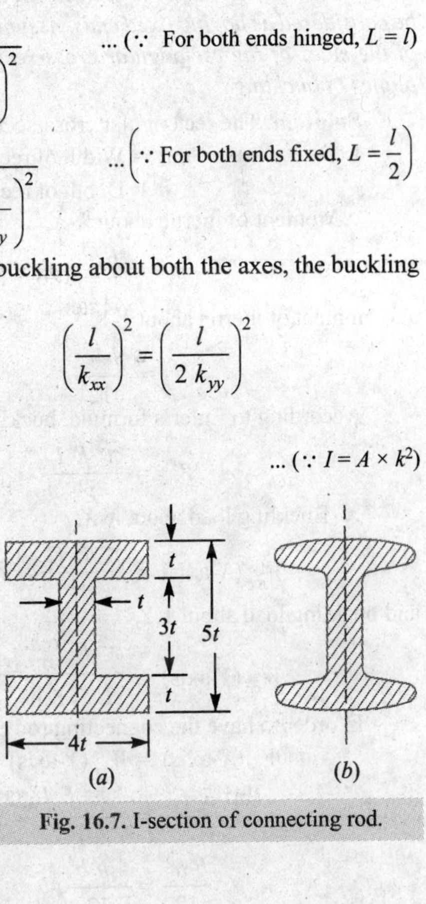

Fig. 16.7. I-section of connecting rod.

The most suitable section for the connecting rod is I-section with the proportions as shown in Fig. 16.7 (*a*).

Area of the section

$$= 2\,(4\,t \times t) + 3\,t \times t = 11\,t^2$$

$\therefore$ Moment of inertia about X-axis,

$$I_{xx} = \frac{1}{12}\left[4\,t\,(5\,t)^3 - 3\,t\,(3\,t)^3\right] = \frac{419}{12}\,t^4$$

and moment of inertia about Y-axis,

$$I_{yy} = \left[2 \times \frac{1}{12}\,t \times (4\,t)^3 + \frac{1}{12}\,(3\,t)\,t^3\right] = \frac{131}{12}\,t^4$$

$$\therefore \quad \frac{I_{xx}}{I_{yy}} = \frac{419}{12} \times \frac{12}{131} = 3.2$$

Since the value of $\frac{I_{xx}}{I_{yy}}$ lies between 3 and 3.5, therefore *I*-section chosen is quite satisfactory.

Notes : 1. The *I*-section of the connecting rod is used due to its lightness and to keep the inertia forces as low as possible. It can also withstand high gas pressure.

2. Sometimes a connecting rod may have rectangular section. For slow speed engines, circular sections may be used.

3. Since connecting rod is manufactured by forging, therefore the sharp corners of *I*-section are rounded off as shown in Fig. 16.7 (*b*) for easy removal of the section from the dies.

Example 16.5. *A connecting rod of length l may be considered as a strut with the ends free to turn on the crank pin and the gudgeon pin. In the directions of the axes of these pins, however, it may be considered as having fixed ends. Assuming that Euler's formula is applicable, determine the ratio of the sides of the rectangular cross-section so that the connecting rod is equally strong in both planes of buckling.*

Solution. The rectangular cross-section of the connecting rod is shown in Fig. 16.8.

Let b = Width of rectangular cross-section, and

h = Depth of rectangular cross-section.

∴ Moment of inertia about *X-X*,

$$I_{xx} = \frac{b.h^3}{12}$$

and moment of inertia about *Y-Y*,

$$I_{yy} = \frac{h.b^3}{12}$$

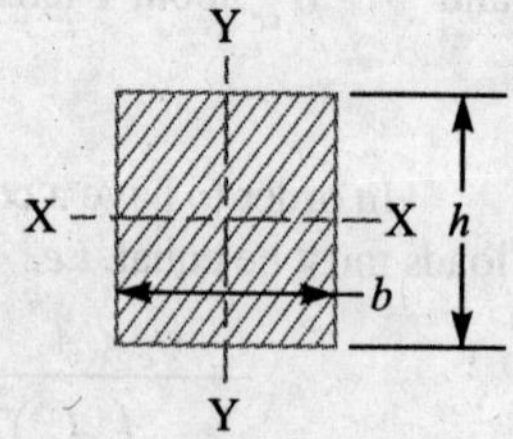

Fig. 16.8

According to Euler's formula, buckling load,

$$W_{cr} = \frac{\pi^2 E I}{L^2}$$

∴ Buckling load about *X-X*,

$$W_{cr}\,(X\text{-axis}) = \frac{\pi^2 E I_{xx}}{l^2} \quad \text{... } (\because L = l, \text{ for both ends free to turn})$$

and buckling load about *Y-Y*,

$$W_{cr}\,(Y\text{-axis}) = \frac{\pi^2 E I_{yy}}{(l/2)^2} = \frac{4\pi^2 E I_{yy}}{l^2} \quad \text{... } (\because L = l/2, \text{ for both ends fixed})$$

In order to have the connecting rod equally strong in both the planes of buckling,

$$W_{cr}\,(X\text{-axis}) = W_{cr}\,(Y\text{-axis})$$

$$\frac{\pi^2 E I_{xx}}{l^2} = \frac{4\pi^2 E I_{yy}}{l^2} \quad \text{or} \quad I_{xx} = 4 I_{yy}$$

$$\therefore \quad \frac{b\,h^3}{12} = \frac{4\,h\,b^3}{12} \quad \text{or} \quad h^2 = 4\,b^2$$

and $h^2 / b^2 = 4$ or $h / b = 2$ **Ans.**

16.16 Forces Acting on a Connecting Rod

A connecting rod is subjected to the following forces :

1. Force due to gas or steam pressure and inertia of reciprocating parts, and
2. Inertia bending forces.

We shall now derive the expressions for the forces acting on a horizontal engine, as discussed below:

1. ***Force due to gas or steam pressure and inertia of reciprocating parts***

Consider a connecting rod *PC* as shown in Fig. 16.9.

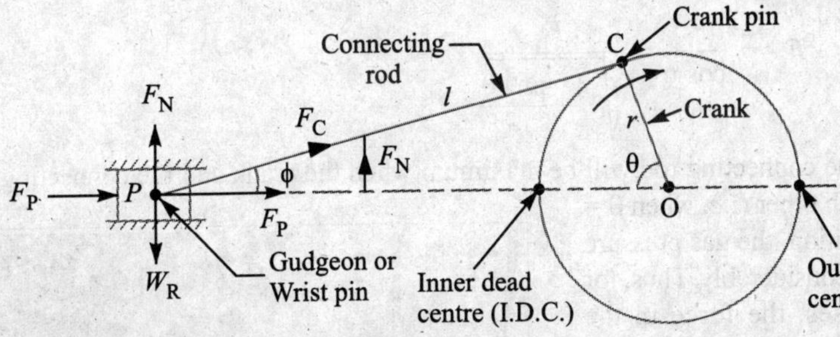

Fig. 16.9. Forces on a connecting rod.

Let p = Pressure of gas or steam,

A = Area of piston,

m_R = Mass of reciprocating parts,

= Mass of piston, gudgeon pin etc. + $\frac{1}{3}$ rd mass of connecting rod,

ω = Angular speed of crank,

ϕ = Angle of inclination of the connecting rod with the line of stroke,

θ = Angle of inclination of the crank from inner dead centre,

r = Radius of crank,

l = Length of connecting rod, and

n = Ratio of length of connecting rod to radius of crank = l / r.

We know that the force on the piston due to pressure of gas or steam,

$$F_L = \text{Pressure} \times \text{Area} = p \times A$$

and inertia force of reciprocating parts,

$$F_I = \text{Mass} \times {}^*\text{Acceleration} = m_R \times \omega^2 \times r \left(\cos\theta + \frac{\cos 2\theta}{n}\right)$$

It may be noted that in a horizontal engine, reciprocating parts are accelerated from rest during the first half of the stroke (*i.e.* when the piston moves from inner dead centre to outer dead centre). It is then retarted during the latter half of the stroke (*i.e.* when the piston moves from outer dead centre to inner dead centre). The inertia force due to the acceleration of reciprocating parts, opposes the force on the piston. On the other hand, the inertia force due to retardation of the reciprocating parts, helps the force on the piston.

∴ Net force acting on the piston pin (or gudgeon or wrist pin),

$$F_P = \text{Force due to pressure of gas or steam} \pm \text{Inertia force}$$
$$= F_L \pm F_I$$

The –ve sign is used when the piston is accelerated and +ve sign is used when the piston is retarted.

---

* Acceleration of reciprocating parts = $\omega^2 r \left(\cos\theta + \frac{\cos 2\theta}{n}\right)$

The force F_P gives rise to a force F_C in the connecting rod and a thrust F_N on the sides of the cylinder walls (or normal reaction on crosshead guides). From Fig. 16.9, we see that force in the connecting rod at any instant.

$$F_C = \frac{F_P}{\cos\phi} = \frac{{}^{*}\ F_P}{\sqrt{1 - \frac{\sin^2\theta}{n^2}}}$$

The force in the connecting rod will be maximum when the crank and the connecting rod are perpendicular to each other (*i.e.* when θ = 90°). But at this position, the gas pressure would be decreased considerably. Thus, for all practical purposes, the force in the connecting rod (F_C) is taken equal to the maximum force on the piston due to pressure of gas or steam (F_L), neglecting piston inertia effects.

Connecting rod.

2. *Inertia bending forces*

Consider a connecting rod *PC* and a crank *OC* rotating with uniform angular velocity ω rad /s. In order to find the acceleration of various points on the connecting rod, draw the Klien's acceleration diagram *CQNO* as shown in Fig. 16.10 (*a*). *CO* represents the acceleration of *C* towards *O* and *NO* represents the acceleration of *P* towards *O*. The acceleration of other points such as *D*, *E*, *F* and *G* etc. on the connecting rod *PC* may be found by drawing horizontal lines from these points to intersect *CN* at *d*, *e*, *f* and *g* respectively. Now *dO*, *eO*, *fO* and *gO* represents the acceleration of *D*, *E*, *F* and *G* all towards *O*. The inertia force acting on each point will be as follows :

Inertia force at $C = m \times \omega^2 \times CO$

Inertia force at $D = m \times \omega^2 \times dO$

Inertia force at $E = m \times \omega^2 \times eO$, and so on.

The inertia forces will be opposite to the direction of acceleration or centrifugal forces. The inertia forces can be resolved into two components, one parallel to the connecting rod and the other perpendicular to the rod. The parallel (or longitudinal) components adds up algebraically to the force acting on the connecting rod (F_C) and produces thrust on the pins. The perpendicular (or transverse) components produces bending action (also called whipping action) and the stress induced in the connecting rod is called ***whipping stress.***

* For derivation, please refer to author's popular book on **'Theory of Machines'**.

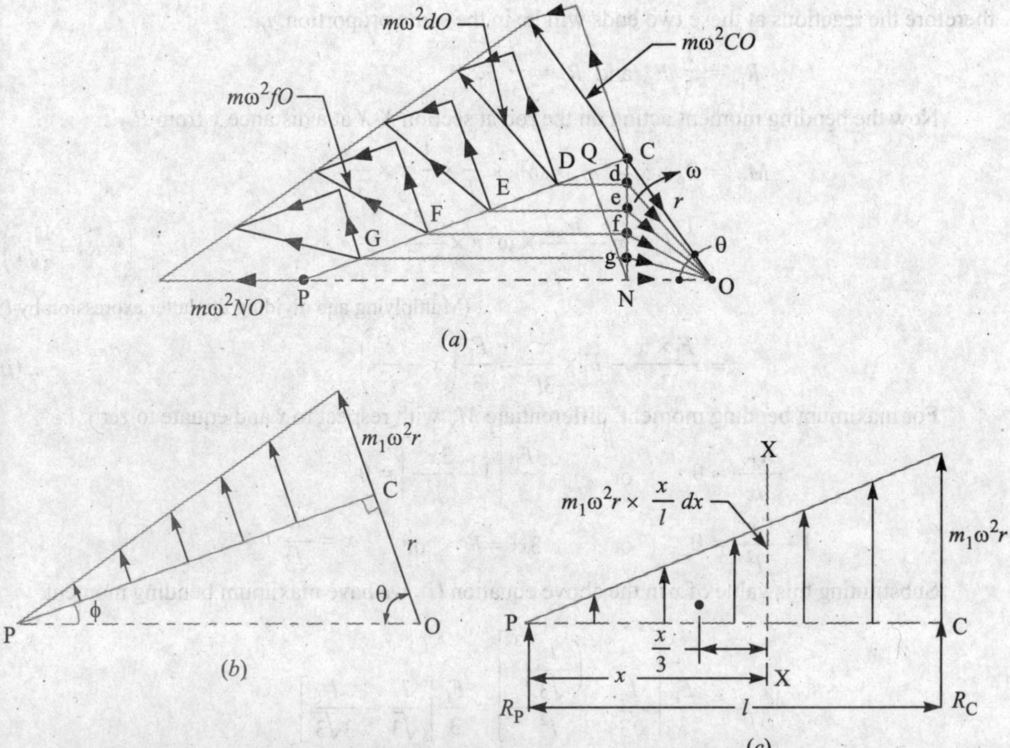

Fig. 16.10. Inertia bending forces.

A little consideration will show that the perpendicular components will be maximum, when the crank and connecting rod are at right angles to each other.

The variation of the inertia force on the connecting rod is linear and is like a simply supported beam of variable loading as shown in Fig. 16.10 (*b*) and (*c*). Assuming that the connecting rod is of uniform cross-section and has mass m_1 kg per unit length, therefore

Inertia force per unit length at the crank pin

$$= m_1 \times \omega^2 r$$

and inertia force per unit length at the gudgeon pin

$$= 0$$

Inertia forces due to small element of length dx at a distance x from the gudgeon pin P,

$$dF_I = m_1 \times \omega^2 r \times \frac{x}{l} \times dx$$

∴ Resultant inertia force,

$$F_I = \int_0^l m_1 \times \omega^2 r \times \frac{x}{l} \times dx = \frac{m_1 \times \omega^2 r}{l}\left[\frac{x^2}{2}\right]_0^l$$

$$= \frac{m_1 \times l}{2} \times \omega^2\, r = \frac{m}{2} \times \omega^2\, r \qquad \text{... (Substituting } m_1.l = m)$$

This resultant inertia force acts at a distance of $2l / 3$ from the gudgeon pin P.

Since it has been assumed that $\frac{1}{3}$ rd mass of the connecting rod is concentrated at gudgeon pin P (*i.e.* small end of connecting rod) and $\frac{2}{3}$ rd at the crank pin (*i.e.* big end of connecting rod), therefore the reactions at these two ends will be in the same proportion, *i.e.*

$$R_P = \frac{1}{3}F_I, \text{ and } R_C = \frac{2}{3}F_I$$

Now the bending moment acting on the rod at section X-X at a distance x from P,

$$M_X = R_P \times x - {}^*m_1 \times \omega^2 r \times \frac{x}{l} \times \frac{1}{2} \times x \times \frac{x}{3}$$

$$= \frac{1}{3}F_I \times x - \frac{m_1 . l}{2} \times \omega^2 r \times \frac{x^3}{3\,l^2} \qquad \ldots\left(\because R_P = \frac{1}{3}F_I\right)$$

... (Multiplying and dividing the latter expression by l)

$$= \frac{F_I \times x}{3} - F_I \times \frac{x^3}{3l^2} = \frac{F_I}{3}\left(x - \frac{x^3}{l^2}\right) \qquad \ldots(i)$$

For maximum bending moment, differentiate M_X with respect to x and equate to zero, *i.e.*

$$\frac{d_{MX}}{dx} = 0 \quad \text{or} \quad \frac{F_I}{3}\left[1 - \frac{3x^2}{l^2}\right] = 0$$

$$\therefore \quad 1 - \frac{3x^2}{l^2} = 0 \quad \text{or} \quad 3x^2 = l^2 \quad \text{or} \quad x = \frac{l}{\sqrt{3}}$$

Substituting this value of x in the above equation (*i*), we have maximum bending moment,

$$M_{max} = \frac{F_I}{3}\left[\frac{l}{\sqrt{3}} - \frac{\left(\frac{l}{\sqrt{3}}\right)^3}{l^2}\right] = \frac{F_I}{3}\left[\frac{l}{\sqrt{3}} - \frac{l}{3\sqrt{3}}\right]$$

$$= \frac{F_I}{3} \times \frac{2l}{3\sqrt{3}} = \frac{2\,F_I \times l}{9\sqrt{3}}$$

$$= 2 \times \frac{m}{2} \times \omega^2 r \times \frac{l}{9\sqrt{3}} = m \times \omega^2 r \times \frac{l}{9\sqrt{3}} \qquad \ldots\left(\because F_I = \frac{m}{2} \times \omega^2 r\right)$$

and the maximum bending stress, due to inertia of the connecting rod,

$$\sigma_{max} = \frac{M_{max}}{Z}$$

where Z = Section modulus.

From above we see that the maximum bending moment varies as the square of speed, therefore, the bending stress due to high speed will be dangerous. It may be noted that the maximum axial force and the maximum bending stress do not occur simultaneously. In an I.C. engine, the maximum gas load occurs close to top dead centre whereas the maximum bending stress occurs when the crank angle $\theta = 65°$ to $70°$ from top dead centre. The pressure of gas falls suddenly as the piston moves from dead centre. In steam engines, even though the pressure is maintained till cut off occurs, the speed is low and therefore the bending stress due to inertia is small. Thus the general practice is to design a connecting rod by assuming the force in the connecting rod (F_C) equal to the maximum force on the piston due to pressure of gas or steam (F_L), neglecting piston inertia effects and then checked for bending stress due to inertia force (*i.e.* whipping stress).

---

* B.M. due to variable loading from $\left(0 \text{ to } m_1\,\omega^2 r \times \frac{x}{l}\right)$ is equal to the area of triangle multiplied by distance of C.G. from X-X $\left(i.e.\ \frac{x}{3}\right)$.

Example 16.6. *Determine the dimensions of an I-section connecting rod for a petrol engine from the following data :*

Diameter of the piston	= *110 mm*
Mass of the reciprocating parts	= *2 kg*
Length of the connecting rod from centre to centre	= *325 mm*
Stroke length	= *150 mm*
R.P.M.	= *1500 with possible overspeed of 2500*
Compression ratio	= *4 : 1*
Maximum explosion pressure	= *2.5 N/mm²*

Connecting rod of a petrol engine.

Solution. Given : $D = 110$ mm $= 0.11$ m ; $m_R = 2$ kg ; $l = 325$ mm $= 0.325$ m ; Stroke length = 150 mm = 0.15 m ; $N_{min} = 1500$ r.p.m. ; $N_{max} = 2500$ r.p.m. ; *Compression ratio = 4 : 1 ; $p = 2.5$ N/mm²

We know that the radius of crank,

$$r = \frac{\text{Stroke length}}{2} = \frac{150}{2} = 75 \text{ mm} = 0.075 \text{ m}$$

and ratio of the length of connecting rod to the radius of crank,

$$n = \frac{l}{r} = \frac{325}{75} = 4.3$$

We know that the maximum force on the piston due to pressure,

$$F_L = \frac{\pi}{4} \times D^2 \times p = \frac{\pi}{4}(110)^2\, 2.5 = 23\,760 \text{ N}$$

and maximum angular speed,

$$\omega_{max} = \frac{2\pi \times N_{max}}{60} = \frac{2\pi \times 2500}{60} = 261.8 \text{ rad/s}$$

We know that maximum inertia force of reciprocating parts,

$$F_I = m_R\,(\omega_{max})^2 r \left(\cos\theta + \frac{\cos 2\theta}{n}\right) \qquad \text{... (i)}$$

The inertia force of reciprocating parts is maximum, when the crank is at inner dead centre, *i.e.* when $\theta = 0°$.

$$\therefore \quad F_I = m_R\,(\omega_{max})^2 r \left(1 + \frac{1}{n}\right) \qquad \text{... [From equation (i)]}$$

$$= 2(261.8)^2\, 0.075 \left(1 + \frac{1}{4.3}\right) = 12\,672 \text{ N}$$

Since the connecting rod is designed by taking the force in the connecting rod (F_C) equal to the maximum force on the piston due to gas pressure (F_L), therefore

Force in the connecting rod,

$$F_C = F_L = 23\,760 \text{ N}$$

---

* Superfluous data.

Consider the *I*-section of the connecting rod with the proportions as shown in Fig. 16.11. We have discussed in Art. 16.15 that for such a section

$$\frac{I_{xx}}{I_{yy}} = 3.2$$

or $$\frac{k^2_{xx}}{k^2_{yy}} = 3.2, \text{ which is satisfactory.}$$

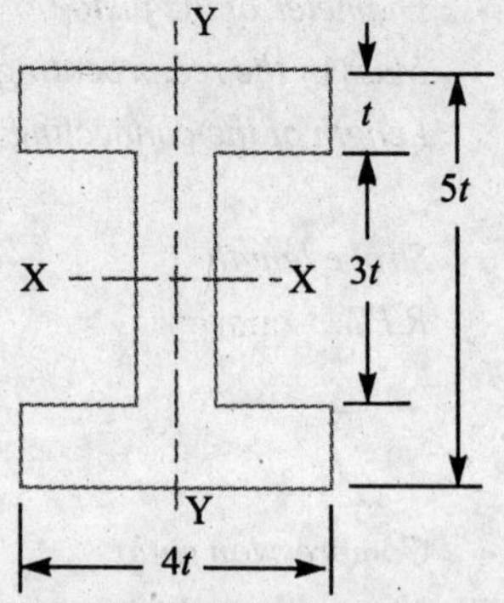

Fig. 16.11

We have also discussed that the connecting rod is designed for buckling about *X*-axis (*i.e.* in a plane of motion of the connecting rod), assuming both ends hinged. Taking a factor of safety as 6, the buckling load,

$$W_{cr} = F_C \times 6 = 23\,760 \times 6 = 142\,560 \text{ N}$$

and area of cross-section,

$$A = 2\,(4t \times t) + t \times 3t = 11\, t^2 \text{ mm}^2$$

Moment of inertia about *X*-axis,

$$I_{xx} = \left[\frac{4\,t\,(5\,t)^3}{12} - \frac{3\,t\,(3\,t)^3}{12}\right] = \frac{419\,t^4}{12} \text{ mm}^4$$

∴ Radius of gyration,

$$k_{xx} = \sqrt{\frac{I_{xx}}{A}} = \sqrt{\frac{419\,t^4}{12} \times \frac{1}{11\,t^2}} = 1.78\,t$$

We know that equivalent length of the rod for both ends hinged,

$$L = l = 325 \text{ mm}$$

Taking for mild steel, $\sigma_c = 320$ MPa = 320 N/mm² and $a = 1 / 7500$, we have from Rankine's formula,

$$W_{cr} = \frac{\sigma_c \times A}{1 + a\left(\dfrac{L}{k_{xx}}\right)^2}$$

$$142\,560 = \frac{320 \times 11\,t^2}{1 + \dfrac{1}{7500}\left(\dfrac{325}{1.78\,t}\right)^2}$$

$$40.5 = \frac{t^2}{1 + \dfrac{4.44}{t^2}} = \frac{t^4}{t^2 + 4.44}$$

or $t^4 - 40.5\,t^2 - 179.8 = 0$

$$\therefore \quad t^2 = \frac{40.5 \pm \sqrt{(40.5)^2 + 4 \times 179.8}}{2} = \frac{40.5 \pm 48.6}{2} = 44.55$$

... (Taking +ve sign)

or $t = 6.67$ say 6.8 mm

Therefore, dimensions of cross-section of the connecting rod are

Height $= 5\,t = 5 \times 6.8 = 34$ mm **Ans.**

Width $= 4\,t = 4 \times 6.8 = 27.2$ mm **Ans.**

Thickness of flange and web

$= t = 6.8 \text{ mm} = 0.0068 \text{ m}$ **Ans.**

Now let us find the bending stress due to inertia force on the connecting rod.

We know that the mass of the connecting rod per metre length,

$$m_1 = \text{Volume} \times \text{density} = \text{Area} \times \text{length} \times \text{density}$$

$$= A \times l \times \rho = 11\, t^2 \times l \times \rho \qquad ..(\because A = 11\, t^2)$$

$$= 11(0.0068)^2\, 1 \times 7800 = 3.97 \text{ kg} \qquad \text{... (Taking } \rho = 7800 \text{ kg/m}^3)$$

$\therefore$ Maximum bending moment,

$$M_{max} = m\, \omega^2\, r \times \frac{l}{9\sqrt{3}} = m_1\, \omega^2\, r \times \frac{l^2}{9\sqrt{3}} \qquad \text{... } (\because m = m_1 . l)$$

$$= 3.97\, (261.8)^2\, (0.075) \times \frac{(0.325)^2}{9\sqrt{3}} = 138.3 \text{ N-m}$$

and section modulus,

$$Z_{xx} = \frac{I_{xx}}{5\, t/2} = \frac{419\, t^4}{12} \times \frac{2}{5\, t} = \frac{419}{30}\, t^3$$

$$= \frac{419}{30}\, (0.0068)^3 = 4.4 \times 10^{-6} \text{ m}^3$$

$\therefore$ Maximum bending or whipping stress due to inertia bending forces,

$$\sigma_{b(max)} = \frac{M_{max}}{Z_{xx}} = \frac{138.3}{4.4 \times 10^{-6}} = 31.4 \times 10^6 \text{ N/m}^2$$

$= 31.4$ MPa, which is safe

Note : The maximum compressive stress in the connecting rod will be,

$$\sigma_{c(max)} = \text{Direct compressive stress} + \text{Maximum bending stress}$$

$$= \frac{320}{6} + 31.4 = 84.7 \text{ MPa}$$

EXERCISES

1. Compare the ratio of strength of a solid steel column to that of a hollow column of internal diameter equal to 3/4th of its external diameter. Both the columns have the same cross-sectional areas, lengths and end conditions. **[Ans. 25/7]**
2. Find the Euler's crippling load for a hollow cylindrical steel column of 38 mm external diameter and 35 mm thick. The length of the column is 2.3 m and hinged at its both ends. Take $E = 200$ GN/m^2. Also determine the crippling load by Rankine's formula, using

   $$\sigma_c = 320 \text{ MPa} ; \text{ and } a = \frac{1}{7500}$$ **[Ans. 17.25 kN ; 17.4 kN]**
3. Determine the diameter of the pistion rod of the hydraulic cylinder of 100 mm bore when the maximum hydraulic pressure in the cylinder is limited to 14 N/mm^2. The length of the piston rod is 1.2 m. The factor of safety may be taken as 5 and the end fixity coefficient as 2. **[Ans. 45 mm]**
4. Find the diameter of a piston rod for an engine of 200 mm diameter. The length of the piston rod is 0.9 m and the stroke is 0.5 m. The pressure of steam is 1 N/mm^2. Assume factor of safety as 5. **[Ans. 31 mm]**

5. Determine the diameter of the push rod made of mild steel of an I.C. engine if the maximum force exerted by the push rod is 1500 N. The length of the push rod is 0.5 m. Take the factor of safety as 2.5 and the end fixity coefficient as 2. **[Ans. 10 mm]**

6. The eccentric rod to drive the D-slide valve mechanism of a steam engine carries a maximum compressive load of 10 kN. The length of the rod is 1.5 m. Assuming the eccentric rod hinged at both the ends, find

   (*a*) diameter of the rod, and

   (*b*) dimensions of the cross-section of the rod if it is of rectangular section. The depth of the section is twice its thickness.

   Take factor of safety = 40 and E = 210 kN/mm^2. **[Ans. 60 mm ; 30 × 60 mm]**

Screwjacks

7. Determine the dimensions of an *I*-section connecting rod for an internal combustion engine having the following specifications :

Diameter of the piston	=	120 mm
Mass of reciprocating parts piston	=	350 kg/m^2 of area
Length of connecting rod	=	350 mm
Engine revolutions per minute	=	1800
Maximum explosion pressure	=	3 N/mm^2
Stroke length	=	180 mm

   The flange width and the depth of the *I*-section rod are in the ratio of 4 t : 6 t where t is the thickness of the flange and web. Assume yield stress in compression for the material as 330 MPa and a factor of safety as 6. **[Ans. t = 7.5 mm]**

8. The connecting rod of a four stroke cycle Diesel engine is of circular section and of length 550 mm. The diameter and stroke of the cylinder are 150 mm and 240 mm respectively. The maximum combustion pressure is 4.7 N/mm^2. Determine the diameter of the rod to be used, for a factor of safety of 3 with a material having a yield point of 330 MPa. Find also the maximum bending stress in the connecting rod due to whipping action if the engine runs at 1000 r.p.m. The specific weight of the material is 7800 kg/m^3.

   [Ans. 33.2 mm ; 48 MPa]

QUESTIONS

1. What do you understand by a column or strut ? Explain the various end conditions of a column or strut.
2. State the assumptions used in Euler's column theory.
3. Define 'slenderness ratio'. How it is used to define long and short columns ?
4. What is equivalent length of a column ? Write the relations between equivalent length and actual length of a column for various end conditions.
5. Explain Johnson's formula for columns. Describe the use of Johnson's formula and Euler's formula.
6. Write the formula for obtaining a maximum stress in a long column subjected to eccentric loading.

7. How the piston rod is designed ?
8. Explain the design procedure of valve push rods.
9. Why an *I*-Section is usually preferred to a round section in case of connecting rods?

OBJECTIVE TYPE QUESTIONS

1. A machine part is designed as a strut, when it is subjected to
   (*a*) an axial tensile force (*b*) an axial compressive force
   (*c*) a tangential force (*d*) any one of these
2. Slenderness ratio is the ratio of
   (*a*) maximum size of a column to minimum size of column
   (*b*) width of column to depth of column
   (*c*) effective length of column to least radius of gyration of the column
   (*d*) effective length of column to width of column
3. A connecting rod is designed as a
   (*a*) long column (*b*) short column
   (*c*) strut (*d*) any one of these
4. Which of the following formula is used in designing a connecting rod ?
   (*a*) Euler's formula (*b*) Rankine's formula
   (*c*) Johnson's straight line formula (*d*) Johnson's parabolic formula
5. A connecting rod subjected to an axial load may buckle with
   (*a*) *X*-axis as neutral axis (*b*) *Y*-axis as neutral axis
   (*c*) *X*-axis or *Y*-axis as neutral axis (*d*) *Z*-axis
6. In designing a connecting rod, it is considered like .... for buckling about *X*-axis.
   (*a*) both ends hinged (*b*) both ends fixed
   (*c*) one end fixed and the other end hinged (*d*) one end fixed and the other end free
7. A connecting rod should be
   (*a*) strong in buckling about *X*-axis
   (*b*) strong in buckling about *Y*-axis
   (*c*) equally strong in buckling about *X*-axis and *Y*-axis
   (*d*) any one of the above
8. The buckling will occur about *Y*-axis,if
   (*a*) $I_{xx} = I_{yy}$ (*b*) $I_{xx} = 4I_{yy}$
   (*c*) $I_{xx} > 4I_{yy}$ (*d*) $I_{xx} < 4I_{yy}$
9. The connecting rod will be equally strong in buckling about X-axis and Y-axis, if
   (*a*) $I_{xx} = I_{yy}$ (*b*) $I_{xx} = 2I_{yy}$
   (*c*) $I_{xx} = 3I_{yy}$ (*d*) $I_{xx} = 4I_{yy}$
10. The most suitable section for the connecting rod is
    (*a*) *L*-section (*b*) *T*-section
    (*c*) *I*-section (*d*) *C*-section

ANSWERS

1. (*b*)	2. (*c*)	3. (*c*)	4. (*b*)	5. (*c*)
6. (*a*)	7. (*c*)	8. (*c*)	9. (*d*)	10. (*c*)

CHAPTER

17

Power Screws

17.1 Introduction

The power screws (also known as ***translation screws***) are used to convert rotary motion into translatory motion. For example, in the case of the lead screw of lathe, the rotary motion is available but the tool has to be advanced in the direction of the cut against the cutting resistance of the material. In case of screw jack, a small force applied in the horizontal plane is used to raise or lower a large load. Power screws are also used in vices, testing machines, presses, etc.

In most of the power screws, the nut has axial motion against the resisting axial force while the screw rotates in its bearings. In some screws, the screw rotates and moves axially against the resisting force while the nut is stationary and in others the nut rotates while the screw moves axially with no rotation.

17.2 Types of Screw Threads used for Power Screws

Following are the three types of screw threads mostly used for power screws :

1. ***Square thread.*** A square thread, as shown in Fig. 17.1 (*a*), is adapted for the transmission of power in either direction. This thread results in maximum efficiency and minimum radial or bursting

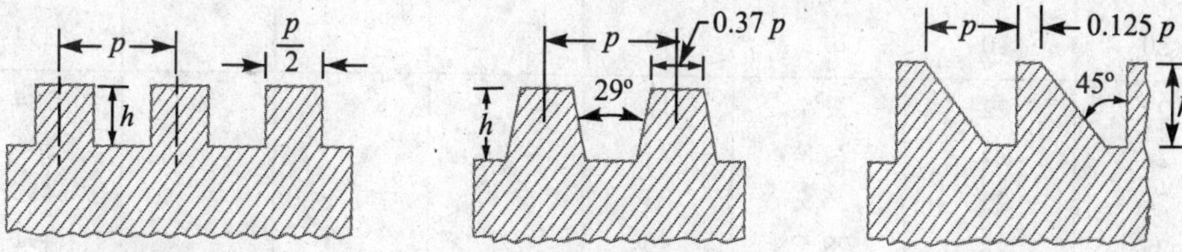

$h = 0.5\,p$
(*a*) Square thread.

$h = 0.5\,p + 0.25$ mm
(*b*) Acme thread.

$h = 0.75\,p$
(*c*) Buttress thread.

Fig. 17.1. Types of power screws.

pressure on the nut. It is difficult to cut with taps and dies. It is usually cut on a lathe with a single point tool and it can not be easily compensated for wear. The square threads are employed in screw jacks, presses and clamping devices. The standard dimensions for square threads according to IS : 4694 – 1968 (Reaffirmed 1996), are shown in Table 17.1 to 17.3.

Screw jacks

2. ***Acme or trapezoidal thread.*** An acme or trapezoidal thread, as shown in Fig. 17.1 (*b*), is a modification of square thread. The slight slope given to its sides lowers the efficiency slightly than square thread and it also introduce some bursting pressure on the nut, but increases its area in shear. It is used where a split nut is required and where provision is made to take up wear as in the lead screw of a lathe. Wear may be taken up by means of an adjustable split nut. An acme thread may be cut by means of dies and hence it is more easily manufactured than square thread. The standard dimensions for acme or trapezoidal threads are shown in Table 17.4 (Page 630).

3. ***Buttress thread.*** A buttress thread, as shown in Fig. 17.1 (*c*), is used when large forces act along the screw axis in one direction only. This thread combines the higher efficiency of square thread and the ease of cutting and the adaptability to a split nut of acme thread. It is stronger than other threads because of greater thickness at the base of the thread. The buttress thread has limited use for power transmission. It is employed as the thread for light jack screws and vices.

Table 17.1. Basic dimensions for square threads in mm (Fine series) according to IS : 4694 – 1968 (Reaffirmed 1996)

Nominal diameter (d_1)	*Major diameter*		*Minor diameter* (d_c)	*Pitch* (p)	*Depth of thread*		*Area of core* (A_c) mm^2
	Bolt (*d*)	*Nut* (*D*)			*Bolt* (*h*)	*Nut* (*H*)	
10	10	10.5	8	2	1	1.25	50.3
12	12	12.5	10				78.5

d_1	d	D	d_c	p	h	H	A_c
14	14	14.5	12				113
16	16	16.5	14	2	1	1.25	154
18	18	18.5	16				201
20	20	20.5	18				254
22	22	22.5	19				284
24	24	24.5	21				346
26	26	26.5	23				415
28	28	28.5	25				491
30	30	30.5	27				573
32	32	32.5	29				661
(34)	34	34.5	31				755
36	36	36.5	33	3	1.5	1.75	855
(38)	38	38.5	35				962
40	40	40.5	37				1075
42	42	42.5	39				1195
44	44	44.5	41				1320
(46)	46	46.5	43				1452
48	48	48.5	45				1590
50	50	50.5	47				1735
52	52	52.5	49				1886
55	55	55.5	52				2124
(58)	58	58.5	55				2376
60	60	60.5	57				2552
(62)	62	62.5	59				2734
65	65	65.5	61				2922
(68)	68	68.5	64				3217
70	70	70.5	66				3421
(72)	72	72.5	68				3632
75	75	75.5	71				3959
(78)	78	78.5	74				4301
80	80	80.5	76				4536
(82)	82	82.5	78				4778
(85)	85	85.5	81	4	2	2.25	5153
(88)	88	88.5	84				5542
90	90	90.5	86				5809
(92)	92	92.5	88				6082
95	95	95.5	91				6504
(98)	98	98.5	94				6960

d_1	d	D	d_c	p	h	H	A_c
100	100	100.5	96				7238
(105)	105	105.5	101	4	2	2.25	8012
110	110	110.5	106				8825
(115)	115	115.5	109				9331
120	120	120.5	114				10207
(125)	125	125.5	119				11 122
130	130	130.5	124				12 076
(135)	135	135.5	129				13 070
140	140	140.5	134				14 103
(145)	145	145.5	139	6	3	3.25	15 175
150	150	150.5	144				16 286
(155)	155	155.5	149				17437
160	160	160.5	154				18 627
(165)	165	165.5	159				19 856
170	170	170.5	164				21124
(175)	175	175.5	169				22 432

Note : Diameter within brackets are of second preference.

Table 17.2. Basic dimensions for square threads in mm (Normal series)according to IS : 4694 – 1968 (Reaffirmed 1996)

Nominal diameter (d_1)	*Major diameter*		*Minor diameter*	*Pitch*	*Depth of thread*		*Area of core* (A_c) mm^2
	Bolt (d)	*Nut* (D)	(d_c)	(p)	*Bolt* (h)	*Nut* (H)	
22	22	22.5	17				227
24	24	24.5	19				284
26	26	26.5	21	5	2.5	2.75	346
28	28	28.5	23				415
30	30	30.5	24				452
32	32	32.5	26	6	3	3.25	531
(34)	34	34.5	28				616
36	36	36.5	30				707
(38)	38	38.5	31				755
40	40	40.5	33	7	3.5	3.75	855
(42)	42	42.5	35				962
44	44	44.5	37				1075

d_1	d	D	d_c	p	h	H	A_c
(46)	46	46.5	38				1134
48	48	48.5	40	8	4	4.25	1257
50	50	50.5	42				1385
52	52	52.5	44				1521
55	55	55.5	46				1662
(58)	58	58.5	49	9	4.5	5.25	1886
(60)	60	60.5	51				2043
(62)	62	62.5	53				2206
65	65	65.5	55				2376
(68)	68	68.5	58	10	5	5.25	2642
70	70	70.5	60				2827
(72)	72	72.5	62				3019
75	75	75.5	65				3318
(78)	78	78.5	68				3632
80	80	80.5	70				3848
(82)	82	82.5	72				4072
85	85	85.5	73				41.85
(88)	88	88.5	76				4536
90	90	85.5	78	12	6	6.25	4778
(92)	92	92.5	80				5027
95	95	95.5	83				5411
(98)	98	98.5	86				5809
100	100	100.5	88				6082
(105)	105	105.5	93				6793
110	110	110.5	98				7543
(115)	115	116	101				8012
120	120	121	106				882
(125)	125	126	111	14	7	7.5	9677
130	130	131	116				10 568
(135)	135	136	121				11 499
140	140	141	126				12 469
(145)	145	146	131				13 478
150	150	151	134				14 103
(155)	155	156	139	16	8	8.5	15 175
160	160	161	144				16 286

d_1	d	D	d_c	p	h	H	A_c
(165)	165	166	149				17 437
170	170	171	154	16	8	8.5	18 627
(175)	175	176	159				19 856

Note : Diameter within brackets are of second preference.

Table 17.3. Basic dimensions for square threads in mm (Coarse series) according toIS : 4694 – 1968 (Reaffirmed 1996)

Nominal diameter	*Major diameter*		*Minor diameter*	*Pitch*	*Depth of thread*		*Area of core*
(d_1)	*Bolt* (d)	*Nut* (D)	(d_c)	(p)	*Bolt* (h)	*Nut* (H)	(A_c) mm^2
22	22	22.5	14				164
24	24	24.5	16	8	4	4.25	204
26	26	26.5	18				254
28	28	28.5	20				314
30	30	30.5	20				314
32	32	32.5	22				380
(34)	34	34.5	24	10	5	5.25	452
36	36	36.5	26				531
(38)	38	38.5	28				616
40	40	40.5	28				616
(42)	42	42.5	30				707
44	44	44.5	32				804
(46)	46	46.5	34	12	6	6.25	908
48	48	48.5	36				1018
50	50	50.5	38				1134
52	52	52.5	40				1257
55	55	56	41				1320
(58)	58	59	44	14	7	7.25	1521
60	60	61	46				1662
(62)	62	63	48				1810
65	65	66	49				1886
(68)	68	69	52	16	8	8.5	2124
70	70	71	54				2290
(72)	72	73	56				2463
75	75	76	59				2734
(78)	78	79	62				3019
80	80	81	64				3217
(82)	82	83	66				3421

d_1	d	D	d_c	p	h	H	A_c
85	85	86	67				3526
(88)	88	89	70				3848
90	90	91	72				4072
(92)	92	93	74	18	9	9.5	4301
95	95	96	77				4657
(96)	96	99	80				5027
100	100	101	80				5027
(105)	105	106	85	20	10	10.5	5675
110	110	111	90				6362
(115)	115	116	93				6793
120	120	121	98				7543
(125)	125	126	103	22	11	11.5	8332
130	130	131	108				9161
(135)	135	136	111				9667
140	140	141	116	24	12	12.5	10 568
(145)	145	146	121				11 499
150	150	151	126				12 469
(155)	155	156	131				13 478
160	160	161	132				13 635
(165)	165	166	137				14 741
170	170	171	142	28	14	14.5	15 837
(175)	175	176	147				16 972

Note : Diameters within brackets are of second preference.

Table 17.4. Basic dimensions for trapezoidal/Acme threads.

Nominal or major diameter (d) mm.	*Minor or core diameter (d_c) mm*	*Pitch (p) mm*	*Area of core (A_c) mm²*
10	6.5	3	33
12	8.5		57
14	9.5		71
16	11.5	4	105
18	13.5		143
20	15.5		189
22	16.5		214
24	18.5	5	269
26	20.5		330
28	22.5		389
30	23.5		434
32	25.5	6	511
34	27.5		594
36	29.5		683

d	d_c	p	A_c
38	30.5		731
40	32.5	7	830
42	34.5		935
44	36.5		1046
46	37.5		1104
48	39.5	8	1225
50	41.5		1353
52	43.5		1486
55	45.5		1626
58	48.5	9	1847
60	50.5		2003
62	52.5		2165
65	54.5		2333
68	57.5		2597
70	59.5	10	2781
72	61.5		2971
75	64.5		3267
78	67.5		3578
80	69.5		3794
82	71.5		4015
85	72.5		4128
88	75.5		4477
90	77.5		4717
92	79.5		4964
95	82.5	12	5346
98	85.5		5741
100	87.5		6013
105	92.5		6720
110	97.5		7466
115	100		7854
120	105		8659
125	110		9503
130	115	14	10 387
135	120		11 310
140	125		12 272
145	130		13 273
150	133		13 893
155	138		14 957
160	143		16 061
165	148	16	17 203
170	153		18 385
175	158		19 607

17.3 Multiple Threads

The power screws with multiple threads such as double, triple etc. are employed when it is desired to secure a large lead with fine threads or high efficiency. Such type of threads are usually found in high speed actuators.

17.4 Torque Required to Raise Load by Square Threaded Screws

The torque required to raise a load by means of square threaded screw may be determined by considering a screw jack as shown in Fig. 17.2 (*a*). The load to be raised or lowered is placed on the head of the square threaded rod which is rotated by the application of an effort at the end of lever for lifting or lowering the load.

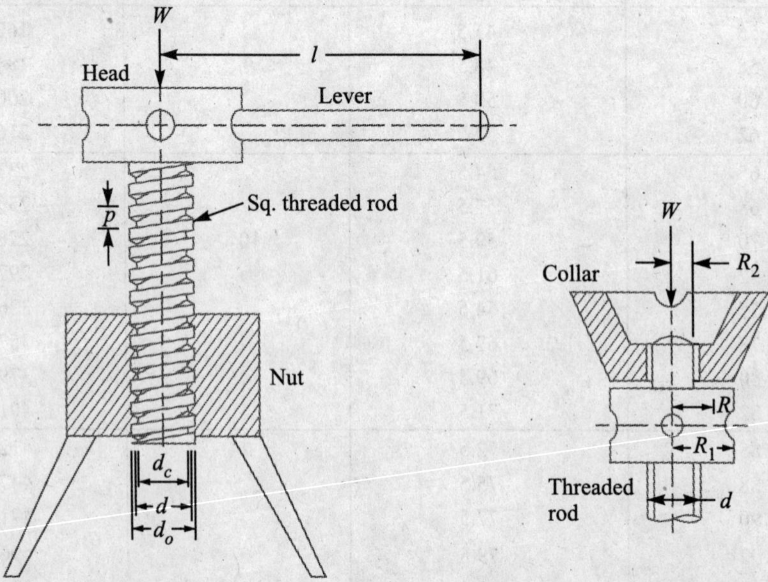

(*a*) Screw jack. (*b*) Thrust collar.

Fig. 17.2

A little consideration will show that if one complete turn of a screw thread be imagined to be unwound, from the body of the screw and developed, it will form an inclined plane as shown in Fig. 17.3 (*a*).

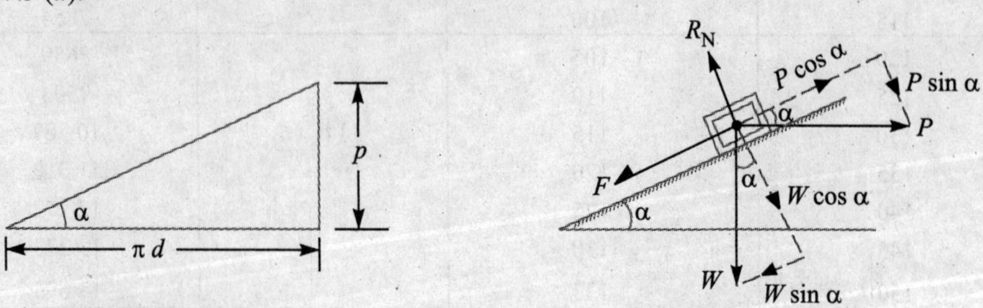

(*a*) Development of a screw. (*b*) Forces acting on the screw.

Fig. 17.3

Let p = Pitch of the screw,

d = Mean diameter of the screw,

α = Helix angle,

P = Effort applied at the circumference of the screw to lift the load,

W = Load to be lifted, and

μ = Coefficient of friction, between the screw and nut

= $\tan \phi$, where ϕ is the friction angle.

From the geometry of the Fig. 17.3 (*a*), we find that

$$\tan \alpha = p / \pi d$$

Since the principle, on which a screw jack works is similar to that of an inclined plane, therefore the force applied on the circumference of a screw jack may be considered to be horizontal as shown in Fig. 17.3 (*b*).

Since the load is being lifted, therefore the force of friction ($F = \mu.R_N$) will act downwards. All the forces acting on the body are shown in Fig. 17.3 (*b*).

Screw jack

Resolving the forces along the plane,

$$P \cos \alpha = W \sin \alpha + F = W \sin \alpha + \mu.R_N \quad \text{...}(i)$$

and resolving the forces perpendicular to the plane,

$$R_N = P \sin \alpha + W \cos \alpha \quad \text{...}(ii)$$

Substituting this value of R_N in equation (*i*), we have

$$P \cos \alpha = W \sin \alpha + \mu (P \sin \alpha + W \cos \alpha)$$
$$= W \sin \alpha + \mu P \sin \alpha + \mu W \cos \alpha$$

or $\quad P \cos \alpha - \mu P \sin \alpha = W \sin \alpha + \mu W \cos \alpha$

or $\quad P (\cos \alpha - \mu \sin \alpha) = W (\sin \alpha + \mu \cos \alpha)$

$$\therefore \quad P = W \times \frac{(\sin \alpha + \mu \cos \alpha)}{(\cos \alpha - \mu \sin \alpha)}$$

Substituting the value of $\mu = \tan \phi$ in the above equation, we get

or $$P = W \times \frac{\sin \alpha + \tan \phi \cos \alpha}{\cos \alpha - \tan \phi \sin \alpha}$$

Multiplying the numerator and denominator by $\cos \phi$, we have

$$P = W \times \frac{\sin \alpha \cos \phi + \sin \phi \cos \alpha}{\cos \alpha \cos \phi - \sin \alpha \sin \phi}$$
$$= W \times \frac{\sin (\alpha + \phi)}{\cos (\alpha + \phi)} = W \tan (\alpha + \phi)$$

$\therefore$ Torque required to overcome friction between the screw and nut,

$$T_1 = P \times \frac{d}{2} = W \tan (\alpha + \phi) \frac{d}{2}$$

When the axial load is taken up by a thrust collar as shown in Fig. 17.2 (*b*), so that the load does not rotate with the screw, then the torque required to overcome friction at the collar,

$$T_2 = \frac{2}{3} \times \mu_1 \times W \left[\frac{(R_1)^3 - (R_2)^3}{(R_1)^2 - (R_2)^2} \right]$$

... (Assuming uniform pressure conditions)

$$= \mu_1 \times W \left(\frac{R_1 + R_2}{2} \right) = \mu_1 \, W R$$

....(Assuming uniform wear conditions)

where R_1 and R_2 = Outside and inside radii of collar,

R = Mean radius of collar = $\dfrac{R_1 + R_2}{2}$, and

μ_1 = Coefficient of friction for the collar.

∴ Total torque required to overcome friction (*i.e.* to rotate the screw),

$$T = T_1 + T_2$$

If an effort P_1 is applied at the end of a lever of arm length l, then the total torque required to overcome friction must be equal to the torque applied at the end of lever, *i.e.*

$$T = P \times \frac{d}{2} = P_1 \times l$$

Notes: 1. When the *nominal diameter (d_o) and the **core diameter (d_c) of the screw is given, then

Mean diameter of screw, $$d = \frac{d_o + d_c}{2} = d_o - \frac{p}{2} = d_c + \frac{p}{2}$$

2. Since the mechanical advantage is the ratio of the load lifted (W) to the effort applied (P_1) at the end of the lever, therefore mechanical advantage,

$$\text{M.A.} = \frac{W}{P_1} = \frac{W \times 2l}{P \times d} \quad \ldots \left(\because P \times \frac{d}{2} = P_1 \times l \text{ or } P_1 = \frac{P \times d}{2l} \right)$$

$$= \frac{W \times 2l}{W \tan(\alpha + \phi)\, d} = \frac{2l}{d \tan(\alpha + \phi)}$$

17.5 Torque Required to Lower Load by Square Threaded Screws

A little consideration will show that when the load is being lowered, the force of friction ($F = \mu.R_N$) will act upwards. All the forces acting on the body are shown in Fig. 17.4.

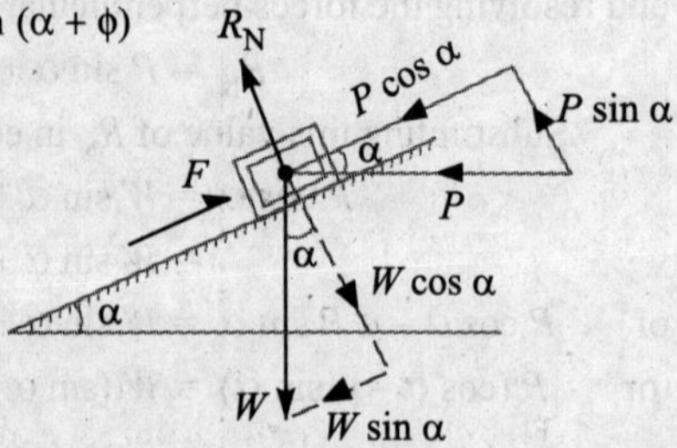

Fig. 17.4

Resolving the forces along the plane,

$$P \cos \alpha = F - W \sin \alpha$$

$$= \mu\, R_N - W \sin \alpha \quad \ldots(i)$$

and resolving the forces perpendicular to the plane,

$$R_N = W \cos \alpha - P \sin \alpha \quad \ldots(ii)$$

Substituting this value of R_N in equation (*i*), we have,

$$P \cos \alpha = \mu\,(W \cos \alpha - P \sin \alpha) - W \sin \alpha$$

$$= \mu\, W \cos \alpha - \mu\, P \sin \alpha - W \sin \alpha$$

or $$P \cos \alpha + \mu\, P \sin \alpha = \mu W \cos \alpha - W \sin \alpha$$

$$P\,(\cos \alpha + \mu \sin \alpha) = W\,(\mu \cos \alpha - \sin \alpha)$$

or $$P = W \times \frac{(\mu \cos \alpha - \sin \alpha)}{(\cos \alpha + \mu \sin \alpha)}$$

Substituting the value of $\mu = \tan \phi$ in the above equation, we have

$$P = W \times \frac{(\tan \phi \cos \alpha - \sin \alpha)}{(\cos \alpha + \tan \phi \sin \alpha)}$$

Multiplying the numerator and denominator by $\cos \phi$, we have

$$P = W \times \frac{(\sin \phi \cos \alpha - \cos \phi \sin \alpha)}{(\cos \phi \cos \alpha + \sin \phi \sin \alpha)}$$

$$= W \times \frac{\sin(\phi - \alpha)}{\cos(\phi - \alpha)} = W \tan(\phi - \alpha)$$

* The nominal diameter of a screw thread is also known as ***outside diameter*** or ***major diameter***.

** The core diameter of a screw thread is also known as ***inner diameter*** or ***root diameter*** or ***minor diameter***.

∴ Torque required to overcome friction between the screw and nut,

$$T_1 = P \times \frac{d}{2} = W \tan(\phi - \alpha)\frac{d}{2}$$

Note : When $\alpha > \phi$, then $P = W \tan(\alpha - \phi)$.

17.6 Efficiency of Square Threaded Screws

The efficiency of square threaded screws may be defined as the ratio between the ideal effort (*i.e.* the effort required to move the load, neglecting friction) to the actual effort (*i.e.* the effort required to move the load taking friction into account).

We have seen in Art. 17.4 that the effort applied at the circumference of the screw to lift the load is

$$P = W \tan(\alpha + \phi) \qquad ...(i)$$

where W = Load to be lifted,

α = Helix angle,

ϕ = Angle of friction, and

μ = Coefficient of friction between the screw and nut = $\tan \phi$.

If there would have been no friction between the screw and the nut, then ϕ will be equal to zero. The value of effort P_0 necessary to raise the load, will then be given by the equation,

$$P_0 = W \tan \alpha \qquad \text{[Substituting } \phi = 0 \text{ in equation } (i)]$$

$$\therefore \text{ Efficiency,} \quad \eta = \frac{\text{Ideal effort}}{\text{Actual effort}} = \frac{P_0}{P} = \frac{W \tan \alpha}{W \tan(\alpha + \phi)} = \frac{\tan \alpha}{\tan(\alpha + \phi)}$$

This shows that the efficiency of a screw jack, is independent of the load raised.

In the above expression for efficiency, only the screw friction is considered. However, if the screw friction and collar friction is taken into account, then

$$\eta = \frac{\text{Torque required to move the load, neglecting friction}}{\text{Torque required to move the load, including screw and collar friction}}$$

$$= \frac{T_0}{T} = \frac{P_0 \times d/2}{P \times d/2 + \mu_1.W.R}$$

Note: The efficiency may also be defined as the ratio of mechanical advantage to the velocity ratio.

We know that mechanical advantage,

$$\text{M.A.} = \frac{W}{P_1} = \frac{W \times 2l}{P \times d} = \frac{W \times 2l}{W \tan(\alpha + \phi)\, d} = \frac{2l}{d \tan(\alpha + \phi)} \qquad \text{...(Refer Art .17.4)}$$

and velocity ratio,

$$V.R. = \frac{\text{Distance moved by the effort } (P_1) \text{ in one revolution}}{\text{Distance moved by the load } (W) \text{ in one revolution}}$$

$$= \frac{2\pi l}{p} = \frac{2\pi l}{\tan \alpha \times \pi d} = \frac{2l}{d \tan \alpha} \qquad \text{... } (\because \tan \alpha = p/\pi d)$$

$$\therefore \text{ Efficiency,} \quad \eta = \frac{M.A.}{V.R.} = \frac{2l}{d \tan(\alpha + \phi)} \times \frac{d \tan \alpha}{2l} = \frac{\tan \alpha}{\tan(\alpha + \phi)}$$

17.7 Maximum Efficiency of a Square Threaded Screw

We have seen in Art. 17.6 that the efficiency of a square threaded screw,

$$\eta = \frac{\tan \alpha}{\tan(\alpha + \phi)} = \frac{\sin \alpha / \cos \alpha}{\sin(\alpha + \phi)/\cos(\alpha + \phi)} = \frac{\sin \alpha \times \cos(\alpha + \phi)}{\cos \alpha \times \sin(\alpha + \phi)} \qquad ...(i)$$

Multiplying the numerator and denominator by 2, we have,

$$\eta = \frac{2 \sin \alpha \times \cos (\alpha + \phi)}{2 \cos \alpha \times \sin (\alpha + \phi)} = \frac{\sin (2\alpha + \phi) - \sin \phi}{\sin (2\alpha + \phi) + \sin \phi} \qquad ...(ii)$$

$$\ldots \left[\begin{array}{l} \because 2 \sin A \cos B = \sin (A + B) + \sin (A - B) \\ 2 \cos A \sin B = \sin (A + B) - \sin (A - B) \end{array}\right]$$

The efficiency given by equation (*ii*) will be maximum when $\sin (2\alpha + \phi)$ is maximum, *i.e.* when

$$\sin (2\alpha + \phi) = 1 \quad \text{or} \quad \text{when} \quad 2\alpha + \phi = 90°$$

$$\therefore \quad 2\alpha = 90° - \phi \quad \text{or} \quad \alpha = 45° - \phi / 2$$

Substituting the value of 2α in equation (*ii*), we have maximum efficiency,

$$\eta_{max} = \frac{\sin (90° - \phi + \phi) - \sin \phi}{\sin (90° - \phi + \phi) + \sin \phi} = \frac{\sin 90° - \sin \phi}{\sin 90° + \sin \phi} = \frac{1 - \sin \phi}{1 + \sin \phi}$$

Example 17.1. *A vertical screw with single start square threads of 50 mm mean diameter and 12.5 mm pitch is raised against a load of 10 kN by means of a hand wheel, the boss of which is threaded to act as a nut. The axial load is taken up by a thrust collar which supports the wheel boss and has a mean diameter of 60 mm. The coefficient of friction is 0.15 for the screw and 0.18 for the collar. If the tangential force applied by each hand to the wheel is 100 N, find suitable diameter of the hand wheel.*

Solution. Given : $d = 50$ mm ; $p = 12.5$ mm ; $W = 10$ kN $= 10 \times 10^3$ N ; $D = 60$ mm or $R = 30$ mm ; $\mu = \tan \phi = 0.15$; $\mu_1 = 0.18$; $P_1 = 100$ N

We know that $\tan \alpha = \dfrac{p}{\pi d} = \dfrac{12.5}{\pi \times 50} = 0.08$

and the tangential force required at the circumference of the screw,

$$P = W \tan (\alpha + \phi) = W \left(\frac{\tan \alpha + \tan \phi}{1 - \tan \alpha \tan \phi}\right)$$

$$= 10 \times 10^3 \left[\frac{0.08 + 0.15}{1 - 0.08 \times 0.15}\right] = 2328 \text{ N}$$

We also know that the total torque required to turn the hand wheel,

$$T = P \times \frac{d}{2} + \mu_1 W R = 2328 \times \frac{50}{2} + 0.18 \times 10 \times 10^3 \times 30 \text{ N-mm}$$

$$= 58\,200 + 54\,000 = 112\,200 \text{ N-mm} \qquad ...(i)$$

Let D_1 = Diameter of the hand wheel in mm.

We know that the torque applied to the handwheel,

$$T = 2 P_1 \times \frac{D_1}{2} = 2 \times 100 \times \frac{D_1}{2} = 100 D_1 \text{ N-mm} \qquad ...(ii)$$

Equating equations (*i*) and (*ii*),

$$D_1 = 112\,200 / 100 = 1122 \text{ mm} = 1.122 \text{ m} \quad \textbf{Ans.}$$

Example 17.2. *An electric motor driven power screw moves a nut in a horizontal plane against a force of 75 kN at a speed of 300 mm / min. The screw has a single square thread of 6 mm pitch on a major diameter of 40 mm. The coefficient of friction at screw threads is 0.1. Estimate power of the motor.*

Solution. Given : $W = 75$ kN $= 75 \times 10^3$ N ; $v = 300$ mm/min ; $p = 6$ mm ; $d_o = 40$ mm ; $\mu = \tan \phi = 0.1$

We know that mean diameter of the screw,

$$d = d_o - p/2 = 40 - 6/2 = 37 \text{ mm}$$

and
$$\tan\alpha = \frac{p}{\pi d} = \frac{6}{\pi \times 37} = 0.0516$$

We know that tangential force required at the circumference of the screw,

$$P = W\tan(\alpha + \phi) = W\left[\frac{\tan\alpha + \tan\phi}{1 - \tan\alpha\tan\phi}\right]$$

$$= 75 \times 10^3\left[\frac{0.0516 + 0.1}{1 - 0.0516 \times 0.1}\right] = 11.43 \times 10^3 \text{ N}$$

and torque required to operate the screw,

$$T = P \times \frac{d}{2} = 11.43 \times 10^3 \times \frac{37}{2} = 211.45 \times 10^3 \text{ N-mm} = 211.45 \text{ N-m}$$

Since the screw moves in a nut at a speed of 300 mm / min and the pitch of the screw is 6 mm, therefore speed of the screw in revolutions per minute (r.p.m.),

$$N = \frac{\text{Speed in mm/min.}}{\text{Pitch in mm}} = \frac{300}{6} = 50 \text{ r.p.m.}$$

and angular speed,
$$\omega = 2\pi N/60 = 2\pi \times 50/60 = 5.24 \text{ rad/s}$$

∴ Power of the motor $= T.\omega = 211.45 \times 5.24 = 1108 \text{ W} = 1.108 \text{ kW}$ **Ans.**

Example. 17.3. *The cutter of a broaching machine is pulled by square threaded screw of 55 mm external diameter and 10 mm pitch. The operating nut takes the axial load of 400 N on a flat surface of 60 mm and 90 mm internal and external diameters respectively. If the coefficient of friction is 0.15 for all contact surfaces on the nut, determine the power required to rotate the operating nut when the cutting speed is 6 m/min. Also find the efficiency of the screw.*

Solution. Given : $d_o = 55$ mm ; $p = 10$ mm $= 0.01$ m ; $W = 400$ N ; $D_2 = 60$ mm or $R_2 = 30$ mm ; $D_1 = 90$ mm or $R_1 = 45$ mm ; $\mu = \tan\phi = \mu_1 = 0.15$; Cutting speed = 6 m / min

Power required to operate the nut

We know that the mean diameter of the screw,

$$d = d_o - p/2 = 55 - 10/2 = 50 \text{ mm}$$

∴
$$\tan\alpha = \frac{p}{\pi d} = \frac{10}{\pi \times 50} = 0.0637$$

and force required at the circumference of the screw,

$$P = W\tan(\alpha + \phi) = W\left[\frac{\tan\alpha + \tan\phi}{1 - \tan\alpha\tan\phi}\right]$$

$$= 400\left[\frac{0.0637 + 0.15}{1 - 0.0637 \times 0.15}\right] = 86.4 \text{ N}$$

We know that mean radius of the flat surface,

$$R = \frac{R_1 + R_2}{2} = \frac{45 + 30}{2} = 37.5 \text{ mm}$$

∴ Total torque required,

$$T = P \times \frac{d}{2} + \mu_1 W R = 86.4 \times \frac{50}{2} + 0.15 \times 400 \times 37.5 \text{ N-mm}$$

$$= 4410 \text{ N-mm} = 4.41 \text{ N-m}$$

We know that speed of the screw,

$$N = \frac{\text{Cutting speed}}{\text{Pitch}} = \frac{6}{0.01} = 600 \text{ r.p.m}$$

and angular speed, $\omega = 2\pi N/60 = 2\pi \times 600/60 = 62.84$ rad / s

∴ Power required to operate the nut

$$= T.\omega = 4.41 \times 62.84 = 277 \text{ W} = 0.277 \text{ kW} \quad \textbf{Ans.}$$

Efficiency of the screw

We know that the efficiency of the screw,

$$\eta = \frac{T_0}{T} = \frac{W \tan\alpha \times d/2}{T} = \frac{400 \times 0.0637 \times 50/2}{4410}$$

$$= 0.144 \quad \text{or} \quad 14.4\% \quad \textbf{Ans.}$$

Example 17.4. *A vertical two start square threaded screw of a 100 mm mean diameter and 20 mm pitch supports a vertical load of 18 kN. The axial thrust on the screw is taken by a collar bearing of 250 mm outside diameter and 100 mm inside diameter. Find the force required at the end of a lever which is 400 mm long in order to lift and lower the load. The coefficient of friction for the vertical screw and nut is 0.15 and that for collar bearing is 0.20.*

Solution. Given : d = 100 mm ; p = 20 mm ; W = 18 kN = 18 × 10^3N ; D_1 = 250 mm or R_1 = 125 mm ; D_2 = 100 mm or R_2 = 50 mm ; l = 400 mm ; $\mu = \tan\phi = 0.15$; $\mu_1 = 0.20$

Force required at the end of lever

Let P = Force required at the end of lever.

Since the screw is a two start square threaded screw, therefore lead of the screw

$$= 2p = 2 \times 20 = 40 \text{ mm}$$

We know that $\tan\alpha = \frac{\text{Lead}}{\pi d} = \frac{40}{\pi \times 100} = 0.127$

1. ***For raising the load***

We know that tangential force required at the circumference of the screw,

$$P = W\tan(\alpha + \phi) = W\left[\frac{\tan\alpha + \tan\phi}{1 - \tan\alpha\tan\phi}\right]$$

$$= 18 \times 10^3 \left[\frac{0.127 + 0.15}{1 - 0.127 \times 0.15}\right] = 5083 \text{ N}$$

and mean radius of the collar,

$$R = \frac{R_1 + R_2}{2} = \frac{125 + 50}{2} = 87.5 \text{ mm}$$

∴ Total torque required at the end of lever,

$$T = P \times \frac{d}{2} + \mu_1 WR$$

$$= 5083 \times \frac{100}{2} + 0.20 \times 18 \times 10^3 \times 87.5 = 569\,150 \text{ N-mm}$$

We know that torque required at the end of lever (T),

$$569\,150 = P_1 \times l = P_1 \times 400 \quad \text{or} \quad P_1 = 569\,150/400 = 1423 \text{ N} \quad \textbf{Ans.}$$

2. ***For lowering the load***

We know that tangential force required at the circumference of the screw,

$$P = W\tan(\phi - \alpha) = W\left[\frac{\tan\phi - \tan\alpha}{1 + \tan\phi\tan\alpha}\right]$$

$$= 18 \times 10^3 \left[\frac{0.15 - 0.127}{1 + 0.15 \times 0.127}\right] = 406.3 \text{ N}$$

and the total torque required the end of lever,

$$T = P \times \frac{d}{2} + \mu_1 \, W R$$

$$= 406.3 \times \frac{100}{2} + 0.20 \times 18 \times 10^3 \times 87.5 = 335\,315 \text{ N-mm}$$

We know that torque required at the end of lever (T),

$$335\,315 = P_1 \times l = P_1 \times 400 \quad \text{or} \quad P_1 = 335\,315 / 400 = 838.3 \text{ N } \textbf{Ans.}$$

Example 17.5. The *mean diameter of the square threaded screw having pitch of 10 mm is 50 mm. A load of 20 kN is lifted through a distance of 170 mm. Find the work done in lifting the load and the efficiency of the screw, when*

1. *The load rotates with the screw, and*
2. *The load rests on the loose head which does not rotate with the screw.*

The external and internal diameter of the bearing surface of the loose head are 60 mm and 10 mm respectively. The coefficient of friction for the screw and the bearing surface may be taken as 0.08.

Solution. Given : p = 10 mm ; d = 50 mm ; W = 20 kN = 20 × 10^3 N ; D_1 = 60 mm or R_2 = 30 mm ; D_2 = 10 mm or R_2 = 5 mm ; $\mu = \tan \phi = \mu_1 = 0.08$

We know that $$\tan \alpha = \frac{p}{\pi d} = \frac{10}{\pi \times 50} = 0.0637$$

∴ Force required at the circumference of the screw to lift the load,

$$P = W \tan (\alpha + \phi) = W \left[\frac{\tan \alpha + \tan \phi}{1 - \tan \alpha \tan \phi} \right]$$

$$= 20 \times 10^3 \left[\frac{0.0637 + 0.08}{1 - 0.0673 \times 0.08} \right] = 2890 \text{ N}$$

and torque required to overcome friction at the screw,

$$T = P \times d / 2 = 2890 \times 50 / 2 = 72\,250 \text{ N-mm} = 72.25 \text{ N-m}$$

Since the load is lifted through a vertical distance of 170 mm and the distance moved by the screw in one rotation is 10 mm (equal to pitch), therefore number of rotations made by the screw,

$$N = 170 / 10 = 17$$

1. *When the load rotates with the screw*

We know that workdone in lifting the load

$$= T \times 2 \pi N = 72.25 \times 2\pi \times 17 = 7718 \text{ N-m} \quad \textbf{Ans.}$$

and efficiency of the screw,

$$\eta = \frac{\tan \alpha}{\tan (\alpha + \phi)} = \frac{\tan \alpha \,(1 - \tan \alpha \tan \phi)}{\tan \alpha + \tan \phi}$$

$$= \frac{0.0637\,(1 - 0.0637 \times 0.08)}{0.0637 + 0.08} = 0.441 \text{ or } 44.1\% \qquad \textbf{Ans.}$$

2. *When the load does not rotate with the screw*

We know that mean radius of the bearing surface,

$$R = \frac{R_1 + R_2}{2} = \frac{30 + 5}{2} = 17.5 \text{ mm}$$

and torque required to overcome friction at the screw and the collar,

$$T = P \times \frac{d}{2} + \mu_1 \, W R$$

$$= 2890 \times \frac{50}{2} + 0.08 \times 20 \times 10^3 \times 17.5 = 100\ 250 \text{ N-mm}$$

$$= 100.25 \text{ N-m}$$

∴ Workdone by the torque in lifting the load

$$= T \times 2\pi N = 100.25 \times 2\pi \times 17 = 10\ 710 \text{ N-m } \textbf{Ans.}$$

We know that torque required to lift the load, neglecting friction,

$$T_0 = P_0 \times d/2 = W \tan \alpha \times d/2 \qquad \text{... } (P_o = W \tan \alpha)$$

$$= 20 \times 10^3 \times 0.0637 \times 50 / 2 = 31\ 850 \text{ N-mm} = 31.85 \text{ N-m}$$

∴ Efficiency of the screw,

$$\eta = \frac{T_0}{T} = \frac{31.85}{100.25} = 0.318 \text{ or } 31.8\% \quad \textbf{Ans.}$$

17.8 Efficiency Vs Helix Angle

We have seen in Art. 17.6 that the efficiency of a square threaded screw depends upon the helix angle α and the friction angle φ. The variation of efficiency of a square threaded screw for raising the load with the helix angle α is shown in Fig. 17.5. We see that the efficiency of a square threaded screw increases rapidly upto helix angle of 20°, after which the increase in efficiency is slow. The efficiency is maximum for helix angle between 40 to 45°.

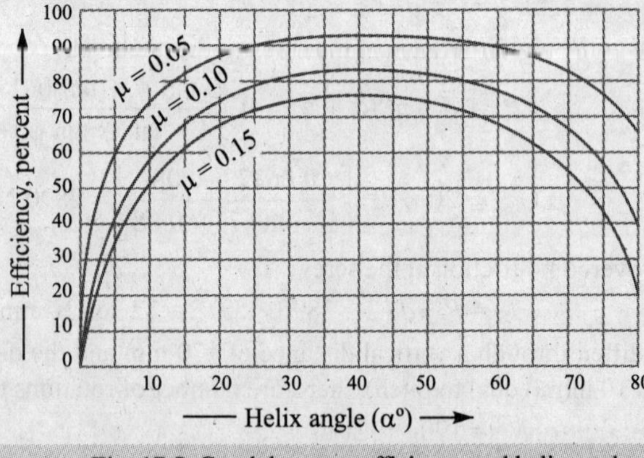

Fig. 17.5. Graph between efficiency and helix angle.

When the helix angle further increases say 70°, the efficiency drops. This is due to the fact that the normal thread force becomes large and thus the force of friction and the work of friction becomes large as compared with the useful work. This results in low efficiency.

17.9 Over Hauling and Self Locking Screws

We have seen in Art. 17.5 that the effort required at the circumference of the screw to lower the load is

$$P = W \tan (\phi - \alpha)$$

and the torque required to lower the load,

$$T = P \times \frac{d}{2} = W \tan (\phi - \alpha) \frac{d}{2}$$

In the above expression, if φ < α, then torque required to lower the load will be ***negative.*** In other words, the load will start moving downward without the application of any torque. Such a condition is known as ***over hauling of screws.*** If however, φ > α, the torque required to lower the load will be ***positive,*** indicating that an effort is applied to lower the load. Such a screw is known as

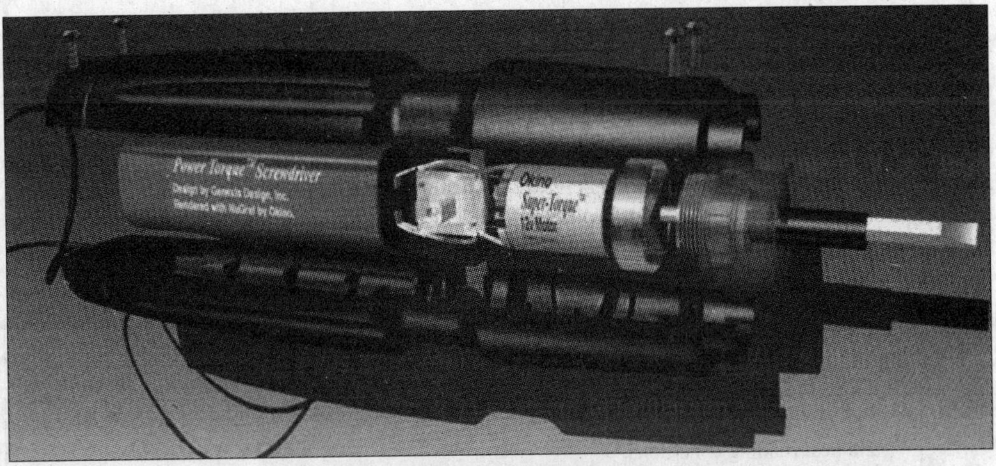

Mechanical power screw driver

self locking screw. In other words, a screw will be self locking if the friction angle is greater than helix angle or coefficient of friction is greater than tangent of helix angle *i.e.* μ or $\tan \phi > \tan \alpha$.

17.10 Efficiency of Self Locking Screws

We know that the efficiency of screw,

$$\eta = \frac{\tan \phi}{\tan (\alpha + \phi)}$$

and for self locking screws, $\phi \geq \alpha$ or $\alpha \leq \phi$.

$\therefore$ Efficiency for self locking screws,

$$\eta \leq \frac{\tan \phi}{\tan (\phi + \phi)} \leq \frac{\tan \phi}{\tan 2\phi} \leq \frac{\tan \phi\,(1 - \tan^2 \phi)}{2 \tan \phi} \leq \frac{1}{2} - \frac{\tan^2 \phi}{2}$$

$$\ldots \left(\because \tan 2\phi = \frac{2 \tan \phi}{1 - \tan^2 \phi} \right)$$

From this expression we see that efficiency of self locking screws is less than $\frac{1}{2}$ or 50%. If the efficiency is more than 50%, then the screw is said to be overhauling.

Note: It can be proved as follows:

Let W = Load to be lifted, and

h = Distance through which the load is lifted.

$\therefore$ Output $= W.h$

and Input $= \dfrac{\text{Output}}{\eta} = \dfrac{W.h}{\eta}$

$\therefore$ Work lost in overcoming friction

$$= \text{Input} - \text{Output} = \frac{W.h}{\eta} - W.h = W.h\left(\frac{1}{\eta} - 1\right)$$

For self locking,

$$W.h\left(\frac{1}{\eta} - 1\right) \leq W.h$$

$$\therefore \quad \frac{1}{\eta} - 1 \leq 1 \quad \text{or} \quad \eta \leq \frac{1}{2} \quad \text{or} \quad 50\%$$

17.11 Coefficient of Friction

The coefficient of friction depends upon various factors like *material of screw and nut, workmanship in cutting screw, quality of lubrication, unit bearing pressure and the rubbing speeds. The value of coefficient of friction does not vary much with different combination of material, load or rubbing speed, except under starting conditions. The coefficient of friction, with good lubrication and average workmanship, may be assumed between 0.10 and 0.15. The various values for coefficient of friction for steel screw and cast iron or bronze nut, under different conditions are shown in the following table.

Table 17.5. Coefficient of friction under different conditions.

S.No.	Condition	Average coefficient of friction	
		Starting	Running
1.	High grade materials and workmanship and best running conditions.	0.14	0.10
2.	Average quality of materials and workmanship and average running conditions.	0.18	0.13
3.	Poor workmanship or very slow and in frequent motion with indifferent lubrication or newly machined surface.	0.21	0.15

If the thrust collars are used, the values of coefficient of friction may be taken as shown in the following table.

Table 17.6. Coefficient of friction when thrust collars are used.

S.No.	Materials	Average coefficient of friction	
		Starting	Running
1.	Soft steel on cast iron	0.17	0.12
2.	Hardened steel on cast iron	0.15	0.09
3.	Soft steel on bronze	0.10	0.08
4.	Hardened steel on bronze	0.08	0.06

17.12 Acme or Trapezoidal Threads

We know that the normal reaction in case of a square threaded screw is

$$R_N = W \cos \alpha,$$

where α is the helix angle.

But in case of Acme or trapezoidal thread, the normal reaction between the screw and nut is increased because the axial component of this normal reaction must be equal to the axial load (W).

Consider an Acme or trapezoidal thread as shown in Fig. 17.6.

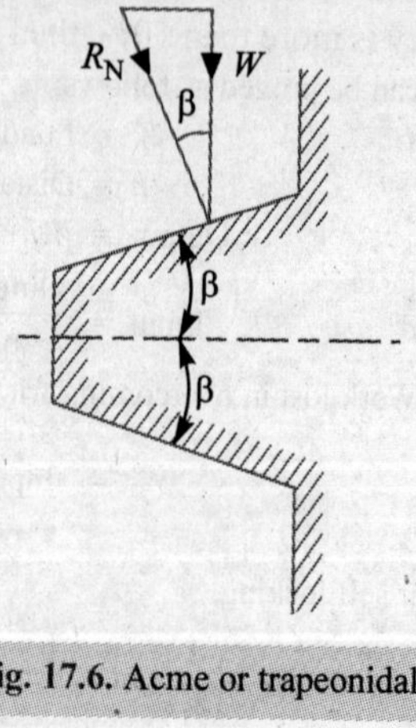

Fig. 17.6. Acme or trapeonidal threads.

Let **2β = Angle of the Acme thread, and

β = Semi-angle of the thread.

* The material of screw is usually steel and the nut is made of cast iron, gun metal, phosphor bronze in order to keep the wear to a mininum.

** For Acme threads, $2\beta = 29°$, and for trapezoidal threads, $2\beta = 30°$.

$$\therefore \quad R_N = \frac{W}{\cos \beta}$$

and frictional force, $F = \mu . R_N = \mu \times \frac{W}{\cos \beta} = \mu_1 . W$

where $\mu / \cos \beta = \mu_1$, known as ***virtual coefficient of friction.***

Notes : 1. When coefficient of friction, $\mu_1 = \frac{\mu}{\cos \beta}$ is considered, then the Acme thread is equivalent to a square thread.

2. All equations of square threaded screw also hold good for Acme threads. In case of Acme threads, μ_1 (*i.e.* $\tan \phi_1$) may be substituted in place of μ (*i.e.* $\tan \phi$). Thus for Acme threads,

$$P = W \tan (\alpha + \phi_1)$$

where ϕ_1 = Virtual friction angle, and $\tan \phi_1 = \mu_1$.

Example 17.6. *The lead screw of a lathe has Acme threads of 50 mm outside diameter and 8 mm pitch. The screw must exert an axial pressure of 2500 N in order to drive the tool carriage. The thrust is carried on a collar 110 mm outside diameter and 55 mm inside diameter and the lead screw rotates at 30 r.p.m. Determine (a) the power required to drive the screw; and (b) the efficiency of the lead screw. Assume a coefficient of friction of 0.15 for the screw and 0.12 for the collar.*

Solution. Given : $d_o = 50$ mm ; $p = 8$ mm ; $W = 2500$ N ; $D_1 = 110$ mm or $R_1 = 55$ mm ; $D_2 = 55$ mm or $R_2 = 27.5$ mm ; $N = 30$ r.p.m. ; $\mu = \tan \phi = 0.15$; $\mu_2 = 0.12$

(a) Power required to drive the screw

We know that mean diameter of the screw,

$$d = d_o - p/2 = 50 - 8/2 = 46 \text{ mm}$$

$$\therefore \quad \tan \alpha = \frac{p}{\pi d} = \frac{8}{\pi \times 46} = 0.055$$

Since the angle for Acme threads is $2\beta = 29°$ or $\beta = 14.5°$, therefore virtual coefficient of friction,

$$\mu_1 = \tan \phi_1 = \frac{\mu}{\cos \beta} = \frac{0.15}{\cos 14.5°} = \frac{0.15}{0.9681} = 0.155$$

We know that the force required to overcome friction at the screw,

$$P = W \tan (\alpha + \phi_1) = W \left[\frac{\tan \alpha + \tan \phi_1}{1 - \tan \alpha \tan \phi_1} \right]$$

$$= 2500 \left[\frac{0.055 + 0.155}{1 - 0.055 \times 0.155} \right] = 530 \text{ N}$$

and torque required to overcome friction at the screw.

$$T_1 = P \times d/2 = 530 \times 46/2 = 12\,190 \text{ N-mm}$$

We know that mean radius of collar,

$$R = \frac{R_1 + R_2}{2} = \frac{55 + 27.5}{2} = 41.25 \text{ mm}$$

Assuming uniform wear, the torque required to overcome friction at collars,

$$T_2 = \mu_2 W R = 0.12 \times 2500 \times 41.25 = 12\,375 \text{ N-mm}$$

$\therefore$ Total torque required to overcome friction,

$$T = T_1 + T_2 = 12\,190 + 12\,375 = 24\,565 \text{ N-mm} = 24.565 \text{ N-m}$$

We know that power required to drive the screw

$$= T.\omega = \frac{T \times 2\,\pi\,N}{60} = \frac{24.565 \times 2\,\pi \times 30}{60} = 77\text{ W} = 0.077\text{ kW} \qquad \textbf{Ans.}$$

... ($\because\ \omega = 2\pi N / 60$)

(b) Efficiency of the lead screw

We know that the torque required to drive the screw with no friction,

$$T_o = W \tan\alpha \times \frac{d}{2} = 2500 \times 0.055 \times \frac{46}{2} = 3163\text{ N-mm} = 3.163\text{ N-m}$$

∴ Efficiency of the lead screw,

$$\eta = \frac{T_o}{T} = \frac{3.163}{24.565} = 0.13 \text{ or } 13\% \qquad \textbf{Ans.}$$

17.13 Stresses in Power Screws

A power screw must have adequate strength to withstand axial load and the applied torque. Following types of stresses are induced in the screw.

1. ***Direct tensile or compressive stress due to an axial load.*** The direct stress due to the axial load may be determined by dividing the axial load (W) by the minimum cross-sectional area of the screw (A_c) *i.e.* area corresponding to minor or core diameter (d_c).

∴ Direct stress (tensile or compressive)

$$= \frac{W}{A_c}$$

This is only applicable when the axial load is compressive and the unsupported length of the screw between the load and the nut is short. But when the screw is axially loaded in compression and the unsupported length of the screw between the load and the nut is too great, then the design must be based on column theory assuming suitable end conditions. In such cases, the cross-sectional area corresponding to core diameter may be obtained by using Rankine-Gordon formula or J.B. Johnson's formula. According to this,

$$W_{cr} = A_c \times \sigma_y \left[1 - \frac{\sigma_y}{4\,C\,\pi^2\,E}\left(\frac{L}{k}\right)^2\right]$$

$$\therefore \qquad \sigma_c = \frac{W}{A_c}\left[\frac{1}{1 - \dfrac{\sigma_y}{4\,C\,\pi^2\,E}\left(\dfrac{L}{k}\right)^2}\right]$$

where W_{cr} = Critical load,

σ_y = Yield stress,

L = Length of screw,

k = Least radius of gyration,

C = End-fixity coefficient,

E = Modulus of elasticity, and

σ_c = Stress induced due to load W.

Note : In actual practice, the core diameter is first obtained by considering the screw under simple compression and then checked for critical load or buckling load for stability of the screw.

2. ***Torsional shear stress.*** Since the screw is subjected to a twisting moment, therefore torsional shear stress is induced. This is obtained by considering the minimum cross-section of the screw. We know that torque transmitted by the screw,

$$T = \frac{\pi}{16} \times \tau\,(d_c)^3$$

or shear stress induced,

$$\tau = \frac{16\,T}{\pi (d_c)^3}$$

When the screw is subjected to both direct stress and torsional shear stress, then the design must be based on maximum shear stress theory, according to which maximum shear stress on the minor diameter section,

$$\tau_{max} = \frac{1}{2}\sqrt{(\sigma_t \text{ or } \sigma_c)^2 + 4\,\tau^2}$$

It may be noted that when the unsupported length of the screw is short, then failure will take place when the maximum shear stress is equal to the shear yield strength of the material. In this case, shear yield strength,

$$\tau_y = \tau_{max} \times \text{Factor of safety}$$

Friction between the threads of screw and nut plays important role in determining the efficiency and locking properties of a screw

3. ***Shear stress due to axial load.*** The threads of the screw at the core or root diameter and the threads of the nut at the major diameter may shear due to the axial load. Assuming that the load is uniformly distributed over the threads in contact, we have

Shear stress for screw,

$$\tau_{(screw)} = \frac{W}{\pi\, n\,.d_c\,.\,t}$$

and shear stress for nut,

$$\tau_{(nut)} = \frac{W}{\pi\, n\,.d_o\,.\,t}$$

where W = Axial load on the screw,

n = Number of threads in engagement,

d_c = Core or root diameter of the screw,

d_o = Outside or major diameter of nut or screw, and

t = Thickness or width of thread.

4. ***Bearing pressure.*** In order to reduce wear of the screw and nut, the bearing pressure on the thread surfaces must be within limits. In the design of power screws, the bearing pressure depends upon the materials of the screw and nut, relative velocity between the nut and screw and the nature of lubrication. Assuming that the load is uniformly distributed over the threads in contact, the bearing pressure on the threads is given by

$$p_b = \frac{W}{\frac{\pi}{4}\left[(d_o)^2 - (d_c)^2\right] n} = \frac{{}^*W}{\pi\, d.t.n}$$

where d = Mean diameter of screw,

t = Thickness or width of screw = $p/2$, and

n = Number of threads in contact with the nut

$$= \frac{\text{Height of the nut}}{\text{Pitch of threads}} = \frac{h}{p}$$

Therefore, from the above expression, the height of nut or the length of thread engagement of the screw and nut may be obtained.

The following table shows some limiting values of bearing pressures.

* We know that $\frac{(d_o)^2 - (d_c)^2}{4} = \frac{d_o + d_c}{2} \times \frac{d_o - d_c}{2} = d \times \frac{p}{2} = d.t$

Table 17.7. Limiting values of bearing pressures.

Application of screw	*Material*		*Safe bearing pressure in N/mm²*	*Rubbing speed at thread pitch diameter*
	Screw	*Nut*		
1. Hand press	Steel	Bronze	17.5 - 24.5	Low speed, well lubricated
2. Screw jack	Steel	Cast iron	12.6 – 17.5	Low speed < 2.4 m / min
	Steel	Bronze	11.2 – 17.5	Low speed < 3 m / min
3. Hoisting screw	Steel	Cast iron	4.2 – 7.0	Medium speed 6 – 12 m / min
	Steel	Bronze	5.6 – 9.8	Medium speed 6 – 12 m / min
4. Lead screw	Steel	Bronze	1.05 – 1.7	High speed > 15 m / min

Example 17.7. *A power screw having double start square threads of 25 mm nominal diameter and 5 mm pitch is acted upon by an axial load of 10 kN. The outer and inner diameters of screw collar are 50 mm and 20 mm respectively. The coefficient of thread friction and collar friction may be assumed as 0.2 and 0.15 respectively. The screw rotates at 12 r.p.m. Assuming uniform wear condition at the collar and allowable thread bearing pressure of 5.8 N/mm², find: 1. the torque required to rotate the screw; 2. the stress in the screw; and 3. the number of threads of nut in engagement with screw.*

Solution. Given : d_o = 25 mm ; p = 5 mm ; W = 10 kN = 10 × 10³ N ; D_1 = 50 mm or R_1 = 25 mm ; D_2 = 20 mm or R_2 = 10 mm ; $\mu = \tan\phi = 0.2$; $\mu_1 = 0.15$; N = 12 r.p.m. ; p_b = 5.8 N/mm²

1. Torque required to rotate the screw

We know that mean diameter of the screw,

$$d = d_o - p/2 = 25 - 5/2 = 22.5 \text{ mm}$$

Since the screw is a double start square threaded screw, therefore lead of the screw,

$$= 2p = 2 \times 5 = 10 \text{ mm}$$

$$\therefore \quad \tan\alpha = \frac{\text{Lead}}{\pi d} = \frac{10}{\pi \times 22.5} = 0.1414$$

We know that tangential force required at the circumference of the screw,

$$P = W\tan(\alpha + \phi) = W\left[\frac{\tan\alpha + \tan\phi}{1 - \tan\alpha\tan\phi}\right]$$

$$= 10 \times 10^3 \left[\frac{0.1414 + 0.2}{1 - 0.1414 \times 0.2}\right] = 3513 \text{ N}$$

and mean radius of the screw collar,

$$R = \frac{R_1 + R_2}{2} = \frac{25 + 10}{2} = 17.5$$

∴ Total torque required to rotate the screw,

$$T = P \times \frac{d}{2} + \mu_1 W R = 3513 \times \frac{22.5}{2} + 0.15 \times 10 \times 10^3 \times 17.5 \text{ N-mm}$$

$$= 65\,771 \text{ N-mm} = 65.771 \text{ N-m } \textbf{Ans.}$$

2. ***Stress in the screw***

We know that the inner diameter or core diameter of the screw,

$$d_c = d_o - p = 25 - 5 = 20 \text{ mm}$$

∴ Corresponding cross-sectional area of the screw,

$$A_c = \frac{\pi}{4}(d_c)^2 = \frac{\pi}{4}(20)^2 = 314.2 \text{ mm}^2$$

We know that direct stress,

$$\sigma_c = \frac{W}{A_c} = \frac{10 \times 10^3}{314.2} = 31.83 \text{ N/mm}^2$$

and shear stress,
$$\tau = \frac{16\,T}{\pi (d_c)^3} = \frac{16 \times 65771}{\pi (20)^3} = 41.86 \text{ N/mm}^2$$

We know that maximum shear stress in the screw,

$$\tau_{max} = \frac{1}{2}\sqrt{(\sigma_c)^2 + 4\tau^2} = \frac{1}{2}\sqrt{(31.83)^2 + 4(41.86)^2}$$

$$= 44.8 \text{ N/mm}^2 = 44.8 \text{ MPa } \textbf{Ans.}$$

3. ***Number of threads of nut in engagement with screw***

Let n = Number of threads of nut in engagement with screw, and

t = Thickness of threads = $p / 2 = 5 / 2 = 2.5$ mm

We know that bearing pressure on the threads (p_b),

$$5.8 = \frac{W}{\pi d \times t \times n} = \frac{10 \times 10^3}{\pi \times 22.5 \times 2.5 \times n} = \frac{56.6}{n}$$

∴ $n = 56.6 / 5.8 = 9.76$ say 10 **Ans.**

Example 17.8. *The screw of a shaft straightener exerts a load of 30 kN as shown in Fig. 17.7. The screw is square threaded of outside diameter 75 mm and 6 mm pitch. Determine:*

1. Force required at the rim of a 300 mm diameter hand wheel, assuming the coefficient of friction for the threads as 0.12;

2. Maximum compressive stress in the screw, bearing pressure on the threads and maximum shear stress in threads; and

3. Efficiency of the straightner.

Solution. Given : W = 30 kN = 30 × 10^3 N ; d_o = 75 mm ; p = 6 mm ; D = 300 mm ; $\mu = \tan \phi = 0.12$

1. ***Force required at the rim of handwheel***

Let P_1 = Force required at the rim of handwheel.

We know that the inner diameter or core diameter of the screw,

$$d_c = d_o - p = 75 - 6 = 69 \text{ mm}$$

Mean diameter of the screw,

$$^*d = \frac{d_o + d_c}{2} = \frac{75 + 69}{2}$$

$$= 72 \text{ mm}$$

and

$$\tan \alpha = \frac{p}{\pi d} = \frac{6}{\pi \times 72} = 0.0265$$

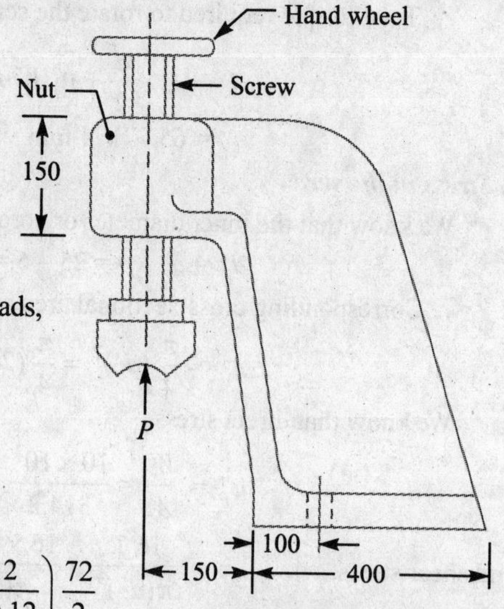

Fig. 17.7

∴ Torque required to overcome friction at the threads,

$$T = P \times \frac{d}{2}$$

$$= W \tan (\alpha + \phi) \frac{d}{2}$$

$$= W \left(\frac{\tan \alpha + \tan \phi}{1 - \tan \alpha \tan \phi} \right) \frac{d}{2}$$

$$= 30 \times 10^3 \left(\frac{0.0265 + 0.12}{1 - 0.0265 \times 0.12} \right) \frac{72}{2}$$

$$= 158\ 728 \text{ N-mm}$$

We know that the torque required at the rim of handwheel (T),

$$158\ 728 = P_1 \times \frac{D}{2} = P_1 \times \frac{300}{2} = 150\ P_1$$

$$\therefore \quad P_1 = 158\ 728 / 150 = 1058 \text{ N} \quad \textbf{Ans.}$$

2. ***Maximum compressive stress in the screw***

We know that maximum compressive stress in the screw,

$$\sigma_c = \frac{W}{A_c} = \frac{W}{\frac{\pi}{4}(d_c)^2} = \frac{30 \times 10^3}{\frac{\pi}{4}(69)^2} = 8.02 \text{ N/mm}^2 = 8.02 \text{ MPa} \quad \textbf{Ans.}$$

Bearing pressure on the threads

We know that number of threads in contact with the nut,

$$n = \frac{\text{Height of nut}}{\text{Pitch of threads}} = \frac{150}{6} = 25 \text{ threads}$$

and thickness of threads, $t = p/2 = 6/2 = 3$ mm

We know that bearing pressure on the threads,

$$p_b = \frac{W}{\pi\, d\,.\,t\,.\,n} = \frac{30 \times 10^3}{\pi \times 72 \times 3 \times 25} = 1.77 \text{ N/mm}^2 \quad \textbf{Ans.}$$

Maximum shear stress in the threads

We know that shear stress in the threads,

$$\tau = \frac{16\ T}{\pi (d_c)^3} = \frac{16 \times 158\ 728}{\pi\ (69)^3} = 2.46 \text{ N/mm}^2 \quad \textbf{Ans.}$$

---

* The mean diameter of the screw (d) is also given by

$$d = d_o - p/2 = 75 - 6/2 = 72 \text{ mm}$$

∴ Maximum shear stress in the threads,

$$\tau_{max} = \frac{1}{2}\sqrt{(\sigma_c)^2 + 4\,\tau^2} = \frac{1}{2}\sqrt{(8.02)^2 + 4\,(2.46)^2}$$

$$= 4.7 \text{ N/mm}^2 = 4.7 \text{ MPa} \quad \textbf{Ans.}$$

3. ***Efficiency of the straightener***

We know that the torque required with no friction,

$$T_0 = W \tan\alpha \times \frac{d}{2} = 30 \times 10^3 \times 0.0265 \times \frac{72}{2} = 28\,620 \text{ N-mm}$$

∴ Efficiency of the straightener,

$$\eta = \frac{T_0}{T} = \frac{28\,620}{158\,728} = 0.18 \text{ or } 18\% \quad \textbf{Ans.}$$

Example 17.9. *A sluice gate weighing 18 kN is raised and lowered by means of square threaded screws, as shown in Fig. 17.8. The frictional resistance induced by water pressure against the gate when it is in its lowest position is 4000 N.*

The outside diameter of the screw is 60 mm and pitch is 10 mm. The outside and inside diameter of washer is 150 mm and 50 mm respectively. The coefficient of friction between the screw and nut is 0.1 and for the washer and seat is 0.12. Find :

1. The maximum force to be exerted at the ends of the lever raising and lowering the gate, 2. Efficiency of the arrangement, and 3. Number of threads and height of nut, for an allowable bearing pressure of 7 N/mm².

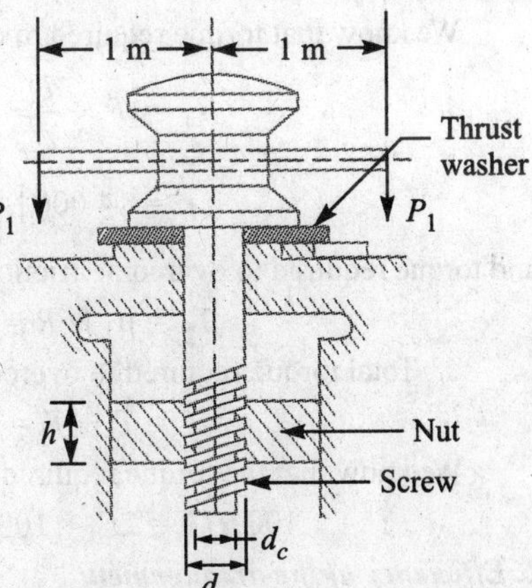

Fig. 17.8

Solution. Given : W_1 = 18 kN = 18 000 N ; F = 4000 N ; d_o = 60 mm ; p = 10 mm ; D_1 = 150 mm or R_1 = 75 mm ; D_2 = 50 mm or R_2 = 25 mm ; $\mu = \tan\phi$ = 0.1 ; μ_1 = 0.12 ; p_b = 7 N/mm²

1. ***Maximum force to be exerted at the ends of lever***

Let P_1 = Maximum force exerted at each end of the lever 1 m (1000 mm) long.

We know that inner diameter or core diameter of the screw,

$$d_c = d_o - p = 60 - 10 = 50 \text{ mm}$$

Mean diameter of the screw,

$$d = \frac{d_o + d_c}{2} = \frac{60 + 50}{2} = 55 \text{ mm}$$

and

$$\tan\alpha = \frac{p}{\pi\,d} = \frac{10}{\pi \times 55} = 0.058$$

(a) ***For raising the gate***

Since the frictional resistance acts in the opposite direction to the motion of screw, therefore for raising the gate, the frictional resistance (F) will act downwards.

∴ Total load acting on the screw,

$$W = W_1 + F = 18\,000 + 4000 = 22\,000 \text{ N}$$

and torque required to overcome friction at the screw,

$$T_1 = P \times \frac{d}{2} = W \tan(\alpha + \phi)\,\frac{d}{2} = W\left(\frac{\tan\alpha + \tan\phi}{1 - \tan\alpha\,\tan\phi}\right)\frac{d}{2}$$

$$= 22\ 000 \left(\frac{0.058 + 0.1}{1 - 0.058 \times 0.1} \right) \frac{55}{2} = 96\ 148 \text{ N-mm}$$

Mean radius of washer,

$$R = \frac{R_1 + R_2}{2} = \frac{75 + 25}{2} = 50 \text{ mm}$$

∴ Torque required to overcome friction at the washer,

$$T_2 = \mu_1 \, W R = 0.12 \times 22\ 000 \times 50 = 132\ 000 \text{ N-mm}$$

and total torque required to overcome friction,

$$T = T_1 + T_2 = 96\ 148 + 132\ 000 = 228\ 148 \text{ N-mm}$$

We know that the torque required at the end of lever (T),

$$228\ 148 = 2P_1 \times \text{Length of lever} = 2P_1 \times 1000 = 2000\ P_1$$

$$\therefore \quad P_1 = 228\ 148 / 2000 = 141.1 \text{ N } \textbf{Ans.}$$

(*b*) ***For lowering the gate***

Since the gate is being lowered, therefore the frictional resistance (F) will act upwards,

∴ Total load acting on the screw,

$$W = W_1 - F = 18\ 000 - 4000 = 14\ 000 \text{ N}$$

We know that torque required to overcome friction at the screw,

$$T_1 = P \times \frac{d}{2} = W \tan(\phi - \alpha) \frac{d}{2} = W \left(\frac{\tan\phi - \tan\alpha}{1 + \tan\phi \tan\alpha} \right) \frac{d}{2}$$

$$= 14\ 000 \left(\frac{0.1 - 0.058}{1 + 0.1 \times 0.058} \right) \frac{55}{2} = 16\ 077 \text{ N-mm}$$

and torque required to overcome friction at the washer,

$$T_2 = \mu_1 \, W R = 0.12 \times 14\ 000 \times 50 = 84\ 000 \text{ N-mm}$$

∴ Total torque required to overcome friction,

$$T = T_1 + T_2 = 16\ 077 + 84\ 000 = 100\ 077 \text{ N-mm}$$

We know that the torque required at the end of lever (T),

$$100\ 077 = 2P_1 \times 1000 = 2000\ P_1 \quad \text{or} \quad P_1 = 100\ 077/2000 = 50.04 \text{ N } \textbf{Ans.}$$

2. ***Efficiency of the arrangement***

We know that the torque required for raising the load, with no friction,

$$T_0 = W \tan\alpha \times \frac{d}{2} = 22\ 000 \times 0.058 \times \frac{55}{2} = 35\ 090 \text{ N-mm}$$

∴ Efficiency of the arrangement,

$$\eta = \frac{T_0}{T} = \frac{35\ 090}{228\ 148} = 0.154 \quad \text{or} \quad 15.4\% \qquad \textbf{Ans.}$$

3. ***Number of threads and height of nut***

Let n = Number of threads in contact with the nut,

h = Height of nut = $n \times p$, and

t = Thickness of thread = $p / 2 = 10 / 2 = 5$ mm.

We know that the bearing pressure (p_b),

$$7 = \frac{W}{\pi . d . t . n} = \frac{22000}{\pi \times 55 \times 5 \times n} = \frac{25.46}{n}$$

$$\therefore \quad n = 25.46 / 7 = 3.64 \text{ say 4 threads } \textbf{Ans.}$$

and $\quad h = n \times p = 4 \times 10 = 40 \text{ mm } \textbf{Ans.}$

Example 17.10. *The screw, as shown in Fig. 17.9 is operated by a torque applied to the lower end. The nut is loaded and prevented from turning by guides. Assume friction in the ball bearing to be negligible. The screw is a triple start trapezoidal thread. The outside diameter of the screw is 48 mm and pitch is 8 mm. The coefficient of friction of the threads is 0.15. Find :*

1. Load which can be raised by a torque of 40 N-m ;

2. Whether the screw is overhauling ; and

3. Average bearing pressure between the screw and nut thread surface.

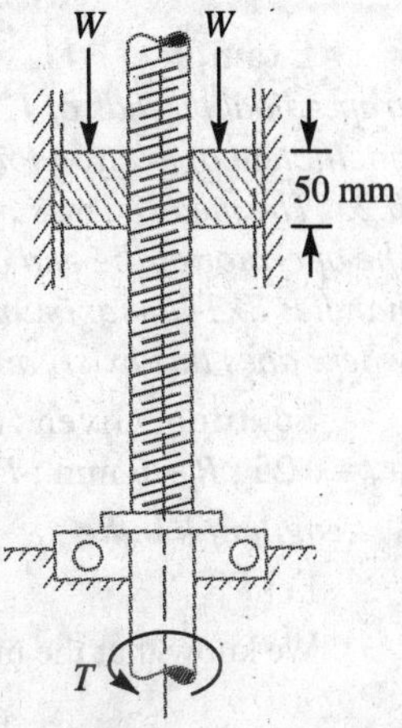

Fig. 17.9

Solution. Given : $d_o = 48$ mm ; $p = 8$ mm ; $\mu = \tan \phi = 0.15$; $T = 40$ N-m $= 40\ 000$ N-mm

1. *Load which can be raised*

Let W = Load which can be raised.

We know that mean diameter of the screw,

$$d = d_o - p/2 = 48 - 8/2 = 44 \text{ mm}$$

Since the screw is a triple start, therefore lead of the screw

$$= 3\,p = 3 \times 8 = 24 \text{ mm}$$

$$\therefore \quad \tan \alpha = \frac{\text{Lead}}{\pi d} = \frac{24}{\pi \times 44} = 0.174$$

and virtual coefficient of friction,

$$\mu_1 = \tan \phi_1 = \frac{\mu}{\cos \beta} = \frac{0.15}{\cos 15°} = \frac{0.15}{0.9659} = 0.155$$

... ($\because$ For trapezoidal threads, $2\,\beta = 30°$)

We know that the torque required to raise the load,

$$T = P \times \frac{d}{2} = W \tan(\alpha + \phi_1)\frac{d}{2} = W\left[\frac{\tan \alpha + \tan \phi_1}{1 - \tan \alpha \tan \phi_1}\right]\frac{d}{2}$$

$$40\ 000 = W\left(\frac{0.174 + 0.155}{1 - 0.174 \times 0.155}\right)\frac{44}{2} = 7.436\,W$$

$$\therefore \quad W = 40\ 000 / 7.436 = 5380 \text{ N} \quad \textbf{Ans.}$$

2. *Whether the screw is overhauling*

We know that torque required to lower the load,

$$T = W \tan(\phi_1 - \alpha)\frac{d}{2}$$

We have discussed in Art. 17.9 that if ϕ_1 is less than α, then the torque required to lower the load will be ***negative,*** *i.e.* the load will start moving downward without the application of any torque. Such a condition is known as overhauling of screws.

In the present case, $\tan \phi_1 = 0.155$ and $\tan \alpha = 0.174$. Since ϕ_1 is less than α, therefore the screw is overhauling. **Ans.**

3. *Average bearing pressure between the screw and nut thread surfaces*

We know that height of the nut,

$$h = n \times p = 50 \text{ mm} \qquad \text{...(Given)}$$

$\therefore$ Number of threads in contact,

$$n = h/p = 50/8 = 6.25$$

and thickness of thread, $t = p/2 = 8/2 = 4$ mm

We know that the average bearing pressure,

$$p_b = \frac{W}{\pi . d . t . n} = \frac{5380}{\pi \times 44 \times 4 \times 6.25} = 1.56 \text{ N/mm}^2 \textbf{ Ans.}$$

Example 17.11. *A C-clamp, as shown in Fig. 17.10, has trapezoidal threads of 12 mm outside diameter and 2 mm pitch. The coefficient of friction for screw threads is 0.12 and for the collar is 0.25. The mean radius of the collar is 6 mm. If the force exerted by the operator at the end of the handle is 80 N, find: 1. The length of handle; 2. The maximum shear stress in the body of the screw and where does this exist; and 3. The bearing pressure on the threads.*

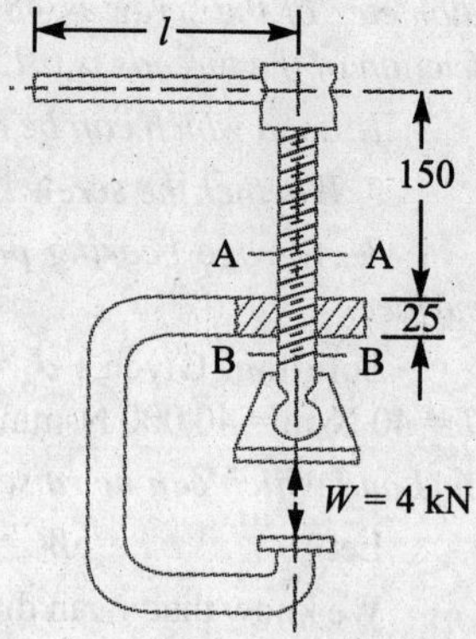

All dimensions in mm.

Fig. 17.10

Solution. Given : $d_o = 12$ mm ; $p = 2$ mm ; $\mu = \tan \phi = 0.12$; $\mu_2 = 0.25$; $R = 6$ mm ; $P_1 = 80$ N ; $W = 4$ kN $= 4000$ N

1. *Length of handle*

Let l = Length of handle.

We know that the mean diameter of the screw,

$$d = d_o - p/2 = 12 - 2/2 = 11 \text{ mm}$$

$$\therefore \quad \tan \alpha = \frac{p}{\pi d} = \frac{2}{\pi \times 11} = 0.058$$

Since the angle for trapezoidal threads is $2\beta = 30°$ or $\beta = 15°$, therefore virtual coefficient of friction,

$$\mu_1 = \tan \phi_1 = \frac{\mu}{\cos \beta} = \frac{0.12}{\cos 15°} = \frac{0.12}{0.9659} = 0.124$$

We know that the torque required to overcome friction at the screw,

$$T_1 = P \times \frac{d}{2} = W \tan (\alpha + \phi_1) \frac{d}{2} = W \left(\frac{\tan \alpha + \tan \phi_1}{1 - \tan \alpha \tan \phi_1} \right) \frac{d}{2}$$

$$= 4000 \left(\frac{0.058 + 0.124}{1 - 0.058 \times 0.124} \right) \frac{11}{2} = 4033 \text{ N-mm}$$

Assuming uniform wear, the torque required to overcome friction at the collar,

$$T_2 = \mu_2 \, W R = 0.25 \times 4000 \times 6 = 6000 \text{ N-mm}$$

$\therefore$ Total torque required at the end of handle,

$$T = T_1 + T_2 = 4033 + 6000 = 10\ 033 \text{ N-mm}$$

We know that the torque required at the end of handle (T),

$$10\ 033 = P_1 \times l = 80 \times l \quad \text{or} \quad l = 10\ 033 / 80 = 125.4 \text{ mm } \textbf{Ans.}$$

2. *Maximum shear stress in the body of the screw*

Consider two sections *A-A* and *B-B*. The section *A-A* just above the nut, is subjected to torque and bending. The section *B-B* just below the nut is subjected to collar friction torque and direct compressive load. Thus, both the sections must be checked for maximum shear stress.

Considering section A-A

We know that the core diameter of the screw,

$$d_c = d_o - p = 12 - 2 = 10 \text{ mm}$$

and torque transmitted at *A-A*,

$$T = \frac{\pi}{16} \times \tau \, (d_c)^3$$

$$\therefore \quad \text{Shear stress, } \tau = \frac{16\,T}{\pi\,(d_c)^3} = \frac{16 \times 10\,033}{\pi \times 10^3} = 51.1 \text{ N/mm}^2$$

Bending moment at *A-A*,

$$M = P_1 \times 150 = 80 \times 150 = 12\,000 \text{ N-mm}$$

$$= \frac{\pi}{32} \times \sigma_b\,(d_c)^3$$

$$\therefore \text{ Bending stress, } \sigma_b = \frac{32\,M}{\pi\,(d_c)^3} = \frac{32 \times 12\,000}{\pi\,(10)^3} = 122.2 \text{ N/mm}^2$$

We know that the maximum shear stress,

$$\tau_{max} = \frac{1}{2}\sqrt{(\sigma_b)^2 + 4\tau^2} = \frac{1}{2}\sqrt{(122.2)^2 + 4(51.1)^2} = 79.65 \text{ N/mm}^2$$

$$= 79.65 \text{ MPa}$$

Considering section B-B

Since the section *B-B* is subjected to collar friction torque (T_2), therefore the shear stress,

$$\tau = \frac{16\,T_2}{\pi\,(d_c)^3} = \frac{16 \times 6000}{\pi \times 10^3} = 30.6 \text{ N/mm}^2$$

and direct compressive stress,

$$\sigma_c = \frac{W}{A_c} = \frac{4W}{\pi (d_c)^2} = \frac{4 \times 4000}{\pi \times 10^2} = 51 \text{ N/mm}^2$$

$\therefore$ Maximum shear stress,

$$\tau_{max} = \frac{1}{2}\sqrt{(\sigma_c)^2 + 4\,\tau^2} = \frac{1}{2}\sqrt{(51)^2 + 4(30.6)^2} = 39.83 \text{ N/mm}^2 = 39.83 \text{ MPa}$$

From above, we see that the maximum shear stress is 79.65 MPa and occurs at section *A-A*. **Ans.**

3. *Bearing pressure on the threads*

We know that height of the nut,

$$h = n \times p = 25 \text{ mm} \qquad \text{...(Given)}$$

$\therefore$ Number of threads in contact,

$$n = h / p = 25 / 2 = 12.5$$

and thickness of threads, $t = p / 2 = 2 / 2 = 1$ mm

We know that bearing pressure on the threads,

$$p_b = \frac{W}{\pi\, d_t . n} = \frac{4000}{\pi \times 11 \times 1 \times 12.5} = 9.26 \text{ N/mm}^2 \textbf{ Ans.}$$

Example 17.12. *A power transmission screw of a screw press is required to transmit maximum load of 100 kN and rotates at 60 r.p.m. Trapezoidal threads are as under :*

Nominal dia, mm	*40*	*50*	*60*	*70*
Core dia, mm	*32.5*	*41.5*	*50.5*	*59.5*
Mean dia, mm	*36.5*	*46*	*55.5*	*65*
Core area, mm²	*830*	*1353*	*2003*	*2781*
Pitch, mm	*7*	*8*	*9*	*10*

The screw thread friction coefficient is 0.12. Torque required for collar friction and journal bearing is about 10% of the torque to drive the load considering screw friction. Determine screw dimensions and its efficiency. Also determine motor power required to drive the screw. Maximum permissible compressive stress in screw is 100 MPa.

Solution. Given : $W = 100$ kN $= 100 \times 10^3$ N ; $N = 60$ r.p.m. ; $\mu = 0.12$; $\sigma_c = 100$ MPa $= 100$ N/mm^2

Dimensions of the screw

Let A_c = Core area of threads.

We know that the direct compressive stress (σ_c),

$$100 = \frac{W}{A_c} = \frac{100 \times 10^3}{A_c}$$

or $$A_c = 100 \times 10^3 / 100 = 1000 \text{ mm}^2$$

Since the core area is 1000 mm^2, therefore we shall use the following dimensions for the screw (for core area 1353 mm^2).

Nominal diameter, $d_o = 50$ mm ;

Core diameter, $d_c = 41.5$ mm ;

Mean diameter, $d = 46$ mm ;

Pitch, $p = 8$ mm. **Ans.**

This screw press was made in 1735 and installed in the Segovia Mint to strike a new series of coper coins which began in 1772. This press is presently on display in the Alcazar castle of Segovia.

Efficiency of the screw

We know that $\tan \alpha = \frac{p}{\pi d} = \frac{8}{\pi \times 46} = 0.055$

and virtual coefficient of friction,

$$\mu_1 = \tan \phi_1 = \frac{\mu}{\cos \beta} = \frac{0.12}{\cos 15°}$$

$$= \frac{0.12}{0.9659} = 0.124 \qquad \text{... } (\because \text{ For trapezoidal threads, } 2\beta = 30°)$$

$\therefore$ Force required at the circumference of the screw,

$$P = W \tan (\alpha + \phi_1) = W \left[\frac{\tan \alpha + \tan \phi_1}{1 - \tan \alpha \tan \phi_1}\right]$$

$$= 100 \times 10^3 \left[\frac{0.055 + 0.124}{1 - 0.055 \times 0.124}\right] = 18\ 023 \text{ N}$$

and the torque required to drive the load,

$$T_1 = P \times d / 2 = 18\ 023 \times 46 / 2 = 414\ 530 \text{ N-mm}$$

We know that the torque required for collar friction,

$$T_2 = 10\%\ T_1 = 0.1 \times 414\ 530 = 41\ 453 \text{ N-mm}$$

$\therefore$ Total torque required,

$$T = T_1 + T_2 = 414\ 530 + 41\ 453 = 455\ 983 \text{ N-mm} = 455.983 \text{ N-m}$$

We know that the torque required with no friction,

$$T_0 = W \tan \alpha \times \frac{d}{2} = 100 \times 10^3 \times 0.055 \times \frac{46}{2} = 126\ 500 \text{ N-mm}$$

$\therefore$ Efficiency of the screw,

$$\eta = \frac{T_0}{T} = \frac{126\ 500}{455\ 983} = 0.278 \text{ or } 27.8\% \textbf{ Ans.}$$

Power required to drive the screw

We know that the power required to drive the screw,

$$= T \times \omega = \frac{T \times 2\,\pi N}{60} = \frac{455.683 \times 2\pi \times 60}{60} = 2865 \text{ W}$$

$$= 2.865 \text{ kW } \textbf{Ans.}$$

Example 17.13. *A vertical two start square threaded screw of 100 mm mean diameter and 20 mm pitch supports a vertical load of 18 kN. The nut of the screw is fitted in the hub of a gear wheel having 80 teeth which meshes with a pinion of 20 teeth. The mechanical efficiency of the pinion and gear wheel drive is 90 percent. The axial thrust on the screw is taken by a collar bearing 250 mm outside diameter and 100 mm inside diameter. Assuming uniform pressure conditions, find, minimum diameter of pinion shaft and height of nut, when coefficient of friction for the vertical screw and nut is 0.15 and that for the collar bearing is 0.20. The permissible shear stress in the shaft material is 56 MPa and allowable bearing pressure is 1.4 N/mm².*

Solution. Given : $d = 100$ mm ; $p = 20$ mm ; $W = 18$ kN $= 18 \times 10^3$N ; No. of teeth on gear wheel $= 80$; No. of teeth on pinion $= 20$; $\eta_m = 90\% = 0.9$; $D_1 = 250$ mm or $R_1 = 125$ mm ; $D_2 = 100$ mm or $R_2 = 50$ mm ; $\mu = \tan\phi = 0.15$; $\mu_1 = 0.20$; $\tau = 56$ MPa $= 56$ N/mm² ; $p_b = 1.4$ N/mm²

Minimum diameter of pinion shaft

Let $\qquad D =$ Minimum diameter of pinion shaft.

Since the screw is a two start square threaded screw, therefore lead of the screw

$$= 2\,p = 2 \times 20 = 40 \text{ mm}$$

$$\therefore \qquad \tan\alpha = \frac{\text{Lead}}{\pi d} = \frac{40}{\pi \times 100} = 0.127$$

and torque required to overcome friction at the screw and nut,

$$T_1 = P \times \frac{d}{2} = W \tan(\alpha + \phi)\frac{d}{2} = W\left(\frac{\tan\alpha + \tan\phi}{1 - \tan\alpha \tan\phi}\right)\frac{d}{2}$$

$$= 18 \times 10^3\left(\frac{0.127 + 0.15}{1 - 0.127 \times 0.15}\right)\frac{100}{2} = 254\,160 \text{ N-mm}$$

$$= 254.16 \text{ N-m}$$

We know that, for uniform pressure conditions, torque required to overcome friction at the collar bearing,

$$T_2 = \frac{2}{3} \times \mu_1 W\left[\frac{(R_1)^3 - (R_2)^3}{(R_1)^2 - (R_2)^2}\right]$$

$$= \frac{2}{3} \times 0.20 \times 18 \times 10^3\left[\frac{(125)^3 - (50)^3}{(125)^2 - (50)^2}\right] \text{ N-mm}$$

$$= 334\,290 \text{ N-mm} = 334.29 \text{ N-m}$$

Since the nut of the screw is fixed in the hub of a gear wheel, therefore the total torque required at the gear wheel,

$$T_w = T_1 + T_2 = 254.16 + 334.29 = 588.45 \text{ N-m}$$

Also the gear wheel having 80 teeth meshes with pinion having 20 teeth and the torque is proportional to the number of teeth, therefore torque required at the pinion shaft,

$$= \frac{T_w \times 20}{80} = 588.45 \times \frac{20}{80} = 147.11 \text{ N-m}$$

Since the mechanical efficiency of the pinion and gear wheel is 90%, therefore net torque required at the pinion shaft,

$$T_p = \frac{147.11 \times 100}{90} = 163.46 \text{ N} - \text{m} = 163\,460 \text{ N-mm}$$

We know that the torque required at the pinion shaft (T_p),

$$163\ 460 = \frac{\pi}{16} \times \tau \times D^3 = \frac{\pi}{16} \times 56 \times D^3 = 11\ D^3$$

$\therefore$ $D^3 = 163\ 460 / 11 = 14\ 860$ or $D = 24.6$ say 25 mm **Ans.**

Height of nut

Let h = Height of nut,

n = Number of threads in contact, and

t = Thickness or width of thread $= p/2 = 20/2 = 10$ mm

We know that the bearing pressure (p_b),

$$1.4 = \frac{W}{\pi\, d.t.n} = \frac{18 \times 10^3}{\pi \times 100 \times 10 \times n} = \frac{5.73}{n}$$

$\therefore$ $n = 5.73 / 1.4 = 4.09$ say 5 threads

and height of nut, $h = n \times p = 5 \times 20 = 100$ mm **Ans.**

Example 17.14. *A screw press is to exert a force of 40 kN. The unsupported length of the screw is 400 mm. Nominal diameter of screw is 50 mm. The screw has square threads with pitch equal to 10 mm. The material of the screw and nut are medium carbon steel and cast iron respectively. For the steel used take ultimate crushing stress as 320 MPa, yield stress in tension or compression as 200 MPa and that in shear as 120 MPa. Allowable shear stress for cast iron is 20 MPa and allowable bearing pressure between screw and nut is 12 N/mm². Young's modulus for steel = 210 kN/mm². Determine the factor of safety of screw against failure. Find the dimensions of the nut. What is the efficiency of the arrangement? Take coefficient of friction between steel and cast iron as 0.13.*

Solution. Given : $W = 40$ kN $= 40 \times 10^3$ N ; $L = 400$ mm $= 0.4$ m ; $d_o = 50$ mm ; $p = 10$ mm ; $\sigma_{cu} = 320$ MPa $= 320$ N/mm² ; $\sigma_y = 200$ MPa $= 200$ N/mm² ; $\tau_y = 120$ MPa $= 120$ N/mm² ; $\tau_c = 20$ MPa $= 20$ N/mm² ; $p_b = 12$ N/mm² ; $E = 210$ kN/mm² $= 210 \times 10^3$ N/mm² ; $\mu = \tan\phi = 0.13$

We know that the inner diameter or core diameter of the screw,

$$d_c = d_o - p = 50 - 10 = 40 \text{ mm}$$

and core area of the screw,

$$A_c = \frac{\pi}{4}(d_c)^2 = \frac{\pi}{4}(40)^2 = 1257 \text{ mm}^2$$

$\therefore$ Direct compressive stress on the screw due to axial load,

$$\sigma_c = \frac{W}{A_c} = \frac{40 \times 10^3}{1257} = 31.8 \text{ N/mm}^2$$

We know that the mean diameter of the screw,

$$d = \frac{d_o + d_c}{2} = \frac{50 + 40}{2} = 45 \text{ mm}$$

and $$\tan\alpha = \frac{p}{\pi d} = \frac{10}{\pi \times 45} = 0.07$$

$\therefore$ Torque required to move the screw,

$$T = P \times \frac{d}{2} = W \tan(\alpha + \phi)\frac{d}{2} = W\left[\frac{\tan\alpha + \tan\phi}{1 - \tan\alpha\tan\phi}\right]\frac{d}{2}$$

$$= 40 \times 10^3 \left[\frac{0.07 + 0.13}{1 - 0.07 \times 0.13}\right]\frac{45}{2} = 181.6 \times 10^3 \text{ N-mm}$$

We know that torque transmitted by the screw (T),

$$181.6 \times 10^3 = \frac{\pi}{16} \times \tau\,(d_c)^3 = \frac{\pi}{16} \times \tau\,(40)^3 = 12\ 568\ \tau$$

$\therefore$ $\tau = 181.6 \times 10^3 / 12\ 568 = 14.45$ N/mm²

According to maximum shear stress theory, we have

$$\tau_{max} = \frac{1}{2}\sqrt{(\sigma_c)^2 + 4\tau^2} = \frac{1}{2}\sqrt{(31.8)^2 + 4\,(14.45)^2} = 21.5 \text{ N/mm}^2$$

Factor of safety

We know that factor of safety

$$= \frac{\tau_y}{\tau_{max}} = \frac{120}{21.5} = 5.58$$

Now considering the screw as a column, assuming one end fixed and other end free. According to J.B. Johnson's formula, critical load,

$$W_{cr} = A_c \times \sigma_y \left[1 - \frac{\sigma_y}{4C\,\pi^2\,E}\left(\frac{L}{k}\right)^2\right]$$

For one end fixed and other end free, $C = 0.25$.

$$\therefore \quad W_{cr} = 1257 \times 200 \left[1 - \frac{200}{4 \times 0.25 \times \pi^2 \times 210 \times 10^3}\left(\frac{400}{10}\right)^2\right] \text{N}$$

... ($\because$ $k = d_c / 4 = 40 / 4 = 10$ mm)

$$= 212\,700 \text{ N}$$

$$\therefore \quad \text{Factor of safety} = \frac{W_{cr}}{W} = \frac{212\,700}{40 \times 10^3} = 5.3$$

We shall take larger value of the factor of safety.

$\therefore$ Factor of safety = 5.58 say 6 **Ans.**

Dimensions of the nut

Let n = Number of threads in contact with nut, and

h = Height of nut = $p \times n$

Assume that the load is uniformly distributed over the threads in contact.

We know that the bearing pressure (p_b),

$$12 = \frac{W}{\frac{\pi}{4}\left[(d_o)^2 - (d_c)^2\right] n} = \frac{40 \times 10^3}{\frac{\pi}{4}\left[(50)^2 - (40)^2\right] n} = \frac{56.6}{n}$$

$\therefore$ $n = 56.6 / 12 = 4.7$ say 5 threads **Ans.**

and $h = p \times n = 10 \times 5 = 50$ mm **Ans.**

Now let us check for the shear stress induced in the nut which is of cast iron. We know that

$$\tau_{nut} = \frac{W}{\pi n . d_o t} = \frac{40 \times 10^3}{\pi \times 5 \times 50 \times 5} = 10.2 \text{ N/mm}^2 = 10.2 \text{ MPa}$$

...($\because$ $t = p / 2 = 10 / 2 = 5$ mm)

This value is less than the given value of $\tau_c = 20$ MPa, hence the nut is safe.

Efficiency of the arrangement

We know that torque required to move the screw with no friction,

$$T_0 = W \tan\alpha \times \frac{d}{2} = 40 \times 10^3 \times 0.07 \times \frac{45}{2} = 63 \times 10^3 \text{ N-mm}$$

$\therefore$ Efficiency of the arrangement

$$\eta = \frac{T_0}{T} = \frac{63 \times 10^3}{181.6 \times 10^3} = 0.347 \text{ or } 34.7\% \textbf{ Ans.}$$

17.14 Design of Screw Jack

A bottle screw jack for lifting loads is shown in Fig. 17.11. The various parts of the screw jack are as follows:

1. Screwed spindle having square threaded screws,
2. Nut and collar for nut,
3. Head at the top of the screwed spindle for handle,
4. Cup at the top of head for the load, and
5. Body of the screw jack.

In order to design a screw jack for a load W, the following procedure may be adopted:

1. First of all, find the core diameter (d_c) by considering that the screw is under pure compression, *i.e.*

$$W = \sigma_c \times A_c = \sigma_c \times \frac{\pi}{4} (d_c)^2$$

The standard proportions of the square threaded screw are fixed from Table 17.1.

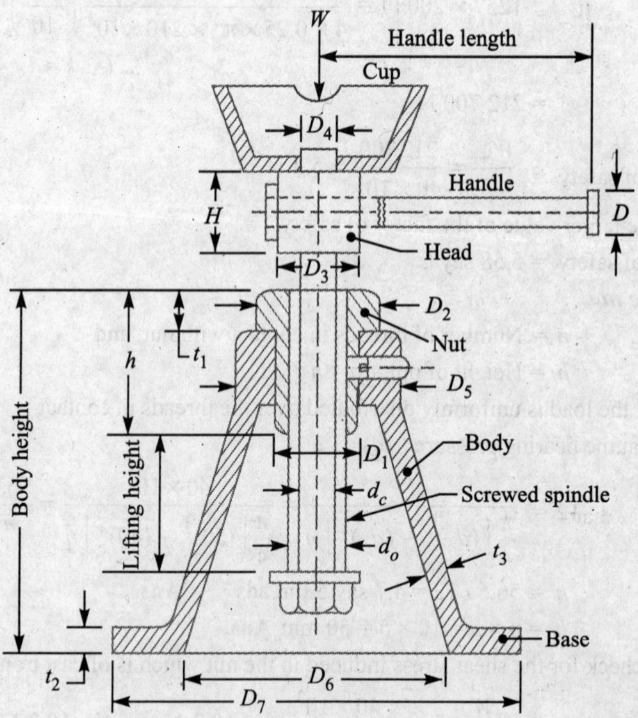

Fig. 17.11. Screw jack.

2. Find the torque (T_1) required to rotate the screw and find the shear stress (τ) due to this torque.

We know that the torque required to lift the load,

$$T_1 = P \times \frac{d}{2} = W \tan(\alpha + \phi) \frac{d}{2}$$

where P = Effort required at the circumference of the screw, and

d = Mean diameter of the screw.

∴ Shear stress due to torque T_1,

$$\tau = \frac{16\, T_1}{\pi (d_c)^3}$$

Also find direct compressive stress (σ_c) due to axial load, *i.e.*

$$\sigma_c = \frac{W}{\frac{\pi}{4}(d_c)^2}$$

3. Find the principal stresses as follows:

   Maximum principal stress (tensile or compressive),

$$\sigma_{c(max)} = \frac{1}{2}\left[\sigma_c + \sqrt{(\sigma_c)^2 + 4\tau^2}\right]$$

   and maximum shear stress,

$$\tau_{max} = \frac{1}{2}\sqrt{(\sigma_c)^2 + 4\,\tau^2}$$

   These stresses should be less than the permissible stresses.

4. Find the height of nut (h), considering the bearing pressure on the nut. We know that the bearing pressure on the nut,

$$p_b = \frac{W}{\frac{\pi}{4}\left[(d_o)^2 - (d_c)^2\right] n}$$

   where n = Number of threads in contact with screwed spindle.

   ∴ Height of nut, $h = n \times p$

   where p = Pitch of threads.

5. Check the stressess in the screw and nut as follows :

$$\tau_{(screw)} = \frac{W}{\pi\, n.d_c.t}$$

$$\tau_{(nut)} = \frac{W}{\pi\, n.d_o.t}$$

   where t = Thickness of screw $= p/2$

6. Find inner diameter (D_1), outer diameter (D_2) and thickness (t_1) of the nut collar.

   The inner diameter (D_1) is found by considering the tearing strength of the nut. We know that

$$W = \frac{\pi}{4}\left[(D_1)^2 - (d_o)^2\right] \sigma_t$$

   The outer diameter (D_2) is found by considering the crushing strength of the nut collar. We know that

$$W = \frac{\pi}{4}\left[(D_2)^2 - (D_1)^2\right] \sigma_c$$

   The thickness (t_1) of the nut collar is found by considering the shearing strength of the nut collar. We know that

$$W = \pi\, D_1.t_1.\tau$$

7. Fix the dimensions for the diameter of head (D_3) on the top of the screw and for the cup. Take D_3 = 1.75 d_o. The seat for the cup is made equal to the diameter of head and it is chamfered at the top. The cup is fitted with a pin of diameter $D_4 = D_3 / 4$ approximately. This pin remains a loose fit in the cup.

8. Find the torque required (T_2) to overcome friction at the top of screw. We know that

$$T_2 = \frac{2}{3} \times \mu_1 W \left[\frac{(R_3)^3 - (R_4)^3}{(R_3)^2 - (R_4)^2} \right]$$ ... (Assuming uniform pressure conditions)

$$= \mu_1 W \left[\frac{R_3 + R_4}{2} \right] = \mu_1 \, W \, R$$ ... (Assuming uniform wear conditions)

where R_3 = Radius of head, and

R_4 = Radius of pin.

9. Now the total torque to which the handle will be subjected is given by

$$T = T_1 + T_2$$

Assuming that a person can apply a force of 300 – 400 N intermittently, the length of handle required

$$= T / 300$$

The length of handle may be fixed by giving some allowance for gripping.

10. The diameter of handle (D) may be obtained by considering bending effects. We know that bending moment,

$$M = \frac{\pi}{32} \times \sigma_b \times D^3$$ ... ($\because \sigma_b = \sigma_t$ or σ_c)

11. The height of head (H) is usually taken as twice the diameter of handle, *i.e.* $H = 2D$.

12. Now check the screw for buckling load.

Effective length or unsupported length of the screw,

$$L = \text{Lift of screw} + \frac{1}{2} \text{ Height of nut}$$

We know that buckling or critical load,

$$W_{cr} = A_c . \sigma_y \left[1 - \frac{\sigma_y}{4 \, C \, \pi^2 \, E} \left(\frac{L}{k} \right)^2 \right]$$

where σ_y = Yield stress,

C = End fixity coefficient. The screw is considered to be a strut with lower end fixed and load end free. For one end fixed and the other end free, $C = 0.25$

k = Radius of gyration = $0.25 \, d_c$

The buckling load as obtained by the above expression must be higher than the load at which the screw is designed.

13. Fix the dimensions for the body of the screw jack.

14. Find efficiency of the screw jack.

Example 17.15. *A screw jack is to lift a load of 80 kN through a height of 400 mm. The elastic strength of screw material in tension and compression is 200 MPa and in shear 120 MPa. The material for nut is phosphor-bronze for which the elastic limit may be taken as 100 MPa in tension, 90 MPa in compression and 80 MPa in shear. The bearing pressure between the nut and the screw is not to exceed 18 N/mm². Design and draw the screw jack. The design should include the design of 1. screw, 2. nut, 3. handle and cup, and 4. body.*

Screw jack

Solution. Given : $W = 80$ kN $= 80 \times 10^3$N ; $H_1 = 400$ mm $= 0.4$ m ; $\sigma_{et} = \sigma_{ec} = 200$ MPa $= 200$ N/mm^2 ; $\tau_e = 120$ MPa $= 120$ N/mm^2 ; $\sigma_{et(nut)} = 100$ MPa $= 100$ N/mm^2 ; $\sigma_{ec\ (nut)} = 90$ MPa $= 90$ N/mm^2 ; $\tau_{e(nut)} = 80$ MPa $= 80$N/mm^2 ; $p_b = 18$ N/mm^2

The various parts of a screw jack are designed as discussed below:

1. Design of screw for spindle

Let d_c = Core diameter of the screw.

Since the screw is under compression, therefore load (W),

$$80 \times 10^3 = \frac{\pi}{4}(d_c)^2 \times \frac{\sigma_{ec}}{F.S.} = \frac{\pi}{4}(d_c)^2 \frac{200}{2} = 78.55\ (d_c)^2$$

... (Taking factor of safety, $F.S. = 2$)

$$\therefore \quad (d_c)^2 = 80 \times 10^3 / 78.55 = 1018.5 \quad \text{or} \quad d_c = 32 \text{ mm}$$

For square threads of normal series, the following dimensions of the screw are selected from Table 17.2.

*Core diameter, $d_c = 38$ mm **Ans.**

Nominal or outside diameter of spindle,

$d_o = 46$ mm **Ans.**

Pitch of threads, $p = 8$ mm **Ans.**

Now let us check for principal stresses:

We know that the mean diameter of screw,

$$d = \frac{d_o + d_c}{2} = \frac{46 + 38}{2} = 42 \text{ mm}$$

and

$$\tan \alpha = \frac{p}{\pi d} = \frac{8}{\pi \times 42} = 0.0606$$

Assuming coefficient of friction between screw and nut,

$$\mu = \tan \phi = 0.14$$

$\therefore$ Torque required to rotate the screw in the nut,

$$T_1 = P \times \frac{d}{2} = W \tan(\alpha + \phi)\frac{d}{2} = W\left[\frac{\tan \alpha + \tan \phi}{1 - \tan \alpha \tan \phi}\right]\frac{d}{2}$$

$$= 80 \times 10^3 \left[\frac{0.0606 + 0.14}{1 - 0.0606 \times 0.14}\right]\frac{42}{2} = 340 \times 10^3 \text{ N-mm}$$

Now compressive stress due to axial load,

$$\sigma_c = \frac{W}{A_c} = \frac{W}{\frac{\pi}{4}(d_c)^2} = \frac{80 \times 10^3}{\frac{\pi}{4}(38)^2} = 70.53 \text{ N/mm}^2$$

and shear stress due to the torque,

$$\tau = \frac{16\ T_1}{\pi (d_c)^3} = \frac{16 \times 340 \times 10^3}{\pi\ (38)^3} = 31.55 \text{ N/mm}^2$$

$\therefore$ Maximum principal stress (tensile or compressive),

$$\sigma_{c(max)} = \frac{1}{2}\left[\sigma_c + \sqrt{(\sigma_c)^2 + 4\,\tau^2}\right] = \frac{1}{2}\left[70.53 + \sqrt{(70.53)^2 + 4\ (31.55)^2}\right]$$

$$= \frac{1}{2}[70.53 + 94.63] = 82.58 \text{ N/mm}^2$$

---

* From Table 17.2, we see that next higher value of 32 mm for the core diameter is 33 mm. By taking $d_c = 33$ mm, gives higher principal stresses than the permissible values. So core diameter is chosen as 38 mm.

The given value of σ_c is equal to $\frac{\sigma_{ec}}{F.S.}$, *i.e.* $\frac{200}{2} = 100$ N/mm^2

We know that maximum shear stress,

$$\tau_{max} = \frac{1}{2}\left[\sqrt{(\sigma_c)^2 + 4\tau^2}\right] = \frac{1}{2}\left[\sqrt{(70.53)^2 + 4 \cdot (31.55)^2}\right]$$

$$= \frac{1}{2} \times 94.63 = 47.315 \text{ N/mm}^2$$

The given value of τ is equal to $\frac{\tau_e}{F.S.}$, *i.e.* $\frac{120}{2} = 60$ N/mm^2.

Since these maximum stresses are within limits, therefore design of screw for spindle is safe.

2. Design for nut

Let n = Number of threads in contact with the screwed spindle,

h = Height of nut = $n \times p$, and

t = Thickness of screw = $p / 2 = 8 / 2 = 4$ mm

Assume that the load is distributed uniformly over the cross-sectional area of nut.

We know that the bearing pressure (p_b),

$$18 = \frac{W}{\frac{\pi}{4}\left[(d_o)^2 - (d_c)^2\right] n} = \frac{80 \times 10^3}{\frac{\pi}{4}\left[(46)^2 - (38)^2\right] n} = \frac{151.6}{n}$$

$\therefore$ $n = 151.6 / 18 = 8.4$ say 10 threads **Ans.**

and height of nut, $h = n \times p = 10 \times 8 = 80$ mm **Ans.**

Now, let us check the stresses induced in the screw and nut.

We know that shear stress in the screw,

$$\tau_{(screw)} = \frac{W}{\pi\, n.d_c.t} = \frac{80 \times 10^3}{\pi \times 10 \times 38 \times 4} = 16.15 \text{ N/mm}^2$$

... ($\because t = p / 2 = 4$ mm)

and shear stress in the nut,

$$\tau_{(nut)} = \frac{W}{\pi\, n.d_o.t} = \frac{80 \times 10^3}{\pi \times 10 \times 46 \times 4} = 13.84 \text{ N/mm}^2$$

Since these stresses are within permissible limit, therefore design for nut is safe.

Let D_1 = Outer diameter of nut,

D_2 = Outside diameter for nut collar, and

t_1 = Thickness of nut collar.

First of all considering the tearing strength of nut, we have

$$W = \frac{\pi}{4}\left[(D_1)^2 - (d_o)^2\right]\sigma_t$$

$$80 \times 10^3 = \frac{\pi}{4}\left[(D_1)^2 - (46)^2\right]\frac{100}{2} = 39.3\left[(D_1)^2 - 2116\right] \quad ...\left[\because \sigma_t = \frac{\sigma_{et(nut)}}{F.S.}\right]$$

or $(D_1)^2 - 2116 = 80 \times 10^3 / 39.3 = 2036$

$\therefore$ $(D_1)^2 = 2036 + 2116 = 4152$ or $D_1 = 65$ mm **Ans.**

Now considering the crushing of the collar of the nut, we have

$$W = \frac{\pi}{4}\left[(D_2)^2 - (D_1)^2\right]\sigma_c$$

$$80 \times 10^3 = \frac{\pi}{4}\left[(D_2)^2 - (65)^2\right]\frac{90}{2} = 35.3\left[(D_2)^2 - 4225\right] \quad \ldots\left[\sigma_c = \frac{\sigma_{ec(nut)}}{F.S.}\right]$$

or $(D_2)^2 - 4225 = 80 \times 10^3 / 35.3 = 2266$

$\therefore$ $(D_2)^2 = 2266 + 4225 = 6491$ or $D_2 = 80.6$ say 82 mm **Ans.**

Considering the shearing of the collar of the nut, we have

$$W = \pi D_1 \times t_1 \times \tau$$

$$80 \times 10^3 = \pi \times 65 \times t_1 \times \frac{80}{2} = 8170\, t_1 \quad \ldots\left[\tau = \frac{\tau_{e(nut)}}{F.S.}\right]$$

$\therefore$ $t_1 = 80 \times 10^3/8170 = 9.8$ say 10 mm **Ans.**

3. *Design for handle and cup*

The diameter of the head (D_3) on the top of the screwed rod is usually taken as 1.75 times the outside diameter of the screw (d_o).

$\therefore$ $D_3 = 1.75\, d_o = 1.75 \times 46 = 80.5$ say 82 mm **Ans.**

The head is provided with two holes at the right angles to receive the handle for rotating the screw. The seat for the cup is made equal to the diameter of head, *i.e.* 82 mm and it is given chamfer at the top. The cup prevents the load from rotating. The cup is fitted to the head with a pin of diameter D_4 = 20 mm. The pin remains loose fit in the cup. Other dimensions for the cup may be taken as follows :

Height of cup = 50 mm **Ans.**

Thickness of cup = 10 mm **Ans.**

Diameter at the top of cup = 160 mm **Ans.**

Now let us find out the torque required (T_2) to overcome friction at the top of the screw.

Assuming uniform pressure conditions, we have

$$T_2 = \frac{2}{3} \times \mu_1 W \left[\frac{(R_3)^3 - (R_4)^3}{(R_3)^2 - (R_4)^2}\right]$$

$$= \frac{2}{3} \times 0.14 \times 80 \times 10^3 \left[\frac{\left(\frac{82}{2}\right)^3 - \left(\frac{20}{2}\right)^3}{\left(\frac{82}{2}\right)^2 - \left(\frac{20}{2}\right)^2}\right] \quad \ldots (\text{Assuming } \mu_1 = \mu)$$

$$= 7.47 \times 10^3 \left[\frac{(41)^3 - (10)^3}{(41)^2 - (10)^2}\right] = 321 \times 10^3 \text{ N-mm}$$

$\therefore$ Total torque to which the handle is subjected,

$$T = T_1 + T_2 = 340 \times 10^3 + 321 \times 10^3 = 661 \times 10^3 \text{ N-mm}$$

Assuming that a force of 300 N is applied by a person intermittently, therefore length of handle required

$$= 661 \times 10^3 / 300 = 2203 \text{ mm}$$

Allowing some length for gripping, we shall take the length of handle as 2250 mm.

A little consideration will show that an excessive force applied at the end of lever will cause bending. Considering bending effect, the maximum bending moment on the handle,

$$M = \text{Force applied} \times \text{Length of lever}$$
$$= 300 \times 2250 = 675 \times 10^3 \text{ N-mm}$$

Let D = Diameter of the handle.

Assuming that the material of the handle is same as that of screw, therefore taking bending stress $\sigma_b = \sigma_t = \sigma_{et} / 2 = 100 \text{ N/mm}^2$.

We know that the bending moment (M),

$$675 \times 10^3 = \frac{\pi}{32} \times \sigma_b \times D^3 = \frac{\pi}{32} \times 100 \times D^3 = 9.82\ D^3$$

$$\therefore \quad D^3 = 675 \times 10^3 / 9.82 = 68.74 \times 10^3 \quad \text{or} \quad D = 40.96 \text{ say } 42 \text{ mm} \quad \textbf{Ans.}$$

The height of head (H) is taken as $2D$.

$$\therefore \quad H = 2\,D = 2 \times 42 = 84 \text{ mm} \quad \textbf{Ans.}$$

Now let us check the screw for buckling load.

We know that the effective length for the buckling of screw,

$$L = \text{Lift of screw} + \frac{1}{2}\text{ Height of nut} = H_1 + h / 2$$
$$= 400 + 80 / 2 = 440 \text{ mm}$$

When the screw reaches the maximum lift, it can be regarded as a strut whose lower end is fixed and the load end is free. We know that critical load,

$$W_{cr} = A_c \times \sigma_y \left[1 - \frac{\sigma_y}{4C\,\pi^2\,E}\left(\frac{L}{k}\right)^2\right]$$

For one end fixed and other end free, $C = 0.25$.

Also $k = 0.25\,d_c = 0.25 \times 38 = 9.5$ mm

$$\therefore \quad W_{cr} = \frac{\pi}{4}(38)^2\ 200\left[1 - \frac{200}{4 \times 0.25 \times \pi^2 \times 210 \times 10^3}\left(\frac{440}{9.5}\right)^2\right]$$

... (Taking $\sigma_y = \sigma_{et}$)

$$= 226\,852\ (1 - 0.207) = 179\,894 \text{ N}$$

Since the critical load is more than the load at which the screw is designed (*i.e.* 80×10^3 N), therefore there is no chance of the screw to buckle.

4. *Design of body*

The various dimensions of the body may be fixed as follows:

Diameter of the body at the top,

$$D_5 = 1.5\,D_2 = 1.5 \times 82 = 123 \text{ mm} \quad \textbf{Ans.}$$

Thickness of the body,

$$t_3 = 0.25\,d_o = 0.25 \times 46 = 11.5 \text{ say } 12 \text{ mm} \quad \textbf{Ans.}$$

Inside diameter at the bottom,

$$D_6 = 2.25\,D_2 = 2.25 \times 82 = 185 \text{ mm} \quad \textbf{Ans.}$$

Outer diameter at the bottom,

$$D_7 = 1.75\,D_6 = 1.75 \times 185 = 320 \text{ mm} \quad \textbf{Ans.}$$

Thickness of base, $t_2 = 2\,t_1 = 2 \times 10 = 20$ mm **Ans.**

Height of the body = Max. lift + Height of nut + 100 mm extra

= 400 + 80 + 100 = 580 mm **Ans.**

The body is made tapered in order to achieve stability of jack.

Let us now find out the efficiency of the screw jack. We know that the torque required to rotate the screw with no friction,

$$T_0 = W \tan \alpha \times \frac{d}{2} = 80 \times 10^3 \times 0.0606 \times \frac{42}{2} = 101\,808 \text{ N-mm}$$

∴ Efficiency of the screw jack,

$$\eta = \frac{T_0}{T} = \frac{101808}{661 \times 10^3} = 0.154 \text{ or } 15.4\% \textbf{ Ans.}$$

Example 17.16. *A toggle jack as shown in Fig. 17.12, is to be designed for lifting a load of 4 kN. When the jack is in the top position, the distance between the centre lines of nuts is 50 mm and in the bottom position this distance is 210 mm. The eight links of the jack are symmetrical and 110 mm long. The link pins in the base are set 30 mm apart. The links, screw and pins are made from mild steel for which the permissible stresses are 100 MPa in tension and 50 MPa in shear. The bearing pressure on the pins is limited to 20 N/mm².*

Assume the pitch of the square threads as 6 mm and the coefficient of friction between threads as 0.20.

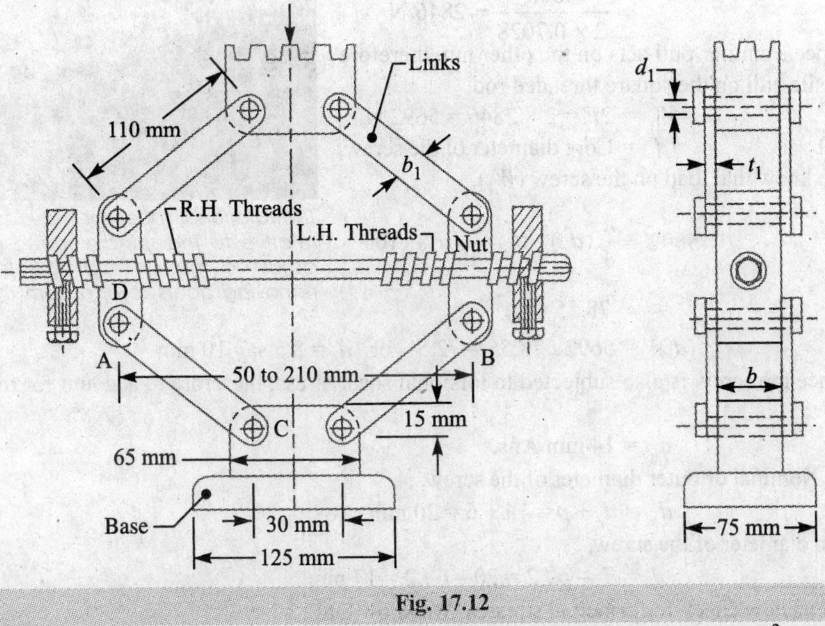

Fig. 17.12

Solution. Given : $W = 4$ kN = 4000 N ; $l = 110$ mm ; $\sigma_t = 100$ MPa = 100 N/mm² ; $\tau = 50$ MPa = 50 N/mm² ; $p_b = 20$ N/mm² ; $p = 6$ mm ; $\mu = \tan \phi = 0.20$

The toggle jack may be designed as discussed below :

1. ***Design of square threaded screw***

A little consideration will show that the maximum load on the square threaded screw occurs when the jack is in the bottom position. The position of the link *CD* in the bottom position is shown in Fig. 17.13 (*a*).

Let θ be the angle of inclination of the link *CD* with the horizontal.

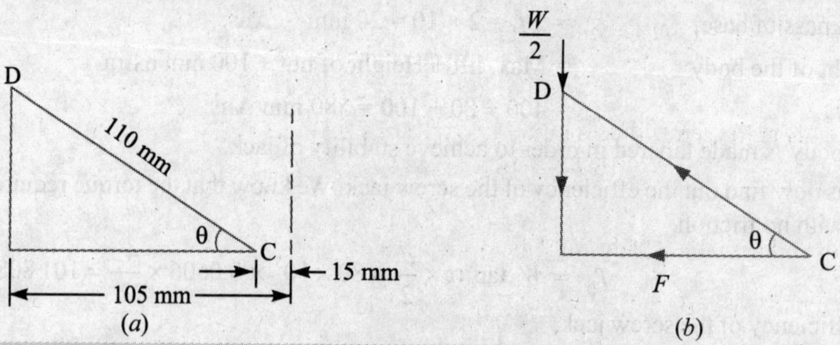

Fig. 17.13

From the geometry of the figure, we find that

$$\cos \theta = \frac{105 - 15}{110} = 0.8112 \text{ or } \theta = 35.1°$$

Each nut carries half the total load on the jack and due to this, the link *CD* is subjected to tension while the square threaded screw is under pull as shown in Fig. 17.13 (*b*). The magnitude of the pull on the square threaded screw is given by

$$F = \frac{W}{2 \tan \theta} = \frac{W}{2 \tan 35.1°}$$

$$= \frac{4000}{2 \times 0.7028} = 2846 \text{ N}$$

The rotational speed of the lead screw relative to the spindle speed can be adjusted manually by adding and removing gears to and from the gear

Since a similar pull acts on the other nut, therefore total tensile pull on the square threaded rod,

$$W_1 = 2F = 2 \times 2846 = 5692 \text{ N}$$

Let d_c = Core diameter of the screw,

We know that load on the screw (W_1),

$$5692 = \frac{\pi}{4}(d_c)^2 \sigma_t = \frac{\pi}{4}(d_c)^2 \, 100$$

$$= 78.55 \, (d_c)^2$$

$\therefore$ $(d_c)^2 = 5692 / 78.55 = 72.5$ or $d_c = 8.5$ say 10 mm

Since the screw is also subjected to torsional shear stress, therefore to account for this, let us adopt

$$d_c = 14 \text{ mm } \textbf{Ans.}$$

$\therefore$ Nominal or outer diameter of the screw,

$$d_o = d_c + p = 14 + 6 = 20 \text{ mm } \textbf{Ans.}$$

and mean diameter of the screw,

$$d = d_o - p/2 = 20 - 6/2 = 17 \text{ mm}$$

Let us now check for principal stresses. We know that

$$\tan \alpha = \frac{p}{\pi d} = \frac{6}{\pi \times 17} = 0.1123 \quad \text{... (where } \alpha \text{ is the helix angle)}$$

We know that effort required to rotate the screw,

$$P = W_1 \tan(\alpha + \phi) = W_1 \left(\frac{\tan \alpha + \tan \phi}{1 - \tan \alpha \tan \phi} \right)$$

$$= 5692 \left(\frac{0.1123 + 0.20}{1 - 0.1123 \times 0.20} \right) = 1822 \text{ N}$$

∴ Torque required to rotate the screw,

$$T = P \times \frac{d}{2} = 1822 \times \frac{17}{2} = 15\ 487 \text{ N-mm}$$

and shear stress in the screw due to torque,

$$\tau = \frac{16\,T}{\pi\,(d_c)^3} = \frac{16 \times 15\ 487}{\pi\,(14)^3} = 28.7 \text{ N/mm}^2$$

We know that direct tensile stress in the screw,

$$\sigma_t = \frac{W_1}{\frac{\pi}{4}(d_c)^2} = \frac{W_1}{0.7855\,(d_c)^2} = \frac{5692}{0.7855\,(14)^2} = 37 \text{ N/mm}^2$$

∴ Maximum principal (tensile) stress,

$$\sigma_{t(max)} = \frac{\sigma_t}{2} + \frac{1}{2}\sqrt{(\sigma_t)^2 + 4\,\tau^2} = \frac{37}{2} + \frac{1}{2}\sqrt{(37)^2 + 4\,(28.7)^2}$$

$$= 18.5 + 34.1 = 52.6 \text{ N/mm}^2$$

and maximum shear stress,

$$\tau_{max} = \frac{1}{2}\sqrt{(\sigma_t)^2 + 4\tau^2} = \frac{1}{2}\sqrt{(37)^2 + 4(28.7)^2} = 34.1 \text{ N/mm}^2$$

Since the maximum stresses are within safe limits, therefore the design of square threaded screw is satisfactory.

2. *Design of nut*

Let n = Number of threads in contact with the screw (*i.e.* square threaded rod).

Assuming that the load W_1 is distributed uniformly over the cross-sectional area of the nut, therefore bearing pressure between the threads (p_b),

$$20 = \frac{W_1}{\frac{\pi}{4}\left[(d_o)^2 - (d_c)^2\right] n} = \frac{5692}{\frac{\pi}{4}\left[(20)^2 - (14)^2\right] n} = \frac{35.5}{n}$$

$$\therefore \quad n = 35.5 / 20 = 1.776$$

In order to have good stability and also to prevent rocking of the screw in the nut, we shall provide $n = 4$ threads in the nut. The thickness of the nut,

$$t = n \times p = 4 \times 6 = 24 \text{ mm} \quad \textbf{Ans.}$$

The width of the nut (b) is taken as $1.5\,d_o$.

$$\therefore \quad b = 1.5\,d_o = 1.5 \times 20 = 30 \text{ mm } \textbf{Ans.}$$

To control the movement of the nuts beyond 210 mm (the maximum distance between the centre lines of nuts), rings of 8 mm thickness are fitted on the screw with the help of set screws.

∴ Length of screwed portion of the screw

$$= 210 + t + 2 \times \text{Thickness of rings}$$

$$= 210 + 24 + 2 \times 8 = 250 \text{ mm } \textbf{Ans.}$$

The central length (about 25 mm) of screwed rod is kept equal to core diameter of the screw *i.e.* 14 mm. Since the toggle jack is operated by means of spanners on both sides of the square threaded rod, therefore the ends of the rod may be reduced to 10 mm square and 15 mm long.

∴ Toal length of the screw

$$= 250 + 2 \times 15 = 280 \text{ mm } \textbf{Ans.}$$

Assuming that a force of 150 N is applied by each person at each end of the rod, therefore length of the spanner required

$$= \frac{T}{2 \times 150} = \frac{15487}{300} = 51.62 \text{ mm}$$

We shall take the length of the spanner as 200 mm in order to facilitate the operation and even a single person can operate it.

3. ***Design of pins in the nuts***

Let d_1 = Diameter of pins in the nuts.

Since the pins are in double shear, therefore load on the pins (*F*),

$$2846 = 2 \times \frac{\pi}{4}(d_1)^2 \tau = 2 \times \frac{\pi}{4}(d_1)^2 \, 50 = 78.55\,(d_1)^2$$

$$\therefore \quad (d_1)^2 = 2846 / 78.55 = 36.23 \text{ or } d_1 = 6.02 \text{ say } 8 \text{ mm } \textbf{Ans.}$$

The diameter of pin head is taken as 1.5 d_1 (*i.e.* 12 mm) and thickness 4 mm. The pins in the nuts are kept in position by separate rings 4 mm thick and 1.5 mm split pins passing through the rings and pins.

4. ***Design of links***

Due to the load, the links may buckle in two planes at right angles to each other. For buckling in the vertical plane (*i.e.* in the plane of the links), the links are considered as hinged at both ends and for buckling in a plane perpendicular to the vertical plane, it is considered as fixed at both ends. We know that load on the link

$$= F / 2 = 2846 / 2 = 1423 \text{ N}$$

Assuming a factor of safety = 5, the links must be designed for a buckling load of

$$W_{cr} = 1423 \times 5 = 7115 \text{ N}$$

Let t_1 = Thickness of the link, and

b_1 = Width of the link.

Assuming that the width of the link is three times the thickness of the link, *i.e.* $b_1 = 3\,t_1$, therefore cross-sectional area of the link,

$$A = t_1 \times 3t_1 = 3(t_1)^2$$

and moment of inertia of the cross-section of the link,

$$I = \frac{1}{12} \times t_1\,(b_1)^3 = \frac{1}{12} \times t_1\,(3t_1)^3 = 2.25\,(t_1)^4$$

We know that the radius of gyration,

$$k = \sqrt{\frac{I}{A}} = \sqrt{\frac{2.25\,(t_1)^4}{3(t_1)^2}} = 0.866\,t_1$$

Since for buckling of the link in the vertical plane, the ends are considered as hinged, therefore equivalent length of the link,

$$L = l = 110 \text{ mm}$$

and Rankine's constant, $a = \dfrac{1}{7500}$

According to Rankine's formula, buckling load (W_{cr}),

$$7115 = \frac{\sigma_c \times A}{1 + a\left(\dfrac{L}{k}\right)^2} = \frac{100 \times 3(t_1)^2}{1 + \dfrac{1}{7500}\left(\dfrac{110}{0.866\,t_1}\right)^2} = \frac{300\,(t_1)^2}{1 + \dfrac{2.15}{(t_1)^2}}$$

or
$$\frac{7115}{300} = \frac{(t_1)^4}{(t_1)^2 + 2.15}$$

$$(t_1)^4 - 23.7\,(t_1)^2 - 51 = 0$$

$$\therefore \quad (t_1)^2 = \frac{23.7 \pm \sqrt{(23.7)^2 + 4 \times 51}}{2} = \frac{23.7 + 27.7}{2} = 25.7$$

or $t_1 = 5.07$ say 6 mm ... (Taking + ve sign)

and $b_1 = 3\,t_1 = 3 \times 6 = 18$ mm

Now let us consider the buckling of the link in a plane perpendicular to the vertical plane.

Moment of inertia of the cross-section of the link,

$$I = \frac{1}{12} \times b_1 (t_1)^3 = \frac{1}{12} \times 3t_1\,(t_1)^3 = 0.25\,(t_1)^4$$

and cross-sectional area of the link,

$$A = t_1 . b_1 = t_1 \times 3\,t_1 = 3\,(t_1)^2$$

$\therefore$ Radius of gyration,

$$k = \sqrt{\frac{I}{A}} = \sqrt{\frac{0.25\,(t_1)^4}{3\,(t_1)^2}} = 0.29\,t_1$$

Since for buckling of the link in a plane perpendicular to the vertical plane, the ends are considered as fixed, therefore

Equivalent length of the link,

$$L = l/2 = 110/2 = 55 \text{ mm}$$

Again according to Rankine's formula, buckling load,

$$W_{cr} = \frac{\sigma_c \times A}{1 + a\left(\dfrac{L}{k}\right)^2} = \frac{100 \times 3(t_1)^2}{1 + \dfrac{1}{7500}\left(\dfrac{55}{0.29\,t_1}\right)^2} = \frac{300(t_1)^2}{1 + \dfrac{4.8}{(t_1)^2}}$$

Substituting the value of $t_1 = 6$ mm, we have

$$W_{cr} = \frac{300 \times 6^2}{1 + \dfrac{4.8}{6^2}} = 9532 \text{ N}$$

Since this buckling load is more than the calculated value (*i.e.* 7115 N), therefore the link is safe for buckling in a plane perpendicular to the vertical plane.

$\therefore$ We may take $t_1 = 6$ mm ; and $b_1 = 18$ mm **Ans.**

17.15 Differential and Compound Screws

There are certain cases in which a very slow movement of the screw is required whereas in other cases, a very fast movement of the screw is needed. The slow movement of the screw may be obtained by using a small pitch of the threads, but it results in weak threads. The fast movement of the screw may be obtained by using multiple-start threads, but this method requires expensive machning and the loss of self-locking property. In order to overcome these difficulties, differential or compound screws, as discussed below, are used.

1. *Differential screw.* When a slow movement or fine adjustment is desired in precision equipments, then a differential screw is used. It consists of two threads of the same hand (*i.e.* right handed or left handed) but of different pitches, wound on the same cylinder or different cylinders as shown in Fig. 17.14. It may be noted that when the threads are wound on the same cylinder, then two nuts are employed as shown in Fig. 17.14 (*a*) and when the threads are wound on different cylinders, then only one nut is employed as shown in Fig. 17.14 (*b*).

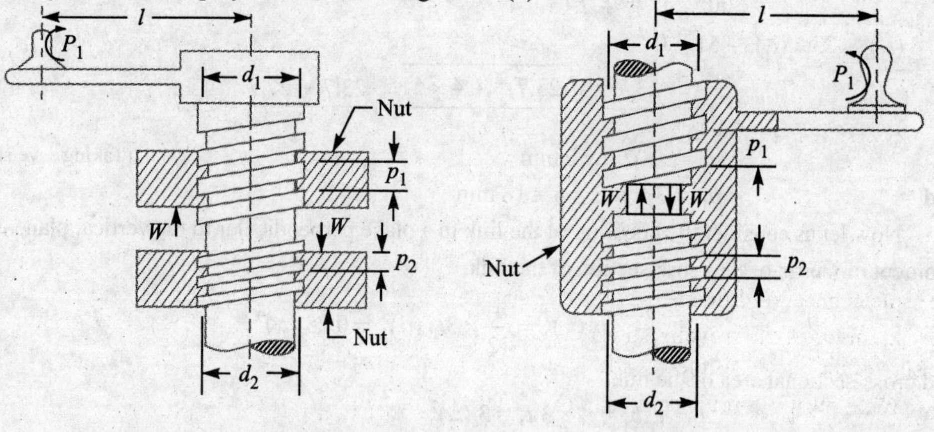

(*a*) Threads wound on the same cylinder. (*b*) Threads wound on the different cylinders.

Fig. 17.14

In this case, each revolution of the screw causes the nuts to move towards or away from each other by a distance equal to the difference of the pitches.

Let p_1 = Pitch of the upper screw,

d_1 = Mean diameter of the upper screw,

α_1 = Helix angle of the upper screw, and

μ_1 = Coefficient of friction between the upper screw and the upper nut = $\tan \phi_1$, where ϕ_1 is the friction angle.

p_2, d_2, α_2 and μ_2 = Corresponding values for the lower screw.

We know that torque required to overcome friction at the upper screw,

$$T_1 = W \tan(\alpha_1 + \phi_1)\frac{d_1}{2} = W\left[\frac{\tan\alpha_1 + \tan\phi_1}{1 - \tan\alpha_1 \tan\phi_1}\right]\frac{d_1}{2} \quad ...(i)$$

Similarly, torque required to overcome friction at the lower screw,

$$T_2 = W \tan(\alpha_2 + \phi_2)\frac{d_2}{2} = W\left[\frac{\tan\alpha_2 + \tan\phi_2}{1 - \tan\alpha_2 \tan\phi_2}\right]\frac{d_2}{2} \quad ...(ii)$$

∴ Total torque required to overcome friction at the thread surfaces,

$$T = P_1 \times l = T_1 - T_2$$

When there is no friction between the thread surfaces, then $\mu_1 = \tan\phi_1 = 0$ and $\mu_2 = \tan\phi_2 = 0$. Substituting these values in the above expressions, we have

$$\therefore \quad T_1' = W \tan\alpha_1 \times \frac{d_1}{2}$$

and

$$T_2' = W \tan\alpha_2 \times \frac{d_2}{2}$$

∴ Total torque required when there is no friction,

$$T_0 = T_1' - T_2'$$

$$= W \tan\alpha_1 \times \frac{d_1}{2} - W \tan\alpha_2 \times \frac{d_2}{2}$$

$$= W\left[\frac{p_1}{\pi d_1}\times\frac{d_1}{2}-\frac{p_2}{\pi d_2}\times\frac{d_2}{2}\right]=\frac{W}{2\pi}(p_1-p_2)$$

$$\left[\because \tan\alpha_1=\frac{p_1}{\pi d_1};\text{ and } \tan\alpha_2=\frac{p_2}{\pi d_2}\right]$$

We know that efficiency of the differential screw,

$$\eta=\frac{T_0}{T}$$

2. ***Compound screw.*** When a fast movement is desired, then a compound screw is employed. It consists of two threads of opposite hands (*i.e.* one right handed and the other left handed) wound on the same cylinder or different cylinders, as shown in Fig. 17.15 (*a*) and (*b*) respectively.

In this case, each revolution of the screw causes the nuts to move towards one another equal to the sum of the pitches of the threads. Usually the pitch of both the threads are made equal.

We know that torque required to overcome friction at the upper screw,

$$T_1 = W\tan(\alpha_1+\phi_1)\frac{d_1}{2}=W\left[\frac{\tan\alpha_1+\tan\phi_1}{1-\tan\alpha_1\tan\phi_1}\right]\frac{d_1}{2} \quad ...(i)$$

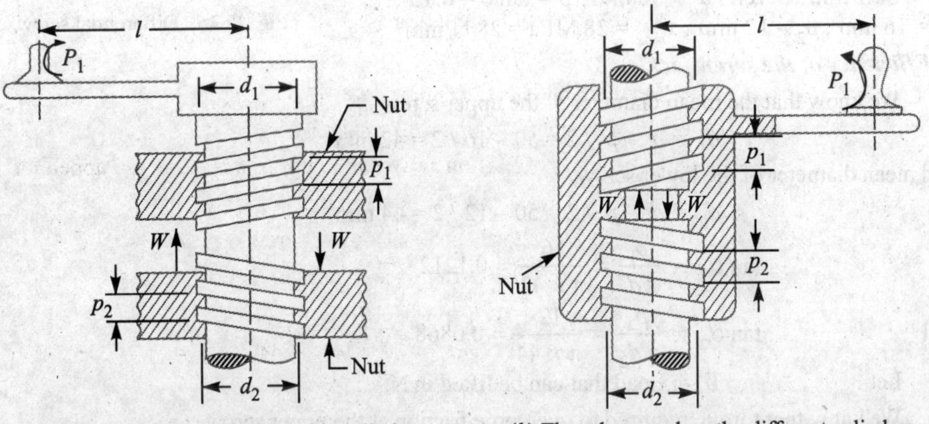

(*a*) Threads wound on the same cylinder. (*b*) Threads wound on the different cylinders.

Fig. 17.15

Similarly, torque required to overcome friction at the lower screw,

$$T_2 = W\tan(\alpha_2+\phi_2)\frac{d_2}{2}=W\left[\frac{\tan\alpha_2+\tan\phi_2}{1-\tan\alpha_2\tan\phi_2}\right]\frac{d_2}{2} \quad ...(ii)$$

∴ Total torque required to overcome friction at the thread surfaces,

$$T = P_1\times l = T_1+T_2$$

When there is no friction between the thread surfaces, then $\mu_1=\tan\phi_1=0$ and $\mu_2=\tan\phi_2=0$. Substituting these values in the above expressions, we have

$$T_1' = W\tan\alpha_1\times\frac{d_1}{2}$$

$$T_2' = W\tan\alpha_2\times\frac{d_2}{2}$$

∴ Total torque required when there is no friction,

$$T_0 = T_1' + T_2'$$

$$= W \tan \alpha_1 \times \frac{d_2}{2} + W \tan \alpha_2 \times \frac{d_2}{2}$$

$$= W \left[\frac{p_1}{\pi d_1} \times \frac{d_1}{2} + \frac{p_2}{\pi d_2} \times \frac{d_2}{2} \right] = \frac{W}{2\pi}(p_1 + p_2)$$

We know that efficiency of the compound screw,

$$\eta = \frac{T_0}{T}$$

Example 17.17. *A differential screw jack is to be made as shown in Fig. 17.16. Neither screw rotates. The outside screw diameter is 50 mm. The screw threads are of square form single start and the coefficient of thread friction is 0.15.*

Determine : 1. Efficiency of the screw jack; 2. Load that can be lifted if the shear stress in the body of the screw is limited to 28 MPa.

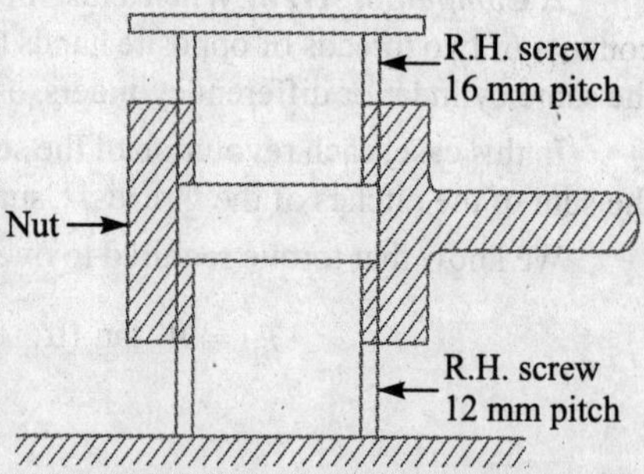

Fig. 17.16. Differential screw.

Solution. Given : $d_o = 50$ mm ; $\mu = \tan \phi = 0.15$; $p_1 = 16$ mm ; $p_2 = 12$ mm ; $\tau_{max} = 28$ MPa $= 28$ N/mm^2

1. ***Efficiency of the screw jack***

We know that the mean diameter of the upper screw,

$$d_1 = d_o - p_1 / 2 = 50 - 16 / 2 = 42 \text{ mm}$$

and mean diameter of the lower screw,

$$d_2 = d_o - p_2 / 2 = 50 - 12 / 2 = 44 \text{ mm}$$

$$\therefore \quad \tan \alpha_1 = \frac{p_1}{\pi d_1} = \frac{16}{\pi \times 42} = 0.1212$$

$$\text{and} \quad \tan \alpha_2 = \frac{p_2}{\pi d_2} = \frac{12}{\pi \times 44} = 0.0868$$

Let W = Load that can be lifted in N.

We know that torque required to overcome friction at the upper screw,

$$T_1 = W \tan(\alpha_1 + \phi) \frac{d_1}{2} = W \left[\frac{\tan \alpha_1 + \tan \phi}{1 - \tan \alpha_1 \tan \phi} \right] \frac{d_1}{2}$$

$$= W \left[\frac{0.1212 + 0.15}{1 - 0.1212 \times 0.15} \right] \frac{42}{2} = 5.8 \, W \text{ N-mm}$$

Similarly, torque required to overcome friction at the lower screw,

$$T_2 = W \tan(\alpha_2 - \phi) \frac{d_2}{2} = W \left[\frac{\tan \alpha_2 - \tan \phi}{1 + \tan \alpha_2 \tan \phi} \right] \frac{d_2}{2}$$

$$= W \left[\frac{0.0868 - 0.15}{1 + 0.0868 \times 0.15} \right] \frac{44}{2} = -1.37 \, W \text{ N-mm}$$

∴ Total torque required to overcome friction,

$$T = T_1 - T_2 = 5.8 \, W - (-1.37 \, W) = 7.17 \, W \text{ N-mm}$$

We know that the torque required when there is no friction,

$$T_0 = \frac{W}{2\pi}(p_1 - p_2) = \frac{W}{2\pi}(16 - 12) = 0.636 \, W \text{ N-mm}$$

∴ Efficiency of the screw jack,

$$\eta = \frac{T_0}{T} = \frac{0.636\,W}{7.17\,W} = 0.0887 \text{ or } 8.87\% \textbf{ Ans.}$$

2. *Load that can be lifted*

Since the upper screw is subjected to a larger torque, therefore the load to be lifted (W) will be calculated on the basis of larger torque (T_1).

We know that core diameter of the upper screw,

$$d_{c1} = d_o - p_1 = 50 - 16 = 34 \text{ mm}$$

Since the screw is subjected to direct compressive stress due to load W and shear stress due to torque T_1, therefore

Direct compressive stress,

$$\sigma_c = \frac{W}{A_{c1}} = \frac{W}{\frac{\pi}{4}(d_{c1})^2} = \frac{W}{\frac{\pi}{4}(34)^2} = \frac{W}{908} \text{ N/mm}^2$$

and shear stress,

$$\tau = \frac{16\,T_1}{\pi\,(d_{c1})^3} = \frac{16 \times 5.8\,W}{\pi(34)^3} = \frac{W}{1331} \text{ N/mm}^2$$

We know that maximum shear stress (τ_{max}),

$$28 = \frac{1}{2}\sqrt{(\sigma_c)^2 + 4\,\tau^2} = \frac{1}{2}\sqrt{\left(\frac{W}{908}\right)^2 + 4\left(\frac{W}{1331}\right)^2}$$

$$= \frac{1}{2}\sqrt{1.213 \times 10^{-6}\,W^2 + 2.258 \times 10^{-6}\,W^2} = \frac{1}{2}\,1.863 \times 10^{-3}\,W$$

$$\therefore \quad W = \frac{28 \times 2}{1.863 \times 10^{-3}} = 30\ 060 \text{ N} = 30.06 \text{ kN} \quad \textbf{Ans.}$$

EXERCISES

1. In a hand vice, the screw has double start square threads of 24 mm outside diameter. If the lever is 200 mm long and the maximum force that can be applied at the end of lever is 250 N, find the force with which the job is held in the jaws of the vice. Assume a coefficient of friction of 0.12. **[Ans. 17 420 N]**

2. A square threaded bolt of mean diameter 24 mm and pitch 5 mm is tightened by screwing a nut whose mean diameter of bearing surface is 50 mm. If the coefficient of friction for the nut and bolt is 0.1 and for the nut and bearing surfaces 0.16, find the force required at the end of a spanner 0.5 m long when the load on the bolt is 10 kN. **[Ans. 120 N]**

3. The spindle of a screw jack has a single start square thread with an outside diameter of 45 mm and a pitch of 10 mm. The spindle moves in a fixed nut. The load is carried on a swivel head but is not free to rotate. The bearing surface of the swivel head has a mean diameter of 60 mm. The coefficient of friction between the nut and screw is 0.12 and that between the swivel head and the spindle is 0.10. Calculate the load which can be raised by efforts of 100 N each applied at the end of two levers each of effective length of 350 mm. Also determine the efficiency of the lifting arrangement. **[Ans. 9945 N ; 22.7%]**

Lead screw supported by collar bearing.

4. The cross bar of a planner weighing 12 kN is raised and lowered by means of two square threaded screws of 38 mm outside diameter and 7 mm pitch. The screw is made of steel and a bronze nut of 38 mm thick. A steel collar has 75 mm outside diameter and 38 mm inside diameter. The coefficient of friction at the threads is assumed as 0.11 and at the collar 0.13. Find the force required at a radius of 100 mm to raise and lower the load. **[Ans. 402.5 N ; 267 N]**
5. The lead screw of a lathe has square threads of 24 mm outside diameter and 5 mm pitch. In order to drive the tool carriage, the screw exerts an axial pressure of 2.5 kN. Find the efficiency of the screw and the power required to drive the screw, if it is to rotate at 30 r.p.m. Neglect bearing friction. Assume coefficient of friction of screw threads as 0.12. **[Ans. 37.76% ; 16.55 W]**
6. The lead screw of a lathe has Acme threads of 60 mm outside diameter and 8 mm pitch. It supplies drive to a tool carriage which needs an axial force of 2000 N. A collar bearing with inner and outer radius as 30 mm and 60 mm respectively is provided. The coefficient of friction for the screw threads is 0.12 and for the collar it is 0.10. Find the torque required to drive the screw and the efficiency of the screw. **[Ans. 18.5 N-m ; 13.6%]**
7. A cross bar of a planer weighing 9 kN is raised and lowered by means of two square threaded screws of 40 mm outside diameter and 6 mm pitch. The screw is made of steel and nut of phosphor bronze having 42 mm height. A steel collar bearing with 30 mm mean radius takes the axial thrust. The coefficient of friction at the threads and at the collar may be assumed as 0.14 and 0.10 respectively. Find the force required at a radius of 120 mm of a handwheel to raise and lower the load. Find also the shear stress in the nut material and the bearing pressure on the threads. **[Ans. 495 N, 346 N ; 1.7 MPa ; 1.84 N/mm^2]**
8. A machine slide weighing 3000 N is elevated by a double start acme threaded screw at the rate of 840 mm/min. If the coefficient of friction be 0.12, calculate the power to drive the slide. The end of the screw is carried on a thrust collar of 32 mm inside diameter and 58 mm outside diameter. The pitch of the screw thread is 6 mm and outside diameter of the screw is 40 mm. If the screw is of steel, is it strong enough to sustain the load? Draw a neat sketch of the system. **[Ans. 0.165 kW]**
9. A sluice valve, used in water pipe lines, consists of a gate raised by the spindle, which is rotated by the hand wheel. The spindle has single start square threads. The nominal diameter of the spindle is 36 mm and the pitch is 6 mm. The friction collar has inner and outer diameters of 32 mm and 50 mm respectively. The coefficient of friction at the threads and the collar are 0.12 and 0.18 respectively. The weight of the gate is 7.5 kN and the frictional resistance to open the valve due to water pressure is 2.75 kN. Using uniform wear theory, determine : 1. torque required to raise the gate; and 2. overall efficiency. **[Ans. 136.85 N-m ; 7.1%]**
10. A vertical square threads screw of a 70 mm mean diameter and 10 mm pitch supports a vertical load of 50 kN. It passes through the boss of a spur gear wheel of 70 teeth which acts as a nut. In order to raise the load, the spur gear wheel is turned by means of a pinion having 20 teeth. The mechanical efficiency of pinion and gear wheel drive is 90%. The axial thrust on the screw is taken up by a collar bearing having a mean radius of 100 mm. The coefficient of friction for the screw and nut is 0.15 and that for collar bearing is 0.12. Find:
    (*a*) Torque to be applied to the pinion shaft,
    (*b*) Maximum principal and shear stresses in the screw ; and
    (*c*) Height of nut, if the bearing pressure is limited to 12 N/mm^2.
    [Ans. 299.6 N-m ; 26.6 N/mm^2, 19 N/mm^2 ; 40 mm]
11. A single start square threaded screw is to be designed for a C-clamp. The axial load on the screw may be assumed to be 10 kN. A thrust pad is attached at the end of the screw whose mean diameter may be taken as 30 mm. The coefficient of friction for the screw threads and for the thrust pads is 0.12 and 0.08 respectively. The allowable tensile strength of the screw is 60 MPa and the allowable bearing pressure is 12 N/mm^2. Design the screw and nut. The square threads are as under :

Nominal diameter, mm	16	18	20	22
Core diameter, mm	13	15	17	19
Pitch, mm	3	3	3	3

[Ans. d_c = 17 mm ; n = 10, h = 30 mm]

12. Design a screw jack for lifting a load of 50 kN through a height of 0.4 m. The screw is made of steel and nut of bronze. Sketch the front sectional view. The following allowable stresses may be assumed

For steel : Compressive stress = 80 MPa ; Shear stress = 45 MPa

For bronze : Tensile stress = 40 MPa ; Bearing stress = 15 MPa

Shear stress = 25 MPa.

The coefficient of friction between the steel and bronze pair is 0.12. The dimensions of the swivel base may be assumed proportionately. The screw should have square threads. Design the screw, nut and handle. The handle is made of steel having bending stress 150 MPa (allowable).

13. A screw jack carries a load of 22 kN. Assuming the coefficient of friction between screw and nut as 0.15, design the screw and nut. Neglect collar friction and column action. The permissible compressive and shear stresses in the screw should not exceed 42 MPa and 28 MPa respectively. The shear stress in the nut should not exceed 21 MPa. The bearing pressure on the nut is 14 N/mm^2. Also determine the effort required at the handle of 200 mm length in order to raise and lower the load. What will be the efficiency of screw? **[Ans. d_c = 30 mm ; h = 36 mm ; 381 N ; 166 N ; 27.6%]**

14. Design and draw a screw jack for lifting a safe load of 150 kN through a maximum lift of 350 mm.

The elastic strength of the material of the screw may be taken as 240 MPa in compression and 160 MPa in shear. The nut is to be made of phosphor bronze for which the elastic strengths in tension, compression and shear are respectively 130, 115 and 100 MPa. Bearing pressure between the threads of the screw and the nut may be taken as 18 N/mm^2. Safe crushing stress for the material of the body is 100 MPa. Coefficient of friction for the screw as well as collar may be taken as 0.15.

15. Design a toggle jack to lift a load of 5 kN. The jack is to be so designed that the distance between the centre lines of nuts varies from 50 to 220 mm. The eight links are symmetrical and 120 mm long. The link pins in the base are set 30 mm apart. The links, screw and pins are made from mild steel for which the stresses are 90 MPa in tension and 50 MPa in shear. The bearing pressure on the pin is 20 N/mm^2. Assume the coefficient of friction between screw and nut as 0.15 and pitch of the square threaded screw as 6 mm.

[Ans. d_c = 10 mm : d_o = 22 mm ; d = 19 mm ; n = 4; t = 24 mm ; b = 33 mm ; d_1 = 10 mm ; t_1 = 7 mm ; b_1 = 21 mm]

QUESTIONS

1. Discuss the various types of power threads. Give atleast two practical applications for each type. Discuss their relative advantages and disadvantages.
2. Why are square threads preferable to V-threads for power transmission?
3. How does the helix angle influence on the efficiency of square threaded screw?
4. What do you understand by overhauling of screw?
5. What is self locking property of threads and where it is necessary?
6. Show that the efficiency of self locking screws is less than 50 percent.
7. In the design of power screws, on what factors does the thread bearing pressure depend? Explain.
8. Why is a separate nut preferable to an integral nut with the body of a screw jack?
9. Differentiate between differential screw and compound screw.

Screw jack building-block system

OBJECTIVE TYPE QUESTIONS

1. Which of the following screw thread is adopted for power transmission in either direction?
   (*a*) Acme threads (*b*) Square threads
   (*c*) Buttress threads (*d*) Multiple threads
2. Multiple threads are used to secure
   (*a*) low efficiency (*b*) high efficiency
   (*c*) high load lifting capacity (*d*) high mechanical advantage
3. Screws used for power transmission should have
   (*a*) low efficiency (*b*) high efficiency
   (*c*) very fine threads (*d*) strong teeth
4. If α denotes the lead angle and ϕ, the angle of friction, then the efficiency of the screw is written as
   (*a*) $\frac{\tan(\alpha - \phi)}{\tan \alpha}$ (*b*) $\frac{\tan \alpha}{\tan(\alpha - \phi)}$
   (*c*) $\frac{\tan(\alpha + \phi)}{\tan \alpha}$ (*d*) $\frac{\tan \alpha}{\tan(\alpha + \phi)}$
5. A screw jack has square threads and the lead angle of the thread is α. The screw jack will be self-locking when the coefficient of friction (μ) is
   (*a*) $\mu > \tan \alpha$ (*b*) $\mu = \sin \alpha$
   (*c*) $\mu = \cot \alpha$ (*d*) $\mu = \operatorname{cosec} \alpha$
6. To ensure self locking in a screw jack, it is essential that the helix angle is
   (*a*) larger than friction angle (*b*) smaller than friction angle
   (*c*) equal to friction angle (*d*) such as to give maximum efficiency in lifting
7. A screw is said to be self locking screw, if its efficiency is
   (*a*) less than 50% (*b*) more than 50%
   (*c*) equal to 50% (*d*) none of these
8. A screw is said to be over hauling screw, if its efficiency is
   (*a*) less than 50% (*b*) more than 50%
   (*c*) equal to 50% (*d*) none of these
9. While designing a screw in a screw jack against buckling failure, the end conditions for the screw are taken as
   (*a*) both ends fixed (*b*) both ends hinged
   (*c*) one end fixed and other end hinged (*d*) one end fixed and other end free.
10. The load cup of a screw jack is made separate from the head of the spindle to
   (*a*) enhance the load carrying capacity of the jack
   (*b*) reduce the effort needed for lifting the working load
   (*c*) reduce the value of frictional torque required to be countered for lifting the load
   (*d*) prevent the rotation of load being lifted

ANSWERS

1. (*b*)	2. (*b*)	3. (*b*)	4. (*d*)	5. (*a*)
6. (*b*)	7. (*a*)	8. (*b*)	9. (*d*)	10. (*d*)

CHAPTER

18

Flat Belt Drives

18.1 Introduction

The belts or *ropes are used to transmit power from one shaft to another by means of pulleys which rotate at the same speed or at different speeds. The amount of power transmitted depends upon the following factors :

1. The velocity of the belt.
2. The tension under which the belt is placed on the pulleys.
3. The arc of contact between the belt and the smaller pulley.
4. The conditions under which the belt is used.

It may be noted that

(*a*) The shafts should be properly in line to insure uniform tension across the belt section.

(*b*) The pulleys should not be too close together, in order that the arc of contact on the smaller pulley may be as large as possible.

* Rope drives are discussed in Chapter 20.

(*c*) The pulleys should not be so far apart as to cause the belt to weigh heavily on the shafts, thus increasing the friction load on the bearings.

(*d*) A long belt tends to swing from side to side, causing the belt to run out of the pulleys, which in turn develops crooked spots in the belt.

(*e*) The tight side of the belt should be at the bottom, so that whatever sag is present on the loose side will increase the arc of contact at the pulleys.

(*f*) In order to obtain good results with flat belts, the maximum distance between the shafts should not exceed 10 metres and the minimum should not be less than 3.5 times the diameter of the larger pulley.

18.2 Selection of a Belt Drive

Following are the various important factors upon which the selection of a belt drive depends:

1. Speed of the driving and driven shafts,
2. Speed reduction ratio,
3. Power to be transmitted,
4. Centre distance between the shafts,
5. Positive drive requirements,
6. Shafts layout,
7. Space available, and
8. Service conditions.

18.3 Types of Belt Drives

The belt drives are usually classified into the following three groups:

1. ***Light drives.*** These are used to transmit small powers at belt speeds upto about 10 m/s as in agricultural machines and small machine tools.

2. ***Medium drives.*** These are used to transmit medium powers at belt speeds over 10 m/s but up to 22 m/s, as in machine tools.

3. ***Heavy drives.*** These are used to transmit large powers at belt speeds above 22 m/s as in compressors and generators.

18.4 Types of Belts

Though there are many types of belts used these days, yet the following are important from the subject point of view:

1. ***Flat belt.*** The flat belt as shown in Fig. 18.1 (*a*), is mostly used in the factories and workshops, where a moderate amount of power is to be transmitted, from one pulley to another when the two pulleys are not more than 8 metres apart.

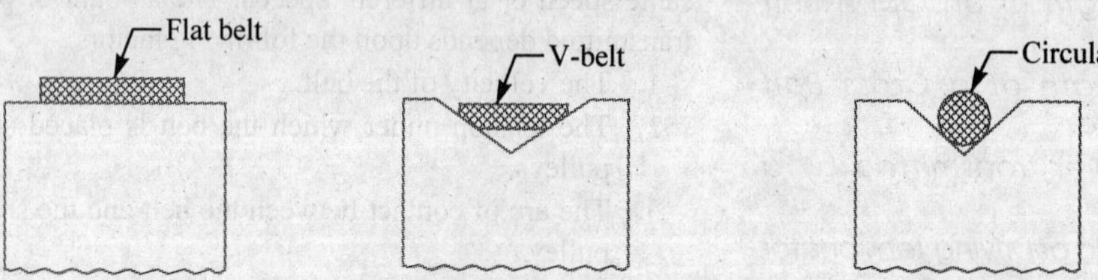

(*a*) Flat belt. (*b*) V-belt. (*c*) Circular belt.

Fig. 18.1. Types of belts

2. ***V- belt.*** The V-belt as shown in Fig. 18.1 (*b*), is mostly used in the factories and workshops, where a great amount of power is to be transmitted, from one pulley to another, when the two pulleys are very near to each other.

3. ***Circular belt or rope.*** The circular belt or rope as shown in Fig. 18.1 (*c*) is mostly used in the factories and workshops, where a great amount of power is to be transmitted, from one pulley to another, when the two pulleys are more than 8 metres apart.

If a huge amount of power is to be transmitted, then a single belt may not be sufficient. In such a case, wide pulleys (for V-belts or circular belts) with a number of grooves are used. Then a belt in each groove is provided to transmit the required amount of power from one pulley to another.

Note : The V-belt and rope drives are discussed in Chapter 20.

18.5 Material used for Belts

The material used for belts and ropes must be strong, flexible, and durable. It must have a high coefficient of friction. The belts, according to the material used, are classified as follows:

1. ***Leather belts.*** The most important material for flat belt is leather. The best leather belts are made from 1.2 metres to 1.5 metres long strips cut from either side of the back bone of the top grade steer hides. The hair side of the leather is smoother and harder than the flesh side, but the flesh side is stronger. The fibres on the hair side are perpendicular to the surface, while those on the flesh side are interwoven and parallel to the surface. Therefore for these reasons the hair side of a belt should be in contact with the pulley surface as shown in Fig. 18.2. This gives a more intimate contact between belt and pulley and places the greatest tensile strength of the belt section on the outside, where the tension is maximum as the belt passes over the pulley.

The leather may be either oak-tanned or mineral salt-tanned *e.g.* chrome-tanned. In order to increase the thickness of belt, the strips are cemented together. The belts are specified according to the number of layers *e.g.* single, double or triple ply and according to the thickness of hides used *e.g.* light, medium or heavy.

(*a*) Single layer belt. (*b*) Double layer belt.

Fig. 18.2. Leather belts.

The leather belts must be periodically cleaned and dressed or treated with a compound or dressing containing neats foot or other suitable oils so that the belt will remain soft and flexible.

2. ***Cotton or fabric belts.*** Most of the fabric belts are made by folding convass or cotton duck to three or more layers (depending upon the thickness desired) and stitching together. These belts are woven also into a strip of the desired width and thickness. They are impregnated with some filler like linseed oil in order to make the belt water-proof and to prevent injury to the fibres. The cotton belts are cheaper and suitable in warm climates, in damp atmospheres and in exposed positions. Since the cotton belts require little attention, therefore these belts are mostly used in farm machinery, belt conveyor etc.

3. ***Rubber belt.*** The rubber belts are made of layers of fabric impregnated with rubber composition and have a thin layer of rubber on the faces. These belts are very flexible but are quickly destroyed if allowed to come into contact with heat, oil or grease. One of the principle advantage of these belts is that they may be easily made endless. These belts are found suitable for saw mills, paper mills where they are exposed to moisture.

4. ***Balata belts.*** These belts are similar to rubber belts except that balata gum is used in place of rubber. These belts are acid proof and water proof and it is not effected by animal oils or alkalies. The balata belts should not be at temperatures above 40°C because at this temperature the balata begins to soften and becomes sticky. The strength of balata belts is 25 per cent higher than rubber belts.

18.6 Working Stresses in Belts

The ultimate strength of leather belt varies from 21 to 35 MPa and a factor of safety may be taken as 8 to 10. However, the wear life of a belt is more important than actual strength. It has been shown by experience that under average conditions an allowable stress of 2.8 MPa or less will give a reasonable belt life. An allowable stress of 1.75 MPa may be expected to give a belt life of about 15 years.

18.7 Density of Belt Materials

The density of various belt materials are given in the following table.

Table 18.1. Density of belt materials.

Material of belt	*Mass density in kg / m³*
Leather	1000
Convass	1220
Rubber	1140
Balata	1110
Single woven belt	1170
Double woven belt	1250

18.8 Belt Speed

A little consideration will show that when the speed of belt increases, the centrifugal force also increases which tries to pull the belt away from the pulley. This will result in the decrease of power transmitted by the belt. It has been found that for the efficient transmission of power, the belt speed 20 m/s to 22.5 m/s may be used.

18.9 Coefficient of Friction Between Belt and Pulley

The coefficient of friction between the belt and the pulley depends upon the following factors:

1. The material of belt;
2. The material of pulley;
3. The slip of belt; and
4. The speed of belt.

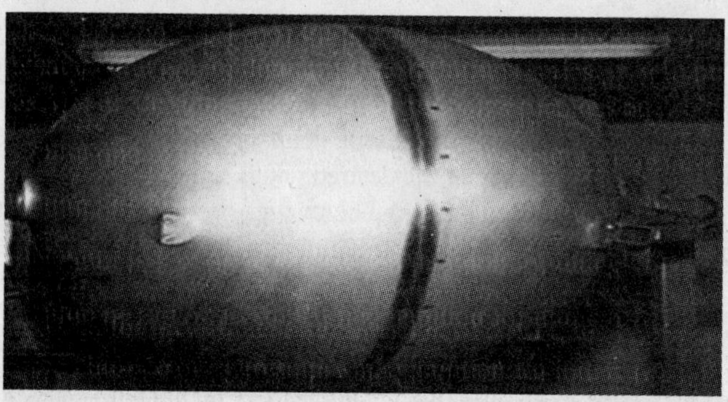

Belts used to drive wheels

According to C.G. Barth, the coefficient of friction (μ) for oak tanned leather belts on cast iron pulley, at the point of slipping, is given by the following relation, *i.e.*

$$\mu = 0.54 - \frac{42.6}{152.6 + v}$$

where v = Speed of the belt in metres per minute.

The following table shows the values of coefficient of friction for various materials of belt and pulley.

Table 18.2. Coefficient of friction between belt and pulley.

Belt material	Pulley material						
	Cast iron, steel			*Wood*	*Compressed paper*	*Leather face*	*Rubber face*
	Dry	*Wet*	*Greasy*				
1. Leather oak tanned	0.25	0.2	0.15	0.3	0.33	0.38	0.40
2. Leather chrome tanned	0.35	0.32	0.22	0.4	0.45	0.48	0.50
3. Convass-stitched	0.20	0.15	0.12	0.23	0.25	0.27	0.30
4. Cotton woven	0.22	0.15	0.12	0.25	0.28	0.27	0.30
5. Rubber	0.30	0.18	—	0.32	0.35	0.40	0.42
6. Balata	0.32	0.20	—	0.35	0.38	0.40	0.42

18.10 Standard Belt Thicknesses and Widths

The standard flat belt thicknesses are 5, 6.5, 8, 10 and 12 mm. The preferred values of thicknesses are as follows:

(*a*) 5 mm for nominal belt widths of 35 to 63 mm,

(*b*) 6.5 mm for nominal belt widths of 50 to 140 mm,

(*c*) 8 mm for nominal belt widths of 90 to 224 mm,

(*d*) 10 mm for nominal belt widths of 125 to 400 mm, and

(*e*) 12 mm for nominal belt widths of 250 to 600 mm.

The standard values of nominal belt widths are in R10 series, starting from 25 mm upto 63 mm and in R 20 series starting from 71 mm up to 600 mm. Thus, the standard widths will be 25, 32, 40, 50, 63, 71, 80, 90, 100, 112, 125, 140, 160, 180, 200, 224, 250, 280, 315, 355, 400, 450, 500, 560 and 600 mm.

18.11 Belt Joints

When the endless belts are not available, then the belts are cut from big rolls and the ends are joined together by fasteners. The various types of joints are

1. Cemented joint, 2. Laced joint, and 3. Hinged joint.

The **cemented joint,** as shown in Fig. 18.3 (*a*), made by the manufacturer to form an endless belt, is preferred than other joints. The **laced joint** is formed by punching holes in line across the belt, leaving a margin between the edge and the holes. A raw hide strip is used for lacing the two ends together to form a joint. This type of joint is known as ***straight-stitch raw hide laced joint,*** as shown in Fig. 18.3 (*b*).

Metal laced joint as shown in Fig. 18.3 (*c*), is made like a staple connection. The points are driven through the flesh side of the belt and clinched on the inside.

Sometimes, **metal hinges** may be fastened to the belt ends and connected by a steel or fibre pin as shown in Fig. 18.3 (*d*).

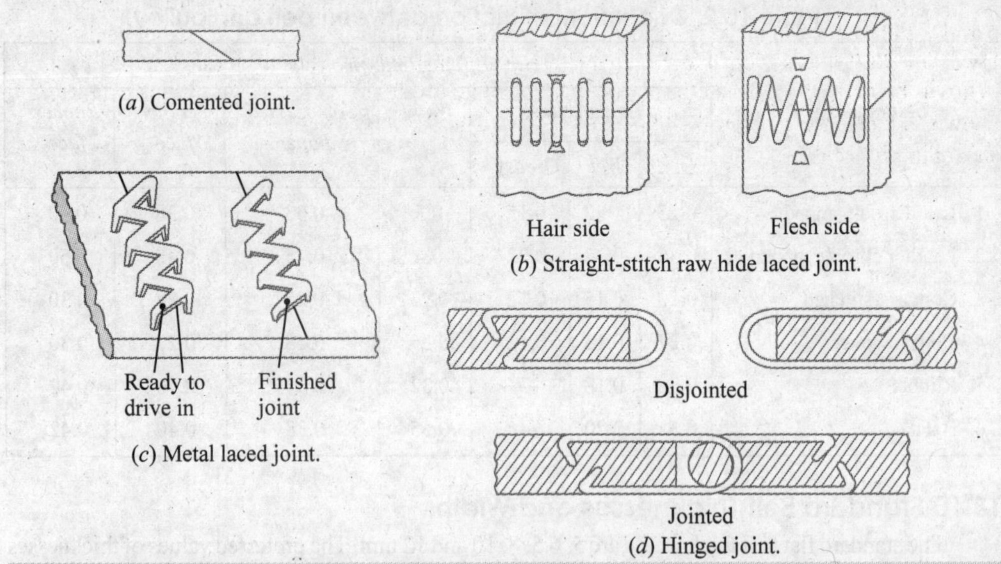

Fig. 18.3. Belt joints.

The following table shows the efficiencies of these joints.

Table 18.3. Efficiencies of belt joints.

Type of joint	*Efficiency* (%)	*Type of joint*	*Efficiency* (%)
1. Cemented, endless, cemented at factory	90 to 100	4. Wire laced by hand	70 to 80
2. Cemented in shop	80 to 90	5. Raw-hide laced	60 to 70
3. Wire laced by machine	75 to 85	6. Metal belt hooks	35 to 40

18.12 Types of Flat Belt Drives

The power from one pulley to another may be transmitted by any of the following types of belt drives.

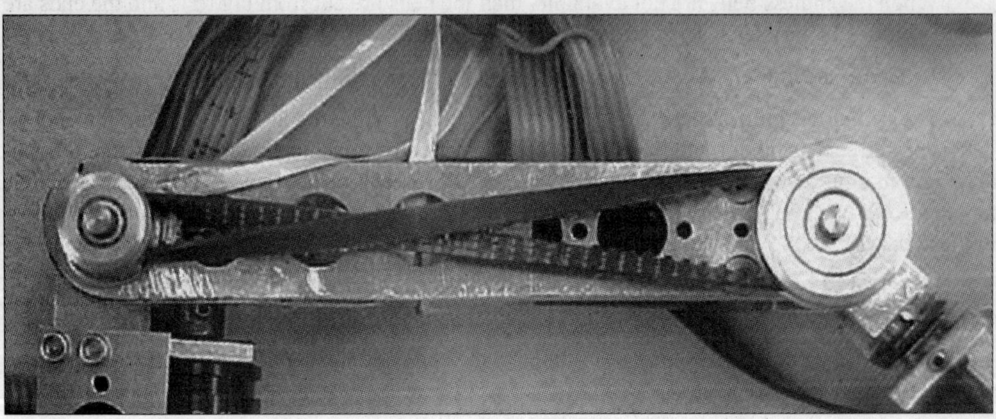

Cross or twist belt drive

1. *Open belt drive.* The open belt drive, as shown in Fig. 18.4, is used with shafts arranged parallel and rotating in the same direction. In this case, the driver *A* pulls the belt from one side (*i.e.* lower side *RQ*) and delivers it to the other side (*i.e.* upper side *LM*). Thus the tension in the lower side belt will be more than that in the upper side belt. The lower side belt (because of more tension) is known as ***tight side*** whereas the upper side belt (because of less tension) is known as ***slack side,*** as shown in Fig. 18.4.

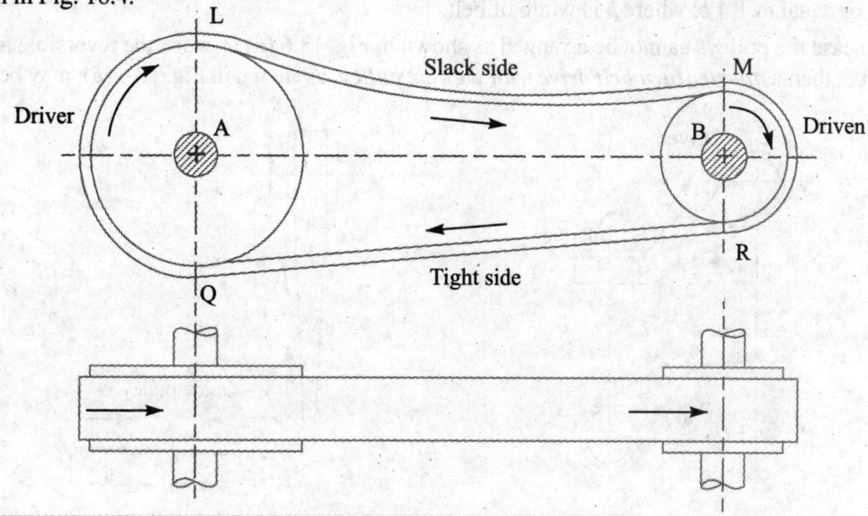

Fig. 18.4. Open belt drive.

2. *Crossed or twist belt drive.* The crossed or twist belt drive, as shown in Fig. 18.5, is used with shafts arranged parallel and rotating in the opposite directions. In this case, the driver pulls the belt from one side (*i.e. RQ*) and delivers it to the other side (*i.e. LM*). Thus, the tension in the belt *RQ* will be more than that in the belt *LM.* The belt *RQ* (because of more tension) is known as ***tight side,*** whereas the belt *LM* (because of less tension) is known as ***slack side,*** as shown in Fig. 18.5.

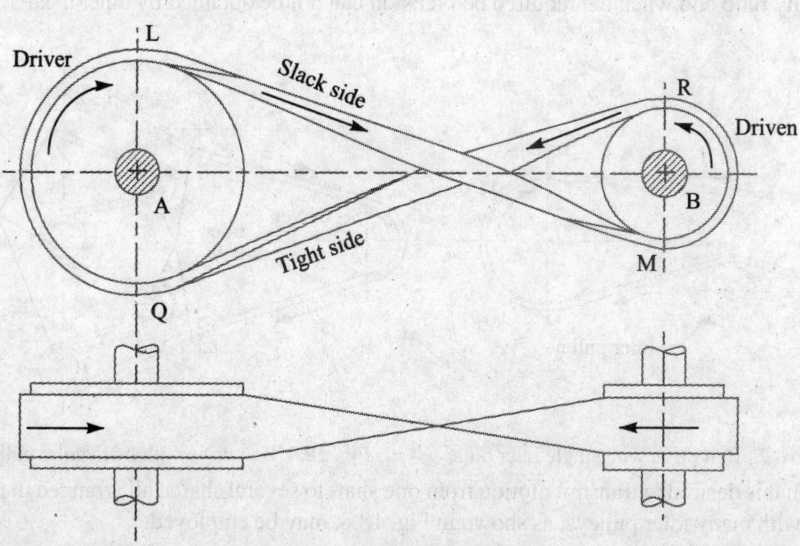

Fig. 18.5. Crossed or twist belt drive.

A little consideration will show that at a point where the belt crosses, it rubs against each other and there will be excessive wear and tear. In order to avoid this, the shafts should be placed at a maximum distance of 20 *b*, where *b* is the width of belt and the speed of the belt should be less than 15 m/s.

3. ***Quarter turn belt drive.*** The quarter turn belt drive (also known as ***right angle belt drive***) as shown in Fig. 18.6 (*a*), is used with shafts arranged at right angles and rotating in one definite direction. In order to prevent the belt from leaving the pulley, the width of the face of the pulley should be greater or equal to 1.4 *b*, where *b* is width of belt.

In case the pulleys cannot be arranged as shown in Fig. 18.6 (*a*) or when the reversible motion is desired, then a ***quarter turn belt drive with a guide pulley,*** as shown in Fig. 18.6 (*b*), may be used.

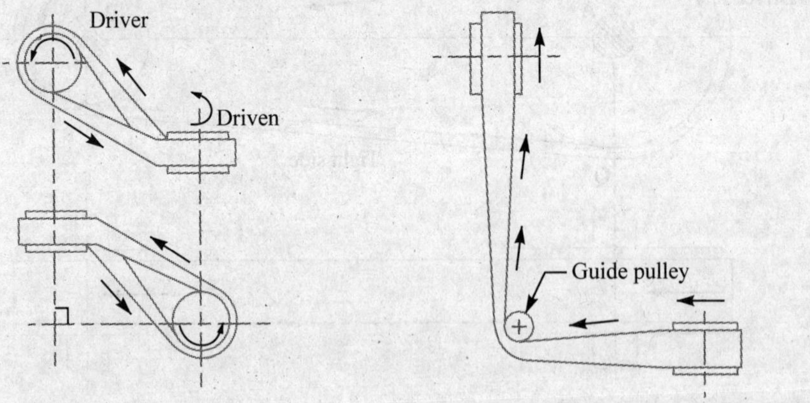

(*a*) Quarter turn belt drive. (*b*) Quarter turn belt drive with guide pulley.

Fig. 18.6

4. ***Belt drive with idler pulleys.*** A belt drive with an idler pulley (also known as ***jockey pulley drive***) as shown in Fig. 18.7, is used with shafts arranged parallel and when an open belt drive can not be used due to small angle of contact on the smaller pulley. This type of drive is provided to obtain high velocity ratio and when the required belt tension can not be obtained by other means.

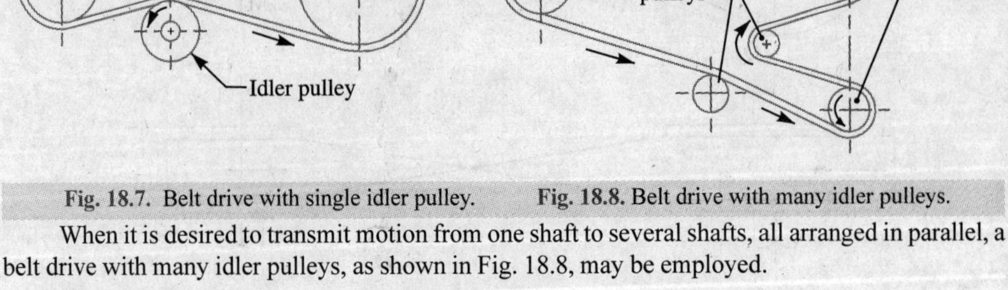

Fig. 18.7. Belt drive with single idler pulley. **Fig. 18.8.** Belt drive with many idler pulleys.

When it is desired to transmit motion from one shaft to several shafts, all arranged in parallel, a belt drive with many idler pulleys, as shown in Fig. 18.8, may be employed.

5. ***Compound belt drive.*** A compound belt drive as shown in Fig. 18.9, is used when power is transmitted from one shaft to another through a number of pulleys.

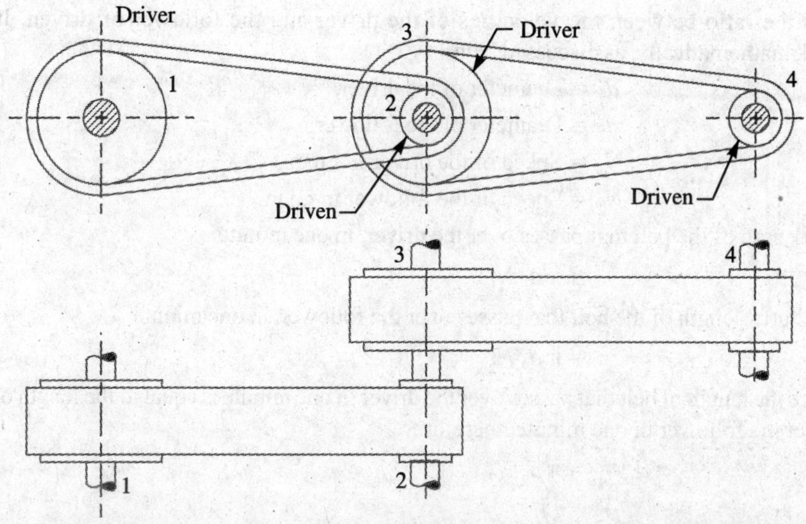

Fig. 18.9. Compound belt drive.

6. ***Stepped or cone pulley drive.*** A stepped or cone pulley drive, as shown in Fig. 18.10, is used for changing the speed of the driven shaft while the main or driving shaft runs at constant speed. This is accomplished by shifting the belt from one part of the steps to the other.

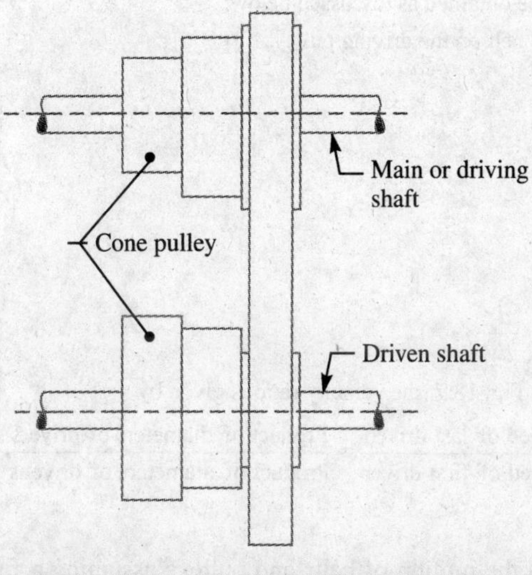

Fig. 18.10. Stepped or cone pulley drive.

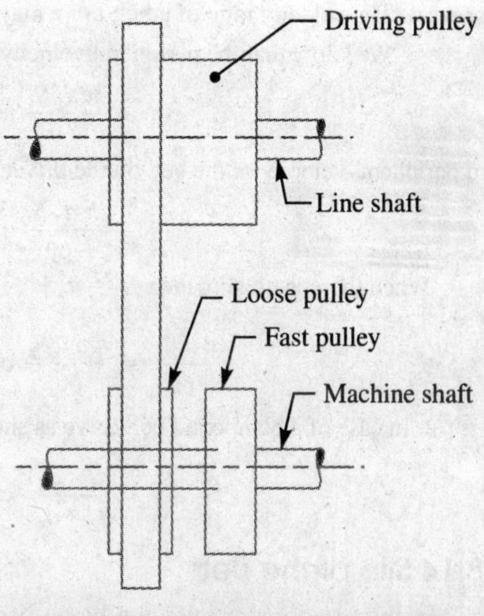

Fig. 18.11. Fast and loose pulley drive.

7. ***Fast and loose pulley drive.*** A fast and loose pulley drive, as shown in Fig. 18.11, is used when the driven or machine shaft is to be started or stopped whenever desired without interferring with the driving shaft. A pulley which is keyed to the machine shaft is called fast pulley and runs at the same speed as that of machine shaft. A loose pulley runs freely over the machine shaft and is incapable of transmitting any power. When the driven shaft is required to be stopped, the belt is pushed on to the loose pulley by means of sliding bar having belt forks.

18.13 Velocity Ratio of a Belt Drive

It is the ratio between the velocities of the driver and the follower or driven. It may be expressed, mathematically, as discussed below:

Let d_1 = Diameter of the driver,

d_2 = Diameter of the follower,

N_1 = Speed of the driver in r.p.m.,

N_2 = Speed of the follower in r.p.m.,

∴ Length of the belt that passes over the driver, in one minute

$$= \pi\, d_1\, N_1$$

Similarly, length of the belt that passes over the follower, in one minute

$$= \pi\, d_2\, N_2$$

Since the length of belt that passes over the driver in one minute is equal to the length of belt that passes over the follower in one minute, therefore

$$\because \quad \pi\, d_1\, N_1 = \pi\, d_2\, N_2$$

and velocity ratio,

$$\frac{N_2}{N_1} = \frac{d_1}{d_2}$$

When thickness of the belt (t) is considered, then velocity ratio,

$$\frac{N_2}{N_1} = \frac{d_1 + t}{d_2 + t}$$

Notes : 1. The velocity ratio of a belt drive may also be obtained as discussed below:

We know that the peripheral velocity of the belt on the driving pulley,

$$v_1 = \frac{\pi\, d_1\, N_1}{60} \text{ m/s}$$

and peripheral velocity of the belt on the driven pulley,

$$v_2 = \frac{\pi\, d_2\, N_2}{60} \text{ m/s}$$

When there is no slip, then $v_1 = v_2$.

$$\therefore \quad \frac{\pi\, d_1\, N_1}{60} = \frac{\pi\, d_2\, N_2}{60} \quad \text{or} \quad \frac{N_2}{N_1} = \frac{d_1}{d_2}$$

2. In case of a compound belt drive as shown in Fig. 18.7, the velocity ratio is given by

$$\frac{N_4}{N_1} = \frac{d_1 \times d_3}{d_2 \times d_4} \quad \text{or} \quad \frac{\text{Speed of last driven}}{\text{Speed of first driver}} = \frac{\text{Product of diameters of drivers}}{\text{Product of diameters of drivens}}$$

18.14 Slip of the Belt

In the previous articles we have discussed the motion of belts and pulleys assuming a firm frictional grip between the belts and the pulleys. But sometimes, the frictional grip becomes insufficient. This may cause some forward motion of the driver without carrying the belt with it. This is called ***slip of the belt*** and is generally expressed as a percentage.

The result of the belt slipping is to reduce the velocity ratio of the system. As the slipping of the belt is a common phenomenon, thus the belt should never be used where a definite velocity ratio is of importance (as in the case of hour, minute and second arms in a watch).

Let s_1 % = Slip between the driver and the belt, and

s_2 % = Slip between the belt and follower,

$\therefore$ Velocity of the belt passing over the driver per second,

$$v = \frac{\pi\, d_1\, N_1}{60} - \frac{\pi\, d_1\, N_1}{60} \times \frac{s_1}{100}$$

$$= \frac{\pi\, d_1\, N_1}{60}\left(1 - \frac{s_1}{100}\right) \quad \text{...}(i)$$

and velocity of the belt passing over the follower per second

$$\frac{\pi\, d_2\, N_2}{60} = v - v\left(\frac{s_2}{100}\right) = v\left(1 - \frac{s_2}{100}\right)$$

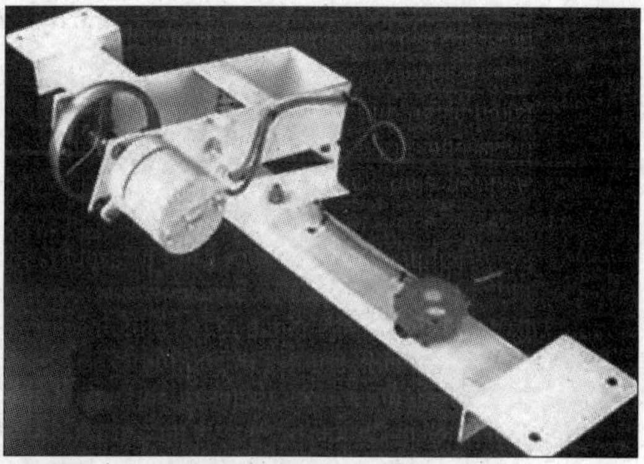

Belt slip indicator is used to indicate that the belt is slipping.

Substituting the value of v from equation (*i*), we have

$$\frac{\pi\, d_2\, N_2}{60} = \frac{\pi\, d_1\, N_1}{60}\left(1 - \frac{s_1}{100}\right)\left(1 - \frac{s_2}{100}\right)$$

$$\therefore \quad \frac{N_2}{N_1} = \frac{d_1}{d_2}\left(1 - \frac{s_1}{100} - \frac{s_2}{100}\right) \quad \text{...}\left(\text{Neglecting } \frac{s_1 \times s_2}{100 \times 100}\right)$$

$$= \frac{d_1}{d_2}\left[1 - \left(\frac{s_1 + s_2}{100}\right)\right] = \frac{d_1}{d_2}\left(1 - \frac{s}{100}\right)$$

...(where $s = s_1 + s_2$ *i.e.* total percentage of slip)

If thickness of the belt (t) is considered, then

$$\frac{N_2}{N_1} = \frac{d_1 + t}{d_2 + t}\left(1 - \frac{s}{100}\right)$$

18.15 Creep of Belt

When the belt passes from the slack side to the tight side, a certain portion of the belt extends and it contracts again when the belt passes from the tight side to the slack side. Due to these changes of length, there is a relative motion between the belt and the pulley surfaces. This relative motion is termed as ***creep***. The total effect of creep is to reduce slightly the speed of the driven pulley or follower. Considering creep, the velocity ratio is given by

$$\frac{N_2}{N_1} = \frac{d_1}{d_2} \times \frac{E + \sqrt{\sigma_2}}{E + \sqrt{\sigma_1}}$$

where σ_1 and σ_2 = Stress in the belt on the tight and slack side respectively, and

E = Young's modulus for the material of the belt.

Note: Since the effect of creep is very small, therefore it is generally neglected.

Example 18.1. *An engine running at 150 r.p.m. drives a line shaft by means of a belt. The engine pulley is 750 mm diameter and the pulley on the line shaft is 450 mm. A 900 mm diameter pulley on the line shaft drives a 150 mm diameter pulley keyed to a dynamo shaft. Fine the speed of dynamo shaft, when 1. there is no slip, and 2. there is a slip of 2% at each drive.*

Solution. Given : N_1 = 150 r.p.m. ; d_1 = 750 mm ; d_2 = 450 mm ; d_3 = 900 mm ; d_4 = 150 mm ; $s_1 = s_2$ = 2%

The arrangement of belt drive is shown in Fig. 18.12.

Let N_4 = Speed of the dynamo shaft.

1. *When there is no slip*

We know that

$$\frac{N_4}{N_1} = \frac{d_1 \times d_3}{d_2 \times d_4} \text{ or } \frac{N_4}{150} = \frac{750 \times 900}{450 \times 150} = 10$$

$\therefore$ $N_4 = 150 \times 10 = 1500$ r.p.m. **Ans.**

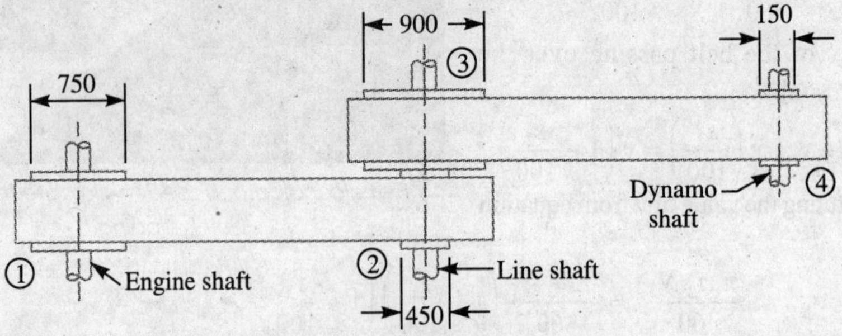

All dimensions in mm.

Fig. 18.12

2. *When there is a slip of 2% at each drive*

We know that

$$\frac{N_4}{N_1} = \frac{d_1 \times d_3}{d_2 \times d_4}\left(1 - \frac{s_1}{100}\right)\left(1 - \frac{s_2}{100}\right)$$

or

$$\frac{N_4}{150} = \frac{750 \times 900}{450 \times 150}\left(1 - \frac{2}{100}\right)\left(1 - \frac{2}{100}\right) = 9.6$$

$\therefore$ $N_4 = 150 \times 9.6 = 1440$ r.p.m. **Ans.**

18.16 Length of an Open Belt Drive

We have discussed in Art. 18.12, that in an open belt drive, both the pulleys rotate in the same direction as shown in Fig. 18.13.

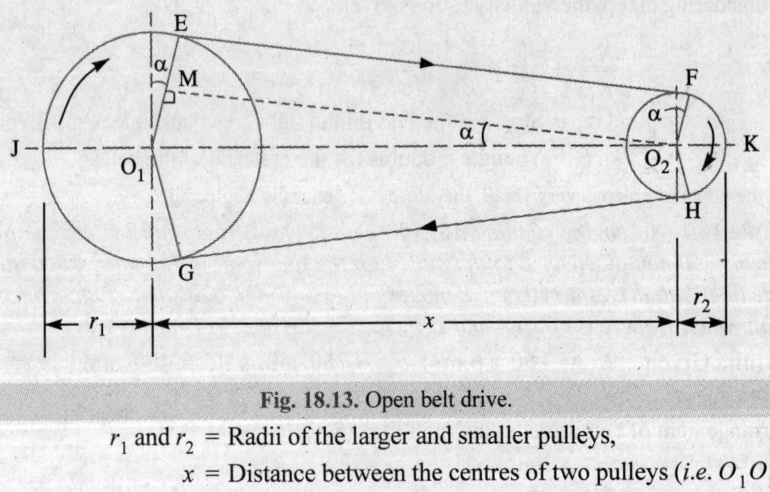

Fig. 18.13. Open belt drive.

Let r_1 and r_2 = Radii of the larger and smaller pulleys,

x = Distance between the centres of two pulleys (*i.e.* O_1O_2), and

L = Total length of the belt.

Let the belt leaves the larger pulley at *E* and *G* and the smaller pulley at *F* and *H* as shown in Fig. 18.13. Through O_2 draw O_2M parallel to *FE*.

From the geometry of the figure, we find that O_2M will be perpendicular to O_1E.

Let the angle $MO_2O_1 = \alpha$ radians.

We know that the length of the belt,

$$L = \text{Arc } GJE + EF + \text{Arc } FKH + HG$$

$$= 2\,(\text{Arc } JE + EF + \text{Arc } FK) \qquad ...(i)$$

From the geometry of the figure, we also find that

$$\sin\alpha = \frac{O_1M}{O_1O_2} = \frac{O_1E - EM}{O_1O_2} = \frac{r_1 - r_2}{x}$$

Since the angle α is very small, therefore putting

$$\sin\alpha = \alpha \text{ (in radians)} = \frac{r_1 - r_2}{x} \qquad ...(ii)$$

$$\therefore \qquad \text{Arc } JE = r_1\left(\frac{\pi}{2} + \alpha\right) \qquad ...(iii)$$

Similarly,

$$\text{arc } FK = r_2\left(\frac{\pi}{2} - \alpha\right) \qquad ...(iv)$$

and

$$EF = MO_2 = \sqrt{(O_1O_2)^2 - (O_1M)^2} = \sqrt{x^2 - (r_1 - r_2)^2}$$

$$= x\sqrt{1 - \left(\frac{r_1 - r_2}{x}\right)^2}$$

Expanding this equation by binomial theorem, we have

$$EF = x\left[1 - \frac{1}{2}\left(\frac{r_1 - r_2}{x}\right)^2 + ...\right] = x - \frac{(r_1 - r_2)^2}{2x} \qquad ...(v)$$

Substituting the values of arc *JE* from equation (*iii*), arc *FK* from equation (*iv*) and *EF* from equation (*v*) in equation (*i*), we get

$$L = 2\left[r_1\left(\frac{\pi}{2} + \alpha\right) + x - \frac{(r_1 - r_2)^2}{2x} + r_2\left(\frac{\pi}{2} - \alpha\right)\right]$$

$$= 2\left[r_1 \times \frac{\pi}{2} + r_1.\alpha + x - \frac{(r_1 - r_2)^2}{2x} + r_2 \times \frac{\pi}{2} - r_2.\alpha\right]$$

$$= 2\left[\frac{\pi}{2}(r_1 + r_2) + \alpha\,(r_1 - r_2) + x - \frac{(r_1 - r_2)^2}{2x}\right]$$

$$= \pi\,(r_1 + r_2) + 2\alpha\,(r_1 - r_2) + 2x - \frac{(r_1 - r_2)^2}{x}$$

Substituting the value of $\alpha = \dfrac{(r_1 - r_2)}{x}$ from equation (*ii*), we get

$$L = \pi\,(r_1 + r_2) + 2 \times \frac{(r_1 - r_2)}{x}\,(r_1 - r_2) + 2x - \frac{(r_1 - r_2)^2}{x}$$

$$= \pi\,(r_1 + r_2) + \frac{2\,(r_1 - r_2)^2}{x} + 2x - \frac{(r_1 - r_2)^2}{x}$$

$$= \pi (r_1 + r_2) + 2x + \frac{(r_1 - r_2)^2}{x}$$ ... (in terms of pulley radii)

$$= \frac{\pi}{2} (d_1 + d_2) + 2x + \frac{(d_1 - d_2)^2}{4x}$$ ... (in terms of pulley diameters)

18.17 Length of a Cross Belt Drive

We have discussed in Art. 18.12 that in a cross belt drive, both the pulleys rotate in the opposite directions as shown in Fig. 18.14.

Let r_1 and r_2 = Radii of the larger and smaller pulleys,

x = Distance between the centres of two pulleys (*i.e.* O_1O_2), and

L = Total length of the belt.

Let the belt leaves the larger pulley at E and G and the smaller pulley at F and H as shown in Fig. 18.14.

Through O_2 draw O_2M parallel to FE.

From the geometry of the figure, we find that O_2M will be perpendicular to O_1E.

Let the angle $MO_2O_1 = \alpha$ radians.

We know that the length of the belt,

$$L = \text{Arc } GJE + EF + \text{Arc } FKH + HG$$

$$= 2 (\text{Arc } JE + FE + \text{Arc } FK) \quad ...(i)$$

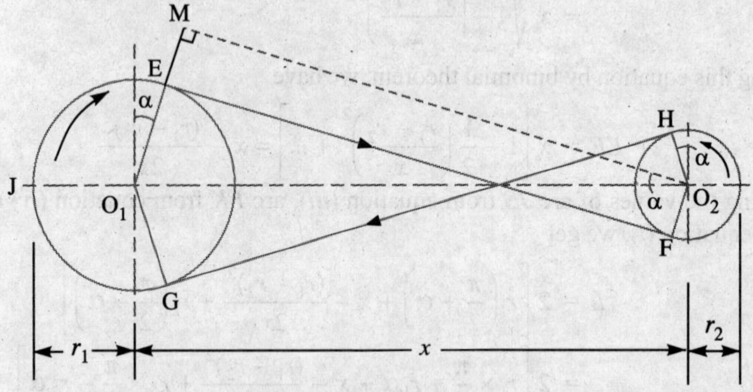

Fig. 18.14. Crossed belt drive.

From the geometry of the figure, we find that

$$\sin \alpha = \frac{O_1M}{O_1O_2} = \frac{O_1E + EM}{O_1O_2} = \frac{r_1 + r_2}{x}$$

Since the angle α is very small, therefore putting

$$\sin \alpha = \alpha \text{ (in radians)} = \frac{r_1 + r_2}{x} \quad ...(ii)$$

$$\therefore \quad \text{Arc } JE = r_1 \left(\frac{\pi}{2} + \alpha\right) \quad ...(iii)$$

Similarly,

$$\text{arc } FK = r_2 \left(\frac{\pi}{2} + \alpha\right) \quad ...(iv)$$

and $$EF = MO_2 = \sqrt{(O_1O_2)^2 - (O_1M)^2} = \sqrt{x^2 - (r_1 + r_2)^2}$$

$$= x\sqrt{1 - \left(\frac{r_1 + r_2}{x}\right)^2}$$

Expanding this equation by binomial theorem, we have

$$EF = x\left[1 - \frac{1}{2}\left(\frac{r_1 + r_2}{x}\right)^2 + ...\right] = x - \frac{(r_1 + r_2)^2}{2x} \quad ...(v)$$

In the above conveyor belt is used to transport material as well as to drive the rollers

Substituting the values of arc *JE* from equation (*iii*), arc *FK* from equation (*iv*) and *EF* from equation (*v*) in equation (*i*), we get,

$$L = 2\left[r_1\left(\frac{\pi}{2} + \alpha\right) + x - \frac{(r_1 + r_2)^2}{2x} + r_2\left(\frac{\pi}{2} + \alpha\right)\right]$$

$$= 2\left[r_1 \times \frac{\pi}{2} + r_1.\alpha + x - \frac{(r_1 + r_2)^2}{2x} + r_2 \times \frac{\pi}{2} + r_2.\alpha\right]$$

$$= 2\left[\frac{\pi}{2}(r_1 + r_2) + \alpha(r_1 + r_2) + x - \frac{(r_1 + r_2)^2}{2x}\right]$$

$$= \pi(r_1 + r_2) + 2\alpha(r_1 + r_2) + 2x - \frac{(r_1 + r_2)^2}{x}$$

Substituting the value of $\alpha = \frac{(r_1 + r_2)}{x}$ from equation (*ii*), we get

$$L = \pi(r_1 + r_2) + 2 \times \frac{(r_1 + r_2)}{x}(r_1 + r_2) + 2x - \frac{(r_1 + r_2)^2}{x}$$

$$= \pi(r_1 + r_2) + \frac{2(r_1 + r_2)^2}{x} + 2x - \frac{(r_1 + r_2)^2}{x}$$

$$= \pi (r_1 + r_2) + 2x + \frac{(r_1 + r_2)^2}{x} \qquad \text{... (in terms of pulley radii)}$$

$$= \frac{\pi}{2} (d_1 + d_2) + 2x + \frac{(d_1 + d_2)^2}{4x} \qquad \text{... (in terms of pulley diameters)}$$

It may be noted that the above expression is a function of $(r_1 + r_2)$. It is thus obvious, that if sum of the radii of the two pulleys be constant, length of the belt required will also remain constant, provided the distance between centres of the pulleys remain unchanged.

18.18 Power Transmitted by a Belt

Fig. 18.15 shows the driving pulley (or driver) *A* and the driven pulley (or follower) *B*. As already discussed, the driving pulley pulls the belt from one side and delivers it to the other side. It is thus obvious that the tension on the former side (*i.e.* tight side) will be greater than the latter side (*i.e.* slack side) as shown in Fig. 18.15.

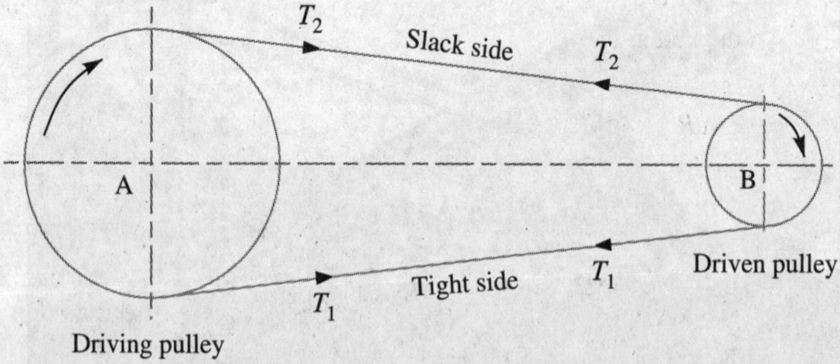

Fig. 18.15. Power transmitted by a belt.

Let T_1 and T_2 = Tensions in the tight side and slack side of the belt respectively in newtons,

r_1 and r_2 = Radii of the driving and driven pulleys respectively in metres,

and v = Velocity of the belt in m/s.

The effective turning (driving) force at the circumference of the driven pulley or follower is the difference between the two tensions (*i.e.* $T_1 - T_2$).

This massive shaft-like pulley drives the conveyor belt.

∴ Work done per second = $(T_1 - T_2)\,v$ N-m/s

and power transmitted = $(T_1 - T_2)\,v$ W ... (∵ 1 N-m/s = 1W)

A little consideration will show that torque exerted on the driving pulley is $(T_1 - T_2)\,r_1$. Similarly, the torque exerted on the driven pulley is $(T_1 - T_2)\,r_2$.

18.19 Ratio of Driving Tensions for Flat Belt Drive

Consider a driven pulley rotating in the clockwise direction as shown in Fig. 18.16.

Let T_1 = Tension in the belt on the tight side,

T_2 = Tension in the belt on the slack side, and

θ = Angle of contact in radians (*i.e.* angle subtended by the arc *AB*, along which the belt touches the pulley, at the centre).

Now consider a small portion of the belt *PQ*, subtending an angle $\delta\theta$ at the centre of the pulley as shown in Fig. 18.16. The belt *PQ* is in equilibrium under the following forces:

1. Tension *T* in the belt at *P*,
2. Tension $(T + \delta T)$ in the belt at *Q*,
3. Normal reaction R_N, and
4. Frictional force $F = \mu \times R_N$, where μ is the coefficient of friction between the belt and pulley.

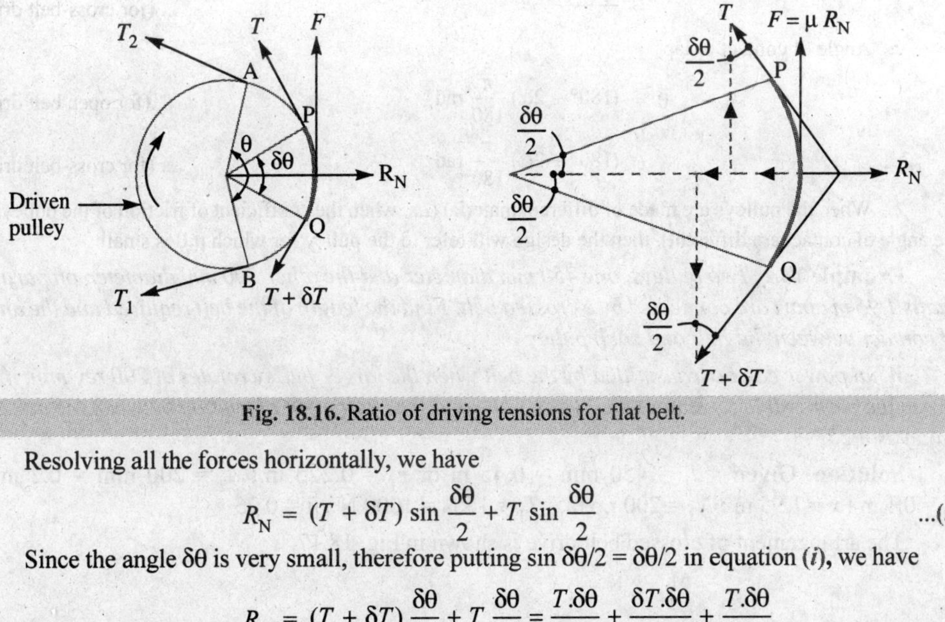

Fig. 18.16. Ratio of driving tensions for flat belt.

Resolving all the forces horizontally, we have

$$R_N = (T + \delta T)\sin\frac{\delta\theta}{2} + T\sin\frac{\delta\theta}{2} \quad \text{...}(i)$$

Since the angle $\delta\theta$ is very small, therefore putting $\sin\delta\theta/2 = \delta\theta/2$ in equation (*i*), we have

$$R_N = (T + \delta T)\frac{\delta\theta}{2} + T\frac{\delta\theta}{2} = \frac{T.\delta\theta}{2} + \frac{\delta T.\delta\theta}{2} + \frac{T.\delta\theta}{2}$$

$$= T.\delta\theta \qquad \text{...}\left(\text{Neglecting } \frac{\delta T.\delta\theta}{2}\right) \quad \text{...}(ii)$$

Now resolving the forces vertically, we have

$$\mu \times R_N = (T + \delta T)\cos\frac{\delta\theta}{2} - T\cos\frac{\delta\theta}{2} \quad \text{...}(iii)$$

Since the angle $\delta\theta$ is very small, therefore putting $\cos\delta\theta/2 = 1$ in equation (*iii*), we have

$$\mu \times R_N = T + \delta T - T = \delta T \quad \text{or} \quad R_N = \frac{\delta T}{\mu} \quad \text{...}(iv)$$

Equating the values of R_N from equations (*ii*) and (*iv*), we get

$$T.\delta\theta = \frac{\delta T}{\mu} \quad \text{or} \quad \frac{\delta T}{T} = \mu.\delta\theta$$

Integrating the above equation between the limits T_2 and T_1 and from 0 to θ, we have

$$\int_{T_2}^{T_1} \frac{\delta T}{T} = \mu \int_0^{\theta} \delta\theta$$

$$\therefore \quad \log_e \left(\frac{T_1}{T_2}\right) = \mu.\theta \quad \text{or} \quad \frac{T_1}{T_2} = e^{\mu.\theta} \qquad ...(v)$$

The equation (*v*) can be expressed in terms of corresponding logarithm to the base 10, *i.e.*

$$2.3 \log \left(\frac{T_1}{T_2}\right) = \mu.\theta$$

The above expression gives the relation between the tight side and slack side tensions, in terms of coefficient of friction and the angle of contact.

Notes : 1. While determining the angle of contact, it must be remembered that it is the angle of contact at the smaller pulley, if both the pulleys are of the same material. We know that

$$\sin\alpha = \frac{r_1 - r_2}{x} \qquad \text{... (for open belt drive)}$$

$$= \frac{r_1 + r_2}{x} \qquad \text{... (for cross-belt drive)}$$

∴ Angle of contact or lap,

$$\theta = (180° - 2\alpha)\frac{\pi}{180} \text{ rad} \qquad \text{...(for open belt drive)}$$

$$= (180° + 2\alpha)\frac{\pi}{180} \text{ rad} \qquad \text{... (for cross-belt drive)}$$

2. When the pulleys are made of different material (*i.e.* when the coefficient of friction of the pulleys or the angle of contact are different), then the design will refer to the pulley for which μ.θ is small.

Example 18.2. *Two pulleys, one 450 mm diameter and the other 200 mm diameter, on parallel shafts 1.95 m apart are connected by a crossed belt. Find the length of the belt required and the angle of contact between the belt and each pulley.*

What power can be transmitted by the belt when the larger pulley rotates at 200 rev/min, if the maximum permissible tension in the belt is 1 kN, and the coefficient of friction between the belt and pulley is 0.25?

Solution. Given : d_1 = 450 mm = 0.45 m or r_1 = 0.225 m ; d_2 = 200 mm = 0.2 m or r_2 = 0.1 m ; x = 1.95 m ; N_1 = 200 r.p.m. ; T_1 = 1 kN = 1000 N ; μ = 0.25

The arrangement of crossed belt drive is shown in Fig. 18.17.

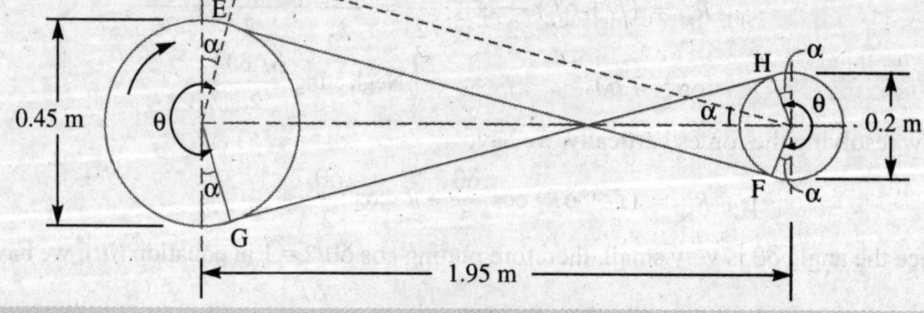

Fig. 18.17

Length of the belt

We know that length of the belt,

$$L = \pi (r_1 + r_2) + 2x + \frac{(r_1 + r_2)^2}{x}$$

$$= \pi (0.225 + 0.1) + 2 \times 1.95 + \frac{(0.225 + 0.1)^2}{1.95}$$

$$= 1.02 + 3.9 + 0.054 = 4.974 \text{ m} \quad \textbf{Ans.}$$

Angle of contact between the belt and each pulley

Let θ = Angle of contact between the belt and each pulley.

We know that for a crossed belt drive,

$$\sin \alpha = \frac{r_1 + r_2}{x} = \frac{0.225 + 0.1}{1.95} = 0.1667$$

$\therefore$ $\alpha = 9.6°$

and $\theta = 180° + 2\alpha = 180 + 2 \times 9.6 = 199.2°$

$$= 199.2 \times \frac{\pi}{180} = 3.477 \text{ rad} \quad \textbf{Ans.}$$

Power transmitted

Let T_1 = Tension in the tight side of the belt, and

T_2 = Tension in the slack side of the belt.

We know that

$$2.3 \log \left(\frac{T_1}{T_2}\right) = \mu.\theta = 0.25 \times 3.477 = 0.8693$$

$$\log \left(\frac{T_1}{T_2}\right) = \frac{0.8693}{2.3} = 0.378 \quad \text{or} \quad \frac{T_1}{T_2} = 2.387 \qquad \text{... (Taking antilog of 0.378)}$$

$$\therefore \quad T_2 = \frac{T_1}{2.387} = \frac{1000}{2.387} = 419 \text{ N}$$

We know that the velocity of belt,

$$v = \frac{\pi\, d_1 N_1}{60} = \frac{\pi \times 0.45 \times 200}{60} = 4.713 \text{ m/s}$$

$\therefore$ Power transmitted,

$$P = (T_1 - T_2)\, v = (1000 - 419)\, 4.713 = 2738 \text{ W} = 2.738 \text{ kW} \quad \textbf{Ans.}$$

18.20 Centrifugal Tension

Since the belt continuously runs over the pulleys, therefore, some centrifugal force is caused, whose effect is to increase the tension on both the tight as well as the slack sides. The tension caused by centrifugal force is called ***centrifugal tension.*** At lower belt speeds (less than 10 m/s), the centrifugal tension is very small, but at higher belt speeds (more than 10 m/s), its effect is considerable and thus should be taken into account.

Consider a small portion PQ of the belt subtending an angle $d\theta$ at the centre of the pulley, as shown in Fig. 18.18.

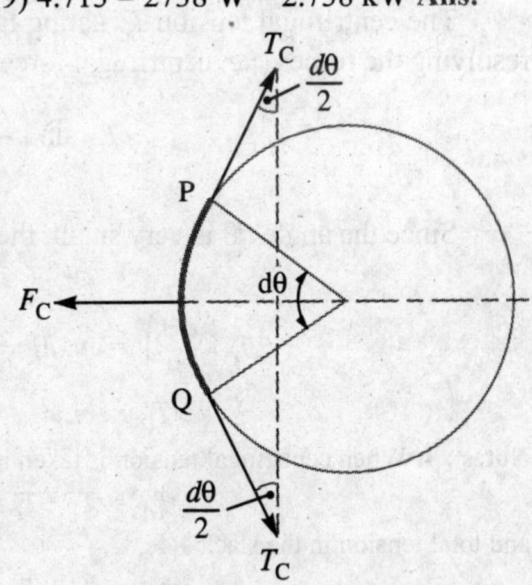

Fig. 18.18. Centrifugal tension.

Let m = Mass of belt per unit length in kg,

v = Linear velocity of belt in m/s,

r = Radius of pulley over which the belt runs in metres, and

T_C = Centrifugal tension acting tangentially at P and Q in newtons.

We know that length of the belt PQ

$$= r.d\theta$$

and mass of the belt PQ $= m.r.d\theta$

∴ Centrifugal force acting on the belt PQ,

$$F_C = m.r.d\theta \times \frac{v^2}{r} = m.d\theta.v^2$$

Belt drive on a lathe

The centrifugal tension T_C acting tangentially at P and Q keeps the belt in equilibrium. Now resolving the forces (*i.e.* centrifugal force and centrifugal tension) horizontally, we have

$$T_C \sin\left(\frac{d\theta}{2}\right) + T_C \sin\left(\frac{d\theta}{2}\right) = F_C = m.d\theta.v^2 \quad ...(i)$$

Since the angle $d\theta$ is very small, therefore putting $\sin\left(\frac{d\theta}{2}\right) = \frac{d\theta}{2}$ in equation (*i*), we have

$$2T_C\left(\frac{d\theta}{2}\right) = m.d\theta.v^2$$

$$\therefore \quad T_C = m.v^2$$

Notes : 1. When centrifugal tension is taken into account, then total tension in the tight side,

$$T_{t1} = T_1 + T_C$$

and total tension in the slack side,

$$T_{t2} = T_2 + T_C$$

2. Power transmitted,

$$P = (T_{t1} - T_{t2})\, v \qquad \text{...(in watts)}$$

$$= [(T_1 + T_C) - (T_2 + T_C)]v = (T_1 - T_2)\, v \qquad \text{... (same as before)}$$

Thus we see that the centrifugal tension has no effect on the power transmitted.

3. The ratio of driving tensions may also be written as

$$2.3 \log \left(\frac{T_{t1} - T_C}{T_{t2} - T_C} \right) = \mu.\theta$$

where T_{t1} = Maximum or total tension in the belt.

18.21 Maximum Tension in the Belt

A little consideration will show that the maximum tension in the belt (T) is equal to the total tension in the tight side of the belt (T_{t1}).

Let σ = Maximum safe stress,

b = Width of the belt, and

t = Thickness of the belt.

We know that the maximum tension in the belt,

$$T = \text{Maximum stress} \times \text{Cross-sectional area of belt} = \sigma.b.t$$

When centrifugal tension is neglected, then

$$T \text{ (or } T_{t1}) = T_1, \textit{ i.e. } \text{Tension in the tight side of the belt.}$$

When centrifugal tension is considered, then

$$T \text{ (or } T_{t1}) = T_1 + T_C$$

18.22 Condition for the Transmission of Maximum Power

We know that the power transmitted by a belt,

$$P = (T_1 - T_2)\, v \qquad \text{...}(i)$$

where T_1 = Tension in the tight side in newtons,

T_2 = Tension in the slack side in newtons, and

v = Velocity of the belt in m/s.

From Art. 18.19, ratio of driving tensions is

$$\frac{T_1}{T_2} = e^{\mu\theta} \quad \text{or} \quad T_2 = \frac{T_1}{e^{\mu\theta}} \qquad \text{...}(ii)$$

Substituting the value of T_2 in equation (*i*), we have

$$P = \left(T_1 - \frac{T_1}{e^{\mu\theta}} \right) v = T_1 \left(1 - \frac{1}{e^{\mu\theta}} \right) v = T_1.v.C \qquad \text{...}(iii)$$

where

$$C = \left(1 - \frac{1}{e^{\mu\theta}} \right)$$

We know that

$$T_1 = T - T_C$$

where T = Maximum tension to which the belt can be subjected in newtons, and

T_C = Centrifugal tension in newtons.

Substituting the value of T_1 in equation (*iii*), we have

$$P = (T - T_C)\, v \times C$$

$$= (T - mv^2)\, v \times C = (T.v - m.v^3)\, C \qquad \text{... (Substituting } T_C = m.v^2)$$

For maximum power, differentiate the above expression with respect to v and equate to zero, *i.e.*

$$\frac{dP}{dv} = 0 \quad \text{or} \quad \frac{d}{dv}(T.v - m.v^3)\, C = 0$$

or $\quad T - 3\, m.v^2 = 0 \qquad$ ...(*iv*)

$\therefore \quad T - 3\, T_C = 0 \quad \text{or} \quad T = 3T_C \qquad$ ... ($\because$ $m.v^2 = T_C$)

It shows that when the power transmitted is maximum, 1/3rd of the maximum tension is absorbed as centrifugal tension.

Notes : 1. We know that $T_1 = T - T_C$ and for maximum power, $T_C = \frac{T}{3}$.

$$\therefore \quad T_1 = T - \frac{T}{3} = \frac{2T}{3}$$

2. From equation (*iv*), we find that the velocity of the belt for maximum power,

$$v = \sqrt{\frac{T}{3m}}$$

Example 18.3. *A leather belt 9 mm × 250 mm is used to drive a cast iron pulley 900 mm in diameter at 336 r.p.m. If the active arc on the smaller pulley is 120° and the stress in tight side is 2 MPa, find the power capacity of the belt. The density of leather may be taken as 980 kg/m³, and the coefficient of friction of leather on cast iron is 0.35.*

Solution. Given: t = 9 mm = 0.009 m ; b = 250 mm = 0.25 m; d = 900 mm = 0.9 m ; N = 336 r.p.m ; $\theta = 120° = 120 \times \frac{\pi}{180} = 2.1$ rad ; σ = 2 MPa = 2 N/mm² ; ρ = 980 kg/m³ ; μ = 0.35

We know that the velocity of the belt,

$$v = \frac{\pi\, d.N}{60} = \frac{\pi \times 0.9 \times 336}{60} = 15.8 \text{ m/s}$$

and cross-sectional area of the belt,

$$a = b.t = 9 \times 250 = 2250 \text{ mm}^2$$

$\therefore$ Maximum or total tension in the tight side of the belt,

$$T = T_{t1} = \sigma.a = 2 \times 2250 = 4500 \text{ N}$$

We know that mass of the belt per metre length,

$$m = \text{Area} \times \text{length} \times \text{density} = b.t.l.\rho = 0.25 \times 0.009 \times 1 \times 980 \text{ kg/m}$$
$$= 2.2 \text{ kg/m}$$

$\therefore$ Centrifugal tension,

$$*T_C = m.v^2 = 2.2\,(15.8)^2 = 550 \text{ N}$$

and tension in the tight side of the belt,

$$T_1 = T - T_C = 4500 - 550 = 3950 \text{ N}$$

Let $\quad T_2$ = Tension in the slack side of the belt.

We know that

$$2.3 \log\left(\frac{T_1}{T_2}\right) = \mu.\theta = 0.35 \times 2.1 = 0.735$$

$$\log\left(\frac{T_1}{T_2}\right) = \frac{0.735}{2.3} = 0.3196 \quad \text{or} \quad \frac{T_1}{T_2} = 2.085 \qquad \text{... (Taking antilog of 0.3196)}$$

---

* $T_C = m.v^2 = \frac{\text{kg}}{\text{m}} \times \frac{\text{m}^2}{\text{s}^2} = \text{kg-m/s}^2$ or N ...($\because$ 1 N = 1 kg-m/s²)

and $$T_2 = \frac{T_1}{2.085} = \frac{3950}{2.085} = 1895 \text{ N}$$

We know that the power capacity of the belt,

$$P = (T_1 - T_2)\, v = (3950 - 1895)\, 15.8 = 32\,470 \text{ W} = 32.47 \text{ kW} \textbf{ Ans.}$$

Notes : The power capacity of the belt, when centrifugal tension is taken into account, may also be obtained as discussed below :

1. We know that the maximum tension in the tight side of the belt,

$$T_{t1} = T = 4500 \text{ N}$$

Centrifugal tension, $T_C = 550$ N

and tension in the slack side of the belt,

$$T_2 = 1895 \text{ N}$$

∴ Total tension in the slack side of the belt,

$$T_{t2} = T_2 + T_C = 1895 + 550 = 2445 \text{ N}$$

We know that the power capacity of the belt,

$$P = (T_{t1} - T_{t2})\, v = (4500 - 2445)\, 15.8 = 32\,470 \text{ W} = 32.47 \text{ kW} \textbf{ Ans.}$$

2. The value of total tension in the slack side of the belt (T_{t2}) may also be obtained by using the relation as discussed in Art. 18.20, *i.e.*

$$2.3 \log \left(\frac{T_{t1} - T_C}{T_{t2} - T_C} \right) = \mu.\theta$$

Example 18.4. *A flat belt is required to transmit 30 kW from a pulley of 1.5 m effective diameter running at 300 r.p.m. The angle of contact is spread over* $\frac{11}{24}$ *of the circumference. The coefficient of friction between the belt and pulley surface is 0.3. Determine, taking centrifugal tension into account, width of the belt required. It is given that the belt thickness is 9.5 mm, density of its material is 1100 kg / m³ and the related permissible working stress is 2.5 MPa.*

Solution. Given : $P = 30$ kW $= 30 \times 10^3$ W ; $d = 1.5$ m ; $N = 300$ r.p.m. ; $\theta = \frac{11}{24} \times 360 = 165°$ $= 165 \times \pi / 180 = 2.88$ rad ; $\mu = 0.3$; $t = 9.5$ mm $= 0.0095$ m ; $\rho = 1100$ kg/m³ ; $\sigma = 2.5$ MPa $= 2.5 \times 10^6$ N/m²

Let T_1 = Tension in the tight side of the belt in newtons, and

T_2 = Tension in the slack side of the belt in newtons.

We know that the velocity of the belt,

$$v = \frac{\pi\, d\, N}{60} = \frac{\pi \times 1.5 \times 300}{60} = 23.57 \text{ m/s}$$

and power transmitted (P),

$$30 \times 10^3 = (T_1 - T_2)\, v = (T_1 - T_2)\, 23.57$$

$$\therefore \quad T_1 - T_2 = 30 \times 10^3 / 23.57 = 1273 \text{ N} \qquad ...(i)$$

We know that

$$2.3 \log \left(\frac{T_1}{T_2} \right) = \mu.\theta = 0.3 \times 2.88 = 0.864$$

$$\therefore \quad \log \left(\frac{T_1}{T_2} \right) = \frac{0.864}{2.3} = 0.3756 \quad \text{or} \quad \frac{T_1}{T_2} = 2.375 \qquad ...(ii)$$

... (Taking antilog of 0.3756)

From equations (*i*) and (*ii*), we find that

$$T_1 = 2199 \text{ N}; \quad \text{and } T_2 = 926 \text{ N}$$

Let b = Width of the belt required in metres.

We know that mass of the belt per metre length,

$$m = \text{Area} \times \text{length} \times \text{density} = b \times t \times l \times \rho$$
$$= b \times 0.0095 \times 1 \times 1100 = 10.45\ b \text{ kg/m}$$

and centrifugal tension, $T_C = m.v^2 = 10.45\ b\ (23.57)^2 = 5805\ b$ N

We know that maximum tension in the belt,

$$T = T_1 + T_C = \text{Stress} \times \text{Area} = \sigma.b.t$$

or $2199 + 5805\ b = 2.5 \times 10^6 \times b \times 0.0095 = 23\ 750\ b$

$\therefore\ 23\ 750\ b - 5805\ b = 2199$ or $b = 0.122$ m or 122 mm

The standard width of the belt is 125 mm. **Ans.**

Example 18.5. *An electric motor drives an exhaust fan. Following data are provided :*

	Motor pulley	*Fan pulley*
Diameter	*400 mm*	*1600 mm*
Angle of warp	*2.5 radians*	*3.78 radians*
Coefficient of friction	*0.3*	*0.25*
Speed	*700 r.p.m.*	—
Power transmitted	*22.5 kW*	—

Calculate the width of 5 mm thick flat belt. Take permissible stress for the belt material as 2.3 MPa.

Solution. Given : $d_1 = 400$ mm or $r_1 = 200$ mm ; $d_2 = 1600$ mm or $r_2 = 800$ mm ; $\theta_1 = 2.5$ rad ; $\theta_2 = 3.78$ rad ; $\mu_1 = 0.3$; $\mu_2 = 0.25$; $N_1 = 700$ r.p.m. ; $P = 22.5$ kW $= 22.5 \times 10^3$ W ; $t = 5$ mm $= 0.005$ m ; $\sigma = 2.3$ MPa $= 2.3 \times 10^6$ N/m^2

Fig. 18.19 shows a system of flat belt drive. Suffix 1 refers to motor pulley and suffix 2 refers to fan pulley.

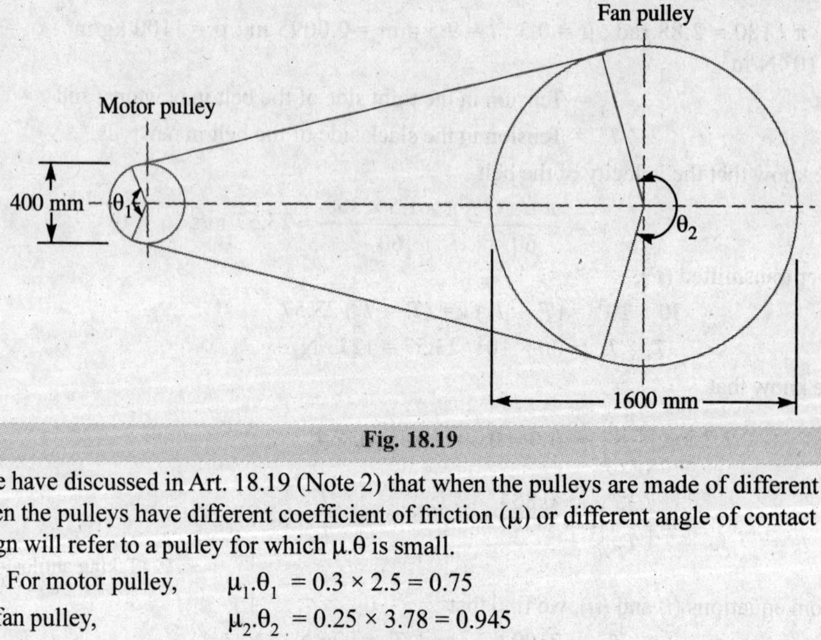

Fig. 18.19

We have discussed in Art. 18.19 (Note 2) that when the pulleys are made of different material [*i.e.* when the pulleys have different coefficient of friction (μ) or different angle of contact (θ), then the design will refer to a pulley for which $\mu.\theta$ is small.

$\therefore$ For motor pulley, $\mu_1.\theta_1 = 0.3 \times 2.5 = 0.75$

and for fan pulley, $\mu_2.\theta_2 = 0.25 \times 3.78 = 0.945$

Since $\mu_1.\theta_1$ for the motor pulley is small, therefore the design is based on the motor pulley.

Let T_1 = Tension in the tight side of the belt, and

T_2 = Tension in the slack side of the belt.

We know that the velocity of the belt,

$$v = \frac{\pi\, d_1.N_1}{60} = \frac{\pi \times 0.4 \times 700}{60} = 14.7 \text{ m/s} \qquad \text{... } (d_1 \text{ is taken in metres})$$

and the power transmitted (P),

$$22.5 \times 10^3 = (T_1 - T_2)\, v = (T_1 - T_2)\, 14.7$$

$$\therefore \quad T_1 - T_2 = 22.5 \times 10^3 / 14.7 = 1530 \text{ N} \qquad \text{...}(i)$$

We know that

$$2.3 \log\left(\frac{T_1}{T_2}\right) = \mu_1.\theta_1 = 0.3 \times 2.5 = 0.75$$

$$\therefore \quad \log\left(\frac{T_1}{T_2}\right) = \frac{0.75}{2.3} = 0.3261 \text{ or } \frac{T_1}{T_2} = 2.12 \qquad \text{...}(ii)$$

... (Taking antilog of 0.3261)

From equations (*i*) and (*ii*), we find that

$$T_1 = 2896 \text{ N}; \quad \text{and } T_2 = 1366 \text{ N}$$

Let b = Width of the belt in metres.

Since the velocity of the belt is more than 10 m/s, therefore centrifugal tension must be taken into consideration. Assuming a leather belt for which the density may be taken as 1000 kg / m³.

∴ Mass of the belt per metre length,

$$m = \text{Area} \times \text{length} \times \text{density} = b \times t \times l \times \rho$$

$$= b \times 0.005 \times 1 \times 1000 = 5\, b \text{ kg/m}$$

and centrifugal tension, $T_C = m.v^2 = 5\, b\, (14.7)^2 = 1080\, b$ N

We know that the maximum (or total) tension in the belt,

$$T = T_1 + T_C = \text{Stress} \times \text{Area} = \sigma.b.t$$

or $$2896 + 1080\, b = 2.3 \times 10^6\, b \times 0.005 = 11\,500\, b$$

$$\therefore \quad 11\,500\, b - 1080\, b = 2896 \text{ or } b = 0.278 \text{ say } 0.28 \text{ m or } 280 \text{ mm } \textbf{Ans.}$$

Example 18.6. *Design a rubber belt to drive a dynamo generating 20 kW at 2250 r.p.m. and fitted with a pulley 200 mm diameter. Assume dynamo efficiency to be 85%.*

Allowable stress for belt = *2.1 MPa*

Density of rubber = *1000 kg / m³*

Angle of contact for dynamo pulley = *165°*

Coefficient of friction between belt and pulley = *0.3*

Solution. Given : P = 20 kW = 20 × 10³ W ; N = 2250 r.p.m. ; d = 200 mm = 0.2 m ; η_d = 85% = 0.85 ; σ = 2.1 MPa = 2.1 × 10⁶ N/m² ; ρ = 1000 kg/m³ ; θ = 165° = 165 × π/180 = 2.88 rad ; μ = 0.3

Let T_1 = Tension in the tight side of the belt, and

T_2 = Tension in the slack side of the belt.

We know that velocity of the belt,

$$v = \frac{\pi\, d.N}{60} = \frac{\pi \times 0.2 \times 2250}{60} = 23.6 \text{ m/s}$$

and power transmitted (P),

$$20 \times 10^3 = (T_1 - T_2)\, v.\eta_d$$
$$= (T_1 - T_2)\, 23.6 \times 0.85$$
$$= 20.1\,(T_1 - T_2)$$

$\therefore \quad T_1 - T_2 = 20 \times 10^3 / 20.1 = 995$ N ...(*i*)

Rubber belt

We know that

$$2.3 \log\left(\frac{T_1}{T_2}\right) = \mu.\theta = 0.3 \times 2.88 = 0.864$$

$$\therefore \quad \log\left(\frac{T_1}{T_2}\right) = \frac{0.864}{2.3} = 0.3756$$

or $\quad \dfrac{T_1}{T_2} = 2.375$ ... (*ii*)

... (Taking antilog of 0.3756)

From equations (*i*) and (*ii*), we find that

T_1 = 1719 N ; and T_2 = 724 N

Let b = Width of the belt in metres, and

t = Thickness of the belt in metres.

Assuming thickness of the belt, t = 10 mm = 0.01 m, we have

Cross-sectional area of the belt

$$= b \times t = b \times 0.01 = 0.01\, b \text{ m}^2$$

We know that mass of the belt per metre length,

$$m = \text{Area} \times \text{length} \times \text{density} = 0.01\, b \times 1 \times 1000 = 10\, b \text{ kg / m}$$

$\therefore$ Centrifugal tension,

$$T_C = m.v^2 = 10\, b\, (23.6)^2 = 5570\, b \text{ N}$$

We know that maximum tension in the belt,

$$T = \sigma.b.t = 2.1 \times 10^6 \times b \times 0.01 = 21\,000\, b \text{ N}$$

and tension in the tight side of belt (T_1),

$$1719 = T - T_C = 21\,000\, b - 5570\, b = 15\,430\, b$$

$$\therefore \quad b = 1719 / 15\,430 = 0.1114 \text{ m} = 111.4 \text{ mm}$$

The standard width of the belt (b) is 112 mm. **Ans.**

Example 18.7. *Design a belt drive to transmit 110 kW for a system consisting of two pulleys of diameters 0.9 m and 1.2 m, centre distance of 3.6 m, a belt speed 20 m / s, coefficient of friction 0.3, a slip of 1.2% at each pulley and 5% friction loss at each shaft, 20% over load.*

Solution. Given : P = 110 kW = 110 × 10^3 W ; d_1 = 0.9 m or r_1 = 0.45 m ; d_2 = 1.2 m or r_2 = 0.6 m ; x = 3.6 m ; v = 20 m/s ; μ = 0.3 ; $s_1 = s_2$ = 1.2%

Fig 18.20 shows a system of flat belt drive consisting of two pulleys.

Let N_1 = Speed of the smaller or driving pulley in r.p.m., and

and N_2 = Speed of the larger or driven pulley in r.p.m.

We know that speed of the belt (v),

$$20 = \frac{\pi\, d_1.N_1}{60}\left(1 - \frac{s_1}{100}\right) = \frac{\pi \times 0.9\, N_1}{60}\left(1 - \frac{1.2}{100}\right) = 0.0466\, N_1$$

$$\therefore \quad N_1 = 20 / 0.0466 = 430 \text{ r.p.m.}$$

and peripheral velocity of the driven pulley,

$$\frac{\pi\, d_2 . N_2}{60} = \text{Belt speed in m/s}\left(1-\frac{s_2}{100}\right) = v\left(1-\frac{s_2}{100}\right)$$

or
$$\frac{\pi \times 1.2 \times N_2}{60} = 20\left(1-\frac{1.2}{100}\right) = 19.76$$

$$\therefore \quad N_2 = \frac{19.76 \times 60}{\pi \times 1.2} = 315 \text{ r.p.m.}$$

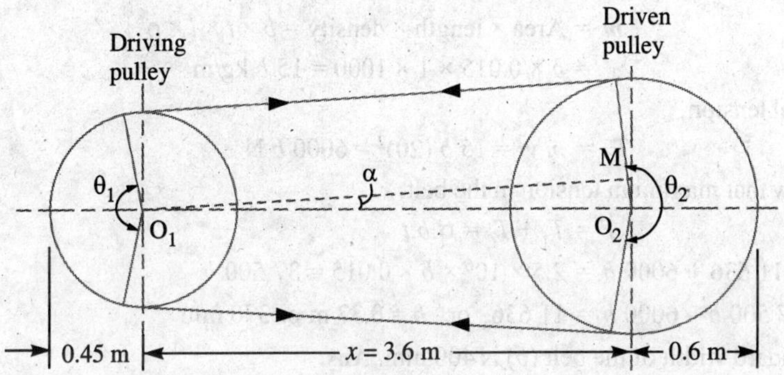

Fig. 18.20

We know that the torque acting on the driven shaft

$$= \frac{\text{Power transmitted} \times 60}{2\,\pi\, N_2} = \frac{110 \times 10^3 \times 60}{2\,\pi \times 315} = 3334 \text{ N-m}$$

Since there is a 5% friction loss at each shaft, therefore torque acting on the belt

$$= 1.05 \times 3334 = 3500 \text{ N-m}$$

Since the belt is to be designed for 20% overload, therefore design torque

$$= 1.2 \times 3500 = 4200 \text{ N-m}$$

Let T_1 = Tension in the tight side of the belt, and

T_2 = Tension in the slack side of the belt.

We know that the torque exerted on the driven pulley

$$= (T_1 - T_2)\, r_2 = (T_1 - T_2)\, 0.6 = 0.6\,(T_1 - T_2) \text{ N-m}$$

Equating this to the design torque, we have

$$0.6\,(T_1 - T_2) = 4200 \quad \text{or} \quad T_1 - T_2 = 4200 / 0.6 = 7000 \text{ N} \qquad ...(i)$$

Now let us find out the angle of contact (θ_1) of the belt on the smaller or driving pulley.

From the geometry of the Fig. 18.20, we find that

$$\sin\alpha = \frac{O_2M}{O_1O_2} = \frac{r_2 - r_1}{x} = \frac{0.6 - 0.45}{3.6} = 0.0417 \quad \text{or} \quad \alpha = 2.4°$$

$$\therefore \quad \theta_1 = 180° - 2\alpha = 180 - 2 \times 2.4 = 175.2° = 175.2 \times \frac{\pi}{180} = 3.06 \text{ rad}$$

We know that

$$2.3 \log\left(\frac{T_1}{T_2}\right) = \mu.\theta_1 = 0.3 \times 3.06 = 0.918$$

$$\therefore \quad \log\left(\frac{T_1}{T_2}\right) = \frac{0.918}{2.3} = 0.3991 \quad \text{or} \quad \frac{T_1}{T_2} = 2.51 \text{ ... (Taking antilog of 0.3991)} \quad ...(ii)$$

From equations (*i*) and (*ii*), we find that

$$T_1 = 11\,636 \text{ N} ; \text{ and } T_2 = 4636 \text{ N}$$

Let σ = Safe stress for the belt = 2.5 MPa = 2.5×10^6 N/m² ...(Assume)

t = Thickness of the belt = 15 mm = 0.015 m, and ...(Assume)

b = Width of the belt in metres.

Since the belt speed is more than 10 m/s, therefore centrifugal tension must be taken into consideration. Assuming a leather belt for which the density may be taken as 1000 kg / m³.

∴ Mass of the belt per metre length,

$$m = \text{Area} \times \text{length} \times \text{density} = b \times t \times l \times \rho$$

$$= b \times 0.015 \times 1 \times 1000 = 15\, b \text{ kg/m}$$

and centrifugal tension,

$$T_C = m.v^2 = 15\, b\, (20)^2 = 6000\, b \text{ N}$$

We know that maximum tension in the belt,

$$T = T_1 + T_C = \sigma.b.t$$

or $$11\,636 + 6000\, b = 2.5 \times 10^6 \times b \times 0.015 = 37\,500\, b$$

∴ $$37\,500\, b - 6000\, b = 11\,636 \quad \text{or} \quad b = 0.37 \text{ m or } 370 \text{ mm}$$

The standard width of the belt (b) is 400 mm. **Ans.**

We know that length of the belt,

$$L = \pi\,(r_2 + r_1) + 2x + \frac{(r_2 - r_1)^2}{x}$$

$$= \pi\,(0.6 + 0.45) + 2 \times 3.6 + \frac{(0.6 - 0.45)^2}{3.6}$$

$$= 3.3 + 7.2 + 0.006 = 10.506 \text{ m } \textbf{Ans.}$$

Example 18.8. *A belt 100 mm wide and 10 mm thick is transmitting power at 1000 metres/min. The net driving tension is 1.8 times the tension on the slack side. If the safe permissible stress on the belt section in 1.6 MPa, calculate the maximum power, that can be transmitted at this speed. Assume density of the leather as 1000 kg/m³.*

Calculate the absolute maximum power that can be transmitted by this belt and the speed at which this can be transmitted.

Solution. Given : b = 100 mm = 0.1 m ; t = 10 mm = 0.01 m ; v = 1000 m/min = 16.67 m/s ; $T_1 - T_2 = 1.8\, T_2$; σ = 1.6 MPa = 1.6 N/mm² ; ρ = 1000 kg/m³

Power transmitted

Let T_1 = Tension in the tight side of the belt, and

T_2 = Tension in the slack side of the belt.

We know that the maximum tension in the belt,

$$T = \sigma.b.t = 1.6 \times 100 \times 10 = 1600 \text{ N}$$

Mass of the belt per metre length,

$$m = \text{Area} \times \text{length} \times \text{density} = b \times t \times l \times \rho$$

$$= 0.1 \times 0.01 \times 1 \times 1000 = 1 \text{ kg/m}$$

∴ Centrifugal tension,

$$T_C = m.v^2 = 1\,(16.67)^2 = 278 \text{ N}$$

We know that

$$T_1 = T - T_C = 1600 - 278 = 1322 \text{ N}$$

and $$T_1 - T_2 = 1.8\ T_2 \qquad \text{...(Given)}$$

$$\therefore \quad T_2 = \frac{T_1}{2.8} = \frac{1322}{2.8} = 472 \text{ N}$$

We know that the power transmitted.

$$P = (T_1 - T_2)\ v = (1322 - 472)\ 16.67 = 14\ 170 \text{ W} = 14.17 \text{ kW } \textbf{Ans.}$$

Speed at which absolute maximum power can be transmitted

We know that the speed of the belt for maximum power,

$$v = \sqrt{\frac{T}{3m}} = \sqrt{\frac{1600}{3 \times 1}} = 23.1 \text{ m/s } \textbf{Ans.}$$

Absolute maximum power

We know that for absolute maximum power, the centrifugal tension,

$$T_C = T/3 = 1600/3 = 533 \text{ N}$$

∴ Tension in the tight side,

$$T_1 = T - T_C = 1600 - 533 = 1067 \text{ N}$$

and tension in the slack side,

$$T_2 = \frac{T_1}{2.8} = \frac{1067}{2.8} = 381 \text{ N}$$

∴ Absolute maximum power transmitted,

$$P = (T_1 - T_2)\ v = (1067 - 381)\ 23.1 = 15\ 850 \text{ W} = 15.85 \text{ kW } \textbf{Ans.}$$

18.23 Initial Tension in the Belt

When a belt is wound round the two pulleys (*i.e.* driver and follower), its two ends are joined together, so that the belt may continuously move over the pulleys, since the motion of the belt (from the driver) and the follower (from the belt) is governed by a firm grip due to friction between the belt and the pulleys. In order to increase this grip, the belt is tightened up. At this stage, even when the pulleys are stationary, the belt is subjected to some tension, called ***initial tension.***

When the driver starts rotating, it pulls the belt from one side (increasing tension in the belt on this side) and delivers to the other side (decreasing tension in the belt on that side). The increased tension in one side of the belt is called tension in tight side and the decreased tension in the other side of the belt is called tension in the slack side.

Let T_0 = Initial tension in the belt,

T_1 = Tension in the tight side of the belt,

T_2 = Tension in the slack side of the belt, and

α = Coefficient of increase of the belt length per unit force.

A little consideration will show that the increase of tension in the tight side

$$= T_1 - T_0$$

and increase in the length of the belt on the tight side

$$= \alpha\ (T_1 - T_0) \qquad \text{...}(i)$$

Similarly, decrease in tension in the slack side

$$= T_0 - T_2$$

and decrease in the length of the belt on the slack side

$$= \alpha\ (T_0 - T_2) \qquad \text{...}(ii)$$

Assuming that the belt material is perfectly elastic such that the length of the belt remains constant, when it is at rest or in motion, therefore increase in length on the tight side is equal to decrease in the length on the slack side. Thus, equating equations (*i*) and (*ii*), we have

$$\alpha\ (T_1 - T_0) = \alpha\ (T_0 - T_2)$$

or $$T_1 - T_0 = T_0 - T_2$$

$$\therefore \quad T_0 = \frac{T_1 + T_2}{2} \quad \text{... (Neglecting centrifugal tension)}$$

$$= \frac{T_1 + T_2 + 2T_C}{2} \quad \text{... (Considering centrifugal tension)}$$

Note: In actual practice, the belt material is not perfectly elastic. Therefore, the sum of the tensions T_1 and T_2, when the belt is transmitting power, is always greater than twice the initial tension. According to C.G. Barth, the relation between T_0, T_1 and T_2 is given by

$$\sqrt{T_1} + \sqrt{T_2} = 2\sqrt{T_0}$$

Example 18.9. *Two parallel shafts whose centre lines are 4.8 m apart, are connected by an open belt drive. The diameter of the larger pulley is 1.5 m and that of smaller pulley 1 m. The initial tension in the belt when stationary is 3 kN. The mass of the belt is 1.5 kg / m length. The coefficient of friction between the belt and the pulley is 0.3. Taking centrifugal tension into account, calculate the power transmitted, when the smaller pulley rotates at 400 r.p.m.*

Solution. Given : $x = 4.8$ m ; $d_1 = 1.5$ m ; $d_2 = 1$ m ; $T_0 = 3$ kN $= 3000$ N ; $m = 1.5$ kg/m ; $\mu = 0.3$; $N_2 = 400$ r.p.m.

We know that the velocity of the belt,

$$v = \frac{\pi\, d_2.N_2}{60} = \frac{\pi \times 1 \times 400}{60} = 21 \text{ m/s}$$

$\therefore$ Centrifugal tension,

$$T_C = m.v^2 = 1.5\,(21)^2 = 661.5 \text{ N}$$

Let T_1 = Tension in the tight side of the belt, and

T_2 = Tension in the slack side of the belt.

We know that the initial tension (T_0),

$$3000 = \frac{T_1 + T_2 + 2T_C}{2} = \frac{T_1 + T_2 + 2 \times 661.5}{2}$$

$$\therefore \quad T_1 + T_2 = 3000 \times 2 - 2 \times 661.5 = 4677 \text{ N} \quad \text{...}(i)$$

For an open belt drive,

$$\sin\alpha = \frac{r_1 - r_2}{x} = \frac{d_1 - d_2}{2\,x} = \frac{1.5 - 1}{2 \times 4.8} = 0.0521 \text{ or } \alpha = 3°$$

$\therefore$ Angle of lap on the smaller pulley,

$$\theta = 180° - 2\alpha = 180 - 2 \times 3 = 174°$$

$$= 174 \times \frac{\pi}{180} = 3.04 \text{ rad}$$

We know that

$$2.3 \log\left(\frac{T_1}{T_2}\right) = \mu.\theta = 0.3 \times 3.04 = 0.912$$

$$\log\left(\frac{T_1}{T_2}\right) = \frac{0.912}{2.3} = 0.3965 \text{ or } \frac{T_1}{T_2} = 2.5 \quad \text{... (Taking antilog of 0.3965) ...}(ii)$$

From equations (*i*) and (*ii*), we have

$$T_1 = 3341 \text{ N} ; \text{and } T_2 = 1336 \text{ N}$$

We know that the power transmitted,

$$P = (T_1 - T_2)\,v = (3341 - 1336)\,21 = 42\,100 \text{ W} = 42.1 \text{ kW} \textbf{ Ans.}$$

Example 18.10. *In a horizontal belt drive for a centrifugal blower, the blower is belt driven at 600 r.p.m. by a 15 kW, 1750 r.p.m. electric motor. The centre distance is twice the diameter of the larger pulley. The density of the belt material = 1500 kg/m³; maximum allowable stress = 4 MPa; μ_1 = 0.5 (motor pulley); μ_2 = 0.4 (blower pulley); peripheral velocity of the belt = 20 m/s. Determine the following:*

1. Pulley diameters; 2. belt length; 3. cross-sectional area of the belt; 4. minimum initial tension for operation without slip; and 5. resultant force in the plane of the blower when operating with an initial tension 50 per cent greater than the minimum value.

Solution. Given : N_2 = 600 r.p.m. ; P = 15 kW = 15 × 10^3 W ; N_1 = 1750 r.p.m. ; ρ = 1500 kg/m³ ; σ = 4 MPa = 4 × 10^6 N/m² ; μ_1 = 0.5 ; μ_2 = 0.4 ; v = 20 m/s

Fig. 18.21 shows a horizontal belt drive. Suffix 1 refers to a motor pulley and suffix 2 refers to a blower pulley.

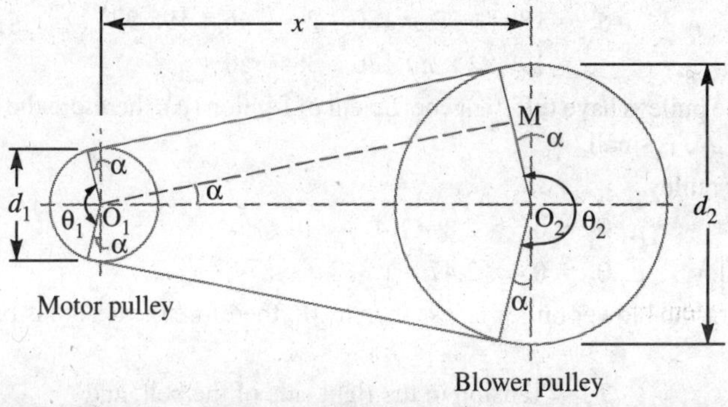

Fig. 18.21

1. ***Pulley diameters***

Let d_1 = Diameter of the motor pulley, and

d_2 = Diameter of the blower pulley.

We know that peripheral velocity of the belt (v),

$$20 = \frac{\pi\, d_1\, N_1}{60} = \frac{\pi d_1 \times 1750}{60} = 91.64\, d_1$$

$$\therefore \quad d_1 = 20 / 91.64 = 0.218 \text{ m} = 218 \text{ mm } \textbf{Ans.}$$

We also know that $\dfrac{N_2}{N_1} = \dfrac{d_1}{d_2}$

$$\therefore \quad d_2 = \frac{d_1 \times N_1}{N_2} = \frac{218 \times 1750}{600} = 636 \text{ mm } \textbf{Ans.}$$

2. ***Belt length***

Since the centre distance (x) between the two pulleys is twice the diameter of the larger pulley (*i.e.* 2 d_2), therefore centre distance,

$$x = 2\, d_2 = 2 \times 636 = 1272 \text{ mm}$$

We know that length of belt,

$$L = \frac{\pi}{2}(d_1 + d_2) + 2\,x + \frac{(d_1 - d_2)^2}{4x}$$

$$= \frac{\pi}{2}(218 + 636) + 2 \times 1272 + \frac{(218 - 636)^2}{4 \times 1272}$$

$$= 1342 + 2544 + 34 = 3920 \text{ mm} = 3.92 \text{ m } \textbf{Ans.}$$

3. ***Cross-sectional area of the belt***

Let a = Cross-sectional area of the belt.

First of all, let us find the angle of contact for both the pulleys. From the geometry of the figure, we find that

$$\sin \alpha = \frac{O_2 M}{O_1 O_2} = \frac{r_2 - r_1}{x} = \frac{d_2 - d_1}{2\,x} = \frac{636 - 218}{2 \times 1272} = 0.1643$$

$\therefore$ $\alpha = 9.46°$

We know that angle of contact on the motor pulley,

$$\theta_1 = 180° - 2\alpha = 180 - 2 \times 9.46 = 161.08°$$
$$= 161.08 \times \pi / 180 = 2.8 \text{ rad}$$

and angle of contact on the blower pulley,

$$\theta_2 = 180° + 2\alpha = 180 + 2 \times 9.46 = 198.92°$$
$$= 198.92 \times \pi / 180 = 3.47 \text{ rad}$$

Since both the pulleys have different coefficient of friction (μ), therefore the design will refer to a pulley for which $\mu.\theta$ is small.

$\therefore$ For motor pulley,

$$\mu_1.\theta_1 = 0.5 \times 2.8 = 1.4$$

and for blower pulley, $\mu_2.\theta_2 = 0.4 \times 3.47 = 1.388$

Since $\mu_2.\theta_2$ for the blower pulley is less then $\mu_1.\theta_1$, therefore the design is based on the blower pulley.

Let T_1 = Tension in the tight side of the belt, and

T_2 = Tension in the slack side of the belt.

We know that power transmitted (P),

$$15 \times 10^3 = (T_1 - T_2)\,v = (T_1 - T_2)\,20$$

$\therefore$ $T_1 - T_2 = 15 \times 10^3 / 20 = 750 \text{ N}$ ...(*i*)

We also know that

$$2.3 \log \left(\frac{T_1}{T_2}\right) = \mu_2.\theta_2 = 0.4 \times 3.47 = 1.388$$

$\therefore$ $\log \left(\frac{T_1}{T_2}\right) = \frac{1.388}{2.3} = 0.6035$ or $\frac{T_1}{T_2} = 4$ ...(*ii*)

... (Taking antilog of 0.6035)

From equations (*i*) and (*ii*),

$$T_1 = 1000 \text{ N}\,; \text{ and } T_2 = 250 \text{ N}$$

Mass of the belt per metre length,

$$m = \text{Area} \times \text{length} \times \text{density} = a \times l \times \rho$$
$$= a \times 1 \times 1500 = 1500\,a \text{ kg/m}$$

$\therefore$ Centrifugal tension,

$$T_C = m.v^2 = 1500\,a\,(20)^2 = 0.6 \times 10^6\,a \text{ N}$$

We know that maximum or total tension in the belt,

$T = T_1 + T_C = 1000 + 0.6 \times 10^6\,a$ N ...(*iii*)

We also know that maximum tension in the belt,

$T = \text{Stress} \times \text{area} = \sigma \times a = 4 \times 10^6\,a$ N ...(*iv*)

From equations (*iii*) and (*iv*),

$$1000 + 0.6 \times 10^6 a = 4 \times 10^6\, a \quad \text{or} \quad 3.4 \times 10^6 a = 1000$$

$$\therefore \quad a = 1000 / 3.4 \times 10^6 = 294 \times 10^{-6}\ \text{m}^2 = 294\ \text{mm}^2\ \textbf{Ans.}$$

4. ***Minimum initial tension for operation without slip***

We know that centrifugal tension,

$$T_C = 0.6 \times 10^6\, a = 0.6 \times 10^6 \times 294 \times 10^{-6} = 176.4\ \text{N}$$

∴ Minimum initial tension for operation without slip,

$$T_0 = \frac{T_1 + T_2 + 2T_C}{2} = \frac{1000 + 250 + 2 \times 176.4}{2} = 801.4\ \text{N}\ \textbf{Ans.}$$

5. ***Resultant force in the plane of the blower when operating with an initial tension 50 per cent greater than the minimum value***

We have calculated above that the minimum initial tension,

$$T_0 = 801.4\ \text{N}$$

∴ Increased initial tension,

$$T_0' = 801.4 + 801.4 \times \frac{50}{100} = 1202\ \text{N}$$

Let T_1' and T_2' be the corresponding tensions in the tight side and slack side of the belt respectively.

We know that increased initial tension (T_0'),

$$1202 = \frac{T_1' + T_2' + 2T_C}{2} = \frac{T_1' + T_2' + 2 \times 176.4}{2}$$

$$\therefore \quad T_1' + T_2' = 1202 \times 2 - 2 \times 176.4 = 2051.2\ \text{N} \qquad \text{...}(v)$$

Since the ratio of tensions will be constant, *i.e.* $\frac{T_1'}{T_2'} = \frac{T_1}{T_2} = 4$, therefore from equation (*v*), we have

$$4T_2' + T_2' = 2051.2 \quad \text{or} \quad T_2' = 2051.2 / 5 = 410.24\ \text{N}$$

and

$$T_1' = 4\, T_2' = 4 \times 410.24 = 1640.96\ \text{N}$$

∴ Resultant force in the plane of the blower

$$= T_1' - T_2' = 1640.96 - 410.24 = 1230.72\ \text{N}\ \textbf{Ans.}$$

Example 18.11. *An open belt connects two flat pulleys. The pulley diameters are 300 mm and 450 mm and the corresponding angles of lap are 160° and 210°. The smaller pulley runs at 200 r.p.m. The coefficient of friction between the belt and pulley is 0.25. It is found that the belt is on the point of slipping when 3 kW is transmitted. To increase the power transmitted two alternatives are suggested, namely (i) increasing the initial tension by 10%, and (ii) increasing the coefficient of friction by 10% by the application of a suitable dressing to the belt.*

Which of these two methods would be more effective? Find the percentage increase in power possible in each case.

Solution. Given : $d_1 = 300$ mm $= 0.3$ m ; $d_2 = 450$ mm $= 0.45$ m ; $\theta_1 = 160° = 160 \times \frac{\pi}{180} = 2.8$ rad ; $\theta_2 = 210° = 210 \times \frac{\pi}{180} = 3.66$ rad ; $N_1 = 200$ r.p.m.; $\mu = 0.25$; $P = 3$ kW $= 3000$ W

Let T_1 = Tension in the tight side of the belt, and

T_2 = Tension in the slack side of the belt.

We have discussed in Art 18.19 (Note 2) that when the pulleys are made of different material [*i.e.* when the pulleys have different coefficient of friction (μ) or different angle of contact (θ)], then the design will be refer to a pulley for which $\mu.\theta$ is small.

∴ For smaller pulley, $\mu.\theta_1 = 0.25 \times 2.8 = 0.7$

and for larger pulley, $\mu.\theta_2 = 0.25 \times 3.66 = 0.915$

Since $\mu.\theta_1$ for the smaller pulley is less than $\mu.\theta_2$, therefore the design is based on the smaller pulley.

We know that velocity of the belt,

$$v = \frac{\pi\, d_1.N_1}{60} = \frac{\pi \times 0.3 \times 200}{60} = 3.142 \text{ m/s}$$

and power transmitted (P),

$$3000 = (T_1 - T_2)\, v = (T_1 - T_2)\, 3.142$$

$$\therefore \quad T_1 - T_2 = 3000 / 3.142 = 955 \text{ N} \qquad ...(i)$$

We know that

$$2.3 \log\left(\frac{T_1}{T_2}\right) = \mu.\theta_1 = 0.25 \times 2.8 = 0.7$$

$$\therefore \quad \log\left(\frac{T_1}{T_2}\right) = \frac{0.7}{2.3} = 0.3043 \quad \text{or} \quad \frac{T_1}{T_2} = 2.015 \qquad ...(ii)$$

... (Taking antilog of 0.3043)

From equations (*i*) and (*ii*), we find that

$$T_1 = 1896 \text{ N, and } T_2 = 941 \text{ N}$$

(i) Power transmitted when initial tension is increased by 10%

We know that the initial tension,

$$T_0 = \frac{T_1 + T_2}{2} = \frac{1896 + 941}{2} = 1418.5 \text{ N}$$

∴ Increased initial tension,

$$T_0' = 1418.5 + 1418.5 \times \frac{10}{100} = 1560.35 \text{ N}$$

Let T_1 and T_2 be the corresponding tensions in the tight side and slack side of the belt respectively.

$$\therefore \quad T_0' = \frac{T_1 + T_2}{2}$$

or

$$T_1 + T_2 = 2\, T_0' = 2 \times 1560.35 = 3120.7 \text{ N} \qquad ...(iii)$$

Since the ratio of the tensions is constant, *i.e.* $T_1 / T_2 = 2.015$ or $T_1 = 2.015\, T_2$, therefore from equation (*iii*),

$$2.015\, T_2 + T_2 = 3120.7 \quad \text{or} \quad T_2 = 3120.7 / 3.015 = 1035 \text{ N}$$

and

$$T_1 = 2.015\, T_2 = 2.015 \times 1035 = 2085.7 \text{ N}$$

∴ Power transmitted,

$$P = (T_1 - T_2)\, v = (2085.7 - 1035)\, 3.142 = 3300 \text{ W} = 3.3 \text{ kW}$$

(ii) Power transmitted when the coefficient of friction is increased by 10%

We know that the coefficient of friction,

$$\mu = 0.25$$

∴ Increased coefficient of friction,

$$\mu' = 0.25 + 0.25 \times \frac{10}{100} = 0.275$$

Let T_1 and T_2 be the corresponding tensions in the tight side and slack side of the belt respectively. We know that

$$2.3 \log \left(\frac{T_1}{T_2} \right) = \mu'.\theta_1 = 0.275 \times 2.8 = 0.77$$

$$\therefore \quad \log \left(\frac{T_1}{T_2} \right) = \frac{0.77}{2.3} = 0.3348 \quad \text{or} \quad \frac{T_1}{T_2} = 2.16 \qquad \text{...}(iv)$$

... (Taking antilog of 0.3348)

Here the initial tension is constant, *i.e.*

$$T_0 = \frac{T_1 + T_2}{2}$$

$$\therefore \quad T_1 + T_2 = 2\,T_0 = 2 \times 1418.5 = 2837 \text{ N} \qquad \text{...}(v)$$

From equations (*iv*) and (*v*), we find that

$$T_1 = 1939 \text{ N, and } T_2 = 898 \text{ N}$$

∴ Power transmitted,

$$P = (T_1 - T_2)\,v = (1939 - 898)\,3.142 = 3271 \text{ W} = 3.217 \text{ kW}$$

Since the power transmitted by increasing the initial tension is more, therefore in order to increase the power transmitted, we shall adopt the method of increasing the initial tension. **Ans.**

Percentage increase in power

Percentage increase in power when the initial tension is increased

$$= \frac{3.3 - 3}{3} \times 100 = 10\% \textbf{ Ans.}$$

Percentage increase in power when coefficient of friction is increased,

$$= \frac{3.271 - 3}{3} \times 100 = 9.03\% \textbf{ Ans.}$$

EXERCISES

1. An engine shaft running at 120 r.p.m. is required to drive a machine shaft by means of a belt. The pulley on the engine shaft is of 2 m diameter and that of the machine shaft is 1 m diameter. If the belt thickness is 5 mm; determine the speed of the machine shaft, when
   1. there is no slip; and 2. there is a slip of 3%. **[Ans. 239.4 r.p.m. ; 232.3 r.p.m.]**
2. A pulley is driven by a flat belt running at a speed of 600 m/min. The coefficient of friction between the pulley and the belt is 0.3 and the angle of lap is 160°. If the maximum tension in the belt is 700 N; find the power transmitted by a belt. **[Ans. 3.974 kW]**
3. Find the width of the belt necessary to transmit 10 kW to a pulley 300 mm diameter, if the pulley makes 1600 r.p.m. and the coefficient of friction between the belt and the pulley is 0.22.

   Assume the angle of contact as 210° and the maximum tension in the belt is not to exceed 8N/mm width. **[Ans. 90 mm]**
4. An open belt 100 mm wide connects two pulleys mounted on parallel shafts with their centres 2.4 m apart. The diameter of the larger pulley is 450 mm and that of the smaller pulley 300 mm. The coefficient of friction between the belt and the pulley is 0.3 and the maximum stress in the belt is limited to 14 N/mm width. If the larger pulley rotates at 120 r.p.m., find the maximum power that can be transmitted. **[Ans. 2.387 kW]**
5. A rough rule for leather belt is that effective tension in it, shall not exceed 15 N/mm of width for a belt of 10 mm thickness. This rule is applied to determine width of belt required to transmit 37 kW, under the following conditions :

   Angle of lap = 165°; Coefficient of friction = 0.3; Velocity of belt = 1500 m/min; Density of leather = 950 kg/m^3.

Find the width of belt required.

Assuming limiting friction between belt and pulley rim, find the stress in the belt.

[Ans. 140 mm ; 1.48 MPa]

6. A leather belt, 125 mm wide and 6 mm thick, transmits power from a pulley 750 mm diameter which runs at 500 r.p.m. The angle of lap is 150° and $\mu = 0.3$. If the mass of 1 m^3 of leather is 1 Mg and the stress in the belt is not to exceed 2.75 MN/m^2, find the maximum power that can be transmitted. **[Ans. 18.97 kW]**

7. An exhaust fan fitted with 900 mm diameter pulley is driven by a flat belt from a 30 kW, 950 r.p.m. squirrel cage motor. The pulley on the motor shaft is 250 mm in diameter and the centre distance between the fan and motor is 2.25 m. The belt is 100 mm wide with a coefficient of friction of 0.25. If the allowable stress in the belt material is not to exceed 2 MPa, determine the necessary thickness of the belt and its total length. Take centrifugal force effect into consideration for density of belt being 950 kg/m^3. **[Ans. 26 mm ; 6.35 m]**

8. A cross belt arrangement has centre distance between pulleys as 1.5 m. The diameter of bigger and smaller pulleys are '*D*' and '*d*' respectively. The smaller pulley rotates at 1000 r.p.m. and the bigger pulley at 500 r.p.m. The flat belt is 6 mm thick and transmits 7.5 kW power at belt speed of 13 m/s approximately. The coefficient of belt friction is 0.3 and the density of belt material is 950 kg/m^3. If the permissible tensile stress for the belt material is 1.75 MPa, calculate: 1. Diameters of pulleys; 2. Length and width of belt.

[Ans. 500 mm, 250 mm ; 4.272 m, 90 mm]

9. A blower is driven by an electric motor through a belt drive. The motor runs at 450 r.p.m. For this power transmission, a flat belt of 8 mm thickness and 250 mm width is used. The diameter of the motor pulley is 350 mm and that of the blower pulley is 1350 mm. The centre distance between these pulleys is 1850 mm and an open belt configuration is adopted. The pulleys are made of cast iron. The coefficient of friction between the belt and pulley is 0.35 and the permissible stress for the belt material can be taken as 2.5 N/mm^2. The mass of the belt is 2 kg/metre length. Find the maximum power transmitted without belt slipping in any one of the pulley. **[Ans. 38 kW]**

Blower driven by electric motor

10. A 18 kW, 900 r.p.m. motor drives a centrifugal pump at 290 r.p.m. by means of a leather belt. The pulleys are of cast iron and are 1.2 metre centre distance. The pulleys of diameter less than 150 mm should not be used. The coefficient of friction between the leather belt and the cast iron pulley is 0.35, and the mass of the belt is 9 kg/m width/m length. The maximum permissible tension per mm width of the belt is 10 N. The drive is to be designed for 20% overload.

Determine the pulley diameters, the required width and length of the belt. Also find the initial tension with which the belt is to be mounted on the pulleys. **[Ans. 460 mm ; 270 mm ; 3.4 m ; 2970 N]**

11. A flat belt, 8 mm thick and 100 mm wide transmits power between two pulleys, running at 1600 m/min. The mass of the belt is 0.9 kg/m length. The angle of lap in the smaller pulley is 165° and the coefficient of friction between the belt and pulleys is 0.3. If the maximum permissible stress in the belt is 2 MN/m^2, find (*i*) Maximum power transmitted, and (*ii*) Initial tension in the belt.

[Ans. 14.821 kW; 1.322 kN]

12. Design a flat belt drive to transmit 110 kW at a belt speed of 25 m/s between two pulleys of diameters 250 mm and 400 mm having a pulley centre distance of 1 metre. The allowable belt stress is 8.5 MPa and the belts are available having a thickness to width ratio of 0.1 and a material density of 1100 kg/m^3. Given that the coefficient of friction between the belt and pulleys is 0.3, determine the minimum required belt width.

    What would be the necessary installation force between the pulley bearings and what will be the force between the pulley bearings when the full power is transmitted?

13. A 8 mm thick leather open belt connects two flat pulleys. The smaller pulley is 300 mm diameter and runs at 200 r.p.m. The angle of lap of this pulley is 160° and the coefficient of friction between the belt and the pulley is 0.25. The belt is on the point of slipping when 3 kW is transmitted. The safe working stress in the belt material is 1.6 N/mm^2. Determine the required width of the belt for 20% overload capacity. The initial tension may be taken equal to the mean of the driving tensions. It is proposed to increase the power transmitting capacity of the drive by adopting one of the following alternatives :

    1. by increasing initial tension by 10%, and
    2. by increasing the coefficient of friction to 0.3 by applying a dressing to the belt.

    Examine the two alternatives and recommend the one which will be more effective. How much power would the drive transmit adopting either of the two alternatives?

QUESTIONS

1. Discuss the different types of belts and their material used for power transmission.
2. Discuss the various important parameters necessary for the selection of a particular drive for power transmission.
3. What are the factors upon which the coefficient of friction between the belt and the pulley depends?
4. How are ends of belts joined? For horizontal belts which side (tight or slack) of the belt should run on the top and why?
5. Explain, with the help of neat sketches, the types of various flat belt drives.
6. List and discuss briefly the factors that control the power transmission capacity of a belt.
7. Prove that the ratio of the driving tensions on the two sides of a pulley is

$$\frac{T_1}{T_2} = e^{\mu\theta}$$

where T_1 = Tension in the tight side of the belt,

T_2 = Tension in the slack side of the belt,

μ = Coefficient of friction between the belt and the pulley, and

θ = Angle of contact in radians.

8. In a belt drive, how will you decide the pulley governing design?
9. It is stated that the speed at which a belt should be run to transmit maximum power is that at which the maximum allowable tension is three times the centrifugal tension in the belt at that speed. Prove the statement.

OBJECTIVE TYPE QUESTIONS

1. The material suitable for the belts used in agricultural equipments is
   (a) cotton (b) rubber
   (c) leather (d) balata gum
2. The power transmitted by means of a belt depends upon
   (a) velocity of the belt
   (b) tension under which the belt is placed on the pulleys
   (c) arc of contact between the belt and the smaller pulley
   (d) all of the above
3. When the speed of belt increases,
   (a) the coefficient of friction between the belt and pulley increases
   (b) the coefficient of friction between the belt and pulley decreases
   (c) the power transmitted will decrease
   (d) the power transmitted will increase
4. In a crossed belt drive, the shafts are arranged parallel and rotate in the .... directions.
   (a) same (b) opposite
5. The tension in the slack side of the belt is .... the tension in the tight side of the belt.
   (a) equal to (b) less than (c) greater than
6. In a flat belt drive, the belt can be subjected to a maximum tension (T) and centrifugal tension (T_C). The condition for transmission of maximum power is given by
   (a) $T = T_C$ (b) $T = 2\,T_C$
   (c) $T = 3\,T_C$ (d) $T = \sqrt{3}\,T_C$
7. When a belt drive is transmitting maximum power,
   (a) effective tension is equal to the centrifugal tension
   (b) effective tension is half of the centrifugal tension
   (c) driving tension in slack side is equal to the centrifugal tension
   (d) driving tension in tight side is twice the centrifugal tension
8. All stresses produced in a belt are
   (a) compressive stresses (b) tensile stresses
   (c) both tensile and compressive stresses (d) shear stresses
9. For maximum power, the velocity of the belt will be
   (a) $\sqrt{\dfrac{T}{m}}$ (b) $\sqrt{\dfrac{T}{2m}}$ (c) $\sqrt{\dfrac{T}{3m}}$
10. The centrifugal tension in the belt
    (a) increases the power transmitted (b) decreases the power transmitted
    (c) has no effect on the power transmitted (d) is equal to maximum tension on the belt

ANSWERS

1. (b)	2. (d)	3. (d)	4. (b)	5. (b)
6. (c)	7. (d)	8. (b)	9. (c)	10. (c)

CHAPTER

19

Flat Belt Pulleys

19.1 Introduction

The pulleys are used to transmit power from one shaft to another by means of flat belts, V-belts or ropes. Since the velocity ratio is the inverse ratio of the diameters of driving and driven pulleys, therefore the pulley diameters should be carefully selected in order to have a desired velocity ratio. The pulleys must be in perfect alignment in order to allow the belt to travel in a line normal to the pulley faces.

The pulleys may be made of cast iron, cast steel or pressed steel, wood and paper. The cast materials should have good friction and wear characteristics. The pulleys made of pressed steel are lighter than cast pulleys, but in many cases they have lower friction and may produce excessive wear.

19.2 Types of Pulleys for Flat Belts

Following are the various types of pulleys for flat belts :

1. Cast iron pulleys, **2.** Steel pulleys, **3.** Wooden pulleys, **4.** Paper pulleys, and **5.** Fast and loose pulleys.

We shall now discuss, the above mentioned pulleys in the following pages.

19.3 Cast Iron Pulleys

The pulleys are generally made of *cast iron, because of their low cost. The rim is held in place by web from the central boss or by arms or spokes. The arms may be straight or curved as shown in Fig. 19.1 (*a*) and (*b*) and the cross-section is usually elliptical.

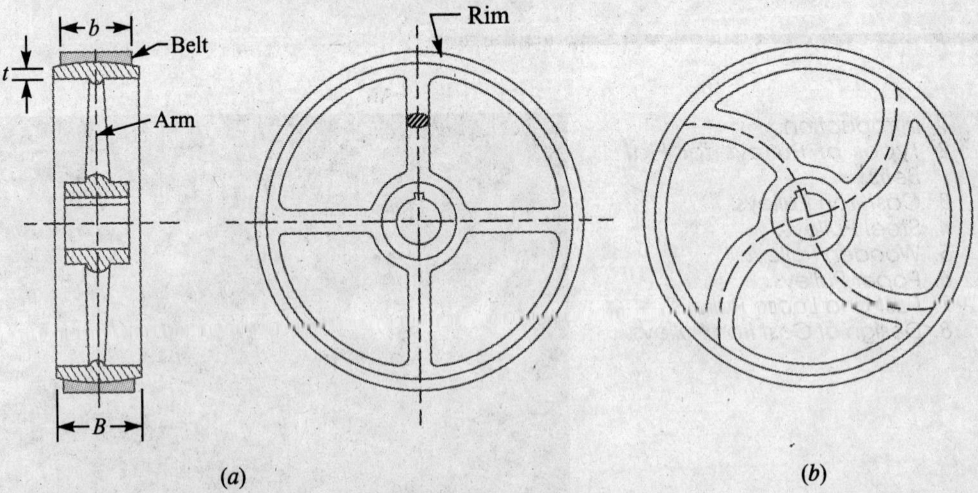

Fig. 19.1. Solid cast iron pulleys.

When a cast pulley contracts in the mould, the arms are in a state of stress and very liable to break. The curved arms tend to yield rather than to break. The arms are near the hub.

The cast iron pulleys are generally made with rounded rims. This slight convexity is known as ***crowning.*** The crowning tends to keep the belt in centre on a pulley rim while in motion. The crowning may be 9 mm for 300 mm width of pulley face.

The cast iron pulleys may be solid as shown in Fig. 19.1 or split type as shown in Fig. 19.2. When it is necessary to mount a pulley on a shaft which already carrying pulleys etc. or have its ends swelled, it is easier to use a split-pulley. There is a clearance between the faces and the two halves are readily tightened upon the shafts by the bolts as shown in Fig. 19.2. A sunk key is used for heavy drives.

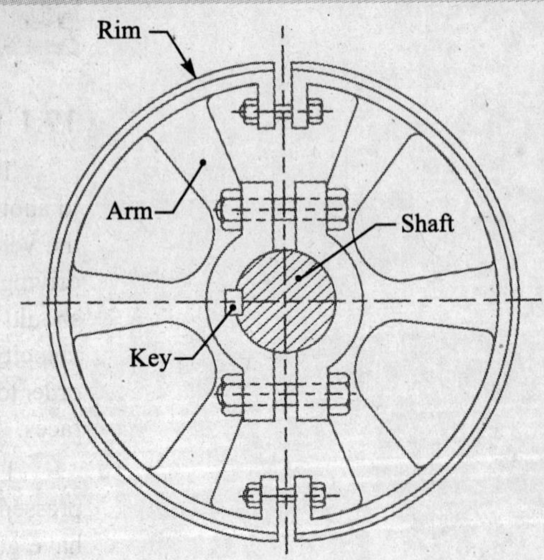

Fig. 19.2. Split cast iron pulley.

* For further details, please refer IS : 1691 – 1980 (Reaffirmed 1990).

19.4 Steel Pulleys

Flat belt drive in an aircraft engine.

Steel pulleys are made from pressed steel sheets and have great strength and durability. These pulleys are lighter in weight (about 40 to 60% less) than cast iron pulleys of the same capacity and are designed to run at high speeds. They present a coefficient of friction with leather belting which is atleast equal to that obtained by cast iron pulleys.

Steel pulleys are generally made in two halves which are bolted together. The clamping action of the hub holds the pulley to its shaft, thus no key is required except for most severe service. Steel pulleys are generally equipped with interchangeable bushings to permit their use with shafts of different sizes. The following table shows the number of spokes and their sizes according to Indian Standards, IS : 1691 – 1980 (Reaffirmed 1990).

Table 19.1. Standard number of spokes and their sizes according to IS : 1691 - 1980 (Reaffirmed 1990).

Diameter of pulley (mm)	*No. of spokes*	*Diameter of spokes (mm)*
280 – 500	6	19
560 – 710	8	19
800 – 1000	10	22
1120	12	22
1250	14	22
1400	16	22
1600	18	22
1800	18	22

Other proportions for the steel pulleys are :

$$\text{Length of hub} = \frac{\text{Width of face}}{2}$$

The length of hub should not be less than 100 mm for 19 mm diameter spokes and 138 mm for 22 mm diameter of spokes.

Thickness of rim = 5 mm for all sizes.

A single row of spokes is used for pulleys having width upto 300 mm and double row of spokes for widths above 300 mm.

19.5 Wooden Pulleys

Wooden pulleys are lighter and possesses higher coefficient of friction than cast iron or steel pulleys. These pulleys have 2/3rd of the weight of cast iron pulleys of similar size. They are generally made from selected maple which is laid in segments and glued together under heavy pressure. They are kept from absorbing moisture by protective coatings of shellac or varnish so that warping may not occur. These pulleys are made both solid or split with cast iron hubs with keyways or have adjustable bushings which prevents relative rotation between them and the shaft by the frictional resistance set up. These pulleys are used for motor drives in which the contact arc between the pulley face and belt is restricted.

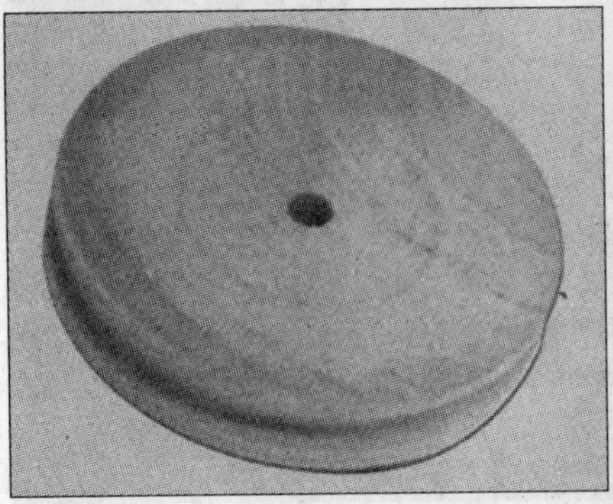

Wooden pulleys.

19.6 Paper Pulleys

Paper pulleys are made from compressed paper fibre and are formed with a metal in the centre. These pulleys are usually used for belt transmission from electric motors, when the centre to centre shaft distance is small.

19.7 Fast and Loose Pulleys

A fast and loose pulley, as shown in Fig. 19.3, used on shafts enables machine to be started or stopped at will. A fast pulley is keyed to the machine shaft while the loose pulley runs freely. The belt runs over the fast pulley to transmit power by the machine and it is shifted to the loose pulley when the machine is not required to transmit power. By this way, stopping of one machine does not interfere with the other machines which run by the same line shaft.

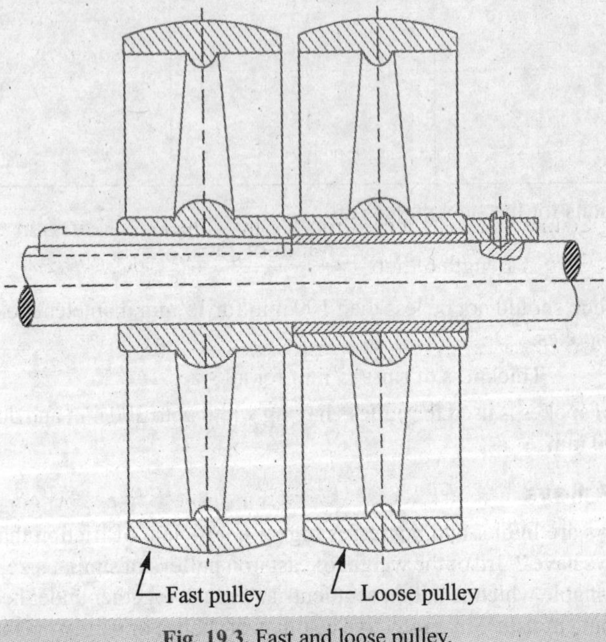

Fig. 19.3. Fast and loose pulley.

The loose pulley is provided with a cast iron or gun-metal bush with a collar at one end to prevent axial movement.

The rim of the fast pulley is made larger than the loose pulley so that the belt may run slackly on the loose pulley. The loose pulley usually have longer hub in order to reduce wear and friction and it requires proper lubrication.

19.8 Design of Cast Iron Pulleys

The following procedure may be adopted for the design of cast iron pulleys.

1. ***Dimensions of pulley***

(*i*) The diameter of the pulley (*D*) may be obtained either from velocity ratio consideration or centrifugal stress consideration. We know that the centrifugal stress induced in the rim of the pulley,

$$\sigma_t = \rho . v^2$$

where ρ = Density of the rim material

= 7200 kg/m^3 for cast iron

v = Velocity of the rim = $\pi DN / 60$, D being the diameter of pulley and N is speed of the pulley.

The following are the diameter of pulleys in mm for flat and *V*-belts.

20, 22, 25, 28, 32, 36, 40, 45, 50, 56, 63, 71, 80, 90, 100, 112, 125, 140, 160, 180, 200, 224, 250, 280, 315, 355, 400, 450, 500, 560, 630, 710, 800, 900, 1000, 1120, 1250, 1400, 1600, 1800, 2000, 2240, 2500, 2800, 3150, 3550, 4000, 5000, 5400.

The first six sizes (20 to 36 mm) are used for *V*-belts only.

(*ii*) If the width of the belt is known, then width of the pulley or face of the pulley (*B*) is taken 25% greater than the width of belt.

$\therefore \quad B = 1.25\, b$; where b = Width of belt.

According to Indian Standards, IS : 2122 (Part I) – 1973 (Reaffirmed 1990), the width of pulley is fixed as given in the following table :

Table 19.2. Standard width of pulley.

Belt width in mm	*Width of pulley to be greater than belt width by (mm)*
upto 125	13
125-250	25
250-375	38
475-500	50

The following are the width of flat cast iron and mild steel pulleys in mm :

16, 20, 25, 32, 40, 50, 63, 71, 80, 90, 100, 112, 125, 140, 160, 180, 200, 224, 250, 315, 355, 400, 450, 560, 630.

(*iii*) The thickness of the pulley rim (*t*) varies from $\dfrac{D}{300} + 2$ mm to $\dfrac{D}{200} + 3$ mm for single belt and $\dfrac{D}{200} + 6$ mm for double belt. The diameter of the pulley (*D*) is in mm.

2. ***Dimensions of arms***

(*i*) The number of arms may be taken as 4 for pulley diameter from 200 mm to 600 mm and 6 for diameter from 600 mm to 1500 mm.

Note : The pulleys less than 200 mm diameter are made with solid disc instead of arms. The thickness of the solid web is taken equal to the thickness of rim measured at the centre of the pulley face.

(ii) The cross-section of the arms is usually elliptical with major axis (a_1) equal to twice the minor axis (b_1). The cross-section of the arm is obtained by considering the arm as cantilever *i.e.* fixed at the hub end and carrying a concentrated load at the rim end. The length of the cantilever is taken equal to the radius of the pulley. It is further assumed that at any given time, the power is transmitted from the hub to the rim or ***vice versa***, through only half the total number of arms.

Let T = Torque transmitted,

R = Radius of pulley, and

n = Number of arms,

∴ Tangential load per arm,

$$W_T = \frac{T}{R \times n / 2} = \frac{2\,T}{R \cdot n}$$

Maximum bending moment on the arm at the hub end,

$$M = \frac{2\,T}{R \times n} \times R = \frac{2\,T}{n}$$

and section modulus,

$$Z = \frac{\pi}{32} \times b_1\,(a_1)^2$$

Now using the relation,

σ_b or σ_t = M / Z, the cross-section of the arms is obtained.

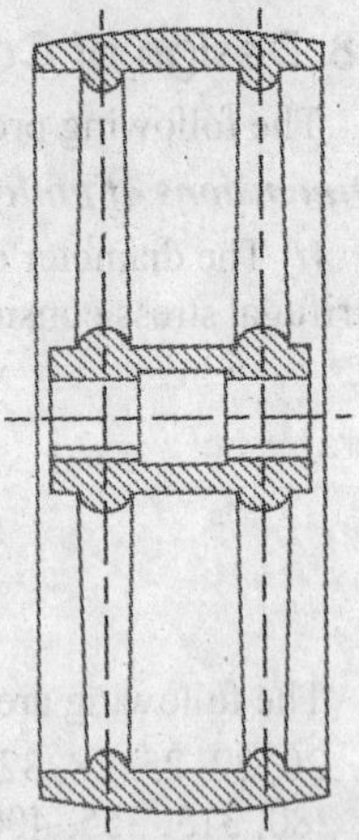

Fig. 19.4. Cast iron pulley with two rows of arms.

(iii) The arms are tapered from hub to rim. The taper is usually 1/48 to 1/32.

(iv) When the width of the pulley exceeds the diameter of the pulley, then two rows of arms are provided, as shown in Fig. 19.4. This is done to avoid heavy arms in one row.

3. ***Dimensions of hub***

(i) The diameter of the hub (d_1) in terms of shaft diameter (d) may be fixed by the following relation :

$$d_1 = 1.5\,d + 25 \text{ mm}$$

The diameter of the hub should not be greater than 2 d.

(ii) The length of the hub,

$$L = \frac{\pi}{2} \times d$$

The minimum length of the hub is $\frac{2}{3}$ B but it should not be more than width of the pulley (B).

Example 19.1. *A cast iron pulley transmits 20 kW at 300 r.p.m. The diameter of pulley is 550 mm and has four straight arms of elliptical cross-section in which the major axis is twice the minor axis. Find the dimensions of the arm if the allowable bending stress is 15 MPa. Mention the plane in which the major axis of the arm should lie.*

Solution. Given : P = 20 kW = 20 × 10³ W ; N = 300 r.p.m. ; *d = 550 mm ; n = 4 ; σ_b = 15 MPa = 15 N/mm²

Let b_1 = Minor axis, and

a_1 = Major axis = $2b_1$ ...(Given)

We know that the torque transmitted by the pulley,

$$T = \frac{P \times 60}{2\,\pi\,N} = \frac{20 \times 10^3 \times 60}{2\,\pi \times 300} = 636 \text{ N-m}$$

* Superfluous data.

∴ Maximum bending moment per arm at the hub end,

$$M = \frac{2T}{n} = \frac{2 \times 636}{4}$$

$$= 318 \text{ N-m} = 318 \times 10^3 \text{ N-mm}$$

and section modulus,

$$Z = \frac{\pi}{32} \times b_1 (a_1)^2 = \frac{\pi}{32} \times b_1 (2b_1)^2$$

$$= \frac{\pi (b_1)^3}{8}$$

Cast iron pulley.

We know that the bending stress (σ_b),

$$15 = \frac{M}{Z} = \frac{318 \times 10^3 \times 8}{\pi (b_1)^3} = \frac{810 \times 10^3}{(b_1)^3}$$

∴ $(b_1)^3 = 810 \times 10^3 / 15 = 54 \times 10^3$ or $b_1 = 37.8$ mm **Ans.**

and $a_1 = 2\,b_1 = 2 \times 37.8 = 75.6$ mm **Ans.**

The major axis will be in the plane of rotation which is also the plane of bending.

Example 19.2. *An overhung pulley transmits 35 kW at 240 r.p.m. The belt drive is vertical and the angle of wrap may be taken as 180°. The distance of the pulley centre line from the nearest bearing is 350 mm. μ = 0.25. Determine :*

1. *Diameter of the pulley ;*
2. *Width of the belt assuming thickness of 10 mm ;*
3. *Diameter of the shaft ;*
4. *Dimensions of the key for securing the pulley on to the shaft ; and*
5. *Size of the arms six in number.*

The section of the arm may be taken as elliptical, the major axis being twice the minor axis.

The following stresses may be taken for design purposes :

Shaft }	*Tension and compression*	— *80 MPa*
Key }	*Shear*	— *50 MPa*
Belt : Tension		— *2.5 MPa*
Pulley rim : Tension		— *4.5 MPa*
Pulley arms : Tension		— *15 MPa*

Steel pulley.

Solution. Given : $P = 35$ kW $= 35 \times 10^3$ W ; $N = 240$ r.p.m. ; $\theta = 180° = \pi$ rad ; $L = 350$ mm $= 0.35$ m ; $\mu = 0.25$; $t = 10$ mm ; $n = 6$; $\sigma_{ts} = \sigma_{tk} = 80$ MPa $= 80$ N/mm² ; $\tau_s = \tau_k = 50$ MPa $= 50$ N/mm² ; $\sigma = 2.5$ MPa $= 2.5$ N/mm² ; $\sigma_t = 4.5$ MPa $= 4.5$ N/mm² ; $\sigma_b = 15$ MPa $= 15$ N/mm²

1. *Diameter of the pulley*

Let D = Diameter of the pulley,

σ_t = Centrifugal stress or tensile stress in the pulley rim
$= 4.5 \text{ MPa} = 4.5 \times 10^6 \text{ N/m}^2$ ...(Given)

ρ = Density of the pulley material (*i.e.* cast iron) which may be taken as 7200 kg/m³.

We know that centrifugal stress (σ_t),

$$4.5 \times 10^6 = \rho.v^2 = 7200 \times v^2$$

$$\therefore \quad v^2 = 4.5 \times 10^6 / 7200 = 625 \quad \text{or} \quad v = 25 \text{ m/s}$$

and velocity of the pulley (v),

$$25 = \frac{\pi\, D\, .\, N}{60} = \frac{\pi\, D \times 240}{60} = 12.568\, D$$

$$\therefore \quad D = 25 / 12.568 = 2 \text{ m } \textbf{Ans.}$$

2. ***Width of the belt***

Let b = Width of the belt in mm,

T_1 = Tension in the tight side of the belt, and

T_2 = Tension in the slack side of the belt.

We know that the power transmitted (P),

$$35 \times 10^3 = (T_1 - T_2)\, v = (T_1 - T_2)\, 25$$

$$\therefore \quad T_1 - T_2 = 35 \times 10^3 / 25 = 1400 \text{ N} \quad \text{...}(i)$$

We also know that

$$2.3 \log \left(\frac{T_1}{T_2}\right) = \mu.\theta = 0.25 \times \pi = 0.7855$$

$$\therefore \quad \log \left(\frac{T_1}{T_2}\right) = \frac{0.7855}{2.3} = 0.3415 \quad \text{or} \quad \frac{T_1}{T_2} = 2.195 \quad \text{...}(ii)$$

... (Taking antilog of 0.3415)

From equations (*i*) and (*ii*), we find that

$$T_1 = 2572 \text{ N} \; ; \text{ and } T_2 = 1172 \text{ N}$$

Since the velocity of the belt (or pulley) is more than 10 m/s, therefore centrifugal tension must be taken into consideration. Assuming a leather belt for which the density may be taken as 1000 kg/m³.

We know that cross-sectional area of the belt,

$$= b \times t = b \times 10 = 10\, b \text{ mm}^2 = \frac{10\, b}{10^6} \text{ m}^2$$

Mass of the belt per metre length,

$$m = \text{Area} \times \text{length} \times \text{density}$$

$$= \frac{10\, b}{10^6} \times 1 \times 1000 = 0.01\, b \text{ kg / m}$$

We know that centrifugal tension,

$$T_C = m.v^2 = 0.01\, b\, (25)^2 = 6.25\, b \text{ N}$$

and maximum tension in the belt,

$$T = \sigma.b.t = 2.5 \times b \times 10 = 25\, b \text{ N}$$

We know that tension in the tight side of the belt (T_1),

$$2572 = T - T_C = 25\, b - 6.25\, b = 18.75\, b$$

$$\therefore \quad b = 2572 / 18.75 = 137 \text{ mm}$$

The standard width of the belt (b) is 140 mm. **Ans.**

3. ***Diameter of the shaft***

Let d = Diameter of the shaft.

We know that the torque transmitted by the shaft,

$$T = \frac{P \times 60}{2\, \pi\, N} = \frac{35 \times 10^3 \times 60}{2\, \pi \times 240} = 1393 \text{ N-m} = 1393 \times 10^3 \text{ N-mn}$$

and bending moment on the shaft due to the tensions of the belt,

$$M = (T_1 + T_2 + 2T_C)\,L = (2572 + 1172 + 2 \times 6.25 \times 140) \times 0.35 \text{ N-m}$$

$$= 1923 \text{ N-m} \qquad \ldots (\because T_C = 6.25\,b)$$

We know that equivalent twisting moment,

$$T_e = \sqrt{T^2 + M^2} = \sqrt{(1393)^2 + (1923)^2} = 2375 \text{ N-m}$$

$$= 2375 \times 10^3 \text{ N-mm}$$

We also know that equivalent twisting momnt (T_e),

$$2375 \times 10^3 = \frac{\pi}{16} \times \tau_s \times d^3 = \frac{\pi}{16} \times 50 \times d^3 = 9.82\,d^3$$

$$\therefore \quad d^3 = 2375 \times 10^3 / 9.82 = 242 \times 10^3 \quad \text{or} \quad d = 62.3 \text{ say } 65 \text{ mm } \textbf{Ans.}$$

4. ***Dimensions of the key***

The standard dimensions of the key for 65 mm diameter shaft are :

Width of key, w = 20 mm **Ans.**

Thickness of key = 12 mm **Ans.**

Let l = Length of the key.

Considering shearing of the key. We know that the torque transmitted (T),

$$1393 \times 10^3 = l \times w \times \tau_k \times \frac{d}{2} = l \times 20 \times 50 \times \frac{65}{2} = 32\,500\,l$$

$$\therefore \quad l = 1393 \times 10^3 / 32\,500 = 42.8 \text{ mm}$$

The length of key should be atleast equal to hub length. The length of hub is taken as $\frac{\pi}{2} \times d$.

$$\therefore \quad \text{Length of key} = \frac{\pi}{2} \times 65 = 102 \text{ mm } \textbf{Ans.}$$

5. ***Size of arms***

Let b_1 = Minor axis, and

a = Major axis = $2b_1$ ... (Given)

We know that the maximum bending moment per arm at the hub end,

$$M = \frac{2\,T}{n} = \frac{2 \times 1393}{6} = 464.33 \text{ N-m} = 464\,330 \text{ N-mm}$$

and section modulus,

$$Z = \frac{\pi}{32} \times b_1\,(a_1)^2 = \frac{\pi}{32} \times b_1\,(2b_1)^2 = 0.393\,(b_1)^3$$

We know that bending stress (σ_b),

$$15 = \frac{M}{Z} = \frac{464\,330}{0.393 \times (b_1)^3} = \frac{1.18 \times 10^6}{(b_1)^3}$$

$$\therefore \quad (b_1)^3 = 1.18 \times 10^6 / 15 = 78.7 \times 10^3 \quad \text{or} \quad b_1 = 42.8 \text{ say } 45 \text{ mm } \textbf{Ans.}$$

and $a_1 = 2b_1 = 2 \times 45 = 90$ mm **Ans.**

Example 19.3. *A pulley of 0.9 m diameter revolving at 200 r.p.m. is to transmit 7.5 kW. Find the width of a leather belt if the maximum tension is not to exceed 145 N in 10 mm width. The tension in the tight side is twice that in the slack side. Determine the diameter of the shaft and the dimensions of the various parts of the pulley, assuming it to have six arms. Maximum shear stress is not to exceed 63 MPa.*

Solution. Given : D = 0.9 m ; N = 200 r.p.m. ; P = 7.5 kW = 7500 W ; T = 145 N in 10 mm width ; $T_1 = 2T_2$; $n = 6$; τ = 63 MPa = 63 N/mm²

We know that velocity of the pulley or belt,

$$v = \frac{\pi D.N}{60} = \frac{\pi \times 0.9 \times 200}{60} = 9.426 \text{ m/s}$$

Let T_1 = Tension in the tight of the belt, and

T_2 = Tension in the slack side of the belt.

We know that the power transmitted (*P*),

$$7500 = (T_1 - T_2)\, v$$
$$= (T_1 - T_2)\, 9.426$$
$$T_1 - T_2 = 7500 / 9.426 = 796 \text{ N}$$

or $2T_2 - T_2 = 796$ N ... ($\because T_1 = 2T_2$)

$\therefore$ $T_2 = 796$ N ;

and $T_1 = 2T_2 = 2 \times 796 = 1592$ N

Note : Since the velocity of belt is less than 10 m/s, therefore the centrifugal tension need not to be considered.

Width of belt

Let b = Width of belt.

Since the maximum tension is 145 N in 10 mm width or 14.5 N/mm width, therefore width of belt,

$$b = T_1 / 14.5 = 1592 / 14.5 = 109.8 \text{ mm}$$

The standard width of the belt (*b*) is 112 mm. **Ans.**

Diameter of the shaft

Let d = Diameter of the shaft,

We know that the torque transmitted by the shaft,

$$T = \frac{P \times 60}{2\,\pi\, N} = \frac{7500 \times 60}{2\,\pi \times 200} = 358 \text{ N-m} = 358\,000 \text{ N-mm}$$

We also know the torque transmitted by the shaft (*T*),

$$358\,000 = \frac{\pi}{16} \times \tau \times d^3 = \frac{\pi}{16} \times 63 \times d^3 = 12.4\, d^3$$

$\therefore$ $d^3 = 358\,000 / 12.4 = 28\,871$ or $d = 30.67$ say 35 mm **Ans.**

Dimensions of the various parts of the pulley

1. ***Width and thickness of pulley***

Since the width of the belt is 112 mm, therefore width of the pulley,

$$B = 112 + 13 = 125 \text{ mm } \textbf{Ans.}$$

and thickness of the pulley rim for single belt,

$$t = \frac{D}{300} + 2 \text{ mm} = \frac{900}{300} + 2 = 5 \text{ mm } \textbf{Ans.}$$

2. *Dimensions of arm*

Assuming the cross-section of the arms as elliptical with major axis equal to twice the minor axis.

Let b_1 = Minor axis, and

a_1 = Major axis = $2b_1$

We know that maximum bending moment on the arm at the hub end,

$$M = \frac{2T}{n} = \frac{2 \times 358\,000}{6} = 119\,333 \text{ N-mm}$$

and section modulus, $Z = \frac{\pi}{32} \times b_1 (a_1)^2 = \frac{\pi}{32} \times b_1 (2b_1)^2 = 0.393\,(b_1)^3$

Assume the arms of cast iron for which the tensile stress may be taken as 15 N/mm². We know that the tensile stress (σ_t),

$$15 = \frac{M}{Z} = \frac{119\,333}{0.393 \times (b_1)^3} = \frac{303\,646}{(b_1)^3}$$

$\therefore$ $(b_1)^3 = 303\,646 / 15 = 20\,243$ or $b_1 = 27.3$ say 30 mm **Ans.**

and $a_1 = 2\,b_1 = 2 \times 30 = 60$ mm **Ans.**

Dimensions of the hub

Diameter of the hub $= 2\,d = 2 \times 35 = 70$ mm **Ans.**

$$\text{Length of the hub} = \frac{\pi}{2} \times d = \frac{\pi}{2} \times 35 = 55 \text{ mm}$$

Since the length of the hub should not be less than $\frac{2}{3}$ B, therefore the length of hub

$$= \frac{2}{3} \times B = \frac{2}{3} \times 125 = 83.3 \text{ say } 85 \text{ mm } \textbf{Ans.}$$

EXERCISES

1. Design the elliptical cross-section of a belt pulley arm near the hub for the following specifications: The mean pulley diameter is 300 mm and the number of pulley arms are 4. The elliptical section has major axis twice the minor axis length. The tight and slack sides tension in the belt are 600 N and 200 N respectively. Assume half number of arms transmit torque at any time and the load factor of 1.75 to account for dynamic effects on the pulley while transmitting torque. The permissible tensile stress for cast iron pulley material is 15 MPa. The pulley hub diameter is 60 mm.

   [Ans. a_1 = 40 mm, b_1 = 20 mm]

2. Design a cast iron driven pulley to transmit 20 kW at 300 r.p.m. The diameter of the pulley is 500 mm and the angle of lap is 180°. The pulley has four arms of elliptical cross-section with major axis twice the minor axis. The coefficient of friction between the belt and the pulley surface is 0.3. The allowable tension per metre width of the belt is 2.5 N. The following allowable stresses may be taken :

   Shear stress for the shaft material = 50 MPa, and

   Bending stress for the pulley arms = 15 MPa.

3. An overhung cast iron pulley transmits 7.5 kW at 400 r.p.m. The belt drive is vertical and the angle of wrap may be taken as 180°. Find :

   (*a*) Diameter of the pulley. The density of cast iron is 7200 kg/m³.

   (b) Width of the belt, if the coefficient of friction between the belt and the pulley is 0.25.

   (*c*) Diameter of the shaft, if the distance of the pulley centre line from the nearest bearing is 300 mm.

(*d*) Dimensions of the key for securing the pulley on to the shaft.

(*e*) Size of the arms six in number.

The section of the arms may be taken as elliptical, the major axis being twice the minor axis. The following stresses may be taken for design purposes :

Shaft and key :	Tension – 80 MPa
	Shear – 50 MPa
Belt :	Tension – 2.5 MPa
Pulley rim :	Tension – 4.5 MPa
Pulley arms :	Tension – 15 MPa

QUESTIONS

1. Discuss the different types of pulleys used in belt drives.
2. Why the face of a pulley is crowned?
3. When a split pulley is used and how it is tightened on a shaft?
4. Explain the 'fast and loose pulley' with the help of a neat sketch.
5. Discuss the procedure used in designing a cast iron pulley.

OBJECTIVE TYPE QUESTIONS

1. The crowning on a 300 mm width of pulley face should be

   (*a*) 9 mm (*b*) 12 mm

   (*c*) 15 mm (*d*) 18 mm

2. The steel pulleys are .... in weight than cast iron pulleys of the same capacity.

   (*a*) heavier (*b*) lighter

3. For a steel pulley of 500 mm, the recommended number of spokes are

   (*a*) 2 (*b*) 4

   (*c*) 6 (*d*) 8

4. The thickness of rim for all sizes of steel pulleys should be

   (*a*) 5 mm (*b*) 10 mm

   (*c*) 15 mm (*d*) 20 mm

5. The width of the pulley should be

   (*a*) equal to the width of belt (*b*) less than the width of belt

   (*c*) greater than the width of belt

ANSWERS

1. (*a*) **2.** (*b*) **3.** (*c*) **4.** (*a*) **5.** (*c*)

CHAPTER

20

V-Belt and Rope Drives

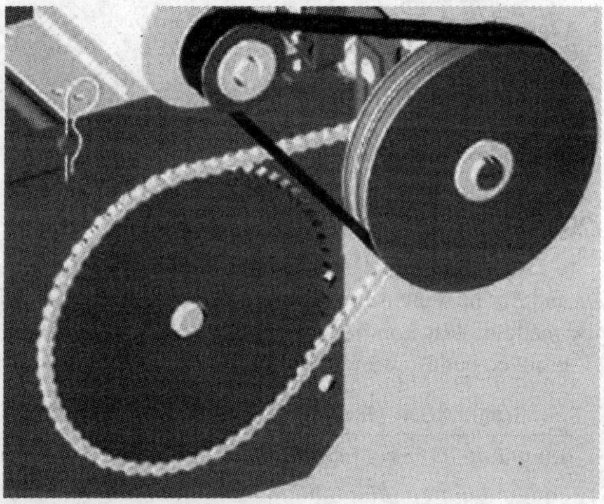

20.1 Introduction

We have already discussed that a *V*-belt is mostly used in factories and workshops where a great amount of power is to be transmitted from one pulley to another when the two pulleys are very near to each other.

The *V*-belts are made of fabric and cords moulded in rubber and covered with fabric and rubber as shown in Fig. 20.1 (*a*). These belts are moulded to a trapezoidal shape and are made endless. These are particularly suitable for short drives. The included angle for the *V*-belt is usually from 30° to 40°. The power is transmitted by the *wedging

* The wedging action of the *V*-belt in the groove of the pulley results in higher forces of friction. A little consideration will show that the wedging action and the transmitted torque will be more if the groove angle of the pulley is small. But a small groove angle will require more force to pull the belt out of the groove which will result in loss of power and excessive belt wear due to friction and heat. Hence the selected groove angle is a compromise between the two. Usually the groove angles of 32° to 38° are used.

action between the belt and the *V*-groove in the pulley or sheave. A clearance must be provided at the bottom of the groove as shown in Fig. 20.1 (*b*), in order to prevent touching of the bottom as it becomes narrower from wear. The *V*-belt drive may be inclined at any angle with tight side either at top or bottom. In order to increase the power output, several *V*-belts may be operated side by side. It may be noted that in multiple *V*-belt drive, all the belts should stretch at the same rate so that the load is equally divided between them. When one of the set of belts break, the entire set should be replaced at the same time. If only one belt is replaced, the new unworn and unstretched belt will be more tightly stretched and will move with different velocity.

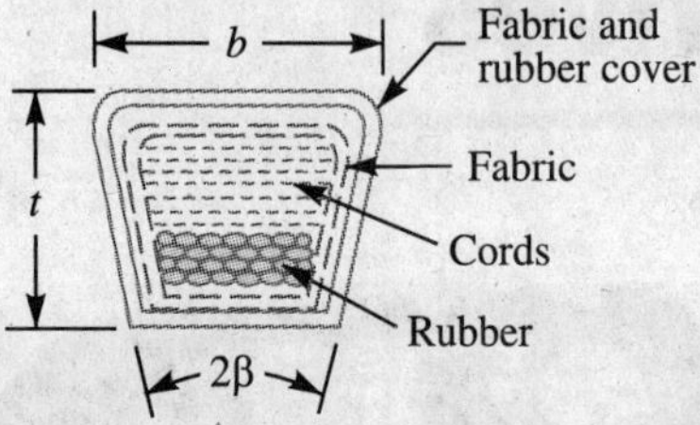

(*a*) Cross-section of a V-belt.

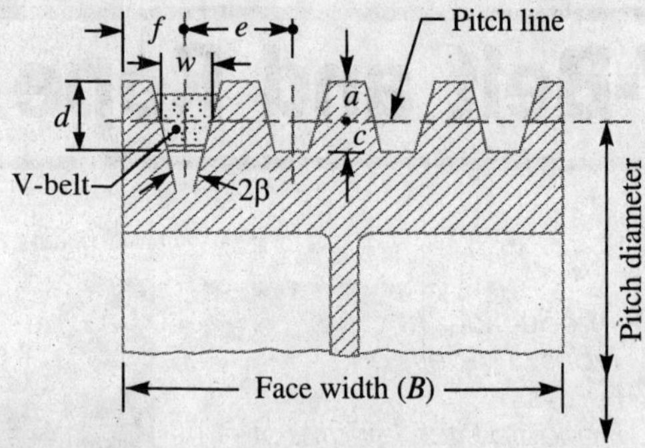

(*b*) Cross-section of a V-grooved pulley.

Fig. 20.1. V-Belt and V-grooved pulley.

20.2 Types of V-belts and Pulleys

According to Indian Standards (IS: 2494 – 1974), the *V*-belts are made in five types *i.e. A*, *B*, *C*, *D* and *E*. The dimensions for standard *V*-belts are shown in Table 20.1. The pulleys for *V*-belts may be made of cast iron or pressed steel in order to reduce weight. The dimensions for the standard *V*-grooved pulley according to IS: 2494 – 1974, are shown in Table 20.2.

Table 20.1. Dimensions of standard V-belts according to IS: 2494 - 1974.

Type of belt	*Power ranges in kW*	*Minimum pitch diameter of pulley (D) mm*	*Top width (b) mm*	*Thickness (t) mm*	*Weight per metre length in newton*
A	0.7 – 3.5	75	13	8	1.06
B	2 – 15	125	17	11	1.89
C	7.5 – 75	200	22	14	3.43
D	20 – 150	355	32	19	5.96
E	30 – 350	500	38	23	–

Table 20.2. Dimensions of standard V-grooved pulleys according to IS : 2494–1974. (All dimensions in mm)

Type of belt	*w*	*d*	*a*	*c*	*f*	*e*	*No. of sheave grooves (n)*	*Groove angle (2β) in degrees*
A	11	12	3.3	8.7	10	15	6	32, 34, 38
B	14	15	4.2	10.8	12.5	19	9	32, 34, 38
C	19	20	5.7	14.3	17	25.5	14	34, 36, 38
D	27	28	8.1	19.9	24	37	14	34, 36, 38
E	32	33	9.6	23.4	29	44.5	20	–

Note : Face width $(B) = (n - 1)\,e + 2f$

20.3 Standard Pitch Lengths of V-belts

According to IS: 2494-1974, the V-belts are designated by its type and nominal inside length. For example, a *V*-belt of type *A* and inside length 914 mm is designated as *A* 914–IS: 2494. The standard inside lengths of *V*-belts in mm are as follows :

610, 660, 711, 787, 813, 889, 914, 965, 991, 1016, 1067, 1092, 1168, 1219, 1295, 1372, 1397, 1422, 1473, 1524, 1600, 1626, 1651, 1727, 1778, 1905, 1981, 2032, 2057, 2159, 2286, 2438, 2464, 2540, 2667, 2845, 3048, 3150, 3251, 3404, 3658, 4013, 4115, 4394, 4572, 4953, 5334, 6045, 6807, 7569, 8331, 9093, 9885, 10 617, 12 141, 13 665, 15 189, 16 713

Material handler.

According to IS: 2494-1974, the pitch length is defined as the circumferential length of the belt at the pitch width (*i.e.* the width at the neutral axis) of the belt. The value of the pitch width remains constant for each type of belt irrespective of the groove angle.

The pitch lengths are obtained by adding to inside length: 36 mm for type *A*, 43 mm for type *B*, 56 mm for type *C*, 79 mm for type *D* and 92 mm for type *E*. The following table shows the standard pitch lengths for the various types of belt.

Table 20.3. Standard pitch lengths of V-belts according to IS: 2494-1974.

Type of belt	*Standard pitch lengths of V-belts in mm*
A	645, 696, 747, 823, 848, 925, 950, 1001, 1026, 1051, 1102 1128, 1204, 1255, 1331, 1433, 1458, 1509, 1560, 1636, 1661, 1687, 1763, 1814, 1941, 2017, 2068, 2093, 2195, 2322, 2474, 2703, 2880, 3084, 3287, 3693.
B	932, 1008, 1059, 1110, 1212, 1262, 1339, 1415, 1440, 1466, 1567, 1694, 1770, 1821, 1948, 2024, 2101, 2202, 2329, 2507, 2583, 2710, 2888, 3091, 3294, 3701, 4056, 4158, 4437, 4615, 4996, 5377.
C	1275, 1351, 1453, 1580, 1681, 1783, 1834, 1961, 2088, 2113, 2215, 2342, 2494, 2723, 2901, 3104, 3205, 3307, 3459, 3713, 4069, 4171, 4450, 4628, 5009, 5390, 6101, 6863, 7625, 8387, 9149.
D	3127, 3330, 3736, 4092, 4194, 4473, 4651, 5032, 5413, 6124, 6886, 7648, 8410, 9172, 9934, 10 696, 12 220, 13 744, 15 268, 16 792.
E	5426, 6137, 6899, 7661, 8423, 9185, 9947, 10 709, 12 233, 13 757, 15 283, 16 805.

Note: The *V*-belts are also manufactured in non-standard pitch lengths (*i.e.* in oversize and undersize). The standard pitch length belt is designated by grade number 50. The oversize belts are designated by a grade number more than 50, while the undersize belts are designated by a grade number less than 50. It may be noted that one unit of a grade number represents 2.5 mm in length from nominal pitch length. For example, a *V*-belt marked *A* – 914 – 50 denotes a standard belt of inside length 914 mm and a pitch length 950 mm. A belt marked *A* – 914 – 52 denotes an oversize belt by an amount of (52 – 50) = 2 units of grade number. Since one unit of grade number represents 2.5 mm, therefore the pitch length of this belt will be 950 + 2 × 2.5 = 955 mm. Similarly, a belt marked *A* – 914 – 48 denotes an undersize belt, whose pitch length will be 950 – 2 × 2.5 = 945 mm.

20.4 Advantages and Disadvantages of V-belt Drive over Flat Belt Drive

Following are the advantages and disadvantages of the *V*-belt drive over flat belt drive :

Advantages

1. The *V*-belt drive gives compactness due to the small distance between centres of pulleys.
2. The drive is positive, because the slip between the belt and the pulley groove is negligible.
3. Since the *V*-belts are made endless and there is no joint trouble, therefore the drive is smooth.
4. It provides longer life, 3 to 5 years.
5. It can be easily installed and removed.
6. The operation of the belt and pulley is quiet.
7. The belts have the ability to cushion the shock when machines are started.
8. The high velocity ratio (maximum 10) may be obtained.
9. The wedging action of the belt in the groove gives high value of limiting *ratio of tensions. Therefore the power transmitted by *V*-belts is more than flat belts for the same coefficient of friction, arc of contact and allowable tension in the belts.
10. The *V*-belt may be operated in either direction, with tight side of the belt at the top or bottom. The centre line may be horizontal, vertical or inclined.

Disadvantages

1. The *V*-belt drive can not be used with large centre distances, because of larger weight per unit length.
2. The *V*-belts are not so durable as flat belts.
3. The construction of pulleys for V-belts is more complicated than pulleys of flat belts.
4. Since the *V*-belts are subjected to certain amount of creep, therefore these are not suitable for constant speed applications such as synchronous machines and timing devices.
5. The belt life is greatly influenced with temperature changes, improper belt tension and mismatching of belt lengths.
6. The centrifugal tension prevents the use of *V*-belts at speeds below 5 m / s and above 50 m / s.

20.5 Ratio of Driving Tensions for V-belt

A *V*-belt with a grooved pulley is shown in Fig. 20.2.

Let R_1 = Normal reactions between belts and sides of the groove.

R = Total reaction in the plane of the groove.

μ = Coefficient of friction between the belt and sides of the groove.

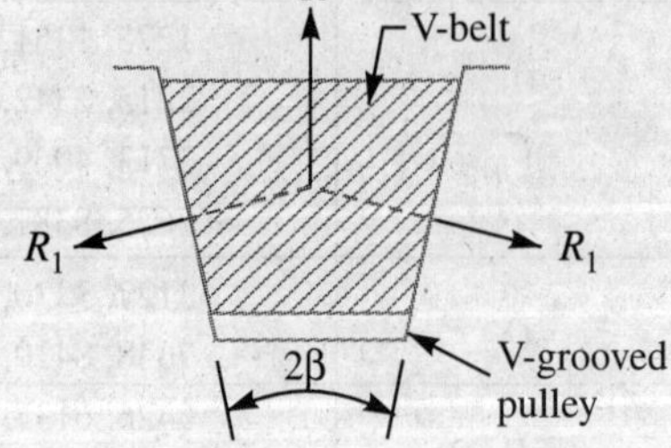

Fig. 20.2. V-belt with pulley.

Resolving the reactions vertically to the groove, we have

$$R = R_1 \sin \beta + R_1 \sin \beta = 2R_1 \sin \beta$$

* The ratio of tensions in *V*-belt drive is cosec β times the flat belt drive.

or $$R_1 = \frac{R}{2 \sin \beta}$$

We know that the frictional force

$$= 2\,\mu.R_1 = 2\,\mu \times \frac{R}{2 \sin \beta} = \frac{\mu.R}{\sin \beta} = \mu.R.\text{cosec}\,\beta$$

Consider a small portion of the belt, as in Art. 18.19, subtending an angle $\delta\theta$ at the centre, the tension on one side will be T and on the other side $(T + \delta T)$. Now proceeding in the same way as in Art. 18.19, we get the frictional resistance equal to $\mu\,R.\ \text{cosec}\ \beta$ against $\mu.R$. Thus the relation between T_1 and T_2 for the V-belt drive will be

$$2.3 \log (T_1 / T_2) = \mu.\theta \text{ cosec } \beta$$

20.6 V-flat Drives

In many cases, particularly, when a flat belt is replaced by V-belt, it is economical to use flat-faced pulley, instead of large grooved pulley, as shown in Fig. 20.3. The cost of cutting the grooves is thereby eliminated. Such a drive is known as ***V-flat drive.***

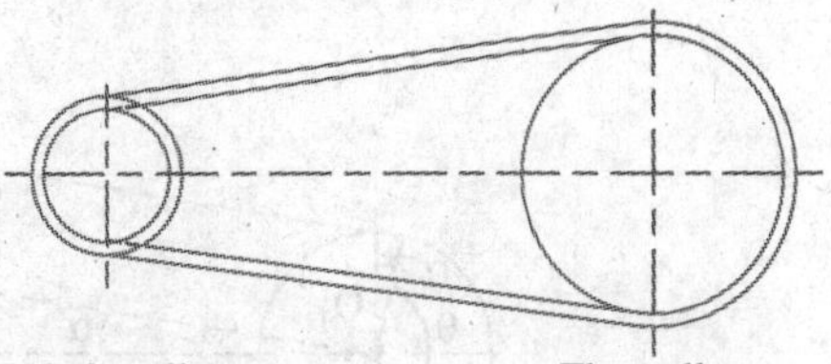

Fig. 20.3. V-flat drive.

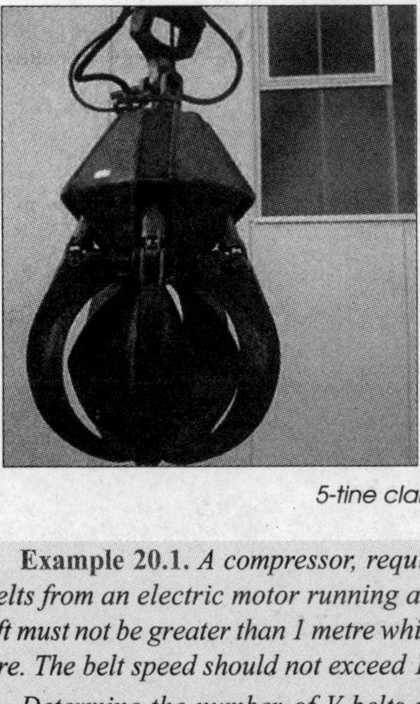

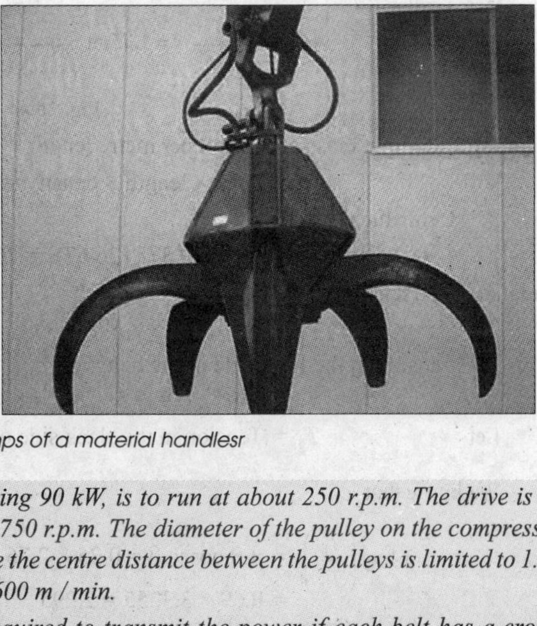

5-tine clamps of a material handlesr

Example 20.1. *A compressor, requiring 90 kW, is to run at about 250 r.p.m. The drive is by V-belts from an electric motor running at 750 r.p.m. The diameter of the pulley on the compressor shaft must not be greater than 1 metre while the centre distance between the pulleys is limited to 1.75 metre. The belt speed should not exceed 1600 m / min.*

Determine the number of V-belts required to transmit the power if each belt has a cross-sectional area of 375 mm², density 1000 kg / m³ and an allowable tensile stress of 2.5 MPa. The groove angle of the pulleys is 35°. The coefficient of friction between the belt and the pulley is 0.25. Calculate also the length required of each belt.

Solution. Given : $P = 90$ kW $= 90 \times 10^3$ W ; $N_2 = 250$ r.p.m. ; $N_1 = 750$ r.p.m. ; $d_2 = 1$ m ; $x = 1.75$ m ; $v = 1600$ m/min $= 26.67$ m/s ; $a = 375$ mm² $= 375 \times 10^{-6}$ m² ; $\rho = 1000$ kg / m³ ; $\sigma = 2.5$ MPa $= 2.5$ N/mm² ; $2\beta = 35°$ or $\beta = 17.5°$; $\mu = 0.25$

First of all, let us find the diameter of pulley on the motor shaft (d_1). We know that

$$\frac{N_1}{N_2} = \frac{d_2}{d_1} \quad \text{or} \quad d_1 = \frac{d_2 N_2}{N_1} = \frac{1 \times 250}{750} = 0.33 \text{ m}$$

For an open belt drive, as shown in Fig. 20.4,

$$\sin \alpha = \frac{O_2M}{O_1O_2} = \frac{r_2 - r_1}{x} = \frac{d_2 - d_1}{2x} = \frac{1 - 0.33}{2 \times 1.75} = 0.1914$$

$$\therefore \quad \alpha = 11.04°$$

and angle of lap on the smaller pulley (*i.e.* pulley on the motor shaft),

$$\theta = 180° - 2\alpha = 180 - 2 \times 11.04 = 157.92°$$

$$= 157.92 \times \frac{\pi}{180} = 2.76 \text{ rad}$$

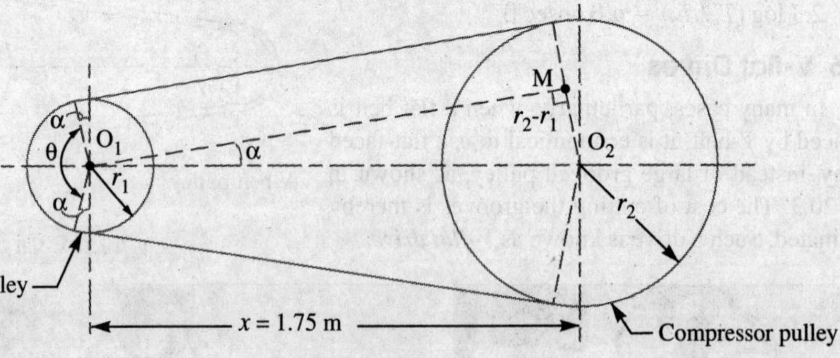

Fig. 20.4

We know that mass of the belt per metre length,

$$m = \text{Area} \times \text{length} \times \text{density} = 375 \times 10^{-6} \times 1 \times 1000 = 0.375 \text{ kg / m}$$

∴ Centrifugal tension,

$$T_C = m.v^2 = 0.375\,(26.67)^2 = 267 \text{ N}$$

and maximum tension in the belt,

$$T = \sigma \times a = 2.5 \times 375 = 937.5 \text{ N}$$

∴ Tension in the tight side of the belt,

$$T_1 = T - T_C = 937.5 - 267 = 670.5 \text{ N}$$

Let T_2 = Tension in the slack side of the belt.

We know that

$$2.3 \log \left(\frac{T_1}{T_2}\right) = \mu.\theta \operatorname{cosec} \beta = 0.25 \times 2.76 \times \operatorname{cosec} 17.5°$$

$$= 0.69 \times 3.3255 = 2.295$$

$$\therefore \quad \log \left(\frac{T_1}{T_2}\right) = \frac{2.295}{2.3} = 0.9976 \quad \text{or} \quad \frac{T_1}{T_2} = 9.95 \quad \text{...(Taking antilog of 0.9976)}$$

and $T_2 = T_1 / 9.95 = 670.5 / 9.95 = 67.4$ N

Number of V-belts

We know that the power transmitted per belt,

$$= (T_1 - T_2)\, v = (670.5 - 67.4)\, 26.67 = 16\,085 \text{ W} = 16.085 \text{ kW}$$

∴ Number of *V*-belts

$$= \frac{\text{Total power transmitted}}{\text{Power transmitted per belt}} = \frac{90}{16.085} = 5.6 \text{ say } 6$$ **Ans.**

Length of each belt

We know that radius of pulley on motor shaft,

$$r_1 = d_1 / 2 = 0.33 / 2 = 0.165 \text{ m}$$

and radius of pulley on compressor shaft,

$$r_2 = d_2 / 2 = 1 / 2 = 0.5 \text{ m}$$

We know that length of each belt,

$$L = \pi (r_2 + r_1) + 2x + \frac{(r_2 - r_1)^2}{x}$$

$$= \pi (0.5 + 0.165) + 2 \times 1.75 + \frac{(0.5 - 0.165)^2}{1.75}$$

$$= 2.09 + 3.5 + 0.064 = 5.654 \text{ m } \textbf{Ans.}$$

Example 20.2. *A belt drive consists of two V-belts in parallel, on grooved pulleys of the same size. The angle of the groove is 30°. The cross-sectional area of each belt is 750 mm² and μ = 0.12. The density of the belt material is 1.2 Mg / m³ and the maximum safe stress in the material is 7 MPa. Calculate the power that can be transmitted between pulleys of 300 mm diameter rotating at 1500 r.p.m. Find also the shaft speed in r.p.m. at which the power transmitted would be a maximum.*

Solution. Given : $n = 2$; $2\beta = 30°$ or $\beta = 15°$; $a = 750 \text{ mm}^2 = 750 \times 10^{-6} \text{ m}^2$; $\mu = 0.12$; $\rho = 1.2 \text{ Mg/m}^3 = 1200 \text{ kg/m}^3$; $\sigma = 7 \text{ MPa} = 7 \times 10^6 \text{ N/m}^2$; $d = 300 \text{ mm} = 0.3 \text{ m}$; $N = 1500$ r.p.m.

We know that mass of the belt per metre length,

$$m = \text{Area} \times \text{length} \times \text{density} = 750 \times 10^{-6} \times 1 \times 1200 = 0.9 \text{ kg/m}$$

and speed of the belt, $v = \dfrac{\pi d N}{60} = \dfrac{\pi \times 0.3 \times 1500}{60} = 23.56 \text{ m/s}$

∴ Centrifugal tension,

$$T_C = m.v^2 = 0.9 (23.56)^2 = 500 \text{ N}$$

and maximum tension, $T = \sigma \times a = 7 \times 10^6 \times 750 \times 10^{-6} = 5250 \text{ N}$

We know that tension in the tight side of the belt,

$$T_1 = T - T_C = 5250 - 500 = 4750 \text{ N}$$

Let T_2 = Tension in the slack side of the belt.

Since the pulleys are of the same size, therefore angle of lap (θ) = 180° = π rad.

We know that

$$2.3 \log \left(\frac{T_1}{T_2}\right) = \mu.\theta \operatorname{cosec} \beta = 0.12 \times \pi \times \operatorname{cosec} 15° = 0.377 \times 3.8637 = 1.457$$

$$\therefore \quad \log \left(\frac{T_1}{T_2}\right) = \frac{1.457}{2.3} = 0.6335 \quad \text{or} \quad \frac{T_1}{T_2} = 4.3 \qquad \text{...(Taking antilog of 0.6335)}$$

and $T_2 = T_1 / 4.3 = 4750 / 4.3 = 1105 \text{ N}$

Power transmitted

We know that power transmitted,

$$P = (T_1 - T_2)\, v \times n = (4750 - 1105)\, 23.56 \times 2 = 171\ 750 \text{ W}$$

$$= 171.75 \text{ kW } \textbf{Ans.}$$

Shaft speed

Let N_1 = Shaft speed in r.p.m., and

v_1 = Belt speed in m/s.

We know that for maximum power, centrifugal tension,

$$T_C = T/3 \quad \text{or} \quad m\,(v_1)^2 = T/3$$

or $$0.9\,(v_1)^2 = 5250/3 = 1750$$

$\therefore$ $$(v_1)^2 = 1750/0.9 = 1944.4 \quad \text{or} \quad v_1 = 44.1 \text{ m/s}$$

We know that belt speed (v_1),

$$44.1 = \frac{\pi d\, N_1}{60} = \frac{\pi \times 0.3 \times N_1}{60} = 0.0157\, N_1$$

$\therefore$ $$N_1 = 44.1/0.0157 = 2809 \text{ r.p.m.} \textbf{ Ans.}$$

Example 20.3. *Two shafts whose centres are 1 metre apart are connected by a V-belt drive. The driving pulley is supplied with 95 kW power and has an effective diameter of 300 mm. It runs at 1000 r.p.m. while the driven pulley runs at 375 r.p.m. The angle of groove on the pulleys is 40°. Permissible tension in 400 mm² cross-sectional area belt is 2.1 MPa. The material of the belt has density of 1100 kg / m³. The driven pulley is overhung, the distance of the centre from the nearest bearing being 200 mm. The coefficient of friction between belt and pulley rim is 0.28. Estimate: 1. The number of belts required ; and 2. Diameter of driven pulley shaft, if permissible shear stress is 42 MPa.*

Solution. Given : $x = 1$ m ; $P = 95$ kW $= 95 \times 10^3$ W ; $d_1 = 300$ mm $= 0.3$ m ; $N_1 = 1000$ r.p.m. ; $N_2 = 375$ r.p.m ; $2\beta = 40°$ or $\beta = 20°$; $a = 400$ mm² $= 400 \times 10^{-6}$ m² ; $\sigma = 2.1$ MPa $= 2.1$ N/mm² ; $\rho = 1100$ kg/m³; $\mu = 0.28$; $\tau = 42$ MPa $= 42$ N/mm²

First of all, let us find the diameter of the driven pulley (d_2). We know that

$$\frac{N_1}{N_2} = \frac{d_2}{d_1} \quad \text{or} \quad d_2 = \frac{N_1 \times d_1}{N_2} = \frac{1000 \times 300}{375} = 800 \text{ mm} = 0.8 \text{ m}$$

For an open belt drive,

$$\sin\alpha = \frac{r_2 - r_1}{x} = \frac{d_2 - d_1}{2x} = \frac{0.8 - 0.3}{2 \times 1} = 0.25$$

$\therefore$ $$\alpha = 14.5°$$

and angle of lap on the smaller or driving pulley,

$$\theta = 180° - 2\,\alpha = 180° - 2 \times 14.5 = 151°$$

$$= 151 \times \frac{\pi}{180} = 2.64 \text{ rad}$$

We know that the mass of the belt per metre length,

$$m = \text{Area} \times \text{length} \times \text{density} = 400 \times 10^{-6} \times 1 \times 1100 = 0.44 \text{ kg/m}$$

and velocity of the belt,

$$v = \frac{\pi d_1 . N_1}{60} = \frac{\pi \times 0.3 \times 1000}{60} = 15.71 \text{ m/s}$$

$\therefore$ Centrifugal tension,

$$T_C = m.v^2 = 0.44\,(15.71)^2 = 108.6 \text{ N}$$

and maximum tension in the belt,

$$T = \sigma \times a = 2.1 \times 400 = 840 \text{ N}$$

$\therefore$ Tension in the tight side of the belt,

$$T_1 = T - T_C = 840 - 108.6 = 731.4 \text{ N}$$

We know that

$$2.3 \log\left(\frac{T_1}{T_2}\right) = \mu.\theta \operatorname{cosec}\beta = 0.28 \times 2.64 \operatorname{cosec} 20° = 0.74 \times 2.9238 = 2.164$$

$\therefore \quad \log\left(\frac{T_1}{T_2}\right) = \frac{2.164}{2.3} = 0.9407$ or $\frac{T_1}{T_2} = 8.72$ ...(Taking antilog of 0.9407)

and $\quad T_2 = \frac{T_1}{8.72} = \frac{731.4}{8.72} = 83.9$ N

1. *Number of belts required*

We know that the power transmitted per belt

$$= (T_1 - T_2)\, v = (731.4 - 83.9)\, 15.71 = 10\,172 \text{ W} = 10.172 \text{ kW}$$

$\therefore$ Number of belts required

$$= \frac{\text{Total power transmitted}}{\text{Power transmitted per belt}} = \frac{95}{10.172} = 9.34 \text{ say 10 } \textbf{Ans.}$$

2. *Diameter of driven pulley shaft*

Let $\quad D$ = Diameter of driven pulley shaft.

We know that torque transmitted by the driven pulley shaft,

$$T = \frac{P \times 60}{2\pi N_2} = \frac{95 \times 10^3 \times 60}{2\pi \times 375} = 2420 \text{ N-m} = 2420 \times 10^3 \text{ N-mm}$$

Since the driven pulley is overhung and the distance of the centre from the nearest bearing is 200 mm, therefore bending moment on the shaft due to the pull on the belt,

$$M = (T_1 + T_2 + 2T_C)\, 200 \times 10 \qquad ...(\because \text{No. of belts} = 10)$$

$$= (731.4 + 83.9 + 2 \times 108.6)\, 200 \times 10 = 2065 \times 10^3 \text{ N-mm}$$

$\therefore$ Equivalent twisting moment,

$$T_e = \sqrt{T^2 + M^2} = \sqrt{(2420 \times 10^3)^2 + (2065 \times 10^3)^2} \text{ N-mm}$$

$$= 3181 \times 10^3 \text{ N-mm}$$

We know that equivalent twisting moment (T_e),

$$3181 \times 10^3 = \frac{\pi}{16} \times \tau \times D^3 = \frac{\pi}{16} \times 42\, D^3 = 8.25\, D^3$$

$\therefore \quad D^3 = 3181 \times 10^3 / 8.25 = 386 \times 10^3$

or $\quad D = 72.8$ say 75 mm **Ans.**

Example 20.4. *Power of 60 kW at 750 r.p.m. is to be transmitted from an electric motor to compressor shaft at 300 r.p.m. by V-belts. The approximate larger pulley diameter is 1500 mm. The approximate centre distance is 1650 mm, and overload factor is to be taken as 1.5. Give a complete design of the belt drive. A belt with cross-sectional area of 350 mm² and density 1000 kg / m³ and having an allowable tensile strength 2 MPa is available for use. The coefficient of friction between the belt and the pulley may be taken as 0.28. The driven pulley is overhung to the extent of 300 mm from the nearest bearing and is mounted on a shaft having a permissible shear stress of 40 MPa with the help of a key. The shaft, the pulley and the key are also to be designed.*

Solution. Given : $P = 60$ kW ; $N_1 = 750$ r.p.m. ; $N_2 = 300$ r.p.m. ; $d_2 = 1500$ mm ; $x = 1650$ mm ; Overload factor = 1.5 ; $a = 350 \text{ mm}^2 = 350 \times 10^{-6} \text{ m}^2$; $\rho = 1000 \text{ kg/m}^3$; $\sigma = 2 \text{ MPa} = 2 \text{ N/mm}^2$; $\mu = 0.28$; $\tau = 40 \text{ MPa} = 40 \text{ N/mm}^2$

1. *Design of the belt drive*

First of all, let us find the diameter (d_1) of the motor pulley. We know that

$$\frac{N_1}{N_2} = \frac{d_2}{d_1} \quad \text{or} \quad d_1 = \frac{d_2 \times N_2}{N_1} = \frac{1500 \times 300}{750} = 600 \text{ mm} = 0.6 \text{ m}$$

and $\quad \sin\alpha = \frac{r_2 - r_1}{x} = \frac{d_2 - d_1}{2x} = \frac{1500 - 600}{2 \times 1650} = 0.2727$ or $\alpha = 15.83°$

We know that the angle of contact,

$$\theta = 180° - 2\alpha = 180 - 2 \times 15.83 = 148.34°$$

$$= 148.34 \times \pi / 180 = 2.6 \text{ rad}$$

Let T_1 = Tension in the tight side of the belt, and

T_2 = Tension in the slack side of the belt.

Assume the groove angle of the pulley, $2\,\beta = 35°$ or $\beta = 17.5°$. We know that

$$2.3 \log\left(\frac{T_1}{T_2}\right) = \mu.\theta \operatorname{cosec} \beta = 0.28 \times 2.6 \times \operatorname{cosec} 17.5° = 2.42$$

$$\therefore \quad \log\left(\frac{T_1}{T_2}\right) = 2.42 / 2.3 = 1.0526 \quad \text{or} \quad \frac{T_1}{T_2} = 11.28 \quad \text{...(i)}$$

...(Taking antilog of 1.0526)

We know that the velocity of the belt,

$$v = \frac{\pi\, d_1\, N_1}{60} = \frac{\pi \times 0.6 \times 750}{60} = 23.66 \text{ m/s}$$

and mass of the belt per metre length,

$$m = \text{Area} \times \text{length} \times \text{density} = 350 \times 10^{-6} \times 1 \times 1000 = 0.35 \text{ kg / m}$$

∴ Centrifugal tension in the belt,

$$T_C = m.v^2 = 0.35\,(23.66)^2 = 196 \text{ N}$$

and maximum tension in the belt,

$$T = \text{Stress} \times \text{area} = \sigma \times a = 2 \times 350 = 700 \text{ N}$$

∴ Tension in the tight side of the belt,

$$T_1 = T - T_C = 700 - 196 = 504 \text{ N}$$

and

$$T_2 = \frac{T_1}{11.28} = \frac{504}{11.28} = 44.7 \text{ N} \quad \text{...[From equation (i)]}$$

We know that the power transmitted per belt

$$= (T_1 - T_2)\,v = (504 - 44.7)\,23.66 = 10\,867 \text{ W} = 10.867 \text{ kW}$$

Since the over load factor is 1.5, therefore the belt is to be designed for 1.5 × 60 = 90 kW.

∴ Number of belts required

$$= \frac{\text{Designed power}}{\text{Power transmitted per belt}} = \frac{90}{10.867} = 8.3 \text{ say } 9 \textbf{ Ans.}$$

Since the *V*-belt is to be designed for 90 kW, therefore from Table 20.1, we find that a '*D*' type of belt should be used.

We know that the pitch length of the belt,

$$L = \pi\,(r_2 + r_1) + 2x + \frac{(r_2 - r_1)^2}{x} = \frac{\pi}{2}\,(d_2 + d_1) + 2x + \frac{(d_2 - d_1)^2}{4x}$$

$$= \frac{\pi}{2}\,(1500 + 600) + 2 \times 1650 + \frac{(1500 - 600)^2}{4 \times 1650}$$

$$= 3300 + 3300 + 123 = 6723 \text{ mm}$$

Subtracting 79 mm for '*D*' type belt, we find that inside length of the belt

$$= 6723 - 79 = 6644 \text{ mm}$$

According to IS: 2494 – 1974, the nearest standard inside length of *V*-belt is 6807 mm.

∴ Pitch length of the belt,

$$L_1 = 6807 + 79 = 6886 \text{ mm } \textbf{Ans.}$$

Now let us find out the new centre distance (x_1) between the two pulleys. We know that

$$L_1 = \frac{\pi}{2}(d_2 + d_1) + 2x_1 + \frac{(d_2 - d_1)^2}{4x_1}$$

$$6886 = \frac{\pi}{2}(1500 + 600) + 2x_1 + \frac{(1500 - 600)^2}{4x_1}$$

$$= 3300 + 2x_1 + \frac{810\ 000}{4x_1} \quad \text{....}(d_1 \text{ and } d_2 \text{ are taken in mm})$$

$$6886 \times 4x_1 = 3300 \times 4x_1 + 2x_1 \times 4x_1 + 810\ 000$$

or
$$3443\,x_1 = 1650\,x_1 + x_1^2 + 101\ 250$$

or
$$x_1^2 - 1793\,x_1 + 101\ 250 = 0$$

$$\therefore \quad x_1 = \frac{1793 \pm \sqrt{(1793)^2 - 4 \times 101\ 250}}{2}$$

$$= \frac{1793 \pm 1677}{2} = 1735 \text{ mm } \textbf{Ans.} \quad \text{...(Taking + ve sign)}$$

2. ***Design of shaft***

Let D = Diameter of the shaft.

We know that the torque transmitted by the driven or compressor pulley shaft,

$$T = \frac{\text{Designed power} \times 60}{2\pi N_2} = \frac{90 \times 10^3 \times 60}{2\pi \times 300} = 2865 \text{ N-m}$$

$$= 2865 \times 10^3 \text{ N-mm}$$

Since the overhang of the pulley is 300 mm, therefore bending moment on the shaft due to the belt tensions,

$$M = (T_1 + T_2 + 2T_C)\,300 \times 9 \quad \text{...}(\because \text{No. of belts} = 9)$$

$$= (504 + 44.7 + 2 \times 196)\,300 \times 9 = 2540 \times 10^3 \text{ N-mm}$$

∴ Equivalent twisting moment,

$$T_e = \sqrt{T^2 + M^2} = \sqrt{(2865 \times 10^3)^2 + (2540 \times 10^3)^2}$$

$$= 3830 \times 10^3 \text{ N-mm}$$

We also know that equivalent twisting moment (T_e),

$$3830 \times 10^3 = \frac{\pi}{16} \times \tau \times D^3 = \frac{\pi}{16} \times 40 \times D^3 = 7.855\,D^3$$

$$\therefore \quad D^3 = 3830 \times 10^3 / 7.855 = 487.6 \times 10^3 \quad \text{or} \quad D = 78.7 \text{ say } 80 \text{ mm } \textbf{Ans.}$$

3. ***Design of the pulley***

The dimensions for the standard V-grooved pulley (Refer Fig. 20.1) are shown in Table 20.2, from which we find that for 'D' type belt

w = 27 mm, d = 28 mm, a = 8.1 mm, c = 19.9 mm, f = 24 mm, and e = 37 mm.

We know that face width of the pulley,

$$B = (n - 1)\,e + 2f = (9 - 1)\,37 + 2 \times 24 = 344 \text{ mm } \textbf{Ans.}$$

4. ***Design for key***

The standard dimensions of key for a shaft of 80 mm diameter are

Width of key = 25 mm **Ans.**

and thickness of key = 14 mm **Ans.**

Example 20.5. *A V-belt is driven on a flat pulley and a V-pulley. The drive transmits 20 kW from a 250 mm diameter V-pulley operating at 1800 r.p.m. to a 900 mm diameter flat pulley. The centre distance is 1 m, the angle of groove 40° and μ = 0.2. If density of belting is 1110 kg / m³ and allowable stress is 2.1 MPa for belt material, what will be the number of belts required if C-size V-belts having 230 mm² cross-sectional area are used.*

Solution. Given : $P = 20$ kW ; $d_1 = 250$ mm $= 0.25$ m ; $N_1 = 1800$ r.p.m. ; $d_2 = 900$ mm $= 0.9$ m ; $x = 1$ m $= 1000$ mm ; $2\beta = 40°$ or $\beta = 20°$; $\mu = 0.2$; $\rho = 1110$ kg/m³ ; $\sigma = 2.1$ MPa $= 2.1$ N/mm² ; $a = 230$ mm² $= 230 \times 10^{-6}$ m²

Fig. 20.5 shows a *V*-flat drive. First of all, let us find the angle of contact for both the pulleys. From the geometry of the Fig. 20.5, we find that

$$\sin\alpha = \frac{O_2M}{O_1O_2} = \frac{r_2 - r_1}{x} = \frac{d_2 - d_1}{2x} = \frac{900 - 250}{2 \times 1000} = 0.325$$

$$\therefore \quad \alpha = 18.96°$$

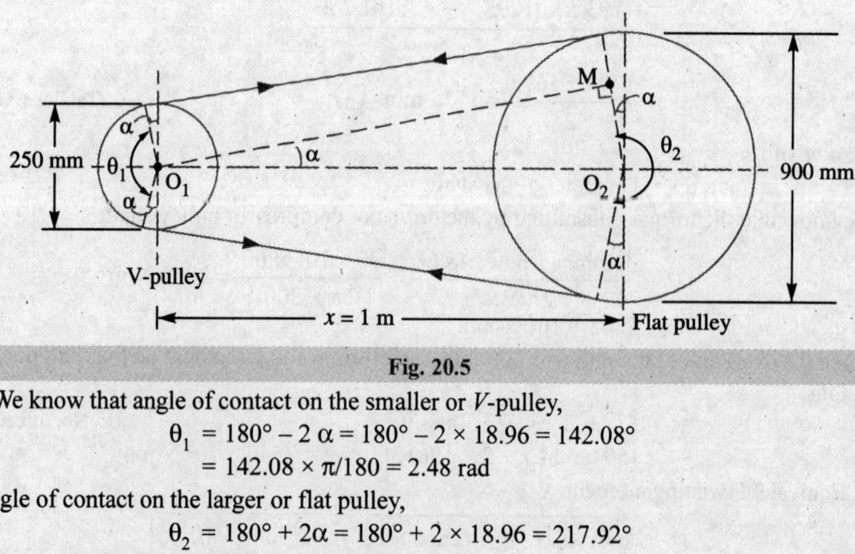

Fig. 20.5

We know that angle of contact on the smaller or *V*-pulley,

$$\theta_1 = 180° - 2\,\alpha = 180° - 2 \times 18.96 = 142.08°$$
$$= 142.08 \times \pi/180 = 2.48 \text{ rad}$$

and angle of contact on the larger or flat pulley,

$$\theta_2 = 180° + 2\alpha = 180° + 2 \times 18.96 = 217.92°$$
$$= 217.92 \times \pi \;/\; 180 = 3.8 \text{ rad}$$

We have already discussed that when the pulleys have different angle of contact (θ), then the design will refer to a pulley for which μ.θ is small.

We know that for a smaller or *V*-pulley,

$$\mu.\theta = \mu.\theta_1 \operatorname{cosec}\beta = 0.2 \times 2.48 \times \operatorname{cosec} 20° = 1.45$$

and for larger or flat pulley,

$$\mu.\theta = \mu.\theta_2 = 0.2 \times 3.8 = 0.76$$

Since (μ.θ) for the larger or flat pulley is small, therefore the design is based on the larger or flat pulley.

We know that peripheral velocity of the belt,

$$v = \frac{\pi\, d_1\, N_1}{60} = \frac{\pi \times 0.25 \times 1800}{60} = 23.56 \text{ m/s}$$

Mass of the belt per metre length,

$$m = \text{Area} \times \text{length} \times \text{density} = a \times l \times \rho$$
$$= 230 \times 10^{-6} \times 1 \times 1100 = 0.253 \text{ kg / m}$$

∴ Centrifugal tension,

$$T_C = m.v^2 = 0.253\ (23.56)^2 = 140.4 \text{ N}$$

Let T_1 = Tension in the tight side of the belt, and

T_2 = Tension in the slack side of the belt.

We know that maximum tension in the belt,

$$T = \text{Stress} \times \text{area} = \sigma \times a = 2.1 \times 230 = 483 \text{ N}$$

We also know that maximum or total tension in the belt,

$$T = T_1 + T_C$$

$$\therefore \quad T_1 = T - T_C = 483 - 140.4 = 342.6 \text{ N}$$

We know that

$$2.3 \log\left(\frac{T_1}{T_2}\right) = \mu.\theta_2 = 0.2 \times 3.8 = 0.76$$

$$\log\left(\frac{T_1}{T_2}\right) = 0.76 / 2.3 = 0.3304 \quad \text{or} \quad \frac{T_1}{T_2} = 2.14 \quad \text{...(Taking antilog of 0.3304)}$$

and

$$T_2 = T_1 / 2.14 = 342.6 / 2.14 = 160 \text{ N}$$

∴ Power transmitted per belt

$$= (T_1 - T_2)\ v = (342.6 - 160)\ 23.56 = 4302 \text{ W} = 4.302 \text{ kW}$$

We know that number of belts required

$$= \frac{\text{Total power transmitted}}{\text{Power transmitted per belt}} = \frac{20}{4.302} = 4.65 \text{ say } 5 \textbf{ Ans.}$$

20.7 Rope Drives

The rope drives are widely used where a large amount of power is to be transmitted, from one pulley to another, over a considerable distance. It may be noted that the use of flat belts is limited for the transmission of moderate power from one pulley to another when the two pulleys are not more than 8 metres apart. If large amounts of power are to be transmitted, by the flat belt, then it would result in excessive belt cross-section.

The ropes drives use the following two types of ropes :

1. Fibre ropes, and **2.** *Wire ropes.

The fibre ropes operate successfully when the pulleys are about 60 metres apart, while the wire ropes are used when the pulleys are upto 150 metres apart.

20.8 Fibre Ropes

The ropes for transmitting power are usually made from fibrous materials such as hemp, manila and cotton. Since the hemp and manila fibres are rough, therefore the ropes made from these fibres are not very flexible and possesses poor mechanical properties. The hemp ropes have less strength as compared to manila ropes. When the hemp and manila ropes are bent over the sheave, there is some sliding of the fibres, causing the rope to wear and chafe internally. In order to minimise this defect, the rope fibres are lubricated with a tar, tallow or graphite. The lubrication also makes the rope moisture proof. The hemp ropes are suitable only for hand operated hoisting machinery and as tie ropes for lifting tackle, hooks etc.

The cotton ropes are very soft and smooth. The lubrication of cotton ropes is not necessary. But if it is done, it reduces the external wear between the rope and the grooves of its sheaves. It may be noted that the manila ropes are more durable and stronger than cotton ropes. The cotton ropes are costlier than manila ropes.

---

* Wire ropes are discussed in Art. 20.12.

Notes : 1. The diameter of manila and cotton ropes usually ranges from 38 mm to 50 mm. The size of the rope is usually designated by its circumference or '***girth***'.

2. The ultimate tensile breaking load of the fibre ropes varies greatly. For manila ropes, the average value of the ultimate tensile breaking load may be taken as $500\ d^2$ kN and for cotton ropes, it may be taken as $350\ d^2$ kN, where d is the diameter of rope in mm.

20.9 Advantages of Fibre Rope Drives

The fibre rope drives have the following advantages :

1. They give smooth, steady and quiet service.
2. They are little affected by out door conditions.
3. The shafts may be out of strict alignment.
4. The power may be taken off in any direction and in fractional parts of the whole amount.
5. They give high mechanical efficiency.

20.10 Sheave for Fibre Ropes

The fibre ropes are usually circular in cross-section as shown in Fig. 20.6 (*a*). The sheave for the fibre ropes, is shown in Fig. 20.6 (*b*). The groove angle of the pulley for rope drives is usually 45°.

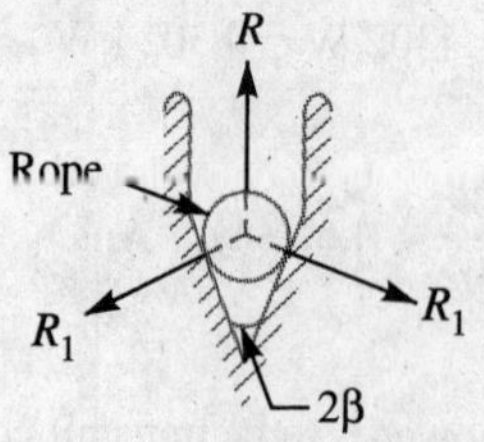

(*a*) Cross-section of a rope.

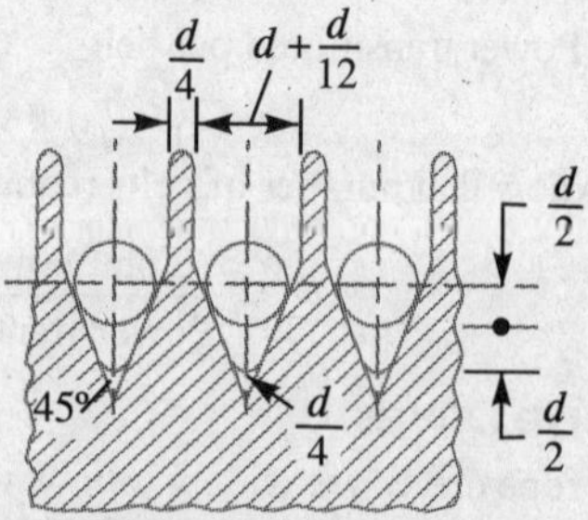

(*b*) Sheave (grooved pulley) for ropes.

Fig. 20.6. Rope and sheave.

The grooves in the pulleys are made narrow at the bottom and the rope is pinched between the edges of the *V*-groove to increase the holding power of the rope on the pulley. The grooves should be finished smooth to avoid chafing of the rope. The diameter of the sheaves should be large to reduce the wear on the rope due to internal friction and bending stresses. The proper size of sheave wheels is 40 *d* and the minimum size is 36 *d*, where *d* is the diameter of rope in cm.

Note : The number of grooves should not be more than 24.

Rope drives

20.11 Ratio of Driving Tensions for Fibre Rope

A fibre rope with a grooved pulley is shown in Fig. 20.6 (*a*). The fibre ropes are designed in the similar way as *V*-belts. We have discussed in Art. 20.5, that the ratio of driving tensions is

$$2.3 \log \left(\frac{T_1}{T_2}\right) = \mu.\theta \operatorname{cosec} \beta$$

where μ, θ and β have usual meanings.

Example 20.6. *A pulley used to transmit power by means of ropes has a diameter of 3.6 metres and has 15 grooves of 45° angle. The angle of contact is 170° and the coefficient of friction between the ropes and the groove sides is 0.28. The maximum possible tension in the ropes is 960 N and the mass of the rope is 1.5 kg per metre length. Determine the speed of the pulley in r.p.m. and the power transmitted if the condition of maximum power prevail.*

Solution. Given : $d = 3.6$ m ; $n = 15$; $2\beta = 45°$ or $\beta = 22.5°$; $\theta = 170° = 170 \times \pi / 180$ $= 2.967$ rad ; $\mu = 0.28$; $T = 960$ N ; $m = 1.5$ kg / m

Speed of the pulley

Let N = Speed of the pulley in r.p.m.

We know that for maximum power, speed of the pulley,

$$v = \sqrt{\frac{T}{3m}} = \sqrt{\frac{960}{3 \times 1.5}} = 14.6 \text{ m/s}$$

We also know that speed of the pulley (v),

$$14.6 = \frac{\pi d . N}{60} = \frac{\pi \times 3.6 \times N}{60} = 0.19\, N$$

$\therefore$ $N = 14.6 / 0.19 = 76.8$ r.p.m. **Ans.**

Power transmitted

We know that for maximum power, centrifugal tension,

$$T_C = T / 3 = 960 / 3 = 320 \text{ N}$$

$\therefore$ Tension in the tight side of the rope,

$$T_1 = T - T_C = 960 - 320 = 640 \text{ N}$$

Let T_2 = Tension in the slack side of the rope.

We know that

$$2.3 \log \left(\frac{T_1}{T_2}\right) = \mu.\theta \operatorname{cosec} \beta = 0.28 \times 2.967 \times \operatorname{cosec} 22.5° = 2.17$$

$$\therefore \quad \log \left(\frac{T_1}{T_2}\right) = \frac{2.17}{2.3} = 0.9435 \quad \text{or} \quad \frac{T_1}{T_2} = 8.78$$ ...(Taking antilog of 0.9435)

and $T_2 = T_1 / 8.78 = 640 / 8.78 = 73$ N

$\therefore$ Power transmitted,

$$P = (T_1 - T_2)\, v \times n = (640 - 73)\, 14.6 \times 15 = 124\,173 \text{ W}$$
$$= 124.173 \text{ kW}$$ **Ans.**

Example 20.7. *A rope pulley with 10 ropes and a peripheral speed of 1500 m / min transmits 115 kW. The angle of lap for each rope is 180° and the angle of groove is 45°. The coefficient of friction between the rope and pulley is 0.2. Assuming the rope to be just on the point of slipping, find the tension in the tight and slack sides of the rope. The mass of each rope is 0.6 kg per metre length.*

Solution. Given : $n = 10$; $v = 1500$ m/min $= 25$ m/s ; $P = 115$ kW $= 115 \times 10^3$ W ; $\theta = 180°$ $= \pi$ rad ; $2\beta = 45°$ or $\beta = 22.5°$; $\mu = 0.2$; $m = 0.6$ kg / m

Let T_1 = Tension in the tight side of the rope, and

T_2 = Tension in the slack side of the rope.

We know that total power transmitted (P),

$$115 \times 10^3 = (T_1 - T_2)\, v \times n = (T_1 - T_2)\, 25 \times 10 = 250\, (T_1 - T_2)$$

$$\therefore \quad T_1 - T_2 = 115 \times 10^3 / 250 = 460$$ ...(*i*)

We also know that

$$2.3 \log \left(\frac{T_1}{T_2}\right) = \mu.\theta \operatorname{cosec} \beta = 0.2 \times \pi \times \operatorname{cosec} 22.5° = 1.642$$

$$\therefore \quad \log \left(\frac{T_1}{T_2}\right) = \frac{1.642}{2.3} = 0.714 \quad \text{or} \quad \frac{T_1}{T_2} = 5.18 \quad \text{...(Taking antilog of 0.714) ...}(ii)$$

From equations (*i*) and (*ii*), we find that

$$T_1 = 570 \text{ N, and } T_2 = 110 \text{ N}$$

We know that centrifugal tension,

$$T_C = m\,v^2 = 0.6\,(25)^2 = 375 \text{ N}$$

∴ Total tension in the tight side of the rope,

$$T_{t1} = T_1 + T_C = 570 + 375 = 945 \text{ N } \textbf{Ans.}$$

and total tension in the slack side of the rope,

$$T_{t2} = T_2 + T_C = 110 + 375 = 485 \text{ N } \textbf{Ans.}$$

Example 20.8. *A rope drive transmits 600 kW from a pulley of effective diameter 4 m, which runs at a speed of 90 r.p.m. The angle of lap is 160°; the angle of groove 45°; the coefficient of friction 0.28; the mass of rope 1.5 kg / m and the allowable tension in each rope 2400 N. Find the number of ropes required.*

Solution. Given : $P = 600$ kW ; $d = 4$ m ; $N = 90$ r.p.m. ; $\theta = 160° = 160 \times \pi/180 = 2.8$ rad ; $2\beta = 45°$ or $\beta = 22.5°$; $\mu = 0.28$; $m = 1.5$ kg / m ; $T = 2400$ N

We know that velocity of the pulley or rope,

$$v = \frac{\pi d N}{60} = \frac{\pi \times 4 \times 90}{60} = 18.85 \text{ m/s}$$

∴ Centrifugal tension,

$$T_C = m.v^2 = 1.5\,(18.85)^2 = 533 \text{ N}$$

and tension in the tight side of the rope,

$$T_1 = T - T_C = 2400 - 533 = 1867 \text{ N}$$

Let $\quad T_2$ = Tension in the slack side of the rope.

We know that

$$2.3 \log \left(\frac{T_1}{T_2}\right) = \mu.\theta \operatorname{cosec} \beta = 0.28 \times 2.8 \times \operatorname{cosec} 22.5° = 0.784 \times 2.6131 = 2.0487$$

$$\log \left(\frac{T_1}{T_2}\right) = \frac{2.0487}{2.3} = 0.8907 \quad \text{or} \quad \frac{T_1}{T_2} = 7.78 \quad \text{...(Taking antilog of 0.8907)}$$

$$\therefore \quad T_2 = T_1 / 7.78 = 1867 / 7.78 = 240 \text{ N}$$

We know that power transmitted per rope

$$= (T_1 - T_2)\,v = (1867 - 240)\,18.85 = 30\,670 \text{ W} = 30.67 \text{ kW}$$

∴ Number of ropes required

$$= \frac{\text{Total power transmitted}}{\text{Power transmitted per rope}} = \frac{600}{30.67} = 19.56 \text{ say } 20 \textbf{ Ans.}$$

Example 20.9. *A rope drive is to transmit 250 kW from a pulley of 1.2 m diameter, running at a speed of 300 r.p.m. The angle of lap may be taken as π radians. The groove half angle is 22.5°. The ropes to be used are 50 mm in diameter. The mass of the rope is 1.3 kg per metre length and each rope has a maximum pull of 2.2 kN, the coefficient of friction between rope and pulley is 0.3. Determine the number of ropes required. If the overhang of the pulley is 0.5 m, suggest suitable size for the pulley shaft if it is made of steel with a shear stress of 40 MPa.*

Solution. Given : $P = 250$ kW $= 250 \times 10^3$ W ; $d = 1.2$ m ; $N = 300$ r.p.m ; $\theta = \pi$ rad ; $\beta = 22.5°$; $d_r = 50$ mm ; $m = 1.3$ kg / m ; $T = 2.2$ kN $= 2200$ N ; $\mu = 0.3$; $\tau = 40$ MPa $= 40$ N/mm²

We know that the velocity of belt,

$$v = \frac{\pi d.N}{60} = \frac{\pi \times 1.2 \times 300}{60} = 18.85 \text{ m/s}$$

and centrifugal tension, $T_C = m.v^2 = 1.3\,(18.85)^2 = 462$ N

∴ Tension in the tight side of the rope,

$$T_1 = T - T_C = 2200 - 462 = 1738 \text{ N}$$

Let T_2 = Tension in the slack side of the rope.

We know that

$$2.3 \log\left(\frac{T_1}{T_2}\right) = \mu.\theta.\text{cosec}\,\beta = 0.3 \times \pi \times \text{cosec}\,22.5° = 0.9426 \times 2.6131 = 2.463$$

$$\therefore \quad \log\left(\frac{T_1}{T_2}\right) = \frac{2.463}{2.3} = 1.071 \quad \text{or} \quad \frac{T_1}{T_2} = 11.8 \quad \text{...(Taking antilog of 1.071)}$$

and

$$T_2 = \frac{T_1}{11.8} = \frac{1738}{11.8} = 147.3 \text{ N}$$

Number of ropes required

We know that power transmitted per rope

$$= (T_1 - T_2)\,v = (1738 - 147.3) \times 18.85 = 29\,985 \text{ W} = 29.985 \text{ kW}$$

∴ Number of ropes required

$$= \frac{\text{Total power transmitted}}{\text{Power transmitted per rope}} = \frac{250}{29.985} = 8.34 \text{ say } 9$$ **Ans.**

Diameter for the pulley shaft

Let D = Diameter for the pulley shaft.

We know that the torque transmitted by the pulley shaft,

$$T = \frac{P \times 60}{2\pi N} = \frac{250 \times 10^3 \times 60}{2\pi \times 300} = 7957 \text{ N-m}$$

Since the overhang of the pulley is 0.5 m, therefore bending moment on the shaft due to the rope pull,

$$M = (T_1 + T_2 + 2T_C)\,0.5 \times 9 \quad \text{...(}\because \text{ No. of ropes = 9)}$$
$$= (1738 + 147.3 + 2 \times 462)\,0.5 \times 9 = 12\,642 \text{ N-m}$$

∴ Equivalent twisting moment,

$$T_e = \sqrt{T^2 + M^2} = \sqrt{(7957)^2 + (12\,642)^2} = 14\,938 \text{ N-m}$$
$$= 14.938 \times 10^6 \text{ N-mm}$$

We know that the equivalent twisting moment (T_e),

$$14.938 \times 10^6 = \frac{\pi}{16} \times \tau \times D^3 = \frac{\pi}{16} \times 40 \times D^3 = 7.855\,D^3$$

$$\therefore \quad D^3 = 14.938 \times 10^6 / 7.855 = 1.9 \times 10^6 \quad \text{or} \quad D = 123.89 \text{ say } 125 \text{ mm}$$ **Ans.**

20.12 Wire Ropes

When a large amount of power is to be transmitted over long distances from one pulley to another (*i.e.* when the pulleys are upto 150 metres apart), then wire ropes are used. The wire ropes are widely used in elevators, mine hoists, cranes, conveyors, hauling devices and suspension bridges. The wire ropes run on grooved pulleys but they rest on the bottom of the *grooves and are not wedged between the sides of the grooves.

The wire ropes are made from cold drawn wires in order to have increase in strength and durability. It may be noted that the strength of the wire rope increases as its size decreases. The various materials used for wire ropes in order of increasing strength are wrought iron, cast steel, extra strong cast steel, plough steel and alloy steel. For certain purposes, the wire ropes may also be made of copper, bronze, aluminium alloys and stainless steels.

20.13 Advantages of Wire Ropes

The wire ropes have the following advantages as compared to fibre ropes.

1. These are lighter in weight,
2. These offer silent operation,
3. These can withstand shock loads,
4. These are more reliable,
5. These are more durable,
6. They do not fail suddenly,
7. The efficiency is high, and
8. The cost is low.

20.14 Construction of Wire Ropes

The wire ropes are made from various grades of steel wire having a tensile strength ranging from 1200 to 2400 MPa as shown in the following table :

Table 20.4. Grade and tensile strength of wires.

Grade of wire	120	140	160	180	200
Tensile strength range (MPa)	1200 – 1500	1400 – 1700	1600 – 1900	1800 – 2100	2000 – 2400

The wires are first given special heat treatment and then cold drawn in order to have high strength and durability of the rope. The steel wire ropes are manufactured by special machines. First of all, a number of wires such as 7, 19 or 37 are twisted into a strand and then a number of strands, usually 6 or 8 are twisted about a core or centre to form the rope as shown in Fig. 20.7. The core may be made of hemp, jute, asbsestos or a wire of softer steel. The core must be continuously saturated with lubricant for the long life of the core as well as the entire rope. The asbestos or soft wire core is used when ropes are subjected to radiant heat such as cranes operating near furnaces. However, a wire core reduces the flexibility of the rope and thus such ropes are used only where they are subjected to high compression as in the case of several layers wound over a rope drum.

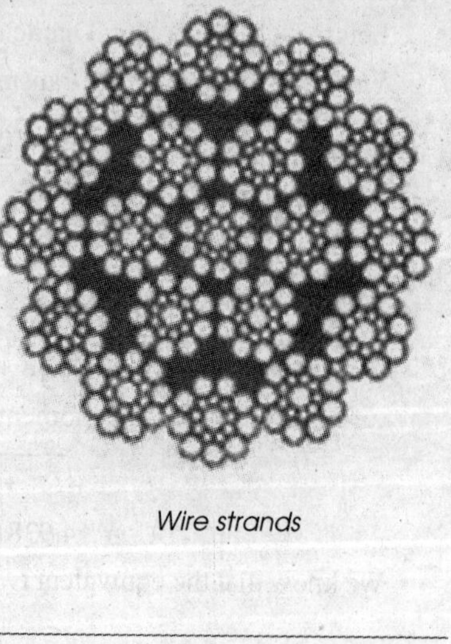

Wire strands

* The fibre ropes do not rest at the bottom of the groove.

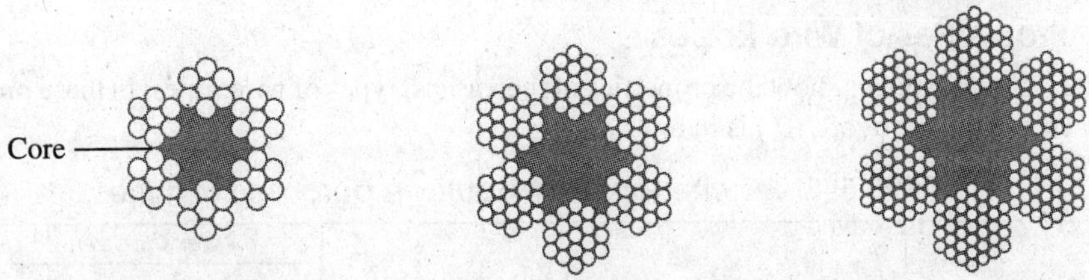

(*a*) 6 × 7 rope. (*b*) 6 × 19 rope. (*c*) 6 × 37 rope.

Fig. 20.7. Cross-sections of wire rope.

20.15 Classification of Wire Ropes

Wire rope

According to the direction of twist of the individual wires and that of strands, relative to each other, the wire ropes may be classified as follows :

1. ***Cross or regular lay ropes.*** In these types of ropes, the direction of twist of wires in the strands is opposite to the direction of twist of the stands, as shown in Fig. 20.8 (*a*). Such type of ropes are most popular.
2. ***Parallel or lang lay ropes.*** In these type of ropes, the direction of twist of the wires in the strands is same as that of strands in the rope, as shown in Fig. 20.8 (*b*). These ropes have better bearing surface but is harder to splice and twists more easily when loaded. These ropes are more flexible and resists wear more effectively. Since such ropes have the tendency to spin, therefore these are used in lifts and hoists with guide ways and also as haulage ropes.

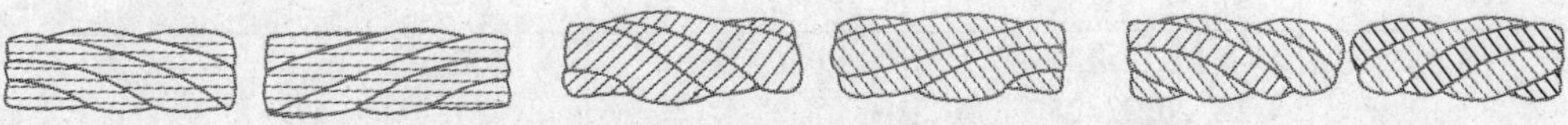

(*i*) right handed (*ii*) left handed (*i*) right handed (*ii*) left handed

(*a*) Cross or regular lay ropes. (*b*) Parallel or lang lay ropes. (*c*) Composite or reverse laid ropes.

Fig. 20.8. Wire ropes classified according to the direction of twist of the individual wires.

3. ***Composite or reverse laid ropes.*** In these types of ropes, the wires in the two adjacent strands are twisted in the opposite direction, as shown in Fig. 20.8 (*c*).

Note: The direction of the lay of the ropes may be right handed or left handed, depending upon whether the strands form right hand or left hand helixes, but the right hand lay ropes are most commonly used.

20.16 Designation of Wire Ropes

The wire ropes are designated by the number of strands and the number of wires in each strand. For example, a wire rope having six strands and seven wires in each strand is designated by 6 × 7 rope. Following table shows the standard designation of ropes and their applications :

Table 20.5. Standard designation of ropes and their applications.

Standard designation	*Application*
6 × 7 rope	It is a standard coarse laid rope used as haulage rope in mines, tramways, power transmission.
6 × 19 rope	It is a standard hoisting rope used for hoisting purposes in mines, quarries, cranes, dredges, elevators, tramways, well drilling.
6 × 37 rope	It is an extra flexible hoisting rope used in steel mill laddles, cranes, high speed elevators.
8 × 19 rope	It is also an extra flexible hoisting rope.

20.17 Properties of Wire Ropes

The following tables show the properties of the various types of wire ropes. In these properties, the diameter of the wire rope (d) is in mm.

Table 20.6. Steel wire ropes for haulage purposes in mines.

Type of rope	*Nominal diameter (mm)*	*Average weight (N/m)*	*Tensile strength (N)*	
			Tensile strength of wire	
			1600 MPa	*1800 MPa*
6 × 7	8, 9, 10, 11, 12, 13, 14, 16 18, 19, 20, 21, 22, 24, 25 26, 27, 28, 29, 31, 35	$0.0347\ d^2$	$530\ d^2$	$600\ d^2$
6 × 19	13, 14, 16, 18, 19, 20, 21 22, 24, 25, 26, 28, 29, 32 35, 36, 38	$0.0363\ d^2$	$530\ d^2$	$595\ d^2$

Table 20.7. Steel wire suspension ropes for lifts, elevators and hoists.

Type of rope	*Nominal diameter (mm)*	*Average weight (N/m)*	*Tensile strength (N)*	
			Tensile strength of wire	
			1100–1250 MPa	*1250–1400 MPa*
6 × 19	6, 8, 10, 12, 14, 16 18, 20, 22, 25	$0.0383\ d^2$	$385\ d^2$	$435\ d^2$
8 × 19	8, 10, 12, 14,16 18, 20, 22, 25	$0.034\ d^2$	$355\ d^2$	$445\ d^2$

Table 20.8. Steel wire ropes used in oil wells and oil well drilling.

Type of rope	*Nominal diameter (mm)*	*Approximate weight (N/m)*	*Ultimate tensile strength (N)*		
			Tensile strength of wire		
			1600 – 1800 MPa	*1800 – 2000 MPa*	*2000 – 2250 MPa*
6 × 7	10, 11, 13, 14, 16, 19, 22, 25	$0.037\ d^2$	$550\ d^2$	$610\ d^2$	–
6 × 19	13, 14, 16, 19 22, 25, 29, 32, 35, 38,	$0.037\ d^2$	$510\ d^2$	$570\ d^2$	$630\ d^2$
6 × 37	13, 14, 16, 19, 22, 25, 26, 32, 35, 38	$0.037\ d^2$	$490\ d^2$	$540\ d^2$	$600\ d^2$
8 × 19	13, 14, 16, 19, 22, 25, 29	$0.0338\ d^2$	–	$530\ d^2$	–

Table 20.9. Steel wire ropes for general engineering purposes such as cranes, excavators etc.

Type of rope	*Nominal diameter (mm)*	*Average weight (N/m)*	*Average tensile strength (N)*	
			Tensile strength of wire	
			1600–1750 MPa	*1750–1900 MPa*
6 × 19	8, 9, 10, 11, 12, 13, 14, 16, 18, 20, 22, 24, 26, 28, 32, 36, 38, 40	$0.0375\ d^2$	$540\ d^2$	$590\ d^2$
6 × 37	8, 9, 10, 11, 12, 13, 14 16, 18, 20, 22, 24, 26, 28, 32, 36, 40, 44, 48, 52, 56	$0.038\ d^2$	$510\ d^2$	$550\ d^2$

20.18 Diameter of Wire and Area of Wire Rope

The following table shows the diameter of wire (d_w) and area of wire rope (A) for different types of wire ropes :

Table 20.10. Diameter of wire and area of wire rope.

Type of wire rope	6 × 8	6 × 19	6 × 37	8 × 19
Wire diameter (d_w)	$0.106\ d$	$0.063\ d$	$0.045\ d$	$0.050\ d$
Area of wire rope (A)	$0.38\ d^2$	$0.38\ d^2$	$0.38\ d^2$	$0.35\ d^2$

20.19 Factor of Safety for Wire Ropes

The factor of safety for wire ropes based on the ultimate strength are given in the following table.

Table 20.11. Factor of safety for wire ropes.

Application of wire rope	*Factor of safety*	*Application of wire rope*	*Factor of safety*
Track cables	4.2	Derricks	6
Guys	3.5	Haulage ropes	6
Mine hoists : Depths		Small electric and air hoists	7
upto 150 m	8	Over head and gantry cranes	6
300 – 600 m	7	Jib and pillar cranes	6
600 – 900 m	6	Hot ladle cranes	8
over 900 m	5	Slings	8
Miscellaneous hoists	5		

20.20 Wire Rope Sheaves and Drums

The sheave diameter should be fairly large in order to reduce the bending stresses in the ropes when they bend around the sheaves or pulleys. The following table shows the sheave diameters for various types of wire ropes :

Table 20.12. Sheave diameters (*D*) for wire ropes.

Type of wire rope	*Recommended sheave diameter (D)*		*Uses*
	Minimum sheave diameter	*Preferred sheave diameter*	
6 × 7	42 *d*	72 *d*	Mines, haulage tramways.
6 × 19	30 *d*	45 *d*	Hoisting rope.
	60 *d*	100 *d*	Cargo cranes, mine hoists
	20 *d*	30 *d*	Derricks, dredges, elevators, tramways, well drilling.
6 × 37	18 *d*	27 *d*	Cranes, high speed elevators and small shears.
8 × 19	21 *d*	31 *d*	Extra flexible hoisting rope.

However, if the space allows, then the large diameters should be employed which give better and more economical service.

Sheave or pulleys for winding ropes

The sheave groove has a great influence on the life and service of the rope. If the groove is bigger than rope, there will not be sufficient support for the rope which may, therefore, flatten from its normal circular shape and increase fatigue effects. On the other hand, if the groove is too small, then the rope will be wedged into the groove and thus the normal rotation is prevented. The standard rim of a rope sheave is shown in Fig. 20.9 (*a*) and a standard grooved drum for wire ropes is shown in Fig. 20.9 (*b*).

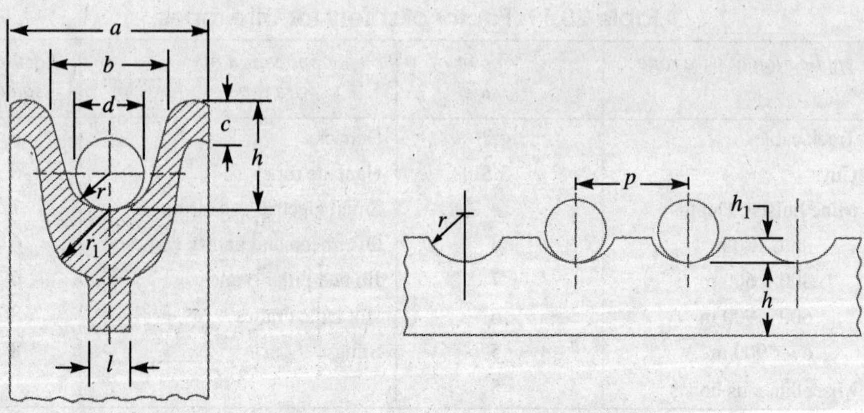

$r = 0.53\ d$; $r_1 = 1.1\ d$; $a = 2.7\ d$; $b = 2.1\ d$; $c = 0.4\ d$; $h = 1.6\ d$; $l = 0.75\ d$

(*a*) Wire rope sheave rim.

$p = 1.15\ d$; $h_1 = 0.25\ d$; $r = 0.53\ d$; $h = 1.1\ d$

(*b*) Grooved rope drum.

Fig. 20.9

For light and medium service, the sheaves are made of cast iron, but for heavy crane service they are often made of steel castings. The sheaves are usually mounted on fixed axles on antifriction bearings or bronze bushings.

The small drums in hand hoists are made plain. A hoist operated by a motor or an engine has a drum with helical grooves, as shown in Fig. 20.9 (*b*). The pitch (*p*) of the grooves must be made slightly larger than the rope diameter to avoid friction and wear between the coils.

20.21 Wire Rope Fasteners

The various types of rope fasteners are shown in Fig. 20.10. The splices in wire ropes should be avoided because it reduces the strength of the rope by 25 to 30 percent of the normal ultimate strength.

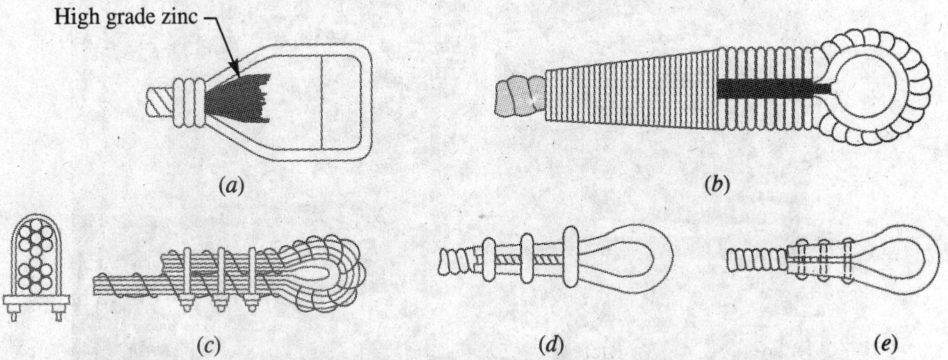

Fig. 20.10. Types of wire rope fastners.

The efficiencies of various types of fasteners are given in the following table.

Table 20.13. Efficiencies of rope fasteners.

Type of fastening	*Efficiency (%)*
(*a*) Wire rope socket with zinc, Fig. 20.10 (*a*)	100
(*b*) Thimble with four or five wire tucks, Fig. 20.10 (*b*)	90
(*c*) Special offset thimble with clips, Fig. 20.10 (*c*)	90
(*d*) Regular thimble with clips, Fig. 20.10 (*d*)	85
(*e*) Three bolt wire clamps, Fig. 20.10 (*e*)	75

20.22 Stresses in Wire Ropes

A wire rope is subjected to the following types of stresses :

1. ***Direct stress due to axial load lifted and weight of the rope***

Let W = Load lifted,

w = Weight of the rope, and

A = Net cross-sectional area of the rope.

$\therefore$ Direct stress, $$\sigma_d = \frac{W + w}{A}$$

2. ***Bending stress when the rope winds round the sheave or drum.*** When a wire rope is wound over the sheave, then the bending stresses are induced in the wire which is tensile at the top and compressive at the lower side of the wire. The bending stress induced depends upon many factors such as construction of rope, size of wire, type of centre and the amount of restraint in the grooves. The approximate value of the bending stress in the wire as proposed by Reuleaux, is

$$\sigma_b = \frac{E_r \times d_w}{D}$$

A heavy duty crane. Cranes use rope drives in addition to gear drives

and equivalent bending load on the rope,

$$W_b = \sigma_b \times A = \frac{E_r \times d_w \times A}{D}$$

where E_r = Modulus of elasticity of the wire rope,

d_w = Diameter of the wire,

D = Diameter of the sheave or drum, and

A = Net cross-sectional area of the rope.

It may be noted that E_r is not the modulus of elasticity for the wire material, but it is of the entire rope. The value of E_r may be taken as 77 kN/mm^2 for wrought iron ropes and 84 kN/mm^2 for steel ropes. It has been found experimentally that $E_r = 3/8\ E$, where E is the modulus of elasticity of the wire material.

If σ_b is the bending stress in each wire, then the load on the whole rope due to bending may be obtained from the following relation, *i.e.*

$$W_b = \frac{\pi}{4}(d_w)^2\, n \times \sigma_b$$

where n is the total number of wires in the rope section.

3. ***Stresses during starting and stopping.*** During starting and stopping, the rope and the supported load are to be accelerated. This induces additional load in the rope which is given by

$$W_a = \frac{W + w}{g} \times a \qquad \text{...(}W\text{ and }w\text{ are in newton)}$$

and the corresponding stress,

$$\sigma_a = \frac{W + w}{g} \times \frac{a}{A}$$

where a = Acceleration of the rope and load, and

g = Acceleration due to gravity.

If the time (t) necessary to attain a speed (v) is known, then the value of 'a' is given by

$$a = v / 60\,t$$

The general case of starting is when the rope has a slack (h) which must be overcome before the rope is taut and starts to exert a pull on the load. This induces an impact load on the rope.

The impact load on starting may be obtained by the impact equation, *i.e.*

$$W_{st} = (W + w)\left[1 + \sqrt{1 + \frac{2a \times h \times E_r}{\sigma_d \times l \times g}}\right]$$

and velocity of the rope (v_r) at the instant when the rope is taut,

$$v_r = \sqrt{2a \times h}$$

where a = Acceleration of the rope and load,

h = Slackness in the rope, and

l = Length of the rope.

When there is no slackness in the rope, then $h = 0$ and $v_r = 0$, therefore

Impact load during starting,

$$W_{st} = 2\,(W + w)$$

and the corresponding stress,

$$\sigma_{st} = \frac{2\,(W + w)}{A}$$

4. ***Stress due to change in speed.*** The additional stress due to change in speed may be obtained in the similar way as discussed above in which the acceleration is given by

$$a = (v_2 - v_1) / t$$

where $(v_2 - v_1)$ is the change in speed in m/s and t is the time in seconds.

It may be noted that when the hoist drum is suddenly stopped while lowering the load, it produces a stress that is several times more than the direct or static stress because of the kinetic energy of the moving masses is suddenly made zero. This kinetic energy is absorbed by the rope and the resulting stress may be determined by equating the kinetic energy to the resilience of the rope. If during stopping, the load moves down a certain distance, the corresponding change of potential energy must be added to the kinetic energy. It is also necessary to add the work of stretching the rope during stopping, which may be obtained from the impact stress.

5. ***Effective stress.*** The sum of the direct stress (σ_d) and the bending stress (σ_b) is called the effective stress in the rope during normal working. Mathematically,

Effective stress in the rope during normal working

$$= \sigma_d + \sigma_b$$

Effective stress in the rope during starting

$$= \sigma_{st} + \sigma_b$$

and effective stress in the rope during acceleration of the load

$$= \sigma_d + \sigma_b + \sigma_a$$

While designing a wire rope, the sum of these stresses should be less than the ultimate strength divided by the factor of safety.

Ropes on a pile driver

20.23 Procedure for Designing a Wire Rope

The following procedure may be followed while designing a wire rope.

1. First of all, select a suitable type of rope from Tables 20.6, 20.7, 20.8 and 20.9 for the given application.
2. Find the design load by assuming a factor of safety 2 to 2.5 times the factor of safety given in Table 20.11.
3. Find the diameter of wire rope (d) by equating the tensile strength of the rope selected to the design load.
4. Find the diameter of the wire (d_w) and area of the rope (A) from Table 20.10.
5. Find the various stresses (or loads) in the rope as discussed in Art. 20.22.
6. Find the effective stresses (or loads) during normal working, during starting and during acceleration of the load.
7. Now find the actual factor of safety and compare with the factor of safety given in Table 20.11. If the actual factor of safety is within permissible limits, then the design is safe.

Wheel that winds the metal rope.

Example 20.10. *Select a wire rope for a vertical mine hoist to lift a load of 55 kN from a depth 300 metres. A rope speed of 500 metres / min is to be attained in 10 seconds.*

Solution. Given : $W = 55$ kN = 55 000 N ; Depth = 300 m ; $v = 500$ m/min ; $t = 10$ s

The following procedure may be adopted in selecting a wire rope for a vertical mine hoist.

1. From Table 20.6, we find that the wire ropes for haulage purposes in mines are of two types, *i.e.* 6 × 7 and 6 × 19. Let us take a rope of type 6 × 19.

2. From Table 20.11, we find that the factor of safety for mine hoists from 300 to 600 m depth is 7. Since the design load is calculated by taking a factor of safety 2 to 2.5 times the factor of safety given in Table 20.11, therefore let us take the factor of safety as 15.

∴ Design load for the wire rope

$$= 15 \times 55 = 825 \text{ kN} = 825\ 000 \text{ N}$$

3. From Table 20.6, we find that the tensile strength of 6 × 19 rope made of wire with tensile strength of 1800 MPa is 595 d^2 (in newton), where d is the diameter of rope in mm. Equating this tensile strength to the design load, we get

$$595\, d^2 = 825\ 000$$

$$\therefore \quad d^2 = 825\ 000 / 595 = 1386.5 \quad \text{or} \quad d = 37.2 \text{ say } 38 \text{ mm}$$

4. From Table 20.10, we find that for a 6 × 19 rope,

Diameter of wire, $\quad d_w = 0.063\, d = 0.063 \times 38 = 2.4$ mm

and area of rope, $\quad A = 0.38\, d^2 = 0.38\, (38)^2 = 550 \text{ mm}^2$

5. Now let us find out the various loads in the rope as discussed below :

(*a*) From Table 20.6, we find that weight of the rope,

$$w = 0.0363\, d^2 = 0.0363\, (38)^2 = 52.4 \text{ N/m}$$

$$= 52.4 \times 300 = 15\ 720 \text{ N} \qquad \text{...}(\because \text{Depth} = 300 \text{ m})$$

(*b*) From Table 20.12, we find that diameter of the sheave (D) may be taken as 60 to 100 times the diameter of rope (d). Let us take

$$D = 100\, d = 100 \times 38 = 3800 \text{ mm}$$

∴ Bending stress,

$$\sigma_b = \frac{E_r \times d_w}{D} = \frac{84 \times 10^3 \times 2.4}{3800} = 53 \text{ N/mm}^2$$

...(Taking $E_r = 84 \times 10^3$ N/mm²)

and the equivalent bending load on the rope,

$$W_b = \sigma_b \times A = 53 \times 550 = 29\ 150 \text{ N}$$

(*c*) We know that the acceleration of the rope and load,

$$a = v / 60t = 500 / 60 \times 10 = 0.83 \text{ m / s}^2$$

∴ Additional load due to acceleration,

$$W_a = \frac{W + w}{g} \times a = \frac{55\ 000 + 15\ 720}{9.81} \times 0.83 = 5983 \text{ N}$$

(*d*) We know that the impact load during starting (when there is no slackness in the rope),

$$W_{st} = 2\,(W + w) = 2(55\ 000 + 15\ 720) = 141\ 440 \text{ N}$$

6. We know that the effective load on the rope during normal working (*i.e.* during uniform lifting or lowering of the load)

$$= W + w + W_b = 55\ 000 + 15\ 720 + 29150 = 99\ 870 \text{ N}$$

∴ Actual factor of safety during normal working

$$= \frac{825\,000}{99\,870} = 8.26$$

Effective load on the rope during starting

$$= W_{st} + W_b = 141\,440 + 29\,150 = 170\,590 \text{ N}$$

∴ Actual factor of safety during starting

$$= \frac{825\,000}{170\,590} = 4.836$$

Effective load on the rope during acceleration of the load (*i.e.* during first 10 seconds after starting)

$$= W + w + W_b + W_a$$

$$= 55\,000 + 15\,7\,20 + 29\,150 + 5983 = 105\,853 \text{ N}$$

∴ Actual factor of safety during acceleration of the load

$$= \frac{825\,000}{105\,853} = 7.8$$

Since the actual factor of safety as calculated above are safe, therefore a wire rope of diameter 38 mm and 6 × 19 type is satisfactory. **Ans.**

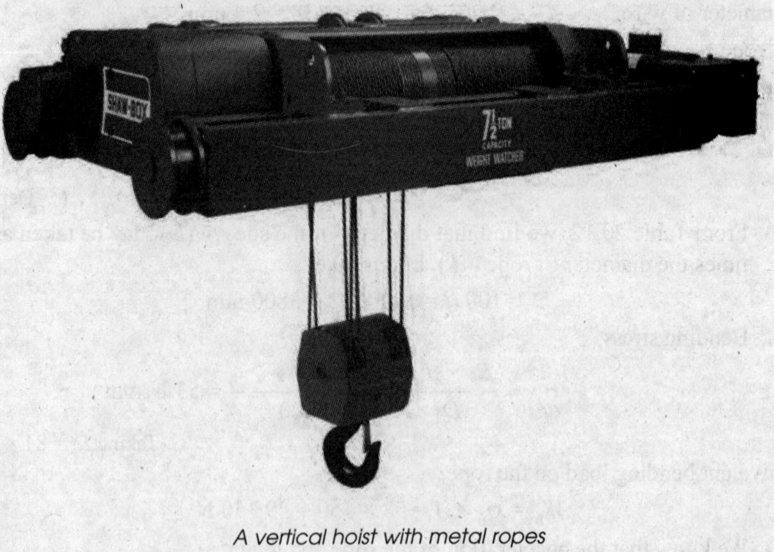

A vertical hoist with metal ropes

Example 20.11. *An extra flexible 8 × 19 plough steel wire rope of 38 mm diameter is used with a 2m diameter hoist drum to lift 50 kN of load. Find the factor of safety (ratio of the breaking load to the maximum working load) under the following conditions of operation :*

The wire rope is required to lift from a depth of 900 metres. The maximum speed is 3 m / s and the acceleration is 1.5 m / s², when starting under no slack condition. The diameter of the wire may be taken as 0.05 d, where d is the diameter of wire rope. The breaking strength of plough steel is 1880 N/mm² and modulus of elasticity of the entire rope is 84 × 10³ N/mm². The weight of the rope is 53 N/m length.

Solution. Given : $d = 38$ mm ; $D = 2$ m $= 2000$ mm ; $W = 50$ kN $= 50\,000$ N ; Depth = 900 m ; $v = 3$ m/s ; $a = 1.5$ m/s² ; $d_w = 0.05\ d$; Breaking strength = 1880 N/mm² ; $E_r = 84 \times 10^3$ N/mm² ; $w = 53$ N/m $= 53 \times 900 = 47\,700$ N

Since the wire rope is 8 × 19, therefore total number of wires in the rope,

$$n = 8 \times 19 = 152$$

We know that diameter of each wire,

$$d_w = 0.05\, d = 0.05 \times 38 = 1.9 \text{ mm}$$

∴ Cross-sectional area of the wire rope,

$$A = \frac{\pi}{4}(d_w)^2\, n = \frac{\pi}{4}(1.9)^2\, 152 = 431 \text{ mm}^2$$

and minimum breaking strength of the rope

$$= \text{Breaking strength} \times \text{Area} = 1880 \times 431 = 810\ 280 \text{ N}$$

We know that bending stress,

$$\sigma_b = \frac{E_r \times d_w}{D} = \frac{84 \times 10^3 \times 1.9}{2000} = 79.8 \text{ N/mm}^2$$

and equivalent bending load on the rope,

$$W_b = \sigma_b \times A = 79.8 \times 431 = 34\ 390 \text{ N}$$

Additional load due to acceleration of the load lifted and rope,

$$W_a = \frac{W + w}{g} \times a = \frac{50\ 000 + 47\ 700}{9.81} \times 1.5 = 14\ 940 \text{ N}$$

Impact load during starting (when there is no slackness in the rope),

$$W_{st} = 2\,(W + w) = 2(50\ 000 + 47\ 700) = 195\ 400 \text{ N}$$

We know that the effective load on the rope during normal working

$$= W + w + W_b = 50\ 000 + 47\ 700 + 34\ 390 = 132\ 090 \text{ N}$$

∴ Factor of safety during normal working

= 810 280 / 132 090 = 6.13 **Ans.**

Effective load on the rope during starting

$$= W_{st} + W_b = 195\ 400 + 34\ 390 = 229\ 790 \text{ N}$$

∴ Factor of safety during starting

= 810 280 / 229 790 = 3.53 **Ans.**

Effective load on the rope during acceleration of the load (*i.e.* during the first 2 second after starting)

$$= W + w + W_b + W_a = 50\ 000 + 47\ 700 + 34\ 390 + 14\ 940$$
$$= 147\ 030 \text{ N}$$

∴ Factor of safety during acceleration of the load

= 810 280 / 147 030 = 5.51 **Ans.**

Example 20.12. *A workshop crane is lifting a load of 25 kN through a wire rope and a hook. The weight of the hook etc. is 15 kN. The rope drum diameter may be taken as 30 times the diameter of the rope. The load is to be lifted with an acceleration of 1 m/s². Calculate the diameter of the wire rope. Take a factor of safety of 6 and Young's modulus for the wire rope 80 kN/mm². The ultimate stress may be taken as 1800 MPa. The cross-sectional area of the wire rope may be taken as 0.38 times the square of the wire rope diameter.*

Solution. Given : $W = 25$ kN = 25 000 N ; $w = 15$ kN = 15 000 N ; $D = 30\, d$; $a = 1$ m/s² ; $E_r = 80$ kN/mm² = 80×10^3 N/mm² ; $\sigma_u = 1800$ MPa = 1800 N/mm² ; $A = 0.38\, d^2$

Let d = Diameter of wire rope in mm.

We know that direct load on the wire rope,

$$W_d = W + w = 25\ 000 + 15\ 000 = 40\ 000 \text{ N}$$

Let us assume that a 6 × 19 wire rope is used. Therefore from Table 20.10, we find that the diameter of wire,

$$d_w = 0.063\ d$$

We know that bending load on the rope,

$$W_b = \frac{E_r \times d_w}{D} \times A = \frac{80 \times 10^3 \times 0.063\,d}{30\,d} \times 0.38\ d^2 = 63.84\ d^2 \text{ N}$$

and load on the rope due to acceleration,

$$W_a = \frac{W + w}{g} \times a = \frac{25\ 000 + 15\ 000}{9.81} \times 1 = 4080 \text{ N}$$

∴ Total load acting on the rope

$$= W_d + W_b + W_a = 40\ 000 + 63.84\ d^2 + 4080$$

$$= 44\ 080 + 63.84\ d^2 \qquad \text{...(i)}$$

We know that total load on the rope

= Area of wire rope × Allowable stress

$$= A \times \frac{\sigma_u}{F.S.} = 0.38\ d^2 \times \frac{1800}{6} = 114\ d^2 \qquad \text{...(ii)}$$

From equations (*i*) and (*ii*), we have

$$44\ 080 + 63.84\ d^2 = 114\ d^2$$

$$d^2 = \frac{44\ 080}{114 - 63.84} = 879 \quad \text{or} \quad d = 29.6 \text{ mm}$$

From Table 20.9, we find that standard nominal diameter of 6 × 19 wire rope is 32 mm. **Ans.**

EXERCISES

1. A *V*-belt drive consists of three *V*-belts in parallel on grooved pulleys of the same size. The angle of groove is 30° and the coefficient of friction 0.12. The cross-sectional area of each belt is 800 mm^2 and the permissible safe stress in the material is 3 MPa. Calculate the power that can be transmitted between two pulleys 400 mm in diameter rotating at 960 r.p.m. **[Ans. 101.7 kW]**

2. Power is transmitted between two shafts by a *V*-belt whose mass is 0.9 kg/m length. The maximum permissible tension in the belt is limited to 2.2 kN. The angle of lap is 170° and the groove angle 45°. If the coefficient of friction between the belt and pulleys is 0.17; find 1. velocity of the belt for maximum power; and 2. power transmitted at this velocity. **[Ans. 28.54 m/s ; 30.66 kW]**

3. A *V*-belt drive system transmits 100 kW at 475 r.p.m. The belt has a mass of 0.6 kg/m. The maximum permissible tension in the belt is 900 N. The groove angle is 38° and the angle of contact is 160°. Find minimum number of belts and pulley diameter. The coefficient of friction between belt and pulley is 0.2. **[Ans. 9 ; 0.9 m]**

4. A-V belt is to transmit 20 kW from a 250 mm pitch diameter sheave to a 900 mm diameter pulley. The centre distance between the two shafts is 1000 mm. The groove angle is 40° and the coefficient of friction for the belt and sheave is 0.2 and the coefficient of friction between the belt and flat pulley is 0.2. The cross-section of the belt is 40 mm wide at the top, 20 mm wide at the bottom and 25 mm deep. The density of the belt is 1000 kg / m^3 and the allowable tension per belt is 1000 N. Find the number of belts required. **[Ans. 3]**

5. Determine the number of V-belts required to transmit 30 kW power under the following conditions :

	Smaller pulley	*Larger pulley*
Speed	1120 r.p.m.	280 r.p.m.
Pitch diameter	225 mm	900 mm
Pulley groove angle	34°	34°

Maximum working load per belt = 560 N
Coefficient of friction = 0.15
Centre distance between pulleys = 875 mm
Mass of belt = 0.3 kg/m **[Ans. 7]**

6. Determine the percentage increase in power capacity made possible in changing over from a flat belt drive to a V-belt drive. The diameter of the flat pulley is same as the pitch diameter of the grooved pulley. The pulley rotates at the same speed as the grooved pulley. The coefficient of friction for the grooved and flat belt is same and is 0.3. The V-belt pulley groove angle is 60°. The belts are of the same material and have same cross-sectional area. In each case, the angle of wrap is 150°. **[Ans. 217.52 %]**

7. A rope drive is required to transmit 750 kW from a pulley of 1 m diameter running at 450 r.p.m. The safe pull in each rope is 2250 N and the mass of the rope is 1 kg / m length. The angle of lap and the groove angle is 150° and 45° respectively. Find the number of ropes required for the drive if the coefficient of friction between the rope and the pulley is 0.3. **[Ans. 22]**

8. Following data is given for a rope pulley transmitting 24 kW :

Diameter of pulley = 400 mm ; Speed = 110 r.p.m ; Angle of groove = 45° ; Angle of lap = 160° ; Coefficient of friction = 0.28 ; Number of ropes = 10 ; Mass in kg / m length of ropes = 53 C^2.

The working tension is limited to 122 C^2; where C = girth (*i.e.* circumference) of rope in metres. Find the initial tension and diameter of each rope. **[Ans. 675.55 N ; 31.57 mm]**

9. Select a suitable wire rope to lift a load of 10 kN of debris from a well 60 m deep. The rope should have a factor of safety equal to 6. The weight of the bucket is 5 kN. The load is lifted up with a maximum speed of 150 metres/min which is attained in 1 second.

Find also the stress induced in the rope due to starting with an initial slack of 250 mm. The average tensile strength of the rope may be taken as 590 d^2 newtons (where d is the rope diameter in mm) for 6 × 19 wire rope. The weight of the rope is 18.5 N/m.

Take diameter of the wire (d_w) = 0.063 d, and area of the rope (A) = 0.38 d^2.

[Ans. 20 mm, 412 MPa]

10. Suggest the suitable size of 6 × 19 hoisting steel wire rope for an inclined mine shaft of 1000 m length and inclination of the rails 60° with the horizontal. The weight of the loaded skip is 100 kN. The maximum acceleration is limited to 1.5 m/s^2. The diameter of the drum on which the rope is being wound may be taken as 80 times the diameter of the rope. The car friction is 20 N / kN of weight normal to the incline and friction of the rope on the guide roller is 50 N / kN of weight normal to the incline. Assume a factor of safety of 5. The following properties of 6 × 19 flexible hoisting rope are given :

The diameter of the rope (d) is in mm. The weight of the rope per metre = 0.0334 d^2 N; breaking load = 500 d^2 N; wire diameter = 0.063 d mm; area of wires in rope = 0.38 d^2 mm^2; equivalent elastic modulus = 82 kN/mm^2. **[Ans. 105 mm]**

QUESTIONS

1. Sketch the cross-section of a *V*-belt and label its important parts.
2. What are the advantages and disadvantages of *V*-belt drive over flat belt drive?
3. Derive the relation for the ratio of driving tensions of a *V*-belt.
4. Describe the fibre ropes. What are its advantages? Draw a neat proportionate sketch of a sheave for fibre ropes.
5. Under what circumstances a fibre rope and a wire rope is used ? What are the advantages of a wire rope over fibre rope ?
6. Discuss the uses and construction of wire ropes. How are wire-rope ends fastened ?
7. Give the application of the following wire ropes :

   (*a*) 6 × 7 rope (*b*) 6 × 19 rope, and (*c*) 6 × 37 rope.

OBJECTIVE TYPE QUESTIONS

1. The included angle for the *V*-belt is usually
   (*a*) 20° – 30°
   (*b*) 30° – 40°
   (*c*) 40° – 60°
   (*d*) 60° – 80°
2. The *V*-belts are particularly suitable for .... drives.
   (*a*) short
   (*b*) long
3. The groove angle of the pulley for *V*-belt drive is usually
   (*a*) 20° – 25°
   (*b*) 25° – 32°
   (*c*) 32° – 38°
   (*d*) 38° – 45°
4. A *V*-belt designated by *A*-914-50 denotes
   (*a*) a standard belt
   (*b*) an oversize belt
   (*c*) an undersize belt
   (*d*) none of these
5. The wire ropes make contact at
   (*a*) bottom of groove of the pulley
   (*b*) sides of groove of the pulley
   (*c*) sides and bottom of groove of the pulley
   (*d*) any where in the groove of the pulley

This heavy duty crane moves within shopfloor on fixed rails.

ANSWERS

1. (*b*)	2. (*a*)	3. (*c*)	4. (*a*)	5. (*a*)

21 Chain Drives

21.1 Introduction

We have seen in previous chapters on belt and rope drives that slipping may occur. In order to avoid slipping, steel chains are used. The chains are made up of number of rigid links which are hinged together by pin joints in order to provide the necessary flexibility for wraping round the driving and driven wheels. These wheels have projecting teeth of special profile and fit into the corresponding recesses in the links of the chain as shown in Fig. 21.1. The toothed wheels are known as ****sprocket wheels or simply sprockets.*** The sprockets and the chain are thus constrained to move together without slipping and ensures perfect velocity ratio.

---

* These wheels resemble to spur gears.

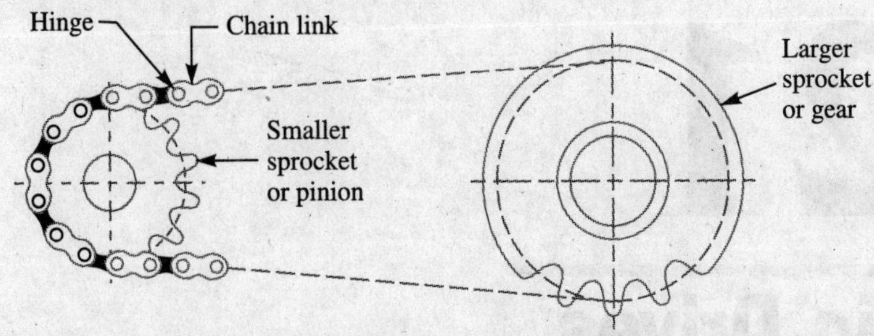

Fig. 21.1. Sprockets and chain.

The chains are mostly used to transmit motion and power from one shaft to another, when the centre distance between their shafts is short such as in bicycles, motor cycles, agricultural machinery, conveyors, rolling mills, road rollers etc. The chains may also be used for long centre distance of upto 8 metres. The chains are used for velocities up to 25 m / s and for power upto 110 kW. In some cases, higher power transmission is also possible.

21.2 Advantages and Disadvantages of Chain Drive over Belt or Rope Drive

Following are the advantages and disadvantages of chain drive over belt or rope drive:

Advantages

1. As no slip takes place during chain drive, hence perfect velocity ratio is obtained.
2. Since the chains are made of metal, therefore they occupy less space in width than a belt or rope drive.
3. It may be used for both long as well as short distances.
4. It gives a high transmission efficiency (upto 98 percent).
5. It gives less load on the shafts.
6. It has the ability to transmit motion to several shafts by one chain only.
7. It transmits more power than belts.
8. It permits high speed ratio of 8 to 10 in one step.
9. It can be operated under adverse temperature and atmospheric conditions.

Disadvantages

1. The production cost of chains is relatively high.
2. The chain drive needs accurate mounting and careful maintenance, particularly lubrication and slack adjustment.
3. The chain drive has velocity fluctuations especially when unduly stretched.

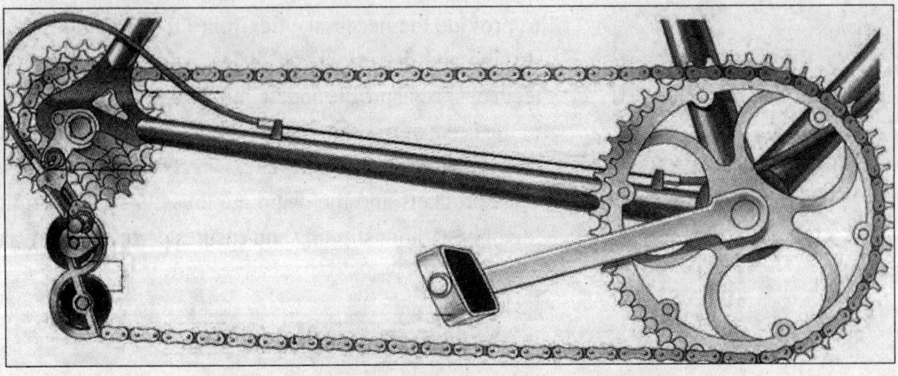

Sports bicycle gear and chain drive mechanism

21.3 Terms Used in Chain Drive

The following terms are frequently used in chain drive.

1. ***Pitch of chain.*** It is the distance between the hinge centre of a link and the corresponding hinge centre of the adjacent link, as shown in Fig. 21.2. It is usually denoted by *p*.

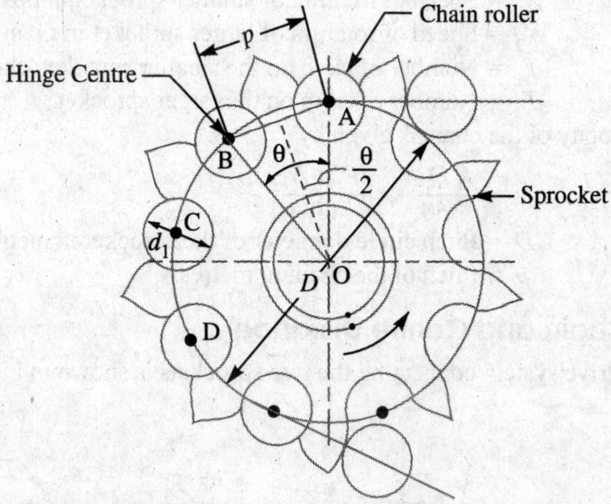

Fig. 21.2. Terms used in chain drive.

2. ***Pitch circle diameter of chain sprocket.*** It is the diameter of the circle on which the hinge centres of the chain lie, when the chain is wrapped round a sprocket as shown in Fig. 21.2. The points *A*, *B*, *C*, and *D* are the hinge centres of the chain and the circle drawn through these centres is called pitch circle and its diameter (*D*) is known as pitch circle diameter.

21.4 Relation Between Pitch and Pitch Circle Diameter

A chain wrapped round the sprocket is shown in Fig. 21.2. Since the links of the chain are rigid, therefore pitch of the chain does not lie on the arc of the pitch circle. The pitch length becomes a chord. Consider one pitch length *AB* of the chain subtending an angle θ at the centre of sprocket (or pitch circle),

Let D = Diameter of the pitch circle, and

T = Number of teeth on the sprocket.

From Fig. 21.2, we find that pitch of the chain,

$$p = AB = 2\,AO \sin\left(\frac{\theta}{2}\right) = 2 \times \left(\frac{D}{2}\right) \sin\left(\frac{\theta}{2}\right) = D \sin\left(\frac{\theta}{2}\right)$$

We know that $\theta = \dfrac{360^\circ}{T}$

$$\therefore \quad p = D \sin\left(\frac{360^\circ}{2\,T}\right) = D \sin\left(\frac{180^\circ}{T}\right)$$

or

$$D = p \operatorname{cosec}\left(\frac{180^\circ}{T}\right)$$

The sprocket outside diameter (D_o), for satisfactory operation is given by

$$D_o = D + 0.8\,d_1$$

where d_1 = Diameter of the chain roller.

Note: The angle θ/2 through which the link swings as it enters contact is called **angle of articulation.**

21.5 Velocity Ratio of Chain Drives

The velocity ratio of a chain drive is given by

$$V.R. = \frac{N_1}{N_2} = \frac{T_2}{T_1}$$

where N_1 = Speed of rotation of smaller sprocket in r.p.m.,

N_2 = Speed of rotation of larger sprocket in r.p.m.,

T_1 = Number of teeth on the smaller sprocket, and

T_2 = Number of teeth on the larger sprocket.

The average velocity of the chain is given by

$$v = \frac{\pi\, D\, N}{60} = \frac{T\, p\, N}{60}$$

where D = Pitch circle diameter of the sprocket in metres, and

p = Pitch of the chain in metres.

21.6 Length of Chain and Centre Distance

An open chain drive system connecting the two sprockets is shown in Fig. 21.3.

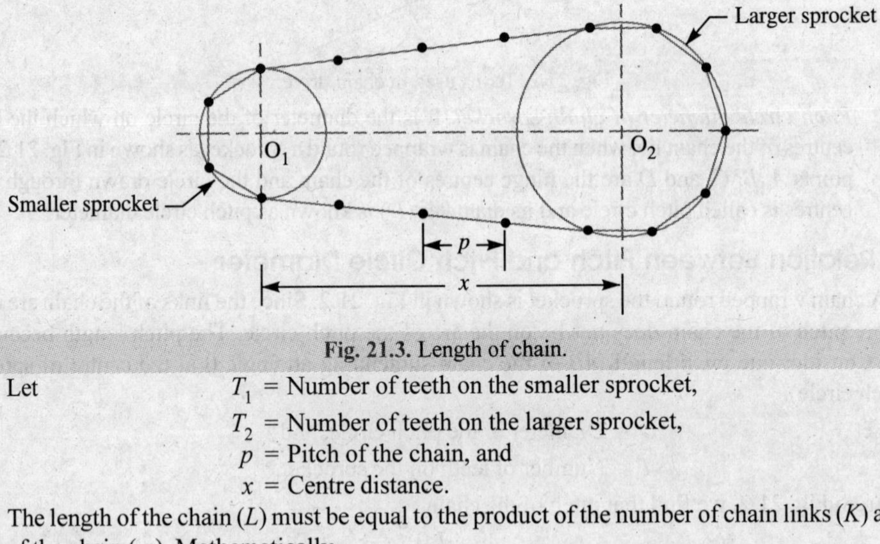

Fig. 21.3. Length of chain.

Let T_1 = Number of teeth on the smaller sprocket,

T_2 = Number of teeth on the larger sprocket,

p = Pitch of the chain, and

x = Centre distance.

The length of the chain (L) must be equal to the product of the number of chain links (K) and the pitch of the chain (p). Mathematically,

$$L = K.p$$

The number of chain links may be obtained from the following expression, *i.e.*

$$K = \frac{T_1 + T_2}{2} + \frac{2\,x}{p} + \left[\frac{T_2 - T_1}{2\,\pi}\right]^2 \frac{p}{x}$$

The value of K as obtained from the above expression must be approximated to the nearest even number.

The centre distance is given by

$$x = \frac{p}{4}\left[K - \frac{T_1 + T_2}{2} + \sqrt{\left(K - \frac{T_1 + T_2}{2}\right)^2 - 8\left(\frac{T_2 - T_1}{2\,\pi}\right)^2}\,\right]$$

In order to accommodate initial sag in the chain, the value of the centre distance obtained from the above equation should be decreased by 2 to 5 mm.

Notes: 1. The minimum centre distance for the velocity transmission ratio of 3, may be taken as

$$x_{min} = \frac{d_1 + d_2}{2} + 30 \text{ to } 50 \text{ mm}$$

where d_1 and d_2 are the diameters of the pitch circles of the smaller and larger sprockets.

2. For best results, the minimum centre distance should be 30 to 50 times the pitch.

3. The minimum centre distance is selected depending upon the velocity ratio so that the arc of contact of the chain on the smaller sprocket is not less than 120°. It may be noted that larger angle of arc of contact ensures a more uniform distribution of load on the sprocket teeth and better conditions of engagement.

21.7 Classification of Chains

The chains, on the basis of their use, are classified into the following three groups:

1. Hoisting and hauling (or crane) chains,
2. Conveyor (or tractive) chains, and
3. Power transmitting (or driving) chains.

These chains are discussed, in detail, in the following pages.

21.8 Hoisting and Hauling Chains

These chains are used for hoisting and hauling purposes and operate at a maximum velocity of 0.25 m / s. The hoisting and hauling chains are of the following two types:

1. ***Chain with oval links.*** The links of this type of chain are of oval shape, as shown in Fig. 21.4 (*a*). The joint of each link is welded. The sprockets which are used for this type of chain have receptacles to receive the links. Such type of chains are used only at low speeds such as in chain hoists and in anchors for marine works.

(*a*) Chain with oval links. (*b*) Chain with square links.

Fig. 21.4. Hoisting and hauling chains.

2. ***Chain with square links.*** The links of this type of chain are of square shape, as shown in Fig. 21.4 (*b*). Such type of chains are used in hoists, cranes, dredges. The manufacturing cost of this type of chain is less than that of chain with oval links, but in these chains, the kinking occurs easily on overloading.

21.9 Conveyor Chains

These chains are used for elevating and conveying the materials continuously at a speed upto 2 m / s. The conveyor chains are of the following two types:

1. Detachable or hook joint type chain, as shown in Fig. 21.5 (*a*), and
2. Closed joint type chain, as shown in Fig. 21.5 (*b*).

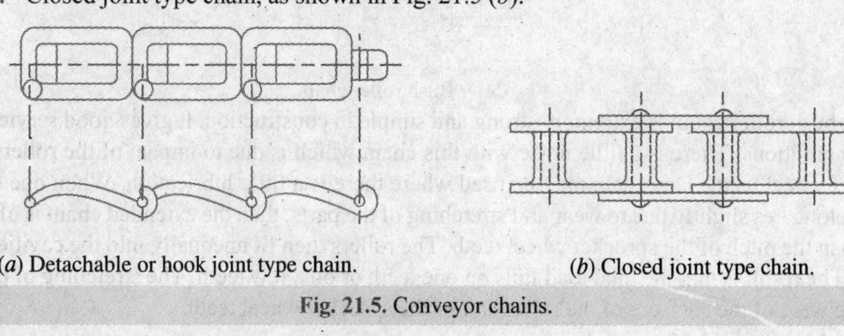

(*a*) Detachable or hook joint type chain. (*b*) Closed joint type chain.

Fig. 21.5. Conveyor chains.

The conveyor chains are usually made of malleable cast iron. These chains do not have smooth running qualities. The conveyor chains run at slow speeds of about 0.8 to 3 m / s.

21.10 Power Transmitting Chains

These chains are used for transmission of power, when the distance between the centres of shafts is short. These chains have provision for efficient lubrication. The power transmitting chains are of the following three types.

1. ***Block or bush chain.*** A block or bush chain is shown in Fig. 21.6. This type of chain was used in the early stages of development in the power transmission.

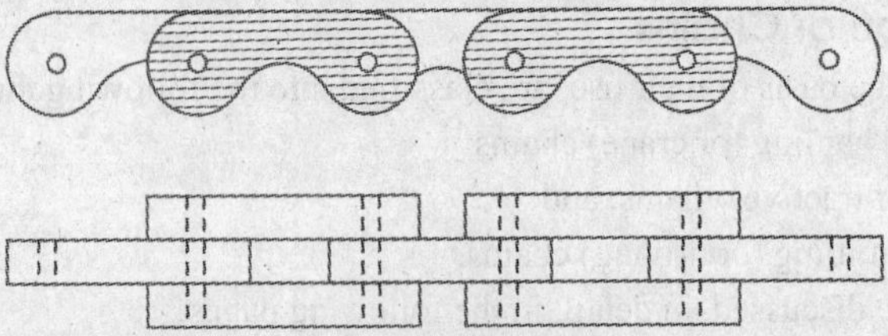

Fig. 21.6. Block or bush chain.

It produces noise when approaching or leaving the teeth of the sprocket because of rubbing between the teeth and the links. Such type of chains are used to some extent as conveyor chain at small speed.

2. ***Bush roller chain.*** A bush roller chain as shown in Fig. 21.7, consists of outer plates or pin link plates, inner plates or roller link plates, pins, bushes and rollers. A pin passes through the bush which is secured in the holes of the roller between the two sides of the chain. The rollers are free to rotate on the bush which protect the sprocket wheel teeth against wear. The pins, bushes and rollers are made of alloy steel.

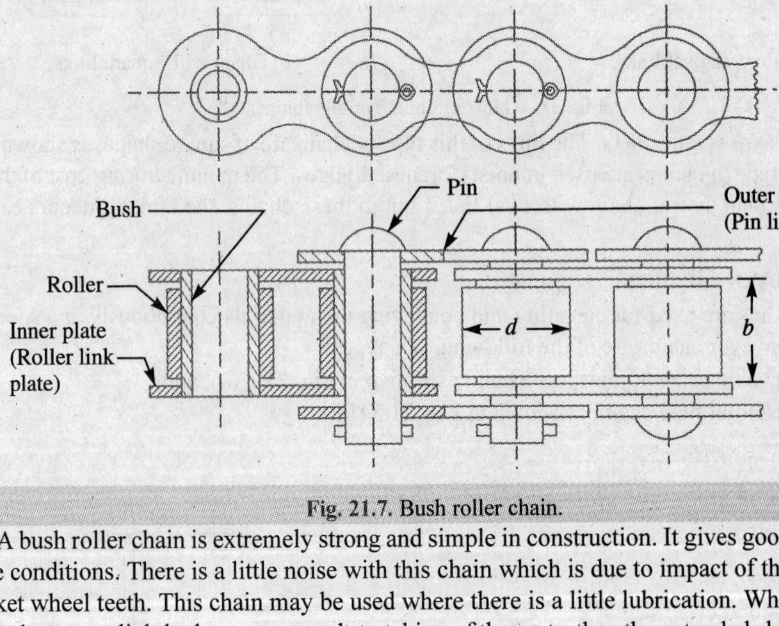

Fig. 21.7. Bush roller chain.

A bush roller chain is extremely strong and simple in construction. It gives good service under severe conditions. There is a little noise with this chain which is due to impact of the rollers on the sprocket wheel teeth. This chain may be used where there is a little lubrication. When one of these chains elongates slightly due to wear and stretching of the parts, then the extended chain is of greater pitch than the pitch of the sprocket wheel teeth. The rollers then fit unequally into the cavities of the wheel. The result is that the total load falls on one teeth or on a few teeth. The stretching of the parts increase wear of the surfaces of the roller and of the sprocket wheel teeth.

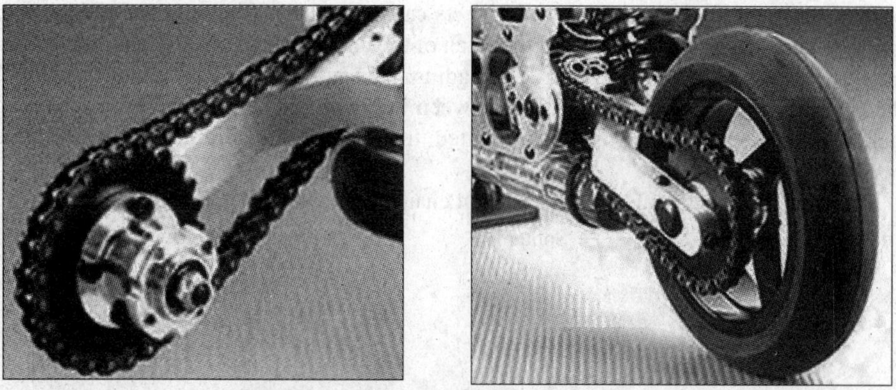

Rear wheel chain drive of a motorcycle

The roller chains are standardised and manufactured on the basis of pitch. These chains are available in single-row or multi-row roller chains such as simple, duplex or triplex strands, as shown in Fig. 21.8.

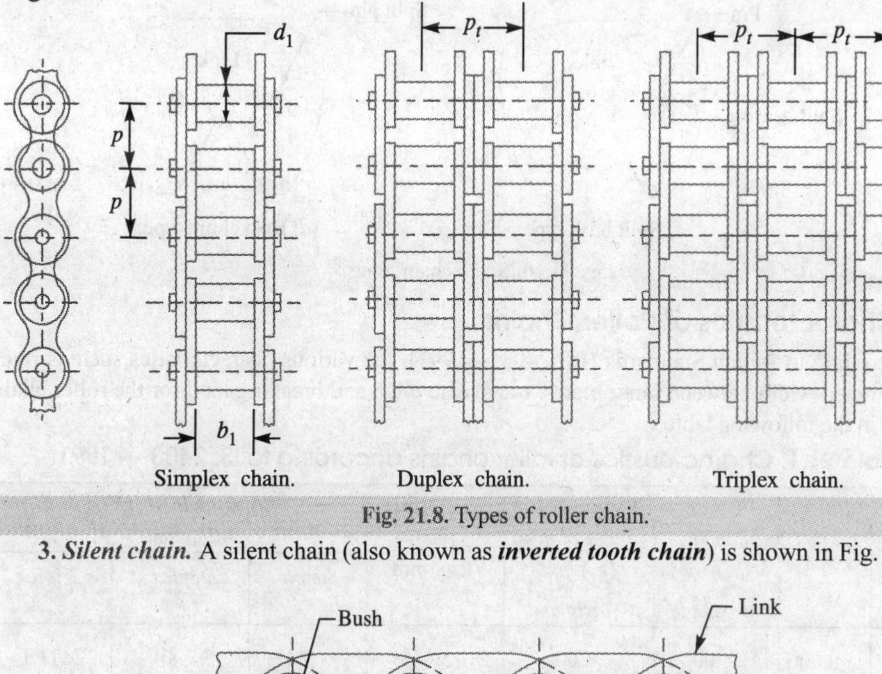

Fig. 21.8. Types of roller chain.

3. *Silent chain.* A silent chain (also known as ***inverted tooth chain***) is shown in Fig. 21.9.

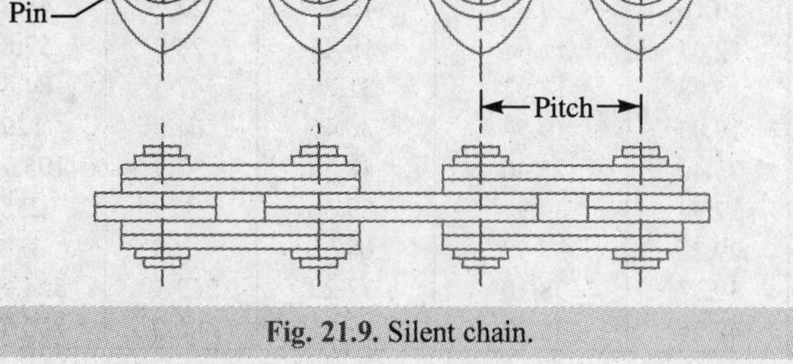

Fig. 21.9. Silent chain.

It is designed to eliminate the evil effects caused by stretching and to produce noiseless running. When the chain stretches and the pitch of the chain increases, the links ride on the teeth of the sprocket wheel at a slightly increased radius. This automatically corrects the small change in the pitch. There is no relative sliding between the teeth of the inverted tooth chain and the sprocket wheel teeth. When properly lubricated, this chain gives durable service and runs very smoothly and quietly.

The various types of joints used in a silent chain are shown in Fig. 21.10.

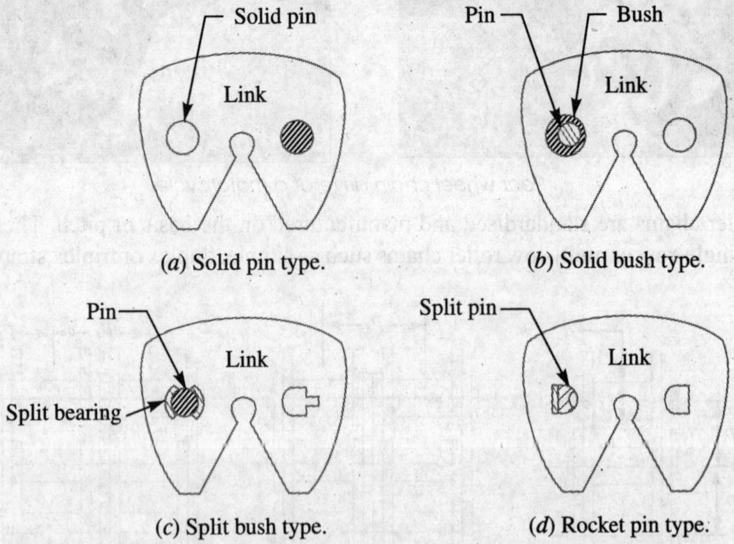

Fig. 21.10. Silent chain joints.

21.11 Characteristics of Roller Chains

According to Indian Standards (IS: 2403 —1991), the various characteristics such as pitch, roller diameter, width between inner plates, transverse pitch and breaking load for the roller chains are given in the following table.

Table 21.1. Characteristics of roller chains according to IS: 2403 — 1991.

ISO Chain number	*Pitch (p) mm*	*Roller diameter (d_1) mm Maximum*	*Width between inner plates (b_1) mm Maximum*	*Transverse pitch (p_1)mm*	*Breaking load (kN) Minimum*		
					Simple	*Duplex*	*Triplex*
05 B	8.00	5.00	3.00	5.64	4.4	7.8	11.1
06 B	9.525	6.35	5.72	10.24	8.9	16.9	24.9
08 B	12.70	8.51	7.75	13.92	17.8	31.1	44.5
10 B	15.875	10.16	9.65	16.59	22.2	44.5	66.7
12 B	19.05	12.07	11.68	19.46	28.9	57.8	86.7
16 B	25.4	15.88	17.02	31.88	42.3	84.5	126.8
20 B	31.75	19.05	19.56	36.45	64.5	129	193.5
24 B	38.10	25.40	25.40	48.36	97.9	195.7	293.6
28 B	44.45	27.94	30.99	59.56	129	258	387
32 B	50.80	29.21	30.99	68.55	169	338	507.10
40 B	63.50	39.37	38.10	72.29	262.4	524.9	787.3
48 B	76.20	48.26	45.72	91.21	400.3	800.7	1201

21.12 Factor of Safety for Chain Drives

The factor of safety for chain drives is defined as the ratio of the breaking strength (W_B) of the chain to the total load on the driving side of the chain (W). Mathematically,

$$\text{Factor of safety} = \frac{W_B}{W}$$

The breaking strength of the chain may be obtained by the following empirical relations, *i.e.*

$$W_B = 106\, p^2 \text{ (in newtons) for roller chains}$$

$$= 106\, p \text{ (in newtons) per mm width of chain for silent chains.}$$

where p is the pitch in mm.

The total load (or total tension) on the driving side of the chain is the sum of the tangential driving force (F_T), centrifugal tension in the chain (F_C) and the tension in the chain due to sagging (F_S).

We know that the tangential driving force acting on the chain,

$$F_T = \frac{\text{Power transmitted (in watts)}}{\text{Speed of chain in m/s}} = \frac{P}{v} \text{ (in newtons)}$$

Centrifugal tension in the chain,

$$F_C = m.v^2 \text{ (in newtons)}$$

and tension in the chain due to sagging,

$$F_S = k.mg.x \text{ (in newtons)}$$

where m = Mass of the chain in kg per metre length,

x = Centre distance in metres, and

k = Constant which takes into account the arrangement of chain drive

= 2 to 6, when the centre line of the chain is inclined to the horizontal at an angle less than 40°

= 1 to 1.5, when the centre line of the chain is inclined to the horizontal at an angle greater than 40°.

The following table shows the factor of safety for the bush roller and silent chains depending upon the speed of the sprocket pinion in r.p.m. and pitch of the chains.

Table 21.2. Factor of safety (n) for bush roller and silent chains.

Type of chain	*Pitch of chain (mm)*	*Speed of the sprocket pinion in r.p.m.*								
		50	200	400	600	800	1000	1200	1600	2000
Bush roller chain	12 – 15	7	7.8	8.55	9.35	10.2	11	11.7	13.2	14.8
	20 – 25	7	8.2	9.35	10.3	11.7	12.9	14	16.3	–
	30 – 35	7	8.55	10.2	13.2	14.8	16.3	19.5	–	–
Silent chain	12.7 – 15.87	20	22.2	24.4	28.7	29.0	31.0	33.4	37.8	42.0
	19.05 – 25.4	20	23.4	26.7	30.0	33.4	36.8	40.0	46.5	53.5

21.13 Permissible Speed of Smaller Sprocket

The following table shows the permissible speed of the smaller sprocket or pinion (in r.p.m.) for the bush roller and silent chain corresponding to different pitches.

Common bicycle is the best example of a chain drive

Table 21.3. Permissible speed of smaller sprocket or pinion in r.p.m.

Type of Chain	*Number of teeth on sprocket pinion*	*Pitch of chain (p) in mm*				
		12	15	20	25	30
Bush roller	15	2300	1900	1350	1150	1000
chain	19	2400	2000	1450	1200	1050
	23	2500	2100	1500	1250	1100
	27	2550	2150	1550	1300	1100
	30	2600	2200	1550	1300	1100
Silent chain	17 – 35	3300	2650	2200	1650	1300

Note: The chain velocity for the roller chains may be as high as 20 m / s, if the chains are properly lubricated and enclosed, whereas the silent chain may be operated upto 40 m / s.

21.14 Power Transmitted by Chains

The power transmitted by the chain on the basis of breaking load is given by

$$P = \frac{W_B \times v}{n \times K_S} \quad \text{(in watts)}$$

where W_b = Breaking load in newtons,

v = Velocity of chain in m/s

n = Factor of safety, and

K_S = Service factor = $K_1.K_2.K_3$

The power transmitted by the chain on the basis of bearing stress is given by

$$P = \frac{\sigma_b \times A \times v}{K_S}$$

where σ_b = Allowable bearing stress in MPa or N/mm^2,

A = Projected bearing area in mm^2,

v = Velocity of chain in m/s, and

K_S = Service factor.

The power rating for simple roller chains depending upon the speed of the smaller sprocket is shown in the following table.

Table 21.4. Power rating (in kW) of simple roller chain.

Speed of smaller sprocket or pinion (r.p.m.)	*Power (kW)*				
	06 B	08 B	10 B	12 B	16 B
100	0.25	0.64	1.18	2.01	4.83
200	0.47	1.18	2.19	3.75	8.94
300	0.61	1.70	3.15	5.43	13.06
500	1.09	2.72	5.01	8.53	20.57
700	1.48	3.66	6.71	11.63	27.73
1000	2.03	5.09	8.97	15.65	34.89
1400	2.73	6.81	11.67	18.15	38.47
1800	3.44	8.10	13.03	19.85	–
2000	3.80	8.67	13.49	20.57	–

The service factor (K_S) is the product of various factors, such as load factor (K_1), lubrication factor (K_2) and rating factor (K_3). The values of these factors are taken as follows:

1. Load factor (K_1) = 1, for constant load
   = 1.25, for variable load with mild shock
   = 1.5, for heavy shock loads
2. Lubrication factor (K_2) = 0.8, for continuous lubrication
   = 1, for drop lubrication
   = 1.5, for periodic lubrication
3. Rating factor (K_3) = 1, for 8 hours per day
   = 1.25, for 16 hours per day
   = 1.5, for continuous service

21.15 Number of Teeth on the Smaller or Driving Sprocket or Pinion

Consider an arrangement of a chain drive in which the smaller or driving sprocket has only four teeth, as shown in Fig. 21.11 (*a*). Let the sprocket rotates anticlockwise at a constant speed of *N* r.p.m. The chain link *AB* is at a distance of *d* / 2 from the centre of the sprocket and its linear speed is given by

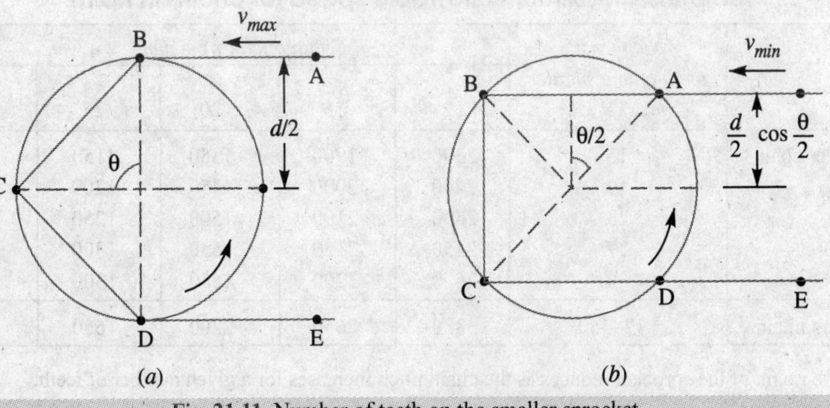

Fig. 21.11. Number of teeth on the smaller sprocket.

$$v_{max} = \frac{\pi \, d \, N}{60} \text{ m/s}$$

where d = Pitch circle diameter of the smaller or driving sprocket in metres.

When the sprocket rotates through an angle $\theta/2$, the link *AB* occupies the position as shown in Fig. 21.11 (*b*). From the figure, we see that the link is now at a distance of $\left(\frac{d}{2} \times \cos\frac{\theta}{2}\right)$ from the centre of the sprocket and its linear velocity is given by

$$v_{min} = \frac{\pi \, d \, N \cos \theta/2}{60} \text{ m/s}$$

From above, we see that the linear velocity of the sprocket is not uniform but varies from maximum to minimum during every cycle of tooth engagement. This results in fluctuations in chain transmission and may be minimised by reducing the angle θ or by increasing the number of teeth on the sprocket. It has been observed that for a sprocket having 11 teeth, the variation of speed is 4 percent and for the sprockets having 17 teeth and 24 teeth, the variation of speed is 1.6 percent and 1 percent respectively.

In order to have smooth operation, the minimum number of teeth on the smaller sprocket or pinion may be taken as 17 for moderate speeds and 21 for high speeds. The following table shows the number of teeth on a smaller sprocket for different velocity ratios.

Table 21.5. Number of teeth on the smaller sprocket.

Type of chain	*Number of teeth at velocity ratio*					
	1	2	3	4	5	6
Roller	31	27	25	23	21	17
Silent	40	35	31	27	23	19

Note: The number of teeth on the smaller sprocket plays an important role in deciding the performance of a chain drive. A small number of teeth tends to make the drive noisy. A large number of teeth makes chain pitch smaller which is favourable for keeping the drive silent and reducing shock, centrifugal force and friction force.

21.16 Maximum Speed for Chains

The maximum allowable speed for the roller and silent chains, depending upon the number of teeth on the smaller sprocket or pinion and the chain pitch is shown in the following table.

Table 21.6. Maximum allowable speed for chains in r.p.m.

Type of chain	*Number of teeth on the smaller sprocket (T_1)*	*Chain pitch (p) in mm*				
		12	15	20	25	30
Roller chain	15	2300	1900	1350	1150	1100
	19	2400	2000	1450	1200	1050
	23	2500	2100	1500	1250	1100
	27	2550	2150	1550	1300	1100
	30	2600	2200	1550	1300	1100
Silent chain	17–35	3300	2650	2200	1650	1300

Note: The r.p.m. of the sprocket reduces as the chain pitch increases for a given number of teeth.

21.17 Principal Dimensions of Tooth Profile

The standard profiles for the teeth of a sprocket are shown in Fig. 21.12. According to Indian Standards (IS: 2403 – 1991), the principal dimensions of the tooth profile are as follows:

1. Tooth flank radius (r_e)

$$= 0.008\, d_1\, (T^2 + 180) \quad \text{...(Maximum)}$$

$$= 0.12\, d_1\, (T + 2) \quad \text{...(Minimum)}$$

where d_1 = Roller diameter, and

T = Number of teeth.

2. Roller seating radius (r_i)

$$= 0.505\, d_1 + 0.069\, \sqrt[3]{d_1} \quad \text{...(Maximum)}$$

$$= 0.505\, d_1 \quad \text{...(Minimum)}$$

3. Roller seating angle (α)

$$= 140^\circ - \frac{90^\circ}{T} \quad \text{...(Maximum)}$$

$$= 120^\circ - \frac{90^\circ}{T} \quad \text{...(Minimum)}$$

4. Tooth height above the pitch polygon (h_a)

$$= 0.625\, p - 0.5\, d_1 + \frac{0.8\, p}{T} \quad \text{...(Maximum)}$$

$$= 0.5\, (p - d_1) \quad \text{...(Minimum)}$$

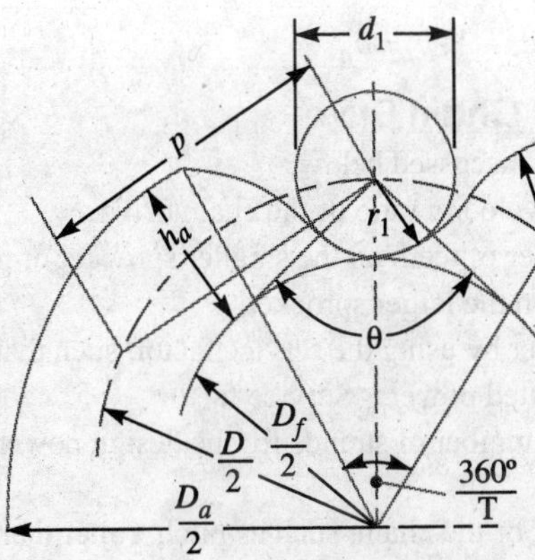

(*a*) Tooth profile of sprocket.

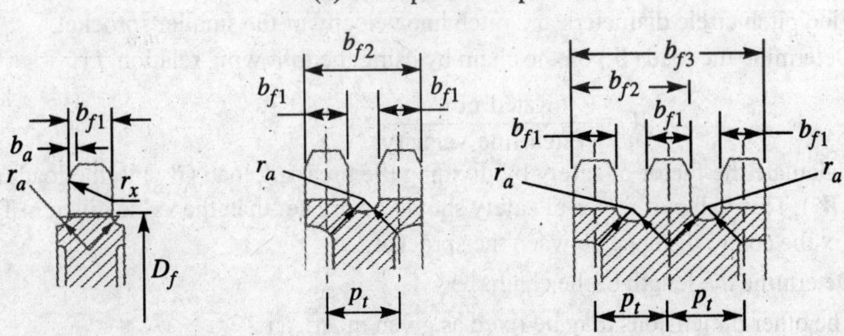

(*b*) Rim profile of sprocket.

Fig. 21.12

Chain drive of an automobile

5. Pitch circle diameter (D)

$$= \frac{p}{\sin\left(\frac{180}{T}\right)} = p \operatorname{cosec}\left(\frac{180}{T}\right)$$

6. Top diameter (D_a)

$$= D + 1.25\, p - d_1 \quad \text{...(Maximum)}$$

$$= D + p\left(1 - \frac{1.6}{T}\right) - d_1 \quad \text{...(Minimum)}$$

7. Root diameter (D_f)

$$= D - 2\, r_i$$

8. Tooth width (b_{f1})

$$= 0.93\, b_1 \text{ when } p \le 12.7 \text{ mm}$$

$$= 0.95\, b_1 \text{ when } p > 12.7 \text{ mm}$$

9. Tooth side radius (r_x) = p

10. Tooth side relief (b_a)

$$= 0.1\, p \text{ to } 0.15\, p$$

11. Widths over teeth (b_{f2} and b_{f3})

$$= (\text{Number of strands} - 1)\, p_t + b_{f1}$$

21.18 Design Procedure of Chain Drive

The chain drive is designed as discussed below:

1. First of all, determine the velocity ratio of the chain drive.
2. Select the minimum number of teeth on the smaller sprocket or pinion from Table 21.5.
3. Find the number of teeth on the larger sprocket.
4. Determine the design power by using the service factor, such that

   Design power = Rated power × Service factor
5. Choose the type of chain, number of strands for the design power and r.p.m. of the smaller sprocket from Table 21.4.
6. Note down the parameters of the chain, such as pitch, roller diameter, minimum width of roller etc. from Table 21.1.
7. Find pitch circle diameters and pitch line velocity of the smaller sprocket.
8. Determine the load (W) on the chain by using the following relation, *i.e.*

$$W = \frac{\text{Rated power}}{\text{Pitch line velocity}}$$

9. Calculate the factor of safety by dividing the breaking load (W_B) to the load on the chain (W). This value of factor of safety should be greater than the value given in Table 21.2.
10. Fix the centre distance between the sprockets.
11. Determine the length of the chain.
12. The other dimensions may be fixed as given in Art. 21.17.

Example 21.1. *Design a chain drive to actuate a compressor from 15 kW electric motor running at 1000 r.p.m., the compressor speed being 350 r.p.m. The minimum centre distance is 500 mm. The compressor operates 16 hours per day. The chain tension may be adjusted by shifting the motor on slides.*

Solution. Given : Rated power = 15 kW ; N_1 = 1000 r.p.m ; N_2 = 350 r.p.m.

We know that the velocity ratio of chain drive,

$$V.R. = \frac{N_1}{N_2} = \frac{1000}{350} = 2.86 \text{ say } 3$$

From Table 21.5, we find that for the roller chain, the number of teeth on the smaller sprocket or pinion (T_1) for a velocity ratio of 3 are 25.

∴ Number of teeth on the larger sprocket or gear,

$$T_2 = T_1 \times \frac{N_1}{N_2} = 25 \times \frac{1000}{350} = 71.5 \text{ say } 72 \textbf{ Ans.}$$

We know that the design power

$$= \text{Rated power} \times \text{Service factor } (K_S)$$

The service factor (K_S) is the product of various factors K_1, K_2 and K_3. The values of these factors are taken as follows:

Load factor (K_1) for variable load with heavy shock

$$= 1.5$$

Lubrication factor (K_2) for drop lubrication

$$= 1$$

Rating factor (K_3) for 16 hours per day

$$= 1.25$$

∴ Service factor, $K_S = K_1.K_2.K_3 = 1.5 \times 1 \times 1.25 = 1.875$

and design power $= 15 \times 1.875 = 28.125$ kW

From Table 21.4, we find that corresponding to a pinion speed of 1000 r.p.m. the power transmitted for chain No. 12 is 15.65 kW per strand. Therefore, a chain No. 12 with two strands can be used to transmit the required power. From Table 21.1, we find that

Pitch, $p = 19.05$ mm

Chain drive

Roller diameter, $d = 12.07$ mm

Minimum width of roller,

$$w = 11.68 \text{ mm}$$

Breaking load, $W_B = 59 \text{ kN} = 59 \times 10^3$ N

We know that pitch circle diameter of the smaller sprocket or pinion,

$$d_1 = p \operatorname{cosec}\left(\frac{180}{T_1}\right) = 19.05 \operatorname{cosec}\left(\frac{180}{25}\right) \text{ mm}$$

$$= 19.05 \times 7.98 = 152 \text{ mm} = 0.152 \text{ m } \textbf{Ans.}$$

and pitch circle diameter of the larger sprocket or gear

$$d_2 = p \operatorname{cosec}\left(\frac{180}{T_2}\right) = 19.05 \operatorname{cosec}\left(\frac{180}{72}\right) \text{ mm}$$

$$= 19.05 \times 22.9 = 436 \text{ mm} = 0.436 \text{ m } \textbf{Ans.}$$

Pitch line velocity of the smaller sprocket,

$$v_1 = \frac{\pi\, d_1\, N_1}{60} = \frac{\pi \times 0.152 \times 1000}{60} = 7.96 \text{ m/s}$$

∴ Load on the chain,

$$W = \frac{\text{Rated power}}{\text{Pitch line velocity}} = \frac{15}{7.96} = 1.844 \text{ kN} = 1844 \text{ N}$$

and factor of safety $= \dfrac{W_B}{W} = \dfrac{59 \times 10^3}{1844} = 32$

This value is more than the value given in Table 21.2, which is equal to 11.

The minimum centre distance between the smaller and larger sprockets should be 30 to 50 times the pitch. Let us take it as 30 times the pitch.

∴ Centre distance between the sprockets,

$$= 30\,p = 30 \times 19.05 = 572 \text{ mm}$$

In order to accomodate initial sag in the chain, the value of centre distance is reduced by 2 to 5 mm.

∴ Correct centre distance

$$x = 572 - 4 = 568 \text{ mm}$$

We know that the number of chain links

$$K = \frac{T_1 + T_2}{2} + \frac{2\,x}{p} + \left[\frac{T_2 - T_1}{2\,\pi}\right]^2 \frac{p}{x}$$

$$= \frac{25 + 72}{2} + \frac{2 \times 568}{19.05} + \left[\frac{72 - 25}{2\,\pi}\right]^2 \frac{19.05}{568}$$

$$= 48.5 + 59.6 + 1.9 = 110$$

∴ Length of the chain,

$$L = K.p = 110 \times 19.05 = 2096 \text{ mm} = 2.096 \text{ m } \textbf{Ans.}$$

EXERCISES

1. Design a roller chain to transmit power from a 20 kW motor to a reciprocating pump. The pump is to operate continuously 24 hours per day. The speed of the motor is 600 r.p.m. and that of the pump is 200 r.p.m. Find: 1. number of teeth on each sprocket; 2. pitch and width of the chain.
2. Design a chain drive to run a blower at 600 r.p.m. The power to the blower is available from a 8 kW motor at 1500 r.p.m. The centre distance is to be kept at 800 mm.

3. A chain drive using bush roller chain transmits 5.6 kW of power. The driving shaft on an electric motor runs at 1440 r.p.m. and velocity ratio is 5. The centre distance of the drive is restricted to 550 ± 2% mm and allowable pressure on the pivot joint is not to exceed 10 N/mm^2. The drive is required to operate continuously with periodic lubrication and driven machine is such that load can be regarded as fairly constant with jerk and impact. Design the chain drive by calculating leading dimensions, number of teeth on the sprocket and specify the breaking strength of the chain. Assume a factor of safety of 13.

QUESTIONS

1. State the advantages and disadvantages of the chain drive over belt and rope drive.
2. Explain, with the help of a neat sketch, the construction of a roller chain.
3. What do you understand by simplex, duplex and triplex chains?
4. Write in brief on
   (*a*)Hoisting and hauling chains,
   (*b*)Conveyor chais, and
   (*c*)Silent chains.
5. Write the design procedure for a chain drive.

OBJECTIVE TYPE QUESTIONS

1. Which one of the following is a positive drive?
   (*a*) Crossed flat belt drive (*b*) Rope drive
   (*c*) *V*-belt drive (*d*) Chain drive
2. The chain drive transmits .... power as compared to belt drive.
   (*a*) more (*b*) less
3. The relation between the pitch of the chain (p) and pitch circle diameter of the sprocket (D) is given by
   (*a*) $p = D \sin\left(\frac{90°}{T}\right)$ (*b*) $p = D \sin\left(\frac{120°}{T}\right)$
   (*c*) $p = D \sin\left(\frac{180°}{T}\right)$ (*d*) $p = D \sin\left(\frac{360°}{T}\right)$
   where T = Number of teeth on the spoocket.
4. In order to have smooth operation, the minimum number of teeth on the smaller sprocket, for moderate speeds, should be
   (*a*) 15 (*b*) 17
   (*c*) 21 (*d*) 25
5. The speed of the sprocket reduces as the chain pitch .... for a given number of teeth.
   (*a*) increases (*b*) decreases

ANSWERS

1. (*d*) 2. (*a*) 3. (*c*) 4. (*b*) 5. (*a*)

C
H
A
P
T
E
R

22

Flywheel

22.1 Introduction

A flywheel used in machines serves as a reservior which stores energy during the period when the supply of energy is more than the requirement and releases it during the period when the requirement of energy is more than supply.

In case of steam engines, internal combustion engines, reciprocating compressors and pumps, the energy is developed during one stroke and the engine is to run for the whole cycle on the energy produced during this one stroke. For example, in I.C. engines, the energy is developed only during power stroke which is much more than the engine load, and no energy is being developed during suction, compression and exhaust strokes in case of four stroke engines and during compression in case of two stroke engines. The excess energy developed during power stroke is absorbed by the flywheel and releases it to the crankshaft during other strokes in which no energy is developed, thus rotating the crankshaft at a uniform speed. A little consideration will show that when the flywheel absorbs energy, its speed increases and when it releases, the speed decreases. Hence a flywheel does not maintain a constant speed, it simply reduces the fluctuation of speed.

In machines where the operation is intermittent like punching machines, shearing machines, riveting machines, crushers etc., the flywheel stores energy from the power source during the greater portion of the operating cycle and gives it up during a small period of the cycle. Thus the energy from the power source to the machines is supplied practically at a constant rate throughout the operation.

Note: The function of a governor in engine is entirely different from that of a flywheel. It regulates the mean speed of an engine when there are variations in the load, *e.g.* when the load on the engine increases, it becomes necessary to increase the supply of working fluid. On the other hand, when the load decreases, less working fluid is required. The governor automatically controls the supply of working fluid to the engine with the varying load condition and keeps the mean speed within certain limits.

Flywheel stores energy when the supply is in excess, and releases energy when the supply is in deficit.

As discussed above, the flywheel does not maintain a constant speed, it simply reduces the fluctuation of speed. In other words, a flywheel controls the speed variations caused by the fluctuation of the engine turning moment during each cycle of operation. It does not control the speed variations caused by the varying load.

22.2 Coefficient of Fluctuation of Speed

The difference between the maximum and minimum speeds during a cycle is called the ***maximum fluctuation of speed.*** The ratio of the maximum fluctuation of speed to the mean speed is called ***coefficient of fluctuation of speed.***

Let N_1 = Maximum speed in r.p.m. during the cycle,

N_2 = Minimum speed in r.p.m. during the cycle, and

N = Mean speed in r.p.m. = $\dfrac{N_1 + N_2}{2}$

∴ Coefficient of fluctuation of speed,

$$C_S = \frac{N_1 - N_2}{N} = \frac{2(N_1 - N_2)}{N_1 + N_2}$$

$$= \frac{\omega_1 - \omega_2}{\omega} = \frac{2(\omega_1 - \omega_2)}{\omega_1 + \omega_2} \qquad \text{...(In terms of angular speeds)}$$

$$= \frac{v_1 - v_2}{v} = \frac{2(v_1 - v_2)}{v_1 + v_2} \qquad \text{...(In terms of linear speeds)}$$

The coefficient of fluctuation of speed is a limiting factor in the design of flywheel. It varies depending upon the nature of service to which the flywheel is employed. Table 22.1 shows the permissible values for coefficient of fluctuation of speed for some machines.

Note: The reciprocal of coefficient of fluctuation of speed is known as **coefficient of steadiness** and it is denoted by *m*.

$$\therefore \qquad m = \frac{1}{C_S} = \frac{N}{N_1 - N_2} = \frac{\omega}{\omega_1 - \omega_2} = \frac{v}{v_1 - v_2}$$

Table 22.1. Permissible values for coefficient of fluctuation of speed (C_S).

S.No.	*Type of machine or class of service*	*Coefficient of fluctuation of speed (C_S)*
1.	Crushing machines	0.200
2.	Electrical machines	0.003
3.	Electrical machines (direct drive)	0.002
4.	Engines with belt transmission	0.030
5.	Gear wheel transmission	0.020
6.	Hammering machines	0.200
7.	Pumping machines	0.03 to 0.05
8.	Machine tools	0.030
9.	Paper making, textile and weaving machines	0.025
10.	Punching, shearing and power presses	0.10 to 0.15
11.	Spinning machinery	0.10 to 0.020
12.	Rolling mills and mining machines	0.025

22.3 Fluctuation of Energy

The fluctuation of energy may be determined by the turning moment diagram for one complete cycle of operation. Consider a turning moment diagram for a single cylinder double acting steam engine as shown in Fig. 22.1. The vertical ordinate represents the turning moment and the horizontal ordinate (abscissa) represents the crank angle.

A little consideration will show that the turning moment is zero when the crank angle is zero. It rises to a maximum value when crank angle reaches 90° and it is again zero when crank angle is 180°. This is shown by the curve *abc* in Fig. 22.1 and it represents the turning moment diagram for outstroke. The curve *cde* is the turning moment diagram for instroke and is somewhat similar to the curve *abc*.

Since the work done is the product of the turning moment and the angle turned, therefore the area of the turning moment diagram represents the work done per revolution. In actual practice, the engine is assumed to work against the mean resisting torque, as shown by a horizontal line *AF*. The height of the ordinate *aA* represents the mean height of the turning moment diagram. Since it is assumed that the work done by the turning moment per revolution is equal to the work done against the mean resisting torque, therefore the area of the rectangle *aA Fe* is proportional to the work done against the mean resisting torque.

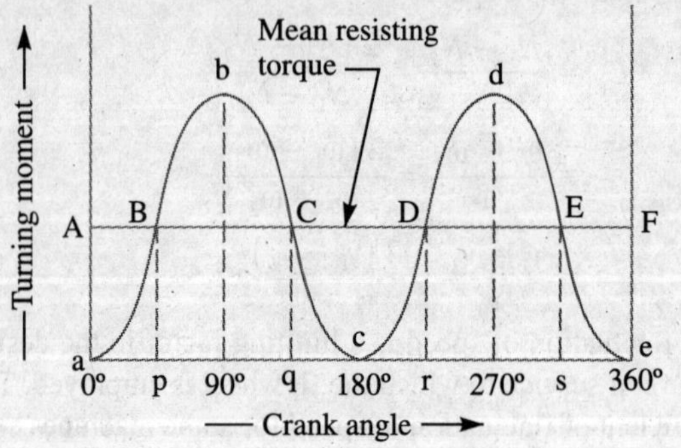

Fig. 22.1. Turning moment diagram for a single cylinder double acting steam engine.

We see in Fig. 22.1, that the mean resisting torque line *AF* cuts the turning moment diagram at points *B*, *C*, *D* and *E*. When the crank moves from '*a*' to '*p*' the work done by the engine is equal to the area *aBp*, whereas the energy required is represented by the area *aABp*. In other words, the engine has done less work (equal to the area *aAB*) than the requirement. This amount of energy is taken from the flywheel and hence the speed of the flywheel decreases. Now the crank moves from *p* to *q*, the work done by the engine is equal to the area *pBbCq*, whereas the requirement of energy is represented by the area *pBCq*. Therefore the engine has done more work than the requirement. This excess work (equal to the area *BbC*) is stored in the flywheel and hence the speed of the flywheel increases while the crank moves from *p* to *q*.

Similarly when the crank moves from *q* to *r*, more work is taken from the engine than is developed. This loss of work is represented by the area *CcD*. To supply this loss, the flywheel gives up some of its energy and thus the speed decreases while the crank moves from *q* to *r*. As the crank moves from *r* to *s*, excess energy is again developed given by the area *DdE* and the speed again increases. As the piston moves from *s* to *e*, again there is a loss of work and the speed decreases. The variations of energy above and below the mean resisting torque line are called ***fluctuation of energy.*** The areas *BbC*, *CcD*, *DdE* etc. represent fluctuations of energy.

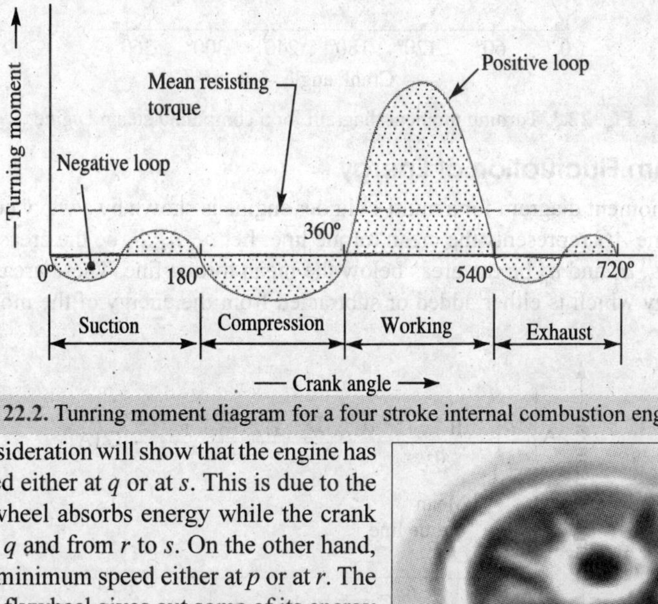

Fig. 22.2. Tunring moment diagram for a four stroke internal combustion engine.

A little consideration will show that the engine has a maximum speed either at *q* or at *s*. This is due to the fact that the flywheel absorbs energy while the crank moves from *p* to *q* and from *r* to *s*. On the other hand, the engine has a minimum speed either at *p* or at *r*. The reason is that the flywheel gives out some of its energy when the crank moves from *a* to *p* and from *q* to *r*. The difference between the maximum and the minimum energies is known as ***maximum fluctuation of energy.***

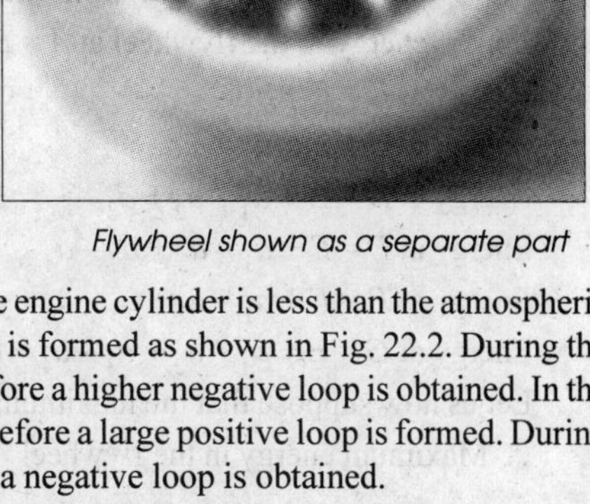

Flywheel shown as a separate part

A turning moment diagram for a four stroke internal combustion engine is shown in Fig. 22.2. We know that in a four stroke internal combustion engine, there is one working stroke after the crank has turned through 720° (or 4π radians). Since the pressure inside the engine cylinder is less than the atmospheric pressure during suction stroke, therefore a negative loop is formed as shown in Fig. 22.2. During the compression stroke, the work is done on the gases, therefore a higher negative loop is obtained. In the working stroke, the fuel burns and the gases expand, therefore a large positive loop is formed. During exhaust stroke, the work is done on the gases, therefore a negative loop is obtained.

A turning moment diagram for a compound steam engine having three cylinders and the resultant turning moment diagram is shown in Fig. 22.3. The resultant turning moment diagram is the sum of the turning moment diagrams for the three cylinders. It may be noted that the first cylinder is the high pressure cylinder, second cylinder is the intermediate cylinder and the third cylinder is the low pressure cylinder. The cranks, in case of three cylinders are usually placed at 120° to each other.

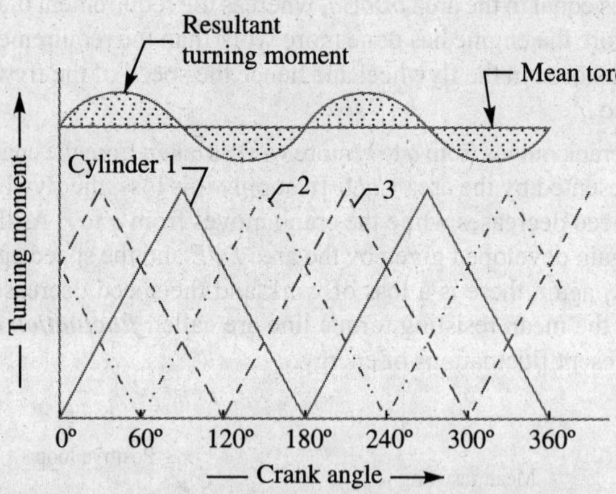

Fig. 22.3. Turning moment diagram for a compound steam engine.

22.4 Maximum Fluctuation of Energy

A turning moment diagram for a multi-cylinder engine is shown by a wavy curve in Fig. 22.4. The horizontal line *AG* represents the mean torque line. Let a_1, a_3, a_5 be the areas above the mean torque line and a_2, a_4 and a_6 be the areas below the mean torque line. These areas represent some quantity of energy which is either added or subtracted from the energy of the moving parts of the engine.

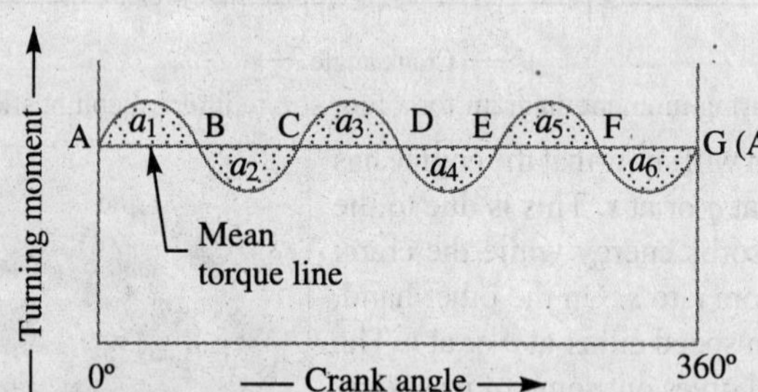

Fig. 22.4. Turning moment diagram for a multi-cylinder engine.

Let the energy in the flywheel at $A = E$, then from Fig. 22.4, we have

Energy at $B = E + a_1$

Energy at $C = E + a_1 - a_2$

Energy at $D = E + a_1 - a_2 + a_3$

Energy at $E = E + a_1 - a_2 + a_3 - a_4$

Energy at $F = E + a_1 - a_2 + a_3 - a_4 + a_5$

Energy at $G = E + a_1 - a_2 + a_3 - a_4 + a_5 - a_6$ = Energy at A

Let us now suppose that the maximum of these energies is at *B* and minimum at *E*.

∴ Maximum energy in the flywheel

$$= E + a_1$$

and minimum energy in the flywheel

$$= E + a_1 - a_2 + a_3 - a_4$$

∴ Maximum fluctuation of energy,

$$\Delta E = \text{Maximum energy} - \text{Minimum energy}$$

$$= (E + a_1) - (E + a_1 - a_2 + a_3 - a_4) = a_2 - a_3 + a_4$$

22.5 Coefficient of Fluctuation of Energy

It is defined as the ratio of the maximum fluctuation of energy to the work done per cycle. It is usually denoted by C_E. Mathematically, coefficient of fluctuation of energy,

$$C_E = \frac{\text{Maximum fluctuation of energy}}{\text{Work done per cycle}}$$

The workdone per cycle may be obtained by using the following relations:

1. Workdone / cycle $= T_{mean} \times \theta$

where T_{mean} = Mean torque, and

θ = Angle turned in radians per revolution

= 2π, in case of steam engines and two stroke internal combustion engines.

= 4π, in case of four stroke internal combustion engines.

The mean torque (T_{mean}) in N-m may be obtained by using the following relation *i.e.*

$$T_{mean} = \frac{P \times 60}{2\pi N} = \frac{P}{\omega}$$

where P = Power transmitted in watts,

N = Speed in r.p.m., and

ω = Angular speed in rad/s = $2\pi N / 60$

2. The workdone per cycle may also be obtained by using the following relation:

$$\text{Workdone / cycle} = \frac{P \times 60}{n}$$

where n = Number of working strokes per minute.

= N, in case of steam engines and two stroke internal combustion engines.

= $N/2$, in case of four stroke internal combustion engines.

The following table shows the values of coefficient of fluctuation of energy for steam engines and internal combustion engines.

Table 22.2. Coefficient of fluctuation of energy (C_E) for steam and internal combustion engines.

S.No.	*Type of engine*	*Coefficient of fluctuation of energy* (C_E)
1.	Single cylinder, double acting steam engine	0.21
2.	Cross-compound steam engine	0.096
3.	Single cylinder, single acting, four stroke gas engine	1.93
4.	Four cylinder, single acting, four stroke gas engine	0.066
5.	Six cylinder, single acting, four stroke gas engine	0.031

22.6 Energy Stored in a Flywheel

A flywheel is shown in Fig. 22.5. We have already discussed that when a flywheel absorbs energy its speed increases and when it gives up energy its speed decreases.

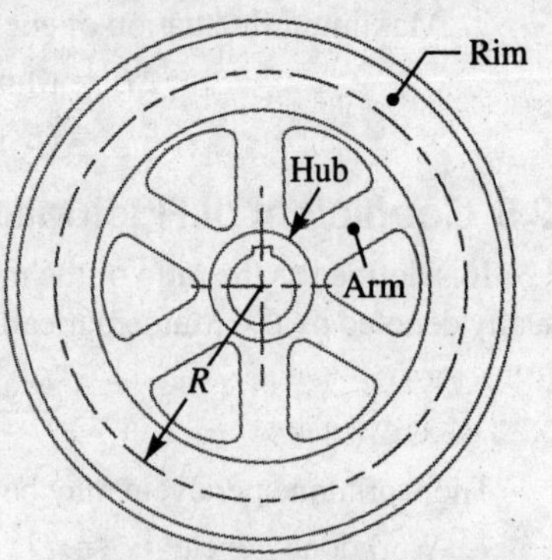

Fig. 22.5. Flywheel.

Let m = Mass of the flywheel in kg,

k = Radius of gyration of the flywheel in metres,

I = Mass moment of inertia of the flywheel about the axis of rotation in kg-m^2 $= m.k^2$,

N_1 and N_2 = Maximum and minimum speeds during the cycle in r.p.m.,

ω_1 and ω_2 = Maximum and minimum angular speeds during the cycle in rad / s,

N = Mean speed during the cycle in r.p.m. $= \dfrac{N_1 + N_2}{2}$,

ω = Mean angular speed during the cycle in rad / s $= \dfrac{\omega_1 + \omega_2}{2}$

C_S = Coefficient of fluctuation of speed $= \dfrac{N_1 - N_2}{N}$ or $\dfrac{\omega_1 - \omega_2}{\omega}$

We know that mean kinetic energy of the flywheel,

$$E = \frac{1}{2} \times I.\omega^2 = \frac{1}{2} \times m.k^2.\omega^2 \text{ (in N-m or joules)}$$

As the speed of the flywheel changes from ω_1 to ω_2, the maximum fluctuation of energy,

$$\Delta E = \text{Maximum K.E.} - \text{Minimum K.E.} = \frac{1}{2} \times I(\omega_1)^2 - \frac{1}{2} \times I(\omega_2)^2$$

$$= \frac{1}{2} \times I \left[(\omega_1)^2 - (\omega_2)^2\right] = \frac{1}{2} \times I\,(\omega_1 + \omega_2)\,(\omega_1 - \omega_2)$$

$$= I.\omega\,(\omega_1 - \omega_2) \qquad \ldots\left(\because \omega = \frac{\omega_1 + \omega_2}{2}\right) \quad \ldots(i)$$

$$= I.\omega^2 \left(\frac{\omega_1 - \omega_2}{\omega}\right) \qquad \ldots\text{[Multiplying and dividing by } \omega]$$

$$= I.\omega^2.C_S = m.k^2.\omega^2.C_S \qquad \ldots(\because I = m.k^2) \quad \ldots(ii)$$

$$= 2\,E.C_S \qquad \ldots\left(\because E = \frac{1}{2} \times I.\omega^2\right) \quad \ldots(iii)$$

The radius of gyration (k) may be taken equal to the mean radius of the rim (R), because the thickness of rim is very small as compared to the diameter of rim. Therefore substituting $k = R$ in equation (*ii*), we have

$$\Delta E = m.R^2.\omega^2.C_S = m.v^2.C_S \qquad \ldots(\because v = \omega.R)$$

From this expression, the mass of the flywheel rim may be determined.

Notes: 1. In the above expression, only the mass moment of inertia of the rim is considered and the mass moment of inertia of the hub and arms is neglected. This is due to the fact that the major portion of weight of the flywheel is in the rim and a small portion is in the hub and arms. Also the hub and arms are nearer to the axis of rotation, therefore the moment of inertia of the hub and arms is very small.

2. The density of cast iron may be taken as 7260 kg / m^3 and for cast steel, it may taken as 7800 kg / m^3.

3. The mass of the flywheel rim is given by

$$m = \text{Volume} \times \text{Density} = 2\,\pi\,R \times A \times \rho$$

From this expression, we may find the value of the cross-sectional area of the rim. Assuming the cross-section of the rim to be rectangular, then

$$A = b \times t$$

where b = Width of the rim, and

t = Thickness of the rim.

Knowing the ratio of b / t which is usually taken as 2, we may find the width and thickness of rim.

4. When the flywheel is to be used as a pulley, then the width of rim should be taken 20 to 40 mm greater than the width of belt.

Example 22.1. *The turning moment diagram for a petrol engine is drawn to the following scales:*

Turning moment, 1 mm = 5 N-m; Crank angle, 1 mm = 1°.

The turning moment diagram repeats itself at every half revolution of the engine and the areas above and below the mean turning moment line, taken in order are 295, 685, 40, 340, 960, 270 mm².

Determine the mass of 300 mm diameter flywheel rim when the coefficient of fluctuation of speed is 0.3% and the engine runs at 1800 r.p.m. Also determine the cross-section of the rim when the width of the rim is twice of thickness. Assume density of rim material as 7250 kg / m³.

Solution. Given : D = 300 mm or R = 150 mm = 0.15 m ; C_S = 0.3% = 0.003 ; N = 1800 r.p.m. or $\omega = 2\pi \times 1800 / 60 = 188.5$ rad/s ; $\rho = 7250$ kg / m³

Mass of the flywheel

Let m = Mass of the flywheel in kg.

First of all, let us find the maximum fluctuation of energy. The turning moment diagram is shown in Fig. 22.6.

Since the scale of turning moment is 1 mm = 5 N-m, and scale of the crank angle is 1 mm = 1° = π / 180 rad, therefore 1 mm² on the turning moment diagram.

$$= 5 \times \pi / 180 = 0.087 \text{ N-m}$$

Let the total energy at $A = E$. Therefore from Fig. 22.6, we find that

$$\text{Energy at } B = E + 295$$
$$\text{Energy at } C = E + 295 - 685 = E - 390$$
$$\text{Energy at } D = E - 390 + 40 = E - 350$$
$$\text{Energy at } E = E - 350 - 340 = E - 690$$
$$\text{Energy at } F = E - 690 + 960 = E + 270$$
$$\text{Energy at } G = E + 270 - 270 = E = \text{Energy at } A$$

From above we see that the energy is maximum at B and minimum at E.

$\therefore$ Maximum energy $= E + 295$

and minimum energy $= E - 690$

We know that maximum fluctuation of energy,

$$\Delta E = \text{Maximum energy} - \text{Minimum energy}$$
$$= (E + 295) - (E - 690) = 985 \text{ mm}^2$$
$$= 985 \times 0.087 = 86 \text{ N-m}$$

We also know that maximum fluctuation of energy (ΔE),

$$86 = m.R^2.\omega^2.C_S = m\,(0.15)^2\,(188.5)^2\,(0.003) = 2.4\,m$$

$$\therefore \quad m = 86 / 2.4 = 35.8 \text{ kg } \textbf{Ans.}$$

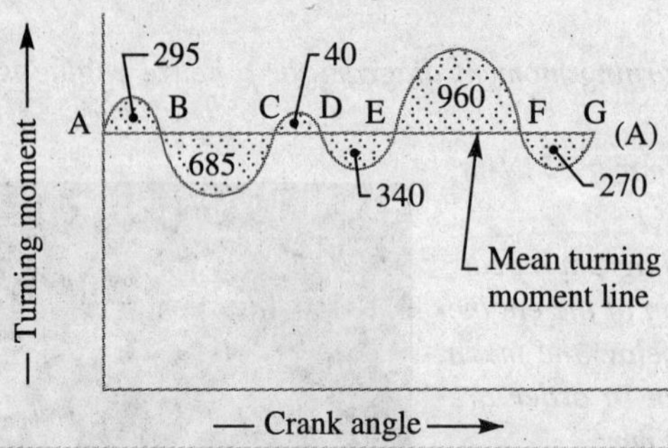

Fig. 22.6

Cross-section of the flywheel rim

Let t = Thickness of rim in metres, and

b = Width of rim in metres = $2\,t$ ...(Given)

$\therefore$ Cross-sectional area of rim,

$$A = b \times t = 2\,t \times t = 2\,t^2$$

We know that mass of the flywheel rim (m),

$$35.8 = A \times 2\pi R \times \rho = 2t^2 \times 2\pi \times 0.15 \times 7250 = 13\,668\,t^2$$

$$\therefore \quad t^2 = 35.8 / 13\,668 = 0.0026 \quad \text{or} \quad t = 0.051 \text{ m} = 51 \text{ mm } \textbf{Ans.}$$

and $b = 2\,t = 2 \times 51 = 102$ mm **Ans.**

Example 22.2. *The intercepted areas between the output torque curve and the mean resistance line of a turning moment diagram for a multicylinder engine, taken in order from one end are as follows:*

$$-35, +410, -285, +325, -335, +260, -365, +285, -260 \text{ mm}^2.$$

The diagram has been drawn to a scale of 1 mm = 70 N-m and 1 mm = 4.5°. The engine speed is 900 r.p.m. and the fluctuation in speed is not to exceed 2% of the mean speed.

Find the mass and cross-section of the flywheel rim having 650 mm mean diameter. The density of the material of the flywheel may be taken as 7200 kg / m³. The rim is rectangular with the width 2 times the thickness. Neglect effect of arms, etc.

Solution. Given : $N = 900$ r.p.m. or $\omega = 2\pi \times 900 / 60 = 94.26$ rad/s ; $\omega_1 - \omega_2 = 2\%\ \omega$ or

$\dfrac{\omega_1 - \omega_2}{\omega} = C_S = 2\% = 0.02$; $D = 650$ mm or $R = 325$ mm $= 0.325$ m ; $\rho = 7200$ kg / m³

Mass of the flywheel rim

Let m = Mass of the flywheel rim in kg.

First of all, let us find the maximum fluctuation of energy. The turning moment diagram for a multi-cylinder engine is shown in Fig. 22.7.

Since the scale of turning moment is 1 mm = 70 N-m and scale of the crank angle is 1 mm = 4.5° = π / 40 rad, therefore 1 mm² on the turning moment diagram.

$$= 70 \times \pi / 40 = 5.5 \text{ N-m}$$

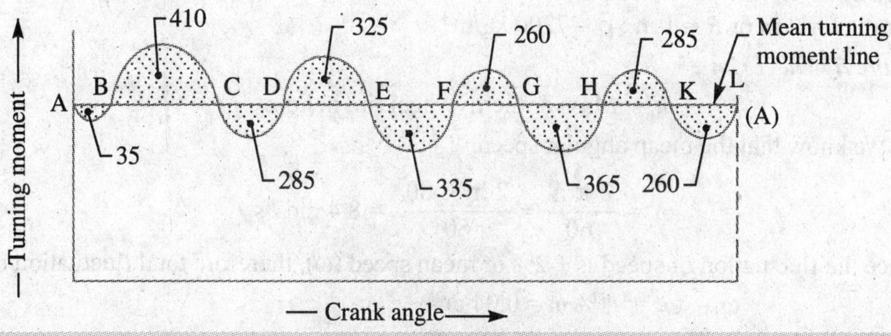

Fig. 22.7

Let the total energy at $A = E$. Therefore from Fig. 22.7, we find that

$$\text{Energy at } B = E - 35$$
$$\text{Energy at } C = E - 35 + 410 = E + 375$$
$$\text{Energy at } D = E + 375 - 285 = E + 90$$
$$\text{Energy at } E = E + 90 + 325 = E + 415$$
$$\text{Energy at } F = E + 415 - 335 = E + 80$$
$$\text{Energy at } G = E + 80 + 260 = E + 340$$
$$\text{Energy at } H = E + 340 - 365 = E - 25$$
$$\text{Energy at } K = E - 25 + 285 = E + 260$$
$$\text{Energy at } L = E + 260 - 260 = E = \text{Energy at } A$$

From above, we see that the energy is maximum at E and minimum at B.

$\therefore$ Maximum energy $= E + 415$

and minimum energy $= E - 35$

We know that maximum fluctuation of energy,

$$= (E + 415) - (E - 35) = 450 \text{ mm}^2$$
$$= 450 \times 5.5 = 2475 \text{ N-m}$$

We also know that maximum fluctuation of energy (ΔE),

$$2475 = \text{m}.R^2.\omega^2.C_S = m\,(0.325)^2\,(94.26)^2\,0.02 = 18.77\,m$$

$\therefore$ $m = 2475 / 18.77 = 132$ kg **Ans.**

Cross-section of the flywheel rim

Let t = Thickness of the rim in metres, and

b = Width of the rim in metres = $2\,t$ ...(Given)

$\therefore$ Area of cross-section of the rim,

$$A = b \times t = 2\,t \times t = 2\,t^2$$

We know that mass of the flywheel rim (m),

$$132 = A \times 2\,\pi\,R \times \rho = 2\,t^2 \times 2\,\pi \times 0.325 \times 7200 = 29\,409\,t^2$$

$\therefore$ $t^2 = 132 / 29\,409 = 0.0044$ or $t = 0.067$ m = 67 mm **Ans.**

and $b = 2t = 2 \times 67 = 134$ mm **Ans.**

Example 22.3. *A single cylinder double acting steam engine develops 150 kW at a mean speed of 80 r.p.m. The coefficient of fluctuation of energy is 0.1 and the fluctuation of speed is ± 2% of mean speed. If the mean diameter of the flywheel rim is 2 metres and the hub and spokes provide 5 percent of the rotational inertia of the wheel, find the mass of the flywheel and cross-sectional area of the rim. Assume the density of the flywheel material (which is cast iron) as 7200 kg / m³.*

Solution. Given : $P = 150$ kW $= 150 \times 10^3$ W ; $N = 80$ r.p.m. ; $C_E = 0.1$; $\omega_1 - \omega_2 = \pm 2\%\ \omega$; $D = 2$ m or $R = 1$ m ; $\rho = 7200$ kg/m^3

Mass of the flywheel rim

Let m = Mass of the flywheel rim in kg.

We know that the mean angular speed,

$$\omega = \frac{2\pi N}{60} = \frac{2\pi \times 80}{60} = 8.4 \text{ rad/s}$$

Since the fluctuation of speed is ± 2% of mean speed (ω), therefore total fluctuation of speed,

$$\omega_1 - \omega_2 = 4\%\ \omega = 0.04\ \omega$$

and coefficient of fluctuation of speed,

$$C_S = \frac{\omega_1 - \omega_2}{\omega} = 0.04$$

We know that the work done by the flywheel per cycle

$$= \frac{P \times 60}{N} = \frac{150 \times 10^3 \times 60}{80} = 112\ 500 \text{ N-m}$$

We also know that coefficient of fluctuation of energy,

$$C_E = \frac{\text{Maximum fluctuation of energy}}{\text{Workdone / cycle}}$$

∴ Maximum fluctuation of energy,

$$\Delta E = C_E \times \text{Workdone / cycle}$$
$$= 0.1 \times 112\ 500 = 11\ 250 \text{ N-m}$$

Since 5% of the rotational inertia is provided by hub and spokes, therefore the maximum fluctuation of energy of the flywheel rim will be 95% of the flywheel.

∴ Maximum fluctuation of energy of the rim,

$$(\Delta E)_{rim} = 0.95 \times 11\ 250 = 10\ 687.5 \text{ N-m}$$

We know that maximum fluctuation of energy of the rim $(\Delta E)_{rim}$,

$$10\ 687.5 = m.R^2.\omega^2.C_S = m \times 1^2\,(8.4)^2\,0.04 = 2.82\ m$$

∴ $m = 10\ 687.5 / 2.82 = 3790$ kg **Ans.**

Cross-sectional area of the rim

Let A = Cross-sectional area of the rim.

We know that the mass of the flywheel rim (m),

$$3790 = A \times 2\pi R \times \rho = A \times 2\pi \times 1 \times 7200 = 45\ 245\ A$$

∴ $A = 3790 / 45\ 245 = 0.084$ m^2 **Ans.**

Example 22.4. *A single cylinder, single acting, four stroke oil engine develops 20 kW at 300 r.p.m. The workdone by the gases during the expansion stroke is 2.3 times the workdone on the gases during the compression and the workdone during the suction and exhaust strokes is negligible. The speed is to be maintained within ± 1%. Determine the mass moment of inertia of the flywheel.*

Solution. Given : $P = 20$ kW $= 20 \times 10^3$ W ; $N = 300$ r.p.m. or $\omega = 2\pi \times 300 / 60 = 31.42$ rad / s ; $\omega_1 - \omega_2 = \pm 1\%\ \omega$

First of all, let us find the maximum fluctuation of energy (ΔE). The turning moment diagram for a four stroke engine is shown in Fig. 22.8. It is assumed to be triangular during compression and expansion strokes, neglecting the suction and exhaust strokes.

We know that mean torque transmitted by the engine,

$$T_{mean} = \frac{P \times 60}{2\,\pi\,N} = \frac{20 \times 10^3 \times 60}{2\,\pi \times 300} = 636.5 \text{ N-m}$$

and *workdone per cycle $= T_{mean} \times \theta = 636.5 \times 4\,\pi = 8000$ N-m ...(*i*)

Let W_C = Workdone during compression stroke, and

W_E = Workdone during expansion stroke.

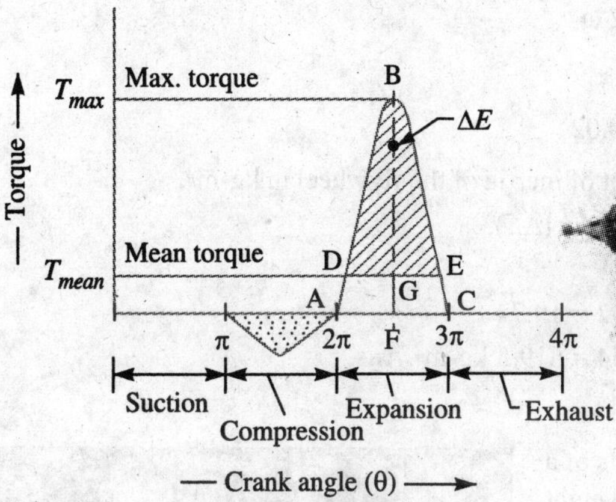

Fig. 22.8

Since the workdone during suction and exhaust strokes is negligible, therefore net work done per cycle

$$= W_E - W_C = W_E - W_E / 2.3 = 0.565\, W_E \qquad ...(ii)$$

From equations (*i*) and (*ii*), we have

$$W_E = 8000 / 0.565 = 14\,160 \text{ N-m}$$

The workdone during the expansion stroke is shown by triangle *ABC* in Fig. 22.8, in which base $AC = \pi$ radians and height $BF = T_{max.}$

∴ Workdone during expansion stroke (W_E),

$$14\,160 = \frac{1}{2} \times \pi \times T_{max} = 1.571\, T_{max}$$

or

$$T_{max} = 14\,160 / 1.571 = 9013 \text{ N-m}$$

We know that height above the mean torque line,

$$BG = BF - FG = T_{max} - T_{mean}$$

$$= 9013 - 636.5 = 8376.5 \text{ N-m}$$

Since the area *BDE* shown shaded in Fig. 22.8 above the mean torque line represents the maximum fluctuation of energy (ΔE), therefore from geometrical relation,

$$\frac{\text{Area of } \Delta\, BDE}{\text{Area of } \Delta\, ABC} = \frac{(BG)^2}{(BF)^2}\text{, we have}$$

---

* The workdone per cycle may also be calculated as follows :

We know that for a four stroke engine, number of working strokes per cycle

$$n = N / 2 = 300 / 2 = 150$$

∴ Workdone per cycle $= P \times 60 / n = 20 \times 10^3 \times 60 / 150 = 8000$ N-m

Maximum fluctuation of energy (*i.e.* area of Δ *BDE*),

$$^*\Delta E = \text{Area of } \Delta ABC \left(\frac{BG}{BF}\right)^2 = W_E \left(\frac{BG}{BF}\right)^2$$

$$= 14\ 160 \left(\frac{8376.5}{9013}\right)^2 = 12\ 230 \text{ N-m}$$

Since the speed is to be maintained within ± 1% of the mean speed, therefore total fluctuation of speed

$$\omega_1 - \omega_2 = 2\ \%\ \omega = 0.02\ \omega$$

and coefficient of fluctuation of speed,

$$C_S = \frac{\omega_1 - \omega_2}{\omega} = 0.02$$

Let I = Mass moment of inertia of the flywheel in kg-m^2.

We know that maximum fluctuation of energy (ΔE),

$$12\ 230 = I.\omega^2.C_S$$

$$= I\,(31.42)^2\ 0.02 = 19.74\ I$$

∴ $I = 12\ 230 / 19.74 = 619.5$ kg-m^2 **Ans.**

22.7 Stresses in a Flywheel Rim

A flywheel, as shown in Fig. 22.9, consists of a rim at which the major portion of the mass or weight of flywheel is concentrated, a boss or hub for fixing the flywheel on to the shaft and a number of arms for supporting the rim on the hub.

The following types of stresses are induced in the rim of a flywheel:

1. Tensile stress due to centrifugal force,
2. Tensile bending stress caused by the restraint of the arms, and
3. The shrinkage stresses due to unequal rate of cooling of casting. These stresses may be very high but there is no easy method of determining. This stress is taken care of by a factor of safety.

We shall now discuss the first two types of stresses as follows:

1. ***Tensile stress due to the centrifugal force***

The tensile stress in the rim due to the centrifugal force, assuming that the rim is unstrained by the arms, is determined in a similar way as a thin cylinder subjected to internal pressure.

Let b = Width of rim,

t = Thickness of rim,

* The maximum fluctuation of energy (ΔE) may also be obtained as discussed below :
From similar triangles *BDE* and *BAC*,

$$\frac{DE}{AC} = \frac{BG}{BF} \quad \text{or} \quad DE = \frac{BG}{BF} \times AC = \frac{8376.5}{9013} \times \pi = 2.92 \text{ rad}$$

∴ Maximum fluctuation of energy (*i.e.* area of Δ *BDE*),

$$\Delta E = \frac{1}{2} \times DE \times BG = \frac{1}{2} \times 2.92 \times 8376.5 = 12\ 230 \text{ N-m}$$

A = Cross-sectional area of rim = $b \times t$,

D = Mean diameter of flywheel

R = Mean radius of flywheel,

ρ = Density of flywheel material,

ω = Angular speed of flywheel,

v = Linear velocity of flywheel, and

σ_t = Tensile or hoop stress.

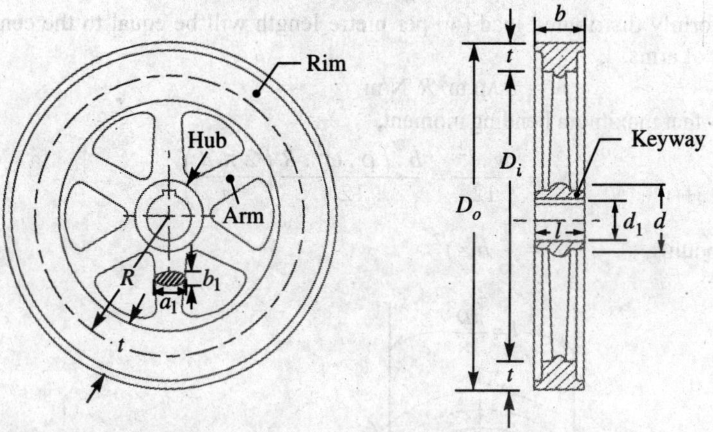

Fig. 22.9. Flywheel.

Consider a small element of the rim as shown shaded in Fig. 22.10. Let it subtends an angle $\delta\theta$ at the centre of the flywheel.

Volume of the small element

$$= A.R.\delta\theta$$

∴ Mass of the small element,

$$dm = \text{Volume} \times \text{Density}$$
$$= A.R.\delta\theta.\rho = \rho.A.R.\delta\theta$$

and centrifugal force on the element,

$$dF = dm.\omega^2.R = \rho.A.R.\delta\theta.\omega^2.R$$
$$= \rho.A.R^2.\omega^2.\delta\theta$$

Vertical component of dF

$$= dF.\sin\theta$$
$$= \rho.A.R^2.\omega^2.\delta\theta \sin\theta$$

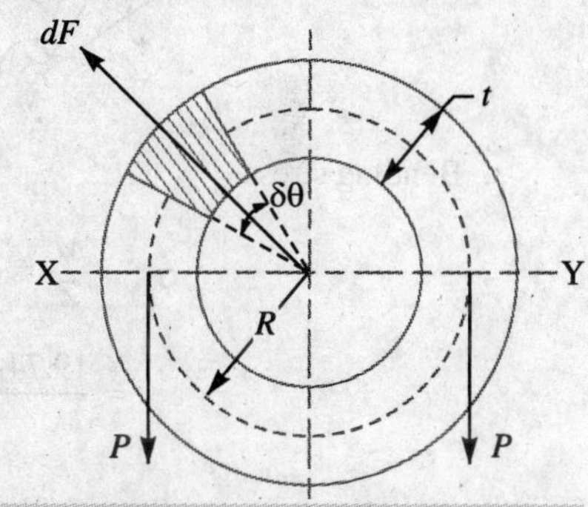

Fig. 22.10. Cross-section of a flywheel rim.

∴ Total vertical bursting force across the rim diameter X-Y,

$$= \rho.A\, R^2.\omega^2 \int_0^{\pi} \sin\theta\, d\theta$$

$$= \rho.A.R^2.\omega^2 \left[-\cos\theta\right]_0^{\pi} = 2\,\rho.A.R^2.\omega^2 \qquad ...(i)$$

This vertical force is resisted by a force of $2P$, such that

$$2P = 2\sigma_t \times A \qquad ...(ii)$$

From equations (*i*) and (*ii*), we have

$$2\rho A.R^2.\omega^2 = 2\,\sigma_t \times A$$

$$\therefore \quad \sigma_t = \rho.R^2.\omega^2 = \rho.v^2 \qquad ...(\because v = \omega.R) \qquad ...(iii)$$

when ρ is in kg / m^3 and v is in m / s, then σ_t will be in N / m^2 or Pa.

Note : From the above expression, the mean diameter (D) of the flywheel may be obtained by using the relation,

$$v = \pi D.N / 60$$

2. *Tensile bending stress caused by restraint of the arms*

The tensile bending stress in the rim due to the restraint of the arms is based on the assumption that each portion of the rim between a pair of arms behaves like a beam fixed at both ends and uniformly loaded, as shown in Fig. 22.11, such that length between fixed ends,

$$l = \frac{\pi D}{n} = \frac{2 \pi R}{n}, \text{ where } n = \text{Number of arms.}$$

The uniformly distributed load (w) per metre length will be equal to the centrifugal force between a pair of arms.

$$\therefore \quad w = b.t.\rho.\omega^2.R \text{ N/m}$$

We know that maximum bending moment,

$$M = \frac{w \,.\, l^2}{12} = \frac{b \,.\, t \,\rho \,.\, \omega^2 \,.\, R}{12}\left(\frac{2 \pi R}{n}\right)^2$$

and section modulus, $\quad Z = \frac{1}{6} b \times t^2$

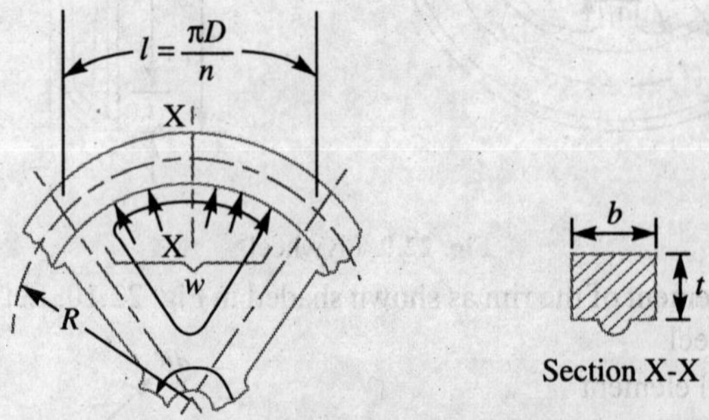

Fig. 22.11

$\therefore$ Bending stress,

$$\sigma_b = \frac{M}{Z} = \frac{b \,.\, t \,\rho \,.\, \omega^2 \,.\, R}{12}\left(\frac{2 \pi R}{n}\right)^2 \times \frac{6}{b \times t^2}$$

$$= \frac{19.74 \,\rho \,.\, \omega^2 \,.\, R^3}{n^2 \,.\, t} = \frac{19.74 \,\rho \,.\, v^2 \,.\, R}{n^2 \,.\, t} \quad \text{...(iv)}$$

...(Substituting $\omega = v/R$)

Now total stress in the rim,

$$\sigma = \sigma_t + \sigma_b$$

If the arms of a flywheel do not stretch at all and are placed very close together, then centrifugal force will not set up stress in the rim. In other words, σ_t will be zero. On the other hand, if the arms are stretched enough to allow free expansion of the rim due to centrifugal action, there will be no restraint due to the arms, *i.e.* σ_b will be zero.

It has been shown by G. Lanza that the arms of a flywheel stretch about $\frac{3}{4}$ th of the amount necessary for free expansion. Therefore the total stress in the rim,

$$= \frac{3}{4}\sigma_t + \frac{1}{4}\sigma_b = \frac{3}{4}\rho.v^2 + \frac{1}{4} \times \frac{19.74 \,\rho \,.\, v^2 \,.\, R}{n^2 \,.\, t} \quad \text{...(v)}$$

$$= \rho.v^2\left(0.75 + \frac{4.935 \, R}{n^2 \,.\, t}\right)$$

Example 22.5. *A multi-cylinder engine is to run at a constant load at a speed of 600 r.p.m. On drawing the crank effort diagram to a scale of 1 m = 250 N-m and 1 mm = 3°, the areas in sq mm above and below the mean torque line are as follows:*

+ 160, – 172, + 168, – 191, + 197, – 162 sq mm

The speed is to be kept within ± 1% of the mean speed of the engine. Calculate the necessary moment of inertia of the flywheel.

Determine suitable dimensions for cast iron flywheel with a rim whose breadth is twice its radial thickness. The density of cast iron is 7250 kg / m³, and its working stress in tension is 6 MPa. Assume that the rim contributes 92% of the flywheel effect.

Solution. Given : = N = 600 r.p.m. or

$\omega = 2\pi \times 600 / 60 = 62.84$ rad / s ; $\rho = 7250$ kg / m³ ; $\sigma_t = 6$ MPa $= 6 \times 10^6$ N/m²

Moment of inertia of the flywheel

Let I = Moment of inertia of the flywheel.

First of all, let us find the maximum fluctuation of energy. The turning moment diagram is shown in Fig. 22.12.

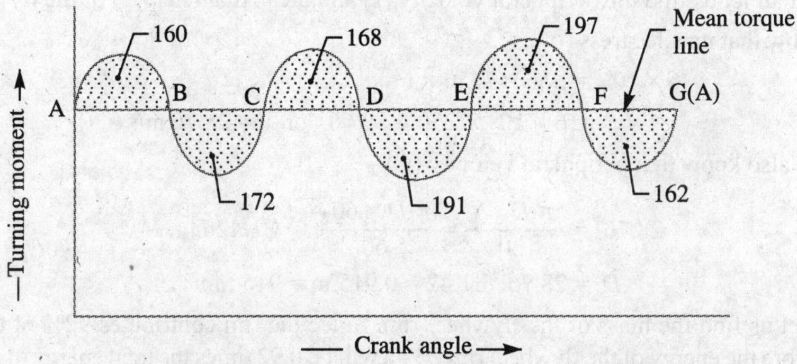

Fig. 22.12

Since the scale for the turning moment is 1 mm = 250 N-m and the scale for the crank angle is

1 mm = 3° = $\dfrac{\pi}{60}$ rad, therefore

1 mm² on the turning moment diagram

$$= 250 \times \frac{\pi}{60} = 13.1 \text{ N-m}$$

Let the total energy at $A = E$. Therefore from Fig. 22.12, we find that

Energy at $B = E + 160$

Energy at $C = E + 160 - 172 = E - 12$

Energy at $D = E - 12 + 168 = E + 156$

Energy at $E = E + 156 - 191 = E - 35$

Energy at $F = E - 35 + 197 = E + 162$

$$\text{Energy at } G = E + 162 - 162 = E = \text{Energy at } A$$

From above, we find that the energy is maximum at F and minimum at E.

$\therefore$ Maximum energy $= E + 162$

and minimum energy $= E - 35$

We know that the maximum fluctuation of energy,

$$\Delta E = \text{Maximum energy} - \text{Minimum energy}$$
$$= (E + 162) - (E - 35) = 197 \text{ mm}^2 = 197 \times 13.1 = 2581 \text{ N-m}$$

Since the fluctuation of speed is ± 1% of the mean speed (ω), therefore total fluctuation of speed,

$$\omega_1 - \omega_2 = 2\% \ \omega = 0.02 \ \omega$$

and coefficient of fluctuation of speed,

$$C_S = \frac{\omega_1 - \omega_2}{\omega} = 0.02$$

We know that the maximum fluctuation of energy (ΔE),

$$2581 = I.\omega^2.C_S = I\,(62.84)^2\ 0.02 = 79\ I$$

$$\therefore \quad I = 2581 / 79 = 32.7 \text{ kg-m}^2 \ \textbf{Ans.}$$

Dimensions of a flywheel rim

Let t = Thickness of the flywheel rim in metres, and

b = Breadth of the flywheel rim in metres = $2\,t$ ...(Given)

First of all let us find the peripheral velocity (v) and mean diameter (D) of the flywheel.

We know that tensile stress (σ_t),

$$6 \times 10^6 = \rho.v^2 = 7250 \times v^2$$

$$\therefore \quad v^2 = 6 \times 10^6 / 7250 = 827.6 \quad \text{or} \quad v = 28.76 \text{ m/s}$$

We also know that peripheral velocity (v),

$$28.76 = \frac{\pi\, D\,.\,N}{60} = \frac{\pi\, D \times 600}{60} = 31.42\ D$$

$$\therefore \quad D = 28.76 / 31.42 = 0.915 \text{ m} = 915 \text{ mm } \textbf{Ans.}$$

Now let us find the mass of the flywheel rim. Since the rim contributes 92% of the flywheel effect, therefore the energy of the flywheel rim (E_{rim}) will be 0.92 times the total energy of the flywheel (E). We know that maximum fluctuation of energy (ΔE),

$$2581 = E \times 2\ C_S = E \times 2 \times 0.02 = 0.04\ E$$

$$\therefore \quad E = 2581 / 0.04 = 64\,525 \text{ N-m}$$

and energy of the flywheel rim,

$$E_{rim} = 0.92\ E = 0.92 \times 64\,525 = 59\,363 \text{ N-m}$$

Let m = Mass of the flywheel rim.

We know that energy of the flywheel rim (E_{rim}),

$$59\,363 = \frac{1}{2} \times m \times v^2 = \frac{1}{2} \times m\,(28.76)^2 = 413.6\ m$$

$$\therefore \quad m = 59\,363 / 413.6 = 143.5 \text{ kg}$$

We also know that mass of the flywheel rim (m),

$$143.5 = b \times t \times \pi\, D \times \rho = 2\,t \times t \times \pi \times 0.915 \times 7250 = 41\,686\ t^2$$

$\therefore \quad t^2 = 143.5 / 41\ 686 = 0.003\ 44$

or $\quad t = 0.0587$ say 0.06 m $= 60$ mm **Ans.**

and $\quad b = 2\,t = 2 \times 60 = 120$ mm **Ans.**

Notes: The mass of the flywheel rim may also be obtained by using the following relations. Since the rim contributes 92% of the flywheel effect, therefore using

1. $I_{rim} = 0.92\ I_{flywheel}$ or $m.k^2 = 0.92 \times 32.7 = 30$ kg–m²

Since radius of gyration, $k = R = D / 2 = 0.915 / 2 = 0.4575$ m, therefore

$$m = \frac{30}{k^2} = \frac{30}{(0.4575)^2} = \frac{30}{0.209} = 143.5 \text{ kg}$$

2. $(\Delta E)_{rim} = 0.92\ (\Delta E)_{flywheel}$

$$m.v^2.C_S = 0.92\ (\Delta E)_{flywheel}$$

$$m\ (28.76)^2\ 0.02 = 0.92 \times 2581$$

$$16.55\ m = 2374.5 \quad \text{or} \quad m = 2374.5 / 16.55 = 143.5 \text{ kg}$$

Flywheel of a printing press

Example 22.6. *The areas of the turning moment diagram for one revolution of a multi-cylinder engine with reference to the mean turning moment, below and above the line, are*

– 32, + 408, – 267, + 333, – 310, + 226, – 374, + 260 and – 244 mm².

The scale for abscissa and ordinate are: 1 mm = 2.4° and 1 mm = 650 N-m respectively. The mean speed is 300 r.p.m. with a percentage speed fluctuation of ± 1.5%. If the hoop stress in the material of the rim is not to exceed 5.6 MPa, determine the suitable diameter and cross-section for the flywheel, assuming that the width is equal to 4 times the thickness. The density of the material may be taken as 7200 kg / m³. Neglect the effect of the boss and arms.

Solution. Given : N = 300 r.p.m. or $\omega = 2\,\pi \times 300/60 = 31.42$ rad/s ; σ_t = 5.6 MPa = 5.6×10^6 N/m^2 ; ρ = 7200 kg/m^3

Diameter of the flywheel

Let D = Diameter of the flywheel in metres.

We know that peripheral velocity of the flywheel,

$$v = \frac{\pi\, D\,.N}{60} = \frac{\pi\, D \times 300}{60} = 15.71\, D \text{ m/s}$$

We also know that hoop stress (σ_t),

$$5.6 \times 10^6 = \rho \times v^2 = 7200\,(15.71\,D)^2 = 1.8 \times 10^6\, D^2$$

$$\therefore \quad D^2 = 5.6 \times 10^6 / 1.8 \times 10^6 = 3.11 \quad \text{or} \quad D = 1.764 \text{ m } \textbf{Ans.}$$

Cross-section of the flywheel

Let t = Thickness of the flywheel rim in metres, and

b = Width of the flywheel rim in metres = 4 t ...(Given)

∴ Cross-sectional area of the rim,

$$A = b \times t = 4\,t \times t = 4\,t^2 \text{ m}^2$$

Now let us find the maximum fluctuation of energy. The turning moment diagram for one revolution of a multi-cylinder engine is shown in Fig. 22.13.

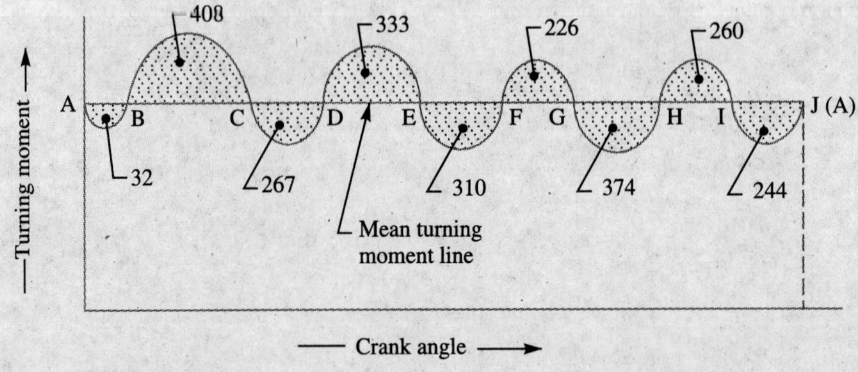

Fig. 22.13

Since the scale of crank angle is 1 mm = 2.4° = $2.4 \times \frac{\pi}{180}$ = 0.042 rad, and the scale of the turning moment is 1 mm = 650 N-m, therefore

1 mm^2 on the turning moment diagram

= 650 × 0.042 = 27.3 N-m

Let the total energy at $A = E$. Therefore from Fig. 22.13, we find that

Energy at $B = E - 32$

Energy at $C = E - 32 + 408 = E + 376$

Energy at $D = E + 376 - 267 = E + 109$

Energy at $E = E + 109 + 333 = E + 442$

Energy at $F = E + 442 - 310 = E + 132$

Energy at $G = E + 132 + 226 = E + 358$

Energy at $H = E + 358 - 374 = E - 16$

Energy at $I = E - 16 + 260 = E + 244$

Energy at $J = E + 244 - 244 = E$ = Energy at A

From above, we see that the energy is maximum at E and minimum at B.

$\therefore$ Maximum energy $= E + 442$

and minimum energy $= E - 32$

We know that maximum fluctuation of energy,

$$\Delta E = \text{Maximum energy} - \text{Minimum energy}$$
$$= (E + 442) - (E - 32) = 474 \text{ mm}^2$$
$$= 474 \times 27.3 = 12\,940 \text{ N-m}$$

Since the fluctuation of speed is ± 1.5% of the mean speed, therefore total fluctuation of speed,

$$\omega_1 - \omega_2 = 3\% \text{ of mean speed} = 0.03\ \omega$$

and coefficient of fluctuation of speed,

$$C_S = \frac{\omega_1 - \omega_2}{\omega} = 0.03$$

Let m = Mass of the flywheel rim.

We know that maximum fluctuation of energy (ΔE),

$$12\,940 = m.R^2.\omega^2.C_S = m\left(\frac{1.764}{2}\right)^2 (31.42)^2\ 0.03 = 23\ m$$

$\therefore$ $m = 12\,940 / 23 = 563$ kg **Ans.**

We also know that mass of the flywheel rim (m),

$$563 = A \times \pi\, D \times \rho = 4\, t^2 \times \pi \times 1.764 \times 7200 = 159\,624\, t^2$$

$\therefore$ $t^2 = 563 / 159\,624 = 0.00353$

or $t = 0.0594$ m $= 59.4$ say 60 mm **Ans.**

and $b = 4\, t = 4 \times 60 = 240$ mm **Ans.**

Example 22.7. *An otto cycle engine develops 50 kW at 150 r.p.m. with 75 explosions per minute. The change of speed from the commencement to the end of power stroke must not exceed 0.5% of mean on either side. Design a suitable rim section having width four times the depth so that the hoop stress does not exceed 4 MPa. Assume that the flywheel stores 16/15 times the energy stored by the rim and that the workdone during power stroke is 1.40 times the workdone during the cycle. Density of rim material is 7200 kg / m³.*

Solution. Given : $P = 50$ kW $= 50 \times 10^3$ W ; $N = 150$ r.p.m. ; $n = 75$; $\sigma_t = 4$ MPa $= 4 \times 10^6$ N/m² ; $\rho = 7200$ kg/m³

First of all, let us find the mean torque (T_{mean}) transmitted by the engine or flywheel. We know that the power transmitted (P),

$$50 \times 10^3 = \frac{2\,\pi\, N \times T_{mean}}{60} = 15.71\ T_{mean}$$

$$\therefore \quad T_{mean} = 50 \times 10^3 / 15.71 = 3182.7 \text{ N-m}$$

Since the explosions per minute are equal to $N/2$, therefore the engine is a four stroke cycle engine. The turning moment diagram of a four stroke engine is shown in Fig. 22.14.

We know that *workdone per cycle

$$= T_{mean} \times \theta = 3182.7 \times 4\,\pi = 40\,000 \text{ N-m}$$

∴ Workdone during power or working stroke

$$= 1.4 \times 40\,000 = 56\,000 \text{ N-m} \quad ...(i)$$

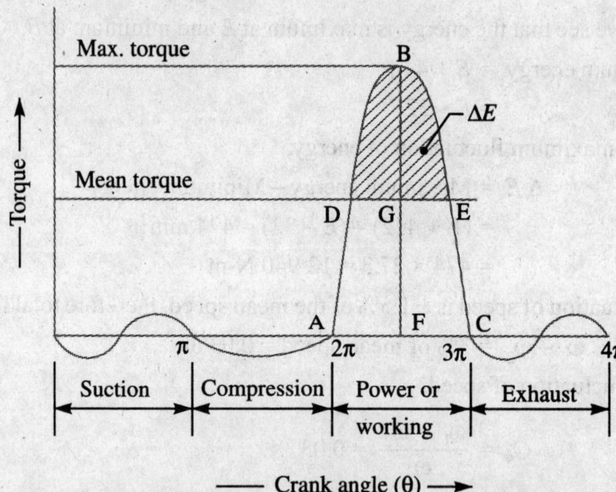

Fig. 22.14

The workdone during power or working stroke is shown by a triangle *ABC* in Fig. 22.14 in which base $AC = \pi$ radians and height $BF = T_{max.}$

∴ Workdone during working stroke

$$= \frac{1}{2} \times \pi \times T_{max} = 1.571\, T_{max} \quad ...(ii)$$

From equations (*i*) and (*ii*), we have

$$T_{max} = 56\,000 / 1.571 = 35\,646 \text{ N-m}$$

Height above the mean torque line,

$$BG = BF - FG = T_{max} - T_{mean} = 35\,646 - 3182.7 = 32\,463.3 \text{ N-m}$$

Since the area *BDE* (shown shaded in Fig. 22.14) above the mean torque line represents the maximum fluctuation of energy (ΔE), therefore from geometrical relation

$$\frac{\text{Area of } \Delta\, BDE}{\text{Area of } \Delta\, ABC} = \frac{(BG)^2}{(BF)^2}, \text{ we have}$$

Maximum fluctuation of energy (*i.e.* area of triangle *BDE*),

$$\Delta E = \text{Area of triangle } ABC \times \left(\frac{BG}{BF}\right)^2 = 56\,000 \times \left(\frac{32\,463.3}{35\,646}\right)^2$$

$$= 56\,000 \times 0.83 = 46\,480 \text{ N-m}$$

Mean diameter of the flywheel

Let D = Mean diameter of the flywheel in metres, and

v = Peripheral velocity of the flywheel in m/s.

---

* The workdone per cycle for a four stroke engine is also given by

$$\text{Workdone / cycle} = \frac{P \times 60}{\text{Number of explosion / min}} = \frac{P \times 60}{n} = \frac{50\,000 \times 60}{75} = 40\,000 \text{ N-m}$$

Flywheel of a motorcycle

We know that hoop stress (σ_t),

$$4 \times 10^6 = \rho.v^2 = 7200 \times v^2$$

$$\therefore\ v^2 = 4 \times 10^6 / 7200 = 556$$

or $$v = 23.58 \text{ m/s}$$

We also know that peripheral velocity (v),

$$23.58 = \frac{\pi\, D\, N}{60} = \frac{\pi\, D \times 150}{60} = 7.855\, D$$

$$\therefore\ D = 23.58 / 7.855 = 3 \text{ m } \textbf{Ans.}$$

Cross-sectional dimensions of the rim

Let t = Thickness of the rim in metres, and

b = Width of the rim in metres = 4 t ...(Given)

$\therefore$ Cross-sectional area of the rim,

$$A = b \times t = 4\,t \times t = 4\,t^2$$

First of all, let us find the mass of the flywheel rim.

Let m = Mass of the flywheel rim, and

E = Total energy of the flywheel.

Since the fluctuation of speed is 0.5% of the mean speed on either side, therefore total fluctuation of speed,

$$N_1 - N_2 = 1\% \text{ of mean speed} = 0.01\,N$$

and coefficient of fluctuation of speed,

$$C_S = \frac{N_1 - N_2}{N} = 0.01$$

We know that the maximum fluctuation of energy ($\Delta\,E$),

$$46\,480 = E \times 2\,C_S = E \times 2 \times 0.01 = 0.02\,E$$

$$\therefore\ E = 46\,480 / 0.02 = 2324 \times 10^3 \text{ N-m}$$

Since the energy stored by the flywheel is $\frac{16}{15}$ times the energy stored by the rim, therefore the energy of the rim,

$$E_{rim} = \frac{15}{16}\,E = \frac{15}{16} \times 2324 \times 10^3 = 2178.8 \times 10^3 \text{ N-m}$$

We know that energy of the rim (E_{rim}),

$$2178.8 \times 10^3 = \frac{1}{2} \times m \times v^2 = \frac{1}{2} \times m\,(23.58)^2 = 278\,m$$

$$\therefore\ m = 2178.8 \times 10^3 / 278 = 7837 \text{ kg}$$

We also know that mass of the flywheel rim (m),

$$7837 = A \times \pi\,D \times \rho = 4\,t^2 \times \pi \times 3 \times 7200 = 271\,469\,t^2$$

$$\therefore\ t^2 = 7837 / 271\,469 = 0.0288 \text{ or } t = 0.17 \text{ m} = 170 \text{ mm } \textbf{Ans.}$$

and $$b = 4\,t = 4 \times 170 = 680 \text{ mm } \textbf{Ans.}$$

Example 22.8. *A shaft fitted with a flywheel rotates at 250 r.p.m. and drives a machine. The torque of machine varies in a cyclic manner over a period of 3 revolutions. The torque rises from 750 N-m to 3000 N-m uniformly during 1 / 2 revolution and remains constant for the following revolution. It then falls uniformly to 750 N-m during the next 1 / 2 revolution and remains constant for one revolution, the cycle being repeated thereafter. Determine the power required to drive the machine.*

If the total fluctuation of speed is not to exceed 3% of the mean speed, determine a suitable diameter and cross-section of the flywheel rim. The width of the rim is to be 4 times the thickness and the safe centrifugal stress is 6 MPa. The material density may be assumed as 7200 kg / m³.

Solution. Given : N = 250 r.p.m. or $\omega = 2\,\pi \times 250 / 60 = 26.2$ rad/s ; $\omega_1 - \omega_2 = 3\%\ \omega$ or $\frac{\omega_1 - \omega_2}{\omega} = C_S = 3\% = 0.03$; $\sigma_t = 6$ MPa $= 6 \times 10^6$ N/m² ; $\rho = 7200$ kg / m³

Power required to drive the machine

The turning moment diagram for the complete cycle is shown in Fig. 22.15.

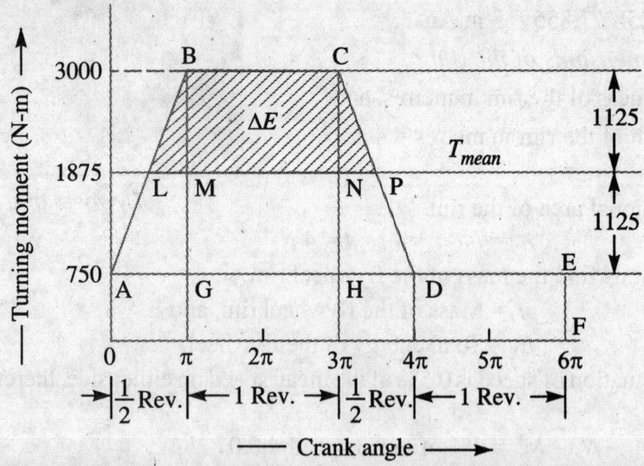

Fig. 22.15

We know that the torque required for one complete cycle

$= \text{Area of figure } OABCDEF$

$= \text{Area } OAEF + \text{Area } ABG + \text{Area } BCHG + \text{Area } CDH$

$$= OF \times OA + \frac{1}{2} \times AG \times BG + GH \times CH + \frac{1}{2} \times HD \times CH$$

$$= 6\,\pi \times 750 + \frac{1}{2} \times \pi\,(3000 - 750) + 2\pi\,(3000 - 750) + \frac{1}{2} \times \pi\,(3000 - 750)$$

$$= 4500\,\pi + 1125\,\pi + 4500\,\pi + 1125\,\pi = 11\,250\,\pi \text{ N-m} \quad \text{...(i)}$$

If T_{mean} is the mean torque in N-m, then torque required for one complete cycle

$$= T_{mean} \times 6\,\pi \text{ N-m} \quad \text{...(ii)}$$

From equations (*i*) and (*ii*),

$$T_{mean} = 11250\,\pi / 6\,\pi = 1875 \text{ N-m}$$

We know that power required to drive the machine,

$$P = T_{mean} \times \omega = 1875 \times 26.2 = 49\,125 \text{ W} = 49.125 \text{ kW}$$ **Ans.**

Diameter of the flywheel

Let D = Diameter of the flywheel in metres, and

v = Peripheral velocity of the flywheel in m/s.

We know that the centrifugal stress (σ_t),

$$6 \times 10^6 = \rho \times v^2 = 7200 \times v^2$$

$$\therefore \quad v^2 = 6 \times 10^6 / 7200 = 833.3 \quad \text{or} \quad v = 28.87 \text{ m/s}$$

We also know that peripheral velocity of the flywheel (v),

$$28.87 = \frac{\pi\, D\, N}{60} = \frac{\pi\, D \times 250}{60} = 13.1\, D$$

$$\therefore \qquad D = 28.87 / 13.1 = 2.2 \text{ m } \textbf{Ans.}$$

Cross-section of the flywheel rim

Let t = Thickness of the flywheel rim in metres, and

b = Width of the flywheel rim in metres = $4\,t$ ...(Given)

∴ Cross-sectional area of the flywheel rim,

$$A = b \times t = 4\, t \times t = 4\, t^2 \text{ m}^2$$

First of all, let us find the maximum fluctuation of energy (ΔE) and mass of the flywheel rim (m). In order to find ΔE, we shall calculate the values of LM and NP.

From similar triangles ABG and BLM,

$$\frac{LM}{AG} = \frac{BM}{BG} \quad \text{or} \quad \frac{LM}{\pi} = \frac{3000 - 1857}{3000 - 750} = 0.5 \quad \text{or} \quad LM = 0.5\,\pi$$

Now from similar triangles CHD and CNP,

$$\frac{NP}{HD} = \frac{CN}{CH} \quad \text{or} \quad \frac{NP}{\pi} = \frac{3000 - 1875}{3000 - 750} = 0.5 \quad \text{or} \quad NP = 0.5\,\pi$$

From Fig. 22.15, we find that

$$BM = CN = 3000 - 1875 = 1125 \text{ N-m}$$

Since the area above the mean torque line represents the maximum fluctuation of energy, therefore maximum fluctuation of energy,

$$\Delta E = \text{Area } LBCP = \text{Area } LBM + \text{Area } MBCN + \text{Area } PNC$$

$$= \frac{1}{2} \times LM \times BM + MN \times \text{BM} + \frac{1}{2} \times NP \times CN$$

$$= \frac{1}{2} \times 0.5\,\pi \times 1125 + 2\,\pi \times 1125 + \frac{1}{2} \times 0.5\,\pi \times 1125$$

$$= 8837 \text{ N-m}$$

We know that maximum fluctuation of energy (ΔE),

$$8837 = m.R^2.\omega^2.C_S = m\left(\frac{2.2}{2}\right)^2 (26.2)^2\, 0.03 = 24.9\, m$$

$$\therefore \qquad m = 8837 / 24.9 = 355 \text{ kg}$$

We also know that mass of the flywheel rim (m),

$$355 = A \times \pi\, D \times \rho = 4\, t^2 \times \pi \times 2.2 \times 7200 = 199\,077\, t^2$$

$$\therefore \qquad t^2 = 355 / 199\,077 = 0.00178 \text{ or } t = 0.042 \text{ m} = 42 \text{ say } 45 \text{ mm } \textbf{Ans.}$$

and $b = 4\,t = 4 \times 45 = 180$ mm **Ans.**

Example 22.9. *A punching machine makes 25 working strokes per minute and is capable of punching 25 mm diameter holes in 18 mm thick steel plates having an ultimate shear strength of 300 MPa.*

The punching operation takes place during 1/10 th of a revolution of the crank shaft.

Estimate the power needed for the driving motor, assuming a mechanical efficiency of 95 per cent. Determine suitable dimensions for the rim cross-section of the flywheel, which is to revolve at 9 times the speed of the crank shaft. The permissible coefficient of fluctuation of speed is 0.1.

The flywheel is to be made of cast iron having a working stress (tensile) of 6 MPa and density of 7250 kg / m³. The diameter of the flywheel must not exceed 1.4 m owing to space restrictions. The hub and the spokes may be assumed to provide 5% of the rotational inertia of the wheel. Check for the centrifugal stress induced in the rim.

Solution. Given : $n = 25$; $d_1 = 25$ mm ; $t_1 = 18$ mm ; $\tau_u = 300$ MPa = 300 N/mm² ; $\eta_m = 95\% = 0.95$; $C_S = 0.1$; $\sigma_t = 6$ MPa = 6 N/mm² ; $\rho = 7250$ kg/m³ ; $D = 1.4$ m or $R = 0.7$m

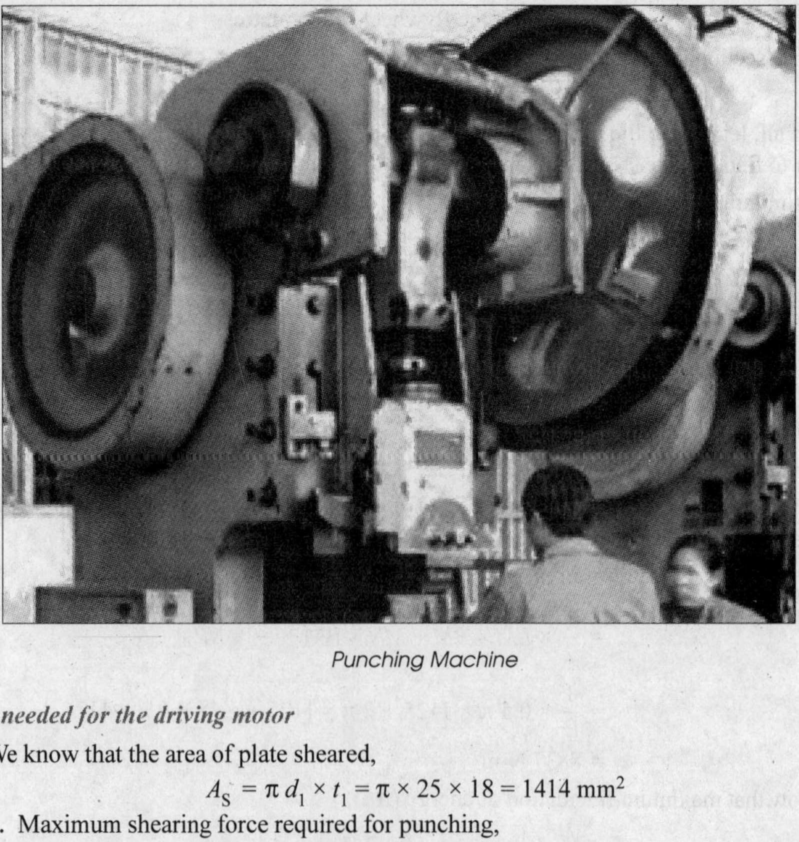

Punching Machine

Power needed for the driving motor

We know that the area of plate sheared,

$$A_S = \pi\, d_1 \times t_1 = \pi \times 25 \times 18 = 1414 \text{ mm}^2$$

∴ Maximum shearing force required for punching,

$$F_S = A_S \times \tau_u = 1414 \times 300 = 424\,200 \text{ N}$$

and energy required per stroke

$$= {}^*\text{Average shear force} \times \text{Thickness of plate}$$

$$= \frac{1}{2} F_S \times t_1 = \frac{1}{2} \times 424\,200 \times 18 = 3817.8 \times 10^3 \text{ N-mm}$$

∴ Energy required per min

$$= \text{Energy / stroke} \times \text{No. of working strokes / min}$$

$$= 3817.8 \times 10^3 \times 25 = 95.45 \times 10^6 \text{ N-mm} = 95\,450 \text{ N-m}$$

We know that the power needed for the driving motor

$$= \frac{\text{Energy required per min}}{60 \times \eta_m} = \frac{95\,450}{60 \times 0.95} = 1675 \text{ W}$$

$$= 1.675 \text{ kW } \textbf{Ans.}$$

* As the hole is punched, it is assumed that the shearing force decreases uniformly from maximum value to zero.

Dimensions for the rim cross-section

Considering the cross-section of the rim as rectangular and assuming the width of rim equal to twice the thickness of rim.

Let t = Thickness of rim in metres, and

b = Width of rim in metres = $2\,t$.

∴ Cross-sectional area of rim,

$$A = b \times t = 2\,t \times t = 2\,t^2$$

Since the punching operation takes place (*i.e.* energy is consumed) during 1/10 th of a revolution of the crank shaft, therefore during 9/10 th of the revolution of a crank shaft, the energy is stored in the flywheel.

∴ Maximum fluctuation of energy,

$$\Delta E = \frac{9}{10} \times \text{Energy/stroke} = \frac{9}{10} \times 3817.8 \times 10^3$$

$$= 3436 \times 10^3 \text{ N-mm} = 3436 \text{ N-m}$$

Let m = Mass of the flywheel.

Since the hub and the spokes provide 5% of the rotational inertia of the wheel, therefore the maximum fluctuation of energy provided by the flywheel rim will be 95%.

∴ Maximum fluctuation of energy provided by the rim,

$$(\Delta E)_{rim} = 0.95 \times \Delta E = 0.95 \times 3436 = 3264 \text{ N-m}$$

Since the flywheel is to revolve at 9 times the speed of the crankshaft and there are 25 working strokes per minute, therefore mean speed of the flywheel,

$$N = 9 \times 25 = 225 \text{ r.p.m.}$$

and mean angular speed, $\omega = 2\,\pi \times 225 / 60 = 23.56$ rad/s

We know that maximum fluctuation of energy (ΔE),

$$3264 = m.R^2.\omega^2.C_S = m\,(0.7)^2\,(23.56)^2\,0.1 = 27.2\,m$$

$$\therefore \quad m = 3264 / 27.2 = 120 \text{ kg}$$

We also know that mass of the flywheel (m),

$$120 = A \times \pi D \times \rho = 2\,t^2 \times \pi \times 1.4 \times 7250 = 63\,782\,t^2$$

$$\therefore \quad t^2 = 120 / 63\,782 = 0.001\,88 \quad \text{or} \quad t = 0.044 \text{ m} = 44 \text{ mm}$$ **Ans.**

and $b = 2\,t = 2 \times 44 = 88$ mm **Ans.**

Check for centrifugal stress

We know that peripheral velocity of the rim,

$$v = \frac{\pi\,D\,.\,N}{60} = \frac{\pi \times 1.4 \times 225}{60} = 16.5 \text{ m/s}$$

∴ Centrifugal stress induced in the rim,

$$\sigma_t = \rho.v^2 = 7250\,(16.5)^2 = 1.97 \times 10^6 \text{ N/m}^2 = 1.97 \text{ MPa}$$

Since the centrifugal stress induced in the rim is less than the permissible value (*i.e.* 6 MPa), therefore it is safe **Ans.**

22.8 Stresses in Flywheel Arms

The following stresses are induced in the arms of a flywheel.

1. Tensile stress due to centrifugal force acting on the rim.
2. Bending stress due to the torque transmitted from the rim to the shaft or from the shaft to the rim.
3. Shrinkage stresses due to unequal rate of cooling of casting. These stresses are difficult to determine.

We shall now discuss the first two types of stresses as follows:

1. ***Tensile stress due to the centrifugal force***

Due to the centrifugal force acting on the rim, the arms will be subjected to direct tensile stress whose magnitude is same as discussed in the previous article.

∴ Tensile stress in the arms,

$$\sigma_{t1} = \frac{3}{4}\,\sigma_t = \frac{3}{4}\,\rho \times v^2$$

2. ***Bending stress due to the torque transmitted***

Due to the torque transmitted from the rim to the shaft or from the shaft to the rim, the arms will be subjected to bending, because they are required to carry the full torque load. In order to find out the maximum bending moment on the arms, it may be assumed as a centilever beam fixed at the hub and carrying a concentrated load at the free end of the rim as shown in Fig. 22.16.

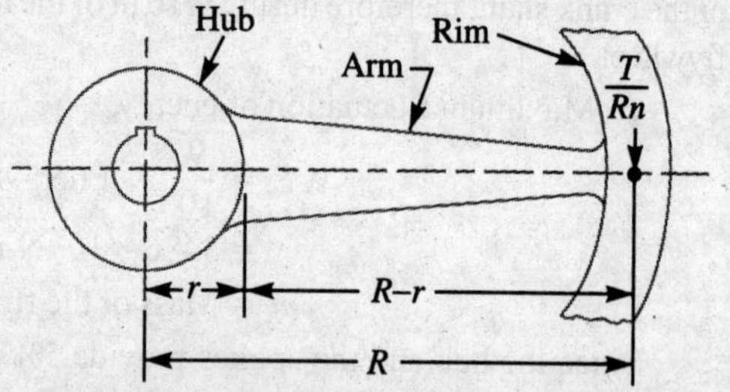

Fig. 22.16

Let T = Maximum torque transmitted by the shaft,

R = Mean radius of the rim,

r = Radius of the hub,

n = Number of arms, and

Z = Section modulus for the cross-section of arms.

We know that the load at the mean radius of the rim,

$$F = \frac{T}{R}$$

∴ Load on each arm $= \dfrac{T}{R\,.\,n}$

and maximum bending moment which lies on the arm at the hub,

$$M = \frac{T}{R\,.\,n}\,(R - r)$$

∴ Bending stress in arms,

$$\sigma_{b1} = \frac{M}{Z} = \frac{T}{R\,.\,n\,.\,Z}\,(R - r)$$

∴ Total tensile stress in the arms at the hub end,

$$\sigma = \sigma_{t1} + \sigma_{b1}$$

Notes: 1. The total stress on the arms should not exceed the allowable permissible stress.

2. If the flywheel is used as a belt pulley, then the arms are also subjected to bending due to net belt tension $(T_1 - T_2)$, where T_1 and T_2 are the tensions in the tight side and slack side of the belt respectively. Therefore the bending stress due to the belt tensions,

$$\sigma_{b2} = \frac{(T_1 - T_2)\,(R - r)}{\frac{n}{2} \times Z}$$

... (∵ Only half the number of arms are considered to be effective in transmitting the belt tensions)

∴ Total bending stress in the arms at the hub end,

$$\sigma_b = \sigma_{b1} + \sigma_{b2}$$

and the total tensile stress in the arms at the hub end,

$$\sigma = \sigma_{t1} + \sigma_{b1} + \sigma_{b2}$$

22.9 Design of Flywheel Arms

The cross-section of the arms is usually elliptical with major axis as twice the minor axis, as shown in Fig. 22.17, and it is designed for the maximum bending stress.

Let a_1 = Major axis, and

b_1 = Minor axis.

∴ Section modulus,

$$Z = \frac{\pi}{32} \times b_1 (a_1)^2 \qquad ...(i)$$

We know that maximum bending moment,

$$M = \frac{T}{R \, . \, n} (R - r)$$

∴ Maximum bending stress,

$$\sigma_b = \frac{M}{Z} = \frac{T}{R.n.Z} (R - r) \qquad ...(ii)$$

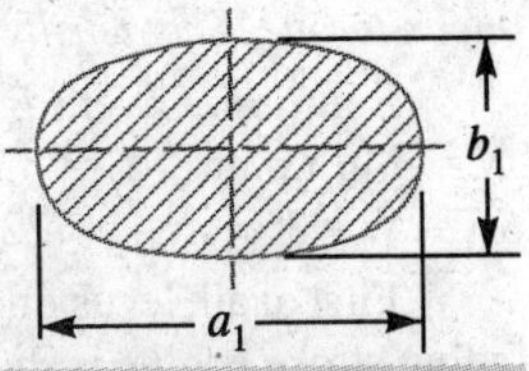

Fig. 22.17. Elliptical cross section of arms.

Assuming $a_1 = 2\,b_1$, the dimensions of the arms may be obtained from equations (*i*) and (*ii*).

Notes: 1. The arms of the flywheel have a taper from the hub to the rim. The taper is about 20 mm per metre length of the arm for the major axis and 10 mm per metre length for the minor axis.

2. The number of arms are usually 6. Sometimes the arms may be 8, 10 or 12 for very large size flywheels.

3. The arms may be curved or straight. But straight arms are easy to cast and are lighter.

4. Since arms are subjected to reversal of stresses, therefore a minimum factor of safety 8 should be used. In some cases like punching machines amd machines subjected to severe shock, a factor of safety 15 may be used.

5. The smaller flywheels (less than 600 mm diameter) are not provided with arms. They are made web type with holes in the web to facilitate handling.

22.10 Design of Shaft, Hub and Key

The diameter of shaft for flywheel is obtained from the maximum torque transmitted. We know that the maximum torque transmitted,

$$T_{max} = \frac{\pi}{16} \times \tau (d_1)^3$$

where d_1 = Diameter of the shaft, and

τ = Allowable shear stress for the material of the shaft.

The hub is designed as a hollow shaft, for the maximum torque transmitted. We know that the maximum torque transmitted,

$$T_{max} = \frac{\pi}{16} \times \tau \left(\frac{d^4 - d_1^4}{d} \right)$$

where d = Outer diameter of hub, and

d_1 = Inner diameter of hub or diameter of shaft.

The diameter of hub is usually taken as twice the diameter of shaft and length from 2 to 2.5 times the shaft diameter. It is generally taken equal to width of the rim.

A standard sunk key is used for the shaft and hub. The length of key is obtained by considering the failure of key in shearing. We know that torque transmitted by shaft,

$$T_{max} = L \times w \times \tau \times \frac{d_1}{2}$$

where L = Length of the key,

τ = Shear stress for the key material, and

d_1 = Diameter of shaft.

Example 22.10. *Design and draw a cast iron flywheel used for a four stroke I.C engine developing 180 kW at 240 r.p.m. The hoop or centrifugal stress developed in the flywheel is 5.2 MPa, the total fluctuation of speed is to be limited to 3% of the mean speed. The work done during the power stroke is 1/3 more than the average work done during the whole cycle. The maximum torque on the shaft is twice the mean torque. The density of cast iron is 7220 kg/m³.*

Solution. Given: $P = 180$ kW $= 180 \times 10^3$ W; $N = 240$ r.p.m. ; $\sigma_t = 5.2$ MPa $= 5.2 \times 10^6$ N/m² ; $N_1 - N_2 = 3\% N$; $\rho = 7220$ kg/m³

First of all, let us find the maximum fluctuation of energy (ΔE). The turning moment diagram of a four stroke engine is shown in Fig. 22.18.

We know that mean torque transmitted by the flywheel,

$$T_{mean} = \frac{P \times 60}{2\,\pi\,N} = \frac{180 \times 10^3 \times 60}{2\,\pi \times 240} = 7161 \text{ N-m}$$

and *workdone per cycle $= T_{mean} \times \theta = 7161 \times 4\,\pi = 90\,000$ N-m

Since the workdone during the power stroke is 1/3 more than the average workdone during the whole cycle, therefore,

Workdone during the power (or working) stroke

$$= 90\,000 + \frac{1}{3} \times 90\,000 = 120\,000 \text{ N-m} \quad ...(i)$$

The workdone during the power stroke is shown by a triangle *ABC* in Fig. 22.18 in which the base $AC = \pi$ radians and height $BF = T_{max.}$

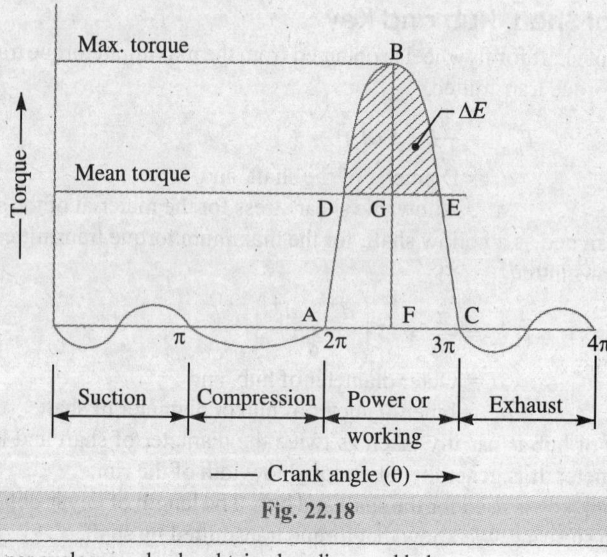

Fig. 22.18

* The workdone per cycle may also be obtained as discussed below :

Workdone per cycle $= \dfrac{P \times 60}{n}$, where n = Number of working strokes per minute

For a four stroke engine, $n = N / 2 = 240 / 2 = 120$

$\therefore$ Workdone per cycle $= \dfrac{180 \times 10^3 \times 60}{120} = 90\,000$ N-m

∴ Workdone during power stroke

$$= \frac{1}{2} \times \pi \times T_{max} \qquad ...(ii)$$

From equations (*i*) and (*ii*), we have

$$\frac{1}{2} \times \pi \times T_{max} = 120\ 000$$

$$\therefore \quad T_{max} = \frac{120\ 000 \times 2}{\pi} = 76\ 384 \text{ N-m}$$

Height above the mean torque line,

$$BG = BF - FG = T_{max} - T_{mean} = 76\ 384 - 7161 = 69\ 223 \text{ N-m}$$

Since the area *BDE* shown shaded in Fig. 22.18 above the mean torque line represents the maximum fluctuation of energy (ΔE), therefore from geometrical relation,

$$\frac{\text{Area of } \Delta\ BDE}{\text{Area of } \Delta\ ABC} = \frac{(BG)^2}{(BF)^2}, \text{ we have}$$

*Maximum fluctuation of energy (*i.e.* area of Δ *BDE*),

$$\Delta E = \text{Area of } \Delta\ ABC \times \left(\frac{BG}{BF}\right)^2 = 120\ 000 \left(\frac{69\ 223}{76\ 384}\right)^2 = 98\ 555 \text{ N-m}$$

1. ***Diameter of the flywheel rim***

Let D = Diameter of the flywheel rim in metres, and

v = Peripheral velocity of the flywheel rim in m/s.

We know that the hoop stress developed in the flywheel rim (σ_t),

$$5.2 \times 10^6 = \rho.v^2 = 7220 \times v^2$$

$$\therefore \quad v^2 = 5.2 \times 10^6 / 7220 = 720 \quad \text{or} \quad v = 26.8 \text{ m/s}$$

We also know that peripheral velocity (v),

$$26.8 = \frac{\pi\ D.N}{60} = \frac{2D \times 250}{60} = 13.1\ D$$

$$\therefore \quad D = 26.8 / 13.1 = 2.04 \text{ m } \textbf{Ans.}$$

2. ***Mass of the flywheel rim***

Let m = Mass of the flywheel rim in kg.

We know that angular speed of the flywheel rim,

$$\omega = \frac{2\ \pi\ N}{60} = \frac{2\ \pi \times 250}{60} = 25.14 \text{ rad / s}$$

and coefficient of fluctuation of speed,

$$C_S = \frac{N_1 - N_2}{N} = 0.03$$

We know that maximum fluctuation of energy (ΔE),

$$98\ 555 = m.R^2.\omega^2.C_S = m\left(\frac{2.04}{2}\right)^2 (25.14)^2\ 0.03 = 19.73\ m$$

$$\therefore \quad m = 98\ 555 / 19.73 = 4995 \text{ kg } \textbf{Ans.}$$

---

* The approximate value of maximum fluctuation of energy may be obtained as discussed below :

Workdone per cycle = 90 000 N-mm ...(as calculated above)

Workdone per stroke = 90 000 / 4 = 22 500 N-m ...(∵ of four stroke engine)

and workdone during power stroke = 120 000 N-m

∴ Maximum fluctuation of energy,

$$\Delta E = 120\ 000 - 22\ 500 = 97\ 500 \text{ N-m}$$

3. ***Cross-sectional dimensions of the rim***

Let t = Depth or thickness of the rim in metres, and

b = Width of the rim in metres = $2\,t$ ...(Assume)

∴ Cross-sectional area of the rim,

$$A = b.t = 2\,t \times t = 2\,t^2$$

We know that mass of the flywheel rim (m),

$$4995 = A \times \pi D \times \rho = 2\,t^2 \times \pi \times 2.04 \times 7220 = 92\,556\,t^2$$

∴ $t^2 = 4995 / 92\,556 = 0.054$ or $t = 0.232$ say 0.235 m = 235 mm **Ans.**

and $b = 2\,t = 2 \times 235 = 470$ mm **Ans.**

4. ***Diameter and length of hub***

Let d = Diameter of the hub,

d_1 = Diameter of the shaft, and

l = Length of the hub.

Since the maximum torque on the shaft is twice the mean torque, therefore maximum torque acting on the shaft,

$$T_{max} = 2 \times T_{mean} = 2 \times 7161 = 14\,322 \text{ N-m} = 14\,322 \times 10^3 \text{ N-mm}$$

We know that the maximum torque acting on the shaft (T_{max}),

$$14\,322 \times 10^3 = \frac{\pi}{16} \times \tau\,(d_1)^3 = \frac{\pi}{16} \times 40\,(d_1)^3 = 7.855\,(d_1)^3$$

...(Taking τ = 40 MPa = 40 N/mm²)

∴ $(d_1)^3 = 14\,322 \times 10^3 / 7.855 = 1823 \times 10^3$

or $d_1 = 122$ say 125 m **Ans.**

The diameter of the hub is made equal to twice the diameter of shaft and length of hub is equal to width of the rim.

∴ $d = 2\,d_1 = 2 \times 125 = 250$ mm = 0.25 m

and $l = b = 470$ mm = 0.47 mm **Ans.**

5. ***Cross-sectional dimensions of the elliptical arms***

Let a_1 = Major axis,

b_1 = Minor axis = $0.5\,a_1$ ...(Assume)

n = Number of arms = 6 ...(Assume)

σ_b = Bending stress for the material of arms = 15 MPa = 15 N/mm² ...(Assume)

We know that the maximum bending moment in the arm at the hub end, which is assumed as cantilever is given by

$$M = \frac{T}{R.n}(R - r) = \frac{T}{D\,.\,n}(D - d) = \frac{14\,322}{2.04 \times 6}(2.04 - 0.25) \text{ N-m}$$

$$= 2094.5 \text{ N-m} = 2094.5 \times 10^3 \text{ N-mm}$$

and section modulus for the cross-section of the arm,

$$Z = \frac{\pi}{32} \times b_1\,(a_1)^2 = \frac{\pi}{32} \times 0.5\,a_1\,(a_1)^2 = 0.05\,(a_1)^3$$

We know that the bending stress (σ_b),

$$15 = \frac{M}{Z} = \frac{2094.5 \times 10^3}{0.05(a_1)^3} = \frac{41\,890 \times 10^3}{(a_1)^3}$$

$\therefore$ $(a_1)^3 = 41\,890 \times 10^3 / 15 = 2793 \times 10^3$ or $a_1 = 140$ mm **Ans.**

and $b_1 = 0.5\,a_1 = 0.5 \times 140 = 70$ mm **Ans.**

6. ***Dimensions of key***

The standard dimensions of rectangular sunk key for a shaft of diameter 125 mm are as follows:

Width of key, $w = 36$ mm **Ans.**

and thickness of key $= 20$ mm **Ans.**

The length of key (L) is obtained by considering the failure of key in shearing.

We know that the maximum torque transmitted by the shaft (T_{max}),

$$14\,322 \times 10^3 = L \times w \times \tau \times \frac{d_1}{2} = L \times 36 \times 40 \times \frac{125}{2} = 90 \times 10^3\,L$$

$\therefore$ $L = 14\,322 \times 10^3 / 90 \times 10^3 = 159$ say 160 mm **Ans.**

Let us now check the total stress in the rim which should not be greater than 15 MPa. We know that total stress in the rim,

$$\sigma = \rho.v^2\left(0.75 + \frac{4.935\,R}{n^2\,.\,t}\right)$$

$$= 7220\,(26.8)^2\left[0.75 + \frac{4.935\,(2.04\,/\,2)}{6^2 \times 0.235}\right] \text{N/m}^2$$

$$= 5.18 \times 10^6\,(0.75 + 0.595) = 6.97 \times 10^6 \text{ N/m}^2 = 6.97 \text{ MPa}$$

Since it is less than 15 MPa, therefore the design is safe.

Example 22.11. *A single cylinder double acting steam engine delivers 185 kW at 100 r.p.m. The maximum fluctuation of energy per revolution is 15 per cent of the energy developed per revolution. The speed variation is limited to 1 per cent either way from the mean. The mean diameter of the rim is 2.4 m. Design and draw two views of the flywheel.*

Solution. Given : $P = 185$ kW $= 185 \times 10^3$ W ; $N = 100$ r.p.m ; $\Delta\,E = 15\%\;E = 0.15\,E$; $D = 2.4$ m or $R = 1.2$ m

1. ***Mass of the flywheel rim***

Let m = Mass of the flywheel rim in kg.

We know that the workdone or energy developed per revolution,

$$E = \frac{P \times 60}{N} = \frac{185 \times 10^3 \times 60}{100} = 111\,000 \text{ N-m}$$

$\therefore$ Maximum fluctuation of energy,

$$\Delta\,E = 0.15\,E = 0.15 \times 111\,000 = 16\,650 \text{ N-m}$$

Since the speed variation is 1% either way from the mean, therefore the total fluctuation of speed,

$$N_1 - N_2 = 2\% \text{ of mean speed} = 0.02\,N$$

and coefficient of fluctuation of speed,

$$C_S = \frac{N_1 - N_2}{N} = 0.02$$

Velocity of the flywheel,

$$v = \frac{\pi\, D.N}{60} = \frac{\pi \times 2.4 \times 100}{60} = 12.57 \text{ m/s}$$

We know that the maximum fluctuation of energy (ΔE),

$$16\,650 = m.v^2.C_S = m\,(12.57)^2\, 0.02 = 3.16\, m$$

$\therefore$ $m = 16\,650 / 3.16 = 5270$ kg **Ans.**

2. *Cross-sectional dimensions of the flywheel rim*

Let t = Thickness of the flywheel rim in metres, and

b = Width of the flywheel rim in metres = $2\,t$ ...(Assume)

$\therefore$ Cross-sectional area of the rim,

$$A = b \times t = 2\,t \times t = 2\,t^2$$

We know that mass of the flywheel rim (m),

$$5270 = A \times \pi\, D \times \rho = 2\,t^2 \times \pi \times 2.4 \times 7200 = 108\,588\, t^2$$

...(Taking $\rho = 7200$ kg / m^3)

$\therefore$ $t^2 = 5270 / 108\,588 = 0.0485$ or $t = 0.22$ m = 220 mm **Ans.**

and $b = 2\,t = 2 \times 220 = 440$ mm **Ans.**

3. *Diameter and length of hub*

Let d = Diameter of the hub,

d_1 = Diameter of the shaft, and

l = Length of the hub,

Steam engine in a Laboratory

We know that mean torque transmitted by the shaft,

$$T_{mean} = \frac{P \times 60}{2\,\pi\,N} = \frac{185 \times 10^3 \times 60}{2\,\pi \times 100} = 17\,664 \text{ N-m}$$

Assuming that the maximum torque transmitted (T_{max}) by the shaft is twice the mean torque, therefore

$$T_{max} = 2 \times T_{mean} = 2 \times 17\,664 = 35\,328 \text{ N-m} = 35.328 \times 10^6 \text{ N-mm}$$

We also know that maximum torque transmitted by the shaft (T_{max}),

$$35.328 \times 10^6 = \frac{\pi}{16} \times \tau\,(d_1)^3 = \frac{\pi}{16} \times 40\,(d_1)^3 = 7.855\,(d_1)^3$$

...(Assuming τ = 40 MPa = 40 N/mm²)

$\therefore$ $(d_1)^3 = 35.328 \times 10^6 / 7.855 = 4.5 \times 10^6$ or d_1 = 165 mm **Ans.**

The diameter of the hub (d) is made equal to twice the diameter of the shaft (d_1) and length of the hub (l) is equal to the width of the rim (b).

$\therefore$ $d = 2\,d_1 = 2 \times 165 = 330$ mm ; and $l = b = 440$ mm **Ans.**

4. *Cross-sectional dimensions of the elliptical arms*

Let a_1 = Major axis,

b_1 = Minor axis = 0.5 a_1 ...(Assume)

n = Number of arms = 6 ...(Assume)

σ_b = Bending stress for the material of the arms = 14 MPa = 14 N/mm² ...(Assume)

We know that the maximum bending moment in the arm at the hub end which is assumed as cantilever is given by

$$M = \frac{T}{R.n}(R - r) = \frac{T}{D\,.\,n}(D - d) = \frac{35\,328}{2.4 \times 6}(2.4 - 0.33) \text{ N-m}$$

$$= 5078 \text{ N-m} = 5078 \times 10^3 \text{ N-mm}$$

...(d is taken in metres)

and section modulus for the cross-section of the arm,

$$Z = \frac{\pi}{32}\,b_1\,(a_1)^2 = \frac{\pi}{32} \times 0.5\,a_1\,(a_1)^2 = 0.05\,(a_1)^3$$

We know that the bending stress (σ_b),

$$14 = \frac{M}{Z} = \frac{5078 \times 10^3}{0.05\,(a_1)^3} = \frac{101\,560 \times 10^3}{(a_1)^3}$$

$\therefore$ $(a_1)^3 = 101\,560 \times 10^3 / 14 = 7254 \times 10^3$

or a_1 = 193.6 say 200 mm **Ans.**

and $b_1 = 0.5\,a_1 = 0.5 \times 200 = 100$ mm **Ans.**

5. *Dimensions of key*

The standard dimensions of rectangular sunk key for a shaft of 165 mm diameter are as follows:

Width of key, w = 45 mm **Ans.**

and thickness of key = 25 mm **Ans.**

The length of key (L) is obtained by considering the failure of key in shearing.

We know that the maximum torque transmitted by the shaft (T_{max}),

$$35.328 \times 10^6 = L \times w \times \tau \times \frac{d_1}{2} = L \times 45 \times 40 \times \frac{165}{2} = 148\,500\,L$$

$\therefore$ $L = 35.328 \times 10^6 / 148\,500 = 238$ mm **Ans.**

Let us now check the total stress in the rim which should not be greater than 14 MPa. We know that the total stress in the rim,

$$= \rho.v^2 \left(0.75 + \frac{4.935\,R}{n^2.\,t}\right)$$

$$= 7200\,(12.57)^2 \left[0.75 + \frac{4.935 \times 1.2}{6^2 \times 0.22}\right] \text{N/m}^2$$

$$= 1.14 \times 10^6\,(0.75 + 0.75) = 1.71 \times 10^6 \text{ N/m}^2 = 1.71 \text{ MPa}$$

Since it is less than 14 MPa, therefore the design is safe.

Example 22.12. *A punching press pierces 35 holes per minute in a plate using 10 kN-m of energy per hole during each revolution. Each piercing takes 40 per cent of the time needed to make one revolution. The punch receives power through a gear reduction unit which in turn is fed by a motor driven belt pulley 800 mm diameter and turning at 210 r.p.m. Find the power of the electric motor if overall efficiency of the transmission unit is 80 per cent. Design a cast iron flywheel to be used with the punching machine for a coefficient of steadiness of 5, if the space considerations limit the maximum diameter to 1.3 m.*

Allowable shear stress in the shaft material = 50 MPa

Allowable tensile stress for cast iron = 4 MPa

Density of cast iron = 7200 kg / m^3

Solution. Given : No. of holes = 35 per min ; Energy per hole = 10 kN-m = 10 000 N-m ; d = 800 mm = 0.8 m ; N = 210 r.p.m. ; η = 80% = 0.8 ; $1/C_S = 5$ or $C_S = 1/5 = 0.2$; $D_{max} = 1.3$ m ; τ = 50 MPa = 50 N/mm^2 ; σ_t = 4 MPa = 4 N/mm^2 ; ρ = 7200 kg / m^3

Power of the electric motor

We know that energy used for piercing holes per minute

= No. of holes pierced × Energy used per hole

= 35 × 10 000 = 350 000 N-m / min

∴ Power needed for the electric motor,

$$P = \frac{\text{Energy used per minute}}{60 \times \eta} = \frac{350\ 000}{60 \times 0.8} = 7292 \text{ W} = 7.292 \text{ kW} \ \textbf{Ans.}$$

Design of cast iron flywheel

First of all, let us find the maximum fluctuation of energy.

Since the overall efficiency of the transmission unit is 80%, therefore total energy to be supplied during each revolution,

$$E_T = \frac{10\ 000}{0.8} = 12\ 500 \text{ N-m}$$

We know that velocity of the belt,

$$v = \pi\, d.N = \pi \times 0.8 \times 210 = 528 \text{ m/min}$$

∴ Net tension or pull acting on the belt

$$= \frac{P \times 60}{v} = \frac{7292 \times 60}{528} = 828.6 \text{ N}$$

Since each piercing takes 40 per cent of the time needed to make one revolution, therefore time required to punch a hole

= 0.4 / 35 = 0.0114 min

and the distance moved by the belt during punching a hole

= Velocity of the belt × Time required to punch a hole

= 528 × 0.0114 = 6.03 m

∴ Energy supplied by the belt during punching a hole,

$$E_B = \text{Net tension} \times \text{Distance travelled by belt}$$
$$= 828.6 \times 6.03 = 4996 \text{ N-m}$$

Thus energy to be supplied by the flywheel for punching during each revolution or maximum fluctuation of energy,

$$\Delta E = E_T - E_B = 12\,500 - 4996 = 7504 \text{ N-m}$$

1. *Mass of the flywheel*

Let m = Mass of the flywheel rim.

Since space considerations limit the maximum diameter of the flywheel as 1.3 m ; therefore let us take the mean diameter of the flywheel,

$$D = 1.2 \text{ m or } R = 0.6 \text{ m}$$

We know that angular velocity

$$\omega = \frac{2\pi \times N}{60} = \frac{2\pi \times 210}{60} = 22 \text{ rad / s}$$

We also know that the maximum fluctuation of energy (ΔE),

$$7504 = m.R^2.\omega^2.C_S = m\,(0.6)^2\,(22)^2\,0.2 = 34.85\,m$$

$$\therefore \quad m = 7504 / 34.85 = 215.3 \text{ kg } \textbf{Ans.}$$

2. *Cross-sectional dimensions of the flywheel rim*

Let t = Thickness of the flywheel rim in metres, and

b = Width of the flywheel rim in metres = $2\,t$ ...(Assume)

∴ Cross-sectional area of the rim,

$$A = b \times t = 2\,t \times t = 2\,t^2$$

We know that mass of the flywheel rim (m),

$$215.3 = A \times \pi D \times \rho = 2\,t^2 \times \pi \times 1.2 \times 7200 = 54.3 \times 10^3\,t^2$$

$$\therefore \quad t^2 = 215.3 / 54.3 \times 10^3 = 0.003\,96$$

or $t = 0.063$ say $0.065 \text{ m} = 65$ mm **Ans.**

and $b = 2\,t = 2 \times 65 = 130$ mm **Ans**

3. *Diameter and length of hub*

Let d = Diameter of the hub,

d_1 = Diameter of the shaft, and

l = Length of the hub.

First of all, let us find the diameter of the shaft (d_1). We know that the mean torque transmitted by the shaft,

$$T_{mean} = \frac{P \times 60}{2\pi N} = \frac{7292 \times 60}{2\pi \times 210} = 331.5 \text{ N-m}$$

Assuming that the maximum torque transmitted by the shaft is twice the mean torque, therefore maximum torque transmitted by the shaft,

$$T_{max} = 2 \times T_{mean} = 2 \times 331.5 = 663 \text{ N-m} = 663 \times 10^3 \text{ N-mm}$$

We know that maximum torque transmitted by the shaft (T_{max}),

$$663 \times 10^3 = \frac{\pi}{16} \times \tau\,(d_1)^3 = \frac{\pi}{16} \times 50\,(d_1)^3 = 9.82\,(d_1)^3$$

$$\therefore \quad (d_1)^3 = 663 \times 10^3 / 9.82 = 67.5 \times 10^3$$

or $d_1 = 40.7$ say 45 mm **Ans.**

The diameter of the hub (d) is made equal to twice the diameter of the shaft (d_1) and length of hub (l) is equal to the width of the rim (b).

$$\therefore \quad d = 2\,d_1 = 2 \times 45 = 90 \text{ mm} = 0.09 \text{ m and } l = b = 130 \text{ mm} \textbf{ Ans.}$$

4. ***Cross-sectional dimensions of the elliptical cast iron arms***

Let a_1 = Major axis,

b_1 = Minor axis = $0.5\,a_1$ ...(Assume)

n = Number of arms = 6 ... (Assume)

We know that the maximum bending moment in the arm at the hub end, which is assumed as cantilever is given by

$$M = \frac{T}{R\,.\,n}(R - r) = \frac{T}{D.n}(D - d) = \frac{663}{1.2 \times 6}(1.2 - 0.09) \text{ N-m}$$

$$= 102.2 \text{ N-m} = 102\ 200 \text{ N-mm}$$

and section modulus for the cross-section of the arms,

$$Z = \frac{\pi}{32} \times b_1\,(a_1)^2 = \frac{\pi}{32} \times 0.5\,a_1\,(a_1)^2 = 0.05\,(a_1)^3$$

We know that bending stress (σ_t),

$$4 = \frac{M}{Z} = \frac{102\ 200}{0.05\,(a_1)^3} = \frac{2044 \times 10^3}{(a_1)^3}$$

$$\therefore \quad (a_1)^3 = 2044 \times 10^3 / 4 = 511 \times 10^3 \quad \text{or} \quad a_1 = 80 \text{ mm} \textbf{ Ans.}$$

and $\quad b_1 = 0.5\,a_1 = 0.5 \times 80 = 40$ mm **Ans.**

5. ***Dimensions of key***

The standard dimensions of rectangular sunk key for a shaft of diameter 45 mm are as follows:

Width of key, w = 16 mm **Ans.**

and thickness of key = 10 mm **Ans.**

The length of key (L) is obtained by considering the failure of key in shearing.

We know that maximum torque transmitted by the shaft (T_{max}),

$$663 \times 10^3 = L \times w \times \tau \times \frac{d_1}{2} = L \times 16 \times 50 \times \frac{45}{2} = 18 \times 10^3\,L$$

$$\therefore \quad L = 663 \times 10^3 / 18 \times 10^3 = 36.8 \text{ say } 38 \text{ mm} \textbf{ Ans.}$$

Let us now check the total stress in the rim which should not be greater than 4 MPa.

We know that the velocity of the rim,

$$v = \frac{\pi D \times N}{60} = \frac{\pi \times 1.2 \times 210}{60} = 13.2 \text{ m/s}$$

$\therefore$ Total stress in the rim,

$$\sigma = \rho.v^2\left(0.75 + \frac{4.935\,R}{n^2\,.\,t}\right) = 7200\,(13.2)^2\left[0.75 + \frac{4.935 \times 0.6}{6^2 \times 0.065}\right]$$

$$= 1.25 \times 10^6\,(0.75 + 1.26) = 2.5 \times 10^6 \text{ N/m}^2 = 2.5 \text{ MPa}$$

Since it is less than 4 MPa, therefore the design is safe.

22.11 Construction of Flywheels

The flywheels of smaller size (upto 600 mm diameter) are casted in one piece. The rim and hub are joined together by means of web as shown in Fig. 22.19 (*a*). The holes in the web may be made for handling purposes.

In case the flywheel is of larger size (upto 2.5 metre diameter), the arms are made instead of web, as shown in Fig. 22.19 (*b*). The number of arms depends upon the size of flywheel and its speed of rotation. But the flywheels above 2.5 metre diameter are usually casted in two piece. Such a flywheel is known as ***split flywheel.*** A ***split flywheel*** has the advantage of relieving the shrinkage stresses in the arms due to unequal rate of cooling of casting. A flywheel made in two halves should be spilt at the arms rather than between the arms, in order to obtain better strength of the joint. The two halves of the flywheel are connected by means of bolts through the hub, as shown in Fig. 22.20. The two halves are also joined at the rim by means of cotter joint (as shown in Fig. 22.20) or shrink links (as shown in Fig. 22.21). The width or depth of the shrink link is taken as 1.25 to 1.35 times the thickness of link. The slot in the rim into which the link is inserted is made slightly larger than the size of link.

Flywheel with web (no spokes)

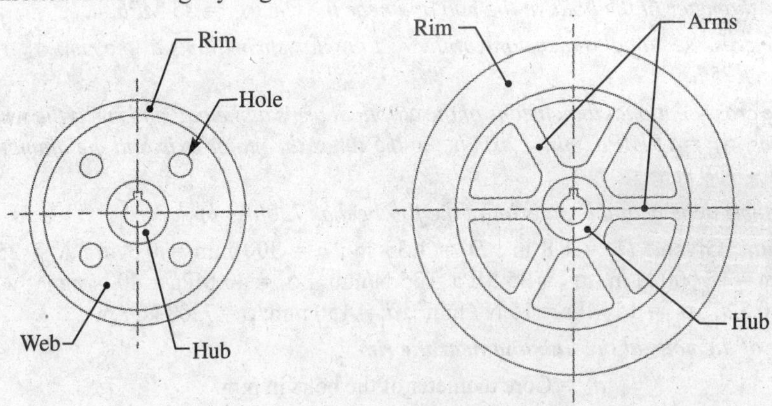

(*a*) Flywheel with web. (*b*) Flywheel with arms.

Fig. 22.19

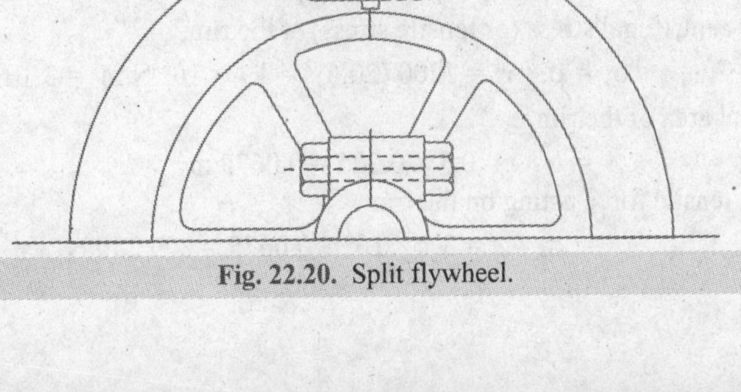

Fig. 22.20. Split flywheel.

Fig. 22.21. Shrink links.

The relative strength of a rim joint and the solid rim are given in the following table.

Table 22.3 Relative strength of a rim joint and the solid rim.

S.No.	*Type of construction*	*Relative strength*
1.	Solid rim.	1.00
2.	Flanged joint, bolted, rim parted between arms.	0.25
3.	Flanged joint, bolted, rim parted on an arm.	0.50
4.	Shrink link joint.	0.60
5.	Cotter or anchor joints.	0.70

Example 22.13. *A split type flywheel has outside diameter of the rim 1.80 m, inside diameter 1.35 m and the width 300 mm. the two halves of the wheel are connected by four bolts through the hub and near the rim joining the split arms and also by four shrink links on the rim. The speed is 250 r.p.m. and a turning moment of 15 kN-m is to be transmitted by the rim. Determine:*

1. *The diameter of the bolts at the hub and near the rim,* σ_{tb} = *35 MPa.*
2. *The cross-sectional dimensions of the rectangular shrink links at the rim,* σ_{tl} = *40 MPa ;* $w = 1.25\ h$.
3. *The cross-sectional dimensions of the elliptical arms at the hub and rim if the wheel has six arms,* σ_{ta} = *15 MPa, minor axis being 0.5 times the major axis and the diameter of shaft being 150 mm.*

Assume density of the material of the flywheel as 7200 kg / m³.

Solution. Given : D_0 = 1.8 m ; D_i = 1.35 m ; b = 300 mm = 0.3 m ; N = 250 r.p.m. ; T = 15 kN-m = 15 000 N-m ; σ_{tb} = 35 MPa = 35 N/mm² ; σ_{tl} = 40 MPa = 40 N/mm² ; $w = 1.25\ h$; $n = 6$; $b_1 = 0.5\ a_1$; σ_{ta} = 15 MPa = 15 N / mm² ; d_1 = 150 mm; ρ = 7200 kg / m³.

1. ***Diameter of the bolts at the hub and near the rim***

Let d_c = Core diameter of the bolts in mm.

We know that mean diameter of the rim,

$$D = \frac{D_o + D_i}{2} = \frac{1.8 + 1.35}{2} = 1.575 \text{ m}$$

and thickness of the rim,

$$t = \frac{D_o - D_i}{2} = \frac{1.8 - 1.35}{2} = 0.225 \text{ m}$$

Peripheral speed of the flywheel,

$$v = \frac{\pi D.N}{60} = \frac{\pi \times 1.575 \times 250}{60} = 20.6 \text{ m / s}$$

We know that centrifugal stress (or tensile stress) at the rim,

$$\sigma_t = \rho \times v^2 = 7200\ (20.6)^2 = 3.1 \times 10^6 \text{ N/m}^2 = 3.1 \text{ N/mm}^2$$

Cross-sectional area of the rim,

$$A = b \times t = 0.3 \times 0.225 = 0.0675 \text{ m}^2$$

∴ Maximum tensile force acting on the rim

$$= \sigma_t \times A = 3.1 \times 10^6 \times 0.0675 = 209\ 250 \text{ N} \quad ...(i)$$

We know that tensile strength of the four bolts

$$= \frac{\pi}{4}(d_c)^2 \sigma_{tb} \times \text{No. of bolts} = \frac{\pi}{4}(d_c)^2\, 35 \times 4 = 110\,(d_c)^2 \quad ...(ii)$$

Since the bolts are made as strong as the rim joint, therefore from equations (*i*) and (*ii*), we have

$$(d_c)^2 = 209\,250 / 110 = 1903 \text{ or } d_c = 43.6 \text{ mm}$$

The standard size of the bolt is M 56 with $d_c = 48.65$ mm **Ans.**

2. ***Cross-sectional dimensions of rectangular shrink links at the rim***

Let h = Depth of the link in mm, and

w = Width of the link in mm = 1.25 h ...(Given)

∴ Cross-sectional area of each link,

$$A_l = w \times h = 1.25\, h^2 \text{ mm}^2$$

We know that the maximum tensile force on half the rim

$$= 2 \times \sigma_t \text{ for rim} \times \text{Cross-sectional area of rim}$$

$$= 2 \times 3.1 \times 10^6 \times 0.0675 = 418\,500 \text{ N} \quad ...(iii)$$

and tensile strength of the four shrink links

$$= \sigma_{tl} \times A_l \times 4 = 40 \times 1.25\, h^2 \times 4 = 200\, h^2 \quad ...(iv)$$

From equations (*iii*) and (*iv*), we have

$$h^2 = 418\,500 / 200 = 2092.5 \text{ or } h = 45.7 \text{ say } 46 \text{ mm } \textbf{Ans.}$$

and $w = 1.25\, h = 1.25 \times 46 = 57.5$ say 58 mm **Ans.**

3. ***Cross-sectional dimensions of the elliptical arms***

Let a_1 = Major axis,

b_1 = Minor axis = 0.5 a_1 ...(Given)

n = Number of arms = 6 ...(Given)

Since the diameter of shaft (d_1) is 150 mm and the diameter of hub (d) is taken equal to twice the diameter of shaft, therefore

$$d = 2\, d_1 = 2 \times 150 = 300 \text{ mm} = 0.3 \text{ m}$$

We know that maximum bending moment on arms at the hub end,

$$M = \frac{T}{R.n}(R - r) = \frac{T}{D.n}(D - d) = \frac{15000}{1.575 \times 6}(1.575 - 0.3)$$

$$= 2024 \text{ N-m} = 2024 \times 10^3 \text{ N-mm}$$

Section modulus, $Z = \frac{\pi}{32} \times b_1 (a_1)^2 = \frac{\pi}{32} \times 0.5\, a_1 (a_1)^2 = 0.05\,(a_1)^3$

We know that bending stress for arms (σ_{ta}),

$$15 = \frac{M}{Z} = \frac{2024 \times 10^3}{0.05\,(a_1)^3} = \frac{40.5 \times 10^6}{(a_1)^3}$$

∴ $(a_1)^3 = 40.5 \times 10^6 / 15 = 2.7 \times 10^6$ or $a_1 = 139.3$ say 140 mm **Ans.**

and $b_1 = 0.5\, a_1 = 0.5 \times 140 = 70$ mm **Ans.**

EXERCISES

1. The turning moment diagram for a multicylinder engine has been drawn to a scale of 1 mm = 1000 N-m and 1 mm = 6°. The areas above and below the mean turning moment line taken in order are 530, 330, 380, 470, 180, 360, 350 and 280 sq.mm.

   For the engine, find the diameter of the flywheel. The mean r.p.m is 150 and the total fluctuation of speed must not exceed 3.5% of the mean.

Determine a suitable cross-sectional area of the rim of the flywheel, assuming the total energy of the flywheel to be $\frac{15}{14}$ that of the rim. The peripheral velocity of the flywheel is 15 m/s.

2. A machine has to carry out punching operation at the rate of 10 holes/min. It does 6 N-m of work per sq mm of the sheared area in cutting 25 mm diameter holes in 20 mm thick plates. A flywheel is fitted to the machine shaft which is driven by a constant torque. The fluctuation of speed is between 180 and 200 r.p.m. Actual punching takes 1.5 seconds. Frictional losses are equivalent to 1/6 of the workdone during punching. Find:

   (*a*) Power required to drive the punching machine, and

   (*b*) Mass of the flywheel, if radius of gyration of the wheel is 450 mm.

3. The turning moment diagram for an engine is drawn to the following scales:

   1 mm = 3100 N-m ; 1 mm = 1.6°

   The areas of the loops above and below the mean torque line taken in order are: 77, 219, 588, 522, 97, 116, 1200 and 1105 mm^2.

   The mean speed of the engine is 300 r.p.m. and the permissible fluctuation in speed is ± 2 per cent of mean speed. The stress in the material of the rim is not to exceed 4.9 MPa and density of its material is 7200 kg/m^3. Assuming that the rim stores $\frac{15}{16}$ of the energy that is stored by the flywheel, estimate

   (*a*) Diameter of rim; and (*b*) Area of cross-section of rim.

4. A single cylinder internal combustion engine working on the four stroke cycle develops 75 kW at 360 r.p.m. The fluctuation of energy can be assumed to be 0.9 times the energy developed per cycle. If the fluctuation of speed is not to exceed 1 per cent and the maximum centrifugal stress in the flywheel is to be 5.5 MPa, estimate the mean diameter and the cross-sectional area of the rim. The material of the rim has a density of 7200 kg / m^3. **[Ans. 1.464 m ; 0.09 m^2]**

5. Design a cast iron flywheel for a four stroke cycle engine to develop 110 kW at 150 r.p.m. The work done in the power stroke is 1.3 times the average work done during the whole cycle. Take the mean diameter of the flywheel as 3 metres. The total fluctuation of speed is limited to 5 per cent of the mean speed. The material density is 7250 kg / m^3. The permissible shear stress for the shaft material is 40 MPa and flexural stress for the arms of the flywheel is 20 MPa.

6. A punching press is required to punch 40 mm diameter holes in a plate of 15 mm thickness at the rate of 30 holes per minute. It requires 6 N-m of energy per mm^2 of sheared area. Determine the moment of inertia of the flywheel if the punching takes one-tenth of a second and the r.p.m. of the flywheel varies from 160 to 140.

7. A punch press is fitted with a flywheel capable of furnishing 3000 N-m of energy during quarter of a revolution near the bottom dead centre while blanking a hole on sheet metal. The maximum speed of the flywheel during the operation is 200 r.p.m. and the speed decreases by 10% during the cutting stroke. The mean radius of the rim is 900 mm. Calculate the approximate mass of the flywheel rim assuming that it contributes 90% of the energy requirements.

8. A punching machine makes 24 working strokes per minute and is capable of punching 30 mm diameter holes in 20 mm thick steel plates having an ultimate shear strength of 350 MPa. The punching operation takes place during $\frac{1}{10}$ th of a revolution of the crankshaft. Find the power required for the driving motor, assuming a mechanical efficiency of 76%. Determine suitable dimensions for the rim cross-section of the flywheel, which revolves at 9 times the speed of crankshaft. The permissible coefficient of fluctuation of speed is 0.4.

   The flywheel is to be made of cast iron having a safe tensile stress of 6 MPa and density 7250 kg/m^3. The diameter of the flywheel must not exceed 1.05 m owing to space restrictions. The hub and spokes may be assumed to provide 5% of the rotational inertia of the wheel. Check for the centrifugal stress induced in the rim.

9. Design completely the flywheel, shaft and the key for securing the flywheel to the shaft, for a punching machine having a capacity of producing 30 holes of 20 mm diameter per minute in steel plate 16 mm thickness. The ultimate shear stress for the material of the plate is 360 MPa. The actual punching operation estimated to last for a period of 36° rotation of the punching machine crankshaft. This crank shaft is powered by a flywheel shaft through a reduction gearing having a ratio 1 : 8. Assume that the mechanical efficiency of the punching machine is 80% and during the actual punching operation the flywheel speed is reduced by a maximum of 10%. The diameter of flywheel is restricted to 0.75 m due to space limitations.

10. A cast iron wheel of mean diameter 3 metre has six arms of elliptical section. The energy to be stored in it is 560 kN-m when rotating at 120 r.p.m. The speed of the mean diameter is 18 m/s. Calculate the following:

(*a*)Assuming that the whole energy is stored in the rim, find the cross-section, if the width is 300 mm.

(*b*)Find the cross-section of the arms near the boss on the assumption that their resistance to bending is equal to the torsional resistance of the shaft which is 130 mm in diameter.

The maximum shear stress in the shaft is to be within 63 MPa and the tensile stress 16 MPa. Assume the minor axis of the ellipse to be 0.65 major axis.

11. A cast iron flywheel is to be designed for a single cylinder double acting steam engine which delivers 150 kW at 80 r.p.m. The maximum fluctuation of energy per revolution is 10%. The total fluctuation of the speed is 4 per cent of the mean speed. If the mean diameter of the flywheel rim is 2.4 metres, determine the following :

(*a*) Cross-sectional dimensions of the rim, assuming that the hub and spokes provide 5% of the rotational inertia of the wheel. The density of cast iron is 7200 kg/m^3 and tensile stress 16 MPa. Take width of rim equal to twice of thickness.

(*b*) Dimensions of hub and rectangular sunk key. The shear stress for the material of shaft and key is 40 MPa.

(*c*) Cross-sectional dimensions of the elliptical arms assuming major axis as twice of minor axis and number of arms equal to six.

12. Design a cast iron flywheel having six arms for a four stroke engine developing 120 kW at 150 r.p.m. The mean diameter of the flywheel may be taken as 3 metres. The fluctuation of speed is 2.5% of mean speed. The workdone during the working stroke is 1.3 times the average workdone during the whole cycle. Assume allowable shear stress for the shaft and key as 40 MPa and tensile stress for cast iron as 20 MPa. The following proportions for the rim and elliptical arms may be taken:

(*a*) Width of rim = 2 × Thickness of rim

(*b*) Major axis = 2 × Minor axis.

13. A multi-cylinder engine is to run at a speed of 500 r.p.m. On drawing the crank effort diagram to scale 1 mm = 2500 N-m and 1 mm = 3°, the areas above and below the mean torque line are in sq mm as below:

+ 160, – 172, + 168, – 191, + 197, – 162

The speed is to be kept within ± 1% of the mean speed of the engine. Design a suitable rim type C.I. flywheel for the above engine. Assume rim width as twice the thickness and the overhang of the flywheel from the centre of the nearest bearing as 1.2 metres. The permissible stresses for the rim in tension is 6 MPa and those for shaft and key in shear are 42 MPa. The allowable stress for the arm is 14 MPa. Sketch a dimensioned end view of the flywheel.

14. An engine runs at a constant load at a speed of 480 r.p.m. The crank effort diagram is drawn to a scale 1 mm = 200 N-m torque and 1 mm = 3.6° crank angle. The areas of the diagram above and below the mean torque line in sq mm are in the following order:

+ 110, – 132, + 153, – 166, + 197, – 162

Design the flywheel if the total fluctuation of speed is not to exceed 10 r.p.m. and the centrifugal stress in the rim is not to exceed 5 MPa. You may assume that the rim breadth is approximately 2.5 times the rim thickness and 90% of the moment of inertia is due to the rim. The density of the material of the flywheel is 7250 kg/m^3.

Make a sketch of the flywheel giving the dimensions of the rim, the mean diameter of the rim and other estimated dimensions of spokes, hub etc.

15. A four stroke oil engine developing 75 kW at 300 r.p.m is to have the total fluctuation of speed limited to 5%. Two identical flywheels are to be designed. The workdone during the power stroke is found to be 1.3 times the average workdone during the whole cycle. The turning moment diagram can be approximated as a triangle during the power stroke. Assume that the hoop stress in the flywheel and the bending stress in the arms should not exceed 25 MPa. The shear stress in the key and shaft material should not exceed 40 MPa. Give a complete design of the flywheel. Assume four arms of elliptical cross-section with the ratio of axes 1 : 2. Design should necessarily include (*i*) moment of inertia of the flywheel, (*ii*) flywheel rim dimensions, (*iii*) arm dimensions, and (*iv*) flywheel boss and key dimensions and sketch showing two views of the flywheel with all the dimensions.

QUESTIONS

1. What is the main function of a flywheel in an engine?
2. In what way does a flywheel differ from that of a governor? Illustrate your answer with suitable examples.
3. Explain why flywheels are used in punching machines. Does the mounting of a flywheel reduce the stress induced in the shafts.
4. Define 'coefficient of fluctuation of speed' and 'coefficient of steadiness'.
5. What do you understand by 'fluctuation of energy' and 'maximum fluctuation of energy'.
6. Define 'coefficient of fluctuation of energy'.
7. Discuss the various types of stresses induced in a flywheel rim.
8. Explain the procedure for determining the size and mass of a flywheel with the help of a turning moment diagram.
9. Discuss the procedure for determining the cross-sectional dimensions of arms of a flywheel.
10. State the construction of flywheels.

OBJECTIVE TYPE QUESTIONS

1. The maximum fluctuation of speed is the
   (*a*) difference of minimum fluctuation of speed and the mean speed
   (*b*) difference of the maximum and minimum speeds
   (*c*) sum of the maximum and minimum speeds
   (*d*) variations of speed above and below the mean resisting torque line
2. The coefficient of fluctuation of speed is the .... of maximum fluctuation of speed and the mean speed.
   (*a*) product (*b*) ratio
   (*c*) sum (*d*) difference

3. In a turning moment diagram, the variations of energy above and below the mean resisting torque line is called
   (*a*) fluctuation of energy
   (*b*) maximum fluctuation of energy
   (*c*) coefficient of fluctuation of energy
   (*d*) none of these
4. If E = Mean kinetic energy of the flywheel, C_S = Coefficient of fluctuation of speed and ΔE = Maximum fluctuation of energy, then
   (*a*) $\Delta E = E / C_S$
   (*b*) $\Delta E = E^2 \times C_S$
   (*c*) $\Delta E = E \times C_S$
   (*d*) $\Delta E = 2\,E \times C_S$
5. The ratio of the maximum fluctuation of energy to the .... is called coefficient of fluctuation of energy.
   (*a*) minimum fluctuation of energy
   (*b*) workdone per cycle
6. Due to the centrifugal force acting on the rim, the flywheel arms will be subjected to
   (*a*) tensile stress
   (*b*) compressive stress
   (*c*) shear stress
   (*d*) none of these
7. The tensile stress in the flywheel rim due to the centrifugal force acting on the rim is given by
   (*a*) $\dfrac{\rho.v^2}{4}$
   (*b*) $\dfrac{\rho.v^2}{2}$
   (*c*) $\dfrac{3\rho.v^2}{4}$
   (*d*) $\rho.v^2$

   where ρ = Density of the flywheel material, and
   v = Linear velocity of the flywheel.
8. The cross-section of the flywheel arms is usually
   (*a*) elliptical
   (*b*) rectangular
   (*c*) *I*-section
   (*d*) L-section
9. In order to find the maximum bending moment on the arms, it is assumed as a
   (*a*) simply supported beam carrying a uniformly distributed load over the arm
   (*b*) fixed at both ends (*i.e.* at the hub and at the free end of the rim) and carrying a uniformly distributed load over the arm.
   (*c*) cantilever beam fixed at the hub and carrying a concentrated load at the free end of the rim
   (*d*) none of the above
10. The diameter of the hub of the flywheel is usually taken
   (*a*) equal to the diameter of the shaft
   (*b*) twice the diameter of the shaft
   (*c*) three times the diameter of the shaft
   (*d*) four times the diameter of the shaft

ANSWERS

1. (*b*)	2. (*b*)	3. (*a*)	4. (*d*)	5. (*b*)
6. (*a*)	7. (*d*)	8. (*a*)	9. (*c*)	10. (*b*)

CHAPTER

23

Springs

23.1 Introduction

A spring is defined as an elastic body, whose function is to distort when loaded and to recover its original shape when the load is removed. The various important applications of springs are as follows :

1. To cushion, absorb or control energy due to either shock or vibration as in car springs, railway buffers, air-craft landing gears, shock absorbers and vibration dampers.
2. To apply forces, as in brakes, clutches and spring-loaded valves.
3. To control motion by maintaining contact between two elements as in cams and followers.
4. To measure forces, as in spring balances and engine indicators.
5. To store energy, as in watches, toys, etc.

23.2 Types of Springs

Though there are many types of the springs, yet the following, according to their shape, are important from the subject point of view.

1. ***Helical springs.*** The helical springs are made up of a wire coiled in the form of a helix and is primarily intended for compressive or tensile loads. The cross-section of the wire from which the spring is made may be circular, square or rectangular. The two forms of helical springs are ***compression helical spring*** as shown in Fig. 23.1 (*a*) and ***tension helical spring*** as shown in Fig. 23.1 (*b*).

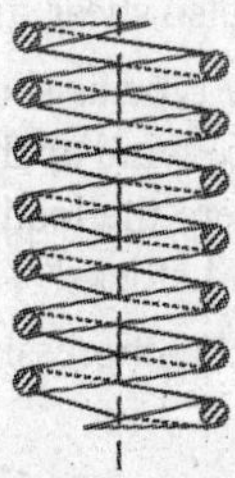

(*a*) Compression helical spring.

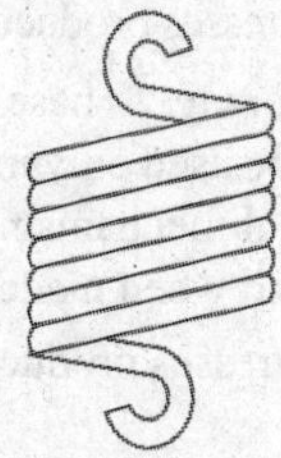

(*b*) Tension helical spring.

Fig. 23.1. Helical springs.

The helical springs are said to be ***closely coiled*** when the spring wire is coiled so close that the plane containing each turn is nearly at right angles to the axis of the helix and the wire is subjected to torsion. In other words, in a closely coiled helical spring, the helix angle is very small, it is usually less than 10°. The major stresses produced in helical springs are shear stresses due to twisting. The load applied is parallel to or along the axis of the spring.

In ***open coiled helical springs,*** the spring wire is coiled in such a way that there is a gap between the two consecutive turns, as a result of which the helix angle is large. Since the application of open coiled helical springs are limited, therefore our discussion shall confine to closely coiled helical springs only.

The helical springs have the following advantages:

(*a*) These are easy to manufacture.

(*b*) These are available in wide range.

(*c*) These are reliable.

(*d*) These have constant spring rate.

(*e*) Their performance can be predicted more accurately.

(*f*) Their characteristics can be varied by changing dimensions.

2. ***Conical and volute springs.*** The conical and volute springs, as shown in Fig. 23.2, are used in special applications where a telescoping spring or a spring with a spring rate that increases with the load is desired. The conical spring, as shown in Fig. 23.2 (*a*), is wound with a uniform pitch whereas the volute springs, as shown in Fig. 23.2 (*b*), are wound in the form of paraboloid with constant pitch

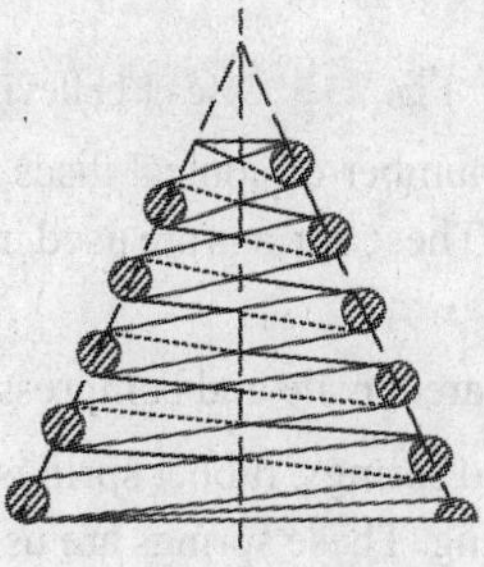

(*a*) Conical spring.

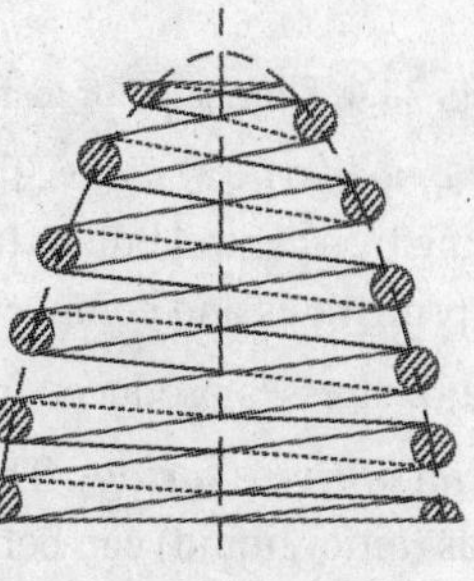

(*b*) Volute spring.

Fig. 23.2. Conical and volute springs.

and lead angles. The springs may be made either partially or completely telescoping. In either case, the number of active coils gradually decreases. The decreasing number of coils results in an increasing spring rate. This characteristic is sometimes utilised in vibration problems where springs are used to support a body that has a varying mass.

The major stresses produced in conical and volute springs are also shear stresses due to twisting.

3. ***Torsion springs.*** These springs may be of ***helical*** or ***spiral*** type as shown in Fig. 23.3. The **helical type** may be used only in applications where the load tends to wind up the spring and are used in various electrical mechanisms. The **spiral type** is also used where the load tends to increase the number of coils and when made of flat strip are used in watches and clocks.

The major stresses produced in torsion springs are tensile and compressive due to bending.

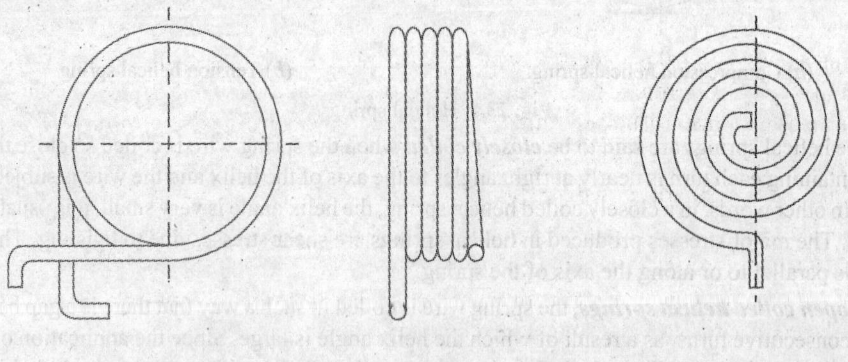

(*a*) Helical torsion spring. (*b*) Spiral torsion spring.

Fig. 23.3. Torsion springs.

4. ***Laminated or leaf springs.*** The laminated or leaf spring (also known as ***flat spring*** or ***carriage spring***) consists of a number of flat plates (known as leaves) of varying lengths held together by means of clamps and bolts, as shown in Fig. 23.4. These are mostly used in automobiles.

The major stresses produced in leaf springs are tensile and compressive stresses.

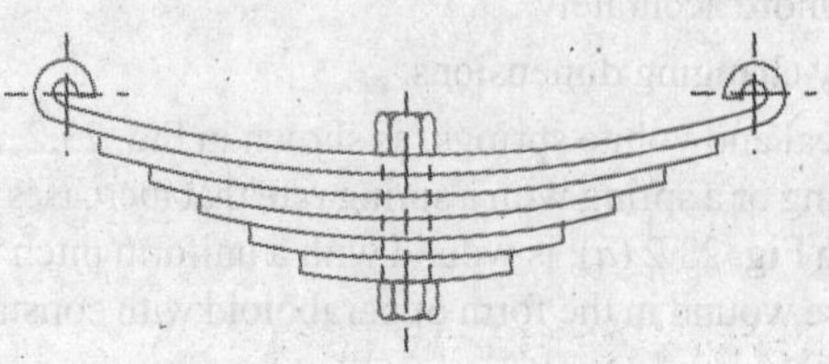

Fig. 23.4. Laminated or leaf springs.

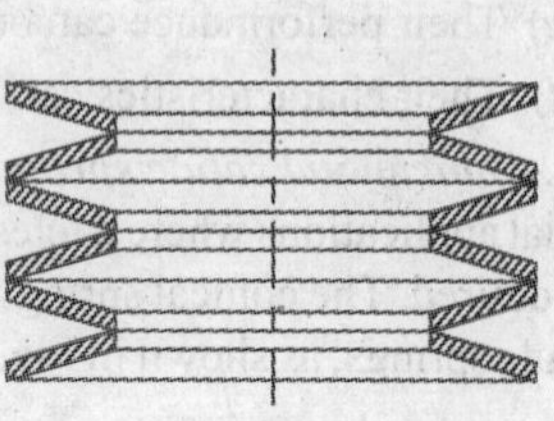

Fig. 23.5. Disc or bellevile springs.

5. ***Disc or bellevile springs.*** These springs consist of a number of conical discs held together against slipping by a central bolt or tube as shown in Fig. 23.5. These springs are used in applications where high spring rates and compact spring units are required.

The major stresses produced in disc or bellevile springs are tensile and compressive stresses.

6. ***Special purpose springs.*** These springs are air or liquid springs, rubber springs, ring springs etc. The fluids (air or liquid) can behave as a compression spring. These springs are used for special types of application only.

23.3 Material for Helical Springs

The material of the spring should have high fatigue strength, high ductility, high resilience and it should be creep resistant. It largely depends upon the service for which they are used *i.e.* severe service, average service or light service.

Severe service means rapid continuous loading where the ratio of minimum to maximum load (or stress) is one-half or less, as in automotive valve springs.

Average service includes the same stress range as in severe service but with only intermittent operation, as in engine governor springs and automobile suspension springs.

Light service includes springs subjected to loads that are static or very infrequently varied, as in safety valve springs.

The springs are mostly made from oil-tempered carbon steel wires containing 0.60 to 0.70 per cent carbon and 0.60 to 1.0 per cent manganese. Music wire is used for small springs. Non-ferrous materials like phosphor bronze, beryllium copper, monel metal, brass etc., may be used in special cases to increase fatigue resistance, temperature resistance and corrosion resistance.

Table 23.1 shows the values of allowable shear stress, modulus of rigidity and modulus of elasticity for various materials used for springs.

The helical springs are either cold formed or hot formed depending upon the size of the wire. Wires of small sizes (less than 10 mm diameter) are usually wound cold whereas larger size wires are wound hot. The strength of the wires varies with size, smaller size wires have greater strength and less ductility, due to the greater degree of cold working.

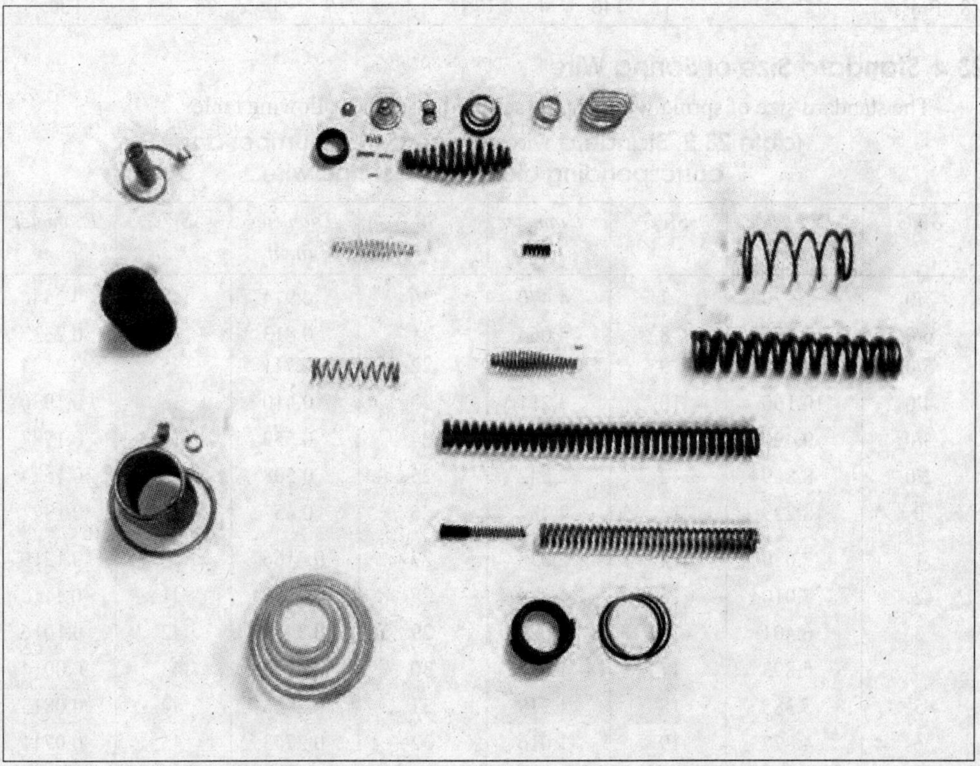

Table 23.1. Values of allowable shear stress, Modulus of elasticity and Modulus of rigidity for various spring materials.

Material	*Allowable shear stress* (τ) *MPa*			*Modulus of rigidity (G)* kN/m^2	*Modulus of elasticity (E)* kN/mm^2
	Severe service	*Average service*	*Light service*		
1. Carbon steel					
(*a*) Upto to 2.125 mm dia.	420	525	651	80	210
(*b*) 2.125 to 4.625 mm	385	483	595		
(*c*) 4.625 to 8.00 mm	336	420	525		
(*d*) 8.00 to 13.25 mm	294	364	455		
(*e*) 13.25 to 24.25 mm	252	315	392		
(*f*) 24.25 to 38.00 mm	224	280	350		
2. Music wire	392	490	612		
3. Oil tempered wire	336	420	525		
4. Hard-drawn spring wire	280	350	437.5		
5. Stainless-steel wire	280	350	437.5	70	196
6. Monel metal	196	245	306	44	105
7. Phosphor bronze	196	245	306	44	105
8. Brass	140	175	219	35	100

23.4 Standard Size of Spring Wire

The standard size of spring wire may be selected from the following table :

Table 23.2. Standard wire gauge (SWG) number and corresponding diameter of spring wire.

SWG	*Diameter (mm)*	*SWG*	*Diameter (mm)*	*SWG*	*Diameter (mm)*	*SWG*	*Diameter (mm)*
7/0	12.70	7	4.470	20	0.914	33	0.2540
6/0	11.785	8	4.064	21	0.813	34	0.2337
5/0	10.973	9	3.658	22	0.711	35	0.2134
4/0	10.160	10	3.251	23	0.610	36	0.1930
3/0	9.490	11	2.946	24	0.559	37	0.1727
2/0	8.839	12	2.642	25	0.508	38	0.1524
0	8.229	13	2.337	26	0.457	39	0.1321
1	7.620	14	2.032	27	0.4166	40	0.1219
2	7.010	15	1.829	28	0.3759	41	0.1118
3	6.401	16	1.626	29	0.3454	42	0.1016
4	5.893	17	1.422	30	0.3150	43	0.0914
5	5.385	18	1.219	31	0.2946	44	0.0813
6	4.877	19	1.016	32	0.2743	45	0.0711

23.5 Terms used in Compression Springs

The following terms used in connection with compression springs are important from the subject point of view.

1. ***Solid length.*** When the compression spring is compressed until the coils come in contact with each other, then the spring is said to be ***solid.*** The solid length of a spring is the product of total number of coils and the diameter of the wire. Mathematically,

Solid length of the spring,

$$L_S = n'.d$$

where n' = Total number of coils, and

d = Diameter of the wire.

2. ***Free length.*** The free length of a compression spring, as shown in Fig. 23.6, is the length of the spring in the free or unloaded condition. It is equal to the solid length plus the maximum deflection or compression of the spring and the clearance between the adjacent coils (when fully compressed). Mathematically,

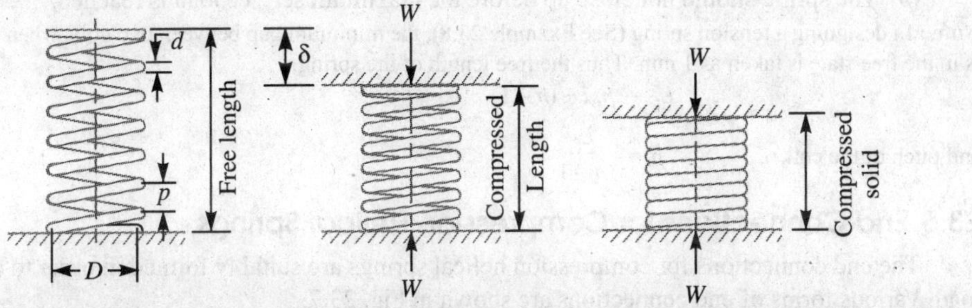

Fig. 23.6. Compression spring nomenclature.

Free length of the spring,

L_F = Solid length + Maximum compression + *Clearance between adjacent coils (or clash allowance)

$$= n'.d + \delta_{max} + 0.15\,\delta_{max}$$

The following relation may also be used to find the free length of the spring, *i.e.*

$$L_F = n'.d + \delta_{max} + (n' - 1) \times 1 \text{ mm}$$

In this expression, the clearance between the two adjacent coils is taken as 1 mm.

3. ***Spring index.*** The spring index is defined as the ratio of the mean diameter of the coil to the diameter of the wire. Mathematically,

Spring index, $C = D / d$

where D = Mean diameter of the coil, and

d = Diameter of the wire.

4. ***Spring rate.*** The spring rate (or stiffness or spring constant) is defined as the load required per unit deflection of the spring. Mathematically,

Spring rate, $k = W / \delta$

where W = Load, and

δ = Deflection of the spring.

---

* In actual practice, the compression springs are seldom designed to close up under the maximum working load and for this purpose a clearance (or clash allowance) is provided between the adjacent coils to prevent closing of the coils during service. It may be taken as 15 per cent of the maximum deflection.

5. ***Pitch.*** The pitch of the coil is defined as the axial distance between adjacent coils in uncompressed state. Mathematically,

Pitch of the coil, $p = \dfrac{\text{Free length}}{n' - 1}$

The pitch of the coil may also be obtained by using the following relation, *i.e.*

Pitch of the coil, $p = \dfrac{L_F - L_S}{n'} + d$

where L_F = Free length of the spring,

L_S = Solid length of the spring,

n' = Total number of coils, and

d = Diameter of the wire.

In choosing the pitch of the coils, the following points should be noted :

(*a*) The pitch of the coils should be such that if the spring is accidently or carelessly compressed, the stress does not increase the yield point stress in torsion.

(*b*) The spring should not close up before the maximum service load is reached.

Note : In designing a tension spring (See Example 23.8), the minimum gap between two coils when the spring is in the free state is taken as 1 mm. Thus the free length of the spring,

$$L_F = n.d + (n - 1)$$

and pitch of the coil, $p = \dfrac{L_F}{n - 1}$

23.6 End Connections for Compression Helical Springs

The end connections for compression helical springs are suitably formed in order to apply the load. Various forms of end connections are shown in Fig. 23.7.

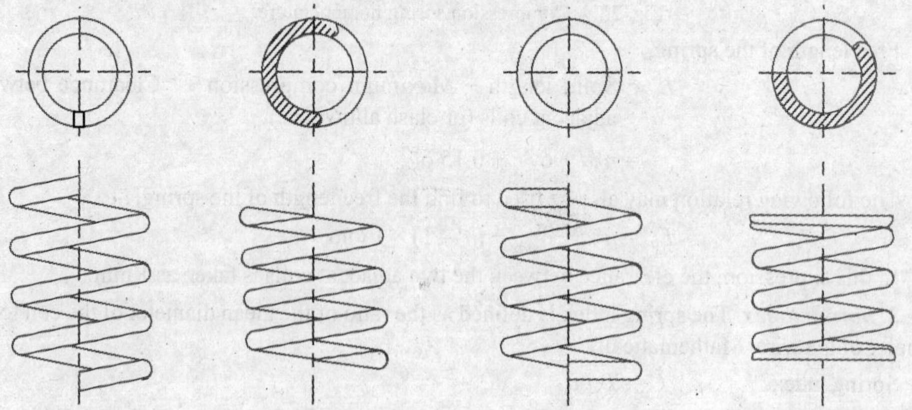

(*a*) Plain ends. (*b*) Ground ends. (*c*) Squared ends. (*d*) Squared and ground ends.

Fig 23.7. End connections for compression helical spring.

In all springs, the end coils produce an eccentric application of the load, increasing the stress on one side of the spring. Under certain conditions, especially where the number of coils is small, this effect must be taken into account. The nearest approach to an axial load is secured by squared and ground ends, where the end turns are squared and then ground perpendicular to the helix axis. It may be noted that part of the coil which is in contact with the seat does not contribute to spring action and hence are termed as ***inactive coils***. The turns which impart spring action are known as ***active turns.*** As the load increases, the number of inactive coils also increases due to seating of the end coils and the amount of increase varies from 0.5 to 1 turn at the usual working loads. The following table shows the total number of turns, solid length and free length for different types of end connections.

Table 23.3. Total number of turns, solid length and free length for different types of end connections.

Type of end	*Total number of turns (n')*	*Solid length*	*Free length*
1. Plain ends	n	$(n + 1)\,d$	$p \times n + d$
2. Ground ends	n	$n \times d$	$p \times n$
3. Squared ends	$n + 2$	$(n + 3)\,d$	$p \times n + 3d$
4. Squared and ground ends	$n + 2$	$(n + 2)\,d$	$p \times n + 2d$

where n = Number of active turns,
p = Pitch of the coils, and
d = Diameter of the spring wire.

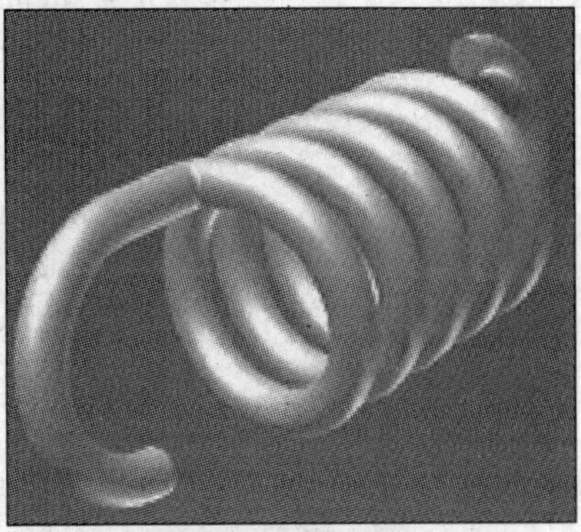

Tension helical spring

23.7 End Connections for Tension Helical Springs

The tensile springs are provided with hooks or loops as shown in Fig. 23.8. These loops may be made by turning whole coil or half of the coil. In a tension spring, large stress concentration is produced at the loop or other attaching device of tension spring.

The main disadvantage of tension spring is the failure of the spring when the wire breaks. A compression spring used for carrying a tensile load is shown in Fig. 23.9.

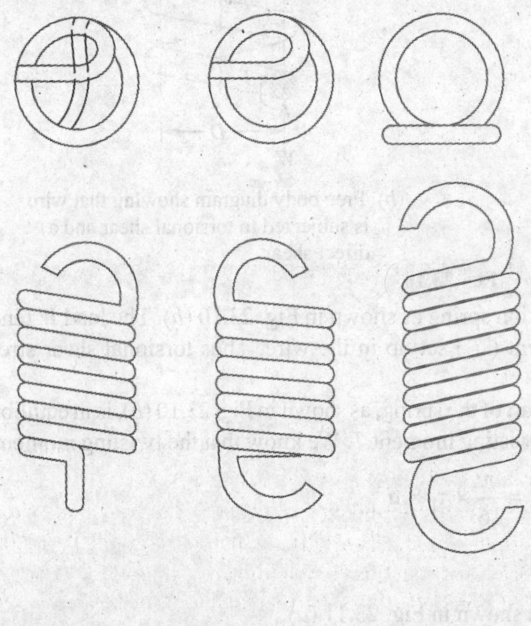

Fig. 23.8. End connection for tension helical springs.

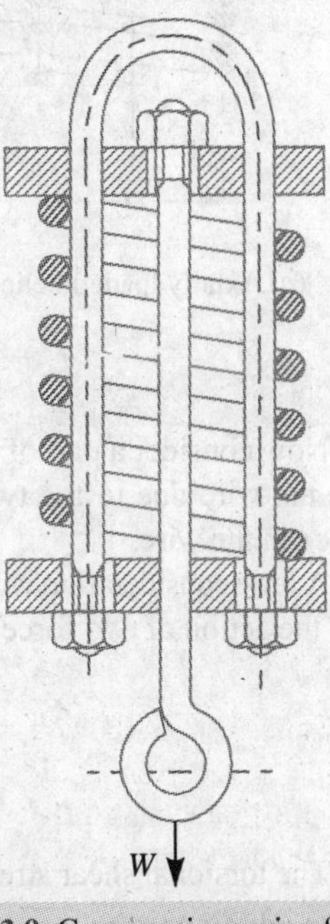

Fig. 23.9. Compression spring for carrying tensile load.

Note : The total number of turns of a tension helical spring must be equal to the number of turns (n) between the points where the loops start plus the equivalent turns for the loops. It has been found experimentally that half turn should be added for each loop. Thus for a spring having loops on both ends, the total number of active turns,

$$n' = n + 1$$

23.8 Stresses in Helical Springs of Circular Wire

Consider a helical compression spring made of circular wire and subjected to an axial load W, as shown in Fig. 23.10 (*a*).

Let D = Mean diameter of the spring coil,

d = Diameter of the spring wire,

n = Number of active coils,

G = Modulus of rigidity for the spring material,

W = Axial load on the spring,

τ = Maximum shear stress induced in the wire,

C = Spring index = D/d,

p = Pitch of the coils, and

δ = Deflection of the spring, as a result of an axial load W.

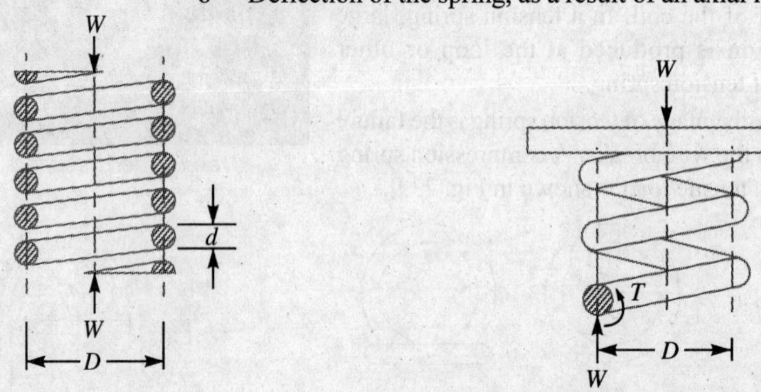

(*a*) Axially loaded helical spring.

(*b*) Free body diagram showing that wire is subjected to torsional shear and a direct shear.

Fig. 23.10

Now consider a part of the compression spring as shown in Fig. 23.10 (*b*). The load W tends to rotate the wire due to the twisting moment (T) set up in the wire. Thus torsional shear stress is induced in the wire.

A little consideration will show that part of the spring, as shown in Fig. 23.10 (*b*), is in equilibrium under the action of two forces W and the twisting moment T. We know that the twisting moment,

$$T = W \times \frac{D}{2} = \frac{\pi}{16} \times \tau_1 \times d^3$$

$$\therefore \quad \tau_1 = \frac{8 W.D}{\pi d^3} \qquad ...(i)$$

The torsional shear stress diagram is shown in Fig. 23.11 (*a*).

In addition to the torsional shear stress (τ_1) induced in the wire, the following stresses also act on the wire :

1. Direct shear stress due to the load W, and
2. Stress due to curvature of wire.

We know that direct shear stress due to the load W,

$$\tau_2 = \frac{\text{Load}}{\text{Cross-sectional area of the wire}}$$

$$= \frac{W}{\frac{\pi}{4} \times d^2} = \frac{4W}{\pi\, d^2} \qquad \text{...}(ii)$$

The direct shear stress diagram is shown in Fig. 23.11 (*b*) and the resultant diagram of torsional shear stress and direct shear stress is shown in Fig. 23.11 (*c*).

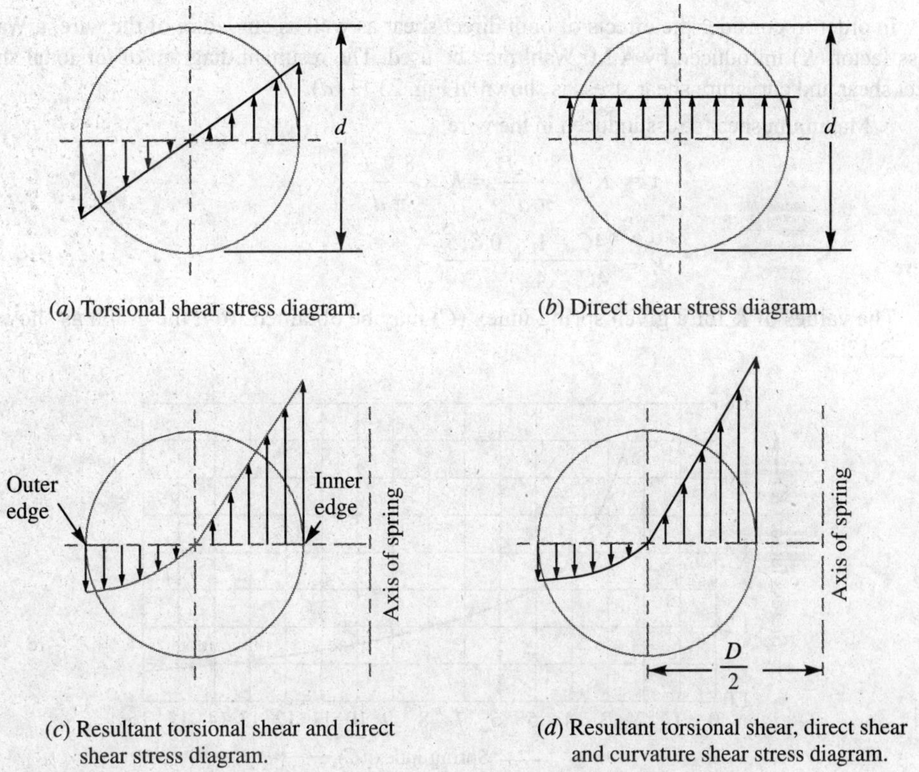

(*a*) Torsional shear stress diagram.

(*b*) Direct shear stress diagram.

(*c*) Resultant torsional shear and direct shear stress diagram.

(*d*) Resultant torsional shear, direct shear and curvature shear stress diagram.

Fig. 23.11. Superposition of stresses in a helical spring.

We know that the resultant shear stress induced in the wire,

$$\tau = \tau_1 \pm \tau_2 = \frac{8W.D}{\pi\, d^3} \pm \frac{4W}{\pi\, d^2}$$

The ***positive*** sign is used for the inner edge of the wire and ***negative*** sign is used for the outer edge of the wire. Since the stress is maximum at the inner edge of the wire, therefore

Maximum shear stress induced in the wire,

$$= \text{Torsional shear stress} + \text{Direct shear stress}$$

$$= \frac{8W.D}{\pi\, d^3} + \frac{4W}{\pi\, d^2} = \frac{8W.D}{\pi\, d^3}\left(1 + \frac{d}{2D}\right)$$

$$= \frac{8\,W.D}{\pi\,d^3}\left(1 + \frac{1}{2C}\right) = K_S \times \frac{8\,W.D}{\pi\,d^3} \qquad \text{...(iii)}$$

... (Substituting $D/d = C$)

where $\qquad K_S$ = Shear stress factor $= 1 + \dfrac{1}{2C}$

From the above equation, it can be observed that the effect of direct shear $\left(\dfrac{8\,WD}{\pi\,d^3} \times \dfrac{1}{2C}\right)$ is appreciable for springs of small spring index C. Also we have neglected the effect of wire curvature in equation (*iii*). It may be noted that when the springs are subjected to static loads, the effect of wire curvature may be neglected, because yielding of the material will relieve the stresses.

In order to consider the effects of both direct shear as well as curvature of the wire, a Wahl's stress factor (K) introduced by A.M. Wahl may be used. The resultant diagram of torsional shear, direct shear and curvature shear stress is shown in Fig. 23.11 (*d*).

∴ Maximum shear stress induced in the wire,

$$\tau = K \times \frac{8\,W.D}{\pi\,d^3} = K \times \frac{8\,W.C}{\pi\,d^2} \qquad \text{...(iv)}$$

where
$$K = \frac{4C - 1}{4C - 4} + \frac{0.615}{C}$$

The values of K for a given spring index (C) may be obtained from the graph as shown in Fig. 23.12.

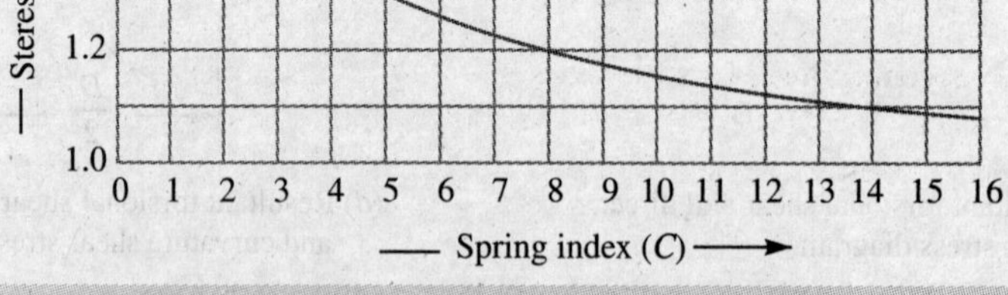

Fig. 23.12. Wahl's stress factor for helical springs.

We see from Fig. 23.12 that Wahl's stress factor increases very rapidly as the spring index decreases. The spring mostly used in machinery have spring index above 3.

Note: The Wahl's stress factor (K) may be considered as composed of two sub-factors, K_S and K_C, such that

$$K = K_S \times K_C$$

where $\qquad K_S$ = Stress factor due to shear, and

$\qquad K_C$ = Stress concentration factor due to curvature.

23.9 Deflection of Helical Springs of Circular Wire

In the previous article, we have discussed the maximum shear stress developed in the wire. We know that

Total active length of the wire,

$$l = \text{Length of one coil} \times \text{No. of active coils} = \pi\, D \times n$$

Let θ = Angular deflection of the wire when acted upon by the torque T.

∴ Axial deflection of the spring,

$$\delta = \theta \times D/2 \qquad ...(i)$$

We also know that

$$\frac{T}{J} = \frac{\tau}{D/2} = \frac{G.\theta}{l}$$

$$\therefore \quad \theta = \frac{T.l}{J.G} \qquad ...\left(\text{considering } \frac{T}{J} = \frac{G.\theta}{l}\right)$$

where J = Polar moment of inertia of the spring wire

$= \frac{\pi}{32} \times d^4$, d being the diameter of spring wire.

and G = Modulus of rigidity for the material of the spring wire.

Now substituting the values of l and J in the above equation, we have

$$\theta = \frac{T.l}{J.G} = \frac{\left(W \times \frac{D}{2}\right)\pi\, D.n}{\frac{\pi}{32} \times d^4\, G} = \frac{16W.D^2.n}{G.d^4} \qquad ...(ii)$$

Substituting this value of θ in equation (*i*), we have

$$\delta = \frac{16W.D^2.n}{G.d^4} \times \frac{D}{2} = \frac{8\,W.D^3.n}{G.d^4} = \frac{8\,W.C^3.n}{G.d} \qquad ... (\because C = D/d)$$

and the stiffness of the spring or spring rate,

$$\frac{W}{\delta} = \frac{G.d^4}{8\,D^3.n} = \frac{G.d}{8\,C^3.n} = \text{constant}$$

23.10 Eccentric Loading of Springs

Sometimes, the load on the springs does not coincide with the axis of the spring, *i.e.* the spring is subjected to an eccentric load. In such cases, not only the safe load for the spring reduces, the stiffness of the spring is also affected. The eccentric load on the spring increases the stress on one side of the spring and decreases on the other side. When the load is offset by a distance e from the spring axis, then the safe load on the spring may be obtained by multiplying the axial load by the factor $\frac{D}{2\,e + D}$, where D is the mean diameter of the spring.

23.11 Buckling of Compression Springs

It has been found experimentally that when the free length of the spring (L_F) is more than four times the mean or pitch diameter (D), then the spring behaves like a column and may fail by buckling at a comparatively low load as shown in Fig. 23.13. The critical axial load (W_{cr}) that causes buckling may be calculated by using the following relation, *i.e.*

$$W_{cr} = k \times K_B \times L_F$$

where k = Spring rate or stiffness of the spring = W/δ,

L_F = Free length of the spring, and

K_B = Buckling factor depending upon the ratio L_F / D.

The buckling factor (K_B) for the hinged end and built-in end springs may be taken from the following table.

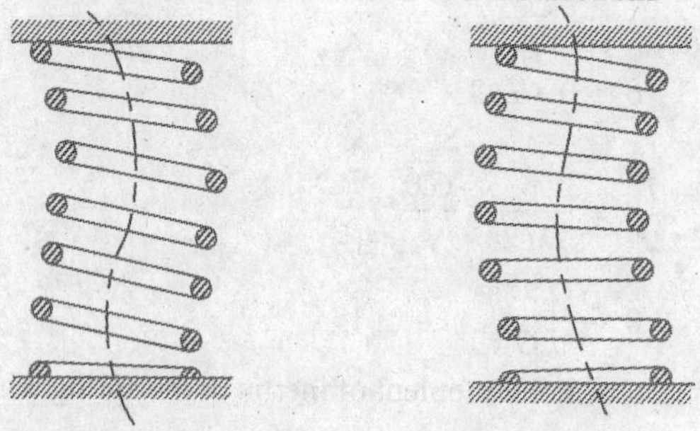

Fig. 23.13. Buckling of compression springs.

Table 23.4. Values of buckling factor (K_B).

L_F/D	*Hinged end spring*	*Built-in end spring*	L_F/D	*Hinged end spring*	*Built-in end spring*
1	0.72	0.72	5	0.11	0.53
2	0.63	0.71	6	0.07	0.38
3	0.38	0.68	7	0.05	0.26
4	0.20	0.63	8	0.04	0.19

It may be noted that a ***hinged end spring*** is one which is supported on pivots at both ends as in case of springs having plain ends where as a ***built-in end spring*** is one in which a squared and ground end spring is compressed between two rigid and parallel flat plates.

It order to avoid the buckling of spring, it is either mounted on a central rod or located on a tube. When the spring is located on a tube, the clearance between the tube walls and the spring should be kept as small as possible, but it must be sufficient to allow for increase in spring diameter during compression.

In railway coaches strongs springs are used for suspension.

23.12 Surge in Springs

When one end of a helical spring is resting on a rigid support and the other end is loaded suddenly, then all the coils of the spring will not suddenly deflect equally, because some time is required for the propagation of stress along the spring wire. A little consideration will show that in the beginning, the end coils of the spring in contact with the applied load takes up whole of the deflection and then it transmits a large part of its deflection to the adjacent coils. In this way, a wave of compression propagates through the coils to the supported end from where it is reflected back to the deflected end. This wave of compression travels along the spring indefinitely. If the applied load is of fluctuating type as in the case of valve spring in internal combustion engines and if the time interval between the load applications is equal to the time required for the wave to travel from one end to the other end, then resonance will occur. This results in very large deflections of the coils and correspondingly very high stresses. Under these conditions, it is just possible that the spring may fail. This phenomenon is called ***surge.***

It has been found that the natural frequency of spring should be atleast twenty times the frequency of application of a periodic load in order to avoid resonance with all harmonic frequencies upto twentieth order. The natural frequency for springs clamped between two plates is given by

$$f_n = \frac{d}{2\pi D^2 . n}\sqrt{\frac{6\,G.g}{\rho}} \text{ cycles/s}$$

where d = Diameter of the wire,
D = Mean diameter of the spring,
n = Number of active turns,
G = Modulus of rigidity,
g = Acceleration due to gravity, and
ρ = Density of the material of the spring.

The surge in springs may be eliminated by using the following methods :

1. By using friction dampers on the centre coils so that the wave propagation dies out.
2. By using springs of high natural frequency.
3. By using springs having pitch of the coils near the ends different than at the centre to have different natural frequencies.

Example 23.1. *A compression coil spring made of an alloy steel is having the following specifications :*

Mean diameter of coil = 50 mm ; Wire diameter = 5 mm ; Number of active coils = 20.

If this spring is subjected to an axial load of 500 N ; calculate the maximum shear stress (neglect the curvature effect) to which the spring material is subjected.

Solution. Given : D = 50 mm ; d = 5 mm ; *n = 20 ; W = 500 N

We know that the spring index,

$$C = \frac{D}{d} = \frac{50}{5} = 10$$

∴ Shear stress factor,

$$K_S = 1 + \frac{1}{2C} = 1 + \frac{1}{2 \times 10} = 1.05$$

and maximum shear stress (neglecting the effect of wire curvature),

$$\tau = K_S \times \frac{8W.D}{\pi\, d^3} = 1.05 \times \frac{8 \times 500 \times 50}{\pi \times 5^3} = 534.7 \text{ N/mm}^2$$
$$= 534.7 \text{ MPa } \textbf{Ans.}$$

* Superfluous data.

Example 23.2. *A helical spring is made from a wire of 6 mm diameter and has outside diameter of 75 mm. If the permissible shear stress is 350 MPa and modulus of rigidity 84 kN/mm², find the axial load which the spring can carry and the deflection per active turn.*

Solution. Given : $d = 6$ mm ; $D_o = 75$ mm ; $\tau = 350$ MPa $= 350$ N/mm² ; $G = 84$ kN/mm² $= 84 \times 10^3$ N/mm²

We know that mean diameter of the spring,

$$D = D_o - d = 75 - 6 = 69 \text{ mm}$$

$\therefore$ Spring index, $\quad C = \dfrac{D}{d} = \dfrac{69}{6} = 11.5$

Let $\quad W$ = Axial load, and

δ / n = Deflection per active turn.

1. ***Neglecting the effect of curvature***

We know that the shear stress factor,

$$K_S = 1 + \frac{1}{2C} = 1 + \frac{1}{2 \times 11.5} = 1.043$$

and maximum shear stress induced in the wire (τ),

$$350 = K_S \times \frac{8\,W.D}{\pi\, d^3} = 1.043 \times \frac{8\,W \times 69}{\pi \times 6^3} = 0.848\,W$$

$\therefore \quad W = 350 / 0.848 = 412.7$ N **Ans.**

We know that deflection of the spring,

$$\delta = \frac{8\,W.D^3.n}{G.d^4}$$

$\therefore$ Deflection per active turn,

$$\frac{\delta}{n} = \frac{8\,W.D^3}{G.d^4} = \frac{8 \times 412.7\,(69)^3}{84 \times 10^3 \times 6^4} = 9.96 \text{ mm } \textbf{Ans.}$$

2. ***Considering the effect of curvature***

We know that Wahl's stress factor,

$$K = \frac{4C - 1}{4C - 4} + \frac{0.615}{C} = \frac{4 \times 11.5 - 1}{4 \times 11.5 - 4} + \frac{0.615}{11.5} = 1.123$$

We also know that the maximum shear stress induced in the wire (τ),

$$350 = K \times \frac{8W.C}{\pi\, d^2} = 1.123 \times \frac{8 \times W \times 11.5}{\pi \times 6^2} = 0.913\,W$$

$\therefore \quad W = 350 / 0.913 = 383.4$ N **Ans.**

and deflection of the spring,

$$\delta = \frac{8\,W.D^3.n}{G.d^4}$$

$\therefore$ Deflection per active turn,

$$\frac{\delta}{n} = \frac{8\,W.D^3}{G.d^4} = \frac{8 \times 383.4\,(69)^3}{84 \times 10^3 \times 6^4} = 9.26 \text{ mm } \textbf{Ans.}$$

Example 23.3. *Design a spring for a balance to measure 0 to 1000 N over a scale of length 80 mm. The spring is to be enclosed in a casing of 25 mm diameter. The approximate number of turns is 30. The modulus of rigidity is 85 kN/mm². Also calculate the maximum shear stress induced.*

Solution. Given : $W = 1000$ N ; $\delta = 80$ mm ; $n = 30$; $G = 85$ kN/mm² $= 85 \times 10^3$ N/mm²

Design of spring

Let D = Mean diameter of the spring coil,

d = Diameter of the spring wire, and

C = Spring index = D/d.

Since the spring is to be enclosed in a casing of 25 mm diameter, therefore the outer diameter of the spring coil ($D_o = D + d$) should be less than 25 mm.

We know that deflection of the spring (δ),

$$80 = \frac{8\,W.C^3.n}{G.d} = \frac{8 \times 1000 \times C^3 \times 30}{85 \times 10^3 \times d} = \frac{240\,C^3}{85\,d}$$

$$\therefore \quad \frac{C^3}{d} = \frac{80 \times 85}{240} = 28.3$$

Let us assume that d = 4 mm. Therefore

$$C^3 = 28.3\,d = 28.3 \times 4 = 113.2 \quad \text{or} \quad C = 4.84$$

and $D = C.d = 4.84 \times 4 = 19.36$ mm **Ans.**

We know that outer diameter of the spring coil,

$$D_o = D + d = 19.36 + 4 = 23.36 \text{ mm } \textbf{Ans.}$$

Since the value of D_o = 23.36 mm is less than the casing diameter of 25 mm, therefore the assumed dimension, d = 4 mm is correct.

Maximum shear stress induced

We know that Wahl's stress factor,

$$K = \frac{4C - 1}{4C - 4} + \frac{0.615}{C} = \frac{4 \times 4.84 - 1}{4 \times 4.84 - 4} + \frac{0.615}{4.84} = 1.322$$

∴ Maximum shear stress induced,

$$\tau = K \times \frac{8\,W.C}{\pi\,d^2} = 1.322 \times \frac{8 \times 1000 \times 4.84}{\pi \times 4^2}$$

$$= 1018.2 \text{ N/mm}^2 = 1018.2 \text{ MPa } \textbf{Ans.}$$

Example 23.4. *A mechanism used in printing machinery consists of a tension spring assembled with a preload of 30 N. The wire diameter of spring is 2 mm with a spring index of 6. The spring has 18 active coils. The spring wire is hard drawn and oil tempered having following material properties:*

Design shear stress = 680 MPa

Modulus of rigidity = 80 kN/mm²

Determine : 1. the initial torsional shear stress in the wire; 2. spring rate; and 3. the force to cause the body of the spring to its yield strength.

Tension springs are widely used in printing machines.

Solution. Given : W_i = 30 N ; d = 2 mm ; $C = D/d$ = 6 ; n = 18 ; τ = 680 MPa = 680 N/mm² ; G = 80 kN/mm² = 80 × 10³ N/mm²

1. ***Initial torsional shear stress in the wire***

We know that Wahl's stress factor,

$$K = \frac{4C-1}{4C-4} + \frac{0.615}{C} = \frac{4\times6-1}{4\times6-4} + \frac{0.615}{6} = 1.2525$$

∴ Initial torsional shear stress in the wire,

$$\tau_i = K \times \frac{8\,W_i \times C}{\pi\,d^2} = 1.2525 \times \frac{8\times30\times6}{\pi\times2^2} = 143.5 \text{ N/mm}^2$$

$$= 143.5 \text{ MPa } \textbf{Ans.}$$

2. ***Spring rate***

We know that spring rate (or stiffness of the spring),

$$= \frac{G.d}{8\,C^3.n} = \frac{80\times10^3\times2}{8\times6^3\times18} = 5.144 \text{ N/mm } \textbf{Ans.}$$

3. ***Force to cause the body of the spring to its yield strength***

Let W = Force to cause the body of the spring to its yield strength.

We know that design or maximum shear stress (τ),

$$680 = K \times \frac{8\,W.C}{\pi\,d^2} = 1.2525 \times \frac{8W\times6}{\pi\times2^2} = 4.78\,W$$

∴ $W = 680 / 4.78 = 142.25$ N **Ans.**

Example 23.5. *Design a helical compression spring for a maximum load of 1000 N for a deflection of 25 mm using the value of spring index as 5.*

The maximum permissible shear stress for spring wire is 420 MPa and modulus of rigidity is 84 kN/mm².

Take Wahl's factor, $K = \frac{4C-1}{4C-4} + \frac{0.615}{C}$, *where C = Spring index.*

Solution. Given : W = 1000 N ; δ = 25 mm ; $C = D/d = 5$; τ = 420 MPa = 420 N/mm² ; G = 84 kN/mm² = 84 × 10³ N/mm²

1. ***Mean diameter of the spring coil***

Let D = Mean diameter of the spring coil, and

d = Diameter of the spring wire.

We know that Wahl's stress factor,

$$K = \frac{4C-1}{4C-4} + \frac{0.615}{C} = \frac{4\times5-1}{4\times5-4} + \frac{0.615}{5} = 1.31$$

and maximum shear stress (τ),

$$420 = K \times \frac{8\,W.C}{\pi\,d^2} = 1.31 \times \frac{8\times1000\times5}{\pi\,d^2} = \frac{16\,677}{d^2}$$

∴ $d^2 = 16\,677 / 420 = 39.7$ or $d = 6.3$ mm

From Table 23.2, we shall take a standard wire of size *SWG* 3 having diameter (d) = 6.401 mm.

∴ Mean diameter of the spring coil,

$D = C.d = 5\,d = 5 \times 6.401 = 32.005$ mm **Ans.** ... ($\because C = D/d = 5$)

and outer diameter of the spring coil,

$D_o = D + d = 32.005 + 6.401 = 38.406$ mm **Ans.**

2. ***Number of turns of the coils***

Let n = Number of active turns of the coils.

We know that compression of the spring (δ),

$$25 = \frac{8W.C^3.n}{G.d} = \frac{8 \times 1000\,(5)^3 n}{84 \times 10^3 \times 6.401} = 1.86\,n$$

$$\therefore \quad n = 25 / 1.86 = 13.44 \text{ say } 14 \textbf{ Ans.}$$

For squared and ground ends, the total number of turns,

$$n' = n + 2 = 14 + 2 = 16 \textbf{ Ans.}$$

3. ***Free length of the spring***

We know that free length of the spring

$$= n'.d + \delta + 0.15\,\delta = 16 \times 6.401 + 25 + 0.15 \times 25$$

$$= 131.2 \text{ mm } \textbf{Ans.}$$

4. ***Pitch of the coil***

We know that pitch of the coil

$$= \frac{\text{Free length}}{n' - 1} = \frac{131.2}{16 - 1} = 8.75 \text{ mm } \textbf{Ans.}$$

Example 23.6. *Design a close coiled helical compression spring for a service load ranging from 2250 N to 2750 N. The axial deflection of the spring for the load range is 6 mm. Assume a spring index of 5. The permissible shear stress intensity is 420 MPa and modulus of rigidity, G = 84 kN/mm².*

Neglect the effect of stress concentration. Draw a fully dimensioned sketch of the spring, showing details of the finish of the end coils.

Solution. Given : $W_1 = 2250$ N ; $W_2 = 2750$ N ; $\delta = 6$ mm ; $C = D/d = 5$; $\tau = 420$ MPa $= 420$ N/mm² ; $G = 84$ kN/mm² $= 84 \times 10^3$ N/mm²

1. ***Mean diameter of the spring coil***

Let D = Mean diameter of the spring coil for a maximum load of $W_2 = 2750$ N, and

d = Diameter of the spring wire.

We know that twisting moment on the spring,

$$T = W_2 \times \frac{D}{2} = 2750 \times \frac{5d}{2} = 6875\,d \qquad \ldots\left(\because C = \frac{D}{d} = 5\right)$$

We also know that twisting moment (T),

$$6875\,d = \frac{\pi}{16} \times \tau \times d^3 = \frac{\pi}{16} \times 420 \times d^3 = 82.48\,d^3$$

$$\therefore \quad d^2 = 6875 / 82.48 = 83.35 \quad \text{or} \quad d = 9.13 \text{ mm}$$

From Table 23.2, we shall take a standard wire of size *SWG* 3/0 having diameter (d) = 9.49 mm.

∴ Mean diameter of the spring coil,

$$D = 5d = 5 \times 9.49 = 47.45 \text{ mm } \textbf{Ans.}$$

We know that outer diameter of the spring coil,

$$D_o = D + d = 47.45 + 9.49 = 56.94 \text{ mm } \textbf{Ans.}$$

and inner diameter of the spring coil,

$$D_i = D - d = 47.45 - 9.49 = 37.96 \text{ mm } \textbf{Ans.}$$

2. ***Number of turns of the spring coil***

Let n = Number of active turns.

It is given that the axial deflection (δ) for the load range from 2250 N to 2750 N (*i.e.* for $W = 500$ N) is 6 mm.

We know that the deflection of the spring (δ),

$$6 = \frac{8\,W.C^3.n}{G.d} = \frac{8 \times 500\,(5)^3\,n}{84 \times 10^3 \times 9.49} = 0.63\,n$$

$$\therefore \qquad n = 6 / 0.63 = 9.5 \text{ say } 10 \textbf{ Ans.}$$

For squared and ground ends, the total number of turns,

$$n' = 10 + 2 = 12 \textbf{ Ans.}$$

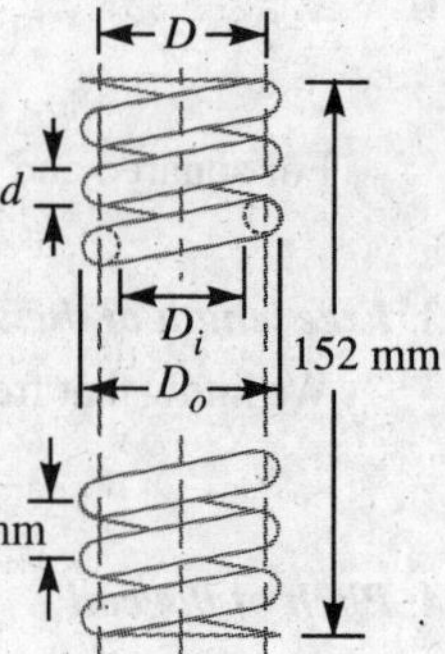

Fig. 23.14

3. ***Free length of the spring***

Since the compression produced under 500 N is 6 mm, therefore maximum compression produced under the maximum load of 2750 N is

$$\delta_{max} = \frac{6}{500} \times 2750 = 33 \text{ mm}$$

We know that free length of the spring,

$$L_F = n'.d + \delta_{max} + 0.15\,\delta_{max}$$
$$= 12 \times 9.49 + 33 + 0.15 \times 33$$
$$= 151.83 \text{ say } 152 \text{ mm } \textbf{Ans.}$$

4. ***Pitch of the coil***

We know that pitch of the coil

$$= \frac{\text{Free length}}{n' - 1} = \frac{152}{12 - 1} = 13.73 \text{ say } 13.8 \text{ mm } \textbf{Ans.}$$

The spring is shown in Fig. 23.14.

Example 23.7. *Design and draw a valve spring of a petrol engine for the following operating conditions :*

Spring load when the valve is open = 400 N
Spring load when the valve is closed = 250 N
Maximum inside diameter of spring = 25 mm
Length of the spring when the valve is open = 40 mm
Length of the spring when the valve is closed = 50 mm
Maximum permissible shear stress = 400 MPa

Petrol engine.

Solution. Given : $W_1 = 400$ N ; $W_2 = 250$ N ; $D_i = 25$ mm ; $l_1 = 40$ mm ; $l_2 = 50$ mm ; $\tau = 400$ MPa $= 400$ N/mm^2

1. ***Mean diameter of the spring coil***

Let d = Diameter of the spring wire in mm, and

D = Mean diameter of the spring coil
= Inside dia. of spring + Dia. of spring wire = $(25 + d)$ mm

Since the diameter of the spring wire is obtained for the maximum spring load (W_1), therefore maximum twisting moment on the spring,

$$T = W_1 \times \frac{D}{2} = 400\left(\frac{25 + d}{2}\right) = (5000 + 200\ d) \text{ N-mm}$$

We know that maximum twisting moment (T),

$$(5000 + 200\ d) = \frac{\pi}{16} \times \tau \times d^3 = \frac{\pi}{16} \times 400 \times d^3 = 78.55\ d^3$$

Solving this equation by hit and trial method, we find that $d = 4.2$ mm.

From Table 23.2, we find that standard size of wire is *SWG* 7 having $d = 4.47$ mm.

Now let us find the diameter of the spring wire by taking Wahl's stress factor (K) into consideration.

We know that spring index,

$$C = \frac{D}{d} = \frac{25 + 4.47}{4.47} = 6.6 \qquad \text{... } (\because D = 25 + d)$$

∴ Wahl's stress factor,

$$K = \frac{4C - 1}{4C - 4} + \frac{0.615}{C} = \frac{4 \times 6.6 - 1}{4 \times 6.6 - 4} + \frac{0.615}{6.6} = 1.227$$

We know that the maximum shear stress (τ),

$$400 = K \times \frac{8\ W_1 . C}{\pi\ d^2} = 1.227 \times \frac{8 \times 400 \times 6.6}{\pi\ d^2} = \frac{8248}{d^2}$$

$$\therefore \quad d^2 = 8248 / 400 = 20.62 \quad \text{or} \quad d = 4.54 \text{ mm}$$

Taking larger of the two values, we have

$$d = 4.54 \text{ mm}$$

From Table 23.2, we shall take a standard wire of size *SWG* 6 having diameter (d) = 4.877 mm.

∴ Mean diameter of the spring coil

$$D = 25 + d = 25 + 4.877 = 29.877 \text{ mm } \textbf{Ans.}$$

and outer diameter of the spring coil,

$$D_o = D + d = 29.877 + 4.877 = 34.754 \text{ mm } \textbf{Ans.}$$

2. ***Number of turns of the coil***

Let n = Number of active turns of the coil.

We are given that the compression of the spring caused by a load of $(W_1 - W_2)$, *i.e.* 400 – 250 = 150 N is $l_2 - l_1$, *i.e.* 50 – 40 = 10 mm. In other words, the deflection (δ) of the spring is 10 mm for a load (W) of 150 N

We know that the deflection of the spring (δ),

$$10 = \frac{8\ W . D^3 . n}{G . d^4} = \frac{8 \times 150\ (29.877)^3\ n}{80 \times 10^3\ (4.877)^4} = 0.707\ n$$

... (Taking $G = 80 \times 10^3$ N/mm²)

$$\therefore \quad n = 10 / 0.707 = 14.2 \text{ say } 15 \textbf{ Ans.}$$

Taking the ends of the springs as squared and ground, the total number of turns of the spring,

$$n' = 15 + 2 = 17 \textbf{ Ans.}$$

3. ***Free length of the spring***

Since the deflection for 150 N of load is 10 mm, therefore the maximum deflection for the maximum load of 400 N is

$$\delta_{max} = \frac{10}{150} \times 400 = 26.67 \text{ mm}$$

An automobile suspension and shock-absorber. The two links with green ends are turnbuckles.

∴ Free length of the spring,

$$L_F = n'.d + \delta_{max} + 0.15\,\delta_{max}$$
$$= 17 \times 4.877 + 26.67 + 0.15 \times 26.67 = 113.58 \text{ mm } \textbf{Ans.}$$

4. *Pitch of the coil*

We know that pitch of the coil

$$= \frac{\text{Free length}}{n' - 1} = \frac{113.58}{17 - 1} = 7.1 \text{ mm } \textbf{Ans.}$$

Example 23.8. *Design a helical spring for a spring loaded safety valve (Ramsbottom safety valve) for the following conditions :*

Diameter of valve seat = 65 mm ; Operating pressure = 0.7 N/mm²; Maximum pressure when the valve blows off freely = 0.75 N/mm²; Maximum lift of the valve when the pressure rises from 0.7 to 0.75 N/mm² = 3.5 mm ; Maximum allowable stress = 550 MPa ; Modulus of rigidity = 84 kN/mm²; Spring index = 6.

Draw a neat sketch of the free spring showing the main dimensions.

Solution. Given : D_1 = 65 mm ; p_1 = 0.7 N/mm² ; p_2 = 0.75 N/mm² ; δ = 3.5 mm ; τ = 550 MPa = 550 N/mm² ; G = 84 kN/mm² = 84 × 10³ N/mm² ; C = 6

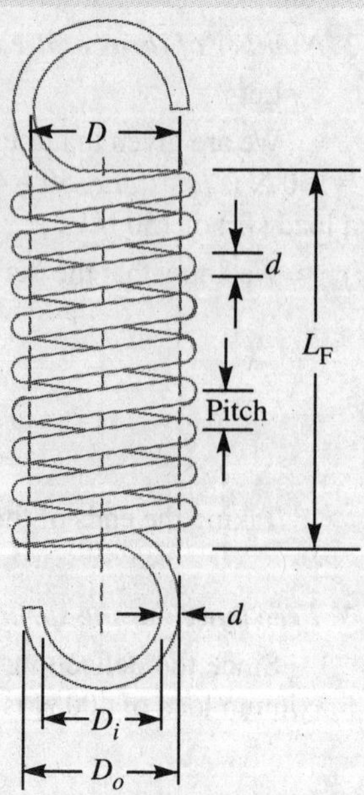

Fig. 23.15

1. *Mean diameter of the spring coil*

Let D = Mean diameter of the spring coil, and

d = Diameter of the spring wire.

Since the safety valve is a Ramsbottom safety valve, therefore the spring will be under tension. We know that initial tensile force acting on the spring (*i.e.* before the valve lifts),

$$W_1 = \frac{\pi}{4}(D_1)^2\, p_1 = \frac{\pi}{4}(65)^2\, 0.7 = 2323 \text{ N}$$

and maximum tensile force acting on the spring (*i.e.* when the valve blows off freely),

$$W_2 = \frac{\pi}{4}(D_1)^2 p_2 = \frac{\pi}{4}(65)^2\, 0.75 = 2489 \text{ N}$$

∴ Force which produces the deflection of 3.5 mm,

$$W = W_2 - W_1 = 2489 - 2323 = 166 \text{ N}$$

Since the diameter of the spring wire is obtained for the maximum spring load (W_2), therefore maximum twisting moment on the spring,

$$T = W_2 \times \frac{D}{2} = 2489 \times \frac{6\,d}{2} = 7467\, d \qquad \text{... } (\because C = D/d = 6)$$

We know that maximum twisting moment (T),

$$7467\, d = \frac{\pi}{16} \times \tau \times d^3 = \frac{\pi}{16} \times 550 \times d^3 = 108\, d^3$$

$$\therefore \quad d^2 = 7467 / 108 = 69.14 \quad \text{or} \quad d = 8.3 \text{ mm}$$

From Table 23.2, we shall take a standard wire of size *SWG* 2/0 having diameter (d) = 8.839 mm **Ans.**

∴ Mean diameter of the coil,

$$D = 6\,d = 6 \times 8.839 = 53.034 \text{ mm } \textbf{Ans.}$$

Outside diameter of the coil,

$$D_o = D + d = 53.034 + 8.839 = 61.873 \text{ mm } \textbf{Ans.}$$

and inside diameter of the coil,

$$D_i = D - d = 53.034 - 8.839 = 44.195 \text{ mm } \textbf{Ans.}$$

2. ***Number of turns of the coil***

Let n = Number of active turns of the coil.

We know that the deflection of the spring (δ),

$$3.5 = \frac{8\, W . C^3 . n}{G . d} = \frac{8 \times 166 \times 6^3 \times n}{84 \times 10^3 \times 8.839} = 0.386\, n$$

$$\therefore \quad n = 3.5 / 0.386 = 9.06 \text{ say } 10 \textbf{ Ans.}$$

For a spring having loop on both ends, the total number of turns,

$$n' = n + 1 = 10 + 1 = 11 \textbf{ Ans.}$$

3. ***Free length of the spring***

Taking the least gap between the adjacent coils as 1 mm when the spring is in free state, the free length of the tension spring,

$$L_F = n.d + (n - 1)\, 1 = 10 \times 8.839 + (10 - 1)\, 1 = 97.39 \text{ mm } \textbf{Ans.}$$

4. ***Pitch of the coil***

We know that pitch of the coil

$$= \frac{\text{Free length}}{n - 1} = \frac{97.39}{10 - 1} = 10.82 \text{ mm } \textbf{Ans.}$$

The tension spring is shown in Fig. 23.15.

Example 23.9. *A safety valve of 60 mm diameter is to blow off at a pressure of 1.2 N/mm². It is held on its seat by a close coiled helical spring. The maximum lift of the valve is 10 mm. Design a suitable compression spring of spring index 5 and providing an initial compression of 35 mm. The maximum shear stress in the material of the wire is limited to 500 MPa. The modulus of rigidity for the spring material is 80 kN/mm². Calculate : 1. Diameter of the spring wire, 2. Mean coil diameter, 3. Number of active turns, and 4. Pitch of the coil.*

Take Wahl's factor, $K = \frac{4C - 1}{4C - 4} + \frac{0.615}{C}$, *where C is the spring index.*

Solution. Given : Valve dia. = 60 mm ; Max. pressure = 1.2 N/mm^2 ; δ_2 = 10 mm ; C = 5 ; δ_1 = 35 mm ; τ = 500 MPa = 500 N/mm^2 ; G = 80 kN/mm^2 = 80 × 10^3 N/mm^2

1. ***Diameter of the spring wire***

Let d = Diameter of the spring wire.

We know that the maximum load acting on the valve when it just begins to blow off,

$$W_1 = \text{Area of the valve} \times \text{Max. pressure}$$

$$= \frac{\pi}{4}(60)^2\ 1.2 = 3394 \text{ N}$$

and maximum compression of the spring,

$$\delta_{max} = \delta_1 + \delta_2 = 35 + 10 = 45 \text{ mm}$$

Since a load of 3394 N keeps the valve on its seat by providing initial compression of 35 mm, therefore the maximum load on the spring when the valve is oepn (*i.e.* for maximum compression of 45 mm),

$$W = \frac{3394}{35} \times 45 = 4364 \text{ N}$$

We know that Wahl's stress factor,

$$K = \frac{4C - 1}{4C - 4} + \frac{0.615}{C} = \frac{4 \times 5 - 1}{4 \times 5 - 4} + \frac{0.615}{5} = 1.31$$

We also know that the maximum shear stress (τ),

$$500 = K \times \frac{8\,W.C}{\pi\,d^2} = 1.31 \times \frac{8 \times 4364 \times 5}{\pi\,d^2} = \frac{72\,780}{d^2}$$

$$\therefore \quad d^2 = 72\,780 / 500 = 145.6 \quad \text{or} \quad d = 12.06 \text{ mm}$$

From Table 23.2, we shall take a standard wire of size *SWG* 7/0 having diameter (d) = 12.7 mm. **Ans.**

2. ***Mean coil diameter***

Let D = Mean coil diameter.

We know that the spring index,

$$C = D/d \quad \text{or} \quad D = C.d = 5 \times 12.7 = 63.5 \text{ mm } \textbf{Ans.}$$

3. ***Number of active turns***

Let n = Number of active turns.

We know that the maximum compression of the spring (δ),

$$45 = \frac{8\,W.C^3.n}{G.d} = \frac{8 \times 4364 \times 5^3 \times n}{80 \times 10^3 \times 12.7} = 4.3\,n$$

$$\therefore \quad n = 45 / 4.3 = 10.5 \text{ say } 11 \textbf{ Ans.}$$

Taking the ends of the coil as squared and ground, the total number of turns,

$$n' = n + 2 = 11 + 2 = 13 \textbf{ Ans.}$$

Note : The valve of n may also be calculated by using

$$\delta_1 = \frac{8\,W_1.C^3.n}{G.d}$$

$$35 = \frac{8 \times 3394 \times 5^3 \times n}{80 \times 10^3 \times 12.7} = 3.34\,n \quad \text{or} \quad n = 35 / 3.34 = 10.5 \text{ say } 11$$

4. ***Pitch of the coil***

We know that free length of the spring,

$$L_F = n'.d + \delta_{max} + 0.15\,\delta_{max} = 13 \times 12.7 + 45 + 0.15 \times 45$$

$$= 216.85 \text{ mm } \textbf{Ans}$$

$$\therefore \quad \text{Pitch of the coil} = \frac{\text{Free length}}{n' - 1} = \frac{216.85}{13 - 1} = 18.1 \text{ mm } \textbf{Ans.}$$

Example 23.10. *In a spring loaded governor as shown in Fig. 23.16, the balls are attached to the vertical arms of the bell crank lever, the horizontal arms of which lift the sleeve against the pressure exerted by a spring. The mass of each ball is 2.97 kg and the lengths of the vertical and horizontal arms of the bell crank lever are 150 mm and 112.5 mm respectively. The extreme radii of rotation of the balls are 100 mm and 150 mm and the governor sleeve begins to lift at 240 r.p.m. and reaches the highest position with a 7.5 percent increase of speed when effects of friction are neglected. Design a suitable close coiled round section spring for the governor.*

Assume permissible stress in spring steel as 420 MPa, modulus of rigidity 84 kN/mm² and spring index 8. Allowance must be made for stress concentration, factor of which is given by

$$\frac{4C - 1}{4C - 4} + \frac{0.615}{C}, \text{ where } C \text{ is the spring index.}$$

Solution. Given : $m = 2.97$ kg ; $x = 150$ mm $= 0.15$ m ; $y = 112.5$ mm $= 0.1125$ m ; $r_2 = 100$ mm $= 0.1$ m ; $r_1 = 150$ mm $= 0.15$ m ; $N_2 = 240$ r.p.m. ; $\tau = 420$ MPa $= 420$ N/mm² ; $G = 84$ kN/mm² $= 84 \times 10^3$ N/mm² ; $C = 8$

The spring loaded governor, as shown in Fig. 23.16, is a *Hartnell type governor. First of all, let us find the compression of the spring.

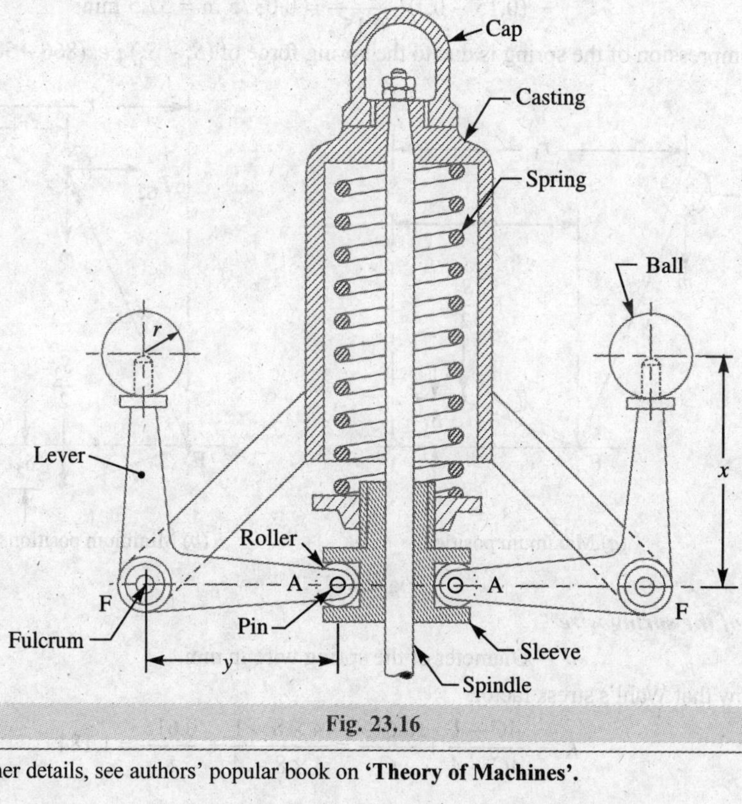

Fig. 23.16

* For further details, see authors' popular book on **'Theory of Machines'**.

We know that minimum angular speed at which the governor sleeve begins to lift,

$$\omega_2 = \frac{2\pi N_2}{60} = \frac{2\pi \times 240}{60} = 25.14 \text{ rad/s}$$

Since the increase in speed is 7.5%, therefore maximum speed,

$$\omega_1 = \omega_2 + \frac{7.5}{100} \times \omega_2 = 25.14 + \frac{7.5}{100} \times 25.14 = 27 \text{ rad/s}$$

The position of the balls and the lever arms at the maximum and minimum speeds is shown in Fig. 23.17 (*a*) and (*b*) respectively.

Let F_{C1} = Centrifugal force at the maximum speed, and

F_{C2} = Centrifugal force at the minimum speed.

We know that the spring force at the maximum speed (ω_1),

$$S_1 = 2\,F_{C1} \times \frac{x}{y} = 2\,m\,(\omega_1)^2\,r_1 \times \frac{x}{y} = 2 \times 2.97\,(27)^2\,0.15 \times \frac{0.15}{0.1125} = 866 \text{ N}$$

Similarly, the spring force at the minimum speed ω_2,

$$S_2 = 2\,F_{C2} \times \frac{x}{y} = 2m\,(\omega_2)^2\,r_2 \times \frac{x}{y} = 2 \times 2.97\,(25.14)^2\,0.1 \times \frac{0.15}{0.1125} = 500 \text{ N}$$

Since the compression of the spring will be equal to the lift of the sleeve, therefore compression of the spring,

$$\delta = \delta_1 + \delta_2 = (r_1 - r)\frac{y}{x} + (r - r_2)\frac{y}{x} = (r_1 - r_2)\frac{y}{x}$$

$$= (0.15 - 0.1)\frac{0.1125}{0.15} = 0.0375 \text{ m} = 37.5 \text{ mm}$$

This compression of the spring is due to the spring force of $(S_1 - S_2)$ *i.e.* (866 – 500) = 366 N.

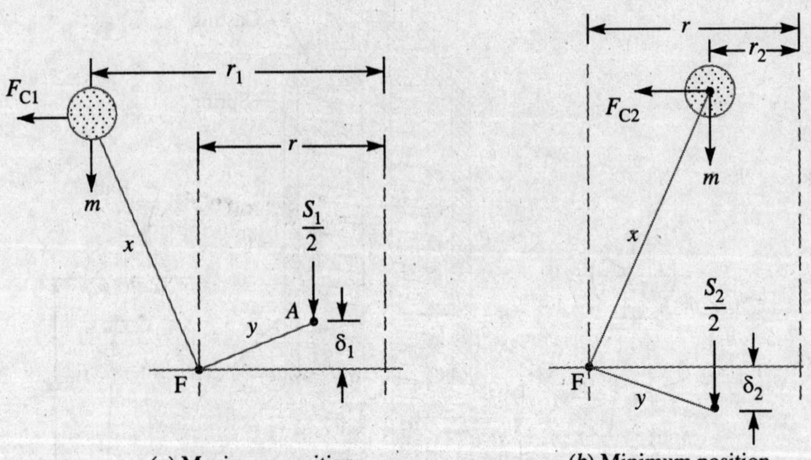

(*a*) Maximum position. (*b*) Minimum position.

Fig. 23.17

1. ***Diameter of the spring wire***

Let d = Diameter of the spring wire in mm.

We know that Wahl's stress factor,

$$K = \frac{4C - 1}{4C - 4} + \frac{0.615}{C} = \frac{4 \times 8 - 1}{4 \times 8 - 4} + \frac{0.615}{8} = 1.184$$

We also know that maximum shear stress (τ),

$$420 = K \times \frac{8\,W.C}{\pi\,d^2} = 1.184 \times \frac{8 \times 866 \times 8}{\pi\,d^2} = \frac{20\,885}{d^2}$$

... (Substituting $W = S_1$, the maximum spring force)

$$\therefore \quad d^2 = 20\,885 / 420 = 49.7 \text{ or } d = 7.05 \text{ mm}$$

From Table 23.2, we shall take the standard wire of size *SWG* 1 having diameter (d) = 7.62 mm **Ans.**

2. ***Mean diameter of the spring coil***

Let D = Mean diameter of the spring coil.

We know that the spring index,

$$C = D/d \text{ or } D = C.d = 8 \times 7.62 = 60.96 \text{ mm } \textbf{Ans.}$$

3. ***Number of turns of the coil***

Let n = Number of active turns of the coil.

We know that compression of the spring (δ),

$$37.5 = \frac{8\,W.C^3.n}{G.d} = \frac{8 \times 366 \times 8^3 \times n}{84 \times 10^3 \times 7.62} = 2.34\,n$$

... (Substituting $W = S_1 - S_2$)

$$\therefore \quad n = 37.5 / 2.34 = 16 \textbf{ Ans.}$$

and total number of turns using squared and ground ends,

$$n' = n + 2 = 16 + 2 = 18$$

4. ***Free length of the coil***

Since the compression produced under a force of 366 N is 37.5 mm, therefore maximum compression produced under the maximum load of 866 N is,

$$\delta_{max} = \frac{37.5}{366} \times 866 = 88.73 \text{ mm}$$

We know that free length of the coil,

$$L_F = n'.d + \delta_{max} + 0.15\,\delta_{max}$$
$$= 18 \times 7.62 + 88.73 + 0.15 \times 88.73 = 239.2 \text{ mm } \textbf{Ans.}$$

5. ***Pitch of the coil***

We know that pitch of the coil

$$= \frac{\text{Free length}}{n' - 1} = \frac{239.2}{18 - 1} = 14.07 \text{ mm } \textbf{Ans.}$$

Example 23.11. *A single plate clutch is to be designed for a vehicle. Both sides of the plate are to be effective. The clutch transmits 30 kW at a speed of 3000 r.p.m. and should cater for an over load of 20%. The intensity of pressure on the friction surface should not exceed 0.085 N/mm² and the surface speed at the mean radius should be limited to 2300 m / min. The outside diameter of the surfaces may be assumed as 1.3 times the inside diameter and the coefficient of friction for the surfaces may be taken as 0.3. If the axial thrust is to be provided by six springs of about 25 mm mean coil diameter, design the springs selecting wire from the following gauges :*

SWG	*4*	*5*	*6*	*7*	*8*	*9*	*10*	*11*	*12*
Dia. (mm)	*5.893*	*5.385*	*4.877*	*4.470*	*4.064*	*3.658*	*3.251*	*2.946*	*2.642*

Safe shear stress is limited to 420 MPa and modulus of rigidity is 84 kN/mm².

Solution. Given : P = 30 kW = 30 × 10³ W ; N = 3000 r.p.m. ; p = 0.085 N/mm² ; v = 2300 m/min ; $d_1 = 1.3\,d_2$ or $r_1 = 1.3\,r_2$; $\mu = 0.3$; No. of springs = 6 ; D = 25 mm ; τ = 420 MPa = 420 N/mm² ; G = 84 kN/mm² = 84 × 10³ N/mm²

First of all, let us find the maximum load on each spring. We know that the mean torque transmitted by the clutch,

$$T_{mean} = \frac{P \times 60}{2 \pi N} = \frac{30 \times 10^3 \times 60}{2 \pi \times 3000} = 95.5 \text{ N-m}$$

Since an overload of 20% is allowed, therefore maximum torque to which the clutch should be designed is given by

$$T_{max} = 1.2\, T_{mean} = 1.2 \times 95.5 = 114.6 \text{ N-m} = 114\,600 \text{ N-mm} \quad ...(i)$$

Let r_1 and r_2 be the outside and inside radii of the friction surfaces. Since maximum intensity of pressure is at the inner radius, therefore for uniform wear,

$$^*p \times r_2 = C \text{ (a constant) or } C = 0.085\, r_2$$

We know that the axial thrust transmitted,

$$W = C \times 2\pi\,(r_1 - r_2) \quad ...(ii)$$

Since both sides of the plate are effective, therefore maximum torque transmitted,

$$T_{max} = \tfrac{1}{2}\,\mu \times W\,(r_1 + r_2)\,2 = 2\pi\,\mu.C\,[(r_1)^2 - (r_2)^2] \quad \text{... [From equation (ii)]}$$

$$114\,600 = 2\pi \times 0.3 \times 0.085\, r_2\,[(1.3\, r_2)^2 - (r_2)^2] = 0.11\,(r_2)^3$$

$$\therefore \quad (r_2)^3 = 114\,600 / 0.11 = 1.04 \times 10^6 \text{ or } r_2 = 101.4 \text{ say } 102 \text{ mm}$$

and $$r_1 = 1.3\, r_2 = 1.3 \times 102 = 132.6 \text{ say } 133 \text{ mm}$$

∴ Mean radius,

$$r = \frac{r_1 + r_2}{2} = \frac{133 + 102}{2} = 117.5 \text{ mm} = 0.1175 \text{ m}$$

We know that surface speed at the mean radius,

$$v = 2\,\pi\, r\, N = 2\,\pi \times 0.1175 \times 3000 = 2215 \text{ m/min}$$

Since the surface speed as obtained above is less than the permissible value of 2300 m/min, therefore the radii of the friction surface are safe.

We know that axial thrust,

$$W = C \times 2\pi\,(r_1 - r_2) = 0.085\, r_2 \times 2\pi\,(r_1 - r_2) \quad \text{... } (\because C = 0.085\, r_2)$$

$$= 0.085 \times 102 \times 2\pi\,(133 - 102) = 1689 \text{ N}$$

Since this axial thrust is to be provided by six springs, therefore maximum load on each spring,

$$W_1 = \frac{1689}{6} = 281.5 \text{ N}$$

1. ***Diameter of the spring wire***

Let d = Diameter of the spring wire.

We know that the maximum torque transmitted,

$$T = W_1 \times \frac{D}{2} = 281.5 \times \frac{25}{2} = 3518.75 \text{ N-mm}$$

We also know that the maximum torque transmitted (T),

$$3518.75 = \frac{\pi}{16} \times \tau \times d^3 = \frac{\pi}{16} \times 420 \times d^3 = 82.48\, d^3$$

$$\therefore \quad d^3 = 3518.75 / 82.48 = 42.66 \text{ or } d = 3.494 \text{ mm}$$

Let us now find out the diameter of the spring wire by taking the stress factor (K) into consideration.

We know that the spring index,

$$C = \frac{D}{d} = \frac{25}{3.494} = 7.155$$

* Please refer Chapter 24 on Clutches.

and Wahl's stress factor,

$$K = \frac{4C - 1}{4C - 4} + \frac{0.615}{C} = \frac{4 \times 7.155 - 1}{4 \times 7.155 - 4} + \frac{0.615}{7.155} = 1.21$$

We know that the maximum shear stress (τ),

$$420 = K \times \frac{8\,W_1.D}{\pi\,d^3} = 1.21 \times \frac{8 \times 281.5 \times 25}{\pi\,d^3} = \frac{21\,681}{d^3}$$

$$\therefore \quad d^3 = 21\,681 / 420 = 51.6 \quad \text{or} \quad d = 3.72 \text{ mm}$$

From Table 23.2, we shall take a standard wire of size *SWG* 8 having diameter $(d) = 4.064$ mm. **Ans.**

Outer diameter of the spring,

$$D_o = D + d = 25 + 4.064 = 29.064 \text{ mm } \textbf{Ans.}$$

and inner diameter of the spring,

$$D_i = D - d = 25 - 4.064 = 20.936 \text{ mm } \textbf{Ans.}$$

2. ***Free length of the spring***

Let us assume the active number of coils $(n) = 8$. Therefore compression produced by an axial thrust of 281.5 N per spring,

$$\delta = \frac{8\,W_1.D^3.n}{G.d^4} = \frac{8 \times 281.5\,(25)^3\,8}{84 \times 10^3\,(4.064)^4} = 12.285 \text{ mm}$$

For square and ground ends, the total number of turns of the coil,

$$n' = n + 2 = 8 + 2 = 10$$

We know that free length of the spring,

$$L_F = n'.d + \delta + 0.15\,\delta = 10 \times 4.064 + 12.285 + 0.15 \times 12.285 \text{ mm}$$

$$= 54.77 \text{ mm } \textbf{Ans.}$$

3. ***Pitch of the coil***

We know that pitch of the coil

$$= \frac{\text{Free length}}{n' - 1} = \frac{54.77}{10 - 1} = 6.08 \text{ mm } \textbf{Ans.}$$

23.13 Energy Stored in Helical Springs of Circular Wire

We know that the springs are used for storing energy which is equal to the work done on it by some external load.

Let W = Load applied on the spring, and

δ = Deflection produced in the spring due to the load W.

Assuming that the load is applied gradually, the energy stored in a spring is,

$$U = \frac{1}{2}\,W.\delta \qquad \text{...}(i)$$

We have already discussed that the maximum shear stress induced in the spring wire,

$$\tau = K \times \frac{8\,W.D}{\pi\,d^3} \quad \text{or} \quad W = \frac{\pi\,d^3.\tau}{8K.D}$$

We know that deflection of the spring,

$$\delta = \frac{8\,W.D^3.n}{G.d^4} = \frac{8 \times \pi\,d^3.\tau}{8K.D} \times \frac{D^3.n}{G.d^4} = \frac{\pi\,\tau.D^2.n}{K.d.G}$$

Substituting the values of W and δ in equation (*i*), we have

$$U = \frac{1}{2} \times \frac{\pi\, d^3 . \tau}{8\, K . D} \times \frac{\pi\, \tau . D^2 . n}{K . d . G}$$

$$= \frac{\tau^2}{4\, K^2 . G} (\pi\, D . n) \left(\frac{\pi}{4} \times d^2 \right) = \frac{\tau^2}{4\, K^2 . G} \times V$$

where V = Volume of the spring wire

= Length of spring wire × Cross-sectional area of spring wire

$$= (\pi\, D . n) \left(\frac{\pi}{4} \times d^2 \right)$$

Note : When a load (say P) falls on a spring through a height h, then the energy absorbed in a spring is given by

$$U = P(h + \delta) = \frac{1}{2}\, W . \delta$$

where W = Equivalent static load *i.e.* the gradually applied load which shall produce the same effect as by the falling load P, and

δ = Deflection produced in the spring.

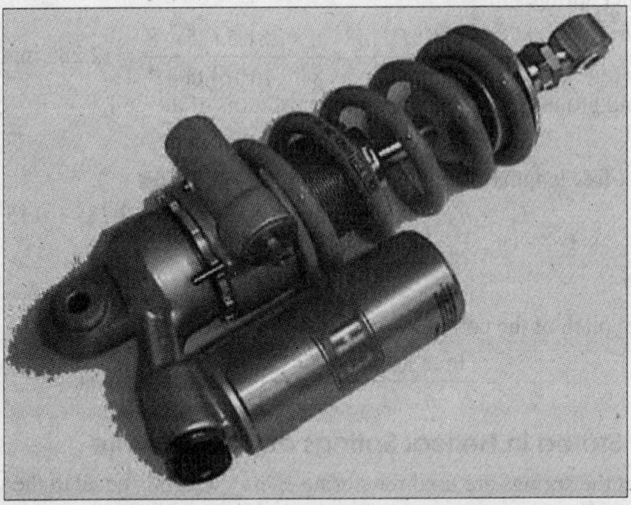

Another view of an automobile shock-absorber

Example 23.12. *Find the maximum shear stress and deflection induced in a helical spring of the following specifications, if it has to absorb 1000 N-m of energy.*

Mean diameter of spring = 100 mm ; Diameter of steel wire, used for making the spring = 20 mm; Number of coils = 30 ; Modulus of rigidity of steel = 85 kN/mm².

Solution. Given : U = 1000 N-m ; D = 100 mm = 0.1 m ; d = 20 mm = 0.02 m ; n = 30 ; G = 85 kN/mm² = 85 × 10⁹ N/m²

Maximum shear stress induced

Let τ = Maximum shear stress induced.

We know that spring index,

$$C = \frac{D}{d} = \frac{0.1}{0.02} = 5$$

∴ Wahl's stress factor,

$$K = \frac{4C - 1}{4C - 4} + \frac{0.615}{C} = \frac{4 \times 5 - 1}{4 \times 5 - 4} + \frac{0.615}{5} = 1.31$$

Volume of spring wire,

$$V = (\pi\, D.n)\left(\frac{\pi}{4} \times d^2\right) = (\pi \times 0.1 \times 30)\left[\frac{\pi}{4}\,(0.02)^2\right] \text{m}^3$$

$$= 0.002\,96 \text{ m}^3$$

We know that energy absorbed in the spring (*U*),

$$1000 = \frac{\tau^2}{4K^2.G} \times V = \frac{\tau^2}{4\,(1.31)^2\; 85 \times 10^9} \times 0.00296 = \frac{5\,\tau^2}{10^{15}}$$

$$\therefore \quad \tau^2 = 1000 \times 10^{15} / 5 = 200 \times 10^{15}$$

or $$\tau = 447.2 \times 10^6 \text{ N/m}^2 = 447.2 \text{ MPa } \textbf{Ans.}$$

Deflection produced in the spring

We know that deflection produced in the spring,

$$\delta = \frac{\pi \tau . D^2\, n}{K.d.G} = \frac{\pi \times 447.2 \times 10^6\; (0.1)^2\; 30}{1.31 \times 0.02 \times 85 \times 10^9} = 0.1893 \text{ m}$$

$$= 189.3 \text{ mm } \textbf{Ans.}$$

Example 23.13. *A closely coiled helical spring is made of 10 mm diameter steel wire, the coil consisting of 10 complete turns with a mean diameter of 120 mm. The spring carries an axial pull of 200 N. Determine the shear stress induced in the spring neglecting the effect of stress concentration. Determine also the deflection in the spring, its stiffness and strain energy stored by it if the modulus of rigidity of the material is 80 kN/mm².*

Solution. Given : $d = 10$ mm ; $n = 10$; $D = 120$ mm ; $W = 200$ N ; $G = 80 \text{ kN/mm}^2 = 80 \times 10^3 \text{ N/mm}^2$

Shear stress induced in the spring neglecting the effect of stress concentration

We know that shear stress induced in the spring neglecting the effect of stress concentration is,

$$\tau = \frac{8\,W.D}{\pi\, d^3}\left(1 + \frac{d}{2D}\right) = \frac{8 \times 200 \times 120}{\pi\,(10)^3}\left[1 + \frac{10}{2 \times 120}\right] \text{N/mm}^2$$

$$= 61.1 \times 1.04 = 63.54 \text{ N/mm}^2 = 63.54 \text{ MPa } \textbf{Ans.}$$

Deflection in the spring

We know that deflection in the spring,

$$\delta = \frac{8\,W.D^3 n}{G.d^4} = \frac{8 \times 200\,(120)^3\; 10}{80 \times 10^3\; (10)^4} = 34.56 \text{ mm } \textbf{Ans.}$$

Stiffness of the spring

We know that stiffness of the spring

$$= \frac{W}{\delta} = \frac{200}{34.56} = 5.8 \text{ N/mm}$$

Strain energy stored in the spring

We know that strain energy stored in the spring,

$$U = \frac{1}{2}\; W.\delta = \frac{1}{2} \times 200 \times 34.56 = 3456 \text{ N-mm} = 3.456 \text{ N-m } \textbf{Ans.}$$

Example 23.14. *At the bottom of a mine shaft, a group of 10 identical close coiled helical springs are set in parallel to absorb the shock caused by the falling of the cage in case of a failure. The loaded cage weighs 75 kN, while the counter weight has a weight of 15 kN. If the loaded cage falls through a height of 50 metres from rest, find the maximum stress induced in each spring if it is made of 50 mm diameter steel rod. The spring index is 6 and the number of active turns in each spring is 20. Modulus of rigidity, G = 80 kN/mm².*

Solution. Given : No. of springs = 10 ; W_1 = 75 kN = 75 000 N ; W_2 = 15 kN = 15 000 N ; h = 50 m = 50 000 mm ; d = 50 mm ; C = 6 ; n = 20 ; G = 80 kN/mm² = 80 × 10³ N/mm²

We know that net weight of the falling load,

$$P = W_1 - W_2 = 75\,000 - 15\,000 = 60\,000 \text{ N}$$

Let W = The equivalent static (or gradually applied) load on each spring which can produce the same effect as by the falling load P.

We know that compression produced in each spring,

$$\delta = \frac{8\,W.C^3.n}{G.d} = \frac{8W \times 6^3 \times 20}{80 \times 10^3 \times 50} = 0.008\,64\,W \text{ mm}$$

Since the work done by the falling load is equal to the energy stored in the helical springs which are 10 in number, therefore,

$$P\,(h + \delta) = \frac{1}{2}\,W \times \delta \times 10$$

$$60\,000\,(50\,000 + 0.008\,64\,W) = \frac{1}{2}\,W \times 0.008\,64W \times 10$$

$$3 \times 10^9 + 518.4\,W = 0.0432\,W^2$$

or $$W^2 - 12\,000\,W - 69.4 \times 10^9 = 0$$

$$\therefore \quad W = \frac{12\,000 \pm \sqrt{(12\,000)^2 + 4 \times 1 \times 69.4 \times 10^9}}{2} = \frac{12\,000 \pm 527\,000}{2}$$

$$= 269\,500 \text{ N} \qquad \text{... (Taking +ve sign)}$$

We know that Wahl's stress factor,

$$K = \frac{4C-1}{4C-4} + \frac{0.615}{C} = \frac{4 \times 6 - 1}{4 \times 6 - 4} + \frac{0.615}{6} = 1.25$$

and maximum stress induced in each spring,

$$\tau = K \times \frac{8W.C}{\pi\,d^2} = 1.25 \times \frac{8 \times 269\,500 \times 6}{\pi\,(50)^2} = 2058.6 \text{ N/mm}^2$$

$$= 2058.6 \text{ MPa } \textbf{Ans.}$$

Example 23.15. *A rail wagon of mass 20 tonnes is moving with a velocity of 2 m/s. It is brought to rest by two buffers with springs of 300 mm diameter. The maximum deflection of springs is 250 mm. The allowable shear stress in the spring material is 600 MPa. Design the spring for the buffers.*

Buffers have springs inside to absorb shock.

Solution. Given : m = 20 t = 20 000 kg ; v = 2 m/s ; D = 300 mm ; δ = 250 mm ; τ = 600 MPa = 600 N/mm²

1. *Diameter of the spring wire*

Let d = Diameter of the spring wire.

We know that kinetic energy of the wagon

$$= \frac{1}{2}\,m\,v^2 = \frac{1}{2} \times 20\,000\,(2)^2 = 40\,000 \text{ N-m} = 40 \times 10^6 \text{ N-mm} \qquad \text{...}(i)$$

Let W be the equivalent load which when applied gradually on each spring causes a deflection of 250 mm. Since there are two springs, therefore

Energy stored in the springs

$$= \frac{1}{2} \times W.\delta \times 2 = W.\delta = W \times 250 = 250\, W \text{ N-mm} \qquad ...(ii)$$

From equations (*i*) and (*ii*), we have

$$\therefore \quad W = 40 \times 10^6 / 250 = 160 \times 10^3 \text{ N}$$

We know that torque transmitted by the spring,

$$T = W \times \frac{D}{2} = 160 \times 10^3 \times \frac{300}{2} = 24 \times 10^6 \text{ N-mm}$$

We also know that torque transmitted by the spring (T),

$$24 \times 10^6 = \frac{\pi}{16} \times \tau \times d^3 = \frac{\pi}{16} \times 600 \times d^3 = 117.8\, d^3$$

$$\therefore \quad d^3 = 24 \times 10^6 / 117.8 = 203.7 \times 10^3 \quad \text{or} \quad d = 58.8 \text{ say } 60 \text{ mm } \textbf{Ans.}$$

2. ***Number of turns of the spring coil***

Let n = Number of active turns of the spring coil.

We know that the deflection of the spring (δ),

$$250 = \frac{8\, W.D^3.n}{G.d^4} = \frac{8 \times 160 \times 10^3\, (300)^3\, n}{84 \times 10^3\, (60)^4} = 31.7\, n$$

... (Taking G = 84 MPa = 84×10^3 N/mm²)

$$\therefore \quad n = 250 / 31.7 = 7.88 \text{ say } 8 \textbf{ Ans.}$$

Assuming square and ground ends, total number of turns,

$$n' = n + 2 = 8 + 2 = 10 \textbf{ Ans.}$$

3. ***Free length of the spring***

We know that free length of the spring,

$$L_F = n'.d + \delta + 0.15\,\delta = 10 \times 60 + 250 + 0.15 \times 250 = 887.5 \text{ mm } \textbf{Ans.}$$

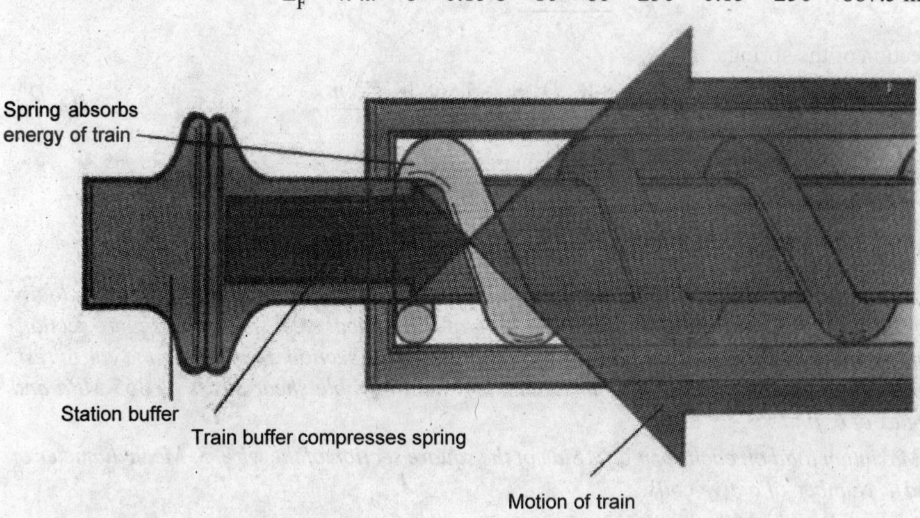

4. Pitch of the coil

We know that pitch of the coil

$$= \frac{\text{Free length}}{n' - 1} = \frac{887.5}{10 - 1} = 98.6 \text{ mm} \quad \textbf{Ans.}$$

23.14 Stress and Deflection in Helical Springs of Non-circular Wire

The helical springs may be made of non-circular wire such as rectangular or square wire, in order to provide greater resilience in a given space. However these springs have the following main disadvantages :

1. The quality of material used for springs is not so good.
2. The shape of the wire does not remain square or rectangular while forming helix, resulting in trapezoidal cross-sections. It reduces the energy absorbing capacity of the spring.
3. The stress distribution is not as favourable as for circular wires. But this effect is negligible where loading is of static nature.

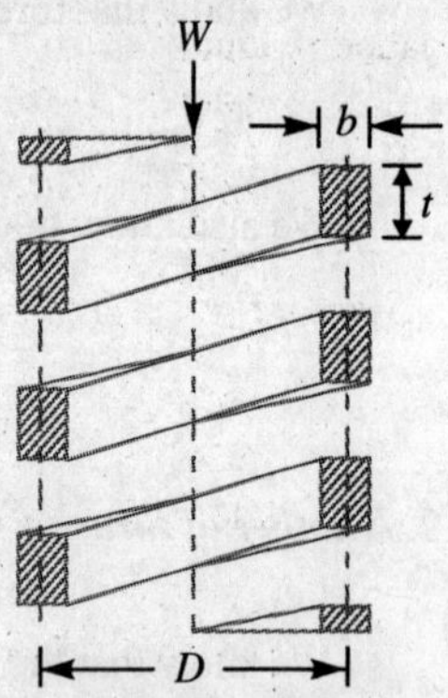

Fig. 23.18. Spring of rectangular wire.

For springs made of rectangular wire, as shown in Fig. 23.18, the maximum shear stress is given by

$$\tau = K \times \frac{W.D\,(1.5\,t + 0.9\,b)}{b^2.t^2}$$

This expression is applicable when the longer side (*i.e.* $t > b$) is parallel to the axis of the spring. But when the shorter side (*i.e.* $t < b$) is parallel to the axis of the spring, then maximum shear stress,

$$\tau = K \times \frac{W.D\,(1.5\,b + 0.9\,t)}{b^2.t^2}$$

and deflection of the spring,

$$\delta = \frac{2.45\,W.D^3.n}{G.b^3\,(t - 0.56\,b)}$$

For springs made of square wire, the dimensions b and t are equal. Therefore, the maximum shear stress is given by

$$\tau = K \times \frac{2.4\,W.D}{b^3}$$

and deflection of the spring,

$$\delta = \frac{5.568\,W.D^3.n}{G.b^4} = \frac{5.568\,W.C^3.n}{G.b} \quad \ldots\left(\because C = \frac{D}{b}\right)$$

where b = Side of the square.

Note : In the above expressions,

$$K = \frac{4C - 1}{4C - 4} + \frac{0.615}{C}, \text{ and } C = \frac{D}{b}$$

Example 23.16. *A loaded narrow-gauge car of mass 1800 kg and moving at a velocity 72 m/min., is brought to rest by a bumper consisting of two helical steel springs of square section. The mean diameter of the coil is six times the side of the square section. In bringing the car to rest, the springs are to be compressed 200 mm. Assuming the allowable shear stress as 365 MPa and spring index of 6, find :*

1. Maximum load on each spring, 2. Side of the square section of the wire, 3. Mean diameter of coils, and 4. Number of active coils.

Take modulus of rigidity as 80 kN/mm².

Solution. Given : m = 1800 kg ; v = 72 m/min = 1.2 m/s ; δ = 200 mm ; τ = 365 MPa = 365 N/mm^2 ; C = 6 ; G = 80 kN/mm^2 = 80 × 10^3 N/mm^2

1. ***Maximum load on each spring,***

Let W = Maximum load on each spring.

We know that kinetic energy of the car

$$= \frac{1}{2} m.v^2 = \frac{1}{2} \times 1800\ (1.2)^2 = 1296 \text{ N-m} = 1296 \times 10^3 \text{ N-mm}$$

This energy is absorbed in the two springs when compressed to 200 mm. If the springs are loaded gradually from 0 to W, then

$$\left(\frac{0 + W}{2}\right) 2 \times 200 = 1296 \times 10^3$$

$$\therefore \quad W = 1296 \times 10^3 / 200 = 6480 \text{ N} \textbf{ Ans.}$$

2. ***Side of the square section of the wire***

Let b = Side of the square section of the wire, and

D = Mean diameter of the coil = 6 b ... ($\because$ $C = D/b = 6$)

We know that Wahl's stress factor,

$$K = \frac{4C - 1}{4C - 4} + \frac{0.615}{C} = \frac{4 \times 6 - 1}{4 \times 6 - 4} + \frac{0.615}{6} = 1.2525$$

and maximum shear stres (τ),

$$365 = K \times \frac{2.4\, W.D}{b^3} = 1.2525 \times \frac{2.4 \times 6480 \times 6\, b}{b^3} = \frac{116\ 870}{b^2}$$

$$\therefore \quad b^2 = 116\ 870 / 365 = 320 \quad \text{or} \quad b = 17.89 \text{ say } 18 \text{ mm } \textbf{Ans.}$$

3. ***Mean diameter of the coil***

We know that mean diameter of the coil,

$$D = 6\, b = 6 \times 18 = 108 \text{ mm } \textbf{Ans.}$$

4. ***Number of active coils***

Let n = Number of active coils.

We know that the deflection of the spring (δ),

$$200 = \frac{5.568\, W.C^3.n}{G.b} = \frac{5.568 \times 6480 \times 6^3 \times n}{80 \times 10^3 \times 18} = 5.4\, n$$

$$\therefore \quad n = 200 / 5.4 = 37 \textbf{ Ans.}$$

23.15 Helical Springs Subjected to Fatigue Loading

The helical springs subjected to fatigue loading are designed by using the *Soderberg line method. The spring materials are usually tested for torsional endurance strength under a repeated stress that varies from zero to a maximum. Since the springs are ordinarily loaded in one direction only (the load in springs is never reversed in nature), therefore a modified Soderberg diagram is used for springs, as shown in Fig. 23.19.

The endurance limit for reversed loading is shown at point A where the mean shear stress is equal to $\tau_e / 2$ and the variable shear stress is also equal to $\tau_e / 2$. A line drawn from A to B (the yield point in shear, τ_y) gives the Soderberg's failure stress line. If a suitable factor of safety ($F.S.$) is applied to the yield strength (τ_y), a safe stress line CD may be drawn parallel to the line AB, as shown in Fig. 23.19. Consider a design point P on the line CD. Now the value of factor of safety may be obtained as discussed below :

* We have discussed the Soderberg method for completely reversed stresses in Chapter 6.

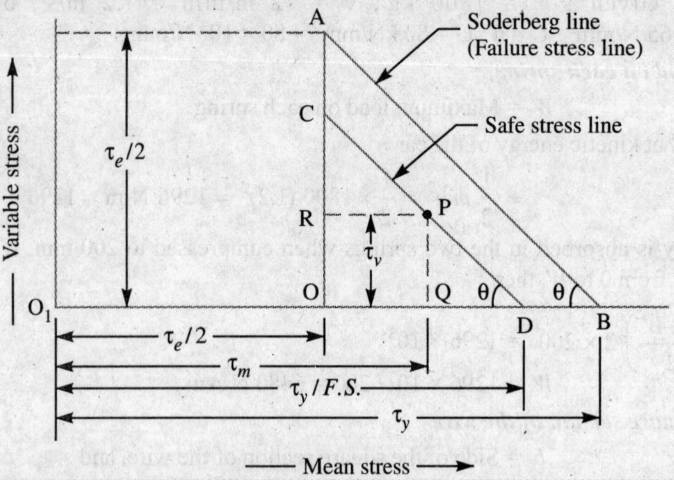

Fig. 23.19. Modified Soderberg method for helical springs.

From similar triangles *PQD* and *AOB*, we have

$$\frac{PQ}{QD} = \frac{OA}{OB} \quad \text{or} \quad \frac{PQ}{O_1D - O_1Q} = \frac{OA}{O_1B - O_1O}$$

$$\frac{\tau_v}{\dfrac{\tau_y}{F.S.} - \tau_m} = \frac{\tau_e/2}{\tau_y - \dfrac{\tau_e}{2}} = \frac{\tau_e}{2\tau_y - \tau_e}$$

or
$$2\,\tau_v \,.\, \tau_y - \tau_v .\, \tau_e = \frac{\tau_e \,.\, \tau_y}{F.S.} - \tau_m .\tau_e$$

$$\therefore \qquad \frac{\tau_e \,.\, \tau_y}{F.S.} = 2\,\tau_v.\tau_y - \tau_v.\tau_e + \tau_m.\tau_e$$

Dividing both sides by $\tau_e.\tau_y$ and rearranging, we have

$$\frac{1}{F.S.} = \frac{\tau_m - \tau_v}{\tau_y} + \frac{2\tau_v}{\tau_e} \qquad ...(i)$$

Notes : 1. From equation (*i*), the expression for the factor of safety (*F.S.*) may be written as

$$F.S. = \frac{\tau_y}{\tau_m - \tau_v + \dfrac{2\,\tau_v.\tau_y}{\tau_e}}$$

2. The value of mean shear stress (τ_m) is calculated by using the shear stress factor (K_S), while the variable shear stress is calculated by using the full value of the Wahl's factor (*K*). Thus

Mean shear stress,

$$\tau_m = K_s \times \frac{8\,W_m \times D}{\pi d^3}$$

where $$K_S = 1 + \frac{1}{2C}; \text{ and } W_m = \frac{W_{max} + W_{min}}{2}$$

and variable shear stress, $$\tau_v = K \times \frac{8\,W_v \times D}{\pi\, d^3}$$

where $$K = \frac{4C - 1}{4C - 4} + \frac{0.615}{C}; \text{ and } W_v = \frac{W_{max} - W_{min}}{2}$$

Example 23.17. *A helical compression spring made of oil tempered carbon steel, is subjected to a load which varies from 400 N to 1000 N. The spring index is 6 and the design factor of safety is 1.25. If the yield stress in shear is 770 MPa and endurance stress in shear is 350 MPa, find : 1. Size of the spring wire, 2. Diameters of the spring, 3. Number of turns of the spring, and 4. Free length of the spring.*

The compression of the spring at the maximum load is 30 mm. The modulus of rigidity for the spring material may be taken as 80 kN/mm².

Solution. Given : $W_{min} = 400$ N ; $W_{max} = 1000$ N ; $C = 6$; $F.S. = 1.25$; $\tau_y = 770$ MPa $= 770$ N/mm² ; $\tau_e = 350$ MPa $= 350$ N/mm² ; $\delta = 30$ mm ; $G = 80$ kN/mm² $= 80 \times 10^3$ N/mm²

1. ***Size of the spring wire***

Let d = Diameter of the spring wire, and

D = Mean diameter of the spring = $C.d = 6\,d$ ... ($\because D/d = C = 6$)

We know that the mean load,

$$W_m = \frac{W_{max} + W_{min}}{2} = \frac{1000 + 400}{2} = 700 \text{ N}$$

and variable load,

$$W_v = \frac{W_{max} - W_{min}}{2} = \frac{1000 - 400}{2} = 300 \text{ N}$$

Shear stress factor,

$$K_S = 1 + \frac{1}{2C} = 1 + \frac{1}{2 \times 6} = 1.083$$

Wahl's stress factor,

$$K = \frac{4C - 1}{4C - 4} + \frac{0.615}{C} = \frac{4 \times 6 - 1}{4 \times 6 - 4} + \frac{0.615}{6} = 1.2525$$

We know that mean shear stress,

$$\tau_m = K_S \times \frac{8\,W_m \times D}{\pi\,d^3} = 1.083 \times \frac{8 \times 700 \times 6\,d}{\pi\,d^3} = \frac{11\,582}{d^2} \text{ N/mm}^2$$

and variable shear stress,

$$\tau_v = K \times \frac{8\,W_v \times D}{\pi\,d^3} = 1.2525 \times \frac{8 \times 300 \times 6\,d}{\pi\,d^3} = \frac{5740}{d^2} \text{ N/mm}^2$$

We know that

$$\frac{1}{F.S.} = \frac{\tau_m - \tau_v}{\tau_y} + \frac{2\,\tau_v}{\tau_e}$$

$$\frac{1}{1.25} = \frac{\dfrac{11\,582}{d^2} - \dfrac{5740}{d^2}}{770} + \frac{2 \times \dfrac{5740}{d^2}}{350} = \frac{7.6}{d^2} + \frac{32.8}{d^2} = \frac{40.4}{d^2}$$

$\therefore$ $d^2 = 1.25 \times 40.4 = 50.5$ or $d = 7.1$ mm **Ans.**

2. ***Diameters of the spring***

We know that mean diameter of the spring,

$D = C.d = 6 \times 7.1 = 42.6$ mm **Ans.**

Outer diameter of the spring,

$D_o = D + d = 42.6 + 7.1 = 49.7$ mm **Ans.**

and inner diameter of the spring,

$D_i = D - d = 42.6 - 7.1 = 35.5$ mm **Ans.**

3. ***Number of turns of the spring***

Let n = Number of active turns of the spring.

We know that deflection of the spring (δ),

$$30 = \frac{8\,W.D^3.n}{G.d^4} = \frac{8 \times 1000\,(42.6)^3\,n}{80 \times 10^3\,(7.1)^4} = 3.04\,n$$

$$\therefore \quad n = 30 / 3.04 = 9.87 \text{ say } 10 \textbf{ Ans.}$$

Assuming the ends of the spring to be squared and ground, the total number of turns of the spring,

$$n' = n + 2 = 10 + 2 = 12 \textbf{ Ans.}$$

4. *Free length of the spring*

We know that free length of the spring,

$$L_F = n'.d + \delta + 0.15\,\delta = 12 \times 7.1 + 30 + 0.15 \times 30 \text{ mm}$$
$$= 119.7 \text{ say } 120 \text{ mm } \textbf{Ans.}$$

23.16 Springs in Series

Consider two springs connected in series as shown in Fig. 23.20.

Let W = Load carried by the springs,

δ_1 = Deflection of spring 1,

δ_2 = Deflection of spring 2,

k_1 = Stiffness of spring 1 = W / δ_1, and

k_2 = Stiffness of spring 2 = W / δ_2

Springs in series.
Fig. 23.20

A little consideration will show that when the springs are connected in series, then the total deflection produced by the springs is equal to the sum of the deflections of the individual springs.

∴ Total deflection of the springs,

$$\delta = \delta_1 + \delta_2$$

or
$$\frac{W}{k} = \frac{W}{k_1} + \frac{W}{k_2}$$

∴
$$\frac{1}{k} = \frac{1}{k_1} + \frac{1}{k_2}$$

where k = Combined stiffness of the springs.

23.17 Springs in Parallel

Consider two springs connected in parallel as shown in Fig 23.21.

Let W = Load carried by the springs,

W_1 = Load shared by spring 1,

W_2 = Load shared by spring 2,

k_1 = Stiffness of spring 1, and

k_2 = Stiffness of spring 2.

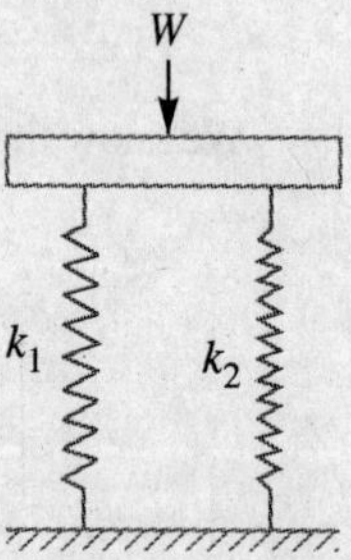

Springs in parallel.
Fig. 23.21

A little consideration will show that when the springs are connected in parallel, then the total deflection produced by the springs is same as the deflection of the individual springs.

We know that $W = W_1 + W_2$

or $\delta.k = \delta.k_1 + \delta.k_2$

∴ $k = k_1 + k_2$

where k = Combined stiffness of the springs, and

δ = Deflection produced.

Example 23.18. *A close coiled helical compression spring of 12 active coils has a spring stiffness of k. It is cut into two springs having 5 and 7 turns. Determine the spring stiffnesses of resulting springs.*

Solution. Given : $n = 12$; $n_1 = 5$; $n_2 = 7$

We know that the deflection of the spring,

$$\delta = \frac{8\,W.D^3.n}{G.d^4} \quad \text{or} \quad \frac{W}{\delta} = \frac{G.d^4}{8\,D^3.n}$$

Since G, D and d are constant, therefore substituting

$$\frac{G.d^4}{8\,D^3} = X, \text{ a constant, we have } \frac{W}{\delta} = k = \frac{X}{n}$$

or $$X = k.n = 12\,k$$

The spring is cut into two springs with $n_1 = 5$ and $n_2 = 7$.

Let k_1 = Stiffness of spring having 5 turns, and

k_2 = Stiffness of spring having 7 turns.

$$\therefore \quad k_1 = \frac{X}{n_1} = \frac{12k}{5} = 2.4\,k \textbf{ Ans.}$$

and $$k_2 = \frac{X}{n_2} = \frac{12k}{7} = 1.7\,k \textbf{ Ans.}$$

23.18 Concentric or Composite Springs

A concentric or composite spring is used for one of the following purposes :

1. To obtain greater spring force within a given space.

2. To insure the operation of a mechanism in the event of failure of one of the springs.

The concentric springs for the above two purposes may have two or more springs and have the same free lengths as shown in Fig. 23.22 (*a*) and are compressed equally. Such springs are used in automobile clutches, valve springs in aircraft, heavy duty diesel engines and rail-road car suspension systems.

Sometimes concentric springs are used to obtain a spring force which does not increase in a direct relation to the deflection but increases faster. Such springs are made of different lengths as shown in Fig. 23.22 (*b*). The shorter spring begins to act only after the longer spring is compressed to a certain amount. These springs are used in governors of variable speed engines to take care of the variable centrifugal force.

A car shock absorber.

The adjacent coils of the concentric spring are wound in opposite directions to eliminate any tendency to bind.

If the same material is used, the concentric springs are designed for the same stress. In order to get the same stress factor (K), it is desirable to have the same spring index (C).

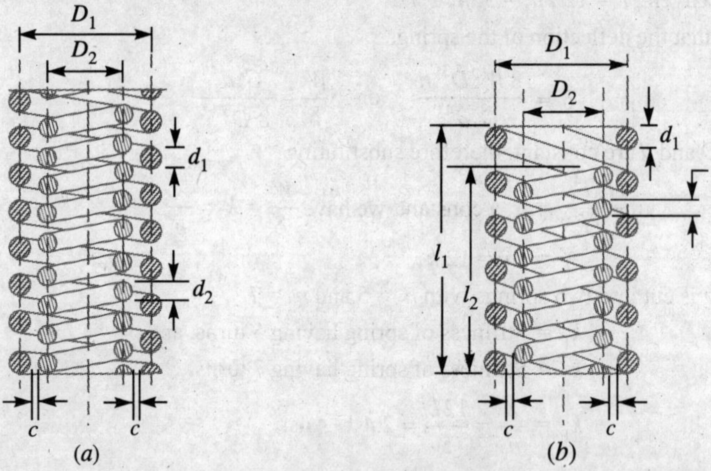

Fig. 23.22. Concentric springs.

Consider a concentric spring as shown in Fig. 23.22 (a).

Let W = Axial load,

W_1 = Load shared by outer spring,

W_2 = Load shared by inner spring,

d_1 = Diameter of spring wire of outer spring,

d_2 = Diameter of spring wire of inner spring,

D_1 = Mean diameter of outer spring,

D_2 = Mean diameter of inner spring,

δ_1 = Deflection of outer spring,

δ_2 = Deflection of inner spring,

n_1 = Number of active turns of outer spring, and

n_2 = Number of active turns of inner spring.

Assuming that both the springs are made of same material, then the maximum shear stress induced in both the springs is approximately same, *i.e.*

$$\tau_1 = \tau_2$$

$$\frac{8\,W_1 . D_1 . K_1}{\pi\,(d_1)^3} = \frac{8\,W_2 . D_2 . K_2}{\pi\,(d_2)^3}$$

When stress factor, $K_1 = K_2$, then

$$\frac{W_1 . D_1}{(d_1)^3} = \frac{W_2 . D_2}{(d_2)^3} \qquad ...(i)$$

If both the springs are effective throughout their working range, then their free length and deflection are equal, *i.e.*

$$\delta_1 = \delta_2$$

or
$$\frac{8W_1\,(D_1)^3\,n_1}{(d_1)^4\,G} = \frac{8W_2\,(D_2)^3\,n_2}{(d_2)^4\,G} \quad \text{or} \quad \frac{W_1\,(D_1)^3\,n_1}{(d_1)^4} = \frac{W_2\,(D_2)^3\,n_2}{(d_2)^4} \qquad ...(ii)$$

When both the springs are compressed until the adjacent coils meet, then the solid length of both the springs is equal, *i.e.*

$$n_1.d_1 = n_2.d_2$$

∴ The equation (*ii*) may be written as

$$\frac{W_1 (D_1)^3}{(d_1)^5} = \frac{W_2 (D_2)^3}{(d_2)^5} \quad ...(iii)$$

Now dividing equation (*iii*) by equation (*i*), we have

$$\frac{(D_1)^2}{(d_1)^2} = \frac{(D_2)^2}{(d_2)^2} \text{ or } \frac{D_1}{d_1} = \frac{D_2}{d_2} = C, \text{ the spring index} \quad ...(iv)$$

i.e. the springs should be designed in such a way that the spring index for both the springs is same.

From equations (*i*) and (*iv*), we have

$$\frac{W_1}{(d_1)^2} = \frac{W_2}{(d_2)^2} \text{ or } \frac{W_1}{W_2} = \frac{(d_1)^2}{(d_2)^2} \quad ..(v)$$

From Fig. 23.22 (*a*), we find that the radial clearance between the two springs,

$$*c = \left(\frac{D_1}{2} - \frac{D_2}{2}\right) - \left(\frac{d_1}{2} + \frac{d_2}{2}\right)$$

Usually, the radial clearance between the two springs is taken as $\frac{d_1 - d_2}{2}$.

$$\therefore \left(\frac{D_1}{2} - \frac{D_2}{2}\right) - \left(\frac{d_1}{2} + \frac{d_2}{2}\right) = \frac{d_1 - d_2}{2}$$

or

$$\frac{D_1 - D_2}{2} = d_1 \quad ...(vi)$$

From equation (*iv*), we find that

$$D_1 = C.d_1, \text{ and } D_2 = C.d_2$$

Substituting the values of D_1 and D_2 in equation (*vi*), we have

$$\frac{C.d_1 - C.d_2}{2} = d_1 \text{ or } C.d_1 - 2\,d_1 = C.d_2$$

$$\therefore \quad d_1 (C - 2) = C.d_2 \text{ or } \frac{d_1}{d_2} = \frac{C}{C - 2} \quad ...(vii)$$

Example 23.19. *A concentric spring for an aircraft engine valve is to exert a maximum force of 5000 N under an axial deflection of 40 mm. Both the springs have same free length, same solid length and are subjected to equal maximum shear stress of 850 MPa. If the spring index for both the springs is 6, find (a) the load shared by each spring, (b) the main dimensions of both the springs, and (c) the number of active coils in each spring.*

Assume G = 80 kN/mm² and diametral clearance to be equal to the difference between the wire diameters.

Solution. Given : W = 5000 N ; δ = 40 mm ; $\tau_1 = \tau_2$ = 850 MPa = 850 N/mm² ; C = 6 ; G = 80 kN/mm² = 80 × 10³ N/mm²

The concentric spring is shown in Fig. 23.22 (*a*).

(a) Load shared by each spring

Let W_1 and W_2 = Load shared by outer and inner spring respectively,

d_1 and d_2 = Diameter of spring wires for outer and inner springs respectively, and

D_1 and D_2 = Mean diameter of the outer and inner springs respectively.

---

* The net clearance between the two springs is given by

$$2\,c = (D_1 - D_2) - (d_1 + d_2)$$

Since the diametral clearance is equal to the difference between the wire diameters, therefore

$$(D_1 - D_2) - (d_1 + d_2) = d_1 - d_2$$

or $$D_1 - D_2 = 2\,d_1$$

We know that $D_1 = C.d_1$, and $D_2 = C.d_2$

$\therefore$ $$C.d_1 - C.d_2 = 2\,d_1$$

or $$\frac{d_1}{d_2} = \frac{C}{C-2} = \frac{6}{6-2} = 1.5 \quad \text{...}(i)$$

We also know that $$\frac{W_1}{W_2} = \left(\frac{d_1}{d_2}\right)^2 = (1.5)^2 = 2.25 \quad \text{...}(ii)$$

and $$W_1 + W_2 = W = 5000 \text{ N} \quad \text{...}(iii)$$

From equations (*ii*) and (*iii*), we find that

$$W_1 = 3462 \text{ N, and } W_2 = 1538 \text{ N } \textbf{Ans.}$$

(*b*) ***Main dimensions of both the springs***

We know that Wahl's stress factor for both the springs,

$$K_1 = K_2 = \frac{4C-1}{4C-4} + \frac{0.615}{C} = \frac{4\times6-1}{4\times6-4} + \frac{0.615}{6} = 1.2525$$

and maximum shear stress induced in the outer spring (τ_1),

$$850 = K_1 \times \frac{8\,W_1.C}{\pi(d_1)^2} = 1.2525 \times \frac{8\times3462\times6}{\pi(d_1)^2} = \frac{66\,243}{(d_1)^2}$$

$\therefore$ $$(d_1)^2 = 66\,243 / 850 = 78 \text{ or } d_1 = 8.83 \text{ say } 10 \text{ mm } \textbf{Ans.}$$

and $$D_1 = C.d_1 = 6\,d_1 = 6\times10 = 60 \text{ mm } \textbf{Ans.}$$

Similarly, maximum shear stress induced in the inner spring (τ_2),

$$850 = K_2 \times \frac{8W_2.C}{\pi(d_2)^2} = 1.2525 \times \frac{8\times1538\times6}{\pi(d_2)^2} = \frac{29\,428}{(d_2)^2}$$

$\therefore$ $$(d_2)^2 = 29\,428 / 850 = 34.6 \text{ or } {}^*d_2 = 5.88 \text{ say } 6 \text{ mm } \textbf{Ans.}$$

and $$D_2 = C.d_2 = 6\times6 = 36 \text{ mm } \textbf{Ans.}$$

(*c*) ***Number of active coils in each spring***

Let n_1 and n_2 = Number of active coils of the outer and inner spring respectively.

We know that the axial deflection for the outer spring (δ),

$$40 = \frac{8\,W_1.C^3.n_1}{G.d_1} = \frac{8\times3462\times6^3\times n_1}{80\times10^3\times10} = 7.48\,n_1$$

$\therefore$ $$n_1 = 40/7.48 = 5.35 \text{ say } 6 \textbf{ Ans.}$$

Assuming square and ground ends for the spring, the total number of turns of the outer spring,

$$n_1' = 6 + 2 = 8$$

$\therefore$ Solid length of the outer spring,

$$L_{S1} = n_1'.\,d_1 = 8\times10 = 80 \text{ mm}$$

Let n_2' be the total number of turns of the inner spring. Since both the springs have the same solid length, therefore,

$$n_2'.d_2 = n_1'.d_1$$

---

* The value of d_2 may also be obtained from equation (*i*), *i.e.*

$$\frac{d_1}{d_2} = 1.5 \text{ or } d_2 = \frac{d_1}{1.5} = \frac{8.83}{1.5} = 5.887 \text{ say } 6 \text{ mm}$$

or $$n_2' = \frac{n_1'.d_1}{d_2} = \frac{8 \times 10}{6} = 13.3 \text{ say } 14$$

and $$n_2 = 14 - 2 = 12 \textbf{ Ans.} \quad \ldots (\because n_2' = n_2 + 2)$$

Since both the springs have the same free length, therefore

Free length of outer spring

$= \text{Free length of inner spring}$

$= L_{S1} + \delta + 0.15\,\delta = 80 + 40 + 0.15 \times 40 = 126$ mm **Ans.**

Other dimensions of the springs are as follows:

Outer diameter of the outer spring

$= D_1 + d_1 = 60 + 10 = 70$ mm **Ans.**

Inner diameter of the outer spring

$= D_1 - d_1 = 60 - 10 = 50$ mm **Ans.**

Outer diameter of the inner spring

$= D_2 + d_2 = 36 + 6 = 42$ mm **Ans.**

Inner diameter of the inner spring

$= D_2 - d_2 = 36 - 6 = 30$ mm **Ans.**

Shock absorbers

Example 23.20. *A composite spring has two closed coil helical springs as shown in Fig. 23.22 (b). The outer spring is 15 mm larger than the inner spring. The outer spring has 10 coils of mean diameter 40 mm and wire diameter 5mm. The inner spring has 8 coils of mean diameter 30 mm and wire diameter 4 mm. When the spring is subjected to an axial load of 400 N, find 1. compression of each spring, 2. load shared by each spring, and 3. shear stress induced in each spring. The modulus of rigidity may be taken as 84 kN/mm².*

Solution. Given : $\delta_1 = l_1 - l_2 = 15$ mm ; $n_1 = 10$; $D_1 = 40$ mm ; $d_1 = 5$ mm ; $n_2 = 8$; $D_2 = 30$ mm ; $d_2 = 4$ mm ; $W = 400$ N ; $G = 84$ kN/mm² $= 84 \times 10^3$ N/mm²

1. ***Compression of each spring***

Since the outer spring is 15 mm larger than the inner spring, therefore the inner spring will not take any load till the outer spring is compressed by 15 mm. After this, both the springs are compressed together. Let P_1 be the load on the outer spring to compress it by 15 mm.

We know that compression of the spring (δ),

$$15 = \frac{8\,P_1\,(D_1)^3\,n_1}{G\,(d_1)^4} = \frac{8\,P_1\,(40)^3\,10}{84 \times 10^3 \times 5^4} = 0.0975\,P_1$$

$$\therefore \quad P_1 = 15 / 0.0975 = 154 \text{ N}$$

Now the remaining load *i.e.* $W - P_1 = 400 - 154 = 246$ N is taken together by both the springs.

Let δ_2 = Further compression of the outer spring or the total compression of the inner spring.

Since for compressing the outer spring by 15 mm, the load required is 154 N, therefore the additional load required by the outer spring to compress it by δ_2 mm is given by

$$P_2 = \frac{P_1}{\delta_1} \times \delta_2 = \frac{154}{15} \times \delta_2 = 10.27\ \delta_2$$

Let W_2 = Load taken by the inner spring to compress it by δ_2 mm.

We know that
$$\delta_2 = \frac{8\ W_2\ (D_2)^3\ n_2}{G\ (d_2)^4} = \frac{8\ W_2\ (30)^3\ 8}{84 \times 10^3 \times 4^4} = 0.08\ W_2$$

$$\therefore \quad W_2 = \delta_2 / 0.08 = 12.5\ \delta_2$$

and
$$P_2 + W_2 = W - P_1 = 400 - 154 = 246 \text{ N}$$

or
$$10.27\ \delta_2 + 12.5\ \delta_2 = 246 \quad \text{or} \quad \delta_2 = 246 / 22.77 = 10.8 \text{ mm } \textbf{Ans.}$$

$\therefore$ Total compression of the outer spring

$$= \delta_1 + \delta_2 = 15 + 10.8 = 25.8 \text{ mm } \textbf{Ans.}$$

2. ***Load shared by each spring***

We know that the load shared by the outer spring,

$$W_1 = P_1 + P_2 = 154 + 10.27\ \delta_2 = 154 + 10.27 \times 10.8 = 265 \text{ N } \textbf{Ans.}$$

and load shared by the inner spring,

$$W_2 = 12.5\ \delta_2 = 12.5 \times 10.8 = 135 \text{ N } \textbf{Ans.}$$

Note : The load shared by the inner spring is also given by

$$W_2 = W - W_1 = 400 - 265 = 135 \text{ N } \textbf{Ans.}$$

3. ***Shear stress induced in each spring***

We know that the spring index of the outer spring,

$$C_1 = \frac{D_1}{d_1} = \frac{40}{5} = 8$$

and spring index of the inner spring,

$$C_2 = \frac{D_2}{d_2} = \frac{30}{4} = 7.5$$

$\therefore$ Wahl's stress factor for the outer spring,

$$K_1 = \frac{4C_1 - 1}{4C_1 - 4} + \frac{0.615}{C_1} = \frac{4 \times 8 - 1}{4 \times 8 - 4} + \frac{0.615}{8} = 1.184$$

and Wahl's stress factor for the inner spring,

$$K_2 = \frac{4C_2 - 1}{4C_2 - 4} + \frac{0.615}{C_2} = \frac{4 \times 7.5 - 1}{4 \times 7.5 - 4} + \frac{0.615}{7.5} = 1.197$$

We know that shear stress induced in the outer spring,

$$\tau_1 = K_1 \times \frac{8\ W_1 . D_1}{\pi\ (d_1)^3} = 1.184 \times \frac{8 \times 265 \times 40}{\pi \times 5^3} = 255.6 \text{ N/mm}^2$$
$$= 255.6 \text{ MPa } \textbf{Ans.}$$

and shear stress induced in the inner spring,

$$\tau_2 = K_2 \times \frac{8\ W_2 . D_2}{\pi\ (d_2)^3} = 1.197 \times \frac{8 \times 135 \times 30}{\pi \times 4^3} = 192.86 \text{ N/mm}^2$$
$$= 192.86 \text{ MPa } \textbf{Ans.}$$

23.19 Helical Torsion Springs

The helical torsion springs as shown in Fig. 23.23, may be made from round, rectangular or square wire. These are wound in a similar manner as helical compression or tension springs but the ends are shaped to transmit torque. The primary stress in helical torsion springs is bending stress whereas in compression or tension springs, the stresses are torsional shear stresses. The helical torsion springs are widely used for transmitting small torques as in door hinges, brush holders in electric motors, automobile starters etc.

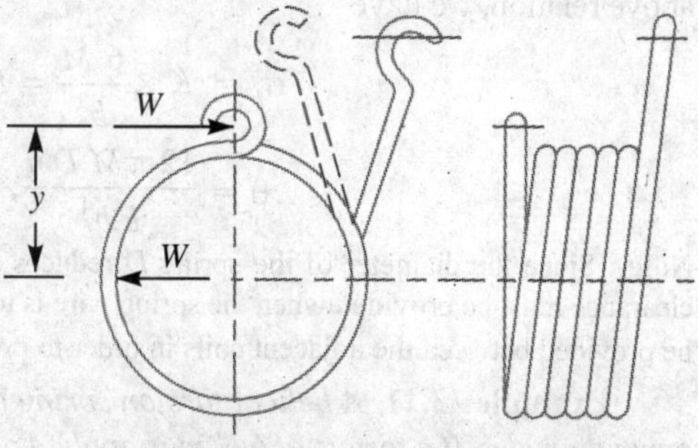

Fig. 23.23. Helical torsion spring.

A little consideration will show that the radius of curvature of the coils changes when the twisting moment is applied to the spring. Thus, the wire is under pure bending. According to A.M. Wahl, the bending stress in a helical torsion spring made of round wire is

$$\sigma_b = K \times \frac{32\,M}{\pi\,d^3} = K \times \frac{32\,W.y}{\pi\,d^3}$$

where K = Wahl's stress factor $= \dfrac{4C^2 - C - 1}{4C^2 - 4C}$,

C = Spring index,

M = Bending moment = $W \times y$,

W = Load acting on the spring,

y = Distance of load from the spring axis, and

d = Diameter of spring wire.

and total angle of twist or angular deflection,

$$^*\theta = \frac{M.l}{E.I} = \frac{M \times \pi D.n}{E \times \pi\,d^4/64} = \frac{64\,M.D.n}{E.d^4}$$

where l = Length of the wire = $\pi.D.n$,

E = Young's modulus,

I = Moment of inertia $= \dfrac{\pi}{64} \times d^4$,

D = Diameter of the spring, and

n = Number of turns.

and deflection, $\delta = \theta \times y = \dfrac{64\,M.D.n}{E.d^4} \times y$

When the spring is made of rectangular wire having width b and thickness t, then

$$\sigma_b = K \times \frac{6\,M}{t.b^2} = K \times \frac{6\,W \times y}{t.b^2}$$

where $K = \dfrac{3C^2 - C - 0.8}{3C^2 - 3C}$

* We know that $M/I = E/R$, where R is the radius of curvature.

$$\therefore \quad R = \frac{E.I}{M} \text{ or } \frac{l}{\theta} = \frac{E.I}{M} \text{ or } \theta = \frac{M.l}{E.I} \quad \ldots\left(\because R = \frac{l}{\theta}\right)$$

Angular deflection, $\theta = \frac{12\,\pi M.D.n}{E.t.b^3}$; and $\delta = \theta.y = \frac{12\,\pi M.D.n}{E.t.b^3} \times y$

In case the spring is made of square wire with each side equal to b, then substituting $t = b$, in the above relation, we have

$$\sigma_b = K \times \frac{6\,M}{b^3} = K \times \frac{6W \times y}{b^3}$$

$$\theta = \frac{12\,\pi M.D.n}{E.b^4}; \quad \text{and} \quad \delta = \frac{12\,\pi M.D.n}{E.b^4} \times y$$

Note : Since the diameter of the spring D reduces as the coils wind up under the applied load, therefore a clearance must be provided when the spring wire is to be wound round a mandrel. A small clearance must also be provided between the adjacent coils in order to prevent sliding friction.

Example 23.21. *A helical torsion spring of mean diameter 60 mm is made of a round wire of 6 mm diameter. If a torque of 6 N-m is applied on the spring, find the bending stress induced and the angular deflection of the spring in degrees. The spring index is 10 and modulus of elasticity for the spring material is 200 kN/mm². The number of effective turns may be taken as 5.5.*

Solution. Given : $D = 60$ mm ; $d = 6$ mm ; $M = 6$ N-m $= 6000$ N-mm ; $C = 10$; $E = 200$ kN/mm^2 $= 200 \times 10^3$ N/mm^2 ; $n = 5.5$

Bending stress induced

We know that Wahl's stress factor for a spring made of round wire,

$$K = \frac{4C^2 - C - 1}{4C^2 - 4C} = \frac{4 \times 10^2 - 10 - 1}{4 \times 10^2 - 4 \times 10} = 1.08$$

∴ Bending stress induced,

$$\sigma_b = K \times \frac{32\,M}{\pi\, d^3} = 1.08 \times \frac{32 \times 6000}{\pi \times 6^3} = 305.5 \text{ N/mm}^2 \text{ or MPa } \textbf{Ans.}$$

Angular deflection of the spring

We know that the angular deflection of the spring (in radians),

$$\theta = \frac{64\,M.D.n}{E.d^4} = \frac{64 \times 6000 \times 60 \times 5.5}{200 \times 10^3 \times 6^4} = 0.49 \text{ rad}$$

$$= 0.49 \times \frac{180}{\pi} = 28° \textbf{ Ans.}$$

23.20 Flat Spiral Spring

A flat spring is a long thin strip of elastic material wound like a spiral as shown in Fig. 23.24. These springs are frequently used in watches and gramophones etc.

When the outer or inner end of this type of spring is wound up in such a way that there is a tendency in the increase of number of spirals of the spring, the strain energy is stored into its spirals. This energy is utilised in any useful way while the spirals open out slowly. Usually the inner end of spring is clamped to an arbor while the outer end may be pinned or clamped. Since the radius of curvature of every spiral decreases when the spring is wound up, therefore the material of the spring is in a state of pure bending.

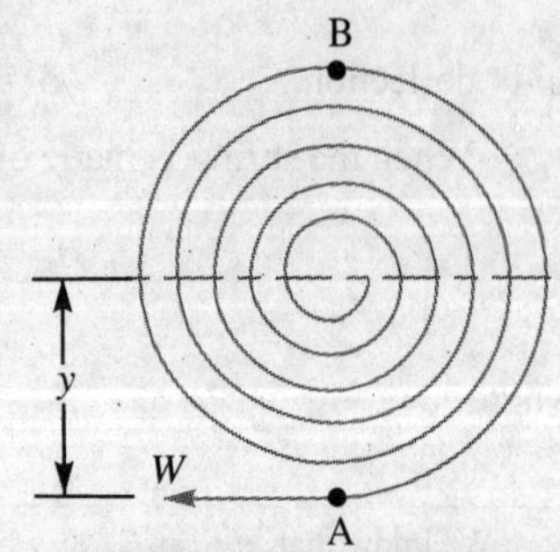

Fig. 23.24. Flat spiral spring.

Let W = Force applied at the outer end A of the spring,

y = Distance of centre of gravity of the spring from A,

l = Length of strip forming the spring,

b = Width of strip,

t = Thickness of strip,

I = Moment of inertia of the spring section = $b.t^3/12$, and

Z = Section modulus of the spring section = $b.t^2/6$

$$\ldots\left(\because Z = \frac{I}{y} = \frac{b.t^3}{12 \times t/2} = \frac{b.t^2}{6}\right)$$

When the end A of the spring is pulled up by a force W, then the bending moment on the spring, at a distance y from the line of action of W is given by

$$M = W \times y$$

The greatest bending moment occurs in the spring at B which is at a maximum distance from the application of W.

Flat spiral spring of a mechanical clock.

∴ Bending moment at B,

$$M_B = M_{max} = W \times 2y = 2W.y = 2M$$

∴ Maximum bending stress induced in the spring material,

$$\sigma_b = \frac{M_{max}}{Z} = \frac{2W \times y}{b.t^2/6} = \frac{12W.y}{b.t^2} = \frac{12M}{b.t^2}$$

Assuming that both ends of the spring are clamped, the angular deflection (in radians) of the spring is given by

$$\theta = \frac{M.l}{E.I} = \frac{12\ M.l}{E.b.t^3} \qquad \ldots\left(\because I = \frac{b.t^3}{12}\right)$$

and the deflection,

$$\delta = \theta \times y = \frac{M.l.y}{E.I}$$

$$= \frac{12\ M.l.y}{E.b.t^3} = \frac{12W.y^2.l}{E.b.t^3} = \frac{\sigma_b.y.l}{E.t} \qquad \ldots\left(\because \sigma_b = \frac{12W.y}{b.t^2}\right)$$

The strain energy stored in the spring

$$= \frac{1}{2}M.\theta = \frac{1}{2}M \times \frac{M.l}{E.I} = \frac{1}{2} \times \frac{M^2 l}{E.I}$$

$$= \frac{1}{2} \times \frac{W^2.y^2.l}{E \times bt^3/12} = \frac{6\ W^2.y^2.l}{E.b.t^3}$$

$$= \frac{6\ W^2.y^2.l}{E.b.t^3} \times \frac{24bt}{24bt} = \frac{144\ W^2 y^2}{Eb^2 t^4} \times \frac{btl}{24}$$

... (Multiplying the numerator and denominator by 24 *bt*)

$$= \frac{(\sigma_b)^2}{24\ E} \times btl = \frac{(\sigma_b)^2}{24\ E} \times \text{Volume of the spring}$$

Example 23.22. *A spiral spring is made of a flat strip 6 mm wide and 0.25 mm thick. The length of the strip is 2.5 metres. Assuming the maximum stress of 800 MPa to occur at the point of greatest bending moment, calculate the bending moment, the number of turns to wind up the spring and the strain energy stored in the spring. Take E = 200 kN/mm².*

Solution. Given : $b = 6$ mm ; $t = 0.25$ mm ; $l = 2.5$ m = 2500 mm ; τ = 800 MPa = 800 N/mm² ; E = 200 kN/mm² = 200 × 10³ N/mm²

Bending moment in the spring

Let M = Bending moment in the spring.

We know that the maximum bending stress in the spring material (σ_b),

$$800 = \frac{12\,M}{b.t^2} = \frac{12\,M}{6\,(0.25)^2} = 32\,M$$

$$\therefore \quad M = 800 / 32 = 25 \text{ N-mm } \textbf{Ans.}$$

Number of turns to wind up the spring

We know that the angular deflection of the spring,

$$\theta = \frac{12\,M.l}{E.b.t^3} = \frac{12 \times 25 \times 2500}{200 \times 10^3 \times 6\,(0.25)^3} = 40 \text{ rad}$$

Since one turn of the spring is equal to 2π radians, therefore number of turns to wind up the spring

$$= 40 / 2\pi = 6.36 \text{ turns } \textbf{Ans.}$$

Strain energy stored in the spring

We know that strain energy stored in the spring

$$= \frac{1}{2}\,M.\theta = \frac{1}{2} \times 24 \times 40 = 480 \text{ N-mm } \textbf{Ans.}$$

23.21 Leaf Springs

Leaf springs (also known as **flat springs**) are made out of flat plates. The advantage of leaf spring over helical spring is that the ends of the spring may be guided along a definite path as it deflects to act as a structural member in addition to energy absorbing device. Thus the leaf springs may carry lateral loads, brake torque, driving torque etc., in addition to shocks.

Consider a single plate fixed at one end and loaded at the other end as shown in Fig. 23.25. This plate may be used as a flat spring.

Let t = Thickness of plate,

b = Width of plate, and

L = Length of plate or distance of the load W from the cantilever end.

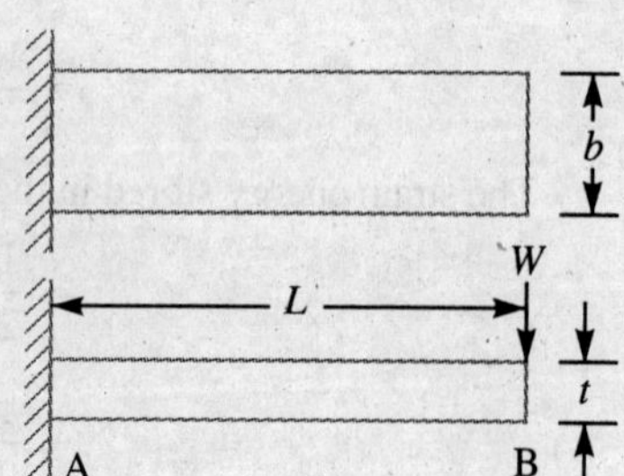

Fig. 23.25. Flat spring (cantilever type).

We know that the maximum bending moment at the cantilever end A,

$$M = W.L$$

and section modulus,

$$Z = \frac{I}{y} = \frac{b\,t^3/12}{t/2} = \frac{1}{6} \times b.t^2$$

∴ Bending stress in such a spring,

$$\sigma = \frac{M}{Z} = \frac{W.L}{\frac{1}{6} \times b.t^2} = \frac{6\,W.L}{b.t^2} \qquad ...(i)$$

We know that the maximum deflection for a cantilever with concentrated load at the free end is given by

$$\delta = \frac{W.L^3}{3E.I} = \frac{W.L^3}{3E \times b.t^3/12} = \frac{4\,W.L^3}{E.b.t^3} \qquad ...(ii)$$

$$= \frac{2\,\sigma.L^2}{3\,E.t} \qquad ...\left(\because \sigma = \frac{6W.L}{b.t^2}\right)$$

It may be noted that due to bending moment, top fibres will be in tension and the bottom fibres are in compression, but the shear stress is zero at the extreme fibres and maximum at the centre, as shown in Fig. 23.26. Hence for analysis, both stresses need not to be taken into account simultaneously. We shall consider the bending stress only.

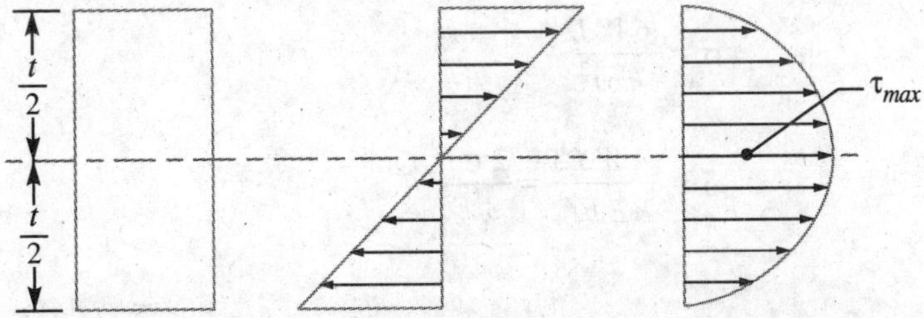

(*a*) Cross-section of plate. (*b*) Bending stress diagram. (*c*) Shear stress diagram.

Fig. 23.26

If the spring is not of cantilever type but it is like a simply supported beam, with length $2L$ and load $2W$ in the centre, as shown in Fig. 23.27, then

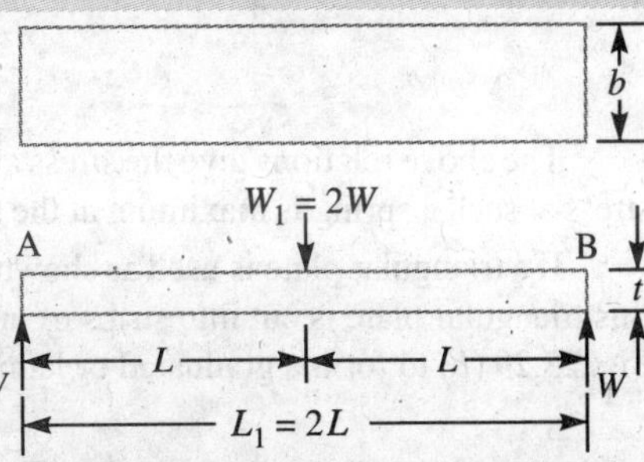

Fig. 23.27. Flat spring (simply supported beam type).

Maximum bending moment in the centre,

$$M = W.L$$

Section modulus, $Z = b.t^2 / 6$

$\therefore$ Bending stress,

$$\sigma = \frac{M}{Z} = \frac{W.L}{b.t^2/6} = \frac{6\,W.L}{b.t^2}$$

We know that maximum deflection of a simply supported beam loaded in the centre is given by

Leaf spring

$$\delta = \frac{W_1\,(L_1)^3}{48\,E.I} = \frac{(2W)\,(2L)^3}{48\,E.I} = \frac{W.L^3}{3\,E.I}$$

...($\because$ In this case, $W_1 = 2W$, and $L_1 = 2L$)

From above we see that a spring such as automobile spring (semi-elliptical spring) with length $2L$ and loaded in the centre by a load $2W$, may be treated as a double cantilever.

If the plate of cantilever is cut into a series of n strips of width b and these are placed as shown in Fig. 23.28, then equations (*i*) and (*ii*) may be written as

$$\sigma = \frac{6\,W.L}{n.b.t^2} \qquad ...(iii)$$

and
$$\delta = \frac{4\,W.L^3}{n.E.b.t^3} = \frac{2\,\sigma.L^2}{3\,E.t} \qquad ...(iv)$$

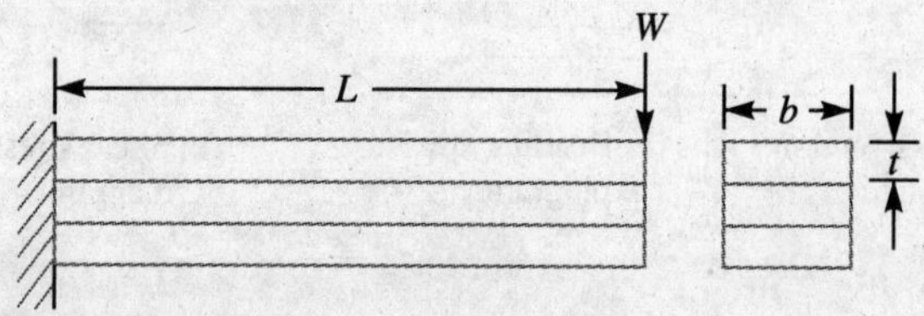

Fig. 23.28

The above relations give the stress and deflection of a leaf spring of uniform cross-section. The stress at such a spring is maximum at the support.

If a triangular plate is used as shown in Fig. 23.29 (*a*), the stress will be uniform throughout. If this triangular plate is cut into strips of uniform width and placed one below the other, as shown in Fig. 23.29 (*b*) to form a graduated or laminated leaf spring, then

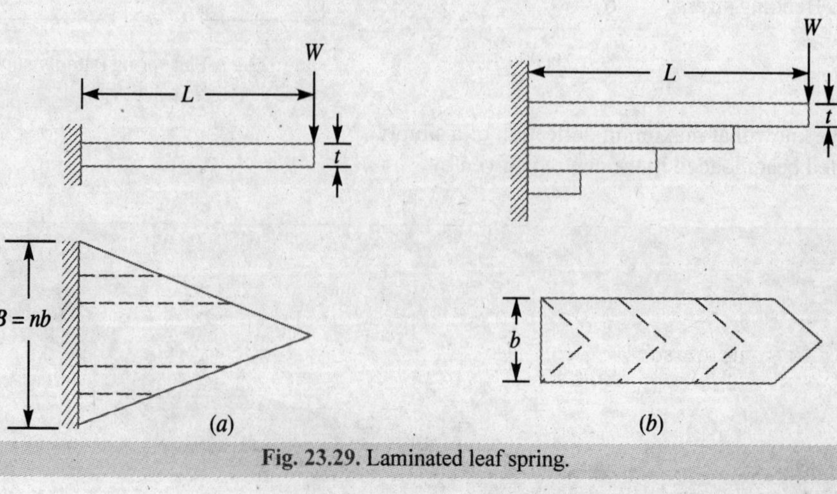

Fig. 23.29. Laminated leaf spring.

$$\sigma = \frac{6\,W.L}{n.b.t^2} \qquad ...(v)$$

and
$$\delta = \frac{6\,W.L^3}{n.E.b.t^3} = \frac{\sigma.L^2}{E.t} \qquad ...(vi)$$

where n = Number of graduated leaves.

A little consideration will show that by the above arrangement, the spring becomes compact so that the space occupied by the spring is considerably reduced.

When bending stress alone is considered, the graduated leaves may have zero width at the loaded end. But sufficient metal must be provided to support the shear. Therefore, it becomes necessary to have one or more leaves of uniform cross-section extending clear to the end. We see from equations (*iv*) and (*vi*) that for the same deflection, the stress in the uniform cross-section leaves (*i.e.* full length leaves) is 50% greater than in the graduated leaves, assuming that each spring element deflects according to its own elastic curve. If the suffixes $_F$ and $_G$ are used to indicate the full length (or uniform cross-section) and graduated leaves, then

$$\sigma_F = \frac{3}{2}\sigma_G$$

$$\frac{6W_F.L}{n_F.b.t^2} = \frac{3}{2}\left[\frac{6\,W_G.L}{n_G.b.t^2}\right] \quad \text{or} \quad \frac{W_F}{n_F} = \frac{3}{2}\times\frac{W_G}{n_G}$$

$$\therefore \quad \frac{W_F}{W_G} = \frac{3\,n_F}{2\,n_G} \qquad ...(vii)$$

Adding 1 to both sides, we have

$$\frac{W_F}{W_G} + 1 = \frac{3\,n_F}{2\,n_G} + 1 \quad \text{or} \quad \frac{W_F + W_G}{W_G} = \frac{3\,n_F + 2\,n_G}{2\,n_G}$$

$$\therefore \quad W_G = \left(\frac{2\,n_G}{3\,n_F + 2\,n_G}\right)(W_F + W_G) = \left(\frac{2\,n_G}{3\,n_F + 2\,n_G}\right)W \qquad ...(viii)$$

where W = Total load on the spring = $W_G + W_F$

W_G = Load taken up by graduated leaves, and

W_F = Load taken up by full length leaves.

From equation (*vii*), we may write

$$\frac{W_G}{W_F} = \frac{2\,n_G}{3\,n_F}$$

or

$$\frac{W_G}{W_F} + 1 = \frac{2\,n_G}{3\,n_F} + 1 \qquad \text{... (Adding 1 to both sides)}$$

$$\frac{W_G + W_F}{W_F} = \frac{2\,n_G + 3\,n_F}{3\,n_F}$$

$$\therefore \quad W_F = \left(\frac{3\,n_F}{2\,n_G + 3\,n_F}\right)(W_G + W_F) = \left(\frac{3\,n_F}{2\,n_G + 3\,n_F}\right)W \qquad ...(ix)$$

∴ Bending stress for full length leaves,

$$\sigma_F = \frac{6\,W_F.L}{n_F.b\,t^2} = \frac{6\,L}{n_F.b.t^2}\left(\frac{3\,n_F}{2\,n_G + 3\,n_F}\right)W = \frac{18\,W.L}{b.t^2\,(2\,n_G + 3\,n_F)}$$

Since $\sigma_F = \frac{3}{2}\sigma_G$, therefore

$$\sigma_G = \frac{2}{3}\sigma_F = \frac{2}{3}\times\frac{18\,W.L}{b.t^2\,(2\,n_G + 3\,n_F)} = \frac{12\,W.L}{b.t^2\,(2\,n_G + 3\,n_F)}$$

The deflection in full length and graduated leaves is given by equation (*iv*), *i.e.*

$$\delta = \frac{2\,\sigma_F \times L^2}{3\,E.t} = \frac{2\,L^2}{3\,E.t}\left[\frac{18\,W.L}{b.t^2\,(2\,n_G + 3\,n_F)}\right] = \frac{12\,W.L^3}{E.b.t^3\,(2\,n_G + 3\,n_F)}$$

23.22 Construction of Leaf Spring

A leaf spring commonly used in automobiles is of semi-elliptical form as shown in Fig. 23.30.

It is built up of a number of plates (known as leaves). The leaves are usually given an initial curvature or cambered so that they will tend to straighten under the load. The leaves are held together by means of a band shrunk around them at the centre or by a bolt passing through the centre. Since the band exerts a stiffening and strengthening effect, therefore the effective length of the spring for bending will be overall length of the spring *minus* width of band. In case of a centre bolt, two-third distance between centres of *U*-bolt should be subtracted from the overall length of the spring in order to find effective length. The spring is clamped to the axle housing by means of *U*-bolts.

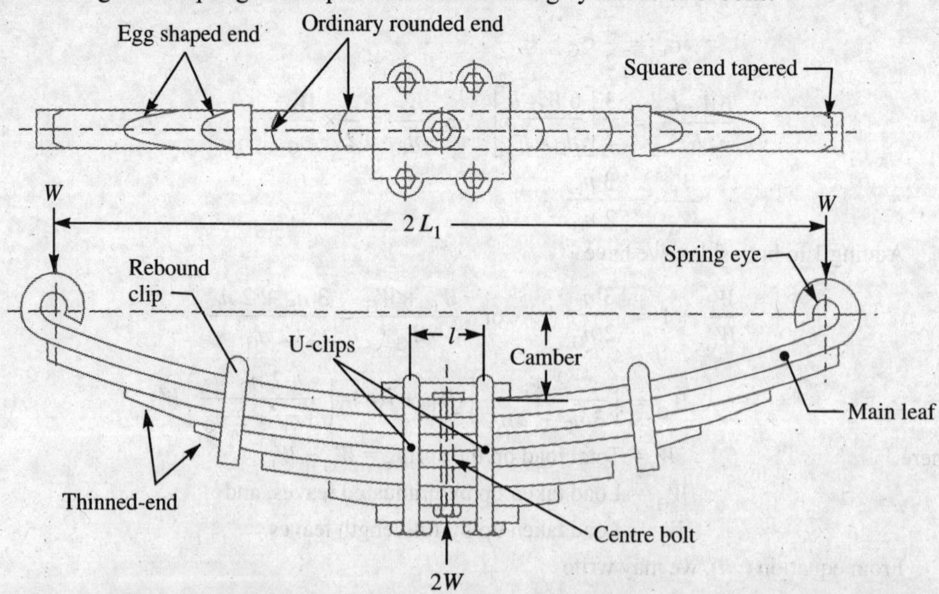

Fig. 23.30. Semi-elliptical leaf spring.

The longest leaf known as ***main leaf*** or ***master leaf*** has its ends formed in the shape of an eye through which the bolts are passed to secure the spring to its supports. Usually the eyes, through which the spring is attached to the hanger or shackle, are provided with bushings of some antifriction material such as bronze or rubber. The other leaves of the spring are known as ***graduated leaves.*** In order to prevent digging in the adjacent leaves, the ends of the graduated leaves are trimmed in various forms as shown in Fig. 23.30. Since the master leaf has to with stand vertical bending loads as well as loads due to sideways of the vehicle and twisting, therefore due to the presence of stresses caused by these loads, it is usual to provide two full length leaves and the rest graduated leaves as shown in Fig. 23.30.

Leaf spring fatigue testing system.

Rebound clips are located at intermediate positions in the length of the spring, so that the graduated leaves also share the stresses induced in the full length leaves when the spring rebounds.

23.23 Equalised Stress in Spring Leaves (Nipping)

We have already discussed that the stress in the full length leaves is 50% greater than the stress in the graduated leaves. In order to utilise the material to the best advantage, all the leaves should be equally stressed. This condition may be obtained in the following two ways :

1. By making the full length leaves of smaller thickness than the graduated leaves. In this way, the full length leaves will induce smaller bending stress due to small distance from the neutral axis to the edge of the leaf.

2. By giving a greater radius of curvature to the full length leaves than graduated leaves, as shown in Fig. 23.31, before the leaves are assembled to form a spring. By doing so, a gap or clearance will be left between the leaves. This initial gap, as shown by *C* in Fig. 23.31, is called ***nip***. When the central bolt, holding the various leaves together, is tightened, the full length leaf will bend back as shown dotted in Fig. 23.31 and have an initial stress in a direction opposite to that of the normal load. The graduated

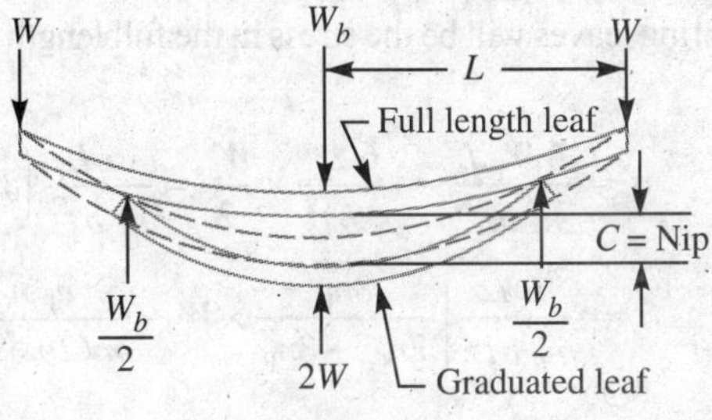

Fig. 23.31

leaves will have an initial stress in the same direction as that of the normal load. When the load is gradually applied to the spring, the full length leaf is first relieved of this initial stress and then stressed in opposite direction. Consequently, the full length leaf will be stressed less than the graduated leaf. The initial gap between the leaves may be adjusted so that under maximum load condition the stress in all the leaves is equal, or if desired, the full length leaves may have the lower stress. This is desirable in automobile springs in which full length leaves are designed for lower stress because the full length leaves carry additional loads caused by the swaying of the car, twisting and in some cases due to driving the car through the rear springs. Let us now find the value of initial gap or nip *C*.

Consider that under maximum load conditions, the stress in all the leaves is equal. Then at maximum load, the total deflection of the graduated leaves will exceed the deflection of the full length leaves by an amount equal to the initial gap *C*. In other words,

$$\delta_G = \delta_F + C$$

$$\therefore \quad C = \delta_G - \delta_F = \frac{6\,W_G \cdot L^3}{n_G\,E.b.t^3} - \frac{4\,W_F.L^3}{n_F.E.b.t^3} \quad \ldots(i)$$

Since the stresses are equal, therefore

$$\sigma_G = \sigma_F$$

$$\frac{6\,W_G.L}{n_G.b.t^2} = \frac{6\,W_F.L}{n_F.b.t^2} \quad \text{or} \quad \frac{W_G}{n_G} = \frac{W_F}{n_F}$$

$$\therefore \quad W_G = \frac{n_G}{n_F} \times W_F = \frac{n_G}{n} \times W$$

$$\text{and} \quad W_F = \frac{n_F}{n_G} \times W_G = \frac{n_F}{n} \times W$$

Substituting the values of W_G and W_F in equation (*i*), we have

$$C = \frac{6W.L^3}{n.E.b.t^3} - \frac{4W.L^3}{n.E.b.t^3} = \frac{2W.L^3}{n.E.b.t^3} \quad \ldots(ii)$$

The load on the clip bolts (W_b) required to close the gap is determined by the fact that the gap is equal to the initial deflections of full length and graduated leaves.

$$\therefore \quad C = \delta_F + \delta_G$$

$$\frac{2W.L^3}{n.E.b.t^3} = \frac{4L^3}{n_F.E.b.t^3} \times \frac{W_b}{2} + \frac{6\,L^3}{n_G.E.b.t^3} \times \frac{W_b}{2}$$

or
$$\frac{W}{n} = \frac{W_b}{n_F} + \frac{3\,W_b}{2\,n_G} = \frac{2\,n_G.W_b + 3\,n_F.W_b}{2\,n_F.n_G} = \frac{W_b\,(2\,n_G + 3\,n_F)}{2\,n_F.n_G}$$

$$\therefore \quad W_b = \frac{2\,n_F.n_G.W}{n\,(2\,n_G + 3\,n_F)} \qquad ...(iii)$$

The final stress in spring leaves will be the stress in the full length leaves due to the applied load ***minus*** the initial stress.

$\therefore$ Final stress,
$$\sigma = \frac{6\,W_F.L}{n_F.b.t^2} - \frac{6\,L}{n_F.b.t^2} \times \frac{W_b}{2} = \frac{6\,L}{n_F.b.t^2}\left(W_F - \frac{W_b}{2}\right)$$

$$= \frac{6\,L}{n_F.b.t^2}\left[\frac{3n_F}{2n_G + 3n_F} \times W - \frac{n_F.n_G.W}{n\,(2n_G + 3n_F)}\right]$$

$$= \frac{6W.L}{b.t^2}\left[\frac{3}{2n_G + 3n_F} - \frac{n_G}{n(2n_G + 3n_F)}\right]$$

$$= \frac{6\,W.L}{b.t^2}\left[\frac{3n - n_G}{n(2n_G + 3n_F)}\right]$$

$$= \frac{6\,W.L}{b.t^2}\left[\frac{3(n_F + n_G) - n_G}{n(2n_G + 3n_F)}\right] = \frac{6\,W.L}{n.b.t^2} \qquad ...(iv)$$

... (Substituting $n = n_F + n_G$)

Notes : 1. The final stress in the leaves is also equal to the stress in graduated leaves due to the applied load *plus* the initial stress.

2. The deflection in the spring due to the applied load is same as without initial stress.

23.24 Length of Leaf Spring Leaves

The length of the leaf spring leaves may be obtained as discussed below :

Let $2L_1$ = Length of span or overall length of the spring,

l = Width of band or distance between centres of *U*-bolts. It is the ineffective length of the spring,

n_F = Number of full length leaves,

n_G = Number of graduated leaves, and

n = Total number of leaves = $n_F + n_G$.

We have already discussed that the effective length of the spring,

$$2L = 2L_1 - l \qquad ...(\text{When band is used})$$

$$= 2L_1 - \frac{2}{3}l \qquad ...(\text{When } U\text{-bolts are used})$$

It may be noted that when there is only one full length leaf (*i.e.* master leaf only), then the number of leaves to be cut will be n and when there are two full length leaves (including one master leaf), then the number of leaves to be cut will be $(n - 1)$. If a leaf spring has two full length leaves, then the length of leaves is obtained as follows :

$$\text{Length of smallest leaf} = \frac{\text{Effective length}}{n-1} + \text{Ineffective length}$$

$$\text{Length of next leaf} = \frac{\text{Effective length}}{n-1} \times 2 + \text{Ineffective length}$$

Similarly, length of $(n-1)$th leaf

$$= \frac{\text{Effective length}}{n-1} \times (n-1) + \text{Ineffective length}$$

The nth leaf will be the master leaf and it is of full length. Since the master leaf has eyes on both sides, therefore

$$\text{Length of master leaf} = 2\,L_1 + \pi\,(d+t) \times 2$$

where d = Inside diameter of eye, and

t = Thickness of master leaf.

The approximate relation between the radius of curvature (R) and the camber (y) of the spring is given by

$$R = \frac{(L_1)^2}{2y}$$

The exact relation is given by

$$y\,(2R + y) = (L_1)^2$$

where L_1 = Half span of the spring.

Note : The maximum deflection (δ) of the spring is equal to camber (y) of the spring.

23.25 Standard Sizes of Automobile Suspension Springs

Following are the standard sizes for the automobile suspension springs:

1. Standard nominal widths are : 32, 40*, 45, 50*, 55, 60*, 65, 70*, 75, 80, 90, 100 and 125 mm. (Dimensions marked* are the preferred widths)
2. Standard nominal thicknesses are : 3.2, 4.5, 5, 6, 6.5, 7, 7.5, 8, 9, 10, 11, 12, 14 and 16 mm.
3. At the eye, the following bore diameters are recommended : 19, 20, 22, 23, 25, 27, 28, 30, 32, 35, 38, 50 and 55 mm.
4. Dimensions for the centre bolts, if employed, shall be as given in the following table.

Table 23.5. Dimensions for centre bolts.

Width of leaves in mm	*Dia. of centre bolt in mm*	*Dia. of head in mm*	*Length of bolt head in mm*
Upto and including 65	8 or 10	12 or 15	10 or 11
Above 65	12 or 16	17 or 20	11

5. Minimum clip sections and the corresponding sizes of rivets and bolts used with the clips shall be as given in the following table (See Fig. 23.32).

Table 23.6. Dimensions of clip, rivet and bolts.

Spring width (B) in mm	*Clip section ($b \times t$) in mm × mm*	*Dia. of rivet (d_1) in mm*	*Dia. of bolt (d_2) in mm*
Under 50	20 × 4	6	6
50, 55 and 60	25 × 5	8	8
65, 70, 75 and 80	25 × 6	10	8
90, 100 and 125	32 × 6	10	10

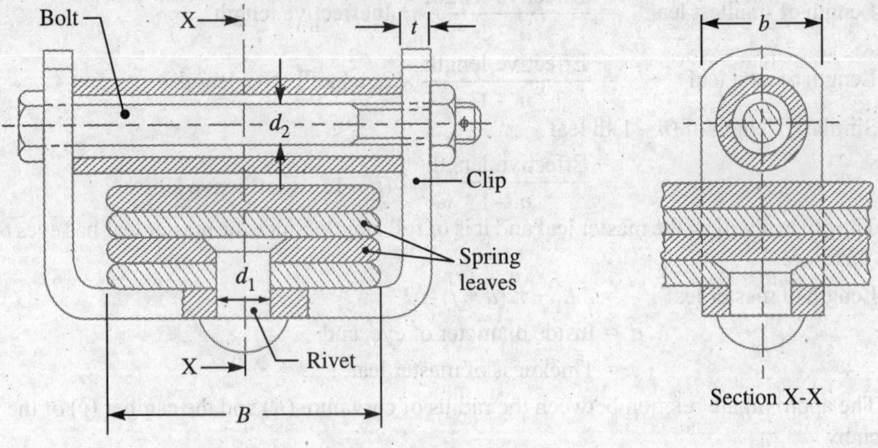

Fig. 23.32. Spring clip.

Notes : **1.** For springs of width below 65 mm, one rivet of 6, 8 or 10 mm may be used. For springs of width above 65 mm, two rivets of 6 or 8 mm or one rivet of 10 mm may be used.

2. For further details, the following Indian Standards may be referred :

(*a*) IS : 9484 – 1980 (Reaffirmed 1990) on 'Specification for centre bolts for leaf springs'.

(*b*) IS : 9574 – 1989 (Reaffirmed 1994) on 'Leaf springs assembly-Clips-Specification'.

23.26 Materials for Leaf Springs

The material used for leaf springs is usually a plain carbon steel having 0.90 to 1.0% carbon. The leaves are heat treated after the forming process. The heat treatment of spring steel produces greater strength and therefore greater load capacity, greater range of deflection and better fatigue properties.

According to Indian standards, the recommended materials are :

1. For automobiles : 50 Cr 1, 50 Cr 1 V 23, and 55 Si 2 Mn 90 all used in hardened and tempered state.
2. For rail road springs : C 55 (water-hardened), C 75 (oil-hardened), 40 Si 2 Mn 90 (water-hardened) and 55 Si 2 Mn 90 (oil-hardened).
3. The physical properties of some of these materials are given in the following table. All values are for oil quenched condition and for single heat only.

Table 23.7. Physical properties of materials commonly used for leaf springs.

Material	*Condition*	*Ultimate tensile strength (MPa)*	*Tensile yield strength (MPa)*	*Brinell hardness number*
50 Cr 1	Hardened	1680 – 2200	1540 – 1750	461 – 601
50 Cr 1 V 23	and	1900 – 2200	1680 – 1890	534 – 601
55 Si 2 Mn 90	tempered	1820 – 2060	1680 – 1920	534 – 601

Note : For further details, Indian Standard [IS : 3431 – 1982 (Reaffirmed 1992)] on 'Specification for steel for the manufacture of volute, helical and laminated springs for automotive suspension' may be referred.

Example 23.23. *Design a leaf spring for the following specifications :*

Total load = 140 kN ; Number of springs supporting the load = 4 ; Maximum number of leaves = 10; Span of the spring = 1000 mm ; Permissible deflection = 80 mm.

Take Young's modulus, E = 200 kN/mm² and allowable stress in spring material as 600 MPa.

Solution. Given : Total load = 140 kN ; No. of springs = 4; n = 10 ; $2L$ = 1000 mm or L = 500 mm ; δ = 80 mm ; E = 200 kN/mm^2 = 200 × 10^3 N/mm^2 ; σ = 600 MPa = 600 N/mm^2

We know that load on each spring,

$$2W = \frac{\text{Total load}}{\text{No. of springs}} = \frac{140}{4} = 35 \text{ kN}$$

$$\therefore \quad W = 35 / 2 = 17.5 \text{ kN} = 17\ 500 \text{ N}$$

Let t = Thickness of the leaves, and

b = Width of the leaves.

We know that bending stress (σ),

$$600 = \frac{6\,W.L}{n.b.t^2} = \frac{6 \times 17\ 500 \times 500}{n.b.t^2} = \frac{52.5 \times 10^6}{n.b.t^2}$$

$$\therefore \quad n.b.t^2 = 52.5 \times 10^6 / 600 = 87.5 \times 10^3 \qquad \text{...}(i)$$

and deflection of the spring (δ),

$$80 = \frac{6\,W.L^3}{n.E.b.t^3} = \frac{6 \times 17\ 500\ (500)^3}{n \times 200 \times 10^3 \times b \times t^3} = \frac{65.6 \times 10^6}{n.b.t^3}$$

$$\therefore \quad n.b.t^3 = 65.6 \times 10^6 / 80 = 0.82 \times 10^6 \qquad \text{...}(ii)$$

Dividing equation (*ii*) by equation (*i*), we have

$$\frac{n.b.t^3}{n.b.t^2} = \frac{0.82 \times 10^6}{87.5 \times 10^3} \quad \text{or} \quad t = 9.37 \text{ say } 10 \text{ mm } \textbf{Ans.}$$

Now from equation (*i*), we have

$$b = \frac{87.5 \times 10^3}{n.t^2} = \frac{87.5 \times 10^3}{10\ (10)^2} = 87.5 \text{ mm}$$

and from equation (*ii*), we have

$$b = \frac{0.82 \times 10^6}{n.t^3} = \frac{0.82 \times 10^6}{10\ (10)^3} = 82 \text{ mm}$$

Taking larger of the two values, we have width of leaves,

$$b = 87.5 \text{ say } 90 \text{ mm } \textbf{Ans.}$$

Example 23.24. *A truck spring has 12 number of leaves, two of which are full length leaves. The spring supports are 1.05 m apart and the central band is 85 mm wide. The central load is to be 5.4 kN with a permissible stress of 280 MPa. Determine the thickness and width of the steel spring leaves. The ratio of the total depth to the width of the spring is 3. Also determine the deflection of the spring.*

Solution. Given : n = 12 ; n_F = 2 ; $2L_1$ = 1.05 m = 1050 mm ; l = 85 mm ; $2W$ = 5.4 kN = 5400 N or W = 2700 N ; σ_F = 280 MPa = 280 N/mm^2

Thickness and width of the spring leaves

Let t = Thickness of the leaves, and

b = Width of the leaves.

Since it is given that the ratio of the total depth of the spring ($n \times t$) and width of the spring (b) is 3, therefore

$$\frac{n \times t}{b} = 3 \quad \text{or} \quad b = n \times t / 3 = 12 \times t / 3 = 4\,t$$

We know that the effective length of the spring,

$$2L = 2L_1 - l = 1050 - 85 = 965 \text{ mm}$$

$$\therefore \quad L = 965 / 2 = 482.5 \text{ mm}$$

and number of graduated leaves,

$$n_G = n - n_F = 12 - 2 = 10$$

Assuming that the leaves are not initially stressed, therefore maximum stress or bending stress for full length leaves (σ_F),

$$280 = \frac{18\ W.L}{b.t^2\ (2n_G + 3n_F)} = \frac{18 \times 2700 \times 482.5}{4\ t \times t^2\ (2 \times 10 + 3 \times 2)} = \frac{225\ 476}{t^3}$$

$\therefore$ $t^3 = 225\ 476 / 280 = 805.3$ or $t = 9.3$ say 10 mm **Ans.**

and $b = 4\ t = 4 \times 10 = 40$ mm **Ans.**

Deflection of the spring

We know that deflection of the spring,

$$\delta = \frac{12\ W.L^3}{E.b.t^3\ (2n_G + 3n_F)}$$

$$= \frac{12 \times 2700 \times (482.5)^3}{210 \times 10^3 \times 40 \times 10^3\ (2 \times 10 + 3 \times 2)} \text{ mm}$$

$= 16.7$ mm **Ans.** ... (Taking $E = 210 \times 10^3$ N/mm²)

Example 23.25. *A locomotive semi-elliptical laminated spring has an overall length of 1 m and sustains a load of 70 kN at its centre. The spring has 3 full length leaves and 15 graduated leaves with a central band of 100 mm width. All the leaves are to be stressed to 400 MPa, when fully loaded. The ratio of the total spring depth to that of width is 2. E = 210 kN/mm². Determine :*

1. The thickness and width of the leaves.

2. The initial gap that should be provided between the full length and graduated leaves before the band load is applied.

3. The load exerted on the band after the spring is assembled.

Solution. Given : $2L_1 = 1$ m $= 1000$ mm ; $2W = 70$ kN or $W = 35$ kN $= 35 \times 10^3$ N ; $n_F = 3$; $n_G = 15$; $l = 100$ mm ; $\sigma = 400$ MPa $= 400$ N/mm² ; $E = 210$ kN/mm² $= 210 \times 10^3$ N/mm²

1. ***Thickness and width of leaves***

Let t = Thickness of leaves, and

b = Width of leaves.

We know that the total number of leaves,

$$n = n_F + n_G = 3 + 15 = 18$$

Since it is given that ratio of the total spring depth ($n \times t$) and width of leaves is 2, therefore

$$\frac{n \times t}{b} = 2 \text{ or } b = n \times t / 2 = 18 \times t / 2 = 9\ t$$

We know that the effective length of the leaves,

$$2L = 2L_1 - l = 1000 - 100 = 900 \text{ mm or } L = 900 / 2 = 450 \text{ mm}$$

Since all the leaves are equally stressed, therefore final stress (σ),

$$400 = \frac{6\ W.L}{n.b.t^2} = \frac{6 \times 35 \times 10^3 \times 450}{18 \times 9\ t \times t^2} = \frac{583 \times 10^3}{t^3}$$

$\therefore$ $t^3 = 583 \times 10^3 / 400 = 1458$ or $t = 11.34$ say 12 mm **Ans.**

and $b = 9\ t = 9 \times 12 = 108$ mm **Ans.**

2. ***Initial gap***

We know that the initial gap (C) that should be provided between the full length and graduated leaves before the band load is applied, is given by

$$C = \frac{2\,W.L^3}{n.E.b.t^3} = \frac{2 \times 35 \times 10^3\,(450)^3}{18 \times 210 \times 10^3 \times 108\,(12)^3} = 9.04 \text{ mm } \textbf{Ans.}$$

3. ***Load exerted on the band after the spring is assembled***

We know that the load exerted on the band after the spring is assembled,

$$W_b = \frac{2\,n_F.n_G.W}{n(2n_G + 3n_F)} = \frac{2 \times 3 \times 15 \times 35 \times 10^3}{18\,(2 \times 15 + 3 \times 3)} = 4487 \text{ N } \textbf{Ans.}$$

Example 23.26. *A semi-elliptical laminated vehicle spring to carry a load of 6000 N is to consist of seven leaves 65 mm wide, two of the leaves extending the full length of the spring. The spring is to be 1.1 m in length and attached to the axle by two U-bolts 80 mm apart. The bolts hold the central portion of the spring so rigidly that they may be considered equivalent to a band having a width equal to the distance between the bolts. Assume a design stress for spring material as 350 MPa. Determine :*

1. Thickness of leaves, 2. Deflection of spring, 3. Diameter of eye, 4. Length of leaves, and 5. Radius to which leaves should be initially bent.

Sketch the semi-elliptical leaf-spring arrangement.

The standard thickness of leaves are : 5, 6, 6.5, 7, 7.5, 8, 9, 10, 11 etc. in mm.

Solution. Given : $2W = 6000$ N or $W = 3000$ N ; $n = 7$; $b = 65$ mm ; $n_F = 2$; $2L_1 = 1.1$ m $= 1100$ mm or $L_1 = 550$ mm ; $l = 80$ mm ; $\sigma = 350$ MPa $= 350$ N/mm^2

1. ***Thickness of leaves***

Let t = Thickness of leaves.

We know that the effective length of the spring,

$$2L = 2L_1 - l = 1100 - 80 = 1020 \text{ mm}$$

$$\therefore \quad L = 1020 / 2 = 510 \text{ mm}$$

and number of graduated leaves,

$$n_G = n - n_F = 7 - 2 = 5$$

Assuming that the leaves are not initially stressed, the maximum stress (σ_F),

$$350 = \frac{18\,W.L}{b.t^2\,(2n_G + 3n_F)} = \frac{18 \times 3000 \times 510}{65 \times t^2\,(2 \times 5 + 3 \times 2)} = \frac{26\,480}{t^2} \quad \text{... } (\sigma_F = \sigma)$$

$$\therefore \quad t^2 = 26\,480 / 350 = 75.66 \text{ or } t = 8.7 \text{ say } 9 \text{ mm } \textbf{Ans.}$$

2. ***Deflection of spring***

We know that deflection of spring,

$$\delta = \frac{12\,W.L^3}{E.b.t^3\,(2n_G + 3n_F)} = \frac{12 \times 3000\,(510)^3}{210 \times 10^3 \times 65 \times 9^3\,(2 \times 5 + 3 \times 2)}$$

$$= 30 \text{ mm } \textbf{Ans.} \quad \text{... (Taking } E = 210 \times 10^3 \text{ N/mm}^2\text{)}$$

3. ***Diameter of eye***

The inner diameter of eye is obtained by considering the pin in the eye in bearing, because the inner diameter of the eye is equal to the diameter of the pin.

Let d = Inner diameter of the eye or diameter of the pin,

l_1 = Length of the pin which is equal to the width of the eye or leaf (*i.e.* b) = 65 mm ...(Given)

p_b = Bearing pressure on the pin which may be taken as 8 N/mm^2.

We know that the load on pin (W),

$$3000 = d \times l_1 \times p_b$$
$$= d \times 65 \times 8 = 520\, d$$
$$\therefore \quad d = 3000 / 520$$
$$= 5.77 \text{ say } 6 \text{ mm}$$

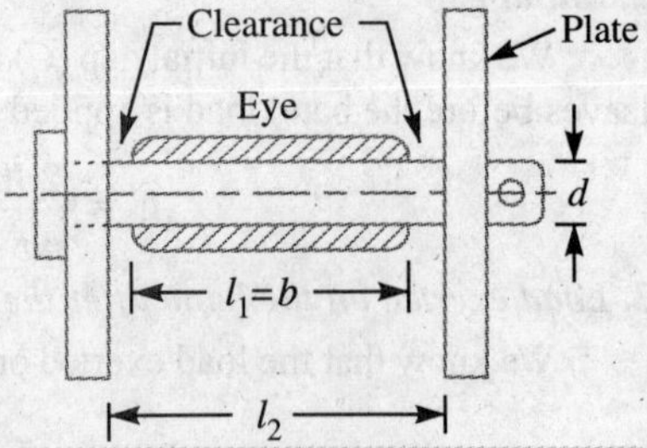

Fig. 23.33

Let us now consider the bending of the pin. Since there is a clearance of about 2 mm between the shackle (or plate) and eye as shown in Fig. 23.33, therefore length of the pin under bending,

$$l_2 = l_1 + 2 \times 2 = 65 + 4 = 69 \text{ mm}$$

Leaf spring.

Maximum bending moment on the pin,

$$M = \frac{W \times l_2}{4} = \frac{3000 \times 69}{4} = 51\,750 \text{ N-mm}$$

and section modulus,
$$Z = \frac{\pi}{32} \times d^3 = 0.0982\, d^3$$

We know that bending stress (σ_b),

$$80 = \frac{M}{Z} = \frac{51\,750}{0.0982\, d^3} = \frac{527 \times 10^3}{d^3} \quad \text{... (Taking } \sigma_b = 80 \text{ N/mm}^2\text{)}$$

$$\therefore \quad d^3 = 527 \times 10^3 / 80 = 6587 \quad \text{or} \quad d = 18.7 \text{ say } 20 \text{ mm } \textbf{Ans.}$$

We shall take the inner diameter of eye or diameter of pin (d) as 20 mm **Ans.**

Let us now check the pin for induced shear stress. Since the pin is in double shear, therefore load on the pin (W),

$$3000 = 2 \times \frac{\pi}{4} \times d^2 \times \tau = 2 \times \frac{\pi}{4} (20)^2 \ \tau = 628.4\ \tau$$

$$\therefore \quad \tau = 3000 / 628.4 = 4.77 \text{ N/mm}^2\text{, which is safe.}$$

4. *Length of leaves*

We know that ineffective length of the spring

$$= l = 80 \text{ mm} \quad \text{... } (\because U\text{-bolts are considered equivalent to a band})$$

$\therefore$ Length of the smallest leaf $= \dfrac{\text{Effective length}}{n-1} + \text{Ineffective length}$

$$= \frac{1020}{7-1} + 80 = 250 \text{ mm } \textbf{Ans.}$$

Length of the 2nd leaf $= \dfrac{1020}{7-1} \times 2 + 80 = 420$ mm **Ans.**

Length of the 3rd leaf $= \dfrac{1020}{7-1} \times 3 + 80 = 590$ mm **Ans.**

Length of the 4th leaf $= \dfrac{1020}{7-1} \times 4 + 80 = 760$ mm **Ans.**

Length of the 5th leaf $= \dfrac{1020}{7-1} \times 5 + 80 = 930$ mm **Ans.**

Length of the 6th leaf $= \dfrac{1020}{7-1} \times 6 + 80 = 1100$ mm **Ans.**

The 6th and 7th leaves are full length leaves and the 7th leaf (*i.e.* the top leaf) will act as a master leaf.

We know that length of the master leaf

$$= 2L_1 + \pi (d + t) 2 = 1100 + \pi (20 + 9)2 = 1282.2 \text{ mm } \textbf{Ans.}$$

5. *Radius to which the leaves should be initially bent*

Let R = Radius to which the leaves should be initially bent, and

y = Camber of the spring.

We know that

$$y(2R - y) = (L_1)^2$$

$$30(2R - 30) = (550)^2 \text{ or } 2R - 30 = (550)^2/30 = 10\ 083 \qquad \ldots (\because y = \delta)$$

$$\therefore \quad R = \frac{10\ 083 + 30}{2} = 5056.5 \text{ mm } \textbf{Ans.}$$

EXERCISES

1. Design a compression helical spring to carry a load of 500 N with a deflection of 25 mm. The spring index may be taken as 8. Assume the following values for the spring material:

   Permissible shear stress = 350 MPa

   Modulus of rigidity = 84 kN/mm^2

   Wahl's factor $= \dfrac{4C-1}{4C-4} + \dfrac{0.615}{C}$, where C = spring index.

   [Ans. d = 5.893 mm ; D = 47.144 mm ; n = 6]

2. A helical valve spring is to be designed for an operating load range of approximately 90 to 135 N. The deflection of the spring for the load range is 7.5 mm. Assume a spring index of 10. Permissible shear stress for the material of the spring = 480 MPa and its modulus of rigidity = 80 kN/mm^2. Design the spring.

   Take Wahl's factor $= \dfrac{4C-1}{4C-4} + \dfrac{0.615}{C}$, C being the spring index.

   [Ans. d = 2.74 mm ; D = 27.4 mm ; n = 6]

3. Design a helical spring for a spring loaded safety valve for the following conditions :

Operating pressure = 1 N/mm^2

Maximum pressure when the valve blows off freely

= 1.075 N/mm^2

Maximum lift of the valve when the pressure is 1.075 N/mm^2

= 6 mm

Diameter of valve seat = 100 mm

Maximum shear stress = 400 MPa

Modulus of rigidity = 86 kN/mm^2

Spring index = 5.5 **[Ans. d = 17.2 mm ; D = 94.6 mm ; n = 12]**

4. A vertical spring loaded valve is required for a compressed air receiver. The valve is to start opening at a pressure of 1 N/mm^2 gauge and must be fully open with a lift of 4 mm at a pressure of 1.2 N/mm^2 gauge. The diameter of the port is 25 mm. Assume the allowable shear stress in steel as 480 MPa and shear modulus as 80 kN/mm^2.

Design a suitable close coiled round section helical spring having squared ground ends. Also specify initial compression and free length of the spring.

[Ans. d = 7 mm ; D = 42 mm ; n = 13]

5. A spring controlled lever is shown in Fig. 23.34. The spring is to be inserted with an initial compression to produce a force equal to 125 N between the right hand end of the lever and the stop. When the maximum force at A reaches to a value of 200 N, the end of the lever moves downward by 25 mm.

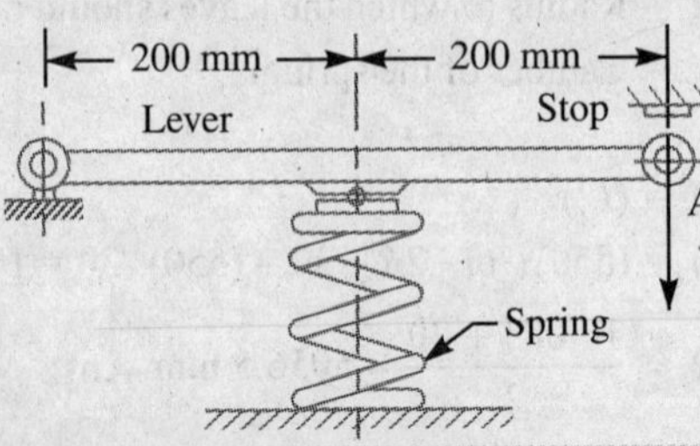

Fig. 23.34

Assuming a spring index as 8, determine: 1. spring rate, 2 size of wire, 3. outside diameter of the spring, 4. number of active coils, and 5. free length, assuming squared and ground ends. The allowable shear stress may be taken as 420 MPa and G = 80 kN/mm^2.

[Ans. 0.33 N/mm ; 3.4 mm ; 27.2 mm; 10 ; 77 mm]

6. It is desired to design a valve spring of I.C. engine for the following details :

(*a*) Spring load when valve is closed = 80 N

(*b*) Spring load when valve is open = 100 N

(*c*) Space constraints for the fitment of spring are :

Inside guide bush diameter = 24 mm

Outside recess diameter = 36 mm

(*d*) Valve lift = 5 mm

(*e*) Spring steel has the following properties:

Maximum permissible shear stress = 350 MPa

Modulus of rigidity = 84 kN/mm^2

Find : 1. Wire diameter; 2. Spring index; 3. Total number of coils; 4. Solid length of spring; 5. Free length of spring; 6. Pitch of the coil when additional 15 percent of the working deflection is used to avoid complete closing of coils.

[Ans. 2.9 mm ; 10.345 ; 9 ; 26.1 mm ; 54.85 mm ; 7 mm]

7. A circular cam 200 mm in diameter rotates off centre with an eccentricity of 25 mm and operates the roller follower that is carried by the arm as shown in Fig. 23.35.

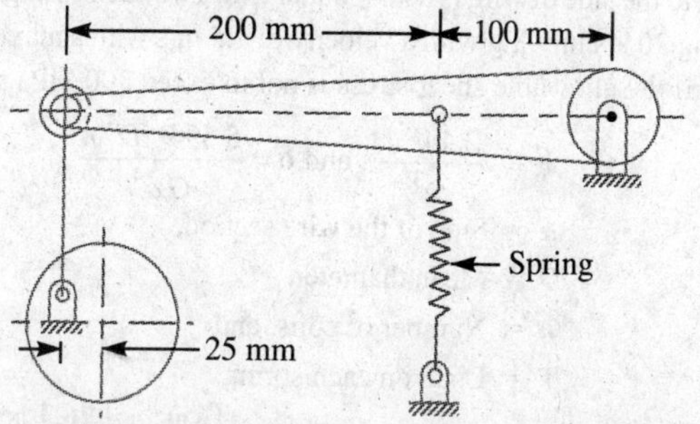

Fig. 23.35

The roller follower is held against the cam by means of an extension spring. Assuming that the force between the follower and the cam is approximately 250 N at the low position and 400 N at the high position.

If the spring index is 7, find the diameter of wire, outside diameter of spring and the number of active coils. The maximum shear stress may be taken as 280 MPa. Use $G = 80$ kN/mm^2.

8. The following data relate to a single plate friction clutch whose both sides are effective:

Power transmitted = 35 kW

Speed = 1000 r.p.m.

Permissible uniform pressure on lining material

= 0.07 N/mm^2

$\frac{\text{Outer radius}}{\text{Inner radius}}$ of plate = 1.5

Coefficient of friction of lining material

= 0.3

Number of springs = 6

Spring index = 6

Stress concentration factor in spring

$$= \frac{4C - 1}{4C - 4} + \frac{0.615}{C}, \quad C = \text{Spring index}$$

Permissible stress in spring steel

= 420 MPa

Modulus of rigidity of spring steel

= 84 kN/mm^2

Compression of spring to keep it engaged

= 12.5 mm

Design and give a sketch of any one of the springs in the uncompressed position. The springs are of round section close coiled helical type and are situated at mean radius of the plate.

[Ans. d = 5.65 mm ; D = 33.9 mm ; n = 6]

9. A railway wagon weighing 50 kN and moving with a speed of 8 km per hour has to be stopped by four buffer springs in which the maximum compression allowed is 220 mm. Find the number of turns in each spring of mean diameter 150 mm. The diameter of spring wire is 25 mm. Take $G = 84$ kN/mm^2.

[Ans. 8 turns]

10. The bumper springs of a railway carriage are to be made of square section wire. The ratio of mean diameter of spring to the side of wire is nearly equal to 6. Two such springs are required to bring to rest a carriage weighing 20 kN moving with a velocity of 1.5 m/s with a maximum deflection of 200 mm. Design the spring if the allowable shear stress is not to exced 300 MPa and $G = 84$ kN/mm^2.

For square wire, $\tau = \dfrac{2.4\, W.D}{a^3}$; and $\delta = \dfrac{5.59\, W.D^3.n}{G.a^4}$

where a = Side of the wire section,

D = Mean diameter,

n = Number of coils, and

W = Load on each spring.

[Ans. a = 26.3 mm ; D = 157.8 mm ; n = 33]

11. A load of 2 kN is dropped axially on a close coiled helical spring, from a height of 250 mm. The spring has 20 effective turns, and it is made of 25 mm diameter wire. The spring index is 8. Find the maximum shear stress induced in the spring and the amount of compression produced. The modulus of rigidity for the material of the spring wire is 84 kN/mm^2.

[Ans. 287 MPa; 290 mm]

12. A helical compression spring made of oil tempered carbon steel, is subjected to a load which varies from 600 N to 1600 N. The spring index is 6 and the design factor of safety is 1.43. If the yield shear stress is 700 MPa and the endurance stress is 350 MPa, find the size of the spring wire and mean diameter of the spring coil. **[Ans. 10 mm ; 60 mm]**

13. A helical spring B is placed inside the coils of a second helical spring A, having the same number of coils and free length. The springs are made of the same material. The composite spring is compressed by an axial load of 2300 N which is shared between them. The mean diameters of the spring A and B are 100 mm and 70 mm respectively and wire diameters are 13 mm and 8 mm respectively. Find the load taken and the maximum stress in each spring.

[Ans. W_A = 1670 N ; W_B = 630 N ; σ_A = 230 MPa ; σ_B = 256 MPa]

14. Design a concentric spring for an air craft engine valve to exert a maximum force of 5000 N under a deflection of 40 mm. Both the springs have same free length, solid length and are subjected to equal maximum shear stress of 850 MPa. The spring index for both the springs is 6.

[Ans. d_1 = 8 mm ; d_2 = 6 mm ; n = 4]

15. The free end of a torsional spring deflects through 90° when subjected to a torque of 4 N-m. The spring index is 6. Determine the coil wire diameter and number of turns with the following data :

Modulus of rigidity = 80 GPa ; Modulus of elasticity = 200 GPa; Allowable stress = 500 MPa.

[Ans. 5 mm ; 26]

16. A flat spiral steel spring is to give a maximum torque of 1500 N-mm for a maximum stress of 1000 MPa. Find the thickness and length of the spring to give three complete turns of motion, when the stress decreases from 1000 to zero. The width of the spring strip is 12 mm. The Young's modulus for the material of the strip is 200 kN/mm^2. **[Ans. 1.225 mm ; 4.6 m]**

17. A semi-elliptical spring has ten leaves in all, with the two full length leaves extending 625 mm. It is 62.5 mm wide and 6.25 mm thick. Design a helical spring with mean diameter of coil 100 mm which will have approximately the same induced stress and deflection for any load. The Young's modulus for the material of the semi-elliptical spring may be taken as 200 kN/mm^2 and modulus of rigidity for the material of helical spring is 80 kN/mm^2.

18. A carriage spring 800 mm long is required to carry a proof load of 5000 N at the centre. The spring is made of plates 80 mm wide and 7.5 mm thick. If the maximum permissible stress for the material of the plates is not to exceed 190 MPa, determine :

1. The number of plates required, 2. The deflection of the spring, and 3. The radius to which the plates must be initially bent.

The modulus of elasticity may be taken as 205 kN/mm^2. **[Ans. 6 ; 23 mm ; 3.5 m]**

19. A semi-elliptical laminated spring 900 mm long and 55 mm wide is held together at the centre by a band 50 mm wide. If the thickness of each leaf is 5 mm, find the number of leaves required to carry a load of 4500 N. Assume a maximum working stress of 490 MPa.

If the two of these leaves extend the full length of the spring, find the deflection of the spring. The Young's modulus for the spring material may be taken as 210 kN/mm^2.

[Ans. 9 ; 71.8 mm]

20. A semi-elliptical laminated spring is made of 50 mm wide and 3 mm thick plates. The length between the supports is 650 mm and the width of the band is 60 mm. The spring has two full length leaves and five graduated leaves. If the spring carries a central load of 1600 N, find :

   1. Maximum stress in full length and graduated leaves for an initial condition of no stress in the leaves.
   2. The maximum stress if the initial stress is provided to cause equal stress when loaded.
   3. The deflection in parts (1) and (2). **[Ans. 590 MPa ; 390 MPa ; 450 MPa ; 54 mm]**

QUESTIONS

1. What is the function of a spring? In which type of spring the behaviour is non-linear?
2. Classify springs according to their shapes. Draw neat sketches indicating in each case whether stresses are induced by bending or by torsion.
3. Discuss the materials and practical applications for the various types of springs.
4. The extension springs are in considerably less use than the compression springs. Why?
5. Explain the following terms of the spring :

   (*i*) Free length; (*ii*) Solid height;

   (*iii*) Spring rate; (*iv*) Active and inactive coils;

   (*v*) Spring index; and (*vi*) Stress factor.
6. Explain what you understand by A.M. Wahl's factor and state its importance in the design of helical springs?
7. Explain one method of avoiding the tendency of a compression spring to buckle.
8. A compression spring of spring constant *K* is cut into two springs having equal number of turns and the two springs are then used in parallel. What is the resulting spring constant of the combination? How does the load carrying capacity of the resulting combination compare with that of the original spring?
9. Prove that in a spring, using two concentric coil springs made of same material, having same length and compressed equally by an axial load, the loads shared by the two springs are directly proportional to the square of the diameters of the wires of the two springs.
10. What do you understand by full length and graduated leaves of a leaf spring? Write the expression for determining the stress and deflection in full length and graduated leaves.
11. What is nipping in a leaf spring? Discuss its role. List the materials commonly used for the manufacture of the leaf springs.
12. Explain the utility of the centre bolt, *U*-clamp, rebound clip and camber in a leaf spring.

OBJECTIVE TYPE QUESTIONS

1. A spring used to absorb shocks and vibrations is
   (a) closely-coiled helical spring (b) open-coiled helical spring
   (c) conical spring (d) torsion spring
2. The spring mostly used in gramophones is
   (a) helical spring (b) conical spring
   (c) laminated spring (d) flat spiral spring
3. Which of the following spring is used in a mechanical wrist watch?
   (a) Helical compression spring (b) Spiral spring
   (c) Torsion spring (d) Bellevile spring
4. When a helical compression spring is subjected to an axial compressive load, the stress induced in the wire is
   (a) tensile stress (b) compressive stress
   (c) shear stress (d) bending stress
5. In a close coiled helical spring, the spring index is given by D/d where D and d are the mean coil diameter and wire diameter respectively. For considering the effect of curvature, the Wahl's stress factor K is given by
   (a) $\frac{4C-1}{4C+4} + \frac{0.615}{C}$ (b) $\frac{4C-1}{4C-4} + \frac{0.615}{C}$
   (c) $\frac{4C+1}{4C-4} - \frac{0.615}{C}$ (d) $\frac{4C+1}{4C+4} - \frac{0.615}{C}$
6. When helical compression spring is cut into halves, the stiffness of the resulting spring will be
   (a) same (b) double
   (c) one-half (d) one-fourth
7. Two close coiled helical springs with stiffness k_1 and k_2 respectively are conected in series. The stiffness of an equivalent spring is given by
   (a) $\frac{k_1 \cdot k_2}{k_1 + k_2}$ (b) $\frac{k_1 - k_2}{k_1 + k_2}$
   (c) $\frac{k_1 + k_2}{k_1 \cdot k_2}$ (d) $\frac{k_1 - k_2}{k_1 \cdot k_2}$
8. When two concentric coil springs made of the same material, having same length and compressed equally by an axial load, the load shared by the two springs will be .... to the square of the diameters of the wires of the two springs.
   (a) directly proportional (b) inversely proportional
   (c) equal to
9. A leaf spring in automobiles is used
   (a) to apply forces (b) to measure forces
   (c) to absorb shocks (d) to store strain energy
10. In leaf springs, the longest leaf is known as
   (a) lower leaf (b) master leaf
   (c) upper leaf (d) none of these

ANSWERS

1. (e)	2. (d)	3. (c)	4. (c)	5. (b)
6. (b)	7. (a)	8. (a)	9. (c)	10. (b)

CHAPTER

24

Clutches

24.1 Introduction

A clutch is a machine member used to connect a driving shaft to a driven shaft so that the driven shaft may be started or stopped at will, without stopping the driving shaft. The use of a clutch is mostly found in automobiles. A little consideration will show that in order to change gears or to stop the vehicle, it is required that the driven shaft should stop, but the engine should continue to run. It is, therefore, necessary that the driven shaft should be disengaged from the driving shaft. The engagement and disengagement of the shafts is obtained by means of a clutch which is operated by a lever.

24.2 Types of Clutches

Following are the two main types of clutches commonly used in engineering practice :

1. Positive clutches, and 2. Friction clutches.

We shall now discuss these clutches in the following pages.

24.3 Positive Clutches

The positive clutches are used when a positive drive is required. The simplest type of a positive clutch is a ***jaw*** or ***claw clutch.*** The jaw clutch permits one shaft to drive another through a direct contact of interlocking jaws. It consists of two halves, one of which is permanently fastened to the driving shaft by a sunk key. The other half of the clutch is movable and it is free to slide axially on the driven shaft, but it is prevented from turning relatively to its shaft by means of feather key. The jaws of the clutch may be of square type as shown in Fig. 24.1 (*a*) or of spiral type as shown in Fig. 24.1 (*b*).

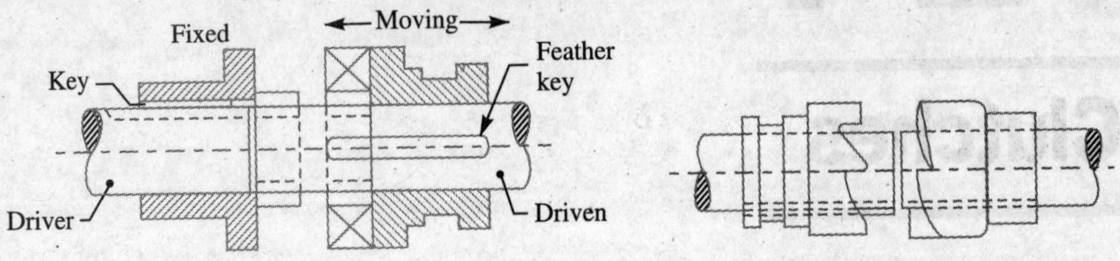

(*a*) Square jaw clutch. (*b*) Spiral jaw clutch.

Fig. 24.1. Jaw clutches.

A square jaw type is used where engagement and disengagement in motion and under load is not necessary. This type of clutch will transmit power in either direction of rotation. The spiral jaws may be left-hand or right-hand, because power transmitted by them is in one direction only. This type of clutch is occasionally used where the clutch must be engaged and disengaged while in motion. The use of jaw clutches are frequently applied to sprocket wheels, gears and pulleys. In such a case, the non-sliding part is made integral with the hub.

24.4 Friction Clutches

A friction clutch has its principal application in the transmission of power of shafts and machines which must be started and stopped frequently. Its application is also found in cases in which power is to be delivered to machines partially or fully loaded. The force of friction is used to start the driven shaft from rest and gradually brings it up to the proper speed without excessive slipping of the friction surfaces. In automobiles, friction clutch is used to connect the engine to the drive shaft. In operating such a clutch, care should be taken so that the friction surfaces engage easily and gradually bring the driven shaft up to proper speed. The proper alignment of the bearing must be maintained and it should be located as close to the clutch as possible. It may be noted that :

1. The contact surfaces should develop a frictional force that may pick up and hold the load with reasonably low pressure between the contact surfaces.
2. The heat of friction should be rapidly *dissipated and tendency to grab should be at a minimum.
3. The surfaces should be backed by a material stiff enough to ensure a reasonably uniform distribution of pressure.

24.5 Material for Friction Surfaces

The material used for lining of friction surfaces of a clutch should have the following characteristics :

* During operation of a clutch, most of the work done against frictional forces opposing the motion is liberated as heat at the interface. It has been found that at the actual point of contact, the temperature as high as 1000°C is reached for a very short duration (*i.e.* for 0.0001 second). Due to this, the temperature of the contact surfaces will increase and may destroy the clutch.

1. It should have a high and uniform coefficient of friction.
2. It should not be affected by moisture and oil.
3. It should have the ability to withstand high temperatures caused by slippage.
4. It should have high heat conductivity.
5. It should have high resistance to wear and scoring.

The materials commonly used for lining of friction surfaces and their important properties are shown in the following table.

Table 24.1. Properties of materials commonly used for lining of friction surfaces.

Material of friction surfaces	*Operating condition*	*Coefficient of friction*	*Maximum operating temperature (°C)*	*Maximum pressure (N/mm^2)*
Cast iron on cast iron or steel	dry	0.15 – 0.20	250 – 300	0.25– 0.4
Cast iron on cast iron or steel	In oil	0.06	250 – 300	0.6 – 0.8
Hardened steel on Hardened steel	In oil	0.08	250	0.8 – 0,8
Bronze on cast iron or steel	In oil	0.05	150	0.4
Pressed asbestos on cast iron or steel	dry	0.3	150 – 250	0.2 – 0.3
Powder metal on cast iron or steel	dry	0.4	550	0.3
Powder metal on cast iron or steel	In oil	0.1	550	0.8

24.6 Considerations in Designing a Friction Clutch

The following considerations must be kept in mind while designing a friction clutch.

1. The suitable material forming the contact surfaces should be selected.
2. The moving parts of the clutch should have low weight in order to minimise the inertia load, especially in high speed service.
3. The clutch should not require any external force to maintain contact of the friction surfaces.
4. The provision for taking up wear of the contact surfaces must be provided.
5. The clutch should have provision for facilitating repairs.
6. The clutch should have provision for carrying away the heat generated at the contact surfaces.
7. The projecting parts of the clutch should be covered by guard.

24.7 Types of Friction Clutches

Though there are many types of friction clutches, yet the following are important from the subject point of view :

1. Disc or plate clutches (single disc or multiple disc clutch),
2. Cone clutches, and
3. Centrifugal clutches.

We shall now discuss these clutches, in detail, in the following pages.

Note : The disc and cone clutches are known as ***axial friction clutches,*** while the centrifugal clutch is called ***radial friction clutch.***

24.8 Single Disc or Plate Clutch

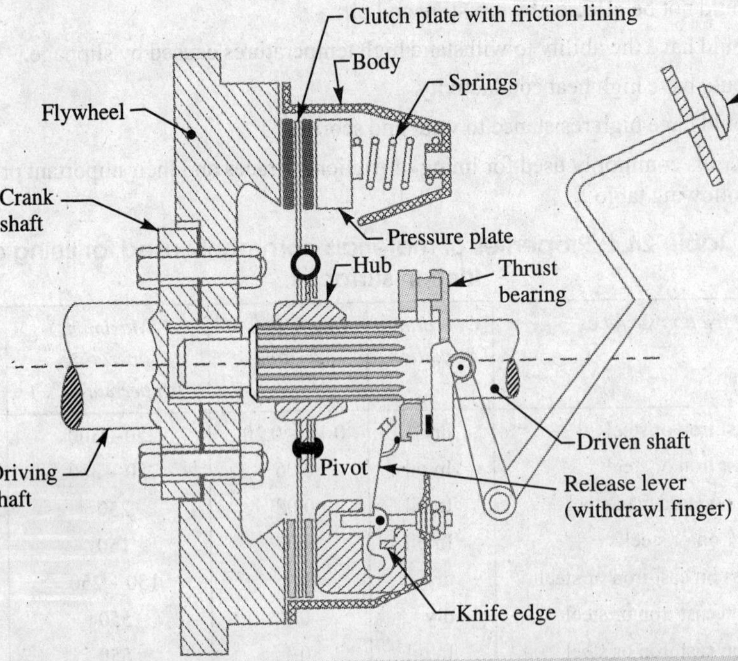

Fig. 24.2. Single disc or plate clutch.

A single disc or plate clutch, as shown in Fig 24.2, consists of a clutch plate whose both sides are faced with a frictional material (usually of Ferrodo). It is mounted on the hub which is free to move axially along the splines of the driven shaft. The pressure plate is mounted inside the clutch body which is bolted to the flywheel. Both the pressure plate and the flywheel rotate with the engine crankshaft or the driving shaft. The pressure plate pushes the clutch plate towards the flywheel by a set of strong springs which are arranged radially inside the body. The three levers (also known as release levers or fingers) are carried on pivots suspended from the case of the body. These are arranged in such a manner so that the pressure plate moves away from the flywheel by the inward movement of a thrust bearing. The bearing is mounted upon a forked shaft and moves forward when the clutch pedal is pressed.

When the clutch pedal is pressed down, its linkage forces the thrust release bearing to move in towards the flywheel and pressing the longer ends of the levers inward. The levers are forced to turn on their suspended pivot and the pressure plate moves away from the flywheel by the knife edges, thereby compressing the clutch springs. This action removes the pressure from the clutch plate and thus moves back from the flywheel and the driven shaft becomes stationary. On the other hand, when the foot is taken off from the clutch pedal, the thrust bearing moves back by the levers. This allows the springs to extend and thus the pressure plate pushes the clutch plate back towards the flywheel.

When a car hits an object and decelerates quickly the objects are thrown forward as they continue to move forwards due to inertia.

The axial pressure exerted by the spring provides a frictional force in the circumferential direction when the relative motion between the driving and driven members tends to take place. If the torque due to this frictional force exceeds the torque to be transmitted, then no slipping takes place and the power is transmitted from the driving shaft to the driven shaft.

24.9 Design of a Disc or Plate Clutch

Consider two friction surfaces maintained in contact by an axial thrust (W) as shown in Fig. 24.3 (a).

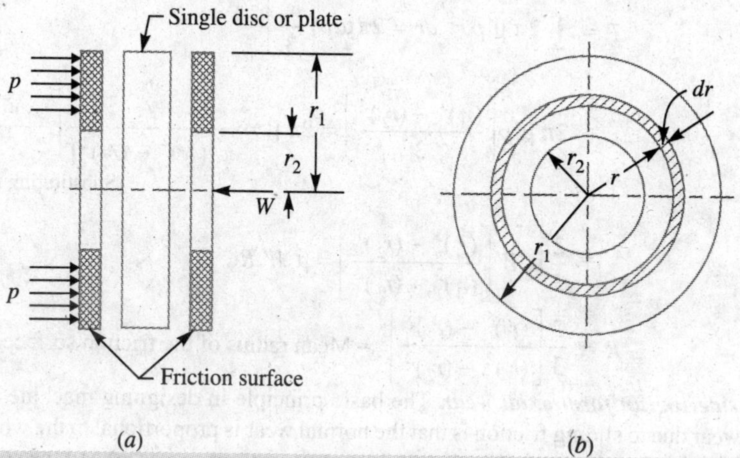

Fig. 24.3. Forces on a disc clutch.

Let T = Torque transmitted by the clutch,

p = Intensity of axial pressure with which the contact surfaces are held together,

r_1 and r_2 = External and internal radii of friction faces,

r = Mean radius of the friction face, and

μ = Coefficient of friction.

Consider an elementary ring of radius r and thickness dr as shown in Fig. 24.3 (b).

We know that area of the contact surface or friction surface

$$= 2\pi\, r.dr$$

∴ Normal or axial force on the ring,

$$\delta W = \text{Pressure} \times \text{Area} = p \times 2\pi\, r.dr$$

and the frictional force on the ring acting tangentially at radius r,

$$F_r = \mu \times \delta W = \mu.p \times 2\pi\, r.dr$$

∴ Frictional torque acting on the ring,

$$T_r = F_r \times r = \mu.p \times 2\pi\, r.dr \times r = 2\,\pi\,\mu\, p.\, r^2.dr$$

We shall now consider the following two cases :

1. When there is a uniform pressure, and
2. When there is a uniform axial wear.

1. *Considering uniform pressure.* When the pressure is uniformly distributed over the entire area of the friction face as shown in Fig. 24.3 (a), then the intensity of pressure,

$$p = \frac{W}{\pi\left[(r_1)^2 - (r_2)^2\right]}$$

where W = Axial thrust with which the friction surfaces are held together.

We have discussed above that the frictional torque on the elementary ring of radius r and thickness dr is

$$T_r = 2\pi\,\mu.p.r^2.dr$$

Integrating this equation within the limits from r_2 to r_1 for the total friction torque.

∴ Total frictional torque acting on the friction surface or on the clutch,

$$T = \int_{r_2}^{r_1} 2\pi\,\mu.p.r^2.dr = 2\pi\mu.p\left[\frac{r^3}{3}\right]_{r_2}^{r_1}$$

$$= 2\pi\,\mu.p\left[\frac{(r_1)^3 - (r_2)^3}{3}\right] = 2\pi\,\mu \times \frac{W}{\pi\,[(r_1)^2 - (r_2)^2]}\left[\frac{(r_1)^3 - (r_2)^3}{3}\right]$$

... (Substituting the value of p)

$$= \frac{2}{3}\,\mu.\,W\left[\frac{(r_1)^3 - (r_2)^3}{(r_1)^2 - (r_2)^2}\right] = \mu\,.W.R$$

where $R = \dfrac{2}{3}\left[\dfrac{(r_1)^3 - (r_2)^3}{(r_1)^2 - (r_2)^2}\right]$ = Mean radius of the friction surface.

2. ***Considering uniform axial wear.*** The basic principle in designing machine parts that are subjected to wear due to sliding friction is that the normal wear is proportional to the work of friction. The work of friction is proportional to the product of normal pressure (p) and the sliding velocity (V). Therefore,

Normal wear ∝ Work of friction ∝ $p.V$

or $p.V = K$ (a constant) or $p = K/V$ ...(*i*)

It may be noted that when the friction surface is new, there is a uniform pressure distribution over the entire contact surface. This pressure will wear most rapidly where the sliding velocity is maximum and this will reduce the pressure between the friction surfaces. This wearing-in process continues until the product $p.V$ is constant over the entire surface. After this, the wear will be uniform as shown in Fig. 24.4.

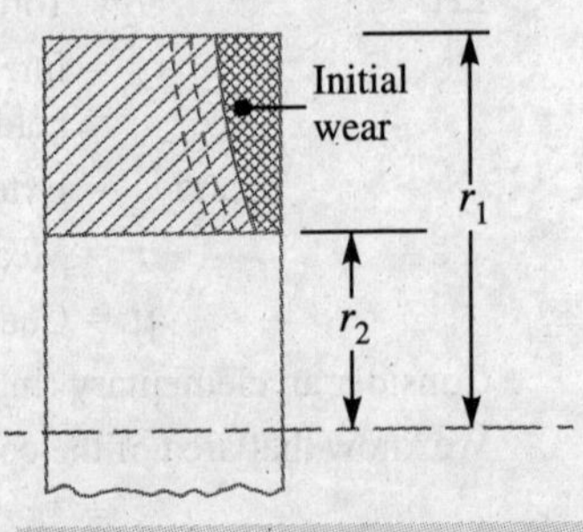

Fig. 24.4. Uniform axial wear.

Let p be the normal intensity of pressure at a distance r from the axis of the clutch. Since the intensity of pressure varies inversely with the distance, therefore

$$p.r = C \text{ (a constant) or } p = C/r \qquad ...(ii)$$

and the normal force on the ring,

$$\delta W = p.2\pi r.dr = \frac{C}{r} \times 2\pi r.dr = 2\pi C.dr$$

∴ Total force acing on the friction surface,

$$W = \int_{r_2}^{r_1} 2\pi\,C\,dr = 2\pi.C\,[r]_{r_2}^{r_1} = 2\pi\,C\,(r_1 - r_2)$$

or $$C = \frac{W}{2\pi\,(r_1 - r_2)}$$

We know that the frictional torque acting on the ring,

$$T_r = 2\pi\,\mu.p.r^2.dr = 2\pi\,\mu \times \frac{C}{r} \times r^2.dr = 2\pi\mu.C.r.dr \qquad ...(\because p = C/r)$$

∴ Total frictional torque acting on the friction surface (or on the clutch),

$$T = \int_{r_2}^{r_1} 2\pi\,\mu\,C.r.dr = 2\pi\,\mu\,C\left[\frac{r^2}{2}\right]_{r_2}^{r_1}$$

$$= 2\pi\mu\,.C\left[\frac{(r_1)^2 - (r_2)^2}{2}\right] = \pi\,\mu.C\,[(r_1)^2 - (r_2)^2]$$

$$= \pi\mu \times \frac{W}{2\pi\,(r_1 - r_2)}\,[(r_1)^2 - (r_2)^2] = \frac{1}{2} \times \mu.W\,(r_1 + r_2) = \mu.W.R$$

where $R = \dfrac{r_1 + r_2}{2}$ = Mean radius of the friction surface.

Notes : 1. In general, total frictional torque acting on the friction surfaces (or on the clutch) is given by

$$T = n.\mu.W.R$$

where n = Number of pairs of friction (or contact) surfaces, and

R = Mean radius of friction surface

$$= \frac{2}{3}\left[\frac{(r_1)^3 - (r_2)^3}{(r_1)^2 - (r_2)^2}\right] \qquad \text{... (For uniform pressure)}$$

$$= \frac{r_1 + r_2}{2} \qquad \text{... (For uniform wear)}$$

2. For a single disc or plate clutch, normally both sides of the disc are effective. Therefore a single disc clutch has two pairs of surfaces in contact (*i.e.* $n = 2$).

3. Since the intensity of pressure is maximum at the inner radius (r_2) of the friction or contact surface, therefore equation (*ii*) may be written as

$$p_{max} \times r_2 = C \quad \text{or} \quad p_{max} = C / r_2$$

4. Since the intensity of pressure is minimum at the outer radius (r_1) of the friction or contact surface, therefore equation (*ii*) may be written as

$$p_{min} \times r_1 = C \quad \text{or} \quad p_{min} = C / r_1$$

5. The average pressure (p_{av}) on the friction or contact surface is given by

$$p_{av} = \frac{\text{Total force on friction surface}}{\text{Cross-sectional area of friction surface}} = \frac{W}{\pi\,[(r_1)^2 - (r_2)^2]}$$

6. In case of a new clutch, the intensity of pressure is approximately uniform, but in an old clutch, the uniform wear theory is more approximate.

7. The uniform pressure theory gives a higher friction torque than the uniform wear theory. Therefore in case of friction clutches, uniform wear should be considered, unless otherwise stated.

24.10 Multiple Disc Clutch

A multiple disc clutch, as shown in Fig. 24.5, may be used when a large torque is to be transmitted. The inside discs (usually of steel) are fastened to the driven shaft to permit axial motion (except for the last disc). The outside discs (usually of bronze) are held by bolts and are fastened to the housing which is keyed to the driving shaft. The multiple disc clutches are extensively used in motor cars, machine tools etc.

A twin disk clutch

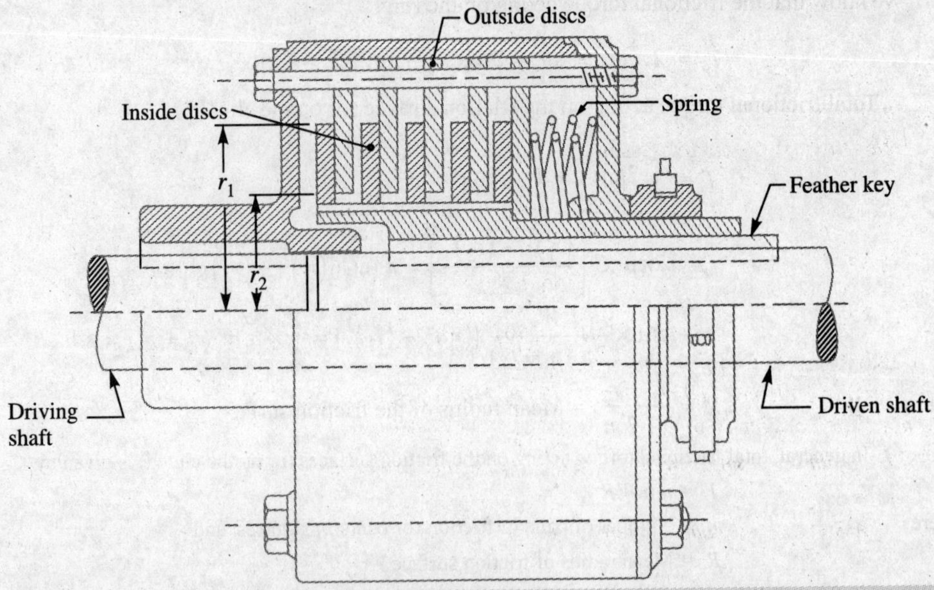

Fig. 24.5. Multiple disc clutch.

Let n_1 = Number of discs on the driving shaft, and

n_2 = Number of discs on the driven shaft.

∴ Number of pairs of contact surfaces,

$$n = n_1 + n_2 - 1$$

and total frictional torque acting on the friction surfaces or on the clutch,

$$T = n.\mu.W.R$$

where R = Mean radius of friction surfaces

$$= \frac{2}{3}\left[\frac{(r_1)^3 - (r_2)^3}{(r_1)^2 - (r_2)^2}\right] \quad \text{... (For uniform pressure)}$$

$$= \frac{r_1 + r_2}{2} \quad \text{... (For uniform wear)}$$

Example 24.1. *Determine the maximum, minimum and average pressure in a plate clutch when the axial force is 4 kN. The inside radius of the contact surface is 50 mm and the outside radius is 100 mm. Assume uniform wear.*

Solution. Given : W = 4 kN = 4000 N ; r_2 = 50 mm ; r_1 = 100 mm

Maximum pressure

Let p_{max} = Maximum pressure.

Since the intensity of pressure is maximum at the inner radius (r_2), therefore

$$p_{max} \times r_2 = C \quad \text{or} \quad C = 50\, p_{max}$$

We also know that total force on the contact surface (W),

$$4000 = 2\pi C\,(r_1 - r_2) = 2\pi \times 50\, p_{max}\,(100 - 50) = 15\,710\, p_{max}$$

$$\therefore \quad p_{max} = 4000 / 15\,710 = 0.2546 \text{ N/mm}^2 \quad \textbf{Ans.}$$

Minimum pressure

Let p_{min} = Minimum pressure.

Since the intensity of pressure is minimum at the outer radius (r_1), therefore,

$$p_{min} \times r_1 = C \quad \text{or} \quad C = 100\, p_{min}$$

We know that the total force on the contact surface (W),

$$4000 = 2\pi C(r_1 - r_2) = 2\pi \times 100\, p_{min}(100 - 50) = 31\,420\, p_{min}$$

$$\therefore \quad p_{min} = 4000 / 31\,420 = 0.1273 \text{ N/mm}^2 \quad \textbf{Ans.}$$

Average pressure

We know that average pressure,

$$p_{av} = \frac{\text{Total normal force on contact surface}}{\text{Cross-sectional area of contact surface}} = \frac{W}{\pi[(r_1)^2 - (r_2)^2]}$$

$$= \frac{4000}{\pi[(100)^2 - (50)^2]} = 0.17 \text{ N/mm}^2 \quad \textbf{Ans.}$$

Example 24.2. *A plate clutch having a single driving plate with contact surfaces on each side is required to transmit 110 kW at 1250 r.p.m. The outer diameter of the contact surfaces is to be 300 mm. The coefficient of friction is 0.4.*

(a) Assuming a uniform pressure of 0.17 N/mm²; determine the inner diameter of the friction surfaces.

(b) Assuming the same dimensions and the same total axial thrust, determine the maximum torque that can be transmitted and the maximum intensity of pressure when uniform wear conditions have been reached.

Solution. Given : $P = 110$ kW $= 110 \times 10^3$ W ; $N = 1250$ r.p.m. ; $d_1 = 300$ mm or $r_1 = 150$ mm ; $\mu = 0.4$; $p = 0.17$ N/mm²

(a) Inner diameter of the friction surfaces

Let d_2 = Inner diameter of the contact or friction surfaces, and

r_2 = Inner radius of the contact or friction surfaces.

We know that the torque transmitted by the clutch,

$$T = \frac{P \times 60}{2\pi N} = \frac{110 \times 10^3 \times 60}{2\pi \times 1250} = 840 \text{ N-m}$$

$$= 840 \times 10^3 \text{ N-mm}$$

Axial thrust with which the contact surfaces are held together,

$$W = \text{Pressure} \times \text{Area} = p \times \pi[(r_1)^2 - (r_2)^2]$$

$$= 0.17 \times \pi[(150)^2 - (r_2)^2] = 0.534[(150)^2 - (r_2)^2] \quad \text{...(i)}$$

and mean radius of the contact surface for uniform pressure conditions,

$$R = \frac{2}{3}\left[\frac{(r_1)^3 - (r_2)^3}{(r_1)^2 - (r_2)^2}\right] = \frac{2}{3}\left[\frac{(150)^3 - (r_2)^3}{(150)^2 - (r_2)^2}\right]$$

$\therefore$ Torque transmitted by the clutch (T),

$$840 \times 10^3 = n.\mu.W.R$$

$$= 2 \times 0.4 \times 0.534[(150)^2 - (r_2)^2] \times \frac{2}{3}\left[\frac{(150)^3 - (r_2)^3}{(150)^2 - (r_2)^2}\right] \quad \text{...}(\because n = 2)$$

$$= 0.285[(150)^3 - (r_2)^3]$$

or $\quad (150)^3 - (r_2)^3 = 840 \times 10^3 / 0.285 = 2.95 \times 10^6$

$\therefore \quad (r_2)^3 = (150)^3 - 2.95 \times 10^6 = 0.425 \times 10^6$ or $r_2 = 75$ mm

and $\quad d_2 = 2r_2 = 2 \times 75 = 150$ mm **Ans.**

(b) Maximum torque transmitted

We know that the axial thrust,

$$W = 0.534\,[(150)^2 - (r_2)^2] \quad \text{... [From equation } (i)]$$

$$= 0.534\,[(150)^2 - (75)^2] = 9011 \text{ N}$$

and mean radius of the contact surfaces for uniform wear conditions,

$$R = \frac{r_1 + r_2}{2} = \frac{150 + 75}{2} = 112.5 \text{ mm}$$

∴ Maximum torque transmitted,

$$T = n.\mu.W.R = 2 \times 0.4 \times 9011 \times 112.5 = 811 \times 10^3 \text{ N-mm}$$

$$= 811 \text{ N-m } \textbf{Ans.}$$

Maximum intensity of pressure

For uniform wear conditions, $p.r = C$ (a constant). Since the intensity of pressure is maximum at the inner radius (r_2), therefore

$$p_{max} \times r_2 = C \quad \text{or} \quad C = p_{max} \times 75 \text{ N/mm}$$

We know that the axial thrust (W),

$$9011 = 2\,\pi\,C\,(r_1 - r_2) = 2\pi \times p_{max} \times 75\,(150 - 75) = 35\,347\,p_{max}$$

$$\therefore \quad p_{max} = 9011 / 35\,347 = 0.255 \text{ N/mm}^2 \quad \textbf{Ans.}$$

Example 24.3. *A single plate clutch, effective on both sides, is required to transmit 25 kW at 3000 r.p.m. Determine the outer and inner diameters of frictional surface if the coefficient of friction is 0.255, ratio of diameters is 1.25 and the maximum pressure is not to exceed 0.1 N/mm². Also, determine the axial thrust to be provided by springs. Assume the theory of uniform wear.*

Solution. Given : $n = 2$; $P = 25$ kW $= 25 \times 10^3$ W ; $N = 3000$ r.p.m. ; $\mu = 0.255$; $d_1 / d_2 = 1.25$ or $r_1 / r_2 = 1.25$; $p_{max} = 0.1$ N/mm²

Outer and inner diameters of frictional surface

Let d_1 and d_2 = Outer and inner diameters (in mm) of frictional surface, and

r_1 and r_2 = Corresponding radii (in mm) of frictional surface.

We know that the torque transmitted by the clutch,

$$T = \frac{P \times 60}{2\,\pi\,N} = \frac{25 \times 10^3 \times 60}{2\,\pi \times 3000} = 79.6 \text{ N-m} = 79\,600 \text{ N-mm}$$

For uniform wear conditions, $p.r = C$ (a constant). Since the intensity of pressure is maximum at the inner radius (r_2), therefore.

$$p_{max} \times r_2 = C$$

or

$$C = 0.1\,r_2 \text{ N/mm}$$

and normal or axial load acting on the friction surface,

$$W = 2\pi\,C\,(r_1 - r_2) = 2\pi \times 0.1\,r_2\,(1.25\,r_2 - r_2)$$

$$= 0.157\,(r_2)^2 \quad \text{... } (\because r_1 / r_2 = 1.25)$$

We know that mean radius of the frictional surface (for uniform wear),

$$R = \frac{r_1 + r_2}{2} = \frac{1.25\,r_2 + r_2}{2} = 1.125\,r_2$$

and the torque transmitted (T),

$$79\,600 = n.\mu.W.R = 2 \times 0.255 \times 0.157\,(r_2)^2\,1.125\,r_2 = 0.09\,(r_2)^3$$

$$\therefore \quad (r_2)^3 = 79.6 \times 10^3 / 0.09 = 884 \times 10^3 \quad \text{or} \quad r_2 = 96 \text{ mm}$$

and

$$r_1 = 1.25\,r_2 = 1.25 \times 96 = 120 \text{ mm}$$

∴ Outer diameter of frictional surface,

$$d_1 = 2r_1 = 2 \times 120 = 240 \text{ mm} \quad \textbf{Ans.}$$

and inner diameter of frictional surface,

$$d_2 = 2r_2 = 2 \times 96 = 192 \text{ mm} \quad \textbf{Ans.}$$

Axial thrust to be provided by springs

We know that axial thrust to be provided by springs,

$$W = 2\pi\, C\,(r_1 - r_2) = 2\pi \times 0.1\, r_2\,(1.25\, r_2 - r_2)$$
$$= 0.157\,(r_2)^2 = 0.157\,(96)^2 = 1447 \text{ N} \quad \textbf{Ans.}$$

Example 24.4. *A dry single plate clutch is to be designed for an automotive vehicle whose engine is rated to give 100 kW at 2400 r.p.m. and maximum torque 500 N-m. The outer radius of the friction plate is 25% more than the inner radius. The intensity of pressure between the plate is not to exceed 0.07 N/mm². The coefficient of friction may be assumed equal to 0.3. The helical springs required by this clutch to provide axial force necessary to engage the clutch are eight. If each spring has stiffness equal to 40 N/mm, determine the dimensions of the friction plate and initial compression in the springs.*

Solution. Given : P = 100 kW = 100 × 10³ W ; *N = 2400 r.p.m. ; T = 500 N-m = 500 × 10³ N-mm ; p = 0.07 N/mm² ; μ = 0.3 ; No. of springs = 8 ; Stiffness/spring = 40 N/mm

Dimensions of the friction plate

Let r_1 = Outer radius of the friction plate, and

r_2 = Inner radius of the friction plate.

Since the outer radius of the friction plate is 25% more than the inner radius, therefore

$$r_1 = 1.25\, r_2$$

For uniform wear conditions, $p.r = C$ (a constant). Since the intensity of pressure is maximum at the inner radius (r_2), therefore

$$p.r_2 = C \quad \text{or} \quad C = 0.07\, r_2 \text{ N/mm}$$

and axial load acting on the friction plate,

$$W = 2\pi\, C\,(r_1 - r_2) = 2\pi \times 0.07\, r_2\,(1.25\, r_2 - r_2) = 0.11\,(r_2)^2 \text{ N} \qquad ...(i)$$

We know that mean radius of the friction plate, for uniform wear,

$$R = \frac{r_1 + r_2}{2} = \frac{1.25 r_2 + r_2}{2} = 1.125\, r_2$$

∴ Torque transmitted (T),

$$500 \times 10^3 = n.\mu.W.R = 2 \times 0.3 \times 0.11\,(r_2)^2\, 1.125\, r_2 = 0.074\,(r_2)^3 \qquad ...(\because n = 2)$$
$$(r_2)^3 = 500 \times 10^3 / 0.074 = 6757 \times 10^3 \quad \text{or} \quad r_2 = 190 \text{ mm} \quad \textbf{Ans.}$$

and

$$r_1 = 1.25\, r_2 = 1.25 \times 190 = 237.5 \text{ mm} \quad \textbf{Ans.}$$

Initial compression in the springs

We know that total stiffness of the springs,

$$s = \text{Stiffness per spring} \times \text{No. of springs} = 40 \times 8 = 320 \text{ N/mm}$$

Axial force required to engage the clutch,

$$W = 0.11\,(r_2)^2 = 0.11\,(190)^2 = 3970 \text{ N} \qquad \text{... [From equation } (i)]$$

∴ Initial compression in the springs

$$= W/s = 3970/320 = 12.4 \text{ mm} \quad \textbf{Ans.}$$

* Superfluous data

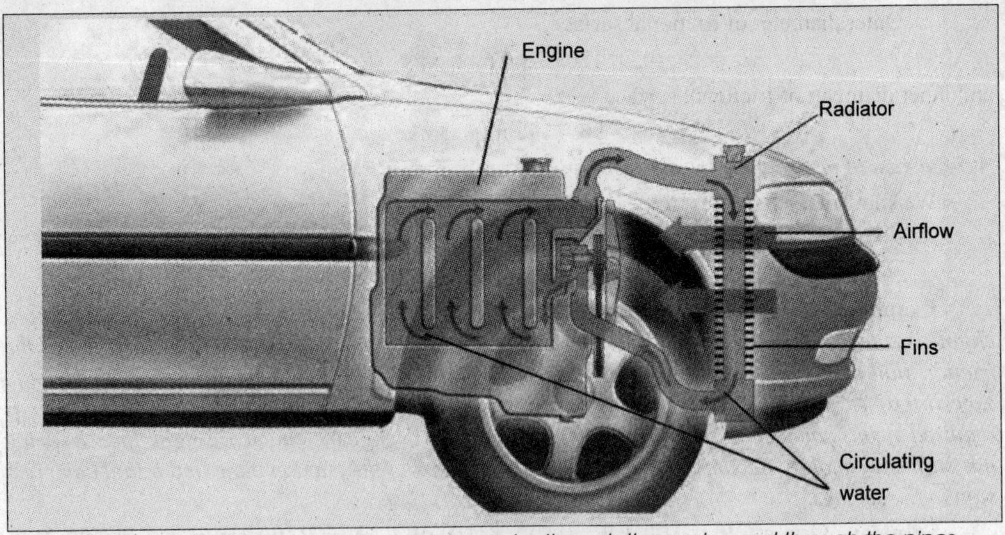

In car cooling system a pump circulates water through the engine and through the pipes of the radiator.

Example 24.5. *A single dry plate clutch is to be designed to transmit 7.5 kW at 900 r.p.m. Find :*

1. *Diameter of the shaft,*
2. *Mean radius and face width of the friction lining assuming the ratio of the mean radius to the face width as 4,*
3. *Outer and inner radii of the clutch plate, and*
4. *Dimensions of the spring, assuming that the number of springs are 6 and spring index = 6. The allowable shear stress for the spring wire may be taken as 420 MPa.*

Solution. Given : P = 7.5 kW = 7500 W ; N = 900 r.p.m. ; r/b = 4 ; No. of springs = 6 ; $C = D/d = 6$; τ = 420 MPa = 420 N/mm²

1. *Diameter of the shaft*

Let d_s = Diameter of the shaft, and

τ_1 = Shear stress for the shaft material. It may be assumed as 40 N/mm².

We know that the torque transmitted,

$$T = \frac{P \times 60}{2\,\pi\,N} = \frac{7500 \times 60}{2\,\pi \times 900} = 79.6 \text{ N-m} = 79\,600 \text{ N-mm} \qquad ...(i)$$

We also know that the torque transmitted (T),

$$79\,600 = \frac{\pi}{16} \times \tau_1\,(d_s)^3 = \frac{\pi}{16} \times 40\,(d_s)^3 = 7.855\,(d_s)^3$$

$$\therefore \quad (d_s)^3 = 79\,600 / 7.855 = 10\,134 \text{ or } d_s = 21.6 \text{ say } 25 \text{ mm} \quad \textbf{Ans.}$$

2. *Mean radius and face width of the friction lining*

Let R = Mean radius of the friction lining, and

b = Face width of the friction lining = $R/4$ ... (Given)

We know that the area of the friction faces,

$$A = 2\pi\,R.b$$

∴ Normal or the axial force acting on the friction faces,

$$W = A \times p = 2\pi\,R.b.p$$

and torque transmitted, $T = \mu\, W.R.n = \mu\,(2\pi\, Rb.p)\, R.n$

$$= \mu\left(2\,\pi R \times \frac{R}{4} \times p\right) Rn = \frac{\pi}{2} \times \mu R^3 .p.n \qquad ...(ii)$$

Assuming the intensity of pressure (p) as 0.07 N/mm^2 and coefficient of friction (μ) as 0.25, we have from equations (*i*) and (*ii*),

$$79\,600 = \frac{\pi}{2} \times 0.25 \times R^3 \times 0.07 \times 2 = 0.055\, R^3$$

... ($\because n = 2$, for both sides of plate effective)

$\therefore \quad R^3 = 79\,600 / 0.055 = 1.45 \times 10^6$ or $R = 113.2$ say 114 mm **Ans.**

and $\quad b = R / 4 = 114 / 4 = 28.5$ mm **Ans.**

3. *Outer and inner radii of the clutch plate*

Let $\quad r_1$ and r_2 = Outer and inner radii of the clutch plate respectively.

Since the face width (or radial width) of the plate is equal to the difference of the outer and inner radii, therefore,

$$b = r_1 - r_2 \text{ or } r_1 - r_2 = 28.5 \text{ mm} \qquad ...(iii)$$

We know that for uniform wear, mean radius of the clutch plate,

$$R = \frac{r_1 + r_2}{2} \text{ or } r_1 + r_2 = 2R = 2 \times 114 = 228 \text{ mm} \qquad ...(iv)$$

From equations (*iii*), and (*iv*), we find that

$r_1 = 128.25$ mm and $r_2 = 99.75$ mm **Ans.**

4. *Dimensions of the spring*

Let $\quad D$ = Mean diameter of the spring, and

d = Diameter of the spring wire.

We know that the axial force on the friction faces,

$$W = 2\pi\, R.b.p = 2\pi \times 114 \times 28.5 \times 0.07 = 1429.2 \text{ N}$$

In order to allow for adjustment and for maximum engine torque, the spring is designed for an overload of 25%.

$\therefore$ Total load on the springs

$$= 1.25\, W = 1.25 \times 1429.2 = 1786.5 \text{ N}$$

Since there are 6 springs, therefore maximum load on each spring,

$$W_s = 1786.5 / 6 = 297.75 \text{ N}$$

We know that Wahl's stress factor,

$$K = \frac{4C - 1}{4C - 4} + \frac{0.615}{C} = \frac{4 \times 6 - 1}{4 \times 6 - 4} + \frac{0.615}{6} = 1.2525$$

We also know that maximum shear stress induced in the wire (τ),

$$420 = K \times \frac{8\, W_s\, C}{\pi\, d^2} = 1.2525 \times \frac{8 \times 297.75 \times 6}{\pi\, d^2} = \frac{5697}{d^2}$$

$\therefore \quad d^2 = 5697 / 420 = 13.56$ or $d = 3.68$ mm

We shall take a standard wire of size *SWG* 8 having diameter (d) = 4.064 mm **Ans.**

and mean diameter of the spring,

$$D = C.d = 6 \times 4.064 = 24.384 \text{ say } 24.4 \text{ mm } \textbf{Ans.}$$

Let us assume that the spring has 4 active turns (*i.e.* $n = 4$). Therefore compression of the spring,

$$\delta = \frac{8\, W_s . C^3 . n}{G.d} = \frac{8 \times 297.75 \times 6^3 \times 4}{84 \times 10^3 \times 4.064} = 6.03 \text{ mm}$$

... (Taking $G = 84 \times 10^3$ N/mm^2)

Assuming squared and ground ends, total number of turns,

$$n' = n + 2 = 4 + 2 = 6$$

We know that free length of the spring,

$$L_F = n'.d + \delta + 0.15\,\delta$$
$$= 6 \times 4.064 + 6.03 + 0.15 \times 6.03 = 31.32 \text{ mm } \textbf{Ans.}$$

and pitch of the coils $= \dfrac{L_F}{n' - 1} = \dfrac{31.32}{6 - 1} = 6.264$ mm **Ans.**

Example 24.6. *Design a single plate automobile clutch to transmit a maximum torque of 250 N-m at 2000 r.p.m. The outside diameter of the clutch is 250 mm and the clutch is engaged at 55 km/h. Find : 1. the number of revolutions of the clutch slip during engagement; and 2. heat to be dissipated by the clutch for each engagement.*

The following additional data is available:

Engine torque during engagement = 100 N-m; Mass of the automobile = 1500 kg; Diameter of the automobile wheel = 0.7 m; Moment of inertia of combined engine rotating parts, flywheel and input side of the clutch = 1 kg-m^2; Gear reduction ratio at differential = 5; Torque at rear wheels available for accelerating automobile = 175 N-m; Coefficient of friction for the clutch material = 0.3; Permissible pressure = 0.13 N/mm^2.

Solution. Given : T = 250 N-m = 250 × 10^3 N-mm ; N = 2000 r.p.m. ; d_1 = 250 mm or r_1 = 125 mm ; V = 55 km/h = 15.3 m/s ; T_e = 100 N-m ; m = 1500 kg ; D_w = 0.7 m or R_w = 0.35 m ; I = 1 kg-m^2 ; T_a = 175 N-m ; Gear ratio = 5 ; μ = 0.3 ; p = 0.13 N/mm^2

1. ***Number of revolutions of the clutch slip during engagement***

First of all, let us find the inside radius of the clutch (r_2). We know that, for uniform wear, mean radius of the clutch,

$$R = \frac{r_1 + r_2}{2} = \frac{125 + r_2}{2} = 62.5 + 0.5\, r_2$$

and axial force on the clutch,

$$W = p.\pi\,[(r_1)^2 - (r_2)^2] = 0.13 \times \pi\,[(125)^2 - (r_2)^2]$$

We know that the torque transmitted (T),

$$250 \times 10^3 = n.\mu.W.R = 2 \times 0.3 \times 0.13\,\pi\,[(125)^2 - (r_2)^2]\,[62.5 + 0.5\,r_2]$$
$$= 0.245\,[\,976.56 \times 10^3 + 7812.5\,r_2 - 62.5\,(r_2)^2 - 0.5\,(r_2)^3]$$

Solving by hit and trial, we find that

$$r_2 = 70 \text{ mm}$$

We know that angular velocity of the engine,

$$\omega_e = 2\pi N / 60 = 2\pi \times 2000 / 60 = 210 \text{ rad/s}$$

and angular velocity of the wheel,

$$\omega_W = \frac{\text{Velocity of wheel}}{\text{Radius of wheel}} = \frac{V}{R_w} = \frac{15.3}{0.35} = 43.7 \text{ rad/s}$$

Since the gear ratio is 5, therefore angular velocity of the clutch follower shaft,

$$\omega_0 = \omega_W \times 5 = 43.7 \times 5 = 218.5 \text{ rad/s}$$

We know that angular acceleration of the engine during the clutch slip period of the clutch,

$$\alpha_e = \frac{T_e - T}{I} = \frac{100 - 250}{1} = -150 \text{ rad/s}^2$$

Let a = Linear acceleration of the automobile.

We know that accelerating force on the automobile,

$$F_a = \frac{T_a}{R} = \frac{175}{0.35} = 500 \text{ N}$$

We also know that accelerating force (F_a),

$$500 = m.a = 1500 \times a \text{ or } a = 500 / 1500 = 0.33 \text{ m/s}^2$$

∴ Angular acceleration of the clutch output,

$$\alpha_0 = \frac{\text{Acceleration} \times \text{Gear ratio}}{\text{Radius of wheel}} = \frac{0.33 \times 5}{0.35} = 4.7 \text{ rad/s}^2$$

We know that clutch slip period,

$$\Delta t = \frac{\omega_0 - \omega_e}{\alpha_0 - \alpha_e} = \frac{218.5 - 210}{4.7 - (-150)} = 0.055 \text{ s}$$

Angle through which the input side of the clutch rotates during engagement time (Δt) is

$$\theta_e = \omega_e \times \Delta t + \frac{1}{2}\alpha_e (\Delta t)^2$$

$$= 210 \times 0.055 + \frac{1}{2}(-150)(0.055)^2 = 11.32 \text{ rad}$$

and angle through which the output side of the clutch rotates during engagement time (Δt) is

$$\theta_0 = \omega_0 \times \Delta t + \frac{1}{2}\alpha_0 (\Delta t)^2$$

$$= 218.5 \times 0.055 + \frac{1}{2} \times 4.7 (0.055)^2 = 12 \text{ rad}$$

∴ Angle of clutch slip,

$$\theta = \theta_0 - \theta_e = 12 - 11.32 = 0.68 \text{ rad}$$

We know that number of revolutions of the clutch slip during engagement

$$= \frac{\theta}{2\pi} = \frac{0.68}{2\pi} = 0.11 \text{ revolutions } \textbf{Ans.}$$

Heat to be dissipated by the clutch for each engagement

We know that heat to be dissipated by the clutch for each engagement

$$= T.\theta = 250 \times 0.68 = 170 \text{ J } \textbf{Ans.}$$

Example 24.7. *A multiple disc clutch has five plates having four pairs of active friction surfaces. If the intensity of pressure is not to exceed 0.127 N/mm², find the power transmitted at 500 r.p.m. The outer and inner radii of friction surfaces are 125 mm and 75 mm respectively. Assume uniform wear and take coefficient of friction = 0.3.*

A twin-disk clutch

Solution. Given : $n_1 + n_2 = 5$; $n = 4$; $p = 0.127$ N/mm² ; $N = 500$ r.p.m. ; $r_1 = 125$ mm ; $r_2 = 75$ mm ; $\mu = 0.3$

We know that for uniform wear, $p.r = C$ (a constant). Since the intensity of pressure is maximum at the inner radius (r_2), therefore,

$$p.r_2 = C \quad \text{or} \quad C = 0.127 \times 75 = 9.525 \text{ N/mm}$$

and axial force required to engage the clutch,

$$W = 2\pi C\,(r_1 - r_2) = 2\pi \times 9.525\,(125 - 75) = 2993 \text{ N}$$

Mean radius of the friction surfaces,

$$R = \frac{r_1 + r_2}{2} = \frac{125 + 75}{2} = 100 \text{ mm} = 0.1 \text{ m}$$

We know that the torque transmitted,

$$T = n.\mu.W.R = 4 \times 0.3 \times 2993 \times 0.1 = 359 \text{ N-m}$$

$\therefore$ Power transmitted, $P = \dfrac{T \times 2\,\pi\,N}{60} = \dfrac{359 \times 2\,\pi \times 500}{60} = 18\,800 \text{ W} = 18.8 \text{ kW}$ **Ans.**

Example 24.8. *A multi-disc clutch has three discs on the driving shaft and two on the driven shaft. The inside diameter of the contact surface is 120 mm. The maximum pressure between the surface is limited to 0.1 N/mm². Design the clutch for transmitting 25 kW at 1575 r.p.m. Assume uniform wear condition and coefficient of friction as 0.3.*

Solution. Given : $n_1 = 3$; $n_2 = 2$; $d_2 = 120$ mm or $r_2 = 60$ mm ; $p_{max} = 0.1$ N/mm² ; $P = 25$ kW $= 25 \times 10^3$ W ; $N = 1575$ r.p.m. ; $\mu = 0.3$

Let r_1 = Outside radius of the contact surface.

We know that the torque transmitted,

$$T = \frac{P \times 60}{2\,\pi\,N} = \frac{25 \times 10^3 \times 60}{2\,\pi \times 1575} = 151.6 \text{ N-m} = 151\,600 \text{ N-mm}$$

For uniform wear, we know that $p.r = C$. Since the intensity of pressure is maximum at the inner radius (r_2), therefore,

$$p_{max} \times r_2 = C \quad \text{or} \quad C = 0.1 \times 60 = 6 \text{ N/mm}$$

We know that the axial force on each friction surface,

$$W = 2\pi C\,(r_1 - r_2) = 2\pi \times 6(r_1 - 60) = 37.7\,(r_1 - 60) \quad \text{...}(i)$$

For uniform wear, mean radius of the contact surface,

$$R = \frac{r_1 + r_2}{2} = \frac{r_1 + 60}{2} = 0.5\,r_1 + 30$$

We know that number of pairs of contact surfaces,

$$n = n_1 + n_2 - 1 = 3 + 2 - 1 = 4$$

$\therefore$ Torque transmitted (T),

$$151\,600 = n.\mu.W.R = 4 \times 0.3 \times 37.7\,(r_1 - 60)\,(0.5\,r_1 + 30)$$

... [Substituting the value of W from equation (*i*)]

$$= 22.62\,(r_1)^2 - 81\,432$$

$$\therefore \quad (r_1)^2 = \frac{151\,600 + 81\,432}{22.62} = 10\,302$$

or $r_1 = 101.5$ mm **Ans.**

Example 24.9. *A multiple disc clutch, steel on bronze, is to transmit 4.5 kW at 750 r.p.m. The inner radius of the contact is 40 mm and outer radius of the contact is 70 mm. The clutch operates in oil with an expected coefficient of 0.1. The average allowable pressure is 0.35 N/mm². Find : 1. the total number of steel and bronze discs; 2. the actual axial force required; 3. the actual average pressure; and 4. the actual maximum pressure.*

Solution. Given : $P = 4.5$ kW $= 4500$ W ; $N = 750$ r.p.m. ; $r_2 = 40$ mm ; $r_1 = 70$ mm ; $\mu = 0.1$; $p_{av} = 0.35$ N/mm²

1. ***Total number of steel and bronze discs***

Let n = Number of pairs of contact surfaces.

We know that the torque transmitted by the clutch,

$$T = \frac{P \times 60}{2\,\pi N} = \frac{4500 \times 60}{2\,\pi \times 750} = 57.3 \text{ N-m} = 57\,300 \text{ N-mm}$$

For uniform wear, mean radius of the contact surfaces,

$$R = \frac{r_1 + r_2}{2} = \frac{70 + 40}{2} = 55 \text{ mm}$$

and average axial force required,

$$W = p_{av} \times \pi\,[(r_1)^2 - (r_2)^2] = 0.35 \times \pi\,[(70)^2 - (40)^2] = 3630 \text{ N}$$

We also know that the torque transmitted (T),

$$57\,300 = n.\mu.W.R = n \times 0.1 \times 3630 \times 55 = 19\,965\,n$$

$$\therefore \quad n = 57\,300 / 19\,965 = 2.87$$

Since the number of pairs of contact surfaces must be even, therefore we shall use 4 pairs of contact surfaces with 3 steel discs and 2 bronze discs (because the number of pairs of contact surfaces is one less than the total number of discs). **Ans.**

2. ***Actual axial force required***

Let W' = Actual axial force required.

Since the actual number of pairs of contact surfaces is 4, therefore actual torque developed by the clutch for one pair of contact surface,

$$T' = \frac{T}{n} = \frac{57\,300}{4} = 14\,325 \text{ N-mm}$$

We know that torque developed for one pair of contact surface (T'),

$$14\,325 = \mu.W'.R = 0.1 \times W' \times 55 = 5.5\,W'$$

$$\therefore \quad W' = 14\,325 / 5.5 = 2604.5 \text{ N} \textbf{ Ans.}$$

3. ***Actual average pressure***

We know that the actual average pressure,

$$p'_{av} = \frac{W'}{\pi[(r_1)^2 - (r_2)^2]} = \frac{2604.5}{\pi\,[(70)^2 - (40)^2]} = 0.25 \text{ N/mm}^2 \textbf{ Ans.}$$

4. ***Actual maximum pressure***

Let p_{max} = Actual maximum pressure.

For uniform wear, $p.r = C$. Since the intensity of pressure is maximum at the inner radius, therefore,

$$p_{max} \times r_2 = C \quad \text{or} \quad C = 40\,p_{max} \text{ N/mm}$$

We know that the actual axial force (W'),

$$2604.5 = 2\pi C\,(r_1 - r_2) = 2\pi \times 40\,p_{max}\,(70 - 40) = 7541\,p_{max}$$

$$\therefore \quad p_{max} = 2604.5 / 7541 = 0.345 \text{ N/mm}^2 \textbf{ Ans.}$$

Example 24.10. *A plate clutch has three discs on the driving shaft and two discs on the driven shaft, providing four pairs of contact surfaces. The outside diameter of the contact surfaces is 240 mm and inside diameter 120 mm. Assuming uniform pressure and μ = 0.3, find the total spring load pressing the plates together to transmit 25 kW at 1575 r.p.m.*

If there are 6 springs each of stiffness 13 kN/m and each of the contact surfaces has worn away by 1.25 mm, find the maximum power that can be transmitted, assuming uniform wear.

Solution. Given : $n_1 = 3$; $n_2 = 2$; $n = 4$; $d_1 = 240$ mm or $r_1 = 120$ mm ; $d_2 = 120$ mm or $r_2 = 60$ mm ; $\mu = 0.3$; $P = 25$ kW $= 25 \times 10^3$ W ; $N = 1575$ r.p.m.

Total spring load

Let $\qquad W$ = Total spring load.

We know that the torque transmitted,

$$T = \frac{P \times 60}{2\,\pi\,N} = \frac{25 \times 10^3 \times 60}{2\,\pi \times 1575} = 151.5 \text{ N-m}$$
$$= 151.5 \times 10^3 \text{ N-mm}$$

Mean radius of the contact surface, for uniform pressure,

$$R = \frac{2}{3}\left[\frac{(r_1)^3 - (r_2)^3}{(r_1)^2 - (r_2)^2}\right] = \frac{2}{3}\left[\frac{(120)^3 - (60)^3}{(120)^2 - (60)^2}\right] = 93.3 \text{ mm}$$

and torque transmitted (T),

$$151.5 \times 10^3 = n.\mu.W.R = 4 \times 0.3 \times W \times 93.3 = 112\,W$$

$\therefore \qquad W = 151.5 \times 10^3 / 112 = 1353$ N **Ans.**

Maximum power transmitted

Given : $\qquad$ No. of springs = 6

$\therefore$ Contact surfaces of the spring = 8

Wear on each contact surface = 1.25 mm

$\therefore \qquad$ Total wear $= 8 \times 1.25 = 10$ mm $= 0.01$ m

Stiffness of each spring $= 13$ kN/m $= 13 \times 10^3$ N/m

$\therefore$ Reduction in spring force

= Total wear × Stiffness per spring × No. of springs

$= 0.01 \times 13 \times 10^3 \times 6 = 780$ N

and new axial load, $\qquad W = 1353 - 780 = 573$ N

We know that mean radius of the contact surfaces for uniform wear,

$$R = \frac{r_1 + r_2}{2} = \frac{120 + 60}{2} = 90 \text{ mm} = 0.09 \text{ m}$$

and torque transmitted, $\qquad T = n\,.\,\mu\,W\,.\,R = 4 \times 0.3 \times 573 \times 0.09 = 62$ N-m

$\therefore$ Power transmitted, $\qquad P = \dfrac{T \times 2\pi\,N}{60} = \dfrac{62 \times 2\pi \times 1575}{60} = 10\,227 \text{ W} = 10.227 \text{ kW}$ **Ans.**

24.11 Cone Clutch

A cone clutch, as shown in Fig. 24.6, was extensively used in automobiles, but now-a-days it has been replaced completely by the disc clutch. It consists of one pair of friction surface only. In a cone clutch, the driver is keyed to the driving shaft by a sunk key and has an inside conical surface or face which exactly fits into the outside conical surface of the driven. The driven member resting on the feather key in the driven shaft, may be shifted along the shaft by a forked lever provided at *B*, in order to engage the clutch by bringing the two conical surfaces in contact. Due to the frictional resistance set up at this contact surface, the torque is transmitted from one shaft to another. In some cases, a spring is placed around the driven shaft in contact with the hub of the driven. This spring holds the clutch faces in contact and maintains the pressure between them, and the forked lever is used only for disengagement of the clutch. The contact surfaces of the clutch may be metal to metal contact, but more often the driven member is lined with some material like wood, leather, cork or asbestos etc. The material of the clutch faces (*i.e.* contact surfaces) depends upon the allowable normal pressure and the coefficient of friction.

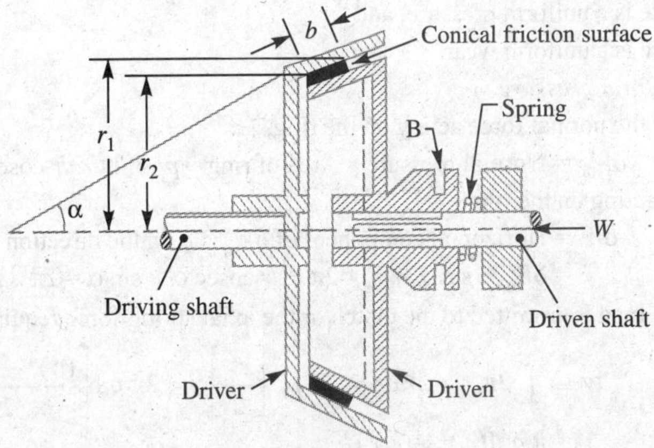

Fig. 24.6. Cone clutch.

24.12 Design of a Cone Clutch

Consider a pair of friction surfaces of a cone clutch as shown in Fig. 24.7. A little consideration will show that the area of contact of a pair of friction surface is a frustrum of a cone.

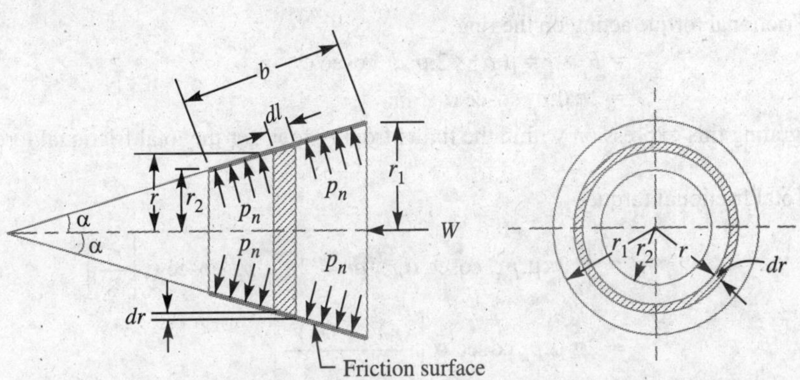

Fig. 24.7. Friction surfaces as a frustrum of a cone.

Let p_n = Intensity of pressure with which the conical friction surfaces are held together (*i.e.* normal pressure between the contact surfaces),

r_1 = Outer radius of friction surface,

r_2 = Inner radius of friction surface,

R = Mean radius of friction surface = $\frac{r_1 + r_2}{2}$,

α = Semi-angle of the cone (also called face angle of the cone) or angle of the friction surface with the axis of the clutch,

μ = Coefficient of friction between the contact surfaces, and

b = Width of the friction surfaces (also known as face width or cone face).

Consider a small ring of radius r and thickness dr as shown in Fig. 24.7. Let dl is the length of ring of the friction surface, such that,

$$dl = dr \text{ cosec } \alpha$$

$\therefore$ Area of ring $= 2\pi\, r.\, dl = 2\pi\, r.dr \text{ cosec } \alpha$

We shall now consider the following two cases :

1. When there is a uniform pressure, and
2. When there is a uniform wear.

1. *Considering uniform pressure*

We know that the normal force acting on the ring,

$$\delta W_n = \text{Normal pressure} \times \text{Area of ring} = p_n \times 2\pi\, r.dr \text{ cosec } \alpha$$

and the axial force acting on the ring,

$$\delta W = \text{Horizontal component of } \delta W_n \text{ (i.e. in the direction of } W)$$

$$= \delta W_n \times \sin \alpha = p_n \times 2\pi\, r.dr \text{ cosec } \alpha \times \sin \alpha = 2\pi \times p_n.r.dr$$

$\therefore$ Total axial load transmitted to the clutch or the axial spring force required,

$$W = \int_{r_2}^{r_1} 2\pi \times p_n.r.dr = 2\pi\, p_n \left[\frac{r^2}{2}\right]_{r_2}^{r_1} = 2\pi\, p_n \left[\frac{(r_1)^2 - (r_2)^2}{2}\right]$$

$$= \pi\, p_n\, [(r_1)^2 - (r_2)^2]$$

and

$$p_n = \frac{W}{\pi\left[(r_1)^2 - (r_2)^2\right]} \qquad ...(i)$$

We know that frictional force on the ring acting tangentially at radius r,

$$F_r = \mu.\delta W_n = \mu.p_n \times 2\pi r.dr \text{ cosec } \alpha$$

$\therefore$ Frictional torque acting on the ring,

$$T_r = F_r \times r = \mu.p_n \times 2\pi r.dr \text{ cosec } \alpha \times r$$

$$= 2\pi\, \mu.p_n \text{ cosec } \alpha.r^2\, dr$$

Integrating this expression within the limits from r_2 to r_1 for the total frictional torque on the clutch.

$\therefore$ Total frictional torque,

$$T = \int_{r_2}^{r_1} 2\pi\mu.p_n.\text{ cosec } \alpha.r^2\, dr = 2\pi\, \mu.p_n \text{ cosec } \alpha \left[\frac{r^3}{3}\right]_{r_2}^{r_1}$$

$$= 2\pi\, \mu.p_n \text{ cosec } \alpha \left[\frac{(r_1)^3 - (r_2)^3}{3}\right]$$

Substituting the value of p_n from equation (*i*), we get

$$T = 2\pi\,\mu \times \frac{W}{\pi\,[(r_1)^2 - (r_2)^2]} \times \text{cosec}\,\alpha \left[\frac{(r_1)^3 - (r_2)^3}{3}\right]$$

$$= \frac{2}{3} \times \mu.W\ \text{cosec}\,\alpha \left[\frac{(r_1)^3 - (r_2)^3}{(r_1)^2 - (r_2)^2}\right] \qquad ...(ii)$$

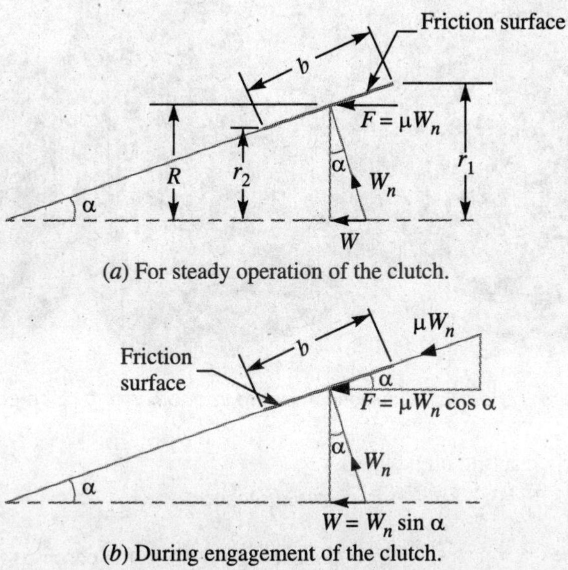

Fig. 24.8. Forces on a friction surface.

2. *Considering uniform wear*

In Fig. 24.7, let p_r be the normal intensity of pressure at a distance r from the axis of the clutch. We know that, in case of uniform wear, the intensity of pressure varies inversely with the distance.

$$\therefore \qquad p_r.r = C \text{ (a constant)} \quad \text{or} \quad p_r = C/r$$

We know that the normal force acting on the ring,

$$\delta W_n = \text{Normal pressure} \times \text{Area of ring} = p_r \times 2\pi r.dr\ \text{cosec}\,\alpha$$

and the axial force acting on the ring,

$$\delta W = \delta W_n \times \sin\alpha = p_r \times 2\pi\, r.dr\ \text{cosec}\,\alpha \times \sin\alpha$$

$$= 2\pi \times p_r.r\ dr$$

$$= 2\pi \times \frac{C}{r} \times r.\ dr = 2\pi\ C.dr \qquad ...\left(\because p_r = \frac{C}{r}\right)$$

$\therefore$ Total axial load transmitted to the clutch,

$$W = \int_{r_2}^{r_1} 2\pi\ C.dr = 2\pi\ C\ [r]_{r_2}^{r_1} = 2\pi\ C\ (r_1 - r_2)$$

or

$$C = \frac{W}{2\pi\,(r_1 - r_2)} \qquad ...(iii)$$

We know that frictional force on the ring acting tangentially at radius r,

$$F_r = \mu.\delta W_n = \mu.p_r \times 2\pi\, r.dr\ \text{cosec}\,\alpha$$

A mammoth caterpillar dump truck for use in quarries and open-cast mines.

∴ Frictional torque acting on the ring,

$$T_r = F_r \times r = \mu . p_r \times 2\pi\, r.dr \text{ cosec } \alpha \times r$$

$$= \mu \times \frac{C}{r} \times 2\pi\, r.dr \text{ cosec } \alpha \times r = 2\pi\, \mu . C \text{ cosec } \alpha \times r\, dr$$

Integrating this expression within the limits from r_2 to r_1 for the total frictional torque on the clutch.

∴ Total frictional torque,

$$T = \int_{r_2}^{r_1} 2\pi\, \mu . C \text{ cosec } \alpha \times r\, dr = 2\pi \mu . C \text{ cosec } \alpha \left[\frac{r^2}{2}\right]_{r_2}^{r_1}$$

$$= 2\pi\, \mu . C \text{ cosec } \alpha \left[\frac{(r_1)^2 - (r_2)^2}{2}\right]$$

Substituting the value of C from equation (*iii*), we have

$$T = 2\pi\mu \times \frac{W}{2\pi\,(r_1 - r_2)} \times \text{cosec } \alpha \left[\frac{(r_1)^2 - (r_2)^2}{2}\right]$$

$$= \mu . W \text{ cosec } \alpha \left[\frac{r_1 + r_2}{2}\right] = \mu\, WR \text{ cosec } \alpha \qquad \text{...}(iv)$$

where $R = \frac{r_1 + r_2}{2}$ = Mean radius of friction surface.

Since the normal force acting on the friction surface, $W_n = W \text{ cosec } \alpha$, therefore the equation (*iv*) may be written as

$$T = \mu\, W_n\, R \qquad \text{...}(v)$$

The forces on a friction surface, for steady operation of the clutch and after the clutch is engaged, is shown in Fig. 24.8 (*a*) and (*b*) respectively.

From Fig. 24.8 (*a*), we find that

$$r_1 - r_2 = b \sin \alpha \quad \text{and} \quad R = \frac{r_1 + r_2}{2} \quad \text{or} \quad r_1 + r_2 = 2R$$

∴ From equation (*i*), normal pressure acting on the friction surface,

$$p_n = \frac{W}{\pi[(r_1)^2 - (r_2)^2]} = \frac{W}{\pi(r_1 + r_2)(r_1 - r_2)} = \frac{W}{2\pi R.b \sin \alpha}$$

or $$W = p_n \times 2\pi R.b \sin \alpha = W_n \sin \alpha$$

where W_n = Normal load acting on the friction surface = $p_n \times 2\pi R.b$

Now the equation (*iv*) may be written as

$$T = \mu(p_n \times 2\pi R.b \sin \alpha) R \operatorname{cosec} \alpha = 2\pi \mu.p_n R^2.b$$

The following points may be noted for a cone clutch :

1. The above equations are valid for steady operation of the clutch and after the clutch is engaged.

2. If the clutch is engaged when one member is stationary and the other rotating (*i.e.* during engagement of the clutch) as shown in Fig. 24.8 (*b*), then the cone faces will tend to slide on each other due to the presence of relative motion. Thus an additional force (of magnitude $\mu.W_n \cos \alpha$) acts on the clutch which resists the engagement, and the axial force required for engaging the clutch increases.

∴ Axial force required for engaging the clutch,

$$W_e = W + \mu.W_n \cos \alpha = W_n . \sin \alpha + \mu W_n \cos \alpha$$
$$= W_n(\sin \alpha + \mu \cos \alpha)$$

It has been found experimentally that the term ($\mu W_n.\cos \alpha$) is only 25 percent effective.

∴ $$W_e = W_n \sin \alpha + 0.25 \mu W_n \cos \alpha = W_n(\sin \alpha + 0.25 \mu \cos \alpha)$$

3. Under steady operation of the clutch, a decrease in the semi-cone angle (α) increases the torque produced by the clutch (*T*) and reduces the axial force (*W*). During engaging period, the axial force required for engaging the clutch (W_e) increases under the influence of friction as the angle α decreases. The value of α can not be decreased much because smaller semi-cone angle (α) requires larger axial force for its disengagement.

If the clutch is to be designed for free disengagement, the value of tan α must be greater than μ. In case the value of tan α is less than μ, the clutch will not disengage itself and axial force required to disengage the clutch is given by

$$W_d = W_n(\mu \cos \alpha - \sin \alpha)$$

Example 24.11. *The contact surfaces in a cone clutch have an effective diameter of 80 mm. The semi-angle of the cone is 15° and coefficient of friction is 0.3. Find the torque required to produce slipping of the clutch, if the axial force applied is 200 N. The clutch is employed to connect an electric motor, running uniformly at 900 r.p.m. with a flywheel which is initially stationary. The flywheel has a mass of 14 kg and its radius of gyration is 160 mm. Calculate the time required for the flywheel to attain full-speed and also the energy lost in slipping of the clutch.*

Solution. Given : D = 80 mm or R = 40 mm ; α = 15° ; μ = 0.3 ; W = 200 N ; N = 900 r.p.m. or $\omega = 2\pi \times 900/60 = 94.26$ rad/s ; m = 14 kg ; k = 160 mm = 0.16 m

Torque required to produce slipping of the clutch

We know that the torque required to produce slipping of the clutch,

$$T = \mu WR \operatorname{cosec} \alpha = 0.3 \times 200 \times 40 \operatorname{cosec} 15° = 9273 \text{ N-mm}$$
$$= 9.273 \text{ N-m } \textbf{Ans.}$$

Time required for the flywheel to attain full-speed

Let t = Time required for the flywheel to attain full speed from the stationary position, and

α = Angular acceleration of the flywheel.

We know that mass moment of inertia of the flywheel,

$$I = m.k^2 = 14\,(0.16)^2 = 0.3584 \text{ kg-m}^2$$

We also know that the torque (T),

$$9.273 = I \times \alpha = 0.3584\,\alpha$$

$$\therefore \quad \alpha = 9.273 / 0.3584 = 25.87 \text{ rad / s}^2$$

and angular speed (ω),

$$94.26 = \omega_0 + \alpha.t = 0 + 25.87 \times t = 25.87\,t \qquad \text{... } (\because \omega_0 = 0)$$

$$\therefore \quad t = 94.26 / 25.87 = 3.64 \text{ s } \textbf{Ans.}$$

Energy lost in slipping of the clutch

We know that angular displacement,

$$\theta = \text{Average angular speed} \times \text{time} = \frac{\omega_0 + \omega}{2} \times t$$

$$= \frac{0 + 94.26}{2} \times 3.64 = 171.6 \text{ rad}$$

$\therefore$ Energy lost in slipping of the clutch,

$$= T.\theta = 9.273 \times 171.6 = 1591 \text{ N-m } \textbf{Ans.}$$

Example 24.12. *An engine developing 45 kW at 1000 r.p.m. is fitted with a cone clutch built inside the flywheel. The cone has a face angle of 12.5° and a maximum mean diameter of 500 mm. The coefficient of friction is 0.2. The normal pressure on the clutch face is not to exceed 0.1 N/mm². Determine : 1. the face width required, and 2. the axial spring force necessary to engage the clutch.*

Solution. Given : $P = 45$ kW $= 45 \times 10^3$ W ; $N = 1000$ r.p.m. ; $\alpha = 12.5°$; $D = 500$ mm or $R = 250$ mm ; $\mu = 0.2$; $p_n = 0.1$ N/mm²

1. *Face width*

Let b = Face width of the clutch in mm.

We know that torque developed by the clutch,

$$T = \frac{P \times 60}{2\,\pi\,N} = \frac{45 \times 10^3 \times 60}{2\,\pi \times 1000} = 430 \text{ N-m} = 430 \times 10^3 \text{ N-mm}$$

We also know that torque developed by the clutch (T),

$$430 \times 10^3 = 2\pi.\,\mu.\,p_n.\,R^2.b = 2\pi \times 0.2 \times 0.1\,(250)^2 b = 7855\,b$$

$$\therefore \quad b = 430 \times 10^3 / 7855 = 54.7 \text{ say } 55 \text{ mm } \textbf{Ans.}$$

2. *Axial spring force necessary to engage the clutch*

We know that the normal force acting on the contact surfaces,

$$W_n = p_n \times 2\pi R.b = 0.1 \times 2\pi \times 250 \times 55 = 8640 \text{ N}$$

$\therefore$ Axial spring force necessary to engage the clutch,

$$W_e = W_n\,(\sin\alpha + 0.25\,\mu\cos\alpha)$$

$$= 8640\,(\sin 12.5° + 0.25 \times 0.2 \cos 12.5°) = 2290 \text{ N } \textbf{Ans.}$$

Example 24.13. *Determine the principal dimensions of a cone clutch faced with leather to transmit 30 kW at 750 r.p.m. from an electric motor to an air compressor. Sketch a sectional front view of the clutch and provide the main dimensions on the sketch.*

Assume : semi-angle of the cone = $12\frac{1}{2}°$; μ = 0.2 ; mean diameter of cone = 6 to 10 d where d is the diameter of shaft; allowable normal pressure for leather and cast iron = 0.075 to 0.1 N/mm²; load factor = 1.75 and mean diameter to face width ratio = 6.

Solution. Given : P = 30 kW = 30 × 10³ W ; N = 750 r.p.m. ; $\alpha = 12\frac{1}{2}°$; μ = 0.2 ; D = 6 to 10 d ; p_n = 0.075 to 0.1 N/mm² ; K_L = 1.75 ; D/b = 6

First of all, let us find the diameter of shaft (d). We know that the torque transmitted by the shaft,

$$T = \frac{P \times 60}{2\pi N} \times K_L = \frac{30 \times 10^3 \times 60}{2\pi \times 750} \times 1.75 = 668.4 \text{ N-m}$$

$$= 668.4 \times 10^3 \text{ N-mm}$$

We also know that the torque transmitted by the shaft (T),

$$668.4 \times 10^3 = \frac{\pi}{16} \times \tau \times d^3 = \frac{\pi}{16} \times 42 \times d^3 = 8.25\, d^3 \quad \text{... (Taking } \tau = 42 \text{ N/mm}^2)$$

$$\therefore \quad d^3 = 668.4 \times 10^3 / 8.25 = 81 \times 10^3 \quad \text{or} \quad d = 43.3 \text{ say 50 mm } \textbf{Ans.}$$

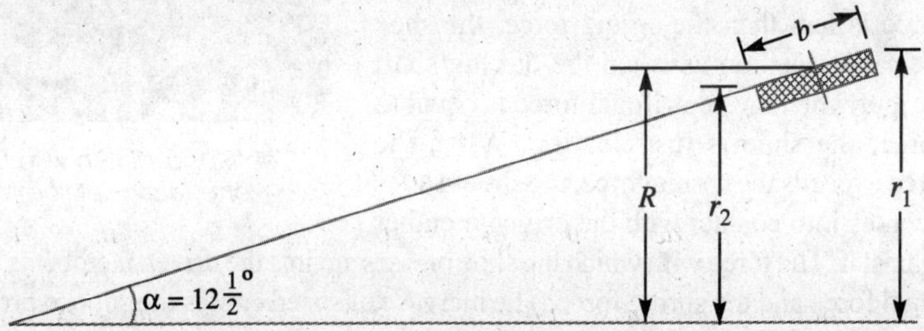

Fig. 24.9

Now let us find the principal dimensions of a cone clutch.

Let D = Mean diameter of the clutch,

R = Mean radius of the clutch, and

b = Face width of the clutch.

Since the allowable normal pressure (p_n) for leather and cast iron is 0.075 to 0.1 N/mm², therefore let us take p_n = 0.1 N/mm².

We know that the torque developed by the clutch (T),

$$668.4 \times 10^3 = 2\pi\mu . p_n . R^2 . b = 2\pi \times 0.2 \times 0.1 \times R^2 \times \frac{R}{3} = 0.042\, R^3$$

$$\text{... } (\because D/b = 6 \text{ or } 2R/b = 6 \text{ or } R/b = 3)$$

$$\therefore \quad R^3 = 668.4 \times 10^3 / 0.042 = 15.9 \times 10^6 \quad \text{or} \quad R = 250 \text{ mm}$$

and $D = 2R = 2 \times 250 = 500$ mm **Ans.**

Since this calculated value of the mean diameter of the clutch (D) is equal to 10 d and the given value of D is 6 to 10d, therefore the calculated value of D is safe.

We know that face width of the clutch,

$$b = D/6 = 500/6 = 83.3 \text{ mm } \textbf{Ans.}$$

From Fig. 24.9, we find that outer radius of the clutch,

$$r_1 = R + \frac{b}{2}\sin\alpha = 250 + \frac{83.3}{2}\sin 12\frac{1}{2}° = 259 \text{ mm } \textbf{Ans.}$$

and inner radius of the clutch,

$$r_2 = R - \frac{b}{2}\sin\alpha = 250 - \frac{83.3}{2}\sin 12\frac{1}{2}° = 241 \text{ mm} \quad \textbf{Ans.}$$

24.13 Centrifugal Clutch

Centrifugal clutch with three discs and four steel float plates.

The centrifugal clutches are usually incorporated into the motor pulleys. It consists of a number of shoes on the inside of a rim of the pulley, as shown in Fig. 24.10. The outer surface of the shoes are covered with a friction material. These shoes, which can move radially in guides, are held against the boss (or spider) on the driving shaft by means of springs. The springs exert a radially inward force which is assumed constant. The weight of the shoe, when revolving causes it to exert a radially outward force (*i.e.* centrifugal force). The magnitude of this centrifugal force depends upon the speed at which the shoe is revolving. A little consideration will show that when the centrifugal force is less than the spring force, the shoe remains in the same position as when the driving shaft was stationary, but when the centrifugal force is equal to the spring force, the shoe is just floating. When the centrifugal force exceeds the spring force, the shoe moves outward and comes into contact with the driven member and presses against it. The force with which the shoe presses against the driven member is the difference of the centrifugal force and the spring force. The increase of speed causes the shoe to press harder and enables more torque to be transmitted.

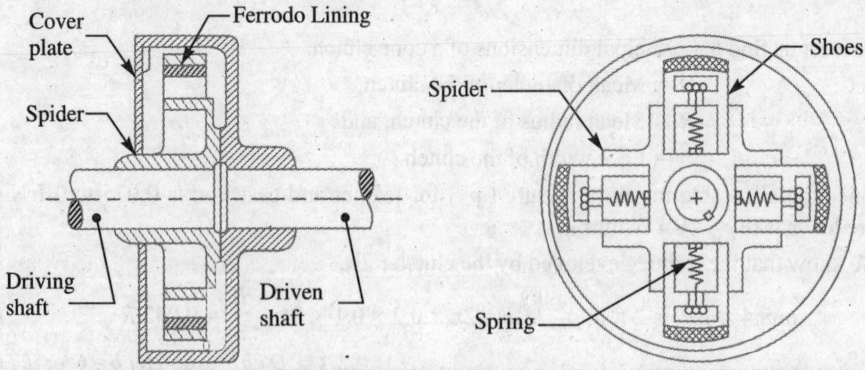

Fig. 24.10. Centrifugal clutch.

24.14 Design of a Centrifugal Clutch

In designing a centrifugal clutch, it is required to determine the weight of the shoe, size of the shoe and dimensions of the spring. The following procedure may be adopted for the design of a centrifugal clutch.

1. ***Mass of the shoes***

Consider one shoe of a centrifugal clutch as shown in Fig. 24.11.

Let m = Mass of each shoe,

n = Number of shoes,

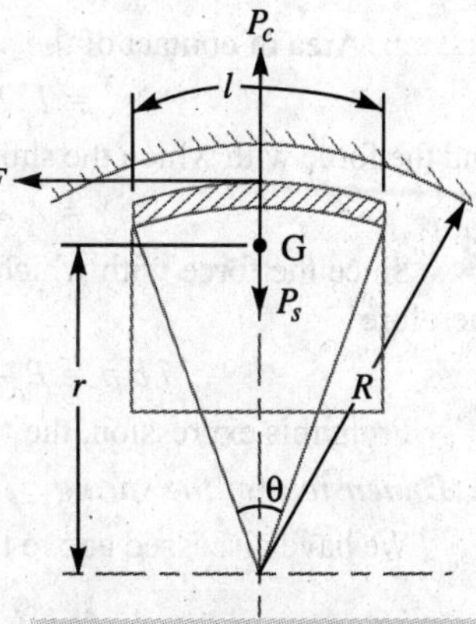

Fig. 24.11. Forces on a shoe of a centrifugal clucth.

r = Distance of centre of gravity of the shoe from the centre of the spider,

R = Inside radius of the pulley rim,

N = Running speed of the pulley in r.p.m.,

ω = Angular running speed of the pulley in rad / s $= 2\pi N / 60$ rad/s,

ω_1 = Angular speed at which the engagement begins to take place, and

μ = Coefficient of friction between the shoe and rim.

We know that the centrifugal force acting on each shoe at the running speed,

$$^*P_c = m.\omega^2.r$$

Since the speed at which the engagement begins to take place is generally taken as 3/4th of the running speed, therefore the inward force on each shoe exerted by the spring is given by

$$P_s = m\,(\omega_1)^2\, r = m\left(\frac{3}{4}\omega\right)^2 r = \frac{9}{16}\, m.\omega^2.r$$

∴ Net outward radial force (*i.e.* centrifugal force) with which the shoe presses against the rim at the running speed

$$= P_c - P_s = m.\omega^2.r - \frac{9}{16}\, m.\omega^2.r = \frac{7}{16}\, m.\omega^2.r$$

and the frictional force acting tangentially on each shoe,

$$F = \mu\,(P_c - P_s)$$

∴ Frictional torque acting on each shoe

$$= F \times R = \mu\,(P_c - P_s)\,R$$

and total frictional torque transmitted,

$$T = \mu\,(P_c - P_s)\,R \times n = n.F.R$$

From this expression, the mass of the shoes (m) may be evaluated.

2. ***Size of the shoes***

Let l = Contact length of the shoes,

b = Width of the shoes,

R = Contact radius of the shoes. It is same as the inside radius of the rim of the pulley,

θ = Angle subtended by the shoes at the centre of the spider in radians, and

p = Intensity of pressure exerted on the shoe. In order to ensure reasonable life, it may be taken as 0.1 N/mm^2.

We know that $\theta = \dfrac{l}{R}$ or $l = \theta.R = \dfrac{\pi}{3}R$ ...(Assuming $\theta = 60° = \pi / 3$ rad)

---

* The radial clearance between the shoe and the rim is about 1.5 mm. Since this clearance is small as compared to r, therefore it is neglected for design purposes. If, however, the radial clearance is given, then the operating radius of the mass centre of the shoe from the axis of the clutch,

$r_1 = r + c$, where c is the radial clearance,

Then $P_c = m.\omega^2\, r_1$ and $P_s = m\,(\omega_1)^2\, r_1$

∴ Area of contact of the shoe

$$= l\,.\,b$$

and the force with which the shoe presses against the rim

$$= A \times p = l.b.p$$

Since the force with which the shoe presses against the rim at the running speed is $(P_c - P_s)$, therefore

$$l.b.p = P_c - P_s$$

From this expression, the width of shoe (b) may be obtained.

3. *Dimensions of the spring*

We have discussed above that the load on the spring is given by

$$P_s = \frac{9}{16} \times m.\omega^2.r$$

The dimensions of the spring may be obtained as usual.

Example 24.14. *A centrifugal clutch is to be designed to transmit 15 kW at 900 r.p.m. The shoes are four in number. The speed at which the engagement begins is 3/4th of the running speed. The inside radius of the pulley rim is 150 mm. The shoes are lined with Ferrodo for which the coefficient of friction may be taken as 0.25. Determine: 1. mass of the shoes, and 2. size of the shoes.*

Solution. Given : P = 15 kW = 15 × 10^3 W ; N = 900 r.p.m. ; n = 4 ; R = 150 mm = 0.15 m ; μ = 0.25

1. *Mass of the shoes*

Let m = Mass of the shoes.

We know that the angular running speed,

$$\omega = \frac{2\pi N}{60} = \frac{2\pi \times 900}{60} = 94.26 \text{ rad/s}$$

Since the speed at which the engagement begins is 3/4 th of the running speed, therefore angular speed at which engagement begins is

$$\omega_1 = \frac{3}{4}\,\omega = \frac{3}{4} \times 94.26 = 70.7 \text{ rad/s}$$

Assuming that the centre of gravity of the shoe lies at a distance of 120 mm (30 mm less than R) from the centre of the spider, *i.e.*

$$r = 120 \text{ mm} = 0.12 \text{ m}$$

We know that the centrifugal force acting on each shoe,

$$P_c = m.\omega^2.r = m\,(94.26)^2\, 0.12 = 1066\, m \text{ N}$$

and the inward force on each shoe exerted by the spring *i.e.* the centrifugal force at the engagement speed, ω_1,

$$P_s = m(\omega_1)^2\, r = m\,(70.7)^2\, 0.12 = 600\, m \text{ N}$$

We know that the torque transmitted at the running speed,

$$T = \frac{P \times 60}{2\pi N} = \frac{15 \times 10^3 \times 60}{2\,\pi \times 900} = 159 \text{ N-m}$$

We also know that the torque transmitted (T),

$$159 = \mu\,(P_c - P_s)\,R \times n = 0.25\,(1066\,m - 600\,m)\,0.15 \times 4 = 70\,m$$

$$\therefore \quad m = 159/70 = 2.27 \text{ kg} \quad \textbf{Ans.}$$

2. *Size of the shoes*

Let l = Contact length of shoes in mm, and

b = Width of the shoes in mm.

Assuming that the arc of contact of the shoes subtend an angle of $\theta = 60°$ or $\pi / 3$ radians, at the centre of the spider, therefore

$$l = \theta.R = \frac{\pi}{3} \times 150 = 157 \text{ mm}$$

Area of contact of the shoes

$$A = l.b = 157\ b \text{ mm}^2$$

Assuming that the intensity of pressure (p) exerted on the shoes is 0.1 N/mm², therefore force with which the shoe presses against the rim

$$= A.p = 157b \times 0.1 = 15.7\ b \text{ N} \quad ...(i)$$

We also know that the force with which the shoe presses against the rim

$$= P_c - P_s = 1066\ m - 600\ m = 466\ m$$

$$= 466 \times 2.27 = 1058 \text{ N} \quad ...(ii)$$

From equations (*i*) and (*ii*), we find that

b = 1058 / 15.7 = 67.4 mm **Ans.**

Special trailers are made to carry very long loads. The longest load ever moved was gas storage vessel, 83.8 m long.

EXERCISES

1. A single disc clutch with both sides of the disc effective is used to transmit 10 kW power at 900 r.p.m. The axial pressure is limited to 0.085 N/mm². If the external diameter of the friction lining is 1.25 times the internal diameter, find the required dimensions of the friction lining and the axial force exerted by the springs. Assume uniform wear conditions. The coefficient of friction may be taken as 0.3.

[Ans. 132.5 mm ; 106 mm ; 1500 N]

2. A single plate clutch with both sides of the plate effective is required to transmit 25 kW at 1600 r.p.m. The outer diameter of the plate is limited to 300 mm and the intensity of pressure between the plates not to exceed 0.07 N/mm^2. Assuming uniform wear and coefficient of friction 0.3, find the inner diameter of the plates and the axial force necessary to engage the clutch.

[Ans. 90 mm ; 2375 N]

3. Give a complete design analysis of a single plate clutch, with both sides effective, of a vehicle to transmit 22 kW at a speed of 2800 r.p.m. allowing for 25% overload. The pressure intensity is not to exceed 0.08 N/mm^2 and the surface speed at the mean radius is not to exceed 2000 m/min. Take coefficient of friction for the surfaces as 0.35 and the outside diameter of the surfaces is to be 1.5 times the inside diameter. The axial thrust is to be provided by 6 springs of about 24 mm coil diameter. For spring material, the safe shear stress is to be limited to 420 MPa and the modulus of rigidity may be taken as 80 kN/mm^2. **[Ans. 120 mm ; 80 mm ; 3.658 mm]**

4. A multiple disc clutch has three discs on the driving shaft and two on the driven shaft, providing four pairs of contact surfaces. The outer diameter of the contact surfaces is 250 mm and the inner diameter is 150 mm. Determine the maximum axial intensity of pressure between the discs for transmitting 18.75 kW at 500 r.p.m. Assume uniform wear and coefficient of friction as 0.3.

5. A multiple disc clutch employs 3 steel and 2 bronze discs having outer diameter 300 mm and inner diameter 200 mm. For a coefficient of friction of 0.22, find the axial pressure and the power transmitted at 750 r.p.m., if the normal unit pressure is 0.13 N/mm^2.

   Also find the axial pressure of the unit normal pressure, if this clutch transmits 22 kW at 1500 r.p.m. **[Ans. 5105 N ; 44.11 kW ; 0.0324 N/mm^2]**

6. A multiple disc clutch has radial width of the friction material as 1/5th of the maximum radius. The coefficient of friction is 0.25. Find the total number of discs required to transmit 60 kW at 3000 r.p.m. The maximum diameter of the clutch is 250 mm and the axial force is limited to 600 N. Also find the mean unit pressure on each contact surface. **[Ans. 13 ; 0.034 N/mm^2]**

7. An engine developing 22 kW at 1000 r.p.m. is fitted with a cone clutch having mean diameter of 300 mm. The cone has a face angle of 12°. If the normal pressure on the clutch face is not to exceed 0.07 N/mm^2 and the coefficient of friction is 0.2, determine :

   (*a*) the face width of the clutch, and

   (*b*) the axial spring force necessary to engage the clutch.

   [Ans. 106 mm ; 1796 N]

8. A cone clutch is to be designed to transmit 7.5 kW at 900 r.p.m. The cone has a face angle of 12°. The width of the face is half of the mean radius and the normal pressure between the contact faces is not to exceed 0.09 N/mm^2. Assuming uniform wear and the coefficient of friction between the contact faces as 0.2, find the main dimensions of the clutch and the axial force required to engage the clutch.

   [Ans. R = 112.4 mm ; b = 56.2 mm ; r_1 = 118.2 mm ; r_2 = 106.6 mm ; W_e = 917 N]

9. A soft cone clutch has a cone pitch angle of 10°, mean diameter of 300 mm and a face width of 100 mm. If the coefficient of friction is 0.2 and has an average pressure of 0.07 N/mm^2 for a speed of 500 r.p.m., find : (*a*) the force required to engage the clutch; and (*b*) the power that can be transmitted. Assume uniform wear. **[Ans. 1470 N ; 10.4 kW]**

10. A cone clutch is mounted on a shaft which transmits power at 225 r.p.m. The small diameter of the cone is 230 mm, the cone face is 50 mm and the cone face makes an angle of 15° with the horizontal. Determine the axial force necessary to engage the clutch to transmit 4.5 kW if the coefficient of friction of the contact surfaces is 0.25. What is the maximum pressure on the contact surfaces assuming uniform wear? **[Ans. 2414 N ; 0.216 N/mm^2]**

11. A soft surface cone clutch transmits a torque of 200 N-m at 1250 r.p.m. The larger diameter of the clutch is 350 mm. The cone pitch angle is 7.5° and the face width is 65 mm. If the coefficient of friction is 0.2, find :

1. the axial force required to transmit the torque;
2. the axial force required to engage the clutch;
3. the average normal pressure on the contact surfaces when the maximum torque is being transmitted; and
4. the maximum normal pressure assuming uniform wear.

[Ans. 764 N ; 1057 N ; 0.084 N/mm^2 ; 0.086 N/mm^2]

12. A centrifugal friction clutch has a driving member consisting of a spider carrying four shoes which are kept from contact with the clutch case by means of flat springs until increase of centrifugal force overcomes the resistance of the springs and the power is transmitted by the friction between the shoes and the case.

Determine the necessary mass and size of each shoe if 22.5 kW is to be transmitted at 750 r.p.m. with engagement beginning at 75% of the running speed. The inside diameter of the drum is 300 mm and the radial distance of the centre of gravity of each shoe from the shaft axis is 125 mm. Assume $\mu = 0.25$. **[Ans. 5.66 kg ; l = 157.1 mm ; b = 120 mm]**

Clutches, brakes, steering and transmission need to be carefully designed to ensure the efficiency and safety of an automobile

QUESTIONS

1. What is a clutch? Discuss the various types of clutches giving at least one practical application for each.
2. Why a positive clutch is used? Describe, with the help of a neat sketch, the working of a jaw or claw clutch.
3. Name the different types of clutches. Describe with the help of neat sketches the working principles of two different types of friction clutches.
4. What are the materials used for lining of friction surfaces?
5. Why it is necessary to dissipate the heat generated when clutches operate?
6. Establish a formula for the frictional torque transmitted by a cone clutch.
7. Describe, with the help of a neat sketch, a centrifugal clutch and deduce an expression for the total frictional torque transmitted. How the shoes and springs are designed for such a clutch?

OBJECTIVE TYPE QUESTIONS

1. A jaw clutch is essentially a
   (a) positive action clutch (b) cone clutch
   (c) friction clutch (d) disc clutch
2. The material used for lining of friction surfaces of a clutch should have .... coefficient of friction.
   (a) low (b) high
3. The torque developed by a disc clutch is given by
   (a) $T = 0.25\ \mu.W.R$ (b) $T = 0.5\ \mu.W.R$
   (c) $T = 0.75\ \mu.W.R$ (d) $T = \mu.W.R$
   where W = Axial force with which the friction surfaces are held together ;
   μ = Coefficient of friction ; and
   R = Mean radius of friction surfaces.
4. In case of a multiple disc clutch, if n_1 are the number of discs on the driving shaft and n_2 are the number of the discs on the driven shaft, then the number of pairs of contact surfaces will be
   (a) $n_1 + n_2$ (b) $n_1 + n_2 - 1$
   (c) $n_1 + n_2 + 1$ (d) none of these
5. The cone clutches have become obsolete because of
   (a) small cone angles (b) exposure to dirt and dust
   (c) difficulty in disengaging (d) all of these
6. The axial force (W_e) required for engaging a cone clutch is given by
   (a) $W_n \sin \alpha$ (b) $W_n (\sin \alpha + \mu \cos \alpha)$
   (c) $W_n (\sin \alpha + 0.25\ \mu \cos \alpha)$ (d) none of these
   where W_n = Normal force acting on the contact surfaces,
   α = Face angle of the cone, and
   μ = Coefficient of friction.
7. In a centrifugal clutch, the force with which the shoe presses against the driven member is the .... of the centrifugal force and the spring force.
   (a) difference (b) sum

ANSWERS

1. (a)	2. (b)	3. (d)	4. (b)	5. (d)
6. (c)	7. (a)			

CHAPTER

25

Brakes

1. Introduction.
2. Energy Absorbed by a Brake.
3. Heat to be Dissipated during Braking.
4. Materials for Brake Lining.
5. Types of Brakes.
6. Single Block or Shoe Brake.
7. Pivoted Block or Shoe Brake.
8. Double Block or Shoe Brake.
9. Simple Band Brake.
10. Differential Band Brake.
11. Band and Block Brake.
12. Internal Expanding Brake.

25.1 Introduction

A brake is a device by means of which artificial frictional resistance is applied to a moving machine member, in order to retard or stop the motion of a machine. In the process of performing this function, the brake absorbs either kinetic energy of the moving member or potential energy given up by objects being lowered by hoists, elevators etc. The energy absorbed by brakes is dissipated in the form of heat. This heat is dissipated in the surrounding air (or water which is circulated through the passages in the brake drum) so that excessive heating of the brake lining does not take place. The design or capacity of a brake depends upon the following factors :

1. The unit pressure between the braking surfaces,
2. The coefficient of friction between the braking surfaces,
3. The peripheral velocity of the brake drum,

4. The projected area of the friction surfaces, and

5. The ability of the brake to dissipate heat equivalent to the energy being absorbed.

The major functional difference between a clutch and a brake is that a clutch is used to keep the driving and driven member moving together, whereas brakes are used to stop a moving member or to control its speed.

25.2 Energy Absorbed by a Brake

The energy absorbed by a brake depends upon the type of motion of the moving body. The motion of a body may be either pure translation or pure rotation or a combination of both translation and rotation. The energy corresponding to these motions is kinetic energy. Let us consider these motions as follows :

1. ***When the motion of the body is pure translation.*** Consider a body of mass (m) moving with a velocity v_1 m / s. Let its velocity is reduced to v_2 m / s by applying the brake. Therefore, the change in kinetic energy of the translating body or kinetic energy of translation,

$$E_1 = \frac{1}{2} m \left[(v_1)^2 - (v_2)^2 \right]$$

This energy must be absorbed by the brake. If the moving body is stopped after applying the brakes, then $v_2 = 0$, and

$$E_1 = \frac{1}{2} m (v_1)^2$$

2. ***When the motion of the body is pure rotation.*** Consider a body of mass moment of inertia I (about a given axis) is rotating about that axis with an angular velocity ω_1 rad / s. Let its angular velocity is reduced to ω_2 rad / s after applying the brake. Therefore, the change in kinetic energy of

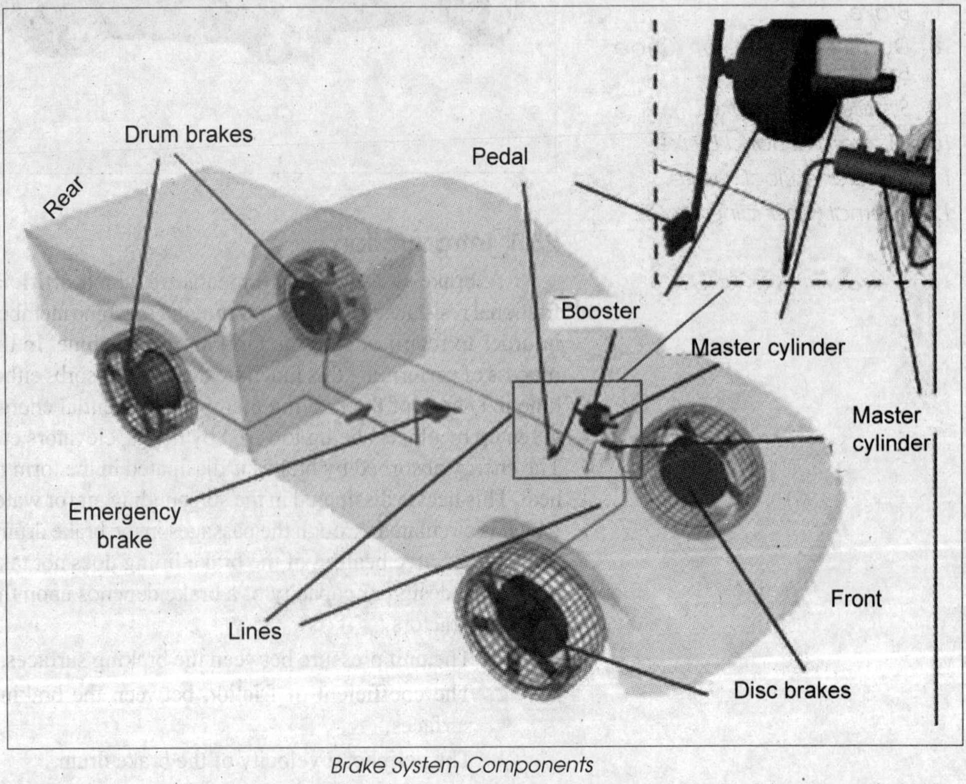

Brake System Components

Note : This picture is given as additional information and is not a direct example of the current chapter.

the rotating body or kinetic energy of rotation,

$$E_2 = \frac{1}{2} I \left[(\omega_1)^2 - (\omega_2)^2\right]$$

This energy must be absorbed by the brake. If the rotating body is stopped after applying the brakes, then $\omega_2 = 0$, and

$$E_2 = \frac{1}{2} I (\omega_1)^2$$

3. ***When the motion of the body is a combination of translation and rotation.*** Consider a body having both linear and angular motions, *e.g.* in the locomotive driving wheels and wheels of a moving car. In such cases, the total kinetic energy of the body is equal to the sum of the kinetic energies of translation and rotation.

∴ Total kinetic energy to be absorbed by the brake,

$$E = E_1 + E_2$$

Sometimes, the brake has to absorb the potential energy given up by objects being lowered by hoists, elevators etc. Consider a body of mass m is being lowered from a height h_1 to h_2 by applying the brake. Therefore the change in potential energy,

$$E_3 = m.g\,(h_1 - h_2)$$

If v_1 and v_2 m / s are the velocities of the mass before and after the brake is applied, then the change in potential energy is given by

$$E_3 = m.g\left(\frac{v_1 + v_2}{2}\right) t = m.g.v.t$$

where v = Mean velocity = $\frac{v_1 + v_2}{2}$, and

t = Time of brake application.

Thus, the total energy to be absorbed by the brake,

$$E = E_1 + E_2 + E_3$$

Let F_t = Tangential braking force or frictional force acting tangentially at the contact surface of the brake drum,

d = Diameter of the brake drum,

N_1 = Speed of the brake drum before the brake is applied,

N_2 = Speed of the brake drum after the brake is applied, and

N = Mean speed of the brake drum = $\frac{N_1 + N_2}{2}$

We know that the work done by the braking or frictional force in time t seconds

$$= F_t \times \pi\, d N \times t$$

Since the total energy to be absorbed by the brake must be equal to the wordone by the frictional force, therefore

$$E = F_t \times \pi\, d N \times t \qquad \text{or} \qquad F_t = \frac{E}{\pi d N.t}$$

The magnitude of F_t depends upon the final velocity (v_2) and on the braking time (t). Its value is maximum when $v_2 = 0$, *i.e.* when the load comes to rest finally.

We know that the torque which must be absorbed by the brake,

$$T = F_t \times r = F_t \times \frac{d}{2}$$

where r = Radius of the brake drum.

25.3 Heat to be Dissipated during Braking

The energy absorbed by the brake and transformed into heat must be dissipated to the surrounding air in order to avoid excessive temperature rise of the brake lining. The *temperature rise depends upon the mass of the brake drum, the braking time and the heat dissipation capacity of the brake. The highest permissible temperatures recommended for different brake lining materials are given as follows :

1. For leather, fibre and wood facing = 65 – 70°C
2. For asbestos and metal surfaces that are slightly lubricated = 90 – 105°C
3. For automobile brakes with asbestos block lining = 180 – 225°C

Since the energy absorbed (or heat generated) and the rate of wear of the brake lining at a particular speed are dependent on the normal pressure between the braking surfaces, therefore it is an important factor in the design of brakes. The permissible normal pressure between the braking surfaces depends upon the material of the brake lining, the coefficient of friction and the maximum rate at which the energy is to be absorbed. The energy absorbed or the heat generated is given by

$$E = H_g = \mu.R_N.v = \mu.p.A.v \text{ (in J/s or watts)} \quad ...(i)$$

where μ = Coefficient of friction,

R_N = Normal force acting at the contact surfaces, in newtons,

p = Normal pressure between the braking surfaces in N/m^2,

A = Projected area of the contact surfaces in m^2, and

v = Peripheral velocity of the brake drum in m/s.

The heat generated may also be obtained by considering the amount of kinetic or potential energies which is being absorbed. In other words,

$$H_g = E_K + E_P$$

where E_K = Total kinetic energy absorbed, and

E_P = Total potential energy absorbed.

The heat dissipated (H_d) may be estimated by

$$H_d = C\,(t_1 - t_2)\,A_r \quad ...(ii)$$

where C = Heat dissipation factor or coefficient of heat transfer in W/m^2/°C

$t_1 - t_2$ = Temperature difference between the exposed radiating surface and the surrounding air in °C, and

A_r = Area of radiating surface in m^2.

The value of C may be of the order of 29.5 W/m^2/°C for a temperature difference of 40°C and increase up to 44 W/m^2/°C for a temperature difference of 200°C.

The expressions for the heat dissipated are quite approximate and should serve only as an indication of the capacity of the brake to dissipate heat. The exact performance of the brake should be determined by test.

It has been found that 10 to 25 per cent of the heat generated is immediately dissipated to the surrounding air while the remaining heat is absorbed by the brake drum causing its temperature to rise. The rise in temperature of the brake drum is given by

$$\Delta t = \frac{H_g}{m.c} \quad ...(iii)$$

where Δt = Temperature rise of the brake drum in °C,

---

* When the temperature increases, the coefficient of friction decreases which adversely affect the torque capacity of the brake. At high temperature, there is a rapid wear of friction lining, which reduces the life of lining. Therefore, the temperature rise should be kept within the permissible range.

H_g = Heat generated by the brake in joules,

m = Mass of the brake drum in kg, and

c = Specific heat for the material of the brake drum in J/kg °C.

In brakes, it is very difficult to precisely calculate the temperature rise. In preliminary design analysis, the product $p.v$ is considered in place of temperature rise. The experience has also shown that if the product $p.v$ is high, the rate of wear of brake lining will be high and the brake life will be low. Thus the value of $p.v$ should be lower than the upper limit value for the brake lining to have reasonable wear life. The following table shows the recommended values of $p.v$ as suggested by various designers for different types of service.

Table 25.1. Recommended values of *p.v*.

S.No.	*Type of service*	*Recommended value of p.v in N-m/m² of projected area per second*
1.	Continuous application of load as in lowering operations and poor dissipation of heat.	0.98×10^6
2.	Intermittent application of load with comparatively long periods of rest and poor dissipation of heat.	1.93×10^6
3.	For continuous application of load and good dissipation of heat as in an oil bath.	2.9×10^6

Example 25.1. *A vehicle of mass 1200 kg is moving down the hill at a slope of 1: 5 at 72 km / h. It is to be stopped in a distance of 50 m. If the diameter of the tyre is 600 mm, determine the average braking torque to be applied to stop the vehicle, neglecting all the frictional energy except for the brake. If the friction energy is momentarily stored in a 20 kg cast iron brake drum, What is average temperature rise of the drum? The specific heat for cast iron may be taken as 520 J / kg°C.*

Determine, also, the minimum coefficient of friction between the tyres and the road in order that the wheels do not skid, assuming that the weight is equally distributed among all the four wheels.

Solution. Given : m = 1200 kg ; Slope = 1: 5 ; v = 72 km / h = 20 m/s ; h = 50 m ; d = 600 mm or r = 300 mm = 0.3 m ; m_b = 20 kg ; c = 520 J / kg°C

Average braking torque to be applied to stop the vehicle

We know that kinetic energy of the vehicle,

$$E_K = \tfrac{1}{2}\, m.v^2 = \tfrac{1}{2} \times 1200\ (20)^2 = 240\ 000 \text{ N-m}$$

and potential energy of the vehicle,

$$E_P = m.g.h \times \text{Slope} = 1200 \times 9.81 \times 50 \times \frac{1}{5} = 117\ 720 \text{ N-m}$$

∴ Total energy of the vehicle or the energy to be absorbed by the brake,

$$E = E_K + E_P = 240\ 000 + 117\ 720 = 357\ 720 \text{ N-m}$$

Since the vehicle is to be stopped in a distance of 50 m, therefore tangential braking force required,

$$F_t = 357\ 720 / 50 = 7154.4 \text{ N}$$

We know that average braking torque to be applied to stop the vehicle,

$$T_B = F_t \times r = 7154.4 \times 0.3 = 2146.32 \text{ N-m} \quad \textbf{Ans.}$$

Average temperature rise of the drum

Let Δt = Average temperature rise of the drum in °C.

We know that the heat absorbed by the brake drum,

$$H_g = \text{Energy absorbed by the brake drum}$$

$$= 357\,720 \text{ N-m} = 357\,720 \text{ J} \qquad \ldots (\because 1 \text{ N-m} = 1 \text{ J})$$

We also know that the heat absorbed by the brake drum (H_g),

$$357\,720 = m_b \times c \times \Delta t = 20 \times 520 \times \Delta t = 10\,400\,\Delta t$$

$$\therefore \quad \Delta t = 357\,720 / 10\,400 = 34.4°\text{C} \quad \textbf{Ans.}$$

Minimum coefficient of friction between the tyre and road

Let μ = Minimum coefficient of friction between the tyre and road, and

R_N = Normal force between the contact surface. This is equal to weight of the vehicle

$$= m.g = 1200 \times 9.81 = 11\,772 \text{ N}$$

We know that tangential braking force (F_t),

$$7154.4 = \mu.R_N = \mu \times 11\,772$$

$$\therefore \quad \mu = 7154.4 / 11772 = 0.6 \quad \textbf{Ans.}$$

25.4 Materials for Brake Lining

The material used for the brake lining should have the following characteristics :

1. It should have high coefficient of friction with minimum fading. In other words, the coefficient of friction should remain constant over the entire surface with change in temperature.
2. It should have low wear rate.
3. It should have high heat resistance.
4. It should have high heat dissipation capacity.
5. It should have low coefficient of thermal expansion.
6. It should have adequate mechanical strength.
7. It should not be affected by moisture and oil.

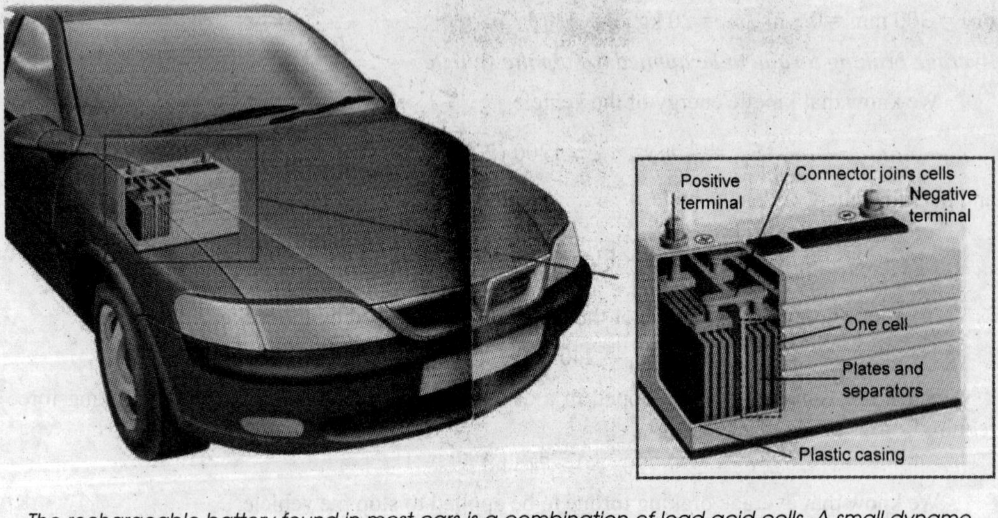

The rechargeable battery found in most cars is a combination of lead acid cells. A small dynamo, driven by the vehicle's engine, charges the battery whenever the engine is running.

The materials commonly used for facing or lining of brakes and their properties are shown in the following table.

Table 25.2. Properties of materials for brake lining.

Material for braking lining	*Coefficient of friction* (μ)			*Allowable pressure (p)*
	Dry	*Greasy*	*Lubricated*	*N/mm²*
Cast iron on cast iron	0.15 – 0.2	0.06 – 0.10	0.05 – 0.10	1.0 – 1.75
Bronze on cast iron	–	0.05 – 0.10	0.05 – 0.10	0.56 – 0.84
Steel on cast iron	0.20 – 0.30	0.07 – 0.12	0.06 – 0.10	0.84 – 1.4
Wood on cast iron	0.20 – 0.35	0.08 – 0.12	–	0.40 – 0.62
Fibre on metal	–	0.10 – 0.20	–	0.07 – 0.28
Cork on metal	0.35	0.25 – 0.30	0.22 – 0.25	0.05 – 0.10
Leather on metal	0.3 – 0.5	0.15 – 0.20	0.12 – 0.15	0.07 – 0.28
Wire asbestos on metal	0.35 – 0.5	0.25 – 0.30	0.20 – 0.25	0.20 – 0.55
Asbestos blocks on metal	0.40 – 0.48	0.25 – 0.30	–	0.28 – 1.1
Asbestos on metal (Short action)	–	–	0.20 – 0.25	1.4 – 2.1
Metal on cast iron (Short action)	–	–	0.05 – 0.10	1.4 – 2.1

25.5 Types of Brakes

The brakes, according to the means used for transforming the energy by the braking element, are classified as :

1. Hydraulic brakes *e.g.* pumps or hydrodynamic brake and fluid agitator,
2. Electric brakes *e.g.* generators and eddy current brakes, and
3. Mechanical brakes.

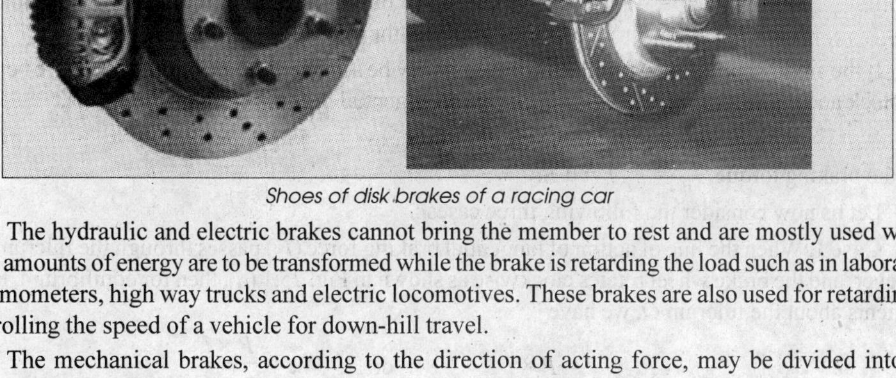

Shoes of disk brakes of a racing car

The hydraulic and electric brakes cannot bring the member to rest and are mostly used where large amounts of energy are to be transformed while the brake is retarding the load such as in laboratory dynamometers, high way trucks and electric locomotives. These brakes are also used for retarding or controlling the speed of a vehicle for down-hill travel.

The mechanical brakes, according to the direction of acting force, may be divided into the following two groups :

(*a*) ***Radial brakes.*** In these brakes, the force acting on the brake drum is in radial direction. The radial brakes may be sub-divided into ***external brakes*** and ***internal brakes.*** According to the shape of the friction element, these brakes may be block or shoe brakes and band brakes.

(*b*) ***Axial brakes.*** In these brakes, the force acting on the brake drum is in axial direction. The axial brakes may be disc brakes and cone brakes. The analysis of these brakes is similar to clutches.

Since we are concerned with only mechanical brakes, therefore, these are discussed in detail, in the following pages.

25.6 Single Block or Shoe Brake

A single block or shoe brake is shown in Fig. 25.1. It consists of a block or shoe which is pressed against the rim of a revolving brake wheel drum. The block is made of a softer material than

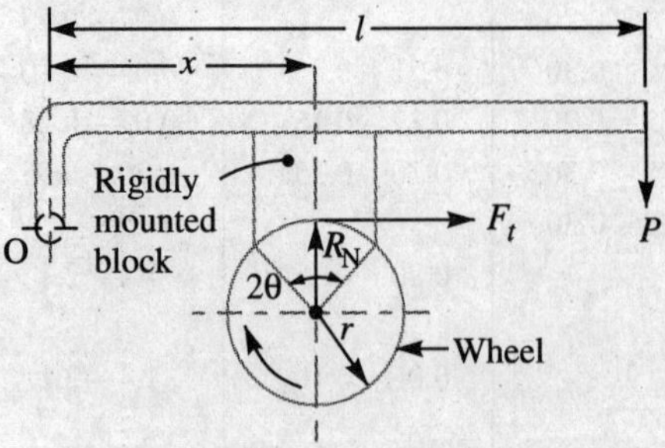

(*a*) Clockwise rotation of brake wheel.

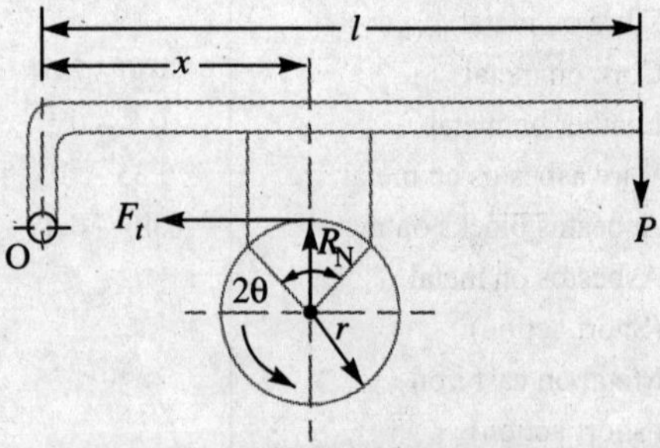

(*b*) Anticlockwise rotation of brake wheel.

Fig. 25.1. Single block brake. Line of action of tangential force passes through the fulcrum of the lever.

the rim of the wheel. This type of a brake is commonly used on railway trains and tram cars. The friction between the block and the wheel causes a tangential braking force to act on the wheel, which retard the rotation of the wheel. The block is pressed against the wheel by a force applied to one end of a lever to which the block is rigidly fixed as shown in Fig. 25.1. The other end of the lever is pivoted on a fixed fulcrum *O*.

Let P = Force applied at the end of the lever,

R_N = Normal force pressing the brake block on the wheel,

r = Radius of the wheel,

2θ = Angle of contact surface of the block,

μ = Coefficient of friction, and

F_t = Tangential braking force or the frictional force acting at the contact surface of the block and the wheel.

If the angle of contact is less than 60°, then it may be assumed that the normal pressure between the block and the wheel is uniform. In such cases, tangential braking force on the wheel,

$$F_t = \mu.R_N \quad ...(i)$$

and the braking torque, $T_B = F_t \cdot r = \mu \, R_N \cdot r$ ...(*ii*)

Let us now consider the following three cases :

Case 1. When the line of action of tangential braking force (F_t) passes through the fulcrum *O* of the lever, and the brake wheel rotates clockwise as shown in Fig. 25.1 (*a*), then for equilibrium, taking moments about the fulcrum *O*, we have

$$R_N \times x = P \times l \quad \text{or} \quad R_N = \frac{P \times l}{x}$$

$\therefore$ Braking torque, $$T_B = \mu.R_N.r = \mu \times \frac{P.l}{x} \times r = \frac{\mu.P.l.r}{x}$$

It may be noted that when the brake wheel rotates anticlockwise as shown in Fig. 25.1 (*b*), then the braking torque is same, *i.e.*

$$T_B = \mu.R_N.r = \frac{\mu.P.l.r}{x}$$

Case 2. When the line of action of the tangential braking force (F_t) passes through a distance '*a*' below the fulcrum *O*, and the brake wheel rotates clockwise as shown in Fig. 25.2 (*a*), then for equilibrium, taking moments about the fulcrum *O*,

$$R_N \times x + F_t \times a = P.l$$

or $$R_N \times x + \mu\, R_N \times a = P.l \qquad \text{or} \qquad R_N = \frac{P.l}{x + \mu.a}$$

and braking torque, $$T_B = \mu\, R_N.r = \frac{\mu.P.l.r}{x + \mu.a}$$

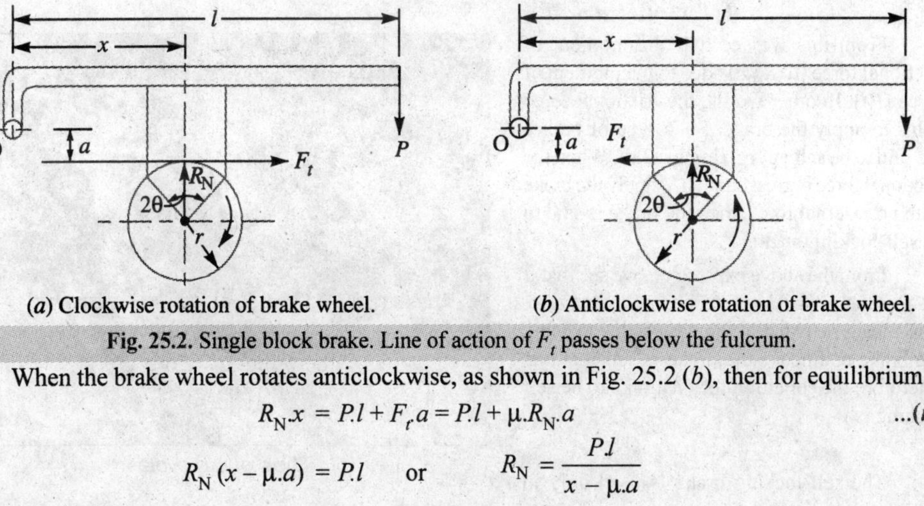

(*a*) Clockwise rotation of brake wheel. (*b*) Anticlockwise rotation of brake wheel.

Fig. 25.2. Single block brake. Line of action of F_t passes below the fulcrum.

When the brake wheel rotates anticlockwise, as shown in Fig. 25.2 (*b*), then for equilibrium,

$$R_N.x = P.l + F_t.a = P.l + \mu.R_N.a \qquad \ldots(i)$$

or $$R_N\,(x - \mu.a) = P.l \qquad \text{or} \qquad R_N = \frac{P.l}{x - \mu.a}$$

and braking torque, $$T_B = \mu.R_N.r = \frac{\mu.P.l.r}{x - \mu.a}$$

Case 3. When the line of action of the tangential braking force passes through a distance '*a*' above the fulcrum, and the brake wheel rotates clockwise as shown in Fig. 25.3 (*a*), then for equilibrium, taking moments about the fulcrum *O*, we have

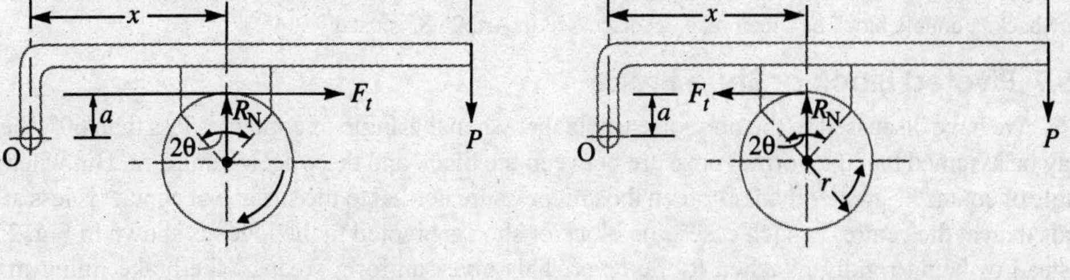

(*a*) Clockwise rotation of brake wheel. (*b*) Anticlockwise rotation of brake wheel.

Fig. 25.3. Single block brake. Line of action of F_t passes above the fulcrum.

$$R_N.x = P.l + F_t.a = P.l + \mu.R_N.a \qquad ...(ii)$$

or $$R_N (x - \mu.a) = P.l \quad \text{or} \quad R_N = \frac{P.l}{x - \mu.a}$$

and braking torque, $$T_B = \mu.R_N.r = \frac{\mu.P.l.r}{x - \mu.a}$$

When the brake wheel rotates anticlockwise as shown in Fig. 25.3 (*b*), then for equilibrium, taking moments about the fulcrum *O*, we have

$$R_N \times x + F_t \times a = P.l$$

or $$R_N \times x + \mu.R_N \times a = P.l \quad \text{or} \quad R_N = \frac{P.l}{x + \mu.a}$$

and braking torque, $$T_B = \mu.R_N.r = \frac{\mu.P.l.r}{x + \mu.a}$$

Notes: 1. From above we see that when the brake wheel rotates anticlockwise in case 2 [Fig. 25.2 (*b*)] and when it rotates clockwise in case 3 [Fig. 25.3 (*a*)], the equations (*i*) and (*ii*) are same, *i.e.*

$$R_N \times x = P.l + \mu.R_N.a$$

From this we see that the moment of frictional force ($\mu.\ R_N.a$) adds to the moment of force (*P.l*). In other words, the frictional force helps to apply the brake. Such type of brakes are said to be **self energizing brakes.** When the frictional force is great enough to apply the brake with no external force, then the brake is said to be **self-locking brake.**

From the above expression, we see that if $x \le \mu.a$, then *P* will be negative or equal to zero. This means no external force is needed to apply the brake and hence the brake is self locking. Therefore the condition for the brake to be self locking is

$$x \le \mu.a$$

The self-locking brake is used only in back-stop applications.

Shoe of a bicycle

2. The brake should be self-energizing and not the self-locking.

3. In order to avoid self-locking and to prevent the brake from grabbing, *x* is kept greater than $\mu.a$.

4. If A_b is the projected bearing area of the block or shoe, then the bearing pressure on the shoe,

$$p_b = R_N / A_b$$

We know that A_b = Width of shoe × Projected length of shoe = $w\ (2r \sin \theta)$

5. When a single block or shoe brake is applied to a rolling wheel, an additional load is thrown on the shaft bearings due to heavy normal force (R_N) and produces bending of the shaft. In order to overcome this drawback, a double block or shoe brake, as discussed in Art. 25.8, is used.

25.7 Pivoted Block or Shoe Brake

We have discussed in the previous article that when the angle of contact is less than 60°, then it may be assumed that the normal pressure between the block and the wheel is uniform. But when the angle of contact is greater than 60°, then the unit pressure normal to the surface of contact is less at the ends than at the centre. In such cases, the block or shoe is pivoted to the lever as shown in Fig. 25.4, instead of being rigidly attached to the lever. This gives uniform wear of the brake lining in the direction of the applied force. The braking torque for a pivoted block or shoe brake (*i.e.* when $2\theta > 60°$) is given by

$$T_B = F_t \times r = \mu'.R_N.r$$

where μ' = Equivalent coefficient of friction $= \dfrac{4\mu \sin\theta}{2\theta + \sin 2\theta}$, and

μ = Actual coefficient of friction.

These brakes have more life and may provide a higher braking torque.

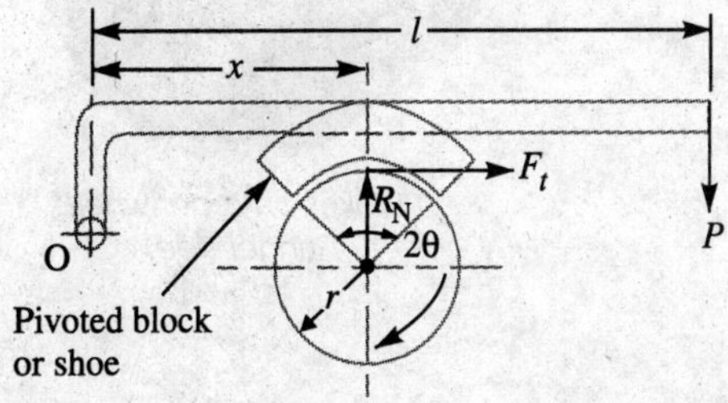

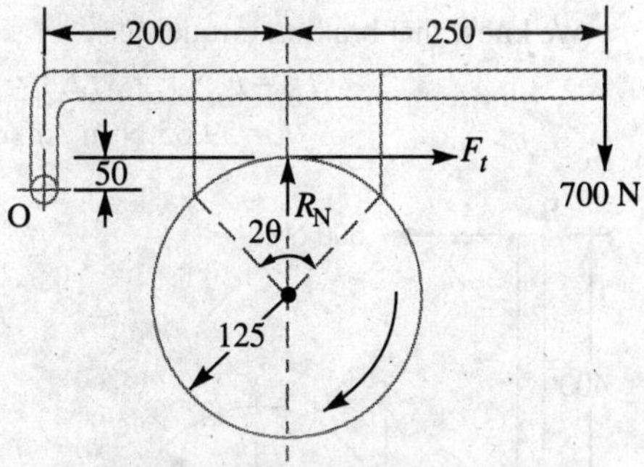

All dimensions in mm.

Fig. 25.4. Pivoted block or shoe brake. Fig. 25.5

Example 25.2. *A single block brake is shown in Fig. 25.5. The diameter of the drum is 250 mm and the angle of contact is 90°. If the operating force of 700 N is applied at the end of a lever and the coefficient of friction between the drum and the lining is 0.35, determine the torque that may be transmitted by the block brake.*

Solution. Given : $d = 250$ mm or $r = 125$ mm ; $2\theta = 90° = \pi / 2$ rad ; $P = 700$ N ; $\mu = 0.35$

Since the angle of contact is greater than 60° , therefore equivalent coefficient of friction,

$$\mu' = \frac{4\mu \sin\theta}{2\theta + \sin 2\theta} = \frac{4 \times 0.35 \times \sin 45°}{\pi/2 + \sin 90°} = 0.385$$

Let R_N = Normal force pressing the block to the brake drum, and

F_t = Tangential braking force = $\mu'.R_N$

Taking moments above the fulcrum O, we have

$$700\,(250 + 200) + F_t \times 50 = R_N \times 200 = \frac{F_t}{\mu'} \times 200 = \frac{F_t}{0.385} \times 200 = 520\, F_t$$

or $520\, F_t - 50\, F_t = 700 \times 450$ or $F_t = 700 \times 450 / 470 = 670$ N

We know that torque transmitted by the block brake,

$$T_B = F_t \times r = 670 \times 125 = 83\ 750 \text{ N-mm} = 83.75 \text{ N-m}$$ **Ans.**

Example 25.3. *Fig. 25.6 shows a brake shoe applied to a drum by a lever AB which is pivoted at a fixed point A and rigidly fixed to the shoe. The radius of the drum is 160 mm. The coefficient of friction of the brake lining is 0.3. If the drum rotates clockwise, find the braking torque due to the horizontal force of 600 N applied at B.*

Solution. Given : $r = 160$ mm = 0.16 m ; $\mu = 0.3$; $P = 600$ N

Since the angle subtended by the shoe at the centre of the drum is 40°, therefore we need not to calculate the equivalent coefficient of friction (μ').

Let R_N = Normal force pressing the shoe on the drum, and

F_t = Tangential braking force = $\mu.R_N$

Taking moments about point A,

$$R_N \times 350 + F_t (200 - 160) = 600 (400 + 350)$$

$$\frac{F_t}{0.3} \times 350 + 40 F_t = 600 \times 750$$

or $$207 F_t = 450 \times 10^3$$

$$\therefore \quad F_t = 450 \times 10^3 / 1207 = 372.8 \text{ N}$$

We know that braking torque,

$$T_B = F_t \times r = 372.8 \times 0.16$$
$$= 59.65 \text{ N-m } \textbf{Ans.}$$

Brakes on a car wheel (inner side)

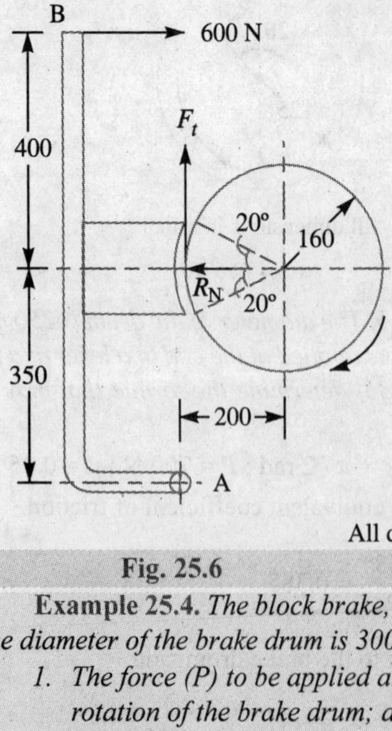

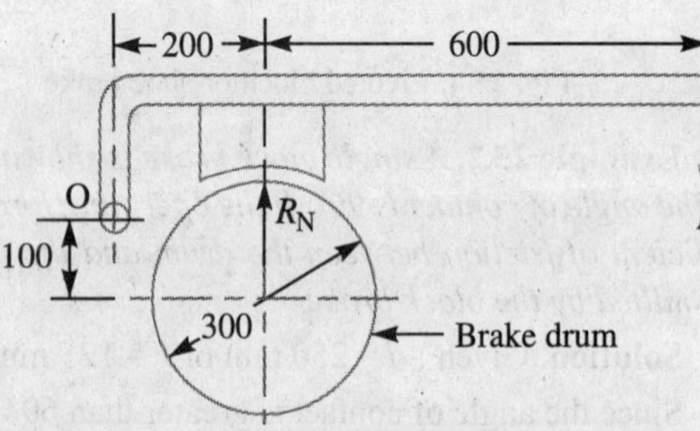

All dimensions in mm.

Fig. 25.6 **Fig. 25.7**

Example 25.4. *The block brake, as shown in Fig. 25.7, provides a braking torque of 360 N-m. The diameter of the brake drum is 300 mm. The coefficient of friction is 0.3. Find :*

1. *The force (P) to be applied at the end of the lever for the clockwise and counter clockwise rotation of the brake drum; and*
2. *The location of the pivot or fulcrum to make the brake self locking for the clockwise rotation of the brake drum.*

Solution. Given : $T_B = 360$ N-m $= 360 \times 10^3$ N-mm ; $d = 300$ mm or $r = 150$ mm $= 0.15$ m ; $\mu = 0.3$

1. ***Force (P) for the clockwise and counter clockwise rotation of the brake drum***

For the clockwise rotation of the brake drum, the frictional force or the tangential force (F_t) acting at the contact surfaces is shown in Fig. 25.8.

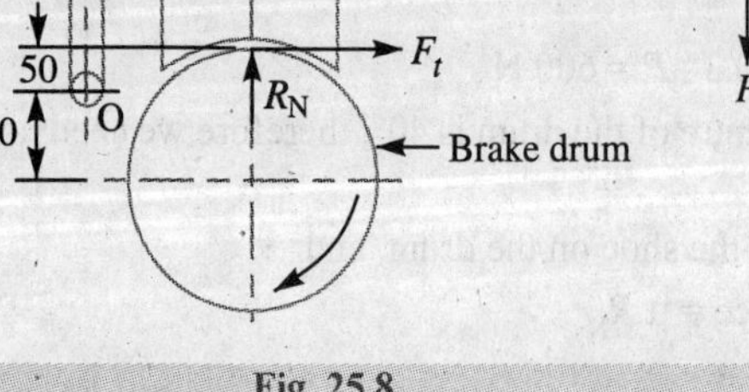

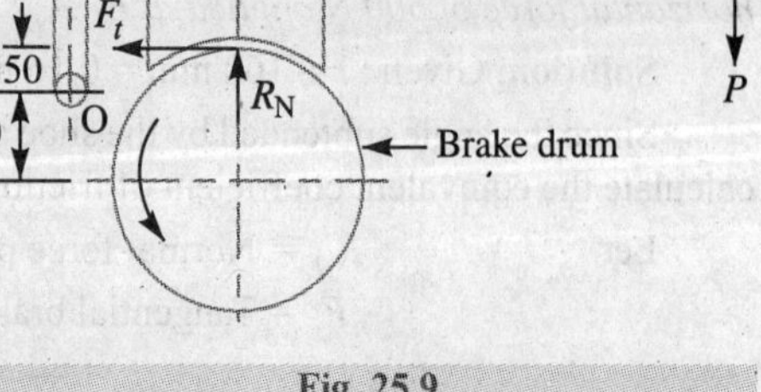

Fig. 25.8 **Fig. 25.9**

We know that braking torque (T_B),

$$360 = F_t \times r = F_t \times 0.15 \quad \text{or} \quad F_t = 360 / 0.15 = 2400 \text{ N}$$

and normal force, $$R_N = F_t / \mu = 2400 / 0.3 = 8000 \text{ N}$$

Now taking moments about the fulcrum O, we have

$$P(600 + 200) + F_t \times 50 = R_N \times 200$$

$$P \times 800 + 2400 \times 50 = 8000 \times 200$$

$$P \times 800 = 8000 \times 200 - 2400 \times 50 = 1480 \times 10^3$$

$$\therefore \quad P = 1480 \times 10^3 / 800 = 1850 \text{ N} \quad \textbf{Ans.}$$

For the counter clockwise rotation of the drum, the frictional force or the tangential force (F_t) acting at the contact surfaces is shown in Fig. 25.9.

Taking moments about the fulcrum O, we have

$$P(600 + 200) = F_t \times 50 + R_N \times 200$$

$$P \times 800 = 2400 \times 50 + 8000 \times 200 = 1720 \times 10^3$$

$$\therefore \quad P = 1720 \times 10^3 / 800 = 2150 \text{ N} \quad \textbf{Ans.}$$

2. *Location of the pivot or fulcrum to make the brake self-locking*

The clockwise rotation of the brake drum is shown in Fig. 25.8. Let x be the distance of the pivot or fulcrum O from the line of action of the tangential force (F_t). Taking moments about the fulcrum O, we have

$$P(600 + 200) + F_t \times x - R_N \times 200 = 0$$

In order to make the brake self-locking, $F_t \times x$ must be equal to $R_N \times 200$ so that the force P is zero.

$$\therefore \quad F_t \times x = R_N \times 200$$

$$2400 \times x = 8000 \times 200 \quad \text{or} \quad x = 8000 \times 200 / 2400 = 667 \text{ mm} \quad \textbf{Ans.}$$

Example 25.5. *A rope drum of an elevator having 650 mm diameter is fitted with a brake drum of 1 m diameter. The brake drum is provided with four cast iron brake shoes each subtending an angle of 45°. The mass of the elevator when loaded is 2000 kg and moves with a speed of 2.5 m / s. The brake has a sufficient capacity to stop the elevator in 2.75 metres. Assuming the coefficient of friction between the brake drum and shoes as 0.2, find: 1. width of the shoe, if the allowable pressure on the brake shoe is limited to 0.3 N/mm²; and 2. heat generated in stopping the elevator.*

Solution. Given : $d_e = 650$ mm or $r_e = 325$ mm $= 0.325$ m ; $d = 1$ m or $r = 0.5$ m $= 500$ mm ; $n = 4$; $2\theta = 45°$ or $\theta = 22.5°$; $m = 2000$ kg ; $v = 2.5$ m / s ; $h = 2.75$ m ; $\mu = 0.2$; $p_b = 0.3$ N/mm²

1. *Width of the shoe*

Let w = Width of the shoe in mm.

First of all, let us find out the acceleration of the rope (a). We know that

$$v^2 - u^2 = 2\,a.h \quad \text{or} \quad (2.5)^2 - 0 = 2a \times 2.75 = 5.5a$$

$$\therefore \quad a = (2.5)^2 / 5.5 = 1.136 \text{ m/s}^2$$

and accelerating force = Mass × Acceleration = $m \times a = 2000 \times 1.136 = 2272$ N

∴ Total load acting on the rope while moving,

W = Load on the elevator in newtons + Accelerating force

$= 2000 \times 9.81 + 2272 = 21\,892$ N

We know that torque acting on the shaft,

$$T = W \times r_e = 21\,892 \times 0.325 = 7115 \text{ N-m}$$

∴ Tangential force acting on the drum

$$= \frac{T}{r} = \frac{7115}{0.5} = 14\ 230 \text{ N}$$

The brake drum is provided with four cast iron shoes, therefore tangential force acting on each shoe,

$$F_t = 14\ 230 / 4 = 3557.5 \text{ N}$$

Since the angle of contact of each shoe is 45°, therefore we need not to calculate the equivalent coefficient of friction (μ′).

∴ Normal load on each shoe,

$$R_N = F_t / \mu = 3557.5 / 0.2 = 17\ 787.5 \text{ N}$$

We know that the projected bearing area of each shoe,

$$A_b = w\,(2r \sin \theta) = w\,(2 \times 500 \sin 22.5°) = 382.7\ w \text{ mm}^2$$

We also know that bearing pressure on the shoe (p_b),

$$0.3 = \frac{R_N}{A_b} = \frac{17\ 787.5}{382.7\ w} = \frac{46.5}{w}$$

$$\therefore \quad w = 46.5 / 0.3 = 155 \text{ mm } \textbf{Ans.}$$

2. ***Heat generated in stopping the elevator***

We know that heat generated in stopping the elevator

$$= \text{Total energy absorbed by the brake}$$

$$= \text{Kinetic energy} + \text{Potential energy} = \frac{1}{2}\,m.v^2 + m.g.h$$

$$= \frac{1}{2} \times 2000\,(2.5)^2 + 2000 \times 9.81 \times 2.75 = 60\ 205 \text{ N-m}$$

$$= 60.205 \text{ kN-m} = 60.205 \text{ kJ } \textbf{Ans.}$$

25.8 Double Block or Shoe Brake

When a single block brake is applied to a rolling wheel, and additional load is thrown on the shaft bearings due to the normal force (R_N). This produces bending of the shaft. In order to overcome this drawback, a double block or shoe brake as shown in Fig. 25.10, is used. It consists of two brake blocks applied at the opposite ends of a diameter of the wheel which eliminate or reduces the unbalanced force on the shaft. The brake is set by a spring which pulls the upper ends of the brake arms together. When a force P is applied to the bell crank lever, the spring is compressed and the brake is released. This type of brake is often used on electric cranes and the force P is produced by an electromagnet or solenoid. When the current is switched off, there is no force on the bell crank lever and the brake is engaged automatically due to the spring force and thus there will be no downward movement of the load.

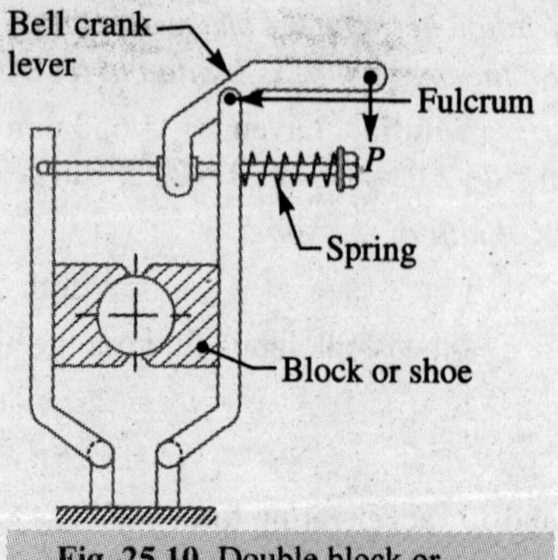

Fig. 25.10. Double block or shoe brake.

In a double block brake, the braking action is doubled by the use of two blocks and the two blocks may be operated practically by the same force which will operate one. In case of double block or shoe brake, the braking torque is given by

$$T_B = (F_{t1} + F_{t2})\,r$$

where F_{t1} and F_{t2} are the braking forces on the two blocks.

Example 25.6. *A double shoe brake, as shown in Fig. 25.11 is capable of absorbing a torque of 1400 N-m. The diameter of the brake drum is 350 mm and the angle of contact for each shoe is 100°. If the coefficient of friction between the brake drum and lining is 0.4; find : 1. the spring force* *necessary to set the brake; and 2. the width of the brake shoes, if the bearing pressure on the lining material is not to exceed 0.3 N/mm².*

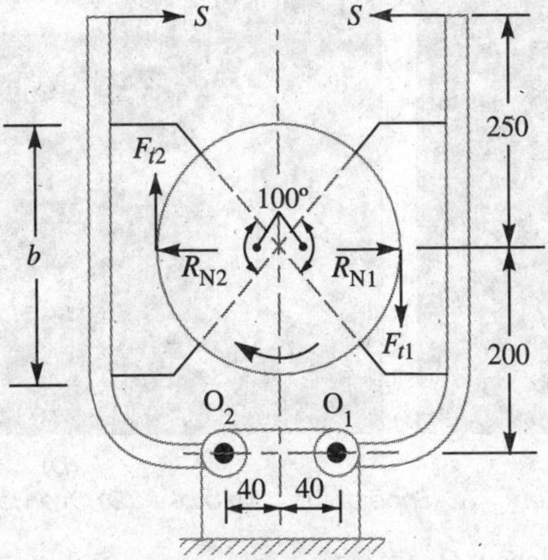

All dimensions in mm.

Fig. 25.11

Solution. Given : T_B = 1400 N-m = 1400 × 10³ N-mm ; d = 350 mm or r = 175 mm ; $2\theta = 100° = 100 \times \pi / 180 = 1.75$ rad ; $\mu = 0.4$; $p_b = 0.3$ N/mm²

1. ***Spring force necessary to set the brake***

Let S = Spring force necessary to set the brake,

R_{N1} and F_{t1} = Normal reaction and the braking force on the right hand side shoe, and

R_{N2} and F_{t2} = Corresponding values on the left hand side shoe.

Since the angle of contact is greater than 60°, therefore equivalent coefficient of friction,

$$\mu' = \frac{4\mu \sin\theta}{2\theta + \sin 2\theta} = \frac{4 \times 0.4 \times \sin 50°}{1.75 + \sin 100°} = 0.45$$

Taking moments about the fulcrum O_1, we have

$$S \times 450 = R_{N1} \times 200 + F_{t1}\,(175 - 40) = \frac{F_{t1}}{0.45} \times 200 + F_{t1} \times 135 = 579.4\,F_{t1}$$

...(Substituting $R_{N1} = F_{t1} / \mu'$)

$$\therefore \quad F_{t1} = S \times 450 / 579.4 = 0.776\,S$$

(A)

(B)

Train braking system : (A) Flexible hose carries the brakepipe between car; (B) Brake hydraulic cylinder and the associated hardware.

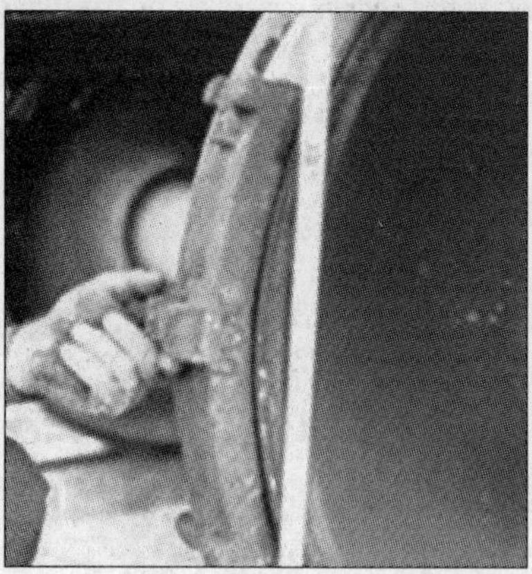
(C)

(D)

Train braking system : (C) Shoe of the train brake (D) Overview of train brake

Again taking moments about O_2, we have

$$S \times 450 + F_{t2} (175 - 40) = R_{N2} \times 200 = \frac{F_{t2}}{0.45} \times 200 = 444.4\ F_{t2}$$

...(Substituting $R_{N2} = F_{t2}/\mu'$)

$$444.4\ F_{t2} - 135\ F_{t2} = S \times 450 \quad \text{or} \quad 309.4\ F_{t2} = S \times 450$$

$$\therefore \quad F_{t2} = S \times 450 / 309.4 = 1.454\ S$$

We know that torque capacity of the brake (T_B),

$$1400 \times 10^3 = (F_{t1} + F_{t2})\ r = (0.776\ S + 1.454\ S)\ 175 = 390.25\ S$$

$$\therefore \quad S = 1400 \times 10^3 / 390.25 = 3587 \text{ N } \textbf{Ans.}$$

2. *Width of the brake shoes*

Let b = Width of the brake shoes in mm.

We know that projected bearing area for one shoe,

$$A_b = b\ (2r \sin \theta) = b\ (2 \times 175 \sin 50°) = 268\ b \text{ mm}^2$$

$\therefore$ Normal force on the right hand side of the shoe,

$$R_{N1} = \frac{F_{t1}}{\mu'} = \frac{0.776 \times S}{0.45} = \frac{0.776 \times 3587}{0.45} = 6186 \text{ N}$$

and normal force on the left hand side of the shoe,

$$R_{N2} = \frac{F_{t2}}{\mu'} = \frac{1.454 \times S}{0.45} = \frac{1.454 \times 3587}{0.45} = 11\ 590 \text{ N}$$

We see that the maximum normal force is on the left hand side of the shoe. Therefore we shall design the shoe for the maximum normal force *i.e.* R_{N2}.

We know that the bearing pressure on the lining material (p_b),

$$0.3 = \frac{R_{N2}}{A_b} = \frac{11\ 590}{268 b} = \frac{43.25}{b}$$

$$\therefore \quad b = 43.25 / 0.3 = 144.2 \text{ mm } \textbf{Ans.}$$

Example 25.7. *A spring closed thrustor operated double shoe brake is to be designed for a maximum torque capacity of 3000 N-m. The brake drum diameter is not to exceed 1 metre and the shoes are to be lined with Ferrodo having a coefficient of friction 0.3. The other dimensions are as shown in Fig. 25.12.*

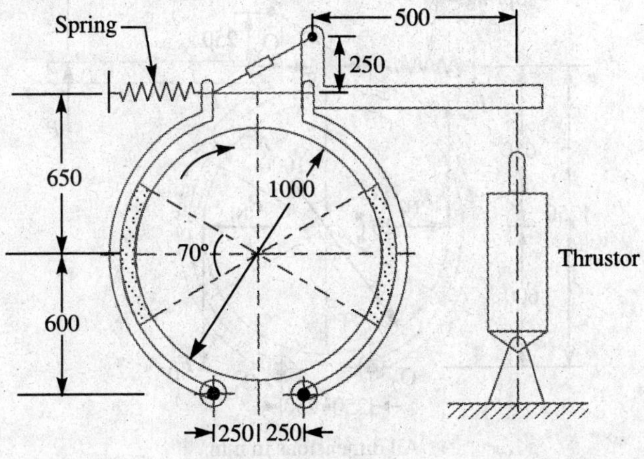

Fig. 25.12

1. *Find the spring force necessary to set the brake.*
2. *If the permissible stress of the spring material is 500 MPa, determine the dimensions of the coil assuming spring index to be 6. The maximum spring force is to be 1.3 times the spring force required during braking. There are eight active coils. Specify the length of the spring in the closed position of the brake. Modulus of rigidity is 80 kN / mm².*
3. *Find the width of the brake shoes if the bearing pressure on the lining material is not to exceed 0.5 N/mm².*
4. *Calculate the force required to be exerted by the thrustor to release the brake.*

Solution. Given : $T_B = 3000$ N-m $= 3 \times 10^6$ N-mm ; $d = 1$ m or $r = 0.5$ m $= 500$ mm ; $\mu = 0.3$; $2\theta = 70° = 70 \times \pi / 180 = 1.22$ rad

1. ***Spring force necessary to set the brake***

Let S = Spring force necessary to set the brake,

R_{N1} and F_{t1} = Normal reaction and the braking force on the right hand side shoe,

and R_{N2} and F_{t2} = Corresponding values for the left hand side shoe.

Since the angle of contact is greater than 60°, therefore equivalent coefficient of friction,

$$\mu' = \frac{4\mu \sin\theta}{2\theta + \sin 2\theta} = \frac{4 \times 0.3 \times \sin 35°}{1.22 + \sin 70°} = 0.32$$

Taking moments about the fulcrum O_1 (Fig. 25.13), we have

$$S \times 1250 = R_{N1} \times 600 + F_{t1}\,(500 - 250)$$

$$= \frac{F_{t1}}{0.32} \times 600 + 250\,F_{t1} = 2125\,F_{t1} \qquad \ldots(\because R_{N1} = F_{t1} / \mu')$$

$$\therefore \quad F_{t1} = S \times 1250 / 2125 = 0.59\,S \text{ N}$$

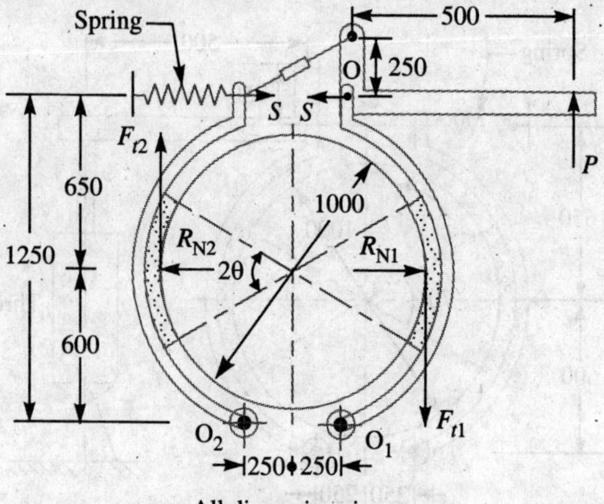

Fig. 25.13

Again taking moments about the fulcrum O_2, we have

$$S \times 1250 + F_{t2}(500 - 250) = R_{N2} \times 600 = \frac{F_{t2}}{0.32} \times 600 = 1875\ F_{t2} \qquad ...(\because R_{N2} = F_{t2}/\mu')$$

or $\quad 1875\ F_{t2} - 250\ F_{t2} = S \times 1250 \quad$ or $\quad 1625\ F_{t2} = S \times 1250$

$\therefore \quad F_{t2} = S \times 1250 / 1625 = 0.77\ S$ N

We know that torque capacity of the brake (T_B),

$$3 \times 10^6 = (F_{t1} + F_{t2})\, r = (0.59\ S + 0.77\ S)\ 500 = 680\ S$$

$\therefore \quad S = 3 \times 10^6 / 680 = 4412$ N **Ans.**

2. ***Dimensions of the spring coil***

Given : $\tau = 500$ MPa $= 500$ N/mm^2 ; $C = D/d = 6$; $n = 8$; $G = 80$ kN/mm$^2 = 80 \times 10^3$ N/mm^2

Let $\quad D$ = Mean diameter of the spring, and

$\quad d$ = Diameter of the spring wire.

We know that Wahl's stress factor,

$$K = \frac{4C - 1}{4C - 4} + \frac{0.615}{C} = \frac{4 \times 6 - 1}{4 \times 6 - 4} + \frac{0.615}{6} = 1.2525$$

Since the maximum spring force is 1.3 times the spring force required during braking, therefore maximum spring force,

$$W_S = 1.3\ S = 1.3 \times 4412 = 5736 \text{ N}$$

We know that the shear stress induced in the spring (τ),

$$500 = \frac{K \times 8 W_S . C}{\pi d^2} = \frac{1.2525 \times 8 \times 5736 \times 6}{\pi d^2} = \frac{109\ 754}{d^2}$$

$\therefore \quad d^2 = 109\ 754 / 500 = 219.5 \quad$ or $\quad d = 14.8$ say 15 mm **Ans.**

and $\quad D = C.d = 6 \times 15 = 90$ mm **Ans.**

We know that deflection of the spring,

$$\delta = \frac{8 W_S . C^3 . n}{G.d} = \frac{8 \times 5736 \times 6^3 \times 8}{80 \times 10^3 \times 15} = 66 \text{ mm}$$

The length of the spring in the closed position of the brake will be its free length. Assuming that the ends of the coil are squared and ground, therefore total number of coils,

$$n' = n + 2 = 8 + 2 = 10$$

∴ Free length of the spring,

$$L_F = n'.d + \delta + 0.15\,\delta$$
$$= 10 \times 15 + 66 + 0.15 \times 66 = 226 \text{ mm } \textbf{Ans.}$$

3. ***Width of the brake shoes***

Let b = Width of the brake shoes in mm, and

p_b = Bearing pressure on the lining material of the shoes.

$= 0.5 \text{ N/mm}^2$ ...(Given)

We know that projected bearing area for one shoe,

$$A_b = b\,(2r.\sin\theta) = b\,(2 \times 500 \sin 35°) = 574\,b \text{ mm}^2$$

We know that normal force on the right hand side of the shoe,

$$R_{N1} = \frac{F_{t1}}{\mu'} = \frac{0.59\,S}{0.32} = \frac{0.59 \times 4412}{0.32} = 8135 \text{ N}$$

and normal force on the left hand side of the shoe,

$$R_{N2} = \frac{F_{t2}}{\mu'} = \frac{0.77\,S}{0.32} = \frac{0.77 \times 4412}{0.32} = 10\,616 \text{ N}$$

We see that the maximum normal force is on the left hand side of the shoe. Therefore we shall design the shoe for the maximum normal force *i.e.* R_{N2}.

We know that bearing pressure on the lining material (p_b),

$$0.5 = \frac{R_{N2}}{A_b} = \frac{10\,616}{574\,b} = \frac{18.5}{b}$$

∴ $b = 18.5 / 0.5 = 37$ mm **Ans.**

4. ***Force required to be exerted by the thrustor to release the brake***

Let P = Force required to be exerted by the thrustor to release the brake.

Taking moments about the fulcrum of the lever O, we have

$$P \times 500 + R_{N1} \times 650 = F_{t1}\,(500 - 250) + F_{t2}\,(500 + 250) + R_{N2} \times 650$$

$$P \times 500 + 8135 \times 650 = 0.59 \times 4412 + 250 + 0.77 \times 4412 \times 750 + 10\,616 \times 650$$

...(Substituting $F_{t1} = 0.59\,S$ and $F_{t2} = 0.77\,S$)

$$P \times 500 + 5.288 \times 10^6 = 0.65 \times 10^6 + 2.55 \times 10^6 + 6.9 \times 10^6 = 10.1 \times 10^6$$

$$\therefore \quad P = \frac{10.1 \times 10^6 - 5.288 \times 10^6}{500} = 9624 \text{ N } \textbf{Ans.}$$

25.9 Simple Band Brake

A band brake consists of a flexible band of leather, one or more ropes, or a steel lined with friction material, which embraces a part of the circumference of the drum. A band brake, as shown in Fig. 25.14, is called a ***simple band brake*** in which one end of the band is attached to a fixed pin or fulcrum of the lever while the other end is attached to the lever at a distance b from the fulcrum.

When a force P is applied to the lever at C, the lever turns about the fulcrum pin O and tightens the band on the drum and hence the brakes are applied. The friction between the band and the drum provides the braking force. The force P on the lever at C may be determined as discussed below :

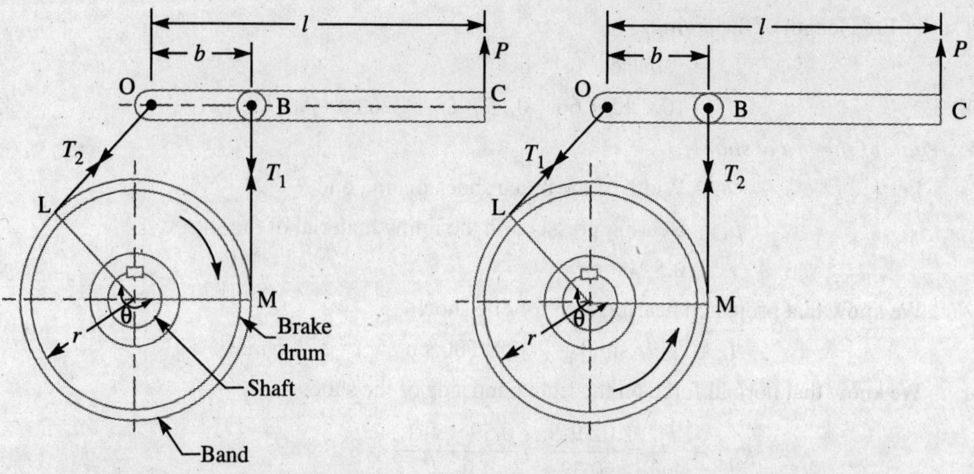

(*a*) Clockwise rotation of drum. (*b*) Anticlockwise rotation of drum.

Fig. 25.14. Simple band brake.

Let T_1 = Tension in the tight side of the band,

T_2 = Tension in the slack side of the band,

θ = Angle of lap (or embrace) of the band on the drum,

μ = Coefficient of friction between the band and the drum,

r = Radius of the drum,

t = Thickness of the band, and

r_e = Effective radius of the drum = $r + t / 2$.

We know that limiting ratio of the tensions is given by the relation,

$$\frac{T_1}{T_2} = e^{\mu.\theta} \quad \text{or} \quad 2.3 \log \left(\frac{T_1}{T_2}\right) = \mu.\theta$$

and braking force on the drum

$$= T_1 - T_2$$

∴ Braking torque on the drum,

$$T_B = (T_1 - T_2)\, r \quad \text{...(Neglecting thickness of band)}$$

$$= (T_1 - T_2)\, r_e \quad \text{...(Considering thickness of band)}$$

Now considering the equilibrium of the lever OBC. It may be noted that when the drum rotates in the clockwise direction as shown in Fig. 25.14 (*a*), the end of the band attached to the fulcrum O will be slack with tension T_2 and end of the band attached to B will be tight with tension T_1. On the other hand, when the drum rotates in the anticlockwise direction as shown in Fig. 25.14 (*b*), the tensions in the band will reverse, *i.e.* the end of the band attached to the fulcrum O will be tight with tension T_1 and the end of the band attached to B will be slack with tension T_2. Now taking moments about the fulcrum O, we have

$$P.l = T_1.b \quad \text{...(for clockwise rotation of the drum)}$$

and

$$P.l = T_2.b \quad \text{...(for anticlockwise rotation of the drum)}$$

where l = Length of the lever from the fulcrum (OC), and

b = Perpendicular distance from O to the line of action of T_1 or T_2.

Notes: 1. When the brake band is attached to the lever, as shown in Fig. 25.14 (*a*) and (*b*), then the force (P) must act in the upward direction in order to tighten the band on the drum.

2. Sometimes the brake band is attached to the lever as shown in Fig. 25.15 (*a*) and (*b*), then the force (P) must act in the downward direction in order to tighten the band. In this case, for clockwise rotation of the drum, the end of the band attached to the fulcrum O will be tight with tension T_1 and band of the band attached to B will be slack with tension T_2. The tensions T_1 and T_2 will reverse for anticlockwise rotation of the drum.

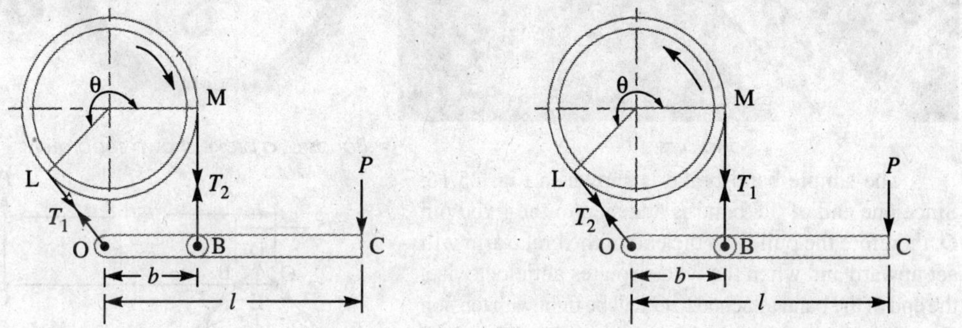

(*a*) Clockwise rotation of drum. (*b*) Anticlockwise rotation of drum.

Fig. 25.15. Simple band brake.

3. If the permissible tensile stress (σ_t) for the material of the band is known, then maximum tension in the band is given by

$$T_1 = \sigma_t \times w \times t$$

where w = Width of the band, and

t = Thickness of the band.

4. The width of band (w) should not exceed 150 mm for drum diameter (d) greater than 1 metre and 100 mm for drum diameter less than 1 metre. The band thickness (t) may also be obtained by using the empirical relation *i.e.* $t = 0.005\ d$

For brakes of hand operated winches, the steel bands of the following sizes are usually used :

Width of band (w) in mm	25 – 40	40 – 60	80	100	140 – 200
Thickness of band (t) in mm	3	3 – 4	4 – 6	4 – 7	6 – 10

Example 25.8. *A simple band brake operates on a drum of 600 mm in diameter that is running at 200 r.p.m. The coefficient of friction is 0.25. The brake band has a contact of 270°, one end is fastened to a fixed pin and the other end to the brake arm 125 mm from the fixed pin. The straight brake arm is 750 mm long and placed perpendicular to the diameter that bisects the angle of contact.*

(*a*) *What is the pull necessary on the end of the brake arm to stop the wheel if 35 kW is being absorbed ? What is the direction for this minimum pull ?*

(*b*) *What width of steel band of 2.5 mm thick is required for this brake if the maximum tensile stress is not to exceed 50 MPa ?*

Solution. Given : d = 600 mm or r = 300 mm ; N = 200 r.p.m. ; μ = 0.25 ; θ = 270° = 270 × π/180 = 4.713 rad ; Power = 35 kW = 35 × 10^3 W ; t = 2.5 mm ; σ_t = 50 MPa = 50 N/mm^2

(a) Pull necessary on the end of the brake arm to stop the wheel

Let P = Pull necessary on the end of the brake arm to stop the wheel.

Band brake

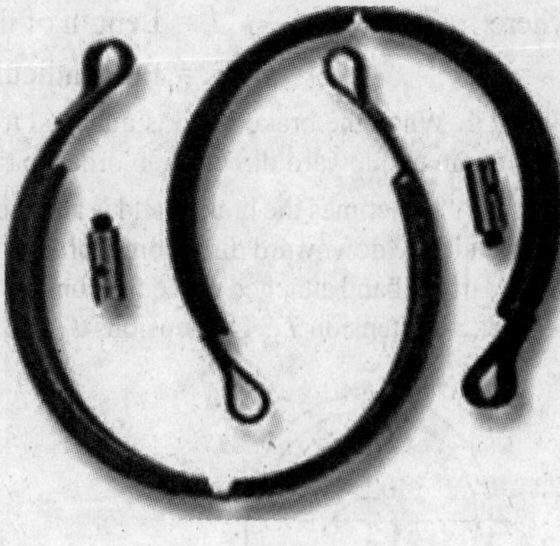

Bands of a brake shown separately

The simple band brake is shown in Fig. 25.16. Since one end of the band is attached to the fixed pin O, therefore the pull P on the end of the brake arm will act upward and when the wheel rotates anticlockwise, the end of the band attached to O will be tight with tension T_1 and the end of the band attached to B will be slack with tension T_2. First of all, let us find the tensions T_1 and T_2. We know that

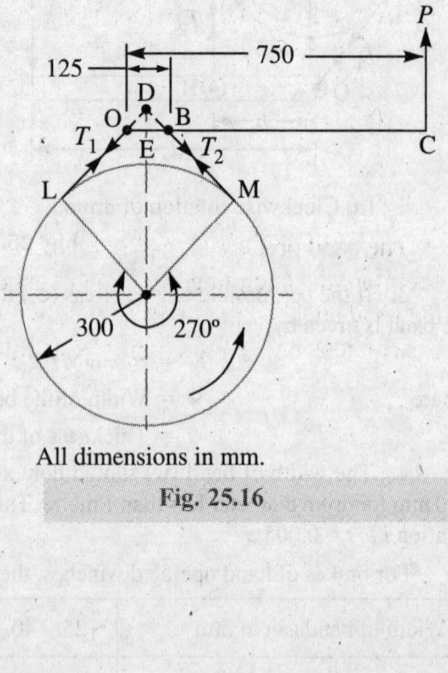

All dimensions in mm.

Fig. 25.16

$$2.3 \log \left(\frac{T_1}{T_2}\right) = \mu.\theta = 0.25 \times 4.713 = 1.178$$

$$\therefore \quad \log \left(\frac{T_1}{T_2}\right) = 1.178 / 2.3 = 0.5123$$

or $$\frac{T_1}{T_2} = 3.25 \qquad ...(i)$$

...(Taking antilog of 0.5123)

Let T_B = Braking torque.

We know that power absorbed,

$$35 \times 10^3 = \frac{2\pi N.T_B}{60} = \frac{2\pi \times 200 \times T_B}{60} = 21\, T_B$$

$$\therefore \quad T_B = 35 \times 10^3 / 21 = 1667 \text{ N-m} = 1667 \times 10^3 \text{ N-mm}$$

We also know that braking torque (T_B),

$$1667 \times 10^3 = (T_1 - T_2)\, r = (T_1 - T_2)\, 300$$

$$\therefore \quad T_1 - T_2 = 1667 \times 10^3 / 300 = 5557 \text{ N} \qquad ...(ii)$$

From equations (*i*) and (*ii*), we find that

$$T_1 = 8027 \text{ N} ; \text{ and } T_2 = 2470 \text{ N}$$

Now taking moments about O, we have

$$P \times 750 = T_2 \times {}^*OD = T_2 \times 62.5\sqrt{2} = 2470 \times 88.4 = 218\,348$$

$$\therefore \quad P = 218\,348 / 750 = 291 \text{ N } \textbf{Ans.}$$

* OD = Perpendicular distance from O to the line of action of tension T_2.

$OE = EB = OB / 2 = 125 / 2 = 62.5$ mm, and $\angle DOE = 45°$

$\therefore \quad OD = OE \sec 45° = 62.5\sqrt{2}$ mm

(b) Width of steel band

Let w = Width of steel band in mm.

We know that maximum tension in the band (T_1),

$$8027 = \sigma_t \times w \times t = 50 \times w \times 2.5 = 125\ w$$

$$\therefore \quad w = 8027/125 = 64.2 \text{ mm } \textbf{Ans.}$$

Example 25.9. *A band brake acts on the $\frac{3}{4}$th of circumference of a drum of 450 mm diameter which is keyed to the shaft. The band brake provides a braking torque of 225 N-m. One end of the band is attached to a fulcrum pin of the lever and the other end to a pin 100 mm from the fulcrum. If the operating force applied at 500 mm from the fulcrum and the coefficient of friction is 0.25, find the operating force when the drum rotates in the anticlockwise direction.*

If the brake lever and pins are to be made of mild steel having permissible stresses for tension and crushing as 70 MPa and for shear 56 MPa, design the shaft, key, lever and pins. The bearing pressure between the pin and the lever may be taken as 8 N/mm².

Solution. Given : $d = 450$ mm or $r = 225$ mm ; $T_B = 225$ N-m $= 225 \times 10^3$ N-mm ; $OB = 100$ mm ; $l = 500$ mm ; $\mu = 0.25$; $\sigma_t = \sigma_c = 70$ MPa $= 70$ N/mm² ; $\tau = 56$ MPa $= 56$ N/mm² ; $p_b = 8$ N/mm²

Operating force

Let P = Operating force.

The band brake is shown in Fig. 25.17. Since one end of the band is attached to the fulcrum at O, therefore the operating force P will act upward and when the drum rotates anticlockwise, the end of the band attached to O will be tight with tension T_1 and the end of the band attached to B will be slack with tension T_2. First of all, let us find the tensions T_1 and T_2.

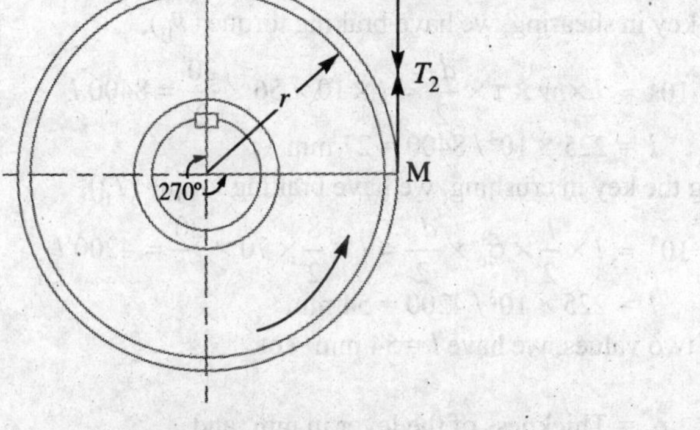

Fig. 25.17

We know that angle of wrap,

$$\theta = \tfrac{3}{4}\text{th of circumference} = \tfrac{3}{4} \times 360° = 270°$$

$$= 270 \times \frac{\pi}{180} = 4.713 \text{ rad}$$

and $$2.3 \log \left(\frac{T_1}{T_2}\right) = \mu.\theta = 0.25 \times 4.713 = 1.178$$

$$\therefore \quad \log \left(\frac{T_1}{T_2}\right) = \frac{1.178}{2.3} = 0.5123 \quad \text{or} \quad \frac{T_1}{T_2} = 3.25 \qquad \text{...}(i)$$

...(Taking antilog of 0.5123)

We know that braking torque (T_B),

$$225 \times 10^3 = (T_1 - T_2)\, r = (T_1 - T_2)\, 225$$

$$\therefore \quad T_1 - T_2 = 225 \times 10^3 / 225 = 1000 \text{ N} \quad \text{...}(ii)$$

From equations (*i*) and (*ii*), we have

$$T_1 = 1444 \text{ N and } T_2 = 444 \text{ N}$$

Taking moments about the fulcrum *O*, we have

$$P \times 500 = T_2 \times 100 = 444 \times 100 = 44\,400$$

$$\therefore \quad P = 44\,400 / 500 = 88.8 \text{ N} \quad \textbf{Ans.}$$

Drums for band brakes.

Design of shaft

Let d_s = Diameter of the shaft in mm.

Since the shaft has to transmit torque equal to the braking torque (T_B), therefore

$$225 \times 10^3 = \frac{\pi}{16} \times \tau\, (d_s)^3 = \frac{\pi}{16} \times 56\, (d_s)^3 = 11\, (d_s)^3$$

$$\therefore \quad (d_s)^3 = 225 \times 10^3/11 = 20.45 \times 10^3 \text{ or } d_s = 27.3 \text{ say } 30 \text{ mm} \quad \textbf{Ans.}$$

Design of key

The standard dimensions of the key for a 30 mm diameter shaft are as follows :

Width of key, w = 10 mm **Ans.**

Thickness of key, t = 8 mm **Ans.**

Let l = Length of key.

Considering the key in shearing, we have braking torque (T_B),

$$225 \times 10^3 = l \times w \times \tau \times \frac{d_s}{2} = l \times 10 \times 56 \times \frac{30}{2} = 8400\, l$$

$$\therefore \quad l = 225 \times 10^3 / 8400 = 27 \text{ mm}$$

Now considering the key in crushing, we have braking torque (T_B),

$$225 \times 10^3 = l \times \frac{t}{2} \times \sigma_c \times \frac{d_s}{2} = l \times \frac{8}{2} \times 70 \times \frac{30}{2} = 4200\, l$$

$$\therefore \quad l = 225 \times 10^3 / 4200 = 54 \text{ mm}$$

Taking larger of two values, we have l = 54 mm **Ans.**

Design of lever

Let t_1 = Thickness of the lever in mm, and

B = Width of the lever in mm.

The lever is considered as a cantilever supported at the fulcrum *O*. The effect of T_2 on the lever for determining the bending moment on the lever is neglected. This error is on the safer side.

∴ Maximum bending moment at *O* due to the force *P*,

$$M = P \times l = 88.8 \times 500 = 44\,400 \text{ N-m}$$

Section modulus,

$$Z = \frac{1}{6}\, t_1.B^2 = \frac{1}{6}\, t_1\, (2t_1)^2 = 0.67\, (t_1)^3 \text{ mm}^3 \qquad \text{...(Assuming } B = 2t_1)$$

We know that the bending stress (σ_t),

$$70 = \frac{M}{Z} = \frac{44\ 400}{0.67\ (t_1)^3} = \frac{66\ 300}{(t_1)^3}$$

$$\therefore \quad (t_1)^3 = 66\ 300 / 70 = 947 \quad \text{or} \quad t_1 = 9.82 \text{ say } 10 \text{ mm } \textbf{Ans.}$$

and $\quad B = 2\,t_1 = 2 \times 10 = 20$ mm **Ans.**

Design of pins

Let $\quad d_1$ = Diameter of the pins at O and B, and

$\quad l_1$ = Length of the pins at O and $B = 1.25\, d_1$ ...(Assume)

The pins at O and B are designed for the maximum tension in the band (*i.e.* T_1 = 1444 N),

Considering bearing of the pins at O and B, we have maximum tension (T_1),

$$1444 = d_1.l_1.p_b = d_1 \times 1.25\, d_1 \times 8 = 10\,(d_1)^2$$

$$\therefore \quad (d_1)^2 = 1444 / 10 = 144.4 \quad \text{or} \quad d_1 = 12 \text{ mm } \textbf{Ans.}$$

and $\quad l_1 = 1.25\, d_1 = 1.25 \times 12 = 15$ mm **Ans.**

Let us now check the pin for induced shearing stress. Since the pin is in double shear, therefore maximum tension (T_1),

$$1444 = 2 \times \frac{\pi}{4}\,(d_1)^2\,\tau = 2 \times \frac{\pi}{4}\,(12)^2\,\tau = 226\,\tau$$

$$\therefore \quad \tau = 1444 / 226 = 6.4 \text{ N/mm}^2 = 6.4 \text{ MPa}$$

This induced stress is quite within permissible limits.

The pin may be checked for induced bending stress. We know that maximum bending moment,

$$M = \frac{5}{24} \times W.l_1 = \frac{5}{24} \times 1444 \times 15 = 4513 \text{ N-mm}$$

... (Here $W = T_1$ = 1444 N)

and section modulus, $\quad Z = \frac{\pi}{32}\,(d_1)^3 = \frac{\pi}{32}\,(12)^3 = 170 \text{ mm}^3$

∴ Bending stress induced

$$= \frac{M}{Z} = \frac{4513}{170} = 26.5 \text{ N-mm}^2 = 26.5 \text{ MPa}$$

This induced bending stress is within safe limits of 70 MPa.

The lever has an eye hole for the pin and connectors at band have forked end.

Thickness of each eye,

$$t_2 = \frac{l_1}{2} = \frac{15}{2} = 7.5 \text{ mm}$$

Outer diameter of the eye,

$$D = 2d_1 = 2 \times 12 = 24 \text{ mm}$$

A clearance of 1.5 mm is provided on either side of the lever in the fork.

A brass bush of 3 mm thickness may be provided in the eye of the lever.

∴ Diameter of hole in the lever

$$= d_1 + 2 \times 3 = 12 + 6 = 18 \text{ mm}$$

The boss is made at pin joints whose outer diameter is taken equal to twice the diameter of the pin and length equal to length of the pin.

The inner diameter of the boss is equal to diameter of hole in the lever.

∴ Outer diameter of boss

$$= 2\,d_1 = 2 \times 12 = 24 \text{ mm}$$

and length of boss $= l_1 = 15$ mm

Let us now check the bending stress induced in the lever at the fulcrum. The section of the lever at the fulcrum is shown in Fig. 25.18.

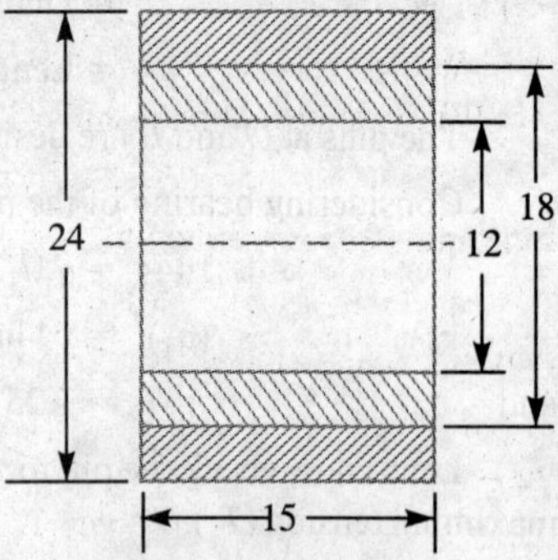

Fig. 25.18

We know that maximum bending moment at the fulcrum,

$$M = P.l = 88.8 \times 500$$

$$= 44\,400 \text{ N-mm}$$

and section modulus,

$$Z = \frac{\frac{1}{12} \times 15\left[(24)^3 - (18)^3\right]}{24/2}$$

$$= 833 \text{ mm}^3$$

∴ Bending stress induced

$$= \frac{M}{Z} = \frac{44\,400}{833} = 53.3 \text{ N/mm}^2$$

$$= 53.3 \text{ MPa}$$

This induced stress is within safe limits of 70 MPa.

25.10 Differential Band Brake

In a differential band brake, as shown in Fig. 25.19, the ends of the band are joined at *A* and *B* to a lever *AOC* pivoted on a fixed pin or fulcrum *O*. It may be noted that for the band to tighten, the length *OA* must be greater than the length *OB*.

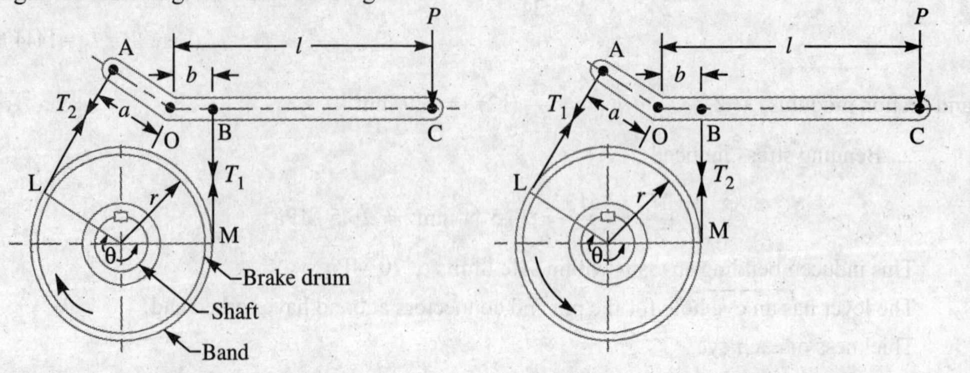

(*a*) Clockwise rotation of the drum. (*b*) Anticlockwise rotation of the drum.

Fig. 25.19. Differencial band brake.

The braking torque on the drum may be obtained in the similar way as discussed in simple band brake. Now considering the equilibrium of the lever *AOC*. It may be noted that when the drum rotates in the clockwise direction, as shown in Fig. 25.19 (*a*), the end of the band attached to *A* will be slack with tension T_2 and end of the band attached to *B* will be tight with tension T_1. On the other hand, when the drum rotates in the anticlockwise direction, as shown in Fig. 25.19 (*b*), the end of the band attached to *A* will be tight with tension T_1 and end of the band attached to *B* will be slack with tension T_2. Now taking moments about the fulcrum *O*, we have

$$P.l + T_1.b = T_2.a \quad \text{...(for clockwise rotation of the drum)}$$

or $$P.l = T_2.a - T_1.b \quad \text{...(i)}$$

and $$P.l + T_2.b = T_1.a \quad \text{...(for anticlockwise rotation of the drum)}$$

or $$P.l = T_1.a - T_2.b \quad \text{...(ii)}$$

We have discussed in block brakes (Art. 25.6), that when the frictional force helps to apply the brake, it is said to be self energizing brake. In case of differential band brake, we see from equations (*i*) and (*ii*) that the moment $T_1.b$ and $T_2.b$ helps in applying the brake (because it adds to the moment $P.l$) for the clockwise and anticlockwise rotation of the drum respectively.

We have also discussed that when the force P is negative or zero, then brake is self locking. Thus for differential band brake and for clockwise rotation of the drum, the condition for self- locking is

$$T_2.a \le T_1.b \quad \text{or} \quad T_2/T_1 \le b/a$$

and for anticlockwise rotation of the drum, the condition for self-locking is

$$T_1.a \le T_2.b \quad \text{or} \quad T_1/T_2 \le b/a$$

Notes: 1. The condition for self-locking may also be written as follows. For clockwise rotation of the drum,

$$T_1.b \ge T_2.a \quad \text{or} \quad T_1/T_2 \ge a/b$$

and for anticlockwise rotation of the drum,

$$T_2.b \ge T_1.a \quad \text{or} \quad T_2/T_1 \ge a/b$$

2. When in Fig. 25.19 (*a*) and (*b*), the length *OB* is greater than *OA*, then the force *P* must act in the upward direction in order to apply the brake. The tensions in the band, *i.e.* T_1 and T_2 will remain unchanged.

3. Sometimes, the band brake is attached to the lever as shown in Fig. 25.20 (*a*) and (*b*). In such cases, when *OA* is greater than *OB*, the force (*P*) must act upwards. When the drum rotates in the clockwise direction, the end of the band attached to *A* will be tight with tension T_1 and the end of the band attached to *B* will be slack with tension T_2, as shown in Fig. 25.20 (*a*). When the drum rotates in the anticlockwise direction, the end of the band attached to *A* will be slack with tension T_2 and the end of the band attached to *B* will be tight with tension T_1, as shown in Fig. 25.20 (*b*).

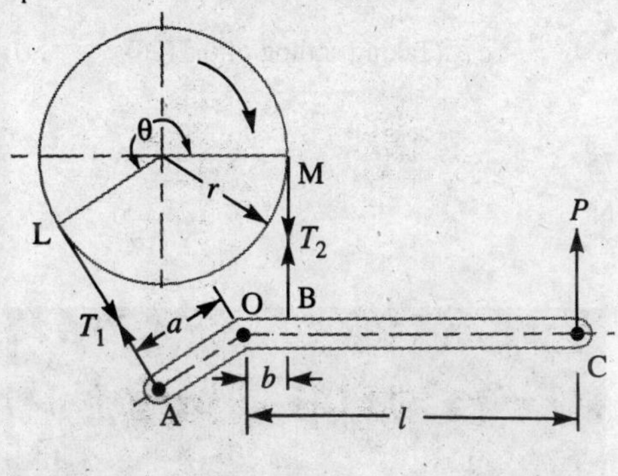

(*a*) Clockwise rotation of the drum.

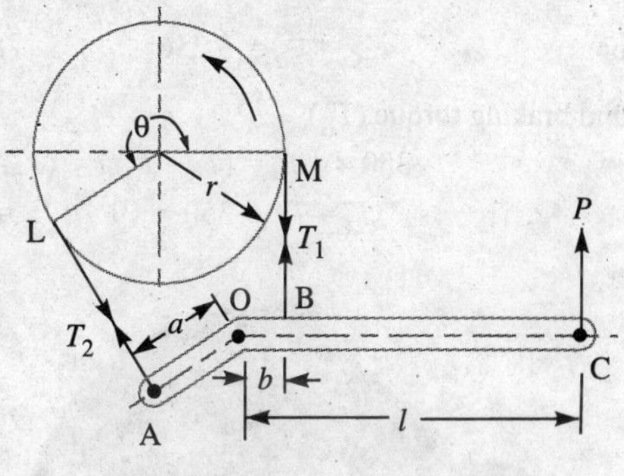

(*b*) Anticlockwise rotation of the drum.

Fig. 25.20. Differential band brake.

4. When in Fig. 25.20 (*a*) and (*b*), the length *OB* is greater than *OA*, then the force (*P*) must act downward in order to apply the brake. The position of tensions T_1 and T_2 will remain unchanged.

Example 25.10. *A differential band brake, as shown in Fig. 25.21, has an angle of contact of 225°. The band has a compressed woven lining and bears against a cast iron drum of 350 mm diameter. The brake is to sustain a torque of 350 N-m and the coefficient of friction between the band and the drum is 0.3. Find : 1. the necessary force (P) for the clockwise and anticlockwise rotation of the drum; and 2. The value of 'OA' for the brake to be self locking, when the drum rotates clockwise.*

Solution. Given : $\theta = 225° = 225 \times \pi / 180 = 3.93$ rad ; $d = 350$ mm or $r = 175$ mm ; $T = 350$ N-m $= 350 \times 10^3$ N-mm ; $\mu = 0.3$

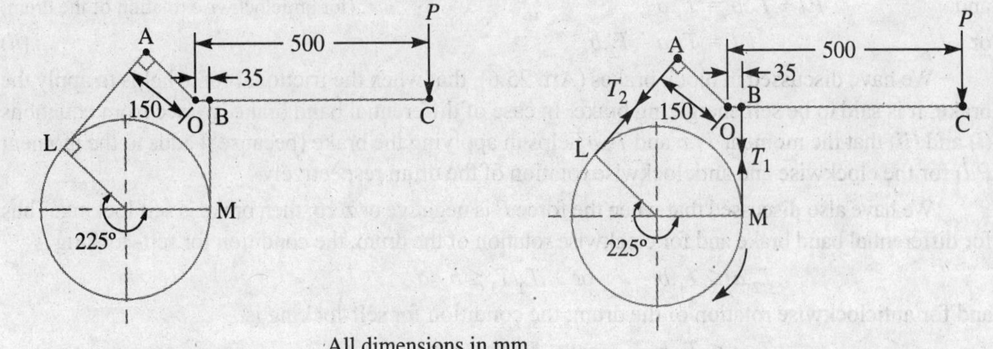

All dimensions in mm.

Fig. 25.21 **Fig. 25.22**

1. ***Necessary force (P) for the clockwise and anticlockwise rotation of the drum***

When the drum rotates in the clockwise direction, the end of the band attached to A will be slack with tension T_2 and the end of the band attached to B will be tight with tension T_1, as shown in Fig. 25.22. First of all, let us find the values of tensions T_1 and T_2.

We know that

$$2.3 \log\left(\frac{T_1}{T_2}\right) = \mu.\theta = 0.3 \times 3.93 = 1.179$$

$$\therefore \quad \log\left(\frac{T_1}{T_2}\right) = \frac{1.179}{2.3} = 0.5126$$

or $$\frac{T_1}{T_2} = 3.256 \qquad \text{...(Taking antilog of 0.5126)} \qquad ...(i)$$

and braking torque (T_B),

$$350 \times 10^3 = (T_1 - T_2)\, r = (T_1 - T_2)\, 175$$

$$\therefore \quad T_1 - T_2 = 350 \times 10^3 / 175 = 2000 \text{ N} \qquad ...(ii)$$

Another picture of car brake shoes

From equations (*i*) and (*ii*), we find that

$$T_1 = 2886.5 \text{ N} ; \text{ and } T_2 = 886.5 \text{ N}$$

Now taking moments about the fulcrum *O*, we have

$$P \times 500 = T_2 \times 150 - T_1 \times 35$$

or
$$P \times 500 = 886.5 \times 150 - 2886.5 \times 35 = 31\,947.5$$

$$\therefore \quad P = 31\,947.5 / 500 = 64 \text{ N } \textbf{Ans.}$$

When the drum rotates in the anticlockwise direction, the end of the band attached to *A* will be tight with tension T_1 and end of the band attached to *B* will be slack with tension T_2, as shown in Fig. 25.23. Taking moments about the fulcrum *O*, we have

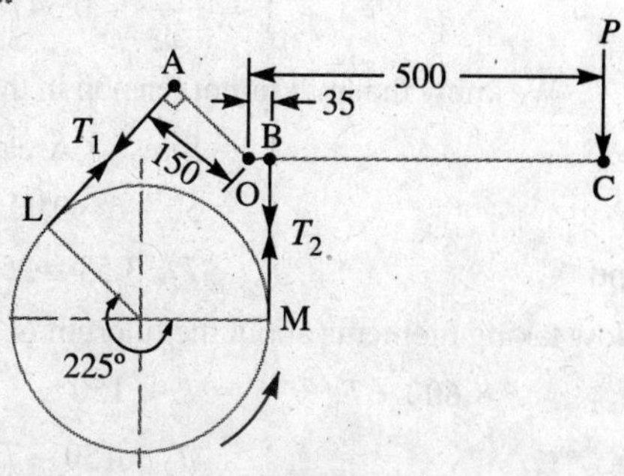

Fig. 25.23

$$P \times 500 = T_1 \times 150 - T_2 \times 35$$

or
$$P \times 500 = 2886.5 \times 150 - 886.5 \times 35$$
$$= 401\,947.5$$

$$\therefore \quad P = 401\,947.5/500 = 804 \text{ N } \textbf{Ans.}$$

2. ***Valve of 'OA' for the brake to be self locking, when the drum rotates clockwise***

The clockwise rotation of the drum is shown in Fig. 25.22.

For clockwise rotation of the drum, we know that

$$P \times 500 + T_1 \times OB = T_2 \times OA$$

or
$$P \times 500 = T_2 \times OA - T_1 \times OB$$

For the brake to be self-locking, *P* must be equal to zero or

$$T_2 \times OA = T_1 \times OB$$

or
$$OA = \frac{T_1 \times OB}{T_2} = \frac{2886.5 \times 35}{886.5} = 114 \text{ mm } \textbf{Ans.}$$

Example 25.11. *A differential band brake, as shown in Fig. 25.24, has a drum diameter of 600 mm and the angle of contact is 240°. The brake band is 5 mm thick and 100 mm wide. The coefficient of friction between the band and the drum is 0.3. If the band is subjected to a stress of 50 MPa, find :*

1. *The least force required at the end of a 600 mm lever, and*
2. *The torque applied to the brake drum shaft.*

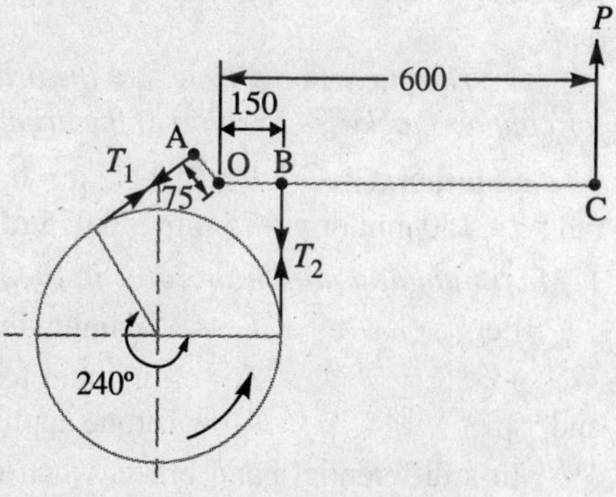

All dimensions in mm.

Fig. 25.24

Solution. Given : $d = 600$ mm or $r = 300$ mm $= 0.3$ m ; $\theta = 240° = 240 \times \pi / 180 = 4.2$ rad ; $t = 5$ mm ; $w = 100$ mm ; $\mu = 0.3$; $\sigma_t = 50$ MPa $= 50$ N/mm^2

1. ***Least force required at the end of a lever***

Let $\quad P$ = Least force required at the end of the lever.

Since the length OB is greater than OA, therefore the force at the end of the lever (P) must act in the upward direction. When the drum rotates anticlockwise, the end of the band attached to A will be tight with tension T_1 and the end of the band attached to B will be slack with tension T_2. First of all, let us find the values of tensions T_1 and T_2. We know that

$$2.3 \log \left(\frac{T_1}{T_2}\right) = \mu.\theta = 0.3 \times 4.2 = 1.26$$

$$\log \left(\frac{T_1}{T_2}\right) = \frac{1.26}{2.3} = 0.5478 \quad \text{or} \quad \frac{T_1}{T_2} = 3.53 \quad \text{... (Taking antilog of 0.5478)} \quad ...(i)$$

We know that maximum tension in the band,

$$T_1 = \text{Stress} \times \text{Area of band} = \sigma_t \times t \times w$$
$$= 50 \times 5 \times 100 = 25\,000 \text{ N}$$

and $T_2 = T_1 / 3.53 = 25\,000 / 3.53 = 7082$ N ...[From equation (*i*)]

Now taking moments about the fulcrum O, we have

$$P \times 600 + T_1 \times 75 = T_2 \times 150$$

$$\therefore \quad P = \frac{T_2 \times 150 - T_1 \times 75}{600} = \frac{7082 \times 150 - 25\,000 \times 75}{600} = -1355 \text{ N}$$

= 1355 N (in magnitude) **Ans.**

Since P is negative, therefore the brake is self-locking.

2. *Torque applied to the brake drum shaft*

We know that torque applied to the brake drum shaft,

$$T_B = (T_1 - T_2)\, r = (25\,000 - 7082)\, 0.3 = 5375 \text{ N-m } \textbf{Ans.}$$

Example 25.12. *A differential band brake has a force of 220 N applied at the end of a lever as shown in Fig. 25.25. The coefficient of friction between the band and the drum is 0.4. The angle of lap is 180°. Find :*

1. The maximum and minimum force in the band, when a clockwise torque of 450 N-m is applied to the drum; and

2. The maximum torque that the brake may sustain for counter clockwise rotation of the drum.

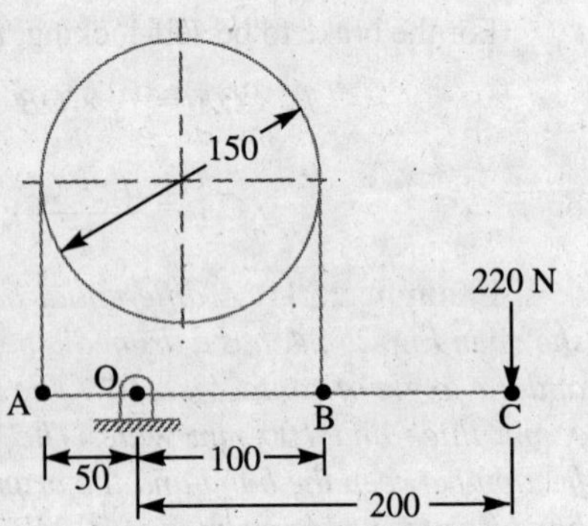

All dimensions in mm.

Fig. 25.25

Solution. Given : $P = 220$ N ; $\mu = 0.4$; $\theta = 180° = \pi$ rad ; $d = 150$ mm or $r = 75$ mm $= 0.075$ m

1. *Maximum and minimum force in the band*

Let T_1 = Maximum force in the band,

T_2 = Minimum force in the band,

and T_B = Torque applied to the drum = 450 N-m ...(Given)

In a differential band brake, when OB is greater than OA and the clockwise torque (T_B) is applied to the drum, then the maximum force (T_1) will be in the band attached to A and the minimum force (T_2) will be in the band attached to B, as shown in Fig. 25.26.

We know that braking torque (T_B),

$$450 = (T_1 - T_2)\, r = (T_1 - T_2)\, 0.075$$

$$\therefore \quad T_1 - T_2 = 450 / 0.075 = 6000 \text{ N}$$

or $T_1 = (T_2 + 6000)$ N ...(*i*)

Now taking moments about the pivot O, we have

$$220 \times 200 + T_1 \times 50 = T_2 \times 100$$

$$44\,000 + (T_2 + 6000)\,50 = T_2 \times 100$$

$$44\,000 + 50\,T_2 + 300\,000 = T_2 \times 100$$

or $\quad T_2 = 6880$ N **Ans.**

and $\quad T_1 = T_2 + 6000$

$= 6880 + 6000 = 12\,880$ N **Ans.** ...[From equation (*i*)]

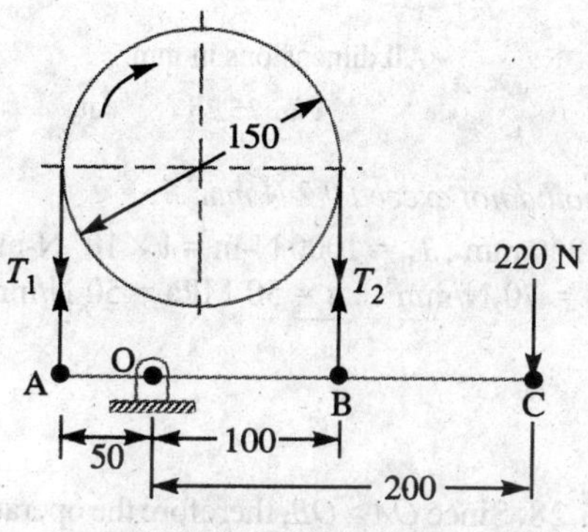
All dimensions in mm.

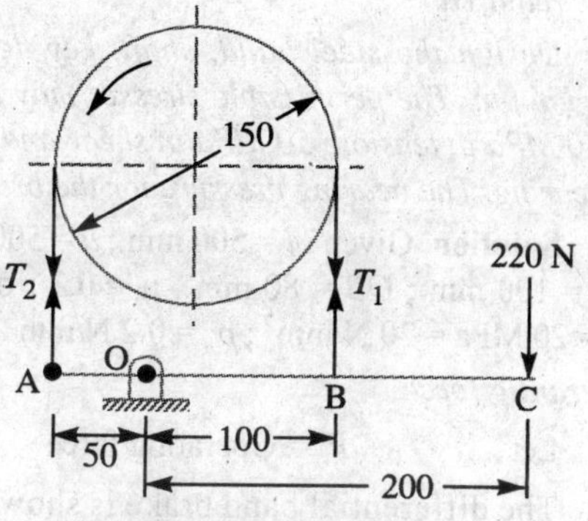
All dimensions in mm.

Fig. 25.26 **Fig. 25.27**

2. ***Maximum torque that the brake may sustain for counter clockwise rotation of the drum.***

When the drum rotates in the counter clockwise direction, the maximum force (T_1) will be in the band attached to B and the minimum force (T_2) will be in the band attached to A, as shown in Fig. 25.27. We know that

$$2.3 \log\left(\frac{T_1}{T_2}\right) = \mu.\theta = 0.4 \times \pi = 1.257$$

$$\log\left(\frac{T_1}{T_2}\right) = \frac{1.257}{2.3} = 0.5465$$

$\therefore \quad \dfrac{T_1}{T_2} = 3.52$ ...(Taking antilog of 0.5465) ...(*i*)

Now taking moments about the pivot O, we have

$220 \times 200 + T_2 \times 50 = T_1 \times 100 = 3.52\,T_2 \times 100 = 352\,T_2$ ...[From equation (*i*)]

or $\quad 44\,000 = 352\,T_2 - 50\,T_2 = 302\,T_2$

$\therefore \quad T_2 = 44\,000 / 302 = 146$ N

and $\quad T_1 = 3.52\,T_2 = 3.52 \times 146 = 514$ N

We know that the maximum torque that the brake may sustain,

$$T_B = (T_1 - T_2)\,r = (514 - 146)\,0.075 = 27.6 \text{ N-m } \textbf{Ans.}$$

Example 25.13. *A differential band brake is operated by a lever of length 500 mm. The brake drum has a diameter of 500 mm and the maximum torque on the drum is 1000 N-m. The band brake embraces 2/3rd of the circumference. One end of the band is attached to a pin 100 mm from the fulcrum and the other end to another pin 80 mm from the fulcrum and on the other side of it when the operating force is also acting. If the band brake is lined with asbestos fabric having a coefficient of friction 0.3, find the operating force required.*

Design the steel band, shaft, key, lever and fulcrum pin. The permissible stresses may be taken as 70 MPa in tension, 50 MPa in shear and 20 MPa in bearing. The bearing pressure for the brake lining should not exceed 0.2 N/mm².

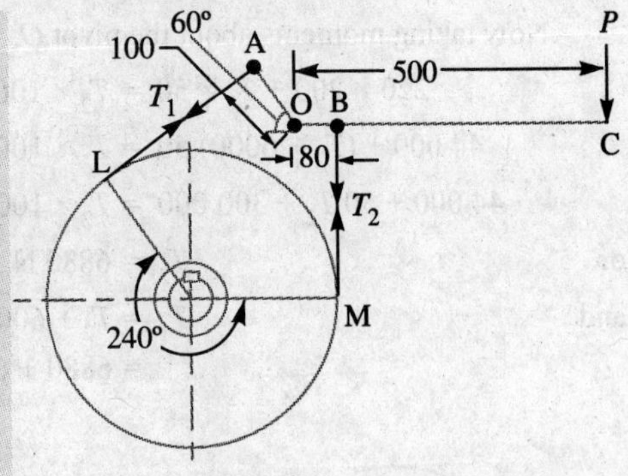

All dimensions in mm.

Fig. 25.28

Solution. Given : l = 500 mm ; d = 500 mm or r = 250 mm ; T_B = 1000 N-m = 1×10^6 N-mm ; OA = 100 mm ; OB = 80 mm ; μ = 0.3 ; σ_t = 70 MPa = 70 N/mm² ; τ = 50 MPa = 50 N/mm² ; σ_b = 20 MPa = 20 N/mm² ; p_b = 0.2 N/mm²

Operating force

Let P = Operating force

The differential band brake is shown in Fig. 25.28. Since $OA > OB$, therefore the operating force (P) will act downward. When the drum rotates anticlockwise, the end of the band attached to A will be tight with tension T_1 and the end of the band attached to B will be slack with tension T_2. First of all, let us find the values of tensions T_1 and T_2.

We know that angle of wrap,

$$\theta = \frac{2}{3}\text{rd of circumference}$$

$$= \frac{2}{3} \times 360° = 240° = 240 \times \frac{\pi}{180} = 4.19 \text{ rad}$$

and $$2.3 \log\left(\frac{T_1}{T_2}\right) = \mu.\theta = 0.3 \times 4.19 = 1.257$$

$\therefore$ $$\log\left(\frac{T_1}{T_2}\right) = \frac{1.257}{2.3} = 0.5465 \quad \text{or} \quad \frac{T_1}{T_2} = 3.52 \quad \text{...(Taking antilog of 0.5465)} \quad ...(i)$$

We know that the braking torque (T_B),

$$1 \times 10^6 = (T_1 - T_2)\, r = (T_1 - T_2)\, 250$$

$\therefore$ $$T_1 - T_2 = 1 \times 10^6 / 250 = 4000 \text{ N} \quad ...(ii)$$

From equations (*i*) and (*ii*), we have

$$T_1 = 5587 \text{ N, and } T_2 = 1587 \text{ N}$$

Now taking moments about the fulcrum O, we have

$$P \times 500 = T_1 \times 100 - T_2 \times 80$$

$$= 5587 \times 100 - 1587 \times 80$$

$$= 431\,740$$

$\therefore$ $$P = 431\,740 / 500 = 863.5 \text{ N} \textbf{ Ans.}$$

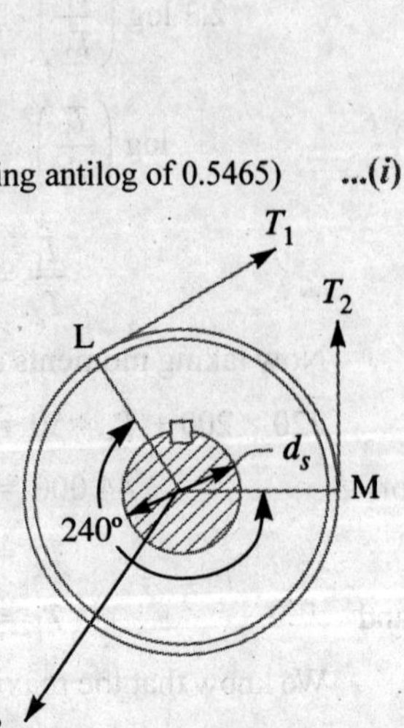

Fig. 25.29

Design for steel band

Let t = Thickness of band in mm, and

b = Width of band in mm.

We know that length of contact of band, as shown in Fig. 25.29,

$$= \pi d \times \frac{240}{360} = \pi \times 500 \times \frac{240}{360} \text{ mm} = 1047 \text{ mm}$$

∴ Area of contact of the band,

$$A_b = \text{Length} \times \text{Width of band} = 1047\, b \text{ mm}^2$$

We know that normal force acting on the band,

$$R_N = \frac{T_1 - T_2}{\mu} = \frac{5587 - 1587}{0.3} = 13\,333 \text{ N}$$

We also know that normal force on the band (R_N),

$$13\,333 = p_b \times A_b = 0.2 \times 1047\, b = 209.4\, b$$

∴ $b = 13\,333/209.4 = 63.7$ say 64 mm **Ans.**

and cross-sectional area of the band,

$$A = b \times t = 64\, t \text{ mm}^2$$

∴ Tensile strength of the band

$$= A \times \sigma_t = 64\, t \times 70 = 4480\, t \text{ N}$$

Since the band has to withstand a maximum tension (T_1) equal to 5587 N, therefore

$4480\, t = 5587$ or $t = 5587 / 4480 = 1.25$ mm **Ans.**

Design of shaft

Let d_s = Diameter of the shaft in mm.

Since the shaft has to transmit torque equal to the braking torque (T_B), therefore

$$1 \times 10^6 = \frac{\pi}{16} \times \tau\, (d_s)^3 = \frac{\pi}{16} \times 50\, (d_s)^3 = 9.82\, (d_s)^3$$

∴ $(d_s)^3 = 1 \times 10^6 / 9.82 = 101\,833$ or $d_s = 46.7$ say 50 mm **Ans.**

Design of key

The standard dimensions of the key for a 50 mm diameter shaft are as follows :

Width of key, $w = 16$ mm

Thickness of key, $t_1 = 10$ mm

Let l = Length of the key.

Considering shearing of the key, we have braking torque (T_B),

$$1 \times 10^6 = l \times w \times \tau \times \frac{d_s}{2} = l \times 16 \times 50 \times \frac{50}{2} = 20 \times 10^3\, l$$

∴ $l = 1 \times 10^6/20 \times 10^3 = 50$ mm **Ans.**

Note : The dimensions of the key may also be obtained from the following relations :

$$w = \frac{d_s}{4} + 3 \text{ mm}; \text{ and } t_1 = \frac{w}{2}$$

Design for lever

Let t_2 = Thicknes. of lever in mm, and

B = Width of the lever in mm.

It is assumed that the lever extends up to the centre of the fulcrum. This assumption results in a slightly stronger lever. Neglecting the effect of T_2 on the lever, the maximum bending moment at the centre of the fulcrum,

$$M = P \times l = 863.5 \times 500 = 431\ 750 \text{ N-mm}$$

and section modulus, $Z = \frac{1}{6} t_2.B^2 = \frac{1}{6} t_2 (2t_2)^2 = 0.67 (t_2)^3$ ...(Assuming $B = 2t_2$)

We know that bending tensile stress (σ_t),

$$70 = \frac{M}{Z} = \frac{431\ 750}{0.67 (t_2)^3} = \frac{644\ 400}{(t_2)^3}$$

$\therefore$ $(t_2)^3 = 644\ 400 / 70 = 9206$ or $t_2 = 21$ mm **Ans.**

and $B = 2\,t_2 = 2 \times 21 = 42$ mm **Ans.**

Design for fulcrum pin

Let d_1 = Diameter of the fulcrum pin, and

l_1 = Length of the fulcrum pin.

First of all, let us find the resultant force acting on the pin. Resolving the three forces T_1, T_2 and P into their vertical and horizontal components, as shown in Fig. 25.30.

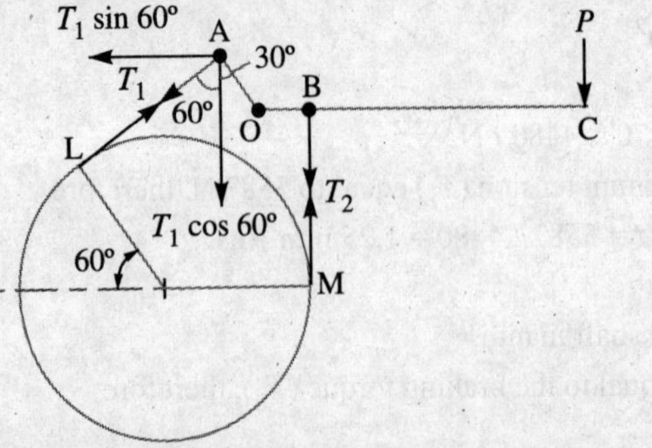

Fig. 25.30

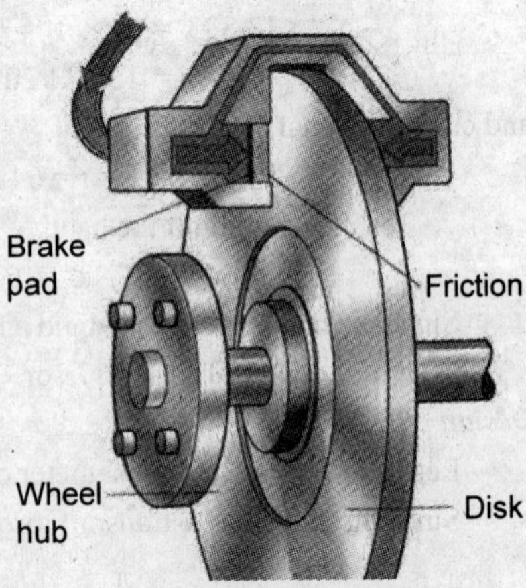

Another type brake disc

We know that sum of vertical components,

$$\Sigma V = T_1 \cos 60° + T_2 + P = 5587 \times \frac{1}{2} + 1587 + 863.5 = 5244 \text{ N}$$

and sum of horizontal components,

$$\Sigma H = T_1 \sin 60° = 5587 \times 0.866 = 4838 \text{ N}$$

$\therefore$ Resultant force acting on pin,

$$R_P = \sqrt{(\Sigma V)^2 + (\Sigma H)^2} = \sqrt{(5244)^2 + (4838)^2} = 7135 \text{ N}$$

Considering bearing of the pin, we have resultant force on the pin (R_P),

$$7\ 135 = d_1.l_1.\sigma_b = d_1 \times 1.25\, d_1 \times 20 = 25 (d_1)^2$$ ...(Assuming $l_1 = 1.25\, d_1$)

$\therefore$ $(d_1)^2 = 7135 / 25 = 285.4$ or $d_1 = 16.9$ say 18 mm **Ans.**

and $l_1 = 1.25\, d_1 = 1.25 \times 18 = 22.5$ mm **Ans.**

Let us now check the pin for induced shear stress. Since the pin is in double shear, therefore resultant force on the pin (R_P),

$$7135 = 2 \times \frac{\pi}{4} (d_1)^2 \tau = 2 \times \frac{\pi}{4} (18)^2 \tau = 509\ \tau$$

$$\therefore \quad \tau = 7135 / 509 = 14 \text{ N/mm}^2$$

This induced shear stress is within permissible limits.

The pin may be checked for induced bending stress. We know that maximum bending moment,

$$M = \frac{5}{24} \times W.l_1 = \frac{5}{24} \times 7135 \times 22.5 = 33\ 445 \text{ N-mm}$$

...(Here $W = R_P = 7135$ N)

and section modulus, $$Z = \frac{\pi}{32}(d_1)^3 = \frac{\pi}{32}(18)^3 = 573 \text{ mm}^3$$

∴ Bending stress induced

$$= \frac{M}{Z} = \frac{33\ 445}{573} = 58.4 \text{ N/mm}^2$$

This induced bending stress in the pin is within safe limit of 70 N/mm^2.

The lever has an eye hole for the pin and connectors at band have forked end. A brass bush of 3 mm thickness may be provided in the eye of the lever. Therefore, diameter of hole in the lever

$$= d_1 + 2 \times 3 = 18 + 6 = 24 \text{ mm}$$

The boss is made at the pin joints whose outer diameter is taken equal to twice the diameter of pin and length equal to the length of pin. The inner diameter of boss is equal to the diameter of hole in the lever.

∴ Outer diameter of boss

$$= 2\ d_1 = 2 \times 18 = 36 \text{ mm}$$

and length of boss $= 22.5$ mm

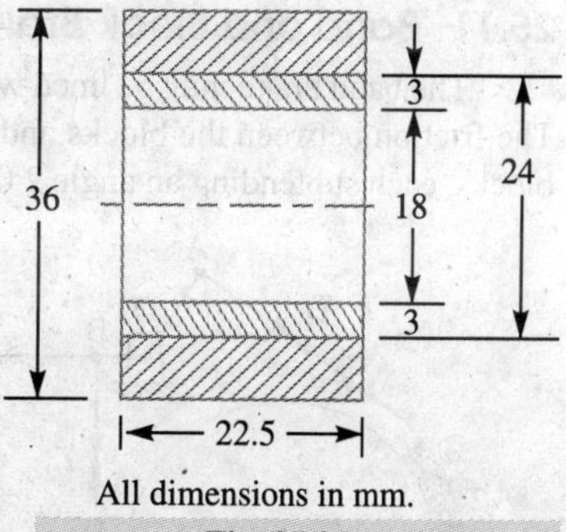

Fig. 25.31

Let us now check the induced bending stress in the lever at the fulcrum. The section of the lever at the fulcrum is shown in Fig. 25.31. We know that maximum bending moment at the fulcrum,

$$M = P \times l = 863.5 \times 500 = 431\ 750 \text{ N-mm}$$

and section modulus,

$$Z = \frac{\frac{1}{12} \times 22.5\left[(36)^3 - (24)^3\right]}{36/2} = 3420 \text{ mm}^3$$

∴ Bending stress induced in the lever

$$= \frac{M}{Z} = \frac{431\ 750}{3420} = 126 \text{ N/mm}^2$$

Since the induced bending stress is more than the permissible value of 70 N/mm^2, therefore the diameter of pin is required to be increased. Let us take

Diameter of pin, $d_1 = 22$ mm **Ans.**

∴ Length of pin, $l_1 = 1.25\ d_1 = 1.25 \times 22 = 27.5$ say 28 mm **Ans.**

Diameter of hole in the lever

$$= d_1 + 2 \times 3 = 22 + 6 = 28 \text{ mm}$$

Outer diameter of boss

$$= 2\,d_1 = 2 \times 22 = 44 \text{ mm}$$

$$= \text{Outer diameter of eye}$$

and thickness of each eye $= l_1 / 2 = 28 / 2 = 14$ mm

A clearance of 1.5 mm is provided on either side of the lever in the fork.

The new section of the lever at the fulcrum will be as shown in Fig. 25.32.

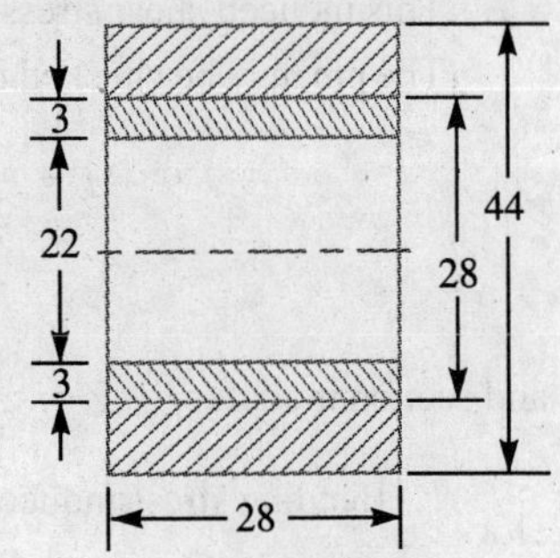

All dimensions in mm.

Fig. 25.32

∴ Section modulus,

$$Z = \frac{\frac{1}{12} \times 28\left[(44)^3 - (28)^3\right]}{44/2} = 6706 \text{ mm}^2$$

and induced bending stress $= \dfrac{431\,750}{6706} = 64.4 \text{ N/mm}^2$

This induced bending stress is within permissible limits.

25.11 Band and Block Brake

The band brake may be lined with blocks of wood or other material, as shown in Fig. 25.33 (*a*). The friction between the blocks and the drum provides braking action. Let there are '*n*' number of blocks, each subtending an angle 2 θ at the centre and the drum rotates in anticlockwise direction.

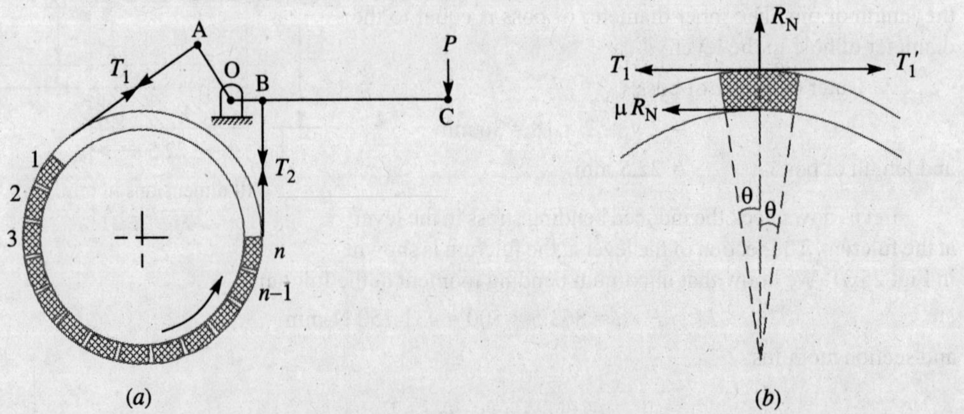

Fig. 25.33. Band and block brake.

Let T_1 = Tension in the tight side,

T_2 = Tension in the slack side,

μ = Coefficient of friction between the blocks and drum,

T_1' = Tension in the band between the first and second block,

T_2', T_3' etc. = Tensions in the band between the second and third block, between the third and fourth block etc.

Consider one of the blocks (say first block) as shown in Fig. 25.33 (*b*). This is in equilibrium under the action of the following forces :

1. Tension in the tight side (T_1),
2. Tension in the slack side (T_1') or tension in the band between the first and second block,
3. Normal reaction of the drum on the block (R_N), and
4. The force of friction ($\mu.R_N$).

Resolving the forces radially, we have

$$(T_1 + T_1') \sin\theta = R_N \quad ...(i)$$

Resolving the forces tangentially, we have

$$(T_1 - T_1') \cos\theta = \mu.R_N \quad ...(ii)$$

Dividing equation (*ii*) by (*i*), we have

$$\frac{(T_1 - T_1')\cos\theta}{(T_1 + T_1')\sin\theta} = \frac{\mu.R_N}{R_N}$$

or

$$(T_1 - T_1') = \mu \tan\theta\,(T_1 + T_1')$$

$$\therefore \quad \frac{T_1}{T_1'} = \frac{1 + \mu\tan\theta}{1 - \mu\tan\theta}$$

Similarly it can be proved for each of the blocks that

$$\frac{T_1'}{T_2'} = \frac{T_2'}{T_3'} = \frac{T_3'}{T_4'} = ... = \frac{T_{n-1}}{T_2} = \frac{1 + \mu\tan\theta}{1 - \mu\tan\theta}$$

$$\therefore \quad \frac{T_1}{T_2} = \frac{T_1}{T_1'} \times \frac{T_1'}{T_2'} \times \frac{T_2'}{T_3'} \times ... \times \frac{T_{n-1}}{T_2} = \left(\frac{1 + \mu\tan\theta}{1 - \mu\tan\theta}\right)^n \quad ...(iii)$$

Braking torque on the drum of effective radius r_e,

$$T_B = (T_1 - T_2)\, r_e$$

$$= (T_1 - T_2)\, r \quad \text{...(Neglecting thickness of band)}$$

Note: For the first block, the tension in the tight side is T_1 and in the slack side is T_1' and for the second block, the tension in the tight side is T_1' and in the slack side is T_2'. Similarly for the third block, the tension in the tight side is T_2' and in the slack side is T_3' and so on. For the last block, the tension in the tight side is T_{n-1} and in the slack side is T_2.

Example 25.14. *In the band and block brake shown in Fig. 25.34, the band is lined with 12 blocks each of which subtends an angle of 15° at the centre of the rotating drum. The thickness of the blocks is 75 mm and the diameter of the drum is 850 mm. If, when the brake is inaction, the greatest and least tensions in the brake strap are T_1 and T_2, show that*

$$\frac{T_1}{T_2} = \left(\frac{l + \mu\tan 7\,^1/_2{}^\circ}{l - \mu\tan 7\,^1/_2{}^\circ}\right)^{12}$$

where μ is the coefficient of friction for the blocks.

With the lever arrangement as shown in Fig. 25.34, find the least force required at C for the blocks to absorb 225 kW at 240 r.p.m. The coefficient of friction between the band and blocks is 0.4.

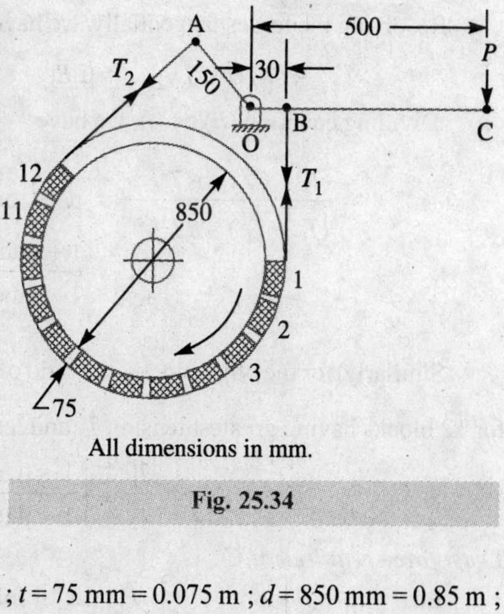

All dimensions in mm.

Fig. 25.34

Solution. Given : $n = 12$; $2\theta = 15°$ or $\theta = 7\,^1/_2°$; $t = 75$ mm = 0.075 m ; $d = 850$ mm = 0.85 m ; Power = 225 kW = 225×10^3 W ; $N = 240$ r.p.m. ; $\mu = 0.4$

Since $OA > OB$, therefore the force at C must act downward. Also, the drum rotates clockwise, therefore the end of the band attached to A will be slack with tension T_2 (least tension) and the end of the band attached to B will be tight with tension T_1 (greatest tension).

Consider one of the blocks (say first block) as shown is Fig. 25.35. This is in equilibrium under the action of the following four forces :

1. Tension in the tight side (T_1),
2. Tension in the slack side (T_1') or the tension in the band between the first and second block,
3. Normal reaction of the drum on the block (R_N), and
4. The force of friction ($\mu.R_N$).

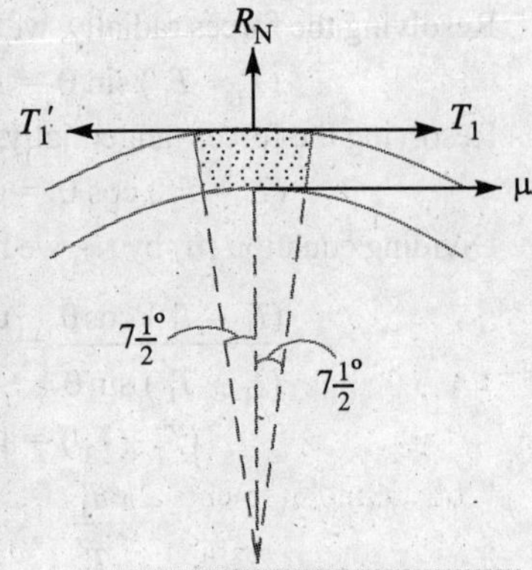

Fig. 25.35

Car wheels are made of alloys to bear high stresses and fatigue.

Resolving the forces radially, we have

$$(T_1 + T_1') \sin 7^1/_2° = R_N \quad \text{...}(i)$$

Resolving the forces tangentially, we have

$$(T_1 - T_1') \cos 7^1/_2° = \mu.R_N \quad \text{...}(ii)$$

Dividing equation (*ii*) by (*i*), we have

$$\frac{(T_1 - T_1') \cos 7^1/_2°}{(T_1 + T_1') \sin 7^1/_2°} = \mu \quad \text{or} \quad \frac{T_1 - T_1'}{T_1 + T_1'} = \mu \tan 7^1/_2°$$

$$\therefore \quad \frac{T_1}{T_1'} = \frac{1 + \mu \tan 7^1/_2°}{1 - \mu \tan 7^1/_2°}$$

Similarly, for the other blocks, the ratio of tensions $\frac{T_1'}{T_2'} = \frac{T_2'}{T_3'}$ etc., remains constant. Therefore for 12 blocks having greatest tension T_1 and least tension T_2 is

$$\frac{T_1}{T_2} = \left(\frac{1 + \mu \tan 7^1/_2°}{1 - \mu \tan 7^1/_2°}\right)^{12}$$

Least force required at C

Let $\quad P$ = Least force required at C.

We know that diameter of band,

$$D = d + 2\,t = 0.85 + 2 \times 0.075 = 1 \text{ m}$$

and power absorbed $= \dfrac{(T_1 - T_2)\pi D.N}{60}$

$$\therefore \quad T_1 - T_2 = \frac{\text{Power} \times 60}{\pi DN} = \frac{225 \times 10^3 \times 60}{\pi \times 1 \times 240} = 17\,900 \text{ N} \quad ...(iii)$$

We have proved that

$$\frac{T_1}{T_2} = \left(\frac{1 + \mu \tan 7\tfrac{1}{2}°}{1 - \mu \tan 7\tfrac{1}{2}°}\right)^{12} = \left(\frac{1 + 0.4 \times 0.1317}{1 - 0.4 \times 0.1317}\right)^{12} = 3.55 \quad ...(iv)$$

From equations (*iii*) and (*iv*), we find that

$$T_1 = 24\,920 \text{ N} \text{ ; and } T_2 = 7020 \text{ N}$$

Now taking moments about O, we have

$$P \times 500 = T_2 \times 150 - T_1 \times 30 = 7020 \times 150 - 24\,920 \times 30 = 305\,400$$

$$\therefore \quad P = 305\,400 / 500 = 610.8 \text{ N} \textbf{ Ans.}$$

25.12 Internal Expanding Brake

An internal expanding brake consists of two shoes S_1 and S_2 as shown in Fig. 25.36 (*a*). The outer surface of the shoes are lined with some friction material (usually with Ferodo) to increase the coefficient of friction and to prevent wearing away of the metal. Each shoe is pivoted at one end about a fixed fulcrum O_1 and O_2 and made to contact a cam at the other end. When the cam rotates, the shoes are pushed outwards against the rim of the drum. The friction between the shoes and the drum produces the braking torque and hence reduces the speed of the drum. The shoes are normally held in off position by a spring as shown in Fig. 25.36 (*a*). The drum encloses the entire mechanism to keep out dust and moisture. This type of brake is commonly used in motor cars and light trucks.

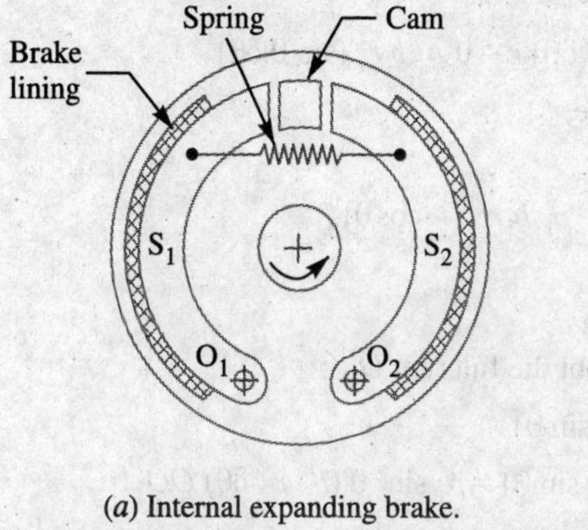

(*a*) Internal expanding brake.

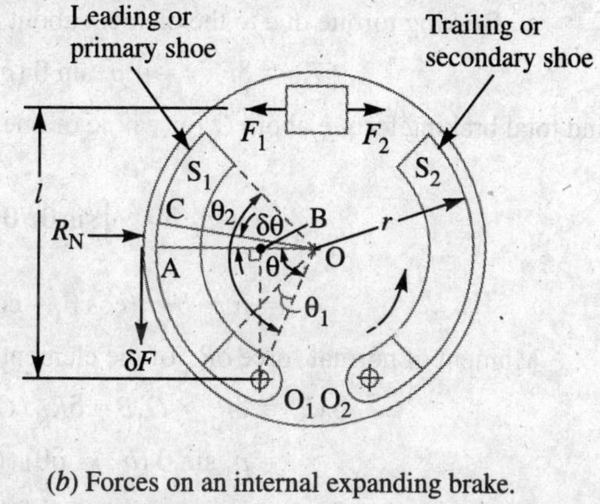

(*b*) Forces on an internal expanding brake.

Fig. 25.36

We shall now consider the forces acting on such a brake, when the drum rotates in the anticlockwise direction as shown in Fig. 25.36 (*b*). It may be noted that for the anticlockwise direction, the left hand shoe is known as ***leading* or *primary shoe*** while the right hand shoe is known as ***trailing* or *secondary shoe.***

Let r = Internal radius of the wheel rim.

b = Width of the brake lining.

p_1 = Maximum intensity of normal pressure,

p_N = Normal pressure,

F_1 = Force exerted by the cam on the leading shoe, and

F_2 = Force exerted by the cam on the trailing shoe.

Inside view of a truck disk brake

Consider a small element of the brake lining AC subtending an angle $\delta\theta$ at the centre. Let OA makes an angle θ with OO_1 as shown in Fig. 25.36 (*b*). It is assumed that the pressure distribution on the shoe is nearly uniform, however the friction lining wears out more at the free end. Since the shoe turns about O_1, therefore the rate of wear of the shoe lining at A will be proportional to the radial displacement of that point. The rate of wear of the shoe lining varies directly as the perpendicular distance from O_1 to OA, *i.e.* O_1B. From the geometry of the figure,

$$O_1B = OO_1 \sin\theta$$

and normal pressure at A, $p_N \propto \sin\theta$ or $p_N = p_1 \sin\theta$

$\therefore$ Normal force acting on the element,

$$\delta R_N = \text{Normal pressure} \times \text{Area of the element}$$

$$= p_N (b \,.\, r \,.\, \delta\theta) = p_1 \sin\theta\, (b \,.\, r \,.\, \delta\theta)$$

and braking or friction force on the element,

$$\delta F = \mu.\delta R_N = \mu\, p_1 \sin\theta\, (b \,.\, r \,.\, \delta\theta)$$

$\therefore$ Braking torque due to the element about O,

$$\delta T_B = \delta F \,.\, r = \mu\, p_1 \sin\theta\, (b \,.\, r \,.\, \delta\theta)\, r = \mu\, p_1\, b\, r^2 (\sin\theta.\delta\theta)$$

and total braking torque about O for whole of one shoe,

$$T_B = \mu\, p_1\, b\, r^2 \int_{\theta_1}^{\theta_2} \sin\theta\, d\theta = \mu\, p_1\, b\, r^2 \left[-\cos\theta\right]_{\theta_1}^{\theta_2}$$

$$= \mu\, p_1\, b\, r^2 (\cos\theta_1 - \cos\theta_2)$$

Moment of normal force δR_N of the element about the fulcrum O_1,

$$\delta M_N = \delta R_N \times O_1B = \delta R_N (OO_1 \sin\theta)$$

$$= p_1 \sin\theta\, (b \,.\, r \,.\, \delta\theta)(OO_1 \sin\theta) = p_1 \sin^2\theta\, (b \,.\, r \,.\, \delta\theta)\, OO_1$$

Total moment of normal forces about the fulcrum O_1,

$$M_N = \int_{\theta_1}^{\theta_2} p_1 \sin^2\theta\, (b \,.\, r\, \delta\theta)\, OO_1 = p_1 \,.\, b \,.\, r \,.\, OO_1 \int_{\theta_1}^{\theta_2} \sin^2\theta\, d\theta$$

$$= p_1 .b.r.OO_1 \int_{\theta_1}^{\theta_2} \tfrac{1}{2}(1 - \cos 2\theta)\, d\theta \qquad \ldots\left[\because \sin^2\theta = \tfrac{1}{2}(1 - \cos 2\theta)\right]$$

$$= \tfrac{1}{2}\, p_1 \,.\, b \,.\, r \,.\, OO_1 \left[\theta - \frac{\sin 2\theta}{2}\right]_{\theta_1}^{\theta_2}$$

$$= \frac{1}{2}\, p_1 \,.\, b \,.\, r \,.\, OO_1 \left[\theta_2 - \frac{\sin 2\theta_2}{2} - \theta_1 + \frac{\sin 2\theta_1}{2} \right]$$

$$= \frac{1}{2}\, p_1 \,.\, b \,.\, r \,.\, OO_1 \, [(\theta_2 - \theta_1) + \tfrac{1}{2} (\sin 2\theta_1 - \sin 2\theta_2)]$$

Moment of frictional force δF about the fulcrum O_1,

$$\delta M_F = \delta F \times AB = \delta F\,(r - OO_1 \cos\theta) \qquad ...(\because AB = r - OO_1 \cos\theta)$$

$$= \mu \,.\, p_1 \sin\theta\,(b \,.\, r \,.\, \delta\theta)\,(r - OO_1 \cos\theta)$$

$$= \mu \,.\, p_1 \,.\, b \,.\, r\,(r \sin\theta - OO_1 \sin\theta \cos\theta)\,\delta\theta$$

$$= \mu \,.\, p_1 \,.\, b \,.\, r \left(r\sin\theta - \frac{OO_1}{2} \sin 2\theta \right) \delta\theta \qquad ...(\because 2\sin\theta\cos\theta = \sin 2\theta)$$

∴ Total moment of frictional force about the fulcrum O_1,

$$M_F = \mu \,.\, p_1 \,.\, b \,.\, r \int_{\theta_1}^{\theta_2} \left(r\sin\theta - \frac{OO_1}{2} \sin 2\theta \right) d\theta$$

$$= \mu \,.\, p_1 \,.\, b \,.\, r \left[-r\cos\theta + \frac{OO_1}{4} \cos 2\theta \right]_{\theta_1}^{\theta_2}$$

$$= \mu \,.\, p_1 \,.\, b \,.\, r \left[-r\cos\theta_2 + \frac{OO_1}{4} \cos 2\theta_2 + r\cos\theta_1 - \frac{OO_1}{4} \cos 2\theta_1 \right]$$

$$= \mu \,.\, p_1 \,.\, b \,.\, r \left[r\,(\cos\theta_1 - \cos\theta_2) + \frac{OO_1}{4} (\cos 2\theta_2 - \cos 2\theta_1) \right]$$

Now for leading shoe, taking moments about the fulcrum O_1,

$$F_1 \times l = M_N - M_F$$

and for trailing shoe, taking moments about the fulcrum O_2,

$$F_2 \times l = M_N + M_F$$

Note: If $M_F > M_N$, then the brake becomes self locking.

Example 25.15. *Fig. 25.37 shows the arrangement of two brake shoes which act on the internal surface of a cylindrical brake drum. The braking force F_1 and F_2 are applied as shown and each shoe pivots on its fulcrum O_1 and O_2. The width of the brake lining is 35 mm. The intensity of pressure at any point A is 0.4 sin θ N/mm², where θ is measured as shown from either pivot. The coefficient of friction is 0.4. Determine the braking torque and the magnitude of the forces F_1 and F_2.*

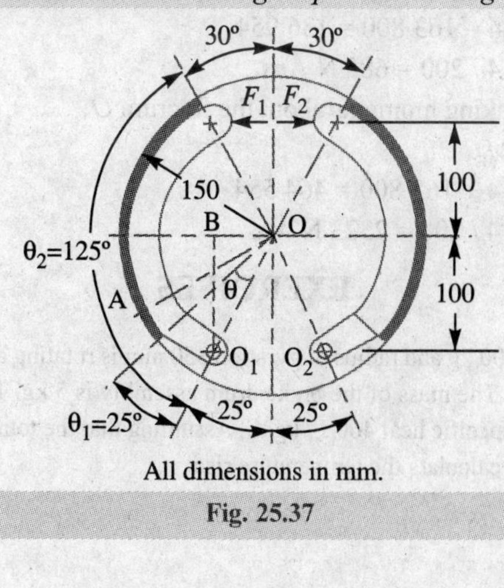

All dimensions in mm.

Fig. 25.37

Solution. Given : $b = 35$ mm ; $\mu = 0.4$; $r = 150$ mm ; $l = 200$ mm ; $\theta_1 = 25°$; $\theta_2 = 125°$

Since the intensity of normal pressure at any point is 0.4 sin θ N/mm², therefore maximum intensity of normal pressure,

$$p_1 = 0.4 \text{ N/mm}^2$$

We know that the braking torque for one shoe,

$$= \mu \cdot p_1 \cdot b \cdot r^2 (\cos \theta_1 - \cos \theta_2)$$
$$= 0.4 \times 0.4 \times 35 (150)^2 (\cos 25° - \cos 125°)$$
$$= 126\ 000\ (0.9063 + 0.5736) = 186\ 470 \text{ N-mm}$$

∴ Total braking torque for two shoes,

$$T_B = 2 \times 186\ 470 = 372\ 940 \text{ N-mm}$$

Magnitude of the forces F_1 and F_2

From the geometry of the figure, we find that

$$OO_1 = \frac{O_1B}{\cos 25°} = \frac{100}{0.9063} = 110.3 \text{ mm}$$

$$\theta_1 = 25° = 25 \times \pi / 180 = 0.436 \text{ rad}$$

and $$\theta_2 = 125° = 125 \times \pi / 180 = 2.18 \text{ rad}$$

We know that the total moment of normal forces about the fulcrum O_1,

$$M_N = \tfrac{1}{2} p_1 \cdot b \cdot r \cdot OO_1 \left[(\theta_2 - \theta_1) + \tfrac{1}{2} (\sin 2\theta_1 - \sin 2\theta_2)\right]$$
$$= \tfrac{1}{2} \times 0.4 \times 35 \times 150 \times 110.3 \left[(2.18 - 0.436) + \tfrac{1}{2} (\sin 50° - \sin 250°)\right]$$
$$= 115\ 815 \left[1.744 + \frac{1}{2} (0.766 + 0.9397)\right] = 300\ 754 \text{ N-mm}$$

and total moment of friction force about the fulcrum O_1,

$$M_F = \mu \cdot p_1 \cdot b \cdot r \left[r (\cos \theta_1 - \cos \theta_2) + \frac{OO_1}{4} (\cos 2\theta_2 - \cos 2\theta_1)\right]$$
$$= 0.4 \times 0.4 \times 35 \times 150 \left[150 (\cos 25° - \cos 125°) + \frac{110.3}{4} (\cos 250° - \cos 50°)\right]$$
$$= 840\ [150\ (0.9063 + 0.5736) + 27.6\ (-0.342 - 0.6428)]$$
$$= 840\ (222 - 27) = 163\ 800 \text{ N-mm}$$

For the leading shoe, taking moments about the fulcrum O_1,

$$F_1 \times l = M_N - M_F$$

or $$F_1 \times 200 = 300\ 754 - 163\ 800 = 136\ 954$$

∴ $$F_1 = 136\ 954 / 200 = 685 \text{ N}$$ **Ans.**

For the trailing shoe, taking moments about the fulcrum O_2,

$$F_2 \times l = M_N + M_F$$

or $$F_2 \times 200 = 300\ 754 + 163\ 800 = 464\ 554$$

∴ $$F_2 = 464\ 554 / 200 = 2323 \text{ N}$$ **Ans.**

EXERCISES

1. A flywheel of mass 100 kg and radius of gyration 350 mm is rotating at 720 r.p.m. It is brought to rest by means of a brake. The mass of the brake drum assembly is 5 kg. The brake drum is made of cast iron FG 260 having specific heat 460 J / kg°C. Assuming that the total heat generated is absorbed by the brake drum only, calculate the temperature rise.

   [Ans. 15.14°C]

2. A single block brake, as shown in Fig. 25.38, has the drum diameter 250 mm. The angle of contact is 90° and the coefficient of friction between the drum and the lining is 0.35. If the torque transmitted by the brake is 70 N-m, find the force P required to operate the brake.

[Ans. 700 N]

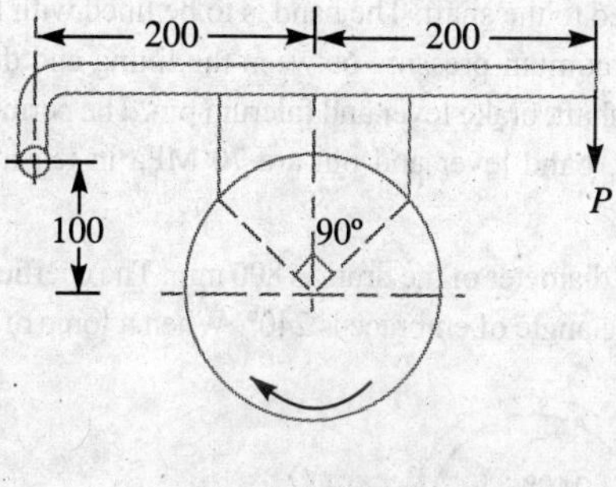

All dimensions in mm.

Fig. 25.38

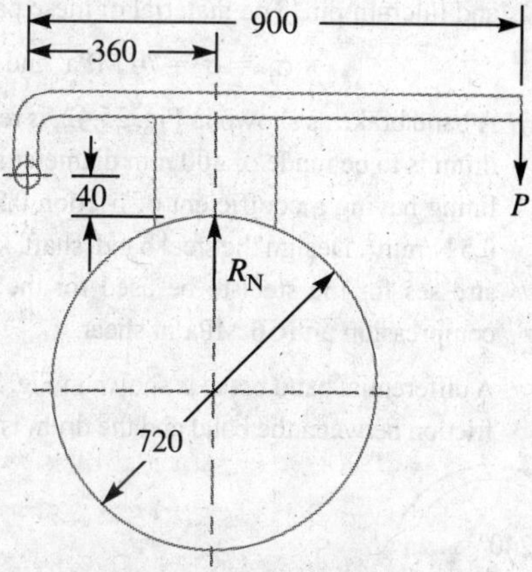

All dimensions in mm.

Fig. 25.39

3. A single block brake, as shown in Fig. 25.39, has a drum diameter of 720 mm. If the brake sustains 225 N-m torque at 500 r.p.m.; find :
   (*a*) the required force (P) to apply the brake for clockwise rotation of the drum;
   (*b*) the required force (P) to apply the brake for counter clockwise rotation of the drum;
   (*c*) the location of the fulcrum to make the brake self-locking for clockwise rotation of the drum; and

   The coefficient of friction may be taken as 0.3. **[Ans. 805.4 N ; 861 N; 1.2 m ; 11.78 kW]**

4. The layout and dimensions of a double shoe brake is shown in Fig. 25.40. The diameter of the brake drum is 300 mm and the contact angle for each shoe is 90°. If the coefficient of friction for the brake lining and the drum is 0.4, find the spring force necessary to transmit a torque of 30 N-m. Also determine the width of the brake shoes, if the bearing pressure on the lining material is not to exceed 0.28 N/mm^2. **[Ans. 99.1 N ; 5 mm]**

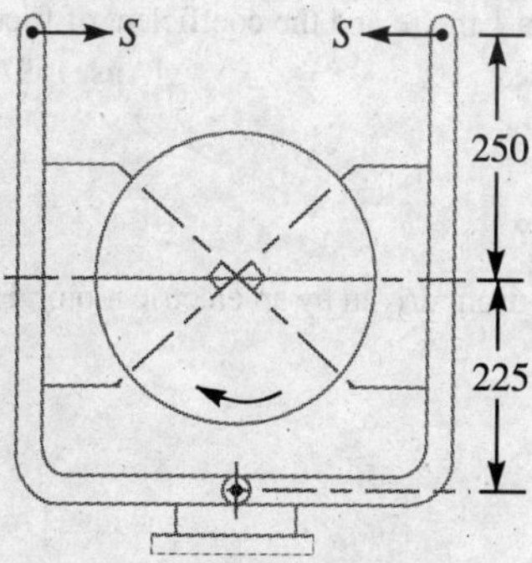

All dimensions in mm.

Fig. 25.40

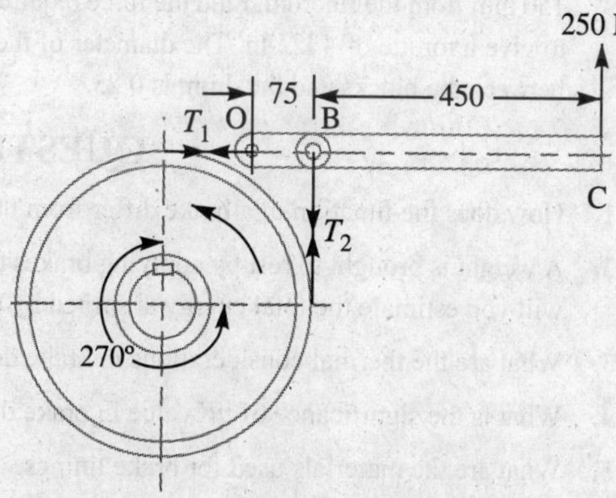

All dimensions in mm.

Fig. 25.41

5. The drum of a simple band brake is 450 mm. The band embraces 3/4th of the circumference of the drum. One end of the band is attached to the fulcrum pin and the other end is attached to a pin *B* as shown in Fig. 25.41. The band is to be lined with asbestos fabric having a coefficient of friction 0.3. The allowable bearing pressure for the brake lining is 0.21 N/mm^2. Design the band shaft, key, lever and fulcrum pin. The material of these parts is mild steel having permissible stresses as follows :

$$\sigma_t = \sigma_c = 70 \text{ MPa, and } \tau = 56 \text{ MPa}$$

6. A band brake as shown in Fig. 25.42, is required to balance a torque of 980 N-m at the drum shaft. The drum is to be made of 400 mm diameter and is keyed to the shaft. The band is to be lined with ferodo lining having a coefficient of friction 0.25. The maximum pressure between the lining and drum is 0.5 N/mm^2. Design the steel band, shaft, key on the shaft, brake lever and fulcrum pin. The permissible stresses for the steel to be used for the shaft, key, band lever and pin are 70 MPa in tension and compression and 56 MPa in shear.

7. A differential band brake is shown in Fig. 25.43. The diameter of the drum is 800 mm. The coefficient of friction between the band and the drum is 0.3 and the angle of embrace is 240°. When a force of 600 N is applied at the free end of the lever, find for the clockwise and anticlockwise rotation of the drum: 1. the maximum and minimum forces in the band; and 2. the torque which can be applied by the brake.

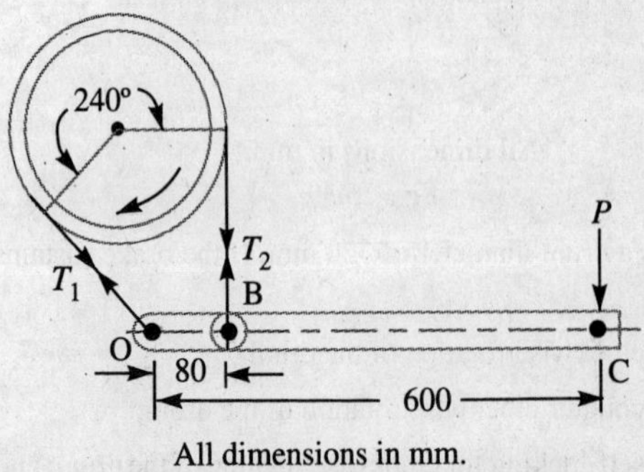

Fig. 25.42

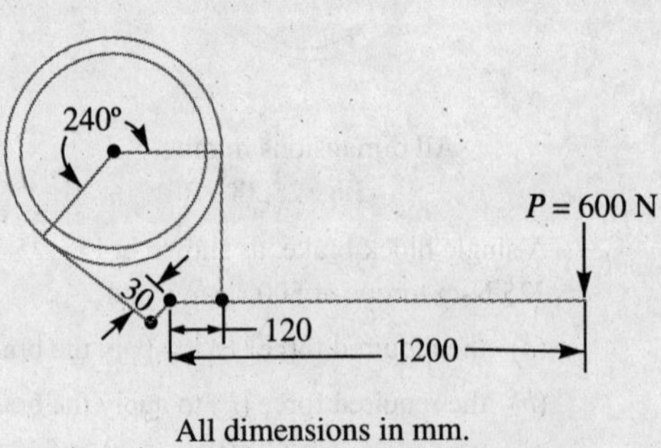

Fig. 25.43

[Ans. 176 kN, 50 kN, 50.4 kN-m ; 6.46 kN, 1.835 kN, 1.85 kN-m]

8. In a band and block brake, the band is lined with 14 blocks, each of which subtends an angle of 20° at the drum centre. One end of the band is attached to the fulcrum of the brake lever and the other to a pin 150 mm from the fulcrum. Find the force required at the end of the lever 1 metre long from the fulcrum to give a torque of 4 kN-m. The diameter of the brake drum is 1 metre and the coefficient of friction between the blocks and the drum is 0.25. **[Ans. 1692 N]**

QUESTIONS

1. How does the function of a brake differ from that of a clutch ?
2. A weight is brought to rest by applying brakes to the hoisting drum driven by an electric motor. How will you estimate the total energy absorbed by the brake ?
3. What are the thermal considerations in brake design ?
4. What is the significance of *pV* value in brake design ?
5. What are the materials used for brake linings.
6. Discuss the different types of brakes giving atleast one practical application for each.
7. List the important factors upon which the capacity of a brake depends.
8. What is a self-energizing brake ? When a brake becomes self-locking.

9. What is back stop action in band brakes ? Explain the condition for it.

10. Describe with the help of a neat sketch the principle of operation of an internal expanding shoe brake. Derive the expression for the braking torque.

Truck suspension system : Front Pivot ball suspension soaks up the bumps and provides unmatched adjustability. Chrome 8 mm CVA joints give added strength.

Note : This picture is given as additional information and is not a direct example of the current chapter.

OBJECTIVE TYPE QUESTIONS

1. A brake commonly used in railway trains is
   (*a*) shoe brake (*b*) band brake
   (*c*) band and block brake (*d*) internal expanding brake
2. A brake commonly used in motor cars is
   (*a*) shoe brake (*b*) band brake
   (*c*) band and block brake (*d*) internal expanding brake
3. The material used for brake lining should have .... coefficient of friction.
   (*a*) low (*b*) high
4. When the frictional force helps to apply the brake, then the brake is said to be
   (*a*) self-energizing brake (*b*) self-locking brake
5. For a band brake, the width of the band for a drum diameter greater than 1 m, should not exceed
   (*a*) 150 mm (*b*) 200 mm
   (*c*) 250 mm (*d*) 300 mm

ANSWERS

1. (*a*) 2. (*d*) 3. (*b*) 4. (*a*) 5. (*a*)

CHAPTER

26

Sliding Contact Bearings

1. Introduction.
2. Classification of Bearings.
3. Types of Sliding Contact Bearings.
4. Hydrodynamic Lubricated Bearings.
5. Assumptions in Hydrodynamic Lubricated Bearings.
6. Important Factors for the Formation of Thick Oil Film.
7. Wedge Film Journal Bearings.
8. Squeeze Film Journal Bearings.
9. Properties of Sliding Contact Bearing Materials.
10. Materials used for Sliding Contact Bearings.
11. Lubricants.
12. Properties of Lubricants.
13. Terms used in Hydrodynamic Journal Bearings.
14. Bearing Characteristic Number and Bearing Modulus for Journal Bearings.
15. Coefficient of Friction.
16. Critical Pressure.
17. Sommerfeld Number.
18. Heat Generated.
19. Design Procedure.
20. Solid Journal Bearing.
21. Bushed Bearing.
22. Split Bearing or Plummer Block.
23. Design of Bearing Caps and Bolts.
24. Oil Grooves.
25. Thrust Bearings
26. Foot-step or Pivot Bearings.
27. Collar Bearings.

26.1 Introduction

A bearing is a machine element which support another moving machine element (known as journal). It permits a relative motion between the contact surfaces of the members, while carrying the load. A little consideration will show that due to the relative motion between the contact surfaces, a certain amount of power is wasted in overcoming frictional resistance and if the rubbing surfaces are in direct contact, there will be rapid wear. In order to reduce frictional resistance and wear and in some cases to carry away the heat generated, a layer of fluid (known as lubricant) may be provided. The lubricant used to separate the journal and bearing is usually a mineral oil refined from petroleum, but vegetable oils, silicon oils, greases etc., may be used.

26.2 Classification of Bearings

Though the bearings may be classified in many ways, yet the following are important from the subject point of view:

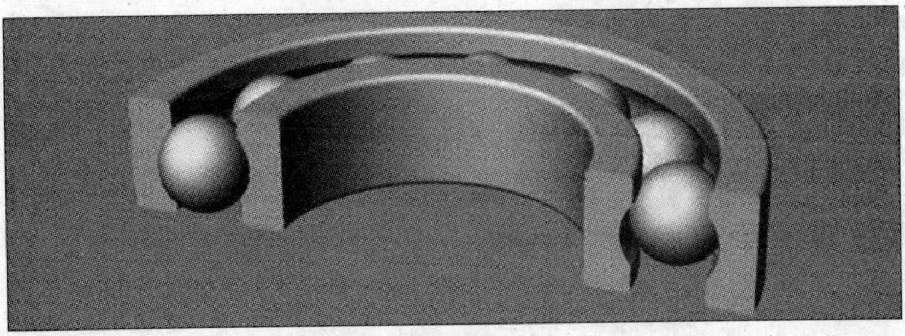

Roller Bearing

1. ***Depending upon the direction of load to be supported.*** The bearings under this group are classified as:

(*a*) Radial bearings, and (*b*) Thrust bearings.

In ***radial bearings***, the load acts perpendicular to the direction of motion of the moving element as shown in Fig. 26.1 (*a*) and (*b*).

In ***thrust bearings,*** the load acts along the axis of rotation as shown in Fig. 26.1 (*c*).

Note : These bearings may move in either of the directions as shown in Fig. 26.1.

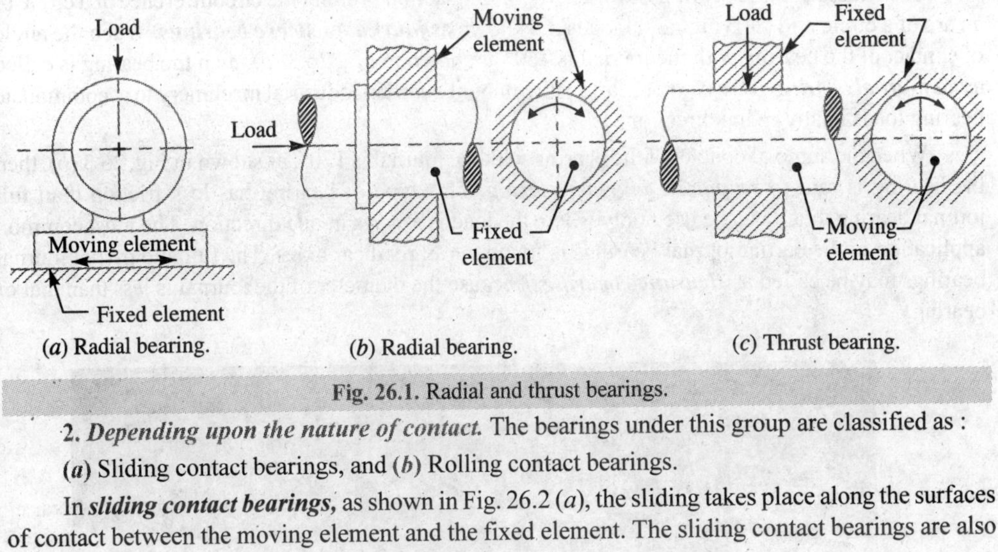

(*a*) Radial bearing. (*b*) Radial bearing. (*c*) Thrust bearing.

Fig. 26.1. Radial and thrust bearings.

2. ***Depending upon the nature of contact.*** The bearings under this group are classified as :

(*a*) Sliding contact bearings, and (*b*) Rolling contact bearings.

In ***sliding contact bearings,*** as shown in Fig. 26.2 (*a*), the sliding takes place along the surfaces of contact between the moving element and the fixed element. The sliding contact bearings are also known as ***plain bearings.***

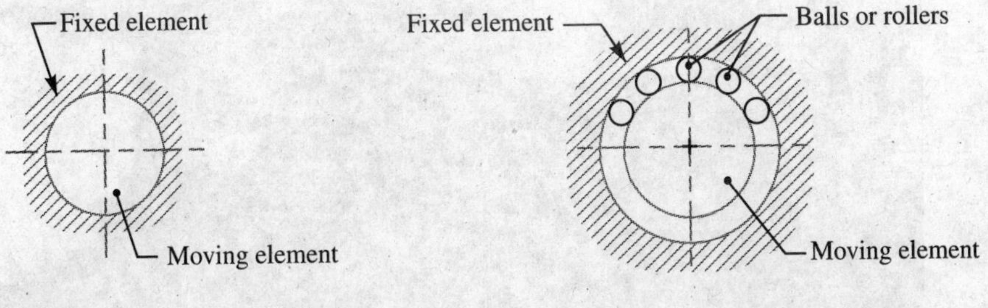

(*a*) Sliding contact bearing. (*b*) Rolling contact bearings.

Fig. 26.2. Sliding and rolling contact bearings.

In ***rolling contact bearings,*** as shown in Fig. 26.2 (*b*), the steel balls or rollers, are interposed between the moving and fixed elements. The balls offer rolling friction at two points for each ball or roller.

26.3 Types of Sliding Contact Bearings

The sliding contact bearings in which the sliding action is guided in a straight line and carrying radial loads, as shown in Fig. 26.1 (*a*), may be called ***slipper*** or ***guide bearings.*** Such type of bearings are usually found in cross-head of steam engines.

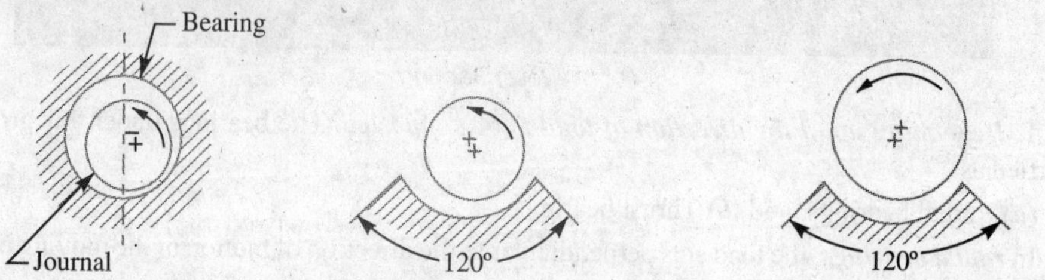

(*a*) Full journal bearing. (*b*) Partial journal bearing. (*c*) Fitted journal bearing.

Fig. 26.3. Journal or sleeve bearings.

The sliding contact bearings in which the sliding action is along the circumference of a circle or an arc of a circle and carrying radial loads are known as ***journal*** or ***sleeve bearings.*** When the angle of contact of the bearing with the journal is 360° as shown in Fig. 26.3 (*a*), then the bearing is called a ***full journal bearing.*** This type of bearing is commonly used in industrial machinery to accommodate bearing loads in any radial direction.

When the angle of contact of the bearing with the journal is 120°, as shown in Fig. 26.3 (*b*), then the bearing is said to be ***partial journal bearing.*** This type of bearing has less friction than full journal bearing, but it can be used only where the load is always in one direction. The most common application of the partial journal bearings is found in rail road car axles. The full and partial journal bearings may be called as ***clearance bearings*** because the diameter of the journal is less than that of bearing.

Sliding contact bearings are used in steam engines

When a partial journal bearing has no clearance *i.e.* the diameters of the journal and bearing are equal, then the bearing is called a ***fitted bearing,*** as shown in Fig. 26.3 (*c*).

The sliding contact bearings, according to the thickness of layer of the lubricant between the bearing and the journal, may also be classified as follows :

1. ***Thick film bearings.*** The thick film bearings are those in which the working surfaces are completely separated from each other by the lubricant. Such type of bearings are also called as ***hydrodynamic lubricated bearings.***
2. ***Thin film bearings.*** The thin film bearings are those in which, although lubricant is present, the working surfaces partially contact each other atleast part of the time. Such type of bearings are also called ***boundary lubricated bearings.***
3. ***Zero film bearings.*** The zero film bearings are those which operate without any lubricant present.
4. ***Hydrostatic or externally pressurized lubricated bearings.*** The hydrostatic bearings are those which can support steady loads without any relative motion between the journal and the bearing. This is achieved by forcing externally pressurized lubricant between the members.

26.4 Hydrodynamic Lubricated Bearings

We have already discussed that in hydrodynamic lubricated bearings, there is a thick film of lubricant between the journal and the bearing. A little consideration will show that when the bearing is supplied with sufficient lubricant, a pressure is build up in the clearance space when the journal is rotating about an axis that is eccentric with the bearing axis. The load can be supported by this fluid pressure without any actual contact between the journal and bearing. The load carrying ability of a hydrodynamic bearing arises simply because a viscous fluid resists being pushed around. Under the proper conditions, this resistance to motion will develop a pressure distribution in the lubricant film that can support a useful load. The load supporting pressure in hydrodynamic bearings arises from either

Hydrodynamic Lubricated Bearings

1. the flow of a viscous fluid in a converging channel (known as ***wedge film lubrication***), or
2. the resistance of a viscous fluid to being squeezed out from between approaching surfaces (known as ***squeeze film lubrication***).

26.5 Assumptions in Hydrodynamic Lubricated Bearings

The following are the basic assumptions used in the theory of hydrodynamic lubricated bearings:

1. The lubricant obeys Newton's law of viscous flow.
2. The pressure is assumed to be constant throughout the film thickness.
3. The lubricant is assumed to be incompressible.
4. The viscosity is assumed to be constant throughout the film.
5. The flow is one dimensional, *i.e.* the side leakage is neglected.

26.6 Important Factors for the Formation of Thick Oil Film in Hydrodynamic Lubricated Bearings

According to Reynolds, the following factors are essential for the formation of a thick film of oil in hydrodynamic lubricated bearings :

1. A continuous supply of oil.
2. A relative motion between the two surfaces in a direction approximately tangential to the surfaces.
3. The ability of one of the surfaces to take up a small inclination to the other surface in the direction of the relative motion.
4. The line of action of resultant oil pressure must coincide with the line of action of the external load between the surfaces.

26.7 Wedge Film Journal Bearings

The load carrying ability of a wedge-film journal bearing results when the journal and/or the bearing rotates relative to the load. The most common case is that of a steady load, a fixed (non-rotating) bearing and a rotating journal. Fig. 26.4 (*a*) shows a journal at rest with metal to metal contact at *A* on the line of action of the supported load. When the journal rotates slowly in the anticlockwise direction, as shown in Fig. 26.4 (*b*), the point of contact will move to *B*, so that the angle *AOB* is the angle of sliding friction of the surfaces in contact at *B*. In the absence of a lubricant, there will be dry metal to metal friction. If a lubricant is present in the clearance space of the bearing and journal, then a thin absorbed film of the lubricant may partly separate the surface, but a continuous fluid film completely separating the surfaces will not exist because of slow speed.

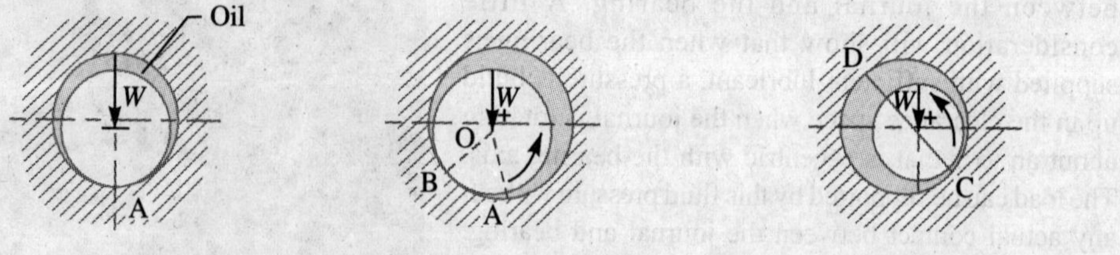

(*a*) At rest. (*b*) Slow speed. (*c*) High speed.

Fig. 26.4. Wedge film journal bearing.

When the speed of the journal is increased, a continuous fluid film is established as in Fig. 26.4 (*c*). The centre of the journal has moved so that the minimum film thickness is at *C*. It may be noted that from *D* to *C* in the direction of motion, the film is continually narrowing and hence is a converging film. The curved converging film may be considered as a wedge shaped film of a slipper bearing wrapped around the journal. A little consideration will show that from *C* to *D* in the direction of rotation, as shown in Fig. 26.4 (*c*), the film is diverging and cannot give rise to a positive pressure or a supporting action.

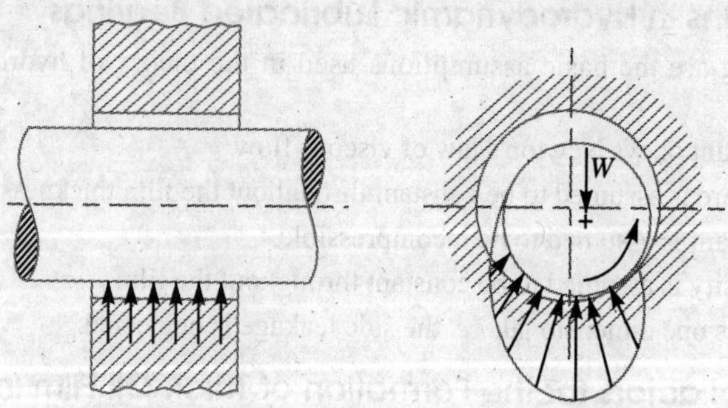

Fig. 26.5. Variation of pressure in the converging film.

Fig. 26.5 shows the two views of the bearing shown in Fig. 26.4 (*c*), with the variation of pressure in the converging film. Actually, because of side leakage, the angle of contact on which pressure acts is less than 180°.

26.8 Squeeze Film Journal Bearing

We have seen in the previous article that in a wedge film journal bearing, the bearing carries a steady load and the journal rotates relative to the bearing. But in certain cases, the bearings oscillate or rotate so slowly that the wedge film cannot provide a satisfactory film thickness. If the load is uniform or varying in magnitude while acting in a constant direction, this becomes a thin film or possibly a zero film problem. But if the load reverses its direction, the squeeze film may develop sufficient capacity to carry the dynamic loads without contact between the journal and the bearing. Such bearings are known as ***squeeze film journal bearing.***

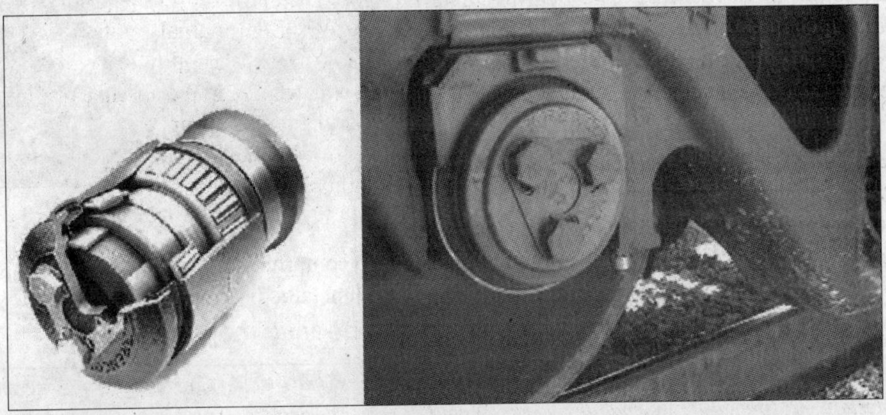

Journal bearing

26.9 Properties of Sliding Contact Bearing Materials

When the journal and the bearings are having proper lubrication *i.e.* there is a film of clean, non-corrosive lubricant in between, separating the two surfaces in contact, the only requirement of the bearing material is that they should have sufficient strength and rigidity. However, the conditions under which bearings must operate in service are generally far from ideal and thus the other properties as discussed below must be considered in selecting the best material.

1. ***Compressive strength.*** The maximum bearing pressure is considerably greater than the average pressure obtained by dividing the load to the projected area. Therefore the bearing material should have high compressive strength to withstand this maximum pressure so as to prevent extrusion or other permanent deformation of the bearing.

2. ***Fatigue strength.*** The bearing material should have sufficient fatigue strength so that it can withstand repeated loads without developing surface fatigue cracks. It is of major importance in aircraft and automotive engines.

3. ***Comformability.*** It is the ability of the bearing material to accommodate shaft deflections and bearing inaccuracies by plastic deformation (or creep) without excessive wear and heating.

4. ***Embeddability.*** It is the ability of bearing material to accommodate (or embed) small particles of dust, grit etc., without scoring the material of the journal.

5. ***Bondability.*** Many high capacity bearings are made by bonding one or more thin layers of a bearing material to a high strength steel shell. Thus, the strength of the bond *i.e.* bondability is an important consideration in selecting bearing material.

6. ***Corrosion resistance.*** The bearing material should not corrode away under the action of lubricating oil. This property is of particular importance in internal-combustion engines where the same oil is used to lubricate the cylinder walls and bearings. In the cylinder, the lubricating oil comes into contact with hot cylinder walls and may oxidise and collect carbon deposits from the walls.

7. ***Thermal conductivity.*** The bearing material should be of high thermal conductivity so as to permit the rapid removal of the heat generated by friction.

8. ***Thermal expansion.*** The bearing material should be of low coefficient of thermal expansion, so that when the bearing operates over a wide range of temperature, there is no undue change in the clearance.

Marine bearings

All these properties as discussed above are, however, difficult to find in any particular bearing material. The various materials are used in practice, depending upon the requirement of the actual service conditions. The choice of material for any application must represent a compromise. The following table shows the comparison of some of the properties of more common metallic bearing materials.

Table 26.1. Properties of metallic bearing materials.

Bearing material	*Fatigue strength*	*Comformability*	*Embeddability*	*Anti scoring*	*Corrosion resistance*	*Thermal conductivity*
Tin base babbit	Poor	Good	Excellent	Excellent	Excellent	Poor
Lead base babbit	Poor to fair	Good	Good	Good to excellent	Fair to good	Poor
Lead bronze	Fair	Poor	Poor	Poor	Good	Fair
Copper lead	Fair	Poor	Poor to fair	Poor to fair	Poor to fair	Fair to good
Aluminium	Good	Poor to fair	Poor	Good	Excellent	Fair
Silver	Excellent	Almost none	Poor	Poor	Excellent	Excellent
Silver lead deposited	Excellent	Excellent	Poor	Fair to good	Excellent	Excellent

26.10 Materials used for Sliding Contact Bearings

The materials commonly used for sliding contact bearings are discussed below :

1. ***Babbit metal.*** The tin base and lead base babbits are widely used as a bearing material, because they satisfy most requirements for general applications. The babbits are recommended where the maximum bearing pressure (on projected area) is not over 7 to 14 N/mm^2. When applied in

automobiles, the babbit is generally used as a thin layer, 0.05 mm to 0.15 mm thick, bonded to an insert or steel shell. The composition of the babbit metals is as follows :

Tin base babbits : Tin 90% ; Copper 4.5% ; Antimony 5% ; Lead 0.5%.

Lead base babbits : Lead 84% ; Tin 6% ; Anitmony 9.5% ; Copper 0.5%.

2. ***Bronzes.*** The bronzes (alloys of copper, tin and zinc) are generally used in the form of machined bushes pressed into the shell. The bush may be in one or two pieces. The bronzes commonly used for bearing material are gun metal and phosphor bronzes.

The ***gun metal*** (Copper 88% ; Tin 10% ; Zinc 2%) is used for high grade bearings subjected to high pressures (not more than 10 N/mm^2 of projected area) and high speeds.

The ***phosphor bronze*** (Copper 80% ; Tin 10% ; Lead 9% ; Phosphorus 1%) is used for bearings subjected to very high pressures (not more than 14 N/mm^2 of projected area) and speeds.

3. ***Cast iron.*** The cast iron bearings are usually used with steel journals. Such type of bearings are fairly successful where lubrication is adequate and the pressure is limited to 3.5 N/mm^2 and speed to 40 metres per minute.

4. ***Silver.*** The silver and silver lead bearings are mostly used in aircraft engines where the fatigue strength is the most important consideration.

5. ***Non-metallic bearings.*** The various non-metallic bearings are made of carbon-graphite, rubber, wood and plastics. The **carbon-graphite bearings** are self lubricating, dimensionally stable over a wide range of operating conditions, chemically inert and can operate at higher temperatures than other bearings. Such type of bearings are used in food processing and other equipment where contamination by oil or grease must be prohibited. These bearings are also used in applications where the shaft speed is too low to maintain a hydrodynamic oil film.

The ***soft rubber bearings*** are used with water or other low viscosity lubricants, particularly where sand or other large particles are present. In addition to the high degree of embeddability and comformability, the rubber bearings are excellent for absorbing shock loads and vibrations. The rubber bearings are used mainly on marine propeller shafts, hydraulic turbines and pumps.

The ***wood bearings*** are used in many applications where low cost, cleanliness, inattention to lubrication and anti-seizing are important.

Industrial bearings.

The commonly used plastic material for bearings is ***Nylon*** and ***Teflon.*** These materials have many characteristics desirable in bearing materials and both can be used dry *i.e.* as a zero film bearing. The Nylon is stronger, harder and more resistant to abrasive wear. It is used for applications in which these properties are important *e.g.* elevator bearings, cams in telephone dials etc. The Teflon is rapidly replacing Nylon as a wear surface or liner for journal and other sliding bearings because of the following properties:

1. It has lower coefficient of friction, about 0.04 (dry) as compared to 0.15 for Nylon.
2. It can be used at higher temperatures up to about 315°C as compared to 120°C for Nylon.
3. It is dimensionally stable because it does not absorb moisture, and
4. It is practically chemically inert.

26.11 Lubricants

The lubricants are used in bearings to reduce friction between the rubbing surfaces and to carry away the heat generated by friction. It also protects the bearing against corrosion. All lubricants are classified into the following three groups :

1. Liquid, 2. Semi-liquid, and 3. Solid.

The ***liquid lubricants*** usually used in bearings are mineral oils and synthetic oils. The mineral oils are most commonly used because of their cheapness and stability. The liquid lubricants are usually preferred where they may be retained.

A grease is a ***semi-liquid lubricant*** having higher viscosity than oils. The greases are employed where slow speed and heavy pressure exist and where oil drip from the bearing is undesirable. The ***solid lubricants*** are useful in reducing friction where oil films cannot be maintained because of pressures or temperatures. They should be softer than materials being lubricated. A graphite is the most common of the solid lubricants either alone or mixed with oil or grease.

Wherever moving and rotating parts are present proper lubrication is essential to protect the moving parts from wear and tear and reduce friction.

26.12 Properties of Lubricants

1. ***Viscosity.*** It is the measure of degree of fluidity of a liquid. It is a physical property by virtue of which an oil is able to form, retain and offer resistance to shearing a buffer film-under heat and pressure. The greater the heat and pressure, the greater viscosity is required of a lubricant to prevent thinning and squeezing out of the film.

The fundamental meaning of viscosity may be understood by considering a flat plate moving under a force P parallel to a stationary plate, the two plates being separated by a thin film of a fluid lubricant of thickness h, as shown in Fig. 26.6. The particles of the lubricant adhere strongly to the moving and stationary plates. The motion is accompanied by a linear slip or shear between the particles throughout the entire height (h) of the film thickness. If A is the area of the plate in contact with the lubricant, then the unit shear stress is given by

$$\tau = P / A$$

According to Newton's law of viscous flow, the magnitude of this shear stress varies directly with the velocity gradient (dV/dy). It is assumed that

(*a*) the lubricant completely fills the space between the two surfaces,

(*b*) the velocity of the lubricant at each surface is same as that of the surface, and

(*c*) any flow of the lubricant perpendicular to the velocity of the plate is negligible.

$$\therefore \quad \tau = \frac{P}{A} \propto \frac{dV}{dy} \quad \text{or} \quad \tau = Z \times \frac{dV}{dy}$$

where Z is a constant of proportionality and is known as ***absolute viscosity*** (or simply viscosity) of the lubricant.

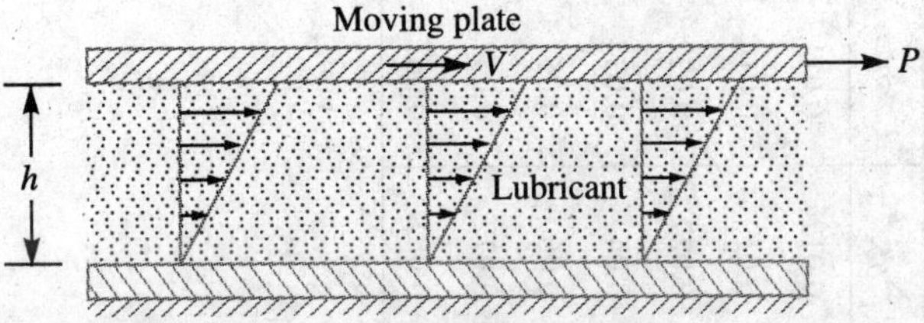

Fig. 26.6. Viscosity.

When the thickness of the fluid lubricant is small which is the case for bearings, then the velocity gradient is very nearly constant as shown in Fig. 26.6, so that

$$\frac{dV}{dy} = \frac{V}{y} = \frac{V}{h}$$

$$\therefore \quad \tau = Z \times \frac{V}{h} \quad \text{or} \quad Z = \tau \times \frac{h}{V}$$

When τ is in N/m^2, h is in metres and V is in m/s, then the unit of absolute viscosity is given by

$$Z = \tau \times \frac{h}{V} = \frac{\text{N}}{\text{m}^2} \times \frac{\text{m}}{\text{m/s}} = \text{N-s/m}^2$$

However, the common practice is to express the absolute viscosity in mass units, such that

$$1 \text{ N-s/m}^2 = \frac{1 \text{kg-m}}{s^2} \times \frac{\text{s}}{\text{m}^2} = 1 \text{ kg/m-s} \qquad \ldots (\because \ 1 \text{ N} = 1 \text{ kg-m/s}^2)$$

Thus the unit of absolute viscosity in S.I. units is kg / m-s.

The viscocity of the lubricant is measured by Saybolt universal viscometer. It determines the time required for a standard volume of oil at a certain temperature to flow under a certain head through a tube of standard diameter and length. The time so determined in seconds is the Saybolt universal viscosity. In order to convert Saybolt universal viscosity in seconds to absolute viscosity (in kg / m-s), the following formula may be used:

$$Z = \text{Sp. gr. of oil} \left(0.000\,22\,S - \frac{0.18}{S}\right) \text{kg/m-s} \qquad \ldots(i)$$

where Z = Absolute viscosity at temperature t in kg / m-s, and

S = Saybolt universal viscosity in seconds.

The variation of absolute viscosity with temperature for commonly used lubricating oils is shown in Table 26.2 on the next page.

2. ***Oiliness.*** It is a joint property of the lubricant and the bearing surfaces in contact. It is a measure of the lubricating qualities under boundary conditions where base metal to metal is prevented only by absorbed film. There is no absolute measure of oiliness.

Table 26.2. Absolute viscosity of commonly used lubricating oils.

S. No.	*Type of oil*	*Absolute viscosity in kg / m-s at temperature in °C*											
		30	*35*	*40*	*45*	*50*	*55*	*60*	*65*	*70*	*75*	*80*	*90*
1.	SAE 10	0.05	0.036	0.027	0.0245	0.021	0.017	0.014	0.012	0.011	0.009	0.008	0.005
2.	SAE 20	0.069	0.055	0.042	0.034	0.027	0.023	0.020	0.017	0.014	0.011	0.010	0.0075
3.	SAE 30	0.13	0.10	0.078	0.057	0.048	0.040	0.034	0.027	0.022	0.019	0.016	0.010
4.	SAE 40	0.21	0.17	0.12	0.096	0.78	0.06	0.046	0.04	0.034	0.027	0.022	0.013
5.	SAE 50	0.30	0.25	0.20	0.17	0.12	0.09	0.076	0.06	0.05	0.038	0.034	0.020
6.	SAE 60	0.45	0.32	0.27	0.20	0.16	0.12	0.09	0.072	0.057	0.046	0.040	0.025
7.	SAE 70	1.0	0.69	0.45	0.31	0.21	0.165	0.12	0.087	0.067	0.052	0.043	0.033

Note : We see from the above table that the viscosity of oil decreases when its temperature increases.

3. ***Density.*** This property has no relation to lubricating value but is useful in changing the kinematic viscosity to absolute viscosity. Mathematically

Absolute viscosity = ρ × Kinematic viscosity (in m^2/s)

where ρ = Density of the lubricating oil.

The density of most of the oils at 15.5°C varies from 860 to 950 kg / m^3 (the average value may be taken as 900 kg / m^3). The density at any other temperature (t) may be obtained from the following relation, *i.e.*

$$\rho_t = \rho_{15.5} - 0.000\ 657\ t$$

where $\rho_{15.5}$ = Density of oil at 15.5° C.

4. ***Viscosity index.*** The term viscosity index is used to denote the degree of variation of viscosity with temperature.

5. ***Flash point.*** It is the lowest temperature at which an oil gives off sufficient vapour to support a momentary flash without actually setting fire to the oil when a flame is brought within 6 mm at the surface of the oil.

6. ***Fire point.*** It is the temperature at which an oil gives off sufficient vapour to burn it continuously when ignited.

7. ***Pour point or freezing point.*** It is the temperature at which an oil will cease to flow when cooled.

26.13 Terms used in Hydrodynamic Journal Bearing

A hydrodynamic journal bearing is shown in Fig. 26.7, in which O is the centre of the journal and O' is the centre of the bearing.

Let D = Diameter of the bearing,

d = Diameter of the journal,

and

l = Length of the bearing.

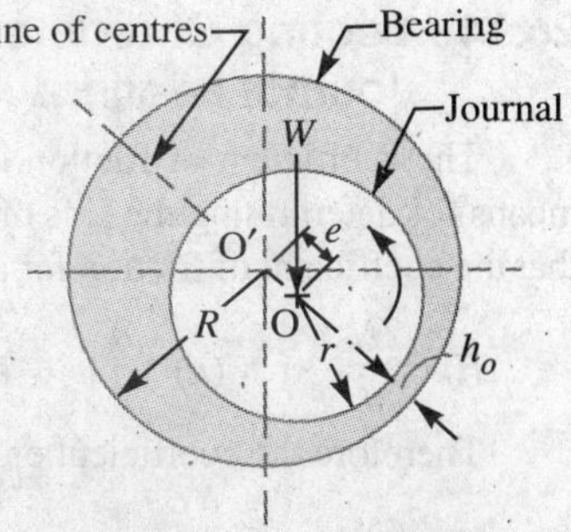

Fig. 26.7. Hydrodynamic journal bearing.

The following terms used in hydrodynamic journal bearing are important from the subject point of view :

1. ***Diametral clearance.*** It the difference between the diameters of the bearing and the journal. Mathematically, diametral clearance,

$$c = D - d$$

Note : The diametral clearance (c) in a bearing should be small enough to produce the necessary velocity gradient, so that the pressure built up will support the load. Also the small clearance has the advantage of decreasing side leakage. However, the allowance must be made for manufacturing tolerances in the journal and bushing. A commonly used clearance in industrial machines is 0.025 mm per cm of journal diameter.

2. ***Radial clearance.*** It is the difference between the radii of the bearing and the journal. Mathematically, radial clearance,

$$c_1 = R - r = \frac{D - d}{2} = \frac{c}{2}$$

3. ***Diametral clearance ratio.*** It is the ratio of the diametral clearance to the diameter of the journal. Mathematically, diametral clearance ratio

$$= \frac{c}{d} = \frac{D - d}{d}$$

4. ***Eccentricity.*** It is the radial distance between the centre (*O*) of the bearing and the displaced centre (*O*′) of the bearing under load. It is denoted by *e*.

5. ***Minimum oil film thickness.*** It is the minimum distance between the bearing and the journal, under complete lubrication condition. It is denoted by h_0 and occurs at the line of centres as shown in Fig. 26.7. Its value may be assumed as *c* / 4.

6. ***Attitude or eccentricity ratio.*** It is the ratio of the eccentricity to the radial clearance. Mathematically, attitude or eccentricity ratio,

$$\varepsilon = \frac{e}{c_1} = \frac{c_1 - h_0}{c_1} = 1 - \frac{h_0}{c_1} = 1 - \frac{2h_0}{c} \qquad \ldots (\because c_1 = c/2)$$

7. ***Short and long bearing.*** If the ratio of the length to the diameter of the journal (*i.e.* *l* / *d*) is less than 1, then the bearing is said to be **short bearing.** On the other hand, if *l* / *d* is greater than 1, then the bearing is known as ***long bearing.***

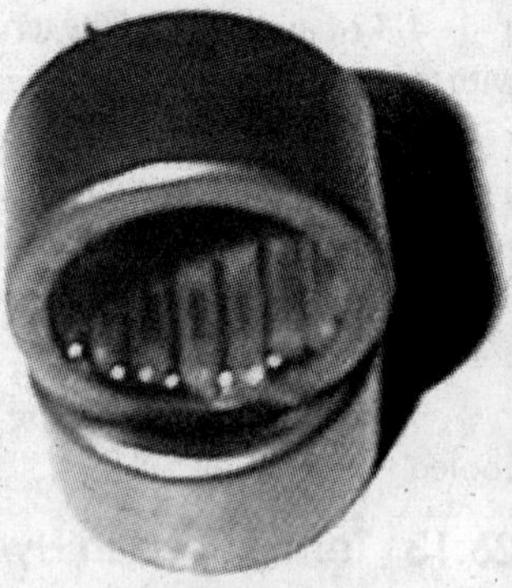

Axle bearings

Notes : 1. When the length of the journal (*l*) is equal to the diameter of the journal (*d*), then the bearing is called ***square bearing.***

2. Because of the side leakage of the lubricant from the bearing, the pressure in the film is atmospheric at the ends of the bearing. The average pressure will be higher for a long bearing than for a short or square bearing. Therefore, from the stand point of side leakage, a bearing with a large *l* / *d* ratio is preferable. However, space requirements, manufacturing, tolerances and shaft deflections are better met with a short bearing. The value of *l* / *d* may be taken as 1 to 2 for general industrial machinery. In crank shaft bearings, the *l* / *d* ratio is frequently less than 1.

26.14 Bearing Characteristic Number and Bearing Modulus for Journal Bearings

The coefficient of friction in design of bearings is of great importance, because it affords a means for determining the loss of power due to bearing friction. It has been shown by experiments that the coefficient of friction for a full lubricated journal bearing is a function of three variables, *i.e.*

(*i*) $\dfrac{ZN}{p}$; (*ii*) $\dfrac{d}{c}$; and (*iii*) $\dfrac{l}{d}$

Therefore the coefficient of friction may be expressed as

$$\mu = \phi\left(\frac{ZN}{p}, \frac{d}{c}, \frac{l}{d}\right)$$

where μ = Coefficient of friction,

ϕ = A functional relationship,

Z = Absolute viscosity of the lubricant, in kg / m-s,

N = Speed of the journal in r.p.m.,

p = Bearing pressure on the projected bearing area in N/mm^2,

= Load on the journal ÷ $l \times d$

d = Diameter of the journal,

l = Length of the bearing, and

c = Diametral clearance.

The factor *ZN* / *p* is termed as ***bearing characteristic number*** and is a dimensionless number. The variation of coefficient of friction with the operating values of bearing characteristic number (*ZN* / *p*) as obtained by McKee brothers (S.A. McKee and T.R. McKee) in an actual test of friction is shown in Fig. 26.8. The factor *ZN/p* helps to predict the performance of a bearing.

The part of the curve PQ represents the region of thick film lubrication. Between Q and R, the viscosity (Z) or the speed (N) are so low, or the pressure (p) is so great that their combination ZN/p will reduce the film thickness so that partial metal to metal contact will result. The thin film or boundary lubrication or imperfect lubrication exists between R and S on the curve. This is the region where the viscosity of the lubricant ceases to be a measure of friction characteristics but the oiliness of the lubricant is effective in preventing complete metal to metal contact and seizure of the parts.

Clutch bearing

It may be noted that the part PQ of the curve represents stable operating conditions, since from any point of stability, a decrease in viscosity (Z) will reduce ZN/p. This will result in a decrease in coefficient of friction (μ) followed by a lowering of bearing temperature that will raise the viscosity (Z).

From Fig. 26.8, we see that the minimum amount of friction occurs at A and at this point the value of ZN/p is known as ***bearing modulus*** which is denoted by K. The bearing should not be operated at this value of bearing modulus, because a slight decrease in speed or slight increase in pressure will break the oil film and make the journal to operate with metal to metal contact. This will result in high friction, wear and heating. In order to prevent such conditions, the bearing should be designed for a value of ZN/p at least three times the minimum value of bearing modulus (K). If the bearing is subjected to large fluctuations of load and heavy impacts, the value of $ZN/p = 15\ K$ may be used.

From above, it is concluded that when the value of ZN/p is greater than K, then the bearing will operate with thick film lubrication or under hydrodynamic conditions. On the other hand, when the value of ZN/p is less than K, then the oil film will rupture and there is a metal to metal contact.

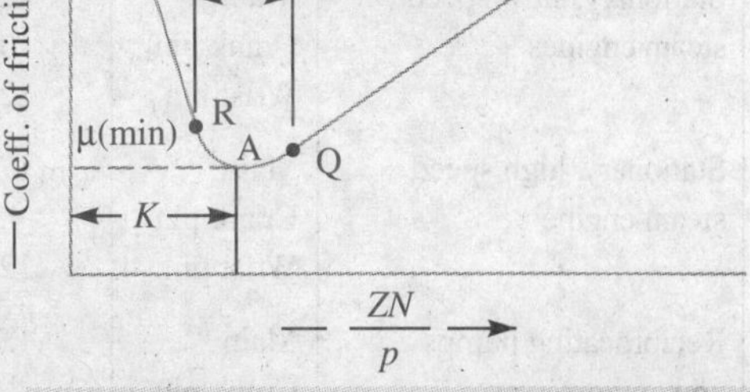

Fig. 26.8. Variation of coefficient of friction with ZN/p.

26.15 Coefficient of Friction for Journal Bearings

In order to determine the coefficient of friction for well lubricated full journal bearings, the following empirical relation established by McKee based on the experimental data, may be used.

*Coefficient of friction,

$$\mu = \frac{33}{10^8}\left(\frac{ZN}{p}\right)\left(\frac{d}{c}\right) + k \quad \text{... (when Z is in kg / m-s and } p \text{ is in N / mm}^2\text{)}$$

where Z, N, p, d and c have usual meanings as discussed in previous article, and

k = Factor to correct for end leakage. It depends upon the ratio of length to the diameter of the bearing (*i.e.* l / d).

= 0.002 for l / d ratios of 0.75 to 2.8.

The operating values of ZN / p should be compared with values given in Table 26.3 to ensure safe margin between operating conditions and the point of film breakdown.

Table 26.3. Design values for journal bearings.

Machinery	*Bearing*	*Maximum bearing pressure (p) in N/mm²*	*Operating values*			
			Absolute Viscosity (Z) in kg/m-s	*ZN/p Z in kg/m-s p in N/mm²*	$\frac{c}{d}$	$\frac{l}{d}$
Automobile and air-craft engines	Main	5.6 – 12	0.007	2.1	—	0.8 – 1.8
	Crank pin	10.5 – 24.5	0.008	1.4		0.7 – 1.4
	Wrist pin	16 – 35	0.008	1.12		1.5 – 2.2
Four stroke-Gas and oil engines	Main	5 – 8.5	0.02	2.8	0.001	0.6 – 2
	Crank pin	9.8 – 12.6	0.04	1.4		0.6 – 1.5
	Wrist pin	12.6 – 15.4	0.065	0.7		1.5 – 2
Two stroke-Gas and oil engines	Main	3.5 – 5.6	0.02	3.5	0.001	0.6 – 2
	Crank pin	7 – 10.5	0.04	1.8		0.6 – 1.5
	Wrist pin	8.4 – 12.6	0.065	1.4		1.5 – 2
Marine steam engines	Main	3.5	0.03	2.8	0.001	0.7 – 1.5
	Crank pin	4.2	0.04	2.1		0.7 – 1.2
	Wrist pin	10.5	0.05	1.4		1.2 – 1.7
Stationary, slow speed steam engines	Main	2.8	0.06	2.8	0.001	1 – 2
	Crank pin	10.5	0.08	0.84		0.9 – 1.3
	Wrist pin	12.6	0.06	0.7		1.2 – 1.5
Stationary, high speed steam engine	Main	1.75	0.015	3.5	0.001	1.5 – 3
	Crank pin	4.2	0.030	0.84		0.9 – 1.5
	Wrist pin	12.6	0.025	0.7		13 – 1.7
Reciprocating pumps and compressors	Main	1.75	0.03	4.2	0.001	1 – 2.2
	Crank pin	4.2	0.05	2.8		0.9 – 1.7
	Wrist pin	7.0	0.08	1.4		1.5 – 2.0
Steam locomotives	Driving axle	3.85	0.10	4.2	0.001	1.6 – 1.8
	Crank pin	14	0.04	0.7		0.7 – 1.1
	Wrist pin	28	0.03	0.7		0.8 – 1.3

* This is the equation of a straight line portion in the region of thick film lubrication (*i.e.* line *PQ*) as shown in Fig. 26.8.

Machinery	Bearing	Maximum bearing pressure (p) in N/mm²	Operating values			
			Absolute Viscosity (Z) in kg/m-s	ZN/p Z in kg/m-s p in N/mm²	$\frac{c}{d}$	$\frac{l}{d}$
Railway cars	Axle	3.5	0.1	7	0.001	1.8 – 2
Steam turbines	Main	0.7 – 2	0.002 – 0.016	14	0.001	1 – 2
Generators, motors, centrifugal pumps	Rotor	0.7 – 1.4	0.025	28	0.0013	1 – 2
Transmission shafts	Light, fixed	0.175	0.025-	7	0.001	2 – 3
	Self -aligning	1.05	0.060	2.1		2.5 – 4
	Heavy	1.05		2.1		2– 3
Machine tools	Main	2.1	0.04	0.14	0.001	1–4
Punching and shearing machines	Main	28	0.10	—	0.001	1–2
	Crank pin	56				
Rolling Mills	Main	21	0.05	1.4	0.0015	1–1.5

26.16 Critical Pressure of the Journal Bearing

The pressure at which the oil film breaks down so that metal to metal contact begins, is known as **critical pressure** or the **minimum operating pressure** of the bearing. It may be obtained by the following empirical relation, *i.e.*

Critical pressure or minimum operating pressure,

$$p = \frac{ZN}{4.75 \times 10^6}\left(\frac{d}{c}\right)^2\left(\frac{l}{d+l}\right) \text{ N/mm}^2 \quad \text{...(when Z is in kg / m-s)}$$

26.17 Sommerfeld Number

The Sommerfeld number is also a dimensionless parameter used extensively in the design of journal bearings. Mathematically,

$$\text{Sommerfeld number} = \frac{ZN}{p}\left(\frac{d}{c}\right)^2$$

For design purposes, its value is taken as follows :

$$\frac{ZN}{p}\left(\frac{d}{c}\right)^2 = 14.3 \times 10^6 \quad \text{... (when } Z \text{ is in kg / m-s and } p \text{ is in N / mm}^2\text{)}$$

26.18 Heat Generated in a Journal Bearing

The heat generated in a bearing is due to the fluid friction and friction of the parts having relative motion. Mathematically, heat generated in a bearing,

$$Q_g = \mu . W . V \text{ N-m/s or J/s or watts} \quad \text{...(}i\text{)}$$

where μ = Coefficient of friction,

W = Load on the bearing in N,

= Pressure on the bearing in N/mm^2 × Projected area of the bearing in mm^2 = $p\,(l \times d)$,

V = Rubbing velocity in m/s = $\frac{\pi d.N}{60}$, d is in metres, and

N = Speed of the journal in r.p.m.

After the thermal equilibrium has been reached, heat will be dissipated at the outer surface of the bearing at the same rate at which it is generated in the oil film. The amount of heat dissipated will depend upon the temperature difference, size and mass of the radiating surface and on the amount of air flowing around the bearing. However, for the convenience in bearing design, the actual heat dissipating area may be expressed in terms of the projected area of the journal.

Heat dissipated by the bearing,

$$Q_d = C.A\,(t_b - t_a) \text{ J/s or W} \quad \ldots (\because 1 \text{ J/s} = 1 \text{ W}) \quad \ldots (ii)$$

where C = Heat dissipation coefficient in W/m^2/°C,

A = Projected area of the bearing in m^2 = $l \times d$,

t_b = Temperature of the bearing surface in °C, and

t_a = Temperature of the surrounding air in °C.

The value of C have been determined experimentally by O. Lasche. The values depend upon the type of bearing, its ventilation and the temperature difference. The average values of C (in W/m^2/°C), for journal bearings may be taken as follows :

For unventilated bearings (Still air)

= 140 to 420 W/m^2/°C

For well ventilated bearings

= 490 to 1400 W/m^2/°C

It has been shown by experiments that the temperature of the bearing (t_b) is approximately mid-way between the temperature of the oil film (t_0) and the temperature of the outside air (t_a). In other words,

$$t_b - t_a = \frac{1}{2}\,(t_0 - t_a)$$

Notes : 1. For well designed bearing, the temperature of the oil film should not be more than 60°C, otherwise the viscosity of the oil decreases rapidly and the operation of the bearing is found to suffer. The temperature of the oil film is often called as the ***operating temperature*** of the bearing.

2. In case the temperature of the oil film is higher, then the bearing is cooled by circulating water through coils built in the bearing.

3. The mass of the oil to remove the heat generated at the bearing may be obtained by equating the heat generated to the heat taken away by the oil. We know that the heat taken away by the oil,

$$Q_t = m.S.t \text{ J/s or watts}$$

where m = Mass of the oil in kg / s,

S = Specific heat of the oil. Its value may be taken as 1840 to 2100 J / kg / °C,

t = Difference between outlet and inlet temperature of the oil in °C.

26.19 Design Procedure for Journal Bearing

The following procedure may be adopted in designing journal bearings, when the bearing load, the diameter and the speed of the shaft are known.

Journal bearings are used in helicopters, primarily in the main rotor axis and in the landing gear for fixed wing aircraft.

1. Determine the bearing length by choosing a ratio of l/d from Table 26.3.
2. Check the bearing pressure, $p = W / l.d$ from Table 26.3 for probable satisfactory value.
3. Assume a lubricant from Table 26.2 and its operating temperature (t_0). This temperature should be between 26.5°C and 60°C with 82°C as a maximum for high temperature installations such as steam turbines.
4. Determine the operating value of ZN/p for the assumed bearing temperature and check this value with corresponding values in Table 26.3, to determine the possibility of maintaining fluid film operation.
5. Assume a clearance ratio c/d from Table 26.3.
6. Determine the coefficient of friction (μ) by using the relation as discussed in Art. 26.15.
7. Determine the heat generated by using the relation as discussed in Art. 26.18.
8. Determine the heat dissipated by using the relation as discussed in Art. 26.18.
9. Determine the thermal equilibrium to see that the heat dissipated becomes atleast equal to the heat generated. In case the heat generated is more than the heat dissipated then either the bearing is redesigned or it is artificially cooled by water.

Example 26.1. *Design a journal bearing for a centrifugal pump from the following data :*

Load on the journal = 20 000 N; Speed of the journal = 900 r.p.m.; Type of oil is SAE 10, for which the absolute viscosity at 55°C = 0.017 kg / m-s; Ambient temperature of oil = 15.5°C ; Maximum bearing pressure for the pump = 1.5 N / mm².

Calculate also mass of the lubricating oil required for artificial cooling, if rise of temperature of oil be limited to 10°C. Heat dissipation coefficient = 1232 W/m²/°C.

Solution. Given : $W = 20\ 000$ N ; $N = 900$ r.p.m. ; $t_0 = 55$°C ; $Z = 0.017$ kg/m-s ; $t_a = 15.5$°C ; $p = 1.5$ N/mm² ; $t = 10$°C ; $C = 1232$ W/m²/°C

The journal bearing is designed as discussed in the following steps :

1. First of all, let us find the length of the journal (l). Assume the diameter of the journal (d) as 100 mm. From Table 26.3, we find that the ratio of l/d for centrifugal pumps varies from 1 to 2. Let us take $l/d = 1.6$.

$$\therefore \quad l = 1.6\, d = 1.6 \times 100 = 160 \text{ mm } \textbf{Ans.}$$

2. We know that bearing pressure,

$$p = \frac{W}{l.d} = \frac{20\ 000}{160 \times 100} = 1.25$$

Since the given bearing pressure for the pump is 1.5 N/mm², therefore the above value of p is safe and hence the dimensions of l and d are safe.

3. $$\frac{Z.N}{p} = \frac{0.017 \times 900}{1.25} = 12.24$$

From Table 26.3, we find that the operating value of

$$\frac{Z.N}{p} = 28$$

We have discussed in Art. 26.14, that the minimum value of the bearing modulus at which the oil film will break is given by

$$3K = \frac{ZN}{p}$$

∴ Bearing modulus at the minimum point of friction,

$$K = \frac{1}{3}\left(\frac{Z.N}{p}\right) = \frac{1}{3} \times 28 = 9.33$$

Since the calculated value of bearing characteristic number $\left(\frac{Z.N}{p} = 12.24\right)$ is more than 9.33, therefore the bearing will operate under hydrodynamic conditions.

4. From Table 26.3, we find that for centrifugal pumps, the clearance ratio (c/d)

$$= 0.0013$$

5. We know that coefficient of friction,

$$\mu = \frac{33}{10^8}\left(\frac{ZN}{p}\right)\left(\frac{d}{c}\right) + k = \frac{33}{10^8} \times 12.24 \times \frac{1}{0.0013} + 0.002$$

$$= 0.0031 + 0.002 = 0.0051 \qquad \text{... [From Art. 26.13, } k = 0.002]$$

6. Heat generated,

$$Q_g = \mu\, W\, V = \mu\, W\left(\frac{\pi d.N}{60}\right) \text{ W} \qquad \text{...}\left(\because V = \frac{\pi d.N}{60}\right)$$

$$= 0.0051 \times 20000\left(\frac{\pi \times 0.1 \times 900}{60}\right) = 480.7 \text{ W}$$

... (d is taken in metres)

7. Heat dissipated,

$$Q_d = C.A\,(t_b - t_a) = C.l.d\,(t_b - t_a) \text{ W} \qquad \text{... } (\because A = l \times d)$$

We know that

$$(t_b - t_a) = \tfrac{1}{2}\,(t_0 - t_a) = \tfrac{1}{2}\,(55° - 15.5°) = 19.75°\text{C}$$

$$\therefore \qquad Q_d = 1232 \times 0.16 \times 0.1 \times 19.75 = 389.3 \text{ W}$$

... (l and d are taken in metres)

We see that the heat generated is greater than the heat dissipated which indicates that the bearing is warming up. Therefore, either the bearing should be redesigned by taking $t_0 = 63°$C or the bearing should be cooled artificially.

We know that the amount of artificial cooling required

$$= \text{Heat generated} - \text{Heat dissipated} = Q_g - Q_d$$

$$= 480.7 - 389.3 = 91.4 \text{ W}$$

Mass of lubricating oil required for artificial cooling

Let m = Mass of the lubricating oil required for artificial cooling in kg / s.

We know that the heat taken away by the oil,

$$Q_t = m.S.t = m \times 1900 \times 10 = 19\,000\, m \text{ W}$$

... [∵ Specific heat of oil (S) = 1840 to 2100 J/kg/°C]

Equating this to the amount of artificial cooling required, we have

$$19\,000\, m = 91.4$$

$$\therefore \qquad m = 91.4 / 19\,000 = 0.0048 \text{ kg / s} = 0.288 \text{ kg / min}$$ **Ans.**

Example 26.2. *The load on the journal bearing is 150 kN due to turbine shaft of 300 mm diameter running at 1800 r.p.m. Determine the following :*

1. Length of the bearing if the allowable bearing pressure is 1.6 N/mm², and

2. Amount of heat to be removed by the lubricant per minute if the bearing temperature is 60°C and viscosity of the oil at 60°C is 0.02 kg/m-s and the bearing clearance is 0.25 mm.

Axle bearing

Solution. Given : $W = 150$ kN $= 150 \times 10^3$ N ; $d = 300$ mm $= 0.3$ m ; $N = 1800$ r.p.m. ; $p = 1.6$ N/mm² ; $Z = 0.02$ kg / m-s ; $c = 0.25$ mm

1. *Length of the bearing*

Let l = Length of the bearing in mm.

We know that projected bearing area,

$$A = l \times d = l \times 300 = 300\, l \text{ mm}^2$$

and allowable bearing pressure (p),

$$1.6 = \frac{W}{A} = \frac{150 \times 10^3}{300\, l} = \frac{500}{l}$$

$\therefore$ $l = 500 / 1.6 = 312.5$ mm **Ans.**

2. *Amount of heat to be removed by the lubricant*

We know that coefficient of friction for the bearing,

$$\mu = \frac{33}{10^8}\left(\frac{Z.N}{p}\right)\left(\frac{d}{c}\right) + k = \frac{33}{10^8}\left(\frac{0.02 \times 1800}{1.6}\right)\left(\frac{300}{0.25}\right) + 0.002$$

$$= 0.009 + 0.002 = 0.011$$

Rubbing velocity,

$$V = \frac{\pi d.N}{60} = \frac{\pi \times 0.3 \times 1800}{60} = 28.3 \text{ m/s}$$

$\therefore$ Amount of heat to be removed by the lubricant,

$$Q_g = \mu.W.V = 0.011 \times 150 \times 10^3 \times 28.3 = 46\,695 \text{ J/s or W}$$

$$= 46.695 \text{ kW } \textbf{Ans.} \qquad \text{... (1 J/s = 1 W)}$$

Example 26.3. *A full journal bearing of 50 mm diameter and 100 mm long has a bearing pressure of 1.4 N/mm². The speed of the journal is 900 r.p.m. and the ratio of journal diameter to the diametral clearance is 1000. The bearing is lubricated with oil whose absolute viscosity at the operating temperature of 75°C may be taken as 0.011 kg/m-s. The room temperature is 35°C. Find : 1. The amount of artificial cooling required, and 2. The mass of the lubricating oil required, if the difference between the outlet and inlet temperature of the oil is 10°C. Take specific heat of the oil as 1850 J / kg / °C.*

Solution. Given : $d = 50$ mm $= 0.05$ m ; $l = 100$ mm $= 0.1$ m; $p = 1.4$ N/mm² ; $N = 900$ r.p.m. ; $d / c = 1000$; $Z = 0.011$ kg / m-s ; $t_0 = 75$°C ; $t_a = 35$°C ; $t = 10$°C ; $S = 1850$ J/kg / °C

1. *Amount of artificial cooling required*

We know that the coefficient of friction,

$$\mu = \frac{33}{10^8}\left(\frac{ZN}{p}\right)\left(\frac{d}{c}\right) + k = \frac{33}{10^8}\left(\frac{0.011 \times 900}{1.4}\right)(1000) + 0.002$$

$$= 0.002\,33 + 0.002 = 0.004\,33$$

Load on the bearing,

$$W = p \times d.l = 1.4 \times 50 \times 100 = 7000 \text{ N}$$

and rubbing velocity,

$$V = \frac{\pi d.N}{60} = \frac{\pi \times 0.05 \times 900}{60} = 2.36 \text{ m/s}$$

∴ Heat generated,

$$Q_g = \mu.W.V = 0.004\ 33 \times 7000 \times 2.36 = 71.5 \text{ J/s}$$

Let t_b = Temperature of the bearing surface.

We know that

$$(t_b - t_a) = \frac{1}{2}(t_0 - t_a) = \frac{1}{2}(75 - 35) = 20°\text{C}$$

Since the value of heat dissipation coefficient (C) for unventilated bearing varies from 140 to 420 W/m²/°C, therefore let us take

$$C = 280 \text{ W/m}^2/°\text{C}$$

We know that heat dissipated,

$$Q_d = C.A\,(t_b - t_a) = C.l.d\,(t_b - t_a)$$

$$= 280 \times 0.05 \times 0.1 \times 20 = 28 \text{ W} = 28 \text{ J/s}$$

∴ Amount of artificial cooling required

$$= \text{Heat generated} - \text{Heat dissipated} = Q_g - Q_d$$

$$= 71.5 - 28 = 43.5 \text{ J/s or W}$$ **Ans.**

2. ***Mass of the lubricating oil required***

Let m = Mass of the lubricating oil required in kg / s.

We know that heat taken away by the oil,

$$Q_t = m.S.t = m \times 1850 \times 10 = 18\ 500\ m \text{ J/s}$$

Since the heat generated at the bearing is taken away by the lubricating oil, therefore equating

$$Q_g = Q_t \text{ or } 71.5 = 18\ 500\ m$$

∴ $m = 71.5 / 18\ 500 = 0.003\ 86 \text{ kg / s} = 0.23 \text{ kg / min}$ **Ans.**

Example 26.4. *A 150 mm diameter shaft supporting a load of 10 kN has a speed of 1500 r.p.m. The shaft runs in a bearing whose length is 1.5 times the shaft diameter. If the diametral clearance of the bearing is 0.15 mm and the absolute viscosity of the oil at the operating temperature is 0.011 kg/m-s, find the power wasted in friction.*

Solution. Given : d = 150 mm = 0.15 m ; W = 10 kN = 10 000 N ; N = 1500 r.p.m. ; l = 1.5 d ; c = 0.15 mm ; Z = 0.011 kg/m-s

We know that length of bearing,

$$l = 1.5\,d = 1.5 \times 150 = 225 \text{ mm}$$

∴ Bearing pressure,

$$p = \frac{W}{A} = \frac{W}{l.d} = \frac{10\,000}{225 \times 150} = 0.296 \text{ N/mm}^2$$

We know that coefficient of friction,

$$\mu = \frac{33}{10^8}\left(\frac{ZN}{p}\right)\left(\frac{d}{c}\right) + k = \frac{33}{10^8}\left(\frac{0.011 \times 1500}{0.296}\right)\left(\frac{150}{0.15}\right) + 0.002$$

$$= 0.018 + 0.002 = 0.02$$

and rubbing velocity,

$$V = \frac{\pi d.N}{60} = \frac{\pi \times 0.15 \times 1500}{60} = 11.78 \text{ m/s}$$

We know that heat generated due to friction,

$$Q_g = \mu.W.V = 0.02 \times 10\,000 \times 11.78 = 2356 \text{ W}$$

∴ Power wasted in friction

$$= Q_g = 2356 \text{ W} = 2.356 \text{ kW } \textbf{Ans.}$$

Example 26.5. *A 80 mm long journal bearing supports a load of 2800 N on a 50 mm diameter shaft. The bearing has a radial clearance of 0.05 mm and the viscosity of the oil is 0.021 kg / m-s at the operating temperature. If the bearing is capable of dissipating 80 J/s, determine the maximum safe speed.*

Solution. Given : l = 80 mm ; W = 2800 N ; d = 50 mm ; = 0.05 m ; c / 2 = 0.05 mm or c = 0.1 mm ; Z = 0.021 kg/m-s ; Q_d = 80 J/s

Let N = Maximum safe speed in r.p.m.

We know that bearing pressure,

$$p = \frac{W}{l.d} = \frac{2800}{80 \times 50} = 0.7 \text{ N/mm}^2$$

and coefficient of friction,

$$\mu = \frac{33}{10^8}\left(\frac{ZN}{p}\right)\left(\frac{d}{c}\right) + 0.002 = \frac{33}{10^8}\left(\frac{0.021\,N}{0.7}\right)\left(\frac{50}{0.1}\right) + 0.002$$

$$= \frac{495\,N}{10^8} + 0.002$$

Front hub-assembly bearing

∴ Heat generated, $$Q_g = \mu.W.V = \mu.W\left(\frac{\pi\, d\, N}{60}\right) \text{J/s}$$

$$= \left(\frac{495\,N}{10^8} + 0.002\right) 2800\left(\frac{\pi \times 0.05\,N}{60}\right)$$

$$= \frac{3628\,N^2}{10^8} + 0.014\,66\,N$$

Equating the heat generated to the heat dissipated, we have

$$\frac{3628\,N^2}{10^8} + 0.014\,66\,N = 80$$

or $N^2 + 404\,N - 2.2 \times 10^6 = 0$

$$\therefore \quad N = \frac{-404 \pm \sqrt{(404)^2 + 4 \times 2.2 \times 10^6}}{2}$$

$$= \frac{-404 \pm 2994}{2} = 1295 \text{ r.p.m } \textbf{Ans.} \quad \text{... (Taking +ve sign)}$$

Example 26.6. *A journal bearing 60 mm is diameter and 90 mm long runs at 450 r.p.m. The oil used for hydrodynamic lubrication has absolute viscosity of 0.06 kg / m-s. If the diametral clearance is 0.1 mm, find the safe load on the bearing.*

Solution. Given : d = 60 mm = 0.06 m ; l = 90 mm = 0.09 m ; N = 450 r.p.m. ; Z = 0.06 kg / m-s ; c = 0.1 mm

First of all, let us find the bearing pressure (p) by using Sommerfeld number. We know that

$$\frac{Z N}{p}\left(\frac{d}{c}\right)^2 = 14.3 \times 10^6$$

$$\frac{0.06 \times 450}{p}\left(\frac{60}{0.1}\right)^2 = 14.3 \times 10^6 \quad \text{or} \quad \frac{9.72 \times 10^6}{p} = 14.3 \times 10^6$$

$$\therefore \quad p = 9.72 \times 10^6 / 14.3 \times 10^6 = 0.68 \text{ N/mm}^2$$

We know that safe load on the bearing,

$$W = p.A = p.l.d = 0.68 \times 90 \times 60 = 3672 \text{ N } \textbf{Ans.}$$

26.20 Solid Journal Bearing

A solid bearing, as shown in Fig. 26.9, is the simplest form of journal bearing. It is simply a block of cast iron with a hole for a shaft providing running fit. The lower portion of the block is extended to form a base plate or sole with two holes to receive bolts for fastening it to the frame. An oil hole is drilled at the top for lubrication. The main disadvantages of this bearing are

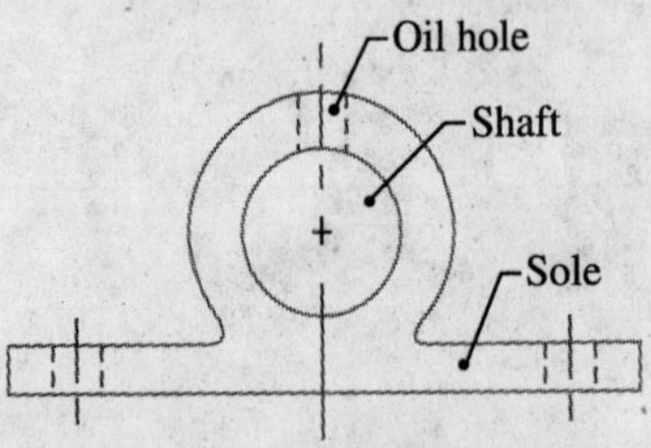

Fig. 26.9. Solid journal bearing.

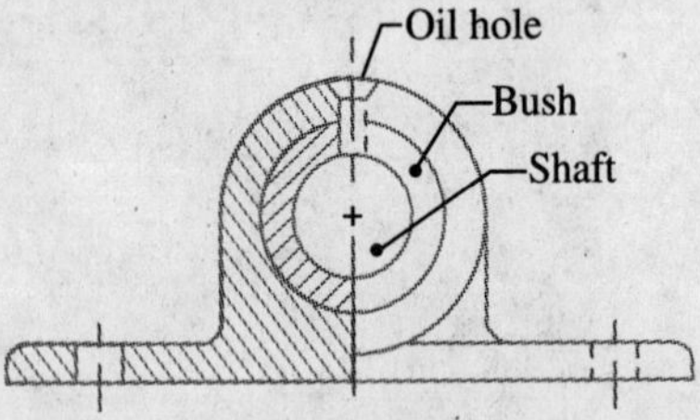

Fig. 26.10. Bushed bearing.

1. There is no provision for adjustment in case of wear, and
2. The shaft must be passed into the bearing axially, *i.e.* endwise.

Since there is no provision for wear adjustment, therefore this type of bearing is used when the shaft speed is not very high and the shaft carries light loads only.

26.21 Bushed Bearing

A bushed bearing, as shown in Fig. 26.10, is an improved solid bearing in which a bush of brass or gun metal is provided. The outside of the bush is a driving fit in the hole of the casting whereas the inside is a running fit for the shaft. When the bush gets worn out, it can be easily replaced. In small bearings, the frictional force itself holds the bush in position, but for shafts transmitting high power, grub screws are used for the prevention of rotation and sliding of the bush.

Bronze bushed bearing assemblies

26.22 Split Bearing or Plummer Block

A split-bearing is used for shafts running at high speeds and carrying heavy loads. A split-bearing, as shown in Fig. 26.11, consists of a cast iron base (also called block or pedestal), gunmetal or phosphor bronze brasses, bushes or steps made in two-halves and a cast iron cap. The two halves of the brasses are held together by a cap or cover by means of mild steel bolts and nuts. Sometimes thin shims are introduced between the cap and the base to provide an adjustment for wear. When the bottom wears out, one or two shims are removed and then the cap is tightened by means of bolts.

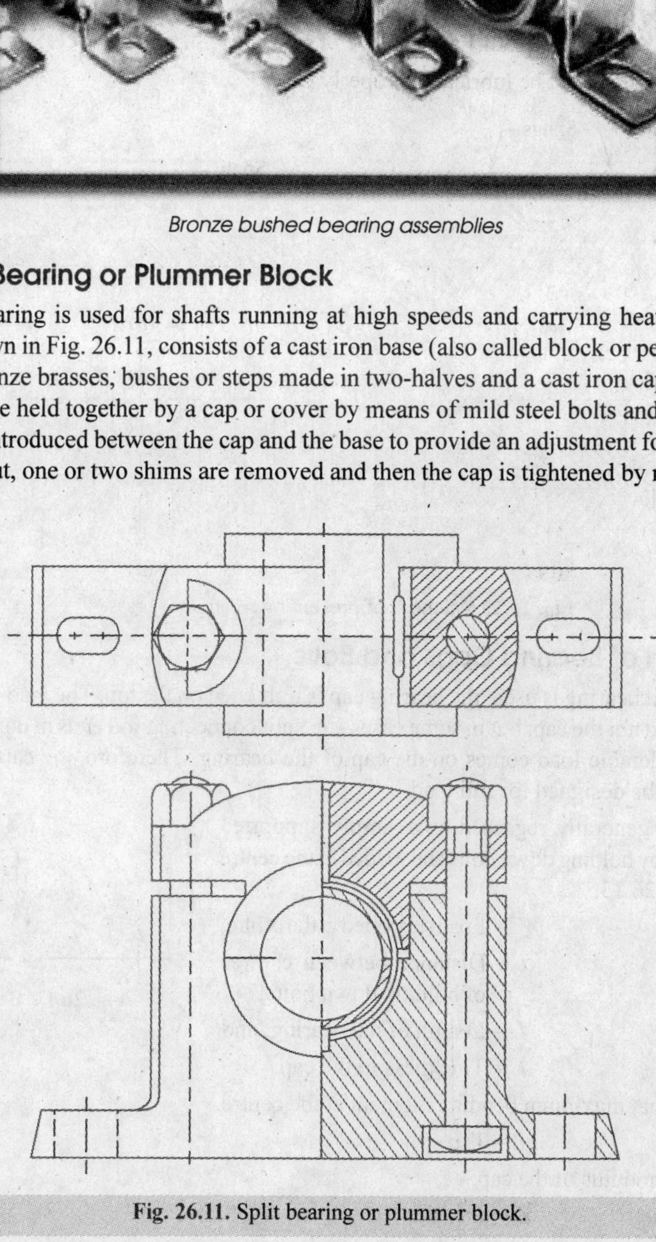

Fig. 26.11. Split bearing or plummer block.

The brasses are provided with collars or flanges on either side in order to prevent its axial movement. To prevent its rotation along with the shaft, the following four methods are usually used in practice.

1. The sungs are provided at the sides as shown in Fig. 26.12 (*a*).
2. A sung is provided at the top, which fits inside the cap as shown in Fig. 26.12 (*b*). The oil hole is drilled through the sung.
3. The steps are made rectangular on the outside and they are made to fit inside a corresponding hole, as shown in Fig. 26.12 (*c*).
4. The steps are made octagonal on the outside and they are made to fit inside a corresponding hole, as shown in Fig. 26.12 (*d*).

The split bearing must be lubricated properly.

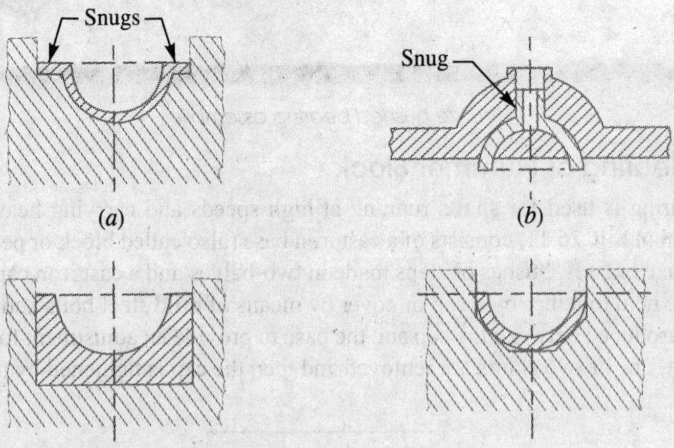

Fig. 26.12. Methods of preventing rotation of brasses.

26.23 Design of Bearing Caps and Bolts

When a split bearing is used, the bearing cap is tightened on the top. The load is usually carried by the bearing and not the cap, but in some cases *e.g.* split connecting rod ends in double acting steam engines, a considerable load comes on the cap of the bearing. Therefore, the cap and the holding down bolts must be designed for full load.

The cap is generally regarded as a simply supported beam, supported by holding down bolts and loaded at the centre as shown in Fig. 26.13.

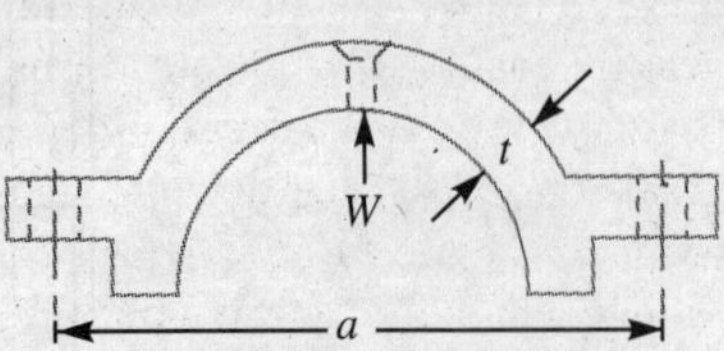

Fig. 26.13. Bearing cap.

Let W = Load supported at the centre,

α = Distance between centres of holding down bolts,

l = Length of the bearing, and

t = Thickness of the cap.

We know that maximum bending moment at the centre,

$$M = W.a / 4$$

and the section modulus of the cap,

$$Z = l.t^2 / 6$$

∴ Bending stress,

$$\sigma_b = \frac{M}{Z} = \frac{W.a}{4} \times \frac{6}{l.t^2} = \frac{3W.a}{2l.t^2}$$

and

$$t = \sqrt{\frac{3\,W.a}{2\sigma_b.l}}$$

Note : When an oil hole is provided in the cap, then the diameter of the hole should be subtracted from the length of the bearing.

The cap of the bearing should also be investigated for the stiffness. We know that for a simply supported beam loaded at the centre, the deflection,

$$\delta = \frac{W.a^3}{48\,E.I} = \frac{W.a^3}{48\,E \times \frac{l.t^3}{12}} = \frac{W.a^3}{4\,E.l\,t^3} \qquad \ldots \left(\because I = \frac{l.t^3}{12}\right)$$

$$\therefore \qquad t = 0.63\,a \left[\frac{W}{E.l.\delta}\right]^{1/3}$$

The deflection of the cap should be limited to about 0.025 mm.

In order to design the holding down bolts, the load on each bolt is taken 33% higher than the normal load on each bolt. In other words, load on each bolt is taken $\frac{4\,W}{3n}$, where n is the number of bolts used for holding down the cap.

Let $\qquad d_c$ = Core diameter of the bolt, and

$\qquad \sigma_t$ = Tensile stress for the material of the bolt.

$$\therefore \qquad \frac{\pi}{4}(d_c)^2\,\sigma_t = \frac{4}{3} \times \frac{W}{n}$$

From this expression, the core diameter (d_c) may be calculated. After finding the core diameter, the size of the bolt is fixed.

26.24 Oil Groves

A self-locking nut used in bearing assemblies.

The oil grooves are cut into the plain bearing surfaces to assist in the distribution of the oil between the rubbing surfaces. It prevents squeezing of the oil film from heavily loaded low speed journals and bearings. The tendency to squeeze out oil is greater in low speed than in high speed bearings, because the oil has greater wedging action at high speeds. At low speeds, the journal rests upon a given area of oil film for a longer period of time, tending to squeeze out the oil over the area of greatest pressure. The grooves function as oil reservoirs which holds and distributes the oil especially during starting or at very low speeds. The oil grooves are cut at right angles to the line of the load. The circumferential and diagonal grooves should be avoided, if possible. The effectiveness of the oil grooves is greatly enhanced if the edges of grooves are chamfered. The shallow and narrow grooves with chamfered edges distributes the oil more evenly. A chamfered edge should always be provided at the parting line of the bearing.

Example 26.7. *A wall bracket supports a plummer block for 80 mm diameter shaft. The length of bearing is 120 mm. The cap of bearing is fastened by means of four bolts, two on each side of the shaft. The cap is to withstand a load of 16.5 kN. The distance between the centre lines of the bolts is*

150 mm. Determine the thickness of the bearing cap and the diameter of the bolts. Assume safe stresses in tension for the material of the cap, which is cast iron, as 15 MPa and for bolts as 35 MPa. Also check the deflection of the bearing cap taking E = 110 kN / mm².

Solution : Given : $d = 80$ mm ; $l = 120$ mm ; $n = 4$; $W = 16.5$ kN $= 16.5 \times 10^3$ N ; $a = 150$ mm ; $\sigma_b = 15$ MPa $= 15$ N/mm²; $\sigma_t = 35$ MPa $= 35$ N/mm² ; $E = 110$ kN/mm² $= 110 \times 10^3$ N/mm²

Thickness of the bearing cap

We know that thickness of the bearing cap,

$$t = \sqrt{\frac{3\,W.a}{2\sigma_b.l}} = \sqrt{\frac{3 \times 16.5 \times 10^3 \times 150}{2 \times 15 \times 120}} = \sqrt{2062.5}$$

$$= 45.4 \text{ say } 46 \text{ mm } \textbf{Ans.}$$

Diameter of the bolts

Let d_c = Core diameter of the bolts.

We know that

$$\frac{\pi}{4}(d_c)^2\,\sigma_t = \frac{4}{3} \times \frac{W}{n}$$

or

$$\frac{\pi}{4}(d_c)^2\,35 = \frac{4}{3} \times \frac{16.5 \times 10^3}{4} = 5.5 \times 10^3$$

$$\therefore \quad (d_c)^2 = \frac{5.5 \times 10^3 \times 4}{\pi \times 35} = 200 \quad \text{or} \quad d_c = 14.2 \text{ mm } \textbf{Ans.}$$

Deflection of the cap

We know that deflection of the cap,

$$\delta = \frac{W.a^3}{4\,E.l.t^3} = \frac{16.5 \times 10^3\ (150)^3}{4 \times 110 \times 10^3 \times 120 (46)^3} = 0.0108 \text{ mm } \textbf{Ans.}$$

Since the limited value of the deflection is 0.025 mm, therefore the above value of deflection is within limits.

26.25 Thrust Bearings

A thrust bearing is used to guide or support the shaft which is subjected to a load along the axis of the shaft. Such type of bearings are mainly used in turbines and propeller shafts. The thrust bearings are of the following two types :

1. Foot step or pivot bearings, and **2.** Collar bearings.

In a ***foot step*** or ***pivot bearing,*** the loaded shaft is vertical and the end of the shaft rests within the bearing. In case of ***collar bearing,*** the shaft continues through the bearing. The shaft may be vertical or horizontal with single collar or many collars. We shall now discuss the design aspects of these bearings in the following articles.

Footstep bearing

26.26 Footstep or Pivot Bearings

A simple type of footstep bearing, suitable for a slow running and lightly loaded shaft, is shown in Fig. 26.14. If the shaft is not of steel, its end must be fitted with a steel face. The shaft is guided in a gunmetal bush, pressed into the pedestal and prevented from turning by means of a pin.

Since the wear is proportional to the velocity of the rubbing surface, which (*i.e.* rubbing velocity) increases with the distance from the axis (*i.e.* radius) of the bearing, therefore the wear will be different at different radii. Due to this wear, the distribution of pressure over the bearing surface is not

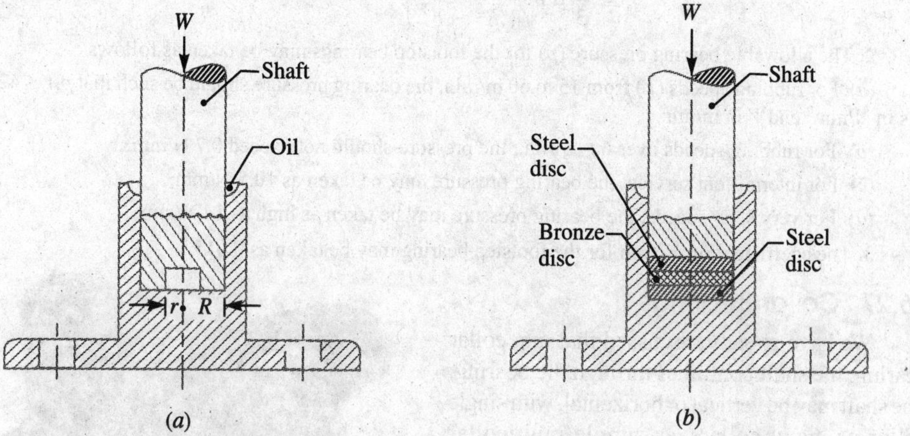

Fig. 26.14. Footstep or pivot bearings.

uniform. It may be noted that the wear is maximum at the outer radius and zero at the centre. In order to compensate for end wear, the following two methods are employed.

1. The shaft is counter-bored at the end, as shown in Fig. 26.14 (*a*).

2. The shaft is supported on a pile of discs. It is usual practice to provide alternate discs of different materials such as steel and bronze, as shown in Fig. 26.14 (*b*), so that the next disc comes into play, if one disc seizes due to improper lubrication.

It may be noted that a footstep bearing is difficult to lubricate as the oil is being thrown outwards from the centre by centrifugal force.

In designing, it is assumed that the pressure is uniformly distributed throughout the bearing surface.

Let W = Load transmitted over the bearing surface,

R = Radius of the bearing surface (or shaft),

A = Cross-sectional area of the bearing surface,

p = Bearing pressure per unit area of the bearing surface between rubbing surfaces,

μ = Coefficient of friction, and

N = Speed of the shaft in r.p.m.

When the pressure in uniformly distributed over the bearing area, then

$$p = \frac{W}{A} = \frac{W}{\pi R^2}$$

and the total frictional torque,

$$T = \frac{2}{3}\mu.W.R$$

∴ Power lost in friction,

$$P = \frac{2\pi N.T}{60} \text{ watts} \qquad \text{... (} T \text{ being in N-m)}$$

Notes : 1. When the counter-boring of the shaft is considered, then the bearing pressure,

$$p = \frac{W}{\pi(R^2 - r^2)}, \text{ where } r = \text{Radius of counter-bore,}$$

and the total frictional torque,

$$T = \frac{2}{3}\mu.W\left(\frac{R^3 - r^3}{R^2 - r^2}\right)$$

2. The allowable bearing pressure (p) for the footstep bearings may be taken as follows :

(*a*) For rubbing speeds (V) from 15 to 60 m/min, the bearing pressure should be such that $p.V \leq 42$, when p is in N/mm² and V in m/min.

(*b*) For rubbing speeds over 60 m/min., the pressure should not exceed 0.7 N/mm².

(*c*) For intermittent service, the bearing pressure may be taken as 10.5 N/mm².

(*d*) For very slow speeds, the bearing pressure may be taken as high as 14 N/mm².

3. The coefficient of friction for the footstep bearing may be taken as 0.015.

26.27 Collar Bearings

We have already discussed that in a collar bearing, the shaft continues through the bearing. The shaft may be vertical or horizontal, with single collar or many collars. A simple multicollar bearing for horizontal shaft is shown in Fig. 26.15. The collars are either integral parts of the shaft or rigidly fastened to it. The outer diameter of the collar is usually taken as 1.4 to 1.8 times the inner diameter of the collar (*i.e.* diameter of the shaft). The thickness of the collar is kept as one-sixth diameter of the shaft and clearance between collars as one-third diameter of the shaft. In designing collar bearings, it is assumed that the pressure is uniformly distributed over the bearing surface.

Collar bearings

Let W = Load transmitted over the bearing surface,

n = Number of collars,

R = Outer radius of the collar,

r = Inner radius of the collar,

A = Cross-sectional area of the bearing surface = $n\,\pi\,(R^2 - r^2)$,

p = Bearing pressure per unit area of the bearing surface, between rubbing surfaces,

μ = Coefficient of friction, and

N = Speed of the shaft in r.p.m.

When the pressure is uniformly distributed over the bearing surface, then bearing pressure,

$$p = \frac{W}{A} = \frac{W}{n.\pi\,(R^2 - r^2)}$$

and the total frictional torque,

$$T = \frac{2}{3}\mu.\,W\left(\frac{R^3 - r^3}{R^2 - r^2}\right)$$

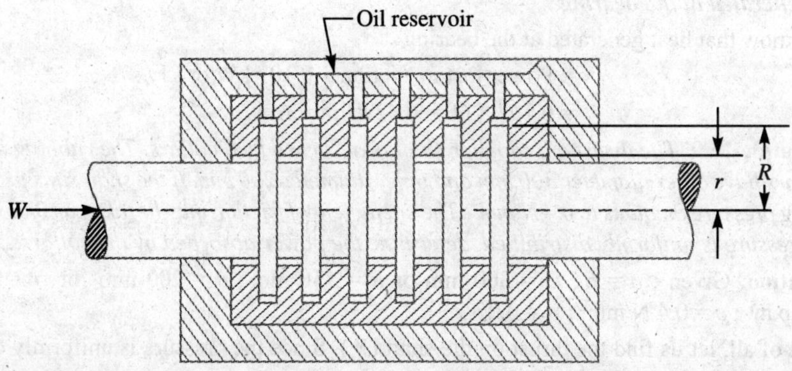

Fig. 26.15. Collar bearing.

∴ Power lost in friction,

$$P = \frac{2\pi\, N.T}{60} \text{ watts} \qquad \text{... (when } T \text{ is in N-m)}$$

Notes : 1. The coefficient of friction for the collar bearings may be taken as 0.03 to 0.05.

2. The bearing pressure for a single collar and water cooled multi-collared bearings may be taken same as for footstep bearings.

Example 26.8. *A footstep bearing supports a shaft of 150 mm diameter which is counter-bored at the end with a hole diameter of 50 mm. If the bearing pressure is limited to 0.8 N/mm² and the speed is 100 r.p.m.; find : 1. The load to be supported; 2. The power lost in friction; and 3. The heat generated at the bearing.*

Assume coefficient of friction = 0.015.

Solution. Given : $D = 150$ mm or $R = 75$ mm ; $d = 50$ mm or $r = 25$ mm ; $p = 0.8$ N/mm² ; $N = 100$ r.p.m. ; $\mu = 0.015$

1. ***Load to be supported***

Let W = Load to be supported.

Assuming that the pressure is uniformly distributed over the bearing surface, therefore bearing pressure (p),

$$0.8 = \frac{W}{\pi (R^2 - r^2)} = \frac{W}{\pi [(75)^2 - (25)^2]} = \frac{W}{15\,710}$$

$$\therefore \quad W = 0.8 \times 15\,710 = 12\,568 \text{ N } \textbf{Ans.}$$

2. ***Power lost in friction***

We know that total frictional torque,

$$T = \frac{2}{3}\mu.W \left(\frac{R^3 - r^3}{R^2 - r^2}\right)$$

$$= \frac{2}{3} \times 0.015 \times 12568 \left[\frac{(75)^3 - (25)^3}{(75)^2 - (25)^2}\right] \text{ N-mm}$$

$$= 125.68 \times 81.25 = 10\,212 \text{ N-mm} = 10.212 \text{ N-m}$$

∴ Power lost in friction,

$$P = \frac{2\,\pi\, N\, T}{60} = \frac{2\,\pi \times 100 \times 10.212}{60} = 107 \text{ W} = 0.107 \text{ kW} \quad \textbf{Ans.}$$

3. *Heat generated at the bearing*

We know that heat generated at the bearing

$$= \text{Power lost in friction} = 0.107 \text{ kW or kJ / s}$$
$$= 0.107 \times 60 = 6.42 \text{ kJ/min } \textbf{Ans.}$$

Example 26.9. *The thrust of propeller shaft is absorbed by 6 collars. The rubbing surfaces of these collars have outer diameter 300 mm and inner diameter 200 mm. If the shaft runs at 120 r.p.m., the bearing pressure amounts to 0.4 N/mm². The coefficient of friction may be taken as 0.05. Assuming that the pressure is uniformly distributed, determine the power absorbed by the collars.*

Solution. Given : $n = 6$; $D = 300$ mm or $R = 150$ mm ; $d = 200$ mm or $r = 100$ mm ; $N = 120$ r.p.m. ; $p = 0.4$ N/mm² ; $\mu = 0.05$

First of all, let us find the thrust on the shaft (W). Since the pressure is uniformly distributed over the bearing surface, therefore bearing pressure (p),

$$0.4 = \frac{W}{n\,\pi\,(R^2 - r^2)} = \frac{W}{6\pi\,[(150)^2 - (100)^2]} = \frac{W}{235\,650}$$

$$\therefore \quad W = 0.4 \times 235\,650 = 94\,260 \text{ N}$$

We know that total frictional torque,

$$T = \frac{2}{3}\,\mu.W\left(\frac{R^3 - r^3}{R^2 - r^2}\right) = \frac{2}{3} \times 0.05 \times 94260\left[\frac{(150)^3 - (100)^3}{(150)^2 - (100)^2}\right] \text{N-mm}$$
$$= 597\,000 \text{ N-mm} = 597 \text{ N-m}$$

∴ Power absorbed by the collars,

$$P = \frac{2\pi.N.T}{60} = \frac{2\,\pi \times 120 \times 597}{60} = 7503 \text{ W} = 7.503 \text{ kW } \textbf{Ans.}$$

Example 26.10. *The thrust of propeller shaft in a marine engine is taken up by a number of collars integral with the shaft which is 300 mm is diameter. The thrust on the shaft is 200 kN and the speed is 75 r.p.m. Taking μ constant and equal to 0.05 and assuming the bearing pressure as uniform and equal to 0.3 N/mm², find : 1. Number of collars required, 2. Power lost in friction, and 3. Heat generated at the bearing in kJ/min.*

Industrial bearings.

Solution. Given : $d = 300$ mm or $r = 150$ mm ; $W = 200$ kN $= 200 \times 10^3$ N ; $N = 75$ r.p.m. ; $\mu = 0.05$; $p = 0.3$ N/mm²

1. *Number of collars required*

Let $\quad n$ = Number of collars required.

Since the outer diameter of the collar (D) is taken as 1.4 to 1.8 times the diameter of shaft (d), therefore let us take

$$D = 1.4\,d = 1.4 \times 300 = 420 \text{ mm} \quad \text{or} \quad R = 210 \text{ mm}$$

We know that the bearing pressure (p),

$$0.3 = \frac{W}{n\,\pi\,(R^2 - r^2)} = \frac{200 \times 10^3}{n\pi\,[(210)^2 - (150)^2]} = \frac{2.947}{n}$$

$$\therefore \quad n = 2.947 / 0.3 = 9.8 \text{ say } 10 \textbf{ Ans.}$$

2. ***Power lost in friction***

We know that total frictional torque,

$$T = \frac{2}{3}\mu W\left(\frac{R^3 - r^3}{R^2 - r^2}\right) = \frac{2}{3} \times 0.05 \times 200 \times 10^3 \left[\frac{(210)^3 - (150)^3}{(210)^2 - (150)^2}\right] \text{ N-mm}$$

$= 1817 \times 10^3$ N-mm = 1817 N-m

∴ Power lost in friction,

$$P = \frac{2\,\pi N.T}{60} = \frac{2\,\pi \times 75 \times 1817}{60} = 14\,270 \text{ W} = 14.27 \text{ kW} \quad \textbf{Ans.}$$

3. ***Heat generated at the bearing***

We know that heat generated at the bearing

= Power lost in friction = 14.27 kW or kJ/s

= 14.27 × 60 = 856.2 kJ/min **Ans.**

EXERCISES

1. The main bearing of a steam engine is 100 mm in diameter and 175 mm long. The bearing supports a load of 28 kN at 250 r.p.m. If the ratio of the diametral clearance to the diameter is 0.001 and the absolute viscosity of the lubricating oil is 0.015 kg/m-s, find : 1. The coefficient of friction ; and 2. The heat generated at the bearing due to friction.

   [Ans. 0.002 77 ; 101.5 J/s]

2. A journal bearing is proposed for a steam engine. The load on the journal is 3 kN, diameter 50 mm, length 75 mm, speed 1600 r.p.m., diametral clearance 0.001 mm, ambient temperature 15.5°C. Oil SAE 10 is used and the film temperature is 60°C. Determine the heat generated and heat dissipated. Take absolute viscosity of SAE10 at 60°C = 0.014 kg/m-s. **[Ans. 141.3 J/s ; 25 J/s]**

3. A 100 mm long and 60 mm diameter journal bearing supports a load of 2500 N at 600 r.p.m. If the room temperature is 20°C, what should be the viscosity of oil to limit the bearing surface temperature to 60°C? The diametral clearance is 0.06 mm and the energy dissipation coefficient based on projected area of bearing is 210 W/m²/°C. **[Ans. 0.0183 kg/m-s]**

4. A tentative design of a journal bearing results in a diameter of 75 mm and a length of 125 mm for supporting a load of 20 kN. The shaft runs at 1000 r.p.m. The bearing surface temperature is not to exceed 75°C in a room temperature of 35°C. The oil used has an absolute viscosity of 0.01 kg/m-s at the operating temperature. Determine the amount of artificial cooling required in watts. Assume d/c = 1000. **[Ans. 146 W]**

5. A journal bearing is to be designed for a centrifugal pump for the following data :

   Load on the journal = 12 kN ; Diameter of the journal = 75 mm ; Speed = 1440 r.p.m ; Atmospheric temperature of the oil = 16°C ; Operating temperature of the oil = 60°C; Absolute viscosity of oil at 60°C = 0.023 kg/m-s.

   Give a systematic design of the bearing.

6. Design a journal bearing for a centrifugal pump running at 1440 r.p.m. The diameter of the journal is 100 mm and load on each bearing is 20 kN. The factor ZN/p may be taken as 28 for centrifugal pump bearings. The bearing is running at 75°C temperature and the atmosphere temperaturic is 30°C. The energy dissipation coefficient is 875 W/m²/°C. Take diametral clearance as 0.1 mm.

7. Design a suitable journal bearing for a centrifugal pump from the following available data :

   Load on the bearing = 13.5 kN; Diameter of the journal = 80 mm; Speed = 1440 r.p.m.; Bearing characterisitic number at the working temperature (75°C) = 30 ; Permissible bearing pressure intensity

= 0.7 N/mm² to 1.4 N/mm²; Average atmospheric temperature = 30°C.

Calculate the cooling requirements, if any.

8. A journal bearing with a diameter of 200 mm and length 150 mm carries a load of 20 kN, when the journal speed is 150 r.p.m. The diametral clearance ratio is 0.0015.

If possible, the bearing is to operate at 35°C ambient temperature without external cooling with a maximum oil temperature of 90°C. If external cooling is required, it is to be as little as possible to minimise the required oil flow rate and heat exchanger size.

1. What type of oil do you recommend ?
2. Will the bearing operate without external cooling?
3. If the bearing operates without external cooling, determine the operating oil temperature?
4. If the bearing operates with external cooling, determine the amount of oil in kg/min required to carry away the excess heat generated over heat dissipated, when the oil temperature rises from 85°C to 90°C, when passing through the bearing.

QUESTIONS

1. What are journal bearings? Give a classification of these bearings.
2. What is meant by hydrodynamic lubrication?
3. List the basic assumptions used in the theory of hydrodynamic lubrication.
4. Explain wedge film and squeeze film journal bearings.
5. Enumerate the factors that influence most the formation and maintenance of the thick oil film in hydrodynamic bearings.
6. Make sketches to show the pressure distribution in a journal bearing with thick film lubrication in axial and along the circumference.
7. List the important physical characteristics of a good bearing material.
8. What are the commonly used materials for sliding contact bearings?
9. Write short note on the lubricants used in sliding contact bearings.
10. Explain the following terms as applied to journal bearings :

    (*a*) Bearing characteristic number ; and (*b*) Bearing modulus.
11. What are the various terms used in journal bearings analysis and design? Give their definitions in brief.
12. Explain with reference to a neat plot the importance of the bearing characteristic curve.
13. What is the procedure followed in designing a journal bearing?
14. Explain with sketches the working of different types of thrust bearing.

OBJECTIVE TYPE QUESTIONS

1. In a full journal bearing, the angle of contact of the bearing with the journal is

   (*a*) 120° (*b*) 180°

   (*c*) 270° (*d*) 360°
2. A sliding bearing which can support steady loads without any relative motion between the journal and the bearing is called

   (*a*) zero film bearing (*b*) boundary lubricated bearing

   (*c*) hydrodynamic lubricated bearing (*d*) hydrostatic lubricated bearing

3. In a boundary lubricated bearing, there is a .... of lubricant between the journal and the bearing.
   (*a*) thick film (*b*) thin film
4. When a shaft rotates in anticlockwise direction at slow speed in a bearing, then it will
   (*a*) have contact at the lowest point of bearing
   (*b*) move towards right of the bearing making metal to metal contact
   (*c*) move towards left of the bearing making metal to metal contact
   (*d*) move towards right of the bearing making no metal to metal contact
5. The property of a bearing material which has the ability to accommodate small particles of dust, grit etc., without scoring the material of the journal, is called
   (*a*) bondability (*b*) embeddability
   (*c*) comformability (*d*) fatigue strength
6. Teflon is used for bearings because of
   (*a*) low coefficient of friction (*b*) better heat dissipation
   (*c*) smaller space consideration (*d*) all of these
7. When the bearing is subjected to large fluctuations of load and heavy impacts, the bearing characteristic number should be .... the bearing modulus.
   (*a*) 5 times (*b*) 10 times
   (*c*) 15 times (*d*) 20 times
8. When the length of the journal is equal to the diameter of the journal, then the bearing is said to be a
   (*a*) short bearing (*b*) long bearing
   (*c*) medium bearing (*d*) square bearing
9. If Z = Absolute viscosity of the lubricant in kg/m-s, N = Speed of the journal in r.p.m., and p = Bearing pressure in N/mm^2, then the bearing characteristic number is
   (*a*) $\frac{Z\,N}{p}$ (*b*) $\frac{Z\,p}{N}$
   (*c*) $\frac{Z}{p\,N}$ (*d*) $\frac{p\,N}{Z}$
10. In thrust bearings, the load acts
   (*a*) along the axis of rotation (*b*) parallel to the axis of rotation
   (*c*) perpendicular to the axis of rotation (*d*) in any direction

ANSWERS

1. (*d*)	2. (*d*)	3. (*b*)	4. (*c*)	5. (*b*)
6. (*a*)	7. (*c*)	8. (*d*)	9. (*a*)	10. (*a*)

CHAPTER

27

Rolling Contact Bearings

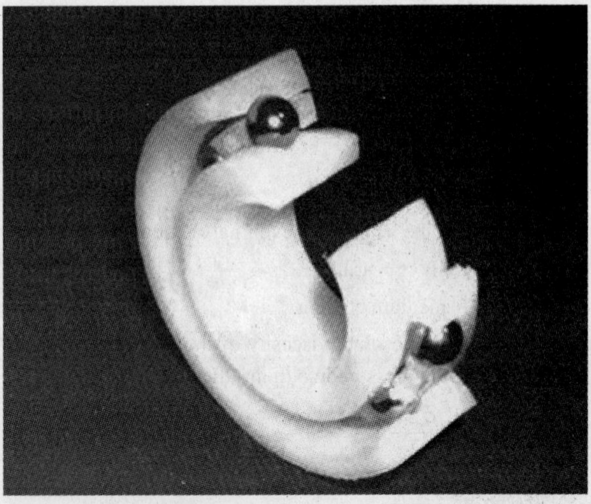

27.1 Introduction

In rolling contact bearings, the contact between the bearing surfaces is rolling instead of sliding as in sliding contact bearings. We have already discussed that the ordinary sliding bearing starts from rest with practically metal-to-metal contact and has a high coefficient of friction. It is an outstanding advantage of a rolling contact bearing over a sliding bearing that it has a low starting friction. Due to this low friction offered by rolling contact bearings, these are called ***antifriction bearings.***

27.2 Advantages and Disadvantages of Rolling Contact Bearings Over Sliding Contact Bearings

The following are some advantages and disadvantages of rolling contact bearings over sliding contact bearings.

Advantages

1. Low starting and running friction except at very high speeds.
2. Ability to withstand momentary shock loads.
3. Accuracy of shaft alignment.
4. Low cost of maintenance, as no lubrication is required while in service.
5. Small overall dimensions.
6. Reliability of service.
7. Easy to mount and erect.
8. Cleanliness.

Disadvantages

1. More noisy at very high speeds.
2. Low resistance to shock loading.
3. More initial cost.
4. Design of bearing housing complicated.

27.3 Types of Rolling Contact Bearings

Following are the two types of rolling contact bearings:

1. Ball bearings; and 2. Roller bearings.

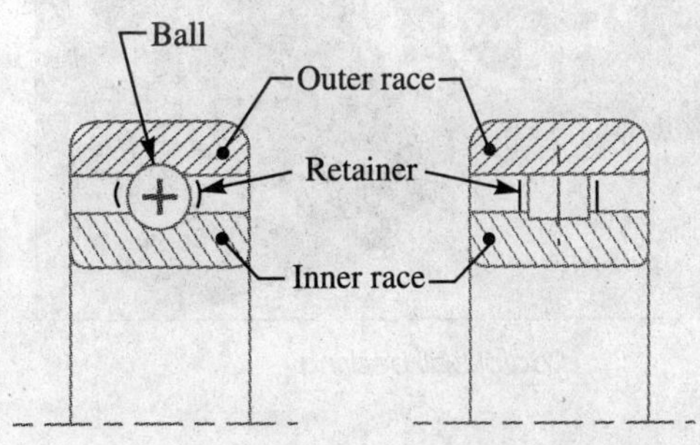

(*a*) Ball bearing. (*b*) Roller bearing.

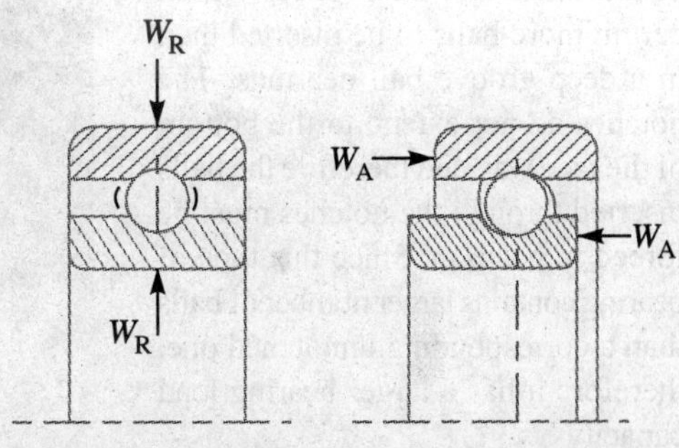

(*a*) Radial ball bearing. (*b*) Thrust ball bearing.

Fig. 27.1. Ball and roller bearings. **Fig. 27.2.** Radial and thrust ball bearings.

The ***ball and roller bearings*** consist of an inner race which is mounted on the shaft or journal and an outer race which is carried by the housing or casing. In between the inner and outer race, there are balls or rollers as shown in Fig. 27.1. A number of balls or rollers are used and these are held at proper distances by retainers so that they do not touch each other. The retainers are thin strips and is usually in two parts which are assembled after the balls have been properly spaced. The ball bearings are used for light loads and the roller bearings are used for heavier loads.

The rolling contact bearings, depending upon the load to be carried, are classified as :

(*a*) Radial bearings, and (*b*) Thrust bearings.

The radial and thrust ball bearings are shown in Fig. 27.2 (*a*) and (*b*) respectively. When a ball bearing supports only a radial load (W_R), the plane of rotation of the ball is normal to the centre line of the bearing, as shown in Fig. 27.2 (*a*). The action of thrust load (W_A) is to shift the plane of rotation of the balls, as shown in Fig. 27.2 (*b*). The radial and thrust loads both may be carried simultaneously.

27.4 Types of Radial Ball Bearings

Following are the various types of radial ball bearings:

1. ***Single row deep groove bearing.*** A single row deep groove bearing is shown in Fig. 27.3 (*a*).

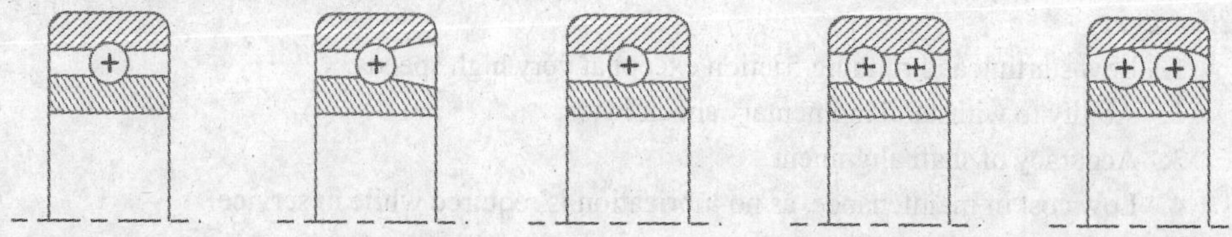

(*a*) Single row deep groove. (*b*) Filling notch. (*c*) Angular contact. (*d*) Double row. (*e*) Self-aligning.

Fig. 27.3. Types of radial ball bearings.

During assembly of this bearing, the races are offset and the maximum number of balls are placed between the races. The races are then centred and the balls are symmetrically located by the use of a retainer or cage. The deep groove ball bearings are used due to their high load carrying capacity and suitability for high running speeds. The load carrying capacity of a ball bearing is related to the size and number of the balls.

Radial ball bearing

2. ***Filling notch bearing.*** A filling notch bearing is shown in Fig. 27.3 (*b*). These bearings have notches in the inner and outer races which permit more balls to be inserted than in a deep groove ball bearings. The notches do not extend to the bottom of the race way and therefore the balls inserted through the notches must be forced in position. Since this type of bearing contains larger number of balls than a corresponding unnotched one, therefore it has a larger bearing load capacity.

3. ***Angular contact bearing.*** An angular contact bearing is shown in Fig. 27.3 (*c*). These bearings have one side of the outer race cut away to permit the insertion of more balls than in a deep groove bearing but without having a notch cut into both races. This permits the bearing to carry a relatively large axial load in one direction while also carrying a relatively large radial load. The angular contact bearings are usually used in pairs so that thrust loads may be carried in either direction.

4. ***Double row bearing.*** A double row bearing is shown in Fig. 27.3 (*d*). These bearings may be made with radial or angular contact between the balls and races. The double row bearing is appreciably narrower than two single row bearings. The load capacity of such bearings is slightly less than twice that of a single row bearing.

5. ***Self-aligning bearing.*** A self-aligning bearing is shown in Fig. 27.3 (*e*). These bearings permit shaft deflections within 2-3 degrees. It may be noted that normal clearance in a ball bearing are too small to accommodate any appreciable misalignment of the shaft relative to the housing. If the unit is assembled with shaft misalignment present, then the bearing will be subjected to a load that may be in excess of the design value and premature failure may occur. Following are the two types of self-aligning bearings :

(*a*) Externally self-aligning bearing, and (*b*) Internally self-aligning bearing.

In an ***externally self-aligning bearing,*** the outside diameter of the outer race is ground to a spherical surface which fits in a mating spherical surface in a housing, as shown in Fig. 27.3 (*e*). In case of ***internally self-aligning bearing,*** the inner surface of the outer race is ground to a spherical surface. Consequently, the outer race may be displaced through a small angle without interfering with the normal operation of the bearing. The internally self-aligning ball bearing is interchangeable with other ball bearings.

27.5 Standard Dimensions and Designations of Ball Bearings

The dimensions that have been standardised on an international basis are shown in Fig. 27.4. These dimensions are a function of the bearing bore and the series of bearing. The standard dimensions are given in millimetres. There is no standard for the size and number of steel balls.

The bearings are designated by a number. In general, the number consists of atleast three digits. Additional digits or letters are used to indicate special features *e.g.* deep groove, filling notch etc. The last three digits give the series and the bore of the bearing. The last two digits from 04 onwards, when multiplied by 5, give the bore diameter in millimetres. The third from the last digit designates the series of the bearing. The most common ball bearings are available in four series as follows :

1. Extra light (100),   2. Light (200),
3. Medium (300),   4. Heavy (400)

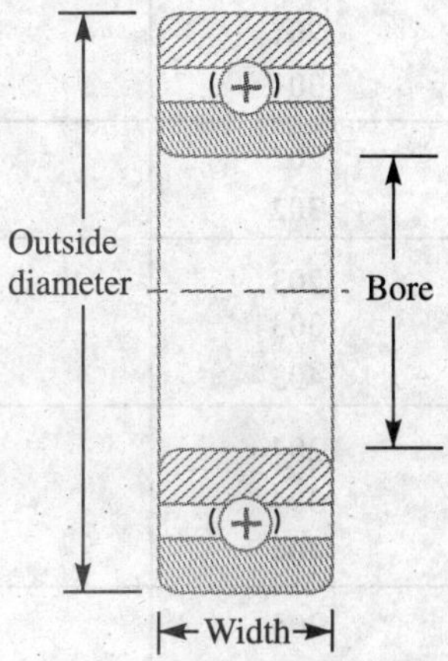

Fig. 27.4. Standard designations of ball bearings.

Notes : 1. If a bearing is designated by the number 305, it means that the bearing is of medium series whose bore is 05 × 5, *i.e.*, 25 mm.

2. The extra light and light series are used where the loads are moderate and shaft sizes are comparatively large and also where available space is limited.

3. The medium series has a capacity 30 to 40 per cent over the light series.

4. The heavy series has 20 to 30 per cent capacity over the medium series. This series is not used extensively in industrial applications.

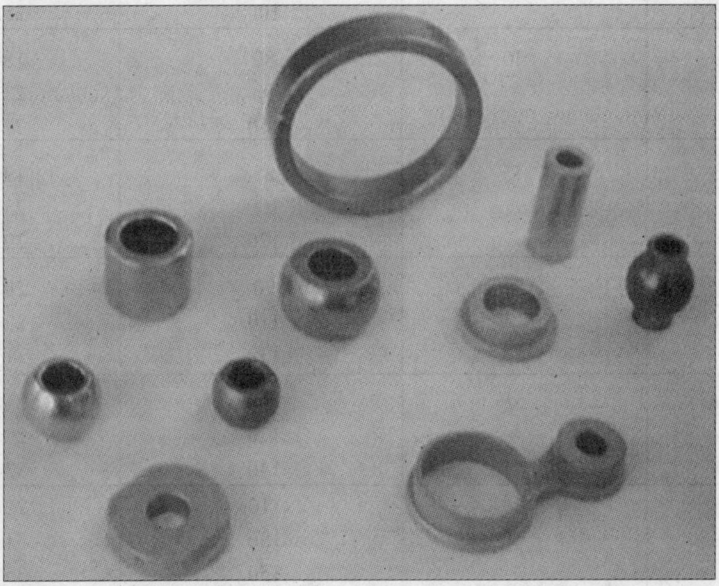

Oilless bearings made using powder metallergy.

The following table shows the principal dimensions for radial ball bearings.

Table 27.1. Principal dimensions for radial ball bearings.

Bearing No.	*Bore (mm)*	*Outside diameter*	*Width (mm)*
200	10	30	9
300		35	11
201	12	32	10
301		37	12
202	15	35	11
302		42	13
203	17	40	12
303		47	14
403		62	17
204	20	47	14
304		52	14
404		72	19
205	25	52	15
305		62	17
405		80	21
206	30	62	16
306		72	19
406		90	23
207	35	72	17
307		80	21
407		100	25
208	40	80	18
308		90	23
408		110	27
209	45	85	19
309		100	25
409		120	29
210	50	90	20
310		110	27
410		130	31
211	55	100	21
311		120	29
411		140	33
212	60	110	22
312		130	31
412		150	35

Bearing No.	Bore (mm)	Outside diameter	Width (mm)
213	65	120	23
313		140	33
413		160	37
214	70	125	24
314		150	35
414		180	42
215	75	130	25
315		160	37
415		190	45
216	80	140	26
316		170	39
416		200	48
217	85	150	28
317		180	41
417		210	52
218	90	160	30
318		190	43
418		225	54

27.6 Thrust Ball Bearings

The thrust ball bearings are used for carrying thrust loads exclusively and at speeds below 2000 r.p.m. At high speeds, centrifugal force causes the balls to be forced out of the races. Therefore at high speeds, it is recommended that angular contact ball bearings should be used in place of thrust ball bearings.

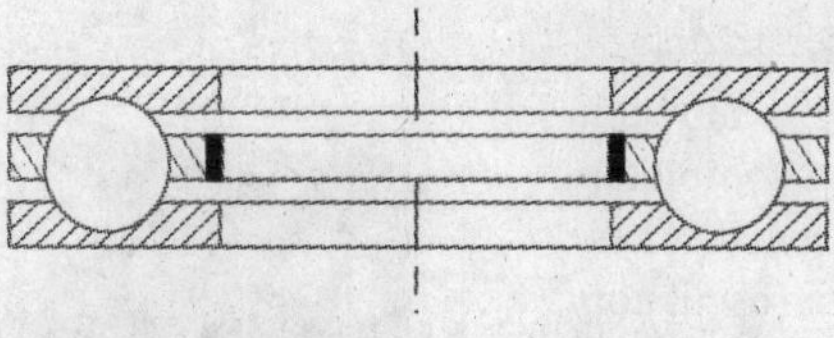

(*a*) Single direction thrust ball bearing.

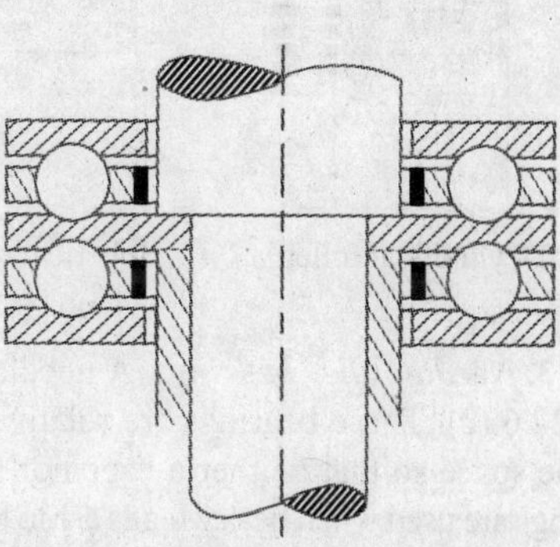

(*b*) Double direction thrust ball bearing.

Fig. 27.5. Thrust ball bearing.

A thrust ball bearing may be a single direction, flat face as shown in Fig. 27.5 (*a*) or a double direction with flat face as shown in Fig. 27.5 (*b*).

27.7 Types of Roller Bearings

Following are the principal types of roller bearings :

1. ***Cylindrical roller bearings.*** A cylindrical roller bearing is shown in Fig. 27.6 (*a*). These bearings have short rollers guided in a cage. These bearings are relatively rigid against radial motion and have the lowest coefficient of friction of any form of heavy duty rolling-contact bearings. Such type of bearings are used in high speed service.

Radial ball bearing

2. ***Spherical roller bearings.*** A spherical roller bearing is shown in Fig. 27.6 (*b*). These bearings are self-aligning bearings. The self-aligning feature is achieved by grinding one of the races in the form of sphere. These bearings can normally tolerate angular misalignment in the order of $\pm 1\frac{1}{2}^{\circ}$ and when used with a double row of rollers, these can carry thrust loads in either direction.

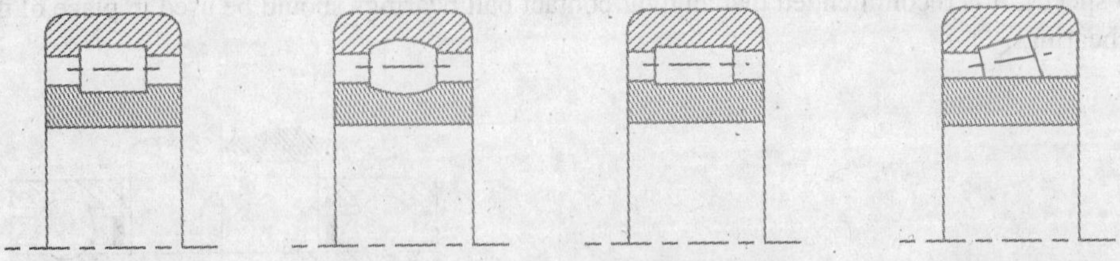

(*a*) Cylindrical roller. (*b*) Spherical roller. (*c*) Needle roller. (*d*) Tapered roller.

Fig. 27.6. Types of roller bearings.

3. ***Needle roller bearings.*** A needle roller bearing is shown in Fig. 27.6 (*c*). These bearings are relatively slender and completely fill the space so that neither a cage nor a retainer is needed. These bearings are used when heavy loads are to be carried with an oscillatory motion, *e.g.* piston pin bearings in heavy duty diesel engines, where the reversal of motion tends to keep the rollers in correct alignment.

4. ***Tapered roller bearings.*** A tapered roller bearing is shown in Fig. 27.6 (*d*). The rollers and race ways of these bearings are truncated cones whose elements intersect at a common point. Such type of bearings can carry both radial and thrust loads. These bearings are available in various combinations as double row bearings and with different cone angles for use with different relative magnitudes of radial and thrust loads.

Cylindrical roller bearings

Spherical roller bearings

Needle roller bearings

Tapered roller bearings

27.8 Basic Static Load Rating of Rolling Contact Bearings

The load carried by a non-rotating bearing is called a static load. The ***basic static load rating*** is defined as the static radial load (in case of radial ball or roller bearings) or axial load (in case of thrust ball or roller bearings) which corresponds to a total permanent deformation of the ball (or roller) and race, at the most heavily stressed contact, equal to 0.0001 times the ball (or roller) diameter.

In single row angular contact ball bearings, the basic static load relates to the radial component of the load, which causes a purely radial displacement of the bearing rings in relation to each other.

Note : The permanent deformation which appear in balls (or rollers) and race ways under static loads of moderate magnitude, increase gradually with increasing load. The permissible static load is, therefore, dependent upon the permissible magnitude of permanent deformation. Experience shows that a total permanent deformation of 0.0001 times the ball (or roller) diameter, occurring at the most heavily loaded ball (or roller) and race contact can be tolerated in most bearing applications without impairment of bearing operation.

In certain applications where subsequent rotation of the bearing is slow and where smoothness and friction requirements are not too exacting, a much greater total permanent deformation can be permitted. On the other hand, where extreme smoothness is required or friction requirements are critical, less total permanent deformation may be permitted.

According to IS : 3823–1984, the basic static load rating (C_0) in newtons for ball and roller bearings may be obtained as discussed below :

1. For radial ball bearings, the basic static radial load rating (C_0) is given by

$$C_0 = f_0 . i . Z . D^2 \cos \alpha$$

where i = Number of rows of balls in any one bearing,

Z = Number of ball per row,

D = Diameter of balls, in mm,

α = Nominal angle of contact *i.e.* the nominal angle between the line of action of the ball load and a plane perpendicular to the axis of bearing, and

f_0 = A factor depending upon the type of bearing.

The value of factor (f_0) for bearings made of hardened steel are taken as follows :

f_0 = 3.33, for self-aligning ball bearings

= 12.3, for radial contact and angular contact groove ball bearings.

2. For radial roller bearings, the basic static radial load rating is given by

$$C_0 = f_0 . i . Z . l_e . D \cos \alpha$$

where i = Number of rows of rollers in the bearing,

Z = Number of rollers per row,

l_e = Effective length of contact between one roller and that ring (or washer) where the contact is the shortest (in mm). It is equal to the overall length of roller ***minus*** roller chamfers or grinding undercuts,

D = Diameter of roller in mm. It is the mean diameter in case of tapered rollers,

α = Nominal angle of contact. It is the angle between the line of action of the roller resultant load and a plane perpendicular to the axis of the bearing, and

f_0 = 21.6, for bearings made of hardened steel.

3. For thrust ball bearings, the basic static axial load rating is given by

$$C_0 = f_0.Z.D^2 \sin \alpha$$

where Z = Number of balls carrying thrust in one direction, and

f_0 = 49, for bearings made of hardened steel.

4. For thrust roller bearings, the basic static axial load rating is given by

$$C_0 = f_0.Z.l_e.D.\sin \alpha$$

where Z = Number of rollers carrying thrust in one direction, and

f_0 = 98.1, for bearings made of hardened steel.

27.9 Static Equivalent Load for Rolling Contact Bearings

The static equivalent load may be defined as the static radial load (in case of radial ball or roller bearings) or axial load (in case of thrust ball or roller bearings) which, if applied, would cause the same total permanent deformation at the most heavily stressed ball (or roller) and race contact as that which occurs under the actual conditions of loading.

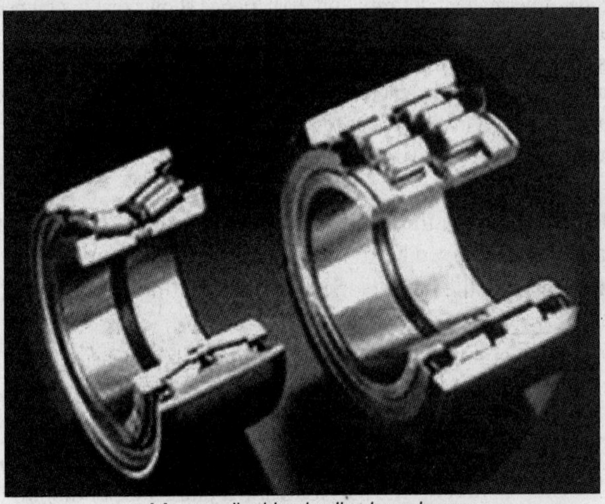

More cylindrical roller bearings

The static equivalent radial load (W_{0R}) for radial or roller bearings under combined radial and axial or thrust loads is given by the greater magnitude of those obtained by the following two equations, *i.e.*

1. $W_{0R} = X_0.W_R + Y_0.W_A$; and   2. $W_{0R} = W_R$

where W_R = Radial load,

W_A = Axial or thrust load,

X_0 = Radial load factor, and

Y_0 = Axial or thrust load factor.

According to IS : 3824 – 1984, the values of X_0 and Y_0 for different bearings are given in the following table :

Table 27.2. Values of X_0 and Y_0 for radial bearings.

S.No.	Type of bearing	Single row bearing		Double row bearing	
		X_0	Y_0	X_0	Y_0
1.	Radial contact groove ball bearings	0.60	0.50	0.60	0.50
2.	Self aligning ball or roller bearings and tapered roller bearing	0.50	0.22 cot θ	1	0.44 cot θ
3.	Angular contact groove bearings :				
	$\alpha = 15°$	0.50	0.46	1	0.92
	$\alpha = 20°$	0.50	0.42	1	0.84
	$\alpha = 25°$	0.50	0.38	1	0.76
	$\alpha = 30°$	0.50	0.33	1	0.66
	$\alpha = 35°$	0.50	0.29	1	0.58
	$\alpha = 40°$	0.50	0.26	1	0.52
	$\alpha = 45°$	0.50	0.22	1	0.44

Notes : 1. The static equivalent radial load (W_{0R}) is always greater than or equal to the radial load (W_R).

2. For two similar single row angular contact ball bearings, mounted 'face-to-face' or 'back-to-back', use the values of X_0 and Y_0 which apply to a double row angular contact ball bearings. For two or more similar single row angular contact ball bearings mounted 'in tandem', use the values of X_0 and Y_0 which apply to a single row angular contact ball bearings.

3. The static equivalent radial load (W_{0R}) for all cylindrical roller bearings is equal to the radial load (W_R).

4. The static equivalent axial or thrust load (W_{0A}) for thrust ball or roller bearings with angle of contact $\alpha \neq 90°$, under combined radial and axial loads is given by

$$W_{0A} = 2.3\ W_R.\tan\alpha + W_A$$

This formula is valid for all ratios of radial to axial load in the case of direction bearings. For single direction bearings, it is valid where $W_R / W_A \leq 0.44 \cot \alpha$.

5. The thrust ball or roller bearings with $\alpha = 90°$ can support axial loads only. The static equivalent axial load for this type of bearing is given by

$$W_{0A} = W_A$$

27.10 Life of a Bearing

The ***life*** of an individual ball (or roller) bearing may be defined as the number of revolutions (or hours at some given constant speed) which the bearing runs before the first evidence of fatigue develops in the material of one of the rings or any of the rolling elements.

The ***rating life*** of a group of apparently identical ball or roller bearings is defined as the number of revolutions (or hours at some given constant speed) that 90 per cent of a group of bearings will complete or exceed before the first evidence of fatigue develops (*i.e.* only 10 per cent of a group of bearings fail due to fatigue).

The term ***minimum life*** is also used to denote the rating life. It has been found that the life which 50 per cent of a group of bearings will complete or exceed is approximately 5 times the life which 90 per cent of the bearings will complete or exceed. In other words, we may say that the average life of a bearing is 5 times the rating life (or minimum life). It may be noted that the longest life of a single bearing is seldom longer than the 4 times the average life and the maximum life of a single bearing is about 30 to 50 times the minimum life.

The life of bearings for various types of machines is given in the following table.

Table 27.3. Life of bearings for various types of machines.

S. No.	*Application of bearing*	*Life of bearing, in hours*
1.	Instruments and apparatus that are rarely used	
	(*a*) Demonstration apparatus, mechanism for operating sliding doors	500
	(*b*) Aircraft engines	1000 – 2000
2.	Machines used for short periods or intermittently and whose breakdown would not have serious consequences *e.g.* hand tools, lifting tackle in workshops, and operated machines, agricultural machines, cranes in erecting shops, domestic machines.	4000 – 8000
3.	Machines working intermittently whose breakdown would have serious consequences *e.g.* auxillary machinery in power stations, conveyor plant for flow production, lifts, cranes for piece goods, machine tools used frequently.	8000 – 12 000
4.	Machines working 8 hours per day and not always fully utilised *e.g.* stationary electric motors, general purpose gear units.	12 000 – 20 000
5.	Machines working 8 hours per day and fully utilised *e.g.* machines for the engineering industry, cranes for bulk goods, ventilating fans, counter shafts.	20 000 – 30 000
6.	Machines working 24 hours per day *e.g.* separators, compressors, pumps, mine hoists, naval vessels.	40 000 – 60 000
7.	Machines required to work with high degree of reliability 24 hours per day *e.g.* pulp and paper making machinery, public power plants, mine-pumps, water works.	100 000 – 200 000

27.11 Basic Dynamic Load Rating of Rolling Contact Bearings

The basic dynamic load rating is defined as the constant stationary radial load (in case of radial ball or roller bearings) or constant axial load (in case of thrust ball or roller bearings) which a group of apparently identical bearings with stationary outer ring can endure for a rating life of one million revolutions (which is equivalent to 500 hours of operation at 33.3 r.p.m.) with only 10 per cent failure.

The basic dynamic load rating (*C*) in newtons for ball and roller bearings may be obtained as discussed below :

1. According to IS: 3824 (Part 1)– 1983, the basic dynamic radial load rating for radial and angular contact ball bearings, except the filling slot type, with balls not larger than 25.4 mm in diameter, is given by

$$C = f_c\,(i \cos \alpha)^{0.7}\, Z^{2/3} \,.\, D^{1.8}$$

and for balls larger than 25.4 mm in diameter,

$$C = 3.647\, f_c\,(i \cos \alpha)^{0.7}\, Z^{2/3} \,.\, D^{1.4}$$

where f_c = A factor, depending upon the geometry of the bearing components, the accuracy of manufacture and the material used.

and i, Z, D and α have usual meanings as discussed in Art. 27.8.

Ball bearings

2. According to IS: 3824 (Part 2)–1983, the basic dynamic radial load rating for radial roller bearings is given by

$$C = f_c\,(i.l_e \cos\alpha)^{7/9}\,Z^{3/4}.\,D^{29/27}$$

3. According to IS: 3824 (Part 3)–1983, the basic dynamic axial load rating for single row, single or double direction thrust ball bearings is given as follows :

(*a*) For balls not larger than 25.4 mm in diameter and $\alpha = 90°$,

$$C = f_c\,.\,Z^{2/3}\,.\,D^{1.8}$$

(*b*) For balls not larger than 25.4 mm in diameter and $\alpha \neq 90°$,

$$C = f_c\,(\cos\alpha)^{0.7}\tan\alpha.\,Z^{2/3}\,.\,D^{1.8}$$

(*c*) For balls larger than 25.4 mm in diameter and $\alpha = 90°$

$$C = 3.647\,f_c\,.\,Z^{2/3}\,.\,D^{1.4}$$

(*d*) For balls larger than 25.4 mm in diameter and $\alpha \neq 90°$,

$$C = 3.647\,f_c\,(\cos\alpha)^{0.7}\tan\alpha\,.\,Z^{2/3}\,.\,D^{1.4}$$

4. According to IS: 3824 (Part 4)–1983, the basic dynamic axial load rating for single row, single or double direction thrust roller bearings is given by

$$C = f_c\,.\,l_e^{\,7/9}\,.\,Z^{3/4}\,.\,D^{29/27} \quad \text{... (when } \alpha = 90°)$$

$$= f_c\,(l_e \cos\alpha)^{7/9}\tan\alpha.Z^{3/4}.\,D^{29/27} \quad \text{... (when } \alpha \neq 90°)$$

27.12 Dynamic Equivalent Load for Rolling Contact Bearings

The dynamic equivalent load may be defined as the constant stationary radial load (in case of radial ball or roller bearings) or axial load (in case of thrust ball or roller bearings) which, if applied to a bearing with rotating inner ring and stationary outer ring, would give the same life as that which the bearing will attain under the actual conditions of load and rotation.

The dynamic equivalent radial load (W) for radial and angular contact bearings, except the filling slot types, under combined constant radial load (W_R) and constant axial or thrust load (W_A) is given by

$$W = X.V.W_R + Y.W_A$$

where V = A rotation factor,

= 1, for all types of bearings when the inner race is rotating,

= 1, for self-aligning bearings when inner race is stationary,

= 1.2, for all types of bearings except self-aligning, when inner race is stationary.

The values of radial load factor (X) and axial or thrust load factor (Y) for the dynamically loaded bearings may be taken from the following table:

Table 27.4. Values of X and Y for dynamically loaded bearings.

Type of bearing	*Specifications*	$\frac{W_A}{W_R} \le e$		$\frac{W_A}{W_R} > e$		e
		X	Y	X	Y	
Deep groove ball bearing	$\frac{W_A}{C_0} = 0.025$				2.0	0.22
	= 0.04				1.8	0.24
	= 0.07				1.6	0.27
	= 0.13	1	0	0.56	1.4	0.31
	= 0.25				1.2	0.37
	= 0.50				1.0	0.44
Angular contact ball bearings	Single row		0	0.35	0.57	1.14
	Two rows in tandem		0	0.35	0.57	1.14
	Two rows back to back	1	0.55	0.57	0.93	1.14
	Double row		0.73	0.62	1.17	0.86
Self-aligning bearings	Light series : for bores					
	10 – 20 mm		1.3		2.0	0.50
	25 – 35	1	1.7	6.5	2.6	0.37
	40 – 45		2.0		3.1	0.31
	50 – 65		2.3		3.5	0.28
	70 – 100		2.4		3.8	0.26
	105 – 110		2.3		3.5	0.28
	Medium series : for bores					
	12 mm		1.0	0.65	1.6	0.63
	15 – 20		1.2		1.9	0.52
	25 – 50		1.5		2.3	0.43
	55 – 90		1.6		2.5	0.39
Spherical roller bearings	For bores :					
	25 – 35 mm		2.1		3.1	0.32
	40 – 45	1	2.5	0.67	3.7	0.27
	50 – 100		2.9		4.4	0.23
	100 – 200		2.6		3.9	0.26
Taper roller bearings	For bores :					
	30 – 40 mm				1.60	0.37
	45 – 110	1	0	0.4	1.45	0.44
	120 – 150				1.35	0.41

27.13 Dynamic Load Rating for Rolling Contact Bearings under Variable Loads

The approximate rating (or service) life of ball or roller bearings is based on the fundamental equation,

$$L = \left(\frac{C}{W}\right)^k \times 10^6 \text{ revolutions}$$

or

$$C = W\left(\frac{L}{10^6}\right)^{1/k}$$

where

L = Rating life,

C = Basic dynamic load rating,

W = Equivalent dynamic load,

and

k = 3, for ball bearings,

= 10/3, for roller bearings.

Roller bearing

The relationship between the life in revolutions (L) and the life in working hours (L_H) is given by

$$L = 60\ N \, . \, L_H \text{ revolutions}$$

where N is the speed in r.p.m.

Now consider a rolling contact bearing subjected to variable loads. Let W_1, W_2, W_3 etc., be the loads on the bearing for successive n_1, n_2, n_3 etc., number of revolutions respectively.

If the bearing is operated exclusively at the constant load W_1, then its life is given by

$$L_1 = \left(\frac{C}{W_1}\right)^k \times 10^6 \text{ revolutions}$$

∴ Fraction of life consumed with load W_1 acting for n_1 number of revolutions is

$$\frac{n_1}{L_1} = n_1\left(\frac{W_1}{C}\right)^k \times \frac{1}{10^6}$$

Similarly, fraction of life consumed with load W_2 acting for n_2 number of revolutions is

$$\frac{n_2}{L_2} = n_2\left(\frac{W_2}{C}\right)^k \times \frac{1}{10^6}$$

and fraction of life consumed with load W_3 acting for n_3 number of revolutions is

$$\frac{n_3}{L_3} = n_3\left(\frac{W_3}{C}\right)^k \times \frac{1}{10^6}$$

But

$$\frac{n_1}{L_1} + \frac{n_2}{L_2} + \frac{n_3}{L_3} + ... = 1$$

or

$$n_1\left(\frac{W_1}{C}\right)^k \times \frac{1}{10^6} + n_2\left(\frac{W_2}{C}\right)^k \times \frac{1}{10^6} + n_3\left(\frac{W_3}{C}\right)^k \times \frac{1}{10^6} + ... = 1$$

$$\therefore n_1(W_1)^k + n_2\,(W_2)^k + n_3\,(W_3)^k + ... = C^k \times 10^6 \quad ...(i)$$

If an equivalent constant load (W) is acting for n number of revolutions, then

$$n = \left(\frac{C}{W}\right)^k \times 10^6$$

$$\therefore \quad n(W)^k = C^k \times 10^6 \qquad \text{...}(ii)$$

where $\quad n = n_1 + n_2 + n_3 + ...$

From equations (*i*) and (*ii*), we have

$$n_1 (W_1)^k + n_2 (W_2)^k + n_3 (W_3)^k + ... = n (W)^k$$

$$\therefore \quad W = \left[\frac{n_1 (W_1)^k + n_2 (W_2)^k + n_3 (W_3)^k + ...}{n} \right]^{1/k}$$

Substituting $n = n_1 + n_2 + n_3 + ...$, and $k = 3$ for ball bearings, we have

$$W = \left[\frac{n_1 (W_1)^3 + n_2 (W_2)^3 + n_3 (W_3)^3 + ...}{n_1 + n_2 + n_3 + ...} \right]^{1/3}$$

Note : The above expression may also be written as

$$W = \left[\frac{L_1 (W_1)^3 + L_2 (W_2)^3 + L_3 (W_3)^3 + ...}{L_1 + L_2 + L_3 + ...} \right]^{1/3}$$

See Example 27.6.

27.14 Reliability of a Bearing

We have already discussed in the previous article that the rating life is the life that 90 per cent of a group of identical bearings will complete or exceed before the first evidence of fatigue develops. The reliability (R) is defined as the ratio of the number of bearings which have successfully completed L million revolutions to the total number of bearings under test. Sometimes, it becomes necessary to select a bearing having a reliability of more than 90%. According to Wiebull, the relation between the bearing life and the reliability is given as

$$\log_e \left(\frac{1}{R} \right) = \left(\frac{L}{a} \right)^b \quad \text{or} \quad \frac{L}{a} = \left[\log_e \left(\frac{1}{R} \right) \right]^{1/b} \qquad \text{...}(i)$$

where L is the life of the bearing corresponding to the desired reliability R and a and b are constants whose values are

$$a = 6.84, \text{ and } \quad b = 1.17$$

If L_{90} is the life of a bearing corresponding to a reliability of 90% (*i.e.* R_{90}), then

$$\frac{L_{90}}{a} = \left[\log_e \left(\frac{1}{R_{90}} \right) \right]^{1/b} \qquad \text{...}(ii)$$

Dividing equation (*i*) by equation (*ii*), we have

$$\frac{L}{L_{90}} = \left[\frac{\log_e (1/R)}{\log_e (1/R_{90})} \right]^{1/b} = {}^*6.85 \, [\log_e (1/R)]^{1/1.17} \qquad \text{... } (\because b = 1.17)$$

This expression is used for selecting the bearing when the reliability is other than 90%.

Note : If there are n number of bearings in the system each having the same reliability R, then the reliability of the complete system will be

$$R_S = R_p$$

where R_S indicates the probability of one out of p number of bearings failing during its life time.

---

* $[\log_e (1 / R_{90})]^{1/b} = [\log_e (1/0.90)]^{1/1.17} = (0.10536)^{0.8547} = 0.146$

$$\therefore \quad \frac{L}{L_{90}} = \frac{[\log_e(1/R)]^{1/b}}{0.146} = 6.85 \, [\log_e (1/R)]^{1/1.17}$$

Example 27.1. *A shaft rotating at constant speed is subjected to variable load. The bearings supporting the shaft are subjected to stationary equivalent radial load of 3 kN for 10 per cent of time, 2 kN for 20 per cent of time, 1 kN for 30 per cent of time and no load for remaining time of cycle. If the total life expected for the bearing is 20 × 10⁶ revolutions at 95 per cent reliability, calculate dynamic load rating of the ball bearing.*

Solution. Given : $W_1 = 3$ kN ; $n_1 = 0.1\ n$; $W_2 = 2$ kN ; $n_2 = 0.2\ n$; $W_3 = 1$ kN ; $n_3 = 0.3\ n$; $W_4 = 0$; $n_4 = (1 - 0.1 - 0.2 - 0.3)\ n = 0.4\ n$; $L_{95} = 20 \times 10^6$ rev

Let L_{90} = Life of the bearing corresponding to reliability of 90 per cent,

L_{95} = Life of the bearing corresponding to reliability of 95 per cent = 20×10^6 revolutions ... (Given)

We know that

$$\frac{L_{95}}{L_{90}} = \left[\frac{\log_e (1/R_{95})}{\log_e (1/R_{90})}\right]^{1/b} = \left[\frac{\log_e (1/0.95)}{\log_e (1/0.90)}\right]^{1/1.17} \quad \text{... } (\because b = 1.17)$$

$$= \left(\frac{0.0513}{0.1054}\right)^{0.8547} = 0.54$$

$\therefore$ $L_{90} = L_{95} / 0.54 = 20 \times 10^6 / 0.54 = 37 \times 10^6$ rev

We know that equivalent radial load,

$$W = \left[\frac{n_1 (W_1)^3 + n_2 (W_2)^3 + n_3 (W_3)^3 + n_4 (W_4)^3}{n_1 + n_2 + n_3 + n_4}\right]^{1/3}$$

$$= \left[\frac{0.1 n \times 3^3 + 0.2\ n \times 2^3 + 0.3 n \times 1^3 + 0.4\ n \times 0^3}{0.1\ n + 0.2 n + 0.3 n + 0.4 n}\right]^{1/3}$$

$$= (2.7 + 1.6 + 0.3 + 0)^{1/3} = 1.663 \text{ kN}$$

We also know that dynamic load rating,

$$C = W \left(\frac{L_{90}}{10^6}\right)^{1/k} = 1.663 \left(\frac{37 \times 10^6}{10^6}\right)^{1/3} = 5.54 \text{ kN } \textbf{Ans.}$$

... ($\because k = 3$, for ball bearing)

Example 27.2. *The rolling contact ball bearing are to be selected to support the overhung countershaft. The shaft speed is 720 r.p.m. The bearings are to have 99% reliability corresponding to a life of 24 000 hours. The bearing is subjected to an equivalent radial load of 1 kN. Consider life adjustment factors for operating condition and material as 0.9 and 0.85 respectively. Find the basic dynamic load rating of the bearing from manufacturer's catalogue, specified at 90% reliability.*

Solution. Given : $N = 720$ r.p.m. ; $L_H = 24\ 000$ hours ; $W = 1$ kN

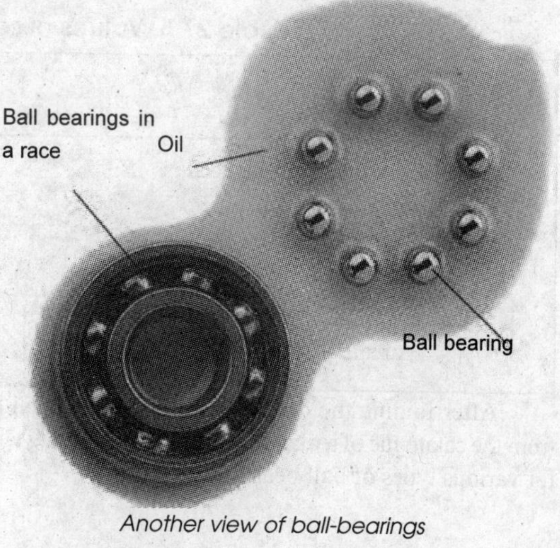

Another view of ball-bearings

We know that life of the bearing corresponding to 99% reliability,

$$L_{99} = 60\, N.\, L_H = 60 \times 720 \times 24\,000 = 1036.8 \times 10^6 \text{ rev}$$

Let L_{90} = Life of the bearing corresponding to 90% reliability.

Considering life adjustment factors for operating condition and material as 0.9 and 0.85 respectively, we have

$$\frac{L_{99}}{L_{90}} = \left[\frac{\log_e (1/R_{99})}{\log_e (1/R_{90})}\right]^{1/b} \times 0.9 \times 0.85 = \left[\frac{\log_e (1/0.99)}{\log_e (1/0.9)}\right]^{1/1.17} \times 0.9 \times 0.85$$

$$= \left[\frac{0.01005}{0.1054}\right]^{0.8547} \times 0.9 \times 0.85 = 0.1026$$

$$\therefore \quad L_{90} = L_{99} / \ 0.1026 = 1036.8 \times 10^6 / \ 0.1026 = 10\,105 \times 10^6 \text{ rev}$$

We know that dynamic load rating,

$$C = W\left(\frac{L_{90}}{10^6}\right)^{1/k}$$

$$= 1\left(\frac{10\,105 \times 10^6}{10^6}\right)^{1/3} \text{kN}$$

$$\ldots (\because k = 3, \text{ for ball bearing})$$

$$= 21.62 \text{ kN } \textbf{Ans.}$$

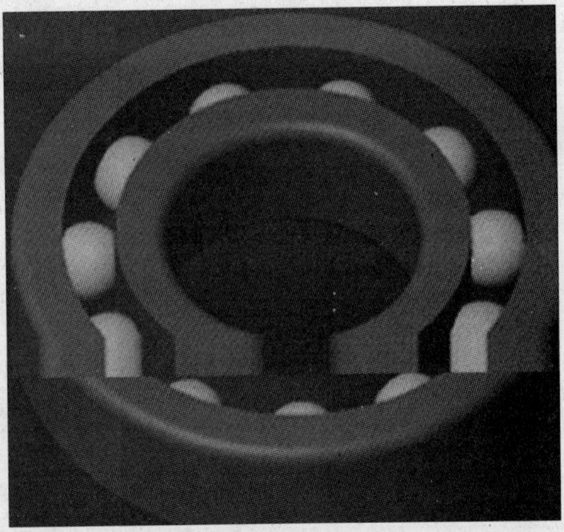

Radial ball bearings

27.15 Selection of Radial Ball Bearings

In order to select a most suitable ball bearing, first of all, the basic dynamic radial load is calculated. It is then multiplied by the service factor (K_S) to get the design basic dynamic radial load capacity. The service factor for the ball bearings is shown in the following table.

Table 27.5. Values of service factor (K_S).

S.No.	*Type of service*	*Service factor (K_S) for radial ball bearings*
1.	Uniform and steady load	1.0
2.	Light shock load	1.5
3.	Moderate shock load	2.0
4.	Heavy shock load	2.5
5.	Extreme shock load	3.0

After finding the design basic dynamic radial load capacity, the selection of bearing is made from the catalogue of a manufacturer. The following table shows the basic static and dynamic capacities for various types of ball bearings.

Table 27.6. Basic static and dynamic capacities of various types of radial ball bearings.

Bearing No.	*Basic capacities in kN*							
	Single row deep groove ball bearing		*Single row angular contact ball bearing*		*Double row angular contact ball bearing*		*Self-aligning ball bearing*	
	Static (C_0)	*Dynamic* (C)	*Static* (C_0)	*Dynamic* (C)	*Static* (C_0)	*Dynamic* (C)	*Static* (C_0)	*Dynamic* (C)
(1)	*(2)*	*(3)*	*(4)*	*(5)*	*(6)*	*(7)*	*(8)*	*(9)*
200	2.24	4	—	—	4.55	7.35	1.80	5.70
300	3.60	6.3	—	—	—	—	—	—
201	3	5.4	—	—	5.6	8.3	2.0	5.85
301	4.3	7.65	—	—	—	—	3.0	9.15
202	3.55	6.10	3.75	6.30	5.6	8.3	2.16	6
302	5.20	8.80	—	—	9.3	14	3.35	9.3
203	4.4	7.5	4.75	7.8	8.15	11.6	2.8	7.65
303	6.3	10.6	7.2	11.6	12.9	19.3	4.15	11.2
403	11	18	—	—	—	—	—	—
204	6.55	10	6.55	10.4	11	16	3.9	9.8
304	7.65	12.5	8.3	13.7	14	19.3	5.5	14
404	15.6	24	—	—	—	—	—	—
205	7.1	11	7.8	11.6	13.7	17.3	4.25	9.8
305	10.4	16.6	12.5	19.3	20	26.5	7.65	19
405	19	28	—	—	—	—	—	—
206	10	15.3	11.2	16	20.4	25	5.6	12
306	14.6	22	17	24.5	27.5	35.5	10.2	24.5
406	23.2	33.5	—	—	—	—	—	—
207	13.7	20	15.3	21.2	28	34	8	17
307	17.6	26	20.4	28.5	36	45	13.2	30.5
407	30.5	43	—	—	—	—	—	—
208	16	22.8	19	25	32.5	39	9.15	17.6
308	22	32	25.5	35.5	45.5	55	16	35.5
408	37.5	50	—	—	—	—	—	—
209	18.3	25:5	21.6	28	37.5	41.5	10.2	18
309	30	41.5	34	45.5	56	67	19.6	42.5
409	44	60	—	—	—	—	—	—
210	21.2	27.5	23.6	29	43	47.5	10.8	18
310	35.5	48	40.5	53	73.5	81.5	24	50
410	50	68	—	—	—	—	—	—

(1)	(2)	(3)	(4)	(5)	(6)	(7)	(8)	(9)
211	26	34	30	36.5	49	53	12.7	20.8
311	42.5	56	47.5	62	80	88	28.5	58.5
411	60	78	—	—	—	—	—	—
212	32	40.5	36.5	44	63	65.5	16	26.5
312	48	64	55	71	96.5	102	33.5	68
412	67	85	—	—	—	—	—	—
213	35.5	44	43	50	69.5	69.5	20.4	34
313	55	72	63	80	112	118	39	75
413	76.5	93	—	—	—	—	—	—
214	39	48	47.5	54	71	69.5	21.6	34.5
314	63	81.5	73.5	90	129	137	45	85
414	102	112	—	—	—	—	—	—
215	42.5	52	50	56	80	76.5	22.4	34.5
315	72	90	81.5	98	140	143	52	95
415	110	120	—	—	—	—	—	—
216	45.5	57	57	63	96.5	93	25	38
316	80	96.5	91.5	106	160	163	58.5	106
416	120	127	—	—	—	—	—	—
217	55	65.5	65.5	71	100	106	30	45.5
317	88	104	102	114	180	180	62	110
417	132	134	—	—	—	—	—	—
218	63	75	76.5	83	127	118	36	55
318	98	112	114	122	—	—	69.5	118
418	146	146	—	—	—	—	—	—
219	72	85	88	95	150	137	43	65.5
319	112	120	125	132	—	—	—	—
220	81.5	96.5	93	102	160	146	51	76.5
320	132	137	153	150	—	—	—	—
221	93	104	104	110	—	—	56	85
321	143	143	166	160	—	—	—	—
222	104	112	116	120	—	—	64	98
322	166	160	193	176	—	—	—	—

Note: The reader is advised to consult the manufacturer's catalogue for further and complete details of the bearings.

Example 27.3. *Select a single row deep groove ball bearing for a radial load of 4000 N and an axial load of 5000 N, operating at a speed of 1600 r.p.m. for an average life of 5 years at 10 hours per day. Assume uniform and steady load.*

Solution. Given : $W_R = 4000$ N ; $W_A = 5000$ N ; $N = 1600$ r.p.m.

Since the average life of the bearing is 5 years at 10 hours per day, therefore life of the bearing in hours,

$$L_H = 5 \times 300 \times 10 = 15\ 000 \text{ hours} \quad \text{... (Assuming 300 working days per year)}$$

and life of the bearing in revolutions,

$$L = 60\ N \times L_H = 60 \times 1600 \times 15\ 000 = 1440 \times 10^6 \text{ rev}$$

We know that the basic dynamic equivalent radial load,

$$W = X.V.W_R + Y.W_A \quad \text{...}(i)$$

In order to determine the radial load factor (X) and axial load factor (Y), we require W_A / W_R and W_A / C_0. Since the value of basic static load capacity (C_0) is not known, therefore let us take $W_A / C_0 = 0.5$. Now from Table 27.4, we find that the values of X and Y corresponding to $W_A / C_0 = 0.5$ and $W_A/W_R = 5000 / 4000 = 1.25$ (which is greater than $e = 0.44$) are

$$X = 0.56 \quad \text{and} \quad Y = 1$$

Since the rotational factor (V) for most of the bearings is 1, therefore basic dynamic equivalent radial load,

$$W = 0.56 \times 1 \times 4000 + 1 \times 5000 = 7240 \text{ N}$$

From Table 27.5, we find that for uniform and steady load, the service factor (K_S) for ball bearings is 1. Therefore the bearing should be selected for $W = 7240$ N.

We know that basic dynamic load rating,

$$C = W\left(\frac{L}{10^6}\right)^{1/k} = 7240\left(\frac{1440 \times 10^6}{10^6}\right)^{1/3} = 81\ 760 \text{ N}$$

$$= 81.76 \text{ kN} \quad \text{... } (\because k = 3, \text{ for ball bearings})$$

From Table 27.6, let us select the bearing No. 315 which has the following basic capacities,

$$C_0 = 72 \text{ kN} = 72\ 000 \text{ N} \quad \text{and} \quad C = 90 \text{ kN} = 90\ 000 \text{ N}$$

Now $\quad W_A / C_0 = 5000 / 72\ 000 = 0.07$

∴ From Table 27.4, the values of X and Y are

$$X = 0.56 \quad \text{and} \quad Y = 1.6$$

Substituting these values in equation (*i*), we have dynamic equivalent load,

$$W = 0.56 \times 1 \times 4000 + 1.6 \times 5000 = 10\ 240 \text{ N}$$

∴ Basic dynamic load rating,

$$C = 10\ 240\left(\frac{1440 \times 10^6}{10^6}\right)^{1/3} = 115\ 635 \text{ N} = 115.635 \text{ kN}$$

From Table 27.6, the bearing number 319 having $C = 120$ kN, may be selected. **Ans.**

Example 27.4. *A single row angular contact ball bearing number 310 is used for an axial flow compressor. The bearing is to carry a radial load of 2500 N and an axial or thrust load of 1500 N. Assuming light shock load, determine the rating life of the bearing.*

Solution. Given : $W_R = 2500$ N ; $W_A = 1500$ N

From Table 27.4, we find that for single row angular contact ball bearing, the values of radial factor (X) and thrust factor (Y) for $W_A / W_R = 1500 / 2500 = 0.6$ are

$$X = 1 \quad \text{and} \quad Y = 0$$

Since the rotational factor (V) for most of the bearings is 1, therefore dynamic equivalent load,

$$W = X.V.W_R + Y.W_A = 1 \times 1 \times 2500 + 0 \times 1500 = 2500 \text{ N}$$

From Table 27.5, we find that for light shock load, the service factor (K_S) is 1.5. Therefore the design dynamic equivalent load should be taken as

$$W = 2500 \times 1.5 = 3750 \text{ N}$$

From Table 27.6, we find that for a single row angular contact ball bearing number 310, the basic dynamic capacity,

$$C = 53 \text{ kN} = 53\ 000 \text{ N}$$

We know that rating life of the bearing in revolutions,

$$L = \left(\frac{C}{W}\right)^k \times 10^6 = \left(\frac{53\ 000}{3750}\right)^3 \times 10^6 = 2823 \times 10^6 \text{ rev} \quad \textbf{Ans.}$$

... ($\because$ $k = 3$, for ball bearings)

Example 27.5. *Design a self-aligning ball bearing for a radial load of 7000 N and a thrust load of 2100 N. The desired life of the bearing is 160 millions of revolutions at 300 r.p.m. Assume uniform and steady load,*

Solution. Given : $W_R = 7000$ N ; $W_A = 2100$ N ; $L = 160 \times 10^6$ rev ; $N = 300$ r.p.m.

From Table 27.4, we find that for a self-aligning ball bearing, the values of radial factor (X) and thrust factor (Y) for $W_A / W_R = 2100 / 7000 = 0.3$, are as follows :

$$X = 0.65 \quad \text{and} \quad Y = 3.5$$

Since the rotational factor (V) for most of the bearings is 1, therefore dynamic equivalent load,

$$W = X.V.W_R + Y.W_A = 0.65 \times 1 \times 7000 + 3.5 \times 2100 = 11\ 900 \text{ N}$$

From Table 27.5, we find that for uniform and steady load, the service factor K_S for ball bearings is 1. Therefore the bearing should be selected for $W = 11\ 900$ N.

We know that the basic dynamic load rating,

$$C = W\left(\frac{L}{10^6}\right)^{1/k} = 11\ 900\left(\frac{160 \times 10^6}{10^6}\right)^{1/3} = 64\ 600 \text{ N} = 64.6 \text{ kN}$$

... ($\because$ $k = 3$, for ball bearings)

From Table 27.6, let us select bearing number 219 having $C = 65.5$ kN **Ans.**

Example 27.6. *Select a single row deep groove ball bearing with the operating cycle listed below, which will have a life of 15 000 hours.*

Fraction of cycle	*Type of load*	*Radial (N)*	*Thrust (N)*	*Speed (R.P.M.)*	*Service factor*
1/10	*Heavy shocks*	*2000*	*1200*	*400*	*3.0*
1/10	*Light shocks*	*1500*	*1000*	*500*	*1.5*
1/5	*Moderate shocks*	*1000*	*1500*	*600*	*2.0*
3/5	*No shock*	*1200*	*2000*	*800*	*1.0*

Assume radial and axial load factors to be 1.0 and 1.5 respectively and inner race rotates.

Solution. Given : $L_H = 15\ 000$ hours ; $W_{R1} = 2000$ N ; $W_{A1} = 1200$ N ; $N_1 = 400$ r.p.m. ; $K_{S1} = 3$; $W_{R2} = 1500$ N ; $W_{A2} = 1000$ N ; $N_2 = 500$ r.p.m. ; $K_{S2} = 1.5$; $W_{R3} = 1000$ N ; $W_{A3} = 1500$ N ; $N_3 = 600$ r.p.m. ; $K_{S3} = 2$; $W_{R4} = 1200$ N ; $W_{A4} = 2000$ N ; $N_4 = 800$ r.p.m. ; $K_{S4} = 1$; $X = 1$; $Y = 1.5$

We know that basic dynamic equivalent radial load considering service factor is

$$W = [X.V.W_R + Y.W_A] K_S \qquad ...(i)$$

It is given that radial load factor (X) = 1 and axial load factor (Y) = 1.5. Since the rotational factor (V) for most of the bearings is 1, therefore equation (i) may be written as

$$W = (W_R + 1.5\, W_A) K_S$$

Now, substituting the values of W_R, W_A and K_S for different operating cycle, we have

$$W_1 = (W_{R1} + 1.5\, W_{A1}) K_{S1} = (2000 + 1.5 \times 1200)\, 3 = 11\,400 \text{ N}$$

$$W_2 = (W_{R2} + 1.5\, W_{A2}) K_{S2} = (1500 + 1.5 \times 1000)\, 1.5 = 4500 \text{ N}$$

$$W_3 = (W_{R3} + 1.5\, W_{A3}) K_{S3} = (1000 + 1.5 \times 1500)\, 2 = 6500 \text{ N}$$

and

$$W_4 = (W_{R4} + 1.5\, W_{A4}) K_{S4} = (1200 + 1.5 \times 2000)\, 1 = 4200 \text{ N}$$

We know that life of the bearing in revolutions

$$L = 60\, N.L_H = 60\, N \times 15\,000 = 0.9 \times 10^6\, N \text{ rev}$$

∴ Life of the bearing for 1/10 of a cycle,

$$L_1 = \frac{1}{10} \times 0.9 \times 10^6\, N_1 = \frac{1}{10} \times 0.9 \times 10^6 \times 400 = 36 \times 10^6 \text{ rev}$$

Similarly, life of the bearing for the next 1/10 of a cycle,

$$L_2 = \frac{1}{10} \times 0.9 \times 10^6\, N_2 = \frac{1}{10} \times 0.9 \times 10^6 \times 500 = 45 \times 10^6 \text{ rev}$$

Life of the bearing for the next 1/5 of a cycle,

$$L_3 = \frac{1}{5} \times 0.9 \times 10^6\, N_3 = \frac{1}{5} \times 0.9 \times 10^6 \times 600 = 108 \times 10^6 \text{ rev}$$

and life of the bearing for the next 3/5 of a cycle,

$$L_4 = \frac{3}{5} \times 0.9 \times 10^6\, N_4 = \frac{3}{5} \times 0.9 \times 10^6 \times 800 = 432 \times 10^6 \text{ rev}$$

We know that equivalent dynamic load,

$$W = \left[\frac{L_1 (W_1)^3 + L_2 (W_2)^3 + L_3 (W_3)^3 + L_4 (W_4)^3}{L_1 + L_2 + L_3 + L_4}\right]^{1/3}$$

$$= \left[\frac{36 \times 10^6 (11\,400)^3 + 45 \times 10^6 (4500)^3 + 108 \times 10^6 (6500)^3 + 432 \times 10^6 (4200)^3}{36 \times 10^6 + 45 \times 10^6 + 108 \times 10^6 + 423 \times 10^6}\right]^{1/3}$$

$$= \left[\frac{1.191 \times 10^8 \times 10^{12}}{621 \times 10^6}\right]^{1/3} = (0.1918 \times 10^{12})^{1/3} = 5767 \text{ N}$$

and

$$L = L_1 + L_2 + L_3 + L_4$$

$$= 36 \times 10^6 + 45 \times 10^6 + 108 \times 10^6 + 432 \times 10^6 = 621 \times 10^6 \text{ rev}$$

We know that dynamic load rating,

$$C = W\left(\frac{L}{10^6}\right)^{1/k} = 5767\left(\frac{621 \times 10^6}{10^6}\right)^{1/3}$$

$$= 5767 \times 8.53 = 49\,193 \text{ N} = 49.193 \text{ kN}$$

From Table 27.6, the single row deep groove ball bearing number 215 having $C = 52$ kN may be selected. **Ans.**

27.16 Materials and Manufacture of Ball and Roller Bearings

Since the rolling elements and the races are subjected to high local stresses of varying magnitude with each revolution of the bearing, therefore the material of the rolling element (*i.e.* steel) should be of high quality. The balls are generally made of high carbon chromium steel. The material of both the balls and races are heat treated to give extra hardness and toughness.

Ball and Roller Bearings

The balls are manufactured by hot forging on hammers from steel rods. They are then heat-treated, ground and polished. The races are also formed by forging and then heat-treated, ground and polished.

27.17 Lubrication of Ball and Roller Bearings

The ball and roller bearings are lubricated for the following purposes :

1. To reduce friction and wear between the sliding parts of the bearing,
2. To prevent rusting or corrosion of the bearing surfaces,
3. To protect the bearing surfaces from water, dirt etc., and
4. To dissipate the heat.

In general, oil or light grease is used for lubricating ball and roller bearings. Only pure mineral oil or a calcium-base grease should be used. If there is a possibility of moisture contact, then potassium or sodium-base greases may be used. Another additional advantage of the grease is that it forms a seal to keep out dirt or any other foreign substance. It may be noted that too much oil or grease cause the temperature of the bearing to rise due to churning. The temperature should be kept below 90°C and in no case a bearing should operate above 150°C.

EXERCISES

1. The ball bearings are to be selected for an application in which the radial load is 2000 N during 90 per cent of the time and 8000 N during the remaining 10 per cent. The shaft is to rotate at 150 r.p.m. Determine the minimum value of the basic dynamic load rating for 5000 hours of operation with not more than 10 per cent failures. **[Ans. 13.8 kN]**

2. A ball bearing subjected to a radial load of 5 kN is expected to have a life of 8000 hours at 1450 r.p.m. with a reliability of 99%. Calculate the dynamic load capacity of the bearing so that it can be selected from the manufacturer's catalogue based on a reliability of 90%. **[Ans. 86.5 kN]**
3. A ball bearing subjected to a radial load of 4000 N is expected to have a satisfactory life of 12 000 hours at 720 r.p.m. with a reliability of 95%. Calculate the dynamic load carrying capacity of the bearing, so that it can be selected from manufacturer's catalogue based on 90% reliability. If there are four such bearings each with a reliability of 95% in a system, what is the reliability of the complete system? **[Ans. 39.5 kN ; 81.45%]**
4. A rolling contact bearing is subjected to the following work cycle :
   (*a*) Radial load of 6000 N at 150 r.p.m. for 25% of the time; (*b*) Radial load of 7500 N at 600 r.p.m. for 20% of the time; and (*c*) Radial load of 2000 N at 300 r.p.m. for 55% of the time.
   The inner ring rotates and loads are steady. Select a bearing for an expected average life of 2500 hours.
5. A single row deep groove ball bearing operating at 2000 r.p.m. is acted by a 10 kN radial load and 8 kN thrust load. The bearing is subjected to a light shock load and the outer ring is rotating. Determine the rating life of the bearing. **[Ans. 15.52×10^6 rev]**
6. A ball bearing operates on the following work cycle :

Element No.	*Radial load (N)*	*Speed (R.P.M.)*	*Element time (%)*
1	3000	720	30
2.	7000	1440	40
3.	5000	900	30

The dynamic load capacity of the bearing is 16 600 N. Calculate 1. the average speed of rotation ; 2. the equivalent radial load ; and 3. the bearing life.

[Ans. 1062 r.p.m. ; 6.067 kN ; 20.5×10^6 rev]

QUESTIONS

1. What are rolling contact bearings? Discuss their advantages over sliding contact bearings.
2. Write short note on classifications and different types of antifriction bearings.
3. Where are the angular contact and self-aligning ball bearings used? Draw neat sketches of these bearings.
4. How do you express the life of a bearing? What is an average or median life?
5. Explain how the following factors influence the life of a bearing:
   (*a*) Load (*b*) Speed (*c*) Temperature (*d*) Reliability
6. Define the following terms as applied to rolling contact bearings:
   (*a*) Basic static load rating (*b*) Static equivalent load
   (c) Basic dynamic load rating (*d*) Dynamic equivalent load.
7. Derive the following expression as applied to rolling contact bearings subjected to variable load cycle

$$W_e = \sqrt[3]{\frac{N_1(W_1)^3 + N_2(W_2)^3 + N_3(W_3)^3 + ...}{N_1 + N_2 + N_3 + ...}}$$

where W_e = Equivalent cubic load,

W_1, W_2 and W_3 = Loads acting respectively for N_1, N_2, N_3 ....

8. Select appropriate type of rolling contact bearing under the following condition of loading giving reasons for your choice.
   1. Light radial load with high rotational speed.
   2. Heavy axial and radial load with shock.
   3. Light load where radial space is very limited.
   4. Axial thrust only with medium speed.

OBJECTIVE TYPE QUESTIONS

1. The rolling contact bearings are known as
   (*a*) thick lubricated bearings (*b*) plastic bearings
   (*c*) thin lubricated bearings (*d*) antifriction bearings
2. The bearings of medium series have capacity .... over the light series.
   (*a*) 10 to 20% (*b*) 20 to 30%
   (*c*) 30 to 40% (*d*) 40 to 50%
3. The bearings of heavy series have capacity .... over the medium series.
   (*a*) 10 to 20% (*b*) 20 to 30%
   (*c*) 30 to 40% (*d*) 40 to 50%
4. The ball bearings are usually made from
   (*a*) low carbon steel
   (*b*) medium carbon steel
   (*c*) high speed steel
   (*d*) chrome nickel steel
5. The tapered roller bearings can take
   (*a*) radial load only
   (*b*) axial load only
   (*c*) both radial and axial loads
   (*d*) none of the above
6. The piston pin bearings in heavy duty diesel engines are
   (*a*) needle roller bearings
   (*b*) tapered roller bearings
   (*c*) spherical roller bearings
   (*d*) cylindrical roller bearings
7. Which of the following is antifriction bearing?
   (*a*) journal bearing
   (*b*) pedestal bearing
   (*c*) collar bearing
   (*d*) needle bearing

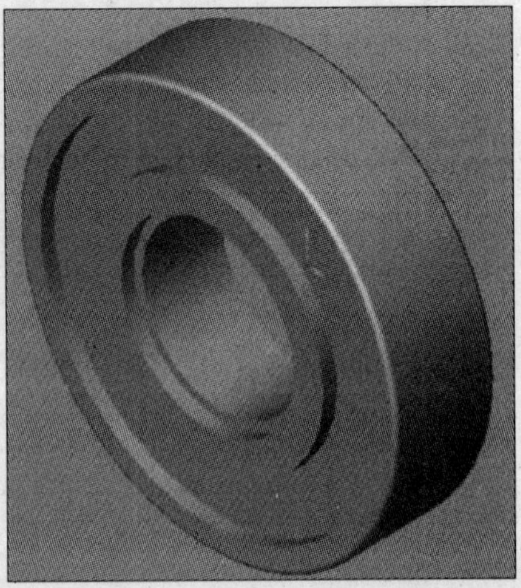

Ball bearing

8. Ball and roller bearings in comparison to sliding bearings have
   (*a*) more accuracy in alignment (*b*) small overall dimensions
   (*c*) low starting and running friction (*d*) all of these
9. A bearing is designated by the number 405. It means that a bearing is of
   (*a*) light series with bore of 5 mm (*b*) medium series with bore of 15 mm
   (*c*) heavy series with bore of 25 mm (*d*) light series with width of 20 mm
10. The listed life of a rolling bearing, in a catalogue, is the
   (*a*) minimum expected life (*b*) maximum expected life
   (*c*) average life (*d*) none of these

ANSWERS

1. (*d*)	2. (*c*)	3. (*b*)	4. (*d*)	5. (*c*)
6. (*a*)	7. (*d*)	8. (*d*)	9. (*c*)	10. (*a*)

CHAPTER

28

Spur Gears

28.1 Introduction

We have discussed earlier that the slipping of a belt or rope is a common phenomenon, in the transmission of motion or power between two shafts. The effect of slipping is to reduce the velocity ratio of the system. In precision machines, in which a definite velocity ratio is of importance (as in watch mechanism), the only positive drive is by ***gears*** or ***toothed wheels.*** A gear drive is also provided, when the distance between the driver and the follower is very small.

28.2 Friction Wheels

The motion and power transmitted by gears is kinematically equivalent to that transmitted by frictional wheels or discs. In order to understand how the motion can be transmitted by two toothed wheels, consider two plain circular wheels *A* and *B* mounted on shafts. The wheels have sufficient rough surfaces and press against each other as shown in Fig. 28.1.

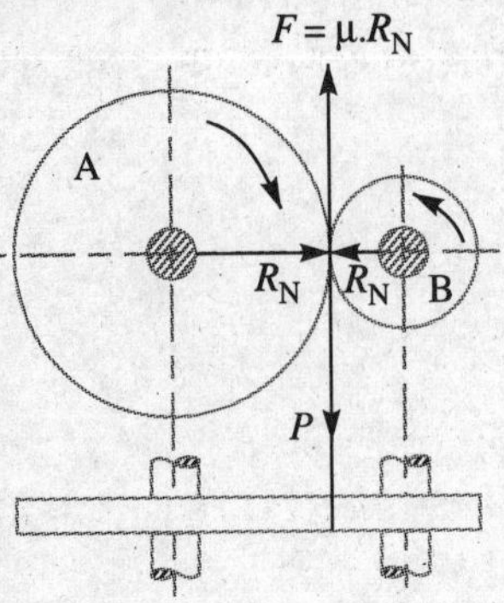

Fig. 28.1. Friction wheels.

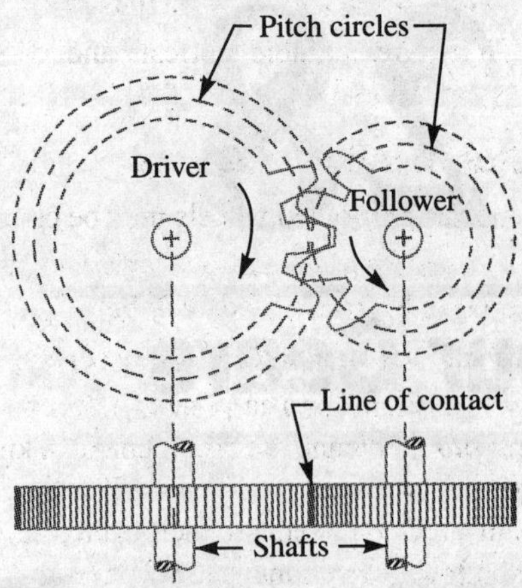

Fig. 28.2. Gear or toothed wheel.

Let the wheel *A* is keyed to the rotating shaft and the wheel *B* to the shaft to be rotated. A little consideration will show that when the wheel *A* is rotated by a rotating shaft, it will rotate the wheel *B* in the opposite direction as shown in Fig. 28.1. The wheel *B* will be rotated by the wheel *A* so long as the tangential force exerted by the wheel *A* does not exceed the maximum frictional resistance between the two wheels. But when the tangential force (P) exceeds the *frictional resistance (F), slipping will take place between the two wheels.

In order to avoid the slipping, a number of projections (called teeth) as shown in Fig. 28.2 are provided on the periphery of the wheel *A* which will fit into the corresponding recesses on the periphery of the wheel *B*. A friction wheel with the teeth cut on it is known as ***gear*** or ***toothed wheel.*** The usual connection to show the toothed wheels is by their pitch circles.

Note : Kinematically, the friction wheels running without slip and toothed gearing are identical. But due to the possibility of slipping of wheels, the friction wheels can only be used for transmission of small powers.

28.3 Advantages and Disadvantages of Gear Drives

The following are the advantages and disadvantages of the gear drive as compared to other drives, *i.e.* belt, rope and chain drives :

Advantages

1. It transmits exact velocity ratio.
2. It may be used to transmit large power.
3. It may be used for small centre distances of shafts.
4. It has high efficiency.
5. It has reliable service.
6. It has compact layout.

In bicycle gears are used to transmit motion. Mechanical advantage can be changed by changing gears.

Disadvantages

1. Since the manufacture of gears require special tools and equipment, therefore it is costlier than other drives.

* We know that frictional resistance, $F = \mu \cdot R_N$

where μ = Coefficient of friction between the rubbing surfaces of the two wheels, and

R_N = Normal reaction between the two rubbing surfaces.

2. The error in cutting teeth may cause vibrations and noise during operation.
3. It requires suitable lubricant and reliable method of applying it, for the proper operation of gear drives.

28.4 Classification of Gears

The gears or toothed wheels may be classified as follows :

1. ***According to the position of axes of the shafts.*** The axes of the two shafts between which the motion is to be transmitted, may be

(*a*) Parallel, (*b*) Intersecting, and (*c*) Non-intersecting and non-parallel.

The two parallel and co-planar shafts connected by the gears is shown in Fig. 28.2. These gears are called ***spur gears*** and the arrangement is known as ***spur gearing.*** These gears have teeth parallel to the axis of the wheel as shown in Fig. 28.2. Another name given to the spur gearing is ***helical gearing,*** in which the teeth are inclined to the axis. The ***single*** and ***double helical gears*** connecting parallel shafts are shown in Fig. 28.3 (*a*) and (*b*) respectively. The object of the double helical gear is to balance out the end thrusts that are induced in single helical gears when transmitting load. The double helical gears are known as ***herringbone gears.*** A pair of spur gears are kinematically equivalent to a pair of cylindrical discs, keyed to a parallel shaft having line contact.

The two non-parallel or intersecting, but coplaner shafts connected by gears is shown in Fig. 28.3 (*c*). These gears are called ***bevel gears*** and the arrangement is known as ***bevel gearing.*** The ***bevel gears,*** like spur gears may also have their teeth inclined to the face of the bevel, in which case they are known as ***helical bevel gears.***

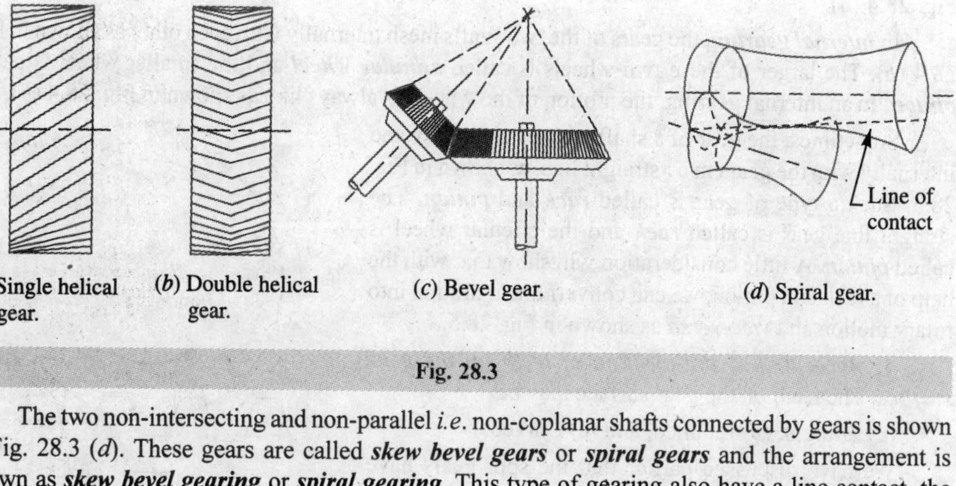

(*a*) Single helical gear. (*b*) Double helical gear. (*c*) Bevel gear. (*d*) Spiral gear.

Fig. 28.3

The two non-intersecting and non-parallel *i.e.* non-coplanar shafts connected by gears is shown in Fig. 28.3 (*d*). These gears are called ***skew bevel gears*** or ***spiral gears*** and the arrangement is known as ***skew bevel gearing*** or ***spiral gearing.*** This type of gearing also have a line contact, the rotation of which about the axes generates the two pitch surfaces known as ***hyperboloids.***

Notes : (*i*) When equal bevel gears (having equal teeth) connect two shafts whose axes are mutually perpendicular, then the bevel gears are known as ***mitres.***

(*ii*) A hyperboloid is the solid formed by revolving a straight line about an axis (not in the same plane), such that every point on the line remains at a constant distance from the axis.

(*iii*) The worm gearing is essentially a form of spiral gearing in which the shafts are usually at right angles.

2. ***According to the peripheral velocity of the gears.*** The gears, according to the peripheral velocity of the gears, may be classified as :

(*a*) Low velocity, (*b*) Medium velocity, and (*c*) High velocity.

The gears having velocity less than 3 m/s are termed as ***low velocity gears*** and gears having velocity between 3 and 15 m / s are known as ***medium velocity gears.*** If the velocity of gears is more than 15 m / s, then these are called ***high speed gears.***

3. *According to the type of gearing.* The gears, according to the type of gearing, may be classified as :

(*a*) External gearing, (*b*) Internal gearing, and (*c*) Rack and pinion.

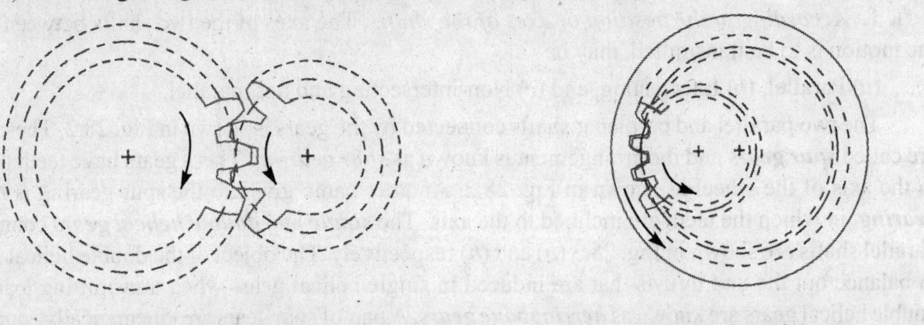

(*a*) External gearing. (*b*) Internal gearing.

Fig. 28.4

In ***external gearing,*** the gears of the two shafts mesh externally with each other as shown in Fig. 28.4 (*a*). The larger of these two wheels is called ***spur wheel*** or ***gear*** and the smaller wheel is called ***pinion.*** In an external gearing, the motion of the two wheels is always unlike, as shown in Fig. 28.4 (*a*).

In ***internal gearing,*** the gears of the two shafts mesh internally with each other as shown in Fig. 28.4 (*b*). The larger of these two wheels is called ***annular wheel*** and the smaller wheel is called ***pinion***. In an internal gearing, the motion of the wheels is always like as shown in Fig. 28.4 (*b*).

Sometimes, the gear of a shaft meshes externally and internally with the gears in a *straight line, as shown in Fig. 28.5. Such a type of gear is called ***rack*** and ***pinion.*** The straight line gear is called ***rack*** and the circular wheel is called ***pinion.*** A little consideration will show that with the help of a rack and pinion, we can convert linear motion into rotary motion and *vice-versa* as shown in Fig. 28.5.

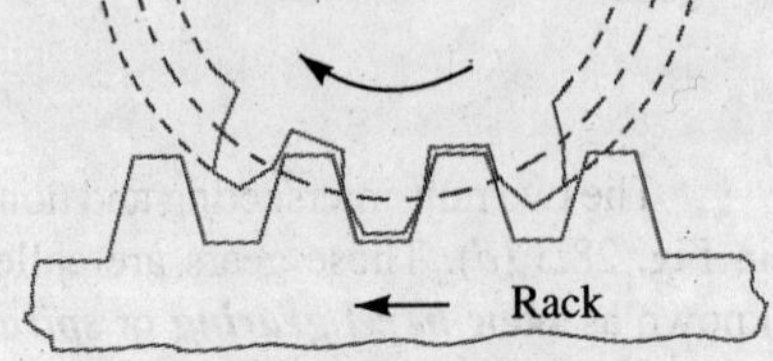

Fig. 28.5. Rack and pinion.

4. *According to the position of teeth on the gear surface.* The teeth on the gear surface may be

(*a*) Straight, (*b*) Inclined, and (*c*) Curved.

We have discussed earlier that the spur gears have straight teeth whereas helical gears have their teeth inclined to the wheel rim. In case of spiral gears, the teeth are curved over the rim surface.

28.5 Terms used in Gears

The following terms, which will be mostly used in this chapter, should be clearly understood at this stage. These terms are illustrated in Fig. 28.6.

1. *Pitch circle.* It is an imaginary circle which by pure rolling action, would give the same motion as the actual gear.

---

* A straight line may also be defined as a wheel of infinite radius.

2. ***Pitch circle diameter.*** It is the diameter of the pitch circle. The size of the gear is usually specified by the pitch circle diameter. It is also called as *pitch diameter*.

3. ***Pitch point.*** It is a common point of contact between two pitch circles.

4. ***Pitch surface.*** It is the surface of the rolling discs which the meshing gears have replaced at the pitch circle.

5. ***Pressure angle or angle of obliquity.*** It is the angle between the common normal to two gear teeth at the point of contact and the common tangent at the pitch point. It is usually denoted by ϕ. The standard pressure angles are $14^1/_2°$ and 20°.

6. ***Addendum.*** It is the radial distance of a tooth from the pitch circle to the top of the tooth.

7. ***Dedendum.*** It is the radial distance of a tooth from the pitch circle to the bottom of the tooth.

8. ***Addendum circle.*** It is the circle drawn through the top of the teeth and is concentric with the pitch circle.

9. ***Dedendum circle.*** It is the circle drawn through the bottom of the teeth. It is also called ***root circle.***

Note : Root circle diameter = Pitch circle diameter × cos ϕ, where ϕ is the pressure angle.

10. ***Circular pitch.*** It is the distance measured on the circumference of the pitch circle from a point of one tooth to the corresponding point on the next tooth. It is usually denoted by p_c. Mathematically,

Circular pitch, $p_c = \pi D/T$

where D = Diameter of the pitch circle, and

T = Number of teeth on the wheel.

A little consideration will show that the two gears will mesh together correctly, if the two wheels have the same circular pitch.

Note : If D_1 and D_2 are the diameters of the two meshing gears having the teeth T_1 and T_2 respectively; then for them to mesh correctly,

$$p_c = \frac{\pi D_1}{T_1} = \frac{\pi D_2}{T_2} \quad \text{or} \quad \frac{D_1}{D_2} = \frac{T_1}{T_2}$$

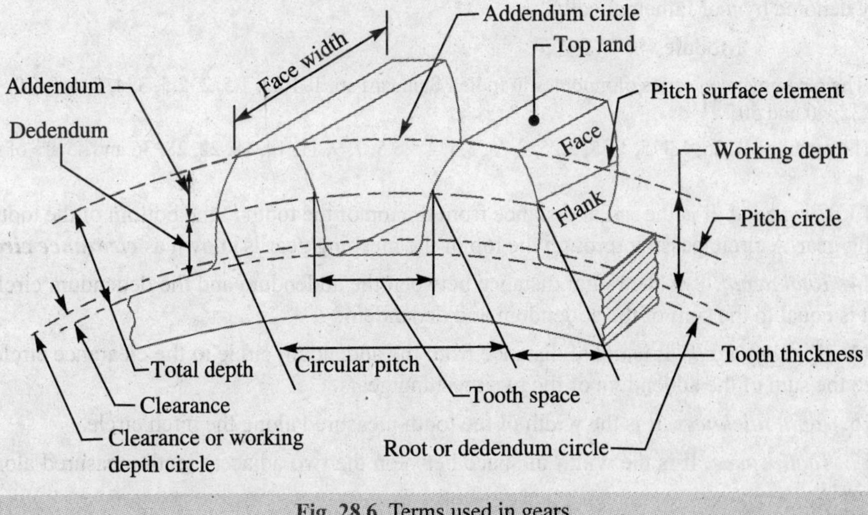

Fig. 28.6. Terms used in gears.

Spur gears

11. ***Diametral pitch.*** It is the ratio of number of teeth to the pitch circle diameter in millimetres. It denoted by p_d. Mathematically,

$$\text{Diametral pitch, } p_d = \frac{T}{D} = \frac{\pi}{p_c} \qquad ...\left(\because p_c = \frac{\pi D}{T}\right)$$

where T = Number of teeth, and

D = Pitch circle diameter.

12. ***Module.*** It is the ratio of the pitch circle diameter in millimetres to the number of teeth. It is usually denoted by m. Mathematically,

$$\text{Module, } m = D / T$$

Note : The recommended series of modules in Indian Standard are 1, 1.25, 1.5, 2, 2.5, 3, 4, 5, 6, 8, 10, 12, 16, 20, 25, 32, 40 and 50.

The modules 1.125, 1.375, 1.75, 2.25, 2.75, 3.5, 4.5,5.5, 7, 9, 11, 14, 18, 22, 28, 36 and 45 are of second choice.

13. ***Clearance.*** It is the radial distance from the top of the tooth to the bottom of the tooth, in a meshing gear. A circle passing through the top of the meshing gear is known as ***clearance circle.***

14. ***Total depth.*** It is the radial distance between the addendum and the dedendum circle of a gear. It is equal to the sum of the addendum and dedendum.

15. ***Working depth.*** It is radial distance from the addendum circle to the clearance circle. It is equal to the sum of the addendum of the two meshing gears.

16. ***Tooth thickness.*** It is the width of the tooth measured along the pitch circle.

17. ***Tooth space.*** It is the width of space between the two adjacent teeth measured along the pitch circle.

18. ***Backlash.*** It is the difference between the tooth space and the tooth thickness, as measured on the pitch circle.

19. ***Face of the tooth.*** It is surface of the tooth above the pitch surface.

20. ***Top land.*** It is the surface of the top of the tooth.

21. ***Flank of the tooth.*** It is the surface of the tooth below the pitch surface.

22. ***Face width.*** It is the width of the gear tooth measured parallel to its axis.

23. ***Profile.*** It is the curve formed by the face and flank of the tooth.

24. ***Fillet radius.*** It is the radius that connects the root circle to the profile of the tooth.

25. ***Path of contact.*** It is the path traced by the point of contact of two teeth from the beginning to the end of engagement.

26. ***Length of the path of contact.*** It is the length of the common normal cut-off by the addendum circles of the wheel and pinion.

27. ***Arc of contact.*** It is the path traced by a point on the pitch circle from the beginning to the end of engagement of a given pair of teeth. The arc of contact consists of two parts, *i.e.*

(*a*) ***Arc of approach.*** It is the portion of the path of contact from the beginning of the engagement to the pitch point.

(*b*) ***Arc of recess.*** It is the portion of the path of contact from the pitch point to the end of the engagement of a pair of teeth.

Note : The ratio of the length of arc of contact to the circular pitch is known as ***contact ratio*** *i.e.* number of pairs of teeth in contact.

28.6 Condition for Constant Velocity Ratio of Gears–Law of Gearing

Consider the portions of the two teeth, one on the wheel 1 (or pinion) and the other on the wheel 2, as shown by thick line curves in Fig. 28.7. Let the two teeth come in contact at point Q, and the wheels rotate in the directions as shown in the figure.

Let $T\,T$ be the common tangent and MN be the common normal to the curves at point of contact Q. From the centres O_1 and O_2, draw O_1M and O_2N perpendicular to MN. A little consideration will show that the point Q moves in the direction QC, when considered as a point on wheel 1, and in the direction QD when considered as a point on wheel 2.

Let v_1 and v_2 be the velocities of the point Q on the wheels 1 and 2 respectively. If the teeth are to remain in contact, then the components of these velocities along the common normal MN must be equal.

$$\therefore \quad v_1 \cos \alpha = v_2 \cos \beta$$

or $$(\omega_1 \times O_1Q) \cos \alpha = (\omega_2 \times O_2Q) \cos \beta$$

$$(\omega_1 \times O_1Q)\frac{O_1M}{O_1Q} = (\omega_2 \times O_2Q)\frac{O_2N}{O_2Q}$$

$$\therefore \quad \omega_1 . O_1M = \omega_2 . O_2N$$

or $$\frac{\omega_1}{\omega_2} = \frac{O_2N}{O_1M} \qquad ...(i)$$

Also from similar triangles O_1MP and O_2NP,

$$\frac{O_2N}{O_1M} = \frac{O_2P}{O_1P} \qquad ...(ii)$$

Combining equations (*i*) and (*ii*), we have

$$\frac{\omega_1}{\omega_2} = \frac{O_2N}{O_1M} = \frac{O_2P}{O_1P} \qquad ...(iii)$$

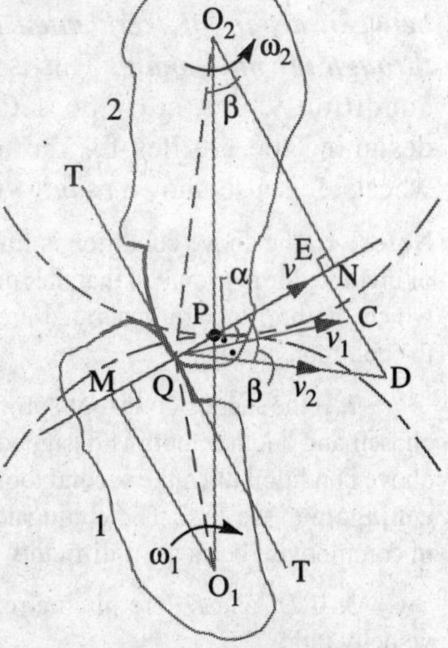

Fig. 28.7. Law of gearing.

We see that the angular velocity ratio is inversely proportional to the ratio of the distance of P from the centres

O_1 and O_2, or the common normal to the two surfaces at the point of contact Q intersects the line of centres at point P which divides the centre distance inversely as the ratio of angular velocities.

Aircraft landing gear is especially designed to absorb shock and energy when an aircraft lands, and then release gradually.

Therefore, in order to have a constant angular velocity ratio for all positions of the wheels, P must be the fixed point (called pitch point) for the two wheels. In other words, ***the common normal at the point of contact between a pair of teeth must always pass through the pitch point.*** This is fundamental condition which must be satisfied while designing the profiles for the teeth of gear wheels. It is also known as ***law of gearing.***

Gear trains inside a mechanical watch

Notes : 1. The above condition is fulfilled by teeth of involute form, provided that the root circles from which the profiles are generated are tangential to the common normal.

2. If the shape of one tooth profile is arbitrary chosen and another tooth is designed to satisfy the above condition, then the second tooth is said to be **conjugate** to the first. The conjugate teeth are not in common use because of difficulty in manufacture and cost of production.

3. If D_1 and D_2 are pitch circle diameters of wheel 1 and 2 having teeth T_1 and T_2 respectively, then velocity ratio,

$$\frac{\omega_1}{\omega_2} = \frac{O_2P}{O_1P} = \frac{D_2}{D_1} = \frac{T_2}{T_1}$$

28.7 Forms of Teeth

We have discussed in Art. 28.6 (Note 2) that conjugate teeth are not in common use. Therefore, in actual practice, following are the two types of teeth commonly used.

1. Cycloidal teeth ; and 2. Involute teeth.

We shall discuss both the above mentioned types of teeth in the following articles. Both these forms of teeth satisfy the condition as explained in Art. 28.6.

28.8 Cycloidal Teeth

A ***cycloid*** is the curve traced by a point on the circumference of a circle which rolls without slipping on a fixed straight line. When a circle rolls without slipping on the outside of a fixed circle, the curve traced by a point on the circumference of a circle is known as ***epicycloid.*** On the other hand, if a circle rolls without slipping on the inside of a fixed circle, then the curve traced by a point on the circumference of a circle is called ***hypocycloid.***

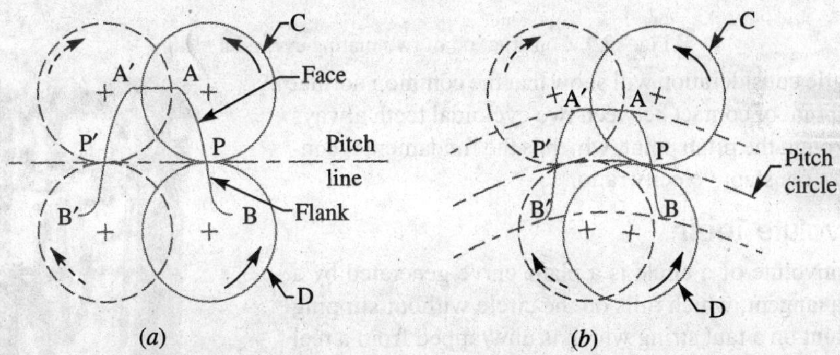

Fig. 28.8. Construction of cycloidal teeth of a gear.

In Fig. 28.8 (*a*), the fixed line or pitch line of a rack is shown. When the circle *C* rolls without slipping above the pitch line in the direction as indicated in Fig. 28.8 (*a*), then the point *P* on the circle traces the epicycloid *PA*. This represents the face of the cycloidal tooth profile. When the circle *D* rolls without slipping below the pitch line, then the point *P* on the circle *D* traces hypocycloid *PB* which represents the flank of the cycloidal tooth. The profile *BPA* is one side of the cycloidal rack tooth. Similarly, the two curves *P' A'* and *P' B'* forming the opposite side of the tooth profile are traced by the point *P'* when the circles *C* and *D* roll in the opposite directions.

In the similar way, the cycloidal teeth of a gear may be constructed as shown in Fig. 28.8 (*b*). The circle *C* is rolled without slipping on the outside of the pitch circle and the point *P* on the circle *C* traces epicycloid *PA*, which represents the face of the cycloidal tooth. The circle *D* is rolled on the inside of pitch circle and the point *P* on the circle *D* traces hypocycloid *PB*, which represents the flank of the tooth profile. The profile *BPA* is one side of the cycloidal tooth. The opposite side of the tooth is traced as explained above.

The construction of the two mating cycloidal teeth is shown in Fig. 28.9. A point on the circle *D* will trace the flank of the tooth T_1 when circle *D* rolls without slipping on the inside of pitch circle of wheel 1 and face of tooth T_2 when the circle *D* rolls without slipping on the outside of pitch circle of wheel 2. Similarly, a point on the circle *C* will trace the face of tooth T_1 and flank of tooth T_2. The rolling circles *C* and *D* may have unequal diameters, but if several wheels are to be interchangeable, they must have rolling circles of equal diameters.

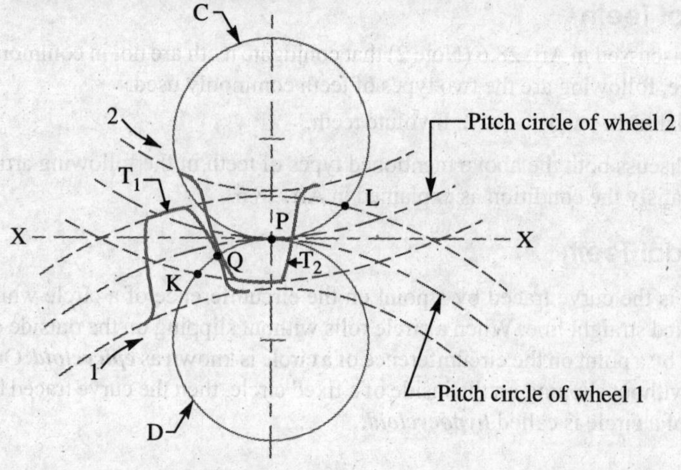

Fig. 28.9. Construction of two mating cycloidal teeth.

A little consideration will show that the common normal *XX* at the point of contact between two cycloidal teeth always passes through the pitch point, which is the fundamental condition for a constant velocity ratio.

28.9 Involute Teeth

An involute of a circle is a plane curve generated by a point on a tangent, which rolls on the circle without slipping or by a point on a taut string which is unwrapped from a reel as shown in Fig. 28.10 (*a*). In connection with toothed wheels, the circle is known as base circle. The involute is traced as follows :

Let *A* be the starting point of the involute. The base circle is divided into equal number of parts *e.g.* AP_1, P_1P_2, P_2P_3 etc.The tangents at P_1, P_2, P_3 etc., are drawn and the lenghts P_1A_1, P_2A_2, P_3A_3 equal to the arcs AP_1, AP_2 and AP_3 are set off. Joining the points A, A_1, A_2, A_3 etc., we obtain the involute curve *AR*. A little consideration will show that at any instant A_3, the tangent A_3T to the involute is perpendicular to P_3A_3 and P_3A_3 is the normal to the involute. In other words, normal at any point of an involute is a tangent to the circle.

Now, let O_1 and O_2 be the fixed centres of the two base circles as shown in Fig. 28.10(*b*). Let the corresponding involutes *AB* and *A'B'* be in contact at point *Q*. *MQ* and *NQ* are normals to the involute at *Q* and are tangents to base circles. Since the normal for an involute at a given point is the tangent drawn from that point to the base circle, therefore the common normal *MN* at *Q* is also the common tangent to the two base circles. We see that the common normal *MN* intersects the line of centres O_1O_2 at the fixed point *P* (called pitch point). Therefore the involute teeth satisfy the fundamental condition of constant velocity ratio.

The clock built by Galelio used gears.

From similar triangles O_2NP and O_1MP,

$$\frac{O_1M}{O_2N} = \frac{O_1P}{O_2P} = \frac{\omega_2}{\omega_1} \quad ...(i)$$

which determines the ratio of the radii of the two base circles. The radii of the base circles is given by

$$O_1M = O_1P\cos\phi, \text{ and } O_2N = O_2P\cos\phi$$

where ϕ is the pressure angle or the angle of obliquity.

Also the centre distance between the base circles

$$= O_1P + O_2P = \frac{O_1M}{\cos\phi} + \frac{O_2N}{\cos\phi} = \frac{O_1M + O_2N}{\cos\phi}$$

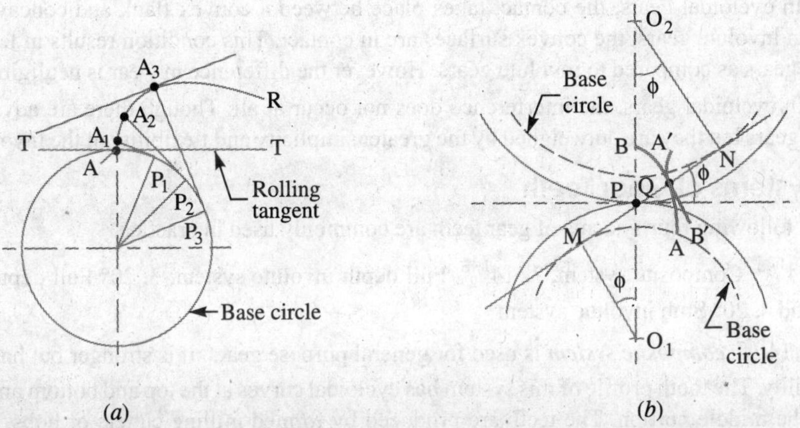

Fig. 28.10. Construction of involute teeth.

A little consideration will show, that if the centre distance is changed, then the radii of pitch circles also changes. But their ratio remains unchanged, because it is equal to the ratio of the two radii of the base circles [See equation (*i*)]. The common normal, at the point of contact, still passes through the pitch point. As a result of this, the wheel continues to work correctly*. However, the pressure angle increases with the increase in centre distance.

28.10 Comparison Between Involute and Cycloidal Gears

In actual practice, the involute gears are more commonly used as compared to cycloidal gears, due to the following advantages :

Advantages of involute gears

Following are the advantages of involute gears :

1. The most important advantage of the involute gears is that the centre distance for a pair of involute gears can be varied within limits without changing the velocity ratio. This is not true for cycloidal gears which requires exact centre distance to be maintained.

2. In involute gears, the pressure angle, from the start of the engagement of teeth to the end of the engagement, remains constant. It is necessary for smooth running and less wear of gears. But in cycloidal gears, the pressure angle is maximum at the beginning of engagement, reduces to zero at pitch point, starts increasing and again becomes maximum at the end of engagement. This results in less smooth running of gears.

3. The face and flank of involute teeth are generated by a single curve whereas in cycloidal gears, double curves (*i.e.* epicycloid and hypocycloid) are required for the face and flank respectively.

* It is not the case with cycloidal teeth.

Thus the involute teeth are easy to manufacture than cycloidal teeth. In involute system, the basic rack has straight teeth and the same can be cut with simple tools.

Note : The only disadvantage of the involute teeth is that the interference occurs (Refer Art. 28.13) with pinions having smaller number of teeth. This may be avoided by altering the heights of addendum and dedendum of the mating teeth or the angle of obliquity of the teeth.

Advantages of cycloidal gears

Following are the advantages of cycloidal gears :

1. Since the cycloidal teeth have wider flanks, therefore the cycloidal gears are stronger than the involute gears for the same pitch. Due to this reason, the cycloidal teeth are preferred specially for cast teeth.

2. In cycloidal gears, the contact takes place between a convex flank and concave surface, whereas in involute gears, the convex surfaces are in contact. This condition results in less wear in cycloidal gears as compared to involute gears. However the difference in wear is negligible.

3. In cycloidal gears, the interference does not occur at all. Though there are advantages of cycloidal gears but they are outweighed by the greater simplicity and flexibility of the involute gears.

28.11 Systems of Gear Teeth

The following four systems of gear teeth are commonly used in practice.

1. $14\frac{1}{2}°$ Composite system, 2. $14\frac{1}{2}°$ Full depth involute system, 3. 20° Full depth involute system, and 4. 20° Stub involute system.

The $14\frac{1}{2}°$ ***composite system*** is used for general purpose gears. It is stronger but has no interchangeability. The tooth profile of this system has cycloidal curves at the top and bottom and involute curve at the middle portion. The teeth are produced by formed milling cutters or hobs. The tooth profile of the $14\frac{1}{2}°$ ***full depth involute system*** was developed for use with gear hobs for spur and helical gears.

The tooth profile of the **20°** ***full depth involute system*** may be cut by hobs. The increase of the pressure angle from $14\frac{1}{2}°$ to 20° results in a stronger tooth, because the tooth acting as a beam is wider at the base. The **20°** ***stub involute system*** has a strong tooth to take heavy loads.

28.12 Standard Proportions of Gear Systems

The following table shows the standard proportions in module (m) for the four gear systems as discussed in the previous article.

Table 28.1. Standard proportions of gear systems.

S. No.	*Particulars*	*$14\frac{1}{2}°$ composite or full depth involute system*	*20° full depth involute system*	*20° stub involute system*
1.	Addendum	1m	1m	0.8 m
2.	Dedendum	1.25 m	1.25 m	1 m
3.	Working depth	2 m	2 m	1.60 m
4.	Minimum total depth	2.25 m	2.25 m	1.80 m
5.	Tooth thickness	1.5708 m	1.5708 m	1.5708 m
6.	Minimum clearance	0.25 m	0.25 m	0.2 m
7.	Fillet radius at root	0.4 m	0.4 m	0.4 m

28.13 Interference in Involute Gears

A pinion gearing with a wheel is shown in Fig. 28.11. *MN* is the common tangent to the base circles and *KL* is the path of contact between the two mating teeth. A little consideration will show, that if the radius of the addendum circle of pinion is increased to O_1N, the point of contact *L* will move from *L* to *N*. When this radius is further increased, the point of contact *L* will be on the inside of base circle of wheel and not on the involute profile of tooth on wheel. The tip of tooth on the pinion will then undercut the tooth on the wheel at the root and remove part of the involute profile of tooth on the wheel. This effect is known as ***interference*** and occurs when the teeth are being cut. In brief, ***the phenomenon when the tip of a tooth undercuts the root on its mating gear is known as interference.***

A drilling machine drilling holes for lamp retaining screws

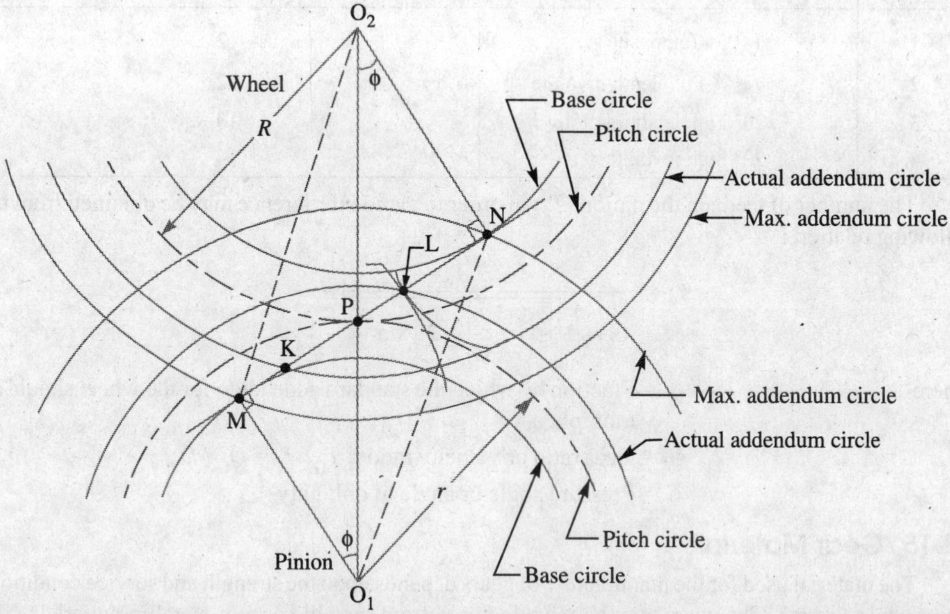

Fig. 28.11. Interference in involute gears.

Similarly, if the radius of the addendum circle of the wheel increases beyond O_2M, then the tip of tooth on wheel will cause interference with the tooth on pinion. The points *M* and *N* are called ***interference points.*** Obviously interference may be avoided if the path of contact does not extend beyond interference points. The limiting value of the radius of the addendum circle of the pinion is O_1N and of the wheel is O_2M.

From the above discussion, we conclude that the interference may only be avoided, if the point of contact between the two teeth is always on the involute profiles of both the teeth. In other words, ***interference may only be prevented, if the addendum circles of the two mating gears cut the common tangent to the base circles between the points of tangency.***

Note : In order to avoid interference, the limiting value of the radius of the addendum circle of the pinion ($O_1 N$) and of the wheel ($O_2 M$), may be obtained as follows :

From Fig. 28.11, we see that

$$O_1N = \sqrt{(O_1M)^2 + (MN)^2} = \sqrt{(r_b)^2 + [(r + R)\sin\phi]^2}$$

where r_b = Radius of base circle of the pinion = $O_1P \cos\phi = r\cos\phi$

Similarly

$$O_2M = \sqrt{(O_2N)^2 + (MN)^2} = \sqrt{(R_b)^2 + [(r + R)\sin\phi]^2}$$

where R_b = Radius of base circle of the wheel = $O_2P\cos\phi = R\cos\phi$

28.14 Minimum Number of Teeth on the Pinion in Order to Avoid Interference

We have seen in the previous article that the interference may only be avoided, if the point of contact between the two teeth is always on the involute profiles of both the teeth. The minimum number of teeth on the pinion which will mesh with any gear (also rack) without interference are given in the following table.

Table 28.2. Minimum number of teeth on the pinion in order to avoid interference.

S. No.	*Systems of gear teeth*	*Minimum number of teeth on the pinion*
1.	$14\frac{1}{2}°$ Composite	12
2.	$14\frac{1}{2}°$ Full depth involute	32
3.	20° Full depth involute	18
4.	20° Stub involute	14

The number of teeth on the pinion (T_P) in order to avoid interference may be obtained from the following relation :

$$T_P = \frac{2A_W}{G\left[\sqrt{1 + \frac{1}{G}\left(\frac{1}{G} + 2\right)\sin^2\phi} - 1\right]}$$

where A_W = Fraction by which the standard addendum for the wheel should be multiplied,

G = Gear ratio or velocity ratio = $T_G / T_P = D_G / D_P$,

ϕ = Pressure angle or angle of obliquity.

28.15 Gear Materials

The material used for the manufacture of gears depends upon the strength and service conditions like wear, noise etc. The gears may be manufactured from metallic or non-metallic materials. The metallic gears with cut teeth are commercially obtainable in cast iron, steel and bronze. The non-metallic materials like wood, rawhide, compressed paper and synthetic resins like nylon are used for gears, especially for reducing noise.

The cast iron is widely used for the manufacture of gears due to its good wearing properties, excellent machinability and ease of producing complicated shapes by casting method. The cast iron gears with cut teeth may be employed, where smooth action is not important.

The steel is used for high strength gears and steel may be plain carbon steel or alloy steel. The steel gears are usually heat treated in order to combine properly the toughness and tooth hardness.

The phosphor bronze is widely used for worm gears in order to reduce wear of the worms which will be excessive with cast iron or steel. The following table shows the properties of commonly used gear materials.

Table 28.3. Properties of commonly used gear materials.

Material (1)	*Condition* (2)	*Brinell hardness number* (3)	*Minimum tensile strength* (N/mm^2) (4)
Malleable cast iron			
(*a*) White heart castings, Grade B	—	217 max.	280
(*b*) Black heart castings, Grade B	—	149 max.	320
Cast iron			
(*a*) Grade 20	As cast	179 min.	200
(*b*) Grade 25	As cast	197 min.	250
(*c*) Grade 35	As cast	207 min.	250
(*d*) Grade 35	Heat treated	300 min.	350
Cast steel	—	145	550
Carbon steel			
(*a*) 0.3% carbon	Normalised	143	500
(*b*) 0.3% carbon	Hardened and tempered	152	600
(*c*) 0.4% carbon	Normalised	152	580
(*d*) 0.4% carbon	Hardened and tempered	179	600
(*e*) 0.35% carbon	Normalised	201	720
(*f*) 0.55% carbon	Hardened and tempered	223	700
Carbon chromium steel			
(*a*) 0.4% carbon	Hardened and tempered	229	800
(*b*) 0.55% carbon	"	225	900
Carbon manganese steel			
(*a*) 0.27% carbon	Hardened and tempered	170	600
(*b*) 0.37% carbon	"	201	700
Manganese molybdenum steel			
(*a*) 35 Mn 2 Mo 28	Hardened and tempered	201	700
(*b*) 35 Mn 2 Mo 45	"	229	800
Chromium molybdenum steel			
(*a*) 40 Cr 1 Mo 28	Hardened and tempered	201	700
(*b*) 40 Cr 1 Mo 60	"	248	900

(1)	(2)	(3)	(4)
Nickel steel			
40 Ni 3	"	229	800
Nickel chromium steel			
30 Ni 4 Cr 1	"	444	1540
Nickel chromium molybdenum steel	Hardness and		
40 Ni 2 Cr 1 Mo 28	tempered	255	900
Surface hardened steel			
(*a*) 0.4% carbon steel	—	145 (core) 460 (case)	551
(*b*) 0.55% carbon steel	—	200 (core) 520 (case)	708
(*c*) 0.55% carbon chromium steel	—	250 (core) 500 (case)	866
(*d*) 1% chromium steel	—	500 (case)	708
(*e*) 3% nickel steel	—	200 (core) 300 (case)	708
Case hardened steel			
(*a*) 0.12 to 0.22% carbon	—	650 (case)	504
(*b*) 3% nickel	—	200 (core) 600 (case)	708
(*c*) 5% nickel steel	—	250 (core) 600 (case)	866
Phosphor bronze castings	Sand cast	60 min.	160
	Chill cast	70 min.	240
	Centrifugal cast	90	260

28.16 Design Considerations for a Gear Drive

In the design of a gear drive, the following data is usually given :

1. The power to be transmitted.
2. The speed of the driving gear,
3. The speed of the driven gear or the velocity ratio, and
4. The centre distance.

The following requirements must be met in the design of a gear drive :

(*a*) The gear teeth should have sufficient strength so that they will not fail under static loading or dynamic loading during normal running conditions.

(*b*) The gear teeth should have wear characteristics so that their life is satisfactory.

(*c*) The use of space and material should be economical.

(*d*) The alignment of the gears and deflections of the shafts must be considered because they effect on the performance of the gears.

(*e*) The lubrication of the gears must be satisfactory.

28.17 Beam Strength of Gear Teeth – Lewis Equation

The beam strength of gear teeth is determined from an equation (known as *Lewis equation) and the load carrying ability of the toothed gears as determined by this equation gives satisfactory results. In the investigation, Lewis assumed that as the load is being transmitted from one gear to another, it is all given and taken by one tooth, because it is not always safe to assume that the load is distributed among several teeth. When contact begins, the load is assumed to be at the end of the driven teeth and as contact ceases, it is at the end of the driving teeth. This may not be true when the number of teeth in a pair of mating gears is large, because the load may be distributed among several teeth. But it is almost certain that at some time during the contact of teeth, the proper distribution of load does not exist and that one tooth must transmit the full load. In any pair of gears having unlike number of teeth, the gear which have the fewer teeth (*i.e.* pinion) will be the weaker, because the tendency toward undercutting of the teeth becomes more pronounced in gears as the number of teeth becomes smaller.

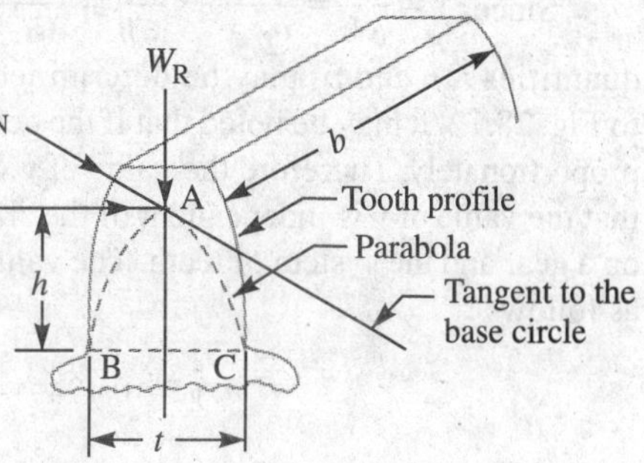

Fig. 28.12. Tooth of a gear.

Consider each tooth as a cantilever beam loaded by a normal load (W_N) as shown in Fig. 28.12. It is resolved into two components *i.e.* tangential component (W_T) and radial component (W_R) acting perpendicular and parallel to the centre line of the tooth respectively. The tangential component (W_T) induces a bending stress which tends to break the tooth. The radial component (W_R) induces a compressive stress of relatively small magnitude, therefore its effect on the tooth may be neglected. Hence, the bending stress is used as the basis for design calculations. The critical section or the section of maximum bending stress may be obtained by drawing a parabola through *A* and tangential to the tooth curves at *B* and *C*. This parabola, as shown dotted in Fig. 28.12, outlines a beam of uniform strength, *i.e.* if the teeth are shaped like a parabola, it will have the same stress at all the sections. But the tooth is larger than the parabola at every section except *BC*. We therefore, conclude that the section *BC* is the section of maximum stress or the critical section. The maximum value of the bending stress (or the permissible working stress), at the section *BC* is given by

$$\sigma_w = M.y / I \qquad ...(i)$$

where M = Maximum bending moment at the critical section $BC = W_T \times h$,

W_T = Tangential load acting at the tooth,

h = Length of the tooth,

y = Half the thickness of the tooth (t) at critical section $BC = t/2$,

I = Moment of inertia about the centre line of the tooth $= b.t^3/12$,

b = Width of gear face.

Substituting the values for *M*, *y* and *I* in equation (*i*), we get

$$\sigma_w = \frac{(W_T \times h)\, t/2}{b.t^3/12} = \frac{(W_T \times h) \times 6}{b.t^2}$$

or

$$W_T = \sigma_w \times b \times t^2 / 6\,h$$

In this expression, *t* and *h* are variables depending upon the size of the tooth (*i.e.* the circular pitch) and its profile.

* In 1892, Wilfred Lewis investigated for the strength of gear teeth. He derived an equation which is now extensively used by industry in determining the size and proportions of the gear.

Let $t = x \times p_c$, and $h = k \times p_c$; where x and k are constants.

$$\therefore \quad W_T = \sigma_w \times b \times \frac{x^2 . p_c^2}{6k . p_c} = \sigma_w \times b \times p_c \times \frac{x^2}{6k}$$

Substituting $x^2 / 6k = y$, another constant, we have

$$W_T = \sigma_w . b . p_c . y = \sigma_w . b . \pi m . y \quad ...(\because p_c = \pi m)$$

The quantity y is known as ***Lewis form factor*** or ***tooth form factor*** and W_T (which is the tangential load acting at the tooth) is called the ***beam strength of the tooth.***

Since $y = \frac{x^2}{6k} = \frac{t^2}{(p_c)^2} \times \frac{p_c}{6h} = \frac{t^2}{6h . p_c}$, therefore in order to find the value of y, the quantities t, h and p_c may be determined analytically or measured from the drawing similar to Fig. 28.12. It may be noted that if the gear is enlarged, the distances t, h and p_c will each increase proportionately. Therefore the value of y will remain unchanged. A little consideration will show that the value of y is independent of the size of the tooth and depends only on the number of teeth on a gear and the system of teeth. The value of y in terms of the number of teeth may be expressed as follows :

$$y = 0.124 - \frac{0.684}{T}, \text{ for } 14\tfrac{1}{2}^\circ \text{ composite and full depth involute system.}$$

$$= 0.154 - \frac{0.912}{T}, \text{ for } 20^\circ \text{ full depth involute system.}$$

$$= 0.175 - \frac{0.841}{T}, \text{ for } 20^\circ \text{ stub system.}$$

28.18 Permissible Working Stress for Gear Teeth in the Lewis Equation

The permissible working stress (σ_w) in the Lewis equation depends upon the material for which an allowable static stress (σ_o) may be determined. The ***allowable static stress*** is the stress at the

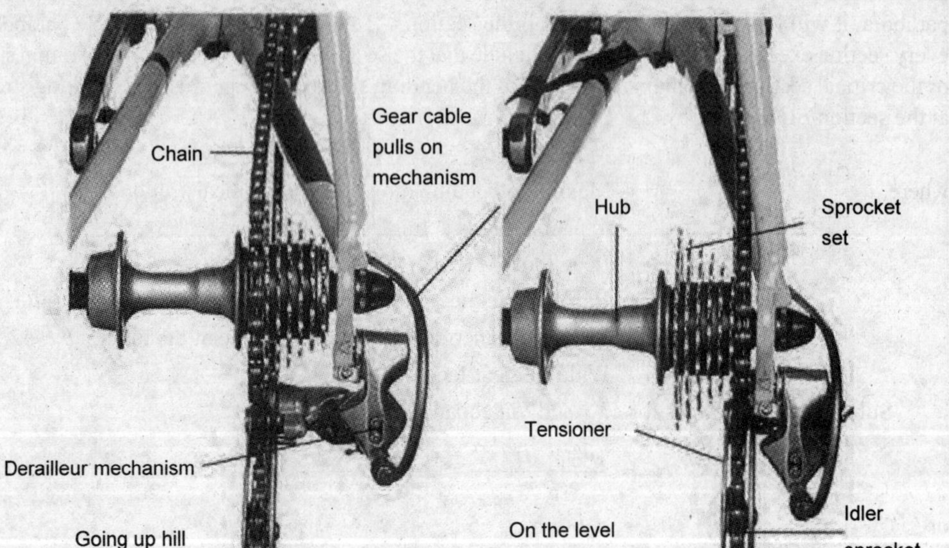

Bicycle gear mechanism switches the chain between different sized sprockets at the pedals and on the back wheel. Going up hill, a small front and a large rear sprocket are selected to reduce the push required for the rider. On the level, a large front and small rear. sprocket are used to prevent the rider having to pedal too fast.

elastic limit of the material. It is also called the ***basic stress.*** In order to account for the dynamic effects which become more severe as the pitch line velocity increases, the value of σ_w is reduced. According to the Barth formula, the permissible working stress,

$$\sigma_w = \sigma_o \times C_v$$

where σ_o = Allowable static stress, and

C_v = Velocity factor.

The values of the velocity factor (C_v) are given as follows :

$$C_v = \frac{3}{3 + v}, \text{ for ordinary cut gears operating at velocities upto 12.5 m / s.}$$

$$= \frac{4.5}{4.5 + v}, \text{ for carefully cut gears operating at velocities upto 12.5 m/s.}$$

$$= \frac{6}{6 + v}, \text{ for very accurately cut and ground metallic gears operating at velocities upto 20 m / s.}$$

$$= \frac{0.75}{0.75 + \sqrt{v}}, \text{ for precision gears cut with high accuracy and operating at velocities upto 20 m / s.}$$

$$= \left(\frac{0.75}{1 + v}\right) + 0.25, \text{ for non-metallic gears.}$$

In the above expressions, v is the pitch line velocity in metres per second.

The following table shows the values of allowable static stresses for the different gear materials.

Table 28.4. Values of allowable static stress.

Material	*Allowable static stress (σ_o) MPa or N/mm²*
Cast iron, ordinary	56
Cast iron, medium grade	70
Cast iron, highest grade	105
Cast steel, untreated	140
Cast steel, heat treated	196
Forged carbon steel-case hardened	126
Forged carbon steel-untreated	140 to 210
Forged carbon steel-heat treated	210 to 245
Alloy steel-case hardened	350
Alloy steel-heat treated	455 to 472
Phosphor bronze	84
Non-metallic materials	
Rawhide, fabroil	42
Bakellite, Micarta, Celoron	56

Note : The allowable static stress (σ_o) for steel gears is approximately one-third of the ultimate tensile stregth (σ_u) *i.e.* $\sigma_o = \sigma_u / 3$.

28.19 Dynamic Tooth Load

In the previous article, the velocity factor was used to make approximate allowance for the effect of dynamic loading. The dynamic loads are due to the following reasons :

1. Inaccuracies of tooth spacing,
2. Irregularities in tooth profiles, and
3. Deflections of teeth under load.

A closer approximation to the actual conditions may be made by the use of equations based on extensive series of tests, as follows :

$$W_D = W_T + W_I$$

where W_D = Total dynamic load,

W_T = Steady load due to transmitted torque, and

W_I = Increment load due to dynamic action.

The increment load (W_I) depends upon the pitch line velocity, the face width, material of the gears, the accuracy of cut and the tangential load. For average conditions, the dynamic load is determined by using the following Buckingham equation, *i.e.*

$$W_D = W_T + W_I = W_T + \frac{21\,v\,(b.C + W_T)}{21\,v + \sqrt{b.C + W_T}} \qquad \text{...}(i)$$

where W_D = Total dynamic load in newtons,

W_T = Steady transmitted load in newtons,

v = Pitch line velocity in m/s,

b = Face width of gears in mm, and

C = A deformation or dynamic factor in N/mm.

A deformation factor (C) depends upon the error in action between teeth, the class of cut of the gears, the tooth form and the material of the gears. The following table shows the values of deformation factor (C) for checking the dynamic load on gears.

Table 28.5. Values of deformation factor (C).

Material		*Involute tooth form*	*Values of deformation factor (C) in N-mm*				
			Tooth error in action (e) in mm				
Pinion	*Gear*		0.01	0.02	0.04	0.06	0.08
Cast iron	Cast iron		55	110	220	330	440
Steel	Cast iron	14½°	76	152	304	456	608
Steel	Steel		110	220	440	660	880
Cast iron	Cast iron		57	114	228	342	456
Steel	Cast iron	20° full depth	79	158	316	474	632
Steel	Steel		114	228	456	684	912
Cast iron	Cast iron		59	118	236	354	472
Steel	Cast iron	20° stub	81	162	324	486	648
Steel	Steel		119	238	476	714	952

The value of C in N/mm may be determined by using the following relation :

$$C = \frac{K.e}{\dfrac{1}{E_P} + \dfrac{1}{E_G}} \qquad \text{...}(ii)$$

where K = A factor depending upon the form of the teeth.

= 0.107, for $14\frac{1}{2}°$ full depth involute system.

= 0.111, for 20° full depth involute system.

= 0.115 for 20° stub system.

E_P = Young's modulus for the material of the pinion in N/mm^2.

E_G = Young's modulus for the material of gear in N/mm^2.

e = Tooth error action in mm.

The maximum allowable tooth error in action (e) depends upon the pitch line velocity (v) and the class of cut of the gears. The following tables show the values of tooth errors in action (e) for the different values of pitch line velocities and modules.

Table 28.6. Values of maximum allowable tooth error in action (e) verses pitch line velocity, for well cut commercial gears.

Pitch line velocity (v) m/s	*Tooth error in action (e) mm*	*Pitch line velocity (v) m/s*	*Tooth error in action (e) mm*	*Pitch line velocity (v) m/s*	*Tooth error in action (e) mm*
1.25	0.0925	8.75	0.0425	16.25	0.0200
2.5	0.0800	10	0.0375	17.5	0.0175
3.75	0.0700	11.25	0.0325	20	0.0150
5	0.0600	12.5	0.0300	22.5	0.0150
6.25	0.0525	13.75	0.0250	25 and over	0.0125
7.5	0.0475	15	0.0225		

Table 28.7. Values of tooth error in action (e) verses module.

Module (m) in mm	*Tooth error in action (e) in mm*		
	First class commercial gears	*Carefully cut gears*	*Precision gears*
Upto 4	0.051	0.025	0.0125
5	0.055	0.028	0.015
6	0.065	0.032	0.017
7	0.071	0.035	0.0186
8	0.078	0.0386	0.0198
9	0.085	0.042	0.021
10	0.089	0.0445	0.023
12	0.097	0.0487	0.0243
14	0.104	0.052	0.028
16	0.110	0.055	0.030
18	0.114	0.058	0.032
20	0.117	0.059	0.033

28.20 Static Tooth Load

The ***static tooth load*** (also called ***beam strength*** or ***endurance strength*** of the tooth) is obtained by Lewis formula by substituting flexural endurance limit or elastic limit stress (σ_e) in place of permissible working stress (σ_w).

∴ Static tooth load or beam strength of the tooth,

$$W_S = \sigma_e . b . p_c . y = \sigma_e . b . \pi\, m . y$$

The following table shows the values of flexural endurance limit (σ_e) for different materials.

Table 28.8. Values of flexural endurance limit.

Material of pinion and gear	*Brinell hardness number (B.H.N.)*	*Flexural endurance limit (σ_e) in MPa*
Grey cast iron	160	84
Semi-steel	200	126
Phosphor bronze	100	168
Steel	150	252
	200	350
	240	420
	280	490
	300	525
	320	560
	350	595
	360	630
	400 and above	700

For safety, against tooth breakage, the static tooth load (W_S) should be greater than the dynamic load (W_D). Buckingham suggests the following relationship between W_S and W_D.

For steady loads, $W_S \geq 1.25\, W_D$

For pulsating loads, $W_S \geq 1.35\, W_D$

For shock loads, $W_S \geq 1.5\, W_D$

Note : For steel, the flexural endurance limit (σ_e) may be obtained by using the following relation :

$$\sigma_e = 1.75 \times \text{B.H.N. (in MPa)}$$

28.21 Wear Tooth Load

The maximum load that gear teeth can carry, without premature wear, depends upon the radii of curvature of the tooth profiles and on the elasticity and surface fatigue limits of the materials. The maximum or the limiting load for satisfactory wear of gear teeth, is obtained by using the following Buckingham equation, *i.e.*

$$W_w = D_P . b . Q . K$$

where W_w = Maximum or limiting load for wear in newtons,

D_P = Pitch circle diameter of the pinion in mm,

b = Face width of the pinion in mm,

Q = Ratio factor

$$= \frac{2 \times V.R.}{V.R. + 1} = \frac{2T_G}{T_G + T_P}, \text{ for external gears}$$

$$= \frac{2 \times V.R.}{V.R. - 1} = \frac{2T_G}{T_G - T_P}, \text{ for internal gears.}$$

$V.R.$ = Velocity ratio = T_G / T_P,

K = Load-stress factor (also known as material combination factor) in N/mm^2.

The load stress factor depends upon the maximum fatigue limit of compressive stress, the pressure angle and the modulus of elasticity of the materials of the gears. According to Buckingham, the load stress factor is given by the following relation :

$$K = \frac{(\sigma_{es})^2 \sin\phi}{1.4}\left(\frac{1}{E_P} + \frac{1}{E_G}\right)$$

where σ_{es} = Surface endurance limit in MPa or N/mm^2,

ϕ = Pressure angle,

E_P = Young's modulus for the material of the pinion in N/mm^2, and

E_G = Young's modulus for the material of the gear in N/mm^2.

The values of surface endurance limit (σ_{es}) are given in the following table.

Table 28.9. Values of surface endurance limit.

Material of pinion and gear	*Brinell hardness number (B.H.N.)*	*Surface endurance limit (σ_{es}) in N/mm^2*
Grey cast iron	160	630
Semi-steel	200	630
Phosphor bronze	100	630
Steel	150	350
	200	490
	240	616
	280	721
	300	770
	320	826
	350	910
	400	1050

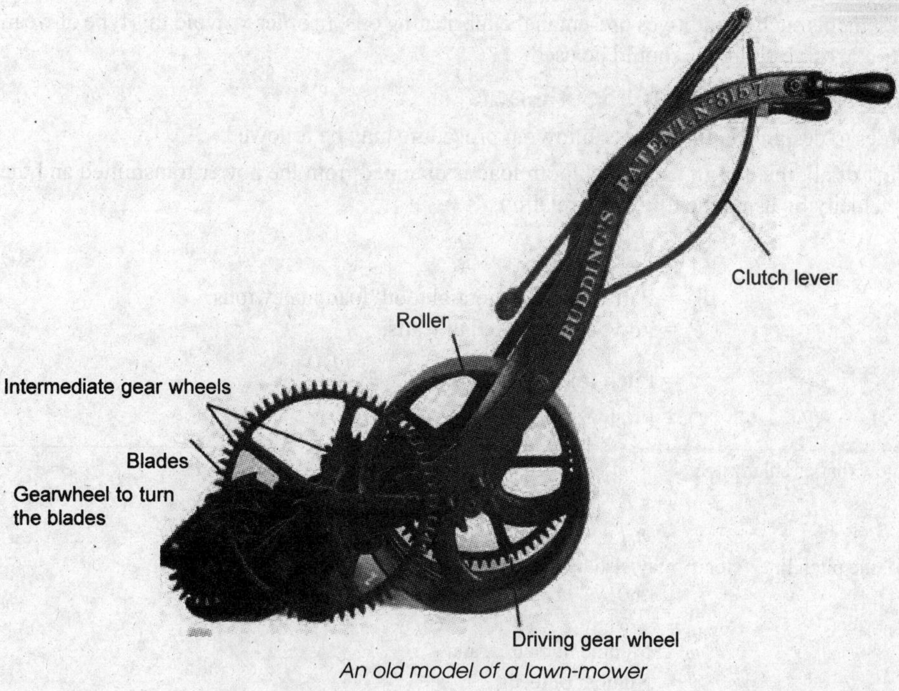

An old model of a lawn-mower

Notes : 1. The surface endurance limit for steel may be obtained from the following equation :

$$\sigma_{es} = (2.8 \times \text{B.H.N.} - 70) \text{ N/mm}^2$$

2. The maximum limiting wear load (W_w) must be greater than the dynamic load (W_D).

28.22 Causes of Gear Tooth Failure

The different modes of failure of gear teeth and their possible remedies to avoid the failure, are as follows :

1. ***Bending failure.*** Every gear tooth acts as a cantilever. If the total repetitive dynamic load acting on the gear tooth is greater than the beam strength of the gear tooth, then the gear tooth will fail in bending, *i.e.* the gear tooth will break.

In order to avoid such failure, the module and face width of the gear is adjusted so that the beam strength is greater than the dynamic load.

2. ***Pitting.*** It is the surface fatigue failure which occurs due to many repetition of Hertz contact stresses. The failure occurs when the surface contact stresses are higher than the endurance limit of the material. The failure starts with the formation of pits which continue to grow resulting in the rupture of the tooth surface.

In order to avoid the pitting, the dynamic load between the gear tooth should be less than the wear strength of the gear tooth.

3. ***Scoring.*** The excessive heat is generated when there is an excessive surface pressure, high speed or supply of lubricant fails. It is a stick-slip phenomenon in which alternate shearing and welding takes place rapidly at high spots.

This type of failure can be avoided by properly designing the parameters such as speed, pressure and proper flow of the lubricant, so that the temperature at the rubbing faces is within the permissible limits.

4. ***Abrasive wear.*** The foreign particles in the lubricants such as dirt, dust or burr enter between the tooth and damage the form of tooth. This type of failure can be avoided by providing filters for the lubricating oil or by using high viscosity lubricant oil which enables the formation of thicker oil film and hence permits easy passage of such particles without damaging the gear surface.

5. ***Corrosive wear.*** The corrosion of the tooth surfaces is mainly caused due to the presence of corrosive elements such as additives present in the lubricating oils. In order to avoid this type of wear, proper anti-corrosive additives should be used.

28.23 Design Procedure for Spur Gears

In order to design spur gears, the following procedure may be followed :

1. First of all, the design tangential tooth load is obtained from the power transmitted and the pitch line velocity by using the following relation :

$$W_T = \frac{P}{v} \times C_S \qquad ...(i)$$

where W_T = Permissible tangential tooth load in newtons,

P = Power transmitted in watts,

*v = Pitch line velocity in m / s $= \frac{\pi D N}{60}$,

D = Pitch circle diameter in metres,

---

* We know that circular pitch,

$$p_c = \pi D / T = \pi m \qquad ...(\because m = D / T)$$

$$\therefore \quad D = m.T$$

Thus, the pitch line velocity may also be obtained by using the following relation, *i.e.*

$$v = \frac{\pi D.N}{60} = \frac{\pi m.T.N}{60} = \frac{p_c.T.N}{60}$$

where m = Module in metres, and

T = Number of teeth.

N = Speed in r.p.m., and

C_S = Service factor.

The following table shows the values of service factor for different types of loads :

Table 28.10. Values of service factor.

Type of load	*Type of service*		
	Intermittent or 3 hours per day	*8-10 hours per day*	*Continuous 24 hours per day*
Steady	0.8	1.00	1.25
Light shock	1.00	1.25	1.54
Medium shock	1.25	1.54	1.80
Heavy shock	1.54	1.80	2.00

Note : The above values for service factor are for enclosed well lubricated gears. In case of non-enclosed and grease lubricated gears, the values given in the above table should be divided by 0.65.

2. Apply the Lewis equation as follows :

$$W_T = \sigma_w.b.p_c.y = \sigma_w.b.\pi\, m.y$$
$$= (\sigma_o.C_v)\, b.\pi\, m.y \qquad ...(\because \sigma_w = \sigma_o.C_v)$$

Notes : (*i*) The Lewis equation is applied only to the weaker of the two wheels (*i.e.* pinion or gear).

(*ii*) When both the pinion and the gear are made of the same material, then pinion is the weaker.

(*iii*) When the pinion and the gear are made of different materials, then the product of $(\sigma_w \times y)$ or $(\sigma_o \times y)$ is the *deciding factor. The Lewis equation is used to that wheel for which $(\sigma_w \times y)$ or $(\sigma_o \times y)$ is less.

A bicycle with changeable gears.

* We see from the Lewis equation that for a pair of mating gears, the quantities like W_T, b, m and C_v are constant. Therefore $(\sigma_w \times y)$ or $(\sigma_o \times y)$ is the only deciding factor.

(*iv*) The product ($\sigma_w \times y$) is called ***strength factor*** of the gear.

(*v*) The face width (*b*) may be taken as 3 p_c to 4 p_c (or 9.5 *m* to 12.5 *m*) for cut teeth and 2 p_c to 3 p_c (or 6.5 *m* to 9.5 *m*) for cast teeth.

3. Calculate the dynamic load (W_D) on the tooth by using Buckingham equation, *i.e.*

$$W_D = W_T + W_I$$

$$= W_T + \frac{21v\,(b.C + W_T)}{21v + \sqrt{b.C + W_T}}$$

In calculating the dynamic load (W_D), the value of tangential load (W_T) may be calculated by neglecting the service factor (C_S) *i.e.*

$W_T = P / v$, where P is in watts and v in m / s.

4. Find the static tooth load (*i.e.* beam strength or the endurance strength of the tooth) by using the relation,

$$W_S = \sigma_e.b.p_c.y = \sigma_e.b.\pi\, m.y$$

For safety against breakage, W_S should be greater than W_D.

5. Finally, find the wear tooth load by using the relation,

$$W_w = D_P.b.Q.K$$

The wear load (W_w) should not be less than the dynamic load (W_D).

Example 28.1. *The following particulars of a single reduction spur gear are given :*

Gear ratio = 10 : 1; Distance between centres = 660 mm approximately; Pinion transmits 500 kW at 1800 r.p.m.; Involute teeth of standard proportions (addendum = m) with pressure angle of 22.5°; Permissible normal pressure between teeth = 175 N per mm of width. Find :

1. The nearest standard module if no interference is to occur;

2. The number of teeth on each wheel;

3. The necessary width of the pinion; and

4. The load on the bearings of the wheels due to power transmitted.

Solution : Given : $G = T_G / T_P = D_G / D_P = 10$; $L = 660$ mm ; $P = 500$ kW $= 500 \times 10^3$ W ; $N_P = 1800$ r.p.m. ; $\phi = 22.5°$; $W_N = 175$ N/mm width

1. ***Nearest standard module if no interference is to occur***

Let m = Required module,

T_P = Number of teeth on the pinion,

T_G = Number of teeth on the gear,

D_P = Pitch circle diameter of the pinion, and

D_G = Pitch circle diameter of the gear.

We know that minimum number of teeth on the pinion in order to avoid interference,

$$T_P = \frac{2\,A_W}{G\left[\sqrt{1 + \frac{1}{G}\left(\frac{1}{G} + 2\right)\sin^2\phi} - 1\right]}$$

$$= \frac{2 \times 1}{10\left[\sqrt{1 + \frac{1}{10}\left(\frac{1}{10} + 2\right)\sin^2 22.5°} - 1\right]} = \frac{2}{0.15} = 13.3 \text{ say } 14$$

... ($\because A_W = 1$ module)

$\therefore \quad T_G = G \times T_P = 10 \times 14 = 140 \quad$ ...($\because T_G / T_P = 10$)

We know that $L = \frac{D_G}{2} + \frac{D_P}{2} = \frac{D_G}{2} + \frac{10\,D_P}{2} = 5.5\,D_P$ ...($\because D_G / D_P = 10$)

$\therefore$ $660 = 5.5\,D_P$ or $D_P = 660 / 5.5 = 120$ mm

We also know that $D_P = m \,.\, T_P$

$\therefore$ $m = D_P / T_P = 120 / 14 = 8.6$ mm

Since the nearest standard value of the module is 8 mm, therefore we shall take

$m = 8$ mm **Ans.**

2. ***Number of teeth on each wheel***

We know that number of teeth on the pinion,

$T_P = D_P / m = 120 / 8 = 15$ **Ans.**

and number of teeth on the gear,

$T_G = G \times T_P = 10 \times 15 = 150$ **Ans.**

3. ***Necessary width of the pinion***

We know that the torque acting on the pinion,

$$T = \frac{P \times 60}{2\pi N_P} = \frac{500 \times 10^3 \times 60}{2\pi \times 1800} = 2652 \text{ N-m}$$

$\therefore$ Tangential load, $W_T = \frac{T}{D_P / 2} = \frac{2652}{0.12/2} = 44\,200$ N ...($\because D_P$ is taken in metres)

and normal load on the tooth,

$$W_N = \frac{W_T}{\cos\phi} = \frac{44\,200}{\cos 22.5°} = 47\,840 \text{ N}$$

Since the normal pressure between teeth is 175 N per mm of width, therefore necessary width of the pinion,

$$b = \frac{47\,840}{175} = 273.4 \text{ mm } \textbf{Ans.}$$

4. ***Load on the bearings of the wheels***

We know that the radial load on the bearings due to the power transmitted,

$W_R = W_N \,.\, \sin\phi = 47\,840 \times \sin 22.5° = 18\,308$ N $= 18.308$ kN **Ans.**

Example 28.2. *A bronze spur pinion rotating at 600 r.p.m. drives a cast iron spur gear at a transmission ratio of 4 : 1. The allowable static stresses for the bronze pinion and cast iron gear are 84 MPa and 105 MPa respectively.*

The pinion has 16 standard 20° full depth involute teeth of module 8 mm. The face width of both the gears is 90 mm. Find the power that can be transmitted from the standpoint of strength.

Solution. Given : $N_P = 600$ r.p.m. ; $V.R. = T_G / T_P = 4$; $\sigma_{OP} = 84$ MPa $= 84$ N/mm^2 ; $\sigma_{OG} = 105$ MPa $= 105$ N/mm^2 ; $T_P = 16$; $m = 8$ mm ; $b = 90$ mm

We know that pitch circle diameter of the pinion,

$D_P = m.T_P = 8 \times 16 = 128$ mm $= 0.128$ m

$\therefore$ Pitch line velocity,

$$v = \frac{\pi D_P . N_P}{60} = \frac{\pi \times 0.128 \times 600}{60} = 4.02 \text{ m/s}$$

Since the pitch line velocity (v) is less than 12.5 m/s, therefore velocity factor,

$$C_v = \frac{3}{3 + v} = \frac{3}{3 + 4.02} = 0.427$$

We know that for 20° full depth involute teeth, tooth form factor for the pinion,

$$y_P = 0.154 - \frac{0.912}{T_P} = 0.154 - \frac{0.912}{16} = 0.097$$

and tooth form factor for the gear,

$$y_G = 0.154 - \frac{0.912}{T_G} = 0.154 - \frac{0.912}{4 \times 16} = 0.14 \qquad \ldots (\because T_G / T_P = 4)$$

$\therefore \qquad \sigma_{OP} \times y_P = 84 \times 0.097 = 8.148$

and $\qquad \sigma_{OG} \times y_G = 105 \times 0.14 = 14.7$

Since $(\sigma_{OP} \times y_P)$ is less than $(\sigma_{OG} \times y_G)$, therefore the pinion is weaker. Now using the Lewis equation for the pinion, we have tangential load on the tooth (or beam strength of the tooth),

$$W_T = \sigma_{wP} . b . \pi\, m . y_P = (\sigma_{OP} \times C_v)\, b . \pi\, m . y_P \qquad (\because \sigma_{WP} = \sigma_{OP} . C_v)$$

$$= 84 \times 0.427 \times 90 \times \pi \times 8 \times 0.097 = 7870 \text{ N}$$

$\therefore$ Power that can be transmitted

$$= W_T \times v = 7870 \times 4.02 = 31\,640 \text{ W} = 31.64 \text{ kW} \textbf{ Ans.}$$

Example 28.3. *A pair of straight teeth spur gears is to transmit 20 kW when the pinion rotates at 300 r.p.m. The velocity ratio is 1 : 3. The allowable static stresses for the pinion and gear materials are 120 MPa and 100 MPa respectively.*

The pinion has 15 teeth and its face width is 14 times the module. Determine : 1. module; 2. face width; and 3. pitch circle diameters of both the pinion and the gear from the standpoint of strength only, taking into consideration the effect of the dynamic loading.

The tooth form factor y can be taken as

$$y = 0.154 - \frac{0.912}{\text{No. of teeth}}$$

and the velocity factor C_v as

$$C_v = \frac{3}{3 + v}, \text{ where } v \text{ is expressed in m / s.}$$

Solution. Given : $P = 20$ kW $= 20 \times 10^3$ W ; $N_P = 300$ r.p.m. ; $V.R. = T_G / T_P = 3$; $\sigma_{OP} = 120$ MPa $= 120$ N/mm^2 ; $\sigma_{OG} = 100$ MPa $= 100$ N/mm^2 ; $T_P = 15$; $b = 14$ module $= 14\, m$

1. ***Module***

Let $\qquad m$ = Module in mm, and

$\qquad D_P$ = Pitch circle diameter of the pinion in mm.

We know that pitch line velocity,

$$v = \frac{\pi D_P N_P}{60} = \frac{\pi m . T_P . N_P}{60} \qquad \ldots (\because D_P = m . T_P)$$

$$= \frac{\pi m \times 15 \times 300}{60} = 236\, m \text{ mm/s} = 0.236\, m \text{ m/s}$$

Assuming steady load conditions and 8-10 hours of service per day, the service factor (C_S) from Table 28.10 is given by

$$C_S = 1$$

We know that design tangential tooth load,

$$W_T = \frac{P}{v} \times C_S = \frac{20 \times 10^3}{0.236\, m} \times 1 = \frac{84\,746}{m} \text{ N}$$

and velocity factor, $$C_v = \frac{3}{3 + v} = \frac{3}{3 + 0.236\, m}$$

We know that tooth form factor for the pinion,

$$y_P = 0.154 - \frac{0.912}{T_P} = 0.154 - \frac{0.912}{15}$$

$$= 0.154 - 0.0608 = 0.0932$$

and tooth form factor for the gear,

$$y_G = 0.154 - \frac{0.912}{T_G} = 0.154 - \frac{0.912}{3 \times 15}$$

$$= 0.154 - 0.203 = 0.1337 \quad \text{... } (\because T_G = 3T_P)$$

$$\therefore \quad \sigma_{OP} \times y_P = 120 \times 0.0932 = 11.184$$

and $$\sigma_{OG} \times y_G = 100 \times 0.1337 = 13.37$$

Since ($\sigma_{OP} \times y_P$) is less than ($\sigma_{OG} \times y_G$), therefore the pinion is weaker. Now using the Lewis equation to the pinion, we have

$$W_T = \sigma_{wP}.b.\pi\, m.y_P = (\sigma_{OP} \times C_v)\, b.\pi\, m\, .\, y_P$$

$$\therefore \quad \frac{84\,746}{m} = 120\left(\frac{3}{3 + 0.236\, m}\right) 14\, m \times \pi\, m \times 0.0932 = \frac{1476\, m^2}{3 + 0.236\, m}$$

or $$3 + 0.236\, m = 0.0174\, m^3$$

Solving this equation by hit and trial method, we find that

$$m = 6.4 \text{ mm}$$

The standard module is 8 mm. Therefore let us take

$$m = 8 \text{ mm } \textbf{Ans.}$$

2. ***Face width***

We know that the face width,

$$b = 14\, m = 14 \times 8 = 112 \text{ mm } \textbf{Ans.}$$

Kitchen Gear : This 1863 fruit and vegetable peeling machine uses a rack and pinion to drive spur gears that turn an apple against a cutting blade. As the handle is pushed round the semi-circular base, the peel is removed from the apple in a single sweep.

3. ***Pitch circle diameter of the pinion and gear***

We know that pitch circle diameter of the pinion,

$$D_P = m.T_P = 8 \times 15 = 120 \text{ mm } \textbf{Ans.}$$

and pitch circle diameter of the gear,

$$D_G = m.T_G = 8 \times 45 = 360 \text{ mm } \textbf{Ans.} \qquad \ldots (\because T_G = 3\,T_P)$$

Example 28.4. *A gear drive is required to transmit a maximum power of 22.5 kW. The velocity ratio is 1:2 and r.p.m. of the pinion is 200. The approximate centre distance between the shafts may be taken as 600 mm. The teeth has 20° stub involute profiles. The static stress for the gear material (which is cast iron) may be taken as 60 MPa and face width as 10 times the module. Find the module, face width and number of teeth on each gear.*

Check the design for dynamic and wear loads. The deformation or dynamic factor in the Buckingham equation may be taken as 80 and the material combination factor for the wear as 1.4.

Solution. Given : $P = 22.5$ kW = 22 500 W ; $V.R. = D_G/D_P = 2$; $N_P = 200$ r.p.m. ; $L = 600$ mm ; $\sigma_{OP} = \sigma_{OG} = 60$ MPa = 60 N/mm^2 ; $b = 10\,m$; $C = 80$; $K = 1.4$

Module

Let m = Module in mm,

D_P = Pitch circle diameter of the pinion, and

D_G = Pitch circle diameter of the gear.

We know that centre distance between the shafts (L),

$$600 = \frac{D_P}{2} + \frac{D_G}{2} = \frac{D_P}{2} + \frac{2\,D_P}{2} = 1.5\,D_P \qquad \ldots (\because D_G = V.R. \times D_P)$$

Arm of a material handler In addition to gears, hydraulic rams as shown above, play important role in transmitting force and energy.

$\therefore \qquad D_P = 600 / 1.5 = 400 \text{ mm} = 0.4 \text{ m}$

and $\qquad D_G = 2\,D_P = 2 \times 400 = 800 \text{ mm} = 0.8 \text{ m}$

Since both the gears are made of the same material, therefore pinion is the weaker. Thus the design will be based upon the pinion.

We know that pitch line velocity of the pinion,

$$v = \frac{\pi D_P \,.\, N_P}{60} = \frac{\pi \times 0.4 \times 200}{60} = 4.2 \text{ m / s}$$

Since v is less than 12 m / s, therefore velocity factor,

$$C_v = \frac{3}{3 + v} = \frac{3}{3 + 4.2} = 0.417$$

We know that number of teeth on the pinion,

$$T_P = D_P / m = 400 / m$$

$\therefore$ Tooth form factor for the pinion,

$$y_P = 0.175 - \frac{0.841}{T_P} = 0.175 - \frac{0.841 \times m}{400} \qquad \text{... (For 20° stub system)}$$

$$= 0.175 - 0.0021\,m \qquad ...(i)$$

Assuming steady load conditions and 8–10 hours of service per day, the service factor (C_S) from Table 28.10 is given by

$$C_S = 1$$

We know that design tangential tooth load,

$$W_T = \frac{P}{v} \times C_S = \frac{22\ 500}{4.2} \times 1 = 5357 \text{ N}$$

We also know that tangential tooth load (W_T),

$$5357 = \sigma_{wP} \,.\, b.\, \pi\, m.y_P = (\sigma_{OP} \times C_v)\, b.\pi\, m.y_P$$

$$= (60 \times 0.417)\ 10\,m \times \pi\, m\ (0.175 - 0.0021\,m)$$

$$= 137.6\,m^2 - 1.65\,m^3$$

Solving this equation by hit and trial method, we find that

$$m = 0.65 \text{ say } 8 \text{ mm } \textbf{Ans.}$$

Face width

We know that face width,

$$b = 10\,m = 10 \times 8 = 80 \text{ mm } \textbf{Ans.}$$

Number of teeth on the gears

We know that number of teeth on the pinion,

$$T_P = D_P / m = 400 / 8 = 50 \textbf{ Ans.}$$

and number of teeth on the gear,

$$T_G = D_G / m = 800 / 8 = 100 \textbf{ Ans.}$$

Checking the gears for dynamic and wear load

We know that the dynamic load,

$$W_D = W_T + \frac{21v\,(b.C + W_T)}{21v + \sqrt{b.C + W_T}}$$

$$= 5357 + \frac{21 \times 4.2\ (80 \times 80 + 5357)}{21 \times 4.2 + \sqrt{80 \times 80 + 5357}}$$

$$= 5357 + \frac{1.037 \times 10^6}{196.63} = 5357 + 5273 = 10\ 630 \text{ N}$$

From equation (*i*), we find that tooth form factor for the pinion,

$$y_P = 0.175 - 0.0021\ m = 0.175 - 0.0021 \times 8 = 0.1582$$

From Table 28.8, we find that flexural endurance limit (σ_e) for cast iron is 84 MPa or 84 N/mm².

∴ Static tooth load or endurance strength of the tooth,

$$W_S = \sigma_e \,.\, b.\, \pi\, m.\, y_P = 84 \times 80 \times \pi \times 8 \times 0.1582 = 26\ 722 \text{ N}$$

We know that ratio factor,

$$Q = \frac{2 \times V.R.}{V.R. + 1} = \frac{2 \times 2}{2 + 1} = 1.33$$

∴ Maximum or limiting load for wear,

$$W_w = D_P \,.\, b.\, Q \,.\, K = 400 \times 80 \times 1.33 \times 1.4 = 59\ 584 \text{ N}$$

Since both W_S and W_w are greater than W_D, therefore the design is safe.

Example 28.5. *A pair of straight teeth spur gears, having 20° involute full depth teeth is to transmit 12 kW at 300 r.p.m. of the pinion. The speed ratio is 3 : 1. The allowable static stresses for gear of cast iron and pinion of steel are 60 MPa and 105 MPa respectively. Assume the following:*

Number of teeth of pinion = 16; Face width = 14 times module; Velocity factor $(C_v) = \dfrac{4.5}{4.5 + v}$, *v being the pitch line velocity in m / s; and tooth form factor (y)* $= 0.154 - \dfrac{0.912}{\text{No. of teeth}}$

Determine the module, face width and pitch diameter of gears. Check the gears for wear; given $\sigma_{es} = 600$ *MPa;* $E_P = 200$ *kN/mm² and* $E_G = 100$ *kN/mm². Sketch the gears.*

Solution : Given : $\phi = 20°$; $P = 12$ kW $= 12 \times 10^3$ W ; $N_P = 300$ r.p.m ; $V.R. = T_G / T_P = 3$; $\sigma_{OG} = 60$ MPa $= 60$ N/mm² ; $\sigma_{OP} = 105$ MPa $= 105$ N/mm² ; $T_P = 16$; $b = 14$ module $= 14\, m$; $\sigma_{es} = 600$ MPa $= 600$ N/mm² ; $E_P = 200$ kN/mm² $= 200 \times 10^3$ N/mm² ; $E_G = 100$ kN/mm² $= 100 \times 10^3$ N/mm²

Module

Let m = Module in mm, and

D_P = Pitch circle diameter of the pinion in mm.

We know that pitch line velocity,

$$v = \frac{\pi D_P \,.\, N_P}{60} = \frac{\pi m \,.\, T_P \,.\, N_P}{60} \qquad \ldots (\because D_P = m.T_P)$$

$$= \frac{\pi m \times 16 \times 300}{60} = 251\ m \text{ mm/s} = 0.251\ m \text{ m/s}$$

Assuming steady load conditions and 8–10 hours of service per day, the service factor (C_S) from Table 28.10 is given by $C_S = 1$.

We know that the design tangential tooth load,

$$W_T = \frac{P}{v} \times C_S = \frac{12 \times 10^3}{0.251\ m} \times 1 = \frac{47.8 \times 10^3}{m} \text{ N}$$

and velocity factor,

$$C_v = \frac{4.5}{4.5 + v} = \frac{4.5}{4.5 + 0.251\ m}$$

We know that tooth form factor for pinion,

$$y_P = 0.154 - \frac{0.912}{T_P} = 0.154 - \frac{0.912}{16} = 0.097$$

and tooth form factor for gear,

$$y_G = 0.154 - \frac{0.912}{T_G} = 0.154 - \frac{0.912}{3 \times 16} = 0.135 \quad \text{... } (\because T_G = 3\,T_P)$$

$$\therefore \quad \sigma_{OP} \times y_P = 105 \times 0.097 = 10.185$$

and $$\sigma_{OG} \times y_G = 60 \times 0.135 = 8.1$$

Since $(\sigma_{OG} \times y_G)$ is less than $(\sigma_{OP} \times y_P)$, therefore the gear is weaker. Now using the Lewis equation to the gear, we have

$$W_T = \sigma_{wG} \cdot b.\pi\, m.y_G = (\sigma_{OG} \times C_v)\, b.\pi\, m.y_G \quad \text{... } (\because \sigma_{wG} = \sigma_{OG}.C_v)$$

$$\frac{47.8 \times 10^3}{m} = 60 \left(\frac{4.5}{4.5 + 0.251\, m} \right) 14\, m \times \pi\, m \times 0.135 = \frac{1603.4\, m^2}{4.5 + 0.251\, m}$$

or $$4.5 + 0.251\, m = 0.0335\, m^3$$

Solving this equation by hit and trial method, we find that

$$m = 5.6 \text{ say } 6 \text{ mm } \textbf{Ans.}$$

Face width

We know that face width,

$$b = 14\, m = 14 \times 6 = 84 \text{ mm } \textbf{Ans.}$$

Pitch diameter of gears

We know that pitch diameter of the pinion,

$$D_P = m.T_P = 6 \times 16 = 96 \text{ mm } \textbf{Ans.}$$

and pitch diameter of the gear,

$$D_G = m.T_G = 6 \times 48 = 288 \text{ mm } \textbf{Ans.} \quad \text{...} (\because T_G = 3\,T_P)$$

This is a close-up photo (magnified 200 times) of a micromotor's gear cogs. Micromotors have been developed for use in space missions and microsurgery.

Checking the gears for wear

We know that the ratio factor,

$$Q = \frac{2 \times V.R.}{V.R. + 1} = \frac{2 \times 3}{3 + 1} = 1.5$$

and load stress factor,

$$K = \frac{(\sigma_{es})^2 \sin \phi}{1.4}\left(\frac{1}{E_P} + \frac{1}{E_G}\right)$$

$$= \frac{(600)^2 \sin 20°}{1.4}\left[\frac{1}{200 \times 10^3} + \frac{1}{100 \times 10^3}\right]$$

$$= 0.44 + 0.88 = 1.32 \text{ N/mm}^2$$

We know that the maximum or limiting load for wear,

$$W_w = D_P.b.Q.K = 96 \times 84 \times 1.5 \times 1.32 = 15\,967 \text{ N}$$

and tangential load on the tooth (or beam strength of the tooth),

$$W_T = \frac{47.8 \times 10^3}{m} = \frac{47.8 \times 10^3}{6} = 7967 \text{ N}$$

Since the maximum wear load is much more than the tangential load on the tooth, therefore the design is satisfactory from the standpoint of wear. **Ans.**

Example 28.6. *A reciprocating compressor is to be connected to an electric motor with the help of spur gears. The distance between the shafts is to be 500 mm. The speed of the electric motor is 900 r.p.m. and the speed of the compressor shaft is desired to be 200 r.p.m. The torque, to be transmitted is 5000 N-m. Taking starting torque as 25% more than the normal torque, determine : 1. Module and face width of the gears using 20 degrees stub teeth, and 2. Number of teeth and pitch circle diameter of each gear. Assume suitable values of velocity factor and Lewis factor.*

Solution. Given : $L = 500$ mm ; $N_M = 900$ r.p.m. ; $N_C = 200$ r.p.m. ; $T = 5000$ N-m ; $T_{max} = 1.25\ T$

1. *Module and face width of the gears*

Let m = Module in mm, and

b = Face width in mm.

Since the starting torque is 25% more than the normal torque, therefore the maximum torque,

$$T_{max} = 1.25\ T = 1.25 \times 5000 = 6250 \text{ N-m} = 6250 \times 10^3 \text{ N-mm}$$

We know that velocity ratio,

$$V.R. = \frac{N_M}{N_C} = \frac{900}{200} = 4.5$$

Let D_P = Pitch circle diameter of the pinion on the motor shaft, and

D_G = Pitch circle diameter of the gear on the compressor shaft.

We know that distance between the shafts (L),

$$500 = \frac{D_P}{2} + \frac{D_G}{2} \quad \text{or} \quad D_P + D_G = 500 \times 2 = 1000 \qquad ...(i)$$

and velocity ratio,

$$V.R. = \frac{D_G}{D_P} = 4.5 \quad \text{or} \quad D_G = 4.5\ D_P \qquad ...(ii)$$

Substituting the value of D_G in equation (*i*), we have

$$D_P + 4.5\ D_P = 1000 \quad \text{or} \quad D_P = 1000 / 5.5 = 182 \text{ mm}$$

and

$$D_G = 4.5\ D_P = 4.5 \times 182 = 820 \text{ mm} = 0.82 \text{ m}$$

We know that pitch line velocity of the drive,

$$v = \frac{\pi D_G . N_C}{60} = \frac{\pi \times 0.82 \times 200}{60} = 8.6 \text{ m/s}$$

∴ Velocity factor,

$$C_v = \frac{3}{3+v} = \frac{3}{3+8.6} = 0.26 \qquad ...(\because v \text{ is less than 12.5 m/s})$$

Let us assume than motor pinion is made of forged steel and the compressor gear of cast steel. Since the allowable static stress for the cast steel is less than the forged steel, therefore the design should be based upon the gear. Let us take the allowable static stress for the gear material as

$$\sigma_{OG} = 140 \text{ MPa} = 140 \text{ N/mm}^2$$

We know that for 20° stub teeth, Lewis factor for the gear,

$$y_G = 0.175 - \frac{0.841}{T_G} = 0.175 - \frac{0.841 \times m}{D_G} \qquad ...\left(\because T_G = \frac{D_G}{m}\right)$$

$$= 0.175 - \frac{0.841\, m}{820} = 0.175 - 0.001\, m$$

and maximum tangential force on the gear,

$$W_T = \frac{2\, T_{max}}{D_G} = \frac{2 \times 6250 \times 10^3}{820} = 15\,244 \text{ N}$$

We also know that maximum tangential force on the gear,

$$W_T = \sigma_{wG}.b.\pi\, m.y_G = (\sigma_{OG} \times C_v)\, b \times \pi\, m \times y_G \qquad ...(\because \sigma_{wG} = \sigma_{OG}.\, C_v)$$

$$15\,244 = (140 \times 0.26) \times 10\, m \times \pi\, m\, (0.175 - 0.001\, m)$$

$$= 200\, m^2 - 1.144\, m^3 \qquad ...(\text{Assuming } b = 10\, m)$$

Solving this equation by hit and trial method, we find that

$$m = 8.95 \text{ say } 10 \text{ mm } \textbf{Ans.}$$

and $b = 10\, m = 10 \times 10 = 100$ mm **Ans.**

2. ***Number of teeth and pitch circle diameter of each gear***

We know that number of teeth on the pinion,

$$T_P = \frac{D_P}{m} = \frac{182}{10} = 18.2$$

$$T_G = \frac{D_G}{m} = \frac{820}{10} = 82$$

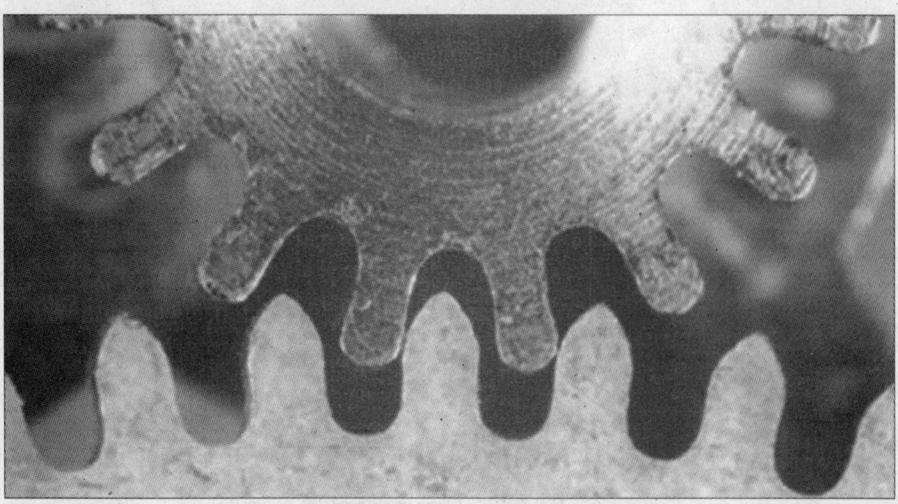

In order to have the exact velocity ratio of 4.5, we shall take

$$T_P = 18 \text{ and } T_G = 81 \textbf{ Ans.}$$

∴ Pitch circle diameter of the pinion,

$$D_P = m \times T_P = 10 \times 18 = 180 \text{ mm } \textbf{Ans.}$$

and pitch circle diameter of the gear,

$$D_G = m \times T_G = 10 \times 81 = 810 \text{ mm } \textbf{Ans.}$$

28.24 Spur Gear Construction

The gear construction may have different designs depending upon the size and its application. When the dedendum circle diameter is slightly greater than the shaft diameter, then the pinion teeth are cut integral with the shaft as shown in Fig. 28.13 (*a*). If the pitch circle diameter of the pinion is less than or equal to 14.75 m + 60 mm (where m is the module in mm), then the pinion is made solid with uniform thickness equal to the face width, as shown in Fig. 28.13 (*b*). Small gears upto 250 mm pitch circle diameter are built with a web, which joins the hub and the rim. The web thickness is generally equal to half the circular pitch or it may be taken as 1.6 m to 1.9 m, where m is the module. The web may be made solid as shown in Fig. 28.13 (*c*) or may have recesses in order to reduce its weight.

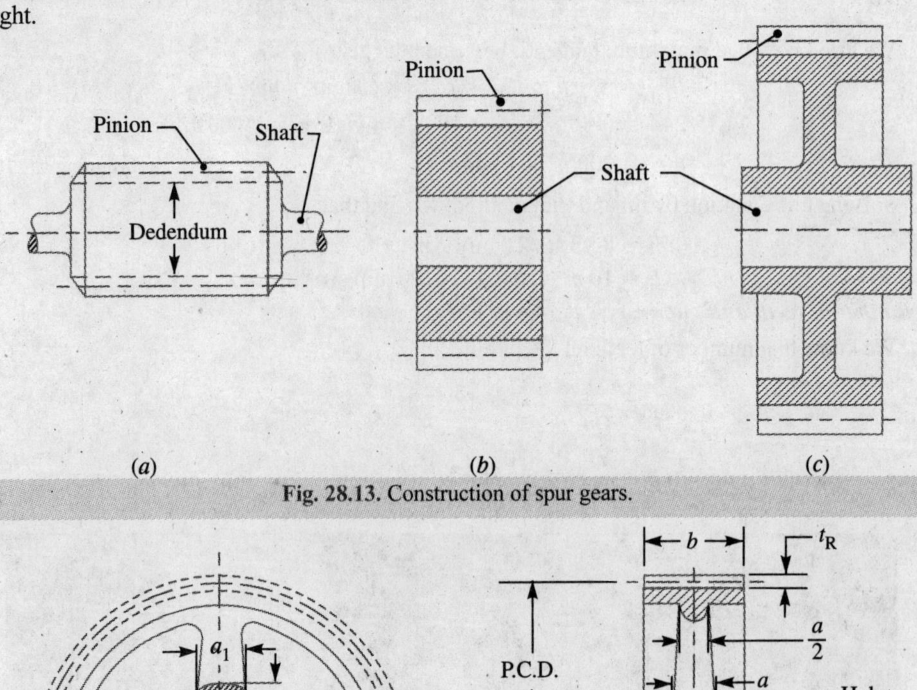

Fig. 28.13. Construction of spur gears.

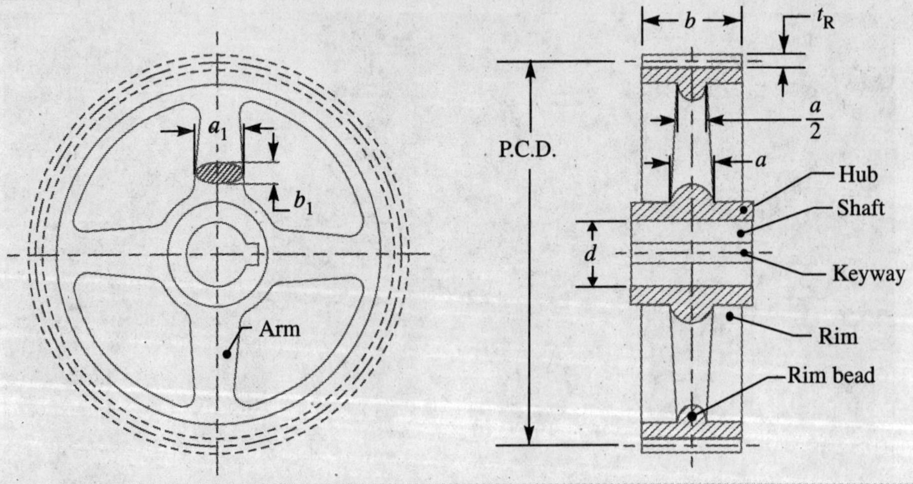

Fig. 28.14. Gear with arms.

Large gears are provided with arms to join the hub and the rim, as shown in Fig. 28.14. The number of arms depends upon the pitch circle diameter of the gear. The number of arms may be selected from the following table.

Table 28.11. Number of arms for the gears.

S. No.	*Pitch circle diameter*	*Number of arms*
1.	Up to 0.5 m	4 or 5
2.	0.5 – 1.5 m	6
3.	1.5 – 2.0 m	8
4.	Above 2.0 m	10

The cross-section of the arms is most often elliptical, but other sections as shown in Fig. 28.15 may also be used.

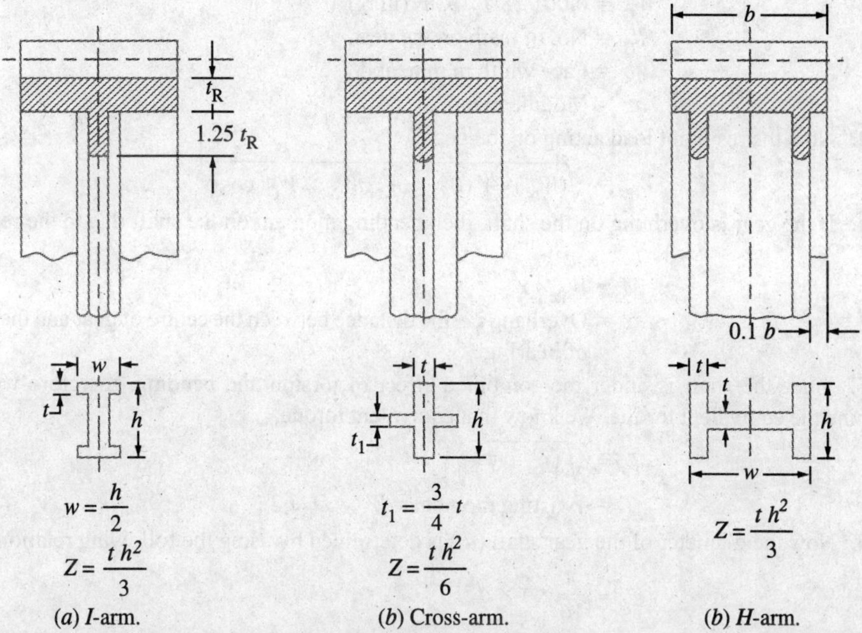

Fig. 28.15. Cross-section of the arms.

The hub diameter is kept as 1.8 times the shaft diameter for steel gears, twice the shaft diameter for cast iron gears and 1.65 times the shaft diameter for forged steel gears used for light service. The length of the hub is kept as 1.25 times the shaft diameter for light service and should not be less than the face width of the gear.

The thickness of the gear rim should be as small as possible, but to facilitate casting and to avoid sharp changes of section, the minimum thickness of the rim is generally kept as half of the circular pitch (or it may be taken as 1.6 *m* to 1.9 *m*, where *m* is the module). The thickness of rim (t_R) may also be calculated by using the following relation, *i.e.*

$$t_R = m\sqrt{\frac{T}{n}}$$

where T = Number of teeth, and

n = Number of arms.

The rim should be provided with a circumferential rib of thickness equal to the rim thickness.

28.25 Design of Shaft for Spur Gears

In order to find the diameter of shaft for spur gears, the following procedure may be followed.

1. First of all, find the normal load (W_N), acting between the tooth surfaces. It is given by

$$W_N = W_T / \cos\phi$$

where W_T = Tangential load, and

ϕ = Pressure angle.

A thrust parallel and equal to W_N will act at the gear centre as shown in Fig. 28.16.

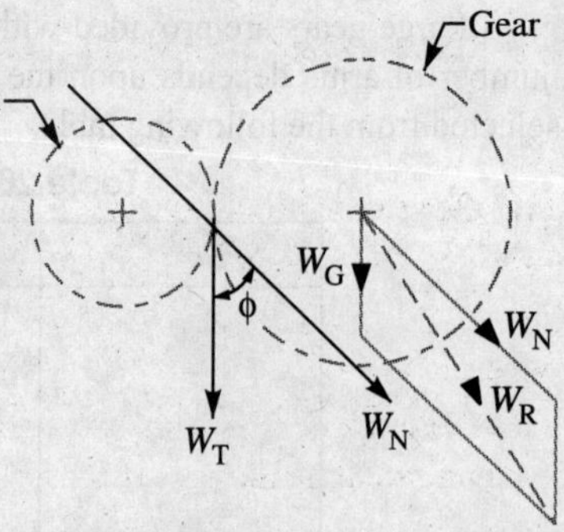

Fig. 28.16. Load acting on the gear.

2. The weight of the gear is given by

$$W_G = 0.001\,18\, T_G . b . m^2 \text{ (in N)}$$

where T_G = No. of teeth on the gear,

b = Face width in mm, and

m = Module in mm.

3. Now the resultant load acting on the gear,

$$W_R = \sqrt{(W_N)^2 + (W_G)^2 + 2\, W_N \times W_G \cos\phi}$$

4. If the gear is overhung on the shaft, then bending moment on the shaft due to the resultant load,

$$M = W_R \times x$$

where x = Overhang *i.e.* the distance between the centre of gear and the centre of bearing.

5. Since the shaft is under the combined effect of torsion and bending, therefore we shall determine the equivalent torque. We know that equivalent torque,

$$T_e = \sqrt{M^2 + T^2}$$

where T = Twisting moment = $W_T \times D_G / 2$

6. Now the diameter of the gear shaft (d) is determined by using the following relation, *i.e.*

$$T_e = \frac{\pi}{16} \times \tau \times d^3$$

where τ = Shear stress for the material of the gear shaft.

Note : Proceeding in the similar way as discussed above, we may calculate the diameter of the pinion shaft.

28.26 Design of Arms for Spur Gears

The cross-section of the arms is calculated by assuming them as a cantilever beam fixed at the hub and loaded at the pitch circle. It is also assumed that the load is equally distributed to all the arms. It may be noted that the arms are designed for the stalling load. The ***stalling load*** is a load that will develop the maximum stress in the arms and in the teeth. This happens at zero velocity, when the drive just starts operating.

The stalling load may be taken as the design tangential load divided by the velocity factor.

Let W_S = Stalling load = $\dfrac{\text{Design tangential load}}{\text{Velocity factor}} = \dfrac{W_T}{C_v}$,

D_G = Pitch circle diameter of the gear,

n = Number of arms, and

σ_b = Allowable bending stress for the material of the arms.

Now, maximum bending moment on each arm,

$$M = \frac{W_S \times D_G / 2}{n} = \frac{W_S \times D_G}{2n}$$

and the section modulus of arms for elliptical cross-section,

$$Z = \frac{\pi\,(a_1)^2\, b_1}{32}$$

where a_1 = Major axis, and b_1 = Minor axis.

The major axis is usually taken as twice the minor axis. Now, using the relation, $\sigma_b = M / Z$, we can calculate the dimensions a_1 and b_1 for the gear arm at the hub end.

Note : The arms are usually tapered towards the rim about 1/16 per unit length of the arm (or radius of the gear).

∴ Major axis of the section at the rim end

$$= a_1 - \text{Taper} = a_1 - \frac{1}{16} \times \text{Length of the arm} = a_1 - \frac{1}{16} \times \frac{D_G}{2} = a_1 - \frac{D_G}{32}$$

Example 28.7. *A motor shaft rotating at 1500 r.p.m. has to transmit 15 kW to a low speed shaft with a speed reduction of 3:1. The teeth are $14\frac{1}{2}°$ involute with 25 teeth on the pinion. Both the pinion and gear are made of steel with a maximum safe stress of 200 MPa. A safe stress of 40 MPa may be taken for the shaft on which the gear is mounted and for the key.*

Design a spur gear drive to suit the above conditions. Also sketch the spur gear drive. Assume starting torque to be 25% higher than the running torque.

Solution : Given : $N_P = 1500$ r.p.m. ; $P = 15$ kW $= 15 \times 10^3$ W ; $V.R. = T_G/T_P = 3$; $\phi = 14\frac{1}{2}°$; $T_P = 25$; $\sigma_{OP} = \sigma_{OG} = 200$ MPa $= 200$ N/mm² ; $\tau = 40$ MPa $= 40$ N/mm²

Design for spur gears

Since the starting torque is 25% higher than the running torque, therefore the spur gears should be designed for power,

$$P_1 = 1.25\,P = 1.25 \times 15 \times 10^3 = 18\,750 \text{ W}$$

We know that the gear reduction ratio (T_G / T_P) is 3. Therefore the number of teeth on the gear,

$$T_G = 3\,T_P = 3 \times 25 = 75$$

Let us assume that the module (m) for the pinion and gear is 6 mm.

∴ Pitch circle diameter of the pinion,

$$D_P = m.T_P = 6 \times 25 = 150 \text{ mm} = 0.15 \text{ m}$$

and pitch circle diameter of the gear,

$$D_G = m.\,T_G = 6 \times 75 = 450 \text{ mm}$$

We know that pitch line velocity,

$$v = \frac{\pi D_P . N_P}{60} = \frac{\pi \times 0.15 \times 1500}{60} = 11.8 \text{ m/s}$$

Assuming steady load conditions and 8–10 hours of service per day, the service factor (C_S) from Table 28.10 is given by

$$C_S = 1$$

∴ Design tangential tooth load,

$$W_T = \frac{P_1}{v} \times C_S = \frac{18\,750}{11.8} \times 1 = 1590 \text{ N}$$

We know that for ordinary cut gears and operating at velocities upto 12.5 m/s, the velocity factor,

$$C_v = \frac{3}{3 + v} = \frac{3}{3 + 11.8} = 0.203$$

Since both the pinion and the gear are made of the same material, therefore the pinion is the weaker.

We know that for $14\frac{1}{2}°$ involute teeth, tooth form factor for the pinion,

$$y_P = 0.124 - \frac{0.684}{T_P} = 0.124 - \frac{0.684}{25} = 0.0966$$

Let b =Face width for both the pinion and gear.

We know that the design tangential tooth load (W_T),

$$1590 = \sigma_{wP}.b.\pi\, m.y_P = (\sigma_{OP}.C_v)\, b.\pi\, m.y_P$$

$$= (200 \times 0.203)\, b \times \pi \times 6 \times 0.0966 = 74\, b$$

$\therefore\ b = 1590 / 74 = 21.5$ mm

In actual practice, the face width (b) is taken as 9.5 m to 12.5 m, but in certain cases, due to space limitations, it may be taken as 6 m. Therefore let us take the face width,

$$b = 6\, m = 6 \times 6 = 36 \text{ mm } \textbf{Ans.}$$

This mathematical machine called difference engine, assembled in 1832, used 2,000 levers, cams and gears.

From Table 28.1, the other proportions, for the pinion and the gear having $14\frac{1}{2}°$ involute teeth, are as follows :

Addendum	=	$1\, m = 6$ mm **Ans.**
Dedendum	=	$1.25\, m = 1.25 \times 6 = 7.5$ mm **Ans.**
Working depth	=	$2\, m = 2 \times 6 = 12$ mm **Ans.**
Minimum total depth	=	$2.25\, m = 2.25 \times 6 = 13.5$ mm **Ans.**
Tooth thickness	=	$1.5708\, m = 1.5708 \times 6 = 9.4248$ mm **Ans.**
Minimum clearance	=	$0.25\, m = 0.25 \times 6 = 1.5$ mm **Ans.**

Design for the pinion shaft

We know that the normal load acting between the tooth surfaces,

$$W_N = \frac{W_T}{\cos\phi} = \frac{1590}{\cos 14^1/_2°} = \frac{1590}{0.9681} = 1643 \text{ N}$$

and weight of the pinion,

$$W_P = 0.00118\, T_P.b.m^2 = 0.001\,18 \times 25 \times 36 \times 6^2 = 38 \text{ N}$$

$\therefore$ Resultant load acting on the pinion,

$$*W_R = \sqrt{(W_N)^2 + (W_P)^2 + 2W_N.W_P.\cos\phi}$$

$$= \sqrt{(1643)^2 + (38)^2 + 2 \times 1643 \times 38 \times \cos 14^1/_2°} = 1680 \text{ N}$$

Assuming that the pinion is overhung on the shaft and taking overhang as 100 mm, therefore

Bending moment on the shaft due to the resultant load,

$$M = W_R \times 100 = 1680 \times 100 = 168\,000 \text{ N-mm}$$

* Since the weight of the pinion (W_P) is very small as compared to the normal load (W_N), therefore it may be neglected. Thus the resultant load acting on the pinion (W_R) may be taken equal to W_N.

and twisting moment on the shaft,

$$T = W_T \times \frac{D_P}{2} = 1590 \times \frac{150}{2} = 119\ 250 \text{ N-mm}$$

∴ Equivalent twisting moment,

$$T_e = \sqrt{M^2 + T^2} = \sqrt{(168\ 000)^2 + (119\ 250)^2} = 206 \times 10^3 \text{ N-mm}$$

Let d_P = Diameter of the pinion shaft.

We know that equivalent twisting moment (T_e),

$$206 \times 10^3 = \frac{\pi}{16} \times \tau\,(d_P)^3 = \frac{\pi}{16} \times 40\,(d_P)^3 = 7.855\,(d_P)^3$$

∴ $(d_P)^3 = 206 \times 10^3 / 7.855 = 26.2 \times 10^3$ or $d_P = 29.7$ say 30 mm **Ans.**

We know that the diameter of the pinion hub

$= 1.8\, d_P = 1.8 \times 30 = 54$ mm **Ans.**

and length of the hub $= 1.25\, d_P = 1.25 \times 30 = 37.5$ mm

Since the length of the hub should not be less than that of the face width *i.e.* 36 mm, therefore let us take length of the hub as 36 mm. **Ans.**

Note : Since the pitch circle diameter of the pinion is 150 mm, therefore the pinion should be provided with a web and not arms. Let us take thickness of the web as 1.8 *m*, where *m* is the module.

∴ Thickness of the web $= 1.8\, m = 1.8 \times 6 = 10.8$ mm **Ans.**

Design for the gear shaft

We have calculated above that the normal load acting between the tooth surfaces,

$$W_N = 1643 \text{ N}$$

We know that weight of the gear,

$$W_G = 0.001\ 18\ T_G.b.m^2 = 0.001\ 18 \times 75 \times 36 \times 6^2 = 115 \text{ N}$$

∴ Resulting load acting on the gear,

$$W_R = \sqrt{(W_N)^2 + (W_G)^2 + 2W_N \times W_G \cos\phi}$$

$$= \sqrt{(1643)^2 + (115)^2 + 2 \times 1643 \times 115 \cos 14\tfrac{1}{2}^\circ} = 1755 \text{ N}$$

Assuming that the gear is overhung on the shaft and taking the overhang as 100 mm, therefore bending moment on the shaft due to the resultant load,

$$M = W_R \times 100 = 1755 \times 100 = 175\ 500 \text{ N-mm}$$

and twisting moment on the shaft,

$$T = W_T \times \frac{D_G}{2} = 1590 \times \frac{450}{2} = 357\ 750 \text{ N-mm}$$

∴ Equivalent twisting moment,

$$T_e = \sqrt{M^2 + T^2} = \sqrt{(175\ 500)^2 + (357\ 750)^2} = 398 \times 10^3 \text{ N-mm}$$

Let d_G = Diameter of the gear shaft.

We know that equivalent twisting moment (T_e),

$$398 \times 10^3 = \frac{\pi}{16} \times \tau\,(d_G)^3 = \frac{\pi}{16} \times 40\,(d_G)^3 = 7.855\,(d_G)^3$$

∴ $(d_G)^3 = 398 \times 10^3 / 7.855 = 50.7 \times 10^3$ or $d_G = 37$ say 40 mm **Ans.**

We know that diameter of the gear hub

$= 1.8\, d_G = 1.8 \times 40 = 72$ mm **Ans.**

and length of the hub $= 1.25\, d_G = 1.25 \times 40 = 50$ mm **Ans.**

Design for the gear arms

Since the pitch circle diameter of the gear is 450 mm, therefore the gear should be provided with four arms. Let us assume the cross-section of the arms as elliptical with major axis (a_1) equal to twice the minor axis (b_1).

∴ Section modulus of arms,

$$Z = \frac{\pi (a_1)^2 b_1}{32} = \frac{\pi (a_1)^2}{32} \times \frac{a_1}{2} = 0.05\,(a_1)^3 \qquad \ldots (\because b_1 = a_1/2)$$

Since the arms are designed for the stalling load and stalling load is taken as the design tangential load divided by the velocity factor, therefore stalling load,

$$W_S = \frac{W_T}{C_v} = \frac{1590}{0.203} = 7830 \text{ N} \qquad \ldots (\because C_v = 0.203)$$

∴ Maximum bending moment on each arm,

$$M = \frac{W_S}{n} \times \frac{D_G}{2} = \frac{7830}{4} \times \frac{450}{2} = 440\,440 \text{ N-mm}$$

We know that bending stress (σ_b),

$$42 = \frac{M}{Z} = \frac{440\,440}{0.05\,(a_1)^3} = \frac{9 \times 10^6}{(a_1)^3} \qquad \ldots (\text{Taking } \sigma_b = 42 \text{ N/mm}^2)$$

∴ $(a_1)^3 = 9 \times 10^6 / 42 = 0.214 \times 10^6$ or $a_1 = 60$ mm **Ans.**

and $b_1 = a_1 / 2 = 60 / 2 = 30$ mm **Ans.**

These dimensions refer to the hub end. Since the arms are tapered towards the rim and the taper is 1 / 16 per unit length of the arm (or radius of the gear), therefore

Major axis of the arm at the rim end,

$$a_2 = a_1 - \text{Taper} = a_1 - \frac{1}{16} \times \frac{D_G}{2}$$

$$= 60 - \frac{1}{16} \times \frac{450}{2} = 46 \text{ mm } \textbf{Ans.}$$

and minor axis of the arm at the rim end,

$$b_2 = \frac{\text{Major axis}}{2} = \frac{46}{2} = 23 \text{ mm } \textbf{Ans.}$$

Design for the rim

The thickness of the rim for the pinion (t_{RP}) may be taken as 1.6 *m* to 1.9 *m*, where *m* is the module. Let us take thickness of the rim for the pinion,

$$t_{RP} = 1.6\, m = 1.6 \times 6 = 9.6 \text{ say } 10 \text{ mm } \textbf{Ans.}$$

The thickness of the rim for the gear (t_{RG}) may be obtained by using the relation,

$$t_{RG} = m \sqrt{\frac{T_G}{n}} = 6 \sqrt{\frac{45}{4}} = 20 \text{ mm } \textbf{Ans.}$$

EXERCISES

1. Calculate the power that can be transmitted safely by a pair of spur gears with the data given below. Calculate also the bending stresses induced in the two wheels when the pair transmits this power.

   Number of teeth in the pinion = 20

   Number of teeth in the gear = 80

Module	= 4 mm
Width of teeth	= 60 mm
Tooth profile	= 20° involute
Allowable bending strength of the material	
	= 200 MPa, for pinion
	= 160 MPa, for gear
Speed of the pinion	= 400 r.p.m.
Service factor	= 0.8
Lewis form factor	$= 0.154 - \frac{0.912}{T}$
Velocity factor	$= \frac{3}{3+v}$

[Ans. 13.978 kW ; 102.4 MPa ; 77.34 MPa]

2. A spur gear made of bronze drives a mid steel pinion with angular velocity ratio of $3^1/_2$: 1. The pressure angle is $14\frac{1}{2}°$. It transmits 5 kW at 1800 r.p.m. of pinion. Considering only strength, design the smallest diameter gears and find also necessary face width. The number of teeth should not be less than 15 teeth on either gear. The elastic strength of bronze may be taken as 84 MPa and of steel as 105 MPa. Lewis factor for $14\frac{1}{2}°$ pressure angle may be taken as

$$y = 0.124 - \frac{0.684}{\text{No. of teeth}}$$

[Ans. m = 3 mm ; b = 35 mm ; D_P = 48 mm ; D_G = 168 mm]

3. A pair of 20° full-depth involute tooth spur gears is to transmit 30 kW at a speed of 250 r.p.m. of the pinion. The velocity ratio is 1 : 4. The pinion is made of cast steel having an allowable static stress, σ_o = 100 MPa, while the gear is made of cast iron having allowable static stress, σ_o = 55 MPa. The pinion has 20 teeth and its face width is 12.5 times the module. Determine the module, face width and pitch diameters of both the pinion and gear from the standpoint of strength only taking velocity factor into consideration. The tooth form factor is given by the expression

$$y = 0.154 - \frac{0.912}{\text{No. of teeth}}$$

and velocity factor is given by

$$C_v = \frac{3}{3+v},$$ where v is the peripheral speed of the gear in m/s.

[Ans. m = 20 mm ; b = 250 mm ; D_P = 400 mm ; D_G = 1600 mm]

4. A micarta pinion rotating at 1200 r.p.m. is to transmit 1 kW to a cast iron gear at a speed of 192 r.p.m. Assuming a starting overload of 20% and using 20° full depth involute teeth, determine the module, number of teeth on the pinion and gear and face width. Take allowable static strength for micarta as 40 MPa and for cast iron as 53 MPa. Check the pair in wear.

5. A 15 kW and 1200 r.p.m. motor drives a compressor at 300 r.p.m. through a pair of spur gears having 20° stub teeth. The centre to centre distance between the shafts is 400 mm. The motor pinion is made of forged steel having an allowable static stress as 210 MPa, while the gear is made of cast steel having allowable static stress as 140 MPa. Assuming that the drive operates 8 to 10 hours per day under light shock conditions, find from the standpoint of strength,

   1. Module; 2. Face width and 3. Number of teeth and pitch circle diameter of each gear.

   Check the gears thus designed from the consideration of wear. The surface endurance limit may be taken as 700 MPa. **[Ans. m = 6 mm ; b = 60 mm ; T_P = 24 ; T_G = 96 ; D_P = 144mm ; D_G = 576 mm]**

6. A two stage reduction drive is to be designed to transmit 2 kW; the input speed being 960 r.p.m. and overall reduction ratio being 9. The drive consists of straight tooth spur gears only, the shafts being spaced 200 mm apart, the input and output shafts being co-axial.

(a) Draw a layout of a suitable system to meet the above specifications, indicating the speeds of all rotating components.

(b) Calculate the module, pitch diameter, number of teeth, blank diameter and face width of the gears for medium heavy duty conditions, the gears being of medium grades of accuracy.

(c) Draw to scale one of the gears and specify on the drawing the calculated dimensions and other data complete in every respect for manufacturing purposes.

7. A motor shaft rotating at 1440 r.p.m. has to transmit 15 kW to a low speed shaft rotating at 500 r.p.m. The teeth are 20° involute with 25 teeth on the pinion. Both the pinion and gear are made of cast iron with a maximum safe stress of 56 MPa. A safe stress of 35 MPa may be taken for the shaft on which the gear is mounted. Design and sketch the spur gear drive to suit the above conditions. The starting torque may be assumed as 1.25 times the running torque.

8. Design and draw a spur gear drive transmitting 30 kW at 400 r.p.m. to another shaft running approximately at 100 r.p.m. The load is steady and continuous. The materials for the pinion and gear are cast steel and cast iron respectively. Take module as 10 mm. Also check the design for dynamic load and wear.

[**Hint** : Assume : $\sigma_{OP} = 140$ MPa ; $\sigma_{OG} = 56$ MPa ; $T_P = 24$; $y = 0.154 - \dfrac{0.912}{\text{No. of teeth}}$;

$C_v = \dfrac{3}{3 + v}$; $\sigma_e = 84$ MPa ; $e = 0.023$ mm ; $\sigma_{es} = 630$ MPa ; $E_P = 210$ kN/mm^2 ; $E_G = 100$ kN/mm^2]

9. Design a spur gear drive required to transmit 45 kW at a pinion speed of 800 r.p.m. The velocity ratio is 3.5 : 1. The teeth are 20° full-depth involute with 18 teeth on the pinion. Both the pinion and gear are made of steel with a maximum safe static stress of 180 MPa. Assume a safe stress of 40 MPa for the material of the shaft and key.

10. Design a pair of spur gears with stub teeth to transmit 55 kW from a 175 mm pinion running at 2500 r.p.m. to a gear running at 1500 r.p.m. Both the gears are made of steel having B.H.N. 260. Approximate the pitch by means of Lewis equation and then adjust the dimensions to keep within the limits set by the dynamic load and wear equation.

QUESTIONS

1. Write a short note on gear drives giving their merits and demerits.
2. How are the gears classified and what are the various terms used in spur gear terminology ?
3. Mention four important types of gears and discuss their applications, the materials used for them and their construction.
4. What condition must be satisfied in order that a pair of spur gears may have a constant velcoity ratio?
5. State the two most important reasons for adopting involute curves for a gear tooth profile.
6. Explain the phenomenon of interference in involute gears. What are the conditions to be satisfied in order to avoid interference ?
7. Explain the different causes of gear tooth failures and suggest possible remedies to avoid such failures.
8. Write the expressions for static, limiting wear load and dynamic load for spur gears and explain the various terms used there in.
9. Discuss the design procedure of spur gears.
10. How the shaft and arms for spur gears are designed ?

OBJECTIVE TYPE QUESTIONS

1. The gears are termed as medium velocity gears, if their peripheral velocity is

(a) 1–3 m / s (b) 3–15 m / s

(c) 15–30 m / s (d) 30–50 m / s

2. The size of gear is usually specified by
   (*a*) pressure angle (*b*) pitch circle diameter
   (*c*) circular pitch (*d*) diametral pitch
3. A spur gear with pitch circle diameter *D* has number of teeth *T*. The module *m* is defined as
   (*a*) $m = d / T$ (*b*) $m = T / D$
   (*c*) $m = \pi D / T$ (*d*) $m = D.T$
4. In a rack and pinion arrangement, the rack has teeth of .... shape.
   (*a*) square (*b*) trepazoidal
5. The radial distance from the .... to the clearance circle is called working depth.
   (*a*) addendum circle (*b*) dedendum circle
6. The product of the diametral pitch and circular pitch is equal to
   (*a*) 1 (*b*) $1/\pi$
   (*c*) π (*d*) $\pi \times$ No. of teeth
7. The backlash for spur gears depends upon
   (*a*) module (*b*) pitch line velocity
   (*c*) tooth profile (*d*) both (*a*) and (*b*)
8. The contact ratio for gears is
   (*a*) zero (*b*) less than one
   (*c*) greater than one (*d*) none of these
9. If the centre distance of the mating gears having involute teeth is increased, then the pressure angle
   (*a*) increases (*b*) decreases
   (*c*) remains unchanged (*d*) none of these
10. The form factor of a spur gear tooth depends upon
    (*a*) circular pitch only (*b*) pressure angle only
    (*c*) number of teeth and circular pitch (*d*) number of teeth and the system of teeth
11. Lewis equation in spur gears is used to find the
    (*a*) tensile stress in bending (*b*) shear stress
    (*c*) compressive stress in bending (*d*) fatigue stress
12. The minimum number of teeth on the pinion in order to avoid interference for 20° stub system is
    (*a*) 12 (*b*) 14
    (*c*) 18 (*d*) 32
13. The allowable static stress for steel gears is approximately .... of the ultimate tensile stress.
    (*a*) one-fourth (*b*) one-third
    (*c*) one-half (*d*) double
14. Lewis equation in spur gears is applied
    (*a*) only to the pinion (*b*) only to the gear
    (*c*) to stronger of the pinion or gear (*d*) to weaker of the pinion or gear
15. The static tooth load should be .... the dynamic load.
    (*a*) less than (*b*) greater than
    (*c*) equal to

ANSWERS

1. (*b*)	2. (*b*)	3. (*a*)	4. (*b*)	5. (*a*)
6. (*c*)	7. (*d*)	8. (*c*)	9. (*a*)	10. (*d*)
11. (*c*)	12. (*b*)	13. (*b*)	14. (*d*)	15. (*b*)

CHAPTER 29

Helical Gears

29.1 Introduction

A helical gear has teeth in form of helix around the gear. Two such gears may be used to connect two parallel shafts in place of spur gears. The helixes may be right handed on one gear and left handed on the other. The pitch surfaces are cylindrical as in spur gearing, but the teeth instead of being parallel to the axis, wind around the cylinders helically like screw threads. The teeth of helical gears with parallel axis have line contact, as in spur gearing. This provides gradual engagement and continuous contact of the engaging teeth. Hence helical gears give smooth drive with a high efficiency of transmission.

We have already discussed in Art. 28.4 that the helical gears may be of ***single helical type*** or ***double helical type.*** In case of single helical gears there is some axial thrust between the teeth, which is a disadvantage. In order to eliminate this axial thrust, double helical gears (*i.e.*

herringbone gears) are used. It is equivalent to two single helical gears, in which equal and opposite thrusts are provided on each gear and the resulting axial thrust is zero.

29.2 Terms used in Helical Gears

The following terms in connection with helical gears, as shown in Fig. 29.1, are important from the subject point of view.

1. ***Helix angle.*** It is a constant angle made by the helices with the axis of rotation.

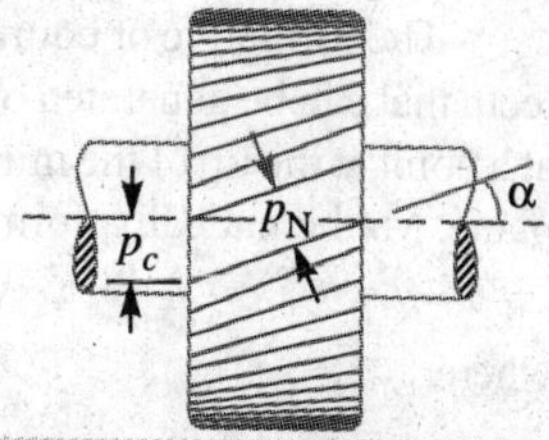

Fig. 29.1. Helical gear (nomenclature).

2. ***Axial pitch.*** It is the distance, parallel to the axis, between similar faces of adjacent teeth. It is the same as circular pitch and is therefore denoted by p_c. The axial pitch may also be defined as the circular pitch in the plane of rotation or the diametral plane.

3. ***Normal pitch.*** It is the distance between similar faces of adjacent teeth along a helix on the pitch cylinders normal to the teeth. It is denoted by p_N. The normal pitch may also be defined as the circular pitch in the normal plane which is a plane perpendicular to the teeth. Mathematically, normal pitch,

$$p_N = p_c \cos \alpha$$

Note : If the gears are cut by standard hobs, then the pitch (or module) and the pressure angle of the hob will apply in the normal plane. On the other hand, if the gears are cut by the Fellows gear-shaper method, the pitch and pressure angle of the cutter will apply to the plane of rotation. The relation between the normal pressure angle (ϕ_N) in the normal plane and the pressure angle (ϕ) in the diametral plane (or plane of rotation) is given by

$$\tan \phi_N = \tan \phi \times \cos \alpha$$

29.3 Face Width of Helical Gears

In order to have more than one pair of teeth in contact, the tooth displacement (*i.e.* the advancement of one end of tooth over the other end) or overlap should be atleast equal to the axial pitch, such that

$$\text{Overlap} = p_c = b \tan \alpha \quad ...(i)$$

The normal tooth load (W_N) has two components ; one is tangential component (W_T) and the other axial component (W_A), as shown in Fig. 29.2. The axial or end thrust is given by

$$W_A = W_N \sin \alpha = W_T \tan \alpha \quad ...(ii)$$

From equation (*i*), we see that as the helix angle increases, then the tooth overlap increases. But at the same time, the end thrust as given by equation (*ii*), also increases, which is undesirable. It is usually recommended that the overlap should be 15 percent of the circular pitch.

$$\therefore \quad \text{Overlap} = b \tan \alpha = 1.15\, p_c$$

or
$$b = \frac{1.15\, p_c}{\tan \alpha} = \frac{1.15 \times \pi m}{\tan \alpha} \quad ...(\because p_c = \pi m)$$

where b = Minimum face width, and

m = Module.

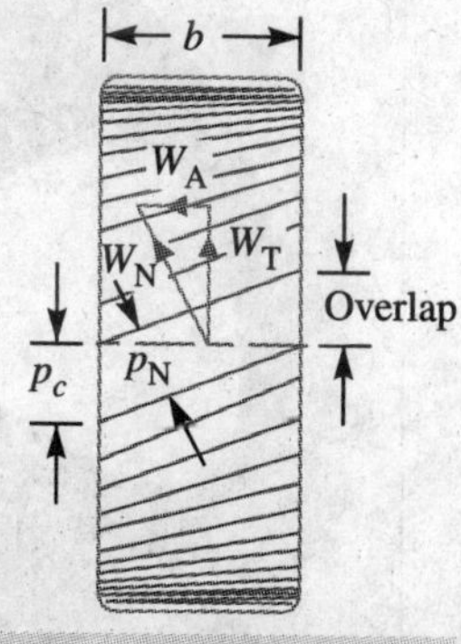

Fig. 29.2. Face width of helical gear.

Notes : 1. The maximum face width may be taken as 12.5 *m* to 20 *m*, where *m* is the module. In terms of pinion diameter (D_P), the face width should be 1.5 D_P to 2 D_P, although 2.5 D_P may be used.

2. In case of double helical or herringbone gears, the minimum face width is given by

$$b = \frac{2.3\, p_c}{\tan \alpha} = \frac{2.3 \times \pi m}{\tan \alpha}$$

The maximum face width ranges from 20 *m* to 30 *m*.

3. In single helical gears, the helix angle ranges from 20° to 35°, while for double helical gears, it may be made upto 45°.

29.4 Formative or Equivalent Number of Teeth for Helical Gears

The formative or equivalent number of teeth for a helical gear may be defined as the number of teeth that can be generated on the surface of a cylinder having a radius equal to the radius of curvature at a point at the tip of the minor axis of an ellipse obtained by taking a section of the gear in the normal plane. Mathematically, formative or equivalent number of teeth on a helical gear,

$$T_E = T / \cos^3 \alpha$$

where T = Actual number of teeth on a helical gear, and

α = Helix angle.

29.5 Proportions for Helical Gears

Though the proportions for helical gears are not standardised, yet the following are recommended by American Gear Manufacturer's Association (AGMA).

Pressure angle in the plane of rotation,

$$\phi = 15° \text{ to } 25°$$

Helix angle, α = 20° to 45°

Addendum = 0.8 m (Maximum)

Dedendum = 1 m (Minimum)

Minimum total depth = 1.8 m

Minimum clearance = 0.2 m

Thickness of tooth = 1.5708 m

In helical gears, the teeth are inclined to the axis of the gear.

29.6 Strength of Helical Gears

In helical gears, the contact between mating teeth is gradual, starting at one end and moving along the teeth so that at any instant the line of contact runs diagonally across the teeth. Therefore in order to find the strength of helical gears, a modified Lewis equation is used. It is given by

$$W_T = (\sigma_o \times C_v)\, b.\pi\, m.y'$$

where W_T = Tangential tooth load,

σ_o = Allowable static stress,

C_v = Velocity factor,

b = Face width,

m = Module, and

y' = Tooth form factor or Lewis factor corresponding to the formative or virtual or equivalent number of teeth.

Notes : 1. The value of velocity factor (C_v) may be taken as follows :

$$C_v = \frac{6}{6+v}, \text{ for peripheral velocities from 5 m / s to 10 m / s.}$$

$$= \frac{15}{15+v}, \text{ for peripheral velocities from 10 m / s to 20 m / s.}$$

$$= \frac{0.75}{0.75+\sqrt{v}}, \text{ for peripheral velocities greater than 20 m / s.}$$

$$= \frac{0.75}{1+v} + 0.25, \text{ for non-metallic gears.}$$

2. The dynamic tooth load on the helical gears is given by

$$W_D = W_T + \frac{21\, v\, (b.C \cos^2 \alpha + W_T) \cos \alpha}{21\, v + \sqrt{b.C \cos^2 \alpha + W_T}}$$

where v, b and C have usual meanings as discussed in spur gears.

3. The static tooth load or endurance strength of the tooth is given by

$$W_S = \sigma_e.b.\pi\, m.y'$$

4. The maximum or limiting wear tooth load for helical gears is given by

$$W_w = \frac{D_P.b.Q.K}{\cos^2 \alpha}$$

where D_P, b, Q and K have usual meanings as discussed in spur gears.

In this case, $$K = \frac{(\sigma_{es})^2 \sin \phi_N}{1.4}\left[\frac{1}{E_P} + \frac{1}{E_G}\right]$$

where ϕ_N = Normal pressure angle.

Example 29.1. *A pair of helical gears are to transmit 15 kW. The teeth are 20° stub in diametral plane and have a helix angle of 45°. The pinion runs at 10 000 r.p.m. and has 80 mm pitch diameter. The gear has 320 mm pitch diameter. If the gears are made of cast steel having allowable static strength of 100 MPa; determine a suitable module and face width from static strength considerations and check the gears for wear, given σ_{es} = 618 MPa.*

Solution. Given : P = 15 kW = 15 × 10^3 W; ϕ = 20° ; α = 45° ; N_P = 10 000 r.p.m. ; D_P = 80 mm = 0.08 m ; D_G = 320 mm = 0.32 m ; $\sigma_{OP} = \sigma_{OG}$ = 100 MPa = 100 N/mm^2 ; σ_{es} = 618 MPa = 618 N/mm^2

Module and face width

Let m = Module in mm, and

b = Face width in mm.

Since both the pinion and gear are made of the same material (*i.e.* cast steel), therefore the pinion is weaker. Thus the design will be based upon the pinion.

We know that the torque transmitted by the pinion,

$$T = \frac{P \times 60}{2\,\pi\,N_P} = \frac{15 \times 10^3 \times 60}{2\,\pi \times 10000} = 14.32 \text{ N-m}$$

∴ *Tangential tooth load on the pinion,

$$W_T = \frac{T}{D_P/2} = \frac{14.32}{0.08/2} = 358 \text{ N}$$

We know that number of teeth on the pinion,

$$T_P = D_P / m = 80 / m$$

and formative or equivalent number of teeth for the pinion,

$$T_E = \frac{T_P}{\cos^3 \alpha} = \frac{80/m}{\cos^3 45°} = \frac{80/m}{(0.707)^3} = \frac{226.4}{m}$$

∴ Tooth form factor for the pinion for 20° stub teeth,

$$y'_P = 0.175 - \frac{0.841}{T_E} = 0.175 - \frac{0.841}{226.4/m} = 0.175 - 0.0037\,m$$

We know that peripheral velocity,

$$v = \frac{\pi\,D_P . N_P}{60} = \frac{\pi \times 0.08 \times 10000}{60} = 42 \text{ m/s}$$

∴ Velocity factor,

$$C_v = \frac{0.75}{0.75 + \sqrt{v}} = \frac{0.75}{0.75 + \sqrt{42}} = 0.104 \quad ...(\because v \text{ is greater than 20 m/s})$$

Since the maximum face width (b) for helical gears may be taken as 12.5 *m* to 20 *m*, where *m* is the module, therefore let us take

$$b = 12.5\,m$$

We know that the tangential tooth load (W_T),

$$358 = (\sigma_{OP} \cdot C_v)\, b.\pi\, m.y'_P$$
$$= (100 \times 0.104)\, 12.5\, m \times \pi\, m\, (0.175 - 0.0037\, m)$$
$$= 409\, m^2\, (0.175 - 0.0037\, m) = 72\, m^2 - 1.5\, m^3$$

Solving this expression by hit and trial method, we find that

$$m = 2.3 \text{ say } 2.5 \text{ mm } \textbf{Ans.}$$

and face width, $b = 12.5\,m = 12.5 \times 2.5 = 31.25$ say 32 mm **Ans.**

Checking the gears for wear

We know that velocity ratio,

$$V.R. = \frac{D_G}{D_P} = \frac{320}{80} = 4$$

∴ Ratio factor,

$$Q = \frac{2 \times V.R.}{V.R. + 1} = \frac{2 \times 4}{4 + 1} = 1.6$$

We know that $\tan\,\phi_N = \tan\phi \cos\alpha = \tan 20° \times \cos 45° = 0.2573$

∴ $\phi_N = 14.4°$

* The tangential tooth load on the pinion may also be obtained by using the relation,

$$W_T = \frac{P}{v}, \text{ where } v = \frac{\pi\,D_P . N_P}{60} \text{ (in m/s)}$$

The picture shows double helical gears which are also called herringbone gears.

Since both the gears are made of the same material (*i.e.* cast steel), therefore let us take

$$E_P = E_G = 200 \text{ kN/mm}^2 = 200 \times 10^3 \text{ N/mm}^2$$

∴ Load stress factor,

$$K = \frac{(\sigma_{es})^2 \sin \phi_N}{1.4}\left(\frac{1}{E_P} + \frac{1}{E_G}\right)$$

$$= \frac{(618)^2 \sin 14.4°}{1.4}\left(\frac{1}{200 \times 10^3} + \frac{1}{200 \times 10^3}\right) = 0.678 \text{ N/mm}^2$$

We know that the maximum or limiting load for wear,

$$W_w = \frac{D_P.b.Q.K}{\cos^2 \alpha} = \frac{80 \times 32 \times 1.6 \times 0.678}{\cos^2 45°} = 5554 \text{ N}$$

Since the maximum load for wear is much more than the tangential load on the tooth, therefore the design is satisfactory from consideration of wear.

Example 29.2. *A helical cast steel gear with 30° helix angle has to transmit 35 kW at 1500 r.p.m. If the gear has 24 teeth, determine the necessary module, pitch diameter and face width for 20° full depth teeth. The static stress for cast steel may be taken as 56 MPa. The width of face may be taken as 3 times the normal pitch. What would be the end thrust on the gear? The tooth factor for 20° full depth involute gear may be taken as* $0.154 - \frac{0.912}{T_E}$, *where* T_E *represents the equivalent number of teeth.*

Solution. Given : $\alpha = 30°$; $P = 35$ kW $= 35 \times 10^3$ W ; $N = 1500$ r.p.m. ; $T_G = 24$; $\phi = 20°$; $\sigma_o = 56$ MPa $= 56$ N/mm^2 ; $b = 3 \times$ Normal pitch $= 3\, p_N$

Module

Let m = Module in mm, and

D_G = Pitch circle diameter of the gear in mm.

We know that torque transmitted by the gear,

$$T = \frac{P \times 60}{2\,\pi\, N} = \frac{35 \times 10^3 \times 60}{2\,\pi \times 1500} = 223 \text{ N-m} = 223 \times 10^3 \text{ N-mm}$$

Formative or equivalent number of teeth,

$$T'_E = \frac{T_G}{\cos^3 \alpha} = \frac{24}{\cos^3 30°} = \frac{24}{(0.866)^3} = 37$$

∴ Tooth factor, $y' = 0.154 - \frac{0.912}{T_E} = 0.154 - \frac{0.912}{37} = 0.129$

We know that the tangential tooth load,

$$W_T = \frac{T}{D_G/2} = \frac{2T}{D_G} = \frac{2T}{m \times T_G} \quad ...(\because D_G = m.T_G)$$

$$= \frac{2 \times 223 \times 10^3}{m \times 24} = \frac{18\ 600}{m} \text{ N}$$

and peripheral velocity,

$$v = \frac{\pi D_G . N}{60} = \frac{\pi . m. T_G . N}{60} \text{ mm/s} \quad ...(D_G \text{ and } m \text{ are in mm})$$

$$= \frac{\pi \times m \times 24 \times 1500}{60} = 1885\ m \text{ mm/s} = 1.885\ m \text{ m/s}$$

Let us take velocity factor,

$$C_v = \frac{15}{15 + v} = \frac{15}{15 + 1.885\ m}$$

We know that tangential tooth load,

$$W_T = (\sigma_o \times C_v)\ b.\pi\ m.y' = (\sigma_o \times C_v)\ 3p_N \times \pi\ m \times y' \quad ...(\because b = 3\ p_N)$$

$$= (\sigma_o \times C_v)\ 3 \times p_c \cos \alpha \times \pi\ m \times y' \quad ...(\because p_N = p_c \cos \alpha)$$

$$= (\sigma_o \times C_v)\ 3\ \pi\ m \cos \alpha \times \pi\ m \times y' \quad ...(\because p_c = \pi\ m)$$

$$\therefore \quad \frac{18\ 600}{m} = 56 \left(\frac{15}{15 + 1.885\ m}\right) 3\ \pi\ m \times \cos 30° \times \pi\ m \times 0.129$$

$$= \frac{2780\ m^2}{15 + 1.885\ m}$$

or $279\ 000 + 35\ 061\ m = 2780\ m^3$

Solving this equation by hit and trial method, we find that

$m = 5.5$ say 6 mm **Ans.**

Pitch diameter of the gear

We know that the pitch diameter of the gear,

$D_G = m \times T_G = 6 \times 24 = 144$ mm **Ans.**

Face width

It is given that the face width,

$$b = 3\ p_N = 3\ p_c \cos \alpha = 3 \times \pi\ m \cos \alpha$$

$= 3 \times \pi \times 6 \cos 30° = 48.98$ say 50 mm **Ans.**

End thrust on the gear

We know that end thrust or axial load on the gear,

$W_A = W_T \tan \alpha = \frac{18\ 600}{m} \times \tan 30° = \frac{18\ 600}{6} \times 0.577 = 1790$ N **Ans.**

Example 29.3. *Design a pair of helical gears for transmitting 22 kW. The speed of the driver gear is 1800 r.p.m. and that of driven gear is 600 r.p.m. The helix angle is 30° and profile is corresponding to 20° full depth system. The driver gear has 24 teeth. Both the gears are made of cast steel with allowable static stress as 50 MPa. Assume the face width parallel to axis as 4 times the circular pitch and the overhang for each gear as 150 mm. The allowable shear stress for the shaft material may be taken as 50 MPa. The form factor may be taken as 0.154 – 0.912 / T_E, where T_E is the equivalent number of teeth. The velocity factor may be taken as $\frac{350}{350 + v}$, where v is pitch line velocity in m / min. The gears are required to be designed only against bending failure of the teeth under dynamic condition.*

Solution. Given : P = 22 kW = 22 × 10^3 W ; N_P = 1800 r.p.m.; N_G = 600 r.p.m. ; α = 30° ; ϕ = 20° ; T_P = 24 ; σ_o = 50 MPa = 50 N/mm^2 ; $b = 4\,p_c$; Overhang = 150 mm ; τ = 50 MPa = 50 N/mm^2

Gears inside a car

Design for the pinion and gear

We know that the torque transmitted by the pinion,

$$T = \frac{P \times 60}{2\,\pi\,N_P} = \frac{22 \times 10^3 \times 60}{2\,\pi \times 1800} = 116.7 \text{ N-m} = 116\ 700 \text{ N-mm}$$

Since both the pinion and gear are made of the same material (*i.e.* cast steel), therefore the pinion is weaker. Thus the design will be based upon the pinion. We know that formative or equivalent number of teeth,

$$T_E = \frac{T_P}{\cos^3 \alpha} = \frac{24}{\cos^3 30°} = \frac{24}{(0.866)^3} = 37$$

$\therefore$ Form factor, $$y' = 0.154 - \frac{0.912}{T_E} = 0.154 - \frac{0.912}{37} = 0.129$$

First of all let us find the module of teeth.

Let m = Module in mm, and

D_P = Pitch circle diameter of the pinion in mm.

We know that the tangential tooth load on the pinion,

$$W_T = \frac{T}{D_P/2} = \frac{2T}{D_P} = \frac{2T}{m \times T_P} \qquad ...(\because D_P = m.T_P)$$

$$= \frac{2 \times 116\,700}{m \times 24} = \frac{9725}{m} \text{ N}$$

and peripheral velocity, $v = \pi D_P.N_P = \pi m.T_P.N_P$

$$= \pi m \times 24 \times 1800 = 135\,735\, m \text{ mm / min} = 135.735\, m \text{ m / min}$$

$$\therefore \text{ Velocity factor, } C_v = \frac{350}{350 + v} = \frac{350}{350 + 135.735\, m}$$

We also know that the tangential tooth load on the pinion,

$$W_T = (\sigma_o.C_v)\, b.\pi\, m.y' = (\sigma_o.C_v)\, 4\, p_c \times \pi\, m \times y' \qquad ... (\because b = 4\, p_c)$$

$$= (\sigma_o.C_v)\, 4 \times \pi\, m \times \pi\, m \times y' \qquad ... (\because p_c = \pi\, m)$$

$$\therefore \quad \frac{9725}{m} = 50 \left(\frac{350}{350 + 135.735\, m} \right) 4 \times \pi^2\, m^2 \times 0.129 = \frac{89\,126\, m^2}{350 + 135.735 m}$$

$$3.4 \times 10^6 + 1.32 \times 10^6\, m = 89\,126\, m^3$$

Solving this expression by hit and trial method, we find that

m = 4.75 mm say 6 mm **Ans.**

Helical gears.

We know that face width,

$b = 4\, p_c = 4\, \pi\, m = 4\, \pi \times 6 = 75.4$ say 76 mm **Ans.**

and pitch circle diameter of the pinion,

$D_P = m \times T_P = 6 \times 24 = 144$ mm **Ans.**

Since the velocity ratio is 1800 / 600 = 3, therefore number of teeth on the gear,

$T_G = 3\, T_P = 3 \times 24 = 72$

and pitch circle diameter of the gear,

$D_G = m \times T_G = 6 \times 72 = 432$ mm **Ans.**

Design for the pinion shaft

Let d_P = Diameter of the pinion shaft.

We know that the tangential load on the pinion,

$$W_T = \frac{9725}{m} = \frac{9725}{6} = 1621 \text{ N}$$

and the axial load of the pinion,

$$W_A = W_T \tan \alpha = 1621 \tan 30° = 1621 \times 0.577 = 935 \text{ N}$$

Since the overhang for each gear is 150 mm, therefore bending moment on the pinion shaft due to the tangential load,

$$M_1 = W_T \times \text{Overhang} = 1621 \times 150 = 243\,150 \text{ N-mm}$$

and bending moment on the pinion shaft due to the axial load,

$$M_2 = W_A \times \frac{D_P}{2} = 935 \times \frac{144}{2} = 67\,320 \text{ N-mm}$$

Since the bending moment due to the tangential load (*i.e.* M_1) and bending moment due to the axial load (*i.e.* M_2) are at right angles, therefore resultant bending moment on the pinion shaft,

$$M = \sqrt{(M_1)^2 + (M_2)^2} = \sqrt{(243\,150)^2 + (67\,320)^2} = 252\,293 \text{ N-mm}$$

The pinion shaft is also subjected to a torque T = 116 700 N-mm, therefore equivalent twisting moment,

$$T_e = \sqrt{M^2 + T^2} = \sqrt{(252\,293)^2 + (116\,700)^2} = 277\,975 \text{ N-mm}$$

We know that equivalent twisting moment (T_e),

$$277\,975 = \frac{\pi}{16} \times \tau\,(d_P)^3 = \frac{\pi}{16} \times 50\,(d_P)^3 = 9.82\,(d_P)^3$$

$$\therefore \quad (d_P)^3 = 277\,975 / 9.82 = 28\,307 \text{ or } d_P = 30.5 \text{ say } 35 \text{ mm } \textbf{Ans.}$$

Let us now check for the principal shear stress.

We know that the shear stress induced,

$$\tau = \frac{16\,T_e}{\pi\,(d_P)^3} = \frac{16 \times 277\,975}{\pi (35)^3} = 33 \text{ N/mm}^2 = 33 \text{ MPa}$$

and direct stress due to axial load,

$$\sigma = \frac{W_A}{\frac{\pi}{4}(d_P)^2} = \frac{935}{\frac{\pi}{4}(35)^2} = 0.97 \text{ N/mm}^2 = 0.97 \text{ MPa}$$

Helical gears

∴ Principal shear stress,

$$= \frac{1}{2}\left[\sqrt{\sigma^2 + 4\,\tau^2}\right] = \frac{1}{2}\left[\sqrt{(0.97)^2 + 4(33)^2}\right] = 33 \text{ MPa}$$

Since the principal shear stress is less than the permissible shear stress of 50 MPa, therefore the design is satisfactory.

We know that the diameter of the pinion hub

$= 1.8\, d_P = 1.8 \times 35 = 63$ mm **Ans.**

and length of the hub $= 1.25\, d_P = 1.25 \times 35 = 43.75$ say 44 mm

Since the length of the hub should not be less than the face width, therefore let us take length of the hub as 76 mm. **Ans.**

Note : Since the pitch circle diameter of the pinion is 144 mm, therefore the pinion should be provided with a web. Let us take the thickness of the web as 1.8 *m*, where *m* is the module.

∴ Thickness of the web $= 1.8\, m = 1.8 \times 6 = 10.8$ say 12 mm **Ans.**

Design for the gear shaft

Let d_G = Diameter of the gear shaft.

We have already calculated that the tangential load,

$W_T = 1621$ N

and the axial load, $W_A = 935$ N

∴ Bending moment due to the tangential load,

$$M_1 = W_T \times \text{Overhang} = 1621 \times 150 = 243\ 150 \text{ N-mm}$$

and bending moment due to the axial load,

$$M_2 = W_A \times \frac{D_G}{2} = 935 \times \frac{432}{2} = 201\ 960 \text{ N-mm}$$

∴ Resultant bending moment on the gear shaft,

$$M = \sqrt{(M_1)^2 + (M_2)^2} = \sqrt{(243\ 150)^2 + (201\ 960)^2} = 316\ 000 \text{ N-mm}$$

Since the velocity ratio is 3, therefore the gear shaft is subjected to a torque equal to 3 times the torque on the pinion shaft.

∴ Torque on the gear shaft,

$$T = \text{Torque on the pinion shaft} \times V.R.$$
$$= 116\ 700 \times 3 = 350\ 100 \text{ N-mm}$$

We know that equivalent twisting moment,

$$T_e = \sqrt{M^2 + T^2} = \sqrt{(316\ 000)^2 + (350\ 100)^2} = 472\ 000 \text{ N-mm}$$

We also know that equivalent twisting moment (T_e),

$$472\ 000 = \frac{\pi}{16} \times \tau \times (d_G)^3 = \frac{\pi}{16} \times 50\,(d_G)^3 = 9.82\,(d_G)^3$$

∴ $(d_G)^3 = 472\ 000 / 9.82 = 48\ 065$ or $d_G = 36.3$ say 40 mm **Ans.**

Let us now check for the principal shear stress.

We know that the shear stress induced,

$$\tau = \frac{16\, T_e}{\pi\,(d_G)^3} = \frac{16 \times 472\,000}{\pi\,(40)^3} = 37.6 \text{ N/mm}^2 = 37.6 \text{ MPa}$$

and direct stress due to axial load,

$$\sigma = \frac{W_A}{\frac{\pi}{4}(d_G)^2} = \frac{935}{\frac{\pi}{4}(40)^2} = 0.744 \text{ N/mm}^2 = 0.744 \text{ MPa}$$

∴ Principal shear stress

$$= \frac{1}{2}\left[\sqrt{\sigma^2 + 4\,\tau^2}\right] = \frac{1}{2}\left[\sqrt{(0.744)^2 + 4\,(37.6)^2}\right] = 37.6 \text{ MPa}$$

Since the principal shear stress is less than the permissible shear stress of 50 MPa, therefore the design is satisfactory.

We know that the diameter of the gear hub

$$= 1.8\, d_G = 1.8 \times 40 = 72 \text{ mm } \textbf{Ans.}$$

and length of the hub $= 1.25\, d_G = 1.25 \times 40 = 50$ mm

We shall take the length of the hub equal to the face width, *i.e.* 76 mm. **Ans.**

Since the pitch circle diameter of the gear is 432 mm, therefore the gear should be provided with four arms. The arms are designed in the similar way as discussed for spur gears.

Design for the gear arms

Let us assume that the cross-section of the arms is elliptical with major axis (a_1) equal to twice the minor axis (b_1). These dimensions refer to hub end.

∴ Section modulus of arms,

$$Z = \frac{\pi b_1 (a_1)^2}{32} = \frac{\pi\,(a_1)^3}{64} = 0.05\,(a_1)^3 \qquad \left(\because b_1 = \frac{a_1}{2}\right)$$

Since the arms are designed for the stalling load and it is taken as the design tangential load divided by the velocity factor, therefore

Stalling load, $$W_S = \frac{W_T}{C_v} = 1621\left(\frac{350 + 135.735m}{350}\right)$$

$$= 1621\left(\frac{350 + 135.735 \times 6}{350}\right) = 5393 \text{ N}$$

∴ Maximum bending moment on each arm,

$$M = \frac{W_S}{n} \times \frac{D_G}{2} = \frac{5393}{4} \times \frac{432}{2} = 291\,222 \text{ N-mm}$$

We know that bending stress (σ_b),

$$42 = \frac{M}{Z} = \frac{291\,222}{0.05\,(a_1)^3} = \frac{5824 \times 10^3}{(a_1)^3} \qquad \text{... (Taking } \sigma_b = 42 \text{ N/mm}^2\text{)}$$

∴ $(a_1)^3 = 5824 \times 10^3 / 42 = 138.7 \times 10^3$ or $a_1 = 51.7$ say 54 mm **Ans.**

and $b_1 = a_1 / 2 = 54 / 2 = 27$ mm **Ans.**

Since the arms are tapered towards the rim and the taper is 1/16 mm per mm length of the arm (or radius of the gear), therefore

Major axis of the arm at the rim end,

$$a_2 = a_1 - \text{Taper} = a_1 - \frac{1}{16} \times \frac{D_G}{2}$$

$$= 54 - \frac{1}{16} \times \frac{432}{2} = 40 \text{ mm } \textbf{Ans.}$$

and minor axis of the arm at the rim end,

$$b_2 = a_2 / 2 = 40 / 2 = 20 \text{ mm } \textbf{Ans.}$$

Design for the rim

The thickness of the rim for the pinion may be taken as 1.6 *m* to 1.9 *m*, where *m* is the module. Let us take thickness of the rim for pinion,

$$t_{RP} = 1.6\ m = 1.6 \times 6 = 9.6 \text{ say } 10 \text{ mm } \textbf{Ans.}$$

The thickness of the rim for the gear (t_{RG}) is given by

$$t_{RG} = m\sqrt{\frac{T_G}{n}} = 6\sqrt{\frac{72}{4}} = 25.4 \text{ say } 26 \text{ mm } \textbf{Ans.}$$

EXERCISES

1. A helical cast steel gear with 30° helix angle has to transmit 35 kW at 2000 r.p.m. If the gear has 25 teeth, find the necessary module, pitch diameters and face width for 20° full depth involute teeth. The static stress for cast steel may be taken as 100 MPa. The face width may be taken as 3 times the normal pitch. The tooth form factor is given by the expression $y' = 0.154 - 0.912/T_E$, where T_E represents the equivalent number of teeth. The velocity factor is given by $C_v = \frac{6}{6 + v}$, where v is the peripheral speed of the gear in m/s. **[Ans. 6 mm ; 150 mm ; 50 mm]**

2. A pair of helical gears with 30° helix angle is used to transmit 15 kW at 10 000 r.p.m. of the pinion. The velocity ratio is 4 : 1. Both the gears are to be made of hardened steel of static strength 100 N/mm^2. The gears are 20° stub and the pinion is to have 24 teeth. The face width may be taken as 14 times the module. Find the module and face width from the standpoint of strength and check the gears for wear. **[Ans. 2 mm ; 28 mm]**

Gears inside a car engine.

3. A pair of helical gears consist of a 20 teeth pinion meshing with a 100 teeth gear. The pinion rotates at 720 r.p.m. The normal pressure angle is 20° while the helix angle is 25°. The face width is 40 mm and the normal module is 4 mm. The pinion as well as gear are made of steel having ultimate strength of 600 MPa and heat treated to a surface hardness of 300 B.H.N. The service factor and factor of safety are 1.5 and 2 respectively. Assume that the velocity factor accounts for the dynamic load and calculate the power transmitting capacity of the gears. **[Ans. 8.6 kW]**

4. A single stage helical gear reducer is to receive power from a 1440 r.p.m., 25 kW induction motor. The gear tooth profile is involute full depth with 20° normal pressure angle. The helix angle is 23°, number of teeth on pinion is 20 and the gear ratio is 3. Both the gears are made of steel with allowable beam stress of 90 MPa and hardness 250 B.H.N.

   (*a*) Design the gears for 20% overload carrying capacity from standpoint of bending strength and wear.

   (*b*) If the incremental dynamic load of 8 kN is estimated in tangential plane, what will be the safe power transmitted by the pair at the same speed?

QUESTIONS

1. What is a herringbone gear? Where they are used?
2. Explain the following terms used in helical gears :
   (*a*) Helix angle; (*b*) normal pitch; and
   (*c*) axial pitch.
3. Define formative or virtual number of teeth on a helical gear. Derive the expression used to obtain its value.
4. Write the expressions for static strength, limiting wear load and dynamic load for helical gears and explain the various terms used therein.

OBJECTIVE TYPE QUESTIONS

1. If T is the actual number of teeth on a helical gear and ϕ is the helix angle for the teeth, the formative number of teeth is written as
   (*a*) $T \sec^3 \phi$ (*b*) $T \sec^2 \phi$
   (*c*) $T/\sec^3\phi$ (*d*) $T \operatorname{cosec} \phi$
2. In helical gears, the distance between similar faces of adjacent teeth along a helix on the pitch cylinders normal to the teeth, is called
   (*a*) normal pitch (*b*) axial pitch
   (*c*) diametral pitch (*d*) module
3. In helical gears, the right hand helices on one gear will mesh .... helices on the other gear.
   (*a*) right hand (*b*) left hand
4. The helix angle for single helical gears ranges from
   (*a*) 10° to 15° (*b*) 15° to 20°
   (*c*) 20° to 35° (*d*) 35° to 50°
5. The helix angle for double helical gears may be made up to
   (*a*) 45° (*b*) 60°
   (*c*) 75° (*d*) 90°

ANSWERS

1. (*a*) 2. (*a*) 3. (*b*) 4. (*c*) 5. (*a*)

CHAPTER

30

Bevel Gears

30.1 Introduction

The bevel gears are used for transmitting power at a constant velocity ratio between two shafts whose axes intersect at a certain angle. The pitch surfaces for the bevel gear are frustums of cones. The two pairs of cones in contact is shown in Fig. 30.1. The elements of the cones, as shown in Fig. 30.1 (*a*), intersect at the point of intersection of the axis of rotation. Since the radii of both the gears are proportional to their distances from the apex, therefore the cones may roll together without sliding. In Fig. 30.1 (*b*), the elements of both cones do not intersect at the point of shaft intersection. Consequently, there may be pure rolling at only one point of contact and there must be tangential sliding at all other points of contact. Therefore, these cones, cannot be used as pitch surfaces because it is impossible to have positive driving and sliding in the same direction at the same time. We, thus, conclude that the elements of bevel gear pitch cones and shaft axes must intersect at the same point.

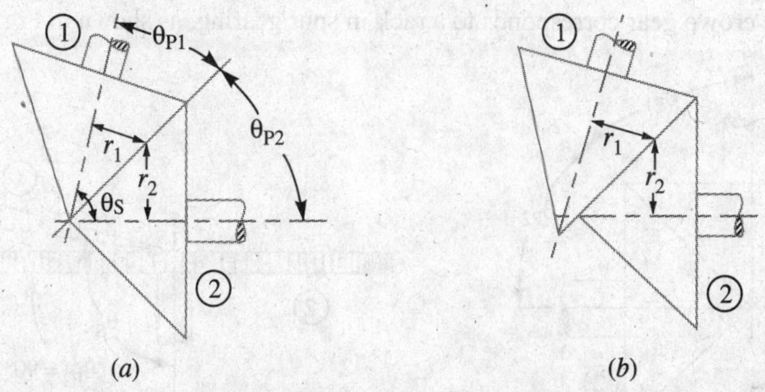

Fig. 30.1. Pitch surface for bevel gears.

The bevel gear is used to change the axis of rotational motion. By using gears of differing numbers of teeth, the speed of rotation can also be changed.

30.2 Classification of Bevel Gears

The bevel gears may be classified into the following types, depending upon the angles between the shafts and the pitch surfaces.

1. *Mitre gears.* When equal bevel gears (having equal teeth and equal pitch angles) connect two shafts whose axes intersect at right angle, as shown in Fig. 30.2 (*a*), then they are known as ***mitre gears.***

2. *Angular bevel gears.* When the bevel gears connect two shafts whose axes intersect at an angle other than a right angle, then they are known as ***angular bevel gears.***

3. ***Crown bevel gears.*** When the bevel gears connect two shafts whose axes intersect at an angle greater than a right angle and one of the bevel gears has a pitch angle of 90°, then it is known as a crown gear. The crown gear corresponds to a rack in spur gearing, as shown in Fig. 30.2 (*b*).

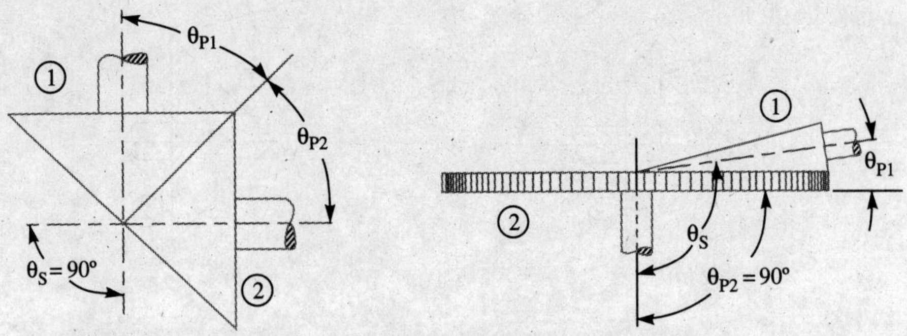

(*a*) Mitre gears.

(*b*) Crown bevel gear.

Fig. 30.2. Classification of bevel gears.

4. ***Internal bevel gears.*** When the teeth on the bevel gear are cut on the inside of the pitch cone, then they are known as ***internal bevel gears.***

Note : The bevel gears may have straight or spiral teeth. It may be assumed, unless otherwise stated, that the bevel gear has straight teeth and the axes of the shafts intersect at right angle.

30.3 Terms used in Bevel Gears

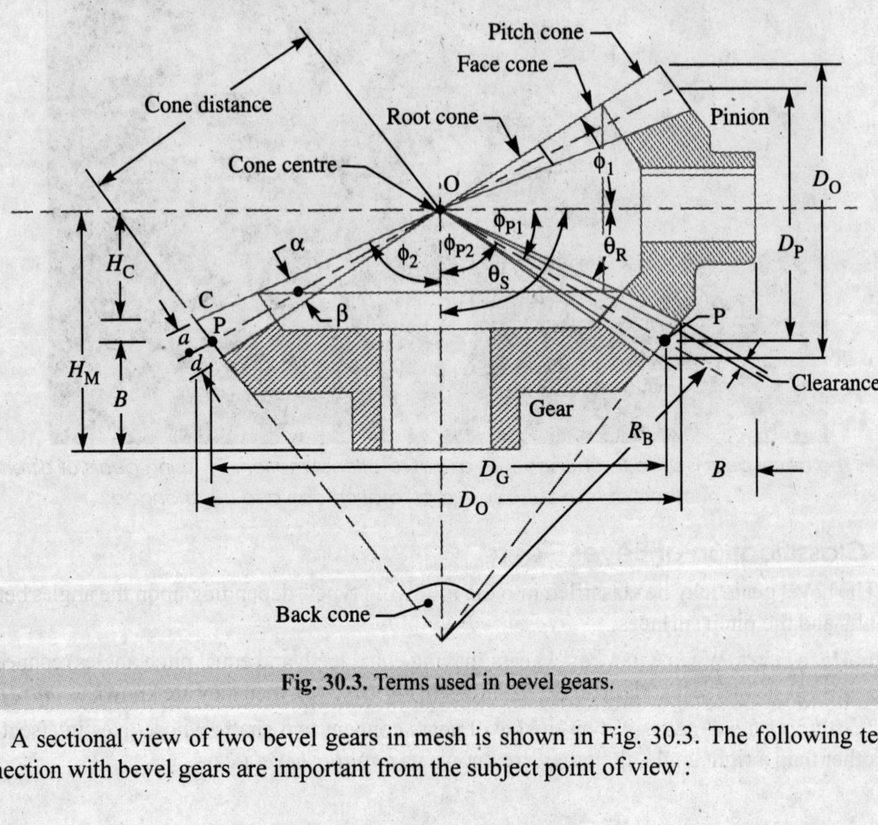

Fig. 30.3. Terms used in bevel gears.

A sectional view of two bevel gears in mesh is shown in Fig. 30.3. The following terms in connection with bevel gears are important from the subject point of view :

1. ***Pitch cone.*** It is a cone containing the pitch elements of the teeth.

2. ***Cone centre.*** It is the apex of the pitch cone. It may be defined as that point where the axes of two mating gears intersect each other.

3. ***Pitch angle.*** It is the angle made by the pitch line with the axis of the shaft. It is denoted by 'θ_P'.

4. ***Cone distance.*** It is the length of the pitch cone element. It is also called as a ***pitch cone radius.*** It is denoted by '*OP*'. Mathematically, cone distance or pitch cone radius,

$$OP = \frac{\text{Pitch radius}}{\sin \theta_P} = \frac{D_P/2}{\sin \theta_{P1}} = \frac{D_G/2}{\sin \theta_{P2}}$$

5. ***Addendum angle.*** It is the angle subtended by the addendum of the tooth at the cone centre. It is denoted by 'α' Mathematically, addendum angle,

$$\alpha = \tan^{-1}\left(\frac{a}{OP}\right)$$

where a = Addendum, and OP = Cone distance.

6. ***Dedendum angle.*** It is the angle subtended by the dedendum of the tooth at the cone centre. It is denoted by 'β'. Mathematically, dedendum angle,

$$\beta = \tan^{-1}\left(\frac{d}{OP}\right)$$

where d = Dedendum, and OP = Cone distance.

7. ***Face angle.*** It is the angle subtended by the face of the tooth at the cone centre. It is denoted by 'ϕ'. The face angle is equal to the pitch angle *plus* addendum angle.

8. ***Root angle.*** It is the angle subtended by the root of the tooth at the cone centre. It is denoted by 'θ_R'. It is equal to the pitch angle *minus* dedendum angle.

9. ***Back (or normal) cone.*** It is an imaginary cone, perpendicular to the pitch cone at the end of the tooth.

10. Back cone distance. It is the length of the back cone. It is denoted by 'R_B'. It is also called back cone radius.

11. ***Backing.*** It is the distance of the pitch point (*P*) from the back of the boss, parallel to the pitch point of the gear. It is denoted by '*B*'.

12. ***Crown height.*** It is the distance of the crown point (*C*) from the cone centre (*O*), parallel to the axis of the gear. It is denoted by 'H_C'.

13. ***Mounting height.*** It is the distance of the back of the boss from the cone centre. It is denoted by 'H_M'.

14. ***Pitch diameter.*** It is the diameter of the largest pitch circle.

15. ***Outside or addendum cone diameter.*** It is the maximum diameter of the teeth of the gear. It is equal to the diameter of the blank from which the gear can be cut. Mathematically, outside diameter,

$$D_O = D_P + 2\,a \cos \theta_P$$

where D_P = Pitch circle diameter,

a = Addendum, and

θ_P = Pitch angle.

16. ***Inside or dedendum cone diameter.*** The inside or the dedendum cone diameter is given by

$$D_d = D_P - 2d \cos \theta_P$$

where D_d = Inside diameter, and

d = Dedendum.

30.4 Determination of Pitch Angle for Bevel Gears

Mitre gears

Consider a pair of bevel gears in mesh, as shown in Fig. 30.3.

Let θ_{P1} = Pitch angle for the pinion,

θ_{P2} = Pitch angle for the gear,

θ_S = Angle between the two shaft axes,

D_P = Pitch diameter of the pinion,

D_G = Pitch diameter of the gear, and

$V.R.$ = Velocity ratio $= \dfrac{D_G}{D_P} = \dfrac{T_G}{T_P} = \dfrac{N_P}{N_G}$

From Fig. 30.3, we find that

$$\theta_S = \theta_{P1} + \theta_{P2} \quad \text{or} \quad \theta_{P2} = \theta_S - \theta_{P1}$$

$$\therefore \quad \sin \theta_{P2} = \sin(\theta_S - \theta_{P1}) = \sin \theta_S \cdot \cos \theta_{P1} - \cos \theta_S \cdot \sin \theta_{P1} \qquad ...(i)$$

We know that cone distance,

$$OP = \frac{D_P/2}{\sin \theta_{P1}} = \frac{D_G/2}{\sin \theta_{P2}} \quad \text{or} \quad \frac{\sin \theta_{P2}}{\sin \theta_{P1}} = \frac{D_G}{D_P} = V.R.$$

$$\therefore \quad \sin \theta_{P2} = V.R. \times \sin \theta_{P1} \qquad ...(ii)$$

From equations (*i*) and (*ii*), we have

$$V.R. \times \sin \theta_{P1} = \sin \theta_S \cdot \cos \theta_{P1} - \cos \theta_S \cdot \sin \theta_{P1}$$

Dividing throughout by $\cos \theta_{P1}$ we get

$$V.R. \tan \theta_{P1} = \sin \theta_S - \cos \theta_S \cdot \tan \theta_{P1}$$

or

$$\tan \theta_{P1} = \frac{\sin \theta_S}{V.R + \cos \theta_S}$$

$$\therefore \quad \theta_{P1} = \tan^{-1}\left(\frac{\sin \theta_S}{V.R + \cos \theta_S}\right) \qquad ...(iii)$$

Similarly, we can find that

$$\tan \theta_{P2} = \frac{\sin \theta_S}{\dfrac{1}{V.R} + \cos \theta_S}$$

$$\therefore \quad \theta_{P2} = \tan^{-1}\left(\frac{\sin \theta_S}{\dfrac{1}{V.R} + \cos \theta_S}\right) \qquad ...(iv)$$

Note : When the angle between the shaft axes is 90° *i.e.* $\theta_S = 90°$, then equations (*iii*) and (*iv*) may be written as

$$\theta_{P1} = \tan^{-1}\left(\frac{1}{V.R}\right) = \tan^{-1}\left(\frac{D_P}{D_G}\right) = \tan^{-1}\left(\frac{T_P}{T_G}\right) = \tan^{-1}\left(\frac{N_G}{N_P}\right)$$

and

$$\theta_{P2} = \tan^{-1}(V.R.) = \tan^{-1}\left(\frac{D_G}{D_P}\right) = \tan^{-1}\left(\frac{T_G}{T_P}\right) = \tan^{-1}\left(\frac{N_P}{N_G}\right)$$

30.5 Proportions for Bevel Gear

The proportions for the bevel gears may be taken as follows :

1. Addendum, $a = 1\,m$

2. Dedendum, $d = 1.2\ m$
3. Clearance $= 0.2\ m$
4. Working depth $= 2\ m$
5. Thickness of tooth $= 1.5708\ m$

where m is the module.

Note : Since the bevel gears are not interchangeable, therefore these are designed in pairs.

30.6 Formative or Equivalent Number of Teeth for Bevel Gears – Tredgold's Approximation

We have already discussed that the involute teeth for a spur gear may be generated by the edge of a plane as it rolls on a base cylinder. A similar analysis for a bevel gear will show that a true section of the resulting involute lies on the surface of a sphere. But it is not possible to represent on a plane surface the exact profile of a bevel gear tooth lying on the surface of a sphere. Therefore, it is important to approximate the bevel gear tooth profiles as accurately as possible. The approximation (known as ***Tredgold's approximation***) is based upon the fact that a cone tangent to the sphere at the pitch point will closely approximate the surface of the sphere for a short distance either side of the pitch point, as shown in Fig. 30.4 (*a*). The cone (known as back cone) may be developed as a plane surface and spur gear teeth corresponding to the pitch and pressure angle of the bevel gear and the radius of the developed cone can be drawn. This procedure is shown in Fig. 30.4 (*b*).

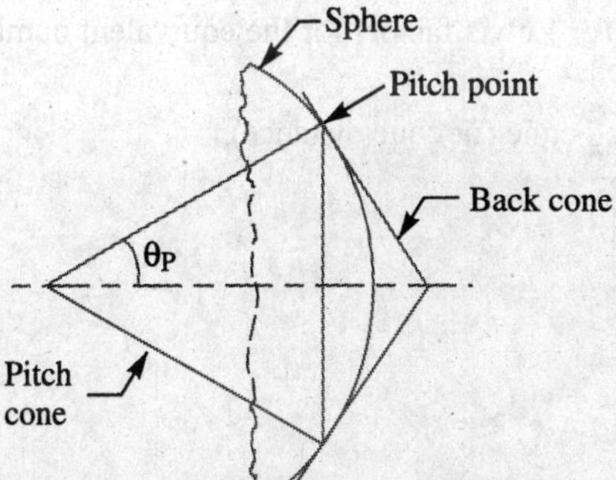

(*a*) Back cone.

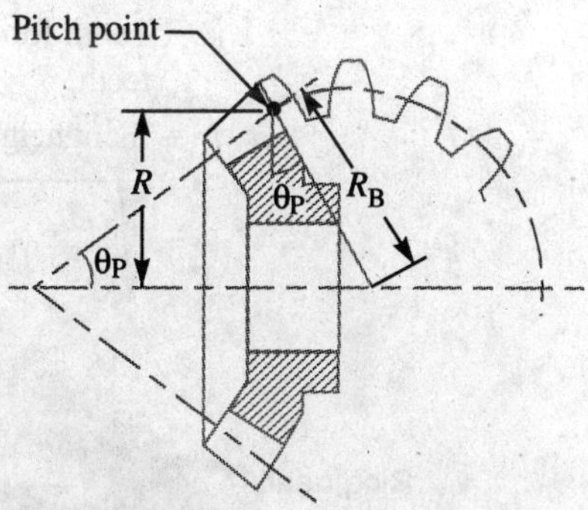

(*b*) Development of back cone.

Fig. 30.4

Let θ_P = Pitch angle or half of the cone angle,

R = Pitch circle radius of the bevel pinion or gear, and

R_B = Back cone distance or equivalent pitch circle radius of spur pinion or gear.

Now from Fig. 30.4 (*b*), we find that

$$R_B = R \sec \theta_P$$

We know that the equivalent (or formative) number of teeth,

$$T_E = \frac{2\ R_B}{m} \quad \ldots \left(\because \text{Number of teeth} = \frac{\text{Pitch circle diameter}}{\text{Module}}\right)$$

$$= \frac{2\ R \sec \theta_P}{m} = T \sec \theta_P$$

where T = Actual number of teeth on the gear.

Notes : 1. The action of bevel gears will be same as that of equivalent spur gears.

2. Since the equivalent number of teeth is always greater than the actual number of teeth, therefore a given pair of bevel gears will have a larger contact ratio. Thus, they will run more smoothly than a pair of spur gears with the same number of teeth.

30.7 Strength of Bevel Gears

The strength of a bevel gear tooth is obtained in a similar way as discussed in the previous articles. The modified form of the Lewis equation for the tangential tooth load is given as follows:

$$W_T = (\sigma_o \times C_v)\, b.\pi\, m.y' \left(\frac{L-b}{L}\right)$$

where σ_o = Allowable static stress,

C_v = Velocity factor,

$= \dfrac{3}{3+v}$, for teeth cut by form cutters,

$= \dfrac{6}{6+v}$, for teeth generated with precision machines,

v = Peripheral speed in m / s,

b = Face width,

m = Module,

y' = Tooth form factor (or Lewis factor) for the equivalent number of teeth,

L = Slant height of pitch cone (or cone distance),

$$= \sqrt{\left(\frac{D_G}{2}\right)^2 + \left(\frac{D_P}{2}\right)^2}$$

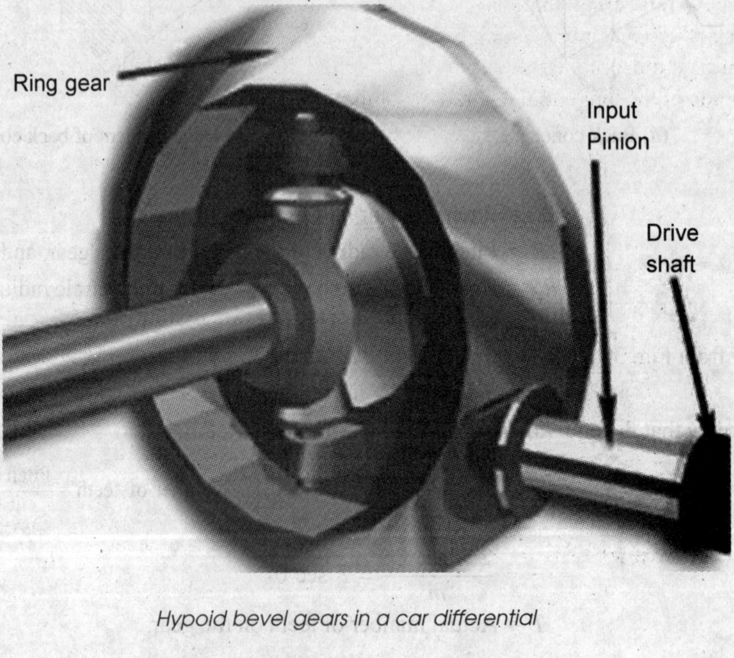

Hypoid bevel gears in a car differential

D_G = Pitch diameter of the gear, and

D_P = Pitch diameter of the pinion.

Notes : 1. The factor $\left(\frac{L-b}{L}\right)$ may be called as ***bevel factor.***

2. For satisfactory operation of the bevel gears, the face width should be from 6.3 *m* to 9.5 *m*, where *m* is the module. Also the ratio *L* / *b* should not exceed 3. For this, the number of teeth in the pinion must not less than $\frac{48}{\sqrt{1+(V.R.)^2}}$, where *V.R.* is the required velocity ratio.

3. The dynamic load for bevel gears may be obtained in the similar manner as discussed for spur gears.

4. The static tooth load or endurance strength of the tooth for bevel gears is given by

$$W_S = \sigma_e.b.\pi\, m.y' \left(\frac{L-b}{L}\right)$$

The value of flexural endurance limit (σ_e) may be taken from Table 28.8, in spur gears.

5. The maximum or limiting load for wear for bevel gears is given by

$$W_w = \frac{D_P.b.Q.K}{\cos \theta_{P1}}$$

where D_P, *b*, *Q* and *K* have usual meanings as discussed in spur gears except that *Q* is based on formative or equivalent number of teeth, such that

$$Q = \frac{2\, T_{EG}}{T_{EG} + T_{EP}}$$

30.8 Forces Acting on a Bevel Gear

Consider a bevel gear and pinion in mesh as shown in Fig. 30.5. The normal force (W_N) on the tooth is perpendicular to the tooth profile and thus makes an angle equal to the pressure angle (ϕ) to the pitch circle. Thus normal force can be resolved into two components, one is the tangential component (W_T) and the other is the radial component (W_R). The tangential component (*i.e.* the tangential tooth load) produces the bearing reactions while the radial component produces end thrust in the shafts. The magnitude of the tangential and radial components is as follows :

$$W_T = W_N \cos \phi, \text{ and } W_R = W_N \sin \phi = W_T \tan \phi \qquad ...(i)$$

These forces are considered to act at the mean radius (R_m). From the geometry of the Fig. 30.5, we find that

$$R_m = \left(L - \frac{b}{2}\right) \sin \theta_{P1} = \left(L - \frac{b}{2}\right) \frac{D_P}{2L} \quad \ldots \left(\because \sin \theta_{P1} = \frac{D_P/2}{L}\right)$$

Now the radial force (W_R) acting at the mean radius may be further resolved into two components, W_{RH} and W_{RV}, in the axial and radial directions as shown in Fig. 30.5. Therefore the axial force acting on the pinion shaft,

$$W_{RH} = W_R \sin \theta_{P1} = W_T \tan \phi \,.\, \sin \theta_{P1} \quad \ldots\text{[From equation (i)]}$$

and the radial force acting on the pinion shaft,

$$W_{RV} = W_R \cos \theta_{P1} = W_T \tan \phi . \cos \theta_{P1}$$

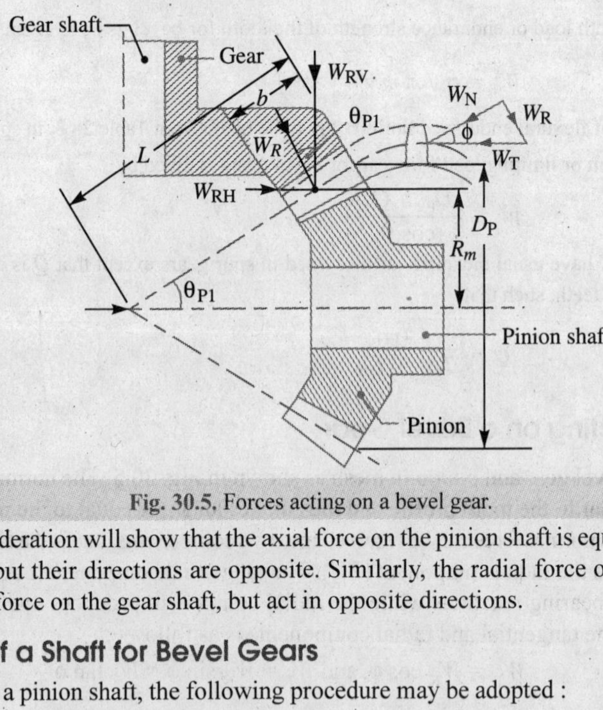

Fig. 30.5. Forces acting on a bevel gear.

A little consideration will show that the axial force on the pinion shaft is equal to the radial force on the gear shaft but their directions are opposite. Similarly, the radial force on the pinion shaft is equal to the axial force on the gear shaft, but act in opposite directions.

30.9 Design of a Shaft for Bevel Gears

In designing a pinion shaft, the following procedure may be adopted :

1. First of all, find the torque acting on the pinion. It is given by

$$T = \frac{P \times 60}{2 \pi N_P} \text{ N-m}$$

where P = Power transmitted in watts, and

N_P = Speed of the pinion in r.p.m.

2. Find the tangential force (W_T) acting at the mean radius (R_m) of the pinion. We know that

$$W_T = T / R_m$$

3. Now find the axial and radial forces (*i.e.* W_{RH} and W_{RV}) acting on the pinion shaft as discussed above.

4. Find resultant bending moment on the pinion shaft as follows :

The bending moment due to W_{RH} and W_{RV} is given by

$$M_1 = W_{RV} \times \text{Overhang} - W_{RH} \times R_m$$

and bending moment due to W_T,

$$M_2 = W_T \times \text{Overhang}$$

∴ Resultant bending moment,

$$M = \sqrt{(M_1)^2 + (M_2)^2}$$

5. Since the shaft is subjected to twisting moment (T) and resultant bending moment (M), therefore equivalent twisting moment,

$$T_e = \sqrt{M^2 + T^2}$$

6. Now the diameter of the pinion shaft may be obtained by using the torsion equation. We know that

$$T_e = \frac{\pi}{16} \times \tau (d_P)^3$$

where d_P = Diameter of the pinion shaft, and

τ = Shear stress for the material of the pinion shaft.

7. The same procedure may be adopted to find the diameter of the gear shaft.

Example 30.1. *A 35 kW motor running at 1200 r.p.m. drives a compressor at 780 r.p.m. through a 90° bevel gearing arrangement. The pinion has 30 teeth. The pressure angle of teeth is* $14\,^1/_2°$. *The wheels are capable of withstanding a dynamic stress,*

$$\sigma_w = 140 \left(\frac{280}{280 + v}\right) \text{ MPa, where } v \text{ is the pitch line speed in m / min.}$$

The form factor for teeth may be taken as $0.124 - \frac{0.686}{T_E}$, *where* T_E *is the number of teeth equivalent of a spur gear.*

The face width may be taken as $\frac{1}{4}$ *of the slant height of pitch cone. Determine for the pinion, the module pitch, face width, addendum, dedendum, outside diameter and slant height.*

High performance 2- and 3 -way bevel gear boxes

Solution : Given : $P = 35$ kW $= 35 \times 10^3$ W ; $N_P = 1200$ r.p.m. ; $N_G = 780$ r.p.m. ; $\theta_S = 90°$; $T_P = 30$; $\phi = 14\,^1/_2°$; $b = L / 4$

Module and face width for the pinion

Let m = Module in mm,

b = Face width in mm

$= L / 4$, and ...(Given)

D_P = Pitch circle diameter of the pinion.

We know that velocity ratio,

$$V.R. = \frac{N_P}{N_G} = \frac{1200}{780} = 1.538$$

∴ Number of teeth on the gear,

$$T_G = V.R. \times T_P = 1.538 \times 30 = 46$$

Since the shafts are at right angles, therefore pitch angle for the pinion,

$$\theta_{P1} = \tan^{-1}\left(\frac{1}{V.R.}\right) = \tan^{-1}\left(\frac{1}{1.538}\right) = \tan^{-1}(0.65) = 33°$$

and pitch angle for the gear,

$$\theta_{P2} = 90° - 33° = 57°$$

We know that formative number of teeth for pinion,

$$T_{EP} = T_P.\sec\theta_{P1} = 30 \times \sec 33° = 35.8$$

and formative number of teeth for the gear,

$$T_{EG} = T_G.\sec\theta_{P2} = 46 \times \sec 57° = 84.4$$

Tooth form factor for the pinion

$$y'_P = 0.124 - \frac{0.686}{T_{EP}} = 0.124 - \frac{0.686}{35.8} = 0.105$$

and tooth form factor for the gear,

$$y'_G = 0.124 - \frac{0.686}{T_{EG}} = 0.124 - \frac{0.686}{84.4} = 0.116$$

Since the allowable static stress (σ_o) for both the pinion and gear is same (*i.e.* 140 MPa or N/mm^2) and y'_P is less than y'_G, therefore the pinion is weaker. Thus the design should be based upon the pinion.

We know that the torque on the pinion,

$$T = \frac{P \times 60}{2\pi N_P} = \frac{35 \times 10^3 \times 60}{2\pi \times 1200} = 278.5 \text{ N-m} = 278\,500 \text{ N-mm}$$

∴ Tangential load on the pinion,

$$W_T = \frac{2T}{D_P} = \frac{2T}{m.T_P} = \frac{2 \times 278\,500}{m \times 30} = \frac{18\,567}{m} \text{ N}$$

We know that pitch line velocity,

$$v = \frac{\pi D_P.N_P}{1000} = \frac{\pi m.T_P.N_P}{1000} = \frac{\pi m \times 30 \times 1200}{1000} \text{ m/min}$$

$$= 113.1\, m \text{ m/min}$$

∴ Allowable working stress,

$$\sigma_w = 140\left(\frac{280}{280 + v}\right) = 140\left(\frac{280}{280 + 113.1m}\right) \text{ MPa or N/mm}^2$$

We know that length of the pitch cone element or slant height of the pitch cone,

$$L = \frac{D_P}{2\sin\theta_{P1}} = \frac{m \times T_P}{2\sin\theta_{P1}} = \frac{m \times 30}{2\sin 33°} = 27.54\, m \text{ mm}$$

Since the face width (b) is 1/4th of the slant height of the pitch cone, therefore

$$b = \frac{L}{4} = \frac{27.54m}{4} = 6.885\, m \text{ mm}$$

We know that tangential load on the pinion,

$$W_T = (\sigma_{OP} \times C_v)\, b.\pi\, m.y'_P\left(\frac{L - b}{L}\right)$$

$$= \sigma_w.b.\pi\, m.y'_P\left(\frac{L - b}{L}\right) \qquad \ldots (\because \sigma_w = \sigma_{OP} \times C_v)$$

or

$$\frac{18\,567}{m} = 140\left(\frac{280}{280 + 113.1m}\right) 6.885\, m \times \pi\, m \times 0.105\left(\frac{27.54m - 6.885m}{27.54m}\right)$$

$$= \frac{66\,780\, m^2}{280 + 113.1m}$$

or

$$280 + 113.1\, m = 66\,780\, m^2 \times \frac{m}{18\,567} = 3.6\, m^3$$

Solving this expression by hit and trial method, we find that

$$m = 6.6 \text{ say } 8 \text{ mm } \textbf{Ans.}$$

and face width, $$b = 6.885\, m = 6.885 \times 8 = 55 \text{ mm } \textbf{Ans.}$$

Addendum and dedendum for the pinion

We know that addendum,

$$a = 1\, m = 1 \times 8 = 8 \text{ mm } \textbf{Ans.}$$

and dedendum, $$d = 1.2\, m = 1.2 \times 8 = 9.6 \text{ mm } \textbf{Ans.}$$

Outside diameter for the pinion

We know that outside diameter for the pinion,

$$D_O = D_P + 2\, a \cos \theta_{P1} = m.T_P + 2\, a \cos \theta_{P1} \quad \ldots (\because D_P = m \,.\, T_P)$$

$$= 8 \times 30 + 2 \times 8 \cos 33^\circ = 253.4 \text{ mm } \textbf{Ans.}$$

Slant height

We know that slant height of the pitch cone,

$$L = 27.54\, m = 27.54 \times 8 = 220.3 \text{ mm } \textbf{Ans.}$$

Example 30.2. *A pair of cast iron bevel gears connect two shafts at right angles. The pitch diameters of the pinion and gear are 80 mm and 100 mm respectively. The tooth profiles of the gears are of 14 $^1/_2$° composite form. The allowable static stress for both the gears is 55 MPa. If the pinion transmits 2.75 kW at 1100 r.p.m., find the module and number of teeth on each gear from the standpoint of strength and check the design from the standpoint of wear. Take surface endurance limit as 630 MPa and modulus of elasticity for cast iron as 84 kN/mm^2.*

Solution. Given : $\theta_S = 90°$; $D_P = 80$ mm = 0.08 m ; $D_G = 100$ mm = 0.1 m ; $\phi = 14\frac{1}{2}°$; $\sigma_{OP} = \sigma_{OG} = 55$ MPa = 55 N/mm^2 ; $P = 2.75$ kW = 2750 W ; $N_P = 1100$ r.p.m. ; $\sigma_{es} = 630$ MPa = 630 N/mm^2 ; $E_P = E_G = 84$ kN/mm^2 = 84×10^3 N/mm^2

Module

Let m = Module in mm.

Since the shafts are at right angles, therefore pitch angle for the pinion,

$$\theta_{P1} = \tan^{-1}\left(\frac{1}{V.R.}\right) = \tan^{-1}\left(\frac{D_P}{D_G}\right) = \tan^{-1}\left(\frac{80}{100}\right) = 38.66°$$

and pitch angle for the gear,

$$\theta_{P2} = 90° - 38.66° = 51.34°$$

We know that formative number of teeth for pinion,

$$T_{EP} = T_P \,.\, \sec \theta_{P1} = \frac{80}{m} \times \sec 38.66° = \frac{102.4}{m} \quad \ldots (\because T_P = D_P / m)$$

and formative number of teeth on the gear,

$$T_{EG} = T_G \,.\, \sec \theta_{P2} = \frac{100}{m} \times \sec 51.34° = \frac{160}{m} \quad \ldots (\because T_G = D_G / m)$$

Since both the gears are made of the same material, therefore pinion is the weaker. Thus the design should be based upon the pinion.

We know that tooth form factor for the pinion having 14 $^1/_2$° composite teeth,

$$y'_P = 0.124 - \frac{0.684}{T_{EP}} = 0.124 - \frac{0.684 \times m}{102.4}$$

$$= 0.124 - 0.006\,68\, m$$

and pitch line velocity,

$$v = \frac{\pi D_P \,.\, N_P}{60} = \frac{\pi \times 0.08 \times 1100}{60} = 4.6 \text{ m/s}$$

Taking velocity factor,

$$C_v = \frac{6}{6+v} = \frac{6}{6+4.6} = 0.566$$

We know that length of the pitch cone element or slant height of the pitch cone,

$$*L = \sqrt{\left(\frac{D_G}{2}\right)^2 + \left(\frac{D_P}{2}\right)^2} = \sqrt{\left(\frac{100}{2}\right)^2 + \left(\frac{80}{2}\right)^2} = 64 \text{ mm}$$

Assuming the face width (b) as 1/3rd of the slant height of the pitch cone (L), therefore

$$b = L/3 = 64/3 = 21.3 \text{ say } 22 \text{ mm}$$

We know that torque on the pinion,

$$T = \frac{P \times 60}{2\pi \times N_P} = \frac{2750 \times 60}{2\pi \times 1100} = 23.87 \text{ N-m} = 23\,870 \text{ N-mm}$$

∴ Tangential load on the pinion,

$$W_T = \frac{T}{D_P/2} = \frac{23\,870}{80/2} = 597 \text{ N}$$

We also know that tangential load on the pinion,

$$W_T = (\sigma_{OP} \times C_v)\, b \times \pi\, m \times y'_P \left(\frac{L-b}{L}\right)$$

or
$$597 = (55 \times 0.566)\, 22 \times \pi\, m\,(0.124 - 0.00\,668\, m)\left(\frac{64-22}{64}\right)$$

$$= 1412\, m\,(0.124 - 0.006\,68\, m)$$

$$= 175\, m - 9.43\, m^2$$

Solving this expression by hit and trial method, we find that

$$m = 4.5 \text{ say } 5 \text{ mm } \textbf{Ans.}$$

The bevel gear turbine

Number of teeth on each gear

We know that number of teeth on the pinion,

$$T_P = D_P / m = 80/5 = 16 \textbf{ Ans.}$$

and number of teeth on the gear,

$$T_G = D_G / m = 100/5 = 20 \textbf{ Ans.}$$

Checking the gears for wear

We know that the load-stress factor,

$$K = \frac{(\sigma_{es})^2 \sin\phi}{1.4}\left[\frac{1}{E_P} + \frac{1}{E_G}\right]$$

$$= \frac{(630)^2 \sin 14\tfrac{1}{2}^\circ}{1.4}\left[\frac{1}{84 \times 10^3} + \frac{1}{84 \times 10^3}\right] = 1.687$$

and ratio factor, $Q = \dfrac{2\,T_{EG}}{T_{EG} + T_{EP}} = \dfrac{2 \times 160/m}{160/m + 102.4/m} = 1.22$

---

* The length of the pitch cone element (L) may also obtained by using the relation

$$L = D_P / 2 \sin \theta_{P1}$$

∴ Maximum or limiting load for wear,

$$W_w = \frac{D_P \cdot b \cdot Q \cdot K}{\cos \theta_{P1}} = \frac{80 \times 22 \times 1.22 \times 1.687}{\cos 38.66^\circ} = 4640 \text{ N}$$

Since the maximum load for wear is much more than the tangential load (W_T), therefore the design is satisfactory from the consideration of wear. **Ans.**

Example 30.3. *A pair of bevel gears connect two shafts at right angles and transmits 9 kW. Determine the required module and gear diameters for the following specifications :*

Particulars	*Pinion*	*Gear*
Number of teeth	*21*	*60*
Material	*Semi-steel*	*Grey cast iron*
Brinell hardness number	*200*	*160*
Allowable static stress	*85 MPa*	*55 MPa*
Speed	*1200 r.p.m.*	*420 r.p.m.*
Tooth profile	$14\frac{1}{2}°$ *composite*	$14\frac{1}{2}°$ *composite*

Check the gears for dynamic and wear loads.

Solution. Given : $\theta_S = 90°$; $P = 9$ kW $= 9000$ W ; $T_P = 21$; $T_G = 60$; $\sigma_{OP} = 85$ MPa $= 85$ N/mm² ; $\sigma_{OG} = 55$ MPa $= 55$ N/mm² ; $N_P = 1200$ r.p.m. ; $N_G = 420$ r.p.m. ; $\phi = 14\,^1/_2°$

Required module

Let m = Required module in mm.

Since the shafts are at right angles, therefore pitch angle for the pinion,

$$\theta_{P1} = \tan^{-1}\left(\frac{1}{V.R.}\right) = \tan^{-1}\left(\frac{T_P}{T_G}\right) = \tan^{-1}\left(\frac{21}{60}\right) = 19.3°$$

and pitch angle for the gear,

$$\theta_{P2} = \theta_S - \theta_{P1} = 90° - 19.3° = 70.7°$$

We know that formative number of teeth for the pinion,

$$T_{EP} = T_P \cdot \sec \theta_{P1} = 21 \sec 19.3° = 22.26$$

and formative number of teeth for the gear,

$$T_{EG} = T_G \cdot \sec \theta_{P2} = 60 \sec 70.7° = 181.5$$

We know that tooth form factor for the pinion,

$$y'_P = 0.124 - \frac{0.684}{T_{EP}} = 0.124 - \frac{0.684}{22.26} = 0.093$$

... (For $14\,^1/_2°$ composite system)

and tooth form factor for the gear,

$$y'_G = 0.124 - \frac{0.684}{T_{EG}} = 0.124 - \frac{0.684}{181.5} = 0.12$$

$$\therefore \quad \sigma_{OP} \times y'_P = 85 \times 0.093 = 7.905$$

and $$\sigma_{OG} \times y'_G = 55 \times 0.12 = 6.6$$

Since the product $\sigma_{OG} \times y'_G$ is less than $\sigma_{OP} \times y'_P$, therefore the gear is weaker. Thus, the design should be based upon the gear.

We know that torque on the gear,

$$T = \frac{P \times 60}{2\pi N_G} = \frac{9000 \times 60}{2\pi \times 420} = 204.6 \text{ N-m} = 204\,600 \text{ N-mm}$$

∴ Tangential load on the gear,

$$W_T = \frac{T}{D_G/2} = \frac{2T}{m.T_G} = \frac{2 \times 204\ 600}{m \times 60} = \frac{6820}{m}\ \text{N} \quad \text{... } (\because D_G = m.T_G)$$

We know that pitch line velocity,

$$v = \frac{\pi D_G . N_G}{60} = \frac{\pi\ m.T_G . N_G}{60} = \frac{\pi\ m \times 60 \times 420}{60}\ \text{mm/s}$$

$$= 1320\ m\ \text{mm/s} = 1.32\ m\ \text{m/s}$$

Taking velocity factor,

$$C_v = \frac{6}{6+v} = \frac{6}{6+1.32\,m}$$

We know that length of pitch cone element,

$$*L = \frac{D_G}{2\sin\theta_{P2}} = \frac{m.T_G}{2\sin 70.7^\circ} = \frac{m \times 60}{2 \times 0.9438} = 32\ m\ \text{mm}$$

Assuming the face width (b) as 1/3rd of the length of the pitch cone element (L), therefore

$$b = \frac{L}{3} = \frac{32\,m}{3} = 10.67\ m\ \text{mm}$$

We know that tangential load on the gear,

$$W_T = (\sigma_{OG} \times C_v)\ b.\pi\ m.y'_G \left(\frac{L-b}{L}\right)$$

$$\therefore \quad \frac{6820}{m} = 55\left(\frac{6}{6+1.32\,m}\right) 10.67\ m \times \pi\ m \times 0.12 \left(\frac{32\ m - 10.67\ m}{32\ m}\right)$$

$$= \frac{885\,m^2}{6+1.32\,m}$$

or $\quad 40\ 920 + 9002\ m = 885\ m^3$

Solving this expression by hit and trial method, we find that

$m = 4.52$ say 5 mm **Ans.**

and $\quad b = 10.67\ m = 10.67 \times 5 = 53.35$ say 54 mm **Ans.**

Gear diameters

We know that pitch diameter for the pinion,

$D_P = m.T_P = 5 \times 21 = 105$ mm **Ans.**

and pitch circle diameter for the gear,

$D_G = m.T_G = 5 \times 60 = 300$ mm **Ans.**

Check for dynamic load

We know that pitch line velocity,

$v = 1.32\ m = 1.32 \times 5 = 6.6$ m/s

and tangential tooth load on the gear,

$$W_T = \frac{6820}{m} = \frac{6820}{5} = 1364\ \text{N}$$

From Table 28.7, we find that tooth error action for first class commercial gears having module 5 mm is

$e = 0.055$ mm

---

* The length of pitch cone element (L) may be obtained by using the following relation, *i.e.*

$$L = \sqrt{\left(\frac{D_G}{2}\right)^2 + \left(\frac{D_P}{2}\right)^2} = \sqrt{\left(\frac{m.T_G}{2}\right)^2 + \left(\frac{m.T_P}{2}\right)^2} = \frac{m}{2}\sqrt{(T_G)^2 + (T_P)^2}$$

Taking $K = 0.107$ for $14\,^1/_2{}^\circ$ composite teeth, $E_P = 210 \times 10^3$ N/mm^2; and $E_G = 84 \times 10^3$ N/mm^2, we have

Deformation or dynamic factor,

$$C = \frac{K.e}{\dfrac{1}{E_P} + \dfrac{1}{E_G}} = \frac{0.107 \times 0.055}{\dfrac{1}{210 \times 10^3} + \dfrac{1}{84 \times 10^3}} = 353 \text{ N / mm}$$

We know that dynamic load on the gear,

$$W_D = W_T + \frac{21\, v\,(b.C + W_T)}{21\, v + \sqrt{b.C + W_T}}$$

$$= 1364 + \frac{21 \times 6.6\,(54 \times 353 + 1364)}{21 \times 6.6 + \sqrt{54 \times 353 + 1364}}$$

$$= 1364 + 10\,054 = 11\,418 \text{ N}$$

From Table 28.8, we find that flexural endurance limit (σ_e) for the gear material which is grey cast iron having B.H.N. = 160, is

$$\sigma_e = 84 \text{ MPa} = 84 \text{ N/mm}^2$$

We know that the static tooth load or endurance strength of the tooth,

$$W_S = \sigma_e.b.\pi\, m.y'_G = 84 \times 54 \times \pi \times 5 \times 0.12 = 8552 \text{ N}$$

Since W_S is less that W_D, therefore the design is not satisfactory from the standpoint of dynamic load. We have already discussed in spur gears (Art. 28.20) that $W_S \geq 1.25\, W_D$ for steady loads. For a satisfactory design against dynamic load, let us take the precision gears having tooth error in action ($e = 0.015$ mm) for a module of 5 mm.

∴ Deformation or dynamic factor,

$$C = \frac{0.107 \times 0.015}{\dfrac{1}{210 \times 10^3} + \dfrac{1}{84 \times 10^3}} = 96 \text{ N/mm}$$

and dynamic load on the gear,

$$W_D = 1364 + \frac{21 \times 6.6\,(54 \times 96 + 1364)}{21 \times 6.6 + \sqrt{54 \times 96 + 1364}} = 5498 \text{ N}$$

From above we see that by taking precision gears, W_S is greater than W_D, therefore the design is satisfactory, from the standpoint of dynamic load.

Check for wear load

From Table 28.9, we find that for a gear of grey cast iron having B.H.N. = 160, the surface endurance limit is,

$$\sigma_{es} = 630 \text{ MPa} = 630 \text{ N/mm}^2$$

∴ Load-stress factor,

$$K = \frac{(\sigma_{es})^2 \sin\phi}{1.4}\left[\frac{1}{E_P} + \frac{1}{E_G}\right]$$

$$= \frac{(630)^2 \sin 14\tfrac{1}{2}{}^\circ}{1.4}\left[\frac{1}{210 \times 10^3} + \frac{1}{84 \times 10^3}\right] = 1.18 \text{ N/mm}^2$$

and ratio factor,
$$Q = \frac{2\, T_{EG}}{T_{EG} + T_{EP}} = \frac{2 \times 181.5}{181.5 + 22.26} = 1.78$$

We know that maximum or limiting load for wear,

$$W_w = D_P.b.Q.K = 105 \times 54 \times 1.78 \times 1.18 = 11\,910 \text{ N}$$

Since W_w is greater then W_D, therefore the design is satisfactory from the standpoint of wear.

Example 30.4. *A pair of 20° full depth involute teeth bevel gears connect two shafts at right angles having velocity ratio 3 : 1. The gear is made of cast steel having allowable static stress as 70 MPa and the pinion is of steel with allowable static stress as 100 MPa. The pinion transmits 37.5 kW at 750 r.p.m. Determine : 1. Module and face width; 2. Pitch diameters; and 3. Pinion shaft diameter.*

Assume tooth form factor, $y = 0.154 - \frac{0.912}{T_E}$, where T_E is the formative number of teeth, width = 1/3 rd the length of pitch cone, and pinion shaft overhangs by 150 mm.

Involute teeth bevel gear

Solution. Given : $\phi = 20°$; $\theta_S = 90°$; $V.R. = 3$; $\sigma_{OG} = 70$ MPa $= 70$ N/mm^2 ; $\sigma_{OP} = 100$ MPa $= 100$ N/mm^2 ; $P = 37.5$ kW $= 37\,500$ W ; $N_P = 750$ r.p.m. ; $b = L/3$; Overhang $= 150$ mm

Module and face width

Let m = Module in mm,

b = Face width in mm = $L/3$, ...(Given)

D_G = Pitch circle diameter of the gear in mm.

Since the shafts are at right angles, therefore pitch angle for the pinion,

$$\theta_{P1} = \tan^{-1}\left(\frac{1}{V.R.}\right) = \tan^{-1}\left(\frac{1}{3}\right) = 18.43°$$

and pitch angle for the gear,

$$\theta_{P2} = \theta_S - \theta_{P1} = 90° - 18.43° = 71.57°$$

Assuming number of teeth on the pinion (T_P) as 20, therefore number of teeth on the gear,

$$T_G = V.R. \times T_P = 3 \times 20 = 60 \qquad \ldots (\because V.R. = T_G / T_P)$$

We know that formative number of teeth for the pinion,

$$T_{EP} = T_P \cdot \sec\theta_{P1} = 20 \times \sec 18.43° = 21.08$$

and formative number of teeth for the gear,

$$T_{EG} = T_G \cdot \sec\theta_{P2} = 60 \sec 71.57° = 189.8$$

We know that tooth form factor for the pinion,

$$y'_P = 0.154 - \frac{0.912}{T_{EP}} = 0.154 - \frac{0.912}{21.08} = 0.111$$

and tooth form factor for the gear,

$$y'_G = 0.154 - \frac{0.912}{T_{EG}} = 0.154 - \frac{0.912}{189.8} = 0.149$$

$$\therefore \quad \sigma_{OP} \times y'_P = 100 \times 0.111 = 11.1$$

and $$\sigma_{OG} \times y'_G = 70 \times 0.149 = 10.43$$

Since the product $\sigma_{OG} \times y'_G$ is less than $\sigma_{OP} \times y'_P$, therefore the gear is weaker. Thus, the design should be based upon the gear and not the pinion.

We know that the torque on the gear,

$$T = \frac{P \times 60}{2\pi N_G} = \frac{P \times 60}{2\pi \times N_P / 3} \quad \text{... } (\because V.R. = N_P / N_G = 3)$$

$$= \frac{37\ 500 \times 60}{2\pi \times 750/3} = 1432 \text{ N-m} = 1432 \times 10^3 \text{ N-mm}$$

∴ Tangential load on the gear,

$$W_T = \frac{2T}{D_G} = \frac{2T}{m.T_G} \quad \text{... } (\because D_G = m.T_G)$$

$$= \frac{2 \times 1432 \times 10^3}{m \times 60} = \frac{47.7 \times 10^3}{m} \text{ N}$$

We know that pitch line velocity,

$$v = \frac{\pi D_G . N_G}{60} = \frac{\pi m . T_G . N_P / 3}{60}$$

$$= \frac{\pi m \times 60 \times 750/3}{60} = 785.5\, m \text{ mm / s} = 0.7855\, m \text{ m / s}$$

Taking velocity factor,

$$C_v = \frac{3}{3 + v} = \frac{3}{3 + 0.7855m}$$

We know that length of the pitch cone element,

$$L = \frac{D_G}{2 \sin \theta_{P2}} = \frac{m.T_G}{2 \sin 71.57°} = \frac{m \times 60}{2 \times 0.9487} = 31.62\, m \text{ mm}$$

Since the face width (b) is 1/3rd of the length of the pitch cone element, therefore

$$b = \frac{L}{3} = \frac{31.62m}{3} = 10.54\, m \text{ mm}$$

We know that tangential load on the gear,

$$W_T = (\sigma_{OG} \times C_v)\, b.\pi\, m.y'_G \left(\frac{L - b}{L}\right)$$

Racks

$$\therefore \quad \frac{47.7 \times 10^3}{m} = 70\left(\frac{3}{3 + 0.7855\,m}\right) 10.54\,m \times \pi\,m \times 0.149 \left(\frac{31.62\,m - 10.54\,m}{31.62\,m}\right)$$

$$= \frac{691\,m^2}{3 + 0.7855\,m}$$

$$143\,100 + 37\,468\,m = 691\,m^3$$

Solving this expression by hit and trial method, we find that

$$m = 8.8 \text{ say } 10 \text{ mm } \textbf{Ans.}$$

and $\qquad b = 10.54\,m = 10.54 \times 10 = 105.4$ mm **Ans.**

Pitch diameters

We know that pitch circle diameter of the larger wheel (*i.e.* gear),

$$D_G = m.T_G = 10 \times 60 = 600 \text{ mm } \textbf{Ans.}$$

and pitch circle diameter of the smaller wheel (*i.e.* pinion),

$$D_P = m.T_P = 10 \times 20 = 200 \text{ mm } \textbf{Ans.}$$

Pinion shaft diameter

Let $\qquad d_P$ = Pinion shaft diameter.

We know that the torque on the pinion,

$$T = \frac{P \times 60}{2\pi \times N_P} = \frac{37\,500 \times 60}{2\pi \times 750} = 477.4 \text{ N-m} = 477\,400 \text{ N-mm}$$

and length of the pitch cone element,

$$L = 31.62\,m = 31.62 \times 10 = 316.2 \text{ mm}$$

∴ Mean radius of the pinion,

$$R_m = \left(L - \frac{b}{2}\right)\frac{D_P}{2\,L} = \left(316.2 - \frac{105.4}{2}\right)\frac{200}{2 \times 316.2} = 83.3 \text{ mm}$$

We know that tangential force acting at the mean radius,

$$W_T = \frac{T}{R_m} = \frac{477\,400}{83.3} = 5731 \text{ N}$$

Axial force acting on the pinion shaft,

$$W_{RH} = W_T \tan\phi.\sin\theta_{P1} = 5731 \times \tan 20° \times \sin 18.43°$$

$$= 5731 \times 0.364 \times 0.3161 = 659.4 \text{ N}$$

and radial force acting on the pinion shaft,

$$W_{RV} = W_T \tan\phi\,.\cos\theta_{P1} = 5731 \times \tan 20° \times \cos 18.43°$$

$$= 5731 \times 0.364 \times 0.9487 = 1979 \text{ N}$$

∴ Bending moment due to W_{RH} and W_{RV},

$$M_1 = W_{RV} \times \text{Overhang} - W_{RH} \times R_m$$

$$= 1979 \times 150 - 659.4 \times 83.3 = 241\,920 \text{ N-mm}$$

and bending moment due to W_T,

$$M_2 = W_T \times \text{Overhang} = 5731 \times 150 = 859\,650 \text{ N-mm}$$

∴ Resultant bending moment,

$$M = \sqrt{(M_1)^2 + (M_2)^2} = \sqrt{(241\,920)^2 + (859\,650)^2} = 893\,000 \text{ N-mm}$$

Since the shaft is subjected to twisting moment (T) and bending moment (M), therefore equivalent twisting moment,

$$T_e = \sqrt{M^2 + T^2}$$

$$= \sqrt{(893\ 000)^2 + (477\ 400)^2}$$

$$= 1013 \times 10^3 \text{ N-mm}$$

We also know that equivalent twisting moment (T_e),

$$1013 \times 10^3 = \frac{\pi}{16} \times \tau\,(d_P)^3$$

$$= \frac{\pi}{16} \times 45\,(d_P)^3 = 8.84\,(d_P)^3$$

... (Taking $\tau = 45$ N/mm²)

$\therefore\quad (d_P)^3 = 1013 \times 10^3 / 8.84 = 114.6 \times 10^3$

or $\quad d_P = 48.6$ say 50 mm **Ans.**

Bevel gears

EXERCISES

1. A pair of straight bevel gears is required to transmit 10 kW at 500 r.p.m. from the motor shaft to another shaft at 250 r.p.m. The pinion has 24 teeth. The pressure angle is 20°. If the shaft axes are at right angles to each other, find the module, face width, addendum, outside diameter and slant height. The gears are capable of withstanding a static stress of 60 MPa. The tooth form factor may be taken as $0.154 - 0.912/T_E$, where T_E is the equivalent number of teeth. Assume velocity factor as $\frac{4.5}{4.5 + v}$, where v the pitch line speed in m/s. The face width may be taken as $\frac{1}{4}$ of the slant height of the pitch cone. **[Ans. $m = 8$ mm ; $b = 54$ mm ; $a = 8$ mm ; $D_O = 206.3$ mm ; $L = 214.4$ mm]**
2. A 90° bevel gearing arrangement is to be employed to transmit 4 kW at 600 r.p.m. from the driving shaft to another shaft at 200 r.p.m. The pinion has 30 teeth. The pinion is made of cast steel having a static stress of 80 MPa and the gear is made of cast iron with a static stress of 55 MPa. The tooth profiles of the gears are of $14\frac{1}{2}$° composite form. The tooth form factor may be taken as $y' = 0.124 - 0.684 / T_E$, where T_E is the formative number of teeth and velocity factor, $C_v = \frac{3}{3 + v}$, where v is the pitch line speed in m/s.

   The face width may be taken as $^1/_3$ rd of the slant height of the pitch cone. Determine the module, face width and pitch diameters for the pinion and gears, from the standpoint of strength and check the design from the standpoint of wear. Take surface endurance limit as 630 MPa and modulus of elasticity for the material of gears is $E_P = 200$ kN/mm² and $E_G = 80$ kN/mm².

   [Ans. $m = 4$ mm ; $b = 64$ mm ; $D_P = 120$ mm ; $D_G = 360$ mm]
3. A pair of bevel gears is required to transmit 11 kW at 500 r.p.m. from the motor shaft to another shaft, the speed reduction being 3 : 1. The shafts are inclined at 60°. The pinion is to have 24 teeth with a pressure angle of 20° and is to be made of cast steel having a static stress of 80 MPa. The gear is to be made of cast iron with a static stress of 55 MPa. The tooth form factor may be taken as $y = 0.154 - 0.912/T_E$, where T_E is formative number of teeth. The velocity factor may be taken as $\frac{3}{3 + v}$, where v is the pitch line velocity in m/s. The face width may be taken as $^1/_4$ th of the slant height of the pitch cone. The mid-plane of the gear is 100 mm from the left hand bearing and 125 mm from the right hand bearing. The gear shaft is to be made of colled-rolled steel for which the allowable tensile stress may be taken as 80 MPa. Design the gears and the gear shaft.

QUESTIONS

1. How the bevel gears are classified ? Explain with neat sketches.
2. Sketch neatly the working drawing of bevel gears in mesh.
3. For bevel gears, define the following :

   (*i*) Cone distance; (*ii*) Pitch angle; (*iii*) Face angle; (*iv*) Root angle; (*v*) Back cone distance; and (*vi*) Crown height.
4. What is Tredgold's approximation about the formative number of teeth on bevel gear?
5. What are the various forces acting on a bevel gear ?
6. Write the procedure for the design of a shaft for bevel gears.

OBJECTIVE TYPE QUESTIONS

1. When bevel gears having equal teeth and equal pitch angles connect two shafts whose axes intersect at right angle, then they are known as

   (*a*) angular bevel gears (*b*) crown bevel gears

   (*c*) internal bevel gears (*d*) mitre gears
2. The face angle of a bevel gear is equal to

   (*a*) pitch angle – addendum angle (*b*) pitch angle + addendum angle

   (*c*) pitch angle – dedendum angle (*d*) pitch angle + dedendum angle
3. The root angle of a bevel gear is equal to

   (*a*) pitch angle – addendum angle (*b*) pitch angle + addendum angle

   (*c*) pitch angle – dedendum angle (*d*) pitch angle + dedendum angle
4. If *b* denotes the face width and L denotes the cone distance, then the bevel factor is written as

   (*a*) b / L (*b*) $b / 2L$

   (*c*) $1 - 2\, b.L$ (*d*) $1 - b / L$
5. For a bevel gear having the pitch angle θ, the ratio of formative number of teeth (T_E) to actual number of teeth (T) is

   (*a*) $\dfrac{1}{\sin \theta}$ (*b*) $\dfrac{1}{\cos \theta}$

   (*c*) $\dfrac{1}{\tan \theta}$ (*d*) $\sin \theta \cos \theta$

ANSWERS

1. (*d*) 2. (*b*) 3. (*c*) 4. (*d*) 5. (*b*)

CHAPTER

31

Worm Gears

31.1 Introduction

The worm gears are widely used for transmitting power at high velocity ratios between non-intersecting shafts that are generally, but not necessarily, at right angles. It can give velocity ratios as high as 300 : 1 or more in a single step in a minimum of space, but it has a lower efficiency. The worm gearing is mostly used as a speed reducer, which consists of worm and a worm wheel or gear. The worm (which is the driving member) is usually of a cylindrical form having threads of the same shape as that of an involute rack. The threads of the worm may be left handed or right handed and single or multiple threads. The worm wheel or gear (which is the driven member) is similar to a helical gear with a face curved to conform to the shape of the worm. The worm is generally made of steel while the worm gear is made of bronze or cast iron for light service.

The worm gearing is classified as non-interchangeable, because a worm wheel cut with a hob of one diameter will not operate satisfactorily with a worm of different diameter, even if the thread pitch is same.

31.2 Types of Worms

The following are the two types of worms :

1. Cylindrical or straight worm, and
2. Cone or double enveloping worm.

The ***cylindrical*** or ***straight worm,*** as shown in Fig. 31.1 (*a*), is most commonly used. The shape of the thread is involute helicoid of pressure angle 14 ½° for single and double threaded worms and 20° for triple and quadruple threaded worms. The worm threads are cut by a straight sided milling cutter having its diameter not less than the outside diameter of worm or greater than 1.25 times the outside diameter of worm.

The ***cone*** or ***double enveloping worm,*** as shown in Fig. 31.1 (*b*), is used to some extent, but it requires extremely accurate alignment.

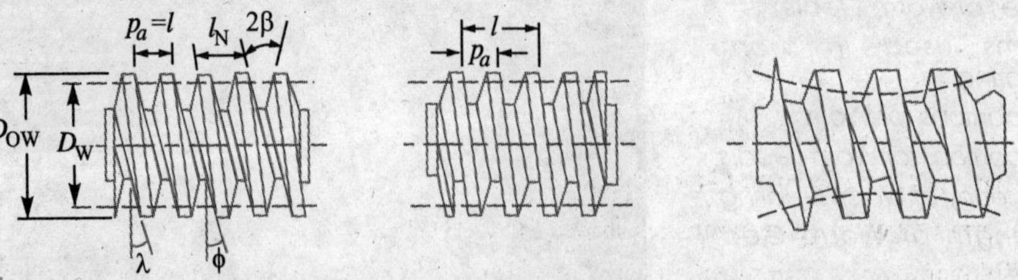

Single threaded. Double threaded.

(*a*) Cylindrical or straight worm. (*b*) Cone or double enveloping worm.

Fig. 31.1. Types of worms.

31.3 Types of Worm Gears

The following three types of worm gears are important from the subject point of view :

1. Straight face worm gear, as shown in Fig. 31.2 (*a*),
2. Hobbed straight face worm gear, as shown in Fig. 31.2 (*b*), and
3. Concave face worm gear, as shown in Fig. 31.2 (*c*).

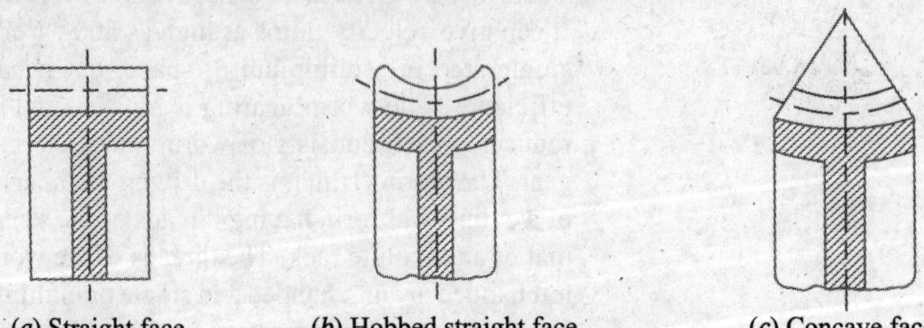

(*a*) Straight face. (*b*) Hobbed straight face. (*c*) Concave face.

Fig. 31.2. Types of worms gears.

The ***straight face worm gear*** is like a helical gear in which the straight teeth are cut with a form cutter. Since it has only point contact with the worm thread, therefore it is used for light service.

The ***hobbed straight face worm gear*** is also used for light service but its teeth are cut with a hob, after which the outer surface is turned.

The ***concave face worm gear*** is the accepted standard form and is used for all heavy service and general industrial uses. The teeth of this gear are cut with a hob of the same pitch diameter as the mating worm to increase the contact area.

Worm gear is used mostly where the power source operates at a high speed and output is at a slow speed with high torque. It is also used in some cars and trucks.

31.4 Terms used in Worm Gearing

The worm and worm gear in mesh is shown in Fig. 31.3.

The following terms, in connection with the worm gearing, are important from the subject point of view :

1. ***Axial pitch.*** It is also known as ***linear pitch*** of a worm. It is the distance measured axially (*i.e.* parallel to the axis of worm) from a point on one thread to the corresponding point on the adjacent thread on the worm, as shown in Fig. 31.3. It may be noted that the axial pitch (p_a) of a worm is equal to the circular pitch (p_c) of the mating worm gear, when the shafts are at right angles.

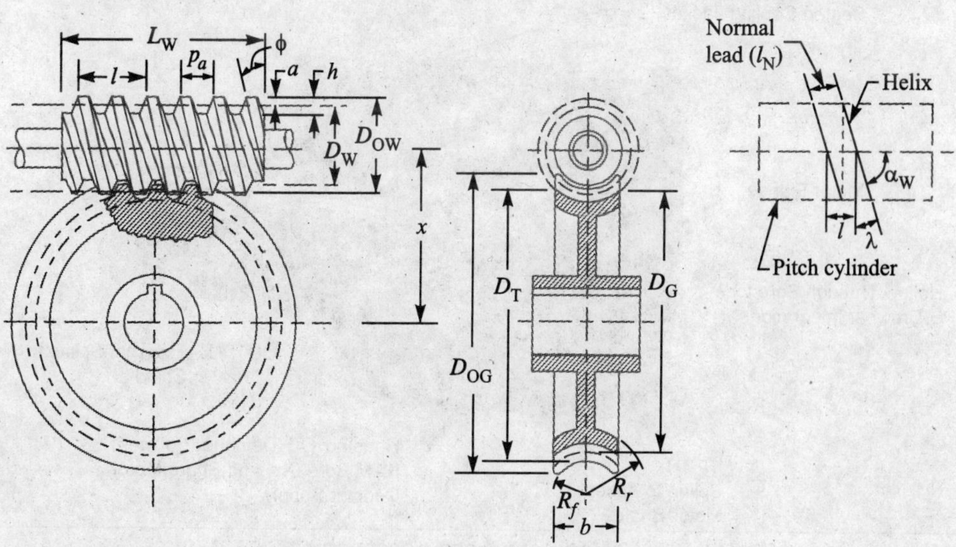

Fig. 31.3 . Worm and Worm gear.

2. ***Lead.*** It is the linear distance through which a point on a thread moves ahead in one revolution of the worm. For single start threads, lead is equal to the axial pitch, but for multiple start threads, lead is equal to the product of axial pitch and number of starts. Mathematically,

Lead, $l = p_a \cdot n$

where p_a = Axial pitch ; and n = Number of starts.

3. ***Lead angle.*** It is the angle between the tangent to the thread helix on the pitch cylinder and the plane normal to the axis of the worm. It is denoted by λ.

A little consideration will show that if one complete turn of a worm thread be imagined to be unwound from the body of the worm, it will form an inclined plane whose base is equal to the pitch circumference of the worm and altitude equal to lead of the worm, as shown in Fig. 31.4.

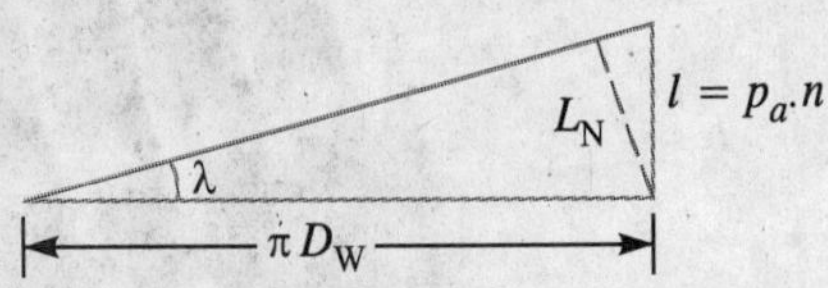

Fig. 31.4. Development of a helix thread.

From the geometry of the figure, we find that

$$\tan \lambda = \frac{\text{Lead of the worm}}{\text{Pitch circumference of the worm}}$$

$$= \frac{l}{\pi D_W} = \frac{p_a \cdot n}{\pi D_W} \qquad ...(\because l = p_a \cdot n)$$

$$= \frac{p_c \cdot n}{\pi D_W} = \frac{\pi m \cdot n}{\pi D_W} = \frac{m \cdot n}{D_W} \qquad ...(\because p_a = p_c \text{ ; and } p_c = \pi m)$$

where m = Module, and

D_W = Pitch circle diameter of worm.

The lead angle (λ) may vary from 9° to 45°. It has been shown by F.A. Halsey that a lead angle less than 9° results in rapid wear and the safe value of λ is 12½°.

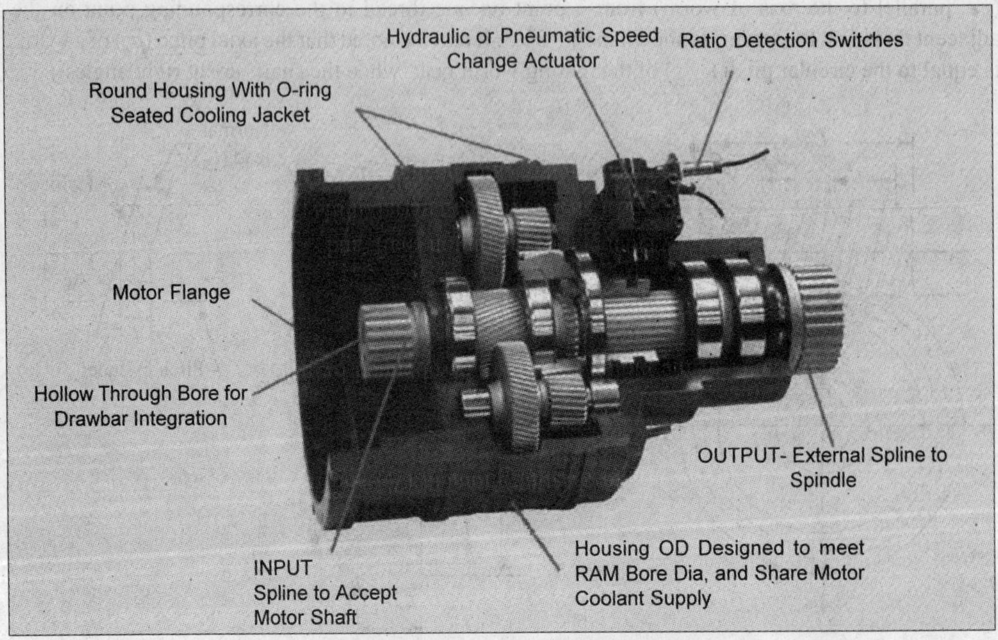

Model of sun and planet gears.

For a compaci design, the lead angle may be determined by the following relation, *i.e.*

$$\tan \lambda = \left(\frac{N_G}{N_W}\right)^{1/3},$$

where N_G is the speed of the worm gear and N_W is the speed of the worm.

4. ***Tooth pressure angle.*** It is measured in a plane containing the axis of the worm and is equal to one-half the thread profile angle as shown in Fig. 31.3.

The following table shows the recommended values of lead angle (λ) and tooth pressure angle (φ).

Table 31.1. Recommended values of lead angle and pressure angle.

Lead angle (λ) in degrees	0 – 16	16 – 25	25 – 35	35 – 45
Pressure angle(φ) in degrees	14½	20	25	30

For automotive applications, the pressure angle of 30° is recommended to obtain a high efficiency and to permit overhauling.

5. ***Normal pitch.*** It is the distance measured along the normal to the threads between two corresponding points on two adjacent threads of the worm. Mathematically,

Normal pitch, $p_N = p_a.\cos \lambda$

Note. The term normal pitch is used for a worm having single start threads. In case of a worm having multiple start threads, the term normal lead (l_N) is used, such that

$$l_N = l\,.\cos \lambda$$

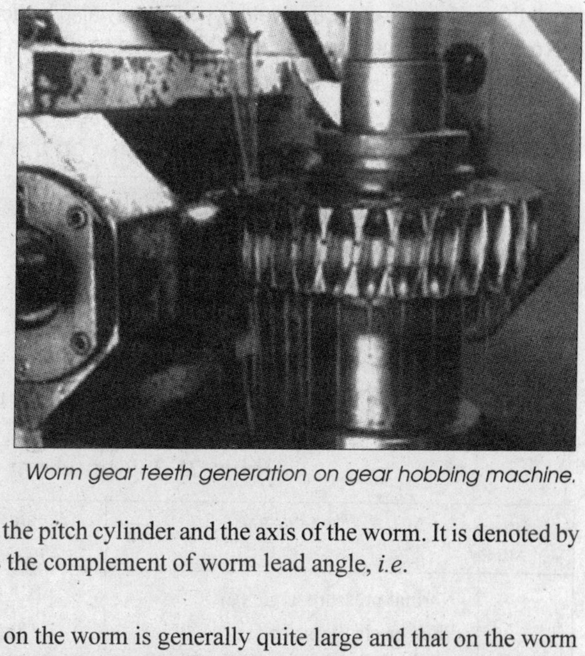

Worm gear teeth generation on gear hobbing machine.

6. ***Helix angle.*** It is the angle between the tangent to the thread helix on the pitch cylinder and the axis of the worm. It is denoted by α_W, in Fig. 31.3. The worm helix angle is the complement of worm lead angle, *i.e.*

$$\alpha_W + \lambda = 90°$$

It may be noted that the helix angle on the worm is generally quite large and that on the worm gear is very small. Thus, it is usual to specify the lead angle (λ) on the worm and helix angle (α_G) on the worm gear. These two angles are equal for a 90° shaft angle.

7. ***Velocity ratio.*** It is the ratio of the speed of worm (N_W) in r.p.m. to the speed of the worm gear (N_G) in r.p.m. Mathematically, velocity ratio,

$$V.R. = \frac{N_W}{N_G}$$

Let l = Lead of the worm, and

D_G = Pitch circle diameter of the worm gear.

We know that linear velocity of the worm,

$$v_W = \frac{l.N_W}{60}$$

and linear velocity of the worm gear,

$$v_G = \frac{\pi \, D_G \, N_G}{60}$$

Since the linear velocity of the worm and worm gear are equal, therefore

$$\frac{l \, . N_W}{60} = \frac{\pi \, D_G \, . N_G}{60} \quad \text{or} \quad \frac{N_W}{N_G} = \frac{\pi \, D_G}{l}$$

We know that pitch circle diameter of the worm gear,

$$D_G = m \, . \, T_G$$

where m is the module and T_G is the number of teeth on the worm gear.

$$\therefore \quad V.R. = \frac{N_W}{N_G} = \frac{\pi \, D_G}{l} = \frac{\pi \, m . T_G}{l}$$

$$= \frac{p_c \, . T_G}{l} = \frac{p_a \, . T_G}{p_a \, . n} = \frac{T_G}{n} \qquad \text{... } (\because p_c = \pi \, m = p_a \text{; and } l = p_a \, . \, n)$$

where n = Number of starts of the worm.

From above, we see that velocity ratio may also be defined as the ratio of number of teeth on the worm gear to the number of starts of the worm.

The following table shows the number of starts to be used on the worm for the different velocity ratios :

Table 31.2. Number of starts to be used on the worm for different velocity ratios.

Velocity ratio (*V.R.*)	36 and above	12 to 36	8 to 12	6 to 12	4 to 10
Number of starts or threads on the worm ($n = T_w$)	Single	Double	Triple	Quadruple	Sextuple

31.5 Proportions for Worms

The following table shows the various porportions for worms in terms of the axial or circular pitch (p_c) in mm.

Table 31.3. Proportions for worm.

S. No.	*Particulars*	*Single and double threaded worms*	*Triple and quadruple threaded worms*
1.	Normal pressure angle (ϕ)	14½°	20°
2.	Pitch circle diameter for worms integral with the shaft	$2.35 \, p_c + 10$ mm	$2.35 \, p_c + 10$ mm
3.	Pitch circle diameter for worms bored to fit over the shaft	$2.4 \, p_c + 28$ mm	$2.4 \, p_c + 28$ mm
4.	Maximum bore for shaft	$p_c + 13.5$ mm	$p_c + 13.5$ mm
5.	Hub diameter	$1.66 \, p_c + 25$ mm	$1.726 \, p_c + 25$ mm
6.	Face length (L_W)	$p_c \, (4.5 + 0.02 \, T_W)$	$p_c \, (4.5 + 0.02 \, T_W)$
7.	Depth of tooth (h)	$0.686 \, p_c$	$0.623 \, p_c$
8.	Addendum (a)	$0.318 \, p_c$	$0.286 \, p_c$

Notes: 1. The pitch circle diameter of the worm (D_W) in terms of the centre distance between the shafts (x) may be taken as follows :

$$D_W = \frac{(x)^{0.875}}{1.416} \qquad \text{... (when } x \text{ is in mm)}$$

2. The pitch circle diameter of the worm (D_W) may also be taken as

$$D_W = 3\,p_c, \text{ where } p_c \text{ is the axial or circular pitch.}$$

3. The face length (or length of the threaded portion) of the worm should be increased by 25 to 30 mm for the feed marks produced by the vibrating grinding wheel as it leaves the thread root.

31.6 Proportions for Worm Gear

The following table shows the various proportions for worm gears in terms of circular pitch (p_c) in mm.

Table 31.4. Proportions for worm gear.

S. No.	*Particulars*	*Single and double threads*	*Triple and quadruple threads*
1.	Normal pressure angle (ϕ)	14½°	20°
2.	Outside diameter (D_{OG})	$D_G + 1.0135\,p_c$	$D_G + 0.8903\,p_c$
3.	Throat diameter (D_T)	$D_G + 0.636\,p_c$	$D_G + 0.572\,p_c$
4.	Face width (b)	$2.38\,p_c + 6.5$ mm	$2.15\,p_c + 5$ mm
5.	Radius of gear face (R_f)	$0.882\,p_c + 14$ mm	$0.914\,p_c + 14$ mm
6.	Radius of gear rim (R_r)	$2.2\,p_c + 14$ mm	$2.1\,p_c + 14$ mm

31.7 Efficiency of Worm Gearing

The efficiency of worm gearing may be defined as the ratio of work done by the worm gear to the work done by the worm.

Mathematically, the efficiency of worm gearing is given by

$$\eta = \frac{\tan\lambda\,(\cos\phi - \mu\tan\lambda)}{\cos\phi\tan\lambda + \mu} \quad ...(i)$$

where ϕ = Normal pressure angle,

μ = Coefficient of friction, and

λ = Lead angle.

The efficiency is maximum, when

$$\tan\lambda = \sqrt{1+\mu^2} - \mu$$

In order to find the approximate value of the efficiency, assuming square threads, the following relation may be used :

$$\text{Efficiency, } \eta = \frac{\tan\lambda\,(1-\mu\tan\lambda)}{\tan\lambda + \mu} = \frac{1-\mu\tan\lambda}{1+\mu/\tan\lambda}$$

$$= \frac{\tan\lambda}{\tan(\lambda + \phi_1)}$$

...(Substituting in equation (*i*), $\phi = 0$, for square threads)

where ϕ_1 = Angle of friction, such that $\tan\phi_1 = \mu$.

A gear-cutting machine is used to cut gears.

The coefficient of friction varies with the speed, reaching a minimum value of 0.015 at a rubbing speed $\left(v_r = \dfrac{\pi\, D_W \,.\, N_W}{\cos \lambda}\right)$ between 100 and 165 m/min. For a speed below 10 m/min, take $\mu = 0.015$. The following empirical relations may be used to find the value of μ, *i.e.*

$$\mu = \frac{0.275}{(v_r)^{0.25}}, \text{ for rubbing speeds between 12 and 180 m/min}$$

$$= 0.025 + \frac{v_r}{18000} \text{ for rubbing speed more than 180 m/min}$$

Note : If the efficiency of worm gearing is less than 50%, then the worm gearing is said to be ***self locking,*** *i.e.* it cannot be driven by applying a torque to the wheel. This property of self locking is desirable in some applications such as hoisting machinery.

Hardened and ground worm shaft and worm wheel pair

Example 31.1. *A triple threaded worm has teeth of 6 mm module and pitch circle diameter of 50 mm. If the worm gear has 30 teeth of 14½° and the coefficient of friction of the worm gearing is 0.05, find 1. the lead angle of the worm, 2. velocity ratio, 3. centre distance, and 4. efficiency of the worm gearing.*

Solution. Given : $n = 3$; $m = 6$; $D_W = 50$ mm ; $T_G = 30$; $\phi = 14.5°$; $\mu = 0.05$.

1. ***Lead angle of the worm***

Let λ = Lead angle of the worm.

We know that $\tan \lambda = \dfrac{m.n}{D_W} = \dfrac{6 \times 3}{50} = 0.36$

∴ $\lambda = \tan^{-1}(0.36) = 19.8°$ **Ans.**

2. ***Velocity ratio***

We know that velocity ratio,

$$V.R. = T_G / n = 30 / 3 = 10 \textbf{ Ans.}$$

3. ***Centre distance***

We know that pitch circle diameter of the worm gear

$$D_G = m.T_G = 6 \times 30 = 180 \text{ mm}$$

∴ Centre distance,

$$x = \frac{D_W + D_G}{2} = \frac{50 + 180}{2} = 115 \text{ mm } \textbf{Ans.}$$

4. ***Efficiency of the worm gearing***

We know that efficiency of the worm gearing.

$$\eta = \frac{\tan \lambda\,(\cos \phi - \mu \tan \lambda)}{\cos \phi \,.\, \tan \lambda + \mu}$$

$$= \frac{\tan 19.8°\,(\cos 14.5° - 0.05 \times \tan 19.8°)}{\cos 14.5° \times \tan 19.8° + 0.05}$$

$$= \frac{0.36\,(0.9681 - 0.05 \times 0.36)}{0.9681 \times 0.36 + 0.05} = \frac{0.342}{0.3985} = 0.858 \text{ or } 85.8\% \textbf{ Ans.}$$

Note : The approximate value of the efficiency assuming square threads is

$$\eta = \frac{1 - \mu \tan \lambda}{1 + \mu / \tan \lambda} = \frac{1 - 0.05 \times 0.36}{1 + 0.05/0.36} = \frac{0.982}{1.139} = 0.86 \text{ or } 86\% \quad \textbf{Ans.}$$

31.8 Strength of Worm Gear Teeth

In finding the tooth size and strength, it is safe to assume that the teeth of worm gear are always weaker than the threads of the worm. In worm gearing, two or more teeth are usually in contact, but due to uncertainty of load distribution among themselves it is assumed that the load is transmitted by one tooth only. We know that according to Lewis equation,

$$W_T = (\sigma_o \cdot C_v)\, b.\, \pi\, m \,.\, y$$

where W_T = Permissible tangential tooth load or beam strength of gear tooth,

σ_o = Allowable static stress,

C_v = Velocity factor,

b = Face width,

m = Module, and

y = Tooth form factor or Lewis factor.

Notes : 1. The velocity factor is given by

$$C_v = \frac{6}{6 + v}, \text{ where } v \text{ is the peripheral velocity of the worm gear in m/s.}$$

2. The tooth form factor or Lewis factor (y) may be obtained in the similar manner as discussed in spur gears (Art. 28.17), *i.e.*

$$y = 0.124 - \frac{0.684}{T_G}, \text{ for } 14\tfrac{1}{2}° \text{ involute teeth.}$$

$$= 0.154 - \frac{0.912}{T_G}, \text{ for } 20° \text{ involute teeth.}$$

3. The dynamic tooth load on the worm gear is given by

$$W_D = \frac{W_T}{C_v} = W_T\left(\frac{6 + v}{6}\right)$$

where W_T = Actual tangential load on the tooth.

The dynamic load need not to be calculated because it is not so severe due to the sliding action between the worm and worm gear.

4. The static tooth load or endurance strength of the tooth (W_S) may also be obtained in the similar manner as discussed in spur gears (Art. 28.20), *i.e.*

$$W_S = \sigma_e . b\, \pi\, m . y$$

where σ_e = Flexural endurance limit. Its value may be taken as 84 MPa for cast iron and 168 MPa for phosphor bronze gears.

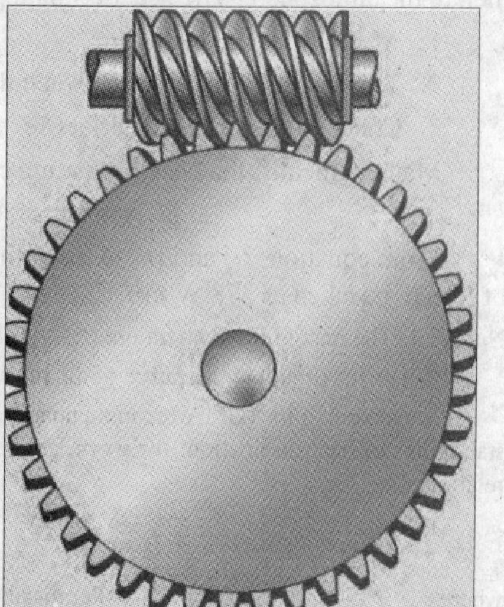

Worm gear assembly.

31.9 Wear Tooth Load for Worm Gear

The limiting or maximum load for wear (W_W) is given by

$$W_W = D_G \,.\, b \,.\, K$$

where D_G = Pitch circle diameter of the worm gear,

b = Face width of the worm gear, and

K = Load stress factor (also known as material combination factor).

The load stress factor depends upon the combination of materials used for the worm and worm gear. The following table shows the values of load stress factor for different combination of worm and worm gear materials.

Table 31.5. Values of load stress factor (K).

S.No.	*Material*		*Load stress factor (K) N/mm²*
	Worm	*Worm gear*	
1.	Steel (B.H.N. 250)	Phosphor bronze	0.415
2.	Hardened steel	Cast iron	0.345
3.	Hardened steel	Phosphor bronze	0.550
4.	Hardened steel	Chilled phosphor bronze	0.830
5.	Hardened steel	Antimony bronze	0.830
6.	Cast iron	Phosphor bronze	1.035

Note : The value of K given in the above table are suitable for lead angles upto 10°. For lead angles between 10° and 25°, the values of K should be increased by 25 per cent and for lead angles greater than 25°, increase the value of K by 50 per cent.

31.10 Thermal Rating of Worm Gearing

In the worm gearing, the heat generated due to the work lost in friction must be dissipated in order to avoid over heating of the drive and lubricating oil. The quantity of heat generated (Q_g) is given by

$$Q_g = \text{Power lost in friction in watts} = P(1 - \eta) \quad \text{...(i)}$$

where P = Power transmitted in watts, and

η = Efficiency of the worm gearing.

The heat generated must be dissipated through the lubricating oil to the gear box housing and then to the atmosphere. The heat dissipating capacity depends upon the following factors :

1. Area of the housing (A),
2. Temperature difference between the housing surface and surrounding air ($t_2 - t_1$), and
3. Conductivity of the material (K).

Mathematically, the heat dissipating capacity,

$$Q_d = A(t_2 - t_1)K \quad \text{...(ii)}$$

From equations (*i*) and (*ii*), we can find the temperature difference ($t_2 - t_1$). The average value of K may be taken as 378 W/m²/°C.

Notes : 1. The maximum temperature ($t_2 - t_1$) should not exceed 27 to 38°C.

2. The maximum temperature of the lubricant should not exceed 60°C.

3. According to AGMA recommendations, the limiting input power of a plain worm gear unit from the standpoint of heat dissipation, for worm gear speeds upto 2000 r.p.m., may be checked from the following relation, *i.e.*

$$P = \frac{3650\, x^{1.7}}{V.R. + 5}$$

where P = Permissible input power in kW,

x = Centre distance in metres, and

$V.R.$ = Velocity ratio or transmission ratio.

31.11 Forces Acting on Worm Gears

When the worm gearing is transmitting power, the forces acting on the worm are similar to those on a power screw. Fig. 31.5 shows the forces acting on the worm. It may be noted that the forces on a worm gear are equal in magnitude to that of worm, but opposite in direction to those shown in Fig. 31.5.

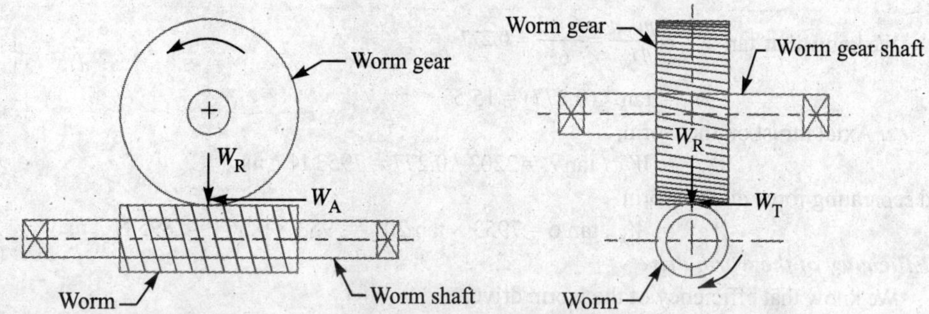

Fig. 31.5. Forces acting on worm teeth.

The various forces acting on the worm may be determined as follows :

1. Tangential force on the worm,

$$W_T = \frac{2 \times \text{Torque on worm}}{\text{Pitch circle diameter of worm } (D_W)}$$

$$= \text{Axial force or thrust on the worm gear}$$

The tangential force (W_T) on the worm produces a twisting moment of magnitude ($W_T \times D_W/2$) and bends the worm in the horizontal plane.

2. Axial force or thrust on the worm,

$$W_A = W_T / \tan \lambda = \text{Tangential force on the worm gear}$$

$$= \frac{2 \times \text{Torque on the worm gear}}{\text{Pitch circle diameter of worm gear} (D_G)}$$

The axial force on the worm tends to move the worm axially, induces an axial load on the bearings and bends the worm in a vertical plane with a bending moment of magnitude ($W_A \times D_W/2$).

3. Radial or separating force on the worm,

$$W_R = W_A \, . \tan \phi = \text{Radial or separating force on the worm gear}$$

The radial or separating force tends to force the worm and worm gear out of mesh. This force also bends the worm in the vertical plane.

Example 31.2. *A worm drive transmits 15 kW at 2000 r.p.m. to a machine carriage at 75 r.p.m. The worm is triple threaded and has 65 mm pitch diameter. The worm gear has 90 teeth of 6 mm module. The tooth form is to be 20° full depth involute. The coefficient of friction between the mating teeth may be taken as 0.10. Calculate : 1. tangential force acting on the worm ; 2. axial thrust and separating force on worm; and 3. efficiency of the worm drive.*

Solution. Given : P = 15 kW = 15 × 10³ W ; N_W = 2000 r.p.m. ; N_G = 75 r.p.m. ; n = 3 ; D_W = 65 mm ; T_G = 90 ; m = 6 mm ; ϕ = 20° ; μ = 0.10

1. ***Tangential force acting on the worm***

We know that the torque transmitted by the worm

$$= \frac{P \times 60}{2 \pi N_W} = \frac{15 \times 10^3 \times 60}{2 \pi \times 2000} = 71.6 \text{ N-m} = 71\,600 \text{ N-mm}$$

∴ Tangential force acting on the worm,

$$W_T = \frac{\text{Torque on worm}}{\text{Radius of worm}} = \frac{71\ 600}{65/2} = 2203 \text{ N } \textbf{Ans.}$$

2. *Axial thrust and separating force on worm*

Let λ = Lead angle.

We know that $\tan \lambda = \frac{m \cdot n}{D_W} = \frac{6 \times 3}{65} = 0.277$

or $\lambda = \tan^{-1}(0.277) = 15.5°$

∴ Axial thrust on the worm,

$$W_A = W_T / \tan \lambda = 2203 / 0.277 = 7953 \text{ N } \textbf{Ans.}$$

and separating force on the worm

$$W_R = W_A \cdot \tan \phi = 7953 \times \tan 20° = 7953 \times 0.364 = 2895 \text{ N } \textbf{Ans.}$$

3. *Efficiency of the worm drive*

We know that efficiency of the worm drive,

$$\eta = \frac{\tan \lambda (\cos \phi - \mu \cdot \tan \lambda)}{\cos \phi \cdot \tan \lambda + \mu}$$

$$= \frac{\tan 15.5° (\cos 20° - 0.10 \times \tan 15.5°)}{\cos 20° \times \tan 15.5° + 0.10}$$

$$= \frac{0.277 (0.9397 - 0.10 \times 0.277)}{0.9397 \times 0.277 + 0.10} = \frac{0.2526}{0.3603} = 0.701 \text{ or } 70.1\% \textbf{ Ans.}$$

31.12 Design of Worm Gearing

In designing a worm and worm gear, the quantities like the power transmitted, speed, velocity ratio and the centre distance between the shafts are usually given and the quantities such as lead angle, lead and number of threads on the worm are to be determined. In order to determine the satisfactory combination of lead angle, lead and centre distance, the following method may be used:

From Fig. 31.6 we find that the centre distance,

$$x = \frac{D_W + D_G}{2}$$

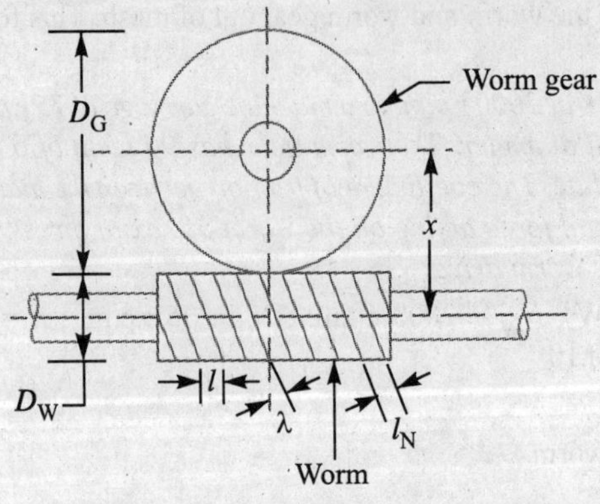

Fig. 31.6. Worm and worm gear.

Worm gear boxes are noted for reliable power transmission.

The centre distance may be expressed in terms of the axial lead (l), lead angle (λ) and velocity ratio ($V.R.$), as follows :

$$x = \frac{l}{2\pi}(\cot \lambda + V.R.)$$

In terms of normal lead ($l_N = l \cos \lambda$), the above expression may be written as :

$$x = \frac{l_N}{2\pi}\left(\frac{1}{\sin \lambda} + \frac{V.R.}{\cos \lambda}\right)$$

or

$$\frac{x}{l_N} = \frac{1}{2\pi}\left(\frac{1}{\sin \lambda} + \frac{V.R.}{\cos \lambda}\right) \quad \text{...}(i)$$

Since the velocity ratio ($V.R.$) is usually given, therefore the equation (i) contains three variables *i.e.* x, l_N and λ. The right hand side of the above expression may be calculated for various values of velocity ratios and the curves are plotted as shown in Fig. 31.7. The lowest point on each of the curves gives the lead angle which corresponds to the minimum value of x / l_N. This minimum value represents the minimum centre distance that can be used with a given lead or inversely the maximum lead that can be used with a given centre distance. Now by using Table 31.2 and standard modules, we can determine the combination of lead angle, lead, centre distance and diameters for the given design specifications.

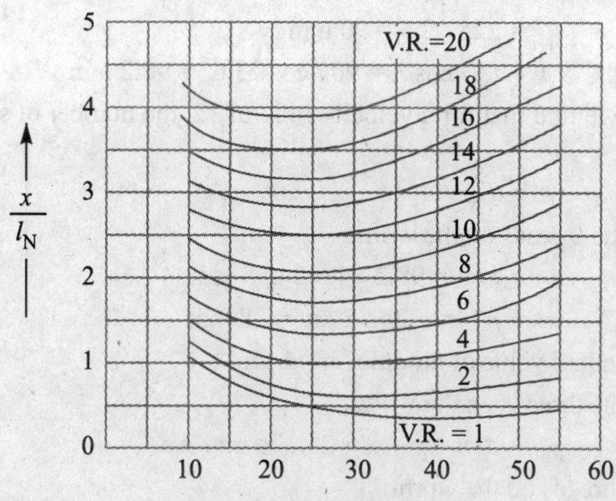

Fig. 31.7. Worm gear design curves.

Note : The lowest point on the curve may be determined mathematically by differentiating the equation (i) with respect to λ and equating to zero, *i.e.*

$$\frac{(V.R.)\sin^3 \lambda - \cos^3 \lambda}{\sin^2 \lambda . \cos^2 \lambda} = 0 \quad \text{or} \quad V.R. = \cot^3 \lambda$$

Example 31.3. *Design 20° involute worm and gear to transmit 10 kW with worm rotating at 1400 r.p.m. and to obtain a speed reduction of 12 : 1. The distance between the shafts is 225 mm.*

Solution. Given : $\phi = 20°$; $P = 10$ kW $= 10\,000$ W ; $N_W = 1400$ r.p.m. ; $V.R. = 12$; $x = 225$ mm

The worm and gear is designed as discussed below :

1. ***Design of worm***

Let l_N = Normal lead, and

λ = Lead angle.

Worm gear of a steering mechanism in an automobile.

We have discussed in Art. 31.12 that the value of x / l_N will be minimum corresponding to

$$\cot^3 \lambda = V.R. = 12 \quad \text{or} \quad \cot \lambda = 2.29$$

$$\therefore \quad \lambda = 23.6°$$

We know that $\dfrac{x}{l_N} = \dfrac{1}{2\pi}\left(\dfrac{1}{\sin \lambda} + \dfrac{V.R.}{\cos \lambda}\right)$

$$\frac{225}{l_N} = \frac{1}{2\pi}\left(\frac{1}{\sin 23.6°} + \frac{12}{\cos 23.6°}\right) = \frac{1}{2\pi}(2.5 + 13.1) = 2.5$$

$$\therefore \quad l_N = 225 / 2.5 = 90 \text{ mm}$$

and axial lead, $l = l_N / \cos \lambda = 90 / \cos 23.6° = 98.2$ mm

From Table 31.2, we find that for a velocity ratio of 12, the number of starts or threads on the worm,

$$n = T_W = 4$$

∴ Axial pitch of the threads on the worm,

$$p_a = l / 4 = 98.2 / 4 = 24.55 \text{ mm}$$

$$\therefore \quad m = p_a / \pi = 24.55 / \pi = 7.8 \text{ mm}$$

Let us take the standard value of module, $m = 8$ mm

∴ Axial pitch of the threads on the worm,

$$p_a = \pi \, m = p \times 8 = 25.136 \text{ mm } \textbf{Ans.}$$

Axial lead of the threads on the worm,

$$l = p_a \,.\, n = 25.136 \times 4 = 100.544 \text{ mm } \textbf{Ans.}$$

and normal lead of the threads on the worm,

$$l_N = l \cos \lambda = 100.544 \cos 23.6° = 92 \text{ mm } \textbf{Ans.}$$

We know that the centre distance,

$$x = \frac{l_N}{2\pi}\left(\frac{1}{\sin \lambda} + \frac{V.R.}{\cos \lambda}\right) = \frac{92}{2\pi}\left(\frac{1}{\sin 23.6°} + \frac{12}{\cos 23.6°}\right)$$

$$= 14.64\,(2.5 + 13.1) = 230 \text{ mm } \textbf{Ans.}$$

Let D_W = Pitch circle diameter of the worm.

We know that $\tan \lambda = \dfrac{l}{\pi D_W}$

$$\therefore \quad D_W = \frac{l}{\pi \tan \lambda} = \frac{100.544}{\pi \tan 23.6°} = 73.24 \text{ mm } \textbf{Ans.}$$

Since the velocity ratio is 12 and the worm has quadruple threads (*i.e.* $n = T_W = 4$), therefore number of teeth on the worm gear,

$$T_G = 12 \times 4 = 48$$

From Table 31.3, we find that the face length of the worm or the length of threaded portion is

$$L_W = p_c (4.5 + 0.02\, T_W)$$

$$= 25.136\,(4.5 + 0.02 \times 4) = 115 \text{ mm} \qquad ...(\because p_c = p_a)$$

This length should be increased by 25 to 30 mm for the feed marks produced by the vibrating grinding wheel as it leaves the thread root. Therefore let us take

$$L_W = 140 \text{ mm } \textbf{Ans.}$$

We know that depth of tooth,

$$h = 0.623\, p_c = 0.623 \times 25.136 = 15.66 \text{ mm } \textbf{Ans.}$$

...(From Table 31.3)

and addendum, $a = 0.286\, p_c = 0.286 \times 25.136 = 7.2$ mm **Ans.**

∴ Outside diameter of worm,

$$D_{OW} = D_W + 2a = 73.24 + 2 \times 7.2 = 87.64 \text{ mm } \textbf{Ans.}$$

2. *Design of worm gear*

We know that pitch circle diameter of the worm gear,

$$D_G = m \,.\, T_G = 8 \times 48 = 384 \text{ mm} = 0.384 \text{ m } \textbf{Ans.}$$

From Table 31.4, we find that outside diameter of worm gear,

$$D_{OG} = D_G + 0.8903\, p_c = 384 + 0.8903 \times 25.136 = 406.4 \text{ mm } \textbf{Ans.}$$

Throat diameter,

$$D_T = D_G + 0.572\, p_c = 384 + 0.572 \times 25.136 = 398.4 \text{ mm } \textbf{Ans.}$$

and face width, $b = 2.15\, p_c + 5 \text{ mm} = 2.15 \times 25.136 + 5 = 59$ mm **Ans.**

Let us now check the designed worm gearing from the standpoint of tangential load, dynamic load, static load or endurance strength, wear load and heat dissipation.

(a) Check for the tangential load

Let N_G = Speed of the worm gear in r.p.m.

We know that velocity ratio of the drive,

$$V.R. = \frac{N_W}{N_G} \quad \text{or} \quad N_G = \frac{N_W}{V.R.} = \frac{1400}{12} = 116.7 \text{ r.p.m}$$

∴ Torque transmitted,

$$T = \frac{P \times 60}{2\,\pi\, N_G} = \frac{10\,000 \times 60}{2\,\pi \times 116.7} = 818.2 \text{ N-m}$$

and tangential load acting on the gear,

$$W_T = \frac{2 \times \text{Torque}}{D_G} = \frac{2 \times 818.2}{0.384} = 4260 \text{ N}$$

We know that pitch line or peripheral velocity of the worm gear,

$$v = \frac{\pi \,.\, D_G \,.\, N_G}{60} = \frac{\pi \times 0.384 \times 116.7}{60} = 2.35 \text{ m/s}$$

∴ Velocity factor,

$$C_v = \frac{6}{6 + v} = \frac{6}{6 + 2.35} = 0.72$$

Gears are usually enclosed in boxes to protect them from environmental pollution and provide them proper lubrication.

and tooth form factor for 20° involute teeth,

$$y = 0.154 - \frac{0.912}{T_G} = 0.154 - \frac{0.912}{48} = 0.135$$

Since the worm gear is generally made of phosphor bronze, therefore taking the allowable static stress for phosphor bronze, σ_o = 84 MPa or N/mm^2.

We know that the designed tangential load,

$$W_T = (\sigma_o \cdot C_v)\, b.\, \pi\, m \,.\, y = (84 \times 0.72)\, 59 \times \pi \times 8 \times 0.135 \text{ N}$$
$$= 12\ 110 \text{ N}$$

Since this is more than the tangential load acting on the gear (*i.e.* 4260 N), therefore the design is safe from the standpoint of tangential load.

(b) Check for dynamic load

We know that the dynamic load,

$$W_D = W_T / C_v = 12\ 110 / 0.72 = 16\ 820 \text{ N}$$

Since this is more than W_T = 4260 N, therefore the design is safe from the standpoint of dynamic load.

(c) Check for static load or endurance strength

We know that the flexural endurance limit for phosphor bronze is

$$\sigma_e = 168 \text{ MPa or N/mm}^2$$

∴ Static load or endurance strength,

$$W_S = \sigma_e \,.\, b.\, \pi\, m \,.\, y = 168 \times 59 \times \pi \times 8 \times 0.135 = 33\ 635 \text{ N}$$

Since this is much more than W_T = 4260 N, therefore the design is safe from the standpoint of static load or endurance strength.

(d) Check for wear

Assuming the material for worm as hardened steel, therefore from Table 31.5, we find that for hardened steel worm and phosphor bronze worm gear, the value of load stress factor,

$$K = 0.55 \text{ N/mm}^2$$

∴ Limiting or maximum load for wear,

$$W_W = D_G \cdot b \cdot K = 384 \times 59 \times 0.55 = 12\ 461 \text{ N}$$

Since this is more than $W_T = 4260$ N, therefore the design is safe from the standpoint of wear.

(e) Check for heat dissipation

First of all, let us find the efficiency of the worm gearing (η).

We know that rubbing velocity,

$$v_r = \frac{\pi\, D_W \cdot N_W}{\cos \lambda} = \frac{\pi \times 0.07324 \times 1400}{\cos 23.6°} = 351.6 \text{ m / min}$$

...(D_W is taken in metres)

∴ Coefficient of friction,

$$\mu = 0.025 + \frac{v_r}{18\ 000} = 0.025 + \frac{351.6}{18\ 000} = 0.0445$$

...(∵ v_r is greater than 180 m/min)

and angle of friction, $\phi_1 = \tan^{-1} \mu = \tan^{-1} (0.0445) = 2.548°$

We know that efficiency,

$$\eta = \frac{\tan \lambda}{\tan (\lambda + \phi_1)} = \frac{\tan 23.6°}{\tan (23.6 + 2.548)} = \frac{0.4369}{0.4909} = 0.89 \text{ or } 89\%$$

Assuming 25 per cent overload, heat generated,

$$Q_g = 1.25\, P\, (1 - \eta) = 1.25 \times 10\ 000\ (1 - 0.89) = 1375 \text{ W}$$

We know that projected area of the worm,

$$A_W = \frac{\pi}{4} (D_W)^2 = \frac{\pi}{4} (73.24)^2 = 4214 \text{ mm}^2$$

and projected area of the worm gear,

$$A_G = \frac{\pi}{4} (D_G)^2 = \frac{\pi}{4} (384)^2 = 115\ 827 \text{ mm}^2$$

∴ Total projected area of worm and worm gear,

$$A = A_W + A_G = 4214 + 115\ 827 = 120\ 041 \text{ mm}^2$$
$$= 120\ 041 \times 10^{-6} \text{ m}^2$$

We know that heat dissipating capacity,

$$Q_d = A\, (t_2 - t_1)\, K = 120\ 041 \times 10^{-6}\, (t_2 - t_1)\, 378 = 45.4\, (t_2 - t_1)$$

The heat generated must be dissipated in order to avoid over heating of the drive, therefore equating $Q_g = Q_d$, we have

$$t_2 - t_1 = 1375 / 45.4 = 30.3°\text{C}$$

Since this temperature difference $(t_2 - t_1)$ is within safe limits of 27 to 38°C, therefore the design is safe from the standpoint of heat.

3. *Design of worm shaft*

Let d_W = Diameter of worm shaft.

We know that torque acting on the worm gear shaft,

$$T_{gear} = \frac{1.25\, P \times 60}{2\, \pi\, N_G} = \frac{1.25 \times 10\,000 \times 60}{2\, \pi \times 116.7} = 1023 \text{ N-m}$$
$$= 1023 \times 10^3 \text{ N-mm}$$

...(Taking 25% overload)

∴ Torque acting on the worm shaft,

$$T_{worm} = \frac{T_{gear}}{V.R. \times \eta} = \frac{1023}{12 \times 0.89} = 96 \text{ N-m} = 96 \times 10^3 \text{ N-mm}$$

Differential inside an automobile.

We know that tangential force on the worm,

$$W_T = \text{Axial force on the worm gear}$$

$$= \frac{2 \times T_{worm}}{D_W} = \frac{2 \times 96 \times 10^3}{73.24} = 2622 \text{ N}$$

Axial force on the worm,

$$W_A = \text{Tangential force on the worm gear}$$

$$= \frac{2 \times T_{gear}}{D_G} = \frac{2 \times 1023 \times 10^3}{384} = 5328 \text{ N}$$

and radial or separating force on the worm

$$W_R = \text{Radial or separating force on the worm gear}$$

$$= W_A \cdot \tan \phi = 5328 \times \tan 20° = 1940 \text{ N}$$

Let us take the distance between the bearings of the worm shaft (x_1) equal to the diameter of the worm gear (D_G), *i.e.*

$$x_1 = D_G = 384 \text{ mm}$$

∴ Bending moment due to the radial force (W_R) in the vertical plane

$$= \frac{W_R \times x_1}{4} = \frac{1940 \times 384}{4} = 186\,240 \text{ N-mm}$$

and bending moment due to axial force (W_A) in the vertical plane

$$= \frac{W_A \times D_W}{4} = \frac{5328 \times 73.24}{4} = 97\,556 \text{ N-mm}$$

∴ Total bending moment in the vertical plane,

$$M_1 = 186\,240 + 97\,556 = 283\,796 \text{ N-mm}$$

We know that bending moment due to tangential force (W_T) in the horizontal plane,

$$M_2 = \frac{W_T \times D_G}{4} = \frac{2622 \times 384}{4} = 251\,712 \text{ N-mm}$$

∴ Resultant bending moment on the worm shaft,

$$M_{worm} = \sqrt{(M_1)^2 + (M_2)^2} = \sqrt{(283\,796)^2 + (251\,712)^2} = 379\,340 \text{ N-mm}$$

We know that equivalent twisting moment on the worm shaft,

$$T_{ew} = \sqrt{(T_{worm})^2 + (M_{worm})^2} = \sqrt{(96 \times 10^3)^2 + (379\ 340)^2} \text{ N-mm}$$

$$= 391\ 300 \text{ N-mm}$$

We also know that equivalent twisting moment (T_{ew}),

$$391\ 300 = \frac{\pi}{16} \times \tau\, (d_W)^3 = \frac{\pi}{16} \times 50\, (d_W)^3 = 9.82\, (d_W)^3$$

...(Taking τ = 50 MPa or N/mm²)

$$\therefore \quad (d_W)^3 = 391\ 300 / 9.82 = 39\ 850 \text{ or } d_W = 34.2 \text{ say } 35 \text{ mm } \textbf{Ans.}$$

Let us now check the maximum shear stress induced.

We know that the actual shear stress,

$$\tau = \frac{16\, T_{ew}}{\pi\, (d_W)^3} = \frac{16 \times 391\ 300}{\pi\, (35)^3} = 46.5 \text{ N/mm}^2$$

and direct compressive stress on the shaft due to the axial force,

$$\sigma_c = \frac{W_A}{\frac{\pi}{4}\, (d_W)^2} = \frac{5328}{\frac{\pi}{4}\, (35)^2} = 5.54 \text{ N/mm}^2$$

$\therefore$ Maximum shear stress,

$$\tau_{max} = \frac{1}{2}\sqrt{(\sigma_c)^2 + 4\,\tau^2} = \frac{1}{2}\sqrt{(5.54)^2 + 4\,(46.5)^2} = 46.6 \text{ MPa}$$

Since the maximum shear stress induced is less than 50 MPa (assumed), therefore the design of worm shaft is satisfactory.

4. *Design of worm gear shaft*

Let d_G = Diameter of worm gear shaft.

We have calculated above that the axial force on the worm gear

= 2622 N

Tangential force on the worm gear

= 5328 N

and radial or separating force on the worm gear

= 1940 N

We know that bending moment due to the axial force on the worm gear

$$= \frac{\text{Axial force} \times D_G}{4} = \frac{2622 \times 384}{4} = 251\ 712 \text{ N-mm}$$

The bending moment due to the axial force will be in the vertical plane.

Let us take the distance between the bearings of the worm gear shaft (x_2) as 250 mm.

$\therefore$ Bending moment due to the radial force on the worm gear

$$= \frac{\text{Radial force} \times x_2}{4} = \frac{1940 \times 250}{4} = 121\ 250 \text{ N-mm}$$

The bending moment due to the radial force will also be in the vertical plane.

$\therefore$ Total bending moment in the vertical plane

$$M_3 = 251\ 712 + 121\ 250 = 372\ 962 \text{ N-mm}$$

We know that the bending moment due to the tangential force in the horizontal plane

$$M_4 = \frac{\text{Tangential force} \times x_2}{4} = \frac{5328 \times 250}{4} = 333\ 000 \text{ N-mm}$$

∴ Resultant bending moment on the worm gear shaft,

$$M_{gear} = \sqrt{(M_3)^2 + (M_4)^2} = \sqrt{(372\ 962)^2 + (333\ 000)^2} \text{ N-mm}$$
$$= 500 \times 10^3 \text{ N-mm}$$

We have already calculated that the torque acting on the worm gear shaft,

$$T_{gear} = 1023 \times 10^3 \text{ N-mm}$$

∴ Equivalent twisting moment on the worm gear shaft,

$$T_{eg} = \sqrt{(T_{gear})^2 + (M_{gear})^2} = \sqrt{(1023 \times 10^3)^2 + (500 \times 10^3)^2} \text{ N-mm}$$
$$= 1.14 \times 10^6 \text{ N-mm}$$

We know that equivalent twisting moment (T_{eg}),

$$1.14 \times 10^6 = \frac{\pi}{16} \times \tau\, (d_G)^3 = \frac{\pi}{16} \times 50\, (d_G)^3 = 9.82\, (d_G)^3$$

$$\therefore \quad (d_G)^3 = 1.14 \times 10^6 / 9.82 = 109 \times 10^3$$

or $\quad d_G = 48.8$ say 50 mm **Ans.**

Let us now check the maximum shear stress induced.

We know that actual shear stress,

$$\tau = \frac{16\, T_{eg}}{\pi\, (d_G)^3} = \frac{16 \times 1.14 \times 10^6}{\pi\, (50)^3} = 46.4 \text{ N/mm}^2 = 46.4 \text{ MPa}$$

and direct compressive stress on the shaft due to the axial force,

$$\sigma_c = \frac{\text{Axial force}}{\frac{\pi}{4}(d_G)^2} = \frac{2622}{\frac{\pi}{4}(50)^2} = 1.33 \text{ N/mm}^3 = 1.33 \text{ MPa}$$

∴ Maximum shear stress,

$$\tau_{max} = \frac{1}{2}\sqrt{(\sigma_c)^2 + 4\,\tau^2} = \frac{1}{2}\sqrt{(1.33)^2 + 4\,(46.4)^2} = 46.4 \text{ MPa}$$

Since the maximum shear stress induced is less than 50 MPa (assumed), therefore the design for worm gear shaft is satisfactory.

Example 31.4. *A speed reducer unit is to be designed for an input of 1.1 kW with a transmission ratio 27. The speed of the hardened steel worm is 1440 r.p.m. The worm wheel is to be made of phosphor bronze. The tooth form is to be 20° involute.*

Solution. Given : $P = 1.1$ kW $= 1100$ W ; $V.R. = 27$; $N_W = 1440$ r.p.m. ; $\phi = 20°$

A speed reducer unit (*i.e.*, worm and worm gear) may be designed as discussed below.

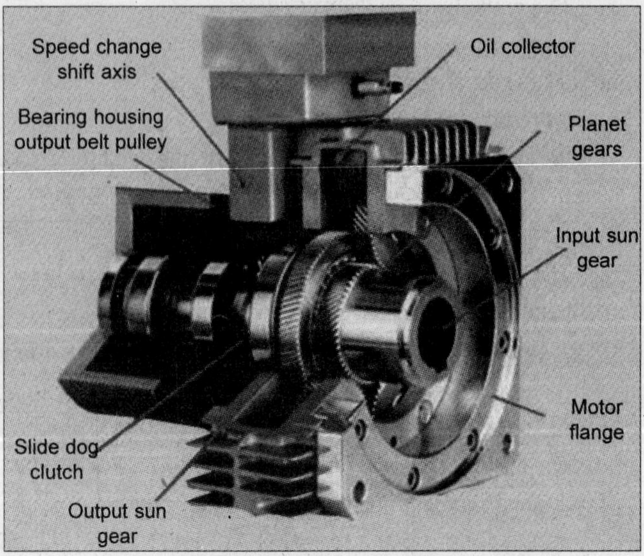

Sun and Planet gears.

Since the centre distance between the shafts is not known, therefore let us assume that for this size unit, the centre distance (x) = 100 mm.

We know that pitch circle diameter of the worm,

$$D_W = \frac{(x)^{0.875}}{1.416} = \frac{(100)^{0.875}}{1.416} = 39.7 \text{ say } 40 \text{ mm}$$

∴ Pitch circle diameter of the worm gear,

$$D_G = 2x - D_W = 2 \times 100 - 40 = 160 \text{ mm}$$

From Table 31.2, we find that for the transmission ratio of 27, we shall use double start worms.

∴ Number of teeth on the worm gear,

$$T_G = 2 \times 27 = 54$$

We know that the axial pitch of the threads on the worm (p_a) is equal to circular pitch of teeth on the worm gear (p_c).

$$\therefore \quad p_a = p_c = \frac{\pi D_G}{T_G} = \frac{\pi \times 160}{54} = 9.3 \text{ mm}$$

and module,

$$m = \frac{p_c}{\pi} = \frac{9.3}{\pi} = 2.963 \text{ say } 3 \text{ mm}$$

∴ Actual circular pitch,

$$p_c = \pi m = \pi \times 3 = 9.426 \text{ mm}$$

Actual pitch circle diameter of the worm gear,

$$D_G = \frac{p_c \cdot T_G}{\pi} = \frac{9.426 \times 54}{\pi} = 162 \text{ mm} \quad \textbf{Ans.}$$

and actual pitch circle diameter of the worm,

$$D_W = 2x - D_G = 2 \times 100 - 162 = 38 \text{ mm} \quad \textbf{Ans.}$$

The face width of the worm gear (b) may be taken as 0.73 times the pitch circle diameter of worm (D_W).

$$\therefore \quad b = 0.73\, D_W = 0.73 \times 38 = 27.7 \text{ say } 28 \text{ mm}$$

Let us now check the design from the standpoint of tangential load, dynamic load, static load or endurance strength, wear load and heat dissipation.

1. *Check for the tangential load*

Let N_G = Speed of the worm gear in r.p.m.

We know that velocity ratio of the drive,

$$V.R. = \frac{N_W}{N_G} \text{ or } N_G = \frac{N_W}{V.R.} = \frac{1440}{27} = 53.3 \text{ r.p.m}$$

∴ Peripheral velocity of the worm gear,

$$v = \frac{\pi D_G \cdot N_G}{60} = \frac{\pi \times 0.162 \times 53.3}{60} = 0.452 \text{ m/s}$$

... (D_G is taken in metres)

and velocity factor,

$$C_v = \frac{6}{6 + v} = \frac{6}{6 + 0.452} = 0.93$$

We know that for 20° involute teeth, the tooth form factor,

$$y = 0.154 - \frac{0.912}{T_G} = 0.154 - \frac{0.912}{54} = 0.137$$

From Table 31.4, we find that allowable static stress for phosphor bronze is

$$\sigma_o = 84 \text{ MPa or N/mm}^2$$

∴ Tangential load transmitted,

$$W_T = (\sigma_o \,.\, C_v)\, b \,.\, \pi m \,.\, y = (84 \times 0.93)\; 28 \times \pi \times 3 \times 0.137 \text{ N}$$
$$= 2825 \text{ N}$$

and power transmitted due to the tangential load,

$$P = W_T \times v = 2825 \times 0.452 = 1277 \text{ W} = 1.277 \text{ kW}$$

Since this power is more than the given power to be transmitted (1.1 kW), therefore the design is safe from the standpoint of tangential load.

2. ***Check for the dynamic load***

We know that the dynamic load,

$$W_D = W_T / C_v = 2825 / 0.93 = 3038 \text{ N}$$

and power transmitted due to the dynamic laod,

$$P = W_D \times v = 3038 \times 0.452 = 1373 \text{ W} = 1.373 \text{ kW}$$

Since this power is more than the given power to be transmitted, therefore the design is safe from the standpoint of dynamic load.

3. ***Check for the static load or endurance strength***

From Table 31.8, we find that the flexural endurance limit for phosphor bronze is

$$\sigma_e = 168 \text{ MPa or N/mm}^2$$

∴ Static load or endurance strength,

$$W_S = \sigma_e \,.\, b. \,\pi m \,.\, y = 168 \times 28 \times \pi \times 3 \times 0.137 = 6075 \text{ N}$$

and power transmitted due to the static load,

$$P = W_S \times v = 6075 \times 0.452 = 2746 \text{ W} = 2.746 \text{ kW}$$

Since this power is more than the power to be transmitted (1.1 kW), therefore the design is safe from the standpoint of static load.

4. ***Check for the wear load***

From Table 31.5, we find that the load stress factor for hardened steel worm and phosphor bronze worm gear is

$$K = 0.55 \text{ N/mm}^2$$

∴ Limiting or maximum load for wear,

$$W_W = D_G \,.\, b \,.\, K = 162 \times 28 \times 0.55 = 2495 \text{ N}$$

and power transmitted due to the wear load,

$$P = W_W \times v = 2495 \times 0.452 = 1128 \text{ W} = 1.128 \text{ kW}$$

Since this power is more than the given power to be transmitted (1.1 kW), therefore the design is safe from the standpoint of wear.

5. ***Check for the heat dissipation***

We know that permissible input power,

$$P = \frac{3650\,(x)^{1.7}}{V.R + 5} = \frac{3650\,(0.1)^{1.7}}{27 + 5} = 2.27 \text{ kW} \quad \text{... (}x\text{ is taken in metres)}$$

Since this power is more than the given power to be transmitted (1.1 kW), therefore the design is safe from the standpoint of heat dissipation.

EXERCISES

1. A double threaded worm drive is required for power transmission between two shafts having their axes at right angles to each other. The worm has 14½° involute teeth. The centre distance is approximately 200 mm. If the axial pitch of the worm is 30 mm and lead angle is 23°, find 1. lead; 2. pitch circle diameters of worm and worm gear; 3. helix angle of the worm; and 4. efficiency of the drive if the coefficient of friction is 0.05. **[Ans. 60 mm ; 45 mm ; 355 mm ; 67° ; 87.4%]**

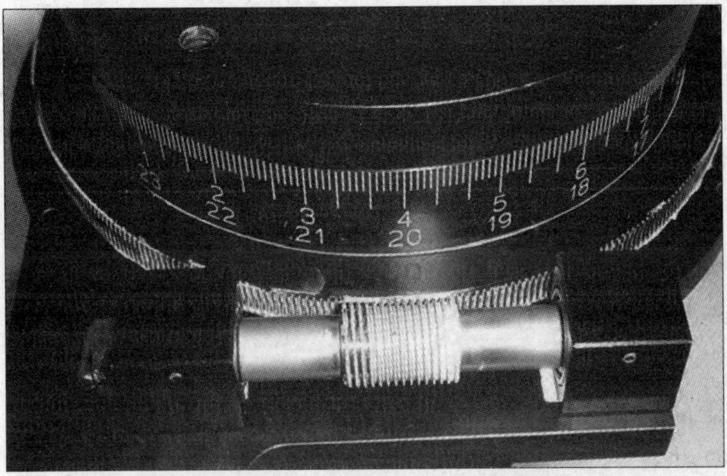

The worm in its place. One can also see the two cubic worm bearing blocks and the big gear.

2. A double threaded worm drive has an axial pitch of 25 mm and a pitch circle diameter of 70 mm. The torque on the worm gear shaft is 1400 N-m. The pitch circle diameter of the worm gear is 250 mm and the tooth pressure angle is 25°. Find : 1. tangential force on the worm gear, 2. torque on the worm shaft, 3. separating force on the worm, 4. velocity ratio, and 5. efficiency of the drive, if the coefficient of friction between the worm thread and gear teeth is 0.04.

   [Ans. 11.2 kN ; 88.97 N-m ; 5220 N ; 82.9%]

3. Design a speed reducer unit of worm and worm wheel for an input of 1 kW with a transmission ratio of 25. The speed of the worm is 1600 r.p.m. The worm is made of hardened steel and wheel of phosphor bronze for which the material combination factor is 0.7 N/mm^2. The static stress for the wheel material is 56 MPa. The worm is made of double start and the centre distance between the axes of the worm and wheel is 120 mm. The tooth form is to be 14½° involute. Check the design for strength, wear and heat dissipation.
4. Design worm and gear speed reducer to transmit 22 kW at a speed of 1440 r.p.m. The desired velocity ratio is 24 : 1. An efficiency of atleast 85% is desired. Assume that the worm is made of hardened steel and the gear of phosphor bronze.

QUESTIONS

1. Discuss, with neat sketches, the various types of worms and worm gears.
2. Define the following terms used in worm gearing :

   (*a*) Lead; (*b*) Lead angle; (*c*) Normal pitch; and (*d*) Helix angle.
3. What are the various forces acting on worm and worm gears ?
4. Write the expression for centre distance in terms of axial lead, lead angle and velocity ratio.

OBJECTIVE TYPE QUESTIONS

1. The worm gears are widely used for transmitting power at .... velocity ratios between non-intersecting shafts.

   (*a*) high (*b*) low

2. In worm gears, the angle between the tangent to the thread helix on the pitch cylinder and the plane normal to the axis of worm is called

   (*a*) pressure angle (*b*) lead angle

   (*c*) helix angle (*d*) friction angle

3. The normal lead, in a worm having multiple start threads, is given by

   (*a*) $l_N = l / \cos \lambda$ (*b*) $l_N = l \cdot \cos \lambda$

   (*c*) $l_N = l$ (*d*) $l_N = l \tan$

   where l_N = Normal lead,

   l = Lead, and

   λ = Lead angle.

4. The number of starts on the worm for a velocity ratio of 40 should be

   (*a*) single (*b*) double

   (*c*) triple (*d*) quadruple

5. The axial thrust on the worm (W_A) is given by

   (*a*) $W_A = W_T \cdot \tan \phi$ (*b*) $W_A = W_T / \tan \phi$

   (*c*) $W_A = W_T \cdot \tan \lambda$ (*d*) $W_A = W_T / \tan \lambda$

   where W_T = Tangential force acting on the worm,

   ϕ = Pressure angle, and

   λ = Lead angle.

ANSWERS

1. (*a*) 2. (*b*) 3. (*b*) 4. (*a*) 5. (*d*)

C H A P T E R

32

Internal Combustion Engine Parts

32.1 Introduction

As the name implies, the internal combustion engines (briefly written as I. C. engines) are those engines in which the combustion of fuel takes place inside the engine cylinder. The I.C. engines use either petrol or diesel as their fuel. In petrol engines (also called ***spark ignition engines*** or ***S.I engines***), the correct proportion of air and petrol is mixed in the carburettor and fed to engine cylinder where it is ignited by means of a spark produced at the spark plug. In diesel engines (also called ***compression ignition engines*** or ***C.I engines***), only air is supplied to the engine cylinder during suction stroke and it is compressed to a very high pressure, thereby raising its temperature from 600°C to 1000°C. The desired quantity of fuel (diesel) is now injected into the engine cylinder in the form of a very fine spray and gets ignited when comes in contact with the hot air.

The operating cycle of an I.C. engine may be completed either by the two strokes or four strokes of the piston. Thus, an engine which requires two strokes of the piston or one complete revolution of the crankshaft to complete the cycle, is known as ***two stroke engine.*** An engine which requires four strokes of the piston or two complete revolutions of the crankshaft to complete the cycle, is known as ***four stroke engine.***

The two stroke petrol engines are generally employed in very light vehicles such as scooters, motor cycles and three wheelers. The two stroke diesel engines are generally employed in marine propulsion.

The four stroke petrol engines are generally employed in light vehicles such as cars, jeeps and also in aeroplanes. The four stroke diesel engines are generally employed in heavy duty vehicles such as buses, trucks, tractors, diesel locomotive and in the earth moving machinery.

32.2 Principal Parts of an Engine

The principal parts of an I.C engine, as shown in Fig. 32.1 are as follows :

1. Cylinder and cylinder liner, 2. Piston, piston rings and piston pin or gudgeon pin, 3. Connecting rod with small and big end bearing, 4. Crank, crankshaft and crank pin, and 5. Valve gear mechanism.

The design of the above mentioned principal parts are discussed, in detail, in the following pages.

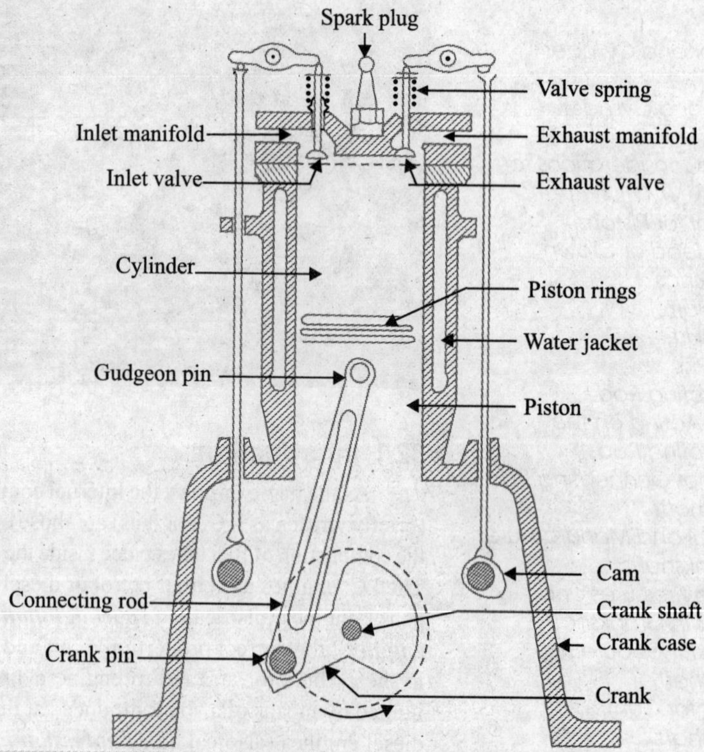

Fig. 32.1. Internal combustion engine parts.

32.3 Cylinder and Cylinder Liner

The function of a cylinder is to retain the working fluid and to guide the piston. The cylinders are usually made of cast iron or cast steel. Since the cylinder has to withstand high temperature due to the combustion of fuel, therefore, some arrangement must be provided to cool the cylinder. The single cylinder engines (such as scooters and motorcycles) are generally air cooled. They are provided with fins around the cylinder. The multi-cylinder engines (such as of cars) are provided with water jackets around the cylinders to cool it. In smaller engines. the cylinder, water jacket and the frame are made as one piece, but for all the larger engines, these parts are manufactured separately. The cylinders are provided with cylinder liners so that in case of wear, they can be easily replaced. The cylinder liners are of the following two types :

1. Dry liner, and 2. Wet liner.

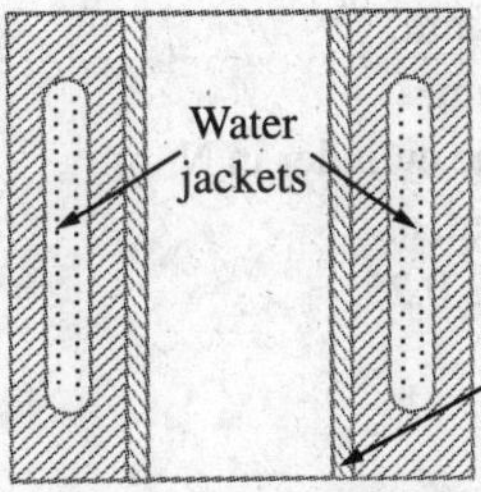

(*a*) Dry liner.

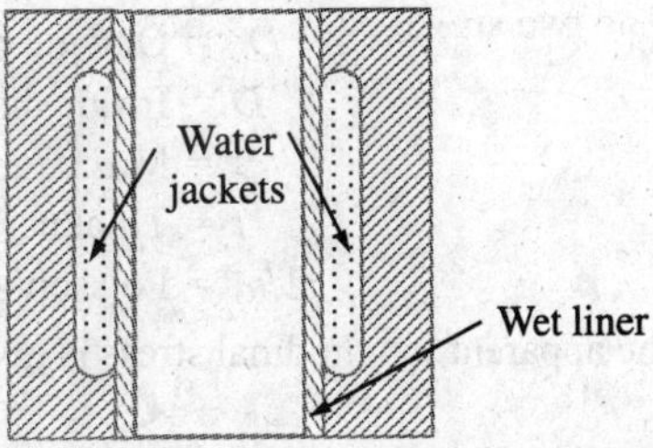

(*b*) Wet liner.

Fig. 32.2. Dry and wet liner.

A cylinder liner which does not have any direct contact with the engine cooling water, is known as ***dry liner,*** as shown in Fig. 32.2 (*a*). A cylinder liner which have its outer surface in direct contact with the engine cooling water, is known as ***wet liner,*** as shown in Fig. 32.2 (*b*).

The cylinder liners are made from good quality close grained cast iron (*i.e.* pearlitic cast iron), nickel cast iron, nickel chromium cast iron. In some cases, nickel chromium cast steel with molybdenum may be used. The inner surface of the liner should be properly heat-treated in order to obtain a hard surface to reduce wear.

32.4 Design of a Cylinder

In designing a cylinder for an I. C. engine, it is required to determine the following values :

1. ***Thickness of the cylinder wall.*** The cylinder wall is subjected to gas pressure and the piston side thrust. The gas pressure produces the following two types of stresses :

(*a*) Longitudinal stress, and (*b*) Circumferential stress.

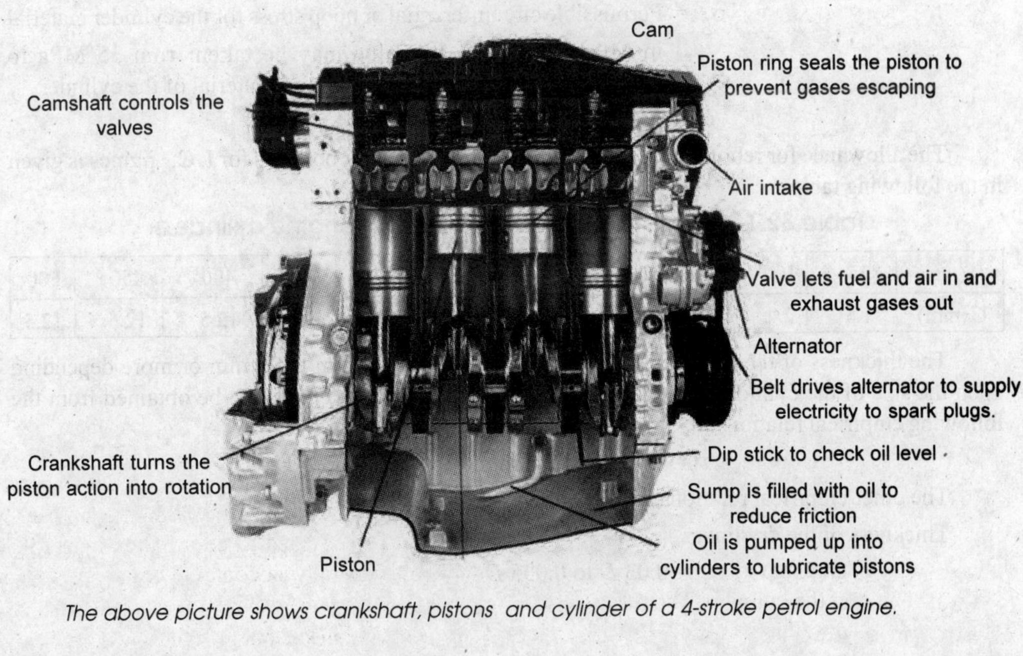

The above picture shows crankshaft, pistons and cylinder of a 4-stroke petrol engine.

Since these two stressess act at right angles to each other, therefore, the net stress in each direction is reduced.

The piston side thrust tends to bend the cylinder wall, but the stress in the wall due to side thrust is very small and hence it may be neglected.

Let D_0 = Outside diameter of the cylinder in mm,

D = Inside diameter of the cylinder in mm,

p = Maximum pressure inside the engine cylinder in N/mm^2,

t = Thickness of the cylinder wall in mm, and

$1/m$ = Poisson's ratio. It is usually taken as 0.25.

The apparent longitudinal stress is given by

$$\sigma_l = \frac{\text{Force}}{\text{Area}} = \frac{\frac{\pi}{4} \times D^2 \times p}{\frac{\pi}{4}[(D_0)^2 - D^2]} = \frac{D^2 \cdot p}{(D_0)^2 - D^2}$$

and the apparent circumferential stresss is given by

$$\sigma_c = \frac{\text{Force}}{\text{Area}} = \frac{D \times l \times p}{2t \times l} = \frac{D \times p}{2t}$$

... (where l is the length of the cylinder and area is the projected area)

$$\therefore \text{ Net longitudinal stress} = \sigma_l - \frac{\sigma_c}{m}$$

$$\text{and net circumferential stress} = \sigma_c - \frac{\sigma_l}{m}$$

The thickness of a cylinder wall (t) is usually obtained by using a thin cylindrical formula, *i.e.*,

$$t = \frac{p \times D}{2\sigma_c} + C$$

where p = Maximum pressure inside the cylinder in N/mm^2,

D = Inside diameter of the cylinder or cylinder bore in mm,

σ_c = Permissible circumferential or hoop stress for the cylinder material in MPa or N/mm^2. Its value may be taken from 35 MPa to 100 MPa depending upon the size and material of the cylinder.

C = Allowance for reboring.

The allowance for reboring (C) depending upon the cylinder bore (D) for I. C. engines is given in the following table :

Table 32.1. Allowance for reboring for I. C. engine cylinders.

D (mm)	75	100	150	200	250	300	350	400	450	500
C (mm)	1.5	2.4	4.0	6.3	8.0	9.5	11.0	12.5	12.5	12.5

The thickness of the cylinder wall usually varies from 4.5 mm to 25 mm or more depending upon the size of the cylinder. The thickness of the cylinder wall (t) may also be obtained from the following empirical relation, *i.e.*

$$t = 0.045\, D + 1.6 \text{ mm}$$

The other empirical relations are as follows :

Thickness of the dry liner

$$= 0.03\, D \text{ to } 0.035\, D$$

Thickness of the water jacket wall

= $0.032\ D + 1.6$ mm or $t / 3\ m$ for bigger cylinders and $3t/4$ for smaller cylinders

Water space between the outer cylinder wall and inner jacket wall

= 10 mm for a 75 mm cylinder to 75 mm for a 750 mm cylinder or $0.08\ D + 6.5$ mm

2. *Bore and length of the cylinder.* The bore (*i.e.* inner diameter) and length of the cylinder may be determined as discussed below :

Let p_m = Indicated mean effective pressure in N/mm^2,

D = Cylinder bore in mm,

A = Cross-sectional area of the cylinder in mm^2,

$= \pi D^2/4$

l = Length of stroke in metres,

N = Speed of the engine in r.p.m., and

n = Number of working strokes per min

= N, for two stroke engine

= $N/2$, for four stroke engine.

We know that the power produced inside the engine cylinder, *i.e.* indicated power,

$$I.P. = \frac{p_m \times l \times A \times n}{60} \text{ watts}$$

From this expression, the bore (D) and length of stroke (l) is determined. The length of stroke is generally taken as 1.25 D to 2D.

Since there is a clearance on both sides of the cylinder, therefore length of the cylinder is taken as 15 percent greater than the length of stroke. In other words,

Length of the cylinder, $L = 1.15 \times \text{Length of stroke} = 1.15\ l$

Notes : (*a*) If the power developed at the crankshaft, *i.e.* brake power (*B. P.*) and the mechanical efficiency (η_m) of the engine is known, then

$$I.P. = \frac{B.P.}{\eta_m}$$

(*b*) The maximum gas pressure (p) may be taken as 9 to 10 times the mean effective pressure (p_m).

3. *Cylinder flange and studs.* The cylinders are cast integral with the upper half of the crank-case or they are attached to the crankcase by means of a flange with studs or bolts and nuts. The cylinder flange is integral with the cylinder and should be made thicker than the cylinder wall. The flange thickness should be taken as 1.2 t to 1.4 t, where t is the thickness of cylinder wall.

The diameter of the studs or bolts may be obtained by equating the gas load due to the maximum pressure in the cylinder to the resisting force offered by all the studs or bolts. Mathematically,

$$\frac{\pi}{4} \times D^2 . p = n_s \times \frac{\pi}{4} (d_c)^2 \sigma_t$$

where D = Cylinder bore in mm,

p = Maximum pressure in N/mm^2,

n_s = Number of studs. It may be taken as $0.01\ D + 4$ to $0.02\ D + 4$

d_c = Core or minor diameter, *i.e.* diameter at the root of the thread in mm,

σ_t = Allowable tensile stress for the material of studs or bolts in MPa or N/mm^2. It may be taken as 35 to 70 MPa.

The nominal or major diameter of the stud or bolt (d) usually lies between 0.75 t_f to t_f, where t_f is the thickness of flange. In no case, a stud or bolt less than 16 mm diameter should be used.

The distance of the flange from the centre of the hole for the stud or bolt should not be less than d + 6 mm and not more than 1.5 d, where d is the nominal diameter of the stud or bolt.

In order to make a leak proof joint, the pitch of the studs or bolts should lie between $19\sqrt{d}$ to $28.5\sqrt{d}$, where d is in mm.

4. *Cylinder head.* Usually, a separate cylinder head or cover is provided with most of the engines. It is, usually, made of box type section of considerable depth to accommodate ports for air and gas passages, inlet valve, exhaust valve and spark plug (in case of petrol engines) or atomiser at the centre of the cover (in case of diesel engines).

The cylinder head may be approximately taken as a flat circular plate whose thickness (t_h) may be determined from the following relation :

$$t_h = D\sqrt{\frac{C.p}{\sigma_c}}$$

where D = Cylinder bore in mm,

p = Maximum pressure inside the cylinder in N/mm^2,

σ_c = Allowable circumferential stress in MPa or N/mm^2. It may be taken as 30 to 50 MPa, and

C = Constant whose value is taken as 0.1.

The studs or bolts are screwed up tightly alongwith a metal gasket or asbestos packing to provide a leak proof joint between the cylinder and cylinder head. The tightness of the joint also depends upon the pitch of the bolts or studs, which should lie between $19\sqrt{d}$ to $28.5\sqrt{d}$. The pitch circle diameter (D_p) is usually taken as $D + 3d$. The studs or bolts are designed in the same way as discussed above.

Example 32.1. *A four stroke diesel engine has the following specifications :*

Brake power = 5 kW ; Speed = 1200 r.p.m. ; Indicated mean effective pressure = 0.35 N / mm^2 ; Mechanical efficiency = 80 %.

Determine : 1. bore and length of the cylinder ; 2. thickness of the cylinder head ; and 3. size of studs for the cylinder head.

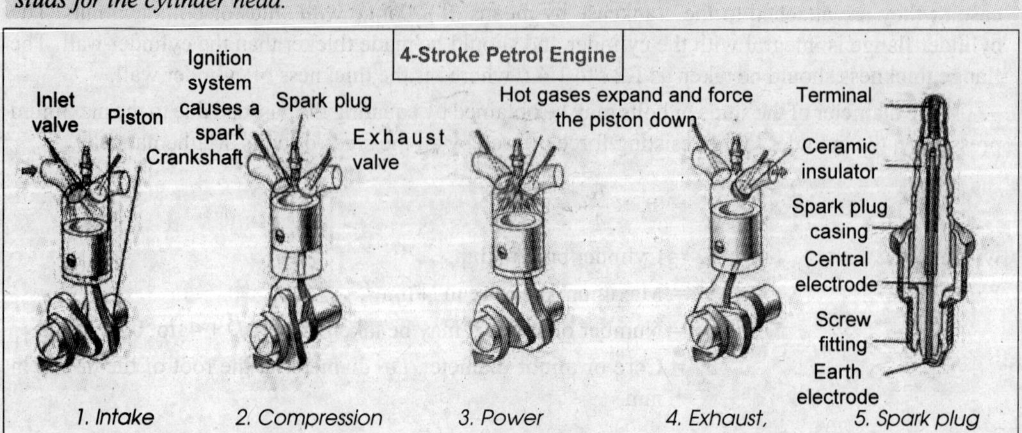

Solution. Given: $B.P. = 5\text{kW} = 5000$ W ; $N = 1200$ r.p.m. or $n = N/2 = 600$; $p_m = 0.35$ N/mm²; $\eta_m = 80\% = 0.8$

1. ***Bore and length of cylinder***

Let D = Bore of the cylinder in mm,

A = Cross-sectional area of the cylinder $= \frac{\pi}{4} \times D^2$ mm²

l = Length of the stroke in m.

$= 1.5\,D$ mm $= 1.5\,D/1000$ m ....(Assume)

We know that the indicated power,

$$I.P = B.P./\eta_m = 5000/0.8 = 6250 \text{ W}$$

We also know that the indicated power ($I.P.$),

$$6250 = \frac{p_m.l.A.n}{60} = \frac{0.35 \times 1.5D \times \pi D^2 \times 600}{60 \times 1000 \times 4} = 4.12 \times 10^{-3}\,D^3$$

...($\because$ For four stroke engine, $n = N/2$)

$\therefore$ $D^3 = 6250/4.12 \times 10^{-3} = 1517 \times 10^3$ or $D = 115$ mm **Ans.**

and $l = 1.5\,D = 1.5 \times 115 = 172.5$ mm

Taking a clearance on both sides of the cylinder equal to 15% of the stroke, therefore length of the cylinder,

$$L = 1.15\,l = 1.15 \times 172.5 = 198 \text{ say } 200 \text{ mm } \textbf{Ans.}$$

2. ***Thickness of the cylinder head***

Since the maximum pressure (p) in the engine cylinder is taken as 9 to 10 times the mean effective pressure (p_m), therefore let us take

$$p = 9\,p_m = 9 \times 0.35 = 3.15 \text{ N/mm}^2$$

We know that thickness of the cyclinder head,

$$t_h = D\sqrt{\frac{C.p}{\sigma_t}} = 115\sqrt{\frac{0.1 \times 3.15}{42}} = 9.96 \text{ say } 10 \text{ mm } \textbf{Ans.}$$

...(Taking $C = 0.1$ and $\sigma_t = 42$ MPa = 42 N/mm²)

3. ***Size of studs for the cylinder head***

Let d = Nominal diameter of the stud in mm,

d_c = Core diameter of the stud in mm. It is usually taken as 0.84 d.

σ_t = Tensile stress for the material of the stud which is usually nickel steel.

n_s = Number of studs.

We know that the force acting on the cylinder head (or on the studs)

$$= \frac{\pi}{4} \times D^2 \times p = \frac{\pi}{4}(115)^2 3.15 = 32\ 702 \text{ N} \quad ...(i)$$

The number of studs (n_s) are usually taken between $0.01\,D + 4$ (*i.e.* $0.01 \times 115 + 4 = 5.15$) and $0.02\,D + 4$ (*i.e.* $0.02 \times 115 + 4 = 6.3$). Let us take $n_s = 6$.

We know that resisting force offered by all the studs

$$= n_s \times \frac{\pi}{4}(d_c)^2\ \sigma_t = 6 \times \frac{\pi}{4}(0.84d)^2\ 65 = 216\ d^2\text{N} \quad ...(ii)$$

...(Taking $\sigma_t = 65$ MPa = 65 N/mm²)

From equations (*i*) and (*ii*),

$$d^2 = 32\ 702/216 = 151 \text{ or } d = 12.3 \text{ say } 14 \text{ mm}$$

The pitch circle diameter of the studs (D_p) is taken $D + 3d$.

$$\therefore \quad D_p = 115 + 3 \times 14 = 157 \text{ mm}$$

We know that pitch of the studs

$$= \frac{\pi \times D_p}{n_s} = \frac{\pi \times 157}{6} = 82.2 \text{ mm}$$

We know that for a leak-proof joint, the pitch of the studs should lie between $19\sqrt{d}$ to $28.5\sqrt{d}$, where d is the nominal diameter of the stud.

∴ Minimum pitch of the studs

$$= 19\sqrt{d} = 19\sqrt{14} = 71.1 \text{ mm}$$

and maximum pitch of the studs

$$= 28.5\sqrt{d} = 28.5\sqrt{14} = 106.6 \text{ mm}$$

Since the pitch of the studs obtained above (*i.e.* 82.2 mm) lies within 71.1 mm and 106.6 mm, therefore, size of the stud (d) calculated above is satisfactory.

$$\therefore \quad d = 14 \text{ mm } \textbf{Ans.}$$

32.5 Piston

The piston is a disc which reciprocates within a cylinder. It is either moved by the fluid or it moves the fluid which enters the cylinder. The main function of the piston of an internal combustion engine is to receive the impulse from the expanding gas and to transmit the energy to the crankshaft through the connecting rod. The piston must also disperse a large amount of heat from the combustion chamber to the cylinder walls.

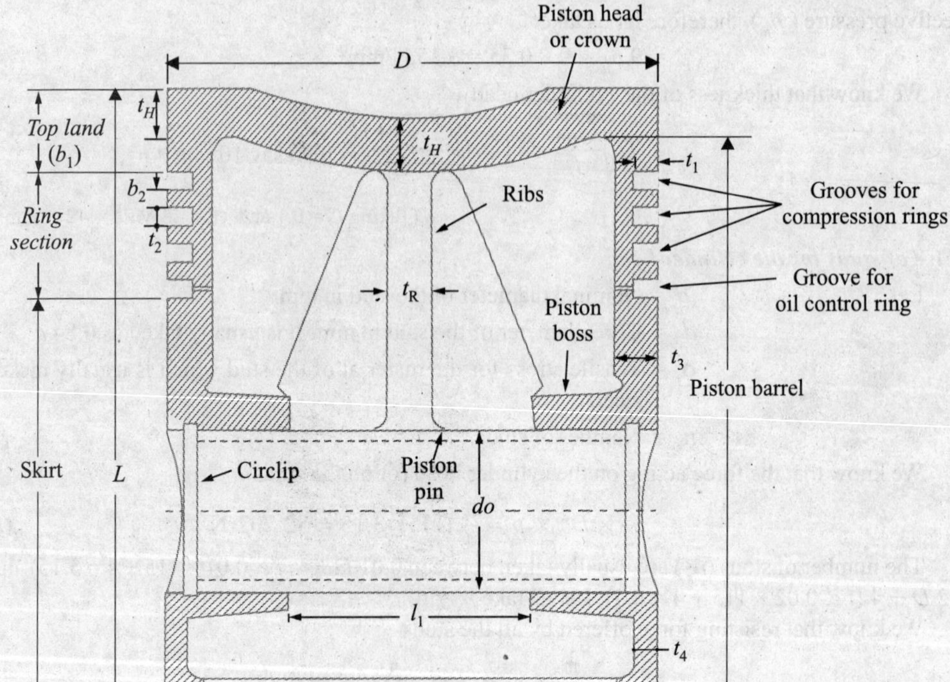

Fig. 32.3. Piston for I.C. engines (Trunk type).

The piston of internal combustion engines are usually of trunk type as shown in Fig. 32.3. Such pistons are open at one end and consists of the following parts :

1. ***Head or crown.*** The piston head or crown may be flat, convex or concave depending upon the design of combustion chamber. It withstands the pressure of gas in the cylinder.

2. ***Piston rings.*** The piston rings are used to seal the cyliner in order to prevent leakage of the gas past the piston.

3. ***Skirt.*** The skirt acts as a bearing for the side thrust of the connecting rod on the walls of cylinder.

4. ***Piston pin.*** It is also called ***gudgeon pin*** or ***wrist pin.*** It is used to connect the piston to the connecting rod.

32.6 Design Considerations for a Piston

In designing a piston for I.C. engine, the following points should be taken into consideration :

1. It should have enormous strength to withstand the high gas pressure and inertia forces.
2. It should have minimum mass to minimise the inertia forces.
3. It should form an effective gas and oil sealing of the cylinder.
4. It should provide sufficient bearing area to prevent undue wear.
5. It should disprese the heat of combustion quickly to the cylinder walls.
6. It should have high speed reciprocation without noise.
7. It should be of sufficient rigid construction to withstand thermal and mechanical distortion.
8. It should have sufficient support for the piston pin.

32.7 Material for Pistons

The most commonly used materials for pistons of I.C. engines are cast iron, cast aluminium, forged aluminium, cast steel and forged steel. The cast iron pistons are used for moderately rated

Twin cylinder airplane engine of 1930s.

engines with piston speeds below 6 m / s and aluminium alloy pistons are used for highly rated engines running at higher piston sppeds. It may be noted that

1. Since the *coefficient of thermal expansion for aluminium is about 2.5 times that of cast iron, therefore, a greater clearance must be provided between the piston and the cylinder wall (than with cast iron piston) in order to prevent siezing of the piston when engine runs continuously under heavy loads. But if excessive clearance is allowed, then the piston will develop ***'piston slap'*** while it is cold and this tendency increases with wear. The less clearance between the piston and the cylinder wall will lead to siezing of piston.

2. Since the aluminium alloys used for pistons have high **heat conductivity (nearly four times that of cast iron), therefore, these pistons ensure high rate of heat transfer and thus keeps down the maximum temperature difference between the centre and edges of the piston head or crown.

Notes: (*a*) For a cast iron piston, the temperature at the centre of the piston head (T_C) is about 425°C to 450°C under full load conditions and the temperature at the edges of the piston head (T_E) is about 200°C to 225°C.

(*b*) For aluminium alloy pistons, T_C is about 260°C to 290°C and T_E is about 185°C to 215°C.

3. Since the aluminium alloys are about ***three times lighter than cast iron, therfore, its mechanical strength is good at low temperatures, but they lose their strength (about 50%) at temperatures above 325°C. Sometimes, the pistons of aluminium alloys are coated with aluminium oxide by an electrical method.

32.8 Piston Head or Crown

The piston head or crown is designed keeping in view the following two main considerations, *i.e.*

1. It should have adequate strength to withstand the straining action due to pressure of explosion inside the engine cylinder, and

2. It should dissipate the heat of combustion to the cylinder walls as quickly as possible.

On the basis of first consideration of straining action, the thickness of the piston head is determined by treating it as a flat circular plate of uniform thickness, fixed at the outer edges and subjected to a uniformly distributed load due to the gas pressure over the entire cross-section.

The thickness of the piston head (t_H), according to Grashoff's formula is given by

$$t_H = \sqrt{\frac{3p.D^2}{16\sigma_t}} \text{ (in mm)} \qquad ...(i)$$

where p = Maximum gas pressure or explosion pressure in N/mm²,

D = Cylinder bore or outside diameter of the piston in mm, and

σ_t = Permissible bending (tensile) stress for the material of the piston in MPa or N/mm². It may be taken as 35 to 40 MPa for grey cast iron, 50 to 90 MPa for nickel cast iron and aluminium alloy and 60 to 100 MPa for forged steel.

On the basis of second consideration of heat transfer, the thickness of the piston head should be such that the heat absorbed by the piston due combustion of fuel is quickly transferred to the cylinder walls. Treating the piston head as a flat ciucular plate, its thickness is given by

$$t_H = \frac{H}{12.56k(T_C - T_E)} \text{ (in mm)} \qquad ...(ii)$$

* The coefficient of thermal expansion for aluminium is 0.24×10^{-6} m / °C and for cast iron it is 0.1×10^{-6} m / °C.

** The heat conductivity for aluminium is 174.75 W/m/°C and for cast iron it is 46.6 W/m /°C.

*** The density of aluminium is 2700 kg / m³ and for cast iron it is 7200 kg / m³.

where H = Heat flowing through the piston head in kJ/s or watts,

k = Heat conductivity factor in W/m/°C. Its value is 46.6 W/m/°C for grey cast iron, 51.25 W/m/°C for steel and 174.75 W/m/°C for aluminium alloys.

T_C = Temperture at the centre of the piston head in °C, and

T_E = Temperature at the edges of the piston head in °C.

The temperature difference ($T_C - T_E$) may be taken as 220°C for cast iron and 75°C for aluminium.

The heat flowing through the positon head (H) may be deternined by the following expression, *i.e.*,

$$H = C \times HCV \times m \times B.P. \text{ (in kW)}$$

where C = Constant representing that portion of the heat supplied to the engine which is absorbed by the piston. Its value is usually taken as 0.05.

HCV = Higher calorific value of the fuel in kJ/kg. It may be taken as 45×10^3 kJ/kg for diesel and 47×10^3 kJ/ kg for petrol,

m = Mass of the fuel used in kg per brake power per second, and

$B.P.$ = Brake power of the engine per cylinder

Notes : 1. The thickness of the piston head (t_H) is calculated by using equations (*i*) and (*ii*) and larger of the two values obtained should be adopted.

2. When t_H is 6 mm or less, then no ribs are required to strengthen the piston head against gas loads. But when t_H is greater then 6 mm, then a suitable number of ribs at the centre line of the boss extending around the skirt should be provided to distribute the side thrust from the connecting rod and thus to prevent distortion of the skirt. The thickness of the ribs may be takes as $t_H / 3$ to $t_H / 2$.

3. For engines having length of stroke to cylinder bore (L / D) ratio upto 1.5, a cup is provided in the top of the piston head with a radius equal to 0.7 D. This is done to provide a space for combustion chamber.

32.9 Piston Rings

The piston rings are used to impart the necessary radial pressure to maintain the seal between the piston and the cylinder bore. These are usually made of grey cast iron or alloy cast iron because of their good wearing properties and also they retain spring characteristics even at high temperatures. The piston rings are of the following two types :

1. Compression rings or pressure rings, and
2. Oil control rings or oil scraper.

The ***compression rings*** or ***pressure rings*** are inserted in the grooves at the top portion of the piston and may be three to seven in number. These rings also transfer heat from the piston to the cylinder liner and absorb some part of the piston fluctuation due to the side thrust.

The ***oil control rings*** or ***oil scrapers*** are provided below the compression rings. These rings provide proper lubrication to the liner by allowing sufficient oil to move up during upward stroke and at the same time scraps the lubricating oil from the surface of the liner in order to minimise the flow of the oil to the combustion chamber.

The compression rings are usually made of rectangular cross-section and the diameter of the ring is slightly larger than the cylinder bore. A part of the ring is cut- off in order to permit it to go into the cylinder against the liner wall. The diagonal cut or step cut ends, as shown in Fig. 32.4 (*a*) and (*b*) respectively, may be used. The gap between the ends should be sufficiently large when the ring is put cold so that even at the highest temperature, the ends do not touch each other when the ring expands, otherwise there might be buckling of the ring.

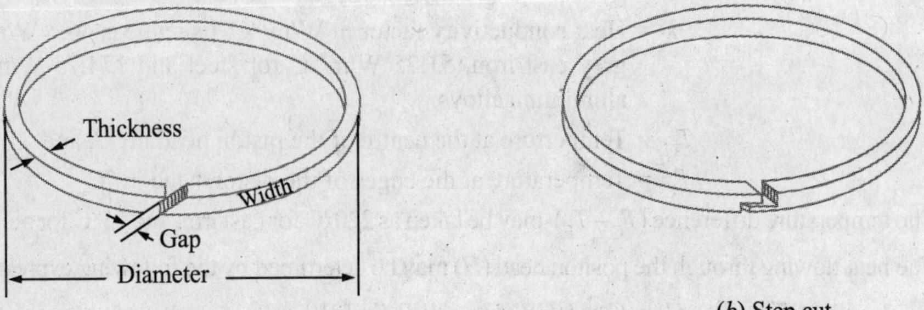

(*a*) Diagonal cut. (*b*) Step cut.

Fig. 32.4. Piston rings.

The radial thickness (t_1) of the ring may be obtained by considering the radial pressure between the cylinder wall and the ring. From bending stress consideration in the ring, the radial thickness is given by

$$t_1 = D\sqrt{\frac{3p_w}{\sigma_t}}$$

where D = Cylinder bore in mm,

p_w = Pressure of gas on the cylinder wall in N/mm^2. Its value is limited from 0.025 N/mm^2 to 0.042 N/mm^2, and

σ_t = Allowable bending (tensile) stress in MPa. Its value may be taken from 85 MPa to 110 MPa for cast iron rings.

The axial thickness (t_2) of the rings may be taken as 0.7 t_1 to t_1.

The minimum axial thickness (t_2) may also be obtained from the following empirical relation:

$$t_2 = \frac{D}{10\,n_R}$$

where n_R = Number of rings.

The width of the top land (*i.e.* the distance from the top of the piston to the first ring groove) is made larger than other ring lands to protect the top ring from high temperature conditions existing at the top of the piston,

∴ Width of top land,

$$b_1 = t_H \text{ to } 1.2\, t_H$$

The width of other ring lands (*i.e.* the distance between the ring grooves) in the piston may be made equal to or slightly less than the axial thickness of the ring (t_2).

∴ Width of other ring lands,

$$b_2 = 0.75\, t_2 \text{ to } t_2$$

The depth of the ring grooves should be more than the depth of the ring so that the ring does not take any piston side thrust.

The gap between the free ends of the ring is given by 3.5 t_1 to 4 t_1. The gap, when the ring is in the cylinder, should be 0.002 D to 0.004 D.

32.10 Piston Barrel

It is a cylindrical portion of the piston. The maximum thickness (t_3) of the piston barrel may be obtained from the following empirical relation :

$$t_3 = 0.03\, D + b + 4.5 \text{ mm}$$

where b = Radial depth of piston ring groove which is taken as 0.4 mm larger than the radial thickness of the piston ring (t_1)

$= t_1 + 0.4$ mm

Thus, the above relation may be written as

$$t_3 = 0.03\, D + t_1 + 4.9 \text{ mm}$$

The piston wall thickness (t_4) towards the open end is decreased and should be taken as $0.25\, t_3$ to $0.35\, t_3$.

32.11 Piston Skirt

The portion of the piston below the ring section is known as ***piston skirt.*** In acts as a bearing for the side thrust of the connecting rod. The length of the piston skirt should be such that the bearing pressure on the piston barrel due to the side thrust does not exceed 0.25 N/mm^2 of the projected area for low speed engines and 0.5 N/mm^2 for high speed engines. It may be noted that the maximum thrust will be during the expansion stroke. The side thrust (R) on the cylinder liner is usually taken as 1/10 of the maximum gas load on the piston.

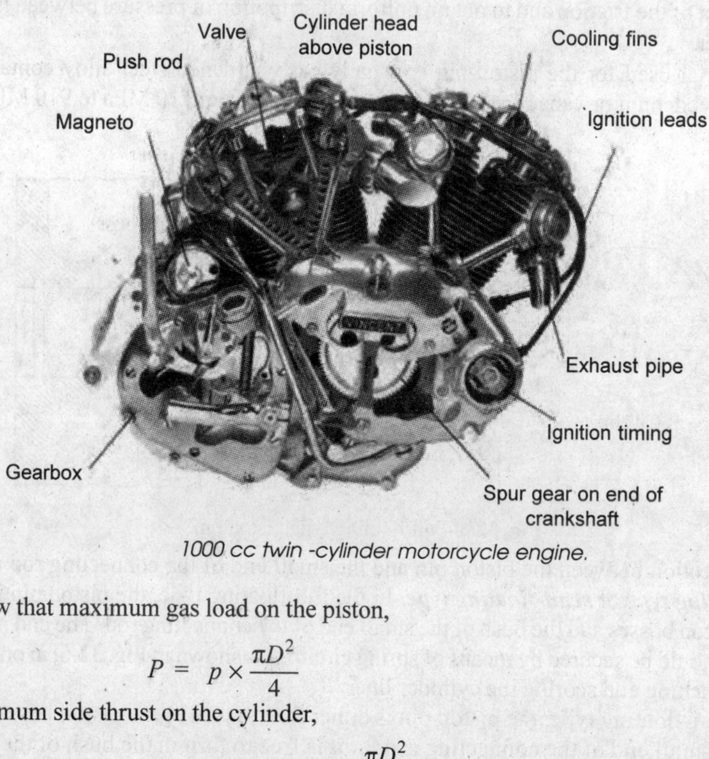

1000 cc twin -cylinder motorcycle engine.

We know that maximum gas load on the piston,

$$P = p \times \frac{\pi D^2}{4}$$

∴ Maximum side thrust on the cylinder,

$$R = P/10 = 0.1\, p \times \frac{\pi D^2}{4} \quad \text{...}(i)$$

where p = Maximum gas pressure in N/mm^2, and

D = Cylinder bore in mm.

The side thrust (R) is also given by

R = Bearing pressure × Projected bearing area of the piston skirt

$= p_b \times D \times l$

where l = Length of the piston skirt in mm. ...(*ii*)

From equations (*i*) and (*ii*), the length of the piston skirt (l) is determined. In actual practice, the length of the piston skirt is taken as 0.65 to 0.8 times the cylinder bore. Now the total length of the piston (L) is given by

$$L = \text{Length of skirt} + \text{Length of ring section} + \text{Top land}$$

The length of the piston usually varies between D and $1.5\ D$. It may be noted that a longer piston provides better bearing surface for quiet running of the engine, but it should not be made unnecessarily long as it will increase its own mass and thus the inertia forces.

32.12 Piston Pin

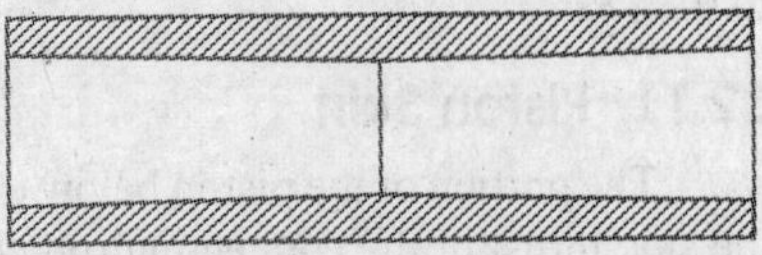

Fig.32.5. Piston pin.

The piston pin (also called gudgeon pin or wrist pin) is used to connect the piston and the connecting rod. It is usually made hollow and tapered on the inside, the smallest inside diameter being at the centre of the pin, as shown in Fig. 32.5. The piston pin passes through the bosses provided on the inside of the piston skirt and the bush of the small end of the connecting rod. The centre of piston pin should be $0.02\ D$ to $0.04\ D$ above the centre of the skirt, in order to off-set the turning effect of the friction and to obtain uniform distribution of pressure between the piston and the cylinder liner.

The material used for the piston pin is usually case hardened steel alloy containing nickel, chromium, molybdenum or vanadium having tensile strength from 710 MPa to 910 MPa.

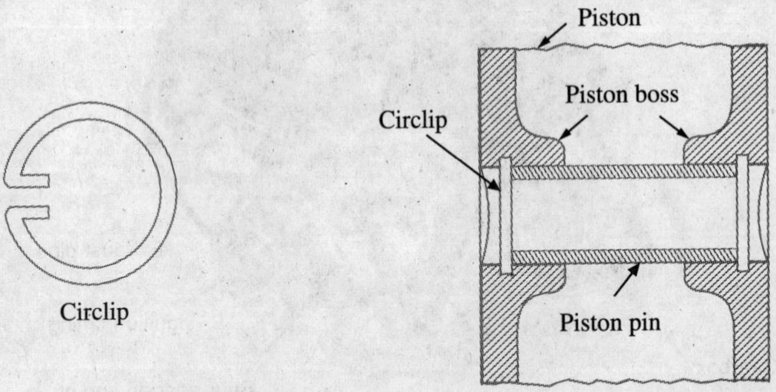

Fig. 32.6. Full floating type piston pin.

The connection between the piston pin and the small end of the connecting rod may be made either ***full floating type*** or ***semi-floating type.*** In the full floating type, the piston pin is free to turn both in the *piston bosses and the bush of the small end of the connecting rod. The end movements of the piston pin should be secured by means of spring circlips, as shown in Fig. 32.6, in order to prevent the pin from touching and scoring the cylinder liner.

In the semi-floating type, the piston pin is either free to turn in the piston bosses and rigidly secured to the small end of the connecting rod, or it is free to turn in the bush of the small end of the connecting rod and is rigidly secured in the piston bosses by means of a screw, as shown in Fig. 32.7

The piston pin should be designed for the maximum gas load or the inertia force of the piston, whichever is larger. The bearing area of the piston pin should be about equally divided between the piston pin bosses and the connecting rod bushing. Thus, the length of the pin in the connecting rod bushing will be about 0.45 of the cylinder bore or piston diameter (D), allowing for the end clearance

* The mean diameter of the piston bosses is made $1.4\ d_0$ for cast iron pistons and $1.5\ d_0$ for aluminium pistons, where d_0 is the outside diameter of the piston pin. The piston bosses are usually tapered, increasing the diameter towards the piston wall.

of the pin etc. The outside diameter of the piston pin (d_0) is determined by equating the load on the piston due to gas pressure (p) and the load on the piston pin due to bearing pressure (p_{b1}) at the small end of the connecting rod bushing.

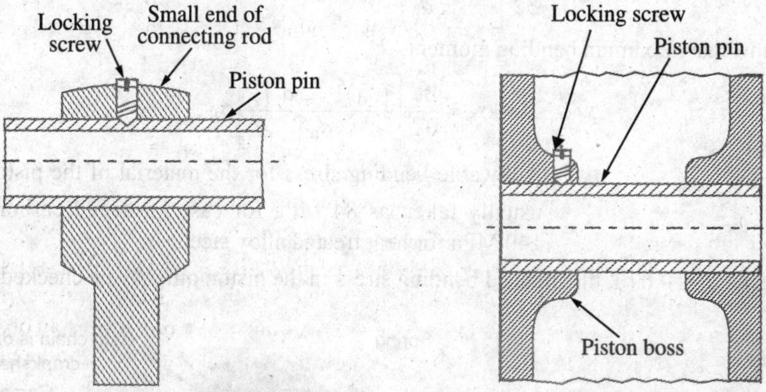

(*a*) Piston pin secured to the small end of the connecting rod.

(*b*) Piston pin secured to the boss of the piston.

Fig. 32.7. Semi-floating type piston pin.

Let d_0 = Outside diameter of the piston pin in mm

l_1 = Length of the piston pin in the bush of the small end of the connecting rod in mm. Its value is usually taken as 0.45 *D*.

p_{b1} = Bearing pressure at the small end of the connecting rod bushing in N/mm². Its value for the bronze bushing may be taken as 25 N/mm².

We know that load on the piston due to gas pressure or gas load

$$= \frac{\pi D^2}{4} \times p \quad ...(i)$$

and load on the piston pin due to bearing pressure or bearing load

$$= \text{Bearing pressure} \times \text{Bearing area} = p_{b1} \times d_0 \times l_1 \quad ...(ii)$$

From equations (*i*) and (*ii*), the outside diameter of the piston pin (d_0) may be obtained.

The piston pin may be checked in bending by assuming the gas load to be uniformly distributed over the length l_1 with supports at the centre of the bosses at the two ends. From Fig. 32.8, we find that the length between the supports,

$$l_2 = l_1 + \frac{D - l_1}{2} = \frac{l_1 + D}{2}$$

Now maximum bending moment at the centre of the pin,

$$M = \frac{P}{2} \times \frac{l_2}{2} - \frac{P}{l_1} \times \frac{l_1}{2} \times \frac{l_1}{4}$$

$$= \frac{P}{2} \times \frac{l_2}{2} - \frac{P}{2} \times \frac{l_1}{4}$$

$$= \frac{P}{2}\left(\frac{l_1 + D}{2 \times 2}\right) - \frac{P}{2} \times \frac{l_1}{4}$$

$$= \frac{P.l_1}{8} + \frac{P.D}{8} - \frac{P.l_1}{8} = \frac{P.D}{8} \quad ...(iii)$$

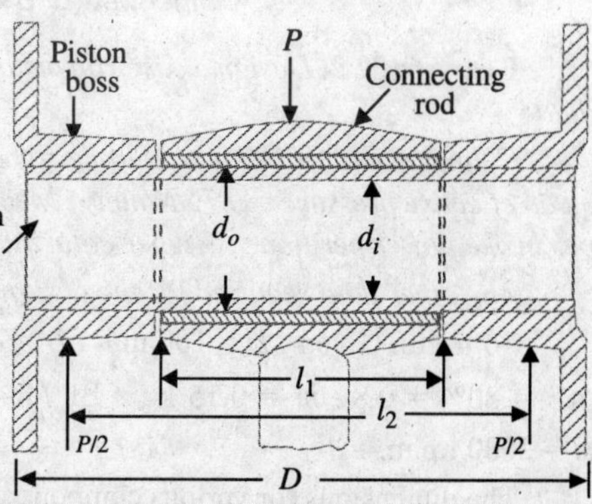

Fig. 32.8

We have already discussed that the piston pin is made hollow. Let d_0 and d_i be the outside and inside diameters of the piston pin. We know that the section modulus,

$$Z = \frac{\pi}{32}\left[\frac{(d_0)^4 - (d_i)^4}{d_0}\right]$$

We know that maximum bending moment,

$$M = Z \times \sigma_b = \frac{\pi}{32}\left[\frac{(d_0)^4 - (d_i)^4}{d_0}\right]\sigma_b$$

where σ_b = Allowable bending stress for the material of the piston pin. It is usually taken as 84 MPa for case hardened carbon steel and 140 MPa for heat treated alloy steel.

Assuming $d_i = 0.6\ d_0$, the induced bending stress in the piston pin may be checked.

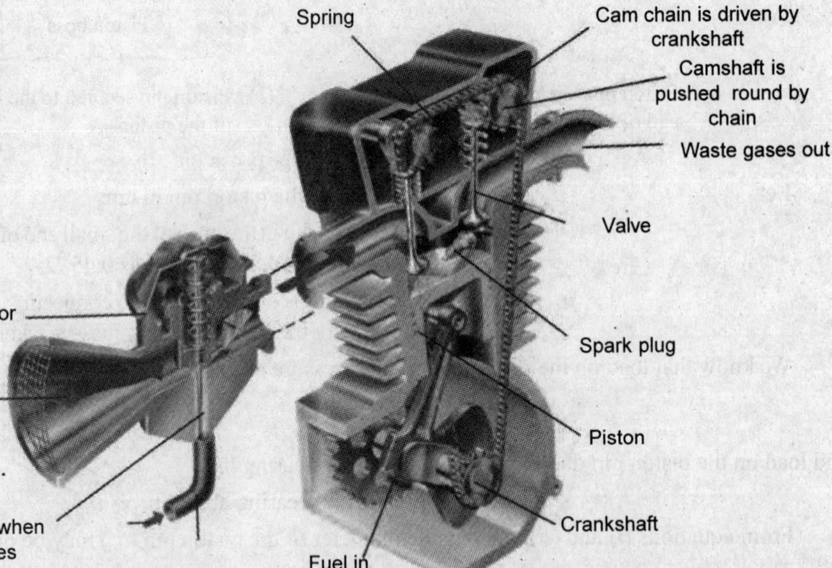

Another view of a single cylinder 4-stroke petrol engine.

Example 32.2. *Design a cast iron piston for a single acting four stroke engine for the following data:*

Cylinder bore = 100 mm ; Stroke = 125 mm ; Maximum gas pressure = 5 N/mm² ; Indicated mean effective pressure = 0.75 N/mm² ; Mechanical efficiency = 80% ; Fuel consumption = 0.15 kg per brake power per hour ; Higher calorific value of fuel = 42 × 10³ kJ/kg ; Speed = 2000 r.p.m.

Any other data required for the design may be assumed.

Solution. Given : $D = 100$ mm ; $L = 125$ mm $= 0.125$ m ; $p = 5$ N/mm² ; $p_m = 0.75$ N/mm²; $\eta_m = 80\% = 0.8$; $m = 0.15$ kg / BP / h $= 41.7 \times 10^{-6}$ kg / BP / s; $HCV = 42 \times 10^3$ kJ / kg ; $N = 2000$ r.p.m.

The dimensions for various components of the piston are determined as follows :

1. ***Piston head or crown***

The thickness of the piston head or crown is determined on the basis of strength as well as on the basis of heat dissipation and the larger of the two values is adopted.

We know that the thickness of piston head on the basis of strength,

$$t_H = \sqrt{\frac{3p.D^2}{16\,\sigma_t}} = \sqrt{\frac{3 \times 5(100)^2}{16 \times 38}} = 15.7 \text{ say } 16 \text{ mm}$$

...(Taking σ_t for cast iron = 38 MPa = 38 N/mm²)

Since the engine is a four stroke engine, therefore, the number of working strokes per minute,

$$n = N/2 = 2000/2 = 1000$$

and cross-sectional area of the cylinder,

$$A = \frac{\pi\, D^2}{4} = \frac{\pi\,(100)^2}{4} = 7855 \text{ mm}^2$$

We know that indicated power,

$$IP = \frac{p_m.L.A.n}{60} = \frac{0.75 \times 0.125 \times 7855 \times 1000}{60} = 12\ 270 \text{ W}$$

$$= 12.27 \text{ kW}$$

∴ Brake power, $BP = IP \times \eta_m = 12.27 \times 0.8 = 9.8$ kW ...($\because \eta_m = BP/IP$)

We know that the heat flowing through the piston head,

$$H = C \times HCV \times m \times BP$$

$$= 0.05 \times 42 \times 10^3 \times 41.7 \times 10^{-6} \times 9.8 = 0.86 \text{ kW} = 860 \text{ W}$$

....(Taking $C = 0.05$)

∴ Thickness of the piston head on the basis of heat dissipation,

$$t_H = \frac{H}{12.56\,k\,(T_C - T_E)} = \frac{860}{12.56 \times 46.6 \times 220} = 0.0067 \text{ m} = 6.7 \text{ mm}$$

...($\because$ For cast iron, $k = 46.6$ W/m/°C, and $T_C - T_E = 220$°C)

Taking the larger of the two values, we shall adopt

$$t_H = 16 \text{ mm} \quad \textbf{Ans.}$$

Since the ratio of L/D is 1.25, therefore a cup in the top of the piston head with a radius equal to 0.7 D (*i.e.* 70 mm) is provided.

2. *Radial ribs*

The radial ribs may be four in number. The thickness of the ribs varies from $t_H/3$ to $t_H/2$.

∴ Thickness of the ribs, $t_R = 16/3$ to $16/2 = 5.33$ to 8 mm

Let us adopt $t_R = 7$ mm **Ans.**

3. *Piston rings*

Let us assume that there are total four rings (*i.e.* $n_r = 4$) out of which three are compression rings and one is an oil ring.

We know that the radial thickness of the piston rings,

$$t_1 = D\sqrt{\frac{3\,p_w}{\sigma_t}} = 100\sqrt{\frac{3 \times 0.035}{90}} = 3.4 \text{ mm}$$

...(Taking $p_w = 0.035$ N/mm², and $\sigma_t = 90$ MPa)

and axial thickness of the piston rings

$$t_2 = 0.7\,t_1 \text{ to } t_1 = 0.7 \times 3.4 \text{ to } 3.4 \text{ mm} = 2.38 \text{ to } 3.4 \text{ mm}$$

Let us adopt $t_2 = 3$ mm

We also know that the minimum axial thickness of the pistion ring,

$$t_2 = \frac{D}{10\, n_r} = \frac{100}{10 \times 4} = 2.5 \text{ mm}$$

Thus the axial thickness of the piston ring as already calculated (*i.e.* $t_2 = 3$ mm)is satisfactory. **Ans.**

The distance from the top of the piston to the first ring groove, *i.e.* the width of the top land,

$$b_1 = t_H \text{ to } 1.2\, t_H = 16 \text{ to } 1.2 \times 16 \text{ mm} = 16 \text{ to } 19.2 \text{ mm}$$

and width of other ring lands,

$$b_2 = 0.75\, t_2 \text{ to } t_2 = 0.75 \times 3 \text{ to } 3 \text{ mm} = 2.25 \text{ to } 3 \text{ mm}$$

Let us adopt $b_1 = 18$ mm ; and $b_2 = 2.5$ mm **Ans.**

We know that the gap between the free ends of the ring,

$$G_1 = 3.5\, t_1 \text{ to } 4\, t_1 = 3.5 \times 3.4 \text{ to } 4 \times 3.4 \text{ mm} = 11.9 \text{ to } 13.6 \text{ mm}$$

and the gap when the ring is in the cylinder,

$$G_2 = 0.002\, D \text{ to } 0.004\, D = 0.002 \times 100 \text{ to } 0.004 \times 100 \text{ mm}$$

$$= 0.2 \text{ to } 0.4 \text{ mm}$$

Let us adopt $G_1 = 12.8$ mm ; and $G_2 = 0.3$ mm **Ans.**

4. ***Piston barrel***

Since the radial depth of the piston ring grooves (b) is about 0.4 mm more than the radial thickness of the piston rings (t_1), therefore,

$$b = t_1 + 0.4 = 3.4 + 0.4 = 3.8 \text{ mm}$$

We know that the maximum thickness of barrel,

$$t_3 = 0.03\, D + b + 4.5 \text{ mm} = 0.03 \times 100 + 3.8 + 4.5 = 11.3 \text{ mm}$$

and piston wall thickness towards the open end,

$$t_4 = 0.25\, t_3 \text{ to } 0.35\, t_3 = 0.25 \times 11.3 \text{ to } 0.35 \times 11.3 = 2.8 \text{ to } 3.9 \text{ mm}$$

Let us adopt $t_4 = 3.4$ mm

5. ***Piston skirt***

Let l = Length of the skirt in mm.

We know that the maximum side thrust on the cylinder due to gas pressure (p),

$$R = \mu \times \frac{\pi D^2}{4} \times p = 0.1 \times \frac{\pi (100)^2}{4} \times 5 = 3928 \text{ N}$$

...(Taking $\mu = 0.1$)

We also know that the side thrust due to bearing pressure on the piston barrel (p_b),

$$R = p_b \times D \times l = 0.45 \times 100 \times l = 45\, l \text{ N}$$

...(Taking $p_b = 0.45$ N/mm²)

From above, we find that

$$45\, l = 3928 \text{ or } l = 3928 / 45 = 87.3 \text{ say } 90 \text{ mm } \textbf{Ans.}$$

∴ Total length of the piston ,

$$L = \text{Length of the skirt} + \text{Length of the ring section} + \text{Top land}$$

$$= l + (4\, t_2 + 3 b_2) + b_1$$

$$= 90 + (4 \times 3 + 3 \times 3) + 18 = 129 \text{ say } 130 \text{ mm } \textbf{Ans.}$$

6. ***Piston pin***

Let d_0 = Outside diameter of the pin in mm,

l_1 = Length of pin in the bush of the small end of the connecting rod in mm, and

p_{b1} = Bearing pressure at the small end of the connecting rod bushing in N/mm². It value for bronze bushing is taken as 25 N/mm².

We know that load on the pin due to bearing pressure

$$= \text{Bearing pressure} \times \text{Bearing area} = p_{b1} \times d_0 \times l_1$$

$$= 25 \times d_0 \times 0.45 \times 100 = 1125\ d_0 \text{ N} \quad \text{...(Taking } l_1 = 0.45\ D\text{)}$$

We also know that maximum load on the piston due to gas pressure or maximum gas load

$$= \frac{\pi D^2}{4} \times p = \frac{\pi (100)^2}{4} \times 5 = 39\ 275 \text{ N}$$

From above, we find that

$$1125\ d_0 = 39\ 275 \quad \text{or} \quad d_0 = 39\ 275 / 1125 = 34.9 \text{ say } 35 \text{ mm } \textbf{Ans.}$$

The inside diameter of the pin (d_i) is usually taken as 0.6 d_0.

$$\therefore \quad d_i = 0.6 \times 35 = 21 \text{ mm } \textbf{Ans.}$$

Let the piston pin be made of heat treated alloy steel for which the bending stress (σ_b)may be taken as 140 MPa. Now let us check the induced bending stress in the pin.

We know that maximum bending moment at the centre of the pin,

$$M = \frac{P.D}{8} = \frac{39\,275 \times 100}{8} = 491 \times 10^3 \text{ N-mm}$$

We also know that maximum bending moment (M),

$$491 \times 10^3 = \frac{\pi}{32}\left[\frac{(d_0)^4 - (d_i)^4}{d_0}\right]\sigma_b = \frac{\pi}{32}\left[\frac{(35)^4 - (21)^4}{35}\right]\sigma_b = 3664\,\sigma_b$$

$$\therefore \quad \sigma_b = 491 \times 10^3 / 3664 = 134 \text{ N/mm}^2 \text{ or MPa}$$

Since the induced bending stress in the pin is less than the permissible value of 140 MPa (*i.e.* 140 N/mm²), therefore, the dimensions for the pin as calculated above (*i.e.* d_0 = 35 mm and d_i = 21 mm) are satisfactory.

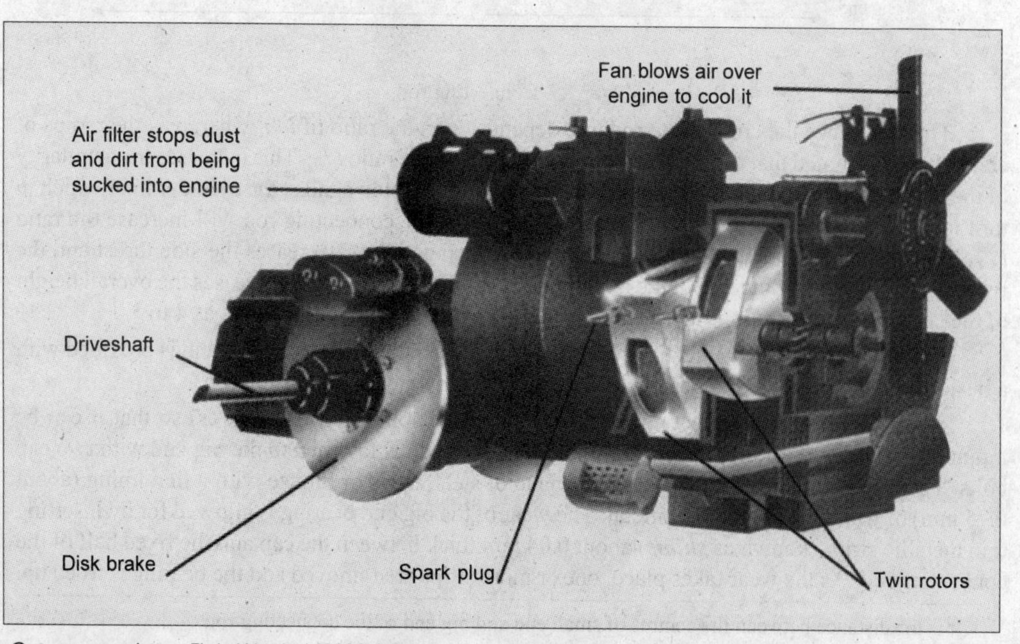

German engineer Fleix Wankel (1902-88) built a rotary engine in 1957. A triangular piston turns inside a chamber through the combustion cycle.

32.13 Connecting Rod

The connecting rod is the intermediate member between the piston and the crankshaft. Its primary function is to transmit the push and pull from the piston pin to the crankpin and thus convert the reciprocating motion of the piston into the rotary motion of the crank. The usual form of the connecting rod in internal combustion engines is shown in Fig. 32.9. It consists of a long shank, a small end and a big end. The cross-section of the shank may be rectangular, circular, tubular, *I*-section or *H*-section. Generally circular section is used for low speed engines while *I*-section is preferred for high speed engines.

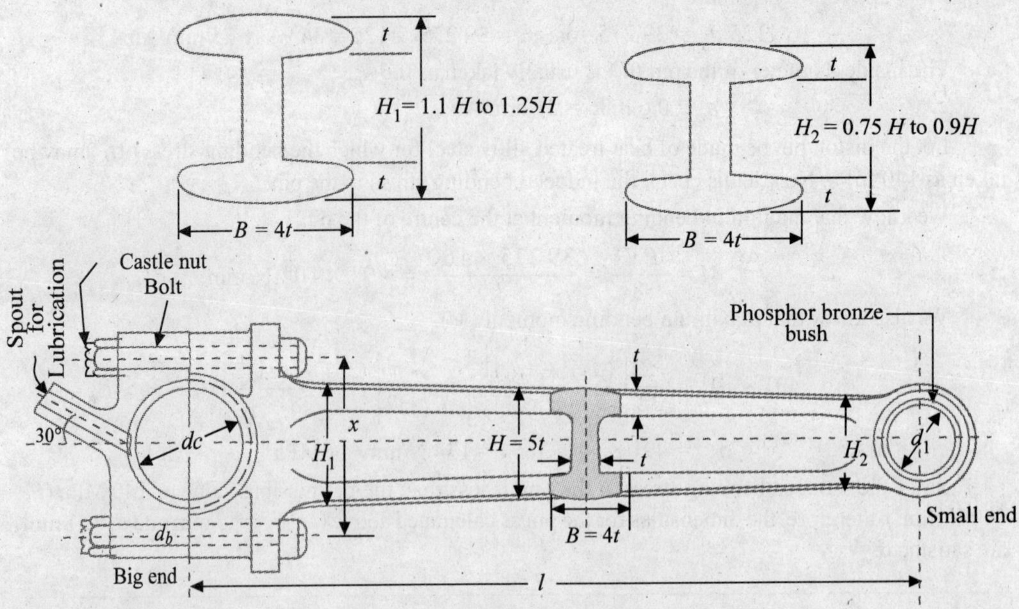

Fig. 32.9. Connecting rod.

The *length of the connecting rod (l) depends upon the ratio of l / r, where r is the radius of crank. It may be noted that the smaller length will decrease the ratio l / r. This increases the angularity of the connecting rod which increases the side thrust of the piston against the cylinder liner which in turn increases the wear of the liner. The larger length of the connecting rod will increase the ratio l / r. This decreases the angularity of the connecting rod and thus decreases the side thrust and the resulting wear of the cylinder. But the larger length of the connecting rod increases the overall height of the engine. Hence, a compromise is made and the ratio l / r is generally kept as 4 to 5.

The small end of the connecting rod is usually made in the form of an eye and is provided with a bush of phosphor bronze. It is connected to the piston by means of a piston pin.

The big end of the connecting rod is usually made split (in two **halves) so that it can be mounted easily on the crankpin bearing shells. The split cap is fastened to the big end with two cap bolts. The bearing shells of the big end are made of steel, brass or bronze with a thin lining (about 0.75 mm) of white metal or babbit metal. The wear of the big end bearing is allowed for by inserting thin metallic strips (known as ***shims***) about 0.04 mm thick between the cap and the fixed half of the connecting rod. As the wear takes place, one or more strips are removed and the bearing is trued up.

* It is the distance between the centres of small end and big end of the connecting rod.

** One half is fixed with the connecting rod and the other half (known as cap) is fastened with two cap bolts.

The connecting rods are usually manufactured by drop forging process and it should have adequate strength, stiffness and minimum weight. The material mostly used for connecting rods varies from mild carbon steels (having 0.35 to 0.45 percent carbon) to alloy steels (chrome-nickel or chrome-molybdenum steels). The carbon steel having 0.35 percent carbon has an ultimate tensile strength of about 650 MPa when properly heat treated and a carbon steel with 0.45 percent carbon has a ultimate tensile strength of 750 MPa. These steels are used for connecting rods of industrial engines. The alloy steels have an ultimate tensile strength of about 1050 MPa and are used for connecting rods of aeroengines and automobile engines.

The bearings at the two ends of the connecting rod are either splash lubricated or pressure lubricated. The big end bearing is usually splash lubricated while the small end bearing is pressure lubricated. In the **splash lubrication system**, the cap at the big end is provided with a dipper or spout and set at an angle in such a way that when the connecting rod moves downward, the spout will dip into the lubricating oil contained in the sump. The oil is forced up the spout and then to the big end bearing. Now when the connecting rod moves upward, a splash of oil is produced by the spout. This splashed up lubricant find its way into the small end bearing through the widely chamfered holes provided on the upper surface of the small end.

In the **pressure lubricating system,** the lubricating oil is fed under pressure to the big end bearing through the holes drilled in crankshaft, crankwebs and crank pin. From the big end bearing, the oil is fed to small end bearing through a fine hole drilled in the shank of the connecting rod. In some cases, the small end bearing is lubricated by the oil scrapped from the walls of the cyinder liner by the oil scraper rings.

32.14 Forces Acting on the Connecting Rod

The various forces acting on the connecting rod are as follows :

1. Force on the piston due to gas pressure and inertia of the reciprocating parts,
2. Force due to inertia of the connecting rod or inertia bending forces,
3. Force due to friction of the piston rings and of the piston, and
4. Force due to friction of the piston pin bearing and the crankpin bearing.

We shall now derive the expressions for the forces acting on a vertical engine, as discussed below.

1. ***Force on the piston due to gas pressure and inertia of reciprocating parts***

Consider a connecting rod *PC* as shown in Fig. 32.10.

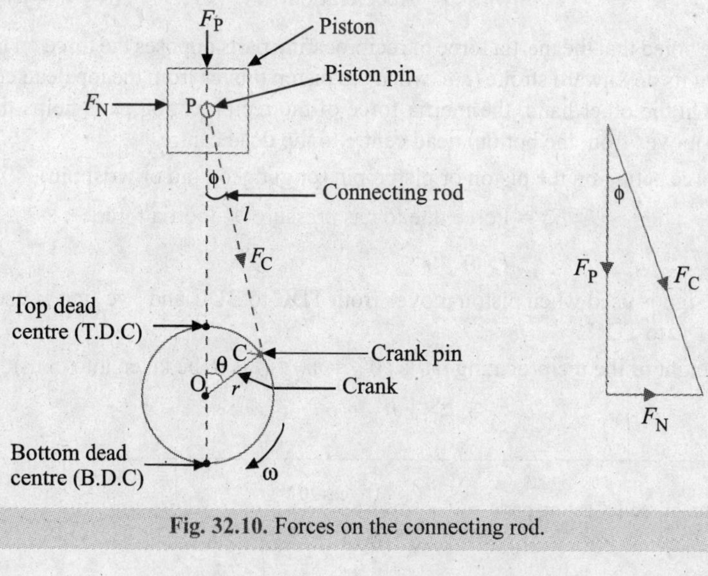

Fig. 32.10. Forces on the connecting rod.

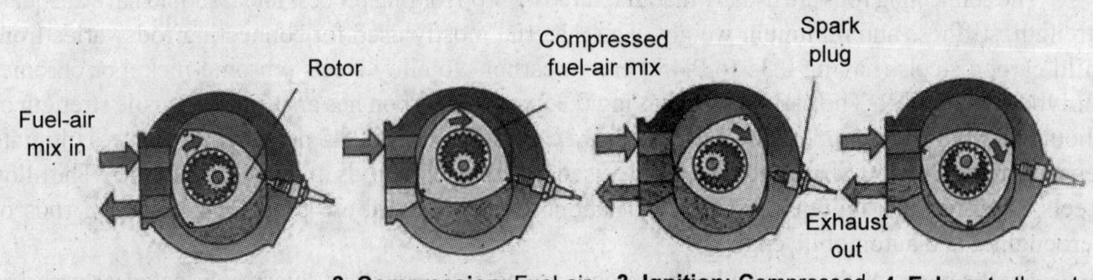

Let p = Maximum pressure of gas,

D = Diameter of piston,

A = Cross-section area of piston $= \dfrac{\pi D^2}{4}$,

m_R = Mass of reciprocating parts,

= Mass of piston, gudgeon pin etc. + $\dfrac{1}{3}$ rd mass of connecting rod,

ω = Angular speed of crank,

ϕ = Angle of inclination of the connecting rod with the line of stroke,

θ = Angle of inclination of the crank from top dead centre,

r = Radius of crank,

l = Length of connecting rod, and

n = Ratio of length of connecting rod to radius of crank = l / r.

We know that the force on the piston due to pressure of gas,

$$F_L = \text{Pressure} \times \text{Area} = p \,.\, A = p \times \pi D^2 /4$$

and inertia force of reciprocating parts,

$$F_I = \text{Mass} \times {}^*\text{Acceleration} = m_R \,.\, \omega^2 \,.\, r \left(\cos\theta + \frac{\cos 2\theta}{n}\right)$$

It may be noted that the inertia force of reciprocating parts opposes the force on the piston when it moves during its downward stroke (*i. e.* when the piston moves from the top dead centre to bottom dead centre). On the other hand, the inertia force of the reciprocating parts helps the force on the piston when it moves from the bottom dead centre to top dead centre.

∴ Net force acting on the piston or piston pin (or gudgeon pin or wrist pin),

$$F_P = \text{Force due to gas pressure} \mp \text{Inertia force}$$

$$= F_L \mp F_I$$

The –ve sign is used when piston moves from TDC to BDC and +ve sign is used when piston moves from BDC to TDC.

When weight of the reciprocating parts ($W_R = m_R \,.\, g$) is to be taken into consideration, then

$$F_P = F_L \mp F_I \pm W_R$$

---

* Acceleration of reciprocating parts $= \omega^2 \cdot r \left(\cos\theta + \dfrac{\cos 2\theta}{n}\right)$

The force F_P gives rise to a force F_C in the connecting rod and a thrust F_N on the sides of the cylinder walls. From Fig. 32.10, we see that force in the connecting rod at any instant,

$$F_C = \frac{F_P}{\cos\phi} = \frac{{}^*F_P}{\sqrt{1 - \frac{\sin^2\theta}{n^2}}}$$

The force in the connecting rod will be maximum when the crank and the connecting rod are perpendicular to each other (*i.e.* when $\theta = 90°$). But at this position, the gas pressure would be decreased considerably. ***Thus, for all practical purposes, the force in the connecting rod*** **(F_C)** ***is taken equal to the maximum force on the piston due to pressure of gas*** **(F_L)**, neglecting piston inertia effects.

2. ***Force due to inertia of the connecting rod or inertia bending forces***

Consider a connecting rod *PC* and a crank *OC* rotating with uniform angular velocity ω rad / s. In order to find the accleration of various points on the connecting rod, draw the Klien's acceleration diagram *CQNO* as shown in Fig. 32.11 (*a*). *CO* represents the acceleration of *C* towards *O* and *NO* represents the accleration of *P* towards *O*. The acceleration of other points such as *D, E, F* and *G* etc., on the connecting rod *PC* may be found by drawing horizontal lines from these points to intresect *CN* at *d, e, f,* and g respectively. Now *dO, eO, fO* and *gO* respresents the acceleration of *D, E, F* and *G* all towards *O*. The inertia force acting on each point will be as follows:

Inertia force at $C = m \times \omega^2 \times CO$

Inertia force at $D = m \times \omega^2 \times dO$

Inertia force at $E = m \times \omega^2 \times eO$, and so on.

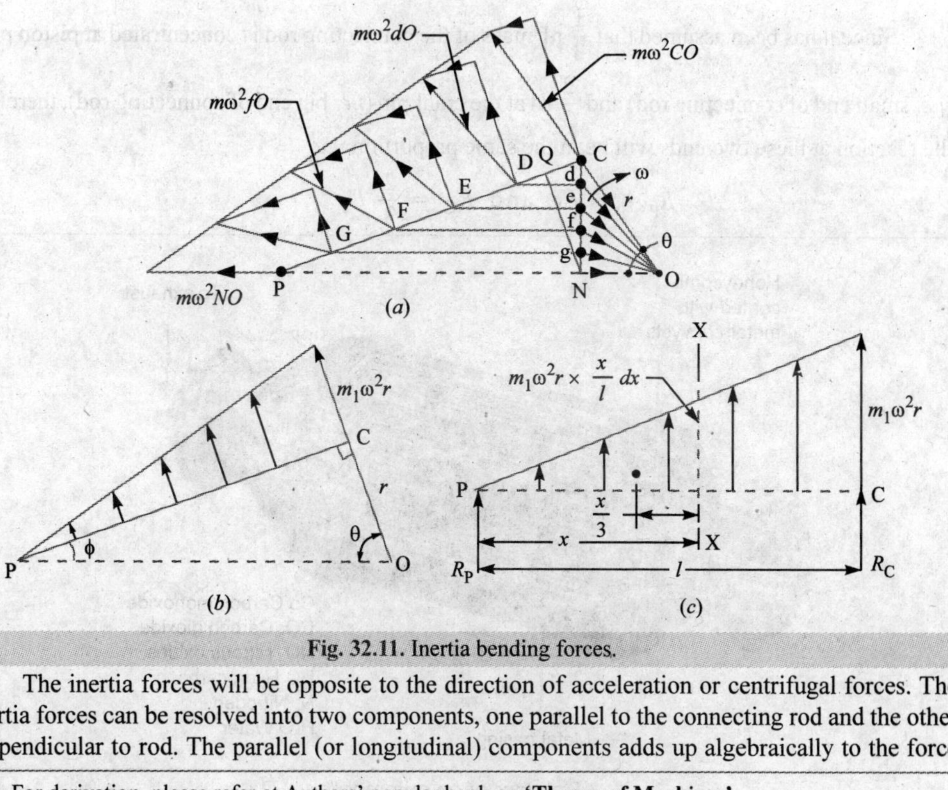

Fig. 32.11. Inertia bending forces.

The inertia forces will be opposite to the direction of acceleration or centrifugal forces. The inertia forces can be resolved into two components, one parallel to the connecting rod and the other perpendicular to rod. The parallel (or longitudinal) components adds up algebraically to the force

* For derivation, please refer ot Authors' popular book on **'Theory of Machines'**.

acting on the connecting rod (F_C) and produces thrust on the pins. The perpendicular (or transverse) components produces bending action (also called whipping action) and the stress induced in the connecting rod is called ***whipping stress.***

It may be noted that the perpendicular components will be maximum, when the crank and connecting rod are at right angles to each other.

The variation of the inertia force on the connecting rod is linear and is like a simply supported beam of variable loading as shown in Fig. 32.11 (*b*) and (*c*). Assuming that the connecting rod is of uniform cross-section and has mass m_1 kg per unit length, therefore,

Inertia force per unit length at the crankpin

$$= m_1 \times \omega^2 r$$

and inertia force per unit length at the piston pin

$$= 0$$

Inertia force due to small element of length dx at a distance x from the piston pin P,

$$dF_1 = m_1 \times \omega^2 r \times \frac{x}{l} \times dx$$

∴ Resultant inertia force,

$$F_I = \int_0^l m_1 \times \omega^2 r \times \frac{x}{l} \times dx = \frac{m_1 \times \omega^2 r}{l}\left[\frac{x^2}{2}\right]_0^l$$

$$= \frac{m_1 . l}{2} \times \omega^2 r = \frac{m}{2} \times \omega^2 r \qquad \text{...(Substituting } m_1 \cdot l = m)$$

This resultant inertia force acts at a distance of $2l / 3$ from the piston pin P.

Since it has been assumed that $\frac{1}{3}$ rd mass of the connecting rod is concentrated at piston pin P (*i.e.* small end of connecting rod) and $\frac{2}{3}$ rd at the crankpin (*i.e.* big end of connecting rod), therefore, the reaction at these two ends will be in the same proportion. *i.e.*

$$R_P = \frac{1}{3} F_I, \text{ and } R_C = \frac{2}{3} F_I$$

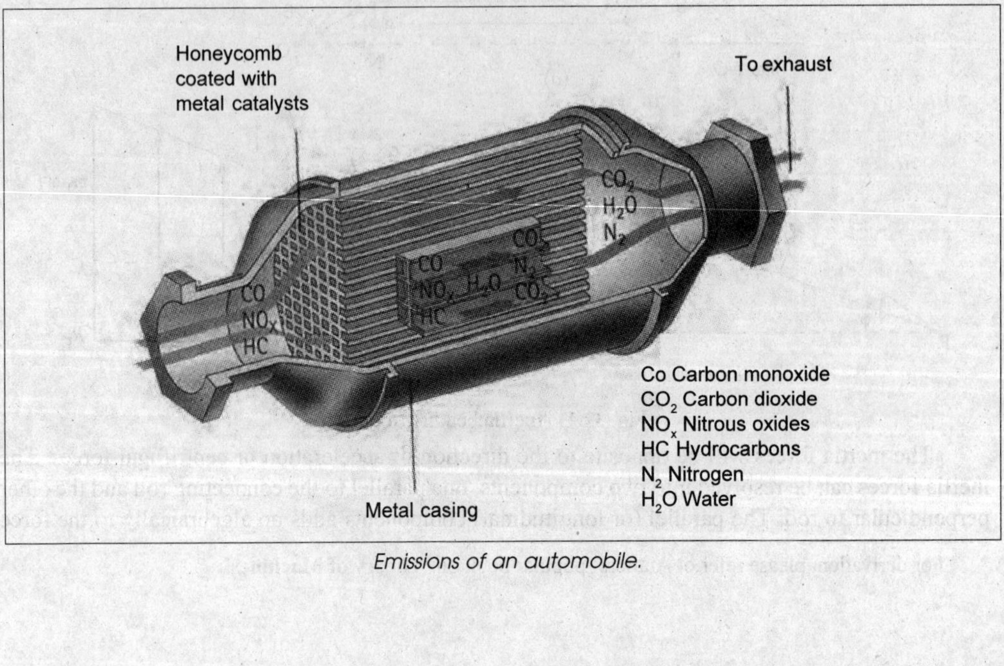

Emissions of an automobile.

Now the bending moment acting on the rod at section $X-X$ at a distance x from P,

$$M_X = R_P \times x - {}^*m_1 \times \omega^2 r \times \frac{x}{l} \times \frac{1}{2} x \times \frac{x}{3}$$

$$= \frac{1}{3} F_I \times x - \frac{m_1 . l}{2} \times \omega^2 r \times \frac{x^3}{3 l^2}$$

...(Multiplying and dividing the latter expression by l)

$$= \frac{F_I \times x}{3} - F_I \times \frac{x^3}{3 l^2} = \frac{F_I}{3}\left(x - \frac{x^3}{l^2}\right) \quad ...(i)$$

For maximum bending moment, differentiate M_X with respect to x and equate to zero, *i.e.*

$$\frac{d_{MX}}{dx} = 0 \text{ or } \frac{F_I}{3}\left[1 - \frac{3x^2}{l^2}\right] = 0$$

$$\therefore \quad 1 - \frac{3x^2}{l^2} = 0 \text{ or } 3x^2 = l^2 \text{ or } x = \frac{l}{\sqrt{3}}$$

Maximum bending moment,

$$M_{max} = \frac{F_I}{3}\left[\frac{l}{\sqrt{3}} - \frac{\left(\frac{l}{\sqrt{3}}\right)^3}{l^2}\right] \quad ...\text{[From equation (i)]}$$

$$= \frac{F_I}{3}\left[\frac{l}{\sqrt{3}} - \frac{l}{3\sqrt{3}}\right] = \frac{F_I \times l}{3\sqrt{3}} \times \frac{2}{3} = \frac{2F_I \times l}{9\sqrt{3}}$$

$$= 2 \times \frac{m}{2} \times \omega^2 r \times \frac{l}{9\sqrt{3}} = m \times \omega^2 r \times \frac{l}{9\sqrt{3}}$$

and the maximum bending stress, due to inertia of the connecting rod,

$$\sigma_{max} = \frac{M_{max}}{Z}$$

where Z = Section modulus.

From above we see that the maximum bending moment varies as the square of speed, therefore, the bending stress due to high speed will be dangerous. It may be noted that the maximum axial force and the maximum bending stress do not occur simultaneously. In an I.C. engine, the maximum gas load occurs close to top dead centre whereas the maximum bending stress occurs when the crank angle $\theta = 65°$ to $70°$ from top dead centre. The pressure of gas falls suddenly as the piston moves from dead centre. ***Thus the general practice is to design a connecting rod by assuming the force in the connecting rod (F_C) equal to the maximum force due to pressure (F_L), neglecting piston inertia effects and then checked for bending stress due to inertia force (i.e. whipping stress).***

3. ***Force due to friction of piston rings and of the piston***

The frictional force (F) of the piston rings may be determined by using the following expression :

$$F = \pi D \cdot t_R \cdot n_R \cdot p_R \cdot \mu$$

where D = Cylinder bore,

t_R = Axial width of rings,

* B.M. due to variable force from $\left(0 \text{ to } m_1 \, \omega^2 r \times \frac{x}{l}\right)$ is equal to the area of triangle multiplied by the distance of C.G. from $X-X\left(i.e. \frac{x}{3}\right)$.

n_R = Number of rings,

p_R = Pressure of rings (0.025 to 0.04 N/mm^2), and

μ = Coefficient of friction (about 0.1).

Since the frictional force of the piston rings is usually very small, therefore, it may be neglected.

The friction of the piston is produced by the normal component of the piston pressure which varies from 3 to 10 percent of the piston pressure. If the coefficient of friction is about 0.05 to 0.06, then the frictional force due to piston will be about 0.5 to 0.6 of the piston pressure, which is very low. Thus, the frictional force due to piston is also neglected.

4. ***Force due to friction of the piston pin bearing and crankpin bearing***

The force due to friction of the piston pin bearing and crankpin bearing, is to bend the connecting rod and to increase the compressive stress on the connecting rod due to the direct load. Thus, the maximum compressive stress in the connecting rod will be

$\sigma_{c\,(max)}$ = Direct compressive stress + Maximum bending or whipping stress due to inertia bending stress

32.15 Design of Connecting Rod

In designing a connecting rod, the following dimensions are required to be determined :

1. Dimensions of cross-section of the connecting rod,
2. Dimensions of the crankpin at the big end and the piston pin at the small end,
3. Size of bolts for securing the big end cap, and
4. Thickness of the big end cap.

The procedure adopted in determining the above mentioned dimensions is discussed as below :

This experimental car burns hydrogen fuel in an ordinary piston engine. Its exhaust gases cause no pollution, because they contain only water vapour.

1. ***Dimensions of cross-section of the connecting rod***

A connecting rod is a machine member which is subjected to alternating direct compressive and tensile forces. Since the compressive forces are much higher than the tensile forces, therefore, the cross-section of the connecting rod is designed as a strut and the Rankine's formula is used.

A connecting rod, as shown in Fig. 32.12, subjected to an axial load *W* may buckle with *X*-axis as neutral axis (*i.e.* in the plane of motion of the connecting rod) or *Y*-axis as neutral axis (*i.e.* in the plane perpendicular to the plane of motion). The connecting rod is considered like both ends hinged for buckling about *X*-axis and both ends fixed for buckling about *Y*-axis.

A connecting rod should be equally strong in buckling about both the axes.

Let A = Cross-sectional area of the connecting rod,

l = Length of the connecting rod,

σ_c = Compressive yield stress,

W_B = Buckling load,

I_{xx} and I_{yy} = Moment of inertia of the section about X-axis and *Y*-axis respectively, and

k_{xx} and k_{yy} = Radius of gyration of the section about *X*-axis and *Y*-axis respectively.

According to Rankine's formula,

$$W_B \text{ about } X-\text{axis} = \frac{\sigma_c . A}{1 + a\left(\frac{L}{k_{xx}}\right)^2} = \frac{\sigma_c . A}{1 + a\left(\frac{l}{k_{xx}}\right)^2} \quad ...(\because \text{For both ends hinged, } L = l)$$

and

$$W_B \text{ about } Y-\text{axis} = \frac{\sigma_c . A}{1 + a\left(\frac{L}{k_{yy}}\right)^2} = \frac{\sigma_c . A}{1 + a\left(\frac{l}{2\,k_{yy}}\right)^2} \quad ...[\because \text{For both ends fixed, } L = \frac{l}{2}]$$

where L = Equivalent length of the connecting rod, and

a = Constant

= 1 / 7500, for mild steel

= 1 / 9000, for wrought iron

= 1 / 1600, for cast iron

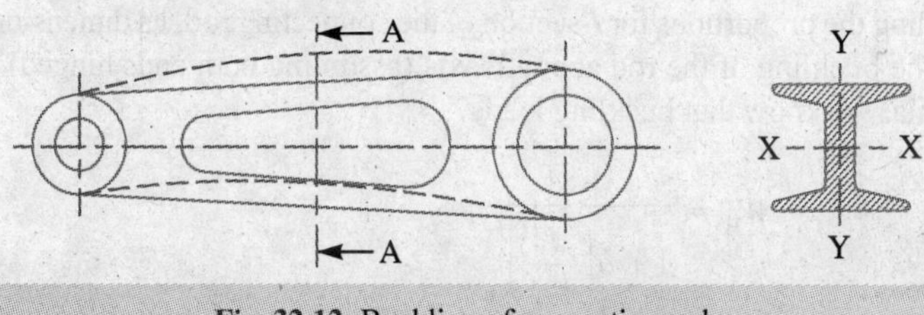

Fig. 32.12. Buckling of connecting rod.

In order to thave a connecting rod equally strong in buckling about both the axes, the buckling loads must be equal, *i.e.*

$$\frac{\sigma_c . A}{1 + a\left(\frac{l}{k_{xx}}\right)^2} = \frac{\sigma_c . A}{1 + a\left(\frac{l}{2k_{yy}}\right)^2} \quad \text{or} \quad \left(\frac{l}{k_{xx}}\right)^2 = \left(\frac{l}{2k_{yy}}\right)^2$$

$$\therefore \quad k_{xx}^2 = 4k_{yy}^2 \quad \text{or} \quad I_{xx} = 4\,I_{yy} \quad ...(\because I = A \cdot k^2)$$

This shows that the connecting rod is four times strong in buckling about *Y*-axis than about *X*-axis. If $I_{xx} > 4\,I_{yy}$, then buckling will occur about *Y*- axis and if $I_{xx} < 4\,I_{yy}$, buckling will occur about *X*-axis. In actual practice, I_{xx} is kept slightly less than $4\,I_{yy}$. It is usually taken between 3 and 3.5 and the connecting rod is designed for bucking about *X*-axis. The design will always be satisfactory for buckling about *Y*-axis.

The most suitable section for the connecting rod is *I*-section with the proportions as shown in Fig. 32.13 (*a*).

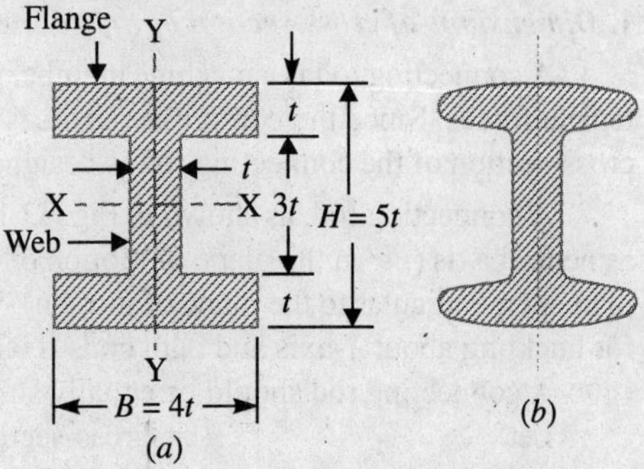

Fig. 32.13. *I*-section of connecting rod.

Let thickness of the flange and web of the section = *t*

Width of the section, $B = 4\,t$

and depth or height of the section,

$$H = 5t$$

From Fig. 32.13 (*a*), we find that area of the section,

$$A = 2\,(4\,t \times t) + 3\,t \times t = 11\,t^2$$

Moment of inertia of the section about *X*-axis,

$$I_{xx} = \frac{1}{12}\left[4\,t\,(5t)^3 - 3t\,(3t)^3\right] = \frac{419}{12}t^4$$

and moment of inertia of the section about *Y*-axis,

$$I_{yy} = \left[2 \times \frac{1}{12}t \times (4t)^3 + \frac{1}{12}(3t)\,t^3\right] = \frac{131}{12}t^4$$

$$\therefore \quad \frac{I_{xx}}{I_{yy}} = \frac{419}{12} \times \frac{12}{131} = 3.2$$

Since the value of $\frac{I_{xx}}{I_{yy}}$ lies between 3 and 3.5, therefore, *I*-section chosen is quite satisfactory.

After deciding the proportions for *I*-section of the connecting rod, its dimensions are determined by considering the buckling of the rod about *X*-axis (assuming both ends hinged) and applying the Rankine's formula. We know that buckling load,

$$W_B = \frac{\sigma_c . A}{1 + a\left(\frac{L}{k_{xx}}\right)^2}$$

The buckling load (W_B) may be calculated by using the following relation, *i.e.*

$$W_B = \text{Max. gas force} \times \text{Factor of safety}$$

The factor of safety may be taken as 5 to 6.

Notes : (*a*) The *I*-section of the connecting rod is used due to its lightness and to keep the inertia forces as low as possible specially in case of high speed engines. It can also withstand high gas pressure.

(*b*) Sometimes a connecting rod may have rectangular section. For slow speed engines, circular section may be used.

(*c*) Since connecting rod is manufactured by forging, therefore the sharp corner of *I*-section are rounded off as shown in Fig. 32.13 (*b*) for easy removal of the section from dies.

The dimensions $B = 4\,t$ and $H = 5\,t$, as obtained above by applying the Rankine's formula, are at the middle of the connecting rod. The width of the section (B) is kept constant throughout the length of the connecting rod, but the depth or height varies. The depth near the small end (or piston end) is taken as $H_1 = 0.75\,H$ to $0.9H$ and the depth near the big end (or crank end) is taken $H_2 = 1.1H$ to $1.25H$.

2. *Dimensions of the crankpin at the big end and the piston pin at the small end*

Since the dimensions of the crankpin at the big end and the piston pin (also known as gudgeon pin or wrist pin) at the small end are limited, therefore, fairly high bearing pressures have to be allowed at the bearings of these two pins.

The crankpin at the big end has removable precision bearing shells of brass or bronze or steel with a thin lining (1 mm or less) of bearing metal (such as tin, lead, babbit, copper, lead) on the inner surface of the shell. The allowable bearing pressure on the crankpin depends upon many factors such as material of the bearing, viscosity of the lubricating oil, method of lubrication and the space limitations.The value of bearing pressure may be taken as 7 N/mm^2 to 12.5 N/mm^2 depending upon the material and method of lubrication used.

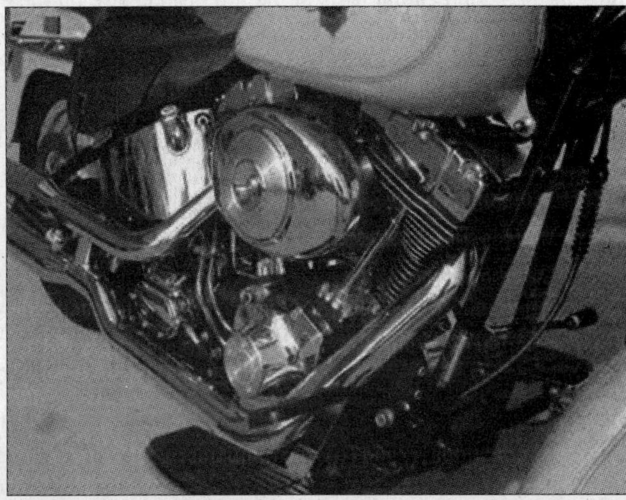

Engine of a motorcyle.

The piston pin bearing is usually a phosphor bronze bush of about 3 mm thickness and the allowable bearing pressure may be taken as 10.5 N/mm^2 to 15 N/mm^2.

Since the maximum load to be carried by the crankpin and piston pin bearings is the maximum force in the connecting rod (F_C), therefore the dimensions for these two pins are determined for the maximum force in the connecting rod (F_C) which is taken equal to the maximum force on the piston due to gas pressure (F_L) neglecting the inertia forces.

We know that maximum gas force,

$$F_L = \frac{\pi\,D^2}{4} \times p \qquad \text{...}(i)$$

where D = Cylinder bore or piston diameter in mm, and

p = Maximum gas pressure in N/mm^2

Now the dimensions of the crankpin and piston pin are determined as discussed below :

Let d_c = Diameter of the crank pin in mm,

l_c = Length of the crank pin in mm,

p_{bc} = Allowable bearing pressure in N/mm^2, and

d_p, l_p and p_{bp} = Corresponding values for the piston pin,

We know that load on the crank pin

= Projected area × Bearing pressure

$$= d_c \,.\, l_c \,.\, p_{bc} \qquad \text{...}(ii)$$

Similarly, load on the piston pin

$$= d_p \,.\, l_p \,.\, p_{bp} \qquad \text{...}(iii)$$

Equating equations (*i*) and (*ii*), we have

$$F_L = d_c \cdot l_c \cdot p_{bc}$$

Taking $l_c = 1.25\, d_c$ to $1.5\, d_c$, the value of d_c and l_c are determined from the above expression.

Again, equating equations (*i*) and (*iii*), we have

$$F_L = d_p \cdot l_p \cdot p_{bp}$$

Taking $l_p = 1.5\, d_p$ to $2\, d_p$, the value of d_p and l_p are determined from the above expression.

3. *Size of bolts for securing the big end cap*

The bolts and the big end cap are subjected to tensile force which corresponds to the inertia force of the reciprocating parts at the top dead centre on the exhaust stroke. We know that inertia force of the reciprocating parts,

$$\therefore \quad F_I = m_R.\omega^2.r\left(\cos\theta + \frac{\cos 2\theta}{l/r}\right)$$

We also know that at the top dead centre, the angle of inclination of the crank with the line of stroke, $\theta = 0$

$$\therefore \quad F_I = m_R.\omega^2.r\left(1 + \frac{r}{l}\right)$$

where m_R = Mass of the reciprocating parts in kg,

ω = Angular speed of the engine in rad / s,

r = Radius of the crank in metres, and

l = Length of the connecting rod in metres.

The bolts may be made of high carbon steel or nickel alloy steel. Since the bolts are under repeated stresses but not alternating stresses, therefore, a factor of safety may be taken as 6.

Let d_{cb} = Core diameter of the bolt in mm,

σ_t = Allowable tensile stress for the material of the bolts in MPa, and

n_b = Number of bolts. Generally two bolts are used.

$\therefore$ Force on the bolts

$$= \frac{\pi}{4}(d_{cb})^2 \sigma_t \times n_b$$

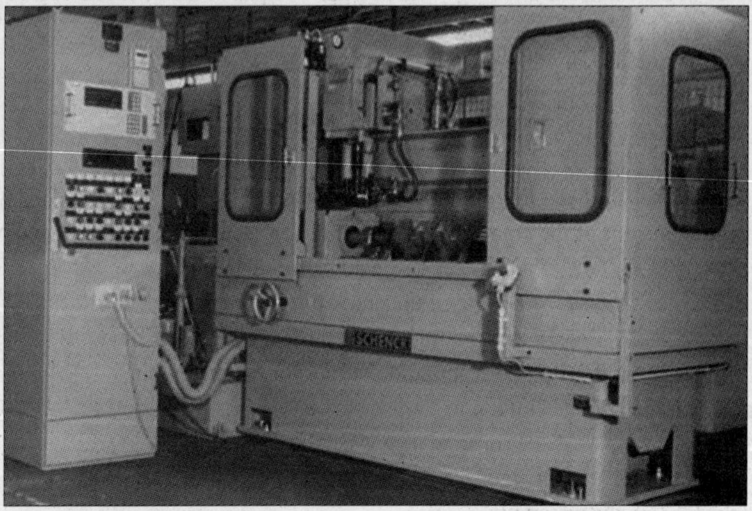

Equating the inertia force to the force on the bolts, we have

$$F_I = \frac{\pi}{4}(d_{cb})^2 \sigma_t \times n_b$$

From this expression, d_{cb} is obtained. The nominal or major diameter (d_b) of the bolt is given by

$$d_b = \frac{d_{cb}}{0.84}$$

4. *Thickness of the big end cap*

The thickness of the big end cap (t_c) may be determined by treating the cap as a beam freely supported at the cap bolt centres and loaded by the inertia force at the top dead centre on the exhaust stroke (*i.e.* F_I when $\theta = 0$). This load is assumed to act in between the uniformly distributed load and the centrally concentrated load. Therefore, the maximum bending moment acting on the cap will be taken as

$$M_C = {}^*\frac{F_I \times x}{6}$$

where x = Distance between the bolt centres.

= Dia. of crankpin or big end bearing (d_c) + 2 × Thickness of bearing liner (3 mm) + Clearance (3 mm)

Let b_c = Width of the cap in mm. It is equal to the length of the crankpin or big end bearing (l_c), and

σ_b = Allowable bending stress for the material of the cap in MPa.

We know that section modulus for the cap,

$$Z_C = \frac{b_c (t_c)^2}{6}$$

$$\therefore \text{ Bending stress, } \quad \sigma_b = \frac{M_C}{Z_C} = \frac{F_I \times x}{6} \times \frac{6}{b_c (t_c)^2} = \frac{F_I \times x}{b_c (t_c)^2}$$

From this expression, the value of t_c is obtained.

Note: The design of connecting rod should be checked for whipping stress (*i.e.* bending stress due to inertia force on the connecting rod).

Example 32.3. *Design a connecting rod for an I.C. engine running at 1800 r.p.m. and developing a maximum pressure of 3.15 N/mm². The diameter of the piston is 100 mm ; mass of the reciprocating parts per cylinder 2.25 kg; length of connecting rod 380 mm; stroke of piston 190 mm and compression ratio 6 : 1. Take a factor of safety of 6 for the design. Take length to diameter ratio for big end bearing as 1.3 and small end bearing as 2 and the corresponding bearing pressures as 10 N/mm² and 15 N/mm². The density of material of the rod may be taken as 8000 kg/m³ and the allowable stress in the bolts as 60 N/mm² and in cap as 80 N/mm². The rod is to be of I-section for which you can choose your own proportions.*

Draw a neat dimensioned sketch showing provision for lubrication. Use Rankine formula for which the numerator constant may be taken as 320 N/mm² and the denominator constant 1 / 7500.

---

* We know that the maximum bending moment for a simply or freely supported beam with a uniformly distributed load of F_I over a length x between the supports (In this case, x is the distance between the cap bolt centres) is $\frac{F_I \times x}{8}$. When the load F_I is assumed to act at the centre of the freely supported beam, then the maximum bending moment is $\frac{F_I \times x}{4}$. Thus the maximum bending moment in between these two bending moments $\left(i.e.\ \frac{F_I \times x}{8} \text{ and } \frac{F_I \times x}{4}\right)$ is $\frac{F_I \times x}{6}$.

Solution. Given : $N = 1800$ r.p.m. ; $p = 3.15$ N/mm^2 ; $D = 100$ mm ; $m_R = 2.25$ kg ; $l = 380$ mm $= 0.38$ m ; Stroke $= 190$ mm ; *Compression ratio $= 6 : 1$; $F.S. = 6$.

The connecting rod is designed as discussed below :

1. ***Dimension of I- section of the connecting rod***

Let us consider an *I*-section of the connecting rod, as shown in Fig. 32.14 (*a*), with the following proportions :

Flange and web thickness of the section $= t$

Width of the section, $B = 4t$

and depth or height of the section,

$$H = 5t$$

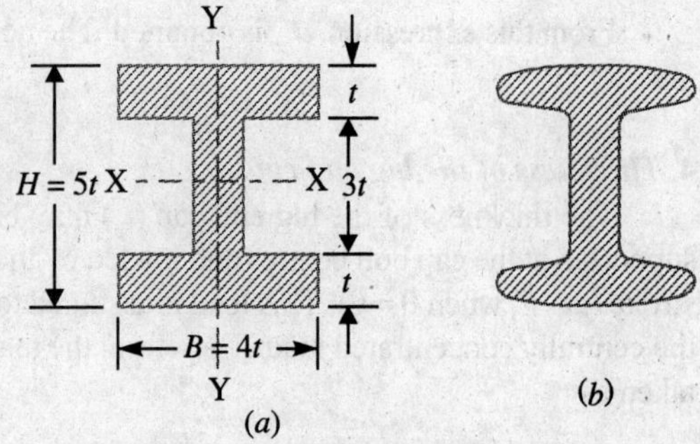

Fig. 32.14

First of all, let us find whether the section chosen is satisfactory or not.

We have already discussed that the connecting rod is considered like both ends hinged for buckling about *X*-axis and both ends fixed for buckling about *Y*-axis. The connecting rod should be equally strong in buckling about both the axes. We know that in order to have a connecting rod equally strong about both the axes,

$$I_{xx} = 4\,I_{yy}$$

where I_{xx} = Moment of inertia of the section about *X*-axis, and

I_{yy} = Moment of inertia of the section about *Y*-axis.

In actual practice, I_{xx} is kept slightly less than $4\,I_{yy}$. It is usually taken between 3 and 3.5 and the connecting rod is designed for buckling about *X*-axis.

Now, for the section as shown in Fig. 32.14 (*a*), area of the section,

$$A = 2\,(4\,t \times t) + 3t \times t = 11\,t^2$$

$$I_{xx} = \frac{1}{12}\left[4t(5t)^3 - 3t \times (3t)^3\right] = \frac{419}{12}t^4$$

and

$$I_{yy} = 2 \times \frac{1}{12} \times t(4t)^3 + \frac{1}{12} \times 3t \times t^3 = \frac{131}{12}t^4$$

$$\therefore \quad \frac{I_{xx}}{I_{yy}} = \frac{419}{12} \times \frac{12}{131} = 3.2$$

Since $\frac{I_{xx}}{I_{yy}} = 3.2$, therefore the section chosen in quite satisfactory.

Now let us find the dimensions of this *I*-section. Since the connecting rod is designed by taking the force on the connecting rod (F_C) equal to the maximum force on the piston (F_L) due to gas pressure, therefore,

$$F_C = F_L = \frac{\pi D^2}{4} \times p = \frac{\pi(100)^2}{4} \times 3.15 = 24\ 740 \text{ N}$$

We know that the connecting rod is designed for buckling about *X*-axis (*i.e.* in the plane of motion of the connecting rod) assuming both ends hinged. Since a factor of safety is given as 6, therefore the buckling load,

$$W_B = F_C \times F.S. = 24\ 740 \times 6 = 148\ 440 \text{ N}$$

* Superfluous data

We know that radius of gyration of the section about X-axis,

$$k_{xx} = \sqrt{\frac{I_{xx}}{A}} = \sqrt{\frac{419t^4}{12} \times \frac{1}{11t^2}} = 1.78\ t$$

Length of crank,

$$r = \frac{\text{Stroke of piston}}{2} = \frac{190}{2} = 95 \text{ mm}$$

Length of the connecting rod,

$$l = 380 \text{ mm} \quad \text{...(Given)}$$

∴ Equivalent length of the connecting rod for both ends hinged,

$$L = l = 380 \text{ mm}$$

Now according to Rankine's formula, we know that buckling load (W_B),

$$148\,440 = \frac{\sigma_c.A}{1 + a\left(\frac{L}{k_{xx}}\right)^2} = \frac{320 \times 11\,t^2}{1 + \frac{1}{7500}\left(\frac{380}{1.78\,t}\right)^2}$$

... (It is given that σ_c = 320 MPa or N/mm² and a = 1 / 7500)

$$\frac{148\,440}{320} = \frac{11\,t^2}{1 + \frac{6.1}{t^2}} = \frac{11\,t^4}{t^2 + 6.1}$$

$$464\,(t^2 + 6.1) = 11\,t^4$$

or

$$t^4 - 42.2\,t^2 - 257.3 = 0$$

∴

$$t^2 = \frac{42.2 \pm \sqrt{(42.2)^2 + 4 \times 257.3}}{2} = \frac{42.2 \pm 53}{2} = 47.6$$

... (Taking +ve sign)

or

$$t = 6.9 \text{ say } 7 \text{ mm}$$

Thus, the dimensions of I-section of the connecting rod are :

Thickness of flange and web of the section

$= t = 7$ mm **Ans.**

Width of the section, $B = 4\,t = 4 \times 7 = 28$ mm **Ans.**

and depth or height of the section,

$H = 5\,t = 5 \times 7 = 35$ mm **Ans.**

Piston and connecting rod.

These dimensions are at the middle of the connecting rod. The width (B) is kept constant throughout the length of the rod, but the depth (H) varies. The depth near the big end or crank end is kept as $1.1H$ to $1.25H$ and the depth near the small end or piston end is kept as $0.75H$ to $0.9H$. Let us take

Depth near the big end,

$$H_1 = 1.2H = 1.2 \times 35 = 42 \text{ mm}$$

and depth near the small end,

$$H_2 = 0.85H = 0.85 \times 35 = 29.75 \text{ say } 30 \text{ mm}$$

∴ Dimensions of the section near the big end

$$= 42 \text{ mm} \times 28 \text{ mm } \textbf{Ans.}$$

and dimensions of the section near the small end

$$= 30 \text{ mm} \times 28 \text{ mm } \textbf{Ans.}$$

Since the connecting rod is manufactured by forging, therefore the sharp corners of I-section are rounded off, as shown in Fig. 32.14 (b), for easy removal of the section from the dies.

2. ***Dimensions of the crankpin or the big end bearing and piston pin or small end bearing***

Let d_c = Diameter of the crankpin or big end bearing,

l_c = length of the crankpin or big end bearing $= 1.3\, d_c$ ...(Given)

p_{bc} = Bearing pressure = 10 N/mm² ...(Given)

We know that load on the crankpin or big end bearing

$$= \text{Projected area} \times \text{Bearing pressure}$$

$$= d_c \cdot l_c \cdot p_{bc} = d_c \times 1.3\, d_c \times 10 = 13\,(d_c)^2$$

Since the crankpin or the big end bearing is designed for the maximum gas force (F_L), therefore, equating the load on the crankpin or big end bearing to the maximum gas force, *i.e.*

$$13\,(d_c)^2 = F_L = 24\,740 \text{ N}$$

$$\therefore \quad (d_c)^2 = 24\,740 / 13 = 1903 \quad \text{or} \quad d_c = 43.6 \text{ say } 44 \text{ mm } \textbf{Ans.}$$

and $$l_c = 1.3\, d_c = 1.3 \times 44 = 57.2 \text{ say } 58 \text{ mm } \textbf{Ans.}$$

The big end has removable precision bearing shells of brass or bronze or steel with a thin lining (1mm or less) of bearing metal such as babbit.

Again, let d_p = Diameter of the piston pin or small end bearing,

l_p = Length of the piston pin or small end bearing $= 2d_p$ ...(Given)

p_{bp} = Bearing pressure = 15 N/mm² ..(Given)

We know that the load on the piston pin or small end bearing

$$= \text{Project area} \times \text{Bearing pressure}$$

$$= d_p \cdot l_p \cdot p_{bp} = d_p \times 2\, d_p \times 15 = 30\,(d_p)^2$$

Since the piston pin or the small end bearing is designed for the maximum gas force (F_L), therefore, equating the load on the piston pin or the small end bearing to the maximum gas force,

i.e.

$$30\,(d_p)^2 = 24\,740 \text{ N}$$

$$\therefore \quad (d_p)^2 = 24\,740 / 30 = 825 \quad \text{or} \quad d_p = 28.7 \text{ say } 29 \text{ mm } \textbf{Ans.}$$

and $$l_p = 2\, d_p = 2 \times 29 = 58 \text{ mm } \textbf{Ans.}$$

The small end bearing is usually a phosphor bronze bush of about 3 mm thickness.

3. ***Size of bolts for securing the big end cap***

Let d_{cb} = Core diameter of the bolts,

σ_t = Allowable tensile stress for the material of the bolts

= 60 N/mm^2 ...(Given)

and n_b = Number of bolts. Generally two bolts are used.

We know that force on the bolts

$$= \frac{\pi}{4}(d_{cb})^2 \sigma_t \times n_b = \frac{\pi}{4}(d_{cb})^2 \; 60 \times 2 = 94.26\,(d_{cb})^2$$

The bolts and the big end cap are subjected to tensile force which corresponds to the inertia force of the reciprocating parts at the top dead centre on the exhaust stroke. We know that inertia force of the reciprocating parts,

$$F_I = m_R \,.\, \omega^2 \,.\, r\left(\cos\theta + \frac{\cos 2\theta}{l/r}\right)$$

We also know that at top dead centre on the exhaust stroke, $\theta = 0$.

$$\therefore \quad F_I = m_R \,.\, \omega^2 \,.\, r\left(1 + \frac{r}{l}\right) = 2.25\left(\frac{2\pi \times 1800}{60}\right)^2 0.095\left(1 + \frac{0.095}{0.38}\right) \text{ N}$$

$$= 9490 \text{ N}$$

Equating the inertia force to the force on the bolts, we have

$$9490 = 94.26\,(d_{cb})^2 \quad \text{or} \quad (d_{cb})^2 = 9490 / 94.26 = 100.7$$

$$\therefore \quad d_{cb} = 10.03 \text{ mm}$$

and nominal diameter of the bolt,

$$d_b = \frac{d_{cb}}{0.84} = \frac{10.03}{0.84} = 11.94$$

say 12 mm **Ans.**

4. ***Thickness of the big end cap***

Let t_c = Thickness of the big end cap,

b_c = Width of the big end cap. It is taken equal to the length of the crankpin or big end bearing (l_c)

= 58 mm (calculated above)

σ_b = Allowable bending stress for the material of the cap

= 80 N/mm^2 ...(Given)

The big end cap is designed as a beam freely supported at the cap bolt centres and loaded by the inertia force at the top dead centre on the exhaust stroke (*i.e.* F_I when $\theta = 0$). Since the load is assumed to act in between the uniformly distributed load and the centrally concentrated load, therefore, maximum bending moment is taken as

$$M_C = \frac{F_I \times x}{6}$$

where x = Distance between the bolt centres

= Dia. of crank pin or big end bearing + 2 × Thickness of bearing liner + Nominal dia. of bolt + Clearance

$= (d_c + 2 \times 3 + d_b + 3)$ mm = 44 + 6 + 12 + 3 = 65 mm

∴ Maximum bending moment acting on the cap,

$$M_C = \frac{F_I \times x}{6} = \frac{9490 \times 65}{6} = 102\ 810 \text{ N-mm}$$

Section modulus for the cap

$$Z_C = \frac{b_c (t_c)^2}{6} = \frac{58 (t_c)^2}{6} = 9.7\ (t_c)^2$$

We know that bending stress (σ_b),

$$80 = \frac{M_C}{Z_C} = \frac{102\ 810}{9.7\ (t_c)^2} = \frac{10\ 600}{(t_c)^2}$$

∴ $(t_c)^2 = 10\ 600 / 80 = 132.5$ or $t_c = 11.5$ mm **Ans.**

Let us now check the design for the induced bending stress due to inertia bending forces on the connecting rod (*i.e.* whipping stress).

We know that mass of the connecting rod per metre length,

m_1 = Volume × density = Area × length × density

$= A \times l \times \rho = 11t^2 \times l \times \rho$ ...($\because A = 11t^2$)

$= 11(0.007)^2 (0.38)\ 8000 = 1.64$ kg

...[$\because \rho = 8\ 000$ kg /m^3 (given)]

∴ Maximum bending moment,

$$M_{max} = m \cdot \omega^2 \cdot r \times \frac{l}{9\sqrt{3}} = m_1 \cdot \omega^2 \cdot r \times \frac{l^2}{9\sqrt{3}} \quad ...(\because m = m_1 \cdot l)$$

$$= 1.64 \left(\frac{2\pi \times 1800}{60} \right)^2 (0.095) \frac{(0.38)^2}{9\sqrt{3}} = 51.3 \text{ N-m}$$

= 51 300 N-mm

and section modulus, $$Z_{xx} = \frac{I_{xx}}{5t/2} = \frac{419\ t^4}{12} \times \frac{2}{5\ t} = 13.97\ t^3 = 13.97 \times 7^3 = 4792 \text{ mm}^3$$

∴ Maximum bending stress (induced) due to inertia bending forces or whipping stress,

$$\sigma_{b(max)} = \frac{M_{max}}{Z_{xx}} = \frac{51\ 300}{4792} = 10.7 \text{ N/mm}^2$$

Since the maximum bending stress induced is less than the allowable bending stress of 80 N/mm^2, therefore the design is safe.

32.16 Crankshaft

A crankshaft (*i.e.* a shaft with a crank) is used to convert reciprocating motion of the piston into rotatory motion or vice versa. The crankshaft consists of the shaft parts which revolve in the main bearings, the crankpins to which the big ends of the connecting rod are connected, the crank arms or webs (also called cheeks) which connect the crankpins and the shaft parts. The crankshaft, depending upon the position of crank, may be divided into the following two types :

1. Side crankshaft or overhung crankshaft, as shown in Fig. 32.15 (*a*), and
2. Centre crankshaft, as shown in Fig. 32. 15 (*b*).

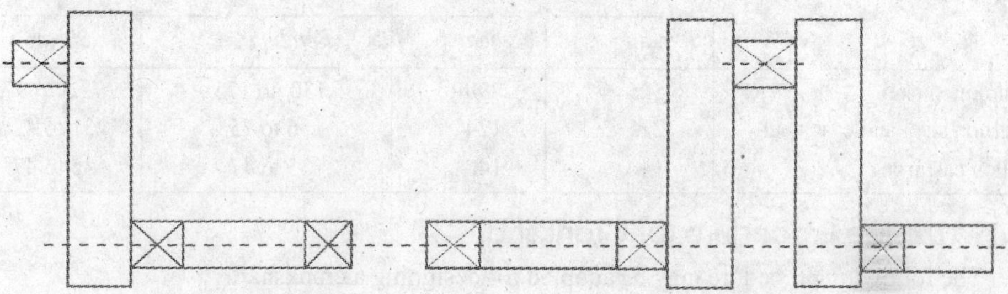

(*a*) Side Crankshaft.

(*a*) Centre Crankshaft.

Fig. 32.15. Types of crankshafts.

The crankshaft, depending upon the number of cranks in the shaft, may also be classfied as single throw or multi-throw crankshafts. A crankhaft with only one side crank or centre crank is called a ***single throw crankshaft*** whereas the crankshaft with two side cranks, one on each end or with two or more centre cranks is known as ***multi-throw crankshaft.***

The side crankshafts are used for medium and large size horizontal engines.

32.17 Material and manufacture of Crankshafts

The crankshafts are subjected to shock and fatigue loads. Thus material of the crankshaft should be tough and fatigue resistant. The crankshafts are generally made of carbon steel, special steel or special cast iron.

In industrial engines, the crankshafts are commonly made from carbon steel such as 40 C 8, 55 C 8 and 60 C 4. In transport engines, manganese steel such as 20 Mn 2, 27 Mn 2 and 37 Mn 2 are generally used for the making of crankshaft. In aero engines, nickel chromium steel such as 35 Ni 1 Cr 60 and 40 Ni 2 Cr 1 Mo 28 are extensively used for the crankshaft.

The crankshafts are made by drop forging or casting process but the former method is more common. The surface of the crankpin is hardened by case carburizing, nitriding or induction hardening.

32.18 Bearing Pressures and Stresses in Crankshaft

The bearing pressures are very important in the design of crankshafts. The *maximum permissible bearing pressure depends upon the maximum gas pressure, journal velocity, amount and method of lubrication and change of direction of bearing pressure.

The following two types of stresses are induced in the crankshaft.

1. Bending stress ; and 2. Shear stress due to torsional moment on the shaft.

* The values of maximum permissible bearing pressures for different types of engines are given in Chapter 26, Table 26.3.

Most crankshaft failures are caused by a progressive fracture due to repeated bending or reversed torsional stresses. Thus the crankshaft is under fatigue loading and, therefore, its design should be based upon endurance limit. Since the failure of a crankshaft is likely to cause a serious engine destruction and neither all the forces nor all the stresses acting on the crankshaft can be determined accurately, therefore a high factor of safety from 3 to 4, based on the endurance limit, is used.

The following table shows the allowable bending and shear stresses for some commonly used materials for crankshafts :

Table 32.2. Allowable bending and shear stresses.

Material	*Endurance limit in MPa*		*Allowable stress in MPa*	
	Bending	*Shear*	*Bending*	*Shear*
Chrome nickel	525	290	130 to 175	72.5 to 97
Carbon steel and cast steel	225	124	56 to 75	31 to 42
Alloy cast iron	140	140	35 to 47	35 to 47

32.19 Design Procedure for Crankshaft

The following procedure may be adopted for designing a crankshaft.

1. First of all, find the magnitude of the various loads on the crankshaft.

2. Determine the distances between the supports and their position with respect to the loads.

3. For the sake of simplicity and also for safety, the shaft is considered to be supported at the centres of the bearings and all the forces and reactions to be acting at these points. The distances between the supports depend on the length of the bearings, which in turn depend on the diameter of the shaft because of the allowable bearing pressures.

4. The thickness of the cheeks or webs is assumed to be from 0.4 d_s to 0.6 d_s, where d_s is the diameter of the shaft. It may also be taken as 0.22*D* to 0.32 *D*, where *D* is the bore of cylinder in mm.

5. Now calculate the distances between the supports.

6. Assuming the allowable bending and shear stresses, determine the main dimensions of the crankshaft.

Notes: 1. The crankshaft must be designed or checked for at least two crank positions. Firstly, when the crankshaft is subjected to maximum bending moment and secondly when the crankshaft is subjected to maximum twisting moment or torque.

2. The additional moment due to weight of flywheel, belt tension and other forces must be considered.

3. It is assumed that the effect of bending moment does not exceed two bearings between which a force is considered.

32.20 Design of Centre Crankshaft

We shall design the centre crankshaft by considering the two crank possitions, *i.e.* when the crank is at dead centre (or when the crankshaft is subjected to maximum bending moment) and when the crank is at angle at which the twisting moment is maximum. These two cases are discussed in detail as below :

1. *When the crank is at dead centre.* At this position of the crank, the maximum gas pressure on the piston will transmit maximum force on the crankpin in the plane of the crank causing only bending of the shaft. The crankpin as well as ends of the crankshaft will be only subjected to bending moment. Thus, when the crank is at the dead centre, the bending moment on the shaft is maximum and the twisting moment is zero.

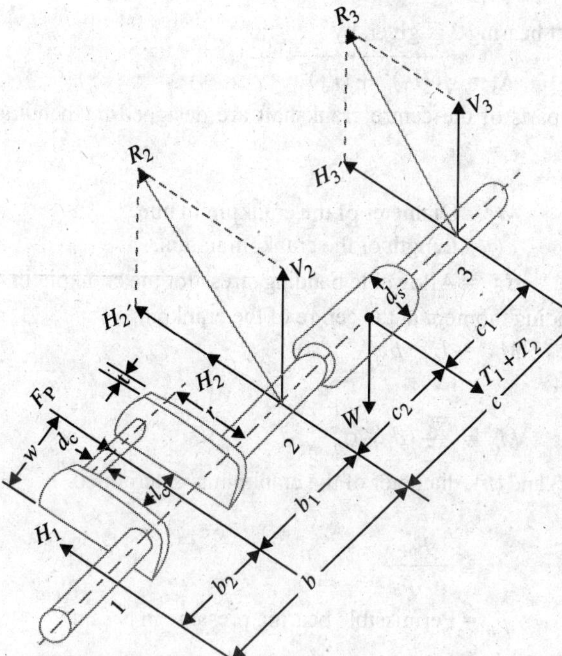

Fig. 32.16. Centre crankshaft at dead centre.

Consider a single throw three bearing crankshaft as shown in Fig. 32.16.

Let D = Piston diameter or cylinder bore in mm,

p = Maximum intensity of pressure on the piston in N/mm^2,

W = Weight of the flywheel acting downwards in N, and

*$T_1 + T_2$ = Resultant belt tension or pull acting horizontally in N.

The thrust in the connecting rod will be equal to the gas load on the piston (F_P). We know that gas load on the piston,

$$F_P = \frac{\pi}{4} \times D^2 \times p$$

Due to this piston gas load (F_P) acting horizontally, there will be two horizontal reactions H_1 and H_2 at bearings 1 and 2 respectively, such that

$$H_1 = \frac{F_P \times b_1}{b}; \quad \text{and} \quad H_2 = \frac{F_P \times b_2}{b}$$

Due to the weight of the flywheel (W) acting downwards, there will be two vertical reactions V_2 and V_3 at bearings 2 and 3 respectively, such that

$$V_2 = \frac{W \times c_1}{c}; \quad \text{and} \quad V_3 = \frac{W \times c_2}{c}$$

Now due to the resultant belt tension ($T_1 + T_2$), acting horizontally, there will be two horizontal reactions H_2' and H_3' at bearings 2 and 3 respectively, such that

$$H_2' = \frac{(T_1 + T_2)\, c_1}{c}; \quad \text{and} \quad H_3' = \frac{(T_1 + T_2)\, c_2}{c}$$

The resultant force at bearing 2 is given by

$$R_2 = \sqrt{(H_2 + H_2')^2 + (V_2)^2}$$

* T_1 is the belt tension in the tight side and T_2 is the belt tension in the slack side.

and the resultant force at bearing 3 is given by

$$R_3 = \sqrt{(H_3)^2 + (V_3)^2}$$

Now the various parts of the centre crankshaft are designed for bending only, as discussed below:

(a) Design of crankpin

Let d_c = Diameter of the crankpin in mm,

l_c = Length of the crankpin in mm,

σ_b = Allowable bending stress for the crankpin in N/mm^2.

We know that bending moment at the centre of the crankpin,

$$M_C = H_1 \cdot b_2 \qquad ...(i)$$

We also know that

$$M_C = \frac{\pi}{32}(d_c)^3 \sigma_b \qquad ...(ii)$$

From equations (*i*) and (*ii*), diameter of the crankpin is determined. The length of the crankpin is given by

$$l_c = \frac{F_P}{d_c \cdot p_b}$$

where p_b = Permissible bearing pressure in N/mm^2.

(b) Design of left hand crank web

The crank web is designed for eccentric loading. There will be two stresses acting on the crank web, one is direct compressive stress and the other is bending stress due to piston gas load (F_P).

Water cooled 4-cycle diesel engine

The thickness (t) of the crank web is given empirically as

$$t = 0.4\, d_s \text{ to } 0.6\, d_s$$
$$= 0.22D \text{ to } 0.32D$$
$$= 0.65\, d_c + 6.35 \text{ mm}$$

where d_s = Shaft diameter in mm,
D = Bore diameter in mm, and
d_c = Crankpin diameter in mm,

The width of crank web (w) is taken as

$$w = 1.125\, d_c + 12.7 \text{ mm}$$

We know that maximum bending moment on the crank web,

$$M = H_1\left(b_2 - \frac{l_c}{2} - \frac{t}{2}\right)$$

and section modulus, $Z = \frac{1}{6} \times w.\, t^2$

$$\therefore \text{ Bending stress, } \quad \sigma_b = \frac{M}{Z} = \frac{6H_1\left(b_2 - \frac{l_c}{2} - \frac{t}{2}\right)}{w\,.\,t^2}$$

and direct compressive stress on the crank web,

$$\sigma_c = \frac{H_1}{w\,.\,t}$$

∴ Total stress on the crank web

$$= \text{Bending stress} + \text{Direct stress} = \sigma_b + \sigma_c$$

$$= \frac{6H_1\left(b_2 - \frac{l_c}{2} - \frac{t}{2}\right)}{w\,.\,t^2} + \frac{H_1}{w\,.\,t}$$

This total stress should be less than the permissible bending stress.

(c) Design of right hand crank web

The dimensions of the right hand crank web (*i.e.* thickness and width) are made equal to left hand crank web from the balancing point of view.

(d) Design of shaft under the flywheel

Let d_s = Diameter of shaft in mm.

We know that bending moment due to the weight of flywheel,

$$M_W = V_3\,.\,c_1$$

and bending moment due to belt tension,

$$M_T = H_3'\,.\,c_1$$

These two bending moments act at right angles to each other. Therefore, the resultant bending moment at the flywheel location,

$$M_S = \sqrt{(M_W)^2 + (M_T)^2} = \sqrt{(V_3 \cdot c_1)^2 + (H_3 \cdot c_1)^2} \quad \text{... } (i)$$

We also know that the bending moment at the shaft,

$$M_S = \frac{\pi}{32}(d_s)^3\, \sigma_b \quad \text{... } (ii)$$

where σ_b = Allowable bending stress in N/mm^2.

From equations (*i*) and (*ii*), we may determine the shaft diameter (d_s).

2. *When the crank is at an angle of maximum twisting moment*

The twisting moment on the crankshaft will be maximum when the tangential force on the crank (F_T) is maximum. The maximum value of tangential force lies when the crank is at angle of 25° to 30° from the dead centre for a constant volume combustion engines (*i.e.*, petrol engines) and 30° to 40° for constant pressure combustion engines (*i.e.*, diesel engines).

Consider a position of the crank at an angle of maximum twisting moment as shown in Fig. 32.17 (*a*). If p' is the intensity of pressure on the piston at this instant, then the piston gas load at this position of crank,

$$F_P = \frac{\pi}{4} \times D^2 \times p'$$

and thrust on the connecting rod,

$$F_Q = \frac{F_P}{\cos \phi}$$

where ϕ = Angle of inclination of the connecting rod with the line of stroke *PO*.

The *thrust in the connecting rod (F_Q) may be divided into two components, one perpendicular to the crank and the other along the crank. The component of F_Q perpendicular to the crank is the tangential force (F_T) and the component of F_Q along the crank is the radial force (F_R) which produces thrust on the crankshaft bearings. From Fig. 32.17 (*b*), we find that

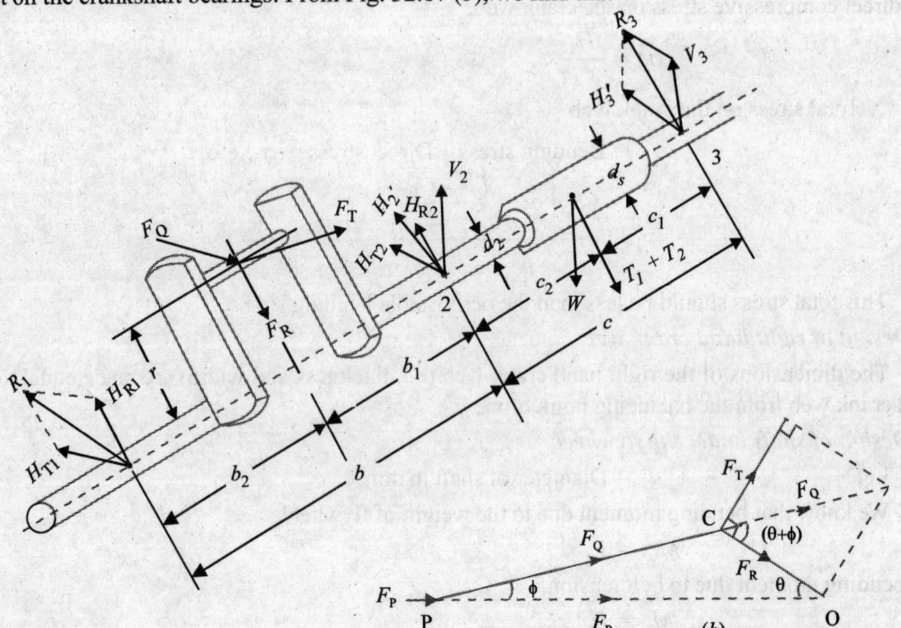

Fig. 32.17. (*a*) Crank at an angle of maximum twisting moment. (*b*) Forces acting on the crank.

$$F_T = F_Q \sin(\theta + \phi)$$

and

$$F_R = F_Q \cos(\theta + \phi)$$

It may be noted that the tangential force will cause twisting of the crankpin and shaft while the radial force will cause bending of the shaft.

* For further details, see Author's popular book on **'Theory of Machines'**.

Due to the tangential force (F_T), there will be two reactions at bearings 1 and 2, such that

$$H_{T1} = \frac{F_T \times b_1}{b}; \quad \text{and} \quad H_{T2} = \frac{F_T \times b_2}{b}$$

Due to the radial force (F_R), there will be two reactions at the bearings 1 and 2, such that

$$H_{R1} = \frac{F_R \times b_1}{b}; \quad \text{and} \quad H_{R2} = \frac{F_R \times b_2}{b}$$

Pull-start motor in an automobile

The reactions at the bearings 2 and 3, due to the flywheel weight (W) and resultant belt pull ($T_1 + T_2$) will be same as discussed earlier.

Now the various parts of the crankshaft are designed as discussed below :

(*a*) *Design of crankpin*

Let d_c = Diameter of the crankpin in mm.

We know that bending moment at the centre of the crankpin,

$$M_C = H_{R1} \times b_2$$

and twisting moment on the crankpin,

$$T_C = H_{T1} \times r$$

∴ Equivalent twisting moment on the crankpin,

$$T_e = \sqrt{(M_C)^2 + (T_C)^2} = \sqrt{(H_{R1} \times b_2)^2 + (H_{T1} \times r)^2} \quad ...(i)$$

We also know that twisting moment on the crankpin,

$$T_e = \frac{\pi}{16}(d_c)^3 \tau \quad ...(ii)$$

where τ = Allowable shear stress in the crankpin.

From equations (*i*) and (*ii*), the diameter of the crankpin is determined.

(b) Design of shaft under the flywheel

Let d_s = Diameter of the shaft in mm.

We know that bending moment on the shaft,

$$M_S = R_3 \times c_1$$

and twisting moment on the shaft,

$$T_S = F_T \times r$$

∴ Equivalent twisting moment on the shaft,

$$T_e = \sqrt{(M_S)^2 + (T_S)^2} = \sqrt{(R_3 \times c_1)^2 + (F_T \times r)^2} \quad ...(i)$$

We also know that equivalent twisting moment on the shaft,

$$T_e = \frac{\pi}{16}(d_s)^3 \tau \quad ...(ii)$$

where τ = Allowable shear stress in the shaft.

From equations (*i*) and (*ii*), the diameter of the shaft is determined.

(c) Design of shaft at the juncture of right hand crank arm

Let d_{s1} = Diameter of the shaft at the juncture of right hand crank arm.

We know that bending moment at the juncture of the right hand crank arm,

$$M_{S1} = R_1\left(b_2 + \frac{l_c}{2} + \frac{t}{2}\right) - F_Q\left(\frac{l_c}{2} + \frac{t}{2}\right)$$

and the twisting moment at the juncture of the right hand crank arm,

$$T_{S1} = F_T \times r$$

∴ Equivalent twisting moment at the juncture of the right hand crank arm,

$$T_e = \sqrt{(M_{S1})^2 + (T_{S1})^2} \quad ...(i)$$

We also know that equivalent twisting moment,

$$T_e = \frac{\pi}{16}(d_{s1})^3 \tau \quad ...(ii)$$

where τ = Allowable shear stress in the shaft.

From equations (*i*) and (*ii*), the diameter of the shaft at the juncture of the right hand crank arm is determined.

(d) Design of right hand crank web

The right hand crank web is subjected to the following stresses:

(*i*) Bending stresses in two planes normal to each other, due to the radial and tangential components of F_Q,

(*ii*) Direct compressive stress due to F_R, and

(*iii*) Torsional stress.

The bending moment due to the radial component of F_Q is given by,

$$M_R = H_{R2}\left(b_1 - \frac{l_c}{2} - \frac{t}{2}\right) \quad ...(i)$$

We also know that $M_R = \sigma_{bR} \times Z = \sigma_{bR} \times \frac{1}{6} \times w.t^2 \quad ...(ii)$

where σ_{bR} = Bending stress in the radial direction, and

$$Z = \text{Section modulus} = \frac{1}{6} \times w \cdot t^2$$

From equations (*i*) and (*ii*), the value of bending stress σ_{bR} is determined.

The bending moment due to the tangential component of F_Q is maximum at the juncture of crank and shaft. It is given by

$$M_T = F_T\left[r - \frac{d_{s1}}{2}\right] \quad \text{... (iii)}$$

where d_{s1} = Shaft diameter at juncture of right hand crank arm, *i.e.* at bearing 2.

We also know that $M_T = \sigma_{bT} \times Z = \sigma_{bT} \times \frac{1}{6} \times t.w^2$ ... (*iv*)

where σ_{bT} = Bending stress in tangential direction.

From equations (*iii*) and (*iv*), the value of bending stress σ_{bT} is determined.

The direct compressive stress is given by,

$$\sigma_d = \frac{F_R}{2w.t}$$

The maximum compressive stress (σ_c) will occur at the upper left corner of the cross-section of the crank.

$$\therefore \quad \sigma_c = \sigma_{bR} + \sigma_{bT} + \sigma_d$$

Now, the twisting moment on the arm,

$$T = H_{T1}\left(b_2 + \frac{l_c}{2}\right) - F_T \times \frac{l_c}{2} = H_{T2}\left(b_1 - \frac{l_c}{2}\right)$$

We know that shear stress on the arm,

$$\tau = \frac{T}{Z_P} = \frac{4.5\,T}{w.t^2}$$

where Z_P = Polar section modulus = $\frac{w.t^2}{4.5}$

∴ Maximum or total combined stress,

$$(\sigma_c)_{max} = \frac{\sigma_c}{2} + \frac{1}{2}\sqrt{(\sigma_c)^2 + 4\tau^2}$$

Snow blower on a railway track

The value of $(\sigma_c)_{max}$ should be within safe limits. If it exceeds the safe value, then the dimension w may be increased because it does not affect other dimensions.

(e) Design of left hand crank web

Since the left hand crank web is not stressed to the extent as the right hand crank web, therefore, the dimensions for the left hand crank web may be made same as for right hand crank web.

(f) Design of crankshaft bearings

The bearing 2 is the most heavily loaded and should be checked for the safe bearing pressure.

We know that the total reaction at the bearing 2,

$$R_2 = \frac{F_P}{2} + \frac{W}{2} + \frac{T_1 + T_2}{2}$$

$$\therefore \text{ Total bearing pressure} = \frac{R_2}{l_2 . d_{s1}}$$

where l_2 = Length of bearing 2.

32.21 Side or Overhung Crankshaft

The side or overhung crankshafts are used for medium size and large horizontal engines. Its main advantage is that it requires only two bearings in either the single or two crank construction. The design procedure for the side or overhung crankshaft is same as that for centre crankshaft. Let us now design the side crankshaft by considering the two crank positions, *i.e.* when the crank is at dead centre (or when the crankshaft is subjected to maximum bending moment) and when the crank is at an angle at which the twisting moment is maximum. These two cases are discussed in detail as below:

1. ***When the crank is at dead centre.*** Consider a side crankshaft at dead centre with its loads and distances of their application, as shown in Fig. 32.18.

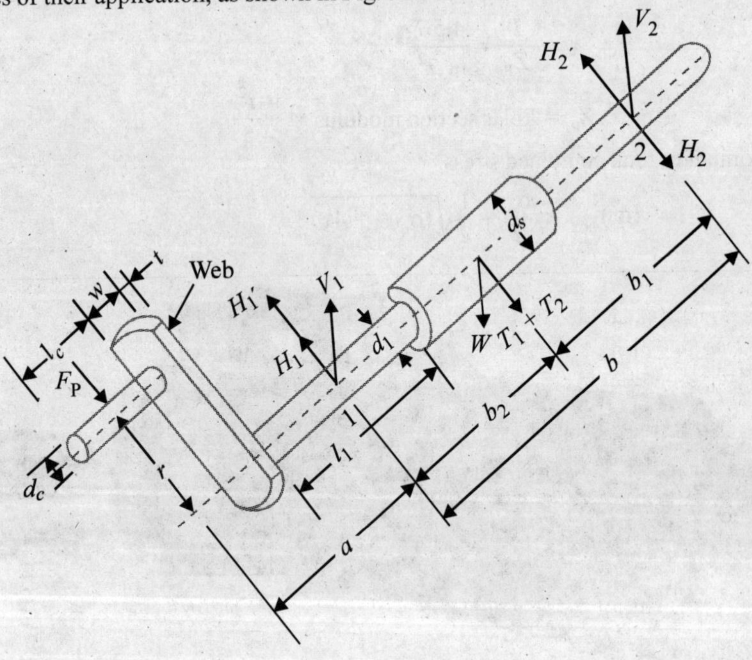

Fig. 32.18. Side crankshaft at dead centre.

Let D = Piston diameter or cylinder bore in mm,

p = Maximum intensity of pressure on the piston in N/mm^2,

W = Weight of the flywheel acting downwards in N, and

$T_1 + T_2$ = Resultant belt tension or pull acting horizontally in N.

We know that gas load on the piston,

$$F_P = \frac{\pi}{4} \times D^2 \times p$$

Due to this piston gas load (F_P) acting horizontally, there will be two horizontal reactions H_1 and H_2 at bearings 1 and 2 respectively, such that

$$H_1 = \frac{F_P(a+b)}{b}; \text{ and } H_2 = \frac{F_P \times a}{b}$$

Due to the weight of the flywheel (W) acting downwards, there will be two vertical reactions V_1 and V_2 at bearings 1 and 2 respectively, such that

$$V_1 = \frac{W.b_1}{b}; \text{ and } V_2 = \frac{W.b_2}{b}$$

Now due to the resultant belt tension ($T_1 + T_2$) acting horizontally, there will be two horizontal reactions H_1' and H_2' at bearings 1 and 2 respectively, such that

$$H_1' = \frac{(T_1+T_2)b_1}{b}; \text{ and } H_2' = \frac{(T_1+T_2)b_2}{b}$$

The various parts of the side crankshaft, when the crank is at dead centre, are now designed as discussed below:

(a) Design of crankpin. The dimensions of the crankpin are obtained by considering the crankpin in bearing and then checked for bending stress.

Let d_c = Diameter of the crankpin in mm,

l_c = Length of the crankpin in mm, and

p_b = Safe bearing pressure on the pin in N/mm^2. It may be between 9.8 to 12.6 N/mm^2.

We know that $F_P = d_c \cdot l_c \cdot p_b$

From this expression, the values of d_c and l_c may be obtained. The length of crankpin is usually from 0.6 to 1.5 times the diameter of pin.

The crankpin is now checked for bending stress. If it is assumed that the crankpin acts as a cantilever and the load on the crankpin is uniformly distributed, then maximum bending moment will be $\frac{F_P \times l_c}{2}$. But in actual practice, the bearing pressure on the crankpin is not uniformly distributed and may, therefore, give a greater value of bending moment ranging between $\frac{F_P \times l_c}{2}$ and $F_P \times l_c$. So, a mean value of bending moment, *i.e.* $\frac{3}{4} F_P \times l_c$ may be assumed.

Close-up view of an automobile piston

∴ Maximum bending moment at the crankpin,

$$M = \frac{3}{4} F_P \times l_c \quad \text{... (Neglecting pin collar thickness)}$$

Section modulus for the crankpin,

$$Z = \frac{\pi}{32}(d_c)^3$$

∴ Bending stress induced,

$$\sigma_b = M / Z$$

This induced bending stress should be within the permissible limits.

(*b*) ***Design of bearings.*** The bending moment at the centre of the bearing 1 is given by

$$M = F_P (0.75\, l_c + t + 0.5\, l_1) \quad \text{...}(i)$$

where l_c = Length of the crankpin,

t = Thickness of the crank web = 0.45 d_c to 0.75 d_c, and

l_1 = Length of the bearing = 1.5 d_c to 2 d_c.

We also know that

$$M = \frac{\pi}{32}(d_1)^3 \sigma_b \quad \text{...}(ii)$$

From equations (*i*) and (*ii*), the diameter of the bearing 1 may be determined.

Note : The bearing 2 is also made of the same diameter. The length of the bearings are found on the basis of allowable bearing pressures and the maximum reactions at the bearings.

(*c*) ***Design of crank web.*** When the crank is at dead centre, the crank web is subjected to a bending moment and to a direct compressive stress.

We know that bending moment on the crank web,

$$M = F_P (0.75\, l_c + 0.5\, t)$$

and section modulus, $Z = \frac{1}{6} \times w \,.\, t^2$

∴ Bending stress, $\sigma_b = \frac{M}{Z}$

We also know that direct compressive stress,

$$\sigma_d = \frac{F_P}{w \,.\, t}$$

∴ Total stress on the crank web,

$$\sigma_T = \sigma_b + \sigma_d$$

This total stress should be less than the permissible limits.

(*d*) ***Design of shaft under the flywheel.*** The total bending moment at the flywheel location will be the resultant of horizontal bending moment due to the gas load and belt pull and the vertical bending moment due to the flywheel weight.

Let d_s = Diameter of shaft under the flywheel.

We know that horizontal bending moment at the flywheel location due to piston gas load,

$$M_1 = F_P (a + b_2) - H_1 \,.\, b_2 = H_2 \,.\, b_1$$

and horizontal bending moment at the flywheel location due to belt pull,

$$M_2 = H_1'.b_2 = H_2'.b_1 = \frac{(T_1 + T_2)\, b_1.b_2}{b}$$

∴ Total horizontal bending moment,

$$M_H = M_1 + M_2$$

We know that vertical bending moment due to flywheel weight,

$$M_V = V_1.b_2 = V_2.b_1 = \frac{W b_1 b_2}{b}$$

∴ Resultant bending moment,

$$M_R = \sqrt{(M_H)^2 + (M_V)^2} \quad ...(i)$$

We also know that

$$M_R = \frac{\pi}{32}(d_s)^3 \sigma_b \quad ...(ii)$$

From equations (*i*) and (*ii*), the diameter of shaft (d_s) may determined.

2. ***When the crank is at an angle of maximum twisting moment.*** Consider a position of the crank at an angle of maximum twisting moment as shown in Fig. 32.19. We have already discussed in the design of a centre crankshaft that the thrust in the connecting rod (F_Q) gives rise to the tangential force (F_T) and the radial force (F_R).

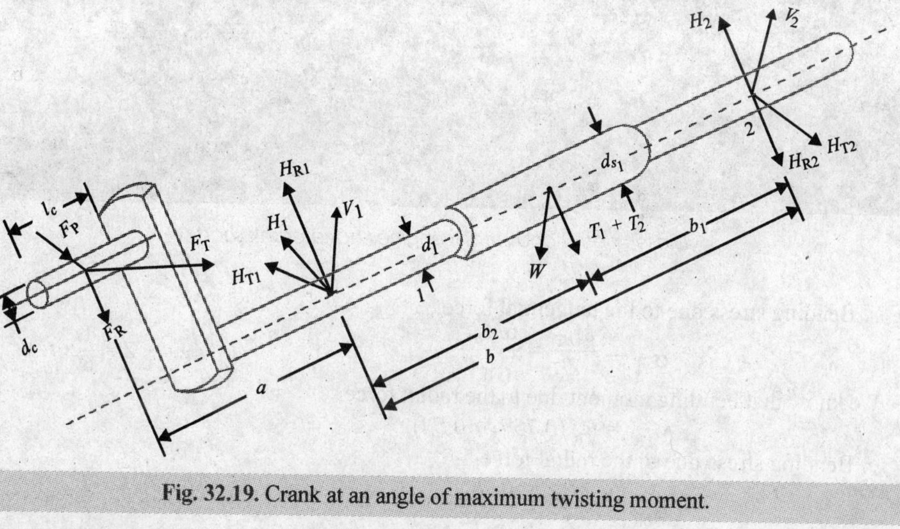

Fig. 32.19. Crank at an angle of maximum twisting moment.

Due to the tangential force (F_T), there will be two reactions at the bearings 1 and 2, such that

$$H_{T1} = \frac{F_T(a+b)}{b}; \quad \text{and} \quad H_{T2} = \frac{F_T \times a}{b}$$

Due to the radial force (F_R), there will be two reactions at the bearings 1 and 2, such that

$$H_{R1} = \frac{F_R(a+b)}{b}; \quad \text{and} \quad H_{R2} = \frac{F_R \times a}{b}$$

The reactions at the bearings 1 and 2 due to the flywheel weight (W) and resultant belt pull ($T_1 + T_2$) will be same as discussed earlier.

Now the various parts of the crankshaft are designed as discussed below:

(*a*) ***Design of crank web.*** The most critical section is where the web joins the shaft. This section is subjected to the following stresses :

(*i*) Bending stress due to the tangential force F_T;

(*ii*) Bending stress due to the radial force F_R ;

(*iii*) Direct compressive stress due to the radial force F_R ; and

(*iv*) Shear stress due to the twisting moment of F_T.

We know that bending moment due to the tangential force,

$$M_{bT} = F_T\left(r - \frac{d_1}{2}\right)$$

where d_1 = Diameter of the bearing 1.

Diesel, petrol and steam engines have crank shaft

∴ Bending stress due to the tangential force,

$$\sigma_{bT} = \frac{M_{bT}}{Z} = \frac{6M_{bT}}{t.w^2} \quad ...(\because Z = \frac{1}{6} \times t \cdot w^2) \quad ...(i)$$

We know that bending moment due to the radial force,

$$M_{bR} = F_R\,(0.75\,l_c + 0.5\,t)$$

∴ Bending stress due to the radial force,

$$\sigma_{bR} = \frac{M_{bR}}{Z} = \frac{6\,M_{bR}}{w.t^2} \quad ...(\text{Here } Z = \frac{1}{6} \times w \cdot t^2) \quad ...(ii)$$

We know that direct compressive stress,

$$\sigma_d = \frac{F_R}{w.t} \quad ...(iii)$$

∴ Total compressive stress,

$$\sigma_c = \sigma_{bT} + \sigma_{bR} + \sigma_d \quad ...(iv)$$

We know that twisting moment due to the tangential force,

$$T = F_T(0.75\,l_c + 0.5\,t)$$

∴ Shear stress, $\tau = \frac{T}{Z_P} = \frac{4.5T}{w.t^2}$

where Z_P = Polar section modulus = $\frac{w.t^2}{4.5}$

Now the total or maximum stress is given by

$$\sigma_{max} = \frac{\sigma_c}{2} + \frac{1}{2}\sqrt{(\sigma_c)^2 + 4\tau^2} \quad ...(v)$$

This total maximum stress should be less than the maximum allowable stress.

(b) ***Design of shaft at the junction of crank***

Let d_{s1} = Diameter of the shaft at the junction of the crank.

We know that bending moment at the junction of the crank,

$$M = F_Q\,(0.75l_c + t)$$

and twisting moment on the shaft

$$T = F_T \times r$$

∴ Equivalent twisting moment,

$$T_e = \sqrt{M^2 + T^2} \quad ...(i)$$

We also know that equivalent twisting moment,

$$T_e = \frac{\pi}{16}(d_{s1})^3\tau \quad ...(ii)$$

From equations (*i*) and (*ii*), the diameter of the shaft at the junction of the crank (d_{s1}) may be determined.

(c) ***Design of shaft under the flywheel***

Let d_s = Diameter of shaft under the flywheel.

The resultant bending moment (M_R) acting on the shaft is obtained in the similar way as discussed for dead centre position.

We know that horizontal bending moment acting on the shaft due to piston gas load,

$$M_1 = F_P(a + b_2) - \left[\sqrt{(H_{R1})^2 + (H_{T1})^2}\right]b_2$$

and horizontal bending moment at the flywheel location due to belt pull,

$$M_2 = H_1'.b_2 = H_2'.b_1 = \frac{(T_1 + T_2)\,b_1.b_2}{b}$$

∴ Total horizontal bending moment,

$$M_H = M_1 + M_2$$

Vertical bending moment due to the flywheel weight,

$$M_V = V_1 \cdot b_2 = V_2 \cdot b_1 = \frac{W\,b_1\,b_2}{b}$$

∴ Resultant bending moment,

$$M_R = \sqrt{(M_H)^2 + (M_V)^2}$$

We know that twisting moment on the shaft,

$$T = F_T \times r$$

∴ Equivalent twisting moment,

$$T_e = \sqrt{(M_R)^2 + T^2} \quad ...(i)$$

We also know that equivalent twisting moment,

$$T_e = \frac{\pi}{16}(d_s)^3\tau \quad ...(ii)$$

From equations (*i*) and (*ii*), the diameter of shaft under the flywheel (d_s) may be determined.

Example 32.4. *Design a plain carbon steel centre crankshaft for a single acting four stroke single cylinder engine for the following data:*

Bore = 400 mm ; Stroke = 600 mm ; Engine speed = 200 r.p.m. ; Mean effective pressure = 0.5 N/mm²; Maximum combustion pressure = 2.5 N/mm²; Weight of flywheel used as a pulley = 50 kN; Total belt pull = 6.5 kN.

When the crank has turned through 35° from the top dead centre, the pressure on the piston is 1N/mm² and the torque on the crank is maximum. The ratio of the connecting rod length to the crank radius is 5. Assume any other data required for the design.

Solution. Given : $D = 400$ mm ; $L = 600$ mm or $r = 300$ mm ; $p_m = 0.5$ N/mm² ; $p = 2.5$ N/mm² ; $W = 50$ kN ; $T_1 + T_2 = 6.5$ kN ; $\theta = 35°$; $p' = 1$N/mm² ; $l / r = 5$

We shall design the crankshaft for the two positions of the crank, *i.e.* firstly when the crank is at the dead centre ; and secondly when the crank is at an angle of maximum twisting moment.

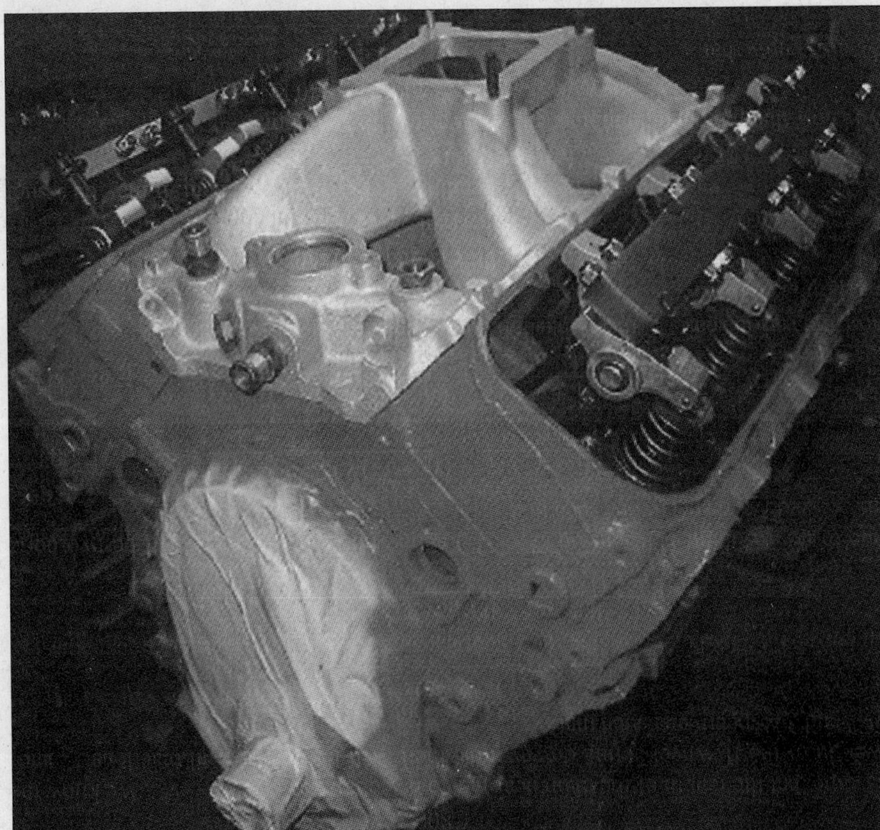

Part of a car engine

1. ***Design of the crankshaft when the crank is at the dead centre*** (See Fig. 32.18)

We know that the piston gas load,

$$F_P = \frac{\pi}{4} \times D^2 \times p = \frac{\pi}{4}(400)^2 2.5 = 314\,200 \text{ N} = 314.2 \text{ kN}$$

Assume that the distance (b) between the bearings 1 and 2 is equal to twice the piston diameter (D).

$$\therefore \qquad b = 2D = 2 \times 400 = 800 \text{ mm}$$

and $$b_1 = b_2 = \frac{b}{2} = \frac{800}{2} = 400 \text{ mm}$$

We know that due to the piston gas load, there will be two horizontal reactions H_1 and H_2 at bearings 1 and 2 respectively, such that

$$H_1 = \frac{F_P \times b_1}{b} = \frac{314.2 \times 400}{800} = 157.1 \text{ kN}$$

and $$H_2 = \frac{F_P \times b_2}{b} = \frac{314.2 \times 400}{800} = 157.1 \text{ kN}$$

Assume that the length of the main bearings to be equal, *i.e.*, $c_1 = c_2 = c / 2$. We know that due to the weight of the flywheel acting downwards, there will be two vertical reactions V_2 and V_3 at bearings 2 and 3 respectively, such that

$$V_2 = \frac{W \times c_1}{c} = \frac{W \times c/2}{c} = \frac{W}{2} = \frac{50}{2} = 25 \text{ kN}$$

and $$V_3 = \frac{W \times c_2}{c} = \frac{W \times c/2}{c} = \frac{W}{2} = \frac{50}{2} = 25 \text{ kN}$$

Due to the resultant belt tension $(T_1 + T_2)$ acting horizontally, there will be two horizontal reactions H_2' and H_3' respectively, such that

$$H_2' = \frac{(T_1 + T_2)\, c_1}{c} = \frac{(T_1 + T_2)\, c/2}{c} = \frac{T_1 + T_2}{2} = \frac{6.5}{2} = 3.25 \text{ kN}$$

and $$H_3' = \frac{(T_1 + T_2)\, c_2}{c} = \frac{(T_1 + T_2)\, c/2}{c} = \frac{T_1 + T_2}{2} = \frac{6.5}{2} = 3.25 \text{ kN}$$

Now the various parts of the crankshaft are designed as discussed below:

(a) Design of crankpin

Let d_c = Diameter of the crankpin in mm ;

l_c = Length of the crankpin in mm ; and

σ_b = Allowable bending stress for the crankpin. It may be assumed as 75 MPa or N/mm².

We know that the bending moment at the centre of the crankpin,

$$M_C = H_1 \cdot b_2 = 157.1 \times 400 = 62\,840 \text{ kN-mm} \quad \text{...}(i)$$

We also know that

$$M_C = \frac{\pi}{32}(d_c)^3 \sigma_b = \frac{\pi}{32}(d_c)^3 75 = 7.364 (d_c)^3 \text{ N-mm}$$

$$= 7.364 \times 10^{-3} (d_c)^3 \text{ kN-mm} \quad \text{...}(ii)$$

Equating equations (*i*) and (*ii*), we have

$$(d_c)^3 = 62\,840 / 7.364 \times 10^{-3} = 8.53 \times 10^6$$

or d_c = 204.35 say 205 mm **Ans.**

We know that length of the crankpin,

$$l_c = \frac{F_P}{d_c \cdot p_b} = \frac{314.2 \times 10^3}{205 \times 10} = 153.3 \text{ say } 155 \text{ mm } \textbf{Ans.}$$

...(Taking p_b = 10 N/mm²)

(b) Design of left hand crank web

We know that thickness of the crank web,

$$t = 0.65\, d_c + 6.35 \text{ mm}$$

$$= 0.65 \times 205 + 6.35 = 139.6 \text{ say } 140 \text{ mm } \textbf{Ans.}$$

and width of the crank web, $w = 1.125\, d_c + 12.7$ mm

$$= 1.125 \times 205 + 12.7 = 243.3 \text{ say } 245 \text{ mm } \textbf{Ans.}$$

We know that maximum bending moment on the crank web,

$$M = H_1\left(b_2 - \frac{l_c}{2} - \frac{t}{2}\right)$$

$$= 157.1\left(400 - \frac{155}{2} - \frac{140}{2}\right) = 39\,668 \text{ kN-mm}$$

Section modulus, $Z = \frac{1}{6} \times w.t^2 = \frac{1}{6} \times 245\,(140)^2 = 800 \times 10^3 \text{ mm}^3$

$\therefore$ Bending stress, $\sigma_b = \frac{M}{Z} = \frac{39\,668}{800 \times 10^3} = 49.6 \times 10^{-3} \text{ kN/mm}^2 = 49.6 \text{ N/mm}^2$

We know that direct compressive stress on the crank web,

$$\sigma_c = \frac{H_1}{w.t} = \frac{157.1}{245 \times 140} = 4.58 \times 10^{-3} \text{ kN/mm}^2 = 4.58 \text{ N/mm}^2$$

$\therefore$ Total stress on the crank web

$$= \sigma_b + \sigma_c = 49.6 + 4.58 = 54.18 \text{ N/mm}^2 \text{ or MPa}$$

Since the total stress on the crank web is less than the allowable bending stress of 75 MPa, therefore, the design of the left hand crank web is safe.

(c) *Design of right hand crank web*

From the balancing point of view, the dimensions of the right hand crank web (*i.e.* thickness and width) are made equal to the dimensions of the left hand crank web.

(d) *Design of shaft under the flywheel*

Let d_s = Diameter of the shaft in mm.

Since the lengths of the main bearings are equal, therefore

$$l_1 = l_2 = l_3 = 2\left(\frac{b}{2} - \frac{l_c}{2} - t\right) = 2\left(400 - \frac{155}{2} - 140\right) = 365 \text{ mm}$$

Assuming width of the flywheel as 300 mm, we have

$$c = 365 + 300 = 665 \text{ mm}$$

Hydrostatic transmission inside a tractor engine

Allowing space for gearing and clearance, let us take $c = 800$ mm.

$$\therefore \quad c_1 = c_2 = \frac{c}{2} = \frac{800}{2} = 400 \text{ mm}$$

We know that bending moment due to the weight of flywheel,

$$M_W = V_3 \cdot c_1 = 25 \times 400 = 10\,000 \text{ kN-mm} = 10 \times 10^6 \text{ N-mm}$$

and bending moment due to the belt pull,

$$M_T = H_3' \cdot c_1 = 3.25 \times 400 = 1300 \text{ kN-mm} = 1.3 \times 10^6 \text{ N-mm}$$

∴ Resultant bending moment on the shaft,

$$M_S = \sqrt{(M_W)^2 + (M_T)^2} = \sqrt{(10 \times 10^6)^2 + (1.3 \times 10^6)^2}$$
$$= 10.08 \times 10^6 \text{ N-mm}$$

We also know that bending moment on the shaft (M_S),

$$10.08 \times 10^6 = \frac{\pi}{32}(d_s)^3 \sigma_b = \frac{\pi}{32}(d_s)^3 42 = 4.12\,(d_s)^3$$

$$\therefore \quad (d_s)^3 = 10.08 \times 10^6 / 4.12 = 2.45 \times 10^6 \text{ or } d_s = 134.7 \text{ say } 135 \text{ mm } \textbf{Ans.}$$

2. ***Design of the crankshaft when the crank is at an angle of maximum twisting moment***

We know that piston gas load,

$$F_P = \frac{\pi}{4} \times D^2 \times p' = \frac{\pi}{4}(400)^2 1 = 125\,680 \text{ N} = 125.68 \text{ kN}$$

In order to find the thrust in the connecting rod (F_Q), we should first find out the angle of inclination of the connecting rod with the line of stroke (*i.e.* angle ϕ). We know that

$$\sin \phi = \frac{\sin \theta}{l/r} = \frac{\sin 35°}{5} = 0.1147$$

$$\therefore \quad \phi = \sin^{-1}(0.1147) = 6.58°$$

We know that thrust in the connecting rod,

$$F_Q = \frac{F_P}{\cos \phi} = \frac{125.68}{\cos 6.58°} = \frac{125.68}{0.9934} = 126.5 \text{ kN}$$

Tangential force acting on the crankshaft,

$$F_T = F_Q \sin(\theta + \phi) = 126.5 \sin(35° + 6.58°) = 84 \text{ kN}$$

and radial force,

$$F_R = F_Q \cos(\theta + \phi) = 126.5 \cos(35° + 6.58°) = 94.6 \text{ kN}$$

Due to the tangential force (F_T), there will be two reactions at bearings 1 and 2, such that

$$H_{T1} = \frac{F_T \times b_1}{b} = \frac{84 \times 400}{800} = 42 \text{ kN}$$

and

$$H_{T2} = \frac{F_T \times b_2}{b} = \frac{84 \times 400}{800} = 42 \text{ kN}$$

Due to the radial force (F_R), there will be two reactions at bearings 1 and 2, such that

$$H_{R1} = \frac{F_R \times b_1}{b} = \frac{94.6 \times 400}{800} = 47.3 \text{ kN}$$

$$H_{R2} = \frac{F_R \times b_2}{b} = \frac{94.6 \times 400}{800} = 47.3 \text{ kN}$$

Now the various parts of the crankshaft are designed as discussed below:

(a) Design of crankpin

Let d_c = Diameter of crankpin in mm.

We know that the bending moment at the centre of the crankpin,

$$M_C = H_{R1} \times b_2 = 47.3 \times 400 = 18\ 920 \text{ kN-mm}$$

and twisting moment on the crankpin,

$$T_C = H_{T1} \times r = 42 \times 300 = 12\ 600 \text{ kN-mm}$$

∴ Equivalent twisting moment on the crankpin,

$$T_e = \sqrt{(M_C)^2 + (T_C)^2} = \sqrt{(18\ 920)^2 + (12\ 600)^2}$$
$$= 22\ 740 \text{ kN-mm} = 22.74 \times 10^6 \text{ N-mm}$$

We know that equivalent twisting moment (T_e),

$$22.74 \times 10^6 = \frac{\pi}{16}(d_c)^3 \tau = \frac{\pi}{16}(d_c)^3 35 = 6.873\,(d_c)^3$$

...(Taking $\tau = 35$ MPa or N/mm²)

∴ $(d_c)^3 = 22.74 \times 10^6 / 6.873 = 3.3 \times 10^6$ or $d_c = 149$ mm

Since this value of crankpin diameter (*i.e.* $d_c = 149$ mm) is less than the already calculated value of $d_c = 205$ mm, therefore, we shall take $d_c = 205$ mm. **Ans.**

(b) Design of shaft under the flywheel

Let d_s = Diameter of the shaft in mm.

The resulting bending moment on the shaft will be same as calculated eariler, *i.e.*

$$M_S = 10.08 \times 10^6 \text{ N-mm}$$

and twisting moment on the shaft,

$$T_S = F_T \times r = 84 \times 300 = 25\ 200 \text{ kN-mm} = 25.2 \times 10^6 \text{ N-mm}$$

∴ Equivalent twisting moment on shaft,

$$T_e = \sqrt{(M_S)^2 + (T_S)^2}$$
$$= \sqrt{(10.08 \times 10^6)^2 + (25.2 \times 10^6)^2} = 27.14 \times 10^6 \text{ N-mm}$$

We know that equivalent twisting moment (T_e),

$$27.14 \times 10^6 = \frac{\pi}{16}(d_s)^3 \tau = \frac{\pi}{16}(135)^3 \tau = 483\ 156\ \tau$$

∴ $\tau = 27.14 \times 10^6 / 483\ 156 = 56.17$ N/mm²

From above, we see that by taking the already calculated value of ds = 135 mm, the induced shear stress is more than the allowable shear stress of 31 to 42 MPa. Hence, the value of d_s is calculated by taking $\tau = 35$ MPa or N/mm² in the above equation, *i.e.*

$$27.14 \times 10^6 = \frac{\pi}{16}(d_s)^3 35 = 6.873\,(d_s)^3$$

∴ $(d_s)^3 = 27.14 \times 10^6 / 6.873 = 3.95 \times 10^6$ or $d_s = 158$ say 160 mm **Ans.**

(c) Design of shaft at the juncture of right hand crank arm

Let d_{s1} = Diameter of the shaft at the juncture of the right hand crank arm.

We know that the resultant force at the bearing 1,

$$R_1 = \sqrt{(H_{T1})^2 + (H_{R1})^2} = \sqrt{(42)^2 + (47.3)^2} = 63.3 \text{ kN}$$

∴ Bending moment at the juncture of the right hand crank arm,

$$M_{S1} = R_1\left(b_2 + \frac{l_c}{2} + \frac{t}{2}\right) - F_Q\left(\frac{l_c}{2} + \frac{t}{2}\right)$$

$$= 63.3\left(400 + \frac{155}{2} + \frac{140}{2}\right) - 126.5\left(\frac{155}{2} + \frac{140}{2}\right)$$

$$= 34.7 \times 10^3 - 18.7 \times 10^3 = 16 \times 10^3 \text{ kN-mm} = 16 \times 10^6 \text{ N-mm}$$

and twisting moment at the juncture of the right hand crank arm,

$$T_{S1} = F_T \times r = 84 \times 300 = 25\ 200 \text{ kN-mm} = 25.2 \times 10^6 \text{ N-mm}$$

∴ Equivalent twisting moment at the juncture of the right hand crank arm,

$$T_e = \sqrt{(M_{S1})^2 + (T_{S1})^2}$$

$$= \sqrt{(16 \times 10^6)^2 + (25.2 \times 10^6)^2} = 29.85 \times 10^6 \text{ N-mm}$$

We know that equivalent twisting moment (T_e),

$$29.85 \times 10^6 = \frac{\pi}{16}(d_{s1})^3 \tau = \frac{\pi}{16}(d_{s1})^3 42 = 8.25\,(d_{s1})^3$$

...(Taking $\tau = 42$ MPa or N/mm²)

∴ $(d_{s1})^3 = 29.85 \times 10^6 / 8.25 = 3.62 \times 10^6$ or $d_{s1} = 153.5$ say 155 mm **Ans.**

(d) Design of right hand crank web

Let σ_{bR} = Bending stress in the radial direction ; and

σ_{bT} = Bending stress in the tangential direction.

We also know that bending moment due to the radial component of F_Q,

$$M_R = H_{R2}\left(b_1 - \frac{l_c}{2} - \frac{t}{2}\right) = 47.3\left(400 - \frac{155}{2} - \frac{140}{2}\right) \text{kN-mm}$$

$$= 11.94 \times 10^3 \text{ kN-mm} = 11.94 \times 10^6 \text{ N-mm} \qquad ...(i)$$

We also know that bending moment,

$$M_R = \sigma_{bR} \times Z = \sigma_{bR} \times \frac{1}{6} \times w.t^2 \qquad ...\left(\because Z = \frac{1}{6} \times w.t^2\right)$$

$$11.94 \times 10^6 = \sigma_{bR} \times \frac{1}{6} \times 245\,(140)^2 = 800 \times 10^3 \sigma_{bR}$$

∴ $\sigma_{bR} = 11.94 \times 10^6 / 800 \times 10^3 = 14.9$ N/mm² or MPa

We know that bending moment due to the tangential component of F_Q,

$$M_T = F_T\left(r - \frac{d_{s1}}{2}\right) = 84\left(300 - \frac{155}{2}\right) = 18\ 690 \text{ kN-mm}$$

$$= 18.69 \times 10^6 \text{ N-mm}$$

We also know that bending moment,

$$M_T = \sigma_{bT} \times Z = \sigma_{bT} \times \frac{1}{6} \times t.w^2 \qquad ...\left(\because Z = \frac{1}{6} \times t.w^2\right)$$

$$18.69 \times 10^6 = \sigma_{bT} \times \frac{1}{6} \times 140(245)^2 = 1.4 \times 10^6 \sigma_{bT}$$

∴ $\sigma_{bT} = 18.69 \times 10^6 / 1.4 \times 10^6 = 13.35$ N/mm² or MPa

Direct compressive stress,

$$\sigma_b = \frac{F_R}{2w \cdot t} = \frac{94.6}{2 \times 245 \times 140} = 1.38 \times 10^{-3} \text{ kN/mm}^2 = 1.38 \text{ N/mm}^2$$

and total compressive stress,

$$\sigma_c = \sigma_{bR} + \sigma_{bT} + \sigma_d$$
$$= 14.9 + 13.35 + 1.38 = 29.63 \text{ N/mm}^2 \text{ or MPa}$$

We know that twisting moment on the arm,

$$T = H_{T2}\left(b_1 - \frac{l_c}{2}\right) = 42\left(400 - \frac{155}{2}\right) = 13\ 545 \text{ kN-mm}$$
$$= 13.545 \times 10^6 \text{ N-mm}$$

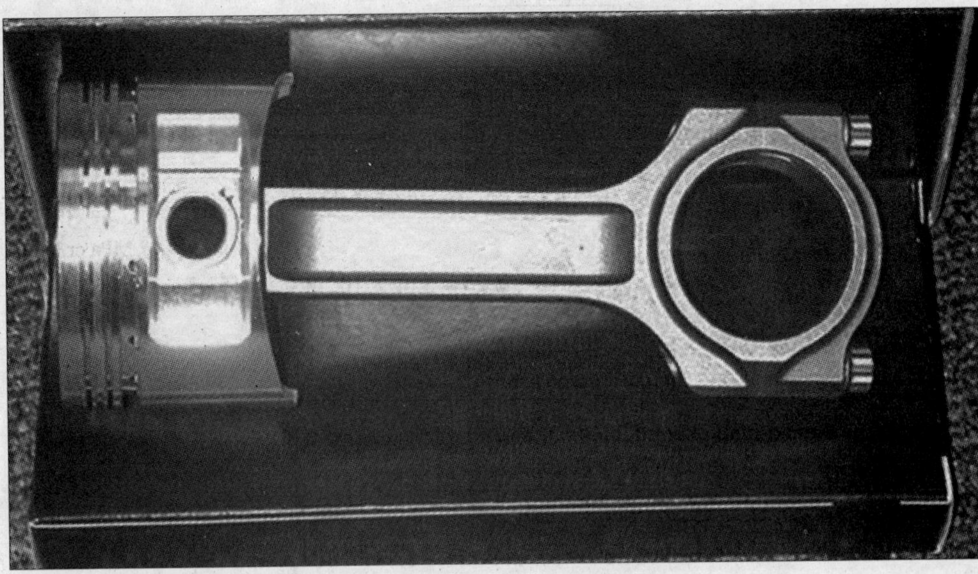

Piston and piston rod

and shear stress on the arm,

$$\tau = \frac{T}{Z_P} = \frac{4.5T}{w.t^2} = \frac{4.5 \times 13.545 \times 10^6}{245\ (140)^2} = 12.7 \text{ N/mm}^2 \text{ or MPa}$$

We know that total or maximum combined stress,

$$(\sigma_c)_{max} = \frac{\sigma_c}{2} + \frac{1}{2}\sqrt{(\sigma_c)^2 + 4\tau^2}$$
$$= \frac{29.63}{2} + \frac{1}{2}\sqrt{(29.63)^2 + 4\ (12.7)^2} = 14.815 + 19.5 = 34.315 \text{ MPa}$$

Since the maximum combined stress is within the safe limits, therefore, the dimension $w = 245$ mm is accepted.

(e) Design of left hand crank web

The dimensions for the left hand crank web may be made same as for right hand crank web.

(f) Design of crankshaft bearings

Since the bearing 2 is the most heavily loaded, therefore, only this bearing should be checked for bearing pressure.

We know that the total reaction at bearing 2,

$$R_2 = \frac{F_P}{2} + \frac{W}{2} + \frac{T_1 + T_2}{2} = \frac{314.2}{2} + \frac{50}{2} + \frac{6.5}{2} = 185.35 \text{ kN} = 185\ 350 \text{ N}$$

∴ Total bearing pressure

$$= \frac{R_2}{l_2 \cdot d_{s1}} = \frac{185\ 350}{365 \times 155} = 3.276 \text{ N/mm}^2$$

Since this bearing pressure is less than the safe limit of 5 to 8 N/mm², therefore, the design is safe.

Example 32.5. *Design a side or overhung crankshaft for a 250 mm × 300 mm gas engine. The weight of the flywheel is 30 kN and the explosion pressure is 2.1 N/mm². The gas pressure at the maximum torque is 0.9 N/mm², when the crank angle is 35° from I. D. C. The connecting rod is 4.5 times the crank radius.*

Solution. Given : $D = 250$ mm ; $L = 300$ mm or $r = L / 2 = 300 / 2 = 150$ mm ; $W = 30$ kN $= 30 \times 10^3$ N ; $p = 2.1$ N/mm², $p' = 0.9$ N/mm² ; $l = 4.5\ r$ or $l / r = 4.5$

We shall design the crankshaft for the two positions of the crank, *i.e.* firstly when the crank is at the dead centre and secondly when the crank is at an angle of maximum twisting moment.

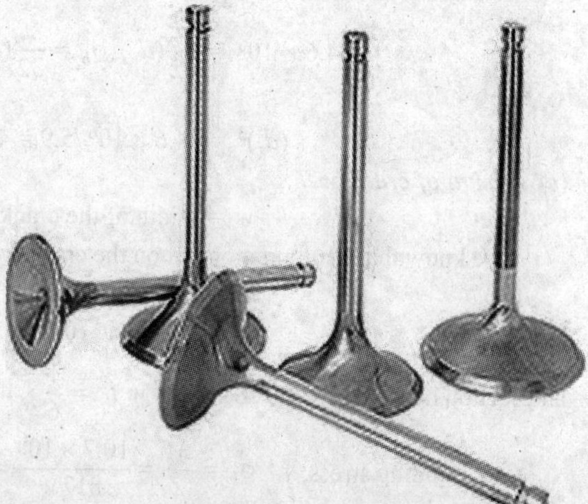

Valve guides of an IC engine

1. *Design of crankshaft when the crank is at the dead centre* (See Fig. 32.18)

We know that piston gas load,

$$F_P = \frac{\pi}{4} \times D^2 \times p$$

$$= \frac{\pi}{4}(250)^2\, 2.1 = 103 \times 10^3 \text{ N}$$

Now the various parts of the crankshaft are designed as discussed below:

(*a*) *Design of crankpin*

Let d_c = Diameter of the crankpin in mm, and

l_c = Length of the crankpin = $0.8\ d_c$ ...(Assume)

Considering the crankpin in bearing, we have

$$F_P = d_c . l_c . p_b$$

$$103 \times 10^3 = d_c \times 0.8\ d_c \times 10 = 8\ (d_c)^2 \quad \text{...(Taking } p_b = 10 \text{ N/mm}^2)$$

$$\therefore \quad (d_c)^2 = 103 \times 10^3 / 8 = 12\ 875 \text{ or } d_c = 113.4 \text{ say } 115 \text{ mm}$$

and $l_c = 0.8\ d_c = 0.8 \times 115 = 92$ mm

Let us now check the induced bending stress in the crankpin.

We know that bending moment at the crankpin,

$$M = \frac{3}{4} F_P \times l_c = \frac{3}{4} \times 103 \times 10^3 \times 92 = 7107 \times 10^3 \text{ N-mm}$$

and section modulus of the crankpin,

$$Z = \frac{\pi}{32}(d_c)^3 = \frac{\pi}{32}(115)^3 = 149 \times 10^3 \text{ mm}^3$$

∴ Bending stress induced

$$= \frac{M}{Z} = \frac{7107 \times 10^3}{149 \times 10^3} = 47.7 \text{ N/mm}^2 \text{ or MPa}$$

Since the induced bending stress is within the permissible limits of 60 MPa, therefore, design of crankpin is safe.

(b) Design of bearings

Let d_1 = Diameter of the bearing 1.

Let us take thickness of the crank web,

$$t = 0.6\,d_c = 0.6 \times 115 = 69 \text{ or } 70 \text{ mm}$$

and length of the bearing, $l_1 = 1.7\,d_c = 1.7 \times 115 = 195.5$ say 200 mm

We know that bending moment at the centre of the bearing 1,

$$M = F_P\,(0.75 l_c + t + 0.5\,l_1)$$
$$= 103 \times 10^3\,(0.75 \times 92 + 70 + 0.5 \times 200) = 24.6 \times 10^6 \text{ N-mm}$$

We also know that bending moment (M),

$$24.6 \times 10^6 = \frac{\pi}{32}(d_1)^3 \sigma_b = \frac{\pi}{32}(d_1)^3\, 60 = 5.9\,(d_1)^3$$

...(Taking σ_b = 60 MPa or N/mm^2)

$\therefore$ $(d_1)^3 = 24.6 \times 10^6 / 5.9 = 4.2 \times 10^6$ or $d_1 = 161.3$ mm say 162 mm **Ans.**

(c) Design of crank web

Let w = Width of the crank web in mm.

We know that bending moment on the crank web,

$$M = F_P\,(0.75 l_c + 0.5\,t)$$
$$= 103 \times 10^3\,(0.75 \times 92 + 0.5 \times 70) = 10.7 \times 10^6 \text{ N-mm}$$

and section modulus, $Z = \frac{1}{6} \times w \cdot t^2 = \frac{1}{6} \times w(70)^2 = 817\,w \text{ mm}^3$

$\therefore$ Bending stress, $\sigma_b = \frac{M}{Z} = \frac{10.7 \times 10^6}{817\,w} = \frac{13 \times 10^3}{w}$ N/mm^2

and direct compressive stress,

$$\sigma_b = \frac{F_P}{w.t} = \frac{103 \times 10^3}{w \times 70} = \frac{1.47 \times 10^3}{w} \text{ N/mm}^2$$

We know that total stress on the crank web,

$$\sigma_T = \sigma_b + \sigma_d = \frac{13 \times 10^3}{w} + \frac{1.47 \times 10^3}{w} = \frac{14.47 \times 10^3}{w} \text{ N/mm}^2$$

The total stress should not exceed the permissible limit of 60 MPa or N/mm^2.

$\therefore$ $60 = \frac{14.47 \times 10^3}{w}$ or $w = \frac{14.47 \times 10^3}{60} = 241$ say 245 mm **Ans.**

(d) Design of shaft under the flywheel.

Let d_s = Diameter of shaft under the flywheel.

First of all, let us find the horizontal and vertical reactions at bearings 1 and 2. Assume that the width of flywheel is 250 mm and $l_1 = l_2 = 200$ mm.

Allowing for certain clearance, the distance

$$b = 250 + \frac{l_1}{2} + \frac{l_2}{2} + \text{clearance}$$
$$= 250 + \frac{200}{2} + \frac{200}{2} + 20 = 470 \text{ mm}$$

and $a = 0.75\,l_c + t + 0.5\,l_1$

$$= 0.75 \times 92 + 70 + 0.5 \times 200 = 239 \text{ mm}$$

We know that the horizontal reactions H_1 and H_2 at bearings 1 and 2, due to the piston gas load (F_P) are

$$H_1 = \frac{F_P\,(a+b)}{b} = \frac{103 \times 10^3\,(239 + 470)}{470} = 155.4 \times 10^3 \text{ N}$$

and $$H_2 = \frac{F_P \times a}{b} = \frac{103 \times 10^3 \times 239}{470} = 52.4 \times 10^3 \text{ N}$$

Assuming $b_1 = b_2 = b/2$, the vertical reactions V_1 and V_2 at bearings 1 and 2 due to the weight of the flywheel are

$$V_1 = \frac{W \cdot b_1}{b} = \frac{W \times b/2}{b} = \frac{W}{2} = \frac{30 \times 10^3}{2} = 15 \times 10^3 \text{ N}$$

and $$V_2 = \frac{W \cdot b_2}{b} = \frac{W \times b/2}{b} = \frac{W}{2} = \frac{30 \times 10^3}{2} = 15 \times 10^3 \text{ N}$$

Since there is no belt tension, therefore the horizontal reactions due to the belt tension are neglected.

We know that horizontal bending moment at the flywheel location due to piston gas load.

$$M_1 = F_P\,(a + b_2) - H_1 \cdot b_2$$

$$= 103 \times 10^3 \left(239 + \frac{470}{2}\right) - 155.4 \times 10^3 \times \frac{470}{2} \qquad \ldots \left(\because\ b_2 = \frac{b}{2}\right)$$

$$= 48.8 \times 10^6 - 36.5 \times 10^6 = 12.3 \times 10^6 \text{ N-mm}$$

Since there is no belt pull, therefore, there will be no horizontal bending moment due to the belt pull, *i.e.* $M_2 = 0$.

∴ Total horizontal bending moment,

$$M_H = M_1 + M_2 = M_1 = 12.3 \times 10^6 \text{ N-mm}$$

We know that vertical bending moment due to the flywheel weight,

$$M_V = \frac{W.b_1.b_2}{b} = \frac{W \times b \times b}{2 \times 2 \times b} = \frac{W \times b}{4}$$

$$= \frac{30 \times 10^3 \times 470}{4} = 3.525 \times 10^6 \text{ N-mm}$$

Inside view of a car engine

∴ Resultant bending moment,

$$M_R = \sqrt{(M_H)^2 + (M_V)^2} = \sqrt{(12.3 \times 10^6)^2 + (3.525 \times 10^6)^2}$$
$$= 12.8 \times 10^6 \text{ N-mm}$$

We know that bending moment (M_R),

$$12.8 \times 10^6 = \frac{\pi}{32}(d_s)^3 \sigma_b = \frac{\pi}{32}(d_s)^3\ 60 = 5.9(d_s)^3$$

$$\therefore \quad (d_s)^3 = 12.8 \times 10^6 / 5.9 = 2.17 \times 10^6 \text{ or } d_s = 129 \text{ mm}$$

Actually d_s should be more than d_1. Therefore let us take

$$d_s = 200 \text{ mm } \textbf{Ans.}$$

2. ***Design of crankshaft when the crank is at an angtle of maximum twisting moment***

We know that piston gas load,

$$F_P = \frac{\pi}{4} \times D^2 \times p' = \frac{\pi}{4}(250)^2\ 0.9 = 44\ 200 \text{ N}$$

In order to find the thrust in the connecting rod (F_Q), we should first find out the angle of inclination of the connecting rod with the line of storke (*i.e.* angle ϕ). We know that

$$\sin\phi = \frac{\sin\theta}{l/r} = \frac{\sin 35°}{4.5} = 0.1275$$

$$\therefore \quad \phi = \sin^{-1}(0.1275) = 7.32°$$

We know that thrust in the connecting rod,

$$F_Q = \frac{F_P}{\cos\phi} = \frac{44\ 200}{\cos 7.32°} = \frac{44\ 200}{0.9918} = 44\ 565 \text{ N}$$

Tangential force acting on the crankshaft,

$$F_T = F_Q \sin(\theta + \phi) = 44\ 565 \sin(35° + 7.32°) = 30 \times 10^3 \text{ N}$$

and radial force,

$$F_R = F_Q \cos(\theta + \phi) = 44\ 565 \cos(35° + 7.32°) = 33 \times 10^3 \text{ N}$$

Due to the tangential force (F_T), there will be two reactions at the bearings 1 and 2, such that

$$H_{T1} = \frac{F_T(a+b)}{b} = \frac{30 \times 10^3 (239 + 470)}{470} = 45 \times 10^3 \text{ N}$$

and

$$H_{T2} = \frac{F_T \times a}{b} = \frac{30 \times 10^3 \times 239}{470} = 15.3 \times 10^3 \text{ N}$$

Due to the radial force (F_R), there will be two reactions at the bearings 1 and 2, such that

$$H_{R1} = \frac{F_R(a+b)}{b} = \frac{33 \times 10^3 \times (239 + 470)}{470} = 49.8 \times 10^3 \text{ N}$$

and

$$H_{R2} = \frac{F_R \times a}{b} = \frac{33 \times 10^3 \times 239}{470} = 16.8 \times 10^3 \text{ N}$$

Now the various parts of the crankshaft are designed as discussed below:

(*a*) ***Design of crank web***

We know that bending moment due to the tangential force,

$$M_{bT} = F_T\left(r - \frac{d_1}{2}\right) = 30 \times 10^3 \left(150 - \frac{180}{2}\right) = 1.8 \times 10^6 \text{ N-mm}$$

∴ Bending stress due to the tangential force,

$$\sigma_{bT} = \frac{M_{bT}}{Z} = \frac{6M_{bT}}{t.w^2} = \frac{6 \times 1.8 \times 10^6}{70\,(245)^2} \qquad \ldots \left(\because Z = \frac{1}{6} \times t.w^2\right)$$

$$= 2.6 \text{ N/mm}^2 \text{ or MPa}$$

Bending moment due to the radial force,

$$M_{bR} = F_R\,(0.75\, l_c + 0.5\, t)$$
$$= 33 \times 10^3\,(0.75 \times 92 + 0.5 \times 70) = 3.43 \times 10^6 \text{ N-mm}$$

∴ Bending stress due to the radial force,

$$\sigma_{bR} = \frac{M_{bR}}{Z} = \frac{6\, M_{bR}}{w.t^2} \qquad \ldots \left(\because Z = \frac{1}{6} \times w.t^2\right)$$

$$= \frac{6 \times 3.43 \times 10^6}{245\,(70)^2} = 17.1 \text{ N/mm}^2 \text{ or MPa}$$

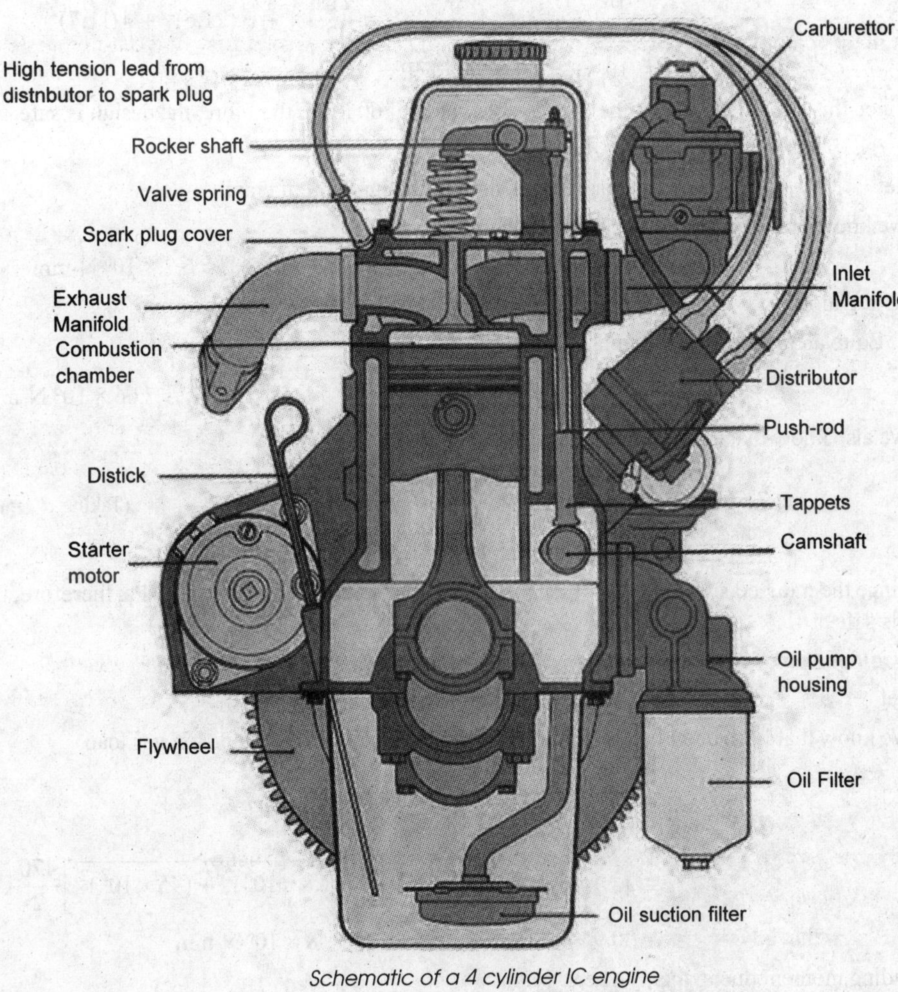

Schematic of a 4 cylinder IC engine

We know that direct compressive stress,

$$\sigma_d = \frac{F_R}{w \cdot t} = \frac{33 \times 10^3}{245 \times 70} = 1.9 \text{ N/mm}^2 \text{ or MPa}$$

∴ Total compressive stress,

$$\sigma_c = \sigma_{bT} + \sigma_{bR} + \sigma_d = 2.6 + 17.1 + 1.9 = 21.6 \text{ MPa}$$

We know that twisting moment due to the tangential force,

$$T = F_T (0.75\, l_c + 0.5\, t)$$

$$= 30 \times 10^3 (0.75 \times 92 + 0.5 \times 70) = 3.12 \times 10^6 \text{ N-mm}$$

∴ Shear stress, $\tau = \dfrac{T}{Z_P} = \dfrac{4.5\, T}{w \cdot t^2} = \dfrac{4.5 \times 3.12 \times 10^6}{245\,(70)^2}$ ... $\left[\because Z_P = \dfrac{w \cdot t^2}{4.5}\right]$

$$= 11.7 \text{ N/mm}^2 \text{ or MPa}$$

We know that total or maximum stress,

$$\sigma_{max} = \frac{\sigma_c}{2} + \frac{1}{2}\sqrt{(\sigma_c)^2 + 4\,\tau^2} = \frac{21.6}{2} + \frac{1}{2}\sqrt{(21.6)^2 + 4(11.7)^2}$$

$$= 10.8 + 15.9 = 26.7 \text{ MPa}$$

Since this stress is less than the permissible value of 60 MPa, therefore, the design is safe.

(b) ***Design of shaft at the junction of crank***

Let d_{s1} = Diameter of shaft at the junction of crank.

We know that bending moment at the junction of crank,

$$M = F_Q (0.75 l_c + t) = 44\,565\,(0.75 \times 92 + 70) = 6.2 \times 10^6 \text{ N-mm}$$

and twisting moment, $T = F_T \times r = 30 \times 10^3 \times 150 = 4.5 \times 10^6$ N-mm

∴ Equivalent twisting moment,

$$T_e = \sqrt{M^2 + T^2} = \sqrt{(6.2 \times 10^6)^2 + (4.5 \times 10^6)^2} = 7.66 \times 10^6 \text{ N-mm}$$

We also know that equivalent twisting moment (T_e),

$$7.66 \times 10^6 = \frac{\pi}{16}(d_{s1})^3 \tau = \frac{\pi}{16}(180)^3 \tau = 1.14 \times 10^6 \tau \quad \text{...(Taking } d_{s1} = d_1\text{)}$$

$$\therefore \quad \tau = 7.66 \times 10^6 / 1.14 \times 10^6 = 6.72 \text{ N/mm}^2 \text{ or MPa}$$

Since the induced shear stress is less than the permissible limit of 30 to 40 MPa, therefore, the design is safe.

(c) ***Design of shaft under the flywheel***

Let d_s = Diameter of shaft under the flywheel.

We know that horizontal bending moment acting on the shaft due to piston gas load,

$$M_H = F_P\,(a + b_2) - \left[\sqrt{(H_{R1})^2 + (H_{T1})^2}\right] b_2$$

$$= 44\,200\left(239 + \frac{470}{2}\right) - \left[\sqrt{(49.8 \times 10^3)^2 + (45 \times 10^3)^2}\right]\frac{470}{2}$$

$$= 20.95 \times 10^6 - 15.77 \times 10^6 = 5.18 \times 10^6 \text{ N-mm}$$

and bending moment due to the flywheel weight

$$M_V = \frac{W.b_1.b_2}{b} = \frac{30 \times 10^3 \times 235 \times 235}{470} = 3.53 \times 10^6 \text{ N-mm}$$

...($b_1 = b_2 = b/2 = 470/2 = 235$ mm)

∴ Resultant bending moment,

$$M_R = \sqrt{(M_H)^2 + (M_V)^2} = \sqrt{(5.18 \times 10^6)^2 + (3.53 \times 10^6)^2}$$

$$= 6.27 \times 10^6 \text{ N-mm}$$

We know that twisting moment on the shaft,

$$T = F_T \times r = 30 \times 10^3 \times 150 = 4.5 \times 10^6 \text{ N-mm}$$

∴ Equivalent twisting moment,

$$T_e = \sqrt{(M_R)^2 + T^2} = \sqrt{(6.27 \times 10^6)^2 + (4.5 \times 10^6)^2}$$

$$= 7.72 \times 10^6 \text{ N-mm}$$

We also know that equivalent twisting moment (T_e),

$$7.72 \times 10^6 = \frac{\pi}{16} (d_s)^3 \tau = \frac{\pi}{16} (d_s)^3\, 30 = 5.9\, (d_s)^3$$

...(Taking τ = 30 MPa)

$$\therefore \quad (d_s)^3 = 7.72 \times 10^6 / 5.9 = 1.31 \times 10^6 \text{ or } d_s = 109 \text{ mm}$$

Actually, d_s should be more than d_1. Therefore let us take

d_s = 200 mm **Ans.**

32.22 Valve Gear Mechanism

The valve gear mechanism of an I.C. engine consists of those parts which actuate the inlet and exhaust valves at the required time with respect to the position of piston and crankshaft. Fig. 32.20 (*a*) shows the valve gear arrangement for vertical engines. The main components of the mechanism are valves, rocker arm, * valve springs, **push rod, ***cam and camshaft.

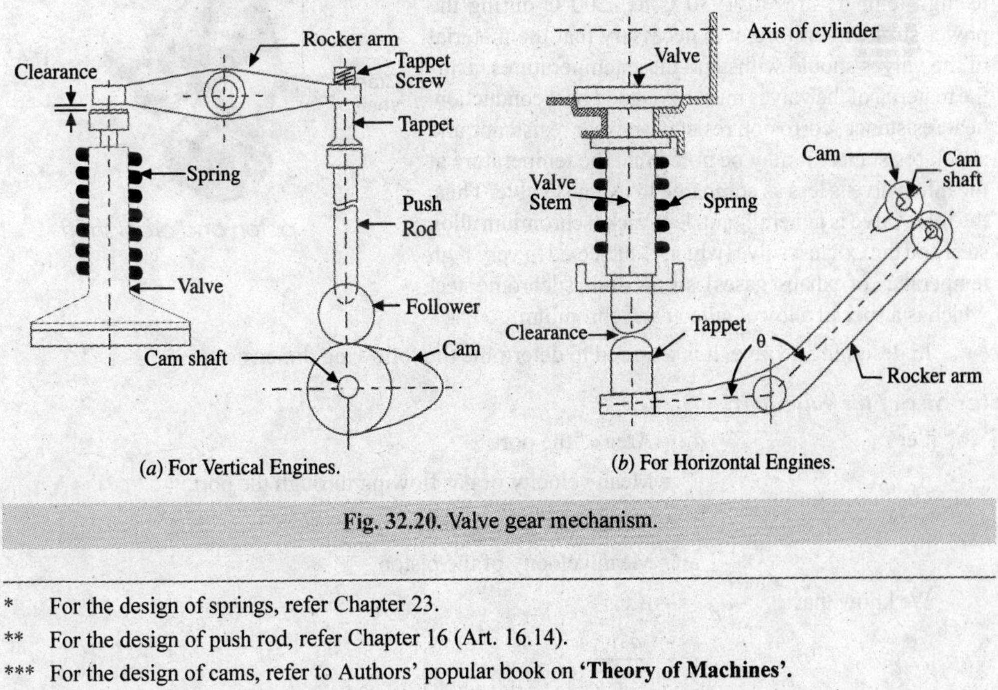

Fig. 32.20. Valve gear mechanism.

* For the design of springs, refer Chapter 23.

** For the design of push rod, refer Chapter 16 (Art. 16.14).

*** For the design of cams, refer to Authors' popular book on **'Theory of Machines'**.

The fuel is admitted to the engine by the inlet valve and the burnt gases are escaped through the exhaust valve. In vertical engines, the cam moving on the rotating camshaft pushes the cam follower and push rod upwards, thereby transmitting the cam action to rocker arm. The camshaft is rotated by the toothed belt from the crankshaft. The rocker arm is pivoted at its centre by a fulcrum pin. When one end of the rocker arm is pushed up by the push rod, the other end moves downward. This pushes down the valve stem causing the valve to move down, thereby opening the port. When the cam follower moves over the circular portion of cam, the pushing action of the rocker arm on the valve is released and the valve returns to its seat and closes it by the action of the valve spring.

In some of the modern engines, the camshaft is located at cylinder head level. In such cases, the push rod is eliminated and the roller type cam follower is made part of the rocker arm. Such an arrangement for the horizontal engines is shown in Fig. 32.20 (*b*).

32.23 Valves

The valves used in internal combustion engines are of the following three types ;

1. Poppet or mushroom valve ; 2. Sleeve valve ; 3. Rotary valve.

Out of these three valves, poppet valve, as shown in Fig. 32.21, is very frequently used. It consists of head, face and stem. The head and face of the valve is separated by a small margin, to aviod sharp edge of the valve and also to provide provision for the regrinding of the face. The face angle generally varies from 30° to 45°. The lower part of the stem is provided with a groove in which spring retainer lock is installed.

Since both the inlet and exhaust valves are subjected to high temperatures of 1930°C to 2200°C during the power stroke, therefore, it is necessary that the material of the valves should withstand these temperatures. Thus the material of the valves must have good heat conduction, heat resistance, corrosion resistance, wear resistance and shock resistance. It may be noted that the temperature at the inlet valve is less as compared to exhaust valve. Thus, the inlet valve is generally made of nickel chromium alloy steel and the exhaust valve (which is subjected to very high temperature of exhaust gases) is made from silchrome steel which is a special alloy of silicon and chromium.

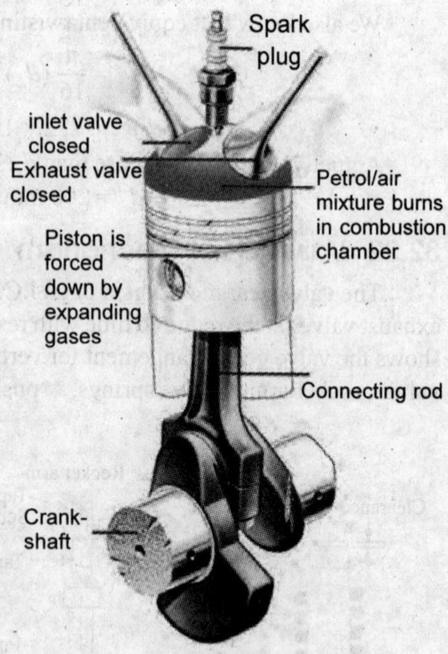

piston and crankshaft

In designing a valve, it is required to determine the following dimensions:

(a) Size of the valve port

Let a_p = Area of the port,

v_p = Mean velocity of gas flowing through the port,

a = Area of the piston, and

v = Mean velocity of the piston.

We know that $a_p.v_p = a.v$

$$\therefore \quad a_p = \frac{a.v}{v_p}$$

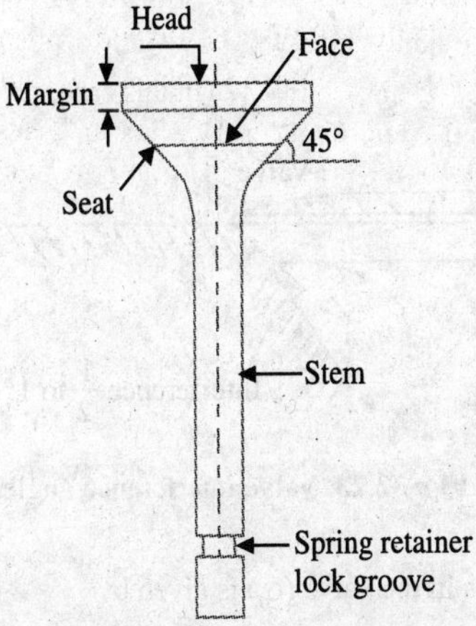

Fig. 32.21. Poppet or mushroom valve.

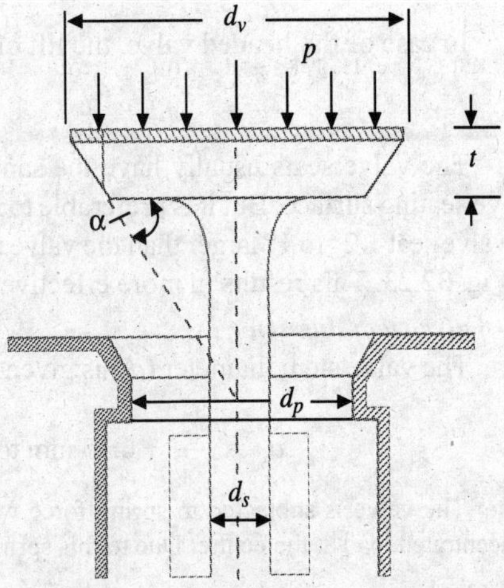

Fig. 32.22. Conical poppet valve in the port.

The mean velocity of the gas (v_p) may be taken from the following table.

Table 32.3. Mean velocity of the gas (v_p)

Type of engine	*Mean velocity of the gas (v_p) m/s*	
	Inlet valve	*Exhaust valve*
Low speed	33 – 40	40 – 50
High speed	80 – 90	90 – 100

Sometimes, inlet port is made 20 to 40 precent larger than exhaust port for better cylinder charging.

(*b*) ***Thickness of the valve disc***

The thickness of the valve disc (t), as shown in Fig. 32.22, may be determined empirically from the following relation, *i.e.*

$$t = k.d_p \sqrt{\frac{p}{\sigma_b}}$$

where k = Constant = 0.42 for steel and 0.54 for cast iron,

d_p = Diameter of the port in mm,

p = Maximum gas pressure in N/mm^2, and

σ_b = Permissible bending stress in MPa or N/mm^2

= 50 to 60 MPa for carbon steel and 100 to 120 MPa for alloy steel.

(*c*) ***Maximum lift of the valve***

h = Lift of the valve.

The lift of the valve may be obtained by equating the area across the valve seat to the area of the port. For a conical valve, as shown in Fig. 32.22, we have

$$\pi\, d_p \cdot h \cos\alpha = \frac{\pi}{4}(d_p)^2 \quad \text{or} \quad h = \frac{d_p}{4\cos\alpha}$$

where α = Angle at which the valve seat is tapered = 30° to 45°.

In case of flat headed valve, the lift of valve is given by

$$h = \frac{d_p}{4} \quad \text{...(In this case, } \alpha = 0°\text{)}$$

The valve seats usually have the same angle as the valve seating surface. But it is preferable to make the angle of valve seat 1/2° to 1° larger than the valve angle as shown in Fig. 32.23. This results in more effective seat.

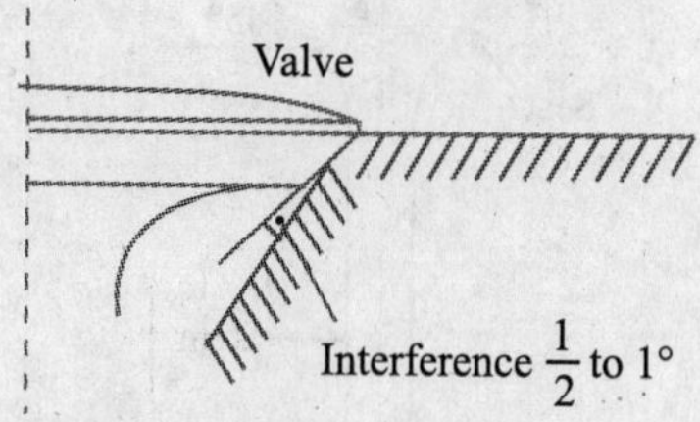

Fig. 32.23. Valve interference angle.

(*d*) ***Valve stem diameter***

The valve stem diameter (d_s) is given by

$$d_s = \frac{d_p}{8} + 6.35 \text{ mm to } \frac{d_p}{8} + 11 \text{ mm}$$

Note: The valve is subjected to spring force which is taken as concentrated load at the centre. Due to this spring force (F_s), the stress in the valve (σ_t) is given by

$$\sigma_t = \frac{1.4\, F_s}{t^2}\left(1 - \frac{2d_s}{3d_p}\right)$$

Example 32.6. *The conical valve of an I.C. engine is 60 mm in diameter and is subjected to a maximum gas pressure of 4 N/mm². The safe stress in bending for the valve material is 46 MPa. The valve is made of steel for which k = 0.42. The angle at which the valve disc seat is tapered is 30°.*

Determine : 1. thickness of the valve head ; 2. stem diameter ; and 3. maximum lift of the valve.

Solution. Given : $d_p = 60$ mm ; $p = 4$ N/mm² ; $\sigma_b = 46$ MPa $= 46$ N/mm² ; $k = 0.42$; $\alpha = 30°$

1. ***Thickness of the valve head***

We know that thickness of the valve head,

$$t = k \cdot d_p \sqrt{\frac{p}{\sigma_b}} = 0.42 \times 60 \sqrt{\frac{4}{46}} = 7.43 \text{ say } 7.5 \text{ mm } \textbf{Ans.}$$

2. ***Stem diameter***

We know that stem diameter,

$$d_s = \frac{d_p}{8} + 6.35 = \frac{60}{8} + 6.35 = 13.85 \text{ say } 14 \text{ mm } \textbf{Ans.}$$

3. ***Maximum lift of the valve***

We know that maximum lift of the valve,

$$h = \frac{d_p}{4 \cos \alpha} = \frac{60}{4 \cos 30°} = \frac{60}{4 \times 0.866} = 17.32 \text{ say } 17.4 \text{ mm } \textbf{Ans.}$$

32.24 Rocker Arm

The * rocker arm is used to actuate the inlet and exhaust valves motion as directed by the cam and follower. It may be made of cast iron, cast steel, or malleable iron. In order to reduce inertia of the rocker arm, an *I*-section is used for the high speed engines and it may be rectangular section for low speed engines. In four stroke engines, the rocker arms for the exhaust valve is the most heavily loaded. Though the force required to operate the inlet valve is relatively small, yet it is usual practice to make the rocker

Roller followers in an engine rocker mechanism

* The rocker arm has also been discussed in Chapter 15 on Levers (Refer Art. 15.9).

arm for the inlet valve of the same dimensions as that for exhaust valve. A typical rocker arm for operating the exhaust valve is shown in Fig. 32.24. The lever ratio a / b is generally decided by considering the space available for rocker arm. For moderate and low speed engines, a / b is equal to one. For high speed engines, the ratio a / b is taken as 1/ 1.3. The various forces acting on the rocker arm of exhaust valve are the gas load, spring force and force due to valve acceleration.

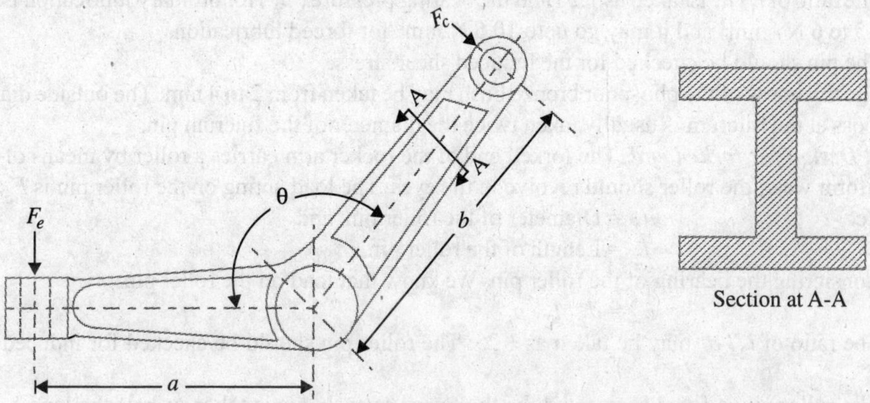

Fig. 32.24. Rocker arm for exhaust valve.

Let m_v = Mass of the valve,

d_v = Diameter of the valve head,

h = Lift of the valve,

a = Acceleration of the valve,

p_c = Cylinder pressure or back pressure when the exhust valve opens, and

p_s = Maximum suction pressure.

We know that gas load,

$$P = \text{Area of valve} \times \text{Cylinder pressure when the exhaust valve opens}$$

$$= \frac{\pi}{4}(d_v)^2 p_c$$

Spring force,

$$F_s = \text{Area of valve} \times \text{Maximum suction pressure}$$

$$= \frac{\pi}{4}(d_v)^2 p_s$$

and force due to valve acceleration,

$$F_{va} = \text{Mass of valve} \times \text{Accleration of valve}$$

$$= m_v \times a$$

∴ Maximum load on the rocker arm for exhaust valve,

$$F_e = P + F_s + F_{va}$$

It may be noted that maximum load on the rocker arm for inlet valve is

$$F_i = F_s + F_{va}$$

Since the maximum load on the rocker arm for exhaust valve is more than that of inlet valve, therefore, the rocker arm must be designed on the basis of maximum load on the rocker arm for exhaust valve, as discussed below :

1. ***Design for fulcrum pin.*** The load acting on the fulcrum pin is the total reaction (R_F) at the fulcrum point.

Let d_1 = Diameter of the fulcrum pin, and

l_1 = Length of the fulcrum pin.

Considering the bearing of the fulcrum pin. We know that load on the fulcrum pin,

$$R_F = d_1 \cdot l_1 \cdot p_b$$

The ratio of l_1 / d_1 is taken as 1.25 and the bearing pressure (p_b) for ordinary lubrication is taken from 3.5 to 6 N / mm² and it may go upto 10.5 N/mm² for forced lubrication.

The pin should be checked for the induced shear stress.

The thickness of the phosphor bronze bush may be taken from 2 to 4 mm. The outside diameter of the boss at the fulcrum is usually taken twice the diameter of the fulcrum pin.

2. ***Design for forked end.*** The forked end of the rocker arm carries a roller by means of a pin. For uniform wear, the roller should revolve in the eyes. The load acting on the roller pin is F_c.

Let d_2 = Diameter of the roller pin, and

l_2 = Length of the roller pin.

Consiering the bearing of the roller pin. We know that load on the roller pin,

$$F_c = d_2 \cdot l_2 \cdot p_b$$

The ratio of l_2 / d_2 may be taken as 1.25. The roller pin should be checked for induced shear stesss.

The roller pin is fixed in eye and the thickness of each eye is taken as half the length of the roller pin.

∴ Thickness of each eye $= l_2 / 2$

The radial thickness of eye (t_3) is taken as $d_1 / 2$. Therefore overall diameter of the eye,

$$D_1 = 2\, d_1$$

The outer diameter of the roller is taken slightly larger (atleast 3 mm more) than the outer diameter of the eye.

A clearance of 1.5 mm between the roller and the fork on either side of the roller is provided.

3. ***Design for rocker arm cross-section.*** The rocker arm may be treated as a simply supported beam and loaded at the fulcrum point. We have already discussed that the rocker arm is generally of *I*-section but for low speed engines, it can be of rectangular section. Due to the load on the valve, the rocker arm is subjected to bending moment.

Let l = Effective length of each rocker arm, and

σ_b = Permissible bending stress.

We know that bending moment on the rocker arm,

$$M = F_e \times l \quad ...(i)$$

We also know that bending moment,

$$M = \sigma_b \times Z \quad ...(ii)$$

where Z = Section modulus.

From equations (*i*) and (*ii*), the value of Z is obtained and thus the dimensions of the section are determined.

4. ***Design for tappet.*** The tappet end of the rocker arm is made circular to receive the tappet which is a stud with a lock nut. The compressive load acting on the tappet is the maximum load on the rocker arm for the exhaust valve (F_e).

Let d_c = Core diameter of the tappet, and

σ_c = Permissible compressive stress for the material of the tappet which is made of mild steel. It may be taken as 50 MPa.

We know that load on the tappet,

$$F_e = \frac{\pi}{4} (d_c)^2 \sigma_c$$

From this expression, the core diameter of the tappet is determined. The outer or nominal diameter of the tappet (d_n) is given as

$$d_n = d_c / 0.84$$

The diameter of the circular end of the rocker arm (D_3) and its depth (t_4) is taken as twice the nominal diameter of the tappet (d_n), *i.e.*

$$D_3 = 2\, d_n \; ; \text{ and } \; t_4 = 2\, d_n$$

5. *Design for valve spring.* The valve spring is used to provide sufficient force during the valve lifting process in order to overcome the inertia of valve gear and to keep it with the cam without bouncing. The spring is generally made from plain carbon spring steel. The total load for which the spring is designed is equal to the sum of initial load and load at full lift.

Let W_1 = Initial load on the spring

= Force on the valve tending to draw it into the cylinder on suction stroke,

W_2 = Load at full lift

= Full lift × Stiffness of spring

∴ Total load on the spring,

$$W = W_1 + W_2$$

Note : Here we are only interested in calculating the total load on the spring. The design of the valve spring is done in the similar ways as discussed for compression springs in Chapter 23 on Springs.

Example 32.7. *Design a rocker arm, and its bearings, tappet, roller and valve spring for the exhaust valve of a four stroke I.C. engine from the following data:*

Diameter of the valve head = 80 mm; Lift of the valve = 25 mm; Mass of associated parts with the valve = 0.4 kg ; Angle of action of camshaft = 110° ; R. P. M. of the crankshaft = 1500.

From the probable indicator diagram, it has been observed that the greatest back pressure when the exhaust valve opens is 0.4 N/mm² and the greatest suction pressure is 0.02 N/mm² below atmosphere.

The rocker arm is to be of I-section and the effective length of each arm may be taken as 180 mm ; the angle between the two arms being 135°.

The motion of the valve may be assumed S.H.M., without dwell in fully open position.

Choose your own materials and suitable values for the stresses.

Draw fully dimensioned sketches of the valve gear.

Solution. Given : d_v = 80 mm ; h = 25 mm ; or r = 25 / 2 = 12.5 mm = 0.0125 m ; m = 0.4 kg ; α = 110° ; N = 1500 r.p.m. ; p_c = 0.4 N/mm² ; p_s = 0.02 N/mm² ; l = 180 mm ; θ = 135°

A rocker arm for operating the exhaust valve is shown in Fig. 32.25.

First of all, let us find the various forces acting on the rocker arm of the exhaust valve.

We know that gas load on the valve,

$$P_1 = \frac{\pi}{4}(d_v)^2\, p_c = \frac{\pi}{4}(80)^2\; 0.4 = 2011 \text{ N}$$

Weight of associated parts with the valve,

$$w = m \cdot g = 0.4 \times 9.8 = 3.92 \text{ N}$$

∴ Total load on the valve,

$$P = P_1 + w = 2011 + 3.92 = 2014.92 \text{ N} \quad ...(i)$$

Initial spring force considering weight of the valve,

$$F_s = \frac{\pi}{4}(d_v)^2\, p_s - w = \frac{\pi}{4}(80)^2\; 0.02 - 3.92 = 96.6 \text{ N} \quad ...(ii)$$

The force due to valve acceleration (F_a) may be obtained as discussed below :

We know that speed of camshaft

$$= \frac{N}{2} = \frac{1500}{2} = 750 \text{ r.p.m.}$$

and angle turned by the camshaft per second

$$= \frac{750}{60} \times 360 = 4500 \text{ deg / s}$$

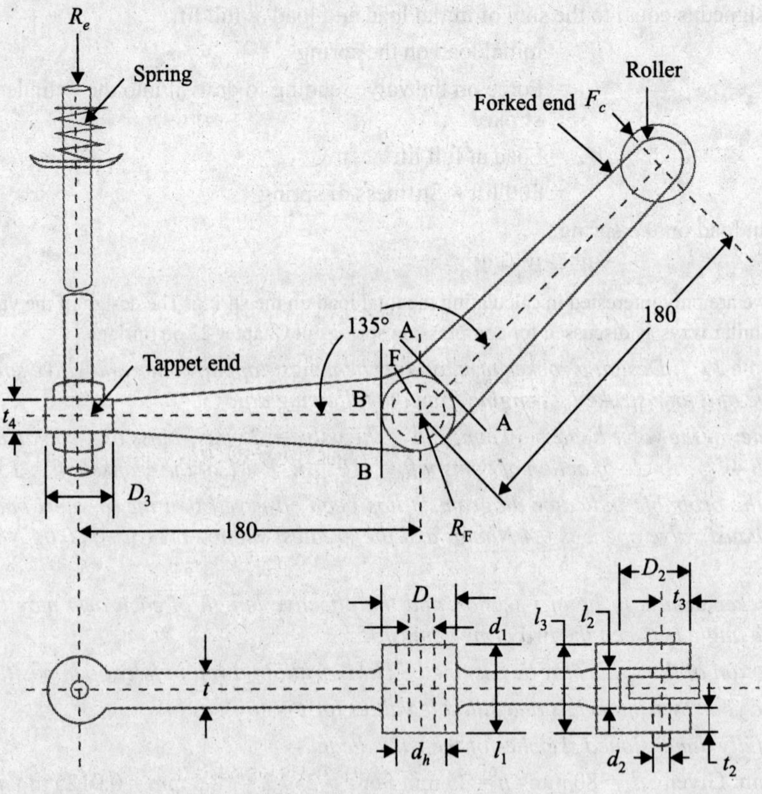

Fig. 32.25

∴ Time taken for the valve to open and close,

$$t = \frac{\text{Angle of action of cam}}{\text{Angle turned by camshaft}} = \frac{110}{4500} = 0.024 \text{ s}$$

We know that maximum acceleration of the valve

$$a = \omega^2 \cdot r = \left(\frac{2\pi}{t}\right)^2 r = \left(\frac{2\pi}{0.024}\right)^2 0.0125 = 857 \text{ m / s}^2 \quad \ldots \left(\because \omega = \frac{2\pi}{t}\right)$$

∴ Force due to valve acceleration, considering the weight of the valve,

$$F_a = m \cdot a + w = 0.4 \times 857 + 3.92 = 346.72 \text{ N} \quad \ldots (iii)$$

and maximum load on the rocker arm for exhaust valve,

$$F_e = P + F_s + F_a = 2014.92 + 96.6 + 346.72 = 2458.24 \text{ say } 2460 \text{ N}$$

Since the length of the two arms of the rocker are equal, therefore, the load at the two ends of the arm are equal, *i.e.* $F_e = F_c = 2460$ N.

Front view of a racing car

We know that reaction at the fulcrum pin *F*,

$$R_F = \sqrt{(F_e)^2 + (F_c)^2 - 2\,F_e \times F_c \times \cos\theta}$$

$$= \sqrt{(2460)^2 + (2460)^2 - 2 \times 2460 \times 2460 \times \cos 135^\circ} = 4545 \text{ N}$$

Let us now design the various parts of the rocker arm.

1. ***Design of fulcrum pin***

Let d_1 = Diameter of the fulcrum pin, and

l_1 = Length of the fulcrum pin = 1.25 d_1 ...(Assume)

Considering the bearing of the fulcrum pin. We know that load on the fulcrum pin (R_F),

$$4545 = d_1 \times l_1 \times p_b = d_1 \times 1.25\,d_1 \times 5 = 6.25\,(d_1)^2$$

...(For ordinary lubrication, p_b is taken as 5 N/mm²)

∴ $(d_1)^2 = 4545 / 6.25 = 727$ or $d_1 = 26.97$ say 30 mm **Ans.**

and $l_1 = 1.25\,d_1 = 1.25 \times 30 = 37.5$ mm **Ans.**

Now let us check the average shear stress induced in the pin. Since the pin is in double shear, therefore, load on the fulcrum pin (R_F),

$$4545 = 2 \times \frac{\pi}{4}(d_1)^2\tau = 2 \times \frac{\pi}{4}(30)^2\tau = 1414\,\tau$$

∴ $\tau = 4545 / 1414 = 3.2$ N/mm² or MPa

This induced shear stress is quite safe.

Now external diameter of the boss,

$$D_1 = 2d_1 = 2 \times 30 = 60 \text{ mm}$$

Assuming a phosphor bronze bush of 3 mm thick, the internal diameter of the hole in the lever,

$$d_h = d_1 + 2 \times 3 = 30 + 6 = 36 \text{ mm}$$

Let us now check the induced bending stress for the section of the boss at the fulcrum which is shown in Fig. 32.26.

Bending moment at this section,

$$M = F_e \times l = 2460 \times 180 = 443 \times 10^3 \text{ N-mm}$$

Section modulus,

$$Z = \frac{\frac{1}{12} \times 37.5\,[(60)^3 - (36)^3]}{60/2} = 17\,640 \text{ mm}^3$$

∴ Induced bending stress,

$$\sigma_b = \frac{M}{Z} = \frac{443 \times 10^3}{17\,640} = 25.1 \text{ N/mm}^2 \text{ or MPa}$$

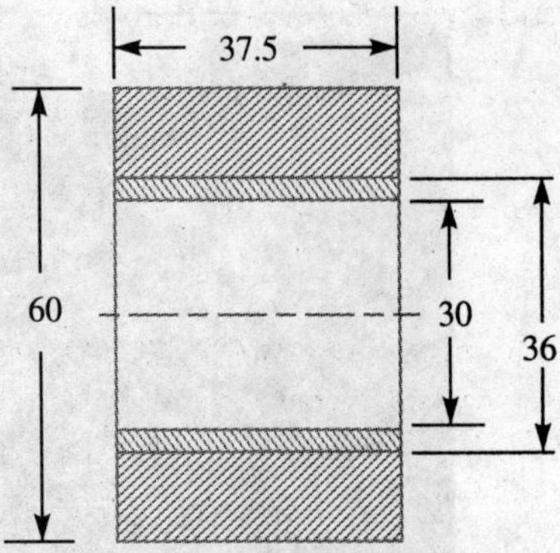

Fig. 32.26

The induced bending stress is quite safe.

2. *Design for forked end*

Let d_2 = Diameter of the roller pin, and

l_2 = Length of the roller pin = $1.25\, d_1$ ...(Assume)

Considering bearing of the roller pin. We know that load on the roller pin (F_c),

$$2460 = d_2 \times l_2 \times p_b = d_2 \times 1.25\, d_2 \times 7 = 8.75\, (d_2)^2$$

... (Taking $p_b = 7 \text{ N/mm}^2$)

∴ $(d_2)^2 = 2460 / 8.75 = 281$ or $d_2 = 16.76$ say 18 mm **Ans.**

and $l_2 = 1.25\, d_2 = 1.25 \times 18 = 22.5$ say 24 mm **Ans.**

Let us now check the roller pin for induced shearing stress. Since the pin is in double shear, therefore, load on the roller pin (F_c),

$$2460 = 2 \times \frac{\pi}{4} (d_2)^2 \tau = 2 \times \frac{\pi}{4} (18)^2 \tau = 509\, \tau$$

∴ $\tau = 2460 / 509 = 4.83 \text{ N/mm}^2$ or MPa

This induced shear stress is quite safe.

The roller pin is fixed in the eye and thickenss of each eye is taken as one-half the length of the roller pin.

∴ Thickness of each eye,

$$t_2 = \frac{l_2}{2} = \frac{24}{2} = 12 \text{ mm}$$

Let us now theck the induced bending stress in the roller pin. The pin is neither simply supported in fork nor rigidly fixed at the end. Therefore, the common practice is to assume the load distrubution as shown in Fig. 32.27.

The maximum bending moment will occur at Y–Y. Neglecting the effect of clearance, we have

Maximum bending moment at Y–Y,

$$M = \frac{F_c}{2}\left(\frac{l_2}{2} + \frac{t_2}{3}\right) - \frac{F_c}{2} \times \frac{l_2}{4}$$

$$= \frac{F_c}{2}\left(\frac{l_2}{2} + \frac{l_2}{6}\right) - \frac{F_c}{2} \times \frac{l_2}{4} \qquad ...(\because t_2 = l_2/2)$$

$$= \frac{5}{24} \times F_c \times l_2 = \frac{5}{24} \times 2460 \times 24$$

$$= 12\,300 \text{ N-mm}$$

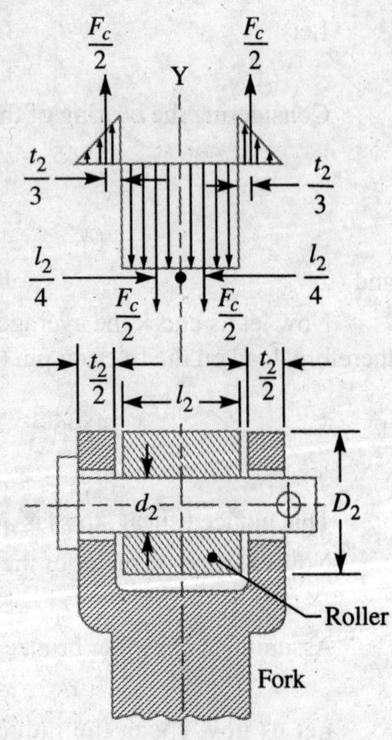

Fig. 32.27

and section modulus of the pin,

$$Z = \frac{\pi}{32}(d_2)^3 = \frac{\pi}{32}(18)^3 = 573 \text{ mm}^3$$

∴ Bending stress induced in the pin

$$= \frac{M}{Z} = \frac{12300}{573} = 21.5 \text{ N/mm}^2 \text{ or MPa}$$

This bending stress induced in the pin is within permissible limits.

Since the radial thickness of eye (t_3) is taken as $d_2 / 2$, therefore, overall diameter of the eye,

$$D_2 = 2\, d_2 = 2 \times 18 = 36 \text{ mm}$$

The outer diameter of the roller is taken slightly larger (atleast 3 mm more) than the outer diameter of the eye.

In the present case, 42 mm outer diameter of the roller will be sufficient.

Providing a clearance of 1.5 mm between the roller and the fork on either side of the roller, we have

$$l_3 = l_2 + 2 \times \frac{t_2}{2} + 2 \times 1.5$$

$$= 24 + 2 \times \frac{12}{2} + 3 = 39 \text{ mm}$$

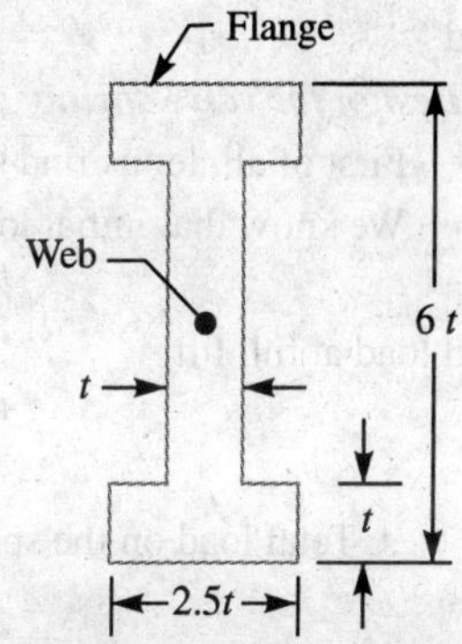

Fig. 32.28

3. ***Design for rocker arm cross-section***

The cross-section of the roker arm is obtained by considering the bending of the sections just near the boss of fulcrum on both sides, such as section $A - A$ and $B - B$.

We know that maximum bending moment at $A - A$ and $B - B$.

$$M = 2460\left(180 - \frac{60}{2}\right) = 369 \times 10^3 \text{ N-mm}$$

The rocker arm is of *I*-section. Let us assume the proportions as shown in Fig. 32.28. We know that section modulus,

$$Z = \frac{\frac{1}{12}\left[2.5t\,(6t)^3 - 1.5t\,(4t)^3\right]}{6t/2} = \frac{37\,t^4}{3\,t} = 12.33\,t^3$$

∴ Bending stress (σ_b),

$$70 = \frac{M}{Z} = \frac{369 \times 10^3}{12.33\,t^3} = \frac{29.93 \times 10^3}{t^3}$$

$$t^3 = 29.93 \times 10^3 / 70 = 427.6 \quad \text{or} \quad t = 7.5 \text{ say } 8 \text{ mm}$$

∴ Width of flange $= 2.5\,t = 2.5 \times 8 = 20$ mm **Ans.**

Depth of web $= 4\,t = 4 \times 8 = 32$ mm **Ans.**

and depth of the section $= 6\,t = 6 \times 8 = 48$ mm **Ans.**

Normally thickness of the flange and web is constant throughout, whereas the width and depth is tapered.

4. ***Design for tappet screw***

The adjustable tappet screw carries a compressive load of $F_e = 2460$ N. Assuming the screw is made of mild steel for which the compressive stress (σ_c) may be taken as 50 MPa.

Let d_c = Core diameter of the tappet screw.

We know that the load on the tappet screw (F_e),

$$2460 = \frac{\pi}{4}(d_c)^2 \sigma_c = \frac{\pi}{4}(d_c)^2 50 = 39.3\,(d_c)^2$$

$$\therefore \quad (d_c)^2 = 2460 / 39.3 = 62.6 \quad \text{or} \quad d_c = 7.9 \text{ say } 8 \text{ mm}$$

and outer or nominal diameter of the screw,

$$d = \frac{d_c}{0.84} = \frac{8}{0.84} = 9.52 \text{ say } 10 \text{ mm } \textbf{Ans.}$$

We shall use 10 mm stud and it is provided with a lock nut. The diameter of the circular end of the arm (D_3) and its depth (t_4) is taken as twice the diameter of stud.

$$\therefore \quad D_3 = 2 \times 10 = 20 \text{ mm } \textbf{Ans.}$$

and $$t_4 = 2 \times 10 = 20 \text{ mm } \textbf{Ans.}$$

5. ***Design for valve spring***

First of all, let us find the total load on the valve spring.

We know that initial load on the spring,

$$W_1 = \text{Initial spring force } (F_s) = 96.6 \text{ N} \qquad \text{...(Already calculated)}$$

and load at full lift,

$$W_2 = \text{Full valve lift} \times \text{Stiffness of spring } (s)$$
$$= 25 \times 10 = 250 \text{ N} \qquad \text{...(Assuming } s = 10 \text{ N/mm)}$$

∴ Total load on the spring,

$$W = W_1 + W_2 = 96.6 + 250 = 346.6 \text{ N}$$

Now let us find the various dimensions for the valve spring, as discussed below:

(a) Mean diameter of spring coil

Let D = Mean diameter of the spring coil, and

d = Diameter of the spring wire.

We know that Wahl's stress factor,

$$K = \frac{4C-1}{4C-4} + \frac{0.615}{C} = \frac{4 \times 8 - 1}{4 \times 8 - 4} + \frac{0.615}{8} = 1.184$$

...(Assuming $C = D/d = 8$)

and maximum shear stress (τ),

$$420 = K \times \frac{8WC}{\pi d^2} = 1.184 \times \frac{8 \times 346.6 \times 8}{\pi d^2} = \frac{8360}{d^2}$$

...(Assuming τ = 420 MPa or N/mm^2)

$$\therefore \quad d^2 = 8360 / 420 = 19.9 \quad \text{or} \quad d = 4.46 \text{ mm}$$

The standard size of the wire is SWG 7 having diameter (d) = 4.47 mm. **Ans.** (See Table 22.2).

∴ Mean diameter of the spring coil,

$$D = C \cdot d = 8 \times 4.47 = 35.76 \text{ mm } \textbf{Ans.}$$

and outer diameter of the spring coil,

$$D_o = D + d = 35.76 + 4.47 = 40.23 \text{ mm } \textbf{Ans.}$$

(b) Number of turns of the coil

Let n = Number of active turns of the coil.

We know that maximum compression of the spring,

$$\delta = \frac{8\,W \cdot C^3 \cdot n}{G \cdot d} \quad \text{or} \quad \frac{\delta}{W} = \frac{8\,C^3 \cdot n}{G \cdot d}$$

Power-brake mechanism of an automobile

Since the stiffness of the springs, $s = W / \delta = 10$ N/mm, therefore, $\delta / W = 1/10$. Taking $G = 84 \times 10^3$ MPa or N/mm^2, we have

$$\frac{1}{10} = \frac{8 \times 8^3 \times n}{84 \times 10^3 \times 4.47} = \frac{10.9\, n}{10^3}$$

$$\therefore \quad n = 10^3 / 10.9 \times 10 = 9.17 \text{ say } 10$$

For squared and ground ends, the total number of the turns,

$$n' = n + 2 = 10 + 2 = 12 \textbf{ Ans.}$$

(*c*) ***Free length of the spring***

Since the compression produced under $W_2 = 250$ N is 25 mm (*i.e.* equal to full valve lift), therefore, maximum compression produced (δ_{max}) under the maximum load of $W = 346.6$ N is

$$\delta_{max} = \frac{25}{250} \times 346.6 = 34.66 \text{ mm}$$

We know that free length of the spring,

$$L_F = n' \cdot d + \delta_{max} + 0.15\ \delta_{max}$$

$$= 12 \times 4.47 + 34.66 + 0.15 \times 34.66 = 93.5 \text{ mm } \textbf{Ans.}$$

(*d*) ***Pitch of the coil***

We know that pitch of the coil

$$= \frac{\text{Free length}}{n' - 1} = \frac{93.5}{12 - 1} = 8.5 \text{ mm } \textbf{Ans.}$$

Example 32. 8. *Design the various components of the valve gear mechanism for a horizontal diesel engine for the following data:*

Bore = 140 mm ; Stroke = 270 mm ; Power = 8.25 kW ; Speed = 475 r.p.m. ; Maximum gas pressure = 3.5 N/mm^2

The valve opens 33° before outer dead cerntre and closes 1° after inner dead centre. It opens and closes with constant acceleration and decleration for each half of the lift. The length of the rocker arm on either side of the fulcrum is 150 mm and the included angle is 160°. The weight of the valve is 3 N.

Solution. Given : $D = 140$ mm $= 0.14$ m ; $L = 270$ mm $= 0.27$ m ; Power $= 8.25$ kW $= 8250$ W ; $N = 475$ r.p.m ; $p = 3.5$ N/mm^2 ; $l = 150$ mm $= 0.15$ m ; $\theta = 160°$; $w = 3$ N

First of all, let us find out dimensions of the valve as discussed below :

Size of the valve port

Let d_p = Diameter of the valve port, and

$$a_p = \text{Area of the valve port} = \frac{\pi}{4}(d_p)^2$$

We know that area of the piston,

$$a = \frac{\pi}{4} D^2 = \frac{\pi}{4}(0.14)^2 = 0.0154\,\text{m}^2$$

and mean velocity of the piston,

$$v = \frac{2\,LN}{60} = \frac{2 \times 0.27 \times 475}{60} = 4.275 \text{ m/s}$$

From Table 32.3, let us take the mean velocity of the gas through the port (v_p) as 40 m/s.

We know that $a_p . v_p = a.v$

$$\frac{\pi}{4}(d_p)^2 40 = 0.0154 \times 4.275 \quad \text{or} \quad 31.42\,(d_p)^2 = 0.0658$$

$$\therefore \quad (d_p)^2 = 0.0658 / 31.42 = 2.09 \times 10^{-3} \quad \text{or} \quad d_p = 0.045 \text{ m} = 45 \text{ mm } \textbf{Ans.}$$

Maximum lift of the valve

We know that maximum lift of the valve,

$$h = \frac{d_p}{4\cos\alpha} = \frac{45}{4\cos 45°} = 15.9 \text{ say } 16 \text{ mm } \textbf{Ans.}$$

...(Taking $\alpha = 45°$)

Thickness of the valve head

We know that thickness of valve head,

$$t = k \cdot d_p \sqrt{\frac{p}{\sigma_b}} = 0.42 \times 45 \sqrt{\frac{3.5}{56}} = 4.72 \text{ mm } \textbf{Ans.}$$

...(Taking $k = 0.42$ and $\sigma_b = 56$ MPa)

Valve stem diameter

We know that valve stem diameter,

$$d_s = \frac{d_p}{8} + 6.35 \text{ mm} = \frac{45}{8} + 6.35 = 11.97 \text{ say } 12 \text{ mm } \textbf{Ans.}$$

Valve head diameter

The projected width of the valve seat, for a seat angle of 45°, may be empirically taken as $0.05\,d_p$ to $0.07\,d_p$. Let us take width of the valve seat as $0.06\,d_p$ *i.e.* $0.06 \times 45 = 2.7$ mm.

$\therefore$ Valve head diameter, $d_v = d_p + 2 \times 2.7 = 45 + 5.4 = 50.4$ say 51 mm **Ans.**

Now let us calculate the various forces acting on the rocker arm of exhaust valve.

We know that gas load on the valve,

$$P_1 = \frac{\pi}{4}(d_v)^2 p_c = \frac{\pi}{4}(51)^2 0.4 = 817 \text{ N} \quad \text{...(Taking } p_c = 0.4 \text{ N/mm}^2\text{)}$$

Total load on the valve, considering the weight of the valve,

$$P = P_1 + w = 817 + 3 = 820 \text{ N}$$

Initial spring force, considering the weight of the valve,

$$F_s = \frac{\pi}{4}(d_v)^2 p_s - w = \frac{\pi}{4}(51)^2 0.025 - 3 = 48 \text{ N}$$

...(Taking $p_s = 0.025$ N/mm²)

The force due to acceleration (F_a) may be obtained as discussed below :

We know that total angle of crank for which the valve remains open

$$= 33 + 180 + 1 = 214°$$

Since the engine is a four stroke engine, therefore the camshaft angle for which the valve remains open

$$= 214 / 2 = 107°$$

Now, when the camshaft turns through 107 / 2 = 53.5°, the valve lifts by a distance of 16 mm. It may be noted that the half of this period is occupied by constant acceleration and half by constant decleration. The same process occurs when the valve closes. Therefore, the period for constant acceleration is equal to camshaft rotation of 53.5 / 2 = 26.75 ° and during this time, the valve lifts through a distance of 8 mm.

We know that speed of camshaft

$$= \frac{N}{2} = \frac{475}{2} = 237.5 \text{ r.p.m.}$$

∴ Angle turned by the camshaft per second

$$= \frac{237.5}{60} \times 360 = 1425 \text{ deg / s}$$

and time taken by the camshaft for constant acceleration,

$$t = \frac{26.75}{1425} = 0.0188 \text{ s}$$

Let a = Acceleration of the valve.

We know that

$$s = u \,.\, t + \frac{1}{2} a.t^2 \quad \text{... (Equation of motion)}$$

$$8 = 0 \times t + \frac{1}{2} a\,(0.0188)^2 = 1.767 \times 10^{-4}\, a \quad \text{...(}\because u = o\text{)}$$

$$\therefore \quad a = 8 / 1.767 \times 10^{-4} = 45\,274 \text{ mm / s}^2 = 45.274 \text{ m / s}^2$$

and force due to valve acceleration, considering the weight of the valve,

$$F_a = m \cdot a + w = \frac{3}{9.81} \times 45.274 + 3 = 16.84 \text{ N} \quad \text{... (}\because m = w/g\text{)}$$

We know that the maximum load on the rocker arm for exhaust valve,

$$F_e = P + F_s + F_a = 820 + 48 + 16.84 = 884.84 \text{ say } 885 \text{ N}$$

Since the length of the two arms of the rocker are equal, therefore, load at the two ends of the arm are equal, *i.e.* $F_e = F_c = 885$ N.

We know that reaction at the fulcrum pin *F*,

$$R_F = \sqrt{(F_e)^2 + (F_c)^2 - 2F_e \times F_c \times \cos\theta}$$

$$= \sqrt{(885)^2 + (885)^2 - 2 \times 885 \times 885 \times \cos 160^\circ} = 1743 \text{ N}$$

The rocker arm is shown in Fig. 32.29. We shall now design the various parts of rocker arm as discussed below:

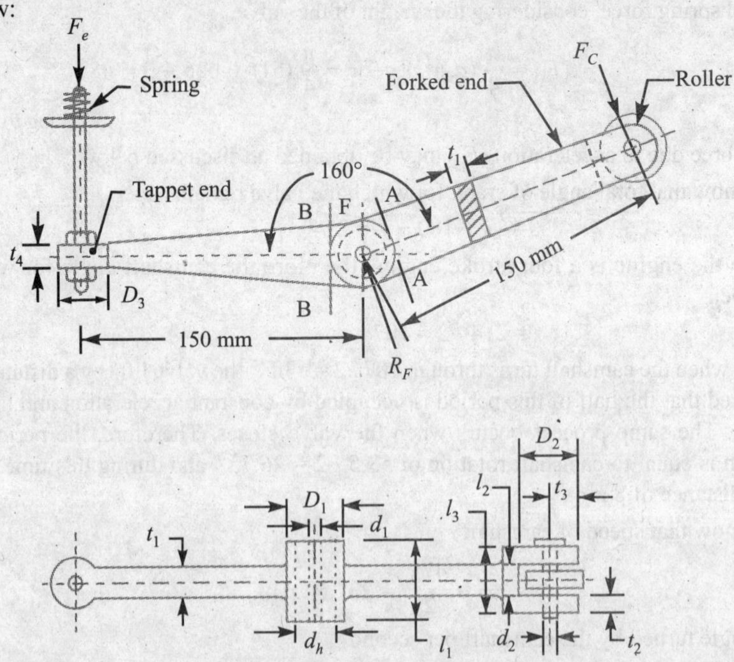

Fig. 32.29

1. ***Design of fulcrum pin***

Let d_1 = Diameter of the fulcrum pin, and

l_1 = Length of the fulcrum pin = 1.25 d_1 ... (Assume)

Considering the bearing of the fulcrum pin. We know that load on the fulcrum pin (R_F),

$$1743 = d_1 \times l_1 \times p_b = d_1 \times 1.25\, d_1 \times 5 = 6.25\,(d_1)^2$$

... (For ordinary lubrication, p_b is taken as 5 N/mm²)

$\therefore$ $(d_1)^2 = 1743 / 6.25 = 279$ or $d_1 = 16.7$ say 17 mm

and $l_1 = 1.25\, d_1 = 1.25 \times 17 = 21.25$ say 22 mm

Now let us check the average shear stress induced in the pin. Since the pin is in double shear, therefore, load on the fulcrum pin (R_F),

$$1743 = 2 \times \frac{\pi}{4}(d_1)^2 \tau = 2 \times \frac{\pi}{4}(17)^2 \tau = 454\,\tau$$

$\therefore$ $\tau = 1743 / 454 = 3.84$ N/mm² or MPa

This induced shear stress is quite safe.

Now external diameter of the boss,

$$D_1 = 2d_1 = 2 \times 17 = 34 \text{ mm}$$

Assuming a phosphor bronze bush of 3 mm thick, the internal diameter of the hole in the lever,

$$d_h = d_1 + 2 \times 3 = 17 + 6 = 23 \text{ mm}$$

Now, let us check the induced bending stress for the section of the boss at the fulcrum which is shown in Fig. 32.30.

Bending moment at this section,

$$M = F_e \times l = 885 \times 150 \text{ N-mm}$$
$$= 132\ 750 \text{ N-mm}$$

Section modulus,

$$Z = \frac{\frac{1}{12} \times 22\,[(34)^3 - (23)^3]}{34/2} = 2927 \text{ mm}^3$$

∴ Induced bending stress,

$$\sigma_b = \frac{M}{Z} = \frac{132\ 750}{2927} = 45.3 \text{ N/mm}^2 \text{ or MPa}$$

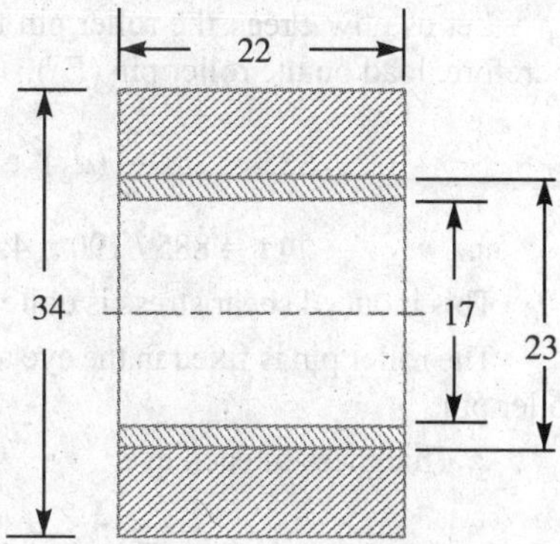

All dimensions in mm

Fig. 32.30

The induced bending stress is quite safe.

2. *Design for forked end*

Let d_2 = Diameter of the roller pin, and

l_2 = Length of the roller pin = 1.25 d_2 ...(Assume)

Considering bearing of the roller pin. We know that load on the roller pin (F_c),

$$885 = d_2 \times l_2 \times p_b = d_2 \times 1.25\, d_2 \times 7 = 8.75\,(d_2)^2$$

...(Taking p_b = 7 N/mm²)

∴ $(d_2)^2 = 885 / 8.75 = 101.14$ or $d_2 = 10.06$ say 11 mm **Ans.**

and $l_2 = 1.25\, d_2 = 1.25 \times 11 = 13.75$ say 14 mm **Ans.**

Power transmission gears in an automobile engine

Let us now check the roller pin for induced shearing stress. Since the pin is in double shear, therefore, load on the roller pin (F_c),

$$885 = 2 \times \frac{\pi}{4} (d_2)^2 \tau = 2 \times \frac{\pi}{4} (11)^2 \tau = 190\ \tau$$

$$\therefore \quad \tau = 885 / 190 = 4.66 \text{ N/mm}^2 \text{ or MPa}$$

This induced shear stress is quite safe.

The roller pin is fixed in the eye and thickness of each eye is taken as one-half the length of the roller pin.

∴ Thickness of each eye,

$$t_2 = \frac{l_2}{2} = \frac{14}{2} = 7 \text{ mm}$$

Let us now check the induced bending stress in the roller pin. The pin is neither simply supported in fork nor rigidly fixed at the end. Therefore, the common practice is to assume the load distribution as shown in Fig. 32.31.

The maximum bending moment will occur at $Y–Y$. Neglecting the effect of clearance, we have

Maximum bending moment at $Y–Y$,

$$M = \frac{F_c}{2}\left(\frac{l_2}{2} + \frac{t_2}{3}\right) - \frac{F_c}{2} \times \frac{l_2}{4}$$

$$= \frac{F_c}{2}\left(\frac{l_2}{2} + \frac{l_2}{6}\right) - \frac{F_c}{2} \times \frac{l_2}{4} \quad ...(\because t_2 = l_2/2)$$

$$= \frac{5}{24} \times F_c \times l_2$$

$$= \frac{5}{24} \times 885 \times 14 = 2581 \text{ N-mm}$$

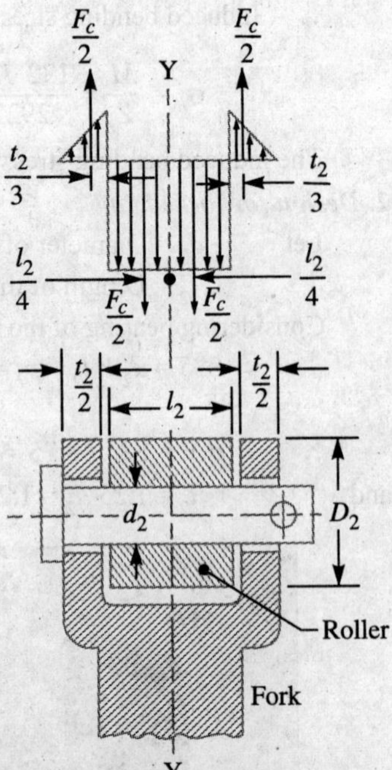

Fig. 32.31

and section modulus of the pin,

$$Z = \frac{\pi}{32} (d_2)^3 = \frac{\pi}{32} (11)^3 = 131 \text{ mm}^3$$

∴ Bending stress induced in the pin

$$= \frac{M}{Z} = \frac{2581}{131} = 19.7 \text{ N/mm}^2 \text{ or MPa}$$

This bending stress induced in the pin is within permissible limits.

Since the radial thickness of eye (t_3) is taken as $d_2 / 2$, therefore, overall diameter of the eye,

$$D_2 = 2\, d_2 = 2 \times 11 = 22 \text{ mm}$$

The outer diameter of the roller is taken slightly larger (at least 3 mm more) than the outer diameter of the eye. In the present case, 28 mm outer diameter of the roller will be sufficient.

Providing a clearance of 1.5 mm between the roller and the fork on either side of the roller, we have

$$l_3 = l_2 + 2 \times \frac{t_2}{2} + 2 \times 1.5 = 14 + 2 \times \frac{7}{2} + 3 = 24 \text{ mm}$$

3. *Design for rocker arm cross-section*

Since the engine is a slow speed engine, therefore, a rectangular section may be selected for the rocker arm. The cross-section of the rocker arm is obtained by considering the bending of the sections just near the boss of fulcrum on both sides, such as section $A–A$ and $B–B$.

Let t_1 = Thickness of the rocker arm which is uniform throughout.

B = Width or depth of the rocker arm which varies from boss diameter of fulcrum to outside diameter of the eye (for the forked end side) and from boss diameter of fulcrum to thickness t_3 (for the tappet or stud end side).

Now bending moment on section $A - A$ and $B - B$,

$$M = 885\left(150 - \frac{34}{2}\right) = 117\ 705 \text{ N-mm}$$

and section modulus at $A - A$ and $B - B$,

$$Z = \frac{1}{6} \times t_1 \cdot B^2 = \frac{1}{6} \times t_1\ (D_1)^2 = \frac{1}{6} \times t_1\ (34)^2 = 193\ t_1$$

...(At sections A–A and B–B, $B = D$)

We know that bending stress (σ_b),

$$70 = \frac{M}{Z} = \frac{117\ 705}{193\ t_1} \quad \text{...(Taking } \sigma_b = 70 \text{ MPa or N/mm}^2\text{)}$$

$\therefore$ $t_1 = 117\ 705 / 193 \times 70 = 8.7$ say 10 mm **Ans.**

4. ***Design for tappet screw***

The adjustable tappet screw carries a compressive load of F_e = 885 N. Assuming the screw to be made of mild steel for which the compressive stress (σ_c) may be taken as 50 MPa.

Let d_c = Core diameter of the tappet screw.

We know that load on the tappet screw (F_e),

$$885 = \frac{\pi}{4}\ (d_c{}^2)\ \sigma_c = \frac{\pi}{4}\ (d_c)^2\ 50 = 39.3(d_c)^2$$

$\therefore$ $(d_c)^2 = 885 / 39.3 = 22.5$ or $d_c = 4.74$ say 5 mm **Ans.**

and outer or nominal diameter of the screw,

$$d = \frac{d_c}{0.84} = \frac{5}{0.8} = 6.25 \text{ say } 6.5 \text{ mm } \textbf{Ans.}$$

We shall use 6.5 mm stud and it is provided with a lock nut. The diameter of the circular end of the arm (D_3) and its depth (t_4) is taken as twice the diameter of stud.

$\therefore$ $D_3 = 2 \times 6.5 = 13$ mm **Ans.**

and $t_4 = 2 \times 6.5 = 13$ mm **Ans.**

5. ***Design for valve spring***

First of all, let us find the total load on the valve spring.

We know that intial load on the spring,

W_1 = Initial spring force (F_s) = 48 N ...(Already calculated)

and load at full lift, W_2 = Full valve lift × Stiffness of spring (s)

= 16 × 8 = 128 N ...(Taking s = 8 N/mm)

$\therefore$ Total load on the spring,

$$W = W_1 + W_2 = 48 + 128 = 176 \text{ N}$$

Now let us find the various dimensions for the valve spring as discussed below:

(*a*) ***Mean diameter of the spring coil***

Let D = Mean diameter of the spring coil, and

d = Diameter of the spring wire.

Inside view of an automobile

We know that Wahl's stress factor,

$$K = \frac{4C - 1}{4C - 4} + \frac{0.615}{C} = \frac{4 \times 6 - 1}{4 \times 6 - 4} + \frac{0.615}{6} = 1.2525$$

...(Assuming $C = D/d = 6$)

and maximum shear stress (τ),

$$420 = K \times \frac{8\,WC}{\pi d^2} = 1.2525 \times \frac{8 \times 176 \times 6}{\pi d^2} = \frac{3368}{d^2}$$

$$\therefore \quad d^2 = 3368 / 420 = 8.02 \text{ or } d = 2.83 \text{ mm}$$

The standard size of the wire is SWG 11 having a diameter (d) = 2.946 mm **Ans.** (see Table 22.2)

$\therefore$ Mean diameter of spring coil,

$$D = C \cdot d = 6 \times 2.946 = 17.676 \text{ mm } \textbf{Ans.}$$

and outer diameter of the spring coil,

$$D_o = D + d = 17.676 + 2.946 = 20.622 \text{ mm } \textbf{Ans.}$$

(*b*) ***Number of turns of the coil***

Let n = Number of turns of the coil,

We know that maximum compression of the spring.

$$\delta = \frac{8\,W \cdot C^3 \cdot n}{G \cdot d} \quad \text{or} \quad \frac{\delta}{W} = \frac{8\,C^3 \cdot n}{G \cdot d}$$

Since the stiffness of the spring, $s = W / \delta = 8$ N / mm, therefore $\delta / W = 1 / 8$. Taking $G = 84 \times 10^3$ MPa or N/mm², we have

$$\frac{1}{8} = \frac{8 \times 6^3 \times n}{84 \times 10^3 \times 2.946} = \frac{6.98\, n}{10^3}$$

$$\therefore \quad n = 10^3 / 8 \times 6.98 = 17.9 \text{ say } 18$$

For squared and ground ends, the total number of turns,

$$n' = n + 2 = 18 + 2 = 20 \textbf{ Ans.}$$

(c) *Free length of the spring*

Since the compression produced under $W_2 = 128$ N is 16 mm, therefore, maximum compression produced under the maximum load of $W = 176$ N is

$$\delta_{max} = \frac{16}{128} \times 176 = 22 \text{ mm}$$

We know that free length of the spring,

$$L_F = n'.\, d + \delta_{max} + 0.15\, \delta_{max}$$
$$= 20 \times 2.946 + 22 + 0.15 \times 22 = 84.22 \text{ say } 85 \text{ mm } \textbf{Ans.}$$

(d) *Pitch of the coil*

We know that pitch of the coil

$$= \frac{\text{Free length}}{n' - 1} = \frac{85}{20 - 1} = 4.47 \text{ mm } \textbf{Ans.}$$

Design of cam

The cam is forged as one piece with the camshaft. It is designed as discussed below :

The diameter of camshaft (D') is taken empirically as

$$D' = 0.16 \times \text{Cylinder bore} + 12.7 \text{ mm}$$
$$= 0.16 \times 140 + 12.7 = 35.1 \text{ say } 36 \text{ mm}$$

The base circle diameter is about 3 mm greater than the camshaft diameter.

∴ Base circle diameter $= 36 + 3 = 39$ say 40 mm

The width of cam is taken equal to the width of roller, *i.e.* 14 mm.

The width of cam (w') is also taken empirically as

$$w' = 0.09 \times \text{Cylinder bore} + 6 \text{ mm} = 0.09 \times 140 + 6 = 18.6 \text{ mm}$$

Let us take the width of cam as 18 mm.

Now the *cam is drawn according to the procedure given below :

First of all, the displacement diagram, as shown in Fig. 32.32, is drawn as discussed in the following steps :

1. Draw a horizontal line *ANM* such that *AN* represents the angular displacement when valve opens (*i.e.* 53.5°) to some suitable scale. The line *NM* represents the angular displacement of the cam when valve closes (*i.e.* 53.5°).
2. Divide *AN* and *NM* into any number of equal even parts (say six).
3. Draw vertical lines through points 0, 1, 2, 3 etc. equal to the lift of valve *i.e.* 16 mm.
4. Divide the vertical lines $3 - f$ and $3' - f'$ into six equal parts as shown by points *a, b, c* ... and a', b', c' .... in Fig. 32.32.
5. Since the valve moves with equal uniform acceleration and decleration for each half of the lift, therefore, valve displacement diagram for opening and closing of valve consists of double parabola.

* For complete details, refer Authors' popular book on **'Theory of Machines'**.

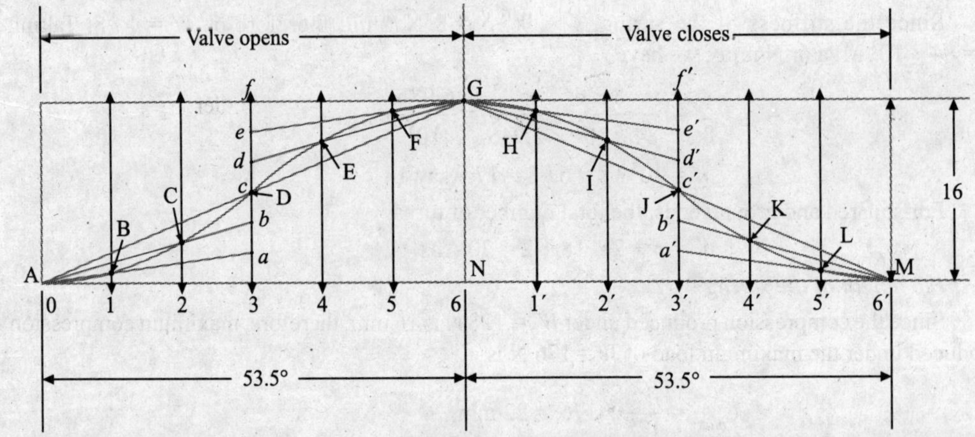

Fig. 32.32. Displacement diagram.

6. Join *Aa, Ab*, *Ac* intersecting the vertical lines through 1, 2, 3 at *B, C, D* respectively.
7. Join the points *B, C, D* with a smooth curve. This is the required parabola for the half of valve opening. Similarly other curves may be drawn as shown in Fig. 32.32.
8. The curve *A, B, C, ..., G, K, L, M* is the required displacement diagram.

Gears keyed to camshafts

Now the profile of the cam, as shown in Fig. 32.32, is drawn as discussed in the following steps:

1. Draw a base circle with centre *O* and diameter equal 40 mm (radius = 40/2 = 20 mm)
2. Draw a prime circle with centre *O* and radius, *OA*= Min. radius of cam + $\frac{1}{2}$ Diameter of roller $= 20 + \frac{1}{2} \times 28$

$$= 20 + 14 = 34 \text{ mm}$$

3. Draw angle *AOG* = 53.5° to represent opening of valve and angle *GOM* = 53.5° to represent closing of valve.
4. Divide the angular displacement of the cam during opening and closing of the valve (*i.e.* angle *AOG* and *GOM*) into same number of equal even parts as in displacement diagram.
5. Join the points 1, 2, 3, etc. with the centre *O* and produce the lines beyond prime circle as shown in Fig. 32.33.
6. Set off points 1*B*, 2*C*, 3*D*, etc. equal to the displacements from displacement diagram.
7. Join the points *A, B, C, ...L, M, A*. The curve drawn through these points is known as ***pitch curve.***
8. From the points *A, B, C, ...K, L*, draw circles of radius equal to the radius of the roller.

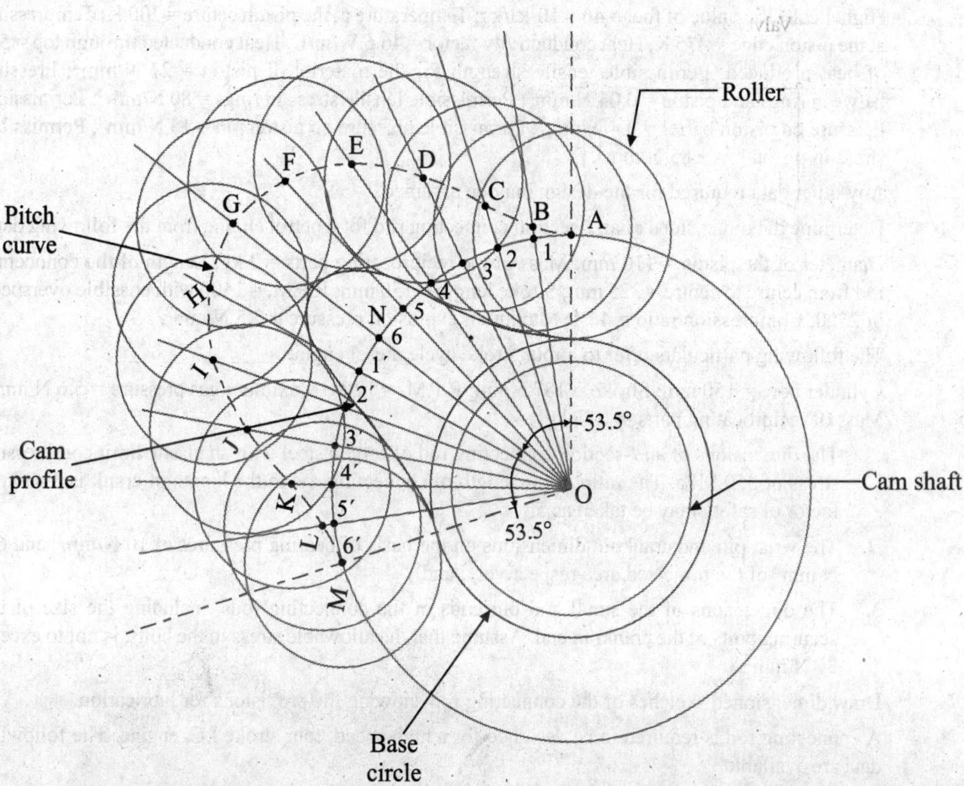

Fig. 32.33

9. Join the bottoms of the circle with a smooth curve as shown in Fig. 32.33. The is the required profile of cam.

EXERCISES

1. A four stroke internal combustion engine has the following specifications:
Brake power = 7.5 kW; Speed = 1000 r.p.m.; Indicated mean effective pressure = 0.35 N/mm^2;Maximum gas pressure = 3.5 N/mm^2; Mechanical efficiency = 80 %.
Determine: 1. The dimesions of the cylinder, if the length of stroke is 1.4 times the bore of the cylinder; 2. Wall thickness of the cylinder, if the hoop stress is 35 MPa; 3. Thickness of the cylinder head and the size of studs when the permissible stresses for the cylinder head and stud materials are 45 MPa and 65 MPa respectively.
2. Design a cast iron trunk type piston for a single acting four stroke engine developing 75 kW per cylinder when running at 600 r.p.m. The other avialable data is as follows:
Maximum gas pressure = 4.8 N/mm^2; Indicated mean effective pressure = 0.65 N/mm^2; Mechanical efficiency = 95%; Radius of crank = 110 mm; Fuel consumption = 0.3 kg/BP/hr; Calorific value of fuel (higher) = 44 × 10^{3}kJ/kg; Difference of temperatures at the centre and edges of the piston head = 200ºC; Allowable stress for the material of the piston = 33.5 MPa; Allowable stress for the material of the piston rings and gudgeon pin = 80 MPa; Allowable bearing pressure on the piston barrel = 0.4 N/mm^2 and allowable bearing pressure on the gudgeon pin = 17 N/mm^2.
3. Design a piston for a four stroke diesel engine consuming 0.3 kg of fuel per kW of power per hour and produces a brake mean effective pressure of the 0.7 N/mm^2. The maximum gas pressure inside the cylinder is 5 N/mm^2 at a speed of 3500 r.p.m. The cylinder diameter is required to be 300 mm with stroke 1.5 times the diameter. The piston may have 4 compression rings and an oil ring. The following data can be used for design:

Higher calorific value of fuel = 46 × 10^3kJ/kg; Temperature at the piston centre = 700 K; Tempressure at the piston edge = 475 K; Heat conductivity factor = 46.6 W/m/K; Heat conducted through top = 5% of heat produced; Permissible tensile strength for the material of piston = 27 N/mm^2; Pressure between rings and piston = 0.04 N/mm^2; Permissible tensile stress in rings = 80 N/mm^2; Permissible Pressure on piston barrel = 0.4 N/mm^2; Permissible pressure on piston pin = 15 N/mm^2; Permissible stress in piston pin = 85 N/mm^2.

Any other data required for the design may be assumed.

4. Determine the dimensions of an *I*-section connecting rod for a petrol engine from the following data:

   Diameter of the piston = 110 mm; Mass of the reciprocating parts = 2 kg; Length of the connecting rod from centre to centre = 325 mm; Stroke length = 150 mm; R.P.M. = 1500 with possible overspeed of 2500; Compression ratio = 4 : 1; Maximum explosion pressure = 2.5 N/mm^2.

5. The following particulars refer to a four stroke cycle diesel engine:

   Cylinder bore = 150 mm; Stroke = 187.5 mm; R.P.M. = 1200; Maximum gas pressure = 5.6 N/mm^2; Mass of reciprocating parts = 1.75 kg.

   1. The dimensions of an *I*-section connecting rod of forged steel with an elastic limit compressive stress of 350 MPa. The ratio of the length of connecting rod to the length of crank is 4 and the factor of safety may be taken as 5;
   2. The wrist pin and crankpin dimensions on the basis of bearing pressures of 10 N/mm^2 and 6.5 N/mm^2 of the projected area respectively; and
   3. The dimensions of the small and big ends of the connecting rods, including the size of the securing bolts of the crankpin end. Assume that the allowable stress in the bolts, is not to exceed 35 N/mm^2.

   Draw dimensioned sketches of the connecting rod showing the provisions for lubrication.

6. A connecting rod is required to be designed for a high speed, four stroke I.C. engine. The following data are available.

   Diameter of piston = 88 mm; Mass of reciprocating parts = 1.6 kg; Length of connecting rod (centre to centre) = 300 mm; Stroke = 125 mm; R.P.M. = 2200 (when developing 50 kW); Possible overspeed = 3000 r.p.m.; Compression ratio = 6.8 : 1 (approximately); Probale maximum explosion pressure (assumed shortly after dead centre, say at about 3°) = 3.5 N/mm^2.

   Draw fully dimensioned drawings of the connecting rod showing the provision for the lubrication.

7. Design a plain carbon steel centre crankshaft for a single acting four stroke, single cylinder engine for the following data:

   Piston diameter = 250 mm; Stroke = 400 mm; Maximum combustion pressure = 2.5 N/mm^2; Weight of the flywheel = 16 kN; Total belt pull = 3 N; Length of connecting rod = 950 mm.

   When the crank has turned through 30° from top dead centre, the pressure on the piston is 1 N/mm^2 and the torque on the crank is maximum.

   Any other data required for the design may be assumed.

8. Design a side crankshaft for a 500 mm × 600 mm gas engine. The weight of the flywheel is 80 kN and the explosion pressure is 2.5 N/mm^2. The gas pressure at maximum torque is 0.9 N/mm^2 when the crank angle. is 30°. The connecting rod is 4.5 times the crank radius.

   Any other data required for the design may be assumed.

9. Design a rocker arm of *I*-section made of cast steel for operating an exhaust valve of a gas engine. The effective length of the rocker arm is 250 mm and the angle between the arm is 135°. The exhaust valve is 80 mm in diameter and the gas pressure when the valve begins to open is 0.4 N/mm^2. The greatest suction pressure is 0.03 N/mm^2 below atmospheric. The initial load may be assumed as 0.05 N/mm^2 of valve area and the valve inertia and friction losses as 120 N. The ultimate strength of cast steel is 750 MPa. The allowable bearing pressure is 8 N/mm^2 and the permissible stress in the material is 72 MPa.

10. Design the various components of a valve gear mechanism for a horizontal diesel engine having the following specifications:

Brake power = 10 kW; Bore = 140 mm; Stroke = 270 mm; Speed = 500 r.p.m. and maximum gas pressure = 3.5 N/mm^2.

The valve open 30° before top dead centre and closes 2° after bottom dead centre. It opens and closes with constant acceleration and deceleration for each half of the lift. The length of the rocker arm on either side of the fulcrum is 150 mm and the included angle is 135°. The mass of the valve is 0.3 kg.

QUESTIONS

1. Explain the various types of cylinder liners.
2. Discuss the design of piston for an internal combustion engine.
3. State the function of the following for an internal combustion engine piston:

   (*a*) Ribs ; (*b*) Piston rings ; (*c*) Piston skirt ; and (*d*) Piston pin
4. What is the function of a connecting rod of an internal combustion engine?
5. Explain the various stresses induced in the connecting rod.
6. Under what force, the big end bolts and caps are designed?
7. Explain the various types of crankshafts.
8. At what angle of the crank, the twisting moment is maximum in the crankshaft?
9. What are the methods and materials used in the manufacture of crankshafts?
10. Sketch a valve gear mechanism of an internal combustion engine and label its various parts.
11. Discuss the materials commonly used for making the valve of an I. C. engine.
12. Why the area of the inlet valve port is made larger than the area of exhaust valve port?

Transmission mechanism in a truck engine

OBJECTIVE TYPE QUESTIONS

1. The cylinders are usually made of
   (*a*) cast iron or cast steel (*b*) aluminium
   (*c*) stainless steel (*d*) copper
2. The length of the cylinder is usually taken as
   (*a*) equal to the length of piston (*b*) equal to the length of stroke
   (*c*) equal to the cylinder bore (*d*) 1.5 times the length of stroke
3. The skirt of piston
   (*a*) is used to withstand the pressure of gas in the cylinder
   (*b*) acts as a bearing for the side thrust of the connecting rod
   (*c*) is used to seal the cylinder in order to prevent leakage of the gas past the piston
   (*d*) none of the above
4. The side thrust on the cylinder liner is usually taken as .... of the maximum gas load on the piston.
   (*a*) 1/5 (*b*) 1/8
   (*c*) 1/10 (*d*) 1/5
5. The length of the piston usually varies between
   (*a*) D and 1.5 D (*b*) 1.5 D and 2 D
   (*c*) 2D and 2.5 D (*d*) 2.5 D and 3 D
   where D = Diameter of the piston.
6. In designing a connecting rod, it is considered like .... for buckling about *X*-axis.
   (*a*) both ends fixed
   (*b*) both ends hinged
   (*c*) one end fixed and the other end hinged
   (*d*) one end fixed and the other end free
7. Which of the following statement is wrong for a connecting rod?
   (*a*) The connecting rod will be equally strong in buckling about *X*-axis, if $I_{xx} = 4\ I_{yy}$
   (*b*) If $I_{xx} > 4\ I_{yy}$, the buckling will occur about *Y*-axis.
   (c) If $I_{xx} < 4I_{yy}$, the buckling will occur about *X*-axis.
   (*d*) The most suitable section for the connecting rod is *T*-section.
8. The crankshaft in an internal combustion engine
   (*a*) is a disc which reciprocates in a cylinder
   (*b*) is used to retain the working fluid and to guide the piston
   (*c*) converts reciprocating motion of the piston into rotary motion and vice versa
   (*d*) none of the above
9. The rocker arm is used to actuate the inlet and exhaust valves motion as directed by the
   (*a*) cam and follower (*b*) crank
   (*c*) crankshaft (*d*) none of these
10. For high speed engines, a rocker arm of........... should be used.
   (*a*) rectangular section (*b*) *I*-section
   (*c*) *T*-section (*d*) circular

ANSWERS

1. (*a*)	2. (*d*)	3. (*b*)	4. (*c*)	5. (*a*)
6. (*b*)	7. (*d*)	8. (*c*)	9. (*a*)	10. (*b*)

INDEX

A

B

C

D

E

F

G

H

I

J

K

L

M

N

O

P

Q

R

S

Micro-V
Micro-V

T

U

V

W

Y

Z